The Technology of Nuclear Reactor Safety

Volume 2

Reactor Materials and Engineering

Theos Jardin Thompson, 1918–1970

The Technology of Nuclear Reactor Safety

VOLUME 2

Reactor Materials and Engineering

EDITORS

T. J.Thompson

J. G. Beckerley

Prepared under the auspices of the
Division of Technical Information
U. S. Atomic Energy Commission

THE M.I.T. PRESS
Massachusetts Institute of Technology
Cambridge, Massachusetts

The Technology of Nuclear Reactor Safety

VOLUME 1

Reactor Physics and Control

VOLUME 2

Reactor Materials and Engineering

ISBN 0 262 20005 8 (hardcover)

Library of Congress catalog card number: 64-24957

Manufactured in the United States of America

Foreword

The period since publication of the first volume of *The Technology of Nuclear Reactor Safety* has seen the construction of many newer and larger nuclear power plants. The impressive safety record of nuclear reactors has continued to grow. It is a record that has been achieved through the efforts of many individuals and groups in industry, government, the national laboratories, and the universities.

One individual who devoted an appreciable fraction of his life to maintaining and improving nuclear reactor safety was Dr. Theos J. Thompson. As head of the reactor project at the Massachusetts Institute of Technology he led that group in more than a decade of valuable accomplishments in nuclear research and engineering. In spite of the burdens of research and teaching, Tommy — as we all knew him — was able to participate in and lead the Advisory Committee on Reactor Safeguards in its important evaluation and promotion of nuclear reactor safety. He continually supported efforts to develop new safety concepts, to perform safety-related experiments, to investigate basic phenomena relevant to reactor safety, and to apply the best scientific and engineering talents to nuclear power reactor design.

For eighteen months we were fortunate to have him with us as a member of the Atomic Energy Commission. His unusual professional competence and dedicated enthusiasm were valuable resources that we drew on. A few months before his untimely death, Tommy completed his work on the final chapter of this volume. It was a long and earnest labor that involved sacrifices of what might have been "spare" time. But he wanted it that way.

As nuclear reactors today and tomorrow generate the energy that enriches our lives, we can be grateful that they perform without danger to plant or public, and we can be grateful for the efforts and leadership of Dr. Thompson and his colleagues in making it possible.

Glenn T. Seaborg

Chairman
U.S. Atomic Energy Commission
1961-1971

Dedication

This volume is dedicated to the memory of Dr. Leslie L. Silverman and Dr. C. Rogers McCullough. Both participated in the planning and writing of this volume. Both were active in the statutory Advisory Committee on Reactor Safeguards, each serving as chairman during the crucial initial period of the national nuclear power program. Both were responsible for developing technical concepts that have contributed, and are continuing to contribute, to the remarkable safety record of nuclear reactors in the United States.

Their many friends, including the authors of this volume and its antecedent volume, will always remember and cherish the balanced judgment, depth of intelligence, sense of humor, honesty, and fairness of these two good men.

Table of Contents

VOLUME 2

Preface

A full account of the history of the project called SIFTOR (Safety Information for the Technology Of Reactors) is given in the preface to Volume 1. Briefly, the project evolved from a suggestion by the Atomic Energy Commission Advisory Committee on Reactor Safeguards that a substantial effort be made to compile the basic body of reactor safety information, including evaluations and generalizations by qualified experts. Following informal discussion within the AEC staff and the Advisory Committee, a tentative outline was developed to identify what such a book might contain. Subsequently, the Massachusetts Institute of Technology agreed to serve as coordinator for the project under contract with the Atomic Energy Commission.

The record of the SIFTOR project is contained in the two volumes of *The Technology of Nuclear Reactor Safety.* The first volume, subtitled *Reactor Physics and Control,* was published late in 1964. With the publication of this volume, subtitled *Reactor Materials and Engineering,* the project is concluded.

Many circumstances, largely beyond the control of the editors, have delayed publication of this second volume. Although most of the chapters presented here were first written in 1963 and 1964, the material has been updated by the authors. Revisions have been made, new material added, and sections inserted to include significant advances made in recent years. Final changes in some chapters were made in the latter half of 1970. The result of the updating process is some variability in coverage from chapter to chapter.

In the intervening years since the publication of Volume 1, there have been substantial advances relevant to reactor safety in the fields of metallurgy, mechanical design, containment, chemical reactions, and heat transfer. The delay in publishing Volume 2 has made it possible to take into account the impact of these new developments on reactor safety. This should increase the useful life of the second volume. At the same time, we must remark, the usefulness of the first volume remains essentially undiminished, since the rate of progress in the technical areas it covers has been relatively slower.

The purpose, intended audience, and scope of *The Technology of Nuclear Reactor Safety* are discussed at the end of Chapter 1 in the first volume. As noted there, the books have been written to assemble in an organized fashion the essence of the safety information concerning reactor technology that has been built up over the past two decades. A second purpose is to provide the reader with an insight into the safety problems that need further investigation.

The volumes are intended for the technically trained who have a basic knowledge of science and engineering. For the expert in a given area, the chapters covering his area of competence should serve as a review and should include some information or viewpoints new to him; the chapters covering other areas should be valuable to him in understanding other aspects of reactor safety that might be related to his own specialized interests.

We cannot emphasize too strongly that many disciplines and many technologies must be drawn on to design and construct safe nuclear reactors and to operate them in a safe manner. For example, heat-transfer systems have materials that can be corroded, transformed by nuclear radiations, mechanically assaulted, metallurgically altered, and so on, so that the design, construction, and operation of heat-transfer systems for a safe nuclear reactor involve many disciplines, including mechanical and chemical engineering, metallurgy, chemistry, and solid-state physics. The SIFTOR books are intended to encourage "cross fertilization" between fields not for academic reasons but because the safety of nuclear reactors *requires* close cooperation and understanding between individuals trained and experienced in many specialties.

In fact, we would have liked to emphasize the interrelationship between disciplines by putting all the chapters into a single volume. Publication in two volumes has the distinct disadvantage of reducing the sense of the interplay between reactor physics and reactor engineering. Consequently, we hope that those who use this volume will also use the first volume and vice versa. To encourage this, we have included cross references to Volume 1 throughout, and we have added as an appendix a complete table of the contents of Volume 1.

The editors wish to express once again their appreciation to those listed on the Acknowledgments page of Volume 1. In addition, we wish to acknowledge the patient skill and help of Mrs. Rae Visminas Driscoll and Miss Rita Falco, who not only made legible much of our scribbling but also maintained the continuity of the project during the production of the two volumes.

T. J. Thompson
J. G. Beckerley

CHAPTER 12

Materials and Metallurgy

THOMAS O. ZIEBOLD*
Massachusetts Institute of Technology, Cambridge, Massachusetts
FRANK G. FOOTE and KARL F. SMITH
Argonne National Laboratory, Argonne, Illinois

CHAPTER CONTENTS

1 INTRODUCTION

Engineering designs are of little use until they are translated into working hardware. Materials engineers—metallurgists and ceramists—participate in this effort by providing designers with information about the behavior of materials in service and by establishing processes for manufacturing hardware which will reliably perform its intended functions. The technology of nuclear reactor systems, as indeed most other modern technology, continually demands not only a search for superior materials when the older ones are found lacking but also improved manufacturing processes and quality control as the established materials are worked harder — that is, they are used under conditions more closely approaching the possible limits. To accommodate this demand, materials engineering ranges from the laboratory development of new alloys and ceramics, based on fundamental knowledge of the properties of solids, to the mill or shop control of product quality, based in many cases on long-established arts.

This chapter focuses attention on the properties of the more commonly used reactor materials and the processes used to fabricate nuclear system components. The text is intended to introduce nuclear engineers with essentially no training in metallurgy to some of the important aspects of this field. Because it is concerned with real materials, whose internal structure is often subject to the flirtatious quirks of nature, metallurgy is largely a descriptive and empirical discipline. In the space of one chapter it is not possible to do much more than present basic terms and concepts. It is the purpose of this chapter, then, to provide the reader with sufficient background knowledge to understand the importance of certain materials properties in the design, construction, and operation of reliable nuclear reactors and to evaluate future developments of new materials and processes.

The properties and fabrication of nuclear fuel materials are discussed in Chapter 13 and are not, therefore, included in this chapter.

1.1 Outline of the Chapter

The text has been organized to present a background of physical metallurgy, a description of engineering materials and manufacturing processes, and some experiences with specific reactor

*Present address: Neutron Products, Inc., Dickerson, Maryland.

components. Section 2 presents several topics from the field of physical and mechanical metallurgy which are important to the design of reliable and safe reactor components. This section is essentially a descriptive outline which points out how materials will behave when subject to stress, high temperatures, and chemically active environments. The illustrations have been selected to show the properties of common reactor materials, but these examples must not be considered sufficiently accurate as to be substitutes for handbook data. For a more detailed study of the subjects presented here, the reader is referred to standard textbooks [1-6].

The brief description of reactor materials given in Sec. 3 is intended to familiarize the reader with characteristic properties of these materials. Further information may be found in handbooks [7-10] and monographs [11-19]. The review of process principles given in Sec. 4 should give the reader an insight into the sources of possible manufacturing defects or deviations from expected mechanical properties. The means for controlling product quality are discussed in Sec. 5.

1.2 Materials Requirements for Nuclear Systems

There are several reasons why nuclear reactor systems deserve more care at all stages of design, fabrication, assembly, and operation than non-nuclear systems. The principal differences can be traced to one source—the existence of radioactive materials in the system.

A major failure of a primary reactor system could conceivably release considerable quantities of radioactive substances to the general public. For this reason, the potential consequences of material failures in a nuclear reactor plant are vastly more serious than similar failures in non-radioactive systems. The three principal barriers between the radioactive source and the outside world are the fuel-element cladding, the primary pressure boundary, and the containment shell. The performance of materials used in these components obviously bears directly on the safety of the reactor plant. Equally important, however, are the materials which constitute the core supports, the control rods, and the rod drive mechanisms.

In addition to requiring containment, radioactivity in the reactor makes it difficult, if not impossible, and costly to inspect and service many of the system components. Consequently, these components must be reliable and safe from the start of operation to the end of their useful life, and this demands the highest possible assurance of quality in design and manufacture. There is no chance for a trial and error selection of materials unless the customer is willing to pay for this with full-scale test assemblies whose failure may be safely accommodated.

Finally, the presence of radiation may alter the properties of materials used in the construction of reactors. This problem has been given particular attention, of course, in the development of nuclear fuel materials, but it is also of importance in the design of all materials which may encounter high radiation fields. Fuel assemblies remain in the reactor for only a relatively short time (a few years), whereas structural components such as core supports and pressure vessels may be exposed to radiation for periods as long as forty years or more. Although the stationary fixtures are in lower radiation fields than the fuel elements, designers must be concerned with effects in these long-life components, since radiation damage generally is a cumulative phenomenon.

Some comments may be made upon the importance of various sections of the reactor plant in regard to materials selection and utilization. The failure of a major component in the primary system, such as a major pipe break or rupture of a bolted closure, will result at the very least in a serious economic loss to the plant. At worst, it is conceivable that radioactivity could be released to the public unless proper precautions are taken. In a similar way, it is conceivable that failure of important components in the control system (the control rods, control-rod channels, and drive mechanisms) could result in the melting of the fuel and release of fission products. It is evident, therefore, that the primary system and control mechanisms are vital to reactor economics and to reactor safety. Other components, such as primary and secondary heat-exchangers and minor systems not directly associated with the primary system, may be less vital to reactor safety, but their malfunction could result in serious economic losses. It is still important to reactor safety, however, to ensure that the failure of some remote secondary system component cannot react back on the primary system and cause a consequential primary system failure.

That accidents such as a major failure of materials somewhere within the primary coolant system can occur is illustrated by the examples given in Chapter 11 of Volume 1, "Accidents and Destructive Tests," and Chapter 20 of this volume. It is also evident from those chapters that failure of certain materials, particularly in control rod mechanisms or other vital parts, can, by itself, lead to nuclear excursions. The necessity for carefully selecting, specifying, fabricating, inspecting, and utilizing materials cannot be overemphasized. Some problems such as radiation damage are unique to reactor applications, while others are of a more conventional nature, similar to those met in many other applications. It is essential that reactor designers consider both the conventional aspects and the unique nuclear aspects involved in the choice of materials. They should never devote so much attention to the materials problems associated with high radiation fields that they neglect the problems associated with high temperatures and great stress, for example.

Materials failures may result from an improper choice of materials for the specified use and condition; they may be caused by lack of proper specifications; they may result from undetected defects such as cracks, inclusions, and voids; they may result from changes in prescribed production practice that can introduce important variations from properties given in handbooks or technical specifications; or they may result from use in a hostile environment not envisioned by the designer. Ensuring that materials are utilized properly in nuclear reactors is a ceaseless battle that starts even before a single blueprint is prepared and is never finished so long as the reactor continues to operate. The reactor designer must select the proper material for each specified use and environmental condition. He must make use of detailed available data on the materials being considered to make the proper initial design choices. If such information does not exist, and the material seems promising, he must experimentally develop the necessary data. He must specify the material, its preparation, and its fabrication methods. He dare not neglect any part of these specifications. Use of published physical and mechanical data without regard to the variables of fabrication and heat treatment of parts to be used is an incorrect and even dangerous procedure, and the original specifications must ensure not only that the correct material is selected but also that it is properly prepared and fabricated. Someone must make certain that materials are carefully supervised and inspected at each step of their preparation and that no inadvertent substitutions occur.

In the sections which follow, it is pointed out that rigid acceptance tests are a vitally important part of the procurement procedures and that specifications are of little value without them. Sometimes, a supplier of materials, and sometimes even the designers themselves, in order to save time or money, will cut corners on the rigid preparation or inspection of the materials in question. Particularly, the supplier may not appreciate the significance of the specifications and may believe that they are more rigid than necessary. A number of cases can be cited where combinations of poor materials below specification coupled with lack of rigid acceptance tests have led to millions of dollars of loss and lengthy delays. Even after construction is complete and the reactor has started operation, surveillance of the materials must continue. Extensive inspection of accessible reactor-system components is required at frequent intervals. Appropriate control of the chemical composition and the environment of the system is essential at all times in order to ensure that the materials behave as anticipated. Vigilance can never cease so long as the reactor remains in operation.

1.3 Environments within a Reactor System

To set the stage for the subsequent discussion of materials and their properties, some of the important environmental conditions in typical power reactors are presented in this section.

Service requirements in a primary reactor system cover a wide spectrum. Because it is desirable to operate the reactor at a high temperature to obtain the best possible thermal efficiency, construction materials should maintain their strength, resistance to creep, ductility, corrosion resistance, and so on, at as high temperatures as possible. Coolant temperatures for many reactors have been tabulated in Chapter 13 and in Appendix 1 of Volume 1. For water-cooled reactors the range of coolant temperatures currently in use in 500 to 650°F (260 to350°C). The coolant temperature in sodium-cooled systems may range up to 1200°F (650°C) and in gas-cooled reactors it may reach 1700°F (927°C). In all these systems it is the diminishing strength of materials at elevated temperatures which to a large extent establishes the upper limit of coolant temperature.

If mechanical strength at high temperature were the only requirement, however, materials design would be fairly simple. But this is not the case by any means. Materials utilized in the reactor core must not only have suitable strength and ductility, resistance to creep and erosion or corrosion by the coolant, fabricability, low cost, and the other more or less conventional characteristics, but they must also have appropriate nuclear characteristics. For example, materials comprising the fuel diluent, fuel cladding, and core structural members must have a low neutron-absorption cross section to permit economic use of neutrons generated by fission. Stainless steel must be ranked below Zircaloy from the neutron economy viewpoint, but Zircaloy costs roughly ten times as much per pound as stainless steel. The choice of one or the other, where a choice is possible, has not always been straightforward.

Selection of materials to be incorporated in or near the reactor core is complicated further by the problems of radiation damage–that is, the alteration of materials properties by radiation bombardment, especially high-energy neutrons. Contemporary power reactors operate with neutron fluxes that are typically in the range of 10^{13} to 10^{14} neutrons/cm^2-sec, and core materials (i.e., materials close to the fuel) will see 10^{22} neutrons/cm^2 in a few years. Projected fast reactors may produce even higher fluxes–up to 10^{15} neutrons/cm^2-sec or more. The permanent damage introduced by fast neutrons in most nonfueled, structural metals is lessened when these materials are used at high temperatures, but the changes may be pronounced for other materials or for components operated at lower temperatures. Graphite, for example, irradiated below about 400°F (200°C) will show substantial anisotropic changes in dimension and in thermal and electrical conductivity and will retain a significant degree of radiation damage as stored internal energy. Irradiation of metals or ceramics containing fissionable isotopes causes a gross disruption of the internal structure of these materials which may lead to swelling or cracking and emission of highly radioactive fission products. Radiation damage significantly alters the brittle-fracture characteristics of pressure-vessel steels, and this subject will be discussed in some detail in this chapter.

Outside the reactor core the service demands on materials are more conventional but still stringent. The requirements for components of a control rod drive mechanism will be similar to other such systems operating in comparable environments. But since the drive mechanism must maintain close dimensional tolerances in order to keep the control rod accurately aligned and positioned, it is necessary that the materials used be resistant to creep, have considerable structural strength, and have sufficient hardness to resist wear at all bearing surfaces. These requirements have resulted in the selection of 17-4PH steel for some rod mechanism parts. It should be noted, however, that this alloy, unless properly heat-treated, is excessively brittle and susceptible to stress corrosion. Failure of 17-4PH components has been encountered in an operating reactor, and this example underscores once more the importance (and difficulty) of controlling materials design and fabriction.

The design and selection of materials for heavy-walled pressure vessels is a field unto itself. It is apparent that the overall unit cost of nuclear power (mills/kw-hr) decreases as the plant capacity increases. Since the power output per unit volume of fuel is limited, generally by the properties of the core materials, an increase in total power can be achieved only by increasing the physical size of the plant. Consequently, there is a continuing demand from reactor engineers for the fabrication of larger and larger pressure vessels. At the present time, American manufacturers are capable of fabricating heavy-walled vessels as large as 30 ft (9.1 m) in diameter, 125 ft (38 m) in length, and as heavy as 1000 tons (450 metric tons) [20]. Fabrication of these massive units involves welding and radiographic inspection of sections as much as a foot or moore in thickness, and the entire vessel must be stress relieved. Field erection of these vessels, as compared with shop fabrication, presents even greater problems. Because of the obvious importance of the integrity of reactor pressure vessels, the properties of boiler steels are discussed in this chapter, and an entire section in Chapter 14 is devoted to the mechanical design of these units.

In concluding these introductory remarks on materials selection, one further comment is in order. All materials within a closed system must be mutually compatible—a choice of one material may well preclude use of another. For example, the presence of copper in an all-aluminum system leads to excessive corrosion by hot water. The presence of even a few parts per million of chloride ion in reactor waters can set up the condition for stress corrosion cracking of stainless steel. Fluorides in small concentrations are detrimental to Zircaloy. The use of certain fluids in a system may be limited by the choice of solid materials or by the choice of other fluids. A serious accident occurred in the Sodium Reactor Experiment (SRE) because an auxiliary hydrocarbon coolant leaked into the primary system (see Chapter 11). A thorough investigation of the compatibility of all materials—solids, liquids, and gases—which will be utilized in any given system must be carried out before final selections are made.

2 ENGINEERING PROPERTIES OF MATERIALS

Modern metallurgical science encompasses many concepts from the fields of physics, chemistry, and mechanical engineering. The characteristics of materials which may be considered as "engineering properties" from the viewpoint of nuclear reactor design are the phenomena of plastic flow, fracture, creep and fatigue, solid-state reactions, and corrosion. In this section each of these topics will be considered, but the discussion will be mainly limited to a summary of observable, macroscopic behavior. It is not the purpose here to review what is known or speculated about the motions of atoms in a metal that are the origins of the gross behavior. Nevertheless, certain atomic concepts will be introduced when they are essential to the understanding of a particular phenomenon.

2.1 Structure-Sensitive Properties

It must be realized at the outset that the properties to be discussed here are all more or less dependent on the metallurgical structure of the material. The term "metallurgical structure" includes the overall composition, the distribution of chemical constituents among various phases in the metal, the crystal structure of the phases, and the size and orientation of the metal grains. Since the metallurgical structure generally depends on the entire history of a metal from the moment it was solidified from a melt, it can be appreciated that a complete characterization of its properties is complicated.

Metals are crystalline solids, but engineering materials never consist of a single, perfect crystal. In the process of solidification, embryonic crystals of the solid metal form at many places in the melt and proceed to grow as more and more of the liquid freezes. When solidification is complete, the metal consists of an aggregate of crystalline grains, each of which is a single crystal, but the crystal lattices are randomly orientated in space. Most of the common engineering metals crystallize to one (or more, depending on the temperature) of the structures illustrated in Fig. 2-1. The networks of surfaces forming interfaces between the grains called grain boundaries, and the grain boundary properties are often as important to the gross properties of the metal as are the properties of the crystalline grains themselves.

The microstructures of metals can be conveniently classified as single phase or multiphase. In single phase pure metals and alloys are the grains have essentially the same chemical composition and crystal structure. Multiphase alloys consist of two or more phases—regions of different chemical composition or crystal structure. In multiphase alloys the "grains" may be single phase or multiphase. The shapes and distributions of metal phases which have been observed are so numerous that it is impossible to classify them in any general manner. Commonly encountered details are the "lamellar" structure (two phases existing as thin platelets so that each phase is contained in every other layer) and the structure showing a precipitated second phase (the precipitate particles are usually much smaller than the matrix grains and may be distributed intragranularly or concentrated in the grain boundaries).

Microstructures may be characterized in another way as "cast," "worked," or "heat treated." A cast microstructure, as its name implies, is the structure which results from solidification of a melt. Depending on the rate of solidification, the size of the casting, and the degree of mixing while freezing, cast structures will show a variable grain size, variable grain shape, different degrees of pre-

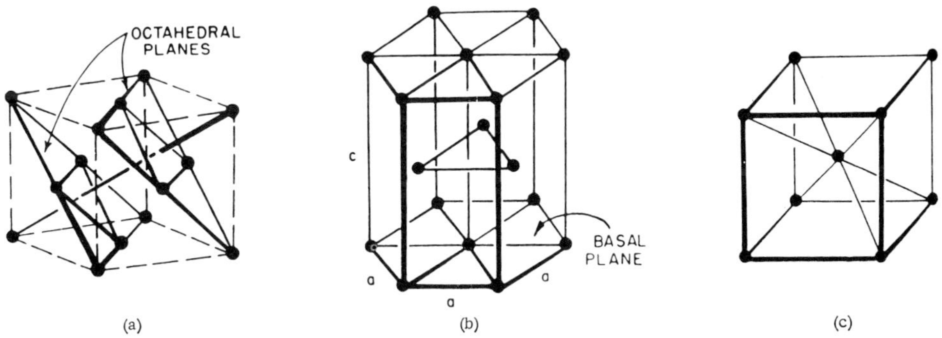

FIG. 2-1 The common crystal structures of metals. (a) Face-centered cubic; (b) close-packed hexagonal; (c) body-centered cubic. (From [141].)

ferred grain orientation, and chemical inhomogeneities within and between grains. If a cast metal is deformed by working (forging, rolling, drawing, etc.), the grain structure will be severely broken up and distorted. Furthermore, if the working produces a primarily unidirectional change in dimensions (as in rolling or drawing, for example), the resulting microstructure may exhibit a preferred orientation wherein most of the grains are aligned so that a certain crystallographic direction lies parallel to the primary working axis. A metal in this condition is said to be "textured" and may exhibit some anisotropy in its gross properties.

The process of heat treating involves holding a metal at some specified high temperature for a certain length of time and then cooling it at a controlled rate. Heat treatments are used to alter the microstructure of a metal so as to obtain a desired grain shape and size or a desired phase distribution, to promote or inhibit the formation of certain phases, and to remove chemical inhomogeneities, internal stresses, and grain textures. A few of the more important heat treatments for reactor materials are discussed in this chapter. A condition which is common to many alloys and which is a starting point for discussing mechanical properties, is the "annealed" structure. Annealing is a somewhat vague term, but it generally implies that a metal has been held at a sufficiently high temperature (somewhere above one-half the melting temperature on the absolute scale) for a long enough time to remove essentially any vestige of prior fabrication history. A single-phase metal in the annealed or "recrystallized" condition will normally consist of large grains of uniform size and composition, random orientation, and essentially equal dimensions in all directions ("equiaxed"). A well-annealed alloy, single-phase or multiphase, is in its softest, or weakest, condition, and is, for all intents and purposes, stable against any further metallurgical changes which might be thermally induced during service.

2.2 Plasticity of Metals

When loaded to sufficiently high stresses, metals will flow plastically and sustain permanent, irreversible strains. At sufficiently low stresses, they will behave elastically—any induced strain will disappear when the load is removed. Roughly stated, the "yield point" delineates the transition from elastic to plastic deformation in the uniaxial tensile test, and this transition property is normally taken as the basis for defining the permissible mechanical loading of metal parts.

The process of plastic flow after the onset of yielding is of interest to metallurgists in deciding how to form metals into desired shapes (as well as providing information about the basic nature of metal plasticity). It is also a region of concern to the design engineer who must know how a metal part will behave in the event that local stresses exceed the yield strength. Metals are said to be "ductile" when they can sustain substantial plastic strain before fracturing. This is manifested by a large elongation or reduction in cross-sectional area in the uniaxial tensile test. Ductility is as important as strength in the selection of materials for load-bearing structural members. Materials which show both high strength and suitable ductility are "tough" metals. "Toughness" reflects the ability of a metal to absorb mechanical energy in the form of plastic strain before it fractures and is indicated by the area beneath a stress-strain curve or by a high energy absorption in an impact test. The concepts of strength, ductility, and toughness are discussed further in the following sections.

2.2.1 The Stress-Strain Curve

The uniaxial tensile test is a common measure of the strength and ductility of metals. In this test a carefully machined bar or rod is pulled to failure at a constant rate of elongation (usually on the order of 10^{-4} to 10^{-2} inches elongation per inch of gage length per second), and the tensile load is recorded as a function of the elongation. "Engineering stress" is defined as the ratio of load on the sample, P, to the initial cross-sectional area, A_0:

$$\sigma_E = P/A_0 \, , \qquad (2\text{-}1)$$

and "engineering strain" is defined as the ratio of elongation or length change, $\Delta\ell$, to the initial length:

$$\epsilon_E = \Delta\ell/\ell_0 . \qquad (2\text{-}2)$$

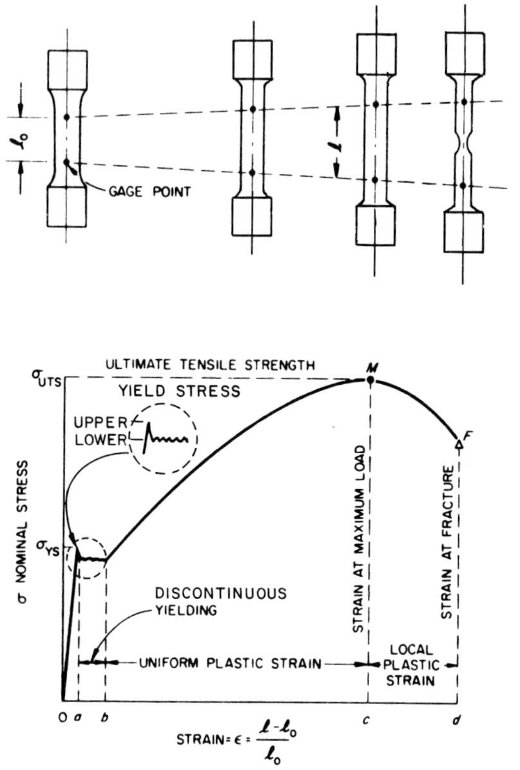

FIG. 2-2 Schematic engineering stress-strain diagram. (After Wechsler [26].)

The behavior of a typical metal during the tensile test is illustrated schematically in Fig. 2-2. At low strains the stress is linearly proportional to the strain, and the sample would return to its original length if unloaded. This is the region of elastic behavior which follows Hooke's law:

$$\epsilon_E = \sigma_E/E,$$

where E is Young's modulus (on the order of 10^6 to 10^7 lb/in.^{2} for most metals). Beyond the "elastic limit" the sample deforms plastically, a permanent strain is induced, and the stress-strain relation is nonlinear. For materials which do not show a sharp transition from elastic to plastic behavior, the "yield strength" is taken as the stress at which the permanent strain (or "offset") reaches a specified value, usually 0.2%.

As a metal is strained beyond the yield point, an increasing stress is required to produce additional plastic strain—that is, the sample becomes effectively stronger as plastic strain increases. This behavior is called "work hardening" or "strain hardening." The point of maximum load on the engineering stress—engineering strain curve is the "ultimate tensile strength," which corresponds to the end of "uniform elongation" of the sample. Beyond this point the sample begins to neck down; the rapidly decreasing cross-sectional area causes a drop in the test load although the actual stress in the necked area continues to increase. Two final definitions of properties taken from the stress-strain curve are the "total elongation," which is the engineering strain at fracture (note the distinction between uniform and total elongation), and the "reduction in area," which expresses the ratio of the cross-sectional area at fracture to the initial cross-sectional area.

Another representation of stress-strain relations is the "true stress-true strain" curve. True stress and true strain are defined in terms of the actual dimensions of the specimen during the tensile test rather than the initial dimensions. The true stress is the ratio of instantaneous load to instantaneous cross-sectional area:

$$\sigma = P/A_i = \sigma_E(A_0/A_i). \quad (2\text{-}3)$$

True strain is defined as the integral of the ratio of an incremental change in length to the instantaneous length (measured in an unstressed state):

$$\epsilon = \int_{\ell_0}^{\ell_i} \frac{d\ell}{\ell} = \log_e(\ell_i/\ell_0) = \log_e(1 + \epsilon_E). \quad (2\text{-}4)$$

After necking begins, the total elongation is no longer a valid measure of the true strain. The true strain during necking may be taken from the instantaneous cross-sectional area:

$$\epsilon = \log_e(A_0/A_i). \quad (2\text{-}5)$$

For many metals the true stress-true strain curve during work hardening may be expressed as

$$\sigma = K\epsilon^n, \quad (2\text{-}6)$$

where K is a constant and n, the "work-hardening exponent," is always less than unity. In terms of this relation, the onset of necking or plastic instability occurs when the true strain is numerically equal to the work-hardening exponent (ϵ = n).

Engineering and true stress-strain curves for representative reactor structural materials are given in Figs. 2-3 and 2-4. These were selected to illustrate the typical behavior of ferritic steels

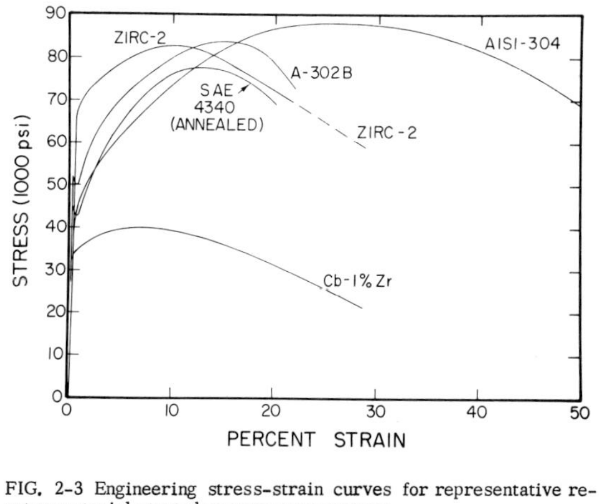

FIG. 2-3 Engineering stress-strain curves for representative reactor materials tested at room temperature.

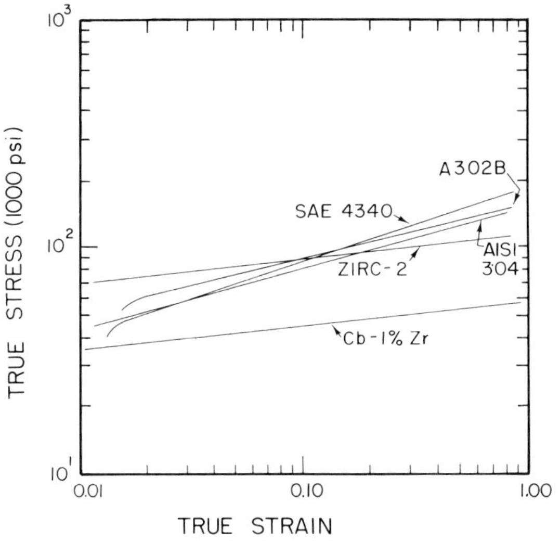

FIG. 2-4 True stress-strain curves for representative reactor materials tested at room temperature.

(ASTM-A302B, SAE 4340), austenitic stainless steels (AISI 304), a zirconium alloy (Zircaloy-2), and a refractory alloy (Cb-1%Zr). These figures reveal two distinctly different modes of yielding. The transition from elastic to plastic strain is rather gradual and indistinct for the austenitic and nonferrous alloys. In marked contrast, the ferritic steels exhibit a distinct "yield drop," and one must distinguish between the "upper yield point" and the "lower yield point." Immediately after yielding, the plastic elongation of the ferritic steels continues with no increase in stress (over a limited range of strain). That is, the metal may be strained to a certain point before work hardening begins. This stage in the process is accompanied by "inhomogeneous" or "discontinuous" yielding as evidenced by local deformation of the sample in the form of bands ("Luders bands") which propagate down the gage section. Note also that the work-hardening characteristics shown in Fig. 2-4 are consistent with Eq. (2-6).

In the following subsections the idiosyncrasies of plastic flow and fracture are discussed in more detail.

2.2.2 The Yield Strength

At room temperature, and slightly above, plastic flow proceeds by "slip" or "twinning" within the metal grains. Slip is a relative sliding of parallel atom planes which results in a relative displacement of the crystal lattice above and below the "slip plane." Twinning is a homogeneous shear which reorients the deformed lattice into a mirror image of the parent lattice across the "twinning plane." Both slip and twinning take place only on certain crystallographic planes and in particular crystallographic directions which are characteristics of the crystal structure. For either process to occur, therefore, the applied load must produce a shear stress component within the metal grain which lies in the slip (or twinning) plane and is parallel with the slip (or twinning) direction, and whose magnitude exceeds some critical value. This "critical resolved shear stress" is determined by uniaxial tensile testing of single crystals.

Slip never occurs by the simultaneous sliding of all the atoms on the slip plane; the shear stress required for this would be enormous. Instead, it is known that slip progresses atom-by-atom across the slip plane, and at any instant of time there is a boundary line between those atoms which have been displaced and those which have not. This line defect is called a "dislocation." Without going into the well-established principles of dislocation behavior, it may be simply stated that the mechanical properties of metals are governed by the elastic properties associated with dislocations and the way in which dislocations interact with each other, with grain boundaries, with impurities, and with other defects as they move through the crystal lattice.

Dislocations are present in all crystals. Ordinary cast and annealed metals will contain on the order of 10^8 cm of dislocation lines per cubic centimeter of material. When a stress is applied, the shear component in a slip plane causes dislocations to glide along the plane, and the passage of a single dislocation results in a unit relative displacement of the crystal above and below the slip plane. The magnitude of the unit slip vector, called the "Burgers vector," is on the order of the interatomic spacing of the atoms in the crystal lattice, and many dislocations must pass across the slip plane before the plastic deformation is observable in any macroscopic sense. Dislocations move relatively easily through perfect crystals, but any perturbation of the crystal by other dislocations, impurity atoms, point defects (vacancies and interstitials), or grosser defects (clusters of point defects, phase boundaries, and grain boundaries) will impede their motion and cause an increase in the resolved shear stress required for slip.

The critical resolved shear stress is experimentally related to the square root of the dislocation density. It is concluded that work hardening, which increases the shear stress necessary to move dislocations, is related in some way to an increase in the number of dislocations caused by plastic strain and their subsequent interactions. That dislocations will multiply under the influence of an applied stress stems from the presence in all metals of "dislocation sources." These sources are themselves dislocations which are fixed in such a way that an applied shear stress of sufficient magnitude will cause them to generate new dislocations. Cold working of metals increases the dislocation density by many orders of magnitude through operation of dislocation sources.

In polycrystalline metals, plastic deformation, which begins as slip within the metal grains, must be transmitted across the grain boundaries. Experimentally, the dependence of the lower yield stress, σ_y, on grain size is known to follow the Petch equation:

$$\sigma_y = \sigma_i + k/\sqrt{d}\,, \tag{2-7}$$

where d is the mean grain diameter. The empirical constants σ_i (the "friction stress") and k (the "source constant") are related respectively to the stress needed to move a dislocation through the lattice and the stress required so that slip in one grain will cause a multiplication in the number of dislocations present in adjacent grains.

In engineering structures the applied loads generally produce a complex stress state. Regardless of the complexity, any set of combined stresses may be resolved to three mutually perpendicular, normal stresses. These "principal stresses," σ_1, σ_2, and σ_3*, result in shear stresses of magnitudes $(\sigma_1 - \sigma_2)/2$, $(\sigma_2 - \sigma_3)/2$, and $(\sigma_3 - \sigma_1)/2$. A widely accepted criterion for yielding under the action of combined stresses is due to Von Mises. By this criterion it is postulated that the "effective stress" is

$$\bar{\sigma} = \left[\tfrac{1}{2}(\sigma_1 - \sigma_2)^2 + \tfrac{1}{2}(\sigma_2 - \sigma_3)^2 + \tfrac{1}{2}(\sigma_3 - \sigma_1)^2\right]^{1/2} \quad (2\text{-}8)$$

and that a randomly oriented polycrystalline metal will yield when the effective stress is equal to the uniaxial tensile yield stress ($\bar{\sigma} = \sigma_y$). Alternatively, the Tresca criterion bases design stresses on the maximum shear stress, which is the difference between the largest and smallest principal normal stresses. (The Tresca criterion is discussed in Chapter 14.) Note that by either criterion the effective stress depends only on the magnitudes of the shear stresses. It follows from the preceding discussion that there can be no plastic flow unless at least one of the shear stresses is nonzero. Hydrostatic loading of a metal, for which $\sigma_1 = \sigma_2 = \sigma_3$, will not produce plastic deformation.

The yield strength is a structure-sensitive property. The addition of alloying constituents can alter yield behavior in several ways. A sharp yield drop is thought to result from the "pinning" of dislocations by foreign atoms in the crystal lattice. Dislocations are favored residence sites for foreign atoms because the foreign atom, being somewhat larger or smaller than the base metal atom, can relax to some degree the lattice strain associated with the dislocation defect. When the alloying atoms "condense" upon dislocations, therefore, the strain energy of the crystal is reduced, and this configuration, called a "Cottrell atmosphere," is thermodynamically more stable than if the foreign atoms were randomly distributed throughout the crystal. The result is an increased yield strength, since a greater resolved shear stress is required to tear the dislocations away from the condensed atmosphere. The yield drop is observed because the "unpinned" dislocations can move and multiply at a lower shear stress than that initially required to free them from their atmospheres.

Alloying may also introduce new phases which impede dislocation movement and thereby increase the yield strength, and the distribution of phases in a multiphase alloy will have a pronounced effect on the yield strength. "Dispersion-hardened" alloys contain a second phase in the form of small particles more or less uniformly distributed through the matrix phase. For example, SAP (sintered aluminum powder) is aluminum which has been dispersion hardened by the addition of aluminum oxide particles. "Precipitation-hardened" alloys are similar, but here the dispersed phase is produced by solid-state transformation—precipitation caused by cooling from a single-phase solid solution. "Martensitic" steels are a further example of hardening by phase transformation (see later discussion). In fact, the control of the yield strength through composition involves so many possibilities that the development of engineering alloys is often a trial-and-error procedure based on past experience. (This is referred to by the uninitiated as "black art.")

For an alloy of fixed composition the yield strength is further dependent on the microstructure. Grain-size dependence is indicated by Eq. (2-7)—"refining" (reducing) the grain size will increase the yield strength. The properties of the grain boundaries themselves will also have an effect. They may be embrittled by grain-boundary precipitates and cause the material to fail "intergranularly" rather than "transgranularly." The grain-boundary strength also depends on temperature, as discussed below. The importance of grain orientation may be understood from the previous discussion of slip. If a large fraction of the grains of a metal are preferentially aligned to a certain crystallographic direction, the magnitude of the resolved shear stress on a particular slip system will depend on the direction of the tensile axis with respect to the texture axis, and the yield strength will therefore depend on the direction of testing. Cold-rolled metals may exhibit a substantially different yield strength in tensile samples cut parallel (longitudinally) or transversely to the rolling direction.

Finally, the yield strength depends on the tensile test conditions, specifically test temperature, strain rate, and sample geometry. Figure 2-5 shows the manner in which the yield strength changes as the test temperature is increased. The materials illustrated range from commerically pure aluminum (1100 Al) to a truly refractory metal (tungsten). It is immediately obvious that each alloy has a temperature limitation on its useful strength and that, conversely, metals can be readily worked if heated to a sufficiently high temperature. These observations are clarified when it is realized that the rate of work hardening also decreases markedly as the temperature increases.

The mode of failure of metals at high temperature is qualitatively different from that at room temperature. The movement of atoms becomes progressively easier as the temperature is raised; dislocations become quite mobile, and grain boundaries are able to slide. Even when loaded to stresses below the yield strength, metals will flow plastically by "creep" at temperatures around one-half the melting temperature (on the absolute scale) or above.* This process is time-dependent—

*By convention, magnitudes are assigned so that $|\sigma_1| > |\sigma_2| > |\sigma_3|$

*As with most "rules of thumb," this one has its limitations. Zirconium, for example, shows significant creep at temperatures well below one-half its absolute melting point. (See Fig. 2-8.)

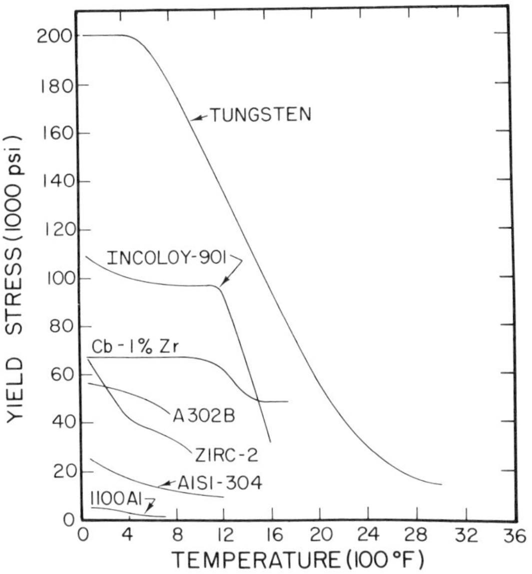

FIG. 2-5 Yield strength of representative materials tested at elevated temperatures.

the material continues to deform with time although the stress is held constant.

The mechanical behavior of metals at low temperatures is again a different story. In Fig. 2-6 the low temperature yield strengths are given for several alloys. Some metals (ferritic steels, for example) exhibit a rapidly increasing yield strength as the temperature is lowered. At some sufficiently low temperature the yield strength equals the ultimate strength, and the sample will fracture before observable (in the usual sense) plastic deformation can occur. In "brittle failure" the material breaks by cleavage rather than by ductile shear. (See Sec. 2.3.)

The yield strength at any temperature depends on the rate of straining. Increasing the strain rate is analogous to decreasing the test temperature. Body-centered cubic metals, which characteristically have a strongly temperature-dependent yield strength, show a significant effect of strain rate. This is particularly so at low temperatures where the change in yield strength with change in temperature is most pronounced. (See Fig. 2-6.) From the standpoint of engineering design, the implication of strain-rate dependence is that materials should be able to withstand shock loads to higher stress levels than static loads. However, this benefit may not be particularly significant except at low temperatures, and there the metal is susceptible to brittle rather than ductile failure. The design engineer should also be alert to the effect of strain rate when comparing published information on the properties of various metals, and when including tensile properties in materials specifications. By testing at a higher strain rate, an unscrupulous supplier could pass an otherwise off-specification lot.

2.2.3 Work Hardening

Work hardening was defined in Section 2.2.1 as the increase in plastic flow stress with increasing plastic deformation. The phenomenon is manifested directly by the shape of a stress-strain curve at strains beyond initial yielding. It is also studied by measuring the tensile properties of a plate or rod which has been progressively cold-worked (by cold rolling, for example) to various degrees of deformation. Figure 2-7 reports the changes in yield strength and uniform elongation for Zircaloy-2 as a function of the reduction in thickness of a cold rolled plate. It is evident from the figure that the strength of a metal can be substantially raised by cold working. Because the ductility is concurrently reduced, the extent of deformation that can be introduced is limited by design or processing considerations.

The effects of cold working can be removed by annealing; that is, by holding the materials for some time at an elevated temperature. The "restoration processes" occur in three stages—"recovery," "recrystallization," and "grain growth"—characterized by microstructural changes associated with the return of mechanical properties to the pre-worked values. A cold-worked metal consists of greatly distorted and deformed crystalline grains containing a high density of tangled dislocations. Considerable energy is stored in the piece in the form of strain in the crystal lattices. Upon annealing, these strains initially relax by movement and rearrangement of the dislocations. The resulting changes in mechanical properties are not large, since most of the dislocations still remain in the piece. This is the recovery stage.

At higher annealing temperatures (or longer annealing times), new crystalline grains will nucle-

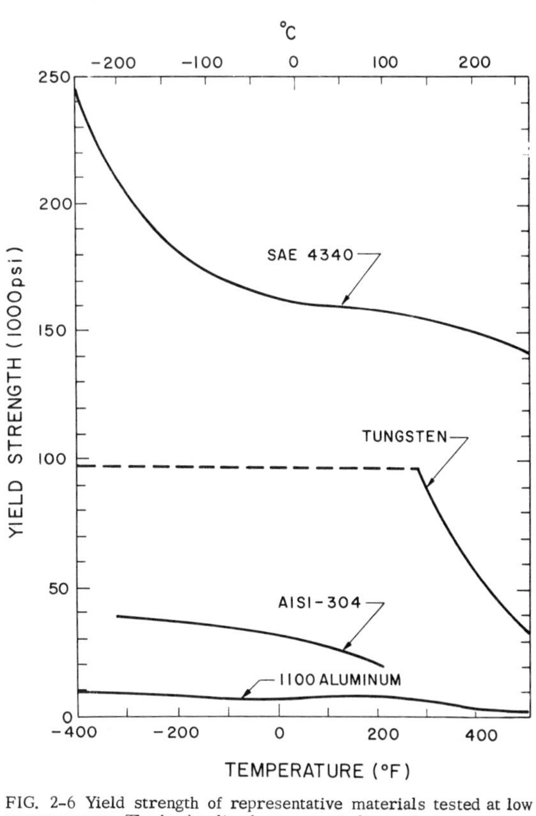

FIG. 2-6 Yield strength of representative materials tested at low temperatures. The broken line for tungsten indicates fracture stress.

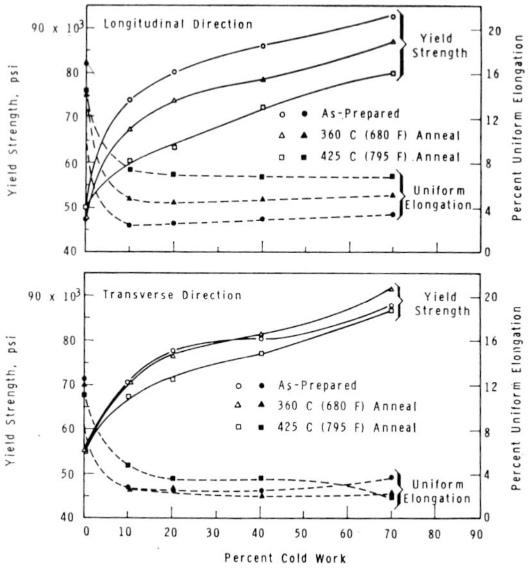

FIG. 2-7 Yield strength and ductility of cold-worked and annealed Zircaloy-2. (From Bement et al. [118].)

ate in the old deformed structure. The new grains grow until the entire piece consists of recrystallized, strain-free grains, and the mechanical properties revert sharply toward pre-worked values. When annealing is at even higher temperatures, recrystallization is followed by grain growth where the larger grains grow at the expense of smaller grains. The driving force comes from the reduction in grain-boundary surface energy.

Both the temperature at which recrystallization occurs and the recrystallized grain size depend on the amount of prior cold work. Increasing the cold work (increasing the reduction in thickness by cold rolling, for instance) results in recrystallization at a lower temperature and a smaller recrystallized grain size. This is one method of controlling the grain size of metals which can be worked. There are two other implications of the recrystallization process from the standpoint of fabrication. First, because recrystallization returns a metal almost to its initial state, substantial cold forming may be accomplished by repeated working and recrystallizing. Second, if a metal is worked at temperatures above the recrystallization range, plastic deformation and recrystallization proceed simultaneously, and there is practically no limit to the amount of deformation that may be accomodated. Thus, the recrystallization range generally separates the "cold-working" and "hot-working" temperatures.

It was mentioned in Sec. 2.1 that cold work can produce a texture in metals and a resulting anisotropy in mechanical properties. Recrystallization does not always remove the anisotropy, because the recrystallized grains in some metals are preferentially oriented with respect to the old grains. A "recrystallization texture" is formed. These effects are illustrated by the properties of Zircaloy-2 in Fig. 2-7. The yield strength of the "as-prepared" material is higher after cold working when tested parallel to the rolling direction than when tested transversely. This is frequently the case for many metals because, even if no strong texture is developed, the effective grain size is different in the two directions. Note, however, that annealing of Zircaloy-2 results in greater recovery of the longitudinal properties, so that the anisotropy is reversed. The data in Fig. 2-7 are actually for low-temperature annealing (not recrystallization), but are included to show that substantial recovery of the properties of Zircaloy-2 may occur at autoclaving temperatures. The anisotropy of hot-worked or recrystallized Zircaloy is indicated in the figure by the intercepts at no cold work. The starting material, presumably recrystallized, has higher transverse than longitudinal strength.

Cold working offers several advantages in addition to increasing the mechanical strength. Rolling or drawing at room temperature produces a smooth surface finish and permits close dimensional control. Consequently, plates or tubes can be finished by cold rolling or drawing, which is inherently less expensive than surface machining. This is particularly advantageous when expensive metals, such as Zircaloy, are being fabricated.

2.2.4 Creep

The progressive elongation with time of a stressed metal is called creep. Generally, three successive stages of creep may be distinguished. In Stage I, "transient" or "primary" creep, the creep rate is initially high but decreases with time. At low temperatures (below about one-half the absolute melting temperature) and stresses below the ultimate strength, the creep rate will diminish to zero, but at higher temperatures a minimum rate is reached and remains constant with time. This period, when the strain rate is constant, is Stage II, "secondary" or "steady-state" creep. In Stage III, "tertiary" creep, the strain rate increases again, until the sample fractures.

The amount and rate of straining are established by the material itself under the imposed stress and temperature conditions. Figure 2-8 shows typical strain-time curves. Creep involves the interplay of strain hardening and recovery. In

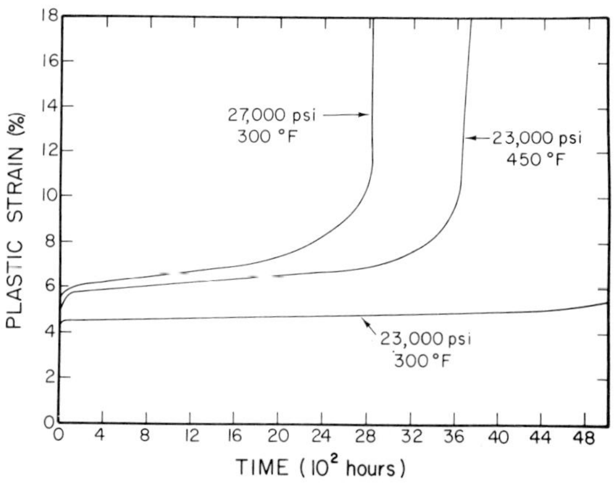

FIG. 2-8 Strain-time curves for Zircaloy-2 at the indicated conditions.

Stage I, work hardening dominates, and the strain rate decreases. In Stage II, the rate of work hardening is exactly compensated by the rate of recovery, and the strain rate is constant (or nearly so). Tertiary creep is caused by necking of the sample or by the accumulation of voids in the grain boundaries. Grain-boundary sliding and progressive recrystallization may also contribute to the high-temperature creep rate.

Several empirical or semitheoretical expressions have been developed to express creep kinetics. The rate of secondary creep is related to the applied stress and temperature by the Dorn-Weertman equation:

$$\dot{\epsilon} = A\sigma^n \exp[-Q_{SD}/RT],$$

in which A and n are constants (n is about 5), Q_{SD} is the activation energy for self-diffusion, R is the gas constant, T is the absolute temperature. The "Larson-Miller parameter" has been widely applied. With this parameter the time, t, for a given strain or for fracture is related to stress and temperature by

$$T(C + \log_{10} t) = f(\sigma).$$

The constant C has the value of 20 (when temperature is in absolute degrees and time is in hours) for many alloys. The function f(σ) must be determined experimentally.

Figure 2-9 reports creep performance for several alloys in terms of the temperature and stress combinations which will produce 1% strain in 10,000 hours. Many of the high-temperature creep-resistant alloys are dispersion hardened. Adding a finely dispersed, hard, second phase not only increases the strength and work-hardening characteristics by impeding dislocation movement; it also makes recovery more difficult.

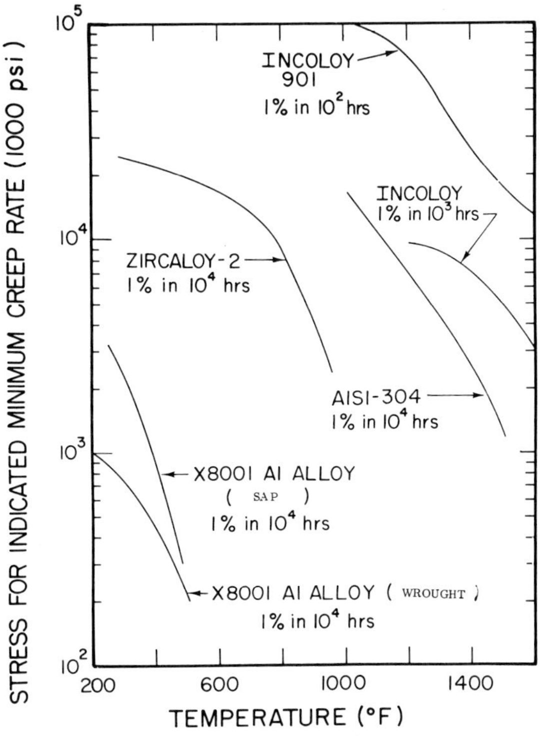

FIG. 2-9 The effect of test temperature and applied stress on the creep rate of representative reactor materials. Creep strain at a given time is indicated on the curves.

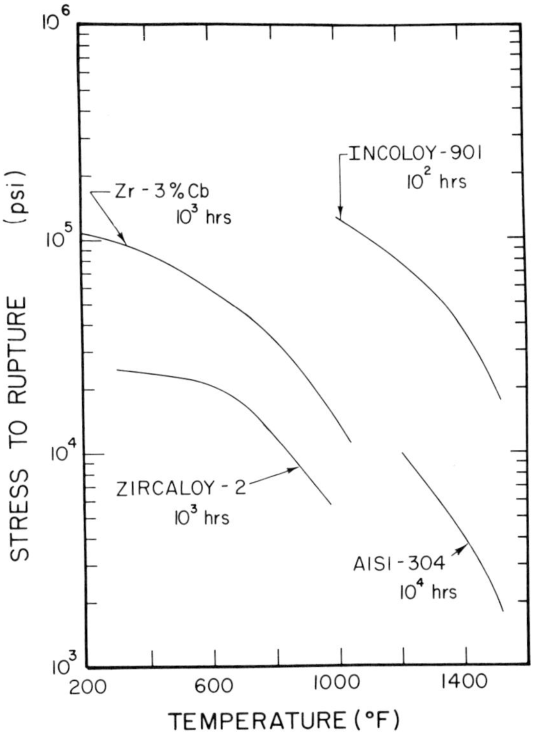

FIG. 2-10 The effect of test temperature and applied stress on the creep-rupture properties of representative reactor materials. Time to rupture is indicated on the curves.

From a reactor design standpoint, creep behavior is most important in connection with fuel cladding materials and pressure tubes. Perhaps the most severe requirements for cladding alloys are those to be met in fast reactors. Operating temperatures in a fast, sodium-cooled reactor will be above 1000°F. The cladding must remain rigid to preserve coolant channel dimensions and to maintain the fuel in its intended position. Stresses will result from the mechanical constraint, the temperature variations, and the buildup of fission products within the fuel element. A significant concern here is the influence of periodic stress changes (by thermal fluctuations, for example) on the overall creep strain. Overloads will, as expected, increase the creep strain and will impose primary creep even though the material has previously reached secondary creep. Cyclic overloading may result in an earlier failure than one overload even though the total time of applying the overload is the same in both cases. Periodic reductions in the load, or unloading, may result in greater ultimate strains if recovery is rapid. Examples of these effects are reported by Kennedy [21].

Pressure-tube reactors, such as the Carolinas-Virginia Test Reactor, the Canadian NPD (see Chapter 10 of Volume 1), and the Hanford "N" reactor, contain the pressurized coolant in tubes running through the reactor. When these tubes are not thermally insulated, as in the NPD, the possibility of creep failure becomes an important design consideration. Tube burst tests, wherein the time to fail a pressurized tube is measured, are useful tests for examining the creep-rupture behavior of alloys for tubular cladding or pressure tubes. Examples of creep-rupture data are shown in Fig. 2-10.

2.3 Fracture

2.3.1 Types of Fracture

Fracture of a stressed metal into two or more parts is characterized as either "ductile" or "brittle." These terms refer to the amount of plastic deformation of the sample prior to fracture. The classification as ductile fracture is often used to imply that a large energy is absorbed in breaking a sample, although this is, strictly speaking, a measure of "toughness." The mode of fracture, as contrasted to deformation preceding fracture, is characterized as "shear" or "cleavage," the former being associated with ductile and the latter with brittle fracture.

Ductile fracture in a simple tensile test occurs in three stages: (1) The sample begins necking down after the "ultimate strength" is passed, and cavities form in the necked region. (2) The cavities begin to coalesce into a crack in the central interior, and the crack spreads slowly outwards. The mean plane of the crack is perpendicular to the tensile axis, but on a local scale the crack proceeds in short segments at 45° to the tensile axis. (3) The crack approaches and reaches the surface in a direction 45° to the tensile axis. This sequence produces the characteristic "cup and cone" configuration of the fracture. The fractured surface appears dull and "fibrous." By contrast, brittle fracture occurs with little overall plastic deformation, and crack propagation is extremely rapid. The surfaces separate by cleavage along characteristic crystal planes within the metal grains, and the fracture surface appears brightly faceted or "granular."

"Intergranular" failure, that is, failure by grain-boundary separation, is observed under some conditions. Brittle films, continuous phases or intermittent precipitates, in the boundaries can lead to brittle, intergranular failure. At sufficiently high temperatures, where the grain boundaries are weaker than the grains themselves, intergranular failure occurs by ductile rupture of the boundaries.

Whether a metal will fail in ductile or brittle fashion depends on a competition between the two processes—cleavage and plastic flow. Cleavage results when there is a sufficiently large tensile stress normal to the cleavage planes. Under conditions where plastic flow can take place easily, the applied tensile stresses cannot attain the values needed to form or propagate a cleavage crack (except, perhaps, on a local scale). Circumstances which hinder plastic flow without appreciably increasing the required cleavage stress will thereby increase the tendency for brittle fracture. From the discussion in the previous sections, it should be apparent that plastic flow becomes more difficult as: (1) the test temperature is lowered, (2) the strain rate is increased, and (3) the differences in principal stresses become smaller (increasing "triaxiality"). Since these factors do not have a strong effect on cleavage, the indicated changes will favor brittle failure. Metallurgical structures which increase the yield strength also inhibit plastic flow and may favor brittle fracture. Decreasing the grain size, however, influences cleavage crack propagation more strongly than it hinders plastic flow, and the net effect of decreasing grain size is to make the metal more resistant to brittle failure.

2.3.2 Ductile-Brittle Transitions

For a given metal in a given metallurgical condition the strongest influence on its plastic behavior is the test temperature. Some metals, namely alloys having a body-centered cubic structure, will exhibit shear fracture at high temperatures but cleavage at low temperatures. The transition from one mode to the other often occurs over a narrow temperature range and defines the "ductile-brittle transition temperature." The transition range and the temperature level at which it occurs depend quite markedly on the way in which it is measured. Impact tests of notched samples are the most severe tests of brittle fracture, since they induce high strain rates and triaxial stresses into the sample. Transition temperatures measured by notched-sample impact tests will be much higher, often 100 or 200°C, than those from unnotched tensile tests at slow strain rates.

Of the several impact tests available, the Charpy V-notch [22] and the drop-weight test [23] have been used most extensively in the evaluation of structural steels. The Charpy sample is a small bar, 1 cm square by 5.5 cm long, in which a transverse notch 2 mm deep is machined. (The notch root radius is 0.25 mm and the included angle of the notch is 45°.) The sample is placed against an anvil and struck by a swinging pendulum with sufficient force to break the bar at the notch. Results of the Charpy test are most commonly reported as the energy required to break the bar, which is determined from the reduced height of swing of the pendulum. Representative Charpy-V results are given in Fig. 2-11. Note that the body-centered cubic, carbon steels, such as A302B and A212B, show a sharp transition from ductile to brittle behavior. The Charpy-V test results may be reported in other ways. For example, examination of the fractured surface will show, in the transition temperature region, a mixture of shear and cleavage fracture. By estimating, or measuring in some way, the fractional area which shows the shear mode, a curve of "% shear" or "% fibrosity" may be plotted against test temperature.

The drop-weight impact test resulted from a program by the Naval Research Laboratory [24] to correlate brittle failures in welded ship structures with service conditions. A relatively large sample (5/8-in. × 2-in. × 5-in. or larger) with a hard-facing weld bead laid down one face, which is slotted to introduce a crack initiation site, is placed on an anvil and struck by a falling weight. When successive samples are tested at decreasing tem-

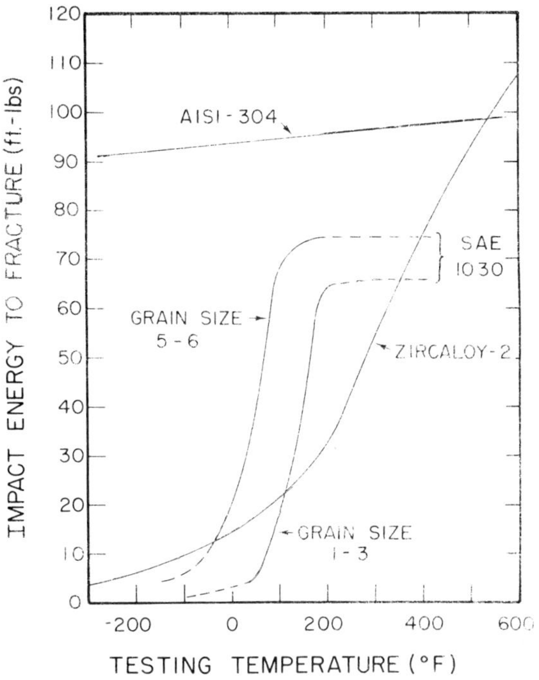

FIG. 2-11 Charpy V-notch impact energy as a function of test temperature for representative reactor materials. Grain size numbers are ASTM designations; higher numbers indicate smaller grain size.

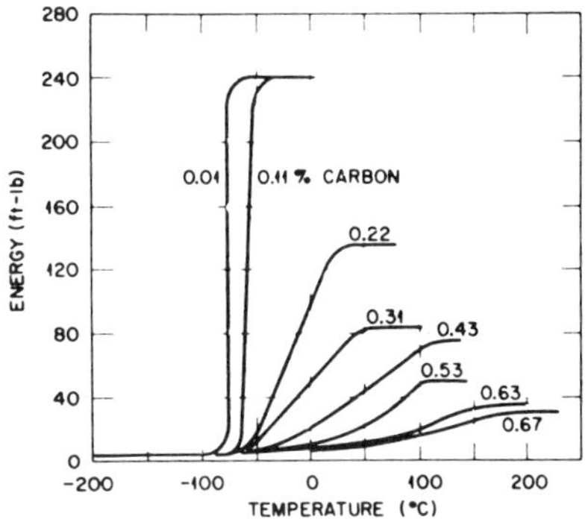

FIG. 2-13 Impact energy for a pearlitic steel as a function of test temperature and carbon content. (After Rinebolt and Harris [142]; taken from Wechsler [26].)

peratures, a sharp transition from no-fracture to complete fracture is observed. The transition temperature determined in this way is defined as the "nil ductility transition," or NDT, temperature. Because the anvil restricts the sample deflection to 0.075 in. in a 4-in. span, the measured NDT temperature is effectively the highest temperature at which the metal will fracture when loaded just to the yield point. It must be remembered, however, that the transition temperature is specific to the test performed. The interpretation of these tests and their correlation with each other and with service failures are discussed in Sec. 2.5.

Several metallurgical variables and processes influence the ductile-brittle transition temperature; namely grain size, composition, phase distribution, cold work, strain aging, quench aging, and irradiation. Most important of these, from the standpoint of design and fabrication practice, are grain size and composition (discussed below) and irradiation (discussed in Sec. 2.4).

A smaller grain size results in a lower transition temperature, as illustrated in Figs. 2-11 and 2-12. As discussed in Sec. 2.4, a smaller grain size also makes carbon steel transition temperatures less sensitive to neutron irradiation, and for these reasons grain-size control by heat treatment is an important part of pressure-vessel specifications.

The effect of composition on the ductile-brittle transition temperature is complex and cannot be generalized at the present time. A few specific conclusions that have been reached [25] are (1) the greater the purity of the metal, the greater the ductility at low temperature (see Fig. 2-13), (2) as silicon is added to carbon steel it first lowers the transition temperature, but after about 0.5% silicon further additions raise the transition, and (3) manganese will decrease the transition temperature of low carbon steel about 5°C for each 0.1% addition. (Manganese also increases the hardenability of steel and permits reduction of the carbon content while the desired yield strength is achieved.)

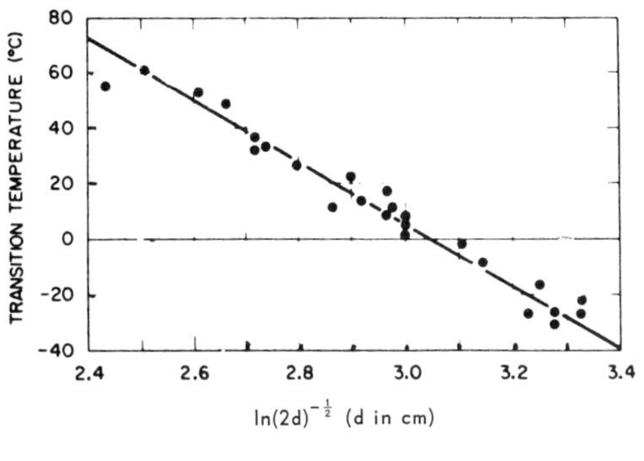

FIG. 2-12 Ductile-brittle transition temperatures as a function of mean grain diameter (2d) for a low carbon steel. (After Heslop and Petch [25]; taken from Wechsler [26].)

2.3.3 Mechanism of Brittle Fracture

As stated in a previous section, whether a metal fails in a ductile or brittle fashion depends on the competition between shear and cleavage. It is appropriate to summarize briefly some theoretical aspects of brittle failure in support of this statement.*

From fundamental considerations it is estimated that the cohesive strength of perfect metal crystals should be on the order of 10^6 to 10^7 lb/in.^{2}, which is several orders of magnitude greater than actual values for engineering materials. It is apparent, from many different experiments, that the discrepancy arises from the presence of structural

*This section is a condensation of the treatment given by Wechsler [26].

irregularities. Griffith proposed that materials contain many fine cracks, which will propagate under an applied stress when the decrease in elastic strain energy upon an incremental increase in crack length just compensates the energy necessary to increase the surface area of the crack. The Griffith criterion gives for the fracture stress in uniaxial tension

$$\sigma_f = \alpha(2\gamma E/c^*)^{1/2}, \qquad (2\text{-}9)$$

where γ is the surface energy, E is Young's modulus, and 2c* is a critical crack length—that is, when the applied stress is σ_f, an internal crack of length 2c* or greater will propagate rapidly, but a crack whose length is less than 2c* will not propagate. The geometrical factor α takes values from 0.8 to 1.3, depending on whether the sample is a thin plate with the crack extending through the thickness (normal to the propagation direction) or a thick plate with an internal crack.

The above relation applies to completely brittle materials. For metals there is evidence that local plastic deformation occurs at the tip of the crack, even under conditions of brittle failure. To account for the energy consumed in plastic flow during crack propagation, Irwin and Orowan modified the Griffith equation to

$$\sigma_f = \alpha[2(\gamma + p)E/c^*]^{1/2} \cong \alpha[2pE/c^*]^{1/2}, \qquad (2\text{-}10)$$

where p is the work of plastic deformation. The approximate expression on the right arises from the observation that the value of p for steels is about a thousand times larger than γ.

To examine the interplay between ductile yielding and brittle fracture, we must introduce some of the concepts of dislocation theory. When an increasing stress is applied to a polycrystalline metal such as steel, it is believed that at some point an avalanche of dislocations is released in the most favorably oriented grains. (This is associated with the yield drop shown in Fig. 2-2.) These dislocations will run together when blocked by a grain boundary and may form a microcrack. If the crack spreads in a time too short to allow stress relaxation to occur by slip in the adjacent grain, brittle fracture will result. Thus, the factor which determines whether yielding or cleavage will occur is whether the stress produced by a dislocation pile-up is sufficiently large to cause a multiplication of dislocations in the next grain.

Bringing together the yield-stress relation as given by the Petch equation [Eq. (2-7)] and the Griffith-Orowan criterion for cleavage propagation [Eq. (2-10)], Cottrell has derived that brittle fracture may occur when the applied stress equals the yield strength if

$$\sigma_y(\sigma_y - \sigma_i)d = \beta\mu(\gamma + p), \qquad (2\text{-}11)$$

or, alternatively,

$$k_y(\sigma_i d^{1/2} + k_y) = \beta\mu(\gamma + p). \qquad (2\text{-}12)$$

In these expressions, σ_y is the lower yield stress, σ_i is the friction stress [see Eq. (2-7)], k_y is the source constant [see Eq. (2-7)], 2d is the grain diameter, μ is the shear modulus, γ and p are surface and plastic-flow energies [see Eq. (2-10)], and the constant β reflects the degree of triaxiality of the stress state. (β decreases with increasing triaxiality—it is taken equal to unity for uniaxial tension, 1/3 for a notched sample, and zero for hydrostatic stress.)

Equations (2-11) and (2-12) express the condition for ductile-brittle transition when the applied stress equals the yield stress. In other words,

$$k_y(\sigma_i d^{1/2} + k_y) < \beta\mu(\gamma + p)$$

implies ductile behavior, and

$$k_y(\sigma_i d^{1/2} + k_y) > \beta\mu(\gamma + p)$$

implies brittle behavior.

Many of the effects of temperature, composition, heat treatment, and so forth, may be interpreted in terms of these relations [5]. Since the right-hand side of Eq. (2-12) varies only slowly with temperature, it is the way in which the terms on the left, which represent the yield strength, change that will determine the transition temperature. Increasing the strain rate increases σ_i and k_y and favors brittle behavior. The friction stress σ_i may be increased by alloying, heat treating (hardenable alloys), cold work, and fast neutron irradiation, and when these treatments do not decrease the effective grain size, they will favor brittle failure. This statement must be qualified somewhat since the various treatments may also affect the shear modulus (μ), surface energy (γ), and plastic-flow energy (p).

The grain-size dependence, as already discussed, is also predicted from Eq. (2-12). Steel deoxidized ("killed") with aluminum has good notch toughness because the small aluminum additions reduce the grain size. Manganese is a particularly beneficial addition to steel because it both reduces the grain size and reduces k_y (by tying up carbon). Nickel and chromium additions also allow fine-grained steels to be produced. Silicon additions influence the grain size but also increase σ_i. Thus, the role of silicon is complicated, as noted earlier.

Although the Griffith surface energy mechanism as discussed above leads to a qualitative picture of crack propagation dependence on material properties, the recent developments of linear elasticity fracture mechanics [27] provide a more useful approach for design purposes. From this point of view* the occurrence of fracture is associated with a critical stress distribution, characteristic of the material, around the tip of a crack. As illustrated in Fig. 2-14, there are three modes of crack surface displacement: mode I, tension normal to the faces of the crack; mode II, shear normal to the leading edge of the crack; mode III, shear parallel to the leading edge of the crack. If we consider mode I deformation and assume that the polar coordinates are centered about the z-axis

*This discussion is based on a review by Weiss and Yukawa [29].

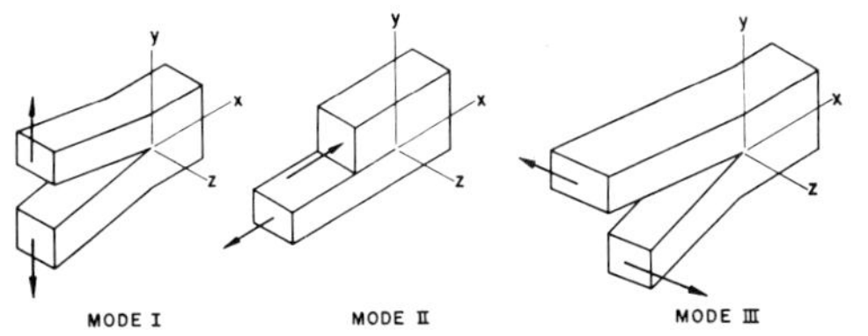

FIG. 2-14 The three basic modes of crack surface displacements. (After Paris and Sih [28].)

with θ measured from the x-z plane, the stress distribution around the crack tip is

$$\left.\begin{aligned} \sigma_x &= (K_I/\sqrt{2\pi r})\cos(\theta/2)[1 - \sin(\theta/2)\sin(3\theta/2)] \\ \sigma_y &= (K_I/\sqrt{2\pi r})\cos(\theta/2)[1 + \sin(\theta/2)\sin(3\theta/2)] \\ \tau_{xy} &= (K_I/\sqrt{2\pi r})\sin(\theta/2)\cos(\theta/2)\cos(3\theta/2) \end{aligned}\right\} \quad (2\text{-}13),$$

with similar expressions for the other deformation modes.

The stress-intensity factor, K, depends on the deformation mode and is a function of the applied stress and crack geometry. For a through-crack of length 2a in an infinite plate with tension σ normal to the crack, $K_I = \sigma\sqrt{\pi a}$; for a disc-shaped crack of diameter 2a inside an infinite solid subjected to uniform tension normal to the crack plane, $K_I = 2\sigma\sqrt{a/\pi}$. (These and the expressions for other geometries have been compiled by Paris and Sih [28].) If the critical-stress system under which fracture occurs is characterized by a stress intensity factor, say K_{Ic}, then a Griffith-type relation is obtained without consideration of the energetics of crack propagation. That is, for the through crack in the infinite plate, the fracture condition is represented by

$$K_{Ic} = \sigma_f\sqrt{\pi a}, \quad (2\text{-}14)$$

which is to be compared with Eq. (2-10). As stated by Weiss and Yukawa [29], the stress-intensity approach is now preferred to the surface-energy approach "primarily because of the straightforwardness of the fracture assumption and the ability to ignore the little-understood surface-energy and plastic-work phenomena accompanying fracture development."

It is clear, from both experiment and analysis, that even "brittle" metals flow plastically at the tip of a crack. From Eqs. (2-13) it is seen that there is a singularity in the elastic stress field just at the crack tip (r = 0). Thus, there is some local volume within which the stress exceeds the yield stress. The size of plastic zone can be estimated by setting σ equal to the yield strength in Eqs. (2-13), which gives

$$r_y = (1/2\pi)(K_I/\sigma_y)^2. \quad (2\text{-}15)$$

(This applies to mode I deformation, but regardless of geometry the parameter $(K_c/\sigma_y)^2$ is a measure of plastic-zone size at the critical point of fast crack propagation.) The application of linear elasticity fracture mechanics is accurate only so long as the plastic-zone size is small in comparison with the crack dimensions and the net remaining cross section, but corrections for plastic-zone size may be applied. (An extension of fracture mechanics to the realm of elastic-plastic stress fields has been given by McKlintock and Irwin [30].)

Another parameter used in fracture mechanics is the force required to extend a crack (or the strain energy release per unit crack length extension). This parameter, $\mathcal{G}$, is related to the stress intensity factor by

$$\left.\begin{aligned} K^2 &= E\mathcal{G} \text{ (plane stress)} \\ K^2 &= E\mathcal{G}/(1-\nu^2) \text{ (plane strain)}, \end{aligned}\right\} \quad (2\text{-}16)$$

where E is Young's modulus and ν is Poisson's ratio. The two parameters K_c and $\mathcal{G}_c$ are entirely equivalent measures of fracture toughness. The parameters for crack propagation by mode I under plane strain loading, K_{Ic} or $\mathcal{G}_{Ic}$, are minimum fracture-toughness values and are used as design properties in the same way as the yield strength [31]. The fracture toughness may be measured in any test for which the defect size and stress distribution are known or calculated. Several tests are discussed by Srawley and Brown [32] and others [27]. It should be noted, however, that there is as yet no agreement on which test or test conditions are most representative of the fracture-toughness properties of actual pressure-vessel structures.

2.3.4 Fatigue Failure

Metals can fail when loaded by alternating stresses with a peak level which would be safe if imposed statically. This is the phenomenon of fatigue. Many different types of fatigue tests have been devised. The stress may be applied by bending, torsion, tension, or compression; the mean stress may or may not be zero; the test constant may be one of several possibilities: peak load, peak stress, peak deflection, or peak strain. The results are usually presented as an "S-N diagram," which shows the failure limit in terms of the number of cycles sustained at each peak stress level.

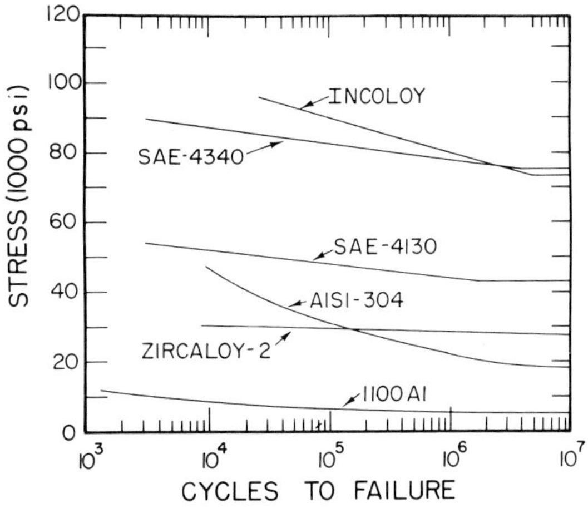

FIG. 2-15 Fatigue properties (S-N curves) of representative reactor materials.

Representative fatigue results are given in Fig. 2-15.

For some metals, the S-N curve is horizontal at a low stress level. This defines the "endurance limit," and cyclical stresses below this limit can be sustained indefinitely. Ferritic steels generally show an endurance limit at a stress of about one-half the ultimate tensile strength. The term "fatigue strength" implies the stress limit for which failure would not be expected until after a specified number of cycles (usually something like 10^7 or 10^8).

From the appearance of typical fatigue fracture surfaces, it is concluded that the fracture starts at some point on the specimen surface; the crack first propagates relatively slowly into the metal, but at some point, when the load-bearing cross section is sufficiently reduced, the crack propagation becomes rapid and the failure is catastrophic. The crack initiation may result from defects or by microscopic slip in the metal grains at the surface, and several mechanisms have been proposed for the initiation of fatigue cracks by slip. Propagation of the crack is probably also controlled by a slip mechanism; thus any treatment which increases the hardness or yield strength of the material may be expected to increase the fatigue strength [5]. (As discussed below, low-cycle fatigue failure may be quite a different phenomenon.)

Since failure is initiated by microscopic cracks at the sample surface, the surface condition is of prime importance in the determination of fatigue life. The removal of machining marks or surface preparations such as nitriding, carburizing, and peening, which produce a surface layer in compression, effectively improves the fatigue properties. Most in-service failures originate at shaft keyways, the roots of screw threads, abrupt changes in section size, and similar points of stress concentration. Corrosive environments markedly reduce the fatigue life—this is usually referred to as "corrosion fatigue." (See Sec. 2.7.)

The temperature dependence of the fatigue limit, at least for high-cycle fatigue failure, follows the dependence of the ultimate strength, and the general effect of temperature is to shift the S-N curve parallel to the stress axis without significantly changing the slope. At sufficiently high temperatures, however, creep and recovery effects can arise, particularly if the mean applied stress is not zero. This is an important engineering problem but beyond the scope of this chapter. The interested reader is referred to Kennedy [21].

The phenomenon of thermal fatigue results from the action of cyclic stresses. The operation of a pressurizer in a pressurized water reactor plant is a pertinent example of a component which is subject to large, cyclic thermal stresses. The effect of thermal stress in combination with pressure cycles on type-347 stainless steel, Inconel-550, and a precipitation-hardened steel (S-816) has received considerable attention [33-38]. For those studies, materials which are relatively stable metallurgically have been used, but most alloys for high-temperature applications are in a metastable condition, with complications of aging, recovery, recrystallization, grain growth, and corrosion needing assessment [38]. Other things being equal, Inconel-550 would seem to provide the greatest freedom from these uncertainities because of its stability at operating temperatures.

For nuclear reactor pressure-vessel design the most important fatigue properties are the low-cycle (10^3 to 10^5 cycles) fatigue resistance and the fatigue

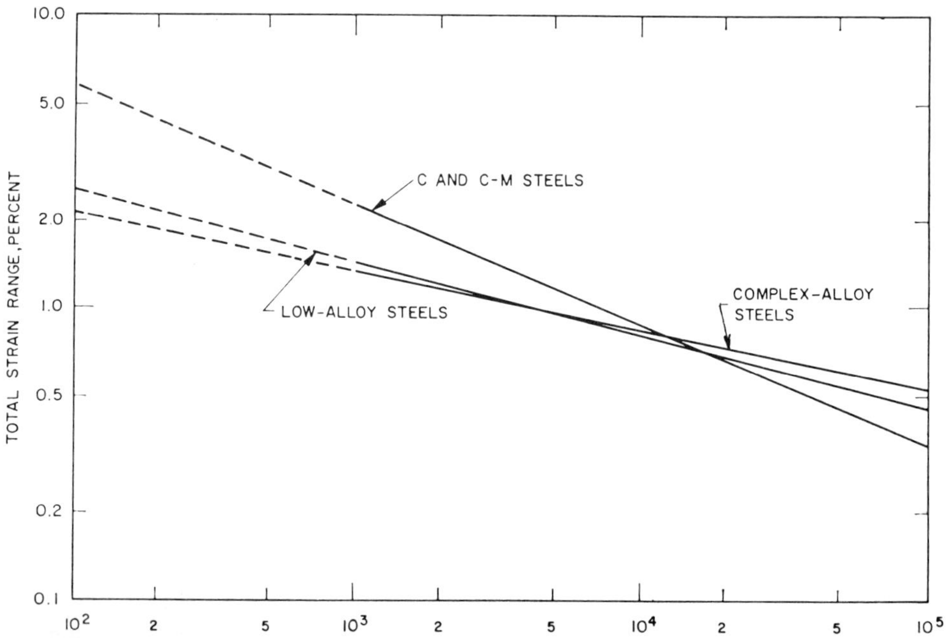

FIG. 2-16 Fatigue properties (strain range vs cycles-to-failure) for three classes of steel. Plain carbon (C), and carbon manganese (C-M) steels, low-alloy steels, and complex-alloy steels with yield strength increasing in that order. (From Langer [39].)

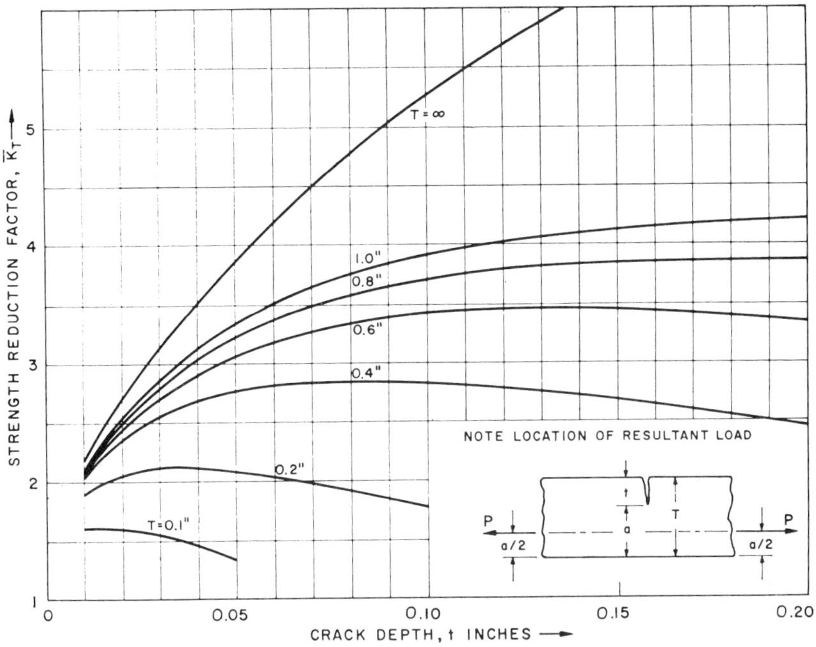

FIG. 2-17 Fatigue strength reduction factors for a crack of arbitrary depth in a section subjected to tension. (After O'Donnell and Purdy [143]; taken from Langer [39].)

notch sensitivity. Langer has proposed, quite reasonably, that "if a catastrophic failure ever occurs in a vessel in a nuclear plant, it will probably originate in an undetected defect which grows under cyclic action to the critical size necessary for brittle fracture" [39]. There are important differences between low-cycle (high loads) and high-cycle (low loads) fatigue. It was stated above that the endurance limit for high-cycle fatigue strength correlates with the ultimate tensile strength. At stresses high enough to produce failure in relatively fewer cycles, however, the important property is the ductility. This is illustrated for three types of steel in Fig. 2-16. Further, it is recognized [28] that the strain range rather than stress range is the significant variable in low-cycle fatigue failure. The most meaningful fatigue tests, therefore, are those in which the strain amplitude at the point of failure is known and held constant.

The presence of sharp notches (cracks or other defects) will reduce the fatigue strength. This is usually stated in terms of "fatigue strength reduction factor," K_f, whose value depends on the notch geometry and dimensions. An example of calculated reduction factors is shown in Fig. 2-17. It should be apparent from this discussion why there is a constant need to guard against imperfections, such as grinding marks or welding electrode arc strikes, in the surfaces of pressure vessels and primary piping.

2.4 Radiation-Induced Changes in Mechanical Properties

Nuclear radiation can significantly, and sometimes dramatically, alter the properties of engineering materials. High-energy radiation impinging on solids will impart most of its energy to the target atoms as heat (lattice and electronic vibrations), but other interactions can occur as summarized in Fig. 2-18. Of the various processes, displacement of target atoms from lattice positions by fast neutrons and fission fragments, and certain transmutations induced by thermal neutrons [e.g., the (n,α) reaction of boron], lead to observable changes in mechanical properties of metals and ceramics. Since the effects of fissioning are discussed in Chapter 13, the following is limited to fast-neutron damage and to the boron transmutation. For further reading several recent reviews are available [40, 41, 42].

2.4.1 Concepts of Radiation Damage

Before presenting the empirical facts of fast neutron damage, it is appropriate to review some of the ideas and the current trends in thinking about the basic nature of radiation damage. There are two questions to be answered: first, how many atom displacements are produced per unit exposure to fast neutrons and what is their spatial distribution; and second, how do various arrangements of displaced atoms influence the macroscopic mechanical properties. Both of these problems are exceedingly complex, and success at quantitative predictions of radiation damage has been quite limited up to the present.

Qualitatively, we may picture the production of displaced atoms as follows. A high-energy neutron traversing a solid undergoes many elastic collisions, which impart kinetic energy to the target atoms. When the recoil energy exceeds a certain threshold value, 25 to 50 ev for most metals, the struck atom will be dislodged from its site and projected through the lattice. Since the recoil

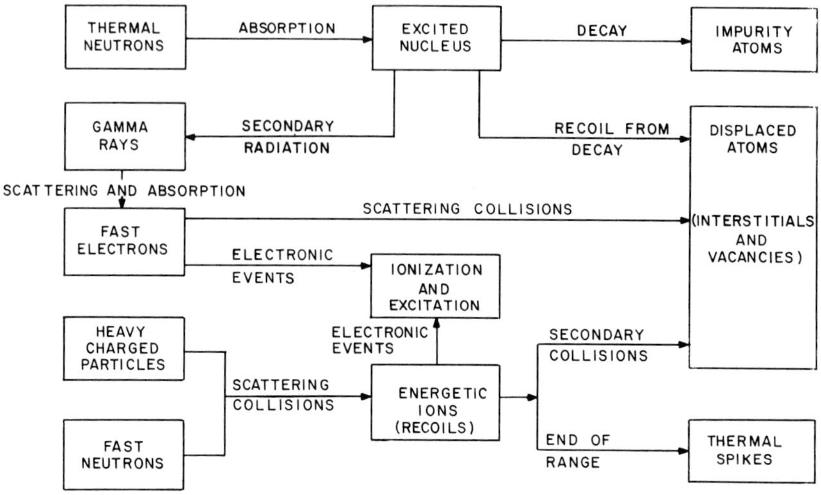

FIG. 2-18 A summary of the interaction resulting from the exposure of solids to nuclear radiation. (From Crawford [144].)

atom energy can be quite high (several times 10^4 for Mev neutrons), these "primary recoils" or "primary knockons" can themselves displace other lattice atoms, which displace others, and so on. Thus, energetic collisions give rise to a shower of displacements, with the kinetic energy being dissipated among many atoms until all have dropped below the threshold for producing displacements.

The rate of production of displacements by monoenergetic neutrons of energy E may be stated as

$$\dot{n}_d = N\sigma_d(E)\phi(E), \tag{2-17}$$

where $\sigma_d(E)$ is the displacement cross section (a function of the neutron energy E), $\phi(E)$ is the neutron flux, and N is the target atom density. The displacement cross section is defined as

$$\sigma_d(E) = \int_{E_d}^{T_m(E)} \nu(T)\kappa(E,T)\,dT, \tag{2-18}$$

where $\kappa(E,T)$ is the differential cross section for the transfer of kinetic energy T from a neutron of energy E, and $\nu(T)$ is the total number of displacements in a cascade originating from a primary recoil whose energy is T. The integration is between the limits E_d, the displacement threshold energy, and $T_m(E)$, the maximum energy which can be transferred by a neutron to a target atom:

$$T_m(E) = \frac{4M}{(M+1)^2}E, \tag{2-19}$$

where M is the atomic mass number of the target atom.

Simplified models (see review by Holmes [43] or Billington and Crawford [44]) yield the following expressions for the differential cross section and displacement yield:

$$\kappa(E,T) = \sigma_s(E)/T_m(E), \tag{2-20}$$

$$\nu(T) = T/2E_d, \tag{2-21}$$

in which $\sigma_s(E)$ is the total elastic scattering cross-section. The latter of these relations is a rather crude approximation for fast-neutron damage but offers at least an order-of-magnitude estimate. When the various equations are combined, the displacement rate produced by monoenergetic neutrons of energy E is

$$\dot{n}_d(E) = N\sigma_s(E)\phi(E)\left(\frac{T_m^2 - E_d^2}{4T_m E_d}\right); \tag{2-22}$$

or, since $T_m \gg E_d$, this is approximately

$$\dot{n}_d(E) \cong N\sigma_s(E)\phi(E)\,E/ME_d. \tag{2-23}$$

[This also assumes heavy targets with $M \gg 1$ in Eq. (2-19)]. Thus, exposure of a piece of iron to 10^{19} fission neutrons (1.5 Mev)/cm^2 can be expected to displace one out of every hundred lattice atoms, and one should certainly expect to see the effect of this disruption on the mechanical properties.

It is known from many experiments [44] that the model used above leads to an overestimate of the number of displacements by a factor of five or more. Obviously, competing processes are removing energy from the recoil cascade without a net displacement of atoms. Inelastic collisions (ionizations and excitations) were ignored in the simple model of the secondary cascade, and these undoubtedly occur for energetic recoils. Furthermore, some secondary collisions result in unstable defects, which immediately relax back to lattice sites; and, more importantly, the regular geometry of metal crystals gives rise to chain events or channeling by which a recoil can be projected along certain crystallographic directions, suffering many small energy transfers without displacing atoms. These events have been recorded experimentally by bombarding single crystals with heavy ions and noting the abnormally long penetration ranges in

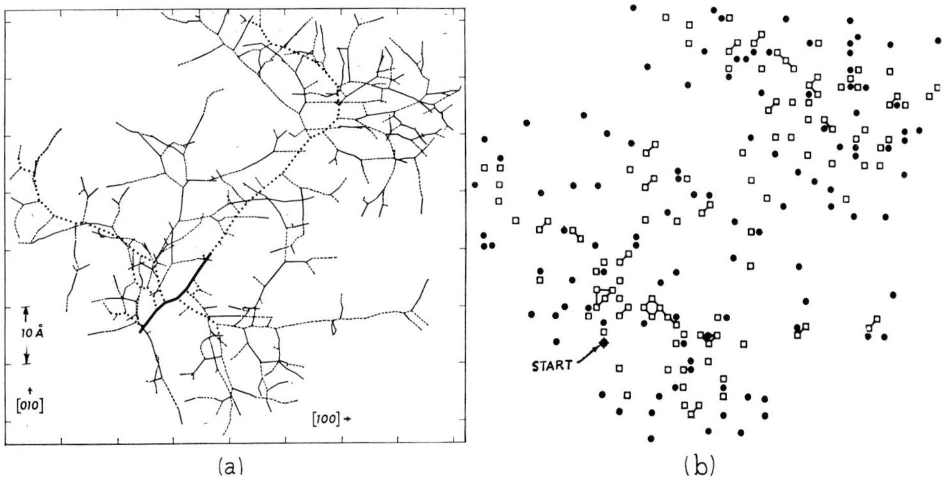

FIG. 2-19 Results of a computer model simulation of the displacement cascade produced by a 5-kev primary recoil in body-centered-cubic ion (a) Projection on a (100) plane of the knock-on atom trajectories. The short heavy track is that of the initiating 5-kev primary; the heavy dotted tracks are secondary knock-ons. Thereafter, alternate broken and solid tracks describe subsequent higher-order knock-ons. (b) Projection on the (100) plane of the damage pattern created by the collisions shown in (a). Open squares are vacancies; filled circles are interstitials. (From Beeler [48].)

certain crystallographic directions. Electron bombardment of single crystals, which leads to preferential ejection of target atoms in specific directions, provides further experimental verification.

Computer simulations of the recoil cascade in three-dimensional crystal arrays have been reported by Vineyard, Erginsoy, et al. [45, 46], by Holmes [47], and by Beeler [48]. These studies have illustrated quite definitely that chain recoils and channeling are probable. Further, they have supported the propositions of Brinkman [49] and Seeger [50] that damage by fast neutrons is characterized by "displacement spikes" or "depleted zones"—local regions in which many displacements have been produced. Figure 2-19 illustrates a simulated cascade in body-centered-cubic iron, initiated by a primary recoil with 5000-ev kinetic energy, and the resulting array of stable point defects.*

After a displacement cascade occurs, the defect array does not remain fixed unless the sample temperature is close to absolute zero. Both vacancies and interstitials diffuse by thermal motion, and the interstitials in particular are thermodynamically quite unstable at normal temperatures. Radiation damage is, therefore, a process involving production of displacements by elastic collision and simultaneous rearrangement and relaxation of the disrupted lattice by thermal diffusion. When the target temperature during bombardment is sufficiently high (roughly one-half the absolute melting temperature), displacement damage anneals out as rapidly as it is produced. As the temperature is lowered, recovery becomes more sluggish, and the net result of radiation damage becomes more observable.

Isolated point defects produced by irradiation would not be expected to influence strongly the mechanical properties of solids. As can be seen from the above, however, fast-neutron bombardment generates highly damaged zones which will relax by thermal motion to more-or-less stable clusters of point defects. Examination of irradiated metals with the electron microscope reveals several types of defect clusters which effectively impede the motion of dislocations and the resulting plastic flow of metals. (Amelinckx [51] has summarized recent work in this area.) As a result of these observations, fast-neutron damage is considered to be roughly analogous to solution strengthening or precipitation hardening of metals.

Neutron irradiation can induce other microstructural changes in addition to those discussed above. Phase transformations can be activated or, in some cases, reversed, but this probably requires the intense damage caused by fission fragments. Thermal-neutron absorption which leads to high-energy capture gamma emission can cause lattice transmutations; for example, the B^{10} (n, α) Li^7 reaction produces gas atoms in the metal which can subsequently agglomerate to form small bubbles. As discussed in Sec. 2.4.2, such reactions can embrittle the metal.

In spite of the fact that there has emerged a reasonably consistent qualitative picture of the nature of radiation damage, progress in the development of quantitative predictions of the induced changes in mechanical properties is slow. At present it is still necessary to base component designs on engineering tests which simulate as closely as possible the expected service environment. The following sections summarize the highlights of experimental knowledge about the effects of fast neutrons on mechanical properties.

*The stable point defects are "interstitials," atoms residing in the spaces between regular crystal lattice sites, "vacancies," unoccupied regular lattice sites, and vacancy "clusters." An associated vacancy-interstitial pair in stable configuration is called a "Frenkel defect."

2.4.2 Tensile Properties

The principal effect of fast-neutron irradiation of most metals is to increase the yield strength and to decrease the rate of work hardening and the uniform and fracture strains. The ultimate strength is increased but not to the same degree as the yield. The temperature dependence of the yield stress is intensified. For the body-centered cubic metals, the ductile-brittle transition temperature is raised, sometimes as much as several hundred degrees Fahrenheit.

Stress-strain curves for a carbon steel, a stainless steel, and Zircaloy-2 are presented in Figs. 2-20 and 2-21 to illustrate the effects of irradiation. Note particularly the substantial decrease in total elongation at fracture. Many irradiated metals show a marked tendency for plastic instability—that is,

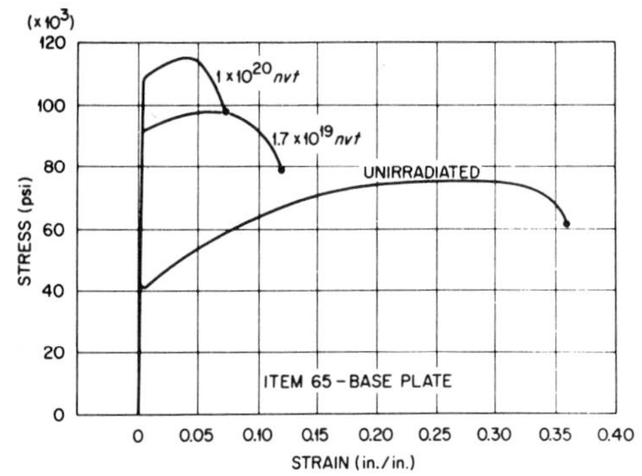

FIG. 2-20 Engineering stress-strain curves for irradiated ASTM A212-B steel. (After Wilson [145]; taken from Wechsler [26].)

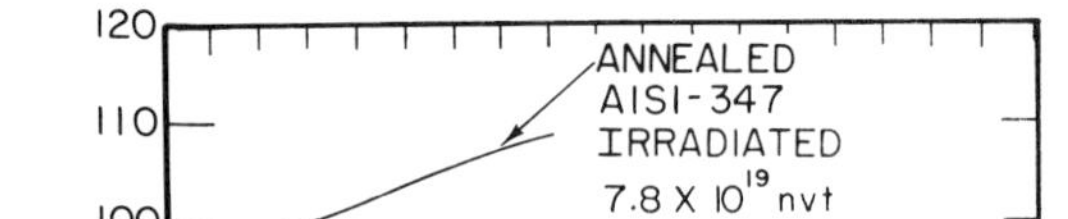

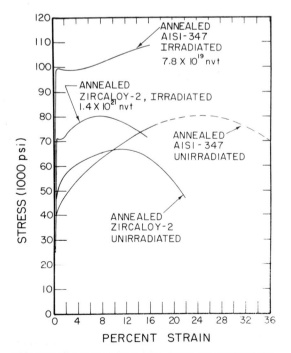

FIG. 2-21 Engineering stress-strain curves for irradiated Zircaloy-2 and AISI 347 stainless steel.

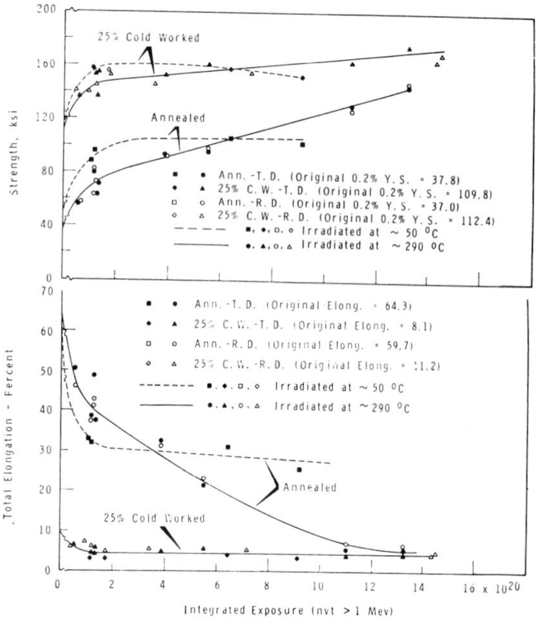

FIG. 2-22 The effect of neutron irradiation on the room-temperature tensile properties of AISI 348 stainless steel. (From Irvin et al. [55].)

necking begins immediately after the yield point is reached. Thus, the uniform and total elongation are severely reduced even though the reduction in area may still be considerable. Commonly, for metals, a load drop appears in the stress-strain curve after irradiation, even for pure single crystals which otherwise would not show a yield drop or even a well-defined yield point.

The rate of increase of yield stress with exposure is a somewhat controversial subject. Holmes [43] has recently reviewed the evidence for a cube-root dependence:

$$\Delta\sigma_y = A(\phi t)^{1/3}. \tag{2-24}$$

A strong case has been presented [50], however, for a square-root dependence at low exposures with saturation of the yield increase at high exposures:

$$\Delta\sigma_y = A\{1 - \exp[-B\phi t]\}^{1/2}. \tag{2-25}$$

This form is more acceptable from a theoretical point of view [50], as well as being consistent with the observed dose dependence of the ductile-brittle transition temperature, as discussed in the next section.

The magnitude of the yield increase is strongly dependent on the metallurgical structure of the starting material, the irradiation temperature, and the tensile test temperature. Figure 2-22 indicates the commonly observed fact that neutron irradiation has a relatively smaller effect on the yield strength of cold-worked metals than on that of annealed metals. It is suggested by these data that the effects of cold work and radiation damage are cumulative, tending toward some saturation value of yield strength. Figure 2-23 shows again the influence of texture on the properties of Zircaloy-2.

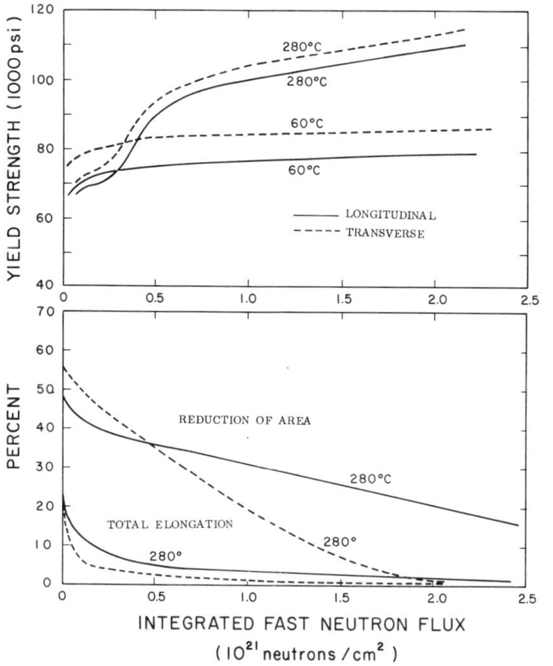

FIG. 2-23 The effect of neutron irradiation on the room-temperature tensile properties of Zircaloy-2. Irradiation temperature is indicated on the curves. (After Bement et al. [118].)

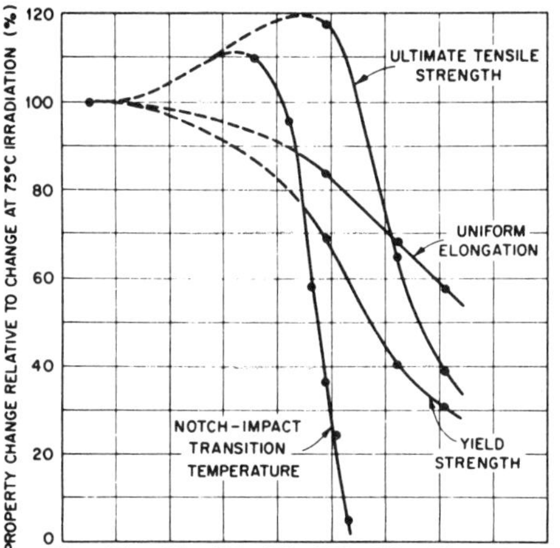

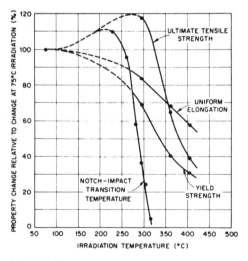

FIG. 2-24 The effect of irradiation temperature on the mechanical properties of ASTM A212-B steel. (After Berggren [146]; taken from Wechsler [26].)

The two figures, for stainless steel and for Zircaloy, illustrate another frequently observed phenomenon — irradiation at somewhat elevated temperature augments the increase in yield strength. Presumably, the induced defects which most strongly influence the mechanical properties must be formed from a rearrangement of the virgin lattice displacements. Hence, diffusion of point defects at moderate temperatures leads to an array which more effectively impedes dislocation movement, but extensive diffusion at higher temeperatures removes defects from the lattice. Therefore, the increase in yield stress should exhibit a maximum at some elevated irradiation temperature. That a maximum occurs for other properties as well is shown in Fig. 2-24.

Test temperature as an important variable is indicated by Fig. 2-25. Note particularly the maximum in postirradiation yield stress around 200°C and the decreasing fracture strain even at test temperatures as high as 800°C. It is concluded from several experiments [52-54] that these effects which are observed in stainless steels, especially the stabilized alloys, are quite probably due to the generation of helium from the $B^{10}(n,\alpha)Li^7$ reaction, even though the steels contain only a few parts per million of boron. The high-temperature embrittlement depends on the thermal-neutron flux and cannot be annealed out at temperatures up to 1150°C [55], which is certainly high enough to remove point defects or even stable clusters. Whether the embrittlement is attributable to helium bubbles in the grain boundaries or to bubble-nucleated precipitations is uncertain, but abnormal grain-boundary inclusions are definitely observed [53].

The damaging effects of fast neutrons in other metals are generally similar to those described above for steels and zirconium alloys. Data for high-temperature alloys have been summarized recently [56-59].

A few recent evaluations of materials irradiated in fast reactors indicate that prolonged exposures—approaching 10^{23} neutrons/cm^2—may cause significant swelling of stainless steel [60-62]. A summary of data from several sources is shown in

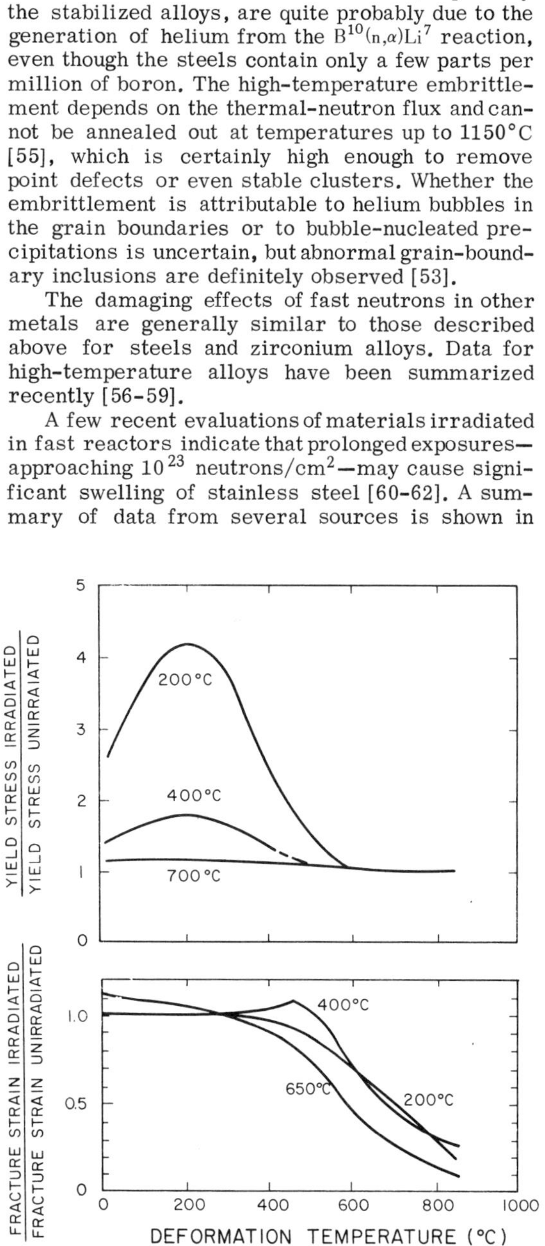

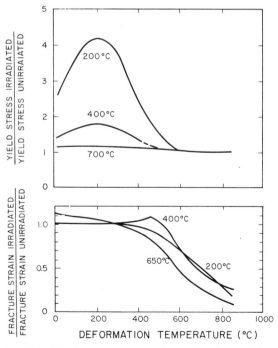

FIG. 2-25 The effect of irradiation on the tensile properties of AISI 304 stainless steel. Irradiation temperatures are indicated on the curves. Total exposure was 7 x 10^{20} n/cm^2 (E > 1 Mev). (From Martin and Weir [52].)

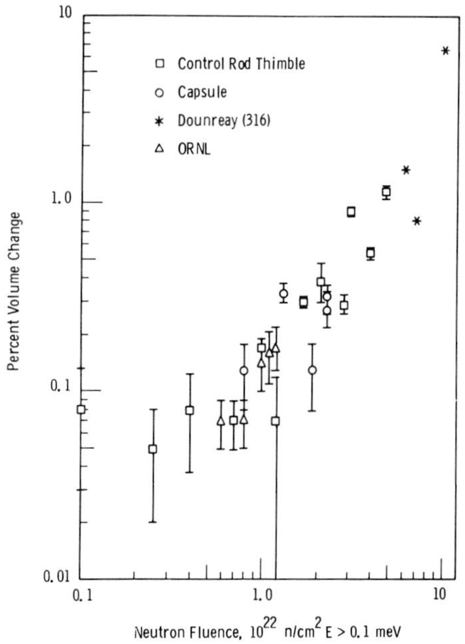

FIG. 2-26 Swelling of austenitic stainless steel (type-304L) by fast-neutron exposure in EBR-2. Neutron exposure (fluence of neutrons with E > 0.1 Mev) is plotted against percent volume change [62].

Fig. 2-26. These data must be treated as still quite preliminary, but they must also be recognized as indicating a serious problem for fast-breeder-cladding design. Further work is needed (and is under way) to define the effects of temperature, stress, prior cold work, and other metallurgical variables on the void-swelling phenomenon.

2.4.3 The Ductile-Brittle Transition

Recalling (Sec. 2.3) that the appearance of a ductile-brittle transition is associated with those metals which have an increased yield strength (relative to changes in the ultimate strength), a sharp yield drop, or a strongly increasing yield strength with decreasing temperature, and recalling further that these three properties are accentuated by fast neutron damage, it should come as no surprise that neutron irradiation has a pronounced effect on the transition temperature. Figure 2-27 illustrates this quite dramatically.

Recent American and British investigations of how neutron irradiation affects the transition temperature have been summarized by Steele and Hawthorne [63] and Harries and Eyre [64]. The dose dependence is indicated in Fig. 2-28. These data are closely proportional to the square root of the integrated flux up to about 10^{19} neutrons/cm^2 and fall below the square-root function for higher exposures. Equation (2-25) appears valid, therefore, for the ΔNDT as well as the incremental yield-strength increase, as proposed by Nichols and Harries [65]. There is evidence [55, 56] that the rate of damage accumulation is unimportant, at least for fast-neutron fluxes in the range 10^{11} to about 10^{14} neutrons/cm^2-sec. This is important in the light of expressed concern that accelerated engineering tests may not be entirely applicable to service conditions for large pressure vessels.

There is a question as to the influence of the neutron energy spectrum on observed changes in the ductile-brittle transition and other mechanical properties. The usual practice has been to report neutron exposures in terms of the integrated spectrum above 1 Mev. This may be valid when comparing test results obtained in similar reactor types, as proposed by the NRL group for tests in light water moderated reactors [66], but it is in doubt when interpreting data from reactors with different spectra. Further, experimental reporting of the integrated flux above 1 Mev is often based on the use of one or two threshold detectors and the assumption that the fast flux distribution corresponds to the fission neutron spectrum. This is a dubious assumption when data from various reactors are compared. Rossin [67] and Shure [68] have proposed that neutron exposures be reported in terms of damage units in which the neutron spectrum is weighted by an appropriate damage cross section. That is, the damage rate is taken as

$$D = \int_0^\infty \sigma_{dam}\, \phi dE ,$$

where σ_{dam} is the energy-dependent damage cross section (in the simplest model this may be taken as proportional to the isotropic elastic scattering cross section) and ϕ is the energy-dependent neutron flux. Experimental verification that this approach clarifies the interpretation of pressure-vessel-damage data is reported by Seman and Pasierb [69].

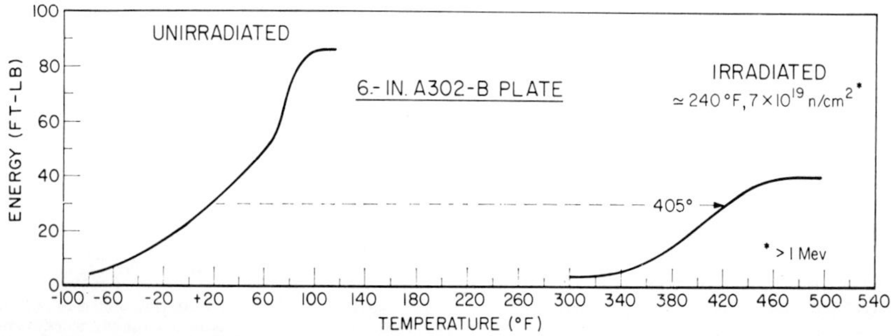

FIG. 2-27. Transition temperature and shear energy characteristics of irradiated A302-B steel. (From Steele and Hawthorne [63].)

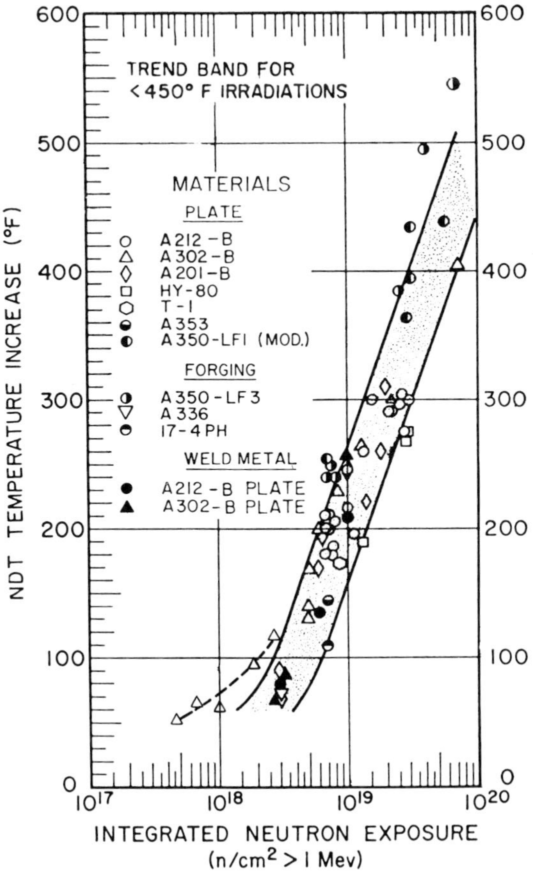

FIG. 2-28 Increase in NDT temperatures of steels irradiated at temperatures below 450°F (232°C). The data are based on drop-weight tests and correlated Charpy V-notch tests. (From Steele and Hawthorne [63]; the five data points at lowest exposure are from Serpan and Steele [66].)

As with the yield strength, higher irradiation temperatures result in less increase in the ductile-brittle transition temperature. Whether or not a maximum is observed, as shown in Fig. 2-24, seems to depend on the metallurgical structure. Nicols and Harries [70] report that an aluminum-killed (i.e., deoxidized) carbon steel exhibits a maximum, whereas a silicon-killed steel does not. It is not clear whether this is a grain-size or a composition effect.

The metallurgical structure has a significant effect on the NDT sensitivity to radiation. Decreasing the grain size results in a smaller NDT increase (as well as lowering the preirradiation NDT). The role of alloying additions is under active investigation, of course, in the hopes of developing a steel which will be insensitive to radiation damage. It is observed that boron (30 ppm or even less) markedly aggravates the embrittling tendency, and it is suggested further that nitrogen may also be responsible for an increased ΔNDT with exposure [64]. Regarding this latter point, if the conclusion is in fact valid, nitrogen content appears to have a more pronounced effect than grain size, which may explain in part why different steelmaking practices lead to a variability in the NDT increase for steels of the same nominal composition. Another possible composition variable is the reported beneficial effect of small uranium additions. Steele and Hawthorne [63] find that 0.06 wt.% uranium decreased and ΔNDT at 1.13×10^{19} n/cm^2 of an A212-B steel by 37% relative to control samples. Beyond these few indications, the effects of alloying additions on radiation embrittlement are not clearly indicated at the present time.

Recently reported work by the Naval Research Laboratory has shown that quite small additions of certain elements can play a large role in the radiation sensitivity of A302-B steel [71, 72]. In particular, if the content of residual elements, sulfur, phosphorus, copper, and vanadium, are kept below 50 ppm, there is no shift in the NDT temperature for exposures to 3×10^{19} neutrons (>1 Mev)/cm^2 at 550°F (288°C). Phosphorus and cop-have been found to be important in this behavior. The work is continuing, and the program includes evaluation of commercial heats of steel with carefully controlled residual elements.

The recovery of ductility by heat treatment is an important phenomenon from the practical standpoint that it may be possible to "repair" reactor pressure vessels by periodic annealing. The results of recovery studies [63, 69] indicate that steels which exhibit the greatest increase in NDT are the most readily relieved and that frequent annealing is quite beneficial even though total recovery is not achieved in any single cycle. In a simulated reactor operation [63] with the vessel at 430°F (221°C) during irradiation and intermediate anneals at 600°F (316°C), the increase in ductile-brittle transition temperature was only 70% of the value it would have reached with no annealing. This represents a possible doubling of the vessel life to reach a specified change in the NDT temperature (presuming that continued cyclic operation would proceed in the same way).

Since most commercial water reactors operate with the pressure vessel at substantially higher temperature than that used in the above example (500-550°F vice 430°F), recovery of the damage would require annealing above 600°F. Nevertheless, the restoration provided by an anneal at even 100°F above the irradiation temperature would be quite significant. For example, the NRL data indicate that 50% recovery can be achieved with a 550°F anneal following irradiation at 430°F. One might expect at least this degree of recovery for a 650°F anneal following a 530°F irradiation, based on the observation that "if steels are irradiated at different elevated temperatures ... and a given temperature increment ... is added for annealing, there appears to be greater recovery for the higher irradiation temperature" [63]. Now, an examination of the NRL trend data for 500-550°F irradiations indicates that a single anneal giving 50% recovery could prolong vessel life by 30 or 40%. (That is, irradiation to 2.7×10^{19} neutrons/cm^2 with one anneal would give the same increase in NDT temperature as irradiation to only 2×10^{19} neutrons/cm^2 with no anneal.) This is not an inconsequential bene-

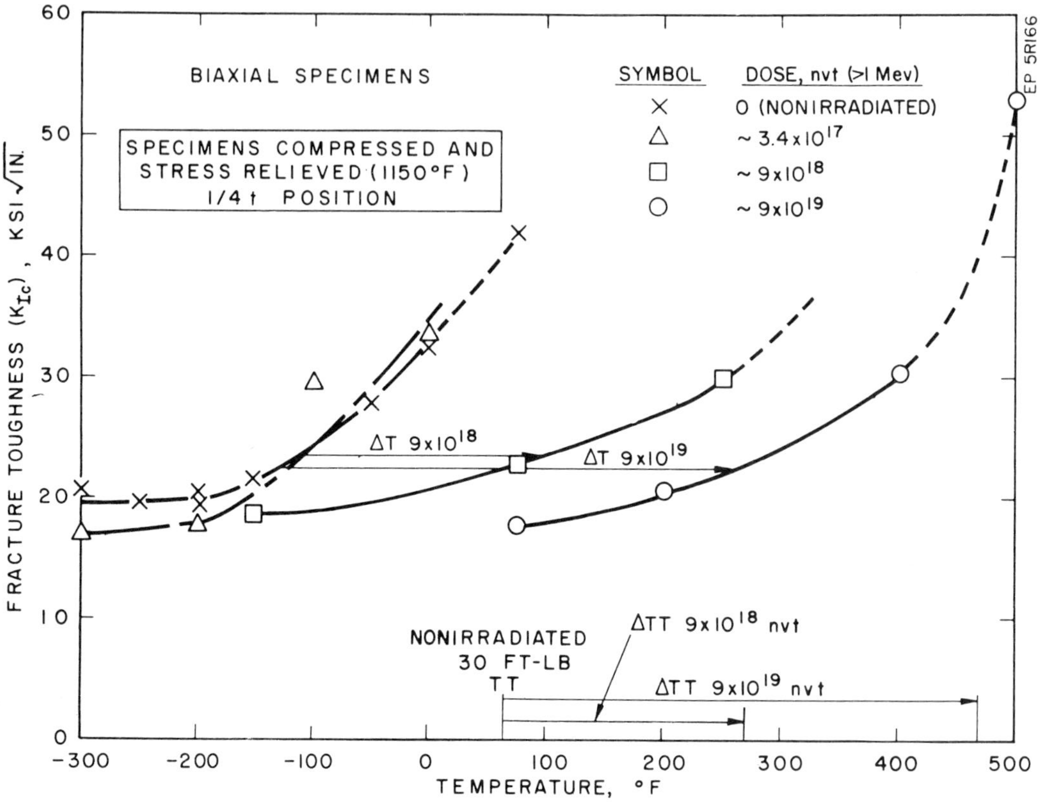

Fig. 2-29 The influence of neutron irradiation on the plane strain fracture toughness of ASTM A302-B steel compared to the shift of the 30 ft-lb Charpy-V transition temperature (ΔTT). (From Johnson [31].)

fit, and although the higher temperature irradiation and annealing data are incomplete at the present time, there seems to be a clear incentive to begin practical consideration of how large reactor vessels might be annealed.

Few measurements have been reported on the influence of neutron irradiation on fracture-toughness parameters. Figure 2-29 presents preliminary results from the Bettis Laboratory for the change in plane strain fracture toughness, K_{Ic}, for A302-B pressure-vessel steel. These data and limited data representing other metallurgical conditions of the same grade of steel suggest that the irradiation-induced temperature shift of the fracture-toughness curves will be no greater than the shift in Charpy transition temperature at the same exposure level [31].

In conclusion, the embrittling effects of fast-neutron irradiation on reactor vessel steels is recognized as a quite serious problem. In spite of the extensive information already developed there are many unresolved questions. Most distressing is the significant variation in radiation sensitivity among various heats of steels of the same nominal composition [70]. The prevailing uncertainties about the influence of prolonged neutron irradiation, as well as considerable uncertainty in the calculated neutron flux at the vessel wall (and possibly in the wall temperature also), have led to the use of surveillance samples in operating reactors to monitor the properties of the actual vessel.

The following have been recommendeded as the most useful features of a surveillance program [73]: (1) The material for the irradiation samples should consist of specimens of the particular heat of steel from which the vessel was constructed. (2) Standard Charpy-V specimens and standard tensile specimens should be tested in the unirradiated condition and after suitable intervals of exposure during the program. (For instance, one or more specimens can be removed during each refueling.) (3) The specimens should be encapsulated or clad to protect them from corrosion (if they are of a corrodable steel that is normally not exposed in the given environment) and exposed at temperatures representative of the vessel in service. The capsules should be individually removable and replaceable. (4) The specimens should be exposed under the same nuclear conditions as those of the pressure vessel and located as near as possible to the vessel interior wall. "If possible, surveillance samples should also be placed at other positions nearer the core of the reactor. This will serve two major purposes. First, if there are significant exposure-rate or spectrum effects, these should show up in a correlation of radiation effects and neutron dosimetry data between accelerated and typical position. These data would also be invaluable for comparison with data from test reactor experiments. Second, if no significant rate

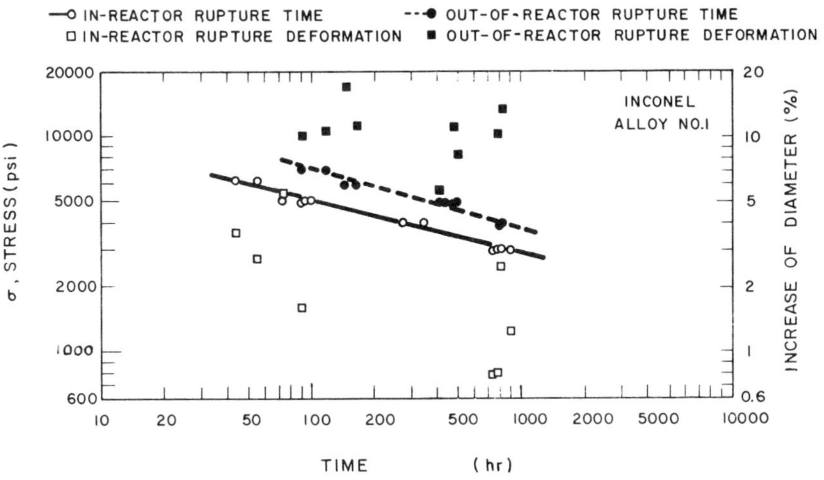

FIG. 2-30 Applied stress vs time-to-rupture for in-reactor tube burst tests on Inconel 600. Test temperature was 815°C (1500°F); neutron flux was 3 x 10^{13} n/cm^2-sec (E > 1 Mev). (After Hinkle [147]; taken from Wechsler [26].)

or spectrum effects are observed, the data from the accelerated position would serve as an extrapolation of the typical position data to a later period in the operating life of the reactor. Utilization of the best neutron dosimetry techniques available in connection with each surveillance unit is implicit in any surveillance program which has significance beyond the immediate monitoring of a component of a particular reactor." [73].

An obvious problem facing the reactor operator who is relying on a surveillance program is the question of what steps to take should the samples in a large reactor plant indicate a critical condition at some future date. It is to be hoped that the several research programs now under way will have arrived at an answer by that time.

2.4.4 Creep, Stress Rupture, and Fatigue

The mechanical properties which involve dynamic processes in metals, as opposed to static or short-time tensile properties, must be considered in a special class as far as their response to radiation damage. Since they involve long test times and since some, such as creep, are high-temperature properties, they will most certainly depend strongly on the rate of damage (neutron flux and neutron energy spectrum) and the irradiation temperature. More importantly, because they are dynamic properties, postirradiation testing will probably reveal a quite different behavior than in-reactor testing. For this reason creep and fatigue properties must be determined by in-reactor experiments. Unfortunately these are difficult and expensive tests, and reported results are limited.

Observations from several experiments [25, 54, 74, 75] indicate that in-reactor creep, stress rupture (tube burst), or fatigue performance is not strikingly different from ex-reactor behavior. Irradiation causes only small decreases in the creep strength and stress-rupture life, although the strain at fracture may be materially reduced. Representative results are shown in Fig. 2-30. Postirradiation tests show a more pronounced increase in creep rate and reduction of rupture life [25, 75]. Grain-boundary fractures have been observed in postirradiation stress-rupture tests, but no mechanism for this behavior has yet been established.

Preliminary in-reactor fatigue tests [76] of A302-B carbon steel are shown in Fig. 2-31. The fatigue life appears to be insensitive to neutron exposure. As discussed in Sec. 2.3.4, fatigue life at low stress (high cycle) correlates with the yield strength, but at high stress (low cycle) it correlates with ductility. Since irradiation increases the yield strength and decreases ductility,

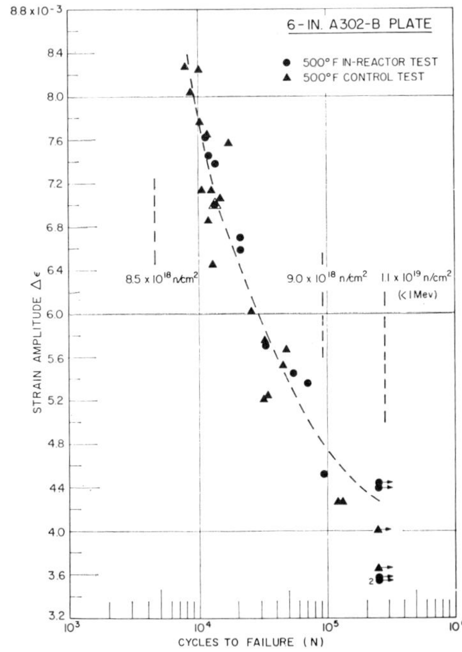

FIG. 2-31 Fatigue behavior of ASTM A302-B steel tested in-reactor at 500°F (260°C). (From Hawthorne and Steele [76].)

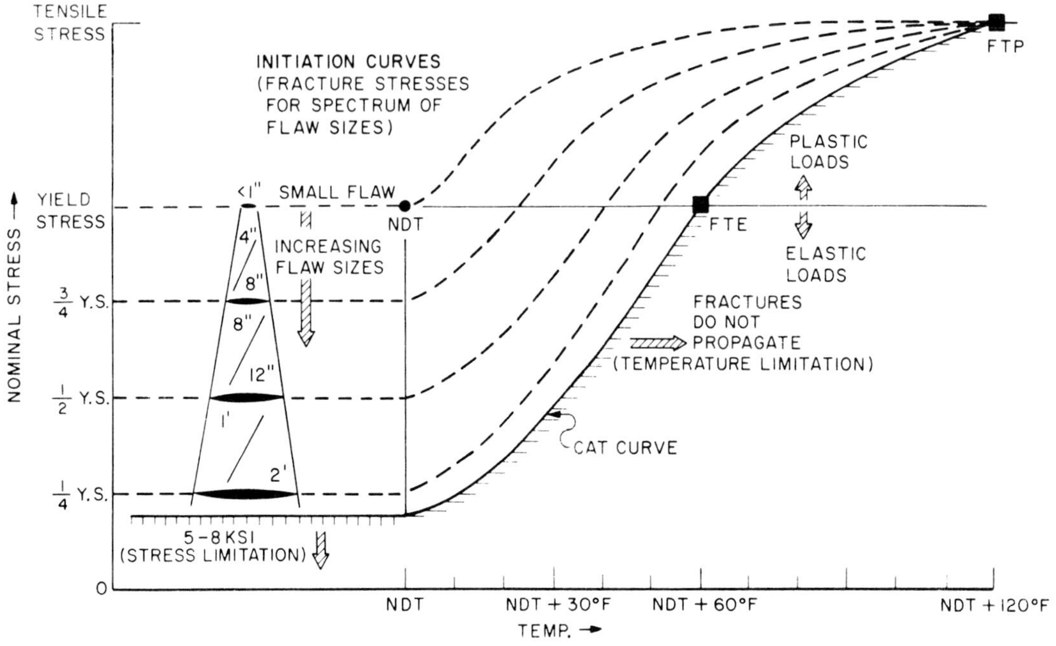

FIG. 2-32 Pellini-Puzak fracture analysis diagram. (From [80].)

one might argue opposing effects for high- and low-cycle fatigue. The point to remember, however, is that in-reactor fatigue processes will undoubtedly be different from postirradiation fatigue, and there is not sufficient knowledge of the basic nature of either fatigue or radiation damage to offer any confident predictions. Designers can only rely on engineering fatigue tests such as those which are currently in progress [76].

2.5 Fracture Analysis in Design

The preceding sections have discussed the mechanical properties of materials without reference to how these properties are applied in the design of structures. In Chapter 14 the mechanical design of reactor components is considered in detail. This section attempts to bridge the gap between the metallurgical testing of materials and the use of measured properties in design. Two questions serve to illustrate the nature of this subject: First, "Is the tensile strength as measured from a small sample, say 1/4 in. × 2 in. in gage dimensions, applicable to the design of a pressure vessel with an 8-in.-thick wall?" Second, "Are the fast-failure characteristics as measured by an impact test representative of the possible behavior of a large vessel with an internal hydrostatic load?"

The first point, namely the scale-up of mechanical properties from tests involving small volumes to large structures, has been reviewed by Irwin [77] and his discussion is outlined here. From a statistical viewpoint, one may consider that the tensile strength depends on the distribution of flaws or local strength inequalities which exist just prior to the onset of unstable fast fracturing. Consider the average tensile fracture stresses, S_1 and S_2, corresponding to two test bars having different test section volumes, V_1 and V_2. A statistical analysis gives the relation

$$S_1/S_2 = (V_2/V_1)^{1/n}, \qquad (2\text{-}26)$$

where n is a parameter characterizing the flaw probability distribution. The value of n is related to the root mean square relative deviation of repeated measurements of the tensile strength, s, by

$$n = 1.2/s, \qquad (2\text{-}27)$$

as long as the scatter in the test results is believed due solely to inherent local strength inequalities rather than to variations in testing procedure. This approach may be useful in demonstrating the performance of a pressure vessel by first constructing and testing a series of smaller prototypes. But the question of the introduction of new classes of flaws in fabricating the full-size vessel does not enter into the statistical analysis, and this could very well overshadow the simple volumetric scale-up.

From another viewpoint, attention should be focused on the "worst flaw" in order to attempt to estimate the strength of a component of a structure in terms of the largest crack which might develop within it during fabrication and use. With this philosophy, failure is seen to occur in the manner described by Irvine et al. [78]:

"The failure behavior of a pressure vessel may be considered to consist of two phases. The first phase is mainly associated with processes leading to vessel damage through cracking or distortion. In a conventional* plant, incipient failure is usually

*Nota bene

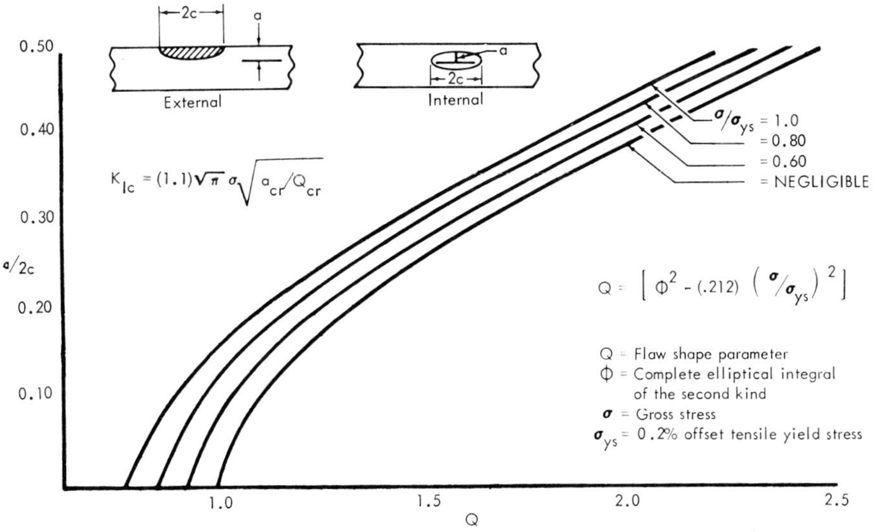

FIG. 2-33 Flaw shape parameter (Q) as a function of flaw geometry (a/2c) and loading (σ/σ_{YS}). (From Tiffany and Masters [81].)

detected during this phase by operational inspection and a plant found to be so affected may be repaired or withdrawn from service.

"If this primary phase remains undetected, the damage increases to a level at which the second phase occurs. This in a pressure vessel is usually characterized by explosive rupture due to the sudden extension of one or more of the cracks arising from the primary phase. It is this possibility of fast crack propagation which is of major interest from a pressure vessel safety viewpoint."

It is usually postulated [79] that the necessary conditions for crack initiation or crack growth to a critical size can exist in a pressure vessel, and interest centers around the conditions under which fast crack propagation can occur. Basically there are two avenues of approach to the problem—phenomenological and analytical. The former is typified by the nil-ductility or transition temperature measurements, and the second is based in fracture mechanics.

Correlations of drop weight and impact tests with in-service failure of ship and boiler plate led to the fracture analysis diagram of Pellini and Puzak [80] which is reproduced in Fig. 2-32. This is a plot of nominal stress against service temperature, which defines several points of interest.

NDT is the nil-ductility transition as determined from a drop-weight test. The correlation of this test with Charpy V-notch transitions is discussed below.

CAT is the crack arrest temperature. This curve represents the temperature of arrest of a propagating brittle fracture for various applied nominal stresses.

FTE is the fracture transition for elastic loading, and the FTE temperature is that temperature at which an explosively loaded plate (explosion-bulge test) will crack in the bulged region but the cracks will not propagate through the elastically loaded zone.

FTP is the fracture transition for plastic loading. Fractures in the explosive test at temperatures above the FTP temperature are entirely by shear.

The diagram shows the correlation between these various transitions, namely that FTE may be expected at NDT + 60°F and FTP at NDT + 120°F. Thus, one approach to pressure-vessel safety is to specify a minimum temperature under full load based on the measured NDT, frequently NDT + 60°F. The usual practice for operation of commercial plants is to reduce stresses (by reduced pressure and controlled heating or cooling) when heating up to or cooling down from the NDT + 60°F temperature of the beltline region of the vessel wall. (The beltline is the region of maximum neutron exposure.) Above NDT + 60°F, the design stress is the only limitation.

Because the drop-weight test, which alone defines the NDT, requires relatively massive samples, the correlation shown in Table 2-1 between Charpy V-notch impact energies and NDT is useful. Caution should be exercised in applying these values, however, since they are based on correlations for particular types of steel. Correlative tests must be performed when new or uninvestigated alloys are involved.

Although the specification of NDT + 60°F places the metal above the CAT for nominally elastic loading, Irvine et al. [78, 79] have pointed out that rapid crack propagation can still occur if the vessel loading produces constant stress (such as a pressure vessel containing a compressible fluid). Furthermore, there is the possibility that a steel may

TABLE 2-1

Charpy V-Notch Impact Energies at a Temperature Corresponding to Nil-Ductility Transition (NDT) Temperature. (From Wylie [83].)

Specific minimum yield strength (psi)	Minimum energy (average of 3 specimens) (ft-lb)	Minimum energy (1 specimen), (ft-lb)
less than 35,000	15	10
35,000-45,000	20	15
45,000-75,000	30	25
75,000-105,000	35	25

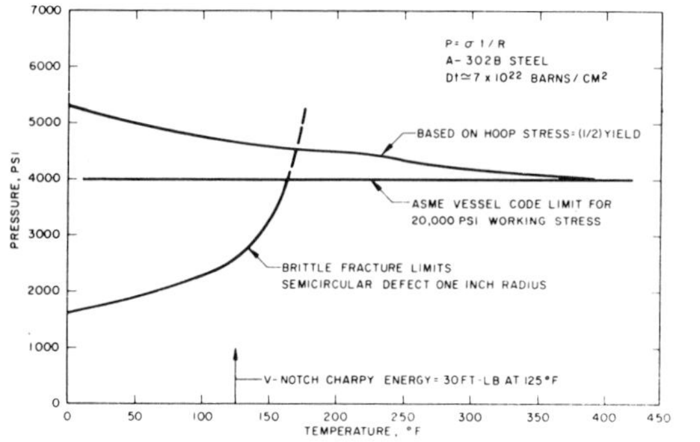

FIG. 2-34 Example of pressure limits for a hypothetical irradiated pressure vessel of A302-B steel based on fracture mechanics principles and two ductile limitations. The lowest curve at any temperature sets the limits. (From Johnson [82].)

exhibit a low impact energy in the fully ductile mode (a low "upper shelf" on the Charpy curve) and may therefore fail at stresses below the yield strength by "ductile tear." Consequently, one must be cautious in using the NDT + T criterion and remember that it is based on an engineering correlation for particular types of steel.*

The application of fracture mechanics to structural design has been reviewed by Tiffany and Masters [81] (from whom this discussion is drawn) and by Johnson [31]. The desired end point is the evaluation of critical flaw sizes required to cause failure at the expected operational stress levels. The necessary parameter is the plane strain fracture toughness, K_{Ic}, which is the minimum stress-intensity factor. For the flaw geometries shown in Fig. 2-33, the applied stress, σ, critical flaw size (expressed as $(a/Q)_c$), and fracture toughness are related by

$$\begin{aligned} (a/Q)_c &= (K_{Ic}/\sigma)^2/1.21\pi \quad \text{(surface flaw)}, \\ (a/Q)_c &= (K_{Ic}/\sigma)^2/\pi \quad \text{(embedded flaw)}. \end{aligned} \qquad (2\text{-}28)$$

(A general discussion of the stress analysis of cracks is given by Paris and Sih [28].) Therefore, if the designer knows the crack geometry and the fracture toughness, he can specify the permissable applied stress. An example from the calculations by Johnson [82] is shown in Fig. 2-34. Note that the NDT + 60°F criterion would be conservative if this vessel were intended for service with 2000 psi internal pressure.

The Pellini-Puzak diagram and fracture mechanics both indicate that allowable stresses depend sensitively on a critical flaw size. Accordingly, the designer is immediately confronted with two questions [81]: "What are the maximum initial flaw sizes that are likely to exist in the vessel prior to its being placed into service?" and "Will these initial flaws grow to critical size to cause failure during the expected service life of the vessel?" With regard to the latter, Langer [39] and Irvine et al. [78] have pointed out that subcritical flaw growth is likely to occur by high strain fatigue or cyclic creep. There is, therefore, an urgent need to develop more information than is currently available on these phenomena.

Regarding the question of flaws existing initially in a fabricated vessel, Sec. 4 of this chapter treats the sources of manufacturing defects, and Sec. 5 deals with procedures for evaluating and controlling manufactured quality. Because structural quality is so intimately tied to fracture analysis of pressure vessels, and correspondingly to their safe design, there are summarized here a few of the important factors in evaluating the fracture toughness of pressure-vessel steels.*

Specifications for pressure-vessel steels generally require determination of the Charpy-V properties to insure that the hydrostatic test and initial pressurization of the vessel will be conducted at or above NDT + 60°F. The critical question is how representative these test values, which cannot at present be determined from the vessel itself, are of the actual material in the finished vessel. Wylie [83] states that "following a large number of tests which compare the impact energy of actual plates with values on the mill test report, it is probable that unless special precautions have been taken any similarity is purely coincidental. The NDT as defined by Charpy impact energy may be in error by as much as 60°F or possibly more. Unfortunately the error is usually toward increased NDT temperatures [i.e., the mill test report is low]." These differences are attributed primarily to the following considerations:

Sample location. Properties are sensitive not only to sample location with respect to depth below the plate surface (usually taken at T/4 for plates of thickness T), but also with respect to location within the ingot.

Cooling rate. Apart from chemistry, cooling rate from the austenitizing temperature is the most important factor in determining notch toughness of carbon and low-alloy steels. In an attempt to assure that a test sample is representative when it is heated apart from a hot-formed vessel, the minimum test-piece size is required to be 3T × 3T × T for plate of thickness T. But the matching of cooling rates between test sample and vessel is still difficult, particularly when accelerated cooling (e.g., by water spray) is used for large vessels.

Forming procedure. Vessels are fabricated by hot forming (at or above the austenitizing temperature), warm forming (at or below the tempering temperature), and cold forming (essentially at ambient temperature). Each of these procedures can, and will, alter the notch toughness properties of the steel.

Other effects, such as prolonged stress relief, nonuniform temperature during stress relief, and nonuniform chemical composition can also cause discrepancies between test-sample and vessel impact properties. As pointed out by Wylie [83], even with the additional requirements now included in

*Note added in proof: For more recent development of approaches to fracture-safe design, see Naval Research Laboratory reports 6300 (June 1965) and 6713 (April 1968).

*This discussion is based on a paper by Wylie [83].

Section III of the ASME Boiler and Pressure Vessel Code, "there is still no guarantee that the final vessel will have the properties shown by the test plates. The problem is magnified in the case of quenched and tempered materials in which the strength is dependent on proper treatment. Where possible, therefore, the final tests should be taken from a cutout from the vessel shell after forming and stress relief." A continuing awareness of this problem and more care in the selection and preparation of test samples can, and should, be expected from designers and fabricators.

2.6 Metallurgical Phase Transformations

In the preceding sections it is pointed out that the properties of engineering alloys, and in fact all metals, are strongly influenced by their metallurgical structure. It is useful, therefore, to review the nature of solid state phase transformations to illustrate how the properties may be affected by various heat treatments during manufacture or by various service environments.

Equilibrium phase diagrams indicate what phases are thermodynamically stable when composition, temperature, and pressure are specified. Although metals are frequently not in thermodynamic equilibrium because of slow transformation kinetics, an understanding of equilibrium diagrams is a prerequisite to predicting whether certain phase changes can possibly occur. Practical examples are the use of phase diagrams in designing precipitation–hardening alloys and in evaluating the compatibility of dissimilar metals to be joined together.

Whereas equilibrium phase diagrams show whether certain transformations are thermodynamically possible, the kinetics of diffusion in the solid state actually determines how far, if at all, such transformations will proceed for specified times, temperatures, and pressures. Thus, in leading up to a discussion of phase-transformation kinetics, it is necessary first to review the basic aspects of equilibrium diagrams and diffusion kinetics.

2.6.1 Equilibrium Phase Diagrams

"A phase is a volume of material which contains no discontinuity in composition or crystal structure." [1] When several phases are present in a solid, each will comprise separate and distinct (by composition or crystal structure or both), homogeneous portions of the alloy. If a system contains a number of components, C, and a number of coexisting phases, P, then the degrees of freedom–that is, the number of state variables that can be independently varied without changing the number of phases in equilibrium–is given by the Gibbs phase rule as

$$F = C - P + 2. \qquad (2\text{-}29)$$

This relation does not, of course, predict how many phases will be present under specified conditions, but it defines certain rules for properly constructing phase diagrams.

Equilibrium diagrams for solid alloys are generally much more complicated than the diagrams for fluids, because a variety of stable crystal structures are observed. Only a few solid systems show complete solubility in a single phase throughout the entire composition range, even when only two components are involved. One pertinent example of complete miscibility is the system UO_2-ThO_2-PuO_2. These three components may be combined in any proportion into a thermodynamically stable single phase. More commonly, metal and ceramic systems show limited terminal solubility–that is, only a limited amount of a solute may be added to a pure element before the single phase solid solution is no longer stable.

A solid solution may be either "substitutional," in which the solute atom is substituted for a solvent atom on a lattice site, or "interstitial," in which the solute atoms reside in the interstices between parent atom lattice positions. Substitutional solid solutions are formed only if the two (or more) kinds of atoms are nearly the same size and if they have similar chemical bonding characteristics. Interstitial solutions are formed when the atom species are greatly different in size or in chemical bonding characteristics.

Many solid systems exhibit "intermediate phases"—single-phase solutions existing over some limited composition range—or "intermetallic compounds," which have a narrow stable composition range about some stoichiometric ratio. Beta brass (about 45-50 wt.% zinc in copper) and delta phase uranium-zirconium (45-55 wt.% zirconium) are examples of intermediate phases. The uranium-aluminum system exhibits three intermetallic compounds, UAl_2, UAl_3, and UAl_4. The standard reference for phase diagrams is Hansen [84].

In addition to the diversity of phases which may appear in solid systems, there are several important "reactions" that occur. "Invariant equilibrium" is any combination of phases and components for which there are no degrees of freedom allowed by the Gibbs phase rule. Thus, for a binary system at constant pressure, the phase rule becomes

$$F = 3 - P.$$

(Note that one degree of freedom has been eliminated by specifying the pressure to be constant.) There can be at most three phases in equilibrium, and if there are indeed three phases, the temperature cannot be altered without changing the number of phases. The term "invariant reaction" refers to the phase changes that occur on heating or cooling through an invariant equilibrium. These reactions for a binary system are summarized in Table 2-2. Each reaction proceeds to the right on cooling and to the left on heating.

The observed combinations of phases and invariant reactions in metal alloy systems are much too numerous for any extended discussion here. We present at this point one diagram, the iron-carbon system in Fig. 2-35, which will illustrate several of the features discussed above and which will be referred to in the subsequent treatment of transformation kinetics. True thermodynamic equilibrium in this system is represented by the boundaries labelled "iron-graphite." Because the transformations proceed by first forming cementite (Fe_3C), and because the transformation from cementite to graphite is exceedingly slow, this

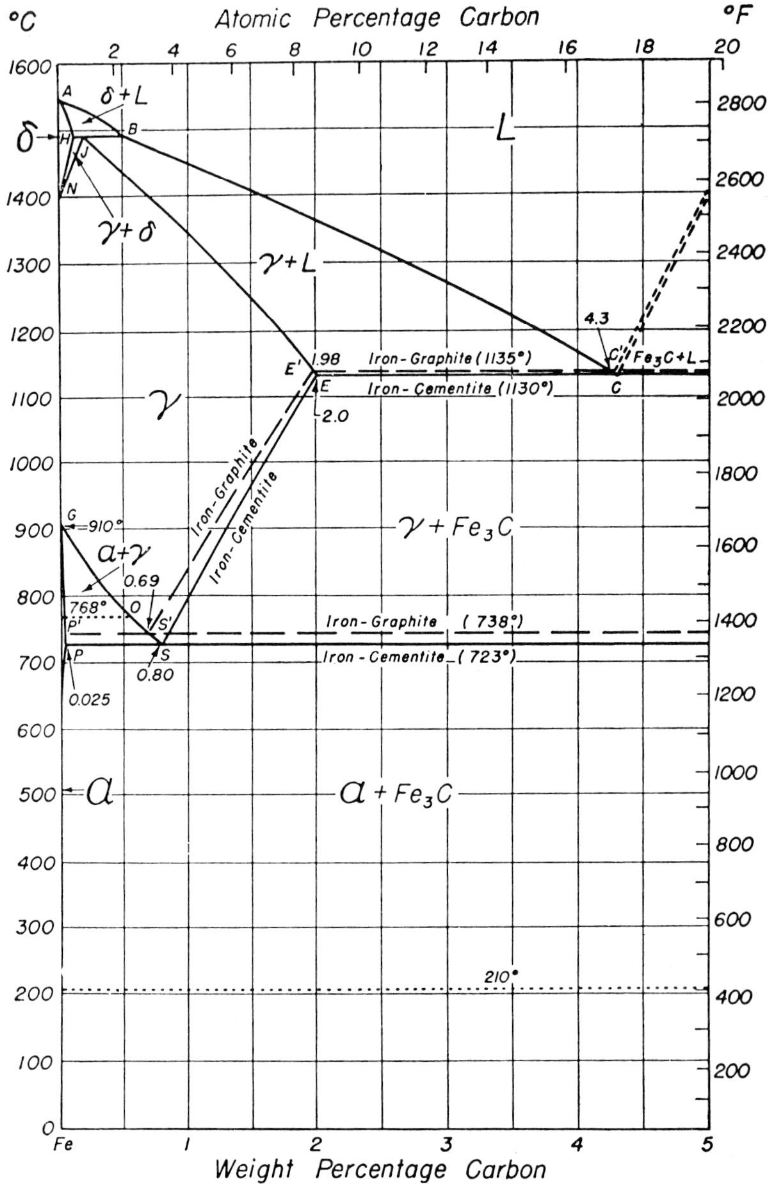

FIG. 2-35 The iron side of the iron-carbon phase diagram. (From Metals Handbook (1948), p. 1182.)

diagram is often shown as the iron-cementite system. In considering the structures formed by ordinary fabrication processes the latter choice is more useful.

Several important features of the diagram should be recognized. Note the invariant reactions: a peritectic at 1490°C, a eutectic at 1130°C, and a eutectoid at 723°C. Note also the comparatively greater solubility for carbon in austenite (γ phase) than in ferrite (α phase). When low-carbon steels are cooled below the eutectoid temperature, carbon is rejected from the austenite and a two-phase structure of ferrite plus cementite is formed. The rate of cooling, the carbon content, and the presence of other alloying constituents all have a pronounced effect on the structure and distribution of these phases, and these in turn determine the mechanical properties of steels. This important transformation is discussed in more detail in a later section.

2.6.2 Diffusion in Solids

Phase transformations in alloys generally involve a redistribution of the various alloying elements, and the rate at which atoms can move through the crystal lattice by diffusion will control the transformation kinetics. (Certain diffusionless transformations are observed, as discussed in the next section, but diffusion-controlled kinetics form the larger class.) Diffusion also controls the kin-

TABLE 2-2

Invariant Reactions in a Binary Alloy System. (Reactions proceed to the right on cooling, to the left on heating.)

Monotectic:	$L \rightarrow L' + \alpha$
Eutectic:	$L \rightarrow \alpha + \beta$
Monotectoid:	$\alpha \rightarrow \alpha' + \beta$
Eutectoid:	$\alpha \rightarrow \beta + \gamma$
Syntectic:	$L + L' \rightarrow \alpha$
Peritectic:	$L + \alpha \rightarrow \beta$
Peritectoid:	$\alpha + \beta \rightarrow \gamma$

L and L′ denote liquid phases of different compositions; α and α′ denote solid phases of different compositions but the same crystal structure; α, β, and γ denote solid phases of different crystal structure.

etics of many other metallurgical processes—homogenization, stress relief, recovery, creep, bonding of dissimilar metals, and gas-metal reactions (oxidation, carburization, nitriding).

Diffusion in solids occurs by thermally induced atom jumps from one lattice site to another, or among interstices in the case of interstitial solutes. Conceptually, one expects that the mean residence time, τ, of an atom at a particular site may be expressed as

$$1/\tau = P\nu \exp[-\Delta H/RT], \tag{2-30}$$

where ν is the vibrational frequency of an atom, ΔH is the activation energy for jumping from one site to an adjacent one, and P is the probability that an atom with sufficient energy can actually reach an adjacent site. The factor P comprises three quantities: (1) the number of equivalent sites that an atom can reach in a single jump, (2) the probability that one of these sites is vacant, and (3) the probability that the gap between neighboring atoms through which the jumping atom must pass is large enough (as a result of thermal vibrations of the surrounding atoms) at the instant that its thermal energy is high enough.

Diffusion may be pictured, then, as a motion from site to site, governed by the fact that some sites are vacant and by the statistical probabilities that at some instant the lattice atoms relax so that one atom will pass from its own site to an adjacent vacant site. Atoms are therefore perambulating at random (at least for cubic crystals) throughout the solid, and their motion will tend to moderate any concentration gradient that may be present. From a straightforward analysis (see Chalmers [3], pages 372 ff) these concepts lead to Fick's law for isothermal diffusion,

$$J = -D\nabla C, \tag{2-31}$$

which defines the diffusion coefficient, D, in terms of net flux of matter, J, driven by a concentration gradient ∇C. It may be shown further that the macroscopic diffusion coefficient is related to the mean residence time, τ, as

$$D = \rho\alpha^2/\tau, \tag{2-32}$$

where α is the unit jump distance and ρ is a geometric factor which depends on how many equivalent directions there are in which the atoms can jump.

From the form of Eqs. 2-30 and 2-32 it is to be expected that, as the temperature is raised and thermal vibration frequencies increase, diffusion becomes more rapid. Experimental diffusion coefficients fall very close to the Arrhenius equation,

$$D = D_0 \exp[-Q/RT], \tag{2-33}$$

in which Q is the diffusion-activation energy. For most metals the constant D_0 lies between 0.01 and 1 cm²/sec. The activation energy is on the order of 20 kcal/mole for interstitial diffusion and in the range of 50 to 90 kcal/mole for substitutional diffusion.

If one considers a small volume element in a solid where the composition is not homogeneous and applies the principle of continuity (rate of accumulation equals difference between inflow and outflow), one obtains an expression for the time rate of change in composition by diffusion flow:

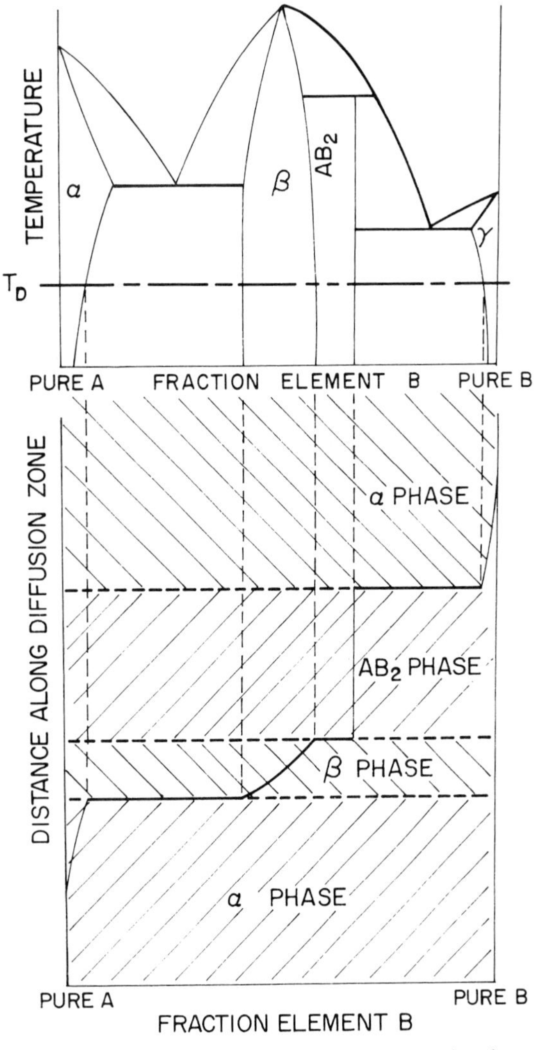

FIG. 2-36 Schematic illustrating the growth of intermediate layers in a diffusion zone and their relation to the equilibrium phase diagram.

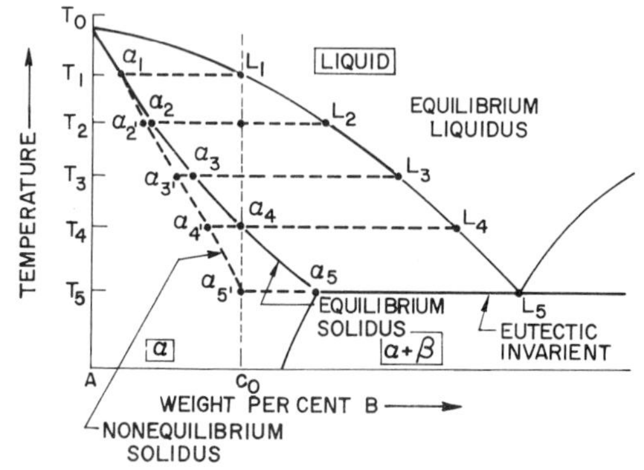

FIG. 2-37 Schematic illustrating nonequilibrium solidification and the development of microsegregation.

$$\partial C/\partial t = -\nabla \cdot J = \nabla \cdot D \nabla C. \qquad (2\text{-}34)$$

Under the proper boundary conditions (which are closely approximated by most practical configurations) the solution to this equation for uniaxial flow is of the form

$$C = f(x/\sqrt{t}). \qquad (2\text{-}35)$$

This is an important relation encountered in many engineering problems. For example, the growth of oxide films often follows a parabolic law,

$$x = k\sqrt{t}, \qquad (2\text{-}36)$$

where x is the film thickness at time t. This implies that diffusion through the film is controlling. (See Chapter 17.)

The concepts of phase equilibrium and diffusion may be applied to the problem of metal-fuel compatibility or, in fact, to the general problem of what happens when dissimilar metals are placed in contact at high temperatures for long times. Considering the simplest case of two pure metals placed in contact and allowed to interdiffuse, one expects that a distinct layer will form in the diffusion zone for each intermediate phase shown on the equilibrium diagram. This is shown schematically in Fig. 2-36. It is assumed that a piece of pure A was placed against a piece a piece of pure B and held at temperature T_D for some time. The diffusion zone exhibits two intermediate layers containing the compound AB_2 and the β phase. The width of each layer will increase proportionally to the square root of the diffusion time, and the width at any time depends on the diffusion coefficients in the layer and in adjacent layers and on the boundary compositions. Formation of bond layers by diffusion is obviously important when a cladding is bonded to a fuel material. Unless properly controlled, intermediate layers may be mechanically weak and brittle; they may impair heat transfer from the fuel; or they may provide a path for rapid corrosion in the event of a penetration through the cladding. Several cases of core-cladding interaction problems have been reviewed in Kaufmann [85].

The consequences of interdiffusion should always be investigated when any two materials are to be held in contact for prolonged times at high temperatures. Even though experiments may indicate little reaction between the base materials, one must still consider the possibility of diffusion paths such as grain boundaries and surfaces. It is quite normal for rates of diffusion down grain boundaries or along surfaces to be several times or even orders of magnitude greater than diffusion through the crystal lattices. Particular attention must be given to metals which form a low melting eutectic. When two such metals are placed in contact and held at a temperature above the eutectic, diffusion will form the eutectic composition, which melts and permits rapid reaction between the two pieces.

Eutectic melting from fuel-cladding interaction during accidental overload has occurred in the EBR-1 (core II), the OMRE, and the SRE reactors. (See Chapter 11 of Volume I.) The EBR-1 accident resulted from extremely abnormal operating conditions which were imposed on the reactor for a stability experiment and which were recognized to involve the risk of core meltdown [86]. The fuel was U-2 wt.% Zr, and the cladding was type-347 stainless steel. Unknown design conditions in the core (possibly bowing of the fuel rods) had promoted a positive temperature coefficient for the reactor reactivity, and the source of the instability was being investigated. When an unexpectedly large temperature excursion took place, an overheated region developed in the core, and the iron in the cladding combined with the uranium in the fuel in a partial meltdown. (The iron-uranium eutectic is at 725°C.)

In the OMRE, near the end of the planned life of the first core, a test element failed as a result of a cooling channel being plugged by inorganic particulate matter. This was believed to be iron rust formed from atmospheric moisture, which entered the core when the reactor was previously opened for cleaning. Some of the aluminum cladding in the failed element had entered into eutectic melting with the fuel. (The uranium-aluminum eutectic is at 640°C.)

Formation of the iron-uranium eutectic in the SRE was also caused by overheating following partial blocking of the coolant channels. The blockage in this case was the result of contamination by decomposition products from tetralin used as an auxiliary coolant for the primary sodium pump seals. Thirteen fuel elements and sixteen moderator elements failed in this incident.

2.6.3 Phase-Transformation Kinetics

This section will now bring together the concepts of phase equilibrium and solid-state diffusion to consider several phase transformation processes which lead to nonequilibrium metallurgical structures. These include coring during solidification, supersaturation and precipitation, inhibited invariant transformation, and diffusionless or martensitic transformations.

2.6.3.1 Nonequilibrium Solidification. When the temperature of a molten alloy is reduced below

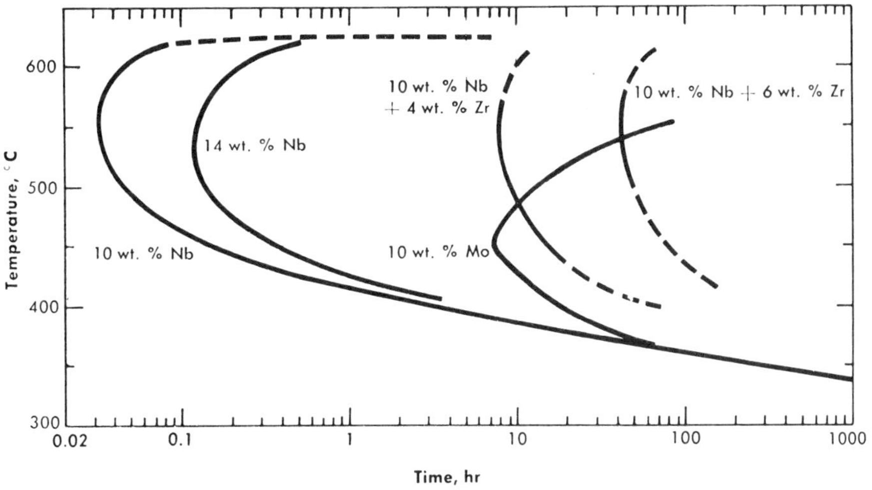

FIG. 2-38 Typical isothermal transformation diagrams (TTT curves). Curves show the time-temperature relations for the initiation of transformation from γ-phase uranium alloys solution-treated at 900°C (1650°F). (From Kaufmann [85].)

the liquidus* temperature, the solid crystallites formed will have a lower content of solute than the liquid, as illustrated in Fig. 2-37. As the temperature is reduced further, the crystallites get larger, and the solid and liquid composition must continuously readjust to higher solute content if equilibrium is to be maintained. Diffusion is too slow in solids for this readjustment to occur during the cooling times experienced in any practical casting procedure, even for very large ingots. Consequently, a composition gradient is established in the crystallites, and the average solid composition at any time is lower in solute than the equilibrium solidus would indicate. Appreciable sub-cooling is required to completely solidify the alloy, and this may produce eutectic freezing of the liquid remaining between the crystallites. Figure 2-37 illustrates a limiting case: liquid compositions to the right of C_0 would show eutectic solidification between the crystallites, and compositions to the left would not.

The structures of cast solids are characteristically "dendritic." The crystallites form as tree-like, branched dendrites with a composition gradient from center to outside of the branches, and such a structure is said to be "cored." The magnitude of the composition variations and the dendrite spacing depend on the partition ratio (ratio of equilibrium liquid to solid composition at any temperature), diffusion rates in the solid, the cooling rate, and other casting conditions which control the rate of nucleation of crystallites in the melt.

Dendritic structures influence the mechanical properties of castings because the regions between dendrite arms tend to be weakened by microsegregation (coring, precipitation of nonequilibrium second phases, and precipitation of equilibrium second phases such as impurities and inclusions) and by microporosity from included gases or shrinkage [87]. It has been reported [87] that for an alloy-steel casting with a yield strength of 185,000 lb/in.^{2} the reduction in area has been increased from 6% to 45%, a factor of 7.5 improvement, by homogenizing* the cast structure.

Dendritic microsegregation in the original cast ingot can influence the properties of wrought materials as well as cast materials. Flemings [87] has summarized work which relates the mechanical properties of wrought materials to segregation in the case ingot. He has reported, for example, that prolonged homogenization of a wrought low-alloy steel increased the transverse reduction in area by 50% and the impact energy from 31-39 ft-lb.

2.6.3.2 Precipitation. An important solid-state transformation for producing alloys with high hardness is the precipitation of a new phase from a solid solution. To obtain a reaction of this type one must select an alloy system whose equilibrium diagram shows that it is possible to cool from a single-phase solid, α, into a two-phase field, $\alpha + \beta$. Many commercial aluminum alloys, usually containing combinations of copper, magnesium, and silicon, and certain stainless steels, such as "17-4 PH" and "17-7 PH," are designed to be precipitation-hardening alloys.

The process for precipitation hardening proceeds as follows. The alloy is solution-treated by holding it at some temperature above the transformation boundary—that is, at a temperature in the solid-solution range—and rapidly quenched from the solutionizing temperature to retain the single phase as a highly supersaturated solution. Hard-

*"Liquidus" refers to the phase boundary between the liquid and the two-phase liquid-solid. "Solidus" refers to the boundary between the solid and the two-phase liquid-solid.

*"Homogenizing" refers to high-temperature annealing intended to reduce microsegregation by diffusion. Homogenizing may also be promoted by cold working and annealing to induce recrystallization.

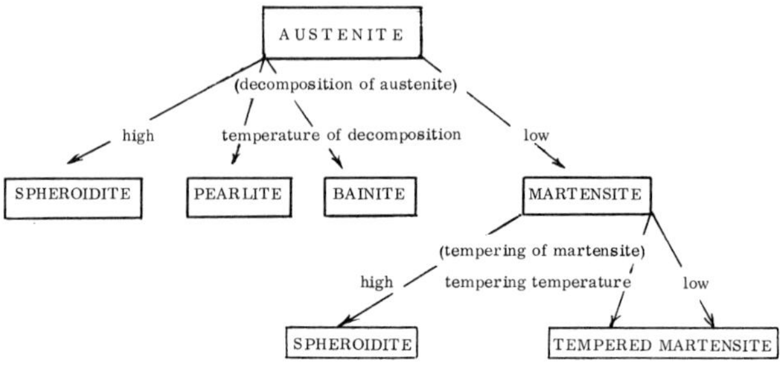

Constituent	Description
AUSTENITE	Face-centered-cubic phase of iron containing carbon in interstitial solid solution and other alloying elements, such as Mn, Ni, Cr, etc., in substitutional solid solution.
FERRITE	Body-centered-cubic phase of iron with very limited solubility for carbon. Like austenite, this phase may have other alloying elements in substitutional solid solution.
SPHEROIDITE	Microconstituent composed of large spheroids of carbide phase in a matrix of ferrite phase.
PEARLITE	Microconstituent composed of alternate plates of carbide phase and of ferrite phase.
BAINITE	Microconstituent composed of carbide phase and ferrite phase.
MARTENSITE	Body-centered-tetragonal phase of iron, produced by entrapping carbon atoms.
TEMPERED MARTENSITE	Microconstituent composed of carbide phase in a ferrite matrix.

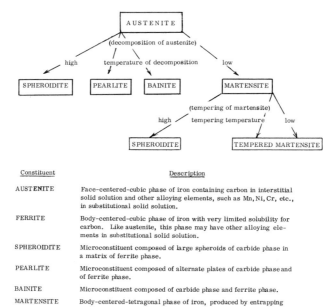

FIG. 2-39 Summary of the constituents produced in steels by the decomposition of austenite. (From Guy [2].)

ening is achieved by subsequently holding the alloy at a temperature below the transformation boundary for a specified time. A fine, uniformly distributed precipitate of the second phase is desired for maximum hardness. Holding these alloys for prolonged times or at higher temperatures in the precipitation range usually produces a coarsening of the precipitate and a corresponding decrease in hardness—a condition referred to as "over-aged."

2.6.3.3 Invariant Transformations. When the temperature of an alloy is suddenly dropped and held at some level below a eutectoid or peritectoid temperature, the indicated transformation does not occur instantaneously. Instead, if one follows the progress of transformation with time (by measuring a sensitive physical property, such as electrical resistivity), it is usually seen to begin at a slow rate, which first increases and then decreases as the reaction nears completion. The kinetics involves two steps: (1) nucleation of the new phase in the parent, and (2) growth of the embryonic nuclei. Nucleation may occur uniformly throughout the material ("homogeneous nucleation") or it may, and most often does, occur on preferred sites such as grain boundaries or inclusions ("heterogeneous nucleation"). Growth of the new constituent requires a redistribution of solute atoms among the various phases (since these have different compositions) and is usually controlled by diffusion.

To a first approximation the process of phase transformation may be pictured as a competition of two forces: (1) a thermodynamic driving force that induces the transformation, and (2) a resistive force that stems primarily from the slow diffusion of atoms through the metal. Since the driving force increases with increased subcooling, but diffusion rates decrease as the temperature is lowered, one expects that there is some point below the equilibrium temperature at which the transformation rate is a maximum. This is indeed the case, as illustrated by Fig. 2-38. These are isothermal "TTT diagrams" (time-temperature-transformation diagrams) for the transformation from γ phase for several uranium alloys. They show the time required to reach an indicated fractional transformation when the alloy is quenched from the γ phase and transformed isothermally at various temperatures.

The practical significance of the transformation kinetics in solids is tremendous. First, curves such as those illustrated in Fig. 2-38, and similar curves for many alloys, indicate that it is possible to retain high-temperature phases in a metastable state at temperatures well below the equilibrium transformation level. In certain cases, uranium alloys for example, the retained metastable phase has more desirable properties than the equilibrium phases. Second, the slow transformations permit a fine control of the reactions, and hence the microstructure, by suitable quenching and heating schedules. In this way, mechanical properties may be tailored to obtain optimum combinations of strength and ductility. The heat treating of steel, an important application of phase-transformation control, is discussed in a later section.

2.6.3.4 Martensitic Transformations. An entirely different class of phase transformations is represented by the formation of martensite from austenite in steel. This is a diffusionless process which occurs instantaneously when the temperature is lowered to a certain level. Martensite is compositionally identical to its parent austenite—that is, no redistribution of solute atoms is involved, so diffusion is not required. In steel the transformation from austenite to martensite is a transformation from a face-centered cubic to a body-centered tetragonal crystal, the latter being highly strained because it is supersaturated with carbon. As a result of the lattice strain, martensite is extremely hard and brittle.

Martensite transformations are characterized by two temperatures: an upper level, the M_s or "martensite start," above which martensite will not form; and a lower level, the M_f or "martensite finish," below which the steel is all martensite. Quenching to any temperature between the M_s and M_f results in only partial transformation from austenite to martensite. Martensitic transformations are observed in iron-nickel and in titanium alloys as well as in carbon steels.

2.6.4 Heat Treatment of Steels

The various procedures for heat treating steels are outlined here for two reasons: they are obviously of direct engineering importance, and they illustrate most of the phenomena which have been discussed in the preceding section on phase transformations.

The mechanical properties of steel may be varied over a wide range not only by altering the composition but also by controlling the decomposition of austenite through suitable heat treatment. Stated in another way, since the mechanical properties of a steel are quite dependent on its thermal history, careful control of fabrication processes is required to guarantee that a finished component has the expected properties. An increase in strength is usually achieved at the expense of a decrease in ductility, and designers should understand these implications of heat treating specifications when ordering steel for reactor plant components.

Referring to the iron-carbon diagram of Fig. 2-35, it is seen that face-centered cubic austenite (γ iron), which has a high solubility for carbon, is unstable below 723°C. Confining attention to the low-carbon, hypoeutectoid steels (i.e., those containing less carbon than the eutectoid composition), the diagram indicates that when the temperature is slowly lowered from the austenite range, the transformation should proceed by first precipitating ferrite (α iron) and then transforming by a eutectoid reaction to ferrite plus cementite (Fe_3C). The eutectoid transformation produces a "lamellar" structure of thin parallel plates which are alternately ferrite and cementite. This lamellar product is called "pearlite."

If the austenite is quenched, however, different structures may be produced by subsequent heat treatment. When transformation proceeds at high temperatures, proeutectoid ferrite plus pearlite result. As the transformation temperature is lowered, the pearlite which forms has a finer structure, and as a result the strength is increased. At low transformation temperatures, "bainite" is formed. This structure contains a dispersion of extremely fine carbides in a strained (carbon

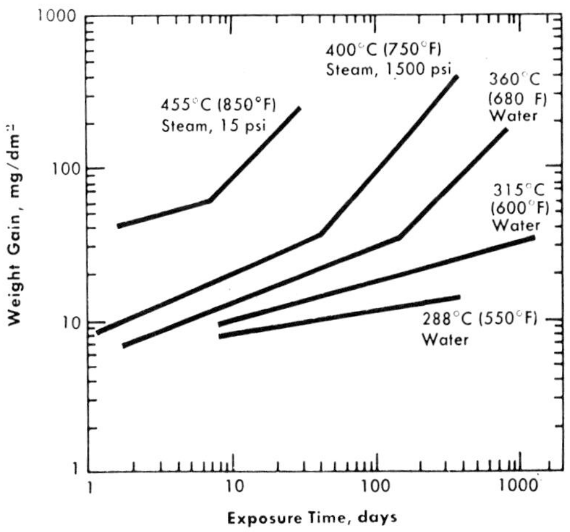

FIG. 2-40 Corrosion weight gains for Zircaloy-2 exposed to high temperature water or steam. (From Kaufmann [85].)

supersaturated) ferrite matrix. Bainitic steels have moderately high yield strengths and are tough.

When a steel is quenched (at a sufficiently high rate to restrict the formation of pearlite or bainite) to a temperature below the M_s, martensite forms. The ultimate hardness of quenched martensitic steels depends almost entirely on the carbon content, but the "hardenability"—that is, the ability to produce martensite by quenching—can be changed by alloying. Manganese, molybdenum, nickel, and chromium are particularly effective in increasing the hardenability of steels. Unless the steel is quenched below the M_f, some austenite is retained. This may subsequently transform to bainite, resulting in a mixed microstructure of bainite and martensite.

Quenched martensite is too brittle for most applications. Its ductility may be improved by "tempering," that is, annealing at a high temperature to precipitate some of the carbon out of the martensite. Tempering at progressively higher temperatures releases more carbon from the martensite and results in more coalescence of the carbide phase particles. It is possible, therefore, to control the final hardness of the steel by controlling the tempering temperature (or time). Tempering is additionally beneficial, particularly for thick sections, in that it reduces residual stresses in the quenched piece.

Figure 2-39 summarizes the various microstructures resulting from the decomposition of austenite. In the above discussion it is implied that these microstructural constituents are produced by quenching and isothermal annealing at selected temperatures below the eutectoid equilibrium. In actual practice different microstructures are produced by controlled cooling rates from the austenitizing temperature, with or without subsequent heat treatment. Commercially used heat treatments are summarized below. These definitions are from the Metals Handbook [7], and the reader is referred to that source or standard textbooks [2, 3] for further details.

Annealing. "Heating to and holding at a suitable temperature, and then cooling at a suitable rate, for such purposes as reducing hardness, improving machinability, facilitating cold working, producing a desired microstructure, or obtaining desired mechanical, physical, or other properties."

Austempering. "Quenching a ferrous alloy from a temperature above the transformation range [rapidly enough] to prevent the formation of high-temperature transformation products, and then holding the alloy until transformation is complete at a temperature below that of pearlite formation and above that of martensite formation."

Austenitizing. "Forming austenite by heating a ferrous alloy into the transformation range (partial austenitizing) or above the transformation range (complete austenitizing)."

Martempering (marquenching). "Quenching an austenitized ferrous alloy in a medium at a temperature in the upper part of the martensite range, or slightly above that range, and holding it in the medium until the temperature throughout the alloy is substantially uniform. The alloy is then allowed to cool in air through the martensite range."

Normalizing. "Heating a ferrous alloy to a suitable temperature above the transformation range and then cooling in air to a temperature substantially below the transformation range."

Spheroidizing. "Heating and cooling to produce a spheroidal or globular form of carbide in steel."

Tempering. "Reheating a quench-hardened or normalized ferrous alloy to a temperature below the transformation range and then cooling at any rate desired."

2.7 Corrosion and Oxidation

At the beginning of Sec. 2 it was stated that engineering metallurgy is by and large an empirical discipline. Of all the branches of metallurgy, this is most surely the case for corrosion technology. Perhaps, if the discussion could be limited to corrosion in a single fluid, a definitive review could be presented here. In fact, nuclear designers use or have considered a staggering variety of coolants—water, steam, sodium and sodium solutions, bismuth-based solution, molten salts, organic fluids, air, helium, carbon dioxide, and various uranium salt solutions in the homogeneous reactors. To discuss corrosion in any less than the few thousand pages to be found in the reference books [6, 88, 89], we shall obviously have to be severely restrictive.

The approach taken in this section is first to outline briefly the various classes of corrosive attack and then to consider two particularly significant corrosion phenomena—stress-corrosion cracking of austenite stainless steels, and hydrogen pickup in zirconium alloys.

2.7.1 Types of Corrosive Attack

The various types of corrosion may be logically considered in three classes: general corrosion, local corrosion, and corrosion involving mechanical stresses. Any one or all of these may occur in a particular system, and designers must necessarily be alert to all of them. Unfortunately (from the cost standpoint) there is no substitute for a dynamic loop test to demonstrate the resistance of actual components or appropriate mock-ups to the several ways in which they may fail by corrosive attack. The variables are so numerous, and seemingly minor changes in such conditions as coolant chemistry or surface chemistry can have so striking an effect on corrosion rates, that it would be foolhardy to predict the performance of unknown or untried systems. Evidence to support this statement is the several instances of extensive plant delay while major components were replaced or rebuilt to circumvent a corrosion problem. It should be apparent, from the magnitude of the costs involved, that the best engineering talent was employed in the design of these components, yet they failed to perform satisfactorily. Corrosion engineering is not simple, and one must often rely on the past mistakes of others.

2.7.1.1 General corrosion includes direct chemical attack, electrochemical attack, solution corrosion, and erosion. Direct attack is essentially an ordinary chemical reaction such as the oxidation of iron to ferrous ions by hydrogen ions in an acid pickling solution, or the oxidation of zirconium by hot water. Metals which form dense, adherent oxide films are self-protective to direct oxidation, but this is not to say that the reaction is stopped absolutely. As illustrated by Fig. 2-40, Zircaloy-2 gains weight continuously when exposed to hot water or steam. The weight gain is due to an ever increasing thickness of oxide film on the metal, which, while retarding corrosion, also impedes heat transfer. The limiting oxide thickness on fuel-element surfaces, from heat-transfer considerations, is something like 100 to 200 mg/dm^2 [85]. The corrosion resistance of the zirconium alloys is sensitive to contaminating impurities, especially nitrogen. Military specifications limit nitrogen content of Zircaloy-2 to a maximum of 100 ppm. Since zirconium so readily "getters" nitrogen from the atmosphere, even when the surface is only overheated by machining, it must be protected during any fabrication process involving high temperatures. (See Sec. 3 for further discussion of the Zircaloys.)

Chromium additions markedly improve the oxidation resistance of iron; hence the development of stainless steels containing around 18% chromium. Corrosion rates of austenitic stainless steels (304, 316, 347) in flowing reactor primary water at 600°F are in the range of 60 to 180 mg/dm^2/year, equivalent to a penetration of 4×10^{-4} in./year. This rate is not appreciably affected by oxygen in the water (up to 5 ppm) or irradiation [85]. A high pH is beneficial, and the primary coolant of water reactors is typically kept at a pH around 8 to 11. (The primary reason for a controlled high pH is to reduce the rate of crud deposition on the fuel-element heat-transfer surfaces.) The stainless steels are also found to have suitable resistance to general corrosion attack in liquid metals and oxidizing gases such as steam and carbon dioxide. There are, however, problems of local attack in some media, as discussed below.

Commercial purity aluminum (1100 alloy) forms a protective film in water below about 250°F, and is used extensively in low-temperature reactors such as the MTR. To cite an example of the dubious nature of nonrepresentative corrosion data, the corrosion rate of X8001 aluminum (1% nickel in commercial aluminum) in static water tests at 300°C (575°F) has been observed to be as low as 5×10^{-4} in./year, a quite acceptable value. However, in 260°C (500°F) water flowing at 20 ft/sec, the penetration rate was 100 times greater [85], probably because corrosion products are not allowed to saturate the water in a dynamic test. Aluminum-alloy claddings have been used successfully in water-cooled reactors at coolant temperatures as high as 420°F (SL-1). This is probably about the upper limit. Aluminum alloys are not at all suitable for sodium-cooled reactors. Even in a water environment, however, the corrosion resistance of aluminum is sensitive to the coolant purity, and reactor systems which have been thoroughly cleaned should be filled initially with demineralized water to avoid pitting. Aluminum alloys have acceptable resistance to hot gases (dry air, nitrogen, and carbon dioxide), but their low strength at high temperatures makes them unattractive for gas-cooled reactors. Aluminum resists general attack by organic coolants up to about 340°C (650°F), and SAP (sintered aluminum powder) is a possible candidate for cladding and pressure tubes in organic-cooled reactors.

Electrochemical, or galvanic, corrosion results whenever dissimilar metals in electrical contact, or even different areas of the same metal, have different electromotive potentials and are immersed in an electrolyte. The metals or areas become anodic and cathodic with respect to each other, and the anode metal dissolves by oxidation to a soluble ion or to an insoluble compound. This type of attack would be expected in aqueous media, but because water-cooled reactors use high resistance water, general galvanic attack is of little consequence. (See later discussion of local corrosion, however.)

General attack in liquid metals is usually by solution corrosion. In the hottest regions of a liquid metal loop the metal from the container is taken into solution. As the coolant reaches colder parts of the loop, the solubility decreases and the solute is deposited on the cold walls. Even though the solubility limits may be quite low, prolonged operation can result in the transport of a significant amount of material. The consequences may be serious either because of penetration of the container wall or because the deposition in cold regions may interfere with moving parts or may radioactively contaminate these parts.

Erosion is a possibility where coolant velocities are high. The effect of erosion may be either to remove protective oxide films, thereby accelerating general attack, or to mechanically remove surface metal. This latter attack is most probable when the coolant has a high-velocity component directed against the wall, at a piping bend or in pumps, for example. Impingement attack is prevented by proper design of coolant passages, but the combination of erosion and corrosion is not easily circumvented. It is for this reason that dynamic loop corrosion tests must always be the final investigation before a new material or component design is accepted for service.

2.7.1.2 Local corrosion connotes pitting, crevice corrosion, and intergranular attack. The rates of penetration by pitting may be several orders of magnitude greater than by general attack. The possibility of pitting, particularly in thin parts, must never be overlooked, but this mode of corrosion would show up in the same tests used to evaluate general corrosion. Pitting is initiated by inhomogeneties in a metal surface and propagates by galvanic action, since the oxygen content at the bottom of a pit is lower than at the surface. Manufacturing quality control, to ensure homogeneous metals without impurity inclusions, and restriction of the oxygen content of the coolant (particularly in liquid metals) should minimize or eliminate pitting.

Crevice corrosion, accelerated attack between close-fitting surfaces, is a galvanic attack similar to pitting except that the "pit" is built in. Crevice corrosion is extremely important where moving parts are involved, since it may lead to excessive wear, high torque, or even complete seizure. As with pitting, crevice corrosion is reduced by restricting the oxygen content of the coolant. In water-cooled reactors, crevice corrosion is not considered to be a problem in systems where the oxygen content of the water is on the order of 0.1 cc/liter, as in hydrogen-bearing water. It must be considered under accidental or temporary high-oxygen conditions (e.g., during refueling), even if the condition exists for only a few days [90]. Crevice corrosion can also arise from residual lithium hydroxide, pickling acids, or cleaning solutions. Cleaning procedures during the fabrication of reactor components must be rigorously followed to preclude the failure of joints or the seizure of moving parts.

Intergranular attack occurs when there is a pronounced difference in reactivity between grain boundaries and the bulk metal. Certain stainless steels, such as type 304, are prone to intergranular attack when improperly heat treated. When these alloys are annealed in the range of 900 to 1400°F they become "sensitized"—chromium carbides are precipitated in the grain boundaries, thereby depleting the adjacent metal of chromium. The depleted zone is anodic with respect to the surrounding alloy, and corrosion proceeds rapidly along the grain boundaries. The heat-affected zones of type-304 stainless steel will be sensitized during welding, and "knife-line" corrosion will occur on either side of the weld. Type-304-stainless-steel-clad fuel elements in superheater experiments in the Vallecitos BWR showed intergranular attack, which was attributed to sensitization of the steel [91]. The maximum cladding temperature was 1200 to 1350°F.

Sensitization of stainless steels may be remedied by annealing above 1600°C to redissolve the carbides. Alternatively, if the carbon content of the steel is kept below 0.03% (as in type 304L) the carbide precipitation is inconsequential. Finally, the steel may be "stabilized" by adding elements which are strong carbide formers, such as titanium (type 321) or columbium (type 347). These elements tie up the carbon to prevent the formation of chromium carbide.

2.7.1.3 Corrosion with mechanical stress includes stress corrosion, corrosion fatigue, and fretting corrosion. The first of these is important

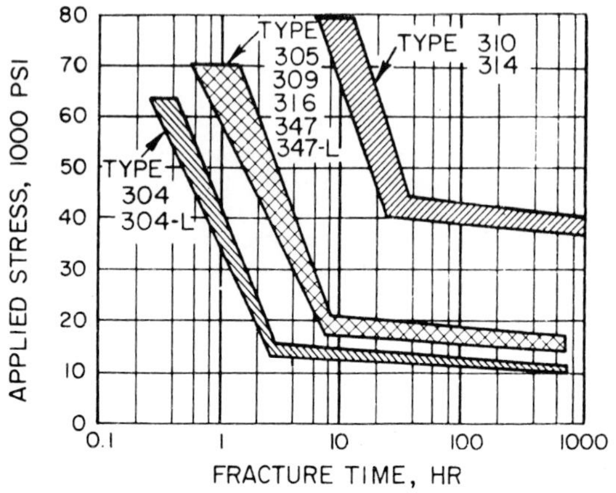

FIG. 2-41 Relative stress-corrosion cracking resistance of commercial stainless steels in boiling 42% magnesium chloride. (From Berry [93].)

enough to warrant extended discussion in the next section.

The consequences of corrosion fatigue are usually stated in terms of a reduction factor which relates the fatigue strength in a particular environment to the "normal" fatigue strength. Certain environments can materially reduce the fatigue strength (as much as a factor of 100 over tests conducted in vacuum [21]) either by pitting or intergranular attack, which effectively produces a notch, or by acceleration of crack propagation. As with fatigue in general, corrosion fatigue can be assessed only by extensive engineering tests.

Fretting corrosion occurs when there is a rubbing or impact motion between two surfaces. The effect of the surface load is to break off protective oxide films and permit accelerated attack. This is a particular problem where components are mechanically joined and subject to vibration—the joints between fuel assemblies and supporting fixtures, for example. The progressive corrosion, which may be accompanied by mechanical deformation of the mating surfaces, will cause the joint to become loosened. This in turn may increase the vibration and the rubbing or impact forces, thereby aggravating the problem. Fretting can be avoided only by proper design of mechanical joints or bearing surfaces. As has been stated several times before, there is no substitute for full-scale, dynamic loop tests of complex components to reveal the possibilities of such insidious processes as fretting corrosion.

2.7.2 Stress-Corrosion Cracking*

"Stainless steel would seem to be the most permanent of the materials we have, and it seemed the thing one could trust most." This opinion, attributed to Eero Saarinen [92] in explaining why he selected stainless steel for construction of the Gateway Arch in St. Louis, Missouri, would probably have been shared by most plant engineers a few years ago. Today, after numerous and often costly failures of stainless steel parts by stress-corrosion cracking, there may be some skepticism about the trustworthiness of this material.

It should first be realized that almost any metal or alloy can be made to fail by stress-corrosion cracking under certain conditions of applied or residual stresses and specific mild corrosives. Stress-corrosion failure in austenitic stainless steels is not peculiar to nuclear reactors; it was recognized and researched prior to their advent. "The novel and often stringent conditions of reactor operation only served to produce a well-known phenomenon in new ways." [93]

Many reported occurences of stress-corrosion cracking in nuclear plants have been associated with primary steam generators or superheaters. It seems, therefore, to occur most frequently where there is intermittent wetting or steam blanketing of heat-transfer surfaces. In addition, a great number of failures involve crevices: tube-to-tube sheet joints, for example.

The mechanism of stress-corrosion cracking of the austenitic steels is not at all clear at the present time. Certainly it results from the combined presence of applied or residual stresses and aqueous solutions containing chlorides or caustics (hydroxyl ions). Some general rules for susceptibility to this type of failure have been stated by Berry [93]:

"1. A pure metal is immune to stress-corrosion cracking.

"2. Alloys made from pure metals may be susceptible to cracking.

"3. There is no universal corrodent that will cause cracking in all alloys, i.e., the corroding conditions that produce cracking are specific to an alloy or alloy system.

"4. Cathodic protection* can prevent stress-corrosion cracking and even stop crack propagation if applied while cracking is in process.

"5. One or more minor impurity elements in a metal or alloy can govern its degree of immunity or susceptibility.

"6. Changes in structure or homogeneity of an alloy by heat treatment can influence its immunity or degree of susceptibility."

The subtle effects implied by the statements in this list, particularly the last two items, indicate the complexity of the problem. There is no agreement as to (1) the mechanism of crack initiation, (2) whether cracks propagate by a two-stage electrochemical-mechanical process or by electrochemical corrosion alone, and (3) whether the electrochemical corrosion is due to highly strained metal, attack at dislocation pileups, cathodic nitride precipitates, anodic hydride precipitates, phase changes in the metal, or some other phenomenon not yet imagined. Berry has reviewed the experimental evidence for the various proposed mechanisms.

More germane to the purposes of this chapter is a discussion of preventive measures which can be taken to eliminate stress-corrosion failures.

*This summary is largely extracted from a review by W. E. Berry [93]. A literature search on this subject is also available [94].

*That is, by making the piece to be protected a cathodic element in the corrosive environment.

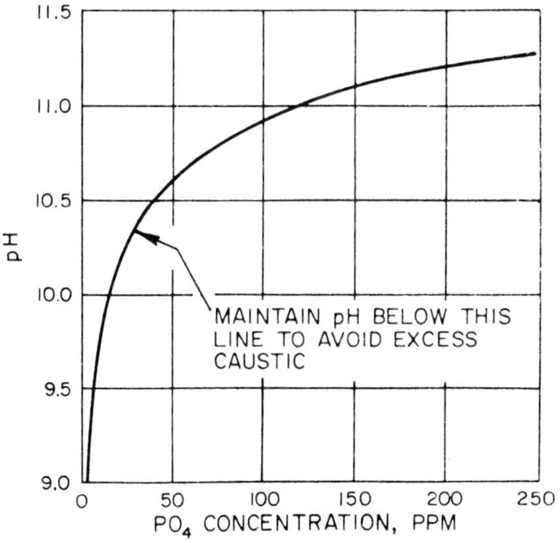

FIG. 2-42 Relation between pH and phosphate concentration defining areas of phosphate vs free caustic residues on evaporating surfaces. (After Singley et al. [95]; taken from Berry [93].)

These include reduction of stress levels, control environment chemistry, cathodic protection, and use of alternate designs or materials.

Figure 2-41 illustrates the influence of applied stress on the time to fracture in boiling 42% magnesium chloride.* Because of the complexity of the interrelation between alloy and environment, it is not possible to specify a universal threshold stress below which cracking will not occur. Failure of type-347 stainless steel has been reported for stresses as low as 2000 lb/in.^{2} in the condensing vapors above a solution containing 50 ppm chloride at 400°F. It would be extremely difficult to reduce residual stresses to such low levels. For many environments, however, solution annealing at 1850 to 1900°F after fabrication can reduce residual stresses sufficiently to preclude cracking, only, of course, if the applied stresses are also low.

An obvious way to eliminate stress-corrosion failures is to remove the offending species from the corroding environment. In boiler water contaminated with chlorides, both oxygen and chloride must be present for cracking to occur. Oxygen is not a requisite for caustic cracking, but this type of attack does not occur as readily as chloride stress corrosion. In comparable tests, for example, greater than 50,000 ppm NaOH was required to crack type-347 stainless steel, whereas only 10 ppm chloride produced rapid failure.

The use of inhibitors in boiler waters has been studied to some extent. Of particular interest is the regulation of phosphate-treated water to prevent free caustic formation on evaporation surfaces, illustrated by Fig. 2-42. Operation above the curve in this figure results in free caustic, and this is believed to be the cause of cracking observed in one of the Shippingport PWR steam generators [95]. Subsequent practice at Shippingport called for operation at 0.2 pH units below this curve.

Cathodic protection, while demonstrably effective in preventing stress-corrosion cracking, is not practical in reactor environments. Reactor primary water has a low conductivity, and sacrificial anodes are undesirable because of the corrosion products they introduce into the coolant.

Although chlorides are not normally present as contaminants in reactor cooling systems, stress corrosion can still occur if oxygen is present and if there are crevices connected with entrapped slag or flux containing chlorides [96]. If residual welding stresses are present, these add to the operational stresses, with the weld acting as a stress raiser, and create the necessary and sufficient conditions for stress corrosion. In several cases, chloride-containing materials in contact with the outside of primary piping or vessels have become wet and formed solutions containing considerable amounts of chloride. For instance, some pipe insulation has been found to contain chlorides which leach out when the insulation becomes wet. In several instances this has led to stress-corosion cracking of these pipes. Trichloroethylene is often used as a machine-shop degreasing agent. There is always a chance that in an improperly cleaned system some of this material could break down under irradiation to release chloride ions.

The effect of fluoride ions, which could be introduced from the use of Teflon seals in the system, is somewhat inconclusive. The effect on type-347

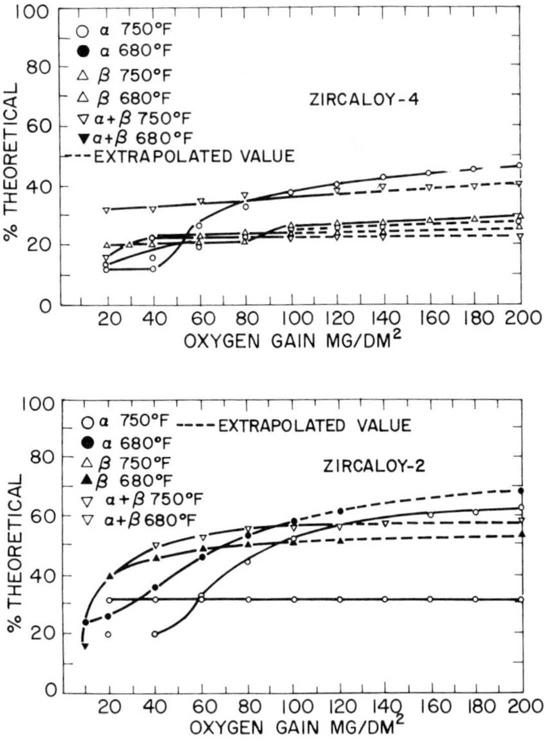

FIG. 2-43 Hydrogen pickup characteristics of the Zircaloys. (From Kass [95].)

*A boiling aqueous solution of $MgCl_2$ causes cracking in susceptible alloys in a few hours. This is commonly used as an accelerated test.

stainless steel has been described as negligible to 230°C (446°F) [97] and as inhibitory to corrosion at a concentration of 190 ppm [98]. Fluoride solutions in strongly acidic or strongly oxidizing media have been shown to be highly corrosive to nickel in alloys. Inconel, however, has shown resistance to molten fluorides in the absence of a highly acidic environment, even under neutron flux and gamma radiation [99].

At present, the most effective way to prevent stress-corrosion failure is by proper design and selection of materials. Almost all failures reported have been related to conditions which concentrate chlorides or caustic. Consequently, stagnant areas and crevices in heat-transfer surfaces should be avoided. Typical of such concentrating areas in combination with high residual stresses is the crevice between rolled-in tubes and tube sheets. The cracking of stainless steel fuel cladding in nuclear superheaters is attributed to concentration of chlorides carried over by wet steam. It has been pointed out that failures do not occur in conventional power plants because the steam is dry and appreciably superheated before it reaches stainless steel surfaces. Furthermore, the steam in conventional power plants contains very little oxygen compared with a superheater receiving steam from a boiling water reactor [93]. There also appears to be at least one other mechanism for concentration of chlorides; namely, that chlorides are concentrated in aluminum corrosion products, presumably by an ion-exchange mechanism [93].

Low-alloy steels and nickel-base alloys, which are not susceptible to chloride stress-corrosion cracking, are the major alternatives to the austenitic stainless steels. However, the low alloy steels exhibit caustic cracking and are subject to severe pitting in the same environments that cause chloride stress corrosion in stainless steels. As a result, the nickel-base alloys appear to be the only suitable materials to use where stress-corrosion environments will be encountered.

2.7.3 Hydrogen Embrittlement of Zirconium Alloys*

During aqueous corrosion, zirconium alloys absorb a certain portion of the nascent hydrogen produced by the corrosion reaction. Because of limited solubility in the alloy, the absorbed hydrogen will form a precipitated second phase of zirconium hydride, which tends to embrittle the metal. Thus, there are two pertinent aspects to this problem: (1) the rate of hydrogen uptake during corrosion, and (2) the effect of hydrogen content on the mechanical properties of the alloy.

As early as 1953 it was observed that Zircaloy-2 absorbs hydrogen during corrosion, that the quantity of hydrogen absorbed was proportional to the amount of corrosion, and that small quantities of hydrogen as a precipitated hydride phase exert embrittling effects upon the mechanical properties of zirconium

*This section is extracted from several articles which appeared in an ASTM monograph [101]. A literature search on this subject is also available [102].

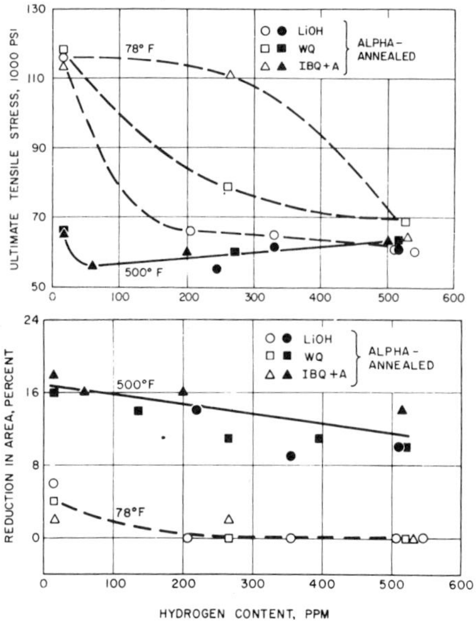

FIG. 2-44 The effect of hybriding on the notched tensile properties of Zircaloy-2. Test temperatures are indicated on the curves. (From Babyak et al. [103].)

alloys [100]. Because the amount of hydrogen taken up was small (some 20 to 40% of the amount released by the corrosion reaction), this was not considered to be a significant problem for short-life elements. In later work, however, longer life Zircaloy components were being anticipated; accelerated hydriding was observed at cladding defects in UO fuel rods; and it was seen that hydrogen could diffuse relatively rapidly down thermal gradients, thereby concentrating and eventually precipitating hydride platelets in cooler regions of a component. As a result, considerable attention was given to reducing the hydrogen-pickup rate. This work led to the development of Zircaloy-4, a low-nickel modification of Zircaloy-2. (See Sec. 3.2.)

The hydrogen-absorption characteristics are usually expressed in terms of "percent theoretical hydrogen," where 100% theoretical would indicate the absorption of all hydrogen from the reaction

$$Zr + 2H_2O = ZrO_2 + 2H_2. \qquad (2\text{-}37)$$

Figure 2-43 summarizes some of the data for Zircaloy-2 and Zircaloy-4. Two autoclave conditions are indicated, 680°F water and 750°F steam. The samples tested were in three metallurgical conditions—(1) α annealed (1250°F), (2) fabricated at 1550°F (α plus β), and (3) quenched from the β range. (These conditions are indicated in the figure as α, $\alpha + \beta$, and β, respectively.)

The superiority of Zircaloy-4 in 680°F water is apparent from Fig. 2-43. Since the corrosion weight gains of Zircaloys-2 and -4 are practically identical [100], the lower fractional hydrogen pickup of Zircaloy-4 indicates also a lower absolute absorption. Of the various metallurgical conditions, the $\alpha + \beta$ fabrication shows the lowest pickup for Zircaloy-4 although the differences are small for the 680°F water test. In the 750°F steam test, only β-quenched Zircaloy-4 has a lower hydrogen pickup when compared with Zircaloy-2.

Babyak et al. [103] have reviewed the influence of hydriding on the tensile properties of Zircaloy. Since the hydrogen embrittlement is attributed to the presence of a second-phase hydride ($ZrH_{1.5}$), the mechanical behavior does not depend on the total hydrogen content per se but rather on the size, distribution, and orientation of the precipitated hydrides. For example, room-temperature test data on samples containing up to 500 ppm hydrogen with the hydrides randomly oriented show little effect on the tensile properties; similar tests on samples containing only 120 ppm hydrogen, but with the hydrides preferentially oriented perpendicular to the stress direction, indicate quite brittle behavior.

Figure 2-44 summarizes the tests reported by Babyak et al. These data are for α-annealed Zircaloy-2 samples that have been hydrogenated in different ways. The symbol "LiOH" refers to samples exposed to concentrated aqueous solutions of lithium hydroxide at 680°F; "WQ" refers to samples equilibrated with zirconium hydride at 995°F and water quenched; "IBQ + A" indicates the same equilibration with zirconium hydride but followed by a quench into iced brine and a subsequent aging heat treatment. The first of these treatments results in grain-boundary precipitation of the hydride, the second produces acicular hydrides primarily at grain boundaries, and the third treatment yields hydrides that are precipitated uniformly on crystallographic planes within the grains. The following conclusions are drawn from this work.

The most pronounced effects of hydriding are the reduction of the notched tensile strength and notched test ductility when the hydrides are in the grain boundaries. The unnotched tensile properties are rather insensitive to either the hydrogen content or the hydride morphology. Metallographic examination of fracture surfaces revealed that, regardless of the hydride morphology, fracture initially occurs in the hydride platelets, and the cracks then propagate through the ductile metal.

The influence of hydriding on the in-reactor performance of Zircaloy cladding has been reported by several groups [100, 102, 104, 105]. Burns and Maffei [105] report that, while the percent hydrogen pickup is unaffected, the corrosion rate of Zircaloy in a fast neutron environment (ETR) is an order of magnitude greater than out-of-pile. Hence the absolute hydrogen absorption is increased by the same factor. The most severe effects of hydriding are seen in the Zircaloy clad of defected UO_2 fuel rods. When a small hole is present in the clad, water enters the tube and reacts rapidly with the hot inner surface. The hydrogen absorbed by the metal diffuses to the outer clad surface, because of the thermal gradient, and precipitates there as hydrides. This effect is severely aggravated by fluorine contamination (a few hundred ppm) of the UO_2 [104]. As a comparative example of the superiority of Zircaloy-4, Kass reports [100] that defected, Zircaloy-2-clad fuel rods failed by hydride embrittlement in 3-1/4 days, whereas identical rods irradiated under the same conditions, but with Zircaloy-4 cladding, did not fail after 40 days exposure.

2.8 Concluding Remarks

The reader must realize that the contents of this section on the engineering properties of materials constitute only the barest introduction to the subject. In closing, we wish to emphasize again the interplay of the various metallurgical factors which have been discussed and to indicate the nature of the real complexities that must be faced by design engineers.

In any practical design problem, many metallurgical properties and processes are quite likely to be superimposed. Temperature gradients, for example, will not only set up thermal stresses but may also induce thermal diffusion (e.g., of hydrogen in zirconium). Redistribution of a constituent by thermal diffusion can, in turn, significantly influence the mechanical behavior of the materials and thereby aggravate the stress design problem. Power cycles, to take another example, may not only cause thermal stresses but may also produce irreversible changes in the materials (thermal fatigue). The combination of a steady stress superimposed on cyclic loads can lead to a more severe degradation of the material than either load can cause separately. And the addition of fast-neutron irradiation to other environmental conditions generally only increases the designer's concern.

Thus, the design engineer cannot consider each metallurgical factor separately but must view the whole environment at once. That this is possible is evidenced by the many successful performances of nuclear reactor components.

3. MATERIALS USED IN NUCLEAR REACTOR SYSTEMS

The properties of important structural materials have been described in the preceding section of this chapter. This section is intended only to outline, by chemical composition and a few physical properties, the various structural, moderator, and control materials which have been used or are seriously considered for use in nuclear reactor systems. For detailed information about these materials the reader is referred to handbooks [7-18]. A list of materials specifications of interest is included in Sec. 5 of this chapter.

3.1 Ferrous Alloys

By far the greatest tonnage of materials used in reactor systems consists of one or more of the structural steel alloys. In many reactor plants the primary system is contained in austenitic stainless steel. In pressurized water and boiling water systems the high primary-system pressure requires heavy-walled reactor vessels, and these are constructed of low-alloy steel with an internal cladding of stainless steel. Gas-cooled reactors, where corrosion is less severe, use unclad carbon steel vessels. In addition to the primary boundary, many reactors utilize stainless steel-core supporting structures and some have stainless-steel-clad fuel elements.

There are two outstanding problems associated with the structural steels—first, the increase in ductile-brittle transition temperature of the boiler steels, which is caused by fast-neutron damage: second, stress-corrosion cracking of austenitic stainless steel in certain aqueous environments. These were discussed in Sec. 2.

3.1.1 Carbon and Low-Alloy Steels

The composition and properties of several carbon and low-alloy steels are listed in Table 3-1 at the end of this section. The ASTM A212-B [106] and A302-B [107] steels are used for heavy-walled pressure vessels in pressurized and boiling water reactors. These vessels currently are as large as 20 ft in diameter and 70 ft high, with 6- to 14-in.-thick walls, and weigh several hundred tons. Vessel fabricators in the United States have tooled up for the fabrication of even larger vessels weighing as much as 1000 tons [20].

The demand for high quality in these vessels requires that the steel be prepared in basic (as opposed to acidic) electric-arc furnaces with rigorous control of the charge materials. The melts are vacuum degassed to remove oxygen, hydrogen, and nitrogen. Cast ingots are rolled, often with substantial cross-rolling, to the desired plate thickness. Quality control of the boiler plate requires visual and ultrasonic inspection and rejection of the piece if the ultrasonic test indicates a defect whose diameter is larger than 3 in. or one-half the plate thickness, whichever is larger [108]. Any defect perpendicular to the plate surface is unacceptable.

Large vessels are fabricated by rolling or press-forging the heavy plate into segments which are welded into cylinders with hemispherical heads. All welds are completely radiographed, usually several times—first after the root pass and then after every several successive passes. The vessels are clad internally with stainless steel weld overlays. Some vessels (steam generators or pressurizers) may have roll-bonded or spot-welded stainless cladding.

Design and material studies for thick-walled pressure vessels are continually being made under

TABLE 3-1

Selected Properties of Representative Reactor Materials[a]

Alloy	Nominal Composition (weight percent)	Room Temperature Tensile Properties[b]: Tensile Str. (1000 psi)	0.2% Yld. Str. (1000 psi)	% Elongation
Part A. Alloy steels				
ASTM A212	0.35 C, 0.9 Mn, 0.23 Si	70-85	38	19
ASTM A302	0.23 C, 0.8 Mn, 0.23 Si, 2.3 Ni, 0.5 Mo	80-100	50	17
SAE 4140	0.40 C, 0.9 Mn, 0.28 Si, 0.95 Cr, 0.2 Mo			
SAE 4340	0.40 C, 0.7 Mn, 0.28 Si, 0.80 Cr, 0.25 Mo, 1.8 Mi			
Part B. Stainless steels				
AISI 304	0.08 C, 18 Cr, 8 Ni	85-180	30-125	50-10
AISI 304L	0.03 C, 18 Cr, 8 Ni	85-180	30-125	60-8
AISI 316	0.08 C, 18 Cr, 12 Ni, 2.5 Mo	80-150	30-135	50-6
AISI 347	0.08 C, 18 Cr, 11 Ni, Cb + Ta = 10× C	91	40	50
AISI 348	0.08 C, 18 Cr, 11 Ni, Cb + Ta = 10× C, Ta < 0.1, Co < 0.2	91	40	50
AISI 430	0.12 C, 16 Cr	70-90	40-55	30-20
17-4 PH (AISI 630)	0.07 C, 1.0 Mn, 1.0 Si, 16.5 Cr, 4 Ni, 0.3 Cb, 4 Cu	-165	-110	-6
Part C. Nickel Alloy				
Inconel 600	77 Ni, 16 Cr, 7 Fe, 0.2 Mn, 0.2 Si	100-160	45-100	43-28
Incoloy 800	32 Ni, 21 Cr, 45 Fe, 0.9 Mn, 0.4 Ti	82	32	46
Hastelloy R-235	60 Ni, 15 Cr, 10 Fe, 5 Mo, 2.5 Co, 2.5 Ti, 2 Al	140	85	33
Part D. Zirconium alloys				
Zircaloy-2	1.5 Sn, 0.15 Fe, 0.10 Cr, 0.05 Ni	71-110	44-97	26-5
Zircaloy-4	1.5 Sn, 0.20 Fe, 0.10 Cr, Ni < 0.007			
Part E. Light-metal alloys				
1100 aluminum	99.0% aluminum ("commercial purity")	13-24	5-22	45-15
6061 aluminum	1.0 Mg, 0.6 Si, 0.3 Cu, 0.3 Cr	18-45	8-40	30-17
X8001 aluminum	1.0 Ni, 0.5 Fe			
Magnox (A12)	1.0 Al, 0.04 Be			

[a]The properties given in this table are nominal and are intended only to illustrate the characteristics of representative materials. The reader is referred to appropriate specifications (see Table 5-1) or suppliers' data for specific information.
[b]Single figures for tensile properties indicate annealed alloys; ranges show properties from annealed to hardest condition.

sponsorship of the Welding Research Council and American Society of Mechanical Engineers [109-112]. Improved heat treatments have been discussed [113] and a compilation of properties of steel sections greater than 4 in. (10 cm) in thickness is in preparation at this writing [114]. An informative treatment on fracture of A302B steel has been given [115].

The alloy steels, such as SAE 4140 or 4340, which have higher yield strengths than the pressure-vessel steels, are used as bolting materials or other members where high stresses are required.

3.1.2 Austenitic Stainless Steels

Compositions and properties of several austenitic stainless steels are given in Table 3-1. Types 304L, 321, 347, and 348 are not subject to sensitization, the first because of low carbon content and the latter three because of stabilizing additions. The austenitic alloys are useful up to about 1000°F before creep must be taken into account. The principal problems associated with using the stainless steels—intergranular and stress corrosion—have been discussed in previous sections of this chapter. Additional information may be found in a review article by Chernock et al. [116].

3.1.3 Other Stainless Steels

The "400 series" stainless steels are ferritic or martensitic (hardenable) alloys, and Table 3-1 lists the properties of the more commonly encountered steels of this type. The non-austenitic grades are used where some corrosion resistance must be traded for lower thermal expansion and magnetic properties. The rotor cans of control rod drive motors or canned circulating pumps use the ferritic stainless steels, for example. The hardenable alloys (types 403, 410, 420) are normally cutlery steels but are used in some applications where high loads are encountered. (See discussion of the failure of type-410 bolts in the SM-1 in Chapter 11 of Volume 1.)

The precipitation-hardening steels, 17-4 PH and 17-7 PH, are used as bearing surfaces and high-strength parts in rod drive mechanisms. It should be emphasized that the high hardness of these alloys is achieved at the expense of ductility and an increased susceptibility to stress-corrosion cracking. (See Sec. 5.1.3.) The precipitation-hardened alloys must be handled with care after they have been heat treated. Even slight bending may introduce high residual stresses or microcracks which could aggravate the stress-corrosion problem.

3.2 Zirconium Alloys

Because of its low neutron-absorption cross-section, zirconium is a useful metal for reactor core components. Early work was directed toward the utilization of crystal bar zirconium from the Van Arkel process (iodide decomposition). It was found, however, that while sufficiently pure crystal bar zirconium has adequate corrosion resistance, its behavior in water is sensitive to impurity content, especially nitrogen. As a result of this observation, and because there was a desire to use the less expensive (and less pure) sponge zirconium from the Kroll process, development efforts were undertaken to improve the corrosion resistance of zirconium by alloying. (See Kass [100] for a review of the alloy development work.) The resulting alloys, which are now widely used, are Zircaloy-2 and Zircaloy-4. The latter is essentially a low nickel modification of Zircaloy-2 which has a lower rate of hydrogen uptake from aqueous corrosion. Properties of these alloys are listed in Table 3-1, and further information is available in a review by Thomas et al. [117].

The corrosion behavior of the Zircaloys in hot water and steam has been extensively characterized [101]. In addition to the effects of nickel content and of metallurgical condition, which were discussed in Sec. 2.7.3, several aspects of coolant chemistry are important. These have been summarized by Kass [100] as follows: (1) "No acclerated corrosion or hydrogen pickup of Zircaloy cladding is to be expected in reactors using LiOH as a means of crud control if the pH is maintained below pH 11.3 and if there are no crevice regions where LiOH can concentrate." (2) Traces of fluoride (about 10 ppm) in the coolant can significantly increase the initial corrosion rates and hydrogen uptake of Zircaloy-2. This contaminant can enter the coolant by leaching of fluorocarbon plastics (Viton A, Teflon) used as seals [104]. (3) Increasing molecular hydrogen (H_2) in the water increases the hydrogen uptake of Zircaloy-2, but Zircaloy-4 is little affected by the hydrogen content in either static or dynamic tests. (4) Hydrogen uptake is markedly decreased by oxygen dissolved in the corroding medium, but it is increased by irradiation [118].

Further evidence of the detrimental nature of fluoride on the corrosion properties of Zircaloy has been reported by Berry [104]. Incomplete removal of pickling acids (aqueous HF-HNO) used to remove contaminated surfaces from Zircaloy components will severely degrade the corrosion behavior. In fact, Berry reports that a 5-min delay in transferring Zircaloy from acid to rinse bath results in a cloudy corrosion film (as opposed to the desired shiny, black film) and accelerated initial corrosion rates. Longer delays in rinsing will lead to spalling (nonprotective) films. While fluorides are extremely harmful, chlorides (up to 10,000 ppm) and iodides (up to 1270 ppm) do not seem to affect the corrosion behavior of Zircaloy in water.

Because of the relatively low strength of Zircaloy at elevated temperatures (see Figs. 2-4 and 2-9), considerable effort has been, and continues to be, devoted to developing higher strength zirconium-base alloys. Modest success has been achieved, the most promising alloys being Zr-2.5%Cb and Zr-3%Cb-1%Sn [119]. The improved high-temperature performance of Zr-2.5%Cb is indicated in Fig. 2-10. In addition to these alloys, which are primarily intended for service in water environments, the Canadians have reported the development of Zr-2.5%Cb-0.5%Cu for use in

CO_2 [120]. It seems apparent at this time that these higher strength, zirconium-base alloys have considerable promise. Their acceptance for operational use awaits further confirmation from long-term, inpile testing and scale-up of fabrication processes for dependable quality control.

3.3 High-Temperature Alloys

There is an ever-present desire to improve the thermal efficiency of nuclear reactors by raising the coolant outlet temperature. Liquid-metal and gas-cooled reactors, of course, operate at much higher temperatures than the present water-cooled reactors, and nuclear superheater stages for water reactors are being pursued. Thus, there is a definite demand for metals with strength and oxidation resistance at high temperatures.

3.3.1 Nickel-Base Alloys

The Inconels and Hastelloys are important nickel-base alloys. Properties of several of these are listed in Table 3-1. Because of the recurrent difficulties with stress corrosion, the nickel-base alloys have been substituted in many components which were fabricated from austenitic stainless steels—steam superheaters, for example—and this trend will undoubtedly continue. The primary obstacles to widespread use of the Inconels or Hastelloys (or other commercial nickel alloys) are their expense and the high neutron-absorption cross section. This latter objection has precluded their use in thermal reactors and may possibly prove unacceptable in fast power reactors as well.

3.3.2 Refractory Metals

The refractory metals include (in order of increasing atomic weight) titanium, vanadium, chromium, zirconium, columbium (niobium), molybdenum, hafnium, tantalum, and tungsten. These metals have melting points ranging from 1688°C (titanium) to 3410°C (tungsten). Hafnium, with its high neutron-absorption cross section, is used as a control material, and zirconium, with its very low cross section, is used as a cladding. The remainder, in spite of their refractory nature, do not have sufficiently attractive nuclear properties to make them generally useful except as minor alloying additions (molybdenum and columbium as γ-phase stabilizers in uranium, for example). Perhaps the most attention has been given to columbium alloys, such as Cb-1% Zr. The capability of this material is indicated in Fig. 2-4.

3.4 Light-Metal Alloys

There are three metals of low atomic mass which have important uses in reactor cores. Two of these, aluminum and magnesium, are utilized for structural members or for fuel cladding or fuel alloys. The third, beryllium or BeO, is used primarily as a solid moderator, but has also been used as a structural material.

3.4.1 Aluminum Alloys

Aluminum has a low thermal-neutron-absorption cross section and is an inexpensive, low-temperature structural material. The neutron absorptions produce only short-lived gamma radioactivity which makes it relatively easy to repair in-core aluminum components. As stated previously, the poor corrosion resistance in water precludes the use of aluminum alloys in water above 250-300°C (500-575°F), but the adaptability of aluminum to research and other low-power water-cooled reactors has been well exploited. It has a low creep strength above 300°C (575°F) and even at somewhat lower temperatures should be used with caution, although its excellent thermal conductivity does much to alleviate local overheating. The various aluminum alloys exhibit low creep strength generally. (See Fig. 2-9.)

An aluminum product with improved creep strength is SAP (sintered aluminum powder). This material is made from fine, partially oxidized aluminum powders which are cold compacted, sintered, and hot worked to produce an aluminum matrix containing a fine dispersion of aluminum oxide particles. Various investigations have shown that the mechanical properties of SAP depend less on the volume fraction of the oxide (usually 6 to 15%) than on the shape, size, and distribution of the dispersed phase [121].

It is reported [121] that SAP retains its strength up to 500°C (930°F), is stable to carbon dioxide and organic coolants, and (with nickel and iron additions) even has attractive corrosion properties in water (to 300°C) and superheated steam (to 350°C). SAP is being considered for pressure tubes in organic-cooled, heavy-water-moderated reactors [122, 123, 124], but there are several outstanding problems at present. The principal difficulties are in the development of processes to fabricate large SAP components or thin-wall tubing with the required homogeneity and reproducibility of mechanical properties, and in the improvement or accommodation (by design) of the very low creep-rupture ductility of this material.

3.4.2 Magnesium Alloys

Magnesium has a low thermal-neutron-absorption cross section and is a good low-temperature structural-alloy base, but because of its poor corrosion resistance to hot water and to liquid metals, its use in nuclear systems is limited essentially to zero-power, water-cooled reactors and to gas-cooled power reactors. The magnesium alloy, Magnox, has been widely used in the United Kingdom gas-cooled power reactors since it does not alloy with uranium and has good oxidation resistance at temperatures up to 350°C (670°F) in carbon dioxide. Beryllium in the amount of 0.01 wt.% addition in some magnesium alloys will protect against oxidation almost up to the melting point.

Alloys of magnesium with 3 to 9 wt.% aluminum and smaller amounts of manganese and zinc are available commercially. The most satisfactory alloys for improving the mechanical properties at increasing temperatures are those containing thorium or lanthanides, but these elements are undesirable for reactor application because of their nuclear characteristics.

3.4.3 Beryllium

Beryllium has an extremely low thermal-neutron-absorption cross section and excellent neutron elastic scattering characteristics which make it valuable as a reflector or moderator material. It has good high-temperature strength, low thermal expansivity, and moderate thermal conductivity. Its moderately high melting point is an advantage in a structural material, but its crystal anisotropy and brittleness make it unreliable as a stress-carrying member except with high safety factors. Furthermore, the generation of helium atoms in beryllium from thermal-neutron absorption can be expected to reduce the already low ductility. The mechanical properties of beryllium are strongly dependent upon purity and working treatment, and it has a tendency to form cracks from stresses set up in machining. These difficulties in fabrication, although not insurmountable, have caused beryllium to be considered principally for use where stresses are low and for use as a bulk material in neutron moderators. It should be mentioned that beryllium dust is highly toxic when inhaled, at least to certain people. Since there is no way to determine who is sensitive, all processes involving handling of beryllium, particularly in finely divided form, must be carried out with rubber gloves, hoods, and proper ventilation.

3.5 Solid Moderators

Two important solid moderators, in addition to beryllium metal, are graphite and beryllia (beryllium oxide). Other moderators are, of course, hydrogen and deuterium in the form of light and heavy water, and the properties of these materials are discussed in Chapter 17. Limited use has been made of zirconium hydride in low-power research reactors and in the SNAP-10A, the first nuclear power source carried on a space satellite. However, we shall confine our discussion here to graphite and beryllia.

3.5.1 Graphite

Graphite was used in the earliest reactors as a neutron moderator and is still popular, partly because of its low cost and ready availability. Graphite has mechanical properties which are quite sensitive to the raw material used and to the mode of preparation [15]. Its strength is low compared to metals but improves somewhat with increasing temperature. Graphite is a peculiarly anisotropic material—its properties vary markedly in different crystallographic directions. It is imperfectly elastic, yielding plastically to some extent under moderate stresses. Oxidation in air becomes significant at about 350°C (670°F). There is a profusion of commercial grades of graphite, differing in texture, grain size, permeability, density, impurity content, and degree of anisotropy [125].

Graphite suffers pronounced damage when exposed to reactor radiations. The principal results are large dimensional changes, increased strength, and sharply decreased thermal and electrical conductivity. All of these changes may show varying degrees of anisotropy. Many of the effects can be partially reversed by annealing at temperatures of 300 to 500°C, but complete recovery, especially in heavily damaged graphite, requires heating to temperatures approaching the graphitization range of the particular type [15]. Irradiation at temperatures above about 250°C greatly decreases the rate at which the damage develops. Because of the property changes associated with irradiation, care must be used in selecting a particular graphite type for large components so as to minimize growth or dimensional distortion and to assure the validity of the original heat-transfer calculation. On annealing, considerable stored energy may be released as heat. This release played an important role in the windscale reactor accident described in Chapter 11.

The properties, design, and performance of graphite as a moderator in nuclear reactors form an extensive body of reported information, and the authors recognize that the above discussion is quite superficial. Space does not permit a suitable expansion, however, and the reader is referred to a recent book by Simmons [126] for a more comprehensive review.

3.5.2 Beryllia

Beryllium oxide is a ceramic with an unusually high thermal conductivity at room temperature; however, the conductivity falls off rapidly up to about 600°F (315°C). Like many ceramics, it has a high compressive strength, depending on composition and manufacturing variables, and it is susceptible to thermal shock. It can be fabricated by standard ceramic techniques to a sintered density of about 3 g/cm³. Except for reactions with the inorganic acids, fused alkalies, and water vapor, beryllia is relatively inert. Beryllia is toxic if inhaled; serious or lethal doses could occur during the firing operation, especially in the presence of atmospheric moisture. Beryllia should be processed only under carefully controlled conditions.

3.6. Control-Rod Materials

The properties of control-rod materials have been compiled in an extensive review by Anderson and Theilacker [17], and there is a monograph on hafnium [13]. The brief comments in this section merely point out the various forms of control materials which have been used in reactor applications.

Boron (natural, or enriched in the isotope B^{10}) has been used in the form of compounds, alloys, or dispersions. It is an effective hardener in steels and most other materials, resulting in embrittlement and difficulties of fabrication. As a result, powdered and sintered compounds (e.g., B_4C) and dispersions (e.g., B_4C in stainless steel) have been favored over alloys. Radiation damage is particularly severe in boron-containing materials, since helium is generated by the $B^{10}(n,\alpha)Li^7$ reaction. Because of this, the design of boron-bearing control or poison materials is in many ways similar to the design of fuel-bearing materials where fission gases (xenon and krypton) are produced.

Hafnium is a chemical homolog of zirconium, and the metallurgical properties of the two are quite similar. Hafnium has superior resistance to aqueous corrosion over zirconium and is used in high purity form in the Shippingport PWR and in the Yankee reactor. It is, however, expensive and not readily available for widespread use.

Cadmium has attractive thermal-neutron-absorption characteristics and is widely used in research reactors and assemblies. Problems have been encountered in the use of cadmium because of its low melting point (321°C or 410°F) and high vapor pressure, but these should be surmountable with proper design. The WTR (Westinghouse Test Reactor) rods were hollow cadmium cylinders canned in stainless steel, and the interior void would accommodate bulging caused by vaporization of the cadmium.

Alloys containing cadmium have been developed for control applications. The MTR uses aluminum-cadmium alloy canned in aluminum and the Yankee power reactor has used silver-indium-cadmium clad with nickel-plate. The latter alloy contains 85% silver, 15% indium, and 5% cadmium, the combination providing complementary neutron absorption properties as well as increased mechanical strength [127]. The strength of this alloy decreases markedly with increasing temperature, however.*

The use of the lanthanides has been limited, with one important exception (europium oxide-stainless steel dispersions in the Army reactors), to development work. Most of this work has been based on the use of lanthanides or their oxides as alloying constituents or as dispersions in stainless steel, aluminum, titanium, or zirconium. Europium, samarium, and gadolinium have certain outstanding advantages with respect to nuclear properties (high cross sections and suitable decay chains), but their metallurgical handling is difficult, and they tend to impair the aqueous corrosion resistance of their matrix materials.

4. METALLURGICAL FABRICATION PROCESSES

This section deals with metallurgical processes used to fabricate reactor system components. An understanding of this background material is necessary for an appreciation of how materials defects may arise, why materials specifications are written as they are, and what acceptance tests or inspection procedures should be used.

Admiral Rickover has said, "Some of the types of difficulties we constantly encounter have to do with faulty welding, faulty radiography and defective castings: that is, with deficiencies in basic conventional processes of present-day technology." [129] This observation emphasizes that whoever is responsible for the procurement of reactor plant components must never presume manufacturing processes to be so well-established that inspection procedures can be relaxed. Only sound specifications, rigid inspection, and rigid acceptance tests can ensure that quality will be satisfactory.

4.1. Melting and Casting

It has been said [130] that probably upwards of 50% of all serious metal difficulties in commercial practice can eventually be traced to the melting and pouring.

The material to be melted, including calculated amounts of the alloying materials (all referred to as the "charge"), is placed in a refractory crucible, in the case of small melts, or in an open hearth for large melts. An open hearth is basically a large open refractory trough heated externally or internally. The source of heat for the charge may be the combustion of oil or gas, electrical resistance, electric arc, electrical induction, or electron bombardment. The crucible or hearth may be either of the tilting type or stationary, some sort of tapping or stopper mechanism being necessary in the latter.

The surface of the melt may be in contact with uncontrolled air, controlled atmosphere, a slag or flux, or vacuum. Gaseous contaminants are a source of many difficulties in the melting process, since many metals have a much higher solubility for atmospheric gases as liquids than as solids. Upon solidification, dissolved gases are rejected to form pores or react to form inclusions. Vacuum degassing, either by vacuum melting or by vacuum treating before casting, is a suitable process for obtaining "clean" ingots or castings. High-grade steels and reactor-grade Zircaloy are treated in this way.

In casting, the alloy melt may be poured into a mold to make an ingot which is subsequently worked into final shape, or the melt may be poured into a mold of essentially the same shape as the finished product. Several methods, as outlined below, are used to produce small castings.

Sand molding is an old process widely used for casting many different alloys. A wooden pattern is made of the required shape and a special molding sand is packed around it in a box-like enclosure. The pattern is removed by "parting" the mold, the mold is reassembled, and the metal is cast by pouring. Dimensional control and surface finish are suitable only for rough parts.

Die castings using permanent molds are made by forcing a given amount of molten metal from a bath into a mold under pressure. These castings are characterized by close tolerances, sharp contours, and smooth surfaces. However, since fast chilling produces a harder case on a softer core, subsequent machining may result in warping of the piece. In addition, the rapid chilling and freezing of the surface tends to produce porosity in the interior of the cast material.

Centrifugal casting involves pouring molten material in a rapidly rotating mold. The rotation may be about an axis through the mold or exterior to the mold with long "sprues" or feeding tubes leading to the mold from the center of rotation. Centrifugal casting gives greater strength to the casting than normal pouring by producing a finer

*Note added in proof: Newer rods to be used in the Yankee reactor consist of silver-cadmium-indium alloy contained in stainless steel cladding [128].

grain size, higher density, improved homogeneity, better detail (at least at the periphery), and elimination of centerline weakness.

Injection casting, particularly as related to processes recently developed for making fuel pins [131], employs gas pressure to force the melt into the mold. The mold, which may be either expendable or permanent, is first evacuated and the open end submerged in the melt. Inert gas pressure on the melt surface forces it into the mold. Long, narrow castings (length/diameter $>$ 100) can be produced by this process because of the rapidity with which the mold is filled before chilling occurs.

Continuous casting is a relatively new process. Molten metal is poured into a mold which is open at top and bottom, and the solidified billet is continuously withdrawn from the bottom. Long pieces, usually intended for hot rolling, are produced, thereby eliminating the need to break down a large ingot.

4.2 Plastic Working

Structural materials which are first produced as ingots are subsequently formed into useful shapes by one or more working operations. Not only can the shape be changed, but many (but by no means all) of the defects found in castings can be eliminated or at least their adverse effects can be minimized. Porosity can be closed up, microsegregation can be reduced, and the grains and grain inclusions can be reoriented to reduce their adverse effects on mechanical properties. The mechanical strength can be increased by cold working.

Forging is a general term used to define the plastic deformation of metals at elevated temperatures into a predetermined shape using compressive forces exerted through some type of die by a hammer, a press, or an upsetting machine. The resulting flow of the forging stock may be either unconfined or confined, depending on the kind of forging die used.

There are several methods of unconfined forging, in which constraint is applied essentially in a single direction, leaving the metal free to flow in the other two dimensions. Flat dies produce a spreading, stretching, and thinning action; edging dies result in a gathering together of material into a desired shape; and still other dies may be used for bending. Punching or piercing is a forging operation severe enough to penetrate a thin section of metal. In confined forging, dies of essentially the shape of the finished piece are used and the piece is struck (drop forged) or pressed (hot pressed) into the desired shape with constraint in all three dimensions.

Rolling of flat plates or sheets is accompanied by running an ingot (or forged billet) between flat rolls. Bars or even more complex shapes are produced by grooved rolls. Successive passes are made, with or without reheating the billet, until the desired reduction is effected.

The temperature range for rolling is selected on the basis of the material characteristics and may vary from red heat down to room temperature. The lower the temperature of rolling, other factors being constant, the greater the tendency toward higher hardness, higher tensile strength, and lower ductility. The amount of cold reduction which a material will withstand is limited by the loss in ductility. The lost ductility can usually be restored by annealing at or above the recrystallization temperature, and the annealed material can then be further cold rolled. Hot rolling is carried out above the recrystallization temperature, and the deformation and recrystallization processes occur more or less simultaneously. Large reductions per pass and large total reductions can be achieved during hot rolling without loss of ductility.

Metals which would be contaminated if exposed to the air while hot are rolled in protective jackets. Zircaloy-clad, metal fuel elements are sealed into welded, mild steel packs. Similarly, Zircaloy and uranium or uranium-alloy billets are hot forged or extruded in copper or steel cans.

Extrusion is a process which applies either hydraulic or mechanical pressure to a heated billet, forcing it through a die to form an elongated shape of greatly reduced cross section. The resulting shape must usually be straightened and trimmed. The process may be executed slowly or rapidly. The latter (impact extrusion) is often used for the manufacture of small aluminum parts. When dissimilar materials, such as a fuel and its cladding, are extruded simultaneously, the process is called coextrusion. By proper shaping of the pieces that form the billet, careful die design, and judicious selection of temperature, ram speed, etc. (processes that comprise a highly sophisticated art), the dimensions of the dissimilar materials can be controlled quite well. Even such "obviously incompatible" materials as uranium dioxide in stainless steel cladding have been successfully coextruded [132]. Coextrusion is used to produce transition joints between dissimilar metals which cannot otherwise be metallurgically joined [133].

In rod and wire drawing, rods are reduced in cross section and wire is manufactured by pulling the metal through a tapered die. The process is almost always at room temperature with the exception of a few metals, such as tungsten and molybdenum, which cannot be worked cold. Swaging is a compression-type process in which a round rod is reduced in diameter by being subjected to a transverse hammering action by radially segmented dies. The hammering action proceeds while the rod is rotated and moved axially. The radial compression is structurally beneficial, hardening the surface more than the core. It is sometimes also beneficial in improving the bonding quality of a cladding on a cylindrical rod.

Many operations, such as rolling, drawing, and extrusion, result in stock which is warped and therefore requires straightening and flattening. The elimination of the irregularities may be achieved either by stretching the piece so as to apply uniform tension to parts, or by a series of bending operations performed by rollers.

Deep drawing (sheet drawing) is best illustrated by the formation of a metal cup from a sheet by means of a punch and die. Materials vary greatly in deep-drawing properties, the brasses, bronzes, and austenitic stainless steels being noted for their adaptability in this process.

4.3 Heat Treating

"A heat treating process includes, broadly speaking, any operation or combination of operations involving heating above room temperature and cooling to room temperature and below for the purpose of obtaining desired properties in a metallic material." [134] One class of heat-treating operations is carried out to effect changes in structure of the material. Within this class are annealing, homogenizing, grain refining, phase transformation, and hardening operations. Another class bears little relation to structural changes. This class includes stress relief, stabilization, and retardation of age hardening. The method of heating may be by furnace, by salt bath, by flame, or by direct electromagnetic induction in the specimen. Cooling may be by water, oil, or air. The method selected depends on the program of required temperatures, their desired distribution, and the degree of control necessary. It should be noted that the cooling rates from a heat treatment are often as important as the high-temperature control itself. For example, an alloy steel may be "austenitized" at any of a range of temperatures, but the cooling rate will almost entirely control the final condition of the alloy. (See Sec. 2.6.4.)

Annealing is a softening process. Most metalworking processes cause internal flow of metal which breaks up the grain structure. This does two important things. One is to make the average grain size smaller and to provide increasing resistance to further deformation (strain-hardening). The other is to give the material directional properties. A complete anneal essentially eliminates both these influences and returns the metal to its state before working. The result is not always desirable, as it may be advantageous to retain the higher strength and hardness of worked material or to preserve certain directional properties. For instance, the initial loading of aluminum fuel at the ETR was annealed during a final finishing operation. As a result, the plates were so soft that the Bernoulli forces established by the coolant distorted the fuel plates. Cold rolling to work-harden the plates solved the problem. Since the irradiation by fast neutrons produces somewhat the same effect as cold working, annealing can be used to partially, if not completely, remove radiation damage.

Homogenization is sometimes necessary if previous operations have left chemical constituents of the alloy inadequately distributed. When the material is held at a temperature near the melting range for a period sufficiently long to allow diffusion to take place, the compositional variations will be reduced. Homogenization is often promoted by recrystallization or phase transformations.

Hardening by heat treating is generally applied to ferrous metals or precipitation-hardening alloys. The heat treatments are specific to each alloy, and important hardening procedures have been discussed in Sec. 2.6.4.

Stress relief is necessary after some forming or welding operations. Severe working or unequal heating generally results in some portions of the metal being in tension while other parts balance it in compression. When the piece is machined or exposed to service conditions, these stresses may be relaxed to some degree, and the piece will warp or perhaps even crack. By heating to a rather moderate temperature, the metal is softened enough to permit local plastic flow and a consequent reduction of the residual stresses. This is an important heat treatment for weldments in massive sections.

4.4 Powder Processes

The use of powder methods for the production of metal parts is attended by certain advantages and certain disadvantages over conventional methods. Important advantages are the minimizing of scrap and machining, better control of composition and structure, and good adaptability to mass production. There are some materials and components which cannot be fabricated in any other manner, such as parts with controlled porosity, parts containing mixtures of metals with nonmetals, and ceramic parts. On the other hand, powder techniques call for expensive dies and raw materials; they may result in diminished mechanical properties, such as the ultimate strength; and they have certain design limitations. A persisting difficulty in the manufacture of large components is the achievement of satisfactory homogeneity and reproducibility of properties, especially the mechanical properties.

The properties of finished parts are sensitive to the shape, size, size distribution, and purity of the starting powders. Powders are made by chemical reduction methods, electrolytic deposition, condensation of metallic vapors, atomizing, grinding, granulation, and stamping. Mixing is usually accomplished with water or organic fluids containing small amounts of organic binders and lubricants.

A common powder process consists of cold pressing and sintering. Blended powders are cold pressed into the desired shape. The green compact is slowly heated to a nominal temperature to decompose and drive off the organic binders. Prolonged heating at temperatures somewhat below the melting point results in densification of the compact by diffusion between and within the powder grains. Densities approaching the true theoretical density can be obtained. Hot pressing, in which the pressing and sintering operations are combined, can be used to produce high-density compacts, but the tooling problems are much more severe.

4.5 Joining

Reactor component designs avoid mechanical joining wherever possible for two reasons: a relatively low reliability of the joint, and the possibility that small pieces such as nuts, bolts, or rivets will come loose and lodge in coolant channels or pumps. Metallurgical joining methods include welding, brazing, and solid-state bonding of various types (roll bonding and pressure bonding). Welding is perhaps the most important of these; the primary system boundary may be a completely welded structure except for the vessel top closure and access ports, and even these may make use of seal welds.

4.5.1 Welding Procedures

In the metal-arc process the heat is produced by an arc formed between the work piece and a metal electrode. The electrode melts and supplies filler metal to the joint. Coated electrodes and electrodes buried in flux (submerged-arc welding) are sometimes used to exclude atmospheric gases from the weld area. The inert-arc process (tungsten-inert gas or T.I.G. welding) substitutes tungsten for the electrode material and provides an inert gas (argon or helium) for the environment of the arc. No flux is necessary in this process. The stainless steels and aluminum are usually welded by this process. A carbon electrode is used in the carbon-arc process, the filler metal being supplied in a separate rod. This process is used mostly for welding copper and its alloys.

Gas welding is accomplished by the combustion of acetylene gas or of hydrogen with oxygen. The filler metal is added separately and fluxes are used. One advantage of the gas process over arc welding is that the gas mixture can be adjusted to provide a reducing, neutral, or oxidizing flame, which in turn will control the oxygen and carbon content of the deposit to some extent. The Thermit process uses the heat-producing reaction of fine aluminum powder and iron oxide. Upon ignition, high-temperature molten iron is produced. In combination with suitable alloying, the metal is conducted into a mold which surrounds the parts to be joined. The process is used principally for heavy repair work. It has also been used to fieldweld the heavy steel reinforcing bars used in concrete reactor containment vessels, the Connecticut Yankee containment, for example.

Electron-beam welding is a recently developed technique wherein a high-energy beam of electrons is used for fusion welding. A primary advantage is the ability to make deep and narrow welds, which reduces the volume of metal affected by the welding.

In resistance welding, the pieces to be joined are pressed between electrodes, and a high electric current is passed through the joint. Resistance heating of the joint interface produces the weld. Advantages of the process are that it is fast, it permits accurate regulation of the heat input, the mechanical pressure usually improves the metal structure, and no filler rods or fluxes are necessary. There are several mechanical variations of resistance-welding methods (including spot, seam, and butt welding) which depend on the type of joint to be made.

The chief weakness of welding processes is their dependence upon human skills and judgments. There is probably no metallurgical operation in which the technique of the operator is more important, and poor welding practice has been the nemesis of many a design. The use of automatic welding machines eliminates many of the human factors but still does not always result in perfect welds, since the machines themselves are subject to faults or to improper operation.

Metallurgical changes in the joint materials are caused by the heating and cooling attendant in all welding processes. Smaller size welded pieces can readily be annealed after joining, substantially returning the work to its original metallurgical state. When size or some other factor (such as welding carried out in situ during repairs) makes it impractical to post-anneal the entire piece, certain advantages can still be obtained by locally heating the work to stress-relief temperatures. This allows local yielding near the weld and reduces the tendency for cracking.

When a weld bead is passed, the adjacent base metal is subjected to a gradient of temperature from ambient (room temperature) to the melting point. This often produces grain growth, phase transformation, or other metallurgical reactions in the base metal. The volume of metal adjacent to a weld in which observable structural changes have occurred is referred to as the "heat-affected zone". Adverse effects in the heat-affected zone, such as sensitization of stainless steel or atmospheric contamination of Zircaloy, must obviously be avoided or removed by subsequent heat treatment. "Stabilized" stainless steels have been developed to avoid sensitization, as discussed in Sec. 2.7.1. The Zircaloys (and other highly reactive metals), are generally welded in dry-boxes filled with helium or argon to prevent atmospheric contamination, which degrades the corrosion resistance of these alloys. If, however, the heat input to the weld can be sufficiently localized (as by resistance welding, for example), the extent of the heat-affected zone may be small enough that the attendant changes are not objectionable, and post-welding heat treatments or protective atmospheres can be largely eliminated.

4.5.2 Other Joining Methods

Brazing uses a lower melting filler material to form the joint; the base metal is not melted. (Soldering is a low-temperature brazing process.) Strength in the joint is provided by surface reaction or by diffusion through the braze metal to alloy it with the base metal. Fluxes are used to dissolve oxide films and allow the braze to wet the joint surfaces. In furnace brazing the pieces are assembled with strips of braze metal placed in the joint, and the entire assembly is heated to melt the braze and promote diffusion. Many points may be brazed simultaneously as, for example, in joining fuel plates into side pieces in the MTR fuel elements. Braze metal may also be applied as a powdered metal suspended in a liquid and brushed onto the joint surface.

Joining methods which do not involve melting of a metal are roll bonding, pressure bonding, explosive bonding, and coextrusion. In all of these processes the bond is formed by solid-state diffusion across the interface, and, when properly made, the joint is indistinguishable from the base metal. Roll bonding is accomplished by hot rolling sandwiched sheets—the rolling deformation breaks down surface resistance to diffusion and promotes recrystallization of grains across the interface. This is a widely used process for cladding plate-type reactor fuel elements, such as the first PWR (Shippingport) seed and the MTR type fuel. It is also an important commercial process used to make "Alclad" (high-strength aluminum alloys clad with pure aluminum).

Pressure bonding differs from roll bonding in that the pieces are not deformed. Components to be bonded are carefully cleaned, assembled into sealed packs, and placed in a furnace under high hydrostatic pressure. The combination of high pressure and high temperature promotes diffusion across the interfaces. Plate-type fuel elements for the second Shippingport core were fabricated by pressure bonding [135]. Explosive bonding is accomplished by the shock wave of a detonated charge. Coextrusion was described in Sec. 4.2.

4.6 Surface Preparations

Coatings or claddings are often applied to finished metal products, sometimes for the sake of appearance, but most often for protection against a corrosive environment. The success of a coating is dependent on its adhesion to the base material; the adhesion may be purely mechanical, by interlocking with surface irregularities on the base metal, or it may be metallurgical, by a limited reaction between the coat and base metal. In any case, the surface to be coated must be free of grease, dirt, scale, or films in order to get good contact between the coating and the base metal. Furthermore, surface cleanliness is vital for metals which are protected from corrosion by the formation of adherent oxide films.

4.6.1 Cleaning Methods

Surface cleaning involves one or more of the following: vapor degreasing; washing with emulsifiable solvents; aqueous alkaline cleaning; electrocleaning; shot, grit, or abrasive blasting; wire brushing; tumbling; and acid pickling. In vapor degreasing, the work piece is hung in the vapor of a nonflammable solvent, which condenses on the surface and washes off the oily contaminants. Parts to be cleaned in an emulsifiable solvent are alternately immersed in the bath and rinsed in running water. In electrocleaning, the metal to be cleaned is used as the cathode (lead, zinc, tin) or as the anode (steel) in an acid or alkaline cleaning electrolytic solution. Hydrogen embrittlement may be an unwelcome side-effect in cathodic treatments. The grit used for the various surface blasting methods may be silica sand, steel grit or shot, or aluminum oxide grit. There is always a danger of unwanted contamination of surfaces by imbedding of the abrasive grit. Steel-shot or wire-brush cleaning with carbon steel materials may result in surface contamination of stainless steels with noncorrosion resistant particles. Tumbling, or "barrel finishing", is useful in reducing surface porosity and in deburring operations. Finally, acid pickling is a treatment by simple immersion in an acid solution to remove oxides and dirt. The choice of acid is of course dependent on the base metal and the type of foreign material to be removed.

It is important to realize that cleaning agents may be a source of contamination which can lead to system failures. Of particular note are chlorinated organic solvents or acids containing fluorine or chlorine used to clean stainless steel parts. These must be thoroughly and completely rinsed away to avoid any possibility of stress corrosion during service. Acids or other cleaning agents left in joints or between close-fitting surfaces will quite likely lead to accelerated crevice corrosion. Stains on any surface may interfere with the formation of protective oxide films and ultimately lead to corrosive failure. Delay in transferring Zircaloy components from hydrofluoric-nitric acid pickling to a rinse can seriously impair the corrosion resistance of these alloys. One cannot emphasize too strongly the necessity for removing all foreign chemicals before placing reactor components in service.

4.6.2 Surface Coatings

Hot dipping is widely used to apply a coating of low melting metal to the surface of a higher melting base metal. Galvanizing (with zinc) and tin plating are examples of this type of surface protection. Electrodeposition is also widely used for applying metal coatings. The work piece is made the cathode in an electrolyte, which is ordinarily an aqueous solution of a salt of the coating element. Electrolytic coatings are mechanically bonded to the base metal, although heating after plating may be used to promote diffusion bonding. Coatings of this type can on occasion peel off and expose the base metal to corrosive attack. For instance, the nickel coating on the Yankee control rods was partially rubbed off by friction and partially peeled off, exposing the cadmium-silver-indium alloy to high-temperature water, and resulting in a serious problem due to deposition of radioactive silver in the primary system.

In vapor deposition the work is placed in the vapor of an easily vaporized metal or in an atmosphere of an easily decomposed volatile compound of the coating metal. For example, nickel carbonyl vapor will deposit metallic nickel on a heated surface and a mixture of tungsten hexachloride and hydrogen will deposit tungsten. Vapor deposition is used in a variety of ways to coat nuclear fuel materials. In flame spraying (or in the more recent variant of plasma gun spraying) the coating material is vaporized or dispersed as droplets (liquid or plastic solid) in the flame or plasma. The flame is directed against the work, which is kept relatively cold. Vapors condense and droplets adhere to the work and to each other to build up a coating. Ceramic materials can also be deposited by this method.

A surface treatment which should be mentioned is the autoclaving of Zircaloy components for water-cooled reactors. After Zircaloy-clad fuel bundles or assemblies have been fabricated, they are degreased, pickled, and cleaned. They are then held in static, water-filled autoclaves at reactor coolant temperatures for approximately 24 hours. This produces an initial corrosion film which, if the pieces have been properly handled, is black, adherent, and protective during service. Autoclaving not only prepares the surface but is a sensitive acceptance test to assure that the components will not fail by excessive corrosion.

4.7 Cleanliness in Reactor Systems

In the preceding section the importance of surface cleanliness has been discussed from the standpoint of corrosion behavior. There is another aspect to cleanliness of reactor components—the exclusion of foreign particles or objects from the reactor system. It is appropriate to point this out in relation to fabrication processes, since these are the source of most contamination.

The consequences of dirt in reactor systems can be serious indeed. A foreign object may lodge in a fuel coolant channel and block or reduce the flow. The starved channel will run hot and may cause the fuel element to swell or blister or even melt. Small objects or particles can certainly do severe damage to pump impellers and may cause erosion in other parts of the system. Particulate matter may foul heat transfer surfaces or may introduce dissolved impurities which will aggravate corrosion somewhere in the system.

"Dirt" which has been found in reactor systems has ranged all the way from small hand tools to fragments of grinding wheels and welding electrodes to bristles from wire or plastic brushes to small particles of paint or just plain dirt. In one instance at the Sanannah River Laboratory, the use of polystyrene brushes to clean off a river water intake screen filter resulted in bristles entering into and lodging in stainless steel heat exchangers. Some of these seemingly inocuous bristles, shaped like a lady's hairpin, straddled stainless steel tubes and, with time and flow conditions, eventually wore their way completely through the tube walls and caused D_2O to be lost to the river water. The initial source and the final result were almost completely unrelated in time and space. One therefore could take the position that any defect in the system is a potential safety or economic hazard for the entire system.

5. MANUFACTURING QUALITY CONTROL

It should be apparent from the foregoing discussion why manufacturing quality control is critically important. First, the failure of any component in a nuclear system is a serious matter. A minor malfunction could conceivably initiate a train of events leading to a major accident. Second, the properties of materials depend sensitively on fabrication procedures. Seemingly insignificant changes in composition, rolling schedules, heat treatments, welding variables, and so forth, may have quite an important, and perhaps undesired, effect on the properties of the finished product.

The required quality of any manufactured item must be communicated to the manufacturer, preferably by contractually binding specifications. There are many ways to make steel plate of given dimensions and nominal composition, but there are few ways to make properly the high-quality plate required for nuclear pressure vessels. Unless quality requirements are made a part of the contracting order, a manufacturer will be free to select the process which costs the least, and this may not produce a satisfactory component.

But specifications cannot stop with a statement of desired quality. That would be like demanding an "excellent" wine from your wineseller without permitting him to taste it or, more importantly, without tasting it yourself—your tastes and his may be different. Inspection procedures and acceptance tests are a necessary part of any quality requirements. In actual fact, it is only through careful specification of the inspections and tests to be carried out that the desired quality can be obtained because most engineering properties are entirely dependent on how the determining tests are carried out. Furthermore, inspection procedures themselves must be subjected to continuous quality control. Demonstration of the validity of sampling plans, qualification of inspection personnel, and calibration of instruments must be rigorously specified and carried out. Otherwise, quality control becomes a meaningless pastime.

Finally, and perhaps most importantly, quality control is a state of mind. Manufacture and inspection depend on people. It is impossible to establish inspection procedures that would absolutely guarantee quality against every conscious or unconscious mistake that a workman might commit. To a great extent, quality depends on the ability and conscientiousness of shop personnel, and there are few substitutes for these virtues. In no place is this more evident than in the quality of weldments. Although there are certain techniques with which to examine welded joints, much of our assurance of quality depends on the ability of the welder. For this reason, critical and difficult welds are made only by men who are specially trained and carefully qualified—not just once, but periodically. Needless to say, inspection personnel, like Caesar's wife, must be above suspicion. The acceptance of a product comes down in the end to a man's initials on an inspection tag.

It is not possible to review here the problems of personnel management. Nevertheless, the intangibles discussed above are such a vital part of quality control that they must at least be pointed out. As for the more tangible aspects of quality control, this section summarizes various sources of manufacturing defects and the tests and technical procedures which are available to guaranteee, as far as possible, that the desired quality has been achieved.

5.1 Sources of Manufacturing Defects

There are certain manufacturing variations which are unavoidable and therefore must be accepted. These are usually subtle differences in grain structure or composition, which give rise to the "heat-to-heat" variation in mechanical or other properties. Figure 5-1 illustrates the variability which one might expect in tensile strength and elongation for a stainless steel sheet product which was specified to have a maximum tensile strength of 100,000 $lb/in.^2$ and a minimum elongation of 40%. Designers must clearly recognize that properties of materials can, and usually do, depart significantly from handbook values.

Beyond these small variations, which are usually acceptable if properly accomodated in the design, there are several classes of manufacturing faults which are clearly not desirable. These are composition variations (off-specification, segregation), physical inhomogeneities (voids, inclusions, cracks, etc.), and improper metallurgical structure (excessive hardness, texturing, etc.) The sources of these defects and means of controlling their occurence are discussed in the following sections.

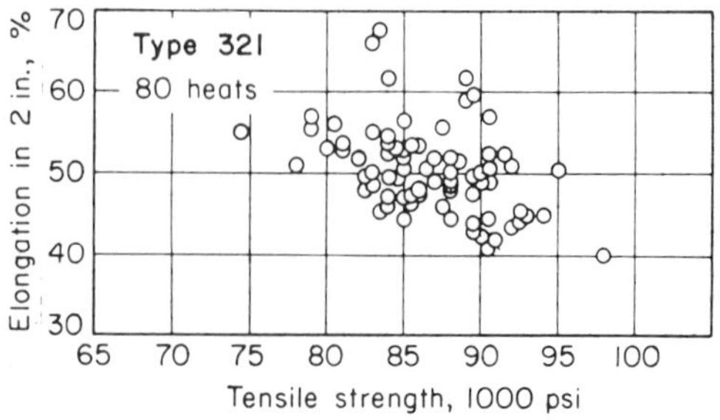

Fig. 5-1 Tensile strength and elongation for eighty heats of AISI 321 stainless steel sheet. (From Metals Handbook (1961), p. 411.)

5.1.1 Composition Variations

5.1.1.1 Off-Specification. Imperfect control of alloying elements implies trouble in the melting and casting process where it is necessary to include certain proportions of certain elements in the material but to exclude certain others as undesirable contaminants. In melting, the ratio of the desired additions in the charge will not necessarily be the ratio in the casting because of different rates of volatilization or oxidation during the melting. If the charge is not properly adjusted to reflect these differences, the final analysis will be off specifications. Without proper process control the quantities of undesirable elements may also rise to an unacceptably high percentage in some melts. Economic pressures on the vendor may persuade him to quote a "ladle" rather than a "check" analysis, or the use of some similar device to avoid scrapping a marginal heat. The responsible reactor builder must be alert for such contingencies or else have representatives whom he trusts to insure that specifications are met.

Interstitial contamination, by carbon, oxygen, nitrogen, and hydrogen, is particularly troublesome. These elements are present in the atmosphere, in scrap charged to the melt, in furnace fuels, and in refractory furnace linings or crucibles. Pickup of the interstitial contaminants can lead to gas voids in the casting, nonmetallic inclusions, general embrittlement, or other detrimental effects specific to certain alloys. Aluminum, iron, nickel, and their alloys in the liquid state are able to dissolve large amounts of hydrogen, which precipitates out on solidification. The result may be gas pockets in the castings or evolution of hydrogen during subsequent processing. One common problem encountered with aluminum-clad fuel elements has been the formation of blisters at the clad/fuel interface because of hydrogen in the metal.

Interstitial contamination is most detrimental to the reactive metals such as zirconium and titanium. Once picked up in the melt, the contaminating elements form stable compounds which cannot be removed by subsequent heat treatments. Oxygen, for example, strengthens zirconium but at the same time reduces its ductility. Nitrogen in excess of a few hundred parts per million seriously impairs the aqueous corrosion resistance of the Zircaloys. Hydrogen has a strong embrittling effect in steels and in the Zircaloys and generally cannot be tolerated.

Vacuum melting or vacuum degassing and rigorous control of the chemistry of charge materials are the most suitable means for minimizing contamination by the gaseous elements. The high grade steels for nuclear pressure vessels are treated by repeated cycling through a vacuum chamber. This has the added benefit of homogenizing the melt and removing small inclusions by flotation [108]. Vacuum induction melting of the Zircaloys produces a "clean" alloy and has eliminated the problem of "stringering"* with its attendant impaired corrosion resistance.

5.1.1.2 Segregation. Inhomogeneities in the chemical composition can result from microsegregation (gradients across dendrites) or from macrosegregation (gradients across or down an ingot). Microsegregation, or "coring," which results from nonequilibrium solidification (see Sec. 2.6.3) is an inherent problem in the casting of alloys. It is removed by homogenizing heat treatments or by subsequent plastic working of the metal. Where extensive heat treatment or working is not possible, the effects of microsegregation may be minimized by solidification control. This is applied, for example, in the production of "premium quality" castings whose mechanical properties are guaranteed [87].

Macrosegregation of chemical constituents oc-curs because ingots do not freeze instantaneously throughout their bulk. Some alloying elements may segregate by gravity separation, and others may be concentrated in the last portion which freezes. Additional segregation may result from incomplete mixing of a melt whose initial charge was inhomogeneous.

In small melts, segregation may be reduced by repeated melting, the ingot being inverted or broken up between melts. This is common practice in the melting of uranium fuel alloys where phase density differences are pronounced and where homogeneity requirements are stringent. In large ingots, macrosegregation must be held within the specified limits by solidification control (careful control of pouring temperatures and pouring rates) and by disposal of material from the top or bottom of the ingot. For example, the plate used for heavy-walled pressure vessels represents only the best material from the ingot; as much as 50% of the original ingot may be discarded (recycled to the melt) [108].

5.1.2 Physical Inhomogeneities

This classification of manufacturing defects includes the broad range of internal and surface voids, inclusions, cracks, and so forth. These defects in bulk metals usually arise from the casting process but may be introduced by subsequent treatment. Defects of this type are quite commonly associated with welds.

5.1.2.1 Casting and Working Defects. "Voids" are caused by the inclusion of gases in the melting

*"Stringers" are thin, nonmetallic inclusions in wrought Zircaloy which result in hairline corrosion.

and casting process which form as bubbles in the interior and occasionally as surface porosity. "Shrinkage porosity" arises from the difference in density between liquid and solid, the solid being more dense than the liquid. "Scabs" are formed by splashing metal while the mold is being filled, the splattered pieces being incorporated as solid inclusions in the casting. "Cold shuts" are caused by entrapment of air in pouring. "Blow-holes" arise from reactions between oxygen in the metal and carbon in the mold, resulting in gas pockets. "Flakes" and "shatter cracks" in forgings are related to gases retained in solid solution. "Laminations" occur as a result of closing up shrinkage porosity during rolling or forging. A "lap" may be produced by neglecting to trim a fin from a billet before rolling; the fin is pressed into the surface, producing a lamination which comes to the surface. "Hot shortness," that is, cracking during hot working, is caused by weak or molten inclusions in the metal. "Shrink cracks" and "hot tears" are caused by different contraction of casting and mold. "Forging burst" arises from an excessively high forging temperature. "Machine tears" and "grinding checks" are the result of overloading the machines.

The various physical defects are mainly eliminated by, first, producing a sound ingot, and second, conditioning the surfaces of billets to be forged or rolled. Ingot porosity is reduced by vacuum treating, as already discussed, and by solidification practices—"hot topping," for example, by feeding molten metal into the shrinkage cavity as freezing proceeds. Surface conditioning involves machining, grinding, flame scarfing, or weld conditioning. The purpose of surface conditioning is to remove all surface defects that might result in cracks, laps, seams, scabs, etc., in or near the surfaces of the finished wrought products.

5.1.2.2 Weld Defects. Welds are subject to a great variety of defects—porosity, entrapped flux, incomplete bonds, atmosphere contamination, and electrode contamination (when nonconsummable electrodes are used). The quality of welds is particularly critical because welded joints are often associated with the highest stresses in a component. Welded nozzles on pressure vessels, angled joints or flanges in piping, tube-to-tube sheet joints, and junctures of fuel plates or rods in assemblies are several examples where the welds are subject to high bending or thermal stresses or to stress concentration because of the joint geometry. Seal welds in a primary coolant boundary, while not subject to high stress, must be absolutely free of porosity, cracks, or corrodable inclusions.

Weld quality control begins with qualification of the weld procedures and the welders. This usually involves the making of several sample joints which are thoroughly evaluated by destructive tests (metallography and mechanical tests). As an example, Canadian practice for the Rolphton, Ontario, aluminum reactor vessel required welders employed on the job to attend a supplemental 6-week training course [136]. A written procedure was required for every welded joint, covering such factors as fluids and times for precleaning, between-pass cleaning, exact electrical (heat input) requirements, bevel angles, and minimum gaps. Disc grinding was done between passes to avoid the slightest porosity. Inspection of welds is generally required after the root pass (first pass) and at completion, but where heavy sections are joined, the inspection may be done at several stages during the build-up of the weld. Defects in welds are usually repairable by grinding out the area and laying in new metal.

Residual stresses, which are always introduced by welding, may be reduced by subsequent stress-relief heat treatments. For small components, this is no particular problem. The normal heat-treating precautions must be observed, of course: prevention of atmospheric contamination of Zircaloy and avoidance of sensitizing stainless steels, for example. When large pressure vessels are stress relieved, the procedure becomes extremely critical. Nonuniform or too rapid heating and cooling must be avoided. Extensive instrumentation with thermocouples and strain gages is usually required. Even the largest, field-erected vessels have been stress relieved after welding. The pressure vessel of the Hunterston Nuclear Generating Station at Ayrshire, Scotland, was stress relieved with internal radiant strip heaters [137]. Seven days were required to reach the desired temperature of 1022°F (550°C). This temperature was held for twelve hours, and then the vessel was slowly cooled over a period of 9 days.

Weld overlays with stainless steel (to clad carbon steel vessels) have been subject to defects of a somewhat different nature. In the manufacture of the pressure vessel for the Elk River Reactor, the internal cladding (type 308L) was applied in two coats. Under the sudden heating and cooling when the second overlay was applied, the austenitic structure of the first pass was partially transformed to martensite. The martensitic phase is brittle and cracked on cooling, although no propagation of cracks into the base metal was observed [138]. Whether or not the extension of cracks into the base metal could occur has been debated at great length. As a result of the condition of the weld overlay, a 5-year (or 250-cycle) operating limit, representing about one-fourth of desired life (or 10% of the desired number of cycles) was set for the Elk River vessel. This limit was established more with the expectation that improved inspection procedures would be established in the interim than on purely technical grounds. To avoid the difficulty in subsequent fabrication, it has been recommended that type-309 stainless be used for the first pass with type-308L for the remaining passes [118].

5.1.3 Improper Metallurgical Structure

Even though a cast or wrought product may have the right chemistry and homogeneity and be free of physical defects, it may still not have the proper grain size, grain orientation, or phase structure. These properties of its internal structure have a pronounced effect on the mechanical properties of a metal. They are controlled through working and heat treating procedures.

Grain size and grain orientation, important to both yield strength and impact properties (see Secs. 2.2 and 2.3), are dependent to a great extent on the entire fabrication history of a piece. Some metals (Zircaloy, for example) have a remarkable "memory" for prior working and heat treating; even extensive annealing will not produce a completely new or randomly arranged grain structure after the metal has been heavily worked.

Impact strength and ductility of a finished piece are functions of the direction of metal flow during the forging or rolling process. The desirable properties are generally better in the direction of flow rather than transverse to the flow direction. (Annealed Zircaloy is an exception to this rule.) This phenomenon is often used to advantage, as in the forging of gears so that the teeth, which bear most of the service stresses, have a tough, fibrous structure. In the case of drawn tubing which is subject to internal pressure, however, the tensile strength is highest in the longitudinal direction, but the greatest stresses will be tangential, hence transverse to the working direction. The anisotropy in mechanical properties produced by working can be reduced, in the case of sheet or plate, by cross rolling.

Hardening heat treatments should be particularly well controlled. The ideal hardness for a piece is that level which will just meet the requirements of the proposed use but little more. Since yield and tensile strengths increase along with hardness, these should also follow the same rule. If higher values of static strength and hardness than necessary are chosen, the user runs the risk that embrittlement and consequent cracking in service will occur. To cite an example, the hydraulic control rod drive piston in the Dresden reactor was made of 17-4 PH steel aged at 900°F. Stress corrosion cracks developed in the piston and drive shaft with an attendant loss of hydraulic fluid and power. The parts in this case had been hardened to a greater degree than necessary for the application. The pistons and drive shafts were subsequently replaced with the same steel hardened at 1100°F, which produced a lower hardness but more ductility to resist the cracking. A similar example is the failure of hardened SM-1 closure bolts by stress-corrosion cracking (see Volume 1, page 693).

The metallurgical structure is controlled generally by a specification of mechanical properties by tensile or impact tests, as discussed in the next section. Representative cross-section samples are heated in a hydrogen atmosphere to form hydride platelets in the metal. The hydrided sections are examined metallographically, and the lot is accepted only if a specified minimum proportion of the hydride platelets are oriented circumferentially around the tube wall. The test assures that the grain structure is favorable to resist cracking when the fuel cladding becomes hydrided by long exposure to hot water.

5.2 Testing and Inspection Methods

5.2.1 Destructive Tests

Four mechanical tests have become standard for metals through the years, and designers are accustomed to writing specifications based on one or more of these tests. These are the tensile, impact, fatigue, and creep tests. Fatigue and creep tests are not normally used for quality control acceptance tests since they involve long test times. The tensile and impact tests, on the other hand, are quite commonly included in materials specifications. The specimen sizes and shapes for these tests are standardized.

The most important requirement for these tests, and usually the most controversial, is to insure that the samples tested are truly representative of the material to be accepted. This is quite generally true of any destructive test where the product itself cannot be tested directly. The location of samples to be taken (usually in adjacent material) and their orientation with respect to the product must be carefully prescribed. Sampling procedures are established by destroying whole pieces and subjecting them to the same tests as the sample pieces. This is an expensive procedure, but one which cannot be avoided until a manufacturer has demonstrated with confidence his ability to turn out a particular product with reproducible properties.

There are many other types of destructive tests—bend tests for welds and bonds, tests for sensitization of stainless steels, metallographic examination and various chemical analyses—too numerous to detail here. For all of these, however, the preceding comments on sampling are pertinent.

5.2.2 Nondestructive Tests

A nondestructive test is one which leaves the usefulness of the object tested unimpaired. The important distinction between destructive and nondestructive tests is that the latter are applied directly to a finished product. One-hundred-percent inspection of certain features (e.g., welds) can be specified with nondestructive tests. Important classes of nondestructive methods include ultrasonic, eddy-current, radiographic, penetrant, and magnetic particle tests. Each of these tests has its own special uses, and its peculiar strengths and weaknesses. Often, if not usually, it is necessary to employ a number of tests on a given fabricated piece of material in order to minimize or eliminate any possibility of serious flaw. Intense inspection is certainly worthwhile on key structural components such as the pressure vessel and its head, the key parts of control rod mechanisms, major core-support units, and other vital in-core structural components. Other major components such as pumps, heat exchanger heaters, and large pipe sections should also be checked extensively.

Ultrasonic tests have been very effective in detecting voids and discontinuities in materials. A sinusoidal or pulse generator of frequencies in the 2.5 to 10 Mc/sec range is connected to a transducer which is coupled to the test piece with a liquid bath or film. The coupling agent is necessary for efficient transmission of acoustic energy from transducer to specimen and from specimen to detector pickup. The output of a pickup transducer (which may be the same as the transmitting transducer) is amplified and transferred to an oscilloscope or recorder. A normal pattern on the oscilloscope (i.e., for a "standard" specimen with no deleterious defects) must have been established by previous exploratory experiments. Any alteration of the recorded pattern is indicative of some type of defect in the specimen. Further investigation is then indicated in the questionable region by either this method or the others below. Different types of flaws and varying orientations of these flaws will have different effects on the patterns. The apparatus can be calibrated using known-defect materials and samples which are subsequently destructively tested, but the interpretation of ultrasonic test results (as with many nondestructive tests) is still largely based on operator experience.

Eddy-current methods can employ either sinusoidal or pulsed electromagnetic radiation for test purposes. Just as with ultrasonic waves, eddy currents may be used in both reflection and transmission arrangements. There are many variants of the eddy-current technique. In one commonly used method the sample is scanned by a probe consisting of two coaxial coils (transmitter and receiver). Small local variations in the electrical and magnetic properties of the material produce variations in the output pattern. The sensitive probe-to-sample spacing can be balanced out. Again, the orientation of the defect and the nature of the defect will give different results. The results are subject to a rough interpretation in terms of the electromagnetic fields induced in the sample if the operator understands the method and has properly calibrated samples. An empirical history must be built up for interpretation just as in the ultrasonic method.

Radiography, or the penetration of the sample by X-rays or γ-rays, is a much older process. Variations in density in the sample caused by irregularities in the interior cause variations in the intensity of the transmitted radiation. The transmitted energy is recorded photographically. Voids or cracks whose major dimensions are parallel to the direction of transmission are recorded as darker areas on the negative. On the other hand, such cracks perpendicular to the direction of transmission may not be detected. Heavy element inclusions, say tungsten contamination in aluminum welds, show as light spots on the radiograph. Results again must be carefully interpreted in terms of empirical test samples.

In the dye penetrant and "Zyglo" methods a liquid of low surface tension, which penetrates into small defects, is applied to the piece. Inspection consists either of direct visual exploration assisted by a bright-colored dye incorporated in the liquid or of examination for a similarly contained fluorescent material by use of ultraviolet light. This method does not disclose subsurface flaws. It can be used to determine roughly the size or total volume of the exposed crack.

The reader will find a more extensive discussion of nondestructive testing methods in Chapter 14.

5.3 Codes and Specifications

Specifications are requirements written by the designer to express what provisions are binding on a supplier. Codes are requirements written by informed and experienced committees, or framed by governmental bodies, to which designers are called upon to conform in the interest of safety. Applicable materials specifications will usually be included in a code.

A material or component specification must be sufficiently complete and binding that the designer is assured of the proper operation of the component covered. Ideally, a specification should so be written as to leave all details of fabrication processes to the judgement of the manufacturer. This is not always possible, however, because of limitations to the existing knowledge of materials properties. Not uncommonly, the specification of certain fabrication procedures (heat treatments, for example) is the only way to guarantee the correct properties of a finished piece.

The vital parts of a component specification are (1) specifications for starting materials, (2) specification of properties of the finished piece, and (3) specification of the inspection and tests which insure that the properties are indeed as specified. This is a simple outline of a document which often is quite difficult to establish. The writing of a specification requires a tremendous technical effort, and may in some cases be satisfactory only after considerable manufacturing experience has been obtained.

Although Chapter 14 contains an extensive discussion of the ASME Boiler and Pressure Vessel Code, it is appropriate to discuss here a few points pertaining to vessel materials.* In addition to design requirements, the code for nuclear pressure vessels (Section III of ASME Code) includes the following significant changes from previous practice [139].

1. Fatigue failure is considered for the first time in a code document.
2. Charpy V-notch impact tests are required for definition of the ductile-brittle transition temperature of ferritic steels.
3. Properties of the materials used must be determined on specimens heat treated to represent the pressure vessel.
4. The fabricator is required to certify inspection results and materials properties.
5. Specific nondestructive tests and test methods are included. The first mention of qualification of inspection personnel is made. A final magnetic particle or liquid penetrant examination is required after hydrostatic test of the vessel.

The first two items point out the importance of mechanical behavior other than uniaxial tensile test properties. Regarding item 3 above, Wylie [83] has discussed the many difficulties that are involved in the handling of test specimens for pressure vessel materials. (See Sec. 2.5 of this chapter for a summary.) The final two items indicate a move to tighter control of inspection procedures demanded by the need for high confidence in the reliability of these components.

It must be emphasized that the ASME Code rules are intended only to provide minimum safety requirements for new construction. They are based on data and results derived from experiment and experience and set a minimum quality level for the design and construction of a pressure vessel

*See Wylie and McGonnagle [139] for further discussion.

which can be stamped with the ASME Code. Although Section III now requires a consideration of thermal stresses and fatigue, the code rules do not consider deterioration that may result during service from corrosion, erosion, radiation effects, or instability of material.*

5.3.1 Standard Materials Specifications

Several technical societies have established specifications for standard engineering materials. These are generally, but not always, "tried and true," and the designer must never assume that just because a specification has been established by an august committee, it will be appropriate for his particular needs. It has been generally true that existing specifications which serve quite well for standard engineering practice have not been definitive or restrictive enough for nuclear applications, and it must always be kept in mind that these standards represent minimum quality requirements.

The ASTM (American Society for Testing and Materials) specifications, dating from about 1902, cover almost every conceivable material, except those which are newly developed. They include methods of testing, recommended practices, and definitions of terms. The ASTM represents a broad variety of industries, and its committees have not tailored the specifications to fit any parochial set of opinions. For this reason, their provisions may be useful to novel type constructions. The ASTM publishes an "Index to Standards" annually. Representative materials specifications are listed in Table 5-1.

TABLE 5-1

ASTM Specifications for Important Components in Water-Cooled Power Reactors*

Component or material	Material	ASTM specifications
Reactor pressure vessel		
Plate	Carbon-silicon steel	A212
	Low-alloy steel	A302
Forgings	Alloy steel	A336
Internal cladding	Austenitic stainless steel	—
	Inconel	—
Flanges and fittings	Carbon steel	A105
Flange bolts and nuts	Alloy steel	A193, A194
Primary piping	Stainless steel	A376, A312, A358
	Carbon steel	A106, A155
Valves, pumps, and fittings	Stainless steel	A182, A351
Pressurizer	Carbon-silicon steel	A212
	Low-alloy steel	A302
	Stainless steel	A240
Core internals, including thermal shield	Stainless steel	A240
Fuel cladding	Austenitic stainless steel (annealed)	A213
	Zirconium alloy	B353

*From Wylie and McGonnagle [139] with additions compiled by M. Bolotsky, USAEC.

The SAE (Society of Automotive Engineers) specifications were initiated in 1911. The SAE handbook contains, in addition to standards and informative reports, sections of detailed recommended practices which are useful to designers. In 1941, the SAE and AISI (American Iron and Steel Institute) jointly revised the composition ranges of the SAE Series. The AISI also issued its own specifications for low-alloy steels. Many of the SAE and AISI series are identical as a result of the work of the joint committee. The SAE specifications cover carbon and alloy steels, methods of testing steels, steel castings, tool and die steels, shot and grit cleaning, aluminum-base alloys, copper-base alloys, magnesium-base alloys, zinc-base alloys, bearing and bushing alloys, threaded parts, springs, and tubings and fittings.

More recently, the SAE issued the AMS series (first meaning "Aeronautical Material Specifications," and later "Aerospace Material Specifications"), which includes aluminum alloys, magnesium alloys, copper alloys, titanium alloys, carbon steels, corrosion and heat-resistant steels, tolerance and processes, nonmetallics, low-alloy steels, accessories, fabricated parts and assemblies.

The AWS (American Welding Society) has issued specifications covering filler metals, welding inspection methods, resistance welding processes, metallizing processes, and brazing and soldering processes.

The Department of Defense issued the Military Handbook in 1958, which amounted to a cross index between the other specifications and the MIL (military) series of specifications. Since the MIL specifications are apt to change abruptly, they are to be used with some reservation. The cross index, however, is valuable in providing approximate equivalents among other standards.

5.3.2 Test and Inspection Specifications

The matter of testing and inspection has been mentioned throughout this chapter, but some repetition is warranted by its importance and by the fact that certain phases are often overlooked. It is easy to assume that the supplier knows his product better than the customer until the results of misplaced confidence have appeared.

There is no real purpose in agreeing upon written specifications for a component if there is no intent to make some kind of acceptance inspection. Errors involving materials, and particularly metals, are not easy to detect. Except for a few metals like gold, copper, silver, and aluminum, which have distinctive colors, most fall in the dull grey category. Differences in density are not apparent without measuring equipment unless the differences are quite large. Magnetic tests are sometimes helpful in distinguishing 400- from 300-series stainless steels, but even the 300 series show some magnetism after cold working. Too many

*Note added in proof: The reader should be aware that several new codes for pressure-component construction are in various stages of preparation. These include construction codes for nuclear piping and for pumps and valves in nuclear systems as well as for installation inspection of pressure components and for periodic in-service surveillance of the reactor pressure vessel in nuclear plants.

cases of the use of a wrong material are on record for this to be considered an unlikely mistake.

Components reasonably capable of disassembly should be inspected internally. Limited means for the rapid identification of metals are available [140]. Chemical analyses should be run on suspect parts when a sample can be obtained. Customer monitoring of processing in the manufacturer's plant is usually permitted and should be effective if the observer is sufficiently informed. Nondestructive testing methods should be used to the fullest extent. It is obviously essential to have particularly high standards on those parts to be sealed into the reactor pressure vessel, including the vessel itself.

There are four necessary parts to any test specification: (1) specification of sampling procedures, (2) specification of the test procedure, (3) qualification of the test method and inspection personnel, and (4) specification of the limits of acceptability of the test results. If any one of these parts is omitted, the others are meaningless. For economic reasons, an additional stipulation is often included—the disposition of rejected material. Is it rejected outright or are repairs permitted? If the latter, then the extent of permissible repairs, repair procedures, and reinspection procedures must be specified.

As with materials specifications, certain inspection procedures have been standardized by the ASTM and other technical groups. Representative ASTM inspection standards have been listed in Table 5-2.

It is relatively easy for a designer to write an unrealistically tight specification. Nuclear power is, after all, a commercial enterprise, and the trick is to write a specification which will guarantee the required safety and reliability at the minimum fabrication cost. The designer must first decide what confidence is required in each component, and these quality levels may all be different. For example, certain properties of fuel elements must be guaranteed with 95% confidence that 95% of the pieces meet the specification. This may be an unrealistic requirement for the nuts-and-bolts industry, yet it is absolutely necessary for critical properties of nuclear fuel components, and the customer must be willing to bear the cost.

Finally, most malfunctions and even accidents which have come to attention have been those which could be classified as oversights rather than as results of abstruse causes. These are maladjustments which could have been avoided by using a little more care and common sense. The use of extra care in cleaning the internals of the system, while permitting the construction crew to leave within it bolts, dirt, and other miscellany is a case in point. Perhaps the best advice is to take care of the details and never to overlook the obvious.

TABLE 5-2

Inspection Methods and Recommended Practices Applicable to Reactor Materials*

Inspection method	Used to detect:	ASTM standard
Magnetic particle	surface flaws	E109, E138, E125, A275
Liquid penetrant	surface flaws	E165
Radiography	internal flaws (for sections up to 2 in. thick)	E52, E71, E94, E99, E142
Ultrasonic	internal flaws	E113, E114, E164, A435, A388
Surveillance and in-reactor tests	effects of radiation	E184, E185
Standards under development		
Eddy current	internal flaws	
Radiography	internal flaws (for sections 3, 6, and 12 in. thick)	

*From Wylie and McGonnagle [139].

REFERENCES

1. John Wulff (ed.), The Structure and Properties of Materials, (4 vols.), John Wiley, N.Y., 1964.
2. A. G. Guy, Elements of Physical Metallurgy (2nd ed.), and Physical Metallurgy for Engineers, Addison-Wesley, Reading, Mass., 1959 and 1962, respectively.
3. Bruce Chalmers, Physical Metallurgy, John Wiley, N.Y., 1959.
4. A. H. Cottrell, The Mechanical Properties of Matter, John Wiley, N.Y., 1964.
5. R. E. Smallman, Modern Physical Metallurgy, Butterworths, London, 1962.
6. H. H. Uhlig, Corrosion and Corrosion Control, John Wiley, N.Y., 1963.
7. T. Lyman (ed.), Metals Handbook, 8th ed., Amer. Soc. for Metals, Novelty, Ohio, 1961.
8. C. R. Tipton (ed.), Reactor Handbook; Vol. 1, Materials, 2nd ed., Interscience, N.Y., 1960.
9. C. A. Hampel, Rare Metals Handbook, 2nd ed., Reinhold, N.Y., 1961.
10. N. G. Woldman (ed.), Engineering Alloys, 4th ed., Reinhold Publ. Corp., N.Y., 1962.
11. D. W. White and J. E. Burke, The Metal Beryllium, Amer. Soc. for Metals, Cleveland, Ohio, 1955.
12. B. Lustman and F. Kerze, The Metallurgy of Zirconium, McGraw-Hill, N.Y., 1955.
13. D. E. Thomas and E. T. Hayes (eds.), The Metallurgy of Hafnium, USAEC, U.S. Gov't. Print. Off., 1960.
14. C. S. Roberts, Magnesium and Its Alloys, John Wiley, N.Y., 1960.
15. R. E. Nightingale (ed.), Nuclear Graphite, Academic Press, N.Y., 1962.
16. J. Barksdale, Titanium; Its Occurrence, Chemistry, and Technology, Ronald Press, N.Y., 1959.
17. W. K. Anderson and J. S. Theilacker (eds.), Neutron Absorber Materials for Reactor Control, USAEC, U.S. Gov't. Print. Off., 1962.
18. —Metallurgy of the Rarer Metals, Butterworth, Inc., Washington, D. C.
    a. A. H. Sully, Chromium, 1954.
    b. G. L. Miller, Zirconium, 2nd ed., 1957.
    c. A. H. Sully, Manganese, 1955.
    d. A. D. McQuillan and M. K. McQuillan, Titanium, 1956.
    e. L. Northcott, Molybdenum, 1956.
    f. G. L. Miller, Tantalum and Niobium, 1959.
19. A. G. Quarrell (ed.), Niobium, Tantalum, Molybdenum, and Tungsten, Elsevier Pub. Co., Amsterdam, 1961.
20. —Nucleonics, 22 (11), 24 (Nov. 1964).
21. A. J. Kennedy, Processes of Creep and Fatigue in Metals, John Wiley, N.Y., 1963.
22. —"Notched Bar Impact Testing of Metallic Materials," ASTM Designation E23-60, 1961 Book of ASTM Standards, Part 3, Amer. Soc. for Test. and Mat'ls., Phila., Pa., p. 79.
23. —"Conducting Drop-Weight Test to Determine Nil-Ductility Transition Temperature of Ferritic Steels," ASTM Designation E208-63T, 1963 Supplement to Book of ASTM Standards, Part 3, Amer. Soc. for Test. and Mat'ls, Phila., Pa., p. 15.
24. P. P. Puzak, M. E. Schuster, and W. E. Pellini, Welding Jour., 33, 481 (1954).
25. J. Heslop and N. J. Petch, "The Ductile-Brittle Transition in the Fracture of Alpha Iron," Phil. Mag., 3, 1128 (1958).
26. M. S. Wechsler, "Radiation Embrittlement of Metals and Alloys," p. 298 of ref. 41.
27. —Fracture Toughness Testing and Its Applications, ASTM-STP 381, Amer. Soc. for Test. and Mat'ls, Phila., Pa., 1965.
28. P. C. Paris and G. C. M. Sih, "Stress-Analysis of Cracks," p. 30 of ref. 27.
29. V. Weiss and S. Yukawa, "Critical Appraisal of Fracture

Mechanics,'' p. 1 of ref. 27.
30. F. A. McClintock and G. R. Irwin, "Plasticity Aspects of Fracture Mechanics," p. 84 of ref. 27.
31. R. E. Johnson, "Fracture Mechanic: A Basis for Brittle Fracture Prevention," AEC report WAPD-TM-505, Nov. 1965.
32. J. E. Srawley and W. F. Brown, "Fracture Toughness Testing," p. 133 of ref. 27.
33. L. F. Coffin, "A Study of the Effects of Cyclic Thermal Stresses on a Ductile Metal," Trans. ASME, J. Basic Eng., 82, 203 (1959).
34. L. F. Coffin, "Strain Cycling and Thermal Stress Fatigue," High Temperature Materials, Proc. Fourth Sagamore Ordnance Mat'ls. Res. Conf., Aug. 1957.
35. L. F. Coffin, "Design Aspects of High-Temperature Fatigue with Particular Reference to Thermal Stresses," Trans. ASME, 78, 527 (1956).
36. L. F. Coffin, "The Problem of Thermal Stress Fatigue in Austenitic Steel at Elevated Temperature," Symp. on Effect of Cyclic Heating and Stressing on Metals at Elevated Temperatures, ASTM-STP 165, Amer. Soc. for Test. and Mat'ls, 1954, p. 31.
37. L. F. Coffin, "The Resistance of Materials to Cyclic Thermal Stresses," ASME paper 57-A-286, 1957.
38. T. C. Yen, "Thermal Fatigue—A Critical Review," Welding Research Council, Bulletin 72, Oct. 1961.
39. B. F. Langer, "The Use of Materials Properties by the Reactor Vessel Designer," p. 418 of ref. 42.
40. D. S. Billington (ed.), Radiation Damage in Solids, Proc. of the Int'l. School of Physics, "Enrico Fermi," Course XVIII, Academic Press, N.Y., 1962.
41. R. Strumane, J. Nihoul, R. Gevers, and S. Amelinckx (eds.), The Interaction of Radiation with Solids, North Holland, Amsterdam, 1964.
42. —Flow and Fracture of Metals and Alloys in Nuclear Environments, ASTM-STP 380, Amer. Soc. for Test. and Mat'ls., Phila., Pa., 1965.
43. D. K. Holmes, "Radiation Damage in Non-fissionable Metals," p. 147 of ref. 41.
44. D. S. Billington and J. H. Crawford, Radiation Damage in Solids, Princeton Univ. Press, Princeton, N. J., 19.
45. G. H. Vineyard, "Dynamic Stages of Radiation Damage," p. 291 of ref. 40.
46. C. Erginsoy, "Dynamics of Atomic Displacement Processes," p. 51 of ref. 41.
47. D. K. Holmes, "Range Calculations and Channeling," p. 33 of ref. 41.
48. J. R. Beeler, "Computer Studies of Neutron Irradiation and Annealing in Finite BCC Iron Specimens," p. 86 of ref. 42.
49. J. Brinkman, "Production of Atomic Displacements by High Energy Particles," Amer. J. of Phys., 24, 246 (1956).
50. A Seegar and U. Essmann, "The Mechanism of Radiation Hardening of Face-Centered Cubic Metals by Fast Neutrons," p. 717 of ref. 40.
51. S. Amelinckx, "The Direct Observation of Lattice Defects by Means of Electron Microscopy," p. 682 of ref. 41.
52. W. R. Martin and J. R. Weir, "The Effect of Irradiation Temperature on the Post-Irradiation Stress-Strain Behavior of Stainless Steels," p. 251 of ref. 42.
53. P. C. L. Pfeil and D. R. Harries, "Effects of Irradiation in Austenitic Steels and Other High Temperature Alloys," p. 202 of ref. 42.
54. J. T. Venard and J. R. Weir, "In-Reactor Stress-Rupture Properties of a 20 Cr-25 Ni Columbium-Stabilized Steel," p. 269 of ref. 42.
55. J. E. Irvin, A. L. Bement, and R. G. Hoagland, "The Combined Effects of Temperature and Irradiation on the Mechanical Properties of Austenitic Stainless Steels," p. 236 of ref. 42.
56. H. E. McCoy and J. R. Weir, "Effect of Irradiation on Bend Transition Temperatures of Molybdenum- and Columbium-Base Alloys," p. 131 of ref. 42.
57. T. T. Clandson and H. J. Pessl, "Irradiation Effects on High-Temperature Reactor Structural Metals," p. 156 of ref. 42.
58. J. Moteff and J. P. Smith, "Recovery of Defects in Neutron-Irradiated Tungsten," p. 171 of ref. 42.
59. B. L. Eyre and A. C. Roberts, "The Nature and Annealing Behavior of Irradiation Damage in Molybdenum," p. 188 of ref. 42.
60. C. Cawthorne and E. J. Fulton, "Voids in Irradiated Stainless Steel," Nature, 216, 575 (1967).
61. E. E. Bloom and J. D. Stiegler, "Void Formation in Stainless Steel" (abstract), J. of Metals, 3, 114A (1968).
62. J. J. Holmes and H. R. Brager, "Irradiation Induced Swelling in Austenitic Stainless Steel," J. of Metals, 3, 115A (1968); also Trans. Amer. Nuclear Soc., 11, 479 (1968).
63. L. E. Steele and J. R. Hawthorne, "New Information on Neutron Embrittlement and Embrittlement Relief of Reactor Pressure Vessel Steels," p. 238 of ref. 42.
64. D. R. Harries and B. L. Eyre, "Effects of Irradiation in Iron and Steels," p. 105 of ref. 42.
65. R. W. Nichols and D. R. Harries, "Radiation Effects on Metals and Neutron Dosimetry," ASTM, STP-341, Amer. Soc. for Test. and Mat'ls., Phila., Pa., 1963, p. 162.
66. C. Z. Serpan and L. E. Steele, "In-Depth Embrittlement of a Simulated Pressure Vessel Wall of A302-B Steel," p. 312 of ref. 42.
67. A. D. Rossin, "Dosimetry for Radiation Damage Studies," USAEC report ANL-6826, Mar. 1964.
68. K. Shure, "Radiation Damage Exposure and Embrittlement of Reactor Pressure Vessels," USAEC report WAPD-TM-471, Nov. 1964.
69. D. J. Seman and E. J. Pasierb, "Considerations on the Annealing of Irradiated Pressure Vessel Steels," Trans. ANS, 8, 416 (Nov. 1965).
70. G. F. Carpenter, N. R. Knopf, and E. S. Byron, "Anomalous Embrittling Effects Observed During Irradiation Studies on Pressure Vessel Steels," Nuc. Sci. and Eng., 19, 18 (1964).
71. Uldis Potapous and J. R. Hawthorne, "The Effect of Residual Elements on the Response of Selected Pressure Vessel Steels and Weldments to Irradiation at 550°F," Nuclear Appl., 6, 27 (1969).
72. —The Effects of Radiation on Structural Metals, STP 426, Amer. Soc. for Test. and Mat'ls, Phila., Pa., 1967.
73. L. E. Steele and J. R. Hawthorne, "Surveillance of Critical Reactor Components to Assess Radiation Damage," Nucl. Eng. and Sci. Cong., Preprint 67, Engineers Joint Council, 1962.
74. M. B. Reynolds, "Strain-Cycle Phenomena in Thin-Wall Tubing," p. 323 of ref. 42.
75. J. J. Holmes, J. A. Williams, D. H. Nyman, and J. C. Tobin, "In-Reactor Creep of Cold-Worked Zircaloy-2," p. 385 of ref. 42.
76. J. R. Hawthorne and L. E. Steele, "In-Reactor Studies of Low-Cycle Fatigue Properties of a Nuclear Pressure Vessel Steel," p. 350 of ref. 42.
77. G. R. Irwin, Dimensional and Geometric Aspects of Fracture of Engineering Materials, Amer. Soc. for Metals, Metals Park, Ohio, 1964, p. 211.
78. W. H. Irvine, A. Quirk, and E. Bevitt, "Fast Fracture of Pressure Vessels," J. Brit. Nucl. Energy Soc., Jan. 1964, p. 31.
79. W. H. Irvine, "The Significance of Recent Developments in Fracture Mechanics in Great Britain in Relation to the Safety of Reactor Pressure Vessels," p. 429 of ref. 42.
80. W. S. Pellini and P. P. Puzak, "Fracture Analysis Diagram Procedures for the Fracture-Safe Engineering Design of Steel Structures," Bulletin 88, Welding Research Council, 1963. Also NRL Report 5920, March, 1963.
81. C. F. Tiffany and J. N. Masters, "Applied Fracture Mechanics," p. 249 of ref. 27.
82. R. E. Johnson, "Fracture Mechanics in Design Against Brittle Failure," Trans. Amer. Nuc. Soc., 8, 367 (Nov. 1965).
83. R. D. Wylie, "The Important Materials Parameters Affecting the Performance of Nuclear Reactor Pressure Vessels," p. 438 of ref. 42.
84. M. Hansen, Constitution of Binary Alloys (2nd ed.), McGraw-Hill, N.Y., 1958.
85. A. R. Kaufmann, Nuclear Reactor Fuel Elements, Interscience, N.Y., 1962.
86. J. L. Roberts, Selected Operating Experience of Commission Power Reactors, USAEC report TID-13305, 1961.
87. M. C. Flemings, "Controlled Solidification," paper presented to Twelfth Sagamore Army Materials Research Conference, Aug. 1965.
88. R. N. Lyon, Liquid Metals Handbook, USAEC report NAVEXOS P-733 (rev.), June 1952.
89. D. J. DePaul, Corrosion and Wear Handbook for Water-Cooled Reactors, USAEC report TID-7006, Mar. 1957.
90. H. Mason, "Selection and Application of Materials for the PWR Reactor Plant," USAEC report WAPD-PWR-971, July 1957.
91. C. N. Spalaris, "Finding a Corrosion Resistant Cladding for Superheater Fuels," Nucleonics, 21, 41 (1963).
92. R. P. Jordan, National Geographic, 128, 614 (1965).
93. W. E. Berry, "Some Facts About Stress Corrosion of Austenitic Stainless Steels in Reactor Systems," Reactor Materials Quarterly, 7 (1), 1 (1964).
94. M. S. Feldman, Stress Corrosion Cracking of Stainless Steel—A Literature Search, USAEC report DP-683, 1963.
95. W. J. Singley, I. H. Welinsky, S. F. Whirl, and H. A. Klein, "Stress Corrosion of Stainless Steel and Boiler Water Treatment at Shippingport Atomic Power Station," Amer. Power Conf., Vol. XXI, 1959, p. 748.
96. F. Turner and H. K. Richardson, Investigation of Cracks in an 18/13/1 Stainless Steel Autoclave, UKAEA report IGR-TN/C-1015, 1958.
97. R. L. Moore and E. Rau, Evaluation of Viton and Other Nonmetallic Gasketing Materials in Contact with Structural Alloys, USAEC report WAPD-BT-22, 1961.
98. J. C. Griess et al, Quarterly Report of the Solution Materials

Section for the Period Ending Jan. 31, 1959, USAEC report CF-59-1-79, 1961.
99. G. W. Keilholtz et al, "Effect of Irradiation on Corrosion of Structural Materials by Molten Fluorides," Nuc. Sci. and Eng., 5, 15 (1959).
100. S. Kass, "The Development of the Zircaloys," p. 3 of ref. 101.
101. –Corrosion of Zirconium Alloys, ASTM-STP 368, Amer. Soc. for Test. and Mat'ls, Phila., Pa., 1964.
102. M. S. Feldman, Hydriding of Zircaloy-2: A Literature Search, USAEC report DP-803, 1963.
103. W. J. Babyak, W. F. Bourgeois, and G. J. Salvaggio, "Effect of Hydrogen and Hydride Morphology on the Tensile Properties of Zircaloy-2," p. 76 of ref. 101.
104. W. E. Berry, "Effect of Fluoride Ions on the Aqueous Corrosion of Zirconium Alloys," p. 28 of ref. 101.
105. W. A. Burns and H. P. Maffei, "Neutron Irradiation and Cold Work Effects on Zircaloy-2 Corrosion and Hydrogen Pickup," p. 101 of ref. 101.
106. –"High Tensile Strength Carbon-Silicon Steel Plates for Boilers and Other Pressure Vessels," ASTM Designation A-212, 1961 Book of ASTM Standards, Part 1, Amer. Soc. for Test. and Mat'ls., Phila., Pa., p. 435.
107. –"Manganese-Molybdenum Steel Plates for Boilers and Other Pressure Vessels," ASTM Designation A-302, 1961 Book of ASTM Standards, Part 1, Amer. Soc. for Test. and Mat'ls, Phila., Pa., p. 482.
108. E. L. Fogelman and R. H. Sterne, "Quality Control and Inspection Procedures for Nuclear Pressure Vessel Plates," Lukens Steel Co., paper presented to Adv. Comm. on Reactor Safeguards, USAEC, Nov. 1965.
109. J. Marin and F. P. J. Rimrott, "Design of Thick-Walled Pressure Vessels Based Upon the Plastic Range," Welding Research Council, Bulletin 41, July 1958.
110. B. Crossland, S. M. Jorgensen, and J. A. Bones, "The Strength of Thick-Walled Cylinders," Trans. ASME, Series B, J. Eng. Ind., 81, 95 (1959).
111. J. Marin and T. Weng, "Strength of Thick-Walled Cylindrical Vessels Under Internal Pressure for Three Steels," Welding Research Council, Bulletin 67, March 1961.
112. J. Marin and T. Weng, "A Critical Evaluation of the Strength of Thick-Walled Cylindrical Pressure Vessels," Welding Research Council, Bulletin 74, Jan. 1962.
113. J. H. Gross, E. H. Kottcamp, and R. D. Stout, "Effect of Heat Treatment on the Microstructure and Low-Temperature Properties of Pressure Vessel Steels," Welding Jour., 37, 160S (1958).
114. C. R. Mayne, "Information and Test Data on Steels Over Four Inches in Thickness," Engineering Foundation, Welding Research Council Bulletin.
115. E. T. Wessel and W. H. Pryle, "Brittle Fracture Characteristics of a Reactor Pressure Vessel Steel," Welding Jour., 40, 415 (1961).
116. W. P. Chernock, R. M. Mayfield, J. R. Weir, "Cladding Materials for Nuclear Fuels," Proc. Third UN Conf. on Peaceful Uses of Atomic Energy, Geneva, 1964, vol. 9, p. 35.
117. W. R. Thomas et al., "Irradiation Experience with Zircaloy-2," Proc. Third UN Int'l Conf. on Peaceful Uses of Atomic Energy, Geneva, 1964, vol. 9, p. 80.
118. A. L. Bement, J. C. Tobin, and R. G. Hoagland, "Effects of Neutron Irradiation on the Flow and Fracture Behavior of Zircaloy-2," p. 364 of ref. 42.
119. W. Jung-Konig, H. Richter, W. Spalthoff, and E. Starke, "Properties and Technology of Zirconium Alloys with Niobium and Tin," Proc. Third UN Int'l Conf. on Peaceful Uses of Atomic Energy, Geneva, 1964, vol. 9, p. 139.
120. C. E. Ellis, S. B. Dalgaard, W. Evans, and W. R. Thomas, "Development of Zirconium-Niobium Alloys," Proc. Third UN Int'l Conf. on Peaceful Uses of Atomic Energy, Geneva, 1964, vol. 9, p. 91.
121. E. A. Block and H. Hug, "The Latest Developments of the SAP Process," Powder Metallurgy, Interscience, 1961, p. 371.
122. D. G. Bozall et al., "Development of Fuel and Coolant Tubes for a Reactor Cooled by Organic Liquid," Proc. Third UN Int'l Conf. on Peaceful Uses of Atomic Energy, Geneva, 1964, vol. 9, p. 102.
123. N. Hansen, P. Knudsen, A. C. Winther, and E. Adolph, "Sintered Aluminum Products for Organic Reactor Application," Proc. Third UN Int'l Conf. on Peaceful Uses of Atomic Energy, Geneva, 1964, vol. 9, p. 122.
124. D. Gualandi and P. Jehenson, "Contribution to the Study of the Technology of $Al-Al_2O_3$ Composites...," Proc. Third UN Int'l Conf. on Peaceful Uses of Atomic Energy, Geneva, 1964, vol. 9, p. 157.
125. –Industrial Graphite Engineering Handbook, Union Carbide Corp., Carbon Products Div., New York, N. Y.
126. J. H. W. Simmons, Radiation Damage in Graphite, Pergamon Press, N. Y., 1965.
127. I. Cohen, E. F. Losco, and J. D. Eichenberg, "Metallurgical Design and Properties of Silver-Indium-Cadmium Alloys for PWR Control Rods," presented at Nuc. Eng. and Sci. Conf., Chicago, 1958.
128. Carl Andognini, Yankee Atomic Electric Company, Rowe, Mass., Personal (TOZ) communication.
129. H. G. Rickover, "Presentation to National Metal Congress and Exposition," Fall Meeting, Amer. Soc. for Metals, 1962.
130. G. Sachs and K. R. Van Horn, Practical Metallurgy, Amer. Soc. for Metals, 1940.
131. F. L. Yaggee, J. E. Ayer, and H. F. Jelinek, "Injection Casting of Uranium-Fission Alloy Pins," Nucl. Met., IV, 51 (1957).
132. J. G. Hunt, D. F. Kaufman, and P. Lowenstein, Hot Extrusion of UO_2 Fuel Elements, AEC Report NMI-1245, Oct. 1961.
133. J. F. Joseph, "Tandem Extruded Joints Undamaged by Irradiation," Nucleonics, 22 (12), 64 (1964).
134. J. F. Young, Materials and Processes, (2nd ed.), John Wiley, N. Y., 1954.
135. F. O. Bingman and C. R. Woods, Gas Pressure Bonding of Production Size PWR Core Plate Type Fuel Elements Containing Ceramic Fuel, USAEC report WAPD-245, May, 1963.
136. –"Reactor Gets Repair-Free Welds," Iron Age, 187, 106 (1961).
137. –"Stress Relieving on Site at Hunterston," Metallurgia, 64, 185 (1961).
138. –"Elk River Reactor Vessel Cladding Defects," USAEC report TID-13083, 1961.
139. R. D. Wylie and W. J. McGonnagle, Quality Control in Fabrication of Nuclear Pressure Vessels, Rowman and Littlefield, N. Y., 1964.
140. –"Rapid Identification (Spot Testing) of Some Metals and Alloys," International Nickel Co., 1952.
141. H. Etherington (ed.), Nuclear Engineering Handbook, McGraw-Hill, N. Y., 1958.
142. J. A. Rinebolt and W. J. Harris, "Effects of Alloying Elements on Notch Toughness of Pearlitic Steels," Trans. ASM, 43, 1175 (1951).
143. W. J. O'Donnell and C. M. Purdy, "The Fatigue Strength of Members Containing Cracks," Trans. ASME, J. Eng. Ind. 86B, 205 (1964).
144. J. H. Crawford, "Radiation Damage in Solids: A Survey," Ceramic Bull., 44, 963 (1965).
145. J. C. Wilson, "Effects of Irradiation on the Structural Materials in Power Reactors," Progress in Nucl. Energy, Series IV Vol. 2, Pergamon Press, 1960, p. 201.
146. R. G. Bergren, Steels for Reactor Pressure Circuits, Special report 69, Iron and Steel Institute, London, 1961, p. 370.
147. N. E. Hinkle, Rad. Effects on Metals and Neutron Dosimetry, ASTM, STP 341, Amer. Soc. for Test. and Mat'l, Phila., Pa., 1963, p. 344.

CHAPTER 13

Nuclear Fuels

D. H. GURINSKY
Metallurgy Division,
Brookhaven National Laboratory, Upton, New York
S. ISSEROW
Nuclear Metals Division,
Textron Inc., West Concord, Massachusetts

CHAPTER CONTENTS*

1 INTRODUCTION

In this chapter the fuel systems currently in use or proposed for use are presented with the view of bringing to the attention of the reader those aspects of design, fabrication and testing which increase the probability of obtaining a more reliable nuclear reactor fuel.

The primary component of a nuclear reactor is the nuclear fuel. When assembled in a reactor in a proper configuration, the fuel fissions and releases heat. The reactor system is built around the fuel for the prime purpose of extracting this heat. In addition to generating heat, the fission process produces radioactive fission products; the escape of these is the primary source of safety problems in a reactor. However, it can be generally stated that with the exception of those reactors (primarily fast) in which bowing of fuel elements or fuel redistribution can lead to an

*This chapter is based on information in the literature or known to the authors prior to early 1964 as revised and rewritten to reflect the principal advances made up to the early spring of 1966.

increase in reactivity,* the failure of fuel elements is not the primary cause of catastrophic accidents. However, unexpected increases in power level, reductions in coolant flow and other abnormal conditions of operation may lead to major fuel ruptures. Such failures may subject operating personnel and even the public to hazards through the excessive release of radioactive fission products. Thus the fuel designer in close cooperation with the nuclear core and heat transfer designers must incorporate into the fuel design a sufficient margin of safety to ensure that major fuel failures will not occur as a result of moderate deviation from the normal operating mode. The goal then is to produce a reliable fuel with a long life, high burnup and low cost which is not subject to failure in the event of moderate system perturbations. Most major fuel element failures result in costly shutdowns to clean up the radioactive fission products. Therefore the discussion in this chapter is concerned with design requirements and test procedures utilized to ensure the retention of radioactive fission products either within the fuel element for solid-fueled reactors or within the primary containment in the case of fluid-fueled reactors.

Nuclear fuel materials may be employed in one of two ways - either as rigid components (fuel elements) of the reactor core or in a fluid form, as for example in the homogeneous reactor. A combination of both of these is used in the LAMPRE (Los Alamos Molten Plutonium Reactor Experiment) where the fluid plutonium-iron alloy is contained in a rigid tantalum fuel tube. As in the case of rigid fuels, many fluid fuels have been suggested. These include the truly homogeneous, the heterogeneous and fluidizable fixed bed. All of these fuel systems have been proposed for use in reactor systems to gain some particular advantage. In some cases the advantage may even be nuclear stability and therefore an increase in safety.

In Secs. 2 and 3 a general discussion of rigid (fuel element) fuels is presented. Fuel elements presently in use and their operational behavior are reviewed in Sec. 4, and the final section discusses the fluid fuel systems. For more detailed information on fuels the reader is referred to a number of books on the subject of nuclear fuels systems and their components [1-13].

2 FUEL SYSTEM COMPONENTS

There are three nuclides, U^{235}, U^{233} and Pu^{239}, which are available in sufficient quantity for use as nuclear fuels. The selection of a specific fuel should be determined only by the feasibility and economics of the type of reactor under design. However, availability, or lack of sufficient development may be overriding reasons in the selection of an initial fuel used in a particular design. For example, most fast reactors built to date have not used plutonium because its cost is high, it is toxic and difficult to work with and the development work on plutonium fuel elements is not at an advanced stage.

2.1 Specific Fuel Systems

Although the number of fissile nuclides is limited there is a large selection of fuel systems. For the rigid-fueled reactors the following are the primary decisions which have to be made in selecting a particular fuel element:

1) Chemical state - metal, ceramic
2) Method of fuel dilution - homogeneous (solid solution), heterogeneous
3) Cladding - material, method of bonding
4) Fuel element configuration - pin, rod, plate, tube

2.1.1 Chemical State

The early reactors utilized metallic uranium because the metallic form provided the highest density of fissile nuclei (only natural uranium was available**). The availability of enriched uranium and the inadequacies (see Sec. 3.2) of natural uranium metal for certain reactors have led to the development of uranium alloys such as: uranium-aluminum, uranium-molybdenum, uranium-columbium,*** uranium-zirconium, uranium-fissium**** and uranium-thorium. Alloys of plutonium are also under development. (All these systems are discussed in reference [1].) Ceramic fuels such as UO_2, U_3O_8, UC, UC_2, UN, US, PuO_2, PuC; cermet fuels such as U_3O_8 in Al, UO_2 in stainless steels, molybdenum, and tungsten; intermetallic fuels such as UAl_3 or UAl_4 in Al, $PuAl_3$ in Al; and ceramals such as UO_2 in BeO, UC_2 in graphite, UO_2 in ThO_2 are now being extensively developed. Intensive testing programs have shown that metallic fuels (uranium base)***** are inferior to the ceramics with respect to burnup potential and high-temperature operational capability.

During the period 1958 to 1965, oxide fuels (UO_2, UO_2-ThO_2) [15, 16, 17, 18] became the most important fuels for water-cooled thermal reactors for economic power. The potential of oxide fuels even for fast reactors is shown by the fact that PuO_2-UO_2 [14] is the fuel which has been selected for SETOR (Southwest Experimental Test Oxide Reactor) and FCR the G. E. fast reactor. Stoichiometric and hypostoichiometric uranium monocarbide [19] and UC-PuC [20] are promising fuels for both fast and thermal reactors because these fuels have thermal conductivities roughly

---

*See discussion of fast reactor core meltdown, etc. in the chapter on Fast Reactor Kinetics.

**CP-1, the very first reactor, contained metal and oxide.

***As recommended by AIME we have used Cb instead of Nb.

****Fissium is a simulated mixture of fission product atoms usually consisting of Mo, Ru, Pd, and Rh, small amounts of Zr and Cb.

*****Uranium-thorium alloys are promising with respect to burnup potential and high temperature operation.

equal to those of corresponding metals, have metal densities greater than their oxides, do not react with liquid sodium and the radiation behavior of UC is comparable to that of UO_2. Improvements in properties such as strength of the fissile carbides are expected by alloying with the carbides of zirconium and columbium. Substantial development and testing still remains to be done to demonstrate that the carbides of the fissile materials are satisfactory as fuel materials. Uranium mononitride is now receiving considerable attention because it has properties similar to those of the carbide and, in addition, it is not attacked by moisture and air as are the carbides. It does, however, require special pressure-melting techniques. Uranium sulfide is also being investigated as a fuel since it too is expected to have good radiation resistance [21].

2.1.2 Fuel Dilution

In most reactors the fuel element must have a large area for heat transfer. To accomplish this the fuel is diluted. The diluent can be either a nuclear inert material, a moderator or a fertile material which serves either as a solvent or a matrix in which the fissile material is dispersed. Examples of homogeneous fuels (solid solution) are U^{235}, U^{233} [4] or Pu in natural uranium; uranium in zirconium [4a]; UO_2 in ThO_2 or ZrO_2; PuO_2 in UO_2, and UC in ZrC. Dispersion fuels which are used are U in Th [4a], UAl_3, UAl_4 or $U(Al, Si)_3$ [22a] in Al; U_3O_8 in Al; UO_2 in stainless steel; U in ZrH_x; UC_2 [23] in graphite; and UO_2 in BeO [24]. (Most of these systems are discussed in references [1] and [4].)

In most cases the fuel is distributed uniformly in the diluent. However in the case of the HIFR the fuel is dispersed according to a pattern devised by the nuclear core designer so that the heat generation can be more uniform and thereby allow for higher heat generation rates. Such fuel elements are made by powder metallurgical techniques or by proper contouring of the fuel-diluent mixtures in the assembly operations.

It can be seen that although there are only two important fissile materials, U and Pu, there are a very large number of fuel systems.

2.1.3 Cladding Materials

Fuels are commonly sealed in cladding which serves to contain fission products. The materials which have most frequently been used as cladding materials are Al, Be, Zr alloys, stainless steel, Mg alloys and graphite. Recently FeAl, Hastelloy [25], Inconel and Incoloy [26], and the refractory metals Cb, Ta [27] and V [4a] have found application.

Pure aluminum and its alloys* are used in low temperature thermal reactors because they are cheap, can be fabricated readily, have good corrosion resistance to water and organic coolants and have low thermal neutron absorption cross section. Beryllium metal because of its extremely low thermal neutron cross section and higher melting point has been studied intensively. Because of its cost, low ductility, poor fabricability and only moderate corrosion resistance to water and moist CO_2 it has not been used as a cladding in any reactor to date.

Zirconium alloys (Zircaloy-2 and -4) [28] have been used in a number of thermal reactors because they have excellent corrosion resistance to water at moderate temperature (600°F or 315°C), low thermal neutron absorption cross section, and good strength. By 1966 economic considerations had led to the selection of zirconium alloy fuel cladding for all water-cooled reactors. Earlier, austenitic stainless steels were used in place of zirconium in many thermal reactors, in spite of their higher thermal neutron cross section, because these steels have moderate to high tensile strengths at temperatures up to 1000°F (540°C), good corrosion resistance to 600°F (315°C) water, are readily fabricated into tubes and plates and are available at reasonable cost. Magnesium alloys (Magnox) are used in the CO_2-cooled reactors of the Calder Hall type mainly because their cross section is low. They possess adequate strength and are corrosion resistant to the CO_2 environment.

The recent development of low permeability graphites has led to their use as a cladding material for high temperature [23] helium-cooled reactors. These graphites have good tensile strengths at elevated temperature (2000-2500°F or 1090 - 1370°C), adequate thermal conductivity, a low coefficient of thermal expansion and a low modulus of elasticity. The latter two properties are important in maintaining low thermal stresses in the material.

Hastelloy [25], Inconel or Incoloy [26] and metals such as Ta [27], Cb [29], and V are being used in special applications because of good corrosion resistance to certain reactor coolants. Hastelloy is the cladding material for the fuel in the closed-cycle nitrogen-cooled reactor [25]. Inconel shows promise as a cladding for fuel elements for the superheat reactor [26]. Tantalum is the containment tube for the LAMPRE PuFe liquid-metal fuel. Columbium (niobium) is being considered for cladding for high temperature reactor fuel elements cooled by lithium [29]. It is used in the Dounreay fast reactor [193, 194, 215].

A promising coating [30, 31] (cladding) which is applied directly to the particles of the fissile material has recently been developed. The coatings studied have been pyrolytic carbon on carbide fuels, ceramics (such as Al_2O_3) on UO_2 and metallic deposits (such as Zr or Cb) on carbide or oxide fuels. Such coatings can be made impervious to gases and can be made thick enough to stop fission fragments generated in the fissile phase. The (coated) particles have varied from tens of microns to hundreds of microns in diameter with the coatings varying in thickness from ten microns to hundreds of microns. It should be noted that to date coated fuel particles are used as the dispersed phase in dispersion-type fuel which in turn is clad. Development work has demonstrated that radiation and temperature stability is increased by applying composite coatings. Pyrolytic carbon coating is used on the fissile phase of the HTGR fuel particles and the Pebble Bed Reactor (AVR).

*These include SAP(APM), a class of alloys dispersion-hardened by Al_2O_3.

2.1.4 Fuel Element Configuration

Although the early reactors used only cylindrical* fuel elements for the most part, it was soon apparent that if higher heat fluxes were to be attained without melting the core of the fuel element, it would be necessary to increase the ratio of surface area to volume. This has led to a reduction in the diameter of the cylindrical fuel elements and many elements can no longer be called rods but are now referred to as "pins" [32, 33]. Most of the UO_2 fuel elements used in water-cooled power reactors are cylindrical in shape because UO_2 fuel element fabrication techniques lend themselves more readily to rod geometry.

Water-cooled research reactors, however, employ either fuel plates or tubular shapes to cope with the higher heat fluxes resulting from the need for higher neutron fluxes for research experiments. Plate elements have been used in certain special purpose power reactors and in the PWR [1a].

3 FUEL ELEMENT DESIGN

The selection and design of fuel elements are influenced by: neutron spectrum (fast, thermal), reactor type, coolant characteristics, neutron flux, power density, linear power density and purpose of reactor. In making a choice among the possibilities, safety considerations play a vital part.

The designer should select a fuel element design which is capable of : (1) transferring the heat generated to the coolant, (2) containing the radioactive fission products, (3) attaining the burnup and corrosion lifetime desired and (4) being evaluated and tested prior to use. The implications of each of these requirements are discussed in detail in the following subsections.

3.1 Thermal Considerations

From a knowledge of the heat transport and heat transfer characteristics of the coolant selected, and the desired power density, the heat transfer surface area of the reactor core can be calculated. With this information it is possible to make the first estimate of the fuel element configuration, i. e., rod, pin, plate. This selection has to take into account such factors as amount of fissionable, fertile, diluent material and cladding permitted by the nuclear core designer as well as metallurgical requirements in the core. For the particular configuration selected, it is then possible to calculate the thermal gradient in the fuel element. This information is of particular importance in that it now becomes possible to determine whether the particular design selected will be adequate. If the element is metallic uranium it is important to know whether the center temperature will exceed the α-β transformation temperature (662°C or 1224°F) or, if the element is UO_2, whether the center temperatures will exceed its melting point (2800°C or 5072°F). Many designers feel that no part of a metallic uranium fuel element should operate above the α-β transformation temperature because the volume change (~ 1%) accompanying this transformation introduces additional stress in the element. Present design practice in UO_2 elements generally avoids central melting** to reduce the release of gaseous fission products from the fuel matrix, to prevent a possible shift of fissile material within the fuel element which could lead to an uncontrolled and unsafe reactivity change, and to avoid the additional cladding stresses produced by the volume increase on melting. Many power reactor designs use the grain-growth temperature (1400 - 1700°C or 2550-3090°F) as the center temperature limit for UO_2.

High temperature gradients in a fuel element lead to high thermal stresses. For metallic elements this can lead either to cracking (if the material is not ductile) or to creep or thermal fatigue if the element is temperature-cycled repeatedly. For ceramic fuel elements (UO_2) the high stresses arising from the thermal gradients are sufficient to crack the UO_2 and even fragment it. This may result in a lower overall thermal conductivity of the elements with a further increase in the gradient.

Cracking of UO_2 due to thermal stress on shutdown has been suggested to explain a loss of reactivity experienced in the Yankee [34] reactor. At Yankee it was observed that on a startup following a shutdown about a 1% loss of reactivity was noted which could not be readily explained. Continued operation showed that this reactivity was recovered over a period of time. One explanation offered was that the shutdown caused the UO_2 to crack sufficiently to reduce the overall thermal conductivity of the elements. Thus, on startup the average temperature of the elements was higher than during the prior (to shutdown) operation of the reactor. A higher average fuel temperature would result in an increased Doppler absorption which would account for the observed loss of reactivity. The observed recovery in reactivity may have been due to sintering of UO_2 during operation which would increase its thermal conductivity. (Sintering of UO_2 occurs readily at high temperatures and quite rapidly in the presence of radiation).

To reduce the center temperature, cylindrical elements have been designed with a central void [35]; alternately, the fissile material in the central region of a fuel element can be replaced by an inert material, as in the case of the HTGR. The HTGR fuel element consists of an annulus of $(U^{235}, Th)C_2$ dispersed in graphite. This design was selected for two reasons: (1) if the U and Th were dispersed throughout the core the temperature gradient could cause uranium and thorium to diffuse from the central region (hottest region) to the periphery (cooler region) of the cylinder and (2) further, the designers realized that if U and Th diffused at different rates, an increase in reactivity could result if the thorium (carbide) particles increased in dimensions as a result of diffusion and precipitation; i. e., further clumping

*For thermal reactors a low ratio of cladding atoms to fuel atoms is highly desirable.

**Fuel elements with molten centers are at present being studied with the view of increasing the thermal performance of oxide elements.

of the thorium carbide would reduce the overall absorption because the original dispersed particles were already black to neutrons. This example is cited to illustrate the importance of close cooperation between the fuel element designer and the reactor physicist if the interplay of the two disciplines is to result in a safe design for a fuel element.

Temperature gradient is important not only to the behavior of the fissile component of a fuel element but can be equally important to the matrix. So called "homogeneous"* fuel elements consisting of a dispersion of uranium in a matrix of zirconium hydride are being used in special purpose reactors. The TRIGA reactor designed by General Atomic utilizes homogeneous fuel elements [36] made by hydriding a uranium-zirconium alloy to the desired Zr/H ratio (see chapter on Kinetics of Solid-Moderator Reactors). When a temperature gradient develops in these elements, hydrogen can migrate to the cooler portion of the fuel and cause a redistribution of the moderator material (H) resulting in a change in reactivity. Here again a reduction in temperature gradient results in a more stable fuel element.

The temperature gradient can also be important for the following reason. Studies [1b, 195] on the swelling rate due to the migration of rare gases in natural uranium metallic elements have shown that the rate increases markedly above 400°C (750°F). Such swelling results in serious dimensional changes and distortion of the fuel elements with a concomitant effect on coolant flow. If the thermal gradient can be kept low as a means of keeping the average temperature everywhere in the element below 400°C (750°F), metallic fuel elements can withstand a higher burnup, thus lowering the fuel cost. Pin, plate and tubular elements are being used by fuel element designers because the central fuel element temperatures can be kept low even while the heat fluxes are being increased.

3.1.1 Mechanical Design

As pointed out previously, the striving for higher heat fluxes leads to the requirement for fuel elements with high ratio of surface area to volume. These elements because of their higher performance must be assured of the required coolant mass flow. Small changes in coolant flow due to changes in coolant channel dimensions can result in unsafe surface temperatures and higher internal temperatures. Higher surface temperatures can lead to higher and unsafe corrosion rates of the cladding materials, to film boiling and possible burnout (discussed in Sec. 2 of Chapter 16, Heat Transfer). To avoid these unsafe operating conditions the fuel element design must incorporate mechanical features that ensure an adequate cross-sectional area of the coolant flow passage under all operating conditions. In the case of rod elements this has resulted in the assembling of rods into clusters or bundles — each element being separated from its neighbors by means of fins, spacers or even by a spacer wire which is wrapped and brazed or welded around each element in the form of a spiral. These construction techniques make the assembly rigid, minimize bowing, ensure that individual rods do not touch anywhere and ensure the adequate flow area requirement. In some designs the clusters or assemblies are joined together to form even larger units to make the structure of the core even more rigid. In this way it has been possible in the case of the PWR elements to minimize the fretting corrosion of Zircaloy. The assemblies are massive enough so that there is very little relative motion between assemblies thereby avoiding the rubbing which removes the protective oxide.

In plate-type elements a number of techniques have been utilized to obtain sound mechanical designs. Since a flat plate is basically an unstable structure and may buckle either way in a thermal gradient, plates have been curved when joined to form an assembly. In this way the bowing direction is predetermined and contacting of plates by opposed random bowing is eliminated during operation. To increase the strength of the individual plate and make it more resistant to the bending caused by pressure differentials (due to unequal pressure drops down the channel) most MTR-type fuel plates now in use are cold worked (hardened). To retain this condition in the final assembly, aluminum-base fuel plates are joined to the side plates by mechanical means (roll bonding or swaging). Brazing, formerly used in joining aluminum fuel plates to the side plates, results in an annealed soft assembly.

In some designs of assemblies [37] of fuel plates it has been found more desirable to allow a certain amount of relative motion between the components of the assembly to avoid buildup of stresses which can result in distortions. In the enriched BGRR element, three V-shaped plates are joined together by means of end rings. Nonuniform cooling results in nonuniform elongation of the components of the assembly with the result that one or both of the end rings breaks during use if the element is joined rigidly to both end rings. By rigidly attaching the bent plates to one end ring and allowing for motion in the other ring, cracking of the rings has been markedly reduced thereby avoiding contact of fuel plates, overheating and release of fission products. To further generalize, the fuel element designer must be aware that fuel elements will be used in a neutron flux gradient and his design must minimize the distortions or high local stresses which result from the nonuniform heat generation.

In this subsection an attempt has been made to point out that heat flux plays an important role in the selection of the configuration of the fuel element. The temperature gradient in the fuel element can produce phase changes, center melting, reorientation of fuel and dimensional changes. Any or all of these consequences can lead to unsafe operating conditions (release of fission products to the coolant stream).

The mechanical design of a fuel element must take into account conditions arising from normal

---

*Such fuel elements have a rapid response to increases in temperature, i.e., a rapid or "prompt" temperature coefficient, since the heat resulting from an increased fission rate is deposited uniformly throughout the fuel matrix.

as well as abnormal conditions of operation. Specifically the fuel element design must incorporate features to minimize instabilities arising from coolant pressure differentials across the fuel element and temperature gradients in the element.

3.2 Economics—Burnup—Radiation Damage

The cost of energy in power reactors, and the cost of operating experimental reactors is intimately tied up with the burnup* attainable in the fuel element. The reactor designer always attempts to select compositions for fuel elements which can be "burned" beyond limits set by the loss of integrity of the fuel element due to radiation damage.

There are three basic processes which give rise to damage in fuel elements:

1) The matrix atoms are displaced from their normal positions when highly energetic particles (neutrons and fission fragments) are stopped in the fuel. The energy absorbed by the matrix atoms which are hit (knocked on) is sufficient to displace the knocked-on atoms from their lattice positions.

2) The fission process produces more atoms than were originally present in the fuel elements; for every fissile atom destroyed approximately two new atoms are created. Since all atoms occupy roughly the same volume, either the fuel element must increase in volume to accommodate the newly created atoms, or adequate void volume must be built into the element to accommodate the new atoms.

3) Fission creates products such as Xe and Kr which behave as gases only after they diffuse and collect in defects (macroscopic, microscopic and submicroscopic) in the fuel element.

3.2.1 Metals

Experience with uranium metal matrix fuel** has demonstrated that proper metallurgical treatment (alloying, heat treatment***) yields a product which shows only moderate distortions due to all three processes up to burnup of 0.5 at.% provided the maximum temperature of the element is kept below 400°C (750°F). However, if the temperature exceeds 400°C, the third process overshadows the effects due to the first two processes. The fission product gases Xe and Kr diffuse, collect in defects (dislocations, grain boundaries) of the fissile material [195], and these gas lense-like regions act as stress raisers. Stresses which arise as a result of anisotropic growth of the α (i.e. the crystal structure is orthorhombic) U are amplified in the regions where the gas has collected. When the amplified stress exceeds the strength of the material, tears and cracks develop in the solid. At higher temperatures >600°C gas bubbles can migrate in α U and coalesce and thereby enlarge which also results in abnormal swelling. If the uranium is in the gamma (γ) condition (body-centered cubic) the allowable burnup and the temperature of operation can be higher before serious distortion of the fuel element sets in.

In the first loading for the Fermi reactor a U-10 wt.% Mo alloy was selected to take advantage of the fact that alloying of uranium with molybdenum results in the retention of the γ-phase of uranium down to room temperature. Heating this alloy [1c] to 400-500°C (750-940°F) causes it to revert to the less radiation-resistant α condition.**** However, in a neutron radiation field, this reverse transformation does not take place and in fact, alloys heat-treated to produce the α-phase material change to the γ-phase. But to maintain the γ-phase it is necessary to irradiate in a minimum critical flux [38], i.e., if the flux is not in excess of the critical flux, reversion to the α condition occurs. It is therefore seen that even in this "promising" alloy type, burnup depends on a number of factors - temperature gradient, flux, flux flattening, cosine distribution, etc. which are difficult to control sufficiently well to permit the burnup to the levels initially thought attainable.

From the information available at this time it seems reasonable to conclude that uranium-base fuels cannot be used for fuel elements which are to operate at moderate to high temperatures and to burnups much above 1 at.%. ***** This type of fuel element****** will, however, continue to be used in low temperature reactors and in plutonium-producing reactors because such elements have a high density of uranium, are relatively easy to fabricate, and have a fair thermal conductivity. It should, however, be pointed out that techniques such as restraint of swelling by a properly selected cladding material may give sufficient improvement in performance to warrant the use of such elements in certain types of moderate temperature reactors.

From the standpoint of radiation stability a very promising fertile-fissile metal base fuel element is the uranium-thorium [42] base fuel system. Although this fuel combination has been studied only sporadically until recently, enough data have been collected to suggest that this fuel combination can be taken to burnups in excess of 1 at.% without serious distortion at average metal temperatures of 600°C (1112°F).

3.2.2 Ceramics

The inadequacy of uranium metal base fuel to meet the need for a fuel material which is "stable"

---

*Burnup is defined as the ratio of atoms fissioned to total atoms of fissile and fertile material in the fuel-matrix composite.

**U-Pu alloys do not behave as well as unalloyed U.

***Anisotropic growth is minimized by a beta heat treatment. See Sec 4.1.5.1.1.

****The alpha condition consists of two phases, alpha and epsilon.

*****It has been shown [39, 40, 41] that a very finely dispersed second phase in U alloys can provide nucleating sites for the fission product gases with an improvement in the burnup behavior. Pugh, UKAE, discusses the marked improvement in burnup response of U-metal-base alloys as a result of adding small amounts of second phase [9,196].

******For example, natural uranium fuel elements are being used successfully in the CO_2-cooled power reactors.

to radiation has led to the development of fissile ceramic base fuels. Test programs have shown that UO_2, when properly encased in a metal container or ceramic, can be operated at high temperatures and to burnup greater than 1 at.%.

The two important functional requirements of a fuel element, radiation resistance and mechanical integrity, are separated in UO_2 elements. Because UO_2 has a low thermal conductivity, large thermal gradients develop during reactor operation. The large thermal stresses which arise crack this brittle material. The UO_2 must therefore be encased in a material which furnishes the containment, the structural shape (surface area) necessary to effect the desired heat removal and to supply the structural strength. In most power reactors UO_2 is encased in tubes although UO_2 is also used in plate [1d] and tubular elements [43] either as a dispersion in a matrix of another ceramic [44] or in a metal or in compartments [45] in a metal plate. One very important and unique property that UO_2 and some other compounds of the fissile materials possess is the ability to contain and absorb a large volume of the generated fission fragments.* Some of the fission products diffuse out of the UO_2, but the rate of diffusion is small below operating temperatures of the order of 1000°C (1830°F). Above this temperature the higher diffusion rate of the rare gases results in larger losses of these gases from the matrix. The fact that the gases diffuse out at the higher temperatures may be exploited in designing the fuel containment to accommodate them, as for example in void volumes at the ends of a fuel rod. Burnup tests on PWR rod elements have been carried to higher than 6 at.% with a volume increase of about 5% and to as high as 46,000 MWD/ton in Yankee without detectable exterior deterioration.

Swelling has been observed in UO_2 dispersed in stainless steel plate elements [46] at high burnup of the fissile phase (40-50% of the contained uranium) when these elements operated at surface temperatures in excess of 700°C (1300°F).

Because UO_2 has such a low thermal conductivity (K 1 to 2 Btu/hr-ft°F or 0.02 to 0.04 watt/cm°C) the thermal gradient is steep and center melting in rod elements may be expected if the power requirements are too high. High temperature causes more rapid evolution of the gaseous fission products which in turn produces high internal gas pressure which can rupture the cladding. The high temperature gradient also increases the tendency for the material to recrystallize and form large grains. This phenomenon is associated with a sweeping-out of the rare gases which, even in the absence of center melting, may give rise to high internal pressures.

Although UO_2 is a much employed material for reactor fuel elements, its successful and safe use in reactors depends on careful design of the fuel element taking into account its physical, mechanical and chemical properties.

The relatively good behavior of UO_2 in power reactors has stimulated intensive radiation testing programs on other ceramic fissile compounds, solid solutions of ceramic fissile compounds in fertile ceramic compounds and dispersions of ceramic fissile compounds in ceramics. Uranium carbide has shown good radiation behavior in tests and was to be used in the second loading of the Hallam reactor and is the reference fuel for HWOCR (Heavy Water Organic Cooled Reactor). It has a thermal conductivity comparable to that of metallic uranium and should therefore show a smaller temperature gradient than UO_2 for comparable heat fluxes. The thermal stresses should also be smaller. Solid solutions of PuC in UC are being radiation-tested and these have shown some promising results [197,198]. Solid solutions of UC in ZrC are also being considered but no test results are available at present.

Irradiation tests of solid solutions of UO_2 in ThO_2 have revealed good properties. This combination is being considered seriously for thorium breeder reactor fuel elements and is used in the Indian Point and Elk River reactors. PuO_2 solid solutions in UO_2 have been irradiation tested and are sufficiently promising for consideration in the fast ceramic reactor [199].

Dispersions of UO_2 in BeO have been irradiation tested and will be used in the EBOR. A dispersion of (U, Th)C_2 (solid solution) in graphite is being successfully tested to a high burnup and will be used in HTGR. The radiation behavior of dispersions (see also Sec. 3.2.3) depends more on radiation behavior of the matrix (dispersion medium) and the fabrication techniques than on the chemical form of the fissioning phase (dispersed phase). This conclusion can be readily reached from the following argument. A dispersion of a fissile material can be fabricated so that the islands of the fissile phase are separated from each other by distances greater than the range of the fission fragments being emitted from the periphery of the dispersed fissile phase. The radiation damage due to the stopping of energetic fission fragments is limited to the fissile particle volume and the small 10-12 micron (0.40-0.48 mils) region around the particle. If a matrix is selected which shows little damage due to the displacements caused by stopping fast neutrons emitted during fission, a stable fuel element can be fabricated. The designer must, however, take into account the swelling that may occur due to accumulation of the fission gases in the dispersed phase.

3.2.3 Cermets

Burnups greater than 50 at.% of the fissile phase have been obtained in many dilute dispersion-type cermet fuel elements. The dispersion of UAl_4 or U(Al, Si)$_3$ in aluminum has become the standard fuel material for the water-cooled research reactors. Dispersions of UO_2 or U_3O_8 ** in aluminum

*The newly created atoms are accommodated in the physical voids of the material. These materials are rarely 100% dense and are actually fabricated to a preselected density.

**Such fuel composites may under certain conditions be subject to the thermite reaction with the concomitant high release of chemical energy as, for example, in a reactor accident [47].

are used where high concentrations of the fissile phase or a high degree of control of composition is required. The latter objective can be obtained by powder metallurgy techniques. The radiation performance of UO_2 dispersed in stainless steel (ferritic and austenitic) has led to its use in the compact water-cooled power reactors. The two-phase alloy U in zirconium [1e]i.e. UZr_2 in a 1 at.% solid solution of U in Zr, is one of the fuel elements employed in naval reactors because of its satisfactory stability in radiation.

The limits of temperature and burnup of the dispersion elements are determined as indicated above in most cases by the mechanical and radiation properties of the matrix. As the burnup is increased the volume of fission gases is increased with a consequent rise in gas pressure in the dispersed phase. When this pressure exceeds the yield strength of the enveloping matrix, swelling or local fracture of the matrix will take place.* In making estimates of the strength of the matrix, it should be remembered that in high pressure water systems, the compressive effect of the coolant on the fuel element can be considered as part of the restraint. Likewise the buildup of internal pressure must be considered during low power operation and during unloading of the fuel.

Although cermets have proven extremely stable under irradiation, element types listed above cannot be used safely above 650°C (1200°F) because the matrices are not strong enough to prevent swelling. The ceramals** UO_2 in BeO and UC_2 in C discussed in the previous section offer the greatest promise for use above 600°C (1112°F). More data are presently available on the behavior of the UC_2-ThC_2 graphite [23, 178] ceramal. This combination has been tested to high burnups at surface temperatures of 1200°C (2190°F). Although fission products are emitted from the fuel compacts, the overall stability is considered very satisfactory. The dimensional changes (contraction) [48] are due to radiation-induced changes in the graphite. The problem of control of fission products for these fuel elements will be discussed in the next section.

3.2.4 Conclusions

In this section it has been pointed out that the economics of reactor operation require a high burnup in fuel elements. This requirement in turn imposes a requirement of radiation stability. In very practical terms a fuel element is said to be stable to radiation if it can be taken to high burnups without serious distortion. Further, it has been pointed out that the single truly limiting factor to the attainment of high burnups is the generation of fission product gases and the management of these in the fuel element rather than the radiation-produced distortions. It has been shown that fissile compounds can be taken to higher burnups because the fission gases can either be stored in the fissile compounds or because they diffuse out of these compounds.

3.3 Fission Product Containment

It has been stated in Sec. 1 that nuclear safety insofar as this term is applied to the fuel element requires only that the radioactive fission products be contained within the fuel element or, stated another way, that the release of radioactive products to the reactor coolant be prevented. It has also been pointed out in the previous sections that to some extent fuel itself or the matrix in which the fuel is embedded offers some containment for the radioactive products. In most fuel element designs the reliance for containment of the fission products is placed on the fuel-encasing material (cladding, can, jacket, matrix, sheath, coating ***).

3.3.1 Criteria for Selection of Cladding

The requisites of cladding materials are: adequate mechanical properties, low neutron cross section, compatibility with the coolant and fuel or fuel matrix, good thermal conductivity, fabricability and radiation stability. The relative importance attached to these requirements depends on the particular reactor fuel and reactor design. It is obvious that in reactors employing natural uranium the selection of cladding material will be based primarily on the thermal neutron absorption cross section because too great a loss of neutrons by a absorption can reduce the effective multiplication factor to less than unity. However, in enriched reactors there is a greater choice, and the factors of reliability, safety, and overall economics may offset the advantage of low absorption.

3.3.1.1 Compatibility with Coolant. Since the cladding must be in contact with the coolant, it must be resistant to corrosion [49]. It is highly desirable that a clad be chosen which is corrosion resistant not only to the coolant but as well to impurities which are inadvertently introduced into the coolant. In the case of water reactors employing stainless steel cladding, the ferritic stainless steels or Inconel or Incoloy might be preferred to the austenitics if the chloride concentration cannot be kept at necessary low levels since the austenitic steels are susceptible to stress-corrosion cracking in a chloride-containing coolant. It is highly desirable to select a clad which does not build up too thick an external layer of corrosion products as the usually poor thermal conductivity of such a layer results in lower heat transfer rates and higher central fuel element temperatures.

In some cases the prevention of the buildup of film **** on the fuel element surface is

*To achieve [200] high burnup (10% of the fissile-fertile atoms) in highly concentrated cermets, e.g. 50 vol.% UO_2 in stainless steel, the UO_2 phase is fabricated to 85-90% theoretical density.

**A "ceramal" is here defined as a solution or dispersion of a ceramic in a ceramic.

***These terms have been used almost interchangeably to denote the materials whose prime function is to prevent the escape of fission products to the coolant.

****In the organic-moderated and -cooled reactor a large fraction of the development effort was accomplished by proper treatment of the coolant, as for example by the addition of LiOH to the PWR coolant or by the control of pH in the case of water-cooled reactors in which aluminum fuel elements are used.

The reactor designer must also be aware of side reactions that may result from the production of the corrosion products. When zirconium or aluminum corrodes in water, hydrogen is formed in addition to the oxides of these metals. In the case of zirconium the hydrogen may be absorbed and the zirconium become embrittled. (Hydrogen pickup is the reason why Zircaloy-2, which was the standard cladding material until recently, is being replaced by Zircaloy-4 [28] in some reactors. The latter alloy is more resistant to hydrogen pickup and embrittlement.) Similarly, the atomic hydrogen evolved when aluminum alloys corrode in water above about 250°C (480°F) is absorbed by the metal. Diffusion of the hydrogen to defects in the lattice, where combination occurs to form hydrogen gas, creates localized high pressures in the metal matrix. The result is a complete disintegration of the aluminum cladding. The development at ANL of the 8001 aluminum alloys which contain small islands of Al-Ni as well as Al-Fe intermetallic compounds permits use up to temperatures of 300°C (572°F). The intermetallic catalyzes the combination of atoms of hydrogen to form hydrogen gas at the corroding surface.

The corrosion of the cladding by pure coolant cannot be the only consideration in the selection of the cladding. The coolant can also act as a carrier or transfer agent for an impurity which can lead to deterioration of the clad. In sodium-cooled reactors, carbon can be transferred from one material of construction of higher carbon activity to one of lower carbon activity. For example, if a sodium coolant circuit contains carbon steel and austenitic stainless steel cladding for fuel element, carbon will be transferred from the carbon steel to the stainless steel. This can lead to an embrittlement of the stainless steel.

In addition, a cladding may be embrittled by reaction with the impurities present in the coolant. Tantalum and columbium (niobium) are embrittled by oxygen in a sodium coolant. To avoid this type of deterioration of cladding it is necessary to reduce the oxygen concentration to low level by suitably treating the coolant with a reactive metal which chemically removes the oxygen (hot-trapping of Na with hot Zr and/or Ti, and/or cold-trapping to precipitate Na_2O).

Cladding failures may also result because a constituent or impurity in the cladding is leached by the coolant. This reaction also takes place because the system is seeking a lower thermodynamic level. The oxides of columbium (niobium) are leached by a lithium coolant because lithium oxide is more stable than the columbium oxide. The failure of the cladding by formation of porosity or by permeation of Li along the grain boundaries or [110] crystallographic planes, with the formation of an oxide phase, can be prevented in columbium by adding zirconium (about 1 wt. %). This alloy when properly heat-treated prevents attack by lithium even in welds.

In some reactors reactivity control is obtained by adding soluble poisons to the coolants. It is obvious that the selection of a cladding material must take into account the possible use of soluble poisons and the effect on corrosion of the trace amounts of these poisons that may be left behind after a flushing operation.

This presentation on compatibility of cladding with coolant would not be complete without mention of one of the very severe "headaches" which plague the reactor designer and corrosion expert, namely, the metal-water reactions postulated in every hazard report in which water is the coolant. This topic is discussed in greater detail in Chapter 17, Chemical Reactions. Although these reactions have been postulated and in fact demonstrated in the laboratory, they have been shown to occur only to a limited extent in the BORAX or SPERT tests performed to date. In the SL-1 accident the energy contribution from the water reaction was small [201a].

3.3.1.2 Compatibility with Fuel. Deterioration of cladding can occur not only as a result of reaction with the coolant but also by reaction with the fuel or matrix. (See also Sec. 4.1.1.2). A reaction which has been extensively studied is the interaction of metallic uranium with claddings of Al and iron-base alloys. The Al-U reaction can be prevented by anodizing the aluminum. However, heat transfer across the U-Al interface is reduced. Anodic coatings were used in the original Brookhaven Graphite Research Reactor fuel elements [50] where a good heat transfer bond was not required. Nickel plate on uranium has also been used or a layer of Al-12 wt.% Si has been interposed to reduce the interaction of U and Al. The reaction of uranium with iron-base alloys can be avoided by keeping the interface temperature below 725°C (1340°F) or by interposing a thin sheet of vanadium between the iron and the uranium.

Little difficulty has been experienced with the interaction of UO_2 with the cladding materials (Zircaloy, stainless steel) used to date. The authors are familiar with some difficulty encountered when Zircaloy was used to encase incompletely dried UO_2. The zirconium was embrittled by the hydrogen produced when the water reacted. Although reaction between UO_2 and Zircaloy does occur at elevated temperatures (700-1000°C or 1300-1850°F), at temperatures normally used in water-cooled power reactors little effect is noted. Difficulty can be encountered with Zircaloy cladding for UO_2 if the UO_2 contains fluorides [51]. The fluoride leads to complete deterioration of the clad in "defected"* elements. Control of fluoride concentration to low values eliminates this difficulty.

---

devoted to the elimination of crud (film) buildup on the fuel element as this film leads to overheating of the fuel and even fuel channel blockage. See references [146] and [151]. See also Sec. 4.2.4.2.

---

*Elements in which a hole has purposely been drilled through the cladding to study the effect of defects in the cladding.

Cladding interaction with carbide fuels presents a problem since carbon can readily migrate into the cladding at elevated temperatures [52]. This effect has been noted particularly where hyperstoichiometric UC is used. The dicarbide present in hyperstoichiometric carbide has a high activity of carbon which readily transfers to many cladding materials * at high temperatures. This transfer of carbon accounts for the choice of hypostoichiometric uranium carbide for the second loading of the Hallam reactor fuel. The fuel element was planned to consist of the carbide encased in stainless steel with a liquid Na bond between the two.

3.3.1.3 Other Criteria. It has been pointed out in the above sections that the selection of a cladding may depend on its neutron cross section because the nuclear requirements (reactivity) dictate such a decision. It should be added that the overall economic considerations lead the designer to the particular selection. A number of economic [53] evaluations have been made comparing the relative merits of stainless steel and zirconium alloys. It seems reasonable to conclude that, as improvements in the quality and uniformity of zirconium are made, it will replace stainless steel in water-cooled power reactors.

An excellent example of selection being based primarily on neutron consideration is the use of Magnox for the cladding of the natural uranium CO_2-cooled reactors of the United Kingdom. Magnox [75, 82] has low creep ductility at the inlet operating temperatures of the Calder Hall reactors and failures due to grain boundary separation caused by uranium elongation have been observed. **

The requirement that a cladding be radiation-resistant is an extremely important one. The cladding is not only subject to deterioration as a result of fast neutron bombardment but is also subject to fission product bombardment at the surface adjacent to the fuel. Most of these recoiling fission fragments end up in the cladding. Although radiation resistance has been recognized as an important requirement, the full implication is only now being recognized in UO_2-fueled reactors because the economics of power reactors require higher operating temperatures for the fuel and higher burnups. Higher burnup and higher temperature result in some swelling of the fuel and the generation of gas pressure within the fuel containment. Also the cladding is subjected to a longer period of irradiation. Irradiation of most cladding materials may result in marked reduction in the uniform elongation *** and a reduction in creep strength. The combination of the swelling of the fuel, gas pressure increase and reduction in ductility of the cladding has resulted in the cracking and tearing of the cladding in test specimens. The reduction in uniform elongation due to irradiation may turn out to be one of the important limitations to high burnups unless the designer can design around this or the metallurgist can develop cladding materials that are more radiation-resistant. Proper heat treatment prior to use may extend the radiation exposure capabilities. A useful cladding material must be fabricable to the shapes required and must be of uniformly high quality, i.e. free of defects. Also a cladding material must be readily weldable since end seals are required to completely encase the fuel.

Although the effect of mechanical properties on the selection of cladding material has not been stressed, it is apparent that the cladding must at least meet the minimum strength requirements, as in the Magnox case. However in those applications **** where the cladding is to be used to restrain the swelling of the fuel, the tensile and creep strength as well as the uniform elongation at operating temperature become very important.

3.3.2 Effect of Bond Type

3.3.2.1 Heat Transfer. Thus far cladding requirements have been discussed abstractly. In any practical application, since the cladding is interposed between the heat source and the heat removal fluid, the method used to effect the transfer of heat across the fuel-cladding interface is of prime importance to the fuel element designer. Two solutions to this problem have been used: either to "bond" the fuel to the cladding or leave it "unbonded" (mechanical bond). ***** These terms have been used rather loosely by the fuel element designer. A bonded fuel element is one in which the fuel makes (1) direct atomic contact with the cladding, (2) is joined to cladding by an intermediate layer of brazing material or (3) is separated from (or joined to) the cladding material by a liquid layer of a good heat conducting material (in most cases liquid Na or NaK). The first two types of bonds are referred to as "metallurgical bonds" and the last type is aptly called a "liquid bond". Note that all the above methods result in good heat transfer across the interface. In the unbonded (mechanically bonded) cladding arrangements, there is no absolute definable area of contact between the cladding and fuel. Unbonded cladding has been used where high heat transfer rates are unnecessary, or where bonding is difficult or impossible. One may therefore redefine these two methods of cladding in heat transfer terms, namely, that in the bonded designs heat transfer is primarily by conduction whereas in the

*Some of the cladding materials considered for carbide fuels form very stable carbides.

**By changing the fabrication procedures used to produce the cans, a finer grain size has been obtained which has better low temperature creep ductility than the earlier Magnox (AL80) cans. Also Magnox (ZR55) has been shown to have greater ductility than Magnox (AL80).

***The uniform elongation of stainless steel in a slow tensile test conducted above 600°C (1112°F) is markedly reduced after an exposure to 8×10^{20} thermal neutrons/cm due to the formation and migration of He to grain boundaries. The He is formed by the (n, α) reaction on the boron impurity present in all stainless steels.

****e.g., ANL use of CB for restraint for Pu alloys.

*****Implies that usually there is mechanical contact between fuel and cladding.

unbonded the heat transfer is by conduction (contacting points), conduction and convection through a gas (where a gas is introduced into interface) and radiation. It is not the objective of this section * to discuss the relative merits of these two methods of cladding from a heat transfer point of view. Instead, the effect of the selection of method on the fission product containment problem will be pointed out.

The advantages of the bonded fuel elements is that the good heat transfer at the interface results in a lower central fuel temperature with all of the implication that this covers. In addition in the metallurgically bonded elements the effect of a defect in the cladding such as a pinhole may be less serious in that coolant access is initially limited to the area of the defect. The fission products which might be carried out of the defect by the coolant can probably be detected by proper instrumentation in the coolant stream before serious contamination results. Where corrosion by the coolant is not a serious problem, it might be argued that use of a liquid-bonded fuel element is a more desirable method of effecting the desired heat transfer; i.e., in those cases where marked distortion of the fuel occurs during use, the distortion of the heat transfer surface (cladding) can be minimized by using undersized fuel material (metal base fuels might be so treated) contained in a cladding which has been provided with free space to accommodate the expansion; the liquid bonding material, the expanding fuel and any gases which are evolved can expand into a free space initially built into the fuel element.

The justification for the use of the unbonded elements is that they work. In many cases it has not been possible to develop adequate and/or cheap methods of bonding the fuel to the cladding. To improve the heat transfer at the interface a gas such as helium with a good thermal conductivity is introduced before the cladding is sealed off. These fuel elements may therefore also be said to be fluid-bonded. The thermal conductivity of this heat transfer fluid does decrease with burnup since the bonding gas is diluted with the gases Kr and Xe of very poor thermal conductivity and the average fuel temperature can be expected to rise with burnup. In these so-called unbonded fuel elements good design calls for a good (almost interference) fit between the fuel and the cladding. Clearance between UO_2 fuel pellets and cladding is a nominal one mil (25 microns) ** in many of the designs now used. Good mechanical fit between fuel and cladding has been obtained in other designs by either drawing the cladding onto the fuel or by hydraulically [54] compressing the cladding onto the fuel. Swaging of the cladding onto the fuel or vibratory compacting of the fuel in the cladding tube are other methods which are being used to obtain good mechanical contact between fuel and clad. The effect of the coolant pressure in effecting contact between clad and fuel cannot be overlooked in evaluating the heat transfer between clad and fuel in high-pressure water-cooled systems.

In most unbonded fuel elements heat transfer across the interface depends not only on the clearance originally built into the fuel element but also on the final clearance or contact (between fuel and cladding) established during operation. The heat transfer across the interface is probably poor on startup of a reactor using unbonded elements. As the fuel heats up it expands until it probably contacts the cladding over a large fraction of the interface area thereby increasing the effective heat transfer coefficient. The average temperature at any point in the fuel depends on the area of contact between cladding and fuel, the contact pressure, the thermal conductivity of the fuel and cladding, and the heat conduction of the gas in the cladding fuel interface. Accurate calculation of central fuel temperatures for unbonded fuel elements is difficult because the over-all heat transfer involves all of these factors. Extensive testing and careful evaluation [55] of the data have made adequate design possible.

3.3.2.2 Other Effects. A special requirement for unbonded fuel element designs is the incorporation of an antiratcheting feature. Ratcheting is defined as the process which results in relative motion of the cladding and fuel accompanied by a deformation of the cladding. Ratcheting can occur if a cladding has a greater linear coefficient of thermal expansion than the fuel. On heating up the cladding expands more than the fuel. On cooling down, the extended portion of the cladding shrinks faster than the fuel and may hang up at two separate points at the clad-fuel interface. Further cooling results in stretching of the cladding between the two points of seizure. Repetition of this process on startup and shutdown is likely to result in failure of the cladding. An example of an antiratcheting design feature is the circumferential groove machined at regular intervals for the full length of the metallic uranium fuel of the Calder Hall reactors. The Magnox cladding is plastically deformed into the groove by hydraulic pressure in one of the final fabrication steps. The ridge formed on the Magnox maintains the same relative position of cladding and fuel on startup and shutdown. In the Yankee reactor fuel element the UO_2 fuel is maintained in compartments by means of discs which are spaced at intervals in the cladding and brazed to it. The design aids in solving the ratcheting problem but also keeps fuel pellets properly spaced throughout core life. Unbonded elements are subject also to failure by deformation of the cladding at the ends of segmented fuel. Most cylindrical UO_2 fuel elements are made by inserting short discs (pellets) into the cladding tube. If gaps develop between the discs, the high hydraulic pressures at which water-cooled power reactors operate may cause a collapse of the cladding into the gaps. Still another mode of failure which has been observed is fragmentation of the outer edges of the UO_2 discs because of poor fit between mating disc surfaces. These fragments may lodge between the cladding and fuel during startup and operation. On shutdown the cladding can plastically deform to failure as it tries to

*See discussion in Chapter 16, Heat Transfer.

**When all tolerances are considered the clearances may be considerably greater than this.

contract around the composite of disc and fragment.

In the above discussion on methods of effecting the heat transfer between cladding and fuel we have shown that from safety considerations metallurgically bonded fuel elements probably have an advantage over unbonded ones. However, in the previous sections it was suggested the burnup considerations might lead to the selection of unbonded or liquid-metal-bonded designs. Safe designs of such fuel elements can be made if the designer is completely aware of all the possible interactions between cladding and fuel.

3.3.3 Effect of Cladding Defects

This subject is discussed in greater detail in Sec. 4.1.5, where operational experience with fuel elements is considered. In general, it may be stated that although such defects are considered undesirable, they do occur, and fuel elements with defects will get into reactors. That the types of failures that can be expected have in fact been experienced is apparent in Sec. 4.1.5.

From an operational point of view cladding defects in metallurgically bonded fuel elements are usually not as serious as the ones in the liquid-bonded or unbonded clad elements. This statement is quite plausible since the corroding liquid (coolant) has immediate access only to a limited area of the fuel element. Fission products appearing (in limited quantities) in the coolant may be detected before serious corrosion of the fuel can take place. Also in principle it is possible to incorporate at the cladding-fuel interface a readily detectable element whose appearance in the coolant is an indication of a cladding failure. If a bonding material is used to join the clad to the fuel, it may be possible to select a brazing material which is corrosion-resistant to the coolant.

In unbonded fuel elements cladding defects can lead to a rapid and large loss of fission products to the coolant with the resultant severe contamination. This certainly is true where uranium-base elements are used in water-cooled and air-cooled reactors. Reaction of uranium with water and air is rapid and produces lower density corrosion products which can cause swelling and progressive failure of the cladding. Where unbonded fuel elements are used it is certainly desirable to consider the use of a fuel which reacts slowly with the coolant. Uranium-zirconium alloys and uranium-aluminum alloys and other more complex uranium alloys react slowly with water. UO_2 fuel is not seriously affected by hydrogenated [56] water. However, waterlogging (see Sec. 4.1.5.3)* can introduce fission products into the coolant. Uranium metal and hypostoichiometric uranium carbide do not react with sodium. From these examples it can be seen that the designer has a number of compatible systems to choose from.

*In the early stages of operation of nonsintered UO_2 fuel elements waterlogging and washout of fuel is a more serious problem. After operation at power the fuel will sinter and densify.

Before concluding the subject of cladding defects, mention should be made of the effect of defects in the bond between the fuel and clad. A bonding defect is primarily of concern in the case of the metallurgically bonded fuel elements. Such defects, if large enough in dimensions, can lead to overheating of the fuel and of the cladding with consequent rupture, etc., of the clad. The fuel element failure in the WTR has been attributed to unbonded areas in some elements (see Sec. 4.3.4.1). Such lack of bond can be detected by the blister test which is a standard test procedure for aluminum and stainless steel elements. In addition, it seems highly desirable to use ultrasonic methods to determine if unbonded areas exist in fabricated elements and to determine the size of the unbonded areas.

In conclusion, it is worth repeating that cladding defects have to be expected and for this reason it seems prudent to incorporate a fission-product-in-coolant detection system ("burst slug detection system") to give adequate warning of impending large-scale failure and to minimize the unpleasant consequences of such failures. (See chapter on Sensing and Control Instrumentation.)

3.4 Development and Validation of Fuel Elements for Reactor Use

In laying out the development program for a new fuel element the designer now has the experience of the last two decades of fuel element development work as a guide. There is ample proof that extensive in-pile and out-of-pile testing is required. In the out-of-pile tests the mechanical, chemical, and physical properties of the components of the fuel elements and fuel element assemblies can be determined. The corrosion behavior of the constituents and components of the fuel elements, as well as the effect of impurities in the coolant, can be evaluated. The effect of temperature and temperature-cycling on the stability and integrity of the fuel element can be studied. The stability of the structures to hydraulic forces can also be assessed.

The relative merits of different fabrication procedures can also be compared in an out-of-pile test program. The hydraulic behavior of a fuel element or fuel element assemblies may be tested. The effect of partial or total blockage on the mechanical behavior can be determined. All of these tests are, however, preliminary to the in-pile tests where the fuel element is subjected to radiation and the resulting internal creation of fission products and generation of heat, to temperature gradients and to thermal stresses. Since the in-pile test is an integrated test, it should approximate as closely as possible the actual conditions of operation of the real reactor. If possible the fabrication of the test fuel element should be by the same processes, using the same material, etc. as planned for the operating reactor.

It has been found very useful in evaluating fuel systems to devise in-reactor tests such that the components can be examined at intervals during the course of irradiation. This might simply involve a visual and dimensional measurement in addition to the more thorough metallographic hot-laboratory tests. Experience has shown the importance of a complete in-pile testing program in which all conditions of thermal, mechanical, hydraulic and flux variables are tested. There are notable examples where "simple" extensions of existing fabrication techniques have resulted in unexpectedly poor performance in reactor of test fuel elements. Most in-pile irradiation programs now include as an essential feature the testing of elements which have a built-in defect (so-called "defected elements") to determine the effect of defects on the behavior of the elements or on the course of the deterioration and disintegration process. In this way it is possible to determine the time constants and sensitivity for the fission-product-in-coolant detection system. Further, it is desirable to know whether a defect in a single element can lead to progressive failures in adjacent elements. Also, in power reactors it is of economic importance to determine how serious it is to continue operation with a defect in an element or elements. Shutdowns are costly and it is highly desirable that a reactor be operated with elements with defects until the the scheduled shutdown.

It seems worth while to point out that the evaluation validation program should preferably be performed on fuel elements made by the selected fabricator since this will permit the simultaneous development of the fuel element and the inspection procedures. The cost of producing fuel elements depends not only on the direct fabrication cost but also on the specifications and inspection procedures. The elimination of unnecessary and onerous specifications and inspection with a well-planned in-pile test program can greatly decrease the cost. Reliability at a minimum cost is the final objective of a sound validation program. Validation of PWR fuel elements is a good example of a thorough development and test program on a new fuel [56a].

4 SPECIFIC FUEL ELEMENTS

The information presented in this section illustrates the variety of ways in which reactor designers have exercised the options available to them in selecting and designing fuel elements for operating reactors. For convenience three categories of fuel elements are considered: (1) solid cylindrical, (2) plates, and (3) tubes. It will be seen that the basic principles of safe fuel element design are the same for the three groups and that some (though not all) methods of fabrication are applicable for any geometry. (Graphite-base elements are considered separately in Sec. 4.4).

The three groups are discussed separately with appropriate tables which include dates to indicate the extent of reactor experience with the particular element. When possible, the discussion draws on the results of actual experience and subsequent out-of-pile examination, both for failures and for successful irradiations. The tables include little on alternative elements that have undergone development although such elements are discussed in the text.

Fuel elements from different classes of reactors are discussed. It should be noted that the fuel element used for a particular reactor is not necessarily the best for that reactor. In some cases a standard, demonstrated fuel element has been used in experimental programs in which other aspects of reactor design were being investigated. While it was recognized that the element used in these programs was probably not the best for ultimate use in a reactor of this type, nevertheless it was considered advisable at that stage to use the fuel element and thereby avoid the expense and risks of developing a more suitable one. In other cases, a power reactor has been used to test a new fuel that was not necessarily optimum for this reactor.

An additional word of caution is inserted here regarding the nomenclature, since it is no more standardized than the fuel elements themselves. In the following discussion, the word "fuel element" is considered in the more limited sense of a small unit of completely contained fuel, i.e., a single entity of fuel enclosed in cladding. The tables indicate how various terms are used for combinations of these simple units into loading units; cluster, bundle, subassembly, assembly. The term "assembly" is used to refer to the smallest group of fuel elements that can be removed from the reactor as a unit. A subassembly consists of a number of elements fastened together more rigidly than the assembly as a whole or in some other way forming a differentiable subgroup of the entire assembly. A "cluster" is often used to refer to a desirable unit in which subassemblies are welded to form a channel for passage of a control rod. The group of elements which make up the fuel in a single pressure tube of a reactor or in any reactor with widely spaced bundles of elements is also often called a cluster. Some confusion also exists in the terminology for the components of the fuel element; that is, the fuel and its cladding. The fuel is sometimes designated as the "core" or the "meat", perhaps in recognition that it may contain not only fissionable material but another material with which it is intermingled, such as an alloying element or a matrix. The term "cladding" is used here to designate the protective material confining the fuel and possessing some mechanical strength; other terms are clad, can, jacket, sheath and coating.

4.1 Solid Cylindrical Elements: Slugs, Rods, Pins

A solid cylinder with external cooling, probably the simplest configuration for a fuel element, was the first configuration to be used. Although this geometry presents limitations in capacity for heat removal, limitations that can be surmounted by appropriate refinements of the cylinder (see below) or by a different configuration, it is attractive because of inherent simplicity. Thus, some of the methods of manufacture are fairly straightforward, well established and amenable to large volume production and even to completely remote operation. The simple geometry also simplifies inspection, since the element can be rotated and translated past an inspecting device.

The various diameters of cylindrical elements are recognized in the use of the terms "slugs", "rods", and "pins", corresponding, respectively, to increasing heat transfer surface per unit fuel

TABLE 4-1

Characteristics of Representative Solid Cylindrical Fuel Elements

(Abbreviations are given at end of book. In this table an asterisk means "See last column for notes".)

Reactor and Coolant	Power (Mw)		Core				Cladding		Fabrication and Bonding			Assembly	Extent of Operating Experience		Notes
	Ther-mal	Elec. (net)	Composition	U^{235} (wt. %)	Dimensions (in.) Diam.	Dimensions (in.) Length	Composi-tion	Thick-ness (mils)	Method	Bond	An-nulus (mils)		Crit.	Full Power	
X-10 Air	3.8		U U	Nat. Nat.	1.1 1.1	4.0 4.0	1100 Al 1100 Al	35 35	Die size Al-Si dip	Mechan-ical Al-Si	None 5-20, nom. 9.5	About 40,000 slugs arrayed in 821 horizontal channels (41 to 54 per channel)	11/43 to 4/52 Since 4/52 Shutdown 1963		U γ-extruded. U α-rolled and β-treated.
BGRR Air	20		U	Nat.	1.1	33 x 4.0	1100 Al	30	Cold hydrostatic collapsing	He	2	Two cartridges in each of 1369 horizontal channels	8/50 On nat. U to 3/48	4/51	U γ-extruded and heat treated at 600°C/12 hr. Al has 6 longitudinal fins. Al is anodized.
NRX* H_2O	42		U UO_2	Nat. Nat.	1.360 1.41 OD 0.60 ID	120.5 136 x 0.88	1060 Al 5052 Al	79 or 40 50	Die size Draw	Mechan-ical	None 10.0 OD 1.5 ID	192 rods hang vertically	7/47	5/48	U α-rolled and β-treated. Al has 3 longitudinal fins. *See entry later in this table for Al-U or Al-Pu booster elements.
Calder Hall CO_2	225	41	U	Nat.	1.150	40	Magnox	60	Hot (500°C) hydrostatic	Mechan-ical	None	Stacked 6 high in each of 1691 channels	5/56	10/56	U vac.-cast, β-treated, α-annealed (550°C/16 hrs.). U has antiratcheting grooves. Cladding has circumferential fins, brace at midsection.
Hunter-ston CO_2	535	150	U	Nat.	1.15	24	Magnox	68				Ten cartridges supported individually in each of 3288 channels	Under construction		Cladding has longitudinal fins with continuous baffle.
EL-2 CO_2	2		U	Nat.	1.05	21.1	Mg or Mg-Zr	60	Hot (400°C) hydrostatic	Mechan-ical	None	Four supported in each of 135 channels	10/52		U cast or γ-extruded. U has helical antiratcheting grooves.
EL-3 D_2O	15		U-1.5 wt. % Mo	1.16	1.35 OD 0.88 ID	12.8	Al (A5)	40	Hot (400°C) hydrostatic or "Diffusion" Ni layer (600-700°C)	Mechan-ical		Four in ribbed Al flow tube	7/57	4/58	Length includes 0.12 in. nat. U end pieces. U has square thread.
G-2 (and G-3) CO_2	250	30	U	Nat.	1.12 or 1.24	11.3	Mg-Zr	60	Hot (400°C) hydrostatic	Mechan-ical	None	28 in each of 1200 horizontal channels	7/58	4/59	Cladding has longitudinal fins, 4 large (supporting) and 12 small.
HWGCR CO_2	590	150	U	Nat.	0.16	160	Mg-Be	18				150 to 200 rods in Mg tube 156 fuel channels			Mg-Be cladding "alloy" - alternate layers by vac. distillation.

Halden D_2O I	6		U	Nat.	1	95	Al	7	Swaged	He				Al has 5 longitudinal fins.
II	20		UO_2	1.5	0.64	0.50	Zy-2	30		He	0.4-1	7 in cluster; 100 clusters		
FR-2 D_2O	12		U	Nat.	1.28	86.4	Al	40	Hot (400°C) hydrostatic	Zn-Sn		158 positions	12/59	
SRE Na I	20	5.7	U	2.75	0.75	12 x 6.0	304 ss	10	Liquid metal	NaK	10	7 in Zircaloy tube for each of 43 subassemblies	4/57 5/58	Helical wire.
II			Th 7.6 wt. % U	93	0.75	12 x 6.0	304 ss	10	Liquid metal	NaK	10	5 around ss tie rod in Zircaloy tube for each of 43 subassemblies	9/60	Helical wire. Clearances increased between fuel elements and tube for coolant flow.
HNPF Na	240	75	U-10 wt. % Mo	3.6	0.59	3, 6, or 9	304 ss	10	Liquid metal	Na	25	18 around hollow spacer tube carrying spacers at 1-ft intervals in Zircaloy tube for 137 subassemblies	1/62 8/62	Spacers. Lower hanger to prevent any parted rod from falling out of process tube. (Reactor shut down 9/64. Na entered canned moderator sections. Swelling and sticking followed.)
			UC	8-3.7 2-4.9	0.892		304 ss	10	Liquid metal	Na	10	8 around graphite in Zircaloy tube		
EBR-I NaK I	1.4		U	93 Nat.	0.364	4 x 1-7/8 8 x 4-3/4	347 ss	22	Liquid metal	NaK	20	Individual rods	8/51 to 12/51	U hot rolled, β-treated and α-annealed (575°C/2 hr.).
II			U-2 wt. % Zr	93 Nat.	0.384	2 x 4-1/4 7-1/2	347 ss	22	Liquid metal	NaK	10	Individual rods	1954 to 11/29/55 (meltdown)	U-2 wt. % Zr slugs centrifugally cast.
III			U-2 wt. % Zr	93	0.364	4-1/4 8-1/2 7-3/4 3-9/16	Zy-2	20	Coextrusion	Metallurgical	0	36 assembled in ss tube about tightening rod with spacer ribs spot welded	11/57	Clad rod heat treated at 800°C and 690°C.
IV			Pu-1.25 wt. % Al		0.232	4 x 2.121	Zy-2	21	Liquid metal	NaK	12.5	60 per subassembly	11/62	Three longitudinal Zircaloy ribs.
EBR-II Na	62.5	16.5	Fissium	49	0.144	14.22	304 ss	9	Liquid metal	Na	6	91 pins per cluster; were wrapped and spot welded	9/61 (dry)	Fissium = 95 wt. % U, 0.2 wt. % Zr, 2.5 wt. % Mo, 1.5 wt. % Ru, 0.3 wt. % Rh, 0.5 wt. % Pd.
Fermi Na	200	60.9	U-10 wt. % Mo (core)	25.6	0.148	30.5	Zr	5	Coextrusion	Metallurgical		140 rods for each of 105 elements		Data are for "Core A".
			U-3 wt. % Mo (blanket)	0.36	0.145	15-1/2 66 total	304 ss	10	Liquid metal	Na	4	16 rods 25 rods	8/63	
PWR* H_2O (2000 psi)	231	60	UO_2	Nat.	0.3575	26 x 0.3493	Zy-2	22.5		He	4-5	120 rods welded into drilled end plates for bundle. 7 bundles on top of each other for each of 113 elements	12/57 12/57 refuelings (new seed) 12/59 to 4/60 8/61 to 10/61	*Blanket of Core I. See Table 4-3 for seed of Core I and for blanket and seed of Core II.
Yankee H_2O (2000 psi)	540	161	UO_2	3.4	0.294	150 x 0.6 (in groups of 25)	348 ss	21		(Air left in)	2	304 or 305 in each of 76 bundles	8/60 1/61* 6/61* refueled 5/62 to 9/62	Ferrules brazed between rods at 8-inch intervals. *Power on 1/61 was 392 Mw(t), on 6/61 485 Mw(t).
Savannah H_2O	69		UO_2	4.2-4.6	0.4255		ss	35		He		164 rods in each of 32 "fuel elements"		Ferrules brazed between rods.

TABLE 4-1 (Continued)

Reactor and Coolant	Power (Mw) Thermal	Power (Mw) Elec., (net)	Core Composition	Core U^{235} (wt. %)	Core Dimensions (in.) Diam.	Core Dimensions (in.) Length	Cladding Composition	Cladding Thickness (mils)	Fabrication and Bonding Method	Fabrication and Bonding Bond	Fabrication and Bonding Annulus (mils)	Assembly	Extent of Operating Experience Crit.	Extent of Operating Experience Full Power	Notes
Saxton H_2O	20	3.2	UO_2	5.7	0.357	0.732	304 ss	15				72 in each of 36 standard assemblies; 9 in each of 9 L-shaped assemblies	4/62	1/63	
Indian Point 1 H_2O	585	151	ThO_2-UO_2 6 diff. comp.	93	0.260	125 x 0.78	304 ss-boron	20.5		He	4 max	195 in each of 120 assemblies	8/62 Remove 1/6 of fuel elements, fall 1963	1/63	Ferrules brazed between rods at 9-inch intervals.
BORAX-IV H_2O	20		ThO_2-6.36 wt. % UO_2	Fully enr.	0.230	0.375-0.750 total 24-24.75	X-8001 Al	16	Dip in molten Pb	Pb	18	8 tubes per plate 6 plates in each of 69 subassemblies	12/56 Shutdown in 6/58	4/57	
BORAX-V H_2O boiler*	20		UO_2	~5	0.375	24	304 ss	15		He		49 removable rods in each of 44-48 subassemblies	12/60 Shutdown 1964	6/61	*Superheater consists of highly enriched UO_2-stainless steel plates similar to SM-1 (APPR) elements. See Table 4-3.
VBWR H_2O	50	10	UO_2	2.3-4.5	~0.5	72 x ~0.5	Zy-2 or 304 ss	11-35				9, 16, or 25 in each of 108 assemblies	8/57 Shutdown 1/60 to 6/60 Dismantled 1963	10/57	
Dresden H_2O	700	208	UO_2	1.5	0.494	4 segments 56 x 0.50	Zy-2	30				36 in each of 452 assemblies	10/59	6/60	
BONUS H_2O	50	16.3 boiler superheater	UO_2 UO_2	2.40 3.25	0.494 0.5		Zy-2 Inconel	25 18		He		32 per bundle; 2 bundles in each of 64 assemblies (4 nat. U) 32 in each of 32 assemblies	4/64		
NPD D_2O	83.3	20	UO_2	Nat.	0.937	22 x 0.832	Zy-2	25				7 rods form "slug"; 9 slugs in each of 132 pressure tubes	4/62	6/62	2 ribs each on 6 outer elements.
CANDU D_2O	693	202	UO_2	Nat.	0.564	24 x 0.792	Zy-2	16		He + A	2 to 5	19 rods form bundle; 12 bundles in each of 306 pressure tubes	(1965)		
CVTR D_2O	64.9	17	UO_2	1/3 1.5 2/3 2.0	0.43		Zy-4	23				19 rods form assembly; 2 assemblies in each of 36 U-shaped pressure tubes	3/63		
PRTR D_2O (MkI)*	70		UO_2 or Al-1.8 wt. % Pu (+ Ni, Si)	Nat.			Zy-2 Zy-2	30 30		He	4 0 at op. temp.	19 rods per cluster in each of 84 shroud tubes		1960	Unbonded wire wrap (72-mil Zy-2). *See Table 4-4 for Mark II, 2 concentric tubes and axial rod. Fabrication and bonding: Insert fuel and resize by swaging or cast Al-Pu into grooved Zircaloy clad.

NRX* H_2O			1) Al-93 wt.% U 2) Al-0.5 to 20 wt. % Pu 3) Al-20 wt. % Pu 4) Al-3.75 wt. % Pu 5) Al-28 wt.% U	93 93	1.360 1.360 1.360 1.360 0.250	8 10 12 9 96	1060 Al 1060 Al 1060 Al 1060 Al 1100 Al	80 80 80 80 45	Sunk and drawn Cast Al-Pu into Al can or machine bore slugs and assemble into Al sheath** Extrusion clad					*See earlier entry in this table for natural U or UO_2. **Al-Pu cast into Al can; Pu conc. varied from 20 wt.% at center to 0.5 wt. % at ends; cans screwed together and assembled into Al sheath which was sunk and drawn over cans.
EGCR He	85	22.3	UO_2	2.46	0.707 OD 0.323 ID	36 x 0.75	304 ss	20				7 rods assembled in ss spider supported in graphite sleeve; 6 assemblies in each of 234 fuel channels	(1965)	
ML-1 N_2	3.3	0.33	UO_2 70 wt. % UO_2 + BeO	93 93	0.176 0.176		Hastel-loy-X Hastel-loy-X	30 30				18 rods (and central Hastelloy-X rod) assembled in Hastel-loy-X liner; 1 sub-assembly in each of 61 pressure tubes	3/61 12/61	MgO pellet at bottom insulates weld; BeO pellet at top disperses heat; 40-mil Hastelloy-X wire wrapped spirally.
HTGR He	115	40	(U^{235}, Th) C_2 dispersed in graphite		2.75 OD 1.75 ID Plugged with graphite	1.5	Graphite (low permeability)				9 cold 11 hot	804	(1965) Reactor shutdown w/o operation	Th/U ~2.5; total 22 wt. % carbide 100-400 micron diam., each coated pyrolytically with 50-60 micron carbon; fuel compacts supported on cylindrical graphite spine.
TREAT air	0.1*		0.245 wt. % U_3O_8 disp. in graphite; 93 blocks, each 3.8 x 3.8 x 8, edges cut at 45°; U enrichment 93 wt. %				Zy-3	25		None			2/59 5/59	*Power = 0.1 Mw(t) steady-state; on transients energy 10^3 Mw-sec.
AGR CO_2	100	27.3	UO_2	2.5	0.4	0.4	ss	15				21 in a cluster; 2 clusters in an assembly; 2 assemblies in an "element"; 2 elements, one on top of the other, in each of 253 channels	8/62 1/63	Stainless steel: 20 Cr; 25 Ni; Cb stabilized. Alumina insulating pellets. UO_2 pellets centrally recessed.
WR-1 OM ter-phenyl	60		UO_2 UO_2	2.1 2.1	0.555 0.555	38 x 0.792 24 x 0.792	SAP ZY-2.5, Cb	25 25			3 to 6 2 to 5	18 elements per bundle; 3 bundles per channel 18 elements per bundle; 5 bundles per channel	(1965) (1966)	
Indian Point 2 H_2O (2250 psi)	2758	875	UO_2	2.23 2.38 2.68 init. Enrich.	0.422	240 x 0.6	Zr	24.3				193 assemblies each of 204 rods	(1969)	Used in conjunction with Boric Acid Solution control and spaced out clusters of small control rods.
Dresden Unit 2 H_2O (1015)	2255	800	UO_2	2.0 init. Enrich.	0.570	252 x 0.57	Zr-2	36				724 assemblies	(1969)	Used in conjunction with cruciform controls and neutron-absorbing curtains.

TABLE 4-2

Fabrication of Solid Cylindrical Fuel Elements [a]

(Footnotes are given at end of table.)

Core	Cladding	Fabrication Method: Simple Insertion: No Subsequent Working	Draw or Die Size: Cold	Draw or Die Size: Hot	Hydrostatic: Cold	Hydrostatic: Hot	Swage	Other	Cast into Cladding	Al-Si	Liquid Metal: Pb	Liquid Metal: Na	NaK	Coextrusion	Others
U	Al		X-10 NRX		BGRR	FR-2 (390°C, with Zn-Sn solder) EL-3	Halden			X-10					
	Mg					Calder Hall (500°C) EL-2 (400°C) G-2									
	SS												SRE (I)		
	Mg-Be														HWGCR alternate layers by vacuum distillation
U-2 wt. % Zr	Zy													EBR-I (III)	
	SS												EBR-I (II)		
U-3 wt. % Mo	SS											Fermi (blanket)			
U-10 wt. % Mo	Zy													Fermi (core)	NRU and NRX extension cladding
	SS											HNPF			
Al-U	Al		NRX												
Th-U	SS												SRE (II)		
UO_2	Al	NRX (5052) WR-1 (SAP)													
	Zy-2 (or Zy-4)	PWR (first blanket) VBWR BONUS (boiler) NPD CANDU [g] CVTR PRTR Sioux Falls (boiler) [d] Halden					PRTR g	PRTR [f]							
	Zr-2.5 wt. % Cb	WR-1													

UO_2	SS	Yankee[c] Savannah Saxton BORAX-V (Superheater) VBWR EGCR AGR Humboldt Bay[b] BR-3[d]						d						e	
	Inconel	BONUS (Superheater)													
	Hast'y-X	ML-1													
ThO_2-UO_2	Al									BORAX-IV					
	SS	Indian Point Elk River[b] Big Rock Point[b]													
UC	Zy	TREAT													
	SS												HNPF		
(U, Th) C_2 in graphite	Graphite	HTGR													
Al-Pu	Al		NRX[h] NRX[i]						NRX[h]					j	
	Zy						PRTR		PRTR						

[a] Reactor abbreviations are given at end of book. In this Table SS = stainless steel, Zy = Zircaloy, Hast'y = Hastelloy.
[b] Not included in table of characteristics (Tables 4-1, 4-3, 4-4).
[c] Cladding crimped into grooved disc and brazed together.
[d] Stretch forming included in earlier BONUS Superheater with stainless steel cladding, also investigated for Yankee and given up.
[e] Irradiation specimen in VBWR.
[f] UO_2 compacted by vibration, may also be subsequently swaged.
[g] Cladding collapses over core during autoclave test.
[h] Al-Pu cast into Al cans, then loaded inside standard Al sheath.
[i] Al-Pu machined to fit inside standard Al sheath.
[j] Al-7.5 wt. % Pu prepared at Hanford for irradiation at SRL (See Report HW-63151, General Electric Co., Hanford Atomic Products Operation).

volume as service requirements increase. Thus, slugs of diameter about 1 in. (2.5 cm) are suitable for reactors with relatively low rates of heat removal: most Pu production reactors, low temperature gas-cooled reactors and the early research reactors. Rods, diameter about 0.5 in. (1.3 cm), are required for more rapid heat removal, as in the various water-cooled power reactors and in EBR-I. Further increase in heat removal rates requires pins, diameter 0.2 in. (5 mm) or less, as in EBR-II and the Fermi Reactor. (In these last cases, a high ratio of cladding to fuel is more tolerable, since the extra cladding is less likely to affect neutron economy.) Another means of increasing the ratio of cladding to fuel and the heat transfer rate is incorporating ribs or fins on the cladding, a technique also applicable to elements of other configurations.

The characteristics of various cylindrical elements are listed in Table 4-1. A condensed representation of methods of fabrication is given in Table 4-2, where, to a rough approximation, the method of fabrication increases in complexity in going from left to right, as does the strength of the fuel-cladding bond.

Included among cylindrical elements are those for BORAX-IV, in which the fuel-containing aluminum tubes are integral with a plate (see references [1f, 57a, and 58]). In some reactors the metal or oxide fuel is hollow but is clad only on the outside as this is the only heat transfer surface.

4.1.1 Features Affecting Design and Fabrication

4.1.1.1 Dimensions and Other Mechanical Aspects. Since many cylindrical fuel elements, notably those with oxide cores, are assembled from components at their final dimensions, clearances must be sufficient to permit assembly and yet not so large as to impair heat transfer. Generally, the gap is filled with helium to improve heat transfer.* The helium may also be used for leak detection either before or during reactor operation. In addition, a void space is left above the fuel to provide a reservoir for the gaseous fission products and also to permit expansion of the fuel core (and any liquid metal bonding material). With such a void, it may be necessary (as shown by BORAX-IV experience) [59] to support the cladding internally against collapse by the coolant pressure; a spring is often used. In some liquid-metal-bonded elements, collapse is not a problem but a stop is provided to prevent the fuel from extending above the liquid metal level. This stop is likely to be an extension on the steel end-plug. An extension of the Zircaloy end-plug has been used (in the CANDU reactor) [60, 61, 62] for another purpose, namely, to insulate the external surface from the hot central portion of the lowermost fuel. This objective can also be achieved with other materials e.g., with pellets of natural uranium (metal or oxide) at the end of a column of enriched pellets (EL-3) or with an inert oxide. *

In a fairly long rod, differences in the thermal expansion of nonbonded or partially bonded fuel and cladding can lead to distortion due to ratcheting (see Sec. 3.3.2.2) as the element undergoes thermal cycling [63]. Metal fuels can be provided with grooves (Calder Hall; EL-2) [22b] or with threads (EL-3) which bind the fuel and cladding so that they both expand and contract together. Oxide elements are made either short and rigid, or are subdivided into compartments. In the Yankee element a series of perforated nickel-plated discs has a circumferential groove into which the cladding is crimped after each group of 25 pellets; the cladding and discs are subsequently brazed together, the "electroless" Ni-P alloy serving as the braze alloy [64, 65].

4.1.1.2 Compatibility of Metallic Uranium with Various Cladding Materials. Interactions of the fuel and the cladding depend on the metallurgical relationships of the components and are hardly unique to any configurations. (See Sec. 3.3.1.2.) Combinations of metallic uranium with aluminum or stainless steel cladding have been used primarily for cylindrical elements and some problems associated with these combinations are reviewed here to show how some fuel elements have evolved. (See also Sec. 4.1.5.1.1.)

Interaction in the uranium-aluminum combination above about 250°C (480°F) leads to the formation of brittle compounds as well as the diffusion of uranium to the aluminum surface [1g]. The nature of the interface is particularly important because it controls the extent of corrosion in case the cladding is penetrated. If it is desired only to prevent the interaction of fuel and cladding, barrier materials may be used on either component, e.g, colloidal graphite on the uranium (BEPO) [54], anodized coating on the aluminum (BGRR, first loading) [50]. If a metallurgical bond to unalloyed uranium is sought, a more severe requirement is placed on the intermediate material, since it must interact with both the uranium and the aluminum to achieve bonding while preventing their interaction. An Al-Si braze can be used for this purpose, forming $U(Al, Si)_3$ and hampering further interaction [1h]. Methods are available for bonding in the solid state under mechanically or hydraulically applied pressure using intermediate layers of a nickel electroplate [1i] or vapor-deposited zinc and tin [1j, 66, 67]. It is worth noting that the combination uranium-zirconium has the advantage of not forming any brittle intermetallics. With this combination, however, impurities or alloying elements in the uranium can introduce difficulties. Thus, carbon in the uranium can lead to the formation of the corrodible zirconium carbide; penetration of the cladding by water is then followed by severe corrosion along the bond line. Preferential corrosion along the bond line can also result from

*Actually the overall thermal conductivity of the He + Xe, Kr is markedly reduced as the rare gases (fission products) are evolved.

**In EGCR, MgO is used at both ends; in ML-1, MgO is used at the bottom and BeO at the top to improve heat dissipation; in AGR, Al_2O_3 is used at both ends and at the center of a stack of pellets.

interaction of the zirconium-base cladding with the alloying elements added to improve the uranium's corrosion resistance. Thus, silicon or molybdenum can be removed from the uranium.

In the combination of metallic uranium fuel with steel cladding, the formation of rather low-melting eutectics (near 700°C or 1290°F) between uranium and the various components of stainless steels is an even more serious problem than the formation of intermetallics. The combination is usually managed by leaving at least a tarnish film on the uranium and assembling it in the steel with an annular gap which is subsequently filled with a liquid metal for heat transfer. It is still necessary to avoid conditions that can lead to intimate contact of the uranium and steel at a temperature where the eutectic can form and cause penetration of the cladding from within. The rate of alloying of fuel alloys with stainless steel has been studied [68]. (Similar considerations apply in the containment of uranium carbide fuel in steel cladding.) It is also necessary to consider possible volatilization of the bonding agents, e.g. sodium, which can generate sufficiently high pressure to expel the fuel and introduce porosity in the molten fuel (EBR-I meltdown and TREAT tests of EBR-II fuel, see Sec. 4.1.5.2).

4.1.2 Fabrication of Cylindrical Elements

Fabrication methods used to combine the fuel and cladding in various cylindrical elements are indicated in Table 4-2. Details regarding the methods are available in report for the individual elements and the various procedures are described elsewhere [1].

The first method listed is the simplest and has been used for many elements, notably for UO_2 in water-cooled power reactors. Essentially, a set of UO_2 pellets is slip-fitted into the cladding, which is then sealed tight. Alternately, the insertion of the metal or oxide fuel is followed by various working procedures to reduce the clearances to promote at least mechanical contact by matching the cladding and fuel contours, and to change the dimensions of the cladding (and in some cases, also of the fuel). In general, standard metal-working procedures are used before or after addition of the second end-closure; the development of some methods, notably hydrostatic canning, has received a strong impetus from the needs of nuclear reactors. In some more recent designs, hydrostatic canning may be considered to be achieved by making the cladding thin enough to collapse onto the fuel during subsequent pressurization, as in the autoclave acceptance test [60, 61, 62]. Such compression of the cladding reduces but does not eliminate the fuel-cladding gap: the cladding springs back after compression; the greater diametral growth of the fuel during operation causes the cladding to yield (i.e. deform plastically). The cladding must be thin enough to collapse uniformly over the fuel rather than collapsing to an ellipse contacting the fuel at essentially only two points. Excessive fuel-cladding clearances—of the order of 7 to 8 mils (0.18 to 0.20 mm) or greater—can lead to wrinkling of the cladding.

Methods involving placement of oxide powders rather than pellets in the cladding are attractive because of their simplicity and applicability to remote handling of materials such as PuO_2. The fuel powder is compacted within the cladding by combinations of methods such as vibration, swaging and gas pressure bonding, whereby densities exceeding 90% of theoretical can be achieved [69]. A hot extrusion technique has been developed for coextrusion of UO_2 within stainless steel cladding [70, 71].

In all of these methods, the tubing used for cladding undergoes rigorous inspection before use. In addition, the first weld may be checked for soundness by hydrostatic and mass spectrometric tests before insertion of the fuel. The closer the dimensions of the assembled combination of cladding and fuel are to the dimensions of the final element (i.e., the less the subsequent working), the closer the clearances have to be made during assembly. Thus, the absence of subsequent working imposes stringent control of the dimensional tolerances of the components.* On the other hand, when subsequent working is applied, further inspection is necessary; this inspection may reveal the need for imposing additional requirements on the starting material, as was found in Hanford's swaging of Zircaloy filled with UO_2 [73]. Cold working without subsequent annealing may lead to cracking under irradiation.

Some of the other methods also leave either the cladding or both components unchanged in their dimensions, but include a step involving metal that is molten during at least part of the operation. Thus, the fuel may be cased into the cladding. On the other hand, in liquid-metal bonding (or "brazing" with Al-Si or Pb) the dimensions of both the fuel and cladding undergo no change,** but liquid metal is allowed to fill the annulus. Different sequences of adding the fuel and the liquid metal to the cladding are possible. In the Al-Si process, the pre-wetted fuel is inserted into the cladding in the Al-12 wt.% Si bath (eutectic) [1h]; lead was added to the BORAX-IV elements by submerging the fuel-filled tubes in the lead [1k, 58]; sodium can be loaded as a solid extruded rod into the steel cladding before the fuel; the NaK eutectic is liquid at room temperature and can be pipetted after insertion of the fuel [11].

Coextrusion, in contrast to simple insertion, involves rather large dimensional changes for both fuel and cladding (usually at least at 10:1 reduction in area) but the ratio of fuel and cladding is maintained except at the ends [1m, 22f, 22g, 74]. Here, appropriate measures (adjustment of compositions and pre-shaping) can be taken not only to maintain the fuel-cladding ratio but also to provide integral end-seals. Coextrusion can also be adapted to provide ribs on the cladding [22g]. Reactor requirements have provided the impetus to the development of this technique, which achieves a metallurgical bond. Coextrusion has been applied to a variety of

*The possible relaxation of dimensional tolerances is being tested [72].

**Of course, it is essential that the bonding process does not cause fuel or cladding to undergo any metallurgical transformation which might alter dimensions.

uranium alloys differing widely in fabrication behavior. Although defective cladding is most likely to occur near the ends, cropping of the ends does not obviate the need for rigorous inspection of all cladding which has been drastically worked subsequent to its earlier inspection.

4.1.3 End Closures

Except when integral end-seals are provided by coextrusion, a separate operation is used to seal the ends of tubes containing fuel [1n]. Coextruded elements need not necessarily incorporate integral end-seals but can be cut to desired length and provided with closures similar to those used in other elements.

Welding, the most common method of sealing the ends, is applicable to various cladding materials. The end plugs are usually of the same material as the cladding. The assurance of absolutely sound welds is provided by process control and inspection, an essential task for reactors containing thousands of individual elements. In some cases, it is necessary to use special procedures to avoid blowout of the molten metal by heated gases that are trapped. The inspection procedure must of course leave no residual effect (or residue) on the weld. For example, if a dye penetrant has been used, its removal is essential, especially for a fast reactor.

Although welded end closures are a potential source of weakness, they have rarely been responsible for fuel element failures. Welds are located at the ends of the elements where handling is most likely to cause damage. Such has been the situation for some of the Calder Hall elements [75].

Brazing has also been used to seal fuel elements. The end plug is sealed as part of the Al-Si canning operation but a weld bead is then run around the exposed braze line, partly as a test of the quality of bond achieved in the canning. The eleven-foot long elements first used for BGRR were induction-brazed with an Al-5 wt.% Si alloy [50]. The braze alloy can be formed in situ by reaction of silicon with aluminum under pressure (BORAX-IV) [1nn, 58].

The ends of the Fermi core pins provide an interesting example of a mechanical seal where the coolant (Na) does not necessitate a metallurgical bond between the fuel and the cladding. Each end of the coextruded fuel is machined and then swaged to a point. A machined zirconium end-cap is slipped over each end and the assembly is cold-swaged [10].

4.1.4 Assembly of Elements

The assembly of individual elements into a loading unit is primarily a mechanical problem and is discussed in the chapter on Mechanical Systems. Nevertheless a brief discussion is warranted here since the achievement of a strong mechanical structure is related to metallurgical changes in the components at various stages, including the joining operation. Furthermore, the inadequacy of mechanical restraints can lead to serious metallurgical consequences as was found in the EBR-I, Mark II meltdown (see Sec. 4.1.5). The importance of constraint in limiting positive reactivity effects has also been shown in SPERT tests of individual and assembled Savannah elements (see Sec. 4.1.5.3.3).

Geometric displacements such as bowing, warping, and vibration can result from the severe conditions encountered during irradiation, including gradients in flux and temperature, or from rapid flow rates and the associated high pressures and pressure differentials. The joining of spacers (ribs, wires, ferrules) can introduce problems in the selection of materials and processes. Similar problems are encountered in joining the elements to end plates. Often the designer finds it best to leave at least one end of the individual elements free to move within a close-fitting hole or slot.

4.1.5 Operational Experience

The preceding sections have cited examples where operating experience revealed difficulties that affected the subsequent design and fabrication of cylindrical fuel elements. This section provides other examples of the consequences of fuel element failures. These failures were not necessarily initiated by inadequacies in the fuel element, but by malfunctions in other parts of the reactor which took the element beyond its design capabilities.* These examples of unplanned incidents are supplemented by information derived from special tests designed to assess the consequences of potential inadequacies in various components of the reactor system. As mentioned earlier, it is necessary for the reactor designer to know of any provisions that have to be made for situations that tax the ability of the fuel element to dissipate heat.

For convenience in the following presentation, the solid cylindrical elements that have been used in reactors are discussed under the following categories: metallic uranium used in relatively low flux in research or power reactors cooled by an oxidizing coolant (air, CO_2, water); metallic uranium used under more severe conditions in power reactors cooled by liquid alkali metals; and UO_2 used in water-cooled reactors.

4.1.5.1 Metallic Uranium in Low-Flux Reactors

4.1.5.1.1 Aluminum Cladding. Aluminum jacketed uranium had earned the right to be called the "classical" fuel element because it has been used so long that its principal limitations are well known. The materials engineer can work with the reactor designer in overcoming or designing around the limitations. Experience has emphasized the importance of the integrity of the cladding and the soundness of the core-cladding bond. In addition, experience with metallic uranium has revealed the anisoptropic behavior of α-uranium under irratiation: if uranium with preferred orientation is irradiated, it undergoes anisotropic dimensional changes, taxing the jacket; if the uranium is

*A fuel element is designed to maintain the proper balance between heat dissipation and heat generation. The design capabilities are exceeded when heat dissipation is reduced below the minimum value set by the designer, e.g. by restriction of coolant flow, or when heat generation is increased above the design maximum, e.g. by a power surge.

heat treated to minimize the texture,* grain size has to be kept small enough to avoid the bumps or wrinkles resulting from differences in the growth of adjacent grains of differing orientation. Hence, it is necessary to β-treat and then quench in a manner that minimizes texture and grain size. Adjustments in composition may facilitate the achievement of fine grain size and also reduce dimensional changes due to gas bubbles (swelling) and/or intergranular cavity formation. These latter changes become significant above about 400°C (750°C).

The evolving recognition of the importance of various features of aluminum-clad uranium is apparent from the modifications that have been made in such elements for various reactors. Thus, experience in X-10 with the initial unbonded elements showed that, in regions of higher temperature, ruptures increased due to increases in the rates of aluminum-uranium interaction and of oxidation of any uranium exposed to air because of a "holiday" in the jacket [1p]. The interaction and the oxidation were facilitated respectively by the absence of a fuel-jacket barrier and by the presence of a gap increasing the area of exposed fuel and thus permitting the volume increase caused by oxidation. Subsequently, Al-Si bonding was used to minimize the fuel-jacket interaction and the gap. As some of the uranium slugs had not received the intended β-treatment, their excessive growth confirmed the effects of the uranium anisotropy. Ruptures in the elements with β-treated slugs initiated at blisters, which apparently originated from defects in the bonding layer. The initial, natural uranium loading for BGRR used γ-extruded uranium that had been heat treated at 600°C (1110°F) for stress relief [50]. Short lengths (4 in. or 10 cm) of uranium were used to avoid warping. Interaction of the aluminum and uranium was avoided by anodizing the aluminum jacket and leaving a gap, which was filled with helium for use in monitoring defects. Failures of these elements were attributed to drastic requirements imposed by high power densities and severe temperature cycling in rapid startup and shutdown. Elements for NRX are unbonded [76]; β-treatment is now standard and is supplemented by slight cold-work to impart to the rods a texture that leads them to increase in length rather than in diameter, since diameter increases have the more undesirable consequence of reducing the flow of coolant [77].

The tendency to go to higher temperatures has increased the importance of combining fuel and cladding in a way that avoids gaps as well as it avoids contact that permits interaction. Improvement over the Al-Si bonding mentioned in the preceding paragraph has been obtained by methods incorporating other elements that achieve a bond while interposing a diffusion barrier. Methods have been developed using electroplated nickel [1i] or vapor-deposited tin-zinc solder [1j, 66, 67]; these methods involve rather high pressures, mechanical or hydrostatic. The effectiveness of bonding is evaluated on the basis of localization of the consequences of exposure of the uranium through a defect in the cladding [78a]. The effectiveness of the nickel in inhibiting interaction of the uranium and the aluminum can be improved by applying the Ni plate in two steps separated by a thin Cr electroplate. The Cr layer separates the Ni into two layers and this prevents interaction of the uranium-bearing and aluminum-bearing nickel compounds [67].

Alloying to improve the aluminum-uranium combination has had only limited use. Modifications that improve the dimensional stability and corrosion resistance of uranium reduce the burden placed on the cladding and the bond. Thus EL-3, with a hollow U-1.5 wt.% Mo core, was reported to have had no cladding failures to 6500 Mwd/metric ton [79]. Because of poor support one fuel rod fell to a position where it was not properly cooled; the resulting rapid melting was not instructive regarding the behavior of such elements.

4.1.5.1.2 Magnesium Cladding. The magnesium cladding adopted for low-temperature gas-cooled reactors avoids the problem of fuel-cladding interaction. The magnesium-uranium combination not only does not form intermetallics but it does not interact in any way that would create a metallurgical bond. Hence, reliance is placed on mechanical bonding, generally with the help of grooves in the uranium. In other respects, the problems with this fuel are quite similar to those with aluminum-clad uranium: ensuring integrity of the cladding and dimensional stability of the fuel.

Calder Hall has provided the principal experience with magnesium-clad uranium. In the relatively few failures that were encountered (62 out of about 160,000 elements) [80] the consequences were mild. Failure was slow enough to allow the elements to be left in the reactor after detection of the failure [75, 81, 82, 83, 84]. Some of the faulty elements had manufacturing defects or handling damage, both generally associated with the welds. Failures due to such defects are observed early in the irradiation. The majority of the failures occurred at higher burnup and are attributed to leakage paths provided by cavity formation ("cavitation") associated with creep at grain boundaries. This effect is related to poor ductility in the low temperature range, where the failures were observed, and is being overcome by modifying the cladding material to refine its grain structure (thereby improving creep ductility) and also by supplementing the support of the individual elements (to reduce the load on the lowermost element) [85].

Several reported failures of EL-2 elements were attributed to attempts to achieve maximum fuel burnup [79]. All but one were due to unspecified failure of the cladding. The exception was due to flow blockage resulting from solidified deposits of polymerized oil. Melting occurred in this incident as well as in an earlier incident where scramming had been followed by cessation of coolant flow

*The texture of a metal is the orientation of its crystallites. This orientation is determined by the metal's history (fabrication procedures and heat treatment) and strongly influences some of the metal's properties, such as the thermal coefficient of expansion and the growth under irradiation. The texture can be determined from X-ray diffraction measurements.

at a time when the uranium was generating enough heat to melt.

4.1.5.2 Metallic Uranium for Liquid-Metal Cooling. Relatively slender metallic uranium rods or pins are used in reactors cooled by liquid alkali metals (Na of NaK) where the heat generation rates are rather high. The high surface temperatures dictate the choice of cladding. The high heat generation rates necessitate good thermal bonding of the uranium to the cladding, either by a direct metallurgical bond (to zirconium) or by a liquid-metal bond (to stainless steel or the refractory metals). Mechanical stability of the assemblies is essential for stable reactivity and for adequate coolant flow. It is also necessary to provide for the possibility that the rate of heat generation may so far exceed the rate of heat dissipation that the fuel will melt, in which case the flow of the molten fuel must be controlled to prevent reassembly into a critical mass.

Both planned and unplanned experiments with metallic elements in liquid-metal-cooled reactors, fast and thermal, are discussed below.

4.1.5.2.1 Fast Reactors. The November 1955 meltdown of the Mark II loading of EBR-I resulted from a situation in which the ratio of heat generation to heat removal was made excessive by concurrent increase in the heat generation and decrease in the heat removal [1g, 86, 87, 201b]. Instabilities (now attributed to bowing of the free standing fuel rods towards the reactor axis) had been noted in prior operation with this core. A series of tests was undertaken to investigate these instabilities, which were not then understood. In the last of these tests the reactor was "pushed" by placing it on a short positive period and cutting off the flow of NaK coolant. Partial meltdown of the core resulted from a delay in the subsequent shutdown. Examination showed that 40 to 50 % of the core had melted and reached temperatures between 850 and 1500°C (1560 - 2550°F). The molten zone had separated into three zones differing in porosity. Molten U-2 wt.% Zr fuel alloy was found to have traveled both downward and upward, presumably under the pressure of volatilized NaK which was also held responsible for the porosity of the molten zones. Metallographic samples showed the formation of the U_6Fe-UFe_2 eutectic (melting point 725°C or 1337°F).

The potential consequences of fuel failure and subsequent meltdown in a fast reactor are serious enough * to have warranted an extensive series of in-pile test in the Transient Reactor Tests Facility (TREAT) [89, 90, 91]. This facility permits brief, rapid power surges during which the specimen fuel temperature rises to a maximum; the specimen can be uncooled or it can be cooled by stagnant or flowing sodium. The test program has included the following types of fuel elements: EBR-II pins with uranium-fission fuel (clad with stainless steel, columbium or tantalum); Fermi elements (U-10 wt.% Mo clad with zirconium); and UO_2 pellets (clad with stainless steel, columbium or tantalum). Preirradiated EBR-II elements are included in the program to establish the effect of fission products, especially the volatile ones. The behavior of the various elements during rapid transients may be summarized in terms of the maximum cladding temperature.

The EBR-II elements have undergone the most extensive tests and their behavior can be summarized as follows. When the cladding temperature exceeds about 800°C (1470°F) voids appear in the sodium bond between the fuel and the cladding but the cladding retains its integrity. When the cladding temperature goes above about 950°C (1740°F), the fuel and the steel cladding form a eutectic, leading to failure of the cladding and permitting expulsion of the molten fuel by the volatilized sodium in directions roughly perpendicular to the axis of the element. Columbium or tantalum cladding retains its integrity to about 1400°C (2550°F) when it ruptures and permits violent ejection of molten fuel.

In the Fermi elements, the zirconium cladding begins to warp at about 1000°C (1830°F) and failure occurs at about 1200°C (2190°F) by deformation and cracking; the ejection of the fuel is less violent than for the EBR-II elements and is apparently influenced only by gravity. The failure of the oxide elements is less violent and is attributable to failure of the cladding when it can no longer contain the gas pressure in the dead space. Thus, Type-304 stainless steel fails at about 1375°C(2507°F) near its melting point of 1445°C (2651°F). Tantalum failed at about 2400°C (4350°F). The extent of the disruption of the oxide fuel is determined by the temperature reached.

In spite of having data from extensive tests such as those in TREAT and in the simulated Fermi meltdown, the reactor designer still faces the problem of obtaining sufficient experimental data to enable him to forecast the consequences of the most serious sequence of credible events. An intriguing example is provided by a sodium analog of waterlogging (see Sec. 4.1.5.3). After an element has undergone sufficient burnup to accumulate a large concentration of fission products, its cladding may fail in a manner that permits the coolant to enter the fuel element while restricting coolant flow between the elements. In the subsequent overheating of the fuel, the internally trapped sodium and the fission products generate enough pressure to expel some of the fuel, probably as a foam. This foamed fuel may or may not increase reactivity but it will almost certainly further restrict flow of coolant. Such a process can repeat itself and progressively lead to complete failure of the reactor core. TREAT tests on preirradiated EBR-II elements should be helpful in indicating the vigor of steps such as the expulsion of the foam.

4.1.5.2.2 Thermal Reactors. The fuel elements used for the first loading of SRE were similar in some respects to the EBR-I elements discussed above (see Table 4-1) but differences in the causes and consequences of failure are worth noting [1r, 92, 93, 94, 95, 96, 201c]. Operation of the SRE

*For the Fermi reactor simulated meltdown tests were performed by pouring the U–10 wt.% Mo fuel alloy over a prototype of the lower portion of the reactor core subassembly [88].

with this unalloyed uranium loading was terminated in July 1959, since evidence of improper cooling had accumulated. Thirteen fuel assemblies were subsequently found to have been substantially damaged from overheating. The fuel had become overheated when the flow channels were blocked by the decomposition products of the organic compound tetralin, which had leaked into the coolant system from the freeze seal of a pump. The temperature became high enough to permit the eutectic reaction of the stainless steel cladding with the uranium fuel (but with the lower specific power of the stockier rods of this thermal reactor, the temperature apparently did not become high enough for independent melting of the uranium). An additional and perhaps earlier cause of failure was the ductile rupture of the cladding as a consequence of numerous rapid cycles of the fuel between the alpha and beta phases as the cooling conditions fluctuated. Differences in deterioration were noted for several experimental fuel compositions. No evidence was found of damage to stainless steel cladding or hardware by carburization or nitriding, but sensitization was seen.

The second core loading of SRE provided an example of the consequences of lack of constraining [97, 98]. Temperature gradients across the thorium-uranium rods led to bowing, which produced positive reactivity effect. This effect was eliminated when the rods were restrained by a wire wrap.

4.1.5.3 Uranium Dioxide (UO_2). UO_2 was developed initially for use as a fuel in water-cooled reactors. Following its use in PWR, it has gained widespread acceptance for such reactors. By now a large body of experience has been accumulated which shows that, aside from being chemically stable in high-temperature hydrogenated water, UO_2 has other features that make it attractive with other coolants (N , He, CO_2, liquid metals, organic compounds) in reactors where chemical stability is a minor consideration. At various levels of enrichment, UO_2 has been used alone or in combination with other oxides or dispersed in metals. Since UO_2 -metal dispersions have been used primarily for plate-type elements, their performance is discussed below with elements of that geometry (see Sec. 4.2.4.1.2). The following discussion of bulk oxide fuel of cylindrical geometry is devoted primarily to in-pile experience in water-cooled reactors since the other various aspects of UO_2 behavior are well documented [7]. It will be seen that, from the safety point of view, UO_2 is close to the ideal fuel, being responsible for few if any fuel element inadequacies and rarely compounding any difficulties introduced by inadequacies in other components [99]. Credit for this excellent performance record can be shared by the material itself and by those who have thoroughly characterized the material to ensure its proper use.

The problems associated with the use of UO_2 as a fuel are similar for various cladding compositions, since in the absence of bonding the fuel-cladding combination is not likely to be a major factor.* Hence, although the following discussion classifies reactor experience on the basis of cladding composition it is apparent that, for the most part, the experience with one cladding is directly applicable to the others; in fact, a rather generalized discussion of the potential problems can be given. These problems derive from the absence of a core-cladding bond; this situation leaves a void and, in case the cladding is penetrated, increases the area of fuel exposed to coolant, thereby leading to release of volatile and soluble fission products. With more porous oxide, both the dead space and the exposed area are increased. In water-cooled reactors the void space makes the element susceptible to "waterlogging" [100] in case of a cladding defect. That is, water entering the void space during low power operation is subsequently converted to steam, which, if trapped, can lead to gross distortion of the cladding (see also Sec. 3.4.4.2). The lack of bonding can combine with the difference in thermal expansion to lead to ratcheting in long elements; although various tests indicate that the problem is not likely to be a serious one [99], it must be recognized that this problem may be aggravated with collapsed cladding (Sec. 4.1.2) because of the more intimate contact. The cladding must be strong enough or must be reinforced to withstand pressure differentials resulting from high pressures either externally (reactor coolant) or internally (fission gases, if burnup becomes high enough for significant gas release). For Zircaloy cladding, corrosion of its internal surface becomes a problem in the event of coolant water penetration (or water presence in the UO_2) because the hydrogen generated by the corrosion reaction is absorbed by the Zircaloy; this hydrogen can, under the influence of a temperature gradient, be transported to a colder zone, where it concentrates to a level sufficient to embrittle the cladding. The hydriding problem can be alleviated by use of Zircaloy-4, which corrodes at approximately the same rate as Zircaloy-2 but absorbs less of the generated hydrogen.

4.1.5.3.1 Aluminum-Clad UO_2. The satisfactory operation of a boiling water reactor with oxide fuel was indicated in BORAX-IV [59, 101]. The fuel element was unique in several respects: a low-temperature cladding was combined with a high temperature fuel (ThO_2-6.36 wt.% UO_2); the core and cladding were bonded with Pb; tube-plates were used for the cladding. On resuming operation of the reactor in February 1958, after a 10-week shutdown, significant fission product release was observed. An in-pile defect test was performed by operating the reactor for two days in March 1958. Subsequent investigation showed one or more ruptured rods in 22 of the 59 subassemblies. The ruptures were attributed to collapse under 300 psi (21 atm) pressure of the aluminum over the void space above the fuel pellets. Cracks and crevices in the collapsed tubing were aggravated by corrosion during the shutdown. It should also be noted that

*This statement has to be qualified for the effect of certain fuel impurities on the cladding. A striking example is the deleterious effect of very small amounts of fluoride on the corrosion behavior of Zircaloy; such fluoride may be present in UO_2[51].

the thickness of cladding was unusually nonuniform. Comfort was derived from the fact that only gaseous fission products were released. (It is not clear whether the Pb ever permitted exposure of the fuel to the water.) The magnitude of the fission product release was attributed to porosity associated with the low bulk density of the fuel (82.3% of theoretical).

4.1.5.3.2 Zircaloy-Clad UO_2. The extensive development that led to the use of Zircaloy-clad UO_2 in PWR and later reactors need not be reviewed here [7, 56]. It is sufficient to point out that the development work has been confirmed by the evaluation of blanket fuel removed from PWR during the first and second refuelings. The following conclusions regarding the performance of the blanket fuel during the lifetime of the first seed are worth quoting [102, 103].*

"1. Crud deposits on the heat transfer surfaces were of negligible thermal importance.

"2. Postirradiation dimensional measurements of the fuel bundles duplicated preirradiation results within measurement error.

"3. The measured performance of the blanket was in good agreement with physics calculations.

"4. The percentages of fission gases released from the fuel were small and in essential agreement with irradiation tests and solid-state diffusion theory.

"5. No microstructural changes were observed in the fuel.

"6. Microstructural examination, coupled with hydrogen analysis, of the Zircaloy showed slowly forming hydride precipitated at the equivalent rate of 5-10 ppm/year. No reaction between fuel and cladding was seen.

"7. Measurements of bursting pressure indicate a 10-20% increase in tensile stress of the Zircaloy-2 cladding as a result of irradiation. The presence of hydrogen did seem to be a factor."

Equally instructive and confirmatory are the results of the evaluation of a fuel rod that was defective when loaded and was operated during the lifetime of the first two seeds [104, 105]. No unusual conditions resulted from reactor operation with complete cladding penetration. The postirradiation examination revealed a transverse crack that started at the inside surface of the Zircaloy. Defects similar in type but differing in number and severity were found in both irradiated and nonirradiated rods fabricated from the same lot of tubing.

Zircaloy-clad UO_2 elements have an excellent record of performance in the Dresden and Kahl reactors [106, 202]. Some of the first assemblies prepared for Dresden were found to have cracks in the wall of the cladding and were replaced before being loaded in the reactor. This experience emphasized the value of thorough testing of the tubing stock used for cladding [107].

Failures of PRTR Mark-I elements, containing mechanically compacted UO_2-PuO_2, appear to be attributable to contaminants in the fuel; postirradiation examination has shown that the corrosion of the Zircaloy cladding started from the inside. Corrosion and hydriding are caused by moisture in the fuel and are accelerated by fluoride. Such failure, progressing from the internal surface of the cladding, has been seen with both vibrationally compacted fuel and swage compacted fuel, but is apparently more rapid with swaged material. The cracking pattern with swaged material suggests that stress induced by the swaging is a factor [108, 109]. A longitudinal slit in an experimental MgO-PuO_2 rod tested in PRTR was attributed to an initiating corrosion on the inside [110].

It is worth emphasizing that the excellent performance of the UO_2 is a consequence of the proper matching of its capabilities with service conditions. When conditions exceed the capabilities of a material, difficulties are bound to ensue, as has been observed in various tests, unintentional as well as those designed to establish the limit of the material's capabilities [111, 112]. It should be obvious that the limits of capabilities have to be defined for material prepared under commercial rather than laboratory conditions.

Experiments are currently underway to test one of the practices regarding the limits of operation with UO_2 fuel [113, 205]. Heretofore UO_2 fuel elements have been designed to assure a maximum operating temperature safely below the melting point. The General Electric experiments are exploring the feasibility of operating a UO_2 rod with gross central melting. Successful operation has been reported with hollow UO_2 fuel pellets, which provide the necessary free volume to accommodate the expansion of UO_2 on melting. Operation of such an element has to allow for the reactivity changes associated with the shift of material on melting and for the possible filling of the central voids in the event of cladding rupture.

4.1.5.3.3 Stainless Steel-Clad UO_2. Yankee was the first reactor to make full-scale application of stainless-steel clad UO_2 rods. These fuel elements have operated so successfully that they have thus far provided no information regarding the consequences of operating with a defective element. As far as is known, no failures occurred in the first five cores [114, 219]. Postirradiation examination of rods from the first core has shown them to be in excellent condition except for local swelling in one rod. Its swollen area was about 0.3-in (7.6 mm) long and had a diameter about 0.010 in. (0.25 mm) greater than that in adjacent areas. The reason for this swelling has not yet been determined [115, 116, 206].

Extensive tests to establish the limits for stainless steel-clad UO_2 were carried out at the VBWR. Thus, the program included thin-walled tubing (5 to 12 mils or 0.127 to 0.305 mm) to establish the limits of non-freestanding cladding, i.e., cladding supported by the UO_2 fuel; such thin cladding is of particular interest for neutron economy. (Similar tests were performed in the

*Later data are worth noting [203]. Results with thin, non-free standing Zircaloy cladding are also available [204].

Saxton reactor [72, 207]. The results of these tests and other experience with stainless steel as cladding for water-cooled reactors have been reviewed [112, 117, 208]. Failures and irradiation tests of intentionally defected rods are summarized. Stresses seem to be a significant factor in the cracking of the cladding [208]. Hence, materials of interest for superheaters are also being considered for water reactors. For superheaters the stress-corrosion cracking of stainless steel is serious enough to hamper its adoption as cladding [118]. GE is giving prime emphasis to Incoloy in its cladding development program, which used a special test loop SADE (Superheat Advance Demonstration Experiment) in the VBWR [26, 119, 218]. For the BONUS reactor Inconel is now specified as the cladding (Zircaloy is used in the boiler).

Experiments have been carried out in SPERT-1 on the behavior of low enrichment UO_2 rod-type elements [120, 121, 201e]. Excursions with periods as short as 3.2 msec were quenched by Doppler effect without core damage. With unconstrained fuel rods, where bowing could occur and cause positive reactivity effects, no cladding failures or damage to the fuel were observed. The benefit of constraint was clearly shown. It also appeared that the low heat transfer rate of such elements would maintain the surface temperature low enough to prevent rapid steam formation. Since steam formation is often a useful shutdown mechanism, this prevention of its formation may adversely affect shutdown of a large nuclear transient. An earlier experiment provided an example of waterlogging, followed by a rupture during a transient.*

4.2 Plate-Type Fuel Elements

Plate-type fuel elements have a high ratio of heat transfer area to element volume, thereby permitting operation at high heat fluxes. This configuration is particularly attractive for the achievement of high neutron fluxes in research reactors and of high specific power in compact power reactors. In some reactors where it is desirable to spread the fuel, it is dispersed in a solvent or, more likely, in a matrix. Usually the high ratio of cladding-plus-fuel diluent to fuel makes it advisable that the fuel be highly enriched. Since plate-type geometry is an effective means of improving heat removal, it reduces the central temperature of a fuel of low conductivity, e.g., the natural UO_2 for the blanket of the second PWR core.

To .exploit properly the heat removal capabilities of flat elements, good thermal bonding between the fuel and the cladding is highly desirable. Metallurgical bonding is facilitated by a strong similarity between the fuel and the cladding. Thus, the cladding (aluminum, Zircaloy, stainless steel) is likely to be identical with the matrix of the fuel. Fuel-cladding combinations are possible in which the cladding and the matrix differ somewhat but have identical constituents.

*Defective welds were responsible for waterlogging-type failures experienced in the superheater section of the BONUS reactor during a transient caused by operator error [209, 216].

Of the plate-type elements the so-called MTR-type, so popular for water-cooled pool and tank reactors, is an outstanding example of a relatively standardized fuel element. And yet, here too it is interesting to note the many variants of this element: reactor designers have exercised in different ways the options available to them in the choice of fuel composition, dimensions of the individual plates, number of plates per assembly, spacing between plates, etc. The designer's freedom includes the right to have the assembly consist of plates differing in composition and geometry. Latitude in the choice of composition for the various plates is provided by the rather versatile techniques of powder metallurgy, which permit a graded fuel composition within a plate and the graded introduction of poison.**

With very few exceptions, plate elements (Table 4-3) have been used only in reactors cooled by water where the heat-transfer capabilities exploit the advantages of plate-type elements. An interesting exception is the SM-1 type element used as a standard and reliable element to test various aspects of organic cooling in the OMRE. Since the organic coolant has lower heat transfer capability than water, these plates had to be assembled in a manner that left them free to move longitudinally in response to differential thermal expansion. The OMRE also included experimental elements with finned aluminum cladding. (These elements are not included in the tabulation, but the section operating experience, Sec. 4.2.4.2, includes the failures of two such elements as a result of coolant blockage).

4.2.1 Features Affecting Design and Fabrication

The principal problem in the design and fabrication of plate-type fuel elements is the mechanical stability of the plates. These plates have to be thin enough, generally 0.1 in. (2.4 mm) or less, to provide the advantages of this geometry for heat removal and yet strong enough to maintain a stable configuration, notably in permitting flow of coolant through the small water channels between plates. This problem is aggravated by the high rates of coolant flow, (e.g., in excess of 30 ft/sec (9 m/sec) for MTR and ETR), which can cause an unbalanced hydrostatic pressure leading to distortion of plates and subsequent blocking of flow channels. The design of such elements — and even seemingly slight deviations and modifications — must take into account the local pressure effects of hydrodynamics, especially as such effects can introduce forces exceeding the strength of individual components. Thus, the first failures in the MTR (in 1954) resulted from a 15% decrease in the flow area of the lower end box which led to lateral pressure differentials between subassemblies and their adjacent channels. These differentials in turn led

**Plate-type elements offer several means of incorporating a poison such as boron carbide in the sides: insertion and sealing in the side plates; dispersion in the side plates; attachment of a poison strip to the side plates.

TABLE 4-3

Characteristics of Representative Plate Fuel Elements

Reactor and Coolant	Power (Mw)		Dimensions of Plate			Core			Cladding		Fabrication Method	Assembly	Extent of Operating Experience		Notes
	Thermal	Elec., (net)	Length (in.)	Width (in.)	Thickness (mils)	Composition	U235 (wt. %)	Thickness (mils)	Composition	Thickness (mils)		No. per Unit & Method	Crit.	Full Power	
NRU D_2O	200		120	2.144 1.962 1.224 (Above are U widths)		U	Nat.	171 171 177	Al (1S)	25	Cable extrusion (extrusion cladding) with oversize cladding (0.1 in.) milled down	5 in 2.5 in. Al. tube. 1 large, 2 medium, 2 small (in width)	11/57	4/58	At outset used cold die sized bars. Had a number of cladding failures, notably one responsible for serious incident in May 1958.
MTR H_2O	40		25-1/8	2.65	17-50 2-65	Al-18 wt.%U	93	20 20	1100 Al	15 22.5	Hot rolling (590°C) in picture frame assembly	19 Roll-swaged into side plates (since 1960) centering combs pinned to each end of fuel plates	3/52	5/52 (30 Mw) 9/55 (40 Mw)	Plate curved about 5-1/2 in. radius. Up to 1960 brazed into side plates.
ETR H_2O (185 psig)	175		37-1/4	2.624	50	Al-23 wt.%U	93		1100 Al		Same	19 Roll-swaged into side plates; centering comb pinned to upper end (with common center)	57		Flat plates. Programmed nonuniformity among coolant channels (outermost channels larger).
ATR H_2O	250		49.5	2.221 to 4.055	17-50 1-80 1-100	Al-34 wt.% U_3O_8 and B	93	20	6061 Al	15	Same	19 in each 45° segment. Roll-swaged or pinned.	(1965)		Plates curved about 3.0 to 5.5 in. radius.
DIDO D_2O	10		23.75 (active)		58	Al-20 wt.% U	93	18	Al	20		10 Brazed to U-shaped spacer bars	11/56		Plates curved about 5-1/2 in. radius.
MITR D_2O	1	5	24.625	3	60	Al-U	93	20	Al	20		18 Brazed into side plates or mechanically pinned or swaged	1/59		Plates curved; 2 outside plates contain no fuel. Center section left open in same plates for beam post; center plates omitted from some assemblies for sample thimbles.
SPERT-1 H_2O	Variable with experiment		24.625	0.923	60	Al-U	93	20	1100 Al	20		51 in 8 rows with 2 side plates and 2 intermediate plates	7/55		
SPERT-2 H_2O (up to 375 psi)			25.125	2.724	60	Al-U	93	20	6061 Al	20		24			
HFBR D_2O	40		23		17-50 2-187	Al-30 wt.% U-3 wt.% Si	93	20 10	6061 Al 6061 Al	15 89		19 Roll swaging	10/65		Plates curved about 6 in. radius.

BGRR air	16		24-1/8	3-1/8	60	Al-8 wt.% U	93	20	Al	20		3 Each bent to V-shape with hooks on top. Slotted rings of 6061 Al (61S) at each end; at one end, ring welded to plates, at other end, rings float in slots, which are closed by welding	Completion of conversion to enriched fuel 3/58	Now using 4043 Al (43S) for rings. 8 elements end to end per channel.
SL-1 (ALPR) H_2O	3	0.3	27.86	4.64	120	Al-17.6 wt.%U -2 wt.% Ni	93	50	X8001 Al	35	Silicon bonding (600°C) of picture frame assembly followed by hot rolling	9 Flanged edges spot welded to side plates	8/58 10/58	Accident 1/61
PWR-Core I H_2O (2000 psi)	231	60	72-3/8	2-1/2	69	Zy-6.7 wt.%U	93	40	Zy-2	15	Hot rolling (780°C) in picture frame assembly	15 per subassembly; 4 subassemblies in each of 32 clusters. Flanged edges welded together (also 2 Zircaloy end plates)	12/57 12/57 Refuelings (new seed) 12/59 to 4/60 8/61 to 10/61	Data refer to seed.
EBWR	100	4.5	54	3.625	280 or 212	U-5 wt.% Zr-1.5 wt.% Nb	Nat. or 1.44	240 or 172	Zy-2	20	Hot rolling (850°C) in picture frame assembly with double vac. seal	6 per assembly; 11 natural assemblies; 104 enriched assemblies. Spot welded intermittently along full length of perforated side plates	12/56 12/56 (20 Mwt) 11/62 (100 Mwt)	
RA-1			24.4	0.20	110	Al-45 wt.% U_3O_8	20		6061 Al		Extrude (380°C)			
HFIR Inner	100		24	3.800	50	Al-26 wt.% U_3O_8 -B_4C		Not uniform	6061 Al (Al-clad)		Hot rolling (500°C) in picture frame assembly	Involute curve: 171 in inner annulus; 369 in outer annulus. Weld on outside. Weld or peen on inside		Formed to curvature by marforming.
Outer				3.500	50	Al-24 wt.% U								
PWR-Core II H_2O (2000 psi)	505	136 Blanket	97.7	3.64		UO_2	Nat.	100	Zy-4	20	Gas pressure bonding 10,000 psi. 1575°F (857°C)	15 in subassembly 4 subassemblies in each of 77 clusters	5/65	Compartmented by internal ribs; longitudinal ribs 1/4 in. apart; transverse ribs 6 in. apart.
		Seed	97.6	3.44		UO_2-ZrO_2	Enr.	36	Zy-4	20	Same	19 in subassembly 4 subassemblies in each of 7 clusters		Compartmented like blanket; 3 different UO_2 concentrations (20 to 38 wt.%). Burnable B^{10} (in stainless steel) poison wafers included in subassemblies.
SM-1 (APPR) and SM-1A H_2O (1200 psi)	10 20.2	1.9 1.6	23	2.76	30	ss (302B)-24 wt.% UO_2	93	20	304L ss	5	Hot rolling (1150°C) in picture frame assembly	18 in each of 38 subassemblies. Brazed into grooved side plates	4/57 Core I shut 4/61 Core II since 6/61	B_4C poison in matrix. Lower Co content specified for ss in Core II.

TABLE 4-3 (Continued)

Reactor and Coolant	Power (Mw)		Dimensions of Plate			Core			Cladding		Fabrication Method	Assembly	Extent of Operating Experience		Notes
	Thermal	Elec., (net)	Length (in.)	Width (in.)	Thickness (mils)	Composition	U^{235} (wt. %)	Thickness (mils)	Composition	Thickness (mils)		No. per Unit & Method	Crit.	Full Power	
PM-2A H_2O (1750 psi)	10	1.6	31.7	2.85	30	ss-26.3 wt. % UO_2	93	20	304L ss	5	Hot rolling (1150°C) in picture frame assembly	18 in each of 32 subassemblies plus 16 in each of 5 subassemblies	10/60 Shutdown 1963	2/61	
OMRE Santowax OM	2.0 to 10.0		37	2.75	30	304 ss-25 wt. % UO_2	93	20	304 ss	5	Hot rolling (1093°C) in picture frame assembly	16 in each of 31 subassemblies. Float freely in grooves in side plates; retained at weld end by bar welded between side plates and engaging notches in fuel plates. Tabs from sides of two fuel plates protrude through side plates and are bent over.	9/57 Dismantled 1963	2/58	2% Si added to ss matrix to facilitate powder manufacture.
SPERT-3 H_2O (2500 psi)			37	1.4 standard, 1.15 small	30	ss-19 wt. % UO_2	93	20	304L ss	5	Hot rolling	38 in 2 rows 32 in 2 rows	12/58		
MTR test of Pu fuel 262 Mwd					60	Al-14 wt. % Pu		20	Al	20	Hot rolling (590°C) in picture frame assembly	18 Brazed into side plates	8/58		Plate curved.

to distortion of the outer fuel plate, displacing the center of the concave plate to give it a convex bulge and bring it in contact with the nearest plate of the adjoining assembly [122, 123a, 124].

Both mechanical and metallurgical means are used to improve the rigidity of the fuel assemblies. In general, sets of parallel plates are attached mechanically to a pair of side plates to give a box-like structure. Spacers usually described as "combs" are often placed across the midwidths of a series of plates to help maintain the spacing. In many designs the plates are curved to improve their stability. Strength of the assemblies is increased by using stronger material for the cladding, fuel matrix and/or the side plates. Thus, for aluminum-clad elements, the ordinary 1100 alloy is being replaced by stronger 6061 alloy, which does not impose a severe penalty in thermal neutron absorption by the principal alloying elements (Mg, Si, Cu and Cr). To retain the strength imparted to the individual plates by cold-rolling, mechanical means of element assembly are favored over means involving heating of the aluminum components. Thus, for the MTR, annealing the cold-worked plates is now avoided by using roll-swaging instead of brazing for attaching the individual fuel plates to the side plates. For Zircaloy and stainless steel, the difference in strength between the worked and the annealed materials is not likely to be a significant consideration; hence, methods involving heating, like welding or brazing, are more acceptable. The lower temperature operation, brazing, may be preferable to avoid warpage.

The importance of the mechanical stability of the assemblies under flow conditions makes flow testing essential not only for validating a fuel design but also as part of the acceptance tests of elements to be inserted into a reactor. The hydraulic test then checks the adequacy of the attachments as well as the individual components. Thus, prior to insertion into the reactor, a substantial fraction (20%) of production assemblies for the MTR is hydraulically tested at 140% of the normal velocity of reactor coolant.

The high specific power at which plate-type elements are usually operated imposes stringent requirements on the homogeneity of the fuel. Areas of high fuel concentration cause hot spots, which can restrict the over-all operating level of the reactor. Thus, the high-density inclusion which was revealed by radiography limited the operating level of the MTR in the test of Al-14 wt.% Pu plates [1s, 210]. The stringent requirement on homogeneity provides an incentive to use powder metallurgy or other expensive techniques involving the blending of particulate matter.

4.2.2 Fabrication of Plate-Type Elements

Hot roll-bonding is the predominant method for fabricating plate-type elements. It is peculiarly suited for and limited to such elements. The following detailed discussion of this method is succeeded by less detailed discussions of several other methods which have had much more limited application.

4.2.2.1 Hot Roll-Bonding. This technique is versatile enough to be applicable to a variety of fuel-cladding combinations [1t, 1u] and to permit the preparation of a series of plates differing in composition and in any of several dimensions. With a change in fuel type the fabrication parameters (e. g., design of billet assembly, temperature, reduction) are adjusted but the basic techniques remain the same and are expected to achieve the following: uniformity of composition and dimensions (over-all, individual) and complete metallurgical bonding of the claddings to the fuel and to the edges and end seals. The paragraphs discussing the following sequence of steps in the roll-bonding of plate elements indicate ways in which the techniques are applied for different compositions: assembly of billet, preparation of components, breakdown rolling and finishing operations.

4.2.2.1.1 Billet Assembly. The preparation of the components is likely to be better appreciated following a description of the billet assembly [1v]. Perhaps the simplest assembly is the one in which the fuel-bearing core is surrounded on four sides by a "picture frame" and this combination is sandwiched between two cover plates (MTR, SM-1). The frame can be eliminated by use of recessed cover plates confining a fuel filler that has beveled ends (PWR seed). The frame can be replaced by two long side plates and two short end plugs preshaped to match the tapers at the core ends (EBWR). The method of assembling these components is determined by the cladding material. Aluminum need not be sealed [3a, 8]. For some elements aluminum is welded around the edges and an evacuation outlet is welded at one end; this outlet is crimped to provide a seal after evacuation [125]. For stainless steel, a seam is welded around the edge, leaving a 0.25 in. (6.4 mm) gap to permit a hydrogen blanket on the internal surfaces for the heating preceding rolling. Zircaloy may or may not be welded but in either case it requires protective sheathing. The mild steel used for this sheathing has to behave properly in rolling and should not release gases (notably nitrogen) during the bonding operation. Any deleterious effect of such released gases can be minimized by sealing the Zircaloy-fuel assembly under vacuum, as was done for EBWR plates by the seal pin technique [1w, 126]. The steel jacket can also contain additional strips of cladding material to provide ribs or flanges on the rolled element. The spacers between the strips are included in the assembly for the PWR seed plates [127]; these strips provide the I-beam structure, which is subsequently welded to obtain a subassembly. (See Sec. 4.2.3). A more recent method of obtaining the I-beam structure in the rolled plate is by use of Zircaloy cladding rolled and machined to a U-shape that avoids the need for separate inserts.

4.2.2.1.2 Preparation of Components. In preparing the various components for billets such as described above, it is the fuel that requires the most attention, since rather conventional methods can be used to prepare and finish the stock for the other components, e. g., the frame for the edge and end claddings, and the cover plates for the face claddings.* Both casting and powder metallurgy techniques are used to prepare fuel cores. In either case, special attention has to be paid to the achievement of homogeneity. It is because of this requirement that powder metallurgy is often used in spite of its higher cost. This technique also provides a means of assuring the exact fuel and poison content of each element if each fuel core is processed individually.

The casting technique selected for each metal fuel composition is determined by the need to achieve homogeneity. For U-Al alloys, reasonably uniform castings can be obtained up to about 25 wt.% U [8]; similar techniques are applicable to Pu-Al alloys [125,128]. As the uranium content is increased, homogeneous binary alloys are more difficult to achieve. In addition the high-uranium binaries become more difficult to fabricate because of the large volume fraction of the compound UAl_4. These difficulties are overcome by additions such as 3 wt.% Si, which suppresses the formation of UAl_4 and gives $U(Al, Si)_3$ as the stable phase, thereby improving the homogeneity, grain size, formability and strength of the alloy [22h, 129]. The Zircaloy-6.7 wt.% U fuel alloy used for the seed of PWR is arc-melted three times with intermediate working and cutting after the first two melts [127]. When only two melts were used, radiography showed segregation of the fissile phase. Whereas the Al-U, Al-Pu and Zr-U castings are worked by forging and/or rolling to obtain the fairly thin bar stock needed for the rolling billet, the U-5wt.% Zr-1.5 wt.% Cb fuel of EBWR was induction-melted and poured from the crucible through a distributor into a 12-cavity mold designed to give bars close to the sizes needed for the rolling assemblies [126].

Standard powder metallurgy operations are used to prepare cores consisting of UO_2 dispersed in metals such as stainless steel or aluminum. Thus, for stainless steel (SM-1) the following sequence is used: screening to obtain powders of the optimum particle-size distribution; weighing of the individual powders; blending; compacting by cold pressing (densities achieved were about 75% of theoretical density); sintering in dry hydrogen at about 1200°C (2190°F) (densities achieved were about 90% of theoretical density); and coining to final size needed for the rolling assembly [130]. For aluminum the sequence is quite similar; however, U_3O_8 is the preferred form of uranium oxide since its reaction with aluminum at temperatures near 600°C (1110°F) (required for fabrication) cause no adverse volume change, whereas UO_2 reacts and causes volume changes that lead to blisters. The relative reactivity with aluminum differs for different grades of UO_2, depending on the method of preparation [22i, 123b]. Stainless steel is inert to UO_2 at temperatures as high as 1400°C (2550°F). Mixtures of U_3O_8 and aluminum are an alternative to silicon-containing U-Al castings (see above) for high-uranium fuels, such as those needed to obtain sufficient U^{235} content when the enrichment is limited to 20 wt.% by export regulation [22a]. (The 1955 Geneva Conference Reactor (GCR) used an Al-UO_2 fuel). Powdered uranium-aluminum compounds such as UAl_3 are amenable to powder metallurgy techniques for the preparation of elements with higher operating limits than elements with dispersed oxide [131].

Achievement of the desired bonding requires that the fuel and cladding components have very clean surfaces. The chemical or mechanical means chosen for the cleaning operation(s) depends of course on the material. Although some prebonding is obtained by prior welding of the edges of the assembly, the initial hot-rolling steps are relied on to effect bonding. An exception is the SL-1 fuel, in which silicon effects the bond before the hot-rolling. Silicon powder is applied to the mating interfaces; the assembly is then pressed briefly near 600°C (1110°F); the Al-Si eutectic (m.p. 577°C or 1071°F) forms in situ and effects bonding. An evacuation tube is unnecessary and the hot rolling becomes a sizing operation [123c].

4.2.2.1.3 Rolling and Finishing. The rolling schedule (temperature, number of passes, reduction per pass) has to be tailored to the specific fuel-cladding combination. The selection of this schedule is guided not only by the requirement to effect bonding but also by the need to avoid end defects which can leave areas in which the cladding is too thin. If not overcome, such defects (well-described as "dogbones") can necessitate a higher nominal cladding thickness to assure the integrity of the cladding. As in coextrusion where the fuel and cladding also undergo extensive deformation together, the end defects can be minimized by matching stiffnesses through adjustment of composition(s) and by preshaping the fuel end and the corresponding end plugs [1x, 125]. Stratagems can be used to supplement the roll bonding to effect bonding. Thus, since 6061 aluminum alloy, which may be desired for stiffer cladding, does not bond as well as 1100 aluminum to the fuel core, the cladding component can consist of Al-clad 6061, i.e., 6061 previously clad with 1100 on the interior surface. The bonding of the Zircaloy end plugs to the Zircaloy side plates in EBWR plates is facilitated by the extra sideward deformation of the end plugs resulting from their proper preorientation, i.e., the roller plate stock is cut to take advantage of its anisotropy by placing its "soft" direction transverse to the rolling direction [1y, 132].

After the achievement of bonding and the preliminary sizing by hot rolling, the plates undergo a series of finishing operations. They are first cold-rolled to the final dimensions. (If a steel jacket was used for the hot rolling, it is removed before the cold rolling). Shearing to the final width and length is usually controlled by radiographic definition of the edges of the fuel. The desired flatness or curvature is imparted with the appropriate forming die, and mechanical or chemical means are used to obtain a suitable surface finish. Any heat treatment required by metallurgical considerations is likely to be incorporated either before or during the above finishing operations. Thus

---

*The cladding thicknesses in Table 4-3 are for the faces. The edge claddings are thicker since they are hardly involved in heat dissipation but are needed for attachment to form an assembly.

MTS-type U-Al plates are annealed to remove hydrogen, which might subsequently cause blistering. EBWR plates were subjected to a heat treatment that improved the dimensional stability of the core alloy under irradiation. The PWR seed elements were annealed at about 600°C (1110°F) for stress relief.

4.2.2.2 Other Methods. Since roll-bonding is not feasible for a brittle fuel such as uranium oxide, two other methods, gas pressure bonding and eutectic bonding, have been developed for the oxide-fueled plates of the PWR second core (seed and blanket). In both methods, the picture frame (or receptacle plate) for the oxide wafers is compartmented by sets of longitudinal and transverse ribs 35 mil (0.89 mm) in width which isolate any failures and thereby minimize the consequent waterlogging (see Sec. 4.1.5.3) and bulging. These ribs also increase the rigidity of the plate. Since the Zircaloy undergoes very little deformation in the bonding operation, the configuration of the compartments as well as the dimensions are preserved. Final hot-flattening may be necessary [22j, 133].

In gas pressure bonding, the combination of temperature (857°C or 1575°F) and pressure (10,000 psi or 700 kg/cm^2) forces the mating Zircaloy surfaces together, which promotes solid-state bonding by diffusion and grain growth across the interface. After the Zircaloy surfaces have been properly roughened by an abrasive belt, the edges of the assembly are welded and the assembly is evacuated. Carbon (from the pyrolysis of methane) is used to coat the oxide wafers with 15 to 40 microinches (0.38 to 1.02 microns) of carbon, which prevents their interaction with the Zircaloy. The bonding operation, which requires several hours, is followed by a heat treatment to restore the optimum corrosion resistance of the Zircaloy. The same basic bonding technique can be used for metal cores. Pressure bonding was used to clad the U-3.5 wt.% Mo plates with finned aluminum for tests in the OMRE [134] and also for the U-3.5 wt.% Mo - 0.1 wt.% Al tubes with spirally finned aluminum for Piqua [135]. The bonding of the cores was facilitated by a nickel electroplate about 1 mil (0.025 mm) thick. These elements illustrate the applicability of pressure bonding to various configurations including ribs.

In eutectic bonding, the mating surfaces of the cladding metal are provided with an intermediate layer of an element that forms a low-melting eutectic [136]. When this type of assembly is heated, melting occurs below the melting point of the cladding. As the assembly is held at temperature, the alloying element diffuses into the cladding metal so that the liquid phase disappears. This technique was described above as silicon bonding for the aluminum plates of the SL-1 elements. For zirconium, several elements are available which can be plated onto a mating surface and form eutectics melting below 1000°C (1830°F). Copper applied by chemical replacement is the preferred bonding agent. Nickel appeared as attractive until it was found to increase severely the absorption of hydrogen by Zircaloy. (The discovery of this effect led to the development of Zircaloy-2.) In the application of eutectic bonding to Zircaloy, the assembly is welded at the edges. Evacuation is considered unnecessary since entrapped air will be gettered by the Zircaloy. During the heating for bonding, a slight positive gas pressure (15 to 30 psig or 2 to 3 atm) is applied to maintain intimate contact between the surfaces being joined.

Extrusion cladding (often designated "cable extrusion" because of its use in the cladding of electrical wire cables) has been adopted for the flat metallic uranium elements of NRU [1z, 22k, 137a]. The technique is not limited to elements of this geometry, but is of interest primarily for rather long elements. In the extrusion cladding process, the aluminum cladding is extruded into a cavity through which the uranium core is simultaneously fed (at lower temperature) as a floating mandrel. The uranium is electroplated with nickel for bonding. The aluminum ends are pressure welded by hot forging, with a fusion weld as a back-up.

4.2.3 Assembly of Elements

Since the thin individual plates have relatively little strength, their assembly into a rigid structure is required. The standard configuration is a rectangular array of parallel plates whose long narrow edges are used for attachment [1aa]. In assembling such a configuration or some of the more complex ones mentioned below, the designer has many options available. Thus, he may maintain equal spacing between plates that differ in composition or geometry (MTR) or he may vary the spacing between identical plates (ETR). In general, side plates are used but intermediate places may be added to hold parallel rows of fuel plates (SPERT-1 and -3). The side plates can include perforations that deliberately reduce their strength to allow free movement of the fuel plates under irradiation (EBWR). The ends of the assemblies may include permanent spacers (combs) or temporary spacers which maintain the spacing during the assembly operation and are subsequently removed.

The individual plates may be modified at the edges to facilitate their assembly. Thus, as indicated above (see Sec. 4.2.2.1.1) the roll-bonding assembly for the PWR seed plates includes strips which form the flanged edges of the I-beam. These flanges then form the side plates when a set of I-beams is welded together. For BORAX-IV and SL-1, the rolled plates were formed to obtain flanged edges parallel to the aluminum side plate to which they were spot welded [123c]. British assemblies are brazed to U-shaped spacer bars.

The attachment of the fuel plates to the side plates may be accomplished mechanically or metallurgically. Thus, roll-swaging is now used for the MTR assemblies, since brazing anneals and softens the cold-worked aluminum. Joining schemes involving pins or tabs are available. The assembly of the SM-1-type plates for OMRE provides an example of rather loose mechanical attachment, designed to leave the plates free to move longitudinally and transversely while held in place by a retaining bar welded between the side plates at the inlet end of the fuel box [1bb, 138, 139]. In the comparison of brazing and welding for metallurgical joining, brazing has the advantage of lower temperature (causing less warping) but welding has the advantage of not introducing another material. Any material used for brazing has to meet the usual requirements of a reactor material, such as adequate strength, corrosion resistance, low neutron absorption cross section.* For aluminum, the Al-12 wt.% Si meets this requirement; it is often introduced as one component of a duplex side plate. For stainless steel, Coast Metals N.P. is applied as a dry powder and fixed in place with an appropriate cement. For Zircaloy, welding is generally used.

The thin plates can be formed to obtain shapes used for assembly of configurations other than a simple box. Thus, for BGRR the plates are bent to V-shape with hooks on the top; three such plates are then assembled with slotted rings at each end; at one end, the ring is welded to the plates; at the other end the ring floats in the slots, which are closed by welding [1cc, 1dd]. Plates can be formed into involutes which are then assembled about a central aperture (HFIR and British designs) [137b]. The assembly of a set of curved plates to form a circle, i.e., a tubular element, is another variant of this technique. These plates can be bent outward at the edges to facilitate joining, or intermediate ribs can be used to hold the curved plates together.

4.2.4 Operational Experience

4.2.4.1 <u>Roll-Bonded Dispersed Fuels.</u> The outstanding performance of various roll-bonded plate-type elements is a consequence of the fortunate combination of fuel element components, bonding method, and operating conditions. The performance of such elements under more severe conditions has been shown in tests where the design capabilities were exceeded either unintentionally because of inadequacies in other components of the system, or intentionally as part of investigations of transient behavior. The following discussion separates aluminum elements (MTR type) and stainless steel elements (SM-1 APPR type).

4.2.4.1.1 <u>Aluminum Cladding and Matrix.</u> Since startup in 1952, the MTR has experienced three instances of fuel assembly failure due to mechanical deficiencies that permitted dimensional changes which could have led to conditions of insufficient cooling [122, 123a, 124]. None of the failures indicated any deficiency in the individual fuel plates. A review of these three failures provides an interesting case history on the design of fuel assemblies with proper consideration of the effects of seemingly minor changes, and on the importance of process control with adequate inspection; relatively little is learned, however, regarding the performance of individual fuel elements under conditions beyond their design capabilities, since only the first failure led to rupture of the cladding with release of activity to the coolant. As noted above (see Sec. 4.2.1), this failure (1954) resulted from decreased flow and the consequent pressure differentials. The later failures (1960 and 1961) were due to poor process control in the assembly to the side plates by brazing and roll-swaging respectively. The poor brazing left some of the fuel plates very poorly bonded to the side plates. The fuel plates deformed and experienced excessive surface temperatures, but no activity was released. As noted earlier, brazing was subsequently replaced by mechanical assembly to retain the strength of cold-worked aluminum. When some of these mechanical joints were not properly made, deformation occurred in the side plates but not in the fuel plates, which showed no sign of excessive temperature.

The initial development work on the flat ETR plates revealed mechanical instabilities, which were subsequently overcome by elimination of brazing for assembly and by adjustment of the different channel spacings. The only in-pile failure of ETR elements occurred in December 1961 because a Lucite sight-box had been left in the reactor after refueling [140, 201f]. The consequent obstruction reduced coolant flow to about 35% of normal in six fuel elements and resulted in the melting of eighteen fuel plates. Activity was released to the water and some of these plates were fused together, but "no evidence of a metal-water reaction was observed or found during or after the incident". In a less serious incident in November 1962, one MTR plate melted partially as a result of restriction of coolant flow by debris from a gasket [141, 201g].**

The BORAX and SPERT tests have shown the behavior of the fuel plates in excursions [201h]; the behavior of the fuel plates in the SL-1 accident is also relevant. With increasingly severe conditions, the plates deform or bow and can then melt. In the destructive experiment on BORAX-I conditions were severe enough to melt the fuel plates, which in previous tests had been plastically deformed by steam pressure. Contact of the molten metal with the coolant generated enough steam pressure to burst the reactor tank. Fuel plate fragments were recovered in different forms, including spongy metallic globules and plates that had been molten only inside, while the cladding had remained solid [57b]. There was little evidence of chemical reaction between the metal and the water. Similarly, in the SL-1 accident some of the fuel elements melted and generated high steam pressures, but chemical reaction does not seem to have contributed significantly to the energy release [142, 201a]. Tests of SPERT-2 elements provide interesting examples of the increasing damage with shorter reactor periods [143, 144]. Thus, with a period near 70 sec, bowing occurred as a result of pressure and/or temperature differentials. With periods near 5 msec, the damage ranged from melting to gross fractures to small fractures revealed by dye penetrant tests. Tests with 5 and 4.6 msec periods show a significant difference in the extent

*The presence of boron or lithium in the braze alloy (or flux) may have important adverse effects on the reactivity of the core. In fact, at least one reactor has failed to reach criticality with its first loading due to an underestimate of such effects.

**A similar incident occurred in ORR on July 1, 1963 [211].

of plate damage. Metallographic examination of plates from the 5 msec period revealed the fuel meat melting and alloying with the clad. The melted fuel shows reduction in thickness with voids and stringers. The 6061 cladding shows intergranular cracking.

Penetration of the X8001 aluminum cladding of a prototype SL-1 plate in an MTR test was attributed to conditions that aggravated the local corrosion of the aluminum: high pH (10) and a scale deposit. The exposed core alloy corroded slowly enough to permit a delay of two days between failure and shutdown. (Compare with CP-5 failure, Sec. 4.3.4.1). Postirradiation examination revealed two blisters filled with corrosion products. The presence of 2 wt.% Ni was believed to have improved the fuel alloy's corrosion resistance and thereby to have reduced the progress of failure [137c, 145].

4.2.4.1.2 Stainless Steel Cladding and Matrix. Thanks to the higher operating temperatures provided by stainless steel and UO_2, few examples are available of the operation of SM-1 (APPR-type) elements under conditions beyond their design capabilities.

APPR-type plates were used for initial operation of the OMRE, since they are "of proven radiation stability and fabrication feasibility" [139]. One of the assemblies failed not because of any inadequacy in its design or fabrication but because of improper latching into the grid plate [146]. The resulting insufficient cooling led to fouling of the surface by the organic coolant, ultimately leaving a massive deposit of carbonaceous material that was able to deform the plates, force them together, and embrittle them. The extent of fission product release due to this sequence could not be ascertained because of the concurrent failure of experimental aluminum-clad plates (see below) and damage to the assembly during forcible removal from the reactor. The damaged plates showed no evidence of swelling or blistering.

Blistering as well as distortions described variously as bowing, warping, and rippling has been observed in SPERT tests of APPR-type fuels [147, 148]. These distortions are attributed to the high temperature differentials resulting from the combination of high rate of heat generation and low conductivity of stainless steel. The cause of the blistering is uncertain but it seems to be related to the oxide stringers of the core, which cause localized heating, leading to high localized thermal stresses and possibly even to local melting of the steel. Since the steel cladding can become susceptible to intergranular corrosion as a result of reaching temperatures of sensitization, the austenitic stainless steel should be stabilized or low in carbon.

4.2.4.2 High-Uranium Fuels. Two examples are available of the failure of a flat fuel element consisting of metallic uranium jacketed in aluminum. Other than the similarities in fuel and cladding, these elements have very little in common. They were jacketed much differently: the NRU element by insertion and sizing (no bond); the OMRE experimental element by gas pressure bonding, giving a metallurgical bond with the help of a 0.5 mil (13 microns) nickel plate. As indicated in the next two paragraphs, the failures differed considerably in cause and consequence.

The NRU incident of May 23, 1958, provides a dramatic example of the consequences of using a fuel element whose integrity is suspect* [149, 150]. It is one of the rare instances in which a fuel element deficiency can be assigned major responsibility for an incident [201i]. The mode of failure of the element shows how lack of bonding of a metallic fuel can magnify the consequences of entry of coolant through a cladding defect (e.g. at a weld). Fuel corrosion compounds the consequences of waterlogging, which permits dead space within the cladding to become filled with water; subsequently this water is converted to steam which can generate high pressures. Thus, in the May 1958 incident, violent failure of the fuel element was indicated by a pressure transient, which was due to rapid formation of steam probably supplemented by generation of hydrogen from aqueous attack on metallic uranium and UD_3 formed as a result of earlier corrosion. Subsequently the distorted state of this element negated proper operation of the mechanism for removal and transfer of the fuel. Consequently part of the element burned and released large quantities of fission products within the reactor building.

The failure of elements in two experimental assemblies in OMRE was due entirely to overheating that resulted from blockage of coolant flow by coarse particulate matter that was strained by the aluminum fins [146, 151]. Such overheating is aggravated by the fouling of the decomposing coolant. In spite of this serious overtaxing of the elements, damage (melting or blisters) was localized. In fact, this test showed that these bonded elements could withstand conditions exceeding the design limits: swelling was not observable in regions where the temperature had considerably exceeded the 750°F (400°C) design temperature.

4.3 Tubular Fuel Elements

Tubular fuel elements can be as effective as plate-type elements for heat transfer and can in fact be visualized as plates that have been bent - one of the methods used for fabrication. While as effective as plates for heat transfer, tubes have the advantage of inherent rigidity and lack of distortion when subjected to a pressure differential as in a pressure tube reactor. The need for extra supporting materials is reduced and longer elements can be used, thereby reducing the number of fuel units and simplifying fuel management. As is seen in Table 4-4, tubular elements are fabricated by a variety of methods. In a sense, tubular elements can be considered a cross between the two previously discussed types: the performance advantages of

*The element whose failure is discussed here was cold-die-sized and did not have a core-cladding bond. Conversion to the extrusion-clad nickel-plated elements was completed in February 1960; fueling since then has been confined to this type of element. (Table 4-3).

TABLE 4-4

Characteristics of Representative Tubular Fuel Elements

(Abbreviations are given at end of book. In this table an asterisk means "See last column for notes".)

Reactor and Coolant	Power (Mw)		Dimensions of Tube (in.)				Core			Cladding		Fabrication Method	Assembly	Extent of Operating Experience		Notes
	Ther-mal	Elec., (net)	Length	OD	ID	Wall (mils)	Composition	U^{235} (wt. %)	Thick-ness (mils)	Com-position	Thick-ness (mils)			Crit.	Full Power	
Piqua Terphenyl	45.5	11.4	Outer 13.5 Inner 13.5	mean 4.63 mean 3.60			U-3.5 wt.% Al-0.1 wt.% Mo	1.94	208	1100 Al	35	Gas pressure bonding. 1 mil Ni plate	4 sections in tandem, each containing 2 concentric tubes, form each of 85 elements	11/63	11/64	Al has twisted fins; contacting fins (from inner and outer tubes) twisted in opposite directions.
CP-5 D_2O	5		31 5/32 27 5/16 27 5/16	A 3.000 B 2.640 C 2.236	2.900 2.516 2.112	62 62 50	Al-16.5 wt.% U Al-24.1 wt.% U Al-24.1 wt.% U	93 93 93	32 32 20	1100 Al	15	Coextrusion with integral end seals	2 inner tubes dimpled outward at 3 points at each end and welded to surrounding tube	2/54		Initially operated with MTR-type plates.
WTR H_2O	60		36 (active)	A 2.500 B 2.063 C 1.625	2.250 1.813 1.375	125	Al-13 wt.% U	93	52	Al	36.5	Assembly of roll-bonded plates bent and brazed at mating edges	Held together by 3-finned end fitting on either end	59 failure 4/4/60 shutdown 3/62		
HWCTR D_2O	~70		Drivers 113 (core)	2.300	1.966	167	Zr-9.3 wt.% U	93	137	Zy-2	15	Coextrusion with integral end seals	Ring of 24 driver tubes surrounding 12 test positions	3/62 Shutdown 11/64		
PRTR D_2O	70 Mark II*		88 (active)	A 3.068 B 1.902	2.208 0.962	430 470	UO_2 UO_2		310 350	Zy-2 Zy-2	60 60		Fastened to bracket at each end, spaced by ribs on cladding			Also have axial fuel rod with 0.668-in. dia. *See Table 4-1 for Mark I.
PM-1 (and PM-3A) H_2O 1300 psia	9.37	1.0	33 1/4	0.506	0.416	45	304L ss-28 wt.% UO_2	93	28	347 ss	8.5	Drawing and isostatic pressing of 3 concentric components	121 tubes in each of 6 peripheral bundles. 15 tubes in one central bundle	2/62 3/62	4/62	
Pathfinder* (Sioux Falls) superheater H_2O	188.9 total	58.5	72 (active)	A 0.839 B 0.630	0.769 0.560	35 35	316L ss-17.5 wt.% UO_2	93	20	316L ss	7.5	Same	Total of 415 superheating elements consisting of 2 concentric tubes and axial poison rod in ss process tube. 3 full length straight spacer wires attached to outside of poison rod and each fuel tube	1964		*Boiler fuel is Zircaloy-clad UO_2 pellets.
GCRE	2.2		28 (active)			45	316 ss-30 wt.% UO_2	93	33	318 ss	6	Assembly of roll-bonded plates	3 plates curved to form tube held by ss clips 4 concentric tubes for each of 61 assemblies	6/60 Shutdown 4/61 after pressure vessel failure dismantled 1962		
Dounreay NaK	72	12	6 x 3 (+ 6 Nat. U)	0.79	0.30		U-9 wt.% Mo	45.5	250	Nb	40	Na bond	Total of 345 core elements, 250 containing enriched U (also 1872 blanket rods surrounding core)	11/59	8/62	

plate-type elements are combined with the fabrication advantages of cylindrical elements.

For test reactors, the axis of the tubular elements provides an accessible position for fast flux irradiation. Both for test reactors and power reactors, a series of concentric tubes can be used; the number has reached six for test reactors (BR-2), [137d] and even more for the direct-cycle reactor studied for aircraft propulsion.

The limited use of tubular elements to date is reflected in the following discussion of various elements.

4.3.1 Features Affecting Design and Fabrication

Tubular geometry introduces few features not already cited in Secs. 4.1.1 and 4.2.1 for other geometries. It is worth noting that an exposed internal surface introduces some problems in inspection, especially in a long tube, but these problems are surmountable.

4.3.2 Fabrication of Tubular Elements

4.3.2.1 Rolling of Plates. The forming of roll-bonded plates into curved segments and their assembly into tubes is an extension of the previously mentioned bending into curved shapes (see Sec. 4.2.3). It is customary to join three plates to form the circle. These plates can be joined to each other directly at the edges (usually bent outward) or through an intermediate length of metal. Joining can be effected mechanically or metallurgically. UO_2 -stainless-steel plates as well as U-Al have been rolled into tubes. For the direct-cycle reactor studied for aircraft propulsion, ribbons of Inconel-clad UO_2-Iconel were rolled into complete circles.

4.3.2.2. Fabrication From Cylindrical Components. Various fabrication methods can be used in assembling the three cylindrical components (inside cladding, core, and outside cladding) of tubular fuel elements. In preparing these assemblies, it is likely to be most convenient to slip the fuel over the inner cladding component. In the following summary of the subsequent processing of an assembly, the methods are separated according to the extent of working that the fuel element undergoes after assembly of the components.

The simplest method involves slip-fitting of the components, e.g., oxide pellets or powder for the PRTR tubes [1ee]. The assembly can be left as assembled or it can be subjected to working, e.g., drawing or swaging. With a metal fuel, metallurgical bonding can be achieved by hot-press bonding or by gas-pressure bonding (see Sec. 4.2.2.2). The aluminum-clad Piqua elements were bonded at 1000°F (540°C) under 7000 psi (490 kg/cm^2) pressure [135]. Liquid-metal bonding is also applicable. Extensive working of the cladding is achieved in extrusion cladding (see Sec. 4.2.2.2); both inside and outside aluminum cladding can be formed by being fed concentrically about the simultaneously fed uranium core, which does not undergo deformation.

Drawing of the assembled elements is one of the features of the process developed by Martin Company for tubular dispersion-type elements assembled with commercial tubing for the cladding [152, 162, 163]. The cylindrical fuel insert can be (1) a tube formed by rolling plate or strip prepared like the flat fuel used for a picture-frame assembly (see Sec. 4.2.2.1), (2) a series of bushings produced by pressing and sintering, or (3) a full length tube prepared by drawing a powder blend in an expendable can and sintering. With powder metallurgy techniques used for these types of fuel core it is possible to obtain a graded fuel composition and to introduce a poison.

The assembly is drawn tightly over a mandrel by cold drawing. The components are in intimate contact and are ready for the diffusion bonding. This one-hour treatment is carried out at 1150°C (2102°F) in dry hydrogen for stainless-steel-UO_2 and 620°C (1148°F) in air for aluminum-UO_2. Sintering and recrystallization across the core-cladding interfaces is facilitated by the prior cold-work. Aluminum elements are strengthened by being subjected to additional cold work after the diffusion anneal.

Coextrusion, involving large dimensional changes for the fuel and the claddings, can be applied to various combinations of fuel and cladding. Aluminum-clad aluminum-uranium alloys are used in CP-5 [1ff, 22f, 221]. Zircaloy-clad zirconium-uranium was used for the driver tubes of HWCTR [43, 153, 217]. High-uranium metal test elements for this reactor were also prepared by coextrusion [154, 212]. Some reactor concepts have also involved long metal elements cooled only internally.

4.3.3 Assembly of Elements

Tubular elements can be used individually or in concentric arrays. In the latter case, appropriate provision has to be made to maintain the spacing. For spacing along the full length of the tubes, the cladding may have integral ribs (Piqua), wrapped wire, or straight wire attached at the end (Pathfinder). For spacing at the ends, inner tubes can be dimpled (if the end seal is relatively soft, like aluminum) and welded to the surrounding tube (CP-5).

4.3.4 Operational Experience

4.3.4.1 Aluminum Cladding and Matrix. A fresh fuel element failed at the WTR on April 3, 1960, during a series of studies undertaken to determine the effect of reduced flow of coolant and incipient boiling on reactor stability [155, 156, 157, 201d]. Some of the fuel melted, so that only the top third of the element could subsequently be removed, but there was apparently no evidence of significant metal-water reaction. The fuel element in question was probably of inferior quality but there is room for debate as to whether this inferior element initiated the failure or compounded any difficulty that may have arisen from other causes. Whatever these other causes and their contribution to the accident, it is hardly likely that an initially satisfactory element would have failed so badly under the conditions at the time of the accident. Subsequent examination of 237 unused cold tubes from the same batch of fuel revealed the presence of various defects "including poor bonding, cracks in the fuel, foreign inclusion, and voids". The key inspection was the ultrasonic test for non-bonding; the results of this test, which had not been used previously, were confirmed by representative sections. The unbonded areas had diameters ranging from about 15 mils (0.37 mm) to over 1 in. (25 mm). The voids were not necessarily between the fuel and the cladding; voids in the fuel alloy could have been caused by hydrogen introduced by poor melting practice. Heat transfer calculations showed that an unbonded area with a diameter greater than 0.5 in. (13 mm) could have accounted for the failure. It should be unnecessary to belabor the lesson regarding the importance of using fuel elements that have been properly fabricated and inspected.

SPERT tests of enriched Al-31 wt.% U tubes clad with ribbed aluminum by coextrusion were less dramatic but provided more useful information about the operation of such elements under abnormal conditions [158]. With different coolant velocities and temperatures, the time after burnout for the fuel elements to reach the melting point (640°C or 1184°F) was varied from 1.3 to 4.9 sec. The reactor was scrammed at different times after the melting, the maximum interval being 7.9 sec. Pressure pulses were produced at a frequency of about 1/sec with peaks of about 150 psi (10 kg/cm^2). Subsequent examination showed that the metal had flowed down the sides of the element; the inside cladding had suffered more damage than the outside cladding. Some metal had been dispersed with the coolant, from which it was recovered as particles; in the most severe meltdown, the particles varied from tiny flakes to a jagged agglomerate. No evidence was seen of any chemical reaction.

One of the coextruded CP-5 elements was found to have developed a defect in 1961. While in the reactor from February 1960 to October 1961, it achieved a burnup of 61 wt.%. During the last four months before removal from the reactor, the presence of a defect was shown by the abnormally high fission product activity in the coolant. Examination after removal from the reactor revealed little damage to the tube: a blister, 0.5 in. x 1 in. (13 mm x 25 mm) on the inside surface of an innermost tube; and a circular ring of white scale on the outside surface directly opposite the blister area. The sequence of events was hard to deduce but it appears reasonable that water penetrated through a defect on the inside and the ensuing water-fuel reaction led to swelling of the cladding. The scale on the outside surface was then a consequence of the high heat flux resulting from the defect on the inside [159, 160, 161].

4.3.4.2 Stainless Steel Cladding and Matrix. A prototype PM-1-type fuel element failed during a 1960 test in SM-1. Subsequent examination revealed that two of the fourteen fuel tubes in this element had blisters on the inside. The extensive evaluation established the failure as being due to waterlogging: water entered through defective end closures and migrated through the permeable cermet core; this water was converted to steam upon startup of the reactor (see Sec. 4.1.5.3) and generated the blisters. In light of these findings, the fabrication process was modified in several ways to increase the resistance to waterlogging by providing more reliable end closures and improving the soundness of the fuel cermet. The reliability of the end closures is improved by 1) welding the dead-end (end plug) seams longitudinally prior to the cladding operation, and 2) welding the ends circumferentially after the cladding operation. The soundness of the fuel cermet is improved by 1) hot rolling the powder strip in a stainless steel envelope before assembly with the dead ends and the claddings and 2) hot isostatically pressing the tubes after the diffusion bonding. The modified process was applied in the fabrication of the fuel elements for both PM-1 and PM-3A. Further improvements in such fuel's performance are expected from precoating the UO_2 particles with stainless steel [152, 163].

4.3.4.3 Zircaloy-Clad Metal Fuel. As part of the investigation of the potential use of high-uranium metal fuels for water-cooled power reactors, extensive defect tests have been performed out-of-pile on coextruded Zircaloy-clad tubes of unalloyed uranium [164] and U-2 wt.% Zr [78b, 165, 166]. These tests have shown that even in the 300 to 350°C (572-622°F) range, the consequences of exposure of the fuel to water are not necessarily catastrophic. Corrosion occurs at a rate that permits detection of fission products (indicating onset of failure) early enough so that the operator is alerted to the need to shut down the reactor, although he need need not do so immediately. He can afford to wait a reasonable period before effectint shutdown. Similar but less extensive out-of-pile tests have been performed on Zircaloy-clad thorium [167, 168].

Little has been reported on the irradiation behavior of Zircaloy-clad high-uranium metal tubes. Two such elements failed in the HWCTR at the end of 1962 after exposure of less than 1000 Mwd/T. These failures were apparently due not to any metallurgical deficiency in the fuel but rather to a mechanical problem: the spiral Zircaloy ribbon used to separate the coaxial tubes was vibrating and wearing through the cladding. Such fretting corrosion was found on the surfaces of elements from out-of-pile flow tests and also on other irradiated elements removed from the HWCTR before failure [169, 170]. The HWCTR test program included irradiation of single tubular elements to burnups in the vicinity of 10,000 Mwd/T [213].

The first loading of driver tubes for the HWCTR, containing a high zirconium fuel like the PWR seed plates, behaved very well [171, 214, 217].

4.3.4.4. Zircaloy-Clad Oxide Fuel. The HWCTR program to evaluate fuels for heavy water reactors included tests of tubular elements of mechanically compacted UO_2. Some of these elements failed at the higher thermal ratings when the coolant pressure (1000 to 1200 psi or 70-85 atm) caused collapse of the outside Zircaloy cladding (2.1 in. or 5.33 cm outer diameter, 30-mil or 0.76 mm wall) and formation of a ridge over vibrationally compacted fuel, which has relatively low density (82 to 87% of theoretical). Such collapse did not occur with swaged fuel (density 89 to 90% of theoretical);

a ridge indicating incipient collapse was observed at the end of one swaged element, where the local density was low. With the more complete collapse in the low-density elements, the lengthwise ridge was able to crack at its apex, exposing the fuel. The fission product release increased with thermal rating but appeared independent of the density or the method of compaction [172].

Earlier in the heavy-water reactor program, before the HWCTR was available, fuel elements were tested in a Savannah River Plant reactor [173]. Four failures of swaged Zircaloy-clad elements were found to involve brittle failure of the outside cladding. In the proposed mechanism, failure is initiated by hydriding of the internal metal surface at a few locations; the hydrogen is present as moisture or another contaminant in the fuel and residual fluoride from the etching may be responsible for localized attack. (The situation is quite similar to that in PRTR Mark-I failures, see Sec. 4.1.5.3.2). The hydrogen diffuses towards the external surface and, during cooling, precipitates as hydride platelets in an oriented pattern determined by the stress pattern in the Zircaloy. This hydride provides a fracture path. Early in the development of the fabrication process, stainless steel was used as a stand-in for the Zircaloy cladding. Failure of some of these elements was apparently initiated at defective weld closures and was followed by cracking of the outside cladding in a manner suggesting stress corrosion.

4.4 Graphite-Base Fuel Elements

Graphite-base fuel elements are here considered separately because they are unique in that it is expected that some fission products will escape from these elements into the coolant stream. In two applications (HTGR, Dragon) control of fission product contamination of the coolant depends on the absorption of fission products either within the fuel element or in external fission product trapping materials [181].

Graphite-base fuel elements will be used in at least four reactors in the late 60's. These are the HTGR [174], the Dragon reactor [175], the AVR [176], and UHTREX [177]. All these reactors have at least three features in common; namely, they will operate at high temperature (exit coolant temperature will be 1400°F or higher), the fuel proper is a uranium dicarbide* and the fuel diluent is graphite. Since more experimental work is available on the HTGR the features of its fuel element will be presented as an example of this type of fuel element.

In the HTGR the fuel consists of a solid solution of UC_2 in ThC_2 dispersed in a graphite matrix. The dispersed carbide particles have a 200-400 micron (8-16 mils) diameter. The mixture of carbide and graphite is formed into compacts (oversized hollow core pellets) by a hot pressing operation. Basically the fuel element consists of these compacts encased in a graphite tube which has been treated (impregnated with carbon) to reduce its permeability to He to about $10^{-3} cm^2/sec$.

*The dicarbide is stable in contact with carbon.

The fissile-fertile solid solution carbide is not dispersed uniformly in the graphite compact. Rather it is concentrated in an annular ring which fits on a central graphite spine, thereby reducing the temperature and temperature gradient to which the carbides are subjected in operation. As pointed out in section 3.1 this is done to reduce migration of the uranium or thorium. Since uranium and thorium carbides react readily with moisture, which would make assembly and handling of the fuel elements difficult, the carbide particles are coated with carbon by decomposing a hydrocarbon in a fluidized bed chemical reactor. It has been shown that properly applied coatings of pyrolytic carbon not only prevent reaction with moisture but also retain the fission products generated in fission. A triplex coating consisting of inner layer of amorphous (sooty) carbon and outer layers of laminar and columnar graphite shows the best retention of fission products. The amorphous inner layer minimizes the effect of fission recoils and reduces the radiation-induced stresses in the outer coatings. In the HTGR fuel, a single layer of coating is used.

In the base of the low-permeability graphite cladding tube there is a section in which fission product trapping materials are contained. Also the bottom of the cladding tube has an opening which on insertion into the reactor is connected to the reactor external fission product trapping system. Helium coolant gas is sucked into the cladding tube by maintaining the trapping system at a pressure lower than the coolant gas. Thus the fuel compacts are continually bathed by helium which flows through the fission product trapping materials contained in the base of the fuel element and thence into the large external fission trap. Thus during reactor operation fission products which are not contained by the pyrolytic coatings are picked up by the helium and carried to the two trapping systems. The permeability of the outer graphite tube has been selected so that the helium can flow into the interior of the tube along its length. Some fission products, however, do get out into the coolant stream by either a process of back diffusion against the He flowing in through the pores or by bulk diffusion through the lattice of the graphite cladding tube.

A quarter-length full-diameter HTGR element [178] has been irradiated to determine the behavior of such an element. It was shown (1) that a low level of radioactivity is maintained in the primary system (loop), (2) that the trapping system cleans up the helium to extremely low levels of both radioactive and chemical contaminants, (3) that there is ready access to the system and (4) that radioactivity in both the main loop and the trapping system is constant as a function of time.

The results of the above test indicate that graphite fuel elements are extremely promising for high temperature reactors. It should be pointed out, however, that they are subject to failure from three main sources. (1) Radiation damage to graphite at elevated temperature can cause serious contraction either in the fuel compacts or the containing tube unless process variables are controlled to minimize this undesirable irradiation behavior. (2) The helium coolant must be controlled at a low oxygen level to avoid oxidation and mass transfer of carbon. (3) Leaks of water or water vapor from the gas to water-steam heat exchanger cannot be tolerated since carbon reacts with water at elevated temperature to give hydrogen and carbon monoxide (water-gas reaction). A water vapor detection system is important to such reactors which transfer their heat directly to the steam-producing system. For further details on the graphite fuel elements of the other reactors, the reader should consult the following references [179-183].

5 FLUID FUELS

5.1 Introduction

Fissionable materials in fluid form at reactor temperatures have been studied primarily because reactor designers would like to couple closely the chemical processing of a fuel with reactor operation. The advantages, disadvantages, features, and development work on such reactors have been completely covered in reference [184], the volume on Fluid Fuel Reactors prepared for the Second U.N. International Conference on Peaceful Uses of Atomic Energy in Geneva in 1958. In the following the presentation is limited to the systems which have received most attention and to the safety problems which have become apparent in the study and operation of these fluid systems.

5.2 Fuel Systems

There are three types of systems which have been studied: (1) water base, (2) metallic, and (3) molten salt. The first group includes the completely homogeneous systems consisting of UO_3 dissolved in water solutions of nitric acid, sulfuric acid, phosphoric acid, UO_2 in concentrated phosphoric and the heterogeneous systems which are uniform dispersions of UO_2 or UO_2-ThO_2 in water.

The Los Alamos water boiler [184a], a solution of UO_3 in HNO_3, was the first fluid fuel reactor. This fuel was replaced at ORNL in their work by UO_3 in H_2SO_4 because of lower parasitic absorption of the latter. This fuel has been the one used in the Homogeneous Reactor Experiments (HRE-I and HRE-II). Because these fluid fuels are unstable above 300°C (572°F), UO_3 dissolved in aqueous H_3PO_4 [184b] (30-60 wt.% H_3PO_4 pressurized with O_2) was studied followed by a study of the UO_2 in concentrated H_3PO_4 [184c] (95 wt.% H_3PO_4 pressurized with H_2). The latter solution has a low vapor pressure, 800 psi (56 kg/cm^2) at 450°C (842°F).

The heterogeneous UO_2-H_2O fuel (slurry) has received a great deal of attention in Holland [185, 186, 187]. UO_2 and ThO_2, H_2O slurries have been studied extensively at ORNL [184d].

The two molten metallic systems* which have been studied are the solution of U in Bi [184e] (also the dispersion of $ThBi_2$ in Bi) and the molten plutonium eutectic fuel [27, 184f, 184g, 188, 189]. Pu forms low temperature eutectics with Fe, Co and Ni. The 9.5 at.% Fe eutectic which melts at 409°C (768°F) was chosen as the fuel for LAMPRE.

The molten salt fuels developed at ORNL [184h] are complex mixtures of solvent fluoride salts usually made up of appropriate amounts of LiF, BeF_2 and ZrF_4 to which the fissile UF_4 or fertile ThF_4 is added. A possible composition of the Molten Salt Reactor (MSR) is 70.7 mole % LiF, 23 mole % BeF_2, 5 mole % ZrF_4, 1 mole % ThF_4, 0.3 mole % UF_4. A unique feature of the molten salt systems is that breeding blanket compositions can be formulated which are completely homogeneous at operating temperatures.

*Dispersions of UO_2 in Na have been studied at ANL.

5.3 Safety Considerations in the Use of Fluid Fuels

As pointed out in Secs. 1 and 3.3, nuclear safety (as we have defined it) is compromised if radioactive material escapes from a fuel element cladding where fuel elements are used, or escapes from the primary containment where fluid fuels are used. It must be recognized that, when fluid fuels are used, one fission product barrier (cladding) is eliminated. The fluid fuel in most designs circulates throughout the reactor system and the hazard associated with a failure anywhere in the primary system is extremely serious. Corrosion problems are no longer primarily centered in the reactor core. Rather, corrosion failure is of primary concern anywhere in the fuel-circulating system. In these systems not only are the β-emitting fission products circulating but also the delayed neutron-emitting fission products circulate. The latter make it possible for the whole circulatory system to become activated. Another type of hazard unique to the circulating fuel reactors is the one which has so aptly been labeled "Where is the uranium?" In a circulating fuel reactor serious reactivity changes can be brought about by a hold-up or deposition of the fissile phase in some part of the circulatory system followed by release of this material and its reintroduction into the reactor core. On the other hand, such reactors do have a large negative temperature coefficient due to thermal expansion of the fluid fuel**; they can be operated with a small excess reactivity because fuel composition can be adjusted during operation and because xenon gas can be removed as it is formed, thus largely eliminating a major reactivity poison effect; and they do not need to operate with a large fission product inventory if continuous chemical processing of the fuel is achieved.

The foregoing relates to the general safety problems of all fluid fuels. Now we shall consider the individual problems associated with each of the fluid fuels presented.

In the uranyl sulfate fuels there are three important problem areas: (1) fuel stability, (2) radiolytic decomposition and (3) corrosion. With respect to (1) it should be noted that if the acidity of the solution falls below a fixed value, UO_3 can precipitate.

**Moreover, this expansion is rapid in view of the fact that the heat transfer from fissile material to moderator-coolant is very rapid.

In addition, if the temperature of such solutions approaches 300°C or 572°F (the temperature depends on composition) the liquid separates into two phases. With respect to (2) the basic problem is that fission fragments are generated throughout the fuel volume and they decompose water into H_2 and O_2. The generation of large volumes of a possibly explosive mixture of gases is a serious problem. This problem has been partially solved by the addition of cupric (Cu^{++}) ion to the fuel solution which acts as a homogeneous catalyst for the recombination of the hydrogen and oxygen. The problem area (3) results from the fact that acidic uranyl sulfate fuel solution is a highly corrosive liquid. Although stainless steel, Zircaloy and titanium are moderately corrosion-resistant to this fuel solution, no really completely satisfactory containment material of low thermal neutron cross section has been developed. The perforation [190] of the core tank of HRE II has been explained as resulting from a localized deposition of UO_3 which melted and/or corroded the Zircaloy tank.*

The uranium-phosphoric-acid fuels have one very serious drawback. They are extremely corrosive and can be contained only in noble metals. Flaws in the gold coating of the stainless steel piping of the LAMPRE I and LAMPRE II required that these experiments be shut down because corrosion perforated the tubing.

Although there has not been much operating experience with the H_2O-UO_2 slurry systems certain problems have been recognized as important. These are: stability against settling in quiescent areas in the fluid circuit, resuspension of the slurry, change in particle size, and erosive-corrosive attack on materials of construction. In addition, it is necessary to maintain a hydrogen atmosphere over UO_2-H_2O slurries to prevent oxidation of UO_2. Since these are water slurries and the energy of fission fragments is dissipated in the water moderator, radiolytic decomposition of H_2O is an additional problem as in all water-base fluid fuels.

In the fluid fuel U-Bi the main problems again are fuel stability, corrosion and impregnation of the graphite moderator. Magnesium was added to the U-Bi fuel to prevent the oxidation of U by oxygen and nitrogen. To prevent the reaction of uranium with the graphite moderator and to reduce the mass transfer and corrosion of iron-base containment materials, zirconium was also added to the U-Bi fuel. Since the fuel permeates the graphite moderator structure and wets it, it was proposed that the moderator be impregnated with a Bi-Mg-Zr molten alloy prior to the introduction of the uranium to the solvent metal alloy.

In the molten plutonium fuels the main problems recognized to date are corrosion of the containment material (tantalum) and variation of the density of the molten fuel. By careful selection of the tantalum, i.e., arc-cast or electron-beam-melted material, the penetration of the tantalum by the molten PuFe alloy is essentially nil below 600°C (1112°F). However, if oxygen gets into the tantalum, for example, at a weld, the fuel alloy can penetrate the tantalum. Another aspect of the oxygen problem is the possible transfer of oxygen from the sodium coolant to the fuel alloy through the tantalum canning material.

The density of molten plutonium fuel varies as a result of the generation of gaseous fission products which gives rise to changes in nuclear reactivity.

Apparently if there is a [191] scum on the molten fuel alloy, the gases form with the scum a low density mass which floats on the surface of the fuel. This effect was noted in the LAMPRE fuel elements in which small additions of carbon were made to the fuel alloy in an effort to reduce attack on the tantalum can by the fuel alloy.

In the molten salt fuels the problems are fuel stability, freezeup of the salt fuels (they have a high melting point), corrosion of containment materials and possibly, generation of F_2. Fuel instability is due primarily to impurities which may be present in the fuel or as a consecutive reaction of corrosion. Impurities in the melt or corrosion may result in the reduction of UF_4 to UF_3. Excess UF_3 in the salt can disproportionate into UF_4 and U. These reactions can be suppressed by proper purification of the salt mixture, pretreatment of the salt with chromium and the use of the alloy INOR 8** [184i] as a construction material. The corrosion problem which was very apparent (chromium removal and the production of voids) when Inconel was used has been solved by the use of the INOR 8 alloy, which is a composite of the Inconel and Hastelloy B compositions. Although the molten salt fuels are very stable to radiation, fluorine is formed in solidified irradiated salt [192], probably as a result of gamma absorption by the solidified salt. Carbon tetrafluoride has been formed in some irradiation tests in which graphite (the proposed moderator for the Molten Salt Reactor) was a component. Whether the production of fluorine by radiolytic decomposition results in increased corrosion has not been resolved to date.

In summary we may say that the fluid fuels are safer in one respect than solid fuel elements since the fission product inventory may be kept lower. However, they must be considered more hazardous than solid fuel elements because the radioactive fission products are distributed throughout the system. What one buys in safety by operating with a smaller excess in reactivity is probably offset by the difficulty in knowing where the fissionable material is at all times. Nevertheless, these reactor fuels are extremely attractive for breeding reactors and development work must continue to be directed at solving the problems which have been recognized. It may be that from such work designs will be evolved which marry the best features of fixed fuels and circulating fuels, i.e., fuels which are fixed during reactor operation and fluid or fluidizable for fission product removal.

*The high concentration of fuel in a localized area results in high local heat generation.

**This alloy is also subject to a marked reduction in uniform elongation due to (n, α) reaction on boron present as an impurity.

ACKNOWLEDGMENTS

Although the opinions and viewpoints in this chapter are those of the authors, we would like to acknowledge the many helpful suggestions and information supplied to us by the following people: Dr. W. K. Barney, General Electric Company, Knolls Atomic Power Laboratory, Schenectady, New York; Dr. B. W. Dunnington and Mr. R. T. Huntoon, du Pont de Nemours Co., Inc., Wilmington, Delaware; Mr. B. R. Hayward, and Dr. C. E. Weber, Atomics International, Canoga Park, California; Dr. A. R. Kaufmann, Nuclear Metals Div., Textron Incorporated, Concord, Massachusetts; Dr. B. Lustman, Bettis Atomic Power Laboratory, Pittsburgh, Pennsylvania; Dr. W. C. Francis, Phillips Petroleum Company, Idaho Falls, Idaho; Mr. J. H. Kittel, Argonne National Laboratory, Argonne, Illinois; Dr. J. S. Kane, Lawrence Radiation Laboratory, Livermore, California; Dr. R. A. Meyer, Dr. D. Ragone, and Mr. D. E. Johnson, General Atomic, San Diego, California; Dr. W. E. Roake, General Electric Company, Richland, Washington; Mr. J. L. Scott, Oak Ridge National Laboratory, Oak Ridge, Tennessee; Mr. J. M. Simmons, U.S.A.E.C., Washington, D. C.; Dr. B. Weidenbaum and Mr. T. J. Pashos, General Electric Company, San Jose, California; and Mr. A. A. Shoudy, Jr., APDA, Detroit, Michigan. We would also like to thank Dr. B. Lustman and Dr. A. R. Kaufmann for having read this chapter and suggested changes.

REFERENCES

1. A. R. Kaufmann (Ed.), Nuclear Reactor Fuel Elements, Metallurgy and Fabrication, Interscience Publishers, Inc., N.Y., 1962.
a) p. 592 and p. 604; b) p. 302; c) p. 331; d) p. 590; e) p. 350; f) p. 472; g) p. 262; h) 466; i) p. 474; j) p. 270; k) p. 471;l) p. 478; m) p. 450; n) p. 481; nn) 486; o) p. 490; p) p. 559; q) p. 569; r) p. 595; s) p. 584; t) p. 442; u) p. 493; v) p. 444; w) p. 495; x) p. 447; y) p. 446; z) p. 435; aa) p. 512; bb) p. 601; cc) p. 518; dd) p. 567; ee) p. 429; ff) p. 458.
2. C. E. Weber and M. Balicki, "Fuel Element Design", Chapter 5 in Reactor Handbook, 2nd ed., Vol. IV, "Engineering", Interscience Publishers, Inc., N.Y., 1963.
3. D. H. Gurinsky and G. J. Dienes (Eds.), Nuclear Fuels, D. Van Nostrand Co., Inc., Princeton, N.J., 1956. a) p. 291.
4. C. R. Tipton, Jr. (Ed.), Reactor Handbook, 2nd ed., Vol. I, "Materials", Interscience Publishers, Inc., N.Y., 1960. a) p. 698.
5. A. N. Holden, Physical Metallurgy of Uranium, Addison-Wesley Publishing Co., Inc., Reading, Mass., 1958.
6. W. D. Wilkinson and W. F. Murphy, Nuclear Reactor Metallurgy, D. Van Nostrand Co., Inc., Princeton, N.J., 1958.
7. J. Belle (Ed.), Uranium Dioxide: Properties and Nuclear Applications, Division of Reactor Development, U.S. Government Printing Office, Washington 25, D. C., 1961.
8. J. E. Cunningham and E. J. Boyle, "MTR-Type Fuel Elements", Proceedings of the First U.N. International Conference on Peaceful Uses of Atomic Energy, Geneva, 1955, Vol. 9, p. 203.
9. S. F. Pugh and B. R. Butcher, "Metallic Fuels", pp. 331-410 in Reactor Technology: Selected Reviews—1964, USAEC TID-8540, Oak Ridge, July 1964.
10. S. Greenberg (Ed.), "Behavior of Cladding Materials in Water and Steam Environments", pp. 213-330 in Reactor Technology: Selected Reviews—1964, USAEC TID-8540, Oak Ridge, July 1964.
11. H. H. Hausner and J. F. Schumar, Nuclear Fuel Elements, Reinhold Publishing Corp., N.Y., 1959.
12. Proceedings of the Symposium on Effects of Irradiation on Fuel and Fuel Elements, Trans. AIME, Nuclear Metallurgy, Vol. VI, 1959.
13. Proceedings of the Symposium on Materials for Gas and Water-Cooled Reactors, Trans. AIME, Nuclear Metallurgy, Vol. VII, 1962.
14. "Reactor News", Nucleonics, 21,5(1963)27.
15. S. V. K. Rao, "Investigation of ThO_2-UO_2 as a Nuclear Fuel", Canadian Report AECL-1785 (CRFD-1154), 1963.
16. L. Weissert, "Development and Testing of Fuel for the Consolidated Edison Thorium Reactor", Report BAW-160, Babcock and Wilcox Co., 1961.
17. L. A. Neimark, J. H. Kittel, and C. L. Hornig, "Irradiation of Metal-Fiber-Reinforced ThO_2-UO_2", USAEC Report ANL-6397, Argonne National Laboratory, 1962.
18. L. Neimark and J. H. Kittel, "Irradiation of Aluminum Alloy Clad Thoria-Urania Pellet", USAEC Report ANL-6538, Argonne National Laboratory, in publication.
19. S. S. Carniglia, "Single Crystal and Dense Polycrystal Uranium Carbide—Thermal, Mechanical and Chemical Properties", Report NAA-SR-Memo 9015, North American Aviation, Inc., 1963; also presented at the International Conference on Carbides in Nuclear Energy held at Harwell, England, 1963.
20. "The Preliminary Irradiation of PuC and UC-PuC", USAEC Report ANL-6678, Argonne National Laboratory.
21. L. A. Neimark and P. D. Shalek, "The Irradiation Behavior of Uranium Monosulfide", Trans. Am. Nucl. Soc., 6(1963)372.
22. Fuel Elements Conference, Paris, November 18-23, 1957, USAEC Report TID-7546.
a. J. E. Cunningham, R. J. Beaver, W. C. Thurber, and R. C. Waugh, "Fuel Dispersions in Aluminum-Base Elements for Research Reactors", Book 1, p. 269.
b. C. Ringot, "Fabrication of the Fourth Set of Fuel Elements for the Experimental Pile EL-2", Book 1, p. 182.
c. J. E. Cunningham and R. E. Adams, "Techniques for Canning and Bonding Metallic Uranium with Aluminum", Book 1, p. 102.
d. M. Gauthron, "Thermopneumatic Cladding", Book 1, p. 18.
e. J. A. Stohr, "Electronic Welding of Metals", Fig. 9, p. 17.
f. R. Montagne and L. Meny, "Coextrusion Applied to the Fabrication of Solid or Disperse Fuel Elements", Book 1, p. 142.
g. A. R. Kaufmann, J. L. Klein, P. Loewenstein, and H. F. Sawyer, "Zirconium Cladding of Uranium and Uranium Alloys by Coextrusion", Fig. 18, p. 175.
h. Book 1, p. 264.
i. J. E. Cunningham, R. J. Beaver, and R. C. Waugh, "Fuel Dispersion in Stainless-Steel Components for Power Reactors", Book 1, p. 243.
j. Book 1, p. 231.
k. A. J. Mooradian, "Aluminum Sheathing of Flat Uranium Plates by Extrusion Cladding", Book 1, p. 121; p. 122.
l. A. R. Kaufmann, J. L. Klein, P. Loewenstein, and H. F. Sawyer, "Zirconium Cladding of Uranium and Uranium Alloys by Coextrusions", p. 157.
23. J. R. Brown, D. C. Pound, and J. B. Sampson, "Hazards Summary Report for the HTGR Critical Facility and Addendum", Report GA-1210 and Add., General Atomic, Division of General Dynamics Corp., 1960.
24. "Experimental Beryllium Oxide Program", Report GA-3053, General Atomic, Division of General Dynamics Corp., 1962.
25. "Aerojet-General Nucleonics, Army Gas-Cooled Reactor Systems Program, Final Hazards Summary for the ML-1 Nuclear Power Plant", Report IDO-28560, 1960 and IDO-28560, Vol. II, Suppl. 1, 1961 Phillips Petroleum Co.
26. C. N. Spalaris, "Finding a corrosion-resistant cladding for superheater fuels", Nucleonics, 21,9(1963)41.
27. G. S. Hanks, R. S. Kirby, and J. M. Taub, "Fabrication of tantalum fuel containers for the LAMPRE-1 Molten Plutonium Reactor", Nucl. Sci. Eng., 14(1962)135.
28. J. N. Chirigos, S. Kass, W. W. Kirk, and G. J. Selvaggio, "Development of Zircaloy-4" in "Fuel Element Fabrication with Special Emphasis on Cladding Materials", Proceedings of an International Atomic Energy Agency Symposium held at Vienna, May 10-13, 1960, Vol. 1, p. 19.
29. E. E. Hoffman, "Corrosion of Materials by Lithium at Elevated Temperatures", USAEC Report ORNL-2674, Oak Ridge National Laboratory, 1959.
30. R. W. Dayton, J. H. Oxley, and C. W. Townley, "Ceramic-Matrix Fuels Containing Coated Particles" in "Battelle Studies of Ceramic-Coated Particle Fuels", Proceedings of a Symposium held at Battelle Memorial Institute, November 5-6, 1962, USAEC Report TID-7654, p. 62, 1963.
31. J. H. Oxley, "Recent developments with coated-particle fuel materials", Reactor Materials, 6,2(1963)1.
32. "Power Reactor Development Company's Enrico Fermi Atomic Power Plant, Revised License Application; Part A, General Information and Request for License; Part B, Technical Information and Hazards Summary Report", Vols. 1 to 6, Report NP-10458; Vol. 7, Report NP-11526, 1961; see also: Directory of Nuclear Reactors, Vol. IV, "Power

Reactors", p. 319, International Atomic Energy Agency, Vienna, 1962.
33. "Enrico Fermi Atomic Power Plant", Report APDA-124, Atomic Power Development Associates, 1959.
34. C. G. Poncelet, "An Analysis of the Reactivity Characteristics of Yankee Core No. 1", Report WCAP-6050, Westinghouse Electric Corp., 1963.
35. "Experimental Gas-Cooled Reactor, Final Hazards Summary Report", Vol. I, "Description and Hazards Evaluation", USAEC Report ORO-586, Oak Ridge Operations Office, 1962.
36. "Hazards Report for Torrey Pines TRIGA Reactor", Report GA-722, General Atomic, Division of General Dynamics Corp., 1959.
37. M. Fox, "The Brookhaven Reactor", Proceedings of the First U.N. International Conference on Peaceful Uses of Atomic Energy, Geneva, 1955, Vol. 2, p. 353.
38. D. E. Thomas et al., "Properties of Gamma-Phase Alloys of Uranium", Proceedings of the Second U.N. International Conference on Peaceful Uses of Atomic Energy, Geneva, 1958, Vol. 5, p. 610.
39. D. Kramer, M. V. Johnston, and C. G. Rhodes, "The influence of freely dispersed precipitation on fission gas bubble formation in uranium-molybdenum ternary alloys", J. Inst. Metals (17 Belgrave Square, London SW1, England).
40. D. Kramer and W. V. Johnston, "Post irradiation annealing of uranium-molybdenum ternary alloys", J. Nucl. Mater., 9(1963) 213.
41. J. A. Brinkman, "Fundamentals of Fission Damage", Proceedings of a Symposium on Effects of Irradiation on Fuel and Fuel Elements held at Chicago, November 4, 1959, Nuclear Metallurgy, Vol. VI, p. 9, 1959.
42. J. H. Kittel et al., "Effects of Irradiation on Thorium and Thorium-Uranium Alloys", USAEC Report ANL-5674, Argonne National Laboratory, 1963.
43. L. M. Arnett et al., "Final Hazards Evaluation of the Heavy Water Components Test Reactor (HWCTR)", Report DP-600, Savannah River Laboratory, E. I. duPont de Nemours and Co., Inc., 1962.
44. W. B. Wright, "Safeguards Report for the EBOR Critical Experiment", Report GA-1100, Suppl. 3, General Atomic, Division of General Dynamics Corp., 1962.
45. S. J. Paprocki, E. S. Hodge, P. J. Gripshover, and D. C. Carmichael, "Fabrication of a Compartmented Flat-Plate Ceramic Fuel Element", in "Fuel Element Fabrication with Special Emphasis on Cladding Materials", Proceedings of an International Atomic Energy Agency Symposium held at Vienna, May 10-13, 1960, Vol. I, p. 283, Academic Press Inc., N.Y., 1961; see also: Directory of Nuclear Reactors, Vol. IV, "Power Reactors", p. 27, International Atomic Energy Agency, Vienna, 1962.
46. W. K. Barney and B. D. Wempler, "Metallography of Irradiated UO_2 Containing Fuel Elements", Report KAPL-1836, Knolls Atomic Power Laboratory, 1958; see also: reference 1, Nuclear Reactor Fuel Elements, Metallurgy and Fabrication, p. 353.
47. J. D. Fleming and J. W. Johnson, "Reactions in Al/U_3O_8 Dispersions", Trans. Am. Nucl. Soc., 6(1963)158.
48. R. E. Nightingale, H. H. Yoshikawa, and E. M. Woodruff, "Radiation-Induced Structural and Dimensional Changes", Chapter 9, p. 286, in Nuclear Graphite, R. E. Nightingale (Ed.), Academic Press Inc., N.Y., 1962.
49. J. N. Wanklyn and P. J. Jones, "The aqueous corrosion of reactor metals", J. Nucl. Mater., 6(1962)291.
50. D. H. Gurinsky, W. T. Warner, J. T. Atherton, C. Binge, H. C. Cook, L. McLean, R. J. Teitel, and B. Turovlin, "The Fabrication of Fuel Elements for the BNL Reactor", Proceedings of the First U.N. International Conference on Peaceful Uses of Atomic Energy, Geneva, 1955, Vol. 9, p. 221.
51. M. F. Notley and J. A. L. Robertson, "Zircaloy-UO_2 failure tied to fluorides", Nucleonics, 19,3(1961)77.
52. B. A. Webb, "Carburization of Austenitic Stainless Steel by Uranium Carbide in Sodium Systems", Report NAA-SR-6246, North American Aviation, Inc., 1963.
53. M. Benedict, "An economic appraisal of stainless steel and zirconium in nuclear power reactors", Metal Progr., 2(1959) 102, 76.
54. R. G. S. Shipper and K. J. Wootton, "Thermal Resistance Between Uranium and Can", Proceedings of the Second U.N. International Conference on Peaceful Uses of Atomic Energy, Geneva, 1958, Vol. 7, p. 684; see also reference 63.
55. J. A. L. Robertson, A. M. Ross et al., "Temperature distribution in UO_2 fuel elements", J. Nucl. Mater., 7(1962) 225.
56. The Shippingport Pressurized Water Reactor, Chapter 5, p. 128, Addison-Wesley Publishing Co., Inc., Reading, Mass., 1958. a) p. 119.
57. A. W. Kramer, Boiling Water Reactors, Addison-Wesley Publishing Co., Inc., Reading, Mass., 1958. a) p. 108; b) p. 77.
58. D. E. Walker et al., "BORAX-IV Reactor: Manufacture of Fuel and Blanket Elements", USAEC Report ANL-5721, Argonne National Laboratory, 1958. a) p. 486.
59. C. F. Reinke et al., "Metallurgical Evaluation of Failed BORAX-IV Reactor Fuel Elements", USAEC Report ANL-6083, Argonne National Laboratory, 1961.
60. J. L. Gray, A. Pon, G. J. Phillips, and D. B. Primeau, "Heavy Water Moderated Natural Uranium Power Reactors", Sixth World Power Conference, Melbourne, October 20-27, 1962, Paper 66 III. 2/2.
61. J. L. Gray, G. A. Pon, G. J. Phillips and D. B. Primeau, "Heavy Water Moderated Natural Uranium Power Reactors", Sixth World Power Conference, Melbourne, October 20-27, 1962, Canadian Report AECL-1646, 1963.
62. A. J. Mooradian, "Economic Fuel for CANDU", Sixth World Power Conference, Melbourne, October 20-27, 1962, Paper 67 II 4/4. See also: AECL-1647.
63. F. Butler, J. Harper, I. H. Morrison, and J. A. Pardoe, "Development of Manufacturing Techniques for the Calder Hall Fuel Element", Proceedings of the Second U.N. International Conference on Peaceful Uses of Atomic Energy, Geneva, 1958, Vol. 6, p. 317; see Figs. 4 and 5 on p. 322.
64. A. G. Thorpe, II, "Design of the Yankee Core I Fuel Assembly", Report YAEC-154, p. 28, Westinghouse Electric Corp., 1960.
65. P. P. King, "Joining Fuel Rods into Subassemblies for YAEC Fuel Elements", Report YAEC-108, Westinghouse Electric Corp., 1958.
66. A. Boettcher et al., "Structure of the Fuel Elements in the Karlsruhe Research Reactor FR-2", Proceedings of the Second U.N. International Conference on Peaceful Uses of Atomic Energy, Geneva, 1958, Vol. 6, p. 449.
67. G. Schneider, "Metallische Bindung zwischen Uran und Aluminum fur Kernreaktor-Brennelemente", Metall, 15(1961) 675.
68. R. S. Neymark, "Rate of Alloying of Uranium Alloys with Stainless Steel (Part I. 1800 to 2300 °F)", Report NAA-SR-3278, North American Aviation, Inc., 1960.
69. Proceedings of the Symposium on Powder Packed Uranium Dioxide Fuel Elements, November 30-December 1, 1961, Report CEND-153, Nuclear Division, Combustion Engineering, Inc., 1962.
70. J. G. Hunt, D. F. Kaufman and P. Loewenstein, "Hot Extrusion of UO_2 Fuel Elements", Report NMI-1245, Nuclear Metals, Inc., 1961.
71. C. J. Baroch, "Design and Fabrication of Coextruded Stainless Steel Clad UO_2 Fuel Rods", Report GEAP-4282, General Electric Co., 1963.
72. D. R. Rees and E. U. Powell, "Saxton tests lead toward better water reactors", Nucleonics, 21,11(1963)72. See p. 75.
73. M. K. Millhollen, "Out-of-Reactor Evaluation of Components and Assemblies - Mark I Swaged UO_2 PRTR Fuel Element", USAEC Report HW-66910, Hanford Atomic Products Operation, General Electric Co., 1960.
74. N. R. Gardner, "Coextruded parts can simplify design", Mater. Design Eng., 48,7(1958)91.
75. V. W. Eldred, A. Stuttard, and J. Skinner, "Changes in Calder fuel elements under irradiation", Nucl. Eng., 5(1960)160.
76. J. F. Palmer and W. M. Barss, "A Study of 50 Sheath Failures in NRX from September 1955 to August 1957", Canadian Report CRR-749 (AECL-650), Chalk River Project, Atomic Energy of Canada, Ltd., 1958.
77. G. C. Garrow, "Summary of the Fabrication and Behavior of Standard Uranium Rods in the NRX Reactor 1957-1960", Canadian Report 101-210 (AECL-1298), Atomic Energy of Canada, Ltd., 1961.
78. AEC-Euratom Conference on Aqueous Corrosion of Reactor Materials held at Brussels, October 14-17, 1959, USAEC Report TID-7587, 1960. (Also European Atomic Energy Society)
a) C. L. Angerman and E. C. Hoxie, "Aqueous Corrosion of Aluminum-Nickel-Uranium Bonds", p. 405. b) S. Isserow and R. G. Jenkins, "Aqueous Corrosion of Zircaloy-Clad Fuel Elements with High Uranium Cores", p. 437.
79. P. Balligand, "Reactor incidents at Saclay", Nucleonics, 18,3(1960)82.
80. V. W. Eldred, "Post-Irradiation Examination - An Essential Part of Fuel Element Assessment", Proceedings of the International Atomic Energy Agency Symposium on Fuel Element Fabrication held at Vienna, May 1960, Vol. 2, p. 157.
81. H. N. Culver, "Calder Hall operating experience", Nucl. Safety, 1,4(1960)90.
82. L. Grainger, "Fuel elements for civil reactors", Nucl. Eng., 6(1961)102.
83. R. V. Moore, H. Kronberger, and L. Grainger, "Advances in the Design of Gas Cooled, Graphite Moderated Power Reactors", Proceedings of the Second U.N. International Conference on Peaceful Uses of Atomic Energy, Geneva, 1958, Vol. 9, p. 104.
84. H. K. Hardy et al., "The development of uranium-Magnox fuel elements for an average irradiation life of 3000 MWD/te", J. Brit. Nucl. Energy Soc., 2,1(1963)33.
85. G. B. Greenough and P. Murray, "Fuel Elements for U.K. Gas-Cooled Reactors", Proceedings of a Symposium on

Materials for Gas- and Water-Cooled Reactors held at New York, October 31, 1962, Nuclear Metallurgy, Vol. VIII, p. 83, 1962.
86. L. A. Mann, "EBR-1 operating experience", Nucl. Safety, 3,2(1961)70.
87. J. H. Kittel, M. Novick, and R. F. Buchanan, "The EBR-1 Meltdown—Physical and Metallurgical Changes in the Core. Final Report", USAEC Report ANL-5731, Argonne National Laboratory, 1957; see also: Nucl. Sci. Eng., 4(1958)180.
88. A. R. Kaufmann, J. J. Pickett, and J. E. Roman "PRDC Meltdown Tests-Pouring of Molten U-10w/o Mo Alloy over Prototype of PRDC Fuel Pin Support and Lower Axial Blanket Rods," Report NMI-4411, Nuclear Metals, Inc., 1959.
89. C. E. Dickerman, E. S. Sowa, and D. Okrent, "Fast reactor safety studies in TREAT - A status report", Nucleonics, 19,4(1961)114.
90. J. R. Weir, Jr., "Argonne transient reactor test factility", Nucl. Safety, 4,2(1962)47.
91. C. E. Dickerman, Argonne National Laboratory, personal communication, April 8, 1963.
92. A. A. Jarrett (Ed.), "SRE Fuel Element Damage, Interim Report", Report NAA-SR-4488, North American Aviation, Inc., 1959.
93. R. L. Ashley et al., "SRE Fuel Element Damage, Final Report", Report NAA-SR-4488 (Suppl.), North American Aviation, Inc., 1961.
94. J. L. Ballif, "Metallurgical Aspects of SRE Fuel Element Damage Episode", Report NAA-SR-4515, North American Aviation, Inc., 1961.
95. B. E. Harper, Jr. (Ed.), "Safeguards Evaluation of Recent SRE Experience Applicable to HNPF", Report NAA-SR-4504, North American Aviation, Inc., 1959.
96. W. B. McDonald and J. H. DeVan, "Sodium reactor experiment incident", Nucl. Safety, 1,3(1960)73.
97. H. F. Donohue and R. W. Keaten, "Fuel Rod Bowing in the SRE", Trans. Am. Nucl. Soc., 5(1962)172.
98. P. N. Haubenreich, "Power-reactor stability experience", Nucl. Safety, 4,3(1963)35.
99. T. D. Anderson, "Integrity of UO_2 fuel elements", Nucl. Safety, 3,1(1961)18.
100. A. J. Mooradian, G. M. Allison, and J. F. Palmer, "Chalk River experience with fuel waterlogging", Nucleonics, 18, 1(1960)81.
101. R. R. S. Robertson and V. C. Hall, Jr., "Fuel Defect Test - BORAX-IV", USAEC Report ANL-5862, Argonne National Laboratory and Atomic Energy of Canada, Ltd., 1959.
102. F. Schwoerer, "Over-All Evaluation of Blanket Fuel Removed from PWR Core 1 During the First Refueling of the Seed", Report WAPD-TM-266, Bettis, Atomic Power Laboratory, 1961.
103. P. A. Fleger, I. H. Mandil, and P. N. Ross, "Shippingport Atomic Power Station Operating Experience, Developments and Future Plans", Report WAPD-T-1429, Bettis, Atomic Power Laboratory, 1961; see also: USAEC Report TID-14480.
104. L. R. Lyman, E. F. Losco, G. J. Selvaggio, and J. G. Goodwin, "Fabrication Defects Observed in PWR Core 1 Blanket Fuel Rods", Report WAPD-TM-321, Bettis, Atomic Power Laboratory, 1962.
105. C. E. Center, H. Feinroth, and J. E. Yingling, "The Second Refueling of Core 1 of the Shippingport Atomic Power Station", Report WAPD-260, Bettis, Atomic Power Laboratory, 1962.
106. R. S. Neymark, "Materials for Dresden and Other Boiling Water Reactors", Proceedings of a Symposium on Materials for Gas- and Water-Cooled Reactors held at New York, October 31, 1962, Nuclear Metallurgy, Vol. VIII, p. 163, 1962.
107. "Dresden Down till January; Zircaloy tube fault sought", Nucleonics, 18,1(1960)18
108. "Ceramics Research and Development Operation, Quarterly Report, April-June, 1963", Report HW-76302, Hanford Atomic Products Operation, General Electric Co., 1963.
109. "Ceramics Research and Development Operation, Quarterly Report, July-September, 1963", Report HW-76303, Hanford Atomic Products Operation, General Electric Co., 1963.
110. "Ceramics Research and Development Operation, Quarterly Report, October-December 1962", Report HW-76300, Hanford Atomic Products Operation, General Electric Co., 1963.
111. R. G. Gray and F. P. Mrazir, "Examination of an In-Pile Failure of a PWR Core 1 Type Fuel Rod", in "Bettis Technical Review", "Reactor Technology", Report WAPD-BT-23, p. 53, Bettis, Atomic Power Laboratory, 1961.
112. S. Naymark and T. J. Pashos, "Fuel Elements for Water Reactors, A Status Report", Report APED-4265, General Electric Co., San Jose, California, 1963.
113. M. F. Lyons et al., "UO_2 Fuel Rod Operation with Gross Central Melting", Trans. Am. Nucl. Soc., 6, (1963)155.
114. "Yankee Core Evaluation Program, Quarterly Progress Report for the Period June 20, 1962 to September 30, 1962", Report WCAP-6052, Westinghouse Electric Corp., 1962.
115. "Yankee Core Evaluation Program Quarterly Progress Report for the Period Ending June 30, 1963", Report WCAP-6055 Westinghouse Electric Corp., 1963.
116. "Yankee Core Evaluation Program Quarterly Progress Report for the Period Ending September 30, 1963", Report WCAP-6056 Westinghouse Electric Corp., 1963.
117. T. J. Pashos, "Experience with Stainless Steel as a Fuel Cladding Material in Water-Cooled Power Reactor Applications", Report APED-4260, General Electric Co., San Jose, California, 1963.
118. "Stress-Corrosion Cracking of Stainless Steel—Fuel Elements", Nucl. Safety, 4,3(1963)33.
119. C. N. Spalaris, "Materials for Superheat Reactors", Proceedings of a Symposium on Materials for Gas- and Water-Cooled Reactors held at New York, October 31, 1962, Nuclear Metallurgy, Vol. VIII, p. 191, 1962.
120. A. H. Spano, "Self-limiting power excursion tests of a water-moderated low-enrichment UO_2 core", Nucl. Sci. Eng., 15(1963)37; see also: Report IDO-16751, Phillips Petroleum Co., 1962.
121. J. E. Houghtaling, T. M. Quigley, and A. H. Spano, "Calculation and Measurement of the Transient Temperature in a Low-Enrichment UO_2 Fuel Rod During Large Power Excursions", Report IDO-16773, Phillips Petroleum Co., 1962.
122. M. H. Bartz, "Performance of Metals During Six Years Service in the Materials Testing Reactor", Proceedings of the Second U.N. International Conference on Peaceful Uses of Atomic Energy, Geneva, 1958, Vol. 5, p. 466.
123. Fuel Elements Conference held at Gatlinburg, Tennessee, May 14-16, 1958, USAEC Report TID-7559, Part 1, 1959.
a) M. H. Bartz, "Fuel Element Problems Associated with Operations of the Engineering Test Reactor and Materials Testing Reactor", p. 28.
b) R. J. Beaver and J. E. Cunningham, "Recent Developments in Aluminum-Base Fuel Elements for Research Reactors", "Discussion", p. 51.
c) R. A. Noland, "Manufacture of the Fuel Plates and Fuel Subassemblies for the Argonne Low-Power Reactor", p. 236 et seq.
d) A. B. Shuck, "Manufacturing Methods for the Experimental Breeder Reactor Mark I and Mark II Fuel Loadings", Fig. 9, "Injection Casting Machine", p. 170.
124. L. J. Harrison, "Mechanical failures of MTR fuel assemblies", Nucleonics, 21,10(1963)78.
125. O. J. Wick, T. C. Nelson, and M. D. Freshley, "Plutonium Fuels Development", Proceedings of the Second U.N. International Conference on Peaceful Uses of Atomic Energy, Geneva, 1958, Vol. 6, p. 700. See Fig. 4 and Table 1.
126. "The Experimental Boiling Water Reactor (EBWR)", USAEC Report ANL-5607, p. 184, Appendix D, "Production of Fuel Subassemblies", Argonne National Laboratory, 1957.
127. H. F. Turnbull et al., "Manufacture of the Seed Fuel Elements for PWR", Proceedings of the Second U.N. International Conference on Peaceful Uses of Atomic Energy, Geneva, 1958, Vol. 6, p. 481.
128. R. E. Tate, A. S. Coffinberry, and W. M. Miner (Eds.), "The Fabrication of Billets Containing Plutonium for MTR Fuel Elements", Chapter 20, p. 23 in The Metal Plutonium, University of Chicago Press, Chicago, Ill., 1961.
129. W. C. Thurber and R. J. Beaver, "Silicon-Modified Uranium-Aluminum Alloys for Foreign Reactor Applications", Am. Inst. Mining Met. Petrol. Engrs., Inst. Metals Div., Spec. Rept., Series 5, No. 7, p. 57.
130. J. E. Cunningham and R. J. Beaver, "APPR Fuel Technology", Proceedings of the Second U.N. International Conference on Peaceful Uses of Atomic Energy, Geneva, 1958, Vol. 6, p. 521.
131. G. W. Gibson and D. R. deBoisblanc, "Uranium-Aluminum Alloy Powders for Use as Nuclear Reactor Fuels", Paper presented at the Fall Meeting of the AIME, Cleveland, October 21-25, 1963.
132. R. E. Macherey, "Fabrication of the Uranium-Base Fuel Plates and Assemblies for the Experimental Boiling Water Reactor", Proceedings of the Second U.N. International Conference on Peaceful Uses of Atomic Energy, Geneva, 1958, Vol. 6, p. 443.
133. M. MacPhee, D. S. Oliver, H. Lloyd, and M. A. Warne, "The Manufacture of Materials Testing Reactor Fuel Elements in the United Kingdom", in "Fuel Element Fabrication with Special Emphasis on Cladding Materials", Proceedings of an International Atomic Energy Agency Symposium held at Vienna, May 10-13, 1960, Vol. 1, p. 419, Academic Press Inc., N.Y., 1961.
134. G. V. Alm, M. H. Binstock, and E. E. Garrett, "Hot Pressure Bonding of OMR Fuel Plates", Report NAA-SR-3583, North American Aviation, Inc., 1959.
135. "Final Safeguards Summary Report for the PIQUA Nuclear Power Facility", Report NAA-SR-5608, North American Aviation, Inc., 1961.
136. J. Glatter et al., "The Manufacture of PWR Blanket Fuel Elements Containing High-Density Uranium Dioxide", Proceedings of the Second U.N. International Conference on Peaceful Uses of Atomic Energy, Geneva, 1958, Vol. 6, p. 630.

137. Research Reactor Fuel Element Conference held at Gatlinburg, Tennessee, September 17-19, 1962, USAEC Report TID-7642, 1962.
a) J. S. Nelles, "Fabrication and Irradiation of Bonded Flat Type Rods for the NRU Reactor", Book 2, p. 398.
b) R. D. Cheverton, "Nuclear Design of the HFIR", Book 1, p. 88.
c) J. H. Kittel, A. P. Gavin, C. C. Crothers, and R. Carlander, "Performance of Aluminum-Uranium Alloy Fuel Plates under High Temperature and High Burnup Conditions", Book 2, p. 425.
d) J. Herpin, "Fuel Elements for the Belgian High Flux Test Reactor BR2", Book 1, p. 152.
138. J. R. Dietrich and W. H. Zinn (Eds.), Solid Fuel Reactors, p. 702, Figs. 7-24 and 7-25, Addison-Wesley Publishing Co., Inc., Reading, Mass., 1958.
139. J. H. Walter, J. F. Leirich, and G. D. Calkins, "Evaluation of Irradiated OMRE Fuel Elements, First Core Loading", Report NAA-SR-4641, North American Aviation, Inc., 1960.
140. F. R. Keller, "Fuel Element Flow Blockage in the Engineering Test Reactor", (Appendices by W. S. Little and J. H. Ronsick), Report IDO-16780, Phillips Petroleum Co., 1962.
141. R. A. Costner, Jr., "MTR fission-break incident", Nucl. Safety, 4,4(1963)144.
142. J. R. Buchanan, "SL-1 final report", Nucl. Safety, 4,3(1963)84.
143. F. Schroeder (Ed.), "Quarterly Technical Report, SPERT Project, April, May, June 1961", Report IDO-16716, Phillips Petroleum Co., 1961.
144. F. Schroeder (Ed.), "SPERT Project Quarterly Technical Report, April, May, June, 1962" Report IDO-16806, Phillips Petroleum Co., 1962.
145. A. P. Gavin and C. C. Crothers, "Irradiation of an Aluminum-Alloy-Clad, Aluminum-Uranium-Alloy-Fueled Plate", USAEC Report ANL-6180, Argonne National Laboratory, 1960.
146. M. W. Rosenthal, "Operating experience with the OMRE", Nucl. Safety, 2,2(1960)76.
147. R. E. Heffner, "Damage to Stainless Steel Fuel in SPERT Reactors", Report IDO-16729, Phillips Petroleum Co., 1961.
148. L. A. Stephan, "Transient Tests of the BSR-II Core in the SPERT I Facility", Report IDO-16768, Phillips Petroleum Co., 1963.
149. J. W. Greenwood, "Contamination of the NRU Reactor in May 1958", Canadian Report CRR-836 (AECL-850), Chalk River Project, Atomic Energy of Canada, Ltd., 1959.
150. A. F. Rupp, "NRU reactor incident", Nucl. Safety, 1,3(1959) 70.
151. J. H. Walter, E. E. Garrett, and J. M. Davis, "Evaluation of Irradiated Experimental OMR Fuel Elements-1" Report NAA-SR-4670, North American Aviation, Inc., 1960.
152. S. Shapiro and M. Galvez, "A New Fabrication Technique for the Production of Stainless-Steel-Oxide-Dispersion Fuel Elements", Proceedings of the Second U.N. International Conference on Peaceful Uses of Atomic Energy, Geneva, Vol. 6, p. 516.
153. A. B. Bremer, E. F. Jordan, and W. MacDonald, "Process Sequence for the Fabrication of HWCTR Driver Tubes", Report NMI-ABB-1, Nuclear Metals, Inc., 1960.
154. A. B. Bremer, "Evaluation of Fifteen Outer Tubes of Zircaloy-Clad Unalloyed Uranium for the HWCTR", NMI-7218-2, Nuclear Metals, Inc., 1962.
155. "Report on WTR Fuel Element Failure, April 3, 1960", Report WTR-49, Westinghouse Electric Corp., 1960.
156. A. N. Tardiff, "Some Aspects of the WTR and SL-1 Accidents", Report IDO-19308, Phillips Petroleum Co., 1962; see also: Reactor Safety and Hazards Evaluation Techniques, Vol. 1, p. 43, International Atomic Energy Agency, Vienna, 1962.
157. R. B. Korsmeyer, "Westinghouse testing reactor incident", Nucl. Safety, 2,2(1960)70.
158. J. R. Seaboch and J. W. Wade, "Fuel Meltdown Experiments", Report DP-314, Savannah River Laboratory, E. I. duPont de Nemours and Co., Inc., 1958.
159. "Reactor Development Program Progress Report, July 1962", USAEC Report ANL-6597, p. 34, Argonne National Laboratory, 1962.
160. "Reactor Development Program Progress Report, August 1962", USAEC Report ANL-6610, p. 47, Argonne National Laboratory, 1962.
161. F. H. Martens, private communication, August 14, 1963.
162. "Evaluation and Results of the Irradiation Testing of the PM-1M Fuel Element", Report MND-M-2583, The Martin Co., 1961.
163. "Status Report on PM Tubular Fuel Element Core Development", Report MND-2825, Aerospace Division, Martin-Marietta Corp., 1962.
164. V. H. Troutner, "Mechanisms and Kinetics of Uranium Corrosion and Uranium Core Fuel Element Ruptures in Water and Steam", Report HW-67370, Hanford Atomic Products Operation, General Electric Co., 1960.
165. S. Isserow, "Corrosion Behavior of Defected Fuel Elements with U-2 w/o Zr Core Clad with Zircaloy-2", Report NMI-4364, Nuclear Metals, Inc., 1958.
166. S. Isserow, "Corrosion Behavior of Fuel Elements with U-2 w/o Zr Core Clad with Zircaloy-2", Report NMI-4388, Nuclear Metals, Inc., 1959.
167. S. Isserow, "The Aqueous Corrosion of Zircaloy-clad-Thorium", Report NMI-1191, Nuclear Metals, Inc., 1957.
168. D. S. Kneppel, "Aqueous Corrosion of Thorium Alloys and Zircaloy-Clad Thorium Alloys", Report NMI-1226, Nuclear Metals, Inc., 1960.
169. L. Isakoff, "Heavy Water Moderated Power Reactors, Progress Report, February 1963", Report DP-835, E. I. duPont de Nemours and Co., 1963.
170. L. Isakoff, "Heavy Water Moderated Power Reactors, Progress Report, March-April 1963", Report DP-845, E. I. duPont de Nemours and Co., 1963.
171. R. R. Hood and L. Isakoff, "Heavy Water Moderated Power Reactors, Progress Report, May-June 1963", Report DP-855, E. I. duPont de Nemours and Co., 1963.
172. G. R. Cole, "Influence of Thermal Rating and Bulk Density on Irradiation Performance of Fused UO_2 Tubes", Presented at Norton Co. Symposium on Powder-Filled UO_2 Elements, Worcester, Mass., November 5-6, 1963, USAEC Report CONF-315-1, 1964.
173. G. R. Caskey, Jr., G. R. Cole, and W. G. Holmes, "Failures of UO_2 Fuel Tubes by Internal Hydriding of Zircaloy-2 Sheathes", Proceedings of the Symposium on Powder Packed Uranium Dioxide Fuel Elements, November 30-December 1, 1961, Vol. II, Report CEND-153, Nuclear Dvisision, Combustion Engineering, Inc., 1962.
174. P. Fortescue, D. Nicoll, C. Rickard, and D. Rose, "HTGR-underlying principles and design", Nucleonics, 18,1(1960)86.
175. "OEEC High Temperature Reactor Project (Dragon) First Annual Report, 1959-1960, (Period covered, April 1, 1959-March 31, 1960)", European Nuclear Energy Agency, Paris, 1960: see also: USAEC Report NP-9161, 1960.
176. O. Mächnig, J. Bincebanck, U. Hennings, and R. Kusher, "Special features of the Brown-Boveri-Krupp reactor", Nucl. Power, 6,59(1961)63.
177. J. H. Russell, "The Turret Experiment at Los Alamos Scientific Laboratory", Proceedings of a Symposium Sponsored Jointly by The Franklin Institute and the American Nuclear Society, Delaware Valley Section, February 10-11, 1960, Franklin Institute Monograph 7, p. 127.
178. R. H. Simon, "In-Pile Loop for the High Temperature Gas Cooled Graphite Moderated Power Reactor", Trans. Am. Nucl. Soc., 5,1(1962)139.
179. "Advanced, Graphite-Matrix, Dispersion Type Fuel Systems", Report GA-4022, Part 1, General Atomic, Division of General Dynamics Corp., 1963.
180. L. R. Zumwalt, E. E. Anderson, and P. E. Gethard, "Fission Product Retention Characteristics of Certain $(ThU)C_2$ Graphite Fuels", Report GA-4551, General Atomic, Division of General Dynamics Corp., 1963.
181. W. V. Goeddel, "HTGR Fuel Materials Development", A chapter in a monograph on "Material and Fuels for High Temperature Nuclear Application". M.I.T. Press, Cambridge, Mass., 1964.
182. Nucl. Safety, 4,4(1963).
a) M. Bender, "Safety in gas-cooled power reactors", p. 21.
b) J. L. Scott, "Fission-product release from uranium-graphite fuels", p. 49.
c) M. H. Fontana, "Loss-of-coolant accidents in gas-cooled reactors", p. 54.
d) S. Peterson, "Integrity of reactor fuels", p. 60.
e) H. N. Culver, "Containment of gas-cooled power reactors", p. 90.
183. S. Peterson, "Chemical reactions of graphite", Nucl. Safety, 5,1(1963)47.
184. J. A. Lane, H. G. MacPherson, and F. Maslan (Eds.), Fluid Fuel Reactors, Addison-Wesley Publishing Co., Inc.,
a) J. A. Lane, S. E. Beall, S. I. Kaplan, and D. B. Hall, "Design and Construction of Experimental Homogenous Reactors", Chapter 7, p. 341.
b) p. 397.
c) p. 404.
d) J. P. McBride, D. G. Thomas, N. A. Krohn, R. N. Lyon, and L. E. Morse, "Technology of Aqueous Suspensions", Chapter 4, p. 128.
e) F. Maslan and F. T. Miles, "Liquid Metal Fuel Reactors", Chapter 18, p. 703.
f) R. M. Kiehn, "Molten Plutonium Fuel Reactor", Chapter 25, p. 939.
g) H. G. MacPherson (Ed.), "Molten-Salt Reactors", Part II, Chapters 11 to 17.
h) p. 867.
i) W. D. Manly et al., "Construction Materials for Molten-Salt Reactors", Chapter 13, p. 595.
185. P. J. Kreyger et al., "Development of a 250 Kw Aqueous Homogeneous Single Region Reactor", Proceedings of the

Second U.N. International Conference on Peaceful Uses of Atomic Energy, Geneva, 1958, Vol. 9, p. 427.

186. K. J. de Jong, "A Subcritical Circulatory Suspension Reactors", Proceedings of the Second U.N. International Conference on Peaceful Uses of Atomic Energy, Geneva, 1958, Vol. 12, p. 525.

187. M. E. A. Hermans, "The Preparation of Uranium Dioxide Fuel for a Suspension Reactor", Proceedings of the Second U.N. International Conference on Peaceful Uses of Atomic Energy, Geneva, 1958, Vol. 7, p. 39.

188. E. O. Swickard (Comp.), "Los Alamos Molten Plutonium Reactor Experiment (LAMPRE) Hazard Report", USAEC Report LA-2327, Los Alamos Scientific Laboratory, 1959.

189. J. W. Anderson, W. D. McNeese, and J. A. Leary, "Preparation and Fabrication of Plutonium Fuel Alloy for Los Alamos Molten Plutonium Reactor Experiment No. 1", USAEC Report LA-2439, Los Alamos Scientific Laboratory, 1960; see also: Nucl. Sci. Eng., 11(1961)434.

190. R. B. Briggs, "Holes in HRE-2 Core Tank", Nucl. Safety, 2,1(1960)99.

191. "Quarterly Status Report on LAMPRE Program for Period Ending May 20, 1962", USAEC Report LAMS-2730, Los Alamos Scientific Laboratory, 1962.

192. "Molten Salt Reactor Program, Semiannual Progress Report, July 31, 196311, USAEC Report ORNL-3529, p. 86-94, Oak Ridge National Laboratory, 1964.

193. R. R. Mathews, J. L. Phillips, K. J. Henry, R. H. Allardice, D. M. Donaldson, and H. E. Tilbe, "Performance and Operation of the Dounreay Fast Reactor", presented at the Third United Nations International Conference on the Peaceful Uses of Atomic Energy, Geneva, 1964, Paper A/Conf. 28/P/130.

194. J. L. Phillips, "The Dounreay Fast Reactor", Nucl. Eng., 10(1965)264.

195. AIME Symposium on Radiation Effects, September 1965, AIME Radiation Effects on Materials, Vol. 26, Gordon and Breach, New York, in press.
a. R. D. Leggett, T. K. Bierlein, B. Mostel, and H. A. Taylor, "Basic Swelling Studies".
b. R. S. Barnes and R. S. Nelson, "Theories of Swelling and Gas Retention in Reactor Materials".

196. R. S. Barnes, R. G. Bellamy, B. R. Butcher, and P. G. Mardon, "The Irradiation Behavior of Uranium and Uranium Alloy Fuels", presented at the Third United Nations International Conference on the Peaceful Uses of Atomic Energy, Geneva, 1964, Paper A/Conf. 28/P/145.

197. A. Strasser, J. Chi, S. Hurwitz, and R. Martin, "Irradiation Behavior of Solid Solution UC-PuC as High Burnup", presented at the Inst. of Metals, Third International Conference on Pu, November 22-26, 1965.

198. C. W. Wheelock, "Carbide Fuels in Fast Reactors", Report NAA-SR-10751, North American Aviation, Inc., 1965.

199. F. Knight et al., "High-Power Operation of PuO -UO Fast Reactor Fuel", Trans. Am. Nucl. Soc., 8(1965)345.

200. B. R. T. Frost et al., "Fabrication and Irradiation Studies of UO Stainless Steel and (UPU)O Stainless Steel Cermets", presented at the Third United Nations International Conference on the Peaceful Uses of Atomic Energy, Geneva, 1964, Paper A/Conf. 28/P/158.

201. T. J. Thompson and J. G. Beckerley (Eds.), Reactor Physics and Control, The Technology of Nuclear Reactor Safety, Vol. 1, The M. I. T. Press, Cambridge, Mass., 1964.
a. p. 675, 678;
b. p. 625;
c. p. 638;
d. p. 645 and Fig. 3-20 on p. 652.
e. p. 684;
f. p. 689;
g. p. 690;
h. p. 684;
i. p. 688.

202. T. J. Pashos, H. E. Williamson, and R. N. Duncan, "Fuel Performance in Boiling-Water Reactors", Trans. Am. Nucl. Soc., 8(1965)362.

203. B. Rubin and L. Lyman, "Behavior of Blanket UO Fuel Rods in PWR Core 1", Trans. Am. Nucl. Soc., 8(1965)359.

204. J. Pawlew, G. H. Chalder, and R. T. Popple, "CANDU Fuel Performance in NPD", Trans. Am. Nucl. Soc., 8(1965)361.

205. M. F. Lyone et al., "Molten UO Fuel Rod Irradiations to High Burnup", Trans. Am. Nucl. Soc., 7(1964)411.

206. W. R. Smalley, "Evaluation of Highly Irradiated Yankee Fuel Cladding", Nuclear Applications, 1(1965)419.

207. D. R. McClintock, E. Paxson, and H. M. Ferrari, "Performance of Thin-walled Stainless-steel-clad Fuel Rods", Nuclear Applications, 1(1965)425.

208. R. N. Duncan et al., "Stainless-steel-clad Fuel Rod Failures", Nuclear Applications, 1(1965)413.

209. E. M. King and E. L. Long, "Bonue Superheater Failure", Nucl. Safety, 7, 1(Fall, 1965) .

210. D. R. de Boisblanc and R. S. Marsden, "Operation of the MTR on a Plutonium Loading", Report IDO-16508, Phillips Petroleum Co., 1958.

211. A. L. Colomb and T. M. Sims, "ORR Fuel Failure Incident", Nucl. Safety, 5, 2(Winter 1963-1964)203.

212. S. R. Nemeth, "Metal Fuel Tube Manufacture", Report DP-976, Savannah River Laboratory, E. I. du Pont de Nemours and Co., Inc., 1965.

213. R. R. Hood (Comp.), Heavy-Water Moderated Power Reactors, Progress Report March-June 1965, Report DP-975, E. I. du Pont de Nemours and Co., Inc., 1965.

214. C. L. Angerman and G. R. Caskey, Jr., "Irradiation Behavior of Zirconium-Uranium Alloy Fuel Tubes", Trans. Am. Nucl. Soc., 8(1965)427.

215. S. A. Cottrell, E. Edmonds, P. Higginson, and W. Oldfield, "Development and Performance of Dounreay Fast Reactor Metal Fuel", presented at the Third United Nations International Conference on the Peaceful Uses of Atomic Energy, Geneva, 1964, Paper A/Conf. 28/P/150.

216. J. E. Cunningham, E. L. Long, Jr., E. A. Franco-Ferreira, and D. G. Harman, "BONUS Reactor Superheater Fuel Assemblies - An Investigation of Failure and Method of Correction", USAEC Report ORNL-3910, Oak Ridge National Laboratory, 1965.

217. C. L. Angerman and G. R. Caskey, "Irradiation Behavior of Zr-U Driver Tubes for HWCTR", Report DP-994, Savannah River Laboratory, E. I. du Pont de Nemours and Co., Inc., 1965.

218. R. N. Duncan, W. H. Arlt and J. S. Atkinson, "Towards Low-Cost High Performance BWR Fuel" Nucleonics 23 (APnl 1965) 50, 62.

219. G. Reed and E. Tarnuzzer, "Examining Yankee Plant Performance in 1965" Nucleonics 24 (March 1966)42.

CHAPTER 14

Mechanical Design of Components for Reactor Systems

N. J. PALLADINO
The Pennsylvania State University
University Park, Pa.

CHAPTER CONTENTS*

*This chapter is based on information in the literature or known to the author prior to early 1964 as revised and rewritten to reflect the principal advances made up until the spring of 1966.

INTRODUCTION

A nuclear power plant consists of a complex array of equipment. Mechanical integrity of this equipment is one of the most important considerations affecting the safety of the plant.

In evaluating the hazards of a nuclear power plant, three levels of safety are identified: safety to the plant, safety to operating personnel and safety to the general public. Usually, hazards evaluations center on questions of safety to the general public. In this chapter, the three levels of safety are considered to be so interrelated that protection of the public begins with safety to the reactor plant and to its operating personnel. Attention will be focused on means for minimizing probabilities of mechanical failures which might affect safety and on methods for limiting the consequences of such failures.

The usual reactor plant provides three barriers against the release of fission products: the fuel matrix and cladding, the primary system envelope, and the plant containment vessel. This chapter is concerned in large part with the second barrier, the primary system envelope.

1.1 The Reactor, a Part of an Integrated Plant

While the potentially severe hazards from a nuclear power plant arise from the reactor core, it must be understood that the core is only one component of an integrated system for the production of power. Reactor integrity is influenced by, and in turn influences, the integrity of the rest of the plant; consequently, in developing the functional requirements of specific components and the systems in which they are to operate, it is exceedingly important to recognize the interactions among them. Because of the subtle effects involved, familiarity with the experiences of the nuclear industry is necessary; unique criteria often must be considered. To illustrate, consider several examples.

One basic consideration in any design is repairability; components must be accessible for repair or replacement. In a conventional power plant, accessibility usually involves questions of arrangement, temperature and pressure. In a nuclear power plant, accessibility also involves consideration of the radioactivity (induced by the core neutrons) of the wear and corrosion products that may deposit on (or near) the part in question. Although, in this case, accessibility might be provided by special tooling or increased shielding, the designer must consider ways to reduce the amount of wear and corrosion products in the coolant stream or to use materials whose wear and corrosion products do not involve hazardous radioactivity following neutron irradiation.

For example, although tests might indicate that a particular cobalt-bearing material is desirable to eliminate wear in a particular reactor core application, the accessibility requirements in other parts of the plant might preclude the use of this material because of the high level of radioactivity of even small amounts of the wear products.

In addition to creating accessibility problems, because of induced radioactivity, corrosion and wear products can interfere with mechanical motions in control drive mechanisms, valves or other moving parts; as a result the quantities of such products must be limited. Moreover, the wear and corrosion products may deposit on the heat transfer surfaces of the fuel elements, impairing heat transfer from the fuel to the coolant; severe local overheating and even fuel element burnout can occur if the quantities of wear and corrosion products are excessive. Obviously then, accessibility to one component or system, as well as the integrity of the core itself, can be influenced by the materials selected for another part.

Another aspect of this same problem is illustrated by the effect of fuel element failure. It is reasonable to assume that, in a core containing many fuel elements, a certain fraction of the fuel elements eventually release fission products to the coolant stream. Some of the fuel elements may have incipient defects eventually leading to release of fission products; other may have initial defects that were not caught upon inspection. Thus, if the amount of fission-product activity in the stream is to be limited to the value on which the reactor design is based, a failed-element detection system in the core may be needed so that failed elements can be identified and removed when necessary. It also is apparent that the fuel element design itself is significantly influenced by the need for inspection to prevent installation of faulty fuel elements. Certain configurations are considerably more difficult and costly to inspect than others. Thus in-core instrumentation, fuel element design and the repairability of those components in the coolant stream likely to be affected by fission products are interrelated, even though the instrumentation, fuel elements and components may be considerably separated in the reactor system.

While nuclear power plants have neither the most intensive or extensive requirements for cleanliness, they are unique in the degree of need for both. Strong motivations exist for cleanliness in a nuclear plant. It is necessary to exclude or remove any superfluous matter that could become radioactive and thereby increase the shielding, decontamination, and accessibility problems. In addition, it is necessary to eliminate particulate matter which could lodge in moving parts and cause mechanical damage or which could clog coolant passages or deposit on heat transfer surfaces and impair heat flow. It is necessary, therefore, to control dirt during assembly of parts and to provide means for cleaning parts during and after manufacture or after incorporation into the reactor system. Such cleaning must be done according to detailed specifications based on the nature of the materials involved in the part. For effective cleaning the component must be designed so that no cleaning solvent or dirt can be entrapped. The solvent used should not be corrosive. Because concentration of the solvent in a cavity could lead to difficulty, back-welding of a welded joint may be required; this will permit use of a cleaning solvent without trapping it in an exposed crevice. Then, after all components are assembled, the system must be cleaned. To reduce the difficulty of this process, assembly under controlled-cleanliness conditions is often required. In addition, the designer is faced with specifying a final clean-up procedure using a solvent that is compatible with all the materials in the system.

Another example of a consideration peculiar to nuclear power plants derives from the fact that moving parts may have to operate in the coolant environment without benefit of normal organic lubricants. Such might be the case in a completely sealed pressurized water system where organic lubricants, if used, could leak into the coolant and be carried to the reactor. Organic lubricants tend to sludge under irradiation and, if the sludge deposits on fuel-element heat transfer surfaces in coolant channels, this could result in a hazardous situation.

As a final example of problems peculiar to nuclear plants, attention is directed to the effects of radiation on properties of materials. These include such effects as increased embrittlement of reactor vessels, increased relaxation of bolts in the core or enhancement of crevice corrosion.

1.2 General Design Requirements

Two general design requirements for achieving safety have been implied in the foregoing section, namely, reliability and maintainability. These are complementary requirements. Reliability is achieved by evaluating performance under assumed operating conditions and providing specifications and drawings to obtain the desire effects. Maintainability, as a design requirement, involves means, procedures and criteria for determining that operating conditions and performance requirements are being met and for taking corrective action when necessary; such corrective action may include adjustment, repair or replacement.

In designing for reliability of mechanical performance of parts, the range of operating conditions must be carefully considered. If a component is to meet its design specifications day in and day out, its operation should not depend upon maintaining some restricted range of variables. For example, the suitability of a flow baffle should not depend upon maintaining a narrow range of pressure drop through the core. If no other alternative is available, then the component must be made fail-safe if prescribed conditions cannot be met. In the case of the baffle, excessive pressure drop should not cause a failure that blocks flow.

The conditions of operation considered in evaluating systems must include those brought about by malfunction of components as well as the so-called normal conditions of steady-state and transient operation. Malfunctions of one part must not lead to progressive failures of other parts. For example, performance of springs in a reactor plant must be evaluated not only on the assumption that they remain whole but also on the assumption that they break. Even if shutdown protection is provided in the event of such breakage, the broken parts may have to be made captive so that they cannot enter the coolant stream and interfere with other operations in the system—such as operation of valves or of the shutdown mechanism itself.

Designing for maintainability in nuclear power plants presents many problems as illustrated by examples in the preceding section. Nevertheless, nuclear power plants must be maintainable. No equipment can be designed that will not require some maintenance during its life. One must not be misled by the fact that certain devices are termed "trouble-free"; they are usually known to be trouble-free for extended periods of time only in tried and proven applications. Adjustments, repairs or replacement of parts eventually become necessary. Large components as well as small can be involved. It is not inconceivable that means for replacing reactor vessels in the field may be required because of radiation effects encountered during their lifetime.

1.3 Specific Criteria for Mechanical Design

The design of a nuclear power plant involves the integration of the efforts of many people, each representing one of a number of technical disciplines. To each of these people fall responsibilities for certain design parameters. In the process of making compromises these parameters remain the prerogative and responsibility of the discipline concerned. For example, regardless of compromises made to achieve a particular design, the mechanical engineer must make sure that the hardware he designs will retain its dimensions, integrity and ability to function in the face of prolonged and short-term mechanical and thermal stress, cycling loads, corrosion, erosion and wear; and that the hardware can be fabricated, cleaned, inspected and maintained [1]. His job is relatively straightforward once the general features and design criteria of a component have been identified. But in getting to this point he, along with his colleagues, must strive for the best possible approach to meeting the objectives of the plants without compromising his design objectives. To achieve this goal, the design of a nuclear power plant evolves by an iterative process.

During the evolution of a design, the mechanical engineer, among others, tries to establish design criteria to guide his own work as well as the work of his colleagues. The criteria include a variety of statements about design features or limiting values of design variables that are intended either to fulfill the design objectives or to provide desired margins of safety. Design criteria are formulated both by the customer and by the persons responsible for each of the technical disciplines involved in the design. Of necessity, many criteria are established early in the design of the plant. Initially these are often general in nature, but they become more specific as work progresses. Criteria must be reviewed periodically. One of the chief sources of difficulty in obtaining a workable system with compatible components results from using a "given" set of design criteria evolved early in a project.

Design criteria provide fixed limits which can severely restrict the benefit of optimization studies. As knowledge is improved, design criteria can sometimes be relaxed; at other times they may have to be made more restrictive. In any event, careful consideration must be given to choosing the optimum time for freezing each of the criteria [1].

Listed below are some of the typical criteria that have been used in the mechanical design of nuclear power plants. No attempt is made here to be either all-inclusive or particularly specific. Details concerning specific components are covered in the ensuing sections of this chapter. Most of the criteria listed below are common to all components. These criteria should not be confused with functional requirements for particular components but rather should be viewed as guidelines for meeting those requirements.

(1) The materials in contact with the reactor coolant must have corrosion and wear rates limited to prescribed values and these values must be as low as knowledge and economics permit.

(2) The reactor plant should be able to operate with a prescribed number of failed fuel elements in the core. A single fuel element failure should not lead to a progressive chain of fuel element failures.

(3) The system components and core internals should be arranged to permit decay-heat removal by natural circulation of the coolant if all other means for handling decay heat should fail.

(4) All reactor parts should be designed so that they can be cleaned without trapping the cleaning solvents in crevices.

(5) All small components such as springs, bolts, nuts, and locking devices must be designed so that, if they break, none of the parts can be released to the coolant stream and carried to regions where they can interfere with mechanical motion, fluid flow or heat transfer or where they can damage bearing, sealing, or other surfaces of importance to the proper function of the plant. These components should be of captive design.

(6) All bolt-locking devices should conform to specifications carefully prepared for the application involved.

(7) Mechanical stresses should be limited to values below the yield point, the promixity to yield varying with specific situations. Creep and radiation effects must be considered in establishing acceptable stress levels.

(8) Combined mechanical and thermal stresses must not produce intolerable dimensional changes and must be below limits leading to acceptable fatigue-life, on the basis of a recognized theory.

(9) All efforts should be made to minimize stress raisers by providing generous fillets in re-entrant corners of stressed parts. If at all possible, fillets of core structural supports should have radii no smaller than 1/8 inch (3 mm).

(10) Welds must be carefully specified with particular attention to obtaining the desired penetration and surface finish and to evaluating and repairing defects.

(11) If appropriate to the application, the plant components must be able to withstand a prescribed shock loading.

(12) Integrity of the pressure barrier or coolant containment portion of the primary system must be provided for by appropriate specification of sealing means.

(13) Irradiation effects on materials must be considered in the design of all components, particularly as related to the brittle fracture characteristics of pressure-containing (stressed) parts.

(14) Small orifices or channels that can become easily clogged should be avoided in the primary loop.

(15) Internal heat generation due to gamma and neutron heating in the reactor vessel and core structures must be recognized as a source of thermal stress and differential expansion so that steps can be taken to prevent undesirable consequences.

(16) Consideration must be given to backups for parts whose failure would have serious consequences.

(17) Devices must be designed so that failure results in a safe situation (fail-safe).

(18) Reliability evaluation should be considered in all designs.

(19) Parts should be designed so that they can be repaired or replaced in a practical manner.

1.4 Scope of the Chapter

This chapter is concerned primarily with the mechanical components of a nuclear power plant not related to fuel element design. The subject matter is treated under four principal topics. First, attention is directed to items associated with the reactor itself. These include reactor vessels, reactor vessel closures, reactor internals other than fuel (the reactor internals, however, do include the core structural supports), control drive mechanisms and auxiliary systems associated with core safety. Second, components external to the reactor and reactor vessel complex are considered. These include piping, pumps and blowers, valves, heat exchangers and steam generators, pressurizing and expansion systems, and components involved in special system problems. The third topic concerns mechanical safety during core handling and refueling. The fourth and last topic relates to nondestructive testing and acceptance inspection.

To avoid duplication, the analytical techniques common to all the sections of this chapter are treated only once, at the point where they are first referred to. As a result, Sec. 2 concerning the reactor vessel forms the reservoir for much of the information on thermal stresses and the effects of radiation on materials. This is not inappropriate because of the prime importance of the reactor vessel to the safety of the nuclear plant. Accordingly, wherever appropriate in the later sections, the reader is referred to the applicable parts of the section on reactor vessels.

Much of the subject matter in this chapter is derived from the author's experience with water-cooled reactors and, therefore, might appear at first glance, to have limited applicability to other reactor types. However, the problems of reactor vessel integrity, thermal shielding, control rod alignment, bolting, locking devices, expansion (both uniform and nonuniform), etc., are common to all reactors. The differences arise chiefly in the relative importance of the various problems in specific reactor types and in the feedbacks that various effects might have in a particular design. One must not conclude, however, that these differences are not important. They most assuredly are, and an attempt has been made to identify some of the more significant differences throughout the chapter. Nevertheless, each problem, whether discussed in terms of one reactor type or another, merits evaluation in any reactor design. For example, in a fast reactor, distortion or deflection of a component in the core might be the cause of a severe reactivity change; in another reactor of the same or different type, the distortion or deflection of a similar component or some other component might impair cooling. The fact that one problem may be more closely associated with one reactor type than with another is not so important for our purpose here as the fact that distortions or deflections can have significant consequences in any reactor and that the consequences must be evaluated for each reactor design regardless of type.

2 REACTOR VESSELS*

The primary functions of a reactor vessel are to house the nuclear core and to contain the primary coolant. In addition, the reactor vessel performs other related and important functions which influence its design: the reactor vessel provides, internally, for directing the flow of coolant to and through the core; it furnishes, externally, anchor points for the primary system piping; it provides supports for control rod mechanisms; and it supplies facilities for fueling and refueling the core.

Because of the important part that the reactor vessel plays in a nuclear plant, it must provide a high degree of dependability throughout its life. It is the one component which cannot be readily replaced nor isolated from the core by valves. In many applications, the reactor vessel must withstand high operating pressures. In all applications it must withstand piping reactions and, in order to contain the coolant, it must provide a high degree of leak-tightness.

As a pressure vessel, the design of the reactor vessel is similar in many respects to the design of other industrial vessels. However, there are a number of special considerations which must be taken into account [2,3]. These play an important part in the establishment of design criteria for reactor vessels.

2.1 Design Criteria and Considerations

Despite the fact that boilers and pressure vessels have been used for more than 150 years, only during the past fifty years have they become regarded as highly reliable devices. During the nineteenth century and the early part of the twentieth century failures took place at an alarming and increasing rate [4]. As steam pressures increased and boilers were used more extensively on land the number of failures mounted to distressing proportions.

Early investigations of boiler accidents revealed an extensive lack of design knowledge and inspection methods as well as widespread carelessness and ignorance in boiler operation. Many conflicting and unsupported theories of failure were being advanced, and in the United States in 1826 the Franklin Institute, among others, began to examine their validity. By 1830 the Institute recognized a need for a basic investigation of the factors involved and organized a committee which instituted an experimental program to study boiling and the problems of boiler design. This work provided the technical basis for the first United States federal legislation in 1833 regarding inspection of boilers on ships. Such inspection is now an activity of the United States Coast Guard.

In 1866 the first of the steam boiler insurance companies in the United States was formed to inspect and insure boilers. These insurance companies were instrumental in developing early rules for design. As the number of accidents increased, many cities and states began to pass laws to regulate the design, manufacture, inspection, and operation of boilers and pressure vessels. By 1911, however, the differences in the various laws were causing widespread confusion.

2.1.1 Code Requirements

To help alleviate the confusion and provide a sound technical basis for regulation, the American Society of Mechanical Engineers (ASME) set up a committee in 1911 to formulate standard specifications for the construction of steam boilers and other pressure vessels and to interpret the specifications when questions arose regarding their intent. Based on this committee's work, the ASME Council adopted its first boiler code as an official Society document in 1915. (This committee is now called the "Boiler and Pressure Vessel Committee" and is often referred to as the "BPVC".) As outlined in section I of reference [5] the objectives of the rules are "to afford reasonably certain protection of life and property and to provide a margin for deterioration in service so as to give a reasonably long safe period of usefulness."

This code has been officially adopted in whole or in part by most states and by many municipalities of the United States as well as by provinces of the Dominion of Canada. Each state, municipality and province that adopts or accepts one or more sections of the ASME Code is invited to appoint a representative to act on the Conference Committee to the main Boiler and Pressure Vessel Committee; the Conference Committee has certain voting privileges on the main BPVC and provides contact with local problems. Since all governmental bodies do not provide for automatic acceptance of revisions of the Code and ASME-approved "Replies" to Code Cases (as the special rulings and interpretations

*The author wishes to acknowledge the valuable assistance provided by B. F. Langer of the Westinghouse Bettis Laboratory in the preparation of Secs. 2 and 3 of this chapter.

are called), it is necessary to confer with local authorities to determine if these modifications are acceptable. It must be recognized that in the last analysis the applicable code requirements are those required by law.

The advent of nuclear reactor vessels brought many questions which were not answered by code rules existing at the time. Recognizing this fact, the BPVC established in 1955 a "Special Committee on Nuclear Power" to consider the special problems associated with nuclear reactor pressure vessels and to make recommendations on the acceptability of new materias, design and inspection rules required for such vessels.

The work of this committee resulted in two series of actions. The first was a series of replies to Code Cases, which were approved by the ASME Board on Codes and Standards, and which have been designated by the suffix N after the case numbers. The second action of the committee has been the preparation of a new section for the Boiler and Pressure Vessel Code, Section III entitled "Rules for Construction of Nuclear Vessels". This section consolidates and amplifies the rulings covered by the Code Cases and was officially approved as part of the Code in November 1963. A 1965 edition was later approved and issued incorporating addenda to the 1963 edition; the 1965 edition has, in turn, been followed by addenda containing important corrections and additions.

In addition to the ASME, a number of other organizations sponsor or publish industrial codes, standards or specifications dealing with pressure vessels and piping [6, 7]. These include the American Standards Association, American Water Works Society, American Welding Society, American Society for Testing Materials, Manufacturers Standardization Society, American Gas Association, American Iron and Steel Institute, American Petroleum Institute, Canadian Standard Association, British Standards Institution, and certain of the military branches of the federal government.

Certain of the standard and specifications developed by the foregoing organizations are made a part of the ASME Boiler and Pressure Vessel Code [8]. For example, the application standards of the American Standards Association, particularly those relating to flanges, bolts and nuts, are either referred to or reproduced by the Code. Similarly, those materials specifications of the American Society for Testing Materials which are approved by the BPVC are incorporated into Section II of the Code entitled "Materials Specifications". As a consequence, in this chapter attention is given primarily to the ASME Code.

There are several basic differences between the general requirements of the new Section III of the Code [10] and Sections I and VIII of the Code [5] on which the nuclear code Cases were based. Some of these are outlined in reference [9]. These and others identified from Section III [10] are summarized below:

1. The approved Section III is essentially self-contained with specific requirements based on more-detailed stress analyses than called for in the Section I and VIII referred to in the Code Cases.
2. Vessels covered by Section III are classified as A, B, C, and Mixed. Reactor vessels fall under Class A. This class covers vessels or tubular elements that are not directly open to the atmosphere, that form part of the reactor coolant system, and within which nuclear fuel is present and a nuclear chain reaction may take place.
3. Special requirements for vessel materials are given in Section III. These are over and above the normal materials requirements for non-nuclear vessels for equivalent temperatures and pressures. The additional requirements relate primarily to further tests and examinations to offset the fact that reactor vessels are inacessible for careful inspection in service.
4. Section III is based on the use of the maximum shear stress theory of failure (also known as the Tresca Criterion) in design instead of the maximum principal stress theory. The maximum shear stress theory was chosen rather than the distortion energy theory (also known as the Mises Criterion) because it is somewhat more conservative, it is relatively easy to apply, and it simplifies fatigue analysis. The maximum shear stress at a point, under the Tresca Criterion, is defined as one-half the algebraic difference between the largest and the smallest of the three principles stresses, a number which then must be compared with one-half the allowable stress. In Section III, however, a new term called "equivalent intensity of combined stress" or simply "stress intensity" has been defined. It is twice the maximum shear stress and can be directly compared to the strength values found from tensile tests.
5. Section III requires detailed calculation and classification of all stresses and provides for application of different limits to the different categories of stress, whereas Sections I and VIII give formulas for minimum allowable wall thickness. Section III requires the calculation of thermal stresses and gives allowable values for them whereas Sections I and VIII do not. Section III also provides for evaluation of cyclic stresses to guard against fatigue failure whereas Sections I and VIII do not.

The five basic stress intensity limits which are to be satisfied are listed below. The first three fall under the heading of primary stresses which are stresses arising from internal resistance to nonself-limiting external forces and moments. The fourth includes secondary stresses which are self-limiting stresses arising from imposed or induced strain patterns. The fifth is the peak stress intensity which is the highest stress in the region under consideration and includes increases in stress brought about by gross and local discontinuities.

a. General Primary Membrane Stress Intensity (P_n). This stress intensity is "derived from the average value, across the thickness of a section, of the general primary stresses produced by design internal pressure or other specified mechanical loads, but excluding all secondary and peak stresses. The allowable value of this stress intensity is S_m, see Tables N-421, N-422, or N-423 of Section III.

b. Local Membrane Stress Intensity P_L). This stress intensity is "derived from the average value across the thickness of a section of the local primary stresses produced by design pressure and specified mechanical loads but excluding all secondary and peak stresses. The allowable value of this stress intensity is 1.5 S_m."

c. Primary Membrane (General P_m, or Local P_L) Plus Bending Stress Intensity P_b. This stress intensity is "derived from the highest value across the thickness of a section of the general or local primary membrane stresses plus primary bending stresses produced by the design pressure and other specified mechanical loads, but excluding all secondary and peak stresses. The allowable value of the stress intensity is 1.5 S_m."

d. Primary, P_m (or P_L) + Pb, Plus Secondary Stress Intensity Q. This stress intensity is derived from the highest value, at any point across the thickness of a section, of the general or local primary membrane stresses plus primary bending stresses plus secondary stresses produced by specified operating pressure and other specified mechanical loads and by general thermal effects. The effects of gross structural discontinuities but not of local structural discontinuities (stress concentrations) shall be included. The allowable value for the maximum range of this stress intensity is $3S_m$."

e. Peak Stress Intensity. [Derived from P_m (or P_L) + P_b + Q + F where F includes the increment added to primary or secondary stress by a concentration such as a notch and certain thermal stresses which may cause fatigue but not distortion of vessel shape.] This stress intensity is "derived from the highest value at any point across the thickness of a section of the combination of all primary, secondary, and peak stresses produced by specified operating pressures and other mechanical loads and by general and local thermal effects and including the effects of gross and local structural discontinuities. The allowable value of this stress intensity is dependent on the range of the stress difference from which it is derived and on the number of times it is to be applied." (See Sec. 2.2.4.)

6. In Section III, the applicable rules from Section VIII for fabrication and inspection of pressure vessels for lethal service was tightened where it was felt necessary, and the special requirements of the Code Cases were incorporated. Some of the special requirements are:

(1) Emphasis has been placed upon procedures for repairs during construction.
(2) Specific material and welding requirements for attachments to pressure parts have been incorporated.
(3) The tolerance rules to Section VIII for pressure vessels were extended to cover spherical and conical shells as well as cylindrical shells and were made applicable to internal pressures as well as external pressures.
(4) For Class A vessels the requirement for 100% radiography of all butt welds was added.
(5) Rules covering methods and evaluation for nondestructive testing, other than radiography, include magnetic particle inspection, liquid penetrant inspection, and ultrasonic inspection.
(6) A requirement was included for pastweld heat treatment of completed vessels made of carbon and low-alloy materials. The preheat rules follow the nonmandatory appendix of Section VIII.

2.1.2 Functional Considerations

While codes form an important source for reactor vessel design criteria, they do not and cannot specify all the criteria which govern design. Many criteria arise from functional considerations. Actually the design must begin with delineation of these criteria and evaluation of their impact on the design before code rules can be applied. This fact is recognized in Section III of the ASME Nuclear Vessel Code; it requires that the owner "provide or cause to be provided for each vessel a specification of functions and design requirements" to serve as a basis for design, construction, and inspection in accordance with the code rules.

The functional considerations must include attention to the interactions between the vessel and the rest of the system in which it operates as well as consideration of the abnormal operating conditions under which the vessel will or may have to perform. Abnormal operating conditions can arise from many causes, such as the malfunction or failure of system components or an unusual operational transient. Although reactor protective systems are provided to limit the consequence of transients, the degree of overpressure or over-temperature imposed on the vessel by such transients must be anticipated and the necessity for the vessel to withstand them must be made a part of the functional requirements of the vessel.

Section III of the code also specifies that the Design Specifications shall be certified by a registered professional engineer "experienced in pressure vessel design." While the Code does not say so, if this function is to be performed adequately, the certifying engineer must be knowledgeable about nuclear plants as well as pressure vessels.

Summarized below are those functional considerations which have become important in reactor vessel design.

1. Structural Loading: The most important criterion for reactor vessel design, of course, is that the vessel must withstand the structural loads imposed upon it. These loads vary from one plant to another and exist in various combinations under different operating conditions for a given plant. Thus analysis and evaluation of reactor vessel reliability often becomes quite complex. The structural loads to be considered have been identified by L. W. Smith [3]. These can be grouped under the following headings:

a. Normal pressure forces imposed by the coolant: An important consideration in applying the Boiler and Pressure Vessel Code is determining the design pressure to be used for a given operating pressure. While the code specifies that safety-valve capacities must prevent the pressure from rising more than a prescribed percent above the maximum allowable working pressure, there is still the question of establishing this pressure. The safety valves must be set high enough so that normal operational swings in pressure can be accommodated without the need for safety-valve action. These swings can be quite important in a nuclear power plant.

b. Mechanical forces inherent to the plant: These may include a variety of effects; significant among these are fluid flow and gravity effects of the coolant, and pipe reactions brought about by thermal expansion of the system. The coolant forces may introduce both steady and transient effects. In liquid-metal systems, where the system pressure may otherwise be small or nonexistent, pressure due to pumping and gravity head may be important. In pressurized water systems, where pumping and gravity effects may be small, water hammer problems may bring about severe mechanical loads.

Pipe reactions due to thermal expansion can be an important consideration in reactor vessel design. As in conventional plants, a careful selection of anchor points must be made to minimize these reactions. However, it should be pointed out that evaluation of the thermal expansion leading to pipe reactions must include consideration of local effects due to internal heat generation (i.e., from gamma radiation and neutrons) in the pipe and vessel in the area of the pipe-to-vessel connection.

c. Thermally induced stresses: Heat generation due to the attenuation of radiation within the pressure vessel walls may introduce stresses as large or larger than the pressure stresses. These stresses will vary with reactor power level and must be evaluated in terms of the numbers and magnitudes of the stress cycles involved. Other thermally induced conditions that must be considered include those brought about by steady-state heat flow through the vessel walls, those encountered during abnormal power changes such as emergency shutdown, those associated with allowable rates of coolant temperature change, those due to removal of decay heat under emergency cooling conditions and those brought about by materials with different coefficients of thermal expansion.

d. Loads brought about by environmental conditions: In naval applications, underwater shock becomes an important design consideration and a complicating factor in identifying loads that the reactor vessel must withstand. In other applications similar types of shock loading may be encountered. These may arise again from military action, earthquakes or reactor accidents.

2. Leak-tightness: As part of its function to contain the primary coolant, the reactor vessel must be designed to limit leakage. This becomes an important criterion in specifying porosity of base and weld metal, particularly in high-pressure gas-cooled reactors. It also becomes an important criterion in selecting reactor vessel closures. Many designers, not satisfied with gasketed closures of any type because they may leak under abnormal temperature swings, have relied only on hermetic seals in the form of either strength welds or seal welds. Other designers have found the use of two concentric gaskets satisfactory to prevent and control leakage by monitoring the space between the gaskets to detect leakage. Gaskets normally used for vessel closures are hollow (pressurized or self-energized) or solid metal "O" rings sometimes coated with a thin layer of silver.

3. Coolant Compatibility: That the vessel materials in contact with coolant must be compatible with the coolant would appear to be a reasonable criterion. However, this criterion becomes difficult to express in meaningful terms. The metal exposed to the coolant must, of course, have acceptable corrosion limits to prevent deterioration over the life of the vessel. In addition, if this material is used as a cladding with a less corrosion-resistant base material for the vessel, the consequences of a leak in the cladding must be examined; such examination may lead to stringent requirements for bonding the cladding to the base metal. So far as the cladding itself is concerned, compatibility with coolant may be largely determined by the need for good corrosion resistance to limit contamination of the coolant rather than to limit cladding deterioration.

4. Radiation Effects: Radiation brings about four effects which create specific design criteria. First, because of heat generation induced by radiation, heavy sections of metal and extensive welding should be avoided where possible in the vessel in the vicinity of the core; particular attention must be given to nozzle reinforcements in this connection. Second, because of the induced radioactivity of corrosion and wear products that could be deposited in regions such as pumps and valves thus interfering with maintenance, it is necessary to minimize the use of materials whose corrosion products could be converted to long-lived ($\gtrsim 10$ hr) gamma emitters by the neutron flux of the core. Third, the effects of neutrons on the properties of materials must be within acceptable limits or (if the properties are dose-dependent) the operating conditions must be changed as the plant gets older; of particular importance in this connection is the influence of radiation on the brittle-fracture characteristics of the vessel. (See Sec. 2.6.3.) Fourth, the presence of residual radioactivity requires provision of special features to permit periodic inspection of the vessel.

5. Dimensional and Alignment Tolerances: In many applications, where the vessel serves as the platform for the control drive mechanisms, the need for accurate alignment imposes severe requirements for close tolerances. Expansions of the vessel during pressure and temperature changes must not destroy this alignment. In addition, it is necessary to make certain that relative expansions between the core, vessel, and control rod systems do not bring about undesirable reactivity effects because of changes in the positions of control rods relative to the core. These requirements may have a bearing on the location of control rods, the configuration of the head, or the materials used in

various parts of the control system. If remotely operated refueling devices are to be used after core depletion, they may introduce additional requirements for close tolerances.

6. Flow Baffles and Thermal Shields: Because the coolant flow distribution influences core performance, the vessel must provide sufficient space for necessary flow baffles and distribution plates to direct the flow; in addition, space must be provided for thermal shields required to limit the radiation impinging on the vessel walls. The vessel designer must consider the supporting arrangements for baffles and thermal shields as well as the problems of rigidity, thermal expansion, thermal stresses and cooling; in addition, the baffle design should be such as to restrict the leakage of coolant around the core.

7. Refueling: As discussed in Sec. 8, refueling can be accomplished by complete removal of the reactor vessel head or by use of fuel ports located in the head. In addition, refueling may require the use of heavily shielded casks or, in the case of water reactors, may be done under flooded-canal conditions. All these requirements introduce design criteria for the reactor vessel. Removal of the head may require cutting and rewelding of a strength weld or a seal weld. The use of multiple fuel ports tends to reduce the ligament efficiency of the head and may require extensive reinforcement of the head. The use of heavy caskets may necessitate special features on the vessel, or in its immediate proximity, for supporting the caskets.

2.2 Analytical Techniques

Detailed stress analyses of reactor vessel components are necessary to show that, under the applicable loads, the appropriate stress limitations in Section 2.1.1 are not exceeded. Section III of the Code requires not only that such stress analyses be made but also that they shall be certified by a registered professional engineer, experienced in pressure vessel design.

In establishing the structural reliability of a reactor vessel, steady-state loads, cyclic loads and transient loads must be considered. Their combined effects must also be evaluated, a problem complicated by the difficulties in predicting the operational history of the vessel. Nevertheless, estimates must be made and the sensitivity of the design to errors in these estimates must be evaluated. In evaluating stresses, radiation effects on properties of materials must be considered.

The loadings that must be taken into account in designing a reactor vessel have been discussed in the preceding section. As indicated in Section III of the Code, these include, but are not limited to, the following:

(1) internal and external design pressure,
(2) weight of the vessel and normal contents under operating or rest conditions, including the additional pressure due to static and dynamic head of liquids,
(3) superimposed loads such as other vessels, operating equipment, insulation, corrosion-resistant or erosion-resistant linings, and piping.
(4) wind loads, snow loads, and earthquake loads where specified,
(5) reactions of supporting lugs, rings, saddles, or other types of support, and
(6) temperature effects.

Evaluation of stresses in a reactor vessel can be quite complex. Many techniques are used, including, in many instances, not only hand calculations, but also computers and experimental techniques. The bases used for such work are discussed in the ensuing parts of this section and include: code thickness formulas, calculation of thermal stresses, influences of stress raisers, design bases for combined stresses, shock and vibration stresses, and experimental stress analysis.

2.2.1 Code Thickness Formulas

The ASME Boiler and Pressure Vessel Code is not intended as a handbook but rather as a set of minimum standards for use by experienced design engineers [8]. Formulas provided by the Code are valid only when used in conjunction with the allowable stress values which form an integral part of the Code. Consequently the formulas as well as allowable stress limits to be used with the proposed Section III of the Code, discussed in Sec. 2.1.1, are in general different from those to be used in conjunction with the Code Cases which refer to Sections I and VIII. It should also be noted that in Sections I and VIII the formulas provide recommended thicknesses on the basis of allowable stresses, whereas in Section III formulas provide stresses to be compared with allowable limits. Nevertheless, Section III does provide formulas (included below) to help in arriving at a tentative thickness for use in design; however, the formulas are not to be construed as formulas for acceptable thicknesses.

1. Definitions and Notations

h = inside depth of head (inches) measured from tangent line.

t = minimum required thickness of shell of head plates (inches).

D = inside diameter of vessel of head skirt (inches).

D_1 = inside diameter (inches) of the conical portion of the head at its points of tangency to the knuckle, measured perpendicular to the axis of the cone. (See Fig. 2-1) [5b].

E = efficiency of longitudinal joints in spherical shells or any heads, or the efficiency of ligaments between openings, whichever is less. For seamless shells or heads, $E = 1.0$. For double-welded butt joints which have been thermally stress-relieved and radiographically inspected, $E = 1.00$ in Section VIII. For other situations the reader is referred to the Code.

L = inside spherical or crown radius (inches).

P = design pressure or maximum allowable working pressure (pounds per square inch). (The design pressure should include a suitable margin above the pressure at which the vessel will be normally operated to allow for probable pressure

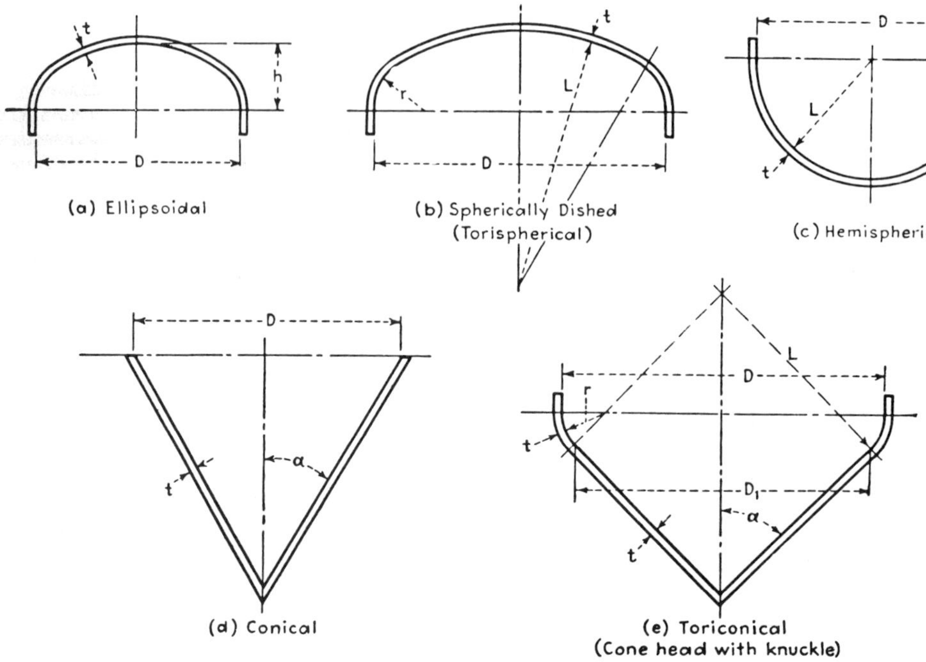

FIG. 2-1 Principal dimensions of typical heads.

surges up to the setting of the pressure-relieving devices).

R = inside radius of the shell course under consideration (inches).

R_o = outside radius of shell (inches).

S = maximum allowable stress value as given in Code tables (pounds per square inch).

S_m = design stress intensity in Section III of Code (pounds per square inch).

α = half the included (apex) angle of the cone at the centerline of the head.

2. Cylindrical Shells*

Section III of the Code gives the following approximate formula for the tentative thickness:

$$t = \frac{PR}{S_m - 0.5P} \quad \text{or} \quad t = \frac{PR_o}{S_m + 0.5P} . \tag{2-1}$$

Equation (2-1) is based on the maximum shear stress theory of failure.

3. Spherical Shells

Section III of the Code provides the following approximate formulas for the tentative thickness:

$$t = \frac{PR}{2S_m - P} \quad \text{or} \quad t = \frac{PR_o}{2S_m} . \tag{2-2}$$

4. Formed Heads under Internal Pressure

Although Section III of the Code is intended to include tentative thickness formulas for formed heads, such formulas have not yet been issued. Section VIII of the Code gives formulas, as outlined below, for the following types of formed heads: (a) ellipsoidal, (b) torispherical, (c) hemispherical, (d) conical, (e) toriconical and (f) flat.

a. Ellipsoidal Heads (See Fig. 2-1a). An ellipsoidal head has the shape of half an ellipsoid of revolution around the axis of the cylindrical vessel to which it is to be attached.

The Section VIII Code formula for the required thickness is

$$t = \frac{PDK}{2SE - 0.2P} , \tag{2-3}$$

where

$$K = \frac{1}{6}\left[2 + \left(\frac{D}{2h}\right)^2\right] .$$

An ellipsoidal head with a minor axis (inside depth of the head minus the skirt) equal to one-half the inside radius of the head skirt is known as the "Manufacturer's Standard Head." For this case $D/2h = 2$, and K in the above formula is unity. This head, like the ASME standard head (see b. below), has a wall thickness approximately the same as that of the cylindrical shell to which it is attached.

*It should be noted that the dimensions of h, t, D, D_1, L, R, R_o, are identical (inches). P, S, S_m also have identical dimensions (psi). E and α are dimensionless. This permits each of the equations to be written in dimensionless form, i.e., as functions of such ratios as t/R, S/P, etc. For example Eq. (2-1) can be written as $t/R_o = (P/S_m)/[1 + 0.5(P/S_m)]$. In this form the equation is independent of the system of units and can be used with cgs or mks units.

b. Torispherical Heads (See Fig. 2-1b). A torispherical head combines a spherical contour with a toroidal knuckle. It is also known as a "spherically dished" head. The spherical end, or crown, has a radius larger than the radius of the cylindrical shell but not greater than the diameter of the shell. The knuckle radius r must be not less than 6% of the outside diameter of the vessel and also not less than 3t. The knuckle must be tangent to both the cylindrical shell and the spherical portion of the head.

The Section VIII Code formula for the required thickness is

$$t = \frac{PLM_1}{2SE - 0.2P}, \tag{2-4}$$

where

$$M_1 = \frac{1}{4}\left[3 + \sqrt{\frac{L}{r}}\right]. \tag{2-5}$$

A torispherical head in which the knuckle radius r is 6% of the inside crown radius ($L/r = 16.67$) is known as the "ASME Standard Head". For this head the required thickness is

$$t = \frac{0.885PL}{SE - 0.1P}. \tag{2-6}$$

This thickness is slightly less than that of the cylindrical shell to which the head is attached, so using the same thickness as the shell is safe.

c. Hemispherical Heads (See Fig. 2-1c). When the thickness of a hemispherical head does not exceed 0.356 LS or P does not exceed 0.665SE, the Section VIII Code formula for the required thickness is

$$t = \frac{PL}{2SE - 0.2P}. \tag{2-7}$$

d. Conical Heads (See Fig. 2-1d). A conical head has a simple conical shape without any transition knuckle. It is a low-cost head because it can be formed by rolling a flat plate without the use of dies. The half apex angle α must be less than 30°; otherwise a transition knuckle is required.

The Section VIII Code formula for the required thickness of the head when $\alpha < 30°$ is

$$t = \frac{PD}{2\cos\alpha\,(SE - 0.6P)}. \tag{2-8}$$

When α exceeds the value for Δ given in Table 2-1 [5b], a conical head without a transition knuckle must be reinforced by a compression ring at the head-shell junction.

The cross-sectional area A of the reinforcing ring in square inches is given by the following formula:

$$A = \frac{P}{SE}\left[\frac{D^2\tan\alpha}{8}\right]\left[1 - \frac{\Delta}{\alpha}\right]. \tag{2-9}$$

e. Toriconical Heads (See Fig. 2-1e). A toriconical head is a conical head with a toroidal transition

TABLE 2-1

Limits for use of reinforcing rings on conical heads

P/(SE)	0.001	0.002	0.003	0.004	0.005	0.006	0.007	0.008	0.009
Δ	11°	15°	18°	21°	23°	25°	27°	28.5°	30°

knuckle. The code requires such a knuckle whenever α exceeds 30° and also requires that the inside knuckle radius r be not less than 6% of the outside diameter of the head skirt nor less than three times the knuckle thickness. The toroidal section must be tangent to both the cylindrical shell and the conical portion of the head.

The required thickness of the knuckle is

$$t = \frac{PM_2D_1}{4\cos\alpha\,(SE - 0.1P)}, \tag{2-10}$$

where

$$M_2 = \frac{1}{4}\left[3 + \sqrt{\frac{D_1}{2r\cos\alpha}}\right]. \tag{2-11}$$

The required thickness of the conical section is

$$t = \frac{PD_1}{2\cos\alpha\,(SE - 0.6P)}. \tag{2-12}$$

f. Flat Heads. The required thickness of a flat head is dependent on the method of its attachment to the shell. The ASME Code shows eighteen typical cases (see Fig. 2-2 [5b]). The required thickness of a flat unstayed circular head is given by the following formula:

$$t = d\sqrt{CP/S}, \tag{2-13}$$

except when the head, cover, or blind flange is attached by bolts causing an edge moment (see Fig. 2-2j and k). The diameter and the constant C are as given for the appropriate design in Fig. 2-2. Formulas for the required thickness of flat heads with edge moments and of noncircular flat heads are given in the Code [5b].

2.2.2 Calculation of Thermal Stresses

Thermal stress problems arise in the design of all reactor vessels. Such stresses are generally due to nonlinear temperature patterns which impose internal constraints on the expansion of adjacent fibers of the vessel wall. As indicated in Sec. 2.1.2, these temperature patterns and thermal stresses may result from a variety of causes, chief among which is the heat generation accompanying gamma-ray and neutron absorption in reactor walls.

Such radiation is attenuated in an approximately exponential manner as it passes through absorbing material such as the steel of thermal shields and reactor vessel walls. Because the heat arising from this attenuation must be removed either from the inside or outside surface or from both surfaces of the material, temperature differences will be developed within the metal. Figure 2-3 shows qualitatively the temperature patterns that might be developed for different circumstances; curve B

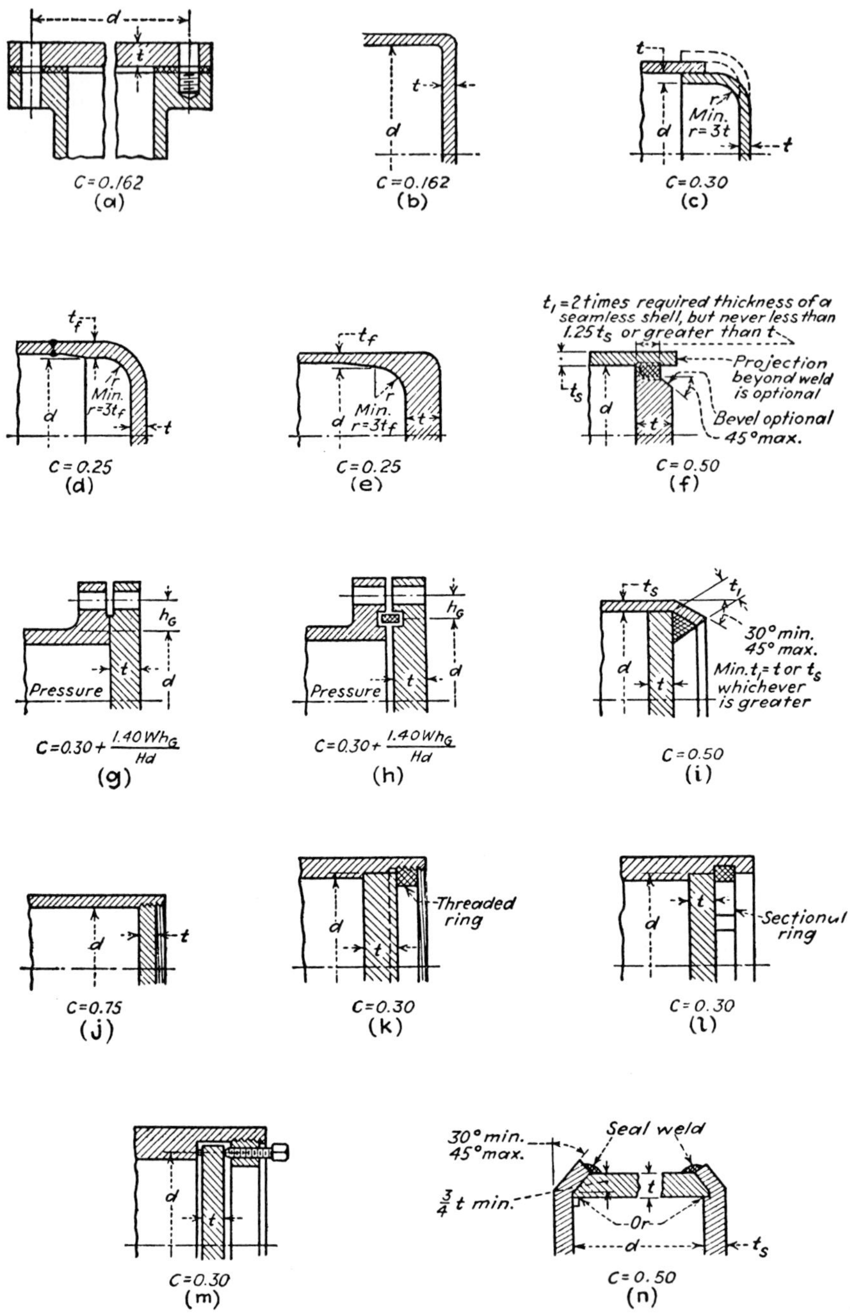

FIG. 2-2 Some acceptable types of unstayed flat heads and covers. [From Section VIII of ASME Code—still applicable under Section III (1965).]

shows the temperature pattern for a vessel cooled on the inside and perfectly insulated on the outside; curve C shows the pattern for a similar situation except that the outside surface is less-than-perfectly insulated so that some heat leakage takes place outwardly as well as inwardly. Curve D shows a temperature pattern more typical of a thermal shield which is cooled well on both sides. It can be seen from these curves that, in all practical situations, there will tend to be a temperature peak within the wall of a pressure vessel and in the thermal shields.

The fact that the heat is deposited in an approximately exponential fashion as a function of the distance through the metal, means that even relatively thin steel thermal shields between the core and the vessel will help greatly in reducing the amount of heat generated in the vessel wall and hence in reducing the temperature differences and thermal stresses within the wall. It also shows that the

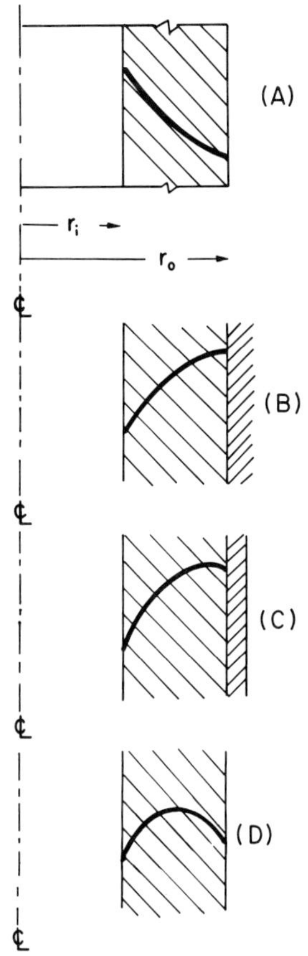

FIG. 2-3 Temperature pattern in cylindrical shells with exponential heat generation. Curve A - heat generation per unit volume. Curve B - temperature pattern with perfect insulator on outside and cooling on the inside. Curve C - temperature pattern with less-than-perfect insulator on outside and cooling on the inside. Curve D - temperature pattern with cooling on both sides.

thermal conductivity of the inner layer of the wall is very important in vessel design both because the deposition of heat occurs largely in that region and because practically all the heat deposited in a reactor vessel wall which is insulated on the outside and cooled on the inside must pass through this inner region. The amount of heat generated and hence the thermal stresses in all regions of the wall vary with power level of the reactor.

Other important sources of thermal stresses are the temperature differences brought about by thermal transients. For this reason, in reactors capable of experiencing wide and rapid swings in coolant temperatures, such as gas-cooled and liquid-metal-cooled reactors, thermal stresses can present more severe problems than in water-cooled reactors. Furthermore, the higher operating temperatures of gas-cooled and liquid-metal-cooled reactors become significant in evaluating combined thermal and mechanical stresses.

In this section, thermal stresses in two geometries of interest to reactor designers are covered along with a discussion of fluid temperature transients. The geometries considered are the hollow cylinder and the hollow sphere.

1. The Hollow Cylinder

The following treatment of thermal stresses in a hollow cylinder is directly applicable to the cylindrical portion of the reactor vessel next to the core and free of penetrations or attachments. The expressions developed for thermal stresses in hollow cylinders are based on the assumption that the distribution of temperature is symmetrical with respect to the axis of the cylinder and is constant along the axis. Under these conditions, the deformation of the cylinder is symmetrical about the axis, and, at some distance from the ends, originally plane cross sections will remain plane. It is also assumed that the material behaves elastically.

In order to compute the thermal stresses it is necessary to know the temperature distribution or the heat generation pattern within the vessel walls. The general case will be given first. Then, two special cases will be considered: the hollow cylinder with steady conduction of heat through the vessel wall and the hollow cylinder with heat generation in the walls brought about by radiation attenuation.

a. General case: The stress components for a given temperature profile, T, away from the ends in a long hollow cylinder are given by the following expressions [13]:

$$\sigma_r = \frac{\alpha E}{1-\nu}\frac{1}{r^2}\left[\frac{r^2 - r_i^2}{r_o^2 - r_i^2}\int_{r_i}^{r_o} T\,r\,dr - \int_{r_i}^{r} T\,r\,dr\right] \quad (2\text{-}14)$$

$$\sigma_t = \frac{\alpha E}{1-\nu}\frac{1}{r^2}\left[\frac{r^2 + r_i^2}{r_o^2 - r_i^2}\int_{r_i}^{r_o} T\,r\,dr + \int_{r_i}^{r} T\,r\,dr - Tr^2\right] \quad (2\text{-}15)$$

$$\sigma_z = \frac{\alpha E}{1-\nu}\left[\frac{2}{r_o^2 - r_i^2}\int_{r_i}^{r_o} T\,r\,dr - T\right], \quad (2\text{-}16)$$

where α = coefficient of linear thermal expansion ($°F^{-1}$ or $°C^{-1}$),
ν = Poisson's ratio,
r_i = inside radius of cylinder (in. or cm),
r_o = outside radius of cylinder (in. or cm),
r = radius to any point in the cylindrical wall (in. or cm),
T = temperature as a function of r measured from an initial uniform temperature at which there are no stresses in the body (°F or °C),
E = modulus of elasticity (psi or kg/cm^2),
σ_r = radial stress (psi or kg/cm^2),
σ_t = tangential stress (psi or kg/cm^2),
σ_z = axial stress (psi or kg/cm^2).

b. Steady heat flow (No internal heat generation): Transfer of heat by conduction through the vessel wall leads to a logarithmic temperature distribution [13].

$$T(r) = \frac{T_i \ln(r_o/r) + T_o \ln(r/r_i)}{\ln(r_o/r_i)} \quad (2\text{-}17)$$

With this temperature pattern the equations of reference [13] may be written as follows:

$\sigma_r =$

$$\frac{\alpha E\theta}{2(1-\nu)\ln(r_o/r_i)}\left[-\ln\frac{r_o}{r} - \left(\frac{r_i^2}{r_o^2 - r_i^2}\right)\left(1 - \frac{r_o^2}{r^2}\right)\ln\frac{r_o}{r_i}\right], \tag{2-18}$$

$\sigma_t =$

$$\frac{\alpha E\theta}{2(1-\nu)\ln(r_o/r_i)}\left[1 - \ln\frac{r_o}{r} - \left(\frac{r_i^2}{r_o^2 - r_i^2}\right)\left(1 + \frac{r_o^2}{r^2}\right)\ln\frac{r_o}{r_i}\right], \tag{2-19}$$

$\sigma_z =$

$$\frac{\alpha E\theta}{2(1-\nu)\ln(r_o/r_i)}\left[1 - 2\ln\frac{r_o}{r} - 2\left(\frac{r_i^2}{r_o^2 - r_i^2}\right)\ln\frac{r_o}{r_i}\right], \tag{2-20}$$

where θ = temperature difference between inside and outside surface of shell.

For very thin cylinder walls the tangential and axial stresses can be approximated by the following:

$$\text{at } r = r_i, \quad \sigma_t = \sigma_z = -\frac{\alpha E\theta}{2(1-\nu)}; \tag{2-21}$$

$$\text{at } r = r_o, \quad \sigma_t = \sigma_z = \frac{\alpha E\theta}{2(1-\nu)}. \tag{2-22}$$

If θ is positive, i.e., if the temperature is higher at the inside radius than at the outside radius, the radial stress σ_r will be compressive at all points and zero at both the inner and outer radii. The tangential stress σ_t and the axial stress σ_z have their largest numerical values at the inner and outer surfaces of the cylinder. If θ is positive, these stresses are compressive at the inner surface and tensile at the outer surface. If θ is negative the reverse is true. (See Fig. 2-4 [12c].)

c. Internal heat generation: Sonneman and Davis [14] have explored the case of a vessel in which the source of radiation is inside the vessel and the radiation is attenuated as it passes through the wall. In order to set forth the equations developed it is necessary to define several new terms in addition to those listed for steady heat-flow through the wall.

Let q_i = rate of heat generation at inner wall (Btu/sec-in.^{3} or cal/sec-cm^3),
β = attenuation factor (in.$^{-1}$ or cm^{-1}),
k = thermal conductivity (Btu/sec-in. °F or cal/sec-cm-°C),
y = r/r_i,
y_o = r_o/r_i.

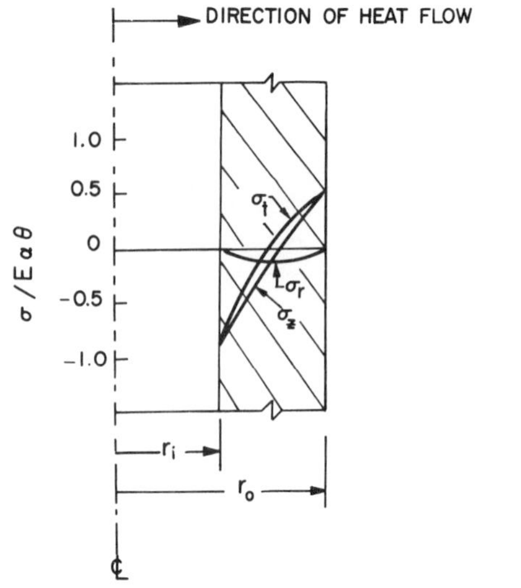

FIG. 2-4 Thermal-stress profiles in cylindrical shell with steady heat flow for $r_o/r_i = 2$. Plotted stress curves are based on these data:

r/r_o	$\sigma_r/E\alpha\theta$	$\sigma_t/E\alpha\theta$	$\sigma_z/E\alpha\theta$
1/2	0	-0.865	-0.865
5/8	-0.113	-0.303	-0.415
3/4	-0.111	+0.047	-0.039
7/8	-0.065	+0.342	+0.278
1	0	+0.555	+0.555

The expression for the temperature distribution T (in °F or °C) in this case is given by Sonneman and Davis as:

$$T = \frac{q_i}{\beta^2 k}\left\{1 - e^{-\beta r_i(y-1)} + e^{\beta r_i} C \ln y + e^{\beta r_i}\left[\mathrm{Ei}(-y\beta r_i) - \mathrm{Ei}(-\beta r_i)\right]\right\}, \tag{2-23}$$

where C is an arbitrary constant and

$$-\mathrm{Ei}(-x) = \int_x^\infty (e^{-t}/t)\,dt. \tag{2-24}$$

The above expression for T can be substituted into Eqs. (2-14), (2-15) and (2-16) to obtain the stresses σ_r, σ_t, and σ_z, using the boundary conditions associated with one of the following cases to evaluate the constant C:

Case I - The wall temperatures on the inside and outside surfaces are equal (typical of thermal shields);

Case II - The inside surface is insulated (typical of inserted thimbles);

Case III - The outside surface is insulated (typical of reactor vessels).

It is quite tedious to calculate the thermal stresses by this technique because the equations lead to expressions involving small differences of large numbers. A computer code was developed for use with the IBM 704 Computer; details are described in reference [15]. The tangential, radial and axial stress distributions resulting from the computer code are given in reference [14] and are plotted in dimensionless form in Figs. 2-5, 2-6 and 2-7 [14], for $\beta r_i = 20$ and a diameter ratio of 1.10 for Cases I, II and III respectively. The maximum tangential stresses are plotted in dimensionless form versus diameter ratio, y_o, with βr_i as a parameter in Figs. 2-8, 2-9 and 2-10 [14] for Cases I, II and III respectively. The tangential stress σ_t is calculated from the following formula by using the value of dimensionless tangential stress σ_T read from the appropriate curve:

$$\sigma_t = \sigma_T \left[\frac{E\alpha q_i}{(1-\nu)k\beta^2} \right]. \qquad (2\text{-}25)$$

2. The Hollow Sphere

The following expressions were derived assuming temperature patterns symmetrical with the center; thus stresses are a function of radius only.

a. General case: The stress components for a given temperature T in a hollow sphere are given by the following expressions [13]:

$$\sigma_r = \frac{2\alpha E}{1-\nu}\left(\frac{1}{r_o^3}\int_0^{r_o} Tr^2 dr - \frac{1}{r^3}\int_0^{r} Tr^2 dr\right), \qquad (2\text{-}26)$$

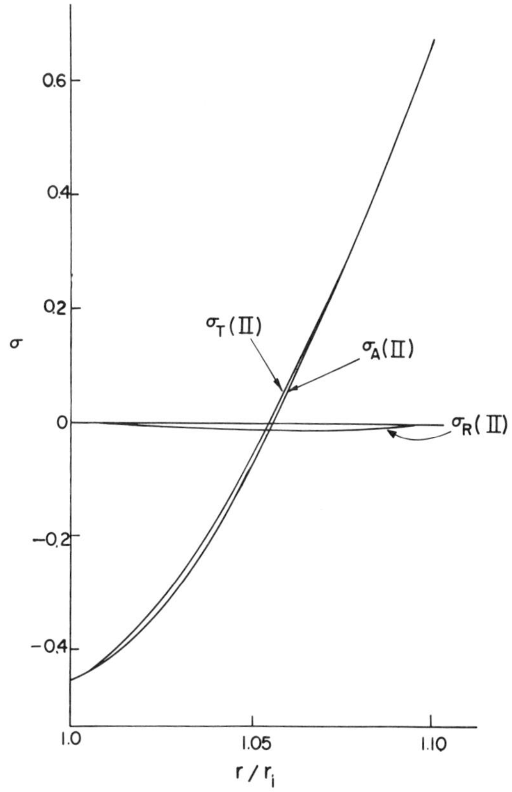

FIG. 2-6 Dimensionless stress (σ - tangential, radial, and axial) versus radial position ($y = r/r_i$) for Case II with $\beta r_i = 20$ and a diameter ratio (y_o) of 1.10.

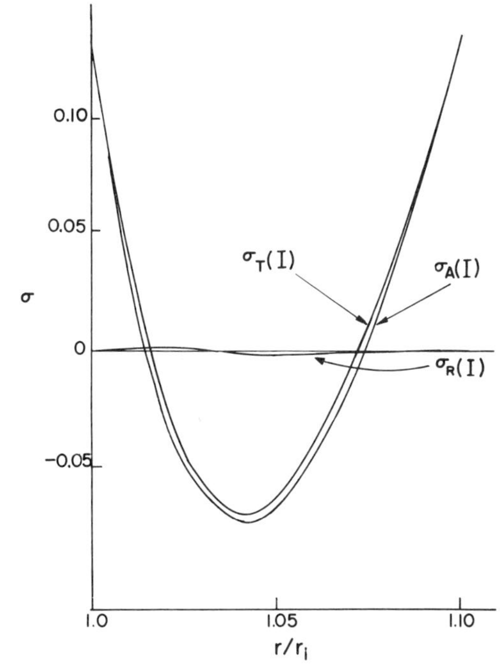

FIG. 2-5 Dimensionless stress (σ - tangential, radial, and axial) versus radial position ($y = r/r_i$) for Case I with $\beta r_i = 20$ and a diameter ratio (y_o) of 1.10.

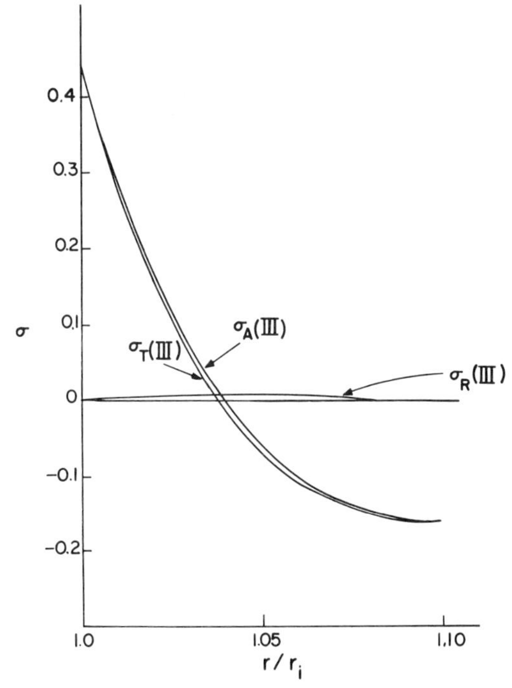

FIG. 2-7 Dimensionless stress (σ - tangential, radial, and axial) versus radial position ($y = r/r_i$) for Case III with $\beta r_i = 20$ and a diameter ratio (y_o) of 1.10.

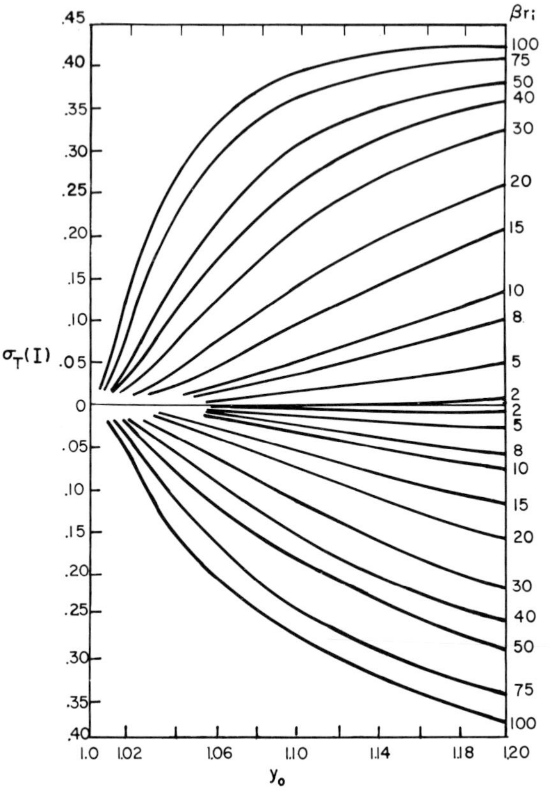

FIG. 2-8 Maximum dimensionless tangential stress σ_T(I) in the wall for Case I versus diameter ratio (y_o) with βr_i as a parameter.

$$\sigma_t = \frac{\alpha E}{1-\nu}\left(\frac{2}{r_o^3}\int_0^{r_o} Tr^2 dr + \frac{1}{r^3}\int_0^{r} Tr^2 dr - T\right). \quad (2\text{-}27)$$

The symbols are defined in the text following Eq. (2-16).

b. Steady heat flow (No internal heat generation): In reference [13] the stress under steady heat flow is given by the following expressions (in which θ is positive if the inner temperature is higher than the outer temperature):

$$\sigma_r = \frac{\alpha E \theta}{(1-\nu)}\,\frac{r_i r_o}{r_o^3 - r_i^3}$$

$$\times\left[r_i + r_o - \frac{1}{r}(r_o^2 + r_i r_o + r_i^2) + \frac{r_i^2 r_o^2}{r^3}\right], \quad (2\text{-}28)$$

$$\sigma_t = \frac{\alpha E \theta}{(1-\nu)}\,\frac{r_i r_o}{r_o^3 - r_i^3}$$

$$\times\left[r_i + r_o - \frac{1}{2r}(r_o^2 + r_i r_o + r_i^2) - \frac{r_i^2 r_o^2}{2r^3}\right]. \quad (2\text{-}29)$$

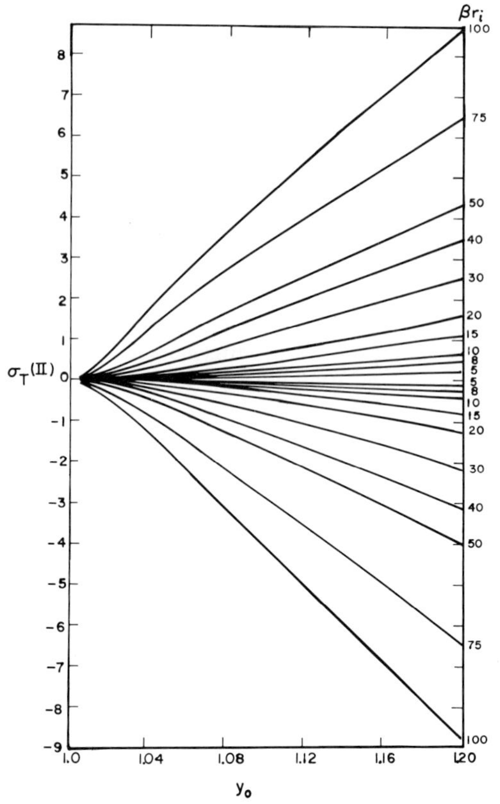

FIG. 2-9 Maximum dimensionless tangential stress σ_T(II) in the wall for Case II versus diameter ratio (y_o) with βr_i as a parameter.

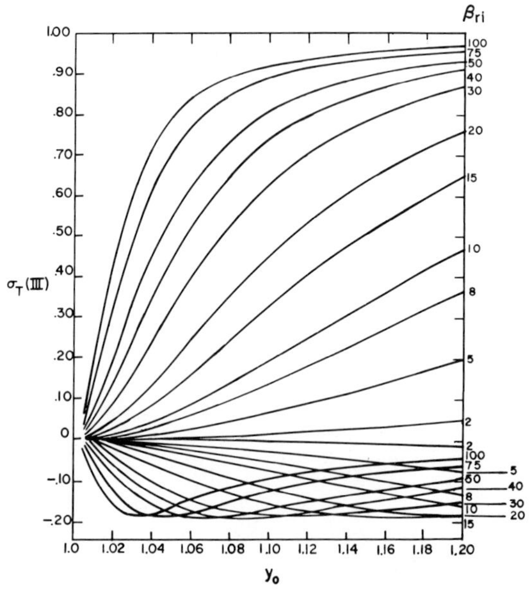

FIG. 2-10 Maximum dimensionless tangential stress σ_T(III) in the wall for Case III versus diameter ratio (y_o) with βr_i as a parameter.

From these, it can be seen that σ_r is zero for $r = r_i$ and $r = r_o$. It becomes a maximum when

$$r^2 = \frac{3r_i^2 r_o^2}{r_i^2 + r_i r_o + r_o^2} \qquad (\sigma_r = \text{max.}). \tag{2-30}$$

The stress σ_t for $\theta > 0$ increases as r increases. When $r = r_i$, one obtains

$$\sigma_t = -\frac{\alpha E \theta}{2(1-\nu)} \left(\frac{r_o(r_o - r_i)(r_i + 2r_o)}{r_o^3 - r_i^3} \right). \tag{2-31}$$

When $r = r_o$, one obtains

$$\sigma_t = \frac{\alpha E \theta}{2(1-\nu)} \left(\frac{r_i(r_o - r_i)(2r_i + r_o)}{r_o^3 - r_i^3} \right). \tag{2-32}$$

For very thin spherical shells, these reduce to the same expressions given above for very thin cylindrical shells.

3. Response to Coolant Temperature Changes

When the temperature of the fluid inside the reactor vessel changes, the walls of the vessel are subjected to thermal stresses. The amount of stress encountered is a function of the rate of temperature change, the extent of the change, the thickness and materials properties of the vessel wall, and the film coefficient between the wall and the fluid. If the fluid temperature within the vessel increases, the walls are subjected to compression on the inside and tension on the outside; if the fluid temperature decreases the reverse is true.

The first question that must be answered is what fluid temperature transients are to be analyzed. Some transients may be rapid but small in magnitude; others may be slower but extend over a wide temperature range. The temperature difference produced in material adjacent to a fluid whose temperature is changing is a function of the rate of fluid temperature change. If this rate is very high, the maximum temperature difference produced in the material is almost immediately equal to the imposed fluid temperature change. If the rate of fluid temperature change is very low, the maximum temperature difference produced in the material will be much smaller than the final total change in fluid temperature because there is time for transfer of heat to take place; the temperatures throughout the material can follow the fluid temperature without the production of large temperature differences in the material. Thus for rapid change of fluid temperature the stress is influenced mostly by the magnitude of the total change; for a slow change of fluid temperature, the rate of change tends to be the important factor. The determination of what is a fast transient and what is a slow transient depends on the wall thickness. Lacking specific experience with a given design, each situation must be analyzed separately. The analysis involves obtaining transient temperature plots and evaluating the associated stresses.

In water-cooled reactors normal fluid temperature changes encountered during load changes are relatively small. In gas-cooled and liquid-metal-cooled reactors the temperature swings can be quite large, and careful analysis of problems brought about by temperature changes is required. These problems are particularly acute in liquid-metal-cooled reactors because the liquid-metal coolant has a larger heat capacity per unit volume and better heat transfer characteristics than a gas. Thus, in a liquid-metal system, the vessel walls may have to be protected by multilayered thermal shields which have provisions for slowly leaking liquid-metal coolant into the regions between the layers and between the outer layer and the wall to provide thermal inertia in the system and thus avoid thermally shocking the vessel wall.

In all reactors, the problems of thermal shock that could be faced by inadvertently bringing a cold loop on the line must be carefully analyzed, as must the reactivity problems brought about by such action. The creep and stress-rupture properties for prolonged exposure at high temperature must be considered in order to assure adequate safety under all conditions of operation. Creep can be particularly important in gas-cooled and liquid-metal-cooled reactors. (See Chapters 9 and 10.)

It is often found that the thermal stresses in the vicinity of the vessel and closure flanges limit the allowable heatup and cooldown rates of the entire plant. To obtain temperatures and stresses in this case, resistance network analogs and digital solutions to transient temperature and stress equations usually must be resorted to. Particular attention must be given to bolted closures where significant time lag is encountered between bolt temperatures and flange temperatures. During heatup the bolts, being colder than the flanges, must elongate to accommodate the differential expansion; yielding may take place if the heatup is too rapid. During cooldown the bolt load decreases; care must be taken to make certain that the bolt load does not drop below that load needed to resist the hydrostatic pressure load without permitting the head to lift. In such a circumstance, if a seal membrane is being used at the head-to-vessel joint, it could be overstressed; if a gasket is being used it would leak.

2.2.3 Influence of Stress Raisers

The formulas presented in Secs. 2.2.1 and 2.2.2 assume ideal materials and ideal geometries. The materials are assumed to be homogeneous, isotropic and free of residual stresses. This may or may not be the case. Safety factors used in establishing allowable working stresses make allowances for departure from ideal material properties; nevertheless, fabrication and inspection processes must be evaluated to assure absence of unusual inhomogeneities or high residual stresses. Where such conditions are identified they must be considered in the design or the defects must be corrected in fabrication. Such defects, as well as conditions of geometry such as sharp fillets and holes that can produce stresses higher than given by the usual stress equations, are called stress raisers.

Geometry cannot always be ideal. Localized stresses considerably higher than those calculated by the formulas presented can occur because of discontinuities such as those which occur at joints between heads, shells and flanges, or because of shell penetrations, or because of changes in section causing stress concentrations. Those stress concentrations relating to shell penetrations are discussed in Sec. 2.3; those relating to flange joints are discussed in Sec. 2.4. In these cases as well as those involving changes of section, further stress concentrations should be avoided by designs that distribute loads without high stress concentration; fillets should be generous. A comprehensive collection of stress concentration data, determined for a large number of relatively simple and common types of discontinuities is given in references [16] and [17].

The choice of appropriate stress concentration factors for use in analyzing cyclic operation is particularly important. Considerations involved in such analyses are discussed as part of the next subsection of this chapter.

2.2.4 Design Bases for Cyclic Operation

Prior to the new Section III, the Boiler Code did not adequately treat methods of evaluating thermal stresses and transient loads. Section III [10] recognizes fatigue as a possible mode of failure arising from cyclic application of thermal and other stresses and specifies design rules to prevent such failure. Inasmuch as the number of stress cycles encountered in reactor vessel work is considerably lower ($< 10^5$ as compared to $\gg 10^7$) than the number encountered in rotating machinery, it was necessary to develop new concepts relating to low-cycle fatigue to prepare this Section [9]. The basic approach used is that developed by B. F. Langer [18]. This approach recognizes that failure in a few thousand cycles can be produced only by strains in excess of the yield strain.

Two separate procedures for cyclic analysis are presented by the Code and are discussed below. One method applies to vessel parts made of carbon or low-alloy steels or of stainless steels or nickel-chrome-iron alloys; the other method applies to high-strength bolting materials (> 100,000 psi or 7000 kg/cm^2 tensile). Both procedures are based on the same theoretical considerations; however, it was determined that the procedures for all fatigue analyses except those associated with high-strength bolting could be simplified. In the case of high-strength bolting the simplification would be overly conservative, so that the more general procedure had to be followed as discussed below.

For reactor vessel components other than high-strength bolts, the procedure for cyclic analysis consists of comparing the amplitude of the alternating peak stresses (see Sec. 2.1.1) expected in service with strain-cycling fatigue data to determine the allowable number of cycles through which these stresses may be oscillated. For convenience the strains involved have been multiplied by the elastic modulus to give fictitious stresses which can be compared directly with stresses calculated on the basis of elastic behavior. These data have been plotted as fatigue strength curves showing the allowable amplitude S_a of the alternating stress component (one-half the alternating stress range) versus the number of cycles. This stress amplitude is based on elastic behavior and hence does not represent a real stress when the elastic range is exceeded. When using these curves for a situation in which thermal transients may be limited by the stresses at several different points of the vessel, each point must be examined.

The fatigue data used for these curves were obtained from tests involving complete stress reversal where the mean stress is zero. The mean stress is defined here as the mean value of the range through which the peak stress fluctuates. Because the presence of a nonzero mean stress detracts from the fatigue resistance of the material, the fatigue curves were adjusted so that the allowable stress amplitude for a given number of cycles was consistent with a mean stress equal to the yield-point stress.

In making the adjustment for mean stress it was assumed that residual stresses due to welding existed, which would produce a mean stress equal to the yield-point stress. Because high-strength bolting material has a much higher yield strength compared to the ultimate strength, the correction as made in the curves is too severe. Furthermore bolts have a much lower probability of containing uncontrolled residual stresses than components subjected to welding. As a result the computed alternating stress intensity for high-strength bolting must be corrected for the effect of mean stress before being used to determine the allowable number of cycles [9].

In the general procedure used for high-strength bolting, both the alternating stress intensity S_{alt}, computed as one-half the range of the alternating peak stresses, and the mean stress S'_{mean}, computed as the average of the minimum and maximum values of the alternating peak stresses, are determined. They are then used to compute an equivalent alternating stress intensity S_{eq}, which is used to find the allowable number of cycles. The value of S_{eq} is determined by the relationship

$$S_{eq} = \frac{S_{alt}}{1 - (S_{mean}/S_u)}, \tag{2-33}$$

where S_u is the ultimate strength and S_{mean} is the adjusted value of mean stress, (the actual mean of the peak stresses taking into account whether or not the yield strength S_y has been exceeded during the cycle). The value of S_{mean} is determined as follows:

(a) if $S_{alt} + S'_{mean} \le S_y$, the mean component is equal to its basic value and $S_{mean} = S'_{mean}$;

(b) if $S_{alt} + S'_{mean} > S_y$ and $S_{alt} < S_y$, then $S_{mean} = S_y - S_{alt}$; or

(c) if $S_{alt} > S_y$, then $S_{mean} = 0$.

For the fatigue analysis of high-strength bolts the design fatigue curve is replaced by the following analytical approximation [9]:

$$N = \left[\frac{10^6 + 0.64\,S_u}{S_{eq} - 0.2\,S_u}\right]^2, \qquad (2\text{-}34)$$

where N = allowable number of cycles, and
S_u = specified minimum tensile strength (psi).

For any part of the vessel, if there are two or more types of stress cycles, the cumulative effect of these stress cycles must be evaluated. Provisions for evaluating cumulative effects are based on a linear damage relationship in which it is assumed that if N_1 cycles would produce failure at stress level S_1, then n_1 cycles at the same stress level would use up the fraction n_1/N_1 of the total life. Failure would occur when $n_1/N_1 + n_2/N_2 + n_3/N_3 +$ ------ is equal to unity. While other hypotheses for estimating accumulated fatigue damage have been shown to be more accurate than the linear-damage assumption, they require detailed knowledge about the sequence of stress cycles [9]. Such information generally cannot be estimated with great accuracy when the vessel is being designed.

An important part of determining the peak stress intensity stresses for cyclic operation is the establishment of appropriate stress concentration factors. In the past it was customary to use lower stress concentration factors for small numbers of cycles than for large numbers of cycles. Langer [9] points out that such practice is not advisable when using strain-fatigue data rather than stress-fatigue data. Thus Section III of the code recommends using the same value of stress concentrations regardless of the number of cycles involved. For geometries such as fillets, grooves and holes, for which theoretical stress concentration factors exist, Langer [9] suggests that it is safe to use the theoretical stress concentration factors even though strain concentrations can exceed the theoretical stress concentration factors; such usage is justified because strain concentrations higher than the stress concentrations occur only when there is gross yielding in the surrounding material which is prevented by the design methods outlined in Section III. Langer further points out that for very sharp notches the theoretical factors grossly overestimate the true weakening effect of the notch in the low and medium strength materials used for pressure vessels. "Therefore," he states, "no factor higher than 5 need ever be used for any configuration allowed by the design rules and an upper limit of 4 is specified for some specific constructions such as fillet welds and screw thread."

Section III of the Code permits the use of experimentally determined fatigue-strength reduction factors, in place of stress concentration factors, provided that they have been determined by tests conducted on a material within the same "P-Number" grouping of Table Q11.1 of Section IX of the Code at a stress level which produces failure in not less than 1000 cycles.

Inasmuch as fatigue analysis of a reactor vessel can be quite time-consuming, rules are set forth in paragraph 415.1 of Section III which may be used a basis for avoiding detailed fatigue analyses certain cases. The rules require knowledge of th pressure fluctuations and the temperature differences that will be encountered in service. Although the fatigue analysis can be waived for certain cases, the other basic stress limits must still be met.

2.2.5 Shock and Vibration Stresses

Reference [3] divides the problem of evaluating shock and vibration stresses into three parts: defining the loads, estimating responses and comparing them to known damage criteria.

The magnitudes of shock loads are dependent, of course, on the application. For most stationary power plants these are estimated on the basis of the earthquake history of the site. Reference [19a], issued in August, 1963, presents pertinent seismological information for engineers in the nuclear field. Included in the document are considerations and computational methods pertinent to the evaluation of seismic forces on reactor components during design. This document indicates that the largest acceleration recorded to date near the epicenters of large magnitude earthquakes in the United States is 0.33g. This was recorded at El Centro, Calif. on May 18, 1940. It is pointed out, however, that higher unrecorded accelerations could have taken place in the past.

For naval applications shock design is based on loads resulting from depth-charge action; analyses must be based on available test data. L. W. Smith [3] points out that a typical and acceptable velocity shock input, in such cases, is 4 ft/sec (1.2 m/sec) for all loading directions and equipment weights. Damage criteria for shock are difficult to establish. Usual stress limits do not necessarily apply. Ductile materials have withstood shock stresses well above the yield point with only a minor amount of permanent set [3].

In evaluating vibrations it is necessary to identify the range of frequencies that might be imposed over the life of the vessel. For example, these may come about from pump or blower impellers, or from electric generators. The particular frequencies which are potentially most damaging must be identified, and the natural frequencies of the vessel or component parts should be made higher than these, if possible, so that resonances can be avoided. If this cannot be done, natural frequencies lower than the forcing frequencies can be used provided that no prolonged operation can take place in the resonance range and provided that stresses involved at other than resonance conditions are below fatigue limits by appropriate margins.

In the following some general rules which summarize dynamic damage criteria are extracted in large part from reference [3].

1. Vibration:
   a. Avoid resonant vibration wherever possible by making designed natural frequencies higher than any forcing frequencies.
   b. If resonance cannot be avoided, ensure that designed natural frequencies and forcing frequencies are in resonance for as short a time as possible and, in any case, for periods well

below fatigue limits; examine higher harmonic frequencies as well as fundamental frequencies. In certain cases structural damping can be incorporated to reduce magnification factors.

c. Add mass to the structure, in such a way that it does not store potential energy, to reduce magnification factors and to lower resonant frequencies.
d. Efforts to strengthen the structure may add mass in such a way that it stores potential energy. This usually stiffens the structure, may increase the resonant frequencies, and may also increase the magnification factors. The net result must be checked to determine if an improvement has been made.
e. Avoid stress raisers and joints that can loosen.
f. Avoid collision of parts.

2. Shock:
   a. Avoid combining parts having similar frequencies to limit resonance excitation.
   b. Avoid brittle materials.
   c. Accept calculated stresses (on an elastic basis) greater than the static yield stress if the first-mode frequency is very low (under 30 or 40 cycles/sec) but keep within allowable deformation limits.
   d. Avoid excessive deformations of parts and avoid collision. Calculated deformations can usually be two or three times the elastic-limit deformation with no large permanent distortion.
   e. Make sure joints are sound and avoid unnecessary stress raisers. At present, judgment and experience must be relied upon to guard aganist failure.

Although the foregoing rules are useful in evaluating reactor plant components, they do not form an adequate basis for evaluating the suitability of a component, such as the reactor vessel or its appurtenances, to withstand shock. A report prepared by NRL in 1960 [19] presents detailed methods for dynamic-shock analysis. This report has contributed significantly to clarifying the considerations involved in such analyses. Analyses are complicated by the fact that the nature of the mounting becomes important so that the entire system of vessel, shield tank (if any), and total vessel support system may have to be evaluated. In such cases, model studies may have to be made.

2.2.6 Evaluation of Brittle Fracture Characteristics

In the design of reactor vessels, consideration must be given to the steps that must be taken to minimize the possibility of catastrophic brittle failure during service. Because of the importance of this subject and the many ramifications involved, including the effects of radiation on brittle fraction characteristics, this subject is treated as a major topic in Section 2.7, after the discussion on general irradiation effects.

2.2.7 Experimental Stress Analysis

Section III of the Code requires that critical or governing stresses in parts for which theoretical stress analysis is inadequate or for which design values are unavailable shall be substantiated by experimental stress analysis. Appropriate results from tests of similar configurations can be used, if available, provided that they are done as prescribed by the Code. In all tests the strength of any material added to the thickness of members for such reasons as corrosion allowance or cladding, must be discounted.

Strain measurement and photoelastic tests are prescribed as permissible for determining the governing stresses. Either two-dimensional or three-dimensional photoelastic techniques may be used provided the model represents the structural effect of the loading. Detailed specifications for strain measurements are prescribed. Brittle-coating tests may be used only to explore for the critical stress region.

The experimental results are to be interpreted on an elastic basis to determine the stresses corresponding to the design loads. Whenever possible, analytical techniques should be used along with experimental methods in an effort to distinguish between primary, secondary and local stresses so that each combination of categories can be controlled by the applicable stress limit.

2.3 Nozzles and Other Vessel Penetrations

The positions of the inlet and outlet nozzles on the reactor vessel influence not only the design of the vessel but also the design of the external piping and the internal core structure. Therefore a number of factors must be considered in locating the nozzles.

Two basic arrangements are usually considered. In one arrangement the inlet and outlet nozzles are all located above the core. In the other arrangement, the inlet nozzles are located below the bottom level of the core and the outlet nozzles above the top of the core. Some of the considerations in selecting nozzle location are the following:

1. If both the inlet and outlet nozzles are at the same elevation above the core, internal seals will be required in the core structure to prevent direct leakage of coolant flow from the inlet nozzles to the outlet nozzles. The baffles required for directing flow may also be complicated. In addition, the space available for installing, welding and inspecting the nozzles is limited particularly for a multiple-loop plant.
2. If the inlet nozzles are at the bottom, the consequence of a major failure in the nozzle or piping at that point must be considered. Such a failure in a liquid-cooled plant could be worse than a failure elsewhere in the piping system in that the vessel could not contain liquid for keeping the core covered during emergency cooling. To help alleviate this problem a leakproof concrete room can be provided beneath the vessel to hold emergency cooling liquid.

3. The location of nozzles should be so selected that piping loads do not induce stresses which add to other discontinuity stresses. In turn, of course, plant layout is affected.
4. Inasmuch as attachment of the nozzle to the vessel involves increased thickness of metal, the problem of shielding the additional metal from gamma heating must be considered. In addition, the thickened portion should be protected against thermal shock that might result if coolant from a cold loop, which was isolated by valves, were to be accidently allowed to enter the vessel. Protection can be provided by use of a thin thermal sleeve which is inserted into the nozzle and is extended over the length of the thickened portion. One end is welded to the inside surface of the nozzle and the other end is left free to expand. Care should be exercised here to avoid resonant vibration of the thermal sleeve. In addition, provision should be made for a small but definite flow outside the sleeve to help prevent crud collection and crevice corrosion.

The ASME Boiler Code (both Section VIII and the new Section III) requires that a penetration in a pressure vessel be reinforced by material of equal cross-sectional area around the periphery of the penetration. This reinforcing material is called compensation material and can be provided either by local increase of shell thickness involving a boss or collar or by a circumferential reinforcement consisting either of an integral enlargement of the thickness all around the vessel or a separate band wrapped around the circumference of the vessel where the penetrations are to be located. Single penetrations or penetrations far-removed from each other are generally reinforced by the first method. If many penetrations are to be accommodated in close proximity to each other, the second method is used, that is, the thickness of the vessel wall is increased all around the circumference in the region of the penetrations. Careful attention must be given in such a case to the behavior of the ligament of metal between the penetrations. The fatigue loading which may be imposed by the action of appurtenances connected to the vessel at the penetration must also be considered.

Section III of the Code states, "The total cross-sectional area of compensation A, required in any given plane for a vessel under internal pressure shall not be less than

$$A = dt_rF$$

where d = diameter in the given plane of the finished opening in its corroded condition, inches; F = 1.00 when the plane under consideration is in the spherical portion of a head or when the given plane contains the longitudinal axis of a cylindrical shell. For other planes through a shell, use the value of F determined from Fig. 452 (of the Code); t_r = thickness in inches which meets the requirements of N 414.1 (of the Code) in the absence of the opening. Not less than half the required material shall be on each side of the centerline".

Limits within which the metal shall be located in order to have value as compensation are pre-

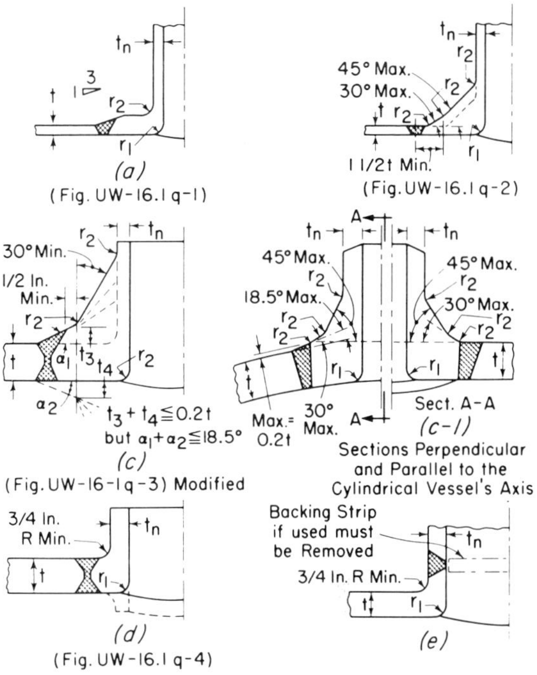

FIG. 2-11 Acceptable welded nozzle attachments readily radiographed to Code standards.
Note: t = thickness of part penetrated;
t_n = thickness of penetrating part;
$t_{c\,min.}$ = 0.7 t_n or 1/4 in., whichever is less;
$r_{1\,min.}$ = 1/4 t or 3/4 in., whichever is less;
r_2 = 1/4 in. minimum.

scribed by the Code. In addition, if a material with a lower design stress value is used, the area provided by such material must be increased in proportion to the inverse ratio of the stress values of the nozzle and the vessel wall material. No reduction in the compensation requirement may be taken for increased strength. The actual strength, however, shall be used in fatigue analysis.

Circular openings need not be provided with compensation material if the hole diameter is less than the shell thickness and less than 5 per cent of the vessel inside diameter and if the arc-distance measured between the center lines of

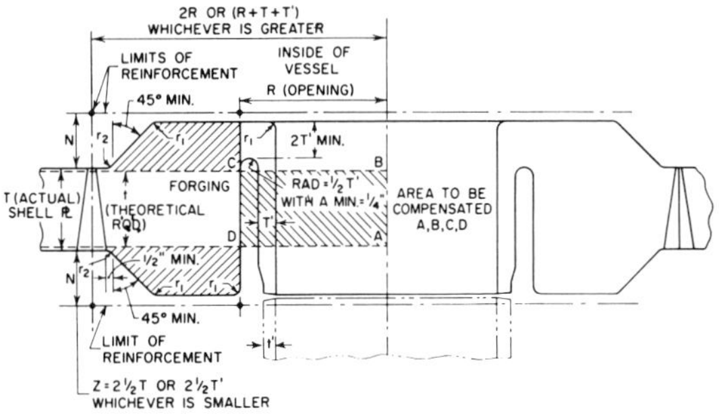

FIG. 2-12 Trepanned nozzle.
Note: r_1 = 1/4 t, 3/4 in. maximum;
r_2 = 1/4 in. minimum.
Reinforcement may be distributed within the limits prescribed by the Code.

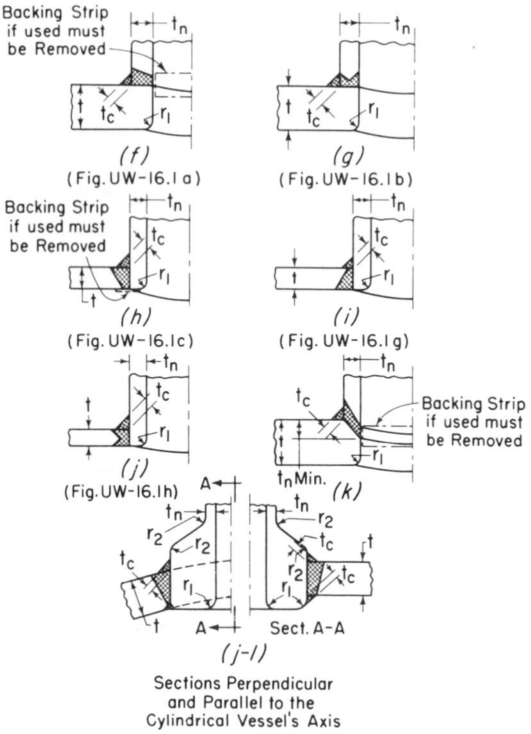

FIG. 2-13 Acceptable full penetration welded nozzle attachments radiographable with difficulty and generally requiring special techniques including multiple exposures to take care of thickness variations.

Note: t = thickness of part penetrated;
t_n = thickness of penetrating part;
$t_c$min. = 0.7 t_n or 1/4 in., whichever is less;
$r_1$min. = 1/4 t or 3/4 in., whichever is less;
r_2 = 1/4 in. minimum.

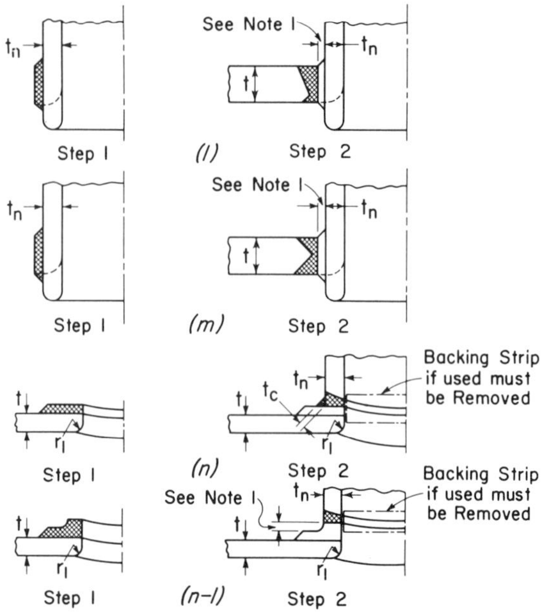

FIG. 2-14 Acceptable full penetration welded nozzle attachments requiring intermediate examination.

Step 1: Examination required before assembly.

Note: t = thickness of part penetrated;
t_n = thickness of penetrating part;
$t_c$min. = 0.7 t_n or 1/4 in., whichever is less;
$r_1$min. = 1/4 t or 3/4 in., whichever is less;
r_2 = 1/4 in. minimum.

Step 2: Radiographic examination made to Figs. 2-11 and 2-13 is required of the attachment used.

Note 1: This dimension shall be at least 3/4 in. when Step 2 radiographic examination is made to Fig. 2-11.

adjacent nozzles along the inside surface of the shelll is not less than three times the sum of their inside radii for openings in a head or along the longitudinal axis of a shell and not less than two times the sum of their radii for openings along the circumference of a cylindrical shell.

At present, suitably reliable methods for calculating the stresses at reinforced openings appear to be lacking. For vessels which meet the requirements for openings and compensation identified in paragraph N-451 of Section III, analyses showing satisfaction of the primary plus secondary stress intensity in the immediate vicinity of the openings is not required. For other situations, analysis or other evaluation of stress is required.

The problems associated with multiple penetrations are particularly difficult. In order to check the validity of assumptions made in design, experimental two-dimensional and/or three-dimensional photoelastic tests are usually made on scale models of the piece.

Section III of the Code identifies two acceptable methods for determining peak stresses around penetrations. One method is the experimental one just referred to. The other is the stress-index method which uses various formulas and available test data for particular configurations and dimensional ratios. This method applies only to single, isolated openings. See reference [10].

Section III of the Code identifies specific methods by which nozzles and other connections must be attached to the shell or head of the vessel. Full-penetration welds, as shown in Figs. 2-11, 2-12, 2-13, and 2-14, are to be used (except as provided in the next paragraph) to achieve continuity of metal and to facilitate radiographic examination [10]. In addition, the Code states that when all or part of the required compensation is attributable to the nozzle, the nozzle shall be attached by full-penetration welds through either the vessel or nozzle thickness or both.

Partial-penetration welds, as shown in Fig. 2-15 [12a], are permitted for making "attachments on which there are substantially no piping reactions, such as control rod housings, pressurizer heater attachments, and openings for instrumentation, and on which there will be no thermal stresses greater than those expected in the vessel itself." The Code states further that, for such attachments, all compensation shall be integral with the part of the vessel penetrated. In addition, partial-penetration welds shall be of sufficient size to develop the full strength of the attachment.

2.4 Flange and Bolt Design

In the design of reactor pressure vessels, careful consideration must be given to discontinuity stresses at the shell-to-flange joint. To reduce

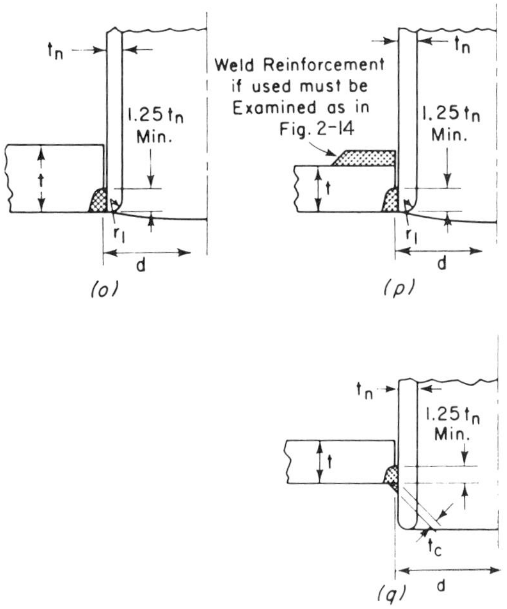

FIG. 2-15 Partial penetration welded nozzle attachments acceptable when independent of reinforcement requirements.

Note:

t = thickness of part penetrated;
t_n = thickness of penetrating part;
$t_c$min. = 0.7 t_n or 1/4 in., whichever is less;
$r_1$min. = 1/4 t_n or 3/4 in., whichever is less;
Weld groove design for oblique nozzles of this type requires special consideration to achieve the 1.25 t_n minimum depth of weld and adequate access for welding and inspection. The welds shown in the sketches may be on either the inside or the outside. With due regard to the requirements in Par. (4)(b) and in the last sentence in Par. (6)(a)(2), the welds shown in the sketches may be on either the inside or the outside.

the stress discontinuity, the thickness of the shell should be gradually increased near the flange to provide a hub beneath the flange, as shown in Fig. 2-16.

Section III of the Code presents more exact formulas for evaluating the behavior of a cylindrical shell when subjected to the action of bending moments of M inch-pounds per inch of circumference, uniformly distributed at the edges and acting at the mean radius of the shell. The stresses

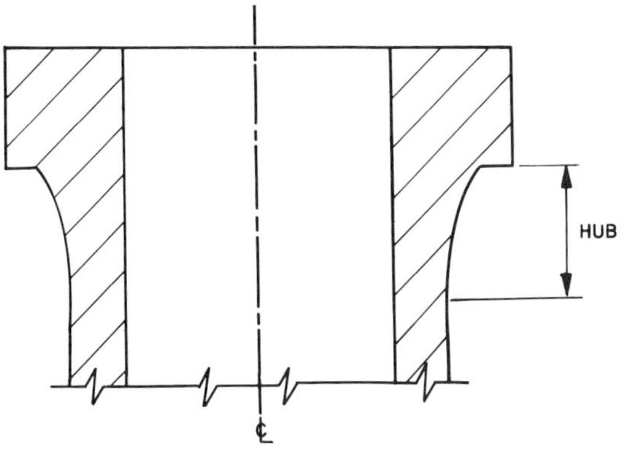

FIG. 2-16 Reduction of stress discontinuity at the shell-to-flange joint.

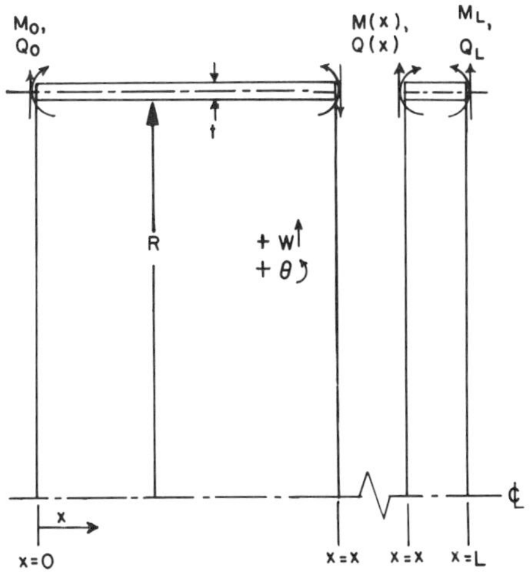

FIG. 2-17 The sign convention arbitrarily chosen for the analysis of cylindrical shells is as indicated. Positive directions assumed for pertinent quantities are indicated.

under such moments must be combined by superposition with stresses independently computed for all other loadings.

The principal stresses (lb/in.^{2} or kg/cm^2) developed at the surfaces of a cylindrical shell at any axial location x, due to uniformly distributed edge-loads as shown in Fig. 2-17 [10a], are given by the following formulas:

$$\sigma_1 = \sigma_t(x) = \{Ew(x)/[R + (t/2)]\} \pm 6\nu M(x)/t^2 , \quad (2\text{-}35)$$

$$\sigma_2 = \sigma_\ell(x) = \pm 6M(x)/t^2 , \quad (2\text{-}36)$$

$$\sigma_3 = \sigma_r = 0 , \quad (2\text{-}37)$$

where in terms preceded by a double sign (±), the upper sign refers to the inside surface of the cylinder and the lower sign refers to the outside surface and

where $w(x)$ = radial displacement of cylinder wall (in. or cm),
x = axial distance measured from the reference end of cylinder (in. or cm),
t = thickness of cylinder (in. or cm),
$M(x)$ = longitudinal bending moment per unit length of circumference (in. lb/in. or cm kg/cm)
E = modulus of elasticity (psi or kg/cm^2),
ν = Poisson's ratio, and
R = inside radius (in. or cm).

The expressions to be used for w(x) and M(x) depend on the loading conditions at the two different edges of the shell and the length of the shell. The interrelations are covered in Section III of the Code. However, if the length of the shell is not less than $3/\beta$, where β is to be evaluated by the formula indicated below, the loading conditions at one edge can be assumed to have no influence on the displacements at the other edge. For this case the following relationships apply:

$$w(x) = (Q_o/2\beta^3 D)f_1(\beta x) + (M_o/2\beta^2 D)f_2(\beta x) , \quad (2\text{-}38)$$

$$\theta(x)/\beta = -(Q_o/2\beta^3 D)f_3(\beta x) - 2(M_o/2\beta^2 D)f_1(\beta x)\,, \quad (2\text{-}39)$$

$$M(x)/2\beta^2 D = (Q_o/2\beta^3 D)f_4(\beta x) + (M_o/2\beta^2 D)f_3(\beta x)\,, \quad (2\text{-}40)$$

$$Q(x)/2\beta^3 D = (Q_o/2\beta^3 D)f_2(\beta x) - 2(M_o/2\beta^2 D)f_4(\beta x)\,, \quad (2\text{-}41)$$

$$f_1(\beta x) = e^{-\beta x}\cos\beta x\,, \quad (2\text{-}42)$$

$$f_2(\beta x) = e^{-\beta x}[\cos\beta x - \sin\beta x]\,, \quad (2\text{-}43)$$

$$f_3(\beta x) = e^{-\beta x}[\cos\beta x + \sin\beta x]\,, \quad (2\text{-}44)$$

$$f_4(\beta x) = e^{-\beta x}\sin\beta x\,, \quad (2\text{-}45)$$

and $Q(x)$ = radial shearing force per unit length of circumference (lb/in. or kg/cm),
θ = dw/dx = rotation of cylinder wall (radians).

The subscript 0 refers to $x = 0$.

The numerical values of functions $f_1(\beta x)$, $f_3(\beta x)$ and $f_4(\beta x)$ are tabulated as $\theta(\beta x)$, $\psi(\beta x)$, $\phi(\beta x)$, and $\xi(\beta x)$ in Table 84 of reference [20]. The values for D and β are obtained from the following expressions:

$$D = \frac{Et^3}{12(1-\nu^2)} \text{ (in./lb or cm/kg)}, \quad (2\text{-}46)$$

$$\beta = \left[\frac{3(1-\nu^2)}{[R + (t/2)]^2 t^2}\right]^{1/4}. \quad (2\text{-}47)$$

When the displacement and rotation are evaluated at $x = 0$, they become

$$w_o = \frac{1}{2\beta^3 D}(\beta M_o + Q_o)\,, \quad (2\text{-}48)$$

$$\theta_o = -\frac{1}{2\beta^2 D}[2\beta M_o + Q_o]\,. \quad (2\text{-}49)$$

The principal stresses developed at any point in the wall of a cylindrical shell due to internal pressure only, excluding the effects of structural discontinuities, are given by:

$$\sigma_1 = \sigma_t = p(1 + Z^2)/(Y^2 - 1)\,, \quad (2\text{-}50)$$

$$\sigma_2 = \sigma_\ell = p(Y^2 - 1)\,, \quad (2\text{-}51)$$

$$\sigma_3 = \sigma_r = p(1 - Z^2)/(Y^2 - 1)\,, \quad (2\text{-}52)$$

where p = internal pressure (psi or kg/cm²),
Y = ratio of outside radius to inside radius,
Z = ratio of outside radius to an intermediate radius.

So far as bolts are concerned, the following rules of Section III of the Code apply, in addition to those requiring analysis for cyclic operation discussed in Sec. 2.2.4 of this chapter.

1. The required bolt load shall be sufficient to resist the hydrostatic end force exerted by the maximum allowable working pressure on the area bounded by the diameter of gasket reaction, and in addition sufficient to maintain, on the gasket or joint contact surface, a compression load which experience has shown to be sufficient to assure a tight joint. A formula for this bolt load is given in Article I-12 of Section III.
2. The gasket or joint contact surface must be seated by applying a minimum initial load at room temperature and without the presence of internal pressure. A separate formula is provided. The bolts must be pretightened sufficiently to satisfy both requirements (1) and (2). When requirement (2) governs, the flange proportions will be a function of bolting instead of internal pressure.
3. The total cross-sectional area of the bolts at the root of thread or section of least diameter under stress shall be governed by the more severe of the above two conditions. The actual cross-sectional area is not to be less than that needed.

In large reactor vessel bolts, precautions must be taken to avoid large torsion stresses due to tightening. This is often accomplished by providing central cavities for the insertion of electric heaters into the bolt so that the bolt can be expanded by heating prior to tightening. After heating the bolts are tightened to a snug position and are then returned to room temperature to provide the desired loading. Care must be exercised in using bolt heaters because of the possibility of overstressing the bolts by overheating them during tightening. Overstressing of reactor vessel closure bolts could lead to a failure of the most severe consequences; failure of a sufficient number of the closure bolts could permit the vessel head to be blown off. This is just about the most serious accident the mechanical designer must guard against. Because of this, hydraulic bolt stretchers are being used more widely for tightening such bolts. A hydraulic bolt stretcher employs a hydraulic cylinder supported on a yoke around the bolt; the piston arm from the cylinder is coupled to the threads of the bolt above the nut. Hydraulic pressure is applied to the underside of the piston to stretch the bolt so that the nut can be run down to its seat. After the hydraulic pressure is released, the desired bolt loading is obtained. This method permits more direct measurement of the extent of bolt stretching during tightening than is possible with bolt heaters, and thereby the chance of overstressing the bolt is reduced [21].

2.5 Pressure-Relief Devices

Pressure-relief devices are important parts of any pressurized installation.

When Section III of the Code was initially issued, Article 9 on this subject was still in preparation.

Since that time, Article 9 has been issued as announced in Mechanical Engineering, August 1964, page 79. Article 9 covers the protection of pressure vessels against overpressure and gives considerable attention to the design, installation, testing and certification of pressure-relief devices.

Each vessel designed to Section III must be protected while in service from consequences arising from the application of steady-state or transient conditions of pressure and temperature which are in excess of the design conditions in the certified design specification called for by the Code. The provisions to meet this requirement and the other applicable rules of Article 9 "are to be the subject of a summary technical report prepared by or on behalf of the owner setting out the degree of overpressure protection provided for each vessel or group of vessels by the pressure-relieving devices, their schematic arrangement in the system, the capacity for relief and dissipation related to the thermal and over-all system characteristics based on both the assumptions made and the analysis of that transient condition which dictated the maximum pressure-relieving requirements, all having regard to the pressure and temperature limitations set out in design specifications."

Acceptable types of pressure-relieving devices include safety and relief valves; rupture disks may be used under special conditions if designed and installed as prescribed by Article 9. While Section III does not specifically define safety and relief valves, the usage is consistent with the following definitions provided in Sections I and VIII:

Safety Valve: An automatic pressure-relieving device actuated by the static pressure upstream of the valve and characterized by full opening pop action. It is used for gas or vapor service.

Relief Valve: An automatic pressure-actuated device actuated by the static pressure upstream of the valve which opens further with the increase in pressure over the opening pressure. It is used primarily for liquid service.

Safety Relief Valve: An automatic pressure-actuated relieving device suitable for use either as a safety valve or relief valve, depending on application.

The following requirements relating to the pressure-relieving devices are extracted from the new Article 9 in Section III of the Code. They are included here for information only and are not all-inclusive; the designer is referred to the latest revision of the Code for compliance purposes.

1. The rated capacity of the pressure-relieving devices including any limitation imposed by the systems connected to the discharge side shall be sufficient to prevent a rise in pressure within the vessels which they protect of more than 10 per cent above the design pressure at the design temperature when the pressure-relieving devices are operating under the condition summarized in the technical report referred to above.
2. The nominal pressure setting of at least one safety or relief valve connected to any vessel or system shall not be greater than the design pressure of the vessel (at design temperature) which it protects. Additional valves required may have higher nominal settings, but in no case shall these settings exceed 105% of the design pressure (at design temperature).

   In determining the setting pressures and discharge capabilities of relief devices required to secure compliance with the general requirements quoted above, full account shall be taken of the pressure drop on both inlet and discharge sides of pressure relief devices at full discharge conditions. In addition, back pressure arising from discharge to closed storage or dissipation systems, or from the use of bursting discs in associations with safety or relief valves, or from the discharge of other devices through common discharge piping shall be considered.

   Where unbalanced safety or relief valves are used, the setting pressure determined by consideration of vessel design pressure and inlet pressure drop under full discharge conditions shall be reduced by total back pressure occasioned by the presence of rupture discs and/or a closed system or other restriction or static pressure on the discharge side.

   If the design or application of a safety or relief valve is such that liquid can collect on the discharge side of the disc, and if such liquid would interfere with proper operation, the valve shall be equipped with a drain at the lowest point where liquid can collect.
3. While pressure-relieving devices need not be installed directly on the vessels which they serve to protect, no stop valve or similar device shall be placed relative to a protective device required for the protection of any vessel so that it could remove the protection afforded to the vessel, except where such stop valves or other devices are shown to be required in the direct interest of system safety or for the purpose of in-service inspection and testing. Any stop valve or similar device on the inlet or discharge side of a protective device provided in conformity with this paragraph shall be so constructed, positively controlled and interlocked that the protection requirement will be complied with under all conditions of operation of the system.
4. Valves operated by pilot control means or devices which are not dependent upon an external energy source are acceptable for meeting the requirements of these rules subject to the following conditions:
   (a) The pilot control device shall be actuated directly by the fluid pressure of the protected vessel or system.
   (b) The main unloading valve operation shall be characterized by full opening pop action in direct response to the operation of the pilot control device.
   (c) The main unloading valve and pilot control device, treated as a combination, shall meet all other requirements of these rules.
   (d) Where operation of the pilot control device is dependent upon integrity of a pressure-sensing element, such as a

bellows, means shall be provided to reveal failure of the sensing element. As an example, a detector or alarm may be provided to sense pressure built up in the chamber as a result of leakage of the sensing element. The chamber shall have a design pressure rating not less than that of chambers on the normally pressurized side of the pressure sensing element.

5. Valves operated by pilot control or other indirect means depending upon an external energy source, such as electrical, pneumatic or hydraulic systems, are not acceptable for meeting the requirements of these rules unless the design is such that the main unloading valve will open automatically by self-actuation even if any essential part of the pilot auxiliary device or the energy source should fail, and provided that:
   (a) the actual set pressure of the valve is reduced to fully compensate for any increase in actuating pressure under self-actuating conditions,
   (b) the capacity rating assigned to the main unloading valve is not greater than the capacity of discharge as limited by such self-actuation.
6. All safety and relief valves and all rupture disks are to be plainly marked by the manufacturer with the data required by Section III. A Code stamp may be applied only upon certification of capacity as specified by the Code.
7. The spring of a safety or relief valve in service for pressures up to and including 250 psig shall not be reset for any pressure more than 10 per cent above or 10 per cent below that for which the valve is marked. For pressure above 250 psig, the spring shall not be reset for any pressure more than 5 per cent above or 5 per cent below that for which the safety or relief valve is marked. If the operating conditions of a valve are changed so as to require a new spring for a different pressure, the valve shall be adjusted by the manufacturer or his authorized representative who shall furnish and install a new name plate.

2.6 Effects of Irradiation

As indicated in Sec. 2.1.2, radiation brings about four effects which must be considered in reactor vessel design. These are heat generation brought about by attenuation of radiation within the metal; radiation effects on the properties of the vessel material, particularly on the brittle-fracture characteristics of the vessel; induced radioactivity in corrosion and wear products from the vessel materials that could be carried by the coolant and deposited in such regions as pumps and valves thus interfering with accessibility for maintenance of such items; and residual radioactivity which influences features for periodic inspection of the vessel.

This section is concerned primarily with the second and third effects listed above, namely radiation effects on the properties of materials and induced radioactivity.

The first of the four effects has already been discussed in connection with thermal stresses. The last is an effect influencing design for which operational procedures and facilities must be provided on the basis of particular plant features.

2.6.1 General Effects of Nuclear Radiations on Metals

This section presents a brief survey of the changes in physical properties brought about by irradiation of reactor-vessel materials. The reader is referred to Chapter 12, Materials and Metallurgy, for further information on this subject.

The nuclear radiations emitted by the reactor core may be characterized as lightly ionizing radiation and heavy particle radiation. In the first category are included beta particles and gamma rays. Gamma rays give up their energy by pair production, Compton scattering or photoelectric effect, all processes which end up with the gamma-ray energy transferred to electrons. Thus all the lightly ionizing radiations result in fast-moving electrons which in turn gradually give up their energy to the surrounding material by ionization. Since the ionization per unit track length is small, the damage produced is small. On the other hand, heavy particles such as protons and fast neutrons which displace atoms from their lattice result in heavy ionizing tracks and cause severe local damage along their track lengths. It is generally agreed that the lightly ionizing radiation produes practically no permanent effect on metals [3,22]. In a reactor vessel, radiation effects result primarily from the action of fast neutrons. The fundamentals of radiation damage in solids as applied to pressure vessels are outlined in Chapter 12, Sec. 2.4 and in several treatments [3, 22, 23, 24].

L. W. Smith [3] has summarized the usual property changes produced by the incidence of neutrons on metals in tabular form reproduced here as Table 2-2. The trends shown in the table are qualitative in nature and are based largely on post-irradiation measurements. Most of the exposures of structural materials to reactor fluxes have been conducted at or near room temperature; relatively few specimens have been exposed to irradiation while being maintained at temperatures approximating temperatures encountered in power reactors. Generally, all changes in physical properties are smaller at elevated temperatures than they are at or near room temperature. Usually the room-temperature changes in properties can be annealed out of pure metals and single-phase alloys to a significant extent, by heating after irradiation; the displaced atoms can diffuse more easily into

TABLE 2-2

Usual property changes produced by the incidence of neutrons on metals

Density decreases	Impact strength decreases
Ductility decreases	Thermal conductivity decreases
Elongation decreases	Electrical conductivity increases
Hardness increases	Modulus of elasticity increases slightly
Yield strength increases	Transition temperature increases (ductile to brittle)
Tensile strength increases	Stress rupture strength decreases*
Creep-rate changes are small	

(Note: *was added to table given in reference [3].)

vacancies at elevated temperatures and thus restore conditions to those existing prior to irradiation. If the metal is an alloy containing an unstable phase, such as austenite in certain steels, exposure to neutrons may cause transformation to a more stable condition at a relatively low temperature. Data obtained to date do not indicate that the behavior of one alloy is appreciably different from another of the same type. L. W. Smith [3] suggests that certain precautions are warranted where similar effects are brought about by neutron exposure and temperature. He points out that precipitation-hardening alloys, for example, should be used in the over-aged condition because under-aged material may reach its maximum hardness during neutron exposure and be more notch-sensitive than in the over-aged condition.

Data indicating irradiation effects are generally correlated against integrated flux (or dose) of neutrons. Some authors differentiate between integrated fast flux and integrated total flux; others do not. In general the errors involved in establishing the fluxes to which specimens have been exposed are high. Nevertheless it appears that neutron changes in steels do not become measureable until nvt, the integrated flux or total neutrons incident per cm^2, reaches values between 10^{17} and 10^{18}/cm^2.

Figure 2-18 [3] shows the effect of irradiation at an nvt of 5×10^{20}/cm^2 on the tensile-load-versus-strain curve for type-347 stainless steel [25]. It will be noted that the yield strength and tensile strength have increased while the elongation has decreased because of the radiation. Figure 2-19 [3] shows the effect of integrated fast flux in excess of 1 Mev on the yield strength (0.2% offset) of austenitic stainless steels at less than 200°F (or 93°C) [26]. This curve tends to support the generalization that effects of irradiation approach saturation levels. A composite picture illustrating the changes in mechanical properties caused by irradiation at temperatures below 200°F (93°C) on carbon-silicon steel ASTM-A-212B has been prepared by M. H. Bartz [27] and is shown in Fig. 2-20 [3].

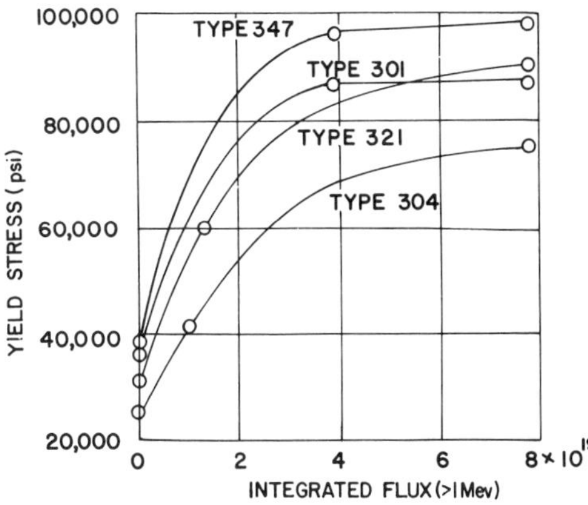

FIG. 2-19 Effect of integrated fast (> 1 Mev) neutron flux (in cm^{-2}) on the yield strength (0.2% offset) of austenitic stainless steels irradiated below 200°F (93°C).

The irradiation effects on steels used for reactor vessels to date have not led to any difficulties in service. Exposures in service, however, are just beginning to enter the range where measurable effects are encountered. Of the numerous effects listed in Table 2-2 the one of most concern is that influencing the transition temperature for brittle fracture. Because of the importance of this subject it is treated separately in the next section.

2.6.2 Induced Radioactivity

In addition to the necessity to avoid corrosion and wear in a vessel because they weaken the material, corrosion and wear must be limited because of the effects that these products may have elsewhere in the system. Not only might they interfere with operation of components requiring close clearances because of the particulate nature of the products, but in addition, they can limit accessibility for maintenance and repair by being deposited in various regions of the system. As a result there is a strong desire to minimize the use of materials

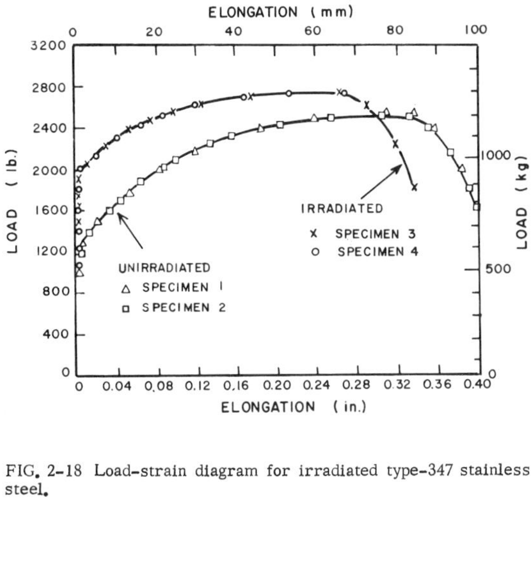

FIG. 2-18 Load-strain diagram for irradiated type-347 stainless steel.

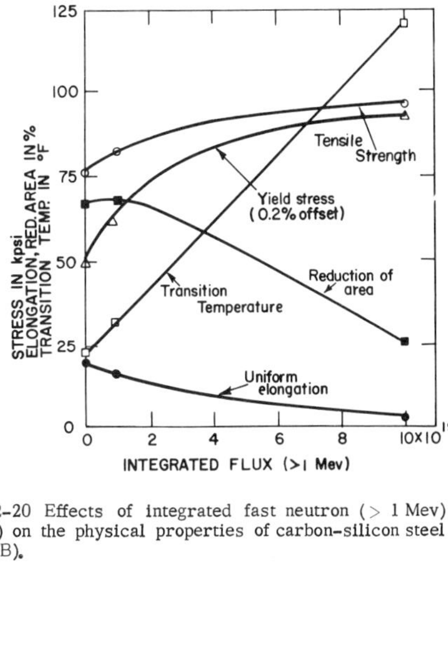

FIG. 2-20 Effects of integrated fast neutron (> 1 Mev) flux (in cm^{-2}) on the physical properties of carbon-silicon steel (ASTM-A-212B).

whose corrosion products can be converted to long-lived gamma emitters in the neutron flux field of the core.

The chief elements of concern in reactor vessel design are cobalt and tantalum. Cobalt in the form of Co^{60} is an undesirable product because it has a half-life of 5.3 years; in addition it emits two gammas per disintegration, one with energy of 1.1 Mev and the other 1.3 Mev. The Co^{60} can be developed either by absorption of neutrons in Co^{59}, the most abundant isotope in natural cobalt, or by the fast neutron reaction $Ni^{60}(n,p)Co^{60}$. By far the more serious concern is that brought about by the cobalt used in the material itself. Thus in thermal reactors, the use of cobalt-base alloys should be limited as much as possible. In addition, however, it must be recognized that cobalt is usually contained in nickel ores, and hence finds its way in varying small amounts into nickel alloys such as austenitic steels. Furthermore, commercial stainless steels may contain as much as 0.4% cobalt because of the scrap used in its manufacture [3]. Tantalum is of concern because it is contained in the Nb stabilizer in stainless steels 309 and 347. When activated, the resulting Ta^{182} emits gammas of about 1.2 Mev and has a half-life of 113 days. The tantalum problem has been alleviated to a large extent with the development (primarily for nuclear applications) of type 348 stainless steel, which is identical to type 347 except that the tantalum content is controlled to 0.1% maximum. The properties of type 348 are quite similar to those of type 347.

Measurements of the gamma activity induced in a number of structural and reactor materials (including those given above) that were irradiated in the ORNL Graphite Reactor have been reported by C. D. Bopp and O. Sisman [38].

2.7 Brittle Fracture Considerations in Vessel Design

Under certain conditions (outlined below) large steel structures can fail by brittle fracture even though they are made of otherwise ductile steel [28]. Service failures of this type have been experienced with all-welded ships during the early part of World War II [29] as well as with bridges, pressure vessels, pipelines, and storage tanks [30]. Studies which were prompted by these incidents showed that the failures could be attributed to the low-temperature brittleness of the materials involved. It was found that the energy which is absorbed by the material before it fractures in low-strain-rate tensile tests can be used as a measure of the ductility or toughness [160]. Various types of tests were devised to determine the temperature at which the transition from ductile to brittle failure takes place under impact loading. Inasmuch as notched specimens were found to have higher and therefore more conservative transition temperatures, most testing has been done with notched specimens.

It appears that three conditions are required to induce brittle fracture [31, 32]. These are:

a stress raiser, such as a minute crack, a sharp notch, or a small flaw as may be caused by an inadvertent arc strike with a welding rod,

a concentration of stress large enough to cause local yielding in the area of the stress raiser, and

a temperature in the material lower than a temperature described as the nil-ductility transition temperature (NDT temperature).

If all of the above conditions are present, a brittle failure can occur. The nil-ductility transition temperature is that temperature below which steel under test does not have to plastically deform prior to fracturing but below which an initiated crack propagates easily through elastically stressed regions. One method to perform this test is to weld a weld bead on a sample of plate material, then to cut a longitudinal notch in the weld and place the plate on an anvil with a slightly curved depression into which the plate can deform elastically under an impact load. After the load is dropped and if the plate does not crack, the temperature is lowered; this is repeated in subsequent tests until a temperature is reached at which the plate breaks. This temperature is the nil-ductility transition temperature. The details of such testing have been written up in a proposed ASTM specification [33].

Two other important temperatures can be defined. First is the FTE temperature (fracture transition temperature for elastic loading) below which the crack will propagate to some extent through an elastically stressed region but will not cause complete cleavage. Above this temperature no propagation of stress occurs in elastically stressed regions. The other temperature is FTP temperature (fracture transition temperature for plastic loading) below which the crack propagates only through the plastically stressed region. Thus, above the FTE temperature but below the FTP temperature a crack started in a plastically stressed region will be arrested when it reaches an elastically stressed region. The temperature differences between the NDT, FTE, and FTP have been found to be about the same for all ferritic materials that have been tested. Thus, FTP = FTE + 60°F (= FTE + 33°C) and FTE = NDT + 60°F. Porse [32] also defines a design transition temperature (DTT) to be used for design purposes. For the present, he proposes the FTE or NDT temperature in Fahrenheit plus 60° (or in Centigrade Plus 33°C) be used as the value for the DTT.

The NDT temperature can also be determined by an explosion test, but the explosion test is not practical for routine use to establish the NDT temperature of steel from a particular heat. Although the drop weight test can be used to routinely establish the NDT of a particular heat, it has been more common practice to use the Charpy V-notch test for this purpose because this test can be performed more easily and on smaller specimens than the drop weight tests require. Correlations have been made between the NDT temperature and the Charpy V-notch test to fix the NDT temperature at the temperature at which one obtains some fixed value of impact without failure (e.g., 30 ft-lbs or 4.6 kg-m for A-302B steel) for the material to be used. However, for thick plates there is considerable scatter in the Charpy V-notch data; a conservative approach would indicate acceptance of the higher values of NDT temperature obtained

[32]. The Charpy V-notch test utilized a bar with a V-notch of prescribed proportions in the center of one side. The bar is supported as a simple beam in the anvil of an impact testing machine; the center of the bar on the side opposite the notch is then struck with a weight on a pendulum to observe the energy necessary to fracture the specimen. The energy of impact is varied by changing the weight or the position from which the pendulum is allowed to fall [34].

In general, it is not sufficient to operate above the NDT temperature in order to avoid brittle-type failure. Consideration must also be given to the state of stress and the size of flaws in the material at the time of failure. This fact is illustrated by the fracture diagram shown in Fig. 2-31 [31] of Chapter 12. As can be seen in this figure, a flaw-free specimen will suffer brittle fracture at the NDT temperature only if the nominal stress in the material is at or slightly above the yield stress. As flaws of increasing size are introduced into the specimen, the nominal stress at which fracture can occur is reduced, until a stress level of 5000 to 8000 psi is reached; below that stress level there does not appear to be enough elastic energy stored in the structure to support propagation of even a very large crack. Above the NDT temperature, a similar trend can be observed, but the stress level at which failure can be expected increases for each flaw size. When the temperature of the material is 60°F higher than the NDT temperature, even a large crack does not propogate under test unless the nominal stress is at the yield stress value. At a temperature 120°F higher than the NDT temperature the probability of brittle fracture is essentially zero at nominal stresses less than the ultimate. Between the NDT temperature and NDT+ 120°F there appears to exist a lower bound on the nominal stress at which brittle fracture might be expected at a given temperature. This is called the crack-arrest-temperature curve (CAT); below and to the right of this curve, brittle fractures are prevented by the crack-arrest properties of the steel. For a given nominal stress level, operation at temperatues above this curve eliminates the question of flaw size [160].

It can be seen from this diagram that temperatures of 60°F to 120°F above the NDT temperature imply substantially greater fracture toughness or resistance to brittle failure than do the NDT temperatures. An extra 60°F above the NDT, therefore, constitutes more than a safety factor as usually used to take into account uncertainties in materials properties or operating conditions; it defines a temperature at which fractures do not propagate under nominal elastic loads. Uncertainties must still be accounted for. Typical of the uncertainties which must be considered are the state of stress, the material structure, the flaw size, the effects of environment including radiation, and the accuracy of the NDT temperature. (See reference [161].)

2.7.1 Radiation Effects on Transition Temperature

As noted in Sec. 2.6, the effect of radiation on structural materials that is of most concern is the effect on the transition temperature for brittle fracture. The effect depends upon the material involved. Austenitic steels behave differently from ferritic steels.

Irvin et al. [162] have investigated the combined effects of temperature and irradiation on the mechanical properties of austenic stainless steels. Notched tensile specimens of AISI 348 stainless steel in the annealed, 10, 20, and 40% cold worked conditions were irradiated at about 290°C and tested at room temperature. The mode of fracture was ductile in nature throughout the range of cold work and neutron exposures studied. A transition from notched strengthening to a notched sensitive behavior was observed for specimens in the 20 and 40% cold work levels after an exposure of about 1.0×10^{20} neutrons/cm^2.

Claudson and Pessl [163] have studied irradiation effects on nickel-based alloys including Inconel 600, Inconel 702, Inconel 718, Inconel 800, Hastelloy C, Hastelloy N, Hastelloy R-235 and Hastelloy X-800. Results showed that little effect upon their mechanical properties could be found for those alloys when irradiated at relatively low temperatures and tested at room temperature. However, when irradiated at high temperatures and tested at 700°C, these alloys exhibited a rapid deterioration of properties occurred, especially the ductility of the material.

For ferritic steels, as indicated in Fig. 2-20, the ductile-to-brittle transition temperature appears to increase linearly with integrated fast neutron flux (> 1 Mev) up to a value of about 10m nvt. This effect causes concern that if the trend were to persist continuously throughout the life of a reactor pressure vessel, the vessel might, late in life, be in jeopardy of failure due to brittle fracture. Pawlicki [35] has indicated that with a properly designed thermal shield inside the reactor vessel, the radiation-damage effects in the vessel are due primarily to neutrons of energy greater than 1 Mev. Pellini [36] has indicated that the increase in the NDT with integrated fast neutron flux (> 1 Mev) is not very sensitive to differences in the chemistry of heat-treatment of the steel or to the initial NDT. He has shown that the results of tests involving a wide variety of steels cluster in a relatively narrow band. (See Fig. 2-22 [23].) Because of differences in irradiation temperature, dose rate and neutron spectrum, more precise correlation could not be made. Nevertheless, it appears that the increase in the NDT temperature, ΔT_{NDT}, can be related to the integrated neutron exposure θ by the expression

$$\Delta T_{NDT} = A\phi^n , \qquad (2\text{-}53)$$

where A is a constant and n, the exponent of θ, is suggested to be 1/3 by some authors and 1/2 by others [23].

Consideration must be given to several complicating factors in evaluating ΔT_{NDT} by the above expression. First, it should be recognized that the influence of having the material under stress during irradiation has not been explored. Recent work by Reynolds [37] is encouraging in this regard because it shows no deleterious effects of the presence of stress during irradiation. Trozera et al.

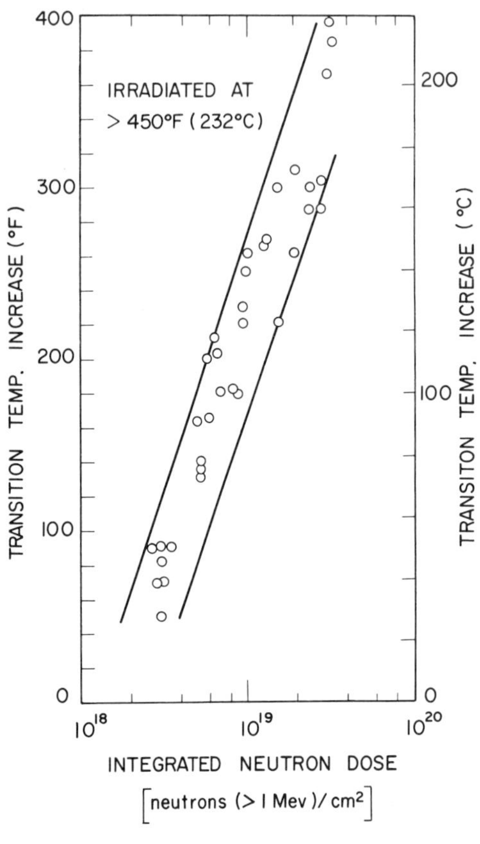

FIG. 2-21. Increase in NDT temperature versus integrated neutron flux for ferritic steels.

[164] present data obtained from tests to determine the effect of irradiation on uniform strain under multiaxial stress. Their results imply a large uncertainty in using simple tensile data to predict properties under complicated stress conditions. More work of this nature is required. Second, and perhaps more important, is the fact that relatively few irradiations have been performed at the temperatures that the materials would encounter in service as a reactor pressure vessel; thus, the effects of annealing are not well-known. As indicated in the preceding section, effects of radiation decrease with increasing temperature. Postirradiation annealing tests have been made to determine the percentage of recovery of notch ductility of pressure vessel steels. The percentage of recovery does not appear to depend on the integrated exposure, but it does depend on the annealing temperature as well as the temperature at which the specimen was irradiated. In the cases tested so far the percentage of recovery decreases as the irradiation temperature increases. In a particular investigation of ASTM A-212 grade B steel, it was found that long-time annealing at 600°F (315°C) following irradiation at 300°F (150°C) produced re-embrittlement [36]. Wechsler and Berggren [23] point out that, based on this fact, the suggestion has been made that the curve relating embrittlement to irradiation temperature passes through a maximum between approximately 300 and 500° F (150° and 260°C). It is clear that more work is required before the effect of temperature on the increase of NDT temperature with irradiation can be completely evaluated.

Porse [32] points out that close attention must be given to the location of structural discontinuities, such as the flange and nozzles, relative to the core when evaluating radiation effects.

Another important factor in the evaluation of the NDT temperature shift is the neutron dosage to which the reactor vessel has been exposed. Bush [165] points out that much of the scatter in Charpy-temperature curves versus neutron exposure can be attributed to inaccurate measurement of the integrated exposure (nvt). Conventional neutron dosimetry with the assumption of a fission spectrum may result in errors in nvt of 300 to 500% because the spectrum actually is strongly perturbed. Current refinements in both dosimetry and reactor physics reduce this uncertainty to 25-50%; however, very few reactor operators use the more sophisticated techniques. For example, application of these techniques to NRL data reduced the scatter in Charpy data vs. nvt from x 50% to x 15%, indicating the value of such methods.

A consideration which, it has been thought, might mitigate the hazard of increased embrittlement of ferritic steel under irradiation is the reduction of neutron flux as it passes through the thickness of a heavy pressure-vessel wall. Serpan and Steele [166] report the results of studies on the in-depth embrittlement of a simulated pressure vessel wall of A-302 B steel. From their work, it appears that the reduction in embrittlement through a 6-inch simulated pressure vessel wall is not large enough to significantly reduce the potential hazard of brittle fracture in a neutron-embrittled reactor pressure vessel.

It is also well to point out that very little information is available concerning the effect of irradiation on fatigue and creep properties of materials; caution should be exercised in the design of equipment where these problems are involved.

Hawthorne and Steele [167] report initial results of dynamic in-reactor reverse bend fatigue tests. Their data did not indicate any pronounced difference in the fatigue strength in the material in the irradiated versus unirradiated condition.

A great deal of additional work on irradiation effects on reactor structural materials is going on at the United States National Laboratories in an effort to provide the data needed. This work is reported periodically in topical reports and in quarterly progress reports [168].

2.7.2 Fracture Mechanics Methods for Evaluating Brittle Fracture

Most of the limitations on operating conditions for reactor vessels, to date, have been based on consideration of ductile-to-brittle fracture transitions. Because of uncertainties, limitations, and qualifications associated with this approach, considerable attention is being given to the use of fracture mechanics to provide an improved basis for evaluating the resistance of material to brittle fracture.

Fracture mechanics is concerned with the effect of stresses near a notch on the integrity of a structural member. As indicated in reference [169], the subject of fracture mechanics can be categorized under three headings: theoretical fracture mechanics, experimental fracture mechanics, and applied fracture mechanics.

2.7.2.1 Theoretical Fracture Mechanics. In the usual treatment of stress analysis, it is assumed that there is continuity of elastic action throughout the structural member being analyzed and that distribution of stresses on a section of the member may be expressed mathematically by relatively simple laws. The condition of stress which exists at a notch or a crack is, however, quite different from that predicted by ordinary stress relationships. If a flaw or a crack exists in a plate, the localized stress adjacent to the crack is considerably higher than the nominal stress in the plate. Thus if a sizeable tensile load, normal to the crack, is applied to the plate, the localized stress, based on elastic action, would be well above the yield point for most structural materials. Consequently, local yielding would take place at the crack tip, and the crack tip radius would increase. This results in a strain hardening near the crack tip as well as an increase in crack tip radius. When the deformation in the material at the crack tip reaches the value associated with the ultimate strength, further increase in load to the member results in growth of the crack. As pointed out in reference [169], if the stress is cyclic instead of monotonic, strain hardening will occur at a lower deformation value. Consequently, crack growth will occur at a lower value of nominal stress under fatigue than it will under monotonic loading.

This concept is consistent with the observed fact that solids containing cracks or other flaws suffer fracture at loads for which the usually calculated stresses are well below the ultimate stress. Sneddon [170] has analyzed the stress field in the vicinity of a disc-shaped crack of radius, a, in an

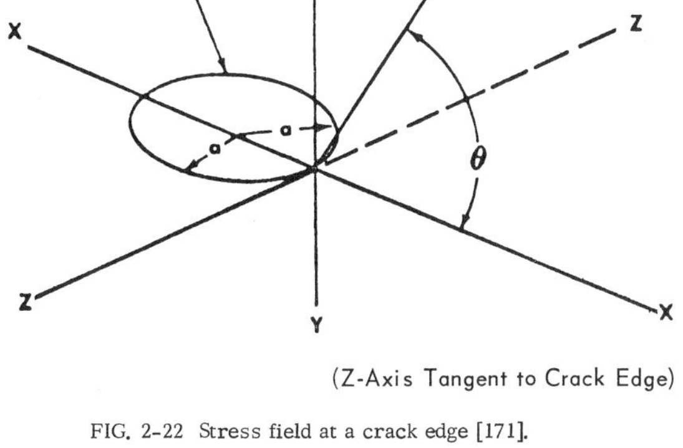

FIG. 2-22 Stress field at a crack edge [171].

infinite isotropic elastic solid under uniaxial tensile stress, σ, in a direction normal to the plane of the crack. See Fig. 2-22. Using this work, Reynolds [171] shows that the tensile stress in the y direction at a point, p, at a distance, r, from the crack edge is given by

$$\sigma_{yy} = \frac{K_I \cos(\theta/2)}{\sqrt{2\pi r}} (1 + \sin(\theta/2)\sin(3\theta/2)) , \qquad (2\text{-}54)$$

where K_I is the stress intensity factor associated with the crack tip radius. The stress intensity factor represents the elevation in stress caused by the crack relative to the unperturbed stress field.

It can be seen that the maximum value for σ_{yy} occurs when $\theta = 0$. Thus the maximum value of σ_{yy} in Fig. 2-23 becomes

$$\sigma_{yy} = \frac{K_I}{\sqrt{2\pi r}} . \qquad (2\text{-}55)$$

This expression implies that, if the material is truly elastic, the stress at the edge of the crack, where $r = 0$, becomes infinite; however, because the material is not elastic for all values of stress, it will yield.

When σ_{yy} in Eq. (2.55) becomes equal to the ultimate stress, it is assumed that the crack will begin to grow and that the stress intensity factor, K_I, has reached a critical value, K_{Ic}. For the crack system shown in Fig. 2-23, it has been shown [172] that K_{Ic} is related to the gross stress, σ, by

$$K_{Ic} = \sigma\sqrt{\pi a} . \qquad (2.56)$$

The critical stress intensity factor, K_{Ic}, is assumed to be a property of the material and is usually evaluated in units of Ksi-$in^{1/2}$.

As pointed out in reference [169], the presence of a sharp defect in a member under uniform uniaxial stress will influence stresses in all principal directions, except for the special area where the defect plane is parallel to the uniaxial stress. Similar situations also exist under more complex loading patterns, particularly those involving three-dimensional problems. Because of the complexity associated with most three-dimensional problems, which often involve triaxial stresses, simplifications are sought. Usually a two-dimensional approximation is used to represent the real system under consideration. As pointed out in reference [169], aside from problems of rotational symmetry, there are two mathematically important two-dimensional approximations, plane stress and plane strain.

In the case of plane stress, it is assumed that the state of stress on every volume element can be described by a set of principal stresses, one of which is zero and two of which are nonzero. The plane stress approximation is usually applied to thin sections such as wall tubing under internal pressure in which the stresses normal to the free surfaces are essentially zero in every volume element.

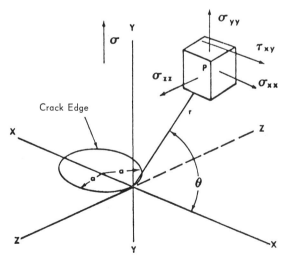

In the case of plane strain it is assumed that the state of stress is such as to produce a zero strain in a particular direction. If this one component of strain is a constant but not zero, the generalized plane strain equations also apply; the requirement is that interior planes remain plane and normal to the consistent strain direction.

In real materials, of course, the stress field is described by neither plane stress nor plane strain. Nevertheless, these approximations are used because of the analytical simplifications which they permit.

The subscript I, which is used to denote the limiting stress intensity or fracture toughness of a material for plane strain, implies a wedge-opening mode of failure. Three principal fracture displacement modes are identified in reference [169]; the wedge-opening mode (Mode I), the forward shear- or edgesliding mode (Mode II), and the parallel shear- or screw-sliding mode (Mode III). For most brittle fractures, Mode I is of major interest because more elastic strain energy is made available under Mode I crack displacement than in the other modes. No subscript is applied to plane stress because, in general, the crack opening displacement modes described above do not apply.

As indicated in reference [171], Irwin [173] showed that the elastic energy, G, released per unit area of crack formed could be given in terms of the stress intensity factor. Thus one can define a critical value of G, namely G_c, corresponding to the critical of K_c at which crack growth occurs. This value, G_c, is known as the fracture toughness of the material and is also assumed to be a property of the material. The relationship between G_c and K_c is given by the following expressions:

$$G_c = \frac{K_c^2}{E} \quad \text{(plane stress)}, \qquad (2\text{-}57)$$

$$G_{Ic} = \frac{K_{Ic}^2}{E}(1 - \nu^2) \quad \text{(plane strain)}. \qquad (2\text{-}58)$$

Substitution of the Eq. (2-56) into the above yields the following expressions for G_c in terms of the gross applied stress σ:

$$G_c = \frac{\pi a \sigma^2}{E} \quad \text{(plane stress)}, \qquad (2\text{-}59)$$

$$G_{Ic} = \frac{\pi a \sigma^2}{E}(1 - \nu^2) \quad \text{(plane strain)}. \qquad (2\text{-}60)$$

In engineering work, G usually has the units in.-lb/in^2 (energy per unit area or force per unit length).

For purposes of predicting or controlling fractures, either of the parameters G and K can be used. The parameter K is often preferred because the principle of superposition can be applied to the stress intensity factor K.

It has been tacitly assumed in the discussion that the fracture stress can be reached by elastic deformations. Reynolds [171] reports that as long as the plastic zone at the end of the crack is small enough so that the elastic stress field around the crack tip is not significantly altered, the fracture toughness concept can be used with reasonable accuracy to predict the behavior of the material containing a flaw of known geometry.

2.7.2.2 Experimental Fracture Mechanics. Fracture toughness tests are used to determine values for K_c or G_c. As indicated in reference [171], these tests are of two types: those in which stress intensity is the quantity measured and those in which energy is measured.

Tests for measuring the stress intensity factor require a specimen containing a crack or crack-initiating notch (of geometry for which the stress intensity factor can be calculated). The specimen is loaded until rapid unstable crack growth occurs. To calculate the critical stress intensity factor, it is necessary to know the gross stress and the position of the crack front (so that the stress intensity factor may be related to gross stress) at the instant of instability. Reynolds [171] points out that instability may be detected acoustically, by discontinuity in the load-deformation curve, or by discontinuity in the electrical conductance of the specimen as a function of load. A wide variety of specimens have been used in such tests, depending on the test equipment available and the material of the specimens used.

Fracture energy tests involve measurements of fracture energy and areas of fracture surfaces. Reynolds points out that G_c, for example, can be obtained by dividing the fracture energy obtained in a Charpy impact test by the area of fracture surface created. The quantity obtained is the total energy per unit fracture area averaged over the specimen net cross section and including both fracture energy and plastic deformation energy. Only by eliminating the plastic deformation energy contribution can data be obtained which are comparable to G_c data obtained from force measurements. In addition, a shear lip indicative of plastic deformation usually surrounds the fracture surface. To measure G_c, it is necessary to minimize the plastic energy contribution to the measured impact fracture energy.

Apart from any differences caused by strain rate effects, the value of G_{Ic} determined from a tensile or slow-bend K_c test should correspond to the value from impact fracture energy measured at the same temperature but corrected for crack starting and shear lip energy absorption. This approach has been pursued by Orner and Hartbower who use Charpy-type impact specimens but sharpen the standard notch by producing a fatigue crack at its base.

From the data obtained in these tests, it is possible to calculate fracture toughness utilizing the appropriate relationship between fracture toughness, fracture stresses, and defect sizes. Care must be exercised to ensure that fracture toughness specimens are tested under conditions truly representative of the service environment of interest [169]. Environment conditions can play an important part in influencing the data obtained, particularly where corrosive reactions may be involved.

It is important to note that fracture toughness as a material parameter is affected by temperature, state of stress in the vicinity of the notch, mode of fracture, and material structure [169].

Fracture toughness can also be influenced by neutron radiation; studies are being made of the effect of neutron radiation on fracture toughness of structural steels. Some preliminary results of such work are reported in reference [169].

2.7.2.3 Applied Fracture Mechanics. Applied fracture mechanics is used to predict design conditions under which brittle fracture might be avoided. Johnson [169] points to two major items that affect the accuracy of such predictions. The first of these is the applicability of the mathematical model which must be used to describe the flaw. While experiments have shown the applicability of equations to artificial defects which were shaped to conform to the geometries assumed in the mathematical models, natural defects which are irregular in shape do not necessarily conform. The second is the accuracy of the parameters used in the calculations.

Subject to these limitations, Johnson illustrates a method by which fracture mechanics may be utilized to set limits on stresses. He assumes a pressure vessel constructed of welded ASTM grade 302B steel with two defects of significant size present in the material. During service life these defects are assumed to grow by slow mechanisms to the sizes indicated below. In the example, no consideration was given to thermal transient stresses, body force stresses, secondary stresses, or other sources which should be recognized in the general case.

The first defect, A, is assumed to be a cluster of closely spaced inhomogeneities in a girth weld which during service coalesce to form an internal disk-shaped crack normal to the vessel axis and of radius $a = 1.0$ inch. The second defect, B, is assumed to be a small semielliptic surface crack, open to the outside surface of the vessel and normal to the hoop stress; this defect is assumed to grow by fatigue to a semicircle (the stable crack boundary), having a radius $a = 1.0$ inch.

Based on the relationships developed by Irwin [173], for part-through elliptical defects, Johnson evaluates the fracture stresses by the following equations.

Normal to the longitudinal stress:

$$\sigma_f(\ell) = \frac{K_{Ic}}{2\sqrt{a/\pi}} \tag{2-61}$$

Normal to the hoop stress:

$$\sigma_f(t) = \frac{(\pi/2)\,K_{Ic}}{[1.21\,\pi a + 0.212\,(K_{Ic}/\sigma_{ys})^2]^{1/2}} \tag{2-62}$$

where σ_{ys} = yield stress.

Utilizing the foregoing information, Johnson then plots curves showing the calculated fracture stress as a function of temperature using fracture toughness data as a function of temperature obtained from reference [174]. He then proceeds to determine and plot the maximum allowable pressure as a function of temperature. The latter curves show that the existence of defect B, oriented normal to the tangential stress, results for this case in a brittle fracture limit on pressure which is lower than the ductile pressure limit for temperatures up to about 315°F; the existence of defect A does not result in any brittle fracture limitation for temperatures at or above room temperature.

Based on the approach outlined by Johnson, Spencer Bush [165] has prepared illustrative calculations which show the critical crack sizes that could lead to catastrophic failure in a postulated reactor vessel. He assumes a vessel of 66 inches I.D. with a wall thickness of 7 inches operating at a pressure of 2500 psig and a temperature of 650°F. The following simplifying assumptions are made.

1. Residual and bending stresses are ignored.
2. The operating pressure is assumed fixed over the operating range of temperatures.
3. Simple thin wall formulas are used to calculate longitudinal and hoop stresses:

$$\sigma_z = \frac{Pr}{2r} = \text{longitudinal stress} = 11{,}750 \text{ psi};$$

$$\sigma_t = \frac{Pr}{t} = \text{transverse or hoop stress} = 23{,}500 \text{ psi}.$$

4. Two defects of the form (described above) postulated by Johnson [169] are assumed, one in a girth weld normal to the vessel axis and one in the vessel wall normal to the hoop stress.
5. The fracture toughness (K_{Ic}) values for an A-302B steel after irradiation are taken from Johnson [169]. Values for nvt's from 9×10^{18} to 9×10^{19} are used as representative of a reasonable range for a PWR pressure vessel after 10-20 years; these cover the potential range, even assuming errors in neutron exposure measurements. This range covers a transition temperature shift of about 150 to 400°F.
6. Yield strengths in the unirradiated state are used at the appropriate temperatures. This is reasonable since changes in yield strength have only a limited effect on the critical crack size.
7. Plane strain failure is assumed which is appropriate for wall thicknesses in the range 5-10 inches.

The critical flaw size for propagation is determined from the equations used by Johnson as quoted above. The results of the calculations are presented in Table 2-3.

From Table 2-3, it can be seen that the most probable failure mode is the splitting of the vessel normal to the hoop stress. Relatively small crack sizes are required for such a failure, and any compositional or additional irradiation effects which lower K_{Ic} values will make the situation more critical.

TABLE 2-3

Critical Crack Sizes in Inches for Catastrophic Failure

Assumed Temp. °F	Neutron Exposure	K_{Ic}	Yield Strength	Crack Size Normal to Hoop Stress: Radius	Hoop Stress: Diameter	Longitudinal Stress: Radius	Longitudinal Stress: Diameter
300	9×10^{19}	24	46,000	0.67	1.34	3.1	6.2
300	9×10^{18}	33	46,000	1.26	2.52	6.0	12.0
400	9×10^{19}	30	44,000	1.02	2.04	4.9	9.8
400	9×10^{18}	43	44,000	2.13	4.26	10.8	21.6

The table also shows that the possibility of splitting the vessel about its girth is exceedingly small, unless it involves a girth weld of poor quality and of inadequate heat treatment which could have drastically lower K_{Ic} values than the base plate.

2.7.3 In-Service Surveillance of Brittle Fracture Characteristics

Most of the civilian reactors now in operation or in advanced states of design provide for in-service surveillance programs whereby samples installed in the reactor vessel can be removed periodically and tested to provide knowledge about properties of materials of critical components, such as the reactor vessel, during service life [175]. The use of such programs is to be highly encouraged; they should be carefully planned to ensure reliable and useful data.

To supplement the knowledge from specimen tests, studies are being made of parts of pressure vessels and other components of decommissioned reactors, including the Army SL-1 and PM-2A and the Organic Moderated Reactor Experiment. These studies will provide significant data on the in-service effects produced in components themselves during the life of the reactor. Such data will be useful for comparison with data from tests on specimens [176].

2.8 Multilayer Vessels

As indicated by the name, multilayer vessels are constructed from multiple layers of thin steel plate, each plate being concentrically wrapped, tightened, and welded longitudinally around an inner, pressure tight cylinder. The longitudinal welds are staggered circumferentially relative to each other and are ground flush so that the next layer can be appropriately fitted around them. Such vessels can be made to have walls of any desired thickness, and the inner cylinder can be fabricated by any weldable material. The wrapping tension and weld shrinkage are used to obtain compressive stresses in the inner layers to compensate for the high tensile stresses imposed on these layers during pressurization. As a result, the pressure stresses in thick wall vessels can be equalized to a large extent during operation.

This method of construction can be applied to vessels of any diameter for which end closures or end bolting flanges can be provided. Increased strength in the wall can be provided by adding more layers or by using steel of higher strength characteristics. Generally only the inner cylinder needs to be pressure-tight, the remaining outer layers being provided with vent holes to protect them against damage from entrapped vapors and to warn of leakage in the inner cylinder. In the event of failure of the vessel, there is less tendency for fragmentation of the wall of the vessel.

Although multilayer vessels have been used extensively for nonnuclear pressure vessels, only two nuclear reactor vessels of this type have been used to date, the vessel for the Saxton Nuclear Experimental Corporation Reactor at Saxton, Pennsylvania and the vessel for SPERT I at NRTS in Idaho. However, the desire for large-diameter vessels in future large nuclear power plants coupled with the experience with the Saxton vessel may very well cause reconsideration of the factors influencing past decisions against them.

Concern about the application of multilayer construction to reactor vessels has centered around the inability to use thermal-stress relief and to obtain meaningful radiographs. In addition, although flanges and nozzles can be welded to these vessels successfully, adequate methods of nondestructive testing have not been completely worked out. The need for vent holes between layers to prevent buildup of pressure between the layers has bothered some nuclear plant designers because of the reduction in the thickness of material providing containment. Other factors for consideration include the action of the vessel under repeated thermal cycling, the influence of irradiation on the brittle fracture characteristics of such vessels and the lack of code coverage for design purposes.

2.9 Thermal Insulation and Canning

To reduce heat losses from the reactor vessel and to permit the use of a biological shield outside the vessel, which operates at a lower temperature than the vessel, thermal insulation must be provided around the vessel. The thermal insulation, in turn, may be canned to prevent the insulating material from coming into contact with water from the biological shield or with reactor coolant during refueling.

Fiberglass in the form of blankets having hexagonal wire mesh facing on the side against the shell has been used successfully to insulate the Shippingport reactor vessel [39]. The blankets were originally 4.5 in. (11.4 cm) and were compressed to 4 in. (10.2 cm) when canned. In the vicinity of the closure bolts the blankets were originally 2.5 in. (6.35 cm) and were compressed to 2 in. (5.08) when canned [39].

The material used for canning the insulation was 20-gage (U. S. Standard)* stainless steel. Inasmuch as access to the lower part of the vessel is not required during refueling, the canning in this region could be permanently welded together. For the closure head, the canned insulation was made in the form of removable sections to permit ready access to the head. Where the head profile was too irregular to permit the use of canned sections, bare Fiberglass blankets were used. The general arrangement of vessel shell insulation and canning is shown in Fig. 2-23 [39].

Other materials can be used for insulation; however, a material should be chosen such as Fiberglass or magnesium oxide that does not suffer loss of insulating capability under irradiation. The canning around the insulation must be effective in protecting the insulation from becoming wet during operation or refueling. If the can contains leaks, waterlogging can take place

*20-gage U. S. Std. sheet is 0.0375 in. (0.9525 mm) thick.

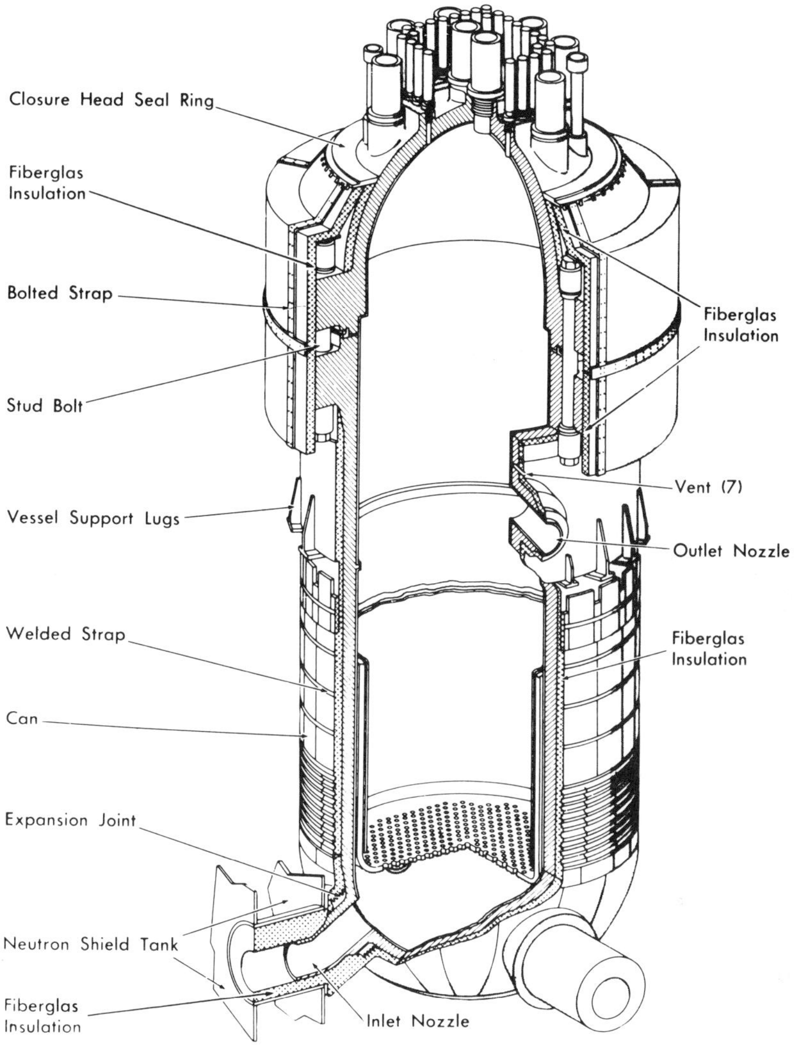

FIG. 2-23. Vessel shell insulation and canning (Shippingport).

during underwater refueling; upon heat-up of the vessel, the water could turn to steam faster than it can escape through the leak and could cause rupture of the can.

Inasmuch as the reactor vessel must be leak-tested prior to being placed in service, the insulation and canning cannot be installed until leak testing of the vessel has been completed. Provisions must be made during design for enough space to install the insulation after such leak testing. Thorough inspection of the completed can itself is also necessary to insure its integrity. Field welds on the can may be inspected by one of the appropriate methods outlines in Sec. 2.11.7.

2.10 Reactor Vessel Supports

A satisfactory method for supporting the reactor is essential to its successful performance in service. The method used will depend in large part upon the particular plant arrangement in which the vessel is to be installed. The vessel support must accommodate expansion of the vessel in both the longitudinal and radial directions and must withstand forces which may be imposed on the vessel by external piping loads on the nozzles and by such mechanical shocks as might be encountered in service.

Two basic types of supports have been used: a cylindrical skirt fastened to the support foundation and extending upward and welded to either the lower or the upper portion of the vessel; and a series of radial brackets welded around the upper portion of the vessel and resting on radial or tapered pins or the support foundation.

The skirt support has been used primarily in applications involving shock, the skirt length, however, being determined primarily by the need to accommodate expansion of the vessel. The bracket-type support is used where the weight of the vessel is sufficient to prevent motion during operation; this type has been used at Shippingport. In both designs, the attachments to the vessel should be kept away from discontinuities in the vessel, if possible, particularly the nozzles, the main flange, and the transition between the bottom head and cylindrical shell.

2.11 Construction and Inspection

Early in the design of a reactor vessel questions of manufacture and inspection must be considered. The size of the vessel as well as sizes of pieces from which the vessel is to be made will be limited by equipment available for forming, welding, heat treating, and machining them within the desired accuracy. The sequence of operation through which manufacturing will progress is influenced by the material selected; stainless steels are handled differently from carbon steels. Accessibility for inspection of welds must be provided in the design. Planning must assure proper control of all manufacturing processes.

Bush [165] points out that while the current generation of nuclear pressure vessels in this country are fabricated in a plant and then shipped to the reactor site for installation, it is quite possible that a substantial portion of the next generation will be field constructed; e.g., the plates will be formed and welded in the field. While plant fabrication offers definite advantages, vessels beyond a certain size are virtually impossible to transport, and excellent results have been noted in many field-fabricated nonnuclear systems. Special attention will have to be given to the problems which will be faced in field fabrication of nuclear reactor vessels.

2.11.1 Standards of Construction

As pointed out in Sec. 2.1.1, many codes exist which apply to reactor vessel design. Almost all of these give attention to minimum requirements that must be met during fabrication. Which of these apply depends upon the applicable laws and the specifications of the purchaser. Efforts are being made to reduce the number of codes, but for the present these include the ASME Boiler and Pressure Vessel Code, the ASA Power Piping Code, the ASTM Material Specifications, General Specifications for Ships of the United States Navy, other military and government specifications and various purchasers' and fabricators' specifications.

To confirm that the vessel complies with the specified requirements, appropriate records must be kept on all materials and parts that go into making the vessel. These include physical and chemical tests, heat number, dimensions, welding procedure, welder qualification records, hydrostatic tests, x-ray films, ultrasonic test reports and all other nondestructive test reports. These must be identifiable with the materials and pieces used in the vessel. These records become a permanent record of the vessel.

Specific details concerning standards of construction are covered in the ensuing discussion where appropriate.

2.11.2 Forming

Fabrication of a vessel begins with the forming of flat plates into appropriately shaped pieces. Thin shells are formed on bending rolls consisting of three rigid cylindrical rolls arranged to form an isosceles triangle of adjustable height; the plate is formed by successive passes through the rolls until the desired radius is obtained. Thick plates are formed into cylindrical shape by large hydraulic press brakes. The dimensions of the formed plates are checked against steel templates as work progresses. Thick shells are generally formed into semi-cylindrical shapes which are welded together along two axial seams. Thin shells could contain only one axial weld. The preparation for welding is accomplished after forming is completed. Components such as head segments and nozzles are hot-formed on a hydraulic press.

2.11.3 Welding

Some vessels require the fabrication of several separate longitudinal sections which may then have to be welded together by circumferential welding. Figure 2-23 [40] shows the fabrication sequence of a thick-walled reactor vessel.

Some idea of the large number of welds involved in reactor vessel fabrication can be obtained from Fig. 2-23. These are not trivial welds; they must withstand the same service as the main body of the vessel. Some of these are made by automatic welding machines; others are made manually, particularly where accessibility is restricted. In either case careful control of the process is required.

All welds of the magnitude involved in reactor vessel walls must have complete crevice-free penetration. Whenever possible, welding from one side only is avoided. The welding is done in a planned series of passes from one side first; this weld is then back-chipped and the welding performed on the other side. After welding is complete, excess material is ground away to produce a continuous surface with the base metal. Tests, as described below, are then made to verify the weld's integrity. Additional destructive tests are made on a separate plate which is welded and heat treated at the same time that the vessel plate is welded and heat treated.

Many special problems must be solved in welding of reactor vessels. Heavy positioning equipment is required. Preheating of large sections must be accomplished. Means must be developed and proved for welding dissimilar materials. Large surfaces must be clad by welding or weld-deposit of a thin corrosion-resistant liner on shapes not available as roll-bonded product. (See Sec. 2.11.4.) These problems must be faced by the designer.

A number of precautionary steps are taken in the welding of large reactor vessels. First, welders who do the welding must pass various welder qualifications tests. They must be instructed to avoid arc strikes on the vessel walls because of the adverse effects such strikes might have on the brittle fracture characteristics of the vessel. To assure good workmanship, quality control records are kept on all welders and on repairs that have to be made on their work. In addition, a welding material control program is necessary

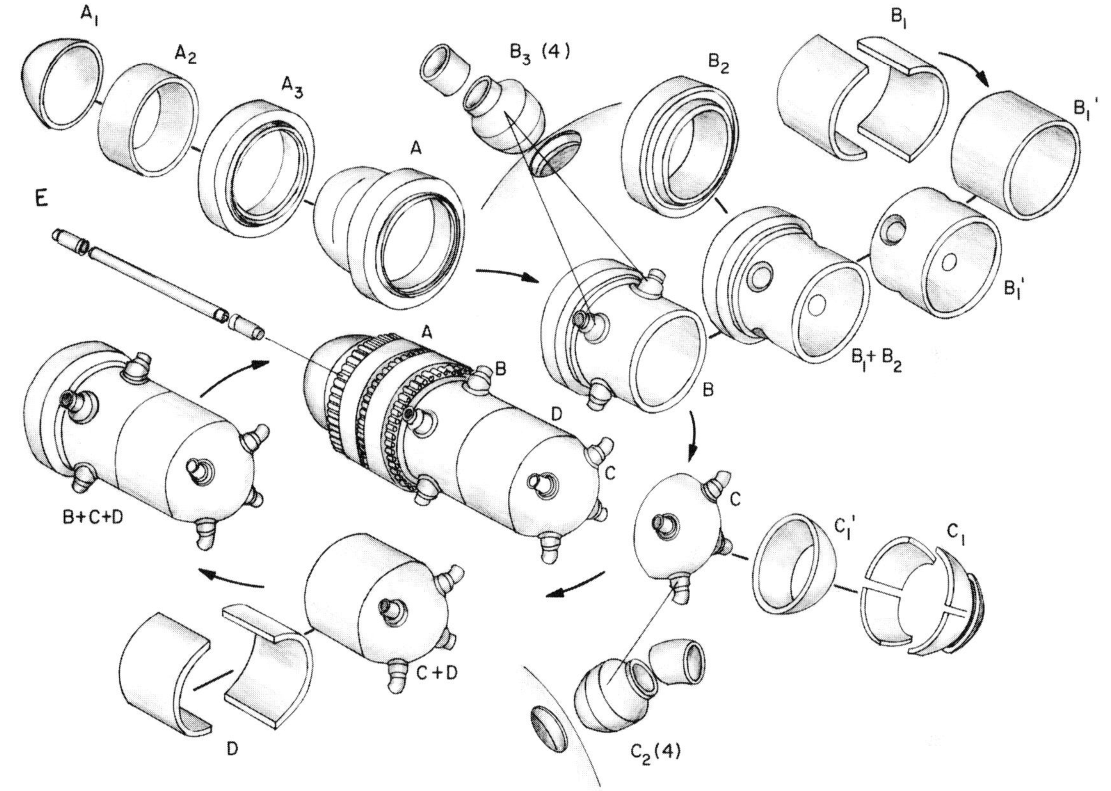

FIG. 2-24. Fabrication sequence of a reactor vessel. A - hydrostatic test head (A_1 - dome, A_2 - shell, A_3 - flange). B - ring 1, complete (B_1 - ring with nozzle holes, B_2 - flange, B_3 - outlet nozzle assembly). C - lower hemispherical head, complete (C_1 - hemispherical shell, C_2 - inlet nozzle assembly). D - ring 2. E - stud bolt assembly (42 required).

to avoid the use of the wrong welding rod and to make sure that welding rods are kept dry. In certain cases, physical and chemical control tests are prescribed on bare rod and fluxes and on the deposited metal.

Standards have been established for conveying welding information from the designers to the manufacturing department; these are in the form of standard welding symbols developed by the American Welding Society and adopted by the American Standards Association [41]. Unqualified notes on drawings such as "To be Welded Throughout"; or "To be Completely Welded" in effect transfer the design of all attachments and connections from the designer to the welding operator, who cannot be expert enough to evaluate the ultimate requirement. This practice is dangerous from many standpoints. Standard welding symbols should be used.

Unacceptable defects as outlined in Section III of the Boiler and Pressure Vessel Code are to be removed by mechanical means or by thermal cropping processes when detected during inspections and tests discussed in Sec. 2.11.7 of this Chapter. The rewelded area must be re-examined by the method specified for the original weld to ensure that it has been satisfactorily repaired. Post-weld heat-treating rules are to apply to all weld repairs.

2.11.4 Cladding

Inasmuch as the corrosion characteristics of most reactor coolants require that only stainless steel be exposed to the coolant, cladding is applied to the interior of carbon steel reactor vessels. The cladding can be applied in any one of several ways. It can be hot-roll bonded to the base metal during manufacturing of the plate; it can be made by weld deposit over the inner surface of the plate prior to, during, or after fabrication; it can be added as a separate plate by a series of overlapping or closely spaced spot welds prior to forming of the plate; small penetrations which are difficult to clad otherwise are clad by the use of somewhat thicker-than-average liners which are bonded to the base metal or other cladding only at the ends. All of these methods have been used in nuclear reactor vessels.

The cladding material usually accounts for 3 to 20% of the total thickness of the plate with the lower percentage characteristic of thick vessels and the larger percentage characteristic of thin vessels. Cladding thickness have ranged between 0.109 and 0.375 in. (0.27 and 0.95 cm). This thickness, however, is not to be relied upon for strength of the vessel. The most important problem encountered in the cladding of vessels is that of assuring a leak-tight well-bonded cladding. Leak-tightness is required so that the reactor coolant does not get to the base metal and cause corrosion. A sound bond is required so that if coolant does get between the cladding and the base metal it cannot cause internal collapse of large sections of the cladding during changes in system operating pressure when the pressure behind the cladding can be higher than the system pressure. Ultrasonic testing is used for detecting unbonded areas. Dye penetrant tests and visual inspection are used for detecting cracks in the cladding.

An important consideration in the design of clad vessels is evaluating the problem of residual stresses in the composite plate in addition to the stresses from other sources. This is one of the reasons that Section III of the Code assumes a mean stress equal to the yield-point stress for cyclic operation. Residual stresses in clad metals can result from initial attachment, from thermal expansion and contraction, and from distortions arising from plastic deformation brought about by high thermal stresses.

2.11.5 Heat Treatment

Heat treatment of welded vessels is required both for obtaining desired service properties and for assuring satisfactory fabrication. The temperatures at which heat treatment is carried out as well as the heating and cooling rates and soaking times vary with the materials and the effects desired. Heat treatment temperatures for reactor vessels are usually in the range from 1150°F to 2150°F (620° to 1175°C). Careful control of temperature and heating and cooling rates are required. Special furnaces are required for large vessels.

In certain cases, a specific atmosphere must be maintained during heat treatment to maintain metallurgical integrity in the material or to maintain cleanliness. Specially fabricated air-tight containers must sometimes be constructed for this purpose. In such cases auxiliary equipment for pumping, manifolding and means for controlling flow of gases are required.

2.11.6 Machining

Machining of reactor vessels presents many problems. Most of these problems arise from the large sizes involved, the tolerances and finishes required and the characteristics of the material.

The large sizes associated with reactor vessels introduce the need for special machining equipment capable of accommodating loads of up to a hundred tons or more; such loads are particularly significant when they must be borne by the machine. In these as well as other cases, special support cradles, jigs, and fixtures must be provided. Handling the vessel and moving it into position on the machine becomes a problem requiring planning. Operations must be observed carefully to insure that the position has not changed because of thermal expansion or the weight of the vessel or the machining forces. Because of the large size and relatively thin sections sometimes involved, dimensional instability may be encountered because of residual stresses in the material; thus it may be necessary to take several rough cuts of progressively reduced depth, alternating from one side to the other. Heat treatment may be required to anneal out stresses developed during machining operations before further machining is done. These problems must be faced and planned for in the preparation of a fabrication procedure and schedule.

In almost every reactor vessel, the tolerances required in the vessel dimensions impose severe problems. Compared to the basic dimensions involved, the tolerances are exceedingly small, so much so that the dimensional range must be specified as applying at a particular temperature; the measurements then must be taken at the temperature. Thus a significant waiting time may be involved in getting the vessel to the desired temperature. In order to obtain close tolerances, the machines used must themselves be quite precise. In view of the fact that some of the tolerances are of a geometric nature, particularly stringent requirements may be imposed on the machine and jigging. Tolerances specified include flatnesses of surfaces, parallelism of related surfaces, finishes of surfaces, perpendicularity of holes, concentricity of holes and positioning of holes.

The materials problems arise out of the fact that machining may have to be done with a special coolant or without a coolant; this is done to avoid contamination of inaccessible areas which are restricted to specific cleaning procedures. It should be noted that chloride corrosion is one of the most serious problems in water-cooled reactors and that many solvents used in machine shops contain chlorine which may be left on the machined part. Thus, thorough final cleaning with a nonhalogenated cleaning agent is very important.

In all of this work it must be appreciated that the vessel becomes progressively more valuable at it nears completion; precautions are necessary to avoid an error that might make useless the many man-hours of work that went into the manufacturing of the vessel up to that point.

2.11.7 Inspection and Testing

Many inspections must be made during and after manufacturing to assure that the vessel is sound and meets all specifications. The inspections include nondestructive tests on the vessel as well as destructive tests made on pieces prepared at the same time and in the same manner as the vessel piece itself. The foregoing discussion mentioned a number of these inspections. These have included the dimension templates used in rolling, the destructive tests on weld specimens, qualification tests of welders, physical and chemical tests on welding rod and fluxes, control of filler material handling, control of preheating and heat treatment temperatures, atmosphere control in heat treatment, and temperature control while making dimensional checks; mention has also been made of accessibility requirements for inspection, the extensive record-keeping involved and the use of special cutting coolants for machining.

Many other inspections and tests are made, however. Nondestructive tests on the vessel include radiography, magnetic particle testing, dye penetrant testing and ultrasonic testing. All welds receive careful inspection by use of X-rays whenever possible. When the X-ray film shows unacceptable defects, the weld must be ground out until the defect is found, and then the weld is repaired by rewelding. The X-ray film must be calibrated by the use of test pieces having known defects. The ASME Boiler Code sets forth particular requirements regarding the sensitivity of X-ray inspections (see reference [10]). High-voltage X-ray machines are required for this purpose in many applications. X-ray machines up to 15 Mev have been used.

Surfaces of welds and base metals are examined carefully for cracks. Magnetic particle testing is used on magnetic materials and dye penetrant testing is used on non-magnetic materials as well as magnetic materials. Ultrasonic testing is used in certain cases to help identify internal defects as well as surface defects. (See Sec. 9 for details of testing methods.)

A number of tests can be and are used for leak testing of the vessel. Soap bubble tests are used externally in certain cases when the vessel is internally pressurized with a gas. If the gas used for pressurizing is freon, the outside of the vessel can be checked for the leakage of freon. A number of designers discourage the use of halides for leak testing because of possible chemical decomposition under irradiation of any halide which remains occluded to the vessel walls and which could lead to corrosion problems. The mass spectrometer leak test is used where great sensitivity is required. The vessel can either be pressurized with helium and examined on the external surfaces with a so-called "helium sniffer," or the vessel can be placed under vacuum internally and surrounded with helium externally; in the latter case the leak detector is attached to the vacuum system. A final hydrostatic test at some number of times design pressure is made on the vessel and closure to "prove" the design and disclose leaks. (See appropriate section of the Boiler and Pressure Vessel Code for the value of the required hydrostatic test pressure.)

Destructive tests are made on test plates welded and heat treated at the same time and in the same way as the vessel. These include tension tests, bend tests, Charpy impact tests, and metallographic examination of grain size.

Because of the critical importance of vessel-plate integrity, inspections must be carefully done and should take advantage of the best available technology. In this connection Bush [165] offers the following commentary:

"While the plates for pressure vessels are inspected according to standard ASTM-ASME standards, these standards represent compromises between the fabricator and the user; too often, they represent obsolete technology because of the time required to make revisions in standards. For example, the most common nondestructive testing technique for inspecting thick steel plate is ultrasonics. The standards reflect the process as it was some years ago; we still use a scan pattern that examines 30-50% of the plate, is limited to detecting defects in in the rolling direction, and uses less than attainable sensitivities. For example, thoroughly tested techniques exist to permit 100% inspection for defects in the rolling direction (laminar), normal to the rolling direction, or at some angle to this direction by a combination of normal incidence and angled beam ultrasonics, and the entire inspection should take no longer than the existing process which can miss both large laminar defects and defects normal to the plate surface. Fortunately modifications in current ASTM Ultrasonic Instpection Standards may help to some degree in a few years.

"While laminar type defects are the most probable, there is a possibility of defects normal to the plate surfaces. The laminar defects, due to distribution of stresses, are inherently less dangerous than defects in the thickness of the plate. The latter may propagate on bending the plate to form the pressure vessel wall, and, generally, no inspection is attempted after the initial tests on the plate. The probability of defects of this nature is not known, but such defects should be detectable by the use of the acoustic emission technique on the finished pressure vessel as described below.

"The acoustic emission process appears quite promising for the detection of flaws in the shell wall or in the primary piping during hydrostatic testing. This technique has been utilized with substantial success, in the testing of missle components and simple vessels. It has been used with aluminum and steel of thicknesses up to four inches to detect flaws which were subsequently repaired. There are no definitive results on reactor pressure vessels using this technique, but it should have an excellent probability of success in detecting flaws [177].

"Basically the acoustic emission process is a dynamic one, depending on locating a growing crack through the noise it emits during loading the pressure vessel. By triangulation or other methods the precise location of the crack can be determined.

"It would be interesting to check a pressure vessel by acoustic emission with permanently mounted pressure transducers. Any defects located during hydrostatic testing could have one or more angled beam transducers mounted so that future growth of the defect could be detected ultrasonically. The fixed mount pressure transducers could also be used if the vessel was pressure tested during future shutdowns.

"While the acoustic emission technique has substantial promise, it is not free from problems. Some of these include: (1) the high noise background requiring sophisticated instrumentation to separate crack growth from general noise; (2) the complexity of detecting defects in nozzles, elbows, etc.; (3) possible deterioration of transducers by temperature and/or irradiation; and (4) the need to cause propagation of the defect if it is to be detected.

"Fixed mount ultrasonic transducers also present problems such as: (1) stability of transducers after extended periods at elevated temperatures in a low neutron and gamma flux; (2) deterioration of transducers by the diffusion of the surrounding shell and pressure block material into the transducer; (3) loss of pressure between the transducer and the shell due to thermal cycling and relaxation of pressure block (the latter is a mechanical problem subject to solution); (4) deterioration of leads by intermediate diffusion; and (5) the proof that one can follow defect growth in a vessel operating at temperature and pressure.

"There is hope that these problems can be solved. Ultrasonic transducers have been opperated at temperatures exceeding those at the outer surface of a reactor pressure vessel for some months without extensive deterioration. Ultrasonics should be satisfactory for detecting defects measuring greater than 5% of the wall, and the method is applicable with metals having the normal range of grain size, welds, plate thickness, and cold work."

2.12 Deterioration of Vessels in Service

2.12.1 Effects Involved

The ASME Boiler Code recognizes that deterioration of vessels can occur by corrosion, erosion, or mechanical abrasion. When vessels or parts of vessels are subject to thinning brought about by these effects, the thickness of the material must be increased over that determined by the appropriate design formulas, or some other suitable method of protection must be provided for the desired life of the vessel. The material added need not be of the same thickness for all parts of the vessel if different rates of attack are expected for the various parts. When the thickness of the material has been increased as provided above, telltale holes may be drilled to provide some positive indication when the thickness has been reduced to a dangerous degree. Specifications for the telltale holes are provided in reference [5b] under paragraph UG-25 for unclad vessels and under paragraph UCL-25(6) for clad vessels. Corrosion-resistant or abrasion-resistant linings whether or not attached to the wall of the vessel shall not be considered as contributing to the strength of the wall except under conditions outlined in Appendix F to Section VIII of the Code.

Appendix F to Section VIII also recognizes another source of vessel deterioration, namely seepage of fluid behind the applied liner. Whenever such seepage is encountered under test, appropriate repairs must be made.

Other sources of vessel deterioration that can be identified include: creep, fatigue, stress corrosion, crevice corrosion, and embrittlement particularly as arising from radiation effects. These effects must be carefully evaluated and provided for during the design of the vessel. Careful inspection during manufacture and use are necessary to guard against failure in service.

The major concern during operation of a nuclear pressure vessel is the possibility, however remote, of a catastrophic failure through rapid crack propagation in a brittle or nonbrittle fashion. Bush [165] points out that, contrary to popular belief, a pressure vessel could fail by fast fracture propagation without the failure of being brittle. The rate of propagation of a fracture is a complex interrelation between fracture toughness, the structural behavior or unloading path of the crack, and the specific size of the crack during loading. Test failures occurring near the brittle-ductile transition temperature will often display a mixture of shear and brittle fracture yet fail catastrophically.

One must always keep in mind one of the inherent differences between nonnuclear and nuclear pressure vessels. Nonnuclear vessels are given extensive periodic inspections; nuclear vessels, at present, cannot be given comparably extensive inspections. Irvine et al [178] of the UAEA express this difference quite clearly.

> "The failure behavior of a pressure vessel may be considered to consist of two phases. The first phase is mainly associated with processes leading to vessel damage through cracking or distortion. In a conventional plant, incipient failure is usually detected during this phase by operational inspection and a plant found to be so affected may be repaired or withdrawn from service.
>
> "If this primary phase remains undetected, the damage increases to a level at which the second phase occurs. This in a pressure vessel is usually characterized by explosive rupture due to the sudden extension of one or more of the cracks arising from the primary phase. It is this possibility of fast crack propagation which is of major interest from a pressure vessel safety standpoint."

The safety record of nonnuclear pressure vessels, particularly of power boilers, has been excellent during the past thirty years. Experience indicates that this excellent record is, at least in part, due to the practice of subjecting such vessels to periodic inspection. Much more effort should be expended in the development of means for conducting periodic inspection of a nuclear pressure vessel during its life. The procedure of fixed-mount pressure transducers described in Sec. 2.11.7 might lend itself to detecting, by acoustical emission, flaws generated in the reactor vessel during operation.

2.12.2 Failure Experience

Fortunately, to date no major failures have been encountered in reactor pressure vessels while in service, although the SL-1 vessel was damaged as a result of a nuclear accident. Nevertheless, three incidents involving nuclear plants that have taken place are of significance to vessel design. These are discussed below.

The first incident involved the Vallecitos Boiling Water Reactor (VBWR) [42]. After approximately five years' operation, the plant was shut down in September, 1962, to look for a steam leak. A circumferential crack was discovered in one of the four stainless steel suction pipes to the reactor recirculation pumps. The crack was about 2.5 in. (6.3 cm) long and was located between the reactor vessel nozzle and a 1 in. (2.54 cm) feedwater injection line. The crack was located well outside the pipe-to-vessel-nozzle weld. Ultrasonic and radiographic examinations revealed indications of cracks in the other three suction lines in corresponding areas.

A detailed and comprehensive thermal and stress analysis of the pipe in the area of failure was performed. The analysis showed that feedwater, which was introduced through a one-inch diameter connection into the suction pipe, caused very high thermal stresses and strain cycling. Unexpected restraint of the piping by a shielding wall resulted in high mechanical stresses and additional plastic strains in the pipe. The mechanism of cracking was believed to be low-cycle fatigue produced by excessive thermal and mechanical strain cycling. The damaged pipe was replaced and the shielding wall modified to avoid restraint on the pipe. To prevent recurrence of the problem, the feedwater injection pipes were removed from the suction lines and the feedwater introduced into the reactor vessel through a pipe (protected by a thermal sleeve) and a water distribution ring.

The second incident concerned the Elk River Reactor. During the process of welding and grinding a weld which connects a 16-in. (40.6-cm) stainless steel pipe to a 16-in. nozzle on the reactor vessel, a crack was discovered in the nozzle adjacent to the weld. Surface and subsurface microfissures were also found in the weld overlay. The remaining three 16-in. nozzles were dye-checked but no other cracks were found. The crack was a result of the field welding process and was repaired in a normal fashion and in no way reflected on the integrity of the vessel.

The foregoing incidents point out that cracks can take place unless great care is taken both in design and fabrication.

The third incident occurred more recently and involved damage to the cladding on the interior surface of the bottom hemispherical head of the Yankee reactor vessel [179]. During refueling of the reactor, it was discovered that the high-flux irradiation-specimen assemblies had broken loose; subsequent inspection of the interior of the vessel disclosed two small areas in the bottom hemispherical head where the 0.109-in.-thick cladding had been penetrated and the alloy steel base metal was exposed. The total area of carbon steel exposed was about 2 square inches [180]. The situation was studied to determine the potential for crevice, galvanic, and general corrosion of the base metal during the service life of the vessel as well as the potential for hydrogen-embrittlement. These studies disclosed no immediate problems; nevertheless, the situation will continue to be monitored.

From this incidence and other experiences with the Yankee plant, Reed and Tarnuzzer [179] offer the following advice about reactor vessels and reactor internals:

> "Until reactor designs become proven and repetitive, provide space so that the reactor vessel can be emptied of its internals, i.e., leave a way to dig yourself out.
>
> "Minimize complex reactor-vessel internals; simple rugged design should be used—the fewer parts the better.
>
> "Avoid inserting measuring devices, irradiation specimens, etc., in long-life power reactors unless these devices can meet the 'simple rugged' criteria or are absolutely necessary.
>
> "Eliminate screwed fastenings with a multiplicity of V-notch stress points from high-flux regions and avoid them elsewhere in the vessel if possible. Our experience indicates an order of preference for fastenings depending upon need for disassembly. Welds should be used wherever possible, then tack-welded screwed fastenings, and finally the lowest order of preference goes to locking-device screwed fastenings.
>
> "Avoid stitch-welded vessel cladding and use primarily weld-deposit cladding.
>
> "Choose the type of reactor vessel wall material to be used behind stainless cladding with corrosion and hydrogen embrittlement in mind, assuming that sometime in the life of the vessel a clad penetrator could occur.
>
> "Owners of new models of reactors and pioneering models must remain alert to the happenings and new discoveries in the industry and act accordingly to investigate and modify (if necessary) their plants.
>
> "Complex refueling-repair problems require some unusual manpower capabilities. Yankee got back into service as early as it did only with the aid of some very artful underwater fishing, feeling, viewing, and gadgeteering."

All three of the foregoing examples confirm the need for continued careful surveillance of vessels during their service life. This is an important and necessary adjunct to maintaining the excellent in-service record brought about by careful attention to design, fabrication, and operating limitations on reactor vessels in service.

Nonnuclear pressure vessels have failed in service and the failures have often been brittle failures as reported by M. E. Shank [30]. Shank points out that most nonship brittle failures occurred under completely static conditions. Most of these occurred, however, before the conditions which might prevent brittle fracture were identified.

Among the more recent failures of interest are the brittle failures of high-pressure gas transmission lines. The pipe for these lines is usually produced under American Petroleum Institute Standards, in several strength grades. The pipe is cold-formed, seam-welded, and often hydrostatically expanded, unheated to obtain the high yield strength desired. Longitudinal failures extending from 180 to 3200 ft (55 m to 980 m) in length have occurred during installation tests. The break always has a sine-wave pattern and does not seem to have any relationship to the weld seam. The U.S. Federal Power Commission reports a number of splits of similar piping (not at the weld) which occurred on test and in service. Shank points out that while some of these undoubtedly represent brittle fractures, others did not.

These experiences, while only briefly discussed here, serve to indicate some of the difficulties faced in the design of steel vessels. Careful design and selection of material as well as good workmanship are of paramount importance in the production of a trouble-free pressure vessel or any steel structure for that matter.

Farmer [43] points out that the British have observed marked differences in the behavior of purposely defected vessels depending upon whether they contain an expandable fluid (gas, steam or pressurized water at high temperature) or a relatively nonexpandable fluid (cold water, sodium, etc.). In preliminary tests they have observed fractures which have the characteristics of brittle fractures at temperatures well above the NDT. This raises a serious question if subsequent tests verify this behavior.

2.13 Prestressed Concrete Reactor Pressure Vessels

The high costs and the potential for embrittlement of steel reactor vessels has stimulated considerable interest in prestressed concrete reactor vessels. However, many factors must be considered in the choice of steels versus concrete vessels; much of the data needed to make this choice is still under development [181].

A prestressed concrete pressure vessel consists of three major components: a leakproof liner, a concrete envelope, and prestressing cables or tendons. The liner is used to prevent leakage of the coolant; it is usually a thin ductile member which is not counted on to provide any structural support and is backed up by the concrete. The concrete serves as the load distribution matrix; prestressing of the concrete prior to operation induces compressive stresses in the vessel walls which tend to be relieved, during operation, by the tensile stresses induced by the internal operating pressures. The prestress is induced by the cables or tendons located at or near the outer extremities of the vessel.

The liner must be designed to withstand the strains expected of the vessel during service. As a consequence of prestressing operations, the liner is usually loaded in compression and therefore must be anchored to the vessel wall to prevent buckling [182]. As the vessel expands in service, the compressive loads in the liner are relieved; thus it becomes necessary to specify not only the allowable stresses and strains but also the allowable number of cycles for stress application. Because of cooling requirements of the vessel, attention must also be given to thermal stresses in the liner. Because of the neutron flow, the degree of embrittlement to be expected in service must also be evaluated.

The concrete used for the vessel must receive careful attention not only to assure adequate strength but also to ensure acceptable shrinkage and creep properties inasmuch as these affect the state of prestress of the vessel at any time after initial load is applied. Provisions must also be made for keeping the temperature of the concrete at acceptable levels during service; at present there exists insufficient data to draw reliable conclusions about the effects of sustained temperatures above 300°F in concrete. Thus the concrete usually must be insulated from high temperatures and must be cooled to avoid excessive gamma heating.

Inasmuch as the strength of the vessel is derived solely from the prestressing cables or tendons, special attention must be given to assuring their reliability. They must be protected against excessive temperatures to prevent undesired creep and against neutron irradiation to prevent embrittlement; this protection is usually provided by the concrete but the design must be such as to ensure adequacy of protection. The tendons must also be protected against corrosion; particular attention must be given to the possibility of galvanic corrosion due to stray currents that may be encountered in electric-power producing plants.

Two methods are used for developing the high tensile stresses in the tendons; these are referred to as pretensioning and post-tensioning. The pretensioning method involves loading of the cable in a fixture, then casting of the concrete and then removal of the pretensioning fixture. The post-tensioning method requires the casting of ducts into the concrete and then the insertion and tensioning of tendons after the concrete has developed its proper strength; the prestress can also be preserved by grouting the cables within the ducts. Since the grouting is done after the vessel has been prestressed for 500 hours or more, most shrinkage effects can be compensated for before the grouting is done.

The post-tensioning method without grouting has the advantage that the cables can be inspected and replaced if necessary. A grouted cable, on the other hand, can be counted upon to maintain its prestress even if local corrosion of the tendons should take place. As a result, some designers believe that a grouted cable would not fail suddenly whereas a tensile failure of ungrouted cable could be instantaneous [182]. The relative merits of grouting versus not grouting constitute a subject of significant controversy at present.

To date, the use of prestressed concrete reactor pressure vessels has been limited to European reactors. They have been used in both England and France [181]. In the United States, the primary effort to date has been by General Atomic on the analysis, development, and testing of a model vessel as a step toward ultimate utilization of a prestressed vessel in a close-coupled reactor plant [183].

As pointed out in reference [184], there is no suitable code of practice for concrete pressure vessels at the present time. In addition to questions associated with the considerations outlined above, many other problems must be overcome in the application of prestressed concrete pressure vessels to nuclear service. Among these are the development of reliable mathematical techniques for analysis of stresses in such vessels, particularly at penetrations and other vessel discontinuities; the development of and confirmatory testing of reliable penetration reinforcement methods; and the development of knowledge about environmental effects on the tendons and the concrete, including temperature, radiation and corrosion effects.

3. REACTOR VESSEL CLOSURES

The chief function of the reactor vessel closure is to provide a demountable, leak-proof head and seal to contain the primary coolant in the reactor vessel during operation. In many applications the closure head may also serve as a platform for the control drive mechanisms and provide a number of ports with their own individual closures for refueling of separate fuel assemblies. In specific applications such as Enrico Fermi, EBR-II, and SGR, which refuel without opening the vessel, the closure assembly incorporates an indexing device to aid with refueling.

3.1 Basic Considerations

Although each reactor design imposes different requirements on the closure, certain basic considerations are common to all of them.

The first consideration is that of providing the strength required for containing the system pressure. The following five characteristics of a closure have a significant influence on its strength: the shape of the closure; the number, size and spacing of penetrations in the head for the refueling ports and/or for attachment of control rod drive mechanisms; the method of attachment of the head to the vessel; bolting loads and flange movements imposed by them; and, in low pressure systems where pressure-strength requirements are small, the loads that might be imposed on the head by heavily shielded casks or by other devices used during refueling through individual ports on the head.

The second consideration, that of leak-tightness, also influences, in considerable detail, the nature of the closure head to be selected. To provide leak-tightness the closure can be either gasketed, seal-welded, or completely welded. Gaskets can range from Flexitallic gaskets to prepressurized or self-pressurizing types. In reactor work gaskets are usually installed in pairs with a bleed off between them to monitor leakage. For applications involving high pressures (> 1000 psi or 70 kg/cm^2) and high temperatures (> 500°F or 260°C), some designers have avoided gasketed closures because of the possibility of leakage under unusual temperature swings; instead, hermetic seals have been used in the form of either strength welds or seal welds. In the Yankee and Consolidated Edison reactors, provisions were made for both gaskets and seal welds; however, the gaskets worked so well that welding of the seal weld has not proved necessary. Leak-tightness is required of course not only for the main closure seal but also for refueling ports and for attachments of control-rod drive mechanisms to the reactor head.

The third design consideration is that of alignment of the closure head with the reactor vessel. This is particularly important where control rods are actuated through penetrations in the head and where separate fuel assemblies are to be refueled through individual ports on the head. Alignment is also important, but generally to lower tolerances, for matching bolts and flange bolt holes in bolted closures and for providing uniform clamping forces on the core structure if the head is used as the clamping means. If the head is to provide penetrations for operating control rods, the materials of construction, the geometry of relative parts, and the tolerances involved must be such that the designed alignments are not destroyed during pressure and temperature changes.

Other considerations involved in the design of the closure, include the speed of opening and securing the closure; the need of facilities for welding, cutting and rewelding seal welds or strength welds; the problems such as galling and bolt replacement caused by repeated opening and securing operations; the limitations on heating and cooling rates which are influenced by flange thickness and bolt arrangement; the problems of attaching penetration nozzles to the head; the problems of cladding the head and the interiors of the penetrations; the problems of keeping the head at or near the temperature of the main vessel; and the requirement for special supports on the head for refueling cask.

3.2 Types of Attachment

The types of attachments for closures that have been considered for application to nuclear reactor vessels can be classified under four headings: strength-weld attachments, bolted attachments, shear block attachments, and threaded rings. By far the most widely used are bolted attachments. Such attachments have involved the use of through bolts in certain applications and stud bolts in others. Figures 3-1 to 3-4, inclusive, show some typical examples of closures.

The most conservative design of attachment one might consider is a combination of two methods for withstanding the internal pressure. For example, such a closure design might incorporate both a strength weld and bolts. Either the weld or the bolts would suffice, but the bolts can be used to keep the weld under compression during operation, thus eliminating problems of tensile fatigue on the vessel. However, such a design is costly and, in view of successful experiences with less complex closure attachments, has not been used for commercial vessels.

Figure 3-1 [44] illustrates the bolted and gasketed closure such as is used on both the Yankee and the Consolidated Edison reactors. This type of attachment is by far the easiest to open and close. In both applications of this design, stud bolts were used for clamping purposes [45].

Figures 3-2 and 3-3 [45] illustrate the design used for Shippingport, incorporating a completed seal weld without gaskets and utilizing through bolts for clamping purposes. The design offers advantages both from the sealing standpoint and from the bolting standpoint. Being hermetic in nature, the seal does not require continuous monitoring of the intermediate bleed point associated with gasket closures. On the other hand, special equipment is required for welding, cutting and rewelding the seal weld. The through bolts offer the advantage that, if seizing of a bolt is encountered because of galling during assembly or disassembly, the bolt can be cut and removed with little risk of damaging the flange or, in the case of shear blocks, without involving contaminated parts.

The shear block type of closure is illustrated in Fig. 3-4 [45]. Its chief advantage is in avoiding the need for heavy flanges. Its disadvantages are the requirement of numerous, carefully machined shear blocks, and the need for handling heavy pieces which might be contaminated with radioactivity if the shear members are interior to the seal; if the shear blocks are exterior to the seal, more re-restrictive limitations on thermal transients must be accepted because of the temperature lag between the closure restraint ring and the vessel.

When bolts are used to provide clamping, they are prestressed during assembly to stresses required at least to resist, without relative axial motion, the hydrostatic test pressure imposed on the vessel after installation. If prestressing were limited to that required for normal operating conditions, lift-off motion of the head during hydrostatic testing could introduce unacceptable flexing of the seal membrane. (See Sec. 2.4 for details concerning required bolt loads.)

Prestressing of large vessel bolts is usually accomplished by the use of electrical heaters or hydraulic belt stretchers as discussed in Sec. 2.4. As pointed out in that section, care must be exercised when using bolt heaters to avoid overstressing the bolts by overheating them during tightening.

Impact wrenches can also be used for tightening bolts without resorting to heating. Impact wrenches, however, are cumbersome for use with bolts much over 3 in. (7.6 cm) in diameter. Furthermore, the torque imposed on the bolt introduces torsional stresses which must be considered in determining

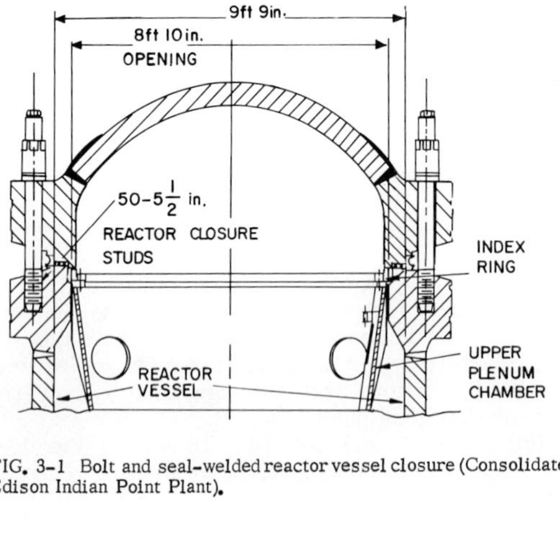

FIG. 3-1 Bolt and seal-welded reactor vessel closure (Consolidated Edison Indian Point Plant).

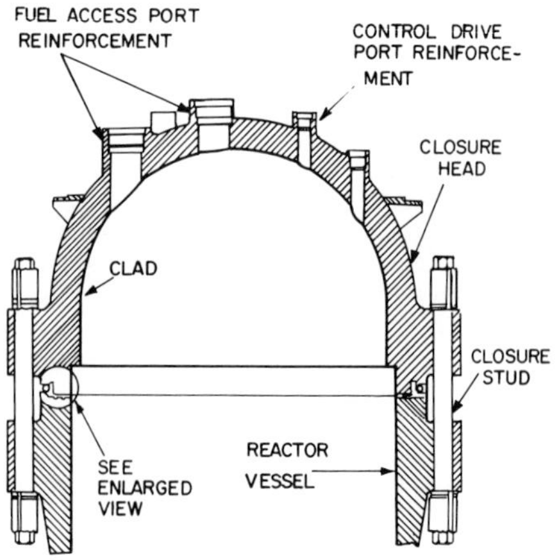

FIG. 3-2 PWR Shippingport reactor plant, reactor vessel closure bolted and seal welded type. Enlarged view of seal weld is shown in Fig. 3-3.

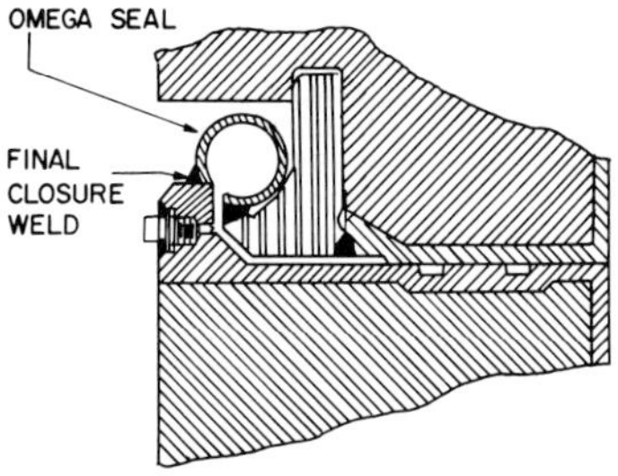

FIG. 3-3 Closure seal weld, PWR, Shippingport. See also Fig. 3-2.

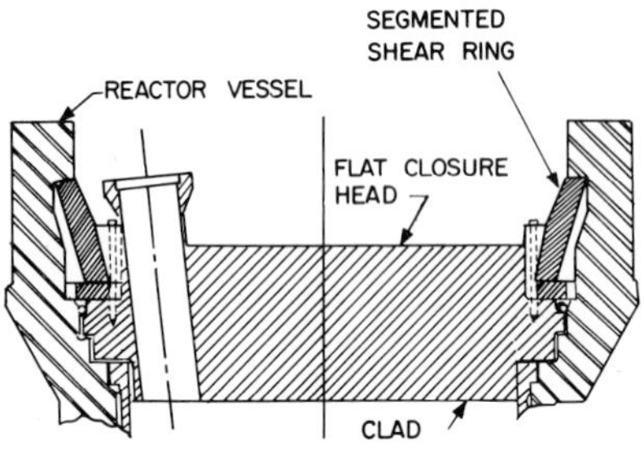

FIG. 3-4 Shear block reactor vessel closure.

the bolt adequacy. In either case, the prestress must be accurately applied and checked by measuring the stretch of the bolt.

When bolts are prestressed, provision must be made for relaxation of the bolts during prolonged operation. In addition, if a gasket is used in the closure, the prestress must be such as to provide gasket pressure under all conditions of operation.

Special care is necessary in designing, specifying, fabricating, and inspecting closure bolts. In addition, periodic inspection of closure bolts should be performed. As outlined in Chapter 11 and in reference [185] two adjacent stud bolts were found to be broken on the SM-1 (APPR) reactor at Ft. Belvoir, Va., during a shutdown inspection. The breaks were determined to have been the result of stress-corrosion cracking. A serious accident could have occurred if additional studs had failed.

Also of particular interest in the design of bolted closures is the rotation of the flange and its influence on bolt and head stresses (see Sec. 2.4). While attention is usually given to this problem, unanticipated characteristics can be encountered in service. For example, on the Humbolt Bay reactor it was found that excessive leakage at the closure was due to the significant amount of flange rotation which occurred when the head closure studs were tensioned [186]. Later during an outage, a modification had to be made to shift the rotational center of the head flange so that improved compression of the 0-ring seal occurred when the studs were tensioned and the vessel pressurized. Unanticipated rotation of a more severe nature could lead to more significant consequences.

3.3 Shape of Closure

Section 2.2.1 describes six formed shapes recognized by the Boiler and Pressure Vessel Code for use as heads on pressure vessels. However, only three of these shapes have found practical application to reactor vessel design. These are the hemispherical head, the flat head and a modified ellipsoidal head.

On the basis of strength consideration, the hemispherical shape provides a head of minimum thickness; however, of the three, it requires the greatest height above the flange. Flat heads require the greatest thickness but generally require the least height above the flange. Ellipsoidal heads are intermediate on both counts but are far more difficult to evaluate than either of the other two.

The chief considerations influencing choice of shape are the system pressure, the diameter of the head, the number, size, spacing and functions of penetrations in the head, and the arrangement of the vessel internals.

For high-pressure systems involving large diameters (> 6 ft or 2 m), hemispherical heads have been used most widely, particularly where numerous penetrations have been required. The Consolidated Edison Indian Point reactor plant, however, used a head which looks more like an ellipsoidal heat than a hemispherical head (see Fig. 3-1). Low-pressure systems such as are used for liquid-metal reactors have used flatheads almost exclusively. Figures 3-5 and 3-6 [46] show the general arrangements used on the Sodium Reactor Experiment (SRE) and the Enrico Fermi reactors respectively.

Figure 3-7 [47] shows a threaded-ring closure utilizing a retained gasket. It is widely used in non-nuclear applications. Figure 3-8 [47] shows the Bridgman joint, named for Dr. P. W. Bridgman. In this joint the pressure load is carried by a large ring bolted to the top of the vessel; leak-tightness is obtained by use of a floating head pulled up against a triangular-shaped gasket by means of bolts [47].

3.4 Seal Membranes

As indicated in Sec. 3.1, a number of vessel heads have been provided with seal-welded closures. For this purpose a seal membrane is used, fastened both to the reactor vessel and to the closure as shown in Fig. 3-3 [45]. These seals are toroidal in shape and are designed to accomodate the expected motions of the head relative to the vessel. They are often referred to as "omega" seals because their shape is similar to the capital Greek letter omega. A groove is provided in the bearing face of the head to assure equalization of pressure between the inside of the vessel and the inside of the seal.

The torus of the seal is generally fastened to the head prior to installation. A separate ring, as shown in Fig. 3-3, is generally used as the base for the torus tube to facilitate both initial installation and replacement, if necessary. The weld between the vessel and the torus tube is made after the head is set in place. The weld champfers are prepared in advance of installation. Usually an automatic welding machine, operating on a track built to suit the vessel, is used to make the final weld. Dye-penetrant checks are made to examine the weld for surface cracks after welding. The seal is also checked for leaks during the hydrostatic tests which are performed after the bolts have been installed.

When the head is to be removed the seal membrane is cut by an automatic cutting machine which cuts the metal at the closure seal weld. The cutting tool at the same time prepares the champfers for the next seal welding operation. Usually a lip is provided behind the seal weld to prevent chips or lubricant from falling into the vessel; the final cut must be very carefully made to control chips and to avoid tearing the membrane. The cutting machine is made to operate on the same circular track as the automatic welding machine.

3.5 Installation and Removal

Inasmuch as the closure head must be demountable to provide access to the core, the problems of installation and removal must be carefully considered. At first glance, this might appear to be a relatively minor problem, but experience has shown that a month or more might be involved in such operations even with careful planning. Considerable time can be consumed in heating and cooling bolts both for removal and reinstallation; time is consumed in removing electrical and cooling connections to control rod drive mechanisms; and in disconnecting the control rods from the mechanisms and in removing the mechanisms themselves. Considerable time is also involved in cutting and rewelding seal welds on the main closure and on the closures associated with the mechanisms.

In addition to the time involved, there are the problems of handling radioactive parts of massive proportions and decontaminating them for easier reinstallation. Also during removal, care is required to make sure that control rods are not inadvertently withdrawn with the head adding reactivity.

The problems involved in installation are concerned primarily with obtaining the desired alignment. Provisions for assuring alignment must be made in the closure design. Tapered guide studs extending upward from the main vessel flange and passing through the bolt holes of the closure flange

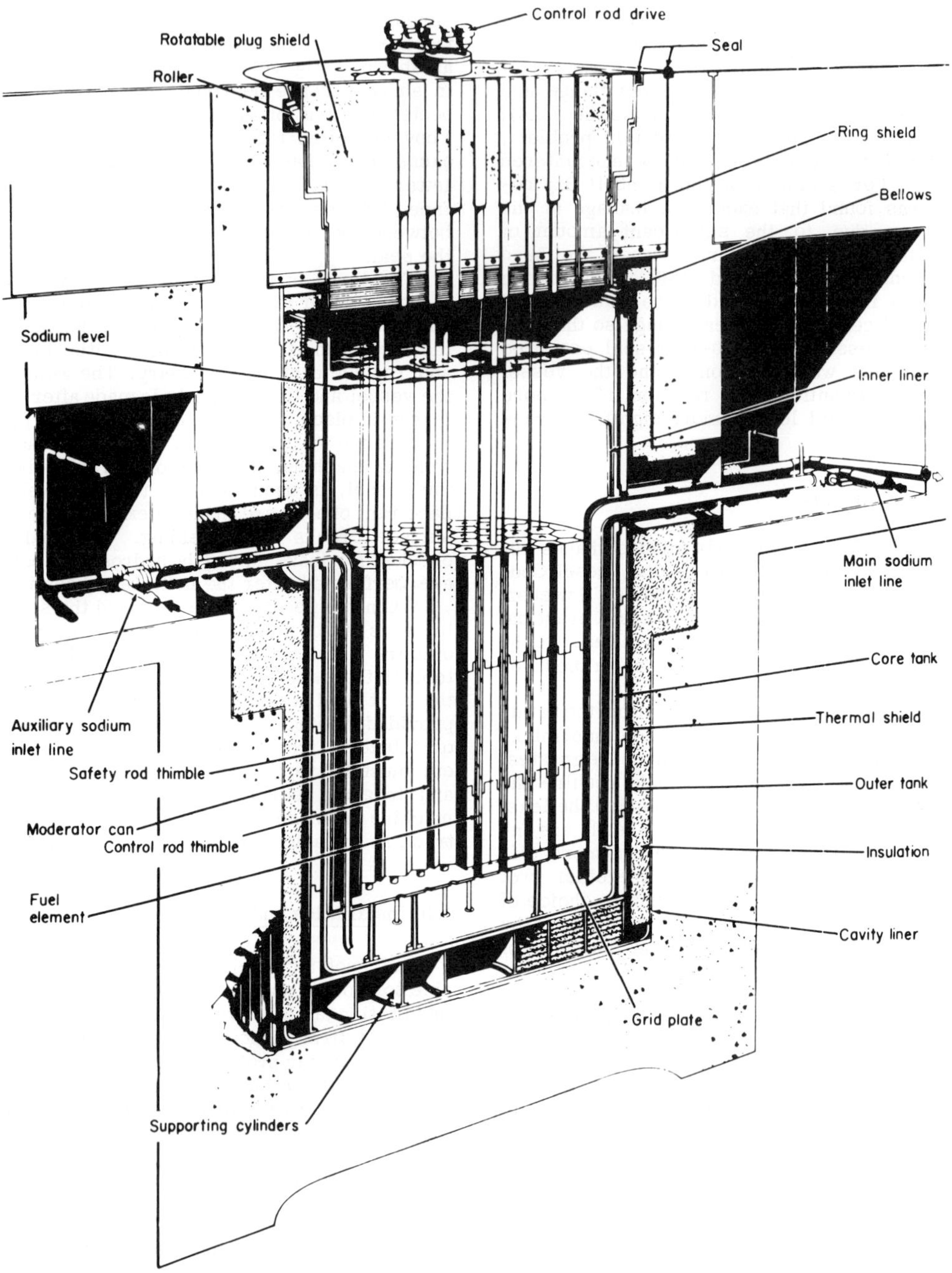

FIG. 3-5 Cutaway view of SRE showing vessel head. Sodium enters above core, is led to lower plenum through double-walled pipes, flows up through core to 6-ft (1.8-m) deep pool in upper plenum, and exits through pipe parallel to inlet.

can be used for initially guiding the head into position as it is lowered by the crane. Special installation jigs can also be used. As the head nears its seating position special installation keys can be used for more accurate alignment. These are attached to the vessel and to the head at locations which are pre-drilled for bolts and locating pins. Such keys provide both circumferential and radial positioning.

The foregoing features are required to facilitate the engagement of the internally provided keys and keyways used for maintaining alignment during operation. In order to provide the required alignment the clearances in the keys and keyways are tight; thus severe impact on the keys cannot be tolerated; such impact could lead to distortion or burring of the keys or the creation of chips that might produce galling and seizing. The design of

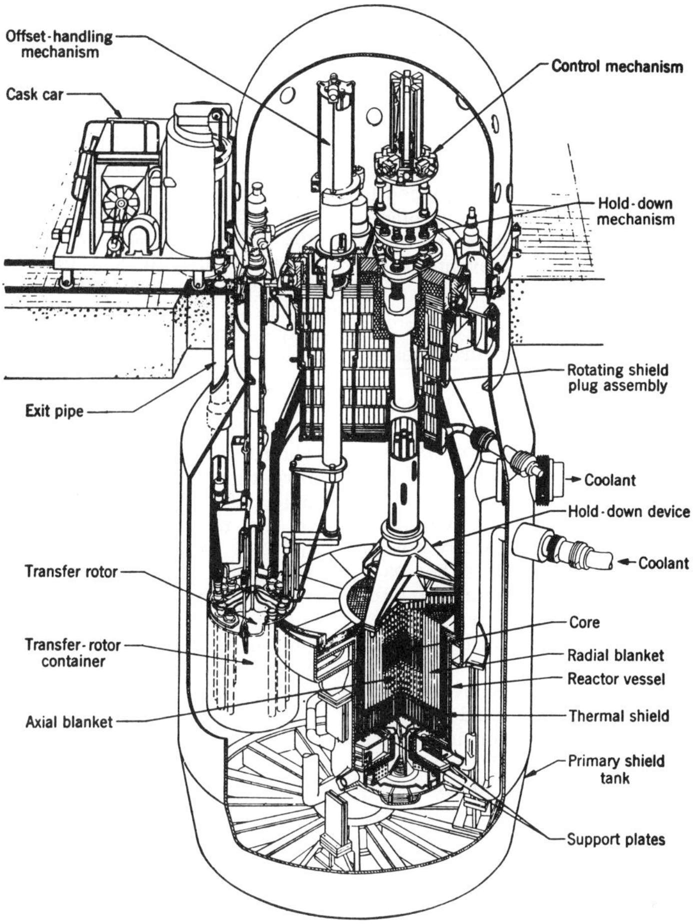

FIG. 3-6 Perspective view of Enrico Fermi reactor.

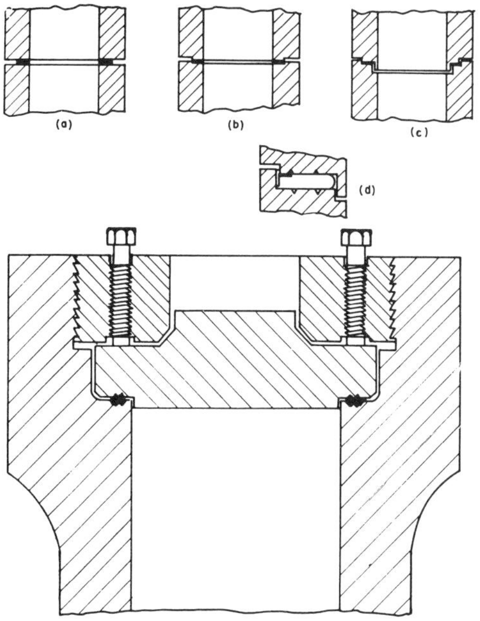

FIG. 3-7 Threaded-ring closure: (a) non-retained, (b) partly retained, (c) fully retained, (d) fully retained with grooves.

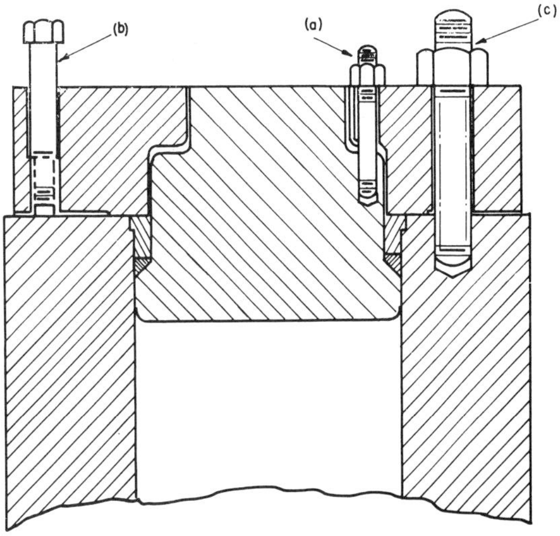

FIG. 3-8 Bridgman closure. Bolts (a) pull floating head against gasket. Bolts (b) are for aid in disassembly. Bolts (c) carry pressure load.

the keys and keyways should provide generous lead-ins; the design might also specify hardened surfaces to reduce the possibility of damage during engagement.

Other devices for maintaining alignment during operation include the use of conical seats between the head and vessel and the use of restraining bolts in the flange joint. Such bolts might be placed in bolt holes 120° apart after the head is in place. They contain expandable collars that can be made to fit the bolt holes tightly and thus prevent lateral motion during operation.

3.6 Influence of Refueling on Closure Design

Some of the influences of refueling arrangements on reactor vessel and closure design are indicated in Sec. 2.1.2. Refueling can be accomplished either by complete removal of the closure head and subsequent removal of the individual units or the entire core assembly, or by opening fuel access ports in the head and removing the fuel assemblies through them. In either case the fuel elements may be withdrawn from the core into special, heavily shielded casks or, in water reactors, they may be withdrawn from the core into a flooded canal.

If refueling involves complete removal of the head, the time required for removing and replacing the head is an important item. The considerations discussed in Sec. 3.5, therefore, require particular attention. In addition, a crane of sufficient capacity must be provided to handle the complete assembly if refueling is based on such procedures. If refueling is accomplished through ports in the head the design of the head becomes more complex because of the added thickness required and the significant amount of reinforcement needed around the penetrations if high pressures are involved (see Fig. 3-9 [48]). If the head contains penetrations for control rod drive mechanisms which are of sufficient size, these might be considered for use as refueling ports. Regardless of which of the foregoing methods is used, enough room must be provided both inside and outside the vessel for the fuel extraction tool to handle the fuel elements.

Aside from the above, attention must be given to the specific problems of using either so-called dry refueling involving casks or wet refueling involving a flooded canal. If shielded casks are used, special facilities may have to be provided for supporting them; in addition, space and facilities must be provided for handling casks, and provisions must be made for cooling the fuel elements while the transfer from the core to the cask is made and while they are in the casks. If wet refueling is used, a seal must be provided between the vessel and canal to avoid flooding those parts of the vessel not involved in refueling; in addition, those items that are covered by water during flooding must be waterproof, or be capable of being made waterproof at refueling (see Fig. 3-10 [49]).

3.7 Strength Considerations

The techniques for evaluating stresses in the closure head are the same as those described for the vessel in Secs. 2.2, 2.3 and 2.4. In designs involving the use of heavy flanges careful investigation of the stresses in the head immediately adjacent to the flange is required. This is generally more difficult to do analytically, with a high degree of reliability, for the head than for the vessel. Hence experimental stress analysis must be resorted to more often for confirming the stresses. In designs involving many penetrations in the head, analysis of ligament stresses may be equally difficult and experimental determination may again be required.

B. F. Langer [50] discusses some of the techniques used for experimental stress analysis in nuclear reactors. A method that has been used successfully for experimental determination of stresses is the use of three-dimensional photoelasticity. Langer presents the results of photoelastic models of reactor vessels and closures to investigate the stresses associated with various hole patterns in flanged hemispherical heads. One of the models is illustrated in Fig. 3-11 [50]. In Fig. 3-12 [50], an overall view is shown of the fringe pattern obtained from a similar model before the model was sliced for detailed examination. The type of information that can be obtained from such tests is illustrated by Fig. 3-13 [50], taken from reference [50]. It shows the stresses on the outer surface of a hemispherical head as measured in a radial slice passing between two holes, and the comparable stresses calculated for an umperforated plate. The calculated stresses shown include those based on shell theory and those based on the Lame formula. In the region away from the holes, theory and experiment agree well; in the region near the holes the stresses peak markedly. Similar results were obtained in other tests of thicker heads. Langer goes on to state that in the ligament between two holes, the theoretical average stress corrected for ligament efficiency checked quite well with the experimental results, but that no good theoretical method was devised for calculating the bending effects produced by the perforations. Work reported in reference [51] indicates that for a given head thickness a simple two-dimensional photoelastic model of a local area can be used to show changes in stresses caused by minor changes in penetration patterns.

Langer [50] describes corroborating tests made using strain gages and brittle lacquer on a metal model of the head (See Fig. 3-14 [50]). Langer also points out that in situations involving complicated geometries, of the sort illustrated in Fig. 3-11, testing of a full-size model may be advisable to determine the thermal stresses produced during warmup and cooldown. The need for such test usually arises in evaluating a new design to which known information cannot be readily applied.

3.8 Temperature Effects

In general the temperature effects on the closure head are similar to those discussed in Sec. 2.2.2 for the reactor vessel, particularly those associated with coolant temperature changes.

The thermal stresses in the head must be carefully evaluated. Those thermal stresses which

FIG. 3-9 PWR final closure head showing reinforcements.

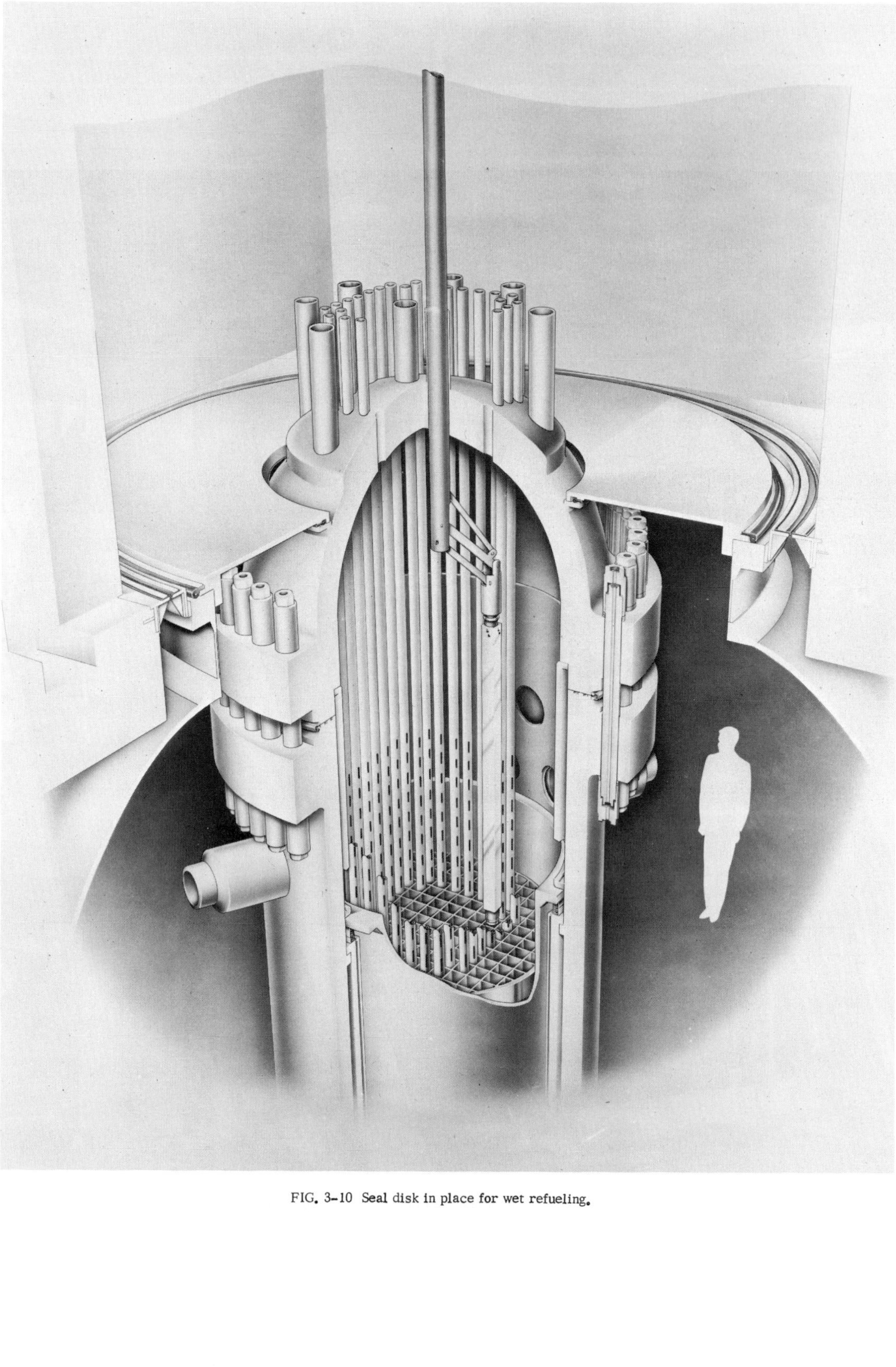

FIG. 3-10 Seal disk in place for wet refueling.

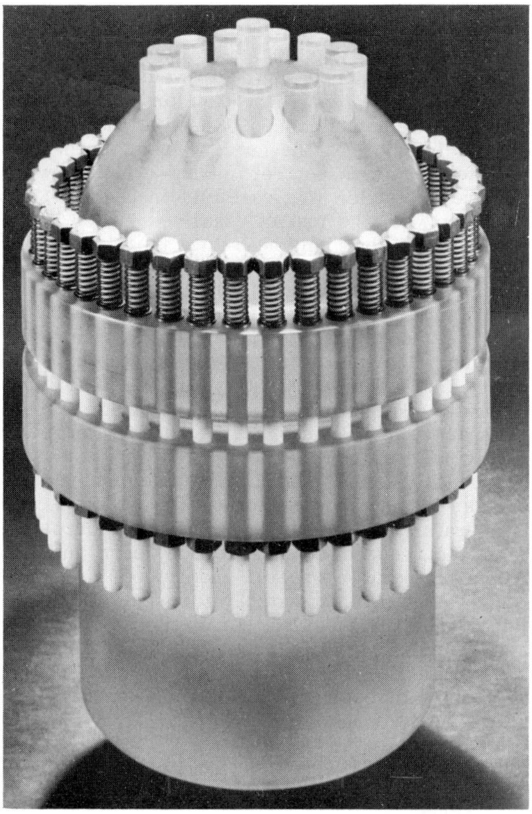

FIG. 3-11 Photoelastic model of perforated hemispherical pressure vessel head.

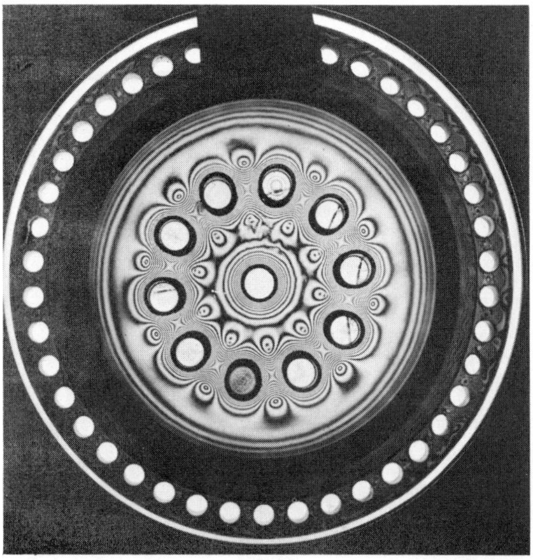

FIG. 3-12 Photoelastic fringe pattern of perforated head.

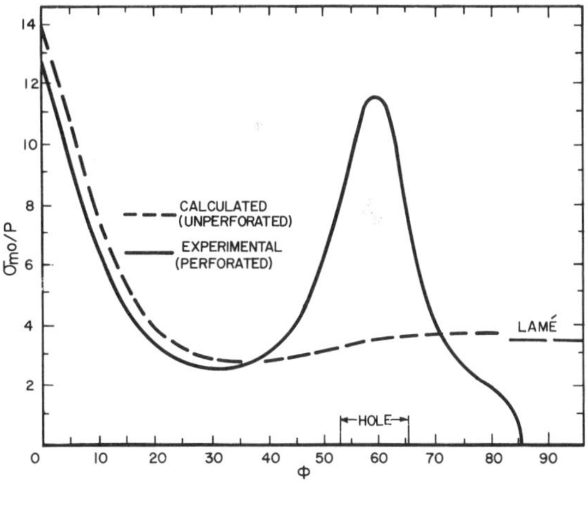

FIG. 3-13 Meridional stress on outer surface of a hemispherical head with a circular hole pattern. $R_o/R_i = 1.12$.

arise from radiation heating are generally smaller in the head than in the vessel adjacent to the core because of the longer path for attenuation of gammas and neutrons as they traverse the coolant between the core and the head. Nevertheless, the head is thicker than the vessel walls and heavily reinforced because of numerous penetrations. The thermal stresses due to radiation heating may be quite significant. If control drive mechanisms are located on the head, thermal stresses may develop because the mechanisms are specially cooled to operate at lower temperatures than the vessel.

Often a thermal barrier is provided in the mechanism housing just above the head to provide the desired temperature distribution pattern in this region of the head. However, in certain designs of mechanisms for water reactors, motion of the control rods produces an exchange of warmer and cooler water between the underside of the head and the inside of the control drive mechanism, which can produce rapid temperature changes. Hence, a thermal sleeve is provided in the penetration which is similar to that described for the reactor vessel nozzles in Sec. 2.3. Despite these precautions, the influence of cyclic operation on the combined thermal, mechanical and pressure stresses in the head must be examined as discussed in Sec. 2.2.4. Failures in this region could lead to ejection of one or more control rods and, therefore, must be considered as especially important.

Because of heat losses through the mechanisms, the head may tend to operate at a lower temperature than the rest of the vessel unless provisions are

FIG. 3-14 Apparatus for brittle-coating and strain gage test of a reactor vessel head.

made for continual passage of warm coolant from the core to the underside of the head. A lower temperature could produce undesirable misalignments and increased stresses in the head.

As indicated above, the considerations involved in evaluating responses to coolant temperature changes for the reactor vessel as discussed in Sec. 2.2.2 apply to the head as well as to the vessel. Particular attention must be given to the closure flanges.

3.9 Construction, Inspection and Testing

The problems of construction, inspection and testing of the closure head are similar to those outlined for the vessel in Sec. 2.10. The chief differences that arise are the need for reinforcements for the penetrations in the head; the necessity for manually applying the cladding by weld deposit because of the complicated shapes involved; the requirement for corrosion-resistant liners for cladding the interior of the penetrations; the needs for careful planning and for accurate location and machining of the penetrations.

Typical of the problems encountered in thick heads with many penetrations are those involved in the manufacture of the head for the Shippingport Reactor. The hemispherical portion of the head was formed from five pieces: four pie sections (sectors) and a central disk. These pieces were 10 in. (25 cm) thick; they were formed to shape and then welded together. The flange was forged and rough machined prior to welding it to the hemispherical dome. The reinforcements were added by weld metal deposit as was the cladding on the interior of the head. Careful preheating and temperature control during welding were required. Welding was carefully inspected in accordance with the methods outlined in Sec. 2.10. Figure 3-9 shows this head after the penetrations were machined.

Two difficult design problems were those of attaching the lining of the penetrations to the cladding of the interior and developing means for attaching control drive mechanism tubes to the head. The tubes were attached by use of large square-toothed threads cut into the cladding in the reinforcement. Because of the need for magnetic properties in the tube, the tube was made out of type-410 stainless steel. It was determined that unless this steel was heat treated at a specific temperature, it was subject to stress corrosion. This fact caused considerable concern until all facets of the problem were understood. A strong-back was used on top of the tubes to back up the joint during part of the operation when considerable uncertainty was encountered. Careful inspections were made to look for incipient cracks in the threads before operation and when the vessel was opened for refueling.

3.10 Special Closure Problems in Specific Applications

The problems of closure design depend in large part upon the particular functional demands of the system. While the foregoing discussion has attempted to be general, there are many applications in which unique problems are encountered. As two examples, attention is given briefly to the problems of pressure tube reactor plants and liquid-metal reactor plants.

3.10.1 The Pressure-Tube Reactor Closure

One of the major problems in the design of a pressure-tube reactor such as the Carolinas Virginia Reactor (CVTR) is that of providing leak-tightness. This problem was particularly important in CVTR because of the cost of the heavy water which it utilizes; the heavy water cost dictated a need for leakage rates below those encountered in light-water moderated plants. A design goal of less than one pound of heavy water per joint per year was established [52].

The CVTR design utilizes a U-tube arrangement for the pressure tubes to permit easy removal and replacement for maintenance and refueling. The pressure tubes are made of Zircaloy and are joined to stainless steel extension pieces which connect with the header system. The heavy-water coolant enters the top of one leg of the pressure tube of the reactor, passes down over a fuel assembly, turns upward through a 180-degree fitting, passes over a second fuel assembly and leaves from the top of the other leg of the tube.

Three separate closure problems had to be faced in the design of the pressure tubes: the development of the Zircaloy-to-stainless steel transition joint; the development of methods for joining the pressure tubes to the feeder lines (known as "jumper tubes") and for making a refueling and jumper-port closure for the top of the pressure tubes; the development of the U-bend joint for the bottom of the pressure tubes.

Two separate approaches were taken in the development of a Zircaloy to stainless steel transition joint [52]. One involved development of a metallurgical joint and the other the development of a gasketed joint. The metallurgical joint was to have been made by either diffusion, eutectic or braze bonding of Zircaloy to stainless steel. Because of the difficulty experienced with a brittle interface in the bond, the joint design was finally based on the use of a gasketed joint. However, a metallurgical joint which appears to be successful has been developed and tested since that time [52].

The problem associated with joining the pressure tubes to the jumper tubes and providing refueling and jumper-port closures arose in large part from the complexity of piping at the top of the pressure tubes. In a pressure tube reactor plant such as CVTR, it is necessary to divide the flow of the coolant from the large primary system piping and distribute it among the pressure tubes which contain the fuel elements. Figure 3-15 shows a model of one-half of a proposed header assembly.

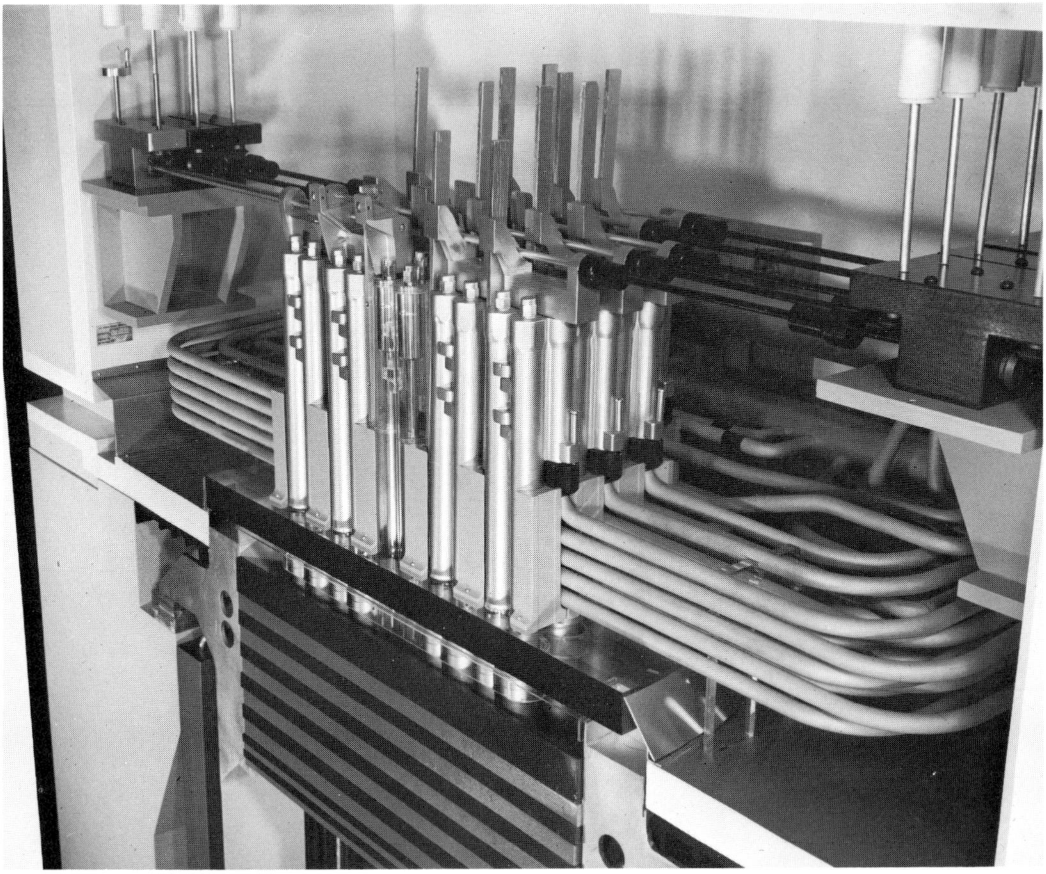

FIG. 3-15 Model of one half of header assembly, CVTR, showing tops of pressure tubes and jumper piping.

FIG. 3-16 Headers of one half of header assembly, CVTR. Pressure tubes go into holes and are joined to jumper piping by gaskets.

Figure 3-16 [53] shows the jumper tube (feeder line) and support structure. Extensive development work was necessary to produce a design of header assembly which met leakage requirements and which was compatible with the operation, maintenance and space requirements. This work involved evaluation of thermal expansion forces and stresses and problems of coolant containment, pressure drop, and fabrication and maintenance feasibility. Both seal-welded and gasketed closures were developed for the refueling and jumper-port closures and for the jumper connections. Gaskets as described below were selected, however, because they were shown to perform satisfactorily and were easier to handle and maintain.

The development of the U-bend joint for CVTR was based on the need for a mechanical joint between the pressure tube and the U-bend to facilitate insertion of the thermal baffle and to simplify fabrication. A welded joint would have made these operations more difficult and would have raised questions about the corrosion life of the joint. In the design of the joint, consideration had to be given to joint integrity, bolt strength, clamp strength, flange strength and radial gasket sealing stress. The mating members of the U-bend joint are all designed for the same material, Zircaloy-4. The bolts and nuts were selected to be of Imonel. Bodies of both 304 stainless steel and Zircaloy were tested and found satisfactory.

All the gasketed joints used the Marman Conoseal. The Conoseal resembles a Belleville spring in appearance. It fits loosely between the male and female flanges at the start of assembly. As the flanges are brought together, the gasket is flattened, resulting in the inside and outside diameters of the gasket decreasing and increasing respectively to make radial contact with the prepared groove in the flanges. Further clamping of the flanges results in an interference fit between gasket and flanges wherein the gasket is radially compressed approximately 2%. Tests showed that the leakage through joints with this gasket design was less than one pound per joint per year even after twelve hundred temperature and pressure cycles [52].

3.10.2 Closures for Liquid-Metal Systems

Liquid-metal systems present a number of special problems involving closures. While such systems do not involve high pressures, they do involve high temperatures. Thus, thermal stresses and thermal shock are important problems. In order to take advantage of the low pressures involved, efforts are made to use more conventional approaches to sealing where compatible with the liquid-metal coolant. Since liquid metals are chemically active, precautions must be taken to avoid contact with air and water [54a]. An inert atmosphere is used above liquid metal surface levels. In addition, as a core coolant, the liquid metal will become radioactive and must be well contained. Also, the thermal insulation used should not react with burning metal. Mineral wool and compacted magnesia have been used for this purpose [54a].

If the liquid metal freezes at room temperature, provisions must be made for heating it to make it liquid. In addition, if two surfaces submerged in Na or NaK are maintained in contact at operating temperature, they may seize or become self-welded. The extent of this effect is dependent on the temperature, the pressure between the parts and the duration of contact.

Because of the importance of preventing leakage of inert gas and liquid-metal vapor, considerable attention must be given to the seals used for control rod drive shafts. Conventional packing glands can be used for shafts having longitudinal motion providing the glands can be located in regions of moderate temperature and can be made accessible for servicing. In high-temperature regions, bellows have been used in applications involving short travel. Bellows have also been used with wobble-plate mechanisms to produce rotating motion. Frozen-sodium seals may be used in certain situations; in such seals sodium is allowed to freeze between a moving and a stationary part with a thin layer of fluid sodium maintained in contact with the moving part by careful control of its temperature. A somewhat similar type of freeze seal has also been used as a nonrotating seal; such as a seal utilizes bismuth or low melting alloy with appropriate heaters to melt the seal material when relative motion of the components is desired. Closures of a pipe can be effected by simply freezing the liquid metal in the pipe. This method has frequently been used for pipe repair or to seal off a section of a system.

4 REACTOR INTERNALS

The reactor internals encompass a wide variety of components. Proceeding inward from the reactor vessel these include the thermal shields and flow guides, the core-cage assembly, the core hold-down system, the fuel-element assemblies, the in-core instrumentation, the control elements and various types of fasteners and locking devices. Each of these components plays an important part in core safety and is discussed in this section. In addition to the design of the components themselves, certain aspects of design common to all of these are considered.

This section discusses in considerable detail the considerations which go into the mechanical design of reactor internals for pressurized-water or boiling-water reactors. While this does not do justice to other reactor types, it serves as the basis for identifying many of the problems common to all reactors, as discussed in Sec. 1, and illustrates the interplay of considerations that must be evaluated in any reactor design.

4.1 Fundamental Considerations

The reader is referred to Secs. 1.3 and 2.1 for a discussion of the general design requirements and specific criteria of mechanical design which apply, in large part, to this section. These are discussed below in terms of the specific components to which they apply. Particular attention is given to those aspects of design which influence safety. These include the fundamental problems of structural integrity, functional reliability, materials compatibility, fabricability, and maintainability. Under structural integrity are covered questions of thermal stresses, creep, fatigue and radiation effects on mechanical properties. For specific details on the computation and evaluation of thermal stresses and on the effect of radiation on metals the reader is referred to Sec. 2.

4.2 Thermal Shields and Flow Guides

Thermal shields are provided interior to the reactor vessel walls to attenuate the nuclear radiations reaching the reactor vessel walls and thereby to reduce the heat generation and accompanying thermal stresses caused by the radiation. The thermal shields generally consist of one or more stainless steel cylindrical shells concentric with the core and supported from a ledge inside the vessel. In addition to attenuating the nuclear radiations, the thermal shields may also furnish the support for the internal coolant flow guides.

Inasmuch as their main purpose is to reduce radiation heating, they themselves must be designed to accommodate such heating. As a result, thermal stresses are an important consideration in their design. After the thickness required to attenuate the radiation has been determined, the number of cylindrical shells and the thickness to be used for each must be established. This is done on the basis that the thermal stresses in each shield cylinder must be kept within allowable limits. The magnitudes of such stresses are determined by the methods outlined in Sec. 2.2.2, paragraph 1 (c), for Case I. The ability of the thermal shields to withstand the thermal stresses is determined by fatigue analysis as outlined in Sec. 2.2.4. For the thermal shields, pressure stresses are small and arise only because of pressure differences required for coolant flow around them. In applications involving high temperatures, the possibility of distortion brought about by uneven cooling of the shields must be considered.

Coolant flow past the thermal shield and the interior of the pressure vessel is required to remove the heat generated. In order to avoid poor flow distribution in the coolant flow passages of the shields and to assure adequate flow in the hottest region of the shields and vessel walls, the average flow velocities are often made higher than indicated by computation. Careful attention must be given to the ultimate path of this flow otherwise it may end up representing a significant bypass flow (often called bypass leakage) around the core. F. A. Grochowsky [55] describes a flow scheme in which such bypass leakage can be avoided. In this arrangement, the pressure drop across the flow distribution baffles is used to drive the flow upward through one passage of the thermal shield and downward through other passages associated with the shields; this coolant flow then joins the main coolant flow at the underside of the core. Seals between the vessel and the thermal shield and between the thermal shields and the core barrel are used to restrict the by-pass leakage (see Fig. 4-1 [55]).

Flow guides are provided interior to the reactor vessel to direct the coolant flow from the inlet nozzles to and through the core and other reactor internals and then to the outlet nozzles. Generally many components of the structural supports, including the core barrel and core holddown structure, play a part in guiding this flow. The flow guides referred to here, however, are those usually attached to the bottom end of the thermal shields in water-cooled and moderated reactors to distribute the flow entering the core. A variety of devices has been used, but among the simplest is the flow baffle shown in Fig. 4-1. It consists of a perforated flat plate which is used to break up the velocity head from the inlet nozzles, to provide sufficient pressure drop for obtaining a reasonably uniform distribution of flow into the core and for forcing flow around the thermal shields. The pressure drop across the baffle should be kept to a minimum, consistent with the flow distribution needs. It should be recognized that the baffles must be capable of taking the required pressure drop; where the diameter of such a baffle is large, this may mean a heavy plate. Attention must also be given to the methods of attaching flow guides; vibration and fretting of parts can be encountered. Breakage of fasteners in the flow guides could eventually lead not only to loss of flow direction but to actual flow blockage by a loose component.

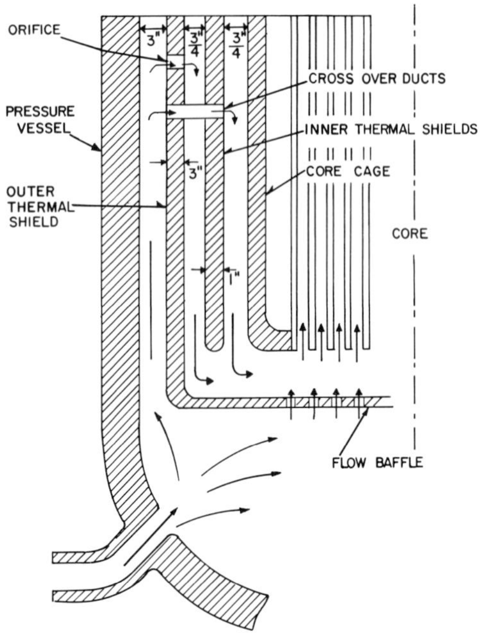

FIG. 4-1 Thermal shield cooling scheme.

Thermal shields and flow guides have been designed either as replaceable assemblies or as integral parts of reactor vessels. If replaceable, they are easier to fabricate, clean and install; their absence during fabrication of the vessel makes it easier to keep the vessel clean. More important, however, is the fact that during the life of the plant they can be removed to facilitate access to the interior surfaces of the vessel for inspection purposes. If replaceable, they also provide greater design flexibility for future cores to be installed in the same vessel; they can be replaced when the core is replaced. The chief disadvantage of replaceable assemblies is that the diameter of the vessel closure must be large enough to accommodate the shield assembly. In addition, means must be provided for clamping the assembly in place. If the thermal shields and flow guides are made integral, the advantages and disadvantages are reversed.

4.3 Core-Cage Assembly

The core cage is the structure used to hold and support the fuel assemblies. It is designed to assure correct alignment for operation of the control rods within the core cage or within the fuel assemblies contained in the cage. Its design must be such as to permit refueling of the core. In addition, much of the in-core instrumentation may be supported by the cage. The core cage is usually supported on a support ledge in the reactor vessel and is held in place by the holddown structure.

While the details differ from one design to another, almost every core cage consists of three basic components: a longitudinal support device (often in the form of a cylindrical barrel), a bottom support plate and a top support grid of some kind. In the design shown in Fig. 4-2 [39], the core-cage barrel holds the bottom support plate and top grid. It also contains grooves and lands on the exterior of a thickened portion to provide a labyrinth seal to limit the flow bypassing the core between the barrel and the thermal shields. The top grid maintains the alignment of the upper portion of the fuel assemblies and carries the hydraulic thrust on the

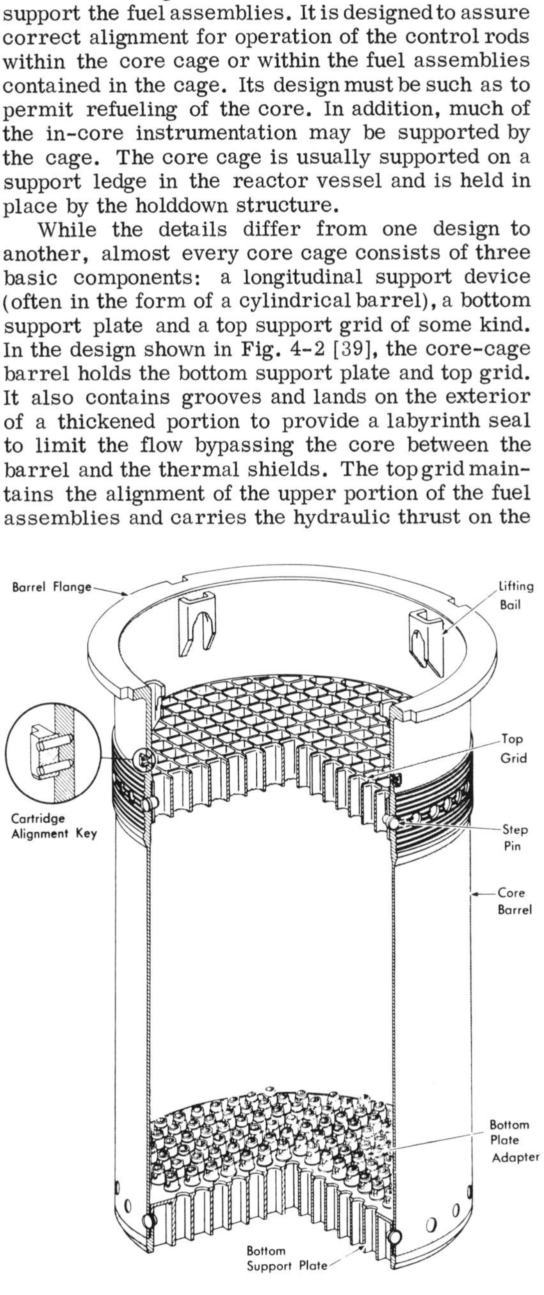

FIG. 4-2 Core-cage assembly.

fuel assemblies through lugs from the assemblies which engage the underside of the grid. In a number of designs the grid supports and/or positions the control rod shroud tubes. The bottom support plate supports the weight of the core and provides stub tubes for engaging the fuel assemblies and directing flow to them. In the case of cores using control rod followers, the bottom plate may also provide support for tubes for the followers extending downward from the bottom plate. The bottoms of these lower shroud tubes are in turn held together by a bottom tie plate. (See Fig. 4-3 [56].)

The design of the core cage involves a number of considerations. First, the core barrel, being immediately adjacent to the core, is subjected to high thermal stresses. Usually these stresses are far less uniform than those in the thermal shields or in the vessel because of variations in the proximity of the fuel assembly to the core barrel. These thermal stresses must be evaluated as indicated in Sec. 2.2; the evaluations generally dictate that the core barrel must be so thin as to raise problems of radial buckling if it is subjected to an external pressure difference. This problem has to be faced in the design of the core-cage barrel for the Shippingport plant. If a thickened section is to be provided for a labyrinth seal, it must be sufficiently far away from the core so that excessive thermal stresses are not introduced.

Other problems encountered in the design of the core-cage barrel are associated with providing for alignment between the core and the vessel and with differential expansion (both radial and axial) between the cage and the vessel and between the cage and the fuel elements. Attachment of the barrel to the top and bottom plates of the cage presents difficulties; radial pins, with their attendant alignment difficulties in manufacture, may have to be used in order to avoid distortions in welding finished top and bottom plates to the barrel. Much finishing work may have to be done to both plates before assembly to the barrel because of lack of accessibility afterward. Problems are also encountered in fabricating the barrel because of its relatively thin section; efforts must be made to avoid untenable distortions.

The top grid presents other design problems. It is close to the core and is subjected to nonuniform heating; it must be designed to avoid distortions that would interfere with its functional requirements. It must be made of sections thin enough to avoid excessive heating and excessive thermal stresses; yet it must be rigid enough to withstand the total hydraulic load imposed upon it during operation, without deflections that would impair control rod alignment. If the grid design is such that it results in spaces between fuel assemblies of water reactors, consideration must be given to the influence of the web thickness of the grid on these spaces and the attendant neutron flux peaks in the water gaps between fuel elements which result from the design. Nevertheless the web must be thick enough to avoid local buckling due to loads imposed by the lugs which transfer the hydraulic loads of the fuel elements to the grid. Furthermore, if the core-cage assembly is designed to withstand the impact of control rod scramming, the effect of such impact must be considered in the design of the top grid.

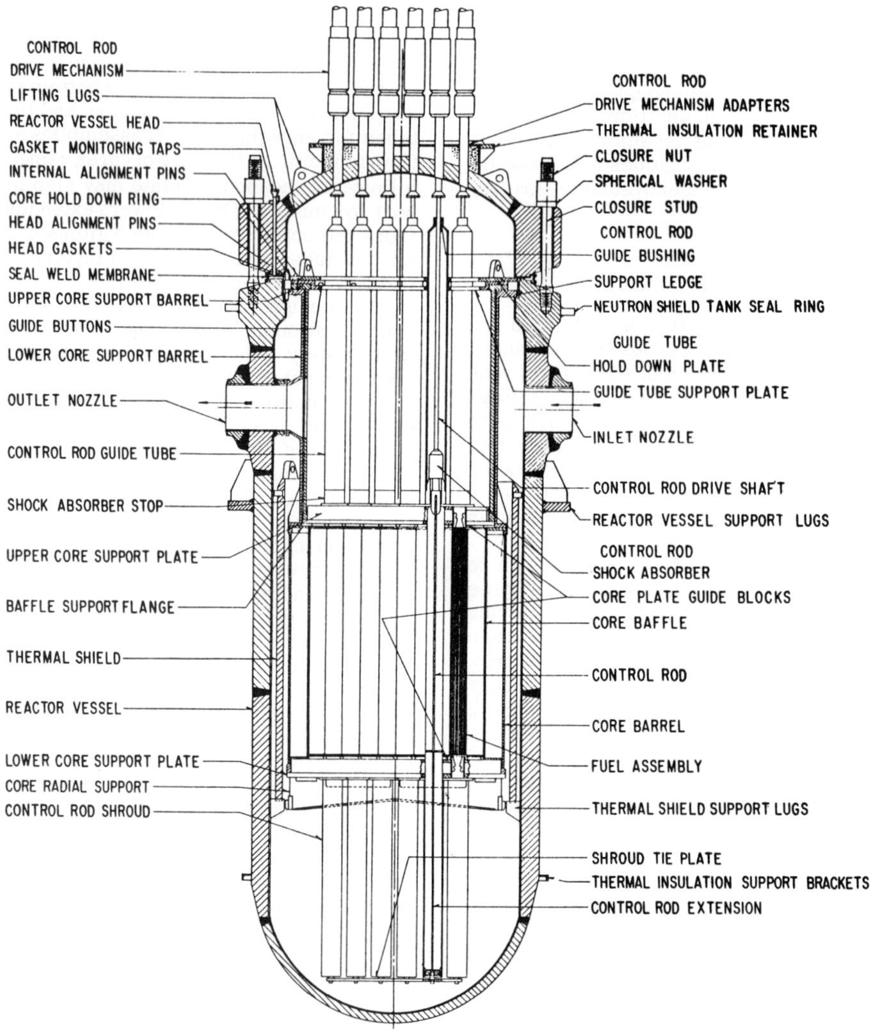

REACTOR VESSEL ASSEMBLY

FIG. 4-3 Reactor vessel assembly, Yankee Nuclear Power Plant.

Computer codes have been developed for determining the adequacy of certain types of grids under the actions described above. In view of the complexities involved and the assumptions implied in such codes, they should be applied with caution unless there has been some experimental verification of their applicability. Sometimes experimental work will show that simpler analytical techniques can be used for evaluating alternative designs. For example, it was determined by test that, for the grid used on Core-I at Shippingport, the maximum stresses from mechanical loads computed by the ligament efficiency modification of solid-plate theory were within ± 5% of measured values and calculated deflections were approximately 20% greater than measured values [51]. In these calculations the maximum stress and deflections were calculated for a solid plate of the same total thickness as the grid and then were divided by a ligament efficiency factor, which was the ratio of the web thickness to the modular pitch; because of the square module and the low ligament efficiencies involved, Poisson's ratio was taken as equal to zero in these calculations. Care should be used in extending this method to other confugurations. Furthermore, it should be recognized that such an approach is useful for preliminary design purposes only; the final design requires careful analytical treatment. The tests referred to above were made on a quarter-scale, carbon steel model of the grid; the actual grid was made of stainless steel. Loads were imposed by hydrostatic pressure on a diaphragm which transmitted forces through pins to simulate the action of the fuel-assembly lugs. Measurements were made by means of strain gages and dial indicators (see Fig. 4-4 [51]).

The development of simple and accurate means

FIG. 4-4 Strain gages shown on quarter-scale model of PWR top grid.

of treating thermal stresses was not as successful. Nevertheless, it was found that in some cases a satisfactory approximation can be achieved with a one-dimensional analysis of the temperature distribution through a solid plate [51].

One of the most difficult problems in grid design is to develop a satisfactory fabrication scheme of reasonable cost. One means which has been successfully used has involved welding together extruded cruciform shapes sufficiently thick to accommodate the distortions and misaligments encountered in fabrication and still to allow sufficient material for final machining; careful jigging was required. The PWR core I grid was fabricated in two steps. After the cruciforms were welded together, a ring was welded around the assembly to provide the required configuration. All welds were made from two sides with back-chipping to provide for 100% penetration. They were carefully dye-checked for cracks. Prior to machining, the complete assembly was annealed with the jigs in place. Figure 4-5 [49] shows a photograph of the grid prior to machining.

The bottom plate involves problems similar to the top grid but with different complications. The bottom plate may also involve a gridwork, but usually this gridwork serves to provide only stiffness for another plate welded on top of it and hence does not require accurate machining as in the top grid. However, welding a plate to the top of such a gridwork is not always a simple matter if crevice-free 100% welds are to be achieved. It may be necessary to make this plate from pieces which are approximately the same size as a module of the gridwork and which are prechampfered for welding. These pieces are then welded to the gridwork and to each other from the top; a seal weld is then applied from the bottom between the plates at the grid. The top welds are then ground smooth and the plate prepared for machining.

The bottom plate must be machined to receive the stub tubes to which the fuel assemblies are engaged during service. If, in addition, these stub tubes also contain flow-measuring devices, provisions must be made for accommodating and cooling the pressure tubes that lead to exterior sensing devices. If thermocouples are to be installed on the underside of the bottom plate, provisions must be made for passage of the lead-wires in a similar fashion (see Sec. 4.6). Figure 4-6 [51] shows the completed PWR bottom plate with instrumentation installed prior to its assembly into the core cage.

The core-cage assembly is used primarily for core designs in which fuel assemblies are to be refueled as individual units. Many modifications to this basic design might be made. For example, the top grid might be eliminated and the fuel elements might be held in place against a peripheral core plate at the top of the assemblies by means of interrupted-screw expander pins. Four of these, one at each corner of a fuel assembly, would have to be turned and removed before the fuel assembly could be removed. Machining tolerances for such a design are quite severe. In other modifications, particularly for core designs involving free radical interchange of coolant, a peripheral baffle plate may be used to confine the coolant flow to the region of the core. (See Fig. 4-7 [56].) In such a design the baffle must be adequately supported to prevent collapse during operation; the support is usually provided by ribs extending to the cage barrel. Actually, in such a design the cage barrel can be replaced by tie bars which hold the top and bottom of the cage together, or the baffle itself can serve as the cage barrel.

Another method of holding a core together can be conceived which does not utilize a core cage at all. This design would consist of a complete array of fuel assemblies permanently attached to an upper support assembly and a bottom plate to form a replaceable cartridge. For large cores in which a planned program of reshuffling fuel is indicated, this cartridge approach is not practical. For small cores, particularly for shipboard use where shock might be an important consideration, the cartridge concept can have merit. A modified cartridge concept has been used for commercial reactors in which the cage is permanently attached to the upper support structure, but provisions are made for individual handling of fuel assemblies.

Other methods of core support have been employed which differ from the core cage or core cartridge concepts described, particularly in reactors other than pressurized-water types. For the most part, however, their problems are similar to those described in this section.

FIG. 4-5 Photograph of PWR top grid prior to machining. (Scale at bottom of photograph is 3 ft in length.)

4.4 Core Holddown Systems

A number of different schemes can be used for holding the core inside the reactor vessel. The most common method is to use the clamping action of the head to hold the core structure in place. A ledge is provided inside the vessel on which the flange of the structure rests; a ledge on the underside of the head is used to provide the clamping action. A Belleville spring, or a clamping ring containing plastically deformable columns, or a hold-down ring consisting of numerous small springs can be used in connection with the clamping action to accommodate machining tolerances and any differential expansions that may be involved. Figure 4-8 [39] shows a core-support spring assembly that has been used.

FIG. 4-6 PWR bottom plate with instrumentation assembled.

If the support ledge in the vessel is located immediately beneath the closure region, the entire core must be held as a cartridge unit extending downward from the ledge. (See Fig. 4-3.) If the support ledge is well down into the vessel, a separate hold-down barrel is used to transmit the clamping action from the head to the core cage flange. In this latter arrangement, the support ledge in the vessel is usually below the upper level of nozzles and is used to support the thermal shields as well as the core cage. (See Figs. 4-9 and 4-10 [39].) If the ledge is near the closure head, separate support lugs are provided for the thermal shields. In either case, if the head is used for clamping the core in the vessel, the added load imposed on the closure must be considered in the design of the closure.

If it is not desired to use the head for clamping, the core cartridge can be bolted to the ledge. If bolting is used, provisions must be made to allow for radial differential expansion between the support flange and the ledge in the vessel. Even if the head is used for clamping, consideration must be given to the requirements for suitable washers between flanges to facilitate sliding.

The structure above the core cage must provide means for guiding the control rod drive shafts and for protecting the control rods from hy-

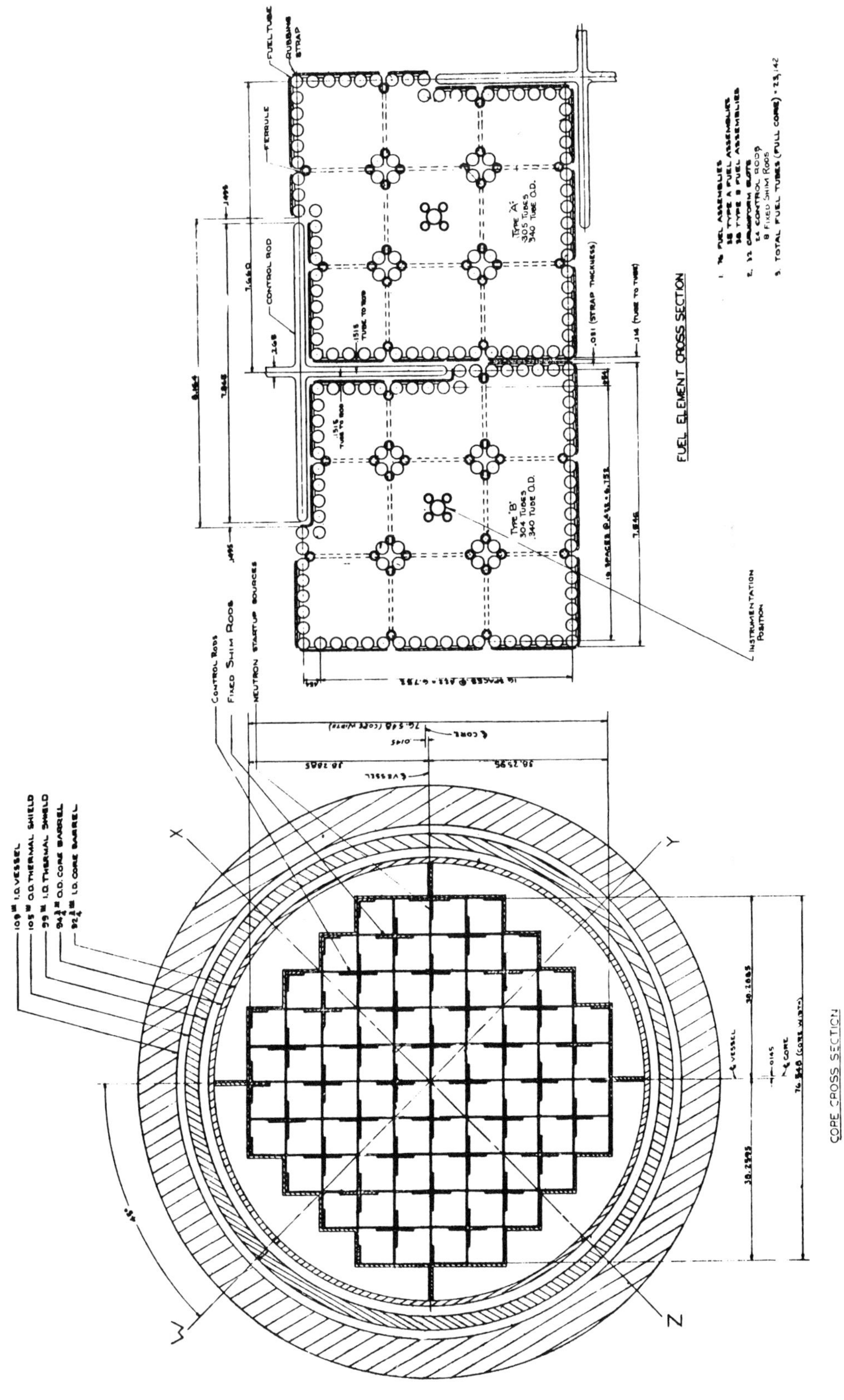

FIG. 4-7 Core cross section, Yankee Nuclear Power Plant.

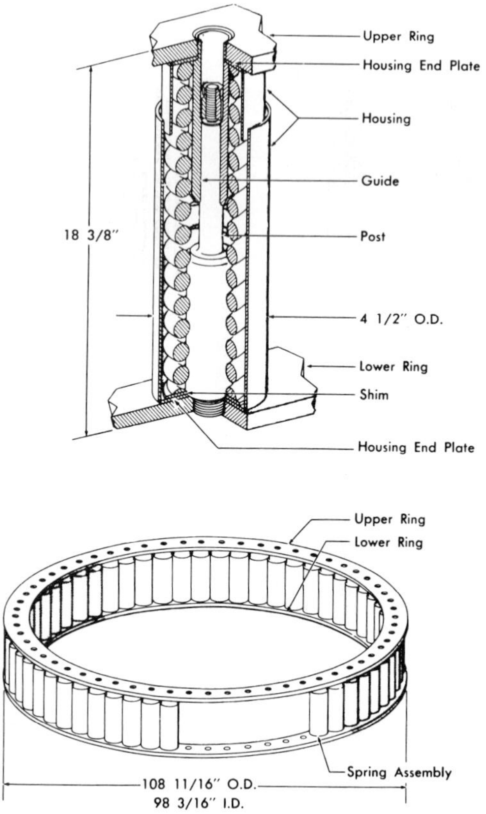

FIG. 4-8 Core support spring assembly.

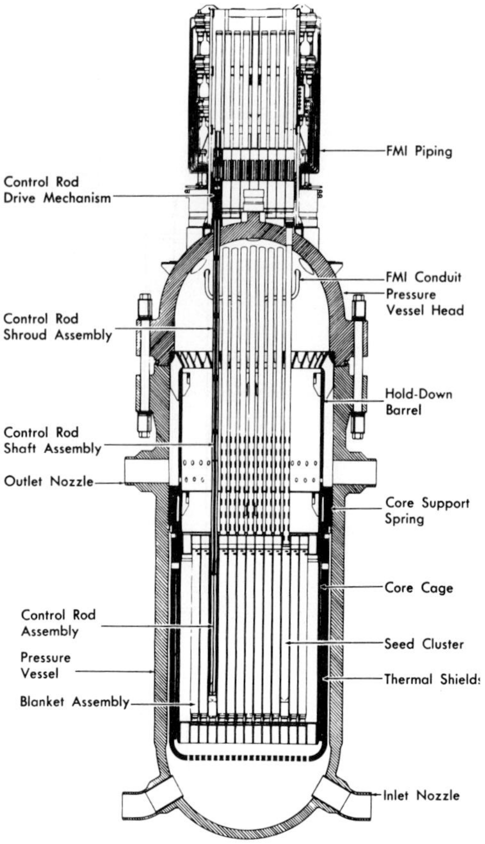

FIG. 4-9 PWR longitudinal section.

draulic forces which may be imposed by cross flow when the rods are partially or completely withdrawn from the core. If the shrouds are not completely effective, the rods can be forced by the full coolant flow against the side of the control rod channel and fail to scram when required. Such action has been experienced in prototype testing. The shrouds must be stiff enough to prevent binding of the control rod assembly when scramming of the rods is called for with full coolant flow in the vessel. The most difficult aspect of this problem is determining the nature of the flow in the region of the shrouds. Fig. 4-11 [51] shows the type of flow pattern obtained in quarter-scale model airflow studies of the PWR reactor.

An important factor which affects the design of the structure above the core is the nature of the flow in the core, i.e., whether the core is designed for one-pass or multi-pass flow of the coolant. The added baffling required by a two-pass design usually adds considerable complexity to the upper core structure. This complexity must be weighed against the gains achieved by a two-pass design; alternative ways of achieving such gains (as, for example, by orificing a one-pass core) should be considered. Other factors that must be considered in the design of the holddown structure are the in-core instrumentation that is to go above the core and the access required for refueling equipment.

As in the design of other structural components of the core, problems of thermal stresses, fatigue, creep and alignment must be considered. Fabrica-

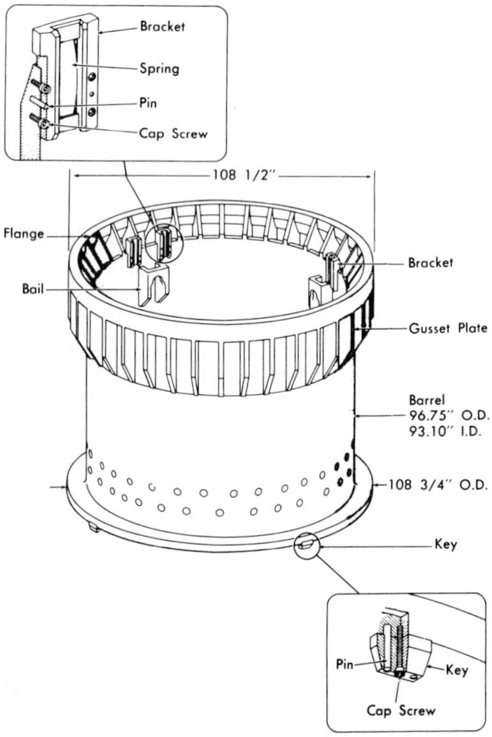

FIG. 4-10 Core hold-down barrel.

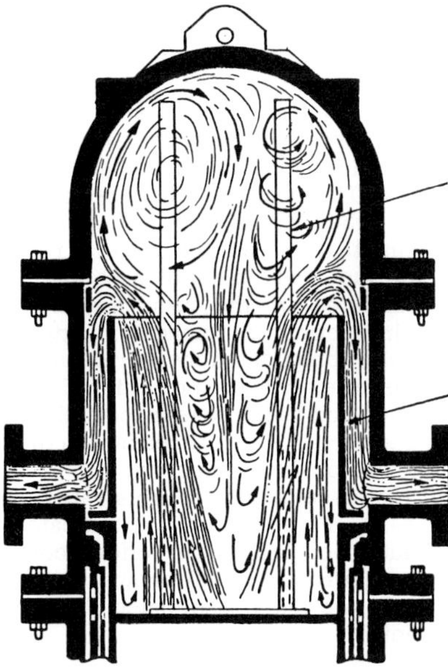

FIG. 4-11 Quarter-scale model airflow study, PWR. (Flow pattern in upper plenum for symmetrical-loop operation.)

tion problems are similar to those discussed for the core cage in Sec. 4.3.

4.5 Fuel-Element Assemblies

This section is concerned primarily with the mechanical aspects of designing fuel elements and fuel element assemblies. The reader is referred to Chapter 3, "The Reactor Core", and Chapter 13, "Fuel Elements", for further details concerning their design.

Fuel elements can take any one of a number of basic shapes and can be assembled into larger units in a variety of ways. The term fuel element is used here to refer to a single element of fuel and cladding such as a single plate or a single rod. Groupings of fuel elements to form larger units will be referred to as assemblies, clusters and/or subassemblies. While these terms can be interchanged without invalidating what is said about them, the terms usually have rather definite meanings. The term "assembly" usually refers to the smallest group of fuel elements that can be removed from the core as a unit and is so used here except where other terms are more appropriate. A "subassembly" is a group of fuel elements which be used to make up a fuel element assembly. The term subassembly is often used to designate a component of a cluster. The term "cluster" is often used to refer to units in which subassemblies are welded in a cluster to form a channel for passage of a control rod. When a subassembly is used as a single replaceable unit, it often retains its name as a subassembly.

The group of rods which make up the fuel in a pressure tube reactor or in any reactor where large gaps exist between rod groups is also usually termed a cluster.

Mechanical considerations in the design of fuel element assemblies involve both the fuel elements themselves and the method of assembly. While the materials, shape, dimension and spacing of fuel elements are selected primarily on the basis of nuclear and heat removal aspects, there are a number of mechanical factors which must be considered. In selecting the shape of a fuel element, consideration must be given to the ease of cladding it to retain fission products, and the ability of the cladding to withstand the internal pressure of gaseous fission products must be evaluated. The shape should also be relatively easy to make with a minimum amount of expensive machining. The shape should be amenable to easy assembly and inspection. Plates, rods, pins, and ribbons are shapes that have been used successfully.

Almost the first consideration in fuel element design is the nature of the fuel material to be used, i.e., whether it is to be uranium metal, an alloy of uranium, a ceramet or a ceramic material. The choice is dependent on a number of factors such as the concentration of uranium desired in the element and the corrosion properties of the resulting matrix. These factors are discussed in detail in Chapter 13, "Fuel Elements".

A number of precautions must be taken in the mechanical design of fuel elements. The detailed nature of the precautions will vary with the shape of the fuel elements, the relative dimensions of cladding and meat, the nature of the fuel, and the degree of bonding or lack of bonding between meat and cladding. Common to all fuel elements is the fact that the fuel element dimensions must be assigned tolerances. These tolerances influence not only the manufacturing procedure to be used and the cost of the elements, but also the performance and even the integrity of the fuel elements if the tolerance range is an appreciable fraction of the cladding thickness. The performance is influenced by the tolerances affecting heat generation rates and dimensional stability. In unbonded fuel elements, attention must be given to variations in diameters and stacked heights of pellets relative to the tube diameter and to the distance between end caps respectively to assure adequate room for expansion.

Attention must also be given to the problems associated with fission-gas release in the fuel matrix. These include the influence of fission gas on the dimensions and integrity of the fuel and the influence of fission-gas pressure on the cladding; the pressure considerations must include effects encountered with the system pressurized and unpressurized. Other problems include waterlogging (see Chapter 13, "Fuel Elements"); vibration characteristics of individual elements; shock-carrying capabilities; hydrogen enbrittlement of cladding; temperature limits for phase changes in fuel and for fuel-cladding reactions; modes of manufacture; locked-in stresses resulting from the manufacturing process or inadequacies of annealing processes; and inspection of individual elements. In addition to these problems are those associated with the fact that fuel elements become part of an assembly; these are discussed later in this section.

An important consideration in fuel element design is the number of inspections to be required and the means to be used for inspecting individual fuel elements. Inspection of all fuel elements can be quite expensive, yet failure to perform adequate inspection can be more costly. It is cheaper to inspect all elements and reject a single defective one early in the manufacturing process than to discover and reject a larger assembly containing a defective element later. It is cheaper to reject an entire assembly than to remove the reactor vessel head at some later time during operations. If the decision is made not to inspect all fuel elements, sufficient statistical planning and inspection must be done to assure that all fabrication processes are in control at all times.

In addition to the above considerations for single fuel elements there exist a number of considerations which apply to assemblies of fuel elements. The grouping of fuel elements into assemblies, subassemblies and clusters facilitates core fabrication and handling. A wide variety of geometries can be used for this purpose. The overall size and geometry of the individual assemblies are usually dictated by the size of the core, the number and location of control rods, the manner in which assemblies are to be held in the core, considerations of thermal distortion or bowing and the manner in which the core is to be refueled. The pattern in which fuel elements are grouped within an assembly is dictated by fuel element shape, spacing to secure appropriate nuclear characteristics, and by inspection requirements. Hexagonal, triangular or rectangular patterns are usually most satisfactory.

Assemblies can be manufactured satisfactorily by welding, by brazing, by mechanical means, or by combinations of these. Welded assemblies have the advantage that they are rugged, require relatively little machining, and fretting problems are not usually encountered in service. Many problems, however, are faced in producing a welded assembly. These include securing good weld penetration without impurity pickup and without extending the heat-affected zone into the uranium-bearing material, obtaining desired dimensions within appropriate tolerances, and avoiding excessive locked-in stresses. Thus in such designs one must face problems of environmental control, accessibility, jigging, and annealing.

Mechanical means for holding subassemblies, however, present other problems. Usually, extensive machining of fuel elements is required. In addition, close tolerances are necessaary so that pieces fit well. Care in design is necessary to avoid problems of vibration, fretting corrosion, and crevice corrosion. However, mechanical assemblies have the important advantage that if a defective fuel element is found during an advanced stage of manufacture, it can be removed and replaced without scrapping the entire assembly. In mechanical assemblies containing control rod passages, designs must be avoided in which it is possible for broken parts of mechanical fasteners to lodge in the control rod channel and interfere with motion of the rod.

Considerations involved in design of fuel element assemblies include those of progressive failure; spacing and tolerances; collapse of adjacent flat plate elements because of Bernoulli forces; pressure stresses and thermal stresses between elements and on assembly shells; differential expansion problems; rubbing and wear; galling and seizing; fretting; bowing; swelling; crevice corrosion; mechanical shock; vibration; integrity of control rod passages; and, again, problems of inspection. In addition to the above, design of fuel element assemblies also usually involves design of latches, springs, orifices for flow control, and bolts and locking devices. (The mechanical features of the last are discussed in Sec. 4.8.)

It is worth noting that in PWR the use of a rectangular array for fuel rods in the blanket was in large measure dictated by the requirement of accessibility for inspection. A triangular pattern would have severely complicated measurement of rod spacing and visual examination of finished assemblies. Each fuel rod bundle was inspected for compliance with specifications on dimensions, corrosion resistance, and cladding integrity.

Some typical fuel assemblies are illustrated in Figs. 4-12 and 4-13. Figure 4-12 [39] indicates a flat plate fuel cluster containing a cruciform control rod channel. Figure 4-13 [44] shows an assembly of fuel rods for the Consolidated Edision (Indian Point) Reactor.

4.6 In-Core Instrumentation

In-core instrumentation can take many forms, including devices to measure inlet coolant tem-

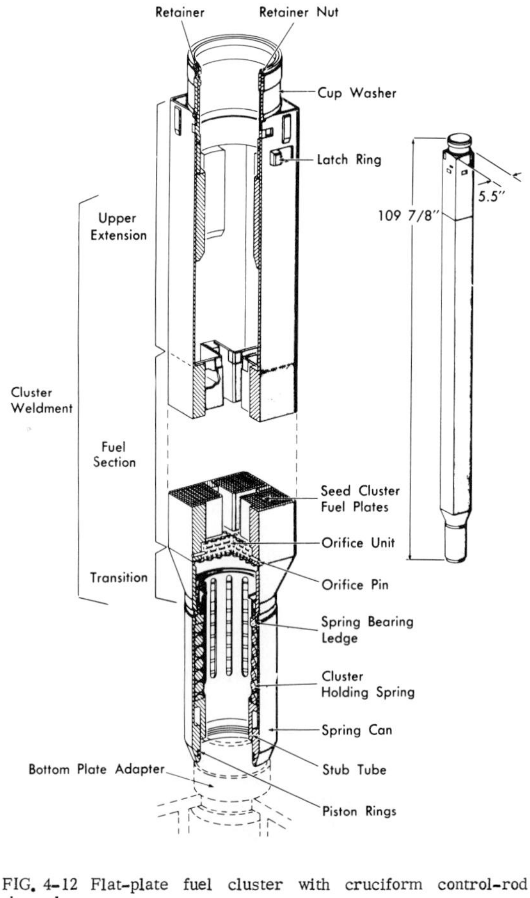

FIG. 4-12 Flat-plate fuel cluster with cruciform control-rod channel.

perature, exit coolant temperature, fuel element temperatures (centerline or surface), coolant flow to individual assemblies, neutron and/or gamma and flux distribution. Devices can also be installed in the core to detect a fuel element failure and to locate the assembly in which the failure has taken place. It is not the intent of this chapter to describe or even discuss each of these instruments but rather to outline some of the mechanical problems which must be faced in using them. More details are discussed in Chapter 6, "Sensing and Control Instrumentation." Because of the compactness of reactor cores, little space is available for instrumentation components and heads. (See Fig. 4-6.) The mechanical problems encountered in developing instrument systems which are part of the core itself can be categorized as follows [51]:

1. Problems associated with accommodating the instruments and leads within the limited space available in a manner compatible with normal operation and also with the refueling scheme. Many of these problems involve difficulties in accomplishing what might be called conventional functions, such as joining and sealing groups of sampling and sensing tubes and leads, and using flow-measuring devices under non-ideal flow conditions.
2. Problems of cooling leads so that they do not cause erroneous readings, particularly tubes for indicating pressure differences in flow-measuring devices; boiling in such leads would give erroneous readings.
3. Problems of specifying and fabricating thermocouple leads and other tubing so that integrity of cladding or tubing can be assured. If, for example, the cladding on a thermocouple lead which goes outside the vessel fails, the primary system is breached. Though this may be a minor breach, it requires attention. The reading of the thermocouple will also be affected.
4. Problems of new devices to perform new functions for which there is little precedence, such as the development of means for inserting and removing flux wires for measuring flux distribution, or the development of means for obtaining a representative mixed sample of effluent from a fuel assembly for failed-element detection and location.

It should be recognized that in-core instrumentation introduces a number of small parts into the core which must be carefully secured. They should be so designed that in the event of failure they cannot get into the coolant stream and be carried to and interfere with the motion of the control rod system or with the flow of coolant.

To meet the requirements for compactness and reliability, appropriate mock-ups of wood or plaster or other materials should be used. Full-scale mock-ups are often desirable. Such mock-ups disclose not only problems of arrangement but also problems to be faced during assembly and disassembly. It must be remembered that disassembly and reassembly of the system will be done in an unfavorable environment after the initial assembly. Often it will have to be done under as much as thirty to forty feet of water. Mock-ups can also assist in determining tooling and template requirements and in training personnel for assembly work.

In-core instrumentation should be considered early in the design of the core because it may very well influence other design features. For example, the installation of a sampling device for detecting and locating fuel elements influences the fuel assembly design. Incorporating a flow-measuring device in the core influences not only the bottom

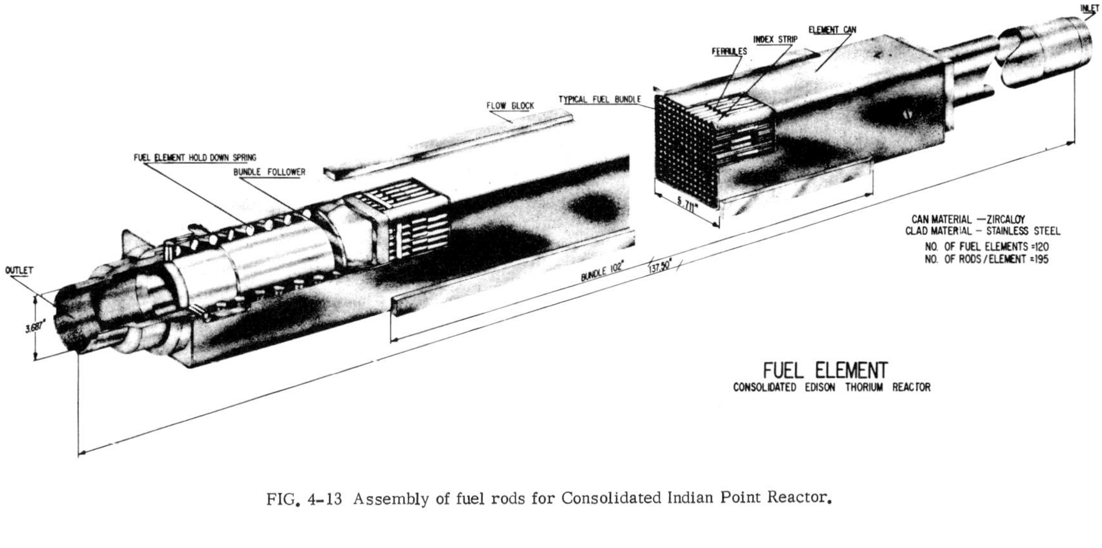

FIG. 4-13 Assembly of fuel rods for Consolidated Indian Point Reactor.

plate of the core cage but also can adversely influence the flow distribution among the individual fuel elements of a fuel assembly. The assembly may have to be made longer to provide for the flow distribution. If use of flux measuring devices involves thimbles in the reactor vessel head, both the core and the head are affected as well as the room above the head.

Materials tests should be made of proposed tubing and leads early in the process of developing manufacturing specifications. For example, at Shippingport it was found that the 0.040 in. (1.02 mm) diameter special metal thermocouple leads contained minute local fissures in the stainless steel cladding; water got into the leads and the thermocouples failed. It was later found that an aging process employed to stabilize thermocouple calibration introduced local areas of brittleness in the sheathing (or cladding) by the precipitation of carbon along grain boundaries. This brittleness coupled with the mechanical handling during installation was believed to be the cause of failure. Reference [51] points out that the difficulties encountered with in-core instrumentation involved those systems for which major compreshensive tests had not been conducted.

4.7 Control Elements

This section deals with the mechanical aspects of various types of control elements used in nuclear reactors. The discussion concerns control rods, installed burnable poisons, chemical control, and back-up systems and last-ditch devices.

4.7.1 Control Rods

By far the most common method of controlling nuclear reactors is by use of control rods. In thermal reactors these contain or consist of materials with high thermal neutron-absorbing properties, e.g., cadmium, boron, and hafnium. Control rods are moved into or out of the core as required. Control rods can be operated from the top or bottom to move vertically through the core or they can be inserted from the side. Rods that are operated from the top must generally be shrouded against cross-flow of exit coolant; they also tend to interfere with refueling. However, top-operated rods can readily utilize gravity for scramming purposes. Rods operated from the side cannot use a gravity scram directly. Rods operated from the bottom can use gravity scram if they are pushed up out of the core as the core goes critical; however, again they require shrouding against coolant cross flow. The use of scram against gravity poses severe safety questions and should be related to the nature of the control drive mechanism use (see Sec. 5). As pointed out in Sec. 5, bottom-mounted mechanisms face problems of crud collection unless special precautions are taken.

The control rods are usually made to have the same length as the core; at the driven end, each rod is attached to a shaft connected to a drive mechanism which moves the rods. This shaft operates in linear bearings contained either in the shroud tubes (discussed previously) or in a specially provided structure. The shaft may or may not extend through the vessel head and, thus, work across the pressure difference between the inside and outside of the vessel. Figures 4-14 and 4-15 [39] show a typical control rod as used in the PWR core. The rod shown is made of hafnium and cruciform in shape. It contains an adapter at the top which provides a remotely operated bayonet joint for uncoupling (during refueling) the rod from the shaft which extends to the control rod drive mechanism on top of the reactor vessel head. The shroud tube and the linear bearings contained in the tube are shown in Fig. 4-15. The shaft from the control rod drive mechanism to the rod is contained in the shroud tube. During refueling this shaft is remotely disconnected from the control rod adapter at the bayonet joint. After this is done the control rod rests in its channel on the spacing buttons which hold the rod in a predetermined position so that the bayonet can be re-engaged after refueling. The entire shroud and shaft assembly is then removed in order to get at the control rod or the fuel cluster in which it is located.

Because of the importance of control rods to the safety of the core, they must function reliably. They are usually the only moving parts in the core. Attention must be given to a number of conflicting requirements in the design of control rods. A number of these are outlined in reference [57] and are included among the following:

1. The control rod must have the structural rigidity and strength to withstand without damage the mechanical loads imposed upon it during operation. Such loads include the impact that accompanies scramming of the control rod. While a buffer mechanism is usually provided to reduce this impact, consideration must be given to the structural adequacy in the event that the buffer sticks, as they sometimes do. Nevertheless, a reliable buffer should be provided to minimize such loads.
2. Control rods must be capable of operating satisfactorily under the worst conditions of distortion that might be expected, specifically in the presence of the maximum thermal gradients that will be experienced.
3. The control rods must be protected against hydraulic forces that would interfere with their operation when partially or fully withdrawn from the core.
4. The linear bearings of the control rod shaft must be protected from becoming fouled by grit, dirt or abrasive wear and corrosion products that might be in the system. Excessive frictional forces arising from such deposit, from hydraulic forces on the rods, or from distortions of the rod must be avoided. Excessively high friction can interfere with the scramming action of the rods or can possibly lead to buckling of the rod when it is being driven into the core by the drive mechanism.
5. In water-moderated reactors, the space left in the core when the control rod is withdrawn may cause excessive peaking of the neutron flux in

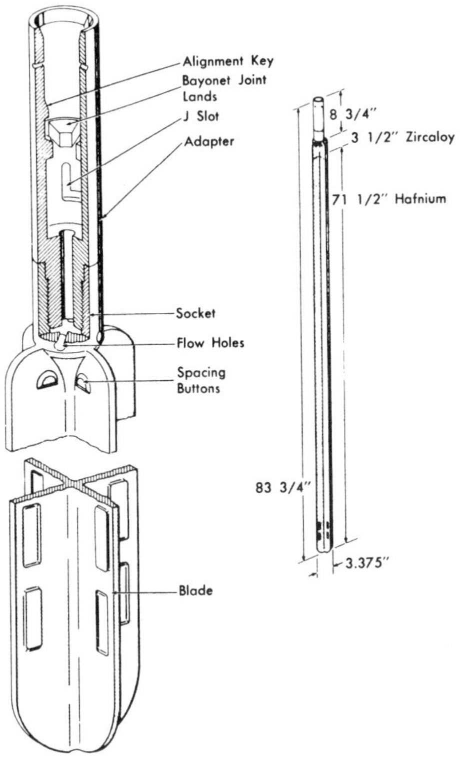

FIG. 4-14 Control rod and adapter, over all and sectional views.

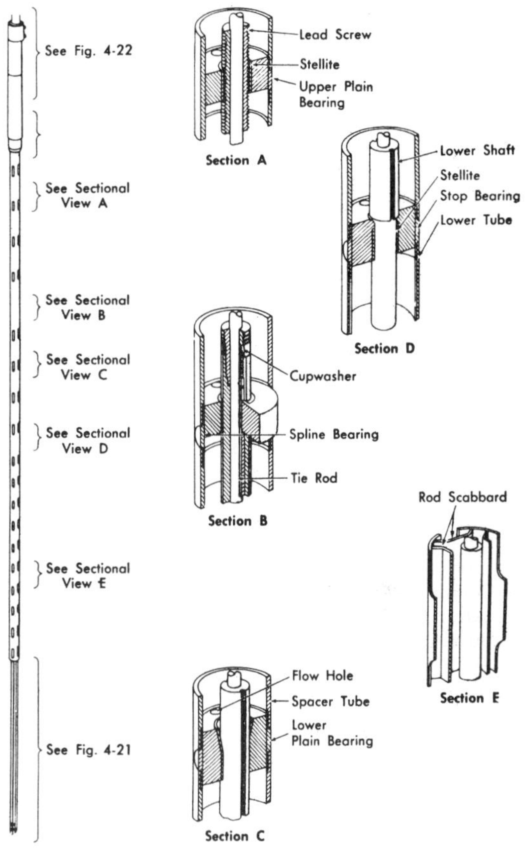

FIG. 4-15 Control-rod shroud tube and linear bearings.

that area. (See Sec. 4 of Chapter 2, "The Reactor Core".) This is sometimes prevented by having a nonmoderating material of low thermal neutron cross section attached to the end of the control rod to fill the space vacated by the control rod as it is withdrawn; this follower can also be made of fuel elements attached to the end of the control rod. Followers of either type require added height of reactor vessels, additional rod guides and provisions for cooling the followers when they are not in their core channels.

6. The number and size of control rods must be compatible with the core layout and with the limited space on the head for the drive mechanisms and the penetrations for the drive shafts. From a maintenance standpoint, it is desirable to be able to remove a control rod through a penetration in the reactor vessel head directly above its operating channel; if advantage is to be taken of this feature, the size, number and location of the rods are further limited by the penetrations that can be provided in the reactor vessel head for this purpose [57].
7. The number, size and arrangement of control rods should be such that the reactor can be shut down with at least one rod stuck in the fully withdrawn position. To meet this requirement sufficient shadowing of rods must be provided so that the sub-region of the core from which the control rod is withdrawn can be made subcritical by the neighboring rods. (See Chapter 2, Sec. 4, "The Reactor Core", and Chapter 11, Sec. 6.) This means that the mechanical designer must yield to the nuclear designer and place the rods closer together than he might otherwise desire. While this criterion is important in assuring safety, it can have far-reaching effects on the mechanical design particularly if the requirement is for cold shutdown; it can easily double the number of rods required and cause the reactor to become larger because of encroachment of control rods on space needed for heat removal and for control rod drive mechanisms [1, 57]. For hot shutdown the one-rod-stuck criterion is not so severe; the use of chemical poison may be considered for permitting cold shutdown with the rod stuck so that corrective action can be taken.
8. The control rods should operate in a well-defined channel so that flow of coolant to the

control rods can be properly controlled. These channels can also provide the surfaces for guiding the rods in the core. The need for rubbing shoes in the form of buttons or dimples on the free end of the rod should be considered in order to assure adequate coolant flow all around the rod. (See Fig. 4-14.) The channel should be large enough to avoid binding of the rod in the channel and should provide a stop at the bottom so that the rod cannot drop out of the bottom of the core in the event of a malfunction of the remote disconnect at the adaptor. Provisions must also be made for resisting any torque imposed upon the rod by the drive mechanism.

9. The control rods should be given corrosion protection. If a protective coating is applied, precautions must be taken against wearing it off. If a separate cladding is used, consideration must be given to the water-logging type failure such as described for fuel elements in Chapter 13, "Fuel Elements"; a bulge brought about by such action could interfere with control rod motion.
10. Consideration should be given to system contamination that can arise from excessive production of wear products from the control rod system.

In evaluating the mechanical aspects of control rods several problems arise which are worthy of further attention. These include materials problems, rod configuration, and rod distortion; each of these is discussed briefly below.

4.7.1.1 Control Rod Materials. While increasing attention has been given to control rod materials, the efforts to date have not been consistent with the needs. In-core tests of new control rod materials in the expected operating environment have been fragmentary when compared to similar tests on fuel element materials [1]. Considerable work remains to be done to provide backup materials for those currently in use.

Hafnium is an excellent control rod material for a number of reasons. In addition to meeting the nuclear requirements, it has good corrosion resistance in many environments; it has good mechanical and physical properties and provides excellent service as an unclad monometallic rod. However, the availability of hafnium is falling behind the needs. Other materials are required. It should be pointed out that hafnium does have radioactive decay products with undesirable half lives and, therefore, wear should be kept to a minimum.

A number of alternate materials have been used or considered. Boron in the form of boron carbide and boron steel has been used. Boron carbide powder contained in aluminum cans has been used for research reactors. The problem of water-logging of this type of control rod is causing it to fall into disrepute. Boron has been used in the SRE control rods; these are made up of stacked rings of 2% boron-in-nickel alloy suspended on a pull tube and operated in a helium-filled stainless steel thimble [58]. However, for applications involving boron steels with high boron concentrations, the brittle nature of the boron steels and the adverse effect of irradiation on them become important. The evolution of helium during neutron irradiation of boron complicates the problem of alloying boron with any material because of both swelling and embrittlement that may be brought about. In rods using a container to hold the boron there is always the concern that the boron may leak out or that water-logging may occur if a leak develops in the container. The most promising approach may be the use of a dispersion of B^{10} particles in a matrix of steel; this incorporates the advantages of encapsulation without the concern of losing the boron or facing severe gross swelling due to radiation damage [59].

A fair amount of work has been done with control rods using cadmium-silver and cadmium-silver-indium alloys. These are not corrosion resistant in the usual reactor environment and must be clad or coated. Experience with stainless-clad control rods has been poor; water-logging, as demonstrated by tests, can be a severe problem. Some methods of bonding now appear promising. The Yankee reactor used cadmium-silver-indium rods coated with nickel; experience has shown that scouring and wear of the coating can lead to a significant amount of corrosion which may plate out radioactive silver on the stainless steel in the system.

Consideration has been given to the use of rare earths as control rod materials [60]. Cermets or clad oxides are proposed. However, much development work and many of the foregoing problems must be faced before these materials find widespread use.

In view of the fact that cladding or other form of containment may be necessary to protect the control rod, careful attention to the manufacture of such cladding is of the utmost importance. In addition, the influence of wear and corrosion on the materials used for cladding must be carefully evaluated.

In addition to the control rod itself, materials problems are faced in connection with the drive shaft and adaptor used to connect the control rod to the drive mechanism. These components must withstand high stresses because of the heavy loads imposed upon them and the limited space which precludes making them as large as might be desired. In addition they are often called up to act as linear bearing surfaces and thus have to be hard enough to resist wear. As a consequence, high-strength materials are required. For such application 17-4 PH stainless steel has been popular among reactor designers. This steel contains 17% chromium, 4% nickel and 3% copper as the carbon precipitating agent. In fabrication, the material is solution heat-treated at 1925°F (1052°C) to insure complete solution of the copper; the material is then cooled to below 90°F (32°C) and then heated again and aged to insure transformation of the austenite to martensite.

The aging temperature of this material turns out to very important. Laboratory tests have shown that 17-4 PH stainless steel is susceptible to stress corrosion if it is aged at the improper temperature. Tests have indicated that an aging temperature of 1050°F (571°C) is satisfactory. Failures were encountered in the Commonwealth Edison Dresden

Reactor with material aged at 900°F (482°C) (see Sec. 4.11). Careful attention to heat treatment is required to avoid problems with special high-strength steels.

4.7.1.2 Control Rod Configuration. Inasmuch as control rods must withstand significant amounts of internal heat generation which must be removed without exceeding allowable temperatures and heat fluxes, they should have large surface-to-volume ratios. It is fortunate that nuclear considerations usually lead to a similar geometrical requirement. This requirement can be met by various shapes, many of them consisting of arrangements of thin plates in geometries which provide structural rigidity to the control rod. Among the shapes used are Y-shaped rods, I-shaped rods, cruciform rods, and thin-walled hollow structures. A thin flat plate control rod is generally not used for power reactors because it responds freely to thermal gradients, can buckle easily under axial loading and has little torsional rigidity; when designed to meet the structural requirements, such a rod does not provide the optimum amount of reactivity that should be associated with a control drive mechanism.

Control rods can also be made in the form of small-diameter pins or rods if properly grouped. Such a design was recently completed for use in pressurized water reactors [61]. In the proposed scheme, each control rod consists of a cluster of a number of cylindrical absorber rods, which are connected at the top end and are coupled to a drive shaft for operating. Each rod in the cluster moves in its own tubular guide which replaces a selected fuel rod in a symmetrical pattern in a single fuel assembly. Inasmuch as the space vacated by each individual rod of the cluster is not immediately adjacent to the space vacated by other rods of the cluster, the water-hole peaking is significantly reduced and the need for control rod followers is just about eliminated. A model of this control rod assembly is shown in Fig. 4-16 [61].

While this design has a number of potential advantages, its susceptibility to jamming in the guides due to the crud or other foreign material in the coolant stream must be guarded against. Questions of alignment and misalignment and problems arising out of tolerances, that must be allowed for in manufacturing, must also be carefully watched.

Another geometry consideration that must be taken into account is that of specifying the control rod and the channel in which it operates. The problems involved are concerned with geometric tolerances as opposed to usual dimensional tolerances. Because of the widespread use of cruciform control rods, this type is used as the basis for discussing the problems involved (see Fig. 4-14). The channel in which such a control rod might operate is shown in Fig. 4-12.

To assist in providing a satisfactory channel for passage of a control rod a mean-free-path envelope of the channel can be defined. It can be considered as the largest unobstructed cruciform passage formed by symmetrical cross sections of the channel which are coincident along the length of the channel. The relationship between the control channel and its mean-free-path envelope is shown in Fig. 4-17. The maximum thickness of the

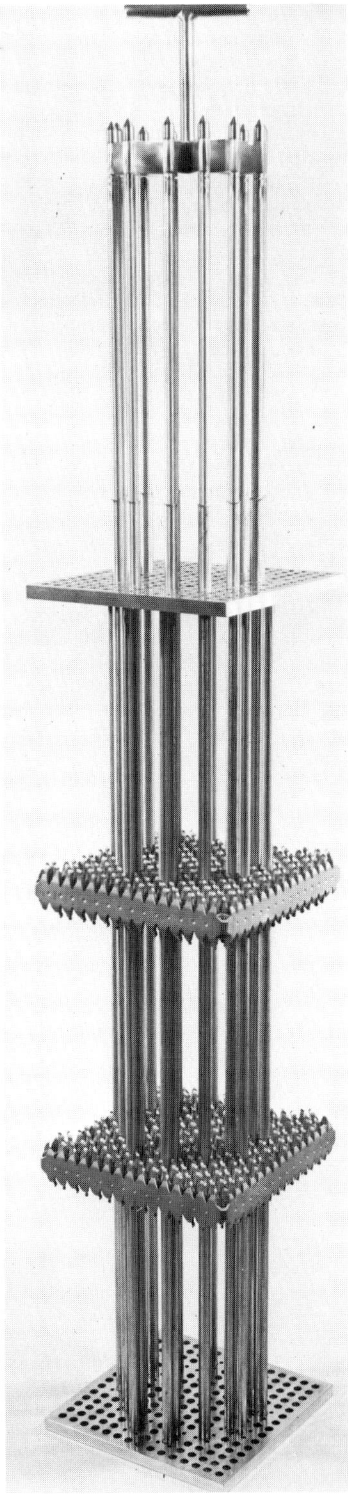

FIG. 4-16 Model of integral control-rod cluster and fuel assembly with absorber rods partially withdrawn.

legs of the envelope is defined by the shortest distance between any points on the two surfaces forming the leg when projected to a common plane and when measured perpendicular to the axial centerplane of the channel. These two points do not necessarily have to be opposite to or even near each other. The maximum thickness for the right-angled cruciform envelope is affected by the selection of the axial centerplanes of the channel; these can be rotated or translated as a unit relative to the channel. The intersection of these centerplanes forms the line of action of the control rod. The centerplanes themselves become the reference planes from which the surfaces for locating the cluster in the core are machined. These centerplanes should be established and marked early in manufacture so that the maximum available mean-free-path envelope can be assured.

A similar envelope can also be defined for a control rod. It should be smaller than the channel envelope by the amount of the desired minimum cold clearance. If shoes are provided on the bottom of the rod, allowance must be made for them. As a matter of fact, a gauge conforming to the control rod envelope can be used for inspecting the rods. A single-plane gauge is not adequate for such inspection.

4.7.1.3 Control Rod Distortion. Heat is generated in control rods by attentuation of capture and fission gamma radiation, by inelastic scattering of neutrons, and by the (n,2) reaction when boron is used. The capture gammas result from neutron absorption by the control rod. The fission gammas produce heat by being attenuated as they

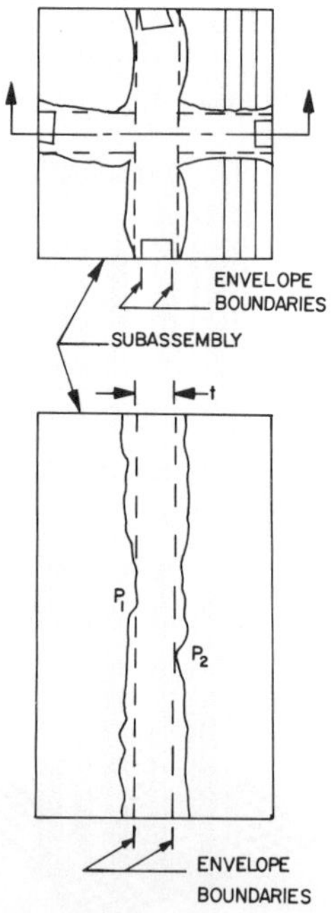

FIG. 4-17 Control-rod channel mean-free-path envelope.
Note: The thickness of the envelope, t, is identical in all four legs and is established by the smallest distance P_1-P_2 in any of the two adjacent or opposite subassemblies.

pass through the rod. The heat from inelastic scattering of neutrons arises from attendant loss in energy of the incident neutron. Typical heat generation rates for hafnium control rods in a high density thermal reactor might be 50-100 watts/cm^3 average and 100-200 watts/cm^3 maximum. (Note: 1 w/cm^3 = 55.9 Btu/hr-in.^{3}.) The rate of heat generation will be even higher in a rod which produces a charged partical reaction upon neutron absorption such as B^{10} (n,α) Li^7. The heat generation rate is influenced by the position of the rod, its effectiveness as a neutron absorber and its mass density.

Provisions must be made for removing the heat generated in the control rods, without exceeding the design criteria. Because of the high heating rates that may be encountered, this can be a difficult problem, particularly in view of the fact that the coolant flow rate on each side of a control rod is dependent upon its attitude and position in its channel.

Because the heating and cooling of the rod is not uniform nor necessarily symmetrical, temperature differentials arise which can lead to distortion of the rod relative to its channel. In designing control rods it is necessary to determine whether or not the resulting deformation will cause the rod to bind in the channel or at the bearings. In the past, the nature of such distortions was determined experimentally on full-scale models of the rods. The general deflection equations and associated boundary conditions could not be solved readily for the temperature patterns encountered. However, by making reasonable simplifying assumptions, Mendes [62] developed methods for obtaining close approximations to the thermal distortion of control rods. His solutions agreed with the test data within the limits of accuracy of the tests.

The approximate equations can be expressed in a closed form which can be solved by hand computation. These equations are summarized below. The deflection of the centroid of a control rod section can be closely approximated by:

$$u_c = \int_0^z \int_0^z \left\{ \frac{1}{I_y} \int_A \alpha(T - T_R)\, x\, dA \right\} dz\, dz\,, \quad (4\text{-}1)$$

$$v_c = \int_0^z \int_0^z \left\{ \frac{1}{I_x} \int_A \alpha(T - T_R)\, y\, dA \right\} dz\, dz\,, \quad (4\text{-}2)$$

where u_c = deflection of centroid in x direction
v_c = deflection of centroid in y direction
x,y = coordinates in a plane perpendicular to the axis of the control rod
z = coordinate along the control rod axis
I_x = area moment of inertial about the x axis
I_y = area moment of inertia about the y axis
$\int_A$ = integration over the control rod cross section
α = linear coefficient of thermal expansion
T = temperature at any point
T_R = room temperature (taken as 70°F or 21°C).

Summation of elements over the rod length gives the axial elongation of the centroid w_c as:

$$w_c = \int_0^z \left\{ \frac{1}{A} \int_A \alpha(T - T_R)\, dA \right\} dz . \qquad (4\text{-}3)$$

Corrections can be applied to the above centroidal deflections to find deflections of noncentroidal points with respect to the principal axis of the cross section. These corrections account for changes in the cross section of the control rod and are not generally significant in practical applications.

In addition to examining the deflections of the control rod, it is prudent to determine if the temperature distribution can lead to a buckling action on portions of the rod. Reference [62] defines two types of buckling—axial and transverse. No equations similar to the approximate equations for determining lateral and longitudinal deflections have been developed for predicting thermal buckling in the general case. Thus, for an arbitrary temperature pattern, the general stress equations may have to be solved, and the stresses obtained from them may have to be examined in order to determine areas of potential buckling. However, for simple temperature distributions and simple cross sections, various standard buckling formulas may be used directly or adapted to fit the needs. The following example is appropriate.

Consider a long cruciform rod with the dimensions and properties indicated in Fig. 4-18 and subjected to a zero temperature in one arm and a constant temperature B in the other arm, both constant over the length and breadth of the flanges. Let it be required to find the temperature difference B which causes the rod to buckle axially and transversely.

1. Axial buckling

   As an example of axial buckling, it will be assumed that the temperature in arm ab is T = 0°F and in arm cd is T = B°F. The cold arm will be assumed to restrain the hot arm so as to produce an average compressive stress in the hot arm equal to the average tensile stress in the cold arm:

$$\sigma_a = (\Delta T/2)\alpha E = B\alpha E/2 . \qquad (4\text{-}4)$$

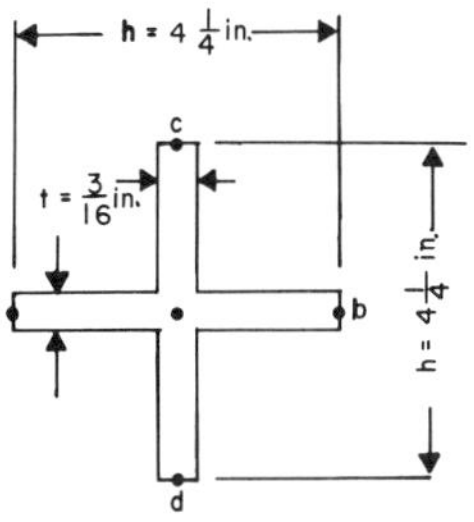

FIG. 4-18 Cross-section and properties of hafnium rod for buckling analysis. Length of rod = 72 in. (1.83 m); y = 0.33, $\alpha = 3.28 \times 10^{-6}$°F ($5.90 \times 10^{-6}$/°C).

This neglects any yielding effects at the junction of the two flanges and assumes uniform elongation over the arm span h.

The L/r_g ratio of the rod shown in Fig. 4-18 (where r_g = least radius of gyration) is > 100, so that Euler's long-column formula may be used:

$$\frac{P}{A} = \sigma_{cr} = \frac{n\pi^2 EI}{AL^2}, \qquad (4\text{-}5)$$

where:

σ_{cr} = critical stress

n = coefficient depending on end conditions (n = 1 for free ends)

$I = \frac{1}{12} th^3 - \frac{1}{12} ht^3 \simeq \frac{1}{12} th^3$, (since $h >> t^3$).

Taking n = 1 will give a conservative value for B. Equating σ_a and σ_{cr} yields the following equation for the magnitude of the temperature B:

$$B = \frac{\pi^2 h^2}{12\alpha L^2} . \qquad (4\text{-}6)$$

Solving this equation using the parameters of Fig. 4-18 yields a conservative temperature difference of 875°F (486°C). This temperature difference is considerably higher than that found in a usual application such as PWR. This type of axial buckling will not usually be encountered. (It must be pointed out that, if such a high temperature difference were to be encountered, the assumptions made in deriving the previous approximate equations for deflection, and hence the equation themselves, would be invalidated.)

2. Transverse buckling

   This type of buckling results from an effective moment applied to the rod due to thermal compressive and tensile stresses within the rod. It is similar to twist-bend buckling of beams as defined by J. B. Den Hartog and is a particularly important consideration in control rods where the span is considerably greater than the thickness. If the temperature pattern produces a bending moment in the stiff plane of such a control rod (as is usually the case), the rod can buckle outward in the flexible direction. Since the portion of the rod in tension remains essentially straight, while the portion buckling in compression deflects laterally, there is a twisting action at the point of buckling. The magnitude of this type of buckling is estimated as follows for the case being considered. The following assumptions are made (refer to Fig. 4-18):

   a. Restraining effect of cross-arm is neglected.
   b. Temperature in arm a is constant and equal to B°F.
   c. Temperature in arm b is equal to 0°F.
   d. Tensile and compressive stresses in arms are uniform.
   e. Shear stresses are neglected.

   Thus,

$$\sigma_a = \frac{\alpha EB}{2}, \qquad (4\text{-}7)$$

$$\sigma_b = \frac{\alpha E B}{2}. \quad (4-8)$$

The moment:

$$M = \sigma_a \frac{h}{2} \cdot \frac{h}{2} \cdot t = \frac{\alpha E B h^2 t}{8}. \quad (4-9)$$

The critical bending moment for a narrow rectangular section under pure bending with ends held vertical, but not fixed in the horizontal plane, is:

$$M' = \frac{\pi t^3 h}{6L} \sqrt{E G(1 - 0.63\, t/h)}. \quad (4-10)$$

Equating the two moments and substituting $E/[2(1 + \nu)]$ for G gives:

$$B = \frac{4}{3} \frac{\pi t^2}{\alpha L h} \sqrt{\frac{(1 - 0.63\, t/h)}{2(1 + \nu)}}. \quad (4-11)$$

Using the dimensions and properties given in Fig. 4-18 gives the temperature difference between ab and cd as: B = 88.8°F (49.3°C). Thus, transverse buckling seems to be the most critical of the two types considered. Whether or not this type of distortion is the one which is controlling in the final design depends on the nature of the applied temperature distribution and on the geometry of the rod. In the case of a rod application similar to that of the PWR control rod in its channel this type of buckling is rather unlikely to occur due to the decrease of the temperature differentials as the rod bows.

4.7.2 Burnable Poisons

One of the major factors influencing safety of power reactors where long life is desired is the amount of excess reactivity that must be carried in the fuel loading (see Chapter 2, "The Reactor Core"). This excess reactivity imposes severe functional requirements on the control system and forms an important source of the inherent positive Δk that can be imposed on the reactor.

To help control this excess reactivity the designer has at his disposal a potent tool, namely, burnable poisons. The poison is built into the reactor in such quantity that little excess reactivity exists during operation at any time during reactor life. The poison burns out as the uranium burns out so that, in principle at least, it should be possible to keep the excess reactivity constant. In actual practice exact correspondence of burnup in the poison and fuel cannot be achieved because of the many factors involved, not the least of which is predicting the various reactivity requirements faced during reactor life such as the need to override transient xenon. The following discussion of burnable poisons is extracted from reference [59].

A number of considerations must be faced in the application of burnable poisons. If the burnable poison is distributed uniformly through the core, the amount that can be accommodated is limited to that which permits the reactor to achieve initial criticality at operating temperature. Because of the difference in microscopic neutron absorption cross-sections of the poison and fuel, they do not burn out at rates proportionate to their respective concentrations. In order to assure criticality throughout life it is necessary to pick a poison that burns out proportionately faster than the fuel rather than the other way around; otherwise the core is left with a high residual amount of poison which shortens life. Furthermore, the burnable poison must have decay products which do not appreciably effect reactivity.

Of the many materials investigated for use as burnable poison, only B^{10} has been extensively used. Its products, Li^7 and He, have acceptable neutron cross sections but the B^{10} tends to burn out rather quickly over the life of a reactor so there is a strong desire to use it with some other poison such as Li^6 which does not burn out as quickly. However, to prevent too rapid a burnout the B^{10} can be lumped to provide self-shielding. Lumping reduces the initial effectiveness of the poison and hence more must be added. This additional inventory results in a higher residual amount in the core later in life and has a slight but not totally insignificant adverse influence on core life.

In addition to reducing the amount of excess reactivity available to the core, burnable poisons can also be used to reduce flux peaks and hence hot spots in the core, thus favorably influencing safety in an auxiliary fashion. This feature is not unimportant because failure of a reactor to ride out a power burst shows up first at the hottest part of the core. It might be worth mentioning that the clad of the fuel elements in the Elk River Reactor contains some boron (of the order of several hundred ppm). This is an unusual application of a burnable poison.

In addition to reactivity, several other factors must be considered in the application of burnable poisons to reactors. These are the problems of incorporating them in the core and the problems that they introduce as a result of radiation damage. If only small amounts of boron are involved, they can be alloyed with the fuel. If concentrations becomes large, radiation damage can be expected in the fuel because of the evolution of helium gas. As a matter of fact, the evolution of helium complicates the problem of alloying B^{10} with any material. Encapsulating the boron is sometimes proposed as a means of introducing it into the reactor. This approach has the drawback that if the cladding of the capsule cracks the boron may be leached out and lead to as much of a safety problem as it helps to solve. The most promising approach may be the dispersion of B^{10} particles in a matrix of steel, thus incorporating the advantages of encapsulation without the possibility of losing the boron or facing severe gross swelling due to radiation damage.

In addition to the above problems, one faces the problem of mechanically incorporating the B^{10} assembly in the reactor. Care must be excercised here to insure that the poison is made an integral part of the core assembly and is not merely fastened by pins, rivets, or bolts which can break, corrode or erode away. The latter methods of attachment must be particularly avoided if the poison assemblies are being placed adjacent to control rod passages, where they sometimes are placed (in water reactors) to reduce water-hole peaking. Even if enough clearance has been provided so that swelling of the boron pieces will not block the rod, failure of the fasteners might cause the boron pieces themselves to block the control rod. The boron pieces should be well buried and sealed within the structural components of fuel assemblies whereever possible to avoid breakage problems.

Despite their problems, burnable poisons can and have been used successfully to augment reactor safety. Further work should be encouraged to develop new methods of applying them and to develop burnable poisons which undergo (n,γ) rather than (n,α) reactions thereby eliminating helium production and reducing radiation damage.

4.7.3 Chemical Control

The use of chemical solutions for control of water-cooled reactors has appealed to reactor designers for some time [59]. Chemical control in general reduces the peak heat flux that must be faced in a reactor design below the value that might be experienced with control rods which distort neutron flux. The chief deterrent to the use of chemical control systems has been the safety requirements that they must be fail-safe, a requirement not yet fully achieved in proposed systems for full chemical control, whether they be integral chemical control systems where the poison is directly in the coolant, or separate chemical systems where the poison flows in tubes at positions comparable to control rod positions.

Nevertheless, chemical control systems are being used in conjunction with control rod systems. Yankee, for example, relies on the use of integral chemical control for shimming the core but gets around the fail-safe requirement by providing enough control rods to shut down the core cold with all rods operating or hot with one rod stuck out of the core. In the latter situation, boric acid solution in the coolant is used for achieving cold shutdown.

In designing reactors which are to use integral chemical control it is necessary to design so that the expulsion of chemical poisons along with moderator from the core (during moderator and fuel element expansion and void formation) does not have a net positive reactivity coefficient. The positive coefficient resulting from the increased thermal utilization should be designed to be substantially less than the negative coefficients due to increased neutron leakage and thermal spectrum change.

A chemical control system separate from the coolant system would, at first glance, appear to be safer than an integral chemical control system, but it actually may not be much better. First, one must consider that in order to minimize the possibility of coolant leakage into the chemical system or chemical leakage into the coolant system, the chemical system will normally be operated at the same pressure as the coolant. Thus one must postulate that a break in the external chemical system is just as probable as a break in the primary coolant system, or possibly even more probable because of the smaller, more easily damaged piping and valves. In the event of such breakage, the control system becomes the biggest source of positive excess reactivity. It would have the advantage over integral chemical control in that inherent shutdown mechanisms in the core would still be present to provide safety. Of course, one might face a double casualty, a pipe break in the external loop of the chemical control system and a pipe break in the reactor portion of the chemical system or conversely a break in the primary coolant system and a pipe break in the reactor portion of the chemical system. Such double breaks are presumed no more probable than a double casualty with control rod systems except that, in the case of the chemical control, unless each chemical rod has its own hook-up, all control can be lost by a single relatively small failure; with mechanical control rods, on the other hand, only one out of many rods is lost, a situation provided for by the stuck-rod criterion.

4.7.4 Back-up Shutdown Systems and Last-Ditch Devices

Because of the important part control rods play in the shutdown of reactors, failure of these rods to act when a shutdown is called for could compromise reactor safety. Such failures might be brought about by blockage or distortion of the control rod passages, mechanical failure or thermal distortion of the control rods themselves or malfunction of the rod-release mechanisms or the signals leading to them. Hence, back-up shutdown systems and/or last-ditch devices are often added when all others fail.

Back-up shutdown systems are usually characterized as being slow-acting. They are used where reactor safety is provided by the inherent shutdown characteristics of the core and it is desired to forestall subsequent power bursts, or where it is desired merely to provide normal shutdown which can no longer be effected by control rods. The time responses of slow devices would be in excess of a second and could be as long as several hours depending on the situation to be handled. Use of such devices should be capable of being initiated manually because the need may well occur at a time when power has failed. Typical of back-up shutdown systems would be manually initiated injection of liquid poisons, a method provided for in the Brookhaven reactor, as well as in Yankee and others; insertion of boron shot, for which provisions are included in the Oak Ridge graphite-moderated reactor, and dumping of moderator or fuel. The Canadian NRX NPD-2 and CANDU reactors have provisions for moderator dumping.

Back-up shutdown systems are sometimes called secondary shutdown systems or alternate shutdown services. Walter in reference [187] summarizes current practices regarding a large sampling of reactors in the United States, the United Kingdom, France, and Canada. Most, but not all, of the reactors examined have secondary shutdown systems, and, in those that do have them, there is a rather large variation in the independence and performance of the systems employed. In addition, basically similar systems, such as those for injecting boron solutions into water-moderated reactors, vary greatly with respect to procedures and time intervals required to complete the shutdown. This is not too surprising because the needs vary from one reactor to another. However, there is no simple way to determine the degree of need in each case. Table 4-1, taken from reference [187] summarizes some of the secondary shutdown systems currently in use.

Last-ditch devices are fast-acting devices and are often referred to as "fuses"; they are used to provide shutdown during power bursts before other shutdown mechanisms come into play. They might be relied upon to shut down a reactor when the inherent shutdown mechanisms are too slow to prevent damage that would result in massive release of fission products. They might also be used to shut down the reactor in time to prevent first-order damage, such as fuel element distortion, when the delay time in the control rod system is too long to do so and the inherent shutdown mechanisms, while preventing release of fission products, would not prevent fuel element distortion. In either case the fuses must be faster than the mechanisms they are to replace.

Fuses can be made so that they respond either to power level or to reactor period. A number of fuse designs of both types have been proposed and tested, some in-core and some out-of-core. In-core tests were made both in the BORAX reactor and in SPERT-I. The basic features of such devices are a trigger mechanism, a propellant and a poison. Because the time delay of a fuse is governed largely by the trigger mechanism, a great deal of attention has been given this item. As a consequence the unique features of one design compared to another usually reside in the trigger mechanism.

Trigger mechanisms generally utilize one of the following elements: a fission-heated fusible rupture disk, a bimetallic element or a radiation-initiated explosive detonator. Propellants that have been used include helium, and explosive gas (hydrogen or propane with air or oxygen), and solid propellants such as smokeless powder. The poison is usually boron in the form of boron powder or boron trifluoride; other poison materials have been studied [63].

Early tests conducted in BORAX reactor and reported by S. N. Stilwell and R. L. Waterfield were instrumental in showing the value of fuses for reactor shutdown [64]. In the fuses tested a fission-heated fusible rupture disk was used to release pressurized helium which in turn drove boron powder into an annular space around a fuel element in the reactor. Data reported included effects with and without fuses present in the reactor. For example, with a reactor period of 14.6 msec the maximum power reached without the fuse was 470 Mw and was attained in 280 msec with a total energy release of 15.52 Mw-sec. With the same reactor period but with a fuse installed, the maximum power reached was determined to be 320 Mw attained at 258 msec with a total energy release of 8.06 Mw-sec.

Evaluations of fuses both before these early tests and following these tests formed the groundwork for a number of different devices [65-69]. A device somewhat similar to that tested in BORAX was developed at Atomics International for use in large power reactors where the time-response requirements might not be as severe as small reactors [70,71]. In this device the heat generated in a fission-heated uranium sensor caused a gas to expand and burst mechanically coupled diaphragms, dispersing boron powder or $B^{10}F_3$ gas into receiver. Simulated out-of-core tests indicated that a total time of 1.10 sec elapsed prior to tripping of the mechanism and that another 0.13 sec was required to fully disperse the poison. These times are slow compared to the response required to deaden a prompt-critical power burst in any reactor.

S. N. Stilwell and R. L. Waterfield report the results of test work performed on fast-period fuses in the SPERT-I reactor [63]. These fuses involved a bimetallic thermostat composed of uranium and zirconium which triggered an electric detonator and fixed pistol powder to propel boron powder along an evacuated passage into the core. Two different geometries were tested. Four of the fuses of one geometry were placed in four quadrants of the core; when the reactor was placed on a 20 msec period these fuses shut down the reactor in about one-third the time required for the inherent mechanisms to operate. The peak power reached with the fuses in place was only one-tenth that reached without the fuse.

Although much interesting work has been done on fuses to date, more work is required before they can be relied on to act as the final shutdown mechanism. All the fuses tested have been developmental in nature and quite sensitive to the environment and to their positions in the core. Some of them have failed to fire in out-of-core tests because of improper assembly. Thus, it is clear that statistical data are required to determine the number and location of fuses to be used. J. R. Tallackson outlined the following areas in which further information regarding fuses is required [72]:

1. Statistical variations and tolerances to be expected in such devices,
2. Stress rupture as affected by time and temperature where materials exhibiting susceptibility to these characteristics are used as critical fuse components,
3. Methods of nondestructive testing which will insure that the device will operate as designed,
4. The general safety problem that is created when any device containing a void or potential void is inserted in the reactor,
5. The behavior of devices at reactor periods less than 10 msec, and
6. Reliability after prolonged storage in the core.

Despite this need for further work, fuses offer the opportunity for enhancing reactor safety.

4.8 Fasteners and Locking Devices

Experience has indicated that fasteners and locking devices used for joining core internals can be a source of difficulty. Care must be taken to make sure that they are not overstressed and that if they break they cannot get into other parts of the system to jam moving parts. The latter can often best be accomplished by making the devices captive. The influence of the failure of small fasteners on the operation of the components which they join should also be examined.

Attention here is directed primarily to threaded fasteners and to the locking devices associated with them. Most problems with fasteners have been associated with bolts or other small threaded devices; the problems faced with them are typical of those of other fasteners.

4.8.1 Bolts and Other Threaded Fasteners

Five problem areas can be identified in the application of threaded fasteners to reactor design: preloading of bolts and the associated stress distribution which must be considered in stress-corrosion problems; relaxation of bolts in service; thread forms; seizing of bolts due to galling during assembly; and breakage of bolt heads.

1. Preloading: Inasmuch as the primary purpose of a bolt is to provide clamping action, preloading of the bolt is an important requirement. Many bolt problems are associated with producing this preloading at assembly and maintaining it throughout the life of the bolt in service [73].

   With a preloaded bolt clamping two completely rigid parts together, a tensile force applied to the bolted structure will not produce separation of the clamped pieces, nor further extension of the bolt, until the resultant tensile force in the bolt is equal to the preloading [73]. If the pieces being clamped are compressible materials of low modulus such as zirconium, or if a gasket is placed between the pieces, the bolt load will be increased beyond the preload value as tensile forces are applied to the structure; this added load is usually small compared to the preloading if the clamped pieces are reasonably rigid. A preloaded joint can also accomodate shear up to the point of slippage; however, because of the inability to observe bolts in a reactor, it is not customary to rely either on the ability of the bolt to resist shear or on the accompanying friction to prevent unscrewing.

   When a bolt is loaded, the stretching action in the threads results in having the threads near the base of the nut, in both the bolt and the nut, carry more load than the other threads. Stewart, in reference [73], presents calculations made by Buckingham on the distribution of load on successive threads for 0.5in.-13 and for 0.5in.-20 threads. These calculations show that for coarse threads the load varies from 53% of the average at the top of the nut to 179% at the base of the nut. In the case of the fine threads, the loads varied from 37% of the average at the top of the nut to 231% of the average at the base. This distribution of stress becomes important in evaluating stress-corrosion problems.

   Inasmuch as most bolts are preloaded by applying a torque, consideration must be given to the torsional loads required to obtain a particular tensile preload. It is difficult to predict the amount of preloading for a given amount of torque. More than 90% of the torque may be used to overcome friction at the threads and at the bearing surfaces on the parts being bolted together. From numerous tests Stewart [73] suggests the following empirical formula* for estimating the torque requirements on unlubricated carbon steel cold-formed bolts:

   $$\text{Torque (lb-in.)} = 0.2 \times \text{bolt diam. (in.)} \times \text{bolt tension (lb)} \qquad (4\text{-}12)$$

   It must be pointed out that there can be considerable variability in the bolt tension produced by a given torque. The torque coefficient in the above equation will vary with surface conditions. Lubricants can be used which will reduce torque coefficients by as much as 50%. Plating of threads will further affect the torque coefficient, the influence depending in large measure on the type and quality of plating applied. Improperly applied plating can cause flaking which could actually increase the torque coefficient. Even without imposed variations in surface conditions, the torque coefficient can vary by 30%, as pointed out by Stewart [73].

   Care must be taken in specifying torque values to be applied to bolts during torquing. The tensile stresses resulting from the combined stresses will be higher than that produced by direct preloading alone. As a result, failure might be expected at a torque less than that indicated by the above formula if the limiting tensile stress were used in the formula. This effect has been demonstrated experimentally by E. A. Davis [74].

2. Relaxation of bolts: Bolts relax their tension to varying degrees in almost all types of service. A number of different factors may contribute to bolt relaxation. A certain amount of relaxation may be due to the extrusion of foreign matter or of plating material from between the contact surfaces. There may also be yielding of surface asperities due to high local stresses. In addition, relaxation may occur because of creep in the bolt shank or bolt threads or because of creep or progressive crushing of the bolt head nut, or bolted material beneath the bolt head or nut due to excessively high bearing stresses. In order to avoid these problems attention must be given to the stresses involved in these areas. For example, the bolt head and nut should be broad enough so that with the maximum hole size through which the bolt passes there is sufficient contact area to avoid high average stresses and high creep rates. (The creep rate, of course, will depend upon the materials being used and the temperatures at which they operate.)

   In nuclear reactors another cause of bolt relaxation exists; that is relaxation due to

---

*Since the coefficient (0.2) is dimensionless, Eq. (4-12) holds for torque (kg-cm), bolt diameter (cm), bolt tension (kg) or other consistent dimensions.

TABLE 4-1
Secondary Shutdown Systems of Nuclear Power Plants

Short name	Type of secondary shutdown system	Negative reactivity of secondary shutdown system	Initiation of secondary shutdown system action	Mechanism for introducing secondary shutdown system negative reactivity	Time interval to introduce secondary shutdown system negative reactivity
Yankee	Boric acid solution	Safety injection, 0.12; chemical shim, 0.15	Safety injection is automatic on low coolant pressure, also manual; chemical shim is manual	Chemical shim, three pumps	~30 min
Saxton	Boric acid solution	Safety injection, 0.275; chemical shim, 0.24	Safety injection is automatic on low coolant pressure, also manual; chemical shim is manual	Chemical shim, two pumps	~12 min
Indian Point	Boric acid solution	One acid tank, 0.084; chemical shim, 0.113	Manual	Two pumps, plus makeup pumps	~3 hr
Savannah	Boric acid solution	>0.108	Solution prepared and added manually when needed	One pump	~5 hr to prepare solution and pump
SM-1	Boric acid solution	> 0.154	Solution prepared and added manually when needed	Two pumps	~30 min to prepare solution
SM-1A	Boric acid solution	> 0.154	Solution prepared and added manually when needed	Two pumps	~30 min to prepare solution
PM-1	Boric acid solution	> 0.125	Solution prepared and added manually when needed	Two pumps	Information not given
PM-2A	Boric acid solution	> 0.143	Solution prepared and added manually when needed	Two pumps	Information not given
PM-3A	Boric acid solution	One batch, 0.114	Solution prepared and added manually when needed	Two pumps	~5 min to pump
HWCTR	Potassium tetraborate solution	0.24	Manual	Gravity drain	~1 min for reactivity reduction of 0.1. ~3 min to complete
Shippingport	Potassium tetraborate solution in portable tank	Batches added as required to assure shutdown	Solution prepared in portable tank and added manually when needed	Two pumps	~1 1/2 hr per batch
CVTR	None so designated, moderator dump could be used	Exceeds loaded core reactivity	Automatic on start up of emergency cooling system after loss-of-coolant accident, also manual	Gravity drain	No information developed
PRTR	None				
Pathfinder	Disodium octoborate solution	Normal reactor water level 0.30; reactor vessel open to pool, 0.07	Manual	One pump, plus one alternate	~40 min
BONUS	Sodium pentaborate solution	~0.27 cold and clean	Manual	Gravity drain	3 to 4 min
Humboldt Bay	Sodium pentaborate solution	Operating 0.20; refueling > 0.15	Manual	High-pressure nitrogen	7 to 8 min
Big Rock Point	Sodium pentaborate solution	0.25	Manual	High-pressure nitrogen initially, then gravity	1 to 10 min
Dresden	Sodium pentaborate solution	From high-pressure tank 0.03; total, 0.15	Manual	High-pressure nitrogen initially, then gravity drain from high-pressure tank, two pumps from high- to low-pressure tank	~2 min for high-pressure tank, 55 min for total solution
Elk River	Sodium pentaborate solution	0.145	Manual	High-pressure air	~15 sec
VBWR	Sodium pentaborate solution	0.10	Manual	High-pressure nitrogen	~10 min
ML-1	Boric acid solution injection into moderator system	0.065 immediately; 0.025 equilibrium	Manual	High-pressure nitrogen	1 to 2 min for procedure; 6 sec for immediate injection, 9 min for total solution injection
Peach Bottom	Electrically driven control rods and thermally released absorbers	Electrically driven rods, 0.11; thermally released absorbers, 0.09	Manual for electrically driven rods	Battery-powered motors on 19 control rods, thermal release and grivity on 55 absorbers	24 sec to 1 min for electrically driven control rods
EGCR Hallam SRE Fermi EBR-II Piqua	None	Not applicable	Not applicable	Not applicable	Not applicable

irradiation. While the exact nature of the phenomenon is not well understood, tests have indicated that approximately 10 to 20% of the preload may be lost over a several-year period because of this effect. Thus some allowance must be made for bolt relaxation due to irradiation.

Because of the nature of most of the factors involved, the rate of relaxation is not constant during bolt life. After extraneous matter has been crushed, relaxation due to these factors proceeds no farther. Furthermore, as bolt tension is reduced by creep and by the above factors to values less than the preloading, the creep rate falls off. Thus, where the bolt tension is of great importance the bolt should be made large enough to permit preloading in excess of that required for normal operation. By this means the required load can be maintained by the bolt even after relaxation has taken place.

The amount of preloading lost by relaxation is a function also of the amount of elastic extension employed in the bolt. This results from the fact that certain of the factors affecting relaxation lead to a fixed amount of change in bolt elongation. Thus in short bolts, where the imposed elongation may be small, the effect of the change in elongation due to relaxation may lead to more severe loss of preloading than in long bolts. Improvement in the situation can be obtained by reducing the diameter of the shank to the same value as the root diameter of the threads. The bolt strength is thereby not reduced below that of the threaded section but the elastic elongation is increased due to the higher tensile stresses imposed on the shank during preloading. This approach is particularly useful where the creep rate of the bolt shank is negligible. Large elastic deformation with its consequent energy absorption capability is advantageous for bolts which may be subjected to impact loading.

3. Thread form: One of the most frequent difficulties encountered in the assembly of bolts in reactor service is galling due to lack of lubricant. This can often be traced back to improper thread forming. Attention should be given to specifying and inspecting for accurate thread forms. Cold-rolled threads are recommended in preference to machined threads because of the more accurately controlled thread forms and surfaces obtained.
4. Seizing due to galling: Because of the desire in certain applications to utilize bolt materials with the same expansion properties as the parts being bolted together in a core, problems of seizing have been encountered, particularly where a stud bolt of type 304 and 347 stainless steel has been threaded into a flange of similar material.

   The problem was foreseen in certain applications and although chrome plating of threads was used to prevent galling, seizing was encountered. In many instances the problem was traced to poor thread forms; in others where chrome plate was used the problem was traced to poor plating, and in others to very close clearances and poor thread surface finishes.

   In order to avoid such problems:
   (a) Attention should be given to carefully specifying and inspecting for accurate thread forms. Cold-rolled threads are preferred to machined threads because of more accurately controlled thread forms and surfaces.
   (b) Wherever possible neolube (a mixture of graphite in alcohol) should be used for assembly of bolted threads.
   (c) Chrome plating should be only a flash chrome of 0.0002 in. to 0.0003 in. (5.1 μ to 7.6 μ) thickness with careful attention to avoiding stringers at the major diameter of the threads.
   (d) Remove or avoid fin-like portions of thread on the uncompleted first or last thread. This fin is caused by the entrance and runout of any pitch thread onto a plane not perpendicular to that thread.
   (e) Use dissimilar materials wherever possible.
   (f) Use generous thread clearances where the application permits.
   (g) Assemble carefully.
5. Breakage of bolt heads: In certain applications breakage of bolt heads has been encountered where the head contained an hexagonal socket for use with a socket wrench. It was found that loose specification of socket depth coupled with equally loose quality control led to insufficient cross sectional area at the bolt neck to resist the loads imposed during operation. An insufficient fillet also contributed to this failure. This particular material was stainless steel type W, an early precipitation-hardening steel similar to 17-4PH stainless steel. Chloride stress-corrosion was also found to be a factor. Careful attention to bolt head design is important to assure satisfactory performance.

Bolt breakage has been encountered in a number of installations. For example, when the Elk River Reactor Vessel was opened to perform the control rod inspection program it was found that one of the bolts holding the emergency spring ring had failed [188]. An engineering evaluation led to the conclusion that the original selection of material was satisfactory and that failure was probably due to the bolt being overstressed during installation. The bolt was replaced with material of the original design. When the vessel was opened for replacement of the central control rod, failure of a different stud was observed. At this time it was decided to replace all three studs with studs of a different material.

When the Yankee Reactor was refueled in 1965, broken bolts were also found on the bottom of the reactor vessel [179]. The first was identified as a thermal-shield-joint cap screw. This discovery led to a detailed examination of the four joints of the thermal shields; all were found partially separated, and repairs, involving a 5-week delay, had to be made. The second bolt found was a shroud-tube-grid cap screw. This was determined to be the result of an isolated locking-up failure which allowed the cap screw to rotate out of position.

4.8.2 Locking Devices for Threaded Fasteners

One of the most important factors that must be considered in the design of a bolted joint used in the core is that of locking the bolt, nut or both. Whenever possible, the locking device should be made so that if it breaks it will prevent pieces of the bolt or the nut from going into the coolant stream.

Because of the wide variety of ways in which bolts can be locked and the variability in the effectiveness of these methods a careful study was made by Langer to identify locking devices that are suitable for core use [75]. As a result of this study, four classes of locking devices were identified. These were designated A, B, C and D. Guidelines were established for both their design and application. These are discussed below. Two important points from these discussions should be kept in mind and their conclusions used in critical areas of the core as identified below. First, it must be obvious whether or not the locking device has been secured. Second, friction devices are never to be counted upon as suitable bolt locking means within the pressure envelope of the reactor.

a. Class "A" locking devices

Fastenings that are exposed to primary or secondary fluid and located where failure could result in a loose fragment being carried to a vital part of the system should be considered as completely inaccessible and must be designed for the highest degree of reliability. (See Figs. 4-19 and 4-20 [76].) The general rules that should be followed are:

1. The locking device must be positive; that is, it must not depend on friction. As a result of this requirement, two operations must be performed in securing the fastening. One is the tightening of the screw thread itself. The other, a separate operation, usually involves plastic deformation or welding of a member to a position where it interferes with the loosening of the main holding member. This rule eliminates the use of jam nuts, spring lockwashers, and self-locking nuts.

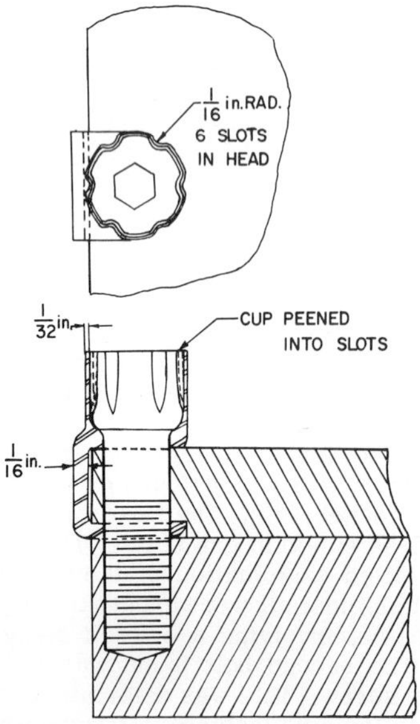

FIG. 4-19 Class A cup-washer locking device.

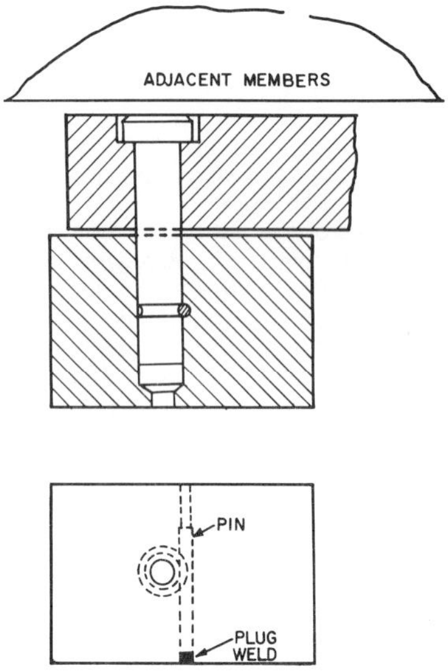

FIG. 4-20 Class A pin-locking device.

2. The deflection used in setting the locking device does not depend on the workman's skill, great precision in the machining of the parts, or a high degree of dimensional stability. Whether or not the lock has been secured must be obvious from cursory inspection. This rule eliminates most peening and upsetting and all prick-punching operations.
3. The locking operation must not involve plastic strain sufficient to crack the materials. Stops or guides should be provided so that the prevention of cracks is not dependent on the skill of the workman. The sharpness of allowable bends must be consistent with the ductility of the material being used. Devices involving plastic deformation must not be used more than once. This rule eliminates most cotter pin designs and makes lock wiring undesirable, except as described below.
4. Locking members must not be so thin that slight corrosion can cause failure. Wires should be avoided in regions of high fluid velocity and should not be less than 0.050 in. (1.3mm) diameter in any case. Sheet members should not be less than 0.015 in. (0.4mm) thick.
5. The design should be such that, if the loaded part of the fastening (e.g., bolt shank) fails, its fragments will be held captive by other members or by the locking device itself.

b. Class "B" locking devices

Fastenings exposed to primary or secondary fluid located in regions of low velocity and in regions where fragments could not be carried into the main fluid stream are also to be considered completely inaccessible and should meet the requirements listed in Part A above, except for paragraph 5. (See Figs. 4-21 and 4-22 [76].)

Lock wiring 0.050 in. (1.3mm) diameter or larger is acceptable, provided the manufacturing drawing specifies the size and material of wire, the size and location of holes, and the configuration of wire. The configuration must be such that loosening of the fastening results in tightening of the wire. Fastenings must be wired in pairs, except that for an odd number of fastenings, one group of three may be used. Lock wiring must be tight. (See Fig. 4-22.)

c. Class "C" locking devices

Fastenings not exposed to primary or secondary fluids but in positions inaccessible for inspection, tightening, or replacement during power operation shall be provided with locking devices that meet the requirements of paragraphs 1, 2, and 3 of Part A above.

d. Class "D" locking devices

Fastenings not exposed to primary or secondary fluids and accessible for frequent inspection, tightening, and replacement during power operation should be provided with locking devices that have proved satisfactory for operation in power plants of the type being designed. These include star washers, spring lockwashers, cotter pins, lock wires, peened bolt ends, and certain proprietary self-locking nuts, as well as any device that meets the requirements of Part C above. Snap rings are acceptable, but should be tested for shockproofness if they are to be used under impact. Set screws and jam-nuts are not recommended.

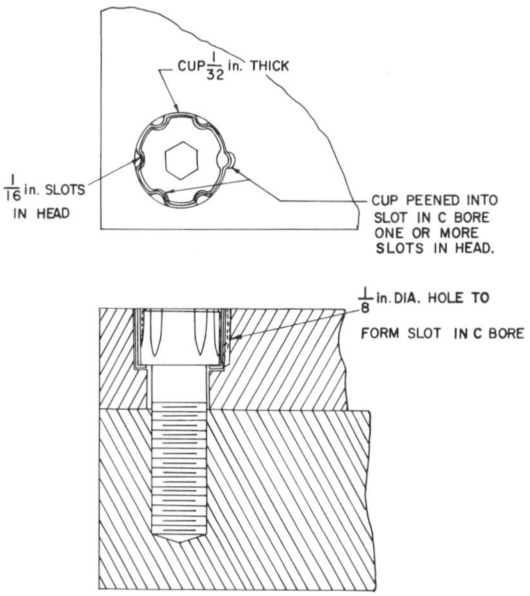

FIG. 4-21 Class B cup-washer locking device. (Allowable for Class A if head is trapped by adjacent member.)

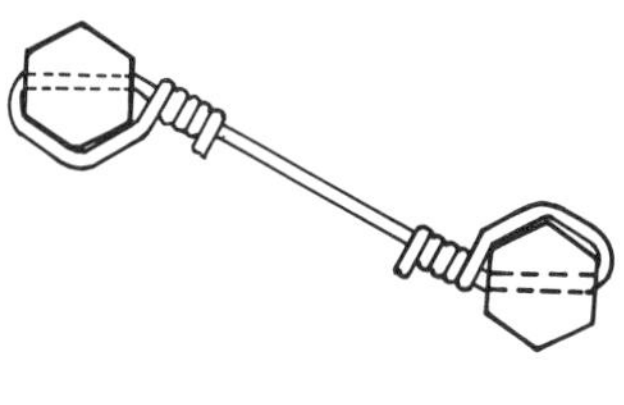

FIG. 4-22 Class B locking wire arrangement. (Class C if wire diameter < 0.050 in. (1.27 mm).)

4.9 Core Assembly

A number of nuclear cores are preassembled dry prior to shipment, in order to mechanically and dimensionally check new designs or design features. Some cores involving a core cage are preassembled dry at the reactor site prior to installation into the reactor vessel. A number of cores, however, are not preassembled particularly where core assembly consists of replacing fuel assemblies in an existing structure in the vessel or where the core design is of a type for which there is considerable experience and for which good control of component dimensions can be provided for proper mating of parts.

Regardless of the method used for assembling the core attention must be given to planning assembly operations, preparing necessary specifications and designing, procuring and debugging assembly equipment. Core assembly should receive careful attention for the following reasons:

1. To guard against accidental damage. At the time of assembly the aggregate of many costly items is brought together. Carelessness in workmanship or handling during this stage can cause damage on a large scale and at a time in the production process where replacement requirements can cause severe delays. This is especially true when core production is a batch process without continuous flow of parts. Damage to fuel elements of course must also be avoided to prevent release of fuel contamination.
2. To guard against sabotage in addition to damage by carelessness.
3. To provide cleanliness control (including use of appropriate tools) and to keep from core components:
   a. matter which could become radiactive and thereby add to the shielding, decontamination, or accessibility problem later,
   b. particulate matter which can lodge in moving parts and cause mechanical damage,
   c. airborne contamination that can cause corrosion of parts (e.g. SO_2 or H_2S), and
   d. dirt that might clog coolant passages or deposit on heat transfer surfaces.
4. To assure correct alignment of components. Special alignment instruments and jigs may be necessary.
5. To avoid criticality hazards. This is necessary because, despite all precautions and an expectation that an assembly will not go critical, assurance can be provided only by carefully planned procedures.

In all cases, the first three of the above items can be readily taken care of by providing a core assembly area where appropriate personnel and tool control can be instituted. Areas for receipt and unpacking of equipment and for cleaning components should be separate from the assembly area. Shops associated with core assembly should also be separate. Change rooms should be available for personnel.

If the air at the assembly site is particularly dirty or if the core is to be exposed for a prolonged period of time, provisions should be made to prevent contamination of the core by the air leaking into the building. All interior ventilation should use air suitably cleaned by filters or precipitrons and, if possible, a slight positive pressure should be maintained to insure that the entrance of outside air is minimized when doors or other access openings are opened and closed.

It must be remembered that while the core is being worked on, it is exposed to contamination. The amount it picks up is a function of how much there is to be picked up and how long the core is exposed to this potential. Since the assembly operations may extend over a number of days or even weeks, means must be provided to limit the amount of dirt brought into the assembly area.

In addition to control of buildings and facilities, attention must be given to the methods of cleaning core parts at assembly and to the definition of cleanliness. A number of specifications have been written for this purpose.

These specifications, either modified or amplified by supplementary information detailed in individual equipment specifications, are used by inspectors to determine whether the subject equipment is adequate from the standpoint of cleanliness and corrosion. Subcontractors must often be supplied with copies of pertinent specifications, and assisted in making certain that these specifications are followed.

In order to provide a guide for inspectors so that they will be in a position either to reject components or to hold them for review, it has become a matter of practical necessity to define further certain phases of cleaning requirements, namely surface finish, surface contamination and general dirt. These rules are, of necessity, very general in nature but they should provide some basis for mutual understanding among the various agencies involved. The ground rules can probably be discussed best on the basis of the type of deviations which have been observed in components produced to date. Past experience in the application of these specifications has shown them to be a very helpful tool in maintaining uniform and consistent cleaning requirements.

In many instances it is possible to permit deviations from the high standards specified, but it is extremely difficult and impractical to list permissible deviations in an equipment specification. For example, if, instead of stating that surfaces shall be "free from all impurities", the specification should state that surfaces shall be "almost" "entirely free of foreign material", there is always a question as to the meaning of the word "almost"; and it becomes necessary to describe in detail all of the surfaces of the component and to state categorically what types of foreign material, in what sizes, distribution and quantities can be permitted on each particular surface. This becomes an almost impossible task, since the types of contamination to which the various components may be subjected during manufacture cannot be anticipated. By stating that surfaces shall be free of all contaminants and that all surfaces shall have at least a signified finish, e.g., 63 microinch (rms) [or 1.6μ (rms)] finish, the task of inspecting components is simplified tremendously. Based on these specifications, the inspector either passes or rejects the item involved, and there is a clear-cut definition between what is acceptable and what is rejectable.

Since it is recognized that the specifications are in some instances unnessarily rigid, it becomes obvious that some of the components which are rejected by the Quality Control Group on the basis of the specifications may actually be satisfactory for operation in reactor systems. It is necessary, therefore, in the interests of economy and time, that components which fail to meet the specified standards be reviewed by some person or persons who are qualified to pass on the deviations which can be accepted in each particular instance. If the deviations are of a dimensional character, such review would normally be made by design engineers; if the deviations concern cleanliness, surface finish, corrosion or metallurgical adequacy, this review should be made by materials application personnel.

The most important reason for attention to core assembly requirements when fuel assembly is involved is that of criticality. This is an important consideration in all operations involving handling of nuclear fuel. Criticality problems are specifically discussed in Chapter 5, "Criticality," of Volume 1 (pages 244 to 284); however, several precautions are listed here to help guide the planning for such work.

1. Keep unneeded water out of assembly area;
2. Have a fire fighting scheme that does not involve moderating or contaminating liquids;
3. Have drain in floor if work is done in a confined area;
4. Have control rods and/or poison strips, which have been suitably calibrated and identified, installed in the core;
5. Perform counts with suitable source in place as each unit is brought up. Plot reciprocal count rate as loading proceeds. Extrapolate the reciprocal count rate to the anticipated critical loading at each step and compare with the existing total plus the next anticipated loading increment. Use plastic to simulate people if the core is accessible to them; people moderate.

4.10 Testing and Inspection

Testing is required in the development of many core components. Some of these tests have been mentioned or implied in the foregoing parts of this section. Reference [57] lists some of the tests required to establish important design information to confirm the adequacy of reactor components. Some of the important non-nuclear characteristics of core components, which should be considered for testing have been extracted from [57] and are listed below. For some reactor plants a number of these might not require testing, especially if the plant represents a minor extension of previously successful designs. However, care is required in eliminating tests if new features are introduced.

a. Vibrational characteristics of the fuel elements and control elements.
b. Thermal distortion characteristics of control elements.
c. Stress concentrations in support structures.
d. Wear of reactor materials in the appropriate environment to establish reliable materials combinations for control rods and drive mechanisms.
e. Life of control rod drive mechanisms under all types of possible operating and environmental conditions.
f. Fatigue limits of new reactor materials at operating temperatures and pressures.
g. Strength and ductility on formed and welded components of fuel elements and subassemblies.
h. Crevice and fretting corrosion of mechanical fasteners.
i. The integrity of bolt-locking devices under hydraulic and vibratory loads.
j. The warping tendencies of fuel elements under severe temperature variations.
k. Adequacy of fuel element shape and fuel element grouping under irradiation.
l. The effectiveness of shrouds for protecting control elements from hydraulic forces.
m. The fuel element and coolant flow channel tolerances that can be maintained during fabrication for use in computation of reactor performance.
n. Flow distribution through the core, to thermal shields and to control elements.
o. Core instrumentation—mechanical, thermal and hydraulic features.
p. The buckling loads of control element drive shafts under various conditions of guidance and alignment.
q. The adequacy of buffer mechanisms to snub the control elements under conditions of emergency insertion.
r. The procedures to be followed and special equipment to be used in correcting faults in the event that parts malfunction during operation.
s. The tool torques required to unlatch parts after exposure to long periods of time in core environment.
t. Deflections under load of the core-supporting structure and reactor vessel head to determine their effect on control rod channel alignments.
u. Natural convection flow characteristics through the core to determine the adequacy of core cooling when no power is available.
v. Pressure drop across subassemblies.
w. Cooling characteristics involved with flow around different fuel element geometries.
x. Inspection procedures.

In order to obtain certain of these data, a complete mock-up of the core serves best, particularly in connection with the problems involving alignments, accessibility and interrelation of hydraulic and mechanical problems. Wherever possible, hydraulic forces, flow distributions, and uneven cooling around fuel element geometries are determined by use of air in the incompressible range. A complete airflow mock-up of the core in wood, plastic and aluminum serves a useful function not only in this connection, but also in demonstrating the mechanical feasibility of the design. In other cases, individual components or features of the components are tested separately.

In addition to mock-ups and tests of development work, means must be established for inspection of core parts. While these may rely on principles outlined in Sec. 4.10, special planning and attention must be given to the design of necessary jigs and fixtures.

Typical of the types of inspection for which planning must be done are those associated with the manufacture of the fuel elements and assemblies for the PWR core. The UO_2 fuel pellets were checked for dimensions and density, and a 4% random sample of pellets was checked for corrosion resistance. Density measurements involved weighing and dimensional checks verified by water displacement measurements. Corrosion resistance was confirmed by tests in 750°F (400°C) steam at 2000 psig (140 kg/cm^2) for 20 hours. The cladding underwent separate tests. The finished mill lengths of Zircaloy-2 tubing were subjected to 5000 psig (350 kg/cm^2) hydrostatic pressure tests and were examined for flaws by eddy current detection. Ultrasonic testing for detecting internal flaws in thin-walled tubing has gradually replaced eddy current testing with most tubing manufacturers. Some designers (for example, at Hanford) believe that for Zircaloy-2 tubing ultrasonic and eddy current testing are complementary since the methods have different sensitivities to defects of certain shapes or orientation. After the tubing was cut to desired lengths the first end cap was welded in place; the integrity of the weld was checked by a radiographic test and a 300 psi (20 atm) helium leak test. The tube was then loaded with fuel pellets, evacuated, filled with high purity helium and then fitted with a second end cap which was welded in place. This weld was tested in the same manner as the first weld. Dimensional checks were again made. The completed rod was then corrosion-tested in 750°F (400°C) steam at 1500 psig (105 kg/cm^2) for 3 days. After the fuel rods were finished, they were welded into bundles. Inspections of rod spacings were then made prior to machining. After final machining, the bundle was cleaned and then corrosion-tested in 680°F (360°C) water at 2700 psig (190 kg/cm^2) for 3 days. A 300 psi (20 atm) helium leak test was then applied as a final check for leaking fuel rods.

As indicated in the text, the fabrication of the Shippingport fuel bundles involved two corrosion tests of the fuel rods (in addition to the tests on tube samples); one test being made on individual rods before assembly; the second being made on the completed bundles after machining and a 1350°F (732°C) vacuum anneal. The anneal probably marred the rod surfaces to an extent necessitating a second corrosion test. Prior to the second test, the bundle surfaces had to be prepared for autoclaving by pickling in hydrofluoric acid and rinsing.

Two disadvantages of this process come to mind—first, the difficulty in removing all of the pickling acid and, secondly, the difficulty in inspecting the corrosion film on the surfaces of 12 fuel rods welded together in an 11 × 11 lattice and 0.055 inch (1.4 mm) spacing. However, the success of the Shippingport blanket fuel design attests to the fact that these problems were successfully overcome by careful attention to manufacturing processes.

4.11 Failure Experience

The importance of the considerations outlined in Sec. 4 is best emphasized by reference to some of the experiences encountered in the testing of reactor components and the operation of reactors. Some of these have been identified in the foregoing parts of Sec. 4. The reader is also referred to Sec. 2.12.2 which presents advice regarding reactor vessels and reactor intervals offered by Reed and Tarnuzzer [179] as a result of their Yankee plant experience.

Typical of the difficulties with core components are those with bolts identified in Sec. 4.8 and those associated with deflection of flow baffle plates. In the Enrico Fermi Reactor the deflection of the baffle plates was a function of the pumping rate and arose out of the higher pressure drop associated with greater flow rates; because of the deflections a flow coefficient of reactivity was introduced into the core which would otherwise not have existed.

Cladding failures have been experienced in a number of reactors. Some of the experience is described in Chapter 11, Sec. 4 and in Chapter 13.

Hughes and Greenwood [77] and Mooradian et al [78] describe the violent failure of a natural uranium metal fuel rod clad with aluminum in the NRU heavy water reactor in Canada in 1958. The failure was attributed to waterlogging of the fuel.

Balligand [79] describes fuel element failures in the EL-2 and EL-3 reactors at Saclay, France. The EL-2 is a heavy-water-moderated reactor cooled by a closed-cycle CO_2 circuit. Several cladding failures were encountered in the period from October 6, 1957 to February 16, 1959, some of them intentional to get experience with such failures. On the other hand the EL-3 reactor, which is a heavy-water-cooled and moderated research reactor, experienced a fuel element failure arising out of a vibration problem in May 1958. The element had been modified slightly and inadvertently located in the wrong place in the reactor. In this position it was poorly supported at the bottom of the calandria and vibrated under the effect of water flow until the welding broke in the aluminum tube (called a cell) in which the fuel rod was mounted. The cell then fell across the bottom of the tank. Because of insufficient cooling, the fuel rod melted.

Dickinson [80] reports the failure of fuel elements in the Sodium Reactor Experiment (SRE) in the United States. A leak of organic material, tetralin, from an auxilliary cooling system to the pump shaft seal, led to partial blocking of flow in several coolant channels. As a result, the fuel element temperatures got high enough to cause the stainless steel cladding to alloy with the uranium metal fuel and, finally, to separate near the bottom with melting of the stainless steel cladding [81].

Colomb and Sims [189] report a similar incident at the Oak Ridge Research Reactor (ORR) in which a partial meltdown of one fuel plate in a 19-plate fuel element occurred. The incident took place with the reactor operating at a power level of 24 Mw during a beginning-of-cycle startup. The meltdown was caused by insufficient cooling as a result of restriction of coolant flow by a large neoprene gasket that had lodged in the upper end box of the fuel element.

A somewhat different situation, causing more core damage, occurred at the Hallam Nuclear Power Facility [190]. Liquid sodium, which removes heat from the uranium fuel in the reactor, leaked into seven of the stainless-steel-clad graphite moderator cans, which had cracked during operation. The sodium caused the graphite to swell and, as a result, the surrounding fuel elements became stuck. Because of this condition, further operation of the plant was discontinued. The nuclear portion of the plant has been shut down since September 1964.

Other fuel element failures have occurred including severe overheating and meltdown of fuel elements in the water-cooled Westinghouse Test Reactor. An indication of a small leak has also been observed at Shippingport. Each of these fuel element failures presents its own lesson.

In addition to cladding failures of fuel elements, fretting failures have occurred in the testing of fuel assemblies. For example, during prototype testing of seed fuel clusters for the Shippingport Reactor, the stainless steel latch used to hold the cluster in the core cage was found to have caused severe fretting of the Zircaloy side plates of the cluster [51]. This cluster had been under test for 7 weeks in a hot-water loop at design flow rate and temperature. Accordingly, the design of the latch mechanism had to be modified to prevent movement between the pieces involved.

Failures have also been experienced in control rods and control rod systems. The scouring of the nickel-clad cadmium-indium-silver control rods has already been mentioned in Sec. 4.7.1. In addition, difficulties have been encountered in the cracking of welds attaching the 2% boron stainless steel blades to the spider of the control rods of the Dresden Reactor. Stress corrosion was believed to be a factor in this failure. A similar type of cracking also took place in the EBWR control rods; these rods were made by spot-welding together formed angles of 2% boron steel. Examination of the EBWR control rods revealed arc-shaped cracks around the spot welds in the boron control section. These cracks were attributed to high stresses in the welds initiated by thermal expansion effects. A series of rivets used to fasten an adapter also showed cracking; stress corrosion has been advanced as the cause for cracking of these rivets.

As indicated in Sec. 4.7.1, failures have been encountered in the testing of 17-4 PH stainless steel drive shafts for control rods. One of the first of the incidents involving such material took place in connection with the Dresden Reactor. On November 14, 1960, during routine procedure prior to startup of the Dresden Reactor, there were indications that one of the 80 control rods was not functioning normally. Subsequent testing indicated that the control blade had become separated from its drive due to a complete break in the index tube of the drive. Later metallurgical examination established that the break was caused by the stress-corrosion cracking of the 17-4 PH stainless steel material. Further testing disclosed that a different aging temperature was required to improve the stress-corrosion resistance of this material. This finding resulted in a review of all reactors built or being built to determine if this material with the same aging treatment was being used in other reactors. Corrective action was taken for all those reactors where there was any question.

A number of incidents involving control rod sticking have been encountered in various reactors from time to time. The causes have been many and varied; in a number of instances the causes have been difficult to find. For example, during a routine shutdown of the Indian Point Reactor, one of the control rods became stuck in the withdrawn position (about one inch from its upper limit) [188]. Upon observing that the rod had not inserted, the operator initiated a scram, but the rod could not be inserted by this action. Investigations subsequent to shutdown clearly showed that the failure to scram was not due to the control rod drive mechanism or hydraulic system. The reactor vessel head had to be removed to find the cause; a foreign object was discovered wedged between the control rod and the upper grid plate. The object was found to be a pin-shaped piece of austenitic steel, about 1 3/4 in. long by 9/32 in. in diameter at the large end and 5/32 in. in diameter at the small end. The origin of the piece of steel could not be traced to anything in the primary circuit nor could it be envisioned as part of any tool that might have inadvertently entered the system.

In another example, one of the control rods of the General Electric Test Reactor (GETR) failed to reach its fully inserted position during a scram [190]. In this instance, investigation showed a 1/4-20 one-inch long socket head cap screw in the shock absorber section of the receiver and a fine metal sliver in the bottom labyrinth of the shock section piston. It is postulated that the control rod failed to drop fully into the core because the piston became slightly offset after contacting the screw and caused the development of the sliver which led to the failure.

The foregoing experiences serve to illustrate that despite the best efforts of designers, failures do take place. It is a credit to those involved that the failures did not lead to more severe situations. This was due in large part to the precautionary tests made during development of components as well as to the careful examination of parts during assembly and to investigation of unusual circumstances during operation.

5 CONTROL ROD DRIVE MECHANISMS

As indicated in Sec. 4.7.1, the most widely used method of controlling reactors is by the use of solid-poison moveable control rods. Motion to these rods is imparted by means of control rod drive mechanisms. Because of their function, control rod drive mechanisms constitute a vital part of the reactor system.

The type of motion required from a control rod drive mechanism will depend upon the function of the control rod in the core. Control rods may be used as regulator rods, shim rods or safety rods. The regulator rod is used in some reactors, particularly research reactors, to maintain the power level of the reactor constant by means of small motions to compensate for small perturbations in power. Such rods are usually operated by an automatic servomechanism; however, facilities for manual control are also usually provided. Shim rods are used to make power level changes and to take care of larger more slowly varying reactivity changes than provided for by the regulator rod; these changes might include those arising from xenon burnout or fuel burnup. Safety rods are used to shut down the reactor quickly. This is usually accomplished by scramming, a process whereby the rods are rapidly inserted into the core either by gravity fall or by the action of highly reliable stored energy devices or by a combination of both means.

In some reactors the various functions are performed by specially designated, separately controlled rods; in other reactors the functions may be shared by all the rods. All rods usually have provisions for scramming despite any other special function they may have. In many applications it is desirable to move rods in groups or gangs; the mechanisms may, therefore, have to have provisions for such operations.

Under normal operating conditions the control rod drive mechanism must provide for controlled motion into or out of the core at a limited rate well below the rate which might lead to an uncontrollable startup accident and in prescribed increments small enough to provide the desired control. In addition, under prescribed circumstances these mechanisms must provide for scramming of control rods. This scramming action must be accomplished in a sufficiently short time to provide protection to the core under the circumstances causing initiation of scram. This includes the time for the appropriate instrumentation to sense the circumstance and transmit a suitable signal to the mechanism as well as the time to release the control rod and accelerate it into the core. Frictional drag on the control rod system due to bearings, hydraulic side forces and similar effects must be considered in determining the accelerating forces required.

The worths of the control rods play an important part in establishing the functional requirements for the control rod drive mechanisms. A control rod moving in the central portion of its travel causes a much larger reactivity change per inch of travel than it does near the beginning or end of its travel. The rod worth near the center of travel thus becomes the basis for establishing the limit on the maximum speed of incremental rod motion. The rod worth near the extremes of travel forms the basis for establishing the maximum time in which scramming of the control rods must be initiated and the rate at which they must be accelerated.

The control rod drive mechanisms are also the means, in many reactors, for providing position indication of the control rods. This indication can be either continuous or stepwise and should include definite indication to show when the rods are fully in or fully out of the core. Other indication might also be provided to identify inoperable control rods.

Many types of mechanisms operating on different principles have been used for driving control rods. The types of mechanisms used depend a great deal upon the reactors to which they are applied and the functions to be performed. The chief criterion in selecting the type of mechanism to be used is that the mechanism must perform reliably in the environment to which it will be exposed. The mechanism selected should be carefully designed and proof-tested. Prototype units should have had performance tests; the production units should be submitted to proof tests of sufficient duration and severity to demonstrate that they meet the design requirements.

In the ensuing parts of this section, attention will be given to more detailed design considerations for control rod drive mechanisms, to the characteristics of principal types, and to the problems of positioning, position indication, scramming, cooling and gas accumulation. Some of the experiences encountered during testing and operation of mechanisms also will be discussed.

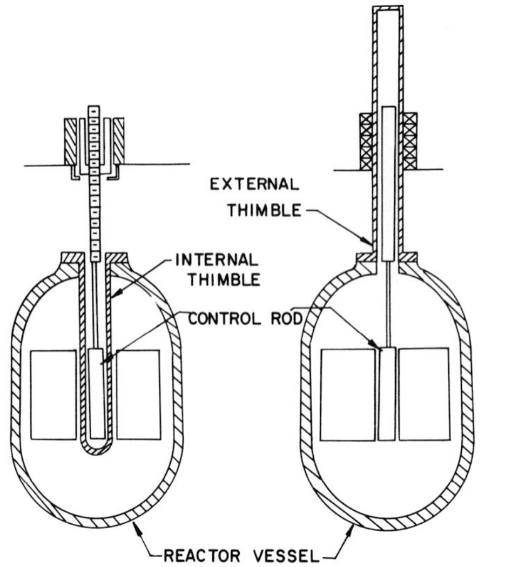

FIG. 5-1 Pressurized-reactor thimbles for control-rod drives. Left sketch shows internal thimble arrangement; right sketch shows external thimble arrangement.

5.1 Design Considerations

Mechanisms for driving control rods can be divided into two groups: those used for nonpressurized reactors and those used for pressurized reactors [82]. The nonpressurized reactors include research reactors operating at low temperatures and at atmospheric pressure as well as power reactors operating at high temperatures with coolants having low vapor pressure, e.g., liquid metals or organic coolants. Such reactors can utilize a wider range of mechanisms and provide a more favorable environment for reliable operation than can the pressurized reactors; the mechanisms for nonpressurized reactors can also be made more accessible for maintenance and repair. The pressurized reactors include the water-cooled and gas-cooled reactors which operate at pressures significantly above atmospheric pressure. The pressures associated with such reactors complicate the design of the control rod drive mechanisms.

If the control rod drive mechanism of a pressurized reactor is to be located outside of the pressure envelope, either a shaft seal must be provided through which the drive shaft connecting the rod and the mechanism must operate, or the control rod must operate in an internal thimble of the vessel as shown in Fig. 5-1 (a). (This figure is adapted from reference [82].) If shaft seals are used, additional drag will be imposed on the rod system and means must be provided for handling the contaminated coolant leakage; provisions must also be made for performing maintenance on such seals. If internal thimbles are used, both the vessel and core design become more complicated; more fissionable material will be required to compensate for the neutron absorption in the thimble shells. Cooling of the rods also becomes more difficult. For these reasons, the mechanisms for pressurized reactors are often enclosed within the pressure envelope, utilizing external thimbles (to accommodate the mechanisms) as shown in Fig. 5-1 (b). In such an arrangement, means must be provided for transmitting power through the pressure envelope. The mechanism in such an application must operate in the environment of the reactor vessel without the benefits of lubricants; such lubricants tend to sludge under irradiation. Cooling problems in the mechanisms and structural problems in the thimbles-to-vessel attachments must be faced in such designs (see Sec. 3). Mechanisms in such an arrangement are not accessible for easy maintenance.

Hydraulic mechanisms can be used for either pressurized or nonpressurized system. Although they have attractive features they have not found widespread use. Except for designs which are to provide fully-in or fully-out motion, such mechanisms can become quite complicated because of the requirements for position-control accuracy. Other problems which are faced in the design of hydraulic mechanisms are concerned with details of critical components such as valves and pumps, particularly where high temperatures are involved (see Sec. 5.2).

There are a number of considerations common to all types of control rod drive mechanisms which apply during design. These can be grouped into two categories: functional requirements and design requirements. Each of these categories is described below. Specific considerations associated with particular designs are discussed in Sec. 5.2.

5.1.1 Functional Requirements

Before a mechanism can be selected or designed, it is necessary to establish the operating or functional requirements of the mechanism. These include the control rod loads that must be handled, the operating speeds desired, the positioning requirements, the position indication requirements, the scramming characteristics desired, the lifetime characteristics required and special requirements. The influence of each of these functional requirements on the design of the mechanism is discussed briefly below.

a. Control Rod Loads: The functional requirements that influence mechanism design first are the loads to be handled by the mechanism during its life, both under normal conditions and emergency insertion. The loads to be handled will influence the drive motor, the power supply and the cooling system of the mechanism. Although the power required to raise a control rod at a speed of several inches per minute may appear small, the power input to the mechanism will have to be many times this value because of electrical and/or mechanical inefficiencies and because of the frictional drag imposed on the rod as mentioned in the introduction to Sec. 5. In addition, in order to have the rods fail-safe in the event that power is lost to the mechanisms, continuous power is required even to hold the control rod at a given position even though the net mechanical output is zero. The holding power is a function of the control rod weight and results in heat which must be removed by the mechanism cooling system. It turns out that for cores with many control rods, the power requirements lead to equipment of substantial size. Often space limitations will be faced in locating the mechanisms on the reactor vessel head, especially if the rod loads to be handled are high. Typical rod loads for water-cooled power reactors, including friction drag will be in the range of 100 to 300 lbs.

b. Operating Speed: The normal operating speeds of reactor control rods are relatively low except for possibly regulator rods of low effectiveness. Typical operating speeds are in the range of 1 to 10 in. per minute. The maximum rod withdrawal speed is based on analyses of the startup accident in which it is postulated that a control rod group is inadvertently but continuously withdrawn at its maximum rate. As a result of these analyses the maximum withdrawal speed is established at a value which will insert reactivity at a rate no larger than can be handled by existing shutdown mechanisms. The startup accident is studied in two parts because of the difference in behavior of the reactor when it is in the subpower range from its behavior when it is operating in the power range. In the low power range, significant heat addition is not immediately available to influence the temperature coefficients of reactivity. In the power range, such coefficients are highly significant.

Schultz [82] and Harrer [83] indicate that for many reactors the rod speed is selected to provide a maximum rate of reactivity change of the order of $10^{-4}\,\Delta k$/sec. Because control rods worths are not accurately known until after final calibration, both authors recommend making provisions for final adjustment of the speed; a range of 4 to 1 is desirable.

c. Positioning Requirements: Because of the influence of the control rods on the reactivity of the core, it is important that they be positioned within prescribed allowable limits. As a result the mechanism cannot have too much backlash. Some braking action is also necessary to stop the movement of the rod without untenable over-travel from the desired position. In addition, the mechanism must be so designed that the control rod cannot move by gravity or other loads imposed on the system after the cessation of the desired motion. Typical specifications require a repeatability of the order of 10^{-5} in reactivity. As a result, the accuracy of actual incremental rod motion may have to be within ± 1/32 inch (~ ± 0.8mm) of the desired incremental motion. These requirements are usually not so straight for power reactors as for research reactors and some power reactors have stepped control rod intermediate positions several inches apart, e.g., Dresden I.

d. Position Indication: Although in general it would be desirable to know the position of a control rod to the same accuracy required for positioning, it is not always easy to do so. In fact many reactors have looser tolerances for position indication than for positioning; this is particularly true when the position indication involves transfer of information across thimble walls into discrete electrical coils. Two sets of tolerances may be specified, one for indicated position and another usually tighter tolerance for incremental motion. For example, the exact position of a control rod may be indicated with an accuracy of only ± 1/2 in. (~ ± 13 mm) while the incremental movement from one position to another might be indicated within ± 1/8 or ± 1/4 in. (~± 0.3 or ± 0.6 mm). Provisions should be made for indicating the extreme ends of travel of the mechanism. A pickup indication should also be provided on mechanisms which releases the rod in scramming; such an indicator scram shows when the control rod is engaged to the mechanism.

e. Scramming Characteristics: As indicated in the introduction to Sec. 5, scramming involves rapid insertion of the control rods in time enough to protect the core against conditions which could lead to damage. The scramming force can be supplied by gravity or by stored energy such as that in a spring or pneumatic device. A clutch or latch, which can be released when power to the device is interrupted, is usually installed between the drive mechanism and the control rod.

In determining the speed with which the rod must be moved into the core during a scram, it is important to recognize that even an infinite speed is not sufficient if the signal to initiate scramming does not reach the mechanism before damage is done to the core. Hence, careful attention must be given to the time delays of the signal-sensing and transmitting equipment as well as of the unlatching device when designing scram circuits. Efforts must also be made to minimize inertial and frictional drag on the control rod system which takes part in the scramming. A typical scram requirement may call for full insertion of the control rod within 3/4 sec following the receipt of a scram signal.

In addition to accelerating a control rod during scram, it is necessary to decelerate it and bring it to a stop. Hence, a buffer device must be provided. Springs and dashpots are commonly used for this purpose. (See Sec. 5.4.) Deceleration must usually be accomplished with the last 10 percent of the stroke.

Because of the high loads involved and the high stresses induced in the control system components, scramming can lead to failure of those components if repeated more often than allowed for in design. Limits exist on repeated impact loadings as they do for other fatigue situations. As a result, control rod drive mechanisms are often designed to insert the rods into the core by driving them inward at a speed faster than allowed on withdrawal but slower than required for scramming. Such action might be called for under circumstances for which power reduction is required but complete shutdown is not. Such action is sometimes referred to as a rod run in or power cutback.

f. Lifetime Characteristics: Inasmuch as corrosion and wear are encountered in service, it is necessary to specify the lifetime requirements that the control-rod drive mechanism will have to meet. Items which influence lifetime would include the number of calendar hours of exposure to operating conditions, the total rod travel expected during the life of the mechanism, the number of scrams expected during the life of a mechanism, and the number of positioning cycles and the associated range of motion expected. For example, it might be estimated that a mechanism will see service for 50,000 hours, will travel about 25,000 ft (~ 7.6 km) during its lifetime, will encounter 500 scrams and will be cycled about 400,000 cycles over an average range of 1/8 inch. (3.2 mm).

g. Special Requirements: These include such items as space limitations, features for remote disconnection of the control rod from the mechanism, location of the buffer, accessibility to the mechanism room, external shock loads, and ability of the mechanism to withstand external wetting during refueling. For shipboard applications the mechanism will also be required to operate under roll, pitch and list conditions.

5.1.2 Design Requirements

Aside from the functional requirements which the control rod drive mechanisms must meet, there are a number of other requirements which will influence the detailed design and performance of the mechanism. The mechanism must be compatible with:

a) The environment in which the mechanism is to operate, including temperature, pressure, and the nature of the reactor coolant which might reach the mechanism. The coolant chemistry, the gas content and the wear and corrosion products in the coolant must be considered. Humidity conditions must be considered for mechanisms or parts of mechanisms exposed to air; lubrication requirement may also be involved. The environment can change at different times; thus environmental effects must be evaluated for normal conditions (startup, operation, shutdown) as well as abnormal conditions (sudden temperature changes, pressure losses, etc.).

   When bottom-mounted drives are used, provisions must be made to prevent sediment from settling onto shafts, bearings and other mechanism parts, or a means for flushing may have to be provided to keep parts free of sediment accumulation.

b) The space available for the mechanisms and the associated cooling and electrical leads.

c) Characteristics of the power source with particular attention to fluctuations, as for example, the fluctuations in voltage and frequency in electrical systems.

d) Availability of and characteristics of cooling means such as air or water. Consideration must be given to quantities required and pressures and temperature involved.

e) The method of attaching the mechanism to the vessel, particularly in pressurized reactors where seals or seal welds may be required.

f) The influence of the core configuration on the design of the mechanism; the control rod spacing in the core may lead to the use of offset shafts and eccentric loading on the mechanism.

g) Requirements for inspection, maintenance and replacement of the mechanisms and/or other nearby components.

h) Refueling requirements.

i) Alignment problems.

j) Frictional and inertial effects during normal operation and scramming.

5.2 Principal Types

Many different types of control-rod drive mechanisms have been used in nuclear reactors to date. Bates [84] has tabulated the nature and characteristics of control rods and drives used in various nuclear power plants in the United States; his table (slightly abbreviated) is reproduced here as Table 5-1 to provide a ready reference.*

---

*Bates' tabulation has been abbreviated by deleting plant identification, location, docket number, status, and fuel type. Note that the data in Table 5-1 are superseded in part by the data of Appendix 1 of this volume. (Appendix 1 tabulates the responses to a questionnaire sent to various reactor installations.)

TABLE 5 - 1

Control Rods and Control Rod Drives in Various Nuclear Power Plants*

Abbreviated Name of Plant	Reactor Type	Loaded Excess Reactivity	Individual Rod Worth	Total Rod Worth	Shutdown Margin	Auxiliary Holddown Means
Shippingport	Pressurized water	15%	Not given; calculated average 0.625%	20%	5%	None
Indian Point	Pressurized water	12.8% (450°F)	Not stated; can stand loss of two rods at 450°F	Movable, 12.5%; fixed, 3% (450°F)	2.4%, worst case	Soluble poison, always for cold clean core
Yankee	Pressurized water	19.8% CC; 6.9% eq.	1% max. for center rod; others less	17%, need 25%	Inadequate by 4.5% CC; adequate HC	Liquid poison required to give 5% cold clean margin
N. S. Savannah	Pressurized water	11.2% CC	1.3% max., 0.93 min. in CC	15% CC	2.8%	Soluble poison
Saxton	Pressurized water	19% CC; 11% HC; 7% eq.	~1.5% at equilibrium	Not listed; insufficient for holddown	None in CC; 2% to 3% eq.	Liquid poison required to hold down cold clean reactor
VBWR	Boiling water	12%	Center rod 5%	18%	3%	Liquid poison, if needed
Dresden	Boiling water	13.1% CC	Not stated; can stand loss of one rod CC	14.1% +2% for end-connector poison	3%	Soluble poison, if needed
Elk River	Boiling water	14.9% CC	4.2% for center rod	16.9% +4.5% burnable poison in fuel	6.5%	Soluble poison, if needed
Pathfinder	Boiling water with superheat	13% CC	1.5% for most effective rod	17%	4%	None
Humboldt Bay	Boiling water	Footnote a	Footnote a	Footnote a	3%	Liquid poison
BONUS	Boiling water with superheat	16.6% CC	1.4% at operating temp.; 1.8% max. CC	19%	2.4%	Fixed poison rods and soluble poison if needed
Big Rock Point	Boiling water	23% CC	Not stated; can stand loss of first rod out CC	28% CC	5%	Soluble poison, if needed
Hallam	Sodium cooled, graphite mod.	Total = 6.3% [b]	Not given; calc. from totals for safeties 0.6% each; shims 0.61% ea.	Safety rods 9%; shim and reg. rods 8%	Total ~10.7% from full power	None
Enrico Fermi	Sodium cooled, fast breeder	40 cents	Said to be \$1.00 for each of 8 safety rods; 46¢ ea. for 2 op. control rods	6.5%	6%	None
EGCR	Helium cooled, graphite mod.	14.88%	Not stated	18.2%	3.32%	None
Peach Bottom	Helium cooled, graphite mod.	29.3% eq.	Most effective rod, 0.01 ± 20% step if suddenly withdrawn	Not stated; 31.3% at equilibrium	2% eq.	None
Piqua	Organic cooled, organic mod.	6.5% HC	7 rods 1.43% each (300°F); 13 rods 0.94% each	Not given; 10% at 300°F	3.4% when core most reactive	None
CVTR	D_2O cooled, D_2O moderated	36% CC	1.4% average	44.8%	8.8%	None

* References to sources of data are given in [84]. For more recent data, see Appendix 1 of this volume as well as other chapters of these volumes. Abbreviations used in table: CC= cold, clean condition; HC= hot, clean condition; eq. = equilibrium condition; ea. =each; reg. =regulating; mod. =moderated.

[a] Stated to be relatively meaningless because of reactor characteristics.

[b] 5.5% at design power plus 0.8% because of fast temperature effects.

TABLE 5 - 1 (continued)

Control Rods and Control-Rod Drives in Various Nuclear Power Plants

Abbreviated Name of Plant	Rod Guides	Type of Rod Drive	Location of Rod Drive	Seals
Shippingport	Guides built into seed fuel elements	Split roller and screw	Top of vessel in thimbles	None required
Indian Point	Guides above core and in bottom grid plate; drive rod guided by its seal; spacer buttons on control rods slide along fuel can surface	Motor-driven stop and oil-hydraulic cylinders	Below pressure vessel attached to buffer-seal flange	Buffer seal, sliding shaft
Yankee	Full length	Magnetic jack and latch	Top of pressure vessel in thimbles	None required
N. S. Savannah	Guides through and above core	Electromechanical and oil-hydraulic drive	Top of vessel above thimbles	Buffer, sliding
Saxton	Follower guides in upper grid plate; guides below core	Magnetic jack	Below pressure vessel in thimbles	None required
VBWR	Fuel channels and guide bushings in vessel head	Electromechanical with pneumatic cylinder and balancing cylinder	On movable platform above reactor; all seals and rods go through single plate assembly	Water seal
Dresden	Guide tubes below core only; fuel assemblies in core	Water-hydraulic drive with mechanical latches	Bottom of pressure vessel in thimbles	None
Elk River	Continuous	Rack and pinion, H_2O lubric.	Bottom of pressure vessel in thimbles	Rotary
Pathfinder	Tubes through core and above; half-way to drives	Rack and pinion	Thimbles above pressure vessel submerged in shield-water pool	Double reactor vessel to seal chain and seal drain to oil-filled drive housing
Humboldt Bay	Guides below core only; fuel assemblies in core	Water-hydraulic drive with mechanical latches	Bottom of pressure vessel in thimbles	None
BONUS	Channels above core; fuel-element boxes in core	Rack and pinion, H_2O lubricated	Thimbles at top of vessel	Rotary
Big Rock Point	Guide tubes below core only; fuel channels act as guides in core	Water-hydraulic drive with mechanical latches	Bottom of pressure vessel in thimbles	None
Hallam	Metal sleeves in graphite moderator	Sprocket and cam for safety rods, magnetic support; ball nut and screw for shim rods	Top of biological shield	None required
Enrico Fermi	Continuous	Lead screw and ball nut	Top	Bellows and metallic O-rings
EGCR	Tube above core, graphite in core	Cable and drum	In top biological shield	None
Peach Bottom	Graphite tubes in core; rod seal in thimbles	Electromechanical with oil-hydraulic and pneumatic cylinders and mechanical latches	Below thimbles on bottom of pressure vessel	Gas seal, sliding
Piqua	Full length	Magnetic jack	Upper part of reactor vessel	Canned and submerged in coolant
CVTR	Core support structure	Rack and pinion, D_2O lubric.	Reactor header cavity	None

TABLE 5 - 1 (concluded)

Control Rods and Control-Rod Drives in Various Nuclear Power Plants

Abbreviated Name of Plant	Scram Direction and Force	Opposition to Scram Force other than Drive and Seal	Scram Time, Objective or Actual	Startup Procedure
Shippingport	Down. Gravity.	Upward water flow	Not given; max. velocity = 8 ft/sec, about 70 in. rod travel	4 rods withdrawn at one time to start with new core; 12 rods are seated at full power; 4 rods withdrawn at one time to compensate for aging
Indian Point	Down. Vessel pressure and gravity.	Back pressure of hydraulic cylinder and coolant flow	Two-thirds insertion in 0.864 sec after receipt of scram signal	Below 10^{-4}% full power, 1 rod or 1 group of 5 rods may be withdrawn at a time; above 10^{-4}% full power, rods may be withdrawn as programed in groups of 4 or 5; above 15% full power, rods may be withdrawn as programed in groups of 4 or 5, manually or automatically
Yankee	Down. Gravity.	Against water flow	Less than 2 sec after initiation of scram signal	Rods withdrawn in groups starting with outside 8 and others in turn; reactor operates with full Boron shim; all rods out except one control group somewhat inserted
N. S. Savannah	Down. Accumulated pressure through hydraul. cyl.	Vessel pressure; 1750 psi	Two-thirds rod insert in 0.8 sec after receipt of scram signal	Manual withdrawal of rods in programed groups up to 15% to 20% power; 9 rods may be put on servo
Saxton	Down. Gravity.	Water flow upward	Not given	One rod withdrawn at a time
VBWR	Down. Air pressure from accumulator[a]	Vessel pressure, 1125psi	0.3 sec for 1/3 of full travel	One rod withdrawn at a time
Dresden	Up. Accumulator pressure[b].	Vessel pressure, 1000 psi; gravity	10% in 0.60 sec; 90% in 2.5 sec	One rod withdrawn at a time
Elk River	Down. Gravity.	None	95% in 1.2 sec	One rod withdrawn at a time above 425°F; motion blocked below this temperature
Pathfinder	Down. Gravity and scram springs, if fully withdrawn	Water flow upward	With signal delays, slightly more than 1 sec full stroke	One rod withdrawn at a time by operator "selection to provide optimum utilization of fuel"
Humboldt Bay	Up. Accumulator pressure[c].	Vessel pressure, 1010 psi		Startup not specified; operate by manual adjustment of control rods; programing to be determined
BONUS	Down. Gravity.	Upward flow of water in core	Not given	One rod withdrawn at a time
Big Rock Point	Up. Accumulator pressure[d].	Vessel pressure, 1050 psi, and gravity	2.5 sec for 70% of full stroke	One rod withdrawn at a time
Hallam	Down. Gravity, only safety rods can scram.	None on safeties	Not given; system delay to effect release in less than 0.30 sec	Safety rods cocked first; cannot be motor-driven to insert when shims are off seats; automatic control operates shims in power range
Enrico Fermi	Down, safeties only. Gravity with spring acceleration.	Upward sodium flow	About 0.6 sec, with 0.3 sec used in getting to core	Rods may be moved singly or as groups; speed reduced once power range reached; safety rods withdrawn first
EGCR	Down. Gravity.	None	About 3 sec for full travel	All rods withdrawn simultaneously up to low power; then by groups of four or five
Peach Bottom	Up. Accumulator pressure[e].	None	Not given; system delay 0.20 sec	One rod withdrawn at a time; programed at power with servo control; some rods cocked before startup
Piqua	Down. Gravity.	Coolant	0.07 sec release; 0.800 sec max. time	One rod withdrawn at a time; manual and automatic switching to make selection
CVTR	Down. Gravity.	Water displacement in guide tube	Not determined	Not established; possibly four-rod group withdrawal

[a] To pneumatic cylinder of drive. Backed up by motor which drives at high speed to follow up insertion.
[b] One for each 3 rods. Or reactor pressure to each cylinder.
[c] One for each 2 rods. Or reactor pressure to each cylinder.
[d] One for each rod. Or reactor pressure.
[e] Accumulator pressure on hydraulic cylinder or gas in pneumatic cylinder.

To facilitate discussion, the principal types of mechanisms used or proposed are classified here under the following headings: The rack and pinion, nut and lead screw, magnetic jack, cable and drum, hydraulic and pneumatic, and harmonic-motion mechanisms. It should be pointed out that, while the mechanisms under a particular heading operate on similar principles, they differ greatly in detail. The reader should not be misled into evaluating mechanisms by type alone; the details of design and the method in which the mechanism is applied may have more to do with the reliability of the mechanism than the general characteristics of its type.

5.2.1 The Rack and Pinion

The rack and pinion provide a relatively simple means for driving control rods, particularly in nonpressurized reactors. They have been used widely for nonpressurized research reactors. Their use is not limited to such reactors, however. Rack-and-pinion drives are being used for the Elk River, Pathfinder, Carolinas-Virginia and BONUS reactors [84]. They have also been used on the Stationary Low Power Reactors No. 1 (SL-1), the Stationary Medium Power Plant No. 1 (SM-1), and the Experimental Boiling Water Reactor EBWR [83]. In such pressurized applications seals are required to limit and control leakage from the reactor (see, for instance, Vol. 1, p. 661).

The rack-and-pinion drive utilizes a motor-driven pinion to operate the rack. The pinion can be driven by almost any type of motor through a gear box to provide the necessary speed reduction. An electromagnetic clutch is usually used to engage and disengage the pinion from the motor for scramming purposes. When power to the electromagnetic clutch is interrupted, the clutch plates separate and permit the rack and control rod to drop into the core by gravity. Acceleration of the rods is augmented by spring action in some applications. Position indication is obtained from the pinion shaft. Limit switches are provided to prevent the motor from driving the rods against the upper or lower mechanical stops. Many rack-and-pinion systems employ a mechanical unidirectional over-running clutch in parallel with the electromechanical clutch so that control rods may be forceably driven into the core by motor, if necessary. Figure 5-2 [82] schematically illustrates the rack-and-pinion drive.

In pressurized systems, the motor, the gear box, and the clutch are usually designed to operate outside the pressure envelope and the rack and pinion inside the pressure envelope. Consequently a shaft seal is provided between the pinion and clutch. The SM-1 reactor utilizes a labyrinth seal consisting of close-fitting rings around the shaft. Clean water is continuously made to flow through the seal from the reactor side to prevent the leakage of reactor water and to cool the seal. In pressurized systems, the rack and pinion are contained in thimbles on the reactor vessel; these thimbles are subjected to the system pressure and environment. Hence, cooling and lubrication must be accomplished by the reactor coolant fluid. In the BONUS reactor the rack and pinion are mounted at the top of the reactor and must operate in a steam environment rather than the water environment of pressurized-water or bottom-mounted boiling-water reactors. In such an application, temperature and lubrication limitations can be quite severe.

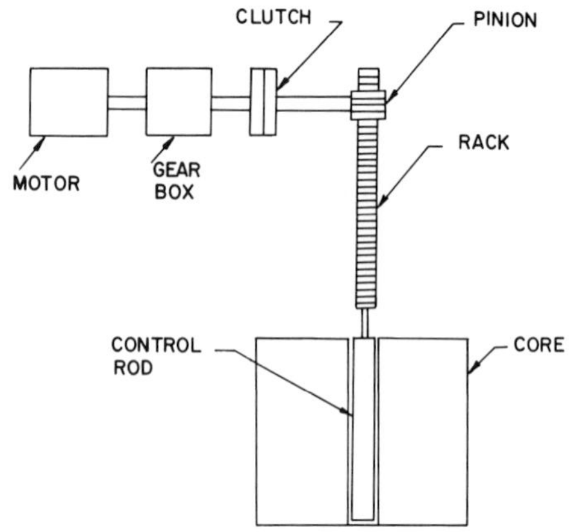

FIG. 5-2 Diagram of rack-and-pinion drive for control rod.

5.2.2 Nut and Lead-Screw Mechanisms

The nut and lead-screw mechanisms utilize either a rotating threaded shaft called a lead screw to impart translation motion to a nonturning nut or a rotating nut to move a nonturning lead screw. These are sometimes referred to as the fixed-screw and fixed-nut designs respectively. However, these same names have been used for specific mechanisms so that it is better to refer to them here as the rotating-screw and the rotating-nut designs respectively.

In the rotating-screw design the rotor of the drive motor is coupled to the lead screw. The nut is usually a ball-nut in which balls circulate through grooves in the nut and the lead screw as the latter rotates. An external ball-return channel is provided on the nut to complete the ball circuit. By this means, ball-bearing action between the nut and the lead screw can be used to reduce friction. The balls actually engage the nut to the lead screw. Once assembled, the nut must not be run off the end of the screw. The nut is in turn attached to a carriage assembly which straddles the screw. The carriage and nut are prevented from turning by guide rollers on the carriage which ride in slots provided on the interior of the mechanism cartridge.

During normal operation, the scram shaft of the control rod is latched to the carriage by a latch assembly which is on the carriage and is pivoted on an axis parallel to the lead screw. This latch assembly is geared to a shaft which extends over the entire length of travel of the carriage. This shaft can rotate over a limited angle but is normally prevented from doing so by an electromagnetically operated armature. When the electromagnet is deenergized the shaft is free to turn under the action of torsion spring; as the shaft turns, it turns the latch arm on the carriage and releases the control rod to initiate a scram. Scramming of the rod is accomplished by the combined action of a scram spring and gravity. To reengage the control rod, the carriage is motor-driven downward until the latch assembly on the carriage passes over the tapered end of the scram shaft and automatically engages the latch if the electromagnet is energized.

When applied to a pressurized system, the armature, lead screw, ball nut, and carriage are contained in a pressure tube on the reactor vessel. The scram magnet is mounted outside the housing. A rod pickup indicator is provided to energize a light on the control panel when the scram shaft on the control rod is fully inserted; this light goes out when it is raised off the bottom. A reluctance-type position indicator coupled directly to the rod motor counts motor revolutions to indicate rod position. When the scram shaft is detached from the mechanism carriage, as it is after scram, the position indicator shows the position of the ball nut and carriage only.

The motor for this use is a canned-rotor synchronous reluctance motor in which the stator is outside the pressure wall and the rotor inside. The pressure wall can be made very thin at the motor because it is backed up by the stator punchings. The rotor is a solid piece with no salient poles or windings; it turns by magnetic attraction as the electric field of the stator rotates with the applied alternating current superposed on top of a direct current. When the frequency of the alternating current is zero the rotor stops but the direct current provides a holding force to prevent slippage.

In the rotating-nut design, the lead screw is attached to the scram shaft on the control rod. The lead screw is provided with translational motion when the nut is turned. The nut can be either a rotating fixed nut or a collapsible rotor nut. The rotating-nut design was used in the Shippingport PWR reactor [85]. The nut was of the collapsible rotor type. The rotor is made of Type 410 stainless steel and is split longitudinally into two arms which are pivoted near the middle and which carry the rollers near their lower ends. When the stator is energized, the top end of each rotor arm moves outward, causing the bottom end of the rollers to move inward and engage the lead screw. The lead screw is prevented from rotating by means of a spline so that when the roller nuts are rotated, the lead screw is moved upward or downward. To scram the rod, the stator field is deenergized, thus permitting the roller nut halves to separate under spring action and allowing the rod to fall by gravity.

The rotor, as well as the lead screw and other moving parts, operates in a pressure tube on the reactor vessel head. As the lead screw moves up or down, its position is indicated by a series of coils outside the pressure tube; a change of inductance is brought about as a magnetic slug on the lead screw moves past the coils.

The motor is a synchronous reluctance motor as described above. The stator is mounted on the outside of the pressure tube, concentric with the mechanism housing so that it can be replaced readily. The magnetic flux passes through the thin wall of the pressure tube which is backed up by the stator windings. Three-phase, variable frequency voltage superposed on a direct current voltage is applied to the stator windings to cause the rotor and roller nut to rotate. The lead screw and control rod are thus raised or lowered as speeds proportional to the a.c. frequency. The direct current is used for holding until a scram is called for. Cooling water from a separate low pressure system flows through a water jacket on the stator to cool the stator windings as

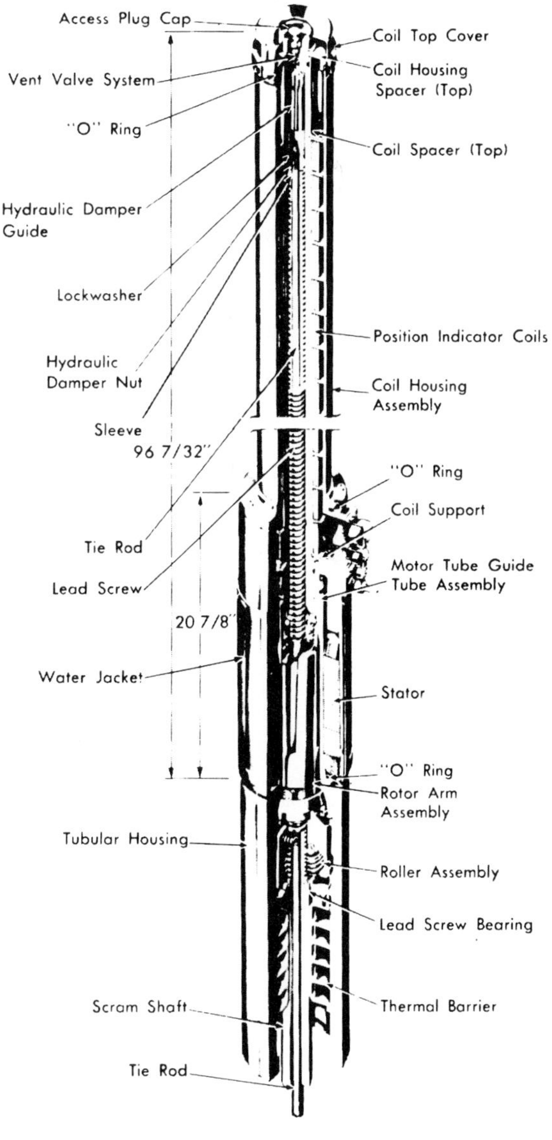

FIG. 5-3 Shippingport PWR control drive mechanism.

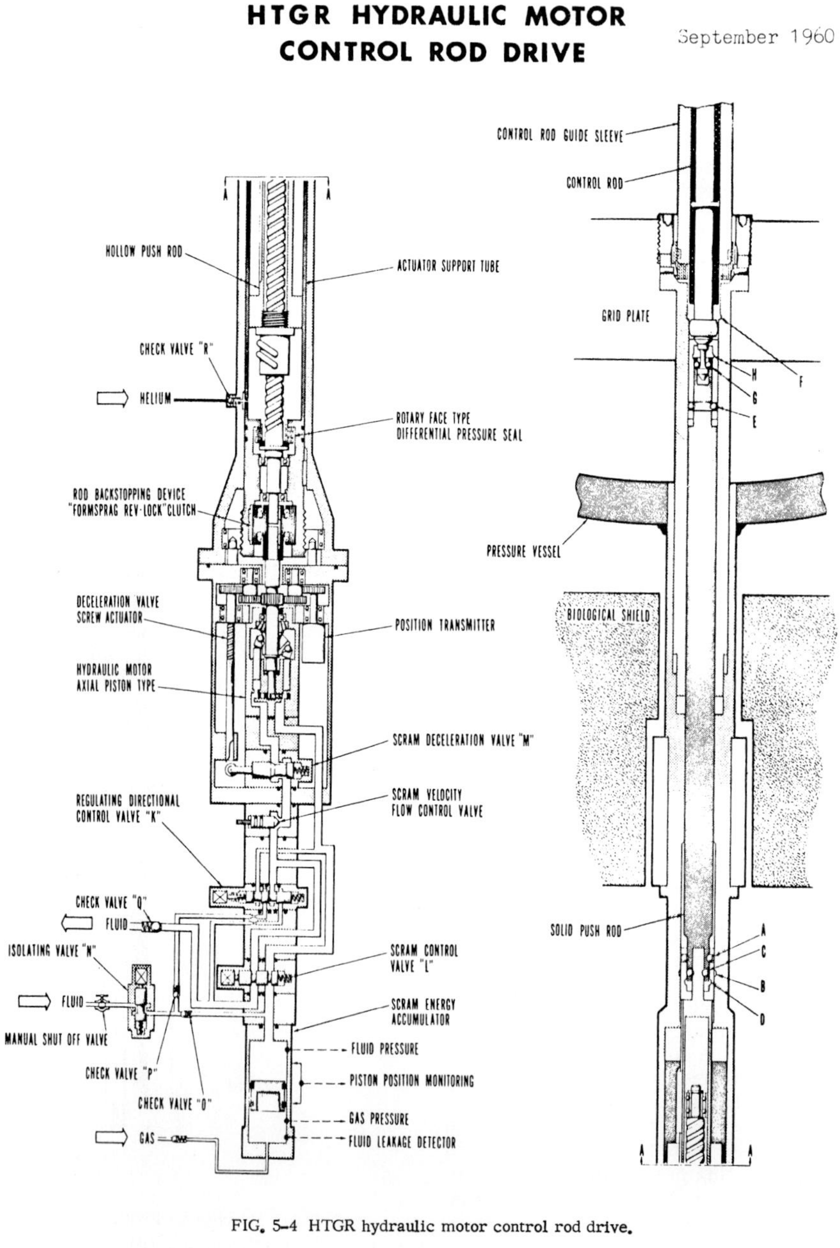

FIG. 5-4 HTGR hydraulic motor control rod drive.

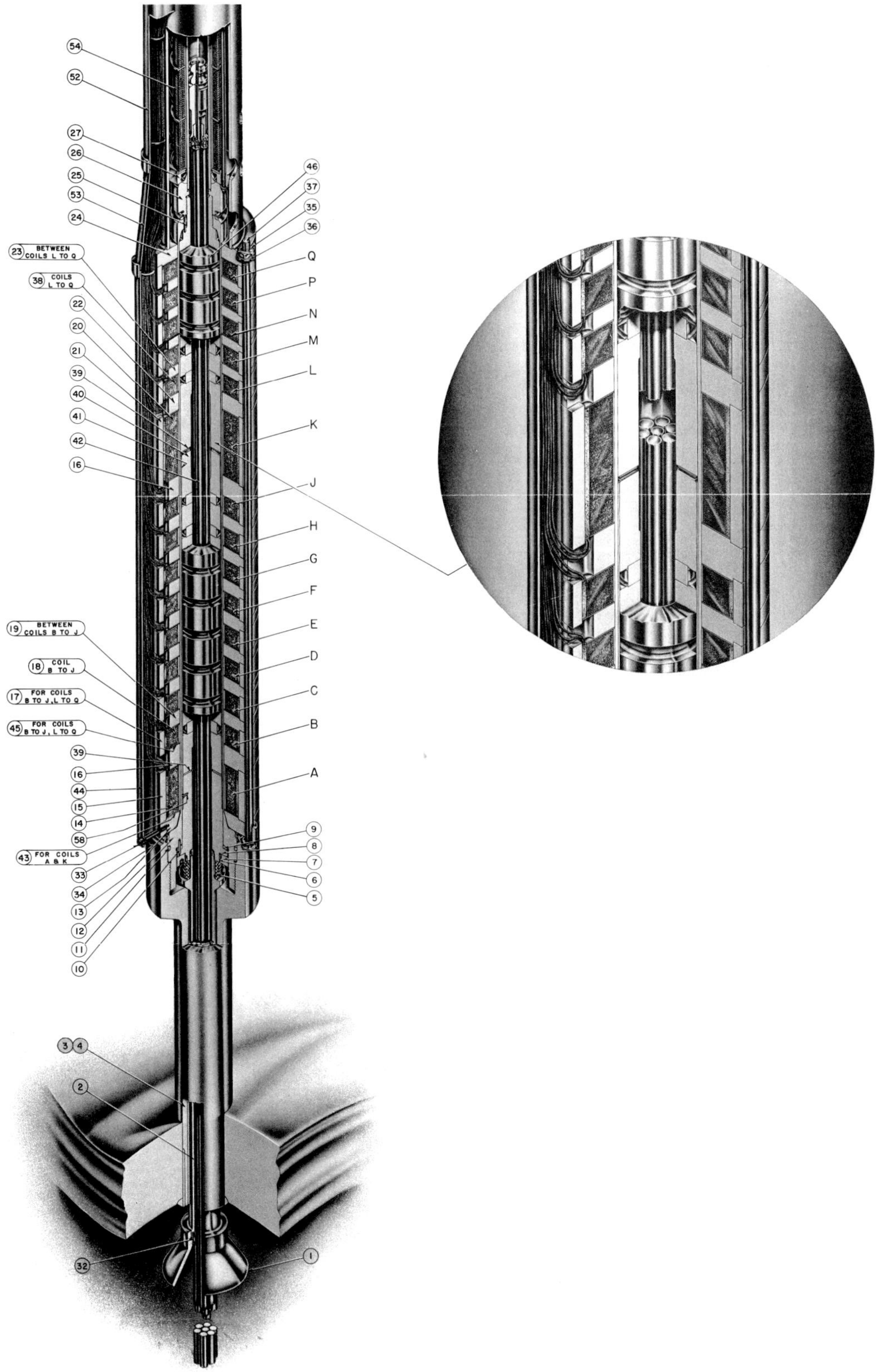

FIG. 5-5 Friction-grip magnetic jack.

well as the primary colant in the mechanism. (Figure 5-3 [39] shows the PWR mechanism.)

The rotating fixed-nut type of mechanism is similar to the split nut type except that the rollers and screw are in continuous engagement. The fixed nut consists of four rollers mounted in a cage which is rotated by a solid four-pole motor. Scramming is accomplished by permitting the lead screw to run down through the nut under the pull of gravity. A mechanism of this type was developed as an alternate unit for the Shippingport reactor.

The nut and lead-screw type of mechanism is used on the Hallam, Enrico Fermi and Peach Bottom reactors in addition to the Shippingport PWR reactor [84]. Each of the mechanism designs used differs considerably.

In the Hallam reactor this type of mechanism is used only on the shim rods which operate in thimbles extending downward through the core. These rods are not scrammed. Scramming is accomplished by a separate set of rods. The thimbles in which the rods operate are sealed and pressurized with helium to prevent contact with sodium or sodium vapor.

The Enrico Fermi reactor uses a rotating-nut design in which the rod is coupled to the lead screw by an electromagnetic latch which is deenergized to drop the rod for scramming purposes [86]. A bellows is used to seal the mechanism from the sodium atmosphere in the reactor. The bellows however, is not accessible for inspection once the reactor is placed into operation. Two types of control rods are used. Safety rods are provided which can be driven into the core at high speed as well as scram. Two motor and speed reducer trains are used with each rod drive. Fast insertion accompanies scram so that rod insertion takes place even if the latch should fail to release or the rods should stick. Helical springs are used to augment gravity action on the rods during scramming. The operating control rods are attached directly to their drives and are not used for scramming.

The Peach Bottom Reactor utilizes a rotating-screw design operated by a hydraulic motor [87]. The control rods are inserted from the bottom and are pushed upward into the core to effect shutdown. The rods remain coupled to the drives at all times during operation and no provisions are made for scramming of rods in the usual way. The scramming action consists of high speed emergency insertion by the hydraulic motor. Thus, the hydraulic motor operates at two separate speeds. Different energy sources and control valves are used for the two speeds. A separate accumulator is provided for emergency insertion.

Figure 5-4 [87] illustrates the unit. Fluid pressure for normal operation is supplied by auxiliary equipment, external to the sub-pile room and not shown here. Upward and downward movement is determined by the spool position of regulating control valve "K". During normal operation the scram control valve "L" is energized to block the ports between the accumulator and motor. When a scram is called for, all control circuits to the drive are de-energized causing the regulating control valve to close and the scram control valve to open. These valves are spring-operated for such actions. By this means the hydraulic fluid from the accumulator is allowed to enter the motor and accelerate the rods to "scram" velocity. Bottled nitrogen is used to pressurize the hydraulic fluid in the accumulator through a floating piston. Deceleration is normally accomplished by the mechanically interlocked deceleration valve "M" during the last two feet of travel. The entire mechanism is housed in a pressure tube attached to the bottom of the reactor vessel.

5.2.3 Magnetic Jack

The magnetic jack mechanism is a step motion linear actuator consisting of a system of coils located around the outside of a pressure tube along its axis, which can jack a control rod up or down magnetically through the wall of the tube. Two types exist, one utilizing a friction grip and the other a positive latch grip.

The friction-grip magnetic jack is illustrated in Fig. 5-5 [88a]. Interior to the pressure tube it contains two grippers, a lower movable gripper and an upper stationary gripper. Six flexible rods, each 1/2-inch (1.27 cm) diameter and made of type-410 stainless steel, pass through axial holes in the center of both grippers to form the driving member of the control rod. By energizing the magnet coils outside the pressure tube in the proper sequence, the driving member may be moved up or down in short steps or held in any position. When the coils around the movable gripper (items B through J in Fig. 5-5) are energized, the magnetic forces cause the drive rod bundle to expand against the bore of the movable gripper. A lifting coil, item K, is then energized to lift the movable gripper approximately 1/8-in. (3.2 mm). After the lift, the stationary gripper coils (items L through Q in Fig. 5-5) are energized to maintain a grip on the drive rod bundle. The movable gripper coils are then deenergized so that the movable gripper can drop to its previous position for repeating the cycle. To reverse the motion the sequence of energizing coils is reversed. Loss of power to the coils permits the control rod to scram either under the action of a spring or gravity or a combination of both. The friction grip magnetic jack has a limited load capacity of less than 200 lbs (92 kg). Its maximum normal speed is about 6 inches/min (15 cm/min). No seals are required but all internal parts must operate in the environment of the reactor coolant.

For large control-rod loads a positive-motion, latch-type magnetic jack mechanism was developed [89]. The general method of operation is similar to that of the friction-grip mechanism except for the method of gripping and transferring load from one gripper to another. Fig. 5-6 [88b] shows a cut-away view of this mechanism; Fig. 5-7 [88b] illustrates the gripping action. In this figure the mechanism is shown with the stationary gripper, item D, engaged to the control rod shaft; coils E and C are both energized. To initiate a lifting cycle the movable gripper coil, item A, is energized to move the latch arm, B, into a groove of the drive rod. The latch arm does not yet engage the drive rod inasmuch as a 1/32-inch (~ 0.8 mm) axial clearance is provided between the latch tooth and corresponding rod tooth at this time. The coil,

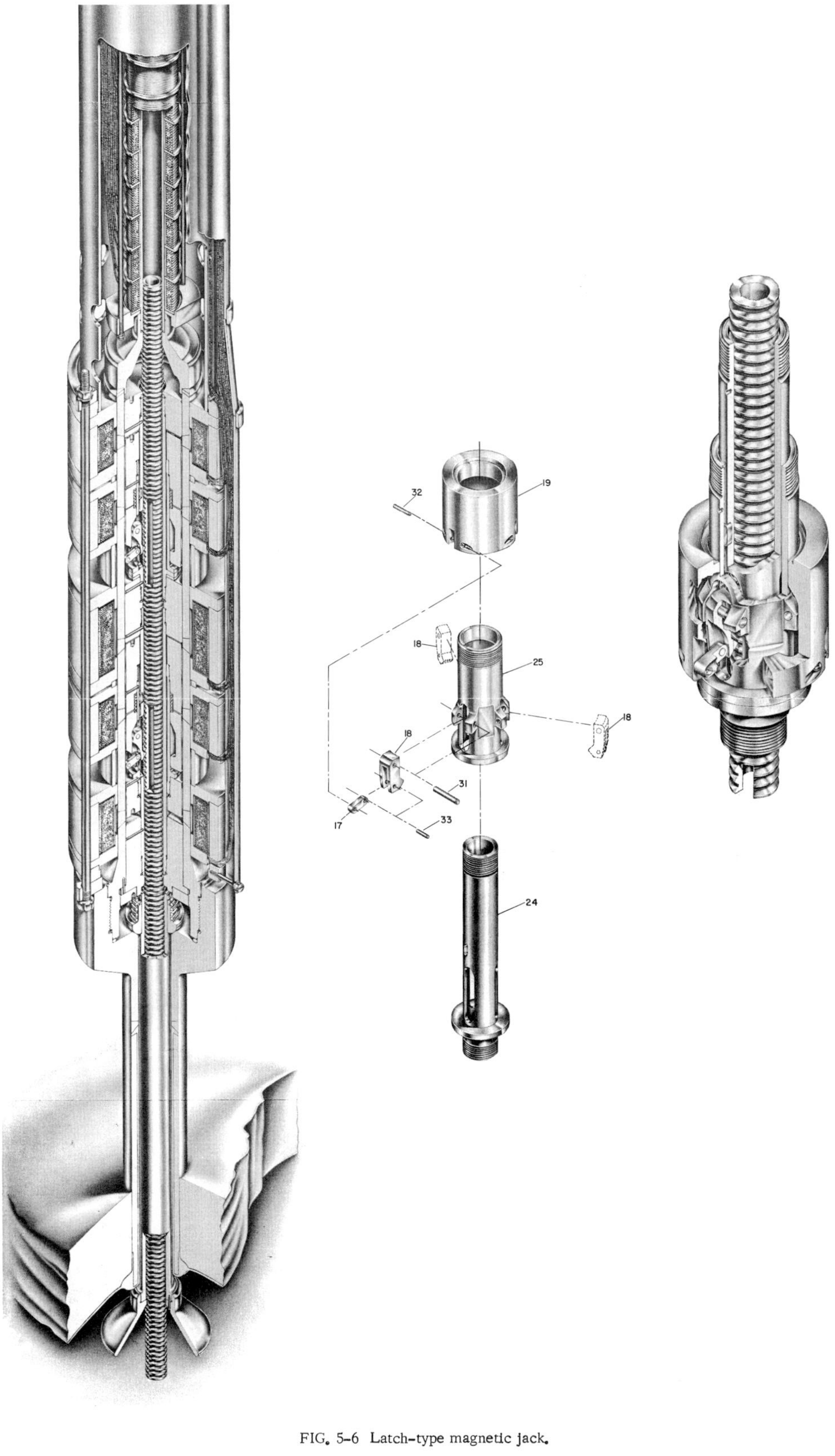

FIG. 5-6 Latch-type magnetic jack.

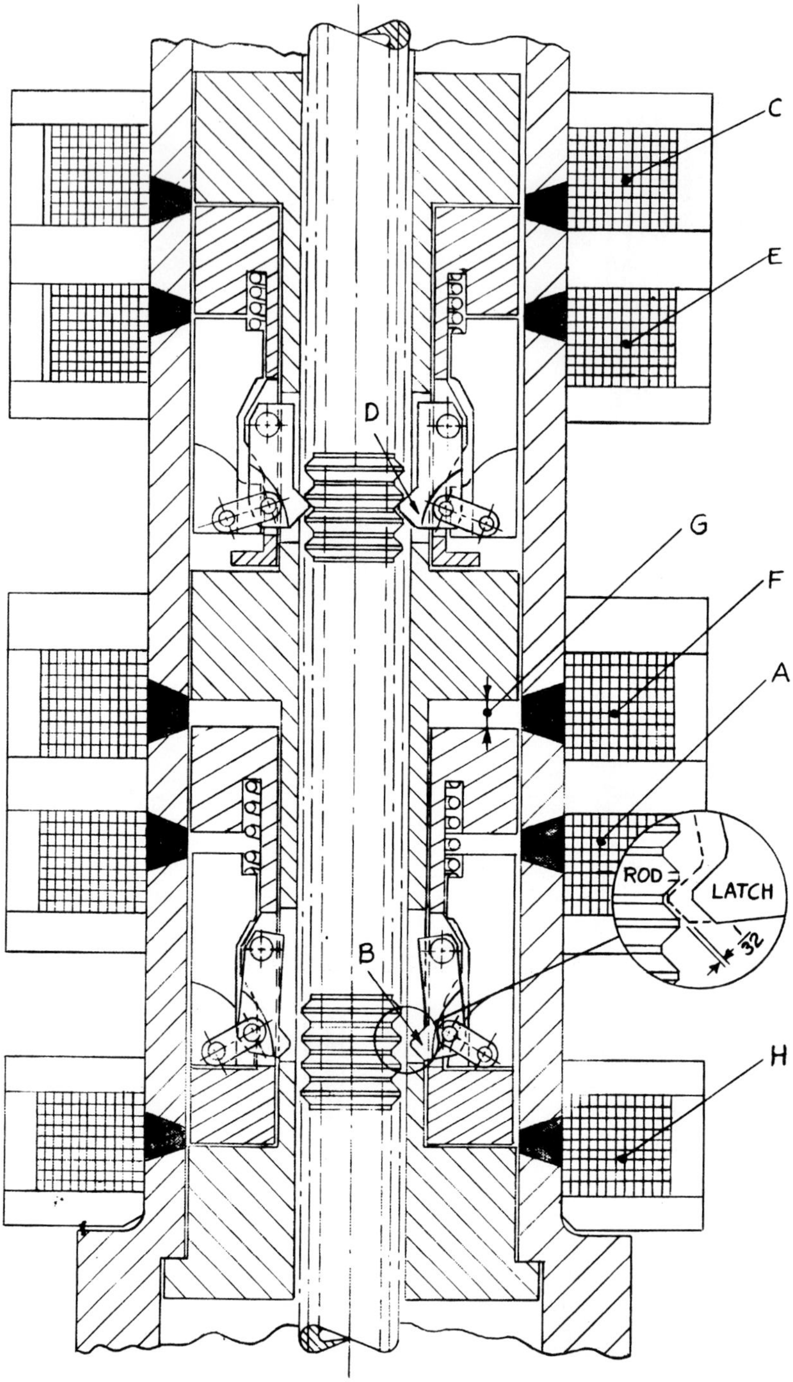

FIG. 5-7 Schematic diagram of latch-type magnetic jack.

C, is then de-energized, allowing the stationary gripper assembly and drive rod to drop until the 1/32-inch clearance is taken up at the movable gripper, B. The stationary gripper continues to drop an additional 1/32-inch to remove the load from the stationary gripper arm, D. The stationary gripper magnet, E, is then de-energized to permit the latch arm, D, to swing out of engagement. Next the lift coil, F, is energized to lift the movable gripper through one step, thereby closing the 3/8-inch (9.5 mm) gap, G. At this point the stationary gripper coil, E, is energized to move the latch arm, D, into a groove of the drive rod; the latch is not yet loaded because of the clearance provided. Coil C is then energized to lift the stationary gripper assembly into contact with the shaft tooth and then moves the drive rod an additional 1/32-inch to remove the load from the movable gripper tooth. The movable gripper coil, A, is then de-energized to let the latch arm B swing clear. The lift coil F is then de-energized and the pull-down coil, H, is energized to return the movable gripper to its initial position ready for the next cycle.

As in the friction-grip jack, scramming is accomplished by interrupting power to all coils. In the latch-time mechanism all tooth angles and linkages must be designed to avoid self-locking.

Magnetic jack mechanisms are employed in the Yankee, Piqua, and Saxton reactors [84]. The Yankee reactor uses the latch type. The Piqua and Saxton utilize the friction-grip type. The Piqua design is different from the other two in that it is completely contained in the pressure vessel with no protruding thimbles. The magnet coils are also inside the vessel. All parts operate in the reactor coolant, Santowax-R, which is at 550°F (200°C).

5.2.4 Cable and Drum Mechanism

Cable-and-drum mechanisms have been used at Argonne on PR-I and ZPR-VII and in England on the Calder Hall graphite-moderated carbon-dioxide-cooled reactor [56, 83, 89]. Cable-and-drum mechanisms are also to be used in the EGCR and the Florida West Coast Nuclear Power Plant (FWCNP) [84].

In this type of drive the control rod hangs from a cable which passes from the control rod upward through a series of pulleys to a motor driven drum. In ZPR, EGCR, and FWCNP, the drum is driven through a series of pulleys to a motor-driven drum. Hall design the rod is permanently coupled to the entire drive mechanism. When the clutch is used, scramming is accomplished by de-energizing the clutch. The ZPR and EGCR utilize gravity to drive the control rod into the core. In FWCNP a spring is used to supply additional acceleration at the start of scram. In the Calder Hall design, with no clutch, scramming is accomplished by de-energizing the three-phase holding windings of the motor to permit the control rod to fall under gravity. All but the Calder Hall design use separate buffers for deceleration of the rod at the end of the scram. The Calder Hall design utilizes breaking action within the drive motor. Fig. 5-8 [90] illustrates the Calder Hall design.

In cable and drum mechanisms, position information is usually taken from the drum shaft and not from the rod. In all such designs there is no way to apply any force to insert a rod that becomes stuck. Problems can also arise from the cable becoming slack. Special provisions are usually made to avoid most slack cable conditions by making sure that the rod never becomes seated unless stuck. To guard against the consequences of a cable failure a positive stop must be provided to prevent the control rod from falling out of the core.

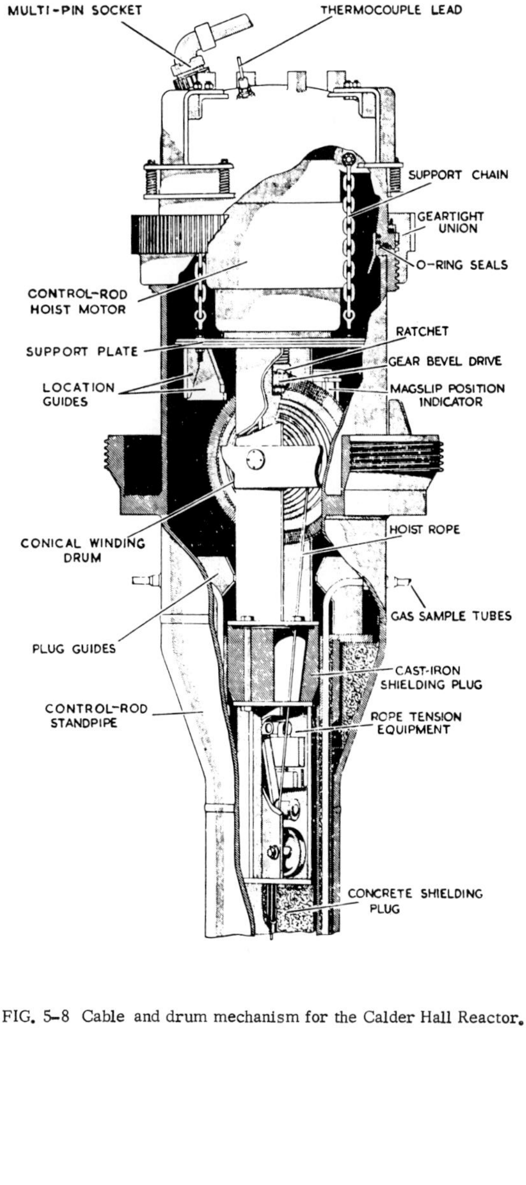

FIG. 5-8 Cable and drum mechanism for the Calder Hall Reactor.

5.2.5 Hydraulic Mechanism

In recent years there has been an increasing use of hydraulic mechanisms for driving control rods in nuclear reactors. Hydraulic mechanisms are being used in the Dresden, Humboldt Bay, Consumers Power Big Rock Point, Consolidated Edision Indian Point, and N. S. Savannah reactors [83]. The first three listed are entirely hydraulic; the Consolidated Edison and N. S. Savannah drives combine hydraulic and electromechanical operation.

The mechanism used for Consumers Power Big Rock Point will be used as the basis for describing a completely hydraulic mechanism [91]. The mechanisms for this plant, as well as for Dresden and Humboldt Bay, are located in protruding thimbles mounted on the bottom head of the pressure vessel; the control rods in all three of these plants are designed to scram upward into the core.

The Big Rock Point drive is shown in Fig. 5-9 [91]. It consists basically of a piston operating in a hydraulic cylinder and uses differential hydraulic pressure on the piston to drive the control rod into or out of the core. A ratchet is used to hold the drive in given positions. The ratchet must be unlocked for further withdrawal of the control rod from any position but need not be unlocked for rod insertion.

As shown in Fig. 5-9 the drive piston operates in an annular space between the inner and outer cylinders. The piston rod extends upward from the piston and, on the outer periphery, contains uniformly spaced grooves, approximately three inches (~ 7.6 cm) apart, which engage the ratchet-type collet lock. The collet lock has six locking fingers for this purpose. As indicated in the figure, the locking groove is shaped to open the fingers when inserting the control rod upwards. In order to withdraw the control rod these fingers must be retracted. This is done by hydraulically raising the collet piston to spread the fingers against an unlocking surface. The collet is held in its lower position by the combination of gravity and the collet return spring.

In the center of the piston rod is a stationary inner cylinder or column attached to the bottom flange of the drive. Water is brought to and from the upper side of the drive piston through this column. A series of small orifices at the top of the column provides a progressive water shutoff to cushion the piston at the upper end of the scram stroke. This column has a piston at the top to seal the upper side of the drive piston from reactor pressure. This column piston also provides a positive stop and uses a series of spring washers to cushion final impact.

Inside the column is a well containing the position indicator probe. The interior of the well is at atmospheric pressure and contains a series of hermetically sealed, magnetically operated switches, each of which indicates a discrete rod position. The switches are operated by a permanent magnet carried by the drive piston. The intervening walls are of nonmagnetic material. Extra switches are provided at each end of the stroke to indicate limits.

The Consolidated Edison drive mechanism operates on a push rod which is attached to the control rods and which extend outside the reactor vessel through sliding seals. The control rods are driven from the bottom of the vessel but are pushed upward out of the core for withdrawal rather than downward so that the pressure in the vessel aids in scramming. The drive mechanism utilizes a hydraulic piston and cylinder to move the push rod collar against a motor-driven stop. This hydraulic force is in opposition to the vessel pressure. Thus to withdraw the rod, the stop must be moved upward toward the reactor; the hydraulic cylinder force causes the push rod to follow. To insert the rod, the stop is driven downward to force the push rod downward. To scram the rod, the pressure in the cylinder is relieved and the vessel pressure drives the control rod to its seat.

The original rod drive system provided for the N. S. Savannah is similar in principle to the foregoing design. In detail, however, a number of differences exist. The control rods are operated through the top head of the pressure vessel rather than through the bottom. The pressure in the vessel tends to withdraw the control rods upward out of the core rather than downward into the core. Scramming is accomplished against vessel pressure by the hydraulic cylinder. A latch is provided to hold the control rod in its seat after a scram until the motor-driven stop follows down. Some consideration has been given to replacing the original rod-drive mechanisms on the N. S. Savannah with mechanisms of a nut and lead-screw design.

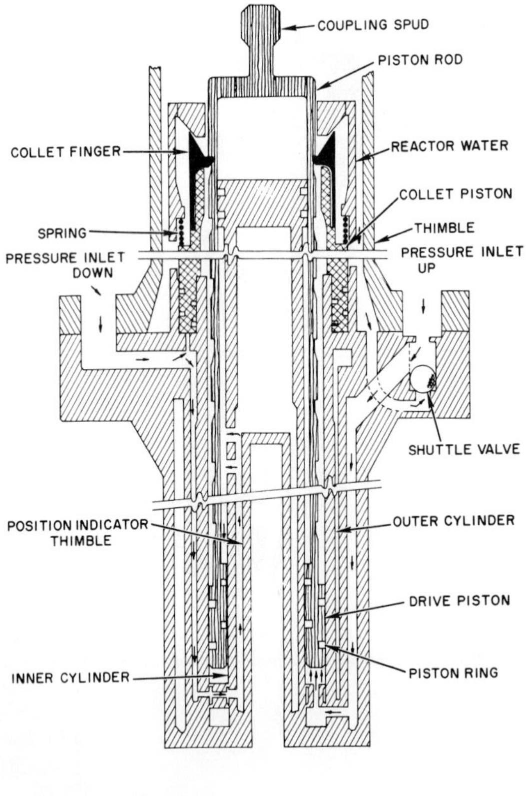

FIG. 5-9 Control rod drive mechanism for Consumers Power Co. Big Rock Point Reactor.

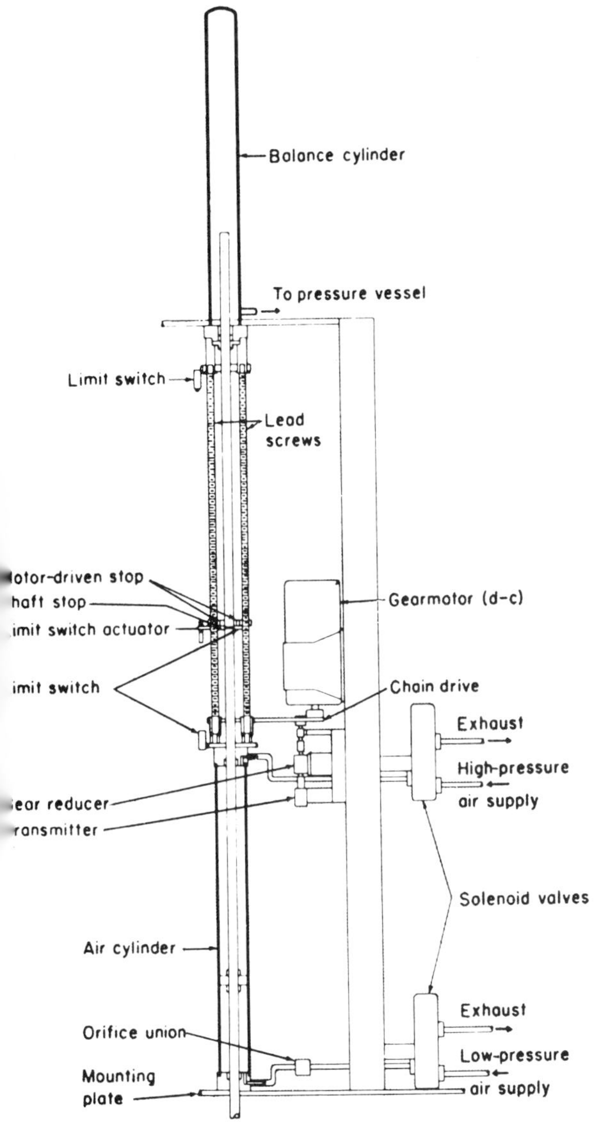

FIG. 5-10 Vallecitos reactor pneumatic control rod drive.

A variation of the hydraulic mechanism was used for the Vallecitos reactor. A double-acting pneumatic piston and assembly cylinder operated by air are used to push the rods up and against the motor-driven stop which travels on lead screws. Fig. 5-10 [83] illustrates the drive used for the Vallecitos reactor.

5.2.6 Harmonic Motion Mechanism

The harmonic motion mechanism is an interesting device for transmitting motion in a control rod drive mechanism [83, 92]. It transmits rotary motion through a flexible wall. As originally developed by United Shoe Machinery Corp., it consists of a circular spline with gear teeth on the inner diameter, a flexible spline with power gear teeth on its outer diameter and a rotating elliptical disk internal to the flexible spline. The elliptical disk is sized to permit engagement of the gear teeth of the splines at the two extremities of the major axis. As the disk rotates it advances the two opposite points at which tooth engagement takes place. Since the flexible spline has fewer teeth than the circular spline, a relative motion is generated between the two splines.

For application to a reactor the relationship of parts would be inverted. The circular spline would become the innermost member, the flexible spline would be the wall of the enclosure and the elliptical component the outer member. A helical gear interior to the mechanism could be used to provide linear motion.

5.3 Positioning Requirements and Position Indication

As indicated in Sec. 5.1.1, it is important to have control rods positioned within prescribed limits because of the importance that rod positions have in determining criticality effects on the reactor and on the ability of the rods to shut down the core. The accuracy with which a control rod must be positioned is dependent upon its worth and upon the accuracy with which reactivity effects involving rod changes must be known. This is a function of the number of control rods in the core, geometry and number of rods, their relative locations, the type of reactor and its use, and extent of withdrawal. Typical requirements for positioning and position indication are set forth in Sec. 5.1.

While it is desirable to know the position of control rods as accurately as they can be set, it is essential to know the positions at least approximately. Accurate positioning information may be attainable at times only at prescribed positions along the path length of the rod motion. These positions become check points. While the exact position of a control rod might be known to rather loose tolerance (e.g., x 1/2 in. or x 1 cm), the incremental motion from one position to another can often be measured with much tighter tolerances. Provisions for indicating when a rod is fully inserted or fully withdrawn should always be made. It must be remembered that fully-inserted rods cannot assist in scramming a reactor; fully-withdrawn rods are not very effective during the early part of a scram. For maximum reliability, the position indication of a rod should be independent of the control drive mechanism if the rod is designed to be separable from the mechanism; in cases such as this, means should be provided to indicate when the rod and mechanism are engaged and when the rod is fully inserted.

In nonpressurized systems, position indication presents few problems. In pressurized systems, however, information for position indication has to be transferred across a pressure thimble wall into a series of separate coils. Several examples of such devices have been described in the preceding section. In most of these designs the position indication is obtained from the drive mechanisms. This method of indication is satisfactory in either pressurized or nonpressurized systems provided that the rod is not separated from the mechanism during scram and provided that the system is in good working order.

In pressurized systems the position indication information is in the form of electrical signals. For example, in the Shippingport design, the inductance charge produced by a magnetic slug near the top of the mechanism is used to indicate the position of the rod. Bridge circuits are usually used to measure the charge in inductance caused by the magnetic slug.

5.4 Problems of Rod Insertion

The problems of rod insertion can be grouped under three general headings; 1) providing emergency power for insertion, 2) assuring a clear and free passage for insertion of the rod, 3) keeping the rod seated. Some of these problems were discussed in Sec. 4.7.1 in connection with core internals. The discussion here will be concerned primarily with the mechanism and its relationship to the control rods and core internals.

In order to function as a safety system, the control rod drive mechanisms must be able to move the control rods into the core during any emergency including loss of power. As a consequence, all control rod drive systems must contain some form of energy storage. The most readily available and most reliable means for energy storage is gravity. It is widely used even when other storage devices are provided. As a matter of fact one should examine critically any control rod scram system that works against the force of gravity. Any devices that are used to augment the action of gravity should be so designed that their failure would not interefere with rod insertion by gravity. Nevertheless, while most mechanisms make use of gravity for scramming or rapid rod insertion, the Dresden Reactor and the Peach Bottom reactor do not. The latter has no scram, in the usual sense of the word, but uses instead a fast insertion accelerated from a pressurized accumulator.

The energy storage devices which can be used to supplement gravity include mechanical springs, flywheels, electrical storage batteries to drive d-c motors and hydraulic accumulators. The type of energy storage system selected for a particular application is dependent upon many factors including the nature of the control rod drive mechanism, the space available, and the method of operating rods, i.e., ganged versus individual rod motion. For scramming purposes springs and accumulators are most common.

Fundamental to the safety of the system is the question whether stored energy is supplied individually for each rod or whether a single storage system is used to serve many rods. It is evident that individual storage provides a higher degree of safety than ganged storage because a failure of a storage device in the first case would affect only one rod whereas a failure in the second case would affect an entire gang of rods. A system utilizing a single energy storage system for all of the control rods of a core would be totally unacceptable for this reason. This consideration becomes particularly important in hydraulic mechanisms where a single pressure accumulation might be proposed for scramming purposes. Bates [84] points out that even if individual accumulators were provided on a hydraulic system the entire system could become inoperable if the piping were placed so that a single accident could rupture all or a large percentage of it. This same observation might also be made of wires involved in an electrical storage system.

If springs are used to augment scrams, care must be taken to make sure that the springs are well contained so that they cannot interfere with gravity scram if whole or broken. Because of their potential for failure, springs shoud not be relied upon solely, for scramming purposes.

In any reactor design consideration must be given to assuring a clear and free passage for insertion of the control rod. Many of the factors that influence free passage of rods have been discussed in Sec. 4.7.1. These include the influences of clearances, hydraulic forces both parallel to and perpendicular to the direction of motion, grit, dirt, corrosion and wear products, and control rod distortion. Related to the question of a free and clear passage for the control rod is the requirement that under no circumstances should the control rod be able to move any of the fuel elements in the core unless such fuel is a part of the control system.

Of more direct concern to the influence of the control drive mechanism on free passage of a control rod is the influence of seals between the mechanisms and the rods. Care must be exercised to make sure that such seals do not impose undue friction on the system or, in the case of seals operating in sodium vapor atmospheres, they do not freeze up because of improper temperature control in the region of the seal. Attention must also be given to the influence of broken parts within the mechanism on the freedom of the control rods to scram as well as the action of close clearances which interfere with free passage of coolant around mechanism parts; such interference could cause vapor binding. The influence of pressure on the unbalance of forces and on accumulation of gases near the top of a mechanism must also be considered.

One might readily assume that once he has provided for successful scramming of a control rod, the mechanical designer has fulfilled his function with regard to control rod insertion. However, such is not the case; he must take steps to make sure that the control rods will stay in position. This requires that the control rods receive no unbalanced force, resulting from any action, that would inadvertently cause the rods to be withdrawn, ejected from or dropped out of the core. Consideration must be given to a wide range of possible actions that might produce such effects. If rods have extensions passing out of a pressure vessel, rod ejection could result from loss of counterbalancing forces. If the flow is upward the rods might be lifted under conditions of high flow. During refueling, control rods could accidently be lifted unless special precautions against such action have been taken. (See SL-1 incident, Sec. 8.1, and Chapter 11, Sec. 3.11.) For shipboard operation, special features may have to be provided to prevent accidental withdrawal of control rods in the event of heavy seas or roll-over.

While no one would argue that a control rod should never be permitted to drop out of the core, mechanisms have been designed and built that would permit such action under special conditions. To avoid this action a positive bottom or seat should be provided for each control rod. One should not rely on a stop in the drive mechansm for this purpose.

5.5 Thermal Problems

The temperature at which a mechanism is to be operated can have a great influence upon its design and its reliability over its lifetime. In designs involving the use of electrical insulation in close proximity to the pressure wall of a mechanism, means must be provided for maintaining the insulation temperature at an acceptable level. This requirement usually provides the most stringent limit on the operating temperature of a mechanism. During short-time latching or emergency conditions the insulation temperature may be permitted to climb above specified limits; however, the integrated life at operating temperature should match the required life with a suitable margin. In making lifetime evaluations of insulation, attention must be given to the hot spot in the windings.

Temperature of the mechanism must also be controlled so that it is well below the saturation temperature of the water if the mechanism is to operate in the water. If the temperature is too close to saturation, vapor formation can take place when rapid motion is demanded of some component of the mechanism. Such vapor could lead to vapor binding or to slowdown of the desired action. Cavitation could also produce severe pitting of parts. Generally the mechanism temperature is held several hundred degrees (200-300°F or ~100 to 200°C) below the coolant temperature in the reactor vessel of pressurized water reactors. To assist in maintaining this temperature differential, thermal barriers are incorporated into the region between the vessel and the mechanism. Such barriers minimize water circulation between the two regions and reduce the heat load that must be normally handled by the mechanism cooling system.

This temperature difference between the water in such a mechanism and the water in the vessel leads to problems even though a thermal barrier is installed. In addition to the steady thermal stress and differential expansions brought about by the temperature difference, there are problems of thermal shock on the vessel attachment every time there is a scram because of the interchange of high temperature and low temperature coolant between thimble and vessel. While this type of problem can be overcome by use of thermal sleeves as outlined in Sec. 3.6, inattention to this need can lead to severe difficulties.

In order to maintain the temperature of a mechanism at the desired level it must be cooled either by appropriate cooling coils or air flow. The heat to be carried away may include both that leaking into the mechanism from the surroundings and that generated within the drive motor. As indicated in Sec. 5.1.1, although the power required to raise a control rod at the required speed may be small, the power input may have to be many times this value because of electrical and/or mechanical inefficiencies and because of frictional drag imposed on the rod. Furthermore, many designs require continuous holding power which has to be dissipated as heat.

5.6 Problems of Gas Accumulation

Although some reference has been made above to the problems brought about by the formation of vapor during operation of a water-cooled system, there are two other gas problems that may have to be faced in the design of control drive mechanisms. The first is that faced in boiling water designs where vapor at the top of the reactor complicates the use of mechanisms on top of the head because of insufficient lubrication of parts; lubrication must be accomplished by pumping pressurized water into critical areas and allowing it to leak into the vessel. The problem can be overcome by locating the mechanisms at the bottom of the vessel where water is available. In such designs attention must be given to utilizing gravity to advantage and to providing definite stops to prevent rods from falling out of the core.

A second problem arose out of having the mechanism buffer operate in the top of the thimble of the Shippingport design. This buffer consists of a piston on the top end of the scram shaft extension that slows down the system by passing into a region of close clearance. The design, however, depends upon the presence of water in the system. But when the system was started up, after a depressurized shutdown, the gas in the system accumulated in the top of the thimble and interfered with the buffer action particularly at low levels of pressurization. As the pressure was increased, the gas was compressed and eventually dissolved so that buffing action could take place. This problem was eventually overcome by providing a venting system at the top of the mechanism tubes. However, the problem points out that gas accumulation can take place and provisions must be made to cope with it. Such gas, if it includes large amounts of oxygen, can contribute to acceleration of corrosion.

5.7 Experiences with Control Rod Drive Mechanisms

Some of the experiences encountered in the designs, testing, and operation of particular control rod drive mechanisms have been discussed in the foregoing parts of Sec. 5. These, as well as other difficulties, serve to indicate the importance of attention to detail in design and the importance of constant surveillance during operation in uncovering difficulties that could lead to hazardous consequences.

Speaking of need for a "bottom" to prevent control rods from falling out of a core, Bates [84] describes two incidents reflecting the subtleties of this rather obvious point.

"The bottom may be inadvertently omitted in many ways. As an example, in the preliminary design of an experimental reactor some years ago, the absorber rods were required to be flexible. One way of providing the necessary flexibility was to arrange the absorber in the form of small cylinders slipped over a central wire support. The whole assembly was suspended from above and arranged to move up and down in a tube through which cooling gas flowed. Not until the experimental assembly was actually under construction was it realized that the cooling-gas tube extended below the reactor core far enough before it changed direction to permit the absorber to fall completely through the core if the support wire failed. The actual loss of a rod in such an instance may appear to be only a nuisance, since its effects would be controllable by moving the remaining rods. The real difficulty becomes apparent, however, when it is observed that the lost rod may have acquired sufficient velocity by the time it is leaving the core to add reactivity at a rate in excess of that which the remaining rods are capable of handling.

"More recently an incident actually occurred in a power reactor which illustrates the validity of the principle. In this reactor the absorber rods are withdrawn downward out of the core by means of a mechanism capable of exerting substantial forces. One of the rods became disconnected from its operating mechanism and failed to follow when withdrawal was attempted. Later, however, while the mechanism was still withdrawn, the rod for some reason freed itself and fell a short distance; in doing so, it effected a step addition of reactivity. Fortunately the result this time was not serious, but it brought up the question of the wisdom of knowingly building into a reactor the potentialities for generating incidents."

The central control rod of the SL-1 reactor could well have experienced such an accident if it had not first been subjected to another more severe one. The details are discussed in a footnote on p. 663 of Vol. 1, and on p. 677, comment (2) in the left-hand column.

Other difficulties with control rods and control rod drives have been reported by S. H. Hanauer in reference [93]. Many of the incidents he reports are concerned with control rods rather than control rod drive mechanisms and have been discussed in Sec. 4.11 of this chapter. A number of the failures reported in that section, however, have to do with the situation in which the poison member becomes detached from its drive. Such failures are worth reemphasizing because they could lead to reactivity excursions having important consequences. For example, Hanauer points out, that in the case of the Dresden reactor, where the control members scram against gravity, if the control element were to become stuck within the core, the control drive could be withdrawn without removing the poison. If this situation were not detected and corrective action taken, the stuck rod might fall out of the core and cause an unexpected increase in reactivity. If only one rod were involved at Dresden the action would not lead to a severe hazard. The situation could be different, however, if a number of rods were simultaneously involved or if the same design were used in another reactor with fewer rods.

Numerous instances of sticking rods have been reported from time to time on a number of reactors. Some of these have been due to binding of the rods themselves, others have been due to seals, and others to unknown causes. In the reactor at the University of Michigan, sticking was found to be due to swelling of one of the shim safety rods within a fuel element. The swelling was attributed to radiolytic dissociation of water that leaked into the rod containing boron carbide powder. New rods were installed but swelling was again encountered before the problem was solved. Although this and other incidents of this type produced no accidents, they were potentially dangerous.

The SL-1 reactor had numerous instances of sticking rods as is reported in detail on pp. 666-668 of Vol. 1. Some of the rods had to be assisted manually during withdrawal; others had to be driven into the core by their motors. Part of the trouble appeared to be friction in the seals through which the drive shaft penetrated the rack-and-pinion housing. Trouble was also encountered by the compression of control shrouds by expanding boron strips, as discussed on p. 665 of Vol. 1. In fact, the SL-1 reactor control rod design, construction, and performance involved problems of almost every variety discussed in this chapter.

Hanauer [93] reports an incident on the General Electric Test Reactor (GETR) which occurred on August 27, 1960, when a scram was caused by an instrument failure in the primary coolant differential pressure-monitoring system. He reports as follows:

"At this time the No. 2 control rod failed to scram and had to be inserted into the reactor by the manual "rundown" equipment. An inspection of the drive mechanism showed that the failure was in the scram latch device. The latch was manually released, and the rod fell. Subsequent inspection showed no visible defects, but lateral pressure applied by hand was found to produce frictional forces sufficient to prevent release. The scram shaft guide bushings were lubricated with high-quality, low-oxidizing grease to correct this condition. The remaining five rods were inspected and lubricated, and tests of performance were made. Since no further difficulties were experienced, operation of the reactor was resumed. Bushing lubrication has been included in the preventive maintenance program, and a program for routine disassembly and overhaul of the drives, on a one-at-a-time basis, has been initiated."

Another experience of related nature is summarized by J. Foster [94] from work by L. C. Oakes [95]. The following excerpt regarding the ball-latch device on the Oak Ridge Research Reactor (ORR) is taken from reference [94].

"There have been instances in which the release of an ORR control rod has been delayed by more than 100 msec following reduction of magnet current to the ball-latch mechanism. There have also been occasions on which the control rods have dropped when no reduction in magnet current took place. In an effort to explain these events, parallel analytical and test programs were carried out. It has been shown analytically and confirmed experimentally that mechanical friction between the balls and plunger surface in the latch can more than offset the force of the release spring and delay release of the control rod following current reduction to the magnet."

L. I. Cobb (191) has summarized the reported malfunctions encountered in the operation of various types of control-rod drive mechanisms at United States AEC-licensed facilities. He groups them under four headings: stuck rods and delayed

TABLE 5-2
Stuck rods and delayed scrams [191]

Description of problem	Cause
1. A rod would not scram from the 98% withdrawn position, but it would scram after the operator drove it to the 98% withdrawn position.	Steel brick piled around the drive shaft caught on some weld spatter and prevented the shaft from turning.
2. To terminate a critical experiment, the operator scrammed the rods. One rod would not scram until he drove it 3% of the way into the core.	A cap on the upper end of the rod hung up on weld spatter and on an unconsumed welding ring inside the rod travel housing.
3. On two different occasions during precritical testing of the scram function, one rod hesitated about 1 or 2 sec before starting to drop into the core.	About 0.01 in. of overlap existed between a rod guide block and the strap against which the block rubbed when the rod moved. Normal vibration of the rod due to coolant flow produced a shallow groove in the strap. When the instrumentation signaled a scram, the guide block remained caught in the groove until the vibration of the rods striking the dashpots set up sufficient force to free the block. Then the rod fell freely.
4. While checking out the reactor before a planned period of operation, the licensee discovered one rod that failed to scram two times out of a series of rod-drop checks.	The upper limit switch did not stop rod motion and occasionally allowed the drive to strike the mechanical stop with sufficient force to cock the magnet carriage. The misallgned carriage caused the rod to bind and prevented free fall when the magnet cureent was turned off by a scram signal.
5. Following a scram signal during reactor operation, one rod stuck about halfway out of the core.	Lack of lubricant on two bearings created enough resistance to finally stop the rod before it fell fully into position.
6. The reactor operator noticed he could move one of the rods without immediately affecting the power level. He scrammed the reactor so that the operability of the rod could be checked. The rod scrammed into the core very slowly.	A teaspoon of metal chips had entered the space between the poison section and the guide tube. The licensee postulated that the chips entered the reactor vessel when a piece of tubing was ground during installation of an experiment.
7. Following a scram from full-power operation, one rod stuck two-thirds of the way out of the core.	A stainless-steel screw had fallen out of the follower section of the rod and had lodged between the rod and the guide tube. The screws in the follower section had only one stake to prevent rotation.
8. Following a planned shutdown of the reactor, the licensee observed that one rod was stuck at the fully withdrawn position.	The licensee had repaired a cam on the rod, and, as a result of the repair, a weldment protruded 40 to 50 mils. This protrusion hung up on a slightly bent ear (used to reposition the guide tube) of the guide tube.
9. During a startup of the reactor, one rod stuck in the core and another moved erratically as it was withdrawn from the core.	Forces of 3.5 g combined with a vibration of 10 cps loosened several screws in a component designed to support fuel assemblies in the core. Most of the loose screws ended up inside the drive mechanism for the two rods.
10. During an attempt to recover from a scram, the operator found he could not withdraw a rod more than a few inches and that he could not insert it each time he turned the control to the "insert" position.	The rod index tube was severely galled. Metal chips from an unknown source had caused the binding pressure.
11. While checking the ability of the rods to scram, the operator noted that one of the rods occasionally would not fall all the way into the core.	The licensee discovered a steel plug 1 in. in diameter and 1 in. in length in the shock tube for the rod. He was unable to find the origin of the plug.
12. During a test involving an unusually high coolant pressure, one rod failed to scram from its fully withdrawn position. It was driven in promptly, however, by the automatic fast-rundown system that backs up the scram system.	A faulty seal allowed sodium to enter the region of the delatching mechanism. Because of the low temperature in the region, the sodium froze and prevented the delatching action required by the scram signal.

TABLE 5-3
Problems with latching and control mechanisms [191]

Description of problem	Cause
1. One rod was selected and given a withdrawal signal. Instead of just one, two rods began moving out of the core.	The second rod that moved had just previously been connected to the selector circuit. When the selector control was moved to the new rod, the ball valve remained open and allowed hydraulic fluid to drive the two rods simultaneously. Misalignment during installation produced a binding condition on the valve.
2. During rod-drive testing, two drives moved out together when the signal required that only one move.	The selector valve for one rod was binding, and it operated very slowly. This kept it open for a short time after the operator turned the switch to "close". A small piece of O-ring was found lodged between the ball and valve body.
3. During operation at about 10% of licensed power, one rod started drifting out of the core following completion of an insert signal. The operator halted the drifting by signaling a rod "insert." The rod would not latch at the fully in position and started to drift out again. The operator scrammed the drive, and it latched properly and remained in the core.	A hard particle about 1/32 in. in diameter had lodged in the latching mechanism; it prevented proper operation.
4. During testing of rod drives, two were found which experienced latching problems.	Failure of two strainers in the reactor coolant demineralizer allowed resins to get into the coolant. A small amount was found in the latching assembly; this prevented the collet finger from engaging the drive index tube.

TABLE 5-4

Instrument problems of control rod systems [191]

Description of problem	Cause
1. During a test of the holding power of the rod magnets, the licensee observed that the magnets would not release the safety rods, although the rods should have scrammed because a condition simulating a power level above the trip point had been imposed on three neutron-flux monitors.	The safety circuit was found to possess a nonfail-safe characteristic. In order to shut off the current to the rod magnets, a negative voltage had to be supplied to the grid of the single vacuum tube that controlled the amount of current being supplied to the rod magnets. If for any reason the proper negative voltage was not supplied to the grid of this tube, the reactor could not be scrammed automatically. Following the discovery of the problem, an instrument technician thought that the grid was drawing current and thus could not maintain the required value of negative voltage. Replacement of the vacuum tube caused the safety circuit to function properly again.
2. During a check-out prior to resuming operation of the reactor, the operator discovered that a simulated high-power trip on one nuclear channel would not scram the reactor as required by the design.	All components of the circuit were operating properly except the spring-loaded push-button type reset switch. The switch must be held in position by spring tension to allow passage of a scram signal. Relaxation of the tension allowed the circuit to open and prevent a scram from a high-power trip as seen by this channel.
3. The reactor coolant low-pressure scram was removed from service for about three weeks during which the reactor was operated.	Instrument technicians incorrectly connected the low-pressure scram inputs to the safety system and thus removed the low-pressure scram feature.
4. To prevent a period scram during a neutron pulsing experiment, the licensee installed a bypass circuit for the period monitor. However, the circuit also inadvertently bypassed two scram trips from the two high-power-level recorders.	The period bypass circuit actually was in parallel with contacts in the safety circuit that would open if either power-monitoring channel reached the trip point.

TABLE 5-5

Materials problems of control rod drive systems [191]

Description of problem	Cause
1. During the inspection of rod components fabricated from 17-4-PH steel, longitudinal cracks were found in two drive shafts. Three cracks were found in one shaft, and two in the other. The longest crack was about 3 in.	Stress corrosion of the 17-4-PH steel that had been heat-treated at about 900°F caused the cracks.
2. As the drive shaft for a rod was being moved from the reactor to the storage pool, one of the fingers from the latching mechanism fell off. The finger was attached to the drive by a pin that was held in place by two small tack welds. Examination of the other rods revealed that half of them had pins with cracked tack welds.	The weld area was too small to support the load placed on it.

scrams; problems with latching and control mechanisms; instrument circuit problems; and materials problems. Tables 5-2, 5-3, 5-4 and 5-5 present a condensed version of the problems encountered in each of these areas.

It should be noted that many of the difficulties encountered in service developed gradually and were detected only by alert surveillance. These experiences show that, even with all the precautions taken in design, continual inspection maintenance is required to assure reliable operation. Feedback from the field to the designer is essential to assure constant improvement of control drive mechanisms.

6 MAIN HEAT REMOVAL SYSTEM

As stated in Sec. 1, the reactor in a nuclear power plant must be viewed as an integral part of the system. Its integrity is influenced by, and in turn influences, the rest of the plant. It is exceedingly important to recognize the interaction between the reactor and the rest of the plant.

Many of these interactions have been implied in the discussions in the previous sections of this chapter. It is the purpose of this section to highlight the principal features of the heat removal systems, outside the reactor vessel, which influence reactor safety. As much could be written about the mechanical design of each of these components in this system as was written about the reactor. However, such treatment would be beyond the intended scope of this book. Emphasis will be placed instead upon the special requirements arising from the fact that these components are in a nuclear power plant. The reader is referred to other sources for treatment of the more standard aspects encountered with these components.

6.1 General

The heat removal system, as used here, includes the major systems and components directly involved in the removal of nuclear heat from the reactor. The auxiliary systems, which are more indirectly involved in heat removal, are discussed in the next section. The heat-removal system consists of the main coolant system with its associated piping, pumps or blowers, valves, heat exchangers, and steam generators; the pressurizer and expansion system; and the secondary steam system.

In the main coolant system, coolant is circulated through the reactor, usually by a pump or blower depending upon the type of coolant, where it picks up heat generated in the core. The coolant then passes either directly to a turbine in the case of certain boiling water or gas-cooled plants, or through tubes in the heat exchanger portion of a steam generator where heat is transferred to a secondary steam system which operates a turbine. In liquid-metal systems the secondary loop may be used as a buffer loop which transfers heat to a tertiary loop which contains the turbine system. In all cases the coolant loses the desired heat and returns to the reactor to complete the cycle. Unless otherwise stated, the discussion that follows will be based upon the use of a heat exchanger between the main coolant system and the secondary system.

The dual-cycle boiling water reactor plant such as used at Dresden, is a combination of the direct and indirect cycles. The dual cycle provides a substantial variation in the amount of subcooling furnished to the reactor and, as such, provides the operator with an extra control device. He can make a considerable change in power level, for example, merely by adjusting the ratio of primary to secondary steam without resorting to control rod movement.

The main coolant system in a nuclear power plant generally involves two or more piping loops connected to a single reactor vessel. Each loop is generally complete in that it contains its own pump, valves and heat exchanger. In many of the plants each of the loops can be isolated from the rest of the plant if need be. In some of the newer plants, however, isolation valves have been eliminated from the primary loops. The use of multiple loops which can be isolated from each other provides an important safety feature to the plant in that one or more loops can remain in operation to cool the reactor even if trouble is encountered in another. Thus they provide "installed spares". In addition they permit the use of smaller, more readily available components than a single loop and ease the problems of plant layout.

Since containment of the coolant is of prime importance to reactor safety, the pressure envelope of all components of the main coolant system should meet the same requirements for system integrity and the same or similar conditions of service as identified for the reactor vessel. Whenever applicable, the pressure-containing components should be designed and fabricated to meet the requirements of the ASME Boiler and Pressure Vessel Code of the ASA Code for Pressure Piping. The main coolant system, as well as the high temperature-rise pressure portions of the auxiliary systems are located within the plant container.

In laying out the main coolant system many factors must be considered. While most of these are usually concerned with the peculiarities of the particular plant involved, there are several general considerations worthy of note. In preparing a nuclear power plant layout consideration must be given to:

1. Differential expansion of components and piping, and the loads and moments caused by such expansion. Careful consideration must be given to the nature and location of anchor points for components.
2. Accessibility for inspection, maintenance, and repair of replacement of components.
3. Vibration sources and mountings to avoid transmission of vibrations.
4. Damage possibilities during construction. Small pipes, fittings or electrical connections should be avoided as much as possible or should be carefully protected to prevent inadvertent damage during constructions.
5. The possibility of progressive damage in the event that one small component fails during service. The layout should attempt to minimize or protect against such possibilities.
6. Radiation effects on gaskets, insulation or other nonmetallic parts in close proximity to the reactor vessel.
7. Means for controlling the spread of radioactivity and providing for cleanup in the event of minor leaks.
8. The problems of making and inspecting field welds on piping and components and controlling the use of welding filler materials.
9. The problems of keeping debris out of the system during construction and cleaning the system upon completion of the plant.
10. Fire protection and fire-fighting means.
11. Accessibility to and provisions for refueling, including provisions for flooding of certain components and protecting other components against flooding where wet refueling is to be done.

6.2 Piping

When speaking of piping, one must include not only the large primary piping in a nuclear power plant but also the myriads of small pipes that are in the auxiliary systems attached to the primary loop. When considered in this light, piping looms as one of the most vulnerable parts of the plant so far as breach of the main coolant loop is concerned. Every piece of pipe in the system is worthy of close attention both during design and during fabrication. A small pipe poorly joined to a large pipe can not only be the source of damage to the small pipe but could also lead to more severe damage of the larger one.

H. Thielsch [96] classifies service failures in piping systems under one or more of the following headings:

1. Design
2. Materials selection
3. Manufacturing and shaping of materials by the mill
4. Final fabrication and welding of the pipe
5. Excessively severe service conditions.

While failures can arise out of a combination of several of these causes, this listing can serve as a basis for discussing the precautions that should be taken with piping in a nuclear power plant. The considerations involved in identifying excessively severe service conditions, however, are discussed under design rather than treated separately.

6.2.1 Piping Design

Piping in a nuclear power plant should be designed, fabricated and welded to the requirements of the ASME Boiler and Pressure Vessel Code and/or the ASA Code for Pressure Piping. The materials specified should meet, in general, the requirements of ASTM and such other supplementary test requirements as deemed necessary by the designer.

In applying these codes and preparing the piping layout, one of the most important considerations becomes that of providing sufficient flexibility in the piping system to absorb the motions due to thermal expansion imposed on the piping during operation without exceeding acceptable stress limits in the piping or in the component to which it is attached. Careful consideration must be given to the size, location, and use of expansion loops and/or expansion joints and the interaction of these systems with each other and with the piping supports. Although formulas are available in handbooks and from suppliers for analyzing the action of individual expansion loops or expansion joints, complex loops may require the use of model studies to obtain piping reactions necessary for calculating the stresses which must be known. When expansion joints are employed, care must be taken to make sure that they are used correctly; expansion joints should not be used in situations leading to double reverse bends unless approved for this purpose by the manufacture of the joint.

The direction of expansion in a pipe run is dependent, of course, on the location of anchor points. Anchor points must be firmly fastened to the pipe and to a rigid part of the plant structure; if they are not, the device used for absorbing expansion cannot serve its function and severe stresses may be encountered elsewhere in the system. In addition to anchor points, piping will require supports to prevent its weight from adding excessive load on components or on the piping itself. The requirement for support becomes particularly important with large pipes. The location and number of supports will vary with the kind of piping and number and size of valves and fittings involved. In general, supports should be provided near changes of direction, near branch lines and near valves. The weight of piping should not be carried through a valve body because, aside from questions of stress, distortions produced by these loads may interfere with operation of the valve.

A good pipe support should have a strong, rigid base properly supported and should be designed to maintain alignment and permit axial motion of the pipe without excessive friction [97]. Figure 6-1 [97] illustrates acceptable methods for supporting pipes.

Equally important in the design of a piping system is the avoidance of notches such as sharp corners, abrupt reinforcements, poor weld forms, or attachments which cause untenable localized stresses. Attention must also be given to thermal fatigue or mechanical shock. Brittle fracture, as discussed in Sec. 2.6.3, must also be considered in piping design. The influence of radiation on brittle fracture characteristics of piping can also be important in some designs.

Of particular concern in piping systems involving nongaseous coolants is the problem of water hammer, generally caused by rapid closing of valves. Careful attention to the action of valves is required to avoid this problem. Some of the conditions leading to excessive trouble with water hammer are analyzed by G. R. Rich in reference [98]. This subject is treated more fully in Chapter 15, "Fluid Flow".

Piping systems should be designed to avoid crud traps which could interfere with flow or which could cause build up of radiation levels in excess of those for which shielding was provided. Piping should also be designed to permit fabrication and inspection in the field. Particular attention should be given to weld-joint design as discussed in Sec. 2. The reader is urged to refer to the entire Secs. 2 and 3 on reactor vessels and closures because most of the details presented in those chapters apply to piping as well as vessels.

6.2.2 Materials Selection

The designer of piping for nuclear power plants faces many of the same problems in the selection

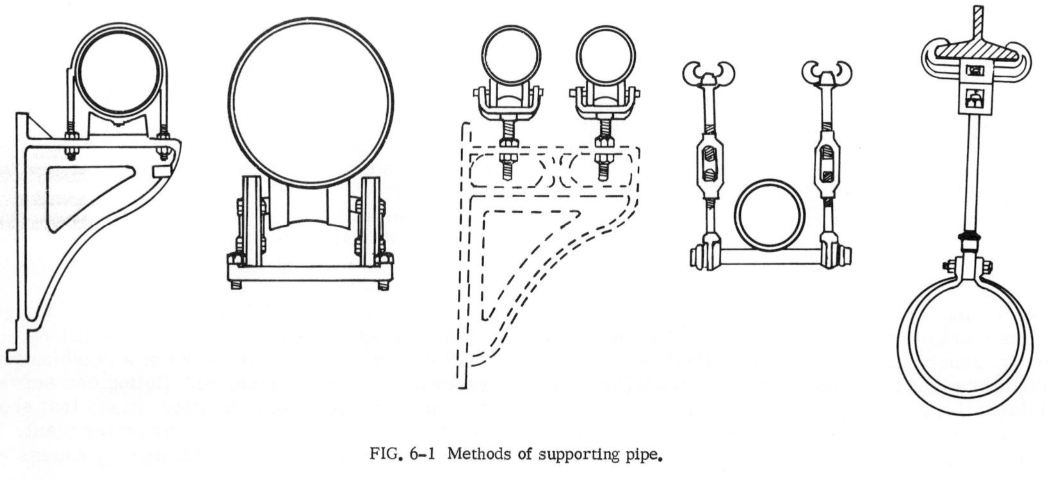

FIG. 6-1 Methods of supporting pipe.

of piping materials as does the reactor vessel designer. The chief problems arise out of the need for compatability with the coolant chemistry to minimize corrosion and the need for adequate strength and fatigue and stress-rupture properties at the temperatures at which the piping is to operate. Thielsch [96] points out that care should be exercised in selecting materials on the basis of accelerated high temperature tests of small-scale laboratory specimens particularly tests for data on creep, stress-rupture, fatigue and corrosion. Failures have been encountered with thick-walled piping made of material which passed such tests. The most widely publicized are the failures of carbon-molybdenum steel piping at temperatures over 800°F because of graphitization adjacent to weld deposits in the heat-affected zone [96]. Likewise, failures have occurred in some type-347 stainless steels at temperatures over 1050°F (566°C) in the pipe base adjacent to weld deposits [96, 99].

Most nuclear power plants built to date have used stainless steel materials for piping. Some loops have been built with unclad carbon steel piping. Where carbon steel piping internally clad with a corrosion resistant material is to be used, consideration must be given to the conditions outlined in Sec. 2.10.4. In all cases careful consideration must be given to the selection of appropriate welding-filler materials. Carbon migration from weld deposits into the base metal can leave decarbonized zones in the weld deposits adjacent to the base metal and lead to crack initiation in service [96].

6.2.3 Manufacture and Shaping of Piping Materials

Defects in piping may have their origin from causes which arise during manufacture. They may begin as far back in the process as the casting of the ingot of material or may be introduced or produced during rolling, forming or extension operations later on. The defects encountered are classified by Thielsch [96] as either notch defects or metallurgical notches. For our purposes here, they will be characterized as mechanical or metallurgical defects. (See Sec. 9.1.2.)

Mechanical defects refer to actual discontinuities or separations in the metal whether they be on the surface or inside the piece. These include cracks, laminations or inclusions in the material; such faults become particularly undesirable when they are perpendicular or have a large component perpendicular to the pipe surface. Those parallel to the surface rarely lead to failures unless they substantially reduce the wall thickness [96]. Mechanical defects often have their origin in mechanical or thermal effects such as encountered in drawing or machining or by shrinkage during casting. Several pipe failures have been traced back to cracks started from a drawing-die groove on the inside of a tube.

Metallurgical defects often involve differences in properties of the material that can lead to failures. Such defects occur most frequently during welding of materials of slightly different composition. Metallurgical defects can also arise where local areas of high hardness appear next to areas of low hardness as a result of localized heat treatment, or where carburization or decarburization has taken place.

Careful attention to procedures is important to producing consistantly satisfactory pipe. New procedures and modified forming equipment should be carefully checked. Radiographic and surface inspection of large pipes should be done routinely.

6.2.4 Final Fabrication and Welding

Any fabrication process that is carelessly performed can lead to failure, but, at times, even those done carefully can be a source of trouble if the design is not well thought out. Although pipe fabrication includes many operations such as bending, extruding, swaging, cutting, welding and heat treating, by far the most common source of trouble is welding.

Many of the problems arising in the welding of pipes are due to the fact that they can be welded from the outside only. The problem here is that of avoiding defects in the first weld pass. In butt welds made by shielded-arc welding, backing rings are used to obtain a good first pass. Nevertheless improper penetration or slag in this pass have led to built-in notches at the inner surface of pipe points. Such effects are particularly encountered where a poor joint fit-up has been obtained. Notches and crevices formed in such cases can lead to stress-corrosion cracking.

As in the case of reactor vessels, arc strikes on piping can lead to cracking or even brittle fracture during service. Localized torch heating may be sufficient to disturb the metallurgical structure of the pipe; such heating may be encountered in flame cutting. It can cause hardness differences if uneven and may effect the corrosion resistance.

Careful consideration must be given to the need for post-weld stress relief. Thielsch [96] points out that under some conditions the various codes permit elimination of such heat-treatment. However, they should not be indiscriminantly eliminated. Failures have occurred, for example, due to stress corrosion in boiler tubes which were cold bent and not stress relieved.

6.2.5 Failure experience

Inasmuch as the major rupture of a primary pipe is usually considered the maximum credible accident in a nuclear power plant, considerable attention has been and is being given by the USAEC to the nature and cause of piping failures which have taken place in nonnuclear industrial piping systems. The USAEC has also been supporting an extensive amount of work on determining the conditions which must be met to prevent piping failures. This work is described in references [192], [193], [194], [195], and [196].

Gibbons and Hackney [192] report the results of an industrial pipe-failure survey which provided 399 case histories involving defective conditions that required repair or replacement. From this study the following qualitative conclusions were derived:

"Piping systems which have been designed and constructed within established code criteria will exhibit high reliability, and a catastrophic, complete severance rupture appears unlikely to occur. Complete ruptures have occurred in nonnuclear piping but with circumstances which are not typical of reactor systems. Less severe failures may be expected. The most probable failure would be cracking due to low-cycle fatigue resulting from the inability of the designer to predict environmental and service conditions on a local basis. The most likely location for failure to occur is in the pipe or pipe-weld heat-affected zone particularly in association with a structural discontinuity. The growth of cracking-type failures can generally be expected to result in leakage prior to gross or major failure. The detection of leakage associated with relatively small cracks should permit repair and prevention of further cracking and aid the study of design inadequacies which allowed cracking to occur. The importance of adequate monitoring and surveillance must therefore be emphasized. The type and amount of monitoring and surveillance conducted must be considered in terms of possible nuclear hazards and the ability of monitoring and surveillance procedures to detect incipient failures."

Kilsby [193] summarizes the range of parameters, conditions and criteria which have been identified for consideration in studying pipe ruptures. These were obtained by a survey of designers, builders, and operators of water-cooled reactor systems. Tagart [194] summarizes the critical known failure parameters which affect the reliability of nuclear pipe systems and discusses the inadequacies of design codes, indicating areas which need additional attention. Klepfer [195] presents recommendations for experimental and analytical work needed to answer the questions raised.

Brothers et al [196] present a review of fracture modes and the state of knowledge about them as related to piping for nuclear plants. The following important questions, which cannot be answered on the basis of present knowledge, are identified:

a. "Whether long, shallow, crack-like defects can develop undetected due to fatigue or localized corrosion, and whether such defects, if developed, could propagate to large-area fractures by unstable ductile rupture.

b. "Whether the time for serious fatigue or corrosion crack (or leak) propagation, in the specific materials of interest, is fast or slow compared to the inherent response time for flaw and leak detection procedures.

c. "Whether service experience thus far, and short-term experimental results showing freedom from localized corrosion fracture, can be taken to imply complete immunity from attack over a 10- to 40-year service exposure."

Thielsch [197] discusses the important part that defects play in the integrity of industrial piping. He points out the difficulty of identifying the conditions under which defects will propagate in piping. Some obvious and apparently serious defects have not propagated at all, even in long service at critically high temperatures and pressures. Other very minor or inconsequential-appearing defects have resulted in failure very rapidly under the same operating conditions.

Several instances of piping cracks have been encountered in nuclear power plants. Smith and Violette [198] report on the cause and repairs of cracking in the recirculation piping of the Vallecitos Boiling Water Reactor (VBWR). A 2 1/2 inch long crack was found in one of the pipes as well as a "crazed" pattern of cracking on the interior surface of the pipe. The cause was thought to be inadvertent restraint on the pipe which led to large strain-cycling during operations. Operating records showed that over 1000 cycles had occurred. (See Vol. 1, pp. 694 and 695.)

Prestele and Froshauer [199] report cracking in the No. 2 main stream lead of the Indian Point Nuclear Power Station:

"Close inspection revealed a circumferential crack about 4-in. long in a 24- by 20-in. reducing fitting in the inlet to the throttle valve. The unit was taken out of service on an emergency basis; and an attempt was made to repair the crack by grinding, rewelding, and stress relieving. X-rays of the repairs revealed additional cracks emanating from the repair, and the attempted repair was repeated unsuccessfully. Finally, a section of the wall, approximately 9 by 12 in., containing all the defects in the fitting was cut out. Internal inspection of the pipe upstream of the window showed a pair of additional cracks radiating from a pressure-impulse nozzle connection. These cracks were ground down to a depth of over 1/2 in. and were repair welded. The nozzle was replaced, and a replacement window was welded into the opening. Test plugs were taken from the other three leads, and the No. 1 lead also showed evidence of cracking. On a later outage this fitting was replaced, as was the No. 2 lead fitting. These failures occurred because the pressure-impulse nozzles just upstream of the fittings are manifolded to an 'average-pressure' transducer for central-control-room information. Under conditions of partial load, the pressure in the leads not yet flowing steam to the turbine is higher than that in the leads flowing steam by an amount equal to the pressure drop from the superheaters to the turbine. This pressure difference caused steam and condensate to flow through the pressure-impulse manifold from the inactive leads to the active leads. On entering the active leads, the steam and condensate splashed against the inside wall of the pipe and reducer, and, because of the large difference between saturation and superheated-steam temperature ($> 550°F$), a thermal stress of a cyclic nature was set up and eventually caused failure. The following steps were taken to prevent a recurrence of the failure: the manifold was disconnected, and a single lead was retained for pressure information."

6.3 Pumps and Blowers

Pumps and blowers can be made in a variety of ways. They can be multistage or single stage;

centrifugal axial, helical, positive displacement or electromagnetic; they can be canned or designed to operate with seals; they can be of single-speed or two-speed design or be capable of operations at any speed up to a prescribed maximum; they may be operated by electric or turbine drive. It is not the intent here to describe all the types and makes of pumps used for nuclear power plants but rather to highlight the features of the principal types which most affect the safety of the plant. In the following discussion the word pump will be used to mean both pumps and blowers except where noted.

6.3.1 Pump and Blower Types

For the purpose of this discussion, pumps and blowers will be grouped as canned pumps, pumps and blowers utilizing seals, and pumps for liquid metals. Each will be described briefly below.

6.3.1.1 Canned Pumps. The canned pumps were developed to provide zero-leakage pumps for use primarily with water-cooled nuclear reactors. The canned pump is an electrically-driven pump in which the pump and the rotor of an electric induction motor are encased in a pressure-tight housing. A thin corrosion-resistant shell or "can" is used to separate the rotor and stator of the motor; the shell withstands the internal operating pressure by being backed up by the stator punchings which are external to the can. The can is completely welded in place to the rest of the pressure-containing housing.

All the moving parts of the pump operate in water including the bearings; hence, only water is available as the lubricant. Bearings usually consist of carbon-graphite sleeves around a hardened stainless material such as stellite. A similar combination of materials is used on a thrust runner. These bearings have required extensive development and still require critical attention for reliable service. Primary cooling water also circulates between the rotor and the can. This circulation helps transmit heat to the cooling coils on the outside of the can. Figure 6-2 [39] shows a cutaway view of a canned rotor pump.

Canned pumps have been built for flows between 1 and 20000 gpm (1 gm = 63.09 cm^3 /min) and pressure heads between 10 and 160 psi (1 psi = 0.0703 kg/cm^2) [100]. Although some pumps have been built for variable-speed operation, most canned pumps being built are limited to two-speed operation. The canned pumps built to date have been primarily single-stage pumps. Multistage pumps, however, can be incorporated in this design.

Although canned pumps were developed primarily for use in water cooled reactor systems, they have been built for and operated in liquid-metal and fused-salt systems. (See Sec. 6.3.2.3.)

6.3.1.2 Pumps and Blowers Using Seals. Because of the lower costs associated with standard pumps and blowers and because of the extensive experience behind them, there is strong motivation to use standard pumps and blowers with suitable seals wherever they are applicable. Pumps of this type have been used successfully in D_2O and H_2O systems and in air systems [54b]. Seals which have been used can be classified as packing-type seals, mechanical seals, water seals, labyrinth seals, and viscosity seals. A single seal assembly may use more than one of these types in a particular application. However, despite the type used, a good seal design will utilize at least two seals in series with a space between them for drain-off of leakage. A still better arrangement involves the use of three seals with a space between each of them; in the space closer to the higher pressure region a clean buffer

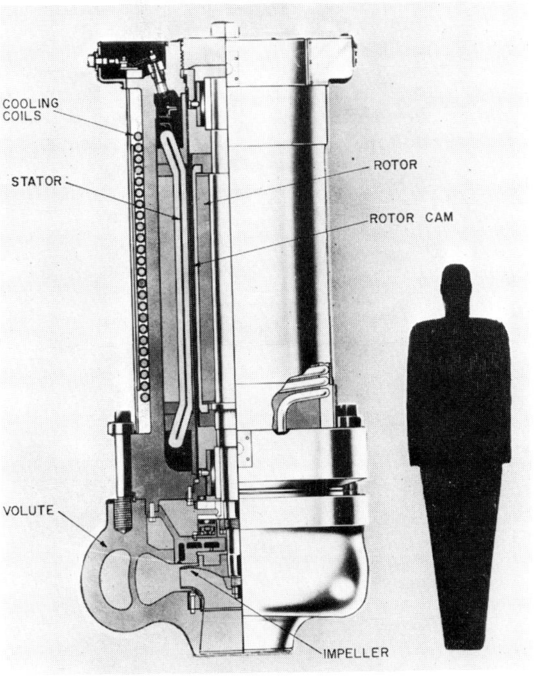

FIG. 6-2 Canned rotor pump.

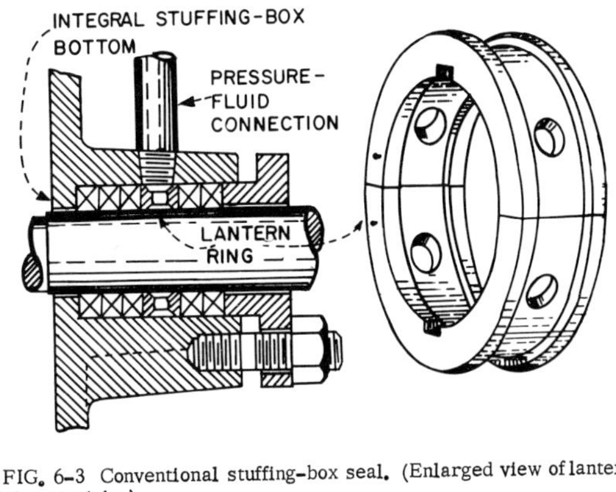

FIG. 6-3 Conventional stuffing-box seal. (Enlarged view of lantern ring at right.)

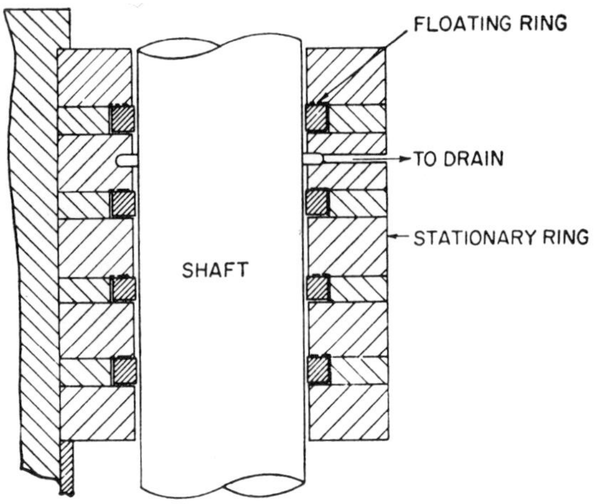

FIG. 6-4 Metallic ring seal.

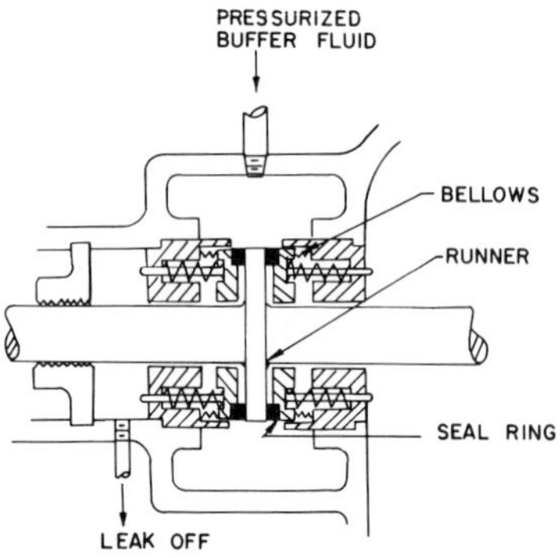

FIG. 6-5 Mechanical seal with plane face runner.

fluid, of the same composition as the fluid being sealed, is introduced at a pressure higher than the internal pump pressure so that the clean fluid leaks into the pump rather than having the contaminated pump fluid leak outward. Some of the clean fluid also leaks outward through the second seal and is collected in the space between the second and third seals.

Figure 6-3 taken from reference [101] shows the stuffing box section of a pump utilizing a lantern ring and conventional type packing. The lantern ring is used to provide the space between packing rings for introduction of the pressurized buffer fluid. The composition packing generally used with such seals is not appropriate for most nuclear applications because of the impurities it tends to introduce into the system and because of embrittlement of the packing by radiation.* Graphite rings or metallic rings separated by floating rings can be used instead as shown in Fig. 6-4 [54a].

The mechanical seal utilizes sealing surfaces located in a plane perpendicular to the shaft and a highly polished rotating disk against which the nose of a stationary ring of a dissimilar material is held in contact by springs. The surfaces form a seal with relatively small frictional losses if the materials are appropriately selected. A seal of this type was tested extensively for use on the Daniels Helium-cooled Pile; in this application a double seal acting as a single runner was planned with clean high pressure helium on the outside as a buffer fluid. A spring-loaded graphite nose ring was used to rub against a stainless steel disc; a bellows was used to seal between the nose ring and spring cage. (See Fig. 6-5.)

In blowers involving relatively low pressures, water-sealed packings are sometimes used to reduce gas leakage outward or, on the section side to prevent any leakage inward. Such a seal is shown in Fig. 6-6 [102]. This seal incorporates a water gland which utilizes the centrifugal action of the gland runner on the water to maintain the water seal between the runner and the casing.

Labyrinth seals are widely used in blowers to restrict leakage. They depend on the pressure drop across the labyrinth to restrict flow. Labyrinth seals can be used either with a smooth shaft or a stepped shaft; the latter provides greater pressure drop but is usually more costly. All labyrinth seals depend upon maintaining close clearances during operation in order to be effective. Rubbing of seals on the shaft will reduce their effectiveness. Labyrinth seals are particularly useful to reduce air leakage into the pump when used in combination with a leak-off connection at a pressure slightly lower than atmospheric. Such use of a labyrinth seal is shown in Fig. 6-5.

The viscosity-type seal can be similar to the mechanical seal with a plane-face runner except that, in place of a stationary nose-ring which rubs against the runner, it utilizes a plate which is separated from the runner by viscous action of the fluid. As in the case of the mechanical seal, a

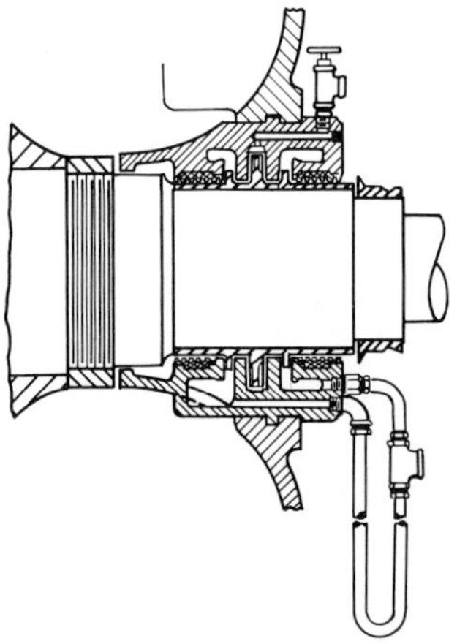

FIG. 6-6 Water-sealed shaft packing.

*In a number of reactor systems including water- and sodium-cooled systems, the gamma radiation levels can be high enough so that radiation damage to seals must be considered. In water systems the high-energy gamma rays from $O^{16}(n,p)n^{16}$ reaction with a seven-second half-life is the chief source.

buffer fluid is used. Another type of viscosity seal consists essentially of a sleeve bearing into which a buffer fluid is bled at the center and allowed to flow toward each end.

Regardless of which type of seal is used, provision must be made for inspection and maintenance. When properly designed, installed and maintained, seals can give good service; improper usuage can lead to either severe increase in leakage or to seizure of the seal. Where a buffer fluid is used, as is generally the case in a high performance seal, precautions must be taken against loss of buffer fluid. Suitable alarms should be provided where loss of the buffer fluid pressure could lead to significant consequences.

6.3.1.3 Pumps for Liquid Metals. Liquid-metal pumps fall into two general categories, mechanical pumps and electromagnetic pumps. Each will be discussed briefly.

Mechanical pumps for use with liquid metals have been largely of the centrifugal type similar to those used on other nuclear plants. However, the problems of sealing such pumps are a bit different. The canned motor pump can be used for liquid metal systems provided special attention is given to the bearing problems. Early designs involving ball bearings submerged in sodium and NaK failed under test. Success was obtained when sleeve bearings were used in combination with forced lubrication by liquid-metal which had been circulated through a specially provided cooling and purification system. Canned rotor pumps for use with liquid metals have been designed for service up to 1600°F (1853°C) with capacities up to 5000 gpm [100].

Reference [103] describes successful tests made on a diaphragm pump for use on liquid metals. It was a positive-displacement, high-temperature pump which operated continuously for approximately 5400 hours without any failure or irregularities in the stainless steel diaphragm.

Another arrangement to avoid bearing problems in mechanical pumps used to pump liquid metals involves the separation of the impeller and bearings so that the bearings can operate in a gas atmosphere. The impeller is attached to the bottom of a long shaft which extends upward to the drive motor and its bearings. A tank surrounds the shaft and extends upward from the pump casing to the motor mounting. The level of liquid metal in this tank is maintained well below the bearings by an overflow line from the side of the tank. An inert gas such as helium is used in the region above the liquid level. A labyrinth seal is provided at the top of the tank to restrict the diffusion of liquid-metal vapor into the motor and bearing regions. Pumps of this type present special problems when considered for high temperature service; these include problems of severe diffusion of liquid metal vapor into the motor and bearings, difficult problems of differential expansion of liquid metal and problems of high-temperature lubrication.

A novel seal which has been under development for use with liquid metal is the so-called frozen-sodium seal. In this seal the outer region is cooled to cause the sodium to freeze in this region to the extent that only a thin layer of liquid sodium is in contact with the rotating parts. Control of temperature in this seal is quite critical; if the temperature is too high, leakage becomes excessive; if it is too low, seizing can take place. Impurities in the sodium have been found to play a part in the seizing of these seals. Further development work is required.

When working with liquid metals that freeze at room temperatures, provisions must be made, in the pump as well as in other parts of the system, to melt the metal prior to start-up of the plant. Other factors that influence pump designs for liquid metals include corrosion effects and the effects of leakage to the surrounding area.

Because of their low electrical resistivity many liquid metals can be pumped by electromagnetic pumps. A number of different types of electromagnetic pumps have been developed for flows up to 20,000 gpm [109]. All of them utilize the fact that a force is produced on an electrical conductor when it is carrying current in the presence of a magnetic field located at right angles to the conductor. These pumps are particularly attractive because they require no seals or bearings. However, it must be remembered that failure of electricity to such pumps not only causes loss of pumping but leaves an open pipe through which flow in the opposite direction is unimpeded. Several of the principal types are described below.

1. D-C Conduction Pump. This pump consists of a thin-walled duct located between the poles of an electromagnet. To two opposite sides of the duct, which contains liquid metal, are attached electrical conductors through which current can be transmitted through the liquid metal in the duct. This current causes the development of thrust on the liquid metal and causes it to flow. (See Fig. 6-7 [54a].) The efficiency of this pump is generally quite low, in part because of the distortion of magnetic field caused by the current flow. To improve efficiency, d-c pumps can be provided with magnetic field compensation. This is achieved by doubling the conductor back through the magnet's gap as shown in Fig. 6-8 [54a]. This arrangement permits a more uniform magnetic field along the length of the pump and results in a larger net force on the fluid. The head developed by a noncompensated pump reaches a peak and then

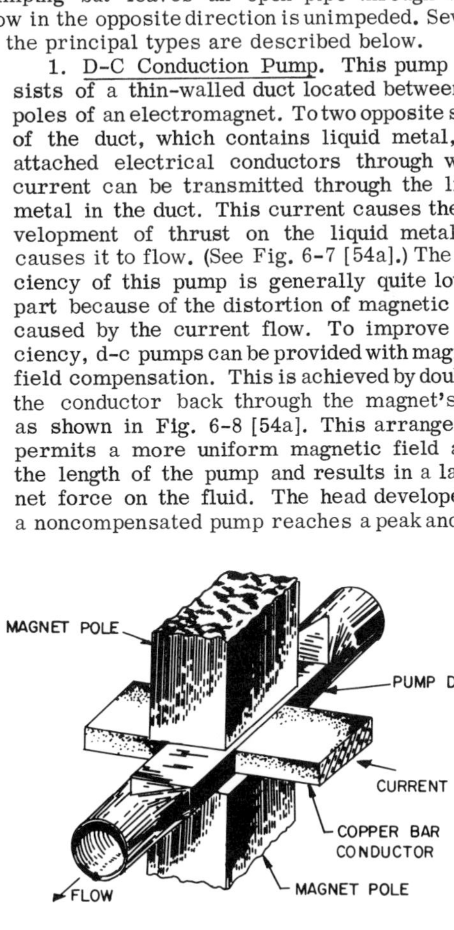

FIG. 6-7 Direct current electromagnetic pump.

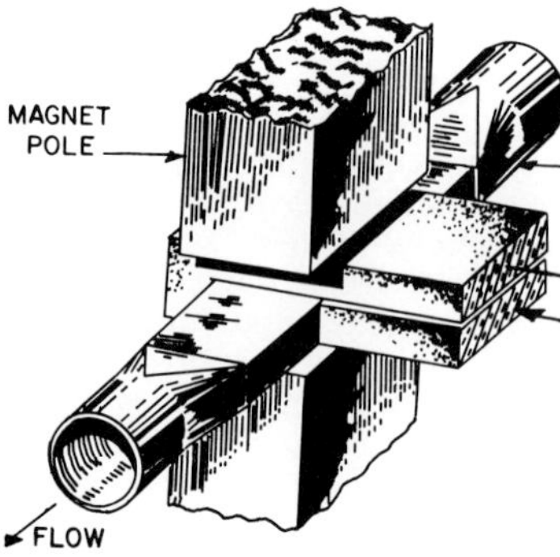

FIG. 6-8 Direct current electromagnetic pump with magnetic field compensation.

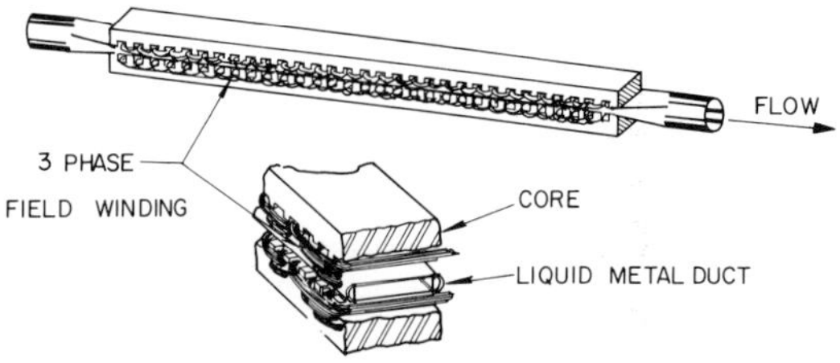

FIG. 6-9 Alternating current linear indirection-type electromagnetic pump.

diminishes as the current is increased; a compensated pump continues to increase the head developed in the fluid as the current is increased [54a]. Other methods of compensation, such as pole shaping, have been used; these generally improve performance at only one particular point of rating [54a].

D-C pumps require high currents at low voltages. These currents can supplied by either rectifiers or generators. Over-all efficiencies from power source to fluid horsepower range from 10 per cent for small pumps (5-10 gpm or 0.3.5-0.630 liter/min) to 40 percent for large pumps (~ 5000 gpm or ~ 300 liter/min) using homopolar generators [54a]. Pumps of this type are simple, rugged, and reliable and are well suited to high temperature service, particularly when the electromagnetic field is generated by one or two turns of large conductors which can be individually insulated by means of high temperature materials.

2. A-C Conduction Pump. This type of pump operates on single-phase a-c in which the current to the electromagnet is properly phased with the duct current. To accomplish this the windings of the electromagnet are usually placed in series with the duct current. This type of pump has the advantage that it uses readily available single-phase a-c without conversion to d-c. It suffers from the disadvantage that eddy current power losses are relatively high and efficiencies are low (< 20%). Hence these pumps have not been used for large installations. However, they are very popular for laboratory use and for use in experimental loops.

3. A-C Linear Induction Pumps. This type of pump operates on the same principle as the squirrel-cage induction motor except that the stator is rolled out flat to produce a linearly traveling magnetic field rather than a rotating magnetic field. It operates from a three-phase power supply and can use 60-cycle alternating current. It can be built in relatively large sizes; pumps of this type have been built with a capacity of 3500 gpm. Efficiencies of pumps in this size range can be made as high as 40 percent. (See Fig. 6-9 [54a].) For high-temperature service this type of pump requires an auxiliary cooling system for the magnet windings.

4. Other Electromagnetic Pumps. A number of variations of the foregoing pumps have been tried with varying degrees of success. Of particular interest are the Einstein-Szilard pump, the helical thin-walled pump, the flat-type rotating field pump and the mechanically driven magnetic pump.

The Einstein-Szilard pump is an adaptation of the linear induction pump in which the fluid flows through an annular duct surrounded by a series of toroidal coils on the outside with a magnetic core on the inside of the annular duct. A moving field is generated in a fashion similar to that of the linear induction pump; this design permits reduction of the edge effects that arise when the conducting fluid enters or leaves a magnetic field [104]. Fabrication problems have retarded development of this type of pump [54a].

The helical thin-walled pump consists of a helical duct in the air gap of an induction motor with a fixed central core. As the field rotates, the field is caused to flow in the helical duct. The flat-type rotating field pump is similar to the helical pump except that it uses a spiral path through which fluid is pumped by a disc-type stator. Both of these pumps have efficiencies which are lower than the efficiency of the linear induction pump, and both are more difficult and costly to fabricate; as a result they have not been used in reactor applications.

The magnetically driven magnetic pump utilizes a rotating set of electromagnets to generate a moving field. This field is used to pump fluid in an adjacent thin-walled chamber. Low efficiency and fabrication difficulties have discouraged its use.

6.3.2 Basic Considerations

As the prime mover of the system coolant, the pump must be a reliable device. It is one of the few mechanical devices in the primary loop that must operate continuously over a long period of time. The loss of coolant flow generally requires immediate reduction of power generation in the core if core design criteria are to be observed. Possible causes of loss-of-coolant flow include mechanical pump failure, loss of power to the pumps or closure of main-piping stop valves. The loss of flow can be complete as, for example, when all pumps cease operating; partial when only one pump out of several cease operating; sequential when several pumps in the system cease functioning sequentially [59]. (See Chapter 15, "Fluid Flow".)

Mechanical pump failure might occur as the result of bearing seizure, breakage of the can in a canned pump or failure of a seal in a sealed pump, blockage by extraneous material, or disengagement of the impeller from the shaft. Pump noise monitoring devices can be used to warn against such mechanical failures. Mechanical failure of a pump would be expected in general to cause only a partial loss of flow in a system having more than one pump because it is improbable that all pumps would fail simultaneously or in such rapid succession as to lead to compounding of effects implied by a sequential loss of flow. Nevertheless, if partial loss of flow that can be accommodated without a scram takes place, the power of the plant should be cut back immediately and investigation of causes made [59].

Although the effect of a single mechanical pump failure in a system with several pumps is not too severe so far as loss of flow is concerned, the loss of power to the pumps can be. Failure of power can be readily envisioned, and, when it occurs, the resulting loss of coolant flow is rapid. Thus the reliability of the power source is important. The loss of power to the pumps can be guarded against by an automatically initiated spare power supply. Yankee uses essentially this method. This guards against flow loss brought about by operator errors and switch-gear failure as well as by power loss. However, in this case the spare power supply must be kept in high state of readiness and must be tested periodically. Various types of scram-initiating devices are also available and used for protecting against loss-of-flow accidents.

Another feature of a pump which affects safety arises from the fact that the pump often requires transmission of power across a pressure envelope or a leak-tight container, either mechanically through a seal or electrically through a "can". Be it a seal or a can, this region of power transmission represents an Achilles heel for the primary loop so far as prolonged loop integrity is concerned. While a can appears better to many designers, it is better only so long as corrosion or stress corrosion does not weaken the can or the balance of pressure is maintained to keep the can backed up. A failure of this type, wherein can leakage led to pump failure, occured on one of the pumps of the Indian Point Nuclear Power Plant [199]. Investigation disclosed that there was direct leakage from the motor gap through the stator can and into the stator winding cavity. The pump was returned to the manufacturer where only by cutting the stator can free of the stator was removal of the rotor possible. Further disassembly revealed severe damage to the lower motor guide bearing which apparently caused failure of the motor.

The design considerations and criteria in matching the pump characteristics to the system needs are important, of course, in assuring satisfactory steady state operation of the system. When designing a pump, the influence of erosion and corrosion must also be taken into account. Furthermore, vanes or other possible pulse-producing components must be designed so that they do not introduce vibrations or fluid pulses whose frequency matches the natural frequency of other components in the core or elsewhere in the system. Of course the speed of operation of the pump must not correspond to the critical speed of the rotor where a rotor is involved; if possible the critical speed should be well above the operating speed. If it is not, means must be provided to pass through this speed rapidly. This aspect of critical speed is particulary important in pumps having a continuously variable speed range.

Care should be taken to avoid problems of cavitation in liquid pumps. Cavitation can rapidly lead to failure of the pump and can produce untenable pressure pulses in the system. Provision must also be made for venting the pump where necessary.

In pumps of pressurized water systems which utilize seals, depressurization of the coolant passing through the seal can lead to evolution of vapor and thereby to erratic seal performance. Such experience was encountered on the pumps of the Canadian Nuclear Power Demonstration (NPD) Reactor (a natural uranium pressure-tube reactor cooled and moderated with heavy water) [200]. Each main coolant pump has two face-type mechanical seals in tandem; sealing is accomplished between a stationary carbon ring and a rotating Stellite ring attached to the shaft. Performance of the seal was initially quite erratic and several outages resulted from seal failures brought about by vapor and gas build-up in the seal. Better venting of the seal and addition of degassing equipment to the coolant circuit led to a much improved situation. Seal failures still occurred, but much less frequently.

6.4 Valves

Large, complex nuclear power plants utilize valves of many types and many sizes for a variety of purposes. They can usually be found in any part of the plant. In a single plant they can range in size from small lift-type check valves in an auxiliary system to large gate valves used as main-stop valves in the primary-coolant loop. They can be power-operated from a remote station or manually operated at the valve. They can be hermetically sealed or sealed by packings. They can be used for throttling, for isolation, for preventing reverse flow or for pressure relief. The problems of designing valves vary greatly with the size, nature and purposes of the valves; in addition, the problems will vary with the fluid in which the valve must work.

Particular attention must be given to the design, construction and maintenance of valves because of the important functions they are called upon to perform. For example, the relief valve of a pressurized system must open when required to prevent overpressure; it must close when the overpressure is relieved so that all the coolant is not lost. Likewise valves which are to isolate a component in the event of leakage are sorely needed when a leak occurs. The problems of providing the reliability required are influenced by many factors, not the least of which the fact that a valve which has not been moved for some period of time may tend to freeze in position and not respond when called upon to do so. In addition, improper or uncalled-for action could lead to severe safety hazards as, for example, when a valve on a cold loop of a pressurized-water reactor is inadvertently opened. The cold water could enter the reactor and lead to a cold water reactivity excursion. In the event of a main steam line rupture in a boiling-water reactor plant, containment must be effected by the closing of fast-action isolation valves in the steam line that leads from the reactor through the containment tube to the turbine. Failure of the isolation valves to close upon rupture of a steam line could lead to major loss of containment and possibly to a major fission product release to the general public. Although two such valves in series are usually provided on each steam line, care must be taken to insure that failure of one valve to close will not lead to failure of the other or that a single signal failure can cause both isolation valves to remain open. The possibility of such action has been identified in some plant designs and in one instance was identified only as the result of a periodic valve test in the plant.

Other malfunctions of valves have also occurred. Reference [201], for example, reports a failure in the main stream bypass valve control system of the Big Rock Point Nuclear Power Plant; this malfunction caused the valve to remain fully open despite attempts by the operator to close the valve by actuating the manual override switch. The reactor scrammed on indicated high flux, but drawdown of the reactor pressure could be stopped only by closing the main steam isolation valve. The malfunction appears to have been caused by sticking of the hydraulic servo valve; this was accompanied by failure of the manual-override closure feature.

In another instance, at the Piqua Nuclear Power Facility, a feedwater valve failed to close on signal from the operator [202]. The failure to close was later determined to be caused by a burr on the valve.

6.4.1 Characteristics of Various Valve Types

Valves may be classified in many ways. For the purpose of this discussion they will be classified as gate, globe, plug, check, control and relief valves. The characteristics of each are discussed below.

6.4.1.1 Gate Valves. This is the most widely used type of valve in many applications both nuclear and non nuclear [105]. This valve utilizes a disc, known as the gate, which moves perpendicular to the axis of the pipe to block the fluid which normally flows directly through the body without turning. Gate valves have relatively low pressure losses when compared to other types. (See Figs. 6-10a [105] and 6-11 [39].) As indicated in Sec. 6.4.2, for nuclear applications the gate should be made of two plates which expand against the seat in order to avoid rubbing action on the mating parts. Gate valves should not be used for throttling because when they are partially opened the bottom of the gate becomes badly eroded.

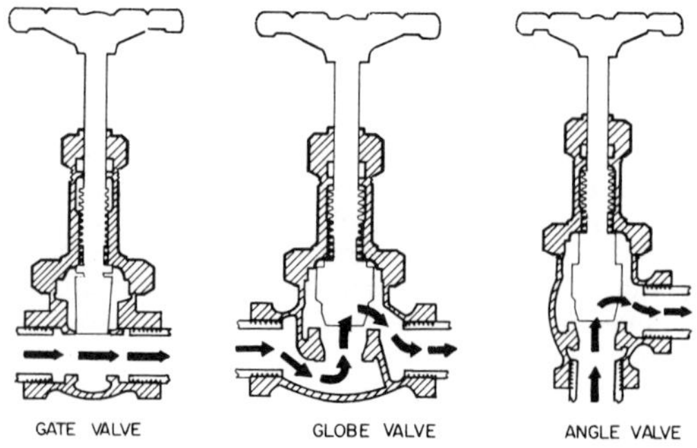

FIG. 6-10a Valves and valve flow characteristics: gate, globe, and angle valves.

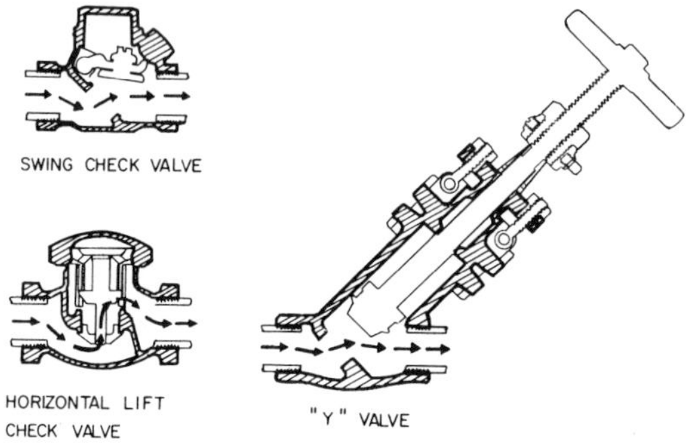

FIG. 6-10b Valves and valve flow characteristics: swing check, horizontal lift check, and "Y" valves.

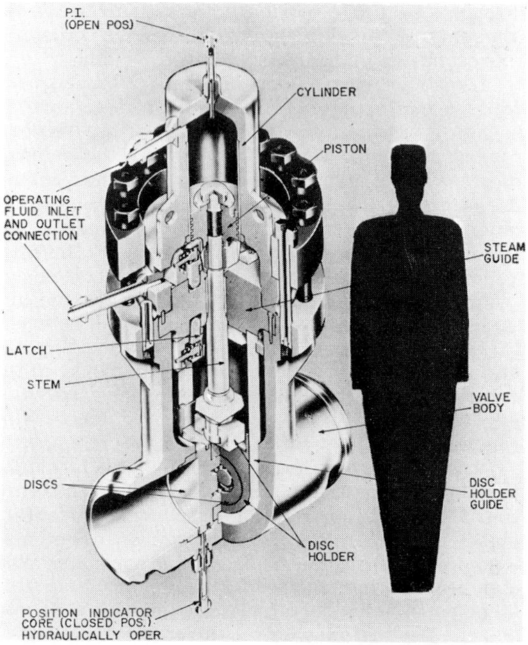

FIG. 6-11 Hydraulically operated gate valve used in Shippingport plant.

6.4.1.2 Globe Valves. Globe valves are particularly well suited to applications which require throttling as well as complete shutting off of flow [105]. The globe valve utilizes a circular orifice which is either parallel to the axis of the valve or at an angle from the axis. (See Fig. 6-10a and the "Y"-valve in Fig. 6-10b.) The fluid flowing through the valve must generally turn approximately 270° to get back into line with the piping. As a result, the pressure drop through a globe valve is much higher than that through a gate valve. Globe valves are generally more expensive than gate valves in large sizes.

6.4.1.3 Plug Valves. The plug valve is also sometimes called a cock valve. It utilizes a rotating plug whose axis is at right angles with the axis of the pipe. The plug contains a carefully bored hole usually of the same diameter as the pipe. The valve is opened or closed by rotating the plug 90° within its seat. The plug and seat are often tapered so that the plug can be lifted slightly to rotate it more easily and is then returned to its seat to provide a tight seal. Wherever possible the plug is provided with a permanent lubricant.

This type of valve is not too well suited to most reactor applications because of the need for a lubricant if the valve is to be operated frequently and because of the possibility that wear and corrosion products could cause the valve to jam. Valves similar to plug valves are sometimes used in refueling devices because they provide a straight through access in a compact assembly.

6.4.1.4 Check Valves. Check valves are used to prevent reversal of flow. In nuclear reactor plants they find their chief application in outlet lines from pumps so that when a pump is shut down the pressure head developed by other pumps does not cause back flow in the line containing the pump which is shut down. Two types of check valves are used, the lift check valve and the swing check valve. (See Fig. 6-10a.) The lift check uses a ball or guided plug which is lifted by the fluid when flow is outward and is seated when the flow is reversed; it has a relatively high pressure drop and so is used primarily for small valves. It does not have a tendency to slam, but care must be exercised to make sure that it is installed properly.

The swing check valve utilizes a pivoted disc appropriately hung and counterweighted to close against a seat if flow is reversed. The chief problems with swing check valves are the tendency to chatter and the tendency to slam and produce water hammer when flow is reversed. Both actions can also cause fatigue failure within the valves. G. M. Fuls [106] presents an analysis of the dynamic action associated with a check valve disk during a flow transient. The methods outlined by Fuls permit evaluation of the impact velocities of the disc and the pressure surges resulting from deceleration of the fluid.

Check valve counterweights and/or springs can be designed so that the valve movement can follow any flow coastdown quite closely; in such a design the valve is essentially closed at zero flow before reversal begins, thereby minimizing water hammer. However, the features which allow this characteristic also impose relatively high resistance to low flows, so that valve design in loops requiring natural circulation flow, for example, must be a compromise between these conflicting requirements.

6.4.1.5 Control Valves. Control valves take many different forms. They can be used for flow control, for pressure control or for controlling motion of hydraulic actuators [107]. Control valves for pressure and flow control usually use a globe-valve-type action or some modification thereof such as a needle valve. Valves for controlling motion of hydraulic actuators often involve sliding spools or rotating spools which are machined with slots, grooves, or holes to yield the desired response. Typical of such valves are those used in hydraulic mechanisms as outlined in Sec. 5.2.5.

The most common type of control valve is the diaphragm valve [105]. The valve plug is moved by an air-operated diaphragm. The valve action can be produced by a single plug operating on one orifice or by a double plug acting on two orifices. Reliability in control valves extends beyond the usual requirement for seat integrity; if they are to operate successfully, the original geometry of the seat and plug must be retained throughout the life of the valve so that the flow response characteristics of the valve are predictable and obtainable.

An excellent example of the application of control valves is the use made of flow-control valves by the Canadians for reactor control. These valves are used for moderator level control; the moderator level is used in turn, to control reactivity. Such control has been applied to the NRX experimental reactor, the NPD-2 Prototype Nuclear Power Station and the CANDU Nuclear Power Station. All three of these use heavy water as the moderator and coolant and are controlled by valves which regulate the level of heavy water that is being pumped into the calandria at a constant rate [108-111]. The heavy water used for moderation is separate from that which is used to cool the reactor so that moderator dumping can be used to shut down the reactor without interfering with the heat removal system.

The NPD-2 and CANDU reactors utilize two separate systems for moderator-level control, the Regulator System and the Protective System [110, 111]. Figure 6-12 [110] shows a flow diagram of the two systems as used in the NPD-2 reactor. The regulator system functions as follows: As shown in the figure, the bottom of the calandria is connected to a moderator dump tank by three 24 inch diameter (61 cm) lines [110]. In the dump tank exists a helium atmosphere; if the pressure of this helium gas (approximately atmospheric) is equal to the pressure equivalent of the head of moderator in the calandria plus the pressure of helium gas above the moderator, the moderator is retained in the calandria. Helium is constantly pumped from the top of the calandria into the dump tank and is returned through control valves to the top of the calandria. In addition, heavy water is circulated at a constant rate from the dump tank through a heat exchanger and back through the dump ports

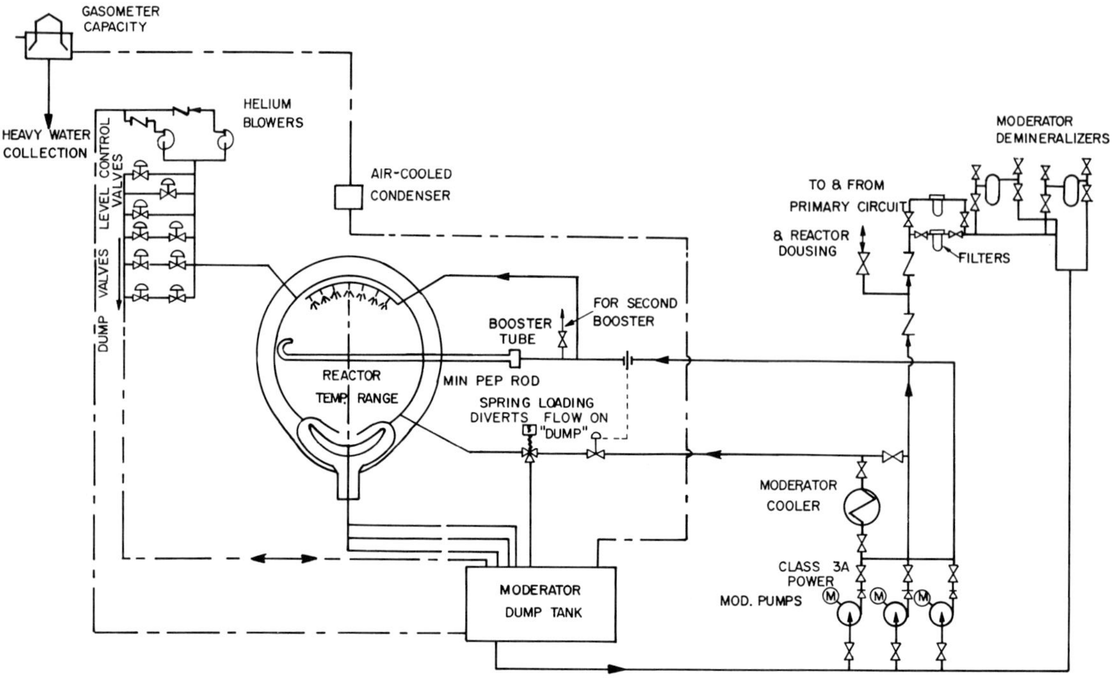

FIG. 6-12 Flow diagram of NPD-2 regulating and protective system.

in the calandria. When the helium control valves are fully closed, the moderator level rises in the calandria because the increased difference between the helium pressure in the dump tank and the helium pressure in the top of the calandria prevents the incoming heavy water to the calandria from returning to the dump tank. When these valves are partially opened the moderator level remains constant; when they are fully opened the moderator level falls. The rates of flow are such that the maximum rate of reactivity change, when the calandria is full, is +0.004% per second if the valves are fully closed and -0.008% per second if the valves are fully opened. The control valves are diaphragm-operated and require air pressure to close.

The protective system consists of six solenoid-operated dump valves connected to large pipes which permit equalization of the helium pressures in the calandria and the dump tank. These valves open when preset limits are exceeded by any of the following variables: reactor period, neutron flux, coolant pressure or temperature, or rate of change of coolant temperature. Complete drainage of the heavy water is accomplished in 30 seconds. All the channels capable of producing a shutdown are triplicated in such a way that trip signals from two channels are required to cause a shutdown; a trip signal from one channel merely sounds an alarm. By this means, any channel can be tested without disturbing operations or losing protection. The dump valves are arranged in three lines to permit utilization of this procedure.

The moderator-level control system offers the important advantage that it avoids the need for moving mechanical components in the reactor where maintenance is difficult to perform. It does, however, transfer the problem of control to valves and to a lesser extent to pump and blowers. The response of this system is governed to a large extent by the response obtained from the control valves. Thus a requirement for fast response is identified, which is different from that encountered in isolation valves. Typical of the response characteristics required are those presented for NRX in reference [109].

6.4.1.6 Relief and Safety Valves. Relief valves and safety valves are used to protect a pressurized system from exceeding a predetermined pressure. Although the names "relief valve" and "safety valve" are used interchangeably, their characteristics are, in general, different. Both valves are designed for normally closed operation, being held closed by a spring acting on the valve plug until the set pressure is reached. The set pressure is the pressure at which the valve disc first begins to lift from its seat. The relief valve, after beginning to lift at the set pressure, is intended to provide a reduction of upstream pressure which is linear with the upstream flow until the rated capacity is reached. The safety valve, on the other hand, begins to lift at the set pressure but is intended to pop fully open as the upstream pressure increases to a slightly higher valve [112]. While a safety valve is sometimes used as a relief valve, the requirement for safety valves regarding set pressure, accumulation pressure and blowdown pressure are usually more stringent [112]. Accumulation pressure is the maximum pressure reached in the equipment before the valve discharges at rated capacity. The blowdown pressure is the pressure in the equipment when the valve reseats. While a relief valve may be a pilot-operated device or a self-actuated device, a safety valve should be self-acutated. The pilot-operated valve uses a pressure-sensing element to open a port which permits pressure to be applied to the operating piston of a hydraulically operated main relief valve. The self-actuated valve opens directly under the action of the internal pressure.

Kosut [112] points out that nuclear power plants place more severe demands on relief and safety valves than do standard fossile-fuel plants. Leakage of nominal amounts of coolant or loss of some of the coolant when a relief or safety valve is lifted in non nuclear plants causes little concern. In nuclear power plants, however, leakage of coolant that is radioactive or may contain radioactive materials constitutes a contamination hazard. Furthermore, he points out that in subsurface craft the treated and highly purified coolant is an item to be conserved. It might also be added that the need for safety valve action when required is more significant in a nuclear power plant because of the potentially severe consequences of a rupture in a pressurized component.

Because of these facts dependability under all conditions of operation becomes a most important factor in safety valve design. All the considerations discussed in Sec. 6.4.2 except possibly those applying to valve stem leakage apply to safety and relief valves. In addition the reader is referred to Sec. 2.5 regarding code requirements for pressure relieving devices. (Also see Sec. 6.6.2.2.) Inasmuch as the same general considerations apply to both safety and relief valves they will be treated jointly in the following discussion.

Pressure-relief valves must have high set-pressure and blowdown-pressure accuracy. The valve must shut tightly on reseating, not only to limit leakage per se but also to prevent excessive erosion of seats brought about by the leakage. In pressurized water reactors, such valves must provide for release of subcooled water from the system, a requirement with which there was little experience prior to the advent of nuclear power plants [113]. In subsurface craft, the problem is further complicated by the fact that the relief valve must operate against variable back pressure.

Kosut [112] points out that, as a result of the extensive development done, relief valves can have blowdowns as low as 3%, an accumulation of approximately zero, and seat pressures as low as ± 10 psig (0.7 kg/cm^2). However, the problem of obtaining and maintaining acceptable seat tightness continues to be a problem. The materials most widely used at present for seats and discs are Stellite #1, 6, 12 and 19 [112]. Stellite #1, 6 and 12 are usually used for hand-facing deposit in pressurized water plants. Because of the poor impact resistance of such materials, relief valve chattering can cause cracking or flaking. Consequently, attention to problems of this type becomes quite important. Test should be made prior to installation to check this point as well as to determine seating characteristics and internal alignment of parts.

In order to assure a good seat, the seating surfaces should be carefully lapped and polished and protected against damage during handling. Because of the erosion of the disc that can be caused by grit, attempts to have a clean system are again called for. Kosut [112] points out that some success has been achieved in eliminating seat erosion, due to foreign particles and low quality steam, by designing the nozzle and disc fluid-flow exit area to have smooth changes, so far as possible. The reader is referred to Sec. 6.4.2.2 for other general considerations regarding valve seats.

Rupture discs may be used as safety devices under special conditions if designed and installed as prescribed by Section III of the ASME Boiler Code. Rupture discs are attractive because they provide a leak-proof seal until blown. A safety relief valve is required in the line to limit leakage after the rupture disc has blown. Tests of proposed rupture-disc designs should be made to assure that they will rupture at the desired pressure.

6.4.2 Basic Design Considerations

May [203] summarizes present limitations of commercially available valves and identifies problem areas in which research and development work has been needed and is being carried on. New trends and techniques are also discussed. While May's report was motivated by space needs, most of the information presented is applicable to valves for nuclear plants. Experience discloses a number of basic design considerations which apply to all valves even though the emphasis may vary from one valve type to another. Such considerations can be grouped under the following headings: sealing valve stems against leakage; providing adequate and reliable valve seats; guarding against corrosion, erosion and wear; assuring valve-body integrity; establishing desired response characteristics; and designing for maintenance and repair. The considerations involved under each of these headings are discussed below.

6.4.2.1 Sealing Valve Stems Against Leakage. The problems of sealing valve stems against leakage have greatly influenced the design of primary-loop valves for a number of nuclear power plants. These problems are also important in many other valves. Aside from the usual packing glands, provisions for sealing against steam leakage have included hermetic sealing by means of bellows, complete nonflexible hermetic sealing, labyrinths of sealing rings, or, in liquid metal systems, frozen seals. (See Sec. 3). Bellow-sealed valves have been used extensively in liquid-metal systems. These bellows are generally made of ausentic stainless steel and are welded to the valve bonnet and to the valve stem. Two bellows in series are often used to provide added protection; these can be designed to be concentric with one another as shown in Fig. 6-13 [54a]. As an added precaution it is good practice to provide a leak-detection device connected to the region between the bellows to

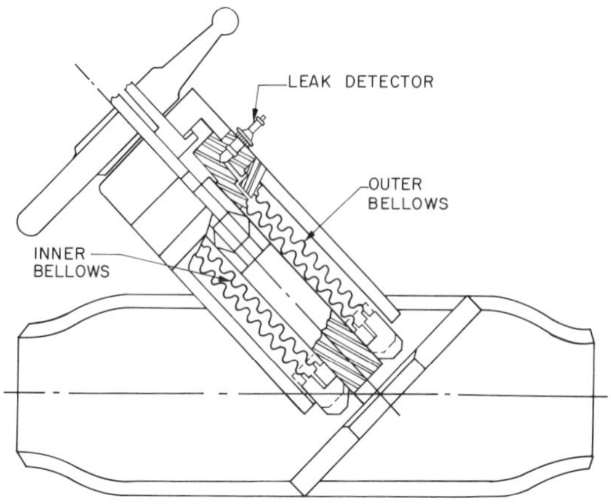

FIG. 6-13 Valve with double-bellows seal.

indicate when the inner bellows develops a leak [54a]. Because the bellows depends upon the elastic range of the bellows materials, the allowable range of movement is limited, particularly at high temperature where the elastic range is lowered. Because of fatigue considerations, the number of cycles of operation may also have to be restricted.

The use of bellows has been avoided in a number of nuclear power plants, particularly pressurized-water plants, both because of the uncertainties in the reliability of bellows for prolonged operation and because of the limitations on the range of motion. In such cases, the desire for hermetic sealing has led to the use of remote operations involving electromagnetic, hydraulic or pneumatic drives. Electromagnetic drives utilize a solenoid around the outside of a canned armature. When the solenoid is energized it creates a magnetic field which moves the armature within the pressure can. Because of the limited forces available with this method, its use is limited to small or balanced pilot valves. For this reason the large remotely-operated valves used for primary coolant loops are hydraulically operated. Figure 6-11 shows such a valve as used in the Shippingport plant. The valve is operated by applying high-pressure water to the piston to open or close it. The piston is solidly fastened to the valve stem. In general, where hydraulic operation is employed, the valve-operating fluid must be the same as that used as the primary coolant so that intermixing of the two will not be detrimental to the reactor operation [113].

Hydraulic valves and solenoid valves are most suitable for applications in which the valves are to be fully opened or fully closed and hence are generally employed for isolation purposes or for on-off control. In the case of hydraulic valves, consideration must be given to minimizing the possibility of broken hydraulic lines to the valve and to the consequences if such lines are broken. The problems arising out of broken hydraulic lines are particularly difficult to plan against because it is not always clear what valve action, if any, should ensue. Maloperation of the same valve could lead to serious consequences if it opened under certain circumstances or closed in others; often it may be best to have it fail "as is". This may be difficult to accomplish with hydraulically-operated valves particularly in pressurized systems. Hydraulic valves operate on a pressure difference across the piston. This pressure difference is supplied either directly from a canned pump or from a system of pressurized flasks with the primary coolant system serving as the low pressure. Loss of a hydraulic line could cause the primary system pressure to actuate the valve even if motion is not desired. In some plants, the primary system pressure serves as the high pressure, with the low pressure side of the valve piston vented to the atmosphere. Inasmuch as the high pressure can be obtained directly from the valve interior without small lines this type of scheme may have more built-in safety. In either case, however, control of the differential pressure is important; this is generally provided by a hermetically sealed solenoid-operated pilot valve which itself can be a source of trouble.

Pneumatically operated valves are similar in many respects to hydraulically operated valves. In such valves compressed air is used to provide the motive power; the air is supplied from pressurized containers whose pressure is maintained by a standard compressor. Many pneumatic valves utilize diaphragms rather than pistons to operate the valve. Valves of this type are used mainly for controlling purposes where the length of travel of the valve is relatively small. The considerations discussed in connection with hydraulically operated valves apply to pneumatic valves.

Valves involving the use of packing rings can be used for nuclear power plants, especially in applications away from the primary piping where they are accessible for observation and maintenance. Kanter [114] points out that since most satisfactory packing rings are generally of organic materials, cooling provisions must be provided to the packing gland where the fluid temperatures are excessive. In applications where the valves are designed to be fully open or fully closed, back seals can be used to restrict stem leakage when the valve is fully open. Such a seal involves an enlargement of the stem which mates against a seat in the bonnet. Care must be exercised in the design of such seats to make sure that they do not bind and give difficulty in closing the valve.

Where the need for opening or closing a valve is infrequent, as might be the case for a valve used as a back-up valve and installed to help isolate a major component if it leaks, a capped valve can be used. This valve is hermetically sealed by a seal-welded cap at all times except when it is to be operated; when the valve is to be operated, the seal-welded cap is removed and a portable motor or hand wheel is used to operate the valve. Stem packing is provided to restrict leakage when the cap is removed.

6.4.2.2 Providing Adequate and Reliable Valve Seats. Valve-seat tightness is an important requirement in all valves for nuclear power plants, but the requirements are particularly stringent for isolation valves. Bake [113] indicates that leakages for such valves in pressurized-water systems have been limited to 2cc of water per hr per in. of nominal pipe size. As a result, careful attention must be given to the design of the seat, to the materials used, and to the surface finishes specified and achieved. Stellite seat materials have been found to be suitable for most applications, provided proper attention has been given to the seat design. The detailed seat designs will vary from one valve type to another but in general the following considerations must be taken into account:

1. The design should be such as to avoid the need for rubbing or wiping action between disc and seat. This is particularly important in gate valves where such action is easily possible unless special steps are taken to avoid it. The valve shown in Fig. 6-11 utilizes a double disc arrangement which is lowered into place in a retracted condition and then expanded to bring the discs against the seat.
2. The valve body must be rigid enough so that planned loads and moments imposed on the valve by piping reactions do not cause distortion of the seat. Even slight distortions can lead to severe leakages. For this reason no severe piping loads or moments should be transmitted through the valve body.
3. Consideration must be given to distortions which might be introduced when the valve is being installed. Welding procedures must be carefully specified. Distortion can also be produced by both steady state and transient thermal stresses.
4. A highly polished finish is required to provide leak-tightness. The finish should be made by rotary motion concentric with the seat so that no radial passages are developed which can permit leakage.
5. Provision must be made to assure that the disc can mate with its seat without imposing severe moments on the valve stem and in turn on the seat. If the stem and valve are integral, careful alignment is necessary; however, a flexible connection should be provided between stem and valve disc if at all possible to avoid this problem.
6. Where the valve seat is tapered, the angle of taper must be such as to eliminate the possibility of binding of the mating parts.
7. For proportional control valves which must operate at times partially closed and for check valves, analyses and tests should be made to assure that no vibration or fluttering is established by coolant flow or small leakage past the valve. When the valve disc is near the seat, severe damage to the seating surfaces can be encountered by this action. Vibrations can also lead to fatigue failure of the valve parts involved.
8. The coolant velocity past the seat when the valve is opened should be limited to a value which will avoid severe erosion. The exact value is dependent not only upon the type of coolant but also upon the nature of the flow path to and through the seat opening.
9. Consideration must be given to the possibility that corrosion and wear products from the valve itself or from other parts of the system may deposit on the seat. The design should be such that the seat tends to shed such deposits. Efforts should also be made to keep the system free of products which can deposit on the valve seats.

6.4.2.3 Guarding Against Corrosion, Erosion and Wear. Although each reactor type presents a different corrosion problem, it has been found that corrosion resistance can be provided by the stainless steels for most applications; types 347, 304, and 316 have all been used [114]. Selection of the type of stainless steel to be used has been based largely upon the welding requirements. Types 347 and 316 have been preferred where extensive welding is required because of their stability against intergranular corrosion. Although these stainless steels have in general given satisfactory results, they have not been shown to be better than some of the 400 series stainless steels in slurry reactors, at least during cold testing. E. A. Goldsmith [115] reports that during a cold test of two valves utilizing both type 416 stainless steel and type 316 stainless steel, the parts made of type 416 were considerably less corroded than those made of type 316. It should also be pointed out that reactor systems using organic coolants do not encounter corrosion problems which require the use of stainless steels; carbon steels have been used satisfactorily for the larger parts of valves in such systems.

The factors affecting erosion in valves are the coolant velocity, the coolant environment and the material [115]. The coolant velocity that can be tolerated depends not only upon the nature of the coolant but also upon the degree of contamination in the coolant and the angle that the fluid must turn through within the valve. Allowable velocities should be selected on the basis of tests. The materials characteristics required for erosion resistance are similar to those required for good wear resistance. Hard-surfaced materials generally provide the best service, particularly where little lubrication is provided by the coolant. The cobalt-base alloys have been used for applications requiring good wear and corrosion resistance. The use of such materials should be limited as much as possible because of the radioactivity that can be induced in the wear products as they pass through the reactor. (See Sec. 2.6.3.) Inconel-X as well as Haynes Alloy No. 25 have been used successfully for valve springs. Haynes-25 has also been used for piston rings in hydraulically operated valves where control of internal leakage is required [113].

6.4.2.4 Assuring Valve-Body Integrity. Inasmuch as valve bodies form an important part of the coolant envelope, their integrity is as important as that of any other part of the system. In fact one might say that the integrity of valves is even more important than that of many other parts because while valves can be used to isolate other components, they are generally not arranged to isolate each other effectively. Valve body integrity is especially important in pressurized systems.

The design of large valve bodies in a number of pressurized applications has been based on stress values indicated for the given material in Section

VIII of the ASME Boiler and Pressure Vessel Code (See Sec. 2 of this chapter). Small nonspecialized valves are often ordered now in accordance with the applicable class designation in ASA B16.5, "Steel Pipe Flanges and Flanged Fittings".

Because of the importance of the integrity of the valve body considerable attention has been given to the various means of fabrication. A number of early valves for nuclear application utilized valve bodies machined from bar stock or forging billets. Forgings were used for the larger valves instead of the more conventional castings in an attempt to avoid porosity in the material. However, because castings are more economical, considerable work was expended by manufacturers to assure sound castings for valve bodies. Admiral Rickover discusses problems of obtaining nonporous castings and forgings in reference [116]. He points out that many of the conventional fabrication procedures are not adequately controlled to provide the sound components required for nuclear power plants. Careful inspection of castings by use of x-rays, magnetic-particle inspection, fluorescent oils, dye stains, or ultrasonics, as appropriate, can permit identifications and location of internal or surface materials discontinuities, voids or inclusions [114]. When these are located they should be removed and repaired by welding.

Among the important considerations in the design of valve bodies are the problems of thermal stress, thermal cycling and thermal shock. The discussions in Sec. 2 apply to valve bodies as well as to reactor vessels and should be referred to by the reader. As indicated in Sec. 2.2.2, the severity of these problems will differ for different systems. Particular attention should be given to avoiding stress raisers (see Secs. 2.2.3 and 6.2.3) because of the importance that they play in determining the service life under thermal cycling. Of particular interest in connection with the evaluation of thermal stress problems in valves is the work of D. J. McDonald on the application of thermal stress theory to the PWR main hydraulic valves [117].

6.4.2.5 Establishing Desired Response Characteristics. Time response requirements for valves will vary greatly. However, experience has shown that there is a tendency to have valves operate too fast rather than too slow. This is especially true on hydraulically operated primary-coolant loop valves. When large 16-in. (40 cm) gate valves go from a fully opened position to a fully closed position in appreciably less than a second, severe acceleration and deceleration loads are imposed on the valve components which can lead to Brinelling or breaking of parts. In addition, fast acting valves, particularly fast-acting check valves, can produce severe water hammer which can lead to damage of the valve or other parts of the system.

It is important in many cases, that valves not be opened or closed inadvertently or that they be opened or closed in appropriate sequence. For example, in water-cooled systems having several loops it is possible to have one or more of the loops shut down and isolated by the main stop valves while the other loops are operating. In such a situation the secondary side of the heat exchanger in the shutdown loop will probably be out of service as well, so that the temperature of the coolant in this loop may be at any temperature down to the ambient temperature in the reactor compartment. If the isolation valves are inadvertently opened during this situation, cold water can be brought into the reactor, particularly if the pump were also started; because of the negative temperature coefficient usually designed into water reactors, this cold water could bring about a severe cold-water reactivity insertion. Because of the seriousness of this situation and the special conditions that lead to shutdown of the loop, valve opening rates should be studied and carefully limited and interlocks should be provided which prevent opening of the main stop valves unless the reactor is shut down. This means that once the stop valves are closed, a loop cannot be placed back in service while the reactor is critical. It should be pointed out that a somewhat similar type of cold water accident could come about if one of the main coolant pumps of one of the loops stopped and later restarted after the temperature of the coolant in the loop has dropped to about or a little below the saturation temperature of the steam in the secondary loop. Protection against this action can be provided by an interlock which prevents startup of the main coolant pump in an idle loop if the temperature difference between the cold loop and the operating loops is greater than a value calculated to be safe. This temperature difference must be established for each particular system design; it is usually in the range of 10°F to 50°F (~5 to 30°C).

Another situation that could lead to serious consequences is the simultaneous closing of stop valves in all loops so that the reactor is isolated from all the heat exchangers. While scramming can be provided to shut down the core in this case, there remains the problem of decay heat removal while the core is isolated. Analyses must be made to determine that the temperature and pressure within the vessel will not reach untenable levels even if it takes several hours to locate and correct the cause of the valve closures.

In addition to the considerations concerning valve response listed above, special mention should be made of the considerations concerning safety or relief valves and those concerning control valves which have been discussed separately in Sec. 6.1.

6.4.2.6 Designing for Maintenance and Repair. Whenever possible valves should be located and shielded so that they are accessible for maintenance. While it may not be possible to gain access to valves in the primary coolant loop during operation, it is generally possible to locate other valves so that they can be observed periodically during operation and repaired where necessary. Consideration should be given to accessibility to the valves for repairs. All valves, particularly large valves, should be designed so that major components can be replaced without removing the valve from the line. Special provisions may have to be made ahead of time to permit such action. Packed valves should always be accessible for tightening or replacing of packing.

6.5 Heat Exchangers and Steam Generators

It is the purpose of this section to describe and discuss some typical designs being used for heat exchangers and steam generators in nuclear power plants in the United States and to summarize the major considerations which apply to the design and operation of such units as indicated by experience. Attention will be concentrated on steam generators as the vehicle for identifying problems. Specific mention of heat exchangers not involving steam generation will be made when different problems are involved. For a more extensive treatment of the subjects discussed in this section the reader is referred to references [118] through [123], all of which are excellently prepared. The reader is also referred to Section 2 of this chapter, concerning reactor vessels; most of the considerations discussed in that section apply to heat exchangers and steam generators.

6.5.1 Types of Steam Generators

Because many of the differences between steam-generator designs arise out of the differences in the reactor coolant used, they will be grouped for our purposes here according to the basic type of coolant with which they are concerned.

6.5.1.1 Steam Generators for Water Systems The authors of reference [119] call attention to the interaction of two traditions in the development of steam generators for water reactors, the traditions of the boilermakers and the traditions of the heat exchanger manufacturers. The boilermakers, as suppliers primarily of boiler equipment for prime movers, are particularly conscious of the need for steam purity so that their designs tend to begin with sizing of steam drums and other steam separating devices. The heat exchanger manufacturers, on the other hand, are more conscious of the needs for economic sizing and servicing of tube bundles as encountered in the chemical industry where the necessity for good steam separation is less stringent. This interaction has played an important part in the development of steam generators for nuclear power plants, for, as immediately apparent, the nuclear steam generator is a heat exchanger which must produce high purity steam.

1. Vertical Steam Generators For Water Reactors

The vertical steam generator generally utilizes an integral drum and has been used extensively for mobile applications. Figure 6-14 [119] illustrates a version used for naval application. It utilizes a bundle of vertical U-tubes, through which the pressurized-water coolant flows. Boiling takes place on the outside of the tubes and the resulting steam-water mixture passes through the riser section to cyclone separators where moisture is removed. The steam leaves through the discharge nozzle and the separated water, along with the incoming feedwater, flows downward through the annulus around the water return baffle.

This design is quite compact but the capacity of the design shown in Fig. 6-14 is limited by the capacity of the steam separators. A vertical steam generator for large steam volumes necessitates enlargement of the top of the shell to accommodate separating equipment of higher capacity.

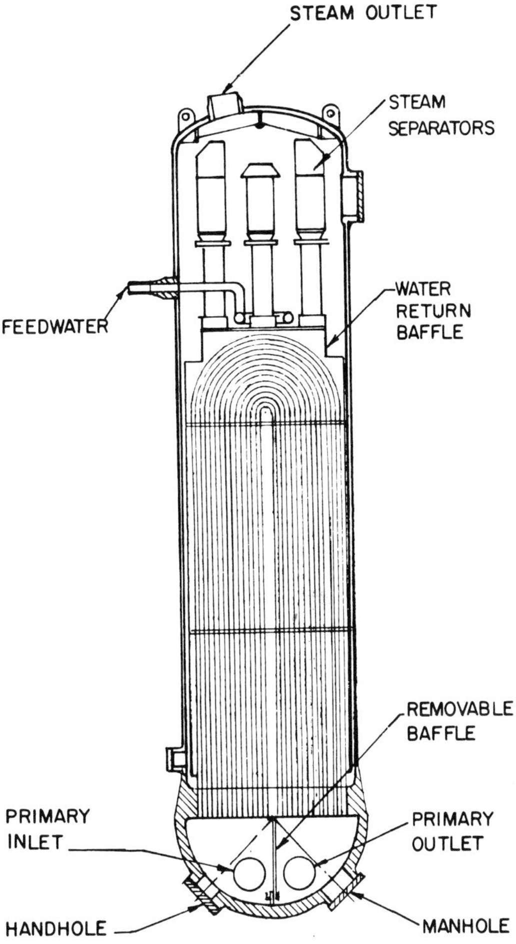

FIG. 6-14 Vertical navy steam generator.

In steam generators of this type, the tubes are usually made of stainless steel and the tubesheet of stainless-steel-clad alloy steel. The shell is usually made of carbon or low-alloy steel; that portion in contact with primary coolant is clad with stainless steel.

A vertical steam generator incorporating an integral superheater was used in the Army Package Power Reactor at Fort Belvoir, but the unit did not provide the desired steam conditions because of the inadequacy of the steam separator. The output steam was wet despite the superheater. This unit was made entirely of stainless steel.

Vertical steam generators are amenable to compact primary piping layouts. In addition, the U-tube arrangement alleviates some of the differential expansion problems between tubes and shell existing in units using two tube sheets. In this type of steam generator, a leakage tube can be plugged through the manholes and handholes provided for this purpose.

Two significant problems are associated with this type of steam generator, however [118]. First are the high combined stresses in the tube-sheet and in the region of the tube-sheet attachment to the shell; these stresses are brought about by the pressure and temperature differences between the primary coolant and the boiling water. Careful analyses of these stresses are necessary for each design. (See reference [123].) Because of these stresses, this type of unit cannot be used for all applications. The second problem is that sludge can readily accumulate on the large flat tube-sheet and can lead to acceleration of stress corrosion; thus, control of boiler water chemistry is very important in this type of unit [118].

2. Horizontal Steam Generators For Water Reactors

Horizontal steam generators are of two general types: those with an integral drum, characteristic of the chemical industry; and those with a separate external drum, more characteristic of the modern water-tube boilers. Figure 6-15 [118] is an illustration of the first type. A unit similar to this was used for the Homogeneous Reactor Experiment (HRE) at Oak Ridge, Tennessee. Figures 6-16 [57] and 6-17 [57] illustrate the second type as used at Shippingport.

In both the integral drum steam generator and the external drum steam generator, primary coolant flows inside of the tubes and boiling takes place outside of the tubes. The integral drum design is generally more compact; however, the space available for efficient steam separation is limited.

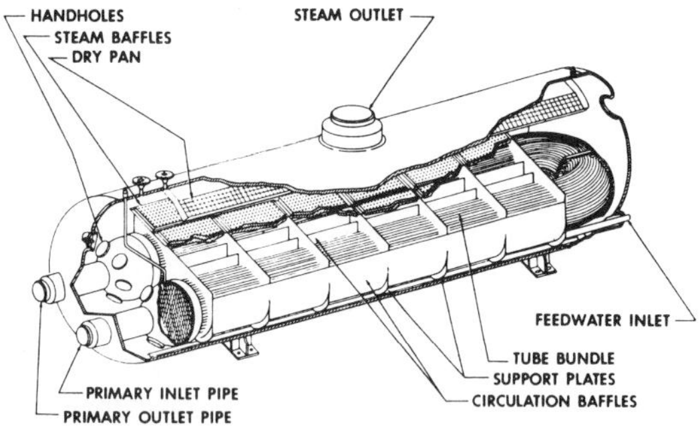

FIG. 6-15 Horizontal integral drum boiler.

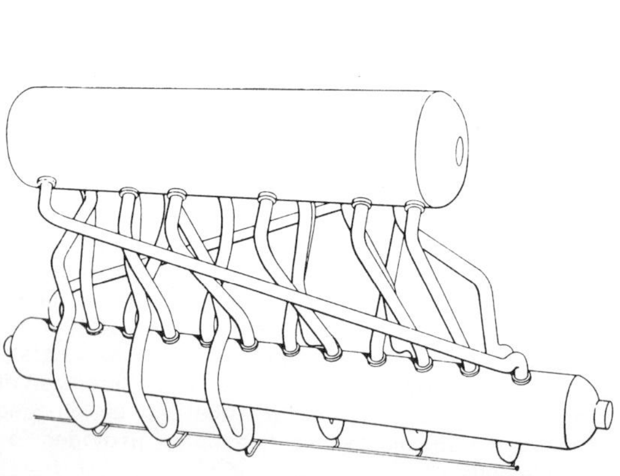

FIG. 6-16 Shippingport steam generator with straight tube heat exchanger.

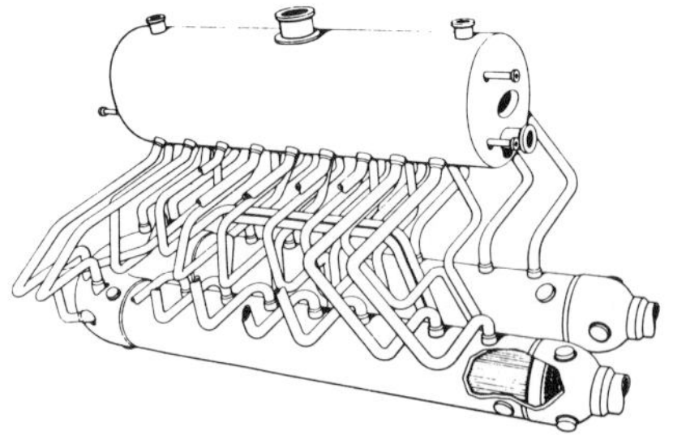

FIG. 6-17 Shippingport steam generator with U-tube heat exchanger.

The Shippingport steam generator shown in Fig. 6-16 employs straight tubes in the heat exchanger portion. The tubes are welded to fixed tube-sheets. In order to avoid excessive differential expansion the tubes, tube sheet, and heat exchanger shell are all made of stainless steel. The steam drum is constructed of carbon steel and is of conventional design. The risers which carry the steam-water mixture from the heat exchanger to the drum and the downcomers which return the cooler water to the heat exchanger are arranged to provide for natural circulation and allow flexibility for expansion. If tubes leak in the heat exchanger, they can be plugged from access holes provided for this purpose.

The Shippingport U-tube design shown in Fig. 6-17 is similar to the straight tube design except that the U-bends in the heat exchanger allow for differential expansion. As a result the shell can be made of carbon steel even though the tubes are made of stainless steel.

The Shippingport plant has four loops, two of them each having a straight tube steam generator and the other two each having a U-tube steam generator. The steam generators cool the nominal 2,000 psi (140 kg/cm^2) pressurized water from 542°F (283°C) to 508 (264°C) and each produces 287,000 lb/hr (132,000 kg/hr) of dry saturated steam at a nominal pressure of 600 psia (20 atm).

3. Direct-Cycle Boiling Water Steam Generation

A direct-cycle boiling water reactor plant utilizes, in the turbine, steam generated directly in the reactor. Steam separation can be accomplished within the reactor vessel itself, or it can be done by means of a separate drum. When steam separation is done within the reactor vessel, the steam-water mixture is directed through cyclone separators above the core. When it is accomplished by means of a separate steam drum, risers and downcomers must be provided between the vessel and drum as is done in a more conventional steam generator.

6.5.1.2 Steam Generators for Liquid-Metal Systems. A number of different types of steam generators have been designed and built for use in liquid-metal systems.

The problems faced in the design of steam generators for liquid-metal systems are somewhat different from those faced in water systems. Inasmuch as most liquid-metal systems use sodium as the coolant, attention must be given to avoiding sodium-water reactions. As a result, a number of steam generators for sodium service use double tubes with an intermediate fluid between them; however, some designs use a simple tube to separate sodium and water. Some systems use an intermediate heat exchanger loop with an appropriate heat exchanger between the primary loop and the steam turbine loop. Such an intermediate loop permits more flexibility in choosing the liquid-metal serving the steam generator and limits the extent of travel of the radioactive primary sodium. However, the intermediate loop interposes an additional loss of temperature between the primary coolant and the steam going to the turbine. In all liquid metal systems the higher temperatures available with liquid metals bring about more severe thermal problems than encountered in water systems.

A number of different types of steam generators have been designed and built for use in liquid-metal systems. Three separate designs each illustrating a different approach will be discussed below.

1. Sodium Reactor Experiment (SRE) Once-Through Steam Generator.

The once-through steam generator for SRE is a horizontally mounted, U-shaped shell and tube heat exchanger containing 199 double tubes. (See Figs. 6-18a and 6-18b [120].) The inner tube contains water and the outer is surrounded by sodium; the annulus between tubes is filled with mercury which serves as a leak detection medium for both tubes. The mercury is maintained at a pressure intermediate between the 625 psi (43 kg/cm^2) steam and 60 psi (4.2 kg/cm^2) sodium. Demineralized and deoxygenated feedwater enters at one end of the unit and flows inside the tubes in counterflow with the sodium on the shell side. The water is preheated, evaporated and superheated while going through the tube. The steam generator produces 88,800 lbs/hr (40,500 kg/hr) of superheated steam at 625 psig (44 kg/cm^2) and 825F (441°C). The 60 psi sodium is cooled from 900F to 440F (227°C) in passing through the unit.

The tube bundles are supported in egg crates; supports are omitted at the U-bend area of the tubes to facilitate thermal expansion. There are no baffles outside the tubes to direct the sodium flow. As a result some stratification of sodium takes place within the shell; this results in an uneven distribution of temperature in the steam

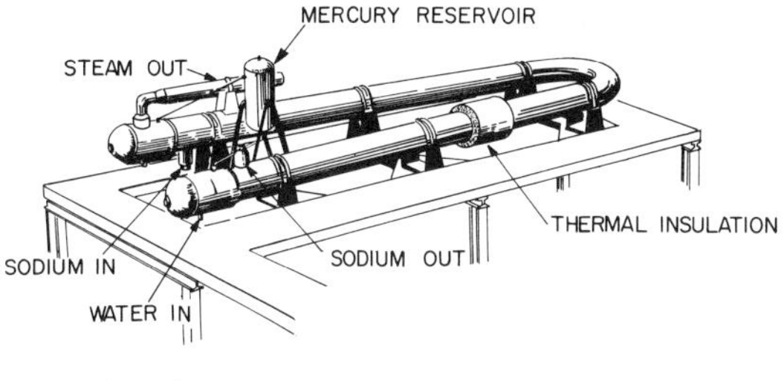

FIG. 6-18a SRE once-through steam generator.

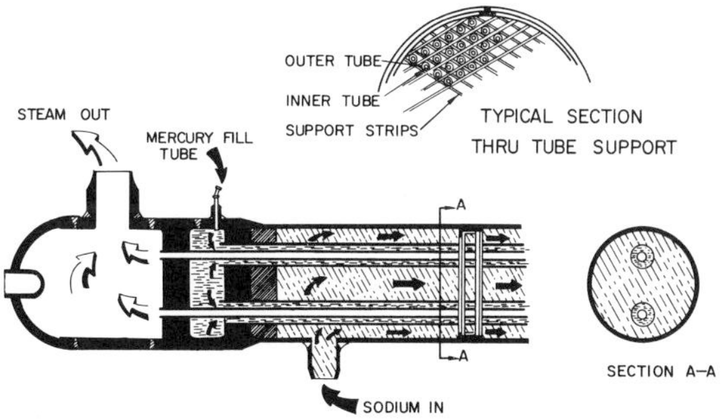

FIG. 6-18b Simplified cut-away of tube arrangement.

coming from the tube bundle. Aside from this, the performance of this unit has been quite satisfactory [120].

2. Enrico Fermi Once-Through Steam Generator.

Each of the steam generators for the Fermi Atomic Power Plant is a vertical-shell once-through unit, but in this design a single tube wall is used rather than the double-tube design used in the SRE. Both systems use intermediate sodium loops between the primary coolant loop and the steam generators.

In the Enrico Fermi design [120] the sodium from the intermediate loop enters the side of the steam generator shell through two 12-inch (30 cm) nozzles and leaves the bottom of the shell through a single 18-inch (46 cm) nozzle. In the shell the sodium passes over serpentine involutes in which steam is generated. These involutes are fabricated from 5/8-inch OD tubes which are 0.042 inch (.107 cm) thick. The tubes and shell are made of 2-1/4% chrome-1% molybdenum stainless steel. The tubes are attached to headers by welding and light rolling.

The feedwater is led into the unit through a 8-inch (20 cm) diameter nozzle which feeds an upper manifold; from this manifold the feedwater passes through the downcomer portion of the tubes which are inside a steam sodium shield. At the bottom of the downcomer section the water turns upward into serpentine involutes where steam generation takes place. The steam leaves the tubes at a lower manifold and is led from the unit through a 12-inch (30 cm) steam outlet nozzle. Each of the two steam generators in the plant produces 476,000 lbs/hr (218 kg/hr) of superheated steam at 900 psig and 780°F (416°C). The sodium is cooled from 820°F (438°C) to 520°F (271°C) in passing through the unit.

During an early hydro test of the unit at 850 psi, a leak of about 200 cc per hour was observed. After some confirming tests, the suspected bundle was removed and examined. Cracks were observed on some of the tubes. Metallographic examination of the cracked tubes indicated that the tubes failed as a result of stress erosion cracking. The defective tubes were replaced; the tube bundles were stress released and cleaned and reinstalled into the unit.

Later difficulties arose because leaks of pressurized water into the hot sodium resulted in a metal-water reaction. This reaction released additional heat which weakened adjacent tubes causing them to expand locally and rupture in turn. It is evident that heat exchangers designed for liquid metal-water systems must take all precautions to avoid leaks and must provide means for insuring that the spread of these leaks and the resultant metal-water reaction is not autocatalytic. Even though radioactive sodium is not involved, a major metal-water reaction could have serious safety implications for such a plant.

3. Steam Generators for Experimental Reactor II - EBR-II.

The steam generator for EBR-II consists of a natural circulation evaporation section, a conventional steam drum and a once-through superheating section. As pointed out in reference [120], the evaporating section is comprised of eight identical shell-and-tube heat exchangers connected in parallel on the tube side to a horizontal steam drum with internal moisture separation. Dry and saturated steam at 580°F (304°C) leaves the top of the steam drum and flows downward through four identical, parallel, vertical shell-and-tube superheaters, from which the steam flows to the turbine generator unit at 840°F (449°C) and 1250 psig (~ 86 kg/cm^2).

The general arrangement of evaporators and superheaters is shown in Fig. 6-19 [120]. Both the evaporators and superheaters are constructed entirely of 2-1/4% chrome-1% molybdenum steel and utilize double-walled tubes. In four of the evaporators the double walls are mechanically bonded, and in four of the evaporators and four of the superheaters the tubes are metallurgically bonded. The units use two tube sheets at each end with the outer tube welded to the sodium sheet and the inner tube welded to the steam tube sheet. The space between the tube sheets communicates directly with the atmosphere [120].

6.5.1.3 Steam Generators for Gas-Cooled Systems. Although the gas-cooled reactor, as represented by the Oak Ridge Reactor, was one of the first reactor types built in the United States, the application of gas-cooled reactors to the production of nuclear power in the United States has lagged behind the water-cooled and liquid-metal-cooled reactors. However, the Gas-Cooled Peach Bottom Atomic Power Station at Peach Bottom, Pennsylvania, represents a departure from this trend. This plant uses pressurized helium as the primary coolant; helium is used because of its excellent heat transport qualities relative to other gases and because it does not react chemically with other materials at high temperatures. The Peach Bottom steam generator design is quite

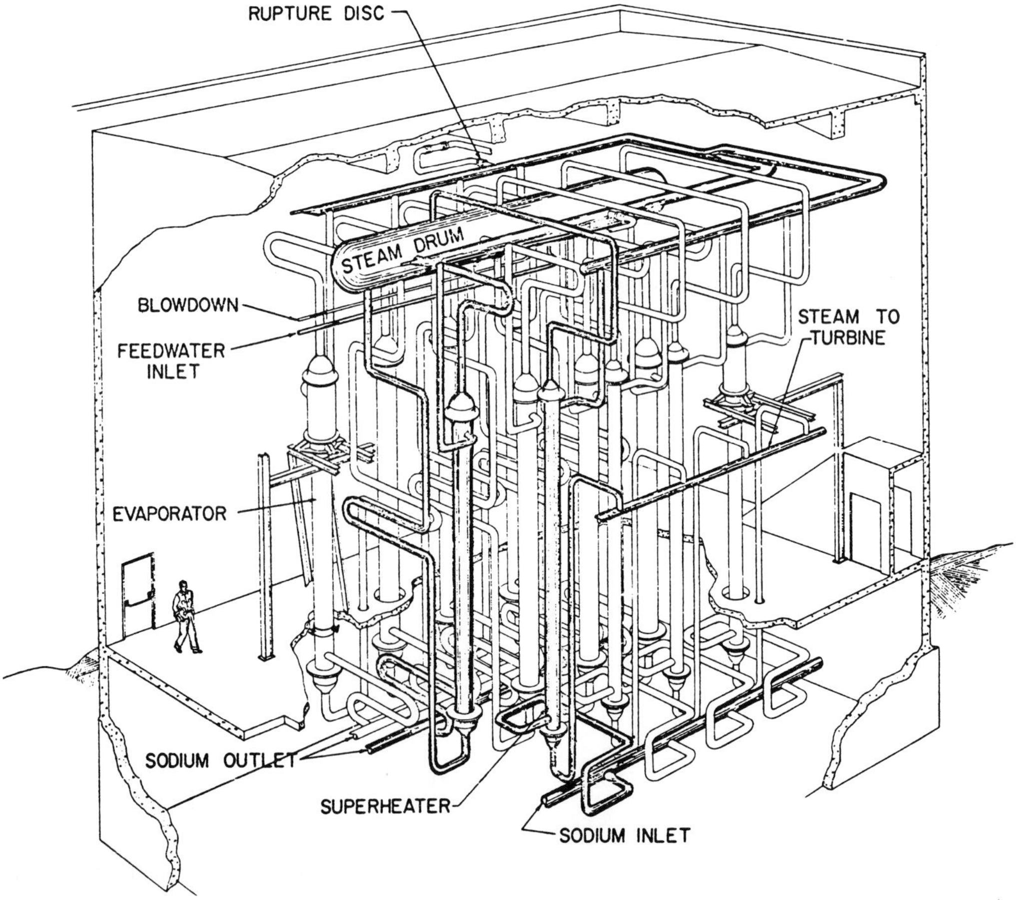

FIG. 6-19 EBR II sodium-heated steam generator.

similar to that developed for the Experimental Gas Cooled Reactor (EGCR) at Oak Ridge, Tennessee. Although further work on EGCR has been discontinued, the description of its steam generator design is presented below because of the pioneering nature of the design and some of its novel features. The similarities and major differences between the Peach Bottom and EGCR steam generators are discussed.

In both designs, one of the objectives has been to provide accessibility to all tube ends for inspection and maintenance without the need for entering the gas side of the unit. As pointed out in reference [118] this avoids the costly and time-consuming operation of removing the helium and then refilling the unit with helium after inspection and perhaps repair operations.

1. Steam Generator for EGCR.

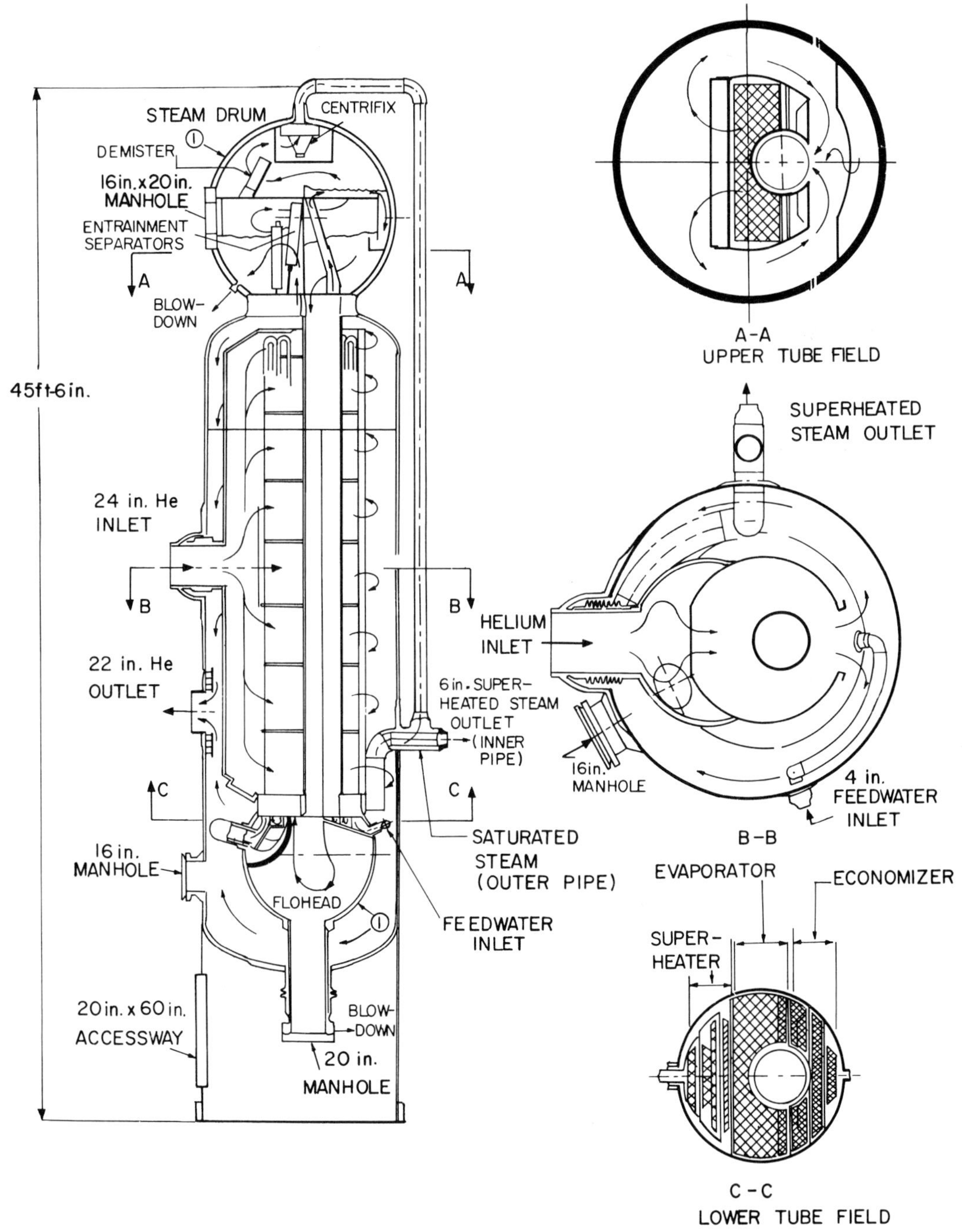

FIG. 6–20 EGCR steam generator.

The EGCR design has two loops with a steam generator in each loop which cools the 350 psia (~ 23 atm) helium from 1043°F to 483°F (562 to 250°C). Each steam generator produces 140,000 lb/hr (64,000 kg/hr) of superheated steam at 1295 psig (91 kg/cm^2) and 903°F (484°C).

The steam generator design has a vertical 108-inch (2.74 m) ID cylinder with a 102-inch (2.59 m) ID spherical steam drum at the top. It contains economizer, evaporator and superheater tube banks in a single shell. (See section C-C of Fig. 6-20 [124].)

As indicated in Fig. 6-20, helium enters the steam generator through a 24-inch (0.61 m) inlet near the middle of one side of the shell. The helium flows into the inlet plenum where it is distributed vertically and then passes horizontally across the nonfinned tubes in the superheater section, the 6 rows of finless and the 13 rows of finned tubes in the evaporator section, and finally across the finned tubes in the economizer section. The helium then flows into an annulus formed between a shroud which encases the tubes and the pressure shell. The cooled helium leaves the steam generator through a 22-inch (0.56 m) nozzle located below the inlet nozzle [124].

As stated in reference [124], feedwater is pumped through the economizer tubes where it is heated to just below the saturation temperature. From the economizer, the heated water enters the steam drum section and flows through the downcomer to the mud drum. The water is heated to the saturation temperature and a portion of the water is converted to steam as it flows upward through the evaporator section to the steam drum. Water which is not converted to steam recirculates back through the 20-inch (0.51 m) diameter downcomer with water from the economizer. The steam is separated from the water in the steam drum and piped from the drum to the superheater inlet.

The following is quoted from reference [124]: "The steam generators contain an integral water trap; that is, all of the water in the steam generator, including that in the steam drum, can be contained in the bottom of the shell at a level just above the bottom of the floating head. The trap prevents large amounts of entrained water from being carried over into the reactor coolant blowers or into the reactor if leaks develop in one or more tubes. In the unlikely event that several tubes rupture in a manner such that the full feedwater flow is pumped into the shell side of one steam generator, the automatic loop isolation on the flood level in that steam generator would prevent filling the shell to the bottom of the helium outlet nozzle.

"In case of fuel element ruptures in the reactor, some radioactive contaminants may be deposited on the tubes or on other helium-side surfaces in the steam generator. Access to the steam-water side of the tube-sheets is permitted in order to locate defective tubes, plug tubes, and perform other maintenance operations. Access to the upper tube-sheet is through a manhole in the steam drum. A manhole in the bottom of the pressure shell allows entrance into the mud drum. During repair, personnel within the steam and mud drums are shielded by the 12-inch (30 cm) thick top and bottom tube-sheets and a shield plug inserted in the downcomer hole at the time of repair.

"Should it become necessary to enter the helium side of the generator, manholes in the pressure shell and the double walled helium shroud permit entrance to the tubefield."

2. Steam Generator for Peach Bottom Atomic Power Station.

The Peach Bottom primary system consists of two loops each with a forced recirculation steam generator which cools the 350 psia (~23 atm) helium from 1352°F to 622°F (734 to 328°C) and which together generate a total of 365,500 lb/hr (166,000 kg/hr) of superheated steam at 1544 psig (109 kg/cm^2) and 1005°F (540°C).

As indicated in Fig. 6-21 [125] the steam generator is a vertical shell-and-tube unit with economizer, evaporator and superheater tubes in a single shell [125]. It differs from the EGCR design in that it has a separate steam drum to provide sufficient water holdup for the evaporator. (See Fig. 6-21.) The helium flow path through this unit is similar to that described for the EGCR unit. The water flow path, while similar in principle, is different in that the tubes are fastened to the tube sheet at the top rather than at the bottom as they are in EGCR and in that external piping is necessary between the external steam drums and the steam generator tubes.

The economizer tubes are manufactured from SA-179 carbon steel, the evaporator tubes from SA-192 carbon steel, and the superheater tubes are made of SA-213-type 304H stainless steel. The steam generator shell is fabricated from SA-212-Grade B steel. The shroud is made of SA-167-type 304 stainless steel [125]. In this design as in the EGCR design the shroud eliminates contact between the hot helium and the steam generator shell.

The design indicates no integral water trap as does the EGCR unit. The most probable source of water in-leakage is thought to be the tube-to-tube-sheet welds. Although they are helium-leak-tested prior to operation, leaks might be brought about by thermal cycling. Reference [125] states, "A baffle running parallel to the main tubesheet but some small distance below it will be installed to enable a continuous helium purge flow to flush out the steam in-leakage. The steam generator tubes will pass through the baffle with a small clearance between the baffle and tubes. Helium will be drawn up through the tube baffle spaces at a rate of 100 lb/hr (45 kg/hr) per baffle. The water-contaminated helium will have the water removed in the helium purification system."

6.5.2 Basic Design Considerations

The considerations involved in the design of heat exchangers and steam generators are grouped under the following headings: design codes; leak-tightness; materials; structural considerations; maintenance; and steam separation. Each of these topics is discussed below.

6.5.2.1 Design Codes. As outlined in Sec. 2.1 the ASME Boiler and Pressure Vessel Code provides specifications for the design and construction of steam boilers and other pressure vessels. Because of the many questions about nuclear pressure-containing vessels which were not answered by the

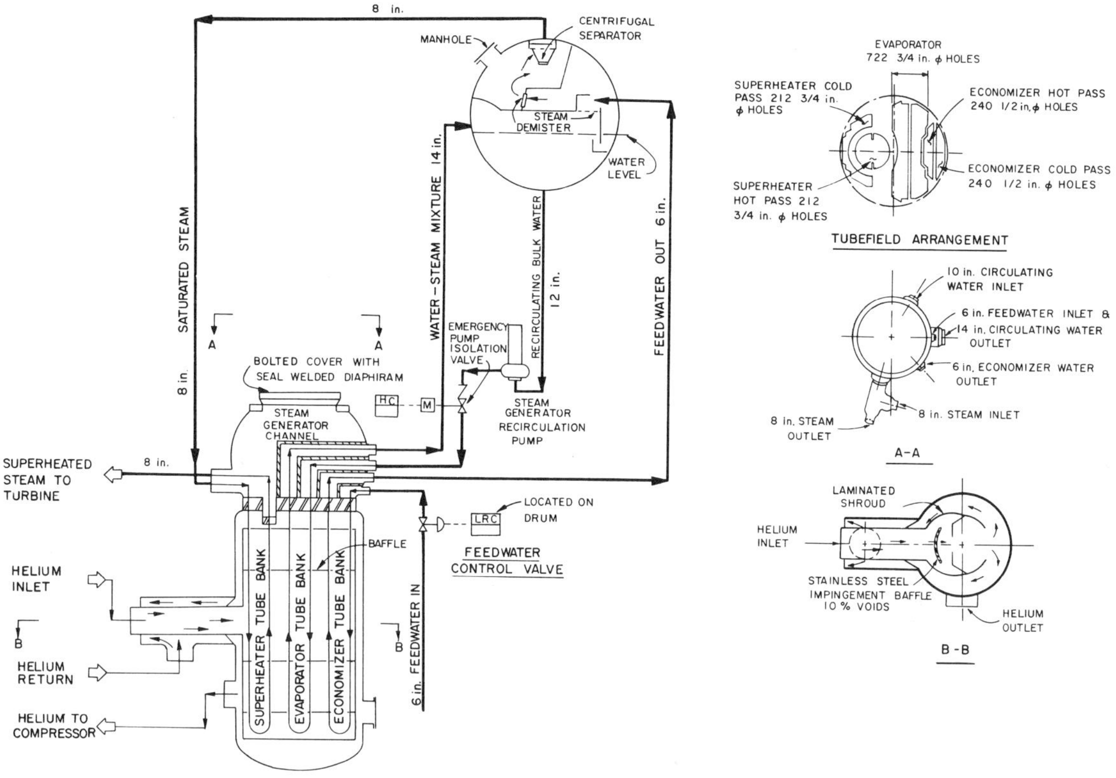

FIG. 6-21 HTGR steam generator circuit.

code, a Special Committee on Nuclear Power was established to consider these questions and make appropriate recommendations. As a result of the committee's work a new Section III of the Code has been prepared as discussed in Sec. 2 of this chapter. In parallel with this work specific modifications have been made to the existing codes as replies to Special Code Cases designated by the suffix N after the case number. These cases establish more stringent design, inspection and fabrication rules for nuclear components than are required for similar nonnuclear equipment [118]. These cases are based on the applicability of Sections I and VIII of the Code; typical of some of the additional requirements are those highlighted in reference [118].

1. Double-welded butt joints or their equivalent for all longitudinal or circumferential seams.
2. Compensation for all vessel penetrations, regardless of diameter.
3. Full penetration welds for all connections except where physically impossible due to spacing.
4. Special requirements regarding the use of safety valves and gauge glasses as well as other requirements of containment vessel design.

The reader should refer to Sec. 2 for more detailed treatment of these code cases as well as the new Section III of the Code. The reader is also referred to references [126] and [127] regarding military specifications applicable to steam generators for pressurized-water systems. Most of the pressurized-water steam generators built to date in the United States have been built in accordance with these references.

6.5.2.2 Leaktightness. Leaktightness in a steam generator, as referred to here, is concerned primarily with control of leakage between the primary-coolant side and secondary-steam side of the heat exchanger. It is an important consideration for a number of reasons some of which apply to all designs and others of which apply only to specific situations. These requirements lead in all cases, however, to the need for care in the fastening of tubes to the tubesheets or headers in a steam generator. Some of the more important reasons for leaktightness are the following:

1. To restrict transport of radioactive contamination and any fission products from the primary system to the steam system.
2. To prevent chemical reactions between the primary fluid and the secondary fluid. In the case of a sodium-to-water heat exchanger, large leaks can produce violent sodium-water reactions. Even if the leaks are small and the reactions nonviolent, water entering the sodium can produce sodium oxide or sodium hydroxide which can lead to fouling of the sodium system.
3. To prevent buildup of high pressure on the low-pressure side of the generator. This

might take place from the primary side to the secondary side of a pressurized water reactor where the primary coolant pressure is considerably higher than the steam pressure. In a liquid-metal or gas-cooled reactor the pressure buildup would tend to be in the opposite direction because the primary system pressure, in these cases, is generally lower than the secondary steam pressure. Such a buildup of pressure could lead to severe failures if the low-pressure side is subjected to significantly higher pressure than that for which it is designed.

4. To prevent introduction of water vapor into a gas-cooled system where the water vapor could react with the moderator.

To obtain leaktightness requires careful attention to design, to materials integrity, and to fabrication and testing. Even where tube-to-tubesheet joints rely on welding for leakproofness, porosity in the welds can be a source of leakage; stress corrosion cracking can also be a source of leakage as discussed in Sec. 6.5.2.3. Where tubes are rolled into the tube sheet, consideration must be given to the consequences of thermal cycling on the integrity of the joint. For the latter two reasons a successful hydrotest does not assure leaktightness if the design is poor.

6.5.2.3 Materials. The general considerations outlined in Secs. 6.2.2, 6.2.3 and 6.2.4 regarding piping apply to steam generators. The considerations involving corrosion and stress-corrosion cracking are generally more demanding when applied to steam generators because the tubing wall thickness involved is significantly thinner and the concentration of dissolved materials more severe in the steam generator than in most of the other piping in the system, particularly on the steam side of the heat exchanger. In addition, local stresses and moments produced by differential temperatures across tube sheets may be significantly higher.

It is pointed out in reference [119] that so far as corrosion resistance is concerned, there are two principle causes for concern; one is the general corrosion rate and the other the resistance of the material to local corrosion after it has undergone fabrication and operation. While the general corrosion rate is normally used as a basis for establishing the corrosion allowance to be used, it gives very little indication of local corrosion behavior. Crevice corrosion and stress-corrosion cracking are local effects which must be considered, as is brought out in The Corrosion and Wear Handbook [128].

As indicated in the descriptions of steam-generator types in Sec. 6.5.1, stainless steels have been used widely because of their good general corrosion resistance. However, stainless steels tend to suffer from stress-corrosion cracking. This tendency has long been recognized as a problem in maintaining the integrity of pressure vessels and heat exchangers in hot water and steam service [122]. Most of the literature is concerned with stress corrosion of stainless steel in water containing chloride which is known as chloride stress corrosion. Factors contributing to chloride stress corrosion include chloride concentration, oxygen concentration and stress concentrations. The authors of reference [122] point out that failures attributable to chloride stress corrosion have been reported at stress levels as low as 5000 psi (3.5 kg/mm^2). Attention to the problems of chloride stress corrosion is particularly important in naval boilers where the feedwater is cooled by sea water, which usually causes contamination by condenser leakage [119]. The presence of two phases within the steam generator may provide a mechanism for concentration of chlorides in local areas such as crevices near the boiling surface.

Austentic stainless steel will also stress crack in solutions containing chemicals other than chlorides, particularly caustics. Quoting from reference [122]:

"It is difficult to tell in many cases whether stress corrosion has occured because of chloride or caustic. The type of attack, transgranular cracking, appears to be identical in both cases. The time for crack initiation appears to be similar for both chloride and caustic stress corrosion, although some evidence has been obtained that a caustic environment may cause more rapid crack propagation, at least under conditions existing in steam generators. Failure of 1/2 in. thick austenitic stainless steel with residual stresses due to cold work has been observed in as little as four hours.

"Conditions which can lead to caustic attack of austenitic stainless steel in steam generators are not as well defined as those for chloride stress corrosion. Results of laboratory investigations indicate that caustic attack can occur in the absence of oxygen and some stress appears to be required. A mechanism for concentrating caustic is also apparently required since cracking did not occur in autoclave tests conducted in solutions containing up to 50,000 ppm sodium hydroxide. Where a mechanism for caustic concentration does exist, severe stress corrosion cracking of stainless steel exposed to solutions containing as little as 40 ppm sodium hydroxide has occurred."

Caustic stress-corrosion cracking is believed to have been the cause of a failure of several tubes in one of the U-tube steam generators at Shippingport, which was discovered during a hydrostatic test in February 1958. At that time the unit had undergone about 150 hours of full power operation plus about 50 hours of low power operation and about two weeks of service at operating temperatures during plant checkout. The tubes were nondestructively tested in place by a Probolog, an eddy current device; the tubes were later removed and examined metallographically. Based on these examinations and laboratory tests on boiler models, the conclusion was made that the failures were due to caustic stress corrosion. Failure by chloride stress corrosion was ruled out because the chloride concentrations had always during steam generation been less than 0.05 ppm which is considerably less than the 500 ppm thought necessary to cause tubing failures of the extent found, in the 250 to 300-hour period during which cracking was thought to occur. Furthermore, the sodium sulfite treatment of the boiler water used in the steam generators had been shown to inhibit chloride stress corrosion [122]. In addition, laboratory tests with model boilers indicated that caustic corrosion of austenitic stainless steel tubing would occur in water conditions similar to those existing in the Shippingport steam generators, provided that a mechanism for concentrating caustic is present. It is believed that the concentration of free caustic which existed in the unit was caused by steam blanketing of the tubes; such steam blanketing was shown to exist in the other duplicate U-tube steam generator in the system. The boiler water treatment at Shippingport consisted of phosphate-pH control of alkalinity and the use of sodium sulfite for oxygen scavenging [122]. Modifications were made to both the water chemistry to control the free caustic in the water and in the steam generators to eliminate steam blanketing. Based on the experience with the Shippingport steam generators, the authors of reference [122] state that it is imperative to observe the following preventive measures in the design and operation of stainless steel-tubed steam generators for nuclear power plant service:

1. The steam generator design should insure the absence of steam blanketed regions in the tube bundle by installation of well distributed steam risers with adequate capacity for steam removal.
2. Sodium hydroxide should not be used for pH control.
3. Only orthophosphates should be used for phosphate control.
4. In order to avoid the possiblility of caustic stress corrosion, the phosphate-pH balance should be maintained at least 0.2 pH units below the Whirl-Purcell curve of stoichiometric neutrality for trisodium phosphate (see Fig. 6-22 [122]).

Cracks were detected in the stainless steel tubes of both steam generators of the Peach Bottom Atomic Power Plant during tests being made in preparation of the plant for its initial start-up. Chloride stress corrosion is again suspected of having been the cause of cracking. Corrective action in a situation such as this could lead to delays of 6 to 9 months in start-up of the plant.

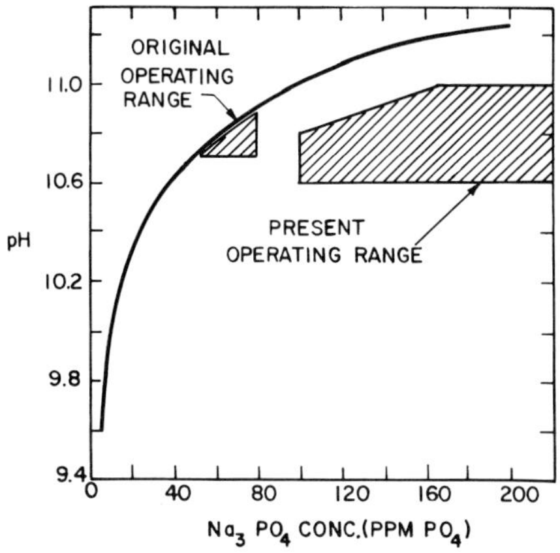

FIG. 6-22 Boiler water phosphate-pH control, Shippingport Atomic Power Station.

A number of stainless steel heat exchangers have failed at the Savannah River AEC Laboratory even at temperatures below the boiling point of water and at low pressure. It is believed these failures are due to the concentration of chloride deposited by evaporation of water from small leaks into gas-filled spaces between the headers of double tube sheet heat exchangers.

6.5.2.4 <u>Structural Considerations</u>. The same careful analyses of mechanical and thermal stresses must be made for steam generators as for the reactor vessel as outlined in Sec. 2 of this chapter. In designs employing tubesheets, the high stresses generally occur at the juncture between the tube sheet and the primary head and in the tube sheet itself. In once-through type boilers the transition zone between nucleate and film boiling may be the area of highest stress [118]. Analyses should include fatigue effects brought about by thermal cycling and thermal shock brought about by severe temperature swings.

6.5.2.5 <u>Inspection and Maintenance</u>. As indicated in the description of various boiler types, careful attention must be given to problems of inspection and maintenance. From the experiences encountered, it is clear that tube failures can be expected with the present state of technology. Maintenance considerations should include attention to problems of tube-and-shell inspection, tube plugging, tube replacement, and tube bundle replacement, as well as to replacement of the complete unit. Attention should be given to providing for inspection of tubes which does not require entering the primary coolant side if possible, particularly where the coolant is helium and the time and cost of removing and replacing helium is a consideration.

Consideration must also be given to cleaning of and blowdown of boilers to reduce the concentration of dissolved impurities and solids in the feedwater.

6.5.2.6 <u>Steam Separation and Purification.</u> Steam separation is an important requirement for steam generators both on units which are to deliver dry and saturated steam and those which use forced recirculation with or without superheating. Moisture must be removed from steam generators which feed saturated steam directly to a turbine in order to avoid excessive erosion of turbine blades, particularly at the low pressure end. Steam must be removed from some recirculated coolants in order to avoid severe cavitation and erosion problems in the recirculating pumps [129].

Steam purification is necessary because the steam produced in the tubes of a steam generator, moving at high velocities, carries with it solid impurities coming from concentration of such impurities in the tubes during boiling. Even relatively low total solids contamination in the steam (0.6 ppm) can result in troublesome blade deposits [119].

Dissolved solids in the boiler interfere with coalescence of bubbles and result in a foaming action in the steam drum, making moisture separation more difficult. In addition, the concentration of the dissolved solids in the unevaporated water increases with time in the boiler tubes, because, except for mechanical carry-over, the solids remain behind with the unevaporated water. As a result it is necessary to blow down the excess solids by appropriate valving.

The separation of steam from water is carried on in the steam drum. Large masses of water can be removed by gravity action. To achieve a high degree of moisture separation and steam purification, special steam separation and purification apparatus is used in the steam drum, such as reversing hoods, baffles, screens and tube-separators whereby gravity is assisted by change in flow direction, impact and centrifugal action. Reference [119] describes and discusses the various devices used. Reference [119] points out that, with adequate water level control in the steam drum and appropriately sized steam separation and purification equipment, the moisture leaving the steam drum can be kept below 0.10% and the solids in the steam below 1 ppm (in some instances as low as 0.04 to 0.08 ppm).

6.6 Pressurizing and Expansion Systems

In pressurized-water-cooled reactors the pressurizer and expansion system forms an important adjunct to the primary loop. The purpose of the pressurizer is to insure that the coolant in the primary system will remain in a subcooled state during steady state and transient operation and to accommodate, without exceeding the allowable pressure range, the contraction and expansion of primary system water as temperatures change during operation. Because of the large volume of water involved, the expansion of water associated with going from room temperature to operating temperature is not accommodated within the pressurizer and expansion system; instead this water is released from the system through a coolant discharge system. When the system is cooled down to room temperature, make-up water must be provided to the system.

6.6.1 Pressurizing Means

Three basic pressurizing means are available to the plant designer of a pressurized water reactor, only two of which are considered practical for large systems and only one of which has been used extensively. These include steam pressurization, gas pressurization, and pressurization by feed pump with controlled bleed and feed of water to and from the system.

The steam pressurization scheme has found the most extensive application. In this system a pressurizer vessel is provided with water in the lower portion and saturated steam, at the desired system pressure, in the upper portion. Electric heaters submerged in the water are provided to generate the steam and to restore the pressure after an outsurge of system water. A spray line for introduction of relatively cool water into the pressurizer is usually provided at the top of the pressurizer to reduce pressure during an insurge of primary system water. The bottom of the pressurizer is connected to the primary coolant system by a surge line which permits flow of water to or from the pressurizer as required by the expansion of water in the primary system. The size of the pressurizer is a function of the magnitude of the volumetric surges, the degree of pressure control required, the heater volume and the variation volume to allow for level-control instrument error. The magnitudes of the surges are a function of the total mass of the coolant, the extent of the imposed power change causing the surge, the rate of power change and the responsiveness of the reactor plant [130].

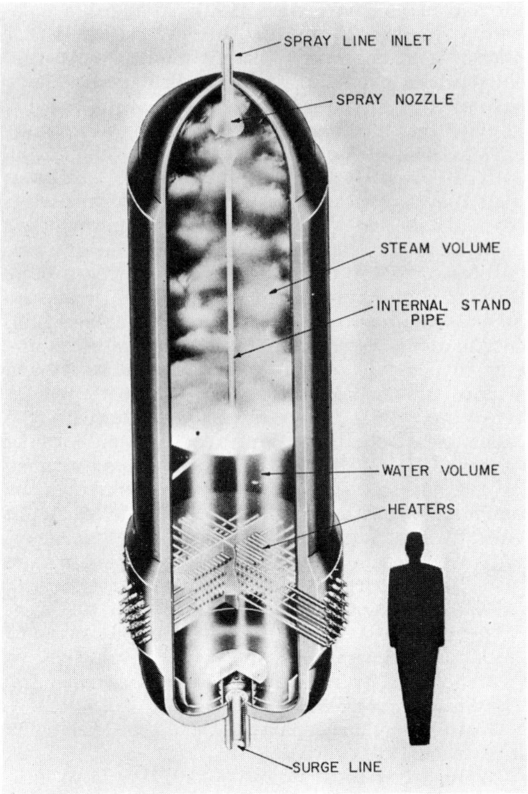

FIG. 6-23 Pressurizer used in Shippingport plant.

Figure 6-23 [131] shows a cutaway view of the pressurizer used in the Shippingport pressurized-water plant. It is a vertical pressure vessel approximately 18 ft (4.6 m) tall and 5 ft (1.27 m) in diameter and is made of carbon steel clad on the inside with a stainless steel liner. It was designed for 2,500 psi (175 kg/cm^2) pressure and operates nominally at 2,000 psi (140 kg/cm^2). The heaters shown near the bottom of the vessel are removable and fit inside stainless steel heater wells which are an integral part of the vessel. The heaters are of the metal sheath type. Each heater is rated at 2,500 watts; 342 heaters are used in the pressurizer. The pressurizer has a spray inlet nozzle at the top for spray cooling and an internal standpipe to provide a reference pressure for the differential pressure cell used to determine the water level in the pressurizer [39].

The use of gas pressurization in a pressurized-water system would utilize a noncondensable pressurized gas such as air, nitrogen, or helium to maintain the desired pressure. Gas pressurization has not found favor for water systems because it raises the question of gas diffusion into the coolant and requires another complex bleed and feed control system involving an additional fluid to maintain pressure. Reference [132] discusses some of the considerations given to steam pressurization vs gas pressurization for the Homogeneous Reactor test at Oak Ridge.

In a pressurized gas-cooled plant, however, pressure is maintained by gas pressurization. This is accomplished by admitting purified gas coolant to the loop from a pressurized storage tank fed from a coolant purification system. Such a system is used for the Peach Bottom Atomic Power Station [125]. A control valve is used automatically to keep pressure approximately constant during all normal load swings. An alarm is sounded if the pressure swing exceeds allowable limits. If the pressure goes below a set value the reactor is automatically scrammed. In the event of an overpressure, a combination of several safety devices protects the system. The first consists of backpressure-operated dump valves which open to blow the helium to the dump tanks in the helium handling system. Should these valves fail to operate and the pressure continue to rise, due, for example, to a steam leak into the helium at the steam generator, ASME Code relief valves would come into play. These valves are arranged in three sets each with a full-sized spare, for the reactor and for each of the two steam generators [125]. All relief valves have an upstream rupture disk with a tell-tale leakoff chamber between it and the relief valve [125].

The third system of pressurization utilizing a feed pump with control bleed and feed of water to and from the system has been used for test loops but has not been used for reactor application.

In the rest of the discussion in this section only the steam pressurizer will be discussed, inasmuch as it is typical of those currently in use in pressurized water systems.

6.6.2 Influence of Pressurizing System on Transient Characteristics of Plant

6.6.2.1 Pressure Control. The pressure of a pressurized-water plant is maintained within prescribed limits by expansion or compression of the steam phase in the steam pressurizer. If the power demand on the nuclear plant is increased suddenly, heat is withdrawn from the steam generator faster than it is supplied by the reactor; as a result the average coolant temperature drops and the water level in the pressurizer drops, thereby causing some of the water in the pressurizer to flash to steam. This flashing helps maintain the pressure until the pressurizer heaters return the pressure to normal. If the power demand on the plant is suddenly decreased, the average temperature of the coolant rises and causes the coolant to expand into the pressurizer, thereby compressing the steam and causing a rise in pressure. This rise in pressure is controlled by introducing reactor coolant through the spray nozzle at the top of the pressurizer. The spray water condenses some of the steam and reduces the pressure to keep it within the allowable limits. In the Shippingport plant the reactor coolant spray into the pressurizer is induced by the pressure drop across the reactor pressure vessel [39].

Maxwell presents, in reference [130], a method for determining the performance characteristics of a saturated-steam pressurizer in a nuclear power plant. He points out that the negative temperature coefficient of reactivity (α_T) can be used as an indication of the temperature responsiveness of a reactor plant to load changes. As the reactor approaches its end-of-core life, the α_T becomes more positive and the plant temperature variations are greater for any given load swing than they were at the beginning of life. Maxwell points out that, for a given reactor plant, surge volumes become greater as α_T becomes more positive so that, to have an adequate pressurizer, it is necessary to base the design on surges resulting at the end-of-core life. Thus the pressurizer may have to be made larger than indicated by beginning-of-life considerations. Drucker and Tong present, in reference [133], experimental verification of analyses of pressure surges in a 200 psia system without sprays. Thomas and Findley present, in reference [134], a computer program for transient analysis of pressurizers for pressurized-water plants; the program is for the Philco 2,000 Computer. The pressurizer model used in reference [134] consists of four distinct fluid regions: steam, spray and condensate, active or main water, and the subcooled lower insurge water.

6.6.2.2 Pressure Relief. In order to protect the system against pressure surges beyond the control capability of the pressure control system, a pressure relief system must be provided. If the pressurizer and reactor vessel can be isolated from each other by means of isolation valves, self-actuated relief valves must be provided for both vessels. (See Sec. 2.5.)

In order to protect against a loss-of-coolant accident in the event that the relief valves do not reseat after relieving a volume surge, the Shippingport plant utilizes two pairs of self-actuated relief valves, each with a motor-operated isolation valve in its discharge line. The isolation valves are interlocked so that one relief valve is always open to the coolant discharge and vent line. As pointed out in reference [39] the use of isolation valves for relief valves required a waiver from the Pennsylvania Code Authorities. This arrangement was used both for the reactor vessel and the pressurizer. In addition, the pressurizer was provided with a pilot-operated steam relief valve of more reliable seating characteristics; it was set at a lower pressure than the self-actuated valves to reduce the number of times that the self-actuated valves had to open. Other pressure-relief devices in the system included a self-actuated water-relief valve in the reactor inlet piping of each loop and pressure-relief valves in a number of the auxiliary systems. In this plant the reactor is scrammed if at any time the system pressure falls below prescribed limits.

6.6.3 Design Considerations

The basic considerations involved in the design of the pressurizer can be grouped under the headings of pressure control, pressure relief, code requirements, structural considerations, determination of pressurizer size, and maintenance. The first two have been discussed in the preceding part of this section. For Code requirements the reader is referred to Secs. 2.1.1 concerning reactor vessels and 6.5.2.1 concerning steam generators. The pressurizer tank is a pressure vessel subjected to conditions similar to those discussed for the reactor vessel with the possible exception of radiation effects. As a result the entire Sec. 2 on reactor vessels should be consulted in the design of the pressurizer. The considerations that will be taken up here are those associated with sizing the vessel and those associated with maintanence. In addition, because of the design implications involved, a brief discussion of the SPERT-III pressurizer-vessel failure is presented in Sec. 6.6.3.3. The reader is also referred to reference [179], which describes cracks found in the cladding of the Yankee pressurizer. The cracks are believed to be associated with the more unfavorable environment in the pressurizer as compared to the rest of the plant; on occasions, the coolant in the pressurizer has been found to contain a considerable amount of oxygen. Detailed investigations of some of the cracks shows that they terminated at the cladding-vessel interface. Continued monitoring is a prudent step in such a situation.

6.6.3.1 Sizing the Pressurizer. As indicated in Sec. 6.6.2 the pressurizer should be sized for the most unfavorable conditions, which in most pressurized water reactors are the end-of-core-life conditions [130]. In order to assure satisfying design requirements it is well to consider the pressurizer as subdivided into a number of distinct volumes, each sized to meet its own requirements. While the number of subdivisions can be varied to suit the particular needs involved, the following volumes should be considered:

1. The Bottom Head and Heater Volume. This volume must be sufficient to keep the top row of heaters covered with water.
2. Level-Instrument-Error Volume. (Sometimes called Variation Volume). The inaccuracies in measurement of water level in the pressurizer can account for an appreciable amount of water volume. Even an error of as small as 2% represented a volume of 3.3 ft.^{3} (.085 m^3) in the Shippingport design. Actually, two level-measuring devices were required, a short scale instrument (125-inches or 3.17 m) for use during normal operation and a wide-seal instrument (190-inches or 4.80 m) for use during start-up and shutdown. Both instruments could be kept within 2% accuracy by manually compensating for density changes throughout a varying pressure and temperature change [39].
3. Flashing-Water Volume. The volume occupied by water which flashes to steam during outsurge of water from the pressurizer must be provided in the pressurizer. This is based on the largest pressure decrease expected during such a surge.
4. Contraction of Primary Coolant. The volume of water which must be provided to the primary coolant when it contracts during a power demand must be provided in the pressurizer. With only temperature coefficient control, this will depend upon the negative temperature coefficient and the rate of load change. A drop of 6°F (3.3°C) in 60 sec. in the Shippingport plant required 17.7 ft^3 0.50 m^3) of water from the pressurizer. The pressure decrease brought about by this action should be checked against the allowable lower pressure limit.
5. Power-level Variation. Because of change in the temperature distribution between zero load and 100% load, additional water may be required in the pressurizer even though the average temperature is the same.
6. Volume Due to Average Temperature Variation. The allowable variation in average loop temperature will produce water volume changes in the loop which must be provided in the pressurizer.
7. Steam Volume. The steam volume must be sufficient to accomodate volume expansion encountered in the primary system during all load transients that might be expected, without exceeding the allowable upper pressure limit. A significant margin should be added for contingency. This volume is required over and above all the others.

In addition to sizing the pressurizer volume, consideration must be given to determining the capacities required for the heaters and the spray coolant. These arise from the surge actions associated with the volume changes indicated above. The heating and cooling capacities assumed in limiting the volume changes to those computed above must be provided for satisfactory operation.

6.6.3.2 Maintenance Considerations. Three specific maintenance considerations should be highlighted for pressurizers. The first is provision for easy replacement of heaters. These are subject to failure after prolonged use. The second is accessibility of relief valves for inspection and repair if necessary. The third involves provision for reducing the pressure in the pressurizer when the system is to be depressurized. If the pressurizer has provisions for spray cooling, the problem is adequately taken care of. However portable plants without such provisions have been built. As a result, depressurization has taken as long as ten days, even with all lagging removed from the pressurizer.

6.6.3.3 SPERT-III Pressurizer-Vessel Failure. On October 26, 1961 a failure occurred in the pressurizer vessel of the SPERT-III reactor project which is used for investigating the kinetic behavior and safety of water reactor systems. The failure occurred while nonnuclear tests were being performed in preparation for a series of experiments to study cold-water accidents [135, 136]. After shutdown of the plant, it was established that a 3/8-inch-wide 2 1/4-inch-long hole had opened in the central girth-seam weld metal. It was also found that a 1-inch bolt used to tie a stabilizing band around the vessel had broken, but closer examination indicated that this bolt had failed earlier.

Examination of the possible causes of the failure indicated that failure was due to overheating of the upper half of the pressurizer. This overheating is believed to have been caused by the uncovering of one or more heaters by lowering the water level in the pressurizer. This belief was based on calculations which showed that, with the temperature distributions measured in the vessel, the water level could be as much as 20% lower than indicated by the instrumentation. This would be enough to expose the one or more heaters to the steam. Inasmuch as the top two heaters were used for control, these heaters, if exposed, would heat the steam but not the water. As the water cooled the error in level indication would increase. It is estimated that temperatures approaching 1,000°F (538°C) were reached in the upper portion of the vessel. Analyses indicated that failure of the vessel could be expected at 2,500 psig (175 kg/cm^2) and 1,000°F (538°C) in as short a time as 10 hours.

The pressurizer vessel was a 2-ft 9-in.-ID, 16-ft 8-in.-high (84 × 510 cm), all-welded vessel of ASTM A-264, grade 3, 0.04 percent maximum carbon steel with type 304L stainless-steel fittings and internal cladding. The backing plate was of ASTM A-212, grade B, firebox-quality carbon steel. The cylindrical vessel walls were 2.95 in. (7.5 cm) thick, including 1/8 in. (.317 cm) of cladding [135].

A number of design changes were in order as a result of this experience. These included alteration of the pressure control circuit. Two approaches were proposed to prevent recurrence of overheating: uniform application of heat to all heaters or use of the bottom heaters as control heaters. In addition, a more reliable liquid-level device was planned. Other action included specification of tests on tensile specimens to be supplied by the fabricator and the requirement for submission of all computations and drawings of the fabricator which reflected the design considerations involved.

6.7 Steam System

Although the mechanical considerations involved in the steam plant are beyond the scope of this book, it is well to point out that steam system reliability is important to the safety of the plant because it represents the ultimate heat sink for the reactor. Its integrity is particularly important where it provides not only for normal cooling but for emergency cooling as well. Availability of cooling water for emergency cooling purposes must be assured where required by the design. The steam system is also particularly important in plants operating separate from major sources of auxiliary power. In such cases the plant must be designed to respond safely in the event of a complete power loss.

7 AUXILIARY SYSTEMS ASSOCIATED WITH REACTOR PLANT SAFETY

Every nuclear power plant contains a number of auxiliary fluid systems which are necessary for proper operation of the plant. LaPointe and Shaw [39] classify such systems into three groups: reactor plant services systems, reactor plant protection systems, and reactor plant information systems. In the first group might be included such systems as charging and volume control systems, coolant vent and drain systems, coolant purification systems, valve operating systems, component cooling systems, and component lubrication systems. In the second group would be included all systems specifically provided for protection of the plant such as pressure control and relief systems, chemical shutdown systems, shutdown cooling systems and safety injection systems. The third group would include such systems as coolant sampling systems, failed-element detection systems and coolant flow indication systems.

Although the auxiliary systems in the second group are the most directly concerned with reactor plant safety, all systems influence safety to some extent. In fact, in certain circumstances some of the so-called reactor plant services systems can be as important to reactor safety as the reactor plant protection systems. For example, failure of a valve operating system, when valve isolation of a broken pipe is required, can lead to severe loss of coolant, a truly significant reactor hazard.

In many power plant designs, one or more of the auxiliary systems is identified as being part of the important safety protection for the plant. Such systems are sometimes described as "engineered safeguards." They are often incorporated to make the plant adequately safe at a location which otherwise might be unsuitable, so far as safety of the public is concerned. In such cases the relationship to safety is clear cut and receives closer scrutiny.

It is the purpose of this section to highlight some of the basic considerations associated with the design of such auxiliary systems and to discuss specifically some of those directly associated with plant safety. Presented first is a review of some of the general considerations which reactor designers and safety reviewers have in mind as they establish the adequacy from the safety viewpoint of the plant design. This is followed by a discussion of specific systems including the coolant charging system, the coolant discharge and vent system, the valve operating system, the chemical shutdown system, the shutdown cooling system and the safety injection system. The pressure control and relief system has been discussed in Sec. 6.6.

7.1 Auxiliary Systems as Engineered Safeguards

In November 1964, the Advisory Committee on Reactor Safeguards (ACRS) reported the results of a review, made at the request of the Atomic Energy Commission, regarding current practices in the use of engineered safeguards [204]. The following paragraph from that review provides a basis for some of the considerations which enter into the design of such safeguards:

"It is important to recognize that engineered safeguards are designed to allow the siting of reactors at locations where, without such safeguards, protection of the public would be exactly balanced by engineered safeguards. On the other hand, the advantages of a remote site may be temporary, if appreciable increases in population density occur near the reactor. Few sites presently in use are such that some engineered safeguards are not desirable. Thus, the protection of the public ultimately depends on a combination of engineered safeguards and adequate distances. Engineered safeguards which can justify decrease of the distances must be extraordinarily reliable and consistent with the best engineering practices as used for applications where failures can be catastrophic. To be worthy of consideration, engineered safeguards must be carefully designed, constructed and installed, equipped with adequate auxiliary power and continuously maintained. Certain designs are based on sound engineering principles supported by materials acceptance tests; others require developmental and proof testing. In any case, provisions for regular and careful testing are required where deterioration may be expected. The acceptance of engineered safeguards to mitigate unfavorable aspects of reactor sites should continue to be based on positive evidence that these design objectives can be attained. In addition, there will probably be a continuing need to develop new devices and design concepts as reactors are proposed for less and less remote sites."

The ACRS report discusses four classes of safeguards: (1) containment and confinement; (2) containment pressure and temperature reducing systems; (3) air cleaning systems to remove fission products; and (4) core spray and safety injection systems. Some of these are clearly outside the scope of this chapter; however, the general requirements discussed below apply to all of them.

As pointed out by Hanauer [205], the engineered safeguard is in the class of "things that have to work", by which is meant that the probability of failure of the safeguard must be acceptably low. On the basis of experiences to date, Smets [206] concludes that a serious accident which will cause a fatality must be expected once in about every 300 reactor-years. If the life of a reactor plant is taken to be 25 to 30 years, then on the average, Hanauer concludes, the probability that the engineered safeguard must operate successfully at least once to avert public danger is of the order of one-tenth. This is significantly higher than probability usually deduced by considering that the component failures will take place randomly. However, Epler [207] points out that experience has shown that nonrandom rather than random component failure dominates safety system reliability.

While the experience with reactor accidents has been outlined largely with reactors designed and operated differently from present power reactors, it is not clear whether present design practices are on the average more safe or less safe than the techniques previously used [205]. While knowledge and competence have increased so have the complexities of the designs. Nevertheless, it should be pointed out that most of the accidents have occurred during shutdown or during experiments; power reactors are shut down less frequently and therefore are exposed less often to shutdown types of accidents. On the other hand, many of the accidents experienced to date involved clean cores. The consequences of such accidents would have been much worse had the cores been full of fission products as in a power reactor that has been in operation for some time.

The following precepts are suggested in reference [205] for use in developing the design of an engineered safeguard:

1. Early identification of those functions which pertain to safety and establishment of appropriate criteria for accomplishing them. This action will help reduce inadvertent compromises on the reliability of engineered safeguards as the design proceeds.
2. Adequate attention to insuring that the engineered safeguard will perform its function when needed.

Most of the examples of inadequate performance of safety systems can be attributed to design errors rather than component failures. These can be summarized in the following categories:

a. "In some instances the system could not work at all. In Heat Transfer Reactor Experiment No. 3 (HTRE-3), for example, a series resistor in a noise filter and a power-supply voltage that was too low combined to ensure that the signal from the ionization chamber could never be as great as the trip level, no matter what the reactor power (Vol. 1, p. 636) [208].

b. "Since accidents are not always obliging enough to occur in the manner foreseen by the designer, the accident sometimes paralyzes the safety system. In Alize I a safety system designed for a slow rise in neutron flux was 'poisoned' by a more rapid rise (Vol. 1, p. 613) [209].

c. "A particularly vulnerable situation arises where a safety system and a nonsafety system (control system) are interdependent in such a way that a single failure affects both; the control system, deprived of its normal signal, induces an uncontrolled change in the process variable; and the safety system, similarly affected, cannot initiate corrective action. In HTRE-3 the regulating-rod servo system and the nuclear safety system shared the same (but not the sole) ionization chamber (Vol. 1, p. 636) [208]. Thus the loss of safety, which is tolerable in most cases because safety action is not often needed, in this and similar situations would surely occur simultaneously with a need for safety action. Such a situation should be avoided."

3. Provision for redundancy and diversity of components in the system. This feature is necessary to avoid system failure as a result of the failure of a single component due either to its inherent characteristics or to outside influences.
4. Provision for monitoring and testing. It is essential that means be provided for confirming initially and periodically (if deterioration can be

expected) that the engineered safeguard as installed will function as intended. Such tests often disclose problems not previously identified. For example, Hake [200] reports the following results of tests on the emergency diesel generation of the Canadian NPD Reactor:

"A single diesel-generator set had been provided to supply essential loads in the event of loss of normal a-c supply. It was intended that the diesel should start automatically when the normal supply was lost. However, because the automatic starting did not prove on test to be reliable enough, it was station policy to run the diesel continuously when the reactor was hot. The diesel was not adequate for this job and necessitated several shutdowns. Two units are now employed to give the starting reliability sought. The cause of the trouble in starting the original unit has recently been traced to a faulty electrical connection in the starting-cycle equipment."

Another example of the type of problem which can be encountered is the situation faced on the core spray system of the BONUS Reactor which led to a proposed change [190]:

"There have been no emergency situations which would make actuation of the core spray mechanism desirable. However, there have been instances in which the core spray has been initiated automatically as a result of instrument faults and/or of special operations being performed during a reactor shutdown which produced erroneous signals that the water level was low and the reactor pressure was low . . . .

"Under the proposed change, the objective of having the core spray mechanism set to come on automatically if reactor operators are not present will still be met . . . .

"The proposed change will allow the operator to make the automatic feature inactive during those times when instrument faults are known to exist or when operations are being performed which could result in an erroneous admission of raw water.

"Since the possibility exists of a failure of the low reactor water level switch or the low reactor pressure switch, two additional methods of introducing core spray water have been provided; the operator can open the core spray valve electrically from the control room by actuating the 'Emergency Core Spray Valve Control Switch' or he can open the core spray manually from the pillbox outside the containment building."

Piper [210] and Gitterman [211] discuss, in some further detail, the problems and considerations which face the designers of reactor safeguards systems. These articles help lend additional insight into the design of such systems.

7.2 Basic Design Considerations

Although the exact nature of reactor plant auxiliary systems will differ from plant to plant, particularly among those plants using different reactor coolants, many of the basic design considerations involved will apply to all auxiliary systems. Chief among these are those discussed below.

7.2.1 System Integrity

The problems of providing for auxiliary system integrity are similar in many respects to those described for the primary coolant system. All auxiliary systems contain piping, valves and pumps or blowers, and in many cases small heat exchangers. The reader should refer to the appropriate sections relating to thse components. Their design should be based on the same careful evaluation of stresses and distortions indicated for the primary components. While considerable experience exists with standard type components, such as the small-diameter piping and valves used in these systems, severe problems can arise from four specific areas.

First, because of the small size involved, there is often difficulty in getting sound and reliable weld attachments, particularly between the primary piping and the smaller auxiliary piping. Not only is a good weld design required, as indicated in Sec. 2, but in addition, accessibility must be provided so that the weld can be properly made.

Second, small piping can be placed under severe stress by the reactions and moments of large components to which it is attached. While the small piping is flexible, difficulty can arise because of inattention to the need for greater flexibility than provided.

Third, small piping can easily be damaged during construction because small pipes are often used as steps by construction workers. Such damage can later be the source of breaching of the primary coolant system. For this reason piping of less than 1/2-inch (1.3cm) size should be avoided unless it is properly protected.

Fourth, it should be recognized that many auxiliary systems are only used intermittently and can cool down to ambient temperatures between times of use. As a consequence, if high temperature coolant is brought into the system suddenly, severe thermal shock can ensue. Analyses of thermal effects and stresses should be made to determine the ability of the system to withstand the effects which will be applied during operation.

7.2.2 System Reliability

To be effective the auxiliary system must be relied upon to perform its function under all the conditions of operation for which performance of the system is required. This means that the considerations of wear, erosion, temperature and radiation damage, which influence the life of the system, must be applied to the design of auxiliary systems as well as the primary system. These can be tempered, where appropriate, by the fact that many of the components of these systems can be and are kept out of high-flux radiation fields and are often accessible for routine inspection and maintenance.

In evaluating the reliability of auxiliary systems, consideration should be given to the consequences of a failure of the system to perform as designed. Examples of the questions which must be answered are the following: If the valve operating system were to fail, should particular valves fail open, closed, or "as is"? If the coolant charging system fails to function, when and how is the reactor power to be reduced? What action or alarm is to be provided if a lubrication system fails? How long do the operating personnel have for corrective action if the shutdown cooling system fails to function; what corrective action should be taken?

When evaluating the effectiveness of a particular auxiliary system which is part of a complex array of auxiliary systems, consideration should be given to the possible interaction of systems. This is particularly important when evaluating the performance of auxiliary systems which are to provide information. For example, flow information based on pressure drop could be influenced by coolant sampling if any interconnection exists. Surges of crud might influence the coolant sampling system if the valve operating system shuts down large valves while a sample is being taken.

7.2.3 Testability and Accessibility for Maintenance and Repair

Because of the many components involved and the desire to use as many standard items as possible in auxiliary systems, considerable attention should be given to providing for testability and accessibility for repair. Whenever possible, components should be kept out of high-flux radiation fields. If inspection is to be done while the system is handling radioactive materials, appropriate shielding should be provided. Means should be provided for isolating the system from the primary system for repair purposes whenever possible. Where draining of coolant samples is required, provisions are required for control of contamination and cleanup of spills.

7.3 The Coolant Charging System

The specific functions to be accomplished by the coolant charging system will depend in large part upon the particular plant involved. It will be the purpose of this section to identify some of the typical functions involved and to highlight some of the design considerations involved.

The coolant charging system generally provides for the following functions:

1. Storage of reactor grade coolant
2. Addition of coolant chemicals, if required, particularly boric acid addition in water-cooled reactors designed for chemical control
3. Coolant charging to the main coolant system
4. Coolant charging to other auxiliary systems and equipment
5. Pressurization with reactor grade coolant for hydrostatic testing, where necessary
6. An emergency coolant supply for safety injection system.

A charging system for a water-cooled reactor might consist of a water storage tank that feeds reactor grade water to a header system to which the suction sides of the charging pumps are con-

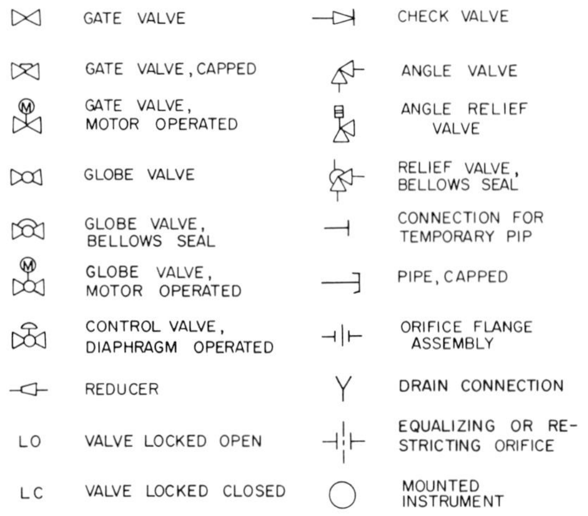

FIG. 7-1a Symbols for used in system diagrams. In addition to the above, instruments are labelled as follows: FI = flow indicator; LC = level controller; LI = level indicator; PI = pressure indicator; PIC = pressure indicator controller; TC = temperature indicator.

nected. (See Fig. 7-1b [39].*) Several charging pumps in parallel would feed a filling header system which contains provision for preheating the water prior to insertion to the system. From the filling header system, lines would lead to each reactor loop, to the pressurizer, and to the other auxiliary systems. Appropriate check valves are necessary in the pump discharge lines, and stop valves in each of the filling lines.

A strainer is usually provided in the suction header piping with a valved by-pass around it for emergency use. In plants having a separate hydrotest charging pump a separate discharge header may be provided for its use. In such an arrangement a valved interconnection should be provided so that either header can by used by any pump if required.

In the design of such a system provisions must be made for make-up water to the storage tank at a rate consistent with maximum needs. If the storage tank is out of doors, provisions are required for heating the water to prevent freezing in cold weather. A small amount of water should be circulated through the storage water tank to prevent excessive buildup of impurities in the water. Provisions should also be made to minimize the oxygen pickup by the water in this tank; Shippingport uses a 0.5 psig (350 g/cm^2) steam blanket over the surface of the water in the storage tank for this purpose [39].

In the Yankee plant where boric acid is used to supplement the control rods for shutdown and operation, provisions are made to pump sufficient boric acid into the main coolant system to establish the necessary boron concentration. A predetermined quantity of 12% boric acid solution is supplied by the chemical shutdown system to the suction header of the high pressure charging pumps for injection into the main coolant system [56].

*Symbols used in system diagrams are defined in Fig. 7-1a.

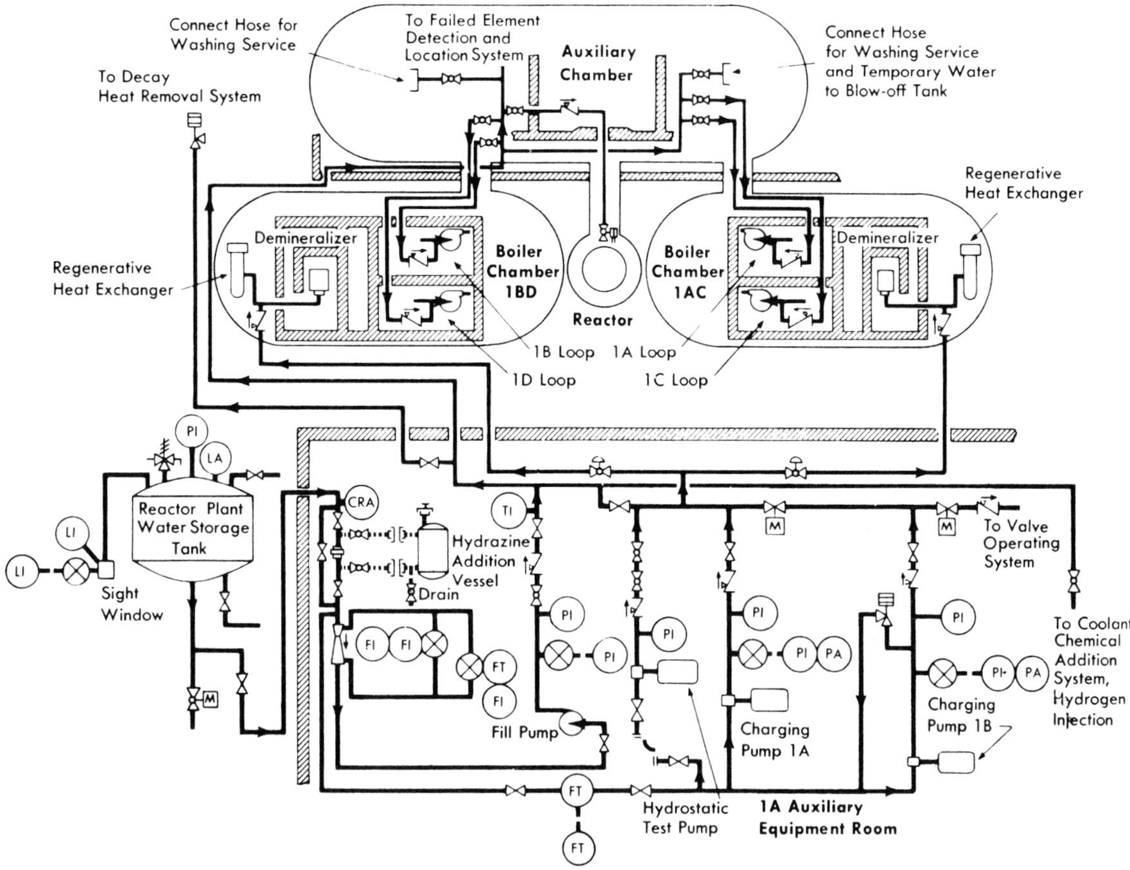

FIG. 7-1b Coolant-charging system, Shippingport plant.

(See Sec. 7.5). In such a system attention must be given to control of the boron concentration and to requirements for proper distribution in the system.

In a liquid metal system, additional problems appear during coolant filling. Because of the influence of metallic oxides such as sodium oxide on the corrosion attack of metals, the oxide concentration must be kept at low levels ($\sim$ 0.005% by wieght). Special precautions must be taken to avoid contact with air. Also, because of reactions with water, both water and water vapor must be excluded. In addition, the system must be kept at a temperature above the melting point of the liquid metal. If electromagnetic pumps are used in the system, gas entrainment in the liquid metal must be avoided if the pumps are to perform satisfactorily. In view of these considerations, the liquid metal is filled directly from the shipping container by use of inert gas pressure. During operation, an expansion tank with an inert gas blanket can be used to accommodate changes in density with temperature [54a].

In a gas-cooled reactor plant such as the Peach Bottom plant a helium handling and storage system serves to provide the coolant charging and discharging system. This system not only provides for storage of helium, but serves as a dump system if excess system pressure builds up in the primary system and serves as a low pressure sink for vent lines, sample lines and relief valves. The principal components of the system are three transfer compressors, two helium dump tanks, and a low pressure helium tank [125]. The functions of the system are accomplished by transfer of helium from one point of the system to another by means of the compressors or pressure equalization.

7.4 The Coolant Discharge and Vent System

The coolant discharge system is sometimes combined with the coolant charging system as indicated above for the gas-cooled reactor and as is the case for the Yankee pressurized-water reactor. Together, the coolant charging and discharge systems are important to reactor safety because they control the insertion and removal of reactor coolant by which heat must be removed from the reactor.

Whether separate or combined with the charging system, the coolant discharge and vent system generally provides the following functions:

1. Coolant removal from the main coolant system for purification or for maintaining pressure in a gas-cooled plant, or for maintaining pressurizer water level during plant heat up.
2. Receiving and containing relief valve discharge from the reactor vessel, the coolant loops, the pressurizer and the purification loop.
3. Removal of noncondensible gases.

As for the charging system, details of design vary with each plant. In a pressurized-water system, normal water removal is accomplished from loop drains. In the Yankee plant all water normally bled from the system flows through the bleed line connected to the cold leg of one of the loops. This is cooled by passing through a feed and bleed heat exchanger before being reduced in pressure. The bleed flow rate is controlled by establishing flow through one or more of the discharge orifices upon opening of the motor-operated gate valves located at the discharge of each orifice. The bleed flows into a low pressure surge tank from which it is led to the waste disposal system. If a low water level exists in the pressurizer vessel, the bleed flow is stopped by closure of the diaphragm-operated valve in the bleed line downstream of the orifices. If the temperature of the bleed exceeds desired limits, an alarm is sounded [56]. The temperature of the water in the low pressure surge tank is maintained within desired limits by recirculating a constant flow of water through the low pressure surge tank cooler.

The coolant discharge system receives relief-valve discharge through a header to which all relief valves are connected. These may discharge to the same surge tank as the normal bleed, as in the Yankee plant or to a separate blow-off tank, as in the Shippingport plant.

Removal of noncondensible gases can be accomplished by several means. First some of the gases dissolved in the main coolant stream come out of solution and accumulate in the low pressure surge tank with the normal bleed water. Second, noncondensible gases can be released through vent valves. The Shippingport plant has valved vents on the primary side of the steam generators, the reactor coolant pumps, the purification demineralizers, and the pressurizer. These are vented to what is called a flash tank, which is comparable to the low pressure surge tank described above. (See Fig. 7-2 [39].)

A number of important considerations are involved in the design of the discharge system. First

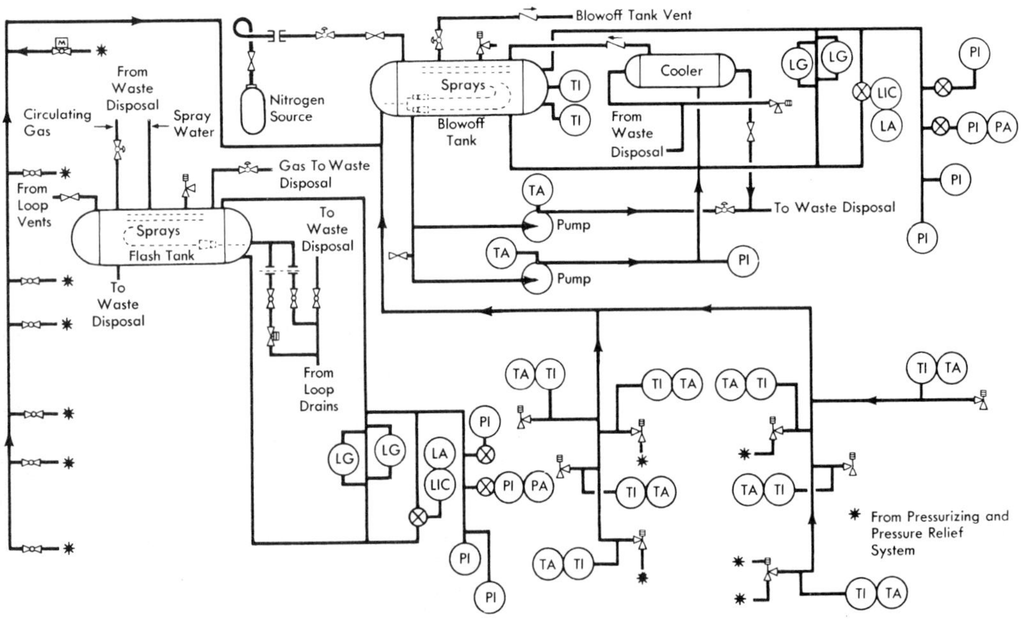

FIG. 7-2 Coolant-discharge and vent system, Shippingport plant.

of all, because this system provides for removal of coolant, it must be conservatively designed to avoid a loss-of-coolant accident. For this reason all drain and vent lines connected directly to the primary coolant system should contain two valves in series (located close to the point at which the lines are attached to the primary system) to reduce the chance of coolant loss. Again, to reduce the possibility of coolant loss, all remotely operated valves in such systems should be backed up by a second such valve, each separated, controlled, if possible, or should be backed up by manually operated valves. The coolant discharge and vent system should be remotely operable from outside the plant container when used for pressurizer control.

To avoid possible contamination of the steam system by reactor coolant, steam system drains and vents should not be connected in any manner to the coolant discharge and vent system piping, even though separated by valves; leakage of such valves could lead to contamination of the steam system.

Consideration should be given to preventing the formation of explosive hydrogen-oxygen pockets in the low-pressure surge tank or any collection tank or flash tank. Circulation of the gases should be provided for. The waste disposal system should be protected from overpressure rising from the flash tank by an automatic valve that isolates the tank from the system if the outlet pressure exceeds a prescribed allowable value.

All flash and blowoff tanks must be carefully sized to receive, cool and mix the influent with the ballast water. Adequate spray water and spray water control is required in these tanks. The blowoff tank should be able to handle the maximum relief valve effluent that may be discharged to it.

In liquid metal systems, a surge or storage tank is often provided both safety and for making adjustments of the amount of coolant in the system [54a]. This storage tank should be able to hold the contents of the entire system. A filter should be provided in the line to the dump tank for normal operation; a separate line without a filter and a fast acting valve is recommended for rapid release of liquid metal from the reactor system to the dump tank [54]. The system should be capable of complete drainage without pockets which can trap oxides and other impurities. However, this tank should be so located that it is not possible to drain accidentally the reactor system during operation, thus causing a loss-of-coolant accident.

7.5 The Valve Operating System

The valve operating system is used to actuate remotely valves which are inaccessible in the

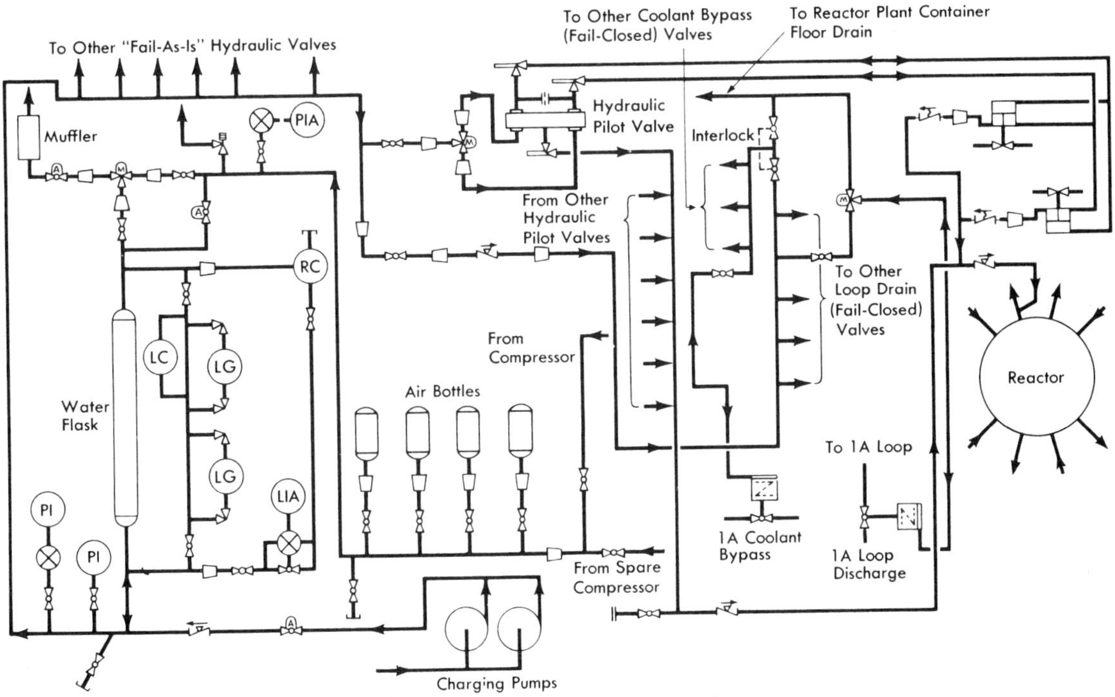

FIG. 7-3 Valve operating system, Shippingport plant.

reactor plant container during plant operation. These valves can be electrically or hydraulically operated as discussed in Sec. 6.4. The discussion in this section is based on an hydraulically operated system for a pressurized-water system; considerations involved in any remote system are similar.

The basic power source for operation of a hydraulic valve-operating system is the hydraulic accumulator which utilizes clean, high purity water supplied by the coolant charging system at reactor system pressure, and which is pressurized during operation by high pressure air from air bottles kept under pressure by means of an air compressor. (See Fig. 7-3 [39]). The pressurized air applies pressure to the water in the accumulator; the water in the accumulator is used to operate the pistons of the valves served by the system. Operation is controlled by selector valves which function to direct the accumulator water to the piston of the hydraulic valve as desired. In order to provide maximum reliability the selector valve is electrically operated by a d-c motor supplied from the plant control batteries. The action of the selector valves is dependent upon whether the hydraulic valve is to fail "as is" or fall closed on loss of hydraulic motive power. On the fail "as is" valves, the selector valves operate hydraulic pilot valves to pressurize one side of the operating piston and vent the other side to the reactor coolant pressure. If accumulator pressure is lost, the pilot valves do not move. On the fail-closed valves the selector valve without the use of a pilot valve admits accumulator water against a spring-loaded piston in the valve body to open the valve; if accumulator pressure is lost, the valve springs close.

The Shippingport plant uses fail-"as-is" action for the eight main stop valves at the inlets and outlets of the reactor vessel, the pressurizer surge-line stop valve, the pressurized spray line globe valve and the failed element detection and location system isolation globe valve. It uses fail-closed action for the four reactor coolant loop bypass valves, four reactor coolant drain valves, and the flash tank inlet valve.

The major design considerations in such a system are to provide a reliable source of stored energy for operating the hydraulic valves (such as provided by accumulators) and to provide a reliable source of power for the selector valves.

The water capacity of the valve operating system must be sufficient to isolate all reactor loops and the pressurizer in the event of a coolant system leak. The air capacity should be such as to have and maintain a prescribed minimum air pressure in the accumulator at the end of the full stroke operation of the valves. The valve operating system should provide for operation of each valve over a partial stroke periodically to reduce the possibility of the valve freezing in position because of corrosion.

7.6 Chemical Shutdown System

The use of neutron-absorbing chemicals for cold shutdown and for control of initial excess reactivity of pressurized-water reactors is finding increased application. The Yankee Nuclear power plant [56] and the Saxton Nuclear Experimental Power Station (Saxton, Pa.) [137] both use boric acid dissolved in the main coolant stream as the neutron-absorbing chemical to provide cold shutdown and to control excess reactivity. It is the function of the chemical shutdown system to prepare, supply, and maintain the boric acid solution which is to be injected into the main coolant.

The chemical shutdown system as used in the Yankee plant consists of a 3000 gallon steam-heated mixing and storage tank, a 100-gpm transfer pump, valves, piping and instrumentation for pressure, temperature and water level control. (See Fig. 7-4 [56].) The solution is prepared in the mixing and storage tank by the addition of technical grade granular boric acid to demineralized water which is heated to 150°F (66°C) by steam heating coils. This temperature is maintained by an automatic temperature-controlled valve; a low temperature alarm is provided to guard against having too low a temperature in the tank. A motor-driven agitator is provided in the tank to aid mixing. The tank is vented to the atmosphere.

The solution prepared in the mixing tank contains 12% boric acid (by weight). A predetermined amount of solution is injected into the main coolant system by the charging pumps (see Sec. 7.3), while the reactor plant is at operating pressure and temperature. The mixing tank is so located as to provide a positive head on the suction header of the charging pumps. Upon injection, the main coolant pumps are used to distribute the solution through the system. Any previously isolated loop is supplied with its predetermined volume of solution before chemical shutdown is accomplished, so that the desired concentration cannot be inadvertently changed by opening of valves in the isolated loop. The entire injection operation can be accomplished by remote control of the motor-operated valve between the mixing tank and the charging-pump suction header.

The boric acid concentration in the main coolant loop is based on providing a 5% $\Delta k/k$ shutdown with all 24 control rods fully inserted in the cold, clean core. The required concentration was determined during initial critical tests of the reactor, although computations had been made previously.

The transfer pump is used for two purposes. The first is to supply boric acid solution from the mixing tank to the safety injection and shield-tank gravity water-storage tank where it is diluted with demineralized water to a 1 wt.% boric acid solution required for safety injection. The second use is to periodically recirculate the boric acid solution in the mixing tank to help maintain a uniform concentration.

The chemical shutdown system also provides for removal of boric acid solution from the main coolant system. This is accomplished by dilution and ion exchange. The boric acid concentration is first reduced by dilution and recirculation for approximately 11 hours after which only 5% of the boric acid remains in the main coolant. The remaining boric acid is then removed by ion exchange requiring approximately nine hours.

So far as safety is concerned, the primary specific considerations involved in the design of the chemical shutdown system are those associated with obtaining and maintaining the proper boric acid

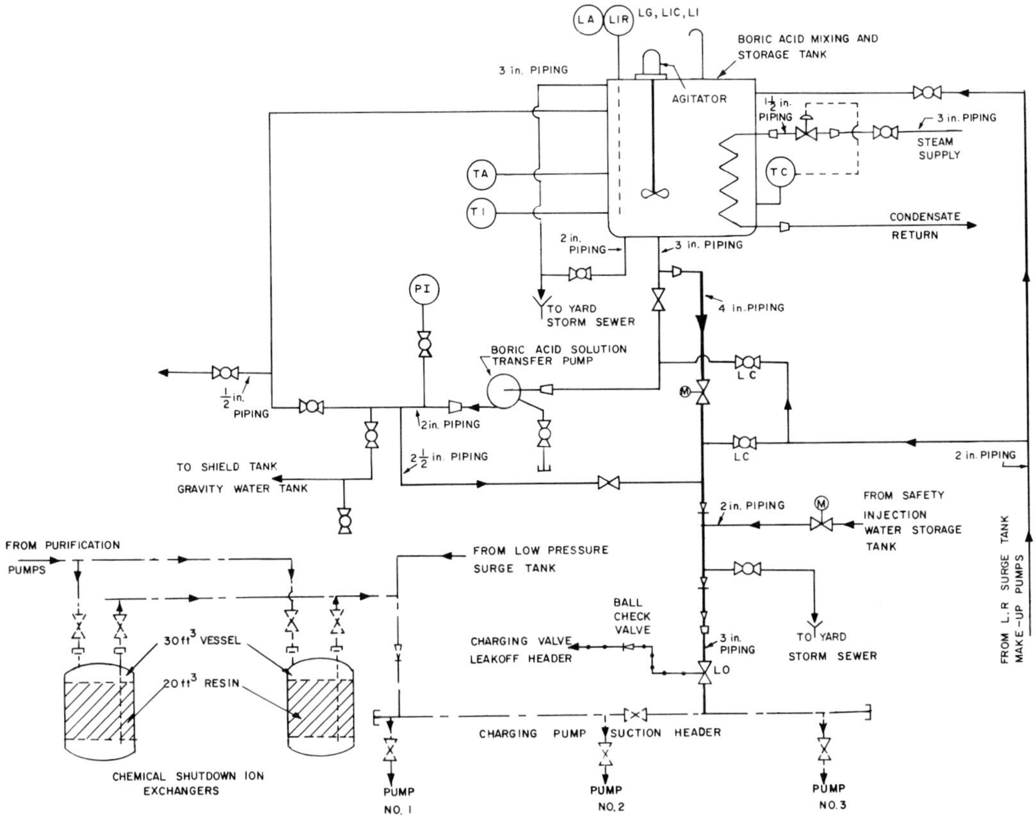

FIG. 7-4 Chemical shutdown system, Yankee plant.

concentration in the system when it is required for shutdown. This means that the solution temperature must be maintained in all parts of the chemical shutdown system so that none of the boric acid comes out of solution prior to getting to the main coolant loops. Attention must also be given to maintaining uniformity of concentration by appropriate recirculation of solution. The system must, of course, be kept tight to avoid leaks.

Appropriate administrative controls are required to assure that the concentration of solution prepared in the mixing tank and the amount of solution introduced into the system are correct. Appropriate operating procedures are also required to assure that an isolated loop is supplied with its appropriate concentration at shutdown for the reason indicated earlier.

During start-up of the reactor the chief concern is that boric acid solution may not be removed uniformly. If for example, some of the boric acid plates out in the core and later flakes off into the coolant stream, it could be the source of a reactivity insertion; if some of the boric acid plates out elsewhere in the system, it could cause a power transient in the core as it passes through. In practice to date these problems have not been a source of trouble; nevertheless they must be considered in each new design.

7.7 Shutdown Cooling System

It is the function of the shutdown cooling system to remove the heat which continues to be generated in the core by the radioactive decay of fission products after core shutdown. For a summary of current information on shutdown heat generation, the reader is referred to reference [138]. The decay heat must be removed to prevent damage to the core and to avoid impairing the safety of the plant. The nature of the shutdown cooling system used for removing this heat will differ from one reactor design to another, but the general considerations involved in providing for shutdown cooling are similar for all plants.

The first requirement of a shutdown cooling system is that it provide for removal of decay heat in the event that all normal power for driving the pumps is lost. In addition, the system must provide for removal of decay heat during normal shutdowns.

Typical of the approaches used in pressurized water-cooled reactors are the shutdown cooling provisions in the Shippingport plant and the provisions in the Yankee plant. In the Shippingport plant, if power is lost to the pumps, the reactor is scrammed and the turbine throttle and auxiliary steam valves are closed. The removal of decay heat, in this circumstance, is accomplished by means of natural circulation of primary coolant from the core to the steam generator where the pressure in the steam drums builds up until it reaches the relief-valve set pressure. The relief-valve then opens and steam is blown to the atmosphere, thus completing removal of the decay heat. The system is so designed that no corrective action is required by plant personnel for at least two hours. By the end of the two-hour period make-up water is required in the steam drum. An emergency diesel generator is provided for this purpose; it supplies power for operating a charging pump to transfer water from the primary coolant water storage tank to the steam drum and to the pressurizer as it cools down. This shutdown cooling system can also be used to remove decay heat following a normal shutdown in which complete cooldown is not required. This method is used to accommodate overnight capacity reduction even with normal a-c power available at the station [39].

The Yankee plant also provides for natural circulation of the primary coolant and operation of the safety and relief valves in the secondary system to dissipate stored and decay heat in the event of a complete loss of power to the pumps. An emergency engine-driven generator is provided to supply make-up water to the secondary system. The probability of complete loss of power to the main coolant pumps is greatly reduced in this plant by having two of the four pump motors fed from a transformer connected to the main generator and each of the other two fed from a transformer connected to separate, incoming 115 kv lines. In addition, separate emergency electrical connections exist to a hydroelectric plant which is within a few hundred yards and operates on the head from the Yankee cooling pond. Provisions are also made to transfer pumps to an energized bus if more than 50% of the flow is lost by power loss to three pumps.

For normal shutdown cooling a separate system is provided in the Yankee plant. It consists of a heat exchanger, circulating pump, piping, valves, and instruments arranged in a low pressure auxiliary loop in parallel with the main coolant loop. (See Fig. 7-5 [56].) The pump in the shutdown cooling system takes primary coolant from the hot leg of the main coolant piping on the reactor side of the loop stop valves and circulates it through the tube side of the shutdown cooler to the cold leg of the main coolant piping where it reenters the system, again on the reactor side of the loop valves. The shell side of the shutdown cooler is supplied with water from the component cooling system, which is in turn cooled by pond water. This arrangement reduces the possibility of having leakage of radioactive main coolant enter the pond water. The shut-

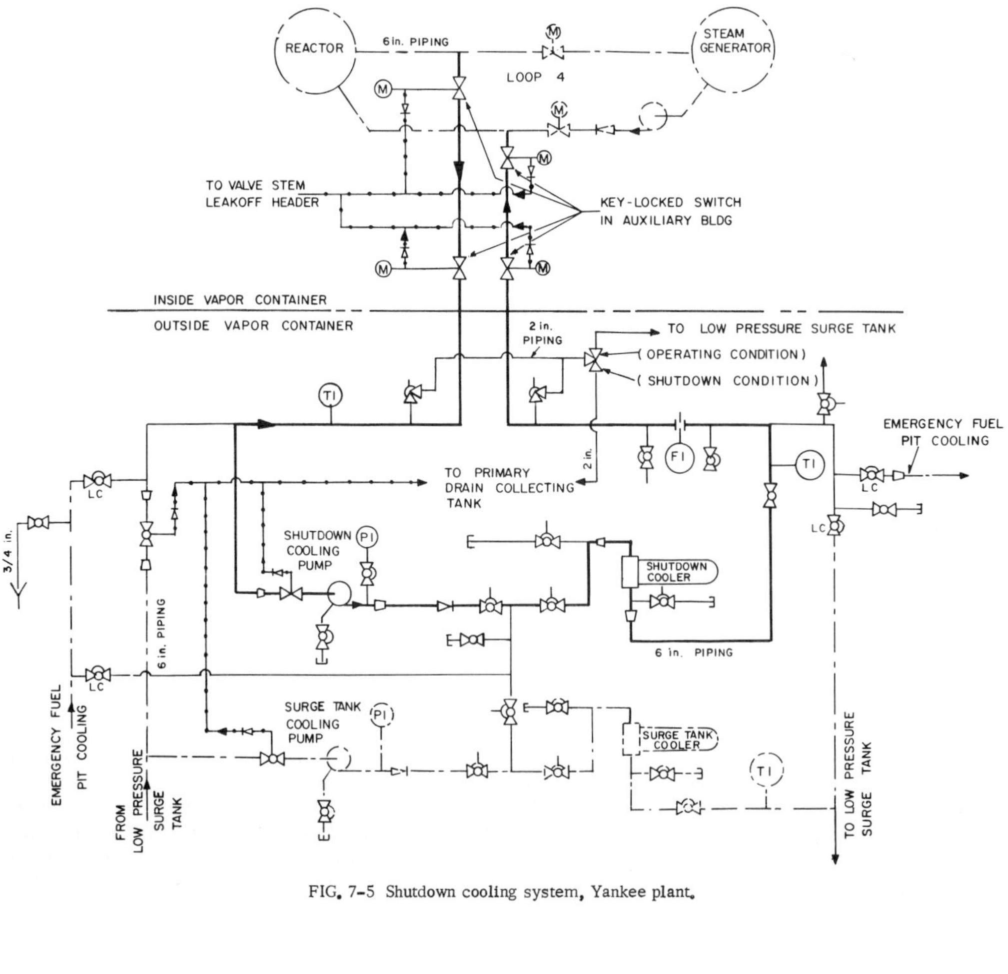

FIG. 7-5 Shutdown cooling system, Yankee plant.

down cooling system for normal operation is placed in service after the main coolant temperature has been reduced to approximately 330°F (166°C), and the pressure to less than 300 psig (20 atm). The system reduces the main coolant temperature to 140°F (60°C) or less and maintains it at that level as long as required. This system is backed up by the low pressure surge tank pump and heat exchanger mentioned in Sec. 7.4.

The Peach Bottom Atomic Power Station, which uses helium as the coolant, has a reactor vessel emergency cooling system for use when normal means of cooling the reactor cannot operate during core shutdown. This system utilizes a steel plate shroud surrounding the pressure vessel to which is welded a series of coils cooled by water from the critical-service water supply. Provisions are made to have the pump in this cooling water supply powered by an emergency diesel generator. The heat lost from the reactor by radiation and convection (mainly radiation) is absorbed and removed by the coils. The coils are located on the outside of the shroud in order to minimize the possibility of water accidentally coming in contact with the vessel. For normal shutdown, decay heat is removed through the steam generator. For prolonged shutdown (> 1 hour), water circulating through the evaporator section of the steam generator is sufficient to carry the decay heat to a separate heat exchanger outside the containment shell [125].

In liquid metal systems a number of different shutdown cooling schemes have been used. These include gravity flow of coolant through the reactor and heat exchanger from an elevated storage tank, natural circulation from core to heat exchanger, and use of auxiliary pumps operated from a stand-by power source [54a].

In addition to the general considerations indicated in Sec. 7.2, attention must be given to the following items in designing the shutdown cooling system:

1. The capacity of the shutdown cooling system must be based on the decay heat associated with many hours of full power operation prior to a need for the system. The length of prior operation used should represent a realistic appraisal. Yankee uses 10,000 full-power hours for this purpose.
2. The system should be capable of handling the decay heat removal for a length of time, following a complete loss of main-coolant pump power, such that corrective action can be taken. The course of action to be taken should be well thought-out and practical. The Shippingport system is based on two hours of such operation.
3. Corrective action following a complete loss of flow should not normally require entering the plant container.
4. The station should be capable of handling decay heat removal without external facilities for a definitely prescribed period of time longer than two hours, in the event that the plant is isolated by a storm. A period of eight hours might be considered a minimum. Therefore an emergency source of power such as a diesel generator must be provided which can supply the needed power for the prescribed length of time.
5. In a pressurized water system it must be remembered that a water level must be maintained in the pressurizer if natural circulation is to take place; if the water level in the system gets too low, vapor locking can take place. Therefore, as the system cools down, make-up water is required in the pressurizer.
6. Provisions must be made for supplying make-up water to the steam drum if flashing of steam is used as part of the heat removal scheme.
7. Although the emergency removal of decay heat should not require immediate action by the plant personnel, the emergency power supply should be placed in service right after the plant power loss, to supply critical power needs.

7.8 Safety Injection System

One of the most severe accidents that could take place in a pressurized-water reactor system is a break in the primary system that could lead to rapid loss of coolant and core meltdown. The safety injection system is provided to supply water to the core to prevent meltdown if a loss of coolant accident were to take place. In such an event the plant is scrammed immediately, either manually or by loss of pressure.

The safety injection system consists of piping, with associated pumps, valves and instrumentation, which connects the reactor plant water storage facilities to the reactor coolant system. The system may also employ one or more components of other systems in its operation. Depending upon the particular plant involved, the safety injection system may be designed, first, to inject large volumes of water into the reactor coolant system to replace water going through the rupture and into the plant container and, second, to circulate and cool the large amount of water that has accumulated in the plant container. If the reactor is designed to use boric acid for cold shutdown, the safety injection pumps draw their water from a storage tank containing borated water; in the Yankee and the Saxton plants, this is the refueling water storage tank. In each of these plants two safety injection pumps are used to inject the borated water into the reactor vessel. In the Saxton plant the injection lines connect to spare inlet and outlet nozzles; in the Yankee plant the lines connect to each of the four coolant loops. Insofar as possible, the entire injection system, except for the injection lines, is located outside the plant container in each plant.

The safety injection system comes into play if the discharge of water from the system is such that the charging pumps cannot supply water at the required rate. In the Yankee plant, safety injection is initiated automatically when the main coolant pressure drops to a set value. (See Fig. 7-6 [56].) The safety injection valves open first and then, as the pressure drops, the injection pumps are started. The system can be operated manually as well as

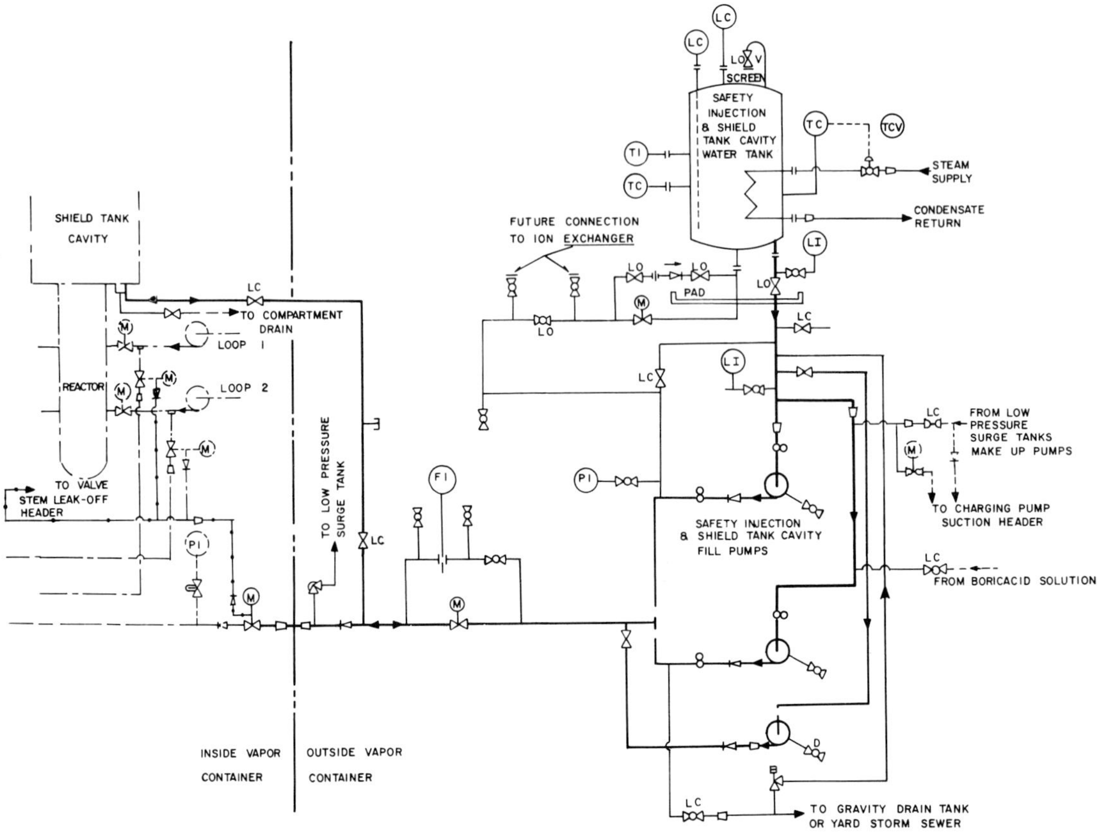

FIG. 7-6 Safety injection system, Yankee plant. Connections to two loops of reactor are shown; connections to other loops are similar.

automatically. With both injection pumps operating, the system can fill the reactor vessel to the top in a little over two minutes if not more than 25% of the total pump discharge is released through the rupture. After the vessel is filled, the flow rate from one pump is to be adjusted by a control valve to replace the water that is boiled off into the plant container. The pumping capacity is 3,600 gpm. The pumps are supplied with power from two separate vital buses. Lights on the control panel are used to indicate that valves have operated as scheduled and that pumps are energized and filling is underway.

A number of specific design considerations apply to the design of the safety injection system. The first of these is tied in closely with the location of the main-coolant pipe connections to the pressure vessel relative to the vertical position of the core. If all pipes are connected to the vessel above the core, a pipe rupture will not impair the ability of the vessel to hold water over the core; if pipe connections exist below the core, water can drain out of the vessel and uncover the core if the pipes are broken. In the latter case, one is faced with the problem of getting the safety-injection water to all parts of the core as it flows into the vessel. If piping must be attached to the bottom of the vessel, it is well for it to have a vertical rise after leaving the vessel so that it can act as a stand-pipe to hold water in the vessel if the rupture is in another part of the loop.

The problem of flow distribution over the core during safety injection should not be dismissed even if all piping is attached to the vessel above the core. A crack in the lower part of the vessel could also cause water to drain right through the core.

Careful analyses are required to determine whether the safety injection system as designed can prevent core meltdown and even water-metal reactions (in certain cores) in the event of a loss-of-coolant accident. Typical of such analyses is that reported for Shippingport in reference [139].

Other factors which must be considered in the design of the safety injection system are the following:

1. A basis for analysis must be established regarding the size rupture that might have to be accommodated. A rupture of a main coolant line should be considered as possible.
2. The prior operating history of the plant should be consistent with that used for the shutdown cooling system as far as decay heat generation is concerned.
3. Adequate missile protection should be provided for any safety injection headers inside the plant container.
4. The plant container must be able to withstand the weight of the large volume of water introduced through the rupture, initially directly from the primary system and later

via the safety injection system. Advantage can be taken of the ability of the plant container to hold water above the core, in the event of a rupture in the primary system below the core, if the container is designed with this in mind. A crude coolant level indicator in the plant container is useful.

5. All external parts of the safety injection system must be protected from freezing where appropriate.
6. Appropriate alarms should be provided to warn of inadvertent initiation of the safety injection system. Indicator lights should be provided to show that valves and pumps have been energized and/or functioning when initiation of the safety injection system is necessary.
7. Thermal shock problems should be investigated to guard against damage of the safety injection system at the time most needed.
8. The pumps and valves in the safety injection system should be individually operated and checked periodically because inactivity could cause them to become bound by corrosion.

8. SAFETY DURING CORE HANDLING, REFUELING, STORAGE AND SHIPMENT

Refueling of reactors and handling, storage, and shipping of reactor fuels present many safety problems. Inattention to the safety precautions necessary in such activities, as well as failure to realize the importance and extent of the considerations involved in the design of core components and the equipment for handling them, can result in nuclear accidents of severe proportions. This was graphically demonstrated by the accident which took place with the SL-1 core at the Stationary Low-Power Reactor Facility of the National Reactor Testing Station in Idaho. A brief review of the circumstances surrounding this accident will serve to indicate the interplay of considerations that can be involved in refueling. (See also discussion in Sec. 3 of Chapter 11, "Accidents and Destructive Tests", Volume 1, pages 608 to 708.)

8.1 The SL-1 Incident

The following summary of findings is extracted from the Interim Report of the AEC's General Manager's Board of Investigation [140]:

1. An explosion occurred in the SL-1 reactor at approximately 9:00 p.m., on January 3, 1961, resulting in the death of three persons, in damage to the reactor and to the reactor room, and in high radiation levels (approximately 500-1000 roentgens per hour).
2. Two members of the crew were killed instantly by the explosion. The third died within two hours as a result of an injury to the head.
3. The explosion involved a nuclear reaction. The thermal neutron flux-time above the reactor is currently estimated to have been approximately 10^{10} neutron/cm^2.
4. Chemical and radioactivity measurements on a single fragment of reactor fuel ejected by the explosion, if representative of the total fuel, suggest that the reaction may have resulted in 1.5×10^{18} fissions. This would have produced 50 Mw-sec of energy.
5. At the time of the explosion, the reactor crew appears to have been engaged in the reassembly of control rod mechanisms and housings on top of the reactor. The pressure generated within the reactor, which may have reached several hundred pounds per square inch, was vented through a number of partially closed nozzles in the head of the reactor, blowing out shield plugs, portions of control rods, and some fuel.
6. The explosive blast was generally upward from the ports in the top of the reactor. Structural damage to the building, principally due to objects projected from the nozzles, was slight.
7. Some gaseous fission products, including radioactive iodine, escaped to the atmosphere outside the building and were carried downwind in a narrow plume. Particulate fission material was largely confined to the reactor building, with slight radioactivity in the immediate vicinity of the building.
8. It is not possible to identify completely or with certainty the causes of the incident. The most likely immediate cause of the explosion appears to have been a nuclear excursion resulting from motion of the central control rod. There is no evidence to support any of several other conceivable initiating mechanisms.
9. It is known that a variety of conditions had developed in the reactor, some having their origin in the design of the reactor and others in the cumulative effects of reactor operation, which may have contributed to the cause and extent of the incident.

As outlined in reference [140], the SL-1 was a direct-cycle boiling-water reactor utilizing enriched uranium fuel clad in aluminum, and moderated and cooled by light water in natural circulation. It was designed to operate at 3 thermal megawatts and to produce 200 kilowatts of electricity and 400 kw of space heat. A saturated steam flow of 9,000 lbs/hr (4100 kg/hr) was generated in the pressure vessel at 300 psi (20 atm) and 420°F (216°C). About 85% of the steam was used to generate electricity. Fifteen percent bypassed the turbine into a heat exchanger which simulated a space-heat load.

The SL-1 was a prototype of a power reactor that could be operated by military personnel in remote arctic areas for three years without refueling. As stated in reference [141], being a prototype unit, the reactor was to contain only such extra safety features as were required for the safety of the plant and its operating personnel. Where feasible, standard commercial components were to be readily available for inspection; however this criterion had to be relaxed for the control rods as indicated below [141]:

> "A special feature of the reactor design was that refueling could take place without removing the head of the pressure vessel. This was accomplished by sizing the control rod

drive nozzles, in the vessel head, to permit the loading of fuel assemblies through the nozzles. It was felt that this concept of fuel handling would simplify significantly changes in the core loading in the ultimate power plants.

"A number of layout studies were made in order to size fuel-element dimensions and control rod strengths and locations. The requirements of convenience of fuel handling were not found to be compatible with rod arrangements under the criteria that the rods by themselves could provide full operating-to-shutdown margin, while simultaneously no single rod could start up the reactor. The decision was therefore made to sacrifice this latter criterion".

In order to achieve a core lifetime of three years, it was necessary to load the core initially with more fuel than could be handled by the control rods. To compensate for the excess reactivity involved, thin strips of aluminum-nickel alloy containing highly enriched boron-10 were spot-welded to the side plates of fuel elements. Early in the life of the reactor, these strips were found to be bowing away from the plates in the regions between weld attachments. This bowing increased with time to the point where great difficulty was encountered in removing elements in the central region of the core. Some of the boron strips were damaged and shattered in the process of removing these elements for inspection during September 1960. It was estimated that about 18% of the boron had been lost from the core by this damage and that, when the core had been reassembled, the shutdown margin was only about 2% [140]. The shutdown margin was increased by the addition of six cadmium strips, welded between two thin aluminum plates, and installed through unused T-shaped control rod shroud tubes on opposite sides of the core.

Considerable difficulty had also been encountered with the movement of control rods. It was frequently found that one or another of the rods would not fall freely when scrammed [140]. Evidence suggests that the control rod shrouds may have been compressed by the expanding boron strips so that rod motion within the shroud was impeded [140]. Difficulty had also been encountered with sticking due to friction in the shaft seals through which the control rod shafts operated. Over the last two months of operation approximately 40 rod stickings were recorded and in some cases rods had to be assisted manually for withdrawal [140].

The crew performing the work on the reactor at the time of the accident had been well trained, and two of the three had been fully qualified by examination. Their previous satisfactory performance as operators, as well as their training, would preclude a deliberate or unknowledgeable withdrawal of the rod unless a sudden emotional instability is postulated [141].

The following conclusion is extracted from reference [142]:

"It is known that certain undesirable conditions had developed with respect to the reactor and its operation, some having their origin in the design of the reactor and others in the cumulative effects of reactor operation, which do not now appear to have had a direct relation to the immediate cause of the incident. The Board observes, however, that the overall effect of these conditions produced an environment in which the possibility of an incident may have been increased beyond that necessary."

8.2 General Safety Considerations During Core Handling, Refueling, and Storage*

The major safety considerations in handling and refueling of cores are providing protection against criticality, avoiding meltdown of spent fuel brought about by inadequate heat removal, and protecting against the spread of contamination from crud from ruptured fuel elements being removed from the core and from fuel elements cut in preparation for inspection. For the safety of operating personnel, protection must also be provided against gamma radiation from spent fuel.

8.2.1 Criticality*

To guard against inadvertent criticality, careful procedural controls are required. Control can be accomplished by controlling the mass of uranium being handled, by controlling the geometry into which fuel may be placed, by using poison rods or solutions installed in the fuel during handling, or by using a combination of all three methods. Regardless of the method used, the limits on mass, distance or minimum poison requirements should be based on the assumption that the amount of hydrogenous material giving maximum reactivity can be present. Allowances must also be made for uncertainties in experiments or calculations in establishing these limits. In order to minimize the criticality harzards involved in shipping fuel elements, both irradiated and unirradiated, the AEC has prepared specific regulations which must be followed as outlined in Sec. 8.5.

Typical of the precautions used in refueling are those reported for the Shippingport plant in reference [143]. The following excerpt is taken from reference [143]. It should be pointed out that the Shippingport core is a seed and blanket core in which the seed assemblies consist of Zircaloy-clad plates of a Zircaloy-uranium alloy and the blanket assemblies consist of round Zircaloy rods containing natural uranium in the form of uranium oxide pellets. This excerpt directs its attention to refueling of the first set of seed assemblies, through penetrations in the reactor vessel head. The refueling was done entirely under water by means of the extraction tool shown in Fig. 8-1 [143].

"A subcritical condition was maintained throughout the refueling operation, primarily by careful control of fuel movement outside of the reactor vessel, careful checking to ensure that each fuel assembly contained a poison or control rod except when switching from a poison to a control rod or vice versa, and cautious insertion of new fuel into the reactor in a prescribed sequence. Fuel handling building safety procedures required that no more than two seed as-

*See Chapter 5, "Criticality," (Volume 1, pp. 244 to 284), especially Sec. 4.

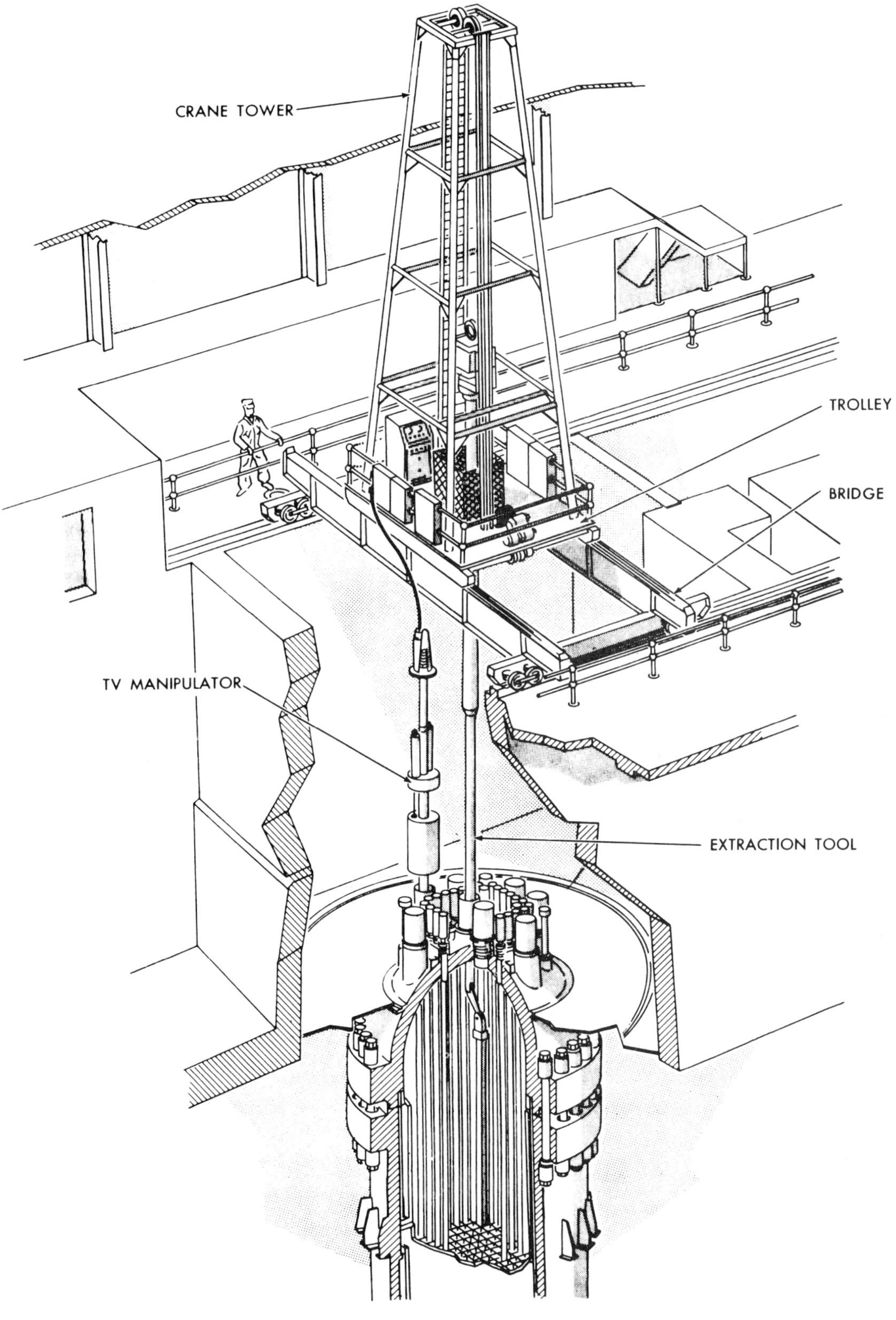

FIG. 8-1 Extraction tool in operation at Shippingport.

semblies be handled without control rods at any time. This limited the parallel work which could be performed on seed assemblies during fuel transfer operations, but was considered necessary since physics calculations indicated that three unrodded Seed 2 assemblies could go critical under water if placed in a certain geometric array.

"A second requirement which prevented the possibility of criticality was the careful checking of each Seed 2 assembly before installation in the reactor to be sure that it contained a hafnium control rod. Presence of the control rod was independently verified by a responsible person from each of the three refueling organizations: the utility, the core manufacturer, and the Atomic Energy Commission. In this way, additional assurance was obtained against the possibility of an inspection error. This control rod verification procedure caused some delay to refueling progress because of the nature of the underwater inspections.

"A third requirement which prevented the possibility of criticality accident was the installation of the Seed 2 clusters in the reactor in increments, corresponding to 0.4% reactivity, and the monitoring of the neutron count rate with the four operational BF_3 neutron detectors installed around the reactor vessel. Thus, any unexpected rise in reactivity could be detected prior to the approach of a critical condition. No unexpected increases in reactivity were detected.

"The safeguard measures were implemented despite the fact that this core was designed to remain subcritical with one rod fully withdrawn under ambient temperature conditions. Despite their duplication and even triplication of purpose, measures such as those described above were considered a necessity in any operation dealing with nuclear fuel during this refueling.

"In addition to the safeguard measures already noted, the fuel handling building was instrumented to detect a criticality condition and to warn occupants immediately if such a condition arose. A boric acid (neutron poison) addition station was maintained in a stand-by condition during all fuel transfer operations and rod testing operations in case the preventive measures failed; and special criticality dosimeters were installed in the fuel handling building to aid in corrective measures in the event of a criticality incident."

8.2.2 Heat Removal During Refueling and Fuel Handling

The chief consideration with regard to heat removal concerns the possible loss of coolant or cooling means when handling spent fuel. Where refueling is accomplished under water this is not a problem. Where refueling involves the withdrawal of fuel into a cask, consideration must be given to the loss of coolant from the cask; if cooling is not normally done in the cask because of the short time involved in using the cask to transfer the fuel to another location, consideration must be given to the action necessary if operations become prolonged for any reason. In either case an emergency means of cooling should be provided. This might be done by making provisions for hose attachments to the cask from building cooling facilities as appropriate. All such provisions should be based on analyses of the time available before the fuel reaches prohibitive temperatures.

Many specific requirements are listed in reference [143] for heat removal from spent fuel under shipment. The reader is referred to Sec. 8.5 for a discussion of these factors.

8.2.3 Contamination Control

Control of radioactive contamination is an important safety consideration during core refueling and handling. Although contamination does not present as severe a hazard as criticality or loss of heat-removal capabilities, it presents a more prevalent hazard. Contamination can arise from crud (radionuclides in corrosion films) on fuel elements or other core components, from ruptured fuel elements removed from the core, from fuel elements damaged in handling, or from fuel elements dismantled or sawed underwater in a refueling canal in preparation for inspection or shipping. To guard against spread of contamination, limitations and procedures for its control must be made a part of the refueling and handling process.

In underwater refueling, particular attention must be given to water-borne contamination during removal of radioactive components. For the Shippingport plant, a plastic bag technique was developed by which non-fuel-bearing components were removed from the reactor directly into a plastic container underwater and transferred to storage in the plastic container. Thus any radioactive crud on the component was prevented from dissolving in the canal water and leading to air-borne contamination when the canal was drained later [143]. A similar technique, involving the use of a metal transfer can, was developed for use with fuel elements, particularly ruptured fuel elements. It was found during the first refueling that the plastic bags and metal transfer can were not normally required. However, devices such as these should be prepared for handling components containing large amounts of crud or for handling ruptured fuel elements. (Plastic bags should not be used for fuel elements as the bag may fuse to the element preventing proper cooling.)

All tools and components used during refueling which come in contact with the primary coolant or reactor internals must be considered contaminated and should be handled with rubber gloves until certified to be within acceptable limits by Health Physics monitoring. Provisions should be made for decontaminating equipment within the confines of the contamination control area. All contaminated regions in and around the refueling area should be decontaminated as soon as practical.

Care must also be taken to guard against air-borne contamination during refueling. Appropriate air monitors should be located at various points in the refueling area. If an area becomes severely contaminated, a radioactive particle detector should be used to indicate if any of the contamination is becoming air-borne prior to clean up. If it is, decontamination procedures should be started. Attention must be given to operations having a high potential for producing air-borne contamination, particularly such operations as cutting, grinding or welding of contaminated material or venting of components containing primary coolant. Appropriate use of respirators should be made by personnel during any such operations. Respirators should also be used in any area where an air particle detector alarm sounds or where Health physicists deem them necessary. All air from the refueling area should be controlled through an appropriate air handling and cleaning system before being released from the building.

By far the most likely means for spreading contamination are the personnel working in the refueling area. Personnel entering the refueling area should be limited to those involved in the operation. They should enter and leave through controlled change rooms where street clothes are exchanged for protective clothing. All personnel leaving the refueling area should be monitored and should divest themselves of all contamination. The change room should be divided so as to separate the contamination side from the street side. Efforts should be made to minimize the amount of contamination brought to the change room; use of shoe cover change stations at the boundary of contamination areas is recommended. A quick-monitoring device should also be provided at such stations to identify any large pickup of contamination. A major factor in limiting personnel contamination during the Shippingport plant refueling was the installation of an aluminum grating on the floor of the reactor pit; this grating separated personnel from the contaminated water which tended to collect on the floor [143].

Controls must also be instituted to prevent the spread of contamination by means of equipment which is brought out of the refueling building. Items of equipment should be decontaminated wherever possible and/or contained in suitable containers or plastic bags. For precautions regarding shipment of fuel see Sec. 8.5.

8.2.4 Radiation Protection for Refueling Personnel

No refueling operations should be planned without an effective Health Physics control procedure. All operations involving removal or handling of radioactive material should be monitored by qualified Health Physicists or cleared by them.

Careful planning is required to make maximum use of personnel. Progress of work can be severely hampered by haphazard exposure of trained personnel to the maximum permissible weekly radiation limit. Temporary shielding should be provided wherever practicable if it appears that radiation levels exceed expected or desired levels. However, the most important factor which will reduce the amount of exposure to high radiation fields is minimization of the number of operations that must be performed in the reactor vessel head area [143]. This is an item for attention by reactor designers as well as by those planning and preparing refueling procedures. In addition, efforts should be made to disassemble and assemble small components which go on the head in some other area if possible. Steps should also be taken to train personnel and supervisors in advance, on mockups, so as to make efficient use of the time spent in the reactor head area.

Reference [143] points out that, during the Shippingport refueling, the radiation field in the vicinity of the head area was considerably higher than anticipated. The radiation field at the work areas on the head during disassembly of completely canned control rod drive mechanisms varied between 150 and 400 mr/hr with all temporary shielding installed and up to 1 r/hr without shielding. This fact is particularly highlighted because it identifies several factors which bear heavily on the reactor systems of this type. As indicated in reference [143], it was found that there are significant unknowns regarding the formation and transport of crud arising from radioactive corrosion and wear products in reactor systems. Nevertheless, it appears clear that efforts should be made to reduce the number of crud traps in areas where work must be done, as in the head area. Trapping of crud in crevices depends not only on the local geometry but also on the distribution of coolant flow in the immediate area. It is postulated that the high crud levels in the crevices of the control rod drive mechanisms were due to thermal circulation of primary coolant during reactor operation and to the influx of crud-bearing water from the core during scramming with settling of crud on the mechanism between scrams. The situation during refueling could have been helped if provisions had been made to flush the mechanism internals with water or steam while still in place during the early stages of refueling.

In view of the fact that the amount of crud formed in the system depends in part upon the wear products formed, efforts should be made to reduce wear by attention to the materials involved, to the surface finished prescribed and to the clearances used. In addition, because most of the crud activity was due to cobalt-60, as little of the cobalt-bearing materials as possible should be used for wearing surfaces. (See Sec. 2.6.3.)

Attention to design problems such as indicated above is important in reducing the radiation exposure encountered by personnel during refueling. Reducing the radiation exposure is important to reactor safety because the number of trained personnel available at any refueling is limited, and when they obtain their maximum allowable weekly dose there is pressure to carry on the work with less qualified personnel.

8.3 Loading and Unloading Fuel

As indicated in Sec. 8.2.1, the chief consideration in loading fuel into a reactor is that of avoiding criticality. In addition, however, care must be taken to avoid damaging the fuel, because this could result in the subsequent loss of fission products [144]. During the unloading or reshuffling of fuel there are the additional considerations of removal of decay heat, control of contamination, and protection of personnel as discussed in Sec. 8.2.

In these operations a number of specific steps must be taken to assure safety. These are the design, procurement and checkout of reliable refueling equipment; preparation of detailed refueling procedures including alternate courses of action in the event that difficulty is encountered during crucial operations; and careful training of personnel.

8.3.1 Fuel Loading and Unloading Equipment

The equipment required for loading and unloading fuel is different for every plant. Its nature will depend upon the type of reactor coolant used in the plant, upon whether the refueling is done wet or dry, upon whether refueling is done through ports in the vessel head or with the head off, and upon whether the refueling is to be accomplished with the reactor shut down or operating. In simple research reactors the tools can be quite simple and can consist of long-handle grapple hooks. In power reactors to be refueled when shut down the tools are far more complex. They can range from the individual extraction tool used at Shippingport as shown in Fig. 8-1 to the built-in refueling machine for the Enrico Fermi plant shown in Fig. 3-6. The gas-cooled power reactors in England will accomplish refueling while the plants are at full power. The design includes a chute-handling machine to install and remove a chute through which the fuel elements are removed and charged. Separate equipment is used for charging and discharging fuel elements. Electrical and mechanical interlocks are provided to insure proper sequencing of operations. Each of the three devices involved constitutes a pressure vessel. The discharge machine has provisions for cleansing and cooling by air or by use of CO_2 reactor coolant.

In the design of all refueling equipment attention must be given to reliability. Materials should be selected to minimize wear or seizure in the environment in which they are to operate. If a lubricant other than the reactor coolant is to be used, it must be sealed to prevent it from entering the primary coolant. Attention must be given to problems of alignment, remote indexing, lead-in and attachment from the fuel elements. The design should be such that the fuel elements cannot be dropped inadvertently while being withdrawn; yet provisions should be provided (See Vol. 1, p. 691, Sec. 5.2.) from the tool if jamming takes place. A separate override on the normal latching or handling means should be provided.

An important part in the development of any refueling equipment is the checkout of the equipment prior to refueling. Such prefueling checks should simulate the accessibility and worst-case tolerance conditions that could be encountered during refueling. Typical of the problems resulting from inadequate checking have been interference between refueling equipment and unrelated structures on or near the reactor vessel head, failure of an extraction tool latch to engage all fuel assemblies because some of the matings of latch and lugs represented the worst tolerance situation, and inadequacy of an uncrimping tool because it was made of too soft a material [143].

Many other reactor installations have experienced difficulty with fuel loading and unloading machines. Hake [200] reports a number of "fueling-machine" failures at the Canadian NPD Reactor of which the following was the most serious:

"By far the most serious failure happened in December 1962 during the first refueling attempt with the reactor at power. Many trials had been carried out on a rig simulating reactor-face conditions, but, on the first attempt on the reactor itself, a substantial heavy-water coolant leak occurred at the connection of the fueling machine to the reactor pressure tube. The leak was large enough to raise the reactor-vault pressure to the point where the vault dousing system operated. This system, by spraying cold heavy water in the vault, suppressed the pressure rise by condensing the steam forming from the leak. The reactor was quickly cooled and depressurized to reduce the leak, and the leak was eventually sealed. Emergency procedures and equipment operated as intended so that cooling of the fuel was not interrupted. The heavy water was seriously downgraded in isotopic concentration and purity by picking up light water, oil, and chemical grout as a result of the leak. The contamination had some deleterious effects on pumps, seals, filters, etc., but the fuel and pressure tubes were unharmed by the incident. Reactivity was regained by rearrangement of the fuel, and station operation was restarted 12 days after the incident. This fueling-machine fault was a serious breach of the main cooling system, but personnel and protective systems responded successfully to the emergency and brought the situation under control without significant release of contamination to the normal accessible areas of the building."

More recent examples of difficulties with fuel loading equipment including jamming of the loom of the fuel transfer machine of the Peach Bottom Atomic Reactor (this occurred during a hot test of the equipment) and the accidental withdrawal and dropping of a fuel element while an adjacent one was being withdrawn from the core of the ESADA-Vallicitos Experimental Superheat Reactor (EVESR). In the latter instance, as the one fuel element was being withdrawn from the core using the cask winch, it caught on an adjacent element, and both were lifted for six feet, at which point the second element dropped back into the core. Apparently the operator had not been observing the cask winch load indicator; but it was also reported that the load-limiting feature was working erratically. Another instance in which the scanning part of the refueling system failed to work in an emergency is given in the Prime Minister's Report to Parliament on the Windscale accident. This part of the incident is quoted on p. 634 of Vol. 1. Malfunctions such as these could lead to severe consequences, and they serve to illustrate the type of considerations that must be reflected in the design of fuel loading and unloading machines.

8.3.2 Preparation of Detailed Refueling Procedures

Prior to refueling, an overall sequence of operations and detailed step-by-step procedures for all work should be prepared. Such procedures are particularly important in complex refueling programs and help materially in streamlining operations, identifying problem areas beforehand, and minimizing radiation exposure of working personnel. Reference [145] is an example of such procedures.

These procedures should indicate the equipment required for each operation so that the equipment can be provided at the site of operations when it is needed. They should also indicate the range of measurements that are acceptable if the data obtained form the basis for proceeding to the next step. Alternative courses of action are to be indicated if the data are outside of the range desired.

The procedures should also indicate actions to be taken if equipment should malfunction or if fuel elements should stick. If field decisions are required, the authority for making such decisions should be clearly defined. Such decisions will undoubtedly have to be made because, although valuable, no set of procedures can be written which will cover all situations. One qualified individual, with appropriate technical backup should be placed in charge of the refueling operation and should be completely responsible for its execution. He should be represented on each shift by a competent shift supervisor whose sphere of action, authority and responsibility has been well defined.

No set of procedures is useful if it is not read and understood by all of the supervisory and working personnel involved in the operation. Hence, the procedures must be disseminated to all concerned. Such dissemination must include all changes made in the procedures. In addition, those who are to carry out the work must be trained for the specific operations they are to perform.

8.3.3 Training of Personnel

All personnel involved in refueling should undergo a familiarization program which includes a description of the various refueling operations and the safety precautions necessary. They should be made to understand the reasons behind such safety precautions and the hazards involved in violating them.

Personnel should be qualified to do all the operations to which they are assigned. This will require extensive training with the actual tools involved and with mockups of the environment in which they must work. These mockups should truly simulate the accessibility conditions to be faced during refueling.

Supervisors should be selected early and undergo training in all aspects of the work which they will be called upon to supervise. They, particularly, should be aware of the hazards involved and the precautions to be taken. They should recognize the need for and be responsible for removing unneeded working personnel from high radiation fields while difficulties are being explored and solutions sought.

All personnel involved in refueling should be aware of the chain of command and the authority of each person in the chain. If the operation is not supposed to involve certain actions on the part of an individual, he should be specifically instructed on that point. For example, if a control rod is not supposed to be withdrawn under certain conditions, each person who would be in position to withdraw the rod must know this fact even if his job is not supposed to involve such action. Knowing what not to do is as important as knowing what to do.

8.4 Spent-Fuel Storage Prior to Shipment

Most spent fuel elements must be stored for a period of time ranging from one to several months prior to shipment to processing plants so that the decay heat to be handled during shipment can decrease to acceptable levels. As a result, facilities must be provided for storing spent fuel elements with provisions made for cooling, monitoring and controlling criticality [144]. Occasionally some alteration of the fuel, such as removal of inert end sections, must be made prior to shipment; such operations at the utility site should be discouraged, however.

Water-filled canals have been widely used for storage purposes. (See Fig. 8-2 [143].) As indicated in reference [146], water provides an excellant shield for both gamma radiation and neutrons and allows good viewing for handling operations. Its disadvantages are that it is a good moderator, thus increasing the criticality problems, and that it does not constitute good containment for flying fragments or gases which might be formed in the event of a chemical or nuclear accident [146]. Water-filled canals also require the use of cans for containing ruptured fuel elements if contamination is to be controlled. Water-filled canals also pose a problem for storage of fuel from liquid-metal plants because of the possibility of violent chemical reaction between the water and any liquid-metal coolant adhering to the fuel. Provisions must be made for removal of any liquid-metal coolant adhering to the fuel prior to placing it in a water canal for storage.

The problems of safety in fuel-storage canals are discussed in references [144, 146] and [127]. A number of these have been identified in Sec. 8.2 and should be referred to in connection with spent-fuel storage. The two added considerations are those of providing appropriate storage racks to ensure criticality control and effective means to cope with fission-gas release.

Storage racks must be rugged to withstand the rough treatment they may undergo. They should be sturdy and rigid to withstand the loads they must handle without distortion or tipping that could impair geometrical control of criticality. They must be so designed as to permit free convective cooling of the fuel assemblies stored in them. They should provide appropriate alignment so that refueling and handling tools can engage and disengage the fuel assemblies in them as required. For added safety storage canals or the portions of canals used for storage should contain borated water, if feasible. Poison rods in the fuel assemblies or cadmium strips in the rack can be used to provide added safety against criticality, particularly where other usage of canal water precludes its being borated. The design of appropriate storage racks need be no problem if the foregoing considerations are applied.

The problem of fission gas release in a storage canal could be a significant problem if the spent fuel surface was contaminated with fissionable material prior to in-core service, or if the fuel was ruptured

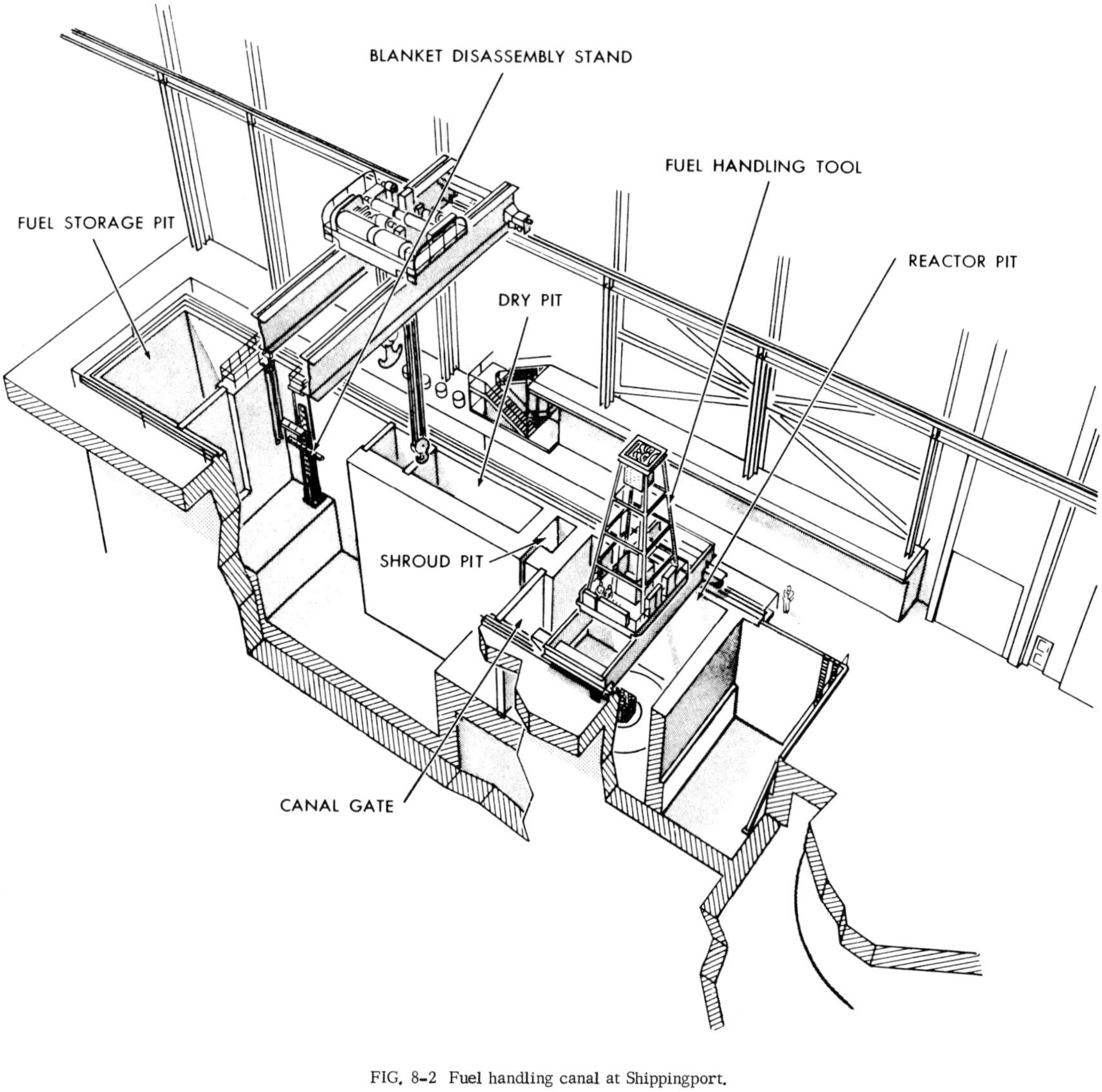

FIG. 8-2 Fuel handling canal at Shippingport.

during operation or severely damaged in handling. Ruptured metallic uranium fuel elements are particularly troublesome because oxidation of the uranium by water will release fission gases [147].

A number of experiences with fuel elements of various types are reported in references [147] and [148]. King in reference [148] reports that if freshly discharged slugs from the ORNL Graphite Reactor are ruptured, sufficient air-borne activity is encountered in the canal room to require the use of air masks for 24 hours unless the elements are immediately canned. Nicholson in reference [147] reports the severe radiation encountered at the General Electric Test Reactor (GETR) as a result of an unexpected gas release from a defective test element, which had been cooling for three days, when the cask containing the element was being cleaned prior to transfer from the building. Based on previous experience with similar uranium-aluminum alloy fuel specimens, no activity was expected. The gas release resulted in exposure of one person to a thyroid dose of 48 rads; six others received an estimated dose of 1875 to 7060 mrem from gaseous I^{131} and I^{133}.

If any fuel elements to be stored in a canal are suspected to contain surface contamination by uranium or are suspected of having been ruptured or of having been damaged to the extent that fuel is exposed, they should be immediately placed in a cask or can which is gas-tight or from which appropriately controlled vents have been provided.

8.5 Shipping Spent Fuel Elements

As the nuclear industry grows, shipping of spent fuel from power reactors will involve a considerable number of ton-miles [149, 150]. Such shipments could easily contain millions of curies [149]. If accidents are to be minimized, an increasing awareness of the hazards involved is required by all of those who play a part in shipping of spent fuel or preparing it for shipment.

Spent fuel elements are transported in heavily shielded containers known as casks. Albrecht in reference [150] likens a loaded spent-fuel cask to a portable nuclear reactor and points out, therefore, that the design considerations are similar. These include prevention of criticality, provisions for adequate heat removal, shielding, and structural integrity. The importance of these considerations cannot be overstressed in designing adequate shipping casks and controlling shipment. If, as a result of a wreck during shipment, the fuel were to go critical or melt because of inadequate cooling and the cask were to rupture, there could be dangerous dispersal of fission products and radioactive particulate matter [150]. A fire could also be involved which would add materially to dispersal of debris.

In order to transport fuel, a license must be obtained from the AEC. Issuance of the license is dependent upon review of information submitted by the applicant concerning criticality, heat transfer, radiation shielding, structural integrity, control of contamination of shipping casks and the handling procedures to be employed.

The USAEC has prepared and published regulations governing the shipment of irradiated solid fuel elements. In 1958, the Commission adopted 10 CFR Part 17, "Regulations to Protect Against Accidental Conditions of Criticality in the Shipment of Special Nuclear Material" [151]. This regulation established procedures for approval of transport of special nuclear material, but set only limited standards.

In 1960 and again in 1961, the Atomic Energy Commission published for public comment its proposed 10 CFR Part 72, "Protection Against Radiation in the Shipment of Irradiated Fuel Elements," to regulate the increasing number of shipments of irradiated solid nuclear fuel. That proposed regulation has never been adopted by the Commission although its provisions have been used as licensing criteria since their publication.

On March 3, 1963, the AEC published a proposed revision of Part 71 (28 Federal Register 2134) incorporating standards developed on the basis of its experience with licensing and shipping operations. As a result of numerous public comments and further study, the notice of proposed rule making revised on March 5, 1963, was withdrawn and was superseded by a new notice published on December 21, 1965 [152].

The new proposed revision of 10 CFR 71 emphasizes performance standards rather than detailed design standards to establish the adequacy of proposed shipping methods. The proposed performance standards are compatible with those developed by the International Atomic Energy Agency. In addition, the new proposal combines Parts 71 and 72 into a single document.

The new proposal also describes a plant for coordinating the activities of the USAEC and the Interstate Commerce Commission in the control of shipments. Since 1948, shipments of radioactive material in interstate and foreign commerce have been regulated by the Interstate Commerce Commission under the Transportation of Explosives and Other Dangerous Articles Act (18 USC and 831-835). The Atomic Energy Commission has provided a safety evaluation of the same shipments in some cases, both as part of its regulatim of the activities of its licensees and its control of its own shipments.

In summary, the revised Part 71 contains: (1) the substance of the earlier Part 71 which covered the shipment of unirradiated fissile materials, as revised to emphasize performance standards, (2) standards and requirements for the shipment of irradiated fissile materials, and (3) standards and procedures for the shipment of "large quantities" of licensed material.

A "large quantity" of licensed material is defined in the regulation in terms of the "transport group" of the radionuclide in question (which is based on relative potential hazard in transport) and in terms of "special form" of the licensed material. A "special form" is a nondispersible form, so that there is no need for further consideration of the hazard of ingestion of the material by a human being. The criteria used to determine whether the material is in "special form" are given in the regulation and depend upon the inherent properties of the material and the properties of a capsule in which it may be transported.

Basically, two aspects must be controlled to provide reasonable assurance of safe transport. First, an individual package of radioactive material must be so designed and its contents so limited as not to cause criticality or intolerable losses in radiation shielding or in containment of the radioactive material. Second, where a number of packages of fissile material are likely to accumulate, the effect of that accumulation must not be sufficient to cause criticality. This problem is unique to fissile material. Even though adequate measures are taken to assure nuclear safety of individual packages or shipments, a criticality incident may be caused by an unsafe accumulation of packages unless a system is established to control the numbers and types of packages which may accumulate in a single vehicle or storage area. Performance standards for an array of packages of fissile material are directed to the avoidance of such an unsafe accumulation.

Part 71 establishes three transport classes, consistent with the "Fissile Classes" developed by the International Atomic Energy Agency, based upon the packaging requirements and controls to be exercised during transport. For Fissile Class I and Fissile Clase II, shipment methods do not depend for safety on control by the shipper during transport, either for individual packages or collection of packages. Fissile Class I packages are so designed that administrative control during shipment is not required for nuclear safety. The design of Fissile Class I packages is required to be such that an unlimited number of such undamaged packages would be subcritical in any arrangement when mixed with any number of other Fissile Class I packages. The possibility that other Fissile Class I packages might be mixed with the particular package requires that consideration be given to moderator present in the other packages interspersed between packages of the design under consideration.

Fissile Class II packages are so designed and labeled that the only control necessary during transport is accomplished by the personnel of the carrier, through application of the "40 unit rule" of the Interstate Commerce Commission, Federal Aviation Agency, and the Coast Guard. Under this rule, carrier personnel, by counting "units" and limiting packages so that no more than 40 "units" are located in one place, would provide effective control over fissile packages.

The number of Fissile Class II packages which may be transported together is so determined that (1) at least five times that number, in any arrangement and undamaged, would not be critical assuming close reflection by water on all sides; and (2) twice that number, following such damage as would result from the hypothetical accident conditions, would not be critical assuming close reflection, optimum interspersed moderation, and any package arrangement. This assures that they will remain subcritical during transport with a sufficient margin of safety to protect against any inadvertent accumulation likely to occur. The number of units assigned to each package is equal to the number forty divided by the number of packages so determined.

If a proposed method of shipment of fissile material is not within Fissile Class I or II, it may only be a Fissile Class III shipment. The nuclear safety of a Fissile Class III shipment depends on control by the shipper to assure that other fissile material is not brought within an unsafe distance of the shipment and that the conditions under which the shipment is authorized are maintained. A Fissile Class III shipment must be transported with a courier accompanying it, by exclusive use of the vehicle or with the use of some other transport control method specifically approved by the Commission.

In order to provide reasonable assurance of adequate radiation shielding, containment of the radioactive material, and absence of nuclear criticality during transport, the performance of the package and the control exercised over it during transport must be evaluated for normal transport conditions and for potential accident conditions. To avoid inconsistencies involved in guarding against every conceivable condition which could be encountered in transport, the transport conditions against which a shipping system must be evaluated can be specified. A set of "normal conditions of transport" intended to represent conditions which may normally occur during transport may be set up. Packages must be designed to withstand these normal conditions. Further, a set of "hypothetical" accident conditions may be specified. Part 71 has specified a set consisting of a thirty-foot drop onto a flat surface, followed by a forty-inch drop onto a six-inch diameter steel bar, followed by exposure to an environment at a temperature of 1475°F for thirty minutes, followed by immersion in water. The hypothetical accident conditions prescribed in the regulations are not intended to represent any one accident, but are so chosen that satisfactory performance of a package exposed to them may be considered to give reasonable assurance of satisfactory performance in accidents likely to occur in transportation.

A package is not expected to withstand without damage the hypothetical accident condiions specified. The extent of allowable damage to a package depends on the effect of that damage on the containment, shielding, and nuclear safety characteristics of the package. It is expected that, in accordance with the Interstate Commerce Commission, Coast Guard, or Federal Aviation Agency regulations, any package which is damaged in transport would not be carried further in normal transport until any necessary repairs were made.

When the performance standards for a single package and for a permissible array of packages have been prescribed, it is necessary that some means be provided to demonstrate the adequacy of a proposed package or array to meet standards. Therefore, a "sample package" must be subjected to the specified transport tests and conditions by actual test, by analysis, or by other assessment. The "sample package" must fairly represent the actual package to be introduced into transport. In some cases of Fissile Class III shipments, where the entire shipment is to be controlled during transport, the "sample package" may be considered to be the entire shipment together with the transporting vehicle. It is the intent that any analytical treatment which has a reasonable degree of certainty may be employed to predict the performance of a package under the specified test conditions. The results of subjecting a package to the test conditions might be determined by engineering analysis, by physical testing of prototype packages or of scale model packages, by testing of package components or by any other method as long as a reasonable degree of certainty is established for the results. A great deal of effort has gone into the establishment of the test conditions to make it possible to use calculational methods of solution. It is hoped that good calculation methods will be developed so as to avoid, at least to some extent, the performance of physical tests which otherwise would be necessary.

In addition to the standard which determine if a package performs adequately when subjected to the normal and accident conditions of transport, there are certain design requirements directed to structural integrity, temperature, radiation shielding, and other general design features of a package.

The following requirements are extracted from the proposed 10 CFR 71. They are not intended to be complete but rather to indicate the nature of the requirements to be met. For complete details the reader is referred to reference [152].

8.5.1 Structural standards for "large quantity" packaging

(a) Load resistance. Regarded as a simple beam supported at its ends along any major axis, packaging shall be capable of withstanding a static load, normal to and uniformly distributed along its length, equal to 5 times its fully loaded weight, without generating stress in any material of the packaging in excess of its yield strength.
(b) Internal pressure. Packaging shall be capable of withstanding an internal pressure within the containment vessel of 20 pounds per square inch gauge or twice the operating gauge pressure, whichever is greater, without generating stress in any material of the packaging in excess of its yield strength.
(c) External pressure. Packaging shall be adequate to assure that the containment vessel will suffer no loss of contents if subjected to an external pressure equivalent to the pressure at a depth of 50 feet of water.
(d) Pressure relief device. Packaging shall be equipped with a means of relieving internal pressure at a gauge pressure at least 20 percent higher than operating pressure and not higher than 75 percent of design pressure.
(e) Lifting and Tie-down systems. Package lifting and tie-down systems are to be such that they are not torn from the package during use, and that stresses delivered to the package through the lifting or tie-down systems would not damage the package.

8.5.2 Temperature standards for large quantity packages

(a) A package used for the transport of a "large quantity" of licensed material shall be so designed and constructed and its contents so limited that with each package exposed to direct sunlight and an ambient temperature of 100°F in still air, and assuming loss of liquid or gaseous coolant in the package except air at atmospheric pressure, and assuming loss of operative mechanical cooling devices, the temperature of the radioactive material will not be higher than the minimum temperature which would result in release of radioactive material from its containment vessel.
(b) Packaging shall be so designed and constructed that primary coolant does not circulate outside of the shielding during transport.

8.5.3 Radiation shielding standards for all packages

(a) A package is to be so designed and constructed and its contents so limited that, except as provided below in paragraph (b), the radiation dose rate originating from the package does not exceed at any time during transport either of the following limits:
(1) 200 milliroentgens per hour or equivalent at any point on the external surface of package; or
(2) 10 milliroentgens per hour or equivalent at a distance of one meter from the center of the package or at the surface of the package if it is farther than one meter from the center.
(b) When a package is transported on a vehicle assigned for the sole use of the licensee, the radiation dose rate originating from the package may exceed the limits specified in paragraph (a) of this section if it does not exceed at any time during transport any of the following limits:
(1) 1000 milliroentgens per hour or equivalent at one meter from any external surface of the package;
(2) 200 milliroentgens per hour or equivalent at any point on the external surface of the transporting vehicle;
(3) 10 milliroentgens per hour or equivalent at two meters from any external surface of the transporting vehicle; and
(4) 2 milliroentgens per hour or equivalent in any area in the vehicle occupied by the driver or any other person.

8.5.4 Standards for normal conditions of transport for a single package

(a) A package used for the transport of fissile material or a "large quantity" of licensed material is to be so designed and constructed and its contents so limited that under the normal conditions of transport:
(1) There will be no release of radioactive material from the containment vessel;
(2) The effectiveness of the packaging will not be substantially reduced;
(3) There will be no mixture of gases or vapors in the package which could, through an increase of pressure or an explosion, significantly reduce the effectiveness of the package;
(4) Radioactive contamination of the liquid or gaseous primary coolant will not exceed 10^{-7} curies of activity of Group I radionuclides per milliliter, 5×10^{-6} curies of activity of Group II radionuclides per milliliter, 3×10^{-4} curies of activity of Group III and Group IV radionuclides per milliliter; and
(5) The internal pressure of the package will not exceed fifty percent of the design gauge pressure.
(b) A package used for transport of fissile material is to be so designed and constructed and its contents so limited that under the normal conditions of transport:
(1) The package will be subcritical;
(2) The geometric and physical form of the package contents would not be substantially altered;
(3) There will be no leakage of water into the containment vessel. This requirement need not be met if, in the evaluation of undamaged packages, it has been assumed that moderation is present to such an extent as to cause maximum reactivity consistent with the chemical and physical form of the material; and

(4) There will be no substantial reduction in the effectiveness of the packaging, as outlined in the regulation.

(c) A package used for the transport of a "large quantity" of licensed material shall be so designed and constructed and its contents so limited that, under the normal conditions of transport, the containment vessel would not be vented directly to the atmosphere.

Much work is going into impact testing of cask prototypes to provide a sound basis for design. Work is also in progress to determine effective means for cask tiedown and to establish the role that such tiedown plays in preventing damage to a cask in the event of a wreck [153].

In addition to the requirements of the AEC, the requirements of other regulatory bodies must be met. These include the current ICC regulations [154] and appropriate state regulations. As the work of various groups involved reaches fruition, it is expected that standard type casks will be developed. However, even the use of such casks will not eliminate the need for evaluating each shipment of fuel for which there is some difference in prior history or in construction. Such precautions are warranted in view of the hazards involved.

9 NONDESTRUCTIVE TESTS AND ACCEPTANCE INSPECTION

It is the purpose of this section to introduce the reader to the field of nondestructive testing and to indicate the important part that it can and does play in the acceptance inspection of components for nuclear power plants. Because of the wide scope of this field, little more can be provided in this section than a glimpse of the field and a background for further reading. A number of excellent articles arising out of symposia on nondestructive testing and several textbooks on the subject are available which provide further details about techniques. These are referred to below.

9.1 General Considerations

McGonnagle [155] defines nondestructive testing as the general name given to all methods of testing and inspection of material which do not impair the future usefulness of the material. The ultimate purpose of such testing and inspection is to determine whether a material or a product is to be accepted or rejected. Nondestructive testing methods provide new tools for making such decisions. In view of the fact that most nondestructive testing involves indirect indication of the desired measurement, a great deal of skill and experience is required to obtain, evaluate, and interpret the data obtained by nondestructive testing methods [155, 156]. Accurate judgment is required in selecting the method most appropriate for making a particular measurement. The techniques used should be well developed and verified by other means where possible; the personnel actually performing the measurements should be well trained. Even with well applied and well performed nondestructive testing, interpretation is difficult because all materials have at least some random structure, forming, in effect, minute faults which will often be identified by nondestructive tests; thus judgment is required in determining whether or not the "relative soundness lies within specified tolerances [155]."

No amount of inspection or testing takes the place of good workmanship in the manufacture of materials and products. Such inspection and testing can have maximum benefits when used in conjunction with a program to identify faults for the purposes of improving processes and keeping them under control.

9.1.1 Steps in Testing

Every nondestructive test involves the application of some form of energy to a piece of material and the measurement of the response of the material to the energy applied. The data obtained must then be evaluated; this usually requires comparison with readings obtained either from other specimens with known characteristics (acceptable and unacceptable) or from reference effects measured simultaneously in or with the test specimen (e.g., use of multiple X-ray films to eliminate film defects). The evaluated data must then be interpreted and appropriate action taken. Correct interpretatior must be based on a knowledge of the probable defects associated with the processing of the material and manufacture of the product. Appropriate action can include acceptance or rejection of the piece or repair of the fault followed by reinspection.

In summary, every nondestructive test involves the following steps:

1. Application of the test energy
2. Measurement of the response of the material to the energy applied
3. Evaluation of the data including comparison with reference readings or standards
4. Interpretation of the data
5. Acceptance, rejection or repair and reinspection

A wide variety of energy forms have been used for nondestructive testing. Chief among these are light energy for visual inspection; pressure energy for leak testing; capillary action for penetrant testing; thermal energy for tests involving heat conduction; acoustic energy for sonic and ultrasonic testing; magnetic energy for magnetic particle tests; electrical energy for a wide variety of tests including electrical resistivity and eddy current tests; and radiation energy for tests utilizing X-rays and gamma rays. The application of these energy methods to nondestructive testing has required extensive development work on both equipment and measurement techniques. Considerable attention has been and continues to be given to the problems associated with sensitivity and resolution. New problems arise as the material or geometry to be tested are changed.

Because the nature of defects and the problems of preparing acceptance specifications play such important parts in nondestructive testing, a brief discussion of these topics is in order before proceeding to a description of some of the test methods and the special requirements involved in applying them to the nuclear field.

9.1.2 Nature of Defects

Following the pattern outlined in Sec. 6.2.3 regarding materials used for piping, defects will be classified as either mechanical or metallurgical defects. Generally, defects of a mechanical nature lend themselves to nondestructive testing; metallurgical defects involving discontinuities in composition are difficult to detect nondestructively unless discernable by surface or subsurface tests.

9.1.2.1 Mechanical Defects. Mechanical defects consist of mechanical discontinuities or separations on the surface or inside of the piece of material. They include defects generally denoted as cracks, tears, flakes, blowholes, pinholes, porosity, laminations, lack-of-bond, and inclusion. They can arise from a variety of sources generally of a mechanical or thermal nature. Some of the characteristics and causes of various types of mechanical defects are discussed below:

Cracks may be transcrystalline or intergranular and may propagate in a variety of directions [155]. Many of them are discernable only under a high-powered microscope; these are often called hairline cracks or microcracks. The space formed by the crack may be clean or filled with oxides or "dirt." Many cracks are formed by thermal effects such as rapid heating, quenching, shrinkage during welding or casting and local overheating during grinding or rough machining. As a result care should be taken in planning thermal operations on large or complex pieces to assure that thermal stresses or distortions caused by poor flow of heat or rapid changes of temperature do not cause cracks as the material passes through temperatures where the strength is very low. Attention should be given to thermal effects also because, even though the material may not crack, it may be left with high residual stresses which could enhance stress corrosion or produce distortions during service. Cracks can also be caused by purely mechanical action; they can result from scratches incurred during drawing or machining; from the impact of sharp edges on forming dies; from dirt or insoluble impurities being included and buried in the material; from blowholes; from the folding of metal into cracks or seams; or from other action similar to the foregoing. Tears are cracks brought about by severe stresses; they often are characterized by jagged edges. Flakes are internal fissures in ferrous metals [155].

Blowholes, pinholes, and porosity are cavities caused during solidification of the material by the entrapment of gases which come out of solution as the material cools. In castings, blowholes may also arise from moisture in the mold. Blowholes are generally oval in shape; their orientation can sometimes be associated with the form of the mold. Their sizes may differ widely. Small blowholes are often termed "pinholes"; however the term "pinholes" may also refer to porosity produced by other forms of gas entrapment like that encountered in welding or brazing. They are microscopic in nature and probably occur during the last stages of solidification when surface tension is high [155]. Numerous pinholes or small blowholes can lead to excessive porosity. The importance of holes is dependent upon their size relative to the size of the piece, their number per unit volume, and their location within the piece.

Laminations refer to discrete layers of materials which fail to bond to each other. They may be caused by discontinuities in the chemical composition of material, by sharp and distributed discontinuities in temperature, and by impurities located along grain boundaries.

Inclusions are particles of impurities which are mechanically held in the material [155]. They not only can act like cracks, but also cause regions of high stress concentration which can markedly reduce the fatigue life of the material. They also introduce inhomogeneity into the material which, when on the surface makes machining, grinding and polishing difficult and leads to acceleration of wear or corrosion during service.

9.1.2.2 Metallurgical Defects. The term "metallurgical defects" is used to refer to unacceptable discontinuities in the chemical composition of material. Differences in properties of the material in the region of the discontinuity could lead to failures during service if severe. Such discontinuities arise out of the fact that the distribution of chemical elements does not remain uniform during solidification of a material. Different rates of precipitation and diffusion of these elements are encountered because of the different rates of cooling of the various parts of the material. This is particularly true during welding where cooling rates are not generally well controlled; discontinuities in properties are most frequently encountered during welding of materials with initially different compositions. Metallurgical defects can also be encountered during localized heat treatment of material if local areas of markedly dissimilar hardness are produced [96]. Carburization or decarburization of steels can also produce metallurgical defects [96].

9.1.3 Acceptance Specifications for Nondestructive Testing

Many of the problems associated with the use of nondestructive testing arise out of the use of improper techniques and the failure to specify clear cut acceptance standards. V. T. Malcolm in reference [156] points out that while ASTM standardizes the methods of nondestructive testing, it does not specify the acceptability of parts by a particular testing method; acceptance standards are a matter of agreement between supplier and purchaser and the applicable codes. Reference data are not "go-no-go" standards, they are only guides.

While nondestructive testing can be said to be an old science dating back to Archimedes [155], its widespread use in industry is relatively young. As a result there is considerable need for improving techniques and standardizing the training of personnel as well as the bases for evaluation and interpretation of results. Malcolm points out in reference [156] that even, after many years of usage, there are no "go-no-go" standards for radiographic, magnetic particle, and dye penetrant inspection.

Because of the important part that nondestructive testing plays in the nuclear industry, special efforts should be expended to assure that specifications for acceptance and rejection are clearly and realistically prepared by persons who are knowledgeable in the field, and that the tests are correctly carried out, and evaluated and interpreted by qualified personnel.

9.2 Testing Methods

McGonnagle [155] classifies all methods of nondestructive testing under the following headings:

1. Visual (oldest and simplest test; can reveal defects not otherwise discernable)
2. Pressure and Leak (uses gas or liquid flow into or through defect)
3. Penetrant (uses low viscosity liquid to detect surface defects by capillary action)
4. Thermal (depends on changes in thermal conductivity brought about by defects)
5. Radiography (utilizes X-ray and gamma-ray radiography)
6. Acoustic (utilizes sonic or ultrasonic vibrations)
7. Magnetic (depends on distortion of magnetic field by defects)
8. Electrical and Electrostatic (utilize influence of defects on electrical resistivity or static electric field)
9. Electromagnetic Induction (based upon influence of defects on eddy currents)
10. Miscellaneous (includes spark tests, surface stress coating, strain gages, etc.)

A comprehensive discussion of each of these methods is beyond the scope of this book. A complete discussion of each method would involve treatment of principles, theory, techniques, equipment, and experiences along with a discussion of the special problems associated with specific applications. For such detailed treatment the reader is referred to references [155] through [159]. It is the purpose of this section to summarize briefly the nature of some of the more important methods used in the nuclear field.

9.2.1 Ultrasonic Tests

This summary is based, in large part, upon "Survey of Ultrasonic Methods and Techniques", by S. A. Wenk in reference [156].

Ultrasonic waves are sound waves having frequencies above the range of human hearing (>20,000 cps). Two characteristics important in ultrasonic testing are that ultrasonic waves travel in well-defined beams and that they are reflected by discontinuities or boundaries of different elastic and physical properties. Ultrasonic waves can be longitudinal waves, transverse waves or surface waves. (See Fig. 9-1 [156].) Longitudinal waves are the most frequently used because they are the easiest to generate and the only type that can be sustained in a liquid. Surface waves are growing in importance for detecting surface defects. Surface waves have been classified into three groups [155]: Rayleigh waves (wavelength small compared to thickness of piece); Lamb waves (wavelength comparable to thickness of piece); and Love waves (waves in thin surface layer whose density is different from that of bulk material).

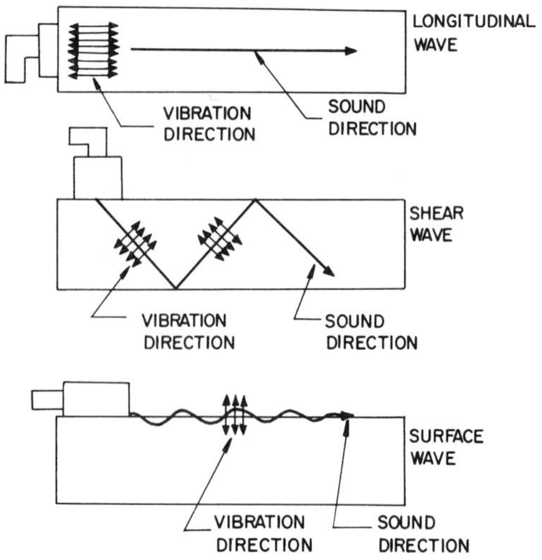

FIG. 9-1 Three basic ultrasonic wave modes in use today.

Ultrasonic waves are produced by piezoelectric crystals, known as transducers, which can transform electrical vibrations into mechanical vibrations and vice versa. Natural quartz crystals are the most commonly used, but barium titanate, lithium sulfate and other piezoelectric materials are also used. For good efficiency the piezoelectric transducer must be coupled to the surface of the test piece. This can be done by direct contact or by immersion of transducer and test piece in a transmitting liquid. In the contact method, a couplant film of water, oil or grease is used between transducer and test piece; in immersion testing the transducer is spaced a short distance from the test piece. Either a single or double transducer system can be used. In the double system one transducer is the transmitting crystal and the other is the receiving crystal; both through-transmission and echo testing can be done with double transducer systems. The single transducer system is the most commonly used system in this country; it is used as an echo-testing device.

The use of a single transducer for echo testing is shown in Fig. 9-2 [156]. Shown in the figure is the signal as displayed by a cathode ray tube. The pips A and B show the crystal-material interface and echo signal respectively. The distance between pips represents both the time and distance of travel of the sound wave. Figure 9-3 [156] shows the resulting signal if a flaw exists in the material; an intermediate pip is produced. If the plane of the defect is not ideally oriented, only a portion of the signal is reflected and the signal is weak. This cannot be readily improved in contact testing but can be easily improved in immersion testing by positioning the crystal to receive the maximum echo signal.

The equipment for producing high-frequency electrical oscillations in ultrasonic testing can be of the continuous radio frequency type (AM or FM) for use with two-crystal operation or of the pulse

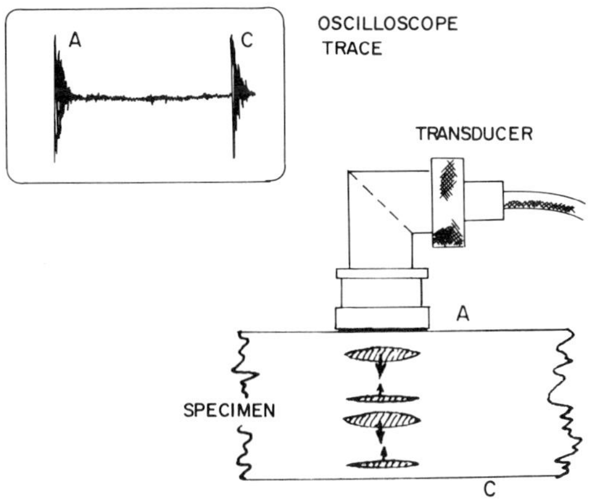

FIG. 9-2 Reflection or echo method of contact scanning with no defect. A is interface between transmitting crystal and specimen. C is the back face of specimen. Shaded ovals represent acoustic pulses from transmitter and the back echo.

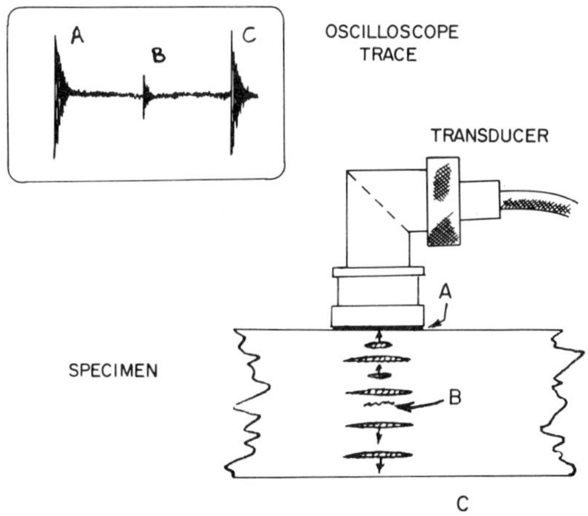

FIG. 9-3 Reflection or echo method of contact scanning with a discontinuity in the material tested. A is the interface between transmitting crystal and specimen. B is the flaw (acoustic discontinuity). Small shaded areas represent acoustic pulses reflected by flaw.

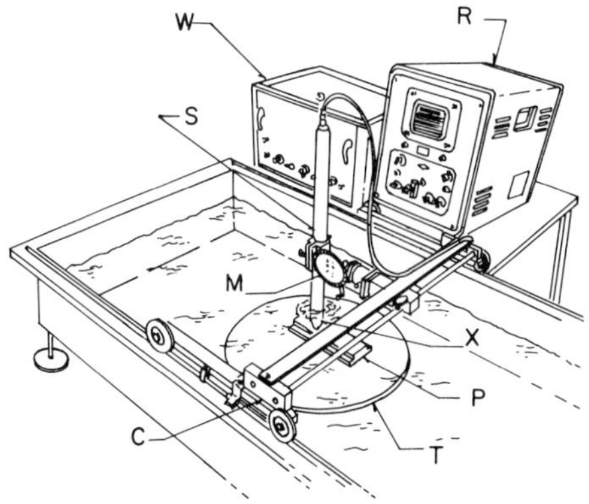

FIG. 9-4 Typical ultrasonic system for immersion testing. P - part being tested, T - support for test material, X - transducer (piezoelectric crystal), S - scanner tube, M - manipulator, C - carriage for scanner, R - reflectoscope, W - wide-band converter used with reflectoscope.

type for use with either one or two-crystal operation. The pulse type is the most widely used in ultrasonic testing. The data can be presented audibly or displayed on a meter, cathode ray tube or recorder. Figure 9-4 [156] shows a typical ultrasonic system for immersion testing.

Ultrasonic tests are used for determining cracks, unbonded areas, porosity and grain size in reactor materials. In addition, as outlined in Sec. 2.11 and reference [177], ultrasonic techniques show promise for detecting cracks in the wall of a pressure vessel during service life. One general approach is to utilize, for detection of cracks in metals, the ultrasonic energy released during growth of such cracks. Ultrasound-sensitive transducers mounted on the vessel could be used to detect these bursts of energy. Such a method might be feasible, not only for detecting crack growth during hydrostatic testing but also for continuous monitoring of the vessel during operation if suitable transducers can be developed which will not deteriorate on prolonged mounting on the vessel. The exact location of the crack growth might be determined by monitoring the output of three or more transducers and using a method of triangulation, for fixing the position. If systems of triangulation can detect the exact location of cracks, then such detectors can be used during refueling periods to check on the vessel condition. Comparison between old and new evaluations may give evidence of vessel defect growth. Further, there is some reason to believe that it may even be possible to actually obtain NDT temperatures for a vessel in service by observing the change in sound velocity in the vessel material as the liquid-filled vessel is carried through the appropriate temperature range.

9.2.2 Penetrant Tests

This summary is based on, "The Use of Penetrants for Inspection of Small Diameter Tubing", by R. G. Oliver, G. M. Tolson and A. Toboada in reference [156].

Penetrants provide a means for supplementing visual inspection of material for surface flaws. A penetrant is used to delineate discontinuities into which the penetrating fluid can enter and be retained. Thus it can be reliable only if the crack or discontinuity reaches to the surface and acts as a capillary and is not covered by dirt or oxides.

Penetrant systems can be of two types: contrast or dye penetrants and fluorescent penetrants. Dye penetrants are viewed by normal light; fluorescent penetrants require the use of ultraviolet light, sometimes called black light. Except for the difference in the light used, the procedures for making penetrant inspections are essentially the same for both types. The steps as illustrated in Fig. 9-5 [156] are as follows:

A. Application of the penetrant. This is done by brushing spraying or dipping. The penetrating time may vary from 1 to 30 min depending on the nature of the flaw and the fluid used.

B. Application of an emulsifier. This is done to render the surface water washable. The

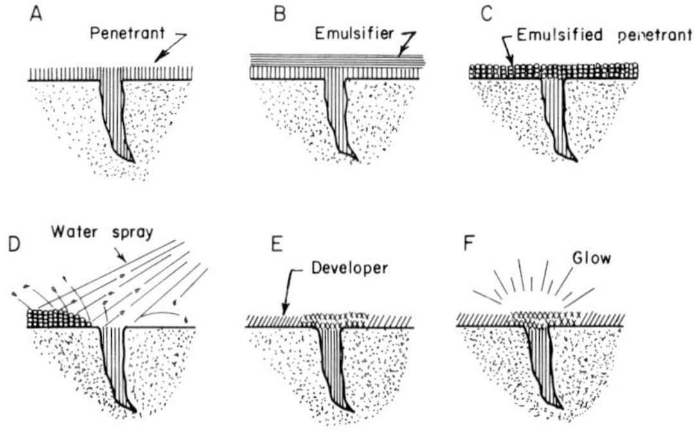

FIG. 9-5 Penetrant inspection procedure using post-emulsification cleaning.

emulsifier could be mixed with the penetrant.

C. Diffusion of the emulsifier into the penetrant. The time allowed for this step influences the ability to identify shallow defects. If the diffusion time is short, shallow defects are more easily identified because the emulsifier will not diffuse deeply into the penetrant.

D. Washing. The emulsified penetrant and excess emulsifier are removed effectively by water spray.

E. Application of a developer. Developer can be either wet or dry. If wet, the developer is applied by dipping or spraying; if dry it is applied by dipping or dusting.

F. Inspection. The time between application of the developer and inspection may vary from 15 min to several hours. The penetrant must diffuse to the surface of the developer to reveal the crack. Thus, if the developer coating is too thick, small cracks will not be revealed because they do not hold enough penetrant to diffuse to the surface.

G. After-cleaning. (Not shown in Fig. 9-5.) As is often the case in reactor systems the residual penetrant may be detrimental to the service of the piece during operation. Therefore the piece must be cleaned by either vapor degreasing or by the use of appropriate solvent cleaner followed by a water wash.

Several factors must be considered in selecting the penetrant system to be used for a particular application. To detect fine cracks in small pieces, such as fuel element tubes, the most sensitive penetrants must be used. Water-washable penetrants requiring the use of emulsifiers are not well suited to this purpose because the emulsifier decreases the penetrant's ability to enter fine cracks. Caution must be used when employing solvents for washing; they are good penetrants and can remove the desired penetrant if used in excess. If a great deal of handling of the test piece is required, the use of a wet developer is discouraged because it may be marred during handling. Fluorescent penetrants have the advantage of good contrast for inspection purposes and do not present as severe a fire hazard, when used in large quantities, as do the low-flash-point dye penetrants. Because of the false indications that can be encountered with fluorescent penetrants, suspected areas should be checked visually under magnification (about 10-power).

All penetrant inspection should be well planned with appropriate marking procedures for indicating flaws. The work should be done by well-trained personnel; interpretation requires careful judgment of qualified people.

9.2.3 X-Ray Tests

This summary is based on "Survey of Radiation Techniques," by G. H. Tenney in reference [156] supplemented by material from reference [155].

The principal methods for radiographic inspection of material are X-ray and gamma radiography. Both are well described and discussed in reference [155]. The X-ray test is used as the example in this section to indicate some of the characteristics of radiographic examination.

X-rays are produced by the impingement of high velocity electrons upon heavy target atoms. The energy of the X-rays is a function of the velocity of the impinging electrons, and the intensity is a function of the number of electrons per unit time bombarding the target. A modern (Coolidge-type) X-ray tube consists of an evacuated tube with a heated spiral tungsten filament inside a cup at one end and a metallic target (often of tungsten) on the other end with a voltage applied between them. The electrons are released by the heated filament (cathode) and are accelerated toward the target (anode) by the applied voltage. X-rays are produced by the transposition of oribital electrons in the target. Inasmuch as only a small fraction of the electron energy appears as X-rays at the target ($\sim$ 3% at 300 kev (135)) the target must be cooled. This can be done by conduction to air-cooled fins or by the use of a coolant flowing through the anode. Inasmuch as the X-ray machine works on direct current, a rectifier is used if the power source is alternating current. The spectrum of radiation from an X-ray tube consists of two parts, spectral peaks superposed on a continuous spectrum. (See Fig. 9-6 [156]).

X-rays and gamma rays are particularly useful in nondestructive testing because they are readily penetrate matter, they are absorbed differently by different media, travel in straight lines, and affect photographic emulsions. As a result, if X-rays are made to pass through a piece of material, they can

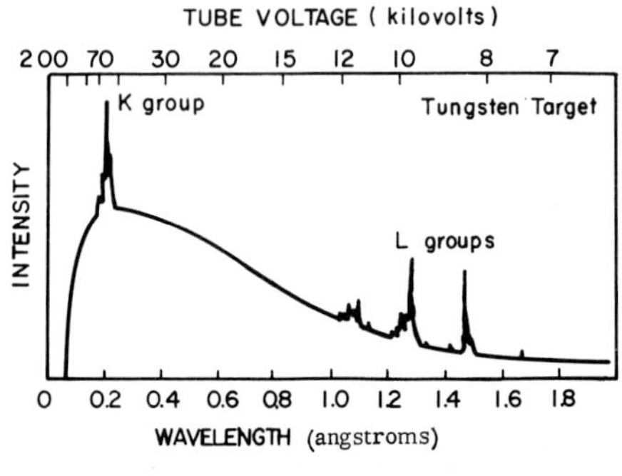

FIG. 9-6 Characteristic X-ray spectrum of tungsten.

reveal defects in the material because of the influence of the defect on the opaqueness of the material to the X-rays.

Several different means are used for determining the amount of radiation passing through a test piece. Of these, the use of radiographic films is the most common in industrial work. Other methods involve the use of fluoroscopes or electronic scanning of an X-ray image. However, because of the preponderant usage of films in normal industrial work, the discussion will be limited to them. They are used widely because they are the most reliable of the X-ray recording methods and provide permanent records for comparison purpose later.

Many factors influence the quality of a radiograph. These include the source size, the object-to film distance, the amount of scatter produced by the piece or surroundings, the type and nature of the film and the amount of and techniques of exposure. Care must be taken in selecting a film appropriate to the job because the film influences the ability to identify flaws.

Industrial radiographers utilize the following terms to describe the quality of a radiograph:

1. Radiographic sensitivity. This expresses the clarity with which images appear on the radiograph. Per cent sensitivity = $s/t \times 100$ where s is the smallest detectable thickness difference and t is the thickness of the specimen.
2. Radiographic contrast. This is the difference in blackening on a radiograph caused by variations in thickness or density in the object.
3. Definition. This refers to the fidelity of the outline of a flaw in the object.
4. Radiographic resolution. This describes the absolute size of the image of the smallest detectable defect.

To determine the radiographic sensitivity, contrast, and definition that can be obtained in a particular application, penetrameters are often used. These are test pieces containing intentional flaws of known characteristics usually in the forms of drilled holes. They are made of the same material as the specimen to be inspected. When the radiograph is being made, the penetrameter is usually placed on the surface facing the source of radiation and adjacent to the area being inspected [155]. If the outline of penetrameter can be seen and if its thickness is known, the upper limit of sensitivity can be computed. Step penetrameters with different thickness can be used for determination of the minimum sensitivity. The outline of the hole can be used to establish definition, and the difference in blackening reveals the contrast. Various types of penetrameters are used. (See Fig. 9-7 [155].) The exact type of penetrameter and the image visibility which is acceptable are usually, and should be, specified for a particular job. When using penetrameters it should be recognized that their use does not preclude the possibility of missing an actual cavity of the same dimensions which is quite diffuse. Penetrameters do not measure resolution.

Another technique of great use in industrial radiography is the use of multiple films. This involves the use of two or more films during a single exposure. Defects which are not present on both films indicate that the flaw is on one of the films and not in the specimen. Multiple films of different speeds can be used to cover a range of thicknesses in a single exposure.

9.2.4 Magnetic-Particle Tests

This summary is based upon information in Chapter 10 of reference [155].

Flaws in a magnetic material can distort an induced magnetic field sufficiently to disclose their presence. A number of the magnetic methods used for nondestructive testing are outlined in reference [155]. Of these, the magnetic particle test is one of the simplest and easiest to use. It can be used on a magnetic specimen of any shape and size to detect surface and subsurface flaws.

The magnetic particle test involves two basic steps: magnetizing the specimen, and applying finely divided magnetic particles. The particles can be applied either wet or dry. If the flaw is on the surface or immediately beneath the surface, it will act to produce a pair of magnetic poles which will attract the magnetic particles by magnetic leakage and identify the location and extent of the flaw. In order for the method to be effective, the surface must be clean and free of oxides. Surface defects are clearly defined by a heavy pile-up of particles; subsurface defects will be less sharply defined. (See Fig. 9-8 [155].) Practically all kinds of mechanical defects can be detected by magnetic particle testing.

A magnetic field can be applied to the specimen by passing a direct current through all or a part of the specimen, by including a magnetic field with a direct-current carrying coil around the specimen, or by applying a permanent or a direct-current electromagnet to the specimen.

Two methods of magnetization in common use are the so-called prod and yoke methods. The prod method is used to produce circular magnetization by current flow through a portion of a specimen. This is done by attaching electrical leads or prods to the specimen; a magnetic field is produced between the contact points. The contact surfaces must be clean to avoid arcing or burning and the magnetization current should be turned on before the prods are positioned on surfaces of the specimen.

The yoke form of magnetization involves the supporting or clamping of the specimen across the poles of a permanent magnet or an electromagnet having a U-shaped core. The specimen completes the magnetization path and thereby permits the flux to return through the specimen rather than through the air.

For greatest sensitivity the specimen should be magnetized so that the magnetic field is perpendicular to the defect. The magnetic particles can be applied when the magnetization is being produced or after the magnization means are removed. The first method is the so-called continuous technique and the second, the residual technique. Normally the continuous technique is used because it is more sensitive; the residual

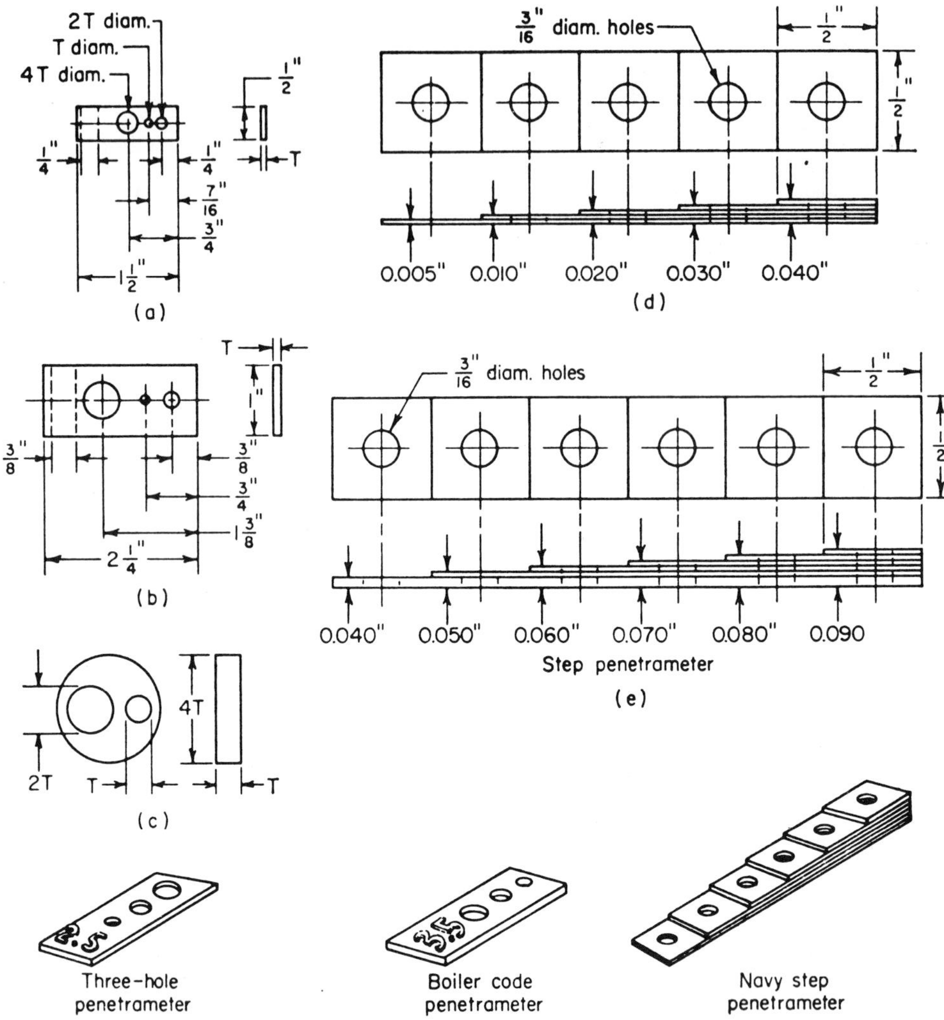

FIG. 9-7 Penetrameters: (a) for metals up to 0.5 in. (1.3 cm) in thickness; penetrameter thickness from 0.005 to 0.050 in. (0.13 to 1.3 mm); (b) for metals up to 8 in. (20 cm) in thickness; penetrameter thickness from 0.060 to 0.160 in. (1.5 to 4.1 mm); (c) for metals over 8 in. (20 cm) in thickness; penetrameter thickness of 0.180 in. (4.6 mm) and over.

technique is used occasionally to help eliminate false readings.

Inspection by means of magnetic particles should be performed before an acid etch, because the etching tends to open narrow defects and round off sharp corners; such action hurts the flaw geometry for good magnetic particle testing. Magnetic particle inspection may produce false indications; therefore only a trained individual can be relied upon to interpret results. The use of magnetic particle inspection should not be counted on to detect flaws more than a half inch below the surface.

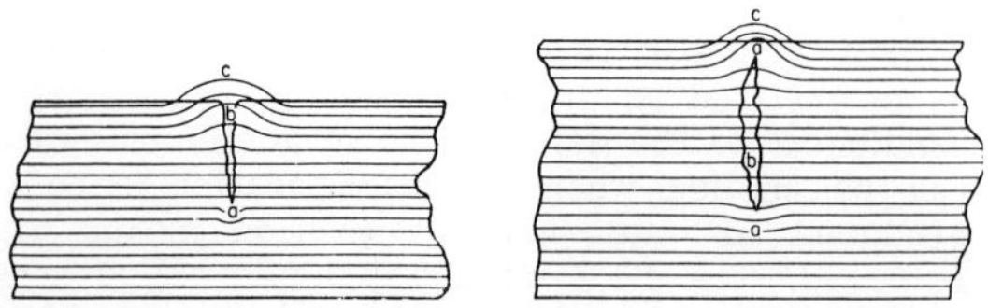

FIG. 9-8 Magnetic leakage due to surface and subsurface flaws.

9.2.5 Eddy-Current Tests

This summary is based on parts of ''Introduction to Eddy Current Methods and Techniques'', by H. L. Libby in reference [156], supplemented by information from reference [155].

The eddy current test utilizes eddy currents produced by electromagnetic induction to detect mechanical and metallurgical defects. Eddy currents induced in the specimen produce a magnetic field which opposes the inducing field; as a result, they change the impedance of the exciting coil or a pickup coil near the specimen. The flow of the eddy currents is influenced by the presence of defects and this influence can be used to identify them. Eddy current tests are most effective for detecting defects at or near the surface. However, great care must be exercised in evaluating and interpreting the results of eddy current tests; many acceptable variations in quality will be indicated which can lead to unnecessary rejection of the specimen unless a correlation has been established between the measurements and the suspected defects.

The eddy current test involves comparison of the impedance change produced in a test coil by a test specimen with that produced by a standard or acceptable specimen under similar conditions. The detection problem is complicated by many factors including coil-to-specimen spacing, coil dimensions, electrical and magnetic properties of the specimens and surface irregularities as well as defects.

The electromagnetic interaction of a coil and specimen results in a "skin effect," or concentration of currents near the surface toward the coil. If a plane conductor is placed in a uniform high frequency magnetic field, the current produced is greatest at the surface and falls off exponentially with depth below the surface. The depth of penetration is defined in this case to be the point at which the current is equal to 1/e times the surface current. While in nonuniform fields the current fall-off does not follow the exponential function, the depth of penetration as defined above is useful for comparing and predicting the behavior of metals under various test conditions.

Test coils for eddy current testing can consist of one winding or several windings. They may use air cores or magnetic cores; magnetic cores increase sensitivity. Sensitivity may also be increased by tuning the coil. Resolution can be improved by shielding the coil with a magnetic or copper material. Although coils can be designed on the basis of theoretical analyses, they should be checked experimentally to determine differences from the ideal assumptions usually made during analysis; mercury-filled tanks with simulated defects can be used for this purpose. Experimental verification of the design will also disclose effects of the leads on inductance and capacitance of the coil. The size of the coil is determined by the type of test specimen and the type and size of defect to be detected. For surface explorations small flat coils are best; they have limited sensitivity for deep explorations. For a coil of a given diameter, the sensitivity of the coil to a defect decreases rapidly as the defect size becomes less than the coil diameter. Different types of coils for testing piping are shown in Fig. 9-9 [155].

A number of different methods are used for read-out of data in eddy current tests. These include amplitude-measuring devices such as a-c voltmeters and ammeters, impedance bridges, amplitude-phase detectors, self-excited oscillator circuits containing the coil and test specimen, and

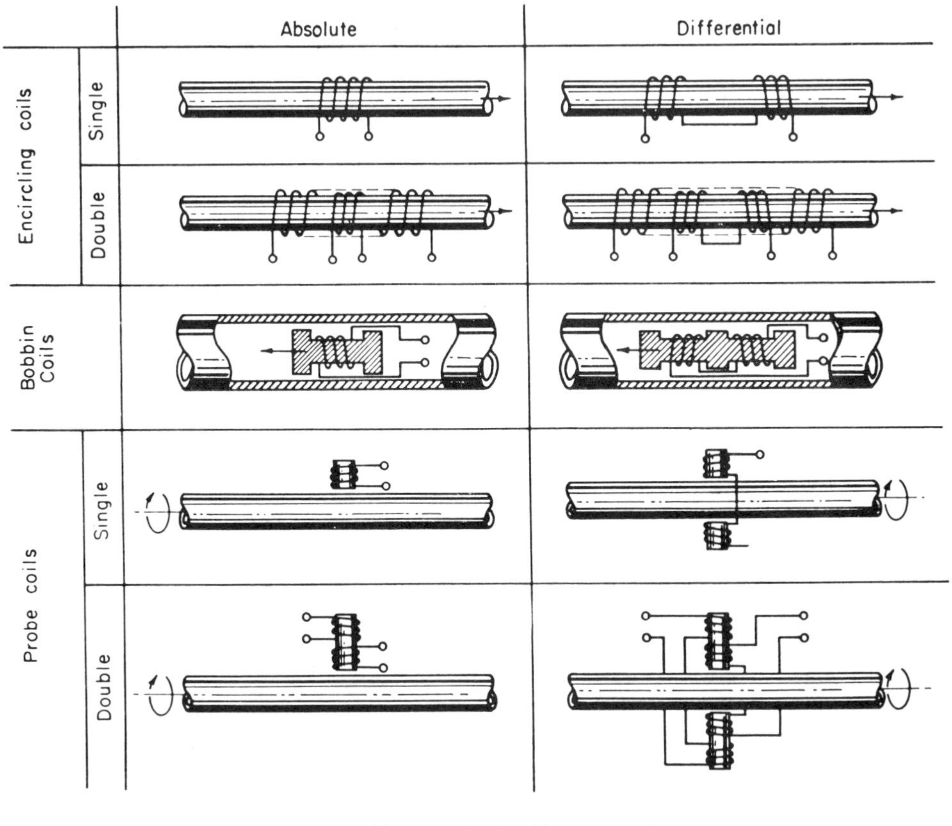

FIG. 9-9 Types of coils for eddy current tests.

phase-measuring circuits. These are discussed in reference [155].

Multiple frequency methods are being developed for use in eddy current testing which give more information than obtained by use of a single frequency. Pulse methods are also under development. (See references [155] and [156].)

9.3 Nuclear Applications

Throughout the chapter, frequent reference has been made to specific applications of nondestructive testing to nuclear components. S. McLain points out in reference [156] that nondestructive testing of fuel elements for heterogeneous reactors and components for various reactor systems differs from ordinary industrial nondestructive testing only in that the specifications for reliability in the nuclear field may be more rigid than those required in other industries. This comes about because of the greater need for component reliability in nuclear systems than exists in many other applications.

Many of the difficulties encountered in the operation of nuclear plants can be traced back to deficiencies which should have been noted and corrected during manufacturing. As a result there has been a widespread increase in the use of nondestructive testing methods in the manufacture of nuclear components. This testing has aided greatly in assuring sound and leak-proof components. This has not come about without a great deal of development work both with regard to measuring methods and to the determination and establishment of acceptable standards. While the inspections and tight standards have added to the cost of nuclear components, they have undoubtedly played an important part in making current reactor plants safe. As more experience is gained in identifying the sources of difficulty in the manufacture of components, it may be possible to reduce the extent of inspection and lower the associated cost without impairing safety.

REFERENCES

1. N. J. Palladino and H. L. Davis, "The engineering design of power reactors", Nucleonics, 18,6(1960)85.
2. "Tentative Structural Design Basis for Reactor Pressure Vessels and Directly Associated Components (Pressurized Water-Cooled System)", Document PB-151987, Office of Technical Services, Department of Commerce, Washington 25, D. C.
3. L. W. Smith, "Structural Analysis", Chapter 6 in Reactor Handbook, Vol. IV, Interscience Publishers, 1964.
4. J. H. Harlow, "The ASME Boiler and Pressure Vessel Code-How the Code originated", Mech. Eng., 81,7(1959)56.
5. ASME Boiler and Pressure Vessel Code, a) Sec. I, "Power Boilers", 1956; b) Sec. VIII "Unfired Pressure Vessels", 1962.
6. "Organizations Sponsoring or Publishing Codes, Standards or Specifications Dealing with Piping and Pressure Vessels", Taylor Forge 2, No. 1, 7, June 1956.
7. "American Standard Code for Pressure Piping", Am. Standards Assn. Bulletin B31.1, 1955.
8. E. M. Kloeblen, "The ASME Boiler and Pressure Vessel Code - The Code in operation", Mech. Eng. 81,7(1959)58.
9. B. F. Langer, "Design Criteria of the ASME Code for Nuclear Vessels", January 17, 1963. Available from: Secretary, ASME Boiler and Pressure Vessel Committee, 345 East 47th St., N.Y. 17, N. Y.
10. "Rules for Construction of Nuclear Vessels", 1965 Edition, Section III of ASME Boiler and Pressure Vessel Code. Available from: Secretary, ASME Boiler and Pressure Vessel Committee, 345 East 47th St., N.Y. 17, N.Y. a) Appendix I.
11. "Case 1270N - ASME Boiler and Pressure Vessel Code, Revisions 4 and 5", Mech. Eng. 83,11(1961)105; 84,1(1962)80.
12. "Case 1273N - ASME Boiler and Pressure Vessel Code", Mech. Eng., a) "Revision 4", 83,2(1961)108, 109, 110; b) "Revision 5", 83,4(1961)107.
13. S. Timoshenko and J. N. Goodier, Theory of Elasticity, McGraw-Hill Book Co., Inc., N.Y., 1951.
14. G. Sonneman and D. M. Davis, "Thermal Stresses in Reactor Vessels", Bettis Technical Review, I,1(1957)47, Report WAPD-BT-1, Bettis, Atomic Power Laboratory, 1957.
15. D. M. David and B. H. Mount, "Calculation of Thermal Stresses in Cylinders with Internal Heat Generation—Description of WBTSGI Code", Report WAPD-TM-59, Bettis, Atomic Power Laboratory, 1957.
16. M. Hetenyi (Ed.), Handbook of Experimental Stress Analysis, John Wiley and Sons, Inc., N. Y., 1950.
17. R. E. Peterson, Stress Concentration Design Factors, John Wiley and Sons, Inc., N.Y., 1953.
18. B. F. Langer, "Design of pressure vessels for low-cycle fatigue", Trans. ASME, Series D, 84(1962)389.
19. "Shock Design of Shipboard Equipment, Part I - Dynamics Analysis Method", USAEC Report NRL-5545, Naval Research Laboratory, 1960.

19a. "Nuclear Reactors and Earthquakes", USAEC Report TID-7024, 1963.

20. S. Timoshenko and S. Woinowsky-Krieger, Theory of Plates and Shells, McGraw-Hill Book Co., Inc., 1959.
21. J. F. Harvey, Pressure Vessel Design: Nuclear and Chemical Applications, D. Van Nostrand Co., Inc., Princeton, N. J., 1963.
22. S. Glasstone and A. Sesonske, "Reactor Structural and Moderator Materials", Chapter 7 in Nuclear Reactor Engineering, D. Van Nostrand Co., Inc., Princeton, N. J., 1963.
23. M. S. Wechsler and R. G. Berggren, "Radiation embrittlement of reactor pressure vessels", Nucl. Safety, 4,1(1962)42.
24. R. H. Ellis, Nuclear Technology for Engineers, McGraw-Hill Book Co., Inc., N.Y., 1959.
25. C. Manual et al., "Atomic radiation changes materials", Power, 99(1955)94.
26. J. C. Wilson and R. G. Berggren, "Effects of neutron irradiation in steel", Trans. Am. Soc. Testing Mater., 55(1955) 689.
27. M. H. Bartz, "Radiation Damage Observations at the MTR", USAEC Report TID-7515, Part 1, 1956.
28. E. R. Parker, Brittle Behavior of Engineering Structures, John Wiley and Sons, N.Y., 1957.
29. D. P. Brown, "Observations on experience with welded ships", Welding J., 31(1952)765.
30. M. E. Shank, "Brittle failure of nonship steel-plate structures", Mech. Eng., 74(1954)23.
31. "Fracture Analysis Diagram Procedure for the Fracture-Safe Engineering Design of Steel Structures", USNRL Report NRL-5920, Naval Research Laboratory, 1963.
32. L. Porse, "Reactor Vessel Design Considering Radiation Effects", Report prepared for presentation at the ASME Convention, Philadelphia, Pa., November 22, 1963.
33. "Drop Weight Tests", Proposed ASTM Specification E208-637.
34. C. A. Keyser, Materials of Engineering, Prenctice-Hall, Inc., Englewood, N. J., 1956.
35. S. S. Pawlicki, "Neutron exposure criteria for reactor vessels", Trans. Am. Nucl. Soc., 6(1963)1949).
36. W. S. Pellini et al., "Analysis of Engineering and Basic Research Aspects of Neutron Embrittlement of Steels", USNRL Report NRL-5780, Naval Research Laboratory, 1962.
37. M. B. Reynolds, "The effect of stress on the radiation stability of ASTM-A-302 Grade B Pressure Vessel Steel", Mater. Res. Std., August 1963, p. 644.
38. C. D. Bopp and O. Sisman, "How to calculate gamma radiation induced in reactor materials", Nucleonics, 14,1(1956) 46.
39. The Shippingport Pressurized Water Reactor, Naval Reactors Branch, USAEC, Westinghouse Electric Corp., and Duquesne Light Co., Addison-Wesley Publishing Co., Inc., Reading, Mass., 1958.
40. Drawing from Combustion Engineering, Inc.
41. "Welding Symbols", ASA Standards Z 32.2.1.
42. "The Cause of Failure and Repair of the VBWR Recirculation Piping", Report APED-4116, Atomic Power Equipment Dept., General Electric Co., San Jose, Calif., 1962.
43. F. R. Farmer, private communication to T. J. Thompson, M.I.T., Cambridge, Mass., 1963.
44. USAEC Docket 50-3, Consolidated Edison Co. of N.Y. Thorium Reactor Facility (Indian Point), HSR, January 1960, Exhibit G-1, (Rev. 3).
45. B. F. Langer, unpublished notes, Bettis, Atomic Power Laboratory, 1963.
46. M. M. El-Wakil, Nuclear Power Engineering, McGraw-Hill Book Co., Inc., N.Y., 1962.

47. W. Coopey, "Holding high-pressure joints", Petroleum Refiner, 35,5(1956)189.
48. Photograph from Combustion Engineering, Inc.
49. Photo from Bettis Atomic Power Laboratory (Westinghouse).
50. B. F. Langer, "Experimental Mechanics of Nuclear Power Reactors", Paper presented at the Annual Meeting of the Society for Experimental Stress Analysis, November 1958, Albany, N.Y. (SESA Proc. Vol. XVIII, No. 2)
51. N. J. Palladino et al., "Experimental Development of the PWR Core and Vessel", Report WAPD-7-711, Bettis Atomic Power Laboratory, 1958.
52. J. R. Reavis, "CVTR Pressure Tube Fitting Design and Development", CVNA-116, Westinghouse Electric Corp., 1962.
53. H. J. Cordle, "Development of the CVTR Pressure Tube Header Assembly", Report CVNA-69, Westinghouse Electric Corp., 1960.
54. H. Etherington (Ed.), Nuclear Engineering Handbook, McGraw-Hill Book Co., Inc., N.Y., 1958.
   a) A. H. Barnes, Section 13-3, "Liquid-Metal Reactor Systems",
   b) S. Untermyer, II, Section 13-1, "Water-Cooled Reactor Systems".
55. F. A. Grochowski, "Thermal shield coolant flow", Bettis Technical Review, I,1(1957)152, Report WAPD-BT-1, Bettis Atomic Power Laboratory, 1957.
56. "Technical Information and Final Hazards Summary Report - Yankee Nuclear Power Station Part B, License Application", Yankee Atomic Electric Co., USAEC Docket 50-29.
57. I. H. Mandil and N. J. Palladino, "Description of the Pressurized Water Reactor (PWR) Power Plant of Shippingport Part C, Core Design", Proceedings of the First U.N. International Conference on Peaceful Uses of Atomic Energy, Geneva, 1955, Vol. 3, p. 211.
58. L. E. Glasgow, "Experience with SRE", Nucleonics, 20, 4(1962)61.
59. N. J. Palladino and A. H. Foderaro, "Intrinsic Reactor Safety Through Design", Reactor Safety and Hazards Evaluation Techniques, Proceedings of a Symposium held at Vienna, May 14-18, 1962, IAEA Vienna, Vol. 1, p. 221.
60. J. A. Ransohoff, "Use of less expensive rare earths as control materials", Nucleonics, 17,7(1959)80.
61. "Presentation to AEC and ACRS of an Improved Control Rod Absorber", Westinghouse Electric Corp., June 5, 1963.
62. W. R. Mendes, "Analytical Analysis of Thermal Distortion of Control Rods", Report WAPD-PWR-(RD2)-229, Bettis, Atomic Power Laboratory, 1958.
63. S. N. Stilwell and R. L. Waterfield, "Fast-Period Reactor Safety Fuse Tests", Report APEX-492, General Electric Co., 1959.
64. S. N. Stilwell and R. L. Waterfield, "BORAX Fuse Tests", Report APEX-441, General Electric Co., 1955.
65. T. W. Donavon, S. N. Stilwell and J. G. Stuart, "Internal Fuses for Low-Power Reactors", Report APEX-397, General Electric Co., 1953.
66. R. J. Spera, S. N. Stilwell, and W. L. Weiss, "Reactor Fuse Propellants and Configurations", Report APEX-388, General Electric Co., 1956.
67. S. H. Fitch and T. H. Springer, "Safety Device Tests in SPERT-I", Report NAA-SR-3045, North American Aviation, Inc., 1958.
68. T. H. Springer and N. C. Miller, "Heat Flow in the Trigger of a Reactor Safety Device", Report NAA-SR-2461, North American Aviation, Inc., 1958.
69. R. L. Waterfield, S. N. Stilwell, and R. F. Spera, "SPERT-I Fuse Trigger Tests", Report APEX-446, General Electric Co., 1959.
70. R. Holland, "Differential Pressure Safety Device", Report NAA-SR-3867, North American Aviation, Inc., 1959.
71. L. E. Johnson, "Coupled Diaphragm Safety Device for Power Reactors", Report NAA-SR-4434, North American Aviation, Inc., 1960.
72. J. R. Tallackson, "Reactor Fuses", Nucl. Safety, 1,1(1959)17.
73. W. C. Stewart, "Bolted Joints", p. 171 in ASME Handbook Metals Engineering Design (1st ed.), McGraw-Hill Book Co., Inc., N. Y., 1953.
74. E. A. Davis, "Combined tension-torsion tests on a 0.34 per cent carbon steel", Trans. ASME, 62(1940)577.
75. B. F. Langer, "Locking Devices", Report WAPD-CE-37, Bettis, Atomic Power Laboratory, 1954.
76. B. F. Langer, unpublished notes, Bettis, Atomic Power Laboratory, 1954.
77. E. O. Hughes and J. W. Greenwood, "Contamination and cleanup of NRU", Nucleonics, 18,1(1960)76.
78. A. J. Mooradian, G. M. Allison, and J. F. Palmer, "Chalk River experience with fuel waterlogging", Nucleonics, 18, 1(1960)81.
79. P. Balligand, "Reactor incidents at Scalay", Nucleonics, 18,3(1960)82.
80. R. W. Dickinson, "Coolant block damages SRE fuel", Nucleonics, 18,1(1960)107.
81. "What happened to SRE's fuel elements?", Nucleonics, 17, 12(1959)23.
82. M. A. Schultz, Control of Nuclear Reactors and Power Plants, (2nd ed.) McGraw-Hill Book Co., Inc., N.Y., 1961.
83. J. M. Harrer, Nuclear Reactor Control Engineering, D. Van Nostrand Co., Inc., Princeton, N. J., 1963.
84. A. E. G. Bates, "Control rods and control rod drives in power reactors", Nucl. Safety, 2,4(1961)17.
85. J. W. Simpson and M. Shaw, "Description of the Pressurized Water Reactor (PWR) Power Plant at Shippingport, Pa. - Part A, Nuclear Power Generation", Proceedings of the First U. N. International Conference on Peaceful Uses of Atomic Energy, Geneva, 1955, Vol. 3, p. 211.
86. Section I.E., "Control and Neutron Source Components", Enrico Fermi Atomic Power Plant, Revised License Application, Technical Information and Hazards Summary Report, Power Reactor Development Co., USAEC NP-10458, 1961.
87. Site Evaluation Report for Peach Bottom Atomic Power Station (HTGR): Part I, (January 1960), Amendment 1, (September 1960), USAEC Docket 50-171, Philadelphia Electric Co.
88. Brochures: a) Latch Type Magnetic Jack Control Rod Drive Mechanism; b) Magnetic Control Rod Drive Mechanism, Model FMJ-200-7.25, Westinghouse Electric Corp, Atomic Equipment Division, Cheswick, Pa. (1964).
89. G. A. Freund, Materials for Control Rod Drive Mechanisms, Rowman and Littlefield, N.Y., 1963.
90. A. E. Harwood et al., "The Design of Electromechanical Auxiliaries Directly Associated with Power Producing Reactors", Proc. Inst. Elec. Engrs. 106A,27(1959)262.
91. "Application to USAEC for Reactor Construction Permit and Operating License", Consumers Power Co., USAEC Docket 50-155 (Big Rock Point), Rev. 1, Amendment 6, 3-19-62.
92. "Modular Package Control Rod Drive Mechanism Incorporating Harmonic Drive", Atomic Power Dept., United Shoe Machinery Corp., Beverly, Massachusetts.
93. S. H. Hanauer, "Control rod and control drive difficulties", Nucl. Safety, 3,1(1961)46.
94. J. Foster, "Latch mechanisms for control rod drives", Nucl. Safety, 3,4(1962)46.
95. L. C. Oakes, "Release Time Investigation of the Oak Ridge Research Reactor Shim Rod Support", USAEC Report CF-53-11-25, Oak Ridge National Laboratory, 1953.
96. H. Thielsch, "Failures in high temperature high pressure piping", Nucl. Safety, 4,3(1963)1.
97. C. W. Ham, "Pipe and Pipe Fittings" in Mark's Handbook (3rd ed.), McGraw-Hill Book Co., Inc. N.Y.., 1930.
98. G. R. Rich, Hydraulic Transients (2nd ed.), Dover Publications, Inc., N.Y., 1963.
99. H. Thielsch, "Summary of experiences and opinions on the use of austenitic stainless steels in steam power piping", Combustion, 57,5(1956)67.
100. C. F. Bonilla, "External Loop Components", Chapter 3 in Reactor Handbook, Vol. IV (in press).
101. I. J. Karassik and R. Carter, "Centrifugal pump question and answers", Monthly series in Power, March 1945-August 1946.
102. DeLaval Steam Turbine Co.
103. D. E. Westerheide, J. C. Clifford, and G. Burnet, "Design and test of a diaphragm pump for liquid metal", Nucl. Sci. Eng., 17(1963)523.
104. E. F. Brill, "Development of Special Pumps, Their Power Supply, Valves, Bearings and Instrumentation for Liquid Metals", Allis-Chalmers Mfg. Co., Milwaukee, Wisconsin.
105. H. F. Rase and M. H. Barrow, Project Engineering of Process Plants, John Wiley and Sons, N.Y., 1957.
106. G. M. Fuls, "Analysis of Check Valve Disc Motion During a Flow Transient", Report WAPD-TM-233, Bettis, Atomic Power Laboratory, 1962.
107. "Fluid Power Book Issue", Machine Design, 35,39(1963).
108. C. G. Lennox and A. Pearson, "NRX Automatic Control System General Description", Canadian Report AECL-434 (CREL-698), 1957.
109. A. Pearson and C. G. Lennox "Experience with the NRX and NRU Control Systems", Proceedings of the Sixth Tripartite Instrumentation Conference Held at Chalk River, Ontario, Canada, April 20-24, 1959, Canadian Report AECL-801, p. 111, 1959.
110. I. N. MacKay, "The Canadian NPD-2 Nuclear Power Station", Proceedings of the Second U. N. International Conference on Peaceful Uses of Atomic Energy, Geneva, 1958, Vol. 8, p. 313.
111. G. C. Laurence, "Reactor safety in Canada", Nucleonics, 8, 10 (1960)73.
112. B. S. Kosut, "Relief and Safety Valves for Nuclear Service", USAEC Report TID-6135, 1960.
113. E. A. Bake, "Primary Coolant Valves for Nuclear Plants", Nuclear Process Instrumentation and Controls Conference Held at Gatlinburg, Tennessee, May 20-22, 1958, USAEC Report ORNL-2695, p. 59, Oak Ridge National Laboratory, 1960.

114. J. J. Kanter, "On the Quality Requirements for Steel Valves for Nuclear Power Plants", Advances in Nuclear Engineering, Vol. I, p. 310, Proceedings of the Second Nuclear Engineering and Science Conference, Pergamon Press, N.Y., 1957.
115. E. A. Goldsmith, "Operational Experiences with Primary Elements and Valves in Slurry Service", Nuclear Process Instrumentation and Controls Conference Held at Gatlinburg, Tennessee, May 20-22, 1958, USAEC Report ORNL-2695, p. 40, Oak Ridge National Laboratory, 1960.
116. H. G. Rickover, "Metallurgy in atomic power", ASME J., (1956)441.
117. D. J. McDonald, "Application of thermal stress theory to the PWR main hydraulic valves", Bettis Technical Review, (1958) , Report WAPD-BT-9, Bettis, Atomic Power Laboratory, 1958.
118. R. C. Barnett and B. N. McDonald, "Nuclear heated boilers", Proc. Am. Power Conf., 12(1960)169.
119. R. E. Vuia, J. A. Paget, and R. D. Winship, "Steam Generators for Pressure Water Reactors", Canadian Report AECL-832, 1959.
120. "Sodium-Heated Steam Generator Summary", USAEC Report TID-18072, 1962.
121. J. A. Paget, "A Survey of Conventional Steam Boiler Experience Applicable to the HTGR Steam Generators", Report GAMD-1208, General Atomic, Division of General Dynamics Corp., 1959.
122. W. J. Singley, I. H. Welinsky, S. F. Whirl, and H. A. Klein, "Stress corrosion of stainless steel and boiler water treatment at Shippingport Atomic Power Station", Proc. Am. Power Conf., 21(1959)748.
123. H. Kraus, "Pressure and thermal stresses in U-tube steam generators for nuclear power plants", Bettis Techncial Review, (1960)13, WAPD-GT-18, Bettis, Atomic Power Laboratory, 1960.
124. "Experimental Gas-Cooled Reactor Final Hazards Summary Report", (EGCR), USAEC Report ORO-586, Vol. I, Oak Ridge Operations Office, 1962.
125. J. R. Brown, D. C. Pound, and J. B. Sampson, "Hazards Summary Report for the HTGR Critical Facility", (Peach Bottom) Report GA-1210, General Atomic, Division of)eneral Atomic, Division of General Dynamics, 1960.
126. "Military Specification Steam Generators, Pressurized Water, Nuclear Naval Ship Propulsion", Bureau of Ships MIL-S-21204, (Ships), 1958.
127. "Military Specification Tube and Pipe, Corrosion Resisting Steel, Seamless and Welded, Radioactive System Service", Bureau of Ships, MIL-T-18063, (Ships) 1954.
128. D. J. De Paul (Ed.), Corrosion and Wear Handbook, USAEC Report TID-7006, 1957.
129. "Boiling reactors: Steam-water separation", Power Reactor Technol., 6,2(1963)69.
130. J. R. Maxwell, "Determining performance characteristics of a saturated steam pressurizer for nuclear power applications", Bettis Technical Review, __(1957)92, Report WAPD-13T-5, Bettis, Atomic Power Laboratory, 1957.
131. Westinghouse Electric Corp., Bettis Atomic Power Laboratory, West Mifflin, Pa.
132. H. F. McDuffie and V. K. Hill, "Homogenious Reactor Project Quarterly Progress Report for the Period Ending January 31, 1956" USAEC Report ORNL-2057 (Del.), Oak Ridge National Laboratory, 1956.
133. E. R. Drucker and K. N. Tong., "Behavior of a steam-pressurizer surge tank", Trans. Am. Nucl. Soc., 5(1962)144.
134. A. W. Thomas and J. A. Findlay, "PRE - A Pressurizer Transient Analysis Program for the Philico 2000 Computer", Report KAPL-M-EC-7, Knolls Atomic Power Laboratory, 1961.
135. "SPERT-III pressurizer vessel failure", Nucl. Safety, 3,4 (1962)91.
136. R. E. Heffner et al., "SPERT-II Pressurizer Vessel Failure", Report IDO-16743, Phillips Petroleum Co., 1962.
137. "Final Safeguards Report", Saxton Nuclear Experimental Corporation, General Public Utilities System, Application to the U.S. Atomic Energy Commision for Reactor Construction Permit and Operating License, AEC Docket 50-146, 1961.
138. J. G. Delene, "Shutdown heat generation", Nucl. Safety, 5,1(1963)40.
139. L. M. Swartz, A. W. Lemmon, Jr., and L. E. Hulbert, "PWR Loss-of-Coolant Accident - Core Meltdown Calculations", Report WAPD-56-544, Bettis, Atomic Power Laboratory, 1957. (See also: TID-4500)
140. "Interim Report on SL-1 Incident, January 3, 1961", USAEC General Manager's Board of Investigation under the Chairmanship of C. A. Nelson, January 27, 1961. (Released February 2, 1961, Release No. D-33.)
141. J. R. Buchanan, "SL-1 final report", Nucl. Safety, 4, 3 (1963)83.
142. Letter Transmitting Final Report of the SL-1 Board of Investigation, Curtis A. Nelson, Chairman, to A. Luedecke, General Manager, September 5, 1962.
143. H. Feinroth and T. D. Sutter, Jr., "The First Refueling of the Shippingport Atomic Power Station", Report WAPD-233, Bettis, Atomic Power Laboratory, 1960. (See also: TID-4500, 15th ed.).
144. W. R. Casto, "Fuel element handling", Nucl. Safety, 2,2 (1960)7.
145. "Procedures for Assembly Installation and Testing of PWR Core 1 Reactor Components and Handling Equipment", Report WAPD-PWR-RD-566, Bettis, Atomic Power Laboratory, 1958.
146. W. H. Carr, "Fuel-processing hazards", Nucl. Safety, 2,2 (1960)45.
147. E. L. Nicholson, "Safety in fuel storage canals", Nucl. Safety, 3,4(1962)63.
148. L. J. King, "Contamination of Shipping Cask and Storage Canal Water by Fuels Irrdiated in Pressurized-Water Reactors - A Review of Pertinent Subjects", USAEC Report CF-60-3-49, 1960.
149. L. B. Shappert, "Accidents in shipping radioactive materials", Nucl. Safety, 5,1(1963)123.
150. W. L. Albrecht, "Shipment of spent fuel elements", Nucl. Safety, 2,1(1960)3.
151. "Regulations to Protect Against Accidental Conditions of Criticality in the Shipment of Special Nuclear Material", Code of Federal Regulations, Title 10, Part 71, Effective 11-2-58.
152. "Regulations to Protect Against Accidental Conditions of Criticality in the Shipment of Special Nuclear Material—Proposed Rule Making", Code of Federal Regulations, Title 10, Part 71, Published for Comment 12-21-65.
153. K. W. Hoff and L. B. Shappert, "Standards and tests for radioactive materials shipping casks", Nucl. Safety, 4,4 (1963)18.
154. Agent T. C. George's Tariff No. 13, Interstate Commerce Commission Regulation No. 4, "Transportation of Explosives and Other Dangerour Articles by Land and Water and Rail Freight and by Motor Vehicle (Highway) and Water Including Specifications for Shipping Containers", issued September 15, 1960 and effective September 25, 1960, T. C. George, Agent, N.Y.
155. W. J. McGonnagle, Nondestructive Testing, McGraw-Hill Book Co., Inc., N.Y., 1961.
156. "Symposium on Nondestructive Tests in the Field of Nuclear Energy, April 16-18, 1957", Am. Soc. Testing Materials (ASTM) Special Technical Publication 223, Philadelphia, Pa., 1958.
157. "Second Symposium on Physics and Nondestructive Testing Held at Argonne National Laboratory, October 3-5, 1961", USAEC Report ANL-6515, Argonne National Laboratory, 1961.
158. "Proceedings of the Symposium on Nondestructive Testing Trends in the AEC Reactor Program, May 20, 1960", USAEC Report TID-7600, 1961.
159. R. C. McMaster (Ed.), Nondestructive Testing Handbook, Vols. I and II, The Ronald Press Co., N.Y., 1959.
160. J. R. Hawthorne, H. B. Piper, and L. E. Steele, "Brittle Fracture of Steel", U.S. Reactor Containment Reactor Technology, Vol. II, USAEC Report ORNL-NSIC-5, 1965.
161. W. S. Pellini et al., "Review of Concepts and Status of Procedures for Fracture-Safe Design of Complex Welded Structures Involving Metals of Low to Ultra-High Strength Levels", USNRL Report NRL-6300, 1965.
162. J. E. Irvin, A. L. Bement, and R. G. Hoagland, "The Combined Effects of Temperature and Irradiation on the Mechanical Properties of Austenitic Stainless Steels", Report BNWL-1, Battelle-Northwest Laboratory, 1965.
163. T. T. Claudson and H. J. Pessl, "Irradiation Effects on High Temperature Reactor Structural Metals", Report BNWL-23, Battelle-Northwest Laboratory, 1965.
164. T. A. Trozera, P. W. Flynn, and G. Buzzelli, "Effects of Neutron Irradiation on Materials Subjected to Multiaxial Stress Distribution", Report GA-5636, General Atomic, 1964.
165. S. H. Bush, Private Communication, 1966.
166. C. Z. Serpan, Jr. and L. E. Steel, "In-Depth Embrittlement of a Simulated Pressure Vessel Wall of A302-B Steel", USNRL Report NRL-6151, 1964.
167. J. R. Hawthorne and L. E. Steele, "In-Reactor Studies of Low Cycle Fatigue Properties of a Nuclear Pressure Vessel Steel", USNRL Report NRL-6127, 1964.
168. "Quarterly Progress Report: Irradiation Effects on Reactor Structural Materials, May, June, July, 1965", Report BNWL-128, Battelle-Northwest Laboratory, 1965.
169. R. E. Johnson, "Fracture Mechanics: A Basis for Brittle Fracture Prevention", Report WAPD-TM-505, Bettis Atomic Power Laboratory, 1965.
170. I. N. Sneddon, "The Distribution of Stress in the Neighborhood of a Crack in an Elastic Solid", Proc. Roy. Soc., London, A187, 229 (1946).
171. M. B. Reynolds, "Fracture Mechanics and the Stability of Engineering Structures", Report GEAP-4678, General Electric Atomic Power Equipment Dept., 1965.

172. P. C. Paris and G. C. Sih, "Stress Analysis of Cracks", Report, Department of Mechanics, Lehigh University, 1964.
173. G. R. Irwin, "Fracture", Handbuch der Physik, Springer-Verlad, Berlin, Vol. 6, pp. 551-590, 1958.
174. E. T. Wessel presentation to the ASTM Subcommittee for Fracture Testing of Medium Strength Metals meeting held at Washington, D. C., Jan. 12, 1965.
175. J. J. DiNunno and A. B. Holt, "Radiation Embrittlement of Reactor Vessels", Nuclear Safety, 4, 2, 34-47, 1962.
176. L. E. Steele, C. Z. Serpan, Jr. and J. R. Hawthorne, "Irradiation Effects on Reactor Structural Materials 1 August–31 October 1965", USNRL Report, NRL Memorandum Report 1663, 1965.
177. M. D. Eriekson, "Containment Vessel Flaw Detection Test", Report BNWC-94, Battelle-Northwest Laboratory, 1965.
178. W. H. Irvine, A. Quirk, and E. Bevitt, "Fast Fracture of Pressure Vessels", J. British Nuclear Energy Society, January 1964.
179. G. Reed and E. Tarnuzzer, "Examining Yankee Plant Performance in 1965", Nucleonics, 24, 3 (1966) 42.
180. "Evaluation of Yankee Vessel Cladding Penetrations", Report WACP-2855, Westinghouse Atomic Power Division, 1965.
181. M. Bender, "A Status Report on Prestressed Concrete Reactor Pressure Vessel Technology", Nuclear Structural Engineering, 1, 83-90, and 202-223, 1965.
182. J. F. Sullivan, "An Evaluation of Prestressed Concrete Pressure Vessels for Nuclear Reactors", Master of Engineering Paper, The Pennsylvania State University, 1965.
183. R. O. Marsh and W. Rockenhauser, "Prestressed Concrete Structures for Large Power Reactors", Report GA-6505, General Atomic, 1965.
184. T. C. Waters and N. T. Barrett, "Prestressed Concrete Pressure Vessels for Nuclear Reactors", J. Brit. Nucl. Energy Soc., July 1963.
185. J. R. Coombe, "SM-1 Pressurized Water Reactor Operating Experience", Nuclear Safety, 4, 2, 100, 1962.
186. J. C. Carroll and J. O. Schuylert, "Humbolt Bay Reactor Operating Experience", Nucl. Safety, 6, 4, 441-451, 1965.
187. C. S. Walker, "Secondary Shutdown Systems of Nuclear Power Reactors", Nucl. Safety, 7, 1, 45-52, 1965.
188. "Action on Reactor Projects Undergoing Regulatory Review", Nucl. Safety, 6, 3, 301-311, 1965.
189. A. L. Colomb and T. M. Sims, "ORR Fuel Failure Incident", Nucl. Safety, 5, 2, 203-207, 1964.
190. "Action on Reactor Projects Undergoing Regulatory Review", Nucl. Safety, 7, 1, 124-132, 1965.
191. L. I. Cobb, "Experience with Control-Rod Systems at Licensed Reactor Facilities", Nucl. Safety, 6, 1, 49-54, 1964.
192. W. S. Gibbons and B. D. Hackney, "Survey of Piping Failures for the Reactor Primary Coolant Pipe Rupture Study", Report GEAP-4574, General Electric Atomic Power Equipment Dept., 1964.
193. E. R. Kilsby, Jr., "A Survey of Water Reactor Primary System Conditions Pertinent to the Study of Pipe Rupture", Report GEAP-4445, General Electric Atomic Power Equipment Dept., 1964.
194. S. W. Tagart, Jr., "Mechanical Design Considerations in Primary Nuclear Piping", Report GEAP-4578, General Electric Atomic Power Equipment Dept., 1964.
195. H. H. Klepfer, "Experimental and Analytical Program Recommendations—Reactor Pipe Rupture Study," Report GEAP-4474, General Electric Atomic Power Equipment Dept., 1965.
196. A. J. Brothers et al., "A Review of Fracture Modes as Related to Reactor Primary Coolant Pipe Rupture", Report GEAP-4446, General Electric Atomic Power Equipment Dept., 1964.
197. H. Thielsch, "Prevention of Pipe Failures", Heating, Piping and Air Conditioning, Series of Articles, July, October, and November, 1964.
198. W. R. Smith, Sr., and J. B. Violette, "The Cause of Failure and the Repair of the VBWR Recirculation Piping", Report GE-APD-4116, General Electric Co., 1962.
199. J. A. Prestele ana G. F. Froshauer, "Operating Experience at Indian Point Nuclear Power Station", Nucl. Safety, 6, 3, 292-298, 1965.
200. G. Hake, "NPD Reactor Operating Experience", Nucl. Safety, 7, 1, 104-113, 1965.
201. "Action on Reactor Projects Undergoing Regulatory Review", Nucl. Safety, 6, 2, 219-227, Winter 1964-65.
202. "Action on Reactor Projects Undergoing Regulatory Review", Nucl. Safety, 5, 4, 410-419, 1964.
203. K. D. May, "Advanced Valve Technology", NASA Report, NASA SP-5019, 1965.
204. H. Kouts, "Report on Engineered Safeguards", letter dated Nov. 18, 1964, to Honorable Glenn T. Seaborg, AEC Press Release No. 6-293, Dec. 17, 1964.
205. S. H. Hanauer, "Design Precepts for Engineered Safeguards", Nucl. Safety, 6, 4, 408-411, 1965.
206. H. B. Smets, "A Review of Nuclear Incidents", Reactor Safety and Hazards Evaluation Techniques, Symposium Proceedings, Vienna, 1962, Vol. I, pp. 89-110, International Atomic Energy Agency, Vienna, 1962.
207. E. P. Epler, "Safety-System Reliability vs. Performance", Nucl. Safety, 6, 4, 411-414, 1965.
208. E. P. Epler, "HTRE-3 Excursion", Nucl. Safety, 1, 2, 57-59, 1959.
209. E. P. Epler, "Failure of Alize I Reactor Safety System", Nucl. Safety, 5, 2, 172-174, Winter 1963-64.
210. H. B. Piper, "Engineered Safeguards", Nucl. Safety, 6, 2, 201-207, Winter 1964-65.
211. H. Gitterman, "Reactor Safeguard Systems", Nucleonics, 23, 10, 56-59 and 63-64, 1965.

CHAPTER 15

Fluid Flow

S. LEVY
General Electric Company, Atomic Products Division,
San Jose, California

CHAPTER CONTENTS*

1 INTRODUCTION

In a nuclear plant, fluid flow affects many components which are intimately tied to plant safety. For instance, the heat generated within the reactor core is removed by circulation of a coolant and an appropriate distribution of the coolant flow is needed to avoid excessive temperatures. Similarly, the satisfactory design of a large fraction of the equipment such as pumps, circulators or vapor separators, is dependent upon an accurate knowledge of the flow conditions within the plant. Fluid flow also plays an important role in many related areas. The rates of corrosion, erosion, vibration,

*Except for a few changes and additions made in proof, this chapter is based on information in the literature or known to the author prior to August 1963.

or mass transfer vary with fluid velocity. Or again, in those reactors where the coolant has moderating properties, changes and oscillations in flow produce changes and oscillations in reactivity and power. Finally, in many serious accidents such as loss of flow or loss of coolant or a sudden energy release from the reactor core, the course of the accident and its consequences depend in part on fluid flow phenomena.

From a fluid flow viewpoint, the safety of the plant can be evaluated if the fluid velocity at every point in the plant is known within a specified margin. This knowledge, to be complete, would have to encompass all operating modes of the plant and cover not only steady state or normal plant conditions, but also transient and abnormal operating circumstances.

This is the fundamental fluid flow problem in nuclear plant design. Its solution requires experimental data and calculation methods which are capable of specifying, within a prescribed accuracy, the fluid velocity at every point of the flow or recirculation systems shown in Fig. 1-1. This figure is a simplified schematic flow diagram typical of most nuclear applications. Only single line connections are shown between the components of the loop. Actually, all lines in the flow loop may consist of multiple parallel lines or systems. By appropriately labeling the various components in Fig. 1-1, it is possible to make it representative of the two power plant flow loops shown in Fig. 1-2.

Examination of the flow around the loops of Figs. 1-1 and 1-2 indicates that the fundamental fluid flow problem in nuclear plants can be subdivided into three distinct areas:

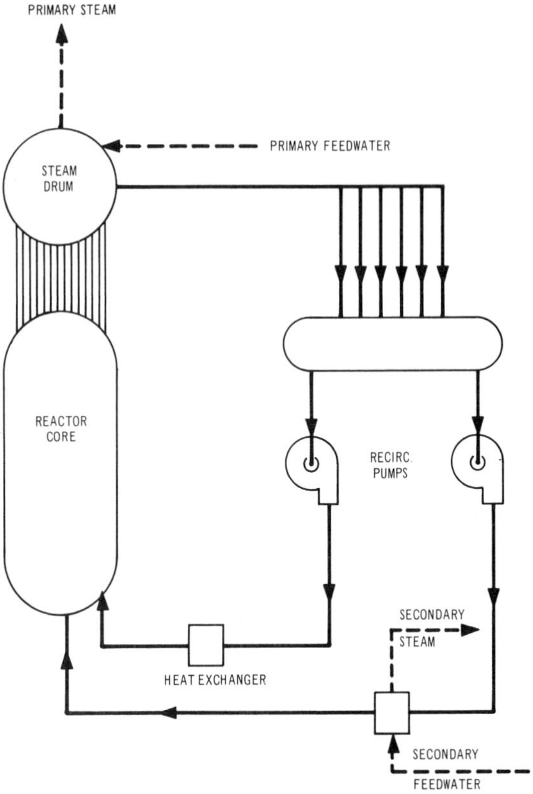

FIG. 1-2 Dresden nuclear power plant flow system.

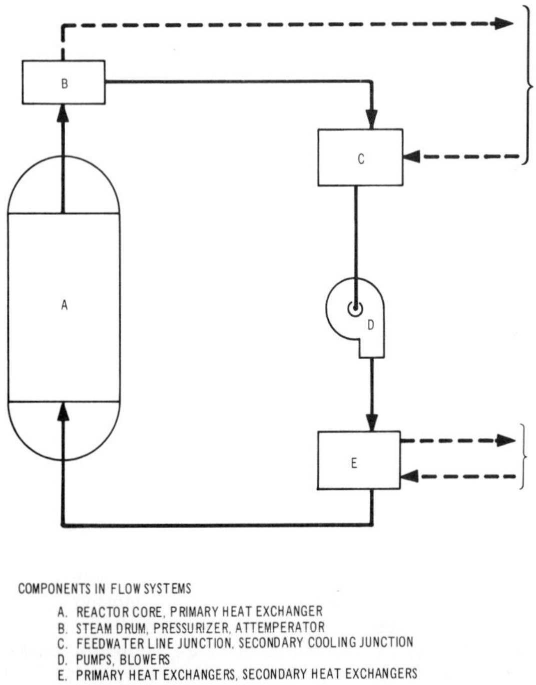

FIG. 1-1 Typical flow system.

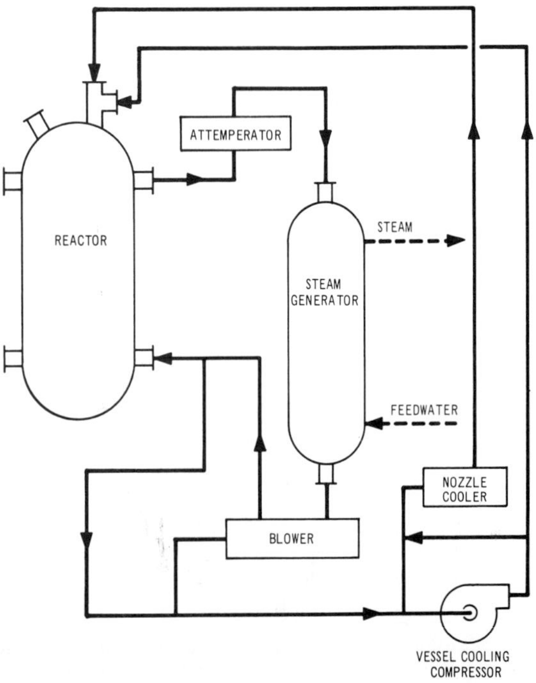

FIG. 1-2 (cont.) EGCR flow diagram.

1. In parts of the loop where the rate of heat exchange is not complex or critical, the main interest is in the total fluid flow or recirculation flow rate. This covers all components of Figs. 1-1 and 1-2 with the exception of the nuclear reactor.

2. In the nuclear reactor itself, the flow distribution becomes important. Here one is concerned about the subdivision of flow between the various parallel flow paths through the reactor. It is necessary to know the flow rate through the various local cross-sectional areas (for instance, through the fuel assemblies) and to determine the bypass flows through the reactor such as leakage through the control rod shrouds or outside the fuel assembly channels, or again, the amounts of flow diverted to cool thermal shields, pressure vessel or other reactor components.

3. Once the flow distribution within the reactor is specified, the next item of interest is the local flow condition. Local fluid velocities must be specified, and particularly those within the limiting or "hottest" core region or fuel assembly. These values are needed to define such local fluid properties as density or enthalpy and to specify the local rates of heat transfer, corrosion, and other velocity-dependent processes.

In the sections that follow, the above three key problem areas are considered. The problems of recirculation flow rate, flow distribution within the reactor, and local fluid distribution within the limiting core regions or fuel assemblies are first examined under steady-state conditions. Methods and bases for calculating these parameters are discussed and their safety implications from a design and operation viewpoint reviewed. The same three problems areas are then examined under transient conditions.

2 STEADY-STATE RECIRCULATION FLOW RATE

2.1 Basic Approach

In each of the flow loops of Figs. 1-1 and 1-2, the recirculation flow rate is obtained from a pressure balance around the loop. The pressure losses in each circuit are equal to the pressure rises around the same circuit. Mathematically speaking, this can be expressed by writing that the integral of the local pressure differential dP around the loop is equal to zero.

$$\oint dP = 0 . \qquad (2\text{-}1)$$

In order to solve Eq. (2-1), it is customary to break the integral of the pressure around the loop into various components of pressure loss and gain. A typical breakdown is:

$$\oint dP = \Delta P_R + \Delta P_p + \Delta P_{C\&E} + \Delta P_B + \Delta P_{V\&F} + \Delta P_H + \Delta P_P = 0 . \qquad (2\text{-}2)$$

In Eq. (2-2), the subscripts R, p, C&E, B, V&F, H, and P have been used, respectively, to represent the pressure loss or gain across the reactor, piping, contraction and expansion geometries, bends, valves and fittings, heat exchangers, and pumps.

Extensive analytical and experimental correlations can be found in the literature for predicting the various terms in Eq. (2-2). Many of the correlations were derived before the development of the nuclear field and are covered in classical hydrodynamics books (see references [1] through [8]). Where the information is available in these references, only summary information is given here. Detailed results are presented only when the fluid flow problems are peculiar to nuclear plant design. Many of these results are found to depend on the type of fluid being considered, and the material described here is subdivided in four groups corresponding to the four different types of fluid used in reactor technology. They are: incompressible single-phase flow, compressible single-phase flow, liquid-vapor flow, and fluid-solid flow.

2.2 Methods of Calculation

2.2.1 Incompressible Single-Phase Flow

Incompressible flow is characterized by negligible density variation. Most single-phase liquid-cooled reactors fall within this category. For all practical purposes, water, liquid metals, and organic coolants are incompressible. For such incompressible fluids, the pressure losses or gains in Eq. (2-2) are predicted by the methods outlined in the following subsections.

2.2.1.1 Reactor Pressure Loss. This term is considered in Sec. 3.2.

2.2.1.2 Piping Pressure Loss. Under steady-state conditions the pressure change along an elemental length dL of piping is made up of acceleration, head, and frictional losses. For a channel of constant cross section a pressure balance equation can be written:

$$\text{pressure change} + \text{acceleration change} + \text{head or hydrostatic change} + \text{frictional loss} = 0$$

or

$$dP + (\rho/2g_c)\, d(V^2) + (g/g_c)\, \rho\, dz + dP_f = 0 . \qquad (2\text{-}3)$$

In Eq. (2-3), ρ is the fluid density and V its average velocity. The symbol z represents the distance measured vertically, g the gravitational acceleration, and g_c a proportionality factor obtained from Newton's law of motion.

For a channel of constant cross-sectional area, the mass flow per unit area is constant and

$$\rho V = G = \text{constant} . \qquad (2\text{-}4)$$

For an incompressible fluid, the density ρ is constant and so is the velocity V, and the acceleration loss vanishes from Eq. (2-3). The hydrostatic term is calculated from the channel length and its angle of inclination from the vertical. If

θ is the angle of inclination of the channel from the horizontal direction, $dz = dL \sin\theta$. The last component of pressure change in Eq. (2-3) is the frictional pressure loss; and it is obtained from

$$dP_f = f\,(dL/D_e)\,\rho\,(V^2/2g_c)\,. \tag{2-5}$$

In Eq. (2-5), D_e is an equivalent channel diameter to be defined later and f is the friction factor.

The friction factor f has been found to depend upon two parameters: a relative channel roughness number (ϵ/D_e) and a dimensionless grouping called the Reynolds number (GD_e/μ). The symbol ϵ represents the average height of the roughness projections from the wall, and μ is the absolute viscosity of the fluid. The Reynolds number is one of the most important parameters in fluid flow. It represents the ratio of acceleration (inertial) to viscous forces. It is also used to determine whether the flow is laminar or turbulent. (Laminar flow is stratified flow with no mixing of fluid particles in the transverse direction, while turbulent flow is characterized by eddying and momentum transport perpendicular to the flow direction.)

Moody [9] has presented the most complete set of curves of the friction factor f in terms of the Reynolds number and relative channel roughness. His curves are reproduced on Fig. 2-1 [9]. The zones of laminar and turbulent flow are marked on the figure.

Analytical expressions have been developed to describe the various curves shown in Fig. 2-1. [9, 10]. The laminar friction factor can be predicted from the relation

$$f = 64/(GD_e/\mu)\,. \tag{2-6}$$

This relation is valid only for circular pipes. For laminar flow in other geometries see Sec. 3.2.1. The friction factor f for turbulent flow in a smooth channel is given by:

$$1/\sqrt{f} = 2.0 \log_{10} [(GD_e/\mu)\sqrt{f}] - 0.80\,. \tag{2-7}$$

For rough channels, the friction factor can be calculated within $\pm$ 5% from

$$f = 0.0055 \left\{ 1 + \left[20000 \frac{\epsilon}{D_e} + 10^6 \frac{\mu}{GD_e} \right]^{\frac{1}{3}} \right\}. \tag{2-8}$$

For the latter channels, the relative roughness parameter (ϵ/D_e) can be obtained from Fig. 2-2 [9]. Figure 2-2 gives value of the ratio (ϵ/D_e) in terms of the channel material and its equivalent hydraulic diameter D_e.

For a given channel geometry, the equivalent hydraulic diameter D_e is defined from

$$D_e = 4\ (\text{flow area})/\text{wetted perimeter}.$$

According to this definition, the pipe diameter and hydraulic diameter are one and the same for a circular channel. For other channels, the definition is an attempt to substitute an equivalent circular pipe diameter for the given channel geometry. The approach is most useful as it extends the use of round pipe data to noncircular channels. It has given very good results especially under turbulent flow conditions.

In the case of heat addition and removal at the channel wall, such properties as fluid density and viscosity vary in the direction transverse to the fluid motion. The values shown in Fig. 2-1 are no longer valid unless they are modified to account for the change in fluid properties. This is usually done by utilizing the isothermal property curves of Fig. 2-1 and evaluating the properties at an intermediate or "film" temperature between the wall and bulk fluid temperatures. Such a "film" temperature can be expected to depend upon the fluid temperature distribution within the channel. Its definition varies with the Reynolds number, channel roughness, and the fluid Prandtl modulus. (The Prandtl modulus represents the ratio of thermal to viscous diffusivity and is equal to $\mu C_p/k$ where C_p and k are the fluid specific heat at constant pressure and the thermal conductivity.) For a given heat transfer rate at the channel wall, it has been shown [11] that increasing the Reynolds number, channel roughness, or Prandtl number increases the heat transfer coefficient and brings the film temperature closer to the bulk fluid temperature. The exact variation of film temperature with all these parameters is rather difficult to predict and it is approximated by taking the film temperature as the arithmetic mean of wall and bulk fluid temperature, or

$$T_a = (T_w + T_b)/2. \tag{2-9}$$

Equation (2-9) has been found satisfactory over a wide range of test conditions [12,13].

Another approach has sometimes been proposed for liquid-cooled reactors. It relies upon forming the ratio of the friction factor with heat addition or removal to the friction factor for isothermal conditions. For liquids, where the temperature dependence of viscosity is much greater than that of other properties, this ratio can be expressed as

$$f/f_{iso} = (\mu_w/\mu_b)^m, \tag{2.10}$$

In Eq. (2-10), μ is the absolute viscosity of the fluid and the subscripts b and w represent its value at the bulk fluid temperature and at wall temperature. Here again, as in the case of the film temperature, the exponent m can be expected to vary with the channel roughness and the fluid Reynolds and Prandtl numbers. Typical analytical predictions obtained by Wiederecht and Sonnemann for turbulent flow in a smooth channel of high pressure water and diphenyl coolant are shown in Fig. 2-3 [11]. It is observed that the ratio f/f_{iso} and the exponent m decrease as the Reynolds and Prandtl number increase. The variation of m is not easy to predict, and it is common practice to select a single value of m for all flow conditions. The most accepted relation is that proposed by Seider and Tate [14] for turbulent flow:

$$f/f_{iso} = (\mu_w/\mu_b)^{0.14}. \tag{2-11}$$

Equation (2-11) has been found acceptable by many investigators [15, 16] and is plotted in Fig. 2-3 for comparison.

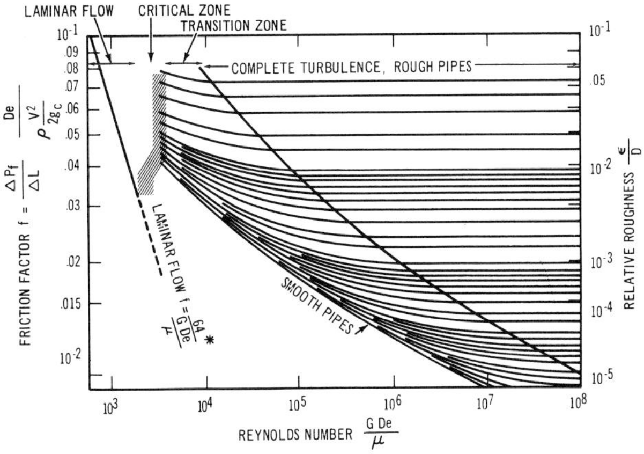

FIG. 2-1 Friction factor for single phase channel flow
*Circular pipe flow only.

Note that in order to use Eqs. (2-9) and (2-11) it is necessary to know the wall temperature. This temperature is obtained from heat transfer considerations, some of which are discussed in the chapter on Heat Transfer.

Equations (2-9) and (2-11) are most commonly used in nuclear designs. In a few instances extensive experimental data exist and empirical calculations are utilized instead. Water at high pressure and high temperature (400 to 630°F or 200 to 330°C) is one fluid for which considerable test results have been obtained [17] and correlated by the following equation:

$$f/f_{iso} = 1 - 0.0025\,(T_w - T_b)\,, \qquad (2\text{-}12)$$

where T_w, T_b are in °R.

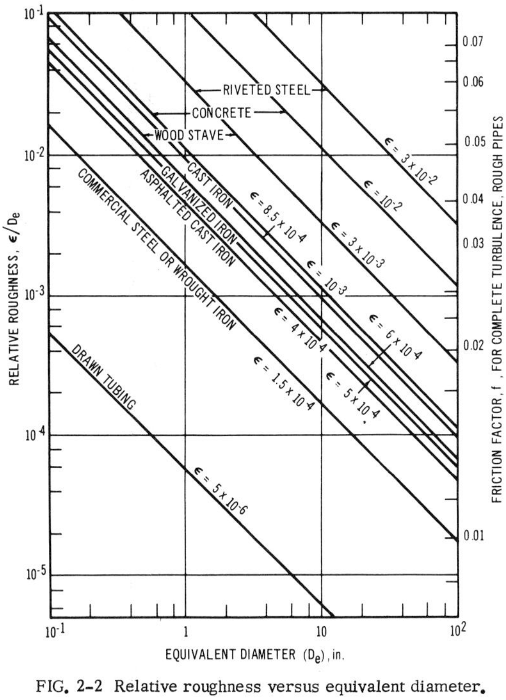

FIG. 2-2 Relative roughness versus equivalent diameter.

2.2.1.3 Contraction-Expansion Losses. Contraction-expansion pressure losses are usually defined in terms of the number of velocity heads lost across the geometry (one velocity head equals $V^2/2g_c$). The pressure change due to a contraction or expansion geometry is made up of two components: the pressure change due to flow area change alone; and the irreversible component of the pressure drop across the contraction or expansion geometry. For sudden contractions and expansions, one can write

$$\Delta P_c = \rho(V^2/2g_c)\,(1-\beta^2) + K_c\rho(V^2/2g_c)\,,$$
$$\Delta P_e = \rho(V^2/2g_c)\,(1-\beta^2) + K_e\rho(V^2/2g_c)\,. \qquad (2\text{-}13)$$

In Eq. (2-13), V is the velocity in the smaller flow area, β the contraction or expansion area ratio, and K_c and K_e the irreversible contraction and expansion loss coefficients.

The coefficients K_c and K_e have been found to depend upon the contraction-expansion geometry and in some cases upon the flow Reynolds number. Typical values are shown in Figs. 2-4, 2-5, and

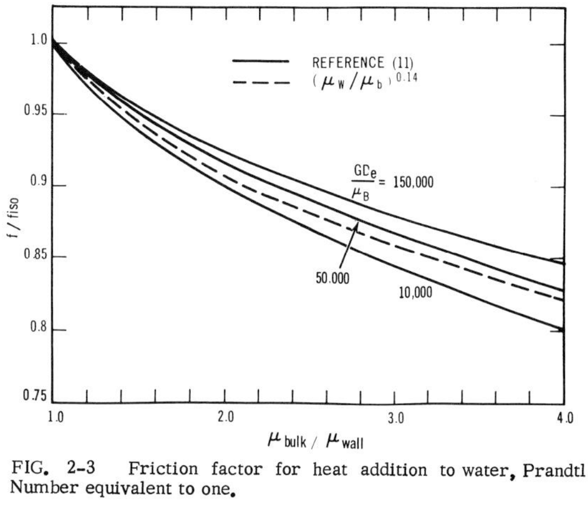

FIG. 2-3 Friction factor for heat addition to water, Prandtl Number equivalent to one.

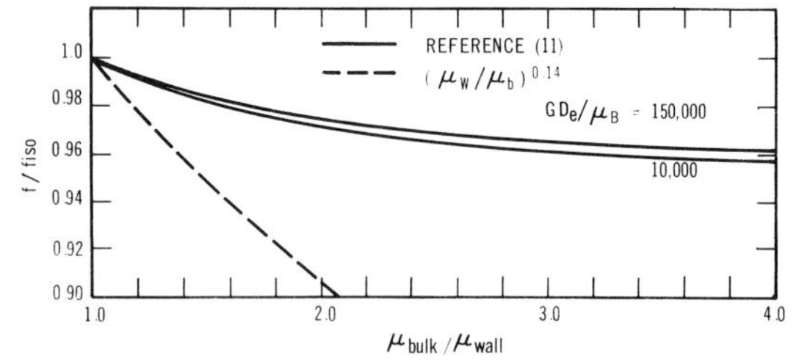

FIG. 2-3 (cont.) Friction factor for heat addition to diphenyl, Prandtl Number equal to six.

2-6. Figure 2-4 shows the loss coefficients K_c and K_e for sudden circular and rectangular contraction-expansion geometries [18]. Figure 2-5 [19,20] shows the loss coefficients for gradual reducers and the variation of the coefficient K_c with rounding and tapering of the entrance geometry. Finally, Fig. 2-6 [21] gives the loss coefficients K_c and K_e for the entrance and exit from the core of a tubular heat exchanger. Additional curves for other heat-exchanger geometries are given in reference [21].

2.2.1.4 Bend Losses. Bend pressure drops are also expressed in terms of the number of velocity heads lost across the bend:

$$\Delta P_B = K_B \rho(V^2/2g_c) \,. \qquad (2\text{-}14)$$

The loss coefficient K_B is a function of geometry, channel roughness and Reynolds number. Typical values are shown in Fig. 2-7 [22] for laminar flow through a one-half inch (1.27 cm) nominal diameter copper pipe using solder type fittings. Values for turbulent flow are given in Figs. 2-8 [22, 24] and 2-9 [26]. Figure 2-8 shows the loss coefficient for a 90° circular bend at a Reynolds number of 100,000 to 225,000 [23]. It also gives a plot of the coefficient K_B for other angles of bend in a circular pipe of uniform diameter and

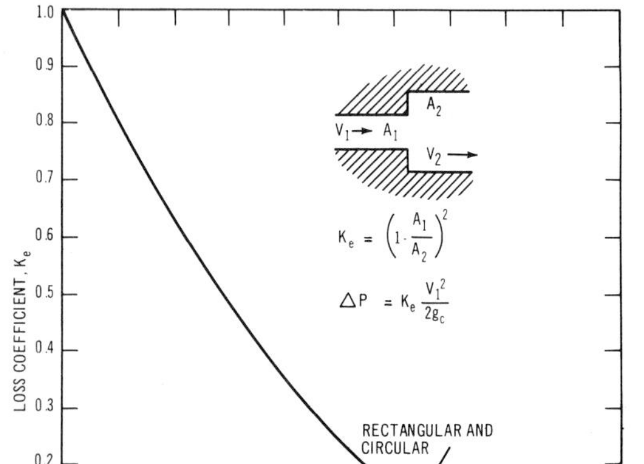

FIG. 2-4 Loss coefficient for sudden enlargement.

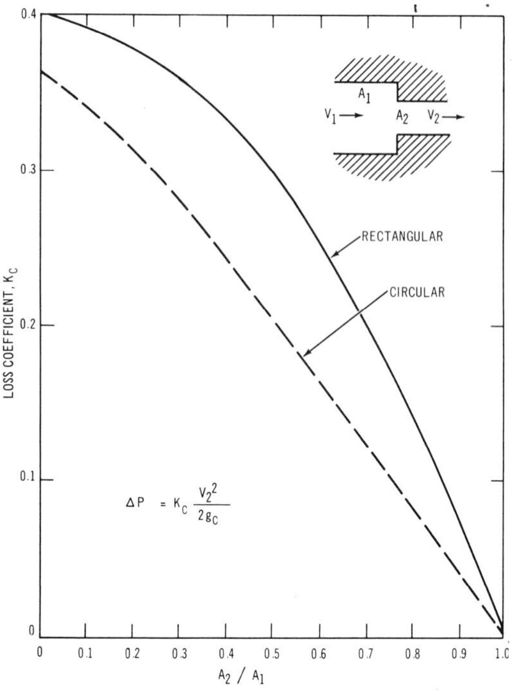

FIG. 2-4 (cont.) Loss coefficient for sudden contraction.

smooth surface at a Reynolds number of 225,000 [24]. Shown on the same plot is the loss coefficient for a miter or simple elbow bend which can be approximated [25] by a formula $(1 - \cos \xi)$ where ξ is the angle of bend. Figures 2-9a and 2-9b give the loss coefficient K_B for a 90° bend made up of a series of smooth or rough, simple elbows [26]. Additional values of the coefficient K_B for other miter bends are given in reference [24]. It should be noted that some bend loss such as the loss in a miter bend can be reduced by the introduction of deflecting vanes.

2.2.1.5 Valve and Fitting Losses. The pressure losses across valves and fittings are expressed by means of a loss coefficient $K_{V\&F}$ or by means of an equivalent frictional pressure drop. The equivalent friction loss is defined in terms of a length L_e that must be added to the channel equipped with valves and fittings to obtain the same pressure drop as that produced by the valves or fittings.

$$\Delta P_{V\&F} = K_{V\&F}\, \rho(V^2/2g_c) = f\,(L_e/D_e)\, \rho(V^2/2g_c) \,. \qquad (2\text{-}15)$$

Typical values of (L_e/D_e) are given in Table 2-1 [27] for turbulent flow through various types of valves and fittings.

2.2.1.6 Heat-Exchanger Losses. For liquid-cooled reactors, heat exchangers are usually of the shell and tube type. When the liquid flows inside the tubes, the pressure drop is made up of an exit and entrance loss and of a frictional pressure drop. These are obtained from Figs. 2-1, 2-2, 2-4, 2-5, and 2-6. When the liquid flows out-

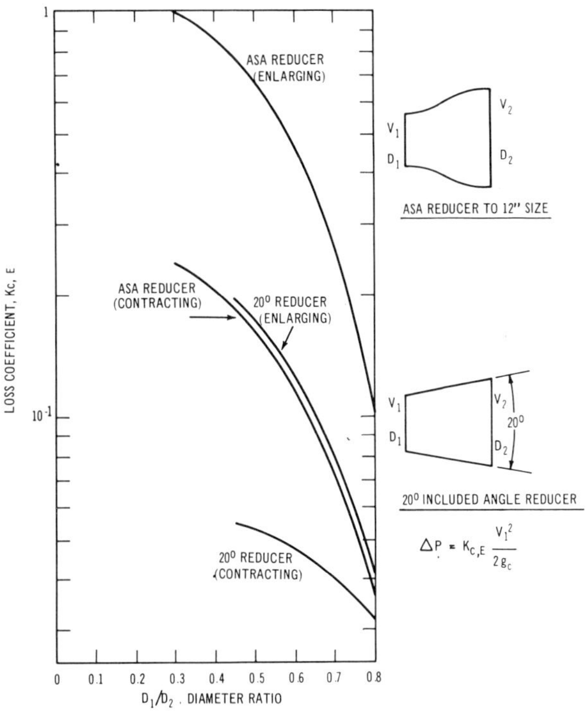

FIG. 2-5 Gradual reducer loss factor.

side and parallel to the tubes, the same figures can again be used if an appropriate equivalent diameter is calculated for the flow inside the shell. Finally, when the liquid flows on the shell side but perpendicular to the tubes, Figs. 2-4 to 2-6 still apply for inlet and exit losses, but the frictional pressure drop is obtained from

$$\Delta P_f = n f \rho (V^2/2g_c) . \qquad (2\text{-}16)$$

In Eq. (2-16), n is the number of rows in the flow direction and V the average fluid velocity between tubes. According to Grimison, the friction factor f is obtained from Fig. 2-10 [28] for the case where the tubes are in line or staggered. Additional data for cross flow to tubes are also found in reference [21].

2.2.1.7 Pump Pressure Rise. The pressure rise ΔP_p across the pump is obtained from manufacturers' supplied curves of pressure head versus flow rate. The curves depend upon the design and type of pump (for instance, see reference [29]). A brief mention will, however, be made here of three significant points:

1. In loss-of-flow accidents, it is important to know not only the pressure head developed by

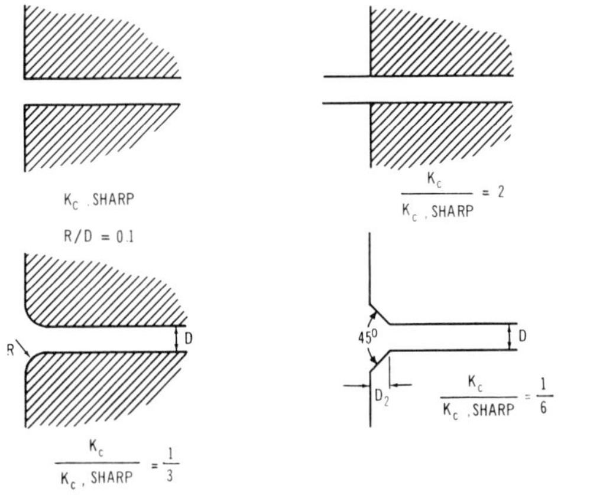

FIG. 2-5 (cont.) Effects of entrance conditions upon contraction coefficient.

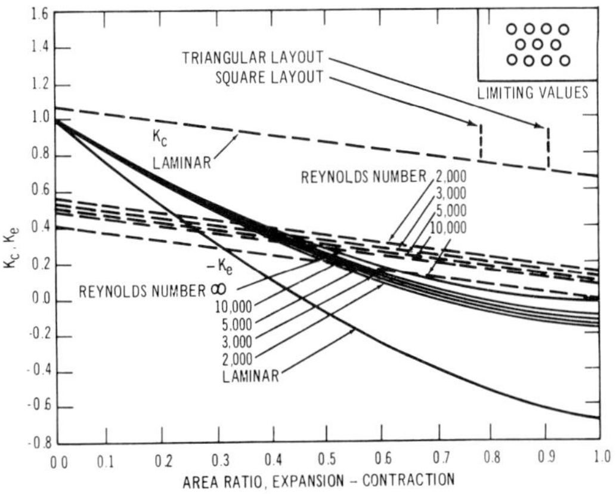

FIG. 2-6 Entrance and exit loss from heat exchanger core.

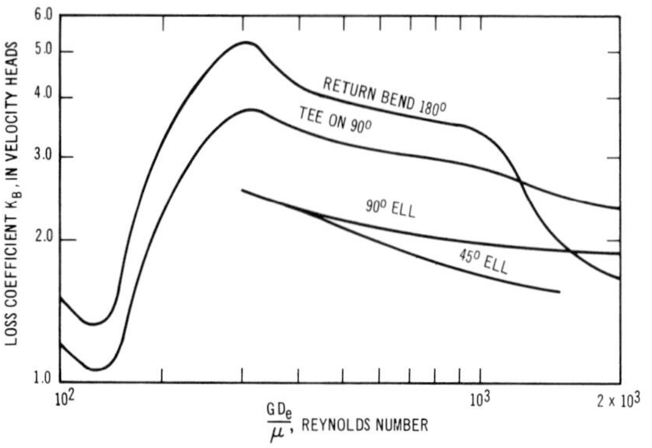

FIG. 2-7 Bend losses in various solder type fittings for streamline flow in a 1/2 inch pipe.

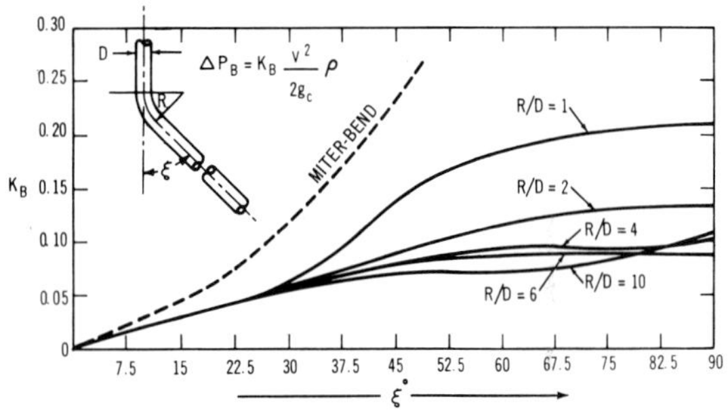

FIG. 2-8 (cont.) Resistance coefficients for bends of uniform diameter.

the pump at rated flow and speed, but also at other flow and speed conditions. Usually, if the head curve is available at one speed, curves at other speeds can be obtained from affinity laws [29]. Figure 2-11 shows a typical set of performance curves for a single suction radial flow pump.

2. It is sometimes necessary to add rotative inertia through a flywheel to a pump in order to obtain sufficient cooling during a loss-of-flow accident. The amount of rotative inertia required may determine the type of pump that can be used. For instance, in a water-cooled reactor, where a large flywheel (above 2000 lb-ft^2 or ~ 80 kg-m^2) is needed, a mechanical seal pump may be preferred if the added flywheel mass causes too large a loss of efficiency in a canned rotor pump.

3. In some reactor applications no pump is used, and the pressure rise required to overcome losses in the loop is obtained from the hydrostatic changes or fluid density variations around the loop. Such loops are referred to as natural circulation loops.

The calculation methods for natural circulation loops are the same as for forced circulations systems. The pressure drops in a natural circulation loop are, however, smaller and must be known more accurately.

2.2.2 Compressible Single-Phase Flow

In compressible flow the fluid density changes are large enough that they must be considered. Compressibility effects become important when the density variation exceeds 10 to 15 percent or when the velocity becomes high. One means of evaluating the importance of compressibility due to high fluid velocity is to calculate the local Mach number. The Mach number M is defined as the ratio of local velocity to the velocity of sound.

The velocity of sound, a, in a fluid is given by

$$a = \sqrt{g_c (dP/d\rho)_S}\,, \tag{2-17}$$

where S represents the entropy.

Compressibility effects can be neglected when the Mach number calculated from the definition of the Mach number and Eq. (2-17) is below 0.3 to 0.4 or when the density variation due to temperature distribution falls below 10 to 15 percent. Beyond these conditions the equations presented in

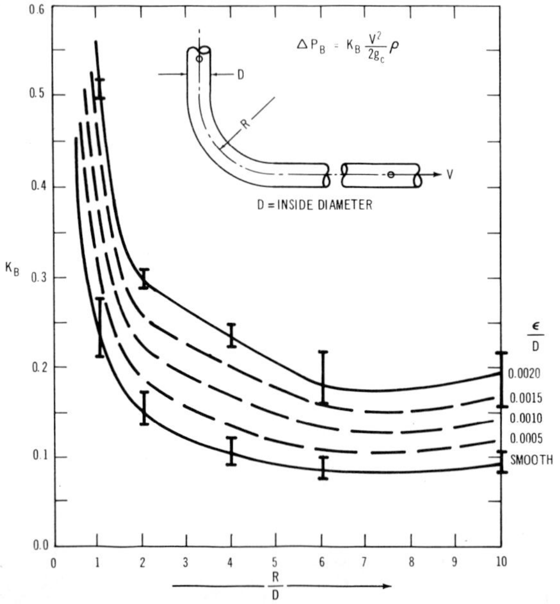

FIG. 2-8 Resistance coefficients for 90° bends of uniform diameter.

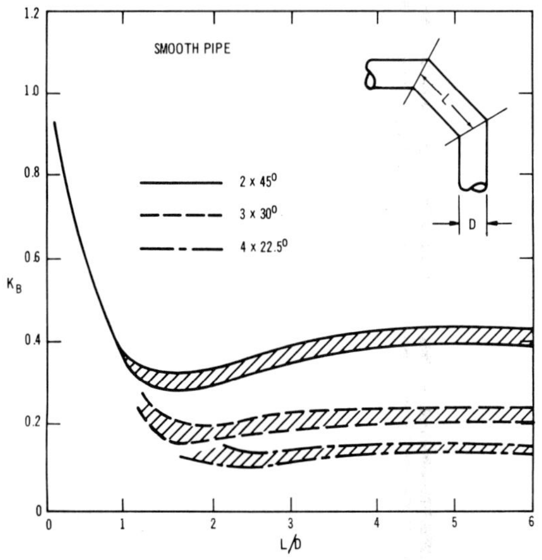

FIG. 2-9a Miter bend losses: smooth pipe.

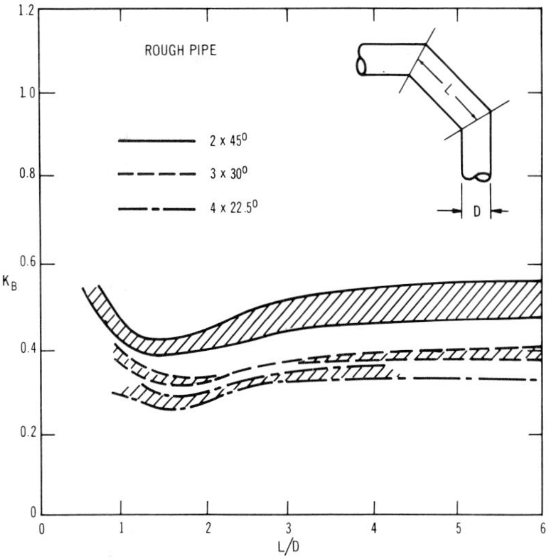

FIG. 2-9b Miter bend losses: rough pipe.

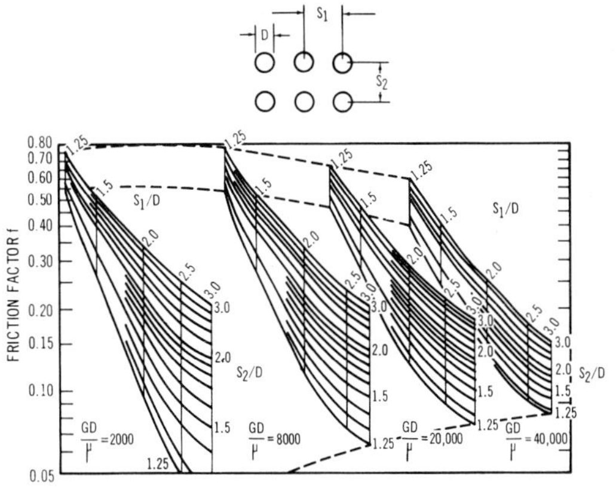

FIG. 2-10 Resistance coefficient for in-line tube bundles.

Section 2.2.1 must be modified. In nuclear applications, this circumstance arises only for high performance gas-cooled reactors. For such reactors, the pressure losses or gains listed in Eq. (2-2) are obtained as noted in the following paragraphs.

2.2.2.1 Reactor Pressure Loss. This term is considered in detail in Sec. 3.2.

2.2.2.2 Piping Pressure Loss. For piping of constant cross-sectional area, Eqs. (2-3) and (2-4) are still valid. However, since the fluid density varies, the acceleration (inertial) term in Eq. (2-3) is not equal to zero and must be retained. On the other hand, the head or hydrostatic term can be neglected because the density of gases is so low. Integration of Eq. (2-3) from position 1 to position 2 and substitution of Eq. (2-5) give, for the case of pressure changes that are small compared with the static pressure level:

$$\Delta P = \frac{G^2}{g_c}\left[\frac{1}{\rho_2} - \frac{1}{\rho_1}\right] + \frac{f}{D_e}\,[L_2 - L_1]\,\frac{G^2}{2g_c\rho_a} \qquad (2\text{-}18)$$

In Eq. (2-18) the friction factor f is defined in terms of the arithmetic average fluid density, ρ_a from position 1 to position 2. In order to calculate the pressure loss ΔP, the variation of the density along the channel must be known. This variation is obtained from an equation of state which, for most gases, can be written as follows:

$$P = \rho R T\,. \qquad (2\text{-}19)$$

In Eq. (2-19), R is an appropriate gas constant and T is the absolute temperature. Equation (2-19) makes it possible to calculate the density ρ in terms of the pressure P if the absolute fluid temperature T is known. The temperature T is obtained from heat transfer considerations (for instance, see

TABLE 2-1

Pressure Drop for Turbulent Flow Through Various Valves and Fittings

	Equivalent length in pipe diameters, L_e/D_e
Fittings	
45-deg. standard elbow	12
90-deg. standard elbow	30
90-deg. long radius elbow	20
Square corner elbow	57
Standard tee with flow through run	20
Standard tee with flow through branch	60
Close pattern return bend	50
Valves	
Conventional gate valve, fully open	13
" " " 3/4 open	35
" " " 1/2 open	160
" " " 1/4 open	900
Conventional globe valve, fully open	340
Conventional angle valve, fully open	145
Conventional swing check valve, fully open	135
In-line ball check valve, fully open	150

chapter on Heat Transfer). Solutions to Eq. (2-18) can then be obtained by trial and error if a relation to predict the friction factor f is available. For a gas which satisfies Eq. (2-19), it can be shown that the friction factor grouping $(\gamma_1 f/2D_e)$ $(L_2 - L_1) = \gamma_1 \Delta P_f/(G^2/\rho_a)$ is uniquely determined by specifying the conditions at position 1 and the fluid end state ratios. Such solutions are usually expressed in terms of the fluid total (impact or stagnation) temperature and pressure, T_S and P_S:

$$T_s = T\,[1 + (1/2)\,(\gamma - 1)\,M^2]\,, \qquad (2\text{-}20)$$

$$P_s = P\,[1 + (1/2)\,(\gamma - 1)\,M^2]^{\gamma/(\gamma-1)}\,. \qquad (2\text{-}21)$$

According to Eqs. (2-20) and (2-21), the total temperature and pressure are those which are obtained by adiabatic deceleration to the rest of the fluid at the local velocity. The above definitions presume that a gas equation of state of the type (2-19) can be written and that the velocity of sound is given by

$$a = \sqrt{(g_c \gamma P/\rho)}\,. \qquad (2\text{-}22)$$

The symbol γ represents the ratio of specific heat at constant pressure to the specific heat at constant volume.

Charts relating the quantities γ_1, M_1, T_{S_2}/T_{S_1} and $\Delta P_S/P_{S_1}$ which satisfy Eqs. (2-18) to (2-22), can be constructed. Typical charts obtained by Manson for air are shown in Figs. 2-12 [30] and 2-13 [30] for T_{S_2}/T_{S_1} = 0.5, 1.0, and 2.0. The application of these charts to other gases over diverse operating conditions is illustrated in Fig. 2-13.

For constant flow area channels, the frictional pressure loss ΔP_f or $(\Delta P_S)_f$ can then be calculated from Figs. 2-12 and 2-13 if the friction factor f is known. The variation of the friction factor f with channel roughness and Reynolds number is nearly the same as for incompressible flow, and the curves of Figs. 2-1 and 2-2 can be used.

In many instances, the fluid temperature varies due to heat addition or removal. When this happens, the curves in Figs. 2-13 and 2-14 [34, 35] are valid if the properties entering the calculation of the friction factor are evaluated at an appropriate film temperature. The use of Eq. (2-9) has been found especially satisfactory for gases, even at very high temperatures [31,32].

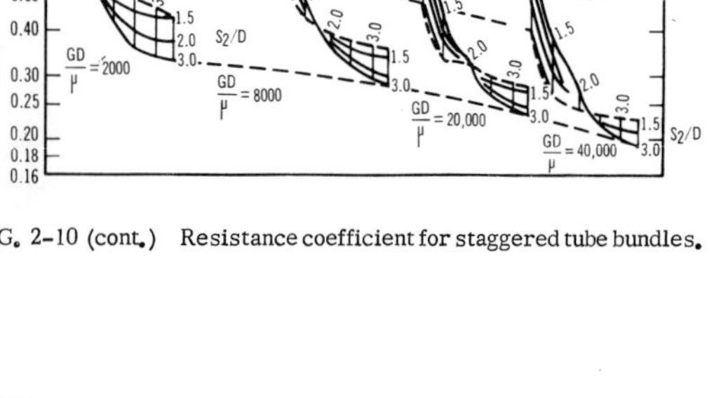

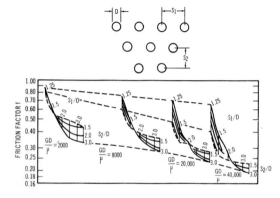

FIG. 2-10 (cont.) Resistance coefficient for staggered tube bundles.

Another method of accounting for property variations has been proposed by Humble, Lowdermilk and Desmon [32]. It uses a modified version of Eq. (2-6), or

$$\frac{1}{\sqrt{f T_a/T_b}} = 2.0\,\log_{10}\left[\frac{GD_e}{\mu_B}\,\frac{\mu_B \rho_f}{\mu_f \rho_B}\left(\frac{f T_a}{T_b}\right)^{1/2}\right] - 0.80\,. \qquad (2\text{-}23)$$

Solutions of Eqs. (2-18) and (2-23) for air similar to those presented in Figs. 2-12 and 2-13 have been obtained by Johnson, Lowdermilk, and Rom [33].

An interesting aspect of the curves shown in Figs. 2-12 and 2-13 is the limiting line labeled "choke line." Choking occurs when the downstream Mach number M_2 equals one, and any attempt to increase the parameter $(\gamma_1\ f/2D_e)\ (L_2 - L_1)$ beyond the values given by the choke line will act to decrease the upstream Mach number M_1 until the downstream Mach number M_2 is again equal to one. This results in a reduction or choking of the flow rate due to friction.

2.2.2.3 Contraction-Expansion Losses. The effects of compressibility upon sudden contraction-expansion geometries are shown in Fig. 2-14 [34, 35]. The contraction-expansion coefficients are expressed in terms of total pressure loss, and compressibility effects are seen to be negligible in most practical cases. At low values of the area ratio, the change in the loss coefficient is small. For area ratios approaching one, the change in the coefficient is large, but the coefficient itself is low and unimportant in fluid flow calculations. The effects of compressibility upon more complex contraction-expansion geometries have not been completely established. The results can, however, be expected to follow the trends of Fig. 2-14. The curves shown in Figs. 2-4 to 2-6 can be utilized if judicious corrections for compressibility effects are applied based upon Fig. 2-14.

The existence of a limiting "choke line" is again noted in the case of a sudden contraction. For subsonic flow, typical of reactor applications, the possibility of flow choking exists whenever the flow area decreases.

The simplest case of flow choking produced by an area reduction is that of frictionless flow through a converging nozzle with fixed upstream conditions and without heat addition. Continuity principles require that

$$(dA/A) + (dG/G) = (dA/A) + (d\rho/\rho) + (dV/V) = 0\,. \qquad (2\text{-}24)$$

If the friction and head losses are neglected, Eq. (2-3) yields

$$V dV + (g_c dP/\rho) = V dV + g_c (dP/d\rho)(d\rho/\rho)$$

$$= V dV + a^2 (d\rho/\rho) = 0\,. \qquad (2\text{-}25)$$

Substitution of Eq. (2-24) gives

$$dA/A = (dV/V)(M^2 - 1)\,. \qquad (2\text{-}26)$$

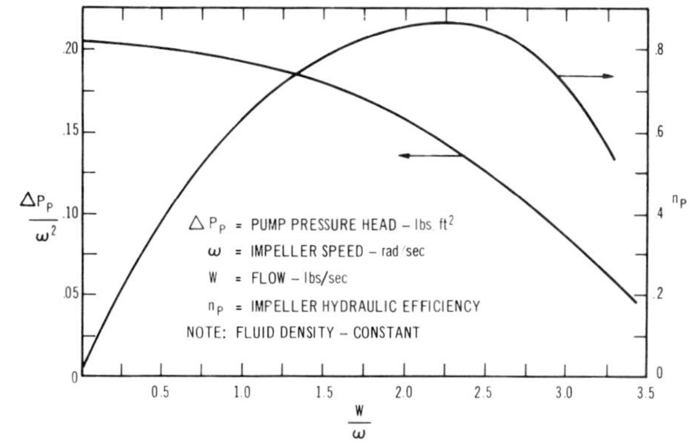

FIG. 2-11 Typical performance curves of single suction radial flow centrifugal pump.

According to Eq. (2-26), a reduction in flow area causes an increase in velocity if the Mach number is below one. A further increase in velocity at M = 1 requires dA/A = 0. From Eq. (2-24), dG/G = 0 or the mass flow rate reaches a maximum corresponding to attaining the sonic velocity at the point of minimum flow cross section or throat of the nozzle. This maximum flow rate per unit area, G, can be computed from the pressure p_o and density ρ_o of the upstream reservoir:

$$G = [2\gamma/(\gamma-1)]\, g_c\, p_o\, \rho_o\, (p/p_o)^{2/\gamma} \times [1-(p/p_o)^{(\gamma-1)/\gamma}]^{1/2} . \qquad (2\text{-}27)$$

The corresponding critical ratio of throat to upstream reservoir pressure is:

$$p/p_o = [2/(\gamma+1)]^{\gamma/(\gamma-1)} . \qquad (2\text{-}28)$$

Pressure ratios equal to or below this critical value will produce sonic velocity at the nozzle throat.

2.2.2.4 Bend Losses. The effects of compressibility upon pressure losses around bends are illustrated in Fig. 2-15 [36] for a 90° bend in a circular pipe. It is seen that compressibility becomes important if the inlet Mach number is large and if the bend radius is sharp. A "choke band" is again shown in Fig. 2-15 corresponding to reaching local sonic velocity around the bend. Here again, data are not available for more complex bend geometries, and the results of 2.2.1.4 may be used with judicious corrections for high Mach number effects based upon Fig. 2-15.

2.2.2.5 Valves and Fitting Losses. The effects of compressibility upon the pressure losses across valves and fittings have not been studied in detail. In most practical designs, a correction factor obtained from Fig. 2-14 may be applied to the incompressible relations.

The use of screens is common in gas circulating systems and the pressure loss across a single screen made up of round wires and a single sheet perforated with round holes can be predicted

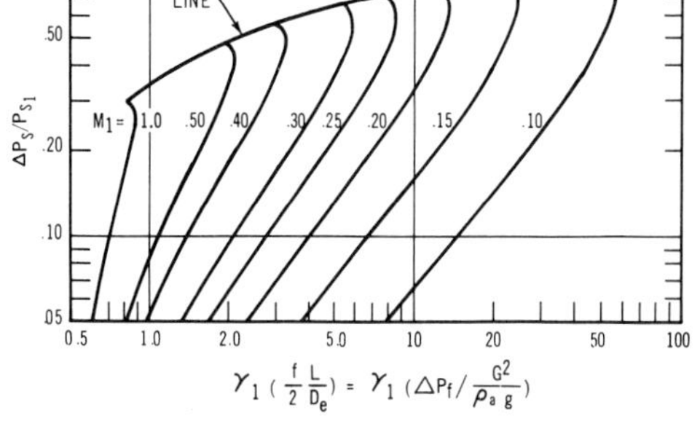

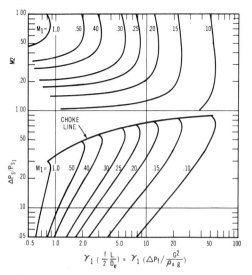

FIG. 2-12 Air flow resistance chart. $T_{S_2}/T_{S_1} = 0.50$, $\gamma_1 = 1.32$, $\gamma_2 = 1.331$.

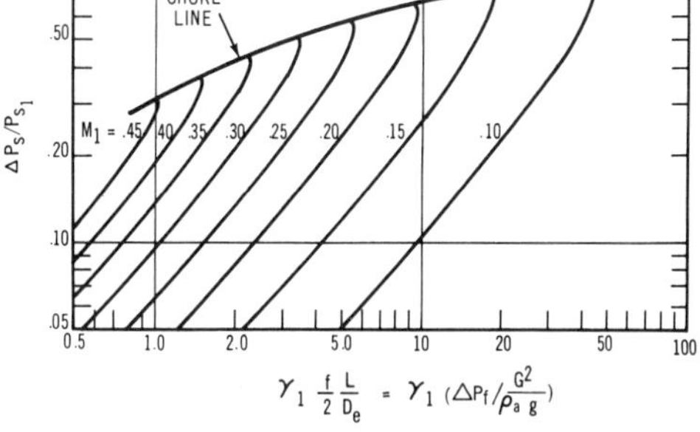

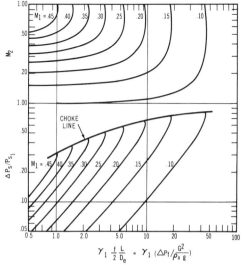

FIG. 2-12 (cont.) Air flow resistance chart. $T_{S_2}/T_{S_1} = 1.00$, $\gamma_1 = 1.38$, $\gamma_2 = 1.38$.

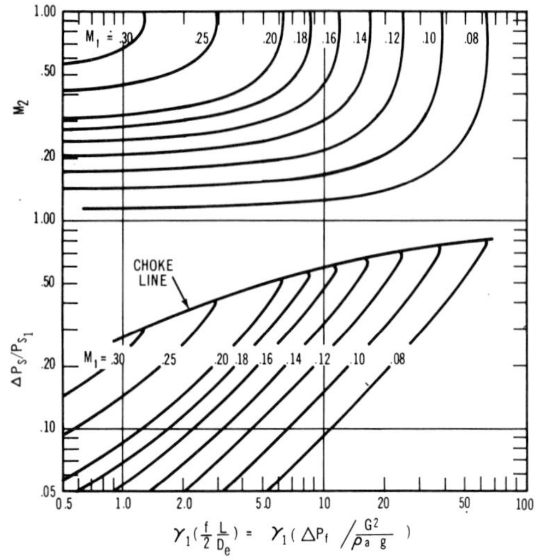

FIG. 2-13 Air flow resistance chart. $T_{S_2}/T_{S_1} = 2.0$, $\gamma_1 = 1.38$, $\gamma_2 = 1.366$.

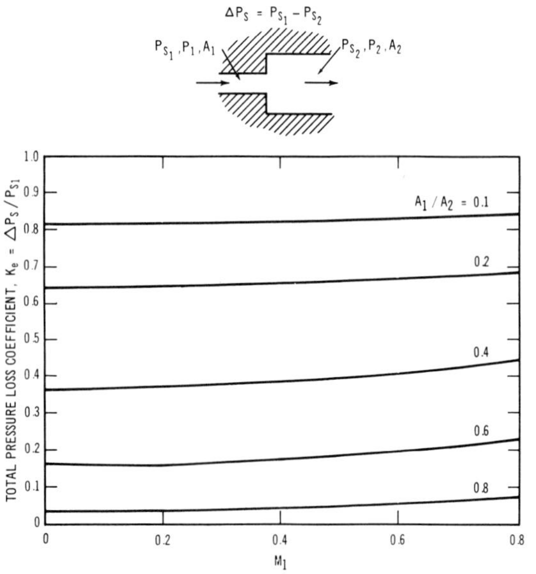

FIG. 2-14 Compressible flow across sudden expansion for $\gamma = 1.4$.

from Fig. 2-16 [37,38]. Data for incompressible pressure drop across packed screens are given in reference [21] and can be extended to compressible flow through appropriate adjustments from Fig. 2-16.

2.2.2.6 Heat-Exchanger Losses. In gas-cooled reactors, extended surfaces and cross flow are dominant in heat-exchanger designs. The extended surfaces vary considerably from one design to another, but usually consist of welding on the tubes of studs, plates, or helical strips to increase the heat transfer area. Extensive pressure loss data for gas-cooled heat exchangers are given in reference [21] for incompressible flow. Recently, new results have been reported for cross flow to studded tubular surfaces and helically finned tubes [39,40]. A few attempts have also been made to develop generalized correlations. The most successful one was presented by Gunther and Shaw [41] and utilizes a volumetric hydraulic diameter D_V defined as

$$D_v = 4\text{ (net free volume)/friction surface.} \quad (2\text{-}29)$$

According to Gunther and Shaw, Eq. (2-5) can be used if the friction factor is obtained from

$$f = [180/(GD_v/\mu)]\,(D_v/S_T)^{0.4}\,(S_L/S_T)^{0.6}$$

$$\text{for } GD_v/\mu < 200,$$

$$f = [1.92/(GD_v/\mu)^{0.145}]\,(D_v/S_T)^{0.4}\,(S_L/S_T)^{0.6}$$

$$\text{for } GD_v/\mu > 200. \quad (2\text{-}30)$$

In Eq. (2-30), S_L and S_T are the longitudinal and transverse pitch in the tube bank.

Jameson [41] proposed a modification of Eq. (2-30) in the turbulent regime. The Jameson relation is

$$f = 3.38\,(GD_v/\mu)^{-0.25}\,(D_v/S_L)^{0.4}\,(S_T/S_L)^{0.6}. \quad (2\text{-}31)$$

Experiments have shown that Eqs. (2-30) and (2-31) are not too accurate. They still yield an approximate and useful guideline for evaluating designs with extended surfaces. More accurate predictions of pressure drop can only be obtained from tests of the type described in references [21], [39], and [40].

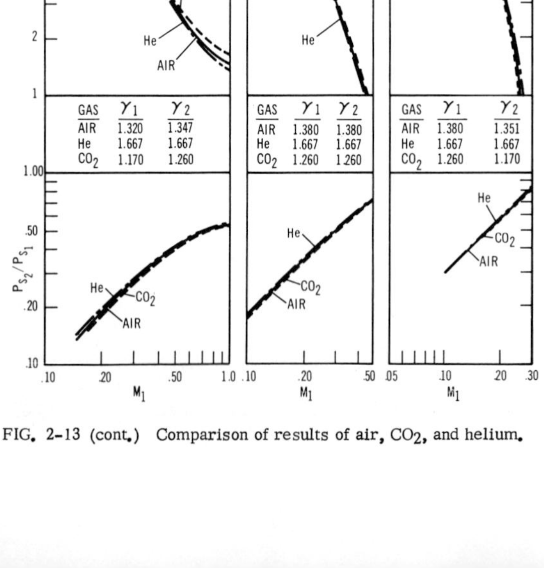

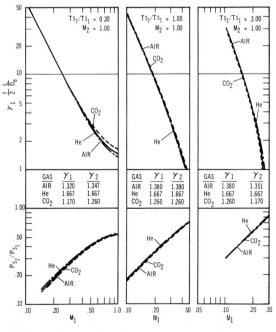

FIG. 2-13 (cont.) Comparison of results of air, CO_2, and helium.

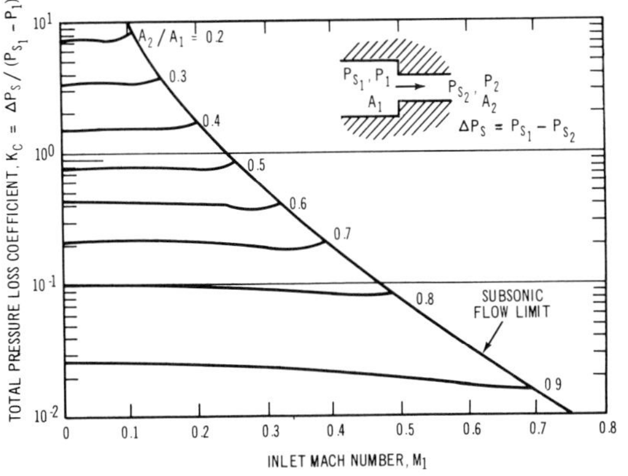

FIG. 2-14 (cont.) Compressible flow across sudden contraction.

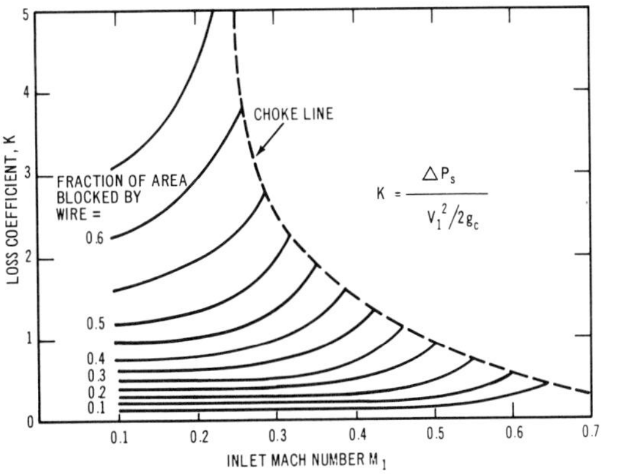

FIG. 2-16 Compressible flow in single round wire screens.

In most reactor designs, the flow velocities in heat exchangers are kept low to reduce pumping losses, and compressibility effects are negligible. If compressibility effects must be included, they can be estimated from the trends described in 2.2.2.3 and 2.2.2.5.

2.2.2.7 Circulator Pressure Rise. The design aspects of circulators for gas-cooled reactors are discussed briefly in the chapter on Mechanical Systems. An important requirement of gas circulator design is that it be able to supply a variable flow to the reactor. With variable flow, it is necessary to solve Eqs. (2-1) and (2-2) for a variety of operating conditions and to specify the circulator pressure rise curve over the same range of performance. A typical set of design curves for a CO_2-cooled reactor described by Goudy and Teire in their review of the development of compressors and drivers is reproduced in Fig. 2-17 [42].

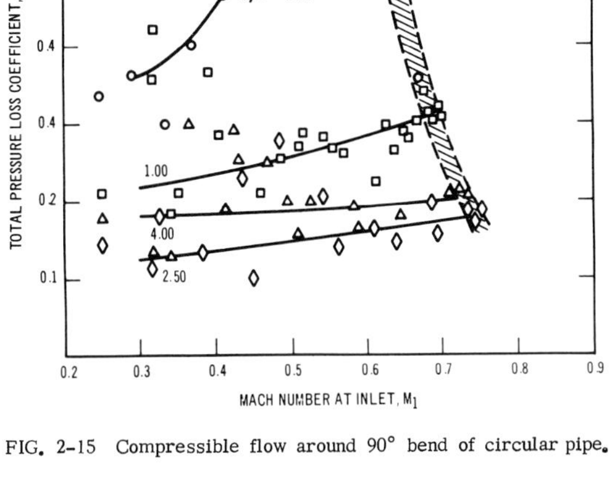

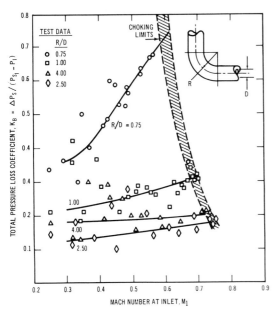

FIG. 2-15 Compressible flow around 90° bend of circular pipe.

2.2.3 Liquid-Vapor Flow

Liquid-vapor flow is characterized by the simultaneous presence of a liquid and a gas phase in the coolant stream. Liquid-vapor flow is found in boiling reactors, evaporators and condensers. The simultaneous presence of liquid and vapor complicates the solutions of pressure drop problems because the liquid and vapor can distribute themselves in an infinite number of flow patterns. This new complication adds to the uncertainties of the calculations and partly accounts for the empiricism and inaccuracies which are prevalent in liquid-vapor flow. We shall limit our attention in this section to the case of liquid-vapor flow without heat addition or removal and discuss the latter case in Sec. 3.

2.2.3.1 Reactor Pressure Loss. This loss is considered in Sec. 3.2.

2.2.3.2 Piping Pressure Loss. Equation (2-3) is still valid and the total pressure drop is made up of three components: hydrostatic, acceleration (inertial), and frictional pressure loss. In order to calculate the head and acceleration pressure drop, the two-phase fluid density must be known. For the special case where the liquid and vapor have the same local velocity, the flow is called homogeneous, and the density is obtained from

$$1/\rho = (1/\rho_L)(1 - x) + (1/\rho_G)\,x, \qquad (2\text{-}32)$$

where x is the weight fraction of vapor in the flowing stream and the subscripts L and G denote the liquid and gas or vapor. Equation (2-32) is valid over a limited range of conditions, i.e., when the liquid density approaches that of the vapor, or for fluids near the critical pressure.

In most cases, the mean liquid and vapor velocity are not equal and the ratio of vapor to liquid velocity, called the slip ratio, is different from one. This slip ratio is usually obtained from experimental tests. The most recent empirical correlation proposed by Marchaterre and Hoglund for

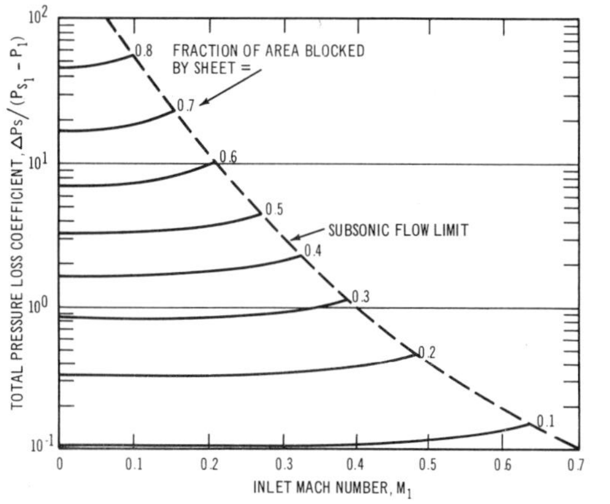

FIG. 2-16 (cont.) Compressible flow across single perforated sheets.

vertical flow is shown in Fig. 2-18 [43]. According to Fig. 2-18, the slip ratio increases as the vapor weight fraction or the ratio of liquid to gas density increases. It is also observed that as the total mass flow rate G goes up or as the hydraulic diameter D_e is reduced, the total liquid Froude number $(G^2/g\rho_L{}^2 D_e)$ increases and the slip ratio goes down.

Once the slip ratio is known, the vapor volume fraction α in the channel can be calculated from a mass balance

$$V_G/V_L = [(1-\alpha)/\alpha]\,(\rho_L/\rho_G)\,[x/(1-x)]\,, \qquad (2\text{-}33)$$

and the local fluid density ρ is equal to

$$\rho = \rho_L\,(1-\alpha) + \rho_G\alpha\,. \qquad (2\text{-}34)$$

The approximate nature of the curves shown in Fig. 2-18 must be recognized. As indicated in reference [43], the plotted slip ratios deviate from experimental data by ± 20 percent. The curves are valid for only vertical flow, total liquid velocity (G/ρ_L) greater than 2880 ft/hr (880 m/hr), and equivalent hydraulic diameter D_e below 3 inches (7.6 cm).

A multitude of other correlations for two-phase fluid density have been proposed in the literature. Particularly worthy of note are the semi-empirical

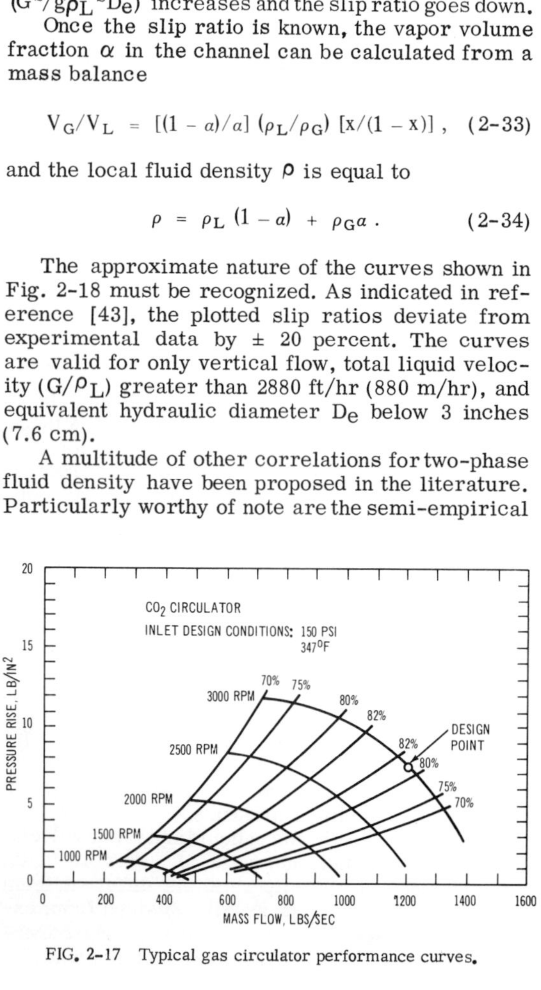

FIG. 2-17 Typical gas circulator performance curves.

curves of Martinelli-Lockhart [44], Martinelli-Nelson [45], Bankoff [46], Hughmark [47], Petrick [48], and the analytical predictions of Griffith [49] and Levy [50, 51].

Martinelli and Lockhart [44] proposed the first significant correlation of vapor volume fraction. Their correlation was based mostly on adiabatic horizontal two-component flow at essentially atmospheric pressure. Hewitt [52] derived the following fit to the Martinelli-Lockhart curve:

$$\ln(1-\alpha) = -1.482 + 4.915\ln X - 5.955(\ln X)^2 + 2.765(\ln X)^3 + 6.399(\ln X)^4 - 8.768(\ln X)^5. \qquad (2\text{-}35)$$

In Eq. (2-35), X is the ratio of the single-phase liquid pressure drop to that of the gas calculated on the basis that the liquid and gas are flowing at the respective flow rates of $G(1 - x)$ and Gx.

Martinelli-Nelson extended the above correlation to steam-water mixtures at various pressures and proposed the curves shown in Fig. 2-19 [45]. More recently, Sher [17] noted that at low qualities the Martinelli-Nelson curves gave slip ratios lower than unity and offered minor modifications to the curves. The modified plots were found in satisfactory agreement with the experimental data of Isbin, et al [53].

In the last few years, Bankoff [46] has suggested that the slip ratio was equal to

$$V_G/V_L = (1-\alpha)/(K'-\alpha)\,, \qquad (2\text{-}36)$$

where K' was an empirical constant. For vertical flow of steam-water mixtures, K' was taken to be a function only of pressure:

$$K' = 0.71 + 0.0001\,P\,, \qquad (2\text{-}37)$$

where P is the fluid pressure in psia. (If P is expressed in kg/cm^2, the coefficient of P should be 0.00142.)

Hughmark [47] extended the work of Bankoff and showed that Eq. (2-36) could be used to predict the slip ratio for horizontal and vertical flow if K' was taken to be a function of the parameter Ψ, with

$$\Psi = \left[\frac{D_e G}{\mu_L(1-\alpha)+\mu_G\alpha}\right]^{1/6}\left[\frac{G^2}{\rho_L{}^2 g D_e}\right]^{1/8}\frac{(1+x)\rho_L}{(1-x)\rho_G}\,. \qquad (2\text{-}38)$$

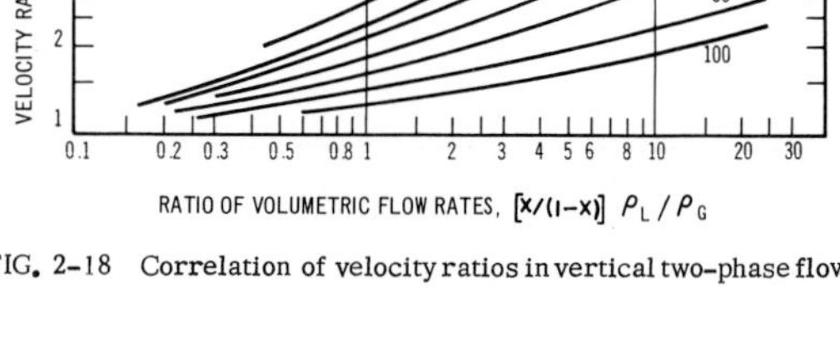

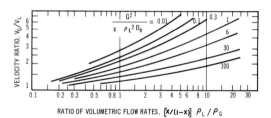

FIG. 2-18 Correlation of velocity ratios in vertical two-phase flow.

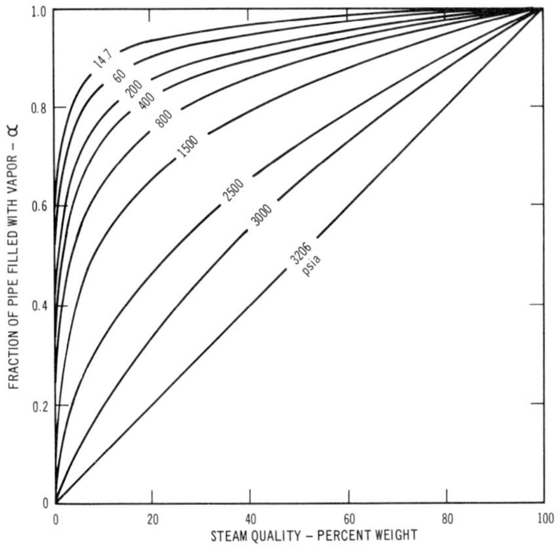

FIG. 2-19 Steam volume fraction.

The proposed relation between Ψ and K' was:

$\Psi =$	1.3	1.5	2.0	3.0	4.0	5.0		
$K' =$	0.185	0.225	0.325	0.49	0.605	0.675		
$\Psi =$	6.0	8.0	10	15	20	40	70	130
$K' =$	0.72	0.767	0.78	0.808	0.83	0.88	0.93	0.98

For downward flow, the two-phase fluid density is not well known. The number of empirical equations for two-phase fluid density and flow in the downwards direction is limited. Petrick has performed some experiments and proposed the following correlation of available slip ratio data:

$$V_G/V_L = 0.63\,(G^2/\rho_L^2\, g\, D_e)^{0.4} \times [x\rho_L/\rho_G(1-x)]^{0.2}\,. \qquad (2\text{-}39)$$

Several inconsistencies can be noted among the above correlations. Not all the equations recognize a flow or hydraulic diameter effect upon the slip ratio, and some correlations even exhibit reverse trends with hydraulic diameter. To date, analytical studies have not been able to resolve these discrepancies or to reduce the experimental uncertainty which ranges from ± 15 to 20 percent. The simplifications which must be introduced in the analytical models have often led to poor agreement with test results. For instance, the momentum exchange model of reference [51] gives too high a slip ratio particularly at high ratios of liquid to gas density. In other analytical solutions, the predictions are only valid for idealistic flow models. In reference [49], a relationship is derived for the velocity of a single gas bubble in vertical slug flow. Similarly, in reference [50], the slip ratio for annular flow with no liquid entrainment, smooth interface, and negligible gravity effects is calculated. The analytical prediction is [50]

$$\frac{V_G}{V_L} = 1 + \left\{\sqrt{\frac{\Delta P_{LTP}}{\Delta P_{LF}}}\;\frac{(1-\alpha)\sqrt{f_{LF}}}{1.6\sqrt{2}}\right\} \times \left\{\frac{3+\sqrt{\alpha}}{1+\sqrt{\alpha}} - \left(\frac{\rho_L}{\rho_G}\right)^{1/2}\left[\frac{2}{\alpha}\ln(1-\sqrt{\alpha}) + 1 + \frac{2}{\sqrt{\alpha}}\right]\right\}. \qquad (2\text{-}40)$$

The friction factor f_{LF} in Eq. (2-40) is calculated on the basis of liquid flow in the channel at the flow rate G(1-x). Eq. (2-40) is, unfortunately, valid only over a limited range of test conditions, i.e., high vapor volume fraction.

For steam-water mixtures, Figs. 2-18 and 2-19 are recommended for vertical upwards flow. For downwards flow of steam-water, Eq. (2-39) is suggested. For other fluids, the correlation of Hughmark is recommended for vertical and horizontal flow conditions. From a practical viewpoint, it should, however, be recognized that the correlation used for the prediction of two-phase density is not as important as the full realization of its inaccuracy and limitations. Furthermore, most of the experimental tests have been performed with water and steam or air mixtures and extrapolation of the available correlations to liquid metal or organic vapors should be done cautiously until further data are available.

C. J. Baroczy [54] has extended the Martinelli approach and has developed a generalized correlation to predict the liquid fraction of all fluids in two-phase flow.

Once the vapor volume fraction is known, the acceleration pressure drop can be calculated. The acceleration loss is obtained by integrating the liquid and gas momentum changes from position 1 to position 2 or

$$\Delta P_{acc} = \frac{G^2}{g_c\rho_L}\left[\frac{(1-x_2)^2}{1-\alpha_2} - \frac{(1-x_1)^2}{1-\alpha_1}\right] + \frac{G^2}{g_c\rho_G}\left[\frac{x_2^2}{\alpha_2} - \frac{x_1^2}{\alpha_1}\right]. \qquad (2\text{-}41)$$

The last pressure drop component of two-phase flow in a channel is the frictional pressure loss. This pressure drop can be predicted from the correlation of Martinelli-Lockhart [44], Martinelli-Nelson [45] or Chenoweth and Martin [55].

Martinelli and Lockhart [44] offered the first comprehensive correlation of two-phase frictional losses. Their correlation was based mostly on adiabatic two-component flow in horizontal channels at atmospheric pressure. Algebraic fits to their curve have been proposed by Hewitt [52] and Chisholm and Laird [56]. In the turbulent flow regime typical of reactor conditions, the following equations have been suggested:

Hewitt:

$$\Delta P_{TPF}/\Delta P_{GF} = 1.445 + 4.957 \ln X + 5.762 (\ln X)^2 - 1.170 (\ln X)^3 - 4.288 (\ln X)^4 + 3.150 (\ln X)^5\,; \qquad (2\text{-}42)$$

Chisholm-Laird:

$$\Delta P_{TPF} / \Delta P_{LF} = 1 + (21/X) + (1/X^2) \,. \quad (2\text{-}43)$$

In Eqs. (2-42) and (2-43), ΔP_{TPF} is the two-phase frictional loss and ΔP_{GF} and ΔP_{LF} are the single-phase gas and liquid frictional loss in the same channel calculated for mass flow rates of Gx and G(1-x), respectively.

Martinelli-Nelson extended the above correlation to steam-water mixtures and their curves are shown in Fig. 2-20 [45]. The pressure drop $(dP/dL)_{Lo}$ appearing in the ordinate of Fig. 2-20 is obtained on the basis of liquid flow in the channel at the total mass flow rate G.

More recently, Chenoweth and Martin presented the correlation shown in Fig. 2-21 [55] for horizontal two-phase flow. The pressure $(dP/dL)_{Go}$ appearing in Fig. 2-21 is again calculated by assuming that gas flows in the channel at the total mass flow rate G. The Chenoweth-Martin correlation has been found particularly accurate for two-phase steam flow in large diameter channels [57].

The uncertainties in the curves of Figs. 2-20 and 2-21 and Eqs. (2-42) and (2-43) cannot be overstressed. The equations have been found to correlate two-phase frictional pressure drop within ±50 percent. The inability of the curves to account for a reduction in the ratio $\Delta P_{TPF}/\Delta P_{Lo}$ with increased mass flow rates as reported by Moen [58] and Petrick [59] must be noted. Similarly, no channel diameter effect is included in the above correlations, contrary to trends found in many experimental investigations [59-62]. Finally, the curves are valid only for smooth channels and must be appropriately modified for roughened wall conditions [57].

As in the case of the vapor volume fraction, analytical models have not proven too useful in two-phase friction calculations. Nevertheless, they have pinpointed the fact that the frictional pressure drop varies with flow pattern. Also, the following relation, obtained for annular flow with a smooth interface, no liquid entrainment, and negligible gravity effects, has been found partly compatible with experimental results [63, 64, 50]:

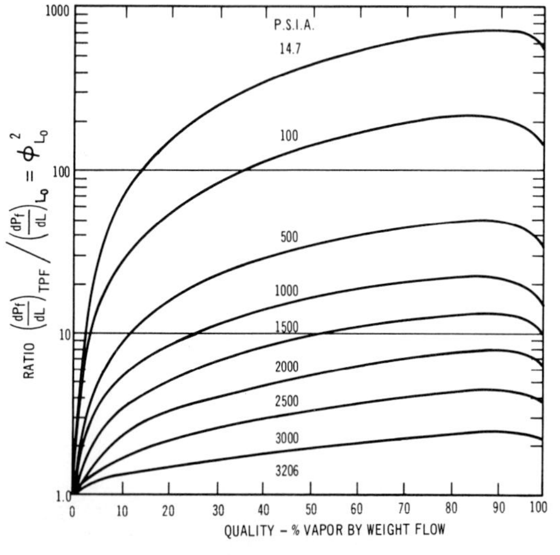

FIG. 2-20 Two-phase friction pressure drop for steam-water mixtures.

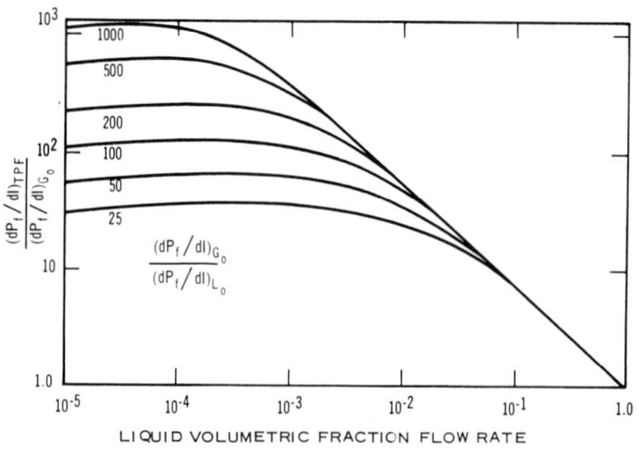

FIG. 2-21 Two-phase friction pressure drop in large horizontal pipes.

$$\Delta P_{TPF} / \Delta P_{LF} = 1/(1-\alpha)^2 \,. \quad (2\text{-}44)$$

Similar expressions with a proportionality constant ranging from 0.8 to 2.0 and exponents from 1.75 to 2.0 have been used to correlate test results [57, 58, 65].

Useful results have also been obtained from the homogeneous model. This model becomes especially accurate as the ratio of liquid to gas density approaches one [66]. In the homogeneous model, the pressure drop is obtained from Eq. (2-5) using the homogeneous density of Eq. (2-32). The friction factor is taken from Figs. 2-1 and 2-2 based upon a total mass flow rate G and a homogeneous viscosity defined as $(1-x)/\mu_L + x/\mu_G$. Other methods of weighing the two-phase viscosity which give equivalent accuracy have also been proposed.

2.2.3.3 Contraction-Expansion Losses. In two-phase flow, the contraction-expansion loss ΔP_{TP} is defined in terms of the corresponding single-phase loss ΔP_{Lo} based upon the total liquid flow rate G in the same geometry. Several models have been suggested to calculate contraction and expansion losses in two-phase flow. Lottes [67] describes several possible models for flow across a sudden expansion.

One of the most useful models is the homogeneous model which for a contraction or expansion geometry gives:

$$\Delta P_{TP} / \Delta P_{L_o} = (1 - x) + x\rho_L/\rho_G \,. \quad (2\text{-}45)$$

Another model singled out in reference [66] for flow across a sudden expansion is that of Romie which yields

$$\frac{\Delta P_{TP}}{\Delta P_{L_o}} = \frac{1}{1-\beta}\left[\left(\frac{x^2\rho_L}{\rho_G}\right)\left(\frac{1}{\alpha_1} - \frac{\beta}{\alpha_2}\right) + (1-x)^2\left(\frac{1}{1-\alpha_1} - \frac{\beta}{1-\alpha_2}\right)\right] . \quad (2\text{-}46)$$

The values of the vapor volume fraction α_1 in the upstream channel and of the volume fraction α_2 in the expanded channel can be calculated from the correlations noted in subsection 2.2.3.2. Test results of Richardson [65] and Petrick [59] with air-water mixtures indicate that α_1 is nearly equal to α_2 and Eq. (2-46) reduces to

$$\Delta P_{TP}/\Delta P_{L_o} = (1 - x)^2/(1 - \alpha) + x^2 \rho_L/\rho_G \alpha \,. \tag{2-47}$$

Available experimental results indicate that Eq. (2-45) predicts slightly high values while Eq. (2-47) is in closer agreement with the data [66].

For other contraction-expansion geometries, the number of experimental and analytical derivations is limited. The two-phase pressure loss can be approximately predicted by calculating the single-phase pressure loss based upon the total liquid flow rate G and multiplying it by the ratios given by Eqs. (2-45) and (2-47). The validity of this approach is illustrated in Fig. 2-22 [68], based upon recent measurements across an orifice geometry with high-pressure steam-water mixtures.

2.2.3.4 Bend Losses. Experimental data or analyses are not available for two-phase flow in bends. Equations (2-45) and (2-47) combined with corresponding single-phase predictions are often used in design calculations. Another accepted approach utilizes the curves of Figs. 2-20 and 2-21 instead of Eqs. (2-45) and (2-47). Some data on two-phase pressure drop in piping components have been reported in reference [69].

2.2.3.5 Valve and Fitting Losses. The methods described in the preceding paragraph are most commonly used.

2.2.3.6 Heat-Exchanger Losses. Since these losses include heat addition or removal, they will be covered in Sec. 3.

2.2.3.7 Pump Pressure Rise. Pumps are designed so that vapor is not present in the flowing stream. If vapor inadvertently occurs in the pump, the pump cavitates, and this condition is discussed in Sec. 2.3.

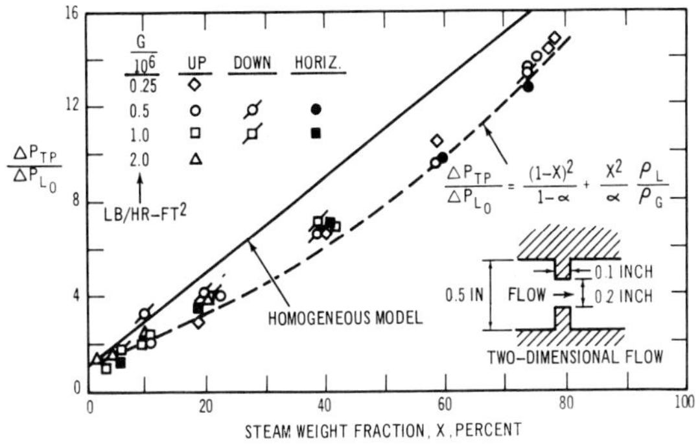

FIG. 2-22 Contraction, expansion loss for steam-water at 1000 psia (70 kg/cm^2).

2.2.4 Fluid-Solid Flow

This kind of flow is characterized by the simultaneous presence of a fluid and solid particles in the moving stream. There are a variety of fluid-solid systems. The most important types can be identified by considering the motion of a fluid (liquid or gas) through a packed or fixed bed of solid particles. When fluid flow starts through the bed, the bed remains relatively fixed and the fluid passes through the interstices between the particles. This condition is typical of pebble type reactors. As the fluid flow rate is increased, the particles start to move upward and remain suspended when the buoyant and drag forces acting on them equal their weight. The bed, at this stage, is fluidized and is representative of fluidized bed reactors. Two types of fluidization have been noted. When the fluid particles are evenly distributed, particulate fluidization occurs. This type of fluidization is dominant with liquid coolants. When the particle distribution is not uniform, the fluidization is called aggregate. Aggregate conditions are usually associated with gas flow. As the fluid flow is increased even more, the particles are carried away in the flowing stream. The moving fluid is now a slurry and is typical of slurry reactors.

For both the pebble bed and fluidized reactors, single phase incompressible and compressible flow prevail outside the reactor. These conditions were treated at length in Secs. 2.2.1 and 2.2.2 and do not require coverage here. The discussion that follows will, therefore, be limited to slurry flow.

2.2.4.1 Reactor Pressure Loss. This pressure drop is discussed in Sec. 3.2.

2.2.4.2 Piping Pressure Loss. The flow of fluid-solid suspensions or slurries in a pipe or channel can be either Newtonian or non-Newtonian. For Newtonian fluids, the shear stress τ is directly proportional to the gradient of the local velocity u, or

$$\tau = (\mu/g_c)(du/dy) = -(\mu/g_c)(du/dr) \,. \tag{2-48}$$

In Eq. (2-48), the viscosity μ is independent of shear stress, and r is the radial distance measured from the center of the pipe.

For non-Newtonian fluids, the viscosity μ varies not only with shear stress, but sometimes even with time. Two types of non-Newtonian fluids are recognized [70]: shear-thinning fluids for which the ratio $\tau/(du/dy)$ decreases with decreased shear rate and shear-thickening fluids which exhibit the opposite trend.

Both shear-thinning and shear-thickening fluids can have time-dependent or time-independent viscosities. When the ratio $\tau/(du/dy)$ decreases with time, the material is sometimes called "thixotropic," while the term "rheopectic" is used to describe materials with the opposite behavior.

We shall concern ourselves here only with uranium oxide and thorium oxide slurries. The properties of these slurries do not vary with time, but they have been found to have both Newtonian and non-Newtonian characteristics. Their behavior depends upon the slurry composition and its degree of flocculation (flocculation is identified with sticking of the original solid particles in loose irregular clusters or flocs).

Consider first the flow of Newtonian slurries. Suspensions of uranium trioxide hydrate ($UO_3 \cdot H_2O$) fall within this category. Also, with appropriate electrolyte addition, thoria (ThO_2) slurries can sometimes have Newtonian properties [71]. The electrolyte behavior of thoria slurries is not entirely understood except for the fact that it occurs over a specific range of electrolyte concentrations [71]. For such Newtonian fluids, the methods developed for single-phase flow in Secs. 2.2 and 2.2.2 are applicable if the slurry density and viscosity are appropriately defined. The density of a suspension can be calculated from Eq. (2-34) if ρ_G is replaced by the density ρ_S of the solid particles, ρ_L by the density ρ_f of the suspending medium and if α is taken to represent the volume fraction occupied by the solid particles. In most fluid-solid suspensions, the local solid and fluid velocity are equal and the homogeneous model of Eq. (2-32) is valid with appropriate changes to the subscripts.

The viscosity of a Newtonian suspension depends not only upon the volume fraction of solids in the stream, but also upon the shape of the particles. Einstein [72] first proposed the relation

$$\mu = \mu_f (1 + 2.5\,\alpha) \qquad (2\text{-}49)$$

to describe the viscosity of dilute suspensions of spherical particles in terms of the viscosity μ_f of the suspending medium. This equation has been extended by various authors to account for higher solid concentration and the interaction and shape of the solid particles. Most of the proposed relations can be written as

$$\mu = \mu_f [1 + A_1\alpha + A_2\alpha^2 + \dots], \qquad (2\text{-}50)$$

where the constants A_1, A_2, ... must be determined empirically. In the case of aqueous uranium trioxide hydrate suspensions (with uranium concentrations up to 250 g/liter), the viscosity has been found to be close to that of water. There was also, up to this concentration, no detectable difference between rods or platelets suspensions [73].

In the case of non-Newtonian fluids, the relationship between shear stress and velocity gradient is much more complicated. Typical data for a ThO_2 suspension are shown in Fig. 2-23 [74]. Shown on the same figure is the curve for the Newtonian suspending medium. It is seen that, at high rates of shear (region AB), the slurry curve approaches that of the suspending medium. However, at lower rates of shear (region BCD), the slurry curve deviates more and more from it. At very low shear rates, the slurry data fall in the region DEF with the data usually falling nearer curve DE than curve DF.

Several equations have been proposed to describe the curve shown in Fig. 2-23. Most uranium oxide and thorium oxide slurries can be handled by means of the Bingham plastic model [75]. According to this model

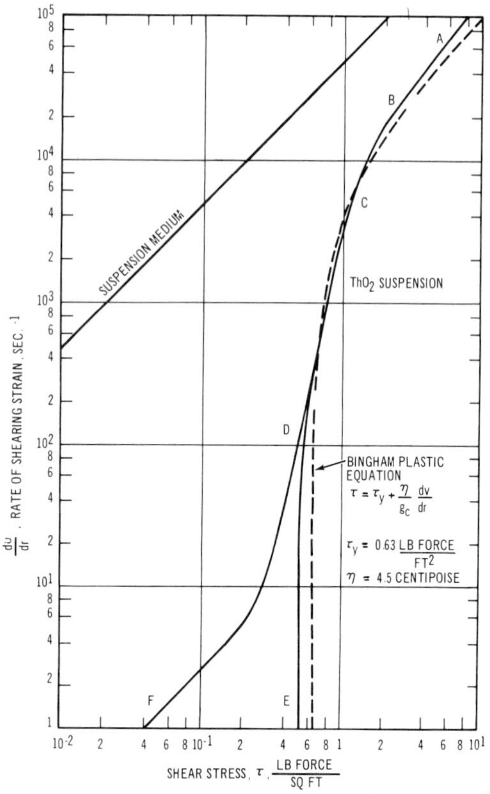

FIG. 2-23 Typical laminar flow shear diagram for ThO_2 suspension.

$$\begin{aligned} du/dr &= (g_c/\eta)(\tau - \tau_y), \quad (\tau > \tau_y), \\ du/dr &= 0 \qquad\qquad\quad (\tau \leqq \tau_y). \end{aligned} \qquad (2\text{-}51)$$

In Eq. (2-51), τ_y is the yield stress, and η is the coefficient of rigidity of the slurry. Other shear stress relations of the power law type [76,77] or of a more generalized nature have been proposed [78,79]. Equation (2-51) has, however, been found to approximately fit the data for uranium oxide and thoria slurries. This is especially true, as shown in Fig. 2-23, at the high shear rates that prevail near the wall of the channel. Departure of the slurry shear curves from Eq. (2-51) is noted at low shear rates and particularly at high concentrations of the solid particles. This departure is not serious as it only affects the low shear zone or central region of the channel. This region plays a small role in determining the pressure drop, and the pressure loss of uranium and thoria slurries can be predicted accurately from Eq. (2-51).

The yield stress τ_y and coefficient of rigidity η have been found to depend upon the solid particles' diameter, their shape, and the volume fraction they occupy. Thomas [76] proposed that

$$\begin{aligned} \tau_y &= (K_1\,\alpha^3/D_p^2)\exp\{0.7\,[(F/F_o - 1)]\}, \\ \eta &= \mu_f \exp[2.5 + (14/\sqrt{D_p}\,\sqrt{F/F_o})]\,\alpha. \end{aligned} \qquad (2\text{-}52)$$

In Eq. (2-52) D_p is the particle diameter, μ_f is the viscosity of the suspending medium, and F/F_o is the platelike particle surface area divided by the equiaxial particle surface area. The constant K_1 usually has the value of 2.27×10^{-9} lb-force (1.01×10^{-3} dyne), but can vary from 1.00 to 3.50×10^{-9} lb-force (0.445 to 1.56×10^{-3} dyne) due to electrolyte effects in thoria slurries. Furthermore, for small particles of interest to reactor design, the ratio F/F_o tends to approach one.

With τ_y and η defined, it is possible to integrate Eq. (2-51) to obtain the laminar friction loss. For flow in cylindrical tubes, Buckingham [80] obtained

$$8G/\rho g_c D = (\tau_w/\eta)[1 - (4\tau_y/3\tau_w) + (1/3)(\tau_y/\tau_w)^4] . \tag{2-53}$$

Since the wall shear stress τ_w is equal to

$$\tau_w = (D/4)(dP_f/dL) , \tag{2-54}$$

an effective viscosity μ_e can be defined for the slurry so that

$$\mu_e = \eta/[1 - (4\tau_y/3\tau_w) + (1/3)(\tau_y/\tau_w)^4] . \tag{2-55}$$

Substitution of Eq. (2-55) into Eq. (2-53) reduces it to the usual Newtonian equation and the slurry laminar frictional pressure drop can be calculated from the laminar friction factor curve shown in Fig. 2-2 if the effective viscosity is used in the definition of the Reynolds number. In other terms, $f = 64/(GD/\mu_e)$. This method has been found capable of correlating slurry data within ±17%. It is illustrated on Fig. 2-24 [74] where a "pseudo shear" diagram is plotted showing the parameter $8G/\rho g_c D$ versus the wall shear stress τ_w. The slope of the straight lines at low shear values corresponds to the effective viscosity μ_e. The main deficiency of this technique has been found to be its inability to account accurately for hydraulic diameter effects.

An alternate method of correlating laminar friction loss in slurry flow is to use the Reynolds number based on the limiting viscosity, η [81]. The friction factor f is then equal to

$$1/(GD/\eta) = (f/64) - (N_{He}/6)/(GD/\eta)^2 + (64N_{He}^4/3f^3)/(GD/\eta)^8 . \tag{2-56}$$

In Eq. (2-56), N_{He} is the Hedstrom number defined by

$$N_{He} = g_c \rho \tau_y D^2/\eta^2 . \tag{2-57}$$

Typical friction factor data obtained with an aqueous thorium oxide slurry are shown in Fig. 2-25 [74]. The test results in the laminar region agree satisfactorily with the prediction of Eq. (2-56).

Examination of Fig. 2-25 shows that the laminar zone for slurries extends beyond the accepted Reynolds number range of Newtonian fluids. It has been found that transition to turbulent flow occurs at Reynolds numbers, based upon the effective viscosity μ_e, of between 2000 and 6000. The transi-

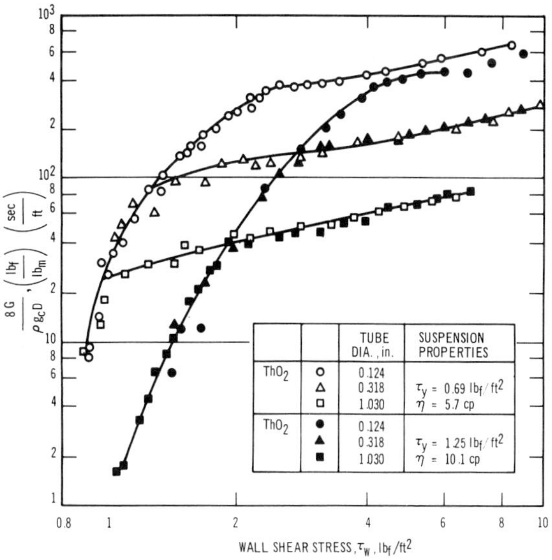

FIG. 2-24 Psuedo shear diagram for ThO_2 suspensions.

tion point is readily identified on Figs. 2-24 and 2-25. The sharp break in slope in the plot of $8G/\rho g_c D$ versus τ_w on Fig. 2-24 at various pipe diameters indicates the start of turbulent flow. Similarly, the intersection of the laminar friction factor curves at constant Hedstrom number with the turbulent curve gives another indication of the transition point on Fig. 2-25. Such an intersection yields an accurate method of calculating the transition Reynolds number if the turbulent friction factor curve is known.

As shown in Fig. 2-25, the turbulent friction factor can be approximately predicted from the single-phase correlation if the coefficient of rigidity η is used instead of the viscosity μ. More accurate expressions for the turbulent friction factor f have been proposed by Thomas [82]. They are

$$f = B(GD/\eta)^{-b} ,$$
$$B = 0.079\ [(\mu_f/\eta)^{0.48} + (g_c\rho\tau_y/\mu_f^2)\ 3.6 \times 10^{-11}]^2 ,$$
$$b = 0.25\ \ [(\mu_f/\eta)^{0.15} + (g_c\rho\tau_y/\mu_f^2)\ 3.6 \times 10^{-11}]^2 . \tag{2-58}$$

Equations (2-58) correlate the available data within ± 10%.

With the friction losses defined from Eqs. (2-53), (2-46), and (2-58), the piping pressure drop can be obtained from Eq. (2-3). For a liquid slurry, the density change and acceleration losses are small and can be neglected. The hydrostatic head loss is computed as for single-phase flow except that the slurry density is calculated from the homogeneous model.

2.2.4.3 Other Fluid-Solid Pressure Losses or Gains. Detailed experimental data for slurries for all other conditions considered in single-phase

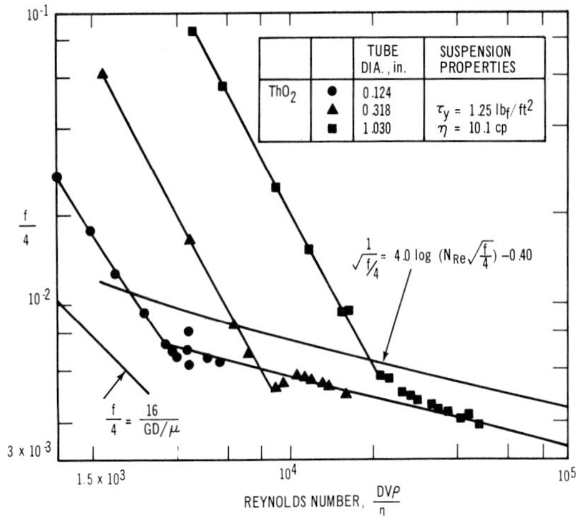

FIG. 2-25 Friction factor data for ThO_2 suspensions.

flow are lacking. The methods developed for piping flow can, however, be readily extended to predict the required pressure drop losses or gains. The curves developed in Secs. 2.2.1 and 2.2.2 are applicable if the homogeneous slurry density is used instead of the single-phase density. In those equations or figures which require a Reynolds number, the slurry Reynolds number GD/μ_e should be substituted for laminar flow and the grouping GD/η used for turbulent flow.

2.3 Design Safety Consideration

This section deals with potential design safety problems arising from the calculation of the steady state recirculation flow rate. One problem that should be considered is the accuracy with which the flow rate can be calculated. Another problem that deserves close scrutiny is the possibility of a reduction in flow rate from unexpected sources. For instance, pump cavitation and vapor carry-under (see Sec. 2.3.3) must be analyzed to show

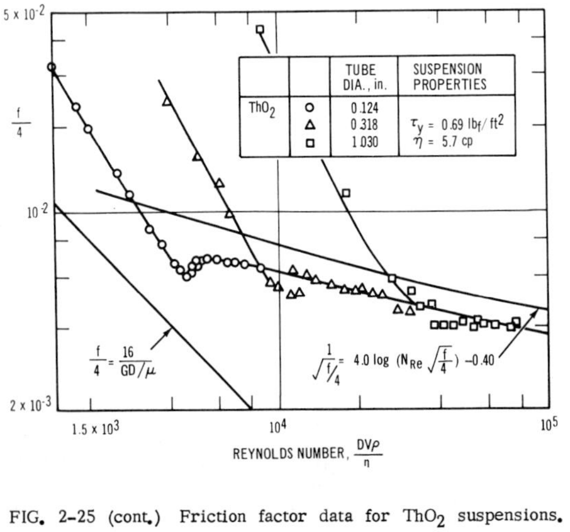

FIG. 2-25 (cont.) Friction factor data for ThO_2 suspensions.

that they do not adversely affect the design predictions. A third source of potential problems includes processes which have safety implications and which depend on the recirculation flow rate. In this group, one must consider the possibility of liquid carryover (see Sec. 2.3.4), mass transfer, corrosion, and erosion and their relation to the design of the recirculation system.

2.3.1 Accuracy of Calculations

As indicated in Sec. 2.2, the calculation of total recirculation flow rate depends on the prediction of a series of pressure losses and gains. The degree of accuracy with which each of these terms can be computed not only varies, but also changes with the type of fluid under investigation. Table 2-2 gives the author's estimate of the possible errors for the four types of fluids previously covered. The errors are listed for each pressure term considered and an overall estimate of the accuracy is given. It should be realized that the values given in Table 2-2 are only approximate and can be expected to change with each nuclear plant design.

Examination of Table 2-2 reveals that the accuracy of pressure drop calculations decreases rapidly as the geometry complexity increases or as a second phase is introduced into the flowing stream. In spite of the large errors associated with some of the pressure terms, the operating point along the pump curve can be selected so as to reduce the effects of inaccuracies in pressure drop calculations. Furthermore, in most practical designs, appropriate conservatism is incorporated in the calculations to insure that the flow rate does not fall below the design value.

2.3.2 Pump Cavitation

Cavitation occurs in a pump whenever the local pressure falls below the vapor pressure corresponding to the fluid temperature at that point. Cavitation is accompanied by the formation of vapor bubbles in the moving stream and can result in serious pump vibration, loss of performance, and pitting and failure of pump materials. The main causes of pump cavitation are too great a suction lift, too high velocities within the pump, and fluid temperatures too close to saturation values. The occurrence of cavitation is usually defined in terms of a cavitation constant θ introduced by Thoma [83,84]. The start of cavitation can be specified from a pressure balance, or

$$H_a + H_s = H_v + H_L + H\theta . \qquad (2\text{-}59)$$

In Eq. (2-59), H_a is the absolute pressure in feet of fluid prevailing at the surface of the pump suction supply; H_s is the static head in the suction vessel above the pump centerline (H_s is negative for a suction lift); H_v is the vapor pressure at the fluid temperature in the pump; H_L is the head loss in the pipe and impeller approach; and H is the pump pressure rise or head. The cavitation constant θ is determined experimentally and is particularly dependent upon the pump specific speed.

The specific speed of a pump N_s is defined as the speed in revolutions per minute at which the pump will develop a head H of 1 ft (0.305m) at a capacity of 1 gpm ($63.1 cm^3/sec$). It is given by the formula

$$N_s = N \sqrt{Q}/H^{3/4} , \qquad (2\text{-}60)$$

where N is the pump speed in rpm and Q the volumetric flow rate in gpm.

Typical values of the cavitation constant θ versus pump specific speed are shown in Fig. 2-26 for single and double suction pumps. Figure 2-26 [29] shows a band of values obtained from the Hydraulic Institute charts. The higher points in these bands correspond to the lower head pumps.

The use of Eq. (2-59) combined with the data of Fig. 2-26 makes it possible to specify the minimum required pump suction head to avoid cavitation. In nuclear plant design, it is also important to verify that cavitation conditions do not develop during system pressure and fluid temperature variations resulting from transient operation of the plant.

While the above discussion has dealt with the occurrence of cavitation in a pump, it should be noted that cavitation can develop in any high velocity region of the recirculation system. If at any time the local pressure falls below the saturation pressure due to high fluid velocity, bubbles will be generated and their collapse due to subsequent pressure increases can produce noise and vibrations harmful to the plant equipment. Conditions for cavitation in butterfly valves are, for instance, discussed in reference [85]. Such conditions are readily avoided by calculating or measuring the local pressure distribution in the flow loops and maintaining it above the corresponding saturation pressure.

2.3.3 Vapor Carryunder

In liquid vapor flow systems, separation of the vapor from the liquid is often necessary to utilize the vapor in the generating of power. Perfect separation is, however, not always possible and the liquid leaving the separating plane often contains vapor. In a forced circulation plant, the presence of vapor or "vapor carryunder" in the liquid returning to the pump aggravates the problem of cavitation covered in Sec. 2.3.2. In a natural circulation reactor, the problem becomes even more acute since the presence of vapor bubbles in the return or downcomer lines seriously reduces the available natural convection driving head. Carryunder has been reported in many natural circulation reactors (EBWR, Elk River, and Humboldt Bay). In most reactor applications, the problem of carryunder is partly alleviated by utilizing the returning and cooler feed fluid to collapse vapor bubbles that may be present in the downcomer. A recent analysis by Bankoff [86] which attempts to predict the rate of vapor bubble collapse in downcomer geometries is worth noting.

In many instances, carryunder can still prevail, and an accurate knowledge of its magnitude is required to assure satisfactory operation of the plant. Two cases will be examined: cooling by natural circulation coupled with gravity separation of the fluid and vapor; and forced coolant circulation combined with high performance steam separators.

Extensive studies of carryunder phenomena for air-water and steam-water mixtures have been performed by Petrick [48] in natural circulation loops. His proposed dimensionless correlation is shown on Fig. 2-27 [48]. A satisfactory correlation of all available carryunder data, including

TABLE 2-2

Accuracy of Pressure Drop and Total Recirculation Flow Rate Calculations

Pressure drop components	Single-phase flow incompressible	Single-phase flow compressible	Liquid-vapor flow	Fluid-solid flow
A. Piping pressure loss				
1. Hydrostatic term	1/2%	1/2%	20%	5%
2. Acceleration term	Negligible	Negligible	20%	5%
3. Friction term	7%	9%	30%	15%
B. Contraction-expansion loss	15%	17%	40%	20%
C. Bend loss	20%	22%	50%	30%
D. Valve and fitting losses	20%	22%	50%	30%
E. Heat exchanger losses	12%	20%	20%	20%
F. Total pressure losses (A to E)	10%	12%	30%	20%
G. Pump pressure rise	3%	3%	3%	3%
H. Total recirculation flow	5%	7%	10%*	10%

*Makes allowance for part of the recirculation loop being single phase.

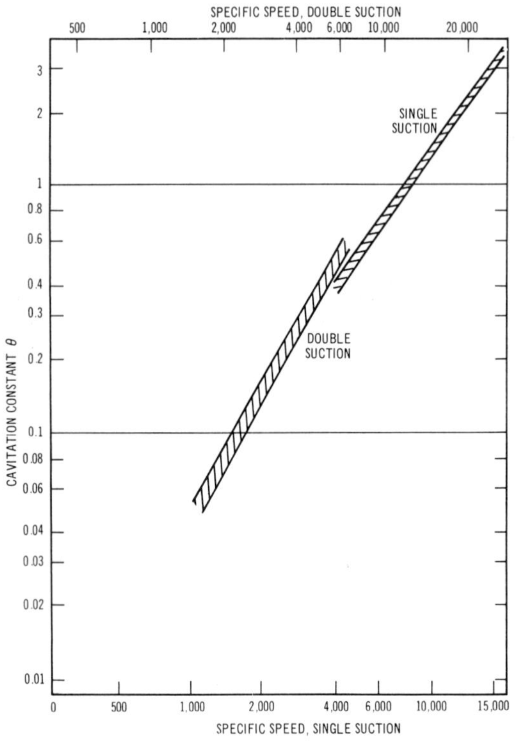

FIG. 2-26 Cavitation constant versus pump specific speed.

some EBWR test points, is obtained in Fig. 2-27. The correlation proposed by Petrick involves two main groupings. The first grouping X_D/X_R gives the ratio of vapor weight fraction in the downcomer to that in the riser. The second grouping ζ is equal to

$$\zeta = (V_{GR}/V_{GD})\left[(\sigma g_c)^{2/3} \Big/ G_L^2 \mu_L\right]\sqrt{\rho_L/\rho_G} \times \left\{\sqrt{(D_R/h_i)} + \sqrt{(h_i/D_R)}\right\} . \qquad (2\text{-}61)$$

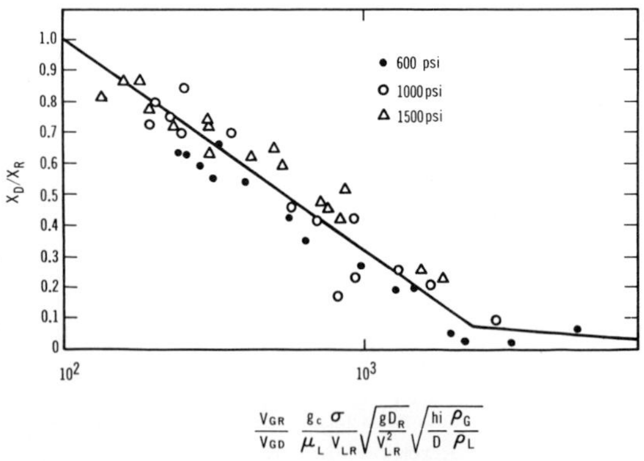

FIG. 2-27 Correlation of steam-water carryunder data.

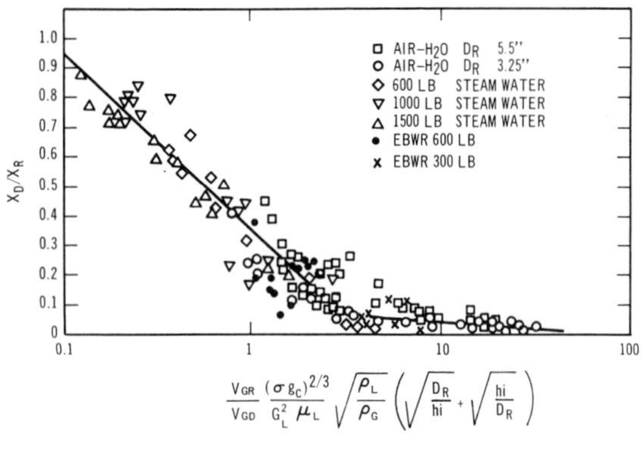

FIG. 2-27 (cont.) Correlation of carryunder data.

The terms V_{GR} and V_{GD} represent the actual vapor velocity in the riser and downcomer and are calculated from the methods described in Sec. 2.2.3.2; σ is the surface tension of the fluid; G_L is the mass flow rate of liquid in the riser; D_R is the riser diameter; and h_i is the actual interface height above the top of the riser. A preliminary criterion suggested by Petrick to avoid carryunder is to fix ζ at a value of 3 or greater. A more recent investigation by Miller and Armstrong [87] shows that this criterion can be relaxed by installing an inverted cone made up of mesh screen atop the riser.

In forced circulation systems utilizing mechanical vapor separators, the carryunder is much smaller. Yet carryunder has been reported in some high performance separators [88]. The carryunder can be expected to vary with the design configuration and the requirements of the vapor separator. No general correlation has been developed to date, and prototype tests are usually performed to measure the carryunder [88]. In such mechanical separators, the residual carryunder is held below that amount which can be suppressed by cold feed addition to the liquid leaving the separator.

2.3.4 Liquid Carryover

Another problem in liquid-vapor separation is the possibility of carrying liquid droplets in the vapor after it leaves the separator. The presence of such droplets can generate hammers in the flow channels as well as cause damage to rapidly rotating components such as turbine blades. In gravity separation, carryover occurs due to the ejection of water drops in the vapor dome by bursting vapor bubbles and the transport of such drops by excessive vapor velocities above the separation plane. In mechanical separators, carryover is caused by vortices and very high vapor velocities shearing away and entraining liquid from the liquid film.

Typical entrainment data for gravity separation of steam-water mixtures in a 19-inch (48.3 cm) diameter test section is shown in Fig. 2-28 [88]. Entrainment increases sharply whenever the steam flow rate exceeds a prescribed value. An extensive

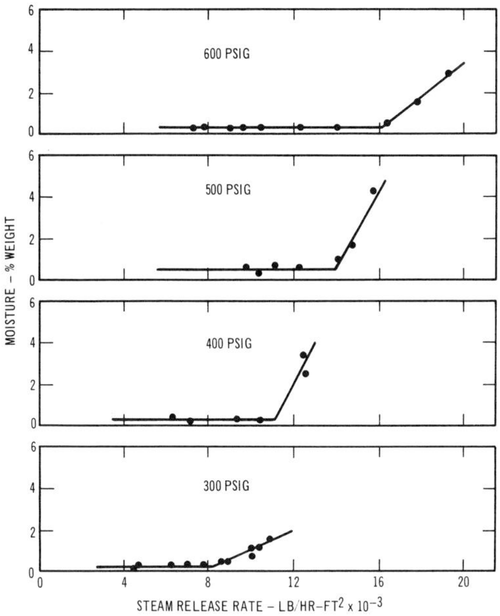

FIG. 2-28 Moisture entrained 33 inches (84 cm) above interface in a 19-inch (48 cm) diameter test section for gravity separation of steam-water.

literature survey of gravity separation data and proposed empirical equations is presented in reference [89]. The most successful correlation of test results was developed by Sterman [89]. According to Sterman, the vapor entrainment or liquid weight fraction in the vapor leaving the interface can be predicted approximately from

$$\mathrm{ENT} = 2.75 \times 10^{8} \left(\frac{V_{GV}^{2}}{\alpha_i g h_v}\right)^{2.3} \left(\frac{\rho_G}{\rho_L - \rho_G}\right)^{-0.25} \times \left(\frac{g \rho_L^{2} (\sigma g_c)^{3/2}}{g^{3/2} (\rho_L - \rho_G)^{3/2} \mu_L^{2}}\right)^{-1.1} . \tag{2-62}$$

In Eq. (2-62), V_{GV} is the superficial gas velocity in the vapor dome (obtained by dividing the total gas volumetric flow by the total flow area) and h_V is the height above the separating interface. The void fraction α_i represents the volume fraction occupied by the vapor below the interface and can be calculated from the data presented in Sec. 2.2.3.2.

In mechanical separators, liquid entrainment is due to liquid vapor interface instability and excessive shear forces at that plane [50,90]. The occurrence of excessive entrainment can be expected to depend upon the separator design and the operating conditions. Idealized predictions offered in references [50] and [90] compare favorably with the correlation proposed by Mozharov [91]. Mozharov predicts that the maximum allowable steam flow rate through a separator is

$$\frac{\rho_G V_{GS}^{2}}{g(\rho_L - \rho_G) D_e} \propto \left(\frac{\rho_G V_{GS} D_e}{\mu_G}\right)^{5/7} \left(\frac{\mu_L}{4\Gamma_L}\right)^{3/7} \times \left(\frac{\sigma g_c}{g(\rho_L - \rho_G) D_e^{2}}\right)^{6/7} . \tag{2-63}$$

In Eq. (2-63), V_{GS} is the vapor velocity through the separator, D_e is the equivalent separator diameter, and Γ_L is the liquid flow rate per unit length of liquid-wetted perimeter.

2.3.5 Other Safety Design Considerations

Flow rate plays a role in many processes with potential safety implications. A brief description of typical problem areas will be given here with special emphasis placed upon the effects of flow rate.

Consider first the problem of mass transfer in a non-isothermal circulating system. In such a system, solution or diffusion of particles from the solid surface to the coolant can take place in the high temperature zones of the loop. The dissolved material is transported by the coolant to the cooler areas where it precipitates on the surfaces. This type of transfer can weaken if not fail structural materials in the high temperature regions. It can also lead to flow blockage in those places where the precipitated material accumulates. Furthermore, if the solute material being transported is radioactive, shielding and maintenance problems can result. The role of flow rate and coolant velocity in this mass transfer process has been clearly established. The rate of dissolution of material into the coolant is directly dependent upon the velocity. Experimental data have been satisfactorily correlated using the analogy between mass and heat transfer [92,93]. For instance, for turbulent flow, the rate of solution is proportional to the flow velocity raised to the 0.8 power. Furthermore, reference [93] states that the rate of precipitation, even though not completely understood, varies with velocity.

Similarly, the rate of surface erosion depends on the velocity in the recirculation system. High velocities produce large kinetic and impact forces and erode baffles, sharp turn surfaces, valves and pipe bends. The effects of erosion are identical to those of mass transfer. Activated particles are transported and deposited where they can produce maintenance, shielding and flow blockage problems. The rate of erosion not only increases with the velocity and density of the fluid, but it also depends upon the material under attack. In most designs, limits are imposed upon local flow velocities to avoid excessive erosion. For instance, the linear velocity of bismuth is held below about 10 ft/sec (~3 m/sec), due to its high density [94]. Limiting velocities for water or sodium are about 25 to 40 ft/sec (7.6 to 12.2 m/sec). For steam, the velocity is usually kept below 200 to 300 ft/sec (61 to 90m/sec). For thoria slurries, the erosion of stainless steel was found negligible for velocities below 20 to 25 ft/sec (6.1 to 7.6

m/sec). The attack, in the latter case, was also found to depend upon the method of slurry manufacture and the reagents added to the suspending medium [73]. Most of the above limits are obtained from erosion tests. They are flexible and vary with the particular application.

Another velocity-dependent process is the chemical corrosion of structural materials by the recirculating coolant. Different corrosion rates have been obtained in static and dynamic studies of corrosion in nuclear plants. Another aspect of the interplay between flow and corrosion is the deposition of corroded particles. Corroded particles accumulate in "dead flow" zones, and the resulting activation can be minimized by appropriate design and layout of the recirculating loops.

2.4 Operating Safety Considerations

The important consideration here is the measurement of flow rate and its comparison with the predicted value. Various methods of measuring flow rates are being used in nuclear plants. They range from direct reading of the flow to indirect calculations from other variables.

Direct measurements of flow are covered in detail in classical fluid flow books [1,2]. Design layouts and calibration curves for such devices as orifices, nozzles, and Venturi meters are given, for example, in "Fluid Meters—Their Theory and Application" [95], or "Flow Measurements of Instruments and Apparatus" [96]. The use of any of these techniques will give an accurate measurement of flow for comparison with the calculated value.

Several other and more indirect methods of measuring flow rate have also been used. The following are worth noting:

(a) Determination of the curve of pump head versus flow in a test loop and measurement of the plant-installed pump head to obtain the flow.
(b) Measurements of pressure drop in geometries for which predictions are available and calculations of the flow from the pressure drop data.
(c) Use of heat balance. A knowledge of the heat input together with test data or fluid enthalpy change gives the flow.
(d) Use of electromagnetic flow meters.
(e) Decay of N^{16} activity between two detector stations and relating the decay to the seven-second half-life of N^{16}.

The accuracy of all the above methods varies. It can range from ± 1% for orifices and nozzles to ± 5 to 10% for the other techniques.

3 STEADY-STATE FLOW DISTRIBUTION WITHIN REACTOR

3.1 Basic Approach

The coolant flow entering the reactor core can take a multitude of parallel flow paths. The flow will subdivide itself among the paths so that the pressure drop is the same across all the paths and the sum of all the flows equals the inlet flow. The flow distribution within the reactor core is computed in two steps: first, the flow in each parallel path is established for an assumed pressure drop across the path; next, the flow rates in all the paths are added to determine the total flow. The correct flow distribution is obtained when the calculated total flow rate by a series of iterations is found to be the same as the available flow. A most important element in these calculations is the prediction of pressure drop versus flow rate in each of the parallel paths.

The most common flow path in a nuclear reactor is a fuel element or assembly, a large number of which are arranged in parallel to form the reactor core. A typical fuel assembly is shown in Fig. 3-1, and the pressure drop across it can be broken down into the following components:

$$\Delta P = \Delta P_{\text{friction}} + \Delta P_{\text{hydrostatic}} + \Delta P_{\text{acceleration}} + \Delta P_{\text{local losses}} . \qquad (3\text{-}1)$$

For the geometry illustrated in Fig. 3-1, the local pressure losses include an inlet, an outlet, a bottom tie plate, an upper tie plate, and a finite number of fuel spacer losses. In many respects, Eq. (3-1) is similar to Eq. (2-2) and many of the methods and solutions described in Sec. 2 can be used. One important difference is that, due to heat addition along the fuel assembly, the fluid properties and enthalpy vary in the flow direction. For

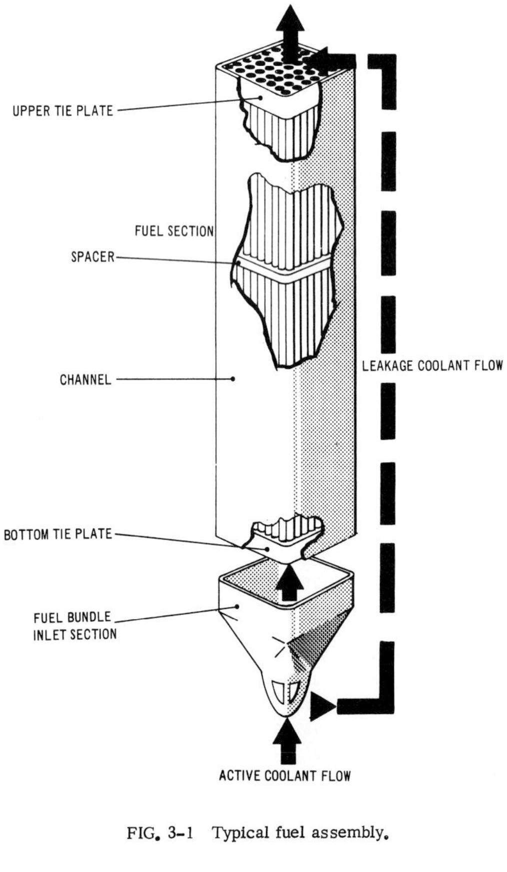

FIG. 3-1 Typical fuel assembly.

this reason, the calculation of flow rate needs to be performed simultaneously with a thermal or heat transfer analysis. Furthermore, the fluid properties and enthalpy change so rapidly that the computations must take into account these changes at frequent intervals or at a large number of "nodes" or points in the flow direction. In most reactors, the heat production is different in each fuel assembly, and the calculations have to be repeated for as many fuel elements as required by symmetry to cover the entire core. In addition, it is necessary to carry out calculations for such leakage flow paths as those around the fuel elements or past the control blades.

The number of flow paths and nodes that must be considered within the reactor core is usually so large and the number of required iterations so great that the calculations are very lengthy and tedious. Fortunately, machine codes have been developed to facilitate the computation of the flow distribution. For instance, STDY-3 [97] and COFFI [98] codes predict the flow distribution in a water cooled reactor core with and without steam generation. Similarly, the HECTIC [99] code predicts the flow distribution in a gas-cooled reactor.

We shall not concern ourselves here with the structure and details of available computer codes, but rather with the accuracy of the calculations and the pressure drop relations on which they are based. The available equations and test results will be presented in four groups corresponding to the four different types of fluid considered in Sec. 2.

3.2 Methods of Calculations

The calculation methods are essentially those presented in Sec. 2.2. For instance, the flow leakage between fuel assemblies can be computed using the information presented in Sec. 2.2. Only in fuel assemblies do the computations become different, and the reason is that some nuclear fuel elements use special flow geometries not covered in the previous sections. We shall deal here with the pressure drop through such special geometries and reserve for Sec. 3.3 any discussion of the overall accuracy of flow distribution calculations.

3.2.1 Incompressible Single-Phase Flow

Fuel geometries used to date with incompressible single-phase fluids have been of the circular, rectangular, annular, and multirod type and variations thereof. Typical designs are illustrated in Fig. 3-2. In the case of multirod fuel elements, a variety of spacers have been utilized to keep the fuel rods apart and some of them are reproduced in Fig. 3-2.

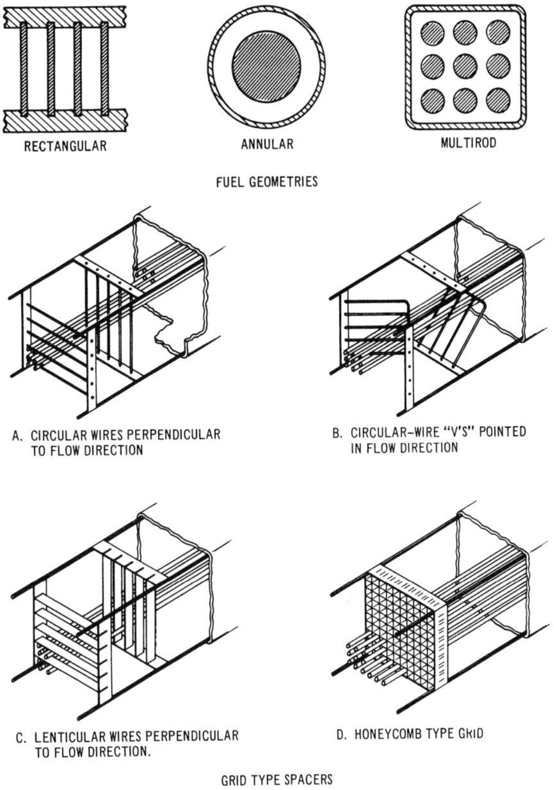

FIG. 3-2 Typical fuel cell geometries.

The pressure drop in the fuel elements shown in Fig. 3-2 is made up of hydrostatic, acceleration, inlet, exit, frictional, spacer, and tie plate pressure losses. The hydrostatic head loss and the acceleration pressure drop are obtained from the methods described in Sec. 2.2.1. Similarly, the inlet and exit losses are calculated from the data and curves presented in Figs. 2-3 and 2-4.

The frictional pressure loss can be predicted from Figs. 2-1 and 2-2 if an appropriate equivalent diameter is used. The effectiveness of this approach is shown in Fig. 3-3 [100]. Figure 3-3 is a plot of experimental data obtained by Hartnett, Koh, and McComas [100] in rectangular ducts of various aspect ratios (the aspect ratio is obtained by dividing the long side of the rectangle by the small side). Shown on the same figure are the theoretical predictions of Deissler and Taylor [101] in the turbulent regime and those of Knudson and Katz [102] in the laminar region. As noted in reference [100], good agreement is obtained in the turbulent regime between the test data and the smooth channel curve of Fig. 2-1, using the equivalent hydraulic diameter. Measurements of friction factor in rectangular channels 1.0 by 0.097 in. (2.54 by 0.246cm) and 1.0 by 0.050 in. (2.54 by 0.127 cm) have also been reported by W. H. Esselman et al. [103], and are plotted in Fig. 3-4 [10]. Good correlation is noted with the curve of Fig. 2-1 for smooth configurations; for the rough channels, the friction factors fall slightly below those of Fig. 2-1 and are about 20 percent low at a Reynolds number of 10^5. Shown on the same figure are some recent experimental data obtained by Janssen and Kervinen [68,69] in a 0.5 by 1.75 in. (1.27 by 4.44 cm) and a 0.25 by 1.75 in. (0.635 by 4.44 cm) rectangular channel. Their test results fall about 10 percent below the smooth channel curve of Moody.

Similar trends have been obtained with other noncircular geometries. The agreement with the Moody curves is within ± 10 percent. Test data

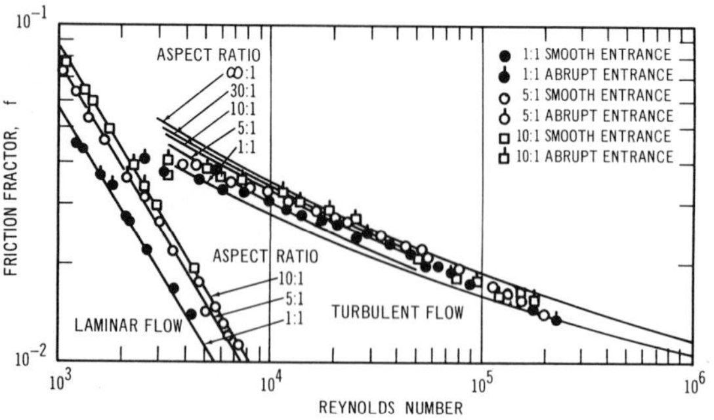

FIG. 3-3 Comparison of experimental and predicted friction factors in rectangular channels.

obtained by Deisler and Taylor [101] for flow past various bundles of rods are shown in Fig. 3-5 [10]. Their experimental results with a triangular array having a rod-clearance-to-diameter ratio of 0.12 fall about 3 percent below the Moody smooth curve. Their square rod pattern data with a 0.12 and 0.2 rod-clearance-to-diameter ratio are about 10 percent below the Moody curve. Plotted on the same figure are test results obtained for the Maritime Gas-Cooled Reactor, EBRII, Yankee, and three configurations of the Fermi reactor [10]. The friction factor values for the latter configurations are well above the curves of Fig. 2-1 and the reason is that the points plotted in Fig. 3-5 include both the friction and spacer loss along the fuel assembly.

A more realistic way of calculating the pressure drop in the latter configurations would be to compute the frictional loss along the fuel rods without spacers and to add to it the spacer pressure drop. De Stordeur [104] expressed the spacer losses in terms of a drag coefficient C_S, or

$$\Delta P_s = \rho C_s V_s^2 s/2gA . \qquad (3-2)$$

In Eq. (3-2), C_S is the drag coefficient, V_S the velocity in the spacer region, A the unrestricted flow area away from the spacer, and s the spacer projected frontal area. A correlation of the drag coefficient C_S in terms of Reynolds number was proposed by De Stordeur for spiral wire spacers

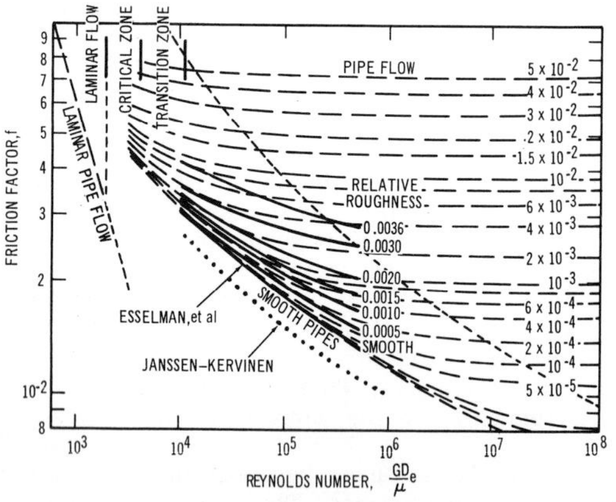

FIG. 3-4 Friction factor in rough rectangular channels.

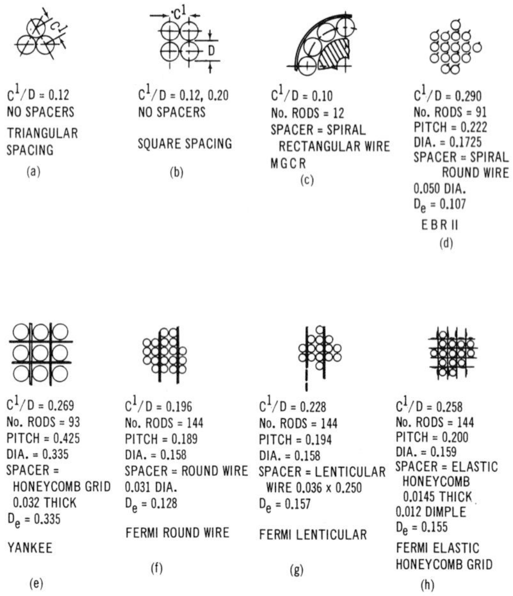

FIG. 3-5 Typical multirod designs.

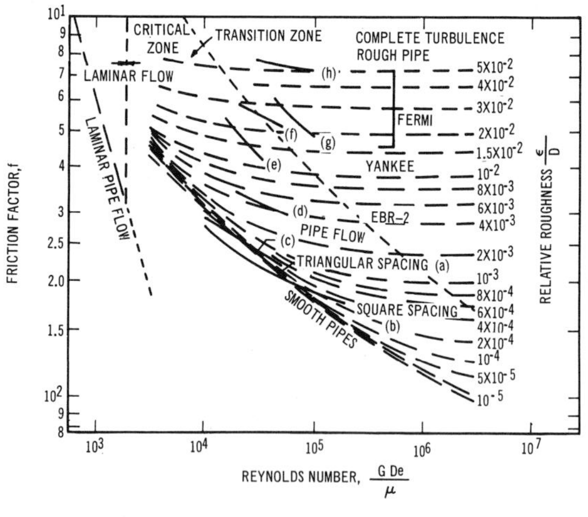

FIG. 3-5 (cont.) Friction factor in multirod designs.

and is shown in Fig. 3-6* [104]. In this instance, s was taken equal to the number of spiral turns over the bundle length times the area of an annulus of thickness equal to the spacer wire diameter. A mean curve was drawn through the experimental data and it correlated the test results within ±40 percent. De Stordeur also examined the drag coefficients for grid-type spacers. His correlations are reproduced on Fig. 3-7 [104]. The proposed drag coefficient curves correlate the test data within ±15 percent.

---

*Measurements just completed by E. D. Waters (U.S. Atomic Energy Commission Report HW-65173 Rev.) on large rods with wire wraps of different pitches fall considerably below the correlation of De Stordeur.

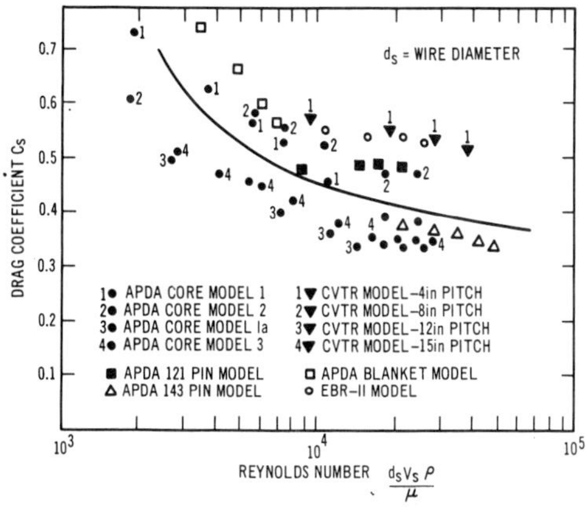

FIG. 3-6 Drag coefficients for spiral wire spacers.

The curves of Fig. 3-7 can be applied to predict the pressure loss across the bottom and upper tie plate when the tie plates are grid-type structures. In some fuel designs the tie plates are made up of solid castings with holes drilled between the fuel rods for the coolant to flow through. For such designs, the pressure drop is predicted by adding the contraction and expansion losses obtained from Figs. 2-3 and 2-4.

3.2.2 Compressible Single-Phase Flow

Several high-performance gas-cooled reactors with compressible flow utilize circular, rectangular, annular, and multirod geometries of the type shown in Fig. 3-2. The pressure drop in such fuel elements is obtained from the data given in Sec. 3.2.1 while the compressibility effects are accounted for by the methods prescribed in Sec. 2.2.2.

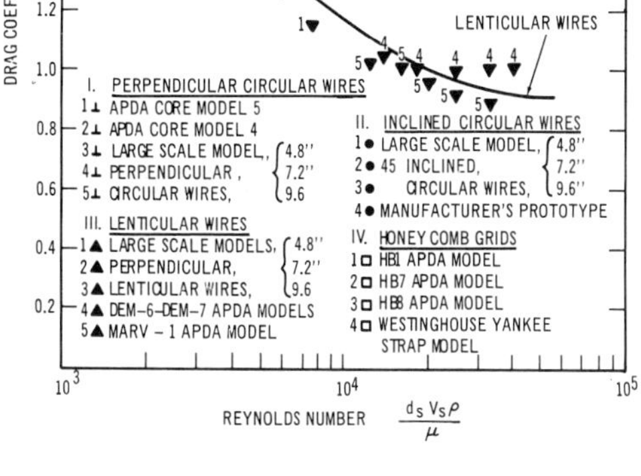

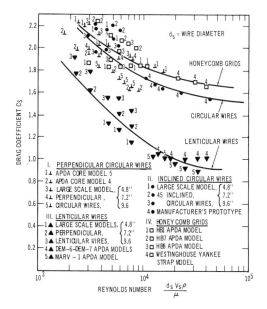

FIG. 3-7 Drag coefficients for transverse grid spacers.

The frictional pressure drop, inlet, exit, spacer, and tie plate losses can be calculated from Figs. 2-1 to 2-5 and Figs. 3-3 to 3-7. The effects of compressibility are estimated from Figs. 2-12 to 2-16.

Other gas-cooled reactors rely upon fuel elements with extended surfaces. Several such geometries have been used to date and some are sketched in Fig. 3-8. They include fuel elements with simple longitudinal, transverse or helical fins or more complex geometries formed from a combination of longitudinal fins and helical splitters (polyzonal axial) or helical fins and axial splitters (polyzonal helical). The calculation of pressure drop in the fuel geometries of Fig. 3-8 is based mostly upon experimental tests. Friction factor values for the transverse fins used in the Calder Hall reactor are given in reference [105]. Test results for polyzonal axial and polyzonal helical fuel elements are given in references [106], [107], and [108]. Particularly worthy of note is the correlation of Wise [106] for polyzonal axial configurations. He was able to correlate the data of references [107] and [108] by means of the relation,

$$(n_{sp}/fl)^{1/2} = 7.25 + 1.53\,(D_{CH} - D_r)/(D_F - D_r)\ , \qquad (3\text{-}3)$$

where n_{sp} is the number of splitters, l the lead of the helix, D_{CH} the outer flow channel diameter, D_F the fin tip diameter, and D_r the root diameter. The friction factor f in Eq. (3-3) is based upon the flow area and hydraulic diameter of the annulus between the fin tips and channels.

A late fuel geometry used in advanced gas-cooled reactors consists of circular rods equipped with very small, closely spaced transverse fins. The fins act as turbulence promoters and hamper the formation of a laminar sublayer close to the heater surface. A substantial increase in heat-transfer performance results but at the expense of an increased pressure drop. Typical pressure drop results for such geometries are presented by Burgoyne, Burnett, and Wilkie [109].

3.2.3 Liquid-Vapor Flow

Liquid-vapor-cooled reactors can have as many as three distinct regimes of flow. If the coolant entering the core is in the subcooled state, single-phase incompressible flow prevails and the associated pressure drop is obtained from the methods of Sec. 3.2.1. As the coolant continues to flow along the fuel elements, a point is reached where boiling starts at the fuel coolant interface. Vapor is then present in the flow channel even though the bulk fluid is still subcooled. The prediction of vapor volume fraction and two-phase pressure drop under subcooled boiling conditions has not been discussed previously and is covered in this section. Further along the fuel assembly, the fluid reaches saturation conditions and net vapor generation starts. Many of the methods advanced in Sec. 2.2.3 apply to this region except for the effects of heat addition which are treated here.

The three possible regimes of liquid vapor flow are shown in Fig. 3-9 [110] which is a plot of

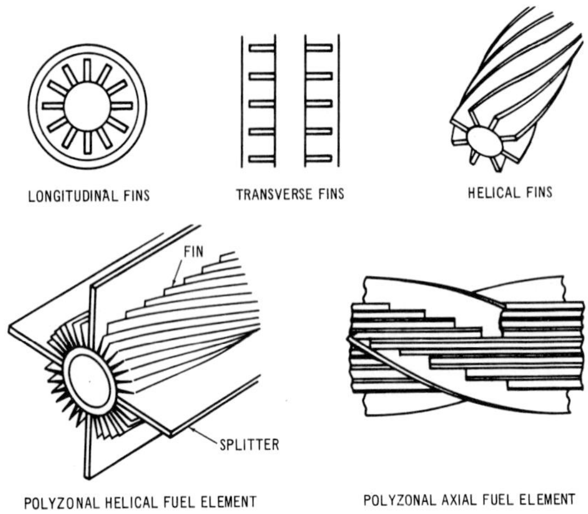

FIG. 3-8 Typical fuel elements with extended surfaces.

vapor volume fraction along a fuel assembly. Single-phase exists in the region OA. Subcooled boiling extends from point A to D, and liquid vapor flow prevails from point D to E. The vapor volume fraction in section DE can be predicted from the information given in Sec 2.2.3 and is shown as curve DFE. In the subcooled region, two distinct regions are shown as proposed by Bowring [110]. In Region I, labeled "highly subcooled", bubbles grow and collapse while sliding along the fuel element surface. In Region II, bubbles leave the heated surface and the amount of voids rises sharply as the bulk fluid temperature increases. Experimental data in both regions have been reported for only steam-water flow. Griffith et al. [111], Marchaterre et al. [111a], Egen et al. [112], Maurer [113], Foglia et al. [114], and Christensen [115] have measured the void fraction for subcooled water flowing vertically upwards.

Predictions of subcooled void fractions, based upon the available test results, have also been proposed for upward flow by Griffith et al, Maurer, and Bowring. The model of Griffith et al. is recommended in region I and that of Bowring for region II. In region I, the vapor volume or void fraction α is equal to

$$\alpha = P_H \delta / A , \qquad (3\text{-}4)$$

where P_H is the heated perimeter and A the flow area. The thickness of the vapor layer δ is the lesser of

$$\delta = 0.033\, D_b , \text{ or}$$

$$\delta = \frac{[\dot{q} - H_t(T_w - T_B)]\mu_B (C_p)_B}{1.07\, H_t^2 (T_{sat} - T_B)} , \qquad (3\text{-}5)$$

with D_b = steam bubble diameter (see Fig. 3-10 [110]) and

$$T_w = T_{sat} + [60\, (\dot{q}/10^6)^{1/4}]/e^{P/900} .$$

In the above equation, H_t is the local heat transfer coefficient obtained from standard correlations, and $\dot{q}$ is the local heat flux.

Region II starts when the temperature difference $\Theta = T_{sat} - T_B$ is the least of :

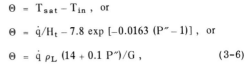

$$\Theta = T_{sat} - T_{in} , \text{ or}$$

$$\Theta = \dot{q}/H_t - 7.8 \exp[-0.0163\, (P'' - 1)] , \text{ or}$$

$$\Theta = \dot{q}\, \rho_L\, (14 + 0.1\, P'')/G , \qquad (3\text{-}6)$$

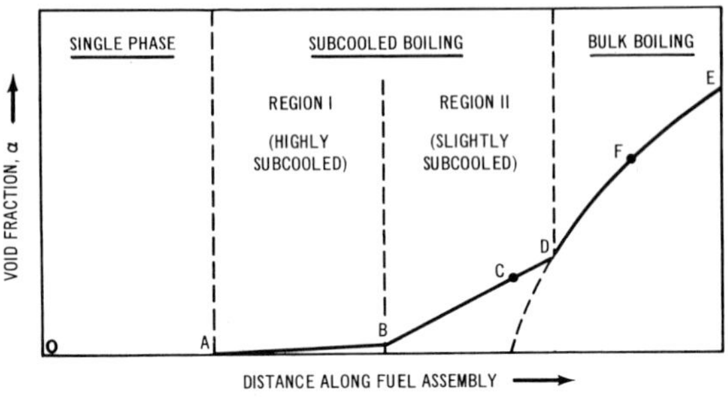

FIG. 3-9 Void fraction buildup along fuel assembly.

where P'' is the pressure in atmospheres. (This specifies B in Fig. 3-9.) The vapor volume fraction in Region II is obtained by defining a fictitious vapor weight fraction x at position C of Fig. 3-9. The value of x at point C is calculated from

$$x = (P/GAE_{fg}) \int_B^C [\dot{q} - H_t\, (T_w - T_{sat})]\, dz/(1 + B') ,$$

$$1 + B' = 1 + 3.2\, (\rho_L\, Cp_L/\rho_G\, E_{fg}) \quad \text{for } 1 < P'' < 9.5 ,$$

$$1 + B' = 2.3 \quad \text{for } P'' > 9.5 , \qquad (3\text{-}7)$$

where P'' is the pressure in atmospheres and E_{fg} is the heat of vaporization. Once x is calculated from Eq. (3-7), the slip ratio V_G/V_L is obtained from Fig. 2-18, and the corresponding void fraction is computed from Eq. (2-33).

Various measurements of frictional pressure drop in the subcooled boiling region have also been made. Reynolds [116], Buchberg et al. [117], Sher

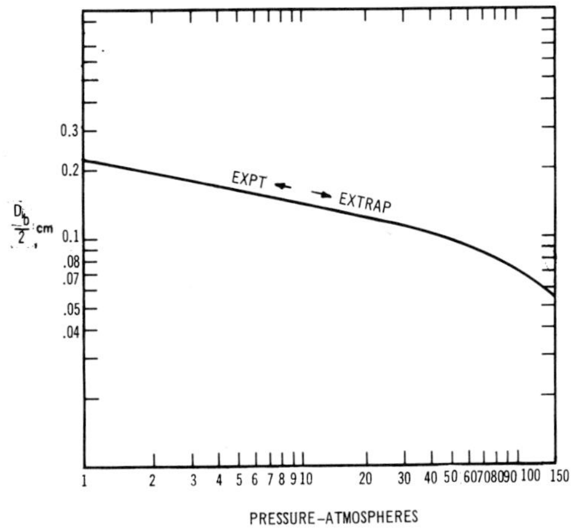

FIG. 3-10 Bubble diameter in steam-water mixtures.

[17], Weatherhead [118], and Jicha and Frank [119] have obtained data for subcooled water boiling at various pressures. In each of these investigations, correlations are given which are valid over the range of the reported experiments. No attempts have been made to date to explain the apparent discrepancies between the proposed correlations or to derive empirically an equation based upon a comprehensive review of the available data. It is recommended that Eq. (2-44) be combined with the prediction of void fraction from the Griffith and Bowring models to predict the two-phase frictional pressure drop for subcooled boiling conditions. This method has the advantage of using proven equations for the void fraction and of being applicable to a large variety of flow conditions and geometries.

The acceleration pressure loss in the subcooled boiling region is obtained from Eq. (2-41) while the inlet, exit, tie plate, and spacer pressure losses are computed from the methods described in Secs. 2.2.3.3 and 3.2.1. All calculations must be based upon the steam weight and volume fraction obtained from Eqs. (3-4) to (3-7).

In the net vapor generation region, DE of Fig. 3-9, the methods and results of Sec. 2.2.3 are generally valid. A few additional comments are in order, however, because of the heat addition along the flow channel.

1. Some investigators have reported that mass velocity has an effect upon the frictional pressure drop in upwards vertical flow with heat addition. The correlations proposed by Sher for steam-water in rectangular channels 0.097 by 1 in. (0.246 by 2.54 cm) and 0.050 by 1 in. (0.127 by 2.54 cm) at 2000 psia (141 kg/cm^2) are reproduced in Fig. 3-11 [17]. Shown in the same figure is a correction factor [120] that can be applied at other pressures to the steam-water curves of Martinelli-Nelson shown in Fig. 2-20 [120].

2. The role of heat addition upon the frictional pressure drop is not completely understood. The most extensive studies have been made by Becker [121] for steam-water mixtures at various pressures, and they show a small, if not negligible, effect of local heat flux upon two-phase pressure drop. Typical results are shown in Fig. 3-12 [121].

3. The enthalpy of the coolant, i.e., its vapor volume and weight fraction, change in the flow direction. The changes in coolant properties can be calculated from a heat balance along the flow channel. One must recognize, however, that the frictional and hydrostatic head loss from point D to point F of Fig. 3-9 must be obtained from an integration process. For instance, the average density $\bar{\rho}_{DF}$ from point D to F is equal to

$$\bar{\rho}_{DF} = \frac{1}{z_F - z_D} \int_D^F \rho \, dz \, . \qquad (3\text{-}8)$$

Similarly, the friction loss $(\Delta P_{TPF})_{DF}$ over the same distance is given by

$$(\Delta P_F)_{DF} = (\Delta P_{L_o})_{DF} \frac{1}{z_F - z_D} \int_D^F \frac{\Delta P_{TPF}}{\Delta P_{L_o}} \, dz. \qquad (3\text{-}9)$$

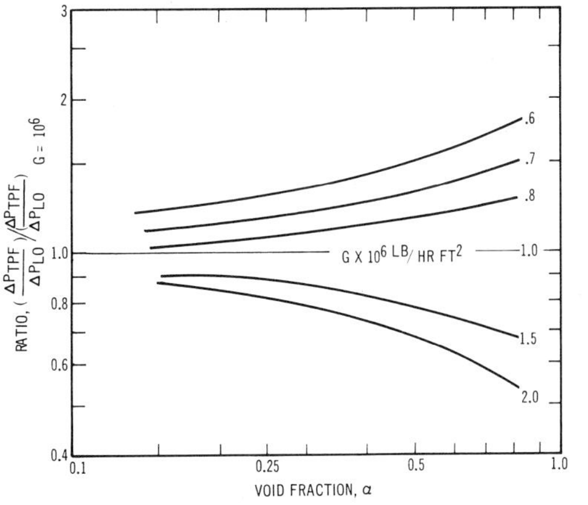

FIG. 3-11 (cont.) Effect of flow rate on steam-water frictional pressure drop.

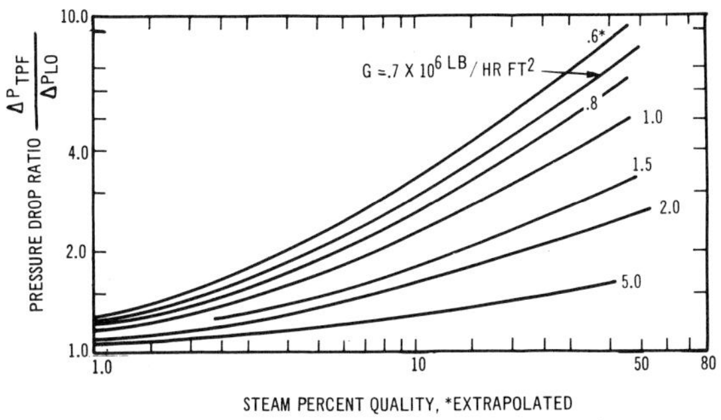

FIG. 3-11 Two-phase frictional pressure drop for steam-water at 2000 psia (141 kg/cm^2).

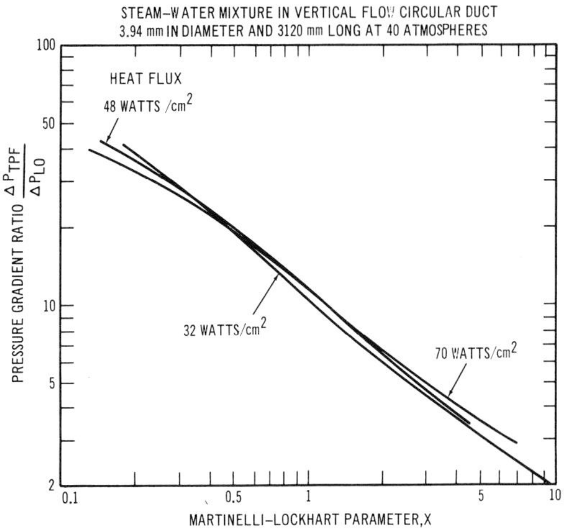

FIG. 3-12 Effect of heat flux on frictional pressure drop.

In Eq. (3-8), ρ is the local fluid density along the fuel assembly. It is obtained from the data given in Sec. 2.2.3.2. Similarly, the ratio ($\Delta P_{TPF}/\Delta P_{Lo}$) needs to be calculated at every position z of the flow channel from the curves of Fig. 2-20 or 2-21.

3.2.4 Liquid-Solid Flow

For slurry type reactors, the methods and results of Sec. 2.2.4 are valid. For pebble bed and fluidized bed reactors, the following correlations suggested by Lyons, Frank, and Scheve [122] are recommended for reactor core pressure drop calculations.

For a fixed bed or pebble bed reactor of length L the frictional pressure loss is calculated from:

$$\Delta P_f = 2fG_s^2 L (1 - \Lambda^2)/g D_p \rho_f \Lambda^3 ;$$

$$f = 100/N_{RE} \text{ (for } N_{RE} < 10) ,$$

$$\Delta P_f = 2fG_s^2 L (1 - \Lambda)/g D_p \rho_f \Lambda^3 ;$$

$$f = 1.75/N_{RE} \text{ (for } N_{RE} > 300) ,$$

$$\Delta P_f = 2fG_s^2 L (1 - \Lambda)/g D_p \rho_f \Lambda^3 ;$$

$$f = [100 (1 - \Lambda)(300 - N_{RE})/290 N_{RE}] + (1.75/N_{RE}^{0.1})[1 - (300 - N_{RE})/290]$$

$$\text{(for } 10 < N_{RE} < 300) . \qquad (3\text{-}10)$$

In the above equations, D_p is the fuel pellet diameter, ρ_f is the fluid density, Λ is the porosity, and N_{RE} is the Reynolds number defined as

$$N_{RE} = \rho_f U_s D_p/\mu = G_s D_p/\mu , \qquad (3\text{-}11)$$

where U_s is the superficial fluid velocity.

For a fluidized bed reactor of length L, the pressure drop is that required to lift the bed and is equal to

$$\Delta P = [\rho_p - \rho_f][1 - \Lambda] L , \qquad (3\text{-}12)$$

where ρ_p is the density of the fuel pellets.

The porosity Λ is defined as the fluid volume divided by total bed volume. It can be specified in terms of the terminal velocity U_t:

$$\Lambda = 0.99 [U_s/U_t]^{0.118} ,$$

$$U_t = (g D_p^2/18 \mu) [\rho_p - \rho_f] , \quad N_{RE}^t < 2 ,$$

$$\Lambda = 1.04 [U_s/U_t]^{0.337} ,$$

$$U_t = 0.153 D_p^{1.14} (\rho_f/\mu)^{0.43} \{g [(\rho_p/\rho_f) - 1]\}^{0.71} , \quad 2 < N_{RE}^t < 500 ,$$

$$\Lambda = 1.15 [U_s/U_t]^{0.43} ,$$

$$U_t = 1.74 \{gD_p [(\rho_p/\rho_f) - 1]\}^{1/2}$$

$$500 < N_{RE}^t < 2 \times 10^5 . \qquad (3\text{-}13)$$

In the above equations, N_{RE}^t is the Reynolds number based upon the terminal velocity U_t or

$$N_{RE}^t = \rho_f U_t D_p/\mu . \qquad (3\text{-}14)$$

3.3 Design Safety Considerations

At the design stage, the safety considerations are the same as those discussed in Sec. 2. First, there is the question of how accurate are the calculations and the predictions of flow distribution. Next, one must recognize the need of evaluating those phenomena which inadvertently reduce the flow rate in the fuel assemblies. Finally, one must take into account flow-induced problems which affect the structural or physical integrity of the reactor core.

3.3.1 Accuracy of Calculations

There are two possible sources of error in flow distribution calculations. The first one is due to uncertainties in the basic pressure drop relations used in the predictions. The second one comes about from uncertainties in the design and reactor variables utilized in the computations.

The accuracy of the pressure drop relations used in flow distribution calculations is approximately the same as that previously estimated in Sec. 2 and summarized in Table 2.2. This means that the flow distribution within a reactor can be predicted within ± 5 to 10 percent depending upon the reactor type and coolant. In many instances, reactor core geometries are more complicated than those covered in Table 2.2, and one might anticipate a larger error; prototype testing is, however, often used to hold the error down.

Several reactor variables and design parameters are used in flow distribution calculations. For instance, the power distribution and the fuel geometry dimensions are needed to carry out the computations. An inaccuracy in some of these variables produces a corresponding error in the flow calculations. Such sources of error are taken into account by means of hot-spot factors as discussed in Sec. 4 of the chapter on Heat Transfer. Some of the more important factors which affect flow distribution are as follows:

1. In pressure vessel type reactors, the coolant flows through a plenum chamber before entering the fuel assemblies. The flow distributions at the inlet of the fuel assemblies can be uneven and will depend upon the design of the plenum chamber. Scale model tests of the reactor and its plenum chamber have often been performed to determine the inlet flow distribution. The tests (see Fig. 4-5 in the chapter on Heat Transfer for typical results) indicate that the flow maldistribution varies from

1 to 18 percent, and that, in most cases, it averages from 4 to 7 percent.

2. Variations in fuel geometry can occur from manufacturing tolerances, or expansion and bowing of the fuel elements in the course of irradiation. Geometrical changes can also be induced by pressure gradients. Pressure differentials across fuel plates or fuel channels can act to reduce the hydraulic diameter or fuel to channel spacings. Similarly, in an open-lattice core, some flow redistribution can be expected among the various fuel assemblies due to uneven heat production. All design features which can adversely affect the coolant flow must be appropriately evaluated and included in the calculations. For instance, reference [123] reports that computer calculations were used to study the flow redistribution in an open lattice. The computations which are based upon constant pressure perpendicular to the flow direction indicate that the loss of flow from a hot channel to the surrounding assemblies is of the order of 5 percent.

3. Deposition of corrosion products or other particles in fuel assemblies, or changes in fuel surface finish due to corrosion or crud buildup can affect the pressure drop and thereby the flow distribution. The effects must be estimated at the design stage and included in the predictions.

4. The calculations of pressure drop and flow distribution depend upon the local fluid properties. The fluid properties in turn are obtained from heat transfer considerations and a knowledge of the power distribution within the reactor core. For this reason, uncertainties in heat transfer or power distribution data can be expected to produce corresponding errors in flow calculations.

3.3.2 Flow Obstruction

In the course of design, several accidents can be postulated which produce a flow reduction in one or several fuel assemblies. Potential reductions of flow must be evaluated and their consequences determined. For instance, in an orificed reactor core it is possible to utilize the wrong orifice or to place an orificed fuel bundle in an inappropriate position. Similarly, a flow reduction would result if some reactor components (nuts, bolts, clips, etc.) failed mechanically and lodged themselves in fuel assemblies. Flow blockage can also come about from unexpected sources. For instance, a sight box was recently overlooked and left inadvertently atop the core of the Engineering Test Reactor. The box was broken up by the coolant flow, and the resulting fragments restricted the flow in eight fuel elements and caused the fuel cladding to fail [124].

It is recommended that a flow blockage margin be calculated for each reactor design. This margin would define the flow reduction that can be tolerated in the hottest fuel assembly before a serious cladding failure happens. In those reactor designs where the margin is found small, design modifications should be considered, or the need for careful monitoring of the flow distribution should be recognized. Of course, an accident like the one produced by the sight box at the Engineering Test Reactor could still obviate such a design margin since the flow was stopped to some elements.

3.3.3 Fuel Failure from Pressure Unbalance and Vibrations

The problems of erosion, corrosion, and mass transfer and their dependence upon flow conditions were mentioned in Sec. 2. Two new problem areas induced by flow are worth noting here. They are the collapse of parallel flat plate elements from a pressure unbalance across the plate and vibrations of fuel rods within a fuel bundle or vibration of the entire bundle. D. R. Miller [125] determined the flow velocity at which a pressure unbalance causes the collapse of long, parallel flat plate fuel assemblies. For instance, for uniform spacing of fuel plates with their long edge attached to side plates, the critical velocity V_c was found to be

$$V_c = [15\, g_c\, E'\, a_1^3\, h_1/\rho\, b_1^4\, (1-\nu^2)]^{1/2}, \quad (3\text{-}15)$$

where a_1 is the plate thickness, b_1 the plate width, E' Young's modulus of elasticity, h_1 the flow channel thickness at midspan, and ν Poisson's ratio. Similar expressions are given in reference [125] for other fuel plate geometries and designs. In a recent experimental study, Scavuzzo [126] points out that fuel plate motion starts at much lower velocities than those calculated from Miller's equations and that such an early motion is caused by the presence of initial plate deflections. Remick [127] also showed that redistribution of flow and plate imperfections must be included in the analysis to obtain agreement with the test results.

Fuel rod vibration can become a serious problem when the vibration amplitude is large. It can lead to mechanical wear at fuel support points or spacers; or it can reduce the rod-to-rod or rod-to-channel clearance; or again, it can produce reactivity oscillations through fuel rod motions; or even result in fuel rod failure from mechanical fatigue. Burgreen et al. [128], measured the vibrations of a single rod and multirod assemblies produced by water near room temperature. They found that the vibration occurs near the rod natural frequency and that the amplitude grows rapidly with water velocity. Dimensionless correlation of their data gave

$$(A'/D_e)^{1/3} = 0.83 \times 10^{-10}\, k_1\, \Gamma^{1/2}\, \Omega, \quad (3\text{-}16)$$

where A' is the amplitude of vibration; k_1 is related to the end fixity of the rod and is equal to 5 when the rod is pin-ended; for rigidly held rods, k_1 is 1, and when one end of the rod is pin-ended and the other end rigidly held, k_1 is 2.08. The parameters Γ and Ω are given by

$$\Gamma = \rho v^2 L^4/g_c E' I, \quad \Omega = v^2/\mu w, \quad (3\text{-}17)$$

where V is the flow average velocity, I is the moment of inertia, and w is the fundamental frequency of vibration.

More recently, E. P. Quinn [129] studied the vibration of a single rod and multirod assemblies in high temperature water and high pressure steam-water mixtures. He found that the amplitude of vibration is much smaller than that predicted by Eq. (3-16) and that the effect of fluid viscosity is small in turbulent flow. Quinn postulates that the mechanism of vibration is a self-sustained one and is caused by the interaction of fluid forces along the curved rods and the variation in local pressure with rod motion. His model indicates that the vibration frequency is near the rod natural frequency and that the amplitude varies inversely with the cube of frequency and directly with the cube of velocity. The amplitude also depends upon the rod geometry, its curvature, and eccentricity.

3.4 Operating Safety Considerations

The principal consideration here is the ability to measure the flow distribution in the reactor core during operation and the comparison of the measurements with the design predictions. In tube type reactors this need is readily met. Each tube can be equipped with a flow measuring device and even a valve to regulate it. In tank type reactors, the measurements become much more difficult and complicated. Several methods have been used to date, including the following:

1. Scale model tests have been used to measure the inlet flow distribution. These model tests can also be used to check the reactor flow distribution, especially in reactors where heat addition and variation of fluid property are not too important [130].
2. Pressure drop tests in prototype fuel assemblies are often performed (for instance, see references [131] and [132]). These tests establish the relation between flow and pressure drop, and thus eliminate one of the important sources of error.
3. Instrumentation has sometimes been used inside the reactor to determine the flow distribution. Instrumented fuel assemblies with turbine type flow meters were, for example, installed in the EBWR [133] and VBWR [134]. Orifices and nozzles have also been utilized in the Shippingport pressurized water reactor. Another technique relies upon a heat balance and measures the inlet and outlet enthalpy of the coolant in the fuel assemblies. Finally, pressure drop measurements in reactor core have been attempted [133]. One difficulty with this last method is in keeping the pressure lines filled with a fluid of a known temperature and density. This means that in high-temperature, liquid-cooled reactors, the pressure lines must be cooled while taking the test data.

4 STEADY-STATE LOCAL FLOW DISTRIBUTION WITHIN FUEL ASSEMBLY

4.1 Problem Areas

Nuclear reactors are being designed so that specific limits such as maximum fuel clad temperature, critical heat flux, or "burnout" margin are not exceeded at any point in the reactor core. In order to satisfy these criteria, one must be able to calculate not only the total flow within each fuel assembly, but also the local flow conditions. This section deals with the computation of three local flow properties: fluid velocity, density distribution, and fluid mixing. All three of the above parameters are important to thermal and safety analyses.

For instance, consider the problem of removing heat from a 7-rod fuel cluster of the type used in the Experimental Gas-Cooled Reactor (EGCR). A cross-sectional view of this fuel element is shown in Fig. 4-1. It consists of 7 rods arranged in a triangular pattern and held together at their upper and lower ends. The 7 rods are contained within a process tube and coolant flows inside the tube and parallel to the rods. The heat generation is not uniform in the fuel rods and varies both along their axis and circumference.

Examination of Fig. 4-1 reveals that the size of the flow passage changes as we traverse arc AB along the central rod. Similarly, the dimension of the coolant path is not constant along arc CD of the outer rods. The velocity distribution will vary along both arcs, and so will the local heat transfer coefficient. The velocity distribution must, therefore, be known to calculate the local rate of heat removal.

If the coolant being used in Fig. 4-1 is of the liquid-vapor or solid-fluid type, the vapor and solid content will change with position in the fuel assembly. When the coolant has moderating properties, the distribution of vapor affects the local power generation. A similar coupling exists between power and fluid-solid density if the solid particles contain fissionable or moderating material. This coupling cannot be taken into account without knowledge of the local density distribution.

The coolant passage in Fig. 4-1 can be subdivided into two regions: an inner zone bounded by the central rod and the inner surface of the outer rods; and an outer zone formed by the process tube and the external surface of the peripheral rods. The average heat generation, coolant flow, and local fluid enthalpy are not the same in these two channels, and some mixing will take place between them. In fact, in some reactors a definite amount of mixing is necessary to insure adequate cooling of the fuel rods. For instance, in the Sodium Reactor Experiment [135] the ratio of heat generation to coolant flow is greater in the

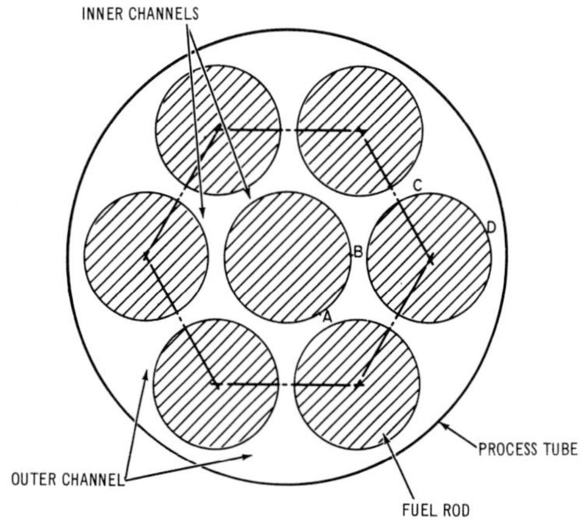

FIG. 4-1 Seven-rod cluster.

inner than in the outer zone of Fig. 4-1, and a stainless steel wire is wrapped around the outer 6 fuel rods to promote mixing between them. The degree of local mixing is important because it determines the local enthalpy conditions.

The above three problem areas will be considered in detail in the following sections. Available test results and calculation methods will first be reviewed. Their implications from a design and operating safety viewpoint will then be discussed.

4.2 Local Fluid Velocity Distribution

Theoretical and analytical predictions of velocity distribution have been obtained for single-phase flow for most geometries of interest to nuclear reactors. For circular pipes the local velocity u is known to have a parabolic shape in laminar flow [1] or

$$u = 2u_m \left[1 - (r - y)^2/r^2\right] . \qquad (4\text{-}1)$$

In Eq. (4-1), y is the distance measured from the wall; r the pipe radius; and u_m the mean fluid velocity. For turbulent flow, the velocity is of the logarithmic type, that is [5]

$$u^+ = 2.5 \ln y^+ + 5.5 . \qquad (4\text{-}2)$$

The dimensionless velocity u^+ and distance y^+ are defined in terms of the wall shear τ_w:

$$u^+ = u/\sqrt{\tau_w/\rho} ,$$

$$y^+ = \rho y\sqrt{\tau_w/\rho}/\mu .$$

Eq. (4-2) is valid except very close to the wall where viscous effects are dominant. In this region, a buffer and laminar sublayer are introduced and their velocities are given by [5]

$$u^+ = -3.05 + 5.00 \ln y^+ \qquad 5 < y^+ < 30 ,$$

$$u^+ = y^+ \qquad 0 < y^+ < 5 . \qquad (4\text{-}3)$$

For rectangular channels the velocity profiles are similar to those in a circular pipe except for corrections at the corners. As previously noted, theoretical predictions of velocity distribution in laminar flow and turbulent flow in rectangular channels are given in references [101] and [102]. Test data have been reported by various investigators [136,137], and they confirm the expected reduction in velocity near the corner. A typical profile obtained by Bell and LeTourneau [137] along the centerline of a 5/8-inch (1.59cm) channel is reproduced in Fig. 4-2 [137]. A decrease in velocity is noted up to one channel thickness from the corner.

Predictions and measurements of velocity distributions in an annular geometry have been reported in various papers [138] to [144]. For laminar flow, the local velocity is

$$u = 6\, u_m \left\{[y/(r_o - r_i)] - y^2/(r_o - r_i)^2\right\} , \qquad (4\text{-}4)$$

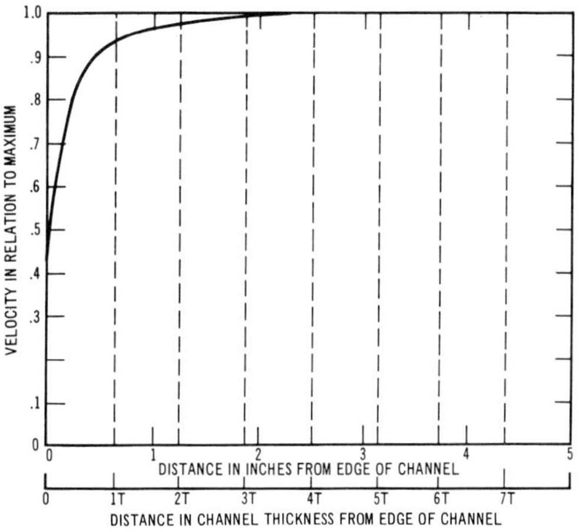

FIG. 4-2 Velocity distribution for turbulent flow in 5/8-inch (15.8 mm) channel.

where y represents the distance measured from the outer wall, and r_o and r_i are the outside and inside radii of the flow annulus. For turbulent flow, the predictions yield logarithmic profiles of the type of Eq. (4-2) which extend from both sides of the annulus. Typical velocity distributions presented in reference [138] are reproduced in Fig. 4-3 [138]. Results for a concentric and an eccentric geometry are shown on the same Figure. A possible discrepancy in the profiles of Fig. 4-3 has been noted at the inner wall [139]. Leung, Kays, and Reynolds [139] used logarithm profiles but chose to determine the radius of zero shear in the annulus from available experimental results. The

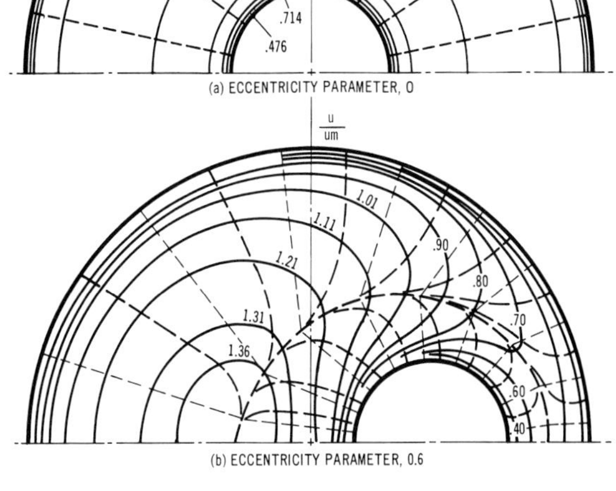

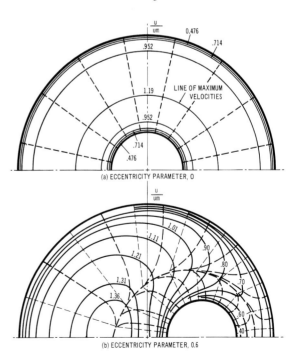

FIG. 4-3 Velocity distribution for turbulent flow in annulus.

location of the radius of zero shear (or maximum velocity) is shown in Fig.4-4 [139]. Test results of various investigators are plotted on Fig. 4-4 together with predictions of the zero shear plane from a laminar flow solution or from assuming equal shear on both walls of the annulus.

For multirod geometries, analytical solutions have been developed by Deissler and Taylor [101] for turbulent flow. Recent velocity measurements obtained by Samuels [145] in a 7-rod cluster are plotted in Fig. 4-5 [145].

Velocity profiles become much more complicated when the fuel elements have extended surfaces. Analytical predictions are practically nonexistent for the complex geometries shown in Fig. 3-8, but some experimental results have been reported lately. A shadowgraph method was used to obtain the velocity profiles shown in Fig. 4-6 for flow along a cylindrical rod with transverse fins. Flow patterns obtained by Cunningham and Slack for a polyzonal axial fuel element are reproduced in Fig. 4-7 [107]. According to Cunningham and Slack, "Fluid entrained between the fins at the upstream end of the helical flute between two fins (point A, Fig. 4-7) moves in a spiral path within the flute until it meets a longitudinal splitter crossing its path (point B). The fluid particles are then thrown roughly radially outwards and recirculate over the channel until they strike the splitter marking the outer boundary of a zone C, whereupon they reenter another helical flute." One can expect that with the more complicated profiles shown in Figs. 4-6 and 4-7, the velocity distributions will be more irregular and, indeed, variations of the heat transfer coefficient have been reported in the axial and radial directions of fuel elements with extended surfaces.

For two-phase flow systems, very few velocity distributions have been reported. Some recent test results reported by CISE [146], Hewitt [147], and Neal [148], and the theoretical solution of Levy [149] are worth noting.

4.3 Local Fluid Density Distribution

The number of experimental and analytical studies of two-phase fluid density distributions is limited. Petrick [48] summarizes the data obtained in vertical flow of air-water mixtures at atmospheric pressure and steam-water at 600 psia

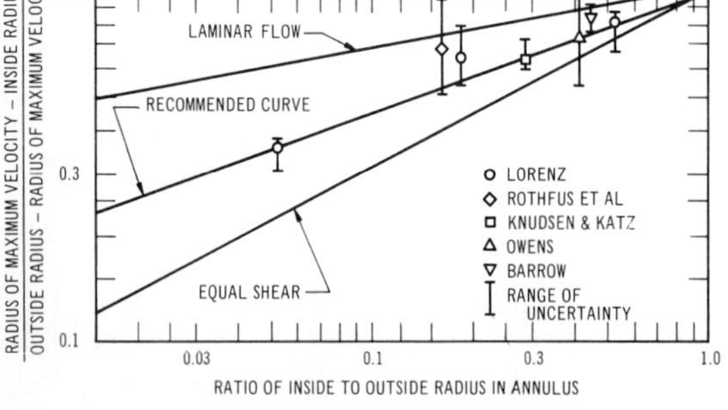

FIG. 4-4 The radius of maximum velocity in an annular passage.

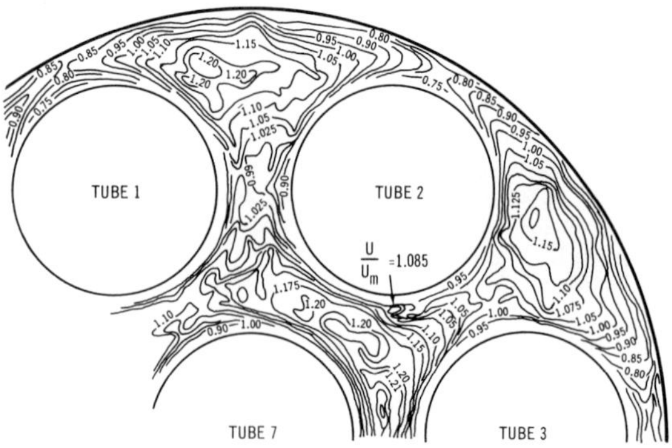

FIG. 4-5 Velocity profile in seven-element cluster at $L/D_e = 3.9$.

($42.2\ kg/cm^2$). Additional test results for mercury and air and argon and water are given in references [146] and [148]. Also worthy of note is the first measurement of voids in parallel rod arrays by Condon and Sher [150]. The test results generally show that the density profiles become more skewed as the gas content is increased and that the density distribution is more curved than the typically flat, single-phase, turbulent velocity profile. These trends are in agreement with the predictions of Bankoff [46] and Levy [149]. Reference [149] proposes that for liquid-gas flow in a pipe, the density profile can be computed from

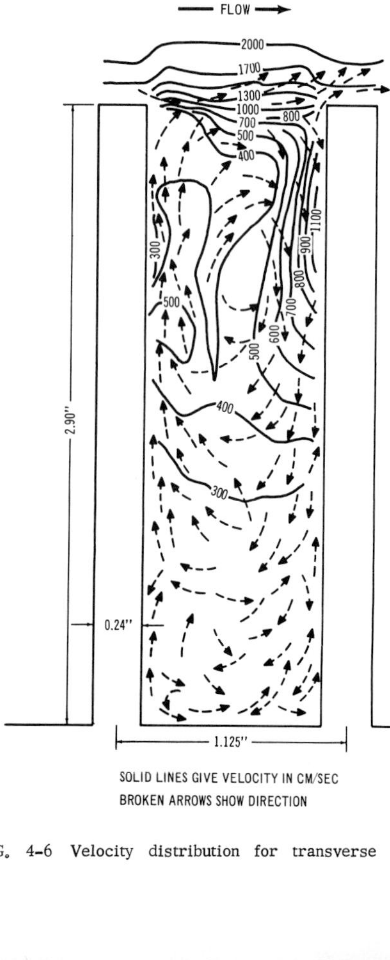

FIG. 4-6 Velocity distribution for transverse fin geometry.

$$\rho = \rho_a (D/2y)^{\beta'} (1/2)(1-\beta')(2-\beta') = \rho_m (D/2y)^{\beta'} , \qquad (4\text{-}5)$$

where ρ_a and ρ_m are the average and minimum (center) density of the fluid in the flow channel. The parameter β' can be related to the average density ρ_a, total flow rate G per unit area in the channel, and the two-phase frictional pressure drop per unit length of pipe, $\Delta P_{TPF}/\Delta L$, or

$$\beta' = 2.5 (\rho_L/G) [1 - (\rho_a/\rho_L)] \sqrt{(\Delta P_{TPF}/\Delta L)(D/\rho_L)} . \qquad (4\text{-}6)$$

The terms $\Delta P_{TPF}/\Delta L$ and ρ_a are calculated from the information given in Secs. 2 and 3.

4.4 Local Fluid Mixing

Local fluid mixing reduces the variation of coolant properties in the direction normal to flow motion. When the flow is laminar, there is by definition no fluid mixing. On the other hand, when the flow is turbulent, turbulent eddies are present and promote local fluid interchange. For single-phase flow, the exchange of momentum by turbulent mixing is expressed by means of an eddy diffusivity ϵ_M which is related to the local shear stress τ [3,5]:

$$\tau = (\mu + \rho\epsilon_M)\, du/dy . \qquad (4\text{-}7)$$

In most cases the turbulent diffusivity term is several times greater than the viscosity and the viscous effects can be neglected. By substituting the appropriate velocity profile in Eq. (4-7) and relating the local shear stress τ to the wall shear stress τ_w through a force balance, the eddy diffusivity can be calculated at every position y in the channel. For instance, for flow in a pipe, ϵ_M (beyond the laminar sublayer) is equal to [152]

$$\epsilon_M = 0.4\, y\, [1 - (y/r)] \sqrt{\tau_w/\rho} . \qquad (4\text{-}8)$$

Equation (4-8) determines the local transport of momentum. It can be used to calculate the local exchange of other properties. For example, if similarity is assumed between the turbulent exchange of momentum and energy, Eq. (4-8) can be used to predict the exchange of enthalpy between two single-phase flow passages in direct contact with each other but at two different bulk temperatures. The exchange of energy per unit area $\dot{q}$ is inversely proportional to the average distance l_a between the passages and directly proportional to the momentum diffusivity at the interface and the difference in energy of the two streams ($\rho C_p T_2 - \rho C_p T_1$). For diffusivity at the interface, one uses the average of the diffusivity coefficients in the two streams, or

$$\dot{q} = [\rho C_p T_2 - \rho C_p T_1)/l_a] (\epsilon_{M_1} + \epsilon_{M_2})/2 . \qquad (4\text{-}9)$$

The diffusivities ϵ_{M_1} and ϵ_{M_2} can be obtained from equations of the type (4-8). For simplification, one can assume that the eddy diffusivities at the interface are equal to the maximum value calculated from Eq. (4-8) or

$$\epsilon_M = (1/40)(G\, D_e/\mu)(\mu/\rho)\sqrt{f/2} . \qquad (4\text{-}10)$$

Knowledge of G, D_e, and f in the two passages specifies ϵ_{M_1} and ϵ_{M_2}.

Several experiments have been performed to date to determine the value of ϵ_M in multirod geometries. The tests are carried out by injecting dye or salt in a moving stream and measuring the concentration or distribution downstream of the injection point. Bell and LeTourneau [152] obtained a value of 0.003 for $\epsilon_M\rho/G\, D_e$ in two multirod fuel assemblies with pitch-to-rod-diameter ratios of 1.13 and 1.20. Similar values of 0.0030 to 0.0037 were obtained in a 12 x 12 rod bundle with a pitch-to-rod-diameter ratio of 1.24 by Nelson, Bishop, and Tong [153]. It is interesting to note that the results of reference [152] and [153] are in agreement with the values computed from Eq. (4-10).

Once ϵ_M is defined, an overall mixing effectiveness in the bundle can be calculated. A typical plot for fuel rods of diameter d and length L given in references [152] and [153] for 4 hot rods centrally located or for a row of hot fuel rods located at the edge of the fuel channel is shown in Fig. 4-8 [153].

A considerable increase in mixing results when the fuel rods are wrapped with a wire to promote local fluid interchange with a corresponding increase in pressure drop. McNown, et al. [154], obtained values of 0.017 for $\epsilon_M\rho/G\, D_e$ in a 25-rod bundle, while Shimasaki and Freede [135] measured a mixing effectiveness of 95 percent in a 7-rod arrangement. Similarly, considerable mixing occurs in the geometries shown in Fig. 3-8. The fins are added not only to increase the heat transfer surface, but also to avoid channeling of cold gas at the tip of the fuel fins.

Increased mixing also takes place at fuel spacers. Some measurements have been reported in reference [152] for a tube sheet type spacer and an equivalent length of mixing was defined. From

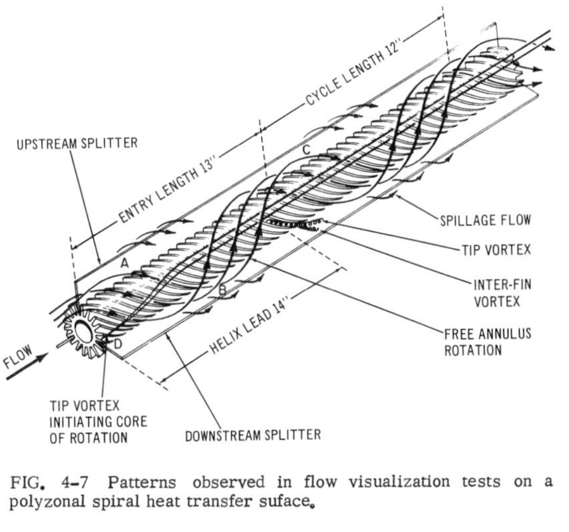

FIG. 4-7 Patterns observed in flow visualization tests on a polyzonal spiral heat transfer suface.

the obtained test data, this length of mixing, L (that can be substituted in the abscissa grouping of Fig. 4-8), was found equal to 91 D_e and 33 D_e for ratios of rod flow area to spacer flow area of 1.72 and 1.44, respectively. Extrapolation of these two data points to other spacer geometries on a log-log plot was suggested in reference [153].

Measurements or predictions of local fluid mixing in a liquid vapor stream are lacking. One can infer from the large pressure drops associated with such systems that the local fluid mixing is considerably greater than in single-phase flow. This trend is confirmed by tests in an annular geometry [155] with the results reported showing that considerable eccentricity was needed before the burnout or critical heat flux values were affected. In the absence of quantitative data, it is suggested that liquid vapor mixing be calculated from Eq. (4-10) by replacing f by the corresponding two-phase friction factor or

$$\epsilon_{TP} = \frac{1}{40} \frac{G D_e}{\mu_L} \frac{\mu_L}{\rho_L} \left(\frac{f_{L_o}}{2} \frac{\Delta P_{TPF}}{\Delta P_{L_o}} \right)^{1/2} . \qquad (4\text{-}11)$$

The friction factor f_{L_0} is based on total liquid flow rate G in the channel. If an analogy between momentum and density is postulated as in reference [149], the interchange of density between two moving two-phase streams with different vapor contents can be calculated from Eq. (4-11) and a relation similar to Eq. (4-9). The corresponding enthalpy exchange between the two streams is then computed from the density values. The same approximate technique can be applied to a fluid-solid system.

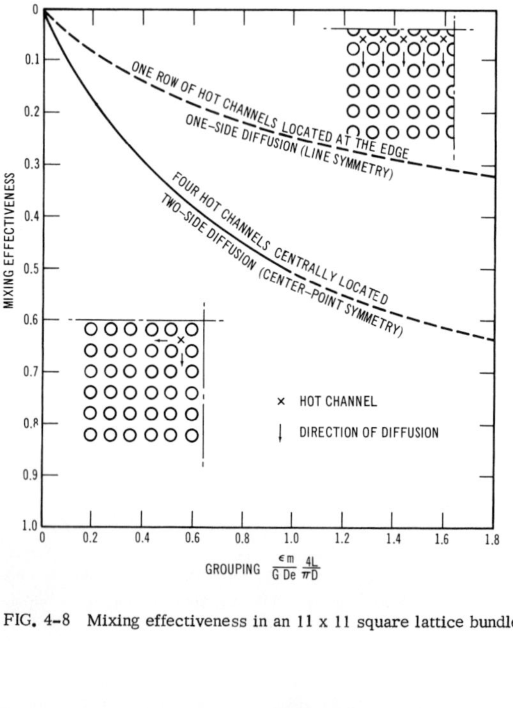

FIG. 4-8 Mixing effectiveness in an 11 x 11 square lattice bundle.

4.5 Design Safety Considerations

These are the same as in Sec. 3.3, namely, accuracy of calculations, possibility of local flow blockage, and local safety problems induced by flow.

4.5.1 Accuracy of Calculations

There are uncertainties in the data and calculation methods presented in the previous sections. It is estimated by the author that for simple fuel geometries and single-phase flow, the local fluid velocity and local fluid mixing can be calculated within 10 and 20 percent, respectively. For more complex geometries, the errors are expected to increase to 15 and 30 percent, and for two-phase flow, they could be as high as 20 to 40 percent. The local density distribution within a two-phase stream can probably be specified within 20 to 30 percent.

There are also uncertainties in the design parameters and reactor variables used in the local velocity, mixing, and density distribution calculations. Some of the more important factors which should be taken into account at the design stage are:

1. Local variations in fuel geometry can occur due to manufacturing tolerances or due to fuel bowing or bulging caused by irradiation. The possible effects of such variations in fuel geometry are illustrated in Fig. 4-3 where an eccentric annular geometry is seen to produce large changes in the local single-phase velocity distribution. These and similar effects are best determined from experimental measurements. In most cases, the tests are concerned with the ultimate effects upon the variable of concern, such as temperature or critical heat flux, rather than the cause of the problem, i.e., distortions of the velocity or density distribution. For instance, critical heat flux tests in an eccentric annular geometry have been performed to study the effects of manufacturing tolerance [152]. Similarly, mass transfer tests in a multirod geometry are reported in reference [153] for offset rod locations and for a local bulge in a fuel rod.

2. Calculations of local flow conditions are dependent upon heat transfer enthalpy and local power distributions. Uncertainties in these variables can be expected to affect the local velocity distribution, fluid mixing, and density profile.

4.5.2 Flow Obstructions

Design calculations must make allowance for potential local flow reductions. Flow obstructions can occur in the course of fuel fabrication or irradiation. Loose manufacturing tolerances or deposition of corrosion products along the fuel rods or spacers will distort the local flow conditions.

4.5.3 Local Safety Problems Induced by Flow

Rod vibrations or fuel plate and channel displacements produced by pressure unbalance have already been noted in Sec. 3.3.3. Similarly, the interaction between flow and local fuel bowing and bulging must be understood. Special attention must be paid to problems with autocatalytic tendencies such as fuel rod bowing which leads to flow reduction which in turn means more bowing of the fuel rod, etc.

4.6 Operating Safety Considerations

The important consideration here is the ability to measure local flow conditions and to compare them with the design predictions. Measurements of local flow properties are, however, very difficult in nuclear reactors and out-of-pile prototype tests are often substituted. Studies performed for the EGCR reactor are typical of such investigations [145]. In a few instances, local variables such as fuel rod temperature have been measured in reactors and the accuracy of the local flow calculations deduced from the in-pile measurements.

5 TRANSIENT FLUID FLOW

An important aspect of nuclear plant design is the calculation of transient flow conditions. In many instances, the transient behavior limits the performance of the plant; in other instances, it determines the sequence of events and consequences of major reactor accidents. This section deals with the calculational models available for studying fluid flow transients and their application to important problems in nuclear reactor technology.

5.1 Calculational Models

In transient flow calculations, fluid velocities vary with time. With the addition of time as a variable, the solutions of fluid flow problems are much more difficult. Even in the simple case of one-dimensional flow of a single-phase or two-phase homogeneous fluid in a channel of constant cross-sectional area, it is necessary to simultaneously solve three partial differential equations. The equations are the equations of continuity, momentum, and energy [157].

$$(\partial\rho/\partial t) + (\partial G/\partial z) = 0\ , \tag{5-1}$$

$$(\partial G/\partial t) + \partial(G^2/\rho)/\partial z = -(\partial P/\partial z) - (f/\rho)\,(|G|G/2D_e) - \rho g\ , \tag{5-2}$$

$$\rho(\partial E/\partial t) + G(\partial E/\partial z) = q + [(\partial P/\partial t) + (G/\rho)\,(\partial P/\partial z)] + (f/\rho)|G|G^2/2\rho\, D_e\ . \tag{5-3}$$

The above equations coupled with initial boundary conditions define the variation of the pressure P(z,t), mass flow rate per unit area G(z,t), and enthalpy E(z,t) in terms of position z and time t. In Eq. (5-3) q is the heat transferred to the coolant per unit volume of coolant and is assumed to be known as a function of time and position.

It should be noted that Eqs. (5-1) to (5-3) neglect the variation of fluid properties, velocity, and pressure in the direction normal to the flow direction. It should also be noted that in such practical systems as a multichannel nuclear reactor, the pressures at the inlet and exit plenums, P_I and P_E are specified or are desired as a function of time. These pressures can be related to the pressure P_o just inside the channel inlet and the pressure P_n just inside the channel exit by using the methods described in Secs. 2 and 3. For instance, the change in pressure from inlet plenum to channel and from channel to exit plenum can be specified in terms of inlet and outlet loss coefficients K_o and K_n and the corresponding area ratios β_o and β_n:

$$\left.\begin{aligned} P_I - P_o &= (1/2)\,(G_o/\rho_o)\,[K_o|G_o| + (1-\beta_o^2)\,G_o]\ ,\\ H_o &= H_I\ ,\\ P_E - P_n &= (1/2)\,(G_n/\rho_n)\,[-K_n|G_n| + (1-\beta_n^2)\,G_n]\ . \end{aligned}\right\} \tag{5-4}$$

In two-phase flow systems, slip flow between the gas and liquid must be taken into account. The corresponding transient equations are more complicated and can be written as follows [152]:

$$(\partial\bar{\rho}/\partial t) + (\partial G/\partial z) = 0\ , \tag{5-5}$$

$$(\partial G/\partial t) + (\partial/\partial z)\,(\rho' G^2) = -(\partial P/\partial z) - (f/\rho)\,(|G|G/2D_e) - \bar{\rho} g\ , \tag{5-6}$$

$$\rho''\,(\partial E/\partial t) + G(\partial E/\partial z) = q\ . \tag{5-7}$$

In Eq. (5-7), the dissipation and the pressure change terms which appeared on the right side of Eq. (5-3) are neglected and the slip effects are represented by appropriate densities $\bar{\rho}$, ρ', and ρ''. These density terms can be expressed in terms of the volume fraction α and the gas weight content x:

$$\bar{\rho} = \rho_L\,(1-\alpha) + \rho_G\,\alpha\ , \tag{5-8}$$

$$(1/\rho') = [(1-x)^2/\rho_L\,(1-\alpha)] + x^2/\rho_G\,\alpha\ , \tag{5-9}$$

$$\rho'' = [\rho_L\,x + \rho_G\,(1-x)]\,(d\alpha/dx)\ . \tag{5-10}$$

The frictional resistance to flow is represented by the grouping (f/ρ) in Eq. (5-6), and it is assumed that this function can be obtained from steady state measurements (see Secs. 2 and 3) and depends only upon the flow rate, enthalpy, and rate of heat transfer:

$$f/\rho = f/\rho\,(E, G, q)\ . \tag{5-11}$$

Equations (5-1) to (5-11) describe the transient flow of a single-phase or two-phase fluid with or without slip in a channel of constant cross-sectional area. In order to solve these equations, various numerical and physical approximations are often necessary. The degree of simplification introduced and the type of model used in the calculations depend on the type of transient under consideration. For very fast transients, Eq. (5-1) to (5-11) must be solved without further approximations and a stepwise time and distance (multi-node) compressible model is used to obtain a finite difference solution to the controlling equations. For fast transients, a momentum integral model and a channel integral model may be used. These models rely upon integrated forms of the momentum or of all the conservation laws to solve such problems as hydrodynamic instabilities in boiling channels. For intermediate and slow transients, further simplifications may be made by neglecting the variation of flow G with position z. This single mass velocity model is effective in treating some of the loss-of-flow accidents or other plant transients of a similar speed.

A description of these four models and of their limitations is given in reference [157]. The following discussion is largely based upon the information presented in that reference.

5.1.1 Multinode Compressible Model

This model gives the most exact representation of transient flow in a channel. For single-phase or two-phase homogeneous flow (corresponding equations can be derived for two-phase slip flow), Eqs. (5-1) and 5-3) can be rewritten as

$$\frac{\rho}{a^2}\frac{\partial P}{\partial t} + \rho\frac{\partial G}{\partial z} + R_h\frac{G}{\rho}\frac{\partial P}{\partial z} - R_h G\frac{\partial E}{\partial z} = -R_h\left[Q + \frac{f}{\rho}\frac{|G|G^2}{2\rho D_e}\right], \tag{5-12}$$

$$\frac{\rho}{a^2}\frac{\partial E}{\partial t} + \frac{\partial G}{\partial z} - R_p\frac{G}{\rho}\frac{\partial P}{\partial z} + R_p G\frac{\partial E}{\partial z} = R_p\left[Q + \frac{f}{\rho}\frac{|G|G^2}{2\rho D_e}\right], \tag{5-13}$$

where

$$R_h = (\partial\rho/\partial E)_{p\,=\,\text{constant}},$$

$$R_p = (\partial\rho/\partial p)_{E\,=\,\text{constant}}, \tag{5-14}$$

$$a = [R_p + (R_h/\rho)]^{-1/2}. \tag{5-15}$$

Equations (5-2), (5-12), and (5-13) each contain the time derivative of only one of the three dependent variables G, P, and E. The equations can be solved by means of a finite difference method with respect to time t and position z. Solutions of the equations reveal that disturbances are propagated at the isentropic sonic velocity a.

Because it can deal with disturbances of the order of sonic velocity, the multinode compressible model is useful for very fast transient calculations. However, in order to avoid numerical instability, it is necessary to take very small time increments (of the order of the time for a sonic wave to pass through one space step) and the model requires excessive computer time.

5.1.2 Momentum Integral Model

There are two new basic assumptions made in this model. First, all fluid properties are considered independent of pressure and are evaluated at some reference pressure which does not vary in the course of the transient. Second, the momentum equation is integrated over the channel length L and this integral form is substituted for the differential equation.

Integration of Eq. (5-2) over the length L gives

$$d\hat{G}/dt = (1/L)(\Delta p - F), \tag{5-16}$$

where $\hat{G}$ represents the average channel mass velocity and is equal to

$$\hat{G} = (1/L)\int_0^L G\,dz. \tag{5-17}$$

For the more general case of slip flow, the symbols ΔP and F are

$$\Delta P = P_I - P_E, \tag{5-18}$$

$$F = \left[\frac{1}{\rho_n'} - \frac{1-\beta_n^2}{2\rho_n}\right]G_n^2 + \frac{K_n|G_n|G_n}{2\rho_n} + \int_0^L \frac{f|G|G}{2\rho D_e}dz - \left[\frac{1}{\rho_o'} - \frac{1-\beta_o^2}{2\rho_o}\right]G_o^2 + \frac{K_o|G_o|G_o}{2\rho_o} + \int_0^L \bar{\rho}g\,dz. \tag{5-19}$$

For single-phase or two-phase homogeneous flow, Eq. (5-19) can be simplified by setting $\rho' = \bar{\rho} = \rho$.

Equation (5-16) gives the time variation of the average channel mass velocity $\hat{G}$ while Eq. (5-7) describes the transient behavior of E and their solution by finite difference methods gives $\hat{G}$ and E as a function of time. The local variations of the mass velocity G with position z are obtained from Eq. (5-5). Equation (5-5) gives the derivative $\partial G/\partial z$ in terms of $\bar{\rho}$ which, according to the first assumption made in the model, is independent of the pressure P and is a known function of the enthalpy E. For single-phase or two-phase homogeneous flow, Eqs. (5-5) and (5-7) can be combined to give a simple expression for $\partial G/\partial z$, or

$$\partial G/\partial z = -(1/\rho)(d\rho/dE)[q - G(\partial E/\partial z)]. \tag{5-20}$$

An important advantage of the momentum integral model is that it neglects sonic effects. In other words, pressure and velocity disturbances are assumed to be propagated at an infinite velocity. This greatly simplifies the machine calculations as time steps of the order of the time for fluid to be transported through one space increment can be utilized. The model is, however, valid only for transients where the changes take place in periods several times that of the sonic transit time in the channel. Furthermore, it cannot be used for large changes in pressure with respect to system absolute pressure.

As previously noted, the model has been found especially useful in predicting the occurrence of hydraulic oscillations [158].

5.1.3 Channel Integral Model

In addition to the momentum integral Eq. (5-16), the continuity and energy equation are integrated in this model over the channel length L. For single-phase or two-phase homogeneous flow, Eqs. (5-1) and (5-3) become

$$d\,\hat{\rho}/dt = G_o - G_n \,, \tag{5-21}$$

$$d\hat{E}/dt = \hat{q} - G_n\,(E_n - E_o) \,, \tag{5-22}$$

where

$$\hat{\rho} = \int_o^L \rho\, dz \,,$$

$$\hat{E} = \int_o^L \rho\,(E - E_o)\, dz \,,$$

$$\hat{q} = \int_o^L q\, dz \,. \tag{5-23}$$

Equations (5-16), (5-21), and (5-22) define the variation of the average channel variables, $\hat{G}$, $\hat{\rho}$, and $\hat{E}$ as a function of time. [Similar results can be obtained with slip flow starting from Eqs. (5-5) to (5-10)]. In order to calculate the variables, $\hat{G}$, $\hat{\rho}$, and $\hat{E}$, it is necessary to first carry out the integrations of (5-23) and these integrations can be performed only if the variation of E with z is known. It is customary to assume that the variation of enthalpy with position remains unchanged during the transient and this approximation allows the solution to proceed.

The latter assumption means that the channel integral model must be used with care in those transients where the enthalpy profile is important or changes rapidly with time. This disadvantage is partly compensated by the ability of the model to use even larger time steps, of the order of the coolant transit time in the entire channel.

5.1.4 Single Mass Velocity Model

This model assumes that the mass velocity G is independent of position z so that

$$G = G(t) \,. \tag{5-24}$$

In order to satisfy Eq. (5-24), $\partial G/\partial z \approx 0$ and according to Eq. (5-5), $\partial\bar{\rho}/\partial t \approx 0$. In other words, variations of density with respect to position are included while variations with respect to time are neglected. With the use of Eq. (5-24), the momentum equation can be integrated for the more general case of slip flow and

$$\begin{aligned} dG/dt = {} & [(P_I - P_E)/L] - (1/L)\,[(1/\rho_n') - (1/\rho_o') \\ & - (1/2)\,(1 - \beta_n^2)\,(1/\rho_n) + 1/2\,(1 - \beta_o^2)\,(1/\rho_o)]\,G^2 \\ & - (1/L)\,(1/2)\,[(K_n/\rho_n) + (K_o/\rho_o)]\,G^2 \\ & - (G^2/2D_eL)\int_o^L (f/\rho)\,dz - (1/2)\int_o^L \bar{\rho}\,g\,dz \,. \end{aligned} \tag{5-25}$$

The method of solution consists of simultaneously handling Eq. (5-25) and (5-7).

The single mass velocity model has the advantage of simplifying the calculations by neglecting the variation of G with respect to z. The model, however, cannot be expected to be accurate in those transients where the density $\bar{\rho}$ varies rapidly with time, i.e., two-phase systems with large changes in vapor content.

5.2. Normal Plant Transients

The first application of the models described in Sec. 5.1 is to the study of normal plant transients. We can subdivide normal plant transients into two categories. The first group is concerned with plant system performance rather than detailed reactor calculations. For such transients, the variations in reactor performance from nominal steady state values are small and a simple simulation of the reactor flow behavior is acceptable. Most of these studies are performed on an analog computer using a single point representation of the core and a channel integral or a single mass velocity flow model. The second category of transients deals with time variations of temperature and heat flux within the core and requires a more detailed evaluation of reactor behavior. A multinode and multichannel representation of the core is usually coupled to a plant model. Such transients can again be examined on an analog computer of large capacity [159]. Another technique is to use an analog model with a simplified reactor representation to calculate transient values of plant pressure and total flow and to utilize the latter results in a digital computer code [160] which describes the detailed behavior of the reactor core. In both approaches, the integral momentum flow model will yield sufficiently accurate answers. However, for simplicity, many of the studies are often performed with a single mass velocity flow model with some potential loss of accuracy.

Normal plant operation includes almost an infinite variety of transients. The important transient conditions to be studied also vary from one reactor type to another. The detailed description of such calculations is beyond the scope of this chapter. The principles of analog computers and typical nuclear power plant simulations are described in nuclear engineering books. Calculated results from such simulations are also available in almost all safeguards reports of nuclear reactors. Finally, it should be noted that the calculation methods and results parallel those discussed for a loss-of-flow accident and this accident is covered at length in the next section.

5.3 Loss-of-Flow Accident

In nuclear plants, loss of flow comes about from failure of power to the coolant pumps and circulators or failure of a component in the recirculating equipment (for instance, bearing failure or seizure). The net result is a sudden reduction in flow rate through the reactor and possible excessive fuel temperatures. The prediction of coolant flow rate during such a loss-of-flow accident is important because it not only affects the consequences of the accident, but also determines the design of the equipment and the type and method of corrective action required in the course of the accident.

5.3.1 Calculation Method

The calculation method can be broken down into two steps. First, the total recirculation rate and inlet and exit reactor pressures are determined as a function of time. Next, these values are used in a detailed transient model of the reactor to calculate local values of the critical variables such as fuel temperature.

The basis for the transient calculation of total recirculation flow rate is the principle of conservation of energy. During normal operation, kinetic energy is stored in the pumps or circulators, and the fluid. When the loss-of-flow accident starts, this kinetic energy is used to overcome frictional losses in the recirculating loop and electrical losses in the pumps or circulators. The rate at which the available kinetic energy is dissipated determines the recirculation flow rate as a function of time.

Let us consider a recirculation system consisting of n loops and let us assume that the flow in each of these loops is $W_1, W_2, \ldots, W_n$. The kinetic energy stored in a channel of uniform cross-sectional A and length L in which fluid of density ρ flows at a velocity V is

$$\text{Channel Kinetic Energy} = (1/2g_c)\,(L\,A\,\rho)V^2\,, \quad (5\text{-}26)$$

and the kinetic energy stored outside the reactor in the entire loop n, made up of m different channels, is

$$(\text{Loop Kinetic Energy})_n = (1/2g_c)\left(\sum_{j=1}^{m} (L_j\,A_j\,\rho_j)V_j^2\right)_n = W_n^2/2g_c\left(\sum_{j=1}^{m} L_j/\rho_j\,A_j\right)_n . \quad (5\text{-}27)$$

If we add to this energy the kinetic energy stored in the reactor and that developed by the pump, and if we equate the rate of available kinetic energy change to the rate at which it is being dissipated in the system, there results for loop n [161]

$$(1/g_c)\left(\sum_{j=1}^{m} L_j/A_j\right)_n dW_n/dt + (1/g_c)\sum (L_R/A_R)\,(d/dt)\,(W_1 + W_2 + \ldots + W_n) = (\Delta P_p)_n - (\Delta P_R) - (\Delta P_L)_n\,. \quad (5\text{-}28)$$

In Eq. (5-28) the subscript R represents the reactor and ΔP_R is the pressure drop across it. Similarly, $(\Delta P_L)_n$ is the total pressure drop across the loop n. Both ΔP_R and $(\Delta P_L)_n$ can include natural convection terms due to fluid density variations which help reduce the rate of coolant loss. For a specified loop flow and total flow, the two terms are calculated from the methods described in Secs. 2 and 3. The grouping $(\Delta P_P)_n$ represents the head developed by the pump for a given loop flow and pump speed. It is obtained from curves of the type shown in Fig. 2-11. The pump speed can be calculated by equating the rate of change of kinetic energy stored in the rotating parts of the pump to the rate at which it is being dissipated across the pump [160]:

$$(1/g_c)\,(I_p)_n\,d\omega_n/dt = (T_E)_n - (T_H)_n - (T_w)_n\,. \quad (5\text{-}29)$$

In Eq. (5-29), I_p is the moment of inertia of the rotating parts, ω the pump rotational speed, and T_E, T_H, and T_W are the electrical, fluid, and windage torques.

The fluid torque is

$$(T_H)_n = W_n\,(\Delta P_p)_n/\omega_n\,\eta'\,\rho\,, \quad (5\text{-}30)$$

where η' is the impeller efficiency and is equal to the pump efficiency if the design point efficiency is independent of pump speed [161]. The electrical and windage torques are obtained from empirical correlations or test data on the pump-motor assembly.

With the use of Eqs. (5-29) and (5-30) and the data given in Secs. 2 and 3 and curves of the type shown in Fig. 2-11, Eq. (5-28) and similar equations for the other loops become first-order differential equations which can be solved for the flow rates W , $W_2, \ldots, W_n$ and for total flow as a function of time. The solutions also specify ΔP_R or the difference $P_I - P_E$ in terms of the time t.

In many cases, a simplified solution of Eq. (5-28) can be obtained by assuming that the pressure losses ΔP_R and $(\Delta P_L)_n$ vary as the square (or some other exponent 1.8 to 2.0) of the flow rate so that

$$\Delta P_R = (\Delta P_R)_i\,(W/W_i)^2 \quad \text{and}$$

$$(\Delta P_L)_n = (\Delta P_L)_{n_i}\,(W_n/W_{n_i})^2\,, \qquad (5\text{-}31)$$

where the subscript i is used to represent the initial conditions before the transient starts.

Once the total flow has been determined as a function of time, the detailed reactor calculations follow. Conditions in the reactor can be determined by means of a thermal and hydraulic digital computer code of the ART [160] type. The ART code is based upon the single mass velocity model described in Sec. 5.1. It also uses a single nodular model to represent fuel element temperature.

The input to the ART code consists of a description of the reactor geometry and a specification of the power and total flow as a function of time. With this input information, the code is capable of calculating the pressure drop across the average channel as a function of time and the flows in the hot channels by imposing this same pressure drop across them. Computations can be performed for as many as thirty axial sections of equal length.

Similar results can be obtained by using an analog model [159].

5.3.2 Typical Results

The results obtained from loss-of-flow calculations vary with the reactor type. For instance, in water-cooled reactors, one is mainly concerned about the occurrence of "burnout" or critical heat flux conditions. In gas-cooled reactors, the rise in fuel surface temperature is of utmost importance during the transient.

The results also depend upon the kind of loss-of-flow accident. There are three broad types of accidents:

1. Simultaneous loss of all pumps or circulators. This could come about from failure of the power supply to the pumps or circulators.
2. Loss of one pump only, the other pumps remaining in operation. This might be caused, for instance, by seizure of the bearing in one pump.
3. Loss of one pump, followed by the loss of the other pumps. In this situation, power in one pump fails, and before this pump has stopped rotating, power fails in the other pumps.

Typical results for pressurized water reactors are shown in Fig. 5-1 [161]. The top curves give the flow coastdown curves for simultaneous failure of power to all pumps. A comparison between test data and analysis is shown. The lower curve of

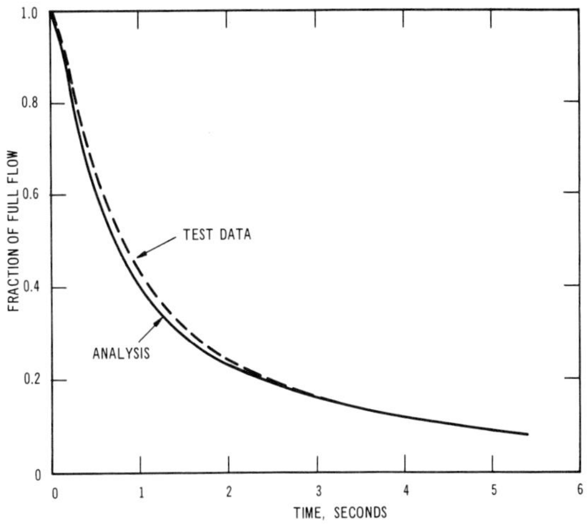

FIG. 5-1 Loss of all pumps in pressurized water reactor.

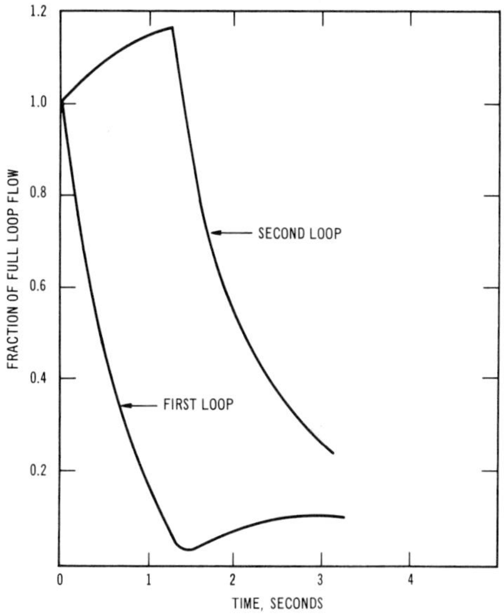

FIG. 5-1 (cont.) Sequential loss of pumps in pressurized water reactor.

Fig. 5-1 shows the calculated flow rates during a sequential pump failure accident in a two-loop system. An increase of flow in the second loop is noted after power failure to the first pump. The increase in flow is due to a reduced reactor pressure drop brought about by a decrease in total flow through the core. In the sequential pump failure illustrated on Fig. 5-1, power to the second pump is assumed to fail after 1.3 sec at which time flow in this loop starts to decay rapidly. The possibility of flow reversal exists in reactors with downwards coolant flow. In such reactors heat added to the coolant produces a buoyancy force which acts to further slow down and even reverse the flow in the course of a loss-of-flow accident.

The buoyancy term becomes much larger if some of the liquid is converted to vapor and the possibility of flow reversal is increased.

Similar results for a boiling water reactor system are shown in Fig. 5-2 [162]. Flow coastdown curves are plotted for power failures to all pumps and for different values of pump flywheel inertia J ranging from 1,000 to 4,000 lb-ft^2. An important difference from the curves of Fig. 5-1 is noted in the sense that the flow settles down to a flow rate of about 40 percent of the original value due to natural circulation in the loops and reactor. A dropoff in power due to increased void content in the reactor is also noted in Fig. 5-2.

5.3.3 Consequences of Loss-of-Flow Accident

The consequences again vary with the type of reactor and kind of loss-of-flow accident. The possible consequences in a boiling water reactor are illustrated in Fig. 5-3 [152] for the coastdown curves previously given in Fig. 5-2. It is seen that power failure in all pumps reduces the "burnout" margin from about 2.5 to 1.3 or above, depending upon the pump inertia. For the design illustrated in Fig. 5-3, "burnout" or critical heat flux condition is not even expected to occur and the consequences of the loss-of-flow accident are not serious and may not require a reactor scram. A similar drop in burnout margin results for pressurized water reactors. The reduction in margin is, however, much more rapid as the flow coastdown curves are steeper and tend to level off at lower flow values. Here again "burnout" can be avoided by starting with a high initial margin or by adding inertia to the pumps. It is interesting to note that in pressurized water reactors, the case of sequential failure of power to the pumps is sometimes more severe than the total loss-of-flow accident. This is due to the fact that, as the flow is reduced, reactivity may be added due to increased inlet water subcooling and the reactor power rises after failure of power to the first pump.

The consequences of loss-of-flow in a gas-cooled reactor are shown in Fig. 5-4 [159]. Due to the small kinetic energy stored in the coolant, loss-of-flow accidents are more serious for this type of reactor and corrective action is required

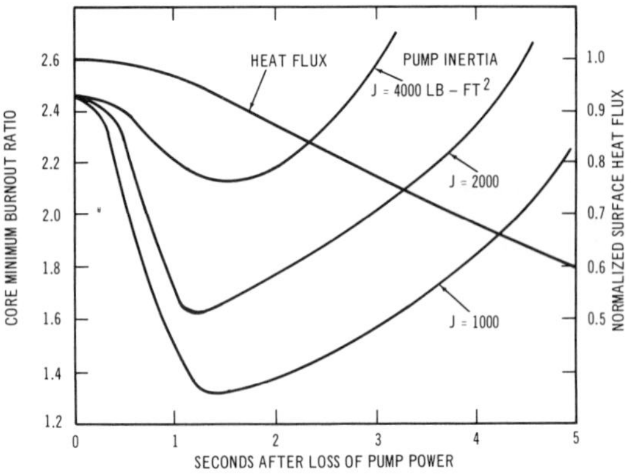

FIG. 5-3 Consequences of loss of flow in boiling water reactors.

as illustrated in Fig. 5-4. Figure 5-4 shows the effect of loss of power to all circulators in the Sizewell Nuclear Power Station [158]. Fuel element thermocouples scram the reactor before the temperatures become excessive. The broken curves in Fig. 5-4 show what would happen if the reactor power were not tripped. In the second portion of Fig. 5-4, power to only one gas circulator is lost and the control rods are inserted as the outlet coolant temperature rises by about 25°C (or 45°F). Continued operation is possible at a reduced power level.

5.4 Flow Oscillations

In most reactor types, there exists a coupling between reactor power and coolant density. This coupling can sometimes lead to power oscillations or "reactor instability" and limit the operating conditions of a given nuclear system. There are two possible reasons for reactor instability. In the first case, typical of the SPERT IA reactor, the power oscillations are caused by the feedback loop between density or vapor voids in the core and the reactor power through the void coefficient of reactivity and the reactor kinetic equations. In the SPERT IA oscillations, it was shown that the feedback loop became unstable even though the reactor flow was constant [163]. This type of reactor in-

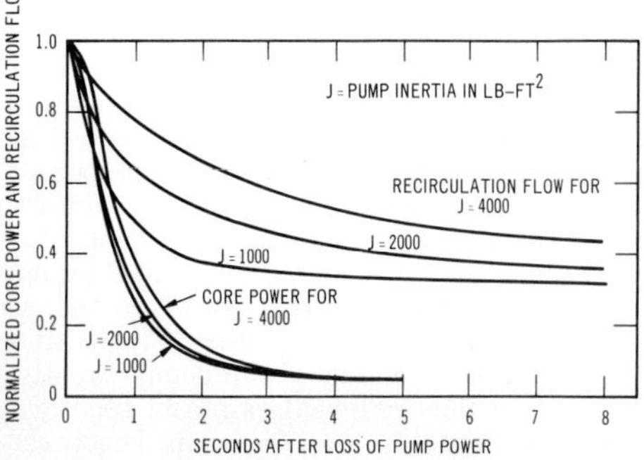

FIG. 5-2 Loss of all pumps in boiling water reactor.

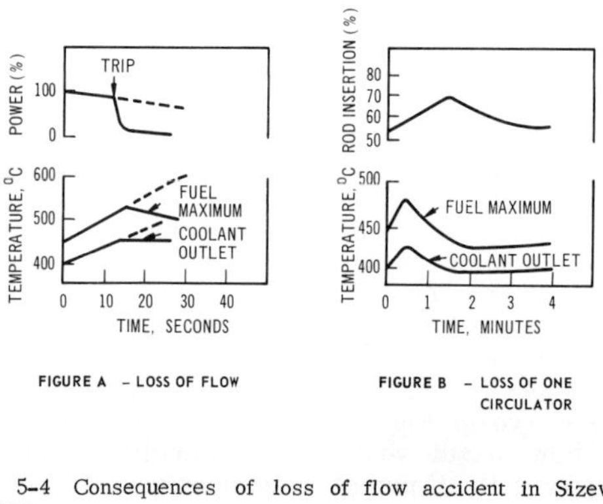

FIG. 5-4 Consequences of loss of flow accident in Sizewell reactor.

stability and its interaction with the feedback loop are discussed in the Water Reactor Kinetics chapter. It will suffice to note that important elements of the feedback loop are the power-to-void and power-to-flow transfer functions. These transfer functions depend on the thermal and hydraulic characteristics of the system, and their predictions rest heavily on hydraulic tests in heated loops. Additional analytical and experimental work is clearly needed in this area. Only two tests have been performed to date, and both of them are with steam-water mixtures. Measurements of the power-to-void and power-to-flow transfer functions for boiling in a SPERT IA channel have been reported [164] and a phenomenological model, based upon steady state measurements of flow rate and void fraction distribution, has been proposed [163]. Test results with water at higher pressures have also been presented by Christensen [115].

The second type of reactor instability is caused by flow oscillations. These flow oscillations can be generated without any feedback loop and have been observed in several out-of-pile tests and loops. We shall concern ourselves here with this purely hydraulic type of instability. Several analytical and experimental studies of hydraulic instability have been published recently [165] to [182]. In the last few years, great strides have been made towards understanding the phenomena involved, but some confusion still prevails in the field. One reason for the confusion is that there are several possible types of hydraulic instability and attempts are made to predict one kind of instability with analytical models which are suitable for the other kinds. Another reason is that the analytical solutions are not simple. The equations become complex and contact with the physical picture is lost. Finally, it would be surprising if good agreement could be reached among investigators on a complex two-phase transient flow problem when their theoretical predictions of steady-state two-phase flow are still at variance. An attempt shall be made here to distinguish between the various possible modes of flow oscillations and to describe their physical cause and basis. Experimental and analytical results will also be briefly presented for each mode of flow oscillations.

5.4.1 Flow Instability Caused by Bubble Nucleation

There are two possible ways for bubble nucleation to produce flow oscillations [165]. The first way is by violent, bumpy boiling. Before bubbles can nucleate, some superheat must be present in the liquid. For clean surfaces, this superheat is sometimes very large, and when a bubble forms, it grows violently and ejects large quantities of liquid from the heated channel. The frequency of the bubble bursts will be related to the time required for superheating of the liquid, forming of the bubble, ejection and runback of the liquid. It can be expected to depend upon the nucleation characteristics of the surface, the system geometry, and the properties which control the bubble growth rate. Instabilities of this type have been noted in tests of SPERT mockup channels of welded construction [163]. These channels had poorer nucleation characteristics than the riveted sections utilized originally. Instability was obtained only in the welded channels and appeared in the form of large bubbles expanding, rising, and leaving the channel at about 7 cycle/sec.

The second type of instability caused by nucleation can be traced to the pressure decrease which follows bubble formation. The process is best illustrated by considering a liquid-filled column opening into a closed vessel. If heat is added to the liquid column, a bubble will eventually form in the column. The formation of the bubble raises the liquid above the bubble and decreases the hydrostatic pressure in the column. The gas in the vessel above the column is simultaneously compressed and tends to counteract the pressure decrease produced by the bubble. If the net effect of the two pressure changes is a pressure reduction in the liquid column, more bubbles are formed and soon enough bubbles are generated to eject the water in the column. This phenomena of geysering in liquid-filled lines has been recently studied by Griffith [165]. As pointed out by Griffith, geysering can be suppressed by using a vessel of small volume above the liquid column. Griffith also noted that the frequency of geysering was a function of the rate of heat addition to the liquid column and the entire process of water ejection and runback took about 7 to 8 sec for tests with water and methanol at one and two atmospheres.

An important aspect of flow instabilities caused by nucleation is that they occur only for large degrees of liquid superheat. Another requirement is that small pressure variations produce large volumetric changes and extensive flashing of liquid to vapor. For steam-water mixtures and most other fluids, these conditions prevail only at low pressures. At low pressures the degree of liquid superheat required for nucleation is considerably greater than at high pressure (see Heat Transfer chapter). Also, at low pressures the susceptibility for flashing increases rapidly and the rate of change in volume with respect to pressure is very large. For a homogeneous steam-water system, Beckjord [166] obtained the following expression for the change Δv in mixture specific volume with respect to the initial specific volume v:

$$\frac{\Delta v}{v} = \left\{\left[\frac{-\Delta v_G + \Delta v_L}{v_L} - \frac{[(v_G/v_L) - 1]\,(\Delta E_L - \Delta E_G)}{E_G - E_L}\right] x - \frac{[(v_G/v_L) - 1]\,\Delta E_L}{E_G - E_L} + \frac{\Delta v_L}{v_L}\right\}\left[\left(\frac{v_G}{v_L} - 1\right) x + 1\right]^{-1}. \tag{5-32}$$

In Eq. (5-32) x is the steam quality, and E the enthalpy. Subscripts L and G are used to represent the liquid water and steam gas. Table 5-1 gives values of $\Delta v/v$ as a function of x and pressure for steam-water mixtures for a pressure decrease of 1 psi (or 0.070 kg/cm^2) [166]. Limiting values of this ratio as the steam quality goes to zero (initiation of first bubble) are also shown. At atmospheric pressure, a change of 1 psi (0.070 kg/cm^2)

TABLE 5-1

Relative Isenthalpic Volume Change of Saturated Steam and Water as a Function of Steam Quality

Pressure psia	Pressure kg/cm^2	Δ v/v	Δ v/v x→0
14.7	1.03	$\frac{101.2x + 5.76}{1600x + 1}$	5.76
25	1.76	$\frac{34.9x + 2.26}{963x + 1}$	2.26
50	3.52	$\frac{8.69x + 0.686}{492x + 1}$	0.686
100	7.03	$\frac{2.16x + 0.210}{249x + 1}$	0.210
200	14.1	$\frac{0.553x + 0.0648}{123.5x + 1}$	0.0648
500	35.2	$\frac{0.0786x + 0.0141}{46x + 1}$	0.0141
1000	70.3	$\frac{0.0196x + 0.00443}{19.6x + 1}$	0.00443

as $x \to 0$ will add 5.76 the original volume by flashing, while at 200 psi (14.1 kg/cm^2) the corresponding change would be only 0.065.

5.4.2 Pressure Drop Instability

This type of instability is due to an increase in pressure drop with decreasing flow in a two-phase system. Under such circumstances, it is possible to operate at more than a single value of flow rate. This is illustrated in Fig. 5-5 [169] where a portion of curve A has been labeled unstable because it gives rise to three possible operating points 1, 2, and 3, at a fixed value of the pressure loss. Operation along curves of the type A can lead to sudden flow reductions or large increases in vapor content in the heated section and, therefore, cause premature burnout to occur. The unstable nature of this phenomena was first recognized by Ledinegg [167] and has been discussed in detail by Markels [168] and Chilton [169].

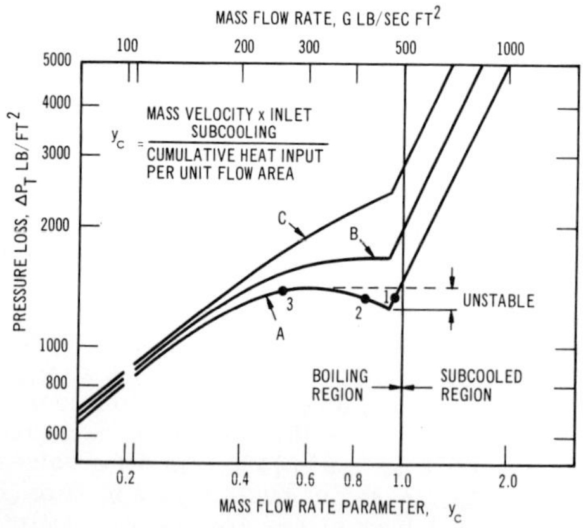

FIG. 5-5 Pressure drop instability model.

A simplified criterion for the occurrence of this phenomena can be developed by considering homogeneous two-phase flow in a channel. If the inlet mass flow rate per unit area is G, the outlet velocity is G/ρ where ρ is the exit density given by Eq. (2-33). The exit steam quality x can be obtained from a heat balance over the channel and is equal to

$$x = [\dot{Q}/AG\,(E_G - E_L)] - (E_L - E_i)/(E_G - E_L)\ , \quad (5\text{-}33)$$

where $\dot{Q}$ is the total heat input, E_i the inlet fluid enthalpy, and A the flow area. Combining Eqs. (2-32) and (5-33) gives

$$G/\rho = [(\rho_L/\rho_G) - 1]\,[\dot{Q}/\rho_L A(E_G - E_L)] + (G/\rho_L)\,\{1 - (E_L - E_i)\,[(\rho_L/\rho_G) - 1]/(E_G - E_L)\}\ . \quad (5\text{-}34)$$

According to Eq. (5-34), for a fixed heat input $\dot{Q}$, the outlet mass velocity can decrease with an increase of inlet mass flow rate if the term in brackets $\{1-(E_L-E_i)\,[(\rho_L/\rho_G)-1]\,/(E_G-E_L)\}$ is negative. This reduced outlet mass velocity leads to a lower two-phase pressure drop even though the inlet flow is increased and is therefore characteristic of curve A shown on Fig. 5-5. A plot of inlet subcooling values which according to Eq. (5-34) will yield unstable pressure drop curves for steam-water mixtures is shown in Fig. 5-6 [171]. Operation above the curve of Fig. 5-6 will give a stable two-phase homogeneous pressure drop curve.

A more exact prediction of conditions which lead to unstable pressure drop curves was proposed by Chilton. Chilton's analysis considers the pressure loss along the entire heated channel and includes acceleration, contraction and expansion losses, nonuniform heat distribution, and subcooled fluid at the inlet of the channel. Chilton neglects the buoyancy effects and assumes that acceleration losses can be calculated from a homogeneous model. The criteria developed by Chilton for uniform and sinusoidal heat input distribution are shown in Fig. 5-7 [169]. They are plotted in terms of the parameters α_1^1 and β_1^1 where

$$\alpha_1^1 = 2(K_T/K_R)\,(E_G - E_L)/a'\,(E_L - E_i)\ ,$$

$$\beta_1^1 = (1/K_h)\,\{[2\,(v_G - v_L)/a'\,v_L] + K_e\}\ . \quad (5\text{-}35)$$

The terms K_h, K_e, and K_T represent single-phase loss coefficients in the heater section, downstream of the heater section, and in the total system.

$$\Delta P_h = K_h\,(G^2/\rho_L g)\ , \quad \Delta P_e = K_e\,(G^2/\rho_L g)\ ,$$

$$\Delta P_T = K_T\,(G^2/\rho_L g)\ . \quad (5\text{-}36)$$

The symbol a$'$ is obtained from the following approximation to the two-phase frictional loss

$$\Delta P_{TPF}/\Delta P_{Lo} = 1 + a'x\ . \quad (5\text{-}37)$$

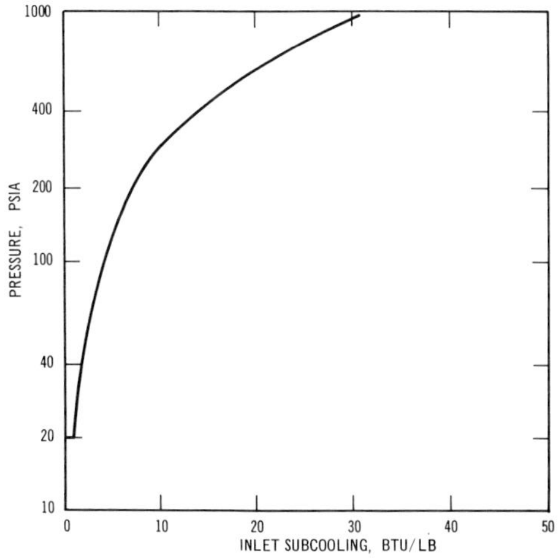

FIG. 5-6 Limit for pressure drop instability for homogeneous steam-water flow.

For given values of the parameters α_1^1, and β_1^1, it is possible to calculate values of the parameter y_c (given by Eq. (5-38) and used in Fig. 5-7) which will avoid pressure drop instability. The grouping y_c is directly related to the mass flow rate G, total heat input $\dot{Q}$, and inlet subcooling, by the relation

$$y_c = G\,(E_L - E_i)\,A/\dot{Q}\,. \tag{5-38}$$

Values of y_c for incipient pressure drop instability are plotted in Fig. 5-7. Figure 5-7 shows that the chances for pressure drop instability are increased when α_1^1 is reduced and β_1^1 increased. From the definition of these two parameters, it can be deduced that pressure drop instability will develop as

1. the inlet subcooling $(E_L - E_i)$ is increased,
2. the resistance beyond the heated section (K_e) is made larger,

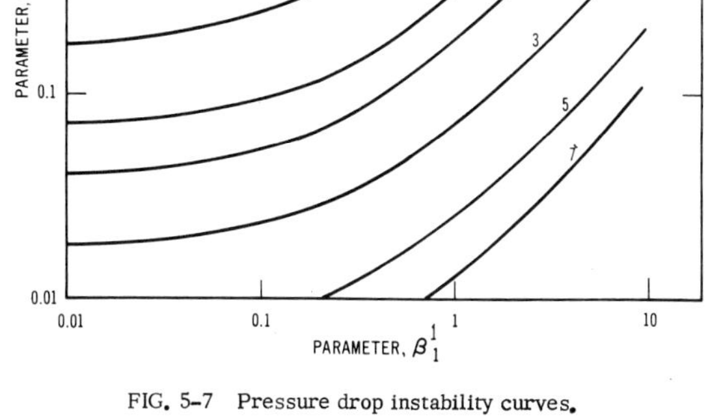

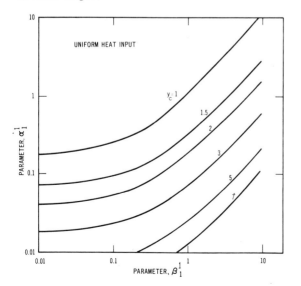

FIG. 5-7 Pressure drop instability curves.

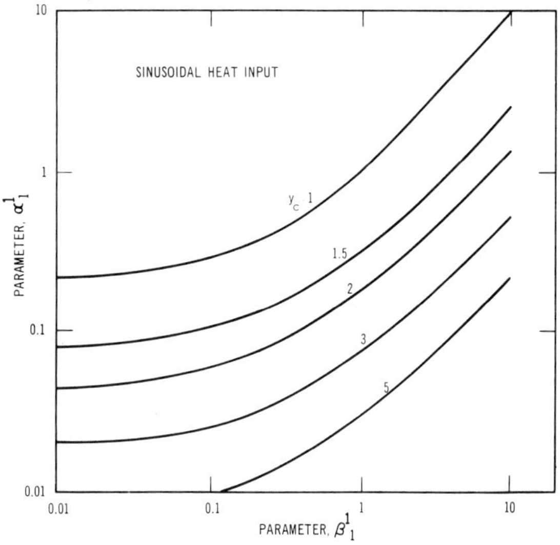

FIG. 5-7 (cont.) Pressure drop instability curves.

3. the resistance at the heater inlet is decreased, and
4. the system pressure is lowered (this is due to a decrease in the parameter a' and the ratio v_G/v_L with pressure).

One final comment should be made about pressure drop instability. It should be recognized that pressure drop instability can lead to a sudden change in flow rate, but it is not the cause of cyclic flow variations. Cyclic flow variations will occur only when the channel or loop under consideration are dynamically unstable. However, as discussed in the next section, operation in a dynamically unstable system along an unstable pressure curve (Curve A, Fig. 5-5) will increase the chances for oscillations.

5.4.3 Closed Loop Flow Oscillations

Flow oscillations can occur in closed loop systems if part of the pressure head driving the flow around the loop comes from unequal density variations in the vertical legs of the loop. Under such circumstances, the hydrostatic pressure unbalance around the loop in part determines the flow rate, and a feedback loop exists between it and the recirculation flow rate. This feedback loop can lead to flow oscillations.

A simple illustration of flow oscillations is obtained by considering a natural circulation loop with a long, large diameter riser, a very short heated section, and a downcomer. Practically all the driving head in such a loop is supplied by the riser, and cyclic variations in riser density can produce periodic variations in the loop flow rate. A pressure balance around the loop gives [170]

$$M'\,[d(G/\rho_L)/dt] + F_f - \hat{a}\,L_R\,(\rho_L - \rho_G)\,g/g_c = 0\,. \tag{5-39}$$

In Eq. (5-39), the first term represents the loop inertia, the second term, the frictional pressure drop around the loop, and the last one, the buoyancy contributed by the riser. The quantity $\hat{\alpha}$ is the time average void fraction in the riser, or

$$\hat{\alpha} = (1/L_R) \int_0^{L_R} \alpha \, dz = (1/T) \int_0^T \alpha \, dt \,. \qquad (5\text{-}40)$$

In Eq. (5-40), L_R represents the vertical riser length and T, the corresponding vapor transit time in the riser. This void fraction $\hat{\alpha}$ acts on the loop flow with a time delay. If the inlet velocity G/ρ_L changes, the riser does not respond until the resulting change in vapor void travels some distance into the riser. The response time is approximately equal to the vapor transit time T, and the loop, if unstable, will oscillate with a period about equal to T.

The quantity M' represents the rate of change of loop pressure drop with respect to the inlet velocity. For flow of a fluid of density ρ_i and velocity V_i in a pipe length l_i and cross-sectional area A_i, M' is equal to

$$M' = (\rho_i A_i l_i/g_c A_i) (dV_i/dt) [dt/d(G/\rho_L)] \,. \qquad (5\text{-}41)$$

Conservation of mass and summation around the loop gives

$$M' = (\rho_L A_h/g_c) \sum l_i/A_i \,, \qquad (5\text{-}42)$$

where A_h is the flow area at the inlet of the heater section.

For a small perturbation $\Delta G/\rho_L$, Eq. (5-39) can be rewritten as

$$M' [d(\Delta G/\rho_L)/dt] + [\partial F_f/\partial(G/\rho_L)] \Delta(G/\rho_L)$$

$$- [\partial\hat{\alpha}/\partial(G/\rho_L] (g/g_c) \; L_R \, \Delta(G/\rho_L) (\rho_L - \rho_G) = 0 \,. \qquad (5\text{-}43)$$

Equation (5-43) is a first-order differential equation which can yield unstable solutions for appropriate combinations of the variables M', $\partial F_f/\partial(G/\rho_L)$, and $\partial\hat{\alpha}/\partial(G/\rho_L)$.

These three terms define the dynamic behavior of a closed loop where the riser plays a dominant role. The flow oscillations are produced by changes in the driving density term $\hat{\alpha}$, and their period will be about equal to the riser void transit time. The chances for oscillation are enhanced as the loop power is raised, since $\partial\hat{\alpha}/\partial(G/\rho_L)$ increases with power. The frictional grouping $\partial F_f/\partial(G/\rho_L)$ acts to damp the oscillations and the system stability improves as this term becomes large. For instance, the addition of an orifice at the inlet of the heater section always increases $\partial F_f/\partial(G/\rho_L)$ and stabilizes the loop. On the other hand, if the two-phase pressure drop curve is of an unstable type, as discussed in subsection 5.4.2, the two-phase contribution to $\partial F_f/\partial(G/\rho_L)$ can act to reduce this term as G/ρ_L increases and the loop becomes less damped. As mentioned previously, high inlet subcooling gives unstable pressure drop curves and increases the chances of instability. Finally, loop stability is increased if the loop inertia constant M' becomes large.

Similar analyses can be developed for other simplified loops. For instance, in test loops where the heated section rather than the riser plays the dominant role, the oscillations will be controlled by the thermal and hydraulic performance of the heater. The frequency of the oscillations will be about equal to the transit time in the heated section, and the chances for loop instability will be again reduced by [170]

1. increasing the loop inertia and the loop throughput velocity, and
2. decreasing the pressure drop associated with the two-phase portion of the heated section and particularly the heated section exit.

The addition of a restriction at the outlet of the heater can in itself produce instabilities. The restriction tends to accumulate vapor in the heated section whenever the flow rate decreases. When this volume of high quality fluid reaches the restriction, it causes a large decrease in pressure drop and the flow rate increases. The cycle can repeat itself and oscillations result.

Analytical solutions and experimental correlations have been developed for more complicated loop geometries. Various models have been proposed by Wallis and Heasley [170], Beckjord [171], Anderson et al. [172], and Wissler [173]. References [170] and [173] postulate homogeneous flow while references [171] and [172] include slip flow. Some of the more refined models described under parallel flow oscillations [177,178] can also be readily modified and applied to calculate loop oscillations. Experimental data on flow oscillations with steam-water mixtures in closed loop systems have been reported in references [172] to [175], and tend to check the analytical results, especially at high pressures. At low pressures (or large ratios of liquid to vapor density), the analysis becomes much more complicated, as it must take into account (see Sec. 5.4.1) the pressure and liquid temperature variation along the heated channel. A low-pressure steam-water model based upon the prediction of the power to void transfer function is developed in references [163] and [164]. This model predicts that instability increases with a reduced boiling length and that the oscillation frequency varies inversely with the nonboiling length of the channel.

5.4.4 Parallel Channel Flow Oscillations

Two possible types of parallel channel flow oscillations are recognized [176]. In the first type, all channels, i.e. the total flow, oscillate in phase with each other. In the second mode, the channel oscillations are out of phase and any flow increase in one channel is compensated by a decrease in other channels so that the total flow is constant and steady. Forced-circulation systems will mostly exhibit the second type of oscillation, while channels in natural circulation can oscillate according to either mode.

The existence of both types of oscillation in a system consisting of n parallel channels can be demonstrated by considering the continuity equation in channel j [176]. Conservation of mass gives

$$dR_j/dt = U_j - F_j\,, \tag{5-44}$$

where R_j is the void volume in the channel and U_j and F_j the exit and inlet liquid volumetric flow rates. Equation (5-44) assumes that the exit void quality in the channel is small and that disturbances are propagated instantaneously through the boiling region.

Conservation of mass specifies the inlet or downcomer volumetric flow rate F_D:

$$F_D = \sum_{j=1}^{n} F_j\,. \tag{5-45}$$

For small sinusoidal variations of the variables, we can write,

$$\delta R_j = \bar{R}_j\, e^{iwt}\,, \tag{5-46}$$

where w is the frequency of the sinusoidal variation and $i = \sqrt{-1}$. By introducing the following definitions of impedance Z and void to flow transfer function H_{RF},

$$Z_D = \delta P_D/\delta F_D\,,\quad Z_E = \delta P_j/\delta U_j\,,$$

$$Z_I = (\delta P_j - \delta P_D)/\delta F_j\,,\quad H_{RF} = \delta F_j/\delta R_j\,, \tag{5-47}$$

Eqs. (5-44) and (5-45) can be combined to give

$$\delta F_j\,(Z_E - Z_I + iw\,Z_E/H_{RF}) = Z_D \sum_{j=1}^{n} \delta F_j\,. \tag{5-48}$$

Equation (5-48) admits two solutions:

$$\delta F_1 = \delta F_2 = \ldots \quad \delta F_n\,,$$

$$Z_E - Z_I + iwZ_E/H_{RF} = nZ_D\,, \tag{5-49}$$

and

$$\sum_{i=1}^{h} \delta F_j = 0\,,$$

$$Z_E - Z_I + iwZ_E/H_{RF} = 0\,. \tag{5-50}$$

Each of the set of Eqs. (5-49) and (5-50) describe one mode of flow oscillation.

Analytical and experimental studies have to date concentrated on the case of Eq. (5-50) with constant total flow. Quandt [177] developed the first analytical solution using small perturbations and a channel integral model. Wallis and Heasley [170] covered the case of homogeneous flow based upon a Lagrangian coordinate approach. More accurate analyses have since been proposed by Meyer and Case [158], Jones [178], and Meyer and Reinhardt [179]. Reference [158] uses an integral momentum model; so does reference [179], except that the equations are linearized to handle small perturbations; reference [178] presents a nodal solution which permits the use of a large number of nodes in the boiling region. Some test results for flow oscillation in parallel channels have been obtained for water-steam mixtures and are listed in references [177] and [158]. The experiments tend to be in good agreement with the analytical predictions.

The conditions which lead to flow oscillations in a parallel channel system are dependent upon a large number of variables. The interrelation of these variables is so complicated that an analysis must be performed in each case to determine the role of any parameter of interest. The physical cause of flow oscillations is, however, easier to explain and will be briefly discussed here [177]. The pressure drop across all the flow channels is constant and for each channel it can be broken down into four components: frictional ΔP_F, elevation ΔP_{EL}, acceleration ΔP_A, and inertial acceleration ΔP_I:

$$\Delta P = \Delta P_F + \Delta P_{EL} + \Delta P_A + \Delta P_I. \tag{5-51}$$

These four components and their sum are shown on Fig. 5-8 [158] for an oscillating flow system. As illustrated in Fig. 5-8, the oscillation amplitude of total pressure drop is different from that of each of the components because they are out of phase. Parallel flow channel oscillations can, therefore, occur when the sum of the amplitudes is zero or total pressure drop constant even though each component, i.e., the flow in the channels, is fluctuating.

5.4.5 Flow Pattern Oscillations

When two phases flow in a channel, it is possible for the two phases to arrange themselves so the flow pattern consists alternately of liquid and gas slugs. This type of flow is called "slug flow" and is usually accompanied by pressure and flow oscillations. The magnitude of the pressure fluctuations is about equal to the hydrostatic pressure associated with the length of each gas slug [180]. The period of the fluctuations is given by the time it takes the gas and liquid slug to travel through the length they occupy [181].

Analyses of slug flow have been presented by Griffith and Wallis [180] for fully developed flow and by Moissis and Griffith [181] for entrance flow conditions. These solutions make it possible to specify the gas and liquid velocities and the average fluid density for slug flow. They, unfortunately, do not determine the operating conditions over which slug flow occurs. Reference [180] proposes an empirical set of curves which is reproduced in Fig. 5-9. Comparison of these curves or other empirical flow regime maps with all the available flow pattern data does not yield good agreement. A very large number of flow pattern maps have been published to date. The curves are valid only for the specific conditions tested and their extrapolation to other conditions is question-

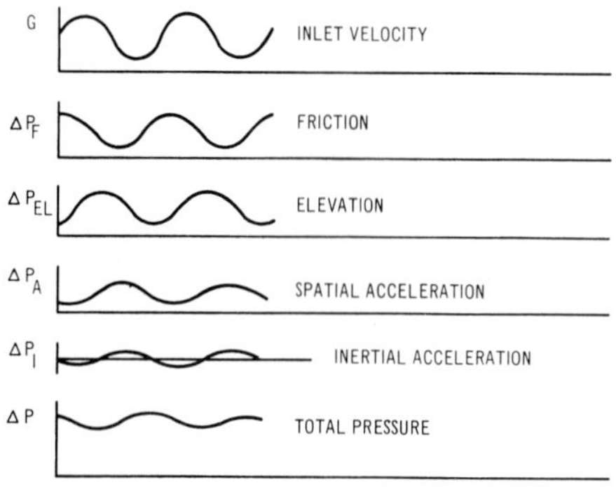

FIG. 5-8 Flow oscillation in parallel channel system.

able. Particularly worthy of note are the flow pattern transition curves of Baker [62] and Kozlov [182].

In reference [50], an attempt was made to predict analytically the boundaries of slug flow. It was shown that in vertical upwards flow, which is of most practical interest to nuclear reactors, slug flow occurs when the energy transfer to the liquid is maximum. This corresponds to a minimum in the measured pressure drop (friction + head loss) as gas is added to the single-phase liquid flow. This is in agreement with a criterion proposed by Govier et al. [183] in their study of flow pattern with air-water mixtures. Reference [50] also defines the transition curve for unsteady upwards annular flow which is characterized by unsteady, large-roll waves at the gas-liquid interface. This transition curve is shown in Fig. 5-10. Plotted on the same figure are the slug flow transition points of reference [183], and a curve is drawn through the test results on the basis that the boundary for slug flow can be correlated by the same variables as unsteady annular flow. It is interesting to note that based upon this correlation, slug flow occurs at lower and lower gas volume fractions as the total flow rate is increased, or the hydraulic diameter is reduced. Slug flow can even be eliminated at high enough values of flow rate or of the frictional parameter $(dP/dL)_{Lo}$. Also, according to Fig. 5-10, the zone between slug and

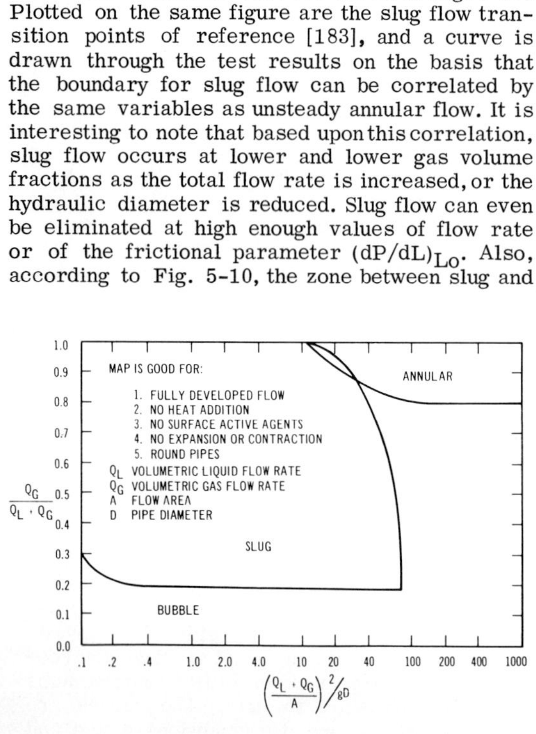

FIG. 5-9 Flow regime map.

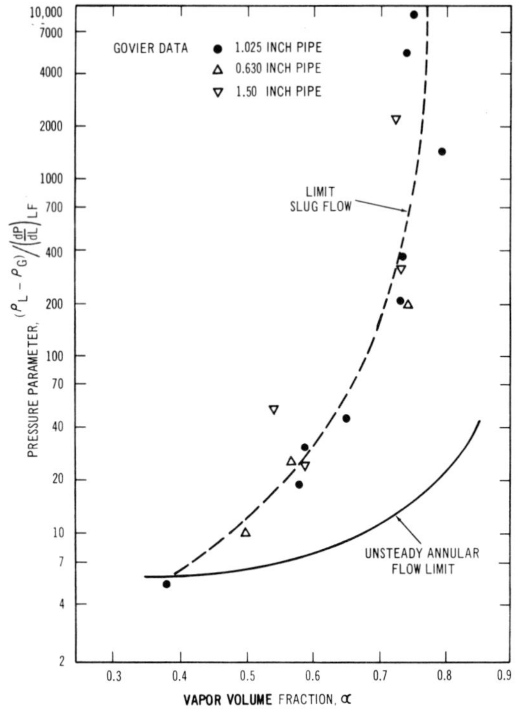

FIG. 5-10 Flow pattern map for upwards two-phase vertical flow.

unsteady annular flow grows as the flow rate is reduced or the hydraulic diameter is increased.

Several studies dealing with flow oscillations have appeared recently in the literature. The following is a brief list of some of the results:

1. Marto and Rohsenow [184] have examined the nucleate boiling instability of alkali metals and have shown that unless artificial nucleation sites are incorporated in the heater surface, the boiling, i.e., the surface temperature, will be unstable. An approximate boiling-stability criterion is presented in the paper.
2. Maulbetsch and Griffith [185] have analyzed system-induced instabilities in forced convection flows with subcooled boiling. By using a linearized, lumped-parameter method, they show that the system is subject to excursive instability (labeled pressure drop instability in this chapter) when the slope of pressure drop versus flow rate of the external system is more strongly negative than that of the heated section. A criterion has also been developed for an oscillatory instability brought about by the presence of a compressible volume either upstream of, or within, the heated section.
3. Several experiments have been performed to study flow oscillations. Some of the tests have been carried out with other fluids besides water (for example, see reference [186] or [187]). Additional tests with steam-water mixtures have also been reported with more emphasis being placed upon measurements of the response of the heated channel to

power modulation (see references [188] and [189]).

4. New models dealing with closed loop and parallel channel flow oscillations have been developed. Most of the recent models have utilized a simplified representation of the two-phase stream in order to obtain non-dimensional groupings or a better understanding of the mechanism responsible for the oscillations (see references [190] and [191]). An interesting evaluation of available analytical models has been made by Neal and Zivi [192], and they concluded that the Jones model [178] gave the best agreement with experimental data. Also worthy of note are the test results and the novel wave propagation model reported by Bijwaard, Staub, and Zuber [193].

5.4.6 Consequences of Flow Oscillations

Flow oscillations produce neutron flux or reactor power oscillations when they are not damped by the reactor feedback loop. In those cases where power oscillations are present, it may be necessary to operate the reactor below the threshold point where the oscillations become large. This can seriously curtail the reactor power level or plant output. It is important to note that when the oscillations develop, the reactor safety is in jeopardy only if the oscillations diverge rapidly. In most practical cases, the oscillation amplitude is low and the period is equal to the controlling fluid transit time. This transit time if of the order of one to several seconds so that the flux trips automatically or the operator can take corrective action before the oscillations diverge too far.

In reactors where power oscillations can occur, tests should be performed at reduced power levels and the test results used to predict the degree of system stability at higher power levels (see chapter on Water Reactor Kinetics). Such careful approach to rated conditions is the best method of minimizing the safety implications.

Another means by which flow oscillations can interplay with reactor safety is through the effect they have on the critical heat flux. Tests performed in natural circulation loops have shown that the critical heat flux decreases when the flow oscillations become excessive [174].

5.5 Special Transient Fluid Flow Problems

This section deals with four special transient fluid flow problems not previously covered: cold fluid addition, loss of coolant, hammer, and shock wave propagation.

5.5.1 Cold Fluid Addition

In many nuclear reactors, it is possible inadvertently to add cold coolant to the reactor recirculating system. If the coolant has moderating properties, its introduction at low temperature can add enough reactivity to produce a nuclear excursion. In water-cooled reactors, this accident is called the "cold water accident." It can occur whenever a loop or piece of equipment has been shut down for a short or long period of time. Inadvertent startup of the loop or component will inject cold fluid into the reactor unless appropriate precautions are taken to avoid it. Should the accident occur, it is important to know the rate of cold fluid addition and the mixed fluid temperature as it enters the reactor core. Calculations of flow rate proceed as in the case of the loss-of-flow accident except that they are performed in the reverse order (see Sec. 5.3). The fluid temperature is obtained by evaluating the degree of mixing between the cold and hot streams. Test results or the methods described in Sec. 4.4 are utilized to calculate the mixed fluid temperature. In most instances, little mixing takes place. For example, experimental investigations show that startup of one cold loop in the Shippingport Atomic Power Station leads to practically no mixing in the reactor inlet plenum [194]. Most of the water entering the vessel from the cold loop travels straight up through a localized (one-quarter) section of the reactor core.

The consequences of cold fluid accidents are directly tied to the amount of reactivity associated with the addition of the cold fluid. Some typical results are discussed in reference [194]. As indicated in that study, the consequences can be held to a tolerable level by means of temperature and valve interlocks. Furthermore, detailed operating procedures are put into effect to avoid occurrence of the accident.

5.5.2 Loss-of-Coolant Accident

A loss-of-coolant accident is defined as one in which the reactor coolant escapes due to loss of integrity of the primary coolant system. The consequences of this accident are dependent upon the size and location of the break in the primary coolant boundary and upon the performance and kind of emergency cooling system provided. We shall concern ourselves here with only the first phase of a loss-of-coolant accident, namely, the calculation of the rate of coolant escape following a rupture in the primary coolant system.

The rate of coolant escape through a break in the primary system depends upon the difference between the coolant pressure within the primary system and the pressure outside the system. This driving pressure differential controls because it is used to overcome the acceleration, frictional, and other pressure losses of the escaping fluid up to the point of the break. In a system design where the volumes inside and outside the primary system are large, the pressure differential remains relatively constant during the accident. In most cases, however, as coolant escapes from the primary system, its pressure decreases while the pressure outside the primary system rises. When this happens, the calculations must take into account the variation of the driving pressure differential with time.

Let us first consider the simplified case of single-phase gas flow with a constant driving pressure differential. The calculations of coolant escape rate follow the methods described in Sec. 2. Typical results for steady flow of a perfect gas in a constant-area, adiabatic channel are shown in

Fig. 5-11. The top curves of Fig. 5-11 give the Mach number at the passage entrance in terms of the duct flow resistance and the ratio of stagnation pressure in the primary system to the back pressure. The corresponding mass rates of coolant escape are shown in the lower graph of Fig. 5-11.

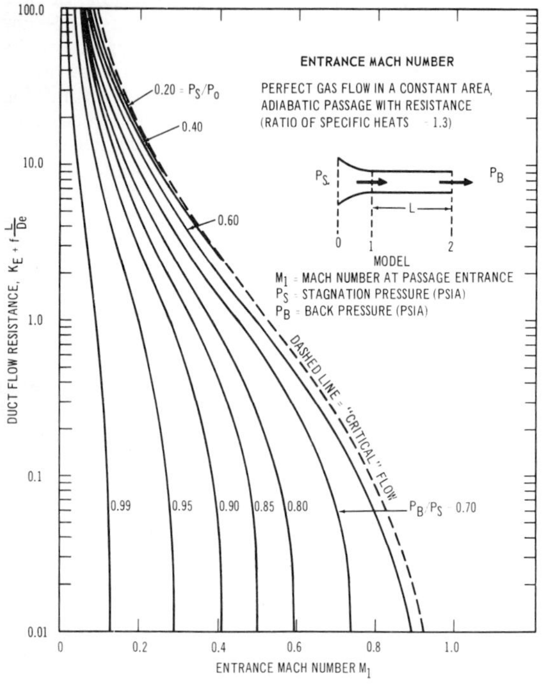

FIG. 5-11 Entrance Mach number for gas escape rates.

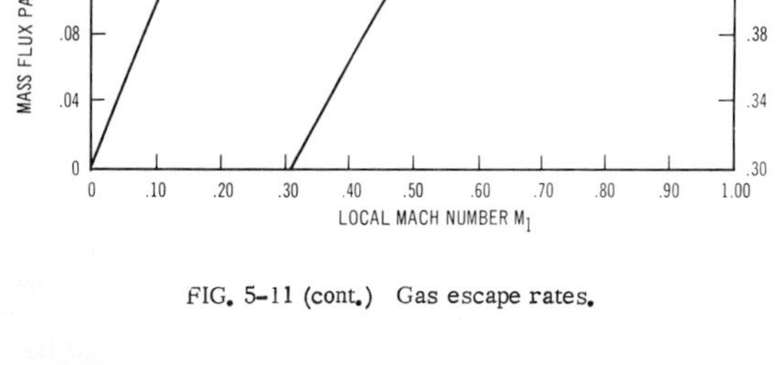

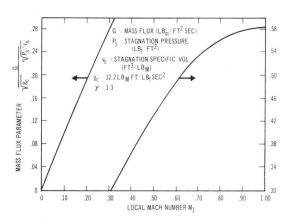

FIG. 5-11 (cont.) Gas escape rates.

The existence of a "critical flow" line corresponding to reaching sonic velocity at the passage exit is noted in Fig. 5-11. When critical conditions exist, the gas coolant escape rate is maximum and remains constant even if the primary system pressure is increased or the back pressure decreased. The occurrence of critical flow is important in loss-of-coolant calculations as it fixes the maximum escape rate of coolant from the primary system. For gas-cooled reactors, the critical flow rate is obtained from Fig. 5-11. It can also be approximated by neglecting the entrance and frictional pressure losses and using Eq. (2-27). Equation (2-27) was derived for critical flow through the throat of a frictionless nozzle.

Let us next consider liquid-cooled reactors. The calculations here depend upon whether the liquid temperature is high enough to produce liquid flashing at or ahead of the break. A typical pressure profile is shown in Fig. 5-12 [195]. In Fig. 5-12 the liquid pressure P_1 decreases due to entrance and friction losses until it reaches saturation pressure at position S. Beyond point S, liquid flashing occurs and two-phase flow exists.

Consider first the case where the liquid temperature is always below saturation. Single-phase flow exists through the entire system of Fig. 5-12. Critical flow seldom occurs under these conditions as liquid sonic velocities are much too high to be generated by the pressure difference $(P_1 - P_3)$. For liquid velocities below the speed of sound, the rate of coolant escape per unit area, G, is calculated by equating the available driving head $(P_1 - P_3)$ to the losses in the flow circuit:

$$G = \sqrt{\frac{2g_c\,(P_1 - P_3)\,\rho_L}{1 + (fL/D_e)}}\,. \tag{5-52}$$

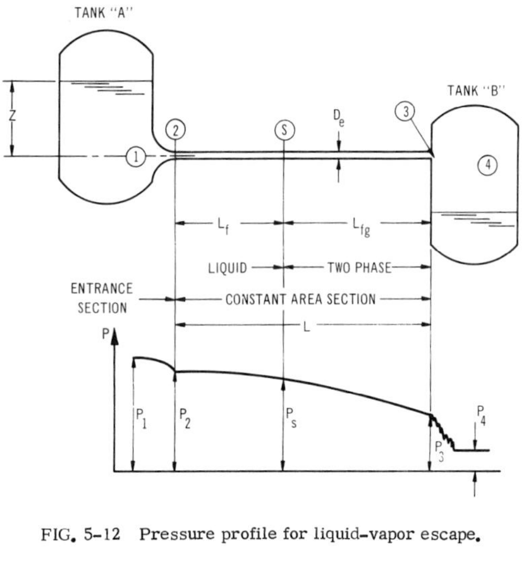

FIG. 5-12 Pressure profile for liquid-vapor escape.

Consider next the case where critical flow and saturation conditions are reached simultaneously at position 3 of Fig. 5-12. Under these circumstances, the fluid reaches acoustic velocity at the same instant that flashing starts. The maximum coolant escape rate per unit area G_0 is obtained from Fig. 5-13 [184]. Figure 5-13 is a plot of G_0 versus the saturation pressure, $P_S = P_3$. It also gives values of the pressure difference $(P_2 - P_S)/[1 + (fL/D_e)]$ required to suppress flashing in the pipe. All the results plotted on Fig. 5-13 are based upon a homogeneous flow model.

In most nuclear systems, saturated conditions are reached before the pipe exit. The flow rate per unit area G between positions 1 and S of Fig. 5-12 is obtained from an equation of the type (5-52):

$$G = \sqrt{\frac{2g_c\,(P_1 - P_S)\,\rho_L}{1 + (fL_f/D_e)}}\,. \qquad (5\text{-}53)$$

Beyond the saturation point, two-phase flow exists and the pressure drop from point S to point 3 is obtained by integrating the continuity, momentum, and energy equations. Typical results calculated by Sajben for homogeneous flow of steam-water mixtures at 1000 and 2000 psia (70.3 and 140.6 kg/cm²) are reproduced in Fig. 5-14 [184]. Figure 5-14 is a plot of the flow ratio G/G_o (where G_o is obtained from Fig. 5-13) in terms of the pressure parameter $(P_S - P_3)/P_S$ and the dimensionless grouping $f_H L_{fg}/D_e$. The symbol f_H represents the two-phase homogeneous friction factor as defined in Sec. 2.

For given values of P_1, P_S, P_3, and L the calculation of coolant escape rate proceeds as follows: The value of G_o is obtained from Fig. 5-13. A value of G is next assumed and Eq. (5-53) is solved for L_f. Similarly, the distance L_{fg} is obtained from the curves of Fig. 5-14. The assumed value of G is correct only if the sum of the distances L_f and L_{fg} adds up to L.

An important shortcoming has been noted in the above calculation. The computed escape coolant rate is low when compared to experimental data and the discrepancy has been traced to the postulate that the flow was homogeneous in the two-phase region. Various models have been proposed in the literature which give better agreement with the test results. The development of such models has evolved from the analysis and measurements of critical flow of a two-phase mixture discharging from a large reservoir into the surroundings.

In such a system, critical flow occurs when the two-phase flow rate stops increasing even though the pressure into which the stream discharges continues to decrease. For homogeneous flow, critical flow conditions are predicted from an energy balance within the flowing fluid and are defined by the relation

$$G_o^2 = -g_c\,[dp/d(1/\rho)]_{\text{constant entropy}}\,. \qquad (5\text{-}54)$$

Values of the critical flow rate G_o obtained for steam-water mixtures from Eq. (5-54) are plotted on Fig. 5-15 in terms of the fluid total energy T.E. The total energy T.E. is equal to

$$\text{T.E.} = E_L(1 - x) + E_G + [(1/\rho_L)(1 - x) + x(1/\rho_G)]\,(G_o^2/2g_cJ)\,, \qquad (5\text{-}55)$$

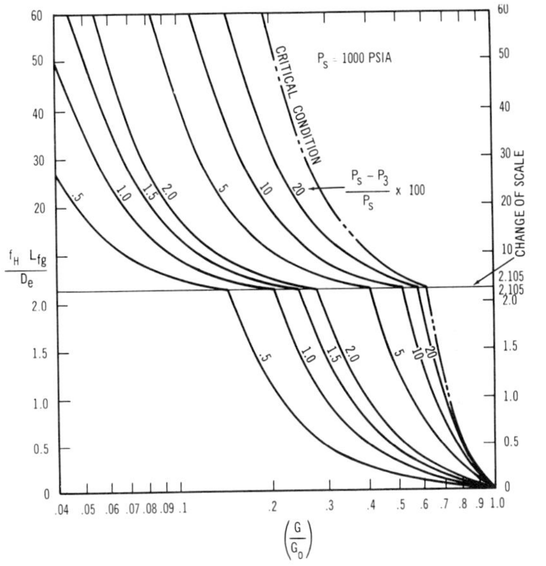

FIG. 5-14 Escape rate of steam-water mixtures.

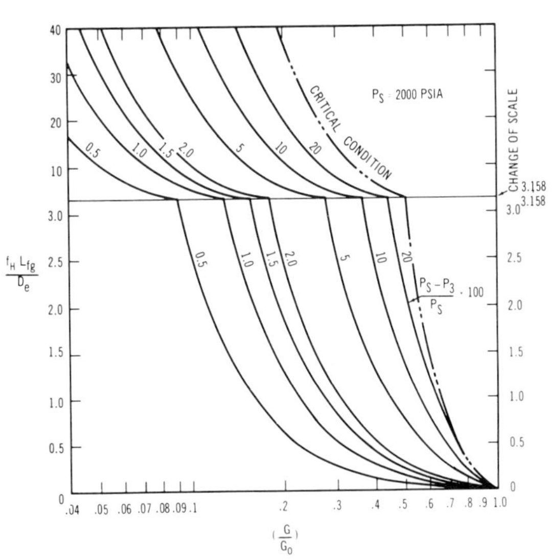

FIG. 5-14 (cont.) Escape rate of steam-water mixtures.

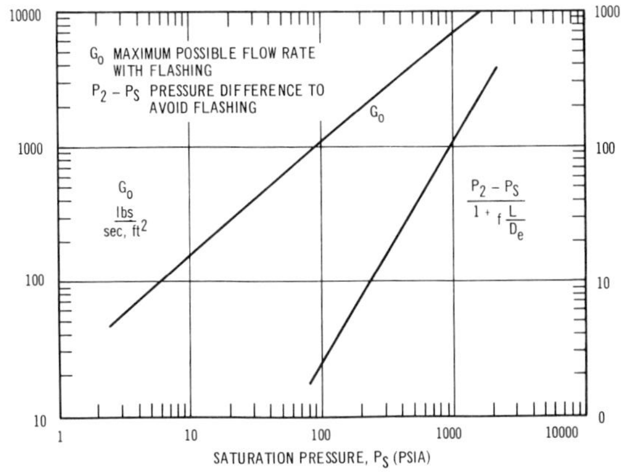

FIG. 5-13 Flow rate to avoid flashing in water escape.

where J is a conversion factor from mechanical to thermal energy.

Deviations of the homogeneous flow predictions from actual measurements with steam-water mixtures are illustrated in Fig. 5-16 [196]. A large discrepancy is noted between tests and analysis at low steam quality, and this explains the recent attempts to develop improved models. Non-homogeneous models have been proposed by Linning [197], Isbin, Moy and DaCruz [196], Massena [198], and Fauske [199]. Fauske's analysis gives the best aggrement with experimental data and his predictions of critical two-phase flow for steam-water mixtures are shown on Fig. 5-17 [199]. The values of steam quality to be used in Fig. 5-17 are obtained from Fig. 5-18 [199] for a given fluid enthalpy E_R in the reservoir.

Many investigations of critical two-phase flow have been reported. Simplified models based upon momentum and energy exchange have been proposed (see reference [200]). The models give adequate agreement with the test data; and except at low vapor content and low pressure, they compare favorably with the predictions of Fauske [199]. More recently, Moody [201] has extended his model to predict critical flow taking into account frictional losses in the discharge pipe. Preliminary experimental measurements of vapor volume fraction at the critical point by Fauske [202] indicate, however, that the ratio of vapor to liquid velocity is lower than postulated by the available models and that nonthermodynamic equilibrium conditions may be prevalent at the critical flow location. These conditions of low vapor slip and nonthermal equilibrium are expected to occur especially when vapor content is low and time for heat exchange between the liquid and vapor is short. For this reason, models have been proposed, based upon homogeneous flow (equal liquid and gas velocity) and no heat transfer between the two phases, to describe the expulsion of liquid-metal coolant in fast reactors (see, for example, reference [203]). The studies which include the effects of superheat show that the expulsion rates may be much higher than originally expected. Experimental studies by Fauske [204] and Zaloudek [205] and a summary paper by Isbin et al. on two-phase critical flow [206] are also worth noting.

All of the previous calculations have dealt with a constant pressure differential driving the fluid out of the primary system. The methods are, however, valid when the pressures within and outside the primary system change with time. For transient pressure conditions, the calculations are performed over small increments of time, and mass, and volume balance equations are written inside and outside the primary system at the end of each time step. The mass and volume balance equations, with appropriate heat transfer assumptions (thermal equilibrium is usually assumed), determine a new value of the driving pressure differential. The coolant escape calculations are then repeated with this new driving head for the next small time interval. Typical results obtained for various size breaks in the water-cooled Shippingport Atomic Power Station are reproduced in Fig. 5-19 [194]. The escape rates shown in Fig. 5-19 are based upon a homogeneous flow model and would tend to underpredict the actual conditions.

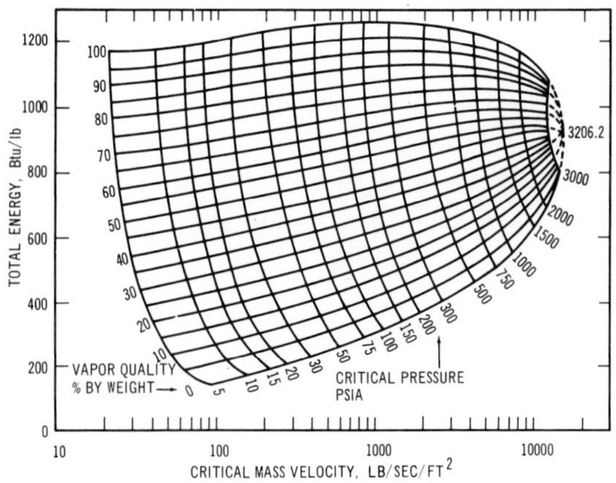

FIG. 5-15 Critical flow prediction from homogeneous model.

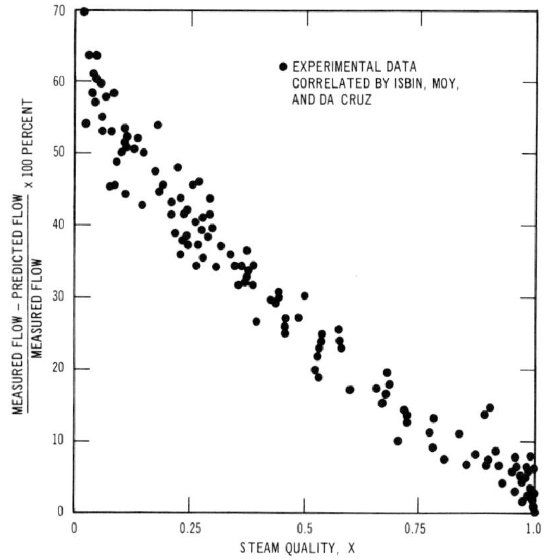

FIG. 5-16 Deviation of critical steam-water flow rates from homogeneous model.

Once the coolant escape rate is specified, the temperatures within the reactor core can be estimated. If the reactor is uncovered in the course of the accident or if the core flow decreases to zero for a sufficiently long time, excessive fuel temperatures result, and some fuel could melt and react chemically with any coolant still present in the primary system. In the case of pipe rupture at operating pressure, there is also the possibility of a large pressure difference which could dislodge the core, the ensuing disarrangement preventing the insertion of control rods. The detailed calculations of the consequences of a loss-of-flow accident are beyond the scope of this chapter, but typical results are, for instance, given in reference [194].

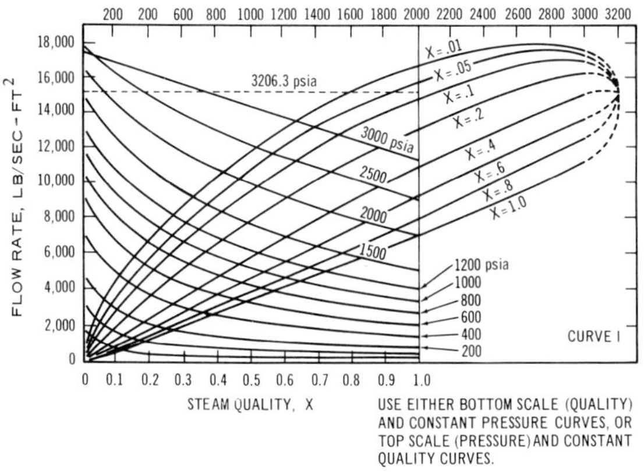

FIG. 5-17 Fauske's critical flow calculations.

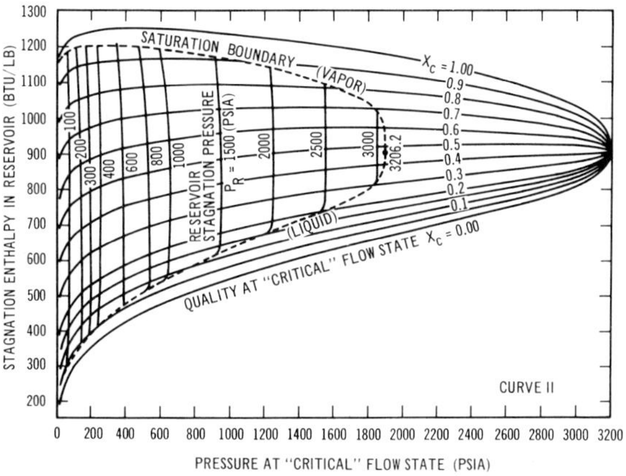

FIG. 5-18 Stagnation properties at "critical" flow states for Fauske's model.

5.5.3 Hammer

When rapid acceleration or deceleration of a fluid occurs in a closed conduit, the change in fluid kinetic energy is converted into a pressure pulse above or below the normal fluid pressure. The pressure pulse is propagated as a wave or "hammer," traveling at acoustic velocity along the conduit or channel containing the fluid. The magnitude of the pressure pulse is determined by equating the change in fluid kinetic energy to the elastic compression or expansion work done on the conduit wall and fluid. High pressure forces are obtained when the change in fluid kinetic energy is large and the fluid incompressible. These forces are sometimes large enough to rupture the conduit.

In nuclear reactors, hammers can be produced by the sudden partial or complete closure of a valve or rapid variations in the speed of a pump. They can also be generated when rapid heat addition to the coolant imparts a sudden acceleration to it.

For a single-phase fluid and a circular conduit of diameter D and wall thickness e_p, the velocity of propagation of the hammer is given by

$$V_H = \{g_c B_f/\rho [1 + (B_f D/B_p e_p)]\}^{1/2}, \qquad (5\text{-}56)$$

where B_f and B_p are the bulk modulus of elasticity of the fluid and pipe material. Typical values of B_f and B_p are listed in Table 5-2 for liquid water and sodium and various pipe materials. When the fluid is a gas, the bulk modulus is considerably smaller due to the compressible nature of the gas. If the gas is compressed or expanded isothermally, B_f is equal to the gas pressure P. For adiabatic conditions, B_f is given by γP where γ is the ratio of the gas specific heat at constant pressure to that at constant volume.

The maximum pressure change associated with a hammer is

$$\Delta P_{max} = \rho \Delta V / V_H g_c . \qquad (5\text{-}57)$$

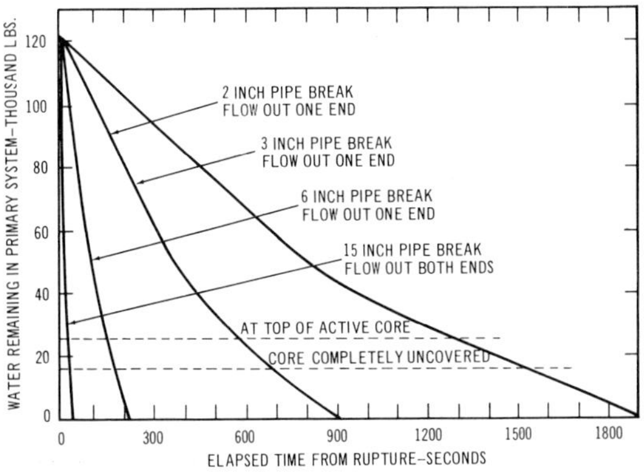

FIG. 5-19 Escape coolant rates from Shippingport reactor.

The maximum pressure rise or decrease will occur if the time in which the velocity change ΔV takes place is less than the time taken by the hammer wave to travel back and forth from the point where ΔV originates to the point of pressure relief. This travel time t_c is equal to

$$t_c = 2L/V_H , \qquad (5\text{-}58)$$

where L is the conduit length measured from the source of the velocity change ΔV to the point of relief pressure such as a reservoir or surge tank.

If the velocity change ΔV occurs in a time t_a greater than t_c, the pressure change produced by the hammer is about equal to ΔP_{max} (t_c/t_a). Similar and more accurate results can be predicted for more complex flow systems and variable rates of change in fluid velocity. Analytical, graphical, and numerical methods can be used to obtain more general solutions as described in references [8] and [207].

Hammers can also occur in two-phase flow. For fluid-solid systems, the pressure change associated with the hammer is greater than in single-phase flow as the solid particles are for all practical purposes incompressible and the pressure forces must be supplied or absorbed in the smaller volume occupied by the fluid. The reverse result is obtained in gas-liquid systems. The size of the

TABLE 5-2

Bulk Modulus of Elasticity

Material	Temperature		Bulk modulus	
	(°F)	(°C)	(lb/ft^2)	(kg/cm^2)
Water (atmospheric pressure)	59	15.0	4.46×10^7	21.8×10^3
	68	20.0	4.58×10^7	22.3×10^3
	100	37.8	4.75×10^7	23.2
	150	65.6	4.72×10^7	23.0
	200	93.3	4.44×10^7	21.6
	212	100	4.35×10^7	21.2
Sodium (atmospheric pressure)	208	97.8	12.4×10^7	60.5×10^3
	300	148.9	11.9×10^7	58.0
	400	204.4	11.5×10^7	56.1
	500	260.0	11.1×10^7	54.0
	600	315.6	10.6×10^7	51.7
Cast Iron	(Relatively independent of temperature)		1.73×10^9	8.44×10^5
Steel			4.18×10^9	20.4
Al alloys			1.44×10^9	7.02
Ni alloys			3.74×10^9	18.2
Wood			0.173×10^9	0.843
Concrete			0.36×10^9	1.76

pressure wave is smaller than for liquid flow, as a large portion of the pressure forces is damped by the compressible gas medium.

The consequences of generating hammers in flow systems vary with the rate and magnitude of the velocity change producing the hammer. They also depend upon the type of fluid and the design of the flow loops. The consequences can sometimes be catastrophic; for instance, in the SL-1 accident, energy generated by the nuclear excursion was rapidly transferred to the reactor water. As steam was formed in the core, it imparted a high velocity to the water above it. It is estimated that the water column displaced from above the core reached a velocity in excess of 100 ft/sec (30 m/sec). Upon its collision with the pressure vessel head, the column's kinetic energy was rapidly converted to compression energy, producing pressures in excess of 10,000 psi (700 kg/cm^2).

The best method for avoiding hammers is to eliminate large and rapid changes in fluid velocity. Slowly-operating valves can be specified or flywheels added to recirculating pumps to reduce the rate of fluid acceleration or deceleration. Relief valves or surge tanks can also be installed to minimize the possible consequences. The surge tank is the simplest device for absorbing or supplying hammer pressure forces. It consists of a static column of fluid and oscillations in its level will rapidly dissipate the change in fluid kinetic energy. Various surge tank designs have been developed. Their principle of operation and the bases for their design are covered in references [8] and [207].

5.5.4 Shock Wave Propagation

Shock waves are flow boundaries across which large and sudden rises of pressure, velocity, and temperature occur. They are usually characterized by a jump from subsonic to supersonic velocity across the wave front. We shall concern ourselves here only with the problem of shock wave propagation.

With the development of supersonic aircrafts, the propagation of shock waves in gases has received considerable attention. The laws governing shock waves in compressible fluids are covered fully in classical hydrodynamic books [6]. Similarly, the propagation of shock waves in water has been carefully studied in relation to underwater explosions. Detailed results are presented by Cole [208]. The status of shock wave propagation in two-phase systems and especially liquid-vapor flow is not as satisfactory. Investigations are presently in progress to measure the speed of sound and the propagation of shock waves in liquid-vapor mixtures. An interesting anomaly occurs with liquid-vapor flow. The speed of sound in liquid-vapor mixtures is lower than that in either component.

A prediction of pseudo-sonic velocity in steam-water based upon a homogeneous model is presented in reference [209]. The calculated values are shown in Fig. 5-20 [209]. The very low sonic velocity calculated at low pressure and low steam quality is caused by the condensation of a large portion of the available vapor, i.e., large change in fluid density, produced by a very small pressure change. The propagation of small and large amplitude waves in a steam-water mixture is also discussed in reference [209]. It is shown in reference [209] that if the liquid weight fraction is low, the liquid is evaporated by the heat of compression; on the other hand, when the steam content is low, the vapor condenses. Preliminary results of pressure wave velocity in boiling water at 10 psi ($\sim$0.7 atm) confirm the predicted trends [209]. Additional work in this area is clearly needed.

A survey of the velocity of sound in two-phase mixtures by Gouse [210] is worth noting.

Additional work is also required in two other problem areas related to shock waves. Investigations should be carried out to determine under what conditions shock waves could be formed in nuclear

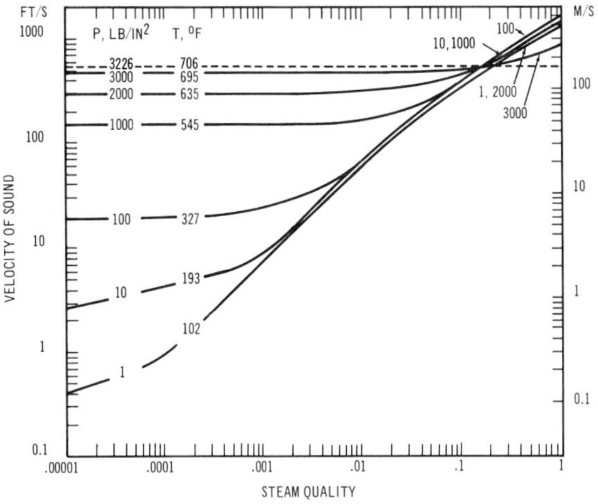

FIG. 5-20 Pseudo-velocity of sound as a function of steam quality.

reactors. It is expected that shock waves can be generated by very large and sudden heat addition to the reactor coolant. Their formation will, however, depend upon the way in which fuel is dispersed in the core following a nuclear excursion and the rate at which heat is transferred from the finely divided fuel to the coolant. Similarly, the consequences of shock waves upon the core internals and pressure vessel integrity need to be studied.

NOMENCLATURE

Dimensions shown are English units. The notations "lb_f," "lb_m" mean "pounds force," "pounds mass" (1 lb_f = 0.0445 x 10^6 dynes, 1 lb_m = 454g).

A	Heat transfer or flow area, ft^2
$A_1, A_2, \ldots$	Constants, dimensionless
A'	Vibration amplitude, ft
a	Speed of sound, ft/sec
a_1	Thickness of fuel plate, ft
B	Constant used in Eq. (2-58)
B_f	Bulk modulus of elasticity of fluid, lb_f/ft^2
B_p	Bulk modulus of elasticity of pipe material, lb_f/ft^2
B'	Constant used in Eq. (3-7)
b	Constant used in Eq. (2-58)
b_1	Width of fuel plate, ft
C_s	Spacer drag coefficient, dimensionless
c_p	Specific heat at constant pressure, Btu/lb-°F
D	Diameter, ft
D_{CH}	Channel diameter, ft
D_F	Fin tip diameter, ft
D_b	Bubble diameter, ft
D_p	Particle or pellet diameter, ft
D_R	Riser diameter, ft
D_r	Root diameter, ft
D_V	Volumetric hydraulic diameter, ft
ENT	Entrainment, dimensionless
E	Fluid enthalpy, Btu/lb
E_{fg}	Heat of vaporization, Btu/lb
E'	Young's modulus of elasticity, lb_f/ft^2
e_p	Thickness of pipe wall, ft
F/F_o	Ratio of platelike surface area to equiaxial particle surface area, dimensionless
F_f	Frictional pressure drop, lb_f/ft^2
F_j	Inlet volumetric flow rate, ft^3/sec
f	Friction factor, dimensionless
f_H	Homogeneous friction factor, dimensionless
G	Mass flow rate per unit area, lb/sec ft^2
G_o	Critical flow rate based on homogeneous flow, lb/sec-ft^2
G_s	Superficial mass flow rate, lb/sec ft^2
g	Gravitational constant, ft/sec^2
g_c	Constant in Newton's law of motion, (lb_m/lb_f) (ft/sec^2)
H	Pump pressure rise, ft
H_a	Absolute pressure, ft
H_s	Static head in the suction vessel above pump, ft
H_{RF}	Transfer function of void to volumetric flow, dimensionless
H_t	Heat transfer coefficient, Btu/sec-ft^2-°R
h	Height, ft
h_1	Flow channel thickness, ft
h_i	Interface height above top of riser, ft
h_v	Height above separating interface, ft
I	Moment of inertia, lb_m/ft^2
I_p	Moment of inertia of rotating parts, lb_m/ft^2
i	Square root of -1
J	Mechanical equivalent of heat, ft-lb_f/Btu
K	Loss coefficient, dimensionless
K_B	Bend loss coefficient, dimensionless
K'	Constant in Eq. (2-36), dimensionless
K_1	Constant in Eq. (2-52)
K_T	Total loss coefficient, dimensionless
K_e	Loss coefficient downstream of test section, dimensionless
K_h	Loss coefficient in heater, dimensionless
k	Thermal conductivity, Btu/ft-sec °R
k_1	Constant related to end fixity of fuel rods
L	Pipe length, ft
L_f	Length of pipe where fluid exists, ft
L_{fg}	Length of pipe with two-phase flow, ft
L_R	Riser length, ft
l	Lead of helix, ft
l_a	Average distance between two channels undergoing mixing, ft
M	Mach number, dimensionless
M'	Rate of change of loop pressure drop with respect to velocity, lb_f-sec^2/ft^3
N	Pump speed, rpm
N_{He}	Hedstrom number, dimensionless
N_{RE}	Reynolds number, dimensionless
$N_{RE}{}^t$	Terminal Reynolds number, dimensionless
N_S	Specific speed, rpm
n	Number of rows or channels

n_{sp}	Number of splitters
P	Pressure, lb_f/ft^2
P_o	Reservoir pressure, lb_f/ft^2
P''	Pressure, atmospheres
P_I	Pressure at reactor inlet plenum, lb_f/ft^2
P_H	Heated perimeter, ft
Q	Volumetric flow rate, gm
$\dot{Q}$	Heat input, Btu/sec
q	Heat input per unit volume, Btu/sec-ft^3
$\dot{q}$	Heat flux, Btu/sec-ft^2
R	Gas constant, ft-lb_f/lb_m-°R
R_h	Derivative of density with respect to enthalpy, (lb/ft^3) (Btu/lb)
R_j	Void volume in channel, ft^3
R_p	Derivative of density with respect to pressure (lb_m/ft^3) (lb_f/ft^3)
r	Radius, ft
r_i	Inside radius, ft
r_o	Outside radius, ft
S	Entropy
S_L	Longitudinal pitch in bundle of rows
S_T	Transverse pitch in bundle of rows
s	Spacer frontal area, ft^2
T	Temperature, °R
T_{sat}	Saturation temperature, °R
T_E	Electrical torque, lb_f/ft
T_H	Fluid torque, lb_f/ft
T_W	Windage torque, lb_f/ft
t	Time, sec
t_a	Actual time for valve closure, sec
t_c	Time for hammer to travel one cycle, sec
U_j	Exit volumetric flow rate, ft^3/sec
U_S	Superficial velocity, ft/sec
U_t	Terminal velocity, ft/sec
u	Velocity, ft/sec
u_m	Mean velocity, ft/sec
V	Average velocity, ft/sec
V_c	Critical velocity, ft/sec
V_{GR}	Gas velocity in riser, ft/sec
V_{GD}	Gas velocity in downcomer, ft/sec
V_{GV}	Superficial velocity in vapor dome, ft/sec
V_S	Velocity in spacer area, ft/sec
v	Specific volume, ft^3/lb_m
W	Mass flow rate, lb/sec
w	Natural frequency of fuel rods, sec^{-1}
X	Martinelli-Lockhart multiplier, dimensionless
x	Vapor weight fraction, dimensionless
x_D	Vapor weight fraction in downcomer, dimensionless
x_R	Vapor weight fraction in riser, dimensionless
y	Distance from channel wall, ft
y_c	Instability parameter, dimensionless
Z	Impedance, $lb_f/sec^2/ft^5$
z	Distance measured in flow direction, ft
α	Volume fraction, dimensionless
α_1^1	Instability parameter, Eq. (5-35)
β	Area ratio, dimensionless
β_1^1	Instability parameter, Eq. (5-35)
β'	Exponent on density profile
Γ	Dimensionless parameter
Γ_L	Liquid flow rate per unit length of film, lb/sec-ft
γ	Ratio of specific heat, dimensionless
δ	Thickness of subcooled void layer, ft
ϵ	Pipe roughness, ft
η	Coefficient of rigidity, lb_m/ft-sec
η'	Impeller efficiency, dimensionless
ζ	Carryunder parameter, Eq. (2-61)
θ	Cavitation constant, dimensionless
Θ	Saturation minus bulk temperature, °F
Λ	Porosity, dimensionless
μ	Absolute viscosity, lb_m/ft-sec
ν	Poisson's ratio, dimensionless
ρ	Density, lb_m/ft^3
ρ_p	Pellet density, lb_m/ft^3
ρ'	Momentum density, Eq. (5-9), lb_m/ft^3
$\underline{\rho}''$	Energy density, Eq. (5-10), lb_m/ft^3
ρ	Average density, lb_m/ft^3
σ	Surface tension, lb_f/ft
τ	Shear stress, lb_f/ft^2
τ_y	Yield stress, lb_f/ft^2
Φ	Dimensionless pressure drop ratio
Ψ	Dimensionless parameter, Eq. (2-38)
Ω	Dimensionless parameter, Eq. (3-17)
ω	Pump rotational speed, rad/sec

Subscripts

a	average
B	bulk
C	contraction
E	expansion
e	equivalent or effective
f	frictional
GF	gas friction
H	heat exchanger
i	interface
L	liquid
LF	liquid friction
LTP	liquid in two-phase flow
Lo	liquid based on total flow
n	position n
p	pipe
R	reactor
S	stagnation or saturation
TP	two-phase
TPF	two-phase friction
W	wall
0,1,2	position

Superscripts

Λ	Integrated over channel length

REFERENCES

1. J. K. Vennard, Elementary Fluid Mechanics, John Wiley and Sons, N.Y., 1947.
2. J. C. Hunsaker and B. G. Rightmire, Engineering Applications of Fluid Mechanics, McGraw-Hill Book Co., Inc., N.Y., 1947.
3. H. Schlichting, Boundary Layer Theory, McGraw-Hill Book Co., N.Y., 1955.
4. H. Lamb, Hydrodynamics, Dover Publications, N.Y., 1945.
5. S. Goldstein, Modern Developments in Fluid Dynamics, Oxford Press, N.Y., 1938.

6. H. W. Liepmann and A. E. Puckett, An Introduction to the Aerodynamics of a Compressible Fluid, John Wiley and Sons, N.Y., 1947.
7. A. H. Shapiro, The Dynamics and Thermodynamics of Compressible Fluid Flow, The Ronald Press, N.Y., 1953.
8. C. Jaeger, Engineering Fluid Mechanics, Blackie and Son, Ltd., London, 1956.
9. L. F. Moody, "Friction factors for pipe flow," Trans. ASME, 66(1944)671.
10. J. P. Waggener, "Friction factors for pressure drop calculations", Nucleonics, 19(1961)145.
11. D. A. Wiederecht and G. Sonnemann, "Investigation of the nonisothermal friction factor in the turbulent flow of liquids", ASME Paper 60-WA-82, 1960.
12. C. S. Keevil and W. H. McAdams, "How heat transmission affects fluid friction in pipes", Chem. Met. Eng., 36(1929) 464.
13. R. G. Diessler, "Analytical investigation of fully developed laminar flow in tubes with heat transfer with fluid properties variable along the radius," Nat. Advisory Comm. Aeronaut. Tech. Note 2410, 1951.
14. E. N. Sieder and G. E. Tate, "Heat transfer and pressure drop of liquids in tubes", Ind. Eng. Chem., 28(1936)1429.
15. F. Kreith and M. Summerfield, "Heat transfer to water at high flux densities with and without surface boiling", Trans. ASME, 71(1949)805.
16. B. S. Petukhov and G. E. Muchnik, "On the problem of the hydraulic resistance in turbulent nonisothermal flow of liquids in tubes", Soviet Physics - Technical Physics, 2(1957)996.
17. H. C. Sher, "Estimations of Boiling and Non-Boiling Pressure Drop in Rectangular Channels at 2000 psi", Report WAPD-TH-300, Bettis, Atomic Power Laboratory, 1957.
18. W. M. Kays, "Loss coefficients for abrupt changes in flow section with low Reynolds number flow in single and multiple-tube systems", Trans. ASME, 72(1950)1067.
19. Tube Turn Catalog - Engineering Data Book No. 211, Tube Turns, Inc., 1949.
20. J. R. Henry, "Design of Power-Plant Installations: Pressure Loss Characteristics of Duct Components", Nat. Advisory Comm. Aeronaut, Wartime Report L-208(ARR No. L4F26), 1944.
21. W. M. Kays and A. L. London, Compact Heat Exchanger, The National Press, Palo Alto, 1955.
22. P. F. Bruins et al., "Friction of fluids in solder type fittings", Trans. Am. Inst. Chem. Eng., 36(1940)721.
23. A. Hofman, "Loss in 90-degree pipe bends of constant circular cross section", Trans. Munich Hydraulic Institute, 3(1929) 29.
24. Hydraulic Institute Standards, Hydraulic Institute, N.Y., 1955.
25. K. Well, Neue Grundlagen der technischen Hydraulik, Oldenbourg, Berlin, 1920.
26. H. Kirchbach, "Loss of energy in miter bends", p. 43 and W. Schubart, "Energy loss in smooth- and rough-surface bends and curves in pipe lines", p. 81, ASME Translation, 1935, Transactions of the Hydraulic Institute of the Munich Technical University, Bulletin 3, R. Oldenbourg, Munich and Berlin, 1929.
27. Crane Valve World, Vol. LIV, No. 1, Crane Co., Chicago, 1956.
28. E. D. Grimison, "Correlation and utilization of new data on flow resistance and heat transfer for cross flow of gases over tube banks", Trans. ASME, 59(1937)583.
29. A. J. Stepanoff, Centrifugal and Axial Flow Pumps, John Wiley and Sons, N.Y., 1957.
30. S. V. Manson, "Gas-friction-heat transfer charts for ducted flows", ASME Paper 57-HT-34,
31. M. E. Davenport, P. M. Magee and G. Leppert, "Heat Transfer and Pressure Drop for a Gas at High Temperature", USAEC Report TID-13485, (Stanford University Report SU-247-2, 1961).
32. L. V. Humble, W. H. Lowdermilk and L. G. Desmon, "Measurements of Average Heat-Transfer and Friction Coefficients for Subsonic Flow of Air in Smooth Tubes at High Surface and Fluid Temperatures", National Advisory Comm. Aeronaut, Rep. 1020, 1951.
33. P. G. Johnson, W. H. Lowdermilk and R. E. Rom, "Temperature and Flow Distribution in Air-Cooled Reactors," Reactor Heat Transfer Conference, 1956.
34. W. B. Hall and E. M. Orme, "Flow of a compressible fluid through a sudden enlargement in a pipe", Proc. Inst. Mech. Engrs., 1969(1955)1007.
35. W. G. Cornell, "The Compressible Flow in a Sudden Contraction in a Channel", General Electric Co., unpublished report.
36. J. T. Higginbotham et al., "A Study of the High Speed Performance Characteristics of 90° Bends in Circular Ducts," Nat. Advisory Comm. Aeronaut Tech. Note 3696, 1956.
37. A. A. Adler, "Variation with Mach Number of Static and Total Pressures through Various Screens", Nat. Advisory Comm. Aeronaut Wartime Rep. L-23, 1946.
38. W. G. Cornell, "Losses in flow normal to plane screens", Trans. ASME, 80(1958)791.
39. N. G. Worley and W. Ross, "The Heat Transfer and Pressure Loss Characteristics of Cross-Flow Tubular Arrangements with Studded Surfaces", Symposium on the Use of Secondary Surfaces for Heat Transfer with Clean Gases, Nov. 9-10, 1960, Inst. Mech. Engr., London, 1961.
40. C. Rounthwaite and N. Cherrett, "Heat Transfer and Pressure Drop Performance of Helically Finned Tubes in Staggered Cross Flow", Symposium on the Use of Secondary Surfaces for Heat Transfer with Clean Gases, Nov. 9-10, 1960, Inst. Mech. Engrs., London, 1961.
41. A. Y. Gunther and W. A. Shaw, "A generalized correlation of friction factors for various types of surfaces in cross flow", Trans. ASME, 67(1945)643.
42. L. J. Goudy and R. D. Teire, "The Development of Compressors and Drives for Gas Cooled Reactors", Franklin Inst., Monograph No. 7, 1960.
43. J. F. Marchaterre and B. M. Hoglund, "Correlation for two-phase flow", Nucleonics, 20, 8(1962)142.
44. R. W. Lockhart and R. C. Martinelli, "Proposed correlation of data for isothermal two-phase two-component flow in pipes", Chem. Eng. Prog., 45(1949)39.
45. R. C. Martinelli and D. B. Nelson, "Prediction of pressure drop during forced circulation boiling of water", Trans. ASME, 78(1948)695.
46. S. G. Bankoff, "A variable density single fluid model for two-phase flow with particular reference to steam water flow," ASME Paper 59-HT-7.
47. G. A. Hughmark, "Holdup in gas-liquid flow", Chem. Eng. Prog. 58(1962)62.
48. M. Petrick, "A Study of Vapor Carryunder and Associated Problems", USAEC Report ANL-6581, Argonne National Laboratory, 1962.
49. P. Griffith and G. B. Wallis, "Two-phase slug flow", ASME Paper 60-HT-28, 1960.
50. S. Levy, "Annular Flow without Liquid Entrainment", Report GEAP-4193, General Electric Co., 1963.
51. S. Levy, "Steam slip theoretical prediction from momentum model", Trans. ASME, Series C, 82(1960)113.
52. G. F. Hewitt, R. D. King and P. C. Lovegrove, "Techniques for Liquid Film and Pressure Drop Studies in Annular Two-Phase Flow", British Report AERE-R-3921, Atomic Research Establishment, Harwell, Berks., England, 1962. See also: J. G. Collier, "Pressure Drop Data for the Forced Convective Flow of Steam/Water Mixtures in Vertical Heated and Unheated Annuli", British Report AERE-R-3808, 1962.
53. H. S. Isbin et al., "Void fractions in two-phase flow", Am. Inst. Chem. Eng., 5(1959)427.
54. C. J. Barocany, A. I. Ch. E. Preprint No. 26, Sixth National Heat Transfer Conference, 1963.
55. J. M. Chenoweth and M. W. Martin, "Turbulent two-phase flows", Petroleum Refiner, 34(1955)151.
56. D. Chisholm and A. D. K. Laird, "Two-phase flow in rough tubes", Trans. ASME 80(1958)276.
57. R. C. Reid et al., "Two-phase pressure drops in large diameter pipes", Am. Inst. Chem. Eng., 3(1957)321.
58. R. H. Moen, Ph.D. Thesis, "An Investigation of the Steam-Water System at High Pressures and High Temperatures", University of Minnesota, 1956.
59. M. Petrick, "Investigation of Two-Phase Air-Water Flow Phenomena", ANL Tech. Memo No. 14, Argonne National Laboratory, 1959.
60. C. J. Hoogendorn, "Gas-liquid flow in horizontal pipes", Chem. Eng. Sci., 9(1959)205.
61. G. W. Govier et al., "The upwards vertical flow of air-water mixtures I. Effect of air and water rates on flow pattern, holdup and pressure drop", Canadian Journal of Chem. Eng., 35(1957)58; and "The upwards vertical flow of air-water mixtures II. Effect of tubing diameter on flow pattern, holdup and pressure drop", Canadian Journal of Chem. Eng., 36(1958)195.
62. O. Baker, "The design of pipe lines for the simultaneous flow of oil and gas", Oil Gas J., 53(1954)185.
63. S. Levy, Second Midwestern Conference on Fluid Mechanics, ASME, Ohio State University, March 1952. Proceedings published by Ohio State Engineering Experiment Station, October 1952.
64. P. A. Lottes and W. S. Flinn, "A method of analysis of natural circulation boiling systems", Nucl. Sci. Eng., 1(1956) 461.
65. B. L. Richardson, "Some Problems in Horizontal Two-Phase Two-Component Flow", USAEC Report ANL-5949, Argonne National Laboratory, 1958.
66. W. H. McAdams et al., "Vaporization inside horizontal tubes ---II. Benzine oil mixtures", Trans. ASME, 64(1942) 193.
67. P. A. Lottes, "Expansion losses in two-phase flow", Nucl. Sci. Eng., 9(1961)26.
68. E. Janssen and J. A. Kervinen, "Two-Phase Pressure Losses", Report GEAP-4202, General Electric Co., 1963.

69. Hanford Report, HW-80970, Rev. 1, March 1964.
70. Committee on Communications Problems in Rheology, Trans. Soc. Rheol., 3(1959)205.
71. H. F. McDuffie and D. C. Kelley, "Homogeneous Reactor Project Quarterly Progress Report for Period Ending October 31, 1954", USAEC Report ORNL-1813 (Decl), Oak Ridge National Laboratory, 1957.
72. A. Einstein, "Contribution on the theory of viscosity of heterogeneous systems", Kolloid-Z, 27(1920)137.
73. A. S. Kitzes and R. N. Lyon, "Aqueous Uranium and Thorium Slurries", Proceedings of First U.N. International Conference on Peaceful Uses of Atomic Energy, Geneva, 1956, Vol. 9, p. 414.
74. D. G. Thomas, "Transport Characteristics of Non-Newtonian Suspensions", in publication.
75. E. C. Gingham, Fluidity and Plasticity, McGraw-Hill Book Co., N.Y., 1922.
76. D. G. Thomas, "Laminar flow properties of flocculated suspension", Am. Inst. Chem. Eng. J., 7(1961)431.
77. A. B. Metzner and J. C. Reed, "Flow on non-newtonian fluids - correlation of the laminar transition and turbulent flow regions", Am. Inst. Chem. Eng. J., 1(1955)434.
78. R. E. Powell and H. Eyring, "Mechanisms for the relaxation theory of viscosity", Nature, 154(1944)427.
79. T. Ree and H. Eyring, The Relaxation Theory of Transport Phenomena, Rheology, Academic Press, N.Y., 1958.
80. E. Buckingham, "On plastic flow through capillary tubes", Am. Soc. Testing Mater. Proc., 21(1921)1154.
81. B. O. A. Hedstrom, "Flow of plastic materials in pipes", Ind. Eng. Chem., 44(1952)561.
82. D. G. Thoma, "Transport characteristics of suspensions, Part IV", Am. Inst. Chem. Eng. J., 8(1962)266.
83. D. Thoma, "Bericht zur Weltkraft Konferenz, London, 1924", Z. Ver. Deut. Ing., 79(1935)239.
84. D. Thoma, "Verhalten einer Kreiselpumpe beim Betrieb in Hohlzog Bereich", Z. Ver. Deut. Ing., 81(1937)972.
85. N. P. Grimm, "Butterfly Valve Cavitation Test", Allis Chalmers Report No. ACNP-62004, 1963.
86. S. G. Bankoff, "Turbulent Liquid Jet Intruding into a Boiling Stream", Proceedings of (Symposium) of 1962 Heat Transfer Fluid Mechanics Institute, Stanford Press, 1962.
87. P. L. Miller and C. P. Armstrong, "Reduction of Vapor Carryunder in Simulated Boiling", USAEC Report ANL-6674, Argonne National Laboratory, 1963.
88. Steam-Water Separation Program, Allis Chalmers Special Report, 1960.
89. G. C. K. Yeh and N. Zuber, "On the Problem of Liquid Entrainment", USAEC Report ANL-6244, Argonne National Laboratory, 1960.
90. N. Zuber, "On the Atomization and Entrainment of Liquid Film in Shear Flow", General Electric Co., Report 62GL153, 1962.
91. N. A. Mozharov, "On the maximum permissible rate of steam flow through a separator", Teploenergetika, 4(1961) 60.
92. C. F. Bonilla, "Mass Transfer in Molten Metal and Molten Salt Systems", Proceedings of First U.N. International Conference on the Peaceful Uses of Atomic Energy", Geneva, 1956, Vol. 9, p. 331.
93. J. N. Taylor, "Mass transfer in liquid metal system", Nuclear Power, 3(1958)53.
94. O. E. Dwyer, "Heat exchange in LMF power reactor systems", Nucleonics, 12, 7(1954)30.
95. "Fluid Meters -- Their Theory and Applications", ASME Research Committee on Fluid Meters, 1959.
96. "Flow Measurement by Means of Thin Plate Orifices, Flow Nozzles and Venturi Tubes", Chapter 4, ASME Power Test Codes, Supplement on Instruments and Apparatus, Part 5, Measurement of Materials, 1959.
97. R. S. Pyle, "Study-3, A Program for the Thermal Analysis of a Pressurized Water Nuclear Reactor, During Steady-State Operation", Report WAPD-TM-213, Bettis, Atomic Power Laboratory, 1960.
98. A. P. Bray, COFFI Code, General Electric Co., San Jose, personal communication.
99. W. C. Reynolds, D. W. Thompson and C. R. Fisher, "HECTIC, An IBM 704 Computer Program for Heat Transfer Analysis of Gas-Cooled Reactors", Report AGN-TM-381, Aerojet-General Nucleonics, 1961.
100. J. P. Hartnett, J. C. Y. Koh, S. T. McComas, "A comparison of predicted and measured friction factors for turbulent flow through rectangular ducts", J. Heat Transfer, 2(1962)82.
101. R. G. Deissler and M. F. Taylor, "Analysis of Axial Turbulent Flow and Heat Transfer through Banks of Rods or Tubes", USAEC Heat Transfer Conference, N.Y., November 1956. (See also: Nat. Advisory Comm. Aeronaut TN-4384.)
102. J. C. Knudson and D. L. Katz, "Fluid Dynamics and Heat Transfer", Engineering Research Institute Bulletin No. 37, University of Michigan, 1954.
103. W. H. Esselman et al., "Thermal and Hydraulic Experiments for Pressurized Water Reactors", Proceedings of Second U.N. International Conference on Peaceful Uses of Atomic Energy, Geneva, 1958, Vol. 7, p. 758.
104. A. N. de Stordeur, "Drag Coefficients for Fuel Element Spacers", Nucleonics, 19,6(1961)74.
105. P. Fortescue and W. B. Hall, "Heat-Transfer Experiments on the Fuel Elements", The British Nuclear Energy Conference, April 1957, Vol. 2, No. 2.
106. A. F. E. Wise, "Investigation of the Polyzonal Spiral Heat Transfer Surface and Its Application to Reactor Design", Proceedings of Symposium on the Use of Secondary Surfaces for Heat Transfer, Institution of Mechanical Engineers, 1961.
107. M. L. Ritz, "Development of the Polyzonal Spiral Fuel Element", Proceedings of Symposium on the Use of Secondary Surfaces for Heat Transfer, Institution of Mechanical Engineers, 1961.
108. C. Cunningham and M. R. Slack, "Heat Transfer and Pressure Drop Performance of Spiral Polyzonal Heat Transfer Surfaces for Gas Cooled Reactors", Proceedings of Symposium on the Use of Secondary Surfaces for Heat Transfer, Institution of Mechanical Engineers, 1961.
109. Burgoyne, Burnett, and Wilkie (initials unknown), UKAEA Report 781(W), 1964.
110. R. W. Bowring, "Physical Model, Based on Bubble Detachment, and Calculation of Steam Voidage in the Subcooled Region of a Heated Channel", OECD Halden Reaktor Prosjekt, Institutt for Atomenergi, Norway, 1962.
111. P. Griffith, J. A. Clark and W. M. Rohsenow, "Void Volumes in Subcooled Boiling Systems", ASME Paper 58-HT-19.
111a. J. F. Marchaterre et al., "Natural and Forced-Circulation Boiling Studies", USAEC Report ANL-5735, Argonne National Laboratory, 1958.
112. R. A. Egen, D. A. Dingee, and J. W. Chastain, "Vapor Formation and Behavior in Boiling Heat Transfer", Report BMI-1163 Battelle Memorial Institute, 1957.
113. G. W. Maurer, "A Method of Predicting Steady-State Boiling Vapor Fractions in Reactor Coolant Channels", Report WAPD-BT-19, Bettis, Atomic Power Laboratory, 1960.
114. J. J. Foglia et al., "Boiling-Water Void Distribution and Slip Ratio in Heated Channels", Report BMI-1517, Battelle Memorial Institute, 1961.
115. H. Christensen, "Power-to-Void Transfer Functions", MIT Sc.D. Thesis, USAEC Report ANL-6385, Argonne National Laboratory, 1961.
116. J. B. Reynolds, "Local Boiling Pressure Drop", USAEC Report ANL-5178, Argonne National Laboratory, 1954.
117. H. F. Buchberg et al., "Heat Transfer, Pressure Drop, and Burnout Studies with and without Surface Boiling for De-Aerated and Gassed Water at Elevated Pressures in Forced Flow System", Proceedings of Heat Transfer and Fluid Mechanics Institute, Stanford University, 1951.
118. R. J. Weatherhead, p. 89, "Reactor Engineering Division Quarterly Report, April 1-June 30, 1955, Section II", USAEC Report ANL-5471, Argonne National Laboratory, 1955.
119. J. J. Jicha and S. Frank, "An experimental Local Boiling Heat Transfer and Pressure Drop Study of a Round Tube", ASME Paper 62-HT-48, 1962.
120. P. A. Lottes et al., "Boiling Water Reactor Technology. Status of the Art Report. Vol. I. Heat Transfer and Hydraulics", USAEC Report ANL-6561, Argonne National Laboratory, 1962.
121. K. M. Becker et al., "An Experimental Study of Pressure Gradients for Flow of Boiling Water in a Vertical Round Duct", Part 3, AE Report, Aktiebolaget Atomenergi, Stockholm, Sweden, April 1962.
122. W. C. Lyons, S. Frank, and M. R. Scheve, "Heat Transfer and Fluid Flow for Fluidized Bed Reactors", Report MND-RP603-31, Martin Nuclear Division, Presented at Fourth National Heat Conference, Buffalo, N.Y., 1960.
123. H. Chelemer and L. S. Tong, "Engineering hot-channel factors for open-lattice cores", Nucleonics, 20,9(1962)68.
124. "Coolant block melts fuel at ETR", Nucleonics, 20,6(1962)86.
125. D. R. Miller, "Critical flow velocities for collapse of reactor parallel plate fuel assemblies", Trans. ASME, 82, Series A, p. 83, 1960.
126. R. J. Scavuzzo, "An Experimental Study of Hydraulically Induced Motion in Flat Plate Assemblies", Report WAPD-BT-25, p. 37, Bettis, Atomic Power Laboratory, 1962.
127. F. J. Remick, "Hydraulically Induced Deflection of Flat Parallel Fuel Plates", Ph.D. Thesis, Pennsylvania State University, 1963.
128. D. Burgreen, J. J. Byrnes and D. M. Benforado, "Vibration of rods inducted by water in parallel flow", Trans. ASME 80(1958)991.
129. E. P. Quinn, "Vibration of Fuel Rods in Parallel Flow", Report GEAP-4059, General Electric Co., 1962.
130. W. J. Taylor, I. Starr and R. L. Baer, "Reactor Flow Studies; Full-Scale Model Flow Tests", Report MND-M-1859, The Martin Co., 1961.
131. E. E. Polomik, "Pressure Drop through 25 Rod Partial NSS Savannah Prototype at 520°F and 1100 psi", Report GEAP-3174, General Electric Co., 1959.

132. E. Janssen and J. A. Kervinen, "Pressure Drop Along a Fuel Cycle Assembly with Various Orifice Configurations", Report GEAP-3655, General Electric Co., 1961.
133. W. A. Sutherland, "Development of In-Core Instrumentation for EBWR", ANL Memorandum.
134. L. K. Holland, "High Power Density Development Project", Fourth Fifth and Sixth Quarterly Reports GEAP-3717, GEAP-3830, and GEAP-3884, General Electric Co., 1961.
135. T. T. Shimasaki and W. J. Freede, "Heat Transfer and Hydraulic Characteristics of the SRE Fuel Element", USAEC Reactor Heat Transfer Conference, 1956.
136. J. Nikuradse, "Geahwindigkeitsverteilung in Turbulenten Stronungen", VDI Zeitschrift, 1926, p. 1229.
137. W. M. Bell and B. W. LeTourneau, "Heat Transfer and Velocity Profiles Near the Corners of Rectangular Channels", Report WAPD-TH-348, Bettis, Atomic Power Laboratory, 1957.
138. R. C. Deissler and M. F. Taylor, "Analysis of Fully Developed Turbulent Heat Transfer and Flow in an Annulus with Various Eccentricities", Nat. Advisory Comm. Aeronaut, TN-3451, 1955.
139. E. Y. Leung, "Heat Transfer with Turbulent Flow in Concentric and Eccentric Annuli with Constant and Variable Heat Flux", Stanford University Thesis (Rep. No. AHT-4), 1962.
140. F. R. Lorenz, "On Turbulent Flow Through Annular Passage", Communications of Institute of Fluid Mechanics, Karlsruhe, 1932.
141. R. R. Rothfus, C. C. Mourad, and V. E. Senecal, "Velocity distribution and fluid friction in smooth concentric annuli", Ind. Eng. Chem., 42,12(1950)2511.
142. W. M. Owens, "Experimental study of water flow in annular pipes", Proc. Am. Inst. Chem. Eng., Vol. 77, Separate No. 88, 1951.
143. H. Barrow, "Fluid Flow and Heat Transfer in an Annulus with a Heated Core Tube", General Discussion on Heat Transfer, IME and ASME, p. 11'3, 1955.
144. J. G. Knudsen and D. L. Katz, "Velocity Profiles in Annuli", Proceedings of Midwestern Conference on Fluid Mechanics, 1950.
145. G. Samuels, "Design and analysis of the experimental gas-cooled reactor fuel assemblies", Nucl. Sci. Eng., 14(1962)37.
146. "A research program in two-phase flow", Centro Informazioni Studii Esperienze, Milano, January 1963.
147. G. F. Hewitt, A. Puffel, and L. E. Gill, "Measurements of the Droplet Size in the Gas Core in Annular Two-Phase Flow", British Report AERE-R-3956, 1962.
148. L. G. Neal, Ph.D. Thesis, Chemical Engineering Department, Northwestern University, Evanston, Illinois, 1962.
149. S. Levy, "Prediction of two-phase pressure drop and density distribution from mixing length theory", ASME Paper 62-HT-6.
150. R. A. Condon and N. C. Sher, "Measurement of void fractions in parallel rod arrays", Nucl. Sci. Eng., 14(1962)327.
151. R. C. Martinelli, "Heat transfer to molten metals", Trans. ASME, 69,8(1947)947.
152. W. M. Bell and B. W. LeTourneau, "Experimental Measurements of Mixing in Parallel Flow Rod Bundles", Report WAPD-TH-381, Bettis Atomic Power Laboratory, 1958.
153. P. A. Nelson, A. A. Bishop, and L. S. Tong, "Mixing in Flow Parallel to Rod Bundles Having a Square Lattice", Report WCAP-1607, Westinghouse Atomic Power Department, 1960.
154. J. S. McNown, C. S. Yih, R. A. Yagle, and W. W. O'Dell, "Tests on Models of Nuclear Reactor Elements, II. Studies on Diffusion", Report APDA-2431-2P, Atomic Power Development Associates, 1957.
155. S. Levy, E. E. Polomik, C. L. Swan and A. W. McKinney, "Eccentric rod burnout at 1000 lbf/in^2 with net steam generation", Int. J. Heat Mass Transfer, 5(1962)595.
156. J. R. Parette and R. E. Grimble, "Average and Local Heat Transfer Coefficients for Parallel Flow Through a Rod Bundle", Report WAPD-TH-180, Bettis, Atomic Power Laboratory, 1956.
157. J. E. Meyer, "Hydronamic models for the treatment of reactor thermal transients", Nucl. Sci. Eng., 10(1961)269.
158. J. E. Meyer and R. P. Case, "Application of a momentum integral model to the study of parallel channel boiling flow oscillations", ASME Paper 62-HT-41, 1961.
159. S. M. Davies, "Simulating sizewell reactor transients with an analog computer", Nucleonics, 21,5(1963).
160. "ART - A Program for the Treatment of Reactor Thermal Transients on the IBM-704", Report WAPD-TM-156, Bettis, Atomic Power Laboratory, 1959.
161. A. J. Arker and D. G. Lewis, "Rapid Flow Transients in Closed Loops", Reactor Heat Transfer Conference, 1956.
162. C. P. Dunlap and A. P. Bray, personal communication.
163. "Kinetic Studies of Heterogeneous Water Reactors", RWD-RL-190, Annual Summary Report for Ramo-Wooldridge, Space Technology Laboratories, Inc., Los Angeles, Dec. 30, 1960.
164. "Kinetic Studies of Heterogeneous Water Reactors", RWD-RL-167, p. 218 Annual Summary Report for Ramo-Wooldridge, Space Technology Laboratories, Inc., Los Angeles, Feb. 29, 1960.
165. P. Griffith, "Geysering in liquid filled lines", ASME Paper 62-HT-39.
166. E. S. Beckjord, "Hydraulic Instability in Reactors", to be published in Nucl. Safety.
167. M. Ledinegg, "Instability of flow during natural and forced circulation", Die Warme, 61(1938)891.
168. M. Markels, "Effects of coolant flow orificing and monitoring on safe pile power", Chem. Eng. Progr. Symp. Ser., 19(1956) 73.
169. H. Chilton, "A theoretical study of stability in water flow through heated passages", J. Nucl. Energy, 5(1947)
170. G. B. Wallis and J. H. Heasley, "Oscillations in two-phase flow systems", J. Heat Transfer, Trans. ASME Series C, 83(363)369.
171. E. S. Beckjord, "The Stability of Two-Phase Flow Loops and Response to Ship's Motion", GEAP-3493 (Rev. 1), General Electric Co., Atomic Products Division, 1960.
172. R. P. Anderson et al., "An Analog Simulation of the Transient Behavior of Two-Phase Natural Circulation Systems", Am. Inst. Chem. Eng., Preprint 27a, 1962.
173. E. H. Wissler, H. S. Isbin, N. R. Amundsen, "Oscillatory behavior of a two-phasenatural circulation loop", Am. Inst. Chem. Eng. J., 3(1956)2.
174. S. Levy and E. S. Beckjord, "Hydraulic instability in a natural circulation loop with net steam generation at 1000 psia", ASME Paper 60-HT-27, 1960.
175. J. Asijee, "Proeven Over De Warmte - A Fooer Van Een Splijtstofelement Vit Een Kern Reactor Van Het Kokeud Water Type", Ph.D. Thesis, Delft University, 1959.
176. S. M. Zivi, "Measurements and Interpretation of Transfer Functions and Hydrodynamic Instabilities in Boiling Loop Experiments", Quarterly Report (Report 18) for Period Ending June 30, 1962, Space Technology Lab. Inc. and Physical Electronics Lab., Canoga Park, Calif., USAEC Report TID-16589. See also: M. Zivi, R. W. Wright, and G. C. K. Yeh, "Kinetic Studies of Heterogeneous Water Reactors", Report STL-6212, Space Technology Laboratories, 1962.
177. E. R. Quandt, "Analysis and Measurements of Flow Oscillations", Chem. Eng. Progr., Monograph and Symposium Series No. 32S, p. 111.
178. A. B. Jones, "Hydrodynamic Stability of a Boiling Channel", Report KAPL-2170, Knolls Atomic Power Laboratory, 1961.
179. J. E. Meyer and E. A. Reinhardt, "A Small Perturbation Approach to the Study of Parallel Channel Boiling Flow Oscillations", Report WAPD-TM-342, Bettis, Atomic Power Laboratory, 1963.
180. P. Griffith and G. B. Wallis, "Two-phase slug flow", J. Heat Transfer, Trans. ASME, Series C, 83(1961)307.
181. R. Moissis and P. Griffith, "Entrance effects in a two-phase slug flow", J. Heat Transfer, Trans, ASME, Series C, 84 (1962)29.
182. B. K. Kozlov, "Types of Gas-Liquid Mixtures and Stability Boundaries in Vertical Tubes", Zh. Tekhn. Fiz., 24,4(1952).
183. G. N. Govier, B. A. Radford and T. S. C. Dunn, "The upwards vertical flow of air-water mixtures", Canada J. Chem. Eng., August 1957, p. 58.
184. P. J. Marto and W. H. Rohsenow, ASME Paper No. 65-HT-22.
185. J. S. Maulbetsch and P. Griffith, MIT Report No. 5382-35.
186. W. Gouse and Andrysiak (initial unknown), MIT Report No. 8973-2.
187. Berenson (initial unknown), Air Force Technical Documentary Report APL TDR G4-117.
188. St. Pierre, Petrick (initials unknown), and S. G. Bankoff, paper presented at the University of Exeter Symposium on Two-Phase Flow, June 1965.
189. G. Possa and J. B. Van Erp, paper presented at the University of Exeter Symposium on Two-Phase Flow, June 1965.
190. Stenning (initial unknown), and T. N. Veziroglu, ASME Paper No. 64 WA-FE-28.
191. J. Boure, Euratom Reports TT 55, Vols. I and II.
192. S. B. H. C. Neal and Zivi (initial unknown), Lecture Series on Boiling and Two-Phase Flow, University of California, 1965.
193. G. Bejwaard, F. W. Staub, and W. Zuber, General Electric Report, GEAP-4778, 1965.
194. L. M. Swartz, A. W. Lemon and L. E. Hulbert, "PWR Loss of Cooland Accident Core Meltdown Calculations", Report WAPD-SC-544, Bettis, Atomic Power Laboratory.
195. M. Sajben, "Adiabatic flow of flashing liquids in pipes", ASME Paper 61-HYD-7.
196. H. S. Isbin, J. E. Moy and A. J. R. DaCruz, "Two-phase steam water critical flow", Am. Inst. Chem. Eng. J., 3,3(1957)261.
197. D. L. Linning, "The adiabatic flow of evaporating fluids in pipes of uniform bore", Proc. Inst. Mech. Eng., IB, 64, 1962.
198. W. A. Massena, "Steam-Water Critical Flow Using the Separated Flow Model", Report HW-75739, General Electric Co., Hanford Atomic Products Operation, 1960.
199. H. Fauske, "Critical Two-Phase Stem-Water Flows", Pro-

ceedings of Symposium of Heat Transfer and Fluid Mechanics Institute, Stanford Press, 1961.
200. S. Levy and F. J. Moody, papers in Journal of Heat Transfer, February 1965.
201. F. J. Moody, ASME Paper No. 65-WA/HT-1.
202. H. K. Fauske, paper presented at the University of Exeter Symposium on Two-Phase Flow, 1965.
203. D. L. Fischer and Häfele (initial unknown), "Shock Front Conditions in Two-Phase Flow including the Case of Desuperheat," paper presented at the Argonne Conference on Safety, Large Fast Reactors, 1965.
204. H. K. Fauske, A. I. Ch. E. Preprint No. 30, Seventh National Heat Transfer Conference.
205. F. R. Zaloudek, Hanford Report, HW-77594, and Battelle Northwest Report, BNWL-34.
206. H. S. Isbin et al., paper presented at the Third International Conference on the Peaceful Use of Atomic Energy, A/CONF. 28/P/232, 1964.
207. G. B. Rich, Hydraulic Transients, McGraw-Hill Book Co., N.Y., 1951.
208. R. H. Cole, Underwater Explosions, Princeton University Press, 1948.
209. H. B. Karplus, "Propagation of Pressure Waves in a Mixture of Water and Steam", Report ARF-4132-12, Armour Research Foundation, 1961.
210. W. Gouse, ASME Paper No. 64-WA FE-35.

CHAPTER 16

Heat Transfer

H. FENECH* and W. M. ROHSENOW
Massachusetts Institute of Technology
Cambridge, Massachusetts

CHAPTER CONTENTS**

*Present address: University of California, Santa Barbara, California.

**This chapter is based on information in the literature or known to the authors prior to January 1969.

1 SOME ASPECTS OF THERMAL CONDUCTION ASSOCIATED WITH NUCLEAR REACTOR SAFETY ANALYSIS

1.1 Introduction

The analysis of the behavior of a nuclear reactor during an accident and the safety evaluation of the accident require the determination of the course of any power excursion which may be involved and the time dependence of the temperature of the various core materials. The heat source is rapidly changing with time. An analysis of the temperature-time behavior introduces problems of unsteady-state conduction in fuel elements whose components may be close to or beyond their melting points and problems of transient heat conduction to a coolant which, in liquid-cooled reactors, may experience a rapid change of phase.

In view of the complexity of these problems and the nonlinear coupling of the neutronic and thermal behavior of the reactor, no analytic solutions have been possible. In most cases one has to resort to a numerical analysis of credible accidents, using digital or analog computers or a combination of both. In this section, special topics associated with the analysis of thermal transients and their safety evaluation are discussed. The thermal conductivity of a mixture of gases and the thermal conductivity of nuclear materials at high temperatures are important in determining the heat conduction in fuel elements as well as the thermal resistance across gaps between fuel and cladding. These topics are treated successively in Secs. 1.2 and 1.3. In Sec. 1.4 the thermal behavior of fuel elements is described. Section 1.5 deals with some special cases of unsteady-state conduction amenable to an analytic treatment and useful in some preliminary hazards evaluation.

1.2 Thermal-Conductivity Coefficients

1.2.1 Pure Gases and Gas Mixtures

Experimental data have been reported in several works, including those listed in references [1] to [8]. Reid and Sherwood [8] outline correlations for calculating the properties of gases and describe methods for interpolation or extrapolation at different pressures and temperatures.

For gases at very high temperatures (1000 to 15,000°K) direct experiments are lacking because of the extreme difficulties associated with these experiments. Amdur and Mason [9] have developed methods and calculated the properties of gases in the temperature range 1000 to 15,000°K; however, they take into account only the molecular translational degrees of freedom and neglect the effects of excitation, dissociation, and ionization. The treatment of gas mixtures is also outlined and developed in detail by Mason and Saxena [10].

In Table 1-1 the calculated values [9] of the thermal-conductivity coefficients of the rare gases in the temperature range 1000 to 15,000°K are given. For the rare gases, the values of the translational thermal conductivity represent the true total thermal conductivity over most of the temperature range. Only for heavier gases at higher temperatures do the effects of electronic excitation need to be taken into account. Mason and Saxena [10] estimate the probable error to be about 5% with an estimated maximum error of 10%.

Cheung [2] has developed formulas for the thermal conductivity and viscosity of gas mixtures which are more accurate than those given in reference [10]; however, the formulas require the diffusion coefficients of each gas in the mixture. The conductivity equation proposed has been compared with 226 binary-mixture conductivities at temperatures from 0 to 774°C (32 to 1425°F). The average deviation is 2.1%. Values of the thermal conductivities of gas mixtures likely to be encountered in fuel elements can be obtained from references [1] and [11]. Some typical values are listed in Tables 1-2a and 1-2b [9, 11].

TABLE 1-1

Translational Thermal Conductivities of Rare Gases at High Temperatures

T (°K)	Units [a]: 10^{-4} cal/cm-sec-°C					
	He	Ne	Ar	Kr	Xe	N_2
1000	8.08	2.59	1.01	0.575	0.344	1.06
1500	10.9	3.39	1.32	0.766	0.457	1.38
2000	14.1	4.13	1.65	0.935	0.556	1.68
2500	17.0	4.80	1.96	1.09	0.647	1.97
3000	19.9	5.43	2.26	1.23	0.732	2.27
3500	22.5	6.02	2.56	1.36	0.811	2.56
4000	25.3	6.61	2.85	1.50	0.885	2.84
4500	27.9	7.20	3.13	1.62	0.959	3.12
5000	30.3	7.79	3.41	1.75	1.03	3.40
5500	32.9	8.38	3.67	1.89	1.11	3.67
6000	35.4	8.93	3.94	2.01	1.17	3.94
6500	38.0	9.49	4.20	2.14	1.24	4.20
7000	40.4	10.0	4.44	2.28	1.32	4.47
7500	42.6	10.6	4.68	2.41	1.38	4.73
8000	45.0	11.1	4.90	2.54	1.45	4.99
8500	47.5	11.6	5.13	2.67	1.52	5.24
9000	49.7	12.1	5.35	2.80	1.59	5.50
9500	52.1	12.7	5.56	2.94	1.66	5.74
10000	54.3	13.2	5.78	3.07	1.74	5.98
11000	58.9	14.2	6.22	3.34	1.88	6.46
12000	63.4	15.0	6.64	3.61	2.01	6.92
13000	67.8	15.9	7.04	3.86	2.13	7.37
14000	72.1	16.8	7.44	4.12	2.25	7.80
15000	76.4	17.6	7.83	4.38	2.37	8.22

[a] Conversion factors: °R = 1.8°K or °K = 0.556°R; 1 cal/cm-sec-°C = 241.9 Btu/hr-ft-°F.

TABLE 1-2a

Atomic Weights of Krypton and Xenon and of Two Mixtures

Gas	In atmosphere	In FP gas
Kr	83.80	85.1
Xe	131.3	134.4
Fission product (FP) gas: 15.3% Kr, 84.7% Xe	—	127.0
Experimental mixture: 15.8% Kr, 84.2% Xe	123.2	—

TABLE 1-2b

Thermal Conductivities in Units of 10^{-7} cal/cm-°C-sec

Gas	Temperature		
	29°C (84°F)	520°C (968°F)	s [c]
He	3670	7360	0.72
Ar	434	914	0.67
Kr	232	534	0.88
Xe	142.7	334	0.88
Experimental gas [a]	149	341	0.86
FP gas [a, b]	147	336	0.86

[a] Compositions as in Table 1-2a.
[b] Conductivity deduced from value of experimental gas.
[c] In the temperature range considered, the thermal conductivity is proportional to T^s. The exponent s is tabulated in this column.

1.2.2 Solids and Aggregates

The fuel elements in heterogeneous reactors can be in the form of pure metals, alloys, ceramics, or cermets. During irradiation the fuel constitution changes and the fuel experiences swelling, cracking, fission-gas evolution, and, in some cases, melting. All these effects are closely associated with the temperature of the fuel, which itself is directly related to the thermal conductivity of the fuel. Generally, the thermal conductivity decreases during irradiation, thus (for fixed fission rate) increasing the fuel temperature. The increase in temperature enhances the release of fission gases and cracking; this further decreases the thermal conductivity and may cause a self-induced worsening of the heat conduction in the fuel, leading in some cases to premature failure of the fuel cladding and direct exposure of the fuel to the coolant.

The theory of heat conduction in solids has not yet reached the point where the thermal conductivity and its temperature dependence can be predicted reliably. Consequently, experimental data should be used whenever available. Data on the thermodynamic properties of solids have been published in numerous forms. One comprehensive compilation of data is given in reference [12]. Powers has made a useful survey of the fundamentals of thermal conductivity of solids at high temperatures [13] and in aggregates [14]. Some interesting aspects of this survey are given below.

The thermal conductivity k of a solid material may be written*

$$k = k_e + k_p = \frac{1}{R_{ev} + R_{ei}} + \frac{1}{R_{pb} + R_{pe} + R_{pd} + R_{pp}} . \qquad (1\text{-}1)$$

The electronic conductivity k_e is the inverse of the electronic thermal resistivity, which has two components: the resistance R_{ev}, caused by scattering of electrons by thermal vibrations of atoms, and the resistance R_{ei}, resulting from scattering of electrons by impurities. The lattice conductivity k_p is the inverse of the sum of four resistances, namely, the thermal resistances resulting from scattering of phonons from boundaries R_{pb}, from electrons R_{pe}, from impurities and defects R_{pd}, and from other phonons R_{pp}. In metals k_e is predominant and in nonmetals k_p is the major conductivity term. The electronic thermal conductivity k_e is related empirically to the electrical conductivity σ by the Wiedemann-Franz ratio,

$$k_e/\sigma T = L . \qquad (1\text{-}2)$$

The constant L is called the Lorentz constant and has the theoretical value of 2.45×10^{-8} when k_e is expressed as watt/cm-°C, σ as ohm^{-1}-cm^{-1}, and T the absolute temperature in °K.

1.2.2.1 Thermal Conductivity of Metallic Materials. Figure 1-1 [13] shows the general trend and magnitude of the thermal conductivity versus temperature for various metallic and nonmetallic materials. For most metals, k either increases or decreases with temperature. For alloys at low temperatures, Fig. 1-1 shows that the conductivities are appreciably lower than those of the base metals, while at high temperatures the conductivity values of alloys and pure metals tend to converge.

Since the electronic thermal conductivity k_e predominates in metals, the Wiedemann-Franz relation, Eq. (1-2), is often used to deduce the thermal conductivity of a metal from electrical conductivity. A more accurate relation, derived by Ewing [15] from curve fitting of numerous data, can be used for predicting the thermal conductivity of liquid and solid metals or alloys as well as organic liquids with a deviation of 5% to 10%. This relation is

$$k = 2.61 \times 10^{-8}(T/r) - 2 \times 10^{-17}(T/r)^2(1/c_p\rho) + 97(c_p\rho^2/MT) , \qquad (1\text{-}3)$$

where k = conductivity in watt/cm-°C,
r = electrical resistivity in ohm-cm,
c_p = specific heat in cal/g-°C,
ρ = density in g/cm^3,
M = molecular weight,
T = absolute temperature in °K.

1.2.2.2 Thermal Conductivity of Oxide Ceramics. For oxide ceramics above room temperature, phonon-phonon interaction by lattice vibration is the principal mode of heat conduction [16]. The thermal conductivity can be related to the mean free path for phonon scattering according to the relation

$$k = \frac{1}{3} C_v v \ell , \qquad (1\text{-}4)$$

where k is the thermal conductivity, C_v the volume heat capacity, v the wave velocity and ℓ the mean free path. The mean velocity can be approximated by $v = (E/\rho)^{1/2}$, where E is the modulus of elasticity and ρ is the density. Since at elevated temperatures both C_v and v are essentially independent of temperature and ℓ varies inversely with temperature, the thermal conductivity of ceramics and nonmetallic materials in general should vary inversely with temperature. Experiment shows that the thermal conductivity of nonmetallic materials varies with temperature as

$$k = A/(T + C) , \qquad (1\text{-}5)$$

where A and C are constants.

Measurements of the thermal conductivity of UO_2 have been reported by Kingery, Hedge, Fieldhouse, and Deem [17]. The data corrected to zero porosity (theoretical density) are given in Table 1-3 [17] and plotted in Fig. 1-2 [17] together with data for ThO_2 and UO_2-ThO_2. Within the claimed experimental errors, the Kingery and Scott data are in good agreement. The low results obtained by Hedge and Fieldhouse (approximately 30% lower than the Kingery values) may be attributed to the fact that the samples had some cracks. Scott presented an empirical expression for the thermal conductivity of unirradiated stoichiometric $UO_{2.0}$:

$$k = \frac{42f}{T' + 400} \text{ (watt/cm-°C)} , \qquad (1\text{-}6)$$

where f is the fraction of theoretical density for UO_2 and T' the temperature in °C. Recently Reiswig [18] gave for the thermal conductivity of unirradiated $UO_{2.0}$ (of theoretical density) from 830 to 2100°C (1526 to 3812°F):

*Nomenclature is listed at end of this chapter.

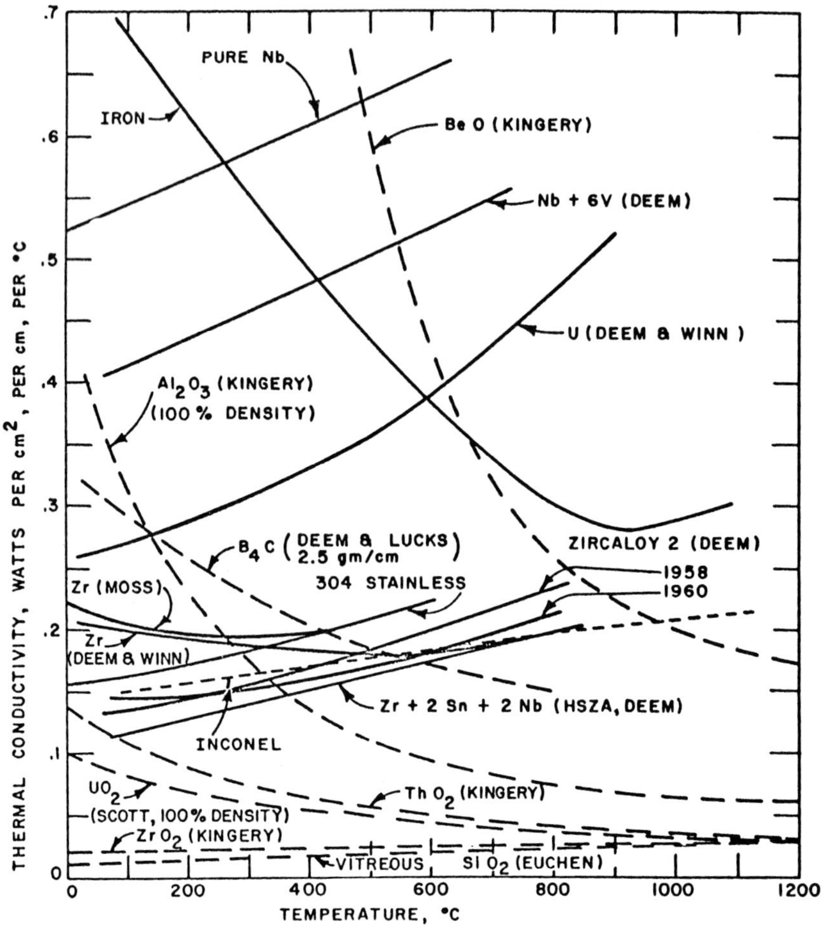

FIG. 1-1. Thermal conductivity versus temperature for various metallic and nonmetallic materials.

$$k = \frac{1}{17.3 + 0.016T} \text{ (watt/cm-°C)}, \qquad (1\text{-}7)$$

where T is in °K. In nonstoichiometric UO_2 the extra oxygen may be accommodated by solid-solution formation or by formation of the second U_4O_9 phase. Measurements of the thermal conductivity of nonstoichiometric UO_2 have been made by Kingery [17], Ross [16]. and Nichols [19] and are plotted in Fig. 1-3 [17]. It is seen that the thermal conductivity decreases with increasing oxygen content in the composition range $UO_{2.0}$ to $UO_{2.1}$. This decrease is probably due to the formation and increasing concentration of the second U_4O_9 precipitate, which has a lower thermal conductivity than UO_2 . Data on uranium dioxide have been reported and discussed in references [20] through [25]. In Ref. [21], Christensen states that "laboratory determinations of thermal conductivity cannot be directly applied to operating fuels. Laboraty specimens are homogeneous and much more complex." The temperature gradients in in-pile fuel are greater than those in laboratory tests. These gradients cause cracks, oxygen migration, and changes in grain structure. Present knowledge does not permit one to say with certainty that the UO_2 fuel center temperature increases, decreases, or remains constant during irradiation [21].

Data on the thermal conductivity of other oxides and ceramics will be found in references [12] and [26].

1.2.2.3 Thermal Conductivity of Heterogeneous Materials. All nuclear solid fuels develop some heterogeneity under irradiation because of cracking, evolution of fission gases, swelling, and other irradiation effects. The subject of heat flow through heterogeneous materials can become very complex when all possible combinations of microstructure are taken into consideration. This subject is treated in detail by Powers [14]. A summary of this work is given in the following paragraphs.

The structure of heterogeneous materials may be classified according to several criteria.

1. Consideration is given to the thermal properties of the medium surrounding a particle.

TABLE 1-3

Thermal-Conductivity Data[a] for Stoichiometric UO_2 Ceramic Specimens

Investigator	Composition	Density (g/cm^3)	Temperature (°C)	Thermal conductivity calculated for nonporous body (watt/cm-°C)
Kingery	$UO_{2.00}$. . .	8.00	200 400 600 800 1000	0.0815 0.0590 0.0452 0.0376 0.0351
Kingery	$UO_{2.00}$. . .	10.08	102 149	0.105 0.0908
Kingery	$UO_{2.00}$. . .[b]	10.55	58 70 82	0.115 0.114 0.114
Hedge and Fieldhouse	$UO_{2.00}$. . .	8.17	202 214 391 638 867 1104 1319 1404 1563 1668	0.0524 0.0510 0.0428 0.0348 0.0264 0.0234 0.0209 0.0199 0.0190 0.0193
Scott	$UO_{2.00}$. . .	10.5	800 900 1000 1100 1150	0.0340 0.0310 0.0275 0.0260 0.0255

[a] Conversion factors: 1 g/cm^3 = 62.43 lb/ft^3; 1 watt/cm-°C = 4.782 Btu/hr-in.-°F.
[b] Reduced from nonstoichiometric $UO_{2.09}$.

a. The concentration of the particles of phase 2 in a medium of phase 1 is sufficiently low so that the perturbation of the heat flux lines due to one particle does not reach the neighboring particles. The conductivity of the medium surrounding a particle may be considered to be that of phase 1 and the equation governing this condition is known as a dilute dispersion equation. Such an equation will yield two conductivity values or two curves, depending on whether the discontinuous phase 2 is of the higher or lower conductivity.

b. If the concentration of phase 2 is increased sufficiently, the conductivity of the medium surrounding a particle may be considered to be equal to that of the mixture. Equations for this condition are called variable-dispersion equations and, like the dilute dispersion equations, yield two curves.

c. When the concentration of phase 2 approaches 50%, particles of phase 2 may touch each other. In this case the mixture is in a state of transition from a dispersion of discontinuous phase 2 to a continuous phase 2 with a discontinous phase 1. The equation describing this transition state is called a mixture equation. Mixture equations yield only one conductivity value or one curve.

2. The shape of the dispersed particles is important. These particles may be spheres or have elongated shapes, resembling, for instance, a rod or a platelet.

3. Another important consideration is the orientation of the particles. Preferred orientation may be in

a. two principal directions (as in a rolled plate),

b. one principal direction and randomness in the other two (as in a mechanically drawn bar), or

c. complete randomness (as in a cast structure or a fluid system).

Some useful equations are presented below for phases consisting of particles which are spherical and nonspherical. A more complete treatment can be found in reference [14].

Heterogeneous materials having a spherical phase. The Rayleigh-Maxwell equation gives the thermal conductivity of a dilute dispersion in the form

$$\frac{k - k_c}{k + 2k_c} = P_d \left[\frac{k_d - k_c}{k_d + 2k_c}\right], \tag{1-8}$$

where k, k_c, k_d are the conductivities of the dispersion, continuous phase, and discontinuous phase, respectively, and P_d is the volume fraction of the discontinuous phase. Curves from this equation are given in Fig. 1-4 [14]. The solid curves are drawn only in the region where they apply. For $k_d/k_c \to 0$, as in porous materials, Eq. (1-8) reduces to

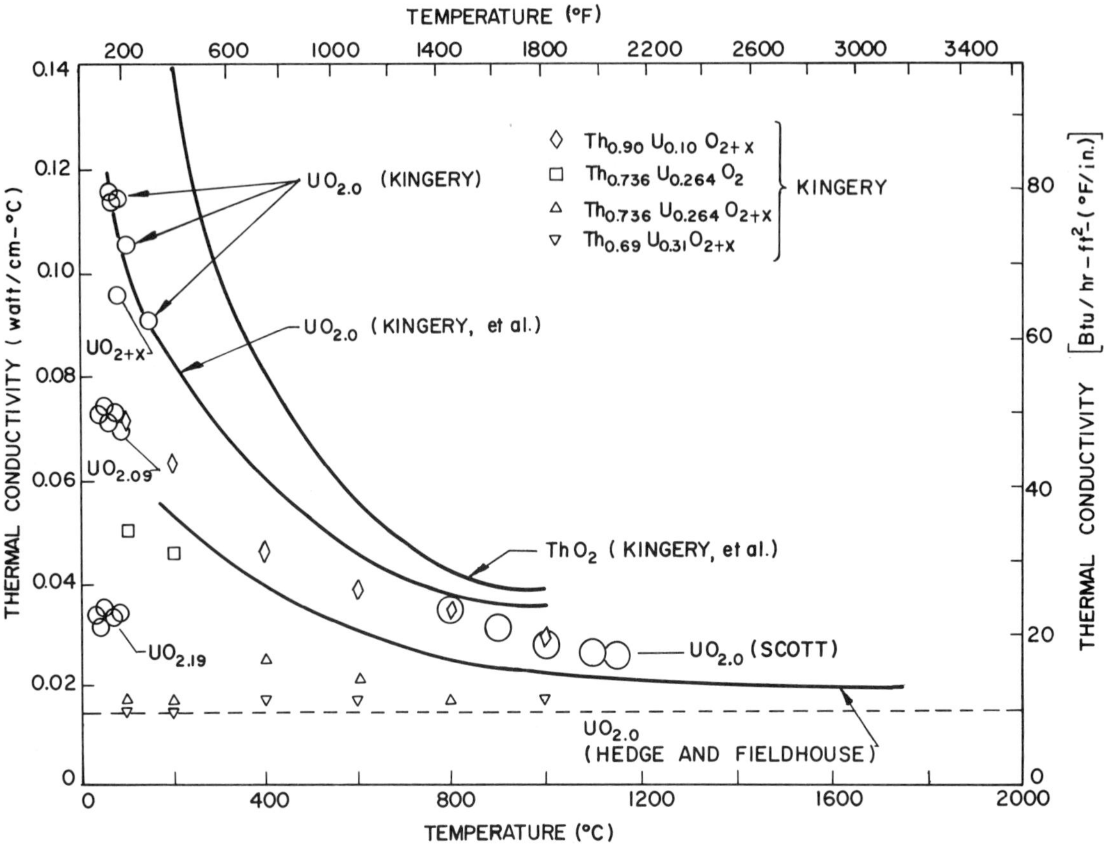

FIG. 1-2. Thermal conductivity of stoichiometric and nonstoichiometric UO_2 of theoretical density as a function of temperature; data for ThO_2 and $UO_2 - ThO_2$ are also shown.

$$k = k_c \frac{1 - P_d}{1 + (P_d/2)} . \quad (1\text{-}9)$$

When $k_d/k_c \to \infty$, as in powder compacts, Eq. (1-8) becomes

$$k = k_c \left(\frac{1 + 2P_d}{1 - P_d}\right), \quad (1\text{-}10)$$

When concentration of the dispersed phase becomes greater than 10% or 15%, the conductivity of the surrounding medium is no longer k_c but can be considered as variable and as k, the conductivity of the total aggregate. An example of a variable dispersion equation is Bruggeman's [27] dispersion equation for spheres:

$$1 - P_d = \frac{k_d - k}{k_d - k_c} \sqrt[3]{\frac{k_c}{k}}, \quad (1\text{-}11)$$

where the symbols have been previously defined. Theoretically, this type of equation may be used for dispersions of high concentration as well as for those that are dilute and for a grain-boundary-film type of structure.

At concentrations of the order of 50%, a discontinous phase no longer remains discontinuous; a mixture-type equation, such as Bruggeman's mixture equation, should be used:

$$P_1 \left(\frac{k_1 - k}{k_1 + 2k}\right) + P_2 \left(\frac{k_2 - k}{k_2 + 2k}\right) = 0 . \quad (1\text{-}12)$$

A plot of the dilute dispersion, variable dispersion, and mixtures equation is given in Fig. 1-5 [14]. Kingery's data for a two-phase system, MgO-BeO, are shown in Fig. 1-6 [28]. The dotted line is the expected thermal conductivity which indicates the transition from a dispersion to a mixture.

Heterogeneous materials with nonspherical phase. When spheres are elongated, the extremes in shape become infinite rods or laminae. Aggregate conductivity for rods and laminae with planes parallel to the heat flux is given by the equation

$$k_p = k_1 P_1 + k_2 P_2 , \quad (1\text{-}13)$$

where k_p is the conductivity of the aggregate, k_1, k_2 and P_1, P_2 are the conductivities and volume fractions of phases 1 and 2, respectively. For laminae perpendicular to the flux, the series equation is applicable:

$$1/k_s = P_1/k_1 + P_2/k_2 . \quad (1\text{-}14)$$

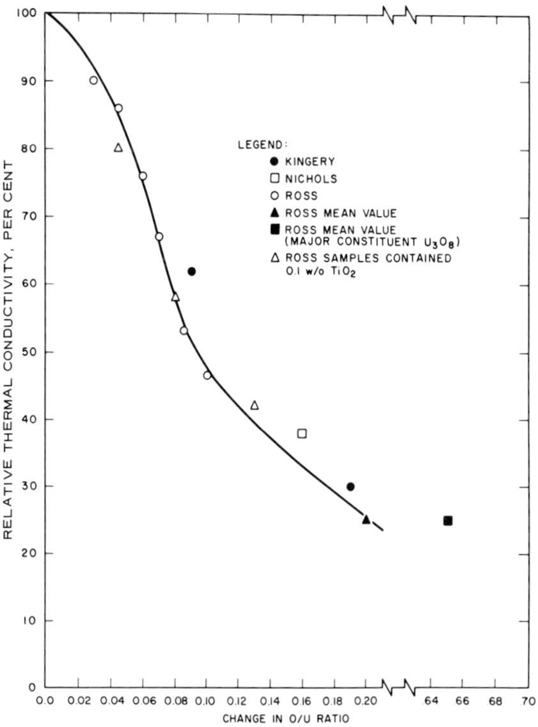

FIG. 1-3. Thermal conductivities of nonstoichiometric uranium oxides relative to stoichiometric UO_2.

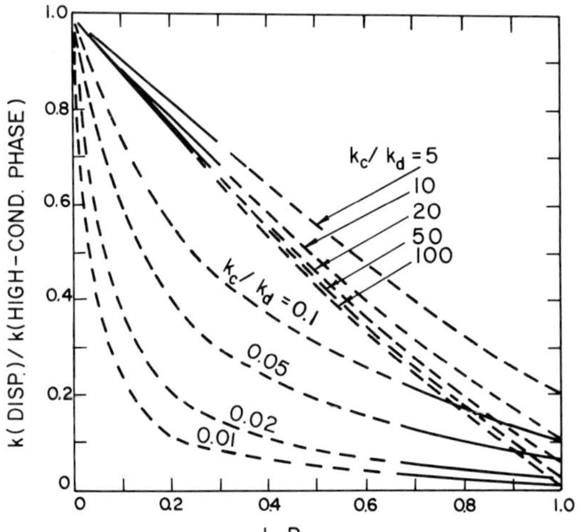

FIG. 1-4. Variation of thermal conductivity (k) of two-phase material (normalized with respect to the conductivity of the high-conductivity phase) as a function of ratio of conductivities of continuous and discontinuous phases (k_c/k_d) and volume fraction of continuous phase (1-Pd). Note that in the upper set of curves $k_c > k_d$, i.e., each of these curves corresponds to a high-conductivity continuous phase. Each curve in the lower set of curves ($k_c < k_d$) corresponds to a low-conductivity continuous phase.

For rods perpendicular to the flux, a Rayleigh-derived [29] equation is used:

$$\frac{k - k_c}{k + k_c} = P_d \left(\frac{k_d - k_c}{k_d + k_c} \right) . \qquad (1\text{-}15)$$

Equations for other types of heterogeneity and phase concentrations are given in reference [14].

Kingery [28] found that cracks within the material can change the conductivity of an aggregate considerably, particularly if the cracks are perpendicular to the flow of heat.

Effect of particle coating on thermal conductivity. Kerner [30] has given a solution to the problem of calculating the effect of coatings on

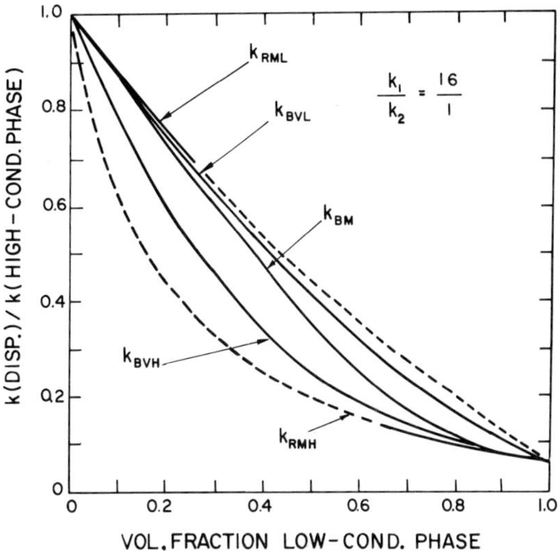

FIG. 1-5. Conductivity of dispersions and mixtures of spheres versus volume fraction of low-conductivity phase according to various equations k_{RML} and k_{RMH} = conductivity according to Rayleigh-Maxwell (dilute dispersion) equation; k_{BVL} and k_{BVH} = conductivity according to Bruggeman's variable dispersion equation; k_{BM} = conductivity according to Bruggeman's mixture equation. The conductivity of the dispersions and mixtures is normalized with respect to the conductivity of the high-conductivity phase. Ratio of conductivities k_1/k_2 = 16.

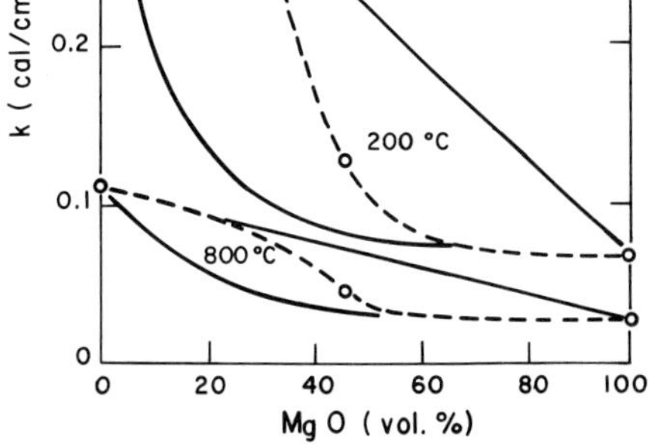

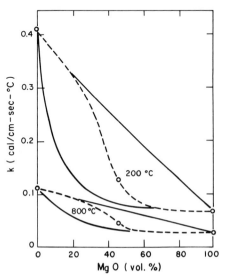

FIG. 1-6. Thermal conductivity in the two-phase system MgO-BeO at 200°C (392°F) and 800°C (1472°F). Broken curves are the thermal-conductivity curves derived from experiment; solid curves are the results expected for one end member as a continuous phase with the other end member dispersed in it. (Note: conversion factor is 1 cal/cm-sec-°C = 20.02 Btu/hr-in.-°F = 2903 Btu/hr-ft^2-(°F/in).)

dispersed particles. The derivation is discussed in reference [14]. For the case of particles 2 coated with material 3 and dispersed in medium 1, the equation for overall conductivity is

$$k = \frac{k_1 P_1 + k_2 P_2 (E_{2z}/E_{1z}) + k_3 P_3 (E_{3z}/E_{1z})}{P_1 + P_2(E_{2z}/E_{1z}) + P_3(E_{3z}/E_{1z})}. \quad (1\text{-}16)$$

The quantities E_{1z}, E_{2z} and E_{3z} are the mean field strengths in the corresponding phases in the direction of the external field and are calculated from the equations:

$$\frac{E_{2z}}{E_{1z}} = \frac{9k_1k_3}{(k_2 + 2k_3)(k_3 + 2k_2) + 2\left(\frac{P_2}{P_2 + P_3}\right)(k_1 - k_3)(k_3 - k_2)}, \quad (1\text{-}17)$$

and

$$\frac{E_{3z}}{E_{1z}} = \frac{3k_1(k_2 + 2k_3)}{(k_2 + 2k_3)(k_3 + 2k_1) + 2\left(\frac{P_2}{P_2 + P_3}\right)(k_1 - k_3)(k_3 - k_2)}. \quad (1\text{-}18)$$

Figure 1-7 [13] gives an example of the effect of particle coating on the conductivity of a dispersion of 15 vol.% of spheres, using a 10/1 ratio between the conductivities of medium to spheres and of coating to medium. For thin coating (below 2% of total volume), the conductivity of the mixture is less than the conductivity of the same mixture without coating, even though the conductivity of the coating is higher than the conductivity of both the continuous medium and the spheres.

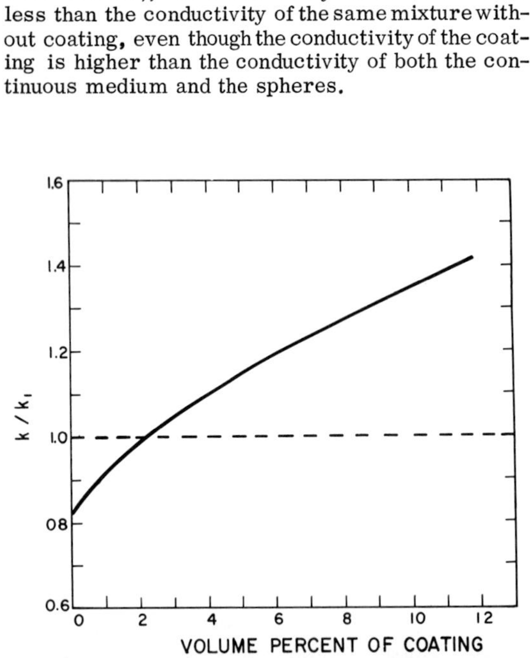

FIG. 1-7. Effect on the overall conductivity of a coating on 15 vol. % spheres in a dispersion. The overall conductivity k is given relative to the conductivity of the continuous medium k_1. The ratio of the conductivity of the coating k_3 to the conductivity of the dispersed spheres k_2 is assumed to be $k_3/k_2 = 100/1$; the ratio k_1/k_2 is assumed = 10/1.

1.2.2.4 Effects of Radiant Heat Transfer on the Thermal Conductivity of Solids at High Temperatures. From measurements of thermal conductivity carried out on polycrystalline, nonmetallic materials at temperatures above 1200°C (~2190°F), it has been found that the conductivities no longer follow the 1/T law at these temperatures [26]. A minimum occurs at about 1500°C (~ 2830°F) and conductivity rises again at higher temperatures as shown in Fig. 1-8 [26]. This effect has been attributed to radiant heat transfer at high temperatures. Heat transferred by radiation from one point at absolute temperature T_1 to another at absolute temperature T_2 in a solid body can be described approximately by the relation

$$q = \frac{AT^3_{avg}}{\alpha}[T_1 - T_2], \quad (1\text{-}19)$$

where α is the absorptivity of the medium in reciprocal length units and A is a constant embodying radiation constants and geometric factors. The constant A, because of complex boundary conditions, cannot be evaluated analytically except for the simplest cases. An overall effective thermal conductivity can be defined as the sum of the thermal and radiant conductivities:

$$k_{eff} = k_t + k_r, \quad (1\text{-}20)$$

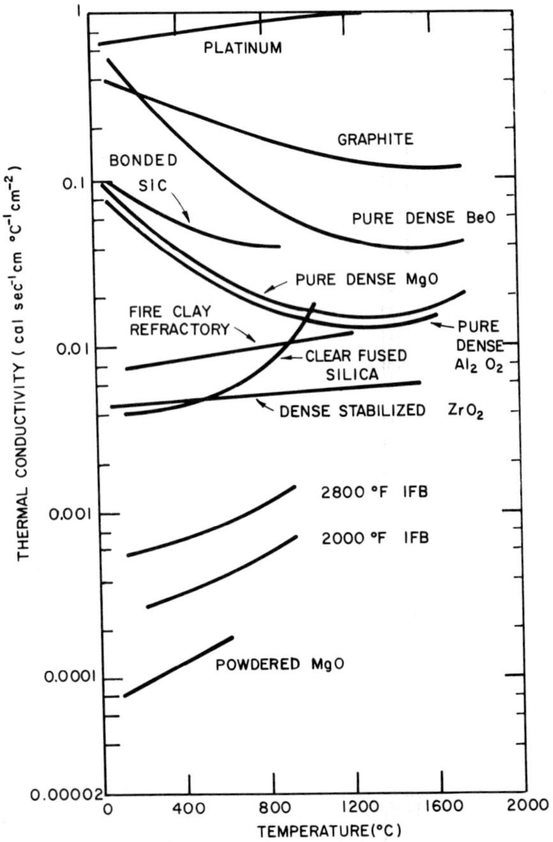

FIG. 1-8. Thermal conductivity of several ceramics over the temperature range ~0 to 1800°C (~30 to 3300°F).

where k_t is the thermal conductivity of the solid and $k_r = cT^3$ (c = constant).

Using Eq. (1-20), Bates [31] predicted an increase in the thermal conductivity of UO_2 for increasing temperature beyond 1500°K. The value of the constant c was estimated and the results shown in Fig. 1-9 [31] were in good agreement with postirradiation examination of some UO_2 fuel rods.

When the material is porous, radiation across the pores can be accounted for by the following equation developed by Loeb [32]:

$$k_m = k_s(1 - P_x) + k_s P_x \left[\frac{P_y k_s}{4\epsilon\gamma d T_m^3} + (1 - P_y) \right]^{-1}, \qquad (1\text{-}21)$$

where k_m is the conductivity of the porous material in the direction of heat flow, k_s is the conductivity of the solid (including radiant conductivity), P_x is the fraction occupied by pores of a cross-sectional area perpendicular to the direction of heat flow in a plane crossing the pores, P_y is the fraction occupied by pores of the length of a line of heat flow that passes through the pores, σ is Stefan's radiation constant (5.670×10^{-5} ergs/cm^2-sec-°K), ϵ the total emissivity of solid surface, γ the geometrical pore factor (2/3 for spherical pores), d the mean pore size, and T_m the mean material temperature.

The effect of radiation on pore conductivity is proportional to the pore size and to the third power of temperature. Pores of large size contribute to increasingly high conductivity at high temperatures, while small-sized pores remain a good barrier to heat flow. The calculated conductivity of different-sized pores over a wide temperature range as calculated by Kingery et al. [26] is illustrated in Fig. 1-10 [26].

1.2.2.5 Effect of Neutron Irradiation on the Thermal Conductivity of Solids and Aggregates. Changes in the thermal conductivity in a fast-neutron, thermal-neutron radiation field warrant close attention both because thermal conductivity is a critical parameter in reactor operation and because the changes produced in this property by neutron irradiation often seem to be particularly large. Although irradiation-produced density and dimensional changes are generally in the range of a few percent, the thermal resistivity may change by as much as a factor of 20 or 30. Changes in thermal conductivity due to neutron interaction are the result of displaced atoms and of the transmutation of atoms of impurities. At temperatures above room temperature, atom displacement is largely removed by annealing. The formation of foreign atoms in metals, soluble or insoluble, continuously reduces conductivity at elevated temperatures.

The thermal conductivity of nonmetallic materials is affected more than that of metals by elevated-temperature irradiation [13].

The effect of low-temperature irradiation on the thermal conductivity and density of a number of nonmetals has been measured for two neutron exposures by Crawford and Wittels. The reported values are listed in Table 1-4 [33a]. Irradiation was carried out in the MTR with a thermal-neutron flux approximately ten times the fast-neutron flux. Therefore, considerable activation of atoms of medium and high thermal-neutron-capture cross section occurred. The thermal conductivities reported were measured at 20°C (68°F) and the densities at 25°C (77°F).

The thermal-conductivity data of BeO and graphite have been reviewed [34a]. The thermal conductivity of graphite changes rapidly during irradiation. Thermal resistivities 40 to 50 times the preirradiation values have been observed [35a, 35b] after fast-neutron irradiations at 30°C (86°F). Measurements were taken in the parallel and transverse directions relative to the extrusion direction of the graphite bar. The neutron-irradiation effect on thermal conductivity saturates as neutron expo-

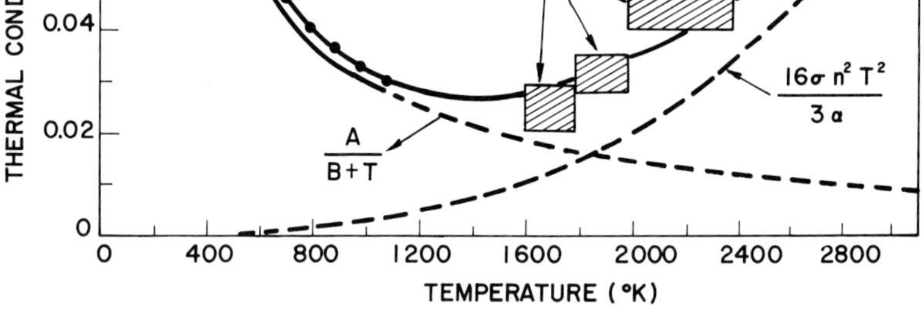

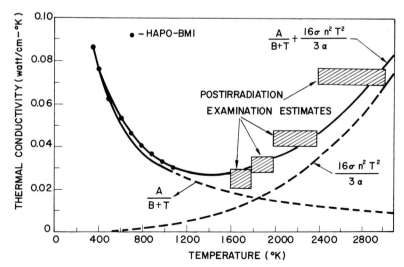

FIG. 1-9. Thermal conductivity of UO_2 versus temperature. Solid curve is predicted sum of lattice [A/(B + T)] and radiation ($16\sigma n^2 T^2/3\alpha$) contributions, where n = refractive index, α = optical absorption coefficient (cm^{-1}), T = temperature (°K). (Conversion factors: 1 watt/cm-°K = 0.2388 cal/ sec-cm-°C = 693 BTU/hr-ft^2-(°F/in.)).

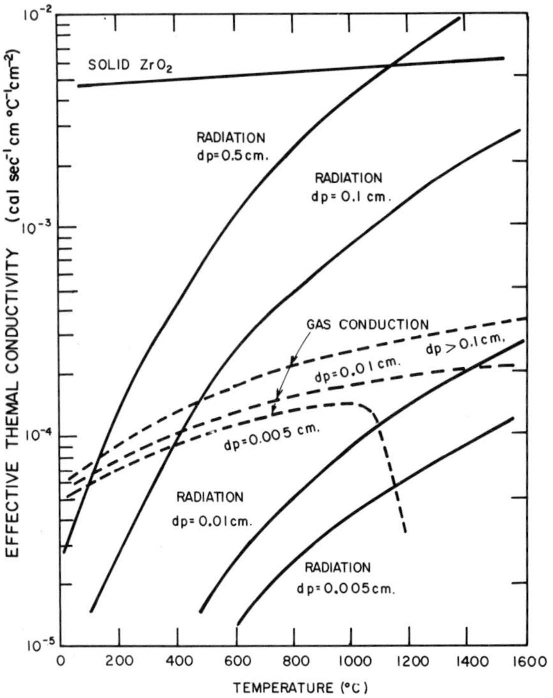

FIG. 1-10. Effective thermal conductivity of pore spaces as a function of temperature, for conduction and radiation heat transfer. In the figure d_p = diameter of pore. (Note: 0.01 cm = 3.937 mil).

sure is continued, and the saturation exposures are about the same for both parallel and transverse cut samples. A larger decrease in conductivity usually occurs in the direction transverse to the direction of extrusion. The change in conductivity in the parallel direction saturates to a value approximately 20% below the saturation value in the transverse direction. The thermal conductivity changes less during irradiation at higher temperatures, as illustrated in Fig. 1-11 [35a]. A similar effect was noted in postirradiation annealing of irradiated graphite, where annealing temperatures of about 400°C (752°F) caused a decrease by a factor of 2 in the change in thermal resistivity.

Hot-pressed BeO specimens were irradiated at exposures ranging from 3×10^9 fast neutrons/cm^2 to about 2×10^{21} fast neutrons/cm^2 and considerable variation was obtained for the thermal conductivities [33a-c, 36, 37, 38]. Depending on the investigator, the postirradiation conductivities ranged from around 0.9 of the preirradiation value down to a low of about 0.2 of the preirradiation value. Typical data illustrating the range of thermal conductivities are presented in Table 1-5 [34a-g]. Results

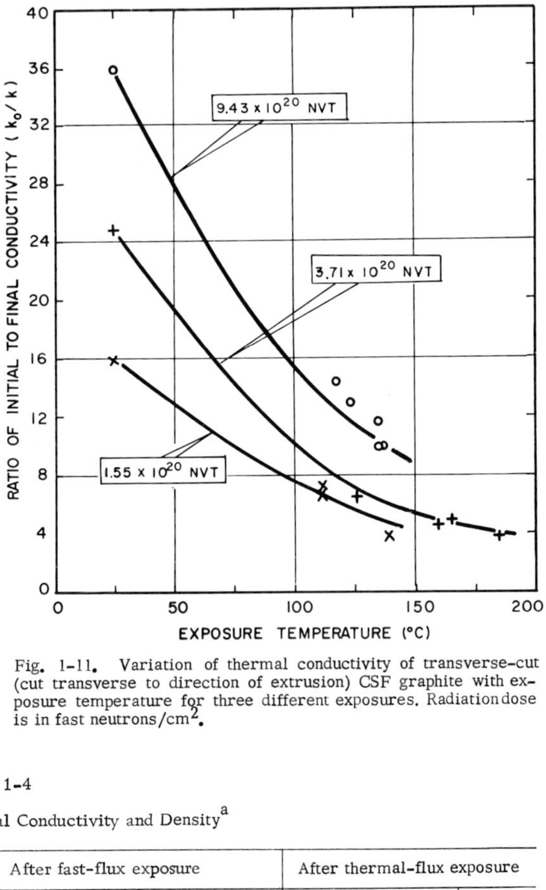

Fig. 1-11. Variation of thermal conductivity of transverse-cut (cut transverse to direction of extrusion) CSF graphite with exposure temperature for three different exposures. Radiation dose is in fast neutrons/cm^2.

TABLE 1-4

Effect of Radiation on Thermal Conductivity and Density[a]

Material	Initial value			After fast-flux exposure			After thermal-flux exposure	
	Thermal conductivity (10^4 cal/°C-cm-sec)	Density (g/cm^3)	Fast neutron-exposure (10^{19}/cm^2)	Thermal conductivity (10^4 cal/°C-cm-sec)	Density (g/cm^3)	Thermal neutron-exposure (10^{19}/cm^2)	Thermal conductivity (10^4 cal/°C-cm-sec)	Density (g/cm^3)
Sapphire	600 ± 200	3.983	6	300 ± 60	3.969	60	200 ± 30	3.944
Al_2O_3 sintered	400 ± 100	3.559	3	230 ± 40	3.553	40	90 ± 5	3.80
BeO	600 ± 200	2.84	7	400 ± 100	2.85	—	—	—
Spinel	250 ± 50	3.60	7	130 ± 10	3.60	40	130 ± 10	3.60
Forsterite	250 ± 50	3.056	6	75 ± 10	3.03	—	—	—
Zircon	120 ± 10	3.73	5	23 ± 1	3.48	30	—	3.38
Steatite	76 ± 5	2.796	7	28 ± 1	2.760	—	—	—
Cordierite	73 ± 5	—	5	20 ± 2	—	30	20 ± 2	—
TiO_2	165 ± 20	4.01	6	110 ± 10	3.99	30	65 ± 5	3.98
Porcelain	270 ± 50	3.41	6	120 ± 10	3.40	40	85 ± 5	3.39
Mica	17 ± 1	2.845	4	12 ± 1	2.738	20	28 ± 3	2.444
Plate glass	25 ± 1	2.509	3	—	2.530	60	—	2.515
Silica glass	35 ± 1	2.204	7	—	2.255	40	—	2.23

[a]Conversion factors: 1 cal/°C-cm-sec = 4.187 watt/cm-°C = 20.02 Btu/hr-in.-°F; 1 g/cm^3 = 62.43 lb/ft^3 = 0.03613 lb/in.^{3}

TABLE 1-5

Radiation-Produced Thermal-Conductivity Changes in BeO (Summary of Results)

Total fast-neutron exposure ($\times 10^{19}$/cm^2)	Temperature during measurement	Original specimen density [a] (g/cm^3)	(Postirradiation thermal conductivity)/(initial condition)
3.6	90°C (194°F)		0.60
7	140°C (284°F)	2.99	0.19
7	140°C (284°F)	2.74	0.31
7	30°C (86°F)	2.8	0.66
50	80°C (176°F)	2.60	0.90
200	550°C (1022°F)	2.8	0.30
	130°C (266°F)	3.0	0.80

[a] Conversion factor: 1 g/cm^3 = 62.43 lb/ft^3.

reported by Elston and Caillat [33c] and Gilbreath and Simpson [33b] are also plotted in Fig. 1-12 [17]. The differences between the results of the various investigators are attributable to the differences in sample densities and measurement techniques.

Fissile fuel materials can undergo a considerable drop in thermal conductivity as a result of increased porosity and fracturing from fission-gas production. The detrimental effect of fission on the conductivity of ceramics is well illustrated in the measurements of Gilbreath and Simpson on the property changes in BeO containing 2 and 10 wt.% of UO_2 [33b]. The property changes were compared with those of BeO (without UO_2) exposed to the same integrated neutron flux. The sample containing 2% UO_2 was irradiated at an estimated temperature of 250°C (480°F) and the 10% UO_2 sample at a temperature of 650 to 700°C (1202 to 1292°F). The thermal conductivity of the fuel samples was observed to decrease drastically upon exposure to fissioning, down to values 15 to 20% of that of the unirradiated samples; whereas the thermal conductivity of the unfueled samples decreased at most to 60% of the unirradiated conductivity value. The effect of fission fragments on conductivity can be separated out from the fast-neutron bombardment by plotting the ratio of the percent change in conductivity in the fueled and unfueled BeO samples. The results are presented in Fig. 1-13 [17]. It is seen that within the scatter of the experimental data, the effect due to fission fragments saturates at an exposure of about 1×10^{18} fissions/cm^3 exposure. This saturation effect is also apparent in UO_2 fuel data.

Data on the neutron-irradiation effects on nuclear fuel are difficult to interpret, as temperature and neutron flux in the samples are not uniform and change with time.

The annealing effect of fast-neutron damage at high temperature affects the change of the thermal conductivity of the fuel as a function of fission density. Ross [16] determined the change in thermal

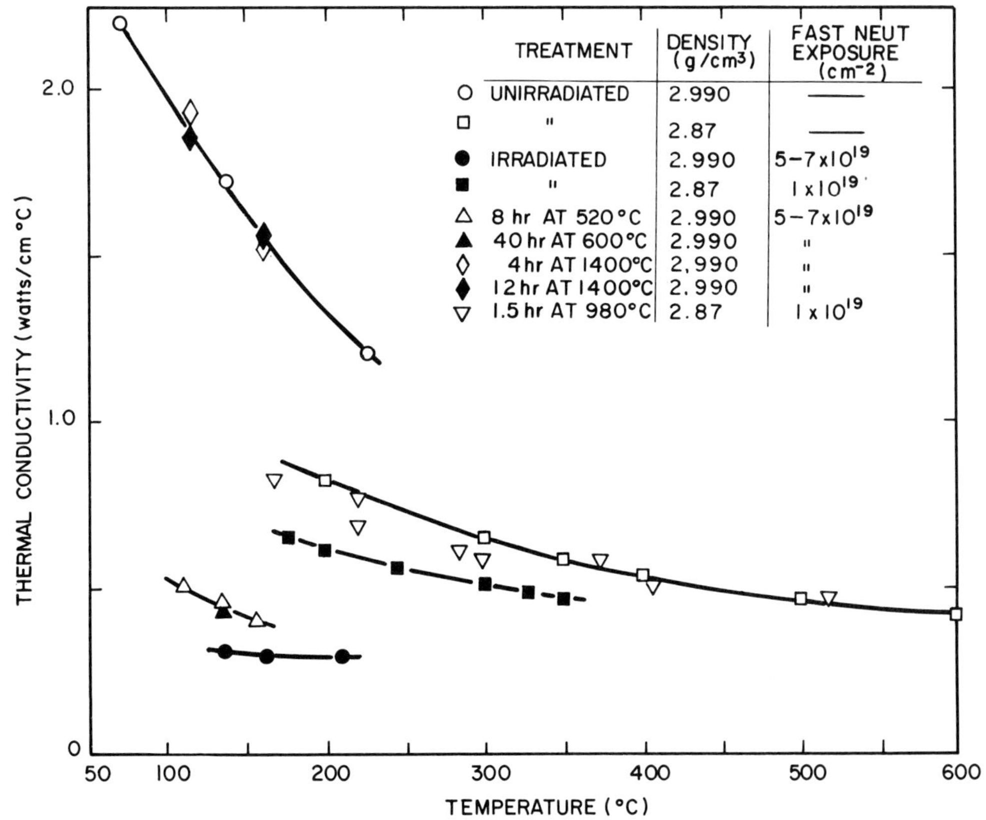

FIG. 1-12. Effect of neutron irradiation on the thermal conductivity of BeO.

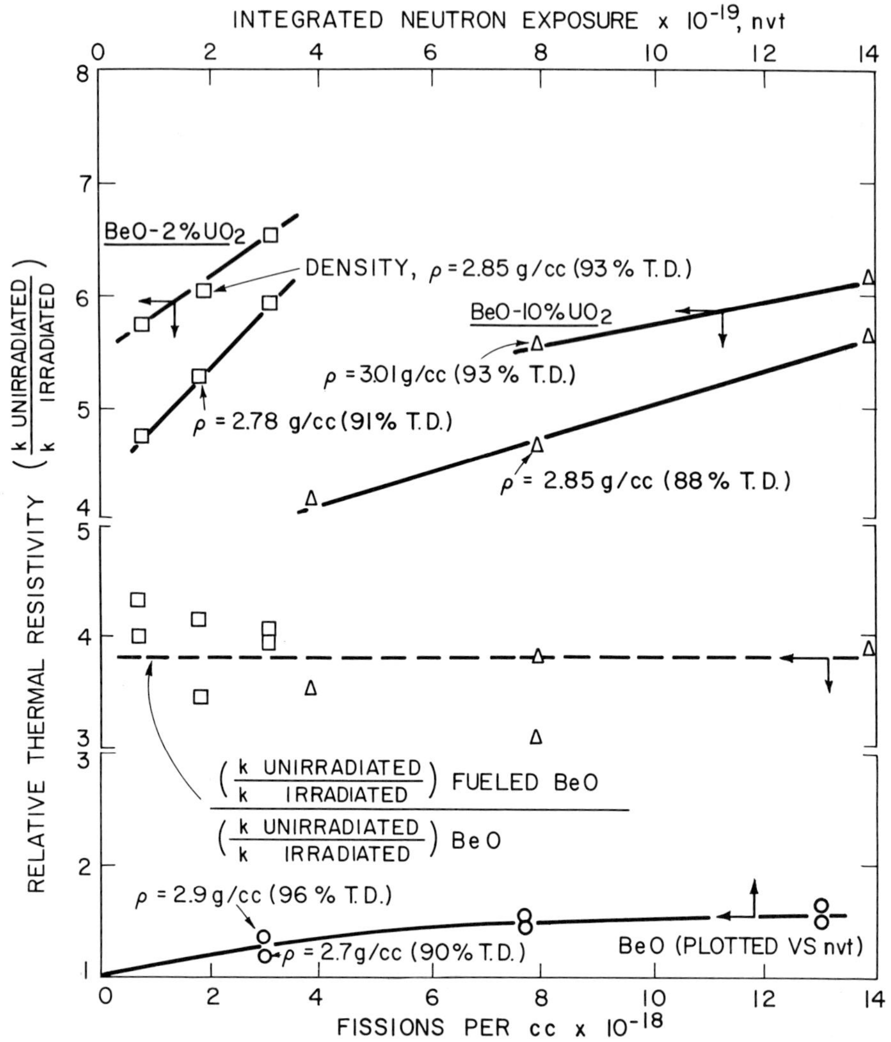

FIG. 1-13. Effect of irradiation on thermal conductivity of fueled and unfueled BeO. Note discontinuity in ordinate axis.

conductivity of 92% dense sintered UO_2 at 60°C (140°F) after exposure levels up to 6.75×10^{18} fissions/cm^3 at low irradiation temperature. His results corrected to 100% theoretical density are plotted in Fig. 1-14 [17]. The thermal conductivity levels off at 75% of its unirradiated value and an exposure of $\sim 3 \times 10^{16}$ fissions/cm^3. The recovery of the irradiation-induced decrease in the thermal conductivity of UO_2 upon annealing was also studied by the same author. It was found that the recovery was complete for irradiation exposures of less than 2.3×10^{16} fissions/cm^3 after 1-hr annealing at temperatures below 1000°C (1832°F). At irradiation exposures greater than 10^{18} fissions/cm^3 the recovery is only about 50% complete at 1000°C and seems to level off at this annealing temperature. Some effects of irradiation on the thermal conductivity of UO_2 have been reported more recently by Clough and Sayers [24]. These workers report no change of the thermal conductivity for exposures of 4.10×10^{19} fissions/cm^3 at temperatures ranging from 500 to 1600°C (932 to 2912°F). No changes are expected by these authors for exposures up to 2.5×10^{20} fissions/cm^3, when impurity concentration becomes appreciable.

1.3 Thermal Resistance at Boundaries between Solids

The fission heat produced in the fuel region of a heterogeneous reactor core has to flow through a fuel-cladding material into the coolant. The thermal resistance to heat flow at the interface between the fuel and the cladding may vary over large limits, depending upon the type of bond between the two regions.

For alloys or cermet types of fuel, a mechanical or metallurgical bond is usually present. Ceramic fuel elements normally have a liquid or gas filler to improve heat conduction. The metallurgical bond provides the highest thermal conductance between a

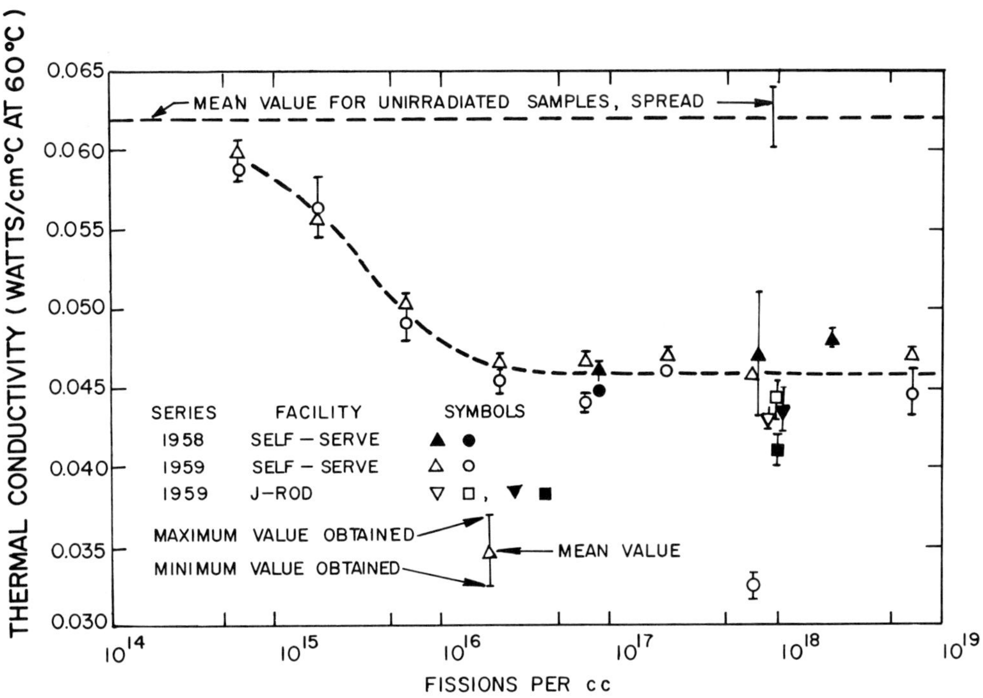

FIG. 1-14. Variation of thermal conductivity of UO_2 at 60°C (140°F) with burnup.

fuel and its cladding; however, such a bond may not stand up under neutron radiation and thermal cycling and may degenerate into a crack filled with gaseous fission products.

An intermediate layer of solid material affords a high-thermal-conductance bond but suffers the danger of being redistributed if the layer melts during a transient. A liquid bond requires additional space in the fuel element for expansion and limits the temperature to well below the boiling point of the bonding liquid. A severe pressure surge in the fuel element would occur as a consequence of the boiling of the liquid during an accident.

Although inert gas bonds (usually helium) have the lowest thermal conductance, they provide expansion space for the gaseous fission products. During a power transient a high thermal resistance at the interface between fuel and cladding is beneficial in two ways: overheating of the cladding is retarded; and the fuel is permitted to increase its temperature more rapidly so that the Doppler temperature coefficient reduces the reactivity earlier during the transient, thus limiting the magnitude of the power transient. (This assumes a negative Doppler coefficient. See also the discussion in Chapters 2 and 4.)

The thermal conductance at the fuel-clad interface is difficult to predict since it depends on the surface state of the two regions, the bonding pressure at the surfaces, and the nature of the bonding solid or gas. All these quantities are difficult to evaluate and are not necessarily uniform either circumferentially or axially in the fuel elements. In the following section the theory of thermal contact as developed to the present is reviewed, and some experimental measurements of the thermal conductance of unirradiated laboratory test specimens and irradiated UO_2 fuel elements are presented.

1.3.1 The Theory of Thermal Conductance of Contact

The theory of thermal contact has advanced to the point where the authors believe that it is possible to predict the thermal conductance within 20 to 30%, provided the surface configuration is known and provided no heat cycling has taken place.

An important hysteresis effect has been observed with time under heat cycling, but no dependable theory has yet been developed for this phenomenon.

Çetinkale and Fishenden [39] have derived an equation for the thermal conductance of metal surfaces in contact by calculation of the isotherms at the contacting points of the two solids. Their expression, in terms of dimensionless numbers, is

$$U = 1 + BC\left\{k \tan^{-1}\left[\frac{1}{C}\left(1-\frac{1}{U}\right)^{1/2} - 1\right]\right\}^{-1} \qquad (1\text{-}22)$$

in which U is the conductance number

$$U = h_c\delta/k_f\,, \qquad (1\text{-}23)$$

h_c the thermal conductance coefficient of the contact, k_f the thermal conductivity of the fluid between the surfaces. δ is the mean distance between the plate surfaces in an idealized model of a contact, as shown in Fig. 1-15a, and is defined as

$$\delta = \epsilon\beta_c\,. \qquad (1\text{-}24)$$

ϵ is a constant less than one which depends on the shape of surface irregularities and is determined experimentally. β_c is the arithmetic mean distance between the surfaces obtained from a magnified profile of the surfaces. The quantities k, C, and B are discussed in the following paragraphs.

The constriction number C in Eq. (1-22) is defined as

$$C = \frac{c}{a} = \left[\frac{A_c \text{ (contact area)}}{A \text{ (total area)}}\right]^{1/2} \tag{1-25}$$

in which c and a are radii of curvature of contacting surface and total surface, respectively, and A_c and A are the areas of solid and of total contact, respectively.

The constriction number is related to the apparent pressure between the surface. For plastic flow

$$C = \sqrt{p/M}\,, \tag{1-26}$$

in which p is the average pressure applied and M is the Meyer hardness. When a pressure p_{max} is applied, leading to plastic flow, and the surfaces are then held together at a pressure p under elastic strain, then

$$C = \sqrt{p_{max}^{1/3}\, p^{2/3}/M}\,. \tag{1-27}$$

The fluid thickness number, B of Eq. (1-22), is given by

$$B = \delta/a \tag{1-28}$$

in which the terms were previously defined. The quantity a may further be related to measured quantities by

$$a = \psi \lambda_c C^{\xi}\,, \tag{1-29}$$

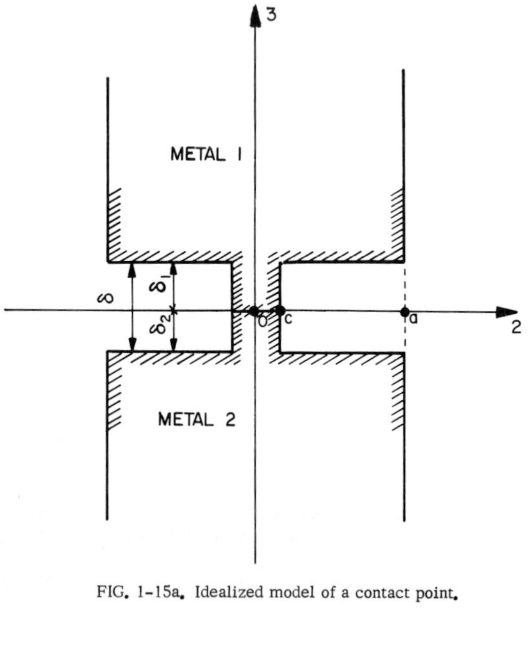

FIG. 1-15a. Idealized model of a contact point.

in which ψ and ξ are constants and λ_c is the sum of the wavelengths of the roughnesses of the two surfaces.

The final quantity of Eq. (1-22), the conductivity number k, is defined by

$$k = k_f/k_s\,, \tag{1-30}$$

in which k_f (neglecting radiation and thermal accommodation at the surfaces) is the fluid conductivity and k_s the geometric mean conductivity of the two metals in contact:

$$\frac{1}{k_s} = \frac{1}{2}\left[\frac{1}{k_1} + \frac{1}{k_2}\right]. \tag{1-31}$$

Thus, in order to compute the thermal conductance h_c from Eqs. (1-22) and (1-23), surface profiles are needed to determine the value of λ_c and β_c. In addition, experimental data of the thermal conductance of the surface similar to the one considered as a function of pressure are needed in order to determine the constants ϵ, ψ, ξ. These constants were found to be independent of the nature of the metal or fluid, but dependent on the surface roughness. For smooth-ground and ground-ground surfaces the constants were found to have the values $\epsilon = 0.61$, $\psi = 0.0048$, and $\xi = -5/3$. No values of these constants were reported for other types of surfaces. By using the above values of the constants with ground surfaces, Eq. (1-22) seems to predict the correct value of the thermal conductance up to a pressure of 800 psi (~50 atm). Fenech and Rohsenow [40] found an approximate analytic solution to the heat conduction at a contact point and derived the following equation for the thermal conductance (h_c) of surfaces in contact:

$$h_c = \frac{\dfrac{k_f}{\delta_1 + \delta_2}\left[(1-\epsilon^2)(\zeta_1 + \zeta_2) + 1.1\epsilon f(\epsilon)\left(\dfrac{1}{k_1} + \dfrac{1}{k_2}\right)\right] + 4.26\epsilon\sqrt{n}}{(1-\epsilon^2)\left[1 - \dfrac{k_f}{\delta_1 + \delta_2}\left(\dfrac{\delta_1}{k_1} + \dfrac{\delta_2}{k_2}\right)\right][\zeta_1 + \zeta_2]} \tag{1-32}$$

where k_1, k_2, and k_f are the conductivity coefficients of metals 1 and 2 in contact and of the fluid between the surfaces; ζ_1 is $4.26\sqrt{n}\,(\delta_1/\epsilon k_1) + (1/k_1)$ and ζ_2 the same expression with δ_1, k_1 replaced by δ_2, k_2; n is the number of contact points per unit area of surface; ϵ is the contact ratio (defined by C earlier):

$$\epsilon^2 = \frac{\text{contact area}}{\text{total area}} = \frac{A_c}{A}\,;$$

$f(\epsilon)$ has a value of 1.0 for all practical purposes, i.e., $\epsilon < 0.1$. The function $f(\epsilon)$ is plotted in Fig. 1-15b [40]. The δ_1 and δ_2 are the shoulder heights of the gaps in the idealized model (see Fig. 1-15a) and are related to the volume average thickness of the fluid on surfaces 1 and 2, ξ_1 and ξ_2, respectively, by

$$\delta_i = \xi_i/[1 - (k_f/k_c)]\,, \quad i = 1, 2, \quad k_f \neq k_i\,. \tag{1-33}$$

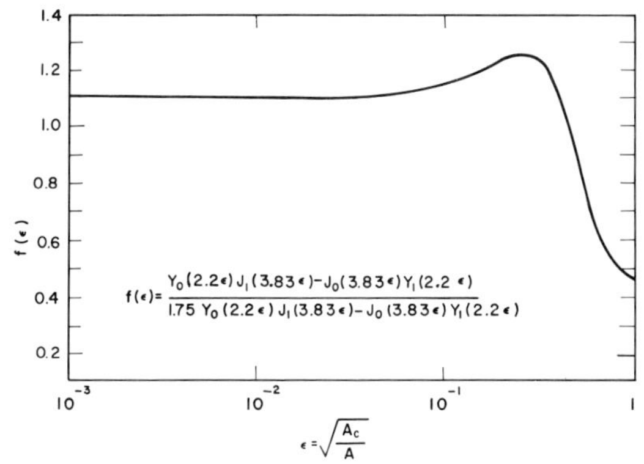

FIG. 1-15b. The function f (ϵ) versus ϵ.

The contact ratio ϵ is related to the load pressure p for plastic flow by the equation

$$\epsilon^2 = p/M\,, \qquad (1\text{-}34)$$

where M is the yield pressure of the softer material. The yield pressure is measured with a Vickers or Knoop testing machine and depends upon the size of indentation. It is therefore necessary to obtain a plot of yield pressure M versus identation area and to use for Eq. (1-34) a value of M corresponding to an indentation area equal to the average contact size, i.e., ϵ^2/n.

The surface parameters n and ξ are obtained from magnified profiles of the surfaces of a length L. After placing the two profiles together in a position of contact, one measures the number of interferences N, the length ℓ_j of the interferences projected on the horizontal, and the areas between surfaces A_1, A_2 above and below a horizontal contact line that passes through most of the contacts.

Then, for this section through the contact, the contact parameters are

$$\xi_{a,i} = A_i/L\,, \quad \sqrt{n} = N/L\,,$$
$$\epsilon = \Sigma_{j=1}^{N} \rho_j/L\,, \quad i = 1, 2\,. \qquad (1\text{-}35)$$

The $\xi_{a,i}$ thus measured is only a two-dimensional void thickness. The volume average void thickness ξ_i is computed approximately from the relation

$$\xi_i = \xi_{a,i}\,(1 + \epsilon), \quad i = 1, 2\,. \qquad (1\text{-}36)$$

An increase in pressure is simulated by overlapping the surface profiles and repeating the measurement of n, ϵ^2, and $\xi_{a,i}$. These measurements can be done graphically or by use of an analog computer after the surface profiles have been recorded on magnetic tape [41]. The number of contact points obtained by this method was found to be in good agreement with the number of contact points measured experimentally for identical surfaces using radioactive gold as a tracer [42]. Equation (1-32) has been extended to the case when an oxide film is present on the surface and when the surfaces have a certain amount of waviness superimposed on the roughness [43]. Equation (1-32) can be simplified in most cases of interest.

When the surfaces have approximately the same roughness or when the two materials in contact have similar conductivity coefficients, the following condition is usually satisfied:

$$\left|\left(\frac{\delta_1 + \delta_2}{\delta_1 - \delta_2}\right)\left(\frac{k_1 + k_2}{k_1 - k_2}\right)\right| > 4\,; \qquad (1\text{-}37)$$

and the thermal conductance equation simplifies to

$$h_c = \frac{k_s}{\delta}\,\frac{1}{1 - (k_f/k_s)}\left[\frac{k_f}{k_s} + \frac{\epsilon^2}{1 - \epsilon^2}\,\frac{1.1(k_f/k_s) + \eta}{\epsilon + \eta}\right], \qquad (1\text{-}38)$$

where

$$\delta = \delta_1 + \delta_2\,, \qquad (1\text{-}39a)$$

$$\frac{1}{k_s} = \frac{1}{2}\left[\frac{1}{k_1} + \frac{1}{k_2}\right] \qquad (1\text{-}39b)$$

$$= 2.13 \quad \delta\sqrt{n}\,. \qquad (1\text{-}39c)$$

When one of the surfaces is very smooth or one of the materials is a much better conductor than the other, such that

$$\sigma_1 k_2/\sigma_2 k_1 > 5\,, \qquad (1\text{-}40)$$

then the thermal-conductance relation simplifies to

$$h_c = \frac{k_s}{\delta}\,\frac{1}{1 - (k_f/k_1)}\left[\frac{k_f}{k_s} + \frac{\epsilon^2}{1 - \epsilon^2}\,\frac{1.1(k_f/k_s + \eta_1)}{\eta_1(k_s/k_1) + \epsilon}\right], \qquad (1\text{-}41)$$

where $\eta_1 = 2.13\ \delta_1\sqrt{n}$.

For contacts under heavy loading or contacts in rarefied atmospheres, the conductance through the fluid film may be neglected. An approximate condition for the applicability of this case is

$$k_f/k_s < \epsilon^2/2\,. \qquad (1\text{-}42)$$

Equation (1-32) with $k_f = 0$ becomes

$$h_c = \frac{\epsilon^2/(1 - \epsilon^2)}{(\delta_1/k_1) + (\delta_2/k_2) + \left(0.46\sqrt{\epsilon^2/n}/k_s\right)}\,. \qquad (1\text{-}43)$$

Measurement of the thermal conductance of Armco Iron-aluminum and stainless steel-stainless steel contacts has shown good agreement with the values calculated from Eqs. (1-32) or (1-41) at pressures up to 20,000 psi (~ 1300 atm). The surface profiles, the number of contact points, and the thermal conductance of the Armco Iron-aluminum contact are shown in Fig. 1-16 (a, b, c) [40]. Calculated values of the contact conductance between fresh uranium metal fuel rods and various canning materials for the surface profile shown in Fig. 1-15a and with NaK, He, or 50% He + 50% (Xe + Kr) as filler are plotted in Fig. 1-17.

The above theoretical development does not account for the important hysteresis effect which takes place over a period of several days under thermal cycling. An effect observed by Cordier

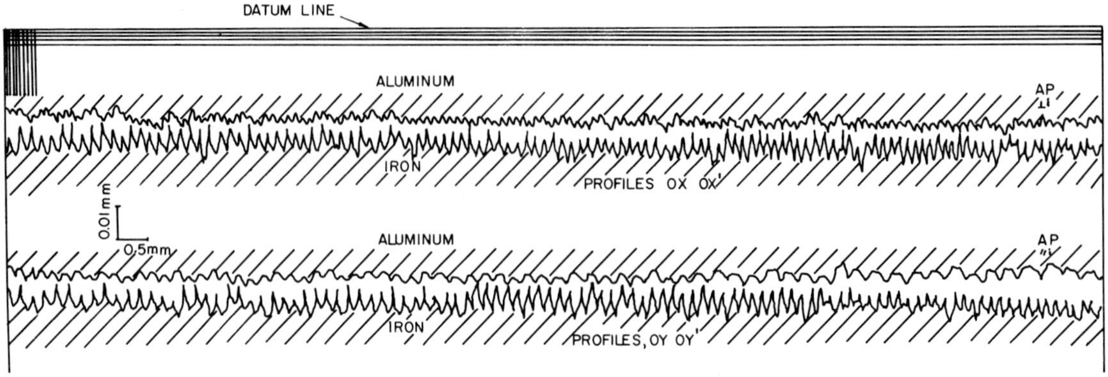

FIG. 1-16a. Recorded profiles of the iron-aluminum contact shown in the no-load position.

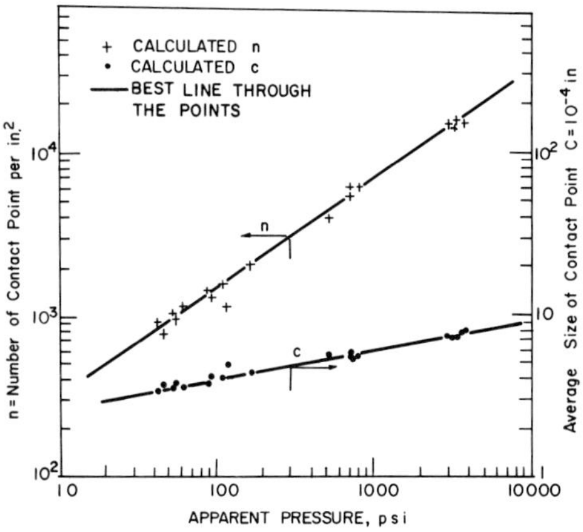

FIG. 1-16b. Iron-aluminum contact; number of contact points per unit area versus pressure. (Conversion factor: 1000 psi = 70.3 kg/cm²).

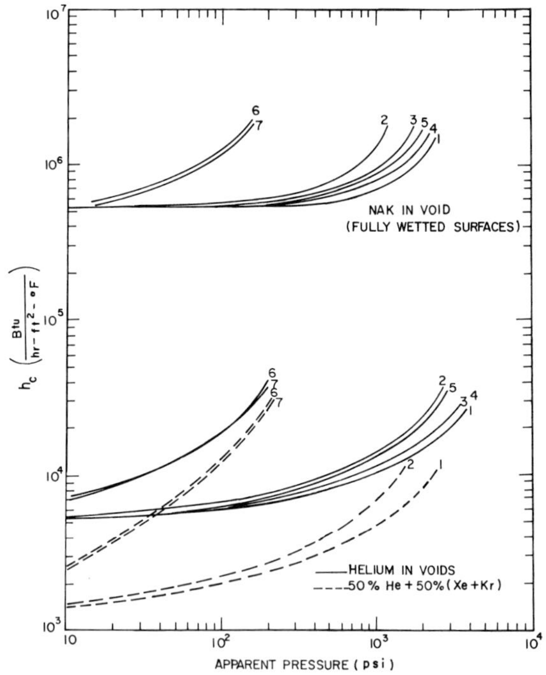

FIG. 1-17. Calculated values of the contact thermal conductance (h_c) versus apparent pressure for fresh uranium metal fuel rods at a mean contact temperature of 400°C (752°F) fora particular surface state configuration (150 rms) with various canning materials: (1) type-304 stainless steel, (2) zirconium, (3) Zircaloy-2, (4) niobium, (5) commercial grade beryllium, (6) 2S-aluminum, (7) Magnox A-12.

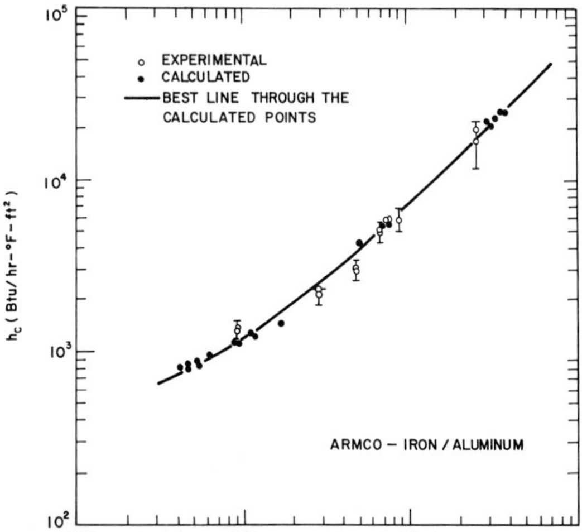

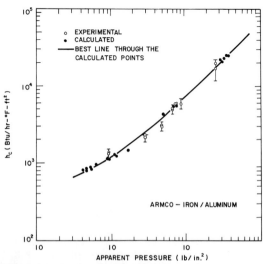

FIG. 1-16c. Pressure dependence of the thermal contact conductance for Armco Iron contact with aluminum.

[44] is illustrated in Fig. 1-18. Under increasing pressure the thermal resistance decreases from A to B. When the pressure at B is maintained, the thermal resistance decreases from B to C over several days. Upon a decrease in pressure the thermal resistance recovers from C to D but continues further to increase from D to F over several days when the pressure at D is maintained constant. The decrease of the resistance from B to C may well be due to a phenomenon of creep, but no explanation has yet been found for the increase from D to F.

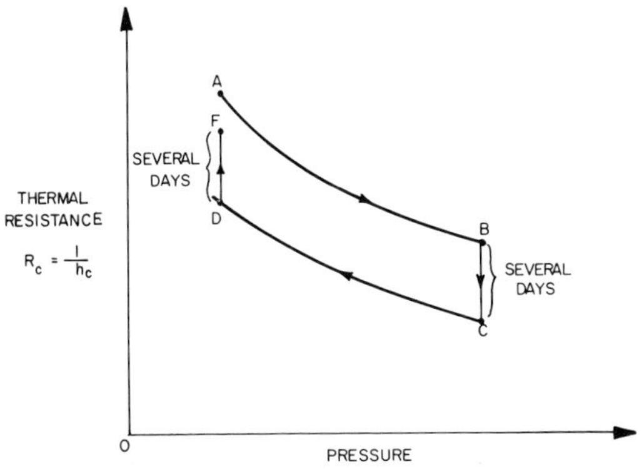

FIG. 1-18. Hysteresis effect on the thermal resistance of a contact under pressure cycling.

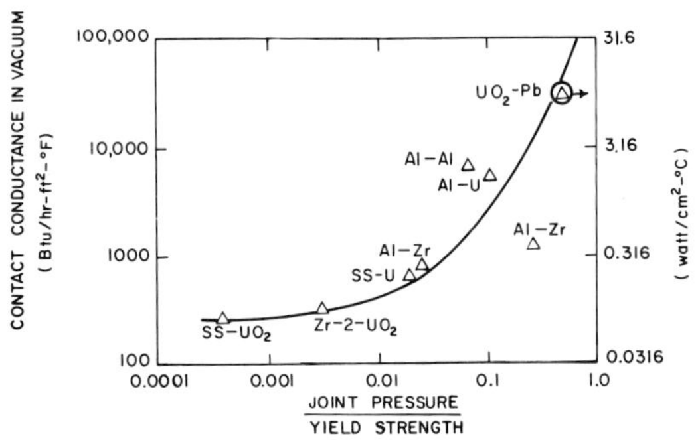

FIG. 1-19. Effect of yield strength of the softer joint material on the contact conductance.

1.3.2 Experimental values

Most measurements of thermal conductance have been made on unirradiated specimens at various pressures and temperatures. Values of the conductance between UO_2 fuel pellets and stainless steel cladding were deduced from overall thermal performance of fuel elements during in-pile tests [45]. Some results of interest to nuclear reactor performance are reported below.

Wheeler [46] measured contact conductance between cladding materials and UO_2 as a function of atmosphere and sheath material. His results for experiments conducted in a vacuum shown in Fig. 1-19 [17] illustrate the pronounced effect of hardness of the softer member of the junction on solid-solid conductance. Skipper and Wooton [33d] measured the thermal resistance between uranium metal and Magnox (magnesium alloy). They reported a significant rise in the interface temperature drop when, as a result of a decrease in He pressure, the mean-free-path length of the gas became larger than 1/40th of the average gap between the uranium surface and the inner surface of the canning material (see Fig. 1-20) [33d]. The effect of oxidation of the uranium surface on the thermal conductance was also investigated and found to be significant, as shown in Fig. 1-21 [33d]. However, it is possible that at higher pressures than that of the experiments (16 psi or about 1 atm), this effect would be reduced due to the cracking of the oxide film and the flow of the metal through the cracks.

Sanderson [47b] also reports measurements of thermal conductance at the junction between unirradiated uranium metal and Magnox cladding. Argon and helium were used as a fluid and the reported thermal resistance as a function of gas pressure is shown in Fig. 1-22 [48]. Fenech and Henry [48] were able to reproduce the general behavior of the contact resistance as a function of the nature and pressure of gas, using a simplified form of Eq. (1-32) and correcting the conductivity coefficient of the gas k_f for the temperature jump at the solid-gas interface by use of the relation

$$\frac{k_f'}{k_f} = \frac{1}{1 + 4\,[(2/a) - 1][\gamma/(\gamma + 1)](1/\text{Pr})(\ell/\delta)}, \quad (1\text{-}44)$$

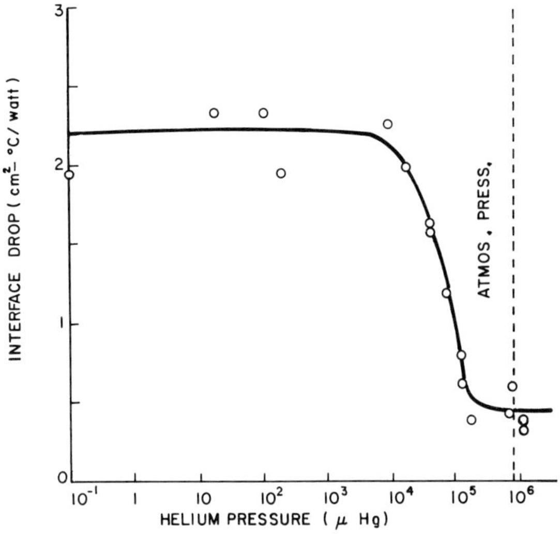

FIG. 1-20. Effect of helium pressure on interface thermal resistance between uranium and Magnox. The uranium was a machined disc with light brown tarnish film. Interface temperature = 272–305°C (522–581°F), interface pressure = 150 psi (10.55 kg/cm^2). [Resistivity conversion factor: 1 cm^2-°C/watt = 0.00568 ft^2-°F-hr/Btu.]

where k_f' is the corrected conductivity coefficient of the gas; Pr the Prandtl number; ℓ the mean-free-path of the gas molecules; δ the average void thickness; γ the ratio of specific heats, c_p/c_v; and a the geometric average of the accommodation coefficient of surfaces 1 and 2, namely,

$$2/a = 1/a_1 + 1/a_2 \, . \quad (1\text{-}45)$$

Sanderson's data for He were well reproduced with an accommodation factor a = 0.5, while the scatter of the Ar data prevented an exact determination of this factor. The calculated values are drawn in full line in Fig. 1-22.

Cohen et al. [45] measured the thermal performance of UO_2, stainless-steel-clad fuel elements as used in a PWR reactor. Specimens were irradiated to burnup up to 5000 Mwd/ton. With the Kingery out-of-pile thermal-conductivity data, the fuel surface temperatures were obtained and used to calculate the clad-fuel contact conductances. The results obtained in several reactor startups are plotted in

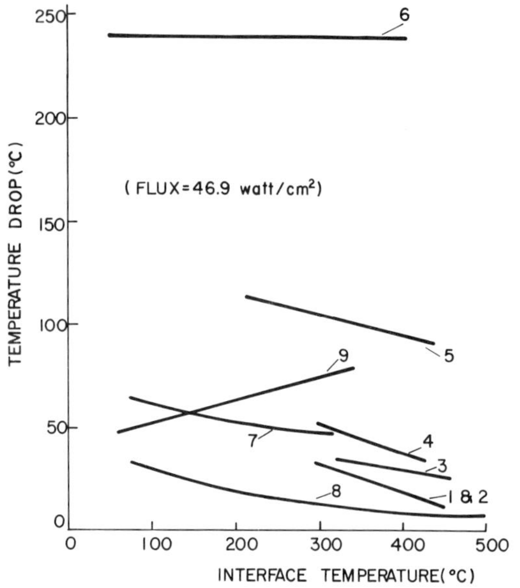

FIG. 1-21. Effect of oxide film (on U-metal surface) on interface temperature drop. The heat flux was 46.9 watt/cm^2 (149000 Btu/hr-ft^2). Numbers refer to: 1, 2—specimen unoxidized in Ar, 18.4 and 17.0 psia respectively; 3, 4, 5—oxidized with 0.4, 0.5, 1.4 mils film and at pressures 16.1, 18.4, 16.9 psia respectively; 6—specimen 8, 2.1 mil oxide film in He, 15.7 psia; 7 - specimen 9, 1.14 mil oxide film in Ar, 15.3 psia; 8 - unoxidized specimens in Ar and He, 14.7-15.7 psia; 9 - specimen 3, oxidized in He, 15.7 psia (no precautions taken against increasing oxidation during test). [Conversion factors: 10 psia = 0.703 kg/cm^2, 1 mil = 0.0254 mm.]

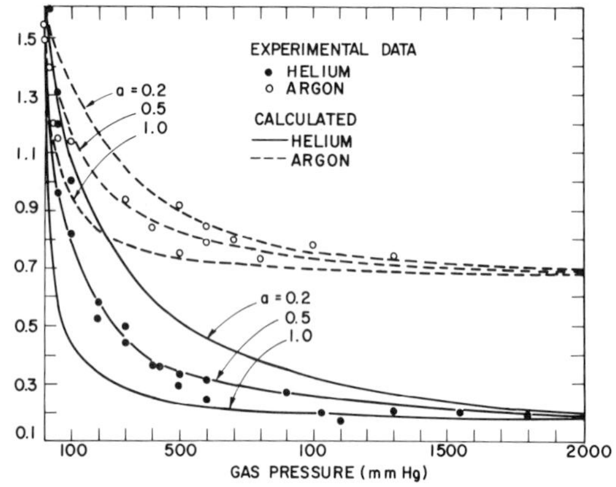

Fig. 1-22. Contact resistance versus interface gas pressure at junction between unirradiated uranium and Magnox cladding. Calculated curves for three values of accomodation coefficient are shown.

Fig. 1-23 [45] and indicate, as anticipated, a marked dependence of contact conductance on pressure. The decrease in thermal conductance after the first startup may well be attributable to a change in the thermal conductivity of the UO_2 due to irradiation or fuel cracking. The dependence of the contact conductance on the initial assembly gas-gap is plotted in Fig. 1-24 [45] and as a function of operating gas-gap in Fig. 1-25 [45]. The calculations of the operating gap at each temperature were made by correcting for thermal expansion of the clad and fuel material and by assuming that the fuel remains centered within the clad. In contradiction to previous work, Cohen et al. observed that the conductance between oxide and sheath in fuel-elements samples is insensitive to the gas medium (see Fig. 1-29, discussion in Sec. 1.4.2). It was postulated by the authors that this insensitivity result from low values of the thermal accommodation coefficients a_1 and a_2 of the gases. Further practical considerations of the thermal conductance of contacts in ceramic fuel elements are discussed in the next section.

1.4 Thermal Performance of Ceramic Fuel Elements

The heat-transfer situation of ceramic fuel elements is complicated by several unique characteristics, each of which, as will be shown below, affects in an important way the thermal capabilities of the fuel elements. In the earlier parts of Sec. 1 it was shown that the thermal-conductivity coefficient of ceramics is strongly dependent on temperature and is affected by neutron irradiation, and that some annealing of the irradiation-induced effects takes place at high temperatures. The low thermal conductivity causes high temperature gradients and, consequently, thermal stress cracking of the relatively brittle ceramics, which in turn changes the configuration of the fuel region and affects the

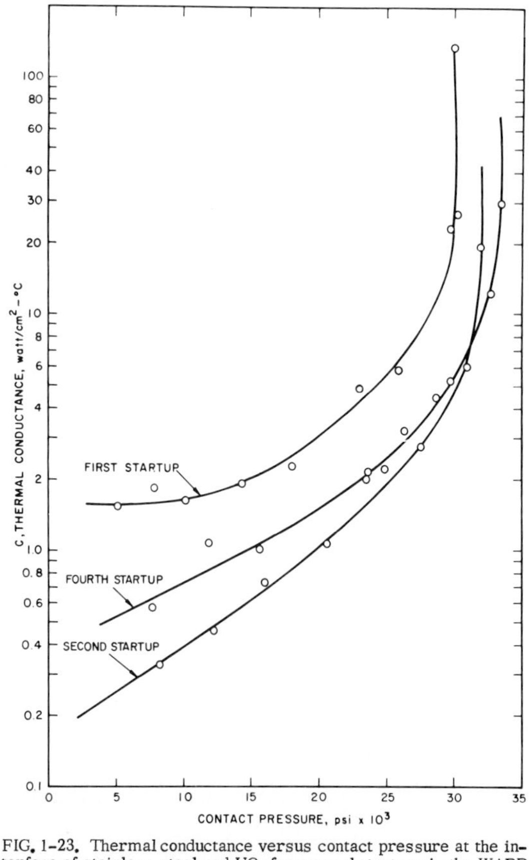

FIG. 1-23. Thermal conductance versus contact pressure at the interface of stainless steel and UO_2 for several startups in the WAPD 22-11 experiment. Clearance zero, Kr+3Xe atmosphere, production grade UO_2. (Note: 10^3psi = 70.3 kg/cm^2).

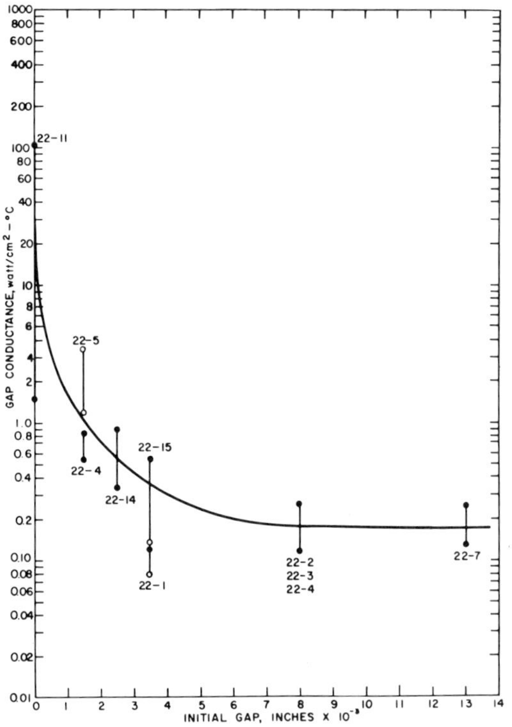

FIG. 1-24. Gap thermal conductance versus initial diametral gap. (Note: 10^{-3} in. = 1 mil = 0.0254 mm.) The numbers on the experimental points refer to WAPD experiments, defined in caption to Fig. 1-29.

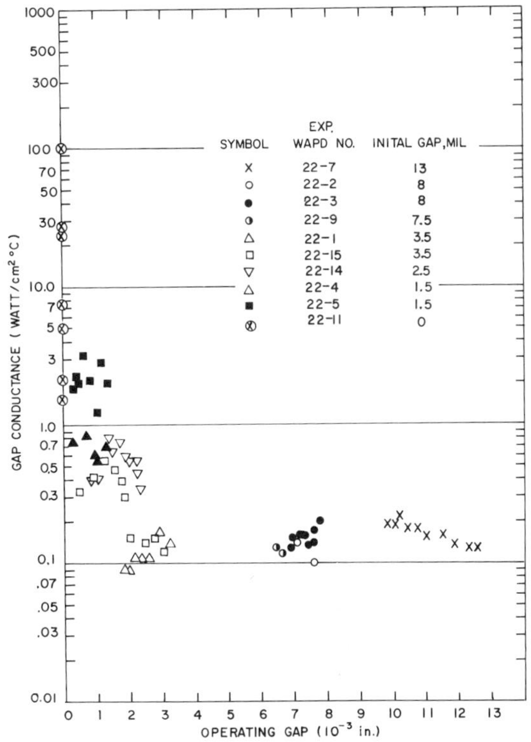

FIG. 1-25. Gap thermal conductance versus operating gap. (Conversion factors: 1 mil = 0.0254 mm, 1 watt/cm^2 -°C = 1761 BTU/hr-ft^2-°F.) "WAPD experiment no.", is defined in caption of Fig. 1-29.

contact conductance at the cladding-fuel interface. Because of the difficulties in predicting how the thermal performance of a fuel element will be affected by each of these characteristics, it is convenient to determine experimentally the overall thermal performance of the fuel elements and then to attempt judicious interpretations of the results. Such a method has been used to study the behavior of UO_2 fuel elements. A very comprehensive study of the results obtained since 1955 has been made by Lustman [45] and is reviewed briefly here.

1.4.1 Mathematical Description of Heat Transfer in Ceramic Fuel Elements

1.4.1.1 The Conductivity Integral and Its Uses. In a cylindrical rod made up of an homogeneous material of conductivity k (T), with a volumetric heat generation rate q (r), the heat conservation equation is

$$\int_0^r q(r)2\pi r dr = [-k(T)2\pi r(dT/dr)]_r \qquad (1\text{-}46)$$

where T is the temperature at r. After rearranging terms and integrating both sides from radius r to the fuel-pellet radius a, or from the centerline of the pellet to a radius r, this equation becomes

$$\int_r^a dr/r \int_0^r q(r)rdr = \int_{T_a}^{T_r} k(T)dT \qquad (1\text{-}47a)$$

or

$$\int_0^a dr/r \int_0^r q(r)rdr = \int_{T_a}^{T_c} k(T)dT\ , \qquad (1\text{-}47b)$$

where T_c and T_a are the temperatures of the fuel pellet at the center and at the outer radius a.

The left side of Eq. (1-47b) is proportional to the total heat-generation rate in the rod. If this value is known, for example, from in-pile measurements or postirradiation burnup analysis, the integral on the right-hand side of the equation can be used to determine the temperature at the rod center, knowing the outside pellet temperature. The outside pellet temperature T_a can be calculated from power output and local coolant-temperature conditions. Alternatively, knowledge of q (r) permits assignment of a value of the integral of k (T) dT between T_a and T_r (i.e., the conductivity integral of Eq. (1-47a)) at any radial position r. When some structural change occurs at a given radius r, a knowledge of this integral from q(r) permits the determination of the temperature T_r at which this structural change (e.g., grain growth) occurs; or, inversely, knowing q(r) and the temperature and location of a structural change, the value of the integral can be deduced and compared with a calculated value obtained from current knowledge of k (T). In Fig. 1-26 [17] the values of

$$\int_{T_1}^{T} k(T)\, dT$$

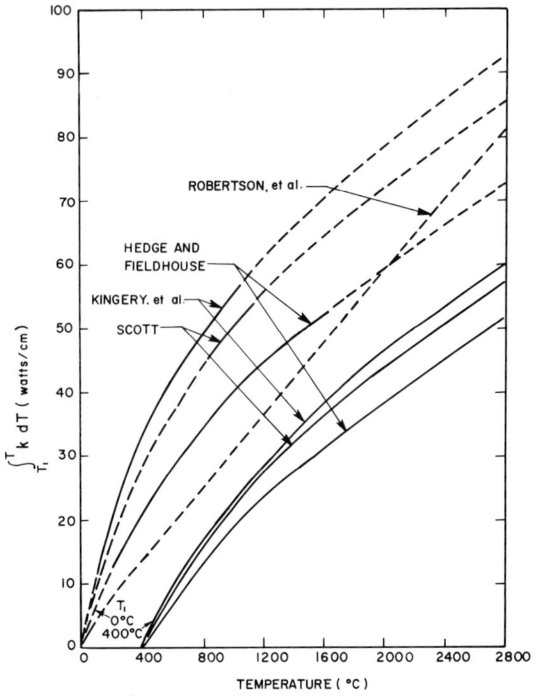

FIG. 1-26 Variation of the conductivity integral as a function of T, the central temperature of the fuel, for two values of the fuel surface temperature and for various measurements of thermal conductivity of UO_2 (95% dense). Solid lines in upper set of curves indicate temperature range over which measurements were performed.

are plotted as a function of T for $T_1 = 0°C$ (32°F) and 400°C (752°F) and for three out-of-pile measurements of the thermal conductivity of UO_2. The data of Kingery et al. were extrapolated by use of the empirical formula k (Btu/hr-ft-°R) = $1.855/T^{0.9}$ (T in degrees Rankine) and those of Scott by the relation k (watt/cm-°C) = 39.9 (T + 400) (T in degrees centigrade) to the temperature at which k = 0.017 watt/cm-°C, beyond which all three sets of data were presumed to show a constant value of k.

It is seen from these curves that for a pellet surface temperature $T_a = 400°C$ (752°F) and a center temperature $T_c = 2750°C$ (4982°F) the predicted power output capabilities vary by about 15%.

1.4.1.2 Effects of Flux Suppression within the Fuel. In Fig. 1-27 [17], variations of q(r), heat-generation rate, are shown for two important cases. One case corresponds to a flux suppression from surface to center of a UO_2 pellet with 7.69 wt.% enrichment, and the other reflects the effect of plutonium buildup on the surface of a natural UO_2 pellet after 10,000 Mwd/ton-UO_2 exposure. The corresponding temperature distribution for these nonuniform heat productions in a pellet with a surface temperature of 400°C (752°F) are compared in Fig. 1-28 [17] with the temperature contours developed in the case of uniform heat generation. In each case, the power output from the pellet was held constant and the data of Kingery et al. in Fig. 1-24 were used to develop the temperature patterns. It is seen that increasing the heat generation at the surface of the fuel reduces the maximum fuel temperature for the same power output.

1.4.1.3 Effects of Fuel-Element Geometry. For cylindrical pellets with a uniform heat-generation rate and radial heat conduction only, Eqs. (1-47) integrate to yield

$$\int_{T_a}^{T_r} k(T)dT = q''(a^2 - r^2)/2a , \qquad (1\text{-}48a)$$

and

$$\int_{T_a}^{T_c} k(T)dT = W/4\pi , \qquad (1\text{-}48b)$$

in which q'' is the surface heat flux and W is the power output per unit length of rod. Equation (1-48b) indicates that for a constant power output per unit length of rod and a constant pellet surface temperature, the central temperature is independent of rod diameter.

For an infinite slab of half-thickness a, the conductivity integral is

$$\int_{T_a}^{T_x} k(T)dT = \frac{q''}{2}\,\frac{(a^2 - x^2)}{a} \qquad (1\text{-}49a)$$

and

$$\int_{T_a}^{T_c} k(T)dT = q''a/2 . \qquad (1\text{-}49b)$$

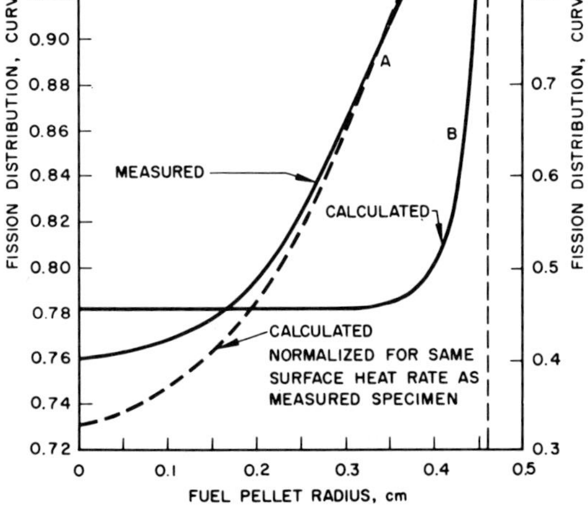

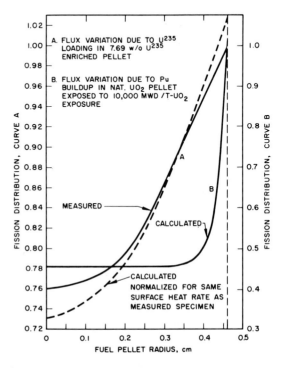

FIG. 1-27. Fission distribution for two cases of nonuniform heat generation.

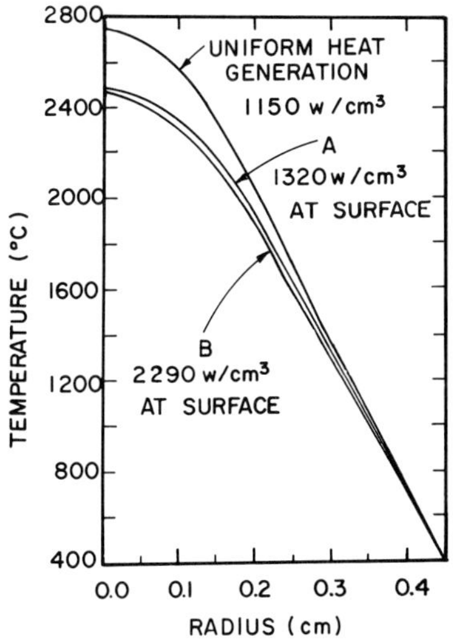

FIG. 1-28. Temperature distribution of fuel pellets as affected by heat-generation pattern. Power output per unit length for each case = 741 w/cm (1.685 Btu/sec-in.). Curves A and B correspond to A and B of Fig. 1-27, i.e. A is for 7.69 wt.% U^{235} enriched pellet and B for the natural UO_2 pellet exposed to 10^4 Mwd/ton UO_2. Extrapolated data of Kingery et al.

When the slab has a half-thickness a and width d, Eq. (1-49b) becomes

$$\int_{T_a}^{T_c} k(T)dT = Wa/4d \; . \qquad (1\text{-}50)$$

For a hollow cylinder of outer radius a, inner radius b, cooled at outer radius only,

$$\int_{T_a}^{T_r} k(T)dT = \frac{q''}{2}\,\frac{a}{a^2-b^2}\left[(a^2-r^2) - 2b^2 \ell n\,\frac{a}{r}\right], \qquad (1\text{-}51a)$$

$$\int_{T_a}^{T_b} k(T)dT = \frac{W}{4\pi}\left[1 - \frac{2b^2}{a^2-b^2}\,\ell n\,\frac{a}{b}\right]. \qquad (1\text{-}51b)$$

For the same hollow cylinder cooled at inner and outer radii such that the surface temperatures T_a and T_b are equal,

$$\int_{T_a}^{T_b} k(T)dT = \frac{q'''}{4}\left[(a^2-R^2) - 2R^2 \ell n\,\frac{a}{R}\right] \qquad R<r<a, \qquad (1\text{-}52a)$$

$$\int_{T_b}^{T_r} k(T)dT = \frac{q'''}{4}\left[(R^2-b^2) - 2R^2 \ell n\,\frac{R}{b}\right] \qquad b<r<R, \qquad (1\text{-}52b)$$

where

$$R^2 = \frac{a^2+b^2}{2[1 + n(ab/R^2)]}$$

and

$$\int_{T_a}^{T_r} k(T)dT = \frac{q_a''a}{2}\left[1 - \frac{2R^2}{a^2-R^2}\,\ell n\,\frac{a}{R}\right], \qquad (1\text{-}53a)$$

$$= \frac{q_b''b}{2}\left[1 - \frac{2R^2}{R^2-b^2}\,\ell n\,\frac{R}{b}\right], \qquad (1\text{-}53b)$$

q_a'' and q_b'' are the heat fluxes through the inner and outer surfaces, respectively.

In fuel elements where the neutron-diffusion approximation is applicable, the power-generation rate for a uniform fuel distribution has the following forms:

$$\text{cylindrical elements}: \; q'''(r) = q_0''' I_0(\kappa r) \qquad (1\text{-}54a)$$

$$\text{slab elements}: \; q'''(x) = q_0''' \cosh(\kappa x)\,, \qquad (1\text{-}54b)$$

where q_0''' is the volumetric heat-generation rate at the center of the element and κ is the inverse of the thermal-neutron-diffusion length. A recommended formula for κ, which gives a more accurate temperature distribution than the one obtained from the use of the thermal-neutron-diffusion length, is given by Adam [49] as

$$\kappa^2 = 3\Sigma_a\Sigma_t\,[1 - (0.8\Sigma_a/\Sigma_t)]\,, \qquad (1\text{-}55)$$

where Σ_a and Σ_t are the macroscopic effective absorption and total cross sections, respectivley. Using the power-generation rate given by Eq. (1-54a) for cylindrical elements, one obtains for the conductivity integral

$$\int_{T_a}^{T_r} k(T)dT = \frac{q''a}{2}\left[\frac{I_0(\kappa a) - I_0(\kappa r)}{\frac{1}{2}\kappa a I_1(\kappa a)}\right] \qquad (1\text{-}56a)$$

$$= \frac{W}{4\pi}\left[\frac{I_0(\kappa a) - I_0(\kappa r)}{\frac{1}{2}\kappa a I_1(\kappa a)}\right] \qquad (1\text{-}56b)$$

and

$$\int_{T_a}^{T_c} k(T)dT = \frac{q''a}{2}\left[\frac{I_0(\kappa a) - 1}{\frac{1}{2}\kappa a I_1(\kappa a)}\right] \qquad (1\text{-}57a)$$

$$= \frac{W}{4\pi}\left[\frac{I_0(\kappa a) - 1}{\frac{1}{2}\kappa a I_1(\kappa a)}\right]. \qquad (1\text{-}57b)$$

Similarly, for an infinite slab of half-thickness a,

$$\int_{T_a}^{T_r} k(T)dT = \frac{q''a}{2}\left[\frac{\cosh(\kappa a) - \cosh(\kappa r)}{\frac{1}{2}\kappa a \sinh(\kappa a)}\right], \qquad (1\text{-}58a)$$

$$\int_{T_a}^{T_c} k(T)dT = \frac{q''a}{2}\left[\frac{\cosh(\kappa a) - 1}{\frac{1}{2}\kappa a \sinh(\kappa a)}\right], \qquad (1\text{-}58b)$$

and for a hollow, externally cooled cylinder of internal diameter b,

$$\int_{T_a}^{T_r} k(T)dT = \frac{q''a}{2}\left[\frac{I_1(\kappa b)\,[J_0(\kappa a) - J_0(\kappa r)] + J_1(\kappa b)\,[I_0(\kappa a) - I_0(\kappa r)]}{\frac{1}{2}\kappa a\,[I_1(\kappa a)J_1(\kappa b) - I_1(\kappa b)J_1(\kappa a)]}\right], \quad (1\text{-}59a)$$

$$\int_{T_a}^{T_c} k(T)dT = \frac{q''a}{2}\left[\frac{\kappa b\,[I_1(\kappa b)J_0(\kappa a) + I_0(\kappa a)J_1(\kappa b)] - 1}{(\kappa^2/2)ab\,[I_1(\kappa a)K_1(\kappa b) - I_1(\kappa b)K_1(\kappa a)]}\right]. \quad (1\text{-}59b)$$

1.4.1.4 Effects of Cladding. The most important effect of cladding on the thermal performance of ceramic fuel is the temperature difference between the inner surface of the cladding and the outer surface of the cladding and the outer surface of the fuel. It has been common practice in the past, for fuel elements with a clearance between the fuel and the cladding at operating conditions, to evaluate the temperature drop between fuel and cladding by assuming that the fuel was centered in the cladding and calculating the thermal resistance of the gas-filled gap. This model, however, does not recognize the fact that the ceramic fuel cracks and that intimate contact is established between the fuel and the cladding. Since in most fuel elements the gas-gap is closed at operating temperatures, it is more realistic to use the concept of thermal contact conductance given in Sec. 1.3.4, when an estimate of the contact resistance is needed.

1.4.2 Thermal Behavior of Sintered High-Density UO_2 Fuel Elements

Cohen et al. [45] conducted experiments on oxide pellets contained within a heavy-walled, stainless steel sheath in which the temperature gradients were measured by thermocouples placed in several radial and azimuthal positions. From these gradients, knowledge of the thermal conductivity of the stainless steel, and measurements of the gamma heating in a capsule adjacent to the sheath, the heat flux and heat-generation rate in the fuel were calculated. A thermocouple in the center of the fuel pellets gave the central temperature attained. Thus, each of the quantities necessary for the measurement of the conductivity integral were directly determined; by making successive measurements as the test reactor power was increased stepwise, the variation of this integral with temperature was determined and compared with the various out-of-pile measurements of this quantity.

The results of these experiments are given in terms of measured conductivity averaged between the center temperature T_c and the inner-sheath wall temperature T_i and are shown in Fig. 1-29 [17]. It may be noted that a continual drop in measured conductivity occurs as the assembled clearance increases from 0 to 3.5 mils (.089 mm). Beyond this value the conductivity remains essentially unaffected.

Another type of experiment, attributed to Hawkings and described by Robertson et al. [50], consists of measurements of the center temperature attained in an aluminum-clad cylindrical uranium-oxide fuel element. The heat-generation rate in the oxide is estimated from the measured heat output of adjacent fuel rods. Robertson assumed a fuel-to-cladding conductance of 2 watt/cm^2-°C (3520 Btu/hr-ft^2-°F) and the values of the conductivity integral (surface temperature 0 °C) obtained at various central temperatures are plotted in Fig. 1-30 [17], together with the data of Cohen et al. [45] and the out-of-pile measurements of Kingery [28]. In the Robertson experiments the assembled fuel-clad clearance is very small and in fact in the case of pellet-rod test "Mark III," the aluminum (57S) sheath was found to have an increased diameter after testing. The data of Cohen et al. for a series of test reactor startups with a press-fit specimen are plotted in the same manner for comparison, again assuming a sheath-fuel conductance of 2 watt/cm^2-°C. The change in the value of the conductivity integral obtained by Cohen at successive startups has already been commented upon in connection with Fig. 1-23.

The performance of swaged or compacted fuel elements is different from sintered fuel elements. A detailed discussion on the thermal behavior of these elements is given by Lustman [17].

The onset of melting at the center of the fuel has been widely accepted as an upper limit to the allowable thermal rating of ceramic fuel elements. However, a number of UO_2 fuel elements have been operated in-pile and center melting has occurred with no apparent damage to the sheath [51]. On the other hand, after 3 hours of operation in a high-temperature loop, a specimen with 45% of the volume of the oxide melted experienced a disastrous failure during operation [52]. The failure consisted of disgorgement of a relatively large amount of fuel into the loop and destruction of the cladding over a section of the fuel rod 1.25 in. (3.118 cm) long. Lyons et al. [53] have since shown that failure due to excessive UO_2 melting was due to severe swelling of the cladding caused by the volumetric expansion of the UO_2 on melting. These authors report successful operation of a UO_2 fuel-rod assembly at extreme thermal performance conditions. The peak surface heat flux* at startup, averaged over the four rods in the assembly, was 1.3×10^6 Btu/hr-ft^2 ($\int_{400}^{T} kdT = 162$ watt/cm) with a calcu-

*Conversion factor for heat flux: 1 Btu/hr-ft^2 = 3.155×10^{-4} watt/cm^2.

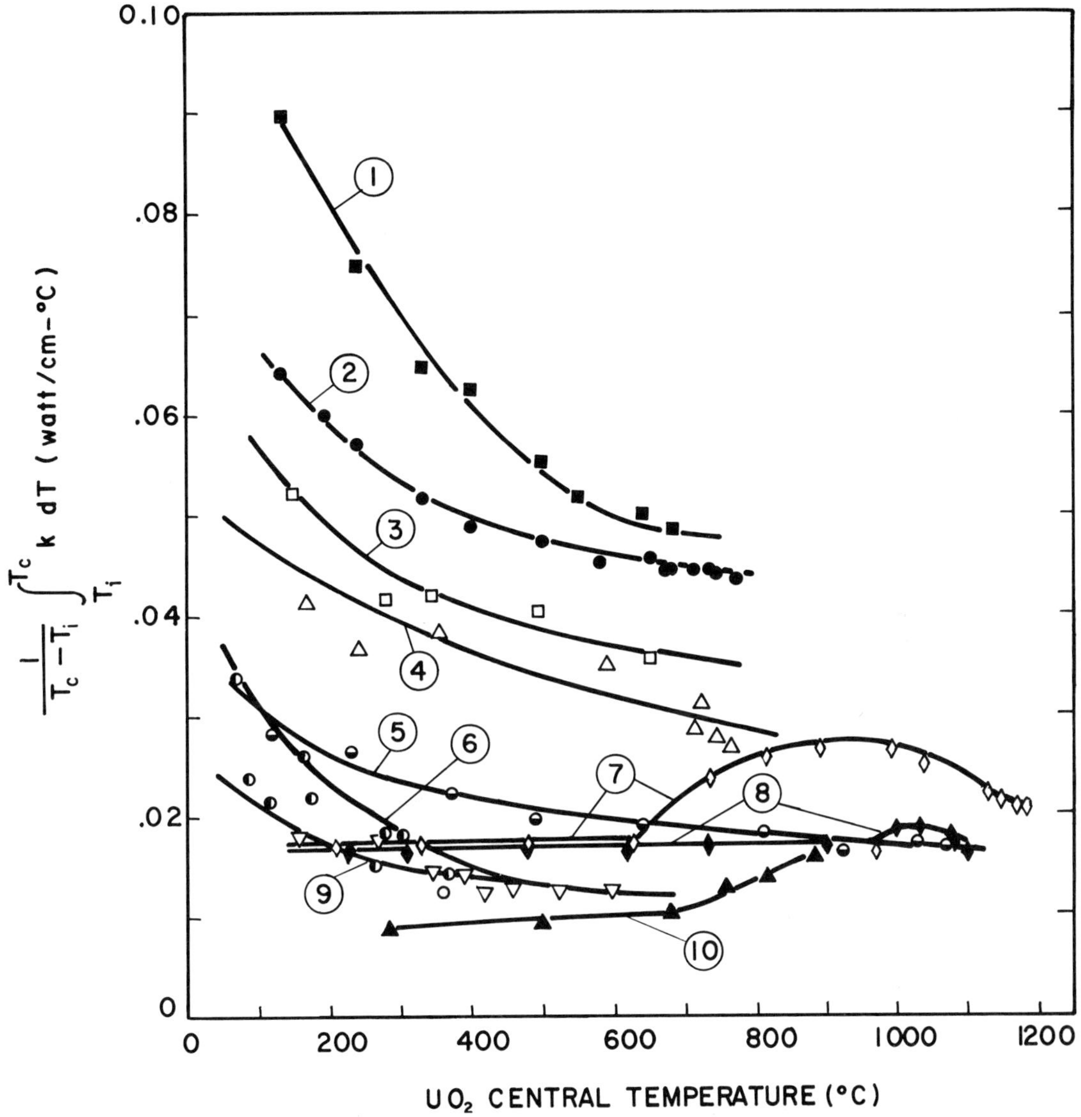

FIG. 1-29. Smeared thermal conductivity (averaged between pellet-center temperature and the inner-sheath wall temperature) of UO_2 as a function of central temperature. The data are from the WAPD-22 Experiment (instrumental capsule irradiation test at the Materials Testing Reactor). Note change in ordinate scale - the figure in reference [17] is in error. The pellet diameter = 0.357 in. (9.07 mm); sheath material was type-304 stainless steel. The various curves correspond to the following: (1) Kingery data, unirradiated UO_2 (95% dense); (2) test WAPD 22-11; zero diametrical clearance, 1 Kr + 3Xe atmosphere, production UO_2; (3) test WAPD 22-4, 1.5 mil (0.038 mm) diametral clearance, He atmosphere, production UO_2 pellets; (4) test WAPD 22-14, 2.5 mil (0.063 mm) diametral clearance, 1 Kr + 3Xe atmosphere, production UO_2 (5) test WAPD 22-6, zero diametral clearance, 1 Kr + 3Xe atmosphere, $UO_{2.15}$ pellets; (6) test WAPD 22-3, 8.0 mil (0.204 mm) diametral clearance, He atmosphere production UO_2; (7) test WAPD 22-15, 3.5 mil (0.089 mm) diametral clearance, 1 Kr + 3Xe atmosphere, UO_2 pellets (1.02% enriched); (8) test WAPD 22-7, 13 mil (0.330 mm) diametral clearance, 1 Kr + 3Xe atmosphere, production UO_2 pellets, second startup; (9) test WAPD 22-1, 3.5 mil (0.089 mm) diametral clearance, He atmosphere, MCW UO_2; (10) test WAPD 22-7, 13 mil (0.330 mm) diametral clearance, 1 Kr + 3Xe atmosphere, production UO_2 pellets, first startup, scrammed at 30 Mw (see also curve 8).

lated peak heat flux of 1.4×10^6 Btu/hr-ft^2 ($\int_{400}^{T} k\,dT$ = 173 watt/cm). At these heat fluxes more than 70% of the UO_2 was melted, and the unmelted rim adjacent to the cladding was approximately 0.04 in. (1.0 mm) thick. Successful operation with this extent of gross central melting was accomplished by using hollow UO_2 fuel pellets. By this means the necessary free volume was provided to accommodate the UO_2 expansion on melting and thereby to eliminate the swelling of the cladding previously experienced with solid pellets. The assembly had attained an average burnup of 2,000 Mwd/ton and endured 19 significant power cycles without difficulty. UO_2 would run down to the bottom of the rod. This would affect power

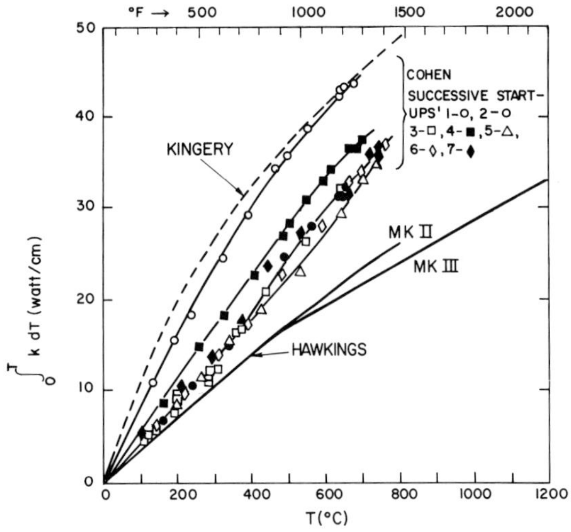

FIG. 1-30. Variation of the conductivity integral with temperature. Assumed interface conductance = $2 w/cm^2$-°C.

distribution and reactivity and would recreate the swelling problem after solidification.

Under certain conditions of fuel cracking and relocation of fuel fragments, Lustman [17] suggests that the molten fuel could come in contact with the cladding and the cooling of the cladding could be insufficient to permit quenching. The cladding may thus be raised to its melting point, creating a burnthrough. This last effect appears the most difficult to cope with in fuel-element design.

Extensive distortion and rupture of a fuel element containing a small defect can occur in water systems by a mechanism known as "waterlogging" [54-56]. A waterlogging failure results when water enters the fuel element during reactor shutdown; on subsequent startup, the water flashes to steam and causes excessive pressure within the element. Eichenberg and coworkers [55] have listed three conditions that could lead to a failure of the cladding of a defective element: 1) a small defect could severely restrict the escape of steam, 2) the original hole could be blocked by a piece of UO_2 that would restrict the escape of steam, and 3) water in the open pores of the UO_2 could cause violent fragmentation of the fuel during heatup. Over a large number of irradiation tests of UO_2 fuel specimens (approximately 30) only four specimens have failed as a result of waterlogging [55, 56]. However, two of these failures occurred as a result of unintentional defects. This appears to demonstrate the difficulty in performing meaningful tests on waterlogging. A defect small enough to produce waterlogging is likely to be a "natural" defect that might be overlooked during inspection. It appears that the conditions which tend to make waterlogging important are small defects, large void volume inside the fuel element, and rapid reactor power changes.

In almost all in-pile tests of fuel elements containing sintered UO_2, it has been observed that the UO_2 fractures as a result of thermal stresses. In addition, there is usually a void space between the oxide and the cladding, although this void space tends to disappear when the element is heated because of the differential expansion between the UO_2 and cladding. Based on the characteristics of UO_2 fuel elements, a mechanism known as "thermal ratcheting," which could result in cladding deformation, has been postulated to occur in the following manner [55]:

1) As the reactor is brought to power, the UO_2 could fracture as the result of thermal stresses, and the clearance between the fuel and cladding would close up.

2) Upon reactor shutdown the void spaces inside the fuel element would be reestablished.

3) Small pieces of the fractured UO_2 could relocate in the void spaces.

4) Further cycling of the fuel element could cause deformation and eventual rupture of the cladding.

The information now available on UO_2 fuel elements indicates that thermal ratcheting is not a serious problem. Further discussions on waterlogging and thermal ratcheting are given in Chapter 13.

1.4.3 Fast-Reactor Fuel-Element Behavior under Transient Heating to Failure

The initial experiments on the transient heating of the fast-reactor fuel element in the Transient Reactor Test Facility (TREAT) have been reported [57, 58]. The mode of failure of elements in fast reactors and the phenomena associated with failure are particularly important because of the possibility that an extensive meltdown of a fast-reactor core could result in the redistribution of fuel to form a supercritical mass.

The experiments are described in reference [57] and also discussed in Chapter 13. These experiments were directed at determining the mode of failure of two typical but markedly different types of fuel elements: those of the Experimental Breeder Reactor No. 2 (EBR-II) and those of the Enrico Fermi Fast Breeder Reactor (EFFBR). The reported experiments concerned the failure of the elements when no coolant was present. The tests involved pulsing the elements to various power levels sufficiently high to cause failure of the elements. The observations included the shape and magnitude of the power pulse, surface and internal temperatures of the fuel elements, and post-irradiation examination of the elements or their remains.

The EBR-II fuel element consists of a uranium - 5 wt.% fissium alloy pin that is 0.144 in. (3.66 mm) in diameter by 14.22 in. (36.12 cm) long, canned in a type-304 stainless steel tube with a 9-mil (0.23-mm) wall and a sodium bond between the fuel pin and the can. The bonding sodium normally fills the tube to a level approximately 0.6 in. (1.52 cm) above the top of the fuel pin and a void space approximately 1 in. (2.54 cm) long is left above the sodium level. Full-length fuel elements were used for the tests. Different typical characteristics of the EBR-II fuel-element behavior were obtained as a function of the maximum temperature reached by the jacket during the test. These tests consisted of power excursions of short duration (0.2 to 30 sec) with various power pulse shapes and values of total energy released. In general, the EBR-II fuel elements did not suffer any damage when the recorded outer-cladding temperatures did not exceed approximately 960°C (1760°F) for the short exposure (< 30 sec) encountered in TREAT. The only action up to this temperature was local boiling and expulsion of the sodium bond (boiling point of sodium at atmospheric pressure is 881°C or 1618°F). The unbonded region might have caused harmful overheating if the element had subsequently been used in an operating reactor.

The temperature region above 960°C (1760°F) and up to 1000°C (1832°F) was characterized by the formation of areas of stainless-steel-uranium alloy (melting point of uranium is 1133°C or 2071°F) (see Chapter 17). The alloy formation was accentuated at pressure contact points, particularly when contact was increased by warping of the fuel element. Penetration of the jacket would sometimes occur depending upon the presence of sufficient pressure contact points between the uranium and cladding surfaces.

Failure of the element always occurred at jacket temperatures between 1000 and 1015°C (1832 and 1859°F). The stainless-steel-uranium alloy would penetrate the cladding and then be forced out (but not violently) by the pressure of vaporizing sodium. Under the effect of gravity and sodium vapor pressure, the alloy tended to flow downward. This increased the contact time between uranium and cladding and led to extensive dissolution of the cladding.

Failure of the fuel element was violent for maximum clad temperatures above 1015°C. The rapidly vaporizing sodium developed a high pressure which forced the uranium out of the element and ejected it laterally and occasionally upward. At very high temperatures (1250 to 1400°C or 2280 to 2550°F) penetrations through the jacket occurred in a number of places along the length of the element. When the element was vented to prevent a pressure build-up at the top, the uranium was driven upward by the force of vaporizing sodium trapped at the base of the element. The overall failure was less violent because of the pressure-release mechanism.

In contrast with the pattern of behavior of the bonded elements, the unbonded elements exhibited quiet meltdown characteristics. The fuel melted and flowed downward (because of gravity and the pressure of inert gas trapped inside the cladding during its assembly) and the molten fuel dissolved large portions of the cladding at the base.

When the refractory metals, tantalum and niobium, were substituted for the unbonded stainless steel cladding, the maximum attainable temperature before failure was raised from 960°C (1760°F) to 1400°C (2552°F). On failure of this type of element, the uranium was ejected from a very localized penetration and the cladding was not extensively dissolved; however, all of the uranium left the confines of the cladding. In two cases in which the failure threshold was not exceeded, although the temperature rose above the melting range of the uranium alloy, voids were found in the uranium after the experiment. These voids were caused by the vaporized sodium, which pushed slugs of uranium alloy upward within the tube. When the fuel element cooled down, the slugs froze in position, leaving voids in the fuel-bearing regions.

The Fermi reactor fuel elements also tested consisted of an 0.158-in. (4.01-mm)-diam uranium-molybdenum-alloy fuel pin, clad with 5-mil (0.13-mm)-thick zirconium. The sample lengths were half the length of the actual fuel elements, which are 32 in. (81.2 cm) long. The fuel and cladding were metallurgically bonded. Considerable fine surface cracking and warping of the element took place during short-duration power pulses when the excursion temperature remained below approximately 1300°C (2372°F). In only one case up to 1300°C did a portion of the alloy dribble from the interior through a crack in the cladding. Actual failure of the elements at higher temperatures seemed to be a case of simple solution of the cladding. The fuel dissolved large portions of the cladding. There was no evidence of any violence. When the fuel was held above the melting point of the uranium alloy for a period of time, the weight of the column of liquid tended to buckle the thin cladding as if the element were pinched between two fingers in several locations. Reference [59] states that the typical high-temperature failure of the sodium-bonded elements is one in which all the fuel is expelled rapidly through a penetration near the hottest point in the fuel pin; this suggests that at these temperatures a supercritical assembly might conceivably be formed by molten-fuel collection in the course of a fast-reactor accident.

In parallel with the TREAT experiments [57] quoted above, sample temperature profiles were calculated as a function of time from experimental data of reactor power, material constants for bond and cladding, and constants derived from uranium data for fuel alloys. Calculations were performed using the IBM 709 code RE-147 ("CYCLOPS"), which was programmed to solve the partial differential equation of heat conduction in cylindrical coordinates by the usual finite difference technique. The assumption of no axial heat conduction, contained in the code, was replaced for these calculations by the assumption of no axial variation in temperature, thus reducing the problem to a one-dimensional analysis. This assumption was based on experimental results which indicated that the axial variation in power was small. The three-region fuel element, e.g., fuel, bond, cladding, was represented by a two-region equivalent by combining the bond and cladding into a simple region of equivalent thermal resistance and volumetric heat capacity. The power generation was assumed to be separable into a product of space and time functions:

$$q = Q_0 R(r) Z(z) n(t) , \qquad (1\text{-}60)$$

where q is the volumetric power-generation rate, R(r) and Z(z) the radial and axial power distributions, and n(t) the relative neutron density as a function of time. The radial power distribution R(r) was calculated by means of an empirical formula, due to Taraba [60], for neutron-flux depression in cylindrical elements. A comparison of the calculated and experimental temperatures at the center of the pin and at the cladding for a fast transient and the EBR-II element are given in Figs. 1-31a and b [57]. As indicated in these figures, the agreement was

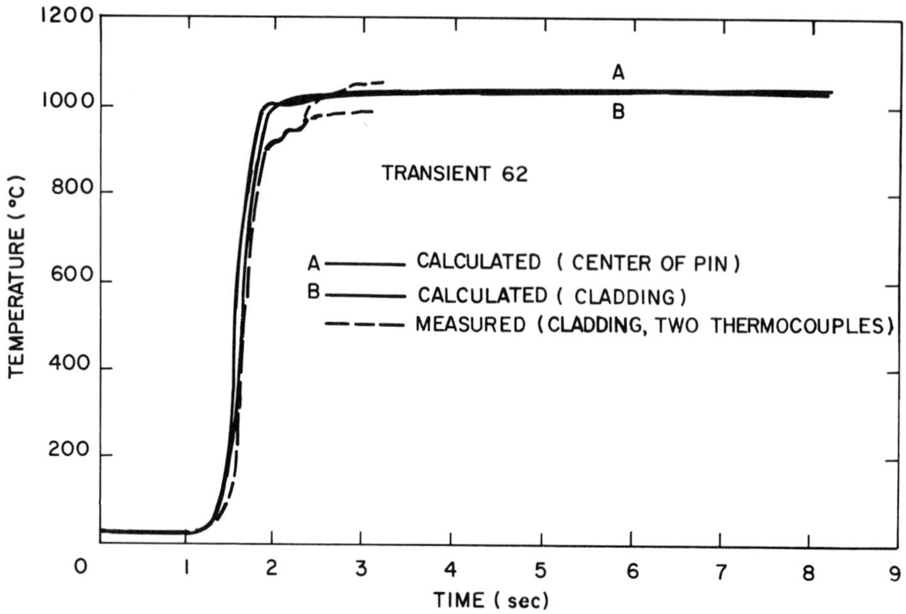

FIG. 1-31a. Comparison of calculated and experimental temperatures at the center of fuel pin and at the cladding in a fast transient. Stainless steel cladding failed at 3.5 sec.

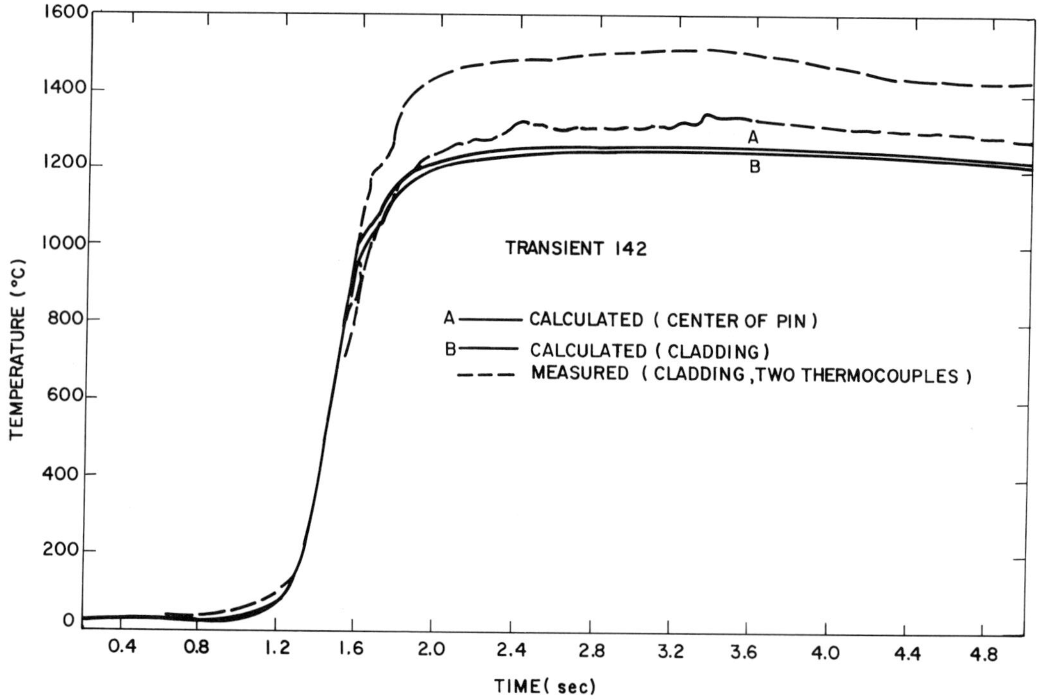

FIG. 1-31b. Comparison of calculated and experimental temperatures at the center of a fuel pin and at the cladding in a fast transient. Difference attributed to malfunction of one thermocouple.

good in some cases, while there was an appreciable error in others. Figure 1-31a is for a stainless steel cladding which failed after 3.5 sec. The recorded temperature shows evidence of appreciable disturbance near failure temperature. The error of 17% between calculated and measured temperatures in Fig. 1-31b was attributed to malfunction of one of the two thermocouples.

Using the temperature-time behavior of the fuel elements obtained during the tests, an analysis of the element failure was attempted in the light of the experimental results. Estimates of relative thermal expansions of fuel and cladding for typical distributions of sample temperatures ruled out failure due to radial expansion. Figure 1-32 [57] is a graph of the calculated internal pressure of the fuel element and bursting pressure of the cladding as a function of temperature for the stainless-steel-clad EBR-II element. At the temperature (1000°C or 1832°F) at which failure was observed, the estimated bursting pressure is about 6 times the total internal pressure. However, the calculated bursting and internal pressures approach each other rapidly above 1000°C and any cladding flaw or appreciable cladding dissolution may shift the bursting pressure D to the left and cause earlier rupture of the cladding. A summary of the results of experiments by the ANL Metallurgy Division [61] on time of penetration of steel and iron by molten uranium and by molten uranium-5 wt.% fissium alloy is given in Table 1-6 [57] If one assumes that in the transient heating of a specimen, local defects occur in the oxide-film coating of the fuel pin, then local attack of the cladding results. No data are available on the effects of contact pressure upon attack of unprotected steel by uranium at the temperature of interest here. However, measurements at 750°C (1382°F) demonstrated rate increases of a factor of about 2 by increasing contact pressure from 0.58 atm to 1.05 atm. An increase from 1.05 atm to 2.07 atm increased the rate by an additional 70%. It was thus concluded that cladding failure of the EBR-II (stainless-steel-clad) element at 1000°C (1832°F) was due to dissolution of cladding unprotected in localized areas by the oxide film on the fuel pin, possibly assisted by internal pressure.

For the EBR-II fuel element when clad with refractory metal (Ta, Nb, Zr), the time for the molten U to penetrate the can is much longer. Some values of penetration time are presented in Fig. 1-33 [61]. On the other hand, the calculated inside pressure and bursting pressure for Ta-clad and Nb-clad elements shown in Fig. 1-34a and b [57] show that at 1500°C or 2732°F (the observed failure temperature) internal pressure is the dominant mechanism of failure.

For the metallurgical-bonded, zirconium-clad Fermi elements, cladding dissolution by a molten fuel core is the cause of failure. This is seen from Fig. 1-33, which indicates that it takes 40 sec for the molten fuel to penetrate the zirconium can of a Fermi-I element at 1294°C (2361°F). This time is reduced to 4 sec at a can temperature of 1370°C (2498°F).

1.5 Time-Dependent Heat Conduction in Fuel Elements

During a power transient the magnitude of the volumetric heat source and its spatial distribution in the fuel are strongly dependent on time. The mode of heat transfer at the boundary between the fuel element and the coolant may change drastically, i.e., one-phase flow convection may develop successively into two-phase flow, subcooled boiling,

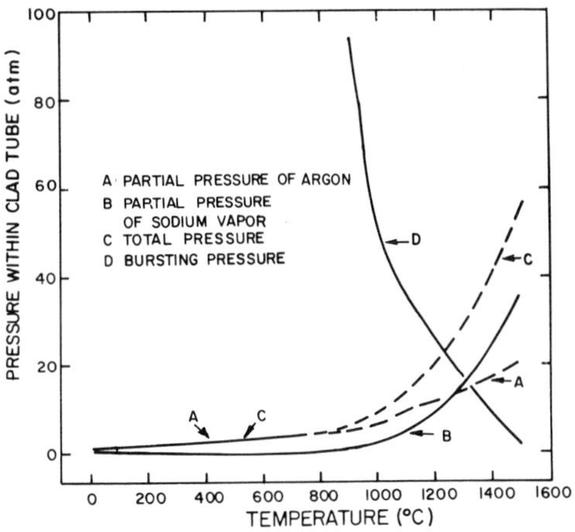

FIG. 1-32 Comparison of calculated cladding bursting pressures and internal pressures for isothermal EBR-II Mark I element.

TABLE 1-6

Time in Seconds for Penetration of Stainless Steel and Iron by Molten Uranium and Uranium-5 wt. % Fissium

Temp.[a] (°C)	Type-304 stainless steel				Type-430 stainless steel		Armco Iron
	in U		in U-5 wt. % Fs		in U	in U-5 wt. % Fs	in U-5 wt. % Fs
	0.025 cm (9.8 mil)	0.102 cm (40.1 mil)	0.025 cm (9.8 mil)	0.102 cm (40.1 mil)	0.102 cm (40.1 mil)	0.102 cm (40.1 mil)	0.102 cm (40.1 mil)
1100	—	—	18.4	64.3	—	—	15.0
1125	—	—	0.61	3.8	—	—	—
1150	0.53	2.7	0.82	3.4	1.93	2.5	1.9
	0.43	2.9	—	3.76	—	—	—
	0.40	2.6	—	—	—	—	—
1187	2.8	11.6	3.8	16.75	3.74	8.8	3.4
	1.92	—	—	—	—	—	—
	1.2	—	—	—	—	—	—
1244	1.7	9.5	3.3	12.7	10.8	—	9.95
1300	—	9.0	—	10.1	—	—	9.6
1350	0.32	6.3	—	8.03	—	—	5.85

[a] The corresponding Fahrenheit temperatures are 2012°F, 2057°F, 2102°F, 2169°F, 2271°F, 2372°F, and 2462°F, respectively.

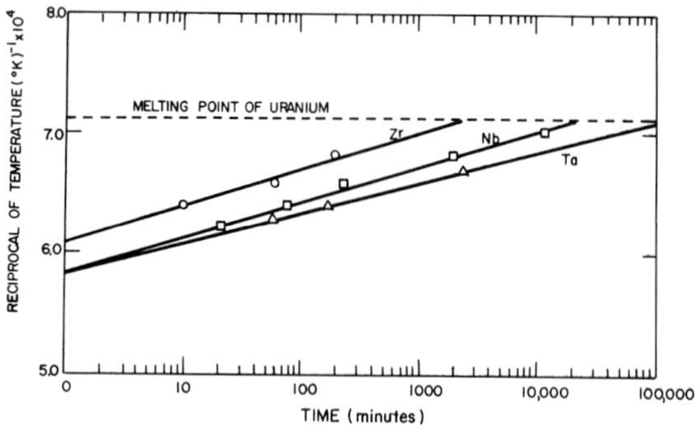

FIG. 1-33. Time required for molten uranium to penetrate 0.051 cm (2 mil) can as a function of temperature for zirconium, niobium, and tantalum.

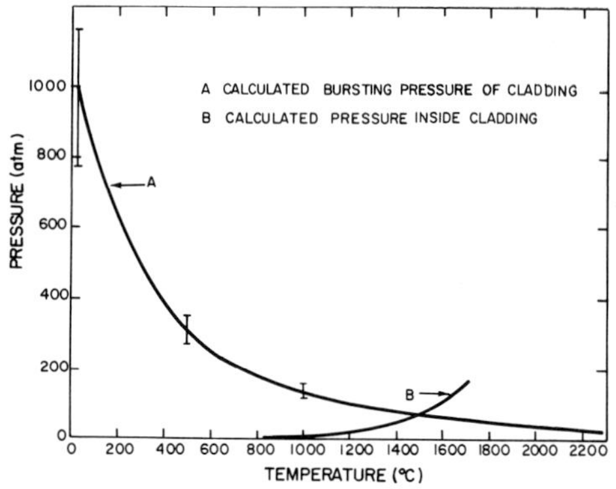

FIG. 1-34a Comparison of calculated bursting pressure of cladding (curve A) and internal pressure (curve B) for isothermal Ta-clad EBR-II pin.

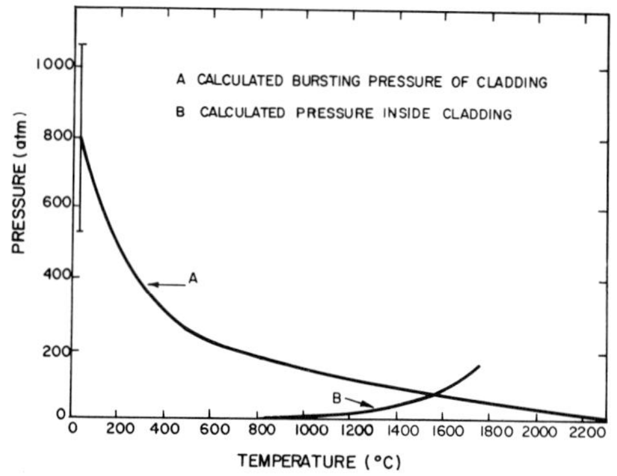

FIG. 1-34b Comparison of calculated bursting pressure of cladding (curve A) and internal pressure (curve B) for isothermal Nb-clad EBR-II pin.

burnout, and film boiling as discussed in Sec. 2 of this chapter. Due to differential expansion and internal pressure buildup, the thermal resistance of the contact between fuel and clad will vary as well as the thermal properties of the fuel-element materials. Further complexities arise due to the dependence of the power density on the fuel temperature and coolant density through the neutron density and neutron energy distribution, which are also time and space dependent. Present methods of analysis of the overall thermal and neutronic behavior of fast reactors and of water-cooled and "dry" or solid-moderated reactors are described in other chapters of this book. Only simple cases of transient heat conduction in one-dimensional geometry are amenable to some simple analytical solution.

Reference [62] treats the case of a clad fuel plate in flowing coolant with constant thermal properties, a constant convection heat-transfer coefficient, and a uniform heat generation increasing exponentially with time. Solutions to other simple cases will be found in references [63, 64, 65] and [66]. Tippets [67] has developed a mathematical procedure using the finite Hankel transform to obtain a solution for the transient radial temperature distribution in a solid cylindrical nuclear fuel element in which the heat-generation rate varies arbitrarily with time. The element is assumed to be a solid cylinder having an initially steady-state temperature distribution. Axial symmetry of temperature and heat generation is assumed, as well as constant material properties, sink temperature, and boundary conductance. The heat-generation rate is assumed to vary radially in accordance with thermal-neutron-diffusion theory and is an arbitrary function of time. It is assumed that the heat-conduction medium is isotropic and homogeneous and that longitudinal conduction is negligible.

Under these conditions the equation of heat conduction can be expressed as

$$\frac{\partial T}{\partial t} = \alpha\left[\frac{\partial^2 T}{\partial r^2} + \frac{1}{r}\frac{\partial T}{\partial r}\right] + \frac{q_0'''}{\rho c_p} m(r)n(t) \text{ ,} \quad (1\text{-}61)$$

where $\alpha = k/\rho c_p$ is the thermal diffusivity, ρ and c_p the density and specific heat, respectively, and $q_0''' m(r)n(t)$ the volumetric heat source separated into space and time functions.

The boundary conditions at the center are

$$\partial T/\partial r = 0 \text{ ,} \quad r = 0 \text{ ,} \quad (1\text{-}62a)$$

and at the fuel-element outer radius a,

$$\partial T/\partial r + UT = 0 \text{ ,} \quad r = a \text{ .} \quad (1\text{-}62b)$$

Equation (1-61) is integrated with respect to time t between $t = t_i - 1$ and $t = t_i$ by using the finite Hankel transform of zero order [68]. The result is

$$\bar{T}(r, t_i) = \bar{T}(r, t_{i-1})\exp(-\alpha\lambda_m^2\Delta t_i)$$

$$+ \frac{q_0'''}{\rho c_p}\bar{m}(r)\exp(-\alpha\lambda_m^2 t_i)\int_{t_{i-1}}^{t_i} n(t)\exp(\alpha\lambda_m^2 t)dt \text{ ,} \quad (1\text{-}63)$$

where

$$\Delta t_i = t_i - t_{i-1} \text{ ,} \quad (1\text{-}64a)$$

$$\bar{T} = \int_0^a rT(r, t)J_0(\lambda_m r)dr \text{ ,} \quad (1\text{-}64b)$$

$$\bar{m}(r) = \int_0^a r m(r) J_0(\lambda_m r)\,dr, \qquad (1\text{-}64c)$$

and λ_m is defined by

$$U J_0(\lambda_m a) - \lambda_m J_1(\lambda_m a) = 0, \quad m = 1, 2, 3. \qquad (1\text{-}64d)$$

Use of the following inversion formula ,

$$f(r) = \frac{2}{a^2}\sum_{m=1}^{\infty} \frac{\lambda_m^2 \bar{f}}{U^2 + \lambda_m^2} \frac{J_0(\lambda_m r)}{J_0^2(\lambda_m a)}, \quad h > 0 \qquad (1\text{-}65)$$

(with $\bar{f}$ being the zero-order Hankel transform of f (r)), gives the desired boundary value solution of Eq. (1-61):

$$T(r, t_i) = \frac{2}{a^2}\sum_{m=1}^{\infty} \frac{J_0(\lambda_m r)}{(1 + \lambda_m^2/n^2) J_1^2(\lambda_m a)} \times \left[\bar{T}(r, t_{i-1}) \exp(-\alpha\lambda_m^2 \Delta t_i) + \frac{q_0'''}{\rho c_p} \bar{m}(r) \exp(-\alpha\lambda_m^2 t_i) \int_{t_{i-1}}^{t_i} n(t) \exp(\alpha\lambda_m^2 t)\,dt\right], \quad h > 0. \qquad (1\text{-}66)$$

The solution (1-66) is used to develop an equation of immediate application to reactor transient heat-transfer problems.

Case I

Consider a long, solid, cylindrical fuel element cooled by the transfer of heat from its surface across a thermal resistance to a uniform sink. Assume that the radial distribution of thermal-neutron flux in the element is in accord with that predicted by thermal-neutron-diffusion theory. Under these conditions the steady-state radial temperature and heat-generation-rate distribution are given respectively by

$$T(r, 0) = T(a, 0) + \frac{W(0)}{2\pi k a \kappa I_1(\kappa a)} [I_0(\kappa a) - I_0(\kappa r)], \qquad (1\text{-}67a)$$

and

$$q'''(r, 0) = \frac{\kappa W(0)}{2\pi a I_1(\kappa a)} I_0(\kappa r), \qquad (1\text{-}67b)$$

where W(t) is the heat-transmission rate at the surface of the element per unit length at time t, κ is the reciprocal of the diffusion length of thermal neutrons, and T (a, 0) is the surface temperature measured above any arbitrary sink or coolant temperature T_c. In general,

$$T(a, 0) = T_c + \frac{W(0)}{2\pi a k U}. \qquad (1\text{-}68)$$

Taking the transform of Eq. (1-67a) and making use of Eqs. (1-67b) and (1-68) and substituting the expression for T (r, 0) in Eq. (1-66) gives the solution:

$$T_1(r, t_1) = \frac{2}{a}\sum_{m=1}^{\infty} M(\lambda, r) \frac{e^{-\alpha\lambda_m^2 t_1}}{1 + \lambda_m^2/U_1^2} \times \left[T_1(a, 0) + \frac{W(0)}{2\pi k a}\phi(\lambda, t)\right] \quad U_1 > 0, \qquad (1\text{-}69)$$

where

$$M(\lambda, r) = \frac{J_0(\lambda_m r)}{\lambda_m J_1(\lambda_m a)}, \quad r < a, \qquad (1\text{-}70a)$$

$$M(\lambda, r) = 1/U_1, \quad r = a, \qquad (1\text{-}70b)$$

$$\phi(\lambda, t) = \frac{1}{\kappa}\left[\frac{I_0(\kappa a)}{I_1(\kappa a)} - \frac{\lambda_m^2}{\kappa^2 + \lambda_m^2}\left(\frac{\kappa}{U_1} + \frac{I_0(\kappa a)}{I_1(\kappa a)}\right) \times \left(1 - \kappa^2 \alpha \int_0^{t_1} n(t) e^{\alpha\lambda_m^2 t}\,dt\right)\right] \quad \kappa > 0 \qquad (1\text{-}71a)$$

$$\phi(\lambda, t) = \left[\frac{2}{a\lambda_m^2} - \frac{1}{U_1} + \frac{2\alpha}{a}\int_0^{t_1} n(t) e^{\alpha\lambda_m^2 t}\,dt\right], \quad \kappa = 0, \qquad (1\text{-}71b)$$

and λ_m is defined by Eq. (1-64d). The subscript 1 has been appended to T(a, 0) and T(r, t_1) to indicate that they are to be measured with respect to the constant sink or coolant temperature existing between $0 < t \le t_1$. The parameter U = h/k (h is over-all heat-transfer coefficient) has the subscript 1 to indicate its correspondence to the time interval $0 < t = t_1$.

Equation (1-69) is applicable to problems for which a steady-state temperature distribution exists at t < 0 and for which it may be assumed that at t = 0 the boundary conductance (as expressed by H = kU) and/or the coolant temperature change instantaneously to new constant values U_1 and T_1, respectively. (Of practical interest is the case of an integral superheat reactor in which the source of superheater coolant is steam generated in the boiler region.)

If the boundary conductance and coolant temperature are uniformly constant from $t<0$ to $t = t_1$, an alternate form of Eq. (1-69) may be obtained by specifying the coolant temperature to be zero:

$$T(r, t_1) = A \sum_{m=1}^{\infty} \frac{M(r, \lambda)}{[1 + (\lambda_m^2/U^2)](\kappa^2 + \lambda_m^2)} \times \left[e^{-\alpha\lambda_m^2 t_1} + \alpha\lambda_m^2 \int_0^{t_1} n(t) e^{-\alpha\lambda_m^2 (t_1 - t)}\,dt\right], \quad U > 0, \qquad (1\text{-}72)$$

where

$$A = \frac{\kappa W(0)}{\pi k a^2}\left[\frac{\kappa}{U} + \frac{I_0(\kappa a)}{I_1(\kappa a)}\right], \quad \kappa > 0, \qquad (1\text{-}73a)$$

or

$$A = \frac{2W(0)}{\pi \kappa a^3}, \quad \kappa = 0, \qquad (1\text{-}73b)$$

and M (r, λ) and λ_m are as previously defined.

Case II

Mathematical solutions of sufficient generality for solving problems involving a variable sink temperature in addition to a space- and time-dependent heat-generation rate are difficult to obtain. In order to provide a method of attack on those problems for which the effect of changes in the coolant temperature cannot be ignored and for which the idealizations necessary to derive an analytic solution would be unacceptable, an adaptation of the solution to Eq. (1-61) suitable for use in a stepwise calculation has been developed. It is assumed as an approximation that the sink or coolant temperature makes instantaneous changes at each time division t_i but is constant throughout each time increment $\Delta t_i = t_i - t_{i-1}$. The difference between the coolant temperature for a time increment Δt_i and the steady-state coolant temperature is ϵ_i. Hence,

$$T_i(r, t_p) - T_{i-1}(r, t_p) = \epsilon_i - \epsilon_{i-1} \qquad (1\text{-}74)$$

where the subscript i means that T refers to the sink temperature existing during the ith time increment. After some algebra, the following solution is derived:

$$T_i(r, t_i) = A \sum_{m=1}^{\infty} \frac{M(\lambda, r)}{[1 + (\lambda_m^2/U^2)]\,(\kappa^2 + \lambda_m^2)} \Bigg[\exp(-\alpha\lambda_m^2 t_i)$$

$$+ (2/aA)(\kappa^2 + \lambda_m^2) \sum_{p=0}^{i-1} (\epsilon_{p+1} - \epsilon_p)\exp[-\alpha\lambda_m^2(t_i - t_p)]$$

$$+ \alpha\lambda_m^2 \int_0^{t_i} n(t)\exp[-\kappa\lambda_m^2(t_i - t)]\,dt\Bigg], \quad U > 0. \qquad (1\text{-}75)$$

The quantities A, M(λ, r), and λ_m are as previously defined. The subscript i appended to $T(r, t_i)$ indicates that T_i is the temperature increase measured above the coolant temperature, existing during the time increment Δt_i.

Equation (1-75) is applicable, for approximate stepwise computation, to problems for which the boundary conductance h remains constant and the coolant temperature varies arbitrarily from a zero steady-state value at t = 0.

For reactor problems the method has the particular advantage of enabling integral solutions to be obtained in which the time function n(t) is arbitrary, which allows immediate application of the solution to a variety of transient problems without further derivations.

Equation (1-72) was programmed on a digital computer and the solutions obtained were compared with results of multi-zone and one-zone lumping numerical methods [69]. The comparison presented in Figs. 1-35a and 1-35b [67] indicate that no appreciable gain in accuracy is obtained with the analytical solution over the multi-lump numerical method. The analytical solution is, however, more convenient to program and requires less machine time.

The assumption of constant and uniform boundary conductance and material properties is necessary for the above analytic method. The consequences of such assumptions were tested by comparing the surface heat flux q″(t) obtained for a base case and similar situations where the boundary conductance, the material properties, and the neutron thermal diffusion length were different. The functions q″(t) and n(t) for the base case are shown in Fig. 1-36 [67]. The following properties were used for the base case (natural uranium fuel):

a = 0.6811 in. (1.73 cm)
c_p = 0.03643 Btu/lb-°F (0.03643 cal/g-°C)
k = 18.82 Btu/hr-ft°F (0.326 watt/cm-°C)
T_m = 400°C (752°F) = mean fuel temperature
h = 5110 Btu/hr-ft²-°F (2.91 watt/cm²-°C)
κ = 1.654 in.⁻¹ (0.651 cm⁻¹)
α = 0.01751 in.²/sec (0.113 cm²/sec)
ρ = 0.6828 lb/in³ (18.9 g/cm³)

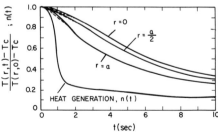

FIG. 1-35a. Cooling of long cylindrical fuel element assuming constant boundary conductance and coolant temperature: heat generation and normalized temperature as a function of time. The solid curves are from analytic solution; the broken curves are computed using the multi-lump numerical method.

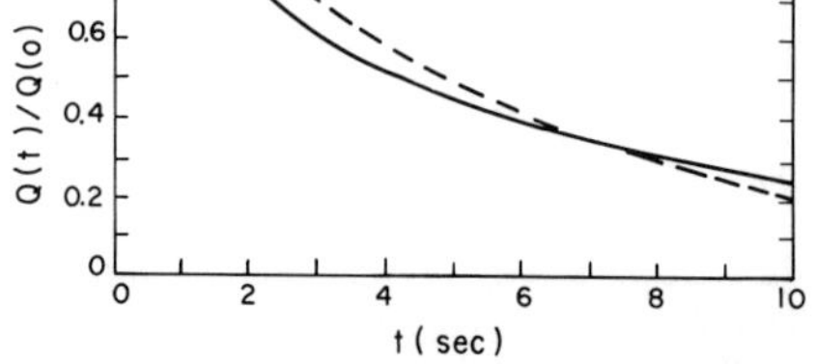

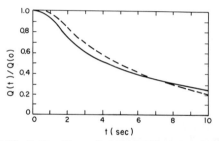

FIG. 1-35b Cooling of long cylindrical fuel element assuming constant boundary conductance and coolant temperature: normalized heat flux as a function of time. The solid curve is from the analytic solution; the broken curve is computed using the one-lump numerical method.

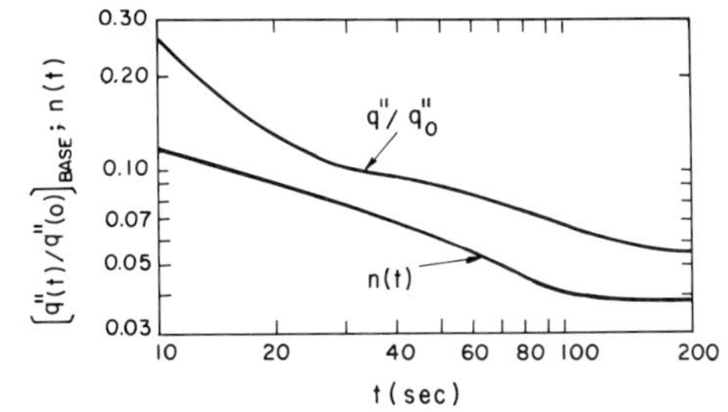

FIG. 1-36. Normalized heat generation and heat flux (base). In the above T_m = 400°C (752°F), U = 5110 Btu/hr-ft^2-°F (2.91 w/cm^2-°C), κ = 1.654/in. (0.651/cm), α = 0.017514 in.^{2}/sec. (0.11298 cm^2/sec).

The effect of variations is shown for two different values of the boundary conductance h (3320 and 6670 Btu/hr-ft^2 °F) for variation of the properties of uranium evaluated at two different temperatures T_m (600°C and 100°C), and for the case of a uniform heat-generation rate in Figs. 1-37a, b, c, respectively [67]. Matsch [70] has made a similar analysis of temperature transients in one-dimensional fuel elements using a finite integral transform method. Matsch considers the case of a two-region element with thermal resistance at the interface, a time-dependent heat generation in the inner region, and a step change in the convective heat-transfer coefficient of the outer surface. In the analysis of severe transients, the accuracy obtained by the above methods is not sufficient. These limitations come from large temperature changes in the fuel element and the change of heat-transfer modes into the coolant as the transient progresses. It therefore becomes necessary to use digital computers. The numerical integration of the time-dependent conduction equation,

$$\nabla \cdot k\nabla T + q''' = \rho c_p \, \partial T/\partial t \, , \qquad (1\text{-}76)$$

is usually performed by lumping the fuel element into discrete space and time intervals and solving numerically by a difference method with appropriate boundary conditions. A detailed description of these methods will be found in References [62] and [67]. A typical IBM-7090 code named SIFT performs such a computation and is described in Reference [71].

2 SPECIAL PROBLEMS OF HEAT TRANSFER BETWEEN SOLIDS AND LIQUIDS

2.1 Introduction

Reactor design involves the ability to predict heat-transfer rates and wall temperatures for a variety of flow geometries and fluid conditions. Many collections of correlation equations have appeared in such places as reactor handbooks. There is relatively good agreement among heat-transfer engineers regarding the prediction of heat-transfer coefficients for fluids in a single phase. However, there is far less agreement on prediction of heat-transfer conditions in two-phase flow.

The major unsettled areas of heat-transfer research are those dealing with the prediction of "burnout" or critical heat flux in forced-convection boiling, prediction of heat-transfer coefficients in film boiling, the influence of transients in boiling heat transfer, and heat transfer to liquid metals. The present discussion is limited to these four areas.

2.2 Critical Heat Flux or "Burnout"

2.2.1 Definition of Critical Heat Flux

The most important limit in the thermal performance of liquid-cooled reactors is the so-called "burnout" condition. It is associated with a sharp reduction in ability to transfer heat from the fuel surface, that is, with a sharply reduced heat flux at a given temperature difference or an increased temperature difference at the same heat flux.

Typical of test results in the vicinity of burnout is the curve shown in Fig. 2-1 [72] with four regions identified. The detailed description of what occurs in these four regions is intimately associated with the details of the two-phase flow regime which exists, the vapor quality, and the total flow rate in a particular geometry.

Imagine the curve of Fig. 2-1 to represent the test results taken at constant inlet enthalpy, constant flow rate, and uniform heat flux in a channel of uniform cross section. Imagine that the system under observation consists of a cylindrical heater unit—say a single fuel rod—cooled by a rising column of a fluid such as water. As the power input to the heater unit is raised (and with it the heat flux q/A), the difference between the surface temperature of the heater wall and the fluid increases in a more or less linear fashion. Region I will usually be a region of no boiling, ordinary single-phase forced convection. Then nucleate boiling and evaporation begin to occur, as shown in the second region of Fig. 2-1. Beyond a critical heat flux the wall surface temperature begins to rise sharply because of the lower heat transfer to the fluid and the wall temperature is observed to oscillate significantly (region III) until the wall is completely dry and film boiling (dry wall) is established (region IV). To understand what happens in regions II and III it is necessary to focus attention on the exit condition. The description of these regions depends largely on the flow rate.

At low flow rates it is possible for vapor to form at stray nucleation sites on the wall and, because the heat flux is low, to establish annular flow where evaporation can continue at the liquid-vapor interface by pure conduction of heat from the wall through the annular liquid film without nucleate boiling. The central core of the channel consists of vapor and liquid droplets. Region II is characterized by an increased heat-transfer coefficient (decrease in slope of the curve) because of the increased velocity due to the presence of the vapor in all cases, including high and low flow, whether or not nucleation occurs. As the quality increases down the channel, the vapor velocity may become large enough to "tear" the liquid off the wall, resulting in the temperature oscillation associated with region III. At sufficiently high heat flux the wall would be dry and a fog flow would exist in the core, resulting in region IV, which might be called "film boiling."

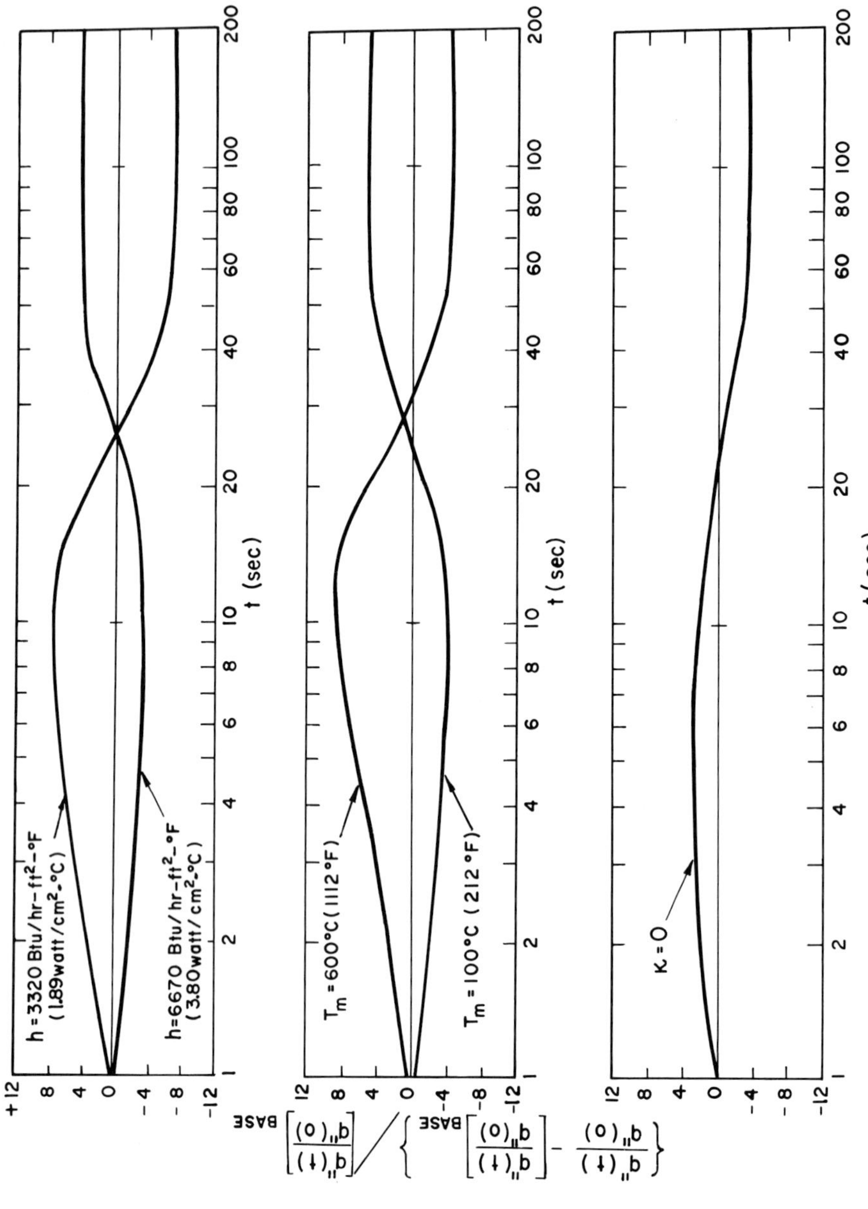

FIG. 1-37. (a) Effect of variation of boundary conductance from base case of Fig. 1-36. (b) Effect of variation of material properties with temperature from base case of Fig. 1-36. (c) Effect of variation of inverse thermal-neutron-diffusion length from base case of Fig. 1-36.

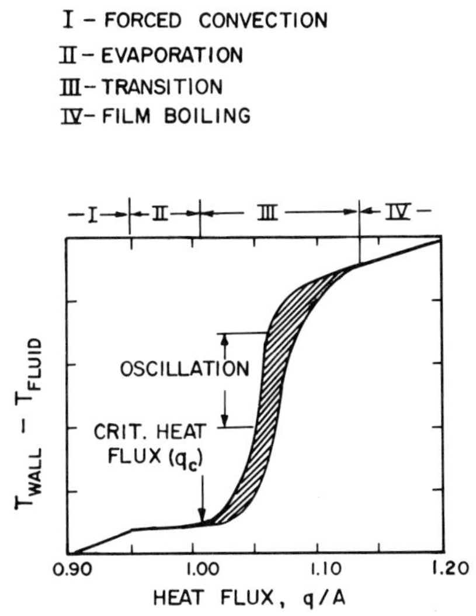

FIG. 2-1. Typical temperature history during burnout.

The term "film boiling" is used to describe the condition when the heater surface is dry and blanketed by a layer, or film, or vapor.

At higher flow rates the condition in region II is identified as nucleate boiling. If the flow rate is sufficiently high, the critical heat flux can occur at very low exit vapor quality, in which case the mechanism in region III is similar to that associated with pool-boiling burnout, where the surface is covered with a liquid. In region IV, the liquid is primarily in bubbly flow with a vapor film separating the liquid from the wall.

These two cases perhaps represent extremes of all possible flowing conditions at the critical heat flux which might occur in nuclear reactors. In the low-flow case it is possible that no nucleate boiling ever really existed and that the critical heat flux was in fact not a departure from nucleate boiling. Several points should be noted. First, the so-called "critical heat flux" in reality is not sharply defined and really represents a more or less narrow region of heat fluxes beyond which a rather sharp, but still ill-defined, wall-temperature change will occur. Second, as shown in Fig. 2-1, the large jump in wall temperature is a function of the vapor quality, the total flow rate, and the geometry. It usually has nothing to do with the metal surface itself. Thus, whether the wall will melt depends entirely on whether the sudden temperature rise of the wall in this phenomenon raises it above the melting point. If a reactor designer finds it desirable to take advantage of high heat fluxes of the order of 10^6 Btu/hr-ft^2 ($\sim$300 watt/cm^2), then melting is sure to occur. (If a heat-transfer coefficient of 50 is assumed, this gives a $T_W - T_f$ of 20,000°F!) Only for very low heat fluxes (less than 10^5 Btu/hr-ft^2) can melting be avoided. In point of fact, superheater elements operate in the region IV regime all the time and some boiling-water-reactor elements operate in regimes somewhere between region III and region IV. Under some conditions, particularly at high exit quality, the heat flux may be low enough to result in a wall temperature in region IV which would be well below the melting point of the wall; hence, no physical burnout would result. At the other extreme, high flow rate, the critical heat flux would represent a condition just below a departure from nucleate boiling. Also the heat flux would be rather high, causing the wall temperature in region IV to be well above the melting point of the wall; hence, physical burnout would result.

In light of the above discussion, terms such as "burnout," "departure from nucleate boiling," "maximum heat flux," etc., should be discarded in favor of the more noncommittal "critical heat flux" defined as the heat flux just below the point, on a curve such as that of Fig. 2-1, where the wall temperature begins to rise sharply.

In the preceding discussion the point which was called the "critical-heat-flux point" was identified. Detecting and measuring the corresponding critical heat flux, $(q/A)_{crit}$, is quite another matter. Some detectors are designed to observe and trip on the amplitude of temperature oscillation of a temperature-measuring device usually placed on the outside surface of the test section and close to the exit. Others are designed to respond to rate of rise of such a temperature indicator. Still others detect the rate of rise of the ratio of the voltage drop over the last quarter of the electrically heated test section to the voltage drop over the entire test section. Clearly, each of these detectors, set to be tripped at various magnitudes of temperature oscillation or various magnitudes of rate of rise of temperature or voltage ratio, can indicate widely differing magnitudes of $(q/A)_{crit}$ for a particular test condition. The possible variation is not known, but the authors believe a ± 10% variation is not unreasonable to expect, particularly when the rate of rise of wall temperature that occurs at low $(q/A)_{crit}$ values in the high-quality region is slow.

2.2.2 Pool Boiling

It is perhaps significant that in even the simplest of geometries—boiling on a heated wire, cylinder, or flat plate submerged in a large pool of liquid—there is disagreement in describing the mechanism or the conditions existing at critical heat flux.

In boiling, bubbles seem to form repeatedly at nucleation sites (cavities) on the heating surface, thus forming a "column" of bubbles. As heat flux increases, visual observations indicate that the number of active nucleation sites per unit area increases. In addition, observations suggest that as heat flux increases to nearly the critical value, the successive bubbles coming from a nucleation site stop appearing as individual bubbles and begin running into each other to from an undulating column of vapor. The following three somewhat different descriptions of the condition at the critical q/A have been offered:

1. As the number of nucleation sites becomes numerous, neighboring bubble or vapor columns coalesce and vapor blanketing occurs at the surface [73].

2. As the bubbles from a nucleation site begin to run into each other, this mutual interference retards the motion of the departing bubble, causing a vapor blanket to cover the surface [74].

3. As the critical heat flux is approached, the appearance of more and more undulating columns of vapor reduces the available area for liquid flow towards the surface. This increases significantly the relative velocity between liquid and vapor, causing the liquid-vapor interface to become unstable, thus essentially starving the heated surface of liquid to cause the formation of the vapor blanket [75].

Each of the above descriptions has led to the formulation of a correlation equation which agrees reasonably well with available data. The equations essentially correlate the average vapor velocity leaving the surface, $(q/A)/\rho_v h_{fg}$, with a function of pressure relative to the pressure at the fluid's critical point. Pursuing the logic of the third description, Zuber [75, 76] suggests the following form of equation for saturated liquids:

$$\frac{(q/A)_{\text{crit. sat.}}}{\rho_v h_{fg}} = 0.13 \left[\frac{\sigma g g_0(\rho_L - \rho_v)}{\rho_v^2}\right]^{1/4}, \quad (2\text{-}1)$$

where $(q/A)_{crit}$ is the critical heat flux, Btu/hr-ft^2; ρ_v, ρ_L is the vapor and liquid density, lb_m/ft^3; h_{fg} is the latent heat, Btu/lb_m; σ is the surface tension, lb_f/ft; g is the acceleration of gravity, ft/hr^2; and g_0 is the constant, 4.17×10^8 lb_m-ft/lb_f-hr^2. This agrees well with a formulation suggested by Kutateladze [77, 78]. Experimental data [76] suggest the coefficient 0.13 is in the range of 0.12 to 0.20.

For boiling of subcooled liquids, Zuber et al. [76] suggest the following equation:

$$\left(\frac{q}{A}\right)_{\text{crit. sub.}} = \left(\frac{q}{A}\right)_{\text{crit. sat.}} + 0.696\sqrt{k_L \rho_L C_L}$$

$$\times \left[\frac{g(\rho_L - \rho_v)}{g_0 \sigma}\right]^{1/4} \left[\frac{\sigma g g_0(\rho_L - \rho_v)}{\rho_v^2}\right]^{1/8}, \quad (2\text{-}2)$$

which agrees well with available data.

Noyes [79] has modified the Zuber-type equation by introducing a Prandtl Number to correlate saturated pool-boiling critical heat flux for sodium at saturation temperature:

$$\frac{(q/A)_{\text{crit. sat.}}}{\rho_v h_{fg}} = 0.144\,\text{Pr}^{-0.245}\left[\frac{\sigma g g_0(\rho_L - \rho_v)}{\rho_v^2}\right]^{1/4}. \quad (2\text{-}3)$$

Data for sodium, water, benzene, methane, carbon tetrachloride, and ethanol have been correlated by the above equation.

Critical-heat-flux data for pool boiling of benzene, diphenyl, and mixtures of the two are presented in reference [80] over a pressure range of 13 to 500 psia (1 to 33 atm).

Most pool-boiling data are taken on submerged wires, vertical flat metal strips, and horizontal plates at the pool bottom. Generally there seems to be little difference in the magnitude of the $(q/A)_{crit}$ for these various cases except when the wire diameters are very small (< 0.005 in. or 0.13 mm) and the strip thicknesses are very small (< 0.006 in. or 0.15 mm). Most data [81] show little or no influence of surface finish on $(q/A)_{crit}$ although there is some disagreement with this conclusion. This conclusion applies to surfaces that can be classed as smooth but not to intentionally grooved or coarse sandblasted surfaces.

2.2.3 Forced Convection

2.2.3.1 General. The present ability to predict and describe the necessary and sufficient conditions for critical heat flux for boiling with forced convection is far inferior to that existing for pool boiling. Forced-convection critical-heat-flux data exist for a variety of test conditions. Most of the data were taken for flow inside of uniformly heated round tubes. Another large body of data exists for uniformly heated thin rectangular passages [82]; data are accumulating for flow in annuli with one or both walls heated; more recently, tests have been performed for flow outside of rod bundles; and a few tests have been performed with nonuniform heat-flux distribution. Most of the data are taken with water as the test fluid. Some data have appeared for a few hydrocarbons and a few liquid metals.

The magnitude of $(q/A)_{crit}$ seems to be seriously affected by the system dynamics of the entire test loop. It has been demonstrated [83] that even with subcooled liquids introduced at the inlet, it is necessary to place a well-throttled valve just ahead of the test section in order to obtain reproducible results for $(q/A)_{crit}$. More recently [84] the quantitative reduction of $(q/A)_{crit}$ due to insufficient inlet throttling has been measured for one particular system. In the late 1950s data appeared for test sections with inlet conditions in the quality regime created by mixing liquid and vapor streams. Also compressible volumes were intentionally introduced at the test-section inlet. In both these latter cases system dynamics or instabilities were found to influence significantly the magnitude of measured $(q/A)_{crit}$.

The heat-transfer mechanism associated with the $(q/A)_{crit}$ is intimately related to the kind of two-phase flow regime existing at its occurrence.

All of these complicating effects have made it virtually impossible to describe adequately the conditions existing when $(q/A)_{crit}$ occurs and to obtain adequate correlations of the data for $(q/A)_{crit}$. Some correlation equations have been suggested for particular fluids (primarily water), for specific geometries, and for limited ranges of operating conditions.

2.2.3.2 Influence of System Instabilities on Critical Heat Flux. The data shown in Fig. 2-2 [85] were taken on a once-through flow system with an expander vessel teed to the flow path just ahead of the test section. The upper curve is obtained when the expander vessel is filled with well-subcooled water or is not in the system. In this case the flow is quite stable. The lower curve is obtained when the expander vessel is filled with a noncondensible gas, saturated vapor, or the test liquid at saturation temperature. In these cases, flow oscillates rather violently, presumably due to the com-

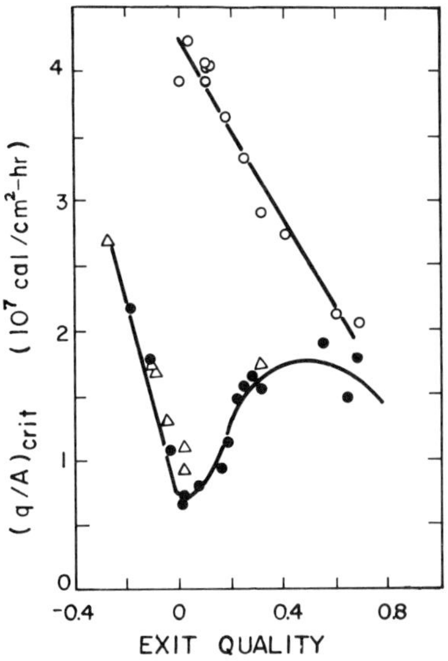

FIG. 2-2. Effect of a compressible volume on the critical heat flux.

pressible volume in the expander. (See Sec. 5.4 of Chapter 15.)

Another operating condition which produces similar results and instabilities occurs when the inlet to the test section is in the vapor-quality regime. The research groups at Harwell (England) and CISE (Italy) have operated test systems with inlet quality determined by mixing liquid and vapor streams ahead of the test section. The data of CISE [86] shown in Figs 2-3a and 2-3b [82] are for a test section L_h/D = 10 and unheated length after mixing of $L_u/D \approx 600$. Here the $(q/A)_{crit}$ is plotted against both inlet quality (or enthalpy) and exit quality (or enthalpy). The critical condition occurs at exit.

In Figs. 2-4a and 2-4b [86] the data come from the same system but with a much longer test section, L_h/D = 300 with $L_u/D \approx 600$. Note the rather unusual crossing of curves.

For each of the above test sections, short and long, the region where the curves dip (represented by A B on the figures) is accompanied by violent flow oscillations. Professor Silvestri suggests that point A occurs where the flow in the unheated length of pipe (just ahead of the test section) is slug flow. Near point B the inlet slug flow changes to annular flow. At the higher total-flow rates the system oscillations do not appear and the dip in the curve disappears.

In this test system the data for $(q/A)_{crit}$ are influenced greatly by the system oscillations and are not representative of the stable forced-convection-boiling processes. To show this, the CISE group ran tests at similar operating conditions but attempted to reduce or eliminate the system oscillations by placing orifices just ahead of the test section. Figure 2-5 [86] shows these results. Curve A shows the data equivalent to that in the preceding two figures when there were no orifices between the test-section inlet and the point of mixing the liquid and vapor streams. Curve B is obtained when the two orifices are placed in series upstream of the heated section. The $(q/A)_{crit}$ in the dip region was increased probably because of the damping of the system oscillations. When five orifices were placed just ahead of the test section, the damping of the system oscillations, particularly as they influenced the flow in the test section, must have been nearly complete, because the resulting curve C is characteristic of data obtained when a system is in stable operation.

We conclude here that data for $(q/A)_{crit}$ taken in a system which is experiencing unstable flow are significantly lower than would be observed if a stable flow condition existed. The amount of the reduction in $(q/A)_{crit}$ will vary from system to system. The only data representative of the test section boiling process which can be correlated or interpreted are data taken in stable operation.

2.2.3.3 Effect of Nonuniform Heat on Critical Heat Flux. The effect of nonuniformity in the power distribution on the critical heat flux is an important problem in the design of nuclear reactors. The results obtained so far are, however, still open to discussion, and they sometimes conflict [82, 86, 87, 88]. It would seem that, where the critical conditions are caused by a steam blanketing of the heating surface only, the local flux condition would be sufficient to determine a local critical $(q/A)_{crit}$. When the critical conditions are caused by the drying out of the wall, the local heat flux should not be signi-

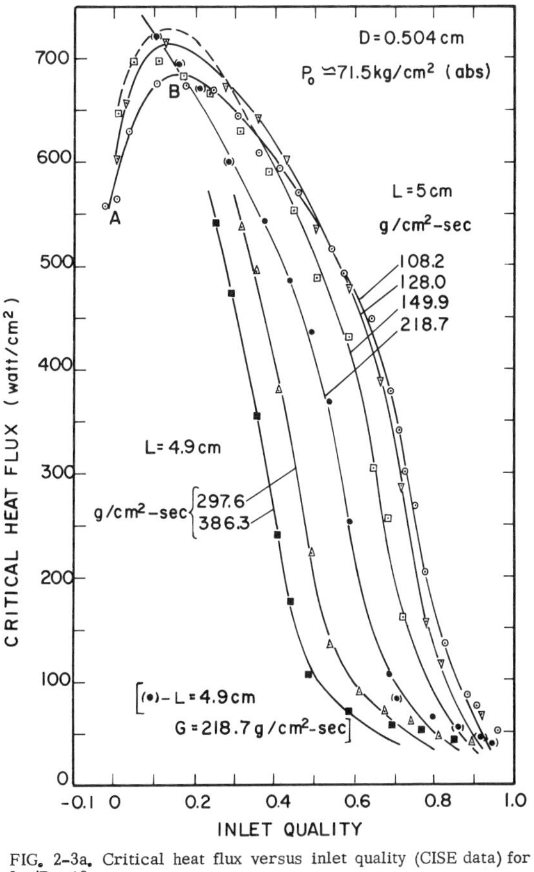

FIG. 2-3a. Critical heat flux versus inlet quality (CISE data) for L_h/D = 10.

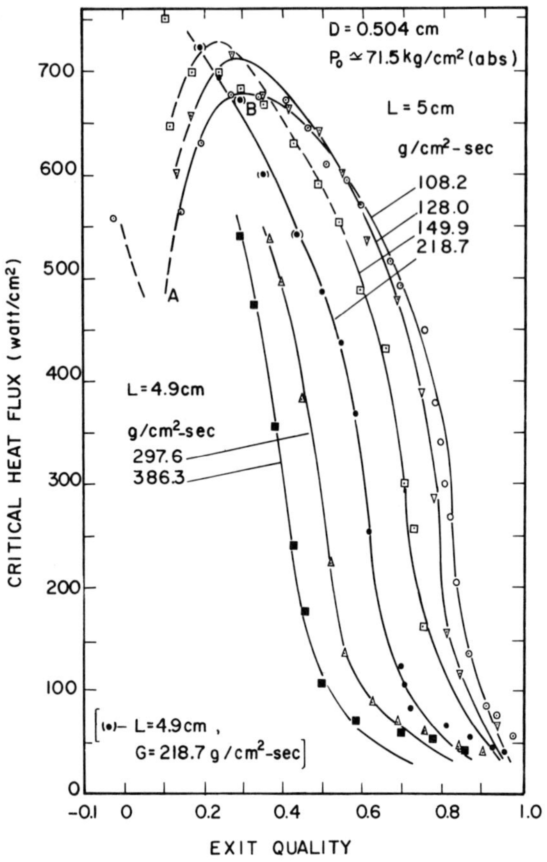

FIG. 2-3b. Critical heat flux versus exit quality (CISE data) for $L_h/D = 10$.

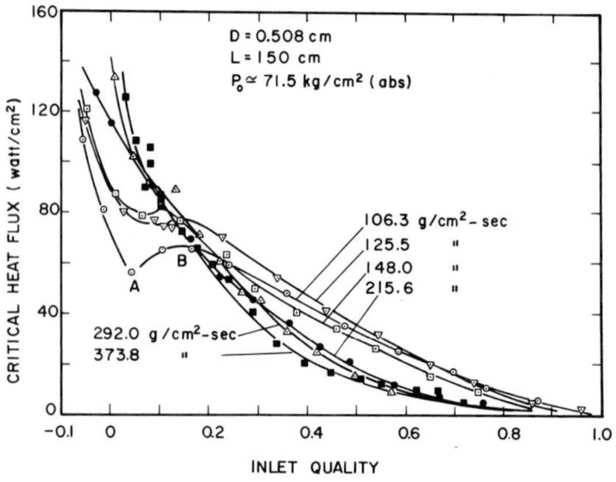

FIG. 2-4a. Critical heat flux versus inlet quality (CISE data) for $L_h/D = 300$.

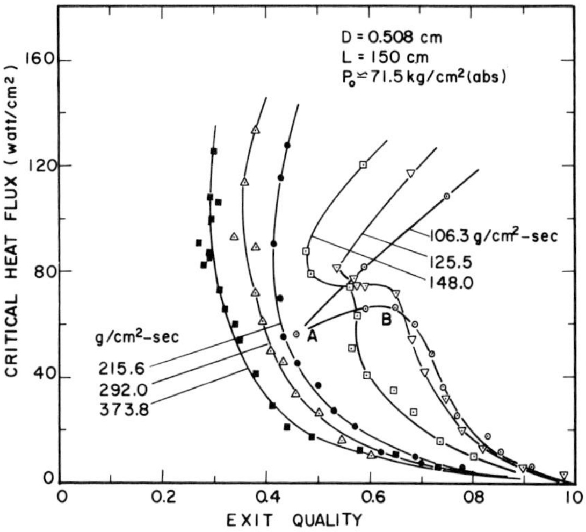

FIG. 2-4b. Critical heat flux versus exit quality (CISE data) for $L_h/D = 300$.

ficant; rather, the total enthalpy of the liquid or the integrated energy added to the system should be important. Unfortunately, in both cases the local flow conditions are dependent on the rate of heat transfer upstream and, therefore, of the axial distribution of the heat flux.

The importance of the axial dependence of the heat flux is best illustrated when the experimental burnout points associated with a given (q/A) distribution are compared with the Macbeth uniform-heat-flux correlation [98]. These results clearly indicate that the local heat flux is not the only criterion. In Fig. 2-6, the same burnout data are plotted in terms of total critical power versus inlet enthalpy. These data are compared with uniform-flux burnout data for the same flow conditions. The scattering indicates that the total enthalpy increase or the concept of average heat flux does not apply in this case either.

As a result of his experimental investigation, Todreas [89] concludes that with uniform and nonuniform axial heat flux, the critical condition is caused either by 1) nucleation-induced disruption of the annular film; or, if such disruption does not occur, then ultimately by 2) dryout resulting from a decrease of the nominal film flow rate to some mean zero value. The authors suggest, on this basis, that a locus of critical conditions be established experimentally, with the nucleation intensity versus local film flow rate used as coordinates. (See Fig. 2-7.) This locus could be obtained from uniform- and nonuniform-axial-flux-distribution data. It could then be used to predict critical conditions for test sections of various flux distributions, under conditions of mass velocity, pressure, and diameter identical to those used to construct the critical locus.

The degree of nucleation intensity is expressed as the ratio of the local critical heat flux to the heat flux necessary to initiate nucleation at the same local conditions of pressure quality and mass velocity, i.e., $(q/A)_{crit}/(q/A)_{inc}$. The incipient heat flux $(q/A)_{inc}$ was determined in this work by the method prescribed by Bergles and Rohsenow [90]. The local film flow rate was taken to be inversely proportional to the enthalpy addition from annular transition location to any point of interest ΔH_{ann-x}. The annular transition location was assumed to be at a quality of 8 wt. % vapor. At low pressures (60 to 200 psia), evaluation of each locus indicated that prediction of the critical power to within ± 10 to 15% and qualitative indication of the critical location could be achieved. At higher pressures, only qualitative agreement of the model with experiments

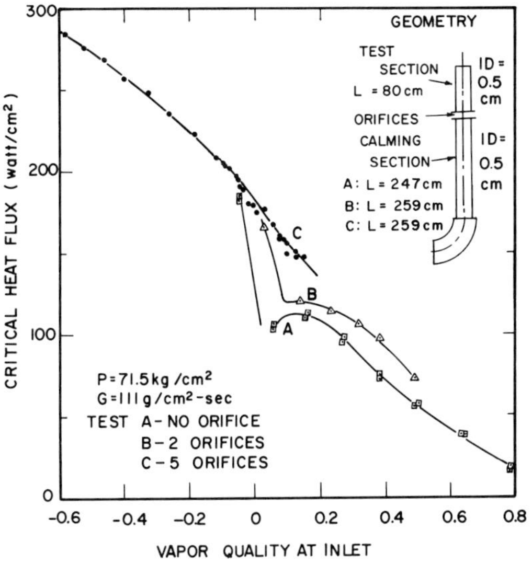

FIG. 2-5. Critical heat flux versus vapor quality at inlet (CISE data) showing effect of orifices at inlet.

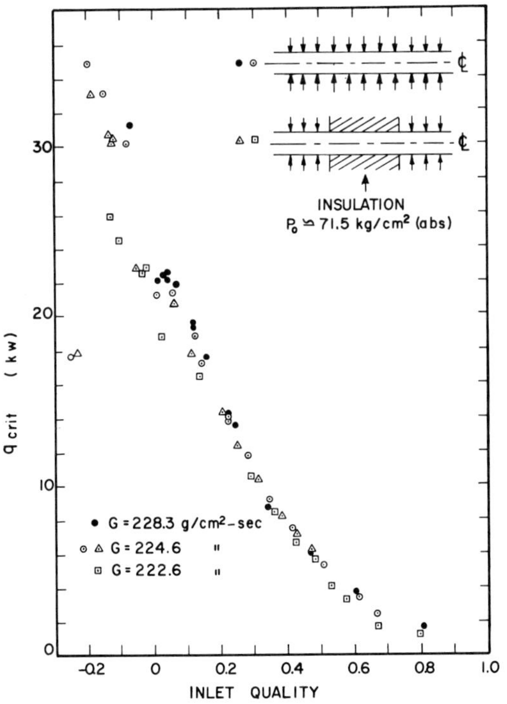

FIG. 2-6. "Hot-patch" test showing effect of nonuniform heating on the critical heat flux (CISE data).

was found. Thus, further experimental work is required.

Tong et al. [91] have compared experiments with nonuniform heat flux and defined a correction factor F as

$$F = \frac{(q/A)_{crit},\ \text{equivalent to uniform flux}}{(q/A)_{crit},\ \text{local nonuniform flux}}. \qquad (2\text{-}4a)$$

The authors postulated that at the wall, the local flow regime consists of a layer of superheated liquid with tiny bubbles separated from the high-velocity main stream by a large bubble layer (Fig. 2-8). It was assumed that the inception of DNB at a distance $z = \ell_c$ downstream is determined by a limiting value of the enthalpy of the superheated layer of liquid. By use of this model, an expression for the numerator of Eq. (2-4a) was found;

$$\begin{matrix}(q/A)\ \text{equi-} \\ \text{valent to} \\ \text{uniform flux}\end{matrix} = \frac{C}{1 - e^{-C\ell_c}} \int_0^{\ell_c} \frac{q(z)}{A} e^{-C(\ell_c - z)} dz \quad (2\text{-}4b)$$

where the constant C was determined empirically from the available data:

$$C = 0.44 \frac{(1 - X_c)^{7.9}}{(G \times 10^{-6})^{1.72}} \text{inch}^{-1}. \qquad (2\text{-}4c)$$

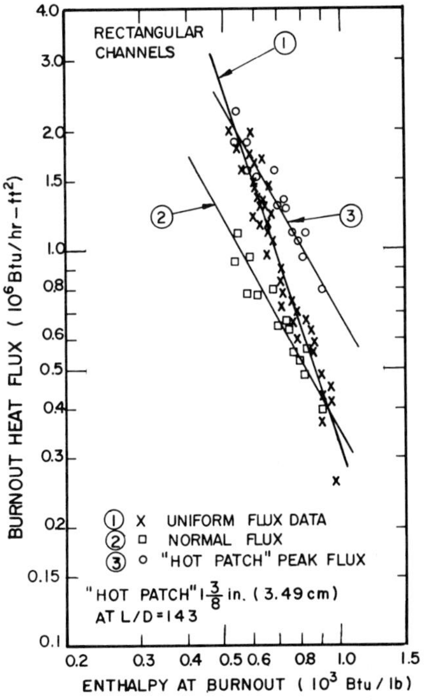

FIG. 2-7. "Hot-patch" test showing effect of nonuniform heating on the critical heat flux (WAPD data).

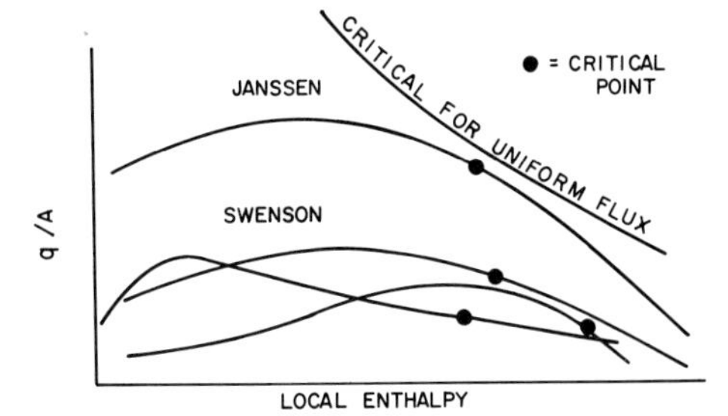

FIG. 2-8. Effect of nonuniform heat flux distribution on critical heat flux.

X_c is the quality at the critical condition and G the flow rate in lb/hr-ft^2.

In the subcooled region, C is large, and the factor F is close to 1.0. Hence, the local $(q/A)_{crit}$ primarily determines the critical conditions. At high qualities, C is small, and the average q/A or Δh primarily determines the critical conditions. It is pointed out that the above formulas should be used only for flux shapes shown on Fig. 2-8 and for the following range of variables:

Channel length	25 to 72 in.
D_e	0.117 to 0.446 in.
p	1,000 to 2,000 psia
G	0.37 to 2.90 × 10^6 lb/hr-ft^2
Inlet enthalpy	76 to 629 Btu/lb
X_c	-0.25 to 0.75

2.2.3.4 Effect of Flow-Cross-Section Shape on Critical Heat Flux. Results for $(q/A)_{crit}$ for one geometry (circular tubes, rectangular passages, flow outside of tube bundles) generally do not agree with those of another geometry. Circular-tube data (see data in references [92] and [93]) generally show higher $(q/A)_{crit}$ for smaller tube diameters at the same mass velocity, pressure, and exit enthalpy with subcooled exit conditions for diameters down to 0.027 in. Similar observations appear to hold for flow in thin rectangular channels [82]. For exit conditions in the quality region the same trend appears, but $(q/A)_{crit}$ goes through a maximum and then decreases as diameter is reduced below 0.10 to 0.15 in.

Some data show a mild effect of channel length and others show very little effect of length. The data reported by Barnett [92] clearly show an effect of D but not of L, and hence not of L/D, when $(q/A)_{crit}$ is compared at the same exit quality, pressure, and mass velocity.

Important to the nuclear reactor designer is knowledge of $(q/A)_{crit}$ for flow along rod clusters. Typical of experimental results for this geometry are those of Becker [94] and Polomik [95].

Figure 2-9 shows some commonly used test sections, where the single annulus is the extreme case of a one-tube rod bundle. Bundles of 3, 7, 19 are formed in an unheated circular shell, and bundles of 4, 9, 16 are formed in square unheated shells. Half rods (dummies) are sometimes placed on the outer shell to attempt to simulate a continuous bundle. Typically, the critical condition usually occurs along the tube facing the large flow area and not where the clearance is smallest; also, in these test sections the critical condition almost always occurs on a tube next to the unheated wall, as shown in Fig. 2-9.

Data for annular flow with only a single inner rod heated fall considerably below the $(q/A)_{crit}$ vs h_{exit} data for flow inside round tubes [94,95]. With both inner and outer walls heated $(q/A)_{crit}$ at a given exit quality is increased to values in the range of those obtained with flow inside round tubes and rectangular passages. Becker [94] suggests that this effect may be a direct function of the ratio of the heated perimeter to the total wetted perimeter of the flow cross section (P_H/P). This suggests that the curve $(q/A)_{crit}$ vs exit quality should rise as the

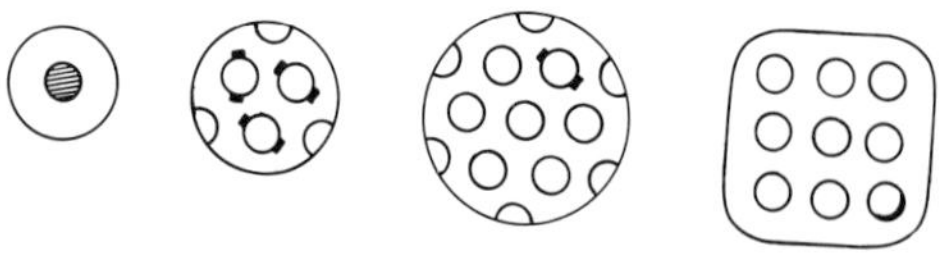

FIG. 2-9. Annular and rod bundle test sections.

number of rods in the cluster increases, since the ratio of heated perimeter to total rod-plus-shell perimeter would approach unity. Becker explains this effect by suggesting that for a given critical condition at the heated surface the flow along the unheated surface forms no vapor; therefore the exit quality would be lower for a particular $(q/A)_{crit}$.

Data [96] with 19 rods in a circular shell ($P_H/P = 0.77$) fell just at the lower edge of the band of data of WAPD [82] for the rectangular channel, suggesting that as P_H/P approaches 1.0, the rod-bundle data would agree with the data for tubes and rectangular channels which are heated around their entire surface.

Becker reports data as exit quality at the critical condition with lines of constant (q/A). Reference [94] shows such a plot as a function of P_H/P embodying most of the annulus, and rod-cluster data. The plot suggests the significance of P_H/P in reducing the critical heat flux.

The above discussion really refers to a kind of average ratio of P_H/P. It is likely that a more critical study of this problem should focus on a "local" P_H/P. To illustrate, a square multirod test section as shown in Fig. 2-9 with uniformly heated rods tends to reach the critical condition in the corner rod on a side facing the unheated wall. An overconservative approach would be to consider a local P_H/P value for the corner rod as an equivalent annulus of the same rod-to-wall spacing.

This effect of P_H/P requires further study and confirmation.

2.2.3.5 Qualitative Description of Critical Condition. There have been many attempts to describe the physical picture of the fluid at the critical condition. While serious disagreement on all of the aspects of the picture still exists, there are certain features upon which there is a little more widespread agreement.

In the subcooled and very low quality regions the picture of the critical condition is probably quite similar to the condition described in pool boiling as discussed in Sec. 2.2.2—continuous streams of bubbles coming from nucleation sites which have become so numerous that liquid cannot flow to the surface to replenish the vapor mass being formed.

In the higher-quality region we visualize a fog flow with a liquid layer on the wall. In the lower-quality regions the q/A may be large enough to produce nucleate boiling in the liquid film, while at higher qualities the critical condition is reached at heat fluxes which are not high enough to produce nucleate boiling at the solid wall surface; here heat is transferred by conduction through the thin liquid layer with evaporation at the liquid-vapor interface. In either event at the critical condition the wall becomes dry — the rate of evaporation exceeds the rate of deposition of liquid from the

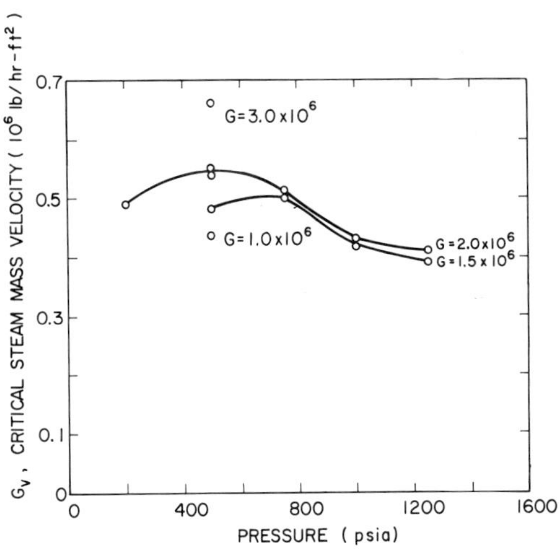

FIG. 2-10. Critical steam mass velocity as a function of pressure for flow in an annulus.

central fog. In the region of higher vapor velocities it is possible that the resulting shear stress may tear the liquid layer from the surface; see Sec. 2.2.4.5 and Fig. 2-10 [97].

In the subcooled region the $(q/A)_{crit}$ increases as subcooling increases (colder liquid) and as total mass flow rate G increases. As the quality region is penetrated, the effect of G is reversed—$(q/A)_{crit}$ decreases as G increases at the same quality. The crossover point is in the low-quality region, 0 to 20%, the exact location depending on pressure as well as G.

In the subcooled region $(q/A)_{crit}$ increases as diameter is decreased to as low as 0.027 in. (0.69 mm). Here the small amount of vapor formed probably produces a greater increase in velocity in the smaller diameter tubes, thus permitting a higher $(q/A)_{crit}$. In the higher-quality region (perhaps greater than 10%) a similar effect is noted over a limited region as D is decreased, but below around 0.10 to 0.15 in. (2.54 to 3.81 mm), as an effect of the resulting increased velocities at these smaller diameters, the shearing effect at the liquid-vapor interface may tear the liquid film from the wall, resulting in a decreasing $(q/A)_{crit}$ as D decreases.

In all regions the effect of pressure is to increase $(q/A)_{crit}$ as pressure is decreased from 2500 psia (170 atm). There appears to be a peak in this effect somewhere in the range of 400 to 800 psia (27 to 54 atm).

For the future it is obvious that a description of the physical process at the critical condition is needed.

2.2.4 Critical-Heat-Flux Correlations

2.2.4.1 General. No forced-convection critical-heat-flux correlation has been evolved which applies universally to data for all fluids and all geometries. This discussion will first direct its attention toward correlations for a particular fluid and a particular flow cross-section shape.

The correlations suggested prior to 1958 are summarized by De Bertoli et al. [82]. Some are purely empirical; others attempt to include some fluid properties by resorting to a description of a postulated mechanism and a limited dimensional analysis. Most of these earlier correlations are limited to rather small ranges of variables and some even show the wrong influence of flow rate at higher qualities. Here the discussion is limited to the more recently proposed correlations.

It should again be emphasized that much of the available data may have been taken with flow oscillations in the system—particularly those taken with inlet quality greater than zero, as shown in Fig. 2-5. These data should not be included in an attempt to correlate successfully the critical-heat-flux data.

For a particular flow geometry (e.g., circular tube) with a given fluid (e.g., water) and heat-flux distribution (e.g., uniform), the following are the independent variables: mass velocity G, length L, diameter D, inlet enthalpy h_i (or inlet subcooling enthalpy Δh_i), pressure p, and, of course, all of the fluid properties. The dependent quantity is then $(q/A)_{crit}$ and, through an energy balance, the exit enthalpy is h_e (or quality x_e). Stated symbolically,

$$(q/A)_{crit} = f_1(G, \Delta h_i, L, D, p) . \qquad (2\text{-}5)$$

From an energy balance, $q = (\pi D^2/4)\, G(h_e - h_i)$ or

$$h_e = h_{sat} + 4\frac{L}{D}\frac{(q/A)}{G} - \Delta h_i . \qquad (2\text{-}6)$$

Combining these last two equations, eliminate Δh_i and, since h_{sat} is a function of p,

$$(q/A)_{crit} = f_2(G, h_e, L, D, p) . \qquad (2\text{-}7)$$

Ideally it would be nice to have a universal correlation with dimensionless combinations of measurable quantities, equally applicable to all fluids and geometries under all operating conditions. To date, this has not been found. In the section that follows a few of the more recent attempts to correlate $(q/A)_{crit}$ will be discussed briefly.

2.2.4.2 Macbeth Correlation. An attempt has been made to correlate the data for one particular fluid, water, using only these independent and dependent variables of Eqs. (2-5) and (2-7) for two specific geometries (Macbeth [98]).

In order to correlate conveniently the data with Eqs. (2-5) to (2-7), it is necessary to seek some simplification of the functional relation. By careful examination of the world's data, excluding all data points taken with inlet conditions in the quality range which are suspected of being associated with flow instabilities, Macbeth found the $(q/A)_{crit}$ to be related linearly with Δh_i. The equation was further simplified by employing the local-condition hypothesis that $(q/A)_{crit}$ depends only on the conditions at the exit in a uniformly heated test section as it approaches the critical condition from the nucleate-boiling or the wetted-wall condition. Then Eq. (2-7) should not include L; so

$$(q/A)_{crit} = f_2(G, h_e, D, p) . \qquad (2\text{-}8)$$

Using data from carefully planned experiments at Winfrith, Barnett [92] shows these equations may be written in the following form:

$$(q/A)_{crit} = \frac{A + (1/4)\,C\,D\,G\,\Delta h_i}{1 + CL}$$

or

$$(q/A)_{crit} = A - (1/4)\,C\,D\,G\,h_{fg}\,x_e\,, \qquad (2\text{-}9)$$

where A and C are functions of D, G, and p.

Macbeth [98] applied these equations to practically all available data for uniformly heated round tubes and rectangular channels, using only data for which the inlet condition was subcooled and no flow oscillations were reported. The coefficients A and C were determined empirically. Macbeth found he could present the results more simply by dividing the representative equations into two zones—high- and low-flow regions, as defined by Fig. 2-11.

In the low-flow region, for uniformly heated round tubes the critical heat flux is given by:

$$(q/A)_{crit} \times 10^{-6} = \frac{(G \times 10^{-6})(h_{fg} + \Delta h_i)}{158\,D^{0.1}(G \times 10^{-6})^{0.49} + 4(L/D)}$$

or

$$(q/A)_{crit} \times 10^{-6} = 0.00633\,h_{fg}\,D^{-0.1}(G \times 10^{-6})^{0.51}(1 - x_e)\,. \qquad (2\text{-}10)$$

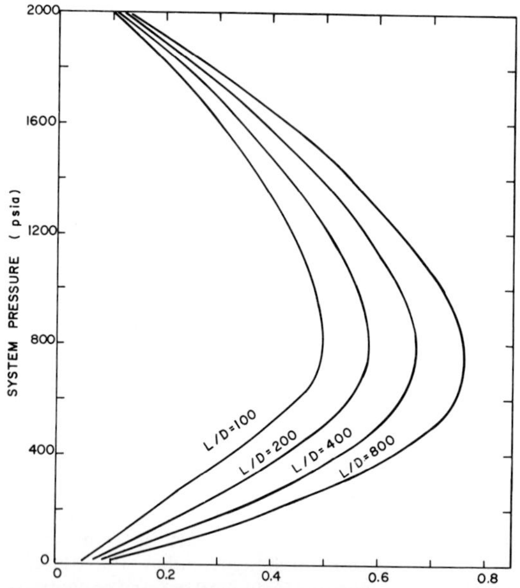

FIG. 2-11. Approximate boundary limits of the low-velocity and high-velocity burnout regimes for round tubes. Note: In the figure the low-velocity region lies to the left of any given curve and the high-velocity region to the right.

In the low-flow region for uniformly heated narrow rectangular channels of width 1 in. (25.4 mm) the critical heat flux is given by:

$$(q/A)_{crit} \times 10^{-6} = \frac{(G \times 10^{-6})(h_{fg} + \Delta h_i)}{3.78\,S^{-1.73}(G \times 10^{-6})^{1.1} + 1.8\,L/S}$$

or

$$(q/A)_{crit} \times 10^{-6} = 0.264\,h_{fg}\,S^{1.73}(G \times 10^{-6})^{-0.1}(1 - x_e)\,, \qquad (2\text{-}11)$$

where S is the internal spacing between the flat heating surfaces in a rectangular channel.

In the high-flow regime, $(q/A)_{crit}$ is given by Eq. (2-9), with A and C given by:

$$A = y_0\,D^{y_1}\,(G \times 10^{-6})^{y_2}\,,$$
$$C = y_3\,D^{y_4}\,(G \times 10^{-6})^{y_5} \qquad (2\text{-}12)$$

where optimized y values are given in Table 2-1 for various pressures for water.

For rectangular channels, replace 1/4 by 0.555 and D by S in Eq. (2-9). The optimized y values are given in Table 2-2.

In the range of reactor design, these equations appear to predict the $(q/A)_{crit}$ to well within 10% RMS error; however, they should not be used beyond the range of the magnitudes of D, G, L, p for which data exists. Also Δh_i must be greater than zero. These equations are not applicable to tube bundles and annuli.

2.2.4.3. Other Correlations. A few of the other correlations are discussed here, but not presented in detail. The reader is referred to the original publications of this material.

Gambill Correlation. Gambill [99] attempted to correlate a large amount of data for seven different fluids over a wide range of values of p, L, G, Δh_i and a variety of flow geometries, limiting the study only to conditions with subcooled exit conditions. He proposed a superposition equation embodying the ordinary forced-convection, nonboiling-type equation and the Zuber equation (2-2). Some data which did not appear to be correct were eliminated. The remainder of the data were correlated to around 40% maximum deviation. Contributing to the spread of results is probably the effect of P_H/P as discussed above.

Westinghouse Correlation. The group at the Westinghouse Atomic Power Dept. proposes correlations [100] of the following form:

Subcooled exit:

$$(q/A)_{crit} = f_4\left(\frac{L}{D}, G, \Delta h_1, p\right); \qquad (2\text{-}13)$$

Exit quality:

$$(q/A)_{crit} = f_5\left(\frac{L}{D}, D, G, \Delta h_i, p\right). \qquad (2\text{-}14)$$

In the latter case the equation is written with $(h_e - h_i)$ as the dependent variable and it is suggested that the same equation should be used for nonuniform-heat-flux cases if $(h_e - h_i)$ is the enthalpy change between inlet and exit even though the critical condition occurs elsewhere in the tube. It was pointed out in conjunction with Fig. 2-8 that this rule of thumb appeared to be applicable when

TABLE 2-1

Optimized y Values and Root-Mean-Square Errors for High-Velocity Regime Burnout Data for Round Tubes

Eq. (2-12)

Pressure[a] (psia)	y_0	y_1	y_2	y_3	y_4	y_5	rms error (%)	No. of expts.
15	1.12	-0.211	-0.324	0.0010	-1.4	-1.05	13.8	88
250 (Nom.)	1.77	-0.553	-0.260	0.0166	-1.4	-0.937	4.7	237
530 (Nom.)	1.57	-0.566	-0.329	0.0127	-1.4	-0.737	5.7	170
1000	1.06	-0.487	-0.179	0.0085	-1.4	-0.555	7.4	405
1570 (Nom.)	0.720	-0.527	0.024	0.0121	-1.4	-0.096	3.4	133
2000	0.627	-0.268	0.192	0.0093	-1.4	-0.343	9.0	362
2700 (Nom.)	0.0124	-1.45	0.489	0.0097	-1.4	-0.529	4.7	37

[a]Conversion factors: 1 psia = 0.07031 kg/cm^2 = 0.06805 atm.

TABLE 2-2

Optimized y Values and Root-Mean-Square Errors for High-Velocity Regime Burnout Data for Rectangular Channels

Eq. (2-12)

Pressure[a] (psia)	y_0	y_1	y_2	y_3	y_4	y_5	rms error %	No. of expts.
600	23.4	-0.472	-3.29	0.123	-1.4	-3.93	6.1	22
800	0.445	-1.01	0.384	0.0096	-1.4	-0.0067	12.9	28
1200	1.88	-0.081	-0.526	0.0035	-1.4	-1.29	4.9	42
2000	0.546	-0.315	-0.056	0.0027	-1.4	-0.725	9.4	359

[a]1 psia = 0.07031 kg/cm^2 = 0.06805 atm.

the flux distribution was a symmetrical cosine distribution but predicted too high a heat flux for nonsymmetrical distributions.

Neither Eq. (2-13) nor Eq. (2-14) suggested in the final form reduces to the form of Eq. (2-8) suggested and shown by Barnett [92] to be valid. Nevertheless data for water over a wide range of variables appear to be correlated by the two equations within 25% at a 95% probability level.

Equations (2-13) and (2-14) do not coincide at the saturation point (zero quality). A third correlation has been proposed to be used in the quality range of ±0.15 [101] but will not be reproduced here because of its complexity.

General Electric Design Curves. The group at the General Electric Atomic Power Equipment Dept. [102] has attempted to draw burnout limit curves. Limit curves have been drawn under all of the data for the annulus with only the central rod heated with $P_H/P < 0.5$; hence they also fall below data for round tubes and rectangular channels. The design curves therefore predict unnecessarily low heat fluxes and result in very conservative designs.

Becker-Persson Correlation. The Swedish group [103] suggests that, as a first approximation, the following simple relation will correlate the data:

$$(q/A)_{crit} \cdot G^{1/2} = f_6(x_e, p) . \qquad (2\text{-}15)$$

Comparing this with Eq. (2-8) shows the variable D missing and the functional relation with G specified. This relation has not been tested against a wide variety of data but for the range studied it seems to represent the data within 15%.

2.2.4.4 World's Data for Critical Heat Flux. Since the authors cannot recommend one single correlation for $(q/A)_{crit}$ for all geometries and flow conditions, the reactor thermal designer will want to go back to original data for a geometry and fluid condition most similar to his particular design. Stein and Lottes [104] have collected the test conditions and ranges of variables covered by various testing groups and have presented this in various tables. An additional early collection of data points has been assembled by Nuclear Development Associates [105] in a single report. All available data points are on IBM cards at Westinghouse [102] and Winfrith [98].

Perhaps these sources will be useful to a designer in attempting to find data taken under conditions which most nearly match his conditions. The use of these direct data must be tempered with the cautions raised in the present discussion in this section.

2.2.4.5 A Limiting Vapor Velocity. There appears to be a limiting vapor velocity in the higher-quality fog-annular-flow regime beyond which the $(q/A)_{crit}$ is greatly reduced [106]. Figure 2-11 [97] shows the magnitude of this critical-mass velocity G_v as a function of pressure at several total-flow-mass velocities G for flow in an annulus. As an approximation $G_v \approx x_e G$. In the range of pressure of 200 to 1200 psia (14 to 84 kg/cm) a reactor design should perhaps try to keep the magnitude of G_V below 4×10^5 lb/hr-ft^2 (195 kg/hr-cm^2) to keep the magnitude of $(q/A)_{crit}$ high.

2.2.4.6 Recommendations for Design. The reactor designer faces the problem of predicting $(q/A)_{crit}$ for his particular geometry, flow conditions, and heat-flux distribution. At present, research has not solved this problem adequately; however, it has provided many tools which are useful.

Initially the designer may refer to the data accumulated on test sections with uniform flux distributions. He should find the data which most nearly match his operating conditions (Sec. 2.2.4.4). In interpreting these data he should be mindful of the possibilities of the existence of flow oscillations which may have been present in the test system but probably do not exist in his reactor.

To help put the test data in proper perspective, he may wish to resort to the Macbeth correlation to calculate the $(q/A)_{crit}$ which would occur in round tubes or rectangular channels, realizing that in rod bundles the magnitude will be reduced as P_H/P is reduced from 1.0 in Fig. 2-10. Other checks may be made by referring to the appropriate correlation discussed in Sec. 2.2.4.3, being mindful that some of these correlations have greater scatter and some are, in fact, rather conservative.

In the final analysis, the test data for near-matching conditions should be used, provided comparison with the correlations indicates that the data are good and provided no flow oscillations were present in the test system.

The influence of nonuniform heat-flux distribution is not well established. With a symmetrical cosine distribution, tests indicate that a critical condition is reached somewhere along the test section, downstream of the peak heat flux, when the integrated average heat flux corresponds to $(q/A)_{crit}$ from uniform-heat-flux tests at the same exit enthalpy in the two cases. With skewed distribution—peaks ahead of or beyond the midpoint—the critical condition occurs in the test section at a somewhat lower integrated average heat flux with the same exit enthalpies (see Sec. 2.2.3.3). The tests with skewed distributions [87] may have flow oscillations, though this was not reported. The amount of this reduction, if real, is not sufficiently well established at present to warrant making a specific recommendation.

In design, consideration should be given to keeping the vapor mass velocity G_V below 4×10^5 lb/hr-ft^2 (195 kg/hr-cm^2) (see Sec. 2.2.4.5).

The most reliable way to obtain suitable information on $(q/A)_{crit}$ for use in designing a particular reactor is to perform tests on a geometrically similar test section with the anticipated axial variation of heat flux at the design conditions of interest.

In the absence of such test data or of previously published test data for conditions similar to the proposed design condtions, the reactor designer may wish to use a conservative design procedure. Then it is recommended that if the geometry is other than flow inside circular tubes or rectangular channels, he use test data for the annular test section with central rod heated and account for nonuniform heat-flux distribution by using the integrated average-heat-flux and exit-enthalpy conditions to match those of the uniform-heat-flux test results. (Swenson's data [87] for highly skewed nonuniform flux distribution are excluded from this recommendation for reasons discussed in Sec. 2.2.3.3.) This procedure will give a very conservative design. The designer may wish to be quite a bit more bold than this in selecting his safety margins. The actual permissible $(q/A)_{crit}$ will be somewhat higher than that predicted by this conservative procedure, as discussed earlier in these recommendations. However, since designs which are less conservative than those outlined above must be based on fragmentary information, the designer should attempt to provide proof in the way of actual tests of his geometry and system conditions. He should also investigate carefully the consequences that would result if the reactor core or a part of it went beyond $(q/A)_{crit}$.

Clearly, further research is needed on the effects of nonuniform heat-flux distribution and fuel-element geometry (P_H/P).

Finally, each reactor should be analyzed for flow instabilities (Sec. 5 of Chapter 15). These instabilities, if present, can cause premature burn-out of fuel elements. If analysis shows them to be present, they must be eliminated by design changes.

2.3 Film Boiling

2.3.1 General

Film boiling occurs beyond the point where $(q/A)_{crit}$ is reached. In this region the heated wall surface is not wet. A vapor blanket film covers the surface and heat is transferred by conduction and radiation across this film to the liquid-vapor interface. Test data are available for film boiling of various fluids on surfaces, particularly horizontal and vertical tubes in an essentially unbounded region. Also, test data are available for various fluids flowing inside tubes in the film-boiling condition.

2.3.2 Unbounded Film Boiling

The following equation was evolved by Bromley [106, 107] for film boiling on the outside of horizontal tubes in a saturated pool of liquid:

$$h_c = 0.62 \left[\frac{k_v^3 \rho_v (\rho_\ell - \rho_v) g (h_{fg} + 0.4 (c_p)_v \cdot \Delta T)}{D \mu_v (T_w - T_{sat})} \right]^{1/4}, \qquad (2\text{-}16)$$

where D is outside tube diameter and μ_V is viscosity of the vapor. Radiation contributes to the heat transfer and increases the vapor-film thickness, reducing the effective contribution of the conduction. The total heat-transfer coefficient is given by

$$h = h_c (h_c/h)^{1/3} + h_r, \qquad (2\text{-}17)$$

where h_c is convective contribution to the heat-transfer coefficient and h_r is calculated for radiation between two parallel planes taking the emissivity of the liquid to be unity. Equation (2-17) has been used to correlate data for a number of hydrocarbons as well as water.

For forced-convection flow of the liquid across the tube, Bromley [107] suggests the following equation when V_∞, the velocity of the liquid far away from the tube wall, is $\geq 2\sqrt{gD}$:

$$h_c = 2.7 \sqrt{\frac{V_\infty k_v \rho_v (h_{fg} + 0.4(c_p)_v \Delta T)}{D(\Delta T)}} \qquad (2\text{-}18)$$

and

$$h = h_c + (7/8)\, h_r \,. \qquad (2\text{-}19)$$

Film boiling on a horizontal surface was analyzed by Berenson [108] and compared with data for pentane, CCl_4, benzene, and ethyl alcohol. The following equation resulted:

$$h = 0.425 \left[\frac{k_{vf}^3 \rho_{vf} (\rho_\ell - \rho_v) g (h_{fg} + 0.4(c_p)_v \Delta T)}{\mu_f (\Delta T) [g_0 \sigma / g(\rho_\ell - \rho_v)]^{1/2}} \right]^{1/4} . \qquad (2\text{-}20)$$

For film boiling on the outside of a vertical tube in a saturated-liquid pool, Hsu and Westwater [109] suggest the following equation:

$$h = 0.0013 \left[\frac{4w}{\pi D \mu} \right]^{0.60} \left[\frac{k_v^3 \rho_v (\rho_L - \rho_v) g}{\mu^2} \right]^{1/3} , \qquad (2\text{-}21)$$

over the range $800 < (4w/\pi D\mu) < 5000$. Here w is the rate of vapor generation, lb/hr.

2.3.3 Forced Convection inside of Tubes

Forced-convection film-boiling data inside tubes has been gathered for water [110, 111] hydrogen and nitrogen [112, 113] and freon-113 [114]. Based on a suggestion of Green [115], Dougall [114] shows that in the higher quality regions beyond 10% these data approach asymptotically the following equation:

$$\frac{h_c D}{k_{vS}} = 0.023 \left[\frac{\rho_{vS} D}{\mu_{vS}} \, \frac{(Q_\ell + Q_v)}{A_p} \right]^{0.8} Pr_{vS}^{0.4} , \qquad (2\text{-}22)$$

where subscript vS refers to saturated vapor, Q_ℓ and Q_v are the volume flow rates of the liquid and vapor, and A_p is $\pi D^2/4$. Then

$$(q/A) = h_c (T_w - T_S) + \epsilon_w \sigma (T_w^4 - T_S^4) , \qquad (2\text{-}23)$$

where ϵ_w is the emissivity of the tube wall surface.

At lower qualities h_c is greater than that predicted by Eq. (2-22). At 1% quality h_c has risen to 4 or 5 times this prediction; hence, Eq. (2-22) should predict "safe" values.

2.4 Transient Boiling

In a variety of accident situations the reactor might experience severe transients. Data directly applicable to these conditions are difficult to obtain. However, a research group has studied the effects of transients on heated ribbon in liquid pools [116-119]; Cole [119] studied the effect of step-change in power input; Rosenthal and Miller [118] and Johnson et al. [116] varied the electrical input exponentially with time; and Howell and Bell [117] studied the effect of a quick release of pressure of the liquid container.

A review of this limited amount of data suggests that starting with cold liquid there is a time delay in nucleation (formation of vapor) which adversely affects the desired decrease in reactivity. On the other hand, a plot of q/A vs $T_{wall} - T_{sat}$ during a transient shows the results of these tests give higher values of (q/A) for steady-state boiling. Furthermore, the peak heat fluxes reached while maintaining nucleate boiling are somewhat higher than those in steady-state boiling. This suggests that predictions of surface temperature rise based on steady-state boiling data will indicate $(q/A)_{crit}$ to occur at lower values than it is likely to in a transient. In one sense this is a conservative method, as burnout is less likely to occur than predicted. In another sense, however, boiling which will tend to shut down the reactor in most cases will occur later than predicted.

In predicting vapor volumes during transients, reference should be made to the data and method of prediction given by Johnson et al. [116], but the results should be used with caution for conditions greatly different from those covered by the tests.

2.5 Heat Transfer to Liquid Metals

2.5.1 General

Liquid-metal-cooled reactors will, for the near future, be designed to operate with single-phase liquids and not in the boiling regime; however, in the case of accidents, the liquid metal may boil. Predictability of heat-transfer performance with liquid metals is still quite poor, particularly with boiling. Liquid-metal heat-transfer systems seem to be plagued with effects of minute amounts of impurities—possibly oxides—causing variation of performance from system to system. The exact nature of these effects is not well established at present.

2.5.2 Forced-Convection, Single-Phase Liquid Metals*

In turbulent flow, expressions for apparent heat transfer and apparent shear stress may be written as follows:

$$q/A = (k + \rho C \epsilon_h)(\partial T/\partial r) ,$$

$$\tau = (\mu + \rho \epsilon_m)(\partial v/\partial r) . \qquad (2\text{-}24)$$

These equations have been integrated across the radius of a tube using the universal velocity

*For definitions of symbols used in this section please refer to the nomenclature list at the end of this chapter.

distribution (see any of many heat transfer texts such as Rohsenow and Choi [120]).

For liquid nonmetals these integrations, when performed for the ratio of energy and momentum diffusivities $E \equiv (\epsilon_h - \epsilon_m) = 1.0$, lead to equations which agree very well with experimental data for turbulent flow of nonmetals, gases, and liquids:

$$\frac{hD}{k_b} = 0.023 \left(\frac{GD}{\mu_b}\right)^{0.8} (Pr_b)^{0.4}, \; Pr \geq 0.5. \qquad (2\text{-}25)$$

For liquid metals, low Prandtl number, the same procedure with $E = 1.0$ leads to the following equation (Lyon [121]):

$$\frac{hD}{k} = 7.0 + 0.025 \, (Re\,Pr)^{0.8}. \qquad (2\text{-}26)$$

Since 1946, data have been gathered by a number of experimenters. Most of the data fall well below the prediction of Eq. (2-26), but a few sets of data agree with it. Most of the test data are taken by measuring inlet and exit temperatures of the liquid metal and wall temperatures along the electrically heated test section. Such measurements lead to values of the heat-transfer coefficient h which inevitably must include a scale or interface resistance at the tube wall, if present.

In these tests, correction for axial conduction upstream must be made, especially at lower Reynolds numbers, in order to obtain the proper liquid temperature corresponding to a particular wall temperature and heat flux. This was pointed out by Trefethen [122]. Unless this correction is made, the deduced data for h at lower Reynolds numbers is too low.

The discrepancy of the data from the prediction of Eq. (2-26) has variously been attributed to scale or deposit at the tube wall, an interface resistance due to nonwetting of the wall, and actual magnitude of E postulated to be below 1.0.

Data have not consistently shown that wetting or nonwetting yield significantly different results. The presence of traces of scale or deposit is difficult to detect. Then assuming the test systems have been "clean" and the fluids "pure" (the meaning of these words is not well defined here!), many investigators have varied the magnitude of E to make the resulting predicted h agree with the data.

Some authors have integrated Eq. (2-24) permitting E to be a function of radial position—Jenkins [123], Deissler [124], Azer and Chao [125], Mizushina and Sasano [126]; others have integrated the equations employing an average value of E, taken as independent of the radial position—Lyon [121], Lykondis and Touloukian [127], Cohen and Rohsenow [120a] and Dwyer [128, 129], to name a few.

Most of the data for uniform heat flux reported prior to 1957 are correlated well with the following equation [120a]:

$$Nu = 6.7 + 0.0041 \, (Re\,Pr)^{0.793} \, e^{41.8\,Pr}. \qquad (2\text{-}27)$$

Most data taken after 1957 fall above this prediction. In analyzing the later data, Dwyer suggests the following relation for $Re\,Pr > 400$:

$$Nu = 7.0 + 0.025 \left[Re\,Pr - \frac{1.82\,Re}{[(\epsilon_m/\nu)_{max}]^{1.4}} \right]^{0.8}, \qquad (2\text{-}28)$$

where $(\epsilon_m/\nu)_{max}$ is given in Fig. 2-12 [128] for the curve labeled "circular pipes."

For concentric annuli with uniform heat flux at the inner wall only, Dwyer [128] recommends

$$Nu_1 = \alpha_1 + \beta_1 \, (E\,Re\,Pr)^{\gamma_1}, \qquad (2\text{-}29)$$

where $\alpha_1 = 4.63 + 0.686\,y$,
$\beta_1 = 0.02154 - 0.000043\,y$,
$\gamma_1 = 0.752 + 0.01657\,y - 0.000883\,y^2$,
$y = (\text{inner radius})/(\text{outer radius})$,

and

$$E = 1 - \frac{1.82}{Pr\,[(\epsilon_m/\nu)_{max}]^{1.4}}, \qquad (2\text{-}30)$$

where $(\epsilon_m/\nu)_{max}$ is given in Fig. 2-12. For parallel plates the above equation applies with $y = 1.0$.

For in-line flow through rod bundles Dwyer recommends

$$Nu = 0.93 + 10.81 \frac{P}{D} - 2.01 \left(\frac{P}{D}\right)^2 + 0.0252 \left(\frac{P}{D}\right)^{0.273} (E\,Re\,Pr)^{0.8}. \qquad (2\text{-}31)$$

Here E is given by Eq. (2-30) and $(\epsilon_m/\nu)_{max}$ by Fig. 2-12.

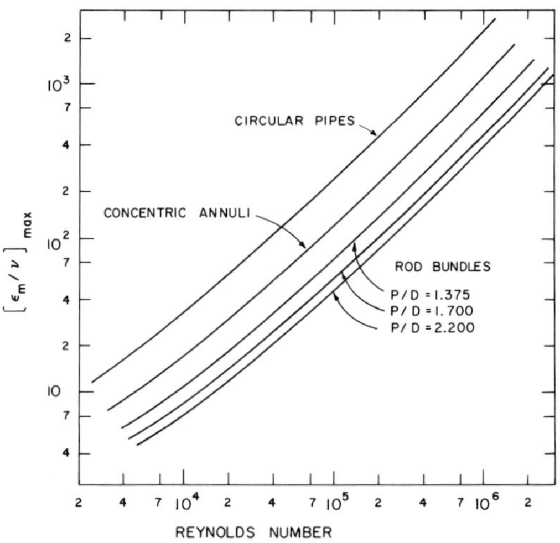

FIG. 2-12. Values of $(\epsilon_m/\nu)_{max}$ for fully developed turbulent flow.

The expression developed by Friedland and Bonilla [130] for constant heat flux emitted by rods spaced on an equilateral triangular pitch,

$$\frac{Nu}{C} = 7.0 + 3.8 \left(\frac{P}{D}\right)^{1.52} + 0.027 \left(\frac{P}{D}\right)^{0.27} Pe^{0.8}, \qquad (2\text{-}32)$$

is preferred because it is simpler and covers a wider range of pitch to diameter ratio, from 1.35

to 10.0. The experimental correlation factor, C, which can be used to modify the theoretical values to bring them into line with the practical results, are as follows, for Pe between 800 and 5,000:

Pe	C
800	0.72
1000	0.76
1500	0.80
2000	0.82
2500	0.84
3000	0.85
4000	0.86
5000	0.86

The corresponding theoretical equation for rods spaced on a square pitch is

$$\frac{Nu}{C} = 7.0 + 4.24\left(\frac{P}{D}\right)^{1.52} + 0.0275\left(\frac{P}{D}\right)^{0.27} Pe^{0.8} \,. \qquad (2\text{-}33)$$

The same correlation factor may be used.

For a P/D ratio significantly below 1.35, the use of an annulus model can lead to serious errors in the determination of the average heat-transfer coefficient, and the less the ratio, the greater the error. There is still much work to be done, both analytical and experimental, on the subject of liquid-metal heat transfer in closely packed rod bundles. For the moment, the only available empirical equation [131] (with slight modification) may be used:

$$Nu = (114 + 0.41Pe)\left[\frac{D_e}{L} + 0.027\left(\frac{P}{D} - 1.1\right)^{0.46}\right],$$

$$\text{for } 1.1 \le \frac{P}{D} \le 1.35, \qquad (2\text{-}34)$$

where $\frac{D_e}{L}$ is the ratio of equivalent flow-channel diameter to rod length.

As the P/D ratio approaches unity, the circumferential variation of surface temperature is greatly increased, and the average heat-transfer coefficient is greatly reduced. For tightly packed rod bundles, i. e., where the rods are touching each other, the correlated result for a triangular array is [132]

$$Nu = 0.15\left(\frac{k_w}{k_f}\right)^{0.02} (Pe)^{0.3} + 0.4\,(k_w/k_f)^{0.5} \qquad (2\text{-}35)$$

Accuracy: ± 15% error

where k_w/k_f is the ratio of the wall and fluid conductivities. A more general expression for triangular and square arrays is [133]

$$Nu = 0.48 + 0.0133(Pe)^{0.70}. \qquad (2\text{-}36)$$

Accuracy: ± 20% error

For the transition region, i.e., $1.0 < P/D < 1.10$, no expression is available; however, a rough answer can be obtained by interpolation of Eqs. (2-36) and (2-34).

The preceding equations apply to Re Pr > 400. For lower flow rates the reader is referred to Dwyer [129].

More recently, results reported by Kirillov et al. [134] and Subbotin et al. [135] were obtained by measuring the radial temperature distribution in the flowing liquid metal, using a temperature probe inserted from the downstream end of the test section. In addition, the actual wall temperature was measured. Knowing the q/A and hence $(\partial T/\partial r)$ at the wall, the temperature distribution could be extrapolated with reasonable precision to determine the temperature at the liquid-metal interface. From the magnitude and the measured actual wall temperature an interface (scale, oxide, etc.) resistance could be determined.

A most interesting result was reported by Kirillov et al. [134] for NaK liquid metal flowing in a copper tube. There existed a significant interface resistance for some time, and after around 500 hours of operation, the interface resistance disappeared or at least became so small it was not measurable. Then the results agreed with Eq. (2-26). One might infer from this that continued purification of the liquid metal to eliminate oxygen and oxides permitted the liquid metal to clean the deposit off the surface.

A later, more puzzling result was reported by Subbotin [135]. In these tests with mercury flowing in a steel tube with uniform heat flux, it was found that the deduced interface resistance varied with Reynolds number as shown in Fig. 2-13 [135]. Surely such a varying resistance could not be associated with a fixed oxide layer at the surface.

The approximate thickness δ of a laminar region near the wall is given by the following relation:

$$\delta/d = k/Re^{\,0.9} \qquad (2\text{-}37)$$

corresponding to $y^+ = 5 = (y/\gamma)\sqrt{\tau_0/\rho}$ [120]. The four points shown plotted on Fig. 2-13 represent the equation $R_c\lambda/a = 630/Re^{0.9}$. There is a surprising agreement. The reported interface resistance varies in the same way as the laminar-layer thickness does with Reynolds number.

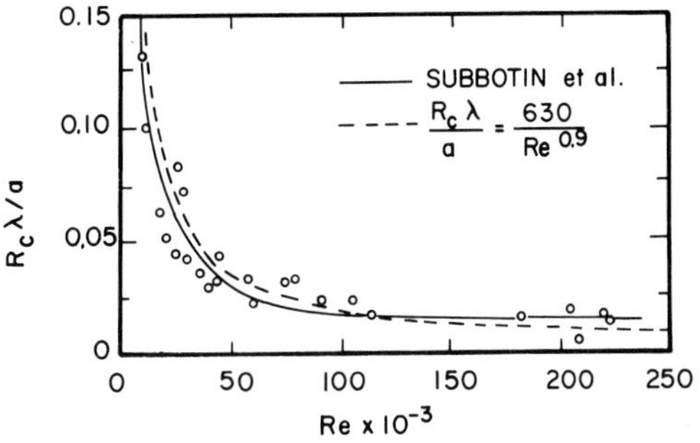

FIG. 2-13. Thermal contact resistance as a function of Reynolds number for flow of heavy liquid metal (Hg) in a tube of steel.

One explanation which suggests itself postulates that oxide-like particles might be finely dispersed throughout the liquid including the nearly laminar region adjacent the wall. Perhaps these particles reduce the effective conductivity of the liquid, particularly in this laminar region. This postulate would suggest little or no actual interface resistance, but rather a decreased conductivity of the liquid. Needless to say, this is only conjecture and should not be considered as fact.

Much more experimental data are needed to understand the effects of impurities and interface resistances. Quite possibly heat-transfer performance in a forced-convection liquid-metal-cooled system which has pure metals, free of oxygen and oxides among other things, would perform according to Eq. (2-26). For the present, however, it is recommended that designs be based on Equations (2-27) through (2-31) and other equations presented by Dwyer [128, 129].

2.5.3 Boiling of Liquid Metals

At present no specific recommendations, generally applicable, can be made regarding the prediction of heat-transfer performance for boiling of liquid metals or for the critical-heat-flux conditions. Active research is in progress by many groups, much of which is reported bimonthly [136]. Although much of the quantitative data reported by the various groups are conflicting or in disagreement, there appears to be emerging some agreement on certain performance features.

Nucleation or bubble initiation in pure liquid metals on clean surface seems to require very much higher temperature difference (hundreds of degrees) above saturation than do nonliquid metals. This leads to significant oscillation in temperature and resultant "bumping" due to rapid flashing of vapor in the highly superheated liquid. In a materials testing reflux capsule with boiling potassium the Oak Ridge group [136a] found that these instabilities (bumping) could be eliminated by drilling a 0.05-in. (1.27-mm) diam well into the heated surface (a "hot finger" of potassium) where vapor would be formed and released into the main body of the liquid.

Similarly, stable flow in forced convection could be obtained by placing an orifice just upstream of the test section to allow some flashing into vapor. In a forced-convection-boiling system it is suspected that, given a small fraction of initial vapor, no actual nucleate boiling will take place. The conductivity of the liquid is probably large enough for the heat transfer to occur by simple conduction in the liquid with surface evaporation at the liquid-vapor interface. Hence, the vapor fraction would increase along the heated flow path without actual nucleate boiling. An annular flow of liquid at the wall and vapor in the central core would probably be established at relatively low qualities for higher flow rates.

A correlation for forced-convection-boiling heat transfer is proposed by Chen [137]. It agrees with some of the available data but not with others. Real confirmation awaits additional good data.

Data on critical heat flux for liquid metals are too sparse to draw conclusions. Orders of magnitudes for $(q/A)_{crit}$ seem to be similar to those for nonliquid metals. No recommendations can be made at present.

3 OVERALL TRANSIENT THERMAL ANALYSIS OF REACTOR CORE

3.1 Introduction

Accidents in nuclear power reactors may be put into two categories: the reactivity accidents in which the core becomes supercritical and the reactor reaches power levels well beyond the design operating power, and loss-of-flow or loss-of-coolant accidents resulting from partial or complete loss of pumping power or from the physical loss of the coolant from the system as, for example, from the rupture of an external pipe. All these accidents are characterized by the fact that the rate of heat production exceeds the rate of heat removal, thereby causing the temperature to increase in the fuel and structural material of the core.

The following section deals with the heat-removal problem during an accident in liquid-cooled reactors. Behavior of gas-cooled reactors in accidents is considered in Chapter 9.

3.2 Heat Removal in Liquid-Cooled Thermal Reactors during an Accident

In this subsection, an outline of some methods of analyzing the behavior of liquid-cooled thermal reactors in accidents is given with particular emphasis on the heat-transfer aspects. Two methods of analysis for the reactivity accident are given. The first one is a simplified one-dimensional treatment, while the second method is more accurate, requires the use of an analog computer, and is more suited to analysis of severe reactivity accidents which could lead to core meltdown and chemical reaction of molten fuel with the coolant.

3.2.1 One-Dimensional Analysis of a Reactivity Accident

The method was developed by Golian et al. [138] and subsequently applied to the safety analysis of the Shippingport reactor [139]. The following is partially abstracted from Reference [139]. A cross section of the Shippingport core showing the seed and blanket arrangement is given in Fig. 3-1 [140]. The accidental increase in reactivity is assumed to occur in the supercritical seed region. Because of the relatively small radial dimensions of the seed, the radial neutron leakage out of the seed substantially affects the criticality of the core. The fact that the power distribution does not significantly vary radially in the narrow annular seed region justifies the following assumptions in this one-dimensional approach.

The volume is considered to be divided into three zones: the fuel, the cladding, and the coolant (see Fig. 3-2 [139]). Since the transient solution of the kinetic equations for the behavior of neutron

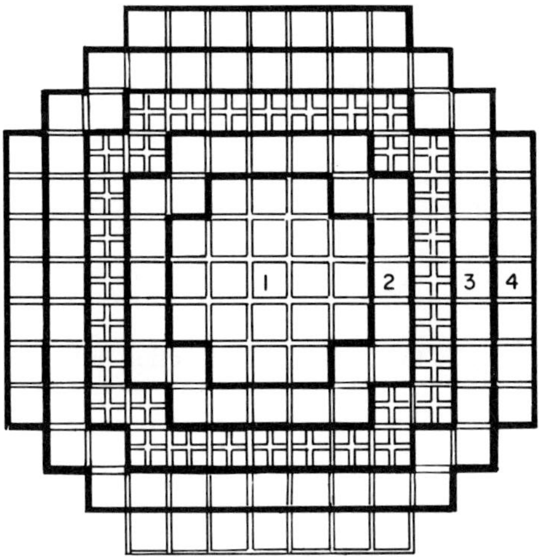

FIG. 3-1. PWR seed and blanket regions. Region 1 has 21 assemblies, region 2 has 24, region 3 has 40, and region 4 has 28. The seed region is between regions 2 and 3 and has 32 assemblies.

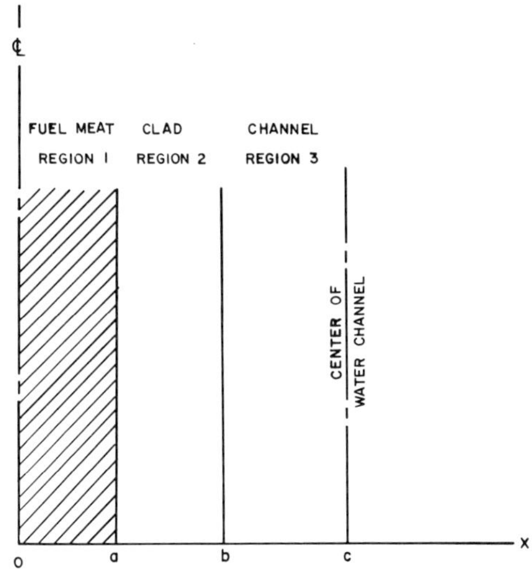

FIG. 3-2. Fuel, cladding, and coolant zones.

flux and power with time dies down in a few periods after the start of the accident, only the asymptotic solution of the form exp(t/T), where T is the period, is considered. For a transient with a period less than 25 msec the transport of heat by coolant flow is insignificant in comparison to the heat transferred to the coolant by conduction. The time-dependent heat source is considered to be localized in the fuel region and distributed uniformly. It is postulated that termination of the excursion occurs when a sufficient moderator-void fraction has been introduced by steam formation to counteract the excess reactivity initially introduced. It is further assumed that the period of the excursion is constant until the excess reactivity is completely compensated for by the shutdown mechanism (see Chapter 7).

The creation of a void volume is taken to be contingent only on the introduction of sufficient energy to raise the temperature and supply the latent heat of vaporization to the coolant. Hydromechanical delays in the void formation are neglected. In view of the low conductivity of water, the temperature gradient in the coolant region is large; it is therefore assumed that steam is formed in a lamina adjacent to the cladding and that heat is transferred to the water through the steam layer by conduction only. The formation of additional steam is considered to be due to conductive heat transfer across the steam film to the remaining water. For short transient periods the energy stored in the fuel element is large compared to the thermal energy conducted to the coolant. It is therefore not unreasonable to keep the thermal conduction as a three-region problem (of fuel, clad, and coolant) and to substitute the thermal properties of water for those of steam in the third region at the onset of boiling. Since the thermal effect on the central region of the addition of the steam layer is negligible, the temperature of the fuel and clad are considered as rising exponentially throughout the transient, i.e., as if the third region consisted entirely of water. The time interval from start to termination of the excursion is divided into two parts: $0 \leq t \leq t_B$ and $t_B \leq t \leq t_B + t^*$ where t_B is the time at which boiling starts and $t_B + t^*$ is the time at which the reactivity insertion is compensated by the effect of void formation (or $t_B + t^*$ is time of maximum transient power).

During the time interval $0 \leq t \leq t_B$, the temperature distributions throughout the three regions satisfy the one-dimensional heat-conduction equations:

$$\frac{\partial \theta_1(x, t)}{\partial t} = \alpha_1^2 \frac{\partial^2 \theta_1(x, t)}{\partial x^2} + \alpha_1^2 S(t), \quad 0 \leq x \leq a, \tag{3-1a}$$

$$\frac{\partial \theta_2(x, t)}{\partial t} = \alpha_2^2 \frac{\partial^2 \theta_2(x, t)}{\partial x^2}, \quad a \leq x \leq b, \tag{3-1b}$$

$$\frac{\partial \theta_3(x, t)}{\partial t} = \alpha_3^2 \frac{\partial^2 \theta_3(x, t)}{\partial x^2}, \quad b \leq x \leq c, \tag{3-1c}$$

where $\theta_i(x, t)$ is the temperature rise in region i at the distance x from the center line at time t. k_i, ρ_i, c_i are the thermal conductivity, density, and thermal capacity of the ith region; $\alpha_i^2 = k_i/\rho_i c_i$; and S(t) is the source term.

The boundary conditions assumed for the three regions shown on Fig. 3-1 are

$$\frac{\partial \theta_1(0, t)}{\partial x} = 0; \quad \frac{\partial \theta_3(c, t)}{\partial x} = 0, \tag{3-2a}$$

$$\theta_1(a, t) = \theta_2(a, t); \quad \theta_2(b, t) = \theta_3(b, t), \tag{3-2b}$$

$$k_1 \frac{\partial \theta_1(a, t)}{\partial x} = k_2 \frac{\partial \theta_2(a, t)}{\partial x};$$

$$k_2 \frac{\partial \theta_2(b, t)}{\partial x} = k_3 \frac{\partial \theta_3(b, t)}{\partial x}. \tag{3-2c}$$

The source term S(t) is related to the power density $q''(t)$ by $S(t) = q''(t)/k_i$.

The asymptotic solutions of Eq. (3-1) have the same time-dependence as the source. Choosing solutions of the form $\theta_i(x, t) = f_i(x)\exp(t/T)$ and satisfying the boundary conditions (3-2) lead to the solutions

$$\theta_1(x, t) = \frac{k_1 T\, S(t)}{\rho_1 c_1}\left[1 + \frac{A_1}{A_0}\cosh \beta_1 x\right], \quad (3\text{-}3a)$$

$$\theta_2(x, t) = \frac{k_1 T\, S(t)}{\rho_1 c_1 A_0}\left[A_2 \cosh \beta_1(x - a) + A_4 \sinh \beta_2(x - a)\right], \quad (3\text{-}3b)$$

$$\theta_3(x, t) = \frac{k_1 T\, S(t)}{\rho_1 c_1 A_0} A_3 \exp\left[-\beta_3(x - b)\right], \quad (3\text{-}3c)$$

where

$$\beta_i = \frac{1}{\alpha_i \sqrt{T}} \qquad i = 1, 2, 3, \quad (3\text{-}4)$$

and $A_0, A_1, \ldots, A_4$ are constants defined on p. 54 of Reference [138]. From Eq. (3-3) it is possible to relate the rise in the center-fuel-plate temperature at the time of boiling to the rise in the clad-coolant-interface temperature at the same time:

$$\frac{\theta_1(0, t_B)}{\theta_3(b, t_B)} = \frac{A_0 + A_1}{A_3}. \quad (3\text{-}5)$$

During the time interval $t_B \le t \le t_B + t^*$, according to the previous assumptions, the temperature rise has the same time dependence as the source; hence,

$$\theta_1(0, t_B + t^*) = \theta_1(0, t_B)\exp(t^*/T). \quad (3\text{-}6)$$

The period T is chosen as a variable and is directly related to the prompt reactivity insertion, $\Delta k_{eff}/k_{eff} = (k-\beta)/k_{eff}$, by the simplified relation

$$\Delta k_{eff}/k_{eff} = \ell^*/T, \quad (3\text{-}7)$$

where ℓ^* is the prompt neutron lifetime and β the delayed neutron fraction. (See Sec. 2 of Chapter 2.)

The time t^* at which the moderator void compensates for the excess reactivity $\Delta k_{eff}/k_{eff}$ is determined by Golian et al. from the relationship between reactivity and moderator-void fraction using the four-factor formula

$$k_{eff} = \eta \epsilon p f P_{NL}, \quad (3\text{-}8)$$

where η is the fertility, ϵ the fast-fission factor, p the resonance-escape probability (p = 1.0 in the seed region), f the thermal utilization, and P_{NL} the non-leakage probability. The result is

$$\exp(t^*/T) = \frac{Q^*}{k_4\beta_4 T\, \theta_3(b, t_B)} + 1, \quad (3\text{-}9)$$

where k_4 is the thermal conductivity of steam, β_4 is given by Eq. (3-4), and Q^* is the minimum quantity of heat required to create the void formation (V_v/V):

$$Q^* = h_{fg} V_v \rho_v. \quad (3\text{-}10)$$

h_{fg} is the latent heat of vaporization and $V_v\rho_v$ the required mass of steam per unit area of fuel element, which is related to the reactivity by

$$V_f = \frac{V_v}{V} = \frac{\ell^*}{T\, k_{eff}\, C_R}. \quad (3\text{-}11)$$

$V_f = V_v/V$ is the void-volume fraction in the moderator and C_R is defined by

$$C_R = \frac{\rho_L - \rho_v}{\rho_A}\left[f\frac{\Sigma_a^{water}}{\Sigma_a^{fuel}} - 2\frac{M^2B^2}{1 + M^2B^2}\right]. \quad (3\text{-}12)$$

ρ_A is the average moderator density obtained from the void fraction V_f and the density of water and steam (ρ_L and ρ_v, respectively):

$$\rho_A = V_f\rho_v + (1 - V_f)\rho_L. \quad (3\text{-}13)$$

M^2 and B^2 are the migration area and the buckling of the equivalent bare reactor.

Combining Eqs. (3-6) and (3-9) gives the fuel temperature at termination of the power excursion:

$$\Theta(0, t_B + t^*) = \Theta_0 + \frac{A_0 + A_1}{A_3}\left[\frac{Q^*}{\beta_4 k_4 T} + \Theta_3(b, t_B) - \Theta_0\right] \quad (3\text{-}14)$$

$$E = V_{fp} T\left\{\frac{\rho_1 c_1}{T}\cdot\frac{A_0}{A_0 + A_1}\left[\Theta(0, t_B + t^*) - \Theta_0\right] - q''(0)\right\}. \quad (3\text{-}15)$$

Θ denotes the actual temperatures and Θ_0 an average fuel-element temperature before the excursion:

$$\Theta(x, t) = \Theta(x, t) + \Theta_0. \quad (3\text{-}16)$$

V_{fp} is the fuel-plate volume and $q''(0)$ the initial power density. The quantity E is the total energy released during the power excursion.

The above method was used in analyzing the reactivity accident of the Shippingport reactor. The following two initial conditions are representative of average operating conditions and typical startup conditions, respectively:

Case I: Initial coolant channel and fuel temperature Θ = 525°F (274°C), water pressure = 2000 psi (136 atm);

Case II: Initial coolant channel and fuel temperature Θ = 190°F (88°C), water pressure = 100 psi (7.7 atm).

For each case, $\Theta_1(0, t_B + t^*)$ and E were computed as functions of reactor period T. The results are plotted on Figs. 3-3 [139] and 3-4 [139]. The calculations indicated that almost all of the heat absorbed by the water was used to heat the water to the boiling point. The remaining thermal energy

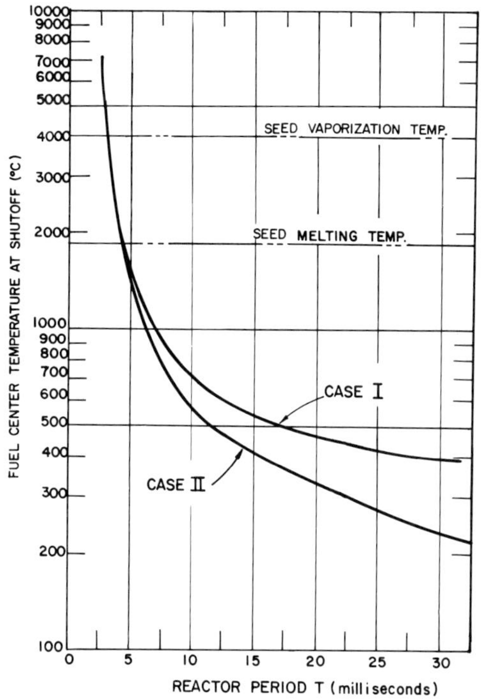

FIG. 3-3. Fuel (seed) center temperature versus reactor period in PWR analysis of two types of reactivity accidents.

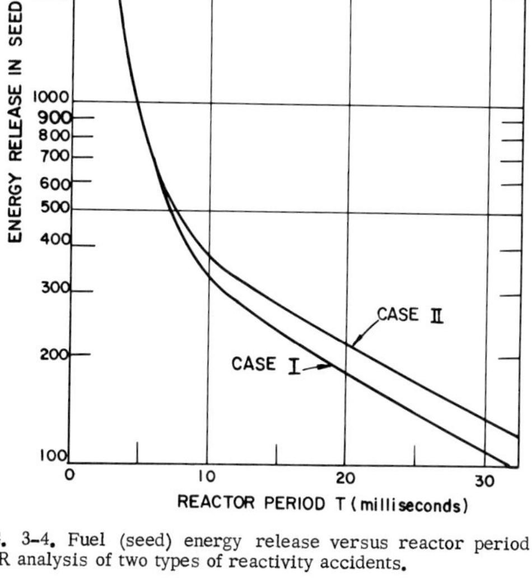

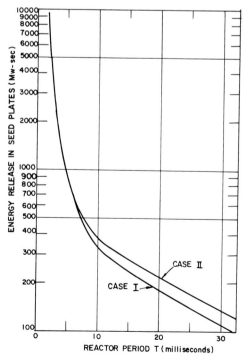

FIG. 3-4. Fuel (seed) energy release versus reactor period in PWR analysis of two types of reactivity accidents.

necessary to vaporize a sufficient volume of steam to terminate the reactivity excursion was a very small fraction of the total energy released. Although radiant heat transfer through the steam was neglected, this effect is significant for excursions of 3 msec or less and would contribute to a reduction of the maximum fuel temperature for those accidents.

3.2.2 Analysis of Severe Reactivity Accidents

The ultimate objective of an analysis of a severe nuclear accident is to determine the amount of metal that can be molten, and, therefore, spatial temperature-time calculations must be made. Several methods of analysis have been developed. Daane and Gartner [141] give an analysis for estimating power excursions in light-water-moderated, stainless-steel-clad reactors for the purpose of evaluating the merit of several configurations. Noordhoff et al. [142] examine two methods previously developed by Golian [138] and Sanford [143] and present a method incorporating some features of both. The method developed by Janssen, Cook, and Hikido [144] incorporates features used in the above references and, in addition, considers the influence of the compressibility of the liquid in the reactor vessel and the effects of the voids formed by radiolytic water decomposition. Owens' method [145] modifies the method of Jannsen et al. to include the effect of direct heating of the water by absorbed nuclear radiation and the effect of a significant Doppler temperature coefficient in slightly-enriched-UO_2 reactors. Details of these methods will be found in the references. The main features of the method of Jannsen et al. are given here, together with Owens' results for the BORAX-1 and SPERT-1 transients.

The core is assumed to consist of two distinct regions, one active and the other inactive. Thermal energy is assumed to be released in the active region of the core, with the heat sources assumed to be distributed uniformly at any instant. The active-core concept is based on the postulate that the actual power distribution in the core may be approximated at every instant and under all conditions of constant or varying power, or presence or absence of voids, by a single step-function in both the radial and axial directions as shown in Fig. 3-5 [144]. Although the principal justification of the approximation is that it simplifies the analyses, its use gives reasonably good quantitative results. The liquid coolant is considered to be confined in a pressure vessel and associated piping. For the purpose of this analysis, any motion of the coolant prior to the beginning of the excursion is disregarded. Two variations of the flow path are shown sketched in Fig. 3-6 [144]. Variation I corresponds to the BORAX-1 and SPERT-1 geometries. Variation II is for a reactor system where the free surface lies several feet from the core at the end of a channel of varying cross section. The initial hydrostatic-pressure distribution is replaced by a constant initial static pressure, equal everywhere to the initial pressure in the core. The pressure outside the reactor enclosure is assumed to be at the initial static

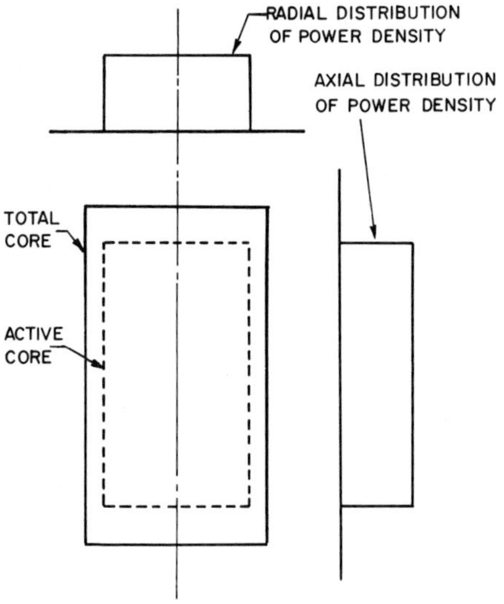

FIG. 3-5. Step-function power distribution in reactor core.

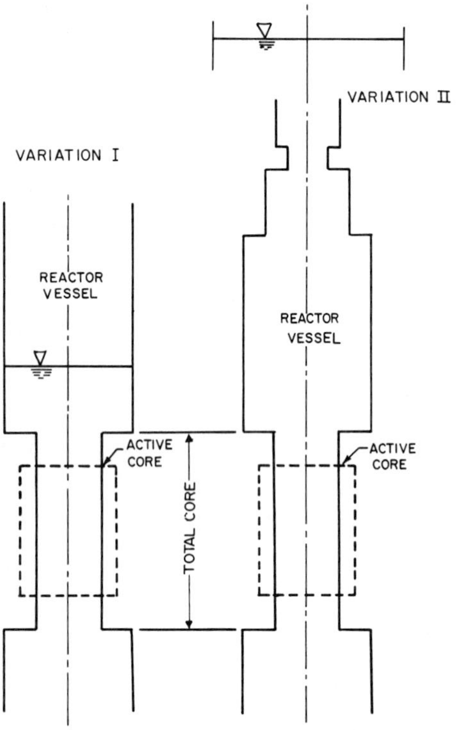

FIG. 3-6. Sketches of two variations of simplified-flow-channel model.

pressure in the core rather than at atmospheric pressure. This should tend to give a conservative prediction for the excursion because the compensating void volume is minimized by use of the higher pressure.

The heat-transfer and steam-void model incorporates the experimental data obtained by Rosenthal [118a] (see Sec. 2) and in particular the time delay in the initiation of boiling and surface-temperature overshoot. Heat transfer to the coolant is given by

$$q'' = (T_w - T_b)/R_T , \tag{3-17}$$

where T_W and T_b are the heating surface and coolant bulk temperature respectively. R_T is the thermal resistance to the flow of heat, which is assumed to be constant in the model. It is expected that the film coefficient would increase to a very high value during nucleate boiling and then decrease to a very low value upon the formation of a steam film. However, even if nucleate boiling became established (contrary to Rosenthal's observations for very short periods), it would last for such a brief time that it may be disregarded. The void volume due to steam formation is found to be

$$V_S = B(T_w - T_{sat})^{1/4} , \tag{3-18}$$

where T_W and T_{sat} are the surface and liquid-saturation temperatures and B a constant which is a function of the geometry and the thermodynamic properties of the fluid. Although the radiolytic gas voids were accounted for in the original analysis, it was found by Owens [145] that the effect of the gas voids contributed only slightly to the transient termination mechanism. Since the assumption was made that the power is uniform in the active core and zero everywhere else, the power is space invariant and one-point neutron kinetics are used in the model. The reactivity is composed of two parts: the initially inserted reactivity ρ_1 and the compensating reactivity ρ_2, so that

$$\rho = \rho_1 + \rho_2 . \tag{3-19}$$

The compensating reactivity ρ_2 is made up of a part due to the presence of void plus a part for the change in density due to temperature changes. To a first-order approximation, the relationship is a linear one:

$$\rho_2 = C_V \frac{V_V}{V} + C_T\Theta , \tag{3-20}$$

where C_V and C_T are constant void and temperature-reactivity coefficients.

An evaluation of the power and temperature versus time for three typical water-cooled and -moderated reactors during a nuclear excursion has been made by Owens [145] using the above method of analysis. The reactors analyzed are characterized by one of the following types of fuel elements: U-Al alloy, highly enriched, Al-jacketed, flat plate; U-Zr alloy, highly enriched, Zr-clad, flat plate; and UO_2, slightly enriched, Zr-clad, rods. The validity of the method of analysis was evaluated by comparing the results with experimental data derived from the BORAX-1 and SPERT-1 tests. The transient equations were solved on an analog computer and the calculated periods were obtained by introducing a step increase of reactivity. The results of the analysis are illustrated in the following

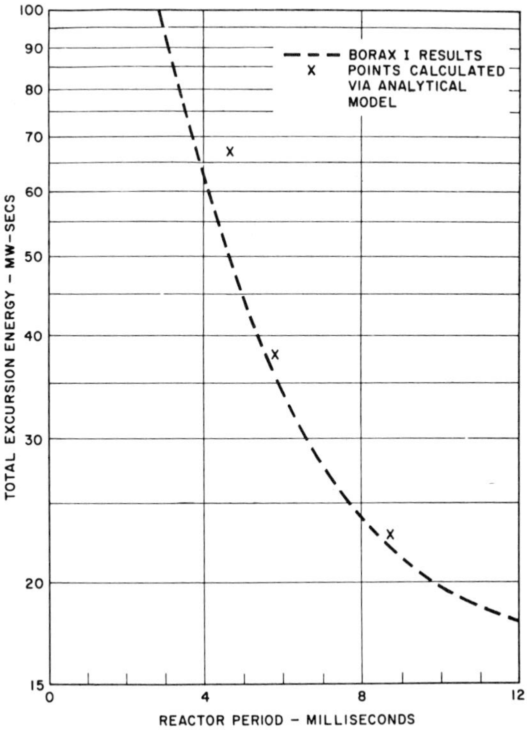

FIG. 3-7. Energy produced in a Borax -1 nuclear excursion versus reactor period.

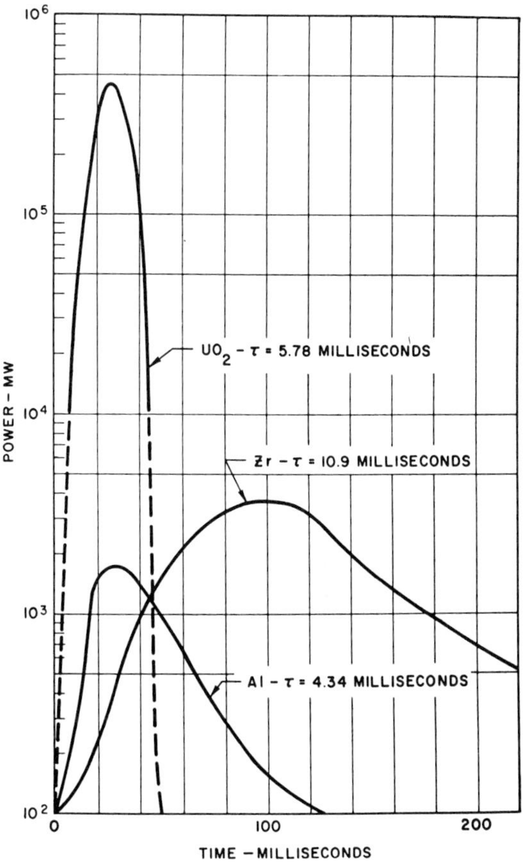

FIG. 3-8. Comparison of power bursts for three reactors. Fuel element are (1) U-Al alloy, highly enriched, Al-jacketed flat plate, (2) U-Zr alloy, highly enriched, Zr-clad flat plate, and (3) UO_2, slightly enriched, Zr-clad rods. Note the difference in initial periods.

figures. Figure 3-7 [145] gives a comparison between the predicted behavior and actual BORAX transient. The deviation (approximately 10% at 5 msec) at very short periods could be due to actual shutdown mechanisms not included in the analysis (such as mechanical deformation of the core) and also could be due to experimental uncertainty. Figures 3-8 [145] and 3-9 [145] illustrate a power excursion for the three reactors. A direct comparison among the three systems is difficult in view of their dissimilar characteristics. In general, the lower thermal diffusivity of Zr compared to Al allows less heat to enter the water, and consequently the Zr reactor experiences larger power excursions due to the greater length of time to shut down.

Figure 3-10 [145] illustrates the effect of period on the peak fuel temperature normalized to the melting temperature and on the percent of molten metal produced. One significant result is the narrow range of accidents between initial and complete melting of the core. It is also interesting to note for the UO_2 core that even in the 4.34-msec excursion the UO_2 and Zr-clad temperatures are below the melting temperature of zirconium. Figure 3-11 [145] supplements the information given in Fig. 3-10 by showing the energy released in these excursions. The relatively lower amount of energy produced in the Al fuel compared to the Zr plate is noteworthy. The broken lines show the total energy release assuming all the metal present reacts chemically with the water. This assumption is not justified since the extent of the chemical reaction will be largely determined by the amount of molten metal available. The dotted lines for the two high-enrichment fuels show the upper limit of the effect, taking this factor into account. Any real energy release due to a metal-water reaction must be in the shaded region between the solid and dotted curves. (See also Chemical Reactions chapter.)

Figures 3-12 [145] and 3-13 [145] give the temperature distribution as a function of period and time for the Zr flat-plate element and the UO_2-Zr-clad fuel. Stress analyses of these fuel elements indicate that the flat plate Zr element will deform but probably not rupture before melting. In the case of the Al fuel, the high thermal conductivity will result in a nearly uniform temperature distribution. In the UO_2-Zr-clad fuel element the yield strength of the cladding will be exceeded long before the oxides become molten.

Figures 3-14 [145] and 3-15 [145] illustrate the importance of various shutdown mechanisms for the low-enrichment UO_2 rod. In Fig. 3-14 the solid curve indicates the power transient with all important shutdown mechanisms in effect. The lower broken curve is the same transient, assuming

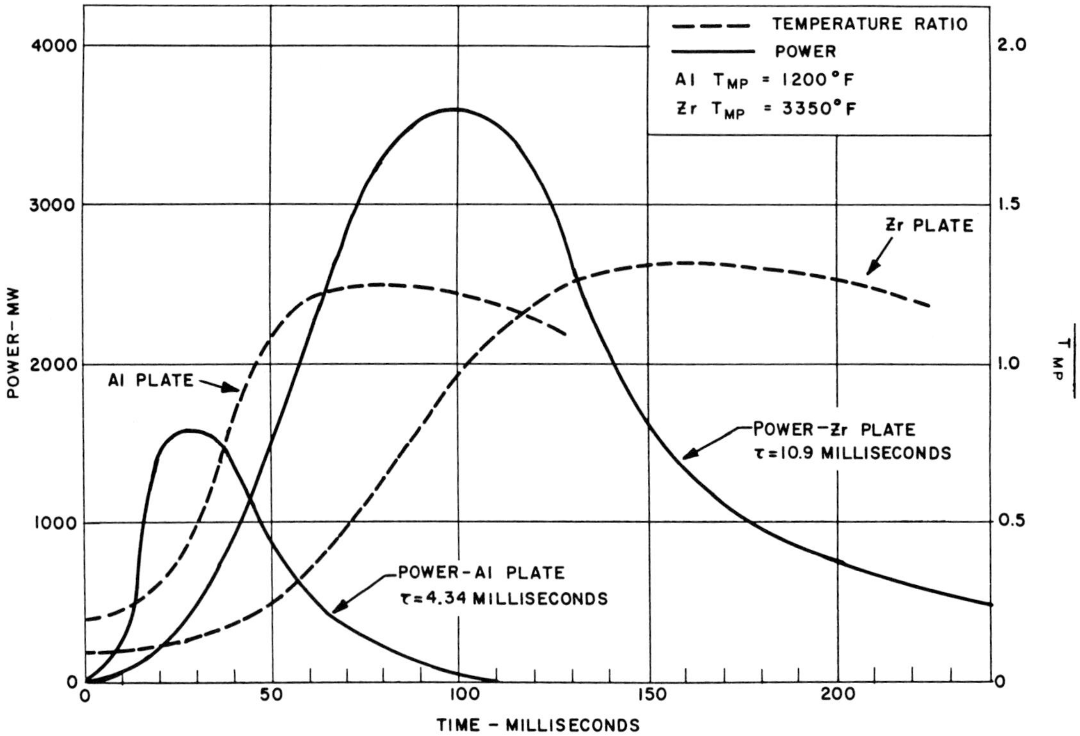

FIG. 3-9. Comparison of power and the ratio of hot-spot to melting temperature for the two flat-plate elements (under same conditions as in Fig. 3-8).

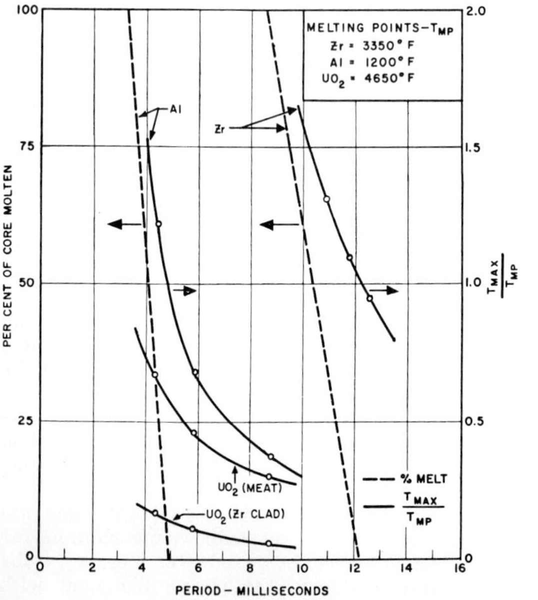

FIG. 3-10. Comparison of hot-spot temperature and percent molten metal for the three reactors versus reactor period.

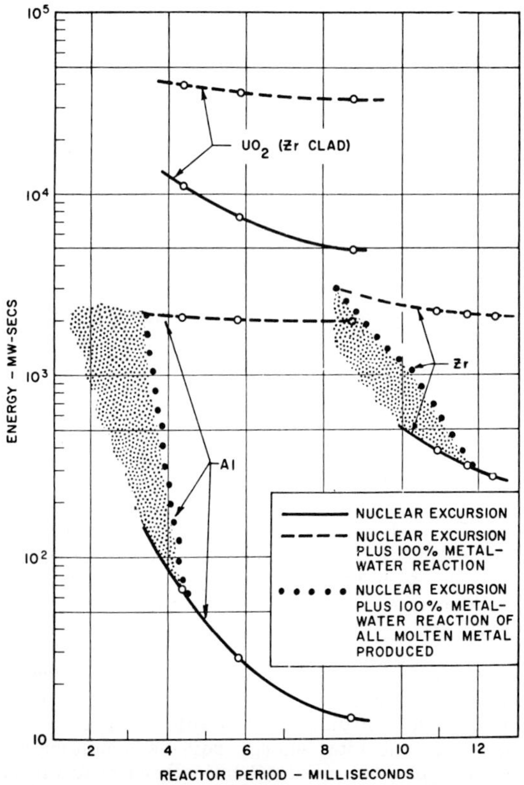

FIG. 3-11. Comparison of energy released in an excursion for the three reactors versus reactor period.

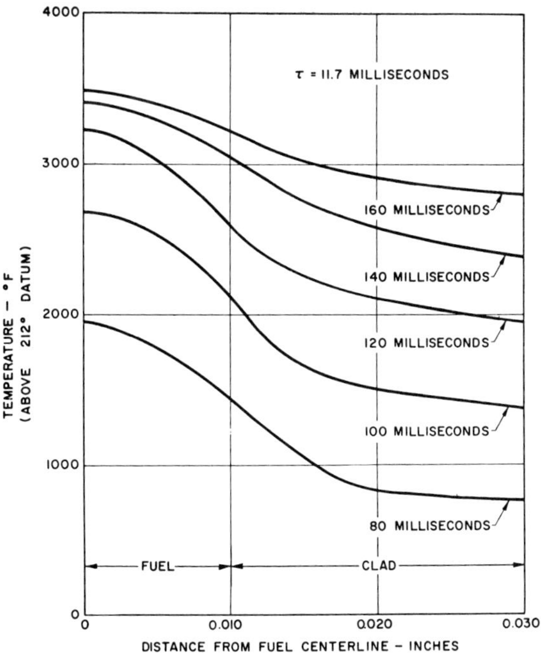

FIG. 3-12. Temperature (at five different time intervals after excursion) versus distance from fuel-element center for Zr flat-plate element.

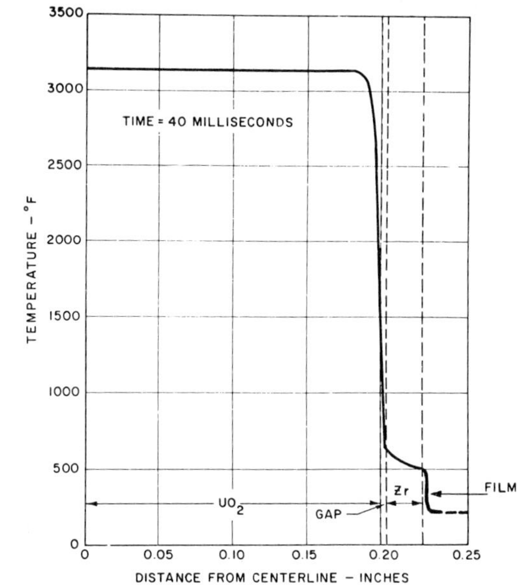

FIG. 3-13. Temperature versus distance from center of fuel element (cylindrical UO_2, Zr-clad) for a representative time (40 msec) after nuclear excursion (initial period = 4.34 msec).

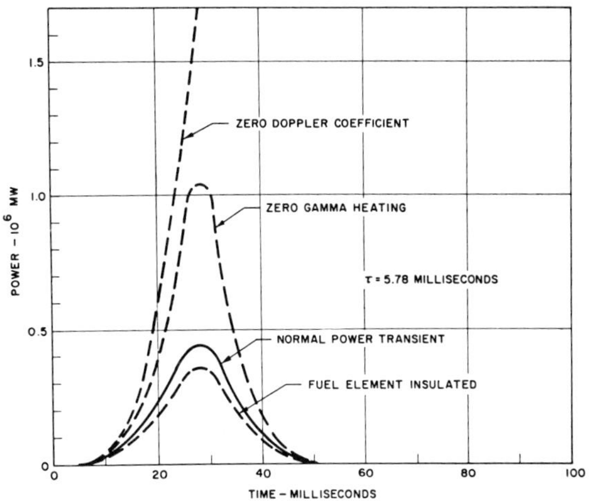

FIG. 3-14. Power excursions for the UO_2, Zr-clad fuel element, showing relative importance of various shutdown mechanisms.

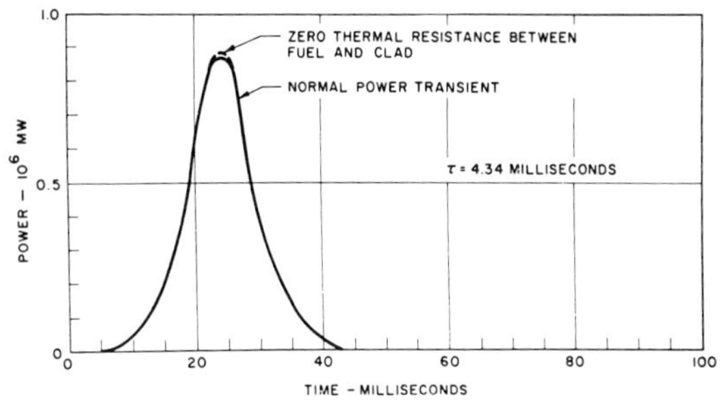

FIG. 3-15. Power excursion for the UO_2, Zr-clad fuel element, showing the effect of the cladding thermal resistance.

the fuel to be perfectly thermally insulated. Insulating the fuel produces a somewhat higher temperature and causes the Doppler effect to be more than sufficient to compensate for the smaller amount of steam void that would have been formed due to reduced heat transfer. The upper broken curve indicates the effect of neglecting the neutron and gamma heating in the water. This seems to be an important mechanism which reduces the peak power by a factor of two or greater. The remaining broken curve illustrates the very important effect of the Doppler coefficient, assumed to be 1.67×10^{-5}/°F (3.0×10^{-5}/°C). In Fig. 3-15 the effect of assuming an infinite heat-transfer coefficient between fuel and clad is shown to be small. This illustrates the weak influence of heat conduction out of the fuel and shows that a slight reduction in fuel temperature has a greater effect on the Doppler coefficient (due to the low enrichment) than it has on the steam-void coefficient.

4 HOT-SPOT AND HOT-CHANNEL FACTORS

4.1 Purpose, General Definitions, and Basic Considerations

An exact knowledge of the temperature distribution and flow conditions in the various regions of a reactor core is of paramount importance in determining the maximum performance allowable over long periods of time or reactor behavior during an accident. Inaccurate thermal analysis may lead to undesirable situations which seriously affect reactor safety. For example:

1) The fuel-element cladding in some region of the core may operate at a temperature where its structural strength is greatly reduced and creep may produce intolerable dimensional changes or deformations.
2) The fuel temperature may be too high for retention of gaseous fission products which, when released, may increase internal cladding pressure to the point where the cladding ruptures.
3) The cladding temperature may become high enough to cause an excessive corrosion rate and ultimate failure.
4) If the fuel-element surface temperature is higher than predicted, fouling due to decomposition of a chemical coolant may occur and subsequently restrict the coolant flow and thereby cause a high fuel temperature.

All these are long-term effects which could occur as the result of unpredicted high temperature or flow conditions during normal operations. In the analysis of the safety of nuclear reactors, it is important to predict accurately the temperature distribution in the core as well as flow conditions throughout an accidental transient. A few percent error in these quantities may be the difference between predicting an operating difficulty or a serious accident (excessive release of fission products due to cladding rupture, metal-water reaction, etc.).

The thermal design of nuclear reactors consists of three successive phases.

1) For a given core configuration, thermal output, and power distribution, the temperature configuration and flow distribution are calculated using the best available correlations and computational techniques. The geometrical configuration in most heterogeneous reactors is too complex for a direct analysis, and the computation is usually made for a simplified geometry and operating condition.
2) Out-of-pile experimental mockups are used to obtain further information concerning flow and temperature close to reactor operating conditions and for a more exact channel geometry. These data permit correction of some shortcomings in the previous analysis.
3) All further unknowns, uncertainties, or simplifications not accounted for in steps 1) and 2) and which may cause higher temperatures or a worsening of the flow conditions in the most critical region of the core are estimated and stated in terms of hot-spot and hot-channel factors. These are used to determine the maximum fuel or cladding temperatures, the maximum heat flux, or the worst flow conditions.

Hot-spot and hot-channel factors are used for both steady-state and transient analyses but are not necessarily independent of the reactor operating conditions. The most critical region of the core is that part which operates at conditions nearest to one or more safe design limits, e.g., near the maximum fuel or cladding temperature, critical heat flux, etc. The location of the critical region depends on many factors, among which the most important are the power distribution for a given control rod position, fuel burnup, and flow distribution. The critical region is therefore not in a fixed location but is likely to shift axially and radially depending on operating conditions and local fuel burnup.

It should be emphasized at this point that the hot-spot and hot-channel factors do not normally include the safety margins used in reactor design and should not be used as a substitute for these margins. The hot-spot and hot-channel factors are introduced to correct the computed or measured temperature, flow, and heat flux in order to obtain as accurate values as possible for these quantities in a specific situation. On the other hand, a well-designed reactor should not operate steadily at conditions too close to some potentially dangerous conditions. Allowances must be made in the design to enable the reactor to undergo some unexpected but possible transient conditions without destroying itself or seriously endangering personnel. Safety margins are therefore introduced to give the needed safe flexibility in operating conditions where these operating conditions are in turn determined by taking into account the hot-spot and hot-channel factors.

For example, the reduction in the design critical-heat-flux ratio* in water-cooled reactors is an indication of the generally improved confidence in predicting the critical heat flux. It is evident, however, that a further substantial decrease in the critical-heat-flux ratio cannot be made unless it is conclusively proved that a specified water-cooled reactor during a transient can withstand heat fluxes equal to or greater than the critical heat flux and can do so without serious core damage requiring major repairs and costly shutdown or endangering the safety of reactor personnel or the general public. Unless these conditions are fulfilled, a safety margin on maximum heat flux (critical-heat-flux ratio) will have to be maintained even though in the future it may be possible to predict much more accurately the maximum heat flux and critical heat flux at operating conditions and during transients.

4.2 Quantities Entering into the Calculation of Hot-Spot and Hot-Channel Factors

Since the hot-spot and hot-channel factors are correction factors which should account for all unknowns, uncertainties, and simplifications made in the thermal analysis, it is virtually impossible to give a comprehensive list of all the quantities which should enter into the determination of these factors. Both the nature and the magnitude of these quantities are dependent on the model chosen in the analysis, the degree of sophistication of the analysis, the amount of experimental work to support and correct the analysis, and the type of reactor under consideration.

In heterogeneous solid-fuel reactors, the quantities which, individually or in combination, limit

*Sometimes called "DNB ratio." See Sec. 2.2.1.

the reactor performance are generally the maximum fuel temperature, cladding temperature, coolant enthalpy, and maximum heat flux. The procedure followed to obtain these quantities is a) to determine the enthalpy and temperature rise in the coolant for the most critical channel (hot-channel) and b) to add up the temperature rises through the laminar film, the oxide layer, the cladding, the contact between cladding and fuel, and the fuel itself as required in order to arrive at the "hot spot" on the cladding or in the fuel. Some quantities, such as temperature rise through the oxide layer and contact between cladding and fuel, may not be present in certain types of reactors.

Each of these quantities is associated with a correction factor, i.e., a "hot-channel factor," if the quantity to be corrected is the enthalpy or coolant temperature, and "hot-spot factors" if the quantity to be corrected is the individual temperature rise through the laminar flow, the oxide layer, the cladding, the surface contact, and the fuel region. Using the notation of Reference [146] the following quantities can be defined:

$$F_H^T = \frac{\Delta H_{max}}{\Delta H_{nom\ av}} = \text{hot channel factor for enthalpy rise}$$

$$F_f^T = \frac{\Delta T_{f\ max\ loc}}{\Delta T_{f\ nom\ av}} = \text{hot spot factor for film temperature rise}$$

$$F_{Ox}^T = \frac{\Delta T_{Ox\ max\ loc}}{\Delta T_{Ox\ nom\ av}} = \text{hot spot factor for oxide layer temperature rise}$$

$$F_C^T = \frac{\Delta T_{C\ max\ loc}}{\Delta T_{C\ nom\ av}} = \text{hot spot factor for cladding temperature rise}$$

$$F_R^T = \frac{\Delta T_{R\ max\ loc}}{\Delta T_{R\ nom\ av}} = \text{hot spot factor for thermal contact temperature rise}$$

$$F_F^T = \frac{\Delta T_{F\ max\ loc}}{\Delta T_{F\ nom\ av}} = \text{hot spot factor for temperature rise in the fuel}$$

$$F_Q^T = \frac{Q_{max\ loc}}{Q_{nom\ av}} = \text{hot spot factor for heat flux} \tag{4-1}$$

The F is used for hot-spot or hot-channel factors; the subscripts H, f, Ox, C, R, F define the quantities to which the factors apply; the superscript T refers to total factors to differentiate these factors from the subfactors defined below.

The hot-spot or hot-channel factors are the ratio of the maximum expected values of the enthalpy rise, temperature rises, or heat flux to their respective uncorrected values (nominal values) obtained from the thermal analysis of the core. The difference between the maximum local values (denoted by subscript "max loc") and the nominal average values (denoted by subscript "nom av") is attributable to simplifications in the analysis and in the experimental model and to uncertainties and unknowns present in the analysis and will be called differences throughout the remainder of this discussion.

Each hot-spot or hot-channel factor includes the contribution from several basic quantities which must be corrected to take into account factors not included in the original simplified model and which, if not corrected for, would introduce errors in the enthalpy and maximum temperatures predicted in the core. The correction factors which account for these differences between the model and the real reactor, when each basic quantity is considered individually, are called the hot-spot or hot-channel "subfactors." These subfactors are denoted by an F, with the same subscripts as the total factors F^T, and with a superscript E or N referring to an engineering or nuclear difference, followed by an additional symbol specifying the basic quantity in error. Thus the notation $F_H^{E,h}$ specifies the engineering hot-channel subfactor for the correction to the coolant enthalphy (H) due to differences in the cladding-to-coolant heat-transfer coefficient (h).

The distinction between engineering and nuclear subfactors is simply a convenient way to indicate the origin of the difference. It is common practice at the present time to assume a uniform heat source in making the basic thermal analysis of a nuclear core. Thus the nuclear subfactors correct not only for inaccurate determination of the neutron fluxes but also for the spatial distribution of the volumetric heat sources. As a result, the nuclear subfactors are much larger than the engineering subfactors. In a real sense this is a simplified bookkeeping method of taking care of this complex situation insofar as it fails to account for the interaction between the separate factors. It should be pointed out that this practice tends to overextend the limits of validity of hot-channel—hot-spot analysis (which assumes small variations of the quantities considered).

In Table 4-1 an extensive list of the subfactors is given. This list does not refer to any reactor in particular and is by no means complete; however, it is intended to give a general idea of the quantities usually in error and their effect on the heat flux, the enthalpy, and temperatures at the local hot spot. The column on the left-hand side of Table 4-1 lists the sources of differences (Nuclear and Engineering). In line with these items are noted the subfactors under the appropriate column (enthalpy, temperature rise, or local heat flux) representing the quantities with which these subfactors are associated. An asterisk is to be interpreted as a subfactor of 1.0.

The individual items in the tabulation are discussed in the following paragraph.

4.2.1 Nuclear Quantities

(A.1, A.2 of Table 4-1) Macroscopic distribution of power in the core. When the thermal analysis is made using a uniform power distribution, it is necessary to correct for the actual spatial distribution. The simplest procedure is to assume

TABLE 4-1

Hot-Spot and Hot-Channel Subfactors

Quantities in error	Enthalpy rise in coolant	Temp. rise thru film	Temp. rise thru oxide	Temp. rise thru clad	Temp. rise thru contact	Temp. rise thru fuel	Local heat flux
A. Nuclear quantities							
1. Radial power distribution (including control rod effect)	$F_H^{N, P\perp}$	$F_f^{N, P\perp}$	$F_{Ox}^{N, P\perp}$	$F_C^{N, P\perp}$	$F_R^{N, P\perp}$	$F_F^{N, P\perp}$	$F_Q^{N, P\perp}$
2. Core: axial distribution (including control rod effect)	$F_H^{N, P_{\parallel}}$	$F_f^{N, P_{\parallel}}$	$F_{Ox}^{N, P_{\parallel}}$	$F_C^{N, P_{\parallel}}$	$F_R^{N, P_{\parallel}}$	$F_F^{N, P_{\parallel}}$	$F_Q^{N, P_{\parallel}}$
3. Fuel assembly: radial power distribution	$F_H^{N, FA}$	$F_f^{N, FA}$	$F_{Ox}^{N, FA}$	$F_C^{N, FA}$	$F_R^{N, FA}$	$F_F^{N, FA}$	$F_Q^{N, FA}$
4. Fuel-element radial and angular power distribution	$F_H^{N, FE}$	$F_f^{N, FE}$	$F_{Ox}^{N, FE}$	$F_C^{N, FE}$	$F_R^{N, FE}$	$F_F^{N, FE}$	$F_Q^{N, FE}$
5. Uncertainties in power distributions due to uncertainties in nuclear analysis (neutron cross section, energy spectrum, method of analysis)	$F_H^{N, \Delta P}$	$F_f^{N, \Delta P}$	$F_{Ox}^{N, \Delta P}$	$F_C^{N, \Delta P}$	$F_R^{N, \Delta P}$	$F_F^{N, \Delta P}$	$F_Q^{N, \Delta P}$
6. Fuel burnup in fuel assembly and fuel element	*	*	*	*	*	$F_F^{N, \Delta BF}$	*
B. Engineering quantities							
1. Flow distribution in plenum chamber	$F_H^{E, W}$	$F_f^{E, W}$	*	*	*	*	*
2. Variations in coolant inlet temperature and pressure, coolant flow rate, and steady-state power fluctuations	$F_H^{E, Er}$	$F_f^{E, Er}$	$F_{Ox}^{E, Er}$	$F_C^{E, Er}$	$F_R^{E, Er}$	$F_F^{E, Er}$	$F_Q^{E, Er}$
3. Fuel-element lattice pitching	$F_H^{E, \Delta p}$	$F_f^{E, \Delta p}$	*	*	*	*	*
4. Variation in fuel-rod diameter	$F_H^{E, \Delta dR}$	$F_f^{E, \Delta dR}$	$F_{Ox}^{E, \Delta dR}$	$F_C^{E, \Delta dR}$	*	*	$F_Q^{E, \Delta dR}$
5. Effects of fuel-element bowing	$F_H^{E, Bow}$	$F_f^{E, Bow}$	*	*	*	*	*
6. Variation in fuel diameter	$F_H^{E, \Delta dF}$	$F_f^{E, \Delta dF}$	$F_{Ox}^{E, \Delta dF}$	$F_C^{E, \Delta dF}$	$F_R^{E, \Delta dF}$	$F_F^{E, \Delta dF}$	$F_Q^{E, \Delta dF}$
7. Eccentricity of fuel material	$F_H^{E, Ecc}$	$F_f^{E, Ecc}$	$F_{Ox}^{E, Ecc}$	$F_C^{E, Ecc}$	$F_R^{E, Ecc}$	$F_F^{E, Ecc}$	$F_Q^{E, Ecc}$
8. Variation in fuel enrichment and density	$F_H^{E, \Delta F}$	$F_f^{E, \Delta F}$	$F_{Ox}^{E, \Delta F}$	$F_C^{E, \Delta F}$	$F_R^{E, \Delta F}$	$F_F^{E, \Delta F}$	$F_Q^{E, \Delta F}$
9. Surface state at the contact between fuel and cladding	*	*	*	*	$F_R^{E, Sc}$	*	*
10. Surface state of the outside surface of cladding	$F_H^{E, So}$	$F_f^{E, So}$	*	*	*	*	*
11. Effect of fission gases released inside cladding	*	*	*	*	$F_R^{E, G}$	*	*
12. Thermal properties of materials and coolant	$F_H^{E, Th}$	$F_f^{E, Th}$	$F_{Ox}^{E, Th}$	$F_C^{E, Th}$	$F_R^{E, Th}$	$F_F^{E, Th}$	*
13. Flow mixing and flow redistribution	$F_H^{E, CF}$	$F_f^{E, CF}$	*	*	*	*	*
14. Heat-transfer correlation between coolant and cladding surface	*	$F_f^{E, h}$	*	*	*	*	*
15. Contact-resistance correlation	*	*	*	*	$F_R^{E, R}$	*	*

* Subfactor = unity.

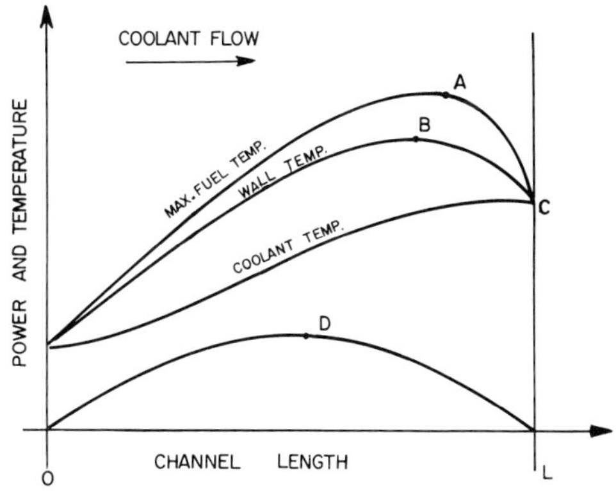

FIG. 4-1. Typical temperature distribution in a hot channel with one-phase flow of coolant. A and B are the hot spots for the fuel and cladding. C is the location of maximum of coolant enthalpy and D the location of maximum heat flux.

that the power distribution function is separable into radial and axial dependence:

$$P(r, z) = R(r)Z(z) . \tag{4-2}$$

The hot-channel factor is thus located at the radius r where the function R(r) is maximum. Therefore, the subfactor $F_Q^{N,P_\perp}$ is the ratio of the maximum value of R(r) divided by its radial average:

$$F_Q^{N,P_\perp} = \frac{R_{max}}{(2/r_C^2)\int_0^{r_C} R(r)rdr}, \tag{4-3}$$

where r_C is the equivalent core radius. The factors $F_H^{N,P_\perp}$, $F_f^{N,P_\perp}$, $F_{Ox}^{N,P_\perp}$, $F_C^{N,P_\perp}$, $F_R^{N,P_\perp}$ and $F_F^{N,P_\perp}$ are identical to $F_Q^{N,P_\perp}$ so long as Eq. (4-2) applies. Along the hot channel the maximum heat flux is normally at a different location from that of the maximum cladding temperature or maximum fuel temperature. (See Fig. 4-1.) The maximum heat flux occurs at the maximum power location and hence where Z(z) has its maximum value. The subfactor $F_Q^{N,P_\parallel}$ is therefore

$$F_Q^{N,P_\parallel} = \frac{Z_{max}}{(1/H_{core})\int_0^{H_{core}} Z(z)dz}, \tag{4-4}$$

where H_{core} is the height of the core.

The remaining subfactors for the axial power distribution in the core, $F_f^{N,P_\parallel}$, $F_{Ox}^{N,P_\parallel}$, $F_C^{N,P_\parallel}$, $F_R^{N,P_\parallel}$, $F_F^{N,P_\parallel}$, are usually smaller than $F_Q^{N,P_\parallel}$, since they apply to a hot spot at a different location. It has been conservatively assumed in the past that each of these subfactors is equal to $F_Q^{N,P_\parallel}$.

The enthalpy subfactor, $F_H^{N,P_\parallel}$, has a different value depending on whether the maximum enthalpy of the coolant is being determined in the hot channel, e.g., in a water-cooled reactor, or the enthalpy of the coolant at the location of the hot spot is used to determine the temperature of that hot spot. In the first choice, the subfactor $F_H^{N,P_\parallel} = 1.0$ since the maximum enthalpy occurs at the channel outlet and has the same value irrespective of the axial power distribution for the same total power in the hot channel. In the second choice, the enthalpy subfactor is a function of the location z of the hot spot and is defined as

$$F_H^{N,P_\parallel} = \frac{\int_0^z Z(z)dz}{\frac{z}{L}\int_0^L Z(z)dz}, \tag{4-5}$$

where L is the total length of the channel.

Since this formulation requires the knowledge of the location of the hot spot, it is now practical to determine the enthalpy and temperature distribution along the hot channel with the worst axial power distribution likely to occur during the core lifetime; see Fig. 4-1. These distributions give directly the maximum enthalpy, maximum temperature, and maximum heat flux in the hot channel, providing Eq. (4-2) holds. In such a case, using the above procedure, the macroscopic power distribution is accounted for in the nominal design and the subfactors A.1, A.2 of Table 4-1 all have a value of 1.0.

The principle of separability of the power breaks down, however, when a hot-spot, hot-channel analysis is made throughout the core lifetime, because of the nonuniformity of fuel burnup and the strong flux distortion introduced by control rods. For accurate power mapping in the core as a function of time, two-dimensional computer codes coupling the neutronic, thermal, and fuel-depletion analysis are required. A typical study of that kind has been made [145] for the Experimental Gas-Cooled Reactor (EGCR). A two-dimensional, two-energy-group code was used to determine the effect of control rod configuration on power peaking. The position of the control rods is shown in Fig. 4-2a [147]. The radial power peaking for several rod configurations is given in Table 4-2 [147] and the axial power peaking is shown in Fig. 4-2b [147] for four different

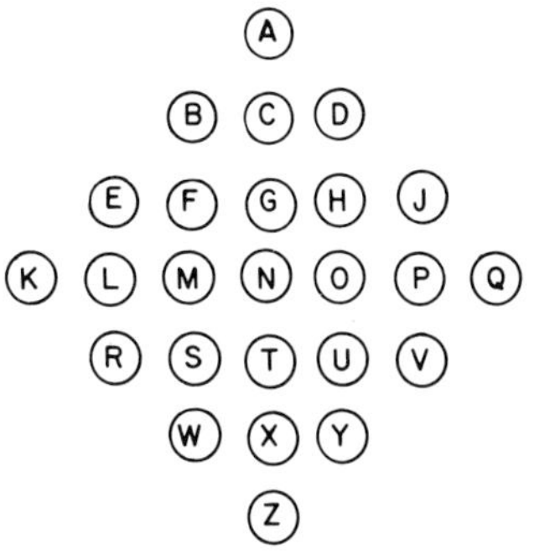

FIG. 4-2a. Identification letters assigned to control rods in the Experimental Gas-Cooled Reactor.

TABLE 4-2

Effect of the Full Insertion of Selected Control Rods on the Radial Power Distribution in the Initial 2.2% Enriched Experimental Gas-Cooled Reactor Core

Control rods inserted	Radial peak-to-average power ratio	Reactivity decrease Δk	Average power per loop (Mw)
None	1.758	0	1.338
N	1.352	0.0246	1.073
N, G, O, T, M	1.201	0.0870	0.682
G, O, T, M	1.188	0.0791	0.866
M, O	1.637	0.0400	1.144
C, P, X, L	2.149	0.0260	1.480
P	1.852	0.0068	1.370
N, O	1.622	0.0380	1.007

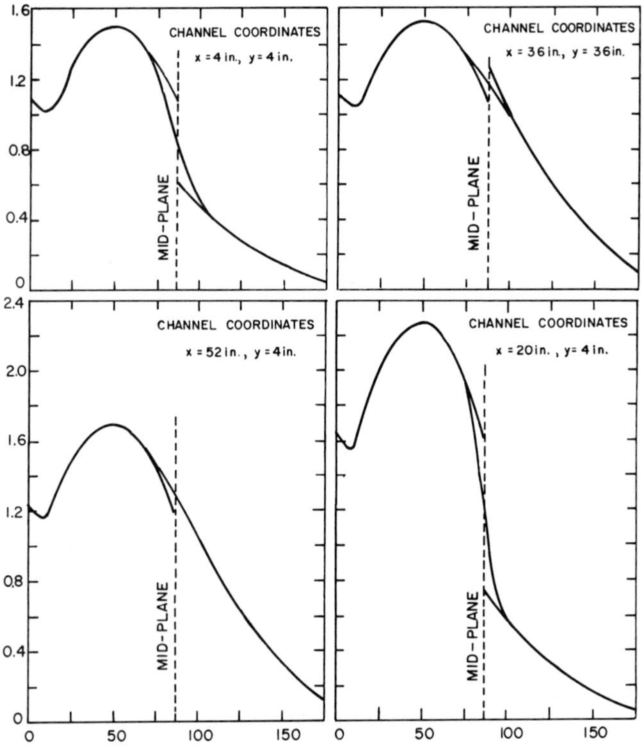

FIG. 4-2b. Axial power distribution in typical fuel channel of the EGCR with the central control rod inserted and rods G, O, T, and M inserted to the core mid-plane. The four central loops contain homogeneous graphite-U^{235} experimental assemblies with C/U = 200. The power-distribution function is normalized to unity for the entire core.

channels. The discontinuity in the power curves at the midplane of the core is due to the simplification in the study, where a three-dimensional power distribution is determined by use of a two-dimensional code. The axial power peaking as a function of controlled reactivity is shown in Fig. 4-3 [147]. If the control rods are employed in a bank, then at the beginning of core life they are inserted from the top well into the core and the power is strongly peaked at the bottom of the core. The power-peaking factor is very large, as is shown in Fig. 4-3. In this case the rods must hold down almost 0.14 Δk at the beginning of core life. Later the power-peaking factor decreases monotonically as the peak power is shifted upward. Power peaking is usually greater in a clean critical core. However, it is important to consider the abnormal power peaking which may result when a change in the control rod configuration exposes some portion of the core which, until that time, has been substantially shielded from the neutron flux. In such a case there is no xenon built up in that region, no fission products, and a maximum fuel inventory.

(A.3, A.4 of Table 4-1) Local radial and angular power distribution. In certain types of reactors the fuel elements are assembled in bundles and inserted into a pressure tube or a moderator sleeve which acts as a guide to the coolant flow. To obtain the hot-spot subfactors for local power peaking in such reactors the power distribution in a homogenized fuel assembly cell and within the hottest element should be considered. These subfactors are obtained by analysis or by experimental measurements of neutron fluxes in a subcritical lattice mockup experiment.

(A.5 of Table 4-1) Differences in power distribution. In the analytical or experimental determination of the power distribution in the core or in a fuel element and fuel assembly, differences are introduced by the simplified methods used or by inaccurate data. An example of how the radial power distribution may be affected by the method of analysis is shown in Fig. 4-4 [148]. The maximum radial power is not only increased but shifted from the center of the core to the core-reflector interface when a more accurate spatial variation of the thermal spectrum is used. Similar cases occur in water-cooled and -moderated lattices. Effects of errors in neutron-cross-section data may also be significant and are considered in the subfactors A.5 of Table 4-1.

(A.6 of Table 4-1) Fuel burnup in fuel assembly and fuel element. The power distribution in the fuel assembly and fuel element change with irradiation time. The principal consequence in a natural- or slightly-enriched-uranium heterogeneous thermal reactor is the buildup of plutonium in the outer region of the fuel slugs, which tends to increase the power depression within the fuel element, thus affecting the maximum fuel temperature. The

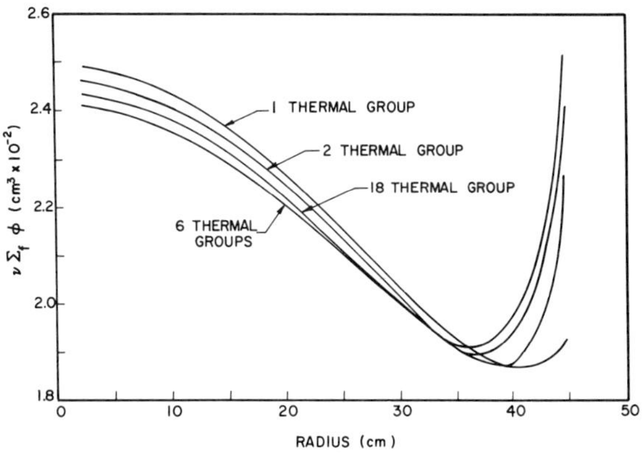

FIG. 4-4. Power distribution versus radius in the High Temperature Gas-Cooled Reactor using different number of thermal groups in calculation. Core temperature = 900°K (627°C or 1161°F); reflector temperature = 300°K (27°C or 81°F).

distribution of heat sources in a highly irradiated fuel is thus more concentrated near the fuel surface so that a more favorable temperature distribution in the fuel results. This gives, for a highly irradiated fuel, a hot-spot subfactor $F_F^{N,\Delta BF}$ smaller than 1.0.

4.2.2 Engineering Quantities

(B.1 of Table 4-1) Flow distribution in plenum chamber. Plenum effects are due to unequal distribution of flow among the channels of a system consisting of a number of identical channels. These effects are accounted for by the subfactor $F_H^{E,W}$ on enthalpy rise obtained from experimental coolant-distribution tests. The results of such a test performed on a quarter-scale model of the Pathfinder reactor are shown in Fig. 4-5a. Air was used as coolant instead of water. Figures 4-5b, c, and d [149] represent the flow-distribution patterns for the same model with two blowers operating instead of three in the three possible configurations. The inner region of the plenum chamber has a flow velocity of 95% of the average velocity. Assuming conservatively that the hot channel is located at the minimum feed velocity, the enthalpy subfactor $F_H^{E,W}$ is 1.05. This value may vary with the design of the plenum chamber. The subfactor $F_f^{E,W}$ is much closer to 1.00 and accounts for the dependence of the heat-transfer coefficient on coolant properties and coolant flow. A similar experiment for the Partially Enriched Gas-Cooled Power Reactor (PEGCPR) has been reported [150]. Using air as a coolant and a simulated one-loop operation (out of two loops), the flow for each channel was within +1.3% and -1.4% of the average flow for the core, indicating a subfactor $F_H^{E,W}$ of 1.014 for this reactor design.

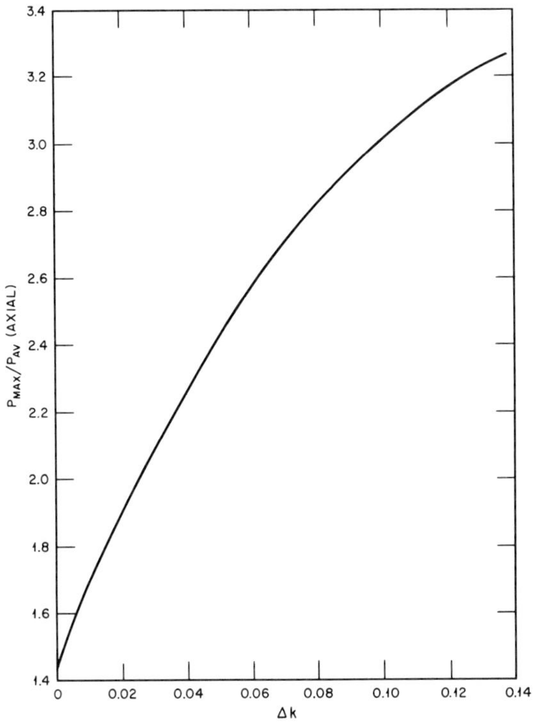

FIG. 4-3. Variation of axial peak-to-average power ratio with Δk for bank insertion of the EGCR control rods

(B.2 of Table 4-1) Variations in coolant inlet temperature and pressure, coolant flow rate, and steady-state power fluctuations. These variations refer to recording errors of the operating conditions of the reactor and the small fluctuations present when running the reactor at normal power. The hot-spot, hot-channel factors for these effects are obtained from the accuracy of the temperatures and flux-measuring devices and the ability of the automatic control system to maintain constant power during normal operation (see Chapter 6).

(B.3, B.4, B.5 of Table 4-1) Variations in lattice pitch, fuel-rod diameter, and effects of fuel-element bowing. All these introduce changes in mechanical dimensions in the core structure. They are attributable to manufacturing tolerances or may be caused by irradiation effects. Fuel-element bowing occurs as a consequence of nonuniform temperature distribution in the fuel element. Variations in lattice pitch and cladding thickness depend on the manufacturing tolerances allowed and on the method of inspection. These tolerances may vary from a fraction of a mil (10^{-3}in.) to several mils, depending on size and on method of fabrication. The amount of bowing should be calculated or measured experimentally.

The basic method for determining the hot-channel and hot-spot factors associated with these changes is to consider a channel of nominal dimensions in a region of average heat flux in parallel with a channel of minimum or maximum average dimensions (and most unfavorable local dimensions) also located in a region of average heat flux. The subfactors are then calculated using the basic assumption that the pressure drops across the nominal and the hot channel are equal [146]. If the channels are interdependent rather than independent (as with parallel-rod fuel elements), then a zero pressure gradient perpendicular to the flow system is usually assumed. The pressure drops across the two channels are equated using available single-phase or two-phase pressure-drop correlations as applicable.

Thus the subfactors $F_H^{E,\Delta p}$, $F_H^{E,\Delta dR}$, $F_H^{E,Bow}$ are equal to the ratio $W_{nom}/W_{min\ av}$ (coolant flow rate in the nominal channel/coolant flow rate in the channel of minimum dimensions) using a change Δp in pitch, ΔdR in fuel-rod diameter, and Bow in bowing effect, respectively. In computing the enthalpy-rise subfactors, several approaches are possible. If it can be shown that the total effect on enthalpy can be conservatively estimated by using the average deviation from the nominal dimension rather than the maximum deviation, this can be used. On the other hand, instances may occur in which such an assumption is nonconservative. For instance, a mistake in milling the slots for an MTR plate-type fuel element will change the coolant-channel width throughout the entire element. In that case the maximum set of unfavorable conditions should be used. The film temperature rise subfactors $F_f^{E,\Delta p}$, $F_f^{E,\Delta dR}$, $F_f^{E,Bow}$ are similarly equal to $h_{nom}/h_{min\ loc}$, the ratio of the nominal heat-transfer coefficient to the minimum heat-transfer coefficient in the minimum channel at the hot spot. In this calculation the deviations Δp, ΔdR, and Bow are the local maximum changes. The subfactor $F_C^{E,\Delta dR}$, on temperature rise through the cladding, is approximately equal to $t_{c\ max\ loc}/t_{c\ nom}$, the ratio of the maximum to nominal average cladding thickness.

A simple example of the computation of these subfactors is given by Letourneau and Grimble [146] for a parallel plate channel using turbulent flow one-phase heat-transfer and pressure-drop correlations of the following form:

friction factor $f \propto (D_eV)^{-0.2}$

heat-transfer coefficient $h \propto V^{0.8}/D_e^{0.2}$,

where V is the average coolant velocity and D_e the hydraulic diameter ($D_e \approx 2d = 2 \times$ channel width). Assuming that the entrance and exit losses are proportional to the velocity head ($V^2/2g$) in the channel the pressure drop is

$$\Delta P = \rho \frac{fL\,V^2}{D_e 2g}, \tag{4-6}$$

with L the total equivalent length of channel. Equating pressure drop in each channel and noting that the cross-flow area $A_f \propto d$, one obtains

$$F_H^{E,\Delta p} \text{ or } F_H^{E,\Delta dR} \text{ or } F_H^{E,Bow} = \left(\frac{d_{nom}}{d_{min\ av}}\right)^{5/3}. \tag{4-7}$$

The minimum film coefficient h occurs in the hottest channel at the local point of maximum channel thickness. Thus

$$F_f^{E,\Delta p} \text{ or } F_f^{E,\Delta dR} \text{ or } F_f^{E,Bow} = \frac{h_{nom}}{h_{min\ loc}} \tag{4-8a}$$

$$= \left(\frac{V_{nom}}{V_{min\ loc}}\right)^{0.8} \left(\frac{d_{max\ loc}}{d_{nom}}\right)^{0.2} \tag{4-8b}$$

$$= \left(\frac{d_{max\ loc}}{d_{min\ av}}\right) \left(\frac{d_{nom}}{d_{min\ av}}\right)^{1/3}. \tag{4-8c}$$

More complicated geometry is not amenable to simple analytic formulation and a thermal analysis of the two channels (nominal and minimum or maximum dimension) should be performed to determine the hot-spot and hot-channel subfactors associated with dimensional variations.

(B.6 of Table 4-1) Variation of fuel diameter. The size of the fuel inserted into the fuel elements may show appreciable variations. An example of this change is given in Table 4-3 [151] for a particular loading in the VBWR core [151]. During fabrication each UO_2-fuel-pellet diameter was measured and then inserted into the cladding tube. The maximum and minimum diameters of the pellets were, respectively, 0.380 in. and 0.372 in. or maximum deviation of ±0.004 in. The main

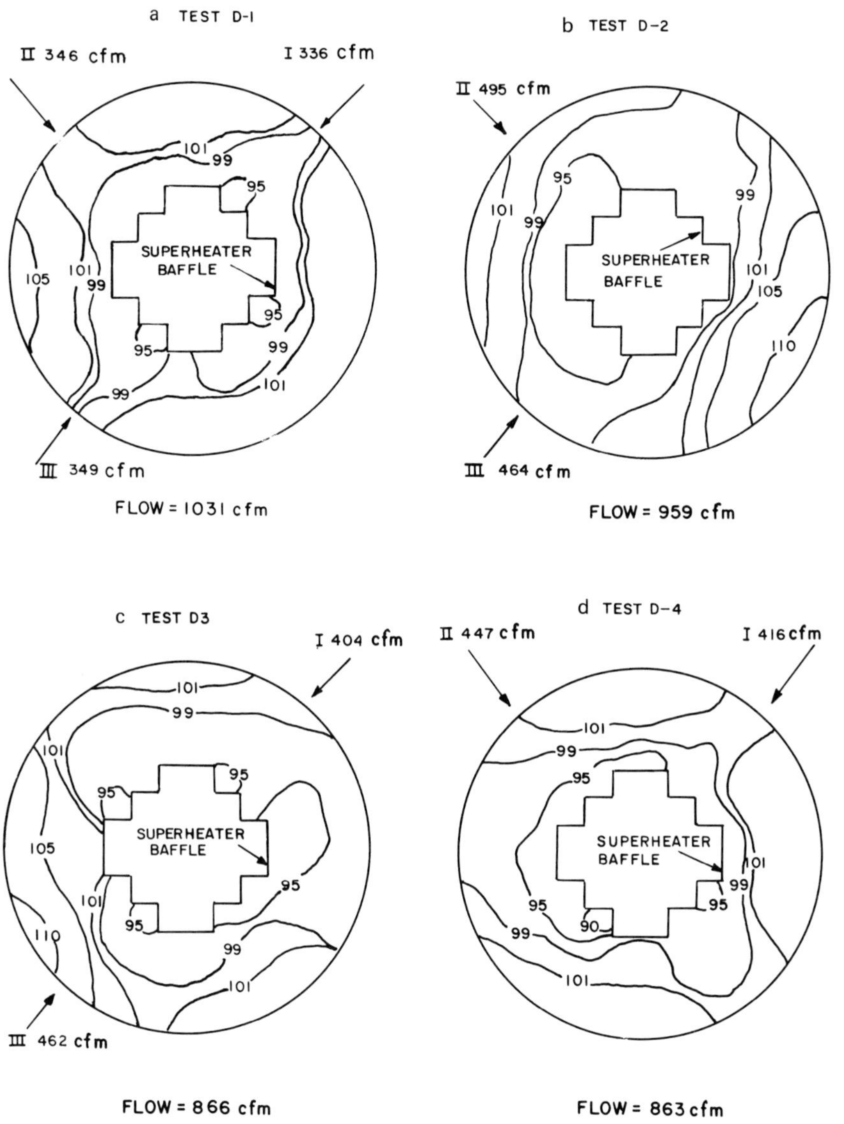

FIG. 4-5. Flow-distribution pattern in the plenum chamber of a quarter-scale model for the Pathfinder Reactor. The contour numbers indicate the average velocity in percent. The upper left pattern is for three blowers in operation, the other patterns for two blowers in operation. Total flow and blower in-flow are given in cubic feet per minute (1 cfm = 0.4720 liter/sec = 0.02832 m^3/min).

effect of a larger fuel diameter is to increase the enthalpy rise of the coolant, the local temperature rises, and the heat-flux increases (due to an increase in fuel volume).

Thus for cylindrical fuel pellets,

$$F_H^{E,\Delta dF} = \left[\frac{d^{pellet}_{max\ av}}{d^{pellet}_{nom}}\right]^2 \tag{4-9}$$

and, conservatively,

$$F_{Ox}^{E,\Delta dF} = F_C^{E,\Delta dF} = F_R^{E,\Delta dF} = F_Q^{E,\Delta dF} = \left[\frac{d^{pellet}_{max\ loc}}{d^{pellet}_{nom}}\right]^2 . \tag{4-10}$$

(B.7 of Table 4-1) Eccentricity of fuel material. If at operating conditions a gap is present in a cylindrically clad fuel element, the pellet may not be centered in the cladding. Plate-type fuel elements may also have different cladding thickness on either side. These variations appreciably affect the heat-flow distribution in the fuel element. The eccentricity subfactors are determined by application of the steady-state heat-conduction equation to a fuel element of nominal dimensions and to a fuel element of maximum eccentricity which will result in a maximum heat flux; changes in the neutron flux are neglected. The enthalpy subfactor $F_H^{E,Ecc}$ is obtained by use of the channel average eccentricity in the ratio $F_H^{E,Ecc} = Q_{max\ av}/Q_{nom}$,

TABLE 4-3

Size Distribution of UO_2 Pellet Loading in the VBWR[a]

Pellet diameter (in.)	Rod number															
	1	2	3	4	5	6	7	8	9	10	11	12	13	14	15	16
0.370	0	0	0	0	0	0	0	0	0	0	0	0	0	0	0	0
0.371	0	0	0	0	0	0	0	0	0	0	0	0	0	0	0	0
0.372	1	1	1	1	1	1	1	1	1	1	1	1	1	1	2	1
0.373	3	3	3	3	3	3	3	3	3	4	2	3	5	3	1	0
0.374	7	7	7	7	7	7	9	10	7	9	23	7	7	13	21	20
0.375	14	13	13	13	13	13	12	12	12	12	13	5	12	15	13	13
0.376	16	16	16	16	16	16	15	17	16	16	8	15	15	15	15	10
0.377	12	13	13	12	11	12	12	12	12	17	12	20	19	12	7	7
0.378	7	7	7	7	8	7	7	3	7	0	1	0	0	0	0	0
0.379	0	0	0	0	0	0	0	0	1	0	0	0	0	0	0	0
0.380	0	0	0	0	0	0	0	0	0	0	0	1	0	0	0	0
Total pellets	60	60	60	59	59	59	59	59	59	59	60	52	59	59	59	51

[a]Numbers entered denote number of pellets having diameter indicated at left.

where Q denotes heat flux. The hot-spot subfactor is obtained using the maximum local value:

$$F_f^{E,Ecc} = F_{Ox}^{E,Ecc} = F_R^{E,Ecc} = F_Q^{E,Ecc} = \left(\frac{Q_{max\ loc}}{Q_{nom}}\right), \qquad (4\text{-}11)$$

$$F_C^{E,Ecc} = \left(\frac{Q_{max\ loc}}{Q_{nom}}\right)\left(\frac{t_{c\ max\ loc}}{t_{c\ nom}}\right), \qquad (4\text{-}12)$$

where t_c stands for cladding thickness. The eccentricity subfactor for the temperature rise in the fuel, $F_F^{E,Ecc}$, is obtained from heat-conduction analysis:

$$F_F^{E,Ecc} = \frac{\Delta T_{max\ loc}^{fuel}}{\Delta T_{nom}^{fuel}}. \qquad (4\text{-}13)$$

For fuel pellets in a cladding with an effective gap between the two, eccentricity may cause surface contact on one side and a wider gap on the opposite side. For the Yankee reactor (fuel pellets 0.3 in. in diameter):

$F_H^{E,Ecc}$ = 1.12 for a gap of 2 mils on one side,

$F_H^{E,Ecc}$ = 1.06 for a 1-mil gap [152].

(B.8 of Table 4-1) Variation in fuel enrichment and density. According to the present (1963) AEC production specifications, the allowable fuel-enrichment variation is 1.53% above the minimum or ±0.76% from the nominal enrichment. It has been conservatively assumed in the past that an increase in fuel enrichment in the hot channel and at the hot spot would not change the neutron-flux level. With this assumption the increase in heat source is proportional to the fissile isotope concentration. Thus, the hot-channel subfactor is

$$F_H^{E,\Delta F} = \left(\frac{e_{max\ av}}{e_{nom}}\right)\left(\frac{\rho_{max\ av}}{\rho_{nom}}\right) \qquad (4\text{-}14)$$

and the hot-spot subfactors are

$$F_f^{E,\Delta F} = F_{Ox}^{E,\Delta F} = F_C^{E,\Delta F} = F_R^{E,\Delta F} = F_F^{E,\Delta F} = F_Q^{E,\Delta F} = \left(\frac{e_{max\ loc}}{e_{nom}}\right)\left(\frac{\rho_{max\ loc}}{\rho_{nom}}\right), \qquad (4\text{-}15)$$

where e stands for enrichment and ρ for fuel density. A variation in neutron flux ϕ due to a change in enrichment alters all the above subfactors by a factor of $\phi_{fuel\ max}/\phi_{fuel\ nom}$. This correction factor can be important and should be obtained from flux calculations in a fuel element or by experiment.

(B.9, B.11, B.15 of Table 4-1) Surface state at the contact between fuel and cladding, effect of fission gases released inside cladding, contact resistance correlation. In heterogeneous fuel elements the thermal resistance at the contact between fuel and cladding or fuel and sleeve appreciably affects the temperature in the fuel. Available correlations for estimating this resistance and the experimental data needed to evaluate the accuracy of a particular correlation are still very scarce (see Sec. 1). For a dry contact with inert gas in the voids and similar surface roughness, a relation for the thermal conductance h_c is obtained from Eq. (1-32) of this chapter:

$$h_c \approx \frac{(1/\delta)(k_f/k_m)\left[(1-\epsilon^2)\left(2.13\sqrt{n}\,\frac{\delta}{\epsilon}+1\right)+1.1\epsilon\right]+2.13\epsilon\sqrt{n}}{(1-\epsilon^2)(k_m-k_f)\left(2.13\sqrt{n}\,\frac{\delta}{\epsilon}+1\right)}, \qquad (4\text{-}16)$$

where n is the number of contact points per unit area, ϵ^2 the contact-area ratio, k_m the mean conductivity of the two solids ($2/k_m = 1/k_1 + 1/k_2$), k_f the gas conductivity, and δ defined by Eq. (1-39a). This correlation was checked with one set of experimental data for a set of helium and argon pressures [53]. The maximum deviation was found to be 40%. Thus for dry contacts, using the above relation for h_c, the hot-spot subfactor is

$$F_R^{E,R} = \frac{\Delta T_{max}}{\Delta T_{correlation}} = \frac{h_{correlation}}{h_{min\ loc}} \cong 1.40 . \tag{4-17}$$

When experimental data are used, the value of $F_R^{E,R}$ is simply equal to (1 + maximum fractional deviation).

The surface state of the contact affects the thermal conductance through the n contact points and the average void thickness as indicated in Eq. (4-16). The pressure and composition of the fission gases affect the thermal conductivity k_f. Therefore, by using Eq. (4-16), one may calculate the hot-spot subfactors $F_R^{E,Sc}$ and $F_R^{E,G}$:

$$F_R^{E,Sc} = F_R^{E,G} = \frac{h_{nom}}{h_{min\ loc}} , \tag{4-18}$$

where the $h_{min\ loc}$ is calculated using the minimum value of k_f to determine $F_R^{E,G}$, and using the maximum value of δ and minimum value of n to determine $F_R^{E,Sc}$.

Values of the thermal conductivities of mixtures of fission-product gases with helium and argon have been published [152, 11].

Values of δ and n are dependent on surface roughness and waviness. For a uranium-Magnox contact of a surface roughness of 5—10 μin. and 1—2 × 10^{-4}-in. waviness, the calculated values are $\delta \cong 1.2 \times 10^{-4}$in. and n ≃ 460 contacts/sq in. for a contact pressure of 100 psi. Uncertainties of the order of 50% may be assumed in computing the $h_{min\ loc}$ and the subfactor $F_R^{E,Sc}$.

(B.10 of Table 4-1) Surface state of the outer surface of fuel-element cladding. The roughness of the heat-transfer surface between fuel element and coolant significantly affects the pressure drop and heat-transfer coefficient in the hot channel. The method used to compute the hot-channel subfactor $F_H^{E,So}$ and the hot-spot subfactor $E_f^{E,So}$ is the same as the one outlined for the change in dimensions (see discussion of subfactors B.3, B.4, and B.5).

For the simple case of a plate fuel element considered above,

$$F_H^{E,So} = \frac{W_{nom}}{W_{min\ av}} = \left(\frac{f_{max\ av}}{f_{nom}}\right)^{1/2} , \tag{4-19}$$

where $f_{max\ av}$ and f_{nom} are the friction factor coefficients for a nominal channel and a channel with a maximum average surface roughness, respectively.

Similarly, making the conservative assumption that $h \propto fW$, W being the flow rate per channel,

$$F_f^{E,So} = \frac{h_{nom}}{h_{min\ loc}} = \left(\frac{f_{nom}}{f_{min\ loc}}\right)\left(\frac{W_{nom}}{W_{min\ av}}\right)$$

$$= \left(\frac{f_{nom}}{f_{min\ loc}}\right)^{1/2}\left(\frac{f_{max\ av}}{f_{min\ loc}}\right)^{1/2} . \tag{4-20}$$

(B.12 of Table 4-1) Thermal properties of materials and coolants. For one-phase cooling, uncertainties in the thermal properties of the coolant introduce errors in the temperature rise of the coolant and in the film heat-transfer coefficient.

To calculate the hot-channel subfactor $F_H^{E,Th}$ for a single-phase flow channel, an estimate of the fractional temperature error corresponding to a given enthalpy must be made ($\epsilon_T = \Delta T/T$). Thus

$$F_H^{E,Th} = 1 + \epsilon_T . \tag{4-21}$$

In a two-phase flow regime, errors in the latent heat of vaporization introduce an error in the vapor-to-liquid ratio. In this case the subfactor $F_H^{E,Th}$ depends on this ratio. A conservative value is obtained by writing

$$F_H^{E,Th} = 1 + (\epsilon_h)_{fg} , \tag{4-22}$$

where $(\epsilon_h)_{fg} = \Delta h_{fg}/h_{fg}$ is the fractional error in latent heat at the channel nominal operating conditions.

The film subfactor $F_f^{E,Th}$ is obtained from the basic definition

$$F_f^{E,Th} = \frac{h_{nom}}{h_{min\ loc}} , \tag{4-23}$$

where $h_{min\ loc}$ is computed using the appropriate correlation and the most pessimistic combination of permissible errors in those thermal properties of the coolant entering the correlation. When the subfactor $F_f^{E,Th}$ is obtained from experimental data at operating conditions, the above errors are included in that factor and a value of 1.0 should be used for $F_f^{E,Th}$.

The subfactors $F_{Ox}^{E,Th}$, $F_C^{E,Th}$, and $F_F^{E,Th}$ are equal to $1 + (\epsilon_k)_i$, where $(\epsilon_k)_i$ is the fractional error in the thermal conductivity of the oxide layer or cladding. The subfactor $F_R^{E,Th}$ is evaluated using the relationship

$$F_R^{E,Th} = \frac{h_{c\ nom\ av}}{h_{c\ min\ loc}} , \tag{4-24}$$

where $h_{c\ nom\ av}$ is the thermal contact conductance for a nominal channel and $h_{c\ min\ loc}$ the one obtained using the minimum values of the thermal conductivities of the material in contact. Equation (4-16) may be used when applicable for this determination.

(B.13 of Table 4-1) Flow redistribution, flow mixing inside assembly. In a vertical open lattice core the static pressure should be uniform in a horizontal plane across the flow. This requirement causes the flow to redistribute itself away from the hot channel. The flow redistribution is determined by the net balance of friction, momentum, and buoyancy of flow in the channels. Computer calculations on the neighboring channels show that the flow which is redistributed away from the hot channel in a conventional large pressurized-water reactor with a uniform inlet flow is about 5% at steady state, neglecting flow mixing between neighboring channels [153]. Thus $F_H^{E,CF} \simeq 1.05$ for this case. Values of $F_h^{E,CF}$ for other reactors should be computed or measured experimentally using dyes or tracers on a mockup model. The subfactor $F_f^{E,CF}$ is attributable to the dependence of the heat-transfer coefficient on coolant properties.

Cross-flow mixing in the hot fuel assembly tends to cool the hottest rod. Results of cross-flow mixing are reported [154] which give a value for $F_H^{E,CF}$ of approximately 0.95. However, this factor for other lattice and fuel-assembly configurations should be determined experimentally.

(B.14 of Table 4-1) Heat-transfer correlation. For one-phase-flow liquid coolant, the film subfactor $F_f^{E,H}$ depends on the correlation used:

$$F_f^{E,H} = \frac{h_{correlation}}{h_{min\ loc}}. \tag{4-25}$$

A comparison of the McAdams, Colburn, and Sieder-Tate correlations with experimental data [155] is shown in Fig. 4-6 [155]. For information on

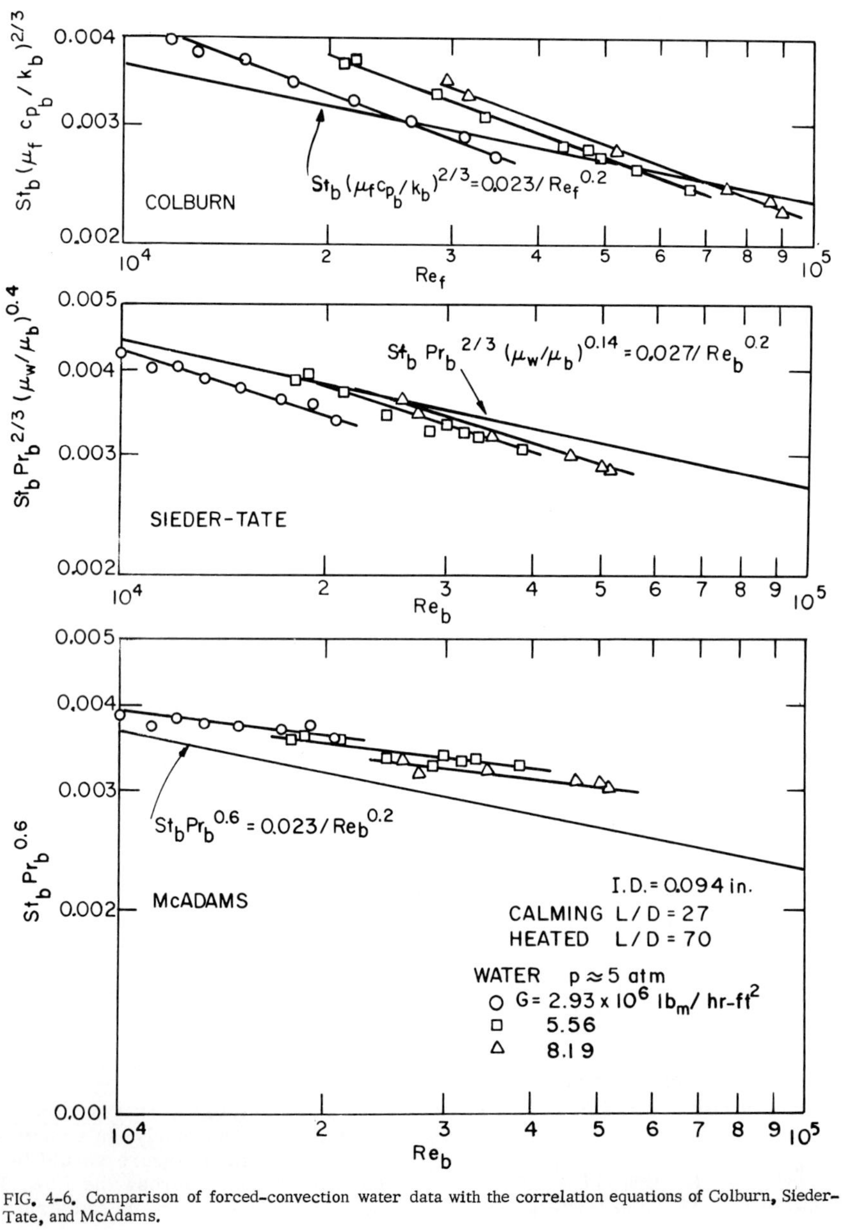

FIG. 4-6. Comparison of forced-convection water data with the correlation equations of Colburn, Sieder-Tate, and McAdams.

gas cooling and axial flow in rod clusters, direct experimental determination of the wall temperature is needed. Experimental data for the EGCR reactor have been reported [155] and the circumferential temperature distribution for the hottest tube at several axial locations is plotted in Fig. 4-7 [156]. A variation of the order of ±10% in circumferential temperature was found, equivalent to a subfactor $F_f^{E,h} \simeq 1.10$.

4.3 Methods Used to Combine the Hot-Spot Hot-Channel Subfactors

An accurate method for combining all the subfactors to arrive at the maximum heat flux, maximum cladding and fuel temperatures should take into consideration that:

a. Some subfactors are deterministic (that is uniquely determined) in nature, e.g., the subfactors A.1 through A.6 or B.1 and B.13 through B.15 of Table 4-1. Other subfactors are stochastic in the sense that a given value of these subfactors has a certain probability of occurrence at a given location and a given time, e.g., B.2 through B.12. Little is known at present about the relevant probability distribution for these subfactors.
b. All the nuclear subfactors and some engineering subfactors (B.6, B.9, B.11, B.12, B.13 of Table 4-1) are functions of irradiation exposure or power transients, and are therefore time dependent.
c. All the subfactors are not completely independent of each other. For example, bowing of a fuel element changes the cross-flow mixing in a fuel bundle, the power generated in the fuel element, the circumferential temperature distribution, etc.

In addition to these considerations, some generalizations can be made which are independent of the method of analysis. Following Lerda and Rossi [157], let w be a thermal parameter defined for each point or for each channel of a given core, e.g., the coolant enthalpy, or the temperature of the cladding, oxide, or fuel. The local correction factor f_w^T associated with w is the ratio of the actual value of w at one point or in one channel to the value of w in a core of nominal design value with a uniform heat source equal to the volumetric average of the actual source. Let the factor F_w^T be the value of f_w^T in the hot channel or at the hot spot. The local correction factor f_w^T is a function of the space coordinates x_1, x_2, x_3 and suitable variables $y_1, \ldots y_n$ characterizing the specific core (lattice pitch, fuel-element diameter, fuel-element bowing, hydrodynamic model, etc.). The variables introducing the spatial dependence are the nuclear quantities A.1 through A.6 and the engineering quantity B.1 listed in Table 4-1. The variables $y_1, \ldots y_n$ are the remaining engineering quantities in the left-hand column of that table. Considering as an example the enthalpy rise in the channel located at (x_1, x_2, x_3), the local correction factor is

$$f_H^T = \frac{\Delta H(x_1, x_2, x_3; y_1, \ldots, y_n)}{\Delta H_{nom\ av}\left(y_1^0, \ldots, y_n^0\right)}, \tag{4-26}$$

where

$\Delta H(x_1, x_2, x_3; y_1, \ldots, y_n)$ = enthalpy rise in an actual channel,

$\Delta H_{nom\ av}\left(y_1^0, \ldots, y_n^0\right)$ = enthalpy rise in the nominal core with a uniform heat source and even flow distribution in the plenum chamber.

The superscript 0 refers to the nominal values of the variables y_i. Note that the $\Delta H_{nom\ av}$ is independent of the space coordinates. Equation (4-26) may be rewritten as

$$f_H^T = \frac{\Delta H(x_1, x_2, x_3; y_1, \ldots, y_n)}{\Delta H\left(x_1, x_2, x_3; y_1^0, \ldots, y_n^0\right)} \times \frac{\Delta H\left(x_1, x_2, x_3; y_1^0, \ldots, y_n^0\right)}{\left(\Delta H_{nom\ av}\ y_1^0, \ldots, y_n^0\right)}. \tag{4-27}$$

If the plenum-chamber flow distribution is arbitrarily grouped with the nuclear factors, it is seen that the first term on the right-hand side of Eq. (4-27) is a result of corrections for engineering errors or approximations; it is designated as the "engineering factor" f_H^E. The second term on the right-hand side of Eq. (4-27) depends only on nuclear factors (with the above restriction) and is referred to as "the nuclear factor" f_H^N. Equation (4-27) thus becomes

$$f_H^T = f_H^E f_H^N, \tag{4-28a}$$

or, generally,

$$f_i^T = f_i^E \cdot f_i^N, \quad i = H, f, Ox, C, R, F, Q. \tag{4-28b}$$

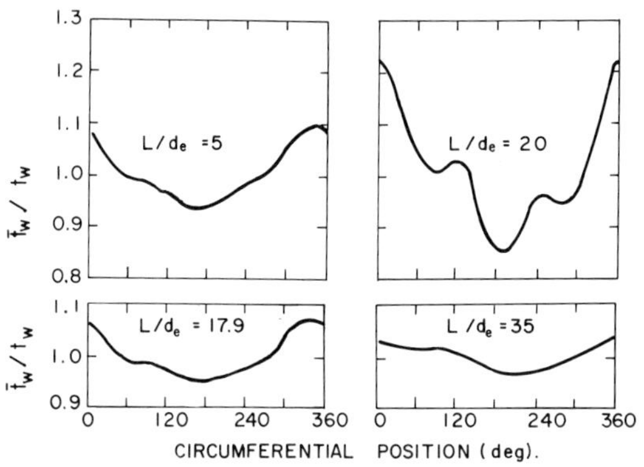

FIG. 4-7. Circumferential variation of the temperature at several axial locations for a peripheral tube of a tube bundle in the Experimental Gas-Cooled Reactor.

The nuclear factors F_i^N are the values of the function f_i^N at the position (x_1, x_2, x_3) of the maximum temperatures, maximum enthalpy, or maximum heat flux, depending on what maximum is considered. All the subfactors which contribute to the nuclear factors are deterministic. The product of these subfactors evaluated at the position (x_1, x_2, x_3) of any one of the maxima will therefore yield the nuclear factors F_i^N. Assuming that the power is separable in space functions and that the minimum flow in the plenum chamber is at the same location as the maximum radial power, we have for the value of F_H^N:

$$F_H^N = F_H^{N,P_\perp} \cdot F_H^{N,P_\parallel} \cdot F_H^{N,\Delta P} \cdot F_H^{N,\Delta BF} \cdot F_H^{E,W} . \quad (4\text{-}29)$$

The notation in Eq. (4-29) is that used in Table 4-1. The remaining nuclear hot-spot factors, $F_f^N, F_{Ox}^N, F_C^N, F_R^N, F_F^N, F_Q^N$, are likewise obtained by multiplying the values of the subfactors at the hot spot, or, generally,

$$F_i^N = \prod_j F_i^{N,j} ,$$
$$i = f, Ox, C, R, F, Q,$$
$$j = P_\parallel, P_\perp, FA, FE, \Delta P, \Delta BF, W, CF, h, R . \quad (4\text{-}30)$$

When the power distribution is not separable into a product of space functions, a power mapping of the core is needed to determine the location of the hot channel and hot spot and an overall power peaking at the hot spot is used instead of the product $F_i^{N,P_\perp} \cdot F_i^{N,P_\parallel}$.

The engineering factors f_i^E are composed of subfactors of a stochastic nature. There is therefore no definite maximum for f_H^E but a given probability for a specified value of f_H^E at the hot spot. The factor f_i^E may be written in two different forms [157], namely, the product form and the sum form.

The product form is

$$f_i^E = \prod_j f_i^{E,j} ,$$
$$i = H, f, Ox, C, R, F, Q,$$
$$j = \Delta p, \Delta dF, \Delta dR, \Delta F, \text{Bow, Ecc, Sc, So, G, Th}, \quad (4\text{-}31)$$

where the $f_i^{E,j}$ are the engineering subfactors in Table 4-1. The product expansion is correct to the first-order derivatives $\partial(i)/\partial j$.

The sum form is

$$f_i^E = 1 + \sum_j \epsilon_{i,j} , \quad (4\text{-}32)$$

where $\epsilon_{i,j}$ is the fractional uncertainty $\Delta_j(i)/i_o$ on the quantities i introduced by a change in j and i_o is the nominal value of i. Thus,

$$\epsilon_{i,j} = f_i^{E,j} - 1 . \quad (4\text{-}33)$$

There are at present two methods of analysis, the conventional or maximum-value method and the statistical method.

4.3.1 Conventional Method

This method of combining the engineering hot-channel and hot-spot factors is the most conservative but is simple and in common use. Equation (4-30) is used to calculate the nuclear factor. Equation (4-31) or (4-32) is used to calculate the engineering factors, and the following is assumed:

1) All the variable quantities j are independent of one another.
2) The maximum values of the subfactors $f_i^{E,j}$ in Eq. (4-31) or maximum deviations $\epsilon_{i,j}$ in Eq. (4-32) all occur in the hot channel or at the hot spot.

The maximum value of the engineering factor F_i^E is obtained by the product of the maximum of the subfactors [Eq. (4-31)] (or the sum of the maximum errors [Eq. (4-32)]). The nuclear factor F_i^N is likewise obtained from Eq. (4-30). The total correction factor is thus

$$F_i^T = F_i^N \cdot F_i^E . \quad (4\text{-}34)$$

For example, the maximum fuel temperature at the hot spot for one-phase coolant flow is

$$T_{\text{fuel max}} = T_{\text{coolant inlet}} + \sum_i F_i^T \Delta T_i , \quad (4\text{-}35)$$

where ΔT_i is the temperature rise in the coolant, through the film, the oxide layer, the cladding, the thermal contact, and the fuel obtained from the nominal design.

4.3.2 Statistical Methods

These methods of combining the engineering hot-channel and hot-spot factors are more accurate but less conservative than the conventional method and take into account the statistical nature of most of the engineering subfactors. Several different approaches to the statistical analysis have been used, depending on the amount of information available on the statistical behavior of the variables and the type of reactor being analyzed. Some of the main features of the methods and assumptions made are outlined below. Details are given in the literature [137, 141, 142, 153, 157, 158, 159, 160, 161, 162].

The simplest and most justifiable method with the present state of knowledge is to assume that all the variables j defined in Eq. (4-31) are independent of one another and have a maximum and minimum value and to combine the maximum deviations statistically to obtain the deviations in the quantities i. The maximum and minimum values may be obtained by quality control (i.e., go—no-go gauges) or by statistical methods where no gauging is possible. The statistical propagation of errors analogous to the deterministic sum of Eq. (4-32) is

$$(\Delta_i)^2 = \sum_j [\Delta_j(i)]^2 . \qquad (4\text{-}36)$$

From these deviations, the values of enthalpy and temperatures at the hot spot are obtained:

$$i = i_0 F_i^N + \Delta_i ,$$

$$i = H, f, Ox, C, R, F, Q . \qquad (4\text{-}37)$$

It has been customary to assume that each of the errors $\Delta_j(i)$ has a mean value of zero and a normal (Gaussian) probability distribution in which the maximum deviation corresponds to a value three times the standard deviation σ. Therefore, the probability that the quantity i has a value given by Eq. (4-37) at the hot spot also has a normal distribution. The value of i corresponding to a deviation of $\Delta_i = 3\sigma$ has thus a 99.87% chance of not being exceeded.

Normally, in addition to the hot-channel factors, a safety margin is allowed above and beyond the corrected value of the parameter i. If there should be no safety margin or a very small one, the situation is different. In the example above, for instance, with no safety margin there is about one chance per thousand that the parameter i will exceed its safe value. If fuel melting were being considered, one element per thousand might fail. This could still be tolerable if the design criteria allowed for 1% fuel failures. On the other hand, if failures or melting of one element led to the progressive failure of others, the situation would be completely intolerable.

An example of this analysis is given in Table 4-4 [163] for the Enrico Fermi reactor. In this instance the centerline temperature of the fuel pin is the limiting consideration. The calculation refers to the pins in the central assembly where the flux is maximum. The left-hand column lists the various physical factors that give rise to uncertainties in this temperature. The next five columns list the extreme uncertainties introduced by these factors in the nominal temperature drop (given at the column head) across each of the five regions in the heat-flow path. Each extreme is expressed in two ways: The columns headed by F list the factors that multiply the nominal ΔT to give the "hot" ΔT; the columns

TABLE 4-4

Typical APDA Hot-Spot Calculation Using Statistical Method

Nominal values	Coolant temp. rise $\Delta T_c = 285°F$		Temp. drop thru film $\Delta T_f = 41.5°F$		Temp. drop thru oxide layer $\Delta T_o = 15.5°F$		Temp. drop thru clad $\Delta T_{Zr} = 39.8°F$		Temp. drop thru fuel $\Delta T_U = 196°F$		Summary	
											$\Sigma(3\sigma)$	$[\Sigma(3\sigma)]^2$
Physical uncertainty	F	3σ(°F)	F	3σ(°F)	F	3σ(°F)	F	3σ(°F)	F	3σ(°F)	(°F)	(°F)^2
Maldistribution of coolant												
(a) to subassemblies	1.15	42.8									42.8	1,831.8
(b) within subassemblies	1.03	8.6									8.6	74.0
Deviation from			1.04	1.7							1.7	2.0
nominal					2.00	15.5					15.5	240.3
dimensions							1.02	0.8			0.8	0.6
									1.03	3.9	3.9	15.2
Maldistribution of U^{235}	1.02	5.7	1.02	0.8	1.02	0.3	1.02	0.8	1.03	3.9	11.5	132.3
Maldistribution of flux	1.05	14.3	1.05	2.1	1.05	0.8	1.05	2.0	1.05	9.8	29.0	841.0
Burnup of core fuel pin	1.10	28.5	1.10	4.2	1.10	1.6	1.10	4.0	1.10	19.6	57.9	3,352.4
Power measurement and control	1.08	22.8	1.08	3.3	1.08	1.2	1.08	3.2	1.08	15.7	46.2	2,134.4
Film-heat-transfer coefficient			1.30	12.5							12.5	156.3
Thermal conductivity of zirconium oxide					1.20	3.1					3.1	9.6
Thermal conductivity of zirconium							1.10	4.0			4.0	16.0
Thermal conductivity of fuel alloy									1.20	39.2	39.2	1,536.6

Total $\Sigma[\Sigma(3\sigma)]^2$ = 10,343.4
(Sq. root of total = 3σ) = 101°F
Total hot-spot deviation = 101°F

Uranium hot spot located at fuel-alloy-length fraction X/L = 0.70
Nominal maximum uranium temperature (without hot-spot factor) = 1,134°F
Hot-spot factor 101

Maximum uranium temperature (99.87% confidence) 1,235°F

headed by 3σ list the difference (in °F) between the "hot" ΔT and nominal ΔT. The table shows that the same physical effect can cause the temperature rises in more than one of the five regions. When these temperatures are not statistically independent, as in the case of maldistribution of flux, the changes are added together to give the total temperature elevation for that particular physical effect in column $\Sigma(3\sigma)$. When the changes are statistically independent, as in the case of variations in fuel-element dimensions, each regional temperature rise is listed in the $\Sigma(3\sigma)$ column separately.

The APDA calculation yields a total hot-spot temperature rise of only 101°F compared to 277°F obtained by simply adding all the 3σ values. In terms of hot-spot factors the statistical method gives 1.2 while the conventional method gives 1.5.

Abernathy [160] calculated the maximum cladding temperature in the EGCR using a similar but more precise method. The maximum deviation, $\theta = T_{max} - T_{nom}$, of the cladding temperature is written to a first-order approximation as a sum of errors:

$$\theta = \sum_i \frac{\overline{\partial T_c}}{\partial u_i} \Delta u_i + \sum_i \frac{\overline{\partial T_c}}{\partial v_i} \Delta v_i , \qquad (4-38)$$

where Δu_i and Δv_i are the deviations from the nominal values of the dimensional and "process" (heat transfer, fission rate, etc.) quantities assumed independent of each other. The partial derivatives of the clad temperature T_c with respect to u_i and v_i are evaluated at the nominal conditions. According to the Central Limit Theorem of probability theory, the distribution of the sum of n independent random variables asymptotically approaches the normal distribution as n approaches infinity. Therefore, if the total number of random variables Δu_i and Δv_i is large, the frequency function for θ is given approximately by the normal frequency function. That is,

$$f_\theta(\theta) \cong (1/\sqrt{2\pi}\,\sigma_\theta)\exp(-\theta^2/2\sigma_\theta^2), \qquad (4-39)$$

where σ_θ^2 is the variance of the distribution. To determine the confidence limits for the maximum cladding temperature it is therefore not necessary, in general, to calculate f_θ explicitly but only the variance σ_θ^2. From Eq. (4-38) the variance σ_θ^2 is given by

$$\sigma_\theta^2 = \sum_i \left[\frac{\overline{\partial T_c}}{\partial u_i}\right]^2 \sigma_{u_i}^2 + \sum_i \left[\frac{\overline{\partial T_c}}{\partial v_i}\right]^2 \sigma_{v_i}^2 , \qquad (4-40)$$

where $\sigma_{u_i}^2$ and $\sigma_{v_i}^2$ are the variance of the frequency distribution of the deviations Δu_i and Δv_i. Since these distributions are not well known at present, it is assumed that the quantities u_i and v_i have equal probability to have values between a maximum and minimum deviation $\bar{u}_i - \Delta u_{i_{max}} \leq u_i \leq \bar{u}_i + \Delta u_{i_{max}}$. Thus, for this rectangular frequency distribution,

$$\sigma_{u_i}^2 = \frac{1}{3}\left[\Delta u_{i_{max}}\right]^2 \quad \text{and} \quad \sigma_{v_i}^2 = \frac{1}{3}\left[\Delta v_{i_{max}}\right]^2 . \qquad (4-41)$$

The probability that the cladding temperature will not exceed $T_{c,nom} + k\sigma_\theta$ is therefore 84.15% for k = 1, 97.72% for k = 2, 99.86% for k = 3, and 99.997% for k = 4.

Another approach is to use the product form [Eq. (4-31)] and combine all the factors statistically. One first determines from measurements or assumes a frequency distribution of the basic stochastic variables, e.g., rod diameter, fuel enrichment, etc. From this information the frequency distribution of each individual subfactor $f_c^{E,j}$ and of their product $f_i^E = \prod_j f_i^{E,j}$ is calculated. This frequency distribution is used to obtain the cumulative probability curve and the confidence limits for a particular hot-spot temperature or coolant enthalpy. The successive steps in this method and the usual approximations made in the analysis are illustrated in the following paragraphs.

a) The frequency distribution of the basic variables. Very few measurements have been made so far on rod diameter, lattice pitch, eccentricity of fuel, cladding thickness, etc. Abernathy [160] has used a rectangular distribution for all stochastic variables (see above). Chelemer and Tong [153] report that sample measurements taken during fabrication of the first Yankee reactor core show that variations in the fuel-rod fabrication tolerances are normally distributed and assume a normal distribution for all stochastic variables. Since the standard deviation is obtained from a finite sample size, a correction is made to the measured standard deviation in order to maintain a high confidence level in the statistical analysis. The confidence level of a statistically determined relationship is the fraction of the time the relationship is expected to be satisfied. Defining m and s as the mean and standard deviation determined from a sample size n taken from a normal distribution, the confidence parameter k is defined by the following probability equation:

$$P(x < m + ks) = P_k , \qquad (4-42)$$

which expresses the probability of the quantity x to be smaller than m + ks. The k values have been listed [163] for various confidence levels γ and the resulting probabilities P_k. For an infinite sample the values of m and s will approach μ and σ respectively, so that

$$P(x < \mu + k_\infty \sigma) = P_k , \qquad (4-43)$$

where k_∞ is the value of k for infinite n.

Comparison of the above two equations gives (using $m \simeq \mu$)

$$\sigma = \frac{m + ks - \mu}{k_\infty} = \frac{ks}{k_\infty} . \qquad (4-44)$$

b) The frequency distribution of the subfactors $f_i^{E,j}$. The subfactors $f_i^{E,j}$ are, in general, complicated functions of the basic variables, e.g., Eqs. 4-8b, c. When the variables are stochastic, it is sometimes permissible to assume that the subfactors are linear functions of these variables. A test for the validity of this assumption is as follows [153].

Let f(x) be a subfactor, a function of the variable x which has a mean value $\bar{x}$ and a standard deviation σ_x. If the following equalities are approximately satisfied,

$$f(\bar{x}) - f(\bar{x} - 2.58\sigma_x) \cong f(\bar{x} + 2.58\sigma_x) - f(\bar{x})$$

$$\cong 2.58\sigma \left[\frac{df}{dx}\right]_{x=\bar{x}} , \qquad (4\text{-}45)$$

the subfactor can be considered linear throughout 99% of the range of x. This test can be applied for functions of more than one variable by considering one variable at a time while holding the others constant. In most cases the standard deviation σ_x is small and the function may be considered linear in the small interval considered.

If a hot-spot or hot-channel subfactor f(x) can be considered as a linear function of its governing physical quantities x_i, each of which is assumed to have a normal distribution, then the subfactor has approximately a normal distribution. The value of the variance σ_f^2 for the subfactor is given by

$$\sigma_f^2 = \sum_i \left[\frac{\partial f}{\partial x_i}\right]^2 \sigma_{x_i}^2 , \qquad (4\text{-}46)$$

where $(\partial f/\partial x)^2$ is the partial derivative of f with respect to x at the nominal conditions and $\sigma_{x_i}^2$ the variance of the independent variables x_i determined earlier in this subsection. The above procedure was used by Chelemer and Tong [153] to determine the hot-channel and hot-spot factors for open-lattice pressurized-water reactors. The results of their calculations are given in Table 4-5 [153]. Although the authors give a confidence limit of 99.86% for each statistical subfactor, they do not determine the confidence limit of the product of the subfactors, i.e., the hot-spot and hot-channel factors. The following paragraph outlines methods to obtain this confidence limit.

c) The frequency distribution of the hot-spot and hot-channel factors. Referring to Eq. (4-28b), the hot-spot or hot-channel factors are the product of deterministic subfactors $f_i^{N,j}$ and stochastic factors $f_i^{E,j}$:

$$f_i^T = \prod_j f_i^{E,j} \cdot \prod_j f_i^{N,j} = f_i^N \cdot \prod_j f_i^{E,j} . \qquad (4\text{-}28b)$$

The deterministic subfactors $f_i^{N,j}$ are simply multiplied together and the frequency distribution of the factors f_i^T is determined from the frequency distribution of the subfactors $f_i^{E,j}$. In combining the subfactors it is necessary to consider correlation between subfactors. That is, it must be determined whether any single factor or group of factors is affected by any other factor, and if so, the manner in which it is affected. For example, consider the two engineering subfactors $F_H^{E,Ecc}$ and $F_H^{E,Bow}$ (given in Table 4-1) for the deviations on coolant enthalpy rise due to eccentricity of the fuel in the rod and due to bowing of the fuel element. The eccentricity of the fuel tends to increase the enthalpy rise on the thin side of the cladding but it also favors bowing of the element on the same side with an additional enthalpy rise due to flow-area restriction. The relationship between the two factors or in general between two dependent quantities, x, y, is described by the covariance σ_{xy} defined by

$$\sigma_{xy} = \frac{1}{n-1} \sum_{i=1}^{n} (x_i - \bar{x})(y_i - \bar{y}) , \qquad (4\text{-}47)$$

where $\bar{x}$ and $\bar{y}$ are the average values and x_i, y_i a set of n observed values. The quantities are said to be positively or negatively correlated if their

TABLE 4-5

Hot-Spot Subfactors for PWR Open-Lattice Rod Bundle

	Subfactor	Heat-flux factor F_q^E	Enthalpy-rise factor $F_{\Delta H}^E$	Film-temp.-drop factor F_θ^E	Probability of not being exceeded (99.9% conf. level)
Statistical	Pellet diameter, density, enrichment and eccentricity	1.041	1.037	1.041	0.9986
	Rod diameter, pitch and bowing[a]	1.004	1.10	1.130	0.9986
Nonstatistical	Inlet flow distribution	—	1.07	1.07	1.0
	Flow redistribution	—	1.05	1.05	1.0
	Flow mixing	—	0.95	0.95	1.0
	Heat-transfer correction	—	—	1.25	1.0
Total engineering hot-channel factor		1.045	1.22	1.57	~1.0

[a]This subfactor was evaluated using expected rather than measured deviations; therefore it has not been combined statistically with the other statistical subfactor.

covariance is positive or negative, respectively. Two quantities stochastically independent have a zero covariance, $\sigma_{xy} = 0$. Dividing the covariance by the product of the standard deviations, the coefficient of correlation is obtained:

$$r_{xy} = \frac{\sigma_{xy}}{\sigma_x \sigma_y}. \quad (4-48)$$

Rude and Nelson [158] suggest several methods for determining the frequency distribution of a product of subfactors applicable when the subfactors are independent of each other and when the subfactors are not independent of each other. These two cases are discussed in the following paragraphs.

c)1. The subfactors are independent of each other. The following multiplication procedure can be used to combine two factors at a time, and successive operations can be made to accommodate more than two factors. The multiplication procedure is illustrated by the following example.

Assume that the two factors F_1, F_2 have the probability distribution given in Table 4-6 [158]. The probability that $F = F_1 \cdot F_2$ exceeds A can be obtained by a sum of probabilities over the region above the curve $F_1 \cdot F_2 = A$ as shown in Fig. 4-8 [158]. That is,

$$P(F_1 \cdot F_2 > A) = \sum_{i=0}^{n-1} P(x_i < F_1 < x_{i+1}) \cdot P\left(F_2 > \frac{A}{x_{i+1}}\right). \quad (4-49)$$

Initially choose $A = 1.041 \cdot 1.0052 = 1.0464$. Plot the curve $F_1 \cdot F_2 = 1.0464$, choose a value for Δx such as $\Delta x = 0.005$. Beginning at the origin, obtain the product of the probability that F_2 exceeds its value on curve A (whose equation is $F_1 \cdot F_2 = A$) corresponding to $F_1 = x_1$ and the probability that F_1 is in the range x_0 to x_1; next obtain the product of probabilities for the strip lying between $F_1 = x_1$ and $F_1 = x_2$, etc. The sum of the products of probabilities obtained for all strips within the region above the curve A_1 is then approximately the probability that $F_1 \cdot F_2$ exceeds A. This procedure can be repeated for other values of A, so that sufficient data can be obtained to describe the variation of probability with increasing values of F.

When the two independent factors F_1 and F_2 have a normal distribution with a frequency function

$$f(x) = \frac{1}{\sqrt{2\pi}\,\sigma_x} \exp\left[-\frac{1}{2}\left(\frac{x-\mu_x}{\sigma_x}\right)^2\right], \quad (4-50)$$

TABLE 4-6

Hot-Channel Factors
Cumulative Probability

Probability of being exceeded	F_1	F_2
0.1	1.041	1.0052
0.01	1.064	1.0092
0.001	1.076	1.0118
0.0001	1.086	1.0138

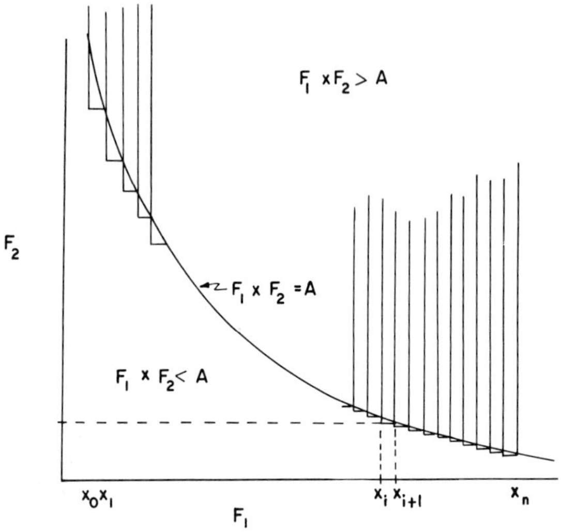

FIG. 4-8. The probability that $F = F_1 \times F_2$ exceeds A. Horizontal broken line represents $F_2(x_{i+1}) = A/x_{i+1}$.

μ_x and σ_x^2 being the mean and the variance of x, the probability

$$P(F_1 \cdot F_2 > A) = \int_0^\infty \int_{A/F_2}^\infty f(x)f(y)\,dx\,dy \quad (4-51)$$

can be written in the following form:

$$P(F_1 \cdot F_2 > A) = \sum_{i=1}^{\infty}\left[1 - \Phi\left(\frac{(A/F_{2,i}) - \mu_1}{\sigma_1}\right)\right] \times \frac{1}{\sigma_2}\,\Phi\left(\frac{(A/F_{2,i}) - \mu_2}{\sigma_2}\right) \cdot \Delta F_{2,i} \quad (4-52)$$

where μ_1, σ_1 and μ_2, σ_2 are the mean and standard deviation of F_1 and F_2 respectively. The range of the subfactor F_2 is divided into small intervals $\Delta F_{2,i}$ where $F_{2,i}$ is the value of F_2 in the interval i.

The functions $\phi(x)$ and $\Phi(x)$ are

$$\phi(x) = (1/\sqrt{2\pi}) \exp(-x^2/2), \quad (4-53)$$

$$\Phi(x) = \int_{-\infty}^{x} \phi(t)\,dt. \quad (4-54)$$

There is an approximate procedure for combining factors which gives very nearly the correct answer. This method uses an extension of Eq. (4-46) to the case of a product of subfactors f_i.

From Eq. (4-28b),

$$f_i^T = f_i^N \cdot \prod_{j=1}^{n} f_i^{E,j}. \quad (4\text{-}28b)$$

To a first-order approximation,

$$\left(\frac{\sigma_T}{\mu_T}\right)^2 = \sum_{i=1}^{n}\left[\frac{\sigma_i}{\mu_i}\right]^2 + 2\sum_{i=1}^{n-1}\sum_{j=i+1}^{n} r_{i,j}\left[\frac{\sigma_i}{\mu_i}\right]\left[\frac{\sigma_j}{\mu_j}\right]. \quad (4-55)$$

σ_T, μ_T and σ_i, μ_i are the standard deviation and mean value of the factor f_i^T and of the n subfactors $f_i^{E,j}$, respectively. $r_{i,j}$ is the coefficient of correlation defined by Eq. (4-48). For a stochastically independent subfactor, $r_{i,j} = 0$.

A product of normally distributed variables may be assumed to be normally distributed for small variations of the variables [164]. Thus, knowing the mean value of f_i^T,

$$\left(\overline{f_i^T}\right) = \mu_T = f_i^N \cdot \prod_{j=1}^{n} \left(\overline{f_i^{E,j}}\right) \tag{4-56}$$

and the variance from Eq. (4-55), the frequency distribution and confidence limits of the hot-spot, hot-channel factors f_i^T can be computed.

When the subfactors and their natural logarithm can be assumed to have a normal distribution, Moody [161] suggests a "lognormal approximation" whereby the product of subfactors is transformed into a sum of log terms and the mean and variance of the sum obtained.

c)2. The subfactors are not independent of each other. If two subfactors $f_i^{E,1}$ and $f_i^{E,2}$ are correlated with a coefficient of correlation $r_{1,2}$, then the probability that the product $f_i^E = f_i^{E,1} \cdot f_i^{E,2}$ exceeds an arbitrary value A is given by a double integral:

$$P\left(f_i^E > A\right) = \int_0^{\infty} \int_{A/y}^{\infty} f(x, y)\,dx\,dy \tag{4-57}$$

where f(x,y) is the joint frequency function for the two variables. If $f_i^{E,1}$ and $f_i^{E,2}$ have a bivariate normal frequency function with means μ_1 and μ_2, standard deviation σ_1, σ_2, and correlation $r_{1,2}$, then, denoting $f_i^{E,1}$ and $f_i^{E,2}$ by x and y, respectively,

$$f(x, y) = \frac{1}{2\pi\sigma_1\sigma_2\sqrt{1 - (r_{1,2})^2}} \times \exp\left\{-\frac{1}{2\left[1 - (r_{1,2})^2\right]}\left[\left(\frac{x - \mu_1}{\sigma_1}\right)^2 - 2r_{1,2}\left(\frac{x - \mu_1}{\sigma_1}\right)\left(\frac{y - \mu_2}{\sigma_2}\right) + \left(\frac{y - \mu_2}{\sigma_2}\right)^2\right]\right\}. \tag{4-58}$$

The integration in Eq. (4-57) can be performed by first integrating with respect to x, then numerically integrating with respect to y.

The approximate method using the coefficient of variation Eq. (4-55) is also applicable when the variations in the subfactors are small.

Rude and Nelson [158] compare the probability for the product F_1 and F_2 (referred above in Table 4-6) to exceed 1.21, using Eq. (4-49) or the approximation Eq. (4-55), with assumption of a normal distribution and different coefficients of correlation ($r_{1,2}$ = 1, 1/2, 0, -1/2). The results are quoted in Table 4-7.

TABLE 4-7

Comparison of Cumulative Probability for the Product of Two Subfactors $F_1 \times F_2 > 1.21$ Using a Different Method and Coefficient of Correlation: The Probability of Each Factor Exceeding 1.10 is 0.025

Equation and correlation	$P(F_1 \times F_2 > 1.21)$
(4-46) $r_{1,2} = 0$	0.0023
(4-46) $r_{1,2} = 1$	0.025
(4-52) $r_{1,2} = 0$	0.0015
(4-52) $r_{1,2} = 1/2$	0.008
(4-52) $r_{1,2} = -1/2$	0.00002

Only in the case of perfect positive correlation $r_{1,2} = 1$ is the probability 0.025 the same as the probability on each factor. In all other cases the probability that the product exceeds 1.21 is smaller. The coefficient of correlation has a great bearing on the confidence limit of the hot-spot and hot-channel factors. But practically no information is yet available on these coefficients for the various subfactors.

Judge and Bohl [165] examined the effects of power flattening in the hot-channel analysis. They suggest that for an equal nonfailure probability a reactor which is operating with a flat power distribution will also have to operate at a lower total power level. H. Guéron [166] has pointed out that the present method of statistical analysis, which determines the probability of failure of only the hot spot or hot channel, is not correct. A more rational analysis should include the probability of failure of all the other channels and all points in those channels. Reference [167] presents a method with an example based on assumed individual-channel-failure probabilities. In References [168] and [169] the Monte Carlo method is used in order to determine the probability distributions of the physical characteristics of interest. A great number of core calculations are performed with values of the basic variables selected at random according to their own distribution. The extreme values of the physical characteristics for each core are then compiled, and their distribution is thus established. These extreme values do not necessarily arise at the nominal hot spots of the design. This method has the advantage of taking into account the probability distribution of the individual points of each channel.

4.3.3 The Synthesis Method of Uncertainty Analysis

Because the conventional method of combining the hot-spot, hot-channel factors is overconservative and the statistical method considers probability only at the hot spot, there was a need for a more rational and generalized statistical method.

This method, developed recently [170], is called the "synthesis method." The procedure is to divide the reactor core into a number of segments (having a natural characteristic length, e.g., pellet height). The statistical method is then applied to each of the characteristic lengths; and the individual probability q_i ($T_i > T^*$) that a certain length i does not exceed a critical temperature T* is determined. The overall core-failure probability is then given by

$$Q(T_{max} > T^*) = 1 - \prod_i [1 - q(T_i > T^*)]$$

This method has been applied successfully to the case of a gas-cooled reactor [170] and is general enough to be adapted to any type of reactor.

4.4 Experimental Determination of Hot-Channel and Hot-Spot Factors

A basic characteristic of the hot-spot and hot-channel analysis is that, in order to trust the results of the computation, the basic information needed for this analysis should be provided at various stages of the analysis from measurements or observations. Following through the method described above, it is clear that many types of information are required.

When the subfactors can be expressed in terms of the basic variables, it is necessary to make measurements over a large sample to determine the variations in those basic variables and compute the corresponding values of the subfactors. For instance, for the subfactor $F_H^{E,\Delta dF}$, Eq. (4-9) requires the knowledge of the maximum pellet diameter and, when a statistical method is used, the probability distribution of the pellet diameter.

When the subfactors describe differences due to situations which cannot be expressed mathematically, a direct measurement of the values of the subfactors needs to be made. In the case of a statistical analysis a large number of measurements is required to simulate all possible situations and to obtain a statistical distribution of the subfactors. See for instance the reported [150] determination of the subfactor $F_H^{E,W}$.

One may also choose to determine experimentally the overall hot-spot or hot-channel factor in a mockup assembly or a full-size reactor. As there is no assurance that the maximum temperature recorded is the maximum possible temperature to be found in the experimental mockup or in any full-size core built subsequently, an analysis is required to determine the confidence limit of these measurements. Some of the work done in the areas mentioned above is outlined in the following section.

4.4.1 Basic Physical Quantities

These include measurements, during fabrication of the fuel elements and fuel assemblies, of the fuel enrichment, fuel diameter, inside and outside cladding thickness, eccentricity of the cladding, pitch between tubes in lattice, etc. Very few of these measurements have been made and reported in the literature. Measurements on pellet diameters, cladding thickness, and fuel weight for a core loading of the VBWR assembly, mentioned earlier in this section (Table 4-3), have been reported [148]. Sample measurements taken during fabrication of the first Yankee core have been described [171]. Measurements of the zirconium-clad uranium, 10 wt.% molybdenum fuel pins of the Enrico Fermi reactor used for irradiation testing in the CP-5 have also been reported and are shown in Table 4-8 [172].

4.4.2 The Subfactors

To determine the distribution of coolant flow in the plenum chamber of water- and gas-cooled reactors, quarter-size models have been built and flow distribution measured.

Results of such experiments for the Pathfinder reactor [146] are shown in Fig. 4-5 of this section. The measured value of $F_H^{E,W}$ was found to be 1.05. A similar experiment was made for the Partially Enriched Gas-Cooled Power Reactor [150] and a value of 1.014 for the same factor is reported.

Heat- and mass-transfer experiments have been made for clusters of parallel gas-cooled fuel rods and values of $F_f^{E,h}$ obtained. Typical results have been reported [153] and commented upon earlier in this chapter (see Fig. 4-7). Results of thermal tests on the AGCR have been given [173]. Values of the ratio h_{loc}/h_{av} on the circumference of an element in the outer circle of the bundle are shown in Fig. 4-9 [173]. Effect on heat transfer of spiral wire fuel elements has been reported [174].

Measurements of flux distributions and their comparison with calculated flux are numerous. Of particular interest is the error introduced by the method used for the physics analysis. Figure 4-10 [175] shows a comparison of the experimental and calculated power distribution in a three-zone core (1.6%, 2.7%, 3.7% enrichments) for two- and four-energy group computations [175]. Figure 4-4 shows also an example of errors introduced by the physics model.

4.4.3 Overall Hot-Spot and Hot-Channel Factors

The overall factors are measured in mockup critical assembly or full-size core. As pointed out earlier, in spite of the directness of these measurements one has to ascertain that a given set of data represents the most likely conditions in the power reactor. Hitchcock [176] analyzed this problem of experimentally determining the maximum possible fuel-element temperature. The analysis is summarized here.

If n wall temperatures, T_{wall}, are measured, the average wall temperature $\overline{T}$ and the variance v^2 can be computed as follows:

$$\overline{T} = \frac{1}{n} \sum_{i=1}^{n} T_{wall}(i), \tag{4-59}$$

TABLE 4-8

Pre- and Postirradiation Diameters of CP-5-1 -2 and -3 Test Pins

(Testing of Fermi Reactor prototype fuel pins in CP-5)

Length (in.) from top of pin	CP-5-1		CP-5-2		CP-5-3	
	Pre	Post	Pre	Post	Pre	Post
0	.1610	.159EC[b]	.163	.163EC	—	—
1	—	—	—	—	.1588	.1595
	—	—	—	—	.1600[a]	.1605[a]
2	—	.159	—	—	.1590	.1622
	—	—	.159	.158	.1588[a]	.1618[a]
3	—	—	—	—	—	—
	.1592	—	—	—	—	—
4	—	.161	—	—	.1591	.1609
	—	—	—	—	.1591[a]	.1605[a]
5	—	—	—	—	—	—
6	—	.171	—	—	.1590	.1605
	.1590	—	.159	.158	.1593[a]	.1608[a]
7	—	—	—	—	—	—
8	—	—	—	—	.1592	.1608
	—	.180	—	—	.1594[a]	.1612[a]
9	—	.181	—	—	—	—
	.1591	—	—	—	—	—
10	—	—	—	—	.1592	.1615
	—	.180	.159	.159	.1592[a]	.1615[a]
11	—	—	—	—	—	—
12	—	.174	—	—	.1587	.1610
	.1591	—	—	—	.1587[a]	.1610[a]
13	—	—	—	—	—	—
14	—	.168	—	—	.1587	.1610
	—	—	.159	.159	.1588[a]	.1610[a]
15	.1590	—	—	—	—	—
16	—	—	—	—	.1588	.1612
	—	.167	—	—	.1588[a]	.1614[a]
17	—	—	—	—	—	—
18	—	.167	—	—	.1590	.1610
	.1590	—	.159	.159	.1588[a]	.1608[a]
19	—	—	—	—	—	—
20	—	.167	—	—	.1587	.1605
	—	—	—	—	.1587[a]	.1605[a]
21	—	—	—	—	—	—
	.1590	—	—	—	—	—
22	—	.166	—	—	.1587	.1601
	—	—	.159	.159	.1588[a]	.1601[a]
23	—	—	—	—	—	—
24	—	.166	—	—	.1588	.1601
	.1591	—	—	—	.1588[a]	.1600[a]
25	—	—	—	—	—	—
26	—	.164	—	—	.1588	.1593
	—	—	.159	.159	.1588[a]	.1600[a]
27	—	—	—	—	—	—
	.1590	—	—	—	—	—
28	—	.163	—	—	.1588	.1595
	—	—	—	—	.1588[a]	.1595[a]

(continued on next page)

TABLE 4-8 (Continued)

Length (in.) from top of pin	CP-5-1		CP-5-2		CP-5-3	
	Pre	Post	Pre	Post	Pre	Post
29	—	.166	—	—	.1590	.1593
30	.1610 .1500EC	— —	.161 —	.162EC —	.1588 [a] .1585	.1595 [a] .1593
31	—	—	—	—	.1589 [a] EC	— EC
32	—	—	—	—	—	—
Length	31	31-5/8	30-13/16	30.9	30-29/32	31
Density	15.66 [c]	13.40	15.6	14.8	15.69	15.33

[a] Measurements 90° apart at same height.
[b] EC denotes eccentricity.
[c] CP-5-1 control pin.

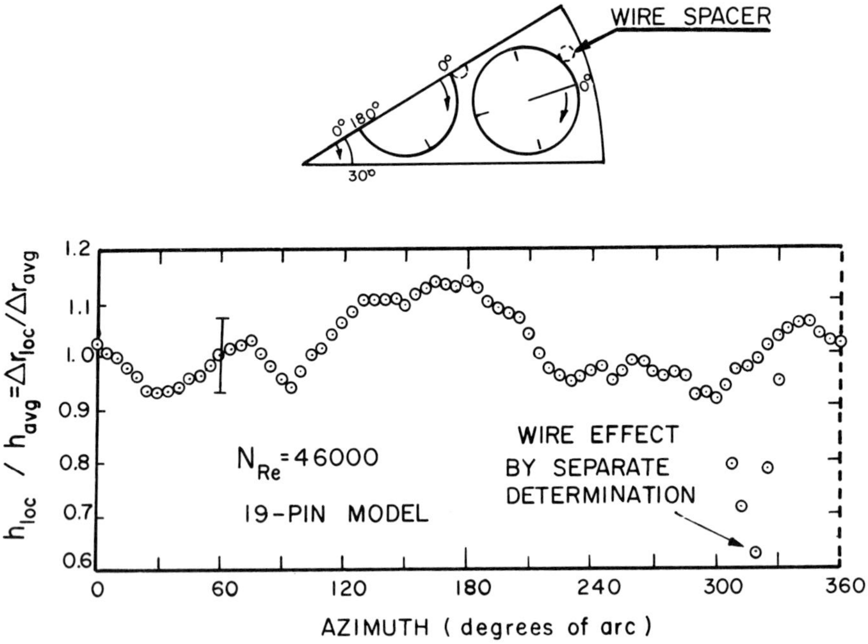

FIG. 4-9. Heat-transfer coefficient for the outer tube of a bundle of parallel tubes for the Army Gas-Cooled Reactor. Upper sketch shows a 30° sector of bundle and indicates positions of origins of azimuthal coordinates.

$$v^2 = \frac{1}{n} \sum_{i=1}^{n} [T_{wall}(i) - \overline{T}]^2 . \qquad (4\text{-}60)$$

In order to estimate the probability that some temperature in the reactor exceeds the measured average $\overline{T}$ by as much as some amount ΔT, one must take into account not only the possible deviations of all the temperatures from the average $\overline{T}$, but also the probable amount by which the measured average, $\overline{T}$, deviates from the true but unknown average and v^2 deviates from the true variance σ^2, which would result from a very large number of observations.

If p is the probability that the maximum temperature in any particular channel exceeds some temperature T_p, then the relation between T_p and p may be written

$$T_p = \overline{T} + f(p, n)v . \qquad (4\text{-}61)$$

A short table of f(p,n) which is reproduced here as Table 4-9 [176] has been published by Hitchcock. Extended tables of the function f may be obtained from the author.

If the total number of fuel channels is N, then the probability P(N,p) that any one of N

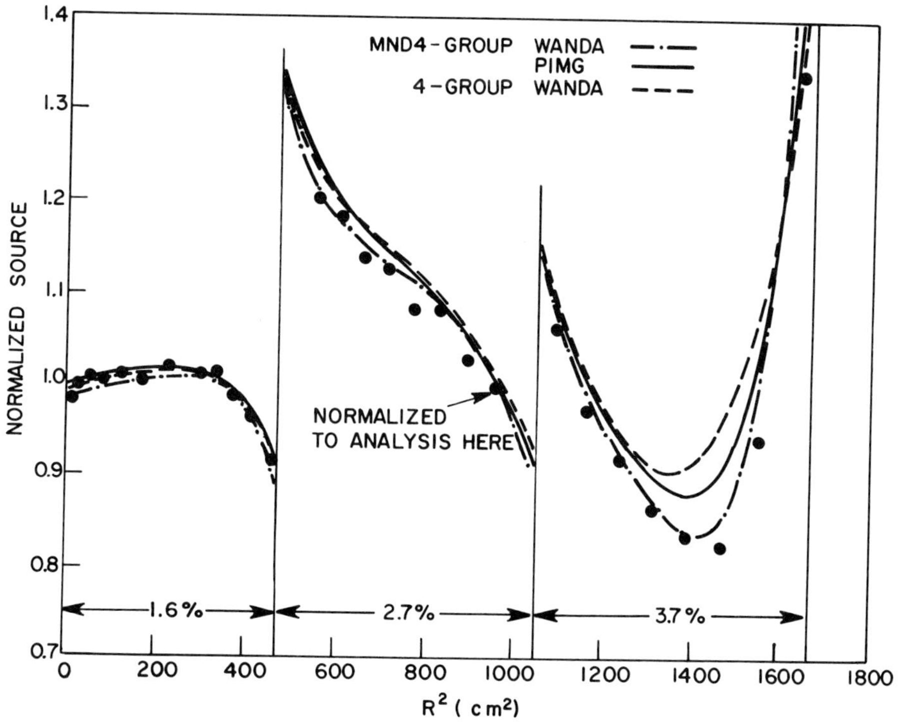

FIG. 4-10. "Core B", three-region cylindrical core, water-to-uranium ratio = 2.5. The stainless-steel-clad UO_2 fuel rods are enriched to 1.6%, 2.7%, 3.7% U^{235}. Loading from center: 1321 - 1.6%, 1604 - 2.7%, 1728 - 3.7%.

TABLE 4-9

Table of Values of f (p, n)

	$p = 10^{-3}$	$p = 10^{-4}$	$p = 10^{-5}$
10	5.28	7.55	10.4
20	3.90	5.02	6.18
30	3.59	4.50	5.39
40	3.45	4.28	5.06
50	3.38	4.16	4.88
60	3.32	4.08	4.77
80	3.26	3.96	4.63
100	3.22	3.92	4.52
Limit	3.09	3.72	4.24

elements has a surface temperature greater than T_p is

$$P(N, p) = 1 - e^{-pN} . \qquad (4\text{-}62)$$

If N were 4000, the following probabilities would be obtained:

p	P(N,p)
10^{-3}	0.982
10^{-4}	0.33
10^{-5}	0.04

In the above discussion, only random errors normally distributed have been considered. In some cases, however, systematic errors in temperature measurement are present and the reference gives a method for the inclusion of these systematic variations. All experimental tests on maximum cladding or fuel temperature have not been subjected to the above analysis.

Several sets of measurement of the fuel surface temperature are reported [177] for the MGCR gas-cooled reactor. The tests include a uniformly heated bundle coated with temperature-sensitive paints, a uniformly heated bundle with thermocouples, and a radially varying heated bundle with thermocouples. The surface-temperature hot-channel factor f was defined as

$$f = \frac{(\text{measured surface temperature}) - (\text{inlet temperature})}{(\text{calculated surface temperature}) - (\text{inlet temperature})}. \qquad (4\text{-}63)$$

The location of the rods A,B,C,D are shown on Fig. 4-11 and the value of f as a function of radial position is given in Table 4-10 [177].

Experimental data have been published on the fuel-plate temperature for a mockup of the Army gas-cooled-reactor systems program [178]. One set of data are shown in Fig. 4-12 [178] together with the calculated temperature using a computer code (TEMP I). The experimental points were measured by thermocouples. The predicted hot-spot temperatures for a range of gas flows were generally 50°F higher than the measured values.

OMRE fuel-plate surface-temperature measurements have been reported [179]. The authors report that the experimentally determined surface-

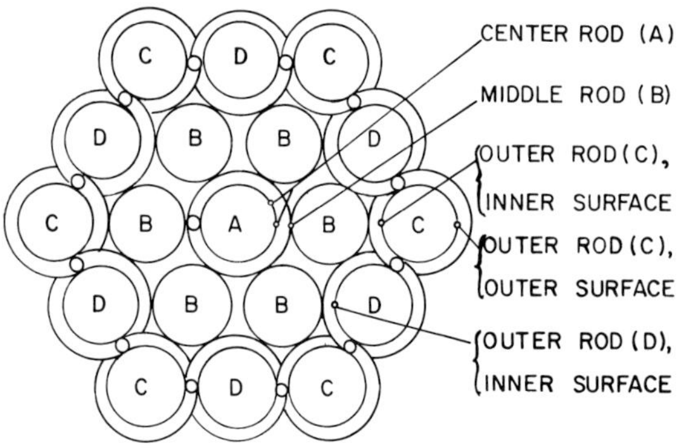

FIG. 4-11. Maritime Gas-Cooled Reactor: temperature measurements in model of rod-bundle element.

TABLE 4-10

Hot-Channel Factor as a Function of Radial Position in a Tube Bundle for the Maritime Gas-Cooled Reactor

Location[b]	Conditions[a] 1	2	3a	3b	3c	Relative power
Center rod, A	1.29	1.21	0.92	1.05	1.19	1.00
Middle rod, B	1.32	(c)	(c)	(c)	(c)	1.14
Outer rod, C						1.60
Inner surface	1.04	1.02	1.14	1.06	1.06	
Outer surface	1.14	0.93	1.03	0.96	0.89	
Outer rod, D						1.42
Inner surface	1.08	1.08	1.10	1.08	1.14	

[a] Notes:

1. Values as determined from temperature-sensitive paint with uniform heat flux.
2. Values as determined from thermocouple measurements with uniform heat flux.
3. Radial power distribution set by physics calculations (temperature measured with thermocouples):

   a. Calculated temperature based on average conditions.
   b. Calculated temperature based on average coolant temperature and local film temperature drop.
   c. Calculated temperature based on no-mixing coolant temperature and local film temperature drop.

[b] See Fig. 4-11.
[c] No measurement.

temperature measurement error was greater than the theoretically predicted error. The temperature error due to the conduction effect in the thermocouple wires can be approximated analytically. Calculation of this error indicates that at reactor operating conditions of 750°F surface temperature and 600°F coolant temperature, the surface temperature indicated by thermocouples would read 24°F too low. The complex of thermal effects attributable to the thermocouple assembly indicated the need for an experimental calibration under reactor heat-transfer conditions.

4.5 Tables of Hot-Spot Hot-Channel Factors for Different Types of Reactors

Tables 4-4, 4-5, and 4-10 through 4-15 of this chapter are given to complete the discussion of

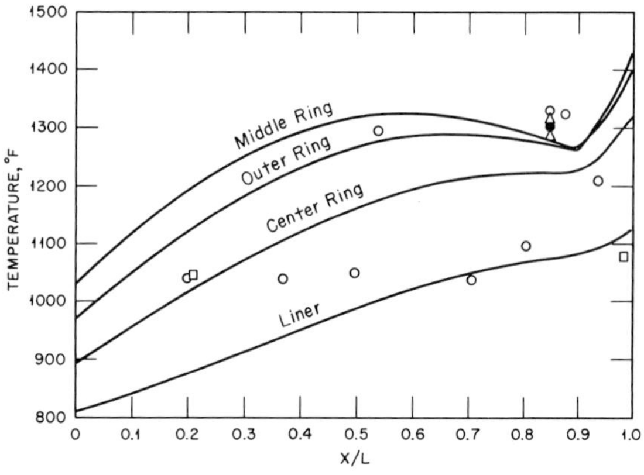

FIG. 4-12. Experimental Gas-Cooled Reactor: fuel-plate surface temperatures (in mockup studies) versus relative axial position for gas flow rate of 1020 lb/hr. Ratio of power generation in fuel pins, central: middle: outer rings is 0.532:1.00:1.44. The open circles indicate experimental data for outer ring, triangles for intermediate ring, filled-in circle for center pin, and squares for inner liner. Test conditions: 70.1 hr operation, 27.0 kw power, 798°F(376°C) inlet gas temperature. Solid curves are computed using TEMP-1 code.

TABLE 4-11

Army Gas-Cooled Reactor Experiment (GCRE-I) Hot-Channel and Hot-Spot Factors

Cause of deviation from nominal conditions	Film factor	Bulk factor	Fuel factor
Fuel considerations			
Enrichment variation	1.002	1.002	1.002
Density, variation	1.015	1.015	1.015
Pellet-diameter variation	1.011	1.011	1.011
Central plug ID variation	1.015	1.015	1.015
Pellet-stack-length variation	—	1.011	—
Reasonable combined effect	1.02	1.027	1.02
Neutron-flux considerations			
Intracell-flux depression	1.05	—	1.05
Control-rod-flux averaging	1.01	1.01	1.01
Uncertainty in flux calc. or meas.	1.03	1.10	1.03
Flow distr. due to control-rod insert	1.008	1.01	—
Flux depression across a pin	1.091	—	—
Combined effect	1.20	1.12	1.092
Metal-fabrication considerations			
Surface-roughness variation	—	1.016	—
Cladding OD variation	1.008	1.030	—
Linear ID variation	1.006	1.030	—
Orifice tolerances	1.033	1.005	—
Spacer tolerances	?	?	—
Pin-spacing variation	1.021	—	—
Reasonable combined effect	1.055	1.040	—
Heat-transfer considerations			
Uncertainty in film correlation	1.10	—	—
Local velocity effects (including intracell thermal conductivity)	1.05	—	—
Insulation-conductance uncertainty	1.008	1.008	1.008
Fuel-conductivity uncertainty (for kal)	—	—	1.200
Gap-conductance uncertainty	—	—	1.030
Combined effect	1.162	1.008	1.245
Fluid-flow considerations			
Uncertainty in pressure loss calcs.	1.02	1.025	—
Combined effect	1.02	1.025	—
Reactor-operation considerations			
Deviation from design power	1.150	1.150	1.150
Reasonable combined effect	1.080	1.080	1.080
Overall combined effect	1.660	1.340	1.500

TABLE 4-12

Different Methods of Calculating Hot-Spot Temperature for the Army Gas-Cooled Reactor Experiment (GCRE-I)

Calculation Method	$F_{\Delta t}$	F_0	$T_{s,\ max}$, °F
a) Product $\prod_{i=1}^{n} F_i$	1.83	1.53	1960
b) Weighted product $\prod_{i=1}^{m} F_i \left[1 + \frac{1}{2} \prod_{i=m+1}^{n} (F_i - 1)\right]$	1.66	1.34	1839
c) Weighted statistical $\prod_{i=1}^{m} F_i \left[1 + \sqrt{\sum_{m+1}^{n} (F_i - 1)^2}\right]$	1.57	1.30	1792
d) Statistical $1 + \sqrt{\Sigma (F_i - 1)^2}$	1.21	1.18	1612
e) Ideal	1.00	1.00	1478

$P_o = 3.0$ Mw $A_{flow} = 44.5$ in.^{2}

$W_o = 26.4$ lb/sec $P_{loss} = 0.88$ Mw

$D_e = 0.1545$ in. $A_{HT} = 133.5$ ft^2

Maximum hot spot occurs at $X/L_H = 0.5$.
Pressure drop is independent of power in the range investigated.
Film hot-spot factor is 1.66.
Bulk hot-spot factor is 1.34.
Friction factor for 48 bearing spacers.

this subject and as general information. However, the reader should not use the values in these tables without discrimination.

The Army Gas-Cooled-Reactor Experiments

The gas-cooled-reactor experiment I (GCRE-I) provides data on fuel-element lifetime, dynamic reactor behavior and control characteristics. It is a water-moderated heterogeneous reactor. The fuel is a dispersion of fully enriched UO_2 in 316 stainless steel with a 6-mil clad of 318 stainless steel. The values of the GCRE-I, hot-spot and hot-channel factors are given in Table 4-11 [178]. Several methods in combining the subfactors lead to a series of estimates of the hot-spot temperatures as shown in Table 4-12 [178]. The first temperature (1960°F) results from the unlikely equal superposition of all factors. The result from the simple statistical propagation of errors (1612°F) is too low because the hot-spot factor includes some effects that definitely superimpose (correlated errors). The preferred choice leads to a hot-spot temperature of 1839°F.

The gas-cooled-reactor experiment II (GCRE-II) is designed to provide more power per pound

TABLE 4-13

Army Gas-Cooled Reactor Experiment (GCRE-II) Hot-Channel Factors for a Clad Core with Circular Coolant Channels

A. Neutron- and gamma-flux distribution			
1. Axial-distribution uncertainty	1.05	1.05	1.05
2. Radial-distribution uncertainty	1.05	1.05	1.05
3. Local flux variation	1.02	1.04	1.04
B. Heat-flux nonuniformities			
1. Fuel homogenity	1.03	1.04	1.04
2. Fuel tolerances	1.001	1.001	1.001
3. Cladding variations	1.002	—	1.01
4. Eccentricity	—	—	—
C. Heat-transfer parameters			
1. Uncertainty in h correlation	—	1.10	—
2. Channel variations on h	—	1.002	—
3. Uncertainty in k	—	—	1.08
D. Mechanical tolerances			
1. External effects on mass flow	1.005	—	—
2. Plenum effects on flow	1.04	—	—
3. Channel variations on flow	1.015	—	—
4. Roughness factor on flow	1.003	—	—
E. Miscellaneous			
1. Irradiation effects	1.01	1.01	1.08
2. Corrosion effects	1.01	1.01	—
Combined factors	1.26	1.34	1.40

TABLE 4-14

Maximum Variation of Variable in Thermal Analysis for the Experimental Gas-Cooled Reactor

Gas temperatures at inlet to core	± 40°F
Macroscopic fission cross section of fuel	± 2%
Radial peak-to-average flux ratio	± 1%
Axial peak-to-average flux ratio	± 2%
Cross-sectional area of fuel pellets	± 1.5%
Graphite sleeve inside diameter	± 0.2%
Stainless steel tubing outside diameter	± 0.8%
Nominal heat-transfer coefficient	± 10%
Observed gas outlet temperature	± 25°F
Thermocouple error	± 25°F

TABLE 4-15

Boiling-Water-Reactor Hot-Channel Factors Comparison of Design Variables

Design variable	By probability analysis			By conventional selection
	Min.	Mean	Max.	
Overall design peaking factor	1.74	2.79	3.53	3.50[a]
Hot-channel factor				
With radial peaking	1.16	1.40	1.64	1.79[a] (including 1.09 overpower)
Without radial peaking	1.04	1.20	1.36	1.38[a] (including 1.09 overpower)
Minimum burnout ratio	1.42	2.12	2.82	1.53

[a] Factors selected for determining the initial core parameters are presented in Table 1 of Reference [145].

of reactor, to have longer fuel-element lifetime, and to operate at higher temperatures and pressures than GCRE-I. It is a semihomogeneous graphite reactor. The fuel consists of a dispersion of UO_2 in a graphite matrix clad with 30-mil Hastelloy-X tubes. The hot-spot and hot-channel factors are given in Table 4-13 [178]. These factors have been used for initial design purposes and will require modification after more experimental data becomes available.

A statistical analysis of hot-spot temperature in EGCR fuel assembly is given in [147] using Tingey's method of error propagation. The maximum deviation of the variables used in the thermal analysis are given in Table 4-14 [147]. The standard deviation of the surface temperature σ was found to be 35.6°F. The corresponding probability that the hottest spot on the fuel-element surface is at a temperature above the indicated value is as follows:

Temperature (°F)	Probability of being above this temperature
Nominal	0.50
Nominal + 20	0.29
" + 40	0.13
" + 60	0.046
" + 80	0.012
" + 100	0.0025
" + 120	0.00038

These probabilities were based on early flux profile determinations and preliminary estimates of tolerances.

Values of hot-spot and hot-channel factors used in pressurized-water reactors and in the Enrico Fermi reactor (fast, sodium-cooled) have been mentioned earlier in this section and given in Tables 4-5 and 4-4 respectively.

A statistical analysis of hot-spot hot-channel factors for boiling-water reactors has been made by Moody [158]. Probability functions and cumulative distributions, using reference core parameters, were computed for the overall design peaking factor, the hot-channel factor, and the minumum burnout ratio. The probability curves of these three variables are shown in Figs. 4-13, 4-14, and 4-15. The corresponding 99% confidence values are listed in Table 4-15 [180] and are compared with the values used in the parametric calculations for the 300 Mw(e) high-power-density conceptual core.

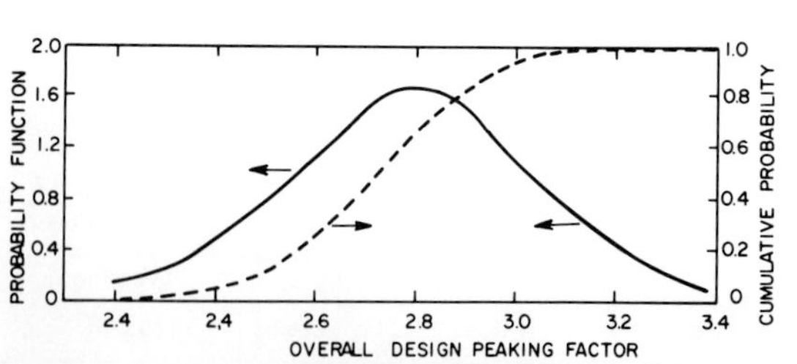

FIG. 4-13. Probability (solid curve) and cumulative probability (broken curve) versus overall design peaking factor (300 Mw, high power density conceptual design study).

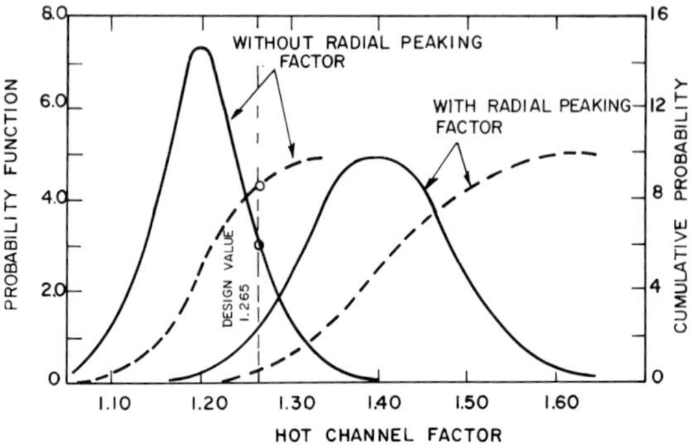

FIG. 4-14. Probability versus hot-channel factor (300 Mw, high power density conceptual design study). Solid curves are probability functions with and without radial peaking factor; broken curves are corresponding cumulative probilities.

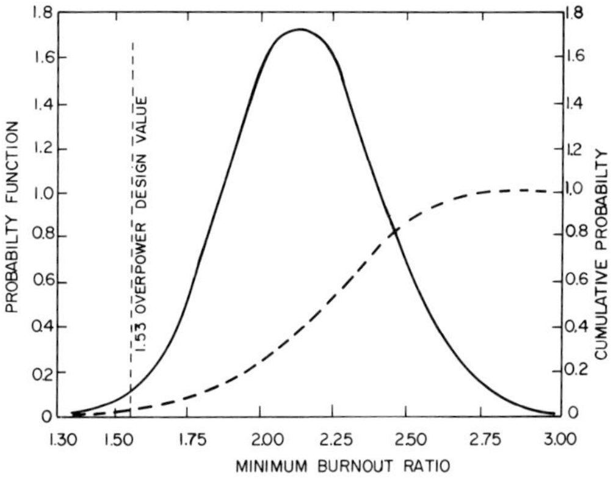

FIG. 4-15. Probability (solid curve) and cumulative probability (broken curve) versus minimum burnout ratio (300 Mw, high power density conceptual design study).

NOMENCLATURE

A	area; constant
A_c	area of solid contact between surfaces
a	radius of idealized heat channel at a contact point; accommodation coefficient (see Eq. (1-45))
B	fluid thickness number (see Eq. (1-28))
Bow	superscript for fuel element bowing (see Table 4-1)
C	constriction number (see Eqs. (1-25) and (1-26)); constant; subscript for clad (see Table 4-1)
C_T	temperature coefficient of reactivity (see Eq. (2-30))
C_V	void coefficient of reactivity (see Eq. (2-30)); volume heat capacity
Cf	superscript for flow mixing and redistribution (see Table 4-1)
c	radius of an idealized contact point
c_p	specific heat at constant pressure

D	diameter
D_e	equivalent diameter
E	modulus of elasticity; function defined by Eq. (2-30); function defined by Eq. (3-15); total energy released during power excursion
Ecc	superscript for fuel eccentricity (see Table 4-1)
Er	superscript for coolant parameter variations (see Table 4-1)
e	fuel enrichment
F	subscript for fuel (see Table 4-1)
F_i^E	engineering hot-spot or hot-channel factor applied to quantity i (equal to sum of appropriate subfactors)
F_i^N	nuclear hot-spot or hot-channel factor applied to quantity i (equal to sum of appropriate subfactors)
$F_i^{E,j}$	engineering hot-spot or hot-channel subfactor applied to the quantity i and caused by a variation of j (see Table 4-1)
$F_i^{N,j}$	nuclear hot-spot or hot-channel subfactor applied to quantity i and caused by a variation of j (see Table 4-1)
F_i^T	total hot-spot or hot-channel factor applied to quantity i
FA, FE	superscripts for fuel assembly and fuel element (see Table 4-1)
f	fraction of theoretical density; friction factor; (as subscript) film; thermal utilization (Eq. (3-8))
f_w^T	local correction factor associated with thermal parameter w (see Sec. 4.3)
f_i^E, f_i^N	engineering and nuclear factors, respectively, associated with thermal parameter i and other variables characterizing specific core (see Sec. 4.3)
$f_i^{E,j}$, $f_i^{N,j}$	subfactors of engineering and nuclear factors, respectively, (see Eqs. (4-30) and (4-21))
G	mass flow rate ($G = w/S = \rho V$)
G_V	gas mass flow rate
g	gravitational acceleration
g_o	constant, 32.17 lb_m ft/lb_f-sec^2 or 4.17×10^8 lb_m ft/lb_f-hr^2
H	enthalpy
h	heat-transfer coefficient; enthalpy
h_c	thermal conductance of contact; convective heat-transfer coefficient
h_{fg}	latent heat of condensation or evaporation
h_r	radiation heat-transfer coefficient
k	thermal conductivity; conductivity number
k_c, k_d	thermal conductivity of continuous and discontinuous phases, respectively
k_e	electronic thermal conductivity
k_{eff}	effective multiplication factor (nuclear chain-reacting system)
k_f	thermal conductivity of fluid; conductivity coefficient of gas
k_L, k_s	thermal conductivity of liquid and solid, respectively
L	length; Lorentz constant
ℓ	mean free path
ℓ^*	prompt neutron lifetime
M	molecular weight; Meyer hardness (load/cross-sectional area of indentation)
N	number of interferences between two surface profiles
Nu	Nusselt number (hD/k)
n	neutron density; number of contact points per unit area
Ox	subscript for oxide (see Table 4-1)
P	wetted perimeter; probability; volume fraction
$P_\perp$, $P_\parallel$	radial and axial power distribution, respectively (see Table 4-1)
Pe	Peclet number
Pr	Prandtl number ($c_p \mu/k$)
P_d	volume fraction of discontinuous phase
P_H	heated perimeter
P_x, P_y	pore fraction
p	pressure; resonance escape probability (Eq. (3-8))
Q	heat quantity; heat flux (Eq. (4-1))
q	rate of heat transfer, Btu/hr
q/A, q''	surface heat flux, Btu/hr-ft^2
q''', q(r)	volumetric heat generation rate, Btu/ft^3
$(q/A)_{crit}$	critical heat flux (see Sec. 2.2.1)
$(q/A)_{inc}$	incipient heat flux (see Sec. 2.2.3.3)
R, R_T, R_D	resistance (electrical, thermal, and mass-diffusional); subscript for thermal contact (see Table 4-1)
R_{ei}, R_{ev}	thermal resistance due to scattering of electrons by impurities and by thermal vibration of atoms, respectively
R_{pb}, R_{pd}, R_{pe}, R_{pp}	thermal resistance due to phonon interactions (see Sec. 1.2.2)
Re	Reynolds number
r	electrical resistivity
Sc	superscript for surface state at fuel-cladding contact (see Table 4-1)
So	superscript for surface state on outside of cladding (see Table 4-1)
St	Stanton number, $h/c_p g$ (see Fig. 4-6)
T	temperature; power transient period; superscript for total (see Eq. (4-1))
T_a, T_c	temperature at outer radius and at center of fuel, respectively
T_b	bulk fluid temperature
T_m	mean absolute temperature
T_o, T_w	wall temperature
T_{sat}	saturation temperature
Th	superscript for thermal properties of materials and coolant (see Table 4-1)
t	time
t_c	cladding thickness
U	overall heat-transfer coefficient; heat-transfer parameter (=h/k); conductance number (Eq. (1-23))
V	volume

V, v	velocity
V_f	void-volume fraction (V_v/V)
V_v	void volume
W	power output per unit length of fuel element; flow rate
w	flow rate; rate of vapor generation
x, X	steam quality (wt.% of steam)
α	absorptivity; thermal diffusivity ($k/\rho c_p$)
β_c	arithmetic mean distance between surfaces (see Eq. (1-24))
γ	ratio of specific heats (c_p/c_v); geometrical pore fraction
ΔBF	superscript for fuel burnup (see Table 4-1)
ΔdF	superscript for variation in fuel diameter (see Table 4-1)
ΔdR	superscript for variation in fuel-rod diameter (see Table 4-1)
ΔP	superscript for uncertainty in power distribution (see Table 4-1)
Δp	superscript for fuel-element lattice pitching (see Table 4-1)
δ	mean distance between plane surfaces of an idealized contact; average void thickness
δ_1, δ_2	equivalent gap thickness in idealized contact (see Fig. 1-15a)
ε	constant; emissivity; contact ratio ($\sqrt{\text{contact area/total area}}$); fast fission factor (Eq. (3-8))
$\varepsilon_h, \varepsilon_m, \varepsilon_D$	eddy diffusivity (energy, momentum, mass)
η	dimensionless number (Eq. (1-41)); fertility (Eq. (3-8))
Θ	absolute temperature
θ	temperature rise
κ	reciprocal of thermal neutron diffusion length
λ_c	surface wavelength
μ	absolute viscosity; mean (Eq. (4-50))
ν	kinematic viscosity (μ/ρ) constant; gap thickness of a contact (Eqs. (1-33), (1-36))
ρ	mass density; reactivity (nuclear reactor)
σ	Stefan-Boltzmann constant ($=0.1815 \times 10^{-8}$ Btu/ft^2-hr-R^4 or 5.67×10^{-5} erg/cm^2-sec); surface tension; electrical conductivity; standard deviation (Eq. 4-50))
Σ	macroscopic neutron cross section (Eq. (1-55))
τ_0	shear stress at the wall
ϕ	angle in spherical coordinates; heat flux (q/A); neutron flux (neutrons/cm^2-sec)
ψ	constant

REFERENCES

1. J. O. Hirschfelder, C. F. Curtiss, and R. B. Bird, Molecular Theory of Gases and Liquids, John Wiley and Sons, Inc., N. Y., 1954.
2. H. Cheung, "Thermal Conductivity and Viscosity of Gas Mixtures", Report UCRL-1830, University of California Lawrence Radiation Laboratory, 1958.
3. L. A. Bromley, "Thermal Conductivity of Gases at Moderate Pressure", Report UCRL-1852, University of California Lawrence Radiation Laboratory, 1952.
4. F. G. Keyes, "Thermal Conductivities for Several Gases with a Description of New Means of Obtaining Data at Low Temperatures and Above 500°C", Project SQUID, Technical Memorandum MIT-1, Massachusetts Institute of Technology, Cambridge, Mass., October 1, 1952; see also: U.S. Government Research Report, Office of Technical Services, PB-13108, 1952.
5. F.G. Keyes, "The Heat Conductivity, Viscosity, Specific Heat and Prandtl Numbers for Thirteen Gases", Project SQUID NP-4621, Technical Rrport MIT-37, April 1, 1952.
6. R. A. Svehla, "Estimated Viscosities and Thermal Conductivities of Gases at High Temperatures", Report NASA-TR-R-132, 1962.
7. S. Chapman, "On the Law of Distribution of Molecular Velocities, and On the Theory of Viscosity and Thermal Conduction, in a Non-Uniform Simple Monatomic Gas", Phil. Trans. Roy. Soc., London, A 216 (1916) 279.
8. R. C. Reid and T. K. Sherwood, The Properties of Gases and Liquids, McGraw-Hill Book Co., Inc., N. Y., 1958.
9. I. Amdur and E. A. Mason, "Properties of gases at very high temperatures", Phys. Fluids, 1 (1958) 370
10. E. A. Mason and S. C. Saxena, "Approximate formula for the thermal conductivity of gas mixtures", Phys. Fluids, 1 (1958) 361.
11. H. von Ubisch, S. Hall, and R. Srivastav, "Thermal Conductivities of Mixtures of Fission Products Gases with Helium and with Argon", Proceedings of the Second U. N. International Conference on Peaceful Uses of Atomic Energy, Geneva, 1958, Vol. 7, p. 697.
12. A. Goldsmith et al., Handbook of Thermophysical Properties of Solid Materials, Armour Research Foundation, The MacMillan Co., N. Y., 1961. (Also Pergamon Press Ltd., London, 1961.)
13. A. E. Powers, "Fundamentals of Thermal Conductivity at High Temperatures", Report KAPL-2143, Knolls Atomic Power Laboratory, 1961.
14. A. E. Powers, "Conductivity in Aggregates", Report KAPL-2145, Knolls Atomic Power Laboratory, 1961.
15. C. T. Ewing et al., "Thermal Conductivity of Metals", "Liquid Metals Technology", Part 1 in Chem. Eng. Progr. Symposium Series No. 20, p. 19, Am. Inst. Chem. Eng., 1957.
16. A. M. Ross, "The Dependence of the Thermal Conductivity of Uranium Dioxide on Density, Microstructure, Stoichiometry and Thermal-neutron Irradiation", Canadian Report CRFD-817, Chalk River, Ontario, 1960.
17. J. Belle (Ed.), Uranium Dioxide: Properties and Nuclear Applications, Division of Reactor Development, USAEC, U. S. Government Printing Office, Washington 25, D. C., 1961.
18. R. D. Reiswig, "Thermal conductivity of UO_2 to 2100°C", J. Am. Ceram. Soc., 44, 1 (1961) 48.
19. R. W. Nichols, "Ceramic fuels - properties and technology", Nucl. Eng., 3 (1958) 327.
20. B. T. Seddon, "Uranium Ceramics Data Manual; Properties of Interest in Reactor Design", British DEG — Report 120 and Addendum, March 7, 1960.
21. J. A. Christensen et al., "Uranium Dioxide Thermal Conductivity", Trans. Am. Nucl. Soc., 7, 2(1964) 391.
22. F. T. Hetzler and E. L. Zebroski, "Thermal Conductivity of Stoichiometric and Hypostoichiometric Uranium Oxide at High Temperatures", Trans. Am. Nucl. Soc., 7, 2 (1964) 392.
23. J. R. MacEwan et al., "An In-Pile Study of the Thermal Conductivity and Electrical Properties of UO_2 and UO_{2-x}", Trans. Am. Nucl. Soc., 8, 2 (1965) 380.
24. D. T. Clough and J. B. Sayers, "The Measurement of the Thermal Conductivity of UO_2 under Irradiation in the Temperature Range 150-1600°C", British Report AERE-R-4670, 1964.
25. M. F. Lyons et al., "UO_2 Pellet Thermal Conductivity from Irradiations with Central Melting", Trans. Am. Nucl. Soc., 7, 1 (1964) 106.
26. J. E. Burke (Ed.), Progress in Ceramic Science, Vol. 2, Pergamon Press, Inc., N. Y., 1962.
27. D. A. G. Bruggeman, "Dielectric constants and conductivities of aggregates of isotropic materials", Ann. Physik, 24 (1935) 636.
28. W. D. Kingery, "Thermal conductivity: XIV, Conductivity of multicomponent systems", J. Am. Ceram. Soc., 42, 12 (1959) 617.
29. Lord Rayleigh, "On the influence of obstacles arranged in rectangular order upon the properties of a medium", Phil. Mag., 34(1892)491.
30. E. H. Kerner, "The Electrical Conductivity of Composite Media", Proc. Phys. Soc. (London), B. 69, 1956, p. 802.
31. J. L. Bates, "Thermal conductivity of UO_2 improves at high temperatures", Nucleonics, 19,6(1961)83.
32. A. L. Leob, "Thermal Conductivity: VIII, A theory of thermal conductivity of porous materials", J. Am. Ceram. Soc., 37(1954)96.

33. Proceedings of the Second U. N. International Conference on Peaceful Uses of Atomic Energy, Geneva, 1958.
a) J. H. Crawford, Jr. and M. C. Wittels, "Radiation Stability of Nonmetals and Ceramics", Vol. 5, p. 300.
b) J. R. Gilbreath and O. C. Simpson, "The Effect of Reactor Irradiation on the Physical Properties of Beryllium Oxide", Vol. 5, p. 367.
c) J. Elston and R. Caillat, "Physical and Mechanical Properties of Sintered Beryllia Under Irradiation", Vol. 5, p. 345.
d) R. G. S. Skipper and K. J. Wootton, "Thermal Resistance Between Uranium and Can", Vol. 7, p. 684.

34a. Reactor Materials, Section 1, 5,4(1962)7.
b. G.W. Arnold and W. D. Comption, "Threshold energy for lattice displacement in - Al_2O_3", Phys. Rev. 4(1960)66.
c. W. Bollman, "Electron-microscopic observations on radiation damage in graphite", Phil. Mag., 5,8(1960)621.
d. P. G. Lucasson and R. M. Walker, "Research Directed Toward the Study of the Radiation Damage Thresholds of the Elements", Report 61-GC-166, General Electric Co., 1961.
e. E. M. Baroody, "Theory of displacement cascades in compounds", Phys. Rev., Series 2, 116(1959)1418.
f. Reactor Core Materials, 3,2(1960)12.
g. K. Weiser, "Theory of diffusion and equilibrium position of interstitial impurities in the diamond lattice", Phys. Rev., Series 2, 126(1962)1427.

35. Proceedings of the First U. N. International Conference on Peaceful Uses of Atomic Energy, Geneva, 1955.
a) W. K. Woods, P. L. Bupp, and J. F. Fletcher, "Irradiation Damage to Artificial Graphite", Vol. 7, p. 455.
b) G. H. Kinchin, "The Effects of Irrradiation on Graphite", Vol. 7, p. 472.
c) G. R. Hennig and J. E. Hove, "Interpretation of Radiation Damage to Graphite", Vol. 7, p. 666.

36. R. C. McGill and J. A. G. Smith, "The Thermal Conductivity of Irradiated and Unirradiated Beryllia", British Report AERE-R-3019, 1959.
37. J. M. Tobin, "Some Effects of Neutron Irradiation on Selected Beryllia Materials", Report GA-2648, General Atomic, Division of General Dynamics Corp., 1962.
38. R. W. Powell, "The Thermal Conductivity of Beryllia", Trans. Brit. Ceram. Soc., 53(1954)389.
39. T. N. Cetinkale and M. Fishenden, "Thermal Conductance of Metal Surfaces in Contact", Proceedings of the General Discussion on Heat Transfer, Sept. 11-13, 1951, p. 271. Institution of Mechanical Engineers, London, 1951.
40. H. Fenech and W. M. Rohsenow, "Prediction of thermal conductance of metallic surfaces in contact", J. Heat Transfer, Trans. ASME, Series C, 85, 1(1963)15; see also: ASME paper 62-HT-32, 1962.
41. J. J. Henry and H. Fenech, "The Use of Analogue Computers for Determining Surface Parameters Required for Prediction of Thermal Contact Conductance", Winter Meeting, ASME, Paper 63-Wa-104.
42. J. J. Henry, "Thermal Conductance of Metallic Surfaces in Contact", USAEC Report NYO-9459, 1963.
43. H. Fenech, "The Thermal Conductance of Metallic Surfaces in Contact", PhD Thesis, Massachusetts Institute of Technology, May 1959.
44. H. Cordier, M. I. T., personal communication, March, 1963.
45. I. Cohen, B. Lustman, and J. D. Eichenberg, "Measurement of the Thermal Conductivity of Metal-Clad Uranium Oxide Rods During Irradiation", Report WAPD-228, Bettis, Atomic Power Laboratory, 1960; see also: J. Nucl. Mater., 3 (1961)331.
46. R. G. Wheeler, "Thermal Contact Conductance of Fuel Element Materials", Report HW-60343, Hanford Atomic Products Operation, General Electric Co., 1959.
47. Internation Developments in Heat Transfer, Part I. Papers presented at the 1961 International Heat Transfer Conference held at the University of Colorado, Boulder, Colo., American Society of Engineers, N. Y., 1961.
a) L. C. Laming, "Thermal Conductance of Machined Metal Contacts", p. 65.
b) P. D. Sanderson, "Heat Transfer from the Uranium Fuel to the Magnox Can in a Gas-Cooled Reactor", p. 53.
48. H. Fenech and J. J. Henry, "An Analysis of Thermal Contact Resistance". Trans. Am. Nucl. Soc., 5,2(1962)476.
49. J. Adam, "Temperature Distribution Within a Fuel Rod with Special Reference to UO_2 Cylindrical Fuel Elements", British Report AERE-M/R-2414, 1957.
50. J. A. L. Robertson et al., "Irradiation Behavior of UO_2 Fuel Elements", Nuclear Metallurgy, VI, p. 45, Am. Inst. Met. Eng., N. Y., 1959.
51. J. L. Bates and W. E. Roake, "Irradiation of Fuel Elements Containing UO_2 Powders". Fifth Nuclear Engineering and Sciences Conference held at Cleveland, Ohio, April 6-9, 1959, Engineers Joint Council, Preprint V-90, N. Y., 1959.
52. J. D. Eichenberg et al., "Effects of Irradiation of Bulk UO_2", Fuel Elements Conference, Paris, Nov. 18-23, 1957, USAEC Report TID-7546, p. 616, 1958.
53. M. F. Lyons et al., "UO_2 Fuel Rod Operation with Gross Central Melting", Trans. Am. Nucl. Soc., 6,1(1963)155.
54. T. D. Anderson, "Integrity of UO_2 fuel elements", Nucl. Safety, 3,1(1961)18.
55. J. D. Eichenberg et al., "Effects of Irradiation on Bulk UO_2", Report WAPD-183, Bettis, Atomic Power Laboratory, 1957.
56. A. J. Mooradian et al., "Chalk River experience with fuel waterlogging", Nucleonics, 18, 1(1960)81.
57. C. E. Dickerman et al., "Studies of Fast Reactor Fuel Element Behavior under Transient Heating to Failure", USAEC Report ANL-6334, 1961.
58. "Melting of fast-reactor fuel elements", Power Reactor Technology, 5,2(1962)29.
59. A. Amorosi and J. G. Yevick, "An Appraisal of the Enrico Fermi Reactor", Proceedings of the Second U. N. International Conference on Peaceful Uses of Atomic Energy, Geneva, 1958, Vol. 9, p. 358.
60. "Metallurgy Division, Quarterly Report, July, August, and September, 1957", USAEC Report ANL-5797, p. 57, Argonne National Laboratory, 1957.
61. R. G. Jenkins, "Metal Interaction Studies", Progress Reports to Argonne National Laboratory, NMI-4815 and NMI-4816, Nuclear Metals, Inc., 1960.
62. The Reactor Handbook, "Engineering", Vol. 2, McGraw-Hill Book Co., Inc., N. Y., 1955; see also: AECD-3656.
63. M. Jakob, Heat Transfer, Vol. 1, John Wiley and Sons, Inc., N. Y., 1949.
64. P. J. Schneider, Conduction Heat Transfer, Addison-Wesley Publishing Co., Inc., Reading, Mass., 1955.
65. H. S. Carslaw and J. C. Jaeger, Conduction of Heat in Solids, Oxford University Press, London, 1947.
66. J. Randles, "Heat Diffusion in Cylindrical Fuel Elements of Water Cooled Reactors", British Report AEEW-R96, 1961.
67. F. E. Tippets, "Transient Heat Conduction Solid Cylindrical Nuclear Fuel Elements", Reactor Heat Transfer Conference of 1956, USAEC Report TID-7529, Part 1, Book 1, p. 66, 1957.
68. I. N. Sneddon, "Finite Hankel Transforms", Phil. Mag., 37(1946)17.
69. G. M. Dusinberre, Numerical Analysis of Heat Flow, McGraw-Hill Book Co., Inc., N. Y., 1949.
70. L. A. Matsch, M. S. Thesis, University of Pittsburgh, 1960.
71. D. Bagwell, "SIFT, an IBM-7090 Code for Computing Heat Distributions", Report K-1528, Gaseous Diffusion Plant, Union Carbide Nuclear Co., 1962.
72. S. Levy and A. P. Bray, "Reliability of burnout calculations in nuclear reactors", Nuclear News, 6,2(1963)3.
73. W. M. Rohsenow and P. Griffith, "Correlation of Maximum Heat Flux Data for Boiling of Saturated Liquids", Am. Inst. Chem. Eng-ASME Heat Transfer Symposium, Louisville, Kentucky, 1955.
74. R. G. Deissler, personal communication. Described by R. Cole in "Photographic study of pool boiling in the region of the critical heat flux". J. Am. Inst. Chem. Eng., 6(1960).
75. N. Zuber, "On the Stability of Boiling Heat Transfer", Trans. ASME, 80(1958)711.
76. N. Zuber, M. Tribus, and J. W. Westwater, "The Hydrodynamic Crisis in Pool Boiling of Saturated and Subcooled Liquids", p. 230, International Heat Transfer Conference held at the University of Colorado, Boulder, Colo., American Society of Engineers, 1961, International Developments in Heat Transfer, Vol. 2, 1961.
77. S. S. Kutateladze, "On the transition to film boiling under natural convection", Kotloturbostroenic, 3(1948)10.
78. S. S. Kutateladze, "On hydrodynamic theory of changes in a boiling process under free convection", Izvestia Akademia Naute Otdelemie Tekhnischeki Naute, 4(1951)529.
79. R. C. Noyes, "Experimental Study of Sodium Pool Boiling Heat Transfer", ASME Paper No. 62-HT-24, National Heat Transfer Conference held at Houston, 1962, ASME-Am. Inst. Chem. Eng.
80. D. A. Huber and J. C. Hoehne, "Pool Boiling of Benzene, Diphenyl, and Benzene-Diphenyl Mixtures Under Pressure", Asme Paper No. 62-HT-30.
81. P. J. Berenson, "Transition Boiling Heat Transfer From a Horizontal Surface", Sc.D Thesis, Mech. Eng. Dept., M. I. T., Feb., 1960.
82. R. A. De Bertoli, S. J. Green, B. W. Le Tourneau, M. Troy, and A. Weiss, "Forced-Convection Heat Transfer Burnout Studies for Water in Rectangular Channels and Round Tubes at Pressures Above 500 psi", Report WAPD-188, Bettis, Atomic Power Laboratory, 1958.
83. W. M. Rohsenow and J. A. Clark, "Heat Transfer and Pressure Drop Data for High Heat Flux Densities to Water at High Sub-Critical Pressures", Fluid Mech. Heat Transfer Inst., Stanford University Press, Stanford, Calif., 1951.
84. W. H. Lowdermilk, C. D. Lanzo, and B. L. Siegel, "Investigation of Boiling Burnout and Flow Stability for Water Flowing in Tubes", Report NACA TN-4382, National Advisory Committee for Aeronautics, 1958.
85. I. T. Aladyev, Z. I. Miropolsky, V. E. Doroshchuk, and M. A.

Styrikovich, "Boiling Crisis in Tubes", p. 237, International Heat Transfer Conference held at the University of Colorado, Boulder, Colo., American Society of Engineers, 1961, International Developments in Heat Transfer, Vol. 2, 1961.

86. M. Silvestri, "Two Phase Flow Problems", Report EUR 352e, Euratom, Proc. Working Group Heat Transfer, Sponsored by Euratom-Unitied States Joint R and D Board held at Brussels, Oct. 29-31, 1963.

87. H. S. Swenson, J. R. Carver, and R. Kakarala, "The Influence of Axial Heat Flux Distribution on the Heat Transfer Performance of a Water-Cooled Reactor", ASME Paper No. 62-WA-297.

88. E. Janssen and J. A. Kervinen, "Burnout Conditions for Rod in Annular Geometry", Report GEAP-3755, General Electric Co., 1963.

89. Neal E. Todreas, "The Effect of Non-Uniform Axial Heat Flux Distribution on the Critical Heat Flux", Sc. D. Thesis, Nuclear Eng. Dept., M. I. T. (September, 1965).

90. A. E. Bergles and W. M. Rohsenow, "The Determination of Forced Convection Surface Boiling Heat Transfer", J. Heat Transfer, ASME Series C, 86, 3 (1964) 365-372.

91. L. S. Tong, "Influence of Axially Non-Uniform Heat Flux on DNB", Eighth National Heat Transfer Conference, A.I. Ch. E. preprint 17, Los Angeles, California (August 8, 1965).

92. P. G. Barnett, "An Investigation into the Validity of Certain Hypotheses Implied by Various Burnout Correlations", British Report AEEW-R-412, Winfrith, Dorset, England, 1963.

93. A. E. Bergles and W. M. Rohsenow, "Forced-Convection Surface-Boiling Heat Transfer and Burnout in Tubes of Small Diameter", Engineering Projects Laboratory Technical Report 8767, Mech. Eng. Dept., M. I. T., 1962.

94. K. Becker, "Burnout Conditions for Flow of Boiling Water in Vertical Rod Clusters", Report AE-74, Aktiebolaget Atomenergi, Stockholm, Sweden, 1962; see also: K. Becker and G. Hernberg, ASME Paper No. 63-HT-25.

95. E. Polomik and E. P. Quinn, "Multi-Rod Burnout at High Pressure", Report GEAP-3940, General Electric Co., 1962.

96. G. M. Hesson, D. E. Fitzsimmons, E. D. Waters, and J. M. Batch, "Preliminary Boiling Burnout Experiments in Axial Flow with a 19-Rod Bundle Geometry", Report HW-73395, Hanford Atomic Products Operation, General Electric Co., 1962.

97. A. W. Bennett, J. G. Collier, and P. M. C. Lacey, "Heat Transfer to Mixtures of High Pressure Steam and Water in an Annulus. III. The Effect of System Pressure on the Burn-Out Heat Flux for an Internally Heated Unit", British Report AERE-R-3934, 1963.

98. R. V. Macbeth, "Burn-Out Analysis", Part II, British Report AEEW-R-167; Part III, British Report AEEW-R-222; Part IV, "World Data for Uniformily Heated Round Tubes and Rectangular Channels", British Report AEEW-R-267, Winfrith, Dorset, England, 1963.

99. W. R. Gambill, "Generalized Prediction of Burnout Heat Flux for Flowing, Subcooled Wetting Liquids", Fifth International Heat Transfer Conference held at Houston, 1962 Am. Inst. Chem. Eng., Chem. Eng. Prog., Series No. 41, 59(1963).

100. L. S. Tong, H. B. Currin, and A. G. Thorp, II, "New DNB Correlation", Report WCAP-1997, Westinghouse Electric Corp., 1962; see also; Nucleonics, 21,5(1963)43.

101. L. S. Tong, "DNB Prediction for an Axially Non-Uniform Heat Flux Distribution," AEC Report WCAP-5584 (September, 1965).

102. E. Janssen and S. Levy, "Burnout Limit Curves for Boiling Water Reactors", Report APED-3892, General Electric Co., 1962.

103. K. M. Becker and P. Persson, "Analysis of Burnout Conditions for Flow of Boiling Water in Vertical Round Ducts", Report AE-113, Aktiebolaget Atomenergi, Stockholm, Sweden, 1963; see also: Report AK-114, 1963.

104. R. P. Stein and P. A. Lottes, "Status of Boiling Burnout for Reactor Design", in Selected Topics in Reactor Technique, Leonard Link (Ed.), USAEC Report TID-8540, 1964.

105. R. P. Stein, "Boiling Burnout for Reactor Design", in Selected Topics in Reactor Technique, L. Link (Ed.), USAEC Report TID-8540, 1964.

106. L. A. Bromley, "Heat transfer in stable film boiling", Chem Eng. Progr., 46,5(1950)221.

107. L. A. Bromley et al., "Heat transfer in forced convection in film boiling", Ind. Eng. Chem., 45(1953)2639.

108. P. Berenson, "Transition Boiling Heat Transfer from a Horizontal Surface", ASME-Am. Inst. Chem. Eng. Heat Transfer Conference held at Buffalo, Aug. 14-17, 1960, Paper No. 18.

109. Y. Y. Hsu and J. W. Westwater, First National Heat Transfer Conference held at Pennsylvania State University, Aug. 11-14, 1957, Am. Inst. Chem. Eng., Paper No. 18.

110. J. B. McDonough, W. Milich, and E. C. King, Technical Report No. 62, MSA Research Corp., 1958.

111. E. E. Polomik, S. Levy, and S. Sawochka, "Film Boiling of Steam-Water Mixtures in Annular Flow at 800, 1100, and 1400 Psi", ASME Paper No. 62-WA-136.

112. J. P. Lewis, J. H. Goodykoontz, and J. F. Kline, "Boiling Heat Transfer to Liquid Hydrogen and Nitrogen in Forced Flow", U.S. Report NASA TN-D-1314, National Aeronautics and Space Administration, 1962.

113. R. C. Hendricks, R. W. Graham, Y. Y. Hsu, and R. Friedman, Experimental Heat Transfer and Pressure Drop of Liquid Hydrogen Flowing Through a Heated Tube", Report NASA TN-D-765, 1961.

114. R. S. Dougall, "Film Boiling on the Inside of Vertical Tubes With Upward Flow of the Fluid at Low Quantities", MIT Report No. 9079-26, 1963.

115. S. J. Green, "Estimated Film Boiling Heat Transfer Coefficients at Burnout", Report WAPD-TH-132, Bettis, Atomic Power Laboratory, 1955.

116. H. A. Johnson, V. E. Schrock, F. B. Selph, J. H. Lienhard, and Z. R. Rosztoczy, "Transient Pool Boiling of Water at Atomospheric Pressure", ASME Paper 29, Part 2, Section A, International Heat Transfer Conference held at the University of Colorado, Boulder Colo., American Society of Engineers, 1961.

117. J. R. Howell and K. J. Bell, "An Experimental Investigation of the Effects of Pressure Transients on Pool Boiling Burnout", Paper 18, Fifth International Heat Transfer Conference held at Houston, 1962, Am. Inst. Chem. Eng.

118. M. W. Rosenthal and R. L. Miller, "An Experimental Study of Transient Boiling", USAEC Report ORNL-2294, Oak Ridge National Laboratory, 1957; see also; Nucl. Sci. Eng., 2(1957)640.

119. R. Cole, "Investigation of Transient Pool Boiling Due to Sudden Large Power Surge", U. S. Report NACA-TN-3885, National Advisory Committee for Aeronautics, 1956.

120. W. M. Rohsenow and H. Y. Choi, Heat, Mass and Momentum Transfer, Chapter XIII, "Radiant Heat Transfer", Prentice-Hall, Inc., N. Y., 1961, p. 189.

121. R. N. Lyon, "Liquid Metal Heat-Transfer Coefficients" Chem. Eng. Prog., 47,(1951)75.

122. L. Trefethen, "Measurement of Mean Fluid Temperatures", Trans. ASME, Aug., 1956.

123. D. R. Jenkins, Heat Transfer of Fluid Mechanics Institute, p. 147, Stanford University Press, Stanford, Calif., 1951.

124. R. G. Deissler, "Analysis of Fully Developed Turbulent Heat Transfer at Low Peclet Numbers in Smooth Tubes with Application to Liquid Metals", National Advisory Committee Aeronaut., Research Memo E52 F05, Washington, D. C., 1952.

125. N. Z. Azer and B. T. Chao, Intern. J. Heat Mass Transfer, 1, 2/3(1960)121.

126. T. Mizushina and T. Sasano, "The Ratio of the Eddy Diffusivities for Heat and Momentum and Its Effect on Liquid Metal Heat Transfer Co-efficients", Paper No. 78, International Heat Transfer Conference held at the University of Colorado, Boulder, Colo., ASME, 1961.

127. P. S. Lykoudis and Y. Touloukian, "Heat Transfer in Liquid Metals", Trans. ASME, 1958.

128. O. E. Dwyer, "Eddy Transport in Liquid-Metal Heat Transfer", USAEC Report BNL-6149, Brookhaven National Laboratory, 1962.

129. O. E. Dwyer, "Heat Transfer to Fluids Flowing Through Pipes, Annuli and Parallel Plates", USAEC Report BNL-6692, Brookhaven National Laboratory, 1963.

130. A. O. Friedland and C. F. Bonilla, J. A. I. Ch. E., 7 (1961) 107.

131. V. I. Subbotin, P. L. Murillov, and M. Y. Suvorov, "Investigation of Heat Transfer in the Intertube Space of Liquid Metal Heat Exchangers", Physics-Power Institute State Committee for the Utilization of Atomic Energy in the USSR (1964).

132. V. I. Subbotin et al., Atomn. Energ., 9 (1960) 584.

133. O. E. Dwyer, Atomic Energy Review, IAEA Vienna, 4, 1 (1966).

134. P. L. Kirillov, V. I. Subbotin, M. Y. Surorov, and M. F. Troyanov, "Heat transfer in pipes to a sodium-potassium alloy and to mercury", J. Nucl. Energy, Part B, 1(1959)123.

135. V. I. Subbotin, M. K. Ibragimov, M. N. Ivanovsky, M. N. Arnoldov, and E. V. Nomofilov, "Turbulent Heat Transfer in Flow of Liquid Metals", Intern. J. Heat Mass Transfer, 4(1961) 79.

136. O. E. Dwyer (Ed.), "High-Temperature Liquid-Metal Technology Review", (Quarterly), Prepared for USAEC by Brookhaven National Laboratory; 1, 4(1963)104.

137. J. C. Chen, "A Proposed Mechanism and Method of Correlating Convective Boiling Heat Transfer with Liquid Metals", USAEC Report BNL-7319, Brookhaven National Laboratory, 1963.

138. S. E. Golian, T. A. Bergstralh, E. G. Harris, and R. C. O'Rourke, "Transient Response of Plane Parallel Fuel Assemblies to Exponential Power Excursions", Report 4495, U. S. Naval Research Laboratory, 1955.

139. D. H. Jones and M. J. Galper, "PWR Reactivity Accidents", Report WAPD-SC-542, Bettis, Atomic Power Laboratory, 1957.

140. L. M. Swartz et al., "PWR Loss-of-Coolant Accident, Core Meltdown Calculations", Report WAPD-SC-544, Bettis, Atomic

Power Laboratory, 1957.
141. R. Daane and L. Gartner, "The Self-Regulation by Moderator Boiling in Stainless Steel-UO_2H_2O Rectors", Report NDA-16, Nuclear Development Corp. of America, 1955.
142. B. H. Noordhoff, C. F. Foremand, and J. W. Martins, "Comparison of Two Methods for Analyzing Accidental Power Transients", Report WAPD-TM-31, Bettis, Atomic Power Laboratory, 1957.
143. E. R. Sanford, "Accidental Power Transients in Metal-Water Reactors", Report WAPD-SC-581, Part 1, Bettis, Atomic Power Laboratory, 1956.
144. E. Janssen, W. H. Cook and K. Hikido, "Metal-Water Reactions. I. A Method for Analyzing a Nuclear Excursion in a Water-Cooled and -Moderated Reactor", Report GEAP-3073, General Electric Co., 1958.
145. J. I. Owens, "Metal-Water Reactions. II. An Evaluation of Severe Nuclear Excursions in Light Water Reactors", Report GEAP-3178, General Electric Co., 1959.
146. B. W. Le Tourneau and R. E. Grimble, "Engineering hot channel factors for nuclear reactor design", Nucl. Sci. Eng., 1(1956)359.
147. "Gas-Cooled Reactor Project, Quarterly Progress Report June 30, 1960", USAEC Report ORNL-2964, Oak Ridge National Laboratory, 1960.
148. "40-Mw(E) Prototype High-Temperature Gas-Cooled Reactor Research and Development Program", Quarterly Progress Report for the Period Ending June 30, 1960, Report GA-1640, General Atomic, Division of General Dynamics Corp., 1961.
149. "Pathfinder Atomic Power Plant Coolant Distribution Tests-Final Report", Report ACNP-5920, Allis Chalmers Mfg. Co., 1959.
150. "Model Studies of Flow and Mixing in the Partially Enriched Gas-Cooled Power Reactor", Report BMI-1397 (USAEC Report TID-4500, 15th Ed.), Battelle Memorial Institute, 1959.
151. W. H. Cook, "Fuel Cycle Program. A Boiling Water Reactor Research Development Program", Second Quarterly Report October-December 1960, Report GEAP-3627, General Electric Co., 1961.
152. "Thermal Design Aspects of the Yankee First Core Fuel Rod", Report YAEC-106 Westinghouse Electric Corp., 1960.
153. H. Chelemer and L. S. Tong, "Engineering hot channel factors for open lattice cores", Nucleonics, 20, 9(1962)68.
154. P. A. Nelson, A. A. Bishop, and L. S. Tong, "Mixing in Flow Parallel to Rod Bundles Having a Square Lattice", Report WCAP-1607, Westinghouse Electric Corp., 1960.
155. A. E. Bergles and W. M. Rohsenow, "The influence of temperature difference on the turbulent forced-convection heating of water", J. Heat Transfer, Trans. ASME, Series C., 84,3(1963)268.
156. H. W. Hoffman et al., "Heat Transfer with Axial Flow in Rod Clusters", Paper 65, International Heat Transfer Conference held at the university of Colorado, Boulder, Colo., ASME, 1961.
157. F. Lerda and C. Rossi, "Contribution to the statistical analysis of hot channel and hot spot factors", Energia Nucleare, 8, (1961)3.
158. P. A. Rude and A. C. Nelson, Jr., "Statistical analysis of hot channel factors", Nucl. Sci. Eng., 7 (1960)157.
159. F. H. Tingey, "Error propagation in hot spot, hot channel analysis", Nucl. Sci. Eng., 9(1961)127.
160. F. H. Abernathy, "The statistical aspects of nuclear reactor fuel element temperature", Nucl. Sci. Eng., 11(1961)290.
161. F. J. Moody, "Probability Theory and Reactor Core Design", Report GEAP-3819, General Electric Co., 1962.
162. Editorial, "APDA introduces statistical hot spot factors", Nucleonics, , 8(1959)92.
163. D. B. Owen, "Tables of Factors or One-Sided Tolerance Limits for a Normal Distribution", Report SCR-13, Sandia Corp., 1958.
164. J. B. Scarborough, Numerical Mathematical Analysis, 3rd Ed., Oxford University Press, London, 1955, and Johns Hopkins Press, Baltimore, 1955.
165. F. D. Judge and L. S. Bohl, "Effective Hot Channel Factors for 'Flat' Power Reactors", Trans. Am. Nucl. Soc., 7, 2 (1964) 497.
166. H. Guéron, Private Communication, M. I. T., November, 1965.
167. A. C. Nelson and W. S. Minkler, "A General Method for Evaluating the Effects of Characteristics in Design Variables on Core Thermal Performance, Nucl. Sci. Eng., 17 (1963) 101-110.
168. A. Borella et al., "Hythest — A Monte Carlo Program for the Evaluation of the Correction Factors in a BWR and BWR Core", EUR 1587. e. (1964). European Atomic Energy Community. Joint Nuclear Research Center, Ispra, Italy.
169. V. L. Businaro and G. P. Pozzi, "A New Approach on Engineering Hot Channel and Hot Spot Statistical Analysis", EUR 1302-e. (1964).
170. H. Fenech and H. Guéron, "The Synthesis Method of Uncertainty Analysis in Nuclear Reactor Thermal Design", Nucl. Sci. Eng., 31 (1968) 505-512.
171. H. Chelemer and L. S. Tong, "Statistically Determined Hot Channel Factors of a Paralled Rod Bundle Core", Report WCAP-1594, Westinghouse Electric Corp., 1960.
172. "Irradiation Testing of Enrico Fermi Prototype Fuel Pins in the CP-5, 1957-1959", Report APDA-130, Atomic Power Development Assoc., 1960.
173. "Army Gas-Cooled Reactor Systems Program, Semiannual Progress Report July 1 - December 31, 1960", Report IDO-28567, Phillips Petroleum Company, 1960.
174. J. G. Knudsen and J. M. Batch, "Heat Transfer Coefficients for a Wire Wrapped 19-Rod Bundle Foil Element" Trans. Am. Nucl. Soc., 4 2,(1961)539,
175. W. J. Eich and W. P. Kovacik, "Reactivity and Spatial Distributions in Multi-Region Lattices: Comparison Between Analysis and Experiment", Trans. Am. Nucl. Soc., 4,1(1961) 101.
176. A. Hitchcock, "Statistical Methods for the Estimation of Maximum Fuel Element Temperature", British Report IGR-TW/R-760, 1958.
177. "Maritime Gas-Cooled Reactor Program, Quarterly Progress Report", Report GA-1183, General Atomic, Division of General Dynamics Corp., 1959.
178. "Army Gas-Cooled Reactor System Program, Semiannual Progress Report", Report IDO-28549, Phillips Petroleum Co., 1959.
179. S. Sudar, "OMRE Fuel Plate Surface Temperature Measurement", Report NAA-SR-4047, North American Aviation, Inc., 1960.
180. V. G. Graynek, "High Power Density Development Project Interim Report, 300 MWe HPD Conceptual Design Study", Report GEAP-3860, General Electric Co., 1962.

CHAPTER 17

Chemical Reactions

LOUIS BAKER, JR. and ROBERT C. LIIMATAINEN*
(Argonne National Laboratory, Argonne, Illinois)

CHAPTER CONTENTS**

1 INTRODUCTION

It was recognized early in the nuclear energy program that many of the elements possessing desirable nuclear properties, i.e., aluminum, zirconium, and uranium, rank high in the electromotive series. The reaction of these elements with oxidizing agents such as water and oxygen is, fortunately, slow at normal temperatures. It was feared, however, that rapid or explosive reactions might be initiated as a result of accidental overheating. Summarizing the present state of knowledge regarding the initiation of rapid or explosive chemical reactions is the principal purpose of this chapter. This discussion is also concerned with slower corrosion-like reactions which can affect the safety of operation of nuclear reactors.

Individual chemical reactions are discussed from a fundamental point of view. It is hoped that this will provide the insight often needed in considerations of safety when existing data do not precisely apply to the situation at hand. Because of

*Present adress: Office of Atomic Energy Affairs, Department of State, Washington, D.C.

**This chapter, excluding the Addendum, is based primarily on information in the literature or known to the authors prior to January 1965. The Addendum summarizes new and significant developments up to February 1969.

the broad scope of the chapter, the coverage is necessarily somewhat abbreviated. The chapter is divided into four major sections. In Sec. 2 slow or corrosion-like reactions between the solid components of reactors (i.e., cladding and structural metals, core metals, solid moderators, control rod materials, and ceramic core materials) and the fluid or environmental components (i.e., water, steam, air, CO_2, H_2, organics, molten salts, liquid metals, etc.) are discussed. In Sec. 3, those reactions which may be rapid or violent are discussed in detail. Four classes of rapid reactions are considered: metal-water reactions, metal-air reactions (ignitions), graphite-gas reactions, and hydrogen-oxygen reactions.

In Sec. 4, fuel meltdowns are discussed to the extent that they constitute an initiating process for a violent metal-water reaction. Information gained from chemical studies of metal-water reactions (described in Sec. 3) is combined with information derived from full-scale meltdown incidents and small-scale meltdown experiments to provide a basis for estimating the extent and rate of reaction to be expected during reactor accidents.

The final part, Sec. 5, concerns miscellaneous chemical problems which include low-melting eutectic alloys, cleanup and analysis of alkali metal coolants, and the aluminum-U_3O_8 thermite reaction.

Most of the chemical reactions of interest in reactor safety occur between a fluid and a solid. The kinetics of fluid-solid reactions are often expressed in the following form:

$$w^n = kt , \qquad (1-1)$$

where w is the extent of reaction, t the time, and n, k constants. The reaction rate law expressed by Eq. (1-1) is called the linear, parabolic, or cubic rate law according to whether n = 1, 2, or 3. It is also common to find experimental results expressed in terms of a nonintegral value of n. The extent of reaction w is usually expressed in the units measured experimentally, often as a weight gain or a quantity of reactant consumed per unit of original area (e.g., mg/cm^2). The extent of reaction can also be expressed as the depth of solid phase penetrated or the thickness of scale generated (e.g., in microns). When a fluid-solid reaction forms a barrier film and the exponent n is greater than unity, the film is said to be protective. This statement refers to the fact that the reaction rate, expressed as the derivative of Eq. (1-1),

$$dw/dt = k/nw^{n-1} , \qquad (1-2)$$

decreases as the thickness of the barrier film increases. The parabolic rate law (n = 2) is observed frequently and describes a reaction controlled by the solid state diffusion of an ionic species through the barrier film.

When a barrier film is so protective that the reaction virtually ceases, it is usually found that the data are best described by a logarithmic rate law,

$$w = k \log (at + t_o). \qquad (1-3)$$

The rate laws are shown qualitatively in Fig. 1-1.

There are many possible reaction mechanisms which can lead to the linear rate law. These include

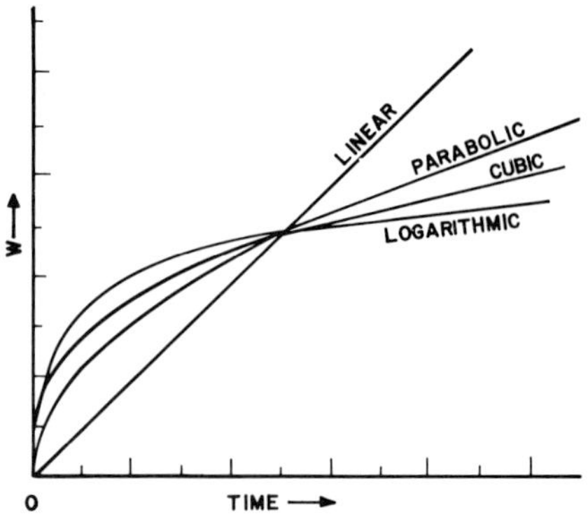

FIG. 1-1 Rate laws for fluid-solid reactions.

reactions whose rates are limited by interfacial processes such as adsorption, ionization, vaporization, etc. The linear rate law implies that the growing barrier film offers either no resistance or a constant resistance to the reaction.

The temperature dependence of chemical reaction rates is generally expressible in terms of the Arrhenius equation,

$$k = A \exp(-\Delta E/RT) , \qquad (1-4)$$

where k is a rate constant, A a pre-exponential factor, ΔE the activation energy, R the gas constant, and T the absolute temperature. A straight line on an Arrhenius plot (log k vs 1/T) with a resulting positive value for the activation energy ΔE is usually taken to mean that the reaction is controlled by one and the same mechanism over the entire temperature range.

Most of the processes which limit the rate are chemical in nature. This would include processes such as ionic diffusion and activated adsorption which generally have activation energies greater than 10 kcal/mole. In addition to chemical processes, reactions may be limited by physical processes such as gas-phase diffusion. Such a limitation might occur, for example, when gaseous products are evolved from a surface such as in metal-water reactions or graphite-air reactions. In these cases it is necessary for the gaseous reactant, either steam or oxygen in the examples cited, to diffuse through the product gases. A gas phase diffusion rate may be formulated as follows:

$$r = h_D \frac{\Delta P}{RT} , \qquad (1-5)$$

where r is the diffusion rate, h_D the mass transfer coefficient, ΔP the difference in partial pressure which drives diffusion, R the gas constant, and T the absolute temperature. A value for h_D may be obtained for small spheres using the fact that the Nusselt number for spheres has a minimum value of 2. The Nusselt number Nu is:

$$Nu = h_D L/D, \qquad (1-6)$$

where L is the characteristic length or diameter and D the diffusion coefficient.

Another example of a reaction rate which is limited by the access of a reactive gas to a solid surface would be the atmospheric oxidation of a reactive metal in the rarified atmosphere of interplanetary space. In this case, the reaction rate could not exceed the rate at which the metal encounters oxygen atoms or molecules.

2 SLOW REACTIONS*

2.1 Corrosion of Cladding and Structural Metals

2.1.1 Zirconium Alloys

Zirconium and its alloys owe their corrosion resistance to the formation of a highly protective barrier film of zirconium dioxide (ZrO_2). Contamination by nitrogen, carbon, and oxygen seriously lowers the corrosion resistance of zirconium and its alloys. Dissolved hydrogen, nitrogen, and oxygen decrease the ductility of the metal and can lead to mechanical failures. Zircaloy-1 and Zircaloy-2 (see reference [1], page 709, for typical analyses) contain a few percent of tin in the alloy to offset the effects of dissolved nitrogen although the alloys are slightly less resistant to corrosion than the pure metal. Zircaloy-3, containing 0.2 to 0.3 percent each of tin and iron, is more corrosion resistant at higher temperatures. Corrosion data for the two most studied forms of zirconium, iodide crystal bar and Zircaloy-2, are generally similar [2]. Both materials have an initial, nominally cubic, reaction rate in water or steam, producing a protective black oxide. The reaction rate is believed to be controlled by the diffusion of oxygen anions, via anion vacancies, through the oxide lattice. Oxidation reactions controlled by ionic diffusion processes usually follow the parabolic law. Much theoretical discussion has been offered to explain the nature of the cubic rate law which applies for both the zirconium-steam reaction and the nearly identical zirconium-oxygen reaction [3, 4, 5, 6].

A transition or "breakaway" to a linear corrosion rate occurs in later stages in the reaction. The post-transition oxide is white in color and marks the beginning of serious corrosion. The protective adherent, black oxide, has been variously reported as cubic, tetragonal, or monoclinic ZrO_2 and the non-protective white oxide has been shown to be monoclinic ZrO_2 [7]. The black oxide apparently owes its color to an excess of metal, while the white oxide is nearly stoichiometric.

Published data [7, 8, 9, 10] were used to compute typical corrosion results given for Zircaloy-2 in Table 2-1. The results indicate that the upper temperature limits for long service life of Zircaloy-2 may approach 454°C (850°F) [11, 12]. Recent studies have indicated that other zirconium alloys containing small quantities of Fe, Cr, or Ni but little or no Sn may be adequate for a few years' use in 1500 psi (100 atm) steam at 500°C (932°F) [13, 14].

The achievement of long service life depends on rigorous adherence to quality control measures. In general, Zircaloy-2 components are subjected to an HNO_3-HF etch to remove cold-worked surface metal. Following removal of the fluoride residue, the parts are autoclaved either in 340°C (650°F) water or 400°C (750°F) steam [15]. The presence of white oxide or any evidence of crack, streaks, blisters, etc., is cause for rejection.

Corrosion results, shown in Table 2-1, probably represent the ideal limits obtainable when no perturbing factors are present to disrupt the protective oxide film. Imperfections and local corrosion can occur as a result of nitrogen contamination or because of copper, iron, and uranium surface inclusions. Local corrosion has been found to occur at gas voids called "stringers." Stringers in the fabricated metal can be avoided by vacuum rather than inert gas melting [16]. Fluoride and chloride contamination are known to increase corrosion rates, making complete removal of etching residues mandatory. Neutron irradiation alone appears to cause little change in corrosion behavior [17, 18], although fission product bombardment can result in serious corrosion. Potentially serious corrosion can result from rubbing of Zircaloy-2 parts under service conditions. Fretting corrosion has been shown to increase greatly normal corrosion rates in flowing water [2].

It is known that hydrogen absorption occurs to a considerable extent during aqueous corrosion [19, 20] and that this can seriously reduce the ductility of the metal. Experimental and theoretical studies [21] have shown that hydrogen undergoes thermal diffusion within zirconium. Hydrogen will migrate to regions of lower temperature. This can lead to high surface concentrations of hydrogen (at, for example, the surface layer of a fuel element cladding) and destructive hydride precipitates.

The reaction of zirconium with oxygen is very similar to its reaction with water or steam. Oxidation in steam can be considered as a reaction with oxygen at a very low concentration since water has a finite but very small equilibrium partial pressure of oxygen associated with it.

The oxidation reaction follows, in general, a cubic rate law [7, 10, 22], although several investigators have been able to correlate their results in terms of a parabolic rate law [23, 24, 25]. The cubic rate law for pure zirconium parallelepipeds between 400 and 900 °C (750-1650°F) is as follows [7]:

$$w^3 = 5.94 \times 10^{16}\, t\, [\exp(-42{,}700/RT)], \quad (2\text{-}1)$$

where w is the weight gain in $\mu g\ O_2/cm^2$, t the time in min, T the temperature in °K, and R the gas constant. There is little or no effect of oxygen pressure over moderate ranges [7, 23, 24]. At oxygen pressures above 300 psi (20 atm), however, ignitions may occur, even at room temperature [26].

The initial cubic (or parabolic) reaction is often followed by a transition or breakaway to a more rapid, usually linear, reaction. However, no breakaway is found [7] with pure zirconium for runs

*Several excellent reviews of the corrosion of reactor materials have been published [255, 343, 344, 345] subsequent to the compilation of this section. The reader is referred to them for additional information.

TABLE 2-1

Corrosion of Cladding Metals in Water and Steam

(Corrosion in mg/dm^2)[a]

Temperature	Zr, Zircaloy-2[b]	Al		Mg AZ-31[e]		Stainless Steels[f]			Hastelloy F
		1100[c]	X8001[d]		302	347	410	17-4PH	
53°C (127°F)	–	–	–	8.6/day	–	–	–	–	–
125°C (257°F)	–	0.1/day	–	–	–	–	–	–	–
150°C (302°F)	–	0.36/day	–	1390/day	–	–	–	–	–
200°C (392°F)	–	2.5/day	–	–	–	–	–	–	–
250°C (482°F)	3 (14d) 12B[h] (1060d)	–	2/day	–	–	5/month[g]	–	–	–
316°C (600°F)	9 (14d) 22B (215d)	–	6/day	–	–	5-15/month[g]	–	–	–
350°C (662°F)	–	–	18/day	–	–	–	–	–	–
427°C (800°F)	48B (12.6d) 175 (50d)	–	–	–	1 (14d) 8 (82d)	3 (14d) 1 (82d)	5 (14d) 10 (82d)	2 (14d) 8 (82d)	2 (14d) 1 (82d)
538°C (1000°F)	78B (1.3d) 175 (50d)	–	–	–	28 (14d) 93 (130d)	8 (14d) 12 (130d)	12 (14d) 12 (130d)	– –	13 (58d) –
732°C (1350°F)	–	–	–	–	116 (14d) 404 (132d)	36 (14d) 83 (132d)	35 (14d) 175 (123d)	36 (14d) 39 (132d)	32 (14d) 52 (132d)

[a] Conversion factors: 1 dm^2 = 100 cm^2 = 15.5 $in.^2$; 1 mg = 2.20 x 10^{-6} lb = 0.0154 grains; 1 mg/dm^2 = $(0.003937/\rho)$ mils with ρ = density in g/cm^3 ($\rho \cong 2.7$ for Al, ~ 1.75 for Mg, ~ 6.4 for Zr, ~ 6.5 for Zircaloy-2, ~ 7.9 for stainless steels).
[b] Reference [2], metal reacted computed from correlating rate equation.
[c] Reference [44], descaled weight loss.
[d] Reference [45], descaled weight loss.
[e] Reference [67], weight loss.
[f] Reference [61], weight gain.
[g] Reference [3], descaled weight loss.
[h] Onset of breakaway oxidation.

ranging from 4000 min (66.7 hr) at 400°C (750°F) to 400 min at 900°C (1650°F). Breakaway oxidations were obtained with many binary alloys, including tin, in from 6 to 650 min at 700°C (1300°F). Of the 20 binary alloy systems studied, only Cu, Ni, Be, and Hf alloys showed any increased resistance to oxidation as compared with pure zirconium. Several percent of either C or Ti in the alloys were particularly deleterious to oxidation resistance.

Zirconium metal can dissolve up to 29 at.% oxygen [27] and, typically, the zirconium-oxygen reaction involves both homogeneous dissolution and the formation of a distinct oxide scale. A recent study estimates that approximately 10 and 20% of the reacted oxygen at 800°C (1475°F) are dissolved in the metal after 100 and 300 min of oxidation, respectively [4]. An initial oxygen content up to ca. 12-15 at.% oxygen markedly shortens the time period preceding breakaway oxidation.

Zirconium reacts with nitrogen to form the golden-yellow zirconium nitride. The nitrogen reaction is much slower than the oxygen reaction; however, in air both zirconium dioxide and zirconium nitride are formed. The reaction with pure nitrogen is described by the following parabolic rate law over the temperature range from 975 to 1640°C (1787°F to 2984°F) at 1 atm pressure [28].

$$V^2 = 5.0 \times 10^3 \, t \, [\exp(-48{,}000/RT)], \qquad (2\text{-}2)$$

where V is the nitrogen consumed in ml/sq cm, t the time in sec, R the gas constant, and T the temperature in °K. The solubility of nitrogen in zirconium over the same temperature range is as follows:

$$\log_{10} C_o = -\frac{2810}{T} + 1.42, \qquad (2\text{-}3)$$

where C_o is solubility in wt.% N_2.

The reaction of zirconium and Zircaloy-2 with air has been studied [25, 29] and the following rate equations [29] are reported:

Zircaloy-2 in dry air

$$w^{2.58} = 1.1 \times 10^9 \, t \, [\exp(-39{,}400/RT)]; \qquad (2\text{-}4)$$

zirconium in dry air

$$w^{2.58} = 1.8 \times 10^9 \, t \, [\exp(-41{,}400/RT)]. \qquad (2\text{-}5)$$

The Zircaloy-2 reaction is somewhat more rapid than the zirconium reaction. The time to breakaway was also found to be longer for zirconium than for Zircaloy-2. It was also noted that high-hafnium-sponge zirconium and graphite-melted zirconium reacted considerably faster in air. It was, therefore, concluded that higher purity improves the corrosion resistance of zirconium in air [29]. In general, oxidation rates in air are slightly greater than in pure oxygen. Unusual dimensional increases have been reported for zirconium scaling in air [29]. It has been shown that these changes require simultaneous presence of oxygen and nitrogen. The dimensional changes apparently result from plastic yielding of the metal under tensile forces generated by the oxide film. This was demonstrated by oxidizing zirconium sheet with one side polished and the other side roughground. The greater stress developed in the unbroken oxide, smooth side, caused the sheet material to be markedly bowed. It was estimated that the compressive stress in the unbroken oxide was between 16,000 and 48,000 psi (11.3 and 33.9 kg/mm^2) [29].

The limited data available on the reaction of zirconium with CO_2 and CO suggests that the reactions are nearly identical to the water and oxygen reactions discussed previously [30]. The corrosion reactions with CO_2 and CO are described approximately by a cubic rate law followed by a transition or breakaway to a more rapid reaction. Indications are that Zircaloy-2 would not be a useful cladding material in carbon dioxide atmospheres above 400°C (750°F). Titanium and aluminum are highly undesirable impurities while small quantities of molybdenum, tungsten, and copper are beneficial [30]. There are also indications that moisture in the CO_2 may increase the reaction rate by a factor of from two to four [2].

The reaction of zirconium with hydrogen is particularly important because of the damaging effects of hydrogen on the ductility of the metal. Deterioration of the mechanical properties of zirconium is detectible in concentrations less than 50 ppm of hydrogen and becomes serious at levels between 500 and 5000 ppm. The solubility of hydrogen in α-zirconium varies between 0.008 wt.% (80 ppm) at 300°C (580°F) and 1.08 wt.% (ca. $ZrH_{1.0}$) at 850°C (1870°F) [8]. The total quantity of hydrogen that can be absorbed by zirconium in one atmosphere of hydrogen varies from about 2.16 wt.% at 20°C (68°F) to 1.71 wt.% at 800°C (1475°F) and has decreased to 1.04 wt.% at 1100°C (2015°F) [31]. The absorbed hydrogen at low temperature exists in precipitated hydride phases. There are at least two hydride phases; a delta, face-centered-cubic phase ($ZrH_{1.40}$ to $ZrH_{1.56}$) and an epsilon, face-centered-tetragonal phase ($ZrH_{1.65}$ to $ZrH_{1.97}$).

The rate of reaction between zirconium and hydrogen is very sensitive to surface pretreatment and the presence of films. The room temperature surface oxide film has a remarkable inhibiting effect on the rate of hydriding [32]. The rate of hydriding of clean metal is described by the following equation:

$$\bar{C} = 2.845 \frac{D^{1/2} t^{1/2}}{h}, \qquad (2\text{-}6)$$

where $\bar{C}$ is the average concentration of hydrogen in the metal in atoms of hydrogen per atom zirconium, D the diffusion coefficient in cm^2/sec, t the time in sec, and h the specimen thickness in cm. The diffusion coefficient varies with temperature over the range 60 to 250°C (140 to 482°F) as follows:

$$\text{hydrogen: } D = 1.09 \times 10^{-3} \exp(-11{,}400/RT)$$
$$\text{deuterium: } D = 0.73 \times 10^{-3} \exp(-11{,}400/RT).$$

Before presenting a brief resumé of the corrosion data for zirconium and its alloys in liquid metals and fused salts, it is appropriate to list some of the principal mechanisms for corrosion in these systems. They are: alloying, solution attack, intergranular penetration, contaminant reactions, erosion, temperature-gradient mass-transfer, and concentration-gradient mass-transfer [33, 34]. Zirconium is considered a good material of construction for Na and NaK up to 600°C (1112°F) [33]. At 500°C (932°F) a one-month exposure of zirconium to Na or NaK gives an average weight change of +0.10 mg/cm^2 which is excellent. The kinetics of the oxidation reaction of zirconium and its alloys in Na at 630°C (1168°F) are given by [35] $\Delta w = kt^n$, where n = 0.37 to 0.44 and k = 0.029 to 0.060 mg/(cm^2) (hr)^n. Zirconium has acceptable resistance to liquid lead and lead-bismuth alloys up to 550°C (1022°F) and limited resistance up to 950°C (1742°F) [33]. Other data on corrosion in sodium gave a parabolic rate law with an energy of activation of 52.9 kcal/mole for zirconium and 45.3 for Zircaloy-2 over the range 400 to 635°C (752 to 1175°F). This rate is insensitive to the oxide content of the sodium over the range 10 to 500 ppm [36]. The addition of aluminum, tin, or molybdenum to the zirconium increases the corrosion by a factor of 2 to 4. Table 2-2 [37] gives some selected corrosion data for unalloyed zirconium and a particular quaternary alloy.

TABLE 2-2

Corrosion of Selected Zirconium Alloys and Zirconium in Sodium at 1000 and 1200°F

Material	Average weight gain, mg/cm^2, for indicated exposure time at 1000°F (538°C)					
	75 hr	163 hr	300 hr	500 hr	1500 hr	2500 hr
Zirconium alloy (wt. %: 1.5 Al, 1.5 Sn, 1.5 Mo)	0.18	0.33	0.40	0.42	0.55	0.58
Unalloyed zirconium	0.14	0.23	0.33	0.30	0.50	0.56

Material	Average weight gain, mg/cm^2, for indicated exposure time at 1200°F (649°C)				
	37.5 hr	130 hr	240 hr	1050 hr	2500 hr
Zirconium alloy (wt. %: 1.25 Al, 1 Sn, 1 Mo)	0.37	0.50	0.76	0.74	0.69
Unalloyed zirconium	0.41	0.53	0.73	0.68	0.69

Zirconium is not compatible with molten uranium; a 9-hr test at 1300°C (2372°F) gave complete penetration of a 0.060-inch (1.52 mm) thick zirconium crucible [38].

In one study zirconium was found to be the only satisfactory metal in contact with dynamic molten lithium at 816°C (1500°F) [39]. Zirconium-niobium alloys are also containment materials of current promise for molten rubidium and lithium in addition to sodium and NaK at temperatures up to 1093°C (2000°F) [40].

When zirconium is in contact with fused salts, protective films are formed as indicated by the decrease in corrosion rates with time, just as is the case for oxidation by water or air. For example, with zirconium in contact with a molten nitrate bath at 500°C (932°F) rates of 41.7×10^{-4}, 3.6×10^{-4}, and 1.7×10^{-4} mg/(mm^2)(hr) were obtained for times of 0.5, 20, and 55 hours, respectively [41]. Zirconium and its alloys readily dissolve in fused fluoride salts such as NaF-ZrF_4-UF_4 at temperatures of 600 to 700°C (1120-1300°F) [42].

2.1.2 Aluminum Alloys

Aluminum in contact with water or steam forms a highly protective barrier film of hydrated aluminum oxides. The composition of the oxides depends on the temperature, and the protectiveness is sensitive to the presence of impurities in the metal and particularly to impurities in the water. Commercially pure aluminum (1100 or 2S Aluminum) is generally satisfactory for use in water below 100°C. Relatively rapid initial corrosion occurs with the evolution of hydrogen and the production of gelatinous aluminum hydroxide. The initial reaction follows a logarithmic rate law and oxide is predominantly Bayerite ($Al(OH)_3$) although there is an amorphous layer and a layer of Boehmite (AlOOH) at temperatures below 100°C [43]. After a few days of exposure, the corrosion reaction reaches a nearly constant rate which continues indefinitely. The reaction rate increases uniformly with temperature above 100°C [44] where the corrosion product is predominantly Boehmite. Uniform attack occurs up to 200°C (392°F), but at still higher temperatures a vigorous penetrating attack may occur.

There has been considerable effort to improve upon the corrosion properties of 1100 aluminum at temperatures above 200°C (392°F). The effects of many alloying additions have been studied in 250 to 350°C (482-662°F) water [45, 46] and in superheated steam of 540°C (1004°F) [47, 48, 49]. Probably the most useful aluminum alloy developed to date has been the X8001 alloy (1 wt.% Ni, 0.5 wt.% Fe). The addition of Ni improves high temperature corrosion resistance by introducing a second phase which is cathodic to the aluminum matrix [45]. Cu and Si additions also improve corrosion resistance above 200°C. A study of fifty commercial aluminum alloys in water showed that only those alloys containing substantial amounts of Cu, Ni, or Si survived six weeks in water at 300 and 350°C (572 and 662°F) [46]. Most of the other alloys failed rapidly by penetrating corrosion. It has been postulated [45] that the rapid penetrating attack is due to a buildup of corrosion-generated hydrogen bubbles which form beneath the oxide film. Mechanical failure of the film then exposes unprotected metal to a hydrogen, steam mixture which in turn produces new blisters at an accelerated pace. Commercial 1100 aluminum was destroyed in about 60 hours in 400°C (725°F) 400 psi (27 atm) steam [47].

Increased corrosion in flowing systems has been attributed to an accelerated dissolution of the protective oxide film. The solubility of aluminum corrosion product in water was determined [50] and the dynamic corrosion rate data were correlated on the basis of parabolic film growth and a linear oxide degradation process acting simultaneously.

Increased aluminum corrosion in water occurs at both extremes of pH. A minimum corrosion rate occurs in the pH 6 to 7 range at 50°C (122°F). Optimum pH is lower at higher temperatures. Corrosion is decreased by the presence of small quantities of oxidizing agents such as dissolved oxygen, hydrogen peroxide, and chromate ion. The presence of phosphate and sulfate has been shown to decrease corrosion rates. Sodium silicate solutions (waterglass) also decrease corrosion rates [51]. The presence of chloride ions even at 1 ppm or cupric ions at 5 ppm can lead to an accelerated pitting corrosion [43]. Electrical coupling of aluminum to Cu, Ni, or graphite parts can lead to galvanic attack under certain circumstances.

Aluminum and its alloys form a very thin (ca. 20 Å) film of amorphous alumina in air at room temperature [52]. The film protects the metal against further oxidation. Thicker, but very protective oxide films form up to about 350°C (662°F). The reaction between about 350 and 450°C (842°F) is parabolic. Between 450 and 600°C (1112°F) the reaction has three branches over which the rate decreases, is constant, and decreases again to very small values, ca. 0.05 μg/cm^2-hr. Weight gains after 170 hr decreased as the temperature increased from 450°C to 600°C [53].

The reaction of aluminum with carbon dioxide was not measurable after a 5-month exposure to 8 atm of CO_2 at 500°C (932°F) [38].

The solubility of hydrogen in aluminum becomes measurable above about 400°C (752°F) and is appreciable in molten aluminum [54]. Hydrogen is not absorbed by the solid metal at lower temperatures although a stoichiometric hydride, AlH_3, can be prepared by indirect methods.

In general, aluminum has poor resistance to liquid metals [33]. However, type 2S and 3S aluminum can be considered for long-time use in contact with Na and NaK alloys at temperatures up to 200°C. At 427°C (800°F) and 500°C (932°F) in static filtered Na and NaK (containing probably 0.005 to 0.01 wt.% oxygen), 99.99% pure aluminum gives a weight change of -10 and -47.3 mg/cm^2-month, respectively, which denotes poor corrosion resistance [33]. Various aluminum alloys such as 3S (3003) and 52S (5052) gave corrosion rates of -9 to -17 mg/cm^2-month also at 427°C [33]. However, other tests have shown much lower corrosion rates for aluminum; for example in distilled Na at 450°C (842°F) high purity aluminum showed a rate of weight change of only -0.25 mg/cm^2-month which indicates excellent resistance. This suggests that the concentration of contaminants in the Na is an important factor. Alumina, Al_2O_3, is acceptable for use in contact with Na and NaK up to 500°C (932°F) provided the ceramic is not porous. A ceramic 1 $Al_2O_3 \cdot$ 1 BeO, shows a corrosion rate of 0.0003 mg/cm^2-month at 500°C in Na [33]. Aluminum has poor resistance to liquid Li and Hg [33]. Therefore, in systems containing aluminum, the inadvertent introduction of mercury, such as from manometers or thermometers, must be avoided. 2S (1100) aluminum can be used in contact with molten Pb and Pb-Bi alloys at temperatures slightly above their melting point, or about 300°C (572°F); at 500°C (932°F) the resistance is limited.

The use of aluminum containment for fused salts such as mixed fluorides is limited simply by the melting point of the salt mixture, which is, usually, so close to that of the aluminum that the aluminum has very little structural strength at the high temperatures required.

Aluminum showed good corrosion resistance during tests in an organic-cooled (Santowax OM, a polyphenyl) loop. Neither the ultimate nor tensile strength of specimens was found to change appreciably during 13 months of operation at 316°C (600°F) [55]. Type 1100 Al showed weight changes of ±0.3 mg/cm^2 during in-pile corrosion tests with a polyphenyl coolant during a 28-month exposure [35].

An APM, aluminum powder metallurgy, alloy has been developed for organic-cooled and moderated reactors. The alloy contains about 5% Al_2O_3 and the balance is essentially aluminum [56]. In themselves, the pure organic fluids of interest in reactor technology are in general not corrosive toward metals, and pile irradiation appears to have little influence on corrosion; however, contaminants such as traces of air, hydrogen, and water vapor can react with metals [57].

2.1.3 Stainless Steels

Stainless steels, or more specifically corrosion-resisting or heat-resisting steels, owe their inertness primarily to the presence of chromium. When the concentration of dissolved chromium in steel exceeds about 11 wt.%, immunity to atmospheric corrosion begins and improves gradually with increasing chromium content [58]. The corrosion resistance of chromium-iron alloys at low temperatures is believed due to the formation of a very thin, transparent chromium oxide film. The film depends for stability on a continuing supply of chromium and oxygen atoms. As a result of this, the chromium must be in a dissolved form and not fixed in a compound. The need for oxygen is readily satisfied in a water or steam environment. At higher temperatures the corrosion film is composed predominantly of magnetite, Fe_3O_4. The diffusion of iron ions through the film is believed to be rate-determining above 232°C (450°F) [3]. The diffusion is inhibited by the chromium oxide at the metal-oxide interface. Many elements can be added to the basic iron-chromium alloy to achieve corrosion resistance and improved physical properties. Nickel is a particularly desirable addition to improve corrosion resistance.

The principal disadvantages of stainless steels in water-cooled thermal reactors are their high thermal neutron cross section and the tendency of the magnetite corrosion film to slough off into flowing water forming activated cruds. The possibility of forming the brittle sigma phase at high operating temperatures, leading to brittle fracture, may also be of concern [59]. The lower temperature limit for its formation depends exponentially on exposure time; it can be placed near 1000°F (538°C) for practical purposes. Excessive intergranular corrosion can occur if the carbon concentration is too high. Pickling in acid solution can also lead to selective attack at grain boundaries [59].

Typical corrosion results for various stainless steels are given in Table 2-1. Usually, there is an initial high rate of weight change followed by a slower reaction.

There are four principal types of stainless steel:

a) austenitic (AISI 300-series) non-hardenable;
b) ferritic (AISI 400-series) non-hardenable;
c) martensitic (AISI 400-series) hardenable;
d) precipitation (17-4-PH, 17-7-PH, AM350, AM355, stainless W, for example) hardenable.

Austenitic steels are generally used in applications where corrosion resistance is the principal factor to be considered. Hydrogen dissolved in the water reduces corrosion. Intergranular corrosion and stress-corrosion cracking are generally not experienced with austenitic stainless steels unless chlorides are present in the water. Therefore, any chlorine containing compounds, even degreasing agents such as trichloroethylene, should be kept away from equipment intended for nuclear service or where radiation can release chlorine from compounds. Ferritic stainless steels such as 430 and 446 contain no nickel and are susceptible to pitting and local galvanic attack. Martensitic stainless steels such as 410 and 416 are the least corrosion resistant of the stainless steels. They tend to develop a thick, black oxide film. Some of the precipitation hardening stainless steels have excellent corrosion resistance as shown in Table 2-1. Experience in the Dresden Reactor has shown, however, that 17-4-PH stainless steel may be very susceptible to stress-corrosion cracking. Present indications are that there is little tendency to crack if the metal is aged at a higher temperature [593°C (1100°F) instead of 482°C (900°F)] after solution heat treating [60]. The previous heat treatment apparently did not sufficiently remove residual stresses.

Certain nickel-chromium alloys, though not strictly steels, are also considered for use in reactor applications where high nickel content and the resulting absorption of neutrons (in thermal reactors) is not a problem. Thus the Inconel (73 Ni, 15 Cr, etc.) alloys and the Hastelloys (45 Ni, 22 Cr, etc.) are among the most corrosion resistant of the non-precious metals known. Results of a corrosion study of stainless steels and nickel-chromium alloys in 800, 1000, and 1350°F (427, 538, and 732°C) 5000 psi (340 atm) steam included in Table 2-1 showed that Hastelloy F, ARMCO 17-4-PH, and ARMCO 17-7-PH were the most resistant alloys tested [61]. The attack of all materials at 732°C (1350°F) was localized in nature.

The reaction of stainless steels with oxygen and air is very similar to the reaction with steam.

The formation of a highly protective layer of chromic oxide (Cr_2O_3) is the principal reason for the extreme inertness of stainless steels. Unalloyed iron reacting with oxygen above 570°C (1058°F) produces the oxide, wustite, FeO, which because of its high concentration of vacancies allows a relatively rapid diffusion of iron ions and a high reaction rate [52]. The presence of nickel, chromium, or manganese in the metal tends to prevent the formation of wustite, producing instead more stable double oxides of the spinel type such as $(FeCr_2)O_4$, $(MnCr_2)O_4$, $(NiCr_2)O_4$, etc.

Oxidation-time curves with 18-8 chromium-nickel and 16%-chromium stainless steels are initially parabolic between 815 and 980°C (1500 and 1796°F). The curves show abrupt increases after certain times which are caused by rupture of the protective spinel layer.

In general, the oxidation resistance of 18-8 chromium-nickel stainless steels is about equivalent to a 25% chromium-iron alloy. Both, however, are inferior to 80-20 chromium-nickel alloys. Oxidation rates in oxygen are somewhat lower than in air at 1050°C (1922°F).

The reaction of stainless steel with carbon dioxide was not measurable after a 5-month exposure to 8 atm of CO_2 at 500°C (932°F) [38].

Iron, chromium, and nickel, separately, are classified as endothermic occluders in regard to their ability to absorb hydrogen. They absorb very small, but measurable, quantities of hydrogen at room temperature and considerably greater quantities at higher temperatures [54]. Nickel has the greatest capacity for hydrogen of the three. At 1200°C (2192°F) the solubility reaches 13 cc H_2/100 g Ni in hydrogen at 1 atm pressure. The solubilities of hydrogen in the three metals are roughly doubled when the metals are melted and continue to increase as the molten metals are heated further. Metals in the category under discussion can be charged with hydrogen considerably beyond their equilibrium capacity. This is accomplished, often inadvertently, by cooling from higher temperature or mechanically working in a hydrogen atmosphere. Atomic hydrogen introduced by aqueous corrosion or other electrolytic process can also accumulate beyond the equilibrium capacity.

Armco iron, 310 stainless steel, and 27 Cr ferritic stainless steel show good corrosion resistance to Na and NaK up to 900°C (1652°F). Nickel, Inconel, Nichrome, and Hastelloys A, B, C, are also good up to 900°C (1652°F) [33]. At 500°C (932°F) the corrosion rates for these ferrous and nickel base alloys are of the order of 0.01 mg/cm²-month which is excellent [33].

Ferrous alloys are satisfactory for use with static mercury but in dynamic systems they are rapidly attacked [58]. The rate of attack of low-carbon steel in a flowing "Harp test" with mercury varied from 5 mils/year (0.13 mm/year) at 500°C (932°F) to 560 mils/year (14.2 mm/year) at 800°C (1472°F) [33]. An evaluation of mercury materials compatibility problems indicated that Haynes 25 was the most resistant and 410 stainless steel was good to 538°C (1000°F) [62].

Steel containing up to 0.5% carbon and stainless steels have good resistance to attack by lead up to 600°C (1112°F) and poor resistance at 1000°C (1832°F) [33]. Of the nonferrous metals, tantalum and niobium appear best and can be used up to 1000°C in contact with molten lead and bismuth alloys.

Iron, ferritic, and austenitic stainless steels can be used in contact with molten lithium, up to 816°C (1500°F) in static systems and up to 593°C (1100°F) in dynamic systems, for 1000 hours with less than 5 mils (0.13 mm) attack [39]. But if the lithium is contaminated by nitrogen, then 35 mil (0.89 mm) penetration of 316 stainless steel will result at 800°C (1472°F) in 100 hours; this is caused by the lithium nitride reaction with the carbides that form the grain boundary materials [34]. Table 2-3 [63] shows data on the rate of removal of type-347 stainless steel in lithium. The 300 series austenitic stainless steels are acceptable materials of construction for molten lithium hydride, at least for 100 hours, up to 718°C (1325°F) [64]. At 1325°F the corrosion rates were about 10^{-5} mg/in²-hr.

TABLE 2-3

Attack and Concentration Data
for Type-347 Stainless Steel in Lithium

Element	Attack rate* (removal of source metal) in Lithium in μg/cm²-month		
	825°C (1517°F)	625°C (1157°F)	425°C (797°F)
Fe	1922	5.8	1.0
Ni	70	6.0	0.07
Mn	62	2.3	0.09
Co	4.1	0.1	0.002
Ta	0.24	0.007	0.006
Zn	0.24		
Ag	0.17	0.0028	0.0007
Sn	0.008	0.0006	<3 x 10^{-5}

Element	Equilibrium concentration in Lithium (μg/g of Lithium)		
	825°C (1517°F)	625°C (1157°F)	425°C (797°F)
Fe	0.31	0.037	0.15
Ni	20		0.071
Mn	56	3.1	0.094
Co	0.0069	0.00015	0.00016
Ta	0.61	0.0058	0.027
Zn	0.99	0.0006	0.026
Ag	0.69	0.012	0.0026
Sn	0.017	0.0012	0.00012

*30-day isothermal exposures. Metal removal determined by radiotracers.

Tantalum containers are used for molten Pu-Fe alloy fuel at 600 to 800°C. The tantalum undergoes some intergranular corrosion but it is not prohibitive. Addition of C, Si, or Al to the tantalum reduces the attack; also, annealing at high temperatures, 1300 to 1450°C (2372 to 2642°F), gives improvement [37].

In fused fluoride salt systems, INOR-8 [Ni, 15-18 Mo, 6-8 Cr, 5 Fe_{max}] showed superior corrosion resistance in 20,000-hour dynamic tests at 704°C (1300°F). Metallographic examination of the metal showed either no attack or pitting up to 2 mils (0.05 mm) [65]. Nickel base alloys give better corrosion resistance than do iron base alloys and the attack for both increases with chromium content [66]. Hastelloy B shows good corrosion resistance to fused fluoride salts, such as molten $ZrF_4 \cdot UF_4 \cdot NaF$ but it has other adverse characteristics such as poor fabricability and brittleness at 650-815°C (1202-1499°F) [66].

Type 304 and 410 stainless steel and 1020 carbon steel showed negligible changes in ultimate tensile strength and yield strength after an in-pile exposure of 8 months to 316°C (600°F) Santowax OM; the neutron irradiation was about 10^{20} nvt and examination of the specimen led to the conclusion that the polyphenyl environment was noncorrosive to these various steels [55]. Other in-pile tests showed the following weight changes for a 28-month exposure at 316°C (600°F) to the polyphenyl moderator-coolant [35]: 1018 C steel, -1.76 mg/cm^2; 410 stainless steel, +0.18 and +0.85 mg/cm^2; 304 stainless steel, +0.22 and +0.01 mg/cm^2.

2.1.4 Magnesium Alloys

The low value of the thermal neutron capture cross section makes magnesium a potential cladding and structural material for use in gas-cooled or water-cooled thermal reactors. Rapid corrosion [67], however, mitigates against long service or use in high temperature water (Table 2-1). Corrosion rates of magnesium alloys increase rapidly with temperature. Pure magnesium disintegrates in a 2.9-day test in pure water at 120°C (248°F). Reactions appear to be linear in studies between 50 and 150°C (122 and 302°F). Corrosion is inhibited somewhat by the accumulation of the corrosion product magnesium hydroxide, $Mg(OH)_2$, in the water. Commerical magnesium alloys such as AZ-31 (27% Al, 1.0% Zr, 0.2-0.4% Mn) show considerably less corrosion than the pure metal. Aluminum, copper, and tin are the most beneficial alloying additions. Minimum corrosion occurs at a pH of 6-7 with a fluoride content in the water of about 10 ppm. Corrosion rates for these alloys are about 1400 mg/dm^2-day (3.2 mils/day) in water at 150°C (302°F).

Slow oxidations of magnesium in air or oxygen are very complex reactions. The reaction with oxygen at room temperature and up to nearly 250°C (482°F) follows a logarithmic rate law in which very thin equilibrium films are formed. Cubic MgO is formed in dry air or oxygen while hydrated oxides are formed when moisture is present. Above 250°C but below 450°C (842°F) in dry oxygen, the oxide film remains protective and a parabolic rate law is followed for considerable lengths of time [52]. Above 475°C (887°F) the oxide loses its protectiveness rapidly and linear attack ensues. The magnesium reaction with dry oxygen at 525°C (977°F) undergoes an induction period during which lateral growth of white oxide occurs across the specimen [68]. The reaction rate then increases and becomes linear. After about 50 hours a transition to a second and still more rapid linear reaction occurs. At 575°C (1067°F) the second transition occurs in about 5 hours and ignition may result. The induction period is believed to correspond to the building up of a protective oxide film and the first transition occurs as this film begins to crack. The linear reaction results from equality between the rates of oxide film growth and of cracking. The thickness of the underlying protective film remains effectively constant. Reaction occurs by the diffusion of magnesium ions (Mg^{++}) through the oxide. The second transition, or breakaway, corresponds to complete failure of the underlying protective film and exposure of bare metal. The high vapor pressure of magnesium, 0.19 mm at 550°C (1022°F), leads to the conclusion that metal is vaporizing and reacting in the vapor phase during the second linear phase of reaction. Vapor phase reaction does not occur below about 515°C (959°F).

The following rate law describes the first linear phase of the reaction between 500 and 575°C (932 and 1067°F):

$$w = 9.5 \times 10^{11}\, t[\exp(-48{,}700/RT)], \quad (2\text{-}7)$$

where w is the quantity of oxygen reacted in mg/cm^2, t the time in hr, R the universal gas constant, and T the temperature in °K. An idea of the magnitude of the rate is obtained by noting that about one atomic layer of magnesium is reacted each second at 525°C (976°F).

The presence of moisture in the reacting gas lowers the temperature at which the oxide film ceases to be protective, shortens the length of the induction period and increases the rate of linear oxidation. Moisture inhibits the development of the second linear (vaporizing) reaction. Details of the reaction, particularly at lower temperatures, are affected markedly by the method of surface preparation.

At 475°C (887°F) rates of oxidation of magnesium were markedly increased by the addition of a few percent of Cu, Ni, Ga, Zn, Sn, and Al [69]. The addition of as little as 0.01% Be reduces the reaction rate in air at 580°C (1076°F) by a factor of 40 or more [70]. The increased oxidation resistance of magnesium-beryllium alloys is the basis of the Magnox series alloys used in British gas-cooled reactors.

Rates of the magnesium oxidation are increased by water vapor, decreased somewhat in the presence of nitrogen and decreased markedly in the presence of SO_2 or CO_2 as shown in Table 2-4. [71]

TABLE 2-4

Rate of Oxidation of Magnesium

Gas	Oxidation Rate of Pure Magnesium at 550°C (1022°F) and 760 mm Hg [mg/cm^2-hr]
O_2	0.24
50% O_2 + 50% N_2	0.059
1% O_2 + 99% N_2	0.026
O_2 + 5% SO_2	negligible
SO_2	negligible
O_2 + 11% CO_2	negligible
CO_2	negligible
O_2 sat'd. with H_2O at 28°C (82°F)	0.61
CO_2 + 8% H_2O	rapid

Results shown in this table indicate that magnesium and its alloys are not rapidly attacked by CO_2 at 550°C (1022°F). This is apparently due to the formation of protective carbonate films[72]. Samples gain about 0.3mg/cm^2 in 3000-hour tests at 550°C.

Magnesium does not react with normal hydrogen [54]. Magnesium absorbs hydrogen only when there is a source of free protons, e.g., in the presence of an electrical discharge or as a result of aqueous corrosion. Hydrogen absorbed in either way cannot be readily expelled by heating. The inability of magnesium to absorb or expel molecular hydrogen is taken as evidence that hydrogen cannot diffuse interstitially or enter into solid solution. Hydrogen introduced into magnesium as protons becomes trapped on a submicroscopic scale.

Magnesium has limited resistance to corrosion by Na and NaK and poor resistance to molten Li and also to Hg [33]. Magnesium cannot be used with mercury because of alloying [38]. However, magnesium has excellent compatibility with uranium.

Magnesium in contact with a polyphenyl moderator for 28 months at 316°C (600°F) gave very high corrosion of -17.8 to -24 mg/cm^2, and anodized magnesium completely corroded away in that time [35]. Magnesium corrosion is very sensitive to water impurities in the organic coolant; therefore, magnesium is considered to be unsatisfactory for practical use in organic-cooled and -moderated reactors [57].

2.2 Corrosion of Core Metals

2.2.1 Uranium

Uranium reacts with water, giving uranium dioxide and hydrogen. At room temperature, or somewhat above, the corrosion rate in water is a function of the dissolved gas as shown in the first two columns of Table 2-5 [73]. However, at 80°C (176°F) or above, the corrosion rate is essentially the same for air-saturated or hydrogen-saturated water and then the temperature becomes an important variable as shown in the third and fourth columns of Table 2-5. Heat treatment of the uranium can reduce its corrosion in water 20 to 30 percent. It is generally necessary to clad

TABLE 2-5

Corrosion of Uranium in Water

At 70°C (158°F) with various dissolved gases		At various temperatures with hydrogen-saturated water (or air-saturated water)	
Dissolved gas	Corrosion rate* (mg/cm^2-hr)	Temperature	Corrosion rate* (mg/cm^2-hr)
H_2	0.35	50°C (122°F)	0.066
N_2	0.25	70°C (158°F)	0.45
He	0.21	90°C (194°F)	1.00
Air	0.01	100°C (212°F)	2.7
O_2	0.003	183°C (361°F)	139
		226°C (439°F)	~800

*Conversion factors: 1 mg/cm^2-hr = 730 mg/cm^2-month = 15 mils/month (assuming 19 g/cm^3 density).

uranium fuel elements to avoid aqueous corrosion and to prevent escape of fission products. Extensive information exists on the various alloys of uranium and their corrosion resistance; among the important alloying elements are Mo, Nb, and Zr (see Fig. 2-1)[74].

The reaction with oxygen below 300°C (572°F) has been shown to occur in two stages [75]. The first-stage reaction appears to be parabolic below about 150°C (302°F) and linear between 150 and 250°C (482°F). The oxidation accelerates to a more rapid linear rate after about 50 μg O_2/cm^2 reaction. Microscopic studies of oxidizing uranium surfaces have shown that the onset of the second stage reaction coincides with the formation of oxide nodules, which suggests that the increase in oxidation rate is associated with an increase of surface area [76]. X-ray diffraction studies have indicated that UO_2 is produced along with higher oxides such as U_3O_7[77]; no U_3O_8 was found in the thin oxide films used. The presence of ten additive elements in uranium metal caused only very small effects on the rate of oxidation [78].

Above 300°C (572F), the reaction of uranium with air and oxygen is rapid with a tendency for the samples to self-heat. Isothermal oxidation rates between 300 and 600°C (1112°F) and probably at higher temperatures are nearly linear. Linear oxidation rates from a variety of sources are plotted in Fig. 2-2. The wide scatter of data reflects both the difficulty of experimental temperature control and the extreme sensitivity of

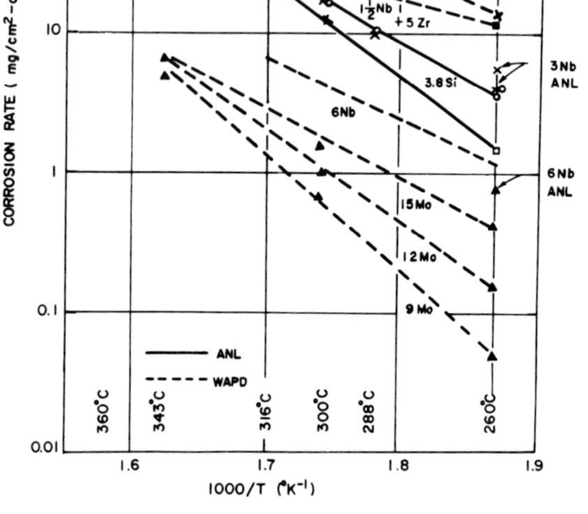

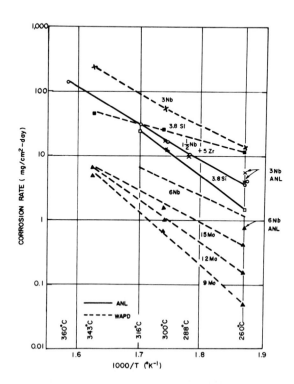

FIG. 2-1 Corrosion of uranium-base alloys in distilled water.

the oxidation reaction to the presence of impurities and to the metallurgical history of specimens. Overall consumption rates of furnace-heated cylindrical uranium specimens in air are given in Fig. 2-3 [79].

The reaction between uranium and nitrogen is slow up to 400°C (752°F); at 750°C (1382°F) the nitridation rate is 12 mg/cm²-hr. From 775 to 900°C (1427 to 1652°F) the reaction rate is quite rapid and the parabolic rate constant is given by:

$$k = 6.2 \times 10^{-6} [\exp(-15{,}500/RT)], \qquad (2\text{-}8)$$

where k is $(g/cm^2)^2/sec$.

Uranium readily reacts with dry hydrogen. Powdered uranium reacts at -80°C (-112°F). At 430°C (806°F) the equilibrium pressure of hydrogen above UH_3 is 1.0 atm. Because of a balance between the rate of formation and decomposition of the UH_3, there is a maximum in the rate of attack at 225°C (437°F).

In Na or NaK the corrosion rate of uranium is quite small (see Table 2-6) [1a]. The solubility of uranium in liquid Na is 50 ppm at 300°C (572°F). The corrosion rate increases markedly if there is oxygen present in the Na or NaK. Uranium does not corrode in molten Pb at temperatures up to

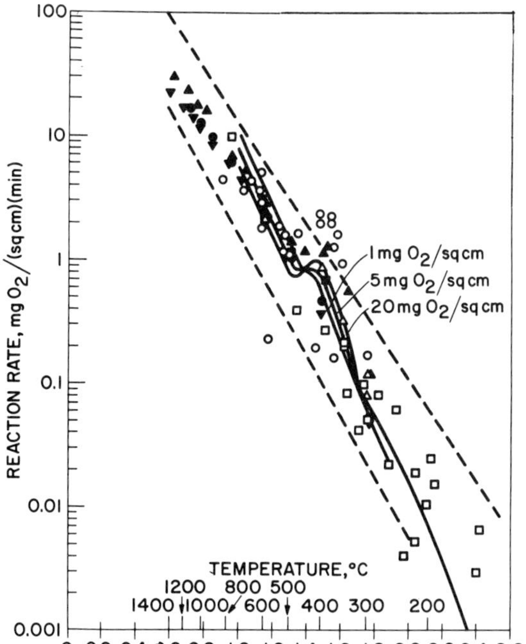

FIG. 2-2 Oxidation rates of uranium in air and in oxygen. The open triangles indicate data of Bessonov and Vlasov in air and oxygen. Open circles are data of Isaacs and Wanklyn in air. The solid circles, squares, triangles (upright and inverted) designate data of Hilliard for various sample sizes in air. The open squares, bounded by broken lines, designate the data of various other workers, measurements made in air and oxygen. The solid curves are ANL data (Baker et al.) after various extents of oxidation in oxygen.

350°C (662°F); however, at 800°C (1472°F) complete penetration of large samples occurs. Uranium does not corrode in contact with molten Li at temperatures up to 600°C (1112°F). The solubility of uranium in molten Bi is very small at 200°C (392°F), but at 1000°C (1832°F) dissolution occurs.

TABLE 2-6

Corrosion Rate of Uranium in Na or NaK

Temperature	Corrosion rate mg/cm²-hr
200°C (392°F)	1.11×10^{-4}
500°C (932°F)	1.39×10^{-4}
600°C (1112°F)	$43. \times 10^{-4}$

The oxidation of uranium by carbon dioxide proceeds as follows [80]:

$$U + 2CO_2 \rightarrow UO_2 + 2CO,$$

$$U + 2CO_2 \rightarrow UO_2 + 2C.$$

The weight gain in CO_2 at 1 atm is shown for various temperatures in Table 2-7. Addition of Ti, Mo, Nb and Cu reduce the attack by CO_2 at 680 to 1000°C (1256 - 1832°F) by factors up to 500, but none of these elements is markedly beneficial at 500°C (932°F) [81]. For additional information on the corrosion of uranium and its alloys a comprehensive summary is available [82].

TABLE 2-7

Oxidation of Uranium by Carbon Dioxide at 1.0 atm

Temperature	Wt. gain in one atm CO_2, mg/cm²-hr
450°C (842°F)	0.1
500°C (932°F)	~ 1
600°C (1112°F)	~ 5
700°C (1292°F)	~ 100
800°C (1472°F)	~ 500

2.2.2 Plutonium

Early plutonium oxidation studies at Los Alamos [83] and in Great Britain [84], conducted in air at temperatures below 100°C, demonstrated a marked increase of oxidation in the presence of moisture (see Fig. 2-4) [86]. In humid air for extended periods of time, oxidation rates of plutonium are quite complicated; for example, rates have been reported to follow logarithmic, parabolic, linear, and accelerating rate laws [85, 86, 87, 88]. In dry air, parabolic behavior is usually observed. Oxidation in dry air at 100°C follows a succession of parabolic-shaped curves [88]. The first parabolic rate continued 9 days during which

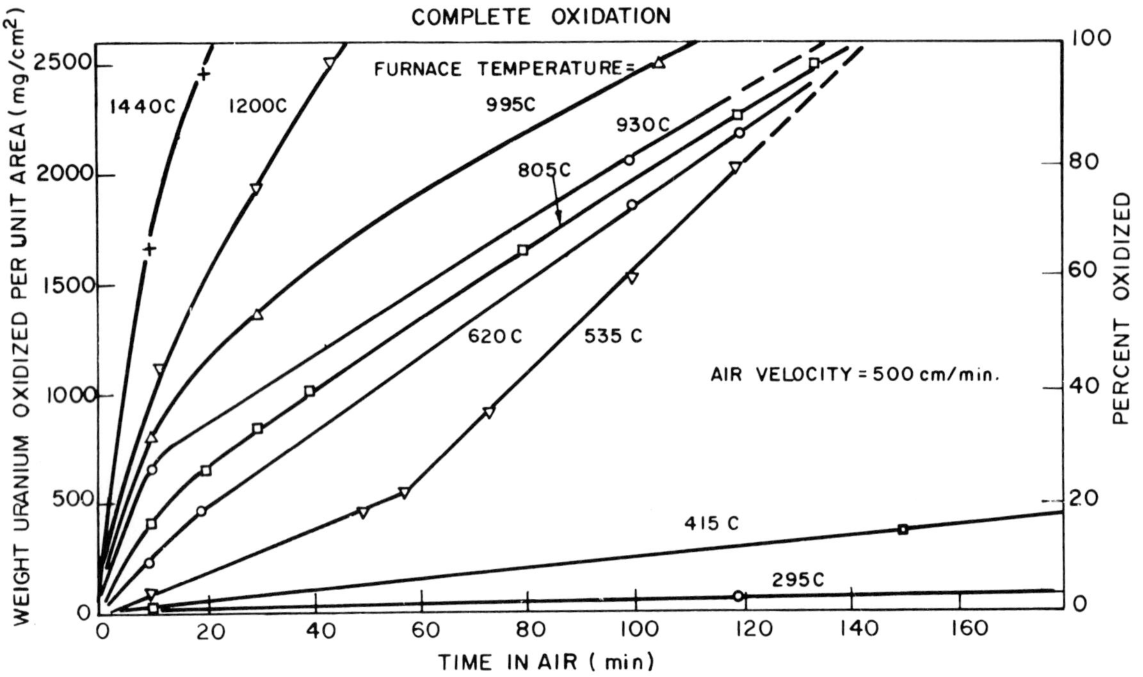

FIG. 2-3. Weight of uranium oxidized vs time for 1/4 in. (6.3 mm) diameter × 3/4 in. (19.0 mm) long specimens at various furnace temperatures.

period the weight gain was 2.4 mg/cm^2. This would correspond to a film of plutonium oxide less than 20 μ (0.8 mils) thick. It was thought that the rate of oxidation is limited by diffusion of oxygen ions through a thin film of PuO or Pu_2O_3. The oxide observed after completion of the experiment was PuO_2.

The oxidation kinetics between 140 and 320°C (284 and 608°F) can be described in terms of a parabolic followed by a linear rate law, i.e., a paralinear rate law [89, 90]. The initially parabolic portion of the reaction continues until approximately 1 μ (0.04 mil) of oxide is built up in the temperature range of 120-220°C (248-428°F). The reaction then obeys the linear rate law as indicated in the Arrhenius plot in Fig. 2-5. The initially parabolic portion of the reaction continues until the oxide film is approximately 50 μ (2 mils) in thickness at 300°C (572°F). The paralinear reaction suggests that the rate is controlled by diffusion of oxygen ions through a thin adherent oxide film. Conversion to the linear rate occurs when the oxide film begins to crack. The adherent interfacial oxide may include both cubic and hexagonal Pu_2O_3 while the readily removable oxide has been shown to be PuO_2.

In the temperature range of stability of the beta phase, after the paralinear law is followed to a weight gain of ~400 μg/cm^2, a faster linear rate of ~20 μg/cm^2-min is observed. The oxide produced in this "third stage" is green instead of gray and is a loose powder. The third stage does not appear to have any temperature

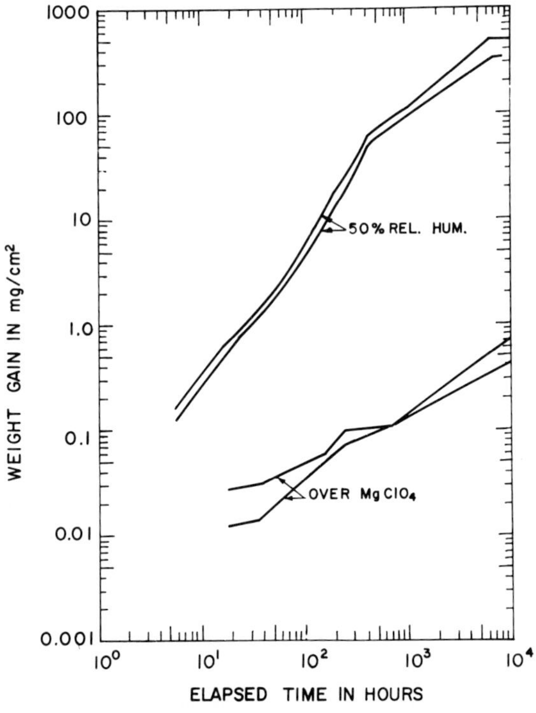

FIG. 2-4 Effect of moisture on the oxidation of plutonium at 75°C (168°F).

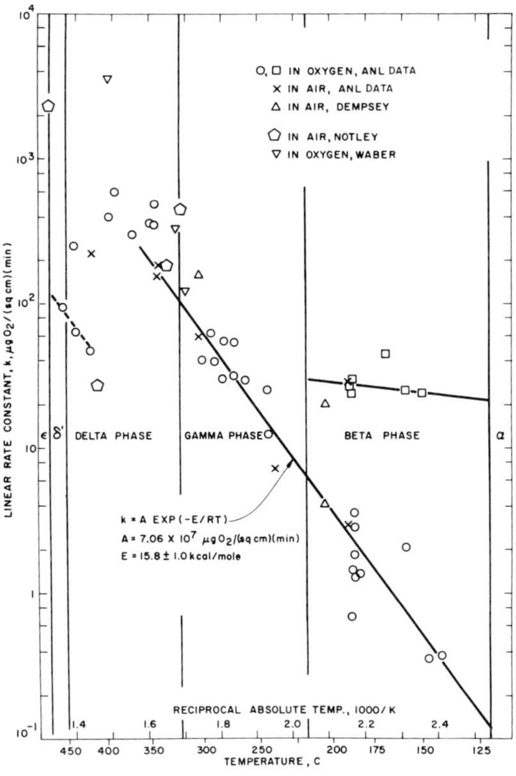

FIG. 2-5 Linear oxidation data for plutonium.

dependence as shown in Fig. 2-5. Very recent experiments have shown the third stage to be associated with moisture content of the oxidizing gas. The third stage is observed at 190°C with 180 ppm moisture in oxygen, but if extreme care is used to remove moisture to below 20 ppm the third stage does not occur.

Several investigators [89, 90, 91] have noted that a different reaction process occurs above 300°C (572°F) which is characterized by an increased protectiveness of the oxide film and the occurrence of a minimum rate in the vicinity of 400°C (752°F). Discontinuities have been noted in thermal expansion and electrical resistivity measurements of plutonium oxides with an oxygen to plutonium ratio between 1.64 and 1.98 at approximately 300°C (572°F) [92, 93]. The plutonium-oxygen phase diagram indicates a phase change in the oxide at 300°C [93]. It would therefore appear more reasonable to attribute the change in the oxidation kinetics near 300°C to a change in the oxide than to attribute it to a change in the metal structure. The oxidation data would be explained in terms of a low diffusion rate of oxygen through the oxide phases that are stable above 300°C (572°F). The minimum rate at 400°C (752°F) would be the net result of this and the increasing diffusion at higher temperatures.

Table 2-8 [94] summarizes the reactions of plutonium with various gases. Plutonium reacts slowly with nitrogen at 250°C (482°F); even at 1000°C (1832°F) complete conversion to the nitride,

TABLE 2-8

Summary of Reactions of Plutonium with Various Gases

Gas	Product of reaction	Rate of reaction
H_2	PuH_2 to PuH_3	Slow at 25-50°C (77-122°F), very rapid at 200°C (392°F). Product normally has composition near to PuH_3.
N_2	PuN	Very slow at low temperatures, slow at temperatures of 800-1000°C (1472-1832°F).
CO_2	PuO_2 (mainly)	Rapid at comparatively low temperatures. When gas is in excess it is reduced to CO but if metal is in excess reduction to C may occur.
O_2	PuO, Pu_2O_3, PuO_2	Reaction with dry O_2 is slow at room temperatures, when partially protective films are formed. Presence of moisture greatly accelerates attack. Ignition of massive Pu may occur at 300°C (572°F).
NH_3	PuN and PuH_2	Reacts at 800-1000°C (1472-1832°F).
Halides		Strongly exothermic reaction with all halides and halogen acid gases.

PuN, does not occur. PuN is quite reactive and is very easily decomposed by moisture or by heating in air.

Plutonium reacts appreciably with hydrogen at 25°C (77°F) and very rapidly at 200°C (392°F) [95]. A hydride of the approximate formula PuH_3 is formed. This compound is stable up to 1000°C (1832°F) under a 1.0 atm hydrogen pressure. However, PuH_3 is oxidized in air at 100°C and reacts smoothly with water. PuH_3 is converted to PuN by heating in N_2 at 230°C (446°F) and to the carbide by contacting with graphite at 800°C (1432°F).

No data are available on the oxidation rates of plutonium by carbon dioxide, but thermodynamic data indicate that a reduction would occur with the formation of CO or C and that the carbon might then form plutonium carbide [95].

Plutonium forms compounds with mercury, lead, and bismuth, but not with lithium, sodium, or potassium. Mercury was used as a coolant in the first fast reactor, Clementine, which was dismantled as a result of fuel element failure. Plutonium should reduce the oxides of sodium and potassium. Thus, severe corrosion would be expected unless these liquid metals were very pure [95]. No information could be found on plutonium in contact with organic coolants.

2.2.3 Thorium

The literature on the reactions of thorium with air, oxygen, nitrogen, and water vapor is somewhat limited. The oxidation of thorium in air [96, 97] and oxygen [98, 99] has received the attention of a number of investigators but comparisons of their results are limited by wide differences in the temperature ranges studied and in the durations of their experiments. The disagreement in observed rate laws, values for the activation energy, and in selected values of experimental rate constants can be seen in Table 2-9 [78] which summarizes the results together with some of the more important experimental details. The reaction in air in the range 100 to 900°C (212-1652°F) was believed to be principally the oxygen reaction, although some evidence of reaction with nitrogen was observed at 400 and 500°C (752 and 932°F) [96].

Rates of reaction of nitrogen with thorium have been determined for the temperature range from 670 to 1490°C (1238-2714°F) at 1.0 atm pressure [100]. The reaction occurs at a slower rate than with oxygen and follows a parabolic law. The value of the energy of activation, 24.3 ±1.3 kcal/mole, is lower than the activation energies reported in the literature for the reaction of nitrogen with other metals.

Information in the literature on the effect of alloying additions on these reactions is limited to a few observations, mainly qualitative in nature, on the reaction in air. A 2 wt.% Be alloy corrodes at a somewhat lower rate in air than unalloyed thorium at temperatures from 425 to 500°C (797 to 932°F) (see Table 2-10) [97]. Alloys of thorium with aluminum (10 wt.%) and with copper (5 to 10 wt.%) were found to have poor resistance to oxidation, disintegrating to fine powders on standing in air [101].

The reaction between thorium and water vapor has been examined in the temperature range 200 to 600°C (392 to 1112°F) and at water vapor pressures between 40 and 100 mm Hg [102]. Thorium dioxide and hydrogen were identified as the main species formed during the reaction along with thorium dihydride which was believed to be a possible side-reaction product. The reaction data were found to follow the logarithmic rate law $w = k \log (1 + 0.45\, t)$, where w is the weight of water reacted per unit area of thorium surface, t the time, and k the rate constant. An average value of 6.44 ± 0.75 kcal/mole was calculated for the activation energy of the reaction.

If exposed to boiling distilled water, thorium becomes covered with an oxide scale and usually gains weight. Considerable variation in corrosion rates has been reported [101] for such tests, from weight gains of 0.03 mg/cm²-hr to weight losses of 0.02 mg/cm²-hr. These rates are high enough to prevent the use of unalloyed or unclad thorium in contact with water in reactor applications. A number of additive elements have been alloyed with thorium to improve its corrosion resistance. The effect of a number of such alloys in 100 to 200°C (212 to 392°F) water [103] is given in Table 2-11 [1b]. The addition of zirconium (see Fig. 2-6 [103]) or small amounts of carbon seemed to be the most beneficial. Alloys with beryllium gave somewhat inconsistent results but some compositions may be good [104].

TABLE 2-9

Kinetic Data for Reactions of Thorium in Air, Oxygen, and Nitrogen

Reference	Source of metal	Gas	Temp. range	Pressure (mm Hg)	Max. time of run (min)	Observed rate law	Rate constant	Activation energy (kcal/mole)
98	Westinghouse[a]	O_2	250-350°C (482-662°F)	450	80	Parabolic	5.0 $\mu g^2/cm^4$-min at 300°C (572°F)	31
			350-450°C (662-842°F)	210	30	Linear	3.6 $\mu g/cm^2$-min at 400°C (752°F)	22
			>450°C (>842°F)			Nonisothermal conditions		
99	Iodide crystal bar	O_2	840-1415°C[b] (1544-2579°F)	760	180	Parabolic	4100 $\mu g^2/cm^4$-min at 850°C (1562°F); 8.6 x $10^7 \mu g^2/cm^4$-min at 1415°C (2579°F)	63
97	Ames	Air	200-500°C (392-932°F)	760	300	Linear	~ 10 $\mu g/cm^2$-min at 400°C (752°F)	--
96	Ames	Air	100-400°C (212-752°F)	760	24 x 10^4	Linear	0.43 $\mu g/cm^2$-min at 100°C (212°F)	13[c]
			400-900°C (752-1652°F)	760	300	Linear	2.8 $\mu g/cm^2$-min at 400°C (752°F); 560 $\mu g/cm^2$-min at 900°C (1652°F)	18
100	Iodide crystal bar	N_2	670-1490°C (1238-2714°F)	760	200	Parabolic	1000 $\mu g^2/cm^4$-min at 670°C (1238°F); $10^6 \mu g^2/cm^4$-min at 1490°C (2714°F)	24

[a] Metal (from Ca reduction of ThO_2) contained 1.0 to 1.5 wt.% ThO_2, 0.4% Ca, 0.03% Fe.
[b] Below 1100°C (2012°F) early data discarded (nonisothermal conditions).
[c] Activation energy applies only to data obtained in range 100 to 200°C (212 to 392°F).

TABLE 2-10

Corrosion of Thorium and a Thorium-Beryllium Alloy in Air at Various Temperatures

	Weight gain (mg/cm^2) after four hours at indicated temperature			
	400°C (752°F)	425°C (797°F)	450°C (842°F)	500°C (932°F)
Th	1.72	9.48	21.40	34.75
Th-Be (2 wt.%)	2.53	3.43	5.82	9.43

An investigation was made of the ability of thorium alloys extrusion-clad with Zircaloy-2 to survive catastrophic rupture when the core alloy was exposed to high temperature water through a small artifical defect in the cladding [105]. At the same time the aqueous corrosion resistance of various thorium alloys was determined at a series of temperatures, and their activation energies were obtained. Thorium-zirconium alloys containing 5-10 wt.% Zr showed the lowest corrosion rate, 25 mg/cm^2-hr, as compared to 150 mg/cm^2-hr for unalloyed thorium in 260°C (500°F) water.

At 1000°C (1832°F) lead and bismuth dissolve thorium rapidly. Lithium, sodium and NaK do not react appreciably with thorium at temperatures in the range from 500 to 600°C (932 to 1112°F), provided they are free from oxygen themselves [101].

2.3 Corrosion of Solid Moderators

2.3.1 Beryllium

Beryllium corrosion tests in water have shown considerable variation (see p. 924 of reference [1]) but expected values can be given as follows: for water temperatures of 30-90°C (86-194°F) the corrosion rate is 4 to 130 mg/dm^2-month; for water at 260°C (500°F) the rate is 40-600 mg/dm^2-month. The corrosion resistance is better in degassed or hydrogenated water than in oxygenated water or in water containing foreign ions. Beryllium undergoes pitting and intergranular, galvanic attack when in contact with type-347 stainless steel and type-1100 aluminum. High purity metal is the most corrosion resistant. Beryllium is quite similar to aluminum in its corrosion behavior.

TABLE 2-11

Effect of Alloy Additions on the Corrosion Behavior of Thorium in High Purity Water

Test temp.	Effect of alloy additions (wt. %) on corrosion resistance			
	Improved	No effect	Slightly harmful	Harmful
100°C (212°F)	0.3 C 0.4-2.5 Be 1-4.3 Si	1.32-5.5 Fe 2-30 Zr	10 Mn 10-30 Mo 2-30 Nb 5-30 Ti 20-50 V	10, 25 Al 5, 10 Cu 1-25 Cr 5 La 25 Pb Up to 10 Mo 25 Ni Up to 1 Si 2 Ag 2 Sn 10 Ta 2 Ti 3 V
178-200°C (352-392°F)	0.4, 2.5 Be 0.3 C 5-30 Ti 5-30 Zr	1-4.3 Si		1, 1.6, 10 Be 1.2 C 2.75, 5.5 Fe 10 Mn 25 Mo 2-30 Nb 10 Ta

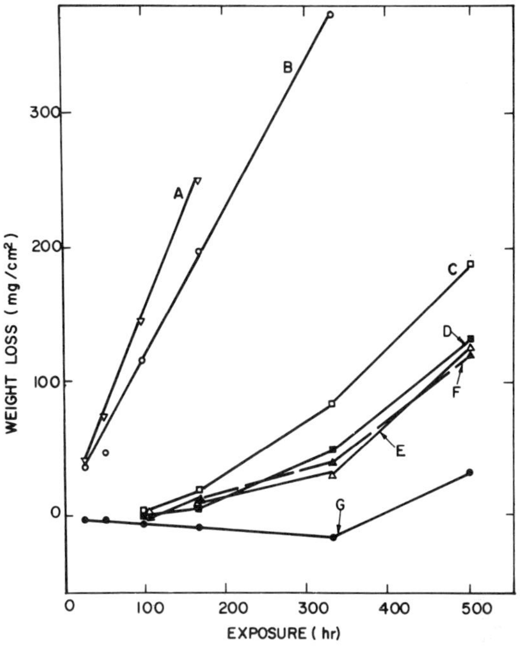

FIG. 2-6 Thorium-zirconium alloys exposed to distilled water at 200°C (392°F). Curve A is weight loss for unalloyed Th; B is for Th-5 wt.% Zr; C for Th-15 wt.% Zr; D for Th-20 wt.% Zr; E for Th-25 wt.% Zr; F for Th-30 wt.% Zr; and G for Th-10 wt.% Zr.

At ordinary temperatures in air, beryllium is not reactive and can retain a polished surface for years. The kinetics of reaction of beryllium with oxygen and nitrogen are given in Figs. 2-7 [106] and 2-8 [106], respectively. Above about 600°C (1112°F) the oxidation rate is appreciable and burning begins at 1200°C (2191°F) with oxygen. The rates for both reactions follow the parabolic law with activation energies of 50 and 75 kcal/mole for oxygen and nitrogen in the range 725 to 950°C (1337 to 1742°F). The reaction products are BeO and Be_3N_2.

Beryllium has good resistance to dry carbon dioxide up to 1000°C (1832°F) [107]. However, if water vapor is present above 600°C (1112°F) then a rapid, breakaway oxidation occurs. The reactions connected with the beryllium-carbon dioxide system are the following [108]:

$$Be + CO_2 = BeO + CO$$
$$2Be + CO_2 = 2BeO + C$$
$$2Be + C = Be_2C.$$

In 1 atm of dry carbon dioxide, the kinetics are as shown in Table 2-12 [107]. Breakaway corrosion of beryllium in moist carbon dioxide can be avoided if the beryllium is fabricated using preoxidized powder [109]. Beryllium does not react significantly with hydrogen at temperatures up to 780°C

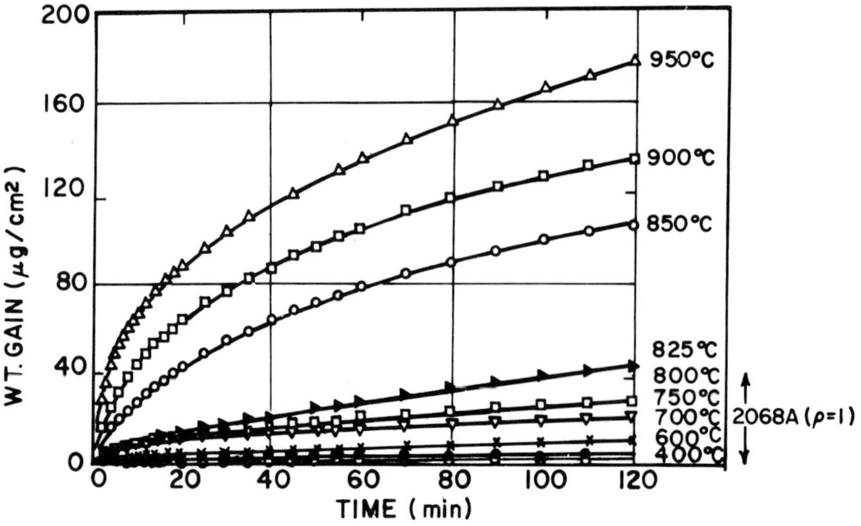

FIG. 2-7 Reaction of Be with O_2. Effect of temperature 400°C (752°F) to 950°C (1742°F), 7.6 cm O_2, abraded through 4/0.

(1436°F) at a hydrogen pressure of 2.6 cm Hg (1/29 atm). However, there is evidence for interstitial absorption of the hydrogen by the beryllium.

Beryllium reacts with fused alkali halides but not with fused alkaline earth halides. Molten beryllium reduces most oxides, nitrides, sulfides, and carbides; at temperatures as low as 900°C (1652°F) the rates are appreciable.

Beryllium has good resistance to corrosion by oxygen-free sodium or NaK under static conditions up to 1000°C (1832°F) (p. 926 of reference [1]). In flowing sodium, beryllium is an acceptable material up to 760°C (1400°F). There is no evidence for stress corrosion of beryllium in either high-temperature sodium or water.

The compound of beryllium that is of most interest in reactor technology is beryllium oxide. BeO is generally very stable and inert to most materials even at high temperatures. There is, however, a loss of beryllium by direct volatilization and by interaction with water vapor at high temperatures as shown in Table 2-13 [1c].

Finally it is appropriate to mention the well-known hazard resulting from the chemical toxicity of beryllium [110]. Be and BeO dusts are especially hazardous. The common precautionary measures are well-ventilated working areas and the use of respiratory masks when necessary.

2.3.2 Graphite

The chemical behavior of graphite with air, oxygen, hydrogen, nitrogen, carbon dioxide, and

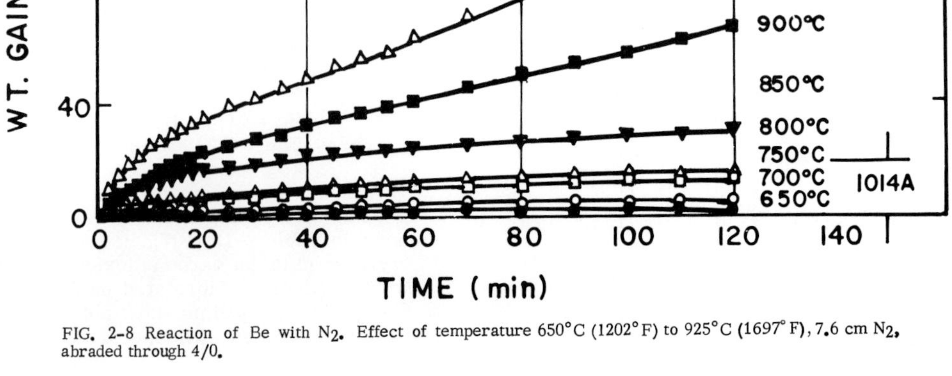

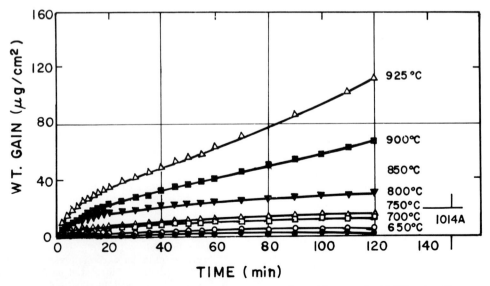

FIG. 2-8 Reaction of Be with N_2. Effect of temperature 650°C (1202°F) to 925°C (1697°F), 7.6 cm N_2, abraded through 4/0.

TABLE 2-12

Reaction Kinetics of Be in 1 Atmosphere of Dry CO_2

Temperature	$(\text{wt. gain})^n = k\,t$	
	n	k
500°C (932°F)	2	5×10^{-16} g^2/cm^4-sec
600°C (1112°F)	2	7×10^{-16} g^2/cm^4-sec
650°C (1202°F)	1	1.4×10^{-10} g/cm^2-sec
700°C (1292°F)	1	4×10^{-10} g/cm^2-sec

steam is covered in the discussion of fast reactions (see Sec. 3.4). The present section is concerned with compatibility problems of graphite in contact with metals, oxides, and fused salts. Table 2-14 [111a] gives a summary of graphite-metal reactions. It is to be noted that graphite can form compounds with K, Li, possibly Na, and with many other metals. In general, reactor grade graphite is not compatible with molten Na or NaK. Graphite is wet by liquid Na above 150°C (302°F); thus Na is absorbed into the pores of graphite by capillary action and rises above the level of the bulk liquid [112]. This absorbed sodium can then cause anisotropic swelling, spalling, and cracking of the graphite. TSP grade graphite upon exposure to sodium undergoes dilations of 0.1, 1.0, and 1.6% elongation at 200°C, 300°C, and 600°C (392°F, 572°F, and 1112°F), respectively (p. 86 of reference [113]). Mass transport of carbon from the graphite into the sodium can occur in graphite-sodium systems and this in turn can cause carbonization of any stainless steel present in the system [114, 115]. In static sodium at 482°C (900°F) graphite shows weight losses of 6% and 60% after exposures of 168 and 841 hours, respectively (p. 299 of reference [116]). If oxygen is present in the sodium, the reaction with graphite is very vigorous as contrasted to the slow reactions discussed above.

Chemical reactions between graphite and various metal oxides are of interest in the development of high-temperature nuclear fuels with ceramic cores. Table 2-15 [111b] gives examples of the reactions which are reductions having the general form,

$$Me_xO_y + yC = yCO + xMe$$

or

$$2Me_xO_y + 3yC = 2yCO + Me_{2x}C_y \ .$$

With an excess of graphite, the carbide is generally formed [111].

TABLE 2-13

BeO Weight Loss at High Temperature Due to Volatilization and Reaction with Water Vapor

Temperature	Weight loss - g/cm^2-hr	
	Volatilization	Reaction with water vapor
1000°C (1832°F)	7.4×10^{-13}	2.013×10^{-3}
1400°C (2552°F)	1.06×10^{-6}	1.11×10^{-1}
2000°C (3632°F)	1.09×10^{-1}	2.32

Graphite is chemically inert to bismuth; however, bismuth may, like sodium, penetrate the pore structure of the graphite [117]. Tests in which the 55% Pb-45% Bi eutectic was circulated for 466 hr at 260°C (500°F) in contact with graphite, showed no evidence for corrosion or erosion of the graphite (p. 445 of reference [111]). The uptake of pure Bi into graphite is a function of the porosity of the graphite, the pressure, and the exposure time; at 550°C (1022°F) and 250 psi (16 atm) values range from 0.5 to 1.5 grams of bismuth/cm^3 of graphite depending on the porosity of the graphite.

In molten salt reactions it is desirable to have the graphite unclad and therefore it must have the ability to resist corrosion. There are three aspects involved in this material problem, namely: salt penetration, compatibility, and effect on other container materials. Tests on the permeation of various grades of graphite at 705°C (1301°F) and 150 psig (10 atm) in contact for 100 hr with a $LiF-BeF_2-ThF_4-UF_4$ molten salt showed from 0 to 1.0% of the graphite was permeated (p. 459 of reference [111]). Reactor design criteria stipulate that the allowable salt absorption should be less than 0.5% of the bulk volume of the graphite. The three grades of graphite which met this specification were B-1, S-4-B, and GT-123-82 with values of 0.0, 0.2, and 0.3% absorption, respectively. The amount of permeation appears to be rather insensitive to the type of fused fluoride salt used. However, CsF, PbF_2, and SnF_2 permeate quite rapidly [118]; this is presumed to be caused by wetting. In circulating loop tests with molten fluoride salts, graphite has been found in the salt after an exposure for 500 hr at 815°C (1499°F) (p. 459 of reference [111]). This graphite will then interact, for example, with nickel-base alloys also present in the system to give carburization. Another chemical problem associated with graphite and molten salts containing UF_4 is the unwanted formation of UO_2 precipitate at 705°C (1301°F) or above. This is thought to be caused by oxygen absorbed in the graphite.

Despite the thermodynamic inertness of graphite and the MSRE (Molten Salt Reactor Experiment) fuel with respect to each other, in the absence of radiation, the unanticipated formation of CF_4 has been observed [119]. An interesting possibility is that elemental fluorine is formed by radiolysis of the salt and that the fluorine then reacts with the graphite [120]. Also, there is a suggestion of subsequent reaction between the gases fluorine and xenon to produce XeF_4 which may account for the observed deficiency of xenon in irradiated cover gas [120]. When HF is sparged through the fused fluoride salt, the graphite corrosion increases markedly; typical values under these conditions range from 0.06 mil/day (1.5 μ/day) to 0.4 mil/day (10 μ/day) [121].

2.4 Corrosion of Control Rod Materials

2.4.1 Boron

Boron is one of the most widely used neutron-absorbing materials for nuclear reactors. It can be incorporated into reactor components as an alloy, as a dispersion in a structural material, or as a clad ceramic body. Common usage includes:

TABLE 2-14

Reactions of Graphite With Metals

Element	Temperature	Product or effect
Al	800°C (1472°F)	Al_4C_3; reaction rapid at 1400°C (2550°F)
B	1600°C (2912°F)	B_4C; commercial preparation carried out at 2400°C (4350°F)
Be	900°C (1652°F)	Be_2C in vacuum or helium
Co	218°C (424°F)	Co_3C; Co_3C is metastable; Co_2C is unstable
Cs		C_8Cs and other lamellar compounds; occurs with liquid or vapor at a few millimeters pressure
Cu	1010°C (1850°F)	No attack when copper-graphite surfaces contacted for 250 hr
Fe	600 to 800°C (1112 to 1472°F)	Fe_3C
Hf	2000°C (3632°F)	HfC
K		C_8K and other lamellar compounds; occurs at temperatures where vapor pressure is a few millimeters of mercury
Li	500°C (932°F)	Li_2C_2 after long periods; intermediate compounds formed after short times
Mg	1100°C (2012°F)	No attack near melting point
Mo	700°C (1292°F)	Mo_2C; MoC forms at >1200°C (>2190°F)
Na	400°C (752°F)	Possibly a $C_{64}Na$ lamellar compound; reaction slow
	400°C (752°F)	No reaction after carefully degassing to exclude O_2
	>450°C (>842°F)	Na_2C_2
Nb	>870°C (>1598°F)	NbC and Nb_2C formed in helium by surface-to-surface contact; some intergranular penetration
Ni	1310°C (2390°F)	No stable carbides formed; solubility of carbon in Ni is 0.65 wt. %
Pb	1090°C (1994°F)	No attack after 24 hr by Pb in Pb-Bi eutectic
Pu	1050°C (1922°F)	PuC from Pu turnings and graphite in vacuum; Pu_2C_3 forms with excess carbon; PuC_2 exists above 1750°C (3180°F)
Rb		C_8Rb and other lamellar compounds; occurs at temperatures where vapor pressure is a few millimeters of mercury
Si	1150°C (2102°F)	β-SiC
Ta	1600°C (2912°F)	Ta_2C in H_2
	2200°C (3992°F)	TaC
Th	2100°C (3812°F)	ThC; ThC_2 in vacuum
U	1150°C (2102°F)	UC; 0.005-cm interface layer in 200 hr

TABLE 2-14 (Cont'd.)

Reactions of Graphite With Metals

Element	Temperature	Product or effect
U	1400°C (2552°F)	UC and trace of UC_2; 0.07-cm interface layer in 200 hr
W	1400°C (2552°F)	W_2C and WC in H_2
	1500°C (2732°F)	Carbide; surface-to-surface contact for 8 min

Boral (B_4C in an Al matrix), Boroxal (B_2O_3 in Al), boron-carbon steel, boron-stainless steel, and a soluble poison such as boric acid.

High-temperature water quickly attacks boron itself, but boron in the form of boron carbide, B_4C, has good resistance to water up to 260°C (500°F) (p. 797 of reference [1]). Boron nitride, BN, is attacked very slowly by boiling water but is quickly hydrolized by steam [122]. Tests on boron containing stainless steels have given conflicting results. Two sets of data [123, 124] show the corrosion resistance of stainless steels containing 1 to 3 wt.% boron in 316°C (600°F) high purity water to be comparable in magnitude to that of the 300-series stainless steels. However, other tests [125] show weight gains of boron stainless steels to be from 4 to 7 times higher than for type-304 stainless in 260°C (500°F) water. Tests of aluminum-base alloys containing boron in boiling distilled water show, in general, corrosion rates comparable to type-1100 aluminum used as a control [126]. Alloys of titanium with boron offer excellent resistance to attack by high-temperature water, while zirconium-boron alloys are quite poor in this respect [124].

When pressed boron powder is heated in air there is no apparent oxidation up to about 750°C (1382°F), but at 800°C (1472°F) a black protective glaze, which is apparently protective to 1000°C (1832°F) coats the specimen [127]. In pure oxygen at 1000°C, the attack is quite rapid. At 900°C (1652°F) and higher, boron combines with nitrogen to form boron nitride [128]. Boron carbide can be used in oxidizing atmospheres to 1000°F [129]. It is slowly etched by hydrogen at temperatures of the order of 1204°C (2200°F) but can be heated to its melting point in CO without change [130]. Boron carbide also has good resistance to nitrogen and CO_2 at 816°C (1500°F) [131]. Boron nitride oxi-

TABLE 2-15

Reactions of Graphite With Selected Metal Oxides

Oxide	Temperature	Remarks
Al_2O_3	1280°C (2336°F)	Al_4C_3 formed
	1307 to 1450°C (2385 to 2642°F)	Reaction follows parabolic rate law with activation energy of 310 kcal/mole CO
BeO	960°C (1760°F)	Be_2C formed
UO_2	1320°C (2408°F)	12% weight lost in 1 hr for -325 mesh UO_2 and graphite flour
	1650 to 2130°C (3000 to 3865°F)	Reaction at interface follows parabolic rate law
PuO_2	850°C (1562°F)	Partial reduction
	1300°C (2372°F)	PuC or Pu_2C_3

dizes slowly in air at temperatures up to 704°C (1300°F) [132]. Alloys of boron with zirconium and titanium, which have been of considerable interest as reactor control and burnable poison materials, are readily attacked by air, oxygen, nitrogen, or hydrogen (p. 796 of reference [1]).

The corrosion resistance of boron-containing materials toward sodium has been studied. Although some attack is observed, boron carbide is capable of being used at temperatures to 593°C (1100°F) as are some boron stainless steels [131]. Some complex nickel-base boron alloys have been considered for application in liquid metals because of the compatibility of nickle with static sodium in the temperature range 760 to 927°C (1400 to 1700°F). Tests for these alloys have shown that although good corrosion resistance is evidenced at 760 and 816°C (1400 and 1500°F) there was considerable second phase depletion [133].

During reactor irradiation, helium is generated in the boron by the reaction B^{10} (n, α) Li^7. This reaction resembles fission and, in the same way, releases considerable heat and ionization energy along paths of particles. At 36% burnup, complete granulation occurs. This reaction can cause embrittlement and deterioriation of the properties of alloys which contain even small quantities of boron [134]. There has been some unfavorable experience with burnable poison (B^{10}) strips in an Al-Ni carrier alloy during in-pile use [135].

2.4.2 Cadmium and Indium

Cadmium does not react with water at temperatures up to 100°C. However, to achieve corrosion resistance at higher temperature it is necessary to use alloys. Silver-cadmium alloys have a nonprotective oxide film and lose weight in water at 288-316°C (550-600°F) [136]. A Ag-30.7 wt.% Cd alloy has a corrosion rate of 1.67 mg/dm^2-day at these temperatures. More severe corrosion occurs in oxygenated than in hydrogenated water. Silver-cadmium-indium alloys show improved corrosion resistance to water (see Fig. 2-9) [136]. However, such alloys, unless fully clad, may permit sufficient silver to be plated out on stainless steel systems to cause a serious problem if the silver is radioactive. The problem is especially acute in borated-oxygenated water and has led to a very difficult maintenance problem at the Yankee Reactor. The half-life of radioactive silver is particularly unfavorable, as it is of the same order of magnitude as the pressurized-water core lifetime.

Moist air oxidizes cadmium (p. 808 of reference [1]). In dry air at 300°C (572°F), the rate constant for the parabolic oxidation is $k = 3.2 \times 10^{-14}$ g^{2}cm^{-4}sec^{-1} for the first 40 hr. For longer times the oxidation slows down and then stops [52]. There are reported negative temperature coefficients for oxidation of cadmium in oxygen in the range 390 to 520°C (734 to 968°F) indicating a progressive sintering of the cadmium oxide scale with increased protectiveness. For pure liquid cadmium in air at 550°C (1022°F) the parabolic rate law applies with a constant of $k = 1.0 \times 10^{-9}$g^{2}cm^{-4} sec^{-1}. Alloying with Mg, Sb, and Sn causes a de-

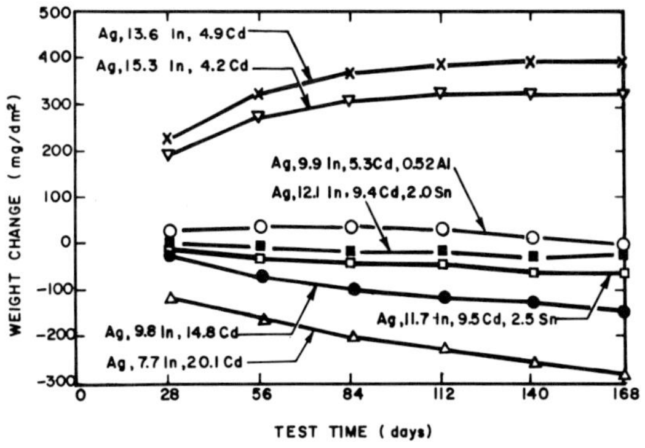

FIG. 2-9 Corrosion behavior of Ag-In-Cd alloys in 600°F (315°C) dynamic water. Numbers refer to weight percent, e.g., 4.9 Cd = 4.9 wt.% Cd. Test conditions: water velocity = 10 ft/sec (3.05 m/sec); pH = 9.5-10.5 with LiOH; H_2 = 25-30 cm^3/kg; $O_2 < 0.14$ ppm.

crease in the oxidation resistance of cadmium, whereas a small amount of zinc is beneficial [52].

2.4.3 Hafnium

Resistance to air oxidation by hafnium (really an alloy with 2-5% Zr) is quite good and it is somewhat better than zirconium. After 2 hr in still air at 950°C (1742°F), a piece of hafnium shows a penetration of only 0.15 mm (6 mils). Hafnium is also superior to zirconium in its corrosion resistance to water and steam (p. 788 of reference [1]); hence no cladding or coating is needed. However, there is a problem with the half-life of the radioactive hafnium produced being comparable to the core lifetime, analogous to silver as mentioned above. Hence, corrosion products must be watched closely. After 294 days in 399°C (750°F) 1500 psi (100 atm) steam, hafnium shows a weight increase of 9 mg/cm^2 [137]. The oxide HfO_2 is very protective. The corrosion behavior of hafnium in contact with hydrogen and nitrogen is similar to that of zirconium.

2.4.4 Rare-Earth Elements

The elements gadolinium, samarium, and europium are promising for reactor control applications but very few corrosion data are available, (see Table 2-16 [138]). In applications of these elements, oxides are usually dispersed in titanium or zirconium. A Ti-10 wt.% Gd_2O_3 alloy was resistant to water at 316°C (600°F) for one month. However, ZrH_2 compacted with 10 wt.% Sm_2O_3 disintegrated in water at 316°C (p. 313 of reference [116]).

2.5 Corrosion of Ceramic Core Materials

Ceramic core materials such as uranium, thorium and plutonium oxides, carbides, and nitrides have not been considered for use in direct contact with water in reactor designs. Interest in their reactions with water and steam arises from the possibility of a cladding failure which would bring the core materials into contact with water.

TABLE 2-16

Corrosion Rates of Rare-Earth Metals in Air as a Function of Temperature

Relative humidity→	Weight gain (mg/cm^2-day) 95°F (35°C) 1%	95°F (35°C) 75%	200°F (93°C) 1%	200°F (93°C) 75%	390°F (200°C)	750°F (400°C)	1110°F (600°C)
La	80	950	510	21,000	30	3,200	13,000
Ce	...	...	...	...	...	200,000	...
Pr	8	76	900	5,500	80	38,000	130,000
Nd	2	7	60	2,000	70	380	4,800
Sm	0	0	0	100	15	17	35
Gd	1	2	0	35	0	210	16,000
Tb	0	0	0	...	0	1,600	40,000
Dy	0	0	0	43	...	350	6,600
Ho	1	1	1	...	11	110	5,400
Er	1	1	0	...	10	90	720
Yb	...	...	...	...	...	170	...
Y	1	1	2	9	4	40	1,900

Studies with UO_2 have indicated that sintered compacts are very resistant to corrosive attack in high-temperature water and steam in the absence of oxygen. The resistance to attack is insensitive to the presence of many impurities. Only the pickup of carbon on sintering in a graphite furnace with an argon atmosphere has been found to cause a loss of fuel stability.

Erosion and corrosion tests have been performed in flowing water (18-28 ft/sec or 5.5-8.5 m/sec) with sintered UO_2 compacts of the Shippingport reactor type (p. 405 of reference [139]). Three kinds of tests were performed: (1) erosion of bare compacts in oxygen-free flowing water, (2) corrosion of bare compacts in oxygenated water, and (3) corrosion of compacts partially protected by defected cladding when exposed to oxygenated water. The extent of erosion in oxygen-free water was greater when sharp edges were exposed to the flow. In all cases, however, erosion was no cause for concern in the Shippingport reactor because of the limited bare fuel surface that could be exposed to the water. Corrosion of bare compacts in oxygenated water (10 cm^3 O_2/kgH_2O) resulted in total weight losses between 10^4 and 10^5 mg/dm^2 in 320-hr tests in 274 to 316°C (525 to 600°F) water. The lower weight losses occurred in water at low pH (4.5 HNO_3) where a thin (1 mil or 25 μ), relatively nonporous film afforded some protection to the underlying oxide. A loose, flaky scale and intermediate reaction rates characterized results in neutral water. The lack of any scale in water of pH 10 resulted in rapid attack and near consumption in 370 hr.

Uranium dioxide, UO_2, undergoes oxidation in air and in oxygen to a tetragonal oxide of composition $UO_{2.3}$-$UO_{2.4}$ (p. 380 of reference [139]). At temperatures above 250°C (482°F), a second oxidation step occurs in which the tetragonal oxide is oxidized to orthorhombic U_3O_8. The stepwise nature of the oxidation to U_3O_8 is indicated in Fig. 2-10 [140] where typical oxidation curves are shown. It is generally believed that the rate of oxidation of UO_2 to a tetragonal oxide is parabolic and that the reaction occurs by the inward diffusion of oxygen. This assumption is based in part on the fact that most investigators have found an increase in oxide density as oxidation proceeds.

Oxidation rates and the nature of the reaction depend on the particle size of the original UO_2. It has been shown, for example, that very finely divided UO_2 prepared by the thermal decomposition of $UO_2C_2O_4$ at 300°C (572°F) in a closed system will oxidize to $UO_{2.5}$ at room temperature. UO_2 powders having a particle size of the order of 0.05 to 0.08 μ or lower, take up considerable oxygen and may ignite spontaneously. These ignitions are probably initiated by an exothermic absorption of oxygen to an extent proportional to the surface. If the specific surface area is large enough, rapid self-heating can bring the temperature into a range where rapid oxidation or burning can occur. Uranium dioxide prepared from precipitates (ammonium diuranate, uranium peroxide, uranium oxalate, etc.) either by direct reduction or by first pyrolyzing and subsequently reducing with hydrogen or alcohol, are particularly likely to be pyrophoric. Longer reduction periods or higher temperatures will usually stabilize such powders. Powders having particle diameters in excess of about 0.2 to 0.3 μ are fairly stable toward oxidation.

ThO_2 and PuO_2 are the highest oxides known for the elements thorium and plutonium. No chemical reactions can, therefore, occur between the oxides and water or oxygen. There are apparently no reported studies of the erosion of ThO_2 and PuO_2 oxide compacts by water.

Uranium mono- and dicarbides (UC and UC_2) react with water to produce a solid uranium oxide and a gaseous mixture of hydrogen and hydrocarbons. The available composition data of the gaseous products are given in Tables 2-17 and

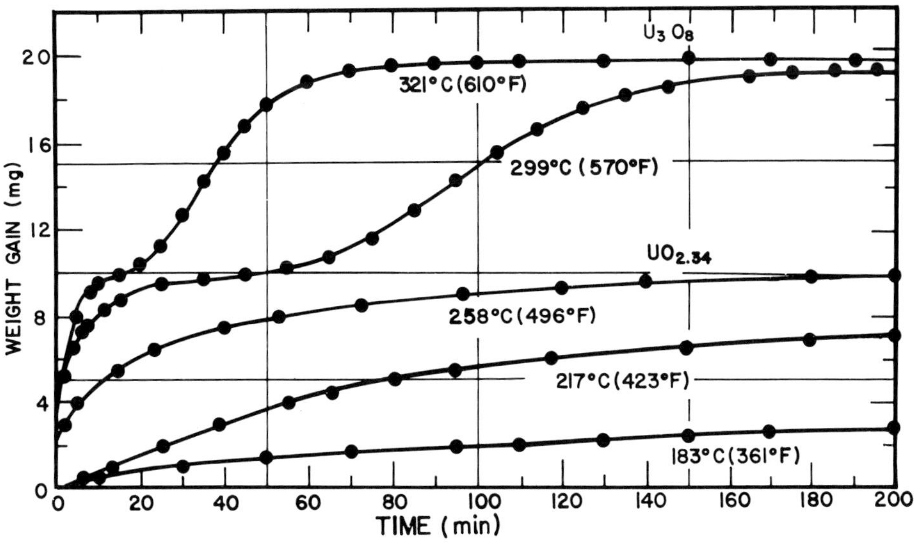

FIG. 2-10 Experimental rate curves for oxidation of UO_2.

2-18 [141]. The results, from a number of investigators, seem consistent and indicate that hydrogen is the primary gaseous product at higher temperatures. Methane is the primary product at room temperature for the UC hydrolysis while relatively large quantities of higher hydrocarbons are generated in the UC_2 hydrolysis. The greater production of higher hydrocarbons, especially those containing even numbers of carbon atoms, was attributed to the fact that the UC_2 lattice contains carbon atoms in pairs [142].

Typical rate data for the reaction of UC_2 with various gases are given in Table 2-19 [143]. Similar data for UC are given in Table 2-20 [144].

The hydrolysis of ThC and PuC were reported to be very similar to that of UC. ThC_2, however, was reported to react about ten times faster than UC_2. The ThC_2 hydrolysis produces about twice as much hydrogen, about one-seventh as much methane, and about ten times as many acetylenic hydrocarbons as UC_2 [142].

It has been reported that uranium mononitride, UN, is unaffected by heating in boiling water for 24 hr [145].

Uranium monocarbide powders were found to ignite between 275 and 250°C (527 and 482°F) in oxygen and at 350°C (662°F) in air [146]. Sample powders capable of ignition were termed "reactive" uranium carbide. This reactivity or sensitivity to ignition developed after arc-melted samples were stored in contact with air. It was surmised that a hydrolysis reaction with the moisture in the air was responsible for the sensitization. Freshly melted samples did not show ignition in air at 426°C (800°F). Results of a study of the ignition of uranium monocarbide powders from three commercial sources were reported in terms of the average specific area [147]. Spheroidized particles ignited in pure oxygen over the temperature range 385 to 405°C (725-761°F) for samples having specific areas from 40 to 6 cm^2/g. Powders composed of irregular particles ignited between 280 and 340°C (536 and 644°F) for a specific area range from 200

TABLE 2-17

Gaseous Products of the Reaction of Uranium Monocarbide With Water

Gas	Gas Conc. Vol. %, at Indicated Temperatures						
	20°C (68°F)	83°C (181°F)	90°C (194°F)	100°C (212°F)	200°C (392°F)	307°C (585°F)	400°C (752°F)
Hydrogen	8	12	22	37	93	96	99
Methane	85	81	72	57	5	3	0.5
Ethane	5	4	4	4			
C_3-C_5 paraffins	1				1.1	0.5	0.0
Olefins		2	0.8	1.0			
Acetylenes		0	0	<0.2			
CO_2		0.3	0.2	0.3	0	0	0
CO		0.4	0.2	0.9	0.4	0.4	0.3

TABLE 2-18

Gaseous Products of the Reactions of Uranium Dicarbide With Water

Gas	Gas Conc. Vol. %, at Indicated Temperatures									
	20°C (68°F)	81°C (178°F)	95°C (203°F)	100°C (212°F)	109°C (228°F)	125°C (257°F)	137°C (279°F)	148°C (298°F)	200°C (392°F)	248°C (479°F)
Hydrogen	10	17	39	47	64	68	75	83	89	96
Methane	60	30	18	10	10	8	7	6	4	1
Ethane	25	30 }	31	30 }		15	11 }		6	1.5
C_3C_5 paraffins	4	8 }								
Olefins		12	9	8 }	25	5	4 }	10	0.6	<0.2
Acetylenes		2 }	1.1	2 }		0.3	2 }		0	<0.2
CO_2		0.2 }								
CO		0.9	0.9	2	0.6	3	0.5	0.2	0.4	1.1

TABLE 2-19

Corrosion of Uranium Dicarbide in Water Vapor, Nitrogen, and Oxygen

Test medium	Temperature	Corrosion rate	
		Linear, mg/(cm²)(sec)	Parabolic, (mg/cm²)²/sec
29 mm H_2O	50°C (122°F)	0.04	—
	150°C (302°F)	0.66	—
	200°C (392°F)	3.2	—
H_2O vapor	250°C (482°F)	—	6520
	300°C (572°F)	—	9210
Nitrogen	400°C (752°F)	—	16
	600°C (1112°F)	—	860
	700°C (1292°F)	—	3300
Oxygen	150°C (302°F)	—	6.1
	200°C (392°F)	—	75
	250°C (482°F)	—	900
	300°C (572°F)	—	Anisothermal

to 20 cm^2/g. Ignitions were very mild. The absence of a sharp temperature rise was attributed to hindered transport of oxygen through the gaseous products of the reaction.

TABLE 2-20

Corrosion of Uranium Monocarbide in Various Gases

Gas	Temperature	Reaction rate for UC (mg of gas reacted/(cm²)(hr))
CO_2	500°C (932°F)	2.4
CO_2	700°C (1292°F)	14
CO_2	850°C (1562°F)	56
CO_2	1,000°C (1832°F)	185
O_2	350°C (662°F)	1.4
N_2	700°C (1292°F)	0.96
N_2	850°C (1562°F)	46
CO	500°C (932°F)	0.007-0.03
CO	600°C (1112°F)	0.007-0.05

3 RAPID OR VIOLENT REACTIONS

3.1 General Considerations

The possible occurrence of rapid or violent chemical reactions within a nuclear reactor has been a subject of continuing concern to reactor designers. Conditions in a cold reactor or in a reactor operating normally are usually such that no violent chemical reactions can be initiated. However, an accident that results in serious overheating of the core, such as a sudden loss of coolant or a nuclear excursion, might initiate violent chemical reactions. Certain other situations can also set the stage for a violent reaction. Thus, the uncontrolled accumulation of radiolytic or electrolytic hydrogen and oxygen in some part of the reactor system could lead to an explosion.

Most of the concern over the initiation of violent reactions has centered on interactions between metals and water and between metals and air. Rapid metal-water reactions, in particular, have been considered a source of energy which could seriously complicate an accident in a water-cooled reactor. As early as 1949 there was concern that chemical reactions could be a serious source of disruptive energy [148]. Table 3-1 lists the heat liberated by the reaction of various structural and cladding metals with water. The quantity of hydrogen liberated by each of the metal-water reactions, also given in the table, indicates that large quantities of gas as well as heat are produced. Moreover, additional energy can be liberated if the hydrogen subsequently reacts with oxygen.

Reactions of sodium of other alkali metal coolants with water were considered a serious hazard during the development of nuclear submarine reactors; however, subsequent studies indicated that the reactions were not as violent as had been feared [149].

The reaction of metals with air must be considered a secondary hazard whenever a reactor accident is under analysis. Primary interest in metal-air reactions, however, stems from fire and explosion incidents which have occurred during processing, fabrication, and storage of metals

TABLE 3-1
Thermodynamic properties of some potentially violent chemical reactions*

Reactant	Temperature (°C)	Oxides formed	ΔH_r Heat of reaction with oxygen[a] (cal/g)	ΔH_r Heat of reaction with water[a] (cal/g)	Hydrogen produced in reaction with water[a] (liters STP/g)
Zr (ℓ)	1852[b]	ZrO_2	-2,883	-1560	0.491
U (ℓ)	1133[b]	UO_2	-1,098	- 596	0.188
Al (ℓ)	660[b]	Al_2O_3	-7,487	-4200	1.246
SS (ℓ)	1370[b]	FeO, Cr_2O_3, NiO	-1330 to - 1430[c]	-144 to - 253[c]	0.441
Mg (ℓ)	651[b]	MgO	-6,000	-3570	0.922
Pu (ℓ)	640[b]	PuO_2	-1,025	- 531	0.187
Na (s)	25	Na_2O	-2,162	—	—
Na (s)	25	NaOH	—	-1466	0.487
K (s)	25	KO_2	-1,729	—	—
K (s)	25	KOH	—	- 857	0.287
Li (s)	25	Li_2O	-10,260	—	—
Li (s)	25	LiOH	—	-7017	1.615
C (s)	1000	CO (g)	-2,267	+2700	1.868
C (s)	1000	CO_2 (g)	-7,867	+2067	3.736
H_2 (g)	1000	H_2O (g)	-29,560	—	—

[a] Numbers are based on one gram of reactant.
[b] Melting point of metal.
[c] Value depends upon whether FeO (m.p. 1377°C) is formed as a solid or as a molten oxide. Higher value is preferred since it is likely that oxides are a complex mixture of solids.
* Data taken from J.P. Coughlin, Bulletin 542, U.S. Bureau of Mines (1954) and from Reactor Handbook, Vol. I, Interscience Publishers, N.Y., 1960.
Abbreviations used: s, ℓ, g = solid, liquid, gas; SS = stainless steel; STP = standard temperature, pressure conditions (0°C, 760 mm Hg).

used in reactors [150]. The pyrophoricity of uranium, zirconium, and plutonium is well known although the detailed sequence of events surrounding many of the incidents is not fully understood.

Reactions of graphite with air and carbon dioxide have been under study in connection with gas-cooled reactors. Neither reaction has been found to be violent; however, at the high temperature that an accident might produce, the reactions are rapid.

3.2 Metal-Water Reactions

High-temperature metal-water reactions have been under almost continuous study since these reactions were first recognized as a hazard. A recent reviewer notes that it has been quite difficult to produce an explosive metal-water reaction deliberately and that this has materially lessened concern over this potential hazard [149]. Fortunately, existing data do suggest that extreme conditions are needed to initiate violent reactions, although some reaction occurs whenever reactive metals are overheated in the presence of water. The serious implications of underestimating reactor hazards, however, have made it imperative to search for dependable methods of estimating the rate and extent of reaction occurring in hypothetical reactor accidents.

An attempt is made in this section to summarize pertinent experimental findings for each of the important metals used in proximity with water in nuclear systems. The data are organized according to the nature of the experiment for the solid metals and details of experimental methods are discussed briefly. Experiments designed to determine isothermal oxidation rates, nonisothermal experiments such as pouring molten metal into water, and experiments in which highly dispersed metal is generated are discussed separately. Finally, an attempt is made to summarize the results of all studies and present a consistent mechanism for each reaction insofar as this can be done. It should be emphasized at this point that past attempts to consider metal-water reactions as fitting one pattern without regard to the individual nature of the metal under consideration are quite unjustified. Recent data make it clear that studies with one metal-water system can be applied to another metal-water system only when there is specific justifying evidence.

Methods of applying the results of metal-water reaction studies to reactor accidents are discussed in Sec. 4, "Fuel Meltdown." The more general problem of what degree of dispersion might be realized in a practical situation as well as testing methods using nuclear heating are also presented in the section on fuel meltdowns.

3.2.1 Zirconium Alloys

Isothermal experiments: Data on the isothermal oxidation of solid Zircaloy-2 [151] has been interpreted [152] by a parabolic rate law having an activation energy of 65.4 kcal/mole and also [153] by an empirical rate law with an exponent varying from n = 2.63 at 1200°C (2192°F) to n = 1.06 at 2200°C (3992°F). These results have been replotted [154] on the assumption that the apparent failure of the parabolic law at high temperature resulted either from a rate limitation due to the gaseous interdiffusion of steam and hydrogen or to the time required to heat the specimen in the presence of steam. Parabolic rate constants computed on this basis are plotted in Fig. 3-1 [154] and show the correlation

$$k = 33.3 \times 10^6 [\exp(-45{,}500/RT)], \qquad (3\text{-}1)$$

where k is in (mg Zr/cm^2)^2/sec. In another study the zirconium-flowing steam reaction between 1000 and 1690°C (1832 and 3074°F) was correlated in terms of the parabolic rate law with an activation energy of 34 kcal/mole [152]. Recalculated parabolic rate constants are plotted in Fig. 3-1. An additional data point obtained by a mathematical analysis of zirconium-water reaction data obtained by an electrical condenser discharge heating method [154] is included in Fig. 3-1. This point was used as the basis for constructing a linear Arrhenius correlation. Isothermal data [155] between 800 and 1200°C (1472 and 2192°F) are in reasonable agreement with the above correlation. In addition, studies indicated that there was no important effect of steam pressure between 200 and 1000 psig (14 and 68 atmospheres).

Studies of the reaction of zirconium and Zircaloy-2 were carried out [156] using induction heating of specimens in the presence of up to 30,000 ppm of water vapor in an atmosphere of flowing helium. The results of the experiments were explained on the premise that the reaction rate was controlled by the transport of the water vapor through the helium. Good agreement was obtained between experimental results and calculations employing mass transfer considerations. In these studies, the reactive metal surface was continuously exposed because of the rapid diffusion of oxygen into the metal.

Nonisothermal experiments: Results of melting zirconium rod and dropping 2- to 10-gram batches into water are summarized briefly in Table 3-2 [157]. From this it appears that the reaction is suppressed by the presence of inert gas and enhanced by the presence of pressurized steam [157].

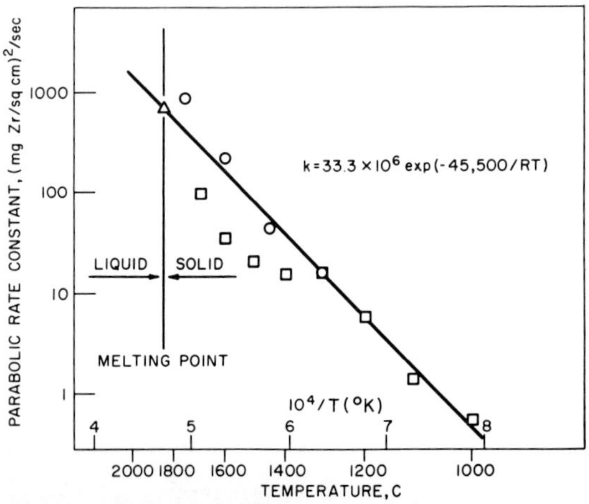

FIG. 3-1 Effect of temperature on the zirconium-water reaction. Open triangle is measurement using Baker's condenser-discharge method; open circles are Bostrom data recalculated; open squares are Lemmon data recalculated.

TABLE 3-2

Zirconium-Water Reaction Studies[a]

Water vapor pressure, psia	Argon pressure, psia	Reaction ml H_2/g Zr	% Reaction[b]
0.5[c]	200-1000	6.7-7.8	1.3-1.6
0.5[c]	15	8.2-12.6	1.7-2.6
0.5[c]	0	28-29	5.7-5.9
40-280	0	110-130	22-26
750	0	280	57

[a] From 2 to 10 g of molten zirconium at 1852°C (3366°F) were dropped into 130 ml of water in a 1400-ml chamber. (In the tabulation 15 psia = 1 atm pressure).
[b] Complete reaction corresponds to 491.2 ml H_2/g Zr.
[c] Room-temperature water (25 mm Hg vapor pressure).

A self-sustaining reaction between Zircaloy and water has been demonstrated [158] by electrical condenser-discharge heating to stimulate exponential periods of 2 to 20 msec. Chemical reaction was initiated very close to the melting point of the metal. Some self-heating was noted in every run, and reaction with foil samples was very extensive. In another study, 25mm-diameter streams of molten zirconium were poured into water; only a thin oxide coating was formed on the resulting metal globules [159].

When molten globules of Zircaloy were dropped into water, the thickness of the oxide layer, measured as a function of water temperature, varied between 24 and 68 μ (corresponding to 2.7 and 5.8% reaction) [152]. Cooling time of the zirconium globules increased markedly with increasing water temperature over the range 33 to 93°C (92 to 200°F).

Experiments with dispersed metal: One of the first demonstrations that finely divided zirconium would react extensively with water was provided by using electrical condenser-discharge heating to melt 2-mil (50-μ) foil strips of zirconium under water [160]. The extent of reaction was determined by collecting the hydrogen generated by the reaction in the absence of air, and it ranged between 20 and 100% for 5 runs with zirconium.

In experiments in which blasting caps were used to disperse molten zirconium streams under water [159, 161], the metal was largely converted into spherical particles which resulted in violent reactions with zirconium, Zircaloy-2, and Zircaloy-B. It was determined that the per cent reaction was a sensitive function of particle size. The reaction could be approximated by assuming that all particles were oxide-coated to a thickness of 25 μ (1 mil).

In another study, 10-20 g of molten zirconium were sprayed under pressure into water in an explosion dynamometer [155]. Spherical particles were formed as in the previous study; the extent of reaction was determined as a function of particle size by measurement of the oxide layer thickness. Mean particle diameters ranged between 250 and 450 μ (10 and 18 mil) and the extent of reaction varied from 27 to 33%. Transient pressures generated within the water column were used to calculate the work, the total impulse, and the mechanical efficiency of the explosion. Explosions appeared to be uniformly violent over the range of metal temperatures from 1900 to 2575°C (3452 to 4667°F), although the investigators claimed that the explosions were more violent above 2400°C (4352°F). The violence of the zirconium-water reaction compared with other metal-water reactions and with black powder is indicated in Fig. 3-2 [155].

The zirconium-water reaction was recently studied extensively by the condenser discharge method [154, 162]. In this study specimen wires were heated almost instantaneously to temperatures between 1100°C (2012°F) and 4000°C (7232°F) in water. Spherical particles were produced at temperatures above the melting point. Because the average particle size decreased continuously as the metal temperature was increased, it was impossible to separate experimentally the effects of metal temperature and particle diameter. The extent of metal-water reaction, determined by measurements of the quantity of hydrogen produced, was measured in water at temperatures from room temperature (water vapor pressure 0.5 psi or 25 mm Hg) to 315°C (600°F) (water vapor pressure 1500 psi or 100 atm).

Experimental results [154], summarized in Fig. 3-3, show that more extensive reaction occurred in heated water than in water at room temperature. The extent of reaction did not increase continuously with water temperature (or vapor pressure) since results in water at 100°C (15 psi or 1 atm water vapor) were nominally identical with results in water at 315°C or 600°F (1500 psi or 100 atm). Extent of reaction is plotted as a function of particle size in Figs. 3-4a [154] and 3-4b [154] disregarding the initial metal temperature. Although the scatter is greater, the results indicate that the reaction may be largely independent of temperature as long as the metal is fully melted. Rates of reaction, judged from transient pressure traces, were explosive (substantial reaction in a few milliseconds) in the more energetic runs indicated in the figure, and much slower (tenths of a second) in less energetic runs. Explosive reactions were found to occur with particles smaller than about 1 mm in heated water and 0.5 mm (500 μ or 20 mils) in water at room temperature (see Fig. 3-4). Explosive reaction appeared to be associated with rapid particle motion.

Discussion: The experimental results overwhelmingly indicate that the quantity and rate of reaction is primarily determined by the size of reacting particles. Only thin oxide films were formed on larger specimens while explosions, self-sustained burning, or very extensive reaction occurred when finely divided metal was used.

Theoretical studies [154] of the reaction of heated zirconium particles with water indicate that the reaction can be described by a two-step scheme. It was assumed that the reaction is initially controlled by the rate of gaseous diffusion of water vapor toward the hot metal particles and of hydrogen, generated by reaction, away from the particles. At a later time the reaction becomes controlled by the parabolic rate law, resulting in rapid cooling of the particle. In these studies the Nusselt number, describing both the gaseous diffusion rate and the rate of convective cooling, was given the theoretical minimum value for spheres, i.e., Nu = 2, whereas the emissivity of the oxide surface was given the theoretical maximum value of unity. The decreased extent of reaction in room-temperature water was described by assuming that in room-temperature water the effective vapor pressure of water, driving diffusion, is one-half of the value in heated water.

Equations were solved on an analog computer using the experimental parabolic rate law, Eq. (3-1) (see Fig. 3-1). Solutions had the form shown in Fig. 3-5 [154] indicating a greatly increased extent and rate of reaction as particle size decreased. The results indicate that extreme self-heating occurs only with the finer particles. An important finding of the study was the surprising result that the reaction was largely independent of the initial metal temperature as long as the particles were fully melted. This resulted from the fact that the reaction rate, in the gas-phase diffusion regime is nearly independent of temperature. Computations of the extent of reaction as a function of particle size are included in Fig. 3-4 and show good agreement with experimental results. The computed extent of reaction is plotted in Fig. 3-6 [154] where it is shown that the curves can be approximated by considering that an oxide film is formed which is 25 μ (1 mil) in depth in room-temperature

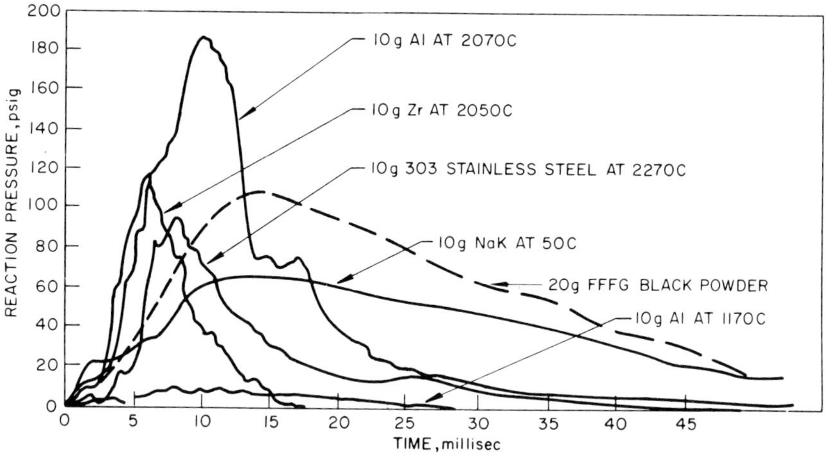

FIG. 3-2 Results of Aeroject explosion dynamometer tests.

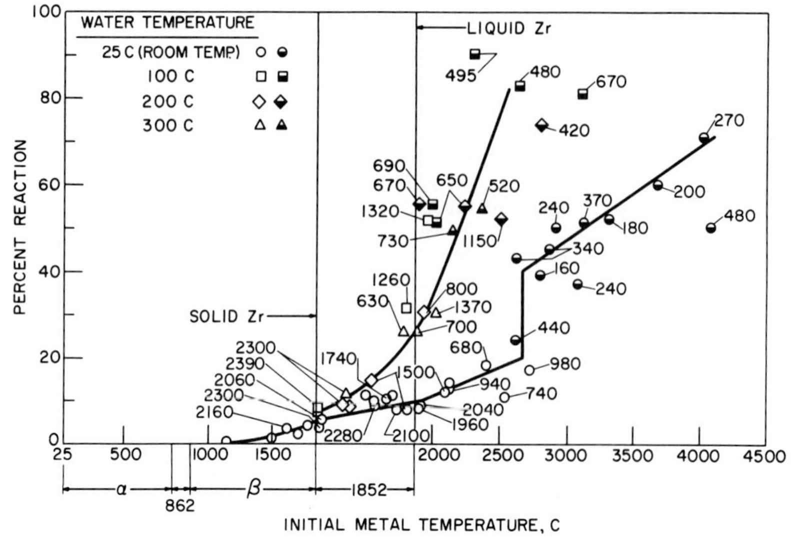

FIG. 3-3 Zirconium-water reactions by the condenser-discharge method. 60-mil (1.52-mm) wires used. The numbers indicate the mean particle diameter of residue in microns. Half-darkened symbols are for runs with an explosive pressure rise.

water and 60 μ (2.4 mil) in depth in heated water. The results are therefore consistent with the Aerojet results [161] and with Battelle falling drop studies [152].

Explosive reactions which occurred with particles smaller than about 1 mm in heated water and 0.5 mm in room-temperature water resulted from the ability of the evolving hydrogen to propel the particles through water at high speed. This had the effect of removing the gaseous diffusion barrier (increasing the Nusselt number) and thereby speeding the reaction through the gas-phase diffusion regime.

The failure to achieve 100% reaction by any experimental method, except for the case of very fine particles, suggests that the protectiveness of the oxide film, reflected in the parabolic rate law, is retained even at temperatures beyond the melting point of the oxide (2700°C or 4892°F). Similar conclusions were reached in connection with the burning of zirconium in oxygen-enriched air [163]. The sharp increases in reaction noted in room-temperature water for metal at 2600°C (4712°F) (see Fig. 3-3) in condenser-discharge studies and at 2400°C (4352°F) by the explosion dynamometer method have been interpreted as ignitions [164].

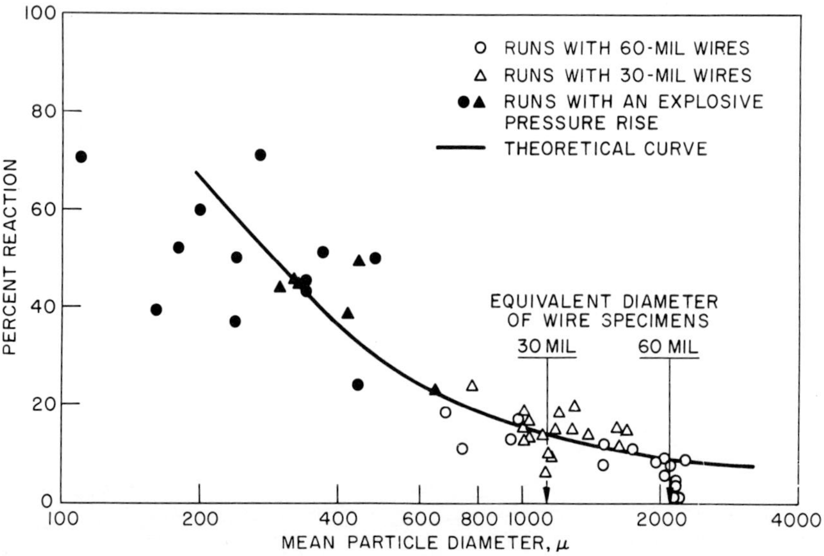

FIG. 3-4a Zirconium-water reaction as function of particle diameter for water at room temperature.

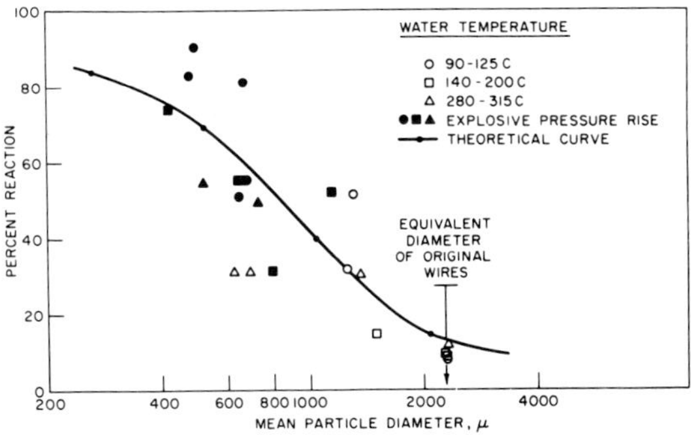

FIG. 3-4b Zirconium-water reaction as function of particle diameter for heated water. Wires used were 60 mil (1.52 mm) in diameter.

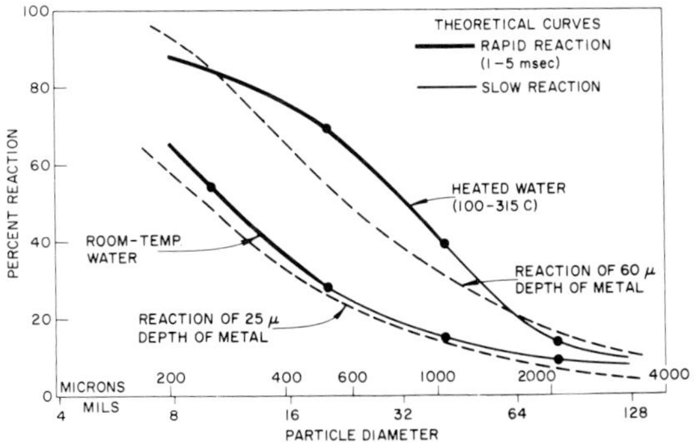

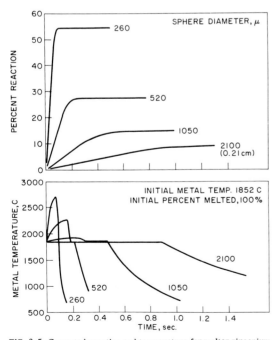

FIG. 3-6 Extent of reaction as a function of particle diameter for molten zirconium spheres formed in water.

These results can be explained on the basis that greater particle subdivision occurs when the oxide is formed in a molten state.

No important differences were reported between pure zirconium and Zircaloy-series alloys in any of the experimental studies of high-temperature metal-water reactions.

3.2.2 Uranium Alloys

Isothermal experiments: An early study of the uranium-steam reaction at 1000°C (1832°F) concluded that the reaction with massive pieces is relatively slow, requiring hours rather than seconds [165]. The oxide film formed at the rate of 0.28 mm/hr (11 mil/hr) in a 15-min test.

Three subsequent studies of the uranium-steam reaction yielded conflicting results. Linear reaction rates were reported up to 880°C (1616°F) in one study [166] and up to 1440°C (2624°F) in another [167]. Limited agreement to a parabolic law was obtained above 880°C in the first case, and over the entire temperature range from 700 to 1280°C (1292 to 2336°F) in the third study [168]. In general, it was found that unalloyed uranium was more reactive with steam than alloyed uranium. A solid U-10 wt.% Nb alloy had the lowest reactivity of several alloys tested [168].

In another study of uranium oxidation [169], uranium cubes, supported on a thermocouple, were heated inductively in flowing steam. The reaction was followed by condensing the effluent steam and determining volumetrically the quantity of hydrogen generated. Results are summarized in Figs. 3-7 and 3-8 [169]. A rapid linear reaction occurred at 400°C (752°F) with a rate nearly identical to that reported in the earlier studies. The reaction was nearly parabolic between 500 and 1200°C (932 and 2192°F) and followed a constant rate law between 600 and 1200°C (1112 and 2192°F) as indicated in Fig. 3-8. The rate law was as follows:

$$w^2 = 9.16 \times 10^4 \, t[\exp(-18{,}600/RT)] , \quad (3\text{-}2)$$

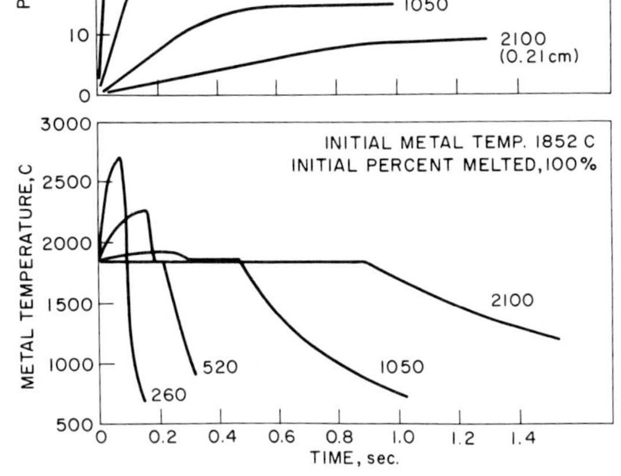

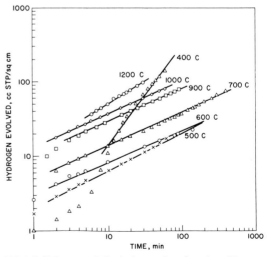

FIG. 3-5 Computed reaction and temperature for molten zirconium spheres in room temperature water.

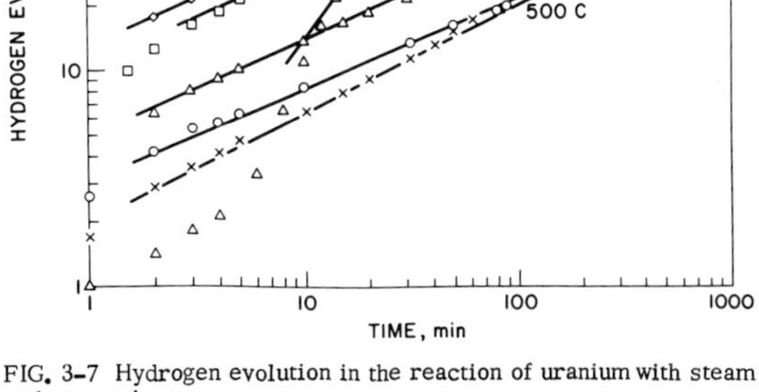

FIG. 3-7 Hydrogen evolution in the reaction of uranium with steam at 1 atmosphere.

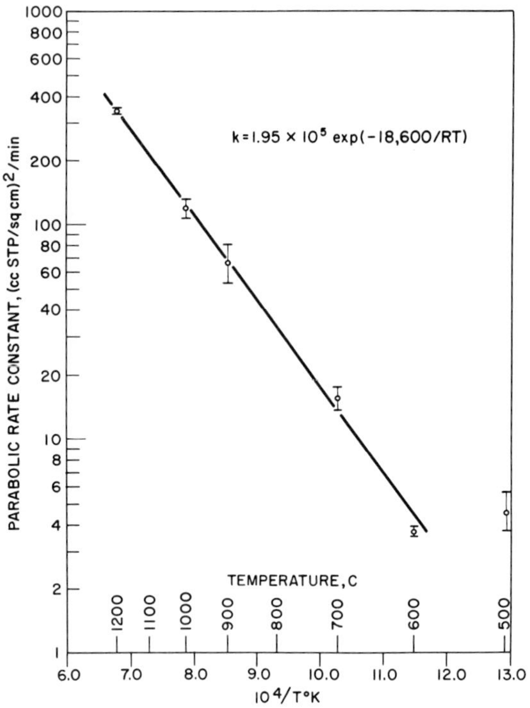

FIG. 3-8 Effect of temperature on the rate of hydrogen evolution for the reaction of steam with uranium.

where w is mg of uranium reacted per cm^2 and t is time in sec. The results of two of the above-mentioned studies of the uranium-steam reaction were in good qualitative agreement with these data [167, 168]. The fact that one earlier study [166] showed wide divergence in results above 500°C (932°F) was probably due to differences in impurity content or metallurgical treatment of the specimens. Ignition and oxidation studies in oxygen show clearly how alloying additions can markedly alter the uranium-oxygen reaction, especially in the temperature range above 500°C (932°F).

Nonisothermal experiments: Natural uranium at 1540°C (2804°F) was poured into cold water in a series of tests [161]. Popping was heard as the evolving hydrogen ignited and burned in air; however, no explosion resulted. Thin plates of metal were found which had about 0.5 mil (13 μ) oxide coating. It was estimated that from 10 to 30% reaction occurred. Similar tests with a 12% Mo alloy gave identical results.

Experiments with dispersed metal: Uranium at 1540°C (2804°F) poured into water and dispersed by a blasting cap [161] was partly atomized but the bulk of the metal remained as comparatively large, irregular platelets, which were covered with surface oxidation. The percentage of reaction varied from 30 to 50%. There were indications of slight explosions in four out of five tests. Results with a 12% Mo alloy were essentially the same, as were the results with a 5.5% Zr, 1.75% Nb ternary alloy. A single run in an explosion dynamometer indicated that uranium was not unusually explosive.

Electrical condenser-discharge results with uranium wires are given in Fig. 3-9 where, as with zirconium, considerably more reaction occurs in heated water than in room-temperature water. Runs were made only in water at 25 and 100°C (77 and 212°F) because of the extensive slow corrosion occurring in water at 200°C (392°F) or higher which would have destroyed the specimen before the run could be completed. Transient pressure traces were similar to those obtained with zirconium. The smaller particles were reacting explosively and larger particles were reacting slowly. The results were consistent with the finding for zirconium that particles smaller than about 0.5 mm (20 mil) in room-temperature water and 1 mm (40 mil) in heated water were in violent motion and therefore reacted very rapidly because of an increased rate of gas-phase diffusion. Several factors complicated the experiments so that it was not possible to ascertain conditions for the explosive behavior of uranium with the certainty that was possible for zirconium. Larger particles (greater than about 1 mm) were not spherical as they were with zirconium and the oxide was not as adherent. This behavior was also reported in the Aerojet explosion studies [155].

Discussion: The uranium-water reaction has been studied much less thoroughly than the zirconium-water reaction. Isothermal studies above 1200°C (2192°F) are in progress. Preliminary studies indicate that it may not be possible to extrapolate the parabolic rate law, Eq. (3-2), to temperatures much above 1200°C (2192°F), as would be required for an analysis of the type described in connection with zirconium.

It is reasonable, on theoretical grounds, to expect that the uranium-water reaction at high temperature is somewhat less extensive than the zirconium-water reaction. The activation energy found for the zirconium reaction (45.5 kcal/mole) is considerably higher than that found for the uranium reaction which is 18.6 kcal/mole for temperatures below 1200°C; above that a value of about 30 kcal/mole is tentatively indicated. It is therefore to be expected that greater reaction occurs with zirconium at very high temperatures. The activation energy difference also implies that uranium reacts more extensively at low temperatures, which is a well-known fact. The rate Eqs. (3-1) and (3-2) indicate that the two metals will follow the same rate law at approximately 1460°C (2660°F) on the basis of hydrogen evolution. The heats of reaction are nearly identical on a molar or a volume basis but, of course, differ on a weight basis.

3.2.3 Aluminum Alloys

Isothermal experiments: In a thermogravimetric study of the isothermal oxidation of molten aluminum in steam over the temperature range 816 to 1260°C (1500 to 2300°F), the reaction followed a linear rate law for pure aluminum and for a 23.4 wt.% uranium-aluminum alloy. Rates reached a maximum at about 1000°C (1832°F) and decreased at higher temperatures [170].

In a "pressure-pulse" study of the isothermal oxidation of aluminum over the range 800 to 1200°C (1472 to 2192°F), the reaction followed a cubic rate law as shown by the broken lines in Fig. 3-10 [171].

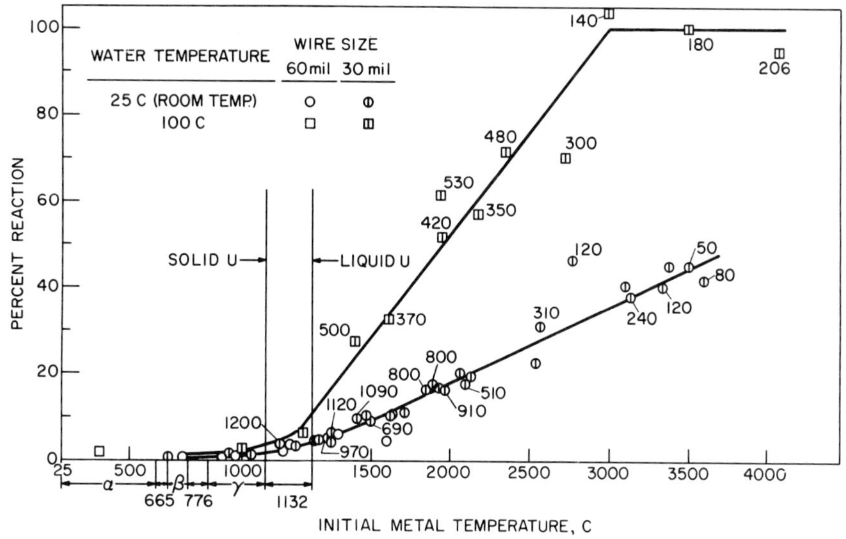

FIG. 3-9 Uranium-water reactions by the condenser-discharge method. 30-mil and 60-mil (0.76-mm and 1.52-mm) wires used. The numbers indicate the mean particle diameter of residue in microns.

The results of the pressure pulse and thermogravimetric studies do not agree in regard to the form of the rate law and especially in regard to the temperature dependence of the reaction rate. Both studies show, however, that at 1200°C (2192°F) and below the reaction is extremely slow and does not constitute a hazard.

A more recent study of the aluminum-steam reaction was made using the levitation-melting method between 1200 and 1750°C (2192 and 3182°F) [172]. One-quarter inch (6.3 mm) pellets of aluminum were supported and heated by a radiofrequency field in the presence of flowing steam. Temperature was measured and controlled by a two-color optical pyrometer. The quantity of aluminum reacting in each run was determined by a gravimetric method in which unreacted aluminum was selectively dissolved in an iodine-methanol solution leaving the oxide intact. The reaction could be described by a cubic rate law at 1200 and 1300°C (2192 and 2372°F) as shown in Fig. 3-10. The results, therefore, agreed with those of the pressure-pulse study although the measured rates differed by a factor or two.

Studies [173] by the levitation method up to 1600°C (2912°F) could be interpreted in terms of an initial cubic-rate law followed by transition to a linear law beyond w = 2.3 mg/cm (see Fig. 3-10):

$$w^3 = 6.7 \times 10^7 t[\exp(-73{,}500/RT)], \qquad (3\text{-}3)$$

where w is in mg of aluminum reacted per cm² and t is time in sec. Delayed ignitions occurred at 1600°C or 2912°F (18 min), at 1650°C or 3002°F (9 min), and at 1700°C or 3092°F (5 min), while immediate ignition occurred at 1750°C or 3182°F. Ignition began with the formation of a small hot spot which rapidly spread across the surface of the sphere. Temperature at the hot spot reached well above 2100°C (3812°F). Burning in steam appeared to be similar to the vapor-phase burning as occurred in air at 1750°C (3182°F) although the voluminous white smoke was less apparent [174].

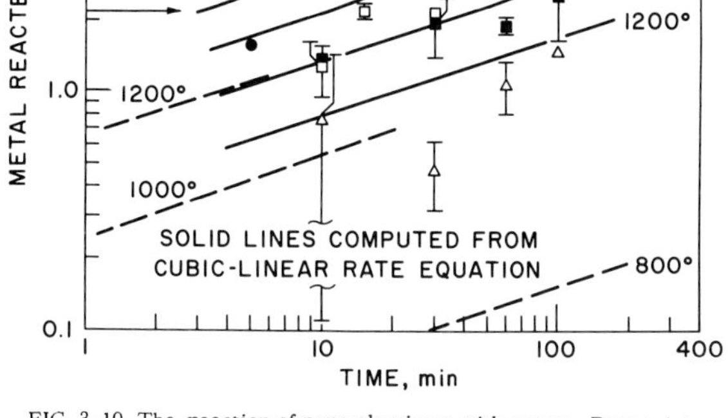

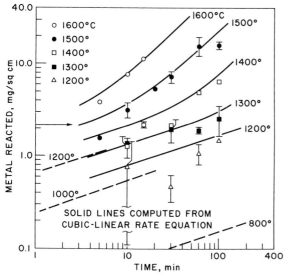

FIG. 3-10 The reaction of pure aluminum with steam. Data points obtained by the levitation-melting method. Broken lines obtained by the pressure-pulse method. Solid lines computed from cubic-linear rate equation based on the data points.

Nonisothermal experiments: A number in investigators have poured molten aluminum into water in an effort to evaluate the conditions needed for explosion. Injections of several pounds of molten aluminum in the form of jets at 750 to 1000°C (1382 to 1832°F) into water gave little or no reaction nor was any appreciable steam pressure buildup noted [175, 176]. Tests in which large crucibles containing molten aluminum at 900°C (1652°F) were smashed under water, showed that pure aluminum did not react, but aluminum containing 1.0 to 7.4% lithium reacted slightly to violently. Aluminum containing 5% uranium reacted violently at 900°C (1652°F) in 3 tests and slightly in 11 tests [159]. Higher temperatures did not increase the activity. Other tests in which molten aluminum and aluminum-lithium alloys were poured into water in 1-in. (25-mm) diameter streams, gave no violent reactions although a thin oxide coating formed on the resulting globules in both cases [159].

The most extensive tests of the effects of contacting large quantities of molten aluminum with water were carried out by taking 50 lb (23 kg) of commercially pure molten aluminum and suddenly discharging the metal into water in steel tanks. Violent explosions occurred in many cases even though the metal temperature was never higher than 900°C (1652°F) [177]. Explosions occurred when the metal was discharged through holes larger than 2.75 in. (7 cm) and only when a certain minimum quantity of metal was used. Explosions occurred when the metal was dropped 18 in. (46 cm) but not when the metal was dropped 10 ft (3.0 m) which suggested that no explosion would occur if the stream were broken up. Breaking the stream with an iron grid prevented explosions even with an 18-in. (46-cm) drop. Depth of water pool and metal temperature were interdependent. Explosions took place in a 3 or 6-in. (7.6 or 15.2-cm) water depth with aluminum at 670°C (1238°F) but required metal at 750°C (1382°F) with a 10-in. (25-cm) water depth. No explosions could be produced when the water level was 20 in. (51 cm). Molten metal spattering without a violent explosion took place when the water level was 2 in. (5 cm) or less. Wetting agents decreased the tendency toward explosion while salt increased it. Coating the bottom of the water container with lime, gypsum, or rust promoted explosiveness while grease, oil, or paint largely eliminated the explosions. Similar but less violent explosions were produced using a molten salt mixture instead of aluminum. This finding made it almost certain that the explosions resulted primarily from steam formation and not from metal-water reaction.

Experiments with dispersed metal: Experiments in which blasting caps were used to disperse molten aluminum streams underwater [159], showed formation of fine metallic granules but no violent explosions. Tests in the explosion dynamometer [155], however, showed that very violent reactions (see Fig. 3-2) could be generated when the initial metal temperature was 1400°C (2552°F) or higher.

Studies in which aluminum wires were electrically exploded by discharging condensers showed that complete reaction forming colloidal dispersions could be obtained with very energetic discharges [178]. Laser beam heating of single small particles in 100°C water [179] showed the existence of an ignition temperature that probably corresponded to the melting point of Al_2O_3, i.e. 2050°C (3722°F). At energies below the ignition, only a few percent reacted. At energies above ignition, reaction was complete. Complete reaction required about 200 msec for a 360-μ particle.

Explosive reactions were initiated in moistened aluminum powder compacts by means of auxiliary explosive charges in one study [180]. Efforts to duplicate the result, however, failed to produce a violent explosion [159].

Discussion: The results of experimental studies suggest very strongly that the temperature reached by the aluminum metal primarily determines the nature of the reaction. Isothermal studies have shown that the oxide film is very protective up to 1300°C (2372°F). This is fully consistent with the repeated failure of many investigators to obtain a vigorous metal-water reaction by pouring molten aluminum into water. Only in the explosion dynamometer tests was the metal temperature high enough to produce a significant reaction.

The explosions produced in the pouring tests were physical in nature and probably did not involve extensive metal-water reaction. The mechanism of these explosions is not fully established. The importance of physical explosions has been emphasized with the suggestion that physical entrapment of water by solidified metal is the initiating mechanism [149]. It appears, nevertheless, that water is momentarily trapped beneath a mound of metal at the bottom of the water tank. The rapid generation of steam then causes violent dispersal of the molten metal. The resulting rapid motion of hot particles greatly increases the heat transfer, forming steam very rapidly, and this process provides the bulk of the energy of the explosion [177].

Aluminum is much less reactive than either zirconium or uranium at temperatures of the order of 1200°C (2190°F). The vigorous burning of aluminum at 1750°C (3182°F), however, makes aluminum more reactive than zirconium or uranium at higher temperatures. This is in accord with the explosion dynamometer tests as indicated by the exceptional violence of the high temperature aluminum-water reaction (see Fig. 3-2). The ignitions noted with aluminum probably occur when the vapor pressure of the aluminum (7 mm Hg at 1750°C or 3182°F) is sufficient to sustain vapor-phase combustion. The ultimate product of this kind of burning is α-alumina which is partly in the form of a very porous mass and partly a finely divided powder. The delayed ignitions occurring in the levitation studies between 1600 and 1750°C (2912 and 3182°F) were very likely due to the ignition of Al_2O, a suboxide formed when aluminum and Al_2O_3 are together at high temperature. Thermodynamic estimates indicate that Al_2O may have a vapor pressure as high as 7-mm Hg at 1600°C (2912°F). The delay in ignition, therefore, is probably related to the time required to generate a significant quantity of Al_2O_3. The fact that the activation energy for the aluminum-steam reaction (73.5 kcal/mole) approximates the heat of vaporization of aluminum (71 kcal/mole) provides further evidence that the reaction is controlled primarily by a vaporization process. A similar coincidence for magnesium was given as evidence for a vaporization-controlled reaction [181].

The ignition reaction observed in the laser experiments with very small particles is markedly different from that observed in levitation experiments with larger pellets of aluminum. With the larger pellets, the product was a finely divided oxide resulting from vapor-phase burning; however, with the small particles, the product was a single transparent crystal of alumina. Both reactions were relatively slow. A levitated pellet, 8 mm in diameter, dropped into liquid water required 30 sec for consumption [182], while the 360-μ particle heated by a laser beam required 200 msec for complete reaction [179].

3.2.4 Stainless Steels

Isothermal experiments: Isothermal studies of the reaction of type-304 stainless steel with flowing steam were performed by direct induction-heating of disc-shaped specimens [183]. Below 1100°C (2012°F) the reaction was too slow for measurement. At 1100° and 1200°C (2192°F) the reaction could be described approximately by a parabolic rate law. The reaction rate at 1300°C (2372°F) was found to depend on the rate at which the sample was heated. Metallographic and electron probe microanalytic studies indicated that a protective coating, high in chromium and nickel content, formed over the surface of slowly heated samples. The coating formed sparsely over rapidly heated samples. Foaming and sample disintegration at 1400°C (2552°F) resulted in a sharp increase in the reaction rate. The foaming probably resulted from the simultaneous melting of the steel and the formation of a molten oxide mixture. At 1400°C, a 1/4-in. thick sample was about 30% reacted in 10 min.

Experiments with dispersed metal and discussion: One test with molten stainless steel (type 303) at 2270°C (4118°F) in the explosion dynamometer, shown in Fig. 3-2, indicated that the explosion was nearly as vigorous as explosions obtained with molten zirconium [155].

Studies of the reaction by the electrical condenser discharge method [184] are summarized in Fig. 3-11. The most significant result of the study was the limited amount of reaction obtained with small particles (260μ or 10 mils) at very high metal temperatures (3700°C or 6692°F) in heated water. Reaction of zirconium and uranium in heated water was more extensive under similar conditions. In addition, the stainless steel-water reaction was not much more extensive in heated water than it was in room-temperature water. These two results indicate that the reaction rate is controlled predominantly by a chemical rate law, even at very high temperatures. Gaseous diffusion of water vapor through the evolving hydrogen, shown to be of major importance in the zirconium-water reaction, appeared to be less important in the high-temperature stainless steel-water reaction.

3.2.5 Magnesium Alloys

Isothermal experiments: A study of the reaction of magnesium with water vapor over the temperature range from 425 to 575°C (797 to 1067°F) using a manometric method showed that the reaction followed a linear rate law after an induction period [181]. Rate constants increased linearly with the water vapor pressure between 31 and 208 mm Hg. Also, there was a gradual increase in activation energy with decreasing water vapor pressure in the 425 to 500°C (797 to 932°F) range. The activation energy in the 500 to 575°C (932 to 1067°F) range reached a limiting value of 33.7 kcal/mole which was nearly identical with the heat of sublimation of magnesium (34.4 kcal/mole). The observations were explained on the basis of a change from a surface to a vapor-phase reaction. The vapor-phase reaction appeared at a higher temperature and lower water-vapor pressures. The change resulted from the very large increase in the vapor pressure of magnesium metal between 500 and 600°C (932 and 1112°F). The rate constants shown in Table 3-3 were reported for a water vapor pressure of 208 mm Hg.

A study of the magnesium-steam reaction between 400 and 600°C (752 and 1112°F) indicated that ignition occurred at 625°C (1157°F) resulting in vapor-phase burning [184a]. Ignition temperature was approximately the same in oxygen and hydrogen peroxide as it was in water vapor. The vapor pressure of magnesium at 625°C (1157°F) is approximately 1.5 mm Hg.

Nonisothermal experiments: Experiments in which molten magnesium was dropped into water

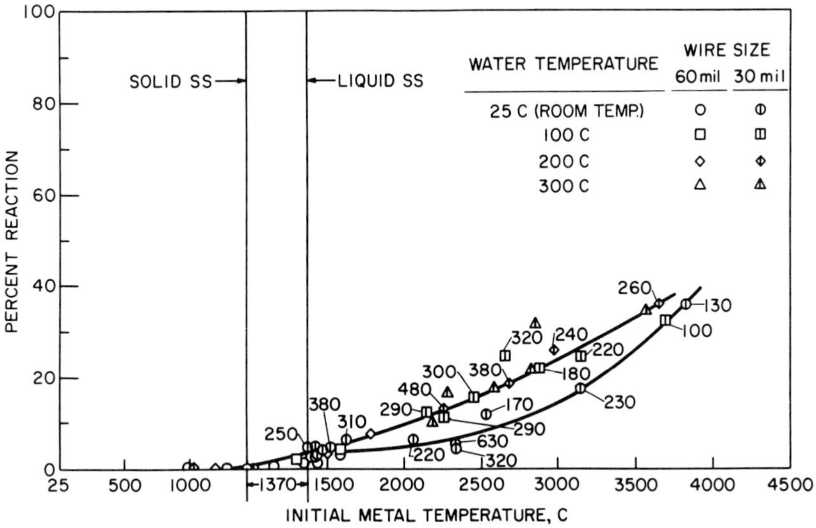

FIG. 3-11 Stainless steel-water reaction by the condenser-discharge method. Type 316 stainless steel used. The numbers indicate the mean particle diameter of residue in microns.

TABLE 3-3
Isothermal Reaction Rates of Magnesium with Water Vapor at 208 mm Hg Pressure

Temperature	Reaction rate $mgH_2O/(cm^2hr)$
425°C (797°F)	0.082
450°C (842°F)	0.141
475°C (887°F)	0.186
500°C (932°F)	0.255
525°C (977°F)	0.359
550°C (1022°F)	0.618
575°C (1067°F)	1.265

and the evolved hydrogen was collected volumetrically gave typical results as shown in Fig. 3-12 [185]. With magnesium initially at 700°C (1292°F) and the water at 20°C (68°F), 49% of the magnesium reacted with the water in about 1.6 sec. The extent of reaction decreases with increasing diameter of the magnesium droplets as indicated in Fig. 3-13. It is interesting to note the relatively short time of reaction which almost corresponds to a quenching; this contrasts to molten aluminum which at 1750°C (3182°F) can burn either submerged or floating in water for considerably longer times [174].

Experiments with dispersed metal: Dispersal of small batches of molten magnesium in water with blasting caps in two qualitative experiments resulted in sharp blasts accompanied by a white flash with 70-80% of the metal reacting [159].

Discussion: The high vapor pressures of magnesium (boiling point 1103°C or 2017°F) and the relatively high heat of reaction indicate that a vigorous vapor-phase burning will occur if magnesium is melted in the presence of water or steam. The course of the magnesium reaction generally parallels that of the aluminum-water reaction if the reactions are compared on the basis of metal vapor pressures.

3.2.6 Liquid Metal Coolants

The metals of interest as coolants for nuclear reactors are Na, NaK, Li^7, Hg, Ga, and Pb-Bi alloy (p. 994 of reference [1]). At the present time Na and NaK are by far the most important; hence this section gives emphasis to these two common coolants. The primary safety problem is that associated with the mixing of the molten metal coolant with steam or water in a steam generator or heat exchanger if a leak or rupture develops in a tube.

A qualitative summary of compatibilities of alkali metals with water and various gases is presented in Table 3-4 [186]. For both water and oxygen there is a general trend of increasing reactivity as the alkali metals proceed from Li to Cs. (See also reference [187].) Mercury is inert to both water and oxygen at ordinary temperatures. Gallium is chemically like aluminum and does not react with water even at 100°C (212°F); gallium up to red heat is only superficially attacked by oxygen. Bismuth at about 700°C (1292°F) decomposes steam and burns to the trioxide in air. At room temperature bismuth is protected by an oxide film. Also, at ordinary temperatures lead is very stable, but at about 300°C (572°F) severe corrosion by air begins. At white heat, lead reacts with steam and burns in air. Lead-bismuth alloys approximate bismuth in their corrosion behavior.

There are three products of the sodium-water reaction. In excess water, sodium hydroxide (NaOH) and hydrogen are formed as indicated in Table 3-1. In excess sodium, sodium monoxide (Na_2O) and hydrogen are formed. The reaction of sodium with water can be considered irreversible. Figure 3-14 [188] shows the results of calculations of equilibrium temperatures and pressures where complete reaction is assumed to occur in a rigid container with the hydrogen produced occupying the net volume between reactants and the oxide or hydroxide. It is apparent that the maximum pressure results from an adiabatic reaction of stoichiometric amounts of sodium and water.

The three reaction products can cause the following difficulties in reactor vessel and piping systems [189]:

Na_2O - plugging of sodium stream;
NaOH - corrosion of water or steam tubes if tube material is austenitic stainless steel or aluminum;
H_2 - excessive pressure buildup, or explosion if oxygen is also present.

Kinetic studies of sodium-water systems may be conveniently divided into "practical" and "basic" experiments. Among the practical experiments is the connecting of a thermal convection NaK loop to a similar water loop through a 0.5-in. (1.25-cm) triggering valve. With the NaK initially at 427°C (800°F) and 30 psig (3 atm) and the water at 177°C (350°F) and 400 psig (28 atm), a 4000 psi sharp peak pressure was noted 20 sec after triggering [189]. However, the addition of a surge volume equivalent to 8% of the NaK, eliminated the pulses and gave pressure fronts of from 400 to 600 psig (28 to 40 atm). Motion pictures have been made of sodium-water reactions both in water and in air. In the argon-blanketed case there was no ignition or explosion, whereas with air there was

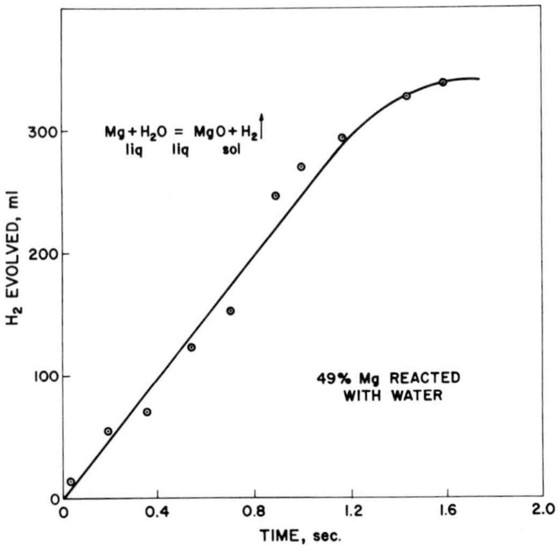

FIG. 3-12. Mg-H_2O reaction data. Initial conditions: Mg at 700°C (1292°F), H_2O at 20°C (68°F).

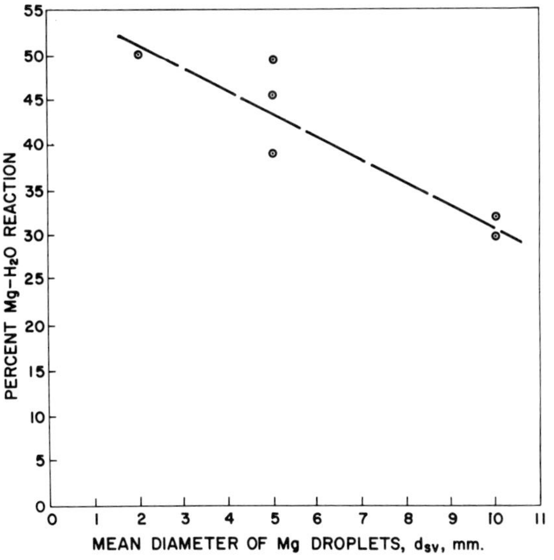

FIG. 3-13. Dependence of extent of Mg-H_2O reaction on particle size.

both ignition of the metal and a hydrogen-oxygen explosion. From these empirical tests the following three safety recommendations were formulated for Na or NaK systems [190] :

1. maintain oxygen content of blanket gas at much less than 5%;
2. provide expansion volume on piping systems; and
3. provide backup surge volumes with a venting device such as a rupture disk.

It is of interest that, in practice, leaks in sodium-water piping systems are usually so small (fine

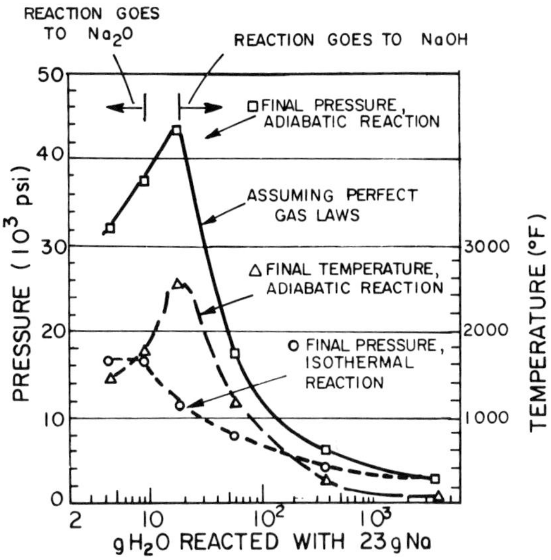

FIG. 3-14. Estimated final pressure and temperature, resulting from the reaction of one mole of sodium with various amounts of water.

TABLE 3-4

Chemical Reactions of the Alkali Metals

With	Lithium	Sodium	Potassium	Rubidium	Cesium
Oxygen	Relatively inert; no reaction below 100°C	Fairly rapid	Fairly rapid	Burns in air	Burns in air
Nitrogen	Reacts. Argon or helium must be used as blanketing gases	No reaction	No reaction	No reaction	No reaction
Hydrogen	Reacts rapidly at m.p. of Li (180°C)	Rapid reaction above 300°C	Rapid reaction above 300°C	Reacts above 600°C	Reacts slowly at 600°C
Water	Slow reaction	Rapid	More rapid	More rapid	Most rapid
Carbon	Reacts at high temps to give Li_2C_2	Reacts at 800-900°C to give Na_2C_2	Dissolves to solid solution; no carbide formed	No carbide formed	No carbide formed
NH_3	Reacts to give $LiNH_2$ (slow)	Reacts to give $NaNH_2$ (slow)	Reacts to give KNH_2 (easy)	Reacts to give $RbNH_2$ (rapid)	Reacts to give $CsNH_2$ (most rapid)
CO	No carbonyl formed (uncertain)	No carbonyl formed, except in liquid NH_2	Forms explosive carbonyl	Carbonyl forms readily (uncertain)	Absorbs CO at room temp.
CO_2	Reacts only at high temps.	Reacts	Reacts	More rapid reaction	Most rapid reaction
Halogens:					
F	Reacts readily emitting light	Ignites	Reacts violently	Reacts, ignites	Reacts most vigorously of all alkali metals
Cl		Reacts	Reacts violently	Reacts, ignites	
Br		Slow reaction	Detonates	Reacts, ignites	
I		Fission-product iodine is probably absorbed (trapped)	Reacts, ignites	Reacts, ignites	
H_2SO_4:					
Cold, conc.	Reacts very slowly	Fairly vigorous	Explosive reaction	Explosive reaction	Explosive reaction
Cold, dilute	Violent reaction	Very vigorous	Explosive reaction	Explosive reaction	Explosive reaction

cracks or pinholes) that the leak is detected not by the direct evidence of the energy or pressure from the metal-water reaction but by subsequent plugging of a pump or valve (p. 62 of reference [191]).

Other experiments have demonstrated that sodium-water reactions are not high-order explosions since peak pressures are attained in milliseconds or seconds as contrasted to microseconds [155]. Results from dispersing 1 to 5 kg of NaK under 10 ft (3 m) of water suggested that less than 0.3% of the exothermic reaction energy was evident in shock waves while 16% went into formation of hydrogen gas bubbles; the remainder was dissipated into the water as heat (p. 126 of reference [188]).

The practical subject of how to extinguish a sodium fire once it has started is discussed in Sec. 3.3.5. Perhaps the most important comment is a negative one: namely, not to apply the conventional materials of water, carbon dioxide, or carbon tetrachloride. Each of these will only aggravate the incident, especially the carbon tetrachloride which can explode when in contact with sodium.

There have been several basic studies of sodium-water reactions which show that the kinetics are usually limited by how the two reactants are contacted (surface area) or by diffusion. The effect of partial pressure of the water vapor, over the range 0 to 9 mm Hg [192], was to increase linearly the rate of reaction. A study based on measuring the hydrogen evolved as a function of time from sodium initially floating in water at 35 to 140°F (2 to 60°C) gave a correlation of volume versus time as follows [193]:

$$V_e^{1/3} - (V_e - V)^{1/3} = kt, \qquad (3\text{-}4)$$

where V_e is the equilibrium or final volume of H_2 evolved, V the instantaneous volume of H_2, t the time, and k the rate constant. This relationship is derived by hypothesizing that the rate of reaction, or hydrogen evolution, is proportional to the surface area of a spherical particle of sodium. For sodium (~ 300 mg/particle) and water at room temperature, k has a value of 0.1 cm/sec. An interesting point encountered in the study was that the reaction of cool, liquid sodium with heavy water was more rapid than with light water. The difference was attributed to the 25% higher viscosity of liquid D_2O relative to H_2O which decreased the cooling action for the case of D_2O. However, there was no observable difference in rates with D_2O or H_2O vapor. At a higher sodium temperature, 300°C (570°F), the molten metal reacted more rapidly with H_2O than D_2O [193].

The characteristics of the sodium-water reaction also change markedly with temperature. Observations have indicated a relatively controlled reaction up to 204°C (400°F). From 204 to 480°C (400 to 900°F) is a fast unstable region [193]. From 540 to 980°C (1000 to 1800°F) is a smooth zone; beyond 980°C the vapor pressure of sodium is great enough to give a vapor phase combustion. The boundaries and even the existence of these zones are open to some question. For example, another investigation [189] has indicated an exponential rate up to 93°C (200°F), a parabolic rate from 93 to 380°C (200 to 700°F) which implies a protective film, and then a very rapid rate with no protective action beyond 380°C.

The reaction rate can be significantly slowed down by adding a film of oil or a foaming agent to the surface of the water on which the sodium droplets are floating [193]. The rate of reaction of sodium with water in isoamyl alcohol is a linear function of the water content over the range 0 to 10% water [194].

The general conclusion that the kinetics are rapid is supported by experiments in which NaK and H_2O were mixed in 2 msec [195] and in experiments where NaK was contacted with a flowing stream of helium containing water vapor [156]. The reaction

$$NaK_2 + 3H_2O = 1.5H_2 + 2KOH + NaOH \qquad (3\text{-}5)$$

yielded the theoretical amount of hydrogen as fast as the reactants were mixed [195]. If oxygen is present, the record of pressure versus time shows first a thermal maximum followed by about a 40 to 80 msec induction period at which time a hydrogen-oxygen explosion occurs. These pressure pulses are shown in Fig. 3-15 [195]. When helium, containing from 100 to 400 ppm of water vapor (total pressure = 1 atm), was passed over molten sodium at 215 to 340°C (419 to 644°F), it was observed that the reaction rate approached the input rate of the water vapor and that the reaction rate also increased with the gas flow velocity. These observations suggest that the apparent reaction rates were limited by the mass transport of water vapor through the inert gas film. Thus, neither the diffusion through the sodium oxide layer (discontinuous) or the inherent chemical kinetics were not rate-determining [156].

One approach to estimating the rate of reaction for the transport-controlled case is to use the relationship

$$R'_m = (D/d)\ Nu\ (p_{H_2O}/RT). \qquad (3\text{-}6)$$

For a typical case [156]:

R'_m = rate of reaction = gmoles/cm²sec
D = diffusion coefficient of water through hydrogen

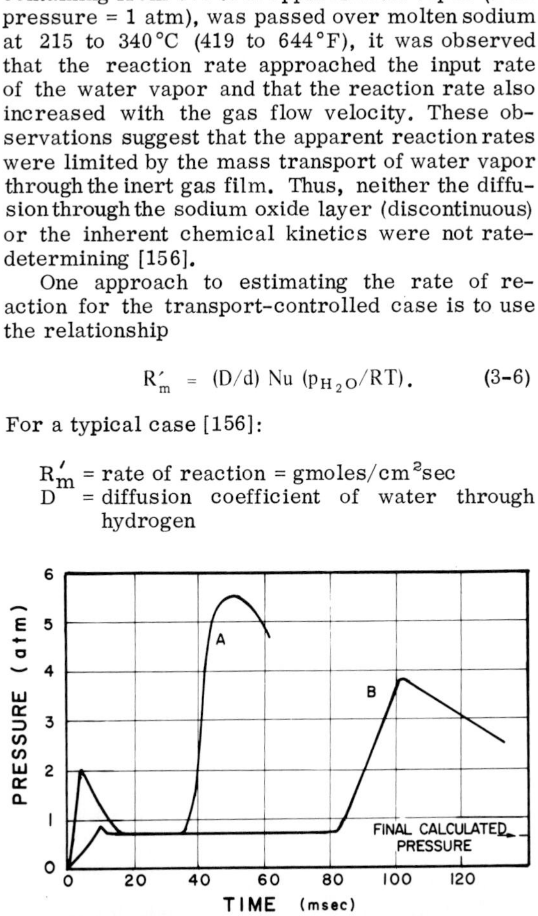

FIG. 3-15 Pressure rise in reaction of NaK with excess H_2O in the presence of air for two runs.

D = 0.175 cm^2/sec at 1000 psi (67 atm)
d = diameter of drop of Na
= 0.1 cm, assumed value
Nu = Nusselt number = 2 for static gas
p = partial pressure of water = 63 atm
T = absolute temperature
= 558 °K (if metal and water are not the same temperature, use mean)
R = gas constant = 82.06 cm^3atm/g mole-°K

One then obtains a value $R'_m = 4.82 \times 10^{-3}$ g moles Na consumed/sec-cm^2. Then from the postulate of $dW/dt = R'_m \times S$ = (g/sec) Na reacted = $R'_m \times cm^2$ surface, it is calculated that 0.87 sec is required to completely react the Na. It is interesting to note that, for this transport-limiting case, the rate is independent of the particular metal being considered.

3.3 Metal-Air Reactions

There have been numerous incidents of metals used in nuclear reactors igniting both in connection with reactor operations and during fabrication and storage [150]. The Windscale incident [196] in England illustrates the hazards of uranium burning in a reactor. The contamination of the NRU Reactor [197] in Canada by the burning of a portion of a fuel rod is an example of the hazards of fuel handling operations. Another incident involving the spontaneous ignition of a nuclear fuel was the Pu metal fire in 1957 at Rocky Flats, Colorado, [198] which resulted in the destruction of a laboratory glove-box train. A serious incident at Bettis [199] involved the ignition of a large quantity of zirconium scrap stored in open, segregated bins. Flames rose to a height of 80 ft (24 m) and were so intensely hot that windows over 100 ft (30 m) away were cracked. An incident at the Y-12 salvage yard at Oak Ridge [200] resulted in two deaths when a drum of finely divided zirconium upset, causing an explosion. A thorium explosion in a metallurgy laboratory [201] resulted in one death and serious financial losses.

Because of such incidents and the increasing number of nuclear reactors with the concomitant processing and handling of chemically reactive metals (often radioactive), studies have been and are being carried out of the factors influencing the ignition of these metals. These studies have concerned themselves with three aspects: isothermal oxidation kinetics, ignition, and burning of metals. Isothermal oxidation studies are considered as slow reactions and are treated in Sec. 2. However, the same reactions which appear slow with relatively massive specimens are largely responsible for the ignitions that occur with finely divided material. In the following discussion, studies of ignition are divided into two groups: those involving single pieces of metal and those dealing with the ignition of metal in aggregate form. A third section briefly outlines the characteristics of the burning of metals.

3.3.1 Zirconium Alloys

Ignition of single pieces: The oxide film formed on zirconium is very protective and adherent. For this reason, it is not possible to study zirconium ignitions by a rising temperature method [78]. Single pieces of zirconium, heated slowly (ca. 10°C/min or 18°F/min) in air or oxygen do not ignite even if heated to 1300°C (2372°F). Apparently a protective film forms at a lower temperature and prevents rapid reaction at higher temperature. Zirconium ignitions have been achieved by another method called the shielded ignition test. This method employs an inert environment, either helium or vacuum, to shield the sample during heating. The inert environment is then rapidly replaced by either air or oxygen and it is determined whether or not ignition occurs. Measured ignition temperatures are plotted in Fig. 3-16 [202] as a function of specific area of sample for foil and wires fabricated from crystal bar zirconium. Observed ignition temperatures in pure oxygen do not differ significantly from those in air. Comparison of results between foil strips with exposed edges and wires with protected ends indicates that the sharp edges have only a minor effect on the observed ignition temperature.

A number of mechanisms have been postulated to account for the ignition of zirconium [203]. Included are mechanisms involving hydrogen generated by metal-water reaction, the presence of lower chlorides, oxynitrates, and peroxides which might detonate. It seems more likely, however, that zirconium ignitions are thermal explosions resulting from more heat being generated than could be dissipated. An attempt to test this hypothesis was made by mathematical simulation of the zirconium ignitions [202]. The cubic rate law Eq. (2-1) was differentiated as follows:

$$\frac{dw}{dt} = \frac{5.94 \times 10^7}{3w^2} \left[\exp(-42{,}700/RT)\right], \qquad (3\text{-}7)$$

where dw/dt is in mg O_2/cm^2-sec. The following heat balance equation was used to describe heat exchange between the sample and the environment:

$$\frac{Cp}{S}\frac{dT}{dt} = \frac{91.2}{32} Q \frac{dw}{dt} - h(T_s - T_o) - \sigma\epsilon(T_s^4 - T_o^4). \qquad (3\text{-}8)$$

Selected temperature-time curves, computed by solving the above equations for specific areas of 5, 50, and 500 cm^2/g, are reproduced in Fig. 3-17 [202] and show both igniting and non-igniting situations for each specific area. The computed curves show that samples of high specific area ignite rapidly if the initial ambient temperature is above a critical value but experience very little self-heating if the temperature is slightly below the critical value. Specimens of low specific area ignite rather slowly above a critical ignition temperature; however, self-heating is considerable in samples that do not ignite. Computed ignition temperatures show good agreement with experimental values as shown in Fig. 3-16 and suggest

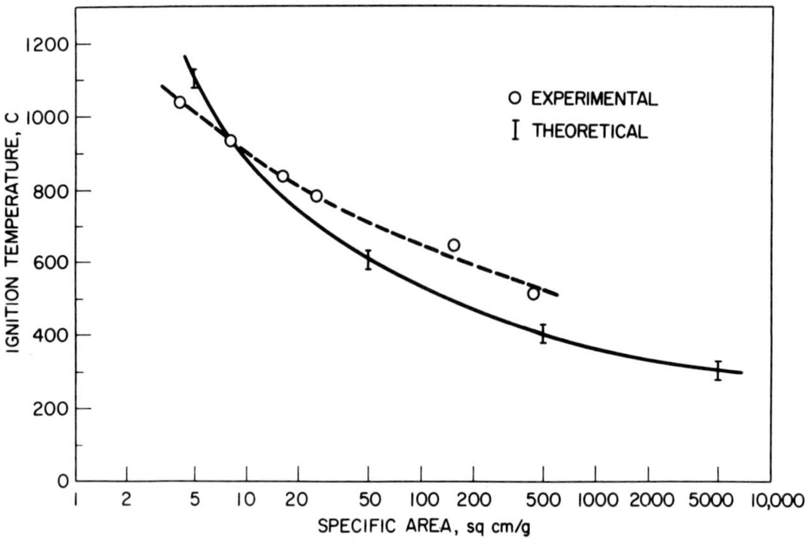

FIG. 3-16 Dependence of zirconium ignition temperature on specific area.

strongly that zirconium ignitions are purely thermal in character and do not require unusual initiating reactions.

The effects of alloying additions on the ignition temperature of zirconium are given in Table 3-5 [78]. Included in Table 3-5 are the cubic-rate constants and the oxidation occurring in 1.0 min at 700°C (1292°F) in 200 mm Hg pure oxygen. The final entry in Table 3-5 illustrates the highly pyrophoric character of zirconium-titanium alloys.

A study of the spontaneous ignition of titanium [204] and zirconium [26] showed that violent ignitions can occur if specimens are suddenly exposed to oxygen at high pressure. Ignition occurs when zirconium rods at room temperature are broken in tension in oxygen at pressures greater than 500 psi (~34 atm). Ignitions occur with zirconium and Zircaloy-2 sheets sealed in glass vials

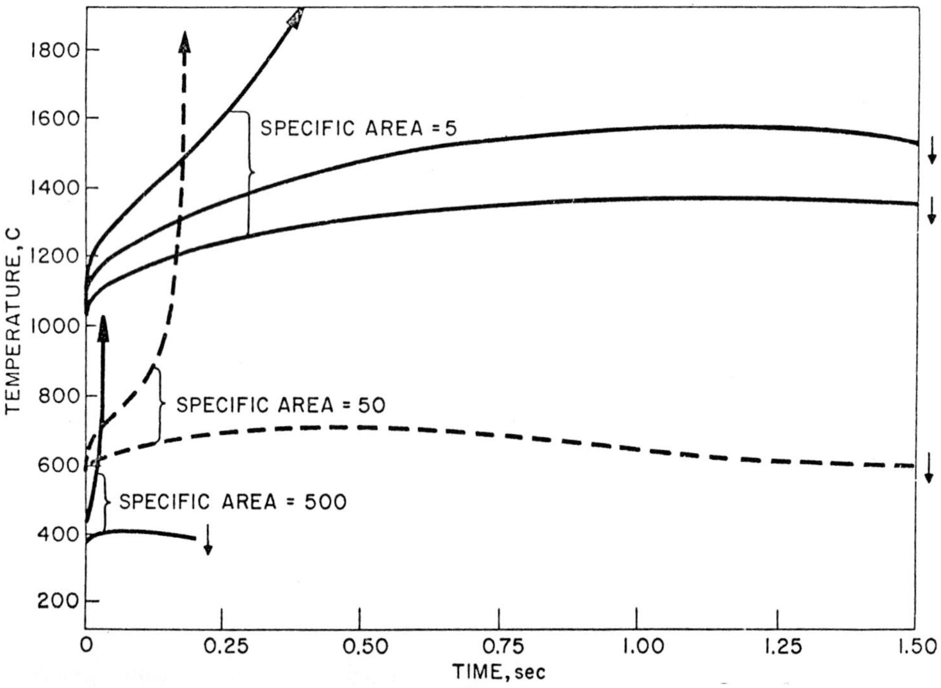

FIG. 3-17 Computed rate of temperature rise of oxidizing zirconium specimens. Specific area values are in cm^2/g; arrows indicate ultimate direction of calculated temperatures.

TABLE 3-5

Shielded Ignition Temperatures of Zirconium and Zirconium Alloys in Oxygen and in Air

All ignition specimens were 0.13 mm (5 mil) foils; all isothermal oxidation specimens were parallelepipeds, 1 x 1.5 x 2 cm (0.39 x 0.59 x 0.79 in.)
Flow rates for oxygen ignition: helium, 5200 cm^3/min; followed by oxygen, 3050 cm^3/min
Flow rates for air ignitions: helium, 1900 cm^3/min; followed by air, 4660 cm^3/min
For conversion to linear flow rates: 1 cm^3/min = 0.20 cm/min
(Other conversion factors: 1000 cm^3 = 61.0 $in.^3$ = 0.0353 ft^3; 1 μg = 2.2 x 10^{-9} lb; 1 cm^2 = 0.1550 $in.^2$ = 0.001076 ft^2)

Alloy composition (at.%)	Oxygen ignition temperature (± 5°C or ± 9°F)	Air ignition temperature (± 5°C or ± 9°F)	Isothermal data at 700°C (1292°F) in 200 mm Hg oxygen	
			Cubic rate constant, K 10^7 $(\mu g/cm^2)^3$/min	Weight gained first minute ($\mu g/cm^2$)
Pure Zr	786°C (1447°F)	784°C (1443°F)	1.6, 0.97[a]	236, 175[a]
0.91 Ni	792°C (1458°F)	794°C (1461°F)	1.5	245
3.60 Cu	784°C (1443°F)	800°C (1472°F)	0.78	195
1.76 U	765°C (1409°F)	766°C (1411°F)	4.1	284
1.84 V	763°C (1405°F)	764°C (1407°F)	Parabolic	332
3.60 Sn	759°C (1398°F)	767°C (1413°F)	1.2	205
2.15 Al	739°C (1362°F)	740°C (1364°F)	12.0	452
3.62 Al	753°C (1387°F)	754°C (1389°F)	8.0	546
1.08 Ti	694°C (1281°F)	698°C (1288°F)	69.7	624
2.12 Ti	634°C (1173°F)	632°C (1170°F)	251.0	946
4.16 Ti	529°C (984°F)	531°C (988°F)	Parabolic	1400
14.9 Ti	372°C (702°F)	374°C (705°F)	Ignited	Ignited

[a] Isothermal run in air at 700°C (1292°F) and 1 atm.

when the vials are broken in the presence of oxygen at 300 psi (~20 atm) or greater or in the presence of an oxygen-helium mixture containing at least 75% oxygen at pressure above 1000 psi (~67 atm). Similar ignitions occur with preoxidized, hydrided, and carbided surfaces although higher pressures are required with carbided surfaces. The ignitions occur only when surfaces are exposed to a rapidly increasing oxygen pressure. One of the most plausible mechanisms given to account for this unusual behavior is that the oxide in equilibrium with the atmosphere is oxygen-deficient and that rapid pressurization causes an exothermic change of composition (oxidation) which results in a rapid temperature rise.

Ignition of aggregates: In general, aggregates and powder compacts of metals ignite at lower temperature than single pieces even when compared at equal specific surface areas. This behavior results from decreased heat losses from particles in the center of the aggregate, caused by the insulating character of the outer regions of the aggregate mass. The phenomenon is illustrated in Fig. 3-18 [205] which shows how the measured ignition temperature of zirconium powder beds decreases monotonically with the depth of powder. Figure 3-18 also shows that finer mesh powder (high specific area) ignites at lower temperatures than a coarser mesh material. Results from studies [206, 207] with very finely divided powders are included in Fig. 3-19 [205] and indicate that the ignition temperature approaches room temperature for sub-micron size particles.

The spark energy required to ignite finely divided zirconium powders has been determined [206]. If the ignition energy of the system is less than 0.01 joule, then this situation presents serious hazards in dry handling because spark energies of this magnitude can be built up and discharged from a human body. On the other hand, commercial magnesium powder, having a spark sensitivity of 0.045 joule, can be handled with relative safety. Zirconium powders having a particle size between 2 and 3 μ (0.08 to 0.12 mil) have an ignition energy of the order of 0.000045 joules and are therefore in the hazardous range. A particle size of ~10 μ (specific area = 920 cm^2/g) is the borderline between safe and hazardous powders on the basis of accidental ignition by static discharge.

A very serious and sometimes hidden hazard can exist in massive zirconium parts prepared by powder metallurgy if sintering has been inadequate. Porous compacts can ignite in the same way as loose powders of an equivalent specific area. Material of this kind having a smooth surface resembling a solid piece of metal has been observed to flash and explode when ignited with a match flame [208].

Burning: Zirconium ignitions in oxygen are characterized by a very bright flash [78] and temperatures above 3000°C (5432°F) have been recorded [163]. Samples are completely consumed in pure oxygen. Ignitions in air are characterized by a more prolonged white glow and, in general, samples are not completely consumed. Maximum burning temperatures in air vary between 1465°C (2669°F) for a zirconium foil 0.002 cm (0.8 mil) thick and 0.5 cm (0.20 in.) wide to 1900°C (3452°F) for a foil 0.002 cm thick and 0.1 cm (0.04 in.) wide [209]. Theoretical values of burning temperatures in air and rates of propagation of burning, computed on the basis that the reaction rate is controlled by the rate of gas-phase diffusion of oxygen through a nitrogen-rich boundary layer [209, 210], are in reasonable agreement with measured values. Propagation of burning in air

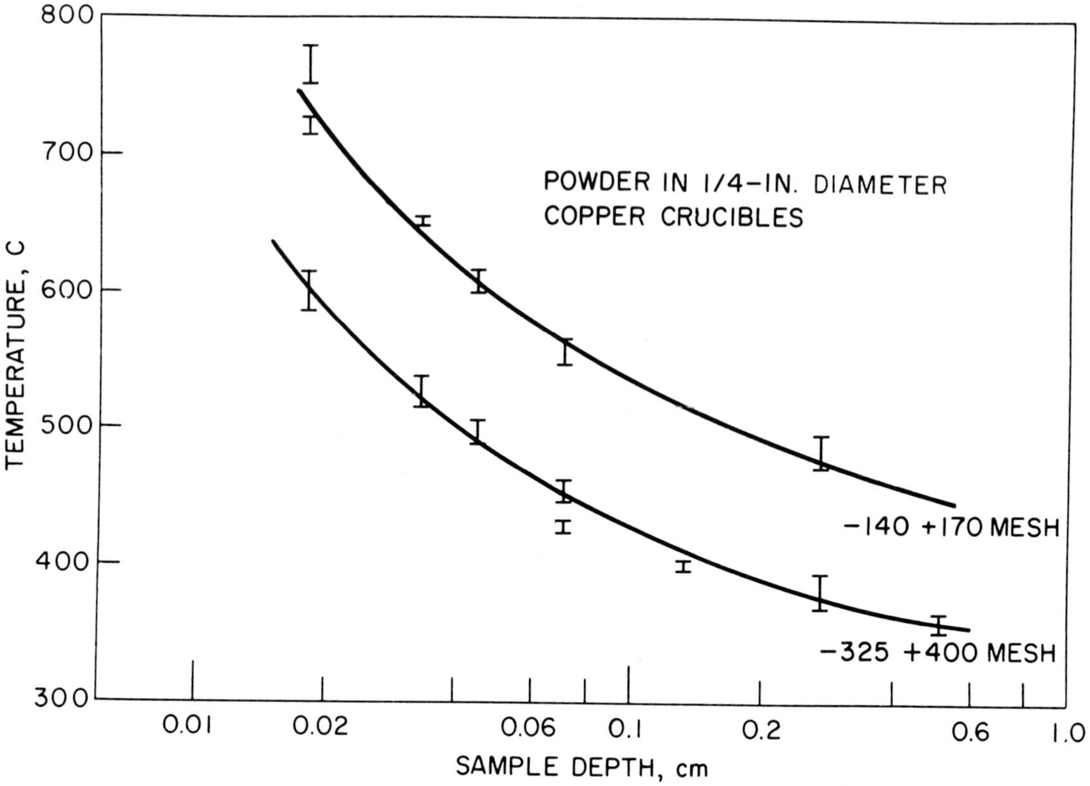

FIG. 3-18 Helium-shielded ignition temperatures in oxygen for various depths of spherical zirconium powder in 0.25 in. (6.3 mm) copper crucibles.

does not occur along pieces over 0.005 in. (0.013 cm) thick and wider than 1/16 in. (0.16 cm) [208].

Propagation of burning in aggregates such as scrap in the form of turnings, shavings or powder and the behavior of larger fires are extremely complex processes. No satisfactory theory exists which describes the course of such fires. Water, carbon dioxide and carbon tetrachloride may intensify zirconium fires while dry chemical powders, sand and certain halogenated hydrocarbons may be effective in decreasing the intensity of burning.

3.3.2 Uranium Alloys

Ignition of single pieces: A number of investigators have heated uranium in air and in oxygen under various experimental conditions [78, 79, 107, 211-214]. Well-defined ignitions occur in oxygen, but, in general, do not occur in air unless foils or wires having a relatively high specific area are used. Irregular thermocycling is frequently observed and is especially characteristic of uranium heated in the range 350 to 450° C (662 to 842° F).

Ignition in oxygen and air has been studied by a rising temperature ignition test [78]. In this method, samples are heated uniformly (usually 10° C/min) in a flowing oxidizing atmosphere. Ignition temperature is defined as the intersection point between linear extensions of the pre-ignition heating rate and the post-ignition self-heating rate taken from the temperature-time record. Ignition temperatures obtained in this way in oxygen and in air are plotted in Fig. 3-20 [215].

Ignition temperatures, plotted in Fig. 3-20, show a marked dependence on specific area. An attempt has been made to compute ignition temperatures (as a function of specific area) from isothermal data in the same manner as for zirconium ignitions. Empirical rate equations, obtained from analysis of isothermal oxidation data between 300 and 625° C (572 and 1157° F) [216] are:

$$300°C < T < 450°C$$
$$(572°F < T < 842°F)$$

$$\frac{dw}{dt} = w^{1/5}\,(3.25 \times 10^7)\;[\exp(-16{,}800/RT)] \tag{3-9}$$

$$T > 450°C\ (842°F)$$

$$\frac{dw}{dt} = \frac{(6.1 \times 10^7)}{w^{1/5}}\;[\exp(-14{,}300/RT)], \tag{3-10}$$

where dw/dt is in mg $O_2/(cm^2)$ (sec). The equations reflect the observation that the reaction accelerates slightly between 300 and 450° C (572 and 842° F) and decelerates slightly above 450° C. The rate equations are solved simultaneously with a heat balance of the form given by Eq. (3-8). Ignition temperatures obtained from computed temperature-time curves [215] are included in Fig. 3-20. The sharp break in computed ignition temperature between samples having specific areas of 5 and 7 cm^2/g results from the change from an accelerating to a decelerating oxidation at 450° C or 842° F (on the assumption that none of the oxide formed below 450° C contributes to

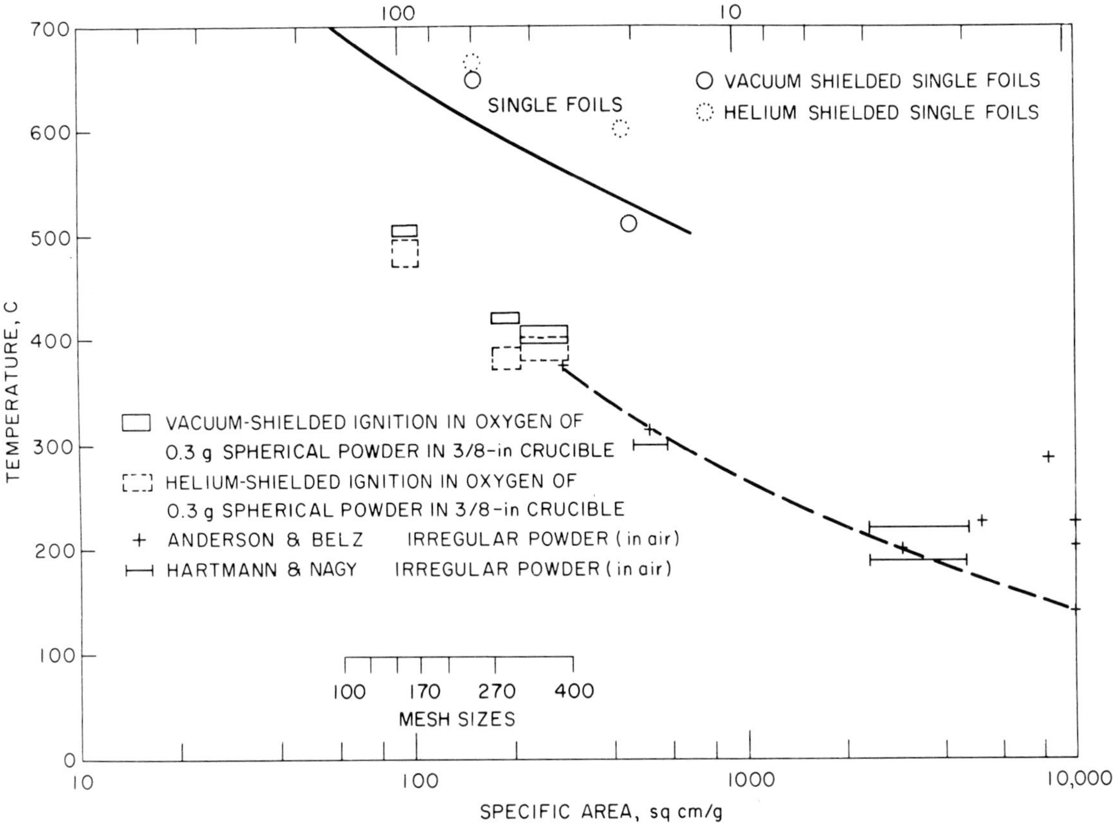

FIG. 3-19 Zirconium ignition temperatures for single foils and powders.

the protectiveness of oxide formed above 450°C). Calculated and experimental ignition temperatures show limited quantitative agreement and suggest that uranium ignitions result simply from an accumulation of heat generated by oxidation.

Alloying additions of many different elements have profound effects on the oxidation and ignition of uranium. There are two major effects on the oxidation reaction [217]. One of these effects, noted particularly with aluminum, is to inhibit the transition from an accelerating reaction to a decelerating reaction at 450°C (842°F). This effect is illustrated in Table 3-6 [217] where the rates of isothermal oxidation at 500°C (932°F) and ignition temperatures are listed for a series of low-level aluminum-uranium alloys. Increased reaction rates in the temperature region around 500°C have also been noted with uranium containing a few atom per cent of Be, Bi, C, Mo, Nb, Pb, Pd, Pt, Ru, Si, Ti, and V. The second effect, produced notably by

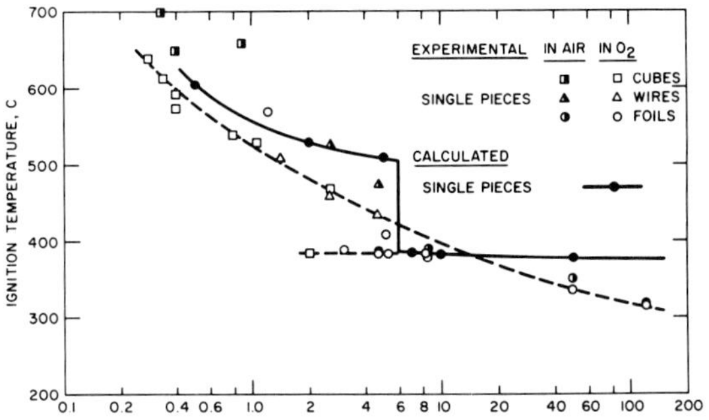

FIG. 3-20 Dependence of uranium ignition on specific area. Half-shaded symbols denote experiments in air; others denote experiments in O_2.

TABLE 3-6

Comparison of Oxidation Rates at 500°C (932°F) and Ignition Temperatures for Selected Low-Level Aluminum Alloys of Uranium

Metal[a]	Ignition temperature[b]	Oxidation rate, $\mu g\ cm^{-2} min^{-1}$ at 500°C after 10,000 $\mu g\ O_2/cm^2$ total oxidation
ANL base metal (15 ppm Al)	595°C (1103°F)	915
BMI base metal (34 ppm Al)	575°C (1067°F)	1510
55 ppm Al	575°C (1067°F)	2120
70 ppm Al	575°C (1067°F)	—
75 ppm Al	420°C (788°F)	2790
235 ppm Al	385°C (725°F)	3060
580 ppm Al	365°C (689°F)	8100
1130 ppm Al (1 at.%)	355°C (671°F)	19800

[a] The aluminum-uranium alloys were prepared from BMI base metal.

[b] Determined by the rising temperature technique in flowing oxygen for 8.5 mm (0.334 in.) cubes.

alloys containing copper, is to render the post-transition oxide more protective and therefore to decrease oxidation rates at temperatures above 500°C (932°F). Some alloying additions produce both effects in that near 500°C (932°F) an increased reaction rate occurs which then decreases markedly at temperatures from 600 to 900°C (1112 to 1652°F). The additions include Bi, Mo, Pd, Pb, Pt, Ru, and V. Additions of Ce, Cr, H, Fe, Ni, Rh, Ag, Ta, Th, Zr were observed to have no significant effect on the uranium oxidation. There is increased resistance to high temperature air oxidation by uranium alloys containing 5 to 15 wt.% molybdenum [211, 107]. The protectiveness of the oxide formed on molybdenum alloy possibly results from the fact that the oxide is more plastic and less porous.

The Chalk River [197] and the Windscale [196] incidents as well as several ignitions at Hanford [218] have been interpreted as indicating an increased pyrophoricity of uranium after irradiation. Studies of fission product release have also indicated an increased oxidation rate with irradiated specimens. It is difficult, however, to determine whether the changes occurred as a result of increased surface area and changed surface texture or whether radiation actually altered the oxidation mechanism [218]. Increased oxidation rates were noted in air only when uranium was irradiated above 10^{20} nvt (10^{-2} at.% burnup) at which point fission gas bubbles caused disruption of the oxide and increased surface area [219]. The available evidence does not indicate a strong influence of irradiation. It is likely that the main effects of irradiation are to increase the active surface through the development of cracks and porosity and self-heating from radioactive decay. The usual effects of radiation on metallic uranium are directional growth, surface roughening, and swelling [220]. These all tend to increase the specific surface area and promote pyrophoricity. Post irradiation swelling has been noted to occur when highly irradiated uranium is heated to 650°C (1202°F) [221, 222]. This swelling would very likely result in ignition if it occurred in air [223]. At the same time it should be pointed out that β-or γ-ray decay heat in irradiated fuel will raise fuel temperatures. Such high fuel temperature situations may be much more likely in nuclear fuels than in other reactor materials.

Ignition of aggregates: The effect of aggregation on the ignition of uranium foils is demonstrated in Table 3-7 [224] as a function of the number of foil specimens in a stack. The ignition temperature decreases sharply in going from one to two foils because of decreased heat losses. In the two-foil case, the heat loss from each foil is only about one-half of that in the single-foil case. Similar behavior was noted [225] in studies of the ignition of spherical uranium powders by a rising temperature ignition test. The effect is shown in Fig. 3-21 [225] where ignition temperatures decrease with increasing height of powder. Ignition temperatures reach nearly constant values at relatively small values of sample heights. These limiting ignition temperatures are plotted in Fig. 3-22 [225] as a function of the specific surface areas of the powders and indicate that submicron size particles would ignite at room temperature. Such

TABLE 3-7

Burning Curve Ignition Temperatures of Stacks of Uranium Foils in Air

Sample[a]	Ignition temperature	
	0.13 mm foil	0.01 mm foil
One foil	400°C (752°F)	320°C (608°F)
Two foils	355°C (671°F)	305°C (581°F)
Four foils	350°C (662°F)	300°C (572°F)
Eight foils	340°C (644°F)	290°C (554°F)

[a] Foils were 16 mm (0.630 in.) square and spaced 2.6 mm (0.098 in.) apart in a ceramic holder.

ignitions have been observed using dust clouds [207]. Uranium dust clouds and settled layers are not as readily ignitable as zirconium dusts. Ignitability of uranium dusts is, however, similar to the ignitability of magnesium and aluminum dusts.

Burning: Uranium ignitions in oxygen are characterized by a bright flash and temperatures probably reach well above 2000°C (3632°F). Ignitions in air are less vigorous. Temperatures reach about 1400°C (2552°F) when 1-cm cubes are ignited in air; however, the temperature drops rapidly and samples are not fully consumed [78]. Theoretical maximum burning temperatures [210] and rates of propagation along foils and wires [209] have been found to agree with experimental values and not to differ much from similar values for zirconium.

3.3.3 Plutonium Alloys

Ignition temperatures in air and oxygen have been measured [226] using the rising temperature method. Figure 3-23 [227] illustrates the results with cube, rod, and foil specimens of pure plutonium and a series of binary alloys in air. The plutonium ignition temperatures fall into two regimes, a high-temperature regime at approximately 500°C (932°F) for samples thicker than 1 mm and a low-temperature regime at approximately 300°C (572°F) for thinner foils. This behavior is consistent with the oxidation behavior of plutonium which changes in the vicinity of 400°C (see Sec. 2.2.2). The effect of certain of the alloying additives is to stabilize the high temperature regime to higher specific areas. Similar data were observed in oxygen.

It has been shown [228] that burning will propagate along plutonium alloy foils (0.24 × 2 mm) in air at 0.2-0.3 cm/sec and in oxygen at 5-8 cm/sec depending on the alloy. Burning characteristics of plutonium appear to be very similar to those described previously for zirconium and uranium.

3.3.4 Magnesium and Aluminum

Magnesium ignition and burning have been studied extensively and could become a serious hazard should magnesium receive wide use in cladding and structural applications. Aluminum is widely used in reactors; however, extreme temperatures are required for ignition.

Ignition temperatures of magnesium and over one hundred magnesium alloys were measured by a

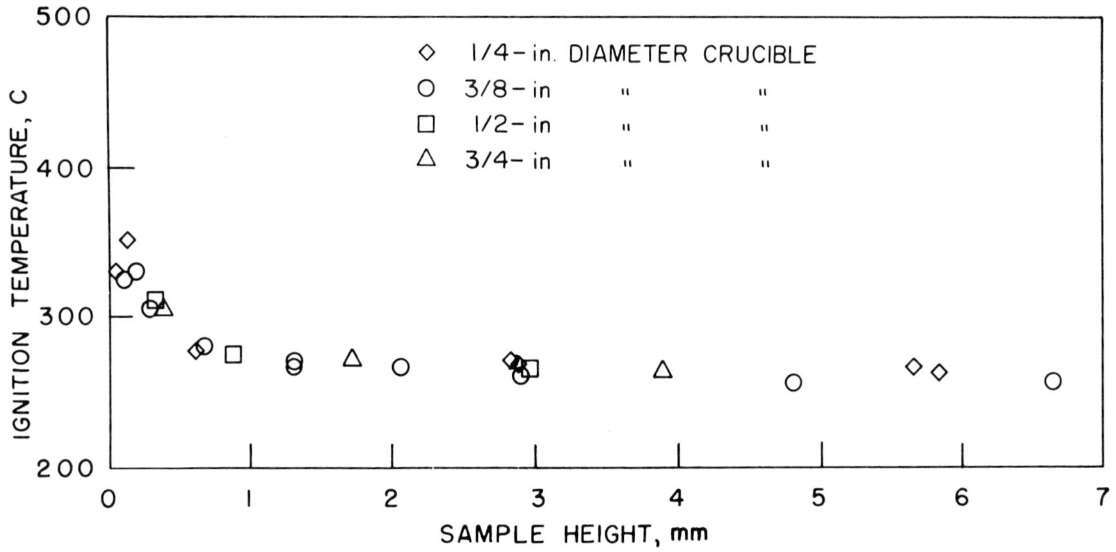

FIG. 3-21 Ignition temperatures of −200 +230 mesh spherical uranium powders as a function of sample height. Crucible diameters varied from 0.25 in. (6.35 mm) to 0.75 (19.05 mm).

rising-temperature method. Ignition temperatures in oxygen are plotted in Fig. 3-24 [229] for a number of these alloys. Ignitions in air result in somewhat erratic temperatures; however, they do not differ much from those indicated in Fig. 3-24. The ignition temperature of pure Mg reached a minimum of 623°C (1153°F) in one atmosphere of oxygen. Higher values are obtained by either decreasing or increasing the oxygen pressure. The burning of magnesium occurs by a vapor phase mechanism and has been likened to the burning of liquid hydrocarbon fuels [230].

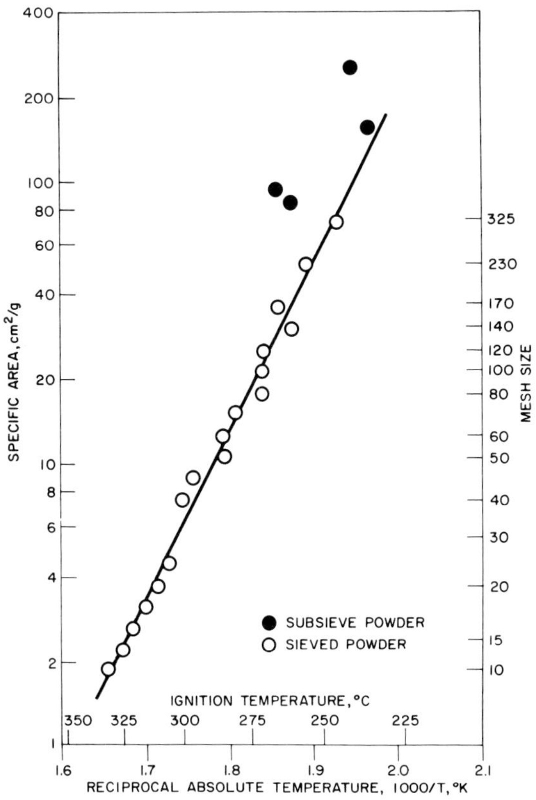

FIG. 3-22 Effect of specific area of spherical uranium powders on experimental ignition temperatures.

Aluminum has been found to ignite at 1750°C (3182°F) in air by intensive heating of levitated pellets [174]. Small aluminum particles ignited in high temperature gases only when the gas temperature was above 1940 to 2090°C (3524 to 3794°F) [231]. It was concluded in this study that ignition occurred when the oxide Al_2O_3 reached its melting point, 2023°C (3673°F). Burning appears to be a complex vapor-phase process involving fragmentation.

3.3.5 Liquid Metal Coolants

The reactions of the alkali metals with the common gases, including oxygen, are summarized in Table 3-4. A brief discussion of the air oxidation of other liquid metal coolants is also included in Sec. 3.2.6. Further discussion in this section is limited to the reactions of sodium and potassium with air and oxygen.

The reaction of sodium with oxygen is very dependent on the quantity of water vapor present and has been found to correlate with the ratio of the oxygen to water vapor partial pressures [232]. When this ratio exceeds 5×10^5, no detectable reaction occurs up to 550°C (1022°F). At ratios between 4×10^5 and 9×10^3, there is no reaction below a certain temperature, the ignition temperature. At and above this temperature a vigorous combustion process develops. At ratios less than 7×10^3, reaction occurs smoothly. At temperatures below 100°C, the reaction rate is relatively slow, but measurable, while at higher temperatures the rate is high enough to result in self-heating. The reaction follows the

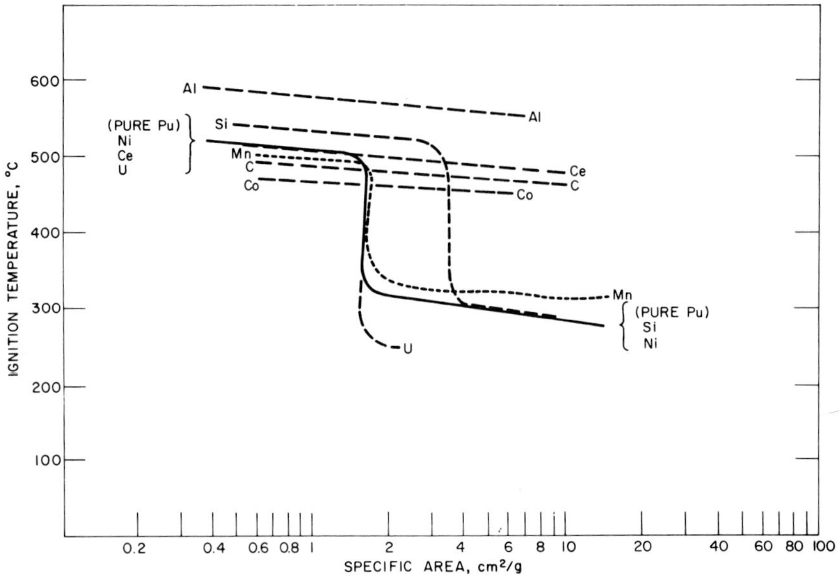

FIG. 3-23 Air ignition of plutonium and plutonium alloys. Each curve is labeled with additive elements. (Alloying element concentration 2 at.%)

parabolic rate law at 105 °C (221 °F) under these conditions.

Similar behavior was reported for potassium. It has been possible to distill potassium in oxygen without reaction under "super-dry" conditions [233]. The reaction in "moist" oxygen, containing 0.25 mg H_2O/liter, is found to be slow and linear with solid potassium between 30 and 60 °C (86 and 140 °F). The reaction with liquid potassium is also relatively slow but parabolic between 70 and 100 °C (158 and 212 °F). Practical situations correspond to the moist category of the foregoing discussion where the oxygen to water ratio is less than 7×10^3.

Ignition: Well-defined ignitions occur with sodium or potassium only when the moisture content of the oxygen is of the order of 0.01 mg H_2O/liter [234]. At higher water vapor concentration, ignitions may or may not occur during oxidation depending upon the nature and thickness of the surface film. Very feeble ignitions occur with potassium at

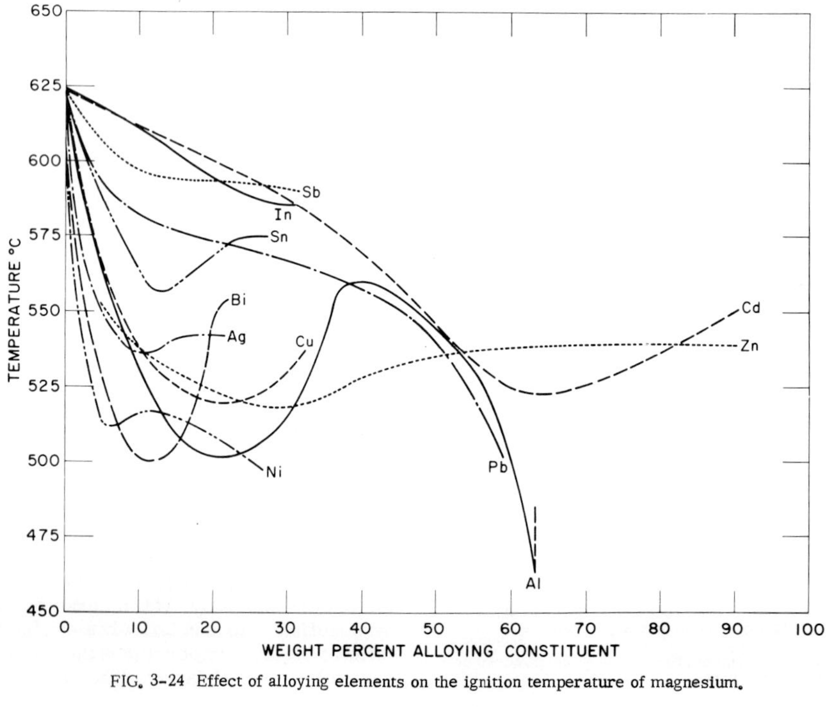

FIG. 3-24 Effect of alloying elements on the ignition temperature of magnesium.

about 150°C (302°F) in moist oxygen. Ignition temperatures of sodium in air may be expressed approximately as follows [235]:

Fine particles condensed from a vapor mist, 25°C (77°F)
Droplets sprayed into air, 120°C (248°F)
Pool with agitated surface, 200°C (392°F)
Pool with undisturbed surface, 300°C (572°F)

Ignition begins as pinhead-size specks of light in a greenish-gray film on the surface of pools of sodium. As self-heating begins, dense clouds of white smoke are evolved. The burning mixture of metal and oxide tends to crawl up the walls of containers and may spill over. Combustion does not occur if the oxygen content of the atmosphere is 5 vol.% or less. There is reaction, however, and there may be considerable smoke even though there is no flame or incandescence. Methods of dealing with the smoke and fume problem have been discussed in detail [236].

Sodium monoxide, Na_2O, and sodium peroxide, Na_2O_2, are both formed when sodium burns in air. The production of monoxide is favored in the presence of excess sodium at lower temperatures. The formation of peroxide is favored by excess air and higher temperature. The superoxide, KO_2, is the stable oxide formed at ordinary temperatures by the oxidation of potassium. The oxide which forms in NaK alloys is Na_2O. The tendency to form KO_2 is prevented by the following reaction (p. 1012 of reference [1]):

$$4Na + KO_2 \longrightarrow 2Na_2O + K, \qquad \Delta H_{298} = -141 \text{ kcal.} \tag{3-11}$$

The above reaction may also have been responsible for several explosions involving NaK. It has been theorized that KO_2 formed in the oxide crust might subsequently have been immersed in the NaK and agitated, resulting in a violent reaction. The superoxide is a strong oxidizing agent and can react violently with organic materials. Oil and grease should not be allowed to contact a system that will contain NaK or K alone.

Many materials are useful in extinguishing sodium fires. These include sodium carbonate, sodium stearate, sodium chloride, large grain vermiculite, graphite, lamp-black, sand, and calcium carbonate [235]. Many commercial formulations of the above materials are available. Water may be applied only if air is excluded and if adequate expansion space is available. Water applied in the presence of air leads to violent hydrogen-oxygen explosions. Carbon tetrachloride can react violently or explode with sodium and should never be used to combat a sodium fire. Carbon dioxide is generally useless on a sodium fire; however, it may be helpful in a combined sodium-hydrocarbon fire by knocking down the high flames of burning hydrocarbon. Fire blankets may also be useful. A blanket of stainless steel foil was found to be particularly effective. It must always be assumed that unreacted sodium is still present even after a fire has been extinguished and that there is always a chance of rekindling until the remaining metal has been destroyed. More detailed discussions of sodium fires are available [235], (p. 1016 of reference [1]). An important safety aspect of sodium-cooled, fast breeder reactors is the sodium-air reaction. Such factors as the burning rate and the peak pressures produced must be taken into account in the design of the containment shell.

Sodium-air accidents in a closed vessel have been classified as follows [237]: stagnant pool, pressurized spray, and explosive ejection. The explosive ejection is probably the most hazardous since the sodium stream is oxidized in flight and the heat of reaction goes mainly into raising the temperature and pressure of the atmosphere in the vessel. In the two slower processes there would be more heat lost to the structure and walls.

Data on the burning rate of a pool of sodium in air are shown in Fig. 3-25 [235]. Typical combustions proceed at rates in the range 0.1 to 0.4 lb Na/ft^2-min (0.5 to 2.0 kg Na/m^2-min). The peak temperatures and pressures produced during tests in which sodium was explosively injected into air are generally lower than the maximum theoretical values; see Fig. 3-26 [238]. The reason for this is primarily the heat losses to walls. Figure 3-27

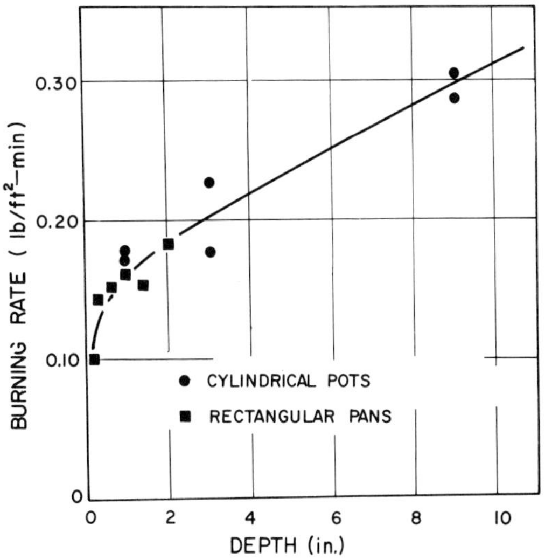

FIG. 3-25a Rate of burning of a pool of sodium in air as a function of pool depth.

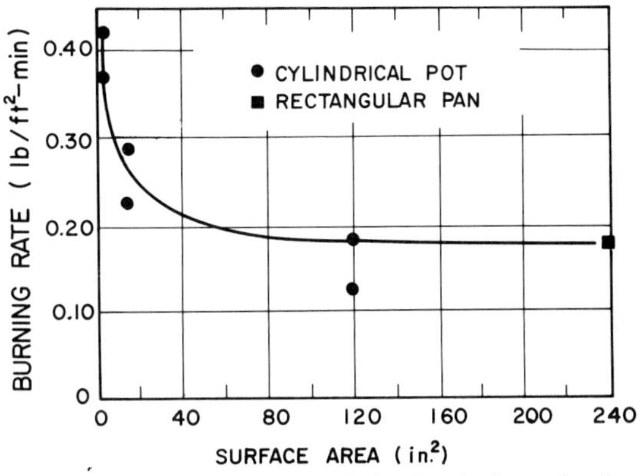

FIG. 3-25b Rate of burning of a pool of sodium in air as a function of pool area.

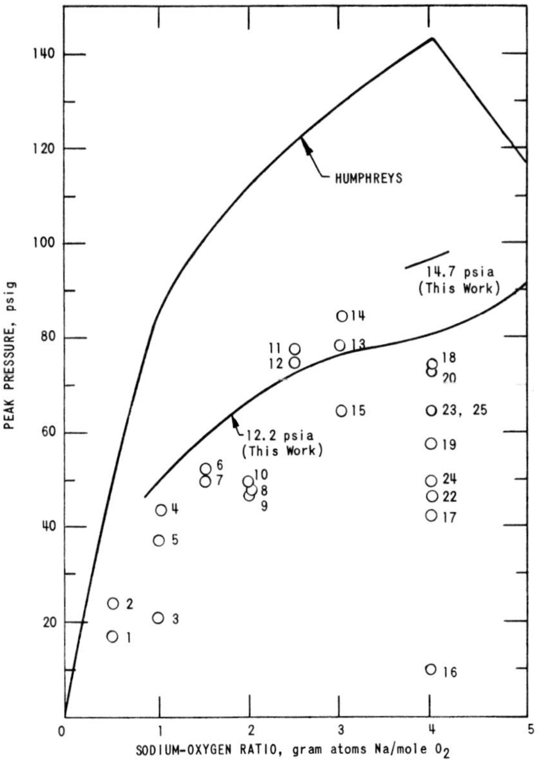

FIG. 3-26 Comparison of maximum theoretical pressures for sodium-air reaction with data of Humphreys. Lines are maximum theoretical pressures; lower lines consider dissociation of reaction products. Numbered circles are experimental data of Humphreys.

[237] shows examples of pressure-time records obtained during these sodium ejection experiments.

A method for estimating the maximum temperature for a sodium-air reaction in a confined volume has been developed [239]. The technique assumes a finely divided stream of sodium burning until all of the oxygen is consumed.

3.4 Graphite-Gas Reactions

An important aspect in the potential hazards of graphite-moderated nuclear reactors is the possibility of ignition and combustion of graphite. This safety problem clearly exists in an air-cooled graphite reactor and also in an inert gas-cooled reactor if there are oxidizing impurities in the gas or if a leak to the atmosphere develops.

There are many variables which influence the kinetics of a heterogenous chemical reaction such as the reaction of graphite and air. Among the variables are: (for the solid) temperature, prior oxidation, irradiation, geometry, purity; (for the gas) temperature, pressure, composition, flow rate [240]. Table 3-8 [241, 242, 243] gives a summary of a few properties of interest in gas-graphite reactions.

The usual operating temperature range for graphite in gas-cooled reactors of various types is as follows:

direct air cooling of pile graphite: ~200°C (390°F)

inert gas cooling in present power reactors: ~300°C to 1300°C (570°F to 2370°F)

gas cooling of space propulsion reactors: up to a maximum of ~3350°C (6060°F)

There are six coolant gases which have received some attention in gas-cooled reactor technology. They are: air, steam, carbon dioxide, nitrogen, hydrogen, and helium.

There is a vast literature on the reactions between the element carbon and various gases. Despite the long history of this topic (for example, in the burning of coal) there are still many unresolved points in the equilibria and kinetics. Some

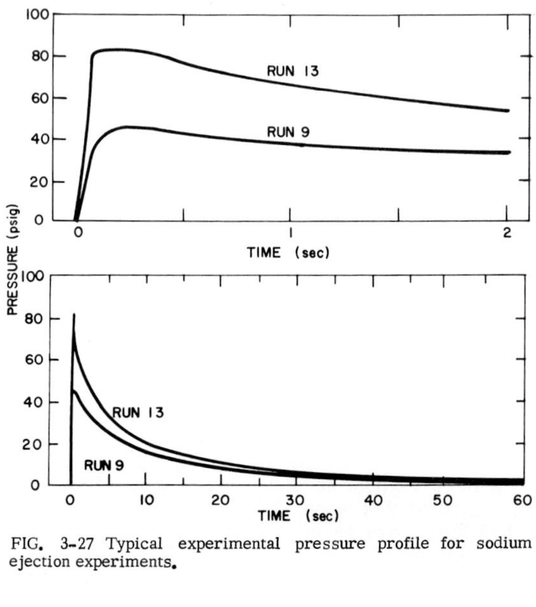

FIG. 3-27 Typical experimental pressure profile for sodium ejection experiments.

TABLE 3-8

Some Properties Pertinent to Gas Graphite Reaction

Reaction	Thermodynamic factors				Kinetic factors	
	ΔH Reaction kcal/gmole	Equilibrium constants			Relative reaction rate	Energy of activation
		Log K_{300}	Log K_{900}	Log K_{1400}		
$C + 1/2\ N_2 = CN$	—	—	—	+ 2.7(2300°K)	Very slow	—
$2C + N_2 = C_2N_2$	+ 73.8	—	—	- 3 (3000°K)	Very slow	78
$C + 2H_2 = CH_4$	- 17.87	+ 8.82	- 0.49	- 2.36	3×10^{-3}	36 (10 to 65)
$C + CO_2 = 2CO$	+ 40.79	- 20.81	- 0.71	+ 2.80	1	86
$C + H_2O = CO + H_2$	+ 31.14	- 15.86	- 0.37	+ 2.44	3	80
$C + O_2 = CO_2$	- 94.03	+ 68.67	+ 22.97	+ 14.78	$1 \times 10^{+5}$	54
$C + 1/2\ O_2 = CO$	- 26.62	+ 23.93	+ 11.13	+ 8.79	—	54
$CO + 1/2\ O_2 = CO_2$	- 67.41	+ 44.74	+ 11.84	+ 5.99	—	—
$CO + H_2O = CO_2 + H_2$	- 9.65	+ 4.95	+ 0.34	- 0.36	—	—

Notes: 1. The heats of reaction are on the basis of solid graphite (ΔH = O) at 18°C (64.4°F) and 1 atm.

2. Log K_{300} is the $\log_{10}$ of K_p at a temperature of 300°K; also given at 900°K and 1400°K.

3. All the K data are for C as β graphite and the other reactants and products as gases. The concentrations in equilibrium are in units of partial pressure in atmospheres. For example,

$$K_p = P_{CO_2}(\text{atm})/[P_{CO}(\text{atm})]\,[P_{O_2}(\text{atm})^{1/2}]$$

4. The relative reaction rate is evaluated at 800°C (1472°F) and 0.1 atm pressure.

5. The energy of activation is in the units kcal/gmole and shows the influence of temperature on the reaction rate in the relationship $r = r_0 \exp(-\Delta E/RT)$, where r = rate, R = universal gas constant, T = absolute temperature, and ΔE = activation energy.

concepts of importance are: [244, 245] energy release or absorption from the reaction, adiabatic flame temperatures, active sites, rates and equilibria of adsorption and desorption, reaction mechanisms (whether mass transport or kinetically controlled), ignition temperature, mathematical correlation models (continuous-reaction model or unreacted-core model with or without a "shrinking core"), protective coatings, inhibitors, and effect of low levels of impurities. Of necessity, all of these subjects cannot be covered in a few pages. The present discussion therefore is greatly abbreviated and reflects what is considered to be of major importance. The presentation is mainly a summary of results which appear reliable and which might be of some use in evaluating reactor safety. Existing references [241, 244, 246] cover the fundamentals very thoroughly. Some insight is being obtained into the nature of oxidation of graphite by microscopic study of the reacted surfaces. These studies have been reviewed [247].

It is, however, appropriate to emphasize again the role of kinetic and diffusion mechanisms in limiting the overall reaction process. Figure 3-28 [248] illustrates a commonly accepted "three-zone" model. At low temperatures, zone I, the chemical reactivity of the graphite is the slow step. In the intermediate range, zone II, the controlling part is diffusion of reactant and product gases in the pores of the graphite together with some influence of graphite chemical reactivity. In the high temperature region, zone III, the limiting factor is the diffusion in the gas-phase (boundary layer around the outer graphite surface) film of products and reactants. For the reaction of graphite with steam, zone II is roughly in the range 700 to 1000°C (1290 to 1830°F). The rate in the kinetic region can be limited by particular aspects such as chemisorption or desorption or rearrangement of species

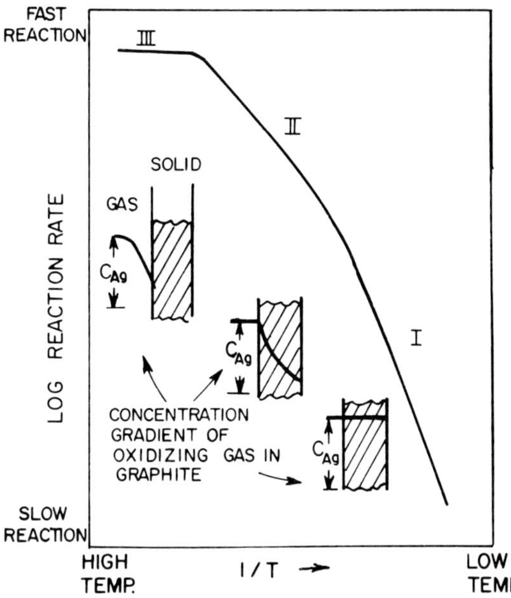

FIG. 3-28 Relative reaction rate of gases with graphite.

on active sites. The diffusion or mass transport region can be governed by pore size in the interior or by turbulence (Reynolds number) on the exterior. As illustrated by the concentration profiles in Fig. 3-28 the whole inner surface of the porous solid (graphite) takes part in the reaction in zone I. Then in zone III, the entire reaction occurs at the outer surface of the graphite and the concentration of the oxidant C_{Ag} is zero in the bulk solid phase.

Finally, in this introductory discussion it is also pertinent to mention briefly the action to be taken during a potential runaway oxidation incident [242]. The reactor operator might, for example, want either to increase the coolant gas (presumably containing, say, air), to try to remove the heat being generated, or to cut back or even shut off the air flow to decrease the supply of oxidant. Unfortunately both of these "remedies" also have drawbacks. Increasing the air flow accelerates the oxidation (Fig. 3-29) [244]; decreasing the flow and hence the cooling might cause melting of the fuel elements from the combined effects of decay and chemical heat generation even after shutdown. An alternate approach is to inject an inhibitor such as chlorine. Adding a gas-phase inhibitor has an advantage in that one can still maintain the coolant stream. However, it is then necessary to cope with the handling and disposal of chlorine; because of this serious limitation, the chlorine method of extinguishing a graphite fire should only be used as a last protective measure.

3.4.1 Air and Oxygen

Reactor practice considers the reaction of graphite and air to be prohibitively rapid above 350°C (662°F), in the context of loss of moderator over a long time scale.

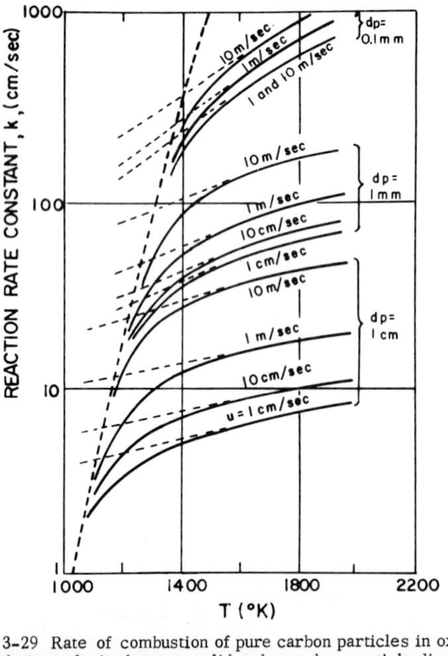

FIG. 3-29 Rate of combustion of pure carbon particles in oxygen. u = relative velocity between solid and gas; d_p = particle diameter. Note: The solid lines give the actual rate constants k. The short broken lines would apply for film-diffusion controlling, that is, k_g. The large broken line shows surface-reaction controlling constant k_s. From the concept of processes in series these three rate constants are related by $(1/k) = (1/k_g) + (1/k_s)$.

Table 3-8 points up that the reaction between graphite and oxygen is very exothermic, rapid, and not influenced by equilibrium considerations in the range of interest.

Experiments conducted to assist in the evaluation of runaway oxidation hazards gave the following reaction rates [249]:

Air-graphite

$$r = 6.97 \times 10^{10} \exp(-50{,}800/RT) \quad (3\text{-}12a)$$

data points in the range ~500 to 700°C (930 to 1290°F)

Oxygen-graphite

$$r = 1.64 \times 10^{12} \exp(-54{,}600/RT), \quad (3\text{-}12b)$$

where r is in grams reacted/gram graphite-hour and r × 0.191 = g/cm²-hr for the particular sample geometry. These results were obtained for EGCR (Experimental Gas-Cooled Reactor) graphite. The kinetics can change markedly from one type and grade of graphite to another (p. 893 of reference [1]). A correlation based on many sources of data gave the following [250]:

$$k = 7.24 \times 10^{9} [\exp(-22{,}000/T)], \quad (3\text{-}13)$$

where k is the reaction rate constant, mass of graphite reacting/[(active mass of graphite sample) (hr)(atm of O_2)], and T is temperature in °K.*

The "active mass" is defined as a fraction η of the total mass of graphite, where η is the ratio of the observed reaction rate to the rate that would be expected if the external gas composition existed uniformly throughout the pores of the sample. The quantity η is estimated from the solution of the differential equations for diffusion and reaction in a porous solid [249, 250]. In a zone I reaction the value of η approaches unity since all of the active sites are theoretically available to the oxidant [249].

When air is passed through a hot graphite channel, the heat produced by the chemical reactions is due to both the C-O_2 and CO-O_2 reactions. Thus, both CO and CO_2 are primary products in the graphite-oxygen reaction. Data over the range 460 to 1420°C (860 to 2588°F) show the equilibrium ratio of:

$$CO/CO_2 = 10^{3.4} [\exp(-12{,}400/RT)], \quad (3\text{-}14)$$

so that more CO is formed at higher temperature [242]. Data show that the largest and most rapid temperature rises are due to the CO-O_2 gas phase reaction.

Serious instability (where the heat generated by the reactions is greater than the heat removed by the air stream) does not occur below 650°C (1202°F) and is confined to flow rates where the Reynolds numbers lie between 2000 and 8000 [252]. Figure 3-30 [252, 253] illustrates the effect of air flow rate; in particular, at flow rates less than 60 liters/min, heat evolved by the graphite oxidation exceeds the heat removed by the air stream and the channel temperature increases. Figure 3-31 [253] shows a typical temperature distribution in an air-cooled graphite channel.

---

*Gulbransen, et al., have reported on the influence of oxygen pressure on the oxidation of graphite [251]. They found a linear pressure dependence up to 800°C and a $P^{0.32}$ dependence from 800 to 1500°C.

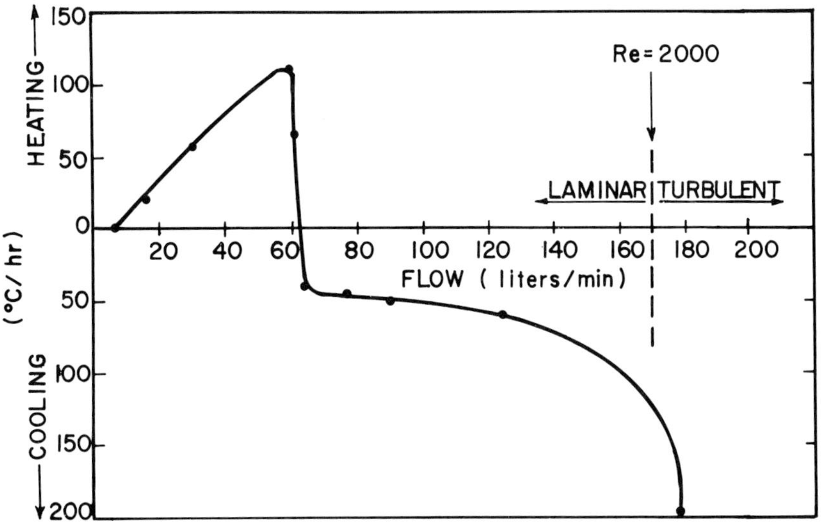

FIG. 3-30 Heating and cooling rates vs. flowrate for an air-cooled graphite channel. Inlet air temperature = 300–380°C (572–716°F). Graphite: 2.75 in. (6.98 cm) diameter, 24 in. (61 cm) long, 670–680°C (1238–1256°F).

An approach which has been suggested for applying rate data to evaluate instability is based on a simple heat balance [254]:

$$G = wHr = \text{rate of heat generation} = \text{cal/hour},$$

where w is the weight of graphite (g), H the heat of reaction (cal/g), and r the reaction rate (g reacted per hour per g graphite).

$$Q = 3600\ Sh(T_{graphite} - T_{gas}) \quad (3\text{-}15)$$
$$= \text{rate of heat loss by convection (cal/hour)},$$

where S is the surface area of graphite (cm^2), h the heat transfer coefficient ($cal/sec\text{-}cm^2\text{-}°C$),

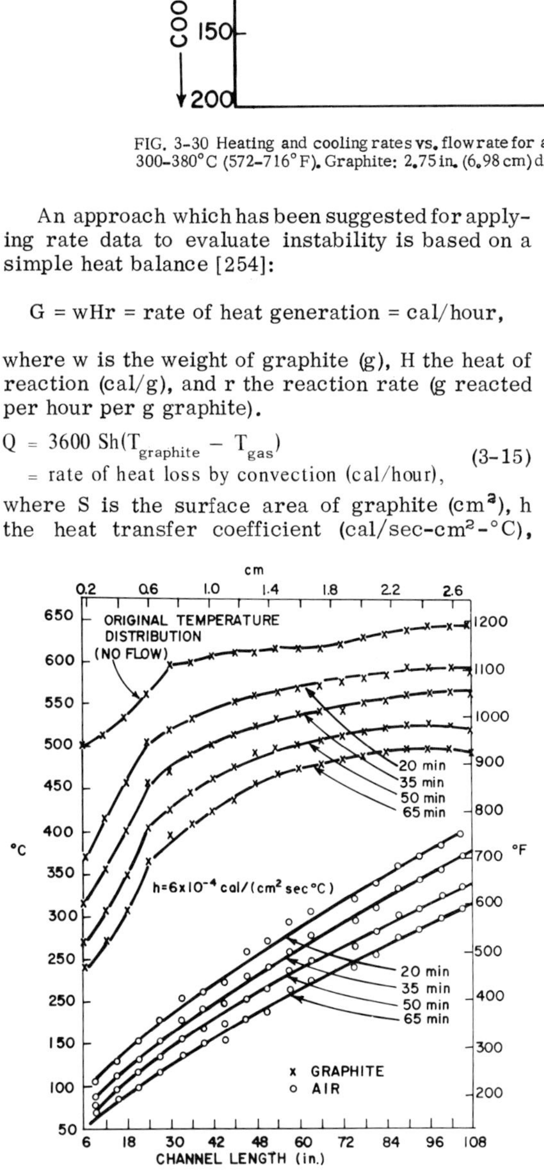

FIG. 3-31 Temperature changes in a 2 in. (5.1 cm) diameter graphite channel with time.

and T the temperature in °C. The criterion for potential thermal instability is that G be greater than Q.

$G < Q$	the temperature decreases
$G = Q$	there is thermal equilibrium
$G > Q$	the temperature rises (radiation losses then come into play).

Many refinements to this basic method are possible, such as the addition of nuclear or decay heating if present to a significant extent, the consideration of temperature gradients within the graphite, etc. Stability plots have been prepared [254] giving safe regions of graphite temperatures and coolant flow with "critical temperatures." It is important to realize the limitation on the above "reaction controlling" approach, namely, that it applies only for temperatures below about 1100°K (827°C or 1521°F). Beyond 1100°K the oxidation process is film diffusion-controlled. Figure 3-29 shows how the rate of combustion of carbon is strongly affected by three parameters: temperature, particle size, and relative velocity between solid and gas. Figure 3-29 points out that, for a given velocity and particle diameter, the high temperature region is gas-film diffusion-controlled and, as the rate curve proceeds downward, the surface reaction begins to be limiting and causes a steep change in rate with temperature.

The rate constants in Fig. 3-29 can be applied by considering, e.g., the case of an unreacted, spherical carbon core shrinking with time because of the combustion reaction. Other geometries such as a cylinder can be similarly treated by using an equivalent sphere which has the same surface-to-volume ratio as the cylinder. For the case of any gas, A, reacting with any solid, B, the chemical equation is:

$$A_{gas} + bB_{solid} = \text{products}. \quad (3\text{-}16)$$

The rate of reaction is considered proportional to

the mass transfer (film) coefficient and the gas concentration driving force [244]:

$$-x^2 \frac{dx}{dt} = \frac{bx_0^2}{\rho_B}(k_g C_A). \qquad (3\text{-}17)$$

Integrating from x_0 to x_c, the radius varies according to:

$$\frac{x_c}{x_o} = \left(1 - \frac{3bk_g C_A}{\rho_B x_o} t\right)^{1/3}, \qquad (3\text{-}18)$$

where x_c is the radius of the reacting particle (carbon) at time t, x_0 the original (t = 0) radius (cm), ρ_B the density of solid (moles/cm^3), C_A the concentration (moles/cm^3) of oxidizing gas (oxygen), and k_g is in cm/sec, t in sec. Note that in this integration the coefficient k_g was assumed constant. However, as indicated in Fig. 3-29 k_g is, in fact, a function of both the particle radius and the temperature. Hence, except for a small range of x_c and T, it is best to obtain a numerical (preferably computer) solution to two simultaneous equations: one for the unsteady state heat balance and the other for the chemical kinetic (mass balance) part. The final result then gives the desired information for the safety evaluation, namely, the rate and amount of the energy release and mass reacted. As a check on the calculations and predictions it is considered highly advisable to mock up the particular system (or a geometrically similar portion thereof) with the particular graphite, gas, flow rates, etc. This should be done because of wide variations in the published kinetic data and gaps still to be filled in the physical chemistry of graphite oxidation. For an example of recent work along this line, see p. 469, Vol. 2 of reference [255].

The effect of temperature and irradiation on the rate of oxidation of graphite is shown in Fig. 3-32 [256]. Lattice defects induced by the neutron irradiation increase the number of active sites thus causing the rate of reaction to increase by a factor of 2 to 6 relative to unirradiated graphite (see [111] p. 391). By annealing the graphite at high temperatures the chemical reactivity goes back to its original value. There is, however,

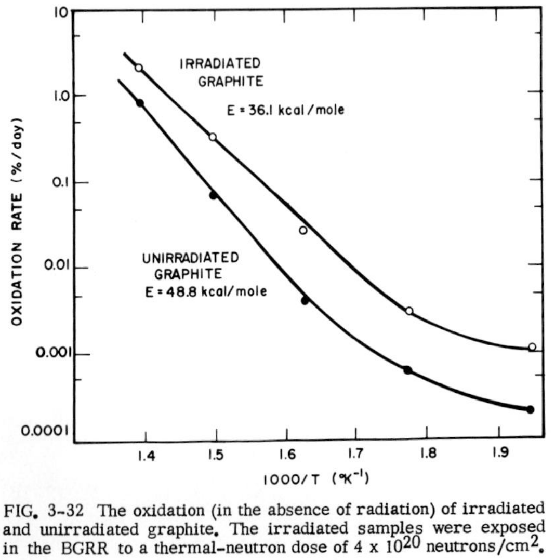

FIG. 3-32 The oxidation (in the absence of radiation) of irradiated and unirradiated graphite. The irradiated samples were exposed in the BGRR to a thermal-neutron dose of 4 x 10^{20} neutrons/cm^2.

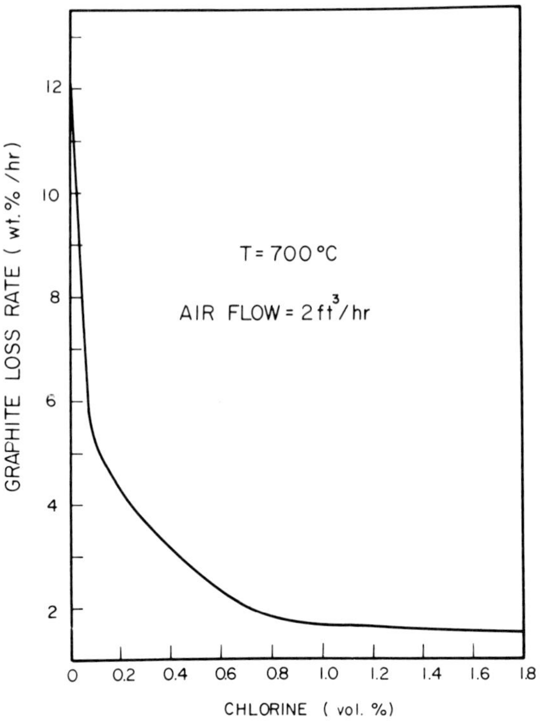

FIG. 3-33 Inhibition by chlorine of the oxidation of graphite in a stream of air.

strong evidence, particularly from United Kingdom work, to indicate that graphite does not undergo oxidation (by air) more rapidly in a gamma radiation field.

Chlorine is an effective inhibitor and extinguisher in the graphite-air system as shown in Figs. 3-33 [242] and 3-34 [242]. The retardation of oxidation rate is thought to occur by chemisorption of chlorine on active sites, thus preventing the oxygen from reaching the carbon. The inhibiting qualities of a number of additives were tested by the UKAEA [257]. These included sulfur dioxide, trimethyl phosphate, carbon tetrachloride, thionyl chloride, and chlorine. All were found inferior in performance to phosphorous oxy-chloride, $POCl_3$, the inhibiting effects of which persist after injection has ceased. They also state that the interruption of chlorine injection results in a rate of oxidation higher than that observed before injection started. There is interest but not much quantitative information on protective coatings for graphite. However, such compounds as silicon carbides [258] and niobium carbide may offer protection and extend the safe operating temperature of graphite in the fuel or moderator. Pyrolytic carbon is also a promising coating material (p. 434 of reference [111]).

When data from experiments on graphite oxidation are plotted as the rate of graphite reacted versus time, it is usual to find first a relatively slow induction period ranging from about 1 to 10 percent "burnoff" of the graphite. This is then followed by a constant rate period in an isothermal test until the area of the sample becomes reduced at about 70 percent burnoff. The interpretation is that the number of active sites increases at first as the pores

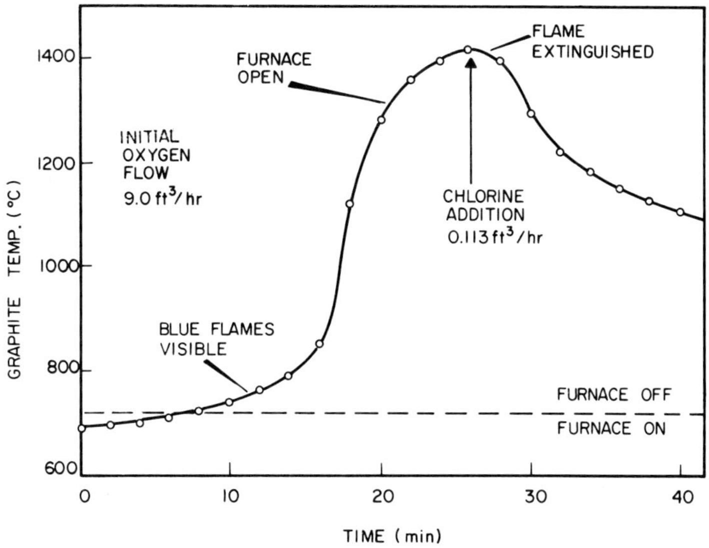

FIG. 3-34 Extinguishing graphite combustion with chlorine. The cylindrical sample was 1 in. (2.54 cm), diameter x 3 in. (7.6 cm) long. Initial oxygen flow = 9 ft^3/hr (0.255 m^3/hr).

are opened by the initial reaction. Then the concentration of available sites remains essentially constant during the bulk of the burnoff [249].

3.4.2 Hydrogen and Nitrogen

From Table 3-8 it is evident that at ambient room temperature the reaction between graphite and hydrogen to produce methane is favored, whereas at high temperature the equilibrium shifts to favor the decomposition of methane into carbon and hydrogen. Thus, in a nonisothermal system carbon tends to be transported from a cool zone and deposited in a hot zone. This migration of carbon can lead to plugging problems. At very high temperatures, 2000 to 3000°K (~3000 to 5000°F), other hydrocarbons such as C_2H_2 are formed. Figure 3-35 [259] shows graphically the methane-hydrogen gas

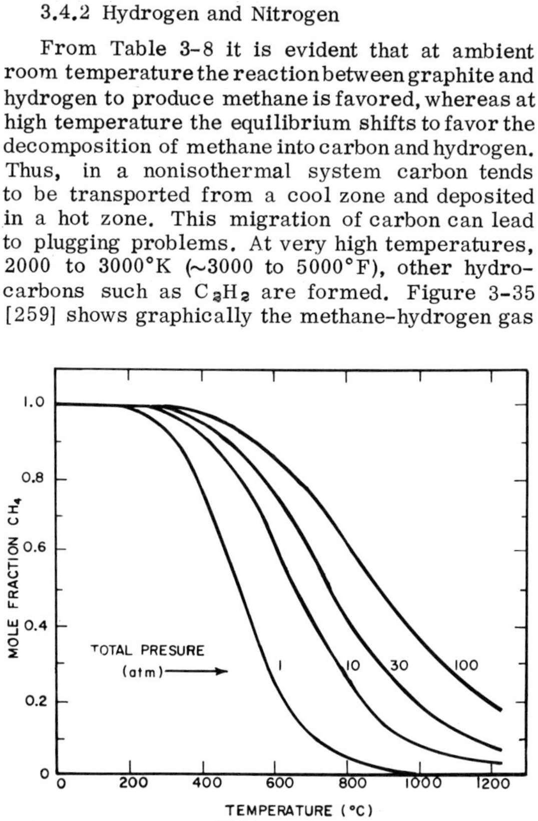

FIG. 3-35 The mole fraction of methane formed at equilibrium for the hydrogen-graphite reaction. Hydrogen and methane are assumed to behave as ideal gases, with no methane present initially. Reaction: $C + 2H_2 = CH_4$.

equilibrium according to the reaction $C + 2H_2 = CH_4$. From mass action, an increase in pressure drives the reaction towards the right, thus increasing the mole fraction of methane at equilibrium at a given temperature. There is not enough information on the kinetic aspects, such as catalysis, adsorption, and radiation effects, to give a correlation suitable even for making good estimates. For carbon char, a rate law of the form

$$\frac{d(CH_4)}{dt} = \frac{k_1(p_{H_2})^2}{1 + k_2(p_{H_2})}, \tag{3-19}$$

where p is the partial pressure of hydrogen and k_1 and k_2 are rate constants, has been used (p. 423 of reference [111]).

The energy of activation changes from 65 kcal/mole above 600°C (1112°F) to 10 kcal/mole at lower temperatures. Below 660°C (1220°F) the adsorption of hydrogen on active sites is probably rate controlling. Qualitatively, it can be said (see Table 3-8) that the reaction between carbon and hydrogen is rather slow compared to the reaction between carbon and carbon dioxide, steam, or oxygen.

Concerning the reaction between graphite and nitrogen, the main feature is that nitrogen can almost be considered an inert gas. Even at 3000°K (4940°F) the equilibrium constant is of the order of 10^{-3} so that not much cyanogen (C_2N_2) can be formed. However, if both hydrogen and nitrogen are present simultaneously, then the following reaction can occur:

$$\frac{1}{2} N_2 + \frac{1}{2} H_2 + C \rightarrow HCN . \tag{3-20}$$

At temperatures of about 1300°C (2370°F), 1500°C (2730°F), and 1700°C (3090°F), reaction rates based on weight loss of graphite are 1.3, 10.0, and 44.0 $\mu g\ cm^{-2} hr^{-1}$, respectively [260], with 1% each of hydrogen and nitrogen in a flowing gas stream of 98% helium. This gives an apparent activation energy of 13.7 kcal/mole. However, when the amount of hydrogen cyanide produced is analyzed, it is always less than 1% of that predicted from the carbon (graphite) weight loss. This fact, plus the observation of sooty carbon deposits on the apparatus, suggests that the back reaction of decomposition of hydrogen cyanide is very rapid. For long-term reactor operation, it has been estimated that hydrogen and nitrogen impurity levels of approximately 100 ppm at 23 atm pressure can be tolerated; then carbon transport across gaps in a high-temperature, helium-cooled graphite reactor should not be a limiting factor. Over the range 2000 to 3500°K (3100 to 5800°F), data on the reaction of nitrogen with graphite from another study [246] are expressed by:

$$\log_{10} r = 5.4 - \left(\frac{2.04 \times 10^4}{T}\right), \qquad (3\text{-}21)$$

where r is the reaction rate in g (carbon) cm^{-2} sec^{-1} and T is in °K. This equation predicts greater rates than the above data.

3.4.3 Carbon Dioxide

The basic equilibrium features of the graphite-CO-CO_2 system are shown in Fig. 3-36 (p. 398 of reference [111]). High temperatures and low pressures favor the formation of carbon monoxide. Since the reaction is endothermic, there is no possibility of an ignition. However, there is the possibility of a subsequent explosion of the product carbon monoxide with oxygen. The characteristics of the CO-O_2 explosion are a function of the concentration of water vapor or nitrogen if present. The lower and upper detonation limits of moist CO-O_2 in air or oxygen are given as 38 and 90%, respectively [261]. The detonation velocity D at room temperature and atmospheric pressure of the stoichiometric mixture $2CO + O_2$ is 1264 m/sec for the dry gases. This is to be compared to 2821 m/sec for $2H_2 + O_2$. However, when the $2CO + O_2$ gas is saturated with water vapor at 35°C (95°F) then the value of D is 1738 m/sec (5700 ft/sec). Also, dry CO-O_2 mixtures are difficult to ignite even with a spark, whereas very little energy is required to ignite a wet mixture. In contrast, the addition of nitrogen decreases the flame velocity considerably and increases the time elapsing until the maximum pressure is reached. As shown in Table 3-8, the oxidation of carbon monoxide to dioxide is highly exothermic (-67.41 kcal/gmole) and not limited by thermodynamics. Experimental data on the $C + CO_2$ reaction fit an equation of the form (p. 40 of reference [111]):

$$-\frac{dc}{dt} = \frac{k_1 P_{CO_2}}{1 + k_2 P_{CO} + k_3 P_{CO_2}}, \qquad (3\text{-}22)$$

where P is partial pressure and k is a rate constant.

This general expression is derived by postulating a mechanism wherein during the reaction $C + CO_2 = 2CO$, carbon monoxide retards the gasification of carbon by decreasing the fraction of the surface which is covered by oxygen atoms. The following kinetic equation can be applied in evaluating the above chemical reaction:

$$r = 2.6 \times 10^9 \, [\exp(-85/RT)], \qquad (3\text{-}23)$$

where r is in cm^3 of CO_2 consumed/cm^2-sec and the energy of activation is given in kcal/gmole [240]. (This is based on experiments with spectroscopically pure carbon at 0.1 atm pressure.) Other investigations with graphite gave the results shown in Table 3-9 and in Fig. 3-37 [241]. (See also reference [262].)

TABLE 3-9

Rate of Reaction of CO_2 (at 0.1 atm Pressure) with Graphite

Temperature	Reaction rate (g/sec-cm^2)	Type of graphite
900°C (1652°F)	6.2×10^{-12}	Wear dust
900°C (1652°F)	0.7×10^{-12}	Graphitized carbon black

Data during operation of a CO_2-cooled reactor are commonly obtained by monitoring the exit gas from the core for carbon monoxide, for example, by mass or infrared spectroscopy. The results indicate a positive dependence of the rate of carbon monoxide production with reactor pressure and power level. The use of CO_2 as a coolant in direct contact with graphite is limited to temperatures less than 600°C (1112°F) by the rate of oxidation (p. 397 of reference [111]). Below 600°C the rate of reaction in the absence of radiation is negligible but at 350°C (662°F) the radiation-induced $C + CO_2$ rate is measurable, although carbon dioxide itself is quite stable in high energy radiation fields. The effects of irradiation on the graphite-CO_2 reaction have been studied by Lind and Wright [263], who report the following rule: "For most practical purposes, the rate of oxidation in pure CO_2, may be determined to a first approximation by calculating the rate of energy absorption by the gas contained within the open pore volume and assuming that 2.35 carbon atoms

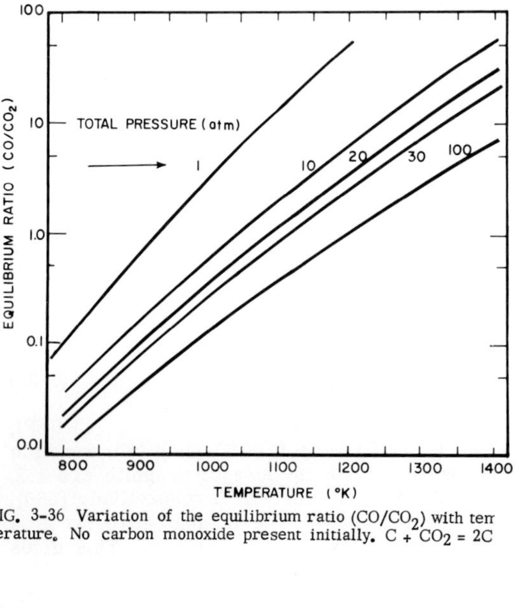

FIG. 3-36 Variation of the equilibrium ratio (CO/CO_2) with ter perature. No carbon monoxide present initially. $C + CO_2$ = 2C

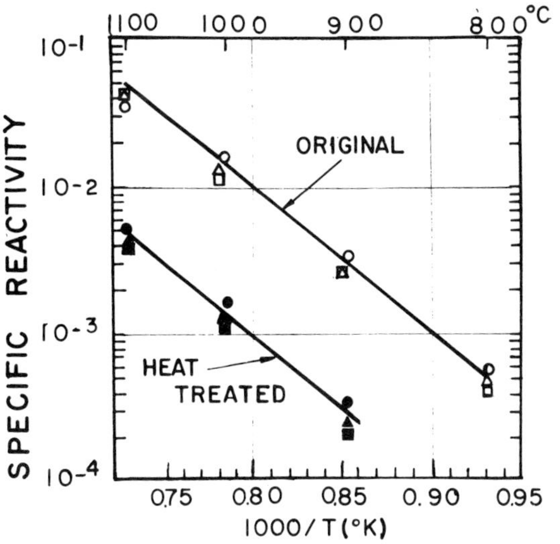

FIG. 3-37 Arrhenius plots for reaction of raw and heat-treated rods of Ceylon graphite with carbon dioxide at 1 atm pressure. The specific reactivity is in g/hr-m^2 per atmosphere of CO_2. The open circles, triangles, and squares are data for the original graphite with apparent densities of 1.61 g/cm^3, 1.83 g/cm^3, and 1.99 g/cm^3, respectively. The solid circles, triangles and squares are data for heat-treated graphite with apparent densities of 1.41 g/cm^3, 1.63 g/cm^3, and 1.76 g/cm^3, respectively. (Conversion factors: 1 g/cm^3 = 0.03613 lb/in^3 = 62.43 lb/ft^3; 1 g/hr-m^3 = 0.2048 x 10^{-3} lb/hr-ft^2.)

are oxidized per 100 ev of energy absorbed." This rule is consistent with results on the effect of pore volume on the radiolytic reaction between CO_2 and graphite reported by Feates [264]. The reaction rate increases with pore volume and is not influenced by previous graphite irradiation. Feates also found that it was not influenced by the H_2O content of CO_2. His measurements were made at 185°C, at various pressures up to 25 atm.

Other variables affecting the graphite oxidation rate are the heat treatment (see Fig. 3-37) and the type of surface [241, 265]. For instance, coating graphite with SiC appears to reduce the rate of oxidation by a factor of approximately six at 700°C (1292°F). By removing impurities, the heat treatment decreases the chemical reactivity of the graphite by a factor of ten. Gas-phase deposition of silicon carbide or trimethyl phosphate on the graphite serves to inhibit the thermal reaction rate, apparently by forming a chemisorbed layer over the active carbon atoms. Unfortunately, the protective coatings do not give much protection during in-pile irradiations; the main benefit is in reducing the pre- and postirradiation reaction rates. Just as the graphite-oxygen reaction can be inhibited by injecting chlorine into the gas phase, studies have shown that the graphite-carbon dioxide rate can be inhibited by adding Br_2, CCl_4, or $POCl_3$ in the gas stream [248]. It is interesting that in the plot of carbon monoxide concentration versus time the rate inhibition experiments show that there is a sharp peak in the amount of CO in the exit gas just when the halogen is added; the CO concentration then quickly decreases to about one-half of its original value [242].

3.4.4 Steam

The equilibria of the steam-graphite system in

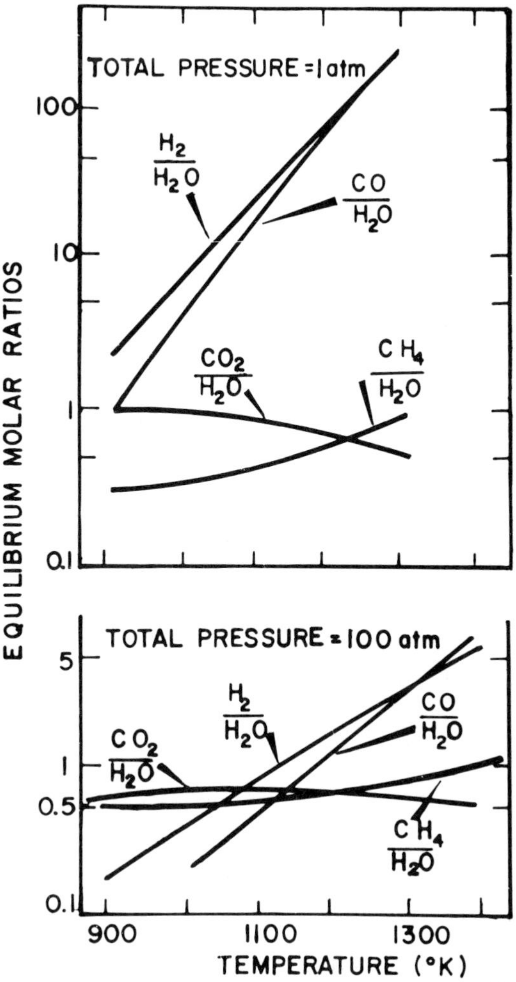

FIG. 3-38 Equilibrium product-steam ratios for the steam-graphite system in which only steam and graphite are present initially. Main reactions are $C + H_2O = CO + H_2$ and $C + 2H_2O = CO_2 + 2H_2$. Secondary reactions are $C + 2H_2 = CH_4$ and $C + CO_2 = 2CO$.

the temperature range of main interest in gas-cooled reactors are shown in Fig. 3-38 [241]. The main features are that (at 1.0 atm) the concentrations of hydrogen and carbon monoxide are of the order of 10 to 100 times that of the water vapor, whereas the carbon dioxide and methane are of the same order of magnitude as the water vapor. Increasing the pressure to 100 atm forces the main "water gas" reactions to the water side, thus making the ratios of components more nearly unity. The additional thermodynamic point of significance is that the reaction of graphite with steam to give hydrogen and carbon monoxide is endothermic, thus resulting in absorption rather than release of energy.

Kinetically, the steam-graphite system is quite comparable to the graphite-carbon dioxide system. The mechanisms, rate equation, and activation energies are similar. The gasification rate equation is (p. 419 of reference [111]):

$$\text{rate} = \frac{k_1 P_{H_2O}}{1 + k_2 P_{H_2} + k_3 P_{H_2O}}, \qquad (3\text{-}24)$$

where the rate constants should be experimentally determined for the particular case. Steam cannot be contacted, as a reactor coolant, directly with graphite because of excessive reaction rates. With steam at 1.0 atm in direct contact with graphite, the reaction rates are 0.15, 4.5, and 56 g/ft^2-hr at 1100, 1500, and 2000°F, respectively (or 0.16, 4.8, and 60 mg/cm^2-hr at 593, 816, and 1093°C, respectively) [266]. The main concern is the water vapor impurities in other coolant gases. Figure 3-39 [267] gives kinetic data for water vapor in argon with graphite. Figure 3-40 [260] gives data on helium-water vapor in contact with graphites containing progressively larger amounts of barium and strontium to simulate fission products. There appears to be essentially a linear, positive dependence of the rate of gasification as a function of the level of impurities in the graphite and of the water vapor content in the gas stream.

The reaction rate between Speer Mod. 2 graphite and steam in a range of 20 to 760 vpm (in helium) and of 825°C to 1025°C is represented by [268]:

$$\text{Reaction rate, mg/g-hr} = 4 \cdot 10^{-4}(C_{H_2O})^{0.72}_{\text{vpm}}.$$

3.5 Hydrogen-Oxygen Reactions

3.5.1 Sources of Hydrogen-Oxygen Mixtures

There are four sources of hydrogen in water-cooled reactor systems and three sources of oxygen, in addition to what might be normally present in the cover gas: the sources of gaseous hydrogen are the aqueous corrosion of core metals, the massive metal-water reaction during an accident, and the electrolysis and radiolytic decomposition of water; and the sources of gaseous oxygen are the in-leakage of air, and the electrolysis and radiolytic decomposition of water.

Aqueous corrosion: The rate of production of hydrogen by aqueous corrosion can be estimated from the nominal corrosion rate data given in Table 2-1 and is normally very small except when produced by accelerated corrosion of fuel material (when, for example, a cladding defect develops).

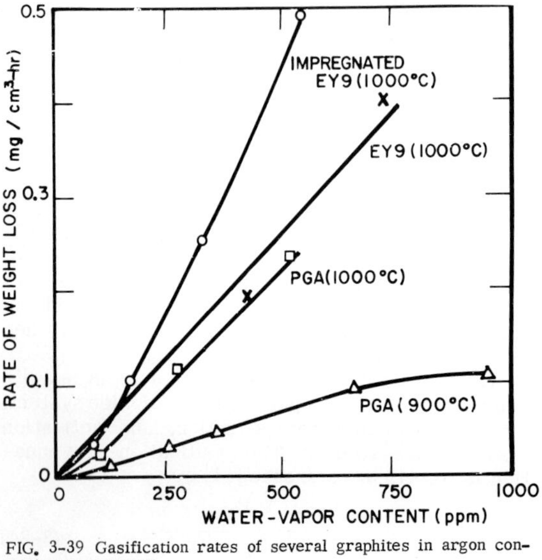

FIG. 3-39 Gasification rates of several graphites in argon containing water vapor.

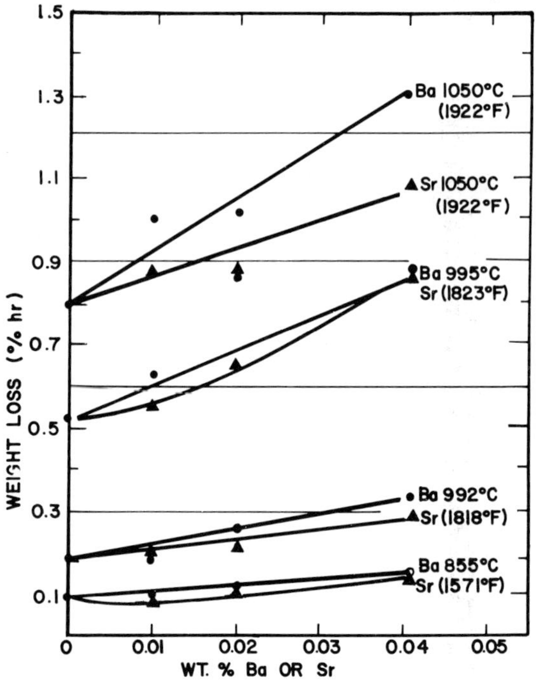

FIG. 3-40 Steam-graphite reaction of barium- or strontium-impregnated graphite compacts exposed to 3.5% H_2O vapor in helium at 1 atm and at a flow rate of 110 ml/min (0.00388 ft^3/min).

Metal-water reaction: The quantity of hydrogen generated during a reactor excursion or a loss-of-coolant accident can be computed from the chemical rate laws given in Sec. 3.2 when the cladding remains intact. It is necessary, in general, to consider the temperature distribution across the reactor in the case of a nuclear excursion accident. The quantity of hydrogen generated by metal-water reaction when a meltdown occurs can best be estimated from the TREAT meltdown experiments described in Sec. 4.

Air leakage: Accumulation of air in a reactor system must be considered both in normal reactor operation and as a result of an accident. Air intake during an accident is not likely to be involved in the primary accident because the pressure within the reactor system is usually greater than atmospheric. The possibility of a subsequent fire or explosion either within or immediately outside of the reactor system should be considered.

Electrolysis: Electrolytic decomposition of water to form a stoichiometric mixture of hydrogen and oxygen might occur inadvertantly whenever a source of electrical power is used either in connection with in-core instrumentation or corrosion control.

Radiolytic decomposition of water: Water is dissociated under the influence of ionizing radiation by two primary processes [269]:

$$H_2O \xrightarrow[\text{radiation}]{\text{ionizing}} H + OH\,,$$
$$H_2O \xrightarrow[\text{radiation}]{\text{ionizing}} H_2O^+ + e_s^-\,. \qquad (3\text{-}25)$$

The H_2O^+ and the secondary electron e_s^- form the

hydroxyl radical and the hydrogen atom, respectively, by the following reactions:

$$H_2O^+ + H_2O \rightarrow H_3O^+ + OH ,$$

$$e_s^- + H_2O \rightarrow e_{aq}^- ,$$

$$e_{aq}^- + H^+ \rightarrow H + H_2O ,$$

$$e_{aq}^- \rightarrow H + OH^- . \qquad (3\text{-}26)$$

The net effect of the radiation is to produce the free radicals H and OH. The above reactions occur within particle tracks. With densely ionizing radiation such as heavy particle recoils, high local concentrations of radicals are produced and radical-radical reactions occur as follows:

$$H + H \rightarrow H_2 , \quad OH + OH \rightarrow H_2O_2 . \qquad (3\text{-}27)$$

The formation of the molecular products, H_2 and H_2O_2, is not as great in the presence of sparsely ionizing radiation such as gamma rays or slow neutrons and the radicals escape the particle tracks intact. The radicals subsequently promote recombination reactions by the following chain mechanisms:

$$H_2 + OH \rightarrow H_2O + H, \quad H_2O_2 + H \rightarrow H_2O + OH. \quad (3\text{-}28)$$

The yields of molecular products [Eq. (3-27)] and the yield of radicals which escape particle tracks are represented by the terms G_F and G_R, respectively, in Table 3-10 [270]. These are defined as the number of molecules or radicals formed per 100 ev of absorbed energy. The rate of recombination reaction, Eq. (3-28), is dependent on the radical yield G_R. The net decomposition of water is favored, therefore, when G_F is large in comparison with G_R and may be nil when G_R is large in comparison with G_F. In the latter case, there is only a low steady-state concentration of H_2 and H_2O_2. The G_F and G_R values depend on the type of radiation to which the water is exposed, as indicated in Table 3-10.

Generally, in heterogeneous reactors there is a metal sheath around the fuel so that no fission fragments and few β-particles enter the water. The radiation, therefore, consists of a mixture of slow and fast neutrons and gamma rays. The slow neutrons contribute negligible energy while the fast neutrons result in energetic proton recoils. A water-moderated heterogeneous reactor such as CP-3′ has been shown [271] to behave as though the mixed radiation were a gamma flux, indicating that the energy deposition from proton or deuteron recoils is small when compared to that from gamma rays. Solutions irradiated in a graphite-moderated reactor (Oak Ridge Pile) show a higher G_F and a lower G_R, as indicated in Table 3-10. This results from a higher ratio of fast neutron to gamma ray flux and a correspondingly greater contribution from proton recoils.

TABLE 3-10

Yields for the Forward and Radical (Recombination) Reactions for Reactor Radiation

	G_F	G_R
Heavy-water reactor (CP-3′)	0.38	2.7
Graphite moderated reactor (Oak Ridge Pile)	1.18	1.57
Homogeneous reactor	1.1 - 1.8	Negligible

In a homogeneous reactor, the solution is exposed to densely ionizing heavy particle recoils which make by far the largest contribution to the absorbed energy and there is a net evolution of the molecular species H_2 and H_2O_2.

Oxygen is apparently not a primary product, but it appears as a result of H_2O_2 decomposition which may occur by several mechanisms; these include free-radical-induced decomposition, thermal decomposition, and impurity-catalyzed decomposition.

Although the rate of production of the species H_2, H_2O_2, OH, and H depends almost exclusively on the type of radiation, the net rate of decomposition of water depends markedly on the nature and concentration of solutes and impurities. Radical scavengers such as chloride, bromide, and iodide ions can decrease the concentration of H and OH and effectively inhibit the recombination reactions, Eq. (3-28). Only a few parts per million of these ions are known to cause rapid decomposition of water under gamma or reactor irradiation. Other ions such as Fe^{++}, Cu^+, and Ce^{+4} in acid solution are also efficient radical scavengers. Some indication of the effects of water impurities on the production of radiolytic gas in heterogeneous reactors is given in Table 3-11 [270].

Excess hydrogen increases the stability of water to irradiation by promoting the recombination reaction, Eq. (3-28). Excess hydrogen has been shown to prevent oxygen evolution entirely during the irradiation of pure water [272]. Hydrogen addition to the feedwater line of BORAX IV greatly decreased the oxygen content of the steam delivered to the turbine as shown in Fig. 3-41 [273]. The specific radiolysis rate in BORAX IV also decreased with increasing reactor power and increasing pH. The decrease in radiolysis rate with increasing pH was attributed to the ability of hydrogen ions to inhibit the recombination reactions. This pH effect is now believed to be due to the reaction of hydrated electrons with hydrogen ions to form hydrogen atoms:

$$e_{aq}^- + H^+ \rightarrow H + H_2O . \qquad (3\text{-}29)$$

The G factor for the radical recombination reaction for homogeneous reactor radiation (see Table 3-10) is negligible so that the net decomposition of the water is unaffected by excess hydrogen or traces of radical scavengers. More sparsely ionizing radiation is produced in low enrichment solution reactors and in slurry reactors; as a result some radical recombination occurs in these systems. The rate of gas generation in a solution reactor may be calculated from the following equation [274]:

$$K = 0.373\,(G)\,(PD) , \qquad (3\text{-}30)$$

where K is the moles H_2 per liter per hour and PD is the reactor power density in kw per liter. The G factor for hydrogen generation in solutions of $U^{235}\,O_2SO_4$ is given as follows [275]:

TABLE 3-11

Chemical Characteristics of Reactors

	Moderator condition						Some impurities in the D_2O				
Reactor	Normal operating power (kw)	Temp.	pH	Electrical conductivity (ohm^{-1} cm^{-1})	D_2 content of gas above D_2O	Evolution of radiolytic gas at max. power (liter/kw-hr)	D_2O_2 µM	Cl^- ppm	NO_3^- ppm	Al ppm	Fe ppm
NRX	40,000	50°C (122°F)	5.8	0.5×10^{-6}	<0.2%	$<5 \times 10^{-5}$	15-120	All less than 0.1 ppm			
CP-5	1,000	40-50°C (104-122°F)	6.5	0.3×10^{-6}		2×10^{-5}					
JEEP	350	50°C (122°F)	5.0	17×10^{-6}	0.5%	0.3	500		10	9	3.5
Zoé	150	35°C (95°F)	~ 4	60×10^{-6}		0.12	700	20	6	12	0.5
P-2	2,000	35°C (95°F)	3.6	130×10^{-6}		0.12	860	30	9	280	1.0
Russian power reactor	30,000	Coolant 270°C (518°F)		(Impurities kept at 3 ppm)		nil					

$$G(H_2) = 1.83 - 0.048(C)^{1/2} , \qquad (3\text{-}31)$$

where C is the uranium concentration in g per liter. Factors for other aqueous homogeneous feed solutions have been tabulated [276].

More detailed information concerning radiolytic problems in water reactors may be found in two recent reviews of the subject [277, 278].

3.5.2 Recombination Methods

It is apparent from the foregoing discussion (see Table 3-11) that the production of radiolytic H_2 and O_2 can be reduced to a very low level in heterogeneous reactors by careful attention to the

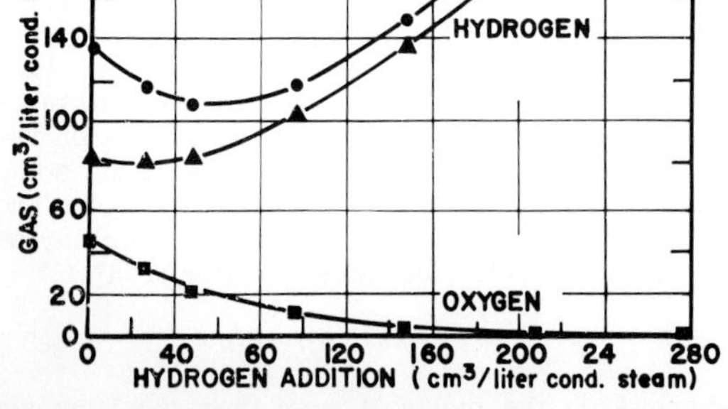

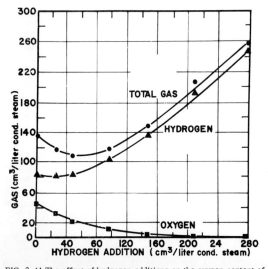

FIG. 3-41 The effect of hydrogen additions on the oxygen content of the steam delivered to the turbine. Reactor power, 4 Mw; pH, 6.1; specific resistance, 0.4×10^6 ohm-cm. Corrected for dissolved oxygen in sample.

purity of the water. This cannot be achieved in a homogeneous reactor.

It has been estimated that the operation of a full-scale homogeneous reactor generating 1000 Mw of energy would be accompanied by the production of as much as 10,000 cfm (280 m^3/min) of explosive gas [279]. Because of this a number of methods has been developed to recombine continuously the H_2 and O_2 to form water, either by internal methods, such as adding recombiner catalysts to the solution, or by external methods involving a tubular burner of a fixed bed of solid recombiner catalyst.

Catalytic recombination in aqueous homogeneous fuel solutions: Recombination of radiolytic hydrogen and oxygen occurs to some extent in UO_2SO_4 solutions even in the absence of added catalytic agents. However, it is not high enough even at 250°C (482°F) to prevent evolution of large quantities of radiolytic gas. The spontaneous recombination reaction is believed to occur as hydrogen slowly reduces the UO_2SO_4. This is followed by a rapid reoxidation of the uranium compounds by dissolved oxygen. It has been discovered that dissolved cupric ion is similarly reduced and rapidly reoxidized. The rate of the cupric-ion-catalyzed recombination, however, may be as much as 10^5 times more rapid than the reaction with UO_2SO_4 solutions alone [280]. The kinetics of the cupric-ion-catalyzed recombination reaction have been studied extensively [279] and the reaction has been shown to be first-order with respect to both the hydrogen pressure and the concentration of cupric sulfate in solution. The reaction occurs only in the liquid phase and between dissolved hydrogen and dissolved oxygen. This is supported by the fact that the reaction rate is limited by the rate of solution of hydrogen in water when there is insufficient agitation. The catalyzed reaction rate increases by a factor of about 20 over the temperature range from 190 to 250° C (374 to 482° F) indicating an activation energy of 24 kcal/mole.

The efficiency of the cupric ion catalyzed reaction has been demonstrated [280]. Sample solutions containing 40 g U and 0.009 moles $CuSO_4$ per liter reached an equilibrium pressure of 1100 psi (75 atm) during irradiation at 229°C (444°F). About 400 psi (27 atm) of this was water vapor and the remainder hydrolytic gas. Equilibrium pressures for uncatalyzed fuel solutions under these conditions were estimated to be between 10,000 and 20,000 psi (~700 to 1400 atm). Tests with a copper recombination catalyst in HRT [281] in the presence of fission and corrosion products indicated that recombination rates were larger than out-of-pile data would have indicated by a factor of 1.5 to 2.

A large number of other materials has been tested for catalytic activity in 0.16 M UO_2SO_4 solution at 250°C (484°F) [279]. Silver and iodide ions were the only ionic species to show significant catalytic activity. At the end of these experiments, however, the silver appeared to have precipitated and the iodide was converted to the elemental form suggesting that these solutions were not stable. The metals Pt, Pd, Os, Ir, Rh, and Ru, added as ionic species, were reduced to the metallic state, in which they acted as extremely effective heterogeneous catalysts, usually initiating explosions of the confined gas.

Noble metals in suspended form have been considered for use as catalysts for operation of homogeneous reactors at 100°C (212°F) where the cupric-ion-catalyzed recombination is too slow. Studies with suspended palladium showed that deposition of the suspended material on walls was a problem [280]. It is also likely to behave erratically, sometimes causing explosions.

Catalytic recombination in slurry fuels: Slurry fuels, particularly those containing thoria, are themselves recombination catalysts achieving recombination rates of the order of 0.005-0.05 moles H_2/hr-liter [280].* An effective catalyst in a slurry fuel, however, must be capable of reacting 5 to 10 moles H_2/hr-liter. Satisfactory performance has been achieved using a $CuSO_4$-H_2SO_4 solution in a thoria slurry. Many slurry fuels of interest, however, contain uranium oxides as well as thoria. It has been shown that $CuSO_4$ solutions are incompatible with slurries containing uranium. The copper may be precipitated as $CuUO_4$ in slurries containing UO_3. Rates as high as 1 or 2 moles H_2/hr-liter have been obtained using heterogeneous catalysts consisting of Ni, Cu, Ag, Pd, Mo, and V as metals or oxides. The highest rates obtained, however, were somewhat less than desired. Heterogeneous catalysts, in general, have the same disadvantages as slurry fuels themselves in that they may tend to settle. They are probably also more susceptible to poisoning by adsorbed impurities than are catalysts present in true solution.

External flame recombiners: A tubular burner has been used to recombine H_2 and O_2 in the off-gas from HRE [274]. The off-gas stream was diluted with steam to produce a nonexplosive mixture; the steam was then condensed ahead of the recombiner. It was found necessary to add a small quantity of steam to the explosive gas mixture just ahead of the recombiner to prevent flashbacks and excessive injector heating. The flame was maintained at a many-holed injector and was continuously ignited by means of a magneto-generated spark. The combustion chamber, designed to handle 15 cfm (0.43 m^3/min) of combustible gas, was a 3-1/2 ft (1.07 m) length of 10 in. (25.4 cm) diameter pipe jacketed by a 12 in. (30.5 cm) pipe through which cooling water flowed. The cooling water removed 70-80% of the heat of combustion although the burner wall temperature reached 620°C (1148°F) during operation at rated load.

The principal hazards associated with the operation of a flame recombiner are the possibility of extinguishment when very little gas is being generated and the possibility of flashbacks through the injector holes. The most serious accident situation was considered to be a flashback all the way through to the reactor core which could result in a high-pressure explosion within the reactor tank. Such a flashback would have to propagate through a region where the steam concentration exceeded 70% and through the throttling valve into the high-pressure core region. Both of these processes were considered highly unlikely.

External catalytic recombiner: An external catalytic recombiner is probably an inherently less hazardous device than a flame recombiner. Consequently, recombiners used in the later homogeneous reactors such as HRT have employed catalytic units. The primary fuel recombiner in HRT was a packed bed 10-1/2 in. (26.7 cm) in diameter by 12 in. (30.5 cm) high consisting of Incoloy or stainless steel ribbon (1/16 in. × 0.005 in. or 1.59 mm × 0.127 mm) plated with 15 g/ft^2 (0.016 g/cm^2) of platinum [283]. The catalyst bed was reactivated periodically by super-heated steam at 500°C (932°F) and 250 psi (17 atm). The bed was mounted at the upper or downstream end of a tank 13 in. (33 cm) in diameter by 5 ft 4 in. (1.63 m) long which also contained a bed of silvered York mesh which served as an iodine trap [284].

A secondary recombiner was also used in the HRT low-pressure system. This was a smaller unit in the form of a removable piping loop and was a cylindrical bed (3 in. or 7.6 cm in diameter by 5-1/2 in. or 14 cm long) containing platinized chromel ribbon. Out-of-pile performance tests [284] indicated that the secondary recombiner alone was sufficient to recombine radiolytic gas from HRT. It had an efficiency of greater than 98% using 26 ℓ/min of O_2 and 1 ℓ/min of H_2. The primary recombiner operated at 99.6% efficiency at a hydrogen flow rate of 80 ℓ/min [286].

The original recombiner used in HRT had to be replaced because of decreased efficiency. The pellet bed used in the original design may have had poor contact between the gas and catalyst. Irreversible poisoning is likely to occur from the accumulation of fission product iodine. The iodine is apparently removed satisfactorily during regeneration at 400°C (752°F) or higher. The catalyst must be free from liquid water during operation. This can be done with an external heater if the steam feed is not sufficiently dry.

*For each 100 ev of fission-recoil energy absorbed, 2.1 molecules of hydrogen are liberated [282].

The gas mixture passing through a catalytic recombiner must be diluted below the inflammability limit, or ignition and flashback will almost certainly occur. Attention must also be given to ensure steady removal of the heat of the recombination reaction in catalytic recombiners.

Recombiners have been extensively used in D_2O-moderated and -cooled reactors because of the high cost of coolant and the need to maintain a tightly closed system.

3.5.3 Explosion Limits

There are two distinct regimes of burning in gaseous systems, namely deflagration and detonation. Although detonation is the more rapid and violent process, deflagration or simple burning can also produce all the effects usually associated with the inexact term, explosion. The general natures of deflagrations and detonations are quite different so that the two regimes are discussed separately.

Deflagration: The mechanism of the hydrogen-oxygen deflagration is a very complicated branching chain reaction in which the principal chain carriers are the free radicals H, OH, and O [287]. Ignition occurs in a hydrogen-oxygen mixture when the rate of production of chain carriers exceeds their rate of destruction. An extremely small quantity of energy contained in an electric spark or introduced into the gas by other means can create a local excess of carriers which results in propagation of the flame throughout the remaining unburned gas. The rate of this propagation is called the laminar flame velocity. Flame velocities are given in Fig. 3-42 [288] for hydrogen, oxygen, and nitrogen mixtures originally at 1 atm and room temperature. The peak value of almost 30 ft/sec (9.15 m/sec) is very high when compared with flame velocities of premixed hydrocarbon fuel gas-air mixtures.

Flame velocities are dependent to some extent upon whether the direction of propagation is up or down. Compositions of limiting mixtures are even more dependent upon direction. The dependence on direction results from the tendency of the burned gas to rise. The convective motion makes upward propagation easier and more rapid. Thus, the lower flammability limits of hydrogen in air saturated with water vapor (room temperature and atmospheric pressure) are 4.1% for horizontal propagation and 9% for downward propagation [289]. In general, the quantity of hydrogen burned increases as the concentration increases above 4%. About 50% of the hydrogen will burn in a 5.6% mixture. Substantially all of the hydrogen is consumed in a 10% mixture. The upper flammability limit for hydrogen-air mixtures may be taken as approximately 74%. Flammability limits widen as the temperature of the unburned gas is increased, as shown in Fig. 3-43 [289]. Increasing the pressure tends to narrow flammability limits somewhat, although the effect is small.

Flammability limits in the presence of steam at 100 psig (7.8 atm) total pressure and a temperature of 146°C (300°F) have been determined [290, 291]. The results were very similar to those obtained at room temperature for hydrogen-air mixtures in the presence of water as shown in Fig. 3-44

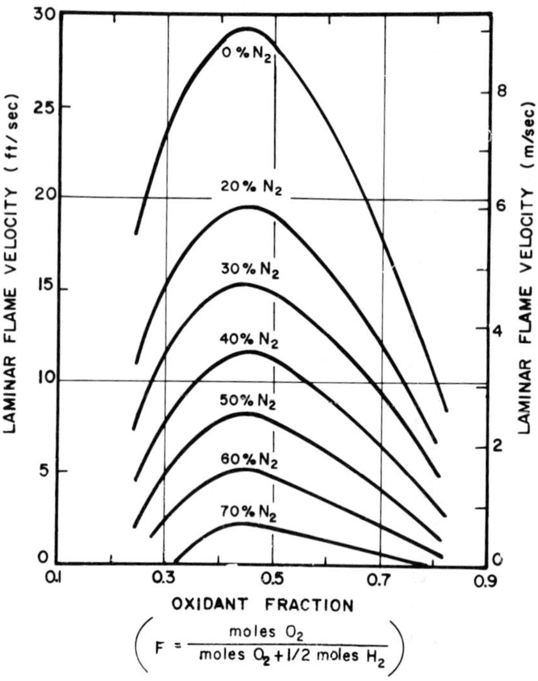

FIG. 3-42 Laminar flame velocities for hydrogen, oxygen, and nitrogen mixtures at atmospheric pressure. (Extrapolated data.)

[289]. Limits have been obtained in autoclave experiments with various hydrogen-oxygen mixtures in the presence of saturated steam up to 260°C (500°F) as shown in Fig. 3-45 [292]. The results can be summarized by noting that mixtures containing partial pressures of hydrogen-oxygen greater than 27% of the total pressure were reactive under all conditions, whereas mixtures containing less than 17% were unreactive under all conditions. As a general rule, results at atmospheric pressure indicate that 7.6 volumes of water vapor per unit volume of the hydrogen-air mixture will render any hydrogen-air mixture completely nonflammable. On the same basis, it requires

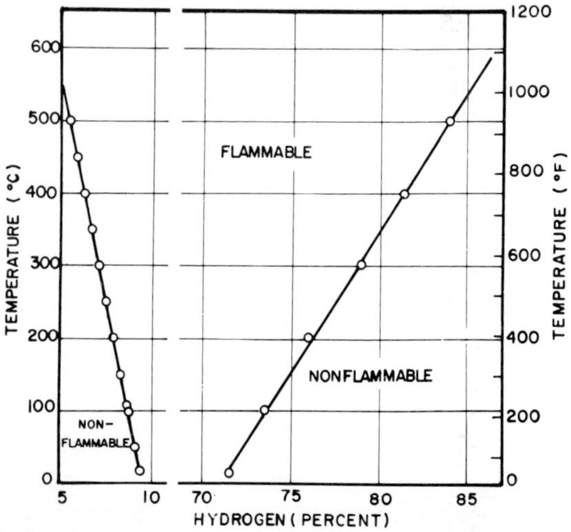

FIG. 3-43 Influence of temperature on limits of flammability of hydrogen in air (for downward propagation of flame).

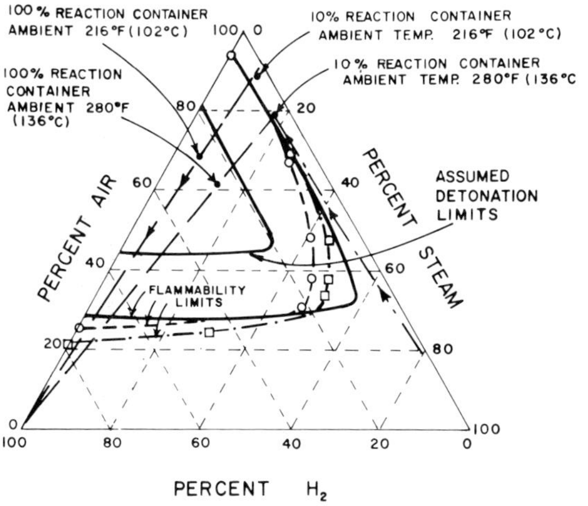

FIG. 3-44 Flammability limits of hydrogen-air-steam mixtures. The solid line is for 75°F (24°C), 0 psig (~1 atm). The broken line, with open circles to indicate measurements, is for 300°F (149°C), 0 psig (~1 atm). The dot-dash line, with open squares to indicate measurements, is for 300°F (149°C), 100 psig (~8 atm).

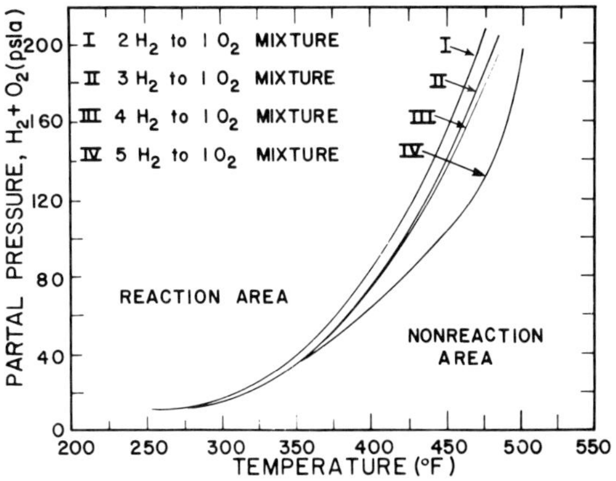

FIG. 3-45 Explosion limits of various H_2 and O_2 mixtures in the H_2-O_2-H_2O system. Tests made in microbomb by spark ignition.

11 volumes of helium or 10.2 volumes of CO_2.

The maximum temperature and pressure reached by the burned gas in hydrogen-oxygen deflagrations is determined largely by the original composition and pressure of the mixture. The adiabatic flame temperature and equilibrium pressure can be calculated precisely from thermodynamic data on the assumption that a constant volume explosion occurs and no external work is performed by the gas. The calculation is made by a reiterative method in which the heat of an assumed isothermal reaction is used to bring the reaction products to the final temperature [287]. The composition of the reaction products must also be computed by reiterative methods; the species present in the burned gas include H_2O, H_2, O_2, OH, H, and O. If nitrogen is present, N_2 and NO must also be considered. Calculated explosion pressures are given in Fig. 3-46 [293] for mixtures of knallgas ($2H_2 + O_2$) with steam. The results are expressed as the ratio of the peak reaction pressure to the initial mixture pressure as a function of initial gas temperature and composition. Measured pressure ratios, also given in Fig. 3-46, show that the calculated pressures are not realized until the onset of detonation. The corresponding extent of reaction is also relatively low until detonating mixtures are produced as shown in Fig. 3-47 [293].

The limits of flammability of deuterium-oxygen-helium mixtures, containing up to 20 vol.% deuterium and 30 vol.% oxygen at initial pressures of 0.5, 1.0 and 2.0 atm and at 25°C (77°F) and 80°C (176°F), have been determined experimentally [294]. Figure 3-48 shows the flammable mixture compositions, in vol. % of the components, at 25°C. In the figure, flammable mixtures are labeled with solid symbols and nonflammable mixtures with hollow symbols. The lower and upper limits of flammability at 25°C were found to be:

Lower limit of flammability at 25°C = 7.8 ± 0.2 vol. % deuterium

for $62 \leqslant [\mathrm{He}] \leqslant 87$ vol. %
and $5 \leqslant [O_2] \leqslant 30$ vol. %,

Upper limit of flammability at 25°C = 96 - [He] ± 0.2 vol. % deuterium

for $70 \leqslant [\mathrm{He}] \leqslant 87$ vol. %
and $[O_2]_{min} = 4.0 \pm 0.2$ vol. %.

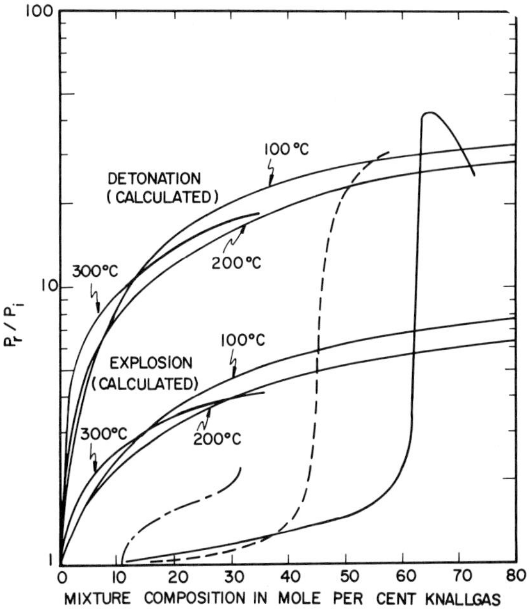

FIG. 3-46 Ratio of peak reaction pressure to initial mixture pressure as a function of composition for knallgas ($2H_2 + O_2$)-steam mixtures (hot wire ignition).
—— 100°C (212°F), steam pressure = 1 atm (14.7 psi)
– – – 200°C (392°F), steam pressure = 15.3 atm (225 psi)
–·–· 300°C (572°F), steam pressure = 85 atm (1250 psi)

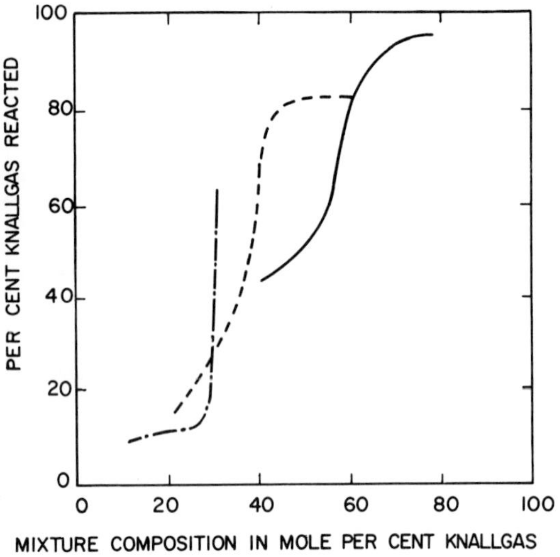

FIG. 3-47 Percent of knallgas ($2H_2 + O_2$) reacted as a function of composition for knallgas-steam mixtures
——— 100°C (212°F), steam pressure = 1 atm (14.7 psi)
- - - - 200°C (392°F), steam pressure = 15.3 atm (225 psi)
—·—·— 300°C (572°F), steam pressure = 85 atm (1250 psi)

These results are valid for each of the three pressures examined. The experimental results indicate that the limits given above are not pressure-dependent within the accuracy and range of the measurements. The limits determined at 80°C are not very different from those reported at 25°C.

The deuterium concentrations are essentially constant at the lower limits and the oxygen concentrations, at the upper limits of flammability.

It is interesting to note that complete flame propagation in a deuterium-oxygen mixture could not be obtained below about 9 vol. % deuterium in oxygen. With helium as a diluent, however, flammability was achieved with 6 vol. % deuterium and 47 vol. % oxygen and helium, respectively. Thus dilution in helium lowers the lower limit of flammability.

The flammability limits were measured in vertical glass tubes, 2 in. in diameter and 48 in. high. Ignition was by an electric spark across a 3/8-in. gap at the bottom of the tube.

The maximum pressures and rates of pressure rise obtained following the ignition of three flammable mixtures have been measured in a 2-ft spherical vessel at 25°C and 1 atm initial pressure, see Fig. 3-49. The maximum recorded pressures, the time to reach these pressures, and the maximum rates of pressure rise are listed in Table 3-12. The usual dependence of pressure on time can be noted. These results are approximately those that would be observed if deuterium were replaced by hydrogen.

Detonations: A detonation is a chemically supported shock wave. Because of this, the velocity of propagation and pressure ratio across the shock are precisely calculable from hydrodynamic equations of momentum, energy, and continuity. The velocity of propagation is the velocity of sound

TABLE 3-12

Maximum Recorded Pressure, Time to Reach Maximum Pressure, and Maximum Rate of Pressure Rise for Three Flammable Mixtures

Mixture	Maximum pressure (psi)	$[\frac{dp}{dt}]_{max}$ (psi./ms)	t_{peak} (ms)
A (30 vol. % D_2, 70 vol. % air)	84	14.3	28
B (17 vol. % D_2, 29 vol. % O_2, 54 vol. % He)	78	9.1	50
C (10 vol. % D_2, 20 vol. % O_2 70 vol. % He)	75	2.0	200

in the burned gas, which is considerably higher than the velocity of sound in the unburned mixture. A detonation wave is therefore said to be supersonic with respect to the unburned mixture. Pressures and temperatures behind the detonation front are higher than those computed for a constant volume explosion. The greater destructiveness of a detonation is due primarily to the much higher velocity and the very sharp nature of the shock front. Detonation wave pressures and the still higher reflected wave pressures were calculated for knallgas-steam mixtures [295, 296] and for heavy knallgas-D_2O vapor mixtures [296, 297]. The reflected wave detonation pressures for knallgas-steam mixtures are included in Fig. 3-46. Experimental pressures, under certain conditions, can exceed calculated pressures somewhat, because of unstable detonation, as shown in the figure.

Composition limits for detonation cannot be expressed precisely since the development of a detonation wave depends greatly on the geometry of the enclosure and the point and method of ignition. Nominal mixture limits for hydrogen-air-steam mixtures for which detonations might occur were included in Fig. 3-42. Detonation limits for knallgas-steam mixtures are shown (in Fig. 3-44) to vary with steam pressure (density). The formation

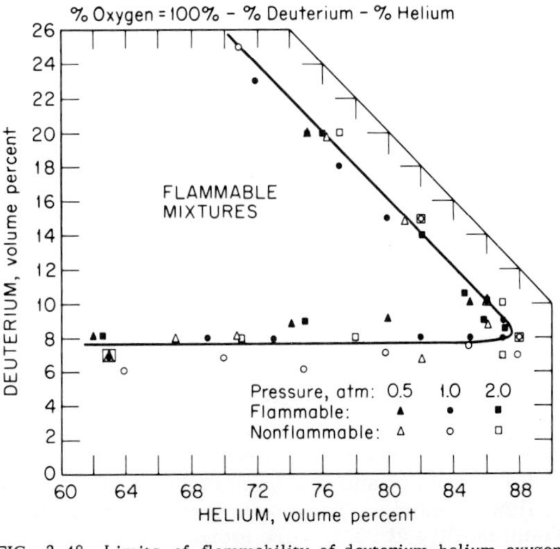

FIG. 3-48 Limits of flammability of deuterium-helium-oxygen mixtures at 0.5, 1.0 and 2.0 atm and 25°C (77°F) initial temperature.

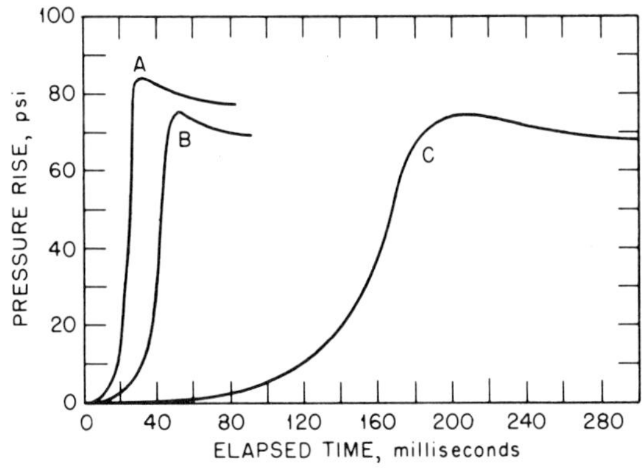

FIG. 3-49 Pressure-versus-time record of deuterium-air (curve A) and deuterium-helium-oxygen mixtures (curves B and C) when ignited in a 2-ft (61-cm) sphere at 25°C (77°F) and 1.0 atm initial pressure. See also Table 3-12.

of a fully developed detonation wave (sometimes called a high-order detonation) is favored by the presence of a long cylindrical channel with ignition occurring at one end. Spark ignition results in a deflagration which propagates down the channel. Compression waves sent out by the subsonic deflagration become successively more rapid and finally become shock-fronted at some point down the tube. The shock front may then rapidly become a detonation wave. Distances required for the appearance of detonation in a 25-mm (1.0-in.) tube are given in Table 3-13 [261]. Distances generally increase with tube diameter and depend on composition and flow condition. Hydrogen-oxygen mixtures saturated with steam are slow detonators as evidenced by weak predetonation reactions (see Figs. 3-44 and 3-45), and as a result exhibit anomalous behavior [298]. Detonations in gaseous mixtures normally do not propagate spherically from a point source since energy attrition is too great.

TABLE 3-13

Distance Traversed by a Flame before Detonation Appearance for Quiescent Stoichiometric H_2-O_2 Mixtures in 1.0-inch Diameter Tube

Initial pressure atm	Distance cm	Distance ft
1	70	2.30
2	60	1.97
3	52	1.71
4	44	1.44
5	35	1.15
6	30	0.98
6.5	27	0.89

4 FUEL MELTDOWN

4.1 General Considerations

Fuel meltdowns are considered here to the extent that they constitute an initiating process for a violent chemical reaction. In addition, an attempt is made to integrate the chemical information with the other processes occurring during a meltdown. The chemical reactions between the cladding metals and water were summarized in Sec. 3.2. The present section considers only meltdowns in water-cooled reactors.

The chemical energy released in a nuclear excursion depends not only on the thermodynamic properties of the system but also on the degree of dispersion of the metal in the water and on the kinetics of the chemical reaction. An indication of the potential magnitude of the chemical energy release is given in Table 4-1, where the fission energy released in the BORAX-1 experiment [299] (135 Mw-sec) is compared with the available chemical energy based upon the reaction of 10% and 100% of the entire 100 lbs (45 kg) of core metal. It is evident from Table 4-1 that the chemical energy release could exceed the fission energy in a massive metal-water reaction. It is also clear that stainless steel is the safest of the core metals from a thermodynamic point of view.

Information on the extent of metal-water reaction to be expected from a postulated reactor incident can be obtained from the results of previous nuclear excursions, small-scale meltdown experiments, and out-of-pile laboratory experiments. Pertinent data from these sources are summarized in the following sections. Although considerable information is now available, significant gaps remain.

4.2 Nuclear Excursions

There have been several reactor excursions in which melting of the core of a water- (or D_2O-) cooled reactor has taken place. Four separate facilities with aluminum-uranium cores have sustained nuclear excursions resulting in extensive damage. Two of these, the NRX [300, 301] and the SL-1 [302, 303, 304] were accidents and two, BORAX-1 [299] and SPERT-1 [305, 306, 307] were full-scale excursion experiments. A summary of three of the excursions which occurred at the NRTS (National Reactor Testing Station), Idaho, USA, is given in Table 4-2. These were the BORAX-1 and the SPERT-1 experiments and the SL-1 accident. All three had similar cores: they were made of aluminum-clad, aluminum-uranium alloy plates. The three incidents were very similar. Each was initiated by the rapid removal of a control rod. In each case a large portion of the center of the core was melted and fragmented. Violent movement of a large portion of the coolant water up and out of the reactor core occurred in each incident.

TABLE 4-1

Calculated Energy Release for Various Metal-Water Reactions (Basis: 100 pounds or 45.4 kg of metal)

Metal	Nuclear Energy Release, Mw-sec	Chemical energy release[a], Mw-sec			
		10 Percent reaction		100 Percent reaction	
		M + H_2O	H_2 + O_2	M + H_2O	H_2 + O_2
Al	135	80	63	800	630
Mg	135	68	46	680	460
Zr	135	30	25	300	250
U	135	11	9.5	110	95
SS-304	135	5	22	50	220

[a] Calculations based on thermodynamic data given in Table 3-1, rounded numbers.

TABLE 4-2

A Summary of Some Destructive Reactor Transients [U-Al Cores, Water-Cooled]

Reactor	BORAX-1	SL-1	SPERT-1
Date of event	July 22, 1954	Jan. 3, 1961	Dec. 5, 1962
Place	NRTS, USA	NRTS, USA	NRTS, USA
Category	Experiment	Accident	Experiment
Reactor period, sec	0.0026	~ 0.004	0.0032
Nuclear energy, Mw-sec	135	130	31
Peak power, Mw	est. 16000	est. 19000	2300
Chemical energy, Mw-sec	Undetermined	24	3.5

Both the BORAX-1 and SPERT-1 were open to the atmosphere (open tank configuration) and motion pictures recorded the rise of the water spout carrying with it core material and instrumentation. Pressure records were made in both experiments; however, they were difficult to interpret due to apparently violent shock waves which destroyed transducers and overloaded recording devices.

In all three incidents the core debris had a similar appearance. Portions of core that had been completely melted were spongy and considerable quantities of metallic oxides were present. No attempt was made to determine the extent of reaction of the core with water in the BORAX experiment. In the postexcursion analysis of both the SL-1 accident and SPERT-1 experiment, portions of the debris were analyzed for alpha aluminia. Alpha alumina is the reaction product formed at temperatures above 600°C (1112°F).

The SL-1 analysis indicated appreciable quantities of alumina in several samples [302, 303]. The results of these analyses and an overall material balance resulted in an estimated chemical energy release of 24 ± 10 Mw-sec compared to the nuclear input of 130 Mw-sec.

A similar analysis of debris from the SPERT-1 destructive test [306] gave an estimated chemical energy input of 3.5 Mw-sec compared with the 31 Mw-sec nuclear energy release.

It is appropriate to point out here that in all three of these analyses, the conclusion was reached that the results of the excursion could be explained in terms of a steam explosion that resulted from the dispersion of molten metal into the water. It is presently believed that hydrogen produced by prompt metal-water reaction contributed little toward the steam explosion.

The accident to the NRX Reactor at Chalk River, Canada, was somewhat different. The aluminum-clad uranium core reactor was D_2O-moderated and primarily H_2O-cooled. An excursion was initiated by the inadvertent removal of a bank of control rods (see Sec. 3.3 of Chapter 11). After the initial power rise to 20 Mw, a sudden increase of reactivity occurred carrying the power to 80 Mw, probably due to expulsion of light water by steam. Eventually, the dumping of the D O moderator ended the excursion. The total nuclear energy release was much greater than the three aforementioned excursions. Because of the long period of the excursion, the peak power was only 80 Mw as compared with 16,000 and 19,000 Mw for the SL-1 and BORAX excursions, respectively.

Severe damage was caused by insufficient cooling of several fuel rods in which experiments were being conducted. Melting of several rods in the NRX core resulted. Appearance of core material indicated that some reaction had obviously occurred. Subsequently, a hydrogen explosion was postulated as an explanation for the lifting of a 4-ton (~4000 kg) dome on the gas holder above the reactor. The investigators stated [301], "The appearance of the residual metal (uranium) in some of the worst cases can only be accounted for by considerable chemical reaction and not just melting of materials. Whether or not the highly exothermic aluminum-water reaction also had a significant role cannot be decided."

In summary, it appears that in each of these incidents the metal-water reactions played somewhat of a role; however, the principal damage resulted from the nuclear energy release. One other point which should be made is that none of these reactors was operating at power at the time of the incident. The water temperatures were in the order of 25°C (80°F) and the pressure in all cases was nearly atmospheric. The possibility of more severe chemical effects at higher initial water temperature and steam pressure cannot be ruled out. It should also be noted that more energetic transients would increase the percentage of the reactor core metal reaching high temperatures. This would increase the chemical energy release relative to the fission energy release because of the exponential dependence of chemical reaction rates on temperature.

4.3 Small-Scale Meltdown Experiments

Extensive experimental data are available on the extent of metal-water reaction and the nature of damage produced by transient nuclear heating of small representative fuel specimens. The technique is to simulate a nuclear excursion or loss-of-coolant accident by using fission heating from a neutron pulse to heat a few grams of fuel in a stainless steel autoclave containing water. Following the transient irradiation, the fraction of metal that reacted with water is computed from the measured quantity of hydrogen produced. The hydrogen is normally determined by a mass spectrometric analysis of the gas phase within the autoclave. Photomicrographs, macrophotographs, and particle size studies are used to characterize the physical damage sustained by the fuel specimens. Thermocouples and pressure transducers are used to determine the temperature and pressure history of the process. Thermocouples, however, did not indicate temperatures beyond the melting or fragmentation point of the specimens because of loss of contact. Pressure records were generally unsatisfactory because of interaction be-

TABLE 4-3

Results of Submerged Fuel Specimen Meltdown Tests in TREAT; Studies with Zirconium

[Water initially at room temperature, except as noted, initial helium overpressure of 20 psia (~1.3 atm).]
[Values with asterisk are estimated.]

Fission energy input (cal/g)	Abiabatic core temperature[a]	CEN run No.	TREAT characteristics		Appearance of specimen after transient	Mean particle diameter	Percent of metal reacted
			Period (msec)	Int. power (Mw-sec)			
Unclad Zirconium-Uranium Alloy Fuel Plates [89.4 wt. % Zr, 10.6 wt. % U, 93% enriched, 1.0 in. (2.54 cm) x 0.5 in. (1.27 cm) x 0.1 in. (0.25 cm)]							
285	—	100	80	208	Melted, one globule	460 mils (11.7 mm)*	6.0
344	—	87	83	251	Melted, one globule	460 mils (11.7 mm)*	5.2
510	—	88	79	372	Melted, one globule	460 mils (11.7 mm)*	9.5
628	—	89	79	458	Melted, one globule	460 mils (11.7 mm)*	11.5
785	—	101	52	573	Spattered	—	67.2
Zircaloy-2 Clad, Mixed Oxide Core Fuel Pins [Cladding: Zircaloy-2, 20 mils (0.51 mm) thick, ends capped. Core: 81.5 wt. % ZrO_2, 9.06 wt. % CaO, 0.74 wt. % Al_2O_3, 8.69 wt. % U_3O_8, 93% enriched, 0.33 in. (0.84 cm) diam., 0.53 in. (1.35 cm) long.]							
254	1600°C (2912°F)	56	105	270	Clad: intact Core: cracked	375 mils (9.53 mm)*	0.3
301	1900°C (3452°F)	28	60	320	Clad: partly fragmented Core: cracked, fragmented	110 mils (2.80 mm)	4.2
362	2300°C (4172°F)	29	63	385	Clad: partly fragmented Core: cracked, fragmented	107 mils (2.72 mm)	8.0
516	3100°C (5612°F)	30	62	550	Clad: partly fragmented Core: fragmented	51 mils (1.30 mm)	14.0
609	3300°C (5972°F)	49	50	648	Clad: partly fragmented Core: fragmented	49 mils (1.25 mm)	24.0
787	3300°C (5972 F)	168	41	985	Clad: fragmented Core: fragmented	15 mils (0.40 mm)	30.0
Zircaloy-2 Clad, UO_2 Core Fuel Pins Cladding: Zircaloy-2 20 mils (0.51 mm) thick, ends capped. Core: UO_2, 11.2% enriched, 0.33 in. (0.84 cm) diam., 0.50 in. (1.27 cm) long.							
215	2700°C (4892°F)	187	74	239	Clad: intact Core: intact	375 mils (9.5 mm)	1.3
290	3000°C (5432°F)	181	72	323	Clad: melted Core: tragmented	55.5 mils (1.4 mm)	24
375	3300°C (5972°F)	214	73	416	Clad: fragmented Core: fragmented	24 mils (0.6 mm)	13
450	3300°C (5972°F)	175	52	505	Clad: fragmented Core: fragmented	29 mils (0.7 mm)	63
595	3300°C (5972°F)	215	49	663	Clad: fragmented Core: fragmented	19 mils (0.5 mm)	21
675	3300°C (5972°F)	216	42	749	Clad: fragmented Core: fragmented	13 mils (0.3 mm)	68
771	3300°C (5972°F)	173	39	867	Clad: fragmented Core: fragmented	26 mils (0.6 mm)	80
Water initially at 285°C, steam overpressure of 1000 psia (68 atm).							
198	2700°C (4892°F)	192H	78	206	Clad: intact Core: intact	375 mils (9.50 mm)*	1.0
282	2700°C (4892°F)	193H	77	301	Clad: intact Core: intact	375 mils (9.5 mm)*	10.7

[a]The value of 3300°C (5972°F) corresponds to the boiling point of UO_2 at one atmosphere.

tween the neutron flux and the transducer circuitry. Meltdown studies with specimens containing zirconium, uranium, aluminum, and stainless steel have been carried out primarily in TREAT [308]. Some studies have been carried out in MTR [309, 310]. These results will be discussed according to the metal.

Zirconium: Studies with zirconium-uranium alloys in MTR indicated 16 to 84% metal-water reaction with dynamic pressure surges to 7000 psi (~470 atm) [310] (p. 867 of reference [1].) Autoclaves were rapidly lowered into MTR and quickly removed to simulate transient nuclear heating. It was concluded that there was no evidence for a self-sustaining chemical reaction of the small alloy specimens below the metal melting point. Beyond the melting point (up to about 3000°C or 5430°F), the results indicated extensive but not explosive reaction.

Studies in TREAT have been performed with zirconium-uranium alloy fuel plates [311] and with Zircaloy-2 clad, mixed oxide [308] and UO_2 [312] core fuel pins. Results are summarized in Table 4-3. Calculated adiabatic core temperatures are given for the clad oxide core specimens. It is believed that oxide core temperatures approach the adiabatic values at temperatures below the oxide melting point because of the poor thermal conductivity of oxides. Adiabatic temperatures are not given for metallic core fuel specimens. The rapid rate of heat loss from metallic specimens to water prevents metallic cores from approaching adiabatic temperatures on the transient periods available in TREAT.

The zirconium-uranium alloy plates simply melted down into the alumina-retaining crucible in four of the five experiments. From 6.0 to 11.5% of the metal reacted with water in these experiments. In the most vigorous experiment of the series, the metal was extensively spattered throughout the autoclave. Hydrogen analysis indicated that 67.2% of the metal reacted and it was likely that the metal temperature reached well above 3000°C (5430°F) in this experiment. Results are plotted in Fig. 4-1 [311].

Experiments conducted with Zircaloy-2 clad, mixed oxide core fuel pins indicated that there was a linear relationship between the extent of metal-water reaction and the nuclear energy input over the range 250 to 600 cal/g of oxide core. The threshold for destruction was about 275 cal/g which corresponded to an adiabatic core temperature of 1800°C (3272°F). Extent of metal-water reaction from these in-pile tests is plotted as a function of adiabatic core temperature in Fig. 4-2 where the apparent threshold for significant metal-water reaction is 1600°C (2912°F). Experiments with Zircaloy-2 clad, pure UO_2 core fuel pins showed that the threshold for fragmentation and for metal-water reaction coincided with UO_2 melting.

Uranium: Meltdown tests in TREAT have been performed with pure uranium wires and rods and with the relatively corrosion-resistant uranium, 5 wt.% zirconium, 1.5 wt.% niobium alloy (this alloy was used in the first core loading of EBWR). Studies with the alloy were performed with bare pins and with Zircaloy-2 clad pins. Results given in Table 4-4 [308] indicate that

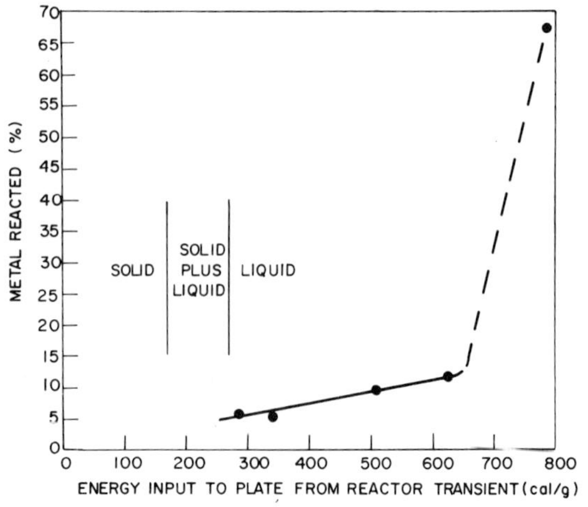

FIG. 4-1 TREAT zirconium-water experiments: percent reaction versus nuclear energy input. [89.35 wt.% zirconium-10.65 wt.% uranium (93% enriched) alloy plates (1.0 in. x 0.5 in. x 0.1 in.)]

three factors are important in governing the extent of metal-water reaction. There are the total fission energy input (see Fig. 4-3) the reactor period (see Fig. 4-4 [308]) and the surface area of the uranium created by fragmentation (see Fig. 4-5 [308]).

For a given energy input, more chemical reaction occurs as the period becomes shorter because there is less time for heat losses to occur. This increases the peak temperature reached by

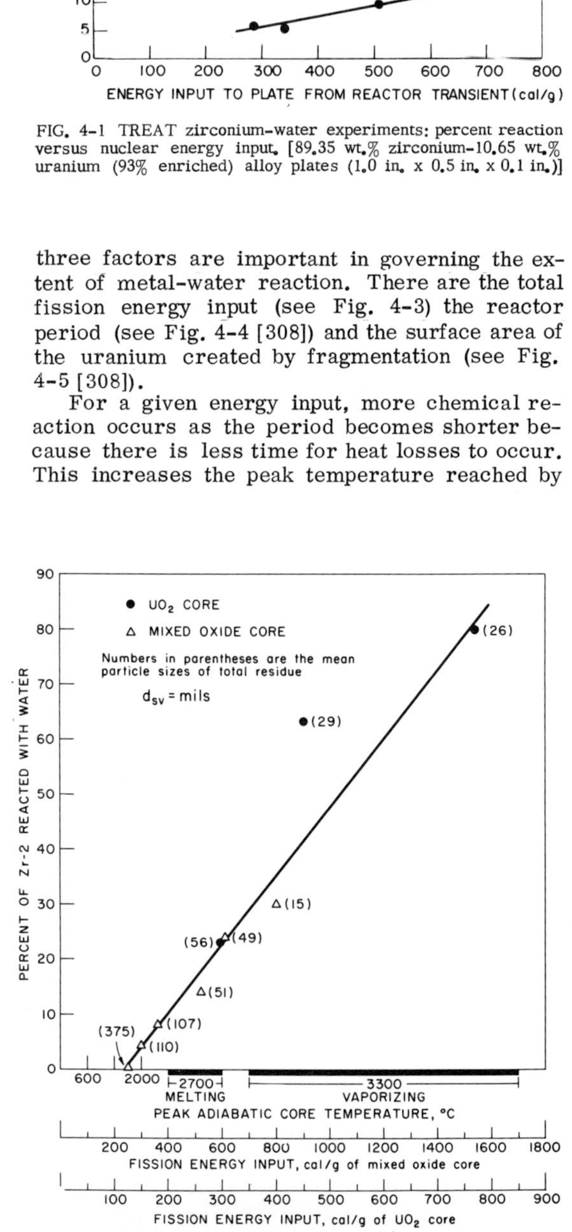

FIG. 4-2 Results of TREAT meltdown tests with Zircaloy-2 clad, urania and mixed oxide core fuel elements submerged in water.

TABLE 4-4

Results of Submerged Fuel Specimen Meltdown Tests in TREAT; Studies with Uranium

[Water initially at room temperature, initial helium overpressure of 20 psia (~ 1.3 atm).]
[Values marked with asterisk are estimated.]

Fission energy input (cal/g)	CEN run No.	TREAT characteristics		Appearance of specimen after transient	Mean particle diameter	Percent of metal reacted
		Period (msec)	Int. power (Mw-sec)			
Zircaloy-2 Clad, Uranium-Zirconium-Niobium Fuel Pins [Cladding: Zircaloy-2, 20 mils (0.51 mm) thick, ends capped. Core: 93.5 wt. % U, 5 wt. % Zr, 1.5 wt. % Nb, 20% enriched, 0.2 in. (0.51 cm) diam., 0.5 in. (1.27 cm) long.]						
59	14	102	56	Clad: ruptured Core: slightly melted	200 mils (5.08 mm)*	0.9
63	15	102	60	Clad: ruptured Core: partly melted	200 mils (5.08 mm)*	1.7
65	13	104	62	Clad: ruptured and bulged Core: partly melted	200 mils (5.08 mm)*	2.0
85	10	112	81	Clad: partly melted Core: melted and fragmented	200 mils (5.08 mm)*	2.0
95	16	96	90	Clad: melted and fragmented Core: melted and fragmented	148 mils (3.76 mm)	9.1
136	18	104	130	Clad: melted and fragmented Core: melted and fragmented	54 mils (1.37 mm)	7.4
136	17	103	130	Clad: melted and fragmented Core: melted and fragmented	—	7.3
179	11	100	170	Clad: melted and fragmented Core: melted and fragmented	91 mils (2.31 mm)	10.0
189	12	77	180	Clad: melted and fragmented Core: melted and fragmented	140 mils (3.56 mm)	8.6
Unclad Uranium-Zirconium-Niobium Fuel Pins [93.5 wt. % U, 5 wt. % Zr, 1.5 wt. % Nb, 20% enriched, 0.2 in. (0.51 cm) diam., 0.5 in. (1.27 cm) long.]						
37	68	100	35	Intact	844 mils (21.4 mm)*	0.1
79	2	103	75	1/2 melted, one globule	844 mils (21.4 mm)*	1.8
84	6	100	80	Melted, one globule	844 mils (21.4 mm)*	2.5
121	8	106	115	Melted, one globule	844 mils (21.4 mm)*	3.2
Unclad Pure Uranium Fuel Pins [20% enriched, 0.2 in. (0.51 cm) diam., 0.5 in. (1.27 cm) long.]						
50	5	125	48	2/3 Melted, one globule	844 mils (21.4 mm)*	0.2
54	3	197	52	3/4 Melted, one globule	844 mils (21.4 mm)*	1.1
Uranium Wire [93% enriched, 64 mil (0.163 cm) diam., 1.0 in. (2.54 cm) long.]						
108	43	515	32	Intact	191 mils (4.86 mm)*	0.4
146	44	304	43	Melted and fragmented	78 mils (1.98 mm)	8.1
146	42	267	43	Melted and fragmented	32 mils (0.81 mm)	7.5
149	45	205	44	Melted and fragmented	33 mils (0.84 mm)	9.3
183	41	96	54	Melted and fragmented	17 mils (0.43 mm)	17.2
305	57	152	90	Melted and fragmented	21 mils (0.53 mm)	50.2
339	58	440	100	Melted and fragmented	23 mils (0.58 mm)	33.2
Uranium Wire [93% enriched, 34 mil (0.086 cm) diam., 1.0 in. (2.54 cm) long.]						
379	35	96	47	Melted and fragmented	101 mils (2.58 mm)	28.3
403	46	96	50	Melted and fragmented	28 mils (0.71 mm)	12.9
410	70	1010	51	Intact	96 mils (2.44 mm)*	0

the metal. The behavior of uranium, as a function of fission energy input, shown in Fig. 4-3, indicated that there are three thermal zones. At energies less than 50 cal/g, the uranium remains in its original solid form with negligible chemical reaction. Above 50 cal/g, meltdown begins and at 100 cal/g the metal has fully melted with about 3% uranium-water reaction. It is important to realize that these meltdown boundaries apply strictly to periods in the vicinity of 100 msec because of heat loss considerations. Increasing the energy progressively to 400 cal/g results in a continual increase in the amount of oxidation until about one-half of the uranium has reacted with the water at a fission heat input of 400 cal/g. At this point the chemical to nuclear energy ratio is about 0.75, which is appreciable. The plot in Fig. 4-3 which expresses the extent of metal-water reaction as a function of fission energy input shows results very similar to those obtained in condenser discharge studies of metal-water reactions in room temperature water (see lower curve in Fig. 3-9) although the exact temperatures reached in the TREAT runs are presently not known.

Aluminum: Meltdown tests in TREAT have been performed with aluminum-uranium alloy plates both clad [313] and unclad [314] and aluminum-clad cermet core plates [315]. The results are given in Table 4-5. The extent of metal-water reaction is plotted as a function of fission energy input in Figs. 4-6 and 4-7. The results indicate that significant metal-water reaction begins when the fission energy reaches about 350 cal/g and increases gradually up to an energy input of about 500 cal/g. In 30°C (86°F) water at energies above 500 cal/g, breakup of both the alloy and cermet materials occurs and extensive chemical reaction takes place. In experiments with heated water (high pressure) at energies above 500 cal/g, breakup does not occur, but very extensive reaction takes place (75 to 95%). At these energies vapor-phase burning of aluminum is thought to begin. Laboratory experiments have indicated that aluminum ignition occurs at 1750°C (3272°F) (see Sec. 3.2.3). The breakup and the subcooled boiling that occurred in the 30°C water experiments apparently resulted in sufficient cooling to limit the duration of burning. In heated water, the combustion of the aluminum proceeded nearly to completion.

The average particle size is plotted as a function of fission energy input for the alloy plates in 30°C water [313, 314] in Fig. 4-8. The cermet plates [308] fragmented in 30°C water at energies above 500 cal/g; however, a single large particle that contained the ceramic portion of the fuel was formed. The aluminum was found in smaller fragments and as the reaction product α-Al_2O_3, in a fine powder. Photographs [313] of the residue from experiments with clad alloy (SPERT 1D) specimens, shown in Figs. 4-9 and 4-10, indicate increasing damage with increasing fission energy. The residues from the experiments with cermet fuel were similar except that the cermet material retained its plate-like shape when the energy was sufficient to cause

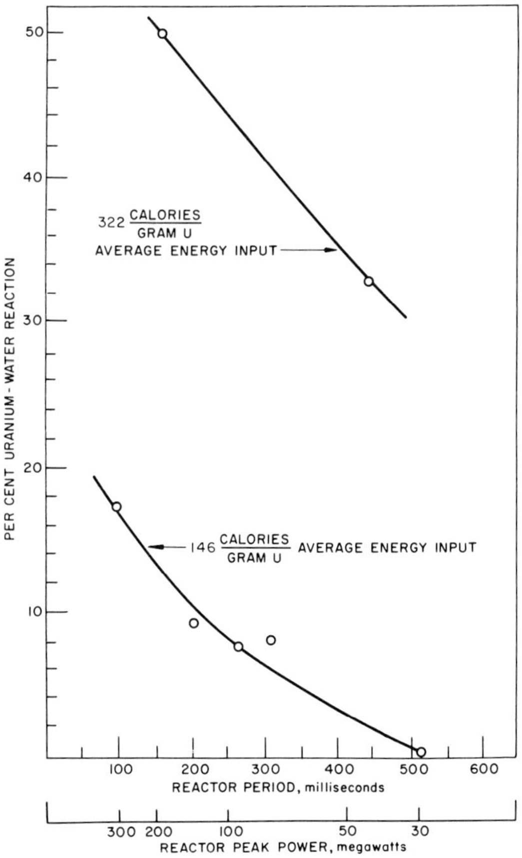

FIG. 4-4 In-pile, metal-water reactions with 64-mil (1.63 mm) diameter uranium wires, 93% enriched.

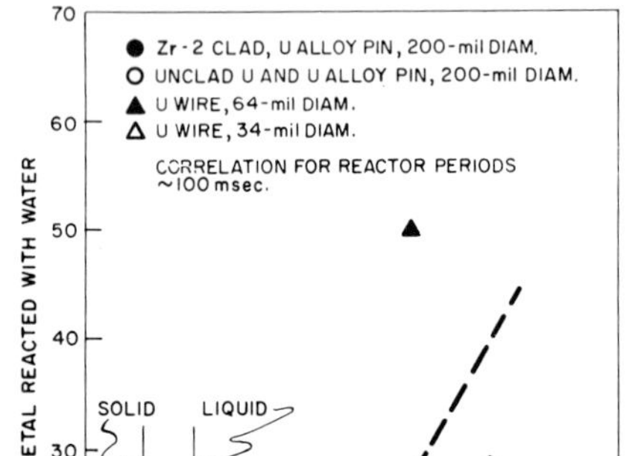

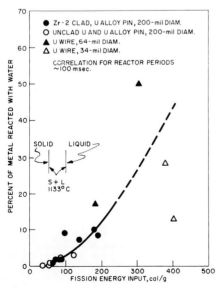

FIG. 4-3 Results of TREAT meltdown tests with uranium specimens.

TABLE 4-5

Results of Submerged Fuel Specimen Meltdown Tests in TREAT; Studies with Aluminum

[Water initially at room temperature, initial helium overpressure of 20 psia (~ 1.3 atm).]
[Values marked with asterisk are estimated]

Fission energy input (cal/g)	CEN run No.	TREAT characteristics: Period (msec)	TREAT characteristics: Int. power (Mw-sec)	Appearance of specimen after transient	Mean particle diameter	Percent of metal reacted
Unclad Aluminum-Uranium Alloy (SL-1 Core) Plates [81 wt. % Al, 17 wt. % U, 2 wt. % Ni, 0.2 in. (0.51 cm) x 0.5 in. (1.27 cm) x 0.5 in. (1.27 cm).]						
288	114	112	174	Melted and bulged	460 mils (11.7 mm)*	0.1
398	97	79	240	Melted, one globule	460 mils (11.7 mm)*	2.8
473	121	79	285	Melted, one globule	460 mils (11.7 mm)*	4.5
530	122	79	320	Melted, one globule	460 mils (11.7 mm)*	7.2
530	123	79	320	Melted, one globule	460 mils (11.7 mm)*	14.2
580	115	77	350	Melted and fragmented	110 mils (2.80 mm)	16.4
672	116	62	405	Melted and fragmented	42 mils (1.07 mm)	19.7
739	98	51	445	Melted and fragmented	—	22.2
880	137	54	530	Melted and fragmented	61 mils (1.55 mm)	27.4
1023	99	50	616	Melted and fragmented	—	54.2
1378	138	41	830	Melted and fragmented	12 mils (0.30 mm)	43.2
Water initially at 285° C, steam overpressure of 1000 psia (68 atm)						
395	141H	87	236	Melted, one globule	460 mils (11.7 mm)*	0.2
550	140H	86	348	One globule	460 mils (11.7 mm)*	76.0
Aluminum Clad, Aluminum-Uranium Alloy Plates (SPERT 1D) [Cladding: 6061 Al, 20 mils (0.51 mm) thick. Core: Al, 24 wt. % U, 93% enriched. Overall dimensions: 0.06 in. (0.153 cm) x 1.4 in. (3.55 cm) x 0.5 in. (1.27 cm).]						
174	109	104	169	Intact, slight melting	320 mils (8.13 mm)*	0.1
245	110	96	238	Melted and bulged	320 mils (8.13 mm)*	0.1
361	111	80	350	Melted, one globule	320 mils (8.13 mm)*	0.3
430	112	50	417	Melted and fragmented	250 mils (6.35 mm)*	3.9
527	117	52	512	Melted and fragmented	45 mils (1.14 mm)	11.0
577	118	50	560	Melted and fragmented	24 mils (0.61 mm)	11.0
652	113	53	633	Melted and fragmented	—	10.4
731	119	42	710	Melted and fragmented	22 mils (0.56 mm)	13.6
794	120	42	770	Melted and fragmented	19 mils (0.48 mm)	36.9
855	136	41	830	Melted and fragmented	83 mils (2.11 mm)	20.0
Water initially at 285° C, steam overpressure of 1000 psia (68 atm)						
390	142H	63	430	Melted, one globule	320 mils (8.13 mm)*	0.4
570	143H	46	650	One globule	320 mils (8.13 mm)*	89
590	177H	43	699	One globule	320 mils (8.13 mm)*	80
Aluminum Clad, Aluminum-Uranium Oxide Cermet Core Fuel Plates (HFIR) Cladding: 6061 Al, 10 mils (0.25 mm) thick. Core: Al-41.45 w/o U_3O_8, 93% enriched, 30 mil (0.76 mm) thick plate: 1.0 in. (2.54 cm) x 0.5 in (3.27 cm) x 0.05 in. (0.13 cm).						
133	198	152	53	Intact	—	0.2
335	200	148	134	Intact, bulged	—	1.2
420	199	109	168	Intact, bulged	—	1.4
425	201	109	170	Intact, ruptured	—	2.5
440	213	100	176	Intact, ruptured	—	2.2
645	204	109	258	Fragmented	—	74.6
1062	205	68	425	Fragmented	—	28.4
Water initially at 120° C, initial helium overpressure of 600 psia (41 atm)						
350	210H	113	140	Intact, bulged	—	0.3
550	211H	61	220	One globule	—	93.9
890	212H	64	360	One globule	—	92.2
Water initially at 285° C, steam overpressure of 1000 psia (68 atm)						
413	206H	106	165	Intact, ruptured	—	11.8
658	207H	104	263	One globule	—	91.3

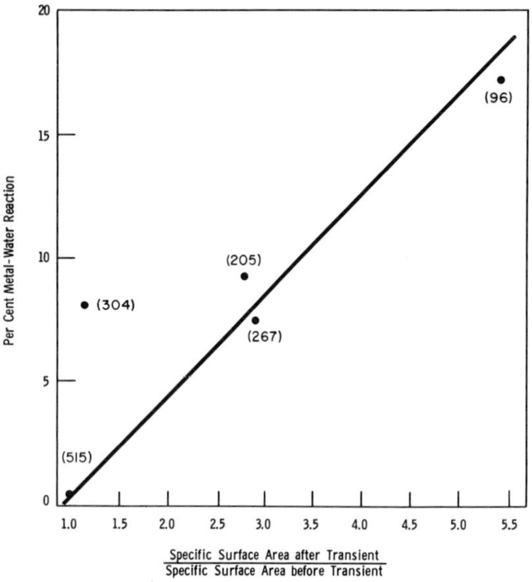

FIG. 4-5 Dependence of extent of uranium-water reaction on surface area for 64-mil (1.63 mm) diameter uranium wires, 93% enriched. Correlation for 146 cal/g or 263 Btu/lb average energy input. Reactor period in msec is indicated by numerals in parentheses.

aluminum melting (~230 cal/g) but not high enough to cause breakup (~500 cal/g). Residues of both materials showed evidence of large spherical pores indicating either interior gas evolution or entrapment of hydrogen or steam. The former appears very likely in the case of the cermet material. Large holes appeared in the surface of the plates at energies just below that required for breakup.

Stainless steel: Meltdown tests in TREAT have been performed [308] with specimens of unclad stainless steel-UO_2 cermet fuel pins and plates and with stainless steel-clad, mixed oxide and pure UO_2 core fuel pins. The results are summarized in Table 4-6.

Results obtained with the oxide core fuel pins are given in Fig. 4-11 and indicate that the stainless steel-water reaction is less extensive than that for zirconium. Comparison between the stainless steel-clad, mixed oxide core fuel pins and the pure urania core fuel pins indicated that there were no significant differences in either the nature of the damage or the extent of metal-water reaction. The threshold for fragmentation of the oxide core specimens corresponded approximately to the melting points of the ceramic core material (2600-2700°C or 4700-4900°F) for both the mixed oxide and the urania cores. A particularly striking example of this is shown by the macrograph of the cross section of a mixed oxide core fuel pin (see Fig. 4-12 [308]). Energy calculations indicated that the core reached a temperature of 2500°C (4532°F) while the macrograph indicated melting of the center region of the core. The stainless steel cladding remained intact during this transient. The nature of the damage produced in a very energetic transient with a urania core specimen is indicated in Fig. 4-13 [308].

Two experiments were performed with urania core specimens in which water was deliberately added to the core to simulate the case of a defected clad. In both cases, a hole developed in the cladding, even though the adiabatic core

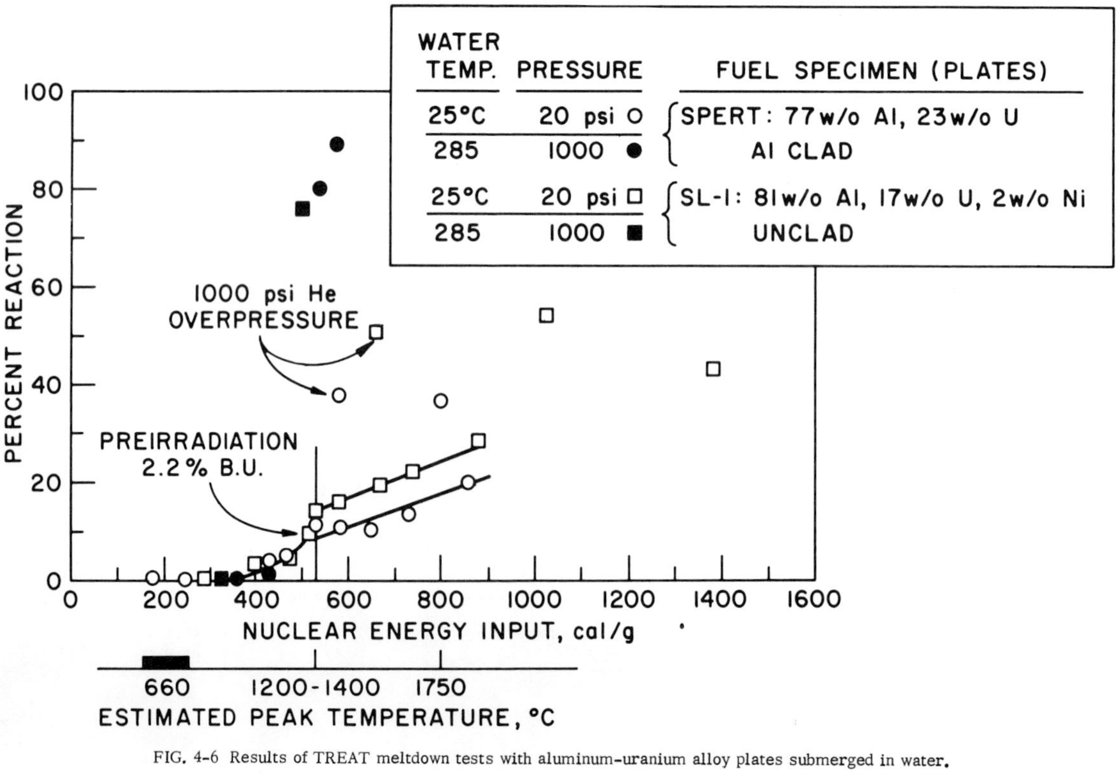

FIG. 4-6 Results of TREAT meltdown tests with aluminum-uranium alloy plates submerged in water.

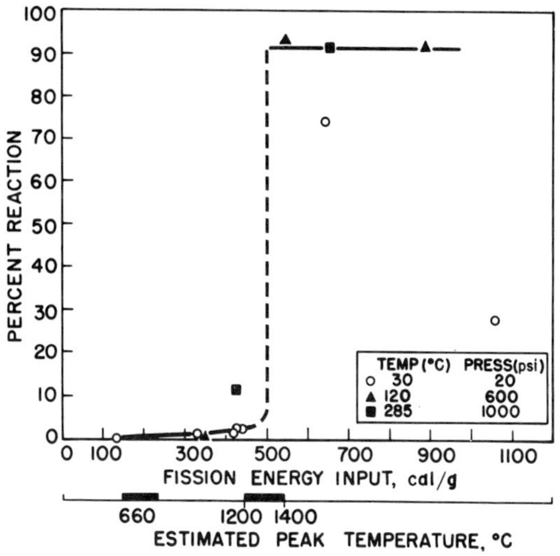

FIG. 4-7 Results of TREAT meltdown tests with aluminum-clad, aluminum-uranium oxide cermet core fuel plates (HFIR) submerged in water.

temperature was only 1300°C (2372°F) in one case.

4.4 Estimation of the Extent of Metal-Water Reaction During Reactor Accidents

The analysis of the rate and extent of metal-water reaction occurring during a reactor accident is ultimately a synthesis of reactor kinetic calculations and chemical reaction calculations. While considerable knowledge exists in both areas, no rigorous means has been developed to analyze an accident involving core meltdown. The principal stumbling block has been the uncertainty regarding the mechanism of fragmentation of the core and, in particular, the particle sizes produced. There appears to be no theoretical approach to the problem of predicting what particle sizes will be produced during a meltdown. Present knowledge in this area seems to be limited to the results obtained in small-scale meltdown experiments, particularly those conducted in TREAT. A direct combination of reactor kinetic and chemical reaction calculations for accidents in which the core remains intact appears feasible. although this case is of minor importance. Two general classes of accidents, the loss-of-coolant and the nuclear excursion accident are discussed in the following paragraphs.

Loss-of-coolant accidents: An attempt has been made to combine reactor kinetic and chemical reaction rate calculations [316]. Calculations have indicated that the heat generated by the oxidation of zirconium cladding significantly increases the cladding temperature as soon as decay heat brings the temperature up to about 1100°C (2000°F). The resulting effect causes varying amounts of clad oxidation prior to melting and a large variation of temperature between the hottest and coolest areas in the core of the hypothetical reactor. These calculations indicated that 5 mils (0.13 mm) out of a total of 30 mils (0.76 mm) of zirconium cladding would react in about 11 min before the cladding began to melt, a result

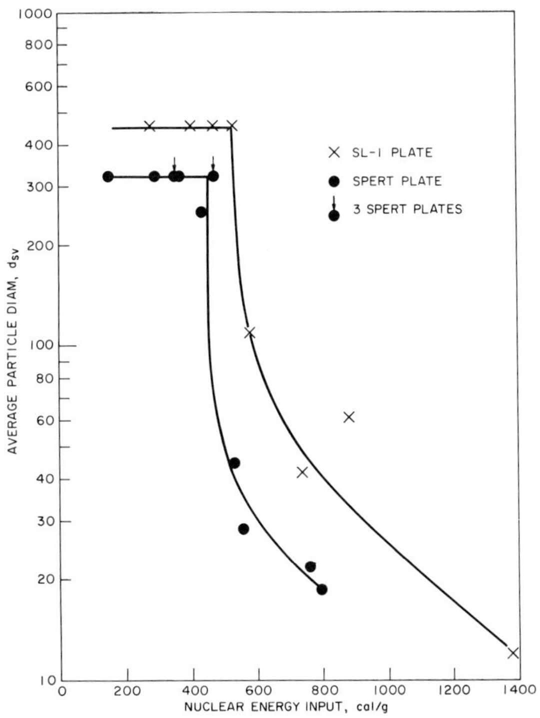

FIG. 4-8 Dependence of average particle diameter on energy of reactor transient for meltdowns of aluminum alloy fuel specimens. (Note 100 mil = 2.54 mm; 100 cal/g = 180 Btu/lb.)

similar to that obtained in analyzing a loss-of-coolant accident in the Shippingport reactor hazards evaluation [153].

A more detailed analysis [317] of a loss-of-coolant accident used the proposed LOFT (Loss of Fluid Test) reactor as a model except that the calculations were made for a 25-mil zirconium-clad core rather than the 15-mil stainless steel-clad core proposed in the LOFT design. The core was divided into ten power sections using the reported power distribution in the LOFT reactor. The rate of the zirconium-water reaction was calculated for each section using the parabolic rate law (Eq. (3-1) in differential form) with the assumption that sufficient steam would be available to react at the rates calculated. A constant rate equal to the supply of steam was used to calculate the reaction when limited amounts of steam were available. An energy balance equating the internal energy of each section was used to calculate the temperature-time history of the ten sections. A flat temperature profile through the fuel pins was assumed and heat losses were ignored. The presence of support structures was also ignored. The extent of reaction of the entire core was calculated by summing the contributions from each core section.

The extents of reaction at the time the core reached average temperatures equal to the melting point of the zirconium clad, 1850°C (3362°F), and of the uranium dioxide core, 2800°C (5072°F), were calculated as follows for the case where the initial core temperature, after blowdown, was uniform at 285°C (545°F).

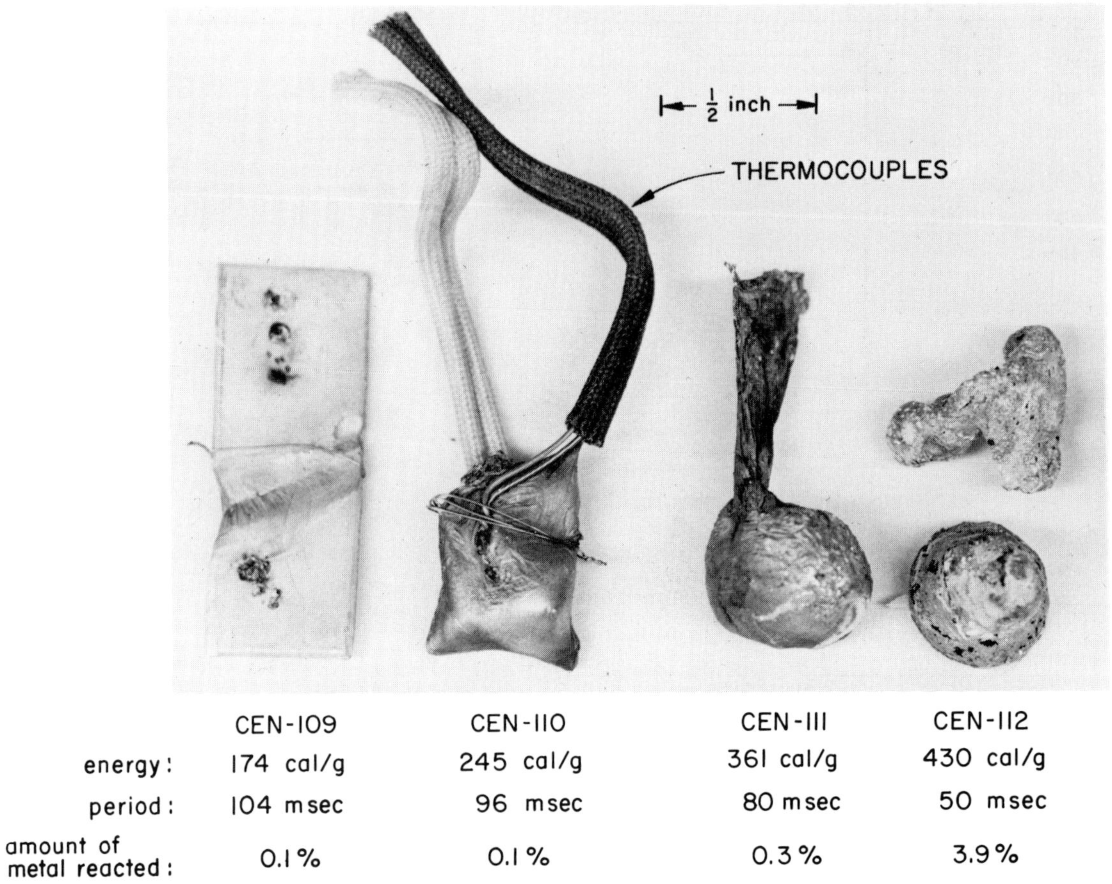

FIG. 4-9 Results of meltdown experiments in TREAT on SPERT fuel samples submerged in water. Original dimensions were 0.06 in. (0.152 cm) by 0.5 in. (1.27 cm) by 1.4 in. (3.56 cm).

1. With unlimited steam, 40 and 62% reaction at 465 and 765 sec, respectively,
2. With a 1000 lb/hr steam flow through the core, 21 and 41% reaction at 740 and 1080 sec, respectively,
3. With a 100 lb/hr steam flow through the core, 6 and 13% reaction at 1200 and 2090 sec, respectively, and
4. With no steam (no metal-water reaction), the average temperatures are reached after 1400 and 2600 sec, respectively.

Calculations were performed for another case where none of the energy stored on the fuel during full power operation was lost to the coolant during blowdown. In this case, portions of the core were at high temperatures immediately after the blowdown. This reduced the times corresponding to anticipated core failure and also reduced the extent of reaction somewhat. The strong interdependence of the core temperature history and the extent of reaction was demonstrated by the calculations.

Calculations of the course of a loss-of-coolant accident in a power reactor indicate that without effective emergency cooling, the core will eventually collapse. As the melted core collapses, further metal-steam reaction will take place as the core melts through the support structures. Additional reaction will also occur as the core falls into the bottom of the reactor vessel if water is present, or during a long cooling period of the consolidated mass at the bottom of the reactor vessel if little or no water is present.

The temperature at which each section of a reactor core will collapse is not known with certainty. Insight into the nature of failure has been provided by recent experiments [318] in a high-temperature steam furnace. Samples (8 in. long by 0.4 in. OD) consisted of a number of high-density UO_2 pellets clad with either type-304 stainless steel or Zircaloy-2 tubing having a 27-mil wall thickness. The samples were elevated into the hot zone of the furnace (maximum temperature 1600°C (2912°F)) over a period of 8 min, held there for 2 min, and rapidly removed. Steam flowed continuously past the samples at a velocity of about 3.2 ft/sec. The appearance of the samples after exposure is shown in Fig. 4-14. The top one-third of the stainless steel cladding (portion reaching between 1400 and 1600°C) appeared to be completely reacted, whereas the remainder had reacted only slightly. The foaming and swelling characteristic of oxidized stainless steel is apparent in Fig. 4-14. Considerable interaction was noted between the stainless steel oxide and the UO_2 pellets. In other studies, it has been shown that UO_2-stainless steel oxide mixtures can form mixtures melting at temperatures as low as 1435°C (2615°F) [319].

For the fuel element clad with Zircaloy-2, the cladding at the hottest end (1600°C) was about one-half reacted, as shown in Fig. 4-14. The extent of reaction was progressively less at lower (cooler) sections of the element. The overall extent of reaction of the Zircaloy-2 in the fuel element was 18% for the 10-min exposure cycle. The hot upper end of the sample was embrittled and had failed by cracking. There was no apparent interaction between the UO_2 and the Zircaloy. It has been shown in other studies, however, that mixtures of ZrO_2, Zr, and UO_2 form solid solutions melting as low as 2150°C (3902°F), well below the melting point of UO_2 itself (2800°C or 5072°F) [319].

Attempts to predict the metal-water reaction occurring after core failure were made for the PWR reactor [153]. The minimum particle size of droplets of zirconium was estimated from the metal surface tension and the smallest thickness of metal in the core. It was assumed that molten

FRAGMENTED PLATE AFTER TRANSIENT

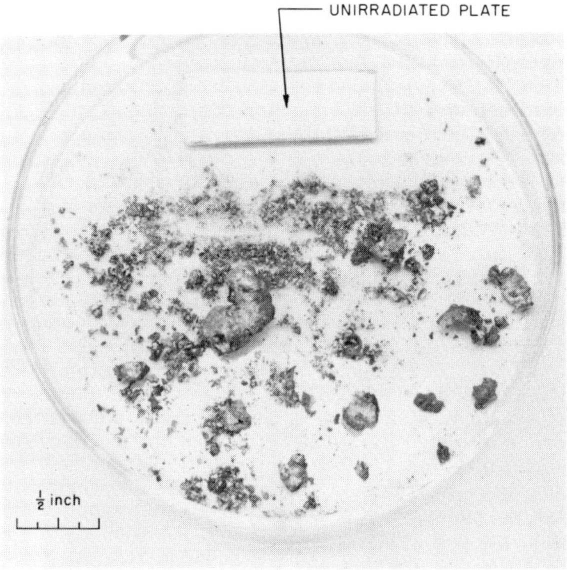

MACROGRAPH OF FRAGMENT

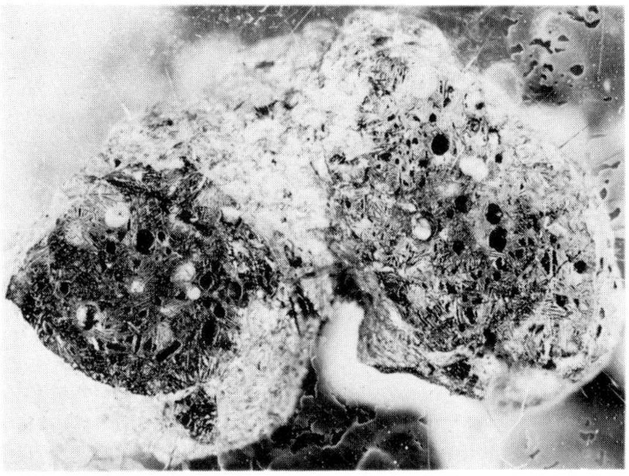

5X

PARTICLE SIZE DISTRIBUTION

d_{sv} = 44.6 mils

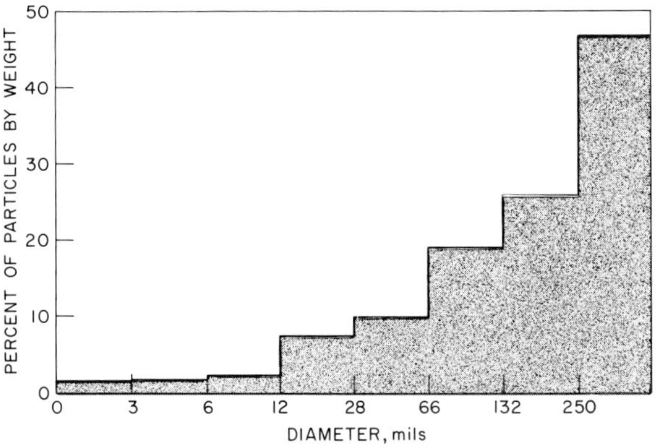

MICROGRAPH OF FRAGMENT

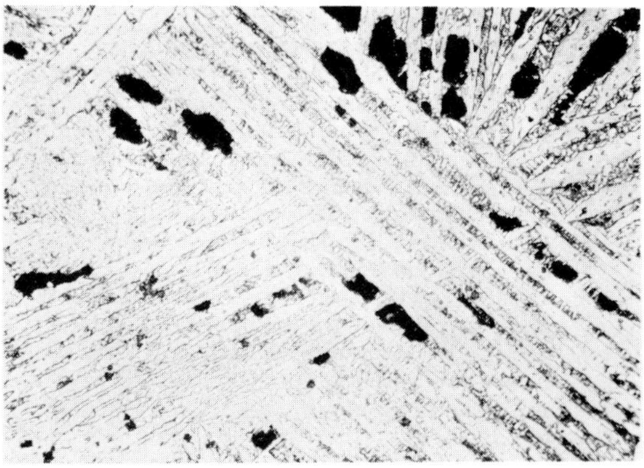

125X

FIG. 4-10 Results of meltdown experiments in TREAT on SPERT fuel samples submerged in water. Energy = 527 cal/g (950 BTU/lb), period = 52 msec, amount of metal reacted = 11%, Test CEN-117. Magnification: macrograph, 4.6x, micrograph, 115x.

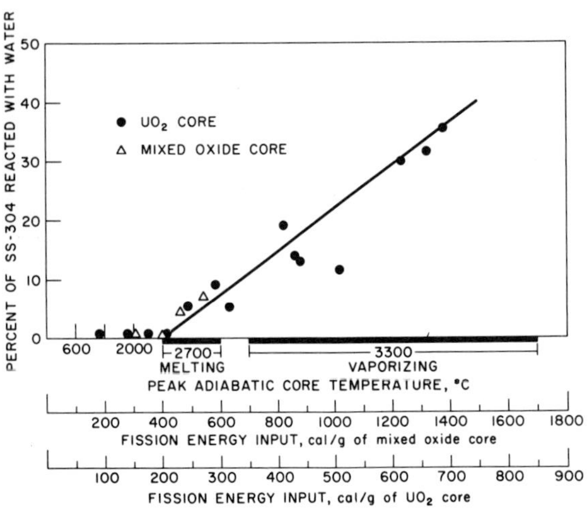

FIG. 4-11 Results of meltdown experiments in TREAT on urania and mixed oxide core, type-304 stainless steel clad fuel pins submerged in water.

droplets disengage at a diameter at which the weight of the droplet equals the surface tension holding the droplet. This gave a minimum droplet diameter of 2.38 mm, which was estimated to yield from 2.7 to 10.6% reaction on quenching in water. More recent calculations of the reaction of molten zirconium spheres with water would indicate about 9% reaction for a 2.38 mm droplet (see Fig. 3-6). The quenching reaction of molten zirconium in water was also simulated to some extent by TREAT experiments with a zirconium-uranium alloy. Reaction was limited to about 12% (see Table 4-3 and Fig. 4-1) unless the metal was heated to beyond about 2600°C (4712°F), where extensive fragmentation and ~70% reaction can occur.

Collapse of a major portion of a reactor core at one time into a water pool at the vessel bottom could lead to a violent steam explosion, if there is any parallel to the explosive reaction when molten aluminum is poured into water [177]. It is unlikely, however, that a large collapse would occur at one time. Because the energy generated at the center of the reactor is greatest, the supporting structures in this region would probably be the first to fail; the failure would then spread radially. In this case, core debris would enter the water over a period of time, minimizing the chance of a serious explosion.

Nuclear excursion accidents: Experience gained from previous full-scale excursions, excluding the NRX incident, is limited to cores containing uranium-aluminum alloys and excursions having periods between 2.6 and 4.0 msec. The results of these excursions (BORAX-1, SL-1, and SPERT-1) seem consistent with the conclusions reached during an evaluation of severe nuclear excursions in light water-cooled reactors [320]. It was concluded that for a flat plate type metallic aluminum fuel, melting would begin to occur only when the period of the excursion was 4.9 msec or less. It was also concluded that complete melting would occur for an excursion having a period of 3.3 msec or less, although this appears to ignore the wide flux variation from the edge of a reactor to the center. For a metallic zirconium core, the limiting periods were 12.2 and 8.6 msec, respectively. The narrow gap in excursion periods between no melting at all and extensive melting implies that extensive melting and the possibility of a massive metal-water reaction must be considered whenever the insertion of sufficient reactivity to achieve these short periods is credible. The differences in period between an aluminum core and a zirconium core resulted from the poorer thermal diffusivity of zirconium in comparison with aluminum. Steam void formation, the principal shutdown mechanism, was slower in the zirconium case.

A low enrichment zirconium-clad, UO_2 core fuel element was shown [320] to have a long thermal relaxation time so that Doppler broadening was the principal shutdown mechanism. Under these conditions it was concluded that the zirconium cladding would not reach the melting temperature and consequently there was little likelihood of a massive metal-water reaction.

These calculations of nuclear excursions did not include terms to describe chemical heating or effects resulting from meltdown or fuel fragmentation [320]. At present, it appears that the nuclear characteristics of an excursion must be calculated separately from the chemical characteristics. Results of small-scale meltdown experiments can then be used to estimate fuel fragmentation patterns and the extent of metal-water reaction. TREAT meltdown studies with Zircaloy, aluminum, and stainless steel-clad oxide core fuel elements were correlated in terms of the adiabatic core temperature. Adiabatic core temperature (calculated from total fission energy input) would result from the nuclear analysis of the maximum credible excursion in a hazards analysis. The average particle sizes of residue and the extent of reaction can be estimated from Tables 4-3, 4-5, and 4-6 and from Fig. 4-2. The TREAT data apply strictly to excursions having periods between 40 and 100 msec. However, more rapid excursions would probably not differ greatly because of the lag in transferring heat from the oxide core to the clad.

TREAT studies with cermet core and metallic core fuel elements were correlated in terms of the energy input to the entire specimen and can be used directly only for excursions having periods in the region of 40-100 msec. More rapid excursions can be treated, approximately, on the assumption that the heating would be nearly adiabatic. In this case, it is necessary to estimate peak temperatures actually reached in TREAT experiments. This is done presently by noting indications of melting and changes in the character of reaction. Regions of solid and melted residue in the TREAT studies are indicated in Figs. 4-1, 4-3 and 4-7. The sharp jump in the zirconium reaction in Fig. 4-1 may be tentatively identified with the melting point of ZrO_2 (2600°C or 4712°F) which has been shown in laboratory studies to result in greater subdivision of particles and more extensive reaction in room-temperature water (see Fig. 3-11). The threshold for extensive aluminum-water reaction at about 750 cal/g in Fig. 4-7 may similarly be identified with the temperature (1750°C or 3182°F) at which aluminum ignites and burns in water as determined in laboratory studies. It should be noted that the TREAT results are currently under study and it is anticipated that a more rigorous estimate of the temperatures reached will be obtained from computer simulation of the experiments. It is also hoped that these studies will result in better methods for estimating the results of postulated reactor accidents.

An important consideration in applying TREAT results to nuclear excursion accidents is the wide variation of integrated flux across a reactor. The reactor must, therefore, be segmented in some way into regions of constant maximum temperature or fission energy input. The extent of metal-water reaction can then be estimated to a first approximation for each segment and the results summed to yield a total for the incident.

The rate of chemical reaction occurring during a nuclear excursion is more difficult to estimate than the total extent of reaction. The overall reaction rate depends primarily upon whether the

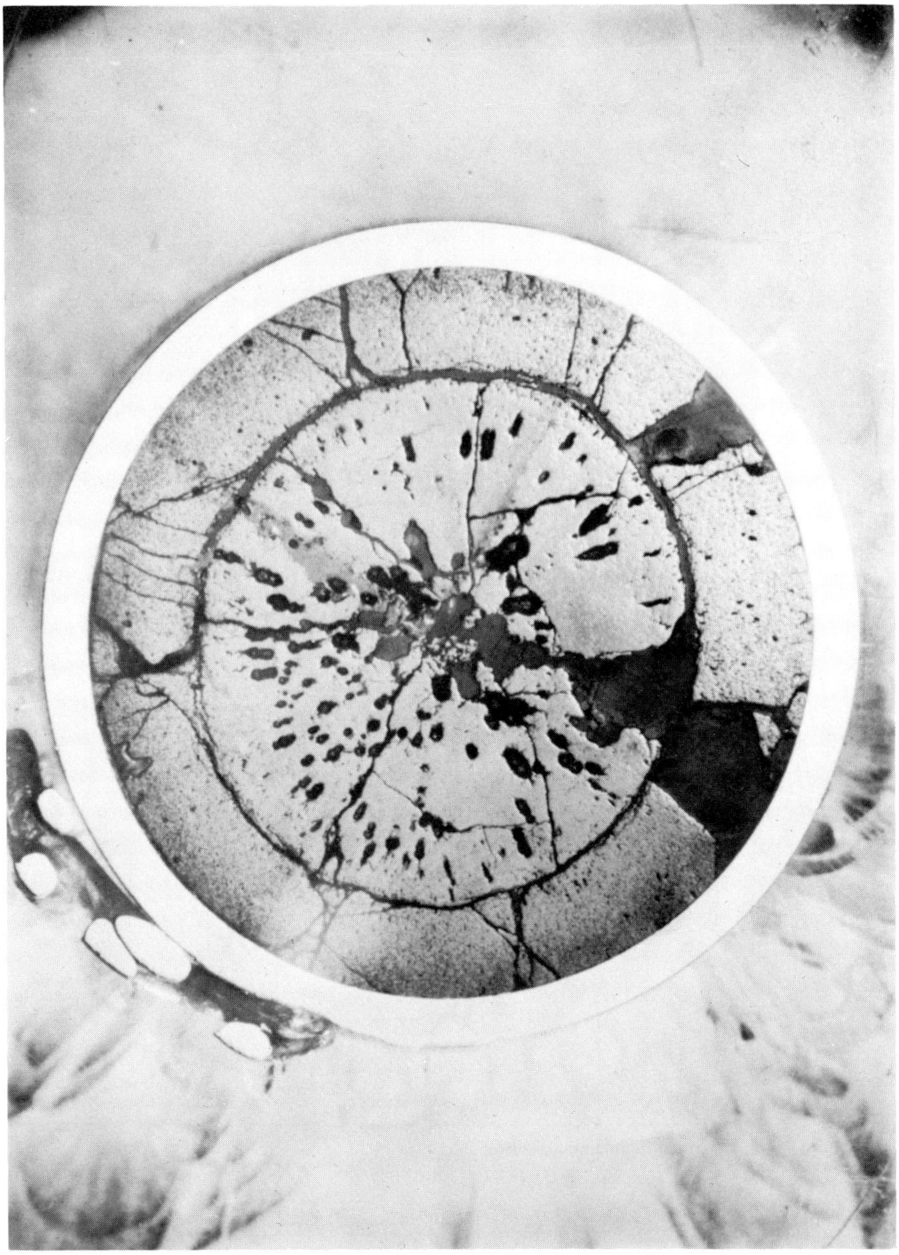

FIG. 4-12 Results of meltdown experiment in TREAT on an oxide core stainless steel-clad fuel pin submerged in water. Cross-section of fuel pin. Energy 395 cal/g (711 Btu/lb), period = 78 msec, and peak adiabatic core temperature = 2500°C (4530°F). Test CEN-21. Magnification 10x.

TABLE 4-6

Results of Submerged Fuel Specimen Meltdown Tests in TREAT; Studies with Stainless Steel

[Water initially at room temperature, except as noted, initial helium overpressure of 20 psia (~ 1.3 atm).]
[Values marked with asterisk are estimated]

Fission energy input (cal/g)	Adiabatic core temperature[a]	CEN run No.	TREAT characteristics: Period (msec)	TREAT characteristics: Int. power (Mw-sec)	Appearance of specimen after transient	Mean particle diameter	Percent of metal reacted[b]
Unclad SS-304 Uranium Oxide Cermet Fuel Pins [90 wt. % SS-304, 10 wt. % UO_2, 93% enriched, 0.34 in. (0.86 cm) diam., 0.50 in. (1.27 cm) long.]							
522	—	51	50	435	Fragmented	33 mils (0.84 mm)	6.6
612	—	50	50	510	Fragmented	24 mils (0.61 mm)	9.6
615	—	52	49	512	Fragmented	40 mils (1.02 mm)	10.2
Unclad SS-304 Uranium Oxide Cermet Fuel Plates [90 wt. % SS-304, 10 wt. % UO_2, 93% enriched, 0.5 in. (1.27 cm) x 1.0 in. (2.54 cm) x 0.1 in. (0.25 cm).]							
550	—	54	51	368	Fragmented	10 mils (0.25 mm)	5.2
725	—	53	52	490	Fragmented	11 mils (0.28 mm)	9.1
732	—	55	51	495	Fragmented	17 mils (0.43 mm)	11.0
SS-304 Clad, Mixed Oxide Core Fuel Pins [Cladding: SS-304, 20 mils (0.51 mm) thick, ends capped. Core: 81.5 wt. % ZrO_2, 9.1 wt. % CaO, 8.7 wt. % U_3O_8, 0.74 wt. % Al_2O_3, 0.33 in. (0.84 cm) diam., 0.50 in. (1.27 cm) long.]							
296	1900°C (3452°F)	40	60	315	Clad: intact Core: cracked	330 mils (8.4 mm)*	0
395	2500°C (4432°F)	21	78	420	Clad: intact Core: partly melted, cracked	330 mils (8.4 mm)*	0
432	2600°C (4712°F)	48	72	460	Clad: partly fragmented Core: fragmented	15 mils (0.38 mm)	4.3
536	3200°C (5792°F)	24	65	560	Clad: partly fragmented Core: fragmented	—	6.6
SS-304 Clad, Urania-Core Fuel Pins [Cladding: SS-304, 20 mils (0.51 mm) thick. Core: 20% enriched UO_2, 0.33 in. (0.84 cm) diam., 0.50 in. (1.27 cm) long.]							
152	2000°C (3632°F)	63	121	120	Clad: intact Core: cracked	330 mils (8.4 mm)*	0.3
215	2700°C (4892°F)	67	290	185	Clad: intact Core: cracked	330 mils (8.4 mm)*	0.3
242	2700°C (4892°F)	65	50	190	Clad: fragmented Core: fragmented	4 mils (0.10 mm)	5.2
292	2900°C (5252°F)	66	97	230	Clad: fragmented Core: fragmented	23 mils (0.58 mm)	9.0
312	3100°C (5612°F)	69	64	245	Clad: fragmented Core: fragmented	2 mils (0.05 mm)	5.4
407	3300°C (5972°F)	61	116	320	Clad: fragmented Core: fragmented	17 mils (0.43 mm)	19.1
432	3300°C (5972°F)	62	49	340	Clad: fragmented Core: fragmented	13 mils (0.33 mm)	14.0
437	3300°C (5972°F)	60	48	344	Clad: fragmented Core: fragmented	11 mils (0.28 mm)	13.5
510	3300°C (5972°F)	71	43	400	Clad: fragmented Core: fragmented	8 mils (0.20 mm)	11.5
(Core saturated with water before test)							
97	1300°C (2372°F)	72	110	76	Clad: partly fragmented Core: cracked	330 mils (8.4 mm)*	0.7
178	2300°C (4172°F)	64	115	140	Clad: partly fragmented Core: cracked	330 mils (8.4 mm)*	0.8

[a] The value 3300°C (5927°F) corresponds to the boiling point of UO_2 at 1 atm.
[b] Percent reaction based on 0.40 liter (STP) of H_2 per gram of stainless steel.

reactive metal in the core is dispersed suddenly or over a period of time. Pressure transducers in the TREAT studies indicated that there were no violent pressure surges. Laboratory studies by the electrical condenser discharge method indicated that very rapid reactions occurred when very fine metal droplets were formed. Since fine particles were also formed in the TREAT studies,

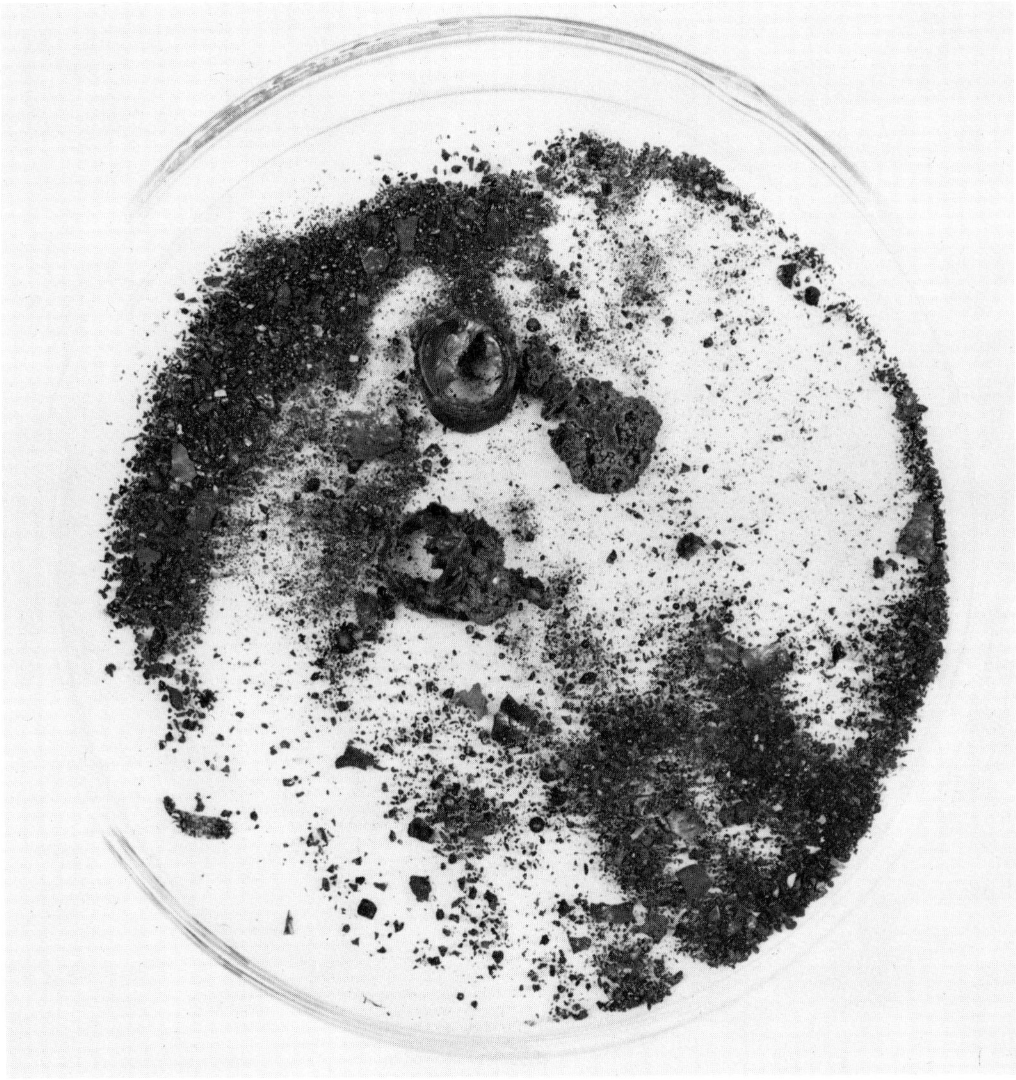

FIG. 4-13 Results of meltdown experiments in TREAT on oxide core stainless steel clad fuel pins submerged in water. Energy = 432 cal/g (779 Btu/lb), period = 49 msec, peak adiabatic core temperature = 4600°C (8300°F), amount of metal reacted = 14.0%. (Test CEN-62).

it appeared that the particles were formed over a period of time so that, while each fine particle reacted rapidly, the overall reaction was relatively slow. It has been suggested that the rate of reaction of the Zircaloy cladding on oxide core fuel elements could be estimated on the assumption that the particle formation rate is equal to the rate at which the cladding is melted [154]. Each particle would then react to the extent indicated in Fig. 3-6. Particles smaller than about 0.5 mm (20 mils) in room-temperature water and about 1.0 mm (39 mils) in heated water would react in a few milliseconds. Larger particles would react at rates indicated in Fig. 3-5.

A comparison between the extent of zirconium-water reaction indicated in Fig. 3-6 and results of TREAT runs with Zircaloy-2 clad, mixed oxide core pins is given in Table 4-7 [154]. It was tentatively assumed that the distribution of particle sizes of oxide and cladding metal together represented the distribution of metal particles alone. The anticipated reaction of each particle size group was then summed to give the total extent of reaction. The comparison of calculated reaction with

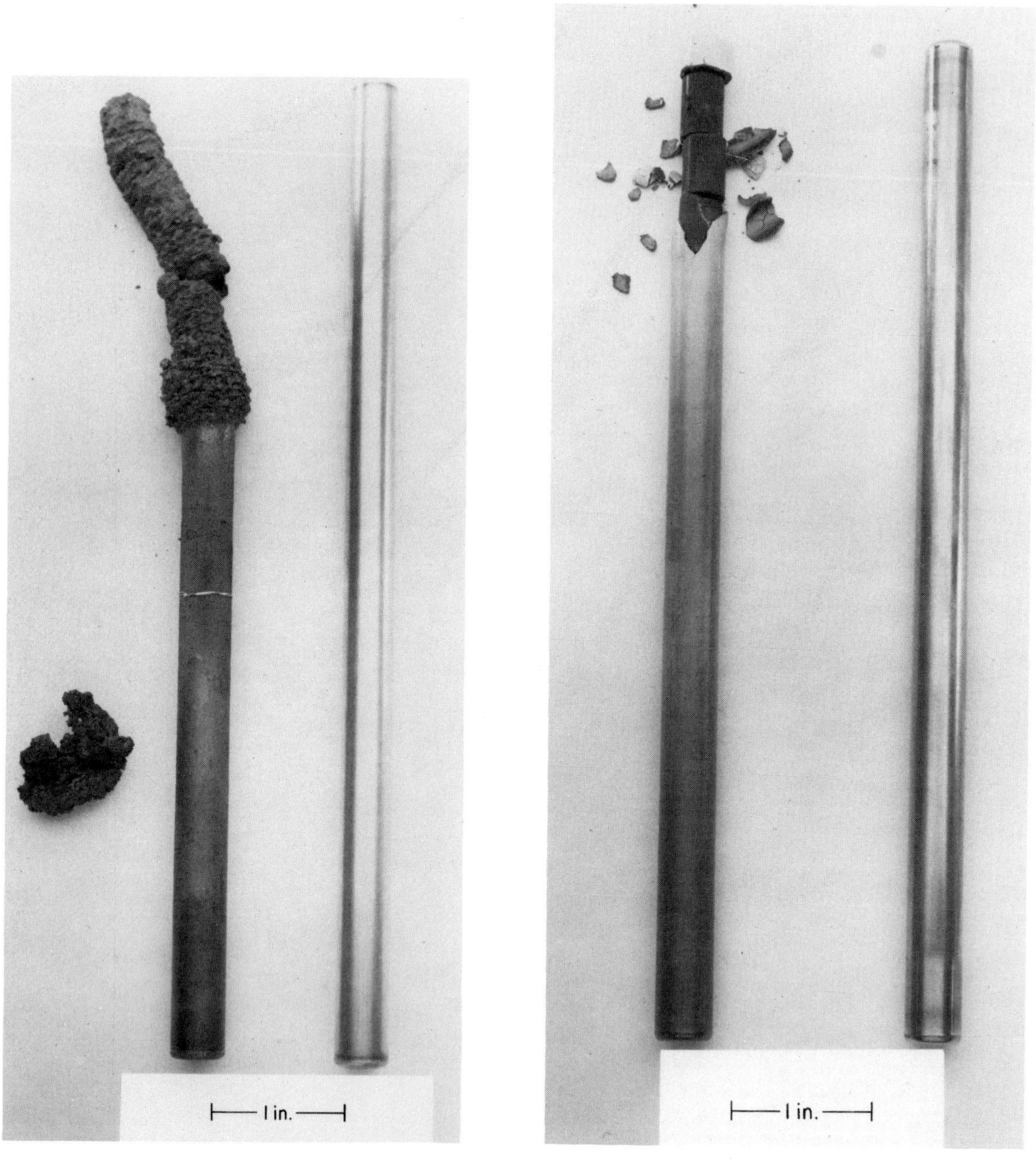

FIG. 4-14 Appearance of simulated fuel pins before and after exposure to 1 atm steam at 1600°C for 10 min. At left: type-304 stainless steel clad, UO_2 core fuel pins. At right: Zircaloy-2 clad, UO_2 core fuel pins.

TABLE 4-7

In-pile Metal-Water Experiments in TREAT

Water at room temperature
Core material: mixed oxide (composition in wt. %: 81.5 ZrO_2, 9.1 CaO, 8.7 U_3O_8, 0.7 Al_2O_3)
Cladding: Zircaloy-2 (20 mils or 0.51 mm thick)
Overall diameter: 0.38 in. (0.965 cm)
Overall length: 1.05 in. (2.67 cm)

Reactor characteristics	CEN Transient No.			
	28	29	30	49
Burst (Mw-sec)	320	385	550	648
Period (msec)	60	63	62	50
Energy (cal/g of oxide core)	301	362	517	610

Particle diameter, reaction data

Size group (mils)	Theoretical % reaction for each size group [a]	Particle size distribution (wt. %) for total sample (metal and oxide)			
1-4	70	0.0005	0.0002	0.1	0.03
4-8	60	0.01	0.024	1.3	0.2
8-16	46	0.2	0.10	6.5	2.8
16-32	28	2.1	3.0	2.0	15.2
32-64	14	11.2	9.7	12.8	25.2
64-128	9	22.3	27.8	77.3	57.6
128-256	6	64.3	59.4	0	0
256-512	-	0	0	0	0

Percent of metal reacted with water

Calculated	8.1	8.3	13.1	14.4
Experimental	4.1	8.0	14.0	24.0

[a] Data taken from Fig. 3-6.

experimental values (based on the measurement of evolved hydrogen) seems very reasonable in view of the many complicating factors.

There is a lack of experience regarding excursions initiated in a reactor operating at full power. Heat and mass transfer effects in saturated water differ markedly from those in subcooled water. Laboratory studies (see Sec. 3.2) and preliminary studies in TREAT [311] have indicated that the zirconium-water reaction and the uranium-water reaction are more extensive in heated water while the stainless steel-water reaction is nearly independent of water temperature. It has been shown that increasing the water temperature (steam pressure) causes a marked increase in the amount of aluminum-water reaction.

The following example is given to illustrate a method of estimating the extent of metal-water reaction which occurred during the SL-1 and SPERT-1 meltdowns. The calculation is based on TREAT meltdown studies summarized in Fig. 4-7. The following two assumptions were made:

1) The portion of core which became molten was assumed to be spherical with x = 0 at the center and x = 1 at the edge. Therefore, the volume = $4\pi/3$.

2) The energy density was assumed to vary linearly with the distance from the center of the sphere. This assumption is probably valid as a first approximation. This relationship was expressed as: $E_x = E_{max} - Ax$.

The maximum energy densities were taken from the literature[302, 321]as: E_{max}(SL-1) = 500 cal/g and E_{max}(SPERT) = 380 cal/g. The energy density at the edge of the sphere (x = 1) was taken as 220 cal/g, the energy required to heat aluminum to its melting point and completely melt it (energies based on E = 0 at 25°C or 72°F). The equations for energy distribution were then determined to be:

$$E_x = 500 - 280x, \text{ for SL-1,} \tag{4-1}$$

$$E_x = 380 - 160x, \text{ for SPERT-1.} \tag{4-2}$$

These equations were then used to calculate the energy distribution in the sphere by calculating the distance from the center of the sphere (x) that achieved energies corresponding to various temperatures. The fraction of sphere attaining the selected temperature or higher was then calculated as x^3. In this way the amount of metal that achieved various energy levels was calculated. This was done in steps of 100°C (180°F).

The extent of reaction was then determined from the TREAT data. This was done by assuming the following:

1) that the TREAT energy input of 220 cal/g corresponds to the complete melting of the material,
2) that the break point of 530 cal/g in the TREAT data corresponds to the temperature range of 1200 to 1400°C (2192 to 2552°F), and

TABLE 4-8

Calculation of Aluminum-Water Reaction in the SL-1 Excursion
(Based on 60 kg of molten core)

T	E(cal/g)	x	x^3	W(kg)	ΔW(kg)	Rx(%)	Metal reacted (kg)
1700°C (3092°F)	500	0	0	0	—	—	—
1600°C (2912°F)	474	0.093	0.0008	0.048	0.048	19.9	0.010
1500°C (2732°F)	448	0.186	0.0066	0.395	0.347	17.2	0.060
1400°C (2552°F)	422	0.279	0.023	1.39	0.99	11.0	0.109
1300°C (2372°F)	396	0.372	0.051	3.06	1.67	11.0	0.184
1200°C (2192°F)	368	0.472	0.105	6.29	3.23	11.0	0.355
1100°C (2012°F)	342	0.564	0.180	10.8	4.5	7.0	0.315
1000°C (1832°F)	316	0.658	0.284	17.0	6.2	3.4	0.211
900°C (1652°F)	290	0.750	0.422	25.3	8.3	1.2	0.100
800°C (1472°F)	264	0.843	0.600	35.9	10.6	0.6	0.064
700°C (1292°F)	238	0.935	0.817	48.9	13.0	0.2	0.026
660°C (1220°F)	220	1.000	1.000	60.0	11.1	0.1	0.011
					Total metal consumed		1.445

Thus: 1.445 kg x 18 Mw-sec/kg = 26.0 Mw-sec chemical energy.

TABLE 4-9

Nuclear Excursions of the SL-1 and SPERT Reactors

	SL-1	SPERT
Excursion Parameters		
Reactor period (msec)	~4	3.2
Integrated power (Mw-sec)	130	31
Peak energy density (cal/g)	500	380
Core Parameters		
Aluminum present in core (kg)	185	51
Uranium present in core (kg)	14	4
Peak core temperature	~1700°C (3092°F)	1200-1300°C (2190-2370°F)
Percent of core melted	32	35
Chemical Parameters		
Observed chemical energy release (Mw-sec)	24 ± 10	3.5
Calculated chemical energy release (Mw-sec)	26	2

3) that the marked increase in extent of reaction at 750 cal/g in the TREAT data corresponds to a temperature of 1750°C (3182°F).

The temperatures between these points were linearly interpolated. These assumptions were based on physical observations and metallographic examinations of the TREAT samples and on laboratory data. The graph in Fig. 4-6 has both TREAT energies and estimated temperatures on the abscissa. These calculations are summarized in Tables 4-8 and 4-9.

The estimates obtained in this manner are in excellent agreement with the values obtained by the postexcursion analyses of the SPERT and SL-1 debris. In both cases estimates of the extent of aluminum-water reaction were made from determinations of the amount of α-alumina present in the debris. The α-alumina analysis method yielded 3.5 Mw-sec of chemical energy release for SPERT [306] and 24 ± 10 Mw-sec for SL-1 [302]. The calculation presented here yielded values of 2 Mw-sec for SPERT and 26 Mw-sec for SL-1.

5 MISCELLANEOUS CHEMICAL PROBLEMS

5.1 Low-Melting Eutectics

One potential hazard associated with the operation of nuclear reactors is the slow alloying of two adjacent solid materials which might result in a low melting solution. The situation is most likely to occur when fuel materials are canned in a metallic container either as part of a reactor or as part of an experimental assembly for use in a test reactor or critical facility. The gradual formation of a low melting solution is likely to lead to a sudden and unexpected failure in the event of a moderate temperature excursion. The possibility of a reaction or mutual dissolution should be considered whenever two solid materials are to be placed in contact with each other for an extended period of time.

Eutectic and minimum melting temperatures for binary metallic alloy systems of potential interest are given in Table 5-1 [322]. Values are given only when the eutectic or minimum melting temperature is significantly lower than the melting point of both of the pure metals. Probably the most important result to be gleaned from Table 5-1 is the fact that iron (or steels) form very low melting eutectic alloys with plutonium, uranium, and thorium. Melting temperatures of mixtures of UO_2 with other oxides are given in Table 5-2 [323] and show that notable depression of the melting point occurs when UO_2 is mixed with oxides of vanadium, molybdenum, or tungsten.

5.2 Aluminum-U_3O_8, Thermite Type Reactions

The need for aluminum-base dispersion fuel elements for high flux reactors has led to the development of aluminum-uranium oxide cermets. The 1955 Geneva Conference Reactor (GCR) used an Al-UO_2 fuel. Great difficulty was experienced

TABLE 5-1

Lowest Eutectic or Minimum Melting Temperatures for Binary Metallic Alloys [a]

(MP = melting point)

Plutonium	(MP 640°C or 1184°F)	
Pu-Be	595°C	(1103°F)
Pu-Cr	615°C	(1139°F)
Pu-Fe	410°C	(770°F)
Pu-Ni	475°C	(887°F)
Magnesium	(MP 649°C or 1200°F)	
Mg-Al	437°C	(819°F)
Mg-Cu	485°C	(905°F)
Mg-Ni	507°C	(945°F)
Mg-Th	596°C	(1105°F)
Aluminum	(MP 660°C or 1220°F)	
Al-Th	632°C	(1170°F)
Al-U	640°C	(1184°F)
Uranium	(MP 1133°C or 2071°F)	
U-Co	734°C	(1353°F)
U-Cr	859°C	(1578°F)
U-Cu	950°C	(1742°F)
U-Fe	725°C	(1337°F)
U-Mn	716°C	(1321°F)
U-Ni	740°C	(1364°F)
U-Th	1086°C	(1987°F)
U-V	1040°C	(1904°F)
Beryllium	(MP 1284°C or 2343°F)	
Be-Cu	866°C	(1591°F)
Be-Fe	1165°C	(2129°F)
Be-Ni	1157°C	(2115°F)
Be-Th	1215°C	(2219°F)
Be-Ti	~950°C	(~1740°F)
Be-Zr	980°C	(1796°F)
Nickel	(MP 1453°C or 2647°F)	
Ni-B	990°C	(1814°F)
Ni-C	1318°C	(2404°F)
Ni-Cr	1345°C	(2453°F)
Ni-Mn	1018°C	(1864°F)
Ni-Mo	1315°C	(2399°F)
Ni-Ta	1360°C	(2480°F)
Ni-Th	1000°C	(1832°F)
Ni-Ti	955°C	(1751°F)
Ni-V	1203°C	(2198°F)
Ni-Zr	961°C	(1762°F)

Iron	(MP 1534°C or 2793°F)	
Fe-B	1149°C	(2100°F)
Fe-C	1147°C	(2097°F)
Fe-Mo	1450°C	(2642°F)
Fe-Nb	1360°C	(2480°F)
Fe-Ta	1410°C	(2570°F)
Fe-Th	860°C	(1580°F)
Fe-Ti	1085°C	(1985°F)
Fe-Zr	934°C	(1713°F)
Titanium	(MP 1668°C or 3034°F)	
Ti-Cr	1390°C	(2434°F)
Ti-Th	1190°C	(2174°F)
Thorium	(MP 1750°C or 3182°F)	
Th-B	1550°C	(2822°F)
Th-Cr	1235°C	(2255°F)
Th-Nb	1435°C	(2615°F)
Th-V	1400°C	(2552°F)
Th-W	1475°C	(2687°F)
Th-Zr	1350°C	(2462°F)
Zirconium	(MP 1852°C or 3366°F)	
Zr-Cr	1300°C	(2372°F)
Zr-Mo	1520°C	(2768°F)
Zr-Nb	1740°C	(3164°F)
Zr-V	1230°C	(2246°F)
Zr-W	1660°C	(3020°F)
Chromium	(MP 1880°C or 3416°F)	
Cr-C	1498°C	(2728°F)
Cr-Ta	1700°C	(3092°F)
Molybdenum	(MP 2620°C or 4748°F)	
Mo-C	2200°C	(3992°F)
Tantalum	(MP 3000°C or 5432°F)	
Ta-C	2800°C	(5072°F)
Tungsten	(MP 3380°C or 6116°F)	
W-C	2475°C	(4487°F)

[a] Alloys are included under the lowest melting parent metals which in turn are listed in order of increasing melting point.

TABLE 5-2

Melting Points of Binary Mixtures of UO_2 and Various Oxides

Composition (mole %)		U/M ratio from chemical analysis	Melting point	Mode of melting
UO_2	Other oxide			
50	50 BeO	1/1.5	2200 ± 50°C (3990 ± 90°F)	Melts with difficulty, droplets
67	33 MgO	1/0.56	1900 ± 50°C (3450 ± 90°F)	Droplets form with difficulty
50	50 MgO	1/1.12	1750 ± 50°C (3180 ± 90°F)	Droplet formation
33	67 MgO	1/1.53	1850 ± 30°C (3360 ± 55°F)	Droplet formation
70	30 CaO	1/0.404	2000 ± 50°C (3630 ± 90°F)	Droplet formation
50	50 Al_2O_3	—	1940 ± 30°C (3520 ± 55°F)	Melts with droplet formation
50	50 SiO_2	1/1	1770 ± 30°C (3220 ± 55°F)	Surface melting
50	50 TiO_2	—	1480 ± 30°C (2700 ± 55°F)	Melts easily
50	50 ZrO_2	—	2600 ± 50°C (4710 ± 90°F)	Same as BeO
50	50 ThO_2	—	— —	Does not melt in arc
50	50 V_2O_3	—	<800°C (<1470°F)	Melted during preliminary roasting
50	50 MoO_3	—	900-1000°C (1650-1830°F)	Same as V_2O_3
50	50 WO_3	—	1000-1200°C (1830-2190°F)	Same as V_2O_3
50	50 Ta_2O_3	—	1850 ± 30°C (3360 ± 55°F)	Melts easily
50	50 Bi_2O_3	—	1800 ± 50°C (3270 ± 90°F)	Melts little, evaporates strongly during heating
50	50 PbO_2	—	1850 ± 50°C (3360 ± 90°F)	Same as Bi_2O_3
50	50 SnO_2	—	— —	Does not melt in arc, strong evaporation
50	50 Cr_2O_3	—	2050 ± 100°C (3720 ± 180°F)	Droplets form with difficulty
50	50 Fe_2O_3	—	1370 ± 30°C (2500 ± 55°F)	Melts easily
50	50 MnO_2	—	1650 ± 30°C (3000 ± 55°F)	Melts easily
100	—	—	2740 ± 100°C (4960 ± 180°F)	Difficult to melt

during manufacture of the GCR fuel elements because of abnormal dimensional growth during the elevated temperature involved in fabrication procedures. The growth of the fuel plate was traced to volume changes accompanying the following reaction [324]:

$$16\ Al + 3\ UO_2 \rightarrow 3\ UAl_4 + 2\ Al_2O_3\ . \qquad (5\text{-}1)$$

The reaction reached 90-100% completion in 10 hr at 600°C (1112°F) in 52 wt.% UO_2 fuel plates.

Attempts to improve upon the UO_2-Al cermet led to the discovery that the reaction of U_3O_8 with aluminum was much slower at the temperatures needed for fabrication [325]. It has been shown that 3000 hr were required for complete reaction at 600°C (1112°F) (p. 369 of [139]). Fuel plates could, therefore, be fabricated from U_3O_8 and aluminum with relatively little difficulty from dimensional changes attending reaction between the components. This process, however, leads to a fuel element in which there remains the potential for an exothermic "thermite" reaction between the components. Several recent studies have been designed to evaluate the safety implications of this potential energy [326, 327, 328].

The U_3O_8-aluminum reaction proceeds in two steps. In the first step, the U_3O_8 is reduced to UO_2 as follows:

$$U_3O_8 + 4/3\ Al \rightarrow 2/3\ Al_2O_3 + 3\ UO_2\ , \qquad \Delta H = -190\ \text{kcal/mole} \qquad (5\text{-}2)$$

In the second step, the UO_2 is reduced to aluminum-uranium intermetallic compounds as indicated by the following unbalanced equation:

$$U_2O + Al \rightarrow Al_2O_3 + UAl_2 + UAl_3 + UAl_4\ . \qquad (5\text{-}3)$$

The heats of formation of the intermetallic compounds were recently determined to be -22.3, -25.2, and -31.2 kcal/mole for UAl_2, UAl_3, and UAl_4 respectively [329]. These relatively high values of the heat of formation imply that the free energies of formation are also large and negative which explains experimental findings that no free uranium is formed in the reaction. It is also apparent that the stability of the intermetallic compounds is the driving force for the reaction represented by Eq. (5-3) since the simple reaction of UO_2 with aluminum to produce uranium metal and aluminum oxide is accompanied by only a small change in free energy and enthalpy.

An estimate of the maximum energy release of aluminum-U_3O_8 compacts at temperatures above the melting point of aluminum (660°C or 1220°F) is given in Fig. 5-1 [326]. The energy indicated by the dotted line would be released by the reduction of U_3O_8 to UO_2, Eq. (5-2), while the solid line indicates the energy release corresponding to complete reaction.

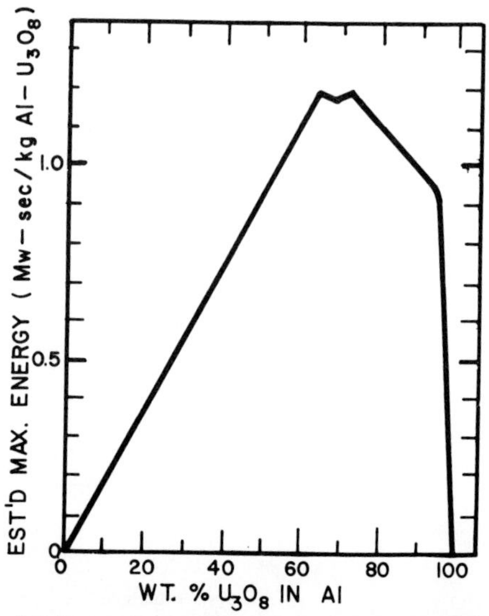

FIG. 5-1 How energy release varies with fuel composition.

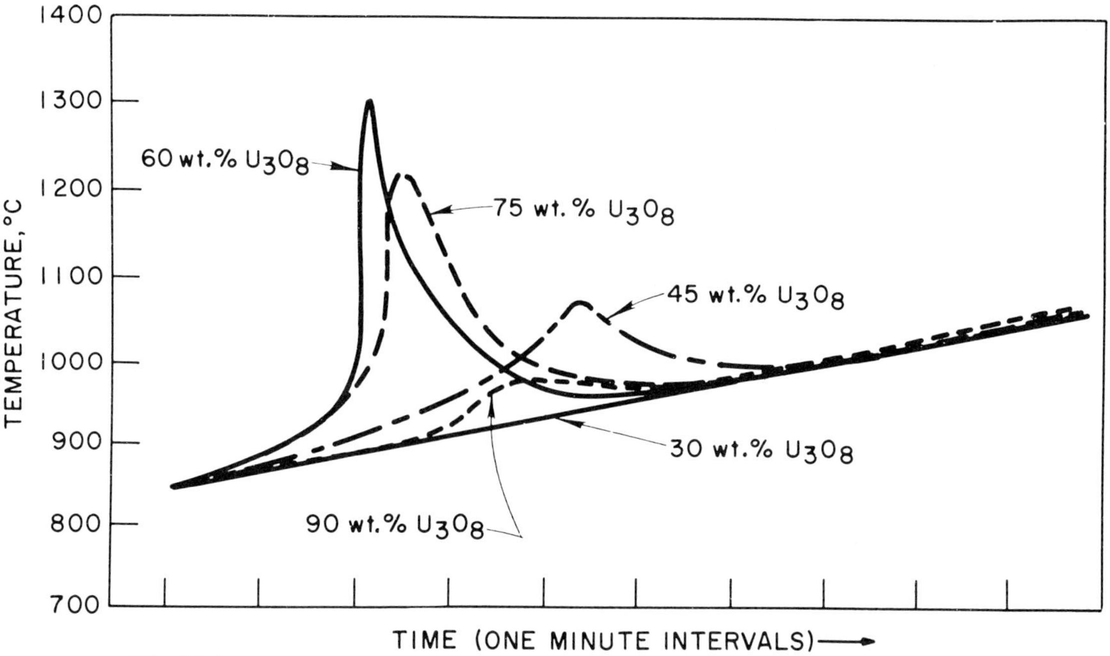

FIG. 5-2 Temperature-time curves for aluminum-U_3O_8 compacts heated at the rate of 25°C/min (45°F/min).

The kinetics of the U_3O_8 reaction have been studied by differential thermal analysis (DTA). In this method, a U_3O_8-Al specimen is heated uniformly along with an inert specimen. The temperature difference between the two specimens is a measure of the rate of heat generation and hence the rate of reaction [326, 327, 328]. Vigorous exothermic reactions were initiated when specimen temperatures reached 815°C (1500°F) to 1049°C (1920°F) depending upon the U_3O_8 content and processing variables. The effect of U_3O_8 content is shown in Table 5-3 [326]. The highest peak height reported, 4.56 mv (Pt-13% Rh vs Pt thermocouple), indicates a spontaneous temperature rise of approximately 333°C (600°F). Similar spontaneous temperature excursions were recorded in another study as shown in Fig. 5-2 [328]. These studies indicated that maximum self-heating occurred with compacts containing 60 to 75 wt.% U_3O_8 and that neglibible self-heating occurred for specimens containing 30 wt.% U_3O_8.

TABLE 5-3

Effect of U_3O_8 Content on Ignition of U_3O_8-Al Specimens

(Differential thermal analysis using Pt-13% Rh vs Pt thermocouples)

U_3O_8 Content (wt. %)	Ignition temperature	Peak area (mv-sec)	Peak height (mv)
85.4	1820°F (993°C)	184	2.98
79.5	1710°F (932°C)	319	4.56
74.5	1500°F (816°C)	401	2.85

The peak temperatures measured are much less than the temperatures that would be expected if the available heat indicated in Fig. 5-1 were released suddenly. This suggests that slow reaction rates prevent a rapid heat release and self-heating to extreme temperatures. High temperatures (>2200°C or 4000°F), however, were reported to occur during ignition of 85.4 wt.% U_3O_8 specimens in one study [326].

The limited information available on the reaction at temperatures above the melting point of aluminum suggests that chemical energy release is slow for compositions in the range used in current reactions, i.e., < 40 wt.% U_3O_8. Significant energy releases and the possibility of ignition during fabrication as well as during nuclear excursion and loss-of-coolant accidents should be considered if higher concentrations of U_3O_8 are used in reactor fuel.

Several metals other than aluminum are thermodynamically capable of reducing UO_2 to uranium. These include lithium, beryllium, magnesium, calcium, strontium, barium, scandium, yttrium, lanthanum, the rare earths, actinium, plutonium, and thorium. Zirconium and hafnium are marginal in this respect; the reaction with zirconium has been shown to occur to a limited extent (p. 342 of reference [139]) although the heat of the reaction is small. Many additional metals are thermodynamically capable of reducing U_3O_8 to UO_2.

5.3 Cleanup and Analysis of Alkali Metal Coolants

The alkali metals, the elements in the first group of the periodic table, have properties such as fairly convenient melting points and relatively low vapor pressures which make them of interest and applicable as liquid metal coolants for reactors. The

TABLE 5-4

Alkali Metals

Metal	Temperature of phase change		Density (g/cm^3) at 25°C (77°F)	
	Melting point	Boiling point	Metal	Oxide
Lithium	186°C (367°F)	1336°C (2437°F)	0.530	2.013 (Li_2O)
Sodium	97.5°C (207.5°F)	880°C (1616°F)	0.963	2.27 (Na_2O)
Potassium	63.7°C (146.7°F)	754°C (1389°F)	0.857	1.836 (K_2O)
Rubidium	38.5°C (101.3°F)	700°C (1292°F)	1.594	3.72 (Rb_2O)
Cesium	28.5°C (83.3°F)	670°C (1238°F)	1.992	4.36 (Cs_2O)

metals are listed in Table 5-4 (except for francium). Physically, these metals are soft and silvery-white. Considerations such as low cost, low neutron absorption cross section, and high thermal conductivity have thus far generally limited the field to sodium and the sodium-potassium alloys, NaK, as heat transfer media for power reactors. This section, therefore, will include only discussions of sodium and NaK. Furthermore, the two coolants will be treated together because of their similarity. Referring again to Table 5-4, it is interesting to note the considerably higher density of the oxide as compared to the parent metal. The NaK eutectic (~ NaK_2), 77 wt.% K, has a melting point of -12.3°C (+9.9°F) and a boiling point of 787°C (1449°F) [188].

Chemically, the alkali metals are quite reactive. Of particular significance in reactor technology is the reaction with oxygen. The sodium or NaK solutions which contain their oxides are very corrosive (for example, toward zirconium) relative to the pure liquid metal. Also, the oxide can cause plugging by precipitating at cool points in the lines. It then follows that two of the major problems in the use of sodium or NaK as reactor coolants are:

1. Determination or monitoring of the amount (or concentration) of impurities in the metal.
2. Removal of the impurities (especially of oxides, hydrides, and carbides).

There are five methods in common use for the analysis of oxide contaminants in sodium and NaK. These methods are briefly summarized in Table 5-5. The classification is based on whether the separation between the metal and oxide is achieved chemically or physically. The concentrations of importance range from a few ppm (< 10) of Na_2O in sodium or NaK at low temperatures to about 0.8 wt.% (8000 ppm) of Na_2O (which corresponds to 0.2 wt.% of O_2) at 538°C (1000°F) as indicated by the solubility curve for Na_2O in sodium as shown in Fig. 5-3 [188].

Since the first step in each of these analytical methods is the accomplishment of a separation of the oxide from the metal, this suggests that the same principle might be used for cleanup operations provided, of course, that the metal it-

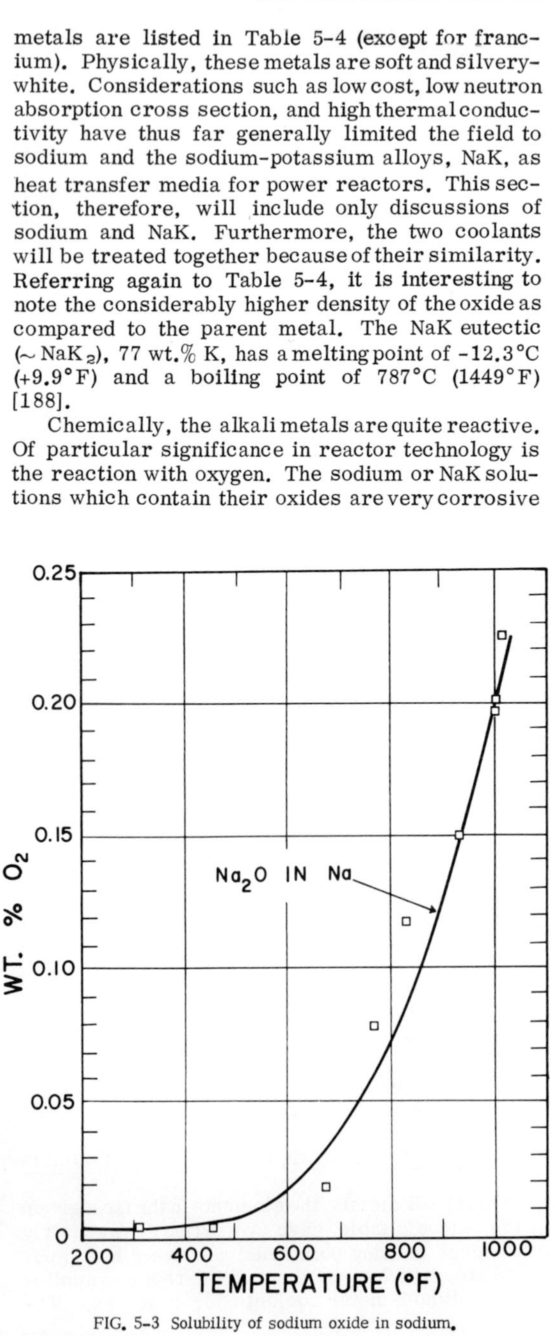

FIG. 5-3 Solubility of sodium oxide in sodium.

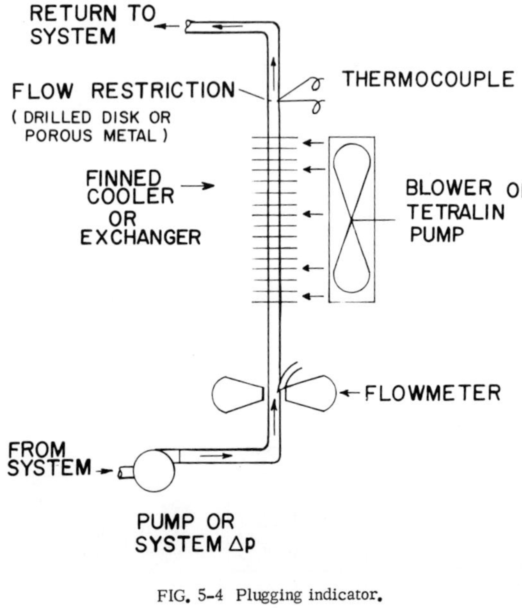

FIG. 5-4 Plugging indicator.

TABLE 5-5

Principal Methods of Analysis for Na_2O in Na and NaK

Method	Reference	Basis for method and comments
Chemical methods		
Amalgamation	188, 338	Na dissolves in Hg but Na_2O does not. After phase separation, the Na_2O is dissolved in H_2O and the solution titrated with HCl to phenolphthalein end point. Applicable to oxide concentration >20 ppm with ± 1.5 ppm accuracy.
Alkyl-Halide (Wurtz)	332, 339, 340	Na contaminated with Na_2O is contacted with butyl bromide (in hexane or xylene) to give NaBr. The unreacted Na_2O is titrated with HCl. Reliability of about ± 20 ppm. [a]
Gettering	341	Hot Ti or Zr (~650°C or ~1200°F) removes oxygen from metal. The weight gain of the getter measures amount of oxide present in Na or Na_2O: ± 10 ppm accuracy.
Physical methods		
Distillation	332, 342	Vacuum distillation of the Na leaving a nonvolatile residue of Na_2O. Oxide content checked by dissolution and titration with H_2O and HCl.
Plugging meter	188	Flowing Na + Na_2O stream is passed through a filter disk whose temperature is lowered until a point where the liquid metal flow rate begins to decrease rapidly. This change is caused by oxide coming out of solution and coating the small holes in the meter. From the solubility data (Fig. 5-3), the Na_2O content is determined. A diagram of a plugging meter installation is given in Fig. 5-4. [188]

[a] A modification is described in ref. 339 where after reacting with n amyl chloride: $2\ C_5H_{11}Cl + 2\ Na \rightarrow C_{10}H_{22} + 2\ NaCl$, then $Na_2O + CO_2 \rightarrow Na_2CO_3$. Analysis is then made spectrophotometrically for Na_2CO_3 at the 11.38 μ absorption band. Lower limit of this method is 20 ppm of Oxygen.

self is not reacted or lost in the process. Thus the two techniques for establishing and maintaining the purity of sodium and NaK, known as cold trapping and hot trapping, can be considered as scale-ups or modifications of the plugging meter and gettering approaches, respectively. The flow restriction can be a porous stainless steel plug or a disk with many small holes drilled in it (for example, 10 of 0.04-in diameter). The cooling can be accomplished with a tetralin heat exchanger or an air blower. (See Fig. 5-4.)

There are two other methods which are of practical interest in checking for the purity of sodium or NaK. One is simply to observe the surface of the liquid metal. A bright, shiny or mirrorlike appearance is indicative of purity whereas a dull or grey or "spotty" surface suggests the presence of oxide or other "crud." The other method makes use of coupons of the metals used in the fabrication of the primary systems, such as stainless steel or zirconium, which are inserted in the coolant loop and periodically removed and examined metallographically for evidence of carbiding, hydriding, or oxidation. If physical degradation is noted, then the coolant must be cleaned up by cold and hot trapping. A cold trap takes out oxides and hydrides from flowing sodium or NaK by a combination of cooling and filtration. Cooling is done either by an external NaK loop or by boiling toluene. These two methods are described schematically in Fig. 5-5 [191, 113]. Filtering is accomplished by a stainless steel wire mesh packed in the central part of the trap [330]. The principle of the cold trap operation is to reduce both the temperature and flow rate of the liquid metal, thus promoting the precipitation and collection of particles of sodium oxide and hydride on the woven mesh. When the cold trap becomes partially or fully plugged, it is removed and disposed of, because of the residual radioactivity in the sodium.

The time required to purify a batch of sodium by cold trapping can be estimated [331]. First a differential material balance of oxide in the sodium is written:

$$W \frac{dX}{dt} = X_c\, w - Xw\,, \tag{5-4}$$

where W is the total weight of sodium, w the flow rate of sodium through the trap, t the time, X the concentration of oxide in sodium at time t, X_o the

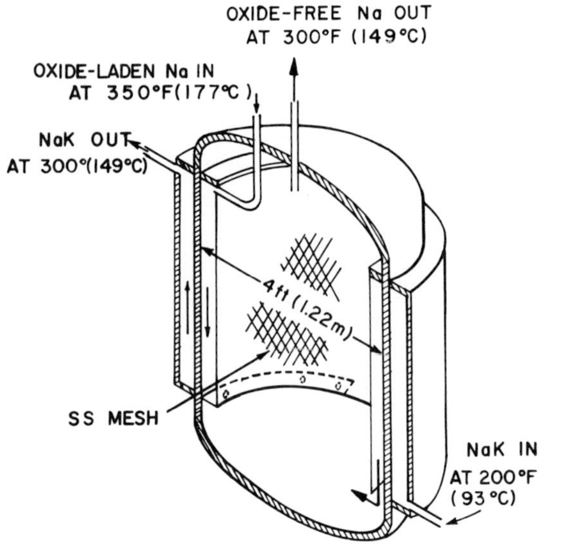

FIG. 5-5a NaK-cooled trap for removal of Na_2O from flowing Na or NaK.

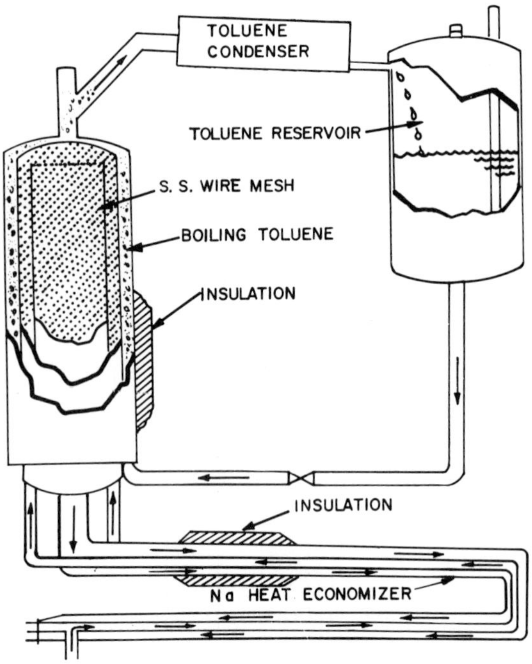

FIG. 5-5b Toluene-cooled trap for removal of Na_2O from flowing Na or NaK.

initial concentration, and X_c is the equilibrium concentration of oxide in sodium corresponding to saturation at the temperature of sodium leaving the cold trap.

Upon integrating Eq. (5-4) the following result is obtained:

$$(X - X_c) / (X_o - X_c) = \exp[-(w/W)t]. \quad (5\text{-}5)$$

After cold trapping, a hot trap is used to achieve additional removal of the oxide down to barely detectable (0 to 10 ppm) concentration levels. Figure 5-6 [113] illustrates a typical hot trap where high surface area Zr sheets at 649°C (1200°F) (av. temp.) react with the oxygen impurity. Another approach is to have some "getter" dissolved in the Na or NaK (such as Ca, Li, Ba, or Mg) which produces an insoluble "getter oxide" by a reduction reaction with the sodium oxide [191]. The calcium in the sodium will also react with nitrogen to produce the insoluble calcium nitride. Table 5-6 [332] gives pertinent material properties.

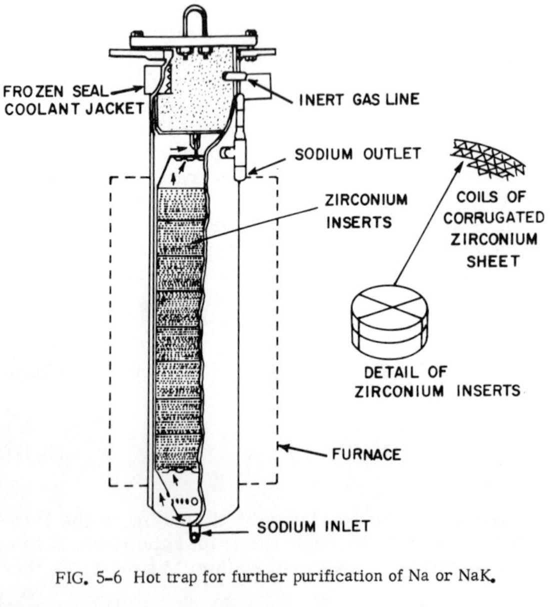

FIG. 5-6 Hot trap for further purification of Na or NaK.

A hot trap is also useful for absorbing hydride and carbide impurities. The hydrogen forms a solid solution in the zirconium and additional reaction yields a second phase of zirconium hydride. The carbon goes into the steel jacket of the hot trap by a carburization reaction [333, 334].

An integral part of the chemical aspects of alkali metal purity control in reactors is the blanket gas that is usually a part of the system. Among the impurities in the gas may be hydrocarbons which will crack when in contact with the high temperature sodium, adding hydrogen to the metal coolant and carburizing the steel. Analysis for hydrogen in sodium or NaK is effected by thermally decomposing the sodium hydride in a vacuum apparatus and then measuring the evolved hydrogen pressure [335, 336]. The dissociation pressure of NaH is 1 atm at 420°C (788°F) [188]. Total carbon is determined by a Van Slyke method with chromic acid, thereby converting the carbon to carbon dioxide, and then measuring the volume of the gas [335, 337]. The carbon, as carbide, is found simply by dissolving the sodium or NaK sample in water and then determining the amount of acetylene formed. A convenient way of purifying the cover gas is to make use of the reactivity of impurities, namely, to bubble the gas through a separate, packed, NaK scrubbing column. The three commonly used blanket gases are: helium, argon, and nitrogen. The main purpose of this inert blanket is to prevent, or minimize, the introduction of impurities such as air, water

TABLE 5-6

Some Thermodynamic and Nuclear Factors Pertinent to Hot Trapping

Metal	Oxide	$-\Delta F$ (500°C or 932°F) kcal.	Solubility of metal in Na at 150°C (302°F) wt. %	Thermal neutron capture cross section (barns)
Ti	TiO_2	17.4	Negligible	5.6
Zr [a]	ZrO_2	38.0	Negligible	0.18
Ba	BaO	40.4	0.56	1.2
Mg	MgO	51.7	1.1	0.06

Typical reduction reactions: $Zr + 2\ Na_2O = ZrO_2 + 4\ Na$

$Mg + Na_2O = MgO + 2\ Na$

[a] Zirconium is a preferable deoxidant to titanium because the zirconium oxide formed is adherent to the parent metal and thus does not get back into the sodium stream as the titanium oxide does.

vapor, etc., into the alkali metal system. It is clear, therefore, that the following three points must be included in any effective program to purify and then maintain the purity of the coolant:

1. Proper specifications and checks on the purity of the incoming (as received) sodium or NaK and inert gas (see reference [188] for recommended specs).
2. Proper cleanliness of the piping system.
3. Proper operation of the equipment to avoid adding contaminants.

The details of these non-chemical aspects are covered in other parts of this book. Suffice it to say here that commercially available inert gases contain 50 to 100 ppm of oxidizing impurities, and that the usual high purity grade of sodium has 60 ppm of oxygen. Because of the deleterious effect of impurities in the sodium on structural materials (such as zirconium and stainless steel) at normal power reactor operating temperatures, it is desirable to keep the oxygen level in the liquid metal to less than 5 ppm.

ADDENDUM*

Richard O. Ivins and Louis Baker, Jr.
(Argonne National Laboratory, Argonne, Illinois)

There has been considerable progress in recent years in understanding the effects of metal-water reactions on the hypothetical accidents (loss of coolant and nuclear excursion) that are considered in evaluating the safety of large water-cooled power reactors. Simulation experiments, where a small number of fuel rods are subjected to an environment characteristic of the accident, have yielded valuable data. New analytical studies, based on computer programs that take into account most of the relevant phenomena, have provided new insights. The following summarizes these developments.

A.1 THE LOSS-OF-COOLANT ACCIDENT

The loss-of-coolant accident has been used in the safety analysis of most power reactors as the design basis accident for design and license considerations. For boiling water reactors (BWR) and pressurized water reactors (PWR), an instantaneous break of one of the large recirculating lines in the primary system is assumed to initiate the event. The pipe break, assumed to be a double-ended rupture, is followed by a blowdown of the primary system. During this period (approximately 10 sec for a PWR and 30 sec for a BWR), coolant leaves the reactor core and a nuclear shutdown is effected by the loss of the moderator-coolant. Typical dimensional and operational parameters are listed in Table A-1 for both a BWR and PWR of current design.

Following the blowdown, the temperature of the cladding rises rapidly, equilibrating with the higher temperature UO_2 fuel, and then continues to rise more slowly as it is heated by fission-product decay. Water remaining in the bottom of the primary vessel supplies a steam flow through the core; the steam then reacts chemically with the cladding.

To prevent the reactor core from reaching a temperature sufficiently high to cause a loss of structural integrity, emergency core cooling systems (ECCS) are provided in both BWR's and PWR's. These systems either spray water into the core from the top or very rapidly fill the primary vessel, flooding the core from the bottom. In either case, water eventually fills the core. Current designs provide redundant systems in both number and design.

*Date of manuscript, February 1969.

TABLE A-1

Typical Parameters for Boiling-Water and Pressurized-Water Reactor Cores

Reactor Model	BWR	PWR
Power, Mw (t)	2255	2758
Fuel rods (UO_2):		
No. of fuel rods	35476	39372
No. of fuel assemblies	724	193
Length of fuel (in.)	144	144
Thickness of Zircaloy cans (in.)	0.085	0.061
OD of Zircaloy cladding (in.)	0.570	0.422
Cladding thickness (in.)	0.036	0.0243
Fuel pellet diam. (in.)	0.488	0.366
Pitch (square array) (in.)	0.738	0.556
Control rods:		
No. of assemblies	177	53
Rods per assembly	1[a]	20[b]
Control rod material	B in SS	Cd in Ag-In
Equivalent core diam. (in.)	182.7	133.7
Vessel inside diam. (in.)	251	173
Reactor pressure (psig)	1000	2250
Coolant outlet temperature (°C)	558	576

[a]Cruciform.
[b]Round.

The chemical reaction of the cladding with steam, which is supplied by the water remaining in the bottom of the primary vessel after the blowdown or introduced by the operation of an ECCS, has three important effects. First, it furnishes energy, which can increase the heating rate of the core. Second, hydrogen, a reaction product, is released to the containment structure. Third, the reaction also changes the character of the cladding (i.e., the metal cladding is converted to an oxide), which can affect its behavior on quenching. Thus, the extent of the reaction occurring during the heating (in steam) and quenching of typical fuel elements has been, and is being, studied extensively.

A.1.1 Simulations of Loss-of-Coolant Accidents

Several experiments have been conducted to simulate the decay heating rate and conditions of steam flow typical of a loss-of-coolant accident with both Zircaloy-2-clad and stainless steel-clad, UO_2-core rods. One technique [A1] involved using a furnace, shown in Fig. A-1, to heat small sample fuel rods in steam. The furnace, which was contained in a steel pressure vessel, had two zones, an internal, steam-filled zone that was surrounded by an alumina tube, and an external zone filled with argon. The argon-filled zone contained molybdenum heater windings and insulation. Water was introduced into the lower part of the steam zone by a positive displacement pump and converted to steam. The hydrogen produced by metal-steam reaction and the unreacted steam were continuously removed from the upper part of the steam zone (high-temperature section) through an outlet valve. The extent of metal-steam reaction was determined by the amount of hydrogen collected. Sample temperatures as high as 1700°C could be tested. In an experiment, the sample was introduced into the lower (cooler) part of the steam-filled zone and then raised by an external crank mechanism into the upper (hotter) part of the steam zone at a rate that would simulate fission-product-decay heating.

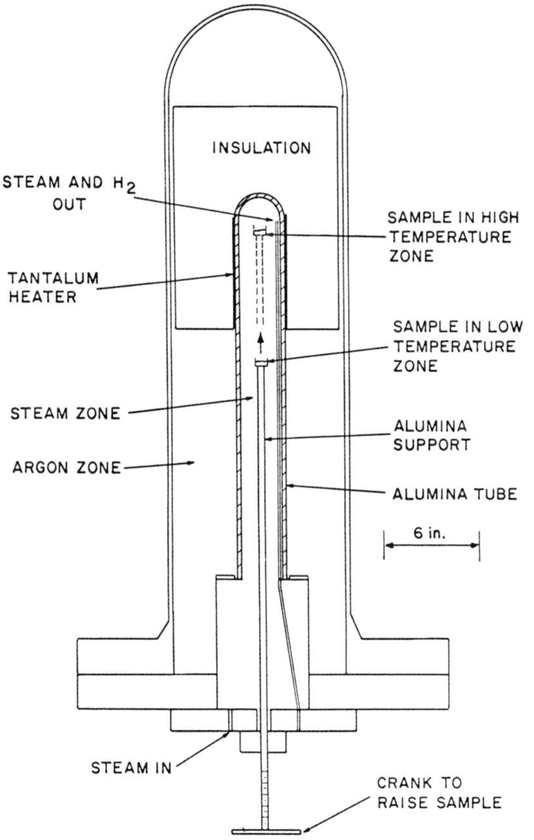

FIG. A-1 High-pressure furnace for metal-steam reaction studies.

A.1.1.1 Zircaloy Cladding

In a number of experiments [A2, A3] conducted with Zircaloy-2-clad samples, the fuel rods consisted of a number of high-density UO_2 pellets sealed within a 27-mil-wall tube forming a rod 3/8 in. OD by 8 in. long.

During one experiment, a sample was raised 8 in. into the high-temperature section of the furnace at the rate of 1 in./min, allowed to remain in position for 2 min, and rapidly withdrawn. In the fully inserted position, the temperature at the upper end of the rod was about 1600°C, while at the lower end the temperature was 900°C. A steam flow equivalent to 10 ml water/min past the fuel rod (3.2-ft/sec linear velocity) was maintained throughout the experiment.

The Zircaloy-2-clad fuel rod after removal from the furnace is shown in Fig. 4-14, along with an unreacted rod. The surface of the upper end of the reacted rod was converted to white ZrO_2, while the lower (cooler) end retained a metallic appearance. The oxidized cladding, uppermost on the rod, was cracked and came apart on handling. The steam flow was sufficient to prevent steam starvation, i.e., the reaction followed the parabolic rate law and was not limited by gas-phase diffusion. The measured temperature history at the top of the rod is shown in the upper portion of Fig. A-2. The temperature of the furnace wall adjacent to the top of the rod and the hydrogen collected during the experiment are also shown in the figure.

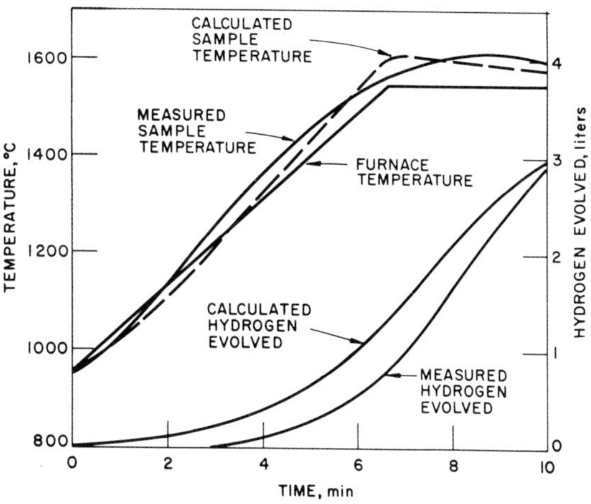

FIG. A-2 Calculated temperature at top of single Zircaloy-2-clad fuel rod and hydrogen evolution from the rod; furnace experiment (steam flow rate: 10 g/min).

A simplified version of the CHEMLOC II program (a digital computer program, discussed later, for calculating a reactor core heatup and steam reaction during a loss of coolant) was used to compute the hydrogen evolution and the rod temperature in this experiment. Results of the calculations are also shown in Fig. A-2. In the calculations, the steam flow and furnace-wall temperature along the fuel rod axis were the input data. The program used the parabolic rate law, Eq. (1-1) with the parabolic rate constant from Eq. (3-1), to calculate the extent of reaction occurring axially along the rod. The measured curve of hydrogen evolved lags the calculated curve because of the delay in collecting the off-gas associated with the experimental technique. Since the total amount of hydrogen collected was 2.96 liters (18.5% cladding reaction) and the calculated amount was 2.94 liters, the agreement was excellent.

Similar calculations based on gas-phase diffusion (i.e., steam diffusing through hydrogen in laminar flow parallel to a rod) predicted that high-reaction zones would occur axially along the rod when the steam flow was lower than that indicated by integration of the parabolic rate law. A photograph [A4] of a fuel rod exposed to steam-flow rates of 0.25 to 2.5 ml/min (compared to 10 ml/min for the rod in Fig. 4-14) is shown in Fig. A-3. The white ring of oxide ring of oxide formed about one-third of the way from the top of the fuel rod was caused by the depletion of steam from the flowing steam-hydrogen mixture at that zone during the latter part of the experiment. The upper band of white oxide occured because the top of the rod first reached temperatures sufficient to support extensive reaction as the rod was raised from the lower (cooler) zone of the furnace to the hot zone.

In one experiment [A-4], a four-rod fuel bundle in a square array with 0.6-in. center-to-center

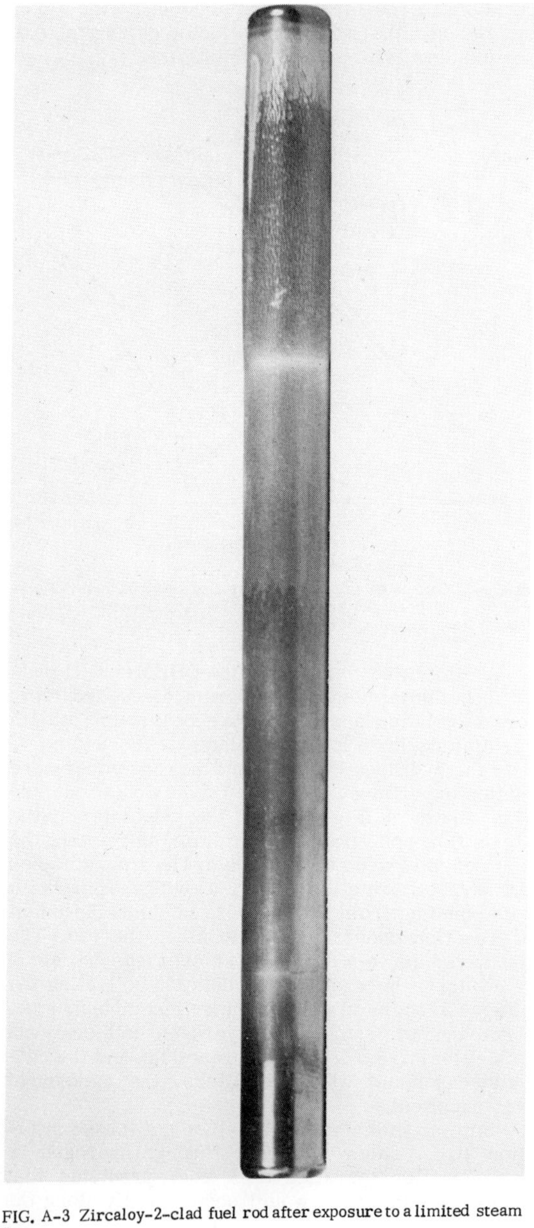

FIG. A-3 Zircaloy-2-clad fuel rod after exposure to a limited steam flow.

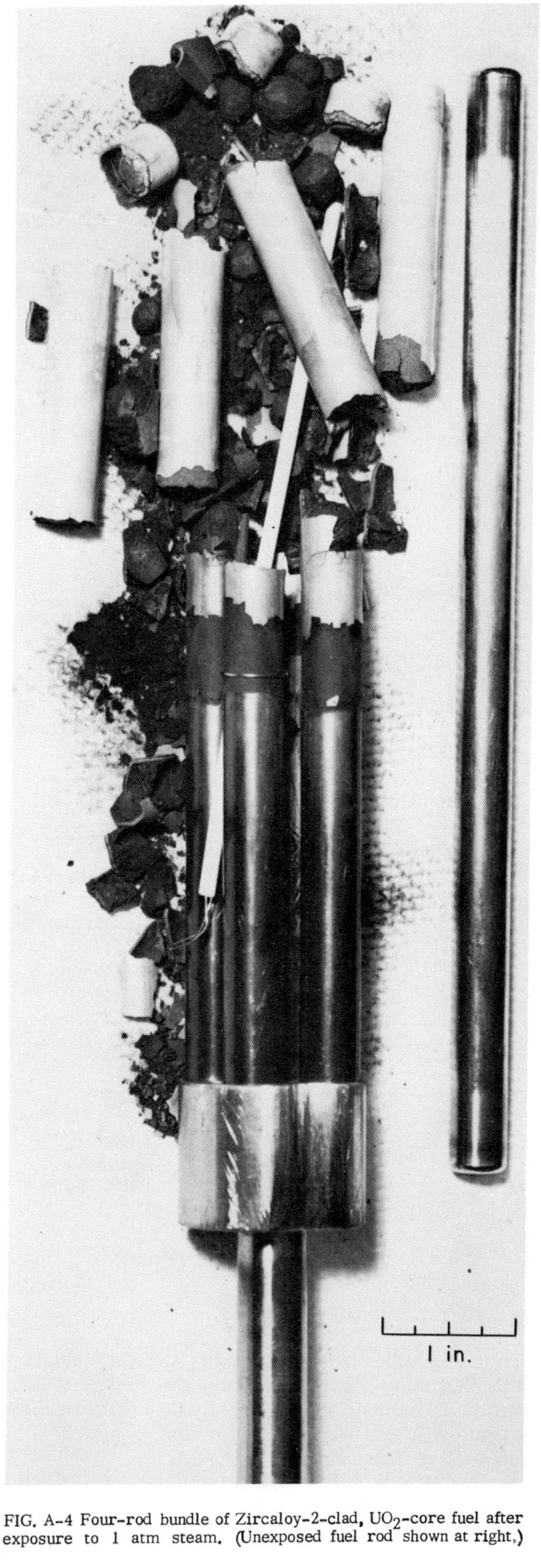

FIG. A-4 Four-rod bundle of Zircaloy-2-clad, UO_2-core fuel after exposure to 1 atm steam. (Unexposed fuel rod shown at right.)

spacing was heated in the same manner as the single rods discussed above. The steam-flow rate was increased to 25 ml/min to ensure that the reaction was not steam-limited. The Zircaloy-2-clad rod bundle is shown in Fig. A-4 after exposure to steam. An unexposed rod is also shown for comparison. Self-heating of the rods was apparent during the experiment; the temperature of the top end of the bundle exceeded 1700°C (more than 200°C above the furnace temperature). The fuel rods were cracked and partly broken up after the experiment, and they broke up further on handling.

An important observation in these experiments was the breakup of the rods after oxidation in steam (see Figs. 4-14 and A-4) even though the rod temperatures were below the melting point of the cladding (Zr m.p. 1852°C).

More recent experiments [A4] have made use of induction heating and the introduction of cold water by either a spray or a flooding system to quench the rods after a period of heatup in steam. A schematic drawing of an apparatus used for induction-heating experiments is shown in Fig. A-5. In these experiments, a 3-in. section of fuel rod was heated in flowing steam to temperatures greater than 2000°C to reveal some of the features of failure in steam. The steam flow was regulated at 1.0 ml/min, a rate typical of flow for a single rod in a loss-of-coolant accident. Temperatures were measured with tungsten-rhenium thermocouples and a two-color pyrometer. A photograph of a fuel rod brought to a peak temperature of 2140°C is shown in Fig. A-6. In the experiment, there was no tendency for the cladding to melt and drip, even though the temperature was above the melting point of the metal cladding. The cross section of the fuel rod, shown in Fig. A-7, indicated that molten cladding was confined between an outer crucible of ZrO and the inner surface of the UO_2. There was also evidence for interaction between the molten Zircaloy and the UO_2. The interaction appeared to be the mechanism for the degradation of the cladding while at temperature, i.e., a low-melting mixture of UO_2 and partly oxidized cladding. Upon cooling, the cladding broke up further, resulting in the two pieces shown in the figure. Experiments at Oak Ridge National Laboratory (ORNL) [A5] have shown that mixtures of ZrO_2, zirconium, and UO_2 form solid solutions that have melting temperatures well below the melting point of UO_2 (2850°C). Samples prepared by melting Zircaloy-clad, UO_2-core fuel in air by induction heating melted at 2150°C in air and 2450°C in vacuum.

Experiments simulating both heating and quenching with a water spray have also been conducted with an induction-heating apparatus [A6]. A steam flow of 1 ml/min was used to simulate a typical loss-of-coolant condition. In three successive experiments, the samples were heated inductively to 2050, 1750, and 2820°C. Water was then sprayed into the top of the chamber at a rate of 190 ml/min, corresponding to 0.05 gal/min per rod for a reactor core. Inductive heating was continued through the spray-cooling period.

In one experiment, denoted 1Z-4, the fuel rod was successfully heated above the melting point of Zircaloy-2 and water was sprayed on the hot rod. The rod cooled from a maximum temperature

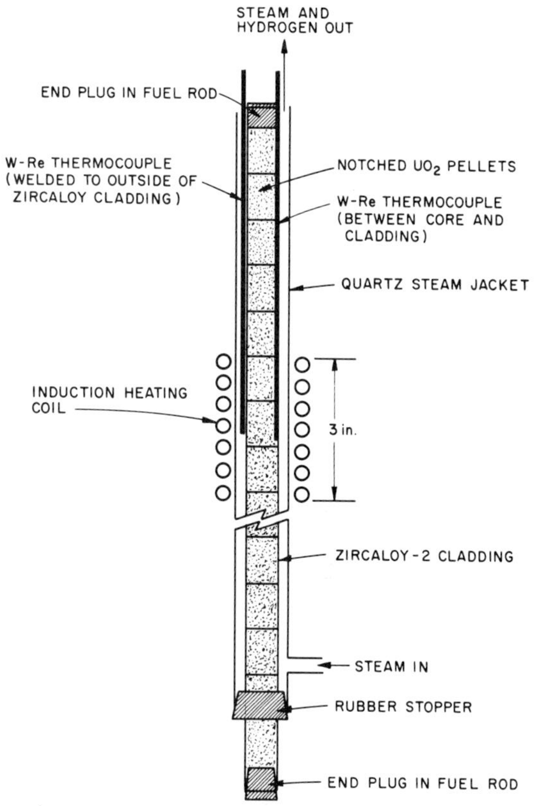

FIG. A-5 Apparatus for induction heating experiment simulating a loss-of-coolant accident.

of 2045°C to 1380°C during the 30-sec spray period. Figure A-8 illustrates the sharp decrease in rod temperature during the spray period and the subsequent increase in temperature after the spray was discontinued, as measured by a thermocouple. Breakup of the cladding after cooldown was noted in this experiment. Figure A-9 shows the heating and cooling curves of the two other experiments. In experiment 1Z-5, a fuel rod heated to a temperature below the melting point of Zircaloy-2 cooled from 1750 to 110°C in 185 sec. In experiment 1Z-6, a fuel rod at a temperature well above the melting point of Zircaloy-2 cooled from ~2820 to 95°C in 180 sec. The rods, shown after cooldown in Fig. A-10, remained intact throughout the experiments; breakup occurred later.

More recently, a series of similar experiments, simulating the heating and quenching of reactor fuel rods following a loss of coolant, have been conducted in which the peak rod temperature reached was varied from 1462 to 2110°C [A7]. In these experiments, a 3-in. section of a 12-in.-long rod was heated inductively with a steam flow of 2 ml/min. The temperature of the heated zone of the rod was raised at a rate of about 5°C/sec to simulate fission-product-decay heating. When the desired peak temperature was reached, water was sprayed on the rod from above or the quartz tube (see Fig. A-6) was flooded from below at a rate of 190 ml/min (25.1 lb/hr). The results of these experiments are tabulated in Table A-2.

TABLE A-2

Results of Out-of-pile Loss-of-coolant Experiments Using Inductively Heated Simulated Fuel Rods[a]

Run No.	Heating rate (°C/sec)	Quench temp (°C)	Bottom or top quench	Power on or off during quench	Hydrogen collected[b] (mol)	ZrO_2 produced[c] (m/o)		Apparent dissolved O_2 in unoxidized metal (m/o)	Equivalent average reaction[d] (m/o)	Rod failed or remained intact
						Avg	Max			
12	9.0	1064	Top	On	0.0	1.6	3.2	—	—	Intact
4	4.8	1462	Bottom	Off	0.025	3.5	6.8	3.45	6.87	Intact
7	6.0	1488	Bottom	On	0.026	3.7	6.1	3.6	7.14	Intact
16[e]	~50	1182	Top	On	0.024	3.8	5.4	2.9	6.59	Intact
15	6.4	1548	Top	Off	0.025	4.0	8.0	3.0	6.87	Intact
8	3.1	1534	Bottom	On	0.029	4.8	8.0	3.34	7.97	Intact
9	5.2	1498	Top	On (3.55)[f]	0.042	7.0	12.0	4.88	11.54	Intact
5	5.0	1663	Bottom	Off	0.062	12.3	21.1	5.0	17.03	Intact
17	4.8	1798	Bottom	Off	0.067	14.0	22.0	4.5	18.40	Failed
18	4.6	1569	Top	On (3)[f]	0.068	14.3	22.3	4.5	18.68	Failed
11	5.7	1849	Top	Off	0.081	16.0	22.1	7.3	22.25	Failed
3	4.7	1646	Bottom	On	0.093	13.0	25.0	—	25.55	Failed
10	4.5	1650	Top	Off	0.080	17.6	29.0	5.18	21.98	Failed
6	5.7	1864	Bottom	Off	0.081	17.4	34.5	5.6	22.25	Failed
13	5.1	1663	Top	On (4.18)[f]	0.102	20.5	34.0	9.3	28.02	Failed
14	6.5	2052	Top	On (2.75)[f]	0.14	30.0	41.0	12.0	38.46	Failed
1	6.3	2110	Top	On	>0.1	—	~100	—	—	Failed
2	5.2	1868	Top	On	>0.1	—	~100	—	—	Failed

[a]Zircaloy-2 clad, UO_2-pellet-core fuel; steam flow, 2 g/min; cooling water spray rate, 190 g/min.
[b]Oxygen produced was taken to be one-half this amount.
[c]From metallurgical examination of 3-in. heated length of cladding.
[d]Equivalent average reaction of 3-in. heated length of cladding based on total oxygen produced.
[e]This run followed a predetermined temperature-time curve simulating a particular ECC (emergency core cooling) situation.
[f]Power turned off at this time (min) after initiation of quench.

Since it was difficult to simulate the heat balance of a fuel rod in a reactor core under quenching conditions in these tests, the induction heating was either continued or stopped during the quenching period. In the experiments in which the induction-heater power was turned off, the samples cooled in a matter of seconds, whereas with power on, the quenching period lasted for several minutes. As would be expected, quenching from the bottom was more effective in cooling the single rods in these tests than spraying water from the top, as shown in Table A-3. The results indicated that the samples failed when the average reaction over the 3-in. sections was greater than about 18%. Thus, the observed breakup of the fuel appears to depend primarily on the extent of oxidation.

Zirconia (ZrO_2) undergoes a phase change at about 1000°C with a 9% volume change. Calcium oxide ZrO_2 (CaO) is added to ZrO_2 to stabilize the material in the manufacture of refractories such as crucibles, etc. The postulated failure mechanism in the fuel is that partial oxidation of the Zircaloy cladding occurs at high temperatures (>1200°C), above the temperature of the phase change for ZrO_2, and when this material is quenched to lower temperatures, the cladding breaks up because of mechanical and thermal stresses. Interaction was noted between the UO_2 pellets and the cladding in these tests, and this may also have contributed to degrading the cladding.

TABLE A-3

Failure Pattern in Out-of-Pile Loss-of-Coolant Experiments Using Single Inductively Heated Simulated Fuel Rods

Quench temperature (°C)	Power on after quench initiated		Power off after quench initiated	
	Top quench	Bottom quench	Top quench	Bottom quench
1064	I[a]			
1182	I			
1462 to 1498	I	I		I
1534 to 1569	F[b]	I	I	
1646 to 1663	F	F	F	I
1798				F
1849 to 1868	F		F	F
2050 to 2110	FF			

[a]Test rod remained intact (I).
[b]Test rod failed (F).

It should be noted that the small samples in these experiments are not mechanically restrained in the test apparatus. Since the mechanical strength of the cladding has been shown to decrease with the degree of oxidation and possibly with interaction with UO_2, the unrestrained rods that did not fail in these tests might have failed if they had been subjected to mechanical restraints.

A.1.1.2 Stainless Steel Cladding

The furnace technique (see Fig. A-1) previously described has also been used to test stainless steel-clad, UO_2-core fuel rods [A2, A3]. The experiments were conducted in a manner similar to those with rods having Zircaloy cladding. In Fig. A-11, the temperature at the top of the rod (hottest zone) and the hydrogen evolution for an experiment with a stainless steel-clad, UO_2-core rod are shown. The appearance of a fuel rod that was exposed to steam under the conditions described above is shown, together with an unexposed element, in Fig. 4-14. The top of the rod reached a maximum temperature of about 1600°C, and the bottom, a maximum of about 900°C, thereby creating an axial temperature gradient of about 700°C. The top one-third of the stainless steel cladding appeared to be completely reacted, whereas the remainder had reacted only slightly. The demarcation between the two reaction zones can be readily seen in Fig. 4-14.

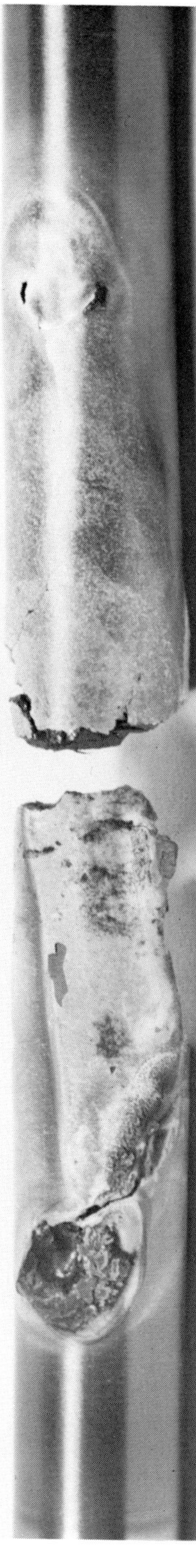

FIG. A-6 Zircaloy-2-clad, UO_2-core rod after induction heating in steam to an indicated cladding temperature of 2140°C.

The maximum temperature at the demarcation was estimated to be about 1400°C. The total hydrogen evolved corresponded to a reaction of 37% of the stainless steel present in the fuel element.

A photomacrograph, shown in Fig. A-12, of a radial cross section near the top of the fuel rod shows areas of interaction between the UO_2 and the stainless steel oxides. The foamy appearance of the oxidized stainless steel is also apparent in the photomacrograph. X-ray analysis of the stainless steel oxides showed a structure of the γ-Fe_3O_4 type.

A four-rod bundle (type-304 stainless steel cladding) that was exposed to high-temperature steam in the same manner as the single rod is shown in Fig. A-13. The holder is shown at the bottom and an unexposed fuel rod on the right. The boundary between the rapid "foaming" reaction and the slow "corrosion" reaction was also at a temperature of about 1400°C. In the area of rapid reaction, the expansion of the reaction products into the steam-flow channel is evident, with bridging between fuel rods in some areas. The amount of hydrogen evolved per rod was about the same as that observed in the experiment with the single fuel rod (i.e., the total amount of hydrogen evolved was approximately four times the amount evolved in the experiments with single rods).

Other experiments were conducted with single rods to determine the effects of both steam flow and steam pressure. In one of these experiments, an increased steam flow of 20 ml water/min was employed with a steam pressure of 1 atm. Steam pressures of 4 and 11 atm, respectively, were used with a steam flow of 10 ml water/min in two other experiments. The results of these experiments are summarized in Table A-4, where it is apparent that neither the rate nor the extent of stainless steel-steam reaction was significantly affected by the changes in steam flow or pressure.

TABLE A-4

Reaction of Stainless Steel-Clad Fuel Elements with Steam in the High-Pressure Furnace (Total exposure time: 10 min[a])

Steam pressure (atm)	Steam-flow rate (ml H_2O/min)	(ft/sec)	Hydrogen generated (liters (STP))	Extent of stainless steel-steam reaction[b] (% of total metal)
1	10	3.2	7.52	37.0
1	20	6.4	7.25	35.7
4	10	0.8	5.57	27.4
11	10	0.3	7.14	35.2

[a]Sample elevated into hottest (~1600°C) part of furnace over an 8-min period, maintained there for 2 min, and then rapidly withdrawn.

[b]Based on 0.51 liter of H_2(STP)/g stainless steel.

A.1.2 Analytical Studies of Loss-of-Coolant Accidents

Many investigators have reported calculations of the core heatup and chemical reaction that occur following a loss of coolant in power reactor. Early analytical efforts [A8-A10], primarily calculations performed by hand, involved the temporal and spatial integration of the parabolic rate law for the zirconium-steam reaction. These calculations, de-

FIG. A-7 Cross section of rod shown in Fig. A-6.

TABLE A-5

Computer Programs that Calculate Core Heatup and Cladding-Steam Reaction Following a Loss of Coolant

	ARC-II Phillips Petroleum Co.	CHEMLOC-II Argonne Nat'l. Lab.	LOCTA-R Westinghouse A.P.D.	MOXY-II Phillips Petroleum Co.	NURLOC Battelle Mem. Inst.	TACT-V General Electric A.P.E.D.
Limited steam flow	Yes	Yes	Yes	Yes	Yes	No
Steam flow	Computed	Input	Input	Input	Computed	No
Melting of cladding	Yes	Yes	Yes	No	No	Yes
Melting of fuel	Yes	Yes	No	No	No	No
Core geometry change (collapse)	No	Yes	Cladding only	No	Yes	Cladding only
Emergency coolant	No	No	Yes	No	Yes[a]	Yes[a]

[a]Cooling during the operation of an emergency system can be calculated by specifying a film heat-transfer coefficient.

scribed earlier, neglected heat losses (i.e., assumed adiabatic heating) when accounting for both fission-product-decay heating and the heat of reaction. The effects of a limited steam flow (i.e., insufficient to allow the reaction rates predicted by the parabolic rate law), materials available for reaction other than fuel cladding, and the eventual loss of geometry or emergency cooling were also neglected.

Recently more refined calculations have been carried out. Waage [A11] has described various computer programs used to calculate heat transfer during blowdown, and core heatup and cladding-steam reaction following the blowdown. Some of the features of programs that calculate core heatup and cladding-steam reaction are listed in Table A-5.

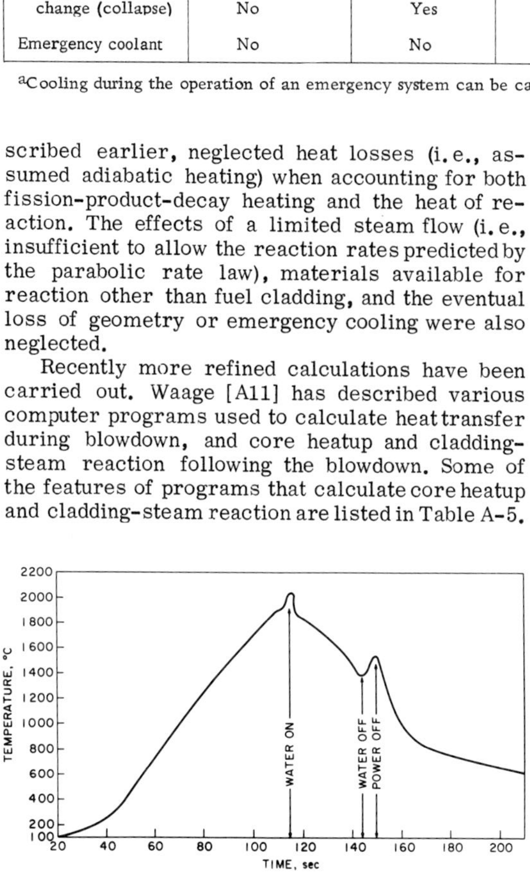

FIG. A-8 Preliminary emergency-coolant-spray experiment with a Zircaloy-2-clad, UO_2-core simulated fuel rod (coolant water spray rate: 190 g/min).

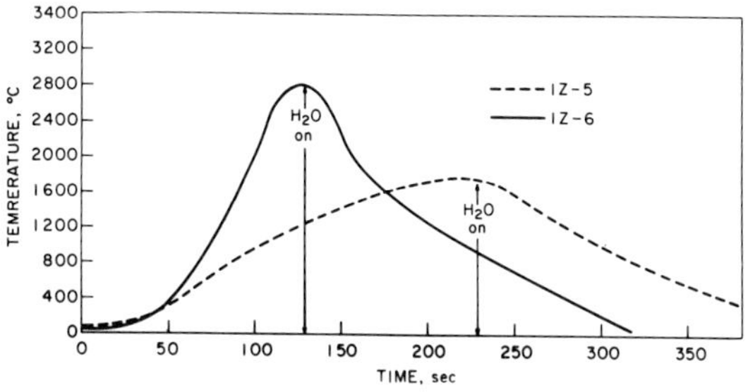

FIG. A-9 Emergency-coolant-spray experiments with Zircaloy-2-clad, UO_2-core simulated fuel rods undergoing nearly total cooldown (coolant water spray rate: 190 g/min).

The computer programs vary greatly in detail. Some programs calculate both the blowdown and core-heatup periods and account for loss of fuel integrity, while others only calculate core heatup and ignore changes in geometry. All the listed programs use the parabolic rate law to describe the cladding-steam reaction, and all but the TACT-V code assume a reaction rate limited by steam availability.

Although the programs vary, the general problem associated with calculating the core heatup and chemical reaction in a loss of coolant can be understood by considering the CHEMLOC-II program [A12].

CHEMLOC-II is a computer program that describes the core heating and chemical reaction from the end of the blowdown period to the time that the core collapses onto the supporting structure, neglecting emergency cooling. The core is assumed to be a parallel array of rods. The array is divided into a number of radial zones, each having the same number of rods. The center zone is a cylinder, while the remaining zones are annuli. Each zone is further divided axially into sections. Steam enters the bottom of the core at a rate that may be constant or may vary with time. The steam flow is required as input information.

The basic calculation for each axial section within the various radial zones is a heat and mass balance made on an elemental section of a fuel rod together with the associated gas-flow area. An element of a fuel rod is shown in Fig. A-14, which includes schematically the heat and mass flows that are accounted for in the program. The program input requires the cladding and fuel temperatures evaluated throughout the core following the blowdown (i.e., at the time the core is bared of coolant), as well as the fission-product-decay energy distribution (both in space and time).

The chemical reaction occurring in an element is calculated either by the parabolic rate law or by assuming the reaction rate to be the rate at which steam diffuses to the cladding surface from the steam-hydrogen gas stream passing each element. In the program, each is calculated, and the lower value is used.

The program calculates the heat and mass transfer throughout the core. Included are heat conduction, convection, and radiation both axially and radially between sections and zones throughout the core, as well as the flow of gas through the core. The average temperatures of fuel, cladding, and gas

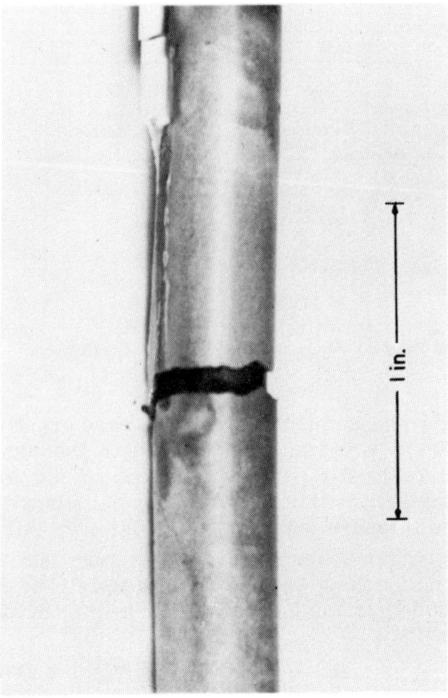

Experiment 1Z-6

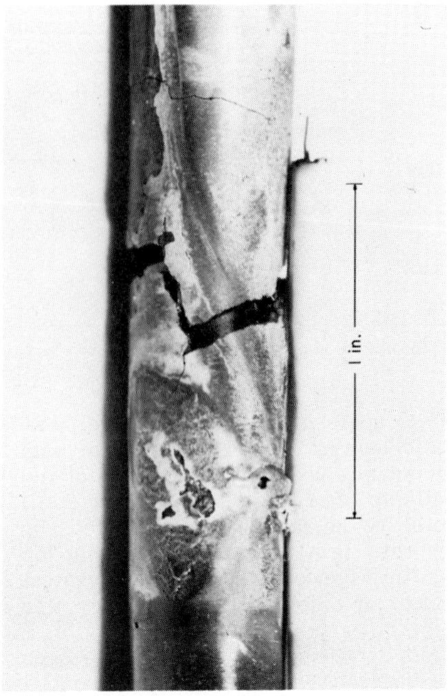

Experiment 1Z-5

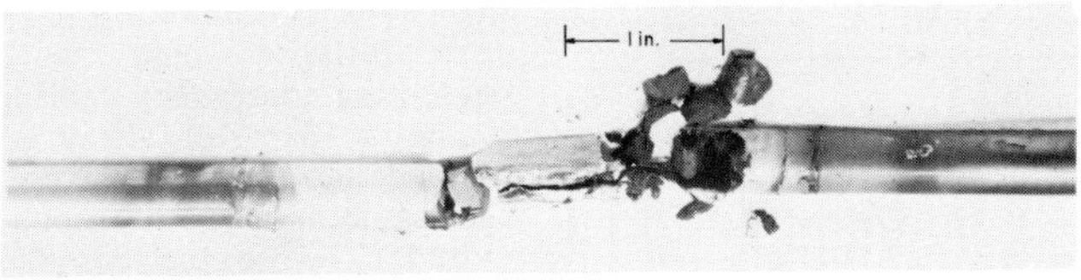

Experiment 1Z-4

FIG. A-10 Zircaloy-2-clad, UO_2-core simulated fuel rods after undergoing an emergency-coolant-spray experiment.

and the bulk gas composition (mol fraction) are calculated for each section in the core.

The program requires materials properties as input. These can be put into the program as functions of temperature.

The loss of geometry within the core is taken into consideration in the program by the use of several input parameters. For example, a section is assumed to change geometry when a preselected temperature (input parameter) is reached, and in the next time interval the fuel and cladding are assumed to slump. An input parameter, defined as the ratio of the length occupied by the slumped material to its original length, is used to define the new geometry. With the slumping of the fuel-cladding mixture, appropriate changes in heat transfer both axially and radially are made in the program. When a predetermined number (also required as input) of axial sections become slumped, steam flow is assumed to cease in the affected zone, and is redistributed into other zones.

In a similar manner, factors are required as input parameters to allow for the dripping of material into lower positions of the core and out of the

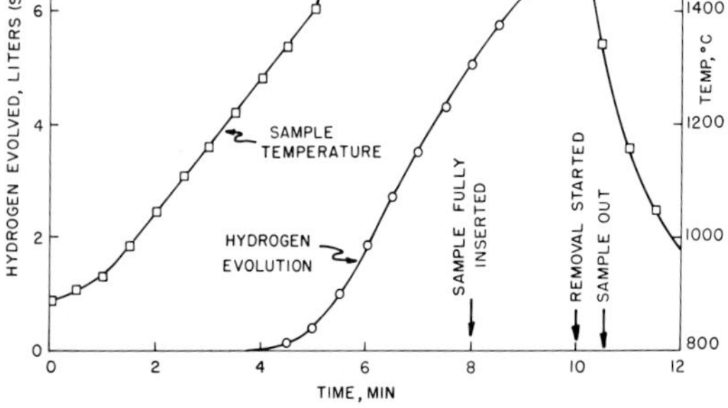

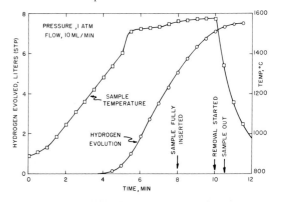

FIG. A-11 Hydrogen evolution and sample temperature during stainless steel-304 clad fuel rod furnace experiment.

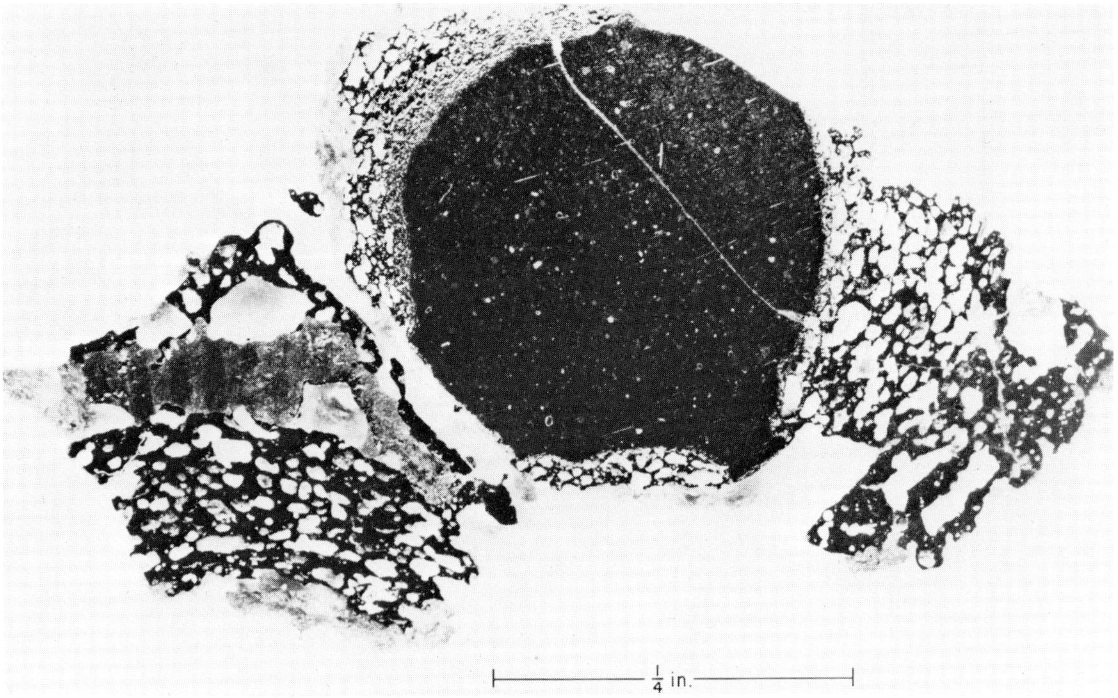

FIG. A-12 Radial cross section near the top of a type-304 stainless steel-clad, UO_2-core simulated fuel element exposed to 1 atm steam at 1600°C for 10 min.

core through the bottom support structure.

Extraneous materials, such as subassembly cans and spacers, are accounted for as both additional surface for reaction and additional mass. Heat lost to the surrounding vessel walls and bottom and top of the core structure is handled by the assumption that the core is surrounded by an annulus with a slab on the top and bottom of the core.

The results of calculations with CHEMLOC for a 3300 Mw(t) boiling water reactor [A13] are shown in Figs. A-15, A-16 and A-17. In the calculations, the failure or slumping temperature was assumed to be 2200°C, an assumption based on the observation of failure at 2140°C in the induction-heating experiments described earlier. It was also assumed that when 24 in. of fuel rod had slumped, the steam flow would be restricted in a zone. The steam flow and decay-heating rate used in the calculations are shown in Fig. A-15. The steam flow was calculated on the assumption that the bottom vessel head was full of water and that the energy required for the

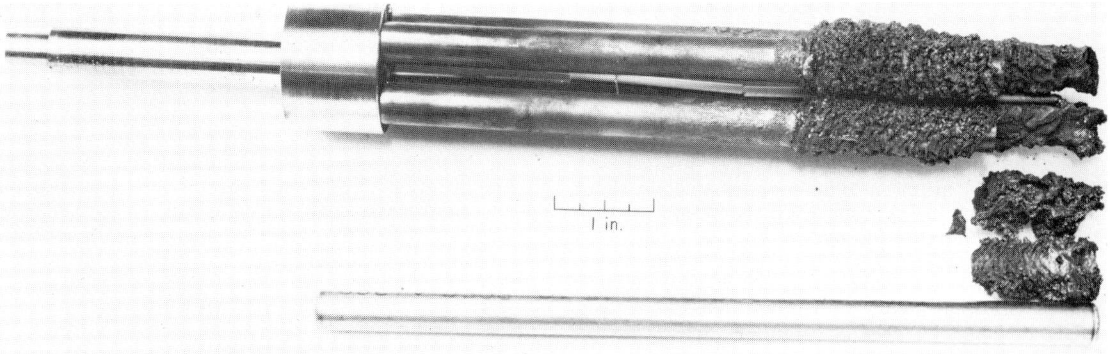

FIG. A-13 Four-rod bundle of type-304 stainless steel-clad, UO_2-core fuel after exposure to 1 atm steam at 1600°C (unexposed fuel rod shown for comparison).

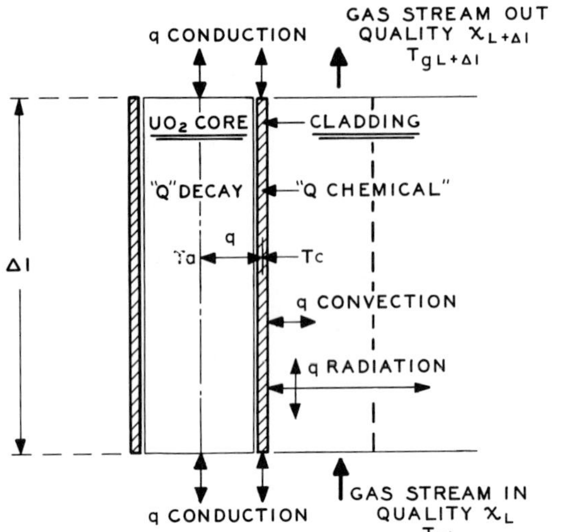

FIG. A-14 Elemental section of a fuel rod, CHEMLOC-II.

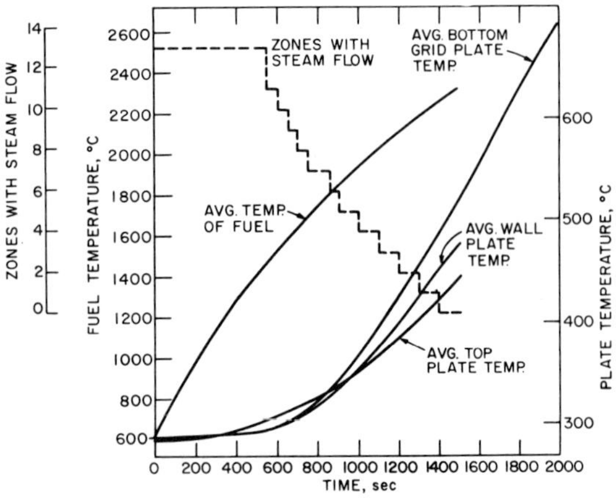

FIG. A-16 Average temperatures of fuel, bottom grid plate, top plate, and wall plate, calculated with CHEMLOC-II program, for a 3300 Mw(t) boiling-water reactor. Slumping temperature assumed to be 2200°C (3828°F). Number of zones with steam flow are also shown as a function of time.

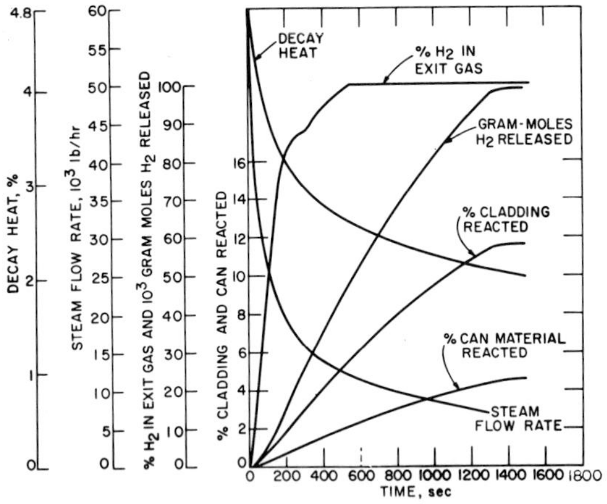

FIG. A-15 Percentage of can and cladding reacted versus time, calculated with CHEMLOC-II program, for a 3300 Mw(t) boiling-water reactor. Slumping temperature assumed to be 2200°C (3828°F). Percent decay heat, steam flow rate, percent hydrogen in exit gas, and gram-moles of hydrogen in exit gas are also shown as a function of time.

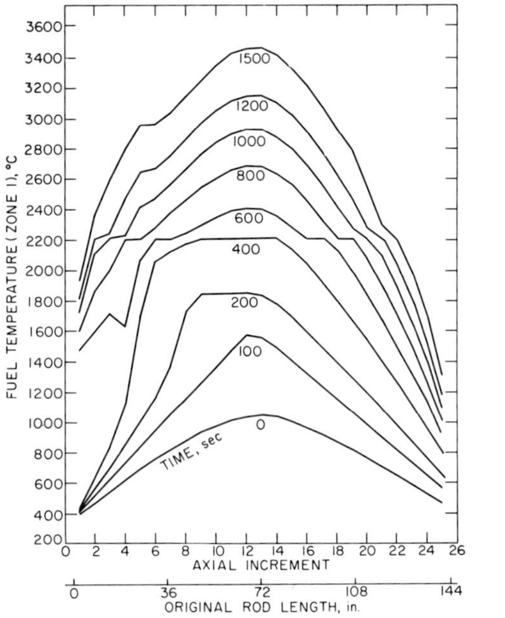

FIG. A-17 Fuel temperature versus axial position, calculated with CHEMLOC-II program, for a 3300 Mw(t) boiling-water reactor. Slumping temperature assumed to be 2200°C (3828°F).

boil-off was supplied by that stored in the vessel head. The notation "% can" in the figure refers to the materials in the core, other than cladding, that react. It can be seen that fuel temperatures reach 2800°C (UO_2 m.p.) in 1000 sec, and that, in about 500 sec after the blowdown, all the steam through the core is converted to hydrogen.

In such calculations, the cladding-steam reaction predicted by the parabolic rate law for zirconium proceeds at a very slow corrosion-like rate up to about 1200°C [A10]. As the temperature increases further, the reaction rate begins to increase rapidly; the heat of reaction, if unlimited steam is available, causes the temperature to increase rapidly and, consequently, results in an even more rapid reaction rate. The possible steam-supply rate to the core, based on the water that might remain in the bottom vessel head, is insufficient to support the reaction rate called for by the rate law. The reaction rates and core-heatup rates are thus controlled by the values of the steam flow postulated for the accident being analyzed.

A.2 THE EXCURSION ACCIDENT

The excursion accident is assumed to result from a rapid reactivity insertion, which causes the temperature of a portion of a reactor core to increase very rapidly. The analysis of such events involves a determination of the reactivity and rate of insertion available from the inadvertent or accidental motion of the control rods. Usually, it is assumed that a control rod falls from the core by gravity. The power increase in the core is terminated by the Doppler effect. From the analysis, the total energy deposited throughout the core can be calculated.

To date, power reactor designs have intentionally limited the available reactivity and insertion rate (and thus the peak temperatures of fuel that can be reached in an excursion) to levels in which little or no loss of fuel integrity is expected to occur. However, to establish the margin of safety, the behavior of rapidly overheated fuel has been examined.

A.2.1 Experimental Simulations of Excursion Accidents

Several experiments have been conducted in order to establish the extent of cladding-steam reaction and the physical behavior of reactor fuel materials during excursions. In these experiments, sample fuel materials typical of those used in test, production, and power reactors have been subjected to transient nuclear heating. Such experiments have been conducted in the TREAT reactor [A14] and the Capsule Driver Core (CDC) facility [A15]. Entire reactor cores have also been subjected to transients in the SPERT program [A15, A16].

The TREAT experiments, concerned primarily with metal-water reactions, are discussed here. The technique involves subjection of fuel samples submerged in a pool of water to a burst of neutrons. In some experiments, samples were contained in opaque high-pressure autoclaves [A14]. Fuel temperatures were obtained from thermocouples spot-welded to sample surfaces. Pressures were recorded by strain-gauge pressure transducers mounted on the autoclaves. To determine the extent of reaction, the gas phase in the autoclave was sampled after the transient irradiation and the amount of hydrogen evolved was determined.

A capsule allowing high-speed motion pictures (in color) to be taken has been used recently in similar experiments [A17]. The events observed are correlated with the reactor power. In the TREAT experiments, calculation of the fission energy developed in the sample is based on calibration experiments. Burnup analyses of test samples are made to relate sample energy to reactor energy.

A.2.1.1 TREAT Experiments with Zircaloy-clad UO_2 Fuel

Samples of Zircaloy-2-clad, UO_2-core (both pelletized and vibrationally compacted) fuel rods of various sizes have been tested in opaque autoclave experiments in TREAT. Results, in terms of the extent of cladding reacted as a function of fission energy input, are shown in Fig. A-18 [A18]. The single pins tested consisted of a single pellet of UO_2 (11.2% enriched), 0.5 in. long by 0.375-in. OD, clad with 25 mils of Zircaloy-2. Tests with single pins of this type were described in the earlier text. The rods were similarly constructed, except that 10 pellets were clad, rather than one. Each vibra-packed UO_2 rod was 9.5 in. long by 0.566 in. OD (30-mil wall) and contained 110 g UO_2 (15% enriched) packed to 80% of theoretical density. One of the vibra-packed fuel rods was pressurized to 1500 psia (helium) during fabrication, rather than to the normal single atmosphere. Experiments were performed with the water initially at 30°C at a pressure of 20 psia and with the water at 285°C at a pressure of 1000 psia.

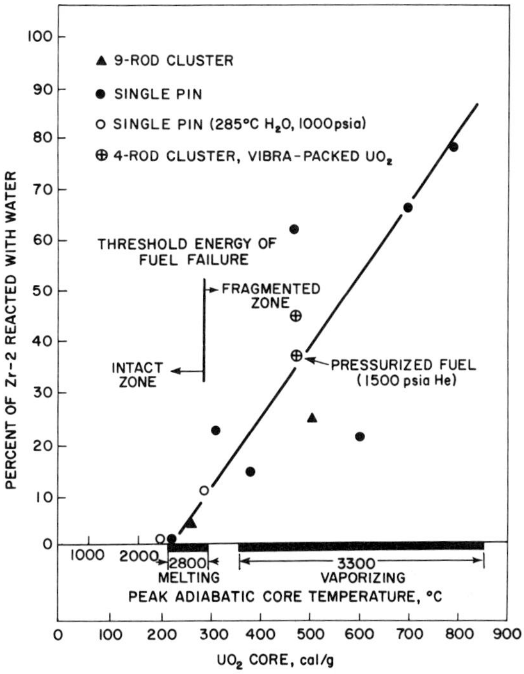

FIG. A-18 Results of meltdown experiments in TREAT on Zircaloy-2 clad, UO_2-core, simulated fuel elements submerged in water (water at 30°C and 30 psia, except as noted).

Post-test observation of the fuel samples established a threshold for failure at a fission energy input of 280-290 cal/g UO_2 (see Fig. A-18). Failure is used here to denote the loss of cladding integrity. In Fig. A-19, photographs of samples from the opaque autoclave tests are shown [A21]. The samples shown were tested at energies above and below the threshold energy.

In recent experiments [A18-A20] using the photographic capsule, several failure modes have been identified. Photographs of two samples after testing are shown in Fig. A-20. In these tests, pellet-core rods were 5.62 in. long and had a 0.42-in. OD. The core consisted of ten 0.5-in.-long UO_2 pellets with a 25-mil cladding wall. The vibra-packed rods were 7.75 in. long and had a 0.56 in. OD with a 30-mil wall. The UO_2 was packed to 86% of theoretical density, compared to an 85% smear density for the pellet-core rods.

In these experiments, the failure of the pellet-fuel rod tested at the lower energy (290 cal/g) occurred after the neutron pulse. The fuel rod appeared to remain intact during the several seconds recorded on film. The fragmented appear-

Intact

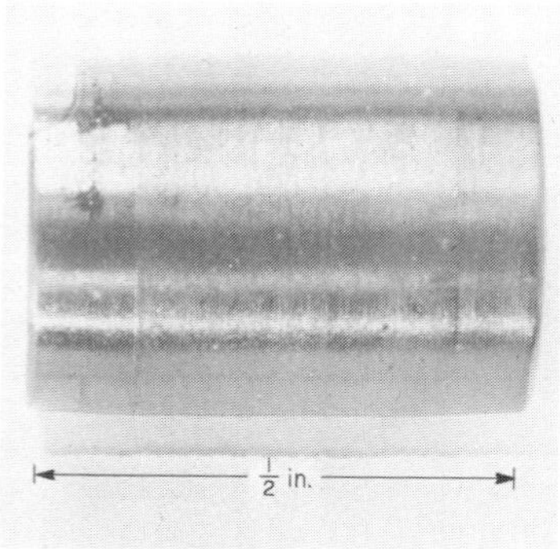

Fission Energy, cal/g	216	280
Reactor Period, ms	74	77
Test Identification No.	CEN 187	CEN 193

Failed

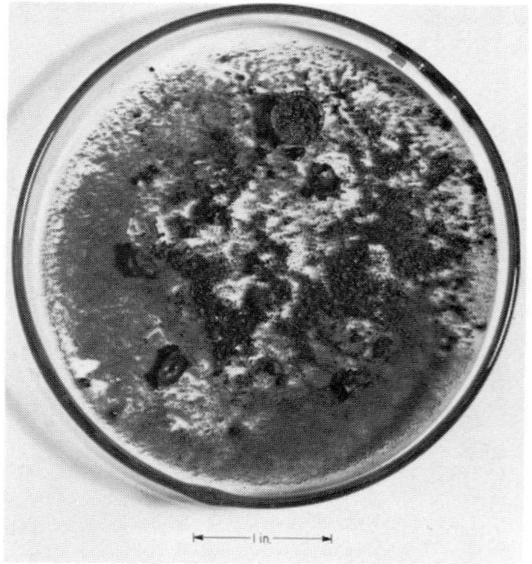

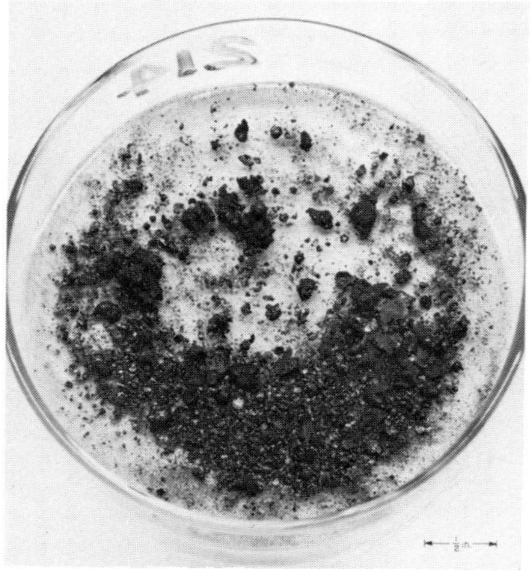

290	375	450
72	73	52
CEN 181	CEN 214	CEN 175

FIG. A-19 Results of opaque autoclave experiments in TREAT with Zircaloy-2-clad, UO_2 pellet core samples.

ance of the rod (see Fig. A-20) was apparently due to cladding oxidation and melting and to fuel-clad interaction, which occurred following the transient.

The pellet-fuel rod tested at the higher energy (330 cal/g) was seen to fail at a time coinciding with the end of the transient, when the rod ruptured and some of the fuel expelled. However, the rod appeared to remain in one large piece, except for the expelled UO_2, during the several seconds following the transient that were recorded on film. The fragmented appearance of the rod is due to the same degradation of the cladding as experienced by the rod described above.

The vibra-packed rod tested at the higher energy (335 cal/g) failed in the same manner as this latter pellet-fuel rod. The cladding rupture occurred when the energy input to the rod was 240 cal/g.

The vibra-packed rod tested at the lower energy (258 cal/g) did not fail during the transient; however, as seen in Fig. A-20, the rod was bulged and the cladding was breached after the test.

It was apparent from the films and temperature records of these tests that the cladding reached its maximum temperature after the transient. The reaction of the cladding with steam occurred primarily during the several seconds of the quenching or cooldown period following the transient rather than during the single second or less of the power burst.

Comparison of failures of Zircaloy-clad UO_2 pellet and powder fuel from TREAT photographic tests suggests the existence of two different failure mechanisms for both fuel types. Both failure mechanisms can occur in a fuel rod, but the one that is predominant depends principally on the total fission energy of the transient. The mechanisms can be defined as:

1. a failure that generally occurs during the power transient and is manifested by rapid pressurization within the fuel specimen, followed by vigorous rupture of the cladding and injection of some of the fuel into the coolant; and
2. a failure that occurs after the transient and is due to excessive cladding temperature, degradation of the cladding by metal-water reaction, and interaction of hot fuel with the cladding.

The first failure mechanism, in which the cladding is breached, can occur without a significant effect of the second deteriorating mechanism. The cladding of some fuel rods subjected to energies between 240 and 290 cal/g were bulged and breached.

The second mechanism, which becomes predominant only after fuel rods have been ruptured at energies above 290 cal/g, results in completely fragmented fuel.

Failures attributable to the first mechanisms, a pressure rupture, can be classified in two ways: (1) prompt — those occurring during the transient, and (2) delayed — those occurring at the end or slightly after the transient. Prompt failures usually result in extensive destruction of the fuel specimens, since fission energy is introduced into the fuel after the cladding has ruptured and significant metal-water reaction occurs following the rupture. The second failure mechanism then becomes predominant and leads to further deterioration of the fuel rod.

Table A-6 summarizes the results of TREAT photographic tests with Zircaloy-clad UO_2 fuels in terms of the failure mechanisms described above.

TABLE A-6

Summary of TREAT Photographic Tests with Zircaloy-clad UO_2 Fuels

Run no. CEN-	Fuel type	Enrichment (wt. % U^{235})	Reactor period (msec)	Total transient energy (cal/g UO_2)	Energy at failure (cal/g UO_2)	Mode of failure
234T	Vibra-Pac	5	43	354	235	Prompt
233T	Vibra-Pac	5	48	335	277	Prompt
239T	Pellet	5	48	331	291	Prompt
240T	Pellet	5	63	263	263	Delayed
238T	Pellet	5	66	244	Did not fail	-
237T	Vibra-Pac	5	67	258	258	Delayed
235T	Vibra-Pac	5	71	233	Did not fail	-
229T	Vibra-Pac	10	74	357	277	Prompt
236T	Vibra-Pac	5	75	223	Did not fail	-
224T	Pellet	10	80	450	295	Prompt
225T	Pellet	10	107	330	330	Prompt
230T	Pellet	10	108	290	290	Delayed

UO_2 Core Pellets		UO_2 Core Vibrationally Compacted	
Clad OD: 0.42 in.		Clad OD: 0.56 in.	
Clad Thickness: 25 mil		Clad Thickness: 30 mil	

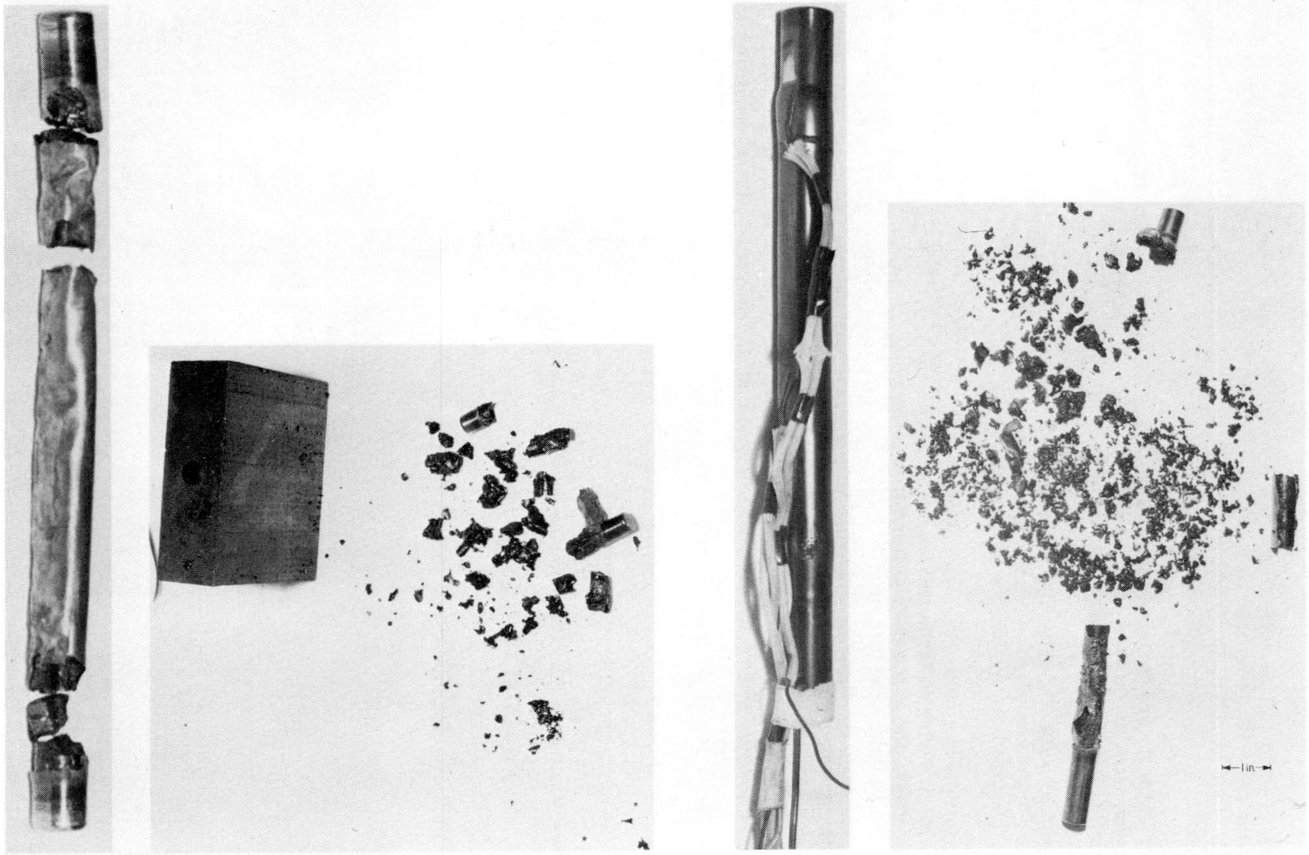

290	330	258	335
108	107	67	48
230T	225T	237T	233T

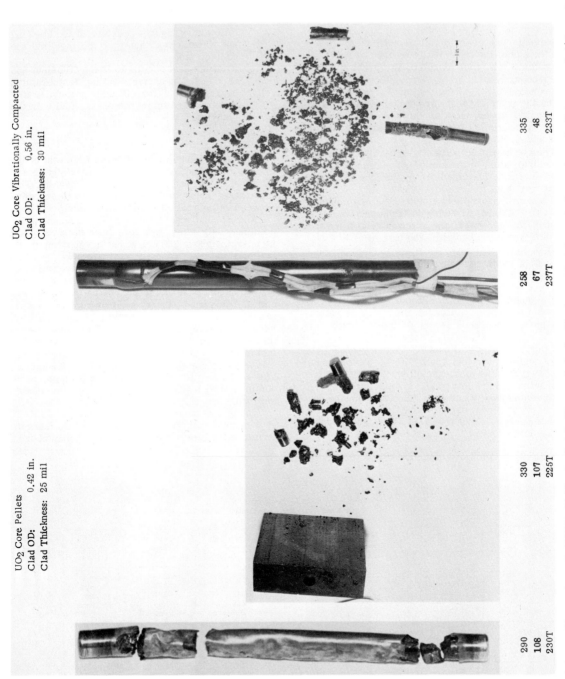

FIG. A-20 Results of photographic experiments in TREAT with Zircaloy-2-clad, UO_2-core fuel rods. The three entries below each photograph are the fission energy (cal/g), reactor period (ms), and test identification number, as in Fig. A-19.

REFERENCES

1. C. R. Tipton, Jr. (Ed.) Reactor Handbook, Vol. 1, "Materials," 2nd Ed., Interscience Publishers, Inc., N.Y., 1960.
   (a) "Corrosion Rate of Uranium in Na or NaK," p. 132.
   (b) "Effect of Alloy Additions on the Corrosion Behavior of Thorium in High-Purity Water," p. 228.
   (c) "BeO Weight Loss at High Temperature Due to Volatilization and Reaction with Water Vapor," p. 938.
2. G. E. Zima, "A Review of the Properties of Zircaloy-2," Report HW-60908, General Electric Co., Hanford Atomic Products Operation, 1959.
3. W. E. Berry, "Corrosion Behavior of Cladding Materials in High-Temperature Water," AEC-EURATOM Conference on Aqueous Corrosion of Reactor Materials, USAEC Report TID-7587, p. 71, 1960.
4. K. Osthagen and P. Kofstad, "Oxidation of zirconium and zirconium-oxygen alloys at 800°C," J. Electrochem. Soc., 109(1962)204.
5. H. H. Uhlig, "Initial oxidation rate of metals and the logarithmic equation," Acta Met., 4(1956)541.
6. N. Cabrera and N. F. Mott, "Theory of the oxidation of metals," Reports on Progress in Physics, Vol. 12, p. 163, The Physical Society, London, 1949.
7. H. A. Porte, J. G. Schnizlein, R. C. Vogel and D. F. Fischer, "Oxidation of Zr and Zr alloys," J. Electrochem. Soc., 107(1960)506.
8. B. Lustman and F. Kerze, Jr., "Metallurgy of Zirconium," National Nuclear Energy Series, Div. VII, Vol. 4, McGraw-Hill Book Co., N.Y., 1955.
9. M. W. Mallett, W. M. Albrecht and R. E. Bennett, "Reaction of zirconium with water vapor at subatmospheric pressures," J. Electrochem. Soc., 104(1957)349.
10. R. G. Charles, S. Barnartt and E. H. Gulbransen, "Prolonged oxidation of zirconium at 350° and 450°," Trans. Met. Soc. AIME, 212(1958)101.
11. J. P. Pemsler, "The Corrosion of Zirconium Alloys in 900°F Steam," Report NMI-1208, Nuclear Metals, Inc., 1958.
12. K. M. Goldman and D. E. Thomas, "Properties of Zircaloy-2," Report WAPD-T-43, Bettis, Atomic Power Laboratory, 1953.
13. J. P. Pemsler, "Corrosion of Zirconium Alloys in 900 and 1000°F Steam," AEC-EURATOM Conference on Aqueous Corrosion of Reactor Materials, USAEC Report TID-7587, p. 96, 1960.
14. J. N. Wanklyn, J. T. Demant and D. Jones, "The Corrosion of Zirconium and Its Alloys by High Temperature Steam," British Report AERE-R-3655, Atomic Energy Research Establishment, Research Group, Harwell, Berks., England, 1961.
15. R. A. Thiede, "Autoclave Testing of Zircaloy-2," USAEC Report HW-65350, General Electric Co., Hanford Atomic Products Operation, 1960.
16. Bettis Technical Review, "Reactor Metallurgy," Report WAPD-BT-10, Bettis, Atomic Power Laboratory, 1958.
17. G. E. Galonian, "Effect of Radiation on the Corrosion of Metallic Materials in 580°F Water," Report KAPL-M-GEC-4, Knolls Atomic Power Laboratory, 1955.
18. K. Alcock and B. Cox, "The Oxidation and Corrosion of Zirconium and Its Alloys," British Report AERE-C/R-2826, Atomic Energy Research Establishment, Research Group, Harwell, Berks., England, 1959.
19. C. M. Schwartz and D. A. Vaughan, "Effect of Hydrogen Pickup on Corrosion Behavior of Zirconium in Water," Report BMI-1120, Battelle Memorial Institute, 1956.
20. J. N. Wanklyn, D. R. Sylvester, J. Dalton and N. J. M. Wilkins, "The Corrosion of Zirconium and Its Alloys in High Temperature Steam. Part II. The Uptake of Hydrogen During Corrosion," British Report AERE-R-3768, Atomic Energy Research Establishment, Risley, Warrington, Lancs., England, 1961.
21. J. M. Markowitz, "Hydrogen Redistribution in Thin Plates of Zr under Large Thermal Gradients," Report WAPD-TM-104, Bettis, Atomic Power Laboratory, 1958.
22. J. Belle and M. W. Mallett, "Kinetics of the high temperature oxidation of zirconium," J. Electrochem. Soc., 101(1954)339.
23. D. Cubicciotti, "The oxidation of zirconium at high temperatures," J. Am. Chem. Soc., 72(1950)4138.
24. M. W. Fassel, "Progress Report No. III. The High Temperature Oxidation of Metals. June 1 - September 26, 1952," USAEC Report NP-4246, DTI Extension, AEC, Oak Ridge, 1952.
25. H. B. Probst, E. B. Evans and W. M. Baldwin, Jr., "Scaling of Zirconium at Elevated Temperatures. Final Report," USAEC Report AECU-4113, DTI Extension, AEC, Oak Ridge, 1959.
26. F. E. Littman, F. M. Church and E. M. Kinderman, "A Study of Metal Ignitions - II. The spontaneous ignition of zirconium," J. Less-Common Metals, 3(1961)379.
27. R. F. Domagala and D. J. McPherson, "System zirconium-oxygen," J. Metals, 6(1954)238.
28. M. W. Mallett, J. Belle and B. B. Cleland, "The reaction of nitrogen with, and the diffusion of nitrogen in, beta zirconium," J. Electrochem. Soc., 101(1954)1.
29. L. F. Kendall, R. G. Wheeler and S. H. Bush, "Reaction kinetics of zirconium and Zircaloy-2 in dry air at elevated temperatures," Nucl. Sci. Eng., 3(1958)171.
30. C. Tyzack, "Zirconium and its alloys," Nucl. Eng., 3(1958) 102.
31. M. N. A. Hall, S. L. H. Martin and A. L. G. Rees, "The solubility of hydrogen in zirconium-oxygen solid solutions," Trans. Faraday Soc., 41(1945)306.
32. E. A. Gulbransen and K. F. Andrew, "Diffusion of hydrogen and deuterium in high purity zirconium," J. Electrochem. Soc., 101(1954)560.
33. E. C. Miller, "Corrosion of Materials by Liquid Metals," Liquid Metals Handbook, 2nd Ed., p. 144, USAEC, U.S. Govt. Printing Office.
34. W. D. Manly, "Fundamentals of liquid metal corrosion," Corrosion, 12,7(1956)46.
35. J. H. Stang, "Corrosion by liquid metals, fused salts and organics," Reactor Materials, 4,1(1961)44; See also: NAA-SR-5350, 1960.
36. J. H. Stang, "Corrosion by liquid metals and fused salts," Reactor Materials, 5,3(1962)42; See also: NAA-SR-6674, 1962.
37. J. H. Stang, "Corrosion by liquid metals," Reactor Materials, 5,2(1962)46; See also: Nucl. Metallurgy, Vol. VII, Sp. Rep. No. 10, 1960; and LAMS-2620, 1961 and Trans. Amer. Nucl. Soc., 4,2(1961)347.
38. A. B. McIntosh and K. Q. Bagley, "Selection of canning materials for reactors cooled by sodium/potassium and carbon dioxide," J. Inst. Metals, 84(1956)251.
39. E. E. Hoffman, "Corrosion of Materials by Lithium at Elevated Temperatures," USAEC Report ORNL-2924, Oak Ridge National Laboratory, 1960.
40. E. S. Bartlett, "Corrosion of Base Metal Alloys," Reactor Materials, 4,2(1961)34; See also: NASA-TN-D769, 1961.
41. L. L. Anderson and G. R. Hill, "Corrosion of Metals in Fused Salt Systems," in "Surface Chemistry Phenomena," Progress Report USAEC TID-14370, p. 47, 1961.
42. W. J. Mecham, R. C. Liimatainen, R. W. Kessie and W. B. Seefeldt, "Decontamination of irradiated uranium by a fluoride volatility process," Chem. Eng. Progr., 53,2(1957)72F.
43. J. E. Draley, "Aqueous Corrosion of 1100 Aluminum and of Aluminum-Nickel Alloys," AEC-EURATOM Conference on Aqueous Corrosion of Reactor Materials, USAEC Report TID-7587, p. 165, 1960.
44. J. E. Draley and W. E. Ruther, "Aqueous Corrosion of 2S Aluminum at Elevated Temperatures," USAEC Report ANL-5001, Argonne National Laboratory, 1953.
45. J. E. Draley and W. E. Ruther, "Corrosion Resistant Aluminum Above 200°C," USAEC Report ANL-5430, Argonne National Laboratory, 1955.
46. R. L. Dillon, R. E. Wilson and V. H. Troutner, "High Temperature Aqueous Corrosion of Commercial Aluminum Alloys," Report HW-37636, General Electric Co., Hanford Atomic Products Operation, 1956.
47. C. Groot and R. E. Wilson, "The Intergranular Corrosion of Aluminum in Super-Heated Steam," Report HW-41797, General Electric Co., Hanford Atomic Products Operation, 1956.
48. J. E. Draley, W. E. Ruther and S. Greenberg, "Corrosion of Aluminum and Its Alloys in Superheated Steam," USAEC Report ANL-6207, Argonne National Laboratory, 1961.
49. J. N. Wanklyn and N. J. M. Wilkins, "The Corrosion of Aluminum, Its Alloys and SAP in High Pressure Steam," AEC-EURATOM Conference on Aqueous Corrosion of Reactor Materials, USAEC Report TID-7587, p. 153, 1960.
50. R. L. Dillon, "Dissolution of Aluminum Oxide as a Regulating Factor in Aqueous Aluminum Corrosion," AEC-EURATOM Conference on Aqueous Corrosion of Reactor Materials, USAEC Report TID-7587, p. 134, 1960.
51. R. M. Haag and F. C. Zyzes, "Corrosion of Aluminum in High-Temperature Water. III. Inhibition of Corrosion by Sodium Silicate," Report KAPL-1741, Knolls Atomic Power Laboratory, 1957.
52. O. Kubaschewski and B. E. Hopkins, Oxidation of Metals and Alloys, (2nd ed.), Academic Press, Inc., N.Y., 1962.
53. D. W. Aylmore, S. J. Gregg and W. B. Jepson, "The oxidation of aluminum in dry oxygen in the temperature range 400-650°C," J. Inst. Metals, 88(1960)205.
54. D. P. Smith, Hydrogen in Metals, The University of Chicago Press, Chicago, Ill., 1948.
55. N. J. Gioseffi and H. E. Kline, "Behavior of structural materials exposed to an organic reactor environment," Trans. Amer. Nucl. Soc., 2,1(1959)25.
56. E. G. Kendall et al., "Fabrication development of APM fuel elements for organic cooled reactors," Nucl. Eng., 6(1961)522.

57. K. Maddocks, "Organic Fluids as Reactor Coolants," British Report AERE-R-3633, Atomic Energy Research Establishment, Risley, 1959.
58. G. Luce, "Stainless Steel, Construction and Engineering Memorandum No. 24," USAEC Report TID-5327, 1956.
59. T. Lyman (Ed.), Metals Handbook, Vol. 1, "Properties and Selection of Metals," (8th ed.), American Society for Metals, Novelty, Ohio, 1961, pp. 419, 429.
60. S. H. Bush, "Trip Report Status of 17-4-PH and Other Precipitation Hardening Stainless Steels," Report HW-68608, General Electric Co., Hanford Atomic Products Operation, 1961.
61. H. A. Pray and W. K. Boyd, "Corrosion of Stainless Steels in Supercritical Water," Report BMI-901, Battelle Memorial Institute, 1954.
62. J. H. Stang, "Corrosion by Molten Metals and Phosphates," Reactor Materials, 5,1(1962)41; See also: TID-11307, 1960.
63. J. H. Stang, "Corrosion by Liquid Metals and Fused Salts," Reactor Materials, 4,4(1961)36; See also: NRL-5572(1960).
64. F. H. Welch, "Properties of Lithium Hydride-V; Corrosion of Austenitic Stainless Steels in Molten LiH," Report APEX-673, General Electric Co., ANP Project, 1961.
65. "Molten-Salt Reactor Program Progress Report for Period from March 1 to August 31, 1961," USAEC Report ORNL-3215, p. 93, Oak Ridge National Laboratory, 1962.
66. W. D. Manley et al., "Metallurgical Problems in Molten Fluoride Systems," Proceedings of the Second U. N. International Conference on Peaceful Uses of Atomic Energy, Geneva, 1958, Vol. 7, p. 223.
67. S. Greenberg and W. E. Ruther, "Aqueous Corrosion of Magnesium Alloys," USAEC Report ANL-6070, Argonne National Laboratory, 1960.
68. S. J. Gregg and W. B. Jepson, "The high-temperature oxidation of magnesium in dry and moist oxygen," J. Inst. Metals, 87(1959)187.
69. T. E. Leontis and F. N. Rhines, "Rates of high-temperature oxidation of magnesium and magnesium alloys," Trans. Amer. Inst. Mining and Met. Engrs., 166(1946)265.
70. K. D. Sinelnikov, V. E. Ivanov and V. F. Zelensky, "Magnesium-Beryllium Alloys as Material for Nuclear Reactors," Proceedings of Second U.N. International Conference on the Peaceful Uses of Atomic Energy, Geneva, 1958, Vol. 5, p. 234.
71. H. Inouye, "The Reactions of Magnesium and Magnesium Alloys with Gases at High Temperatures," Report CF-58-1-93, Oak Ridge National Laboratory, 1958.
72. R. Caillet and R. Darras, "Corrosion of Magnesium and Certain of Its Alloys in Gas-Cooled Reactors," Proceedings of Second U.N. International Conference on the Peaceful Uses of Atomic Energy, Geneva, 1958, Vol. 5, p. 220.
73. W. A. Mollison, G. C. English and F. Nelson, "Corrosion of Uranium in Distilled Water," Report CT-3055, University of Chicago Metallurgical Laboratory, 1945.
74. R. W. Nichols, "Uranium and its alloys," Nucl. Eng., 2(1957)362.
75. L. Leibowitz, J. G. Schnizlein, J. D. Bingle and R. C. Vogel, "The kinetics of oxidation of uranium between 125° and 250°C," J. Electrochem. Soc., 108(1961)1155.
76. L. Leibowitz, J. G. Schnizlein, L. W. Mishler and R. C. Vogel, "A microscopic study of oxide films on uranium," J. Electrochem. Soc., 108(1961)1153.
77. L. Leibowitz, J. D. Bingle and M. Homa, "An x-ray study of oxidized uranium surfaces," J. Electrochem. Soc., 111 (1964)248.
78. J. G. Schnizlein, P. J. Pizzolato, H. A. Porte, J. D. Bingle, D. F. Fischer, L. W. Mishler and R. C. Vogel, "Ignition Behavior and Kinetics of Oxidation of the Reactor Metals, Uranium, Zirconium, Plutonium, and Thorium and Binary Alloys of Each," USAEC Report ANL-5974, Argonne National Laboratory, 1959.
79. R. K. Hilliard, "Oxidation of Uranium in Air at High Temperatures," Report HW-58022, General Electric Co., Hanford Atomic Products Operation, 1958.
80. D. J. Littler (Ed.), "Properties of Reactor Materials and Effects of Radiation," Proceedings of International Conference held at Berkeley Castle, Gloucestershire, England, May 30-June 2, 1961, Butterworths, London, 1962.
81. J. E. Antill and K. A. Peakall, "Oxidation of uranium alloys in carbon dioxide and air," J. Less Common Metals, 3 (1961)239.
82. W. D. Wilkinson, Uranium Metallurgy, Vol. II, "Uranium Corrosion and Alloys," Interscience Publications, Inc., 1962, p. 757.
83. K. W. Covert and M. Kolodney, "Protection of Plutonium Against Atmospheric Oxidation," USAEC Report LA-314, Los Alamos Scientific Laboratory, 1945.
84. E. Dempsey and A. E. Kay, "Some investigations on plutonium metal," J. Inst. Metals, 86(1958)379.
85. J. T. Waber, "The Corrosion Behaviors of Plutonium and Uranium," Proceedings of the Second U.N. International Conference on Peaceful Uses of Atomic Energy, Geneva, 1958, Vol. 6, p. 204.
86. J. T. Waber and E. S. Wright, "The Corrosion of Plutonium" (Presented at the AEC-ASM Conference, Chicago, 1957), The Metal Plutonium, A. S. Coffinberry and W. N. Miner (Eds), p. 194, University of Chicago Press, Chicago, Ill., 1961.
87. J. B. Raynor and J. F. Sachman, "Oxidation of plutonium in moist air and argon," Nature, 197(1963)587.
88. J. F. Sachmann, "The Atmospheric Oxidation of Plutonium Metal," (International Conference on the Metallurgy of Plutonium, Grenoble, 1960), Plutonium 1960, p. 222, Cleaver Hume, Ltd., London, 1961.
89. J. G. Schnizlein and D. F. Fischer, "Oxidation Kinetics of Plutonium between 140 and 450°C," 12th Annual AEC Corrosion Symposium, Pleasanton, Calif., May 20-22, 1963.
90. J. G. Schnizlein and D. F. Fischer, "Metal Oxidation and Ignition Kinetics: Isothermal Oxidation of Plutonium," Chemical Engineering Division Summary Report, Jan., Feb., March 1963, USAEC Report ANL-6687, p. 171, Argonne National Laboratory, 1963.
91. M. J. F. Notley, E. N. Hodkin and J. A. C. Davidson, "The Oxidation of Plutonium and Certain Plutonium Alloys in Air and Carbon Dioxide," British Report UKAEA-AERE-R-4070, United Kingdom Atomic Energy Authority, Atomic Energy Research Establishment, Risley, Warrington, Lancs., England 1962.
92. C. E. Holley, Jr., R. N. R. Mulford, E. J. Huber, Jr., E. L. Head, F. H. Ellinger, and C. W. Bjorklund, "Thermodynamics and Phase Relationships for Plutonium Oxides," Proceedings of the Second U.N. International Conference on Peaceful Uses of Atomic Energy, Geneva, 1958, Vol. 6, p. 215.
93. T. D. Chikalla, C. E. McNeilly and R. E. Skavdahl, "The Plutonium-Oxygen System," Report HW-74802, General Electric Co., Hanford Atomic Products Operation, 1962.
94. E. L. Francis, (Comp.) "Plutonium Data Manual," British Report IGR-161 (RD/R), Industrial Group, Research and Development Branch, Risley, Warrington, Lancs., England, 1959.
95. K. Q. Bagley, "Plutonium and its alloys," Nucl. Eng., 2 (1957)461.
96. M. W. Mallett and W. M. Albrecht, "The Corrosion of Thorium in Air," Report BMI-819, Battelle Memorial Institute, 1953.
97. J. G. Feibig, "Air Corrosion of Thorium," Report CT-2400, Ames Laboratory, Ames, Iowa, 1945.
98. P. Levesque and D. Cubicciotti, "The reaction between oxygen and thorium," J. Am. Chem. Soc., 73(1951)2028.
99. A. F. Gerds and M. W. Mallett, "Surface reaction between oxygen and thorium," J. Electrochem. Soc., 101(1954)171.
100. A. F. Gerds and M. W. Mallett, "Reaction of nitrogen with, and the diffusion of nitrogen in, thorium," J. Electrochem. Soc., 101(1954)175.
101. O. N. Carlson, P. Chiotti, G. Murphy, D. Peterson, B. A. Rogers, J. F. Smith, M. Smutz, M. Voss and H. A. Wilhelm, "The Metallurgy of Thorium and Its Alloys," Proceedings of the First International Conference on Peaceful Uses of Atomic Energy, Geneva, 1955, Vol. 9, p. 74.
102. B. E. Deal and H. J. Svec, "Metal-Water Reactions. III. Kinetics of the reaction between thorium and water vapor," J. Electrochem. Soc., 103(1956)421.
103. W. E. Berry, H. A. Pray and R. S. Peoples, "Corrosion of Thorium and Thorium Binary Alloys in Distilled Water at 100 and 200°C," Report BMI-951, Battelle Memorial Institute, 1954.
104. J. W. Arendt, W. W. Binger, J. Hopkins and F. Nelson, "Aqueous Corrosion of Thorium and Thorium Alloys," Report CT-3036, Ames Laboratory, Ames, Iowa, 1945.
105. D. S. Kneppel, "Aqueous Corrosion of Thorium Alloys and Zircaloy-Clad Thorium Alloys," Report NMI-1226, Nuclear Metals, Inc., 1960.
106. E. A. Gulbransen and K. F. Andrew, "The kinetics of the reactions of beryllium with oxygen and nitrogen and the effect of oxide and nitride films on its vapor pressure," J. Electrochem. Soc., 97(1950)383.
107. J. E. Antill and P. Murray, "Reactions between Fuel Elements and Gaseous Coolant," Progress in Nuclear Energy, Series IV, Vol. 3, p. 65, J. M. Nichols (Ed.), Pergamon Press, N. Y., 1960.
108. A. Draycott, F. D. Nicholson, G. H. Price and W. I. Stuart, "Study of the Variables Affecting the Corrosion of Beryllium in Carbon Dioxide," Report AAEC/E-83, Australian Atomic Energy Commission, 1961.
109. R. B. Adams, G. H. Price and W. J. Stuart, "Reactions of Preoxidized Beryllium Powder in Moist Carbon Dioxide," Report AAEC/E-88, Australian Atomic Energy Commission, 1962.
110. N. I. Sax (Ed.), Dangerous Properties of Industrial Materials, Reinhold Publishing Corp., N. Y., 1963, p. 502.
111. R. E. Nightingale (Ed.), Nuclear Graphite, Academic Press, N.Y., 1962.
    (a) "Reactions of Graphite with Metals," p. 143.

(b) "Reactions of Graphite with Selected Metal Oxides," p. 145.
112. T. A. Coultas and R. Cygan, "Compatibility of Sodium, Graphite, and Stainless Steel," Report NAA-SR-258, North American Aviation, Inc., 1957.
113. C. Starr and R. W. Dickenson, Sodium-Graphite Reactors, Addison-Wesley Publishing Co., Reading, Mass., 1958.
114. A. R. Kaufmann (Ed.), Nuclear Reactor Fuel Elements, Interscience Publishers, Inc., N. Y., 1962, p. 251.
115. J. G. Gratton, "Solubility of Carbon in Sodium", Report KAPL-1807, Knolls Atomic Power Laboratory, 1957. Interscience Publishers, Inc., N.Y., 1962, p. 251.
116. W. D. Wilkinson and W. F. Murphy, Nuclear Reactor Metallurgy, D. van Nostrand Co., N. J., 1960.
117. W. G. O'Driscoll and J. C. Bell, "Graphite: Its properties and behavior," Nucl. Eng., 3(1958)479.
118. "Molten Salt Reactor Program Quarterly Progress Report for Period Ending July 31, 1960," USAEC Report ORNL-3014, p. 81, Oak Ridge National Laboratory, 1960.
119. "Molten Salt Reactor Program Quarterly Progress Report for Period Ending February 28, 1962," USAEC Report ORNL-3282, p. 103, Oak Ridge National Laboratory, 1962.
120. "Molten Salt Reactor Program Quarterly Progress Report for Period Ending August 31, 1962," USAEC Report ORNL-3369, p. 112, Oak Ridge National Laboratory, 1962.
121. "Chemical Engineering Division Summary Report, July, August, September, 1956," USAEC Report ANL-5633, p. 17, Argonne National Laboratory, 1956.
122. W. J. Hallett, "Report on Boron Nitride," Report NEPA-255, Fairchild Engine and Airplane Corporation, 1947.
123. N. R. Grant, "Corrosion of Boron-Stainless Steel," Reactor Engineering Division Summary Report, April, May, June, 1956, USAEC Report ANL-5601, p. 57, Argonne National Laboratory, 1956.
124. D. N. Dunning, W. K. Anderson and P. R. Mertens, "Boron containing control materials," Nucl. Sci Eng. 4(1958)402.
125. L. B. Prus, E. S. Byron and J. F. Thompson, "Boron stainless steel alloys," Nucl. Sci. Eng. 4(1958)415.
126. W. C. Thurber, J. A. Milko and R. J. Beaver, "Boron-Aluminum and Boron-Uranium-Aluminum Alloys for Reactor Application," USAEC Report ORNL-2149, Oak Ridge National Laboratory, 1957.
127. H. S. Cooper, "Boron," Chap 5, Rare Metals Handbook, 2nd Ed., C. A. Hampel (Ed.), Reinhold Publishing Corp., London, 1961.
128. H. F. Rizzo et al., "Refractory compositions based on silicon-boron-oxygen reactions," J. Amer. Ceramic Soc., 43(1960) 498.
129. "Boron carbide," Advanced Materials Technology, 1,3(1958)4.
130. A Handbook on Boron Carbide, Elemental Boron and Other Stable Boron Rich Materials, Norton Co., Worcester, Mass., 1955.
131. D. N. Dunning and W. E. Ray, "Control rod materials," Nucleonics, 16,5(1958)88.
132. K. M. Taylor, "Boron nitride," Materials and Methods, 43(1956)88.
133. G. M. Slaughter, C. F. Leitten, Jr., P. Patriarca, E. E. Hoffman and W. D. Manly, "Sodium corrosion and oxidation resistance of high temperature brazing alloys," Welding J., 36(1957)217.
134. E. S. Byron (Ed.) "Boron Materials," Chap. 4 in Neutron Absorber Materials for Reactor Control, W. K. Anderson and J. S. Theilacker (Eds.), U.S. Govt. Printing Office, 1962.
135. W. J. Kahn and D. H. Shaftman, "A Retrospective Analysis of Aspects of the ALPR (SL-1) Design," USAEC Report ANL-6692, Argonne National Laboratory, 1962.
136. I. Cohen (Ed.), "Silver and Silver-Base Alloys," Chap. 5 in Neutron Absorber Materials for Reactor Control, W. K. Anderson and J. S. Theilacker (Eds.), U.S. Govt. Printing Office, 1962.
137. F. R. Lorenz, Jr., "Hafnium," Chap. 3 in Neutron Absorber Materials for Reactor Control, W. K. Anderson and J. S. Theilacker (Eds.) U.S. Govt. Printing Office, 1962.
138. F. H. Spedding and A. H. Daane (Ed.), The Rare Earths, John Wiley and Sons, Inc., N.Y., 1961, pp. 167, 186.
139. J. Belle (Ed.), Uranium Dioxide: Properties and Nuclear Applications, U.S. Govt. Printing Office, 1961.
140. S. Aronson, R. B. Roof, Jr. and J. Belle, "Kinetic study of the oxidation of uranium dioxide," J. Chem. Phys., 27(1957) 137.
141. M. J. Bradley and L. M. Ferris, "Processing of Uranium Carbide Reactor Fuels, I. Reaction with Water and HCl," USAEC Report ORNL-3101, Oak Ridge National Laboratory, 1961.
142. C. P. Kempter, "Hydrolysis properties of uranium monocarbide and dicarbide," J. Less-Common Metals, 4(1962)419.
143. F. A. Rough and W. Chubb, "An Evaluation of Data on Nuclear Carbides," Report BMI-1441, Battelle Memorial Institute, 1960.
144. K. A. Peakall and J. E. Antill, "Oxidation of uranium monocarbide," J. Less-Common Metals, 4(1962)426.
145. H. W. Newkirk, "Chemical Reactivity of Uranium Monocarbide and Uranium Mononitride with Water at 100°C," Report HW-59408, General Electric Co., Hanford Atomic Products Operation, 1959.
146. E. W. Murbach, "The oxidation of 'reactive' uranium carbide," Trans. Met. Soc., AIME, 227(1963)488.
147. M. Tetenbaum, R. Wagner and J. D. Bingle, "Metal Oxidation and Ignition Kinetics: Ignition Studies of Uranium Monocarbide Powders by the Burning Curve Method," Chemical Engineering Division Summary Report, April, May, June, 1961," USAEC Report ANL-6379, p. 192. Argonne National Laboratory, 1961.
148. M. M. Mills, "A Study of Reactor Hazards," Report NAA-SR-31 (Del.) North American Aviation, Inc., 1949.
149. L. F. Epstein, "Recent Developments in the Study of Metal-Water Reactions," Progress in Nuclear Energy, Series IV, Vol. 4, p. 461, Pergamon Press, N.Y., 1961.
150. R. B. Smith, "Pyrophoricity - A technical mystery under vigorous attack," Nucleonics, 14,12(1956)28.
151. W. A. Bostrom, "The High Temperature Oxidation of Zircaloy in Water," Report WAPD-104, Bettis, Atomic Power Laboratory, 1954.
152. A. W. Lemmon, Jr., "Studies Relating to the Reaction between Zirconium and Water at High Temperatures," Report BMI-1154, Battelle Memorial Institute, 1957.
153. B. Lustman, "Zirconium-Water Reactions," Report WAPD-137, Bettis, Atomic Power Laboratory, 1955.
154. L. Baker, Jr. and L. C. Just, "Studies of Metal-Water Reactions at High Temperatures: III. Experimental and Theoretical Studies of the Zirconium-Water Reaction," USAEC Report ANL-6548, Argonne National Laboratory, 1962.
155. H. M. Higgins and R. D. Schultz, "The Reaction of Metals in Oxidizing Gases at High Temperatures," Report IDO-28000, Aerojet-General Corp., 1957.
156. S. C. Furman, "Metal-Water Reactions: V. The Kinetics of Metal-Water Reactions - Low Pressure Studies," Report GEAP-3208, General Electric Co., 1959.
157. M. Milich and E. C. King, "Molten Metal-Water Reactions," USAEC Report NP-5813, Tech. Rpt. No. 44, DTI Extension, AEC, Oak Ridge, 1955.
158. D. C. Layman and H. L. Mars, "Some Qualitative Observations of Zirconium-Water Reactions," Report KAPL-1534, Knolls Atomic Power Laboratory, 1956.
159. H. M. Higgins, "A Study of the Reaction of Metals and Water," Report AECD-3664, Hanford Atomic Products Operation, 1955.
160. W. C. Ruebsamen, F. J. Shon and J. B. Chrisney, "Chemical Reaction between Water and Rapidly Heated Metals," Report NAA-SR-197, North American Aviation, Inc., 1952.
161. H. M. Higgins, "The Reaction of Molten Uranium and Zirconium Alloys with Water," Report AGC-AE-17, Aerojet-General Corp., 1956.
162. L. Baker, Jr., R. L. Warchal, R. C. Vogel and M. Kilpatrick, "Studies of Metal-Water Reactions at High Temperatures: I. The Condenser Discharge Experiment: Preliminary Results with Zirconium," USAEC Report ANL-6257, Argonne National Laboratory, 1961.
163. P. L. Harrison and A. D. Yoffe, "The Burning of Metals," Proceedings of the Royal Society of London, Series A, 261 (1961)357.
164. L. F. Epstein, "Correlation and prediction of explosive metal-water reaction temperatures," Nucl. Sci. Eng., 10 (1961)247.
165. W. M. Manning, J. J. Katz and H. R. Hoekstra, "Reactor Hazards," USAEC Report ANL-WMM-596, Argonne National Laboratory, 1950.
166. B. E. Hopkinson, "Kinetics of the uranium-steam reaction," J. Electrochem. Soc., 106(1959)102.
167. A. J. Scott, "Fission Product Release by the High Temperature Uranium-Steam Reaction," Report HW-6204, General Electric Co., Hanford Atomic Products Operation, 1959.
168. A. W. Lemmon, Jr., "The Reaction of Steam with Uranium and with Various Uranium-Niobium-Zirconium Alloys at High Temperatures," Report BMI-1192, Battelle Memorial Institute, 1957.
169. R. E. Wilson and P. Martin, "Metal-Water Reactions: Isothermal Studies of the Uranium-Steam Reaction by the Volumetric Method," Chemical Engineering Division Summary Report, April, May, June 1962, USAEC Report ANL-6569, p. 148, Argonne National Laboratory, 1962.
170. W. F. Zelezny, "Metal-Water Reactions: Rates of Reaction of Aluminum and Aluminum-Uranium Alloys with Water Vapor at Elevated Temperatures," Report IDO-16629, Phillips Petroleum Co., 1960.
171. D. Mason and P. Martin, "Metal-Water Reactions: Pressure Pulse Method," Chemical Engineering Division Summary Report, July, Aug., Sept., 1961, USAEC Report ANL-6413, p. 178, Argonne National Laboratory, 1961.
172. R. E. Wilson, L. Mishler and C. Barnes, "Metal-Water Reactions: Studies of the Aluminum-Water Reaction by the

Levitation-Melting Method," Chemical Engineering Division Summary Report, Oct., Nov., Dec., 1962, USAEC Report ANL-6648, p. 196, Argonne National Laboratory, 1963.

173. R. E. Wilson, C. Barnes and L. Baker, "Studies of the Aluminum-Steam Reaction by the Levitation Melting Method," Chemical Engineering Division Semiannual Report, Jan. - June 1964, USAEC Report ANL-6900, p. 233, Argonne National Laboratory, 1964.

174. R. E. Wilson and P. Martin, "Metal-Water Reactions: Levitation Method," Chemical Engineering Division Summary Report, April, May, June, 1961, USAEC Report ANL-6379, p. 208, Argonne National Laboratory, 1961.

175. J. M. West and J. T. Weills, "Reactor Engineering and Services Division Quarterly Report, Sept. 1-Nov. 30, 1950," USAEC Report ANL-4549, p. 5, Argonne National Laboratory, 1950.

176. R. B. Cox, H. M. Higgins and E. G. Lucken, "Research Development and Testing of Underwater Propulsion Devices," Report AGC-AE-464, Aerojet-General Corp., 1950.

177. G. Long, "Explosions of molten aluminum in water—cause and prevention," Metal Progress, 71(1957)107.

178. R. F. Plott, "Reactions Produced by the Electrical Explosion of a Metal Immersed in a Fluid," USAEC Report ANL-5040, Argonne National Laboratory, 1950.

179. L. Leibowitz, L. W. Mishler, and P. W. Krause, "Studies of Metal-Water Reactions by the Laser Heating Method," Chemical Engineering Division Semiannual Report, July-Dec. 1964, USAEC Report ANL-6925, p. 198, Argonne National Laboratory, 1965.

180. A. A. Shidlovskii, "Explosive methyl alcohol-water mixtures with magnesium and aluminum," Zh. Prik. Khim., 19(1946) 371.

181. D. S. Gibbs and H. J. Svec, "Kinetics of the Reaction between Magnesium and Water Vapor," Report ISC-779, Iowa State College, 1956.

182. R. O. Ivins and R. Koonz, "Study of the Combustion of Alumium in Water," Chemical Engineering Division Semiannual Report, July - Dec. 1963, USAEC Report ANL-6800, p. 338, Argonne National Laboratory, 1964.

183. R. E. Wilson and C. Barnes, "Isothermal Studies of the Stainless Steel-Steam Reaction by the Volumetric Method," Chemical Engineering Division Semiannual Report, Jan. - June 1964, USAEC Report ANL-6900, p. 239, Argonne National Laboratory, 1964.

184. L. Baker, Jr., R. Warchal and R. Koonz, "Metal-Water Reactions: Condenser-Discharge Method," Chemical Engineering Division Summary Report, July, Aug., Sept., 1962, USAEC Report ANL-6596, p. 186, Argonne National Laboratory, 1962.

184a. O. Kubaschewski and H. Ebert, "Reaction of water vapor and hydrogen peroxide upon light metals at high temperatures," Z. Metallk., 38(1947)232.

185. R. C. Liimatainen, "Nuclear reactor safety: The meltdown of metals at high temperatures and metal-water reaction," Arkhimedes J. Finnish Phys. Soc, No. 1, 1964.

186. M. Sittig, "Sodium - Its Manufacture, Properties and Uses," ACS Monograph No. 133, p. 188, Reinhold Publishing Co., Inc., 1956.

187. M. M. Markowitz, "Alkali metal-water reactions," J. Chem. Educ., 40(1963)633.

188. C. B. Jackson (Ed.), "Liquid Metals Handbook," Sodium-NaK Supplement, USAEC Report TID-5277, 1955.

189. "Chemical Considerations in the Sodium Cooled D_2O Moderated Reactor (SDR)," Report NDA-84-6, Nuclear Development Corp. of America, 1958.

190. N. R. Adolph, "Recent Test Results of Sodium-Water Systems," Proceeding of the Second U.N. International Conference on Peaceful Uses of Atomic Energy, Geneva, 1958, Vol. 7, p. 119.

191. J. R. Dietrich and W. H. Zinn, Solid Fuel Reactors, Addison-Wesley Publishing Col, Inc., Reading, Mass., 1958.

192. H. M. Saltsburg, "The Kinetics of Molten Metal-Water Reactions: A Report on Na_2K—Water Vapor," Report KAPL-1763, Knolls Atomic Power Laboratory, 1957.

193. L. Corrsin, H. Steinmentz and B. Marano, "Sodium-Water Reaction Rate Studies," Report NDA-84-19, Nuclear Development Corp. of America, 1959.

194. Knolls Atomic Power Laboratory Staff, "Progress Report No. 45," Report KAPL-341, Knolls Atomic Power Laboratory, 1950.

195. M. Kilpatrick, L. Baker and C. McKinney, "Studies of fast reactions which evolve gases. The reaction of sodium-potassium alloy with water in the presence and absence of oxygen," J. Phys. Chem., 57,4(1953)385.

196. Memo to Parliament, "Accident at Windscale No. 1 Pile 10th October 1957," (Cmmd. 302), Her Majesty's Stationery Office, London, 1957. (Copies available through British Information Services, 45 Rockefeller Plaza, N.Y. 20, N.Y.)

197. J. W. Greenwood, "Contamination of NRU Reactor in May 1958," Report CRR-836, AECL-850, Atomic Energy of Canada, Ltd., Chalk River, 1959.

198. AEC Serious Accident Bulletin No. 130, USAEC Nov. 27, 1957.

199. AEC Serious Accident Bulletin No. 84, USAEC Aug. 15, 1955.

200. "Final Report of Explosion in Oak Ridge, Tennessee, Y-12 Salvage Yard Adjacent to Bldg. 9929-1," Report Y-1137A, Carbide and Carbon Chemicals Corp., 1956.

201. AEC Serious Accident Bulletin No. 107, USAEC Aug. 20, 1956.

202. L. Baker, Jr., J. D. Bingle and R. Koonz, "Metal Oxidation and Ignition Kinetics: Theory of Metal Ignition," Chemical Engineering Division Summary Report, July, Aug., Sept., 1961, USAEC Report ANL-6413, p. 152, Argonne National Laboratory, 1961.

203. Interim Report "Zirconium Fire and Explosion Hazard Evaluation," USAEC Report TID-5365, 1956. (Division of Organization and Personnel Safety and Fire Protection Branch, AEC).

204. F. E. Littman, F. M. Church and E. M. Kinderman, "A study of metal ignitions. I. The spontaneous ignition of titanium," J. Less-Common Metals, 3(1961)367.

205. J. G. Schnizlein and J. W. Allen, "Metal Oxidation and Ignition Kinetics: Zirconium Powder Ignition," Chemical Engineering Division Summary Report, April, May, June, 1962, USAEC Report ANL-6569, p. 145, Argonne National Laboratory, 1962.

206. H. C. Anderson and L. H. Belz, "Factors controlling the combustion of zirconium powders," J. Electrochem. Soc., 100(1953)240.

207. I. Hartman, J. Nagy and M. Jacobson, "Explosive Characteristics of Titanium, Zirconium, Thorium, Uranium and and Their Hydrides," Report BM-RI-4835, U.S. Bureau of Mines, 1951.

208. W. W. Allison, "Zirconium, Zircaloy, and Hafnium Safe Practice Guide for Shipping, Storing, Handling, Processing, and Scrap Disposal," Report WAPD-TM-17, Bettis, Atomic Power Laboratory, 1960.

209. L. Leibowitz, L. Baker, Jr., J. G. Schnizlein, L. W. Mishler and J. D. Bingle, "Burning velocities of uranium and zirconium in air," Nucl. Sci. Eng., 15(1963)395.

210. E. M. Mouradian and L. Baker, Jr., "Burning temperatures of uranium and zirconium in air," Nucl. Sci. Eng., 15(1963) 388.

211. J. W. Isaacs and J. N. Wanklyn, "The Reaction of Uranium with Air at High Temperatures," British Report AERE-R-3559, United Kingdom Atomic Energy Authority, Research Group, Atomic Energy Research Establishment, Harwell, Berks., England, 1960.

212. Y. Adda, "Investigation of the Kinetics of the Reactions of Oxidation, Nitridation, and Hydridation of Uranium," French Report CEA-757, Commissariat a l'Energie Atomique, 1958.

213. L. Baker, Jr. and J. D. Bingle, "Metal Oxidation and Ignition Kinetics: Isothermal Oxidation of Uranium at High Temperatures," Chemical Engineering Division Summary Report, Jan., Feb., Mar., 1962, USAEC Report ANL-6543, p. 168, Argonne National Laboratory, 1962.

214. A. F. Bessonov and V. G. Vlasov, "Oxidation mechanism of metallic uranium," Fiz. metal. metalloved, 12(1961)403.

215. L. Baker, Jr. and J. D. Bingle, "Metal Oxidation and Ignition Kinetics: Theory of Uranium Ignition," Chemical Engineering Division Summary Report, Oct., Nov., Dec., 1962, USAEC Report ANL-6648, p. 186, Argonne National Laboratory, 1963.

216. L. Baker, Jr. and J. D. Bingle, "Metal Oxidation and Ignition Kinetics: Theory of Uranium Ignition," Chemical Engineering Division Summary Report, April, May, June, 1962, USAEC Report ANL-6569, p. 136, Argonne National Laboratory, 1962.

217. L. Baker, Jr., J. G. Schnizlein, J. D. Bingle and A. J. Buhl, "Metal Oxidation and Ignition Kinetics: Theory of Uranium Ignition," Chemical Engineering Division Summary Report, July, Aug., Sept., 1962, USAEC Report ANL-6596, p. 175, Argonne National Laboratory, 1962.

218. G. E. Zima, "Pyrophoricity of Uranium in Reactor Environments," USAEC Report HW-62442, Hanford Atomic Products Operation, 1960.

219. R. K. Hilliard, "Fission Product Release from Uranium: Effect of Irradiation Level," USAEC Report HW-72321, General Electric Co., Hanford Atomic Products Operation, 1962.

220. J. H. Kittel and S. H. Paine, "Effects of high burnup on natural uranium," Nucl. Sci. Eng., 3(1958)250.

221. N. R. Chellew and R. K. Steunenberg, "Fission gas release and swelling during heating of irradiated EBR-II type fuel," Nucl. Sci. Eng., 14(1962)1.

222. B. A. Loomis and D. W. Pracht, "Swelling of Uranium and Uranium Alloys on Post-Irradiation Annealing," USAEC Report ANL-6532, Argonne National Laboratory, 1962.

223. J. G. Schnizlein, "Metal Oxidation and Ignition Kinetics: Ignition of Irradiated Uranium," Chemical Engineering Division Summary Report, July, Aug., Sept., 1962, USAEC Report ANL-6596, p. 185, Argonne National Laboratory, 1962.

224. L. Baker, Jr., J. G. Schnizlein and J. D. Bingle, "Metal Oxidation and Ignition Studies: Ignition of Uranium Foil Aggregates," Chemical Engineering Division Summary Report, Jan., Feb., Mar., 1961, USAEC Report ANL-6333, p. 205,

Argonne National Laboratory, 1961.
225. M. Tetenbaum, L. W. Mishler and J. G. Schnizlein, "Uranium powder ignition studies," Nucl. Sci. Eng., 14(1962)230.
226. J. G. Schnizlein and D. F. Fischer, "Metal Oxidation and Ignition Kinetics: Plutonium Ignition Studies," Chemical Engineering Division Summary Report, Oct., Nov., Dec., 1962, USAEC Report ANL-6648, p. 192, Argonne National Laboratory, 1963.
227. J. G. Schnizlein and D. F. Fishcher, "Plutonium Oxidation and Ignition Studies," Chemical Engineering Division Summary Report, July-Dec. 1963, USAEC Report ANL-6800, Argonne National Laboratory, 1964.
228. J. G. Schnizlein and D. F. Fischer, "Metal Oxidation and Ignition Studies: Plutonium Ignition Studies," Chemical Engineering Division Summary Report, April, May, June, 1960, USAEC Report ANL-6183, p. 135, Argonne National Laboratory, 1960.
229. M. W. Fassell, L. B. Gulbransen, J. R. Lewis and J. H. Hamilton, "Ignition temperatures of magnesium and magnesium alloys," J. Metals, 3(1951)522.
230. K. P. Coffin, "Some Physical Aspects of the Combustion of Magnesium Ribbons," Fifth Symposium (International) on Combustion, U. of Pittsburgh, Aug. 30-Sept. 3, 1954, p. 267, Reinhold, N.Y., 1955.
231. R. Friedman and K. A. Mace, "Ignition and combustion of aluminum particles in hot ambient gases," Combust. Flame, 6(1962)9.
232. P. B. Longton, "Alkali Metal-Gas Reactions Part IX; The Influence of Moisture on the Reaction of Sodium with Oxygen," British Report IGR-TN/C-535, United Kingdom Atomic Energy Authority, Industrial Group, Culcheth Laboratories, Culcheth, Lancs., England, 1957.
233. P. B. Longton, "The Reaction of Potassium with Wet Oxygen: Report of Progress to March 31, 1956," British Report IGR-TM/C-039, United Kingdom Atomic Energy Authority, Industrial Group, Culcheth Laboratories, Culcheth, Lancs., England, 1956.
234. P. B. Longton, "The Ignition Temperatures of Sodium and Potassium in Oxygen," British Report RHM(56)/136, 1956.
235. J. D. Gracie and J. J. Droher, "A Study of Sodium Fires," Report NAA-SR-4383, North American Aviation Inc., 1960.
236. H. K. LeMar, "Liquid Metals Smoke Abatement," Report PWAC-235, Pratt and Whitney Aircraft Div., United Aircraft Corp., 1957.
237. J. R. Humphreys, Jr., "Sodium-Air Reactions as They Pertain to Reactor Safety and Containment," Proceedings of the Second U.N. International Conference on Peaceful Uses of Atomic Energy, Geneva, 1958, Vol. 11, p. 177.
238. L. Baker and A. D. Tevebaugh, "Sodium-Air Reaction Calculations," Chemical Engineering Division Semiannual Report, Jan. - June 1965. USAEC Report ANL-7055, Argonne National Laboratory, 1965.
239. E. Hines et al., "How strong must reactor housings be to contain Na-air reactions," Nucleonics, 14,10(1956)38.
240. D. R. de Halas, R. E. Dahl and J. L. Jackson, "Hanford Studies for EGCR Combustion Characteristics - Summary Report," Report HW-71296, General Electric Co., Hanford Atomic Products Operation, 1961.
241. P. L. Walker, Jr., F. Rusinko, Jr. and L. G. Austin, "Gas Reactions of Carbon," p. 133, Advances in Catalysis and Related Subjects, Vol. XI, Academic Press, Inc., N.Y., 1959.
242. R. E. Dahl, "Evaluation of Chlorine Inhibition of Graphite Oxidation as a Gas Cooled Reactor Safeguard," Report HW-67225, General Electric Co., Hanford Atomic Products Operation, 1961.
243. J. Berkowitz, "The Reaction of Graphite with Nitrogen at Elevated Temperatures," Thermodynamics of Reactor Materials, Proceedings of Symposium of the International Atomic Energy Agency, Vienna, June, 1962, Vol. II, p. 345.
244. O. Levenspiel, Chemical Reaction Engineering, p. 338, John Wiley and Sons, N.Y., 1962.
245. T. J. Clark and R. C. Giberson, "Studies of Surface Sorption in Gas-Graphite Systems, Preliminary Report," Report HW-67793, General Electric Co., Hanford Atomic Products Operation, 1961.
246. R. H. Fox, "Nitrogen-Graphite Compatibility Study," Proceedings of the US/UK Meeting on the Compatibility Problems of Gas-Cooled Reactors, held at Oak Ridge, Tennessee, USAEC Report TID-7597, Book 2, p. 545, 1960.
247. S. Peterson, "Chemical reactions of graphite," Nuclear Safety, 5,1(1963); see also J. Inorg. Nucl. Chem. 24,(1962)1129.
248. H. Heddon and E. Wicke, "About Some Influences on the Reactivity of Carbon," p. 249, Proceedings of the Third International Conference on Carbon, Buffalo, N.Y., 1959.
249. R. E. Dahl, "Oxidation of Graphite under High Temperature Reactor Conditions," Report HW-68493, General Electric Co., Hanford Atomic Products Operation, 1961.
250. J. W. Prados, "Graphite oxidation," Nuclear Safety, 2,4(1961)8.
251. E. A. Gulbransen, "The Oxidation of Graphite at Temperatures of 600 to 1500°C," J. Electrochem. Soc., 10, 6 (1963) 476.
252. D. G. Schweitzer, G. C. Hrabak and R. M. Singer, "Oxidation and heat transfer studies in graphite channels, I: The effect of air flow rate on the $C\text{-}O_2$ and $CO\text{-}O_2$ reactions," Nucl. Sci. Eng., 12(1962)39.
253. D. G. Schweitzer, "Thermal properties of air-cooled graphite channels," Nucl. Sci. Eng., 13(1962)275.
254. R. E. Dahl, "Experimental Evaluation of the Combustion Hazard to the Experimental Gas Cooled Reactor - Preliminary Burning Rig Experiments," Report HW-67792, General Electric Co., Hanford Atomic Products Operation, 1961.
255. Corrosion of Reactor Materials, Vols. 1 and 2, International Atomic Energy Agency, Vienna, 1962.
256. W. L. Kosiba and G. J. Dienes, "The Effect of Radiation on the Rate of Oxidation of Graphite," US/UK Graphite Conference held at St. Giles Court, London, Dec. 16-18, 1957, p. 121. USAEC Report TID-7565 (Part I), 1959.
257. J. R. Beattie, J. B. Lewis and R. Lind, "Graphite Oxidation and Reactor Safety," presented at the Third U. N. International Conference on the Peaceful Uses of Atomic Energy, Geneva, 1964, Paper A/Conf. 28/P/ 185.
258. M. C. Brockway, "Graphite," Reactor Materials, 5,1(1962)25; see also HW-68494, 1961.
259. D. D. Wagman et al., "Heats, free energies and equilibrium constants of some reactions involving O_2, H_2, H_2O, C, CO, CO_2 and CH_4," J. Res. Nat. Bur. Std., 34(1945)143.
260. L. R. Zumwalt et al., "Carbon Transport and Corrosion in High Temperature Gas Cooled Reactors," Corrosion of Reactor Materials, Proceedings of the Symposium of the International Atomic Energy Agency, Vienna, June 1962, Vol. II, p. 345.
261. W. Jost, Explosion and Combustion Processes in Gases, pp. 184, 187, McGraw-Hill, N.Y., 1946.
262. F. M. Lang, "Specific rate of oxidation in air and carbon dioxide of various purified graphites," Comptes Rendus, 255(1962)1511.
263. R. Lind and J. Wright, "Factors Controlling Reaction between Graphite and Radiolysed Carbon Dioxide," presented at the Third U. N. International Conference on the Peaceful Uses of Atomic Energy, Geneva, 1964, Paper A/Conf. 28/P/566.
264. F. S. Feates, "The Effects of Water Vapour Carbon Monoxide and Pore Volume on the Radiolytic Reaction between Carbon Dioxide and Graphite in Sealed Tubes," British Report AERE-R-4332, United Kingdom Atomic Energy Authority, Reasearch Group, Atomic Energy Research Establishment, Harwell, Berks., England, 1964.
265. A. R. Anderson et al., "Chemical Studies of Carbon Dioxide and Graphite under Reactor Conditions," Proceedings of Second International Conference on Peaceful Uses of Atomic Energy, Geneva, 1958, Vol. 7, p. 335.
266. J. O. Kolb and W. B. Cottrell, "Graphite Oxidation," Gas-Cooled Reactor Program Semiannual Progress Report for Period Ending Sept. 30, 1962, USAEC Report ORNL-3372, p. 23, Oak Ridge National Laboratory, 1963.
267. J. E. Antill and K. A. Peakall, "Attack of graphite by an oxidizing gas at low partial pressures and high temperatures," J. Nucl. Materials, 2(1960)31.
268. J. P. Blakely, "Rates of Reaction of a 1-Inch-Diameter Graphite Sphere with $He\text{-}CO_2$ Mixtures," USAEC Report ORNL-TM-751, Oak Ridge National Laboratory, 1964.
269. G. C. Clark (Ed.), Encyclopedia of X-rays and Gamma Rays, Reinhold Publishing Corp., N.Y., 1963.
270. R. F. S. Robertson, "The Radiolytic Behavior of Water in a Nuclear Reactor," Progress in Nuclear Energy, Series IV, Vol. 1, p. 265. McGraw-Hill Publishing Co., N.Y., 1956.
271. E. J. Hart, W. R. McDonell and S. Gordon, "The Decomposition of Light and Heavy Water Boric Acid Solutions by Nuclear Reactor Radiations," Proceedings of the First U.N. International Conference on Peaceful Uses of Atomic Energy, Geneva, 1955, Vol. 7, p. 593.
272. A. O. Allen, C. J. Hochanadel, J. A. Ghormley and T. W. Davis, "Decomposition of water and aqueous solutions under mixed fast-neutron and gamma-radiation," J. Phys. Chem., 56(1952)576.
273. G. K. Whitham and R. R. Smith, "Water Chemistry in a Direct Cycle Boiling Water Reactor," Progress in Nuclear Energy, Series IV, Vol. 2, p. 92, Pergamon Press, London and N.Y., 1960.
274. J. A. Ransohoff and C. F. Graham, "Gas Handling Systems: External Recombiners," Reactor Handbook, Vol. 2, "Engineering," USAEC Report AEDC-3646, p. 701, 1955.
275. J. W. Boyle, C. J. Hochanadel, T. J. Sworski, J. A. Ghormley and W. R. Kieffer, "The Decomposition of Water by Fission Recoil Particles," Proceedings of the First U.N. International Conference on Peaceful Uses of Atomic Energy, Geneva, 1955, Vol. 7, p. 576.
276. H. F. McDuffie, "The Systems $UO_2SO_4\text{-}H_2O$ and $UO_2SO_4\text{-}D_2O$: Radiation Stability," The Reactor Handbook, Vol. 2, "Engineering," USAEC Report AECD-3646, p. 560, 1955.
277. R. G. Sowden, "Radiolytic problems in water reactors", J. Nucl. Mater., 8(1963)81.

278. A. O. Allen, The Radiation Chemistry of Water and Aqueous Solutions, D. Van Nostrand Co., Inc., Princeton, N.J., 1961.
279. H. F. McDuffie, E. L. Compere, H. H. Stone, L. F. Woo and C. H. Secoy, "Homogeneous catalysis for homogeneous reactors: catalysis of the reaction between hydrogen and oxygen," J. Phys. Chem., 62(1958)1030.
280. J. R. McCord, "Internal Recombination of Hydrogen and Oxygen: A Literature Review," Report WCAP-120, Bettis, Atomic Power Laboratory, 1956.
281. S. E. Beall, P. N. Haubenreich and J. W. Hill, Jr., "HRT Operations: Results of Internal-Recombination Experiment," Homogeneous Reactor Program Progress Report, Dec. 1, 1960-May 31, 1961, USAEC Report ORNL-3167, p. 4, Oak Ridge National Laboratory, 1961.
282. L. R. Steele, S. Gordon and C. E. Dryden, "Water Decomposition by Fission Fragment Recoil Energy in an Aqueous Slurry of Uranium-Thorium Oxides," Nucl. Sci. Eng., 15, 4 (1963) 458.
283. W. R. Gall and M. I. Lundin, "HRT Design: Recombiners," Homogeneous Reactor Program Progress Report, May 1-October 31, 1959, USAEC Report ORNL-2879, p. 35, Oak Ridge National Laboratory, 1960.
284. W. D. Burch and L. B. Shappert, "Behavior of iodine and xenon in the homogeneous reactor test," Nucl. Sci. Eng., 15(1963)124.
285. S. E. Beall, P. H. Haubenreich and J. W. Hill, Jr., "HRT Operations: Recombiner Performance Tests," Homogeneous Reactor Program Progress Report, Aug. 1-Nov. 30, 1960, USAEC Report ORNL-3061, p. 17, Oak Ridge National Laboratory, 1961.
286. I. Spiewak and F. N. Peebles, "HRT Component Testing and Development: HRT Recombiner Test," Homogeneous Reactor Program Quarterly Progress Report for Period Ending July 31, 1960, USAEC Report ORNL-3004, p. 28, Oak Ridge National Laboratory, 1960.
287. B. Lewis and G. von Elbe, Combustion, Flames, and Explosions of Gases, (2nd ed.), Academic Press, N.Y., 1961.
288. T. H. Pigford, "Explosion and Detonation Properties of Mixtures of Hydrogen, Oxygen, and Water Vapor," USAEC Report ORNL-1322, Oak Ridge National Laboratory, 1952.
289. Z. M. Shapiro and T. R. Moffette, "Hydrogen Flammability Data and Application to PWR Loss-of-Coolant Accident," Report WAPD-SC-545, Bettis, Atomic Power Laboratory, 1957.
290. Bureau of Mines, Division of Explosives Technology, "Research on the Combustion and Explosion Hazards of Hydrogen-Water Vapor-Air Mixtures," Progress Report No. 1, March-June 1956. M. G. Zabetakis for Westinghouse Electric Corp., Report AECU-3326.
291. Bureau of Mines, Division of Explosives Technology, "Research on the Combustion and Explosions Hazards of Hydrogen-Water Vapor-Air Mixtures," Final Report, 1956, M. G. Zabetakis for Westinghouse Electric Corp., Report AECU-3327.
292. H. A. Pray, C. E. Schweickert and E. F. Stephan, "Explosion Limits of the Hydrogen-Oxygen-Water System at Elevated Temperatures," Report BMI-705, Battelle Memorial Institute, 1951.
293. J. A. Luker and E. C. Hobaica, "Effect of initial mixture density on the formation of detonation in knallgas-saturated with water vapor," J. Chem. Eng. Data, 6(1961)253.
294. A. L. Furno, S. R. Harris and M. G. Zabetakis, "Flammability of Deuterium in Oxygen-Helium Mixtures," USAEC Report TID-20898, Explosives Research Center, Bureau of Mines, U. S. Dept. of the Interior, Pittsburgh, Pa., 1964.
295. J. A. Luker, P. L. McGill, and L. B. Adler, "Knallgas and knallgas-steam mixtures at high initial temperature and pressure," J. Chem. Eng. Data, 4(1959)136.
296. L. B. Adler, J. A. Luker, and E. A. Ryan, "Detonation properties of heavy knallgas, 2 D_2 + O_2," J. Chem. Eng. Data, 6(1961)256.
297. J. A. Luker, "Formation of gaseous detonation waves," J. Chem. Eng. Data, 7(1962)209.
298. L. B. Adler, E. C. Hobaica, and J. A. Luker, "The effect of external factors on the formation of detonation in saturated knallgas-steam mixtures," Combustion and Flame, 3(1959) 481.
299. J. R. Dietrich, "Experimental Investigation of the Self-Limitation of Power During Reactivity Transients in a Subcooled Water-Moderated Reactor, BORAX-1 Experiments," Report AECD-3668, 1954.
300. W. B. Lewis, "The Accident to the NRX Reactor on December 12, 1952," Canadian Report DR-32, Atomic Energy of Canada, Ltd., Chalk River Project, Chalk River, Ontario, 1953.
301. D. G. Hurst, "The Accident to the NRX Reactor: Part II," Canadian Report GPI-14, Atomic Energy of Canada, Ltd., Chalk River Project, Chalk River, Ontario, 1953.
302. "Final Report of SL-1 Recovery Operation," Report IDO-19311, General Electric Co., 1962.
303. J. R. Buchanan, "SL-1 final report," Nuclear Safety, 4,3 (1963)83.
304. "Report on the Nuclear Incident at the SL-1 reactor, January 3, 1961, at the National Reactor Testing Station," Report IDO-19302, Phillips Petroleum Co., 1962.
305. T. R. Wilson, "An Engineering Description of the SPERT-1 Reactor Facility," Report IDO-16318, Phillips Petroleum Co., 1957.
306. "Quarterly Technical Report, SPERT Project, Jan., Feb., March, 1963," Report IDO-16893, Phillips Petroleum Co., 1963.
307. "Report on SPERT-1 destructive test results," Trans. Amer. Nucl. Soc., 6,1(1963)137.
308. R. C. Liimatainen, R. O. Ivins, M. F. Deerwester and F. J. Testa, "Studies of Metal-Water Reactions at High Temperatures. II. Treat Experiments: Status Report on Results with Aluminum, Stainless Steel-304, Uranium and Zircaloy-2," USAEC Report ANL-6250, Argonne National Laboratory, 1962.
309. O. J. Elgert and A. W. Brown, "In-Pile Molten Metal-Water Reaction Experiments," Report IDO-16257, Phillips Petroleum Co., 1956.
310. W. N. Lorentz, "The Chemical Reaction of Zirconium-Uranium Alloys at High Temperatures," Report WAPD-PM-22, Bettis, Atomic Power Laboratory, 1955.
311. R. O. Ivins, F. J. Testa and P. Krause, "Metal-Water Reaction Studies in TREAT," Chemical Engineering Division Summary Report, July, Aug., Sept., 1962, USAEC Report ANL-6569, p. 191, Argonne National Laboratory, 1962.
312. R. C. Liimatainen, F. J. Testa and J. Hepperly, "Studies of the the Meltdown of Oxide-Core Metal-Clad Fuel Pin in TREAT," USAEC Report ANL-6900, p. 254, Argonne National Laboratory, 1964.
313. R. O. Ivins, F. J. Testa and P. Krause, "Studies of the Aluminum-Water Reaction in TREAT," Chemical Engineering Division Summary Report, Oct., Nov., Dec., 1962," USAEC Report ANL-6648, p. 201, Argonne National Laboratory, 1963.
314. R. O. Ivins, F. J. Testa and P. Krause, "Studies of the Aluminum-Water Reaction in TREAT," Chemical Engineering Division Summary Report, Jan., Feb., March, 1963, USAEC Report ANL-6687, p. 179, Argonne National Laboratory, 1963.
315. R. O. Ivins and F. J. Testa, "Studies with Aluminum-U_3O_8 Cermet Fuels (HFIR Fuel) in TREAT," Chemical Engineering Division Semiannual Report, July - Dec. 1965, USAEC Report ANL-7125, Argonne National Laboratory, 1966.
316. J. I. Owens, R. W. Lockhart, D. R. Iltis and K. Hikido, "Metal-Water Reactions: VIII. Preliminary Consideration of the Effects of a Zircaloy-Water Reaction During a Loss of Coolant Accident in a Nuclear Reactor," Report GEAP-3279, General Electric Co., Atomic Products Division, 1959.
317. L. Baker and R. O. Ivins, "Analyzing the Effects of a Zirconium-Water Reaction," Nucleonics, 23, 7 (1965) 70.
318. R. E. Wilson and C. Barnes, "Reaction of Flowing Steam with Stainless Steel and Zircaloy-2-Clad UO_2 Fuel Elements," Chemical Engineering Division Semiannual Report, July - Dec. 1965, USAEC Report ANL-7125, Argonne National Laboratory, 1966.
319. G. W. Parker and J. G. Wilhelm, "Melting Points of Mixtures of Oxidized Fuel Cladding with UO_2," Nuclear Safety Program Semiannual Progress Report, July - Dec. 1964, USAEC Report ORNL-3776, p. 37, Oak Ridge National Laboratory, 1965.
320. J. I. Owens, "Metal-Water Reactions: II. An Evaluation of Severe Nuclear Excursions in Light Water Reactors," Report GEAP-3178, General Electric Co., Atomic Products Division, 1959.
321. R. O. Ivins, "A study of the reaction of aluminum/uranium alloy fuel plates with water initiated by a destructive reactor transient," Trans. Amer. Nucl. Soc., 6,1(1963)101.
322. M. Hansen, Constitution of Binary Alloys, McGraw-Hill Book Co., Inc., N.Y., 1958.
323. S. G. Tresvyatskiy and V. I. Kushakovskiy, "The melting point determination of binary mixtures of uranium oxide with some other oxides in air," Atomnaya Energiya, 8,1(1960)56; see also: Nucleonics, 18,7(1960)101.
324. R. C. Waugh, "The Reaction and Growth of Uranium Dioxide - Aluminum Fuel Plates and Compacts," USAEC Report ORNL-2701, Oak Ridge National Laboratory, 1959.
325. R. J. Beaver and J. E. Cunningham, "Recent Developments in Aluminum-Base Fuel Elements for Research Reactors," Fuel Elements Conference held at Gatlinburg, Tenn., May 14-16, 1958, USAEC Report TID-7559 (Part 1), p. 40, 1959.
326. J. D. Fleming and J. W. Johnson, "Aluminum-U_3O_8 exothermic reactions," Nucleonics, 21,5(1963)84.
327. D. F. Mason and J. D. Bingle, "Thermal Analysis of Uranium Oxide-Metal Cermets," Chemical Engineering Division Summary Report, Jan., Feb., March, 1961, USAEC Report ANL-6333, p. 221, Argonne National Laboratory, 1961.
328. L. Baker, Jr., J. D. Bingle, R. Warchal, and C. Barnes, "Aluminum-U_3O_8 Thermite Reaction," Chemical Engineering Summary Report, July-December 1963, USAEC Report ANL-6800, Argonne National Laboratory, 1964.

329. M. I. Ivanov, V. A. Tumbakov and N. S. Podalskaya, "The heats of formation of UAl_2, UAl_3, and UAl_4," Atomnaya Energia 5(1958)166.
330. R. B. Hinze, "Control of Oxygen Concentration in a Large Sodium System," Report NAA-SR-3638, North American Aviation, 1959.
331. B. C. Voorhees and W. H. Bruggeman, "Interim Report on Cold Trap Investigations," Report KAPL-612, Knolls Atomic Power laboratory, 1951.
332. M. Davis and A. Draycott, "Compatibility of Reactor Materials in Flowing Sodium," Proceedings of the Second U.N. International Conference on the Peaceful Uses of Atomic Energy, Geneva, 1958, Vol. 7, p. 91.
333. W. J, Anderson, "Removal of Carbon from Liquid Sodium System," Report NAA-SR-6386, North American Aviation, 1961.
334. "SRE operates again with core 2 - but with 5-Mw(th) power limit," Nucleonics, 20,1(1962)74.
335. R. L. Carter, R. L. Eichelberger and S. Siegel, "Recent Developments in the Technology of Sodium-Graphite Reactor Materials," Proceedings of the Second U. N. International Conference on the Peaceful Uses of Atomic Energy, Geneva, 1958, Vol. 7, p. 72.
336. L. P. Pepkowitz and E. R. Proud, "Determination of hydrogen," Anal. Chem, 21(1949)1000.
337. L. P. Pepkowitz and W. D. Moak, "Precision determination of low concentrations of carbon in metals," Anal. Chem., 26(1954)1022.
338. L. P. Pepkowitz and W. C. Judd, "Determination of sodium monoxide in sodium," Anal. Chem., 22(1950)1283.
339. H. J. deBruin, "Determination of Traces of Oxygen in Sodium Metal by Infrared Spectrophotometry," Anal. Chem., 32 (1960)360.
340. H. Steinmentz and B. Minushkin, "Experimental Determination of Contaminants in Sodium," Report NDA-2154-6, Nuclear Development Corp. of America, 1961.
341. R. W. Dayton and C. R. Tipton, "Determination of Oxygen in Sodium," Report BMI-1213, Battelle Memorial Institute, 1957.
342. J. C. White, "Procedure for the Determination of Oxygen in Na, NaK by the Distillation Method," USAEC Report CF-56-4-31, Oak Ridge National Laboratory (X-10), 1956.
343. C. R. Breden, "Boiling Water Reactor Technology - Status of the Art Report. Vol. II: Water Chemistry and Corrosion," USAEC Report ANL-6562, Argonne National Laboratory, 1963.
344. J. R. DiStefano and E. E. Hoffman, "Corrosion Mechanisms in Refractory Metal-Alkali Metal Systems," USAEC Report ORNL-3424, Oak Ridge National Laboratory, 1963.
345. J. N. Wanklyn and P. J. Jones, "The aqueous corrosion of reactor metals," J. Nucl. Mater., 6(1962)291.

ADDENDUM REFERENCES

A1. R. E. Wilson and C. Barnes, "Studies of Metal-Water Reactions by the High-Pressure Furnace Method", Chemical Engineering Division Semiannual Report, Jan.-June 1964 USAEC Report ANL-6900, p. 204, Argonne National Laboratory, 1964.
A2. R. E. Wilson and C. Barnes, "Reaction of Flowing Steam with Stainless Steel and Zircaloy-2-clad UO_2 Fuel Elements" Chemical Engineering Division Semiannual Report, July-Dec. 1965, USAEC Report ANL-7125, p. 153, Argonne National Laboratory, 1966.
A3. R. E. Wilson and C. Barnes, "Reaction of Flowing Steam with Stainless Steel-Clad and Zircaloy-2-Clad UO_2-Core Simulated Fuel Elements", Chemical Engineering Division Semiannual Report, Jan.-June 1966 USAEC Report ANL-7225, p. 164, Argonne National Laboratory, 1966.
A4. R. E. Wilson et al., "Reaction of Flowing Steam with Zircaloy-2-Clad, UO_2-Core Simulated Fuel Elements", Chemical Engineering Division Semiannual Report, July-Dec. 1966, USAEC Report ANL-7325, p. 142, Argonne National Laboratory, 1967.
A5. G. W. Parker and J. G. Wilhelm, "Melting Point of Mixtures of Oxidized Fuel Cladding with UO_2", Nuclear Safety Program Semiannual Progress Report for Period Ending December 31, 1964, USAEC Report ORNL-3776, p. 37, Oak Ridge National Laboratory, 1965.
A6. R. E. Wilson and C. Barnes, "Laboratory Studies with Zircaloy-2-Clad, UO_2-Core Fuel Rods Simulating the Conditions of a Loss-of-Coolant Accident", Chemical Engineering Semiannual Report, Jan.-June 1967 USAEC Report ANL-7375, p. 144, Argonne National Laboratory, 1967.
A7. J. C. Hesson, "Analysis of Loss-of-Coolant Accidents", Argonne National Laboratory Reactor Development Program Progress Report, USAEC Report ANL-7438, p. 130, Argonne National Laboratory, 1968.
A8. B. Lustman, "Zirconium Water Reaction Data and Application to PWR Loss-of-Coolant Accident," USAEC Report WAPD-SC-543, Bettis Atomic Power Laboratory, 1957.
A9. J. I. Owens, R. W. Lockhart, D. R. Iltis and K. Hikido, "Metal-Water Reactions: VIII. Preliminary Consideration of the Effects of a Zircaloy-Water Reaction during a Loss-of-Coolant Accident in a Nuclear Reactor", General Electric Co., Atomic Products Division, USAEC Report GEAP-3279, 1959.
A10. L. Baker, Jr., and R. O. Ivins, "Analyzing the effects of a zirconium-water reaction," Nucleonics 23, 7(1965) 70.
A11. J. M. Waage, "Description of calculational methods and digital-computer codes for analyzing coolant-blowdown and core-heatup phenomena." Nuclear Safety 8 (1967) 549.
A12. J. C. Hesson, J. L. Anderson and R. O. Ivins, "CHEMLOC-II: A Computer Program Describing the Core Heating and Cladding-Steam Reaction for a Water-Cooled Power Reactor Following a Loss of Coolant," USAEC Report ANL-7361, Argonne National Laboratory, 1968.
A13. J. C. Hesson and R. O. Ivins, "Thermal Reactor Safety Studies," Chemical Engineering Division Summary Report, July-Dec. 1967, USAEC Report ANL-7425, Argonne National Laboratory, 1968.
A14. R. C. Liimatainen, R. O. Ivins, M. F. Deerwester and F. J. Testa, "Studies of Metal-Water Reactions at High Temperatures: II. TREAT Experiments," USAEC Report ANL-6250, Argonne National Laboratory, 1962.
A15. Quarterly Technical Report SPERT Project, Oct., Nov., Dec., 1966, USAEC Report IDO-17245, Phillips Petroleum Co., 1967.
A16. R. Scott, Jr., C. L. Hale and R. N. Hagen, "Transient Tests of the Fully Enriched UO_2 Stainless Steel Plate — Type C — Core in the SPERT III Reactor. Data Summary Report," USAEC Report IDO-17223, Phillips Petroleum Co., 1967.
A17. R. O. Ivins, L. J. Harrison and L. Leibowitz, "Photographic Studies of Aluminum-Clad Fuel Plate Samples in TREAT", Chemical Engineering Division Semiannual Report, Jan.-June 1966, USAEC Report ANL-7225, p. 173, Argonne National Laboratory, 1966.
A18. D. Armstrong, L. Harrison and F. Testa, "In-Pile Experiments with Zircaloy-2-Clad, UO_2-Core Fuel Rods Simulating the Condition of an Excursion Accident," Chemical Engineering Division Semiannual Report, Jan.-June 1967, USAEC Report, ANL-7375, p. 147, Argonne National Laboratory, 1967.
A19. D. R. Armstrong, R. O. Ivins, L. J. Harrison and F. J. Testa, "TREAT Excursion Simulations," Chemical Engineering Division Semiannual Report, July-Dec. 1967, USAEC Report ANL-7425, Argonne National Laboratory, 1967.
A20. R. C. Liimatainen and F. J. Testa, "Scale-Up Experiments in TREAT on UO_2-Core Fuel Clusters," Chemical Engineering Division Semiannual Report, July-Dec. 1965, USAEC Report ANL-7125, p. 170, Argonne National Laboratory, 1966.
A21. L. Harrison, R. C. Liimatainen and F. J. Testa, "Photographic Studies of Metal-Clad, UO_2-Core Fuel Rods in TREAT," Chemical Engineering Division Semiannual Report, July-Dec. 1966, USAEC Report ANL-7325, p. 158, Argonne National Laboratory, 1967.

CHAPTER 18

Fission-Product Release

G. W. Parker and C. J. Barton
Oak Ridge National Laboratory
Oak Ridge, Tennessee

CHAPTER CONTENTS

1 INTRODUCTION

1.1 General Considerations

In the event that an accident leads to extensive failures of fuel element cladding, it is probable that fission products will be released. These fission products may be partially or totally trapped in the primary reactor system, or they may be released to the containment, or they may escape, in part, from the containment.

The quantities of fission products existing in a given reactor core at a given time can be accurately estimated from the history (power level versus time) of that core. However, the fraction of fission products that will be released under a given set of circumstances cannot be accurately estimated in any easy way. Therefore in early evaluations of the safety of nuclear reactors it was assumed that 100% or, at least, a very large percentage of the fission products would be released from the fuel to the containment [1]. Clearly, this assumption was overconservative in many cases. Since those early evaluations were made, considerable effort has been made to improve the estimation of the fraction of fission products that can escape to the containment.

The purpose of this chapter is to provide a review* of what is known about the release of fission products from the fuel to the containment. This is a problem of many parts. Fission products include many atomic species in varying amounts and with very different physical, chemical, and biological properties. Some are definitely more volatile than others; some may deposit or, by chemical interaction, may be selectively trapped in their immediate environment before they can escape into the containment proper. This chapter will be concerned with all the major problems involved when the fraction of fission products released to the containment is estimated.

Theoretical considerations and early observations of fuel melting [2] demonstrated that the fission-product elements could be divided into two groups: those obviously volatile and those obviously nonvolatile. It was shown to be reasonable to consider that only the volatile fraction was released to

*A similar review is presented by M. H. Fontana in Chapter 3, "U. S. Reactor Containment Technology," W. B. Cottrell and A. W. Savolainen (Ed.), USAEC report ORNL-NSIC-5, Oak Ridge National Laboratory, 1965.

the containment. Then, when the particles and the radiations emitted by the fission products and their relative biological effectiveness (RBE) were known, a more realistic evaluation of the hazards involved could be made.

Of course, this division of fission products into two groups, volatile and nonvolatile, is a gross oversimplification of the true situation. Some elements exhibit volatilities in an intermediate range and others move from the volatile to the nonvolatile group when the environmental conditions are changed. Many experiments have been carried out to further narrow the area of ignorance. Simulation of the conditions that might exist as a result of an accident is difficult, and laboratory experiments designed for this purpose are often subject to criticism for lacking a high degree of correlation with many factors associated with the bonafide accident. On the other hand, the accidents of record, which, fortunately have been relatively minor, tend to confirm the value of experimentation, because there has been fair agreement between the available experimental data and the observed release of fission products.

In the following sections the source term, that is, the quantity of fission products produced and available for release at any given time, is considered first. Various types of fuel are then discussed and their differences noted. Then there is a discussion of the mechanisms of release. Section 2 deals with the theoretical considerations of factors influencing the fraction of the fission products of any given type that can be released to the containment. Section 3 discusses available experimental information on factors affecting fission-product release. Section 4 deals with penetrations through fuel-element cladding. Section 5 discusses available data on releases resulting from actual reactor fuel-failure incidents. The last two sections are concerned with the calculation of "prompt" fission-product release and with analytical models of release from molten fuels.

1.2 Radioactive Fission Products in Irradiated Fuel

The first step in an analysis of the release of fission products from core to containment is to determine the amount of fission products present at any given time in core life and at any time after shutdown.

There are nearly 35 elements and over 200 different isotopes formed in the fission process. The most important of these are listed in Table 1-1, approximately in the order of the volatility of the elements but with some changes dictated by observed behavior in fission-product release experiments. It is of interest to note the very large quantities of radioactivity that will exist in a large power reactor after a year of operation. Since most power reactors will reload in a period of the order of 1 to 2 years, this table represents, within perhaps a factor of 2, the maximum fission-product radioactivity per megawatt of power present in a power reactor.

Note that the half-lives of many of the isotopes are sufficiently short so that they do not figure prominently in the amount of radioactivity present after one day's shutdown. A glance at Table 1-1 shows that only Xe^{133}, I^{131}, and I^{133} remain as highly volatile fission products in appreciable quantities after the first 24 hrs.

The quantity of the stable atomic species is also indicated in the table. Weights of the elements in grams per megawatt of thermal power become kilograms for a 1000-Mw reactor. In general, the total weight in grams per megawatt of the stable isotopes in the fuel after 1 year of irradiation, except for a few long-lived nuclides, is many times higher than that of the radioactive isotopes.

If the iodine inventory in a 1000-Mw reactor is uniformly dispersed in a 3,000,000 ft^3 (85,000 m^3) containment shell, the resultant concentration of radioactive iodine (excluding I^{129}) in the atmosphere would be approximately 2.4 mg (200,000/85,000)/m^3. The total concentration, including both radioactive and nonradioactive iodine, is 30 mg (2,600,000/85,000)/m^3. Thus, if all the iodine contained within the fuel elements were to be released from a 1000-Mw reactor, the physical quantities present in the atmosphere would be appreciable. It should be pointed out, however, that such large reactors involve large quantities of fuel and provide much surface area for adsorption of the iodine as it emerges from the fuel. In fact, the determination of the fraction of the total iodine that escapes from the fuel to the containment, the transport of fission-product iodine, is the most difficult problem to solve.

The fission products decay as a sequence of beta and gamma emitters until a stable product is formed. The physical states of the intermediate steps in the chain are important, since some may be volatile and some may be nonvolatile.

The detailed numerical yield values and genetic relationships of the fission products, including decay characteristics, are a major factor in the evaluation of both the stable and the radioactivity inventories associated with the fission process. Since reactors vary in their use of fissile material, separate yield data are available for U^{233}, U^{235}, Pu^{239}, Th^{232}, and U^{238}. In addition, the fission-product distribution varies with change in neutron energy. Two of the most nearly complete compilations of fission-product yields from various fuel materials are given by Katcoff [3] and by Blomeke and Todd [4]. Table 1-2 [3]* summarizes the thermal-neutron fission yields from U^{233}, U^{235}, and Pu^{239}. Figures 1-1 [3], 1-2 [3], 1-3 [3], and 1-4 [3] illustrate the differences in yield data, showing the regions of greatest variation. Table 1-3 [3] summarizes existing data on the fast-neutron and 14-Mev-neutron yields for U^{235} and U^{238}. The chain relationships of all the mass numbers from 73 to 166 are given according to Katcoff in Table 1-4 [3].

It so happens that the rare-gas fission-product elements xenon and krypton and the halogens occur at the beginning of the important decay chains. Since, as shown in Table 1-1, these two groups of elements are the most volatile members of the chain, it is obvious that a large fraction of the inventory of these elements may be released, in certain circumstances, somewhat selectively. For instance, during startup with a new core or after a long shutdown, a larger fraction of the total release would involve

*Additional sources from which Tables 1-2 to 1-4 were derived are given in reference [3].

TABLE 1-1

Yields and Characteristics of Important Fission-Product Isotopes Arranged Approximately in the Order of the Volatility of the Elements[a]

Isotope	Half-life	Activity (kcurie/Mw(t)) after 1 year of irradiation[b]		Total weight (g/Mw(t)) after 1 year of irradiation		Boiling[c] point (°C)	Probable release and transport form[d]	Health physics properties
		At shutdown	1 day after shutdown	Isotope	Element			
HIGH VOLATILITY (GROUP I)								
Kr	Stable	-	-	4.9		-153	Elemental gas	External radiation, slight health hazard.
-83m	114m	3	0					
-85	10.27y	0.1	0.1	0.3				
-85m	4.4h	8	0.2					
-87	78m	15	0					
-88	2.8h	23	0.1					
-89	3m	31	0					
-90	33s	38	0		5.2			
Xe	Stable	-	-	55.4		-108	Elemental gas	External radiation, slight health hazard.
-131m	12d	0.3	0.3	0.003				
-133m	2.3d	1	0.7	0.003				
-133	5.27d	54	47	0.3				
-135m	15.6m	16	0					
-135	9.2h	34	14	0.01				
-137	3.9m	48	0					
-138	17m	53	0					
-139	41s	61	0		55.7			
Br	Stable	-	-	0.03		59	Elemental Br_2 or HBr	External whole-body radiation, moderate health hazard.
-83	2.3h	3	0					
-84	32m	6	0					
-85	3m	8	0					
-87	56s	15	0		0.03			
I	Stable	-	-	0.45		183	Elemental gas I_2 or HI; variable gaseous organic compounds; adsorbed forms on small particles.	External radiation, internal irradiation of thyroid, high radio-toxicity.
-129	1.7×10^7y	10^{-6}	10^{-6}	1.92				
-131	8d	25	23	0.20				
-132[e]	2.3h	38	0	0.004				
-133	21h	55	26	0.05				
-134	52m	63	0	0.002				
-135	6.7h	55	4.4	0.01				
-136	86s	53	0	0.000	2.64			
INTERMEDIATE VOLATILITY (GROUP II)								
Cs	Stable	-	-	22.7		685	Elemental Cs converting to CsOH.	Internal hazard to whole body.
-134	2y	0.029	0.29	0.03				
-136	13d	0.147	0.147	0.002				
-137	26.6y	1.28	1.3	12.2	35.0			
Te	Stable	-	-	5.1		987	Elemental Te above 1000°C, converting to TeO_2	External radiation, moderate health hazard. Te^{132} contributes a health hazard from I^{132} daughter.
-125m	58d	0.005	-	0.0003				
-127m	105d	0.414	0.5	0.04				
-127	9.4h	2.1	0.5					
-129m	34d	2.9	2.3	0.2				
-131m	30h	3.9	2.2	0.0007				
-131	25m	26	0	0.1				
-132	77h	37.3	30.1	0.1				
-133m	52m	54	0					
-134	44m	57	0					
-135	2m	36	0		5.4			
GROUP VOLATILE UNDER HIGHLY OXIDIZING CONDITIONS (GROUP III)								
Ru	Stable	-	-	17.4		(4230)	Volatile oxide, RuO_4	Internal hazard to kidney and GI tract.
-103	41d	25.7	25.7	0.8				
-106	1y	1.54	1.54	0.5	18.7			
Tc								
-99	2.12×10^5y	10^{-2}	10^{-2}	9.3		(4600)	Volatile oxide, Te_2O_7	Internal hazard to GI tract and lung.
-99m	6.04h	5.5	0.1	0.001	9.3			

[a] This table is an expansion of a similar table published by J. R. Beattie in UKAEA Report AHSB(S)R-9.

[b] Assumed thermal neutron flux, $5 \times 10^{12}/cm^2$.

[c] See reference [7], Appendix VI by J. L. Margrave.

[d] Chemical states of released fission-product elements are affected by temperature and by the chemical composition of the ambient atmosphere during and after release. Those states shown in the table are selected to represent "most probable" chemical states during typical accident conditions.

[e] 38.0 kcurie of I^{132} are produced by decay of Te^{132} in the reactor. Te^{132} released in an accident will produce more I^{132}.

TABLE 1-1 (concluded)

Isotope	Half-life	Activity (kcurie/Mw(t)) after 1 year of irradiation[b]		Total weight (g/Mw(t)) after 1 year of irradiation		Boiling[c] point (°C)	Probable release and transport form[d]	Health physics properties
		At shutdown	1 day after shutdown	Isotope	Element			
Mo	Stable			34.3		(4800)		Internal hazard to GI tract and lung.
-99	67h	51.5	40	0.1	34.4		Volatile oxide, MoO_3	
LOW VOLATILITY (GROUP IV)								
Sr	Stable	-	-	6.5		1366	Elemental Sr converting to SrO.	Internal hazard to bone and lung.
-89	54d	39	39	1.5				
-90	28y	1.2	1.2	8.7				
-91	9.7h	51	9.7	0.0	14.6			
Ba	Stable	-	-	14.6		1635	Elemental Ba converting to BaO.	Internal hazard to bone.
-140	12.8d	53	48	0.7	16.1			
Sb								
-125	2.7y	0.04	0.041	0.10	0.10	1640	Elemental Sb or Sm_2O_3	Internal hazard to GI tract, lung, total body, and bone.
REFRACTORY (GROUP V)								
Sm	Stable			5.2		1602	Elemental Sm converting to Sm_2O_3	Internal hazard to bone, lung, and GI tract.
-151	93y	0.01	0.01	0.42				
-153	47h	1.4	1.0	0.003				
-156	~10h	0.1	0.02	0.00	5.6			
Pm								
-147	2.6y	4.8	4.8	5.0	5.0	(2700)	Elemental Pm converting to Pm_2O_3	Internal hazard to GI tract, bone and lung.
-149	54h	11.8	8.7	0.02				
Pr	Stable			11.4		3020	Elemental Pr converting to Pr_2O_3	Internal hazard to GI tract.
-143	13.7d	52	45	0.8				
-145	6h	36	2.34	0.01	12.2			
Y	Stable			4.9		2783	Elemental Y converting to Y_2O_3	Internal hazard to GI tract.
-90	64.5h	1.2	0.9	0.002				
-91	58d	53	52.4	2.2				
-92	3.6h	52.6	0.5	0.006	7.1			
Nd	Stable			39.9	40.2	3090	Elemental Nd converting to Nd_2O_3.	Internal hazard to GI, liver, and lung.
-147	11.3d	23.2	20	0.3				
La	Stable			12.9		3370	Elemental La converting to La_2O_3	Internal hazard to GI tract.
-140[f]	40h	54	36	0.1	13.0			
Ce	Stable	-	-	23.9		3470	Elemental Ce converting to CeO_2.	Internal hazard to bone, liver, and lung
-141	32d	55.3	55.3	1.8				
-143	33h	54	33	0.053				
-144	290d	30.0	30.0	9.5	36.8			
Zr	Stable	-	-	43.5	45.9	4325	Elemental Zr converting to ZrO_2.	Internal hazard to GI tract, total body, and lung.
-95	63d	53	53	2.4				
Nb	Stable			0.0		4930	Elemental Nb converting to Nb_2O_5	Internal hazard to GI, bone, lung, total body.
-95m	90h	0.5	0.5	0.001				
-95	35d	53	53	1.3	1.3			

[b] Assumed thermal neutron flux, $5 \times 10^{12}/cm^2$.
[c] See reference [7], Appendix VI by J. L. Margrave.
[d] Chemical states of released fission-product elements are affected by temperature and by the chemical composition of the ambient atmosphere during and after release. Those states shown in the table are selected to represent "most probable" chemical states during typical accident conditions.
[f] See Ba^{140}.

xenon, krypton, and the halogens than would be the case when the reactor was in equilibrium power operation. Of course, the total release during start-up with a new core or after a long shutdown is smaller than that which would occur under the same circumstances later in core life. The quantity of any fission product available for release builds up monotonically with operating time.

1.2.1 Fission-Product Equations

Although the reader is referred to other sources [3,4] for a complete discussion of the complex process of calculating yields of the various fission products, a few illustrations of the basic methods may be useful here.

TABLE 1-2

Thermal-Neutron Fission Yields (%) from U^{233}, U^{235}, and Pu^{239}

Fission product	U^{233}	U^{235}	Pu^{239}	*Fission product*	U^{233}	U^{235}	Pu^{239}
47-h Zn^{72}		1.6×10^{-5}	1.2×10^{-4}	136-d Sn^{123}		0.0013	
4.9-h Ga^{73}		1.1×10^{-4}		9.6-d Sn^{125}	0.052	0.013	0.071
7.8-m Ga^{74}		3.5×10^{-4}		2.0-y Sb^{125}		0.021	
11.3-h Ge^{77}	0.011	0.0031		91-h Sb^{127}	0.60	0.13	0.39
38.7-h As^{77}	0.021	0.0083		105-d Te^{127m}		0.035	
2.1-h Ge^{78}		0.020		57-m Sn^{128}		0.37	
91-m As^{78}		0.020		25.0-m I^{128}		3×10^{-5}	
9.0-m As^{79}		0.056		37-d Te^{129m}		0.35	
total Br^{80}	3.9×10^{-4}	1.0×10^{-5}		1.7×10^{7}-y I^{129}		0.8	
57-m Se^{81m}		0.0084		2.6-m Sn^{130}		2.0	
18.4-m Se^{81}		0.14		12.6-h I^{130}		5×10^{-4}	
35.9-h Br^{82}	1.1×10^{-3}	4×10^{-5}		30-h Te^{131m}		0.44	
25-m Se^{83}		0.22		8.05-d I^{131}	2.9	~3.1	3.77
2.4-h Br^{83}	0.87	0.51	0.084	stable Xe^{131}	3.39	2.93	3.78
stable Kr^{83}	1.17	0.544	0.29	77-h Te^{132}	4.4	~4.7	5.1
6.0-m Br^{84}		0.019		stable Xe^{132}	4.64	4.38	5.26
31.8-m Br^{84}		0.92		20.8-h I^{133}		~6.9	5.2
stable Kr^{84}	1.95	1.00	0.47	5.27-d Xe^{133}		6.62	6.91
39-s Se^{85}		~1.1		stable Cs^{133}	5.78	6.59	6.91
10.6-y Kr^{85}	0.58	0.293	0.127	52.5-m I^{134}		7.8	
stable Rb^{85}	2.51	1.30	0.539	stable Xe^{134}	5.95	8.06	7.47
stable Kr^{86}	3.27	2.02	0.76	6.7-h I^{135}	5.5	6.1	5.7
18.6-d Rb^{86}	2.3×10^{-4}	2.9×10^{-5}	2.3×10^{-5}	9.2-h Xe^{135}		6.3	
16-s $Se^{(87)}$		~2		2.6×10^{6}-y Cs^{135}	6.03	6.41	7.17
5×10^{10}-y Rb^{87}	4.56	2.49	0.92	86-s I^{136}	1.8	3.1	2.1
stable Sr^{88}	5.37	3.57	1.42	stable Xe^{136}	6.63	6.46	6.63
50.5-d Sr^{89}	5.86	4.79	1.71	13-d Cs^{136}	0.12	0.0068	0.11
28-y Sr^{90}	6.43	5.77	2.25	30-y Cs^{137}	6.58	6.15	6.63
9.7-h Sr^{91}	5.57	5.81	2.43	stable Ba^{133}		5.74	6.31
58-d Y^{91}	5.1	~5.4	2.9	83-m Ba^{139}	6.45	6.55	5.87
stable Zr^{91}	6.43	5.84	2.61	12.8-d Ba^{140}	5.4	6.35	5.4
2.7-h Sr^{92}		5.3		stable Ce^{140}	6.47	6.44	5.60
stable Zr^{92}	6.64	6.03	3.14	3.8-h La^{141}	7.1	6.4	5.7
10.3-h Y^{93}		6.1		33-d Ce^{141}		~6.0	5.1
1.1×10^{6}-y Zr^{93}	6.98	6.45	3.97	stable Pr^{141}	6.4		(4.5)*
stable Zr^{94}	6.68	6.40	4.48	stable Ce^{142}	6.83	6.01	5.01
65-d Zr^{95}	6.1	6.2	5.8	33-h Ce^{143}		5.7	5.3
stable Mo^{95}	6.11	6.27	5.03	stable Nd^{143}	5.99	6.03	4.57
stable Zr^{96}	5.58	6.33	5.17	280-d Ce^{144}	4.5	~6.0	3.79
23-h Nb^{96}	6.5×10^{-3}	6.1×10^{-4}	3.6×10^{-3}	5×10^{15}-y Nd^{144}	4.61	5.62	3.93
17.0-h Zr^{97}		5.9	5.5	stable Nd^{145}	3.47	3.98	3.13
stable Mo^{97}	5.37	6.09	5.65	stable Nd^{146}	2.63	3.07	2.60
52-m Nb^{98}	0.20	0.064	0.20	11.1-d Nd^{147}		~2.7	2.2
stable Mo^{98}	5.15	5.78	5.89	2.6-y Pm^{147}	1.9		1.94
66.5-h Mo^{99}	4.80	6.06	6.10	1.3×10^{11}-y Sm^{147}	1.98	2.36	2.07
stable Mo^{100}	4.41	6.30	7.10	stable Nd^{148}	1.34	1.71	1.73
stable Ru^{101}	2.91	5.0	5.91	53.1-h Pm^{149}			1.4
stable Ru^{102}	2.22	4.1	5.99	stable Sm^{149}	0.76	1.13	1.32
39.7-d Ru^{103}	1.8	3.0	5.67	stable Nd^{150}	0.56	0.67	1.01
stable Ru^{104}	0.94	1.8	5.93	80-y Sm^{151}	0.335	0.44	0.80
4.45-h Ru^{105}		0.9		stable Sm^{152}	0.220	0.281	0.62
36-h Rh^{105}			3.9	47-h Sm^{153}	0.11	0.15	0.37
1.01-y Ru^{106}	0.24	0.38	4.57	stable Eu^{153}	0.13	0.169	
22-m Rh^{107}		0.19		stable Sm^{154}	0.045	0.077	0.29
13.4-h Pd^{109}	0.044	0.030	1.40	24-m Sm^{155}		0.033	0.23
7.6-d Ag^{111}	0.024	0.019	0.23	4-y Eu^{155}		0.033	
21.0-h Pd^{112}	0.016	0.010	0.12	15.4-d Eu^{156}	0.011	0.014	0.11
43-d Cd^{115m}	0.0011	0.0007	0.0031	15.4-h Eu^{157}		0.0078	
53-h Cd^{115}	0.020	0.0097	0.0038	60-m Eu^{158}		0.002	
total 115	0.021	0.0104	0.041	18.0-h Gd^{159}		0.00107	0.021
3.0-h Cd^{117m}		0.011		6.9-d Tb^{161}		7.6×10^{-5}	0.0039
27.5-h Sn^{121}	0.018	0.015	0.043	82-h Dy^{166}			6.8×10^{-5}

*Estimated total chain yield.

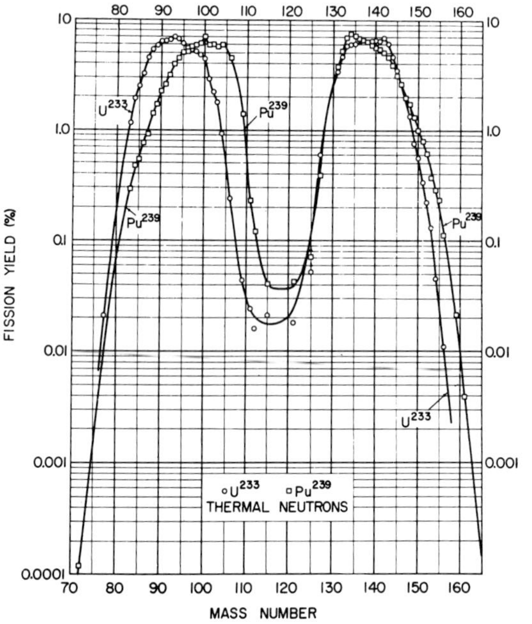

FIG. 1-1 Fission-product distribution from fission of U^{233} and Pu^{239} by thermal neutrons.

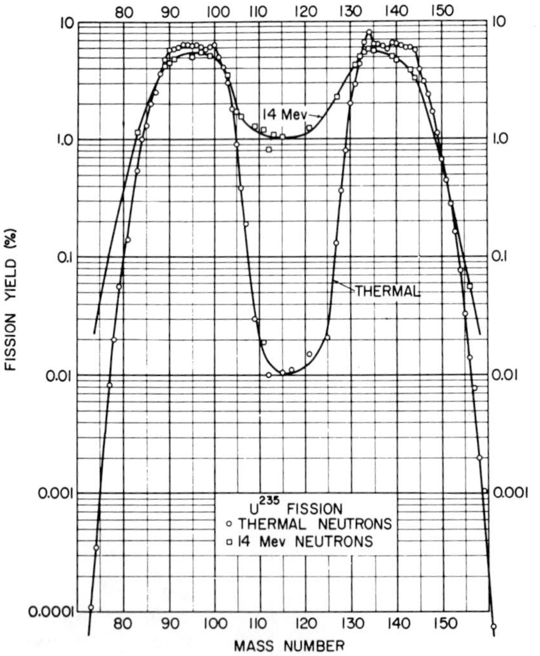

FIG. 1-2 Fission-product distribution from fission of U^{235} by thermal neutrons and by 14-Mev neutrons.

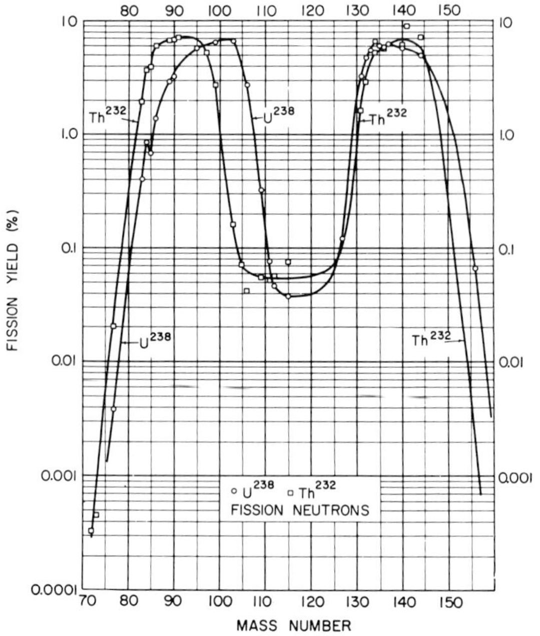

FIG. 1-3 Fission-product distribution from fission of Th^{232} and U^{238} by fission neutrons.

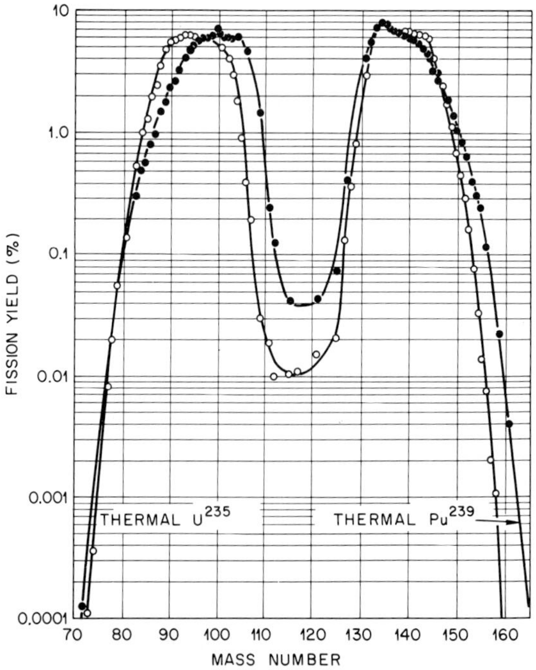

FIG. 1-4 Fission-product distribution from fission of U^{235} and Pu^{239} by thermal neutrons.

TABLE 1-3

Fission-Product Yields from Fission Induced by Fission-Spectrum Neutrons (Fast) and by 14-Mev Neutrons

*Fission product**	U^{235} (*fast*)	Pu^{239} (*fast*)	U^{238} (*fast*)	Th^{232} (*fast*)	U^{235} (*14 Mev*)	U^{238} (*14 Mev*)
Zn^{72}				3.3×10^{-4}		
Ga^{73}				4.5×10^{-4}		
Ge^{77}				0.009		
As^{77}			0.0038	0.020		
Br^{82}					0.004	
Br^{83}				1.9	1.16	0.62
Kr^{83}			0.40	1.99		
Br^{84}						1.1
Kr^{84}			0.85	3.65		
Kr^{85}			0.153	0.87		
Kr^{86}			1.38	6.0		
Sr^{89}			2.9	6.7	4.5	2.7
Sr^{90}	5.0	2.2	3.2	6.8	4.5	3.1
Sr^{91}				7.2	4.9	3.6
Y^{91}						2.8
Y^{93}						4.5
Zr^{95}			5.7		5.0	5.2
Nb^{96}					0.003	
Zr^{97}		5.2		5.2	5.6	5.8
Mo^{99}	6.1	6.0	6.3	2.7	5.17	5.7
Mo^{101}						5.5
Mo^{102}						3.9
Ru^{103}			6.6	0.16	3.5	
Ru^{105}						2.3
Rh^{105}				0.07	1.70	3.4
Ru^{106}			2.7	0.042	1.58	
Pd^{109}	0.146	2.0	0.32	0.055	1.31	1.2
Ag^{111}	0.071		0.076	0.052	1.20	0.96
Pd^{112}	0.041	0.14	0.046	0.057	0.81	0.69
Ag^{113}					1.1	0.85
Cd^{115m}			0.003	0.003	0.062	0.06
Cd^{115}	0.038	0.067	0.034	0.072	1.00	0.64
total 115			0.037	0.075	1.06	0.70

*Fission product**	U^{235} (*fast*)	Pu^{239} (*fast*)	U^{238} (*fast*)	Th^{232} (*fast*)	U^{235} (*14 Mev*)	U^{238} (*14 Mev*)
Sn^{121}					1.23	0.96
Sn^{125}					1.34	0.45
Sb^{127}			0.12		2.28	1.7
Sb^{129}						1.4
I^{131}				1.2	4.3	4.8
Xe^{131}			3.2	1.62	(4.3)†	
Te^{132}			4.7	2.4	4.2	4.7
I^{132}					5.0	
Xe^{132}			4.7	2.87	(5.0)†	
I^{133}					5.4	
Xe^{133}						6.6
Cs^{133}			5.5		(5.6)†	
I^{134}					5.3	
Xe^{134}			6.6	5.38	(5.9)†	
I^{135}					4.5	
Xe^{135}						5.5
Cs^{135}			6.0		(5.7)†	
Xe^{136}			5.9	5.65		
Cs^{136}					0.23	
Cs^{137}	6.3	6.8	6.2	6.3		
Ba^{139}					5.0	4.6
Ba^{140}		5.0	5.7	6.2	4.6	4.6
Ce^{141}				9.0		
Ce^{143}					3.9	3.6
Pr^{143}						3.2
Ce^{144}	5.0		4.5	7.1	3.3	3.3
Nd^{147}	2.3		2.6			2.0
Pm^{149}	1.1		1.8			
Sm^{153}	0.21	0.48	0.41			0.39
Eu^{156}	0.025		0.071		0.055	0.12
Gd^{159}	0.0034		0.0084			
Tb^{161}	4.6×10^{-4}		0.0016			

*For half-lives, see Table 1-2, with the exception of Mo^{102} (11.5 min), Sb^{129} (4.6 hr), I^{132} (2.3 hr) and Pr^{143} (13.7 day).
†Estimated total chain yield.

The following information is required to permit the calculation of a given isotope formed during fission:

1) time of irradiation,
2) flux density of the reactor or facility,
3) relative fission yield of each isotope along the decay chain of the particular mass number of interest and any branching ratios involved,
4) half-lives (or decay constants) of the isotopes along the decay chain, and
5) cross sections for capture by the nuclides, if these are significant, to the degree of accuracy desired.

The Xe^{133} chain, including I^{133}, follows the decay scheme shown in Fig. 1-5. The chain diagram shows that the mass-133 chain has an initial yield of 4.0% in the 4-min. Sb^{133}, which decays by branching to two isomers of tellurium, both of which produce 20.8-hr I^{133}. Additional prompt yield of I^{133} increases the total to 6.9%. In turn, the 20.8-hr I^{133} undergoes beta decay with a branching ratio of 2.4% to 2.3-day metastable Xe^{133}. Isomeric transition then takes place to complete the total yield through 5.27-day Xe^{133}. The end of the chain is seen in the production of stable Cs^{133} by beta decay of Xe^{133}. As an illustration, a few simplified calculations of the growth and decay of 20.8-hr I^{133} and 5.27-day Xe^{133} are given, following the methods of Dillon and Burris [5].

The following calculation simplifies the decay chain to instantaneous production of I^{133} by fissioning, followed by direct decay of I^{133} to Xe^{133}, and assumes no burnup by neutron capture. The decay scheme may be represented as A → B → C, where A is I^{133}, B is Xe^{133}, and C is stable Cs^{133}. With a clean reactor core, the rate of change of isotope A is

$$\frac{dN_A}{dt} = R_A - \lambda_A N_A , \qquad (1\text{-}1)$$

where N_A=number of atoms of isotope A at time t, λ_A= the decay constant for A, and R_A = the rate of production of A by the fission process. Equation (1-1) can be solved for N_A:

$$N_A = \frac{R_A}{\lambda_A}\,(1 - e^{-\lambda_A t}). \qquad (1\text{-}2)$$

TABLE 1-4

Decay Chains and Yields from Thermal-Neutron Fission of U^{225}

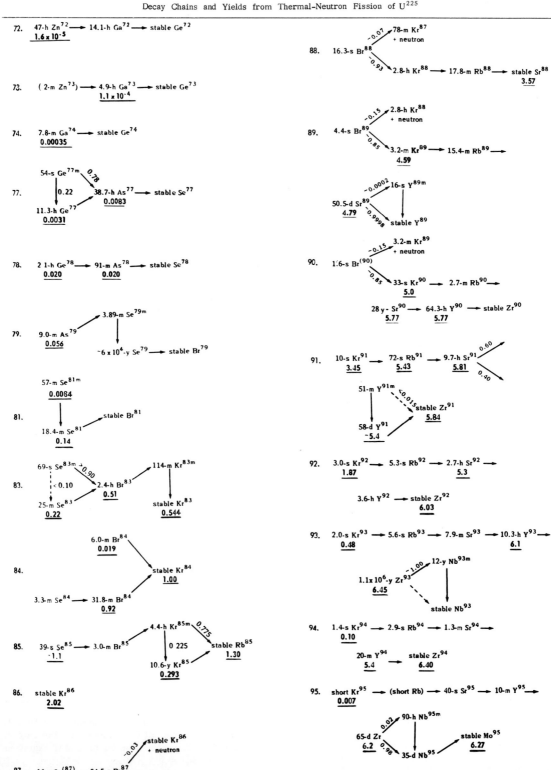

72. 47-h Zn^{72} → 14.1-h Ga^{72} → stable Ge^{72}
47-h Zn^{72}: 1.6×10^{-5}

73. (2-m Zn^{73}) → 4.9-h Ga^{73} → stable Ge^{73}
4.9-h Ga^{73}: 1.1×10^{-4}

74. 7.8-m Ga^{74} → stable Ge^{74}
7.8-m Ga^{74}: 0.00035

77. 54-s Ge^{77m} —0.78→ 38.7-h As^{77} → stable Se^{77}
54-s Ge^{77m} —0.22→ 11.3-h Ge^{77} → 38.7-h As^{77}
38.7-h As^{77}: 0.0083
11.3-h Ge^{77}: 0.0031

78. 2 1-h Ge^{78} → 91-m As^{78} → stable Se^{78}
2 1-h Ge^{78}: 0.020
91-m As^{78}: 0.020

79. 9.0-m As^{79} → 3.89-m Se^{79m} → ~6×10^4-y Se^{79} → stable Br^{79}
9.0-m As^{79}: 0.056

81. 57-m Se^{81m} → 18.4-m Se^{81} → stable Br^{81}
57-m Se^{81m}: 0.0084
18.4-m Se^{81}: 0.14

83. 69-s Se^{83m} —0.90→ 2.4-h Br^{83} → 114-m Kr^{83m} → stable Kr^{83}
69-s Se^{83m} —<0.10→ 25-m Se^{83} → 2.4-h Br^{83}
2.4-h Br^{83}: 0.51
25-m Se^{83}: 0.22
stable Kr^{83}: 0.544

84. 6.0-m Br^{84} → stable Kr^{84}
3.3-m Se^{84} → 31.8-m Br^{84} → stable Kr^{84}
6.0-m Br^{84}: 0.019
stable Kr^{84}: 1.00
31.8-m Br^{84}: 0.92

85. 39-s Se^{85} → 3.0-m Br^{85} → 4.4-h Kr^{85m} —0.775→ stable Rb^{85}
4.4-h Kr^{85m} —0 225→ 10.6-y Kr^{85} → stable Rb^{85}
39-s Se^{85}: ~1.1
stable Rb^{85}: 1.30
10.6-y Kr^{85}: 0.293

86. stable Kr^{86}
2.02

87. 16-s $Se^{(87)}$ → 54.5-s Br^{87} —~0.03→ stable Kr^{86} + neutron
54.5-s Br^{87} —~0.97→ 78-m Kr^{87} → 5×10^{10}-y Rb^{87}
16-s $Se^{(87)}$: ~2
5×10^{10}-y Rb^{87}: 2.49

88. 16.3-s Br^{88} —~0.07→ 78-m Kr^{87} + neutron
16.3-s Br^{88} —~0.93→ 2.8-h Kr^{88} → 17.8-m Rb^{88} → stable Sr^{88}
stable Sr^{88}: 3.57

89. 4.4-s Br^{89} —~0.15→ 2.8-h Kr^{88} + neutron
4.4-s Br^{89} —~0.85→ 3.2-m Kr^{89} → 15.4-m Rb^{89} →
3.2-m Kr^{89}: 4.59

50.5-d Sr^{89} —~0.0002→ 16-s Y^{89m} → stable Y^{89}
50.5-d Sr^{89} —~0.9998→ stable Y^{89}
50.5-d Sr^{89}: 4.79

90. 1.6-s $Br^{(90)}$ —~0.15→ 3.2-m Kr^{89} + neutron
1.6-s $Br^{(90)}$ —~0.85→ 33-s Kr^{90} → 2.7-m Rb^{90} →
33-s Kr^{90}: 5.0

28 y - Sr^{90} → 64.3-h Y^{90} → stable Zr^{90}
28 y - Sr^{90}: 5.77
64.3-h Y^{90}: 5.77

91. 10-s Kr^{91} → 72-s Rb^{91} → 9.7-h Sr^{91} —0.60→ (51-m Y^{91m})
9.7-h Sr^{91} —0.40→ (58-d Y^{91})
10-s Kr^{91}: 3.45
72-s Rb^{91}: 5.43
9.7-h Sr^{91}: 5.81

51-m Y^{91m} —<0.015→ stable Zr^{91}
51-m Y^{91m} → 58-d Y^{91} → stable Zr^{91}
stable Zr^{91}: 5.84
58-d Y^{91}: ~5.4

92. 3.0-s Kr^{92} → 5.3-s Rb^{92} → 2.7-h Sr^{92} →
3.0-s Kr^{92}: 1.87
2.7-h Sr^{92}: 5.3

3.6-h Y^{92} → stable Zr^{92}
stable Zr^{92}: 6.03

93. 2.0-s Kr^{93} → 5.6-s Rb^{93} → 7.9-m Sr^{93} → 10.3-h Y^{93} →
2.0-s Kr^{93}: 0.48
10.3-h Y^{93}: 6.1

1.1×10^6-y Zr^{93} —~1.00→ 12-y Nb^{93m} → stable Nb^{93}
1.1×10^6-y Zr^{93} - - → stable Nb^{93}
1.1×10^6-y Zr^{93}: 6.45

94. 1.4-s Kr^{94} → 2.9-s Rb^{94} → 1.3-m Sr^{94} →
1.4-s Kr^{94}: 0.10

20-m Y^{94} → stable Zr^{94}
20-m Y^{94}: 5.4
stable Zr^{94}: 6.40

95. short Kr^{95} → (short Rb) → 40-s Sr^{95} → 10-m Y^{95} →
short Kr^{95}: 0.007

65-d Zr —0.02→ 90-h Nb^{95m} → 35-d Nb^{95} → stable Mo^{95}
65-d Zr —0.98→ 35-d Nb^{95} → stable Mo^{95}
65-d Zr: 6.2
stable Mo^{95}: 6.27

96. stable Zr^{96}
6.33

Table 1-4 (continued)

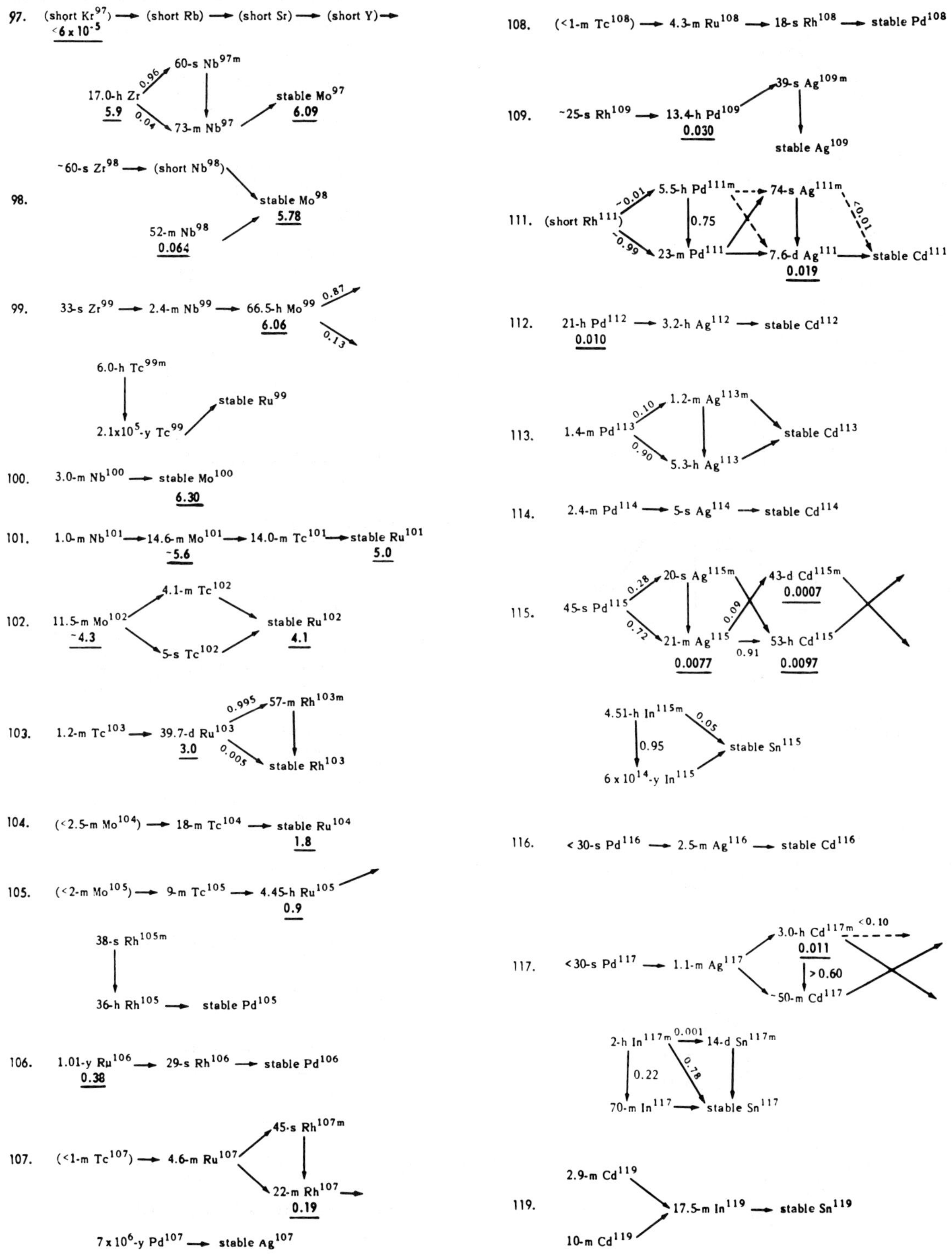

Bold-face underlined numbers give experimental fission yields. Last fission yield along any chain usually represents total chain yield. Lower values for yields of earlier chain members may be caused by (1) direct formation in fission of later chain members, (2) chain branching, (3) experimental uncertainty. Latter accounts for cases where early chain member has higher yield than later one. Where branching occurs, arrows are shown only for decay modes observed experimentally; fraction in each branch is given where known. Parentheses indicate nuclide probably occurs in fission but has not been so observed.

Table 1-4 (continued)

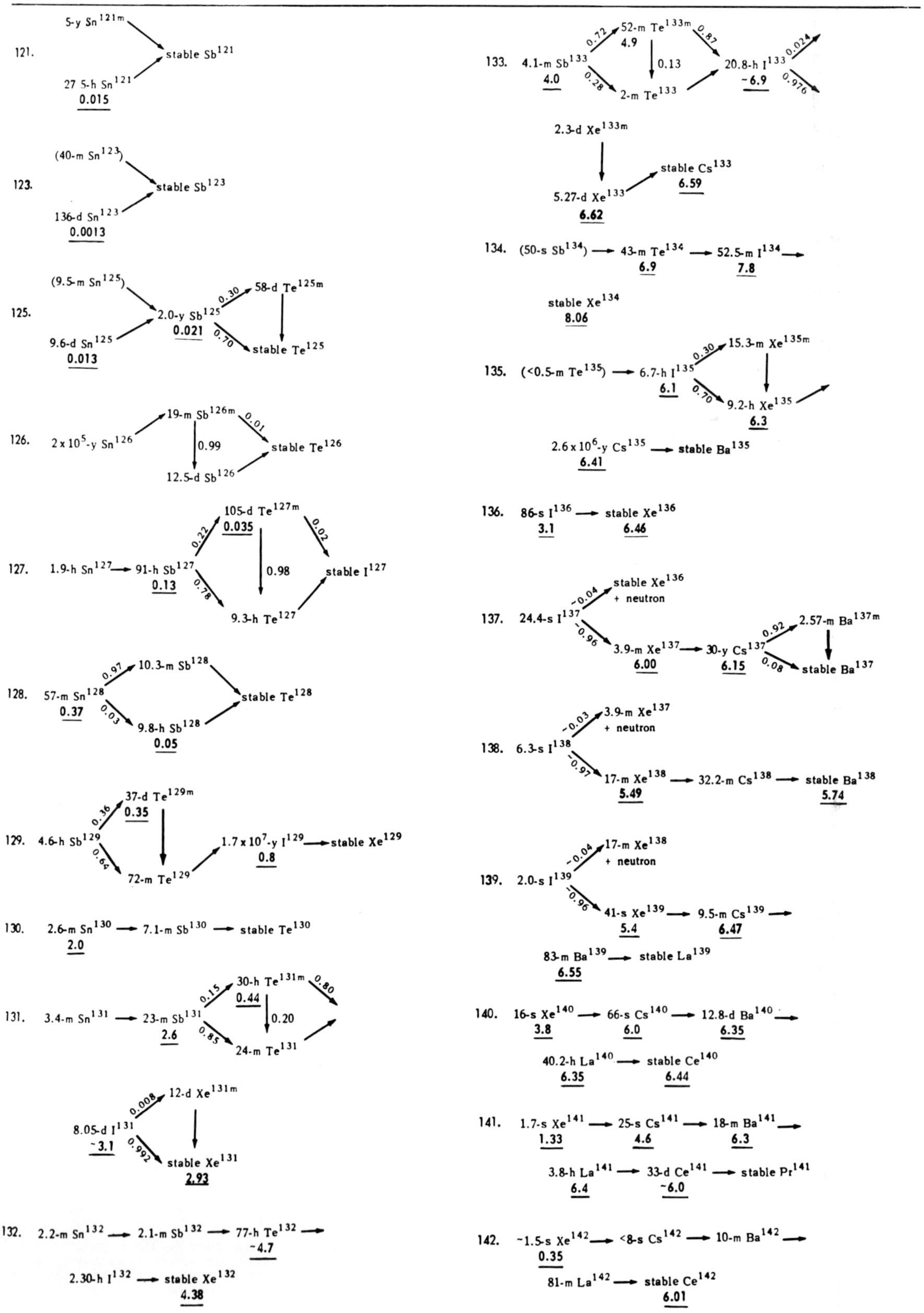

Table 1-4 (concluded)

143. 1-s Xe^{143} → (short Cs) → 13-s Ba → 18-m La^{143} →
0.051
33-h Ce^{143} → 13.7-d Pr^{143} → stable Nd^{143}
6.0 6.03

144. short Xe^{144} → (short Cs) → (short Ba) → (short La) →
0.006
280-d Ce^{144} → 17.4-m Pr^{144} → 5×10^{15}-y Nd^{144}
~6.0 5.62

145. 3.0-m Ce^{145} → 5.96-h Pr^{145} → stable Nd^{145}
3.98

146. 13.9-m Ce^{146} → 24.4-m Pr^{146} → stable Nd^{146}
3.07

147. 1.2-m Ce^{147} → 12.0-m Pr^{147} → 11.1-d Nd^{147} →
~2.7
2.6-y Pm^{147} → 1.3×10^{11}-y Sm^{147}
2.36

148. 40-s Ce^{148} → 1.95-m Pr^{148} → stable Nd^{148}
1.71

149. (2.0-h Nd^{149}) → 53.1-h Pm^{149} → stable Sm^{149}
1.13

150. stable Nd^{150}
0.67

151. (13-m Nd^{151}) → 28.4-h Pm^{151} → 80-y Sm^{151} → stable Eu^{151}
0.44

152. stable Sm^{152}
0.281

153. 47-h Sm^{153} → stable Eu^{153}
0.15 0.169

154. stable Sm^{154}
0.077

155. 24-m Sm^{155} → 4-y Eu^{155} → stable Gd^{155}
0.033 0.033

156. 9-h Sm^{156} → 15.4-d Eu^{156} → stable Gd^{156}
0.013 0.014

157. 15.4-h Eu^{157} → stable Gd^{157}
0.0078

158. 60-m Eu^{158} → stable Gd^{158}
0.002

159. 18.0-h Gd^{159} → stable Tb^{159}
0.00107

161. (3.7-m Gd^{161}) → 6.9-d Tb^{161} → stable Dy^{161}
7.6×10^{-5}

166. 82-h Dy^{166} → 27.3-h Ho^{166} → stable Er^{166}

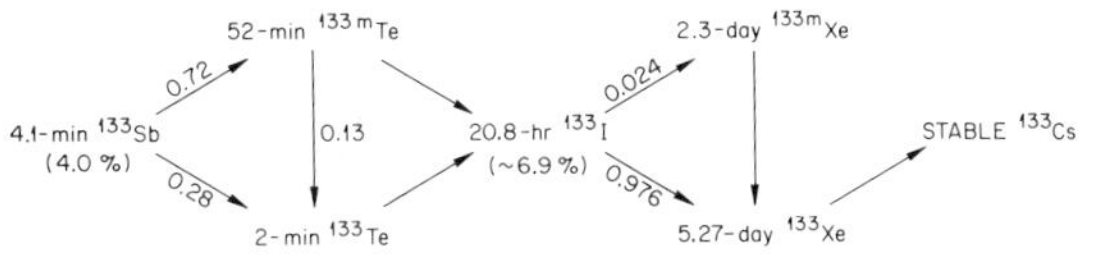

FIG. 1-5 Decay scheme of mass-133 chain.

In Eq. (1-2) it is assumed that $N_A = 0$ at $t = 0$. Similarly, the rate of change of the number of B atoms is

$$\frac{dN_B}{dt} = R_B - \lambda_B N_B + \lambda_A N_A, \qquad (1\text{-}3)$$

and the number of B atoms is

$$N_B = \frac{R_A + R_B}{\lambda_B}(1 - e^{-\lambda_B t}) - \frac{R_A}{\lambda_B - \lambda_A}(e^{-\lambda_A t} - e^{-\lambda_B t}). \qquad (1\text{-}4)$$

In Eq. (1-4) it is also assumed that $N_B = 0$ at $t = 0$. To calculate the disintegration rates of A or B the following equations are used:

$$\text{Disintegration rate of A} = \lambda_A N_A, \qquad (1\text{-}5)$$

$$\text{Disintegration rate of B} = \lambda_B N_B, \qquad (1\text{-}6)$$

where N_A is obtained from Eq. (1-2) and N_B from Eq. (1-4).

For the sample calculation, we desire to compare the disintegration rates of I^{133} and Xe^{133} after 1 day of operating time and to express the activity in terms of reactor operating power so that N_A will be the number of atoms of I^{133} per kilowatt of reactor power and R_A will be the number of atoms of A per second per kilowatt of reactor power. If we use 200 Mev per fission, or 3.1×10^{13} fissions per second equal to 1 kw reactor power, a 6.9% yield for I^{133} and zero direct fission yield for Xe^{133}, and if we note that $\lambda_A = 9.25 \times 10^{-6}$/sec and $\lambda_B = 1.522 \times 10^{-6}$/sec, then $N_A = 1.27 \times 10^{17}$ atoms I^{133}/kw (after 1-day of operation); $N_B = 1.13$

$\times\ 10^{17}$ atoms Xe^{133}/kw (after 1-day of operation). The above results derived from Eqs. (1-2) and (1-4), are based on an initial (t = 0) clean reactor core. The disintegration rates, Eqs. (1-5) and (1-6), may be expressed as curies where 1 curie = 3.7×10^{10} disintegrations per second:

dN_A/dt = 31.9 curies/kw reactor power (after 1 day of operation)

dN_B/dt = 2.16 curies/kw reactor power (after 1 day of operation).

In order to obtain comparative energy-release rates, the disintegration rates may be multiplied by the decay energy per disintegration (Mev/dis). For this calculation, the number of curies of each isotope multiplied by the decay energy in Mev/dis will be expressed in "Mev-curies," where 1 Mev-curie = 3.7×10^{10} Mev of decay energy per second. After 1 day of reactor operation, the activity of I^{133} (decay energy 1.01 Mev total) is 32.2 Mev-curies/kw reactor power, and the activity of Xe^{133}(decay energy 0.196 Mev total) is 0.42 Mev-curies/kw.

After the reactor is shut down, R_A and R_B become zero. For isotope A, Eq. (1-2) reduces to the simple decay formula with the solution:

$$N_A = N_{A_0} e^{-\lambda_A t} \qquad (1\text{-}7)$$

where N_{A_0} = number of atoms of isotope A at shutdown and t = time from shudown. For isotope B, Eq. (1-3) reduces to

$$\frac{dN_B}{dt} = -\lambda_B N_B + \lambda_A N_{A_0} e^{-\lambda_A t}.$$

The solution of this equation is

$$N_B = N_{B_0} e^{-\lambda_B t} + \frac{\lambda_A}{\lambda_B - \lambda_A} \cdot N_{A_0}(e^{-\lambda_A t} - e^{-\lambda_B t}), \qquad (1\text{-}8)$$

where N_{B_0} = number of atoms of B at shutdown and t = time from shutdown. The disintegration rates are determined by substitution of the calculated N_A and N_B in Eqs. (1-5) and (1-6).

For reactor operation of 1 day followed by 10 days of decay, $N_A = 4.3 \times 10^{13}$ atoms I^{133}/kw reactor power and $N_B = 5.6 \times 10^{16}$ atoms Xe^{133}/kw. The activity of I^{133} at this time is 1.08×10^{-2} curies and the activity of Xe^{133} is 2.2 curies. In order to compare the decay power or energy-release rate of the two isotopes, we can multiply these numbers by the energy in Mev/dis to obtain 1.09×10^{-2} Mev-curies for I^{133} and 0.43 Mev-curies for Xe^{133}.

The result of the above illustrative calculation is presented in Fig. 1-6. This diagram is typical of the rate-of-growth and decay relationships between members of the same fission-product chain. The time required for saturation of the 20.8-hr iodine is shown to be only a few days, while that for the 5.27-day xenon is several weeks. Decay of the two nuclides on shutdown illustrates a different behavior

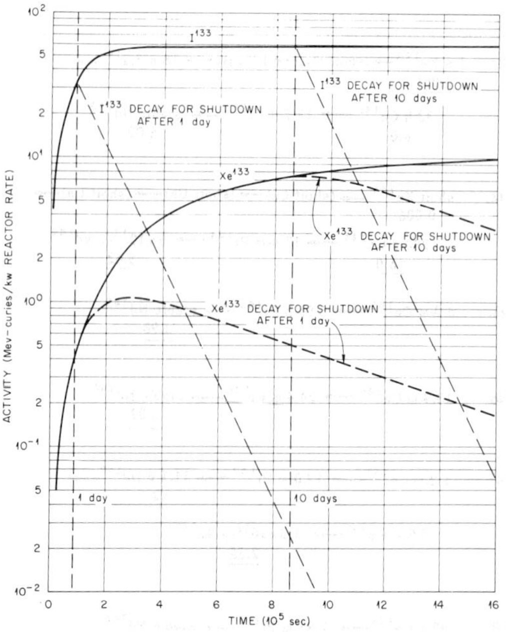

FIG. 1-6 Growth and decay of I^{133} and Xe^{133}.

in that the relatively short-lived precursors of the iodine contribute little to its inventory, and, therefore, exponential decay is all that is seen. For xenon, after a short irradiation period, a significant fraction of the inventory is contributed by the decay of iodine. For example, following shutdown after 1 day of operation, the xenon continues to grow for about 2 days before it begins to decay exponentially. After 10 days of operation the effect is much less important and decay is exponential within a few hours.

Since there are so many elements and radioactive species included in the fission spectrum, it is desirable to use a simplified picture, if such use is realistic, by reducing the number of species under consideration to a number less than 10. One such scheme is shown in Table 1-5. A study of Table 1-1 shows that krypton, xenon, and iodine are very important, because they are volatile and also because they are present in quite large quantities. In addition, iodine concentrates in the thyroid and, consequently, is biologically potent. Therefore, these elements must be included in the table of important fission products. Radioactive cesium must be included, even though its yield is low, because it is moderately volatile and, when ingested, constitutes a hazard to the whole body. Tellurium has isotopes with very short lives and high yields. It also is moderately volatile, and it contributes a health hazard through I^{132}, the daughter product of Te^{132}. Strontium is included because it has relatively high radioactivity and it is a long-lived hazard to bone and lung, even though of relatively low volatility. Barium is listed because of high yield and slight volatility and because of its hazard to bone and lung tissue. Ruthenium is included because of its high volatility under strongly oxidizing conditions. Other isotopes in

Table 1-1 are excluded from Table 1-5 because of low volatility, low fission yield, relatively low toxicity, or very short half-life.

TABLE 1-5

Summary of Important Fission-Product Activities Arranged According to Volatility

Element	Activity (kcurie/Mw(t)) for 1 Year of Operation		Weight (g)
	At shutdown	1 day after shutdown	
Rare gases (Xe, Kr)	385	62	61
I	289	53	2.6
Te, Cs	220	37	40
Ru	27	27	19
Sr, Ba	144	98	31

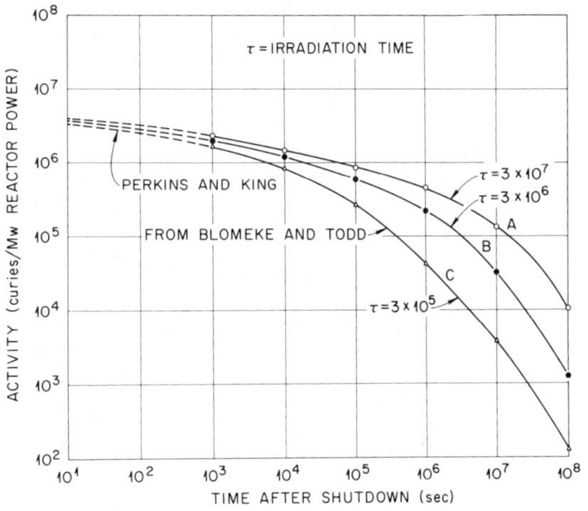

FIG. 1-7 Decay of total fission-product inventory resulting from different irradiation times.

The time in core and the power level of a reactor are major factors in the determination of the inventory of fission products. The short-lived isotopes reach their saturation concentration soon after reactor startup, while the longer lived activities, those with half-lives in the range from months to years, continue to accumulate during an average reactor core life.

Since radioactive iodine constitutes a particularly serious problem, it requires special emphasis. Five isotopes of iodine listed in Table 1-1 exceed I^{131} in the amount of activity at the instant of shutdown. However, a significant amount of only one of these (20.8-hr I^{133}) persists 24 hr after shutdown.

As a rough approximation, the total radioactivity of the fission products will decay by a factor of 10 in 24 hr, even after high fuel burnup. Comparison of the two activity columns in Table 1-1 shows that no significant amount of bromine or krypton activity remains after a 24-hr decay period and, as mentioned above, few of the volatile and semivolatile isotopes are still present in significant quantity at that time. Figure 1-7 [3, 4] shows a graph of the decay of the gross fission products as a function of time after shutdown. Curve A indicates the integrated total activity per megawatt for an operating time of 1 year (3×10^7 sec) and curves B and C for 35 days (3×10^6 sec) and for 3.5 days (3×10^5 sec).

A more detailed illustration of the growth and decay of individual products and groups of products is given in Figs. 1-8 to 1-11. Bolles and Ballou [6] have published detailed abundance and decay curves for fission products resulting from prompt or simultaneous fission which are typical of the spectrum from weapons tests or criticality accidents. In one such curve (Fig. 1-8), the fission elements are divided into nine chemical groups, which somewhat correspond to the volatility groups discussed above. For comparison, the corresponding abundance and decay curves for fission products resulting from reactor operation during periods from 3.5 days to 350 days are given in Figs. 1-9, 1-10, and 1-11. In these figures, the elements are grouped according to their relative volatility. It is clear from these figures that, after extended irradiation, about 50% of the total activity is always associated with the refractory elements (the rare earths, zirconium, and niobium). The halogens, rare gases, alkaline earth metals (barium and strontium), and tellurium are each present initially to the extent of 10% to 15% of the total activity. Tellurium falls below 8% after 24 hr or less, and iodine drops below 6% in not more than 10 days. Only the barium-strontium group, of the four mentioned above, gains in abundance after 10 days. The refractory group continues to increase to between 70% and 80% of the total after this short period. For periods longer than 50 days, Ru^{106}, Cs^{137}, and Sr^{90} become the most significant radioactivities in addition to the refractory group.

1.2.2. Np^{239} and Pu^{239}

Special mention of these two transuranium elements is required with regard to reactor safety. It is particularly important to point out that in low-enrichment reactors, neutron capture by U^{238} continuously produces 2.3-day Np^{239}, which contributes a large (about 30%) fraction of the total gamma activity of the core at shutdown. Pu^{239} is produced by decay of Np^{239} at a rate approximately equal to 80% of the rate that U^{235} is consumed by fission (1.0 g/Mw-day). Therefore, Pu^{239} is formed at a rate of about 0.8 g/Mw-day in low-enrichment reactors. It is partially consumed by fissioning during long operating times but not enough to alter significantly the hazard from this source. Fortunately, neptunium and plutonium both form stable oxides (NpO_2 and PuO_2) which are comparable in volatility to uranium dioxide (UO_2); and a large fraction (see Table 3-4) of these biologically potent species would be expected to remain in the reactor core under most accident conditions.

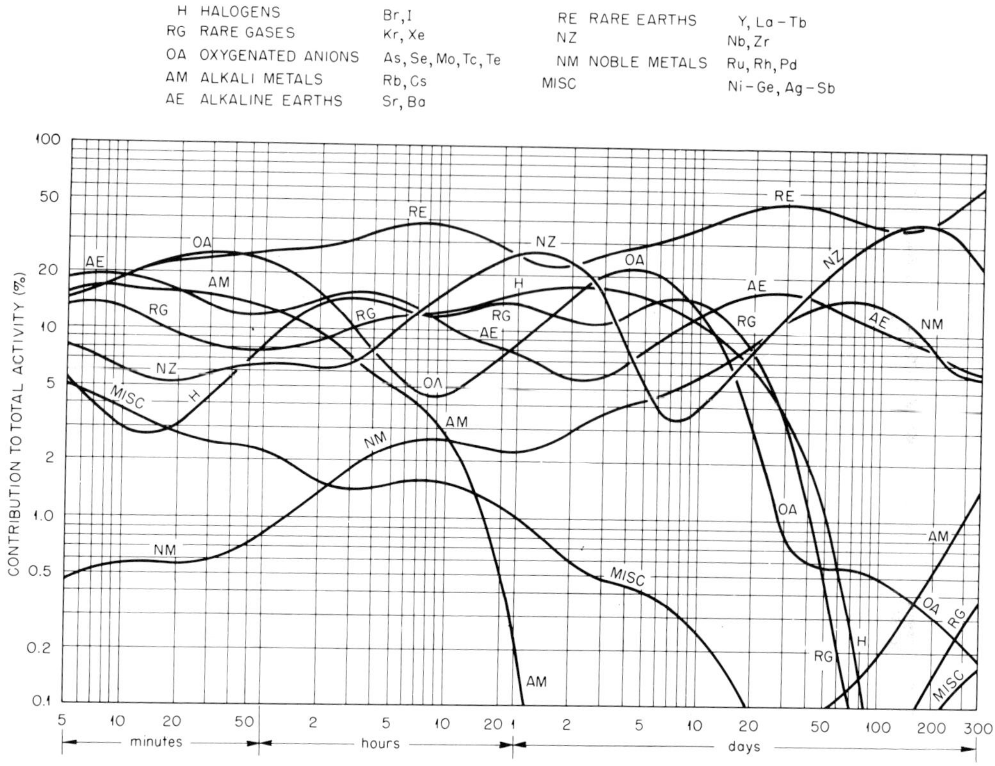

FIG. 1-8 Calculated abundances of groups of U^{235} fission products after instantaneous fission, according to Bolles and Ballou [6].

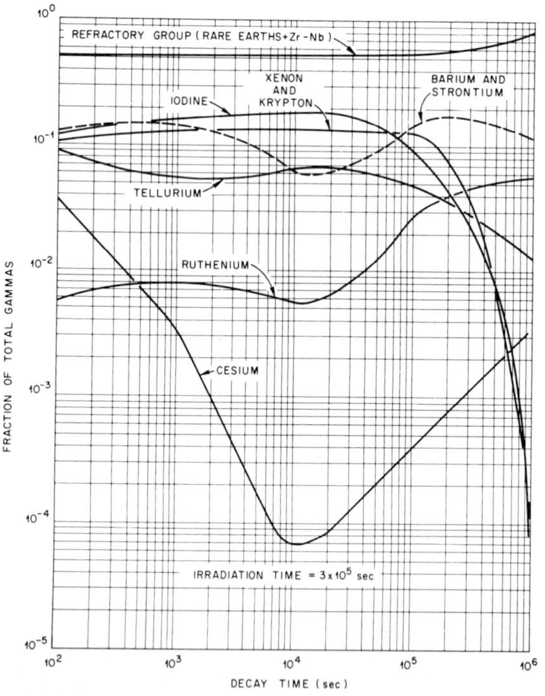

FIG. 1-9 Calculated abundances of groups of U^{235} fission products after a 3.5-day irradiation period.

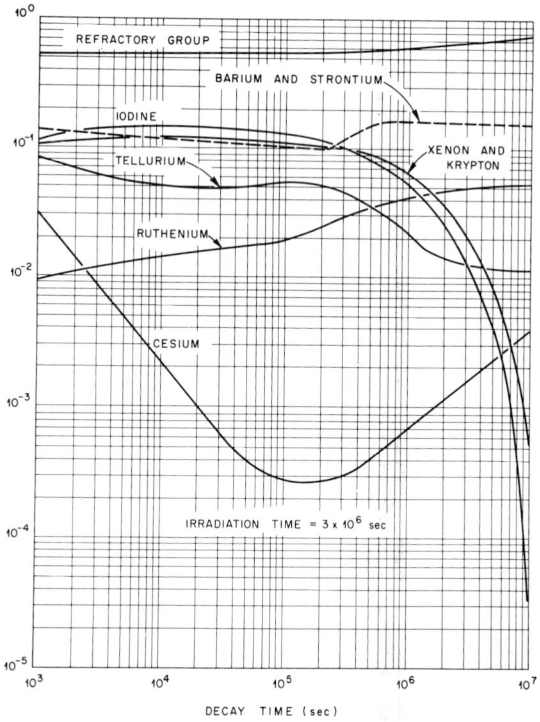

FIG. 1-10 Calculated abundances of groups of U^{235} fission products after a 35-day irradiation period.

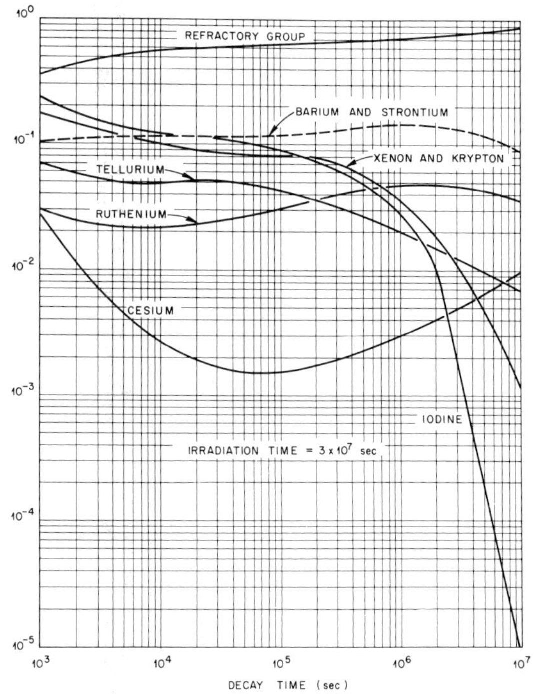

FIG. 1-11 Calculated abundances of groups of U^{235} fission products after a 350-day irradiation period.

1.3 Fuel-Element Classification

The discussion of fission-product behavior in this section will concentrate on those fuels already in use (see also Chapter 13). In a consideration of the release of fission products, it is convenient to divide the fuel types into six different classes, as shown in Table 1-6.

The first class consists of the metals and alloys, which are generally characterized by normal metallic behavior and by melting points normally below 2000°C (3632°F). Although there are some minor variations in the behavior of these metals and alloys on melting (due to differences in their chemical activity), it is convenient to classify them as one group.

The second class of fuel elements includes the oxides dispersed in a metallic matrix. The melting point is still determined by the metal used, but there are some differences in the behavior of the dispersed oxide.

The third class includes the oxide fuel materials. They are characterized by melting points normally above 2000°C (3632°F) and by a relatively high degree of chemical inertness.

Class IV includes the carbides dispersed in graphite or in pyrolytic carbon. These, in general, have high melting points but are characterized by vigorous chemical reaction with a number of materials that could conceivably be part of the reactor system, including air, water, etc.

The fifth group consists of miscellaneous compounds, namely, two uranium sulfides and the nitrides of uranium and thorium; the sixth group consists of three fuel systems that are liquid at reactor operating temperatures.

Melting temperatures have been included in the table because they comprise one of the most important physical properties in determining the extent of fission-product release, as will be discussed in more detail in Sec. 1.4.1. Melting ranges estimated from published phase diagrams are given for some materials that do not have unique melting points.

The term ''ceramic fuel'' is avoided in this classification system because it appears to mean different things to different people. The selection of materials for inclusion in this table is admittedly rather arbitrary. It could be expanded extensively by the addition of many materials that may at some future date become important components of reactor fuel systems or it could be compressed by eliminating all materials that have not actually been employed in reactors or tested extensively under reactor environments. The fluid-fuel section is quite restricted in scope because few experimental reactors have used such fuels to date.

TABLE 1-6

Classification and Melting Point of Fuel Components and Mixtures

Class I: Metals and Alloys (Sec. 3.1)

Fuel component (compositions in wt. %)	Melting point	Reference
	1133°C (2071°F)	[7]
U-Al eutectic (87-13)	640°C (1184°F)	[9]
U-Mo (90-10)	~1150-1250°C (~2100-2280°F)	[9]
U-Zr (98-2)	~1150°C (~2100°F)	[9]
U-Nb (10-90)	2400°C (4352°F)	[22]
U-Th eutectic (72-28)	1370°C (2498°F)	[9]
U-Pu (95-5)	1070-1100°C (1958-2012°F)	[10]
Pu	640°C (1184°F)	[10]
Pu-Al (3.7-96.3)	650°C (~1200°F)	[10]
Th	1750°C (3182°F)	[9]
U-ZrH_2 (8-92)	1170-1300°C (2138-2372°F)	[9]
Zircaloy-2 (98.5 Zr-1.5 Sn) (cladding material)	~1850°C (~3360°F)	[9]

Class II: Oxides Dispersed in a Metallic Matrix (Sec. 3.2)

Fuel component	Matrix melting point	Reference
Al-UO_2	659°C (1218°F)	[7]
Al-U_3O_8	659°C (1218°F)	[7]
Stainless steel-UO_2	1370-1425°C (2498-2597°F)	[9]
Nichrome-V-UO_2	~1390°C (~2530°F)	[9]

TABLE 1-6 (continued)

Class III: Oxide Fuel Materials (Sec. 3.3)		
Fuel component (compositions in wt. %)	Melting point	Reference
UO_2[a]	2860°C (5180°F)	[23]
U_3O_8	2860°C (UO_2)	[11]
PuO_2	2280°C (4136°F)	[13]
(Th, U)O_2 solid solution (93.6-6.4)	~3200°C (~5790°F)	[9]
(U, Pu) O_2 solid solution (95-5)	2700°C (4890°F)	[17]
UO_2 - BeO (5-95)	2150-2500°C (3900-4530°F)	[11]
BeO	2550°C (4622°F)	[7]
ThO_2	3300°C (5972°F)	[9]

Class IV: Carbides in Graphite or Pyrolytic Carbon (Sec. 3.4)		
Fuel component	Melting point	Reference
UC	2500°C (4532°F)	[14]
UC_2	2480°C (4496°F)	[14]
ThC	2625°C (4757°F)	[14]
ThC_2	2655°C (4813°F)	[14]
PuC	1654°C (3009°F)	[13]
PuC_2	2250°C (4082°F)	[13]
ZrC	3535°C (6397°F)	[14]
NbC	3500°C (6332°F)	[14]
UC-UC_2 (ss min.)	2400°C (4352°F)	—
UC_2-C (eutectic)	2450°C (4442°F)	[14]
ThC_2-C (eutectic)	2500°C (4532°F)	[14]
ZrC-C (eutectic)	2430°C (4406°F)	[14]
NbC-C (eutectic)	3250°C (5882°F)	[14]

Class V: Miscellaneous Compounds (Sec. 3.5)		
Fuel component	Melting point	Reference
UN	2850°C (5162°F)	[15]
US	2460°C (4460°F)	[15]
US_2	1680°C (3056°F)	[15]
ThN	2630°C (4766°F)	[15]

Class VI: Fuel Systems Liquid at Reactor Operating Temperatures (Sec. 3.6)		
Fuel component (composition in mole %)	Melting point	Reference
NaF-ZrF_4-UF_4 (53-40.5-6.5)	530°C (986°F)	[9]
LiF-BeF_2-ZrF_4-UF_4 (65-29.1-5.0-0.9)	450°C (842°F)	[16]
Pu-Fe (90.5-9.5)	411°C (772°F)	[13]

[a]A wide range of values for the melting point of UO_2 is found in the literature. The value given here is believed by the authors of this chapter to be accurate within the uncertainty assigned by the investigators (± 45°C or ± 81°F). The melting point of irradiated UO_2 has been reported [12] to vary with burnup from 2790°C (5054°F) at zero burnup to a maximum of 2930°C (5306°F) at 0.09 at. % burnup and then to decrease to 2760°C (5000°F) and 2660°C (4820°F) at 8.4 and 11.3 at. % burnup, respectively.

1.4 Mechanisms of Release

1.4.1 Fuel Melting

The amount of a fission product released when irradiated fuel is heated might be expected to follow a relatively simple relation between the fuel melting temperature and the vapor pressure of the fission product; however, since identical experimental conditions have rarely been maintained in fission-product release experiments with different fuel materials, it has been difficult to establish the nature of the relationship. The release of some fission-product elements would no doubt be affected by chemical reactions and by solubility in the fuel or its environs, as well as by the fuel temperatures reached.

Without fuel melting and subsequent release of fission products, a nuclear reactor accident would involve no more hazard to the general public than any conventional power plant accident of a similar nature. Thus, melting, followed by the release of fission products, is one of the unique features of a serious nuclear reactor accident.

The degree of melting is always difficult to postulate, since much uncertainty exists concerning the rate of heat loss from the reactor core and the possibility of reassembly of a critical mass in the bottom of the primary vessel. The melting of fuels with reasonable burnup levels could lead to high release rates for the volatile elements (xenon iodine, tellurium, cesium). The low melting fuels may also release some ruthenium and cesium in addition to xenon, iodine, and tellurium; while the higher melting ones release all of these and may also release some strontium and barium.

1.4.2 Fuel Oxidation

Burning of either a metallic fuel (Sec. 3.1.1) or of a lower oxide (UO_2) (Sec. 3.3.1.2) greatly enhances fission-product release by increase of the exposed surface area by many orders of magnitude, as well as by local overheating and gas expulsion. Hilliard [17] and Parker et al. [18] have noted that the fraction of most fission products released from metallic uranium is proportional to the extent of oxidation. They have also observed that the rate of release is nearly proportional to temperature up to 1500°C (2732°F).

An oversight in many hazards summaries is the failure to consider the effect that UO_2 burning may contribute to the total release of fission products following meltdown (or cladding rupture) and subsequent exposure of fuel to air at temperatures below 1500°C. The correct procedure would be to sum the effect of melting or high-temperature diffusion with the effect of oxidation on the residual inventory of fission products.

1.4.3 Gaseous Fission-Product Diffusion

Diffusion of gaseous fission products at temperatures near or above their boiling point is the principal mode of escape from an unaltered fuel matrix. This process is enhanced by the effect of burnup. Bubbles or gas pockets become evident as the fuel swells when heated to just below the melting temperature. When the mechanical strength of the fuel is exceeded by internal gas pressure (dissolved gases or excess oxygen from UO_2), the bubbles break through the surface and sweep the collected gases, including halogens and other volatile elements, out of the fuel. At tracer level, the release is often delayed and limited by retention of fission products in lattice defects or voids. At high burnup, the release may begin somewhat below the fuel melting temperature as a result of failure of the cladding due to an increase in pressure of accumulated gases.

The initial phase of the diffusion process (see Sec. 3.3.1.1) invariably consists of a prompt-burst-type release which may account for more than half the total volatile release. The residual fraction is then released more slowly at a steady rate. The fraction released in a given time can be fitted to an equation of the type:

$$f = \frac{6}{a}\sqrt{\frac{Dt}{\pi}} - \frac{3Dt}{a^2},$$

where D = the diffusion coefficient of the particular species that exists in the temperature range of interest, t = time, and a = the radius of the spheres that are assumed to represent the material. This latter process is of relatively little importance in nuclear safety considerations because, at the high temperatures required for rapid diffusion, another mechanism, such as melting or oxidation, is more likely to be the controlling factor. However, grain growth, a process occurring above approximately 1700°C (3092°F) in UO_2, results in a large increase in diffusion rate (see Fig. 3-15, Sec. 3.3.1.1), and in the temperature range 1700 to 2800°C (3092 to 5072°F) diffusion would probably occur at a sufficiently high rate to contribute significantly to the accident hazard.

1.4.4 Migration of Solid-Phase Fission Products

In heterogeneous fuel systems, solubility of the fission products in the fuel matrix and cladding is of little significance as a mechanism prompting release, except perhaps at temperatures approaching the fuel melting point. Even partially melted fuel plates (see reference [19]) of aluminium alloy from the Oak Ridge Reactor (ORR) accident (see Table 5-1) showed no significant migration of fuel or cladding penetration in a part of the fuel plate very close to the melted region, presumably because such a process is strongly time-dependent.

Distinct similarities in physical properties of the fuel and fission-product phases favor dissolution. For example, elements that alloy readily, such as tellurium, ruthenium, tin, antimony, and molybdenum may dissolve in metallic fuel or cladding and thus increase the migration through the cladding. Solubility, in the sense of strong alloy formation, may then favor retention by the fuel until the metallic fuel or cladding containing the alloyed fission products is completely destroyed by oxidation. Oxides of the rare earths or alkaline earths will dissolve in UO_2 at very high temperatures, thereby gaining additional access to the UO_2 surface and to the fuel-void volume. The alkaline earths, however, having relatively unstable oxides at high temperatures, will volatilize rapidly if the system is extremely low in free oxygen as, for example, in the presence of melted Zircaloy cladding (see Sec. 3.3).

1.4.5 Compound Formation

Compound formation between fission products and fuel components is normally not significant; however, in theory it should occur to some extent between volatile elements, such as cesium and the halogens, when fuel rods are operated at temperatures high enough to permit distillation of these fission products and condensation in the cooler parts of the fuel jacket. Such compounds, however, have relatively low stability at fuel temperatures likely to be reached in a loss-of-coolant accident, and they would probably dissociate in the event of high-temperature cladding rupture. Experiments on fission-product release from uranium-aluminum alloy failed to show appreciable compound formation when cesium and iodine were released simultaneously by melting, even though the aerosol of iodine and cesium was allowed to age for an extended period in the same container. This was demonstrated by the diffusion-tube method (see Sec. 3.1.3).

Compound formation may be of somewhat more significance in the case of pyrocarbon-based fuels. The rapid diffusion of soluble but relatively unstable carbides of strontium and barium may account for the observed high rates of diffusion of these elements (see Sec. 4.2). The behavior of cesium, which also diffuses rapidly in graphite-matrix fuels, is unusual because of the formation of an interlamellar compound, such as cesium octacarbide (CsC_8). Other fission products (e.g., zirconium, niobium) are immobilized as carbides because of the high temperature stability of the compound. It seems likely, however, that compounds which form when two fission-product elements are deposited on the same surface may be of more importance than those formed in the fuel.

It has been reported [20a] that there is evidence of formation of a uranium iodide compound when irradiated uranium is melted in pure helium. Since fission-produced iodine atoms are surrounded by uranium atoms, it appears that favorable conditions for reaction exist. A later report [21] gives evidence for the formation of uranium iodide in fuels irradiated to about 200 Mwd/ton and heated in helium. At temperatures above 1700°C, (3092°F), partial dissociation to a lower iodide and elemental iodine was reported to occur. Since uranium iodides are easily oxidized and since it is highly unlikely that fuel materials will continue to be surrounded by pure, nonoxidizing gases in a reactor accident, it seems probable that uranium iodide formation would not significantly affect release of fission-product iodine from uranium or uranium alloys under accident conditions.

2 THEORETICAL CONSIDERATIONS

2.1 Free Energies and Vapor Pressures

2.1.1 Oxide Fuel Systems and the Effect of Oxygen on Fission-Product Release

Processes favoring release of fission products to the environment and those favoring retention in the fuel have been discussed by Parker et al. [24]. Data in this same report indicate a correlation between fuel melting temperature and release of rare gases or iodine. The results of studies performed since that report, especially studies of the release of fission products from high-burnup fuel materials, have altered some of the earlier beliefs on the subject, but there can be little question that the chemical form of the fission products has a profound effect on their release from fuel during reactor accidents and on their subsequent behavior. Although methods to determine directly the chemical form of released fission products remain to be devised, it is possible to draw inferences as to their probable chemical form from observations of the effect of various environments on the extent of fission-product release.

The two physical properties most relevant are the vapor pressure of the elements and compounds that can form under accident conditions and the free energies of formation of the compounds; the latter indicate the stability of the compounds at elevated temperatures. It should be recognized that thermodynamic data apply, strictly speaking, only to equilibrium conditions, which seldom, if ever, exist in reactor accidents. Consequently, it is necessary to be cautious in the use of such data to predict the behavior of fission products under accident conditions. It seems probable that, at elevated temperatures resulting from loss-of-coolant accidents in reactors fueled with high-melting materials, equilibrium will be at least approached, so that conclusions based on thermodynamic considerations are of some value. Kingery and Wygant [25] have discussed other limitations on thermodynamics with special reference to ceramic materials. The low concentrations of fission products in fuels irradiated to the burnup levels expected to be attained in most power reactors make the assumption of ideal behavior seem reasonable. The significance of fugacities with regard to fission-product behavior in accidents is not sufficiently clear to warrant further discussion. At present, there is little evidence that compound formation between fission products or between fission products and fuel components strongly affects the release of fission products. (See Sec. 1.4.5).

The free energy of oxide formation is of considerable importance in the prediction of the form of released fission products, because UO_2 and $(U, Th)O_2$ fuels are being used, or proposed for use, in a large number of nuclear power reactors, and also, oxygen is likely to be present in the environment of accident-ruptured fuel materials. Glassner [26] has assembled a useful compilation of thermodynamic data on oxides, such as given in Table 2-1, as has Coughlin [27]. Similar data for the elements have been tabulated by Stull and Sinke [28] and by Kelley and King [29].

Unfortunately, it is usually necessary to extrapolate available data considerably in order to obtain free-energy values at the melting point of UO_2 (2860°C or 5180°F) or higher temperatures. Extrapolation or estimates are usually sufficiently accurate to tell whether or not a fission-product oxide is stable at temperatures of interest. For example, the fact that the free energy of formation of cesium oxide (Cs_2O) is positive at temperatures above approximately 1300°C (2370°F) shows that cesium is likely to exist in the vapor phase at high temperatures even in air; thus, its release is likely to be relatively unaffected by the presence of oxygen. It will react readily, of course, with oxygen at lower temperatures, so that its behavior subsequent to its release may be affected to a greater extent by the environment than is its release.

Free-energy data also indicate that ruthenium oxide (RuO_4) is stable at temperatures up to 1430°C (2584°F). Hilliard and Reid [30] postulated the formation of this highly volatile oxide (B. P. 135°C or 275°F) to account for ruthenium-release values found when high-burnup uranium specimens were oxidized in air. These investigators say that RuO_4 is apparently not formed in the presence of uranium metal or lower uranium oxides. The behavior of molybdenum released from oxidizing uranium appears [20a, 30] to be similar to that of ruthenium, although free-energy values show that its oxides are more stable than the ruthenium oxides. The reported [30] decreasing release of tellurium with increasing temperature, ascribed to the decreasing

TABLE 2-1

Free Energy of Oxide Formation of Fissile, Fertile, and Fission-Product Elements

Compound	$-\Delta F_T^o$ (kcal/gram-atom of oxygen)				
	500°K	1000°K	1500°K	2000°K	2500°K
ThO_2	135	124	113	101	89
Ce_2O_3	134	123	112	100	90
La_2O_3	132	123	112	101	91
SrO	129	117	106	87	67
BaO	122	111	100	87	70
ZrO_2	119	108	97	86	75
UO_2	119	109	99	89	79
U_3O_8	96	86	76	dec	
PuO_2	128	102	92		
Cs_2O	54	34	7	dec	
MoO_2	55	46	36	26	dec
MoO_3	50	40	32	29	26
TeO_2	28	16	5	dec	
RuO_2	15	5	dec		
RuO_4	11	6	2	dec	

dec = decomposes

Note: 500°K = 227°C = 440°F; 1000°K = 727°C = 1340°F; 1500°K = 1227°C = 2240°F; 2000°K = 1727°C = 3140°F; 2500°K = 2227°C = 4040°F.

stability of tellurium oxide (TeO_2), was not confirmed by the results of similar studies [18]. Recently reported vapor pressure data [31] show that tellurium is more volatile than TeO_2; it should be emphasized that the experimental conditions under which vapor pressure data are obtained do not correspond to those of interest here (see Sec. 6.3).

A compilation of vapor-pressure data similar to that given earlier is shown in Figs. 2-1 [24] and 2-2 [24]. From the information shown in these figures, one could predict, for example, that the release of barium and strontium will be much greater in the absence of oxygen than in its presence, since the elements are more volatile than the oxides and since the free energy values [26] also show that the oxides are quite stable at high temperatures. There is abundant support for this prediction in the literature.

Rosenthal and Cantor [32] have calculated the contribution that fission-product cesium and rubidium would make to the pressure buildup in UO_2 fuel elements.

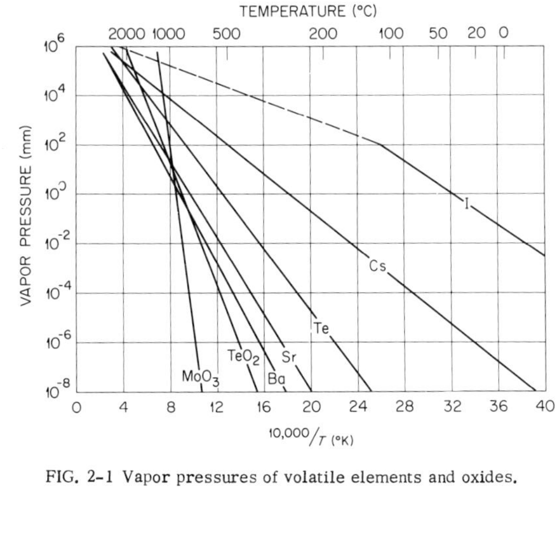

FIG. 2-1 Vapor pressures of volatile elements and oxides.

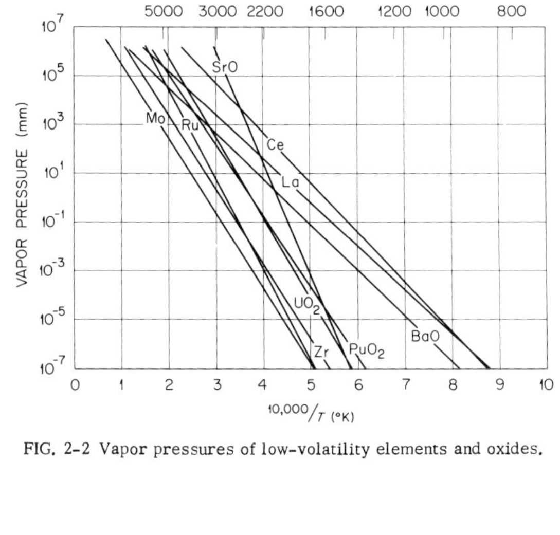

FIG. 2-2 Vapor pressures of low-volatility elements and oxides.

2.1.2 Graphite-Based Fuel Systems

Interest in the use of fuels containing uranium carbide (UC) in a graphite matrix in high-temperature gas-cooled reactors has prompted both theoretical and experimental studies of fission-product behavior in materials and coolant streams. Two theoretical studies have been reported by Brewer [33, 34] at General Atomic. Unfortunately, these brief reports, which were prepared primarily for internal use, do not specify the fuel composition or the burnup level considered, and they also give no information on the methods employed in calculating the reported results; however, the reports do contain some interesting conclusions based on thermodynamic and vapor pressure data. Thermodynamic data on carbides compiled by Brewer et al. [35] have been revised and extended by Krikorian [36, 37]. Very little data on the vapor pressure of carbides exist, but, since most of the fission-product carbides are unstable at high temperatures, the vapor pressures can be assumed to be those of the elements. Cantor [38] has estimated vapor pressures of barium carbide (BaC_2) to be $10^{-4.86}$ and $10^{-2.14}$ atm at 1000 and 1500°K, respectively, while the corresponding estimates for strontium carbide (SrC_2) are $10^{-3.91}$ and $10^{-1.4}$ atm. Cantor states that these values are probably high because they are about 0.1 of the vapor pressures of the metals at these temperatures, while available data for calcium and calcium carbide (CaC_2) show that the vapor-pressure ratio in that case is more nearly 0.01.

Brewer [33, 34] postulates that fission-product bromine and iodine will combine with fission-product cesium (or rubidium). Experimental evidence supporting this belief is lacking at the present time and, in fact, Castleman [21] and Parker et al. [39b] have reported data showing that cesium and iodine released from molten irradiated uranium or aluminum-uranium alloy deposited independently. Brewer [34] predicts that, in a "runaway" reactor containing fuel elements brazed with zirconium carbide (ZrC), the upper temperature limit of fuel-can integrity would probably be set by the melting temperature of the ZrC-C eutectic which has been reported [40] to be 2430°C (4406°F). He also indicates that internal gas pressure (36 atm owing to fission products alone, if no condensation is assumed) would probably cause can failure before it reached this temperature. He states that the vapor pressure inside the fuel can, after "extensive" burnup, will be between 40 and 60 atm, including the helium pressure, at the melting temperature (4500°C or 8130°F) of the carbides.

2.2 Behavior of Released Particles in Reactor Accidents

Any solid material, when heated to a temperature sufficient to induce surface oxidation, will disperse a small quantity of fine particles to the atmosphere. At temperatures approaching the melting, ignition, or boiling point of the material, the process is accelerated.

Reactor accidents can create conditions that could result in vaporization of some reactor core materials, including fuel components and fission products. It is, therefore, important in any safety analysis to postulate the fraction of the fission-product inventory that may be incorporated in solid particles and to estimate (on the basis of both theory and experiment) the size distribution and concentration of the particles forming the aerosol. The transport of this fraction of the radioactive release is then determined by the properties of these particles; their deposition or washout from the containment-vessel atmosphere becomes significant in an assessment of the hazard of a serious reactor accident. To achieve the maximum deposition when deposition is desirable or to minimize the deposition when deposition is detrimental may be within the reach of safeguards devices if the parameters that control deposition under accident conditions are understood.

Schwendiman et al. [41] have summarized the theories of particle deposition in aerosol transport systems and Chamberlain [42] has discussed the transport of particles in terms of deposition velocity to a variety of surfaces, including the environment external to containment systems. According to these studies, the deposition of particles on a surface is controlled by the interactions of the dynamic fluid properties of the gases in the system and the properties of particles moving in the fluid. Chamberlain recognizes the difficulty of the problem and points out that it is necessary to know the surface and flow conditions within 0.01 cm of the surface.

The following mechanisms control deposition: gravitational settling; Brownian diffusion; thermal deposition; electrical depositon; and turbulent deposition in straight and bent pipes. Each of these mechanisms are discussed in the following paragraphs.

2.2.1 Gravitational Settling

Schwendiman et al. [41] treat the problem of gravitational settling in terms of the Stokes relation for suspended particles and for gases flowing under laminar conditions in a horizontal tube. Equations attributed to Thomas [43] are used to derive a simple expression from which the fractional deposition due to gravity can be calculated. The results are experimentally confirmed. The length for 100% deposition is given by Schwendiman et al. as

$$L_{100} = \frac{8Q}{3u_t r\pi} = \frac{8}{3}\frac{rV}{u_t}, \qquad (2\text{-}1)$$

and the length for 50% deposition as

$$L_{50} = 0.354 L_{100}, \qquad (2\text{-}2)$$

where β is a dimensionless parameter equal to $(L/r)(u_t/V)$ and L = distance downstream from inlet, u_t = terminal settling velocity, r = tube radius, V = average velocity in tube.

Some values of lengths for 50% and 100% deposition for some tube diameters, particle sizes (d_p) and densities (ρ_p) are presented in Table 2-2. These are values predicted by the above equations.

2.2.2 Diffusion of Very Small Particles

For low volumetric flow rate, very small particles will deposit on surfaces because they are readily moved by the gas molecules. Browning [44] and others [45] have used this phenomenon to determine the size of very small particles, since the deposition under the proper conditions is an actual separation process. Comparison of the deposition of several radioactive species is possible. The original equations of Townsend [46] were modified by Gormley and Kennedy [47] and later by DeMarcus and Thomas [48]. In Table 2-3 Schwendiman illustrates the diffusional deposition as a function of gas flow rate and particle size for particles in the range 0.001 to 0.1μ. Chamberlain [42] summarizes the predictions of various theories on diffusive transport of very small particles in a diagram (Fig. 2-3) in which the turbulent deposition (see discussion below) is also given for the largest particles.

Thermal effects are also treated by Schwendiman et al. [41]. In Fig. 2-4, the effect of a 100°K/cm

TABLE 2-2

Length of a Horizontal Tube for 100% and 50% Deposition Due to Gravity

	Tube diameter (cm)	Q (cm^3/sec)	$\rho_p = 2$		$\rho_p = 5$		$\rho_p = 10$	
			L_{100} (cm)	L_{50} (cm)	L_{100} (cm)	L_{50} (cm)	L_{100} (cm)	L_{50} (cm)
$d_p = 2\mu$	1	39	2617	927	1048	371	526	186
	2	157	5235	1854	2096	742	1052	372
V = 50 cm/sec	4	628	10,470	3706	4192	1484	2104	744
$K_m = 1.075$	6	1413	15,705	5560	6288	2226	3156	1116
$d_p = 5\mu$	1	39	437	310	136	48	87	31
	2	157	875	619	272	96	175	62
V = 50 cm/sec	4	628	1750	1239	544	192	350	124
$K_m = 1.032$	6	1413	2625	1858	816	288	525	186
$d_p = 10\mu$	1	39	110	39	44	16	22	8
	2	157	221	78	88	31	44	16
V = 50 cm/sec	4	628	442	156	176	62	88	31
$K_m = 1.016$	6	1413	663	234	264	93	132	47

TABLE 2-3

Length of Tube within Which 20%, 50% and 75% of Particles Will Deposit

Q (cm^3/sec)	Length for 20% Deposited (cm)			Length for 50% Deposited (cm)			Length for 75% Deposited (cm)		
	0.001μ	0.01μ	0.1μ	0.001μ	0.01μ	0.1μ	0.001μ	0.01μ	0.1μ
0.25	0.043	3.4	268	0.23	18	1400	0.53	42	3300
0.50	0.085	6.8	536	0.45	36	2800	1.06	83	6600
1	0.17	13.6	1070	0.90	72	5630	2.12	166	13,200
2	0.34	27	2140	1.81	143	11,300	4.24	332	
5	0.85	68	5280	4.5	358	28,150	10.6	830	
10	1.7	136	10,560	9.0	715	56,300	21.2	1660	
20	3.4	272	21,120	18.1	1430		42.4	3320	
40	6.8	544		36.2	2860		85	6640	
100	17	1360		90	7150		212		
200	34	2720		181	14,300		424		
400	68	5440		362	28,600		848		

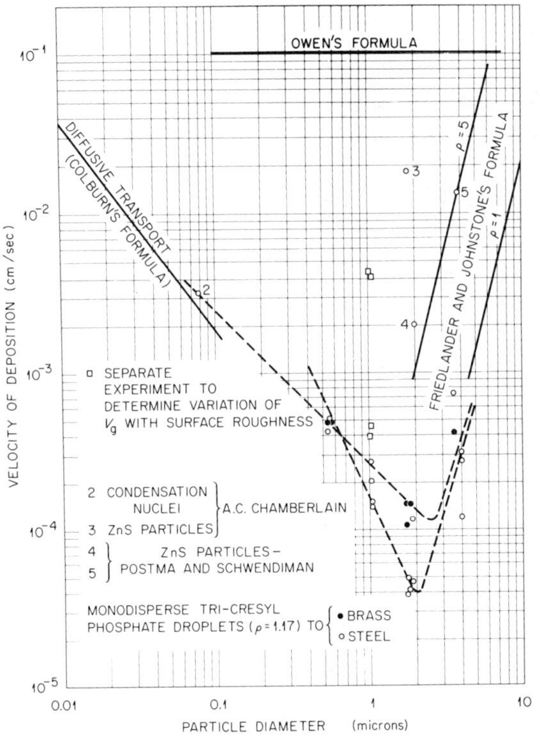

FIG. 2-3 Deposition of particles to vertical surfaces. Velocity of deposition versus particle size.

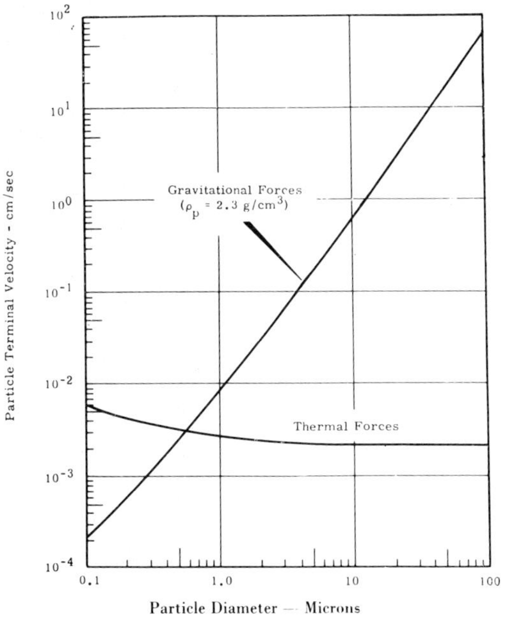

FIG. 2-4 Comparison of settling velocity and velocity due to a 100° K/cm thermal gradient.

thermal gradient on the particle terminal velocity is shown.

Electrical effects are not considered to be significant in accident situations, and they will not be treated here.

2.2.3 Turbulent Deposition in Straight and Bent Pipes

Friedlander and Johnstone [49] have analyzed the mechanism of transport of particles in a stream where turbulence may cause them to be deposited by impaction against the pipe or tube wall. Postma [50] has corrected the Friedlander mass-transfer equations to conform to experimental observations. Data calculated by use of the modified Friedlander Johnstone equation [49] are illustrated by the values in Table 2-4 in terms of particle densities (ρ_p) from 4 to 8 g/cm and for particle diameters (d_p) from 1 to 10μ.

2.2.4 Summary

It is apparent from the preceding brief discussion that particle transport behavior depends in a complex manner on all the properties of an aerosol. The controlling factors are mainly particle size, density, and gas velocity.

TABLE 2-4

Fraction of Entering Particles That Will Not Deposit Within Lengths of Vertical Tubes as Predicted from Equation 26[a] [41]

d_p μ	Tube diameter (cm)	Tube Reynolds number	Flow Rate (cm^3/sec)	$\rho_p = 4$			$\rho_p = 6$			$\rho_p = 8$		
				Tube length (cm)			Tube length (cm)			Tube length (cm)		
				200	500	2000	200	500	2000	200	500	2000
1	0.5	4000	241	0.998	0.994	0.977	0.995	0.987	0.948	0.991	0.977	0.910
	1.0	6000	722	0.999	0.998	0.991	0.998	0.995	0.979	0.996	0.990	0.963
	2.0	8000	1926	1.000	0.999	0.997	0.999	0.999	0.994	0.999	0.997	0.990
	4.0	10,000	4832	1.000	1.000	0.999	1.000	1.000	0.999	1.000	0.999	0.998
2	0.5	4000	241	0.963	0.910	0.686	0.918	0.810	0.431	0.862	0.690	0.226
	1.0	6000	722	0.985	0.963	0.858	0.966	0.918	0.711	0.942	0.860	0.547
	2.0	8000	1926	0.996	0.990	0.960	0.991	0.977	0.912	0.984	0.960	0.850
	4.0	10,000	4832	0.999	0.998	0.991	0.998	0.995	0.980	0.996	0.991	0.966
5.0	0.5	4000	241	0.072	0.001	0.000	0.004	0.000	0.000	(0.000)	(0.000)	(0.000)
	1.0	6000	722	0.344	0.069	0.000	0.111	0.004	0.000	(0.028)	(0.000)	(0.000)
	2.0	8000	1926	0.750	0.486	0.056	0.552	0.226	0.003	0.379	0.089	0.008
	4.0	10,000	4832	0.940	0.856	0.537	0.880	0.726	0.277	0.811	0.592	0.123
10	0.5	4000	241	0.000	0.000	0.000	(0.000)	(0.000)	(0.000)	(0.000)	(0.000)	(0.000)
	1.0	6000	722	0.002	0.000	0.000	(0.000)	(0.000)	(0.000)	(0.000)	(0.000)	(0.000)
	2.0	8000	1926	0.193	0.016	0.000	0.050	0.001	0.000	(0.012)	(0.000)	(0.000)
	4.0	10,000	4832	0.701	0.411	0.029	0.523	0.198	0.002	0.386	0.093	0.000

[a]K/V used in calculation is outside the range of the straight line portion of deposition correlation relation.

In the size range of condensation nuclei in which agglomeration occurs at a significant rate (0.001 to 0.1μ), diffusion to surfaces is most important for laminar flow conditions. At high velocities, deposition of particles is very slight except for the smallest (0.001μ) particles.

In the larger size ranges (2 to 10μ) either gravitational settling or turbulent impaction may be the factor controlling deposition. Either can be relatively efficient or inefficient depending on the particular circumstances.

The intermediate size particles, 0.1 to 2μ, are relatively unaffected by all deposition processes, and they thus become the ones least likely to disappear rapidly from the gas phase.

In order to predict the behavior of a mixture of particles covering the size range 0.001 to 10μ, relatively detailed analysis of the effect of the various processes is necessary. The expected flow rates, pipe sizes and distribution of particles are all needed for a reliable prediction.

2.3 Reactor Accident Models as Guides to Fission-Product Release

Models of reactor accidents can be used as guides to estimating fission-product release. The study of a reactor accident by the use of sequential analytical models has been proposed as part of the American Nuclear Society's Standards Program for some time [51]. The development of models was a vital part of the Loss-of-Fluid Test Program (LOFT) [52], conducted by Phillips Petroleum Company as part of the U. S. Atomic Energy Commission safety program.

Since any credible accident will result in only partial release of the fission products from the core, even the roughest appraisal of fission-product hazard resulting from the release is dependent upon knowledge of the interaction of the many variables closely related to the sequence of failures. The resulting temperatures and fuel environments, the heat capacity of components, the transport velocity of ambient gases etc., are all important.

2.3.1 The Minor Transient

This category is defined to include all accidents in which the fission-product release is less than about 1% of the release expected in a design-basis accident. Specific examples of this category are the SPERT destruction tests, which were relatively innocuous in terms of fission-product release, and the SL-1 accident, in which the kinetic energy produced by the water hammer caused severe damage to both the fuel and the reactor vessel, but fission-product release was limited, at most, to a few percent of the radio-iodine inventory.

Generally, the reason that a "minor transient" does not cause a significant release of fission products is that the period in which the fuel is fluid is too short to permit either diffusion or bubble formation, i. e., too short to permit the action of those mechanisms by which volatile materials are ordinarily separated from overheated fuel. In a minor transient, the fission-product release may also be minimized by the presence of large amounts of water, which dissolve the soluble fission products and speed solidification of the fuel.

In accidents belonging in this category, contaminated water and dispersed fuel residues are likely to create the major problem; these may be entirely limited to the reactor building or, as in the case of the SL-1 accident, they may produce a minor release of iodine to the atmosphere because of open building construction.

2.3.2 The Loss-of-Fluid Meltdown (LOFM)

We define this as the maximum credible accident for UO_2 fuels because the fuel is at a high temperature for much longer times than in transient accidents. The description of a typical LOFM varies with the size and power level of the reactor. Even the fuel cladding in small reactors may not be ruptured by the limited available heat; while in the largest reactors now under consideration, 3000 to 8000 Mw(t), all the oxide fuel, as well as part of the core vessel, could possilby melt if no safety device or heat sink were provided, or if that which had been provided failed to operate properly.

This class of accident in currently operating power reactors is generally thought of as resulting in the sequence of core destruction steps described in the model analysis. Essentially, the entire core will be destroyed, most of the UO_2 will eventually melt, and some may even oxidize after cooling below 1500°C(2730°F). The release of fission products will extend over a long period, and its significance as an external hazard will depend mainly upon the integrity of the containment or confinement system. Under the worst conditions, that is, when atmosphere exchange to the environment is rapid (one change per hour), a large break in the containment shell would be expected to release to the environment, as the result of fuel meltdown, the following approximate portions of the fission-product inventory, based on the authors' evaluation of the data in Sec. 3: 100% of the rare gases; 50% of the iodine, cesium, tellurium; 25% of the ruthenium; 0.5% of the strontium and barium; and 0.1% of the rare earths and other refractory fission-product elements, e. g., zirconium, niobium.

Information on the transport of fission products shows that deposition would tend to diminish fission-product release from the confinement system to a fraction of the above estimates unless a major breach occurs. Leakage from the containment would normally be somewhat selective, provided the openings are very small [10 μ (0.4 mil) or less]. The magnitude of the radiation hazard posed by the above-listed radionuclides is estimated later in this section.

2.3.3 The Major Transient

A major nuclear excursion could conceivably be a more severe accident than the loss-of-coolant meltdown. However, there are inherent limiting phenomena that tend to make the two approximately equal or, at least, tend to limit the major transient in the worst case to about the same net fission-product release as the meltdown.

In TID-14844 [53], the fission-product release is estimated without reference to the accident regime, but it is indicated that it applies to the maximum credible accident. The following broad limits of release from the fuel (not adjusted for retention in the primary vessel) were suggested in this report for three groups of fission-product elements: 100% for the rare gases; 50% for iodine (should include tellurium and cesium); and 1% for solids (rare earths and refractory elements, zirconium, niobium, ruthenium, etc.).

The SL-1 accident vaporized about 5% of the inner 16 elements of the core (see vol. 1, pp. 676 and 704). In the case of UO_2 systems, a large fraction of the reactor core could conceivably be disrupted initially in an accident of this type. However, fuel vaporization has been estimated by Ergen in an unpublished calculation as restricted to not more than 10% of the fuel before dispersal of the components reduces the possibility of further vaporization. Elbaum and Thompson [54] have made quantitative estimates of the magnitude of transients shut down by fuel vaporization for UO_2 rods or uranium-aluminum-alloy plate fuels. Transient melting has been shown by Lorenz et al. [55] to result in a much smaller fission-product release than would occur in a loss-of-coolant meltdown, and on this basis the severity of a major transient is considered to be limited to the sum of that from the initial vaporization step plus that from the slow meltdown of a fraction of the fuel core following the transient. The hazard to the containment system is, of course, not quite the same in these two cases, and, therefore, environmental release in a major transient may exceed that of a contained release from a loss-of-coolant accident.

Since the possibility of the additional hazard from the chemical energy of metal-water reactions is not readily dismissed, it is perhaps necessary to concede that, in a major transient in a metal-fueled reactor system, this effect may possibly contribute sufficiently to the total accident-released energy to exceed the hazard of a loss-of-coolant meltdown (see Chapter 17 also). Again, it would appear that this would probably not imply a higher fission-product release to the containment system but, instead, a greater likelihood of environmental involvement by containment breaks.

2.3.4 Inventory Considerations

From inventory considerations (see Figs. 1-7 to 1-11, inclusive) the amounts of the significant long-lived fission products in a typical reactor can be calculated. Table 2-5 shows the percentage of the total inventory of gamma activity for various elements at shutdown and at 1 day and 1 month after shutdown according to ratios shown in Fig. 1-11. Table 2-6 shows the same values adjusted for the fraction released. Similar tabulations can be developed for all other inventory periods and decay times, including instantaneous release from a clean core. The classification of gamma activities in Tables 1-9 through 1-11 is that adopted by Blomeke and Todd [4].

Decay of the total inventory is given in Fig. 1-7 as a function of power level, so that the percentage values in Tables 2-5 and 2-6 can be converted to curies of gamma activity.

The individual fission products have different relative importances that need to be considered here in order to determine which species merit special attention. The biological significance of the fission products as an ingestion or inhalation hazard have been rated by Beattie [56] and others in the following order: iodine; tellurium (1/10 as significant

TABLE 2-5

Gamma Activities Expressed as Percent of Total Gamma Inventory

Element	At shutdown	1 day after shutdown	1 month after shutdown
Xe-Kr	17	8	3
I	22	9	3.5
Te	7	4	2
Cs	2	0.1	0.2
All other solids	52	79	91.3

TABLE 2-6

Gamma Activities Adjusted for Percent Released

Element	At shutdown	1 day after shutdown	1 month after shutdown
100% Xe-Kr	17	8	3
50% I	11	4.5	1.8
50% Te	3.5	2	1
50% Cs	1	0.05	0.1
1% All Solids	0.5	0.8	0.9
Maximum percent of total gamma activity	33.0	15.3	6.8

as iodine); strontium and cesium (1/10 as significant as tellurium). These values are based on the radiotoxicity of the element and the likelihood of release and ingestion. With respect to the conclusions about relative importances, some reservations may be in order, since the assumed releases were probably based on Windscale experience with a metallic uranium fire.

In addition, when a high retention fraction is credited for iodine, then the limiting radiation hazard may be from the rare gases in spite of their low biological potency. Beattie [56] and others have recognized this limitation and have suggested safeguards such as pumping of the gas into holding tanks.

2.3.5 Particle Sizes of Released Fission Products

Experimental observation of released particles carrying fission products in aged aerosols has indicated the presence of only relatively small particles generated from the melting of UO_2 and other fuel components. The summary in Table 2-7 and the data in Fig. 2-5 are typical of many observations made in nuclear safety experimental facilities. From these data, it is apparent that the average accident-produced particle can be no larger than 0.05μ (0.002 mil); this evidence suggests rather high mobility for a cloud of such released particles.

It can be speculated that these observations possibly do not allow for a more realistic and longer residence time in the high-temperature zone where agglomeration to 1μ or larger could be anticipated. Unfortunately, the agglomeration process tends to lead to a highly flocculent particle of low density, which does not increase greatly in aerodynamic size. On the other hand, particles of diameter 0.1μ (0.004 mil) and smaller are too small to be significantly affected by gravity.

In an assessment of the importance of particles to the dispersion process, it is likely that the condensible elements, excluding halogens, will always be associated with filterable solids. Experience within atomic energy establishments indicates that decontamination factors of about 10^4 or better for contaminated air streams are attainable when the equivalent of an AEC High-Efficiency Particulate Air Filter is used. This subject is discussed in some detail in Chapter 19, Sec. 4.3.2. It is possible to classify the fission products as (1) filterable solids and (2) gaseous molecules. Included in the latter group are the halogens and the rare gases xenon and krypton and, in the former, all other elements. A small overlap occurs, mainly because iodine may adsorb at low temperature on some airborne solids, and, under nonoxidizing conditions, it usually forms iodides with low volatility. The efficiency of the adsorption process appears to be low, however, after a few hours in a containment system. In one set of experiments (Table 2-8) the fraction of the iodine remaining on small particles after a few hours was less than 1%. Admittedly, however, different conditions of time, space, and atmosphere may give a higher adsorption of iodine on fuel particles.

The nature of the particles from reactor fuels has been well documented. Well-formed particles of uranium oxide (U_3O_8) are generally found (see Fig. 2-6) when UO_2 fuels melt in air/ however, the particles are spherical if melting occurs in an inert or reducing environment (see Fig. 2-7). Other spherical particles of cladding metal or metal oxide are also observed. Under specific laboratory conditions, penetration of filters is reported only if the transient time between the high-temperature zone and the filter is short (seconds). This observation can be explained on the basis that some gaseous volatile species, such as ruthenium oxides, do not have an opportunity to condense or agglomerate before reaching the filter.

2.3.6 Chemical States of Fission-Product Elements

2.3.6.1 Under Oxidizing Conditions. The ability to evaluate the hazards of fission products is dependent upon knowledge of their chemical and physical forms in the released state. If the release process is thought of as essentially a smoke- and fume-producing phenomenon, with hot gases mixing with air and steam, it is less difficult to label the forms than if the atmosphere or the process are not described. For most situations, it is likely that an oxidizing condition will prevail either within the reactor vessel or immediately outside it. With this assumption, the expected chemical states of released fission products are as given in Table 2-9 [57].

It may be assumed that the forms of most fission products under these conditions are determined by their high degree of dispersal and small dimensions and by their affinity for oxygen. The form generally most volatile for fission products is the free element—except for ruthenium and possibly tellurium and molybdenum, for which the oxidized forms are predominant. Oxygen also enhances UO_2 vaporization by some mechanism that leads to formation of UO_3 or U_3O_8 as the final form.

TABLE 2-7

Comparison of Particles Produced by Various Methods of Melting UO_2 Fuel Specimens

Method	Time molten	Sample size[a] (g)	Atmosphere	Median particle size[a] (μ)	Limitations
Arc-image	1 - 6 min	0.3	He, air, CO_2	0.02-0.1	Small samples, short time molten
Induction: tungsten crucible	1 > 10 min	30-50	He	0.03-1	Nonoxidizing atmosphere
Induction: direct coupling	~ 1 min	20-80	Air, steam + air	0.07-0.3	Clad samples, short time molten
In-pile	5 - 10 min	6	He, air + steam	0.01-5	Small specimens, expensive
TREAT	1 - 10 sec	30	Ar, air + steam	0.02-0.16	Transients only, expensive
Center resistor	~ 1 min	40	He	0.08-0.1	Nonoxidizing atmosphere, short time molten
Plasma torch	10 - 30 sec	150-300	He, air, steam	~0.02	High gas temperature, short time molten
Arc-discharge	1 sec - > 10 min	5-10	Air, steam	0.03	Uncontrolled temperature, sputtering

[a]1g = 0.0022 lb = 0.0353 oz; 1 μ = 0.001 mm = 0.0394 mil.

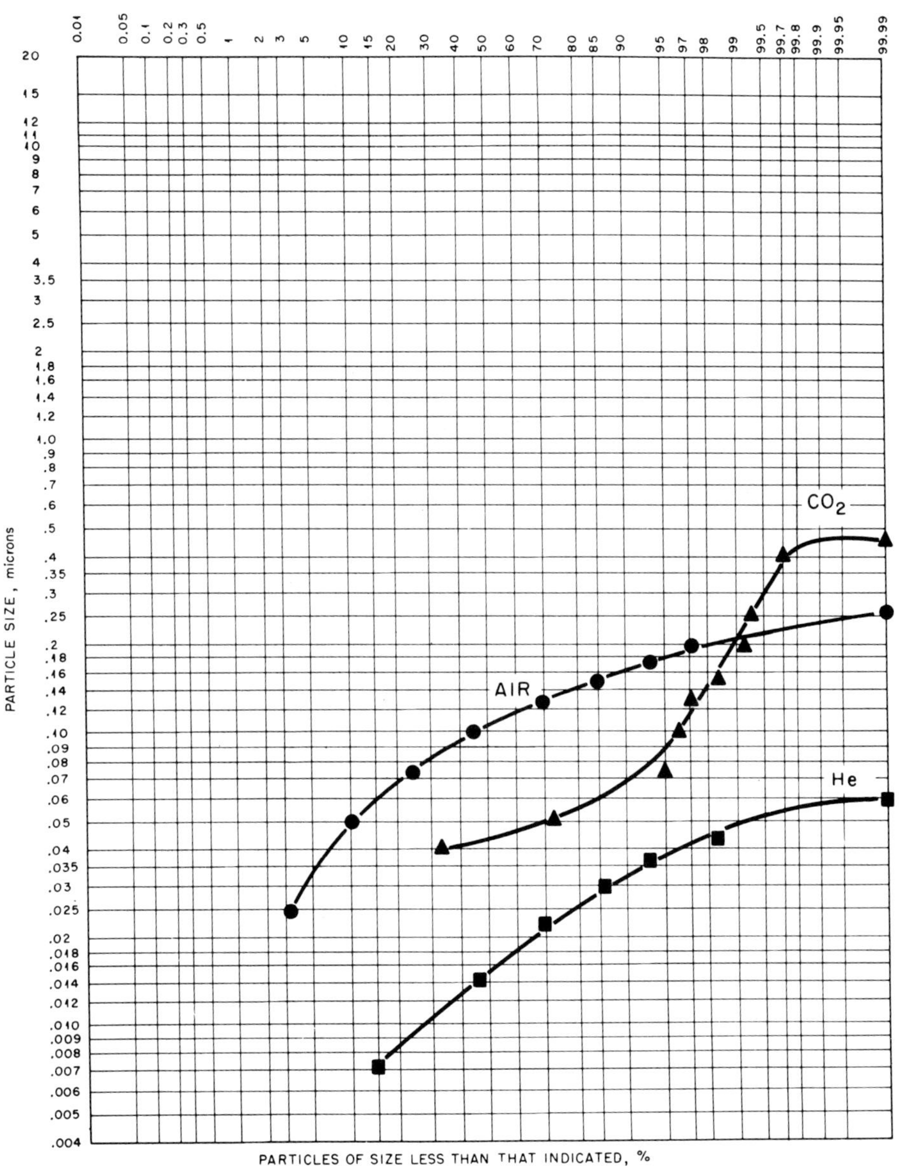

FIG. 2-5 Particle-size evaluation of oxides vaporized from melted UO_2 (arc-image furnace).

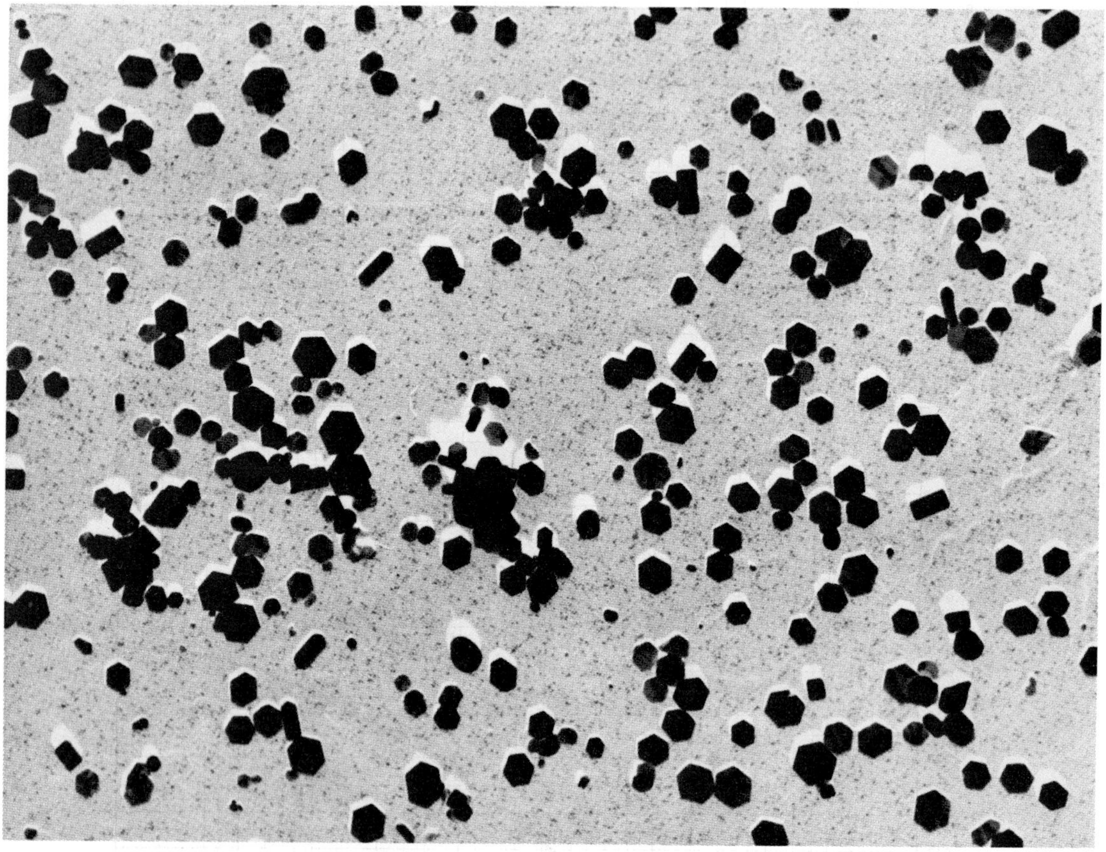

FIG. 2-6 Replica of crystalline U_3O_8 particles vaporized from UO_2 melted in air. 30,000× (particle diameters 0.02 to 0.2μ).

TABLE 2-8

Iodine Deposition and Desorption in the ORNL Containment Mockup Facility

Atmosphere	Air	Steam-air		Steam-air and organics	
Total Time in Containment	5 hr	5 hr	18 hr	5 hr	5 hr
Pressure (psig)[a]	15 air	10 steam 15 air	12 steam 15 air	10 steam 15 air	15 steam 15 air
	Total I in each fraction (%)				
I held in containment tank:					
Retained on tank walls	79.6	60.8	38.2	19.3	20.9
Collected in steam condensate	-	34.9	54.0	52.9	47.3
Total retention	79.6	95.7	92.2	72.2	68.2
I removed from tank:					
By pressure release	3.0	1.1	2.4	12.9	8.5
By Ar displacement	7.4	1.2	2.4	13.3	15.9
By air sweep	2.9	0.3	1.7	0.5	2.6
Total removed, airborne	13.3	2.6	6.5	26.7	27.0
I removed on test samples	6.1	1.7	1.4	0.8	2.2
Distribution of airborne I from tank:					
Retained on filters (particulate I)	4.1[b]	0.5	1.5	0.1	0.3
Retained on Ag/Cu screens	5.3	0.8	2.3	6.4	8.5
Retained on charcoal papers	3.0	0.4	0.4	2.0	15.8
Retained in charcoal cartridges (4-in. bed)	0.3	0.6	2.0	16.4	2.3
Penetration through 1.5 in. of charcoal	0.003	0.0002	0.05	0.3	0.03
Amount of I in "penetrating" form	~0.3	~0.5	1.2	1.6	1.4

[a]Note: 15 psig ≅ 2 atm absolute = 2.066 kg/cm^2.
[b]High value due to use of organic membrane filter.

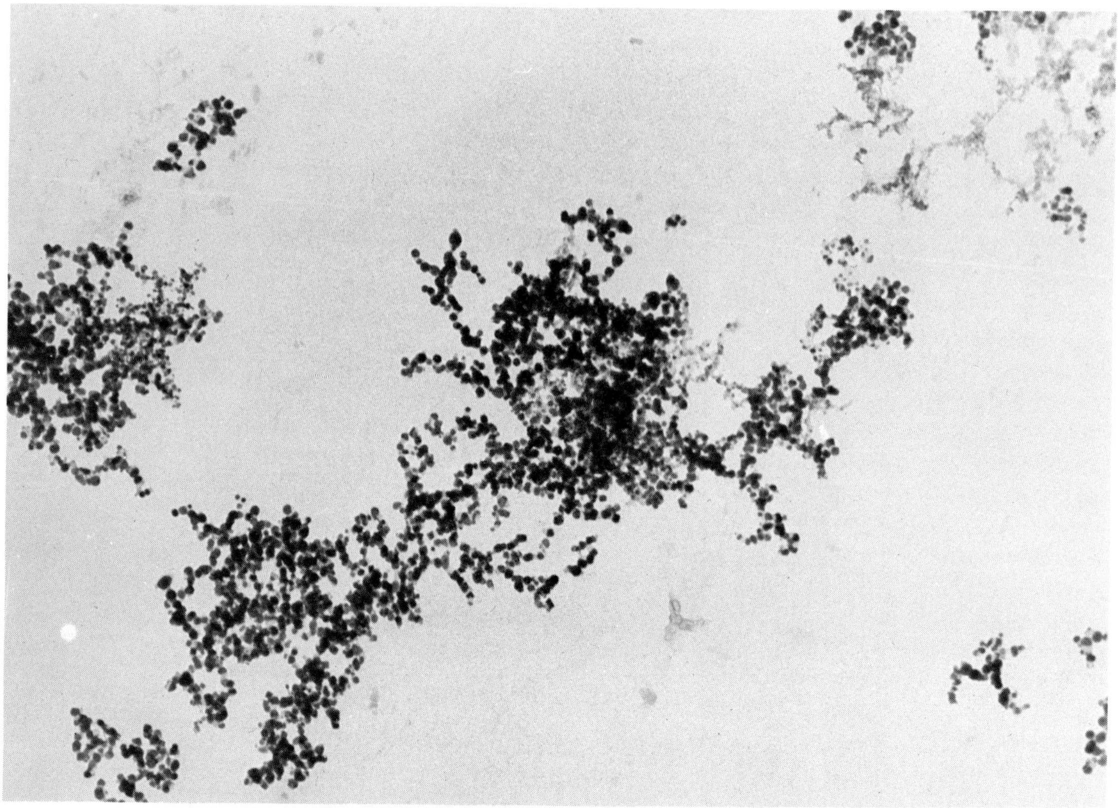

FIG. 2-7 Replica of spherical UO_2 particles vaporized from UO_2 melted in helium. 46,000× (particle diameters 0.01 to 0.05μ).

2.3.6.2 Under Reducing Conditions. The alkaline earths, strontium and barium, are most volatile in strongly reducing or oxygen-free systems; therefore, the elemental form is the volatile species. Vapor-pressure data tend to confirm this conclusion. In the absence of oxygen, however, the release of ruthenium and molybdenum is strongly suppressed.

A further variation in iodine behavior is sometimes noted with steam in the absence of air. This situation may produce a relatively high percentage of iodine in the form of hydrogen iodide (HI). To some extent, this form may be more reactive to metals than the elemental form and may influence transport phenomena.

It is also possible that within the reactor vessel, particularly in the reducing atmosphere, the presence of carbides of the various metals may contribute to the production of iodine in the organic form. The partial conversion of iodine to the form of an organic compound [illustrated in Table 2-9 as methyl iodide (CH_3I)] is a somewhat poorly understood process, which probably involves initial iodine deposition followed by chemical exchange.

For the less volatile fission products, Miller and Browning have proposed a model that may contribute to the study of vaporized species [39a, 58b].

3 EXPERIMENTS ON RELEASE OF FISSION PRODUCTS FROM VARIOUS TYPES OF FUEL

Data on release of fission products form various types of fuel are collected in this section and arranged according to the fuel classification given in Table 1-6. Some information on oxidation of fuels is included for materials that oxidize readily under possible accident conditions. An understanding of

TABLE 2-9

Probable Chemical States of Released Fission Products[a]

Element	Vaporized state	Deposited state
Rare gases	Kr, Xe	None
I	I_2	$I^- \rightleftharpoons I_2$; CH_3I
Te	$TeO_2 + Te$	TeO_2
Cs	Cs	Cs_2O
Ru	RuO_4	$Ru_2O_3 + Ru$
Sr	Sr	SrO

[a]This simplified summary is based partly on observed chemical behavior, including changes in volatility with atmosphere, and partly on physical properties. The equilibration of mixed states assumes an atmosphere containing oxygen and some reducing agent on plate-out surfaces.

factors affecting oxidation rates is essential for an adequate assessment of the hazard of the fission-product release that may result from destruction of fuel elements by oxidation. Where experimental information on fission-product release from a fuel is lacking, another, similar fuel system for which there are data is indicated as the best available source of information.

Data on distribution of released fission products often found in reports on fission-product-release experiments are deliberately omitted from this section because such information is discussed in the chapter on fission-product transport.

3.1 Class I, Metals and Alloys

3.1.1 Metallic Uranium

The importance of experimental studies of the extent of fission-product release accompanying the oxidation or melting of irradiated uranium in different atmospheres was demonstrated by the Windscale incident [59, 60] (see also Vol. 1, p. 633). Reactors fueled with metallic uranium have a minor role in nuclear power production in the United States, but they continue to be of importance in plutonium-producing reactors in this country and abroad, as well as in certain gas-cooled power reactors, principally those in Great Britain. When coolant is lost in reactors of this type, hot metallic uranium may come into contact with air, steam, or CO_2, creating a potential hazard by releasing fission products. The amount of information in this section is indicative of the large quantity of data in the literature rather than of the importance of this fuel.

The principal studies contributing to the evaluation of this hazard have been performed at Harwell, Hanford, Brookhaven, and Oak Ridge. Since it seems reasonable to assume that rates of fission-product release will be proportional to rates of oxidation, and experimental studies have established the validity of this assumption, a discussion of the effect of various parameters on oxidation rates is important for an understanding of this potential reactor hazard. A brief review of this subject* is found in Chapter 17, but further consideration seems justified at this point.

3.1.1.1 Oxidation of Uranium in Air. A number of studies of oxidation rates of uranium in air have been published. Early investigations [62-66] were performed at comparatively low temperatures; the results of these studies are excluded from the present discussion because Windscale experience indicated that temperatures above the melting point of the metal may be attained in reactor accidents involving the oxidation of uranium. Later studies of the oxidation reaction [18, 67-71] were made at temperatures as high as 1440°C (2624°F). Only the more important factors affecting oxidation rates will be covered here.

*Information on the pyrophoricity of uranium in reactor environments has also been reviewed by Zima [61].

Effect of Ratio of Surface Area to Weight. Hilliard [67] has found that overall oxidation rates, defined as the average rate computed from the time required to completely oxidize his small specimens, can be fitted to equations of the form:

Overall rate = $W_0/A_0t_c = Ce^{K}A_0/W_0$, where W_0 is the original weight of the specimen (mg), A_0 the original surface area (cm^2), and t_c the time required for complete oxidation (min). C and K are constants for a particular temperature and, presumably, for a particular air-flow rate (500 cm/min or 16.4 ft/min in this case). Hilliard has also found a linear relation between the weight-to-area ratio and the time required for complete oxidation at 805, 995, 1200, and 1400°C (or 1483, 1825, 2192, and 2624°F). British data [68] obtained with 1-kg samples having a surface-to-volume ratio similar to that of Calder reactor fuel elements show that oxidation was completed in about 90 min at 800°C (1472°F) under their air-flow conditions (higher flow rates than Hilliard's). Comparison of these data with calculations based on Hilliard's data indicates that extrapolation of oxidation data obtained with small samples and low air-flow rates may not give a reliable indication of accident behavior of full-sized fuel elements. Other studies [18], also made with small specimens, show that even a small variation in surface-to-weight ratio produces a significant change in oxidation rate. This investigation covered the range 0.41 to 0.77 cm^2/g* at 1000°C (1832°F) and 0.38 to 0.53 cm^2/g at 1200°C (2192°F) at a single air velocity (120 cm/min or 3.94 ft/min).

Effect of Air Velocity. Hilliard [67] reports that the oxidation rate (for 10-min oxidation periods) increased in proportion to the logarithm of air velocity for the conditions employed in his tests. His data show a 1.6-fold increase in oxidation rate with a 100-fold increase in air velocity and indicate that air velocity is not a critical parameter.

Effect of Furnace Temperature. Data on the variation of oxidation rate with furnace temperature obtained by Hilliard with three specimens of different sizes are shown in Fig. 3-1 [67]. At higher temperatures, it appears from these data that the logarithm of the rate is linearly dependent upon 1/T. More detailed data covering approximately the same temperature range are shown in Fig. 3-2 [18]. These data, obtained by use of a continuously weighing balance with an air velocity of 220 cm/min (7.22 ft/min), show that the oxidation process is rather complicated, especially in the 600 to 900°C (1110 to 1650°F) range, and that the rate of oxidation at each temperature varies over a wide range. Possible explanations of the observed oxidation behavior are considered below in the discussion on the oxidation of uranium in CO_2.

Self-Heating Effect. In most of the uranium oxidation studies, the furnace or furnace gas temperature is measured. Because the heat of oxidation of uranium is quite large (259 and 285 kcal/g-atom

*1 cm^2/g = 0.488 ft^2/lb = 70.3 $in.^2/lb$.

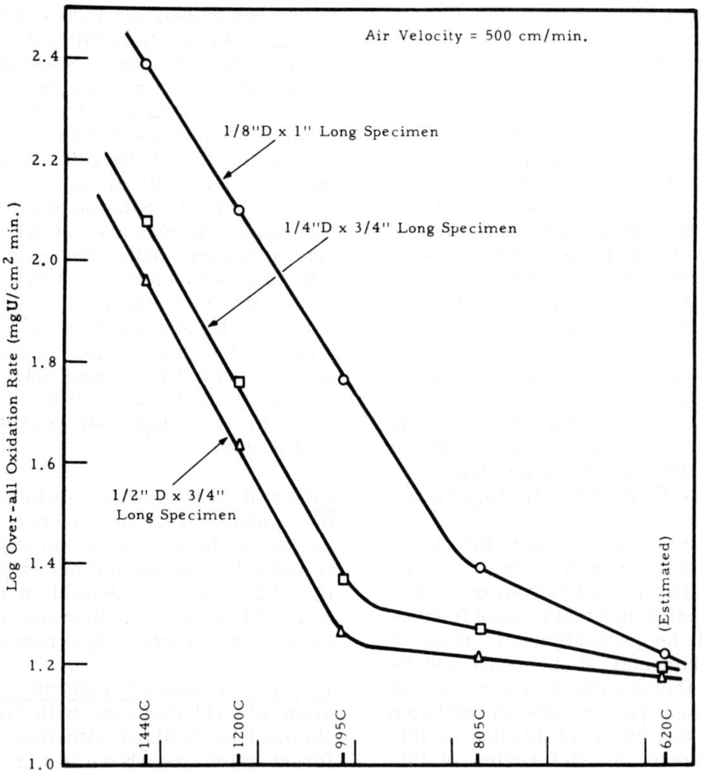

FIG. 3-1 Overall oxidation rate of uranium versus the reciprocal of the absolute temperature for various sizes of cylindrical specimens.

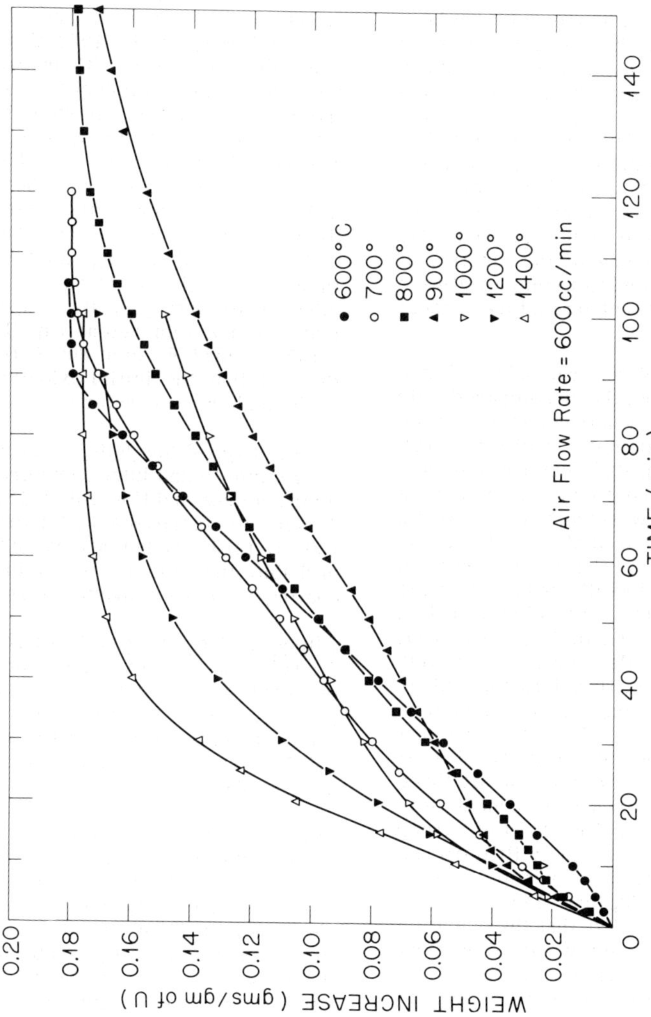

FIG. 3-2 Oxidation of uranium in air.

of uranium* for UO_2 and U_3O_8 formation, respectively) and the oxide layer formed serves as an effective heat barrier, the temperature of the unoxidized metal rises considerably above that of the oxidizing atmosphere. Hilliard [67] reports observation of temperature cycling in the temperature range 400 to 650°C (750 to 1200°F). It should be recognized that uranium temperatures in much of the published work on oxidation of uranium were higher than furnace temperatures at times because of the self-heating effect.

Ignition Studies. The ignition temperature has been defined by Wanklyn [72] as the gas temperature at which the reaction becomes controlled by the supply of gas. Wanklyn has developed a theory for the ignition of uranium that is said [70] to be consistent with most of the existing data. This subject has been investigated rather extensively by Baker and his coworkers at Argonne National Laboratory [73] with various reactor-core materials, and it is covered in Chapter 17. These studies have helped to provide a better understanding of the kinetics of uranium oxidation, but the conditions employed were rather far removed from expected reactor-accident conditions.

Effect of Burnup. The oxidation rate of irradiated uranium (~0.1% burnup) has been compared with that of similar specimens of unirradiated uranium under the same experimental conditions [18]. Uranium cylinders, 0.25 in. (0.635 cm) in diameter and 0.7 in. (1.78 cm) long were contained in porous alumina thimbles. Gas-flow velocity (220 cm/min or 7.22 ft/ min) was measured at room temperature without allowance for the cross-sectional area of the alumina cup. Results obtained at gas temperatures of 800, 1000, and 1200°C (1472, 1832, and 2192°F) are shown in Figs. 3-3, 3-4, and 3-5 [18]; the figures also include a similar comparison of the effects of irradiation on oxidation rates in CO_2 and steam, which will be discussed in the next subsection. The data show that irradiated uranium oxidized more rapidly, at least initially, than unirradiated uranium at all three temperatures, but the difference was most pronounced at 1000°C (1832°F). The effect of burnup on oxidation rate was also studied by Hilliard and Reid [30] over a broad range of burnup values at temperatures of 1000, 1200, and 1440°C (1832, 2192, and 2624°F). They stated that the oxidation rate was independent of burnup to a level of about 10^{18} nvt (5×10^{-4} at. % burnup), but that above this level the specimens were oxidized more extensively. The scatter of their data is too large to provide more than a qualitative indication of the upward trend in oxidation rate at higher burnup (maximum 4×10^{20} nvt = 0.17 at. % burnup). More data on this effect at higher burnup levels would be desirable, particularly with larger specimens that would permit extrapolation to the surface-to-volume ratios of full-size fuel rods.

3.1.1.2 Oxidation of Uranium in CO_2. Interest in oxidation rates of uranium in CO_2 results from the use of this gas as the coolant in reactors of the Calder Hall type. Consequently, most of the publications [18, 68, 70, 74-79] on this subject are of British origin. Only two of the published reports [18, 78] contain data on the oxidation of irradiated uranium in CO_2.

Effect of Temperature. Antill et al. [77] state that the rate of reaction of pure CO_2 with uranium over the temperature range 500 to 780°C (932 to 1436°F) increases with rising temperature, with a marked increase when the metal is near or at the β-γ transition temperature. They state that at higher temperatures, up to 1000°C (1832°F), the rate gradually decreases with increasing temperature because the oxide product forms a barrier between the gas and metal. All studies of this type have given results confirming the importance of the β-γ transition temperature. The interesting oxidation behavior in CO_2 at 800°C (1472°F) shown in Fig. 3-3, which was also observed by Megaw and Bridges [79], appears to be due to formation of a metastable oxide during rapid oxidation. The maximum O:U ratio observed [18] was approximately 2.32. Other data [18] also show that complete oxidation is not achieved as rapidly at 1200 or 1400°C (2192 or 2552°F) as at 800°C (1472°F), in spite of the fact that the initial oxidation rates are higher at the higher temperatures.

Effect of Gas-Flow Rate. Data obtained by exposure of uranium to undiluted commercial CO_2 at different temperatures and flow rates [18] show that the oxidation rate increases with increasing flow rate at 1200°C (2192°F) but not at 1400°C (2552°F). It is probable that the nature of the protective oxide coating is more important than the gas-flow rate.

Effect of Water Vapor. Antill et al. [77] report that 3.1 vol. % water vapor in CO_2 produces an increase in reaction rate over that of pure CO_2 at 500°C (932°F). However, little effect of water vapor was noted at temperatures ranging from 850 to 1000°C (1562 to 1832°F) or with low concentrations (500 ppm) at 500°C.

Effect of Burnup. The effect of irradiation of uranium on its oxidation rate in CO_2 was studied first by Parker et al. [18]. Their data, shown in Figs. 3-3, 3-4, and 3-5, demonstrate that the burnup effect is more pronounced in CO_2 than in air. This effect was also investigated by Diffey and King [78] at irradiation levels of 1250 to 2350 Mwd/ton. They could find no effect within the limits of their experiments, at a furnace temperature of 600°C (1112°F), but at 800°C (1472°F) the irradiated material swelled because of pressure of fission gases and oxidized very rapidly even after furnace heat was shut off. The particle size of the oxide formed was reported to be independent of burnup.

Several investigators have discussed possible reasons for the increased oxidation rate of irradiated uranium and the increased release of fission products at high burnups discussed elsewhere in this chapter. Hilliard and Reid [30] ascribe the increased oxidation rate at temperatures above the melting point of uranium (1133°C or 2071°F) under their experimental conditions to formation of

---

*1 kcal = 3.968 Btu; 1 g-atom uranium = 0.525 lb; 1 kcal/g-atom uranium = 7.56 Btu/lb U.

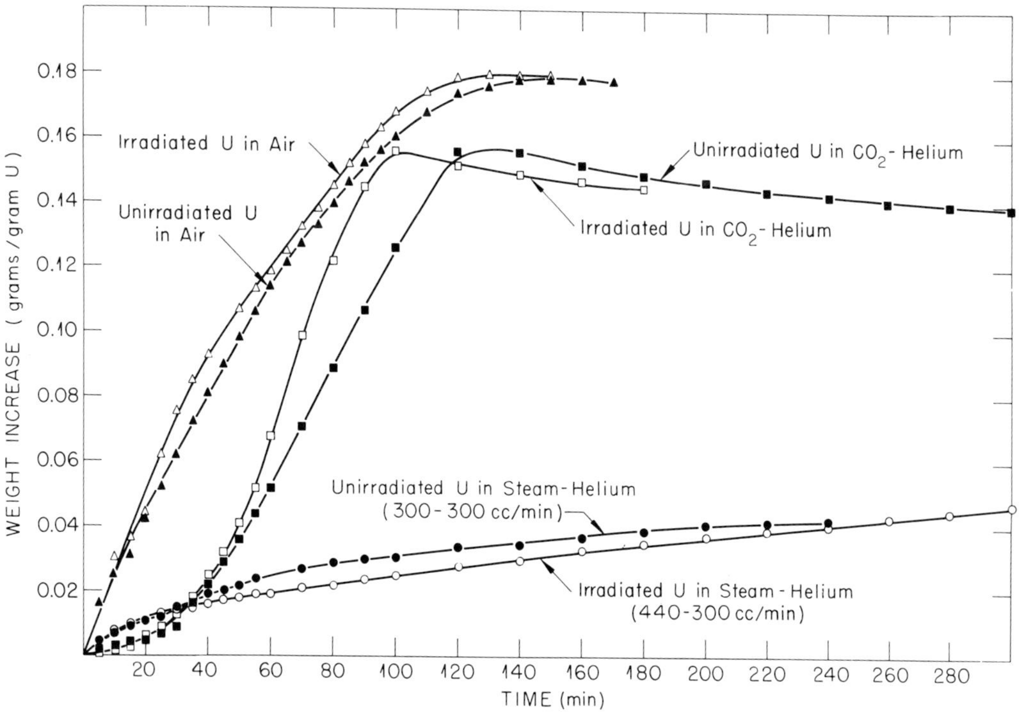

FIG. 3-3 Comparison of oxidation rates of irradiated and unirradiated uranium in air, CO_2-helium mixture, and steam-helium mixture at 800°C (1470 F).

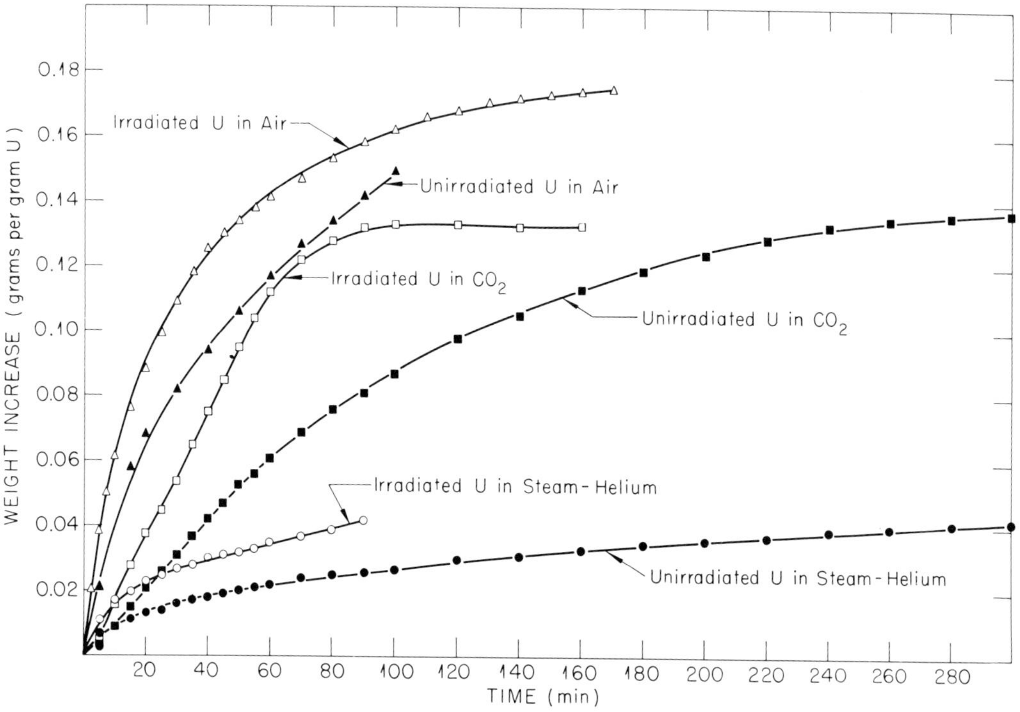

FIG. 3-4 Comparison of oxidation rates of irradiated and unirradiated uranium in steam diluted with helium, and in air or CO_2 at 1000°C (1830°F). Flow rate ≈ 600 cm^3/min.

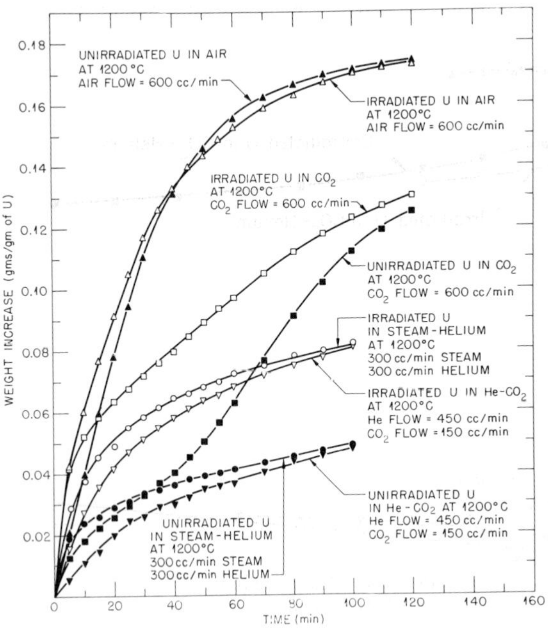

FIG. 3-5 Comparison of oxidation rates of irradiated and unirradiated uranium in steam diluted with helium, air, and CO_2 (or CO_2 diluted with helium at 1200°C (2190°F).

fission-gas bubbles, which burst through the thin oxide layer covering the molten uranium, allowing it to flow and cover the bottom of the crucible. Diffey and King [78] express the belief that the very high oxidation rates of irradiated uranium in CO_2 at 800°C (1472°F) are due to swelling caused by release of fission-product gas within the uranium. Buddery and Scott [80] have studied bubble formation accompanying the melting of irradiated uranium in some detail. It seems reasonable to assume that increased access of oxygen to unoxidized uranium could result from cracks or holes in the oxide coating produced by fission gas. It has also been suggested [18] that imperfections in the oxide coating of irradiated uranium could result from the presence of fission-product atoms and that these imperfections could lead to cracking and diminished protectivity of the oxide coating. Experimental evidence to support this explanation is lacking at present. Further study appears necessary in order to provide a more adequate explanation of the burnup effect.

Self-Heating Effect. The self-heating effect of uranium in CO_2 was studied by Antill et al. [77], who placed a thermocouple in the center of a specimen 2.5 cm (0.984 in.) long and 2 cm (0.787 in.) in diameter. Self-heating was not observed until the gas temperature reached 725°C (1337°F). The specimen temperature then increased gradually until a sudden increase of 150°C (270°F) was noted when it reached 775 to 780°C (1427 to 1436°F). The oxidation rate was not measured in this experiment, but the investigators believe that the sudden increase in temperature coincided with a 10-fold increase in oxidation rate.

3.1.1.3 Oxidation of Uranium in Steam. Early studies [62, 81, 82] of the oxidation of uranium in steam were performed at temperatures below 600°C (1112°F) because of interest in water-cooled reactors fueled with uranium. More recently, the uranium-steam reaction has been investigated [18, 83, 84] at temperatures as high as 1440°C (2624°F).

Effect of Temperature and Steam-Flow Rate. Weight-increase data obtained by exposure of uranium specimens to steam-helium mixtures at different temperatures are included in Figs. 3-3, 3-4, 3-5, and 3-6 [18]. The oxidation rate increases, in general, with increasing temperature in the 800 to 1400°C (1472 to 2552°F) range. It may seem a bit surprising, however, that the oxidation rate shown in Fig. 3-6 decreases with increasing steam-flow rate. This result is explained [18] on the basis that the sintering action of steam on UO_2, reported by Huddle [82], Hopkinson [83], and others, occurs at a rate proportional to the steam-flow rate.

Inspection of the curves in Fig. 3-6 shows that the initial oxidation rate is the maximum rate observed. After the initial rapid stage of oxidation has been completed, the rate becomes essentially constant for some time and then gradually decreases if oxidation proceeds for an extended period. The high initial rate is due to the reaction of essentially unprotected uranium with steam. The linear part of the curve is explained by Hopkinson [83] as follows: When the oxide reaches a certain thickness, the stresses connected with film growth cause cracking, but the oxide layer immediately adjacent to the metal remains intact; the oxidation rate, being controlled by diffusion through a layer of constant thickness, thus remains constant. The final slow decrease in rate is said [18] to be due, possibly, to sintering of the loose oxide around the specimen, which would result in an increased diffusion path for the steam.

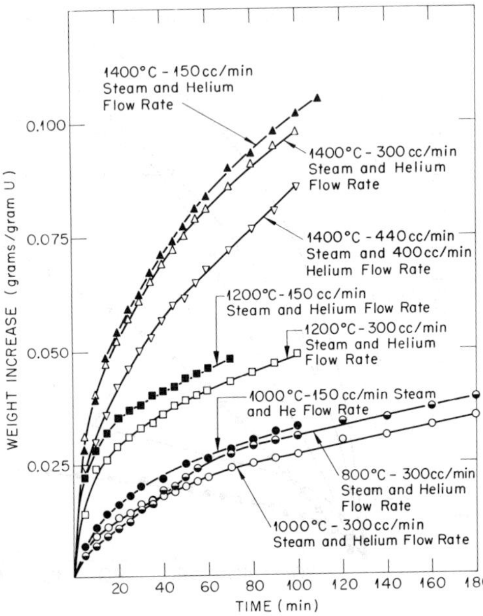

FIG. 3-6 Effect of varying temperature and steam-flow rate on the oxidation of uranium in steam diluted with helium.

Effect of Burnup. A comparison of the oxidation rates of irradiated and unirradiated uranium in steam at 800 to 1000°C (1472 to 1832°F), Figs. 3-3 and 3-4, shows the same type behavior that has been observed in the tests made in air and in CO_2. The irradiation effect is not observed in tests made at 1200°C (2192°F), Fig. 3-5, because of the higher steam flow employed with the irradiated specimens. As noted above, increased steam-flow rates produce lower oxidation rates in steam. Apparently, no other studies of oxidation rates of irradiated uranium in steam have been published.

3.1.1.4 Comparison of Oxidation Rates in Various Atmospheres. The data shown in Fig. 3-3, 3-4, and 3-5 permit ready comparison of the oxidation rates of both irradiated and unirradiated uranium in three different atmospheres. At all three temperatures, the rates are in the same decreasing order: air, CO_2, steam. It is observed that complete oxidation of irradiated uranium occurs about as fast in CO_2 as in air at 800 and 1000°C (1472 and 1832°F), while none of the specimens are completely oxidized in steam at this temperature. These data would have permitted prediction of the observed ineffectiveness of CO_2 to quench the Windscale Pile No. 1 fire (see Vol. I, p. 633) if they had been available at that time. The data also show that steam provides the minimum oxidation rate and, consequently, the successful use of water to quench the Windscale fire could also have been predicted.

3.1.2 Release of Fission Products from Metallic Uranium

Details of release experiments are reported here only where such information appears to be needed to aid the evaluation of the data reported. Experiments to determine the extent of release of fission products from metallic uranium are discussed on the basis of the atmospheric environment of the fuel material.

3.1.2.1 Uranium Melted in an Inert Gas. Studies of the release of fission products accompanying the melting of irradiated uranium were carried out at Ames [2] and at the Argonne National Laboratory. Burris et al. [85] report volatilization of a large fraction of the rare gases, halogens, and cesium. Megaw et al. [68] who melted a 1-kg specimen of tracer-level-irradiated uranium in argon, report that only 0.17% of its I^{131} content was released as compared to a 12 to 15% release in air and CO_2. Hilliard [86] heated 11.5-gm (0.0254-lb) cylinders of tracer-level-irradiated uranium in helium and in air and gave a comparison of fission-product release shown in Table 3-1 [17].

Several experiments involving the melting of irradiated uranium (0.1 at. % burnup) in impure helium were performed at the Oak Ridge National Laboratory. The data are given in Table 3-2 [18]. The lack of correlation between the amount of uranium oxidized and the fraction of fission gases released indicates that oxidation had a minor effect on fission-product release in these experiments.

The discrepancy between the high gas-release values of the ORNL experiments and the low values shown in Table 3-1 (10%) can probably be attributed to the difference in gas concentration (sufficient in the ORNL experiments to form bubbles) in the uranium used in the two investigations. (There was approximately 10^6 greater concentration of xenon in the ORNL experiments.)

In some experiments [18], a sensitive in-stream gamma detector ahead of the cold charcoal trap permitted observation of fission-gas release in more detail than was afforded by the use of traps alone. The data indicate that a large fraction of the fission gases was released very rapidly when the uranium melted. The fraction released, as shown by trapping of the released gas in refrigerated charcoal, increased from only 98% in 16 min after melting began to 99.3% after 38 min. A sharp peak in the counter trace that recorded the rate of gas release during the cooling period was noted at the freezing point of uranium. A smaller peak that probably corresponded to the β-α inversion point (660°C or 1220°F) was noted, and another large peak was noted when the uranium was cooled rapidly from 513°C (955°F) to room temperature. These observations indicate that a part of the fission gas not released while irradiated uranium is in the molten condition will be "squeezed" out during the cooling period, but the fraction released by this process is probably too small to be of significance in hazards analyses.

3.1.2.2 Uranium Oxidized in Air. A number of investigations of fission-product release as the result of exposure of irradiated uranium to air have been reported, because air is a likely atmospheric environment of uranium in loss-of-coolant accidents, regardless of the coolant employed. Among these are the studies conducted at Hanford [17, 30, 86], at ORNL [18] and in England [68]. It has been reported [18, 30] that data on fission-product release obtained with fuel material that has undergone a significant degree of burnup may be quite different from results obtained with tracer-level-irradiated fuel specimens. Consequently, emphasis here will be placed on data obtained with high-burnup material where such data are available.

TABLE 3-1

Comparison of Fission-Product Release from Tracer-Level-Irradiated Uranium Heated 25 Min at 1215°C (2219°F) in Air and Helium Atmospheres

Atmosphere	Percent released from specimen						
	I^{131}	Te^{132}	Xe^{133}	Sr^{89}	Cs	Ru^{103}	Ba^{140}
Air	65	60	50	0.4	1.0	0.65	0.09
He	41	4.2	10	0.18	1.2	0.09	0.08

TABLE 3-2

Fraction of Rare Gases Released on Melting of Irradiated Uranium (0.1 at.% Burnup) in Impure Helium

Maximum furnace temperature	U oxidized (%)	rare gases released (%)
1170°C (2138°F)	0.9	98.9
1180°C (2156°F)	3.9	97.1
1200°C (2192°F)	4.1	97.7
1250°C (2282°F)	15.3	99.6

TABLE 3-3

Fission-Product Release from Irradiated Uranium[a] Incompletely Oxidized in Air[b]

Furnace temp. (°C)[c]	Time (min)	U oxidized (%)	Total activity released (%)							
			Xe-Kr	I	Te	Cs	Ru	Zr	Ce	Sr
1000	< 1	11.1	-	3.1	-	0.01	0.004	0.003		
1000	10	46.9	~ 100	-	-	2.4	0.1	0.0007	0.0001	0.05
1000	20	53.2	~ 100	67.3	-	2.8	0.13	0.005	0.008	0.055
1000	40	86.9	~ 100	79.5	-	18.4	5.2	0.018	0.006	0.05
1200	< 1	25.0	-	12.5	-	1.62	0.019	0.17		
1200	5	43.6	97.7	31.9	8.1	18.5	0.035	-	-	0.024
1200	8	66.2	99.2	23.3	12.7	14.5	0.16	-	-	0.028
1200	10	94.0	100	-	-	11.2	0.51	0.05	0.06	2.7
1200	10	68.0	98.7	39.9	24.8	17.1	0.22	-	-	0.6
1200	12	77.5	99.8	46.4	23.0	-	-	-	-	0.9
1200	15	72.0	99.8	52.8	51.6	28.6	4.3	-	-	3.1
1200	15	64.8	99.6	71.5	68.0	13.65	2.0	-	-	1.1
1200	20	65.4	99.4	57.3	71.3	13.0	1.8	-	-	0.85
1200	30	72.3	-	62.1	77.4	19.2	2.34	-	-	1.77

[a] 0.1 at.% burnup, preheated in helium.
[b] Velocity 120 cm/min (3.9 ft/min), measured at room temperature.
[c] 1000°C = 1832°F; 1200°C = 2192°F.

Data on fission-product release from irradiated uranium (0.1 at % burnup), partially oxidized in horizontal furnace-tube apparatus, are shown in Table 3-3 [18]. The cylindrical specimens, with a diameter of 0.25 in. (6.35 mm) and an approximate length of 0.75 in. (19.05 mm), were contained in shallow boats inside a quartz furnace tube. They were heated and cooled in flowing helium and exposed to air for varying lengths of time at two furnace temperatures. The data confirm the results of Hilliard's [17] studies made with similar specimens irradiated to 2.4×10^{14} nvt, which showed increasing release of the more volatile fission products with increasing fraction of uranium oxidized. However, the release of iodine and tellurium from the low-burnup uranium appeared to level out at about 80% in Hilliard's experiments, whereas the release of these fission products from the oxidized portion of higher burnup material, as shown in Table 3-3, appear to be essentially complete. In general, the release of tellurium was somewhat lower than that of iodine, and less cesium was released than tellurium. Moderate releases of ruthenium and strontium were noted at 1200°C (2192°F), but very little of the cerium and zirconium volatilized.

The effect of burnup on the release of fission products from uranium partially oxidized in high-temperature air was also studied by Hilliard and Reid [30]. The largest effect was noted in the release of cesium and iodine, and their data for these isotopes are shown in Figs. 3-7 and 3-8 [30]. These workers calculated the concentration of various fission products in their uranium specimens at the lower and upper burnup limits employed in their studies, 2.4×10^{14} (10^{-7} at. %) and 4×10^{20} nvt (0.17

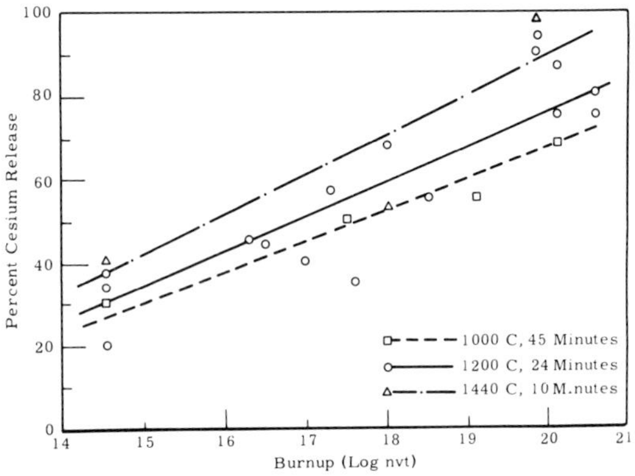

FIG. 3-7 Release of cesium as a function of burnup.

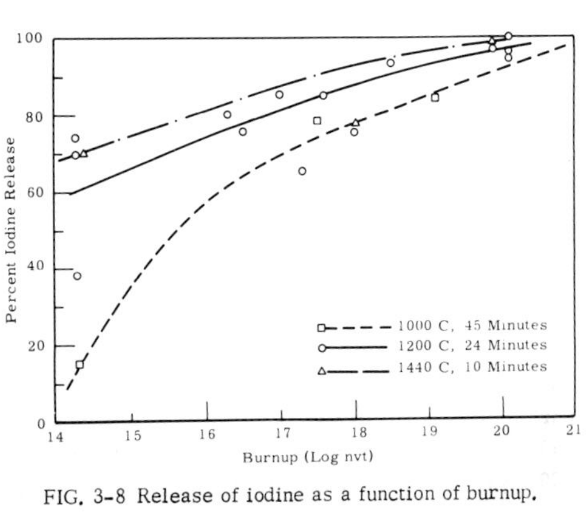

FIG. 3-8 Release of iodine as a function of burnup.

at. %); the results are given in Table 3-4 [30]. They also compared rates of release of xenon from uranium exposed to air at 1200°C (2192°F) at different burnup levels. The data are shown graphically in Fig. 3-9 [30]. It is clear that xenon is released almost instantaneously from uranium on melting in air at 1200°C (2192°F) when the concentration of the gas in uranium is reasonably high.

Data on the distribution of fission products liberated by the complete oxidation of irradiated uranium (0.1 at % burnup) in an air stream are shown in Table 3-5 [18]. These data were obtained by use of a vertical furnace tube made of mullite; the small cylindrical specimens were held in deep (2.25-in. or 5.71-cm) porous alumina extraction thimbles. The air stream had less ready access to the uranium specimen than was afforded by the open-boat apparatus mentioned above. Suspension of the samples in the vertical furnace tube was required to permit continuous weighing of the uranium while it was oxidizing, and the alumina thimble prevented loss of the uranium oxidation products.

Data in Table 3-5 show that release of iodine and ruthenium was quite high even with a furnace temperature of 800°C (1472°F), while the release of cesium and tellurium was rather low. An increase of the furnace temperature to 1000°C (1832°F) produced a moderate increase in cesium release but a large increase in the fraction of iodine and tellurium released. At 1200°C (2192°F), the release of iodine, tellurium, and ruthenium was essentially complete, and a substantial fraction of the cesium also escaped. Very small amounts of the refractory elements cerium, zirconium, and strontium were released.

Iodine was found mostly in the hot (200°C or 392°F) charcoal bed, where it would be expected if it were liberated in the molecular form. Most of the cesium remained in the mullite furnace tube,

TABLE 3-4

Theoretical Concentration of Fission Products in Irradiated Uranium

Fission-product element	Concentration (at %)	
	2×10^{14} nvt	4×10^{20} nvt
Xe	1.9×10^{-8}	3.5×10^{-2}
I	1.8×10^{-9}	1.9×10^{-3}
Cs	1.4×10^{-8}	3.1×10^{-2}
Te	2.0×10^{-9}	3.9×10^{-3}
Sr	1.0×10^{-8}	1.5×10^{-2}
Ba	7.5×10^{-9}	9.3×10^{-3}
Zr	2.4×10^{-8}	5.4×10^{-2}
Ce	1.1×10^{-8}	2.6×10^{-2}
Ru	1.1×10^{-8}	1.5×10^{-2}
Mo	1.5×10^{-8}	2.7×10^{-2}
Gross fission products	1.6×10^{-7}	3.2×10^{-1}

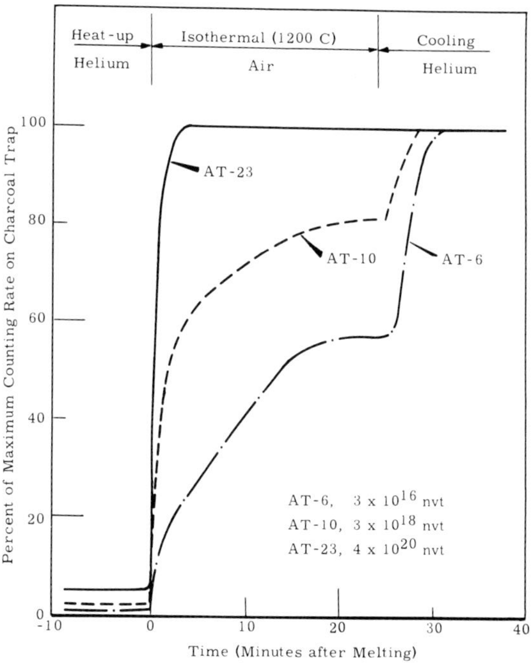

FIG. 3-9 The effect of burnup on xenon release rate.

TABLE 3-5

Activity Released by Complete Oxidation of Irradiated Uranium[a] in Air[b]

Experiment Number	Temp. (°C)[d]	Total Activity Released (%)							
		Xe-Kr	I	Cs	Ce	Te	Ru	Zr	Sr[c]
68	800		48	0.06	0.001	2.9	73	0.05	0.002
81	1000	97.1	89	0.4	0.002	80	77	0.02	0.002
83	1200	99.2	90	14	0.0006	96	85	0.01	0.02
32	1200		99.2	16	0.03	84	78	0.08	0.005

[a]Burnup = 0.1 at. %.

[b]Air flow rate, 220 cm/min (7.21 ft/min) measured at room temperature.

[c]Values probably low owing to chemisorption effect.

[d]800°C = 1472°F; 1000°C = 1832°F; 1200°C = 2192°F.

possibly in the form of Cs_2O. Although the distribution of tellurium varied considerably in the four experiments, it appears that a large fraction of tellurium released by uranium oxidizing in air will be in the form of particulate matter, probably TeO_2. Ruthenium was undoubtedly released as a volatile oxide, RuO_3 or RuO_4, but it probably was quickly converted to a less-volatile lower oxide; this conversion accounts for the fact that a large fraction of this element remained in the furnace tube. The part that was airborne long enough to reach the filters stopped there.

3.1.2.3 Uranium Oxidized in CO_2. There is comparatively little information in the literature on the release of fission products from uranium oxidized in CO_2. Complete oxidation of irradiated uranium (0.1 at % burnup) at 800, 1000, and 1200°C (1472, 1832, 2192°F) in commercial-grade CO_2 (1 to 2 % O_2) or in CO_2 diluted with helium gave release data recorded in Table 3-6 [18]. The data show that cesium and ruthenium release values obtained in CO_2 were very much lower than the corresponding values obtained in air (Table 3-5) except for the anomalous ruthenium release result in experiment 84, discussed below. Iodine and tellurium were released to about the same extent in CO_2 that they were in air at the same temperature. Megaw et al. [68] found much lower I^{131} release rates in CO_2 than in air in studies made with 1-kg slugs of tracer-level-irradiated uranium, but they were unable to provide an explanation for the different behavior in the two oxidizing atmospheres. The induction-heating method that they employed allowed the gas surrounding the specimens and the surface of the oxide coating to remain relatively cool. They found that evidence of surface adsorption of I^{131} on the U_3O_8 and a greater iodine-retention ability of the UO_2 produced in CO_2, as compared to that of the U_3O_8 produced in air, could explain the observed differences. A more likely explanation, however, is that oxygen in air will oxidize iodides more readily than will CO_2.

In experiment 84, Table 3-6, a very large fraction of the ruthenium was released under unusual circumstances. Approximately 10% of the uranium was inadvertently oxidized in air at the beginning of the experiment and, after the remaining uranium was oxidized in CO_2, a high-velocity stream of helium was passed through the apparatus for 30 min. This combination of atmospheric conditions produced higher release values of several nuclides than resulted from oxidation in CO_2 at the same temperature under normal conditions (experiment 31), but the behavior of ruthenium in this experiment is especially noteworthy. All the filter papers used to collect particulate matter in this experiment contained ruthenium, including, in order, a 3.0 μ Millipore, a 0.4 μ Millipore, a coarse cellulose Whatman paper, a 0.4 μ Millipore, and another Whatman cellulose paper. (Note: 1 μ = 0.25 mil) This fact indicates that the ruthenium was in the form of very small particles or a gas and suggests that ruthenium probably volatilized as RuO_3 or RuO_4 for reasons that are not clear. The existence of released ruthenium in the form of an oxide was indicated by the fact that it dissolved easily. Ruthenium oxide is much easier to dissolve than the metal.

3.1.2.4 Uranium Oxidized in Steam-Helium Mixtures. The principal published reports on the release of fission products from irradiated uranium oxidized in steam are those of Scott [84], of Parker et al. [18]. The latter investigators diluted their steam with helium, and results of these studies are summarized in this section.

Experiments on the release of fission products from irradiated uranium (0.1 at. % burnup) oxidized in steam were most conveniently performed through use of helium as the carrier gas. In addition to serving as an inert carrier of steam, the helium swept fission gases into the cold charcoal trap after steam had been removed from the furnace-exit gas mixture by condensation. Release data obtained when irradiated uranium specimens were exposed to a mixture of steam and helium at 800, 1000, and 1200°C (1472, 1832 and 2192°F) are displayed in Table 3-7 [18]. The slow oxidation rates attained in this atmosphere made it impractical to oxidize the uranium completely. Consequently, the

TABLE 3-6

Fission Products Released by Complete Oxidation of Irradiated Uranium[a] in CO_2[b] or in CO_2 Diluted with Helium[c]

Experiment number	Atmosphere	Temp.[f] (°C)	Percent of total activity released								
			Gross γ	I	Cs	Ce	Te	Ru	Zr	Sr	Rare gases
67	CO_2-He	800	0.7	5.8	0.02	0.001	1.9	0.04	0.044		Not determined
82	CO_2	1000	0.7	85	0.002	0.0003	39	0.08	0.002	0.001	75
62	CO_2-He	1200	0.6	68	0.9	0.003	95		0.59	0.012	Not determined
31	CO_2	1200	7.6	53	1.7	0.002	69	2.0	0.09	0.01	Not determined
84	CO_2[d]	1200	14.6	85	1.8	0.004	96	93[e]	0.020	0.3	99.2

[a] Burnup = 0.1 at. %.

[b] CO_2 flow velocity, 200 cm/min (6.56 ft/min), measured at 25°C (72°F).

[c] CO_2 flow rate, 150 cm^3/min (0.00534 ft^3/min); He flow rate, 450 cm^3/min (0.0160 ft^3/min).

[d] Approximately 10% of the uranium was oxidized by accidental admission of air at the beginning of this run. After the completion of the CO_2 oxidation, high-velocity He (125 ft/min or 3810 cm/min) was passed through the apparatus for 30 min.

[e] Average of two analyses. High results probably due to unusual conditions mentioned above.

[f] 800°C = 1472°F; 1000°C = 1832°F; 1200°C = 2192°F.

TABLE 3-7

Fission Products Released by Incomplete Oxidation of Irradiated Uranium[a] in Steam Diluted with Helium[b]

Temp.[c] (°C)	U oxidized (%)	Total activity released (%)								
		Gross γ	I	Cs	Ce	Te	Ru	Zr	Sr	Xe-Kr
800	36	0.002	0.3							0.9
1000	34	0.1	5.2	0.04	0.007	2.0	0.02	0.04	0.02	3.0
1200	65	0.2	15	0.2	0.0006	79	0.01	0.2	0.02	

[a]Burnup= 0.1 at. %; preheated in He.
[b]Steam and He flow rate each approximately 300 cm^3/min (0.0106 ft^3/min), measured or calculated rate at room temperature.
[c]800°C = 1472°F; 1000°C = 1832°F; 1200°C = 2192°F.

release values shown in Table 3-7 should be divided by the fraction of uranium oxidized before they are compared with the data in Tables 3-5 and 3-6. That this correction ignores loss of fission products through diffusion from unoxidized portions of the specimens probably accounts for the fact that adjusted values of more volatile elements such as iodine and tellurium exceeded 100% in some cases. The data in Table 3-7 show that the dense, adherent coating of UO_2 formed around uranium specimens exposed to steam at temperatures in the range 800 to 1200°C (1472 to 2192°F) resulted in marked reduction of the fraction of fission products released, except for tellurium. These data confirm Scott's values [84], showing that iodine and tellurium are released to approximately the same extent as the rare gases in this environment. The high tellurium release value obtained in steam at 1200°C (2192°F), as compared with iodine release observed at this temperature, suggests the possibility that hydrogen released by the uranium-steam reaction may have combined with tellurium to form highly-volatile H_2Te. However, no corroborating evidence for the formation of such a compound was noted in these experiments, and Scott's data, which show the same range of iodine and tellurium release values at 1215°C (2219°F) and higher iodine than tellurium release at 1440°C (2624°F), do not appear to support this hypothesis. The low values for release of cesium, ruthenium, and rare gas observed in a steam atmosphere are especially noteworthy.

3.1.2.5 Uranium Heated in Steam-Air Mixtures. Data on fission-product release from irradiated uranium oxidized in steam mixed with air are contained in Table 3-8 [18]. These data were obtained with tracer-level (~10^{17} nvt or 10^{-4} at. % burnup)-irradiated cylindrical specimens weighing about 11 g. The steam-to-air ratio by volume was calculated to be 12 to 1 at 20°C (68°F), and the air-flow velocity was 120 cm/min (3.94 ft/min) measured at 20°C. Horizontal furnace-tube-open-boat apparatus was employed in these experiments. The data, when adjusted for fraction of uranium oxidized, are comparable to the air-oxidation data shown in Table 3-5 except for the ruthenium results.

3.1.2.6 Uranium Oxidized in Undiluted Steam. Scott [84] measured fission-product release from tracer-level-irradiated uranium (10^{-7} at. % burnup) heated in flowing (406 cm/min or 13.3 ft/min) undiluted steam. The relation found between fission-product release and fraction of uranium oxidized is shown in Fig. 3-10 [84]. The maximum release values of other elements, with a heating period of 120 min at 1440°C (2624°F), were cesium, 1.1%; ruthenium, 0.8%; strontium, barium, and zirconium, 0.3% or less.

Release values obtained when irradiated uranium (0.1 at. % burnup, except for one test) was heated in various atmospheres at 1200°C (2192°F) are compared in Table 3-9 [18]. The high ruthenium release obtained upon complete oxidation of the

TABLE 3-8

Fission-Product Release from Tracer-Level-Irradiated Uranium[a] Heated in Air-Steam Mixture[b]

Furnace[c] Temp. (°C)	Time (min)	U oxidized (%)	Total activity released (%)					
			Xe-Kr	Gross γ	I	Te	Cs	Ru
1000	11	17	4.7	0.5	4.3	1.8	0.8	0.4
1000	40	25	5.5	0.3	4.9	6.2	0.4	0.2
1000	60	42	-	0.3	2.7	4.7	0.2	0.02
1200	10	13	26	8.9	8.6	6.5	1.6	0.04
1200	40	28	65	5.9	46	42	5.8	0.3
1200	61	49	-	4.9	47	60	6.9	4.0

[a]Approximate burnup level = 10^{-4} at. %; preheated in He.
[b]Steam-to-air ratio by volume was 12:1 at 20°C (68°F). Air flow velocity was 120 cm/min, (3.93 ft/min) measured at room temperature.
[c]1000°C = 1832°F; 1200°C = 2192°F.

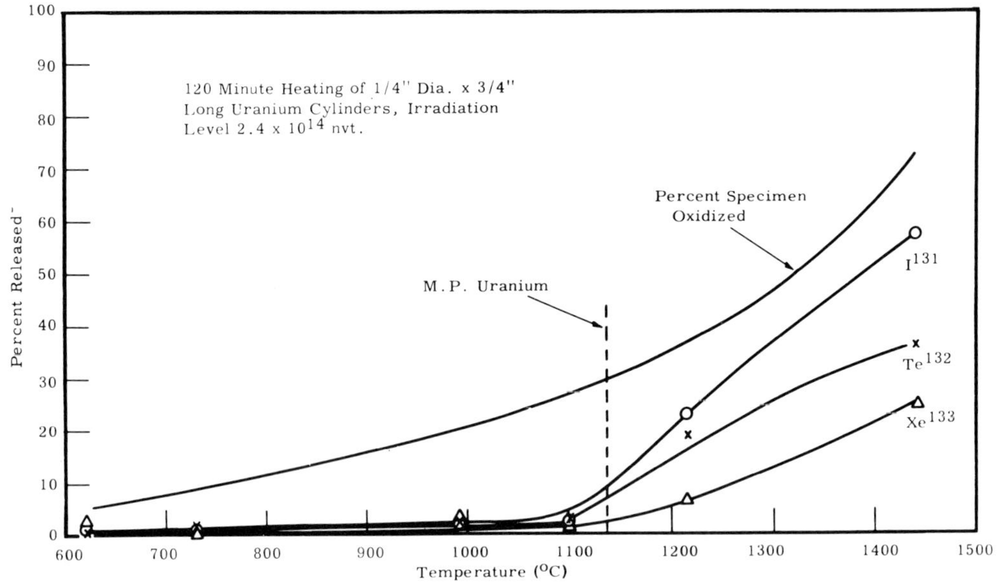

FIG. 3-10 Effect of temperature on fission-product release in steam atmosphere.

TABLE 3-9

Comparison of Fission-Product Release from Irradiated Uranium[a] Heated in Various Atmosphere at 1200°C (2192°F)

Atmosphere	Time heated (min)	U oxidized (%)	Xe-Kr	Total activity released (%)					
				I	Te	Cs	Ru	Sr	Zr
Air	20	65	99.4	57	71	13	1.8	0.09	~0.05
Air	250	100	99.2	90	96	~14	85	-	0.01
CO_2	630	100	~99	53	69	1.7[b]	2.0	0.01[b]	0.1
CO_2-He	410	90		68	95	0.9[b]		0.01[b]	0.6
Steam-He	123	65		15	79	0.2[b]	0.01	0.02[b]	0.2
He	148	4	98	47	0.6	2.0		0.9	
Air[c]	50	92	81	92	66	~1.0	0.6	0.01	0.007
Steam[c]	120	33	4	18	12				

[a] Burnup = 0.1 at. %; preheated in He.
[b] Release values probably low because of chemisorption in mullite furnace tube or alumina crucible.
[c] Tracer-level irradiation.

specimen in air, as compared to partial oxidation in CO_2, is quite noticeable.

3.1.3 Uranium-Aluminum Alloys

Uranium-aluminum alloys clad with aluminum have been employed extensively in research reactors (HFIR, CP-5, MITR, LITR, and ORR) and in materials-testing reactors (MTR and ETR at the NRTS). Early data on fission-product release from fuel elements of this type were obtained [24] with low-burnup fuel materials. The release of fission products from irradiated uranium-aluminum-alloy specimens has been more recently studied over a wide range of temperatures in several atmospheres, with fuels of several degrees of burnup used [39b, 58a]. In most of the recent experiments, the fuel was held at the maximum temperature for 2 min, but the specimens were molten for periods ranging from about 10 to 17 min because of the time required for heating and cooling. Some experiments were performed with longer heating periods because of the possibility that fission-product-decay heat could maintain this low-melting fuel material in the molten state for considerable lengths of time after a loss-of-coolant accident.

Data showing the effect on fission-product release of several variables including temperature, atmosphere, time at temperature, and air-flow rate are given in Table 3-10 [39b]. These data were all obtained with fuel specimens irradiated to 23.6 at. % U^{235} burnup. It is apparent that, at this burnup level, release of rare gases is almost quantitative at any temperature above the melting point of fuel. Release of other elements increases, in general, with increasing temperature, as might be expected. The effect of atmosphere is most noticeable in the tellurium and cesium results. Release of cesium was

TABLE 3-10

Effect of Maximum Temperature, Time at Temperature, and Atmosphere[a] on Fission-Product Release from Uranium-Aluminum-Alloy Specimens[b]

Maximum Temp.	Time at maximum temperature (min)	Atmosphere[a]	Release (%)					
			Gross γ	Rare gases	I	Te	Cs	Ru
800°C (1472°F)	2	He	7.4	99.5	29.8	5.3	13.0	0.18
900°C (1652°F)	2	He	13.5	~100	52.8	4.3	20.8	0.08
1000°C (1832°F)	2	He	23.5	~100	82.1	2.9	47.7	0.19
1105°C (2021°F)	2	He	40.7	~100	82.4	2.9	69.5	0.25
700°C (1292°F)	2	Air	2.3	97.9	37.8	0.3	3.1	0.02
800°C (1472°F)	2	Air	3.1	99.4	78.6	0.2	3.8	<0.1
900°C (1652°F)	2	Air	5.2	100.0	91.9	2.1	6.2	0.1
1000°C (1832°F)	2	Air	6.7	99.8	97.3	<9.7	8.8	0.2
1090°C (1994°F)	2	Air	12.0	100.0	98.4	44.8	12.4	0.6
1145°C (2093°F)	2	Air	16.8	100.0	94.2	62.0	18.6	0.4
700°C (1292°F)	2	Steam-Air	0.9	98.3	27.0	<0.03	0.6	<0.02
800°C (1472°F)	2	Steam-Air	2.5	99.5	76.8	0.3	1.1	0.1
900°C (1652°F)	2	Steam-Air	6.8	99.9	90.6	5.7	6.5	0.5
1000°C (1833°F)	2	Steam-Air	10.6	~100	95.6	22.6	11.0	0.5
1085°C (1985°F)	2	Steam-Air	25.5	~100	96.8	67.9	30.5	0.8
700°C (1292°F)	60	Air	3.3	97.7	58.0	<0.14	3.5	<0.02
800°C (1472°F)	60	Air	4.5	99.5	84.7	0.7	5.9	0.03
900°C (1652°F)	60	Air	6.3	99.95	95.3	2.9	9.2	0.2
1000°C (1832°F)	60	Air	18.1	99.98	92.8	16.6	23.3	0.1
1090°C (1994°F)	60	Air	16.1	99.98	98.3	78.4	37.8	0.03
840°C (1544°F)[c]	60	Air	5.3	~100	94.6	1.5	6.5	0.1
870°C (1598°F)[c]	60	Air	8.1	~100	95.8	4.0	6.9	0.7

[a]Gas flowing at a rate of 250 cm^3/min or 0.00989 ft^3/min (measured at room temperature) equivalent to a gas velocity of approximately 34 cm/min (1.115 ft/min). The steam flow rate in steam-air mixtures was four times that of air.

[b]Burnup level, 23.6 at. % U^{235}. Specimens were in the form of 5/16-in. (0.794-cm)-diam disks punched from MTR-type fuel plates, reirradiated to build up a suitable inventory of short-lived fission products.

[c]Air flow rate in these experiments was increased to 3000 cm^3/min (0.016 ft^3/min) or about 430 cm/min (14.1 ft/min) measured at room temperature.

much higher than that of tellurium in helium, while the order was reversed in air and steam-air atmospheres, at least at high temperatures. The release of iodine was slightly higher in oxidizing atmospheres than in helium and, at 900°C (1652°F) or higher temperatures, more than 90% of this important fission product was released in the presence of oxygen. Mixing steam with air had no significant effect on fission-product release, in marked contrast to the effect of steam on fission-product release from metallic uranium. Increasing the time at maximum temperature in air produced a moderate increase in tellurium and cesium release, but the release of iodine and rare gases was so high with a short heating period that no time effect could be observed for these elements. The release values obtained for ruthenium in air were so low that no trend with either time or temperature could be established from the data obtained. The data in Table 3-10 also show that a drastic increase in air flow rate had little, if any, effect on fission-product release.

Data on the effect of burnup on fission-product release from uranium-aluminum alloys at different temperatures are shown in Table 3-11 [58a]. There is a noticeable burnup effect in the release of all the fission-product elements examined in these experiments, but the largest and most important effect is in the release of the most volatile species, iodine and the rare gases. These data provide an explanation for the low results previously reported [24] for fuel specimens irradiated only to trace level. The burnup effect on fission-gas release from molten uranium-aluminum fuel is shown in more detail in Fig. 3-11 [58a]. The data show little increase in release with increasing burnup above 3.2%, except

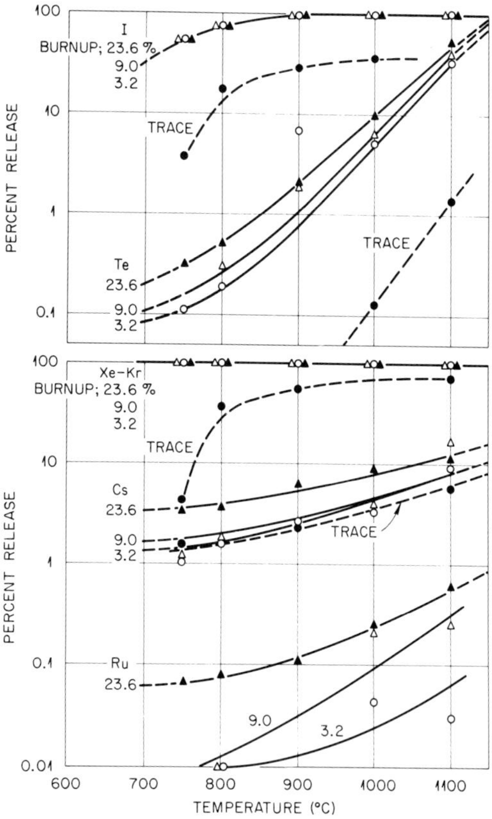

FIG. 3-11 Burnup effect on fission-product release from uranium-aluminum alloy melted in air.

TABLE 3-11

Effect of Temperature and Burnup on Fission-Product Release from Irradiated Uranium-Aluminum Alloys[a]

Burnup Level (%)	Release (%)				
	Iodine	Tellurium	Cesium	Ruthenium	Rare Gas
			At 750°C (1382°F)		
Tracer	3.7	~ 0.01	~ 1.2		4.3
3.2	52.8	0.14		0.0005	98.2
9.0	54.2	0.05	1.3	0.004	
23.6	56	~ 0.3	~ 3.6	0.07	~ 98
			At 800°C (1472°F)		
Tracer	16.7	0.02	1.6	~0.002	37.2
3.2		~ 0.3	1.1	0.01	99.4
9.0	71.9	0.04	1.7	~0.01	99.4
23.6	78.6	~ 0.5	3.8	0.08	99.4
			At 900°C (1652°F)		
Tracer	28.8	0.03	2.6	~0.004	54.0
3.2	97.4	6.1	2.7	0.002	>99.5
9.0	95.0	1.9	7.5		>99.5
23.6	92	2.0	6.2	0.1	>99.9
			At 1000°C (1832°F)		
Tracer	41.2	0.14			
3.2	98.5	5.3	3.2	0.04	~100
9.0	97.2	6.3	3.5	0.2	~100
23.6	97.3	9.7	8.8	0.25	99.8
			At 1000°C (2012°F)		
Tracer	(34.3)	1.3	6		71.8
3.2	99.5	31.7	9.5	0.03	~100
9.0	93.5	37.1	19.6	0.25	~100
23.6	98.4	~50	12.4	0.6	100

[a]Specimens were heated for 2 min at maximum temperature in air flowing at 250 cm^3/min (0.00989 ft^3/min).

for cesium release at the highest burnup level. This fact seems to indicate that the burnup effect, whatever its explanation may be, is saturated at a comparatively low burnup level.

See Table 4-1 of Chapter 11, Vol. 1, which includes mention of a number of incidents involving this type of fuel material. SPERT tests [87-89] of Al 31% uranium-69% aluminum tubes clad with ribbed aluminum provided information on the melting behavior of this type of fuel under abnormal reactor operating conditions (See Chapter 13, Sec. 4.3.4.1 and Chapter 7, references).

3.1.4 Uranium-Molybdenum Alloys

Only limited use has been made of uranium-molybdenum alloys in reactors (See Sec. 4.1.5.1 in Chapter 13). An alloy containing 1.5 wt. % molybdenum is used in the EL-3 (French) reactor and one incident involving a fuel-element failure in this reactor has been reported (See Table 4-1 of Chapter 11).

The small use of uranium-molybdenum alloys is reflected in the lack of published information on fission-product release from such alloys. Information on fission-product release from uranium-zirconium alloys is the best guide available to the probable release from uranium-molybdenum fuels. British investigators [90] studied the oxidation behavior of uranium-molybdenum alloys containing up to 15.5 wt. % molybdenum in CO_2 at temperatures in the range 500 to 1000°C (932 to 1832°F) and in air at 500 to 900°C (932 to 1652°F). The molybdenum additions decreased the oxidation rates of uranium in CO_2 at 800 to 1000°C (1472 to 1832°F) and in air at 500 to 680°C (932 to 1256°F) by factors of 1 to 140 but increased them in CO_2 at 500 to 680° (932 to 1256°F) by factors of 1 to 8. The beneficial effects of molybdenum in reducing the oxidation rate of uranium in air and CO_2 were explained by sintering and by an increase in the plasticity of the product. This theory was supported by particle-size studies of the oxidation products.

Freas et al. [91] studied the meltdown behavior of uranium-10 wt. % molybdenum fuel pins irradiated to 0.2 to 0.9 total at. % burnup. It seems unfortunate that they did not extend their studies to include the determination of fission-product release from the irradiated fuel materials. They observed the behavior of fuel pins during 150-sec excursions to 1010°C (1850°F) and during excursions characterized by temperature rise rates of about 38 to 3000°C/sec (68 to 5400°F/sec) to temperatures above the melting point of the fuel alloy.

Fuel pins with total burnups of about 0.4 to 0.5 at. % were reported to have withstood three 150-sec excursions to 1010°C (1850°F) without rupturing, with a 2 to 3% increase in diameter. Fuel pins with total burnups in excess of 0.9 at. % ruptured and showed diameter increases in excess of 10% after one 150-sec excursion followed by a 1-min exposure at 1010°C (1850°F).

Fuel pins with burnups ranging from 0.2 to 0.9 at. % (total) failed by cladding ruptures during the rapid excursions at temperatures in the approximate range 1035 to 1090°C (1895 to 1994°F). The molten irradiated fuel alloy did not appear to flow readily and tended to form large globules on the side of the pin. This behavior was ascribed to the spongy nature of the molten material.

An incident involving fuel melting that has occurred since publication of Table 4-1 of Chapter 11 is the flow-blockage melt of two elements in the Enrico Fermi sodium-cooled fast reactor on October 5, 1966 [92-94]. The fuel was 10 wt. % molybdenum-uranium enriched to 25.6 wt. % U-235. Considerable plate-out occurred in the primary system and little plate-out occurred in the heat exchangers. The results are summarized also by Keilholtz and Battle [95].

Samples of the argon cover gas and sodium coolant were taken at various locations and at various times after the incident. Readings of gross beta activity as a function of time are shown in Fig. 3-12 [93], labeled coil 34 through coil 38. In addition, gross beta activity in various cover gas samples is shown. Activities of the principal isotopes present (Cs^{137}, Sr^{89}, I^{131} in sodium; Kr^{85}, Xe^{133} in argon) were also determined.

The following general comments on the fission products observed were made:

1) Noble-gas release appears to be larger than obtained from SRE meltdown analysis [96] and various experiments (50% versus 10% or less) [97].
2) Particulate release minus plate-out losses for the first sodium sample taken also appears to be larger than expected ($\approx$ 10%).
3) Considerable plate-out in the primary system has occurred (Fig. 3-12). Less than 1% of the original fission products remain in the sodium.
4) Contrary to expectations, measurements indicate very little plate-out in the heat exchangers. Cold trapping did not remove much of the activity.
5) Although solubility data for noble gases in sodium vary greatly [98, 99], conservative calculations show their retention in sodium under accident conditions is small [$<$ 20%).
6) Measurements show nonuniform distribution of the released fission-product activity in both the argon and the sodium systems (Fig. 3-12).
7) Sodium analysis appears less reliable than cover-gas analysis. For best accuracy, sodium samples should be taken as soon as possible after melting; gas analysis should not be attempted until short-lived activities have died out.

3.1.5 Uranium-Zirconium Alloys

Uranium-zirconium alloys have received more extensive use in reactors than have uranium-molybdenum alloys; therefore, more fission-product release information is available. Fission-product release from this type of fuel and the melting behavior of fuel element subassemblies consisting of nine 70-mil (1.78-mm)-thick Zircaloy-clad fuel plates 15 in. (38.1 cm) long and 2-1/2 in. (6.35 cm) wide, with a total surface area of 4.28 ft^2 (0.398 m^2) were studied by Rodgers and Kennedy [100]. This is one of the few reported investigations of fuel melting and fission-product release by use of full-size fuel elements. The center fuel plate only was removed and irradiated to tracer level (1.4 $\times 10^{16}$ nvt) in fission-product release studies. Melting was accomplished by induction heating. Approximately half the fuel assembly was surrounded by the work coil. An experiment involving melting of a uranium-zirconium subassembly in a steam atmosphere was completed in 45 sec and 654 g (1.44 lb) of melt was released from the subassembly (17% of the total subassembly weight), the maximum observed in the series of experiments. The fraction of the heated zirconium that reacted with steam to generate hydrogen varied from 0.033 to 0.197, with an average of 0.09.

Fission-product-release data obtained in four experiments are given in Table 3-12 [100]. The data show that practically all the xenon in the melted portion of the irradiated fuel plate was released.

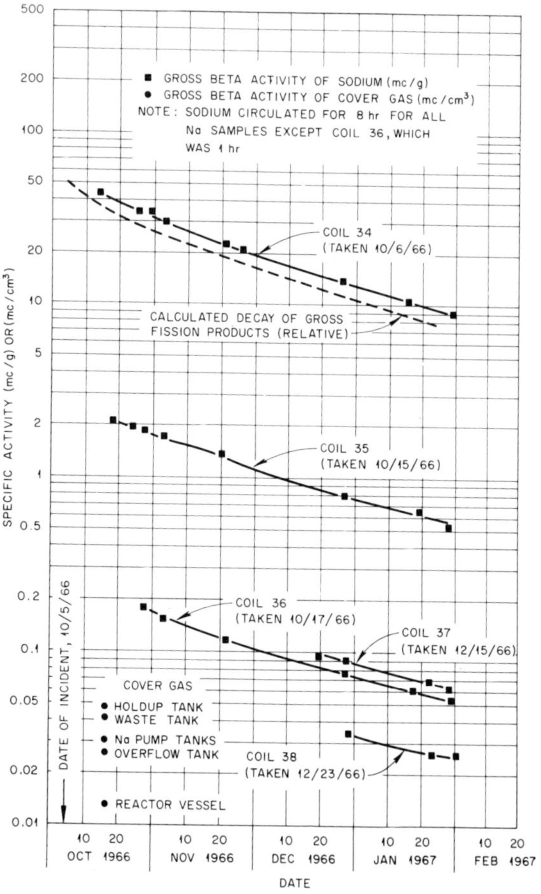

FIG. 3-12 Total sodium and cover-gas activity after incident at Fermi Reactor.

TABLE 3-12

Fission Product Release from PWR-Type Clad Fuel Assemblies[a] Melted in Flowing Steam[b]

Run No.	Total Release from Fuel (%)						
	Xe	I	Sr	Y	Ba	Ce	Cs
6	42	1.3					
7	47	7.6	0.7	0.6	0.5	0.6	
8	58	25.8	1.4		1.6	0.6	0.5
9	50	7.9	1.1		0.8	0.6	0.4

[a]Full-size, 9-plate fuel subassemblies were used in these tests, with the center plate irradiated to tracer level. Approximately half the assembly was surrounded by the work coil of the induction heater.

[b]Steam velocity, 0.06 ft/sec (1.8 cm/sec).

There seems to be little reason to doubt that iodine release values would have been much higher for high-burnup fuel material than the average release observed with this tracer-level-irradiated fuel.

Data obtained on melting fuel specimens irradiated to a significant burnup level (15 at. % U^{235}) in air and steam atmospheres are shown in Table 3-13 [24]. These data confirmed the indication shown in the previous table that the rare gases are quantitatively released, but the iodine and cesium results are much higher on the average than those obtained with low-burnup fuel. The release of less volatile elements such as strontium, barium, and cerium was unaffected by burnup, a fact that agrees with uranium-aluminum alloy experience described in Sec. 3.1.3 of this chapter.

It would be highly desirable, in view of continuing use of zirconium-uranium alloys in reactors, to supplement the early experiments on fission-product release and to examine the distribution of the volatilized fission products in some detail.

3.1.6 Uranium-Niobium Alloys

Niobium-rich uranium alloys have sufficiently high melting points to be attractive for use as high-temperature fuels. The phase diagram for the system and considerable information on physical properties of alloys containing 10 and 20 wt. % uranium are available [22] but fission-product release from these materials remains to be investigated. Limited information on the effect of various coolants on these alloys has been reported [22], but a more detailed study of their oxidation behavior, particularly in steam and steam-air mixtures at high temperatures, is needed in order to permit prediction of the results of exposure to typical loss-of-coolant accident conditions.

Release data obtained with uranium-zirconium alloys most nearly approximate the probable release from uranium-niobium alloys and can be used to predict the extent of fission-product release until data can be obtained.

3.1.7 Uranium-Thorium Alloys

No data on fission-product release or oxidation rate appear to be available for uranium-thorium alloys. There seems to be no established use for such alloys at present in reactor technology and, consequently, no strong incentive to investigate

TABLE 3-13

Fission-Product Volatilization from Melted Encapsulated Zircaloy Punched Disks, 15% Burnup

Run no.	Atmosphere during melting	Heating time (sec)[a]	Max. Temp.[b] (°C)	Total activity released (%)						
				Rare gases	Gross γ	I_2	Cs	Sr	Ba	Ce
10-10	Air	12.5	1705	100	2	28	10			
10-15	Air	16		100	2.2	32	11			
10-16[c]	Air	12	1750	100	6.4		12			
11-25	Air	30	1705	100	2.5	14	7.3	0.9	0.1	0.005
11-26	Air	30	1800	100	4.1	30	13	0.8	0.3	0.004
	Average	20	1740	100	3.4	26	10.6	0.85	0.2	0.004
10-20	Steam	33		100	5.3	13	8.9	2.4	0.3	0.05
10-22	Steam	32	1775	100	5.5	52	23	3.1	1.2	0.01
11-11	Steam	18		100	7.8	66	22	4.8		
11-12	Steam	31	1750	100	7.2	56	24	1.3		
11-14	Steam	35	1750	100	4.8	57	13	0.2		
11-21	Steam	30.5	1750	100	8.0	45	19	0.2	0.05	
11-24	Steam	31.5	1730	100	5.7	36	20	0.8	0.2	
	Average	30	1750	100	6.3	47	19	1.8	0.4	0.03

[a]Sample usually melted in approximately 12 sec.

[b]Optical pyrometer temperature. Note: 1705°C = 3101°F; 1730°C = 3146°F; 1740°C = 3164°F; 1750°C = 3182°F; 1775°C = 3227°F; 1800°C = 3272°F;

[c]Punched disk not reirradiated.

their fission-product release or oxidation behavior. Since only one stable oxide of thorium exists, ThO_2, which forms a continuous series of solid solutions with UO_2, formation of such solid solutions may make the oxidation behavior of alloys of these two metals more like that of thorium than of pure uranium, which readily oxidizes to U_3O_8 in air.

Data on the volatilization of fission products from molten and solid thorium-uranium alloys (tracer-level-irradiated) containing 3 wt. % uranium heated in a vacuum have been reported by Milne and Young [101]. These investigators found that to increase the surface-to-weight ratio (cm^2/g)* of the solid alloy from 0.8 to greater than 10 markedly increased the rate of volatilization of cesium, strontium, and tellurium but increased the volatilization of cerium and rare earths only slightly. Virtually all the cesium, strontium, and tellurium were volatilized after being heated for 120 min at 1650°C (3002°F), while only half the yttrium and rare earths and about one-fourth the cesium volatilized under these conditions. Volatilization rates from the molten alloy increased, as expected, with increasing temperature in the range 1700 to 2070°C (3092 to 3758°F), with increasing time in the range 30 to 120 min, and with increasing surface-to-weight ratio (cm^2/g) in the range 0.2 to 0.7 at 1910°C (3470°F). Rates of volatilization of fission-product xenon and iodine were not included in the studies but probably would have been too large to measure.

While the above data are interesting and serve as a check on the relative volatilities of the elements at elevated temperatures, they are not very helpful in a prediction of fission-product release from thorium-uranium alloys under accident conditions. Data obtained with uranium-zirconium alloys, Sec. 3.1.5, are probably the best guide available at present for the prediction of the extent of fission-product release from uranium-thorium alloys.

3.1.8 Uranium-Plutonium Alloys

The introductory remarks in the previous section appear to apply also to uranium-plutonium alloys. However, because of the increasing importance of plutonium as a fuel component, it is desirable to provide both information on oxidation behavior of these alloys and data on fission-product release.

The published [10a] phase diagrams of the uranium-plutonium system show that these metals form solid solutions with a minimum liquidus value of 610°C (1130°F) at about 10 at. % uranium. All alloys in this system will solidify at temperatures below the melting point of pure uranium. Since PuO_2 is the only stable oxide of plutonium, it seems likely that oxidation of these alloys may, under most conditions, produce UO_2 - PuO_2 solid solutions. Rates of oxidation and particle-size distributions in the oxidation products cannot be predicted with any degree of certainty, and experimental data on this subject may be required in the future.

Because the plutonium-rich alloy compositions in this system have low melting points, it seems likely that melting of irradiated fuels of this type could be accomplished without extensive release of fission products except the rare gases. It is more difficult to predict, however, the release of fission products that would occur on oxidation of the molten alloys.

*1 cm^2/g = 0.488 ft^2/lb.

3.1.9 Plutonium

The numerous phase transitions that pure plutonium undergoes on heating to temperatures below its melting point (640°C or 1184°F) and the large volume change accompanying some of the transitions have discouraged use of the pure metal in reactors. It is included in the table of fuels and fuel components because of its potential importance as a component of various types of fuel and because it is produced in considerable quantities in reactors with low-enrichment-uranium fuels.

Plutonium is known to be more volatile than uranium. Its boiling point is 3235°C (5855°F), compared to 3937°C (7100°F) for uranium (see Appendix VI of reference [7]). However, in view of the fact that uranium will have a higher concentration than plutonium in most fuel materials, it seems unlikely that the difference in volatility of the elements will be important in reactor accidents.* The vapor pressure of PuO_2 is so close to that of UO_2 (Fig. 2-2) that fractionation of the two oxides at high temperatures also seem unlikely. Effects of high-temperature disproportionation of UO_2 and PuO_2 on fractionation need experimental investigation.

High-temperature oxidation-rate data for plutonium are rather meager. Available results [10b] show that the oxidation proceeded linearly between 200°C (392°F) and 300°C (572°F), a fact that indicated the formation of a porous oxide coating. At 416°C (781°F) (delta phase) the oxidation proceeded slowly, as compared to the rate at 303°C (577°F), and the fact that the oxidation relationship was parabolic indicated that the oxide formed a protective coating at this temperature. At 487°C (909°F), the highest temperature at which oxidation of plutonium was studied in this investigation, the specimen oxidized according to the parabolic law initially, then the rate slowly increased and was followed after 4 hr by a period during which the metal oxidized quite rapidly.

A summary of published research on plutonium release during overheating or fires [102a] shows that the oxidation behavior of plutonium is quite complicated because of the various allotropic forms of the metal that exist and because of the effect of moisture. More studies [102b, c, 103] emphasize the effect of higher temperatures and of moisture on particle-size distribution.

Felt and Merritt [103] ignited completely four large (450 to1770 g) samples of plutonium in an ambient air stream with a velocity of about 525 cm/sec. Temperatures recorded by thermocouples embedded in the samples ranged from 500 to 1000°C during the 22-to 90-min ignition time. The particulate oxides entrained in the air stream during the burning were collected and examined. From

*The relative importance of plutonium will be greater in most fast-breeder-reactor designs.

2.9×10^{-4} to 4.9×10^{-2} wt. % of the plutonium was found in the air stream, with the result that release rates ranged from 3.2×10^{-2} to 4.5×10^{-3} wt. %/hr for the bare metal. The amount released by delta stabilized metal was lower than for the alpha-phase metal. The size of particles observed was small, the largest being 8μ in diameter, with a mass median diameter of 4.2μ.

Mishima [102b] and Stewart [104] report that larger particles, from 8 to 200μ were observed in their work. In the Felt and Merritt studies, as well as in those of Schnizlein et al. [73], temperature spikes were observed in the plutonium. Immersion of the burning metal in magnesium oxide (MgO) sand reduced the amount released by over an order of magnitude, to 2.9×10^{-4} wt. %/hr.

Ettinger, Moss, and Bailey [105] have studied aerosols produced during sodium and plutonium fires for various atmospheres from air to 100% nitrogen. Small quantities of sodium, a plutonium-cobalt-cerium alloy, and alpha- and delta-phase plutonium were burned separately. Samples of the oxides formed were collected and sized. The count median diameter (c. m. d.) and geometric standard deviation (σ_g) were determined graphically from log-probability plots of the data, and the mass median diameter (m. m. d.) was calculated by use of the Hatch and Choate equation [106]:

$$\log \text{m.m.d.} = \log \text{c.m.d.} + 6.9 \log^2 \sigma_g .$$

The aerosol from the plutonium-cobalt-cerium alloy (57% plutonium) showed a c. m. d. of 0.04 to 0.09μ; σ_g, 1.24 to 1.54; m. m. d., 0.05 to 0.14μ. The results of the aerosol from the metallic plutonium were nearly the same, giving a c. m. d., of 0.02 to 0.06μ; σ_g, 1.24 to 1.76; m. m. d., 0.03 to 0.13μ. The oxygen content of the atmosphere did not affect the particle size observed.

Fires involving both plutonium alloy and sodium produced airborne particles that gave a plutonium-sodium ratio ranging from 0.35 to less than 0.008%. The high ratio occurred initially and was rapidly reduced as the molten sodium blanketed the plutonium release.

3.1.10 Plutonium-Aluminum Alloys

It was mentioned in the previous section that the phase transitions of plutonium do not encourage use of the pure metal as a fuel material and that plutonium-aluminum alloys were the first plutonium-containing fuel materials to be employed in reacttors. One reactor incident that involved melting of an experimental plutonium-aluminum-alloy fuel element (NRX) has been reported (see Table 5-1). The NRX fuel slugs were reported [10c] to be 9 in. (22.9 cm) long and 1.36 in. (3.45 cm) in diameter. They contained 3.8 wt. % plutonium (0.43 at. %), a limit set by the allowable heat flux at the coolant-sheath interface. Plutonium-aluminum billets fabricated for use in MTR-type fuel elements [10d] contained about 10 wt. % plutonium (1.24 at. %). It appears, therefore, that alloys in this system with melting points slightly lower than that of pure aluminum are considered most promising at present.

No studies of the oxidation behavior of plutonium-aluminum alloys have come to the attention of the authors, but free-energy data and studies of methods of preparing plutonium-aluminum alloys [10e] show that the reaction

$$3PuO_2 + 2Al \rightarrow 3Pu + 2Al_2O_3$$

proceeds in the presence of air at 1000°C (1832°F) and higher temperatures. In fact, this reaction is routinely used at Chalk River for the preparation of these alloys, with the addition of cryolite as a fluxing agent, at a temperature of 1200°C (2192°F). The reduction reaction is also favored by the complexing of plutonium as $PuAl_4$ in the presence of excess aluminum; molten aluminum-rich plutonium alloys may be expected to be little, if any, more susceptible to oxidation than pure aluminum.

There appear to be no published data on fission-product release from irradiated plutonium-aluminum alloys, but the data on uranium-aluminum alloys, Sec. 3.1.3, should serve quite well for prediction of the release of fission products from these plutonium-containing alloys.

3.1.11 Thorium

Thorium is normally considered a fertile rather than a fissionable material. The fission-product distribution in Th^{232} produced by fission neutrons is shown in Fig. 1-3. The future role of metallic thorium in reactor technology is not clear at present, and it is included in Table 1-6 because Th^{232} is the fertile species in most breeder-reactor schemes. Milne and Young have studied the volatilization of fission products from molten and solid thorium-uranium alloys [101]. It seems probable, however, that thorium is more likely to be employed in the form of the oxide or carbide than as the metal; consequently, the oxidation and fission-product release behavior of metallic thorium will not be considered here.

3.1.12 Uranium-Zirconium Hydride (ZrH_2)

The high-temperature stability of ZrH_2 has prompted interest in use of uranium ZrH_2 mixtures in compact, light-weight reactors of the type needed for space applications. There appears to be little published information on fission-product release from fuels of this type or on their oxidation behavior.

Zirconium hydride is similar both in physical appearance and in other physical properties to the metal. Consequently, uranium-zirconium-alloy data (Sec. 3.1.5) may be used to predict the probable extent of fission-product release from these fuels.

3.2 Class II, Oxides Dispersed in Metallic Matrix

3.2.1 Aluminum-UO_2

Very little data seem to be available on the release of fission products from UO_2 dispersed in aluminum. Creek et al. [24] report results obtained with three samples of tracer-level-irradiated Geneva reactor fuel: An average of 5.6% of the rare gases and 0.003% of the iodine was released when the samples were melted in air. The melting point of aluminum (659°C or 1218°F) is low enough so that one would not expect diffusion of fission products from UO_2, even when the material is in the form of small particles, to be great enough to be significant in reactor accidents. (Investigations of the reaction of UO_2 particles with aluminum at temperatures above and below the melting point of aluminum are reviewed in Chapter 17.) The particles will be destroyed and volatile fission products will be released if reduction of UO_2 by aluminum occurs.

3.2.2 Aluminum-U_3O_8

No fission-product-release data on fuel consisting of U_3O_8 dispersed in an aluminum matrix seem to be available. This fuel has been proposed [107] for use in the HFIR at the Oak Ridge National Laboratory, and has been in use in the Puerto Rico (BONUS) Reactor [108]. The possibility of a thermite-type reaction in fuel of this type is discussed in Chapter 17.

3.2.3 Stainless Steel-UO_2

Studies of fission-product release from UO_2 dispersed in stainless steel (Army Power Package Reactor or SM-1) fuel coupons have been made with material irradiated at tracer level and with 20 to 30% burnup fuel (Tables 3-14 to 3-17 [24]. The data in Tables 3-16 and 3-17 show a definite correlation between fission-product release and preheat time (time required to heat the fuel specimen from room temperature to the melting point of the cladding and matrix material). The rare gas and iodine data display considerable scatter, but the correlation is somewhat better for gross-gamma and cesium data (see Fig. 3-13 [24]).

The higher melting point of the stainless steel resulted in a greater release of volatile fission products from the melted fuel, as compared to that from UO_2 dispersed in aluminum. Little difference was noted in release values in air and steam, but the fact that the release of cesium and strontium

TABLE 3-14

Fission-Product Volatilization from APPR Clad Coupons[a] Melted in Air or Steam

Run no.	Atmosphere	Preheat time to melt (sec)	Total activity released (%)				
			Rare gases	Gross γ	I	Cs	Sr
1	Air	125	45	5.4	49		
2	Air	55	59	2.2	25		
3	Air	42	38				
4	Air	94	48	4	15		
5	Air	137	40	1.5	34		
6	Air	90	54	8	31	11	0.001
7	Air	144	61	6	41	13	0.001
8	Air	75	44				
9	Air	151	50	4.5	38		0.0001
10	Air	75	46	4.4	35	3.4	0.1
		Average	48	4.5	34	9.1	0.03
11	Steam		39	2.3	11.2	0.3	0.4

[a]Coupons irradiated in the X-10 graphite reactor for one week (tracer level).

TABLE 3-15

Fission-Product Volatilization by Melting of APPR Punched Disks, 25% Burnup[a]

Run no.	Atmosphere during melting	Max. temp. (°C)	Time to melt (sec)	Total activity released (%)		
				Gross γ	Cs	Sr
1	Air	1575	13	0.4	12	-
2	Air	1650	15	0.2	5.8	-
3	Air	1575	47	17	68	
4	Steam		45	17	75	0.06
5	He	1650	37	26	99.9	6.8

[a]Decayed through long cooling period, not reirradiated.

Note: 1575°C = 2867°F; 1650°C = 3002°F.

TABLE 3-16

Variation in the Amounts of Iodine and Rare Gases Released from APPR Disks[a] with Different Preheat Times

Sample no.	Preheat time (sec)	Pyrometer reading at melting pt.	Total rare gases released (%)	Total I released (%)
1	17	1300°C (2372°F)	16	3.7
2	20	1565°C (2849°F)	31.3	5.7
3	31	1500°C (2732°F)	19.4	15.2
4	59	1618°C (2944°F)	44.6	15.3
5	61	1550°C (2822°F)	41.0	15.9
6	72	1521°C (2770°F)	50.7	17.6
Average	43.3		34.8	12.2

[a]20% burnup.

TABLE 3-17

Variation in Amount of Cesium Released from APPR Disks[a] with Different Preheat Times

Sample no.	Thickness of heater	Preheat time (sec)	Total Cs released (%)
1	0.062 in. (1.57mm)	6	15.7
2	0.125 in. (3.16mm)	11	18.7
3	0.25 in. (6.35mm)	20	34.8
4	0.31 in. (7.87mm)	34	64.0
5	0.31 in. (7.87mm)	42	72.2
Average		22.6	41.0

[a]30% burnup.

was higher in helium than in oxidizing gases, reflected the higher volatility of the element as compared to the oxide. At the maximum temperatures attained in these experiments, 1575 to 1650°C (2867 to 3002°F), Cs_2O would be largely dissociated; however, the fuel specimens were at this temperature for only a fraction of the total heating time.

3.2.4 Nichrome-UO_2

Fuels consisting of UO_2 dispersed in Nichrome V were considered for use in the direct-cycle-reactor system at the Aircraft Nuclear Propulsion Project at one time, and a few experiments were performed to determine the extent of fission products from fuel specimens of this type. The data given in Table 3-18 [24] were obtained with tracer-level-irradiated fuel specimens and the values are not significantly different from those obtained with tracer-irradiated stainless steel-UO_2 dispersions under comparable conditions. Values for release of iodine and cesium are lower than those obtained with high-burnup fuel materials. A more detailed examination of release from this type of fuel may be required if use of Nichrome-UO_2 dispersions in power reactors is contemplated.

3.3 Class III, Oxide Fuel Materials

3.3.1 Uranium Dioxide, UO_2

The advantages and disadvantages of UO_2 as a fuel material and its physical properties are discussed in Chapter 13. A more complete treatment of this subject is found in the book by Belle [11].

Fission products may be released from UO_2 fuels during an accident by diffusion, oxidation, and melting, listed in the approximate order of increasing

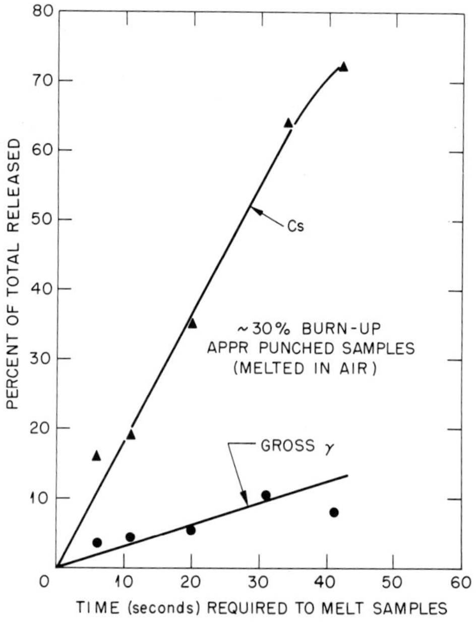

FIG. 3-13 Correlation of percent release of Cs from stainless steel-UO_2 dispersion fuel with total heating time.

TABLE 3-18

Release of Fission Products from Dispersions of UO_2 in Nichrome V When Heated in Air

Heating time	Total fission-product activity released (%)			
	Rare gases	I	Cs	Sr
30 sec	9.7	4.2	0.7	
32 sec	10.2	4.4	0.3	0.0001
4 hr	77	99.1	1.7	0.01

extent of fission-product release. Each release mechanism is affected by many parameters, and to obtain experimental data on all parameters under all conceivable accident conditions would obviously require a large effort. Work in this field has been directed toward determination of the relative importance of the parameters and toward evaluation of their effect on fission-product release under most probable accident conditions. A great deal of information in the literature on release under operating conditions is omitted in order to give more attention to data obtained under accident conditions. The various release mechanisms will be considered in the order indicated above.

3.3.1.1 Release from UO_2 by Diffusion. Release of fission products from UO_2 by diffusion has been studied by more investigators than has release by the other two mechanisms combined. In spite of this fact, the subject is still not thoroughly understood. Data obtained for present or projected reactor operating conditions have been reviewed by Lane et al. [109] and by Cottrell et al. [110]. Information on mechanisms of fission-gas diffusion has been reviewed by Carroll [111]. Only during the past few years have investigators extended their measurements in this field to temperatures above 1500°C (2732°F) and included release rates of fission products other than the rare gases. Another limitation on the usefulness of much published diffusion data for prediction of release of fission products in accidents is the fact that specimens employed were irradiated to very low levels of burnup not representative of the burnup levels expected in operating reactors.

Consideration of the kinetics of fission-gas release from UO_2 has been complicated by the "burst" effect. Davies and Long [112] summarized experimental observations of this effect as follows: "Whenever the temperature of the sample is raised rapidly, either in-pile or out-of-pile, both the initial rate of evolution and the total amount evolved are higher than anticipated by the diffusion laws, and this phenomenon persists even after several anneals at successive temperatures."

Data reported by Stevens and coworkers [113] indicate that an oxidized surface layer is a major cause of the burst effect with lightly irradiated UO_2. Davies and Long [112] have rejected this explanation because they consider it unlikely that this mechanism can account for the subsequent bursts when the temperature of the sample in an atmosphere of hydrogen is raised and for the very low apparent activation energy observed at low temperatures, after high-temperature annealing experiments. They offer a tentative explanation involving emission from pores and other defects in the UO_2 lattice; they suggest that the fission gases enter these imperfections and the free space within the irradiation capsule by a knock-out process (knocked out by fission fragments). Their statement (reference [112], p. 8) that only a small fraction of the total inert-gas content of the UO_2 is evolved by the burst mechanism is not supported by much of the published data and seems to indicate that there are different mechanisms responsible for the burst effect. The more important of these, from the accident-release standpoint, is probably the surface-oxidation effect, because of the comparatively large

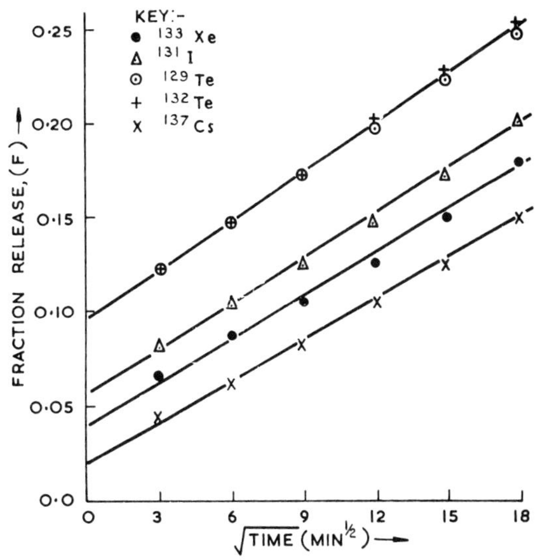

FIG. 3-14 Plots of fraction released versus square root of time for low-density UO_2 at 1600°C (2910°F). UO_2 density = 8.15 g/cm³.

quantity of gas likely to be released by ordinary reactor-grade UO_2.

A plot of fraction of fission products released against square root of time, given by Davies et al. for low-density UO_2 (8.15 g/cm³) heated at 1600°C (2912°F) in purified hydrogen, is reproduced in Fig. 3-14 [114]. One fact that this plot illustrates is that xenon, which one intuitively expects to migrate most rapidly, actually is one of the slower moving elements. Under other conditions examined by the same investigators, high-density UO_2 (10.16 g/cm³) at 1300°C (2372°F), xenon was by far the slowest moving element, and cesium was the fastest.

Data have been reported [114] on the effect of time and temperature on fission-product release from high density UO_2 at 2000 to 2200°C (3632 to 3992°F). At these temperatures, grain growth is rather rapid. Under the most drastic conditions employed, 5-hr heating at 2200°C (3992°F), almost half the small (50 mg), lightly irradiated specimen volatilized, along with almost all the tellurium and xenon and fractions of other isotopes.

Similar studies [20c, 115a] made by heating tracer-irradiated pressurized-water-reactor (PWR)-type UO_2 pellets in a flowing stream of purified helium for 5.5 hr gave the data in Table 3-19 [20c]. The smaller fraction of the sample volatilized in these experiments, compared to the British results obtained at comparable times and temperatures, is due possibly to the difference in sample size (0.05 versus 7 g). It is clear from these data that escape rates are high enough at temperatures of 1700°C (3092°F) and above to permit release of significant quantities of fission products over a period of hours. At 2100°C (3812°F) and above, even such low-volatility elements as barium and zirconium volatilize to a significant extent in 5.5 hr.

A plot of diffusion constant versus the reciprocal of the absolute temperature for the diffusion of rare gas (xenon) from tracer-irradiated PWR-type UO_2

TABLE 3-19

Diffusion of Fission Products from UO_2[a] into Purified Helium[b]

Temp.[c] (°C)	Release (%)									
	Rare gases	Gross γ	I	Te	Cs	Ru	Sr	Ba	Zr	U
1515	1.3	0.9	5.8	2.9	1.4	0.9	0.1			
1610	2.7	2.1	6.5	12	1.7	1.5	0.1			
1710	2.6	6.3	9.6	20	2.7	3.8	0.4	1.3		
1800	3.7	5.2	12	21	3.2	6.9	1.0			
1900	9.7	12	16	48	8.6	8.5	2.3			
1980	12	12	42	76	15	13	4.2	8.7	1.8	
2105	25	23	40	81	24	22	13	21	0.5	0.5
2150	59	18	74	95	53	49	28	40	12	0.5
2200	65	33	75	96	70	50	36	59	18	
2260	87	38	84	96	65	90	55	75	35	1.3

[a]Tracer-level-irradiated PWR UO_2 samples heated 5.5 hrs in tantalum crucibles by RF induction.
[b]Helium purified by contact with hot zirconium sponge; flow rate 50 cm^3/min (0.00177 ft^3/min).
[c]1515°C = 2759°F; 1610°C = 2930°F; 1710°C = 3110°F; 1800°C = 3272°F; 1900°C = 3452°F; 1980°C = 3596°F; 2105°C = 3821°F; 2150°C = 3902°F; 2200°C = 3992°F; 2260°C = 4100°F.

is shown in Fig. 3-15 [115a]. The inflection point at about 1800°C (3272°F) in this plot is believed to be due to grain growth and, since this phenomenon is a function of both time and temperature, the slope probably does not change sharply.

The effect of burnup on release of fission products by diffusion at four temperatures is shown by the data in Table 3-20 [21c]. Data from Table 3-19 are included for comparison, although the samples were not identical even for PWR UO_2. None of the high-burnup UO_2 materials received to date have included intact pellets such as those employed to obtain the data in Table 3-19. Fragments of variable size were, of necessity, employed to study the burnup effect. These fragments ranged in weight roughly from 0.1 to 0.2 g and sample weights varied from about 1 to 2 g. There is no clear-cut burnup effect evident at lower temperatures (1400 to 1610°C) (2552 to 2930°F) or at low burnups (1000 Mwd/ton, but otherwise it is quite evident that diffusion of fission products increases with increasing burnup except, possibly, for the low-volatility elements.

Data on diffusion rates at temperatures between 2260°C (4100°F) and the melting point of UO_2 (2860°C or 5180°F) are very scarce. Release data obtained in a transient reactor experiment (see Table 3-28), where the specimen reached a temperature above 2600°C (4712°F) for a very short time (seconds) but did not melt, indicate that release rates of the more volatile elements are very high in this temperature range.

One type of release from UO_2 that has received relatively little attention is the escape of fission gases resulting from rapid cooling. Rothwell has reported [116] that tracer-irradiated, fully enriched UO_2 specimens (10 to 20 mg) heated to 2000°C (3632°F) in flowing purified helium exhibited a small burst on initial heating, a constant release rate at 2000°C, a rapid fall in count rate with decreasing temperature that reached background at 1600°C (2912)°F), followed by a large burst of activity that began at about 1450°C (2642°F). The quantity of gas released on cooling is said to have exceeded that evolved on heating at 2000°C (3632°F) for 30 min, amounting to about 1/3 of the total Kr^{85} content of the specimen. The release rate on cooling

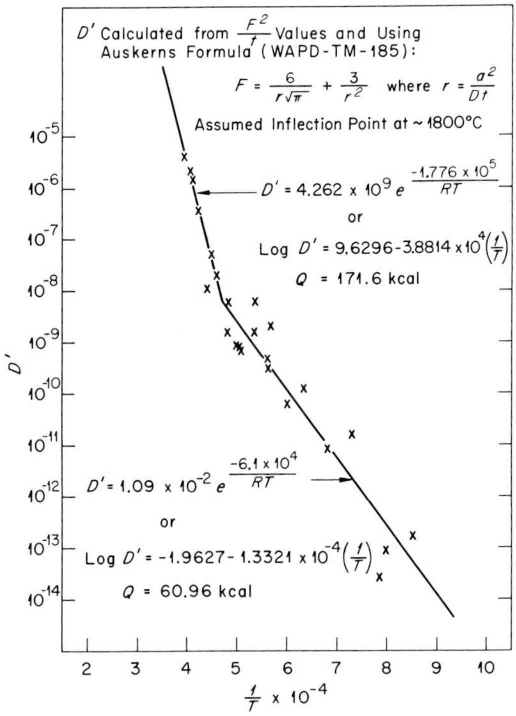

FIG. 3-15 Rare-gas diffusion from PWR UO_2 in helium.

TABLE 3-20

Effect of Burnup Level and Temperature on Diffusion of Fission Products from UO_2 Heated 5.5 hr in Pure Helium

Temp.	Irradiation level (Mwd/ton)	Individual fission products released[a] (%)						
		Xe-Kr	I	Te	Cs	Ru	Sr	Ba
1400°C (2552°F)	~ 1[b]	0.8	4.0	3.9	0.02	0.02	0.001	
	1005[c]	0.8	0.9	0.8	2.6	0.001	0.1	
	1000[b]	0.5	1.6	1.2	0.5	0.001	0.06	1.8
	4000[b]	6.1	23	16	21	0.006	0.08	0.5
1610°C (2930°F)	~ 1[b]	2.7	6.5	12	1.7	1.5	0.1	
	1005[c]	2.6	3.7	12	12	0.1	2.0	17
	1000[b]	6.0	5.5	27	20	0.3	0.2	12
	4000[b]	14	25	48	43	0.2	0.5	15
1780°C (3236°F)	~ 1[b]	3.7	12	21	3.2	6.9	1.0	
	1005[c]	12	24	67	27	0.4	9.0	39
	1000[b]	14	26	35	22	0.4	3.7	21
	4000[b]	42	59	60	40	5.7	5.8	18
1980°C (3596°F)	~ 1[b]	12	41	75	15	13	4.2	8.7
	1005[c]	29	53	74	84	6.0	15	57
	1000[b]	49	63	90	70	4.8	~ 10	51
	4000[b]	71	81	81	98	15	33	60

[a]Includes that portion adsorbed on crucible and reflector parts.

[b]PWR-type UO_2 (93 to 94% of theoretical density). Only the 1 Mwd/ton pellets were full size (7 g). The high-burnup samples were 0.1 to 0.2 g fragments with a total weight of 1 to 2 g.

[c]EGCR-type UO_2 (97% of theoretical density). Samples were similar to PWR samples in total weight and fragment size.

was found to depend strongly on the cooling rate. Rothwell attributes the large cooling release to a phase change in which a substoichiometric oxide that is stable at high temperatures reverts to UO_2 and uranium metal at temperatures of 1450°C (2642°F) and below.

Parker and coworkers have compared the gas release in tracer-irradiated and in highly irradiated UO_2 specimens heated to 1600°C (2912°F) and then cooled. The results are shown in Fig. 3-16 [115a].

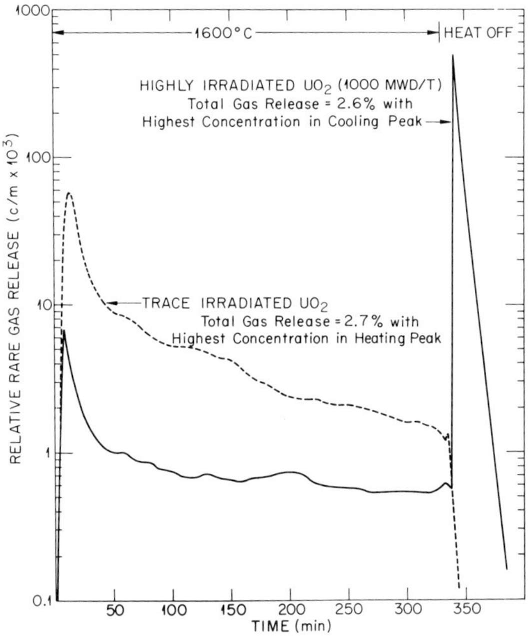

FIG. 3-16 Comparison of rate of release of fission gas from trace irradiated and highly irradiated UO_2 on heating to 1600°C (2910°F) and cooling in helium.

The fact that this temperature is probably too low for formation of appreciable amounts of substoichiometric oxide may account for failure to observe significant release on cooling of the tracer-irradiated material, but the reason for the large release on cooling of the highly irradiated specimen is not clear. A similar release was noted [18] on rapid cooling of irradiated uranium, previously heated above its melting point, from about 520°C (968°F) to room temperature. It has not been established whether other fission products show a "cooling-burst" effect, but the possible contribution of this effect to the overall hazard of loss-of-coolant accidents needs consideration if reactor accident conditions permit rapid cooling of over-heated fuel.

3.3.1.2 Release from UO_2 by Oxidation

Oxidation Rates. Several investigations of the oxidation of UO_2 have been reported [70, 117-120]. Most of the studies at low temperatures were performed with powder materials, while those at higher temperatures employed sintered bodies of varying density. Aronson et al. [119] showed that the low-temperature oxidation (160 to 350°C or 320 to 662°F) of UO_2 proceeded in two steps: the first, the oxidation to tetragonal $UO_{2.34 \pm 0.03}$, followed by conversion to orthorhombic U_3O_8. The temperature range was extended by Antill et al. [117, 120] from 350 to 1000°C (662 to 1832°F). Their data, obtained with cylindrical pellets with a density of about 10.5 g/cm³, showed that at 350 to 600°C (662 to 1112°F), a slow initial rate was followed by rapid oxidation when the U_3O_8 fell away from the specimen as a fine powder. At 650 to 800°C (1202 to 1472°F), the oxide tended to become protective but broke down to a higher rate after a period that increased as the temperature was raised. Above 900°C (1652°F) no "breakaway" was observed, because

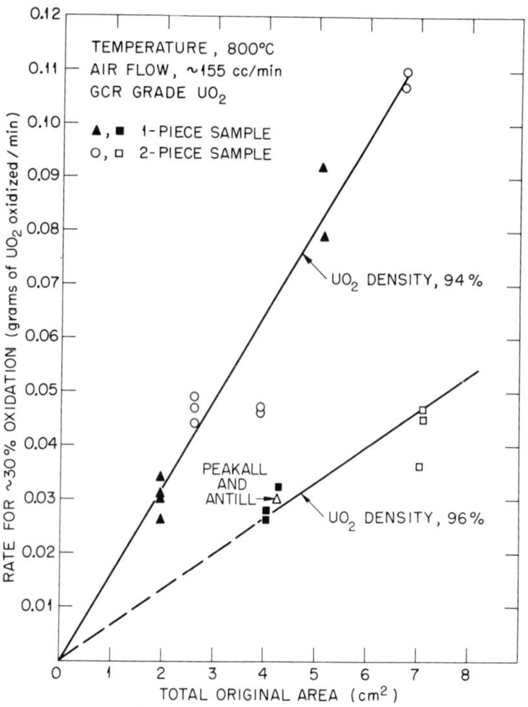

FIG. 3-17 Effect of geometric surface area and density on oxidation rate of UO_2.

the U_3O_8 formed an adherent shell around the specimen. Thermal cycling was shown to break up the protective oxide coating formed at higher temperatures. The maximum linear oxidation rate in this study [120] (51 to 52 mg/cm^2-hr) was observed at 450 and 500°C (842 and 932°F).

A similar study by Livey (p. 81 of reference [70]) made with hot-pressed $UO_{2.13}$ with a density of about 10 g/cm^3, showed a peak oxidation rate of 160 mg/cm^2-hr at 700°C (1292°F). American work [19] on the oxidation of PWR-type UO_2 (93% of theoretical density) in air at temperatures in the range 400 to 1400°C (752 to 2552°F) showed a maximum in the initial oxidation rate of UO_2 at 500°C (932°F) in agreement with the data of Antill et al. [120] but also showed a minimum at about 700 to 800°C (1292 to 1472°F) which does not agree with Livey's results [70].

It was also shown [20c] that the protective effect of an adherent coating of U_3O_8 diminishes at temperatures of 1000°C (1832°F) and above, possibly because of an increased rate of diffusion of oxygen through the oxide. Specimens heated to 1580°C (2876°F) and cooled in helium showed only surface cracking, while U_3O_8, formed by oxidation of uranium at 1400°C (2552°F), lost weight rapidly when heated in air at 1500 to 1580°C (2822 to 2876°F), approaching the UO_2 composition in about 1.5 hr. These observations show that UO_2 will not oxidize in air at temperatures of 1550°C (2822°F) or higher.

A comparison of initial oxidation rates of specimens differing only slightly in density is shown in Fig. 3-17 [115a]. Surface areas plotted are geometrical areas. Agreement with the data of Antill et al. [120], as mentioned above, seems quite good.

It is clear from this figure that oxidation rates are strongly dependent on density and on surface area, a fact that probably explains some of the confusion that has arisen with regard to oxidation rates.

Some studies of the oxidation of UO_2 in CO_2 containing 10 to 600 ppm oxygen at 500 to 900°C (932 to 1652°F) have been reported [70]. Oxidation rates were said to be slow, with weight gains of 4 to 20 mg for 7-g specimens in 50 to 100 hr.

Oxidation of UO_2 by steam apparently does not occur at significant rates. In fact, steam is known to promote sintering of powdered UO_2 at moderately high temperatures (800 to 1000°C or 1472 to 1832°F). Data on rates of oxidation of UO_2 in steam-air mixtures seem to be lacking at present.

The complexity of the oxidation of UO_2 in air makes it desirable to have oxidation rates with full-size, highly irradiated fuel elements and to obtain fission-product-release data concurrently with the oxidation studies. Such investigations can obviously be performed only in well-designed and well-shielded hot-cell facilities.

Fission-Product Release. Fewer studies of fission-product release accompanying the oxidation of UO_2 have been made than of release by diffusion. Parker and coworkers [115a] first studied release from tracer-irradiated PWR-type UO_2 pellets of theoretical density and later [20c] investigated release from the same type of fuel when irradiated to different levels of burnup, up to a maximum of 7000 Mwd/ton. Release values determined with specimens irradiated at tracer level (~1 Mwd/ton) are plotted in Fig. 3-18 [20c], while data obtained at the highest burnup level are given in Table 3-21. It is clear from these data that fission-product release is not the simple function of temperature that might be expected from the oxidation behavior discussed in the previous section.

The effect when the heating time is varied at different temperatures is shown in Table 3-22 [20c]. Specimens employed were PWR-type material with a density of 93 to 94% of theoretical, irradiated to a burnup of 4000 Mwd/ton. In the range investigated, increasing exposure time in air seemed to have no significant effect on fission-product release below 800°C (1472°F), but at this temperature and above, increasing release of some isotopes with increasing exposure was observed, as might be expected.

The effect of burnup on oxidation release of the more volatile fission products is shown graphically in Fig. 3-19 [20c] for two temperatures. The largest effect in the case of iodine and ruthenium came in the first 1000 Mwd/ton of burnup, as was true of the rare gases at 1200°C (2192°F). The release of tellurium appeared to increase more or less regularly with increasing burnup in the range tested. The release of cesium, even at 1200°C, was too low to establish an unequivocal correlation, although the results obtained did indicate a slight increase in release with increasing burnup.

The outstanding feature of these results is the high release of iodine and ruthenium. Especially favorable conditions for the release of the latter indicate that this element is readily converted to a volatile oxide.

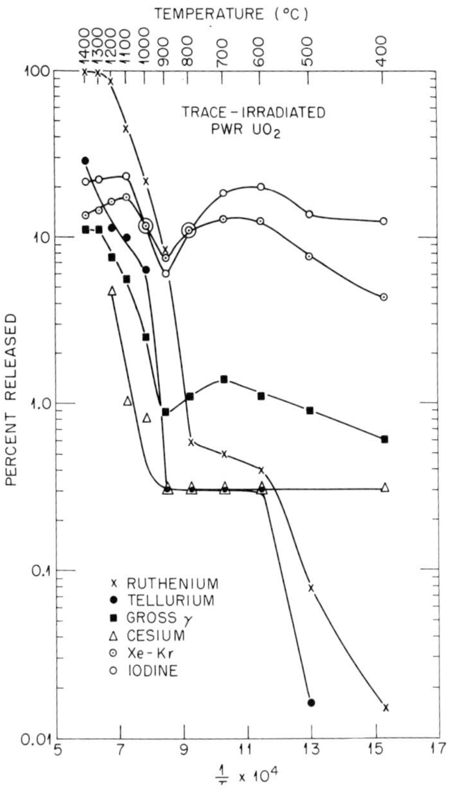

FIG. 3-18 Fission-product release by the oxidation of UO_2 to U_3O_8 in air, showing discontinuity between 600°C (1112°F) and 900°C (1652°F).

3.3.1.3 Release from UO_2 by Melting. Melting of irradiated UO_2 is the most drastic mechanism for release of fission products, but, because of the very high melting point of UO_2, it is also the least likely to occur. The principal parameters investigated have been burnup, sample size, time molten, atmosphere, melting method, and type of cladding. Most of the melting experiments have been performed in helium, because of reactivity of container or heater materials with oxygen, but it seems probable that the atmosphere surrounding molten UO_2 will have little effect on the extent of release. A few experiments have been performed with CO_2, air, and steam-air mixtures. There can be little doubt, however, that the atmosphere will have a drastic effect on postrelease behavior of the more reactive fission-product elements. The first experiments on measurement of fission-product release from molten UO_2 were performed at the Oak Ridge National Laboratory; a variety of methods were used in these investigations.

Results of the first melting experiments with small tracer-irradiated UO_2 specimens melted in an arc-image furnace with its beryllium oxide (BeO) support tube are given in Table 3-23. The only parameters varied were the sample weight and melting time. The data show that, with these small specimens, release of all fission products sought, except for strontium, barium, and rare earths, was very high when the specimens were completely melted. Results of other experiments with different atmospheres and three levels of burnup are shown in Table 3-24 [121]. There appears to be a definite increase in release between tracer-level irradiation and 2800 Mwd/ton but there were no significant differences between the results at 2800 and 11,000 Mwd/ton. There is no evidence in these data that the atmosphere surrounding the molten specimens affected the extent of release.

Data obtained with larger samples of tracer-irradiated UO_2 heated in tungsten crucibles are given in Table 3-25 [122]. The tungsten crucible served as both container and susceptor for radio-frequency induction heating. The time that the UO_2 remained molten varied to some extent in these experiments, and there seems to be a positive correlation between release and time molten for the more volatile elements. In general, comparable release data were obtained by the two melting techniques except for ruthenium. This element is known to be quite oxygen-sensitive, and it is possible that its higher release on melting in the arc-image furnace can be attributed to traces of oxygen in the helium supply and to the absence of the good

TABLE 3-21

Fission-Product Release from PWR-Type UO_2[a] Irradiated to 7000 Mwd/ton and Heated in Air[b] for 90 min

Temp.[c] (°C)	Individual fission products released (%)						
	Rare gases	I	Te	Cs	Ru	Sr	Ba
500	3.1	4.1	<0.5	0.0006	0.1	<0.0007	<0.0004
600	4.2	3.1	<0.1	<0.002	0.7	<0.0004	<0.009
700	6.1	15	<0.08	<0.005	0.1	<0.0005	<0.007
800	9.4	9.0	<0.3	0.002	9.8	<0.0005	0.03
850	15	34	1.4	0.02	35	<0.005	<0.08
900	34	29	80	<0.01	78	<0.03	<0.8
1000	86	78	37	<0.03	93	<0.04	<0.3

[a]Samples 0.5 to 0.9 g of 96% density material in porous alundum cups, preheated for 13 to 16 min in He.
[b]Air flow, 100 cm^3/min (0.00353 ft^3/min).
[c]500°C = 932°F; 600°C = 1112°F; 700°C = 1292°F; 800°C = 1472°F; 850°C = 1562°F; 900°C = 1652°F; 1000°C = 1832°F.

TABLE 3-22

Fission-Product Release from PWR-Type UO_2[a] Irradiated to 4000 Mwd/ton and Heated in Air[b]

Temperature	Time at Temp. (min)		Individual fission products released (%)							
	He	Air	Rare gases	I	Te	Cs	Ru	Sr	Ba	U
500°C	16	23	1.5	3.6	<0.007	<0.0004	<0.005	<0.0004		
(932°F)	18	90	2.9	3.2	<0.01	<0.0007	<0.01	<0.0004	<0.0008	
600°C	14	18	4.4	10	<0.006	0.002	0.08	<0.001		
(1112°F)	15	90	4.5	8.0	8.4	<0.001	1.8	<0.001	<0.004	
700°C	14	12	9.3	9.6	0.01	0.001	1.7	<0.0002	<0.0004	
(1292°F)	13.5	15	7.0	10.	0.004	<0.001	0.4	<0.0003	<0.0006	
	14	90	6.8	6.5	<0.05	<0.0005	2.3	<0.0004	<0.002	
800°C	13	15	14	7.1	0.007	0.015	1.0	<0.0004	<0.0007	
(1472°F)	14	90	14	16	<0.06	<0.01	12	<0.0004	<0.001	
900°C	14	19	21	49	0.4	0.009	17	<0.0001	0.01	
(1652°F)	15	90	22	47	6.0	0.015	53	<0.0008	<0.004	
1000°C	16	15	40	84	12	0.09	72	<0.0003	<0.02	
(1832°F)	13.5	90	44	75	32	0.37	92	0.1	0.08	0.06
1100°C	14	14	66	79	16	<0.02	91	<0.05	<0.003	
(2012°F)	14	90	73	84	39	0.2	99	0.006	0.01	<0.003
1200°C	14	16.5	71	82	37	0.8	99	<0.01	<0.001	
(2092°F)	13	90	80	95	66	6.4	99.6	0.007	0.7	<0.003

[a]Sample approximately 1 g of intermediate-density (93 to 94%) material in porous alundum cups

[b]Air flow, 100 cm^3/min (0.00353 ft^3/min).

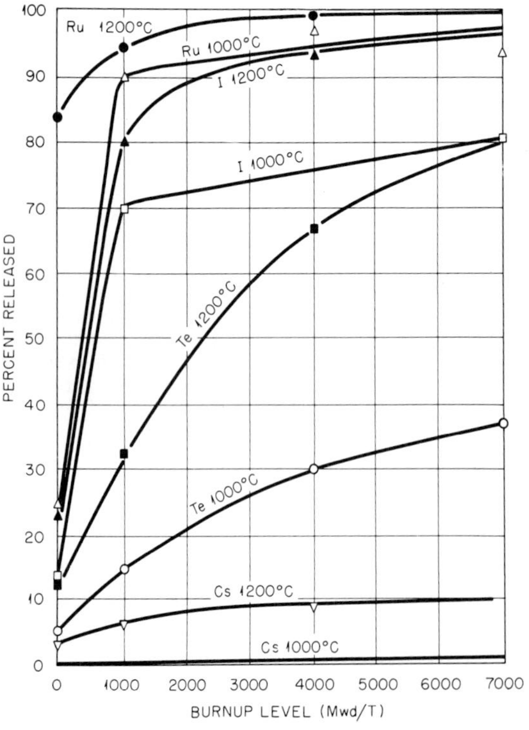

FIG. 3-19 Effect of burnup on fission-product release by oxidation at 1000°C and 1200°C.

oxygen getter (tungsten) that was available in the induction-heating experiments with larger samples.

Results of a third type of melting experiment, in which a tungsten rod resistor passed through cored UO_2 pellets served as the heating element, are shown in Table 3-26 [123]. Both clad and unclad elements were employed, but the experiments were limited to a helium atmosphere. Complete melting of the specimens could not be accomplished before the tungsten rods melted. Nevertheless, the results are quite useful, because the high interior fuel temperature and cooler surface achieved with this heated method more nearly simulate nuclear heating than any other out-of-pile technique. Also, this method can be used with a number of fuel rods arranged to simulate a reactor core, as described in another section of this chapter. The results in Table 3-26 obtained with unclad specimens, after adjustment for the fraction of the fuel melted, are not drastically different from the results of the other melting methods except that release of strontium and barium is rather high; in addition, observation of low ruthenium release indicated absence of free oxygen. The release from the stainless-steel-clad specimen was similar to those from unclad fuel, but the zirconium-clad specimens gave quite different results. The data in Table 3-26, coupled with postmelting examinations, indicate that the molten zirconium wet the UO_2 and spread over the surface. It thus served as an effective oxygen-getter, a fact that accounts for the high strontium and barium escape. Cesium, iodine, and rare gases were apparently unaffected by the cladding.

A number of experiments have been performed [58a, 124, 125] in the Oak Ridge Research Reactor (ORR) in which miniature stainless-steel- or zircaloy-clad UO_2 fuel specimens, irradiated to tracer level, were heated in helium and various other atmospheres by fission heat and then fission-product release was measured. The data shown in Table 3-27 are values for the release from the fuel and from the furnace zone, which had a minimum temperature of 1000°C (1832°F). Part of the

TABLE 3-23

Fission-Product Release from UO_2[a] Melted in Helium[b]

Run no.	Sample weight (g)	Time at high temperature (sec)	Release (%)								
			Rare gases[c]	Gross γ	I	Te	Cs	Ru	Sr	Ba	TRE[d]
1	0.57	120	64	9.4	71	60	59	28	0.18		
2	0.34	120	91	13	70	72	25	60	0.07	0.8	0.2
3	0.56	120	93	6.0	84	86	34	32	0.16	0.9	1.1
4	0.56	180	56	7.7	67	63	24	75	0.11	1.3	0.7
5	0.58	180	63	14	46	54	12	36	0.11	2.6	0.5
6	0.37	120	69	10	51	43	7.1	20	0.26	0.5	0.3
7	0.18	120	99.4	30	84	86	90	72	0.20	2.0	0.7
8	0.25	90	99.6	31	95	96	93	76	3.9	7.3	3.8

[a]Trace-irradiated pellet melted simultaneously with BeO support tube in arc-image furnace.
[b]Helium flow rate, 100 cm^3/min (0.00353 cu ft/min).
[c]Rare-gas release less than 95%, indicates incomplete melting.
[d]Total rare earths.

TABLE 3-24

Effect of Irradiation and Atmosphere on Fission-Product Release Resulting from the Melting of UO_2[a]

Atmosphere	Irradiation level (Mwd/ton)	Wt. of sample (g)	Individual fission products released (%)								UO_2 vaporized (%)
			Xe-Kr	I	Te	Cs	Ru	Sr	Ba	Rare earths	
He (impure)	Tracer	0.22[b]	99.5	90	92	91	61	2.1	4.5	2.2	
	2800	0.03	99.9	92	98	99	60	2.1	6.6	5.1	21
Air	Tracer	0.2[b]	98	95	79	38	68	0.2	0.5	0.5	
	2800	0.04	100	99.7	94	93	95	0.4	1.8	3.0	
CO_2	Tracer	0.2	81	77	71	61	45	0.3	1.1	0.9	14
	2800	0.02	99.9	99	99	90	74	0.5	2.5	2.8	
	11000	0.05	99.9	99.9	99	97	79	0.6	2.9	2.3	

[a]EGCR UO_2, with O/U ratio of 2.04 and density 95% of theoretical (average), melted in arc-image furnace.
[b]Average of two results; all others are averages of three results.

TABLE 3-25

Fission-Product Release from UO_2[a] Melted in Helium[b] by the Tungsten-Crucible Method

Molten time (min)	UO2 vaporized (%)	Gross γ release (%)	Release %							
			Rare gases	I	Te	Cs	Ru	Sr	Ba	Ce
1.0	0.10	23	93	77	90	63	0.45	0.33	4.8	0.05
1.5	0.16	15[c]	98	98	98	66	0.05	0.47	2.6	0.07
2.0	0.16	26	99	99	99	60	0.32	0.41	3.0	0.17
2.5	0.25	13[c]	99	95	99	72	0.33	0.53	2.4	0.13
1.5[d]	-	14[c]	99	88	92	80	0.20	0.26	2.6	0.40
2.5[d]	-	13[c]	99	93	96	89	0.70	0.50	3.6	1.10

[a]Sample: 29 g PWR UO_2 irradiated at tracer level and preheated in helium for 4.5 to 5.0 min.
[b]Atmosphere: purified helium flowing at a rate of 700 cm/min (0.0247 ft^3/min).
[c]Decayed 4 to 7 days longer than previous sample.
[d]UO_2 sample had a slightly higher density than the first four samples.

TABLE 3-26

Fission-Product Release[a] from Tracer-Irradiated PWR-Type UO_2 Melted in a Single-Element Tungsten-Resistor Furnace Filled with Helium[b]

Element	Heat duration (min)	UO_2 vaporized (%)	Gross γ release (%)	Individual fission products released (%)							
				Xe-Kr	I	Te	Cs	Ru	Sr	Ba	Ce/RE
UO_2	5.0	0.8	7.1	63	47	56	44	1.6	1.6	5.3	<0.6
UO_2	4.0	0.2	5.7	50	30	42	41	0.4	0.8	2.9	<0.5
UO_2	4.4	0.3	6.9	34	25	33	>40	0.05	1.2	4.3	0.5
UO_2 (SS clad)	4.7	0.2	5.0	56	52	31	46	0.5	1.0	4.2	0.3
UO_2 (Zr clad)	7.0	0.1	2.6	52	24	1.1	28	0.1	10.1	10.6	0.5
UO_2 (Zr clad)	6.7	0.04	5.2	41	50	0.6	32	0.2	10.0	7.5	0.5

[a]Results are not corrected for the fraction of the sample melted that is approximately equal to the percent rare-gas release. Release is from fuel and cladding.

[b]He flow rate, 400 cm^3/min (0.0141 ft^3/min).

UO_2 release from the fuel is believed to have been due to flowing or spattering of the molten UO_2 rather than to vaporization. Although the first experiments, giving the data shown in Table 3-27, were planned to melt the specimens, uncontrollable variations in the neutron flux and other factors prevented complete melting of the fuel in several experiments. The data show that almost quantitative release of iodine, tellurium, and cesium occured even without complete melting of the specimens. The release of strontium, barium, and ruthenium was variable but rather high, except when none of the specimens melted. There was approximately a one-to-one correspondence between the fraction of zirconium and cerium released and the fraction of UO_2 released, indicating that these refractory fission products probably accompanied the UO_2. The 1968 data showed that almost all of the barium, zirconium, cerium, and uranium remained in the furnace; less than 1% of these fission products and only a trace of uranium reached the aging chamber. However, about one-half of the iodine inventory, one-third of the cesium, and 5% of the tellurium and ruthenium reached the aging chamber [125].

Two experiments on the release of fission products from UO_2 heated under transient reactor conditions in the TREAT reactor have reported. The data are included in Table 3-28 [58c], along with information on the reactor transient conditions employed. The fuel was not irradiated before exposure to neutrons during the reactor transient; consequently, these results are representative only of release behavior in a SPERT-type core or in a reactor core that goes to a high temperature shortly after becoming critical for the first time. The results are interesting, however, because it was possible to relate release, as well as transport behavior, to the chemical species present while the fuel was hot. Thus, Sr^{89} and Ba^{140}, which would normally be among the less volatile fission pro-

TABLE 3-27

Fission Products Released by In-Pile Heating of Tracer-Irradiated Stainless-Steel-Clad UO_2 [a] in Helium[b]

Experiment no.	UO_2 vaporized (%)	UO_2 melted (%)	Release (%)							
			I^{131}	Te^{132}	Cs^{137}	Ru^{106}	Sr^{89}	Ba^{140}	Zr^{95}	Ce^{144}
7	0.72	0	96		96	11	5.2	14	0.9	2.5
11	5.1	70-80	98	99.4	96	48	18	26	12	7.6
5	34	100	99.4	99.2	98	91	57	55	43	44
8	36	80-90	99.0	98	98	57	44	70	58	43
3	43	100				48	93	96	87	85
4	44	100	99.6			81	77	94	56	59
6	47	80-90	99.0	95	99.3	59	55	52	60	49
2[c]	50	100				83	99.8	99.8	87	93
10	56	100	97	96	96		69	72	58	59
9	70	100	99.7	99.1	99	93	71	71	72	70

[a]Sample length, 1 in. (25.4mm) diameter, 0.210 in. (5.33mm); sample in maximum flux zone (presumably molten) 5 min except for experiment 2, 10 min.

[b]Helium velocity (at fuel melting point), 60 ft/min (30.5 cm/sec) except for experiment 10, 125 ft/min (63.5 cm/sec) and experiment 11, 350 ft/min (178 cm/sec).

[c]Sample irradiated unclad in this experiment only.

TABLE 3-28

Fission Products Released by In-Pile Heating of Stainless-Steel-Clad UO_2 Under Transient Reactor Conditions

Experiment no.	Reactor transient conditions		UO_2 vaporized (%)	Total Release (%)[a]							
	Integrated power (Mw-sec)	Period (msec)		Sr^{89} (Br+Kr)	Cs^{137} (I+Xe)	Ba^{140} (Xe+Cs)	I^{131} (Sn+Sb)	Te^{129} (Sn+Sb)	Ce^{144} (Ba+La)	Zr^{95} (Rb+Sr)	Ru^{103} (Tc)
1[b]	320	108	0.005	23	26	8.2	46	21	7.0	0.2	0.1
2[c]	328	87	0.75	58	53	42	31	21	8.8	1.5	0.6

[a]The fission-product species found by radiochemical analysis is listed, while the precursor species that was probably present in the previously unirradiated fuel while it was at high temperatures is shown in parentheses.

[b]A low-power transient was run with experiment 1 the day before the main transient. The fuel employed in this experiment did not melt, but it is believed to have reached a temperature above 2600°C (4712°F).

[c]Approximately 65% of the fuel used in this experiment appeared to have melted.

ducts, were released to a high degree because of their volatile precursors. On the other hand, I^{131} and Te^{129}, which would normally show the maximum release, were released only moderately because of their metallic precursors, which could presumably alloy with the stainless steel cladding. Cerium, zirconium, and ruthenium all had relatively low volatility precursors and consequently exhibited low release values. Further TREAT tests are discussed in Sec. 6 of Chapter 17, which considers loss-of-coolant accidents. Especially significant is the fact that cladding failures occur at an energy input of about 300 calories/g.

A comparison of fission-product-release results obtained by different methods of melting UO_2 (listed in Table 2-9) is given in Table 3-29. It is apparent that a large fraction of the rare gases, iodine, tellurium, and cesium will be released when UO_2 melts, except that tellurium may be retained by molten zirconium cladding. Ruthenium release was large only in the arc-image-furnace experiments and in those ORR in-pile experiments where a large fraction (50%) of the specimen volatilized. In the arc-image-furnace experiments, only ceramic materials (BeO and UO_2) were present in the high-temperature zone; consequently, oxygen liberated from the UO_2 specimen may have contributed to the volatility of the ruthenium rather than being absorbed by hot tungsten or cladding material. The release of strontium and barium was high only in the experiments with zirconium-clad specimens heated by tungsten resistor rods (Table 3-26), (oxygen-getting action of the cladding material), in the ORR experiments (large fraction of sample volatilized), and in the TREAT experiment (volatile precursors). The large fraction of sample volatilized in the ORR experiments indicates that the fuel probably reached a temperature considerably above the melting point of UO_2. It seems questionable whether fission-product-decay heat would be sufficient to heat fuel elements to the UO_2 melting point (2850°C or 5162°F), melt the fuel, and then vaporize half of it.

Further experimentation and possibly theoreti-

TABLE 3-29

A Comparison of Fission-Product Release by Different Methods of Melting UO_2[a] in Helium

Method of melting	Sample wt. (g)	UO2 vaporized (%)	Release (%)								
			Gross γ	Xe-Kr	I	Te	Cs	Ru	Sr	Ba	Ce
Arc-image	0.25	~2.0	~30	99.5	95.2	96.2	92.9	76.3[b]	3.9	7.3	3.8
Tungsten crucible	29.0	0.16	15.1	98.0	98.3	97.6	66.0	0.05	0.47	2.57	0.07
Tungsten resistor (unclad)	39	0.8	14	~100	70	90	82	1.1	2.5	9.0	1.0
Tungsten resistor (SS clad)	39	0.4	9	~100	93	55	83	0.9	1.7	7.4	0.5
Tungsten resistor (Zr clad)	39	0.17	8	~100	83	1.8[c]	66	0.4	23[d]	20[d]	1.2
In-pile (ORR)[e]	6	0.1			90	75	77	4	1.5	1	0.3
In-pile (TREAT)[f]	30	0.01			11	0.9	2	0.04	0.7	1.6	0.003

[a]UO_2 irradiated to tracer level except for dvc-image-furnace specimens.

[b]Impure He, included some air to oxidize Ru.

[c]Zr alloys with Te and depresses release rate.

[d]High Sr and Ba releases, attributed to loss of O by Zr-gettering.

[e]Release from high temperature zone (~1000°C or 1830°F).

[f]Release in 0.1 sec transient.

cal investigations appear to be required to determine whether the in-pile or the out-of-pile experiments provide more realistic accident conditions. Actually, the question seems academic so far as release of the more volatile elements on melting is concerned, since it is obvious that they are released at very rapid rates even below the UO_2 melting temperature.

3.3.2 Uranium Oxide, U_3O_8

The oxide of uranium that is produced when uranium metal or finely divided UO_2 is exposed to air at temperatures in the range of 800 to 1000°C (1472 to 1832°F) is U_3O_8 (See Sec. 3.1.1). This oxide has not been used to any extent in reactor fuels except when dispersed in a metallic matrix (see Sec. 3.2.2), but this use merits limited consideration of the high-temperature behavior of this material.

The uranium-oxygen system is said [11a] to be one of the most complex of the metal-oxide systems, and discussion of its phase relations is beyond the scope of this review. The important point for reactor safety is that U_3O_8 is not stable in air or steam at 1500°C (2732°F) or higher, and decomposition to a lower oxide would undoubtedly accelerate fission-product release from this material when it is overheated in a reactor accident. Available thermodynamic data indicate that UO_2 is approximately 23 kcal/gram-atom of oxygen more stable than U_3O_8 at 2000°K (1727°C or 3141°F). This disadvantage is offset to some extent by the above-mentioned fact that UO_2 oxidizes under some circumstances to give U_3O_8. It appears, however, that because UO_2 is more stable than U_3O_8 under most accident conditions, the lower oxide is preferable for reactor applications—if only from the standpoint of fission-product release in loss-of-coolant accidents. Other considerations have led to the use of U_3O_8 dispersed in aluminum as the fuel in several reactors.

3.3.3 Plutonium Dioxide, PuO_2

There seems to be little interest at present in the use of pure PuO_2 pellets as fuel material in reactors but some interest has been expressed [126] in possible utilization of stainless steel dispersions of PuO_2 in large fast power reactors. Melting of the matrix material may occur in a loss-of-coolant accident, and it is necessary to consider the probable behavior of fission products from small PuO_2 particles in such an event.

There is apparently little published information on fission-product release from massive or particulate forms of PuO_2, and consideration of this subject is necessarily speculative. As a first approximation, data on diffusion rates of fission products in UO_2 (Sec. 3.3.1) may be used to predict release rates from solid, undamaged PuO_2.

In Sec. 3.3.5 it is mentioned that the plutonium in PuO_2 or in (Pu, U)O_2 solid solutions cannot be oxidized to a higher valence than four [127]. Data in the literature [128] indicate that this compound is relatively stable in a neutral atmosphere, but it appears that heating in hydrogen or an inert atmosphere to temperatures of 1100 or 1200°C (2012 or 2192°F) reduces the oxygen-to-plutonium ratio to a composition corresponding to $PuO_{1.98}$. It is possible that a reactive molten metal such as zirconium in contact with PuO_2 would provide reducing conditions that could alter appreciably the oxygen content of stoichiometric PuO_2.

Plutonium dioxide is reported [128] to melt in a helium atmosphere at 2280°C (4136°F), thereby inducing some deterioration to give $PuO_{1.62}$. Under 0.1 to 1.0 atm of oxygen, the melting point is said [128] to be 2400°C (4352°F) for $PuO_{2.00}$. No information is available on the effect of steam on the melting point or on the composition of molten PuO_2, but steam-air mixtures are more likely to be the atmospheric environment of water-cooled reactors in loss-of-coolant accidents than pure steam. It seems probable, however, that melting of stoichiometric PuO_2 would not result in liberation of sufficient oxygen to have any greater effect on the behavior of released fission products than does the release of oxygen on melting UO_2. Because of the lower melting point of the former, release of low- or intermediate-volatility fission products from the melted material is likely to be appreciably lower than that from molten UO_2.

3.3.4 Thorium Dioxide-Uranium Dioxide Solid Solutions, (Th, U)O_2

Mixed thorium-uranium oxide fuels are potentially important for use in power reactors because of the Th^{232}-U^{233} breeding cycle. Little information exists in the literature on the behavior of these fuels under accident conditions. The irradiation behavior of powder-compacted fuels of this type has been examined at the Oak Ridge National Laboratory, and some measurements of fission-gas release were made in connection with these studies [129]. Values of Kr^{85} release during the lifetime of the tests (several months in most cases) ranged from 0.5 to 18.3% of the calculated gas content of the fuel. The highest value was obtained with a fuel surface temperature of 705°C (1301°F). The center of the fuel, of course, reached a much higher temperature.

Some useful data on fission-product release from defective fuel elements in a water-cooled reactor were obtained [130] in the BORAX-4 reactor (Sec. 6) which was operated at low power for 2 days with 22 defective fuel elements out of a total of 69. The fission-product buildup in various parts of the system and the release rates were measured. The only fission products released in significant quantities were daughters of short-lived fission gases. Rates of emission of Xe^{138} and Kr^{89} from the fuel during steady-state operation at 2.5 Mw were found to be 4 and 0.7 curies/min, respectively. No information on fuel temperatures is given in this report, but surface temperatures were undoubtedly low, and the data obtained have only qualitative significance so far as loss-of-coolant accidents are concerned. It is interesting to note that the decontamination factor for I^{131}, defined as activity per unit weight of reactor water to the activity in the same weight of steam, was about 33, indicating that a large fraction of this nuclide was present in a volatile form, presumably I_2.

In general, information on fission-product release from UO_2 fuel materials (Sec. 3.3.1) will be the best guide to probable release values for $(Th, U)O_2$ fuels under loss-of-coolant reactor accident conditions until direct measurements with this type of fuel become available. However, certain differences in these fuels need to be considered. Since the mixed fuels will be predominantly ThO_2, their melting points will be several hundred degrees higher than that of UO_2, so that the probability of melting in an accident will be decreased. The mixed oxides will also be less affected by exposure to oxygen, since thorium has only one stable valence and ThO_2 not only dilutes the UO_2 but tends to stabilize the UO_2 structure. Differences in diffusion release are less obvious. Use of powder-compacted fuels would seem to enhance release by diffusion, but the powder materials sinter in the reactors and the center of the fuel even melts at high heat ratings. Therefore, no generalization concerning release from $(Th, U)O_2$ fuels by diffusion seems justified at present. It seems obvious that fission-product release from these fuels will need to be investigated from the accident viewpoint if present expectations of their increasing importance in power reactor technology are fulfilled.

3.3.5 Uranium Dioxide-Plutonium Dioxide Solid Solutions, $(U, Pu)O_2$

British investigators [131, 132] have studied the oxidation behavior of $(U, Pu)O_2$ solid solutions, as well as that of PuO_2 and other solid solutions, in some detail. Solid solutions of this type are said [10f] to be potentially useful as the fuel material in a fast-reactor core. The available data indicate that, while plutonium in PuO_2 or in solid solutions containing PuO_2 cannot assume a higher valence than four, the uranium in $(U, Pu)O_2$ solid solutions oxidizes in air at 750°C (1382°F) to an average valence of five. Powder samples were employed in these studies and, consequently, changes in fuel-element configurations as the result of exposure of solid solutions to high-temperature air could not be observed.

Brett and Russel [127] have reported that pellets cotaining more than 20 mole % PuO_2 were physically stable on oxidation in air at 750°C (1382°F). The pellets retained a fluorite lattice and oxidized to a limit approximately defined by the equation

$$\frac{O}{M} = 2.00 + 0.02P$$

where P = mole % PuO_2.

These investigators have also reported that pellets containing 0 to 20 mole % PuO_2 remained a singlephase after arc-melting in argon, while those with 40 to 60 mole % PuO_2 contained two phases after this treatment; for pellets containing 80 mole % PuO_2, three phases were present.

From the preceding discussion, it seems that physical changes in $(U, Pu)O_2$ fuel bodies, which will affect the release of fission products at temperatures below the melting point of the fuel, will depend on fuel composition, atmosphere, temperature, and possibly other parameters. It appears, therefore, that experimental investigations of fission product release from specific fuel compositions under typical reactor accident conditions will be required in order to define clearly the hazard of fission-product release from unmelted fuels of this type in reactor accidents.

The release of fission products from melted $(U, Pu)O_2$ solid solutions has apparently not been investigated but it presumably would be more nearly comparable to that from $BeO-UO_2$ fuels (Sec. 3.3.6) than to that from pure UO_2 (Sec. 3.3.1), depending on fuel composition. Plutonium-dioxide-rich fuels melt in approximately the same range as $BeO-UO_2$ mixtures, while the melting point of uranium-rich fuels will not be significantly different from that of pure UO_2. There seems to be no reason to expect that fuel composition will noticeably affect fission-product release from molten oxide fuel mixtures except as it affects the melting point. Vapor pressure data for PuO_2 and UO_2 (Fig. 2.2) seem to indicate that fractionation would probably not create a major hazard except, possibly, in the case of plutonium-rich fuels.

3.3.6 Uranium Dioxide-Beryllium Oxide, UO_2-BeO

The excellent nuclear and physical properties of BeO make it attractive to use as a diluent for UO_2 in ceramic fuel elements. Not only does BeO provide enlarged heat-transfer surface through dilution; in addition, it has been found [133] that sintered mixtures of UO_2 with ZrO_2 or BeO and minor amounts of other oxides containing less than 50 wt. % UO_2 were little affected by exposure to air for 200 hr at 1260°C (2300°F). More recent experience [115b] with $BeO-UO_2$ fuel materials stabilized by addition of Y_2O_3 showed that recoil was the dominant mechanism for in-pile release of fission products from these materials in flowing dry air at temperatures up to 1400°C (2552°F). Diffusion of I^{131} from the fuel was observed at 1650°C (3002°F).

The presence of water vapor in the air stream was reported [21d] to cause slow deterioration of the fuel-element surfaces because of the reaction:

$$BeO + H_2O(gas) \rightleftharpoons Be(OH)_2(gas).$$

Not only was the release of fission products mildly accelerated, as compared to rates measured in dry air; but also, deposits of BeO were found downstream from the maximum temperature zone. These observations are of more concern for long-term reactor operations than for predictions of fission-product release in loss-of-coolant accidents.

Measurements of fission-product release accompanying the melting of fuel specimens of this type have been reported by Conn et al. [20e] and by Parker and coworkers [134]. The former investigators employed a plasma jet to melt specimens irradiated to 0.44% burnup and reirradiated to establish a suitable inventory of I^{131}, Ba^{140}, and Sr^{89}. Their total release data are given in Table 3-30 [20e]. The data indicate that the release of iodine was related to the amount of sample melted. The correlation was poor in the case of Ba^{140} release and nonexistent in the Sr^{89} data.

TABLE 3-30

Release of Fission Products from $BeO-UO_2-Y_2O_3$ Fuel Melted by a Plasma Jet

Test no.	M-1	M-2	M-3	M-4
Time of melt (min)	2	8	2	2
Fraction of sample melted	0.8-0.9	1.0	0.5	1.0
Total I^{131} release (%)	68	88	48	87
Total Ba^{140} release (%)	10	9	2.7	6
Total Sr^{89} release (%)	2	1.6	1.8	2

A more thorough investigation of parameters affecting the release of fission products from small tubular samples of $BeO-UO_2-Y_2O_3$ fuel was made [134] by use of an arc-image furnace, an apparatus in which the image of a carbon arc is focused sharply on the object heated. The fuel specimens were surrounded by a glass envelope in fission-product-release experiments. Air flowing through this envelope carried particles and gases evolved from the heated fuel to a collection train. The length of time molten was not well controlled with this heating arrangement. Each tubular fuel specimen was held in a horizontal position with the light beam focused initially on its front end. When this part of the specimen melted, the molten portion dropped far enough to be out of the high-temperature zone, and the specimen was advanced so that another portion was heated to the melting point. This process was continued until all the specimen and a small part of the solid BeO rod used to support the specimen were melted. The time required to complete the operation varied from about 80 to 95 sec. Data obtained with three types of fuel are recorded in Table 3-31 [134]. Most of the melting experiments were performed with fuel irradiated to 0.43% burnup of U^{235}, but two experiments were made with low-burnup (0.01%) material. Most of the specimens were reirradiated to build up an inventory of short-lived isotopes.

High values for the release of iodine, tellurium, cesium, and ruthenium are noted in Table 3-31. A significant fraction of the uranium content of the fuel and a smaller amount of the BeO also volatilized. This enhanced volatility of both uranium and beryllium could have been an effect of the high (surface-to-volume) ratio of the very small samples. No significant burnup effect was shown over this narrow range of burnup and a tenfold increase in air velocity likewise had little effect on fission-product release, although it had a large effect on fission-product transport, as discussed in Sec. 4. Maximum fuel-melting temperatures measured by means of an optical pyrometer in these experiments were 2550 ± 30°C (462 ± 54°F), and freezing temperatures of 2450 ± 25°C (4442 ± 45°F) were recorded. These values are somewhat higher than the value 2315°C (4200°F) reported elsewhere [20e] for this type of fuel.

Some data were also obtained on the rate of release of fission products from tracer-irradiated ceramic-coated fuel elements (hollow cylinders, ~1 cm diameter, 1 mm wall thickness) heated for 5-hr periods in flowing helium. The data are presented graphically in Fig. 3-20 [134] in a form that permits extrapolation to temperature outside the measurement range (1015 to 1400°C or 1859 to 2552°F). The straight lines through the points fit the equation

$$\% \text{ released} = a \exp(-b/T),$$

where T is the absolute temperature. The constants a and b have the values as follows:

Gross - γ : $a = 1.747 \times 10^7$, $b = 3.252 \times 10^4$ °K;
I - γ : $a = 2.663 \times 10^9$, $b = 3.386 \times 10^4$ °K;
Te - β : $a = 1.655 \times 10^{11}$, $b = 4.1898 \times 10^4$ °K;
Cs - γ : $a = 3.327 \times 10^{13}$, $b = 5.414 \times 10^4$ °K;
Ru - γ : $a = 1.694 \times 10^{13}$, $b = 5.228 \times 10^4$ °K;

3.3.7 Beryllium Oxide, BeO

This material is included in Table 1-6 because of its importance as a moderator and as a fuel

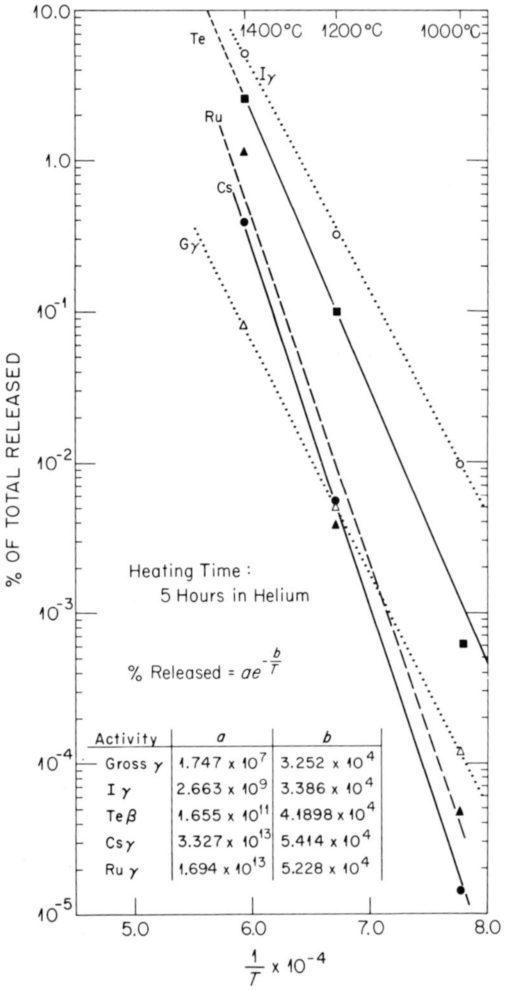

FIG. 3-20 Release of fission products on heating of ceramic-coated $BeO-UO_2-Y_2O_3$ for 5 hr in flowing helium. Data fits exponential: % released = a exp (-b/T), where a and b are constants (see text). In the above figure the symbol Gγ is the abbreviation for gross gamma activity.

TABLE 3-31

Release of Fission Products from BeO-UO_2-Y_2O_3 Fuel Specimens Melted in Air

Fuel type	Total activity or fuel component vaporized (%)												
	Rare gases	Gross γ	I	Te	Cs	Ru	Sr	Zr	Ba	Ce	TRE	Be	U
Uncoated	75.7	11.6	56.2	62.7	61.5	55.8	0.03	0.02	0.3	0.6	0.26	0.86	8.9
Uncoated	71.3	11.	76.7	68.2	62.0	63.7	0.15	0.02	0.54	0.5	0.19	0.58	8.2
Uncoated	77.1	11.6	79.0	75.1	50.1	64.4	0.11	0.03	0.56	0.4	0.45	1.6	7.8
Uncoated[a]	73.8	7.6	70.7	66.3	45.7	54.2	0.02	0.42	0.22	0.5	0.18	0.76	9.3
Uncoated	82.0	10.0	83.4	84.5	41.4	61.9	0.11	0.01	0.13	0.4	0.12	0.58	16.1
Uncoated	69.7	26.0[b]	72.4	71.0	72.9	66.9	0.10	0.02	0.39	0.6	1.4	0.89	12.6
Uncoated[c]	69.5	18.4[b]	78.8	71.7	49.3	57.6	0.05	0.03	0.15	0.5	0.28	1.23	7.8
Coated[e]	53.4	10.6[b]	73.5	63.2	44.6	50.0	0.08	0.004	0.5	0.9	0.19	0.60	9.8
Coated[e]	57.0	13.8[b]	77.3	72.4	32.3	58.3	0.02	0.005	0.5	0.42	0.29	0.69	10.6
Coated	d	4.9	d	64.3	60.9	49.9	0.02	0.009	0.16	0.48	0.27	0.66	13.0
Inside	59.4	3.87	73.5	68.4	33.5	48.3	0.008	0.0034	0.156	0.072	0.2	0.327	5.91
Only	59.2	5.8	66.8	76.6	55.4	55.3	0.013	0.005	0.15	0.18	0.14	0.339	5.76
Coated[e]	60.9	5.7	63.3	78.7	57.6	59.6	0.01	0.004	0.23	0.45	0.2	0.366	6.66

[a]0.01% burnup of U^{235}. All other specimens irradiated to 0.43% burnup.

[b]High release of gross gamma due to short cooling period after reirradiation, resulting in presence of more volatile short-lived gamma emitters.

[c]Air flow velocity in this experiment was 5 cu ft/min (2.36 liter/sec). In all other experiments it was 0.5 cu ft/min (0.236 liter/sec).

[d]Not reirradiated.

[e]Coatings consisted of a few mils (1 mil = 25μ) of ZrO_2 or pure BeO.

component (see the previous section). The effects of irradiation on solid BeO have been studied extensively [135] at the Oak Ridge National Laboratory and elsewhere. Various other chemical and physical properties of BeO have also been measured both in the presence and in the absence of reactor irradiation, but none of the published information seems sufficiently relevant to probable behavior in nuclear reactor accidents to warrant discussion here.

3.3.8 Thorium Oxide, ThO_2

Fission products can occur in ThO_2 as a result of exposure to fission neutrons (see Fig. 1-3) and as a result of fissioning of U^{233} caused by decay of 27.4-day Pa^{233}, which is produced by thermal-neutron capture by Th^{232}. The very high melting point of ThO_2 (3300°C or 5970°F) and its lack of reactivity with atmospheric environments likely to occur in reactor accidents, such as air and steam, point to diffusion as the sole probable mechanism of fission-product release from pure thorium oxide. Since the concentration of fission products in this material is unlikely to be high, the probability that ThO_2 will reach high enough temperatures for fission-product release through diffusion to become significant in reactor accidents seems too low, and the utility of the pure material in reactor cores is too uncertain at present to merit detailed consideration. Mixtures with UO_2 are discussed elsewhere (Sec. 3.3.4).

3.4 Carbides in Graphite or Pyrolytic Carbon

This class of fuel materials will be considered collectively because of the scarcity of useful information on release of fission products. Rates of diffusion of fission products through essentially undamaged coating materials are discussed elsewhere in this chapter (Sec. 4.2), and this section will be confined to release of fission products resulting from the reaction of carbide fuel materials with oxidizing atmospheres.

Results of an investigation of the oxidation kinetics of dense compacts of UC were reported [136]. While the results obtained with the "home-made" fuel compacts employed in these studies are certainly not representative of the oxidation behavior of commercial pyrolytic-carbon-coated uranium carbide fuels, these data do give some indication of reaction rates to be expected from uncoated uranium carbides or from fuel having damaged coatings.

Data were obtained on reaction rates of the UC compacts in CO_2 at 500 to 1000°C (932 to 1832°F), in O_2 at 350 to 1000°C (662 to 1832°F), in N_2 at 700 to 1000°C (1292 to 1832°F), and in CO at 400 to 600°C (752 to 1112°F). Since reaction rates in CO_2 and CO are only of interest in reactors cooled with CO_2 and since pure N_2 is an unlikely accident environment, only the oxidation in pure O_2 is considered here. Weight-gain-versus-time curves are shown in Fig. 3-21 [136]. About one-third or less of the specimens, 1.5-cm (0.591 in.)-diam by 0.5-cm (0.197-in.)-thick discs, was oxidized in these experiments. A comparison of the reaction rate of the UC with that of uranium metal in this environment was made at only one temperature (350°C or 662°F). The metal oxidized much more rapidly than the monocarbide at this temperature.

Graphite-gas reactions are discussed in Chapter 17. Some studies of the combustion of such fuels in oxygen have been conducted as part of an effort to develop a reprocessing scheme [134]. Exposure of a bed of pea-sized lumps of fueled graphite, heated initially to 650 to 700°C (1200 or 1290°F), to flowing oxygen resulted in complete combustion of the material in 3 hr or less. The centerline temperature of the burning bed was said to have reached sufficiently high temperatures (undetermined) on occasion to sinter the ash to a friable solid.

Some data on fission-product release accompanying the combustion of irradiated graphite fuels were also reported [137]. With HTGR fuel samples irradiated to about 10,000 Mwd/metric ton of uranium plus thorium, 6-hr combustion in oxygen at furnace temperatures of 800°C (1472°F) and 1200°C (2192°F)—fuel-bed temperatures were not measured—volatilized 78 to 99% of the ruthenium, 24 to 88% of the cesium, 0.008 to 0.07% of the zirconium, and 0.01 to 0.1% of the rare earths. Release of rare gases, iodine, and tellurium was not studied because their radioactivity had decayed, but volatilization of these elements would presumably be virtually complete under the experimental conditions. Decontamination factors greater than 10^4 were observed for ruthenium and cesium in the off-gas after passage through a 40μ (1.6-mil) porosity nickel filter and a Millipore filter at room temperature.

Data on fission-product release resulting from in-pile burning of fuel specimens made up of pyrolytic-carbon-coated uranium carbide spheres (diameter = 175 to 205μ or 6.9 to 8.1 mil) in a graphite matrix have been reported [39c]. The fuel specimens, irradiated to trace level, were partially burned during a 15-min exposure in the ORR. The fuel temperature, initially at 890°C (1634°F) in helium, rose to approximately 1400°C during burning in a flowing air stream. The results of two experiments are summarized in Table 3-32 [39c]. In considering the data in this table, one should bear in mind that the samples were incompletely burned and, in one experiment (no. 3), 93% of the uranium was in the unburned portion of the fuel because of uneven distribution of uranium in the fuel and, consequently, there was uneven oxidation of the specimen. The results show that significant quantities of all the isotopes studied escaped from the fuel, ranging from 4 or 5% for Zr^{95} to 51 or 55% for Cs^{137}, but only the more volatile elements escaped from the high-temperature zone of the furnace in significant amounts under the conditions of these experiments.

Laboratory studies [138] of the reaction of steam with unirradiated UC showed that 5-g batches of 5-mm (0.2-in.) fragments were converted to UO_2 in 45 min at 750°C (1382°F). It was observed that the reaction proceeded in two steps. The UC was initially converted to UO_2 and free carbon, followed

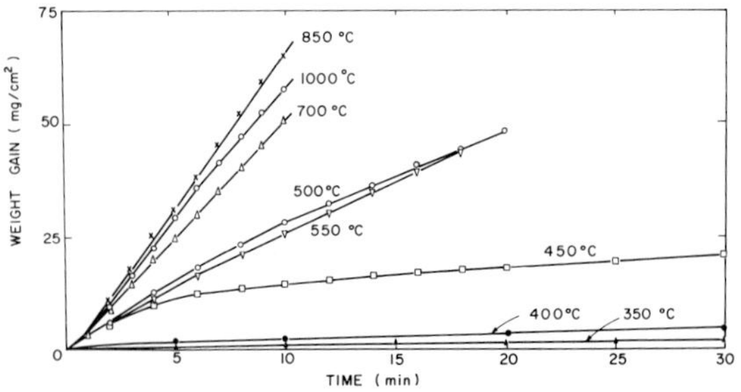

FIG. 3-21. Weight gain as a function of time for UC in O_2 at various temperatures.

by further reaction of the carbon with steam to give CO_2, CO, and hydrogen (5 moles of off-gas per mole of UC consumed).

In experiments with irradiated UC, air could not be excluded, and the uranium was oxidized to U_3O_8. When 4- to 7-g specimens were steamed for about 5 hr at 700 or 800°C (1292 or 1472°F), the condensed steam from the off-gas was found to contain 1.1% (700°C) to 42% (800°C) of the total cesium in the irradiated fuel samples. Cesium was the principal fission product volatilized from these long-cooled fuel specimens.

3.5 Miscellaneous Compounds

This group of compounds will be considered collectively, as in the previous section, in part because of the absence of published information on fission-product release and also because of present uncertainty about the future of these compounds in reactor technology. Available information on the chemical and physical properties of these compounds has been summarized by Dell and Allbutt [15].

The nitrides and sulfides of uranium and thorium have at least one important property in common: they are easily oxidized. Uranium mononitride (UN) powder is reported not to be pyrophoric at room temperature but to ignite on warming with the ignition temperature depending on the surface area of the powder and other undefined factors. Finely divided U_3O_8 and nitrogen gas are produced by ignition of UN in air. This compound is said to be less easily hydrolyzed than UC. Hydrolysis in water vapor pressures of about 200 mm Hg becomes rapid only above 250°C (482°F) and, again, the rate of reaction depends largely on particle size. Nitrogen is converted to ammonia in this reaction.

Less information on the chemical reactivity of thorium mononitride (ThN) appears to be available but ThN and ThC are said to react more readily with water vapor than the corresponding uranium compounds and small sintered compacts were reported to disintegrate completely on exposure to atmospheric moisture for a short time.

Both uranium monosulfide (US) and the disulfide (US_2) are reported to be pyrophoric when ground. Compacts of US apparently do not oxidize at 20°C (68°F), but oxidize slowly to UOS at 160°C (320°F). Similarly, compacts of US_2 suffer only

TABLE 3-32

Fission Products Released by In-Pile Burning of Pyrolytic-Carbon-Coated Uranium Carbide in a Graphite Matrix

	Material found outside fuel residue (% of total in assembly)								
	Sr^{89}	Zr^{95}	Ru^{106}	I^{131}	Te^{132}	Cs^{137}	Ba^{140}	Ce^{144}	U
Experiment 1	13.8	3.7	48.2	28.9	42.7	51.3	14.5	9.1	2.1
Experiment 3	13.1	5.0	8.8	16.5	23.2	54.5	7.0	7.0	0.9

surface oxidation at 20°C but oxidize to UOS at 300°C (572°F). Both compounds presumably will oxidize to UO_2 or U_3O_8 at higher temperatures in the presence of excess oxygen.

Summarizing: The reactivity of the nitrides with oxygen and water vapor would undoubtedly contribute to the hazard of loss-of-coolant accidents in reactors fueled with these materials but no more than for uncoated uranium carbide fuels. The high melting point of the compounds indicates that fuel melting would probably result in release of most fission products to an extent comparable to that from UO_2.

3.6 Fuel Systems Liquid at Reactor Operating Temperatures

Several liquid-fuel systems have been considered for reactor use including uranium nitrate dissolved in water (LASL Water Boiler—see Chapter 11 Sec 3.2), uranium dissolved in molten bismuth, Liquid Metal Fast Reactor (LMFR); uranyl phosphate in phosphoric acid, Los Alamos Power Reactor Experiment, (LAPRE); and uranyl sulfate dissolved in dilute D_2SO_4, Homogeous Reactor Experiment (HRE). The molten fluoride and molten plutonium-alloy reactor systems are the only fluid-fuel concepts being supported sufficiently at present to anticipate future construction or operation of experimental reactors and, consequently, these are the only fuels in this class considered here.

3.6.1 Sodium Fluoride-Zirconium Fluoride-Uranium Tetrafluoride, $NaF-ZrF_4-UF_4$

Mixtures of this type fueled the Aircraft Reactor Experiment (ARE), the first molten-salt reactor constructed and operated [139] (1954 at the Oak Ridge National Laboratory). Chemical problems involved in this operation are discussed in reference [140]. Since the objective of the molten-salt program at that time was to develop a reactor for aircraft propulsion with a 1000-hr-life requirement, operation of the reactor terminated before the longer lived fission products reached their equilibrium concentrations, and only limited information on fission product behavior was obtained in this experiment. This information, together with some data on fission-product behavior in the same type of fuel in in-pile experiments, has been summarized by Robinson et al. [141] and in reference [139], pp. 588-591. Robinson discusses electroneutrality in molten-fluoride fuels in reference [142].

All the available evidence, including operating experience with the ARE, indicates that very little of the fission gases krypton and xenon remain in a circulating fluoride fuel mixture. Therefore, no significant hazard in a reactor accident could result from release of an accumulation of these gases.

The fission products rubidium, cesium, strontium, barium, zirconium, yttrium, and the rare earth elements all form stable fluorides, and it is stated [140] that they should exist in molten-fluoride mixtures in their ordinary valence state and that quite high burnups would be required before a molten-fluoride reactor could saturate its fuel with any of these fission products.

The valence states assumed by the nonmetallic elements selenium, tellurium, bromine, and iodine are said [140] to depend strongly on the oxidation potential defined by the container and the fluoride melt. In general, it appears that, in a reactor employing INOR-8 (an alloy similar to Inconel but with less chromium) as the container material, the chromium in the alloy will reduce a small fraction of the uranium content of the fuel mixture to the trivalent state, so that slightly reducing conditions would be expected to prevail during normal operation. Iodine is consequently expected to exist in the fluoride mixture as the iodide, but it is difficult to predict the extent of its release on exposure to accident environments. Water vapor will hydrolyze molten fluorides, but water or steam are not likely to be employed as operating coolants for the fuel in high-temperature reactors of the molten-salt type. Oxygen will react only slowly with components of the molten-salt mixtures and, when present in small amounts, it is scavenged by materials like UF_4 and ZrF_4 that are present in very high concentrations as compared to the fission-product iodide concentration. Larger amounts of oxygen will liberate iodine at a rate dependent on the rate of diffusion into the liquid. Fission-product gamma energy in solidified fuel may cause slow release of iodine after an accident.

Efforts to determine the behavior of iodine in the ARE were unsuccessful. No iodine was found in off-gas samples nor by analysis of solidified salt samples. The latter had cooled for a month or more before analysis (about four half-lives for I^{131}). The negative results of off-gas analysis could be explained by deposition of iodine in the sample line as well as by complete absence of iodine in the off-gas. Rare gases and daughter products of short-lived rare gases (Rb^{88}, Xe^{135}, and Cs^{138}) were the only isotopes identified in material trapped from the off-gas in cooled charcoal, but unidentified peaks were also observed in the gamma-ray spectrum. Establishment of the presence of Ru^{103}, Ru^{106}, and Zr^{95}-Nb^{95} on the wall of a pipe section from the ARE [141] indicated that the more noble elements are probably reduced to the metals and would not contribute to the hazard of reactor accidents.

It appears that the hazard of fission-product release from NaF-ZrF_4-UF_4 fuels under accident conditions probably would be low but cannot be clearly defined at present.

3.6.2 Lithium Fluoride-Beryllium Fluoride-Zirconium Tetrafluoride-Uranium Tetrafluoride-Uranium Tetrafluoride, LiF-BeF_2-ZrF_4-UF_4

Mixtures of LiF, BeF_2, ZrF_4, and UF_4 fuel the Molten-Salt Reactor Experiment (MSRE), a 7.3-Mw(t) experimental reactor that went critical at Oak Ridge National Laboratory in June 1965 and reached full power (7.3 Mw(t) in May 1966. By March 1968 it had run for 9,005 equivalent full-power hours and it was critical 80% of the time from December 1966 to March 1968. It was shut down in December 1969 after having achieved all its experimental objectives.

Most of the considerations in the previous section apply equally well to this fuel system. A preliminary safety analysis report (PSAR) for this reactor, including a systems description has been published [143], and a good review of molten salt reactor technology that includes experience with the MSRE has been published [144a-h]. Scarcity of data on fission-product release under accident conditions at the time of the writing of the PSAR is reflected in the description of the maximum credible activity release, where it is assumed that 10% of the solid fission products and 10% of the iodine escape the salt and become dispersed in the primary reactor container. The only way for this to occur would be for 10% of the fuel mixture to be dispersed also, and this seems rather unlikely. Even if the fuel mixture were to be dispersed into small droplets by some undefined mechanism, the liquid would probably solidify almost instantaneously, and migration of fission products from the solidified particles would undoubtedly be at such a low rate that the fission-product-containing fuel particles, rather than separated fission products, would constitute the principal hazard. The beryllium in the fuel material, as was pointed out in the hazards report [143], might constitute a significant hazard in this event. Observations of escape of fission products from fuel at temperatures in the range 480 to 820°C (896 to 1508°F) in a helium atmosphere were reported [144c, 145]. Only fission gases were observed in the sweep gas from liquid fuel but Te^{143} and I^{132} were detected after solidification of the fuel. A review of observations of fission-product behavior in the MSRE has been given [144c].

Figure 3-22 [144a] gives a design flowsheet for the system including operating conditions at 7.3 Mw(t) power. Figure 3-23 [144b] shows a drawing of the MSRE core and reactor vessel. Flow of the liquid fuel is into the core vessel through the flow distribution, down in an annulus along the sides of the core, up through the core, and out. The control rods are flexible hollow cylinders of gadolinium oxide-aluminum oxide (Gd_2O_3-Al_2O_3) ceramic canned in Inconel and threaded on flexible stainless steel hose, which serves also as an air duct. There is no real "blanket" region in the MSRE as there would be in a breeder power reactor. The 54-in.-diam core is made up of low-permeability graphite bars, 2 in. square and 64 in. tall, exposed directly to the liquid fuel, which flows in passages machined into the faces of the bars. Specimens of graphite showed no visible

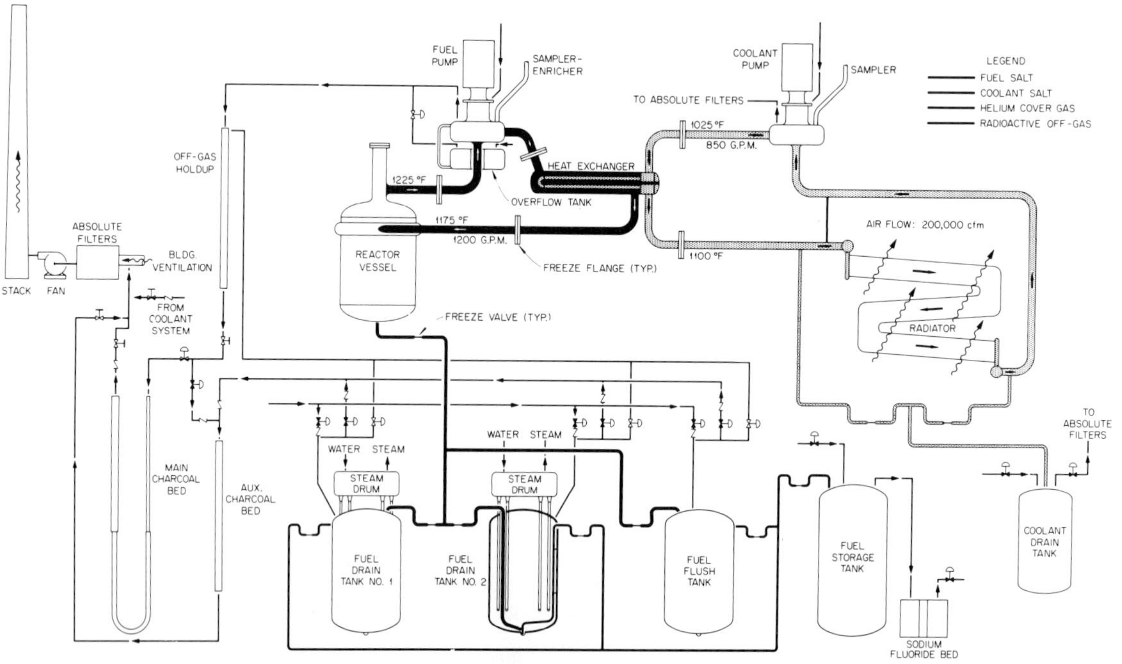

FIG. 3-22 Design flowsheet of the MSRE.

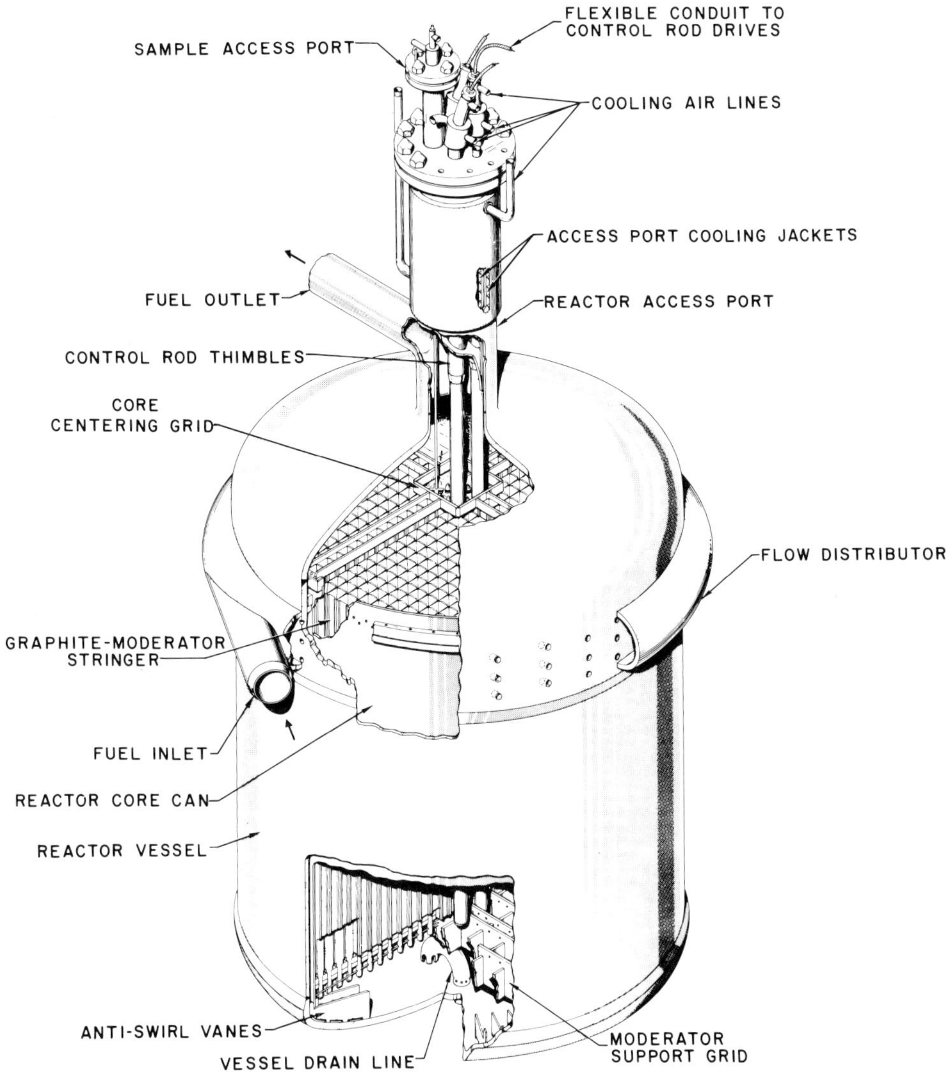

FIG. 3-23 Details of the MSRE core and reactor vessel.

radiation effects after 10,000-hr exposure, but from past radiation-damage experience, it is anticipated that the graphite will first shrink somewhat and then expand to occupy more than its original volume. It is anticipated that the graphite in such a reactor must be removable.

All metal components of the system in contact with molten salt are made of Hastelloy-N (a commercial version of INOR-8), which is stronger than austenitic stainless steels and most nickel-base alloys, but like these metals is subject to loss of high-temperature ductility and shortened stress-rupture life under neutron irradiation. A modified Hastelloy-N with improved resistance to irradiation damage has been developed [144d]. Since no metallic parts of the reactor structure are in high fast-neutron fluxes, this should not be a significant problem. The observed generalized corrosion rate of the metallic parts was less than 0.3 mil (0.0076 cm) in 3 years.

The design is such that the bowl of the fuel pump shown in Fig. 3-22 is utilized as the surge space for the circulating loop. Dry, deoxygenated helium at 5 psig blankets the salt. About 50 gpm of the 1200-gpm pump discharge is sprayed into the gas space of the pump bowl to provide contact between the molten salt and the cover gas, so that the Xe^{135} and other noble gases are allowed to escape. The stripping of xenon by this method has cut the xenon reactivity effect observed in this reactor by a factor of five. The noble metals (molybdenum, niobium, ruthenium, tellurium) persist in metallic form, mostly plating out on the surfaces. Less than 1% remains in the circulation salts, but a significant fraction goes out with the off gas as a smoke or dust of metallic particles. Other fission products, including iodine and the rare earths, remain in the molten salt, as does the plutonium produced when U^{238} is present in the fuel mixture.

Uranium fluoride fuel in capsule form is added to the reactor through a combined sampling and enriching line to the pump bowl. "Freeze valves" (consisting of special sections of pipe that can be cooled to freeze the molten salt and thus to stop flow) are used to isolate various parts of the systems as needed. The reactor liquid fuel can be dumped through such a valve into a subcritical (unmoderated by graphite) drain tank for storage. These tanks are cooled by bayonet-type water coolers. Normally, all salt piping and vessels are electrically heated to prepare for salt filling and to help keep the salt molten when there is no nuclear power. The heat from this experimental reactor eventually passes via a heat exchanger to a secondary coolant fluoride-salt system and is eventually dissipated in an air-flow heat exchanger.

Table 3-33 shows a comparison of some calculated and measured nuclear parameters of the system using U^{235} and U^{233} [144b] fuels. In general, the fact that fuel can be added as needed, together with removal of most of the xenon absorber, makes the reactivity requirements low. The observed negative temperature coefficients of the molten-salt system, coupled with the low reactivity requirements, make the system response acceptable even though the delayed neutron fraction of 0.0017 with U^{233} fuel and circulating salt (which sweeps neutron precursors out of the core) is the lowest of any operating reactor to date. Calculations and experiments indicate that the reactor is inherently stable at all power levels, the degree of stability increasing with increasing power.

This type of reactor is potentially a thermal-breeder reactor operating on the thorium-U^{233} cycle with a significant breeding gain. Since there is no fuel fabrication nor time allowances for fabrication and cooling, the fuel inventory is low. Thus, the doubling time for advanced designs of reactors of this type may be close to those obtainable for sodium-cooled fast breeders (see Chapter 10). Two concepts for this type of reactor have been discussed. One is a two-fluid system with a fertile thorium containing fluoride salt for the blanket and a fissile salt containing U^{233} in the graphite core. The second, and simpler, system is a one-fluid concept, [144g] which uses a single salt stream containing both uranium and thorium fluorides. By adjustment of the degree of graphite moderation between the core and blanket, neutron captures can be favored by uranium (followed by fissioning) in the core and by thorium (followed by decay to U^{233}) in the blanket. Present design effort [144g] is directed toward development of a breeder reactor of the one-fluid type.

Extrapolation of vapor-pressure data for the MSRE fuel (reference [145], p. 117) indicates that the mixture boils at about 1400°C (2550°F). The probability that the fuel mixture will reach this temperature seems rather low. Even if it did, it is probable that reactions with the environment would be more likely to liberate fission products than would direct volatilization.

Summarizing: Although it now appears that some noble metals may be released as dust, rare gases are the only fission products certain to be released from overheated fluoride fuel mixtures; and they are largely removed during ordinary reactor operations so that the inventory available at any time is limited. Further experimentation will be required to determine the extent of release of iodine and other hazardous fission products under accident conditions.

3.6.3 Plutonium-Iron Alloy

The plutonium-iron eutectic composition (90 at. % plutonium-10 at. % iron) was employed as the fuel material in the Los Alamos Molten

TABLE 3-33

Summary of MSRE Nuclear Parameters with U^{235} and U^{233} Fuels

Parameter	Units	U^{235} Fuel		U^{233} Fuel	
		Calculated	Measured	Calculated	Measured
Initial critical concentration in salt	g U/liter	33.06[a]	32.85 ± 0.25[a]	15.30[b]	15.15 ± 0.1[b]
Reactivity loss due to circulation of delayed-neutron precursors	% δk/k	0.222	0.212 ± 0.004	0.093	c
Control-rod worth at initial critical loading[d]	% δk/k				
1 Rod		2.11	2.26	2.75	2.58
3 Rods, banked		5.46	5.59	7.01	6.9
Temperature coefficient of reactivity at operating loading	$\frac{\delta k/k}{°F}$ ($\times 10^5$)				
Total		-8.1	-7.3 ± 0.2	-8.8	-8.5
Fuel		-4.1	-4.9 ± 2.3	-5.7	c
Concentration coefficient of reactivity	$\frac{\% \delta k/k}{\% \delta c/c}$	0.234	0.223	0.389	0.369

[a] U^{235} only.
[b] Uranium of the isotopic composition of the material added during the critical experiment (91% U^{233}).
[c] Measurement obscured by effect of circulating voids.
[d] Normal full travel of rod(s).

Plutonium Reactor Experiment (LAMPRE). The high plutonium content of the plutonium-iron eutectic poses a serious heat-transfer problem. As a result, it has been suggested that ternary alloys with less than 10 at. % plutonium be used in any future high-power reactors of this type.

Sealed tantalum capsules employed in the LAMPRE to contain the alloy fuel prevented escape of fission products as long as the capsules were intact but apparently afforded limited opportunity for observation of fission-product behavior except for fission gases. Some difficulty was experienced [146] with fuel column separation in LAMPRE Core I capsules, but this was an operating problem rather than a hazard. This type of core will probably not be used in power reactors fueled with molten plutonium alloys.

Bidwell [147] has considered the fission-product distribution between a molten plutonium-alloy fuel (7.5 at. % plutonium-25 at. % cobalt-67.5 at. % cerium) and liquid sodium coolant in a "dynamic core" fast reactor where the fuel is pumped through an outside loop by the coolant. Such an arrangement is presumably more suitable for large power reactors than the core configuration used in LAMPRE. Bidwell concludes, on the basis of phase-diagram information and thermodynamic data, that rubidium, cesium, strontium, barium, europium, bromine, iodine, krypton, and xenon would be extracted into the liquid sodium coolant. This list includes two of the most biologically potent fission products, iodine and strontium, and a large amount of highly contaminated liquid-metal coolant would contribute significantly to the hazard of a loss-of-coolant accident with this type of reactor.

A safety analysis report [148] on the Fast Reactor Core Test Facility, designed to extend the testing program started with LAMPRE, assumed a fission-product release of 100% of the inert fission gases and 10% of the remaining fission products in the maximum credible accident. The latter figure seems questionable if a significant fraction of a contaminated sodium inventory or of the fuel itself burns. The design of the facility provides for maintenance of an inert atmosphere over accident-released fuel and coolant materials, but the possibility of failure of such provisions must be recognized.

More detailed consideration of the hazard of molten-plutonium-fuel reactors must necessarily be deferred until details of future reactors are available and experimental data on fission-product behavior in dilute plutonium alloys are obtained. Data on reaction rates of such mixtures with the atmospheres that would exist in credible accidents and on the release of fission products resulting from exposure of molten mixtures to oxidizing atmospheres will certainly be of interest. The behavior of plutonium oxide produced by such exposures will also need to be considered.

4. PENETRATION THROUGH FUEL-SYSTEM CLADDING

The discussion in this section will be limited to penetration of fission products through fault-free cladding materials, since the subject of escape of fission products through fuel-element defects is discussed in Chapter 13, and actual experience is covered in the following section. The release of fission products resulting from melting of the cladding was covered in Sec. 3.

4.1 Effusion through Metals

Fuel-element designers generally assume complete retention of fission products by metallic fuel-cladding materials, and there seems to be little reason to question this assumption on the basis of the scanty information available in the literature. (One important exception is the diffusion of tritium through stainless steel clads.) Although some experimenters have observed escape of fission products from metal-clad specimens, it was found in several cases that this could be attributed either to surface contamination or to defects in the cladding material.

Castleman et al. [149] have reviewed earlier work on the diffusion of rare gases through metals and reported the results of their efforts to measure the diffusion of xenon through aluminum and stainless steel. They found 2S aluminum sheet, 0.010 to 0.035 in. thick, to be impermeable to xenon at temperatures ranging from 23 to 473°C, and they also could detect no penetration of a 0.020-in.-thick sheet of type-304 stainless steel by xenon at 510 to 650°C. The tracer technique employed in this investigation would have permitted measurement of permeabilities as low as 0.55×10^{-12} std cm^3/cm^2-sec/cm Hg gas pressure with a metal specimen 1 mm thick, acording to the authors.

Gordon et al. [150] also attempted to measure the diffusion rate of helium through 2S aluminum at temperatures up to 500°C and found that the rate was below their limit of detection (1.8×10^{-10} std cm^3/cm^2-sec). However, Murray and Pincus [151] report that they were able to measure the diffusivity of helium through aluminum by using "sandwiches" produced by alternately bombarding an aluminum surface with helium ions and then covering the bombarded surface with a layer of vapor-deposited aluminum. They report that the diffusivity increased from about 6×10^{-10} std cm^3/cm^2-sec at 400°C to 2.2×10^{-9} at 570°C. They found lower diffusivities ($\sim 4 \times 10^{-11}$) when the helium was produced in aluminum-lithium-alloy material by irradiation in a reactor. Murray [152] attributes the lower values and the failure to vary with temperature in the expected proportion to the entrapment of the gaseous atoms in bubbles. Murray and Pincus [151] also report values of about 1×10^{-9} for the diffusivity of tritium in irradiated aluminum-lithium alloy in the same temperature range. The conditions employed to produce the specimen in this investigation are certainly not typical of the conditions that exist in most metal-clad fuels. The measured diffusivities, therefore, may be considered as upper limits rather than as actual escape rates, especially since krypton and xenon diffusivities would be expected to be lower than that of helium.

A more thorough study of fission-product diffusion through metal has been performed by Battelle investigators [153]. They used an 80% iron-20% chromium alloy in the form of 0.002-in.-thick foils, 0.010-in. wall tubing, and 0.050-in.-thick plates in their studies. This alloy has a body-centered cubic structure similar to that of commercial ferritic stainless steels from room temperature to its melting point. The tubing was used as cladding for UO_2 pellet material for in-pile tests. Fission products were introduced into the thin foils by recoil impregnation for postirradiation fission-gas-release studies, while the thicker sheets were employed in investigations of postirradiation behavior of solid fission products that were also introduced into the alloy by recoil. Significant results are shown in Figs. 4-1, -2, -3, -4. The authors summarize their observations on postirradiation gas release as follows:

1) Release of Xe^{133} with time gives an initial rapid release followed by release that is approximately linear with the square root of heating time for the remaining 6 to 8 hr of heating.
2) For heating times in excess of 10 hr and up to 260 hr, release is linear with the square root of time but occurs at a still lower rate than is observed at the intermediate time of heating.
3) A burst of Xe^{133} is given off each time the specimen is heated to a higher temperature, but only a small burst is given off when the specimen is subsequently heated to a lower temperature.
4) The total amount of gas given off in the initial bursts when a specimen is heated through a series of temperatures is approximately equal to the amount released in a single heating at the highest temperature.
5) The rate and total amount of gas released in the initial burst are both temperature and concentration dependent.
6) Removal of a surface layer following heating permits a second large burst to occur during heating to the same temperature.
7) Removal of a surface layer from the specimen before heating does not prevent the occurrence of a burst during heating.
8) The release rate of Xe^{133} following the burst may not be considered independently of the fraction of gas released in the burst.

The in-pile fission-gas-release measurements show that

1) Release of fission gases through 10-mil iron-20 wt. % chromium alloy cladding is quite low at temperatures up to 1200°C, with the exception of Xe^{133}. Release of this isotope increases significantly at 1100°C and above.
2) Release fractions of the various gaseous isotopes, with the exception of Xe^{133}, are roughly proportional to their decay constants. On a log-log plot, the proportionality is -0.5.

Studies of diffusion rates of solid fission products show that the diffusion rates for Ce^{141} and $Zr\text{-}Nb^{95}$ in the alloy are similar but significantly less than the diffusion rate for Ru^{103}. Diffusion rates of other solid fission products in this material have apparently not been reported.

The burst effect mentioned above is illustrated by the data in Fig. 4-1 [153], while the temperature dependence of the initial release rate is shown in Fig. 4-2 [153]. A decreased release rate noted with increasing gas concentration and the results of differential heating tests are interpreted by the authors as indicating that the gas "precipitated" or clustered in bubbles.

Results of in-pile release tests are shown in Fig. 4-3 [153]. The wide range of Xe^{133} rates is said to indicate increasing R/B (rate of release/rate of formation) with time. The R/B values are based on equilibrium production rates, and approximately 30 days would be required for Xe^{133} to reach its equilibrium value.

Diffusion of solid fission products in the alloy is shown in Fig. 4-4 [154]. The distribution of fission products was found to be roughly symmetrical about the original interface, and Ru^{103} obviously diffused more rapidly than the other species studied.

In summary, this program at Battelle Memorial Institute produced very interesting information on transport of gaseous and solid fission products through stainless-steel-like materials. Again, it should be noted that tritium is a fission product whose presence outside the clad may not be de-

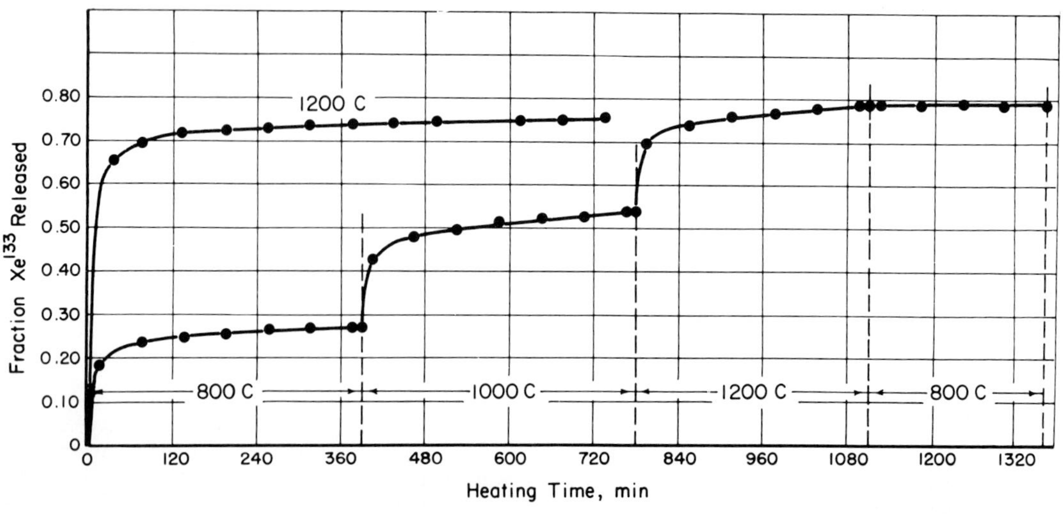

FIG. 4-1 Release of Xe^{133} from Fe-20 wt. % Cr alloy showing additive nature of gas burst.

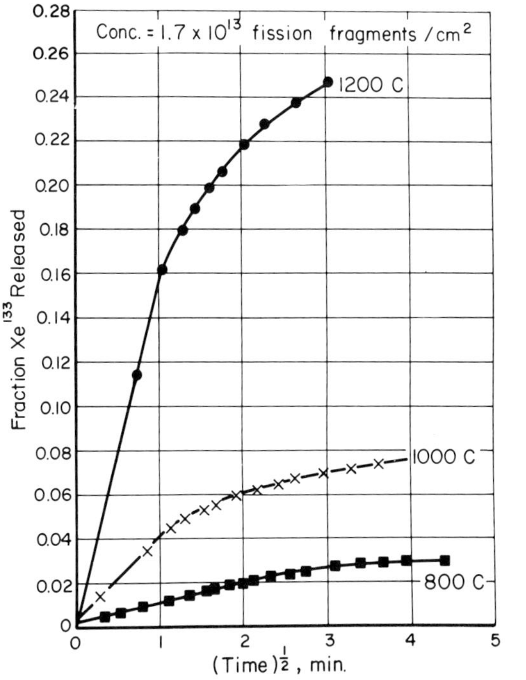

FIG. 4-2 Effect of temperature on the release of Xe^{133} from Fe-20 wt. % Cr alloy during short heating times.

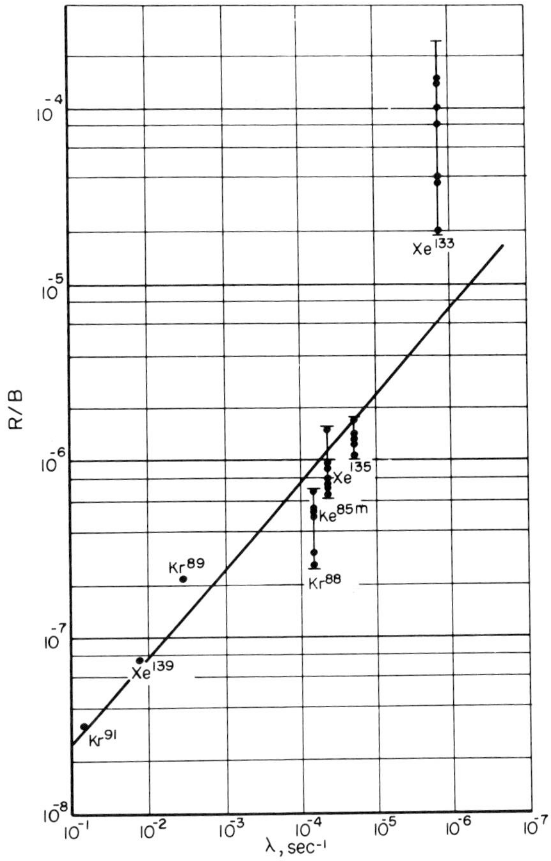

FIG. 4-3 Ratio of release rate to production rate versus decay constant for fission gases produced by a flux of 10^{12} neutrons/cm^2-sec at 1200°C (2192°F). The specimens were Fe-20 wt. % Cr tubes with 0.010 in (0.25 mm) wall thickness.

tected, since it emits a very weak beta ray that requires special detection methods. Recoil-impregnated materials show significant diffusion rates at temperatures of 800 to 1200°C, but a maximum in-pile escape rate (R/B) of 2×10^{-4} was observed for Xe^{133} at 1200°C. It should be pointed out that in the pellet-type fuel configuration used in most power reactors, a very small fraction (approximately 0.1%) of the fission products will be introduced into the cladding by recoil and currently available information indicates that diffusion rates of gases through metals at temperatures up to 1200°C (2192°F) are too low to make a significant contribution to the hazard of a loss-of-coolant accident. It would be desirable to have data on diffusion rates at temperatures up to the melting point of stainless steel. Dispersion of small fuel particles in stainless steel will favor introduction of fission products into the matrix material by the recoil process.

4.2 Effusion through Ceramic, Graphite, and Carbon Fuel Coatings

The economic incentive for high temperature in nuclear reactors has led to the development of all-ceramic fuel elements. It is essential that such elements retain fission products efficiently under operating conditions in most reactors and, consequently, there is a great deal of information in the literature on the effectiveness of various fuel-coating materials. Comparatively little of the published data is relevant to loss-of-coolant accidents in which fuel elements are heated to very high temperatures. The state of the art of preparing various types of coated particles and available information on fission-product release were summarized in a symposium on "Coated Particle Fuels", resulting in publication of six papers [155 a-f].

The high melting points of ceramic fuel elements, especially those composed of carbides, pyrolytic carbon, and graphite, make data on rates of diffusion of fission products through these materials of interest in hazards evaluations, because it appears quite possible that fuel elements of this type could remain intact at high temperatures for fairly long periods of time in loss-of-coolant accidents. "High temperatures" are arbitrarily defined here as 1500°C (2732°F) and above, since this represents the approximate upper limit now being considered for advanced gas-cooled reactors. Some data obtained at lower temperatures are included in this review because of a lack of high-temperature data.

4.2.1 Oxide Fuel Coatings

Results obtained when a graphite sphere fueled with 127μ UO_2 particles coated with a 42μ layer of vapor-deposited Al_2O_3 was irradiated are shown in Fig. 4-5 [156]. These data show that although

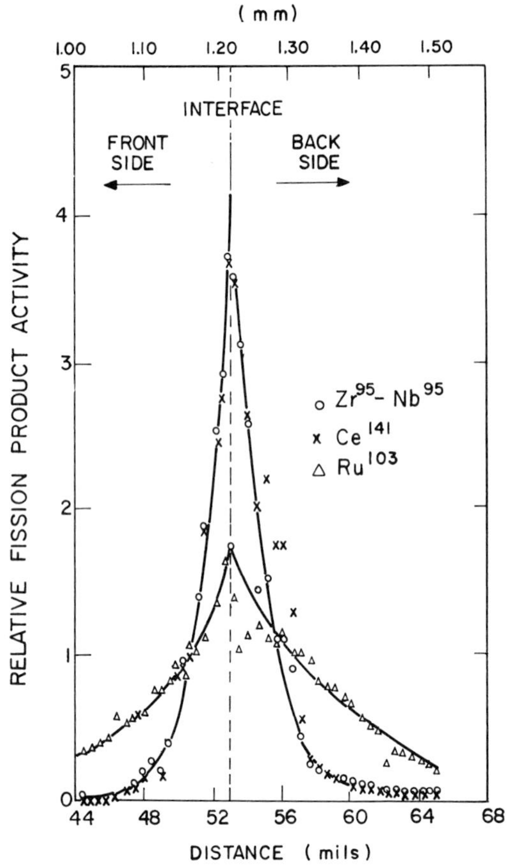

FIG. 4-4 Fission-product distributions in Fe-20 wt. % Cr alloy heat-treated for 98.5 hr at 1000°C (1832°F). Fission-product concentration ≈ 2 x 10^{13} recoils/cm^2.

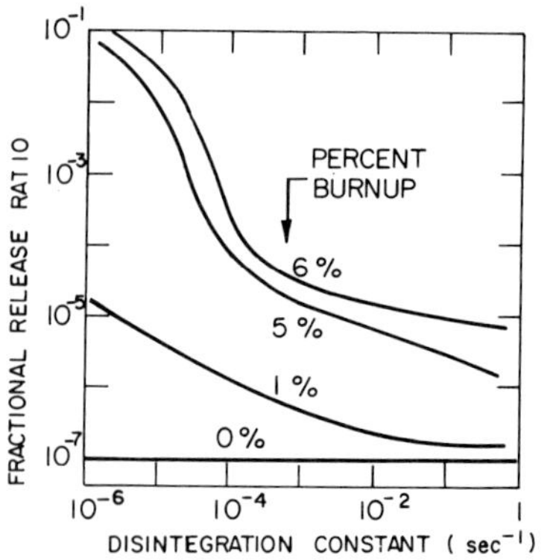

FIG. 4-5 In-pile behavior of fission-product release from alumina-coated UO_2 particles.

the initial fission-product retention of this fuel material was excellent, it obviously deteriorated rather drastically when exposed to reactor irradiation at temperatures between 510 and 730°C. Battelle investigators have reported [157] that alumina coatings on UO_2 fail rapidly at low temperatures. Results obtained with fuel consisting of an inner layer of pyrolytic carbon under the alumina coating were stated [158] to be similar to those shown in Fig. 4-5 when exposed in a reactor at 1050°C. Studies of fission-product behavior in alumina-coated UO_2 fuel particles are reported by Raines and Goldthwaite [159]. Their data show that when particle coatings of the type used in these tests are intact, diffusion of fission gases through them is so slight that it can be neglected and that a major release occurs as a consequence of particle cracking. In these tests, cracking is said [159] to have become important at between 1 and 3 at. % burnup. Later improvements in methods of applying this type of coating [155a] have resulted in production of UO_2 particles that retained their integrity at burnups of at least 10 at. % [155b].

Data on the performance of other oxide fuel coatings at high burnup levels do not appear to be available. Data obtained with uranium-impregnated MgO at tracer-level irradiation [160] permitted a comparison of diffusion rates in this media with those in graphite. Molybdenum and arsenic were found to be more volatile (showed higher diffusion rates) in MgO than in graphite; palladium, tellurium, iodine, and barium showed about the same behavior in both matrices; while cadmium, silver and probably strontium, appeared to be less volatile in MgO than graphite.

The causes of cracking of the oxide fuel coatings have not been adequately elucidated, but it is clear that fuels having this type of coating are not likely to become important in reactor technology unless coatings are developed that can maintain their integrity at high fuel burnup and when exposed to large thermal shocks. Further discussion of safety aspects of oxide-coated fuels should therefore be delayed until detailed descriptions of fuel configurations proposed for actual reactor use are available.

4.2.2 Siliconized Silicon Carbide Coating

Data have been reported [159] on fission-gas release from a sphere fueled with UC particles and coated with siliconized silicon carbide (Si-SiC). Details of the coating and the method employed to apply the coating to the sphere were not presented. Initial fission-gas release from the specimen was below detection level, but increased rather abruptly by a factor of about 10^6, indicating coating failure, after exposure in a reactor to a burnup of about 1% with a surface temperature of 720°C. After the failure occurred, the release fraction was found to be related to half-life, being greater for the longer-lived isotopes. The results are said [159] to be typical of fission-gas release from uncoated fueled graphite, indicating that coatings of this type have little or no effect on fission-gas release after they crack.

4.2.3 Graphite

The term graphite covers a large family of materials. Scott states [161] that ordinary reactor-grade graphite has a permeability of 1 cm^2/sec, while that of low-permeability materials is of the order of 10^{-8} cm^2/sec. Consequently, in a discussion of the penetration of fission products through graphite, it is necessary to specify the type of material involved. Correlation of properties of graphites and their performance requires a detailed knowledge of methods of preparation that, in some cases, is considered proprietary information. Most of the data in the literature on fission-product penetration of graphite and pyrolytic carbon were obtained as part of fuel-element-development efforts, and the emphasis has been on performance under proposed reactor operating conditions rather than under conditions that might be present in potential accidents. However, some tests have been conducted at quite high temperatures because of the above-mentioned interest in high-temperature reactors, especially advanced gas-cooled reactors.

Early studies [162-165] of the diffusion of fission products in uranium-impregnated porous graphite materials (AGOT-KC, EBP, and AUF) showed rather different, and to some extent surprising, behavior of various fission products. The important results of these studies can be summarized as follows: cesium and strontium diffused quite rapidly at 1500°C, while xenon moved relatively slowly at this temperature. Barium, tellurium, and iodine diffused rapidly at 1900°C, while ruthenium, molybdenum, zirconium, and the rare earths, except for praseodymium, diffused slowly; only 2% of the zirconium diffused in 4 hr at 2200°C. Rates of diffusion of iodine and tellurium in uranium-impregnated AUF graphite are shown in Fig. 4-6 [164]. More recent studies have served to confirm, at least qualitatively, the relative rates of diffusion reported earlier. One investigation [160] covered a wide range of fission-product species, as shown by Table 4-1. At 2600°C, more than half of all species tested except palladium had diffused out of the graphite in a 30-sec heating period. These results indicate that biologically potent fission products such as strontium and iodine will penetrate ordinary reactor-grade graphite at a rapid rate in overheated fuels containing this material. Bromley [166] has published a thorough review of transport and diffusion of fission products in graphite, which should be consulted for a more detailed examination of this subject. Bryant et al. [167] have investigated the effect of several variables on rates of escape of fission products from graphite in the 1550 to 2400°C temperature range. Typical data obtained are shown in Table 4-2 and in Fig. 4-7 [167]. The shape of the curve for barium is said to be typical of that for strontium, yttrium, cesium, and the lanthanide elements, while the curve for tellurium is reported to be typical for antimony and iodine. It is concluded that diffusion through the graphite matrix is a rate-limiting process in the loss of some metallic fission products, but the behavior exhibited by tellurium, according to these authors and others, can apparently be explained on the basis that fission products that recoil into graphite grains or crystals will move slowly to pores and rapidly through pores to sample surface.

Gas transport through porous media such as AGOT graphite has been the subject of theoretical and experimental studies at ORNL [168-174] and elsewhere. The model developed to correlate gas-transport phenomena in porous materials has also been shown [172, 173] to be applicable to low-permeability graphite.

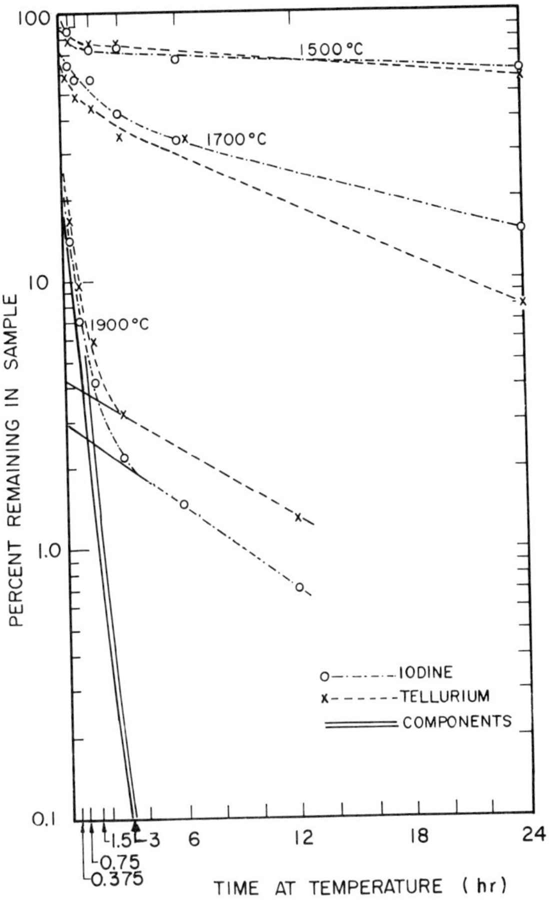

FIG. 4-6 Diffusion of iodine and tellurium from uranium-impregnated AUF graphite.

Data reported by British investigators [175] showing that no correlation exists between diffusion rates of cesium, barium, and strontium in different graphites and their gaseous permeabilities are given in Table 4-3. This lack of correlation was interpreted as indicating that the rate-controlling process in the diffusion is an activated surface migration on the pore walls of the graphite, possibly accompanied by some intergranular diffusion. Arrhenius plots of data obtained with a commercially produced graphite (Morgan Crucible Company, EY9) are shown in Fig. 4-8 [175]. These workers emphasize the fact that the kinetics of the escape of fission products from graphite are complicated and that diffusion coefficients are no more than a convenient way of expressing the rates of escape for purposes of comparison.

Orth [176] has studied diffusion losses of lanthanide and actinide elements from graphite at temperatures from 1600 to 2600°C. He finds a close correlation between diffusion rates of the lanthanide elements from graphite and their boiling points, and similar relationships appear to hold with those actinide elements for which boiling-point information was available. Data on volatilization of uranium, neptunium, and plutonium from

TABLE 4-1

Percent Retention of Fission Products in Graphite at 2400°C as a Function of Time

Element	Percent Remaining			
	30 sec	60 sec	120 sec	240 sec
As^{77}	80	44	48	-
Br^{83}	13	2.2	1.0	-
Sr^{89}	2.8	0.18	0.03	-
Ru^{103}	-	-	-	70
Rh^{105}	-	-	-	92
$Pd^{109,112}$	42	31	34	23
Ag^{111}	<<0.1	-	-	-
Cd^{105}	<<0.1	-	-	-
Sb^{127}	20	-	11	6.2
$Sn^{125,121}$	-	-	-	0.1
Te^{132}	16	5.2	4.8	3.4
$I^{131,133}$	5	3.3	3.0	1.4
Cs^{137}	-	-	-	0.26
Ba^{140}	<<0.1	-	-	-
Ce^{143}	-	-	-	27
Nd^{147}	-	-	-	19
$Eu^{156,157}$	-	-	-	74

TABLE 4-2

Fission-Product Retention in Admixture Samples[a]

Element	Temperature (°C)	Time (min)	Fraction remaining
Kr	1550	10	0.77
Sr	1550	5	0.74
Y	2400	8	0.78
Zr	2400	8	0.98
Mo	2400	8	0.97
Ru	2400	10	0.92
Ag	1550	5	0.015
Sb	1925	10	0.25
Te	1925	180	0.33
I	1925	180	0.30
Xe	1925	10	0.70
Cs	1550	5	0.57
Ba	1925	10	0.48
Ce	2400	8	0.47
Eu	1925	10	0.05

[a]The samples were 0.230 in. diam × 3/8 in long, with 1 to 2μ uranium particles.

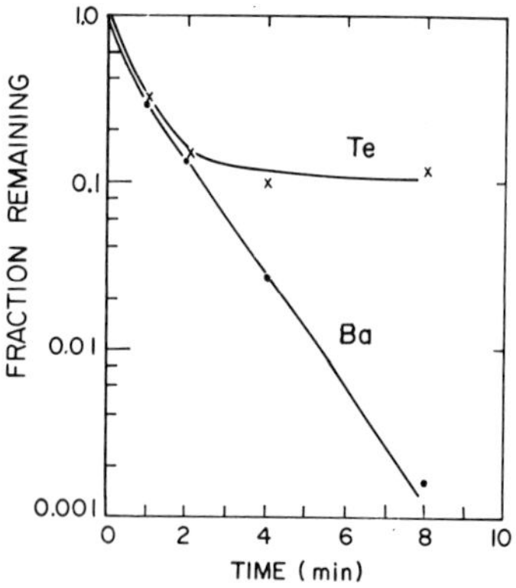

FIG. 4-7 Losses of barium and tellurium from admixture samples as a function of time at 2400°C (4350°F).

solution-impregnated graphite pins given in Table 4-4 indicate greater separation of uranium and plutonium than has been observed in oxide systems.

4.2.4 Pyrolytic Carbon

A great deal of effort has been, and is being, expended to develop pyrolytic carbon coatings for fuel particles, especially for UC particles. The status of these efforts in the latter part of 1962 was reviewed in a symposium [177]. Goeddel [155d] has reviewed work on the development and evaluation of coated-particle fuel for the Peach Bottom high-temperature gas cooled reactor. In 1968, the Fort St. Vrain reactor fuel design adopted an (SiC) coating, which should control the releases of all fission products, not just iodine, cesium, and the rare gases. Therefore, damage to fuel particles becomes more important. One such test on a transiently heated previously irradiated high temperature gas-cooled reactor (HTGR) fuel element was carried out in 1967 at the ORR [178, 179]. A test on a bonded bed of coated particles (UC_2 particles with a 43μ buffer layer, a 22μ SiC layer, a 69μ isotropic carbon layer, and a thin anisotropic sacrificial layer) under high-temperature blocked-channel conditions was carried out in 1968 [180]. About 10% of the coatings failed during the 1700°C irradiation. Theoretical studies have been carried out on fission-product release and transport in HTGR coated-particle fuel [180, 181]. The steam-graphite reaction has been studied theoretically and experimentally in a preliminary way [180, 182-185]. Relatively few of the coatings developed have been adequately tested under proposed service conditions, and even fewer have been tested at temperatures of interest in hazards evaluations.

The fission-product-release data obtained with pyrolytic-carbon-coated particles have been mostly release rates of the rare gases, which are of less interest to nuclear safety people than release rates of iodine and some other fission products. Nevertheless, gas-release data do show the extent of survival of the coatings under some conditions that might exist in reactor accidents and may give an indication of the order of magnitude of release rates to be expected for other fission products.

TABLE 4-3

Comparison of Gas Permeability Constants and Diffusion Data for Four Different Graphites (500 to 1100°C)

10^{12} x Darcy permeability (cm^2)	B_o $(cm)^{2a}$	K_o $(cm)^a$	Pore radius (μ m)	-log D_o		
				Cs	Sr	Ba
200	5×10^{-10}	2×10^{-6}	2	2.1	4.1	-
4	1×10^{-11}	4×10^{-7}	0.6	2.0	3.5	3.8
2	2×10^{-12}	1×10^{-7}	0.4	1.7	2.9	3.5
0.008	1×10^{-13}	1×10^{-8}	0.16	2.4	3.3	3.9

[a] B_o and K_o are terms in the equation $K\ (cm^2/sec) = \frac{B_o P_m}{\mu} + \frac{4}{3} V_z K_o$.

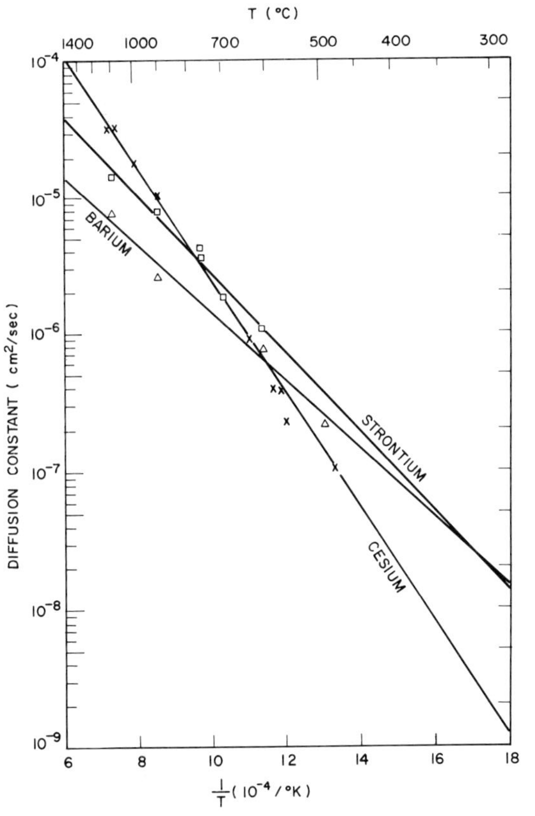

FIG. 4-8 The variation with temperature of the diffusion constants for cesium, strontium, and barium in a commercial graphite.

TABLE 4-4

Loss of Uranium, Neptunium and Plutonium from U^{238}-Impregnated Graphite Pins Heated in Helium

Temperature (°C)	Length of heating (hr)	Percent lost during heating		
		U^{238}	Np^{239}	Pu^{239}
2050	4.0	<1	≤ 1	46
2100	2.5	2.5	-	-
2400	2.0	9	45	98

Postirradiation studies of Xe^{133} release from pyrolytic carbon were reported by Morrison et al. [186]. Their data, showing the time-dependent Xe^{133} release at 1200°C, are displayed in Fig. 4-9 [186].

Data on the half-life dependence of the fission-gas-release rates at 815°C, obtained with uncoated UC particles and with similar particles having different types of pyrolytic carbon coating, are shown in Fig. 4-10 [177a]. These data clearly demonstrate the superiority of duplex coatings (laminar inner layer and columnar outer layer). The fact that uncoated and coated particles showed about the same half-life dependence for release is interpreted [177a] as indicating that the mechanism of release from cracked coatings in these tests was similar to that from uncoated particles. Particles used to obtain the data shown in Fig. 4-10 were heated to higher temperatures after defective particles were removed [187]. A small burst of gas activity was noted when the particles were heated in purified helium, but the release rate dropped in a short time to 10^{-4}%/hr of the Kr^{85} content of the sample. Particles heated to 1900°C showed a rate of release of 0.02%/hr and those heated to 2100°C, a rate of 0.13%/hr. Some small bursts of activity corresponding to a fraction of the Kr^{85} content of a single particle were noted in these experiments, indicating cracking of particle coatings. Data obtained at General Atomics [177b, 188] show that lightly irradiated pyrolytic-carbon-coated (Th, U)C_2 particles with 30μ- to 60μ-thick coatings gave negligible Xe^{133} release on heating at 1700 to 2000°C for 2 or 3 days but that they deteriorated in a matter of minutes if heated to 2400°C or higher.

Data [177b] obtained with the same type of particles in a graphite matrix show temperature-independent release of Xe^{133} of about 0.5% in 48 hr in the temperature range 1400 to 2000°C. This release was ascribed to particles that cracked when the compacts were manufactured. Release values reported for I^{131} and Te^{132} in this temperature range were of about the same order of magnitude, but high Ba^{140} release rates were noted at 1800°C. The technique employed in these experiments unfortunately did not yield reliable data except for xenon at the upper end of the temperature range. The high escape rate for barium as compared to xenon was interpreted [177b] as indicating that barium can dissolve in and diffuse through pyrolytic carbon coatings that are impermeable to xenon.

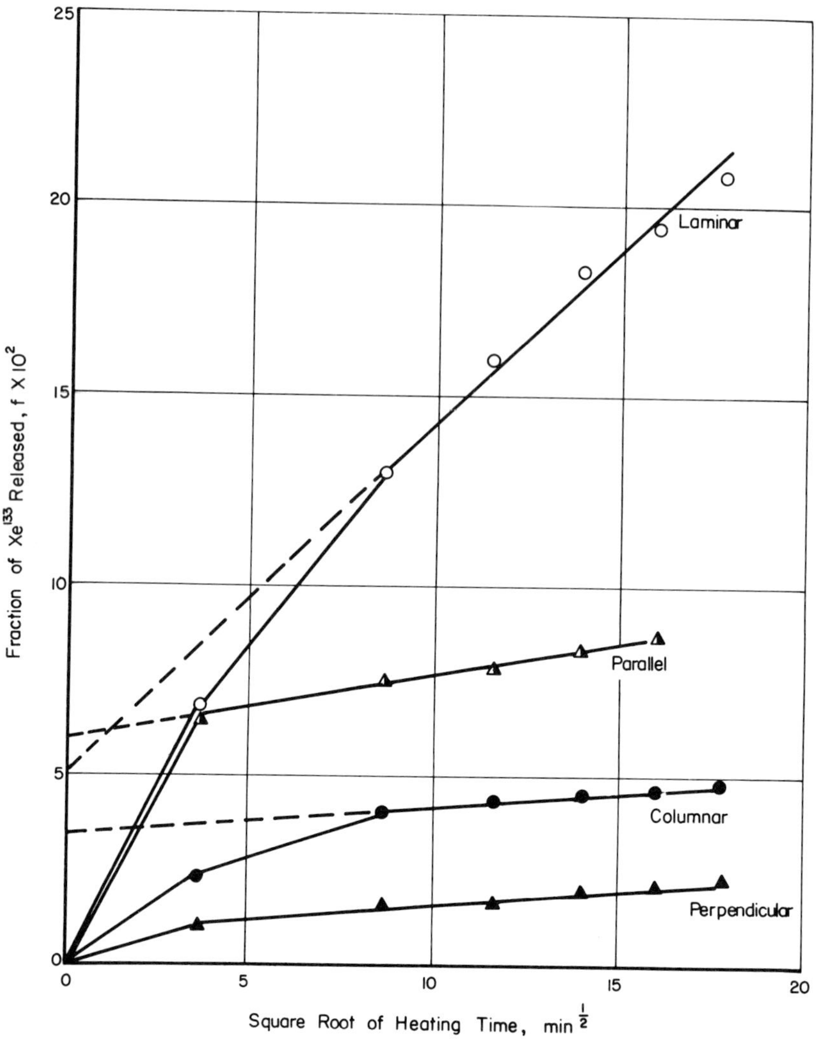

FIG. 4-9 Typical time-dependent Xe^{133} release from pyrolytic carbon at 1200°C (2190°F).

Columnar-type pyrolytic-carbon-coated UC particles that had been irradiated to 10^{17} nvt were heated in a collection system for 100 hr at 1400°C [189]. Preliminary data indicated that 3% of the Ba^{140} content of the particles and a similar fraction of the Te^{132}, I^{131}, and Cs^{137} were released. The fractions of Ce^{141} and Ru^{103} released were about 0.1 of the Ba^{140} value (0.3%), while the fractional release of Zr^{95} was 0.08%. These release values are much higher than those obtained at General Atomics [177b] at this temperature with lightly irradiated pyrolytic-carbon-coated (Th, U)C_2 particles in a graphite matrix. Data on the release of cesium from duplex- and triplex-coated particles, given in Table 4-5 [190], show that release rates are quite low, at least up to approximately 1800°C.

It appears that a great deal more data on fission-product release from pyrolytic-carbon-coated fuel particles will be needed, especially at temperatures above 1500°C.

Several fission products, including xenon, krypton, tellurium, iodine, cesium, and strontium, diffuse through graphite at a significant rate at 1550°C, while others, such as zirconium, molybdenum, and ruthenium, are relatively immobile at 2400°C. Silver, cadmium, and tin are the most mobile elements in this medium studied to date. The available data indicate that when fuels made up of pyrolytic-carbon-coated particles in a graphite matrix are heated to temperatures of 1500 to 2000°C, penetration of prolytic-carbon coating will probably be the rate-controlling factor. Sufficient data are not available at present to permit a comparison of loss rates from oxidized and intact fuels of this type.

5. BEHAVIOR OF FISSION PRODUCTS IN REPORTED FUEL FAILURES*

A review of nuclear reactor incidents is presented in Chapter 11. Sec. 4 of this review consists

*See also Volume 1 (especially Chapter 11) and Chapter 13 in this volume.

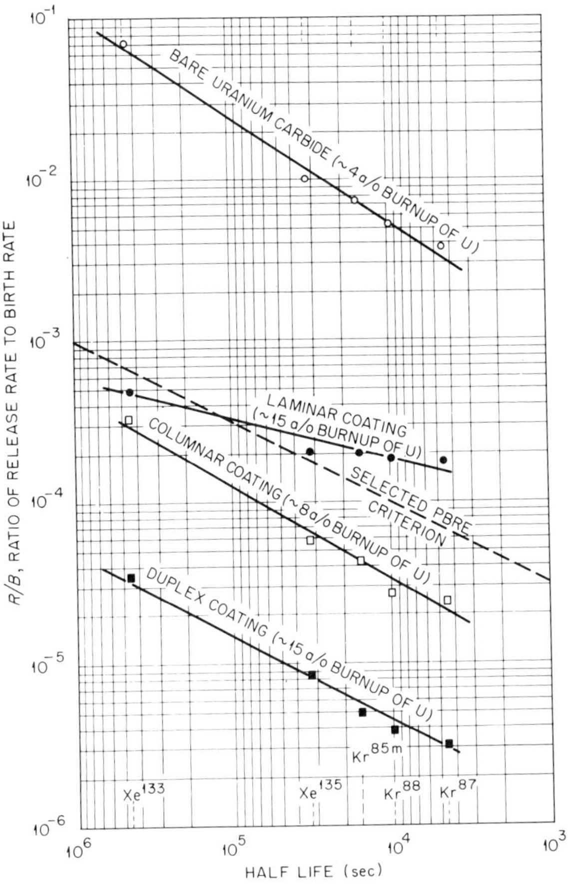

FIG. 4-10 Relation between the ratio (R/B) of release rate to birth rate and the half-life for release of inert gases from pyrolytic-carbon-coated and uncoated UC particles at 816°C (1500°F). The calculation of R/B for Kr^{85m} is based on fission yield of 2.0%.

of incidents involving fuel-element failures, but very little information is reported on fission-product behavior. This appears to be a common occurrence in reports of reactor incidents, and it is interesting to speculate on the reasons for the scarcity of this type of information in the literature. Most of the reports on reactor incidents contain much more of the details of circumstances that led to the accident than of its consequences. The dispersal of fission products in some incidents was of such a nature that only qualitative statements concerning fission-product behavior were possible, but, even in some of these cases, the qualitative information reported leaves much to be desired. It seems that, at most reactor installations that had reportable incidents, accident-released fission products were regarded as an annoyance rather than as a source of information that might be useful to reactor designers and to operators of other reactors of the same type.

The information shown in Table 5-1 was assembled in an effort to draw together widely scattered data on fission-product behavior in reactor accidents. It may be noted that a number of the incidents listed in Chapter 11 are not included in this table, in most cases because of the above-mentioned lack of reported data on release of fission products. Furthermore, inspection of Table 5-1 shows that rather scanty data were available for some of the incidents included. This review will have served a useful purpose if it does nothing more than encourage people responsible for reporting reactor incidents to give more adequate information on the release of fission products.

Several statements seem to be warranted by the data in Table 5-1. The first is that every one of the accidents listed (the BORAX-4 incident, as noted in the table, cannot be considered an accident) involved fuel or clad melting or burning or both. (See also discussion in Sec. 3.3.4.) It is also worthy of note that the fuel in almost all these accidents was either metallic uranium or an alloy of uranium and that the cladding material in most cases was aluminum, a very low-melting material. It is encouraging that no fuel-element failures have been reported in reactors employing high-melting fuel materials (e.g., UO_2, m.p., 2850°C or 5162°F) clad with high-melting materials such as zirconium (m.p., 1900°C or 3450°F) or stainless steel (m.p., ~1500°C or 2730°F), although impressive numbers of megawatt-days of power have been generated by reactors employing this type of fuel (Shippingport, Dresden, Yankee, etc.).

It is interesting to find that evidence of a metal-water reaction was noted in only four of the incidents involving water-cooled reactors, NRX (1952), EL-3, NRU, and SL-1. It seems unlikely that the chemical reaction contributed a significant fraction of the total energy involved in the first three incidents and may have been as much as 20% in the last. Chapter 11, p. 682, describes the SPERT-1 destructive test, which

TABLE 4-5

Release of Cesium from Duplex- and Triplex-Coated Particles during Postirradiation Heating

	Cs^{137} released/hr (%)									
Temp. (°C)	1370	1370	1520	1600	1660	1700	1800	1800	2000	2100
No particles failed	0	1	1	0	0	0	0	1	1	3
NCC Duplex	0.002	0.06	-	0.1	-	0.04	-	0.6	2.0	2.0
GA Triplex	0.005	-	0.002	-	0.004	-	0.006	-	-	-

TABLE

Behavior of Fission Products in

Reactor	Date of incident	Reactor type	Reactor fuel	Type of incident	Cause of incident
ORNL graphite reactor	1947-1948	Graphite moderated, air cooled	Al-clad-natural-U slugs	Fuel burning and melting	Cladding failures resulted in oxidation of U, and, in some cases, the U_3O_8 plugged the channel and caused fuel to melt
NRX	December 1952	D_2O moderated, water cooled	Al-clad U rods	Fuel melting and U-H_2O reaction	Reduced coolant flow caused light water to be boiled out
NRX	July 1955	D_2O moderated, water cooled	Al-clad U rods	Melting of experimental PuAl alloy fuel rod	Water entered sheath and steam pressure lifted sheath out of contact with casting can; alloy melted and penetrated sheaths and reactor wall
G1	October 1956	Graphite moderated, air cooled	Al-clad natural-U rods	Fuel burning and melting	Reduced coolant flow
Windscale No. 1	October 1957	Graphite moderated, air cooled	Magnox-clad natural-U	Fuel burning and melting	Local overheating in reactor during Wigner energy release
EL-3 [a]	April 1958	D_2O cooled and moderated	Al-clad, hollow, U-2% Mg alloy	Fuel melting and oxidation	Vibration caused break in poorly supported fuel element, which dropped to bottom of tank and melted
NRU	May 1958	D_2O moderated and cooled	Al-clad U	Fuel burning following explosive failure of waterlogged fuel element	Waterlogged element burst, lodged in core, and broke during removal efforts; part of element burned
OMRE	October 1958	Organic cooled moderated	Stainless-steel-clad, UO_2-stainless-steel dispersion plates	Melting of Al cladding of	Al fins on an experimental fuel assembly strained out particles in organic coolant and reduced flow
BORAX-4	March 1959	Boiling water	UO_2-ThO_2 pellets, Al-clad, Pb-bonded	Deliberate operation with defective fuel elements	22 of 69 fuel elements had cladding defects
SRE	July 1959	Na cooled, graphite moderated	Stainless-steel-clad U rods bonded with NaK	Fuel melting	Reduced coolant flow caused by decomposition of Tetralin that leaked into system resulted in formation of Fe-U eutectic and cladding failure
WTR	April 1960	Water cooled and moderated	U-Al alloy in Al-clad plates	Fuel melting	Deliberate operation with one-third normal coolant flow or defective fuel bonding or both
SL-1	January 1961	Boiling water	U-Al alloy in Al-clad plates	Fuel melting and high-pressure steam generation	Withdrawal of control rod beyond specified limit
ETR	December 1961	Water cooled and moderated	U-Al alloy in Al-clad plates	Fuel melting	Reduced coolant flow (35% of normal) caused by Lucite sight glass
MTR	November 1962	Water cooled, Be moderated	Al-clad U-Al alloy plates	Fuel melting	Gasket material from roof of seal tank restricted coolant flow
ORR	July 1963	Water cooled and moderated	U-Al alloy in Al-clad plates	Fuel melting	Reduced coolant flow caused by neoprene gasket
Enrico Fermi	October 1966	Na cooled fast breeder	U-10 wt.% Mo clad in stainless steel	Fuel melting (2 elements warped)	Coolant flow blockage by loose metal plate

[a] See also Chapter 11, Table 4-1.

5-1

Reported Reactor Fuel-Failure Incidents[a]

Extent of contamination	Major fission products released	Amounts of fission products released (curies)	Comments and references
U_3O_8 particles and accompanying fission products contaminated a large part of the restricted area adjacent to the stack	Mixed nonvolatile fission products in UO_2 particles found on ground; volatile fission products dispersed to atmosphere through stack	Not reported; can be calculated from irradiation times, estimated flux, and slug weight	Showed need of stack filters; particles found ranged from about 90 to 400 μ [191, 192]
Reactor building badly contaminated; active cloud carried detectable activity 1/4 mile; low-level river contamination	Volatile and gaseous	10,000; long-lived fission products in coolant water; high intensity cloud of airborne short-lived fission products	Nontypical operating conditions existed at time of accident due to tests with reactor [193, 194]
Reactor vessel contaminated with Pu	Apparently not reported; can be assumed that mixed fission products, excluding Xe and Kr, accompanied the Pu	Not reported	Only reported meltdown of a Pu fuel element in an operating reactor, except for the experimental reactor Clementine [10c, 195, 196]
Blocked fuel channel held most of fission products; activity below maximum permissible limit outside station	I and rare gases	20 to 50 (estimated)	Reactor remained at full power for about 20 min after trouble started [197a]
Widespread; I^{131} contamination of milk supply of large area	I^{131}, Te^{132}, Cs^{137}, Sr^{89}, Sr^{90}, Ru^{106}, Ce^{144}	I^{131}, 30,000; Te^{132}, 12,000; Cs^{137} 600; Sr^{89}, 80; Ru^{106}, 80; Ce^{144}, 80; Sr^{90}, 2	Only reactor incident to date that caused significant damage to public [59] (See also Chap. 11, p. 633)
D_2O coolant and He cover gas	Xe and I isotopes	Xe^{133}, 100 curies in He cover gas; I^{131}, I^{133}, Xe^{131}, Xe^{133}, and Xe^{135} in D_2O, amounts not reported	1.6 kg of U (estimated) oxidized; showed that only rare gases transferred to cover gas [198]
Reactor building badly contaminated; detectable contamination in about 100 acres of adjacent land	Mixed fission products found in surface contamination and stack filters; less Ru and Ba^{140} + La^{140} than expected	Total amounts not measured; surface contamination up to 2.5 r/hr; air, 100 yd from building 1 hr after burning, 200,000 dis/(min)(m^3)	This is the most serious incident reported involving a waterlogged fuel element [199, 200] (See also Chap. 11, p. 688)
Mixed fission products in coolant; Xe in cover gas	Xe, Kr, Rb, I, Ba, La, and Te	1000 (total)	Showed need of more efficient filtration of coolant; coolant was purified and reused [201]
Reactor water and air around pool	Fission gases, Xe^{138} and Kr^{88}	Xe^{138}, 4 curies/min; Kr^{88}, 0.7 curies/min	This cannot be considered an accident [130, 201a]
Na coolant only	Zr^{95}+Nb^{95}, Ce^{141}, Cs^{137}, Ba^{140} + La^{140}, Ru^{103}, I^{131}, Cs^{134}	Zr^{95} + Nb^{85}, 306; Ce^{141}, 131; Ba^{140} + La^{140}, 36 each; Cs^{137}, 27.7; Ru^{103}, 20.9; I^{131}, 16.3; Cs^{134}, 0.4	Since all fission products except rare gases remained in coolant, the fraction of fission products released could be accurately determined (0.3% of fuel element activity) [202, 203] (See also Chap. 11, p. 638)
Primary coolant and air in vicinity	Fission gases, Xe, and Kr	5000, mixed fission products to coolant; 261, airborne Xe and Kr just after accident; 800, total	A normal fuel element would probably not have failed [197c, 204, 205] (See also Chap. 11, p. 645)
Building badly contaminated; extensive low-level deposition of I^{131}	I^{131} outside building; mixed fission products inside	I^{131}, 84 curies outside building; no report on mixed fission products	Reactor not operating at time of accident; only reactor incident involving fatalities [197c, 206, 207] (See also Chap. 11, p. 653)
Primary coolant and atmosphere	Mixed fission products to coolant water; fission gases and daughters (particles) to stack	6.0, fission gases; 0.4, particulate; immediate release through stack; 42, mixed fission products to water; continuing release to water and atmosphere from unclad U^{235}	Showed need of more precautions in order to keep extraneous materials out of core [208, 209] (See also Chap. 11, p. 689)
Mixed fission products in coolant water, some airborne activity in building and at area monitors	I^{131}, I^{133}, I^{135}, Ba^{140}, and Sr^{91} found in coolant water	Release as factor increase in activity over normal; total, 13 to 16; I^{131} and I^{133}, 62; I^{135}, 126; Ba^{140}, 25; Sr^{91}, 36	Loss of 0.7 g of U^{235} from one fuel plate (see also ETR and ORR indents [210, 211] (See also Chap. 11, p. 690)
Core and pool water; air in reactor building and some surface contamination from fission-gas daughters	I^{133}, I^{134}, I^{135}, Cs^{138}, Kr^{88}, Xe^{135}, and Xe^{138} found in coolant water; Kr^{88}, Rb^{88}, Xe^{138}, and Cs^{138} airborne	1000 to 3000, in coolant water	See ETR, above; short operating time after restart (~30 min) resulted in nonequilibrium fission-product distribution [212, 213]
Core and primary Na coolant with inert cover gas	Cover gas: Xe^{131}, Xe^{133}, Kr^{85}; Coolant: Cs^{137}, Sr^{90}, I^{131}, Ba^{140}, La^{140}	45, Noble gases in cover gas (1/6 of amt. available from 2 melted elements)	

included approximately 10% metal-water reaction.

Except for the Windscale incident, comparatively small quantities of released fission products other than the rare gases escaped from the confines of the reactor buildings, although they were not designed primarily for containment. None of the incidents included in the table, however, approached the seriousness of the maximum credible or design-basis accident (DBA).

6. CALCULATION OF "PROMPT" FISSION-PRODUCT RELEASE

During the early stages of a severe reactor accident, fuel-rod claddings can rupture from the combined effects of internal gas pressure and loss of cladding strength as the clad overheats. The internal gas pressure is largely due to fission gases that have been released from the fuel during normal reactor operation. Rapid release of these gases and of other fission-product vapors that may have accumulated within the void regions of the fuel rod constitute the initial portion of the fission-product-release component known as "prompt release." The latter portion consists of fission products that may be released from the exposed fuel during the period required to cool the reactor core with emergency cooling methods. Therefore, "prompt release" is properly applicable only under conditions characterized by minimum damage to the core and complete recovery of the system from the accident.

6.1 Release Processes during Irradiation

Several processes contribute to release of fission products from reactor fuels during normal power plant operation. Each has been discussed in detail in numerous reviews [11b, 110, 112, 214, 215], so only brief descriptions will be presented here. It should be noted from the outset that experimental in-pile work is heavily oriented toward release of only the noble fission-product gases, krypton and xenon, from oxide fuel systems, UO_2 in particular. The emphasis on gases arises from a combination of factors that include the importance of internal gas pressure buildup to fuel-element integrity, the high volatility of the gases, and the advantages their chemical inertness provides to experimental collection and analysis. The importance of UO_2 in current power reactor technology explains the emphasis on oxide fuels. Fission-gas release from other fuel systems may be qualitatively quite similar, but will differ in magnitude depending upon the physical properties of the new material. The release of fission products other than noble gases, even from oxide fuels, is extremely difficult to specify because of the lack of experimental data and the complexity of the physicochemical interactions that these more reactive species may undergo both within and outside the fuel body. Therefore, any discussion of release processes is largely confined to noble-gas escape from UO_2.

It is convenient to divide the processes according to the particular temperature range in which each is most effective. The first release process is caused by escape of fission-fragment recoils and by the knockout or sputtering of fuel material from a thin surface layer of the solid. Lustman [11b] quotes an expression that may be used to calculate recoil escape rate, dN/dt atoms/sec, from a body of geometric surface area S cm^2 whose thickness dimension is large compared with the recoil length of the nuclide in the fuel, ζ cm. Thus,

$$\frac{dN}{dt} = \frac{fYS\zeta}{4} \qquad (6\text{-}1)$$

where f is the fissioning rate per unit volume, fissions/sec-cm^3, and Y is the fission yield of the nuclide, atoms/fission. Fission-fragment-recoil distances are typically 5 to 10 μ in fuel materials, so for pellet-type fuel designs, recoil escape amounts to only several hundredths of one percent of the total fission-gas inventory when fuel-element discharge exposures are reached. Also, some of the recoiling fragments will travel across the gas

TABLE

Accumulation

Value of μ	Accumulation formation										
	(-13)	(-12)	(-11)	(-10)	(-9)	(-8)	(-7)	(-6)	(-5)	(-4)	(-3)
(-13)	-	-	-	-	-	-	-	-	-	-	-
(-12)	-	-	-	-	-	-	-	-	-	-	-
(-11)	-	-	-	-	-	-	-	-	-	-	-
(-10)	-	-	-	-	-	-	-	-	-	-	-
(-9)	-	-	-	-	-	-	-	-	-	-	-
(-8)	-	-	-	-	-	-	-	-	-	-	-
(-7)	-	-	-	-	-	-	-	-	-	-	-
(-6)	-	-	-	-	-	-	-	-	-	-	-
(-5)	-	-	-	-	-	-	-	-	-	-	-
(-4)	-	-	-	-	-	-	-	-	-	-	-
(-3)	-	-	-	-	-	-	-	-	-	-	6.9855(-8)
(-2)	-	-	-	-	-	-	-	-	-	2.2418(-8)	6.9855(-7)
(-1)	-	-	-	-	-	-	-	-	7.1215(-3)	2.2417(-7)	6.9861(-6)
(0)	-	-	-	-	-	-	-	2.2553(-9)	7.1215(-8)	2.2416(-6)	6.9823(-5)
(1)	-	-	-	-	-	-	7.1350(-10)	2.2552(-8)	7.1211(-7)	2.2404(-5)	6.9448(-4)
(2)	-	-	-	-	-	2.2566(-10)	7.1350(-9)	2.2551(-7)	7.1172(-6)	2.2284(-4)	6.5823(-3)
(3)	-	-	-	-	7.1363(-11)	2.2556(-9)	7.1348(-8)	2.2539(-6)	7.0789(-5)	2.1120(-3)	3.9772(-2)
(4)	-	-	-	2.2557(-11)	7.1363(-10)	2.2565(-8)	7.1307(-7)	2.2418(-5)	6.7092(-4)	1.2749(-2)	2.9695(-2)
(5)	-	-	7.1365(-12)	2.2557(-10)	7.1359(-9)	2.2553(-7)	7.0923(-6)	2.1247(-4)	1.0486(-3)	9.4552(-3)	9.4558(-3)
(6)	-	2.2553(-12)	7.1364(-11)	2.2556(-9)	7.1321(-8)	2.2431(-6)	6.7218(-5)	1.2820(-3)	2.9965(-3)	2.9970(-3)	-
(7)	7.1365(-13)	2.2567(-11)	7.1361(-10)	2.2554(-8)	7.0937(-7)	2.1259(-5)	4.0557(-4)	9.4822(-4)	9.4838(-4)	-	-
(8)	7.1365(-12)	2.2566(-10)	7.1322(-9)	2.2433(-7)	6.7231(-6)	1.2827(-4)	2.9992(-4)	2.9997(-4)	-	-	-
(9)	7.1361(-11)	2.2554(-9)	7.0935(-8)	2.1251(-6)	4.0564(-5)	9.4849(-5)	9.4865(-5)	-	-	-	-
(10)	7.1322(-10)	2.2433(-8)	6.7232(-7)	1.2828(-5)	2.9995(-5)	3.0000(-5)	-	-	-	-	-
(11)	7.0938(-9)	2.1261(-7)	4.0565(-6)	9.4352(-6)	9.4868(-6)	-	-	-	-	-	-
(12)	6.7232(-8)	1.2828(-5)	2.9995(-6)	3.0000(-6)	-	-	-	-	-	-	-
(13)	4.0565(-7)	9.4852(-7)	9.4863(-7)	-	-	-	-	-	-	-	-
(14)	2.9995(-7)	3.0000(-7)	-	-	-	-	-	-	-	-	-
(15)	9.4868(-8)	-	-	-	-	-	-	-	-	-	-

space and become embedded in the cladding surface. Escape of fission gases via knockout or sputtering events caused by collisions with recoils is difficult to quantify. Rodgers [216] found that 1 in 16 fission fragments leaving the surface of uranium metal ejected about 10^4 atoms of the solid. In addition to the fission gas that may be present in this volume of the solid, some nearby gas may also escape because high local temperatures are generated along recoil tracks. However, these release mechanisms are balanced somewhat by return of gas to the fuel via collisions with fission-fragment recoils outside the fuel surface [217]. Therefore we must rely upon experimental work to provide an estimate of recoil-induced release. Investigations have shown that these temperature-independent mechanisms account for most of the fission-gas release that occurs from UO_2 fuel pellets at temperatures below about 1000°C [218, 219]. Since appreciable fractions of the fuel in a reactor operate at temperatures considerably above 1000°C, recoil-induced release is usually considered unimportant.

At fuel temperatures between 1000 and about 1600°C, thermally induced migration of fission gases to the fuel surface appears to be the dominant release process. Classical diffusion theory [11b, 110] is frequently used to describe the kinetics of gas release in this temperature range.

The equivalent-sphere model originally proposed by Booth [220] and discussed by Lustman [11b] constitutes the basis for calculation of release from the sintered UO_2 characteristic of recent water-reactor fuels. The sintered fuel, although near theoretical density, contains some interconnected porosity, which effectively increases the surface area for release of fission gases. In the equivalent-sphere approach, the fuel body is considered to be composed of spherical particles of uniform size whose surface-to-volume ratio is equivalent to the actual surface-area-to-fuel-volume ratio of the sintered material. Fick's diffusion equations are then formulated by use of spherical coordinates and solutions obtained for the appropriate boundary conditions. The pertinent mathematical relationships arising from the equivalent-sphere diffusion approach have been presented by Lustman [11b] and Beck [221]. For reactor irradiation conditions where fission-product species are being generated, decaying, and diffusing simultaneously, the following expressions may be used to compute fractional release values. For a radioactive isotope:

$$G = \frac{N\lambda}{B} = 3\left(\frac{\coth\sqrt{\mu}}{\sqrt{\mu}} - \frac{1}{\mu}\right) - \exp(-\mu_\tau) + \frac{6\mu\exp(-\mu_\tau)}{\pi^2}\sum_1^\infty \frac{\exp(-n^2\pi^2\tau)}{n^2(n^2\pi^2 + \mu)}; \tag{6-2}$$

and for a stable isotope:

$$F = \frac{N}{Bt} = 1 - \frac{1}{15\tau} + \frac{6}{\pi^4\tau}\sum_1^\infty \frac{\exp(-n^2\pi^2\tau)}{n^4}, \tag{6-3}$$

where G = ratio of the number of external nondecayed atoms at any instant to the total number of nondecayed atoms; F = ratio of the number of external atoms at any instant to the total number of atoms produced; N = accumulation of undecayed, released atoms from a unit volume of sphere ($atoms/cm^3$); B = production rate ($atoms/sec\text{-}cm^3$); $\mu = \lambda a^2/D$; $\tau = Dt/a^2$; λ = radionuclide decay constant (sec^{-1}); a = radius of spherical body (cm); D = diffusion coefficient (cm^2/sec); t = time (sec). Tabulated values of these functions for various values of the dimensionless parameters μ and τ have been given by Beck [221] and are reproduced in Tables 6-1 and 6-2.

It should be pointed out that the derivation of these equations is based upon the assumption that the fission-product production rate and the diffusion coefficient remain constant with time. Diffusion

6-1

Function, G = Nλ/B

for value of τ shown

(-2)	(-1)	(0)	(1)	(2)	(3)	(4)	(5)	(6)	(7)	-
-	-	-	-	-	-	-	-	-	1.0000(-6)	1.0000
-	-	-	-	-	-	-	-	1.0000(-6)	1.0000(-5)	1.0000
-	-	-	-	-	-	-	1.0000(-6)	9.9999(-6)	9.9995(-5)	1.0000
-	-	-	-	-	-	9.9999(-7)	9.9999(-6)	9.9995(-5)	9.9950(-4)	1.0000
-	-	-	-	-	9.9993(-7)	9.9998(-6)	9.9996(-5)	9.9950(-4)	9.9602(-3)	1.0000
-	-	-	-	9.9933(-7)	9.9993(-6)	9.9994(-5)	9.9950(-4)	9.9502(-3)	9.5163(-2)	1.0000
-	-	-	9.9333(-7)	9.9933(-6)	9.9988(-5)	9.9949(-4)	9.9502(-3)	9.5163(-2)	6.3212(-1)	1.0000
-	-	9.3334(-7)	9.9333(-6)	9.9928(-5)	9.9943(-4)	9.9501(-3)	9.5163(-2)	6.3212(-1)	9.9995(-1)	1.0000
	5.6365(-7)	9.3333(-5)	9.9328(-5)	9.9883(-4)	9.9495(-3)	9.5162(-2)	6.3212(-1)	9.9995(-1)	1.0000	1.0000
2.1068(-7)	5.6365(-5)	9.3329(-5)	9.9283(-4)	9.9435(-3)	9.5156(-2)	6.3211(-1)	9.9995(-1)	9.9999(-1)	-	9.9999(-1)
2.1067(-6)	5.6362(-5)	9.3284(-4)	9.8835(-3)	9.5096(-2)	6.3205(-1)	9.9989(-1)	9.9993(-1)	-	-	9.9993(-1)
2.1066(-5)	5.6333(-4)	9.2842(-3)	9.4497(-2)	6.3145(-1)	9.9929(-1)	9.9933(-1)	-	-	-	9.9933(-1)
2.1055(-4)	5.6838(-3)	8.8559(-2)	6.2552(-1)	9.9335(-1)	9.9340(-1)	-	-	-	-	9.9340(-1)
2.8943(-3)	5.3196(-2)	5.7123(-1)	9.3906(-1)	9.3911(-1)	-	-	-	-	-	9.3911(-1)
1.9857(-2)	3.2638(-1)	6.5204(-1)	6.5209(-1)	-	-	-	-	-	-	6.5209(-1)
1.2035(-2)	2.6997(-1)	2.7000(-1)	-	-	-	-	-	-	-	2.7000(-1)
9.1864(-2)	9.1868(-2)	-	-	-	-	-	-	-	-	9.1563(-2)
2.9700(-2)	-	-	-	-	-	-	-	-	-	2.9700(-2)
-	-	-	-	-	-	-	-	-	-	9.4568(-3)
-	-	-	-	-	-	-	-	-	-	2.9970(-3)
-	-	-	-	-	-	-	-	-	-	9.4838(-4)
-	-	-	-	-	-	-	-	-	-	2.9997(-4)
-	-	-	-	-	-	-	-	-	-	9.4865(-5)
-	-	-	-	-	-	-	-	-	-	3.0000(-5)
-	-	-	-	-	-	-	-	-	-	9.4863(-6)
-	-	-	-	-	-	-	-	-	-	3.0000(-6)
-	-	-	-	-	-	-	-	-	-	9.4868(-7)
-	-	-	-	-	-	-	-	-	-	3.0000(-7)
-	-	-	-	-	-	-	-	-	-	9.4863(-8)

coefficients have been determined from in-pile irradiation capsule data but usually are obtained from out-of-pile annealing studies. They follow the familiar Arrhenius expression:

$$D_T = D_0 \exp(-\frac{Q}{RT}), \qquad (6\text{-}4)$$

where D_T = diffusion coefficient at temperature T; D_o = limiting value; Q = activation energy for diffusion; R = gas constant; T = temperature.

Tabulations of diffusion parameters have been given by Lustman [11b] and by Morrison et al. [222]. A collection of values from a selected number of literature references is listed in Table 6-3. The values result in an uncertainty of roughly two orders of magnitude in diffusion coefficient at any particular temperature with Bastrom's data (reported by Lustman [11b]) representative of minimum values and Auskern's [223] data representative of the maximum. In spite of these uncertainties, diffusion theory has been used with some success to correlate gas-release data from irradiation capsule puncture tests. However, certain predictions of the theory have been difficult to reconcile with experimental observations. Specifically, the observed half-life dependence for release often does not conform to that predicted by the theory. Also, burst-type releases of fission gases, which occur during periods of changing reactor conditions, cannot be completely explained by transient diffusion effects.

At temperatures above 1600°C, gross structural changes in UO_2 occur, which influence fission-product migration and release. Between 1600 and 1800°C, equiaxed grain growth predominates, and it is generally acknowledged that the process accelerates escape of fission gases from UO_2 beyond that predicted by classical volume diffusion theory [225]. The mechanism is considered to involve grain-boundary sweeping, but adequate models for gas release under these conditions are not yet available [226]. Most workers resort to the empirical estimate that about one-third of the gas produced in the equiaxed-grain-growth region escapes the fuel body [227] without delay [228].

At temperatures extending from 1800°C to the melting point of UO_2 (2850°C), extensive columnar grain growth occurs [225, 226]. The driving force for this phenomenon comes from the steep thermal gradient in UO_2, combined with the presence of voids or the formation of fission-gas bubbles after irradiation has proceeded for some time. The voids or bubbles migrate up the thermal gradient as fuel that vaporizes from the hotter sides recrystallizes on the cooler sides. The migrating "holes" collect fission gases lying along their path and then release the contents to the axial cavity that forms along the fuel centerline. This relatively rapid, repetitive process results in nearly quantitative release of fission gas from the columnar-grain-growth region [226, 228]. Considerable progress is being made in the theoretical description of fuel and fission-gas behavior under these conditions, but the kinetic models are complex and still in the developmental stage. For the purposes of safety evaluations, it remains best to assume 100% gas release for the fuel-volume fraction that experiences columnar grain growth.

6.1.1 Environmental Factors

The discussion in the foregoing section quite adequately illustrates the importance of temperature to fission-gas release. The poor thermal conductivity of UO_2 results in steep thermal gradients in the fuel during irradiation, and therefore all the release processes described above can be expected to occur in cylindrical UO_2 fuel rods.

For fuels having higher thermal conductivities, such as UC and UN, the higher temperature processes associated with grain growth may not be so important unless the fuels are operated with high surface temperatures [229]. The thermal performance of fuel elements can be characterized by the parameter,

$$\int_{T_{surface}}^{T_{center}} k\, dT, \qquad (6\text{-}5)$$

where k is the thermal conductivity of the fuel [11b]. For a cylindrical fuel pellet (flux depression effects ignored), the relationship between this parameter, called the thermal-conductivity integral, and other quantities of interest is

$$\int_{T_a}^{T_c} k dT = \frac{qa^2}{4} = \frac{Qa}{2} = \frac{W}{4\pi}, \qquad (6\text{-}6)$$

where T_a = surface temperature of pellet; T_c = center temperature of pellet; a = radius of pellet (cm); q = volumetric heat-generation rate (watts/cm³); Q = surface heat flux (watt/cm²); and W = linear power of pellet (watt/cm).

The last equality shows that, for constant power output per unit length, the center temperature of the fuel is independent of pellet diameter. Actually, the temperature at identical fractions of the radius of any pellet will be the same, provided the linear-power rating remains constant. Since fission-gas release was shown to be directly related to temperature in the preceding section, it follows that the fraction of fission gas released at each location, as well as the cumulative fraction released in the total cross section, is independent of the absolute value of the pellet radius for a fixed value of the thermal-conductivity integral. Therefore, it has become customary to define fractional release in fuel elements in terms of the thermal-conductivity integral or the linear-power rating. For irradiated, 95% dense UO_2, White [229] has recommended the following expression for thermal conductivity:

$$k_{UO_2} = \frac{40.14}{T + 174} + 2.675 \times 10^{-13} T^3 (T \text{ in}^\circ K). \qquad (6\text{-}7)$$

This equation results in a value of 86 watts/cm for the thermal-conductivity integral over the temperature range 273°K to the melting point of

TABLE 6-2

Release of Stable Isotopes

τ	$\tilde{F} = \tilde{N}/Bt$
(-12)	2.2568(-6)
(-11)	7.1365(-6)
(-10)	2.2567(-5)
(-9)	7.1363(-5)
(-8)	2.2566(-4)
(-7)	7.1350(-4)
(-6)	2.2553(-3)
(-5)	7.1215(-3)
(-4)	2.2418(-2)
(-3)	6.9865(-2)
(-2)	2.1068(-1)
(-1)	5.6365(-1)
(0)	9.3334(-1)
(1)	9.9333(-1)
(2)	9.9933(-1)
(3)	9.9993(-1)
(4)	9.9999(-1)
(5)	1.0000

UO_2 (3133°K). The cumulative fractional gas release from a cylindrical pellet can then be expressed as a sum of the release fractions characteristic of each temperature region multiplied by the fraction of the fuel volume that lies in each temperature region. Mathematically, the relationship is

$$F = f_r \frac{\int_{T_s}^{T_r} k\,dT}{\int_{T_s}^{T_c} k\,dT} + f_d \frac{\int_{T_r}^{T_d} k\,dT}{\int_{T_s}^{T_c} k\,dT} + f_{ex} \frac{\int_{T_d}^{T_{ex}} k\,dT}{\int_{T_s}^{T_c} k\,dT} + f_{col} \frac{\int_{T_{ex}}^{T_c} k\,dT}{\int_{T_s}^{T_c} k\,dT} \qquad (6\text{-}8)$$

where F = cumulative release from pellet; f_r = release fraction due to recoil; f_d = release fraction due to diffusion; f_{ex} = release fraction due to equiaxed grain growth; f_{col} = release fraction due to columnar grain growth; T_s = fuel surface temperature; T_c = fuel center temperature ($\leqslant$2800°C); T_r = temperature limit for recoil release; T_d = temperature limit for diffusion release; and T_{ex} = temperature limit for release during equiaxed grain growth.

In practice, the thermal history of the fuel depends upon several variables, including the fuel composition and density, the neutron flux distribution throughout the reactor core, and the operating history of the reactor. Changes in the first two variables during irradiation are usually not considered in release calculations. The neutron flux distribution is quite important in the determination of the power output at each location and, hence, the temperature profile within the fuel. The last variable probably accounts for many fluctuations in fission-gas release. The shuffling of fuel assemblies between reactor cycles leads to nonuniform temperature distributions across the fuel during its in-pile lifetime. For the stable fission gases, the most significant release will probably occur during the time the fuel spends at maximum power, but for relatively short-lived radioactive nuclides, the final power-output conditions of the fuel will determine the release fractions. There is scanty published data concerning fission-gas release as a function of fuel burnup, but available results [231] indicate that release in the diffusion-controlled temperature range does increase with burnup, particularly above the 10 to 20,000 Mwd/ton level. The mechanism is not well known, but macroscopic structural changes may be the major cause. Optical and electron microscopy have shown reductions in grain size [232] and the presence of fission-gas bubbles in UO_2 [233] and other fuels [234] having high fission exposures. Most solid fission products are also insoluble in the fuel, and autoradiographic examinations have revealed chemically similar species segregated in specific zones or concentrated along grain boundaries [235, 236]. The accumulation of fission products at localized positions within irradiated fuels leads to stress concentrations around the positions. At elevated temperatures, the fuel matrix yields under the stress, and swelling of the body results [237]. Thermal cycles to low temperatures, combined with the nonuniform stress distribution, may induce cracking and thus increase the surface area for gas release [238]. Only empirical methods are available to estimate the effect of burnup on diffusion-controlled release. Daniel et al. [232] and Evans and Shilling [227] have

TABLE 6-3

Noble-Gas Diffusion Parameters for UO_2

Temperature range (°C)	D_0 (cm^2/sec)	Q (kcal/g-mole)	UO_2 Burnup (Mwd/ton)	Investigators	Reference
800-1600	8.0×10^{-3}	92	Trace	Melehan, et al.	[218]
800-1400	6.6×10^{-6}	71.7	Trace	Bostrom	[11b]
800-1600	3.1×10^{-5}	71	Trace	Long	[224]
800-1600	8.1×10^{-7}	67	Trace	Susko	[11b]
900-1400	4.9×10^{-4}	73.8	Trace	Auskern	[223]
900-1700	4.3×10^{-4}	72	4000	Parker et al Morrison et al.	[115a] [222]

evolved similar approaches based upon work with postirradiation annealing conducted by Lustman [11b]. Each involves multiplication of the "reference" noble gas apparent diffusion coefficient (D/a^2) determined at the tracer burnup level by integer factors that increase with increasing fuel burnup. The factors suggested are shown in Fig. 6-1. The curves, while different, do not result in widely different release estimates, and each should be considered equally valid on the basis of the limited data involved in their development.

6.1.2 Methods for Calculation of Release

When the variety of processes and factors that affect fission-gas release from UO_2 and the current uncertainty regarding the quantitative effect of each are considered, it is not surprising that a number of methods have been suggested to estimate fission-gas release. However, each technique represents a simplification, modification, or detailed treatment of the thermal-conductivity-integral approach outlined in the preceding section. A simple correlation, proposed by Horn et al. [239], states that to a reasonable approximation the fraction of gas released is equal to the fractional volume of fuel with a temperature above 2000°C. Hoffman and Coplin [240] propose a technique based upon correlation of measured Kr^{85} releases in a series of fuel-rod irradiations with the maximum volumetric average temperatures attained by the rods. The maximum volumetric average temperature is defined as the arithmetic average between the fuel

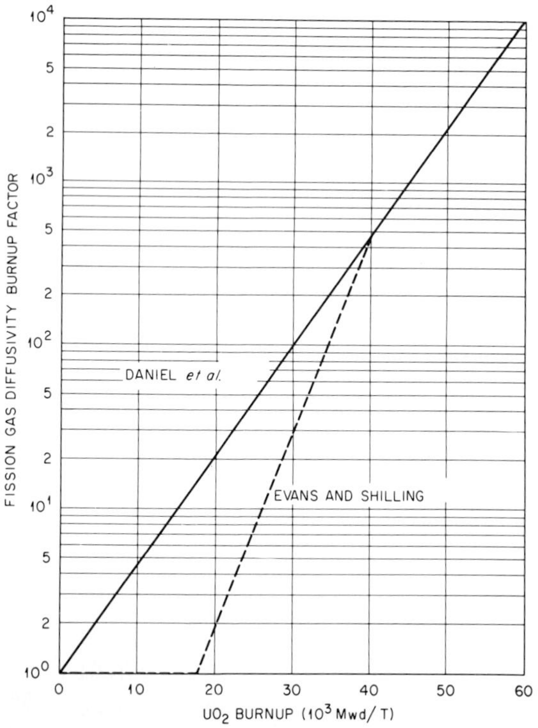

FIG. 6-1 Suggested burnup correction factors for fission-gas diffusion from UO_2.

center and surface temperatures at the peak-power position of the rod. This value may also be normalized to the peak-to-average heat rating over the length of the rod. Lewis [241] has suggested the following correlation for fission-gas release during irradiation of UO_2:

$$R < \left\{ 0.005\int_{T_s}^{1000} k d\theta + 0.10\int_{1000}^{1300} k d\theta + 0.60\int_{1300}^{1600} k d\theta + 0.95\int_{1600}^{T_c < 2800} k d\theta \right\} \Bigg/ \int_{T_s}^{T_c < 2800} k d\theta \quad . \tag{6-9}$$

In this equation, R is the gas-release fraction and $\int k d\theta$ is the thermal-conductivity integral for UO_2 fuel having a density greater than 10.5 g/cm^3. The integration limits are temperatures in degrees Centigrade.

Diffusion theory has been applied to calculate fission-gas release from UO_2 under in-pile irradiation conditions. In order to use the method, the radial temperature distribution in the pellet must be computed, and then the rod must be divided into small, approximately isothermal volume increments. The pertinent fission-gas diffusivity for each volume increment is obtained from the appropriate Arrhenius activation energy equation. These are used in the diffusion expression to calculate the amount of release from each increment, and then the results are summed to obtain total release. This approach has been used by Lustman et al. [11b], who have also showed that calculated fission-gas releases agree reasonably well with measured values obtained from postirradiation examination of prototype fuel rods. The same approach has been incorporated by Morrison et al. [242] in the development of a computer code (REGAP) to calculate fission-product release to void spaces in fuel rods during irradiation on a core-wide basis. In this work, diffusion theory is used to compute release over the entire range of fuel temperatures, that is, even from the regions of equiaxed and columnar grain growth. Since extrapolated diffusion-coefficient values are quite high in these temperature regions, the release estimates one obtains are nearly identical to numbers obtained by assuming 33% release between 1600 and 1800°C and 100% release above 1800°C. Nevertheless, use of diffusion theory in these temperature regions should be considered as only empirical, and extrapolation of the results requires caution. In some cases, workers do not use fission-gas diffusivities obtained from out-of-pile experiments, but, rather, they apply the diffusion model to irradiation-capsule results and extract diffusion parameters by reverse calculation. This approach is the basis of the "D′-empirical" method developed by Parker and Lorenz [243].

Each of the calculational techniques just outlined will produce somewhat different fission-gas-release estimates. An intercomparison of the methods for a common problem is shown in Fig. 6-2. In this figure, fission-gas release as a function of $\int k\,dT$ is plotted for two different fuel-pellet surface temperatures, 400 and 600°C. The thermal performance of the fuel is determined by the thermal-conductivity equation recommended by White et al. [229]. Sensitivity of the diffusion-model approach to basic input diffusion-parameter data is also illustrated. The limiting release values at low $\int k\,dT$ values represent the temperature-independent recoil and knockout contributions as estimated by the separate investigators. It can be seen that the various methods yield gas-release estimates that vary by as much as an order of magnitude for lower fuel temperatures but tend to converge to within a factor of two at high fuel temperatures. Since Lewis' method gives the highest gas-release estimates over the widest range of $\int k\,dT$ values and is easily used, it should be recommended for nuclear reactor safety calculations. However, it should be mentioned that effects of fuel burnup on diffusion release do not appear in Fig. 6-2 results. Therefore, for high fuel burnups, the diffusion approach might yield higher release values than the method developed by Lewis.

Curves like those in Fig. 6-2 are very useful in the calculation of fission-gas release on a core-wide basis, provided the power distribution throughout the reactor is known. The power output of the fuel is directly related to $\int k\,dT$ as shown in Eq. (6-6). Reactor operating conditions dictate the surface temperatures that will be achieved, but the two values used in Fig. 6-2 are representative of the range to be expected in water-cooled power reactors. However, in calculations of this type, it must be remembered that both the distribution of fission gas and its release depend upon the entire power-production history of each reactor fuel assembly. Thus, a familiarity with fuel-management practices is advantageous in the development of total core-gas-release estimates.

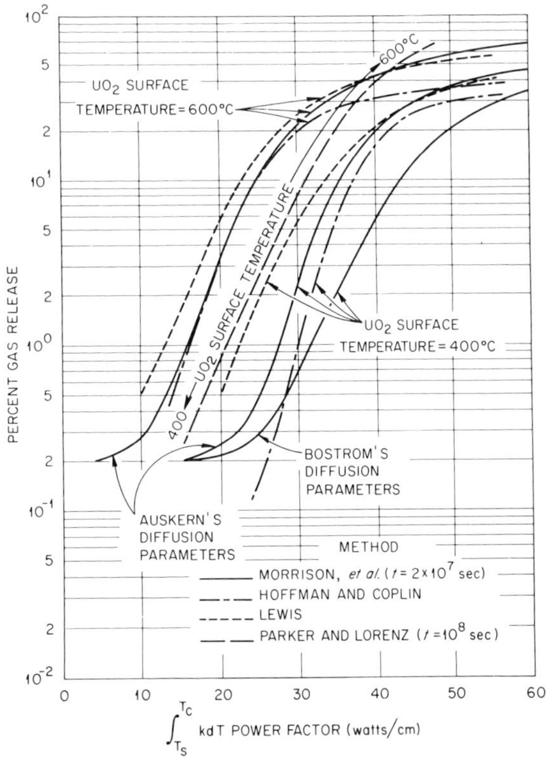

FIG. 6-2 Comparison of methods for calculation of fission-gas release from UO_2.

6.1.3 Other Fission Products

Very little experimental information concerning escape of other fission products from reactor fuels during operation is available. Results of some in-pile experiments indicate that iodine migration rates in UO_2 are approximately equal to that of xenon [218]. Out-of-pile annealing experiments conducted by Parker et al. [115a] suggest that under an inert atmosphere iodine diffuses about twice as fast as does xenon in UO_2. Release data are also presented for other fission-product elements, including tellurium, cesium, strontium, and ruthenium. Similar out-of-pile experiments with UO_2 have been conducted by Davies et al. [114], who found that at temperatures above 1000°C, the emission rates of iodine, tellurium, and cesium were equal to or in excess of the emission rate of xenon, the rates being dependent upon fuel compact density. Oi and Takagi [244] have published diffusion coefficients for strontium, ruthenium, zirconium-niobium, cerium, and lanthanum obtained from out-of-pile anneals of small single-crystal specimens of UO_2 at temperatures up to 2500°C. No theoretical or empirical relationships have been developed to estimate release of fission products other than the noble gases from UO_2 during irradiation, except for use of the diffusion model with changes in diffusion-coefficient values [222]. Fission-product-release values obtained by this method do not take into account effects of chemical reactions that these more reactive species might undergo with the fuel, the cladding, or each other. The existing data on reaction of fission-product iodine with Zircaloy cladding material has been summarized by Allen [228]. The extent of such reactions, the volatility of the reaction products, and the vaporization from the fuel itself must be considered in an attempt to determine the vapor-phase concentration of fission products available for rapid release when cladding rupture occurs during the reactor accident. No quantitative analysis of this type has been done.

6.2 Release during Accident Thermal Transient

The second portion of the "prompt" release component, i.e., escape of fission gases and other products during the thermal transient preceding complete core recovery, has received much less attention than the initial portion. However, this release probably will be small compared with the initial portion if the emergency cooling systems (ECCS) function as expected. Nevertheless, for off-design operation of ECCS the question is still of interest. The conditions for release in this case are similar to those used in postirradiation fission-product-diffusion experiments. The radial temperature profile across individual fuel pins will be nearly flat [222] during the post-loss-of-coolant thermal transient, in contrast to the large thermal gradients that exist during power operation. However, the entire fuel cross section will experience a rapid rise in temperature, followed by some period at the peak temperature, and then a rapid temperature decrease as the ECCS becomes effective.

Numerous solutions to Fick's diffusion equations for release under isothermal conditions have been obtained [11b, 218, 221, 245, 246]. These expressions may be used to estimate release in a time-varying temperature field if an iterative procedure is followed. The technique will be outlined in terms of use with the equivalent-sphere diffusion approach. First, the thermal transient is divided into equal, approximately isothermal, time intervals, and the value for the release parameter (τ) is computed for each interval. Then these can be used in a sequential calculation of the cumulative fractional release throughout the history of the transient. In a large reactor core, the transient will vary with location because of spatial variation of the heat source and heat-transfer characteristics. To account for this in accident analyses, the reactor core of interest should be subdivided into a number of equal volume regions whose individual thermal-transient histories collectively represent the behavior of the whole system. Following the above approach, and approximating a large reactor core as a right-circular cylinder, one obtains the generalized equations for calculation of fission-product release from each core region during the period of thermal transient:

$$F_{i,k,l,n} = \frac{6}{\sqrt{\pi}} \left(\sum_{n=1}^{n} \tau_{i,k,l,n} \right)^{1/2} - 3 \left(\sum_{n=1}^{n} \tau_{i,k,l,n} \right) \quad (6\text{-}10)$$

for $F < 0.9$, and

$$F_{i,k,l,n} = 1 - \frac{6}{\pi^2} \exp \left[-\pi^2 \left(\sum_{n=1}^{n} \tau_{i,k,l,n} \right) \right] \quad (6\text{-}11)$$

for $0.9 \leqslant F \leqslant 1.0$, where $F_{i,k,l,n}$ = the cumulative fractional release of isotope i from axial section k of radial core zone l after n time intervals; and $\tau_{i,k,l,n}$ = the release parameter for isotope i located in axial section k of radial core zone l for the n-th time interval. The release parameter $\tau_{i,k,l,n}$ is further defined as

$$\tau_{i,k,l,n} = \frac{D_{i,k,l}\,\Delta t_n}{a^2}, \quad (6\text{-}12)$$

where $D_{i,k,l}$ = the diffusion coefficient for isotope i located in axial section k of radial core zone l (cm^2/sec); Δt_n = the n-th time interval (sec); a = the equivalent-sphere radius (cm); and the diffusion coefficient values are obtained from the Arrhenius equation

$$D_{i,k,l} = D_{0_i} \exp \left(- \frac{Q_i}{RT_{k,l}} \right), \quad (6\text{-}13)$$

where D_{0i} = the limiting value of the diffusion coefficient (cm^2/sec); Q_i = the Arrhenius activation energy (cal/mole); R = the universal gas constant (cal/(mole)(°K)); and T = the absolute temperature (°K).

It must be remembered that the total cumulative fractional release from each core region in the reactor will consist of the sum of two quantities; the $F_{i,k,l,n}$ values computed by Eq. (6-10) or (6-11), and the values for release that has occurred during the period of normal reactor operation for each region. These latter numbers must be obtained independently by use of one of the methods described earlier. Furthermore, if one desires a summation of release from all regions in order to obtain total core-release values, the nonuniform distribution of the fission-product inventory must be taken into account. It is adequate to assume that the fission products are distributed according to the operating power distribution for the reactor. Thus, each of the release quantities must be multiplied by a volume-weighted power-distribution ratio before the final summation is performed.

Mathematically, these steps can be expressed as follows:

$$F_{T_{i,k,l,n}} = \left[F_{i,k,l,n} + f_{i,k,l} \right] (V_{k,l}), \quad (6\text{-}14)$$

where $F_{T_{i,k,l,n}}$ = cumulative fractional inventory release values; $F_{i,k,n,l}$ = cumulative fractional release due to postaccident thermal transient; $f_{i,k,l}$ = fractional release due to preaccident operation of the fuel; and $V_{k,l}$ = volume fraction of the total fission-product inventory generated in each region. The total core fractional release for each isotope after each time interval ($F_{CORE_{i,n}}$) then will be given by

$$F_{CORE_{i,n}} = \sum_{\ell=1}^{\ell} \sum_{k=1}^{k} F_{K_{i,k,l,n}}. \quad (6\text{-}15)$$

The above technique involves considerable labor if done by hand. Morrison et al. [222] have developed a computer program (FRACREL) that performs the same operations in a matter of minutes. Obviously, in either case, one must have at hand the temperature-versus-time data for the transient in order to begin the calculation.

Many of the factors that affect fission-product release during irradiation also influence release during postirradiation heating. A discussion of postirradiation annealing experiments and the influencing factors that have been examined can be found in Sec. 3 of this chapter. Briefly, fuel composition and structure, burnup level, and composition of the atmosphere above the fuel surface all exert measureable influences on observed releases. Bursts of fission-product release during both heating and cooling of fuel specimens are also reported. Computations of the type outlined above are able to include the first three effects through variation of fission-product-diffusion parameters but do not include treatment of "burst" release phenomena. An empirical method, presented by Landoni and Moffette [247] for computation of postirradiation release from gas-cooled-reactor fuels, includes both the short-term (burst) and the long-term components of release observed in annealing studies of prototype fuel samples [248]. The release parameters for each component are of the Arrhenius equation type but applicable only to the particular uranium carbide uncoated-particle fuel compacts characteristic of gas-cooled-reactor fuels.

6.3 Prompt Release of Radioiodine

Conservative assessments of the public radiologic exposure following a Loss-of-Coolant accident (LOCA) indicate that further steps need to be taken in order to show that public exposure is not excessive. The prompt release of radioiodine from fuel sheathing which ruptures in the course of the accident must be confined to a greater extent than can now be assured in the assumed accident and with the assumed containment concept. The source term, namely the quantity of iodine escaping the core, is believed to be grossly overestimated in the conservative analyses made so far. The principal factors affecting the source term are the conditions leading to rupture of sheating of individual fuel rods and the availability of fission product iodine in the gas contained within each rod.

There are more than 10,000 fuel elements in a typical core, and the conditions to which they are subjected in normal operation or in a LOCA vary widely. The distribution of conditions that apply to a particular reactor and accident is supplied by the reactor designer. This distribution may be expressed in terms of number of elements in the core (or some equivalent thereof) whose burnup, power history, pre-accident pressure, accident-imposed temperature history, etc., fall in various intervals. The further data necessary to carry through an assessment of the quantity of iodine released in the accident are the relationships of these conditions to (1) the incidence of sheath ruptures, and (2) the release of iodine from an element suddenly depressurized.

Sheath rupture thresholds are dependent upon several important variables which can be investigated in the laboratory. Generally, the following variables are involved:

(1) initial (pre-accident) internal pressure;
(2) rate of cladding temperature increase;
(3) maximum clad temperature.

Rupture thresholds are also implicitly dependent upon the geometry of the element, particularly as it affects the pressure loading of the sheath as plastic strain increases.

The power transient incidental to a LOCA introduces a condition that would enhance the gap inventory by an amount that can now be estimated only crudely. Assumptions now used should be verified and, if possible, improved.

Conventional methods for calculating radioiodine release from UO_2 are based mainly on the Booth diffusion model concept; however, the release of radioiodine-131 is slightly more significant since the diffusion parameter, D', for iodine exceeds that for xenon by a factor of at least two. The effect of continued fuel irradiation is also significant and may account for more than an order of magnitude in the equilibrium rate of release. A calculation based on conservative assumptions for radioiodine release to the clad gaps was given in a recent paper [254]. A typical case for water reactor fuel irradiated to 10,000-25,000 Mwd/T is given in Table 6-4. A very important point is made in that at lineal power ratings below 10 kw/ft only a few percent of the iodine is released while doubling the power will increase the release by a factor of ten.

In addition to the steady-state diffusion process, a second contribution of radioiodine may come from the heating burst especially during an overpower transient phase of the accident. In exploratory analyses the release of iodine (i.e., the gap inventory) has been estimated to increase 10% in highly rated fuel and 340% in fuel normally of low rating. These estimates demonstrate the importance of verifying the magnitude of the transient contribution.

Actual release requires rupture of the sheathing and depends as well upon other inherent

TABLE 6-4

Calculated Equilibrium Fuel Gap Iodine^{131} Inventory[a]

Localized UO_2 Fuel Heat Rating (kw/ft)[b]	Iodine^{131} in Fuel Gap[c] (percent of equilibrium)		
	Low Burnup[d]	Intermediate Burnup[e]	High Burnup[f]
9.0	0.045	0.092	0.29
11.6	0.46	0.85	2.5
13.9	3.3	6.0	14.4
16.7	15.5	24.0	35.0
18.9	27.0	36.1	49.0
20.1	33.5	44.0	54.0
21.0	37.0	47.0	57.0

[a]Based on annealing data in Ref. 259 for BWR-size UO_2.
[b]Assumes 94 percent dense UO_2 and 3.5 mil radial gap.
[c]Release from the cladding and fuel plenum may be an order of magnitude lower.
[d]Equivalent to 1,000 Mwd/T.
[e]Equivalent to 10,000 Mwd/T.
[f]Equivalent to 25,000 Mwd/T.

retention processes likely to be in effect in a fuel element. Unfortunately, little credit can be afforded the retention effect until some additional experimental work involving a transient of similar proportions is actually performed.

Other modes of core behavior contributing to iodine release could be postulated and analyzed (such as fuel element fragmentation), but these are not consistent with the LOCA postulated.

In view of the need for a new look at regulatory criteria and standards for the evaluation of potential fission product release consistent with engineered reactor safeguards, comparable assessment studies are currently in progress at BMI [255], ORNL [256], and Phillips Petroleum Company's LOFT Project [257] to devise an acceptable approach for calculating the extent to which fission product radioactivity, especially radioiodine, could be released in a safeguard-limited loss-of-coolant accident.

The mechanistic assessment of the potential inventory of volatile fission products (rare gases and halogens) may be outlined by use of an accident sequence and core response process as follows:

1. Reactor vessel depressurization on rapid loss-of-coolant.
2. Core heatup from sensible heat and radioactive decay.
3. Rod rupture beginning at 1200-1400°F accompanied by prompt release of a fraction of the gap inventory of iodine and rare gases (Fraction I).
4. A thermal transient prior to turn-around by emergency coolant. This is accompanied by an additional release consisting of a heating burst (Fraction II), and a time dependent, slower, transient diffusional releast (Fraction III).
5. Upon cooling of the core, some additional gaseous release is incurred by fuel pellet breakup (Fraction IV).
6. Assuming that fuel rod fragmentation is prevented by the safeguard coolant, the remaining contribution may come only from aqueous leaching of the exposed or expelled fuel including any incurred by a UO_2-water reaction (Fraction V).
7. The maximum clad temperature reached is conservatively taken as 3500°F (no cladding will melt); however 100 percent of the rods will fail and the maximum duration of the transient before turn-around could be 5 minutes.

Six critical points seem to be involved in the justification of this method of evaluation of the loss-of-coolant prompt release fission product source term. These are:

1. The validity of the application of classical diffusion theory or the Booth Model [258] which correlates the experimental release data with temperature, UO_2 density, length of irradiation time, etc. with the equilibrium fuel gap radionuclide inventory.
2. The validity of the higher release rate or diffusion parameter, D′, for iodine compared to xenon from UO_2 to the fuel voids. We cite mainly two references [114, 259] using 70 separate determinations of which 63 give D′ ratios of iodine over xenon ranging from 1.5 to 12. The median ratio is 6.25 and the square root of the ratio (2.5) is approximately the increase in diffusion rate of iodine over xenon in reactor grade 94% dense UO_2.
3. The correlation of burnup with the rate of xenon and iodine release. ORNL diffusion annealing data [259] show a significant burnup effect (Table 6-4), however one cannot assign a quantitative relation from it. Therefore use of the BMI [255] and UK [260] analysis of the Westinghouse data [261] has been proposed. These suggest an increase of one order of magnitude in D′ for each 15,000 Mwd/T of accumulated burnup. This correction however is probably overconservative according to a recent study by Baily, et al. [262].
4. The significance and magnitude of the heating and cooling burst contribution. This effect is recognized by most experimenters as a departure from simple diffusion or grain growth. It is characterized as a rapid adjustment of fission gas inventory remaining in the UO_2 for the transient temperature change and is difficult to assess.
5. Since radioiodine may be retained in the ruptured rod, a prompt cladding and gap release coefficient must be assigned to each release process. At present this value is mainly speculative; however tentative values derived by Collins [263] and Feuerstein [264] may be used. These range from about 0.10 to 0.65 depending upon pressure, rupture temperature, etc.
6. The estimation of Fraction V is dependent upon some conclusions yet to be drawn from current clad-failure programs.

The mechanistic method of evaluating potential volatile fission product release may be useful in the justification of a significant reduction in presently accepted guidelines. For the average fuel rod heat rating, around 8 kw/ft, the potential radioiodine release is lower than for UO_2 melting by perhaps a factor of 100; at power ratings above 18 kw/ft, the potential is probably only a factor of 2 or 3 lower.

7 ANALYTICAL MODELS OF FISSION-PRODUCT RELEASE FROM MOLTEN FUELS

Only limited effort has been directed toward development of analytical models to describe release of fission products from molten reactor fuels. Release under these conditions can be affected by a variety of factors. Reactions that are kinetically hindered at lower temperatures may proceed quite readily at the very high temperatures characteristic of molten ceramic materials. Mass transfer rates within the melt are probably influenced by convection currents that develop because of temperature differences between the interior and the surface. If a crust of solidified material forms on the external surface of the molten mass, this will offer a barrier to release. Vaporization of fission products at the surface can be influenced by vapor-phase mass-transport conditions and by chemical reaction with the atmosphere surrounding the melt. Highly volatile fission-product species that are insoluble in the liquid may form vapor bubbles, while more refractory species may form solid-phase precipitates. Lack of fundamental thermodynamics and kinetic data, combined with uncertainties concerning the fluid dynamics, has severely limited theoretical effort. Currently available approaches for estimation of the rate of fission-product release from molten reactor fuels consist of purely empirical expressions and isolated attempts to model individual aspects of the process.

Morrison et al. [222] adopted a simple empirical approach to provide time-dependent release functions for the calculation of fission-product release from UO_2 under water-reactor loss-of-coolant accident conditions. Results of out-of-pile fuel-melting experiments performed by Parker et al. [121] were approximated by use of a linear function of the square root of time to describe cumulative fractional releases.

However, the experiments were conducted with only gram quantities of UO_2 and, therefore, the release functions described will tend to grossly overestimate rates of release from much larger masses of molten UO_2. Miller [249] has proposed a more mechanistic mathematical model, in which it is assumed that the rate-limiting step for fission-product release from a fuel melt is boundary-layer diffusion in the gas phase above the surface of the melt. The model successfully predicts fuel-vaporization rates and their dependence upon external gas pressure and composition. Application to fission-product release is severely hampered by the lack of fundamental solubility data for fission products in molten-fuel solvents. Without these data, the driving force for release cannot be defined. Some success has been attained in correlation of the model with experiment by use of one experiment of a series to define the Henry's law coefficent, which is then used to predict release from other experiments in the series. However, no verification of the extracted coefficients is available. McKenzie [250] has used a similar approach to correlate measured rates of plutonium volatilization from molten, irradiated uranium with the Langmuir vaporization equation and Raoult's law. Agreement between experimentally measured and calculated rates of plutonium loss under vacuum is interpreted as indicating that plutonium in neutron-irradiated uranium follows the ideal solution laws.

Castleman and Tang [251] have recently formulated a generalized expression for use in the computation of fission-product release from a melt. The equation for the fractional release of a fission product, F_r, is derived by a consideration of diffusion in both the liquid and the gas phase. The rate of transport across the vapor-liquid interface is expressed by the Langmuir equation, generalized for nonideal, multicomponent solutions. The equation obtained is

$$F_r = \sum_{n=1}^{\infty} \frac{1 - \exp[-D_1 t\varphi_n^2/4\ell^2]}{\dfrac{D_1 C_1}{8A\ell}\cdot\dfrac{\varphi_n^3}{\tan\varphi_n} + \dfrac{1}{4}\left(1 + \dfrac{P_T C_1\sqrt{D_1}}{P^0\gamma C_2\sqrt{D_2}} + \dfrac{D_1 C_1}{4A\ell}\right)\varphi_n^2}$$

where $A = \alpha P^\circ\gamma/[2\pi MRT]^{1/2}$; α = vaporization coefficient of vaporizing species, B; P_T = total system pressure; P° = vapor pressure of pure B; γ = activity coefficient of B in condensed phase; C_1 = molar density of condensed phase; C_2 = molar density of vapor phase; t = time at temperature; ℓ = depth of condensed phase; D_1 = diffusion coefficient of B in condensed phase; D_2 = diffusion coefficient of B in vapor phase; T = absolute temperature; R = gas constant; and φ_n = the roots of the equation,

$$\varphi_n \tan\varphi_n = \frac{2\alpha\ell}{[2\pi MRT]^{1/2}}\frac{1}{\cos\varphi_n}\left(\frac{P^\circ\gamma}{C_1 D_1} + \frac{P_T}{C_2\sqrt{D_1 D_2}}\right) + \frac{2\alpha\ell}{[2\pi MRT]^{1/2}}\left(\frac{P^\circ\gamma}{C_1 D_1} + \frac{P_T}{C_2\sqrt{D_1 D_2}}\right).$$

The investigators have used a reduced form of the above expression to correlate observed rates of fission-product-iodine release from molten uranium with depth of the melt. From this correlation and from the experimental observation that release rates are independent of cover-gas flow rate and molecular weight, it is concluded that the rate-controlling process for iodine escape is diffusion in the condensed phase rather than in the gas phase. However, the large iodine diffusion coefficients that this treatment yields indicate that, even in this carefully conducted experiment, convective forces in the uranium melt have influenced mass-transport rates. Fontana and Wantland [253] have been developing analytical methods to describe internal convection flow patterns in molten pools and the effect of enhanced mass transport of dissolved products on their release from the surface. Fontana [230] has also developed calculational techniques to compute thermodynamic property data for compounds in any phase, which may be used to supply thermophysical properties to the mass-transfer model. Initial calculations for yttrium release from a large pool of molten UO_2 indicate that the effect of internal convection on mass transfer is so important that it cannot be neglected where its occurrence may be suspected.

REFERENCES

1. "Theoretical Possibilities and Consequences of Major Accidents in Large Nuclear Power Plants", USAEC Report WASH-740, March 1957. (Staff Brookhaven Natl. Lab.)
2. F. H. Spedding, I. B. Johns et al., "Removal of Fission Products from Molten Uranium by Diffusion and Chemical Reactions", Report MUC-NS-3068, University of Chicago Metallurgical Laboratory, 1942.
3. S. Katcoff, "Fission product yields from neutron-induced fission", Nucleonics 18, 11 (1960) 201.
4. J. O. Blomeke and M. F. Todd, "Uranium^{235} Fission Product Production as a Function of Thermal Neutron Flux, Irradiation Time, and Decay Time", USAEC Report ORNL-2127, Oak Ridge National Laboratory, 1958.
5. I. G. Dillon and L. Burris, Jr., "Estimation of Fission Product Spectra in Fuel Elements Discharged from the Power Breeder Reactor and the Experimental Breeder Reactor No. 2", USAEC Report ANL-5334, Argonne National Laboratory, Oct. 1954 (Decl. Sept. 1955).
6. Bolles R. C. and N. E. Ballou, "Calculated activities and abundances of U^{235} fission products", Nucl. Sci. Eng., Vol. 5 (1959) pp. 156-185.
7. J. O. M. Bockris, J. L. White, and J. O. Mackenzie (Eds.), Physicochemical Measurements at High Temperatures, Academic Press, Inc., N. Y., 1959.

8. A. N. Nesmeyanov, Vapor Pressure of the Chemical Elements (R. Gary, Ed.) Elsevier Pub. Co., N. Y. (1963).
9. C. R. Tipton, Jr. (Ed.), Reactor Handbook (2nd ed), Vol. I, "Materials", Interscience Publishers, Inc., N. Y., 1960.
10. A. S. Coffinberry and W. N. Miner (Eds.), The Metal Plutonium, University of Chicago Press, Chicago, Ill., 1961.
   a) F. W. Schonfeld, "Phase Diagrams Studied at Los Alamos", Chapter XXII.
   b) A. E. Kay, "Some Physical and Physicochemical Properties of Plutonium", Chapter XVII.
   c) K. L. Wauchope, "The Preparation of Plutonium-Aluminum Alloy Fuel Elements for the NRX Reactor", Chapter XXVIII.
   d) R. E. Tate, "The Fabrication of Billets Containing Plutonium for MTR Fuel Elements", Chapter XXIX.
   e) O. J. C. Runnalls, "The Preparation of Plutonium-Aluminum and Other Plutonium Alloys", Chapter XXVI.
   f) W. M. Cashin, "A Mixed-Oxide Concept of a Plutonium-Fueled Power Reactor", Chapter XXXV.
   g) K. L. Wauchope, "The Preparation of Plutonium-Aluminum Alloy Fuel Elements for the NRX Reactor", Chapter XXVIII.
11. J. Belle (Ed.), Uranium Dioxide: Properties and Nuclear Applications, USAEC, Superintendent of Documents, Washington, D. C., 1961. a) H. R. Hoekstra, "Phase Relationships in the Uranium-Oxygen and Binary Oxide Systems", Chapter 6. b) B. Lustman, "Irradiation Effects in Uranium Dioxide," Chapter 9.
12. W. E. Roake, "Irradiation Alteration of Uranium Dioxide", Report HW-73072, General Electric Co., Hanford Atomic Products Operation, Mar. 1962.
13. E. Grison, W. B. H. Lord and R. D. Fowler (Eds.), Plutonium 1960, Clever-Hume Press, Ltd., London, 1961.
14. S. Langer, "Melting Points and Thermodynamic Properties of Some Reactor Core Materials", Report GAMD-2101, General Atomic, Division of General Dynamics Corp., Mar. 1961.
15. R. M. Dell and M. Allbutt, "The Nitrides and Sulphides of Uranium, Thorium and Plutonium: A Review of Present Knowledge", British Report AERE-R-4253, Mar. 1963.
16. R. B. Lindauer, "Revisions to MSRE Design Data Sheets, Issue No. 8", USAEC Report ORNL-CF-63-6-30, Oak Ridge National Laboratory, 1963.
17. R. K. Hilliard, "Fission Product Release from Uranium Heated in Air", Report HW-60689, General Electric Co., Hanford Atomic Products Operation, Aug. 1959. Also HW-SA-2640 T. D. Chikalla, "Melting behavior in the system UO_2-PuO_2." General Electric Co., Hanford Atomic Products Operation, June 13, 1962.
18. G. W. Parker, G. E. Creek, W. J. Martin, and C. J. Barton, "Fuel Element Catastrophe Studies: Hazards of Fission Product Release from Irradiated Uranium", USAEC Report ORNL-CF-60-6-24, Oak Ridge National Laboratory, 1960.
19. Nuclear Safety, Vol. 5, No. 2, p. 203, 1963.
20. The Third Conference on Nuclear Reactor Chemistry held in Gatlinburg; Tennessee, October 9-11, 1962, USAEC Report TID-7641, 1963.
   a) A. W. Castleman, Jr., "The Chemical and Physical Behavior of Released Fission Products", p. 155.
   b) L. C. Schwendiman and L. F. Coleman, "Particulates Generated During the Air Oxidation of Uranium", p. 94.
   c) G. W. Parker, G. E. Creek, R. A. Lorenz, and W. J. Martin, "Parametric Studies of Fission-Product Release from UO_2 Fuels", p. 15.
   d) P. K. Conn, E. A. Aitken, C. C. Browne, and R. E. Fryxell, "Studies of Fission Gas Release from In-Pile Fuel Elements. II. Corrosion-Induced Loss from Be0 Ceramic Elements", p. 61.
   e) P. K. Conn, R. L. Stuart, J. Y. Gerhardt, and P. H. Wilks, "Be0 Ceramic Fuel Meltdown Experiments Using a Plasma Jet", p. 148.
21. A. W. Castleman, Jr. and I. W. Tang, "Vaporization of Fission Products from Irradiated Uranium. II. Some Observations on the Chemical Behavior of Fission Products Iodine and Cesium," USAEC Report BNL 13651, Brookhaven National Laboratory, 1969.
22. J. A. DeMastry, "Niobium-10wt. % Uranium Alloy", Chapter in "Properties of Fuels for High-Temperature Reactor Concepts", R. W. Enderbrock (Ed.), Report BMI-1598, Battelle Memorial Institute, 1962.
23. T. C. Ehlert and J. L. Margrave, "Melting point and spectral emissivity of uranium dioxide", J. Am. Ceram. Soc., 41 (1958) 330.
24. G. E. Creek, W. J. Martin, and G. W. Parker, "Experiments on the Release of Fission Products from Molten Reactor Fuels", USAEC Report ORNL-2616, Oak Ridge National Laboratory, July 1959.
25. W. D. Kingery and J. F. Wygant, "Thermodynamics in ceramics. I. Energy and heat content", Bull. Am. Ceram. Soc., 31, 5 (1952) 165.
26. A. Glassner, "The Thermochemical Properties of the Oxides, Fluorides, and Chlorides to 2500°K", USAEC Report ANL-5750, Argonne National Laboratory, 1957.
27. J. P. Coughlin, "Contributions to the Data on Theoretical Metallurgy: II. Heats and Free Energy of Formation of Inorganic Oxides", U. S. Bureau of Mines, Bulletin 542, 1954.
28. D. R. Stull and G. C. Sinke, "Thermodynamic Properties of the Elements", Am. Chem. Soc., Advan. Chem., 18 (1956).
29. K. K. Kelley and E. G. King, "Contributions to the Data on Theoretical Metallurgy. XIV. Entropies of the Elements and Inorganic Compounds", U. S. Bureau of Mines, Bulletin 592, 1961.
30. R. K. Hilliard and D. L. Reid, "Fission Product Release from Uranium—Effect of Irradiation Level", Report HW-72321, General Electric Co., Hanford Atomic Products Operation, June 1962.
31. R. A. Stinchcombe, "Generation of Tellurium Aerosols and Preliminary Filtration Experiments", British Report AERE-M-1130, Jan 1963.
32. M. W. Rosenthal and S. Cantor, "Some Remarks on the Contribution of Fission Product Cesium to the Pressure Buildup in UO_2 Fuel Elements", USAEC Report ORNL-CF-60-3-81, 1960.
33. L. Brewer, "Fate of Fission Product Gases in the Coolant Stream", Report GAMD-903, General Atomic, Division of General Dynamics Corp., Aug 1959.
34. L. Brewer, "Vaporization Processes in a Runaway Reactor", Report GAMD-919, General Atomic, Division of General Dynamics Corp. Aug. 1959.
35. L. Brewer, L. A. Bromley, P. W. Gilles, and N. L. Lofgren, "Thermodynamic and Physical Properties of Nitrides, Carbides, Sulfides, Silicides, and Phosphides", Paper 4 in Vol. 19B, Chemistry and Metallurgy of Miscellaneous Materials: Thermodynamics, L. L. Quill, (Ed.), Nat. Nucl. Energy Ser., McGraw-Hill Book Co., Inc., N. Y., 1950.
36. O. H. Krikorian, "Thermodynamic Properties of the Carbides", Report UCRL-2888, University of California Lawrence Radiation Laboratory, Apr 1955.
37. O. H. Krikorian, "Estimation of High Temperature Heat Capacity of Carbides", Report UCRL-6785, University of California Lawrence Radiation Laboratory, Feb 1962.
38. S. Cantor, private communication to R. B. Evans III, Oak Ridge National Laboratory, August 1961.
39. "Nuclear Safety Program, Semiannual Progress Report for Period Ending June 30, 1963", USAEC Report ORNL-3483, Oak Ridge National Laboratory, 1963. (Same as TID-4500, 22nd ed.)
   a) W. E. Browning, R. D. Ackley, "Characterization and control of Accident—Released Fission Products", p. 26.
   b) G. W. Parker, G. E. Creek, W. J. Martin, R. A. Lorenz, "Fission-Product Release from Aluminium-Uranium Alloys", p. 9.
   c) W. E. Browning, Jr., R. P. Shields, C. E. Miller, Jr., B. F. Roberts, "Release of Fission Products on In-Pile Melting or Burning of Reactor Fuels", p. 22.
   d) C. J. Barton, G. W. Parker, G. E. Creek, W. J. Martin, "Hot-Cell Containment Mockup Facility for Transport Evaluation", p. 33.
40. C. Agte and K. Moers, Z. Anor. Alleg. Chem., 198 (1931) 236.
41. L. C. Schwendiman, G. A. Schmel, and A. K. Postma, "Radioactive Particle Retention in Aerosol Transport Systems," Proceedings of the International Symposium on Radioactive Pollution of Gasesous Media, Saclay, November 12-16, 1963. Vol. II. p. 373, Presses Universitaires de France, Paris, 1965.
42. A. C. Chamberlain, "Particle Transport Across Boundary Layers," USAEC Report TID-7641, p. 1, 1963.
43. J. W. Thomas, "Gravity Settling of Particles in a Horizontal Tube," Journal of the Air Pollution Control Association 8, p. 132 (1958).
44. W. E. Browning, Jr., R. D. Ackley, "Particle Size Distribution of Radioactive Aerosols by Diffusion Coefficient Measurments," USAEC Report ORNL-3319, p. 44, Oak Ridge National Laboratory, Aug 1962.
45. J. W. Thomas, "The Diffusion Battery Method for Aerosol Particle Size Determination," ORNL-1648, January 1954.
46. J. S. Townsend, "The Diffusion of Ions into Gases," Transactions of the Royal Society, 193-A, pp. 129-158 (1900).
47. P. G. Gormley, M. Kennedy, "Diffusion from a Stream Flowing Through a Cylindrical Tube," Proc. of the Royal Irish Academy, 52-A, pp. 163-169 (1949).
48. W. C. DeMarcus, J. W. Thomas, "Theory of a Diffusion Battery," USAEC Report ORNL-1413, p. 44, Oak Ridge National Laboratory, Oct 1952.
49. S. K. Friedlander, H. F. Johnstone, "Deposition of Particles from Turbulent Gas Streams," Ind. Eng. Chem., 49, p. 1151 (1957).
50. A. K. Postma, "Studies in Micromeritics. Part II.—The Deposition of Particles in Circular Conduits Due to Thermal Gradients," USAEC Report H. W—70791, Hanford Atomic Products Operation, 1961.
51. G. W. Parker, "A Review of Fission-Product Release Re-

search", Trans. Am. Nucl. Soc. 6 (1963) 120.
52. T. R. Wilson, O. M. Hauge, G. B. Matheny,"Feasibility and Conceptual Design for the STEP Loss-of-Coolant Facility", USAEC Report IDO-16833 (Rev. 1) Phillips Petroleum Co., Idaho, 1963. See also G. W. Parker et al., Fission-Product Release under LOFT Conditions, LOFT Assistance programs, W. B. Cottrell Nuclear Safety Progress Report for period ending Dec. 31, 1968. ORNL-4374, June 1969.
53. J. J. DiNunno et al., "Calculation of Distance Factors for Power and Test Reactors", USAEC Report TID-14844, 1962.
54. G. J. Elbaum, T. J. Thompson, Analysis of Rapid Excursions Involving Fuel-Element Rupture. Trans. Am. Nucl. Soc., 9 (1) p. 331 San Diego June 11-15, 1967 (Also MIT Sc. D. Thesis 1967).
55. "Nuclear Safety Semiannual Progress Report", USAEC Report ORNL-3691, Oak Ridge National Laboratory, Nov 1964.
56. J. R. Beattie, "Future Trends in the Assessment of Hazards from Fission Product Releases", presented for the Colloquium on Radioactive Pollution of Gaseous Media at Saclay, France, November 1963.
57. G. W. Parker, G. E. Creek, W. J. Martin, "Fuel Element Decomposition Products", AEC Seventh Air Cleaning Conference, USAEC Report TID-7627, p. 263, 1962.
58. "Nuclear Safety Program Semiannual Progress Report for Period Ending December 31, 1963", USAEC Report ORNL-3547, Oak Ridge National Laboratory, 1964.
    a) G. W. Parker, W. J. Martin, C. J. Barton, G. E. Creek, and R. A. Lorenz, "Fission Product Release from Aluminum-Uranium Alloys", p. 5.
    b) W. E. Browning, Jr., C. E. Miller, Jr., B. F. Roberts, and R. P. Shields, "Behavior of Fission Products Released During In-Pile Destruction of Reactor Fuels", p. 42.
    c) G. W. Parker, R. A. Lorenz, and C. E. Miller, Jr., "Release of Fission Products on In-Pile Melting of Reactor Fuels Under Transient Reactor Conditions", p. 25.
    d) G. W. Parker, W. J. Martin, C. J. Barton, G. E. Creek, and R. A. Lorenz, "Properties of Fission Product Aerosols Produced by Overheated Reactor Fuels", p. 3.
    e) L. F. Parsly, T. H. Row, P. P. Holz, and L. F. Franzen, "Nuclear Safety Pilot Plant", p. 73.
59. W. Penny et al., "Accident at Windscale No. 1 Pile on 10th October 1957", Cmnd. 302, Her Majesty's Stationery Office, London, November 1957.
60. J. F. Loutit, W. G. Marley, and R. S. Russel, "The Nuclear Reactor Accident at Windscale", Medical Research Council, The Hazards to Men of Nuclear and Allied Radiations (Second Report), Appendix H, Her Majesty's Stationery Office, London, 1960.
61. G. E. Zima, "Pyrophoricity of Uranium in Reactor Environments", Report HW-62442, General Electric Co., Hanford Atomic Products Operation, 1960.
62. T. Wathen, "Corrosion of Uranium Metal in Air and Steam at Various Temperatures", British Report BR-223A, 1943.
63. J. T. Waber, "An Analysis of Project Data on the Corrosion of Uranium in Various Media", USAEC Report LA-1381, Los Alamos Scientific Laboratory, Dec. 1948, Decl. Mar. 1957.
64. D. Cubicciotti, "The Reaction Between Uranium and Oxygen", J. Am. Chem. Soc., 74 (1952) 1079.
65. J. Loriers, "On the Oxidation of Metallic Uranium", Compt. Rend., 234 (1952) 91.
66. J. T. Waber, "A Review of the Corrosion of Uranium and Its Alloys", USAEC Report LA-1524, Los Alamos Scientific Laboratory, Nov. 1952. Decl. June 1956.
67. R. K. Hilliard, "Oxidation of Uranium in Air at High Temperatures", Report HW-58022, General Electric Co., Hanford Atomic Products Operation, Dec. 1958.
68. W. J. Megaw, R. C. Chadwick, A. C. Wells, and J. E. Bridges, "The Oxidation and Release of I^{131} from Uranium Slugs Oxidizing in Air and Carbon Dioxide, J. Nucl. Energy Pts. A/B, Reactor Sci. Technol., 15 (1961) 176.
69. Baker, L., J. D. Bingle, G. Klepac, and R. Koonz, "Isothermal Oxidation of Uranium at High Temperature", USAEC Report ANL-6413, p. 160. Argonne National Laboratory, 1961.
70. J. E. Antill and P. Murray, "Reactions Between Fuel Elements and Gaseous Coolants", in Vol. 5, Technology Engineering and Safety, C. M. Nichols (Ed.), Progress in Nuclear Energy, Series IV, Pergamon Press, N. Y. and London, 1960.
71. J. W. Isaacs and J. N. Wanklyn, "The Reaction of Uranium with Air at High Temperatures", British Report AERE-R-3559, 1960.
72. J. N. Wanklyn, "The Ignition of Uranium", British Report AERE-M/M-184, 1957.
73. Schnizlein J. G., P. J. Pizzolato, H. A. Porte, J. D. Bingle, D. F. Fischer, L. W. Mishler, and R. C. Vogel, "Ignition Behavior and Kinetics of Oxidation of the Reactor Metals, Uranium, Zirconium, Plutonium, and Thorium, and Binary Alloys of Each", USAEC Report ANL-5974, Argonne National Laboratory, Apr. 1959.
74. R. A. U. Huddle, "The Oxidation of Uranium by Carbon Dioxide-Temperature Range 175-500°C", British Report TRDC P. 36, 1954.
75. D. R. Silvester, British Report AERE-M/R-2437.
76. J. E. Antill, K. A. Peakall, N. Crick, and E. Smart, "Compatibility of UO_2, UC, and U with CO_2", British Report AERE-M/M-168, 1957.
77. J. E. Antill, K. A. Peakall, N. Crick, and M. gardner, "Kinetics of the Oxidation of Uranium by Carbon Dioxide", British Report AERE-M/R-2524, 1958.
78. H. R. D. Diffey and D. T. King, "The Oxidation of Irradiated Uranium in Carbon Dioxide", British Report AERE-R-3699, 1961.
79. W. J. Megaw and J. E. Bridges, unpublished data quoted by G. W. Dolphin. See also: W. J. Megaw and J. Rundo in Reports on Progress in Physics, Vol. XXV, p. 337, 1962.
80. J. H. Buddery and K. T. Scott, "A Study of the Melting of Irradiated Uranium", J. Nucl. Mater., 5, 1 (1962) 81.
81. J. J. Katz and E. Rabinowitch, The Chemistry of Uranium, Vol. VIII, p. 167, Nat. Nucl. Energy Ser., McGraw-Hill Book Co., Inc., N. Y., 1951.
82. R. A. U. Huddle, "The Uranium-Steam Reaction", British Report AERE-M/R-1281, 1963.
83. B. E. Hopkinson, "The Kinetics of the Uranium-Steam Reaction", British Report AERE-M/R-1281A, 1957.
84. A. J. Scott, "Fission Product Release by the High Temperature Uranium-Steam Reaction", Report HW-62604, General Electric Co., Hanford Atomic Products Operation, 1959.
85. L. Burris, H. M. Feder, S. Lawroski, W. A. Rodger, and R. C. Vogel, "The melt refining of irradiated uranium: Application to EBR-II fast reactor fuels. I. Introduction", Nucl. Sci. Eng., 6 (1959) 493-495.
86. R. K. Hilliard, C. E. Linderoth, and A. J. Scott, "Fission product release from overheated uranium—A laboratory study", Health Phys., 7 (1961) 1.
87. J. R. Seebach and J. W. Wade, "Fuel Meltdown Experiments", Report DP-314, E. I. du Pont de Nemours and Co., Inc., (Decl.) 1958.
88. F. Schroeder, informal report presented at American Nuclear Safety Meeting, November 1963 N. Y., See: Nucleonics, January 1964, p. 25; Nucl. News, December 1963, p. 30; Nucleonics, June 1964, p. 27.
89. W. K. Ergen, "SPERT-I destructive test with UO_2 fuel", Nucl. Safety 5, 3 (1964) 231.
90. J. W. Antill, K. A. Peakall, and M. Gardner, "Oxidation of Uranium-Molybdenum Alloys in Carbon Dioxide and Air at 500-1000°C", British Report AERE-M/R-2805, 1959.
91. D. G. Freas, A. F. Leatherman, and J. E. Gates, "Meltdown Studies of Irradiated Uranium-10 w/o Molybdenum Fuel Pins", Report BMI-PRDC-656, Battelle Memorial Institute, 1960.
92. Atomic Power Development Associates, Inc., "Report on the Fuel Melting Incident in the Enrico Fermi Atomic Power Plant on October 5, 1966", AEC Docket 50-16 (December 15, 1968).
93. R. E. Mueller, R. J. Beaudry, and J. G. Feldes, "Results of Fission Product Activity Analysis to Determine Extent of Fuel Failure in Fermi Accident", Trans. Amer. Nuclear Society, 10 (1) 334-335 (1967).
94. A. E. Klickman and R. C. Callen, "Anomalous Reactivity Effects in the Fermi Incident", Trans. Amer. Nuclear Society, 10 (1) 334 (1967).
95. G. W. Keilholtz and G. C. Battle, Jr., "Fission Product Release and Transport in Liquid Metal Fast Breeder Reactors", ORNL-NSIC-37, March 1969.
96. R. S. Hart, "Distribution of Fission Product Contamination in the SRE", NAA-SR-6890, North American Aviation (1961).
97. K. K. Brown, "Re-examining the Significance of Fission Products in Accident Situations", Nucleonics, 23,6, (1965).
98. "Chemical Engineering Division Semi-Annual Report, January-June, 1964." ANL-6900, Argonne National Laboratory, August 1964.
99. C. R. Mitra and C. F. Bonilla, "Solubility and Stripping of Rare Gases in Molten Metals, Final Report," BNL-3337, Brookhaven National Laboratory (1955).
100. S. J. Rodgers and G. E. Kennedy, "Fission Product Release During a Simulated Meltdown of a PWR Type Core", MSA Research Corp., Technical Report 63, October 1958.
101. T. A. Milne and C. T. Young, "The Volatilization of Fission Products from Molten and Solid Thorium-Uranium Alloy", Report NAA-SR-1680, North American Aviation, Inc., 1956.
102. a) J. Mishima, "A Review of Research on Plutonium Release During Overheating and Fires", USAEC Report HW-83668, Hanford Laboratory, August 1964.
    b) J. Mishima, "Plutonium Release Studies, I. Release from Ignited Metal", BNWL-205, December 1965.
    c) J. Mishima, "Plutonium Release Studies II, Release from Ignited, Bulk Metallic Pieces", BNWL-357, November 1966.
103. R. E. Felt and H. D. Merritt, "Plutonium Metal Fire Hazards Study (Rev. 1) Memorandum dated March 4, 1966" Isochem Inc., Richland, Washington.
104. K. Stewart, "The Particulate Material Formed by the Oxidation of Plutonium", Progress in Nuclear Energy, Pergamon

Press, New York, Series IV, Vol. 5, 1963.
105. H. J. Ettinger, W. D. Moss and H. Busey, "Characteristics of the Aerosol Produced from Burning Sodium and Plutonium," USAEC Report LA-3491, 1966.
106. T. Hatch and S. P. Choate, "Statistical Description of the Size Properties of Non-Uniform Particulate Substances", J. Franklin Inst., 207, 369 (1929).
107. F. T. Binford and E. N. Cramer, "High Flux Intensity Reactor—A Functional Description", USAEC Report ORNL-3572, Oak Ridge National Laboratory, 1962.
108. "BONUS Reactor—Final Hazards Summary Report", Report PRWRA-GNEC-5, General Nuclear Engineering Corp., 1962.
109. J. A. Lane et al., "A Study of Problems Associated with Release of Fission Products from Ceramic Fuels in Gas-Cooled Reactors", USAEC Report ORNL-2851, Oak Ridge National Laboratory, 1959.
110. W. B. Cottrell et al., "Fission Product Release from UO_2", USAEC Report ORNL-2935, Oak Ridge National Laboratory, 1960.
111. R. M. Carroll, "Fission product release from UO_2", Nucl. Safety 4, 1 (1962) 35-42.
112. D. Davies and G. Long, "Abnormal Kinetics in the Release of Inert Gases from Uranium Dioxide", British Report AERE-M-969, 1963.
113. W. H. Stevens, J. R. MacEwan, and A. M. Ross, "The Diffusion Behavior of Fission Xenon in Uranium Dioxide", USAEC Report TID-7610, 1961.
114. D. Davies, G. Long, and W. P. Stanaway, "The Emission of Volatile Fission Products from Uranium Dioxide", British Report AERE-R-4342, 1963.
115. "Nuclear Reactor Chemistry—Second Conference, Gatlinburg, Tennessee, October 10-12, 1961", USAEC Report TID-7622, 1962.
a) G. W. Parker, G. E. Creek, and W. J. Martin, "Fission Product Release from Reactor-Grade UO_2 by Diffusion, Oxidation, and Melting", p. 149. See also: ORNL-3176.
b) E. A. Aitken, P. K. Conn, E. S. Collins, and R. E. Honnell, "Studies of Fission Gas Release from In-Pile Tests: I. Recoil Loss from Ceramic Fuel Elements", p. 193.
c) J. M. Blocher, Jr., M. F. Browning, A. C. Secrest, V. M. Secrest, and J. H. Oxley, "Preparation of Ceramic-Coated Nuclear Fuel Particles", p. 57.
116. E. Rothwell, "The release of Kr^{85} from irradiated uranium dioxide on post-irradiation annealing", J. Nucl. Mater., 5, (1962) 241-249.
117. K. A. Peakall and J. E. Antill, "Oxidation of uranium dioxide in air at 350-1000°C", J. Nucl. Mater, 2 (1960) 194-195.
118. K. T. Scott and K. T. Harrison, "The oxidation of uranium dioxide", J. Nucl. Mater., 8, (1963) 307.
119. S. Aronson, R. B. Roof, Jr., and J. Belle, "Kinetic study of the oxidation of uranium dioxide", J. Chem. Phys. 27 (1957) 137.
120. K. A. Peakall, J. E. Antill, and M. J. Bennett, "The Oxidation of UO_2 and Defect Fuel Elements", British Report AERE-R-3603, not for publication.
121. G. W. Parker, G. E. Creek, and W. J. Martin, "Influence of Irradiation Level on Fission Product Hazards Associated with UO_2-Fueled Reactors", USAEC Report ORNL-3319, p. 11, Oak Ridge National Laboratory, 1962.
122. G. W. Parker, G. E. Creek, and W. J. Martin, "Fission Product Release from Melted UO_2 by the Tungsten Crucible Method", USAEC Report ORNL-3401, Oak Ridge National Laboratory, 1963.
123. G. W. Parker and R. A. Lorenz, "Melting of UO_2 by a Centered Tungsten Resistor", USAEC Report ORNL-3401, p. 11, Oak Ridge National Laboratory, 1963.
124. Browning, W. E., Jr., C. E. Miller, Jr., R. P. Shields, B. F. Roberts "Release of Fission Products During In-Pile Melting of UO_2", Nucl. Sci. Eng., 18, 151-162 (1964).
125. Roberts, B. F. and S. H. Fried, Nuclear Safety Program Annual Progress Report for Period ending Dec. 31, 1968 ORNL-4374, June 1969.
126. J. J. Edwards et al., "Fast Reactor Fuel Cycle Costs and Temperature Coefficients of Reactivity for PuO_2-SS and PuO_2-UO_2", Report APDA-154, Atomic Power Development Associates, Inc., 1963.
127. N. H. Brett and L. E. Russell, "The Sintering Behavior and Stability of (Pu, U)O_2 Solid Solutions", British Report AERE-R-3900, 1962.
128. T. D. Chikalla, C. E. McNeilly, and R. E. Skavdahl, "The Plutonium-Oxygen System", Report HW-74802, General Electric Co., Hanford Atomic Products Operation, 1962.
129. S. A. Rabin, S. D. Clinton, and J. W. Ullmann, "Irradiations of Non-Sintered ThO_2-UO_2 and ThO_2-PuO_2 Fuel Rods for Power Reactor Applications", Proceedings of Powder-Filled UO_2 Fuel Element Symposium, Worcester, Mass., November 5-6, 1963.
130. R. F. S. Robertson and V. C. Hall, Jr., "Fuel Defect Test—BORAX-IV", USAEC Report ANL-5862, Argonne National Laboratory, 1959.
131. E. E. Jackson and M. H. Rand, "The Oxidation Behavior of Plutonium Dioxide and Solid Solutions Containing Plutonium Dioxide", British Report AERE-R-3636, 1963.
132. N. H. Brett and A. C. Fox, "Oxidation Products of Plutonium Dioxide-Uranium Dioxide Solid Solutions in Air at 750°C", British Report AERE-R-3937, 1963.
133. A. S. Sasko, R. S. Roth, and S. M. Lang, "Study of the Stabilization of Uranium Oxide in Binary Combinations", Report WADC-TR-53-449 (Part I), National Bureau of Standards, October 1953 (Classified).
134. G. W. Parker, G. E. Creek, and W. J. Martin, "Preliminary Report on the Release of Fission Products on Melting GE-ANP Fuel", USAEC Report ORNL-CF-60-1-50, Oak Ridge National Laboratory, January 1960, and Suppl. March 1960.
135. G. W. Keilholtz, J. E. Lee, Jr., R. P. Shields, and W. E. Browning, Jr., "Radiation Damage in Beryllium Oxide", Radiation Damage in Reactor Materials, Proceedings of a Symposium held at Venice, May 1962, International Atomic Energy Agency, Vienna, 1963.
136. K. A. Peakall and J. E. Antill, "Oxidation of uranium monocarbide", J. Less-Common Metals, 4 (1962) 426.
137. R. E. Blanco, G. I. Cathers, L. M. Ferris, T. A. Gens, R. W. Horton, and E. L. Nicholson, "Processing of graphite reactor fuels containing coated particles and ceramics", Nucl. Sci. Eng. 20 (1964) 13-22.
138. J. H. Goode et al., in "ORNL Status and Progress Report for Period Ending December 31, 1963", USAEC Report ORNL-3561, p. 7, Oak Ridge National Laboratory, 1964.
139. H. G. MacPherson (Ed.), "Molten Salt Reactors", Part II in Fluid Fuel Reactors, Addison-Wesley Publishing Co., Inc., Reading, Mass., 1958.
140. W. R. Grimes et al., "Chemical Aspects of Molten Fluoride Reactors", Progr. Nucl. Energy Series IV, Vol. 2, Technology, Engineering and Safety, C. M. Nichols (Ed.), Pergamon Press, N. Y. and London, 1960.
141. M. T. Robinson et al., "Some aspects of the behavior of fission products in molten fluoride reactor fuels", Nucl. Sci. Eng., 4 (1958) 288.
142. M. T. Robinson, "On the chemistry of the fission process in reactor fuels containing UF_4 and UO_2", Nucl. Sci. Eng., 4 (1958) 263.
143. S. E. Beall, W. L. Breazeale, and B. W. Kinyon, "Molten Salt Reactor Experiment Preliminary Hazards Report", USAEC Report ORNL-CF-61-2-46, Oak Ridge National Laboratory, 1961.
144a. M. W. Rosenthal, P. R. Kasten, and R. B. Briggs, "Molten-Salt Reactors—History, Status, and Potential," Nucl. Appl. Tech. 8, 107 (1970).
144b. P. N. Haubenreich and J. R. Engel, "Experience with the Molten-Salt Reactor Experiment," Nucl. Appl. Tech. 8, 118 (1970).
144c. W. R. Grimes, "Molten-Salt Reactor Chemistry," Nucl. Appl. Tech. 8, 137 (1970).
144d. H. E. McCoy, et al., "New Developments in Materials for Molten-Salt Reactors," Nucl. Appl. Tech. 8, 156 (1970).
144e. M. E. Whatley, et al., "Engineering Development of the MSBR Fuel Recycle," Nucl. Appl. Tech. 8, 170 (1970).
144f. D. Scott and W. P. Eatherly, "Graphite and Xenon Behavior and their Influence on Molten-Salt Reactor Design," Nucl. Appl. Tech. 8, 179 (1970).
144g. E. S. Bettis and R. C. Robertson, "The Design and Performance Features of a Single-Fluid Molten-Salt Breeder Reactor," Nucl. Appl. Tech. 8, 190 (1970).
144h. A. M. Perry and H. F. Bauman, "Reactor Physics and Fuel-Cycle Analyses," Nucl. Appl. Tech. 8, 208 (1970).
145. R. B. Briggs (Program Director), "Molten Salt Reactor Program Semiannual Progress Report, January 31, 1963", USAEC Report ORNL-3419, p. 97, Oak Ridge National Laboratory, 1963.
146. K Division Personnel, "LAMPRE-I Final Design Status Report", USAEC Report LA-2833, Los, Alamos Scientific Laboratory, 1963.
147. R. M. Bidwell, "Fission-product behavior in direct-contact-core-liquid-metal-fueled reactors", Nucl. Sci. Eng., 18, 4 (1964) 426-434.
148. K Division Personnel "Fast Reactor Core Test Facility Safety Analysis Report", USAEC Report LA-2735, Los Alamos Scientific Laboratory, 1962.
149. A. W. Castleman, F. E. Hoffman, and A. M. Eshaya, "Diffusion of Xenon Through Aluminum and Stainless Steel", USAEC Report BNL-624, Brookhaven National Laboratory, 1960.
150. P. Gordon, J. Atherton, and A. Kauffman, "Study of Helium Diffusion Through Aluminum", USAEC Report AECD-3313, 1952.
151. G. T. Murray and Q. Pincus, "Permeability of Cladding Materials to Inert Gases", USAEC First Annual Report, NYO-9000, 1959.
152. G. T. Murray, "Permeability of Cladding Materials to Inert Gases", USAEC Second Annual Report, MRC-195, 1960.
153. A. A. Bauer et al., "Fission Product Migration in and Release

from Iron-20 w/o Chromium", USAEC Report BMI-1611, January 1963.
154. F. R. Winslow, T. S. Elleman, G. G. Cocks, J. Bugl, and A. A. Bauer, "Fission Product Release from Fuel Element Cladding—Quarterly Progress Report", USAEC Report BMI-X-10039, April 1963.
155. Papers from the symposium on coated particle fuels at the November 1963 meeting of the American Nuclear Society at New York City, Nucl. Sci. Eng. 20 (2) 1964.
   a) J. M. Blocher, Jr., M. F. Browning, W. J. Wilson, V. M. Secrest, A. C. Secrest, R. B. Landrigan, and J. H. Oxley, "Properties of Ceramic-Coated Nuclear-Fuel Particles", pp. 153-170.
   b) C. W. Townley, N. E. Miller, R. L. Ritzman, and R. J. Burian, "Irradiation Studies of Ceramic-Coated Nuclear Fuel particles", pp. 171-179. See also Nucleonics 22 (2) 43 (February 1964).
   c) F. L. Carlsen, Jr., E. S. Bomar, and W. O. Harms, "Development of fueled Graphite Containing Pyrolytic-Carbon-Coated Carbide Particles for Non-Purged Gas-Cooled Reactor Systems", pp. 180-200.
   d) W. V. Goeddel, "Development and Utilization of Pyrolytic-Carbon-Coated Carbide Fuel for the High-Temperature Gas-Cooled Reactor", pp. 201-218.
   e) R. A. Reuter, "Duplex Carbon-Coated Fuel Particles", pp. 219-226.
   f) H. G. Sowman, R. L. Surver, and J. R. Johson, "The Development of Spherical Pyrolytic-Carbon-Coated UC_2 and ThC_2 Fuel Particles", p. 227-234.
156. J. H. Oxley, "Recent Developments with Coated-Particle Fuel Materials", Reactor Mater. 6 (2) 1 (1963).
157. R. W. Dayton and R. F. Dickerson, "Progress Relating to Civilian Applications During July 1962", USAEC Report BMI-1589 (Del.), August 1962.
158. J. H. Oxley, "Coated-Particle Fuel Materials", Reactor Mater. 6 (1) 31 (1963).
159. G. E. Raines and W. H. Goldthwaite, "In-Pile Fission-Gas-Release Behavior of Alumina-Coated UO_2 Particles Irradiated to High Burnup", Report BMI-1552, Battelle Memorial Institute, 1961.
160. G. A. Cowan and C. J. Orth, "Diffusion of Fission Products at High Temperatures from Refractory Matrices", Proceedings of the Second U. N. International Conference on Peaceful Uses of Atomic Energy, Geneva, 1958, Vol. 7, p. 328.
161. J. L. Scott, "Fission product release from uranium-graphite fuels", Nucl. Safety, 4, 4 (1963) 49.
162. C. A. Smith and C. T. Young, "Diffusion of Fission Fragments from Uranium-Impregnated Graphite", Report NAA-SR-72, North American Aviation, Inc., 1951.
163. C. T. Young and C. A. Smith, "Preliminary Experiments on Fission Product Diffusion from Uranium-Impregnated Graphite in the Range 1800-2200°C", Report NAA-SR-232, North American Aviation, Inc., 1953.
164. L. B. Doyle, "High Temperature Diffusion of Individual Fission Elements from Uranium Carbide-Impregnated Graphite", Report NAA-SR-255, North American Aviation, Inc., 1953.
165. D. D. Cubicciotti, "Diffusion of Xenon from Uranium Carbide-Impregnated Graphite at High Temperatures", Report NAA-SR-194, 1952.
166. J. Bromley, "Transport and Diffusion of Fission Products in Graphite", UKAEA Report AERE-R-4004, March 1962.
167. E. A. Bryant, G. A. Cowan, J. E. Sattizahn, and K. Wolfsberg, "Rates and Mechanisms of the Loss of Fission Products from Uranium-Graphite Fuel Materials", Nucl. Sci. and Eng. 15 (1963). 288.
168. R. B. Evans, III, G. M. Watson, and E. A. Mason, "Gaseous diffusion in porous media at uniform pressure", J. Chem. Phys., 35 (1961) 2076.
169. R. B. Evans, III, G. M. Watson, and E. A. Mason, "Gaseous diffusion in porous media. II. Effect of pressure gradients", J. Chem. Phys. 36, (1962) 1894.
170. E. A. Mason, R. B. Evans, III, and G. M. Watson, "Gaseous diffusion in porous media. III. Thermal transpiration", J. Chem. Phys., 38 (1963) 1808.
171. R. B. Evans III, J. Truitt, and G. M. Watson, "Interdiffusion of helium and argon in large-pore graphite", J. Chem. Eng. Data, 6 (1961) 522.
172. R. B. Evans, III, G. M. Watson, and J. Truitt, "Interdiffusion of gases in a low-permeability graphite at uniform pressure", J. Appl. Phys., 33 (1962) 2682.
173. R. B. Evans, III, G. M. Watson, and J. Truitt, "Interdiffusion of gases in a low-permeability graphite. II. Influence of pressure gradients", J. Appl. Phys., 34 (1963) 2020.
174. J. W. Prados and J. L. Scott, "Models for Fission Gas Release from Coated Fuel Particles", USAEC Report ORNL-3421, Oak Ridge National Laboratory, 1963.
175. J. Bromley and M. R. Large, "The Migration of Fission Products in Artificial Graphite", Proceedings of the Fifth Conference on Carbon, Vol. 1, p. 365, Pergamon Press, N. Y. and London, 1962.
176. C. J. Orth, "Diffusion of Lanthanides and Actinides from Graphite at High Temperatures", Nucl. Eng. 9 (1961) 417.
177. "Ceramic-Matrix Fuels Containing Coated Particles", Proceedings of a Symposium held at Battelle Memorial Institute, November 5-6, 1962. USAEC Report TID-7654.
   a) R. R. Sellers, P. E. Reagan, R. M. Carroll, D. F. Toner, J. L. Scott, and E. L. Long, "Irradiation Tests", p. 79.
   b) L. R. Zumwalt, E. E. Anderson, and P. E. Gethard, "Fission-Product Release from (Th, U) C_2-Graphite Fuels", p. 223.
178. J. A. Conlin and G. L. Segaser, "HTGR Fuel Test in the ORR Poolside Capsule Facility," p. 266, Nuclear Safety Program Ann. Progr. Rept. Dec. 31, 1967, USAEC Report ORNL-4228, Oak Ridge National Laboratory.
179. S. H. Freid et al., "Fission-Product Release from an Overheated HTGR Fuel," USAEC Report ORNL-TM-2388, Oak Ridge National Laboratory, 1968.
180. W. B. Cottrell, "Nuclear Safety Progress Report for the Period Ending December 31, 1968" (March 1969). Sec. 5.
181. J. Appel and B. Roos, "A Study of the Release of Radioactive Metallic Isotopes from HTGR's" USAEC Report GA-8399, General Atomic, 1967.
182. A. P. Malinauskas, J. L. Rutherford, and R. B. Evans III, Gas "Transport in MSRE Moderator Graphite. I. Review of Theory and Counter-diffusion Experiments," USAEC Report ORNL-4148, Oak Ridge National Laboratory, September 1967.
183. E. A. Mason, A. P. Malinauskas, and R. B. Evans III, J. Chem. Phys., 46: 3199 (1967).
184. A. P. Malinauskas, J. Chem. Phys., 42: 156 (1965); 45: 4704 (1966).
185. C. M. Blood, G. M. Hebert, and L. G. Overholser, "Oxidation of Bonded Coated-Particle Fuel Compacts by Steam," USAEC Report ORNL-4269, Oak Ridge National Laboratory, July 1968.
186. D. L. Morrison, T. S. Elleman, R. S. Barnes and D. N. Sunderson, "Post-Irradiation Release of Xe^{133} from Pyrolytic Carbon", Report BMI-1634, 1963.
187. M. T. Morgan, R. M. Martin, D. C. Evans, and J. G. Morgan, "Fission-Gas Release from High-Burnup Coated Particles", USAEC Report ORNL-3523, Oak Ridge National Laboratory.
188. L. R. Zumwalt, E. E. Anderson, and P. E. Gethard, "Fission-Product Retention Characteristics of Certain (Th, U) C_2-Graphite Fuels", USAEC Report GA-4551, September 1963. (Chapter in Materials and Fuels for High Temperature Nuclear Applications, MIT Press.)
189. D. L. Morrison T. S. Elleman, and D. N. Saunderman, "Fission-Gas Diffusion Studies", Report BMI-1630 (Del.), Battelle Memorial Institute, 1963.
190. O. Sisman et al., "Irradiation Behavior of High-Temperature Fuel Materials", USAEC Report ORNL-3591, p. 128, May 1963.
191. M. E. Ramsey and C. D. Cagle, "Research Program and Operating Experience on ORNL Reactors", Proceedings of the First U. N. International Conference on Peaceful Uses of Atomic Energy, Geneva, 1955, Vol. 2, p. 281.
192. J. S. Cheka and H. J. McAlduff, "Progress Reports on the Particle Problem", USAEC Reports: ORNL-146, August 1948; ORNL-172, September 1948; ORNL-211, November 1948; and ORNL-319, March 1949.
193. W. B. Lewis, "The Accident to the NRX Reactor on December 12, 1952", Canadian Report AECL-232, 1953.
194. D. G. Hurst, "The Accident to the NRX Reactor, Part II", Canadian Report AECL-233, 1953.
195. G. F. W. Gilbert, "The Operation of Engineering Test Reactors", Proceedings of the Atomic Power Symposium Held at Chalk River, Ontario, May 4-5, 1959. Canadian Report AECL-799, 1959.
196. O. J. C. Runnalls, "Irradiation Histories of Plutonium-Aluminum Alloy Fuel Rods", Canadian Report UK/C/4/114, 1955.
197. Reactor Safety and Hazards Evaluation Techniques, Vol. 1, International Atomic Energy Agency, Vienna, 1962.
   a) D. Martin, J. Bauzit, R. Cante, and L. Hebrard, "The Combustion in Air of Magnesium-Clad Fuel Elements", p. 3.
   b) A. N. Tardiff, "Some Aspects of the WTR and SL-1 Accidents", p. 43.
198. P. Balligand, "Reactor incidents at Saclay", Nucleonics, 18, 3 (1960) 82.
199. J. W. Greenwood, "Contamination of the NRU Reactor in May 1958", Canadian Report CRR-836, 1959.
200. A. F. Rupp, "NRU reactor incident", Nucl. Safety, 1, 3 (1960) 70.
201. M. W. Rosenthal, "Operating experience with the OMRE", Nucl. Safety, 2, 2 (1960) 75.
201a. W. K. Ergen "Reactor accidents", Nucl. Safety, 1, 3 (1960) 22.
202. A. A. Jarrett, "SRE Fuel Damage, Interim Report", Report NAA-SR-4488, North American Aviation, Inc., 1959. See also: NAA-SR-4488 (Suppl.), Final Report 1961.
203. W. B. McDonald and J. H. DeVan, "Sodium Reactor Experiment incident", Nucl. Safety, 1, 3 (1960) 73.
204. Westinghouse Electric Corp., "Report on WTR Fuel-Element Failure, April 3, 1960", Report WTR-49 July 7, 1960.

205. R. B. Korsmeyer, "Westinghouse Testing Reactor incident", Nucl. Safety, 2, 2 (1960) 70.
206. "IDO Report on the Nuclear Incident at the SL-1 Reactor, January 3, 1961, at the National Reactor Testing Station", USAEC Report IDO-19302, Idaho Operations Office, 1962.
207. J. R. Buchanan, "SL-1 final report", Nucl. Safety, 4, 3 (1963) 83.
208. F. R. Keller, "Fuel Element Flow Blockage in the Engineering Test Reactor", Report IDO-16780, Phillips Petroleum Co., 1962.
209. J. R. Buchanan, "Accidents in nuclear energy operations", Nucl. Safety, 3, 4 (1962) 93.
210. R. A. Costner, Jr., "MTR fission break incident", Nucl. Safety, 4, 4 (1963) 144.
211. E. H. Smith (Ed.), MTR Progress Report, Cycle No. 182, October 29-November 19, 1962, Report IDO-16831, Phillips Petroleum Co., 1962.
212. T. M. Sims and A. L. Colomb, "Preliminary Report on Fission Product Release at the Oak Ridge Research Reactor on July 1, 1963", USAEC Report ORNL-TM-627, Oak Ridge National Laboratory, 1963.
213. A. L. Colomb and T. M. Sims, "ORR fuel failure incident", Nucl. Safety, 5, 2 (1963/64) 203.
214. Childs, B. G., "Fission Product Effects in Uranium Dioxide", J. Nucl. Mat. 9 217-244 (1963).
215. Carroll, R. M., "Fission-Gas Behavior in Fuel Materials", Nuclear Safety 8, 345 (1967).
216. Rodgers, M. D., "Mass Transport of Uranium by Fission Fragments", J. Nucl. Mat., 15 (1), 65-72 (1965).
217. Lewis, W. B., "The Return of Escaped Fission Product Gases to UO_2", Canadian Report, DM-58 (January, 1960).
218. Melehan, J. B., Barnes, R. H., Gates, J. E., and Rough, F. A. "Release of Fission Gases From UO_2 During and After Irradiation", USAEC Report BMI-1623, Battelle Memorial Institute, March, 1963.
219. Carroll, R. M., and Sisman, O., "In-Pile Fission-Gas Release From Single-Crystal UO_2", Nucl. Sci. Engng, 21, 147-158 (1965).
220. Booth, A. H., "A Method of Calculating Fission Gas Diffusion From UO_2 and Its Application to the X-2-f Loop Test," AECL Report CRDC-721, (September, 1957).
221. Beck, S. D., "The Diffusion of Radioactive Fission Products From Porous Fuel Elements", USAEC Report BMI-1433 (April, 1960).
222. Morrison, D. L., et al., "An Evaluation of the Applicability of Existing Data to the Analytical Description of a Nuclear-Reactor Accident", USAEC Report BMI-1779 (August, 1966).
223. Auskern, A. B., "The Diffusion of Krypton-85 from Uranium Dioxide Powder", Westinghouse Report WAPD-TM-185 (February, 1960).
224. Long, G., Davies, D., and Findlay, J. R., "Diffusion of Fission Products in Uranium Dioxide and Uranium Monocarbide", First Conference on Nuclear Reactor Chemistry Gatlinburg, Tenn., TID-7610 (1960), pp. 1-2.
225. Notley, M. J. F., and MacEwan, J. R., "The Effect of UO_2 Density on Fission Product Gas Release and Sheath Expansion, Nucl. Appl., 2, 117-122 (1966).
226. Frost, B. R. T., "Behavior of Fuels at High Burnup Levels, Nucl. Eng.
227. Evans, D. M., and Shilling, A. W., "Parameters for Estimating the Release of Fission Gases from UO_2 Fuel", British Report TRG-Report 1240(W), May, 1966.
228. Allen, J., "The Release of Iodine From Uranium Dioxide Fuel", J. Brit. Nucl. Energ. Soc., 6, 127-133, (1967).
229. Brassfield, H. C., White, J. F., Sjodahl, L., and Bittel, J. T., "Recommended Property and Reaction Kinetics Data for Use in Evaluating a Light-Water-Cooled Reactor Loss-of-Coolant Incident", General Electric Report GEMP-482 (April, 1968).
230. Fontana, M. H., "TCDATA: A Fortran Program for Computing Thermodynamic data of Compounds in Any Phase State", USAEC Rept. ORNL-4304 (September, 1968).
231. Anderson, T. D., "Effects of High Burnup on Bulk UO_2 Fuel Elements", Nuclear Safety, 6, 164-169 (1965).
232. Daniel, R. C., Bleiberg, M. L., Meieran, H. B., and Yeniscavich, W., "Effects of High Burnup on Zircaloy-Clad Bulk UO_2 Plate Fuel Element Samples", USAEC Report, WAPD-263, September, 1962.
233. Whapham, A. D., "Electron Microscope Observation of the Fission-Gas Bubble Distribution in UO_2", Nucl. Appl. 2, 123-130 (1966).
234. Ross, A. M., and Rose, D. H., "Replica Electron Microscopy of Irradiated Uranium Carbide Fuel from Phase I of the X-721 Experiment", AECL-2701, Atomic Energy of Canada Ltd. (August, 1967).
235. Boyle, R. F., Peterson, J. P., Von Ketron, W., and Lannin, T. E. "Autoradiography and Burr Drilling Techniques for Fission Product Migration Studies of Irradiated UO_2, USAEC Report GEAP-4452. General Electric Co. (1964).
236. Oi, N., Ohno, T., and Naito, K., "Relocation of Fission Products and Pu in Irradiated UO_2 Pellet", J. Nucl. Sci. Tech. 1, 284-289 (1964).
237. Eyre, B. L., and Bullough, R., "The Formation and Behavior of Gas Bubbles In A Non-Uniform Temperature Environment", J. Nucl. Mat. 26, 249-266 (1968).
238. Notley, M. J. F., and MacEwan, J. R., "Stepwise Release of Fission Gas From UO_2 Fuel", Nucl. Appl. 2, 477-480 (1966).
239. Horn, G. R., Bates, J. L., and de Halas, D. R., "Irradiation Brewing of UO_2, Trans. Am. Nucl. Soc. 8, 24, (1965).
240. Hoffmann, J. P., and Coplin, D. H., "The Release of Fission Gases From UO_2 Pellet Fuel Operated at High Temperatures", USAEC Report GEAP-4596 (September, 1964).
241. Lewis, W. B., "Engineering for the Fission Gas in UO_2 Fuel", Nucl. Appl., 2, 171-181 (1966).
242. Morrison, D. L., et al., "An Evaluation of the Applicability of Existing Data to the Analytical Description of a Nuclear Reactor Accident", Quarterly Progress Report for April through June, 1967, USAEC Report BMI-1810 (July, 1967).
243. Parker, G. W., and Lorenz, R. A. "Calculation of Amount of Volatile Radioactivity in Fuel Rod Void Spaces", in "Nuclear Safety Program Annual Progress Report for Period Ending December 31, 1967, USAEC Report ORNL-4228 (April, 1968), p. 11.
244. Oi, N., and Takagi, J., "Diffusion of Non-Gaseous Fission Products in UO_2 Single Crystals, Z. Naturforsch 19a, 1331 (1964); 20a 673 (1965).
245. Crank, J., "Mathematics of Diffusion", Oxford University Press, London (1956).
246. Raines, G. E., Townley, C. W., Beck, S. D., and Goldthwaite, W. H., "A Method for the Study and Correlation of Fission-Gas-Release Behavior in Fuel Materials During Irradiation," Battelle Memorial Institute report BMI-1548 (October, 1961).
247. Landoni, J. A., and Moffette, T. R., "Peach Bottom HTG Emergency Cooling Accident Temperatures and Thermal Release of Nuclear Poisons and Radioactive Nuclides," General Atomic Report, GAMD-3977 (Rev. 1) (June, 1963).
248. "40-MW (E) Prototype High Temperature Gas-Cooled Reactor Research and Development Program", Quarterly Progress Report for Period Ending June 30, 1961, General Atomic Report, GA-2493 (July, 1961).
249. Miller, C. E., Jr., "A Boundary-Layer Diffusion Model of Fission-Product Release from Reactor Fuels", Nucl. Appl. 5, 198-205 (1968).
250. McKenzie, D. E., "The Volatilization of Plutonium from Neutron Irradiated Uranium", Can. J. Chem. 34, 515 (1956).
251. Castleman, A. W., Jr., and Tang, I. N., personal communication to be published, (1968).
252. Castleman, A. W., Jr., Tang, I., Horn, R., MacKay, R., Lewkowitz, S., et al. "Safety Studies" in Annual Report Nuclear Engineering Department. Brookhaven National Laboratory, USAEC Report, BNL-50082, pp. 164-172, December 31, 1967.
253. Wantland, J. L., and Fontana, M. H., "Enhancement of Heat and Mass Transfer By Internal Convection in Molten Cores", in Nuclear Safety Program Annual Progress Report for Period Ending December 31, 1968, USAEC Report ORNL-4374 (June, 1969). pp. 48-57.
254. G. W. Parker, R. A. Lorenz and G. E. Creek, "The Calculation of Fission Product Release from Core Coolant Safeguarded Reactors," Trans. Am. Nuclear Soc. 12 (1), 902-903, Nov. 1969.
255. D. L. Morrison, W. Carbiener and R. Ritzman, "An Evaluation of the Applicability of Existing Data to the Analytical Description of a Nuclear-Reactor Accident," BMI Report 1856, Jan. 1969.
256. G. W. Parker, R. A. Lorenz, et al., "Prompt Release of Fission Products from Zircaloy-Clad UO_2Fuels," Nuclear Safety Program Annual Progress Report for Period Ending Dec. 31, 1967, ORNL-4228, pp. 3-24, W. B. Cottrell, Ed. April 1968.
257. V. F. Baston, E. G. Good, W. A. Yuill, "Fission Product Release Analysis," IDO Report 17292, May 1969.
258. A. H. Booth and G. T. Rymer, "Determination of the Diffusion Constant of Fission Xenon in UO_2 Crystals and Sintered Compacts," CRDC-720, August 1958.
259. G. W. Parker, G. E. Creek, W. J. Martin, C. J. Barton, R. A. Lorenz, "Out-of-Pile Studies of Fission-Product Release From Overheated Reactor Fuels at ORNL, 1955-1965," USAEC Report ORNL-3981, July 1967.
260. D. M. Evans and A. W. Shilling, "Parameters for Estimating the Release of Fission Gases from UO_2 Fuel," TRG Report 1240 (w), May 1966.
261. R. C. Daniels, et al., "Effects of High Burnup on Zircaloy-Clad Bulk UO_2 Plate Fuel Samples," WAPD-263, Sept, 1962.
262. W. E. Baily, C. N. Spalaris, D. W. Sandusky, E. L. Zebroski, "Effect of Temperature and Burnup on Fission Gas Release in Mixed Oxide Fuel," Paper presented at American Ceramic Society Meeting, May 1969 (to be published).
263. R. D. Collins, J. J. Hillary, and J. C. Taylor "Air Cleaning for Reactors with Vented Containment," British Report TRG 1318(w).
264. H. Feuerstein, "Behavior of Iodine in Zircaloy Capsules", USAEC Report ORNL-4543, Oak Ridge National Laboratory, August 1970.

Chapter 19

Fission Product Behavior and Retention in Containment Systems

Leslie Silverman
Harvard University
Cambridge, Massachusetts
D. L. Morrison and R. L. Ritzman
Battelle Memorial Institute
Columbus, Ohio
T. J. Thompson
Massachusetts Institute of Technology
Cambridge, Massachusetts

CHAPTER CONTENTS

1 INTRODUCTION

The radiation exposure that the public may receive in the event of a serious accident in a nuclear reactor is governed almost totally by the amount of fission products that escape or are released from the containment. This amount depends in turn upon the quantity of fission products that are released from the fuel and are transported to the containment, the fractional deposition and removal by engineered systems within the containment, and the fractional leakage from the containment. Chapter 18 discusses the mechanisms for the release of fission products from the fuel elements and gives estimates of the fractional releases of various types of fission-product nuclides. The concepts of reactor containment and possible leakage from it during an accident are discussed in a later chapter. This chapter is concerned with the transport and deposition of fission products after they leave the fuel element and the means available for fission-product retention within the containment by both natural and engineered processes.

While the primary system surrounding the reactor core is intact, the release of fission products to the containment is very low, only that encountered in normal operation. In the event of an accident where there is a failure of the primary piping system and the primary coolant is lost, fission products, already present in the coolant in small amounts from normal operation or in larger amounts from the initiating event, can be disseminated into the containment. If no standby heat-removal capability is provided, the reactor core can overheat, resulting in the release of additional fission products from the fuel. The quantity of the fission products released to the containment is thus directly influenced by the engineered safety systems for core cooling. Proper operation of the standby coolant system eliminates or reduces the release of fission products from the fuel and alters their transport from the fuel to the containment.

The subsequent behavior of the fission products in the containment is markedly controlled by the operation of engineered safety systems in the containment. For small reactors, natural mechanisms of heat removal to control the pressure and temperature in the containment and to control and retain the fission products are adequate from a safety viewpoint. With large reactors or in sites with small exclusion zones, more positive control of the pressure transient in the containment and of the fission-product disposition is desirable. Engineered systems in the containment that can help to fulfill this dual purpose can be provided. If these engineered systems are incorporated into the containment design, their proper operation can overwhelm many of the natural mechanisms of fission-product behavior. Hence this subject is an important one to discuss in this chapter.

Discussion of the transport of fission products from the fuel and their behavior in the containment involves the knowledge and understanding of significant gas phase and surface processes. The concentration of the fission products, the macroscopic and trace constituent composition of the gas phase, temperature, pressure, and radiation are important in determining the physical and chemical behavior of the fission products in the gas phase. The gas-phase environment changes with time during an accident, so transient processes must be noted. Transport of fission products within the gas phase and changes in their chemical and physical form with time are most important. The composition of surfaces, reactions between the fission products (in vapor and aqueous phases) and the surfaces, temperature, and radiation may all influence the surface processes. The operation of engineered safety systems in the containment may exercise a controlling influence on many of the gas-phase and surface processes, but even if these systems operate, fission-product behavior and retention in the containment can be described principally by gas-phase transport and surface interactions. The mechanism of transport and the nature of the surfaces encountered may be markedly different during the operation of engineered removal systems than they are under natural conditions.

Since the presence of fission products within reactor fuel as a result of the fission chain reaction constitutes the principal potential hazard, it is only natural that a great deal of attention be devoted to this subject. Chapter 18 discusses the quantities of fission products formed, the mechanisms for their release from the fuel, and the fractional releases involved. Beattie [1], Blomeke and Todd [2], and others have listed the principal fission products and their production rates as well as their characteristics. Abstracts of these references are contained in Chap. 18. Chapter 4 of [3] also discusses this material and provides a good review for Chaps. 18 and 19.

Of all of the fission-product elements that could be released during an accident, iodine, and in particular the isotope I^{131}, has been considered to be the most significant due to its fission yield and its chemical and biological behavior. For light-water reactors when the criteria of Part 100 of Title 10 of the Code of Federal Regulations and TID-14844 [4] are applied, the iodine dose in the event of an accident is in most cases controlling, except close to the reactor. Even then, the containment, which gives a higher dose for direct gamma ray exposure, is considered to be unshielded in the calculation. If engineered removal systems can achieve decontamination factors between 50 and 200 for iodine within the containment, the noble gases will then become the controlling factor for doses under the TID-14844 assumptions [5]. Farmer [6,7] has used this iodine dose and the resulting estimates of thyroid cancer as an important part of his probabilistic analysis of safety features. Because of its importance, iodine release and behavior has been the subject of several reviews [3,8,9] and will be emphasized in this chapter. However, the behavior of other fission products cannot be ignored, and, where information on their behavior is available, it is discussed.

Sources of information on fission-product behavior are numerous and include specific inputs from reactor safety programs as well as general literature references. The fission products are isotopes of chemical elements whose properties have been studied in nonnuclear areas, and information on the chemical and physical behavior from these sources is invaluable in understanding fission-product behavior in reactor containments. Much can be borrowed from the chemical engineering literature on heat and mass transfer. Information from basic studies is combined with observations from large-scale experiments to provide the needed understanding of fission-product processes. Sound engineering judgment and analysis must be used to apply this information to the reactor accident situation.

Topics discussed in this chapter include the fission-product source term in the containment, which depends upon the transport of fission products from the fuel to the containment and the physical and the chemical characteristics of the fission products; the natural processes of transport and deposition that affect fission-product behavior in the containment; and the retention of fission products in the containment by the use of engineered removal systems.

2 FISSION PRODUCT SOURCE TERM IN THE CONTAINMENT

2.1 Factors Affecting the Source Term

The extent of the fission-product release from

the reactor core and primary system establishes the concentration limits for fission products in the containment. The chemical composition of the fission products released to the containment depends upon the type of reactor, the type of accident, and upon the time-temperature history of the reactor core following the initiation of the accident. Factors which affect the release of fission products from various fuels are discussed extensively in Chap. 18 and is not repeated here. In this section, attention is directed to a description of the fission products after they are released from the fuel, some of the factors influencing the retention of the fission products within the primary system, and changes in the form and chemical composition of the fission products as they are transported from their point of release to containment and within the containment itself.

Fission-product release from the fuel in a light-water reactor depends strongly upon the time-temperature history during the loss-of-coolant accident, which in turn is influenced by the operation of emergency or standby cooling systems. If these systems perform as they are designed, the release may be very small or at least will be limited to the prompt release fractions described in Chapter 18. The stored gases within the individual fuel pins will be released if the clad is ruptured. The chemical composition of the prompt-release fission products delivered to the containment will consist mainly of the volatiles, the noble gases, and the halogens. If the performances of the standby cooling systems are degraded, greater quantities of fission products will be released from the core. As the temperatures within the core approach the melting point of the fuel, the fission-product composition will include less volatile elements as well. Prompt release can occur late in the blowdown phase of the accident and prior to standby coolant injection. Transport and deposition of the fission products in a steam atmosphere within the core and primary system must be considered to determine what portion of the released fission products reach the containment. As a most conservative estimate, it can be assumed that all the fission products released from the fuel are available in the containment. Any fission products that are present in the primary coolant will be transported to the containment with the coolant during blowdown. The fission products may remain entrained or dissolved in the liquid phase, or may be released from the liquid to the containment atmosphere. If the core has been re-covered by the standby coolant, any subsequent release of fission products would occur to water, and it is expected that behavior similar to that observed in the coolant under normal operation would be encountered. Only small releases are expected under these conditions, since the fuel temperatures are relatively low.

For containment system design and accident evaluation purposes, it has been assumed that guidelines in TID-14844 [4] are applicable to establish the fission-product source term in the containment. In this document it is assumed that 100% of the noble gases, 50% of the halogens, and 1% of all other fission products are released from the core. It is further assumed that one-half of the released halogens plate out, so that only 25% of the core inventory need be considered in the containment. No mechanism is postulated to support these guidelines, but an accident with severe overheating of the core or substantial meltdown could lead to these quantities of fission products in the containment. Lower release fractions for specific cases can be estimated if mechanisms are applied, but prudence will dictate that conservative values should always be selected when safety systems are designed and the consequences of accidents are assessed. It is unlikely that the values in TID-14844 would be greatly changed.

In liquid-metal fast-breeder reactors (LMFBRs), the release of fission products to the containment, is in part conceptually the same as in light-water reactors. High temperatures in the fuel promote the release of fission products. If there is a loss of cooling capability in the core concurrent with a breach in the primary system, fission products can be released from overheated fuel and discharged into the containment following transport through the primary system. Higher power densities planned for oxide-fueled cores aggravate the situation in LMFBRs, and fuels may reach the higher temperatures more quickly than in light-water reactors. Transport and deposition of the released fission products in sodium or inert environments must be considered in LMFBRs (in contrast to steam environments in light-water reactors). The presence of plutonium in the fuel and sodium aerosols in the containment are factors unique to the LMFBR situation. Fission products released to the coolant during normal operation and the radioactivity of the coolant itself are also part of the source term in an accident situation. A major difference in the fission-product availability in the containment in the fast reactor case results from the possibility that a reactivity accident may lead to rapid disassembly of part of the core. In this situation, a significant fraction of the fuel and fission products could be vaporized and transported rapidly to the containment through loss of integrity of the primary system. The fission-product source in this case would consist of nearly all of the chemical elements which occur as fission products, rather than only the most volatile.

Release of fission products to the containment during accidents in gas-cooled reactors depends primarily upon the specific fuel design if a loss-of-coolant accident is assumed [10]. In reactors utilizing metallic fuel with metal cladding, fuel-element cladding-temperature transients during accidents are limited to prevent combustion of the cladding. If this is achieved, only those fission products present in the normal coolant will be released. If combustion of the cladding and oxidation of the graphite moderator occurs, the amount of fission-product release is interrelated with the extent of oxidation of the fuel [11]. In reactors which utilize ceramic fuel with ceramic coatings, all of the fission products (with the exception of cesium) are contained in the coating or the surrounding graphite sleeves during normal operation [12]. Release of fission products during an accident depends upon the temperatures within the core and upon chemical reactions with the graphite and the fuel coatings. Oxidation can significantly increase the release during an accident.

2.2 Chemical and Physical Form of the Fission Products

2.2.1 Fission-Product Vapors

The chemical and physical form of the reactive fission product elements under accident conditions is the result of dynamic processes. Interaction of the fission products with the major constituents of the atmosphere in and around the reactor core, primary system, and containment may result in oxidation or reduction of the susceptible species, depending upon specific combinations. Reactions between the fission products, which in themselves are trace constituents of the atmosphere, with other trace contaminants must also be considered. Adsorption and desorption of fission products on surfaces could also promote changes in the chemical composition of the gaseous fission products. Noble gases are of course unreactive and do not change form during the accident. The other fission products normally considered volatile (i.e., halogens, tellurium, cesium, and, under some conditions, the volatile oxides of metals such as ruthenium) are of concern in this section from a physico-chemical viewpoint.

Due to its biological significance and its high volatility, the physical chemistry of fission-product iodine during an accident has been studied extensively. Keilholtz and Barton [8] have summarized much of the pertinent information. Because a wide variety of environmental conditions in reactor accidents can be anticipated if all the different types of reactors are considered, data on iodine behavior have been obtained under dry conditions, in air, in CO/CO_2, in a helium atmosphere, under high humidity and condensing conditions in steam-air atmospheres, and in steam. Although methods of sampling and identification of the chemical and physical forms of iodine have varied among the experimenters, three methods have been used extensively: diffusion-tube sampling, Maypack sampling, and honeycomb sampling. A brief description of these techniques will give an appreciation of the uncertainties associated with the data obtained using the techniques and will serve as a basis for interpretation of experimental results. In addition, the development of the techniques themselves have provided information on the physical chemistry of the fission-product elements.

Use of diffusion tubes to identify the chemical and physical forms of iodine is based upon the measurement of the distribution of radioactivity deposited on the walls of a cylindrical or rectangular channel from a gas under laminar flow conditions. The equations developed by Gormley and Kennedy [13] which are applicable to cylindrical tubes are:

$$F = 0.819\,e^{-\beta z} + 0.975\,e^{-6.1\beta z} + 0.0325\,e^{-16\beta z}, \quad F \le 0.78, \qquad (2\text{-}1)$$

$$F = 1 - 4.07\,h^{2/3} + 2.4\,h + 0.446\,h^{4/3}, \quad F > 0.78, \qquad (2\text{-}2)$$

where F = fraction penetrating, z = length of tube or channel, $\beta = 3.66\pi D/Q$, $h = \pi Dz/2Q$. D = diffusion coefficent of aerosol usually expressed in cm^2/sec, and Q = volumetric flow rate through the tube.

The value of the diffusion coefficient (D) for gases can be obtained from gas diffusion equations such as that proposed by Gilliland [14]:

$$D = 0.0043 \times \frac{T^{3/2}}{P\left(V_A^{1/3} + V_B^{1/3}\right)^2}\sqrt{\frac{1}{M_A} + \frac{1}{M_B}}, \qquad (2\text{-}3)$$

where V and M are the molecular volume and mass of constituents A and B (for instance, iodine and air) and T is the absolute temperature. Diffusion constants for any gases may be taken from references such as [15] and used in Eqs. (2-1) and (2-3) as well.

Equations (2-1) and (2-2) can be used for determining the average size of particles, if the parameters Q and T are known, by substituting penetration values obtained from radioactive counting methods. This approach is valid for a monodisperse aerosol. Since most aerosols are polydisperse, a modified approach must be taken. Gieseke [16] has derived a relationship assuming a log-normal distribution of particles. The analysis is based upon the calculations of Fuchs, Stechkina, and Starosselskii [17] and the generalized penetration equation is:

$$F = \frac{N_g}{N_0} = 0.915\int_{-\infty}^{+\infty} e^{-\beta z} f(r)\,d(\ln r) + 0.059\int_{-\infty}^{+\infty} e^{-\beta z} f(r)\,d(\ln r), \qquad (2\text{-}4)$$

where

$$\beta = \beta(x) = \frac{3.77\,b}{aQ} D(r),$$

$$f(r) = \frac{1}{\sqrt{2\pi}\,\ln\sigma_g}\exp\left[\frac{-(\ln r - \ln r_g)^2}{2(\ln\sigma_g)^2}\right],$$

r_g = geometric-mean particle radius, r = particle radius, σ_g = geometric standard deviation, D(r) = diffusion coefficient as a function of particle size, z = distance along the channel, 2a = channel width, and b = channel height. It can be seen that the series giving the penetration has been truncated after the first two terms. The fractional penetration, N_g/N_0, was calculated by Fuchs at various values of $\beta z/D$ and for numerous choices of r_g and σ_g. The results were presented [17] in graphical form and can be used to obtain plots of deposition rate versus distance.

Gieseke has compared the particle sizes obtained if the aerosol is assumed to be monodisperse with the values obtained using the polydisperse assumption. A difference on the order of a factor of ten was noted between the mean size of the log-normal distribution and the particle size predicted for a monodisperse aerosol. The mean diameter for the polydisperse assumption was larger than the assumed monodisperse size. This suggests that caution should be employed in using the results of diffusion tube measurements as a determination of particle size. Since most processes affecting particle behavior during an accident depend strongly upon particle size, and since it is difficult to discern from experiments whether sizes are being obtained correctly (or even that the experiment simulates accident conditions), each removal mechanism should be examined to determine if a conservative approach is being followed for the specific accident conditions assumed.

The original concept of the diffusion tube which utilizes a single material of construction and operates under isothermal conditions has been modified to provide additional data. Browning and his colleagues [18] have described a combination of diffusion-tube coatings which can be used to define the chemical nature of the particles present and to obtain an approximation of their sizes. For molecular or elemental iodine, a silver-plated tube is employed. For high-molecular-weight organic iodine compounds, a rubber-lined tube is used. Finally, for compounds of low molecular weight, a tube lined with activated carbon is employed. Studies [19] of the deposition profiles on rubber indicate that at least two iodine compounds are deposited, the behavior is complex, and not solely diffusion-controlled and one of the compounds formed may be unstable. The rubber itself may, however, affect the results. Castleman [20] has used a diffusion tube which had a temperature gradient imposed along it to study fission-product behavior in helium. A chromatographic separation of fission products according to chemical composition was obtained.

The Maypack [21] is another and perhaps more commonly used method designed to classify the various forms in which fission-product iodine may be gas-borne in an accident. It consists of a series of filters and adsorbers which are individually effective in different degrees against different forms of iodine. Obviously, there are many possible variants. The pack shown in Fig. 2-1 [22] has in sequence:

a) a glass-fiber filter for removing solid particles;
b) six silver-plated screens for adsorption of elemental iodine vapor;
c) carbon-loaded filter paper for the adsorption of reactive vapor compounds of iodine;
d) a bed of activated carbon granules (charcoal) for the adsorption of less reactive vapor compounds of iodine.

Other variations in Maypack design and loading

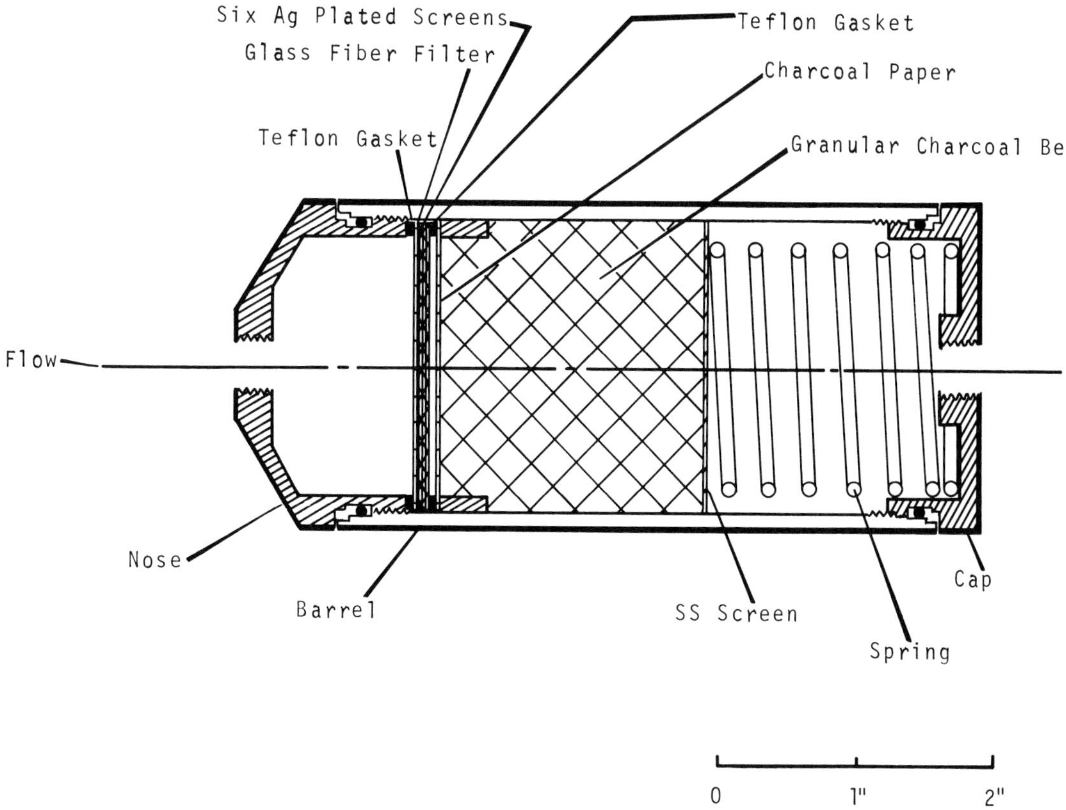

FIG. 2-1 ADF Maypack (Schematic).

have been presented by Collins [23] and Parsly [24].

While the Maypack samplers provide a good indication of the nature of the physical and chemical composition of the iodine released, they are not free from ambiguities. Deposition of one form of iodine on several components can occur. Particular problems [24] are encountered in high humidity (100+% relative humidity) with loss of efficiency by the silver-plated screens and excessive deposition of particulate material in the inlet check valves. It may be necessary to run several packs with different arrangements and combinations of layers to remove the uncertainties. Used properly, the Maypack provides a semiquantitative method of classifying iodine releases into elemental, particulate, and penetrating vapor fractions.

Honeycomb samplers [25] which distinguish more clearly between molecular and particulate iodine have also been used. A honeycomb sampler consists of a silver-plated honeycomb, a high efficiency filter, and charcoal beds. If used in a steam atmosphere, the sampler is heated to prevent condensation and low charcoal efficiency.

In attempting to identify radioactive particles of larger sizes, where diffusion tubes can no longer be used for characterization, Silverman and Browning [26] employed uniform Dacron polyester staple fibers placed in a filter pack. This pack could later be separated into its original sheets or sections and counted to obtain relative activity per layer. By operating separate packs at different gas velocities over a wide velocity range, it is possible to approximate the particle sizes in the diffusion range and to detect the presence of larger sizes beyond this range. Definition of size in the latter case requires instruments such as the cascade impactor, the conifuge Goetz particle size spectrometer, or direct measurement on membrane filters either by radioautography or optical sizing. These instruments and their use are described in detail elsewhere by Silverman, Billings, and First [27].

Early experimenters examining the behavior of fission-product iodine under accident conditions assumed that this fission product existed in the vapor phase as an element (diatomic molecule, I_2) or was attached to particulates. The presence of other non-elemental vapor forms was not suggested until 1951. At that time, Chamberlain and Chadwick [28] observed on occasions a low efficiency for removal of what was presumed to be elemental iodine by caustic soda bubblers. Little attention was paid to nonelemental forms until Morris et al. [29] deduced the presence of an unidentified form (or range of compounds) from the deposition and sorption behavior of trace quantities of I^{132} released in the PLUTO containment vessel. The compound(s) were subsequently fractionated into two groups by Eggleton and Atkins [30]: one fraction was identified as the alkyl iodides, predominantly methyl iodide ($\sim 85\%$), but the other fraction was not identified. Since that time, considerable effort has been devoted to the study of the forms of iodine released from overheated fuel and transported to the containment.

Attention has been directed predominantly toward methyl iodide: this has been a subject of a review article by Mishima [31], and has also been reviewed by Durant et al. [32]. The principal concern over methyl iodide arises from its low solubility in water and its lack of reactivity with containment surfaces. At temperatures below those required for decomposition, methyl iodide can remain airborne for a considerable period of time in a containment under natural response conditions and can eventually become the predominant species. Partial decomposition has been reported at temperatures as low as 90° C but temperatures between 300° and 400° C are required for thermal decomposition of methyl iodide in the absence of a catalyst. The problems associated with removal of methyl iodide with engineered systems is discussed later in this chapter. Under dry conditions, methyl iodide is sorbed well by charcoal but loss of charcoal effectiveness can occur in the presence of moisture. Removal of methyl iodide by containment sprays also presents problems by virtue of the low partition of the methyl iodide into the liquid phase.

Durant [32] has summarized the data on release of methyl iodide from fuel and has examined the formation of methyl iodide in containment. The results of UKAEA estimates of methyl iodide released from cooled irradiated UO_2 as given by Durant are presented in Table 2-1. A survey of literature for data on methyl iodide formation in containment vessels is given in Table 2-2. Mishima [31] has presented additional information on the formation of methyl iodide and its behavior in nuclear systems. It can be concluded from all of this information that the amount of methyl iodide present under any given set of accident conditions is not easy to calculate and is the result of many competing processes. The mechanisms of its formation are not known. The amount of methyl iodide released from overheated reactor fuel has been reported to be from a few thousandths to 20% of the elemental iodine airborne. In general, the proportion of iodine that appears as methyl iodide increases as the total concentration of iodine released to the gas phase decreases. The type of fuel cladding does not appear to affect the amount of methyl iodide released. The composition of the atmosphere affects the amount of methyl iodide with reducing conditions tending to lower the amount. Radiation also can affect methyl iodide. Effects of radiation on the behavior of methyl iodide have been studied in a preliminary manner by Tang and Castleman [41] and by Kircher and Barnes [42]. It is not possible from either of these studies to determine whether radiation will cause significant decomposition of methyl iodide during an accident or promote its formation, especially in light-water reactors.

Castleman [43] has studied the chemical states of iodine released from metallic U, UO_2, and U_3O_8 into steam environments at temperatures of 1000° to 1300° C and has reported the formation of other nonelemental iodine species under these conditions. His results are summarized in Tables 2-3, 2-4, and 2-5. Methyl iodide was determined to be the major constituent of fraction B by the use of gas chromatography. The chemical reactions that have been postulated to occur during and

TABLE 2-1.

UKAEA Estimates of Methyl Iodide Released from Lightly Cooled Irradiated UO_2*

	Experiment	Atmosphere	UO_2 fuel Mwd/ton	Percent of iodine released from fuel	Percent of iodine inventory released as methyl iodide	
					Maypack	Gas chromatograph
a	No. 21	54% CO_2, 35% CO plus air, H_2, H_2O, 200 ppm CH_4	150	2.8 — 3	1	0.2
a	No. 22	CO_2/CO	150	4 — 9	2	0.5
a	No. 23	CO_2/CO Sat. H_2O	150	40	4	5.0
a	No. 24	CO_2/CO plus CH_4 (1000 ppm)	150	33	20	1.0
a	No. 25	CO_2/CO	150	27 — 35	10	Not done
a	No. 26	CO_2/air	100	25 — 47	2	0.7
a	No. 27	CO_2/5% CH_4	150	13 — 27	0.7	0.1
b	Table 3	Steam-air	0.5	6.2	0.02	-
b	Table 3	Steam-air	0.5	6.8	0.02	-
b	Table 3	Steam-air	1.0	2.4	0.01	-
b	Table 3	Steam-air	2.0	2.4	0.02	-
b	Table 3	Steam-air	5.0	2.9	0.001	-
b	Table 3	Steam-air	1.0	13.7	0.01	-
b	Table 3	Steam-air	2.0	15.4	0.003	-
b	Table 3	Steam-air	7.0	40.5	0.03	-
b	Table 4	Steam	100	28	0.04	-
b	Table 4	Steam	100	35	0.5	-
b	Table 4	Steam	100	34	3	-
b	Table 4	Steam	100	9	2	-
b	Table 4	Steam	100	27	0.08	-
b	Table 4	Steam-air	100	60	0.2	-

a - Reference [33]
b - Reference [34]
* From Reference [32]

following the release from metallic uranium into steam are:

$$U(F.P.) + 2H_2O \rightarrow UO_2 + 2H_2 + F.P.,$$

$$I + H_2 \rightarrow HI + H,$$

$$2I \rightleftharpoons I_2,$$

$$I_2 + H_2O \rightleftharpoons HIO + I^- + H^+,$$

$$3IO \xrightarrow{\Delta} IO_3^- + 2I^-.$$

Evidence has also been presented for the existence of hypoiodous acid as a volatile iodine species produced in water-air mixtures [44].

Castleman has obtained data on the state of iodine released into helium and air [45, 46, 47]. In helium either a compound with uranium is formed or elemental iodine is released as indicated by the following equations:

$$U + 2I \rightarrow UI_2, \qquad (T = 1200°C)$$

$$UI_x(ads) + O_2 \rightarrow UO_2I_2 + I_{x-2}, \qquad (Ambient)$$

$$UO_2I_2 \rightarrow UO_2 + I_2.$$

It was also noted that in helium, iodine was transported independently of cesium.

Determination of the chemical state of iodine released from sodium has been studied by Castleman [48] and Pollack [49]. Both groups concluded that iodine would vaporize from sodium as NaI, although there is some disagreement with respect to the value of the distribution coefficient.

Investigations into the chemical state of other fission products that could be in a gaseous state

TABLE 2-2.

Summary of Literature Data on Methyl Iodide Formation*.

Test[a] facility	Surface	I_2[b] deposition, %	Aging Time, hr.	Surface-to-volume ratio, m^2/m^3	I_2[c] conc., μ g/m^3	Temp., °C	Atmosphere	CH_3I,[b] %
ADF	Painted Steel	20-95	5	5.2	0.03-630	20	Air	0.05-0.7
ADF		50	5		0.2-640	80	Steam-Air	0.1-6.0
CMF	Bare Stainless Steel	25-90	3-4	9.1	5-220	25	Air	0.3-3.7
CMF		8-60	4-18		2000-8000	30-110	Steam-Air	0.5-1.3
CMF		20	5-16		2000	30-110	Steam-Air-0.3% CH_4	2.5-21
NSPP			36-48	1.4	2600-25000	30	Air	6-10
Zenith	Painted Concrete & Steel	90	5	1.4	0.4-0.8	25	Air	0.09-0.6
Zenith		1	Ventilated		0.8	25	Air	<0.003[d]
HATR	Painted Steel	10	0.5	11	40,000	400	CO-CO_2	0.25
HATR		95	20	7.8	1300	25	CO-CO_2-Air	5

NOTES:

a ADF - Aerosol Development Facility, Battelle - Northwest [35]
CMF - Containment Mockup Facility, Oak Ridge National Lab. [36, 37, 38]
NSPP - Nuclear Safety Pilot Plant, Oak Ridge National Lab. [36]
Zenith - Reactor Facility, United Kingdom Atomic Energy Authority [39]
HATR - Highly Active Test Facility, United Kingdom Atomic Energy Authority [40]

b Values shown are maximum reported for test in terms of starting iodine inventory and are not necessarily the final equilibrium value.

c Maximum iodine concentration in gas phase at start of test.

d Limit of experimental measurement.

TABLE 2-3

Distribution and chemical states of iodine released from metallic U into steam

Chemical state	Location: Condenser	Location: Boiler	Location: Traps	Over-all
I°	12.2	5.8	49.9	12.0
I-	83.9	78.8	0	83.2
I^{+5}, +7	1.0	15.4	0	1.8
Others* A	1.5	0	0	1.4
Others* B	1.4	0	50.1	1.6
Iodine distribution	94.3	5.2	0.5	100.0

*A = fraction not removed from the aqueous solution; B = fraction not recovered from CCl_4.

TABLE 2-4

Distribution and chemical states of iodine released from UO_2 into steam

Chemical state	Location: Condenser	Location: Boiler	Location: Traps	Over-all
I°	79.1	28.6	60.1	76.6
I-	8.6	68.1	0	10.5
I^{+5}, +7	0.4	3.3	0	0.5
Others* A	2.5	0	0	2.3
Others* B	9.4	0	39.9	10.1
Iodine distribution	92.8	3.8	3.5	100.0

*A = fraction not removed from the aqueous solution; B = fraction not recovered from CCl_4.

TABLE 2-5

Distribution and chemical state of iodine released from UO_2 into steam and hydrogen

Chemical state	Location: Condenser	Location: Boiler	Location: Traps	Over-all
I°	10.0	0	20	11.0
I-	71.6	61.5	0	61.9
I^{+5}, +7	9.5	38.5	0	9.1
Others* A	2.4	0	0	2.0
Others* B	6.5	0	80	16.0
Iodine distribution	83.9	2.9	13.2	100.0

*A = fraction not removed from the aqueous solution; B = fraction not recovered from CCl_4.

during an accident have been quite limited. Tellurium has been studied by Malinauskas [50], who reports the formation of an oxyhydroxide compound at high temperatures if water vapor is present. Otherwise, tellurium release as TeO_2 would be expected. Cesium, due to its chemical reactivity, would be expected to oxidize rapidly upon release, with the formation of water soluble oxides. Higher releases of ruthenium and molybdenum have been observed in experiments under oxidizing conditions (see Sec. 2.3). Formation of the volatile oxides of these metals has been postulated as a mechanism.

2.2.2 Particulates

Fission products released from reactor fuel could be associated with particles, and the behavior of these particulate-associated fission products should be considered along with the gaseous fission products in assessments of reactor safety. Several extensive treatments of particle behavior under accident conditions have been reported [51, 52, 53], and several texts on aerosols in general are available [54, 55, 56]. From a reactor safety viewpoint, the formation of particles, the mechanisms of particle behavior, and reactions occurring at the surfaces are all important to consider. Particle formation involves nucleation, condensation growth, and agglomeration, while thermophoresis, diffusiophoresis, gravitational settling, and transport in convective flows affect the airborne concentration of particulates during an accident. Particle removal by natural effects in a reactor containment will occur as a result of a combination of these phenomena. Reactions of fission products and water at the surfaces of particles must also be considered in predicting particulate behavior. Formation of particulates are briefly considered below, and natural removal mechanisms are discussed in Sec. 3.3.

The dispersion of particulates from ruptured or melted fuel into the circulating liquid or gas-stream within the pressure vessel or containment volume will produce colloidal and larger size particulate suspensions. The larger sizes are attributed to agglomerating and flocculating mechanisms. Microscopic gas bubbles which are insoluble in a circulating coolant can behave as suspended particles. They will have a buoyancy, however, which tends to float them to the upper fluid region, in contrast to denser particles which may stay in equilibrium or tend to settle by gravity. When the fluid in which the particulates are suspended is liquid, the suspension is defined as a hydrosol, and when suspended in a gas or air media, the term aerosol is applied. Fission-product particulates released into a steam atmosphere would also be called aerosols. These two terms describe a particle size range in which certain hydrodynamic or aerodynamic behavior can be expected and which can in some cases be predicted by analytical methods. Hydrosols may involve particle sizes in the colloidal range and hence sedimentation can be neglected. Both hydrosols and aerosols exhibit Brownian motion. Aerosols are considerably less stable than hydrosols. Aerosols have a much greater relative density compared to the medium in which they are suspended than hydrosols. They also have greater convection gradients acting upon them. The aerosol particles exhibit a more lively movement due to constant bombardment by gas molecules. However, ion clouds do not fit the definition of aerosols since the particles are not large enough to scatter visible radiation. Nevertheless, large ions and condensation nuclei may provide targets and surfaces for the adsorption of gases and vapors.

The study of particulate behavior requires some determination or estimation of the number, density, and size of the primary particles available. In reactor accidents, an aerosol formed by condensation is the most important class of mechanisms, but the possibility of the formation of dispersion aerosols by comminution of the coolant should not be neglected [51]. Formation by the latter mechanism may result during a destructive excursion in a LMFBR. Aerosols can also form by homogeneous nucleation (condensation onto nuclei formed of the same material as the condensing vapor or condensation of vapors onto foreign nuclei). The supersaturated vapor conditions can arise by the cooling of an unsaturated vapor, by evaporation from a hot surface, or by a gas-phase reaction. The source of the vapor determines the special arrangement of the condensing material, including temperature and concentration gradients.

Methods for predicting the nucleation rates of aerosols have been reviewed by LaMer [57] and by Dunning [58]. Most of the studies on nucleation have been concerned with homogeneous nucleation of liquid droplets, primarily water droplets. The analysis of most of this work has been based on theoretical treatments by Volmer and Weber [59], Becker and Doring [60], Zeldovich [61], Frenkel [62], and Kuhrt [63]. The basic theory of homogeneous nucleation has also been extended to include the case of heterogeneous nucleation [64]. An extensive review with analysis of both homogeneous and heterogeneous nucleation has been presented by Hirth and Pound [65].

In the case of homogeneous nucleation, a critical supersaturation of the vapor and critical particle size is required before the particle will stabilize and grow [50]. The driving force for the formation of the particle depends upon the ratio of the vapor pressure on the particle to the equilibrium vapor pressure (P/P_0), i.e., $-nRT\ln(P/P_0) = -(4/3)\pi r^3(\rho/M)RT\ln(P/P_0)$, where r is the particle radius, ρ is the density of the vapor, and M is the molecular weight. There is an opposite driving force on the particle due to the increase in surface energy with increasing size (i.e., $4\pi r^2\gamma$, where γ is the surface tension). When these two forces are balanced, the radius of the particle will have attained the critical size (r_c) for stability:

$$r_c = \frac{2\gamma M}{RT} \ln\left(\frac{P}{P_0}\right). \tag{2-5}$$

For calculations involving growth from some initial droplet size, it should be noted that use of the critical size will predict zero growth rate. The initial droplet size must then be larger than the critical size. This is because growth equations apply to an average of all drops, including those which decay, while nucleation theory applies only to those drops which grow [66]. Becker and Doring [60] have shown that the droplets being considered in the two theories become identical at a size of about 1.3 times the critical size, hence this should be used for the initial size in growth equations.

An equation can be derived for the rate of formation of droplets of radius greater than r_c [51]. The derivation is based upon the statistical probability, due to random collisions, of the existence of particles of sizes greater than r_c. The rate of formation, J (nuclei/cm^3-sec), is given by:

$$J = \frac{\alpha\sqrt{2}N^{3/2}}{\sqrt{\pi}R^2}\left(\frac{P_0}{T}\right)^2\frac{\sqrt{\gamma M}}{\rho}\left(\frac{P}{P_0}\right)^2 \times \exp\left\{-\frac{16\pi M^2\gamma^3 N}{3\rho^2 RT[RT\ln(P/P_0)]^2}\right\}, \quad (2\text{-}6)$$

where α is the accommodation or condensation coefficient (the probability of a collision resulting in the molecule sticking) and N is Avogadro's number. A value of unity can usually be assumed for α for monatomic vapors condensing on clean metallic surfaces, but lower values have been observed for different surface-vapor interactions.

Using Eq. (2-6), the information necessary to calculate nucleation rates of particles can be established. The surface tension, γ, is an important factor in the equation and values of γ are needed for droplets with relatively small groups of molecules. Properties for small droplets with all molecules near the surface may differ from bulk properties, leaving considerable uncertainty in the estimate. Another problem is that nucleation as described above is for a single pure component. In an accident situation it is expected that cladding, fuel, coolant, and fission products will all be forming condensation nuclei together. It then becomes not only necessary to estimate single-component properties, but properties of mixtures.

Several attempts have been made to use nucleation-rate theory to predict particle formation in systems with materials of a nature similar to those encountered in a nuclear accident. Stewart [67] and Lavrenchik [68] have considered the case of particle formation in the fireball of a nuclear explosion. They both conclude that realistic calculations of nucleation rate are impossible because of the lack of knowledge concerning the physical properties of condensing material. Stockham and Snow [69] also find that the lack of surface tension data prevents calculation of nucleation rates of Zircaloy and uranium vapors ablated from a space nuclear-energy power plant on reentry into the atmosphere.

The likelihood of obtaining precise calculations of nucleation rates is therefore not very good due to the lack of information available on properties of mixtures. Some limiting assumptions can be made, however, for specific situations, and estimates can be made if considerable uncertainty is tolerable. For reactor safety evaluations, upper limit approaches are often useful to establish conservative bounds. The nucleation theories described above are a means to achieve these estimates. There are experimental data available on particle formation under simulated accident conditions, and these can be used to guide the analyses.

2.3 Fission-Product Transport and Deposition in the Core and Primary System

In a loss-of-coolant accident, the primary system, even though its integrity has been impaired, presents a potential barrier to retard or prevent release of fission products to the containment. While the retention by the primary system depends upon the particular reactor type and the accident, the most important factors are transport of the fission products in the gas phase, reaction rates of the fission products with the surfaces, and the effective surface-area-to-volume ratio encountered along the transport path.

Experimental data on fission-product transport and deposition in a reactor core and primary system under accident conditions have not been obtained directly. Inferences on expected behavior can be obtained from the release and transport studies in the in-pile meltdown facility, the Containment Mockup Facility (CMF), and the Containment Research Installation (CRI) at Oak Ridge National Laboratory; in the Aerosol Development Facility (ADF) at the Pacific Northwest Laboratories; and in the Contamination-Decontamination Experiment (CDE) facility at the Idaho Nuclear Corporation. In these facilities, the fraction of fission products released from the high temperature zone downstream from the molten fuel and/or to the containment vessels is an indication of the behavior that could be expected in the core and primary system of a reactor. It should be recognized that this is a qualitative description only and that the retention will depend strongly upon the temperature and gas flow conditions during accidents. Additional data have been obtained by Genco [70] in laboratory studies that can be used with analytical models to predict accident behavior.

Data on the release of fission products from the high temperature zone (>1,000° C) in in-pile experiments [71] are presented in Table 2-6. The effects of atmosphere can be noted. Atmosphere appears to affect the ruthenium behavior most significantly. Data on the distribution of fission products released from simulated fuel and from high-burnup UO_2 fuel in the CMF [72] are presented in Table 2-7. Iodine release to the containment tank varied from 87% to 100% in these runs. Hilliard [22] has reported data on the deposition of fission products from specimens simulating irradiated UO_2 and from irradiated UO_2 in the ADF. Results from the experiments with irradiated fuel are presented in Table 2-8. The fraction deposited in the delivery line was nearly the same for the simulated fission-product samples as the irradiated fuel samples. Deposition data for fission products in the heated (425° C) transport line in the CDE runs [73] are presented in Table 2-9.

TABLE 2-6

Effects of Atmospheres on Material Released from the High Temperature Zone (>1000° C) of the Furnace in UO_2 Fuel-destruction In-pile Experiments [71]

Atmosphere	Material released (%)					
	I	Te	Cs	Ru	Sr-Ba	Zr-Ce
Helium[a]	86	86	71	4.0	1.4	0.4
Moist helium	86	63	71	1.0[b]	3.9[b]	1.1[b]
87% steam,[a] 12% helium, 1% hydrogen	66	87[c]	47	0.44	1.5	0.18
Air[d]	68	60	53	0.21	1.6	0.11
Moist air[a]	95	72	68	15	1.2	0.44
87% steam,[d] 13% air	67	92	42	0.79	0.83	0.13

[a]Average for three experiments.
[b]Corrected (× 5) for incomplete melting.
[c]One experiment.
[d]Oxygen depleted by cladding before fuel melting.

TABLE 2-7.

Distribution of Fission Products Released from Simulated Fuel and from High-Burnup UO_2 Fuel in the Containment Mockup Facility [72]

Location	Fission Products Found (% of Total Inventory)							
	Cesium				Tellurium			
	Run 3-5[a]	Run 12-9[b]	Run 2-24[c]	Run 3-25[d]	Run 3-5	Run 12-9	Run 2-24	Run 3-25
Furnace tube	6.2	3.0	81.7	26.2	2.5	0.3	70.0	27.5
Aerosol tank	13.9	15.1	0.8	17.6	46.3	6.9	19.9	1.3
Condensate		43.6	0.6	44.2		0.8	0.2	0.04
Roughing filter	16.5	0.8	0.008	0.019	6.2	0.4	0.1	0.009
Absolute filter	0.04	0.002	0.001	0.042	0.008	0.05	0.4	0.009
Total release from fuel	36.6	62.5	83.1	88.1	55.0	8.4	90.6	28.8

Location	Fission Products Found (% of Total Inventory)							
	Ruthenium				Strontium			
	Run 3-5	Run 12-9	Run 2-24	Run 3-25	Run 3-5	Run 12-9	Run 2-24	Run 3-25
Furnace tube	6×10^{-4}	0.07	9.2	0.024	1×10^{-3}	0.01		2.0
Aerosol tank	1×10^{-3}	0.35	0.06	0.66	7×10^{-3}	0.04	0.03	3.2
Condensate		0.12	0.001	0.10		3×10^{-4}	9×10^{-3}	3.4
Roughing filter	8×10^{-4}	4×10^{-4}	0.003	0.08	2×10^{-3}	2×10^{-4}	2×10^{-4}	0.005
Absolute filter	3×10^{-5}	2×10^{-4}	0.002	0.01		2×10^{-4}	2×10^{-4}	0.011
Total release from fuel	2×10^{-3}	0.54	9.3	0.88	0.01	0.05	0.04	8.6

[a] In run 3-5, high burnup (7000 Mwd/ton) UO_2 fuel was melted with air in the furnace tube and in the containment tank; aging time in the containment tank was approximately 3.5 hr.

[b] In run 12-9, simulated stainless-steel-clad UO_2 fuel was melted with air in the furnace tube and with a steam-air atmosphere at 27 psig in the containment tank; aging time in the tank was approximately 5.5 hr.

[c] In run 2-24, simulated Zircaloy-clad UO_2 fuel was heated to or near the melting point of UO_2 with air in the furnace and with a steam-air mixture at 27 psig in the containment tank; complete oxidation of the fuel to a ZrO_2-U_3O_8 mixture occurred as the fuel cooled below 1000° C; aging time in the tank was approximately 5 hr.

[d] In run 3-25, simulated Zircaloy-clad UO_2 fuel was melted with a steam-helium atmosphere in the furnace and with a steam-air mixture at 27 psig in the containment tank; aging time in the tank was approximately 5 hr.

Although temperature- and residence-time considerations are important in applying these results to the reactor accident situation, the data qualitatively indicate that perhaps only 20% of the iodine released from the fuel will be retained by the core and primary system. Tellurium and cesium retention may also be low, while retention of the more refractory elements such as barium, ruthenium, and zirconium will be high.

Extensive laboratory studies of fission-product deposition have been reported by Genco et al. [70]. Iodine deposition on prefilmed Zircaloy and stainless steel surfaces and tellurium deposition and desorption on bare and prefilmed Zircaloy surfaces were studied as related to fission-product behavior in the primary vessel during a loss-of-coolant accident in a water-cooled reactor. The prefilmed surfaces were typical of those found within the primary system of a reactor. They were prepared by exposing the material for 1000 hr in a pressurized water loop with temperature, flow rate, and chemical composition the same as in a power reactor. Iodine deposition experiments were conducted with I-air-steam and HI-hydrogen-steam over the temperature range of 150° to 750° C. Tellurium deposition and desorption experiments were conducted at 200° to 500° C.

Iodine deposition rates were extremely low under primary system conditions and decreased with increasing temperature according to an Arrhenius relationship [70]. Rate coefficents of the order of 10^{-2} and 10^{-3} cm/sec were observed for HI in a temperature range of 150° to 750° C. The deposition coefficients for I_2 were first order, while those for HI exhibited a dependency upon square root of time and square root of HI concentration. First-order rate coefficients obtained at various temperatures for prefilmed type-304 stainless steel exposed to iodine are shown in Fig. 2-2. Because of the corrosive action of iodine, deposition was greater on bare metal surfaces than on prefilmed surfaces. Cracking of the oxide film which led to the exposure of bare metal enhanced deposition on the prefilmed samples.

Prefilmed and bare Zircaloy reacted with tellurium vapor at and above 400° C to form an apparent zirconium telluride. Less than 20% of the deposited tellurium was desorbed during 49 hr of exposure at 400° C under continuous evacuation.

Laboratory data on fission product deposition similar to those obtained by Genco [70] can be applied to the prediction of fission-product behavior during an accident by using appropriate analytical models. A well-mixed volume model has been developed to describe transport and deposition in a plenum of a reactor vessel and in a primary coolant line under loss-of-coolant conditions [74]. Although this model may not apply to all situations, it can be used to illustrate fission-product behavior and to identify the significant factors. It is assumed that the entire region is well mixed and that only a small fraction of the available surface area is covered by fission products. Under these conditions, the equations governing transport and deposition are:

$$\frac{dC}{dt} + \frac{Q}{V}(C - C_0) + \frac{A}{V}\frac{dS}{dt} = 0 , \qquad (2\text{-}7)$$

TABLE 2-8.

Fission-Product Release from Irradiated Specimens [22]

Element	Run (f)	Release from UO_2 (%) (a)	Delivered to containment mg(b)	Delivered to containment % (c)	Deposited in delivery line (%) (d)
Iodine	IA 26	(e)	0.07	72	6
	IA 32	82	0.28	68	14
	IA 33	81	0.25	65	20
	IA 39	99	0.45	89	8
	IA 40	75	0.35	67	0.2
	IB 41	99	0.42	83	13
	IB 42	96	0.42	83	12
Cesium	IA 32	65	1.8	58	11
	IA 33	64	1.6	55	15
	IA 39	99	4.7	78	12
	IA 40	30	1.0	18	30
	IB 41	93	4.3	74	12
	IB 42	94	4.5	74	10
Tellurium	IA 32	51	0.2	47	7
	IA 33	50	0.2	41	16
	IB 41	94	0.7	75	13
	IB 42	96	0.8	83	11
Ruthenium	IA 39	14	0.2	6	55
	IA 40	21	0.5	12	38
	IB 41	53	1.5	38	27
	IB 42	18	0.6	15	13
Barium	IA 39	0.22	0.06	0.18	12
	IA 40	0.13	0.03	0.07	39
	IB 41	0.07	0.02	0.05	29
	IB 42	0.10	0.02	0.06	25

(a) Escaped from fuel-clad UO_2 and granular bed, percent of inventory.
(b) Estimated from irradiation history.
(c) Percent of inventory.
(d) Percent of material entering delivery line.
(e) Insufficient analyses.
(f) Atmosphere: IA 26, air; IA 32, steam; IA 33, IA 39, IA 40, and IB 41, air plus steam; IB 42, steam plus hydrogen.

TABLE 2-9.

Deposition of Fission Products in Transport Line of the CDE Facility [73]

Fission Product	Percentage of Release Deposited in Transport Line [a, b]		Percentage of Fuel Inventory Deposited in Transport Line [b]	
	Run 4	Run 5	Run 4	Run 5
Sr-89	29		0.0085	
Zr-95	37	93	0.00037	0.00011
Mo-99	trace		trace	
Ru-103	2.5	4.3	0.006	0.00013
I-131	20	12.8	4.8	5.8
Te-129		6.7		0.23
Te-132	15		3.3	
Cs-137	8	31.2	0.63	9.9
Ba-140	17		0.019	
Ce-141	100		0.004	

[a] Release considered as sum of nuclide activity in transport line and containment vessel.

[b] Blanks indicate that no data were obtained.

$$\frac{dS}{dt} + \frac{k_g RT}{M}(C - C^*) = 0 , \qquad (2-8)$$

$$\frac{dS}{dt} + k_A C^* - k_D S = 0 , \qquad (2-9)$$

where C = airborne concentration, g/cm^3, S = surface concentration, g/cm^2, C* = airborne concentration at the surface, C_o = inlet airborne concentration, A = surface area, cm^2, V = volume, cm^3, Q = inlet and outlet flow rate, cm^3/sec, k_g = gas-phase mass-transfer coefficient, g/(cm^2-sec-atm), k_A = sorption coefficient, cm/sec, and k_D = desorption coefficient, sec^{-1}. An analytical solution is not available for these equations but they have been programmed for computer solution in a code designated PLENUM [74].

A series of sensitivity calculations has been performed to determine the controlling factors for

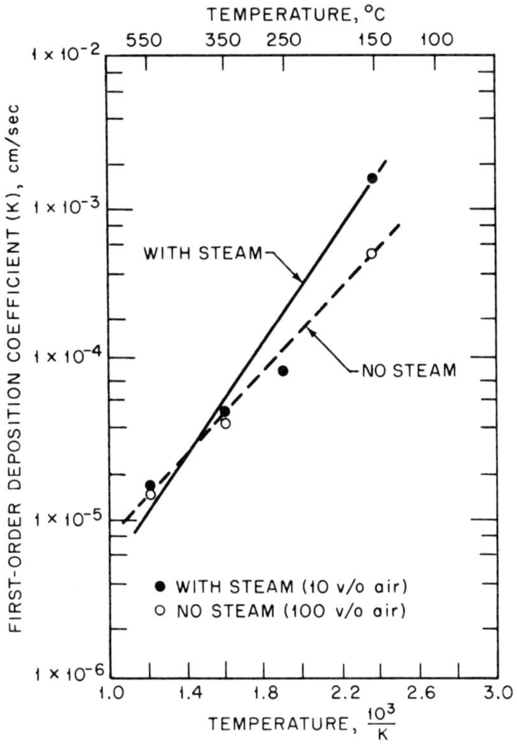

FIG. 2-2 First-order rate coefficients obtained at various temperatures for prefilmed type-304 stainless steel exposed to iodine-air [70].

retention of fission products by the primary system. In these calculations, it was assumed that the flow rate was controlled by the boil-off rate of water in the lower head of the pressure vessel, and that the surface temperatures in the primary system remained constant at 300° C. These conditions may exist if standby cooling is severely degraded and if the source of steam (the driving force for fission-product transport) is the water remaining in the pressure vessel after blowdown. A constant surface temperature was assumed for convenience in the calculations. A value for the adsorption coefficient from the experimental work by Genco [70] for iodine deposition from steam onto stainless steel at 300° C was used. The results of these calculations indicate that at least 50% of the iodine that is released from the core under these conditions will not be retained by the primary system. Five times the geometric surface area, based upon a typical cylindrical pressure vessel design, was assumed to be available for deposition. In some cases, releases to the containment of more than 90% were calculated. The most important factor affecting the retention was the steam-flow rate. Deposition decreased with increasing flow rate or shorter residence time in the vessel. A similar conclusion, that only a small percentage of I_2 and HI (<10%) would deposit on primary system surfaces at temperatures expected in an accident, was also reported by Genco [70], based upon calculations using laboratory deposition data.

Although these calculations were performed for accident conditions with no standby cooling, it is expected that the conclusions on the fraction of iodine retained within the primary system (on the surface of the vessel, piping, and internals) would be nearly the same with standby cooling. Surface temperatures would be lower (which would tend to enhance deposition) while flow rates would be higher due to added steam flow (which decreases the residence time and the deposition). Most of the fission products released from the fuel would be entrained in the coolant.

3. NATURAL REMOVAL PROCESSES IN THE CONTAINMENT VESSEL

3.1 General

There has been extensive study of fission-product behavior under conditions simulating those in reactor containments without engineered fission-product removal systems. The terms "natural effects," "natural conditions," or "natural response" are sometimes used to describe this type of accident environment. Under natural response conditions, heat is removed by being transferred to the containment walls and is subsequently conducted out of the building through the exterior walls. Transfer of heat to the walls promotes natural circulation of the containment atmosphere and condensation of steam at the walls. Fission-product behavior is influenced greatly by these gas-phase mass-transfer processes within the containment. In power reactors utilizing steel-shell containments, natural responses will completely remove the stored heat in the coolant and the decay heat from the containment; hence, the removal of fission products by natural effects is adequate to achieve reactor safety. However, with the increasing size of reactors and the use of concrete containments, which reduces heat transfer through the containment shell, the need for containment-atmosphere cooling and more positive fission-product control has become greater, and so engineered safety systems have been incorporated into the reactor designs. The natural response case is of interest even for these latter systems because of the possibility of a nuclear accident in which the engineered safety systems fail to operate. An understanding of the removal of fission products by natural effects can provide a basis for establishing limiting cases of fission-product release to the environment and evaluating the effectiveness of the engineered removal systems. A number of investigations of the processes governing fission-product behavior under natural response conditions have been conducted in a variety of experimental and analytical studies and a basis is available upon which a conservative but reasonable prediction of fission-product behavior during an accident can be made.

3.2 Experimental Data on Natural Removal Processes

Two basic types of studies have been performed

which yield data on fission-product behavior under natural conditions. The first type are laboratory studies in which the physical chemistry of fission-product behavior at various surfaces or interfaces can be studied. Conditions can be carefully controlled, mechanisms can be elucidated, and rate coefficients for the various processes can be obtained. Data from such studies can be used in analytical models to describe fission-product behavior in reactor containments under accident conditions. Laboratory studies, however, yield little or no information on gas-phase mass-transfer processes and do not reveal the resultant of competing processes. A second type of study of fission-product behavior in model containment vessels or in experimental reactor containment vessels can provide data on this kind of information.

A number of reviews of fission-product behavior under accident conditions have been written [8, 53, 75, and 76]. Iodine has been more extensively studied than the other fission products and this fact is reflected in the reviews. In some of the work, limited data pertaining to the behavior of some of the other fission products have been reported.

3.2.1 Laboratory Studies

Many of the results from laboratory deposition studies are expressed in terms of a deposition velocity, v_d (cm/sec). This quantity is defined as follows:

$$v_d = \frac{\text{rate of deposition on the surface/cm}^2\text{- sec}}{\text{concentration/cm}^3\text{ of gas}} .$$

If the fission product is uniformly dispersed within the gas phase in a closed containment vessel, then the total loss from the gas phase is:

$$Av_d C dt = -V dC ,$$

where A is the surface area of the containment, V is the volume, and C is the concentration of the fission product. The concentration at any time is obtained by integration, and

$$\frac{C}{C_0} = e^{-Av_d t/V} .$$

This approach grossly oversimplifies the deposition processes at the surfaces and implies that the surface processes are rate-limiting in the experiment. It does provide a means for intercomparing data obtained from a variety of sources under many conditions and can be used in a simple manner to apply the laboratory results to reactor containments. The validity of the application depends upon the degree to which the experiments simulate the containment conditions.

Forberg, Westermark, and Holmquist [77] studied iodine deposition on small samples of concrete and other materials. The samples were exposed to iodine vapor over a period of several days at ambient temperature. The rate of sorption on concrete was independent of the iodine loading of the surface for up to 50 $\mu g/cm^2$, and deposition velocities taken from the initial portions of the sorption curves were 2×10^{-2} cm/sec (Table 3-1).

Croft, Davis, and Iles [78] exposed deposition coupons in a 27 m^3 room with high concentrations of iodine. Deposition coefficients obtained from the early portions of these experiments, run under static conditions, are given in Table 3-1. Deposition was found to be dependent on the nature of the surface, with bare and mild steel and concrete surfaces showing higher deposition values than painted surfaces, which were in turn more receptive than the plastic (PTFE and polyethylene) surfaces. Oxidized or polished mild steel surfaces were similar to each other in deposition behavior.

Results from experiments reported by Chamberlain [75] on deposition on small cylinders in a flowing gas stream are included in Table 3-1. Morris and Nicholls [76] have also performed experiments in a flow system to determine iodine deposition on various materials. The data indicated that under the conditions of their experiments, gas-phase mass transfer was not controlling and, hence, the deposition velocity and the sorption coefficient are nearly equal. Results on paints have been given as average values for three different paints: "Air Spun," "Tretol," and "Tretaline." Deposition velocities attributable to Jarman [79] have also been reported by Morris and Nicholls. The complete experimental conditions are unknown, but the results indicate that the deposition velocities approximate the sorption coefficients for most cases.

Iodine deposition on stainless steel surfaces and on commercial paints used within the nuclear power industry was studied by Rosenberg [81]. These experiments covered a range of conditions expected in the containment vessel during a loss-of-coolant accident in a water-cooled reactor. Sorption-desorption experiments were performed with I_2 from both the vapor and aqueous phases as well as under condensing steam conditions. Vapor-phase kinetic studies were performed in steam-air mixtures from 25° to 170° C, while aqueous-phase studies were conducted between 25° and 90° C, using saturated solutions.

For I_2 deposition from both the vapor and aqueous phase under containment system conditions, sorption-desorption experiments indicated that considerable iodine penetrates into the interior of commercial coatings and that a significant portion of the diffusing iodine is irreversibly retained. These observations as well as the experimental data can be explained quite adequately by assuming a diffusion-with-chemical-reaction model. Under condensing steam conditions, I_2 was found to deposit on commercial coatings, and the presence of condensed water did not adversely affect the rate of deposition. Application of the experimental deposition data to loss-of-coolant accident conditions indicated that the effectiveness of commercial coatings in removing airborne I_2 is highly dependent on the particular coating used. The order of decreasing I_2 deposition is inorganic

TABLE 3-1.

Sorption Coefficients for Iodine on Various Surfaces.

Ref.	Exposure time, min	Iodine vapor concentration, μ g/m^3	Atmosphere	Temp., °C	Iodine sorption coefficient on indicated surface, 0.01 cm/sec				
					Aluminum	Mild steel	Stainless steel	Paint	Concrete
80	60	0.35	Moist air	Amb.	7	17	-	-	-
78	20	7.5×10^4	Dry air	20	-	62[b]	-	0.72	6.7
76	20	10 to 150	Dry air	20	16	7.7	4.8	2.1	26[b]
76	120 to 240	10 to 150	Dry air	150	0.73	0.30	0.38	0.11	7.6
76	120 to 240	10 to 150	Air - 40 v/o water vapor	150	6.2	1.7	0.91	2.2	0.081
76	300	1,000 to 2,000	CO_2	20	0.02 -1.0[a]	0.1-1.5 3-10[c]	0.5 -0.8[a]	0.02 -0.2[a, e] 0.03 -0.02[a, f]	-
79	60	40 to 100	Dry air	Amb.	1.8	16.8	-	-	-
75	60	100 to 1,000	Dry air	Amb.	1.7	34	-	-	-
75	60	1,000 to 10,000	Dry air	Amb.	1.4	13	-	-	-
75	60	700	Dry air	200	4.3	16	-	-	-
75	~200	20,000	Steam and air	150	0.51	12	5.4	1.4	-
77	3,000	23	Room air	Amb.	-	-	-	-	2

[a] May have been influenced by high airborne iodine concentration.
[b] May have been influenced by mass transfer.
[c] Acid-etched mild steel.
[d] Corrected for estimated mass transfer.
[e] Gloss paint.
[f] Chlorinated rubber paint.

zinc primer, epoxy, phenolic, acrylic latex, and vinyl base coatings. Some of the vapor-phase deposition data at 115° C in a steam-air atmosphere are shown in Fig. 3-1 [82].

Vapor-phase deposition experiments with CH_3I and liquid-phase deposition experiments with I^-, Cs^+, and Te^{+4} indicated no significant degree of retention of these ions by commercial coatings. In the vapor phase, copious quantities of iodine will deposit on stainless steel provided water vapor is present. The process follows first-order kinetics according to an Arrhenius relationship having an activation energy of 36 kcal/mole-K. However, deposition does not occur under condensing-steam conditions, aqueous-phase conditions, or extremely dry vapor-phase conditions (< 10 ppm H_2O).

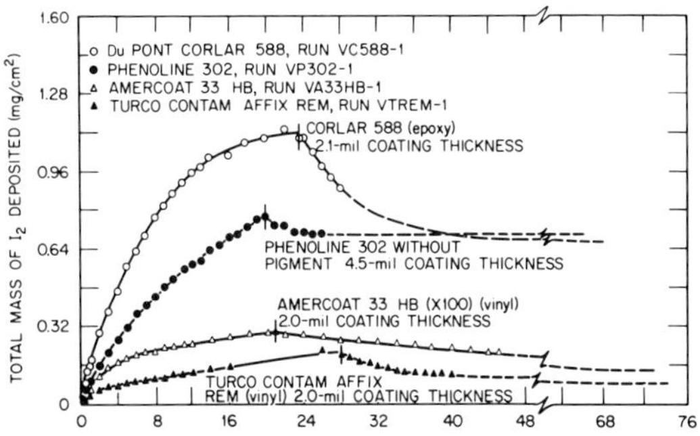

FIG. 3-1 Vapor-phase deposition and desorption of elemental iodine on painted surfaces.

These laboratory data on deposition conditions are useful in several respects for the description of fission-product behavior during an accident. By having a good measurement of the surface reaction rates for iodine, data described below from the model containment-vessel studies (which include both gas-phase and surface processes) can be more readily interpreted. The contribution of each of the processes to the observed iodine behavior can be discerned. Surface deposition coefficients can be used directly with assumed or predicted gas-phase mass-transfer rates to estimate the behavior of fission products in reactor containment vessels. The surface coefficients are more accurately known than the gas-phase mass-transfer rates occurring under accident conditions, so that the accuracy of the estimated iodine deposition rates under natural-response conditions are limited by the latter rather than the former. The accuracy of the techniques employed to make the overall predictions can be tested in model containment vessel studies.

3.2.2 Containment Vessel Studies

As noted above, it is difficult to apply the results of laboratory experiments to large reactor containments under accident conditions. Analytical models are needed for this purpose and before confidence can be attained in the predictions of the mathematical models, experimental validation of the model is required. Experiments on fission-product behavior have been conducted in a number of model containment vessels under simulated accident conditions. These provide additional data for the development of analytical models and as well as a means through which the models can be verified. A few experiments have also been performed in containment shells of experimental reactors, and these can be used to provide further understanding of fission-product behavior under natural response conditions.

The CMF and the CRI facilities have been used primarily to study the effects of aging on the behavior of fission products. Behavior of fission products released from stainless-steel-clad UO_2 irradiated to 7000 Mwd/ton were studied in the CMF [83]. The CMF is a cylindrical stainless steel model containment vessel with a volume of 6.3 ft^3 and a surface-area-to-volume ratio of 3.5 ft^{-1}. The fuel samples were heated in steam and air flowing through the furnace and the fission products were released into the containment vessel pressurized to 29 psig with steam and air. Airborne activities as a function of time following the release from the Zircaloy-clad rod are shown in Fig. 3-2. Less tellurium was released from the Zircaloy-clad rod than from the stainless-steel-clad rod due to alloying effects, while more strontium and barium were released from the Zircaloy-clad specimen due to reducing effects. No effect attributable to differences in burnup could be discerned.

A series of runs with a stainless steel liner were conducted in the Containment Research Installation (CRI) to determine fission-product release and composition over the range of release and environmental conditions of loss-of-coolant accidents [84]. The volume of the CRI is 160 ft^3 and the surface-area-to-volume ratio is 0.89 ft^{-1}. Conditions for the steam-air runs in this sequence are tabulated in Table 3-2, and data on the variations in iodine behavior with time are shown in Fig. 3-3. The total time of experiments was between 20 and 24 hr. The effect of natural deposition processes for the iodine is also summarized in Table 3-3. It was concluded from these tests that iodine is initially depleted from the gas phase by a fast chemisorption process that may leave up to 10% or more of the iodine in a relatively stable airborne state. A particulate form with a mean mass diameter of 0.25μ is postulated. Solid aerosols of cesium and tellurium are not depleted by adsorption or diffusion or by the limited steam condensation rates employed in the tests (maximum 4 × 10^{-6} gm/cm^2-sec), but obey Stokes settling for the particle size distribution according to stirred-settling theory. The fraction of organic iodide formed is low but variable and appears to be formed at a constant rate, probably in a surface reaction on the walls. Comparison of the behavior between the real fission products and the simulated fission products used in the CSE experiments shows no significant difference.

Fission behavior has been studied under natural response conditions made at the Nuclear Safety Pilot Plant (NSPP) [85, 86, 87, 88]. This stainless steel model containment vessel has a volume of 1350 ft^3 and the surface-area-to-volume ratio is 0.52 ft^{-1}. Tests in the NSPP generally fall into three categories according to the environmental conditions in the model containment vessel and the source of the fission products. In the first category are the natural effects tests to study transport and deposition of iodine in air using an elemental iodine source. Stable iodine tagged with I^{131} was swept by air into the containment vessel and the vessel atmosphere was mixed for 10 minutes. The airborne iodine concentration as a function of time for

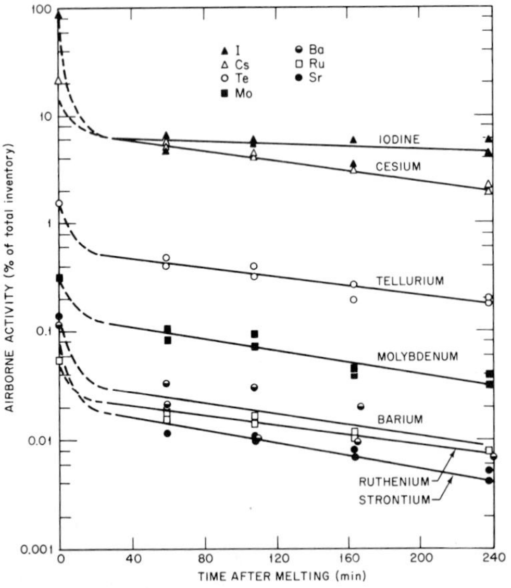

FIG. 3-2 Composition of CMF tank atmosphere as determined by gas samples in Run 4-11.

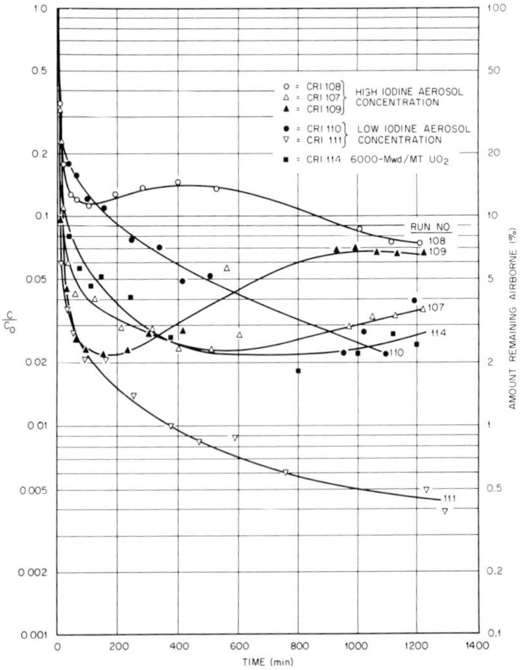

FIG. 3-3 Variation in radioiodine behavior in stainless-steel-lined CRI.

TABLE 3-2.

Conditions of Steam-Air Runs in CRI and Data on Iodine Concentration in Condensates and on Stainless Steel Tank Walls.

Run designation	Initial iodine in tank		Distribution in containment vessel at end of aging period [a] (%)		Amount remaining airborne[b] (%)	Monolayers on wall[b]
	Total (mg)	mg/m^3	In condensate	On walls		
CRI 104 (I_2) tracer	30	6.7	15	84.6		0.64
CRI 105 (HI) tracer	20	4.5	13	86.8		0.43
CRI 107[c] (CSE-Sim) zircaloy-UO_2 melt	9.3	2.1	14.8	81.7	3.5	0.19
CRI 108 zircaloy-UO_2 melt	20	4.5	36.5	56.3	7.2	0.32
CRI 109 zircaloy-cladding melt (1850° C)	8.9	2.0	40.1	53.5	6.4	0.12
CRI 110 zircaloy-cladding rupture (1300° C)	2.2	0.5	20.9	77	2.1	0.044
CRI 111 zircaloy-cladding rupture (1200° C)	0.2	0.04	27.3	72.3	0.4	0.004
CRI 114 (6000-Mwd/Mt UO_2) zircaloy-UO_2 melt	1.0	0.22	13.2	84.2	2.6	0.025
CMF 1007 (CSE Sim)[d]	0.052	0.29	68	30.8		0.0026
CMF 1008 (CSE Sim)[d]	0.34	1.9	46	42		0.024

[a] Difference between sum of condensate and wall plateout and 100% is amount remaining airborne.

[b] Based on 0.3 $\mu g/cm^2$.

[c] Run 107 was conducted as a CSE simulant validation test, with separate vaporization filaments for fission-product groups.

[d] CMF runs listed for comparison.

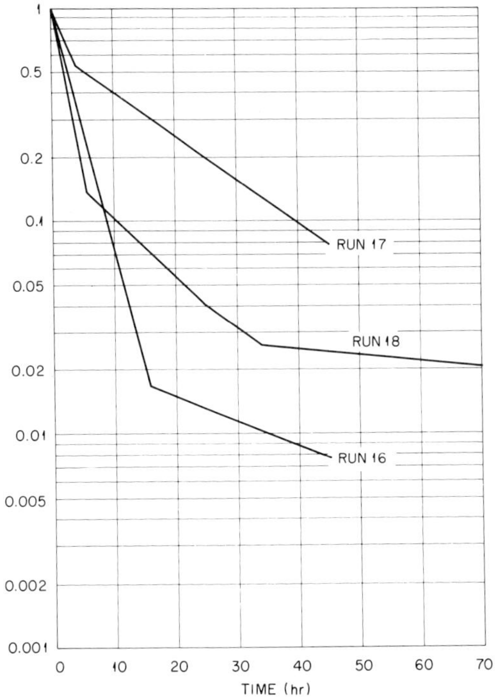

FIG. 3-4 Behavior of airborne I_2 in the model containment vessel of the NSPP.

three runs is shown in Fig. 3-4. Although the three runs were to be identical, some subtle phenomenon was apparently affecting the iodine behavior.

The second category of tests in the NSPP involved the study of the behavior of fission products released from trace-irradiated UO_2 into steam-air atmospheres in the model containment vessel [86]. Two tests were performed in which lightly irradiated, stainless-steel-clad UO_2 pellets were melted with a plasma torch and the fission products were released into the model containment vessel. The vessel atmosphere at the time of meltdown was air plus steam at 107° C and 25 psig. After melting the fuel, the steam was allowed to condense; then the concentration of the fission products in the vessel atmosphere, in the condensate, and on various surfaces was studied. In both runs, the removal of airborne fission products was allowed to take place entirely by natural process. Data on total airborne iodine concentration as a function of time for the two runs are shown in Fig. 3-5. Initial gas-phase concentrations decreased by factors of 5 and 15 during the 48 hr observation period in Run 8 and by a factor of 10 in Run 9 during the first 24 hr observation period prior to spraying. Accumulation of iodine in the wall-runoff-condensate samplers is shown in Fig. 3-6. Iodine followed the steam condensation quite closely under the conditions of this test. Deposition-coupon data revealed that the differences between deposition materials under condensing conditions are less than those under dry con-

TABLE 3-3.

Effect of Natural Deposition Processes for Radioiodine in CRI Steam-Air Runs

Run	Conditions	Initial halftime (min)	Amount deposited (%)	Secondary halftime (min)	Amount deposited (%)	Not deposited[a] (%)
107	Simulant + UO_2 melt	4.5	95.0	260	1.5	3.5
108	I_2 + SnO_2 aerosol + UO_2 melt	4.0	87	132	5.5	7.2
109	I_2 + partial zircaloy melt	3.0	93.5[b]			6.4
110	I_2 + zircaloy rupture	3.9	82	186	15.7	2.1
111	I_2 + zircaloy rupture	3.0	96.5	145	3.0	0.4
114	6000-Mwd/Mt UO_2 melt	4.6	93.0	159	4.3	2.6

[a] Includes all iodine remaining airborne at end of aging period.

[b] Maximum value of 97% was reached, but desorption rate increased later.

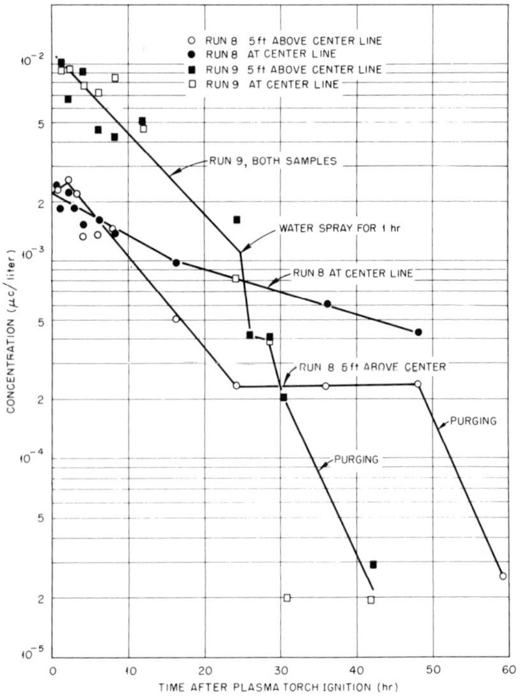

FIG. 3-5 NSPP Runs 8 and 9—Total iodine concentration in Maypack samples.

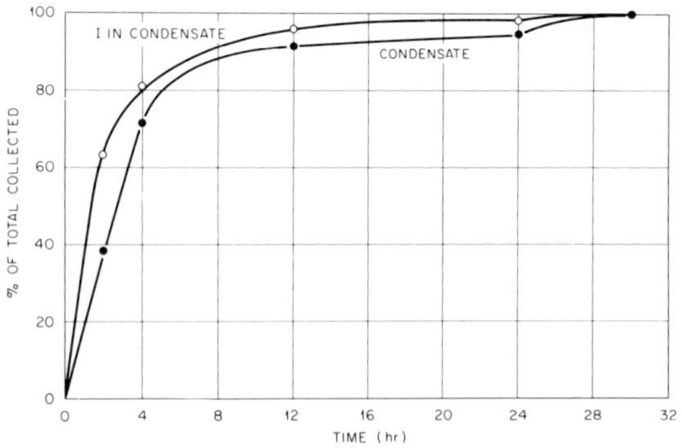

FIG. 3-6 Accumulation of wall runoff condensate and of I in condensate; NSPP Run No. 9.

ditions. Since in condensing conditions, the surfaces are covered with a water film, the deposition process is comprised of a two-step process: absorption from the gas phase into the water film, followed by diffusion to the coupon surface and deposition on the surface; in dry atmosphere runs, the process consists simply of direct deposition on the solid surface.

Since the use of highly irradiated fuel involves experiments at high radiation levels, it is more convenient to use simulated fission products. A third category of experiments was conducted in NSPP to obtain data on the behavior of fission-product simulants for comparison with trace burnup and highly irradiated fuel tests [87]. Fuel for these runs was prepared by adding stable and radioactive tracer isotopes of strontium or barium, iodine, cesium, cerium, ruthenium, and tellurium to the UO_2. The fuel matrix was clad with stainless steel. In two of the runs, a reducing atmosphere was present in the furnace, while in the third an oxidizing atmosphere was maintained. Deposition in the containment vessel was rapid during the early portion of all these runs. Iodine which was transported to the model containment vessel following release under reducing conditions deposited more rapidly than in the oxidizing atmosphere runs. One hour after plasma torch ignition, the percentage of iodine entering the vessel in an airborne state was 0.6 and 2.5 for release into the two reducing atmospheres, while it was 10 for release into the oxidizing atmosphere. The spread in the data indicates a variance in the experimental conditions, such as the number of passes over the fuel made by the plasma torch between runs and the difficulty in obtaining reproducible results in model containment vessel tests. In all runs a draft tube was used to mix the fission products with the containment vessel atmosphere. After the initial deposition, removal of iodine and cesium by natural processes was slow, with half-times of one to several hours.

The Contamination-Decontamination Experiment [73] is a pilot-plant facility in which fission-product behavior under natural-response conditions has also been studied. The volume of the CDE containment vessel is 86 ft^3 and the containment surface-area-to-volume ratio is 2.4 ft^{-1}. The inner surfaces of the stainless steel walls are painted. Five high-level runs from a radioactivity standpoint (burnups up to 2000 Mwt/ton with less than a 30-hr decay time before fuel meltdown) and numerous trace-level runs have been completed. Steam-air atmospheres were present in the containment vessel during the runs. Iodine mass concentrations in the containment vessel were several orders of magnitude below those in the NSPP runs. In the containment vessel, the distribution of the fission products between the condensate and the wall was found to depend upon the amount of condensate formed (or steam injected) and the solubilities of the fission products. Generally, about one-half of the total fission products reaching the containment vessel was removed by the condensate and the other half was deposited on the wall. Increasing the amount of condensate increased the fraction of fission products it removed. Condensation also promoted fission-product movement to the wall, resulting in a higher deposition on condensing surfaces than on noncondensing surfaces. Fission products other than the noble gases and iodine behaved similarly as small particulates (0.001μ to 0.7μ). Tin, a volatile component of Zircaloy, constituted the major mass of the particulates. The airborne iodine concentration decreased with an initial half-time of approximately 8 min. Time-dependent I^{131} behavior is shown in Fig. 3-7. The iodine activity remaining in the atmosphere after environmental conditions stabilized appeared to be hypoiodous acid (HOI) rather than organic iodides.

Experiments have been performed in the Aerosol Development Facility in a manner whereby a direct comparison could be made between the transport of simulant and reference fission-product aerosols

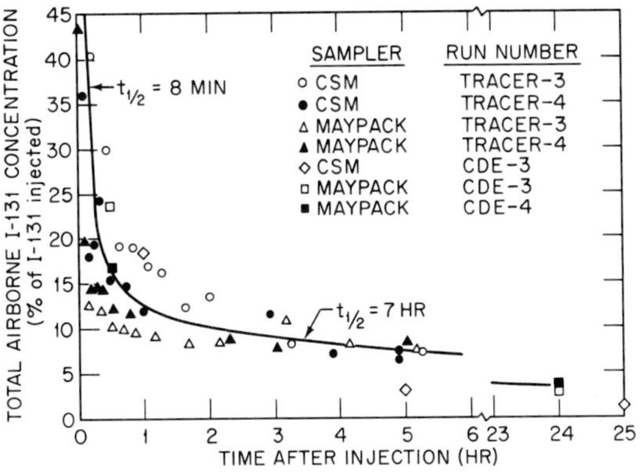

FIG. 3-7 Time-dependent I-131 concentration in containment atmosphere.

in a condensing-steam containment atmosphere [22]. The reference fission-product aerosol was produced by heating stainless-steel-clad UO_2 irradiated to 1 × 10^{17} nvt (~ 0.35 Mwd/ton) in oxidizing and reducing atmospheres at various temperatures up to that of molten UO_2. The simulant aerosol was produced by volatilizating stable fission-product elements or compounds that were radioactively traced for analytical purposes. The aerosol was injected into a containment vessel filled with an equimolar steam and air mixture. Both a stainless steel aerosol tank (SAT) with a volume of 3.2 ft^3 (and a surface-area-to-volume ratio of 1.8 ft^{-1}) and a painted aerosol tank (PAT) with a volume of 5.4 ft^3 were used. Steam condensed at a steady rate by heat transfer through the vessel surfaces and was replaced by a continuous steam addition to maintain isothermal (~ 80° C) and isobaric (~ -1 in. Hg gage) containment conditions. Isothermal and isobaric conditions were maintained to simplify data interpretation, and the addition of steam promoted mixing of the fission products in the gas phase.

Gas-phase concentrations in the model containment vessel as a function of time are shown in Figs. 3-8, 3-9, and 3-10. The effects on the airborne behavior of a continuous release over an extended period of time can be seen in Fig. 3-10. The results of all tests, however, were similar from the time at which release was stopped. All particulate materials decreased exponentially to background levels. Elemental iodine decreased exponentially until either the gas-liquid equilibrium was attained or until background radioactivities obscured this information. The high molecular weight organic iodides (charcoal paper samples) were rapidly removed for a short time, then their concentration diminished very slowly for the duration of the test. Methyl iodide concentration (charcoal bed) remained relatively constant with time.

Two variables of the containment environment, steam-condensation rate and gas velocity past surfaces, were shown to be of major importance to transport from the gas phase. The differences in the behavior of iodine early in the runs shown in Figs. 3-8 and 3-9 are attributable to these effects. Furnace atmosphere during release, UO_2

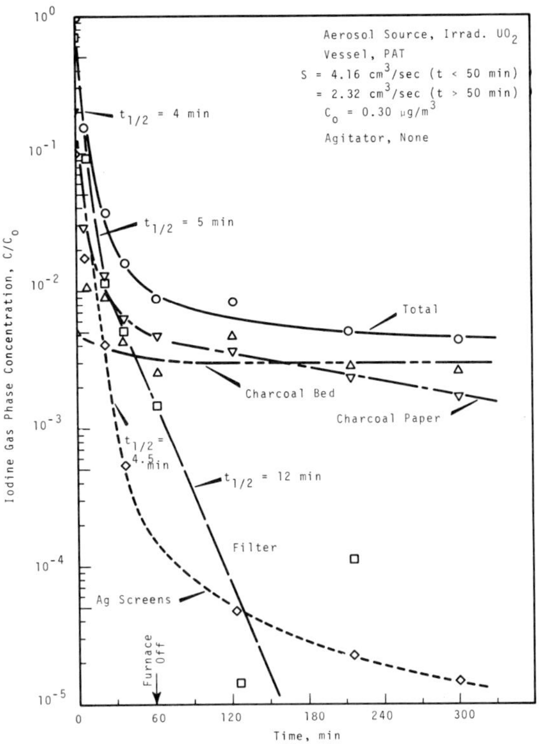

FIG. 3-8 Iodine gas-phase concentration, Run IA 39, Aerosol Development Facility [22].

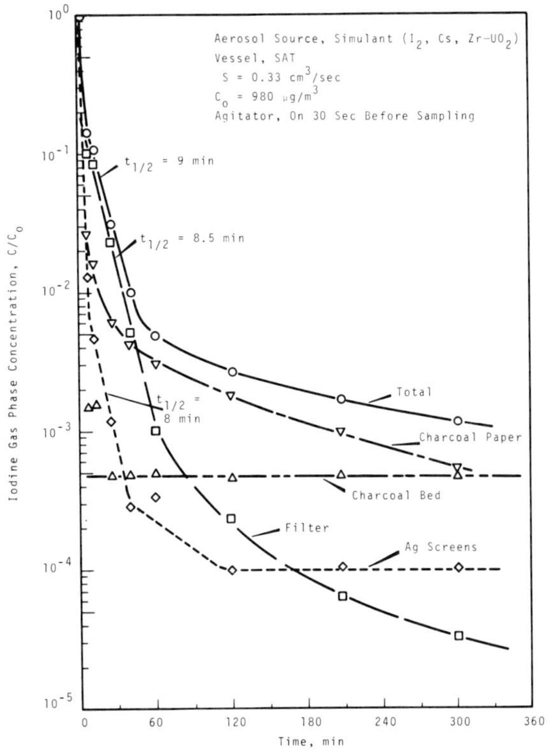

FIG. 3-9 Iodine gas-phase concentration, Run SB 78, Aerosol Development Facility [22].

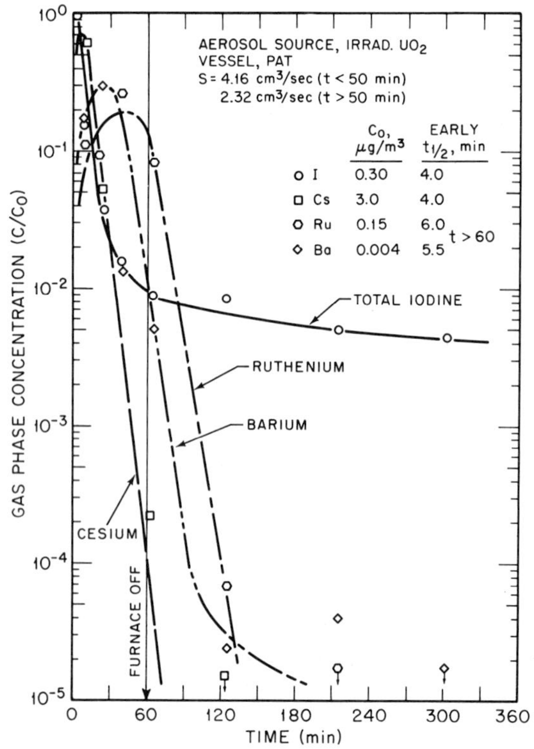

FIG. 3-10 Gas-phase concentrations, Run IA 39, Aerosol Development Facility [22].

cladding type, total mass released, type of containment vessel interior surface, and age of painted surfaces were shown to have minor effects on containment behavior.

Two fission products (iodine and cesium) were water soluble and were largely transported to the condensate pool in the vessel sump. The other fission products (tellurium, ruthenium, barium, and uranium) were insoluble particles which, after transport to the vessel surface, remained there even after extensive steaming of the vessel.

Based on the gas-phase mass-transfer coefficients and on aerosol distribution within the containment vessel, the results of the ADF tests indicated that the containment transport behavior of simulant and reference fission-product aerosols was identical within the limits of experimental error.

Fission-product behavior has also been studied under natural-response conditions in the large Containment Systems Experiment (CSE) containment vessel [89]. The main vessel of the CSE has a volume of 3×10^4 ft^3 and a total surface-area-to-volume ratio of 0.29 ft^{-1}. During Run A-5 in the CSE vessel, simulant fission-product materials were injected into a steam-air atmosphere at 47 psia and 253° F (123° C). Makeup steam was added to the vessel to maintain the temperature and pressure during the run. This enabled direct comparison with ADF runs and promoted mixing of the fission products in the containment vessel. While the pressure and temperature in a containment during an accident decreases with time, results from tests such as CSE are easier to interpret if some of the variables such as pressure and temperature are held constant. Measurement of the gas-phase concentrations of iodine, cesium, and uranium were made as a function of time during the run.

The total airborne iodine concentration is shown in Fig. 3-11 as a function of time. The total iodine concentration reflects the behavior of its constituents. Initially the half-life for the iodine in the gas phase was 16.2 min. As the gas-phase concentration of iodine approached 1% of its initial value, the airborne half-life became 26 hr. Elemental iodine concentration, however, decreased exponentially with a half-life of 13.5 min until it was about 0.1% of its initial value. This rapid decrease is attributed to transfer from gas to surfaces with the rate being gas-film diffusion-controlled. After about two hours, the steam condensate became saturated and further reduction in the gas-phase concentration occurred only slowly (24-hr half-life). Particulate-associated iodine decreased with a 50-min half-life for about an hour, then with a 14-min half-life until its concentration reached a value about one third that of elemental iodine, after which it decreased slowly at about the same rate as elemental iodine. Methyl iodide decreased slowly during the entire test with an average half-life of 30 hr. Distribution of the iodine within the containment during the test is shown in Fig. 3-12.

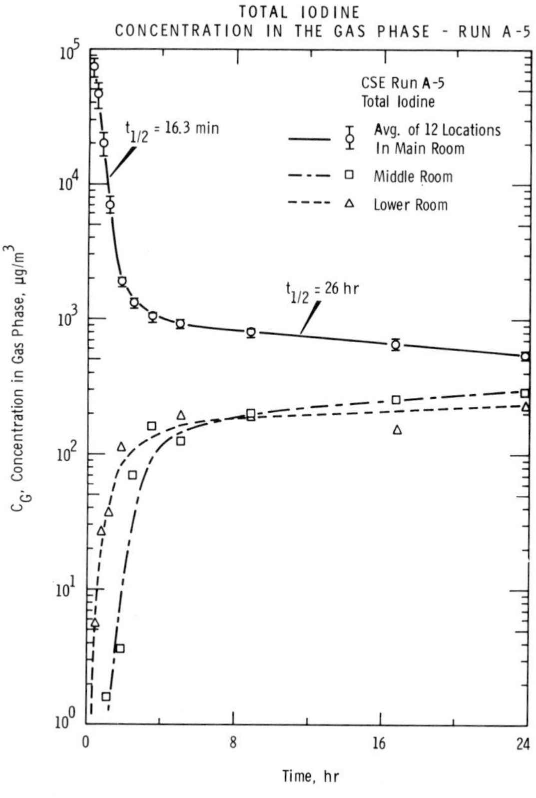

FIG. 3-11 Total iodine concentration in the gas phase, Run A-5, Containment Systems Experiment Facility [89].

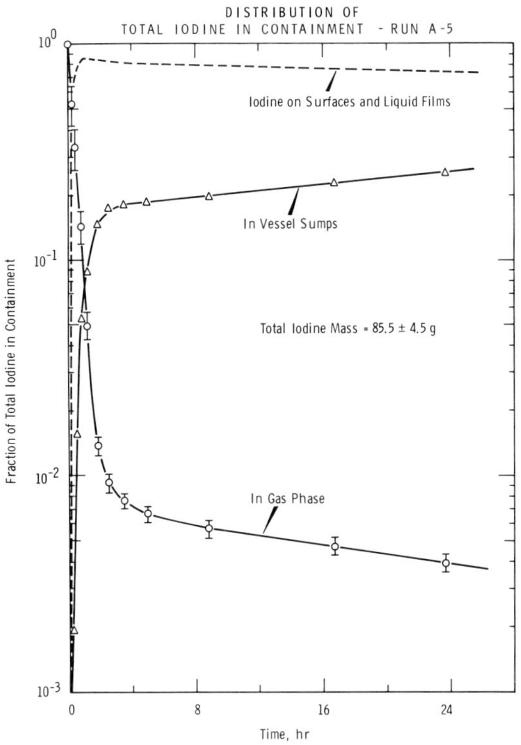

FIG. 3-12 Distribution of total iodine in containment, Run A-5, Containment Systems Experiment Facility [89].

The cesium concentration decreased exponentially with a 42-min half-life for 2 hr, then more slowly with a 1.9-hr half-life until background was reached. The airborne uranium concentration decreased with a half-life of 53 min for about 2 hr, then with a 100-min half-life until background was reached.

This experiment showed that aerosols and vapors released into a vessel containing air and steam are mixed very rapidly in the main room of the CSE vessel. The error bars shown on the data in Fig. 3-11 are indicative of the spread in the samples taken at twelve locations in the room. Makeup steam additions to the main room would also be expected to provide mixing. The experimental results for elemental iodine removal were in excellent agreement with theory based upon natural convection and heat-mass transfer analogy. All materials except methyl iodide were removed from the atmosphere under the conditions of this experiment by natural processes until the concentrations were less than 0.1% of their initial values.

The half-life for elemental iodine removal by natural processes early in other CSE tests has also been determined. The values vary somewhat, depending upon experimental conditions, but range between 10 min and 15 min in the CSE vessel [90].

Deposition experiments in the Zenith, Dido, and Pluto reactor shells have been reported by Megaw and May [80] and by Croft and Iles [91]. Their data showed changing deposition rates during the aging of an iodine aerosol released initially as molecular vapor. The initial airborne iodine concentrations in dry air were 0.01 to 14 $\mu g/m^3$ in the Dido and Pluto experiments and 0.3 to 0.8 $\mu g/m^3$ in the Zenith shell. The maximum iodine concentrations used were lower than possible accident conditions by a factor of 100 or more and the absence of vaporized fuel particles reduced the degree of accident simulation. At these low concentrations, conversion of the molecular iodine into nonelemental forms would be expected which would support the observation of changing deposition rates during aging.

3.3 Analytical Models of Natural Removal Processes

Valid analytical models of fission-product behavior in reactor containments under accident conditions are needed to apply the data obtained in the laboratory and model containment vessel experiments to the much larger, practical systems. A variety of approaches can be taken to analytical modeling of fission-product behavior under natural conditions, and proof of the applicability of any model must rest upon its representation of test results. Models describing fission-product behavior under natural conditions differ in detail and in the treatment of many parameters, but generally the models fall into classifications based upon geometry and gas-phase transport mechanisms. Two geometries are usually considered: tubes and large sections which can be effectively represented as large flat plates. In tubes, laminar or turbulent flow conditions are important, while in large containment vessels, the models can be grouped by diffusion-controlled or well-mixed conditions within the gas phase. The presence of condensing steam in loss-of-coolant accidents in water reactors is important as a gas-phase transport process and in its influence on surface processes.

Under postulated accident conditions for power-reactor containments, the well-mixed-volume models are usually applicable. Models which have been developed to describe fission-product behavior under these conditions have been based upon the data described in the preceding section and reasonable confidence can be placed in their use for accident evaluations. The major uncertainties are associated with the description of the accident conditions or, more importantly, the gas-phase mass-transfer processes. Surface deposition processes for the several forms of fission-product iodine are reasonably well understood and estimates of deposition rates to within at least 20% can be made for specified gas-phase conditions and iodine form. Specification of the gas-phase transport processes is less certain, but even then, an overall prediction of elemental iodine removal rates by natural effects during an accident can be made to within a factor of two. Within some regions of the containment and for some conduits in reactor systems the tube geometry models may be more applicable.

A number of analytical models have been developed to treat fission-product deposition from flowing gas streams [13, 92-101]. A heat-mass analogy approach has been applied to obtain analytical relationships for the distribution of species between the gas stream and the walls of a conduit. Davies has discussed the deposition of particulates from flowing streams [101]. Epstein and Evans [94] have developed a model to describe the deposition processes in the coolant of a water-cooled reactor, but the principles described are applicable more generally. Isothermal conditions are assumed and the bulk fluid phase is assumed to be well mixed. Turbulent heat, mass, and momentum analogies are used. Transport of the fission products across a boundary layer at the wall is described and adsorption and desorption at the wall is treated. The steady-state case has been solved and deposition of both particles and gases is considered by Epstein and Evans. Gormley and Kennedy [13] treat the laminar flow case under isothermal conditions and assume a perfect sink condition at the wall. Ozisik has presented the steady-state case for both gaseous and particulate fission products [100] and has expanded this to include both a transient case [97] and a case with a temperature gradient at the collecting surface [96]. Under laminar flow conditions, the amount of fission products deposited decreases exponentially in the asymptotic region where the mass profiles are fully developed. By comparing the slopes of experimental and theoretical curves, diffusion coefficients for the molecules can be predicted if perfect sink conditions exist at the wall. Conversely, if the diffusion coefficient is known, the curve of penetration of fission products versus distance along a conduit can be predicted.

A model assuming radioactive decay and imperfect sink conditions at the wall has been developed by Venerus [98]. A slug-flow profile is assumed in this model and the equations are solved for steady-state conditions. This approach was modified by Ozisik and Neill [99] to include a parabolic velocity profile, radioactive decay, and imperfect sink conditions at the wall. Deposition as a function of time and distance was calculated. For the imperfect sink conditions, the slope of the deposition curve in the asymptotic region varied with time and radioactive decay. The equilibrium slopes of deposition versus distance are shown in Fig. 3-13.

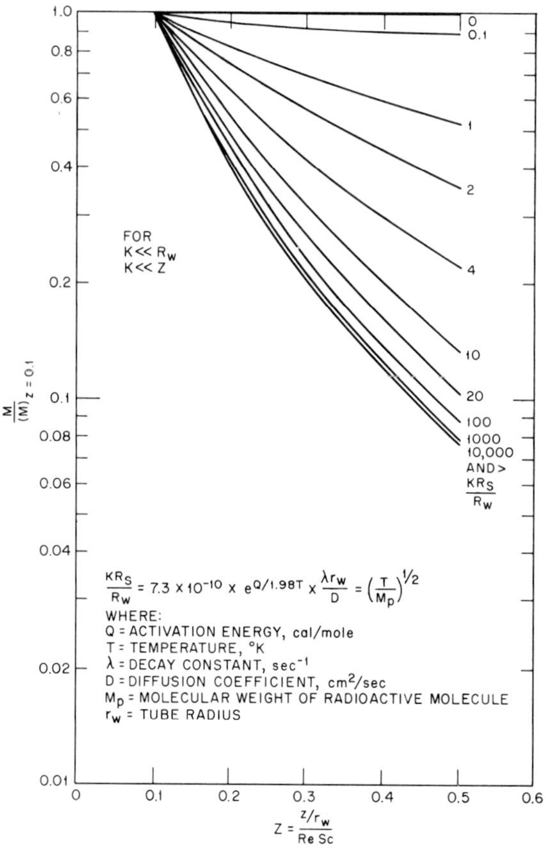

FIG. 3-13 Slopes of deposition curves in the range $0.1 \leq Z \leq 0.5$ for large times.

Raines et al. [93] have considered deposition of gaseous fission products from flowing streams in which the boundary-layer development was included. Perfect sink conditions were assumed at the wall. Deposition profiles that would be predicted under isothermal, turbulent flow conditions are shown in Fig. 3-14. This model has been used in the analysis of data from deposition experiments in helium at temperatures of up to 1200°F (650°C). Tellurium, cesium, zirconium, barium, and cerium appear to follow model predictions well, while iodine does not. At the temperatures of the experiments, the wall apparently is not a perfect sink for iodine.

The more pertinent models for the calculation of fission-product removal by natural processes from the containment under accident conditions are the well-mixed-volume models. In a loss-of-coolant accident in a water reactor, the blowdown itself should mix the containment atmosphere thoroughly and steam condensation should tend to keep it mixed. For these accidents, the assumption of a well-mixed volume appears reasonable. In liquid-metal fast-breeder reactors and in gas-cooled reactors, the case for a well-mixed volume cannot be so clearly drawn. Low-pressure coolant blowdown may not promote as much mixing in the containment of these types of reactors. However, in large containment vessels under accident conditions, temperature gradients would be expected to promote natural convection circulation of the atmosphere. The major difference between the water-cooled reactors and other types is that the description of fission-product behavior in the former type must consider steam condensation, while deposition in the latter types is likely to occur under relatively dry conditions. In either case the analytical treatment of a well-mixed-volume model is basically the same. The primary mechanisms involved are (1) transport of fission products to the neighborhood of a surface, (2) transfer of fission-product material across a boundary layer at the surface, and (3) interactions between the fission products and the surface.

Two well-mixed-volume models have been developed to describe fission-product deposition under dry conditions [53,102]. Both particulate

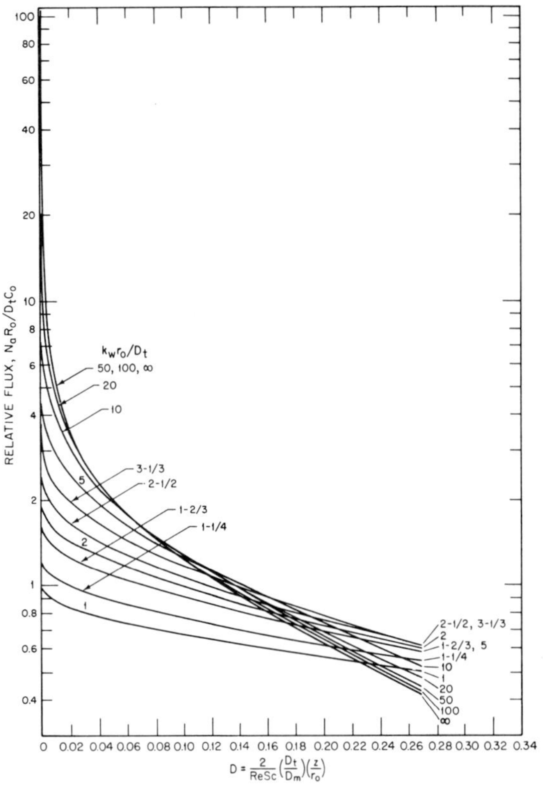

FIG. 3-14 Generalized theoretical deposition profiles with parametric variation of wall factor.

and gaseous fission-product deposition can be described by the appropriate selection of the transport coefficients. The two models are based upon similar assumptions that the fission products are uniformly mixed in the bulk gas and that deposition occurs by diffusion across a stagnant gas film on the boundary layer at the surface. The equations describing the deposition are [53]:

$$\frac{dC}{dt} + \frac{A}{V}\frac{dS}{dt} = 0, \tag{3-1}$$

$$\frac{dS}{dt} = \frac{k_G RT}{M}(C - C^*), \tag{3-2}$$

$$\frac{dS}{dt} = k_A C^* \left(1 - \frac{S}{S_T}\right) - k_D S, \tag{3-3}$$

where S = surface concentration, S_T = equilibrium surface concentration, C = airborne concentration in bulk gas, C* = airborne concentration at air-surface interface, t = time, A = surface area, V = volume of vessel, R = gas constant, T = absolute temperature, M = molecular weight of adsorbing fission product, k_G = mass-transfer coefficient, k_A = adsorption coefficient, and k_D = desorption coefficient. Over a small time interval τ, the ratio S/S_T can be assumed to be constant and the solution gives the ratio of the final airborne concentration (C_2) to the initial concentration (C_1) as

$$\frac{C_2}{C_1} = e^{-\alpha\tau}\left(\frac{\alpha e^{\lambda\tau}}{\alpha - \lambda}\right) - \frac{\lambda}{\alpha - \lambda}, \tag{3-4}$$

where $\alpha = Ak_G k_A^* RT/V(k_A^* M + k_G RT)$, $k_A^* = k_A(1 - \bar{S}/S_T)$, $\bar{S}$ = average value of S over the time interval τ, and $\lambda = -k_D k_G RT/(k_A^* M + k_G RT)$. The surface concentration is given as

$$S = \frac{VC_i}{A}\left(1 - \frac{C}{C_i}\right), \tag{3-5}$$

where C_i = initial airborne concentration and C = airborne concentration at any time t. The ratio S/S_T is then given as

$$\frac{S}{S_T} = \frac{1 - C/C_i}{1 - C_f/C_i}, \tag{3-6}$$

where C_f is the final equilibrium airborne concentration.

A simpler case exists when the surface concentration is very small and $S/S_T \ll 1$. For this condition, desorption can be ignored, $k^*_A = k_A$ = constant, and the calculations for airborne concentration can be made directly using Eq. (3-4) in the abbreviated form:

$$\frac{C}{C_i} = e^{-\alpha\tau}. \tag{3-7}$$

If diffusion across the boundary layer is assumed to be the rate-limiting process, Eq. (3-7) reduces to the solution presented by Watson [102]. Examples of comparisons among Watson's model predictions of deposition under dry conditions with limited surface coverage are given in Figs. 3-15 and 3-16. Figure 3-15 shows the typical behavior of the normalized iodine concentration-time curve at a given position in the NSPP. The mass-transfer coefficient in terms of concentration (k_C) and the dimensionless parameter describing the surface conditions (ξ) used to calculate the theoretical curve are given on the figure. To predict the results to be expected in CRI experiments, the mass-transfer coefficient from the NSPP experiment was scaled with vessel size and pressure, and the surface parameter was scaled using the proper surface-to-volume ratio. Comparison between the experimental data in CRI and the predicted results based upon the adjusted parameters from the NSPP experiments are shown in Fig. 3-16. It can be seen that the general behavior of the concentration-time curve can be predicted. Although the dry-wall deposition models have been applied only to describe iodine behavior, there is no reason why the models cannot be used to predict the behavior of other fission products. The gas-phase mass-transfer coefficients and the surface-reaction coefficients must be adjusted accordingly.

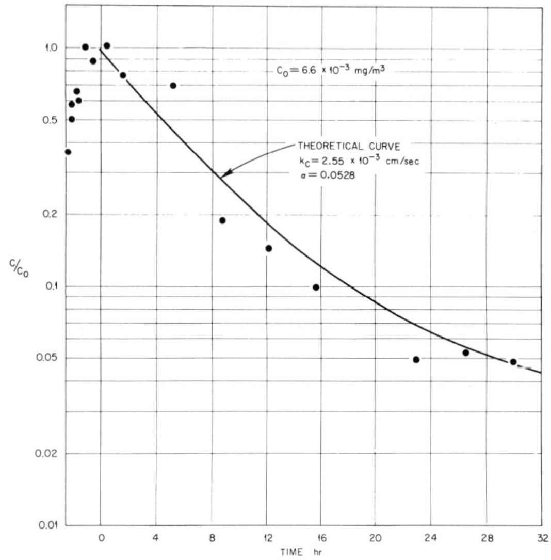

FIG. 3-15 Iodine concentrations vs. time for NSPP Run No. 6 (iodine-dry air) [102]. (The parameter α in the above figure is denoted by ξ in the text.)

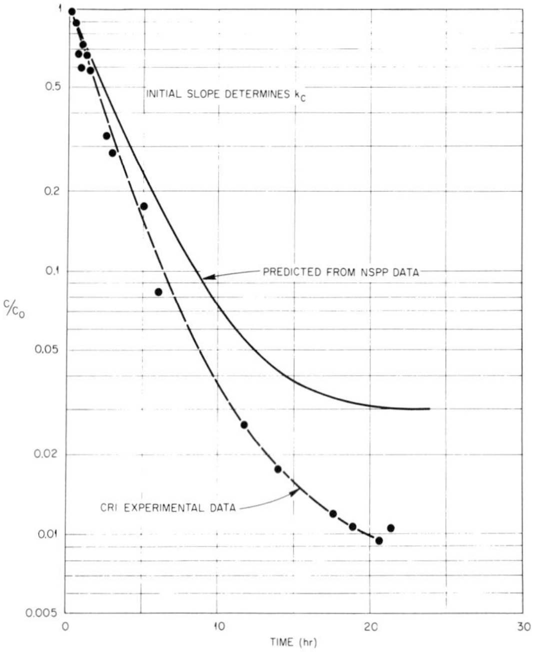

FIG. 3-16 CRI experimental iodine concentration and values predicted from NSPP data [102].

Several analytical models have been developed to describe fission-product deposition in steam-air atmospheres [53, 102-105, 90]. The most general of these models is the well-mixed-volume model derived by Ritzman et al. [103]. The behaviors of one gaseous and one particulate form of iodine can be treated simultaneously with this model in which airborne concentration, the amount deposited on the surface, and the amount in the condensate are predicted as a function of time. Operation of an engineered removal system can also be included through the use of an effective flow rate (Q_f) of containment atmosphere through the system and an efficiency factor, E. The equations describing fission product transport and deposition in this well-mixed-volume model are based upon several assumptions, mass balances, and flux definitions. These are the major assumptions:

(1) The atmosphere in the containment is well mixed.

(2) Steam condensation occurs equally on all surfaces and forms a water film of uniform thickness on all surfaces; this film runs off the surfaces at a rate equal to the condensation rate.

(3) The iodine concentration in the water film is time-dependent but at any time the concentration is everywhere the same.

(4) Iodine uptake by the solid surfaces is irreversible.

(5) The iodine in the airborne particles is immediately released to the water upon contact between the particle and the water film.

(6) Equilibrium exists at the air-water interface, and the iodine distribution at these conditions is given by a partition coefficient.

The general equations describing this system can now be developed. The flux terms are denoted by J_B, J_W, J_P, and J_D, see Eqs. (3-11) to (3-14). The mass balance on gaseous iodine is:

$$\frac{dC}{dt} - \frac{Q_i C_i}{V} + (Q_f E + Q_L)\frac{C}{V} + \frac{A}{V} J_B = 0 ; \qquad (3\text{-}8)$$

mass balance on particulate iodine:

$$\frac{C_i^* \pi d^3}{6}\left[\frac{dN}{dt} - \frac{Q_i N_i}{V} + (Q_f E^* + Q_L)\frac{N}{V}\right] + \frac{A}{V} J_w = 0 ; \qquad (3\text{-}9)$$

mass balance of iodine in liquid:

$$J_B + J_P = J_D + \frac{R' C_L}{A \rho_L} + \frac{V_L}{A}\frac{dC_L}{dt} ; \qquad (3\text{-}10)$$

flux from liquid to wall:

$$J_D = k_x C_L = \frac{dS}{dt} ; \qquad (3\text{-}11)$$

flux to liquid (particles):

$$J_P = \frac{C_i^* \pi d^3 D}{6\delta} N ; \qquad (3\text{-}12)$$

total flux of particles:

$$J_w = \frac{C_i^* \pi d^3}{6A} V\left(\frac{AD}{V\delta} + \frac{v}{h}\right) N ; \qquad (3\text{-}13)$$

flux to liquid (gaseous):

$$J_B = \frac{k_G RT}{M}(C - HC_L) ; \qquad (3\text{-}14)$$

where the symbols not previously defined are C_i = inlet airborne concentration (g/cm^3), C_i^* = inlet concentration in particles, mass/unit volume of particle (g/cm^3), C_L = concentration in liquid (g/cm^3), D = diffusion coefficient of particles (cm^2/sec), d = particle diameter (cm), E = filter or scrubber efficiency for gaseous species, E* = filter efficiency for particles, H = partition coefficient for iodine between air and water, h = effective height of containment vessel (cm), k_x = reaction-rate constant, with wall surface (cm/sec), N = particle concentration (particles/cm^3), N_i = particle concentration in inlet flow (particles/cm^3), Q_f = volumetric flow through filter (cm^3/sec), Q_i = volumetric inlet flow rate (cm^3/sec), Q_L = volumetric leak rate (cm^3/sec), R′ = condensation rate (g/sec), S = surface concentration (g/cm^2), V = volume of containment vessel (cm^3), V_L = volume of liquid film on wall surfaces (cm^3), v = terminal settling velocity of particles (cm/sec), δ = film thickness for particle deposition (cm), ρ_L = liquid density (g/cm^3). In addition to the above equations, initial conditions are needed. At t= 0, they are taken as

$$C = C_L = N = S = 0$$

$$C_i = \epsilon C_T$$

$$C_i^* = (1-\epsilon)C_T \bigg/ \left(\frac{\pi d^3}{6}\right) N_i ,$$

where ϵ = fraction of iodine in gaseous form.

An analytical solution for the differential equations of the model, with appropriate boundary conditions, has not been obtained. The model has been programmed for computer solution and the code is designated COVEDEP [103]. The equations are solved time-incrementally, with the results for the previous time increment setting the initial conditions for computation in the following time increment (unless otherwise specified by additional input data). This approach allows treatment of variable-rate processes for many of the important physical events which are incorporated within the model. Although COVEDEP offers considerable flexibility in being able to describe fission-product behavior in the containment under natural conditions, values for various parameters in the computer code, such as the steam-condensation rate, the liquid volume on the wall, and the mass transport coefficients, must be supplied as input.

Watson, Perez, and Fontana [102] have developed a slightly different approach for describing iodine behavior in a well-mixed volume under condensing conditions. Because there is a net flow of steam toward the walls, it is assumed that the iodine flux, which consists of diffusion and bulk-flow components, may be approximated by the bulk component alone. In this approach it is further assumed that the solubility of iodine in the steam condensate is high enough to permit the iodine and steam to condense together with the same composition as exists in the gas phase. Mathematical relations have been derived for the concentration of iodine as a function of time in terms of the surface-to-volume ratios, condensing steam fluxes, and steam concentrations. For the case of constant steam pressure in a steam atmosphere alone, Watson et al. [102] find that:

$$\frac{C}{C_0} = \exp\left[-\left(\frac{A}{V}\right)\left(\frac{J_0 RT}{P}\right)\right], \qquad (3\text{-}15)$$

where J_0 is the constant steam flux to the wall and P is the vessel pressure. For decaying steam pressure, the steam flux, J(t′), and the total gas concentration, n(t′), must be known or estimated as a function of time. The solution for this case is

$$\ln\left(\frac{C}{C_0}\right) = -\left(\frac{A}{V_0}\right)\int_0^t \frac{J(t')}{n(t')}\,dt' . \qquad (3\text{-}16)$$

Both the COVEDEP-model and the Watson-model predictions have been compared with Nuclear Safety Pilot Plant (NSPP) data [106]. The Watson model adequately predicts the airborne half-life for iodine at later stages of the experiment, but underestimates the removal rate of iodine to the wall at early stages of the experiment. The COVEDEP code, on the other hand, more closely agrees with the early-time data obtained during the NSPP tests, in terms of both the airborne half-life for iodine and its accumulation rate in the condensate. The basic differences between the two models reduce to the physical process that controls the iodine deposition in a water layer. Near the beginning of the experiment (and accident), when the wall is already wetted by condensed steam but prior to the iodine release, the deposition is controlled by the vapor-solution equilibrium rather than the vapor-condensed-phase thermal equilibrium. This early effect is neglected in the Watson model but considered in the COVEDEP model.

In addition to the two previously discussed well-mixed volume models used to describe fission-product behavior under natural-response condensing steam conditions, there have also been developed two models based upon the transport of fission products in gas phase by natural convection [104, 105, 90]. The basic assumption made by Yuill is that the process which limits the rate of deposition of material on the containment wall is the transport of material from somewhere in the center of the building to within the diffusion or collection distance of the wall. Another assumption made was that the gross effects of convectional patterns may be described by representing the containment building as a cylinder having the same radius and volume as the building. It was further assumed that the total deposition rate may be described by considering a plane having the same area as the surfaces in the cylindrical containment building. If the containment air is considered to be passing over the vertical plane representing the containment surface, deposition from the moving gas stream is given by

$$\frac{M_t}{M_\infty} = 1.13\sqrt{\frac{Dt}{l^2}}\,, \qquad (3\text{-}17)$$

where M_t = amount of diffusing species absorbed in time t, M_∞ = amount of diffusing species which would be absorbed in infinite time, D = diffusivity of species considered, t = time the air is in contact with the plane during one cycle past the plane, and l = geometric factor related to the thickness of the moving gas stream. The time t is simply the height of the equivalent cylinder divided by the velocity of the flow past the wall. Equation (3-17) is valid for small values of M_t/M_∞ and applies to laminar flow past a wall which is an infinite sink.

The maximum velocity, v_{max}, for free convection over the wall may be calculated by Eq. (3-18), which was derived by Prandtl:

$$v_{max} = 0.55\sqrt{g\beta\Delta TZ} \qquad (3\text{-}18)$$

where g is the gravitational constant, β is the thermal coefficient of density per degree, ΔT is the temperature difference, and Z is the height of an equivalent cylinder representing the containment building. The calculation of this velocity is based upon heat fluxes to the wall, which are available from containment-heat transfer codes such as CONTEMPT [107]. Yuill [104] tabulates physical constants which can be used to relate the heat flux to the velocity.

The time dependence of the airborne molecular iodine concentration can be calculated after the velocity is determined. If θ is the amount of time elapsed since the iodine was released, the fraction of iodine remaining airborne is

$$\ln\left(\frac{C}{C_0}\right) = -\frac{0.565\,\theta}{Zl}\sqrt{DZ'v}\,, \qquad (3\text{-}19)$$

where $Z' = A/2\pi r$ and A is the total surface area suitable for iodine plateout. Additional equations have been presented by Yuill to account for deposition under turbulent flow conditions and for the deposition of particulates.

Calculations have been made of the amount of iodine remaining airborne in the LOFT containment as a function of time after release, using the natural circulation model and the COVEDEP model. These two models predict similar behavior, especially during the earlier period. Four to five hours after the blowdown, the model predictions tend to diverge. It is significant that the two approaches agree during the early time portion of the accident, since the potential for leakage is high at that time and it is important to know the airborne inventory for safety assessments. Model predictions also have been compared by Yuill to CSE results. Half-lives between 13.0 and 15.7 min for iodine were predicted, while half-lives of 10-11 min were observed.

Hilliard and Knudsen [90] have presented slightly different methods for predicting the rate of transfer of gaseous fission products from steam-air atmospheres to containment-vessel surfaces. The principal advantage of this model is the ability to predict the gas-phase mass-transfer coefficient. Assumptions in addition to those used in earlier models are: 1) flow patterns are caused only by natural convection due to heat transfer to vertical walls, and 2) the steam-air mixture is saturated at all locations. The simplification is made that the walls are at thermal equilibrium, though the model can be extended to cover the unsteady state.

The correlation recommended by McAdams [108] for turbulent heat transfer by natural convection, modified for density gradients caused by mass transfer of steam, is

$$h_i = 0.13\,k_f\left(\frac{g}{\nu^2}\,\mathrm{Pr\,Sc}\right)^{1/3}\left(1 - \frac{M_bT_i}{M_iT_b}\right)^{1/2}. \qquad (3\text{-}20)$$

By analogy,

$$k_c = 0.13\,D_v\left(\frac{g}{\nu^2}\,\mathrm{Pr\,Sc}\right)^{1/3}\left(1 - \frac{M_bT_i}{M_iT_b}\right)^{1/3}. \qquad (3\text{-}21)$$

In the preceding equations, k_f is the thermal conductivity of the gas film, D_v is the diffusion coefficient, g is the acceleration due to gravity, ν is the kinematic viscosity, Pr is the Prandtl number, Sc is the Schmidt number, M is the molecular weight of the gas, and T is the temperature. The subscripts b and i refer to the bulk gas mixture and the inside surface, respectively. All the terms are known except T_i, the temperature at the interface. But T_i can be estimated by an energy balance:

$$\begin{aligned}\frac{g}{A} &= h_i(T_b - T_i) + \frac{k_c}{RT_b}\,\Delta H_{vap}(P_b - P_i)\\ &= h_w(T_i - T_0)\\ &= h_0(T_0 - T_a)\,.\end{aligned} \qquad (3\text{-}22)$$

For stated conditions of bulk gas temperature (T_b), surrounding air temperature (T_a), and wall coefficient (h_w), the value of T_i can be calculated. Equation (3-21) is then used to calculate the mass-transfer coefficient for elemental iodine or any other diffusing species for which the effective diffusivity is known. The gas-phase concentration calculated for a puff release, assuming gas-phase diffusion is limiting, is

$$\frac{C_g}{C_0} = \exp\left[-\left(k_c\frac{A_T}{V} + k_s\frac{A_W}{V}\right)t\right]. \qquad (3\text{-}23)$$

where C_g is the gas-phase iodine concentration at any time t, C_0 is the gas-phase iodine concentration at t = 0, k_c is the gas-phase iodine mass-transfer coefficient due to diffusion, k_s is the iodine mass-transfer coefficient due to condensation, A_T is the total internal surface area, A_W is the heat transfer surface area, and V is the gas volume.

Calculations for different cases of h_w, T_b, and T_a showed that k_c and k_s are relatively independent of these parameters over the range of interest. Also, k_c is about ten times greater than k_s. Thus, the concentration half-time depends chiefly on the total surface-area-to-volume ratio.

Values for the gas-phase mass-transfer coefficients as a function of the temperature difference, $T_b - T_i$, are given in Fig. 3-17 [109].

The predicted values of heat- and mass-transfer rates are compared in Table 3-4, with experimental values from tests in two sizes of vessels. The COVEDEP model also has been used to predict the results of experiment D-1 [110]. The predicted half-life for the disappearance of iodine was 7 min, which is in substantial agreement with the experiment and the model predictions of Hilliard. Hilliard and Knudsen believe that improvement could be made to the natural convection model if either: (1) liquid reaction rates and diffusion were considered, or (2) allowance was made for the portion of surface area which may have laminar boundaries. Prediction using the natural convection model for a large PWR with 4-ft-thick concrete walls and A_T/V of 0.2 ft^{-1} gave 23 min for the concentration half-life of elemental iodine. This rate holds until the concentration is about 1.5% of its initial value.

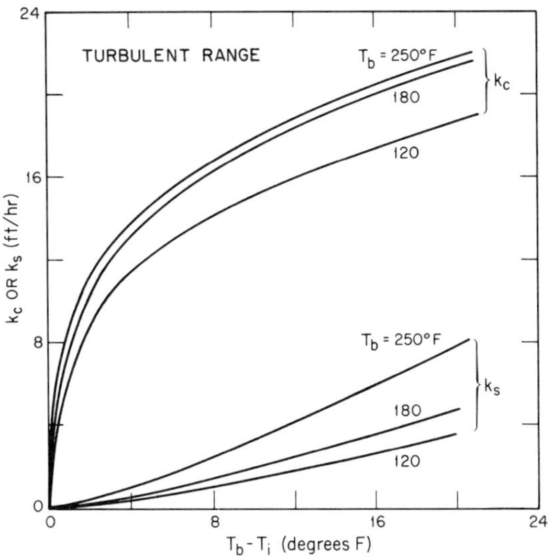

FIG. 3-17 Gas-phase mass-transfer coefficients [109].

Other analyses that are of note in predicting fission-product behavior under natural response conditions are those of Ozisik [111], which predict the steady-state flux of molecular sized matter from a mixture of vapor and noncondensible gas to the walls of a large containment vessel during condensation of the vapor.

The analytical models previously described do not treat particulate behavior in detail. Since some additional factors require consideration in particle transport and deposition, an elaboration on particulate behavior is made in the following paragraphs.

Particle transport and deposition during a reactor accident depend upon the properties of the particles and upon a number of environmental factors. Depending upon the reactor type, the accident, and the point of fission-product release and particle formation, transport and deposition can occur in gas and liquid environments that range from stagnant through well-stirred to turbulent flow conditions. Gaseous environments may include dry as well as condensing conditions. Under this spectrum of environmental conditions, reactions between gaseous fission products and the particles and agglomerations of particles must be considered. The nature of the particles, their sizes, size distributions, and chemical composition will greatly influence the surface reactions. The hydrophobic (water-repelling) or hydrophilic (water-attracting) nature of the particles or their solubility in water will affect the steam condensation on them.

Although a wide range of conditions and a combination of factors influencing particle behavior are possible, two broad categories of conditions appear most important to reactor safety: The behavior of particulates in the condensing-steam environment following a loss-of-coolant accident in a light-water reactor, and the behavior of

TABLE 3-4.

Model Predictions vs. Experiments in the CSE [90]

Run Number	$T_b - T_i$ °F	Steam Flux lb/hr-ft^2	k_c (I_2) ft/hr	k_s ft/hr	$T_{1/2}$ (I_2) min
D-1[a] Prediction	7.2	0.26	16.0	2.2	5.0
Experiment	(c)	0.375	(c)	(c)	8±1
D-2[a] Prediction	7.2	0.25	16.0	2.2	5.0
Experiment	(c)	0.340	(c)	(c)	10±1
A-1[b] Prediction	7.8	0.10	16.2	1.5	8.7
Experiment	(c)	0.096	(c)	(c)	16±2
A-2[b] Prediction	7.8	0.10	16.2	1.5	8.7
Experiment	5.4±2.0	0.11	(c)	(c)	10±1

[a] CSE drywell, $A_T/V = A_W/V = 0.46$ ft^{-1}, $h_w = 20$ Btu h^{-1} ft^{-2} °F^{-1}, $T_b = 250$°F, $T_a = 110$°F.
[b] Main CSE vessel, $A_T/V = 0.293$ ft^{-1}, $A_W/V = 0.19$ ft^{-1}, $h_w = 38$ Btu h^{-1} ft^{-2} °F^{-1}, T_b = 180° F, $T_a = 110$° F.
[c] Not measured or not applicable.

aerosols resulting from an accident in a sodium-cooled reactor system. The former category has been discussed by Davis [51] and Horst [52], and the latter category has been discussed by Reist [112].

In the light-water reactor, the containment system atmosphere consists of a steam-air mixture contaminated by radioactive gases and aerosols. The bulk atmosphere quickly comes to equilibrium with the unevaporated coolant and hence is a saturated mixture. Condensation of steam occurs when this saturated atmosphere has access to a cooler surface, such as the walls of the containment vessel. As soon as a film of liquid condensate forms and a quasi-steady state obtains, the conditions adjacent to the wall are also those of saturation but at a lower temperature than in the bulk of the steam. Consequently gradients of temperature and composition (ratio of steam to air) are established normal to the surface.

These temperature and composition gradients are the direct cause of forces on the aerosol particles. A particle in a temperature gradient experiences a force proportional to the gradient which drives the particle away from the warmer region. Likewise, a particle which is located in a region in which diffusion is taking place is driven with the molecular flux by a force proportional to the partial pressure gradient. These phenomena are known, respectively, as thermophoresis and diffusiophoresis.

The temperature and composition differences between the bulk of the mixture and that close to the wall also imply differences in the density of the mixture. This density variation causes buoyancy forces and gives rise to natural convective flow down the walls. The convective flow contributes to the mixing in the containment vessel and enhances the supply of aerosol near the wall. Further, when this flow becomes turbulent it can be the source of an added deposition mechanism.

It is conceivable that the profiles of temperature and humidity are such that a supersaturated condition can occur in the boundary layer between the wall and the bulk of the system. If supersaturation occurs, condensation might also take place on particles present in this region. Even undersaturation could result in limited condensation on particles, provided that they are soluble. Any increase in particle size and mass due to condensation can have an appreciable effect on the subsequent trajectory of the particle. Its gravitational settling velocity would increase. Due to its increased inertia, the particle would move in a path of smaller curvature when accelerated. Inertial deposition to structures and spray drops would thus become more prominent, particularly if turbulent conditions were present.

In addition to natural condensation on the containment walls, engineered safety systems designed to suppress pressure buildup also cause appreciable condensation. Condensation on spray drops alters their efficiency for collecting aerosol particles by increasing the drop size and by providing additional collection forces. Condensation of steam-air mixtures upon injection into water pools can rapidly demobilize particles carried along with the mixture. Ice surfaces provide a situation similar to that already discussed for the containment walls, the primary difference being a lower wall temperature.

The basic processes in the liquid-metal fast-breeder reactor case are the same with the possible exception of steam condensation. Particles are subjected to the forces of gravity, diffusiophoresis, thermophoresis, and forces of an electrical nature. The major difference between the two categories of accidents is the controlling mechanisms for particle behavior. Diffusiophoresis and thermophoresis as a result of the steam condensation, and coagulation and settling enhanced by the wet conditions are expected to be the controlling factors in the water reactor case, while rapid agglomeration and settling under dry conditions are expected to be the most important factors in the LMFBR case [112].

The motion of a particle due to its diffusion is a random process and becomes ordered only when a large number of particles is considered. When the latter is the case, the force acting on the particles can be attributed to a gradient in the particle concentration. The random motion becomes ordered in the sense that the net movement of particles is away from the region of high concentration, and the net flux of particles due to diffusion is governed by Fick's law. Diffusion coefficients for particles vary substantially with particle size, and diffusion processes are usually insignificant for particles $>0.1\mu$ in radius [112]. Under accident conditions, diffusion processes for removal of particulates at surfaces in the containment are of minor importance compared to the other phenomena. However, Brownian diffusion may be important for the scavenging of submicron particles by spray drops [113].

A review by Waldmann and Schmitt [114] covers the theoretical and experimental investigations of both thermophoresis and diffusiophoresis up to 1965. For reactor accident considerations, the free-molecular regime is the most important dynamic regime for treating diffusiophoresis and thermophoresis. In this regime the particle is so small or its mean free path so large that the particle has no influence on the velocity distribution of the fluid molecules. Both diffusiophoresis and thermophoresis are analyzed in the free molecule regime by calculating the momentum from molecular collisions that is transferred per unit of time to the particle. The gas molecules are assumed to have the Chapman-Enskog [115] velocity distribution. Waldmann [116] and Bakanov and Derjaguin [117] independently performed these calculations. For the special case of a vapor (1) diffusing through resting air (2), the mass flux of the vapor is equal to the condensation rate for a steam-air condensation system in the steady state. The velocity, V_D, of the particles in the concentration gradient is given by [118]:

$$V_D = -\frac{(m_1)^{1/2}}{X_1(m_1)^{1/2} + X_2(m_2)^{1/2}} \frac{D}{p_2} \frac{dp_1}{dz}, \quad (3\text{-}24)$$

where D is the diffusion coefficient for the diffusing vapor in the gas, m_1 and m_2 are the molecular weights of the vapor and the gas, respectively,

X_1 and X_2 the mole fractions of the vapor and air, p_1 and p_2 the partial pressures, and dp_1/dz is the partial pressure gradient of the vapor. It is apparent that velocity is independent of particle size in the free molecular regime.

If the diffusing molecules are water and the concentration gradient dC/dz is expressed in concentration units ($g/cm^3/cm$), then a deposition velocity, V_g (cm/sec), can be derived [118]:

$$V_g = -k' \frac{dC}{dz} = -257 \frac{dC}{dz}. \quad (3\text{-}25)$$

Davis [51] has used the deposition velocity to calculate the total weight of water condensed in the containment, and he further has related the weight of water to the latent heat of condensation, H_l, and the latent heat of condensation per unit mass, H_L, to yield an expression for the fractional loss of particles (n/n_0) from the gas phase:

$$\log\left(\frac{n}{n_0}\right) = -\frac{384\, H_L}{\Delta H_L} V = -0.170 \frac{H_L}{V}, \quad (3\text{-}26)$$

where ΔH_L is 2.26×10^3 watt-sec/g and V is the containment volume in cm^3.

The mathematical descriptions for thermophoresis are similar to those for diffusiophoresis, with the inclusion of a term for the thermophoretic force which depends upon the temperature gradient [113]. The velocity is again independent of particle size and is in the direction of decreasing temperature. For a condensing steam-air system this would be a velocity towards the cool wall. Horst [52] has evaluated the velocity equation for the condition that the steam-air system is everywhere saturated and compared the thermophoretic velocity to the diffusiophoretic velocity. Thermophoresis was found to be negligible for transport of particles to a condensation surface for cases of interest in water-reactor safety. Reist [112] has concluded that thermal deposition may be initially important in an LMFBR containment and could contribute significantly to rapid removal of material. Its effects should then diminish to the point where they can be neglected.

Experiments have confirmed the respective theories for the free molecule regime by agreeing with the predicted results within 10% or less. Schmitt and Waldmann [119] measured the diffusiophoretic velocities of oil drops in several gas mixtures and got good agreement with theory. They used a modified form of Millikan's oil drop apparatus for their experiments. Schmitt [120] also found thermophoretic velocities of oil drops (low conductivity) which agreed with free molecule theory. Goldsmith and May [118] obtained excellent agreement with theory for the diffusiophoretic velocity of an aerosol produced by passing air or oxygen over a heated nichrome wire. They performed experiments for water vapor diffusing through air and helium, measuring the velocity through the deposition it caused on a plate normal to the concentration gradient.

It is necessary to consider coagulation and settling together since settling rates depend strongly upon particle size, and coagulation is the process which determines the particle size. Although the two processes are not separable physically, it is convenient to treat the two separately mathematically. The description of the two processes presented by Davis [51] will be followed.

The basis for settling calculations is Stokes' law, in which the force of gravity on a particle (mg) is balanced by the fractional force of viscous drag ($6\pi\eta r V_f$) For a spherical body of radius r moving in a continuous medium of viscosity η, the velocity of fall, V_f, is given by

$$V_f = \frac{mg}{6\pi\eta r} = \frac{2r^2\rho g}{9\eta}. \quad (3\text{-}27)$$

The assumption of a continuous medium breaks down when the particle diameter is on the order of or smaller than the free path of molecules. This problem is usually handled by a correlation factor to Stokes' law: the Cunningham slip coefficient, Cu. In these cases,

$$V_f = \frac{2r^2\rho g(\mathrm{Cu})}{9\eta} \quad (3\text{-}28)$$

Values for the Cunningham correction factor can be obtained from standard references on aerosols, such as Fuchs [54]. A curve showing the correction factor under room temperature conditions is given in Fig. 3-18. Particles with diameters of 0.1μ fall twice as fast as Stokes' law prescribes; 0.01μ particles fall 15 times as fast.

The assumption in Stokes' law that a falling body is a sphere with density ρ describes an ideal condition. For complicated chain-agglomerate particles, shape corrections should be applied. Some shape factors have been calculated for long rods and elliptical platelets [54], but in general it is not clear how to handle the shape factor and density of chain agglomerates. This factor is ignored and accepted as an uncertainty in calculations of particulate settling rates.

Particle settling in a well-stirred tank (i.e., stirred settling) can readily be calculated from Stokes' law. The concept is that the aerosol is

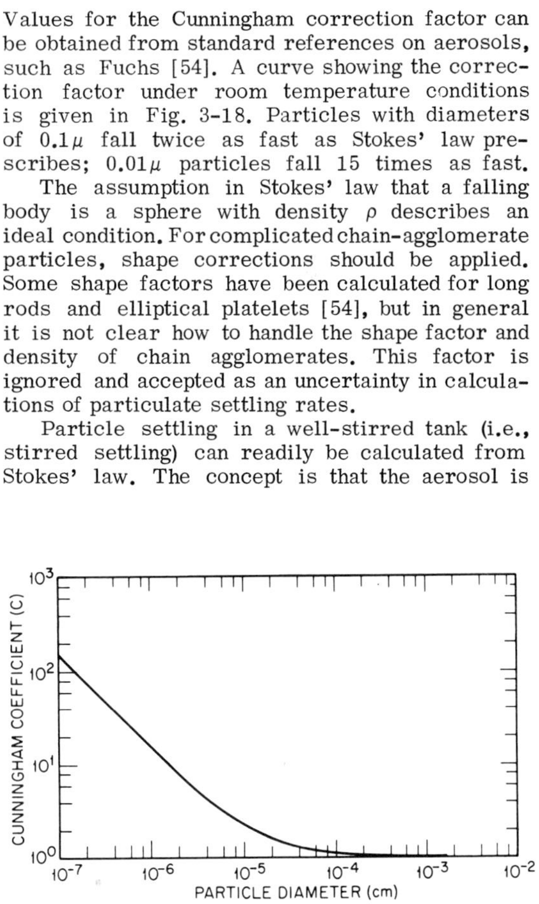

FIG. 3-18 Cunningham coefficient at room conditions [51].

well stirred so that the concentration is everywhere the same, except in the boundary layer along the floor of the tank. There is no mixing in the boundary layer and any particle which crosses into the boundary layer settles. For a cylindrical tank of height h, the fractional loss of particles (n/n_0) from the gas by settling as a function of time satisfies

$$\log \frac{n}{n_0} = -\frac{2r^2 \rho g t\,(Cu)}{2.303\,(9\eta h)}. \qquad (3\text{-}29)$$

Aerosol particles, regardless of their composition, most generally coalesce or coagulate if they come in contact, and the process goes on continuously so that the aerosol particles increase in size. Thermal motion commonly provides the basis for the collisions. As coagulation proceeds, the particle-number concentration decreases. The rate of change of the number concentration, $-dn/dt$, is equal to the rate of collision, which is proportional to n^2. Hence

$$-\frac{dn}{dt} = k'' n^2. \qquad (3\text{-}30)$$

From kinetic theory it can be shown that $k'' = 4kTC/3\eta$ for particles of one size greater than 0.2μ. It follows then that the number concentration n at time t is related to the number concentration at time zero, n_0, by

$$\frac{1}{n} = \frac{1}{n_0} + \frac{4kT(Cu)}{3\eta} t. \qquad (3\text{-}31)$$

Simultaneous consideration of coagulation and settling has been handled from first principles by means of long computer codes [121, 122]. Davis [51] has adopted a simple approach which uses the concept of stirred settling and assumes the self-preserving size distribution function. His idea is that settling occurs according to the stirred settling concept and coagulation occurs in such a way as to reestablish the self-preserving function. The self-preserving size distribution function is given by

$$\frac{dN}{dr} = 0.05\,\phi\, r^{-4}, \qquad (3\text{-}32)$$

where dN/dr is the distribution function, r is the particle radius, and ϕ is the volume fraction (volume of particles per unit volume of aerosol). There are theoretical bases to support the notion that an aged aerosol approaches this size distribution function as a steady state. There are experimental results which substantiate this notion for atmospheric aerosols, and some data indicate that agglomerated stainless steel oxide particles roughly achieve this distribution function, which implies that there are maximum and minimum particle sizes.

Description of particulate behavior when the mechanisms of diffusiophoresis, thermophoresis, coagulation, and settling are all acting simultaneously has been treated by Davis [51]. In this development, it is presumed that each mechanism operates independently and hence that the airborne fraction resulting from all mechanisms is simply the product of the airborne fractions from each individual process. Results of an illustrative calculation of the airborne volume fraction versus time are presented in Fig. 3-19.

Two factors that may also have a marked effect on particle behavior are the adsorption of water onto the particles with a consequent change of shape, and the condensation of water onto particles to the extent that they grow enough to settle substantially faster [51]. Water can affect the shape of agglomerate particles by making them more compact. This could increase penetration through filters. Condensation will occur most readily on soluble particles and hydrophilic particles which can be postulated to form during an accident. Since the settling velocity depends upon the square of the particle radius [see Eq. (3-29)], an effective increase in particle size by condensation could result in a greater contribution of this mechanism to particle removal.

4. ENGINEERED SYSTEMS FOR FISSION-PRODUCT REMOVAL

4.1 Introduction

While all of the fission products, with the exception of the noble gases, will eventually be removed from the containment atmosphere by natural processes as described in the preceding section, the natural removal mechanisms are inherently slow. If a major release of fission products to the containment occurs, it is desirable to reduce the potential for public exposure as rapidly as possible. This can be accomplished by means of two interrelated operations: reduction of the driving force for containment leakage, and reduction of the airborne inventory of fission

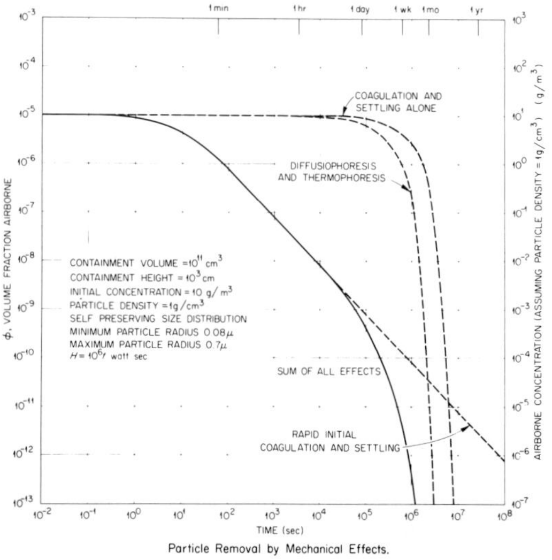

FIG. 3-19 Particle removal by mechanical effects [51].

products. Pressure reduction systems, which provide means to accomplish the former, are a part of current containment designs. With minor modifications, these same systems can be used to remove the fission products (except noble gases) from the containment atmosphere. In addition to reducing the airborne inventory rapidly, the engineered removal systems provide positive control over the fission products. Several engineered removal concepts have been developed and can be used individually or in combination. These include containment spray systems, high efficiency filter and charcoal adsorber systems, pressure suppression pools, reactive containment coatings, foam encapsulation, and diffusion boards.

Air cleaning is used in two ways as an engineered safety feature in water-cooled reactors: (1) recirculating systems in the containment shell reduce the concentration of airborne fission products in the containment atmosphere after an accident and thereby reduce fission-product leakage from the containment system; and (2) once-through (single-pass) systems in the secondary containment building or volume collect and retain any fission products that have leaked from the containment shell (or even directly from the primary cooling system, as in some possible accidents in boiling-water reactors). Use of a once-through system reduces dispersal of fission products to the environment and simultaneously relieves pressure buildup in the secondary containment structure by releasing the air or other gas to the atmosphere.

For the design of air-cleaning systems as an engineered safety feature, a loss-of-coolant accident followed by at least partial core meltdown has been widely used as the design-basis accident, although other types of accidents are also considered. Important considerations in the use of air-cleaning systems are their effectiveness, their reliability, and their resistance to radiation, thermal transients, abnormal pressures, pressure surges, missiles, and corrosion under accident conditions.

The design of the air-cleaning system or systems for a given reactor is integrally related to the design of the reactor's containment system. The containment system can be defined as the reactor containment structure, its associated engineered safety systems, and the components that are provided to maintain its integrity. The basic envelope that surrounds a reactor may be one of many types. Those predominant in the power reactor field in the United States are steel pressure shells and various types of concrete structures with steel liners. These structures are provided with various penetrations, including equipment and personnel air-locks, electrical and instrument penetrations, and piping penetrations, together with their associated isolation valves. The penetrations are carefully designed to maintain the integrity of the system.

While other engineered safety features are provided to attain atmospheric pressure as rapidly as possible, and even though the design leakage rates of containment shells are low, air-cleaning systems further reduce fission-product escape. If other engineered safety features are only partially effective, or if the containment leakage rate is higher than the design value, it is especially important for the air-cleaning systems to function.

Spray cooling systems, recirculating air-cooling systems, and other heat-removal systems included as engineered safety features in containment shells are designed to reduce the pressure and temperature of the postaccident containment atmosphere as quickly as possible and thus minimize the release of fission products to the environment. Currently designed systems for removing fission products from the atmosphere of the containment shell are in most cases combined with the containment shell cooling systems by the addition of filters and adsorbers to the recirculating air-cooling systems and/or the addition of a chemical to the containment spray cooling system.

The U.S. AEC Advisory Committee on Reactor Safeguards has summarized the philosophy on air cleaning systems as engineered safety systems as follows [123]:

"The function of an air cleaning system as an engineered safeguard is to remove and to retain fission products from an unlikely partial or total fuel meltdown. Fission products that are thus retained should be fixed in a form that prevents redispersion. Components of the air cleaning system should be so located that decontamination and essential handling can be accomplished readily and without hazard to the health and safety of the public.

"For cleaning or decontamination purposes, the released fission products from a reactor fuel meltdown may be divided into four groups. These are: the noble gases, krypton and xenon; the halogens, bromine and iodine; volatile solids, such as tellurium, selenium, cesium and ruthenium; and other solids, primarily strontium, yttrium and barium.

"Because of their chemical nature and short half-life, radioactive noble gases can usually be treated only by containment or by controlled release from elevated locations such as tall stacks. Therefore, with noble gases consideration must be given to meteorological dispersion and dilution as influenced by characterization of the surrounding environment.

"It is convenient to divide the remaining fission products into two physical groups: gases and particulates. Gases (essentially iodine and bromine in elemental form) are removed by adsorbents such as activated charcoal, by chemisorption on silver, or by absorption in a reactant solution. Particulates which range in size from several microns down to less than 0.1 can be removed by impingement, scrubbers, electrostatic precipitators or filters. The final device in a particulate cleaning train is usually one based on mechanical filtration principles. Containment spray or 'dousing' systems for condensing steam may also serve as decontamination systems, because of the gas-contacting and impingement action of the spray droplets. The multiple contacting which is possible within a contained gas volume makes a containment vessel with sprays equivalent to a scrubbing chamber.

"Halogen gases may be adsorbed upon the surfaces of released particulates and may react with them. Hence, it is necessary to use a combination of adsorption, absorption and filtration devices to remove halogens. Because iodine may occur in both inorganic and organic states, the gas cleaning system must be capable of removing both. Since the halogens adsorbed on particulates are not irreversibly bound, it is necessary to follow the filter with an adsorber. A liquid scrubber should be followed by both when maximum decontamination is necessary.

"The air cleaning components of a reactor safety system include: a ventilation system, a heat-removal device, air cleaners and an air mover with motor. The system must be capable of working continuously in hot, saturated steam environments for a period of time long enough to remove the required portion of the released fission products from the containment or confinement system. To handle the anticipated release, the air cleaning system must have sufficient capacity in flow, in adsorbent and chemically reactive materials, and in filtration surface. Adsorption and filtration systems must be designed and installed so that the decay heat of collected fission products will not cause combustion or destruction of their media or overheating to the point where collected fission products will be redispersed. The media must also be protected against shock waves, missiles, moisture entrainment, liquid slugging and radiation damage, as well as corrosion and chemical attack. The duct work and filter housing should be protected against mechanical injury or missiles damage to avoid bypassing or leakage of untreated air. The system should be leak tested at the same pressure differential as it would have to endure under accident conditions.

"Because electrical power is necessary for circulating and recirculating both air and water, it is essential that backup power be available for maintaining minimal flow rates.

"The decontamination efficiency required of an air cleaning system will depend on whether the system is once-through or recirculating. The decontamination factors needed will be based on the dose to the environment and the dilution to be assumed for stack dispersion. In recirculating systems, the decontamination factor is related to the number of containment volumes passed through the cleanup system. Decontamination factors of 10 to 1000 or more may be required in most applications.

"Gas leaks which bypass filters or adsorbers in effect decrease air cleaning efficiency. When an iodine removal efficiency of 99% is projected, a bypass or leak around the beds of 1% leads to double the iodine release. Appropriate design and testing of associated gas handling equipment is required.

"The reliable performance of an air cleaning system must be assured by frequent 'in-place' testing which includes monitoring with gases and particulates that simulate the expected fission products. Ease of testing for leaks, and access for inspection of seals, gaskets and clamps are necessary. A continuous monitor of resistance or pressure drop through the cleaning system is desirable where the decontamination unit is always in use. In the case of emergency or standby cleaning systems, at least quarterly operation and checking of both air mover and cleaner is desirable.

"The nature of fission product releases to be expected in the unlikely event of a major accident is not yet well enough known to permit more than conservative lower bounds on the efficiency to be determined for air cleaning. Reliable lower bounds may, however, be assumed when individual cases are reviewed."

The functions of equipment employed in the United Kingdom reactors for filtering gaseous effluents are described by Smith and Bainbridge [124]. The objective is to limit the radioactive material released so that neither the operators nor the public suffer any harm. The major gases to be filtered are air, used for cooling the biological shield, and CO_2, used as coolant in the reactor and reactor service machinery. The shield-cooling air system is such that even if the particulate activity is greater than expected, the filtration equipment ensures that the air can be safely released to the atmosphere. Filtration of the CO_2 in a closed circuit will ensure that the coolant gas is kept clean during normal operation, so that occasional discharges through filters can be safely made. Filtration-absorption equipment is also provided to enable operators to deal with very unlikely, though credible, conditions of the coolant circuit, so that releases of fission products to the atmosphere can be severely limited.

The most important principle to recognize in developing an effective air cleanup system, whether it be a dousing system, recirculating spray, recirculating filter and a charcoal or other adsorbent bed, or even a foam encapsulation system, is that the intent is to provide additional reactive or removal surfaces, beyond the shell and exposed components within the containment, for the collection of fission products. This extra surface area must be provided rapidly after the accident and within a known time period, assuming that surface materials with a predictable behavior are available and that they interact with a gaseous-particulate system in a predictable manner. A system interaction of an unpredictable nature or time scale cannot be analyzed, so it must be assumed in a safety analysis report that its effectiveness in reducing the fission-product content in the gaseous phase is zero.

In order to consider the effectiveness of cleanup or decontamination systems within a containment on an analytical basis, the following assumptions must be made:

1. The cleanup system removes 100% or a known percentage, k, of the contaminant that passes through or contacts it.

2. In the passage of the contaminant through the cleanup system or in the volume of the containment system, an irreversible reaction takes place on contact with the surface. There is no driving mechanism such as vapor pressure or diffusion causing subsequent re-release from the collecting surface. (In the case of iodine vapor, this assumption in practice requires that a reactant be present at the surface to change it to a nonvolatile compound.) This assumption may be violated in some systems by the organic iodide compounds which have been observed to form.

3. The particulates can be considered to be fairly uniform in size, shape, and composition and to remain suspended during the period of recirculation. If this assumption is not obeyed, the particles will be deposited anyway, but the calculational estimates cannot be so accurate.

4. The suspended matter is completely mixed throughout the containment at all times.

The cleaning or removal rate of contaminants from the containment volume will depend upon the deposition rate as well as upon the recirculating-air flow rate; the efficiency of the collector; the settling rate of particles (v_t) (their agglomeration rate may also be considered in the same time interval); the wall loss, deposition or reaction rate with surfaces and particles (v_d); the height of the containment vessel (h) corresponding to the point of highest mixed volume; the rate of leakage to outside or, in a confinement system, the rate of inflow (Q_1), as well as the rate of outflow (Q); the initial particle concentration in the vessel (C_o) (C is concentration/cc); the initial gas concentration (C_v); and the containment volume (V) and its surface area (A).

Based upon the assumptions stated above, some simple analytical relationships can be derived which are useful for calculation of the amount of fission products remaining airborne in the containment and leakage as a function of time. Three cases covering different time dependencies of release to the containment are of interest. If all the release to the containment occurs instantaneously or over a period of time short with respect to the other processes, it is treated as a puff release. If the release occurs continuously throughout the accident and if the removal systems also operate continuously, it is designated as a continuous release case. If a release to the containment occurs over a relatively long period of time and then is terminated but the removal systems continue to operate, it is defined as a discontinued release case. Using the above assumptions, the equation describing the fission-product behavior in the containment is

$$\frac{dC}{dt} = -\lambda_N C - \lambda_R C - \lambda_L C , \qquad (4\text{-}1)$$

where C = airborne fission product concentration at any time in the containment, mass/unit volume; λ_N = rate constant describing natural removal processes, time^{-1}; λ_R = rate constant describing engineered removal processes, time^{-1}; λ_L = rate constant describing the leakage, time^{-1}. The rate constants are assumed to remain constant throughout the accident. The equation is generally applicable to all forms of fission products if the rate constants are defined appropriately and there is no interconversion of one form into another.

For the puff release case, C = C_o at t = 0. Thus,

$$\frac{C}{C_0} = e^{-\lambda t} , \qquad (4\text{-}2)$$

where λ is the sum of the rate constants for all of the processes. Both C and C_o can be multiplied by the volume V to yield the total amount of fission products, M and M_o, where M and M_o may be expressed in moles, grams, curies, etc. So for the puff release case, $M/M_o = e^{-\lambda t}$.

Values for the rate constants for natural removal can be obtained from experiments or from the analytical models described in Sec. 3 (e.g., Hilliard's and Yuill's models). For a spray system using a reactive solution so that gas-film resistance controls the removal rate, the rate constant for removal defined in Eq. (4-17) can be used. In a recirculating filter-charcoal absorber system, $\lambda_R = EQ_f/V$, where E is the trapping efficiency and Q_f is the flow rate through the filter. The rate constant for the leakage can be obtained directly from the leak rate.

For the continuous release case,

$$\frac{M}{M_t} = \frac{1}{\lambda t}[1 - e^{-\lambda t}] . \qquad (4\text{-}3)$$

where M_t = the total amount of fission products injected into the containment. Two time periods must be defined for the discontinued release case. Let θ be the time for a constant fission-product injection rate with the engineered removal system operating, let τ be the time from the end of the injection period, also with the engineered removal system operating, and let T = $\theta + \tau$. Then for the discontinued release case,

$$\frac{M_T}{M_\theta} = \frac{1}{\lambda\theta}[1 - e^{-\lambda\theta}][e^{-\lambda\tau}] . \qquad (4\text{-}4)$$

If the leak rate remains constant during the accident and is negligible compared to the removal processes rate, the amount of the fission product, M_L, that leaks from the containment in time t for the puff release case is

$$M_L = \int_0^t \lambda_L M_0 e^{-\lambda t} dt . \qquad (4\text{-}5)$$

Hence the fractional leakage for the puff release case is

$$\frac{M_L}{M_0} = \frac{\lambda_L}{\lambda}[1 - e^{-\lambda t}] , \qquad (4\text{-}6)$$

the fractional leakage for the continuous release case is

$$\frac{M_L}{M_t} = \frac{\lambda_L}{\lambda^2 t}[\lambda t + e^{-\lambda t} - 1] , \qquad (4\text{-}7)$$

and the fractional leakage for the discontinued release case is

$$\frac{M_L}{M_\theta} = \frac{\lambda_L}{\lambda^2\theta}[\lambda\theta - e^{-\lambda\tau} + e^{-\lambda T}] . \qquad (4\text{-}8)$$

Examination of Eq. (4-6), (4-7), and (4-8) reveals that fractional leakage from a containment vessel is primarily an exponential function of time. For long periods following the release of fission products to the containment, the limiting value of the fractional leakage depends upon the ratio of the leak rate, λ_L, to the sum of the rate constants for all of the removal mechanisms within the containment, λ. The time required to reach the limiting value of the leakage is governed by the removal processes in the containment. A few examples may be considered to obtain an appreciation of the concepts involved in these equations. If a design leak rate of 0.2 vol. %/day is selected as typical for a large power reactor containment, the value for λ_L is 2.3×10^{-8} sec^{-1}. From the information presented in Sec. 3.3, Hilliard [90] has predicted that the half-life for natural removal of iodine in a large containment vessel with a surface-area-to-volume ratio of 0.2 ft^{-1} (which includes internal surfaces in addition to those of the shell) would be 23 min, or $\lambda_N = 5.0 \times 10^{-4}$ sec^{-1}. As is usually the case, $\lambda_L \ll \lambda_N$ and λ is nearly equal to λ_N. For a puff release to the containment, the fractional leakage would be at 95% of its limiting value in slightly less than 2 hr. The limiting value of the fractional leakage for this case would be 4.6×10^{-5}, which means that the amount leaked would be only 4.6×10^{-5} times that released to the containment. Both the natural removal process and the low leakage provide means to reduce the release considerably.

The additional protection provided by the engineered removal systems to be discussed subsequently can also be assessed through these equations. Values for the coefficient describing the engineered removal rate, λ_R, can be obtained from Eq. (4-17), if use of a containment spray system only is considered. For the containment spray system described for Diablo Canyon (Sec. 4.2.4), the volumetric spray flow rate, F, is 2600 g/min for single pump operation. The containment height, h, is 142 ft and the free-gas volume is 2.6×10^6 ft^3. If the spray droplet diameter is assumed to be 1000μ, then from Table 4-2, the terminal velocity, U_t, is 397 cm/sec, and the mass transfer coefficient for a reactive spray at 100° C is 11.6 cm/sec. Through Eq. (4-17), the combination of these values yields $\lambda_S = \lambda_R = 1.7 \times 10^{-2}$ sec^{-1}. In this case, 95% of the limiting value for the fractional leakage would be obtained in about 3 min. The limiting value for the fractional leakage would be 1.4×10^{-6}. It is apparent that the use of a containment spray system such as that described provides additional safety both in terms of substantially shortening the time to reach a limiting value for the leakage and reducing the amount leaked.

Similar analyses could be performed for the other cases, describing the time dependence of the release of fission products to the containment, and also for other combinations of the engineered removal systems. The conclusions, however, would be substantially the same as those for the puff release case and the containment spray. In general, the use of engineered removal systems rapidly reduces the airborne inventory of fission products released to the containment atmosphere, consequently shortening the time to attain a limiting value for the leakage. The limiting value also directly depends upon the removal rates within the containment. A reduction factor of ten for the half-life of fission-product removal in the containment will result in a reduction factor of approximately ten for the leakage.

4.2 Containment Sprays

A containment spray system in a water-cooled power reactor promotes fission-product retention within the containment by two means: reduction of pressure and absorption of fission products. The immediate objective of instituting spray action after a loss-of-coolant accident is to reduce the containment over-pressure by condensing the steam released to the containment from the primary system. A reduction of this pressure will diminish the driving force for leakage from the containment. The sprays are supplied with water from a tank, pool, or pressure-suppression pool with a subsequent recycling of water by the containment sump (or pressure-suppression pool) through coolers and perhaps filters and back to the spray header. The reduction of the temperature (pressure) in the containment may often be aided by cooling from a recirculating-air cleaning system within the containment consisting of a HEPA filter, charcoal adsorber section (usually preceded by moisture deentrainers and prefilters), and a cooler. In fact, almost all of the pressurized-water recirculating-air cleaning systems in the second generation reactors (e.g., Connecticut Yankee, Ginna, Indian Point-2, H.B. Robinson-2, Diablo Canyon, etc.) have such cooling coil systems. Often there is a separate set of cooling coils and fans present in the containment through which the containment air is recirculated during normal operations to cool it and thus prevent overheating of wiring and equipment due to leakage of reactor waste heat to the containment. If these normal operation systems are to be used for accident conditions as well, the fan and its motor and other vulnerable components must be designed for post-accident conditions (perhaps air-steam mixtures at 40 to 50 psig and 250° to 300°F).

Containment sprays also have a second function: the retention of fission products within the containment by absorption. In the event of the release of fission products to the containment, the sprays will tend to "wash out" or "scrub" fission products from the containment atmosphere. The removal of fission products by sprays can be enhanced by the use of chemical additives to react with iodine, methyl iodides, and other iodine compounds. Sodium hydroxide (NaOH) and basic sodium thiosulfate ($Na_2S_2O_3$) solutions have been proposed and, in fact, the use of these reactive solutions is currently accepted design practice. Hydrazine (N_2H_4) has also been investigated [125] for methyl iodide removal, but its use is not recommended because of its toxicity and its possible decomposition to yield hydrogen with concomitant explosion hazard. The remainder of this section will be devoted to the fission-product removal function of containment sprays.

4.2.1 Underlying Principles

Griffiths [126] has reviewed early work on the

use of sprays to enhance the removal rate of fission products from the containment atmosphere. His work is described in Chap. 4 of [3]. In general, the sprays will remove water-soluble gaseous and particulate fission products by absorbing them and dissolving them within the droplets which in turn carry the fission products to the floor and into a sump and piped fission-product removal system. From there the water can be recirculated through the spray system. The nonsoluble fission products suspended in the atmosphere as fine particles will tend to be washed out of the gaseous phase by the sprays. For simplicity, most of the following discussion will center on iodine removal since it is generally regarded to be the fission product which limits the dose during an accident, but the principles are generally applicable.

The mass transfer of the vapor forms of iodine from the gas to the liquid phase takes place in three stages: transport in the gas phase by diffusion or convection to the drop surface, equilibration between the two phases at the surface, and transfer by diffusion or convection to within the drop. The processes involved may be conveniently defined in terms of transfer coefficients. The overall-mass-transfer coefficient, K_G (g/cm^2-sec-atm), is related to the liquid-mass-transfer coefficient, k_L, and the gas film coefficient, k_G, by:

$$\frac{1}{K_G} = \frac{1}{k_G} + \frac{1}{H_S k_L} , \tag{4-9}$$

where H_S is the solubility coefficient (g/cm^3-atm).

If the transport within the liquid is diffusion-controlled, the liquid-mass-transfer coefficient, k_L, is given by:

$$k_L = \frac{dm/dt}{\pi d^2 (C_1 - C)} = \frac{2\pi^2}{3} \frac{D_L}{d} , \tag{4-10}$$

where m = mass of the drop of diameter d at average concentration C at time t, and C_1 = concentration at the drop surface. The diffusivity of iodine in water, D_L, can easily be calculated from [127]:

$$D_L = 7.4 \times 10^{-8} \frac{(XM)^{1/2} T}{\eta V^{0.6}} , \tag{4-11}$$

where T = absolute temperature, V = molecular volume, M = molecular weight of the solvent, X = degree of association of the solvent (2.6 for water), and η = viscosity. At 20°C, D_L = 1.14×10^{-5} cm^2/sec and at 100°C, D_L = 5.15×10^{-5} cm^2/sec. Table 4-1 gives calculated values of k_L as a function of drop diameter and temperature.

Circulation within the drop will tend to enhance the transport within the liquid phase. Theoretical analyses and experimental measurements are available for estimating the circulation within drops confined to flow regimes with low Reynolds numbers [128, 129, 130], but these analyses are only of limited interest to containment spray operation. There presently appears to be no conclusive theory or experimental evidence beyond the Stokes flow regime for use in predicting drop circulation or the role internal circulation plays in producing drop distortion. At low Reynolds numbers the circulation velocity increases with fall velocity [128], but extrapolation of these observations to the high Reynolds numbers of interest is questionable. Surfactants may also affect internal circulation, and Garner and Haycock [128] have shown that small concentrations of surfactants produce caps of immobile interfaces on leading surfaces of drops. The transfer of shear is retarded, reducing internal circulation. Since good information on internal circulation is not available, it is generally conservative to assume it is absent and to use Eq. (4-10) for k_L.

Iodine undergoes a variety of hydrolysis reactions in aqueous solutions that must be considered in deriving a value for the partition coefficient H_S. Eggleton [131] suggests the following:

$$I_2(g) \overset{K1}{\rightleftharpoons} I_2(aq) ,$$

$$I_2(aq) + I^- \overset{K2}{\rightleftharpoons} I_3^- ,$$

$$I_2(aq) + H_2O \overset{K3}{\rightleftharpoons} H^+ + I^- + HIO(aq) ,$$

$$HIO(aq) \overset{K4}{\rightleftharpoons} HIO(g) ,$$

$$I_2(aq) + H_2O \overset{K5}{\rightleftharpoons} HOI + I^- + H^+ ,$$

$$3I_2(aq) + 3H_2O \overset{K6}{\rightleftharpoons} IO_3^- + 5I^- + 6H^+ .$$

It is apparent from the above list of reactions (which does not include all possibilities) that the partition coefficient for elemental iodine depends strongly upon the pH of the solution. A high pH or basic solution favors partition of I_2 into the aqueous phase. A dependence of the partition coefficient on concentration has also been reported [132]. Partition of iodine into the aqueous phase is enhanced at low concentrations. For neutral and alkaline solutions, high partition coefficients for I_2, on the order of 10^3 to 10^5, have been observed. Only limited information is available, however, on the partition of CH_3I into aqueous solutions, and it is generally observed to be low. Schwendiman [133] reports that for a temperature range of 5° to 70°C, the partition coefficient (H) for CH_3I in water can be represented by the following equation:

$$\log(H) = -4.82 + \frac{1597}{T} ,$$

where T is the absolute temperature in degrees Kelvin.

The overall-mass-transfer coefficient may be expressed as the mass of constituent transferred/(sec)-(unit area)/unit partial pressure difference across the transfer zone (k_G), or per unit concentration difference across the transfer zone (k_c). Moreover, k_c is equivalent to the velocity of deposition, V_g, and the relation between the two is:

$$k_c = V_g = \frac{RT}{M} K_G . \tag{4-12}$$

TABLE 4-1.

Mass-transfer Coefficients for Iodine to Water Drops [a]

Mean drop diam (μ)	At 20° C			
	k_L (cm/sec)	k_G (g/sec-cm^2-atm)	K_G (g/sec-cm^2-atm)	V_g (cm/sec)
50	1.5×10^{-2}	4.25×10^{-1}	1.8×10^{-2}	1.7
100	7.5×10^{-3}	2.64×10^{-1}	9.05×10^{-3}	8.56×10^{-1}
200	3.75×10^{-3}	1.93×10^{-1}	4.56×10^{-3}	4.31×10^{-1}
500	1.5×10^{-3}	1.43×10^{-1}	1.84×10^{-3}	1.74×10^{-1}
1000	7.5×10^{-4}	1.23×10^{-1}	8.7×10^{-4}	8.25×10^{-2}
2000	3.75×10^{-4}	1.04×10^{-1}	4.4×10^{-4}	4.31×10^{-2}
Mean drop diam (μ)	At 100° C			
	k_L (cm/sec)	k_G (g/sec-cm^2-atm)	K_G (g/sec-cm^2-atm)	V_g (cm/sec)
50	6.8×10^{-2}	4.22×10^{-1}	4.98×10^{-3}	6.04×10^{-1}
100	3.4×10^{-2}	2.56×10^{-1}	2.5×10^{-3}	3.03×10^{-1}
200	1.7×10^{-2}	1.82×10^{-1}	1.25×10^{-3}	1.51×10^{-1}
500	6.8×10^{-3}	1.29×10^{-1}	5.0×10^{-4}	6.05×10^{-2}
1000	3.4×10^{-3}	1.11×10^{-1}	2.5×10^{-4}	3.03×10^{-2}
2000	1.7×10^{-3}	9.1×10^{-2}	1.25×10^{-4}	1.51×10^{-2}

[a] From Griffiths, [126], p. 8.

R is the gas constant and M the molecular weight. At 20°C, V_g for iodine is 94.7 K_G (cm/sec) and at 100°C, Vg is 121 K_G. At 20°C, H_s is .125 g/cm^3-atm and at 100°C, H_s is 7.4×10^{-2}g/cm^3-atm. Griffiths has calculated the mass transfer coefficients for iodine to water drops as shown in Table 4-1. The effect of drop diameter and temperature are plotted in Fig. 4-1.

If the spray water contains an additive which reacts chemically with iodine, the I_2 can no longer exert its partial pressure and so iodine concentration in the liquid phase is very much favored. Sodium hydroxide solutions (~ 0.2 mole NaOH) and alkaline sodium thiosulfate solutions (0.063 mole $Na_2S_2O_3$, pH = 9) have been suggested for this purpose. The presence of the base promotes the hydrolysis of I_2 in accordance with the chemical reaction scheme described above, and sodium thiosulfate is a well-known reducing agent for I_2. Vapor pressure of iodine above a solution of iodide ions is negligible.

If the drops contain an active solution, the transfer rate of iodine to the drop is controlled by the film diffusion rate at the drop surface, expressed in terms of the gas-mass-transfer coefficient (g-moles/sec-cm-atm.) as [134]:

$$k_G = \frac{D_v \rho}{M_m d P_f} Sh, \tag{4-13}$$

where D_v = diffusion coefficient of iodine in air (cm^2/sec), ρ = density of air, M_m = ratio of the mean molecular weight of air to that of the I_2 mixture in the boundary layer, d = drop diameter, P_f = partial pressure of the air in the gas film (atm), Re = Reynolds number ($\rho vd/\eta$), Sc = Schmidt number ($\eta/\rho D_v$), and Sh = Sherwood number (2 + 0.6 $Re^{1/2}$ $Sc^{1/3}$). At low concentrations of iodine, M_m can be taken as the mean molecular weight of air, and P_f as the air pressure. From the general gas law,

$$P_f = \frac{\rho}{M_m} RT \tag{4-14}$$

and hence

$$k_G = \frac{D_v}{d}\frac{1}{RT}(2 + 0.6\,Re^{1/2}\,Sc^{1/3}). \tag{4-15}$$

Equation (4-15) is used to calculate the mass transfer coefficients given in Table 4-2, assuming that there was always an excess of reactive salt. Additional values for mass transfer coefficients are given by Parsly [135].

In the case of a reactive spray, if the gas phase is assumed to be well mixed at all times and the mass transfer rate to be controlled by diffusion from the bulk gas to the surface of the drop, the time required to reduce the concentration of iodine from C_o to C_t is given by

$$\frac{C_t}{C_0} = e^{-\lambda_s t}. \tag{4-16}$$

The removal rate constant due to sprays, λ_s, is related to the containment volume and the spray system operating characteristics in the following manner:

$$\lambda_s = \frac{6Fk_c\theta}{Vd} \tag{4-17}$$

where F = total volumetric spray rate (ft^3/min), k_c = gas-phase mass-transfer coefficient (ft/min),

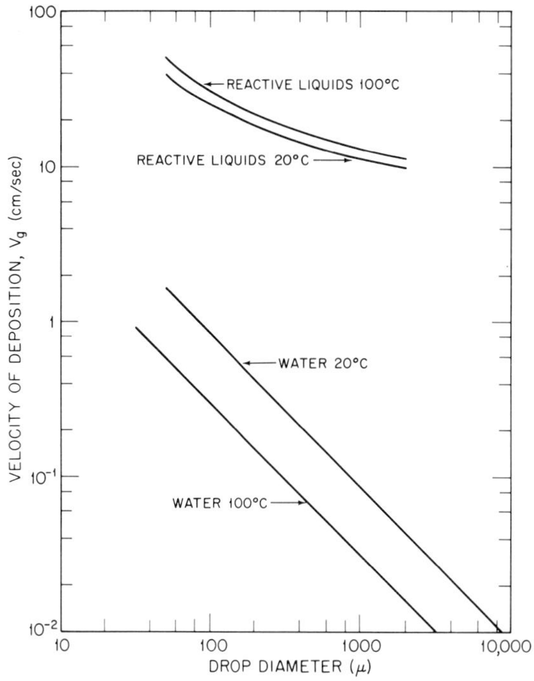

FIG. 4-1 Overall-transfer coefficients for transfer of iodine to liquid drops (from Griffiths, [126] Fig. 2).

θ = time spray drops are airborne = h/V_t (min), h = average height of containment sprayed, V_t = terminal velocity of spray droplet, V = volume of gas being sprayed (ft^3), and 6/d = surface-to-volume ratio for a spherical drop of diameter d, (ft^{-1}). For precise calculations, the effective value of λ_s should be obtained by summing over the spectrum of drop sizes produced by the nozzle. However, it is convenient to use the simplifying assumption that the entire spray can be treated as if it were of uniform size. Row [136] has calculated that if performance estimates are based on mean drop diameters, except the number mean, should be conservative. Thus the simplifying assumption is usually employed.

The time that a spray drop remains airborne, θ, can be estimated by dividing the average vertical distance from the spray nozzles to the intercepting surfaces, h_{ave}, by the average vertical drop velocity. Again a simplifying assumption is made: the average velocity is assumed to be the terminal velocity that a drop attains when it has been slowed by viscous drag from its initial velocity. This assumption is usually valid since the drops will reach their terminal velocities within a few meters of the nozzle and the average fall heights are on the order of 30 to 45 m. For example, a 700-μ drop would travel a distance of 3m before it slowed to 1.02 times its terminal velocity [137].

The problem of methyl iodide removal by containment sprays is more complex to treat theoretically, since it involves liquid-phase mass-transfer and slower chemical reactions between CH_3I and the spray additives. Schwendiman [133] and Parsly [138] have discussed absorption of CH_3I by falling droplets and have presented equations describing absorption under a number of conditions of interest. The relative reaction rates and liquid-phase mass-transfer rates define the specific equation to be used.

Particulate matter may be removed from the atmosphere by diffusion to solid or liquid surfaces, by fallout under gravity (often aided by coagulation), and by impaction on solid or liquid surfaces. The discussion here is limited to impaction. The fraction of particulate matter lying in the path of a falling drop and removed from the atmosphere by impact is the efficiency of impaction, E. Table 4-3 gives E for impaction of aerosol particles by water drops. Griffith states that it is probable that submicron particles of unit density are not removed by sprays. The proportion of aerosol removed per second (λ) will be proportional to the region swept out by the drops travelling with terminal velocity V_w, or:

$$\lambda = \frac{\pi d_w^2}{4} N V_w E, \tag{4-18}$$

where d_w = drop diameter and N = number of drops/cc. In a given spray, N is determined by the

TABLE 4-2.

Mass-transfer Coefficients (k_G) for Transfer of Iodine into Drops of Liquid that Combine Chemically with Iodine [a]

Temperature (°C)	Drop diam (cm)	V_t (cm/sec)	η/ρ (cm^2/sec)	D_v (cm^2/sec)	Re	Sc	$Re^{1/2}Sc^{1/3}$	Sh	k_G (g/sec-cm^2-atm)	V_g (cm/sec)
20	5×10^{-3}	7.75	0.15	0.085	0.26	1.76	0.61	2.37	0.425	40.2
	1×10^{-2}	25	"	"	1.67	"	1.56	2.94	0.264	25.0
	2×10^{-2}	76	"	"	10.1	"	3.84	4.3	0.193	18.2
	5×10^{-2}	200	"	"	67	"	9.9	7.94	0.143	13.5
	1×10^{-1}	397	"	"	264	"	19.6	13.75	0.123	11.6
	2×10^{-1}	649	"	"	866	"	35.3	23.2	0.104	9.9
100	5×10^{-3}	7.75	0.23	0.11	0.17	2.09	0.53	2.32	0.422	51.1
	1×10^{-2}	25	"	"	1.1	"	1.35	2.81	0.256	31.0
	2×10^{-2}	76	"	"	6.6	"	3.31	3.99	0.182	22.1
	5×10^{-2}	200	"	"	43	"	8.45	7.07	0.129	15.6
	1×10^{-1}	397	"	"	172	"	16.9	12.15	0.111	13.5
	2×10^{-1}	649	"	"	564	"	29.9	19.9	0.091	11.0

[a] From Griffiths, [126], p. 11.

TABLE 4-3.

Efficiency of Impaction of Aerosol Particles by Water Drops[a]

Particle diam (μ)	E, Efficiency of impaction: Water drop diam and terminal velocity			
	100 μ 25 cm/sec	200 μ 76 cm/sec	500 μ 200 cm/sec	1000 μ 400 cm/sec
1	0.0003	0.007	0.03	0.04
2	0.009	0.06	0.24	0.37
5	0.02	0.12	0.46	0.70

a From Griffiths [126].

drop diameter (which fixes the terminal velocity) and the mass flow rate of the liquid (M'):

$$N = 6 \times 10^{-4} \frac{M'}{V_w \pi d_w^3}. \qquad (4\text{-}19)$$

Hence

$$\lambda = 1.5 \times 10^{-4} \frac{EM'}{d_w}. \qquad (4\text{-}20)$$

Typical values for particles are given and plotted in Fig. 4-2. The equation above holds for mass flow rates at least ten times as great as shown in the table.

4.2.2 Experimental Data

Some of the early investigations on the behavior of fission products in water droplets were carried out by Mausteller and Campana [139]. They describe the results of tests in which 550°F, 2000 psi water containing simulated fission products was allowed to leak at a controlled rate into a 4000-ft^3 vessel. The fission products used were Na^{24}, added as Na_2Co_3 to simulate soluble fission products; Mo^{99}, added as MoO_3 to simulate insoluble fission products; and I^{131}, added as I_2 to represent iodine and gaseous fission products. Fallout coefficients of 0.1 to 0.8 hr^{-1} (half-lives of 0.9 to 7 hr) were observed, with most of the coefficients in a range of 0.2 to 0.35 hr^{-1}. More than 80% of the activity was associated with particles in the 0.4μ to 0.7μ range. Only 6% to 15% of the activity was found in water droplets. Calculations showed that the life of a 10μ-diam droplet in air at 50% relative humidity is only about 0.1 sec, while that of a 100μ-diam droplet is about 10 sec. Average droplet diameters of fogs and mists were said to be in the range of 5 μ to 100 μ. They also allowed 144 liters of water at 550°F and 2000 psi to expand rapidly into a 1340-ft^3 vessel. A maximum of 2% of I^{131} and Na^{24} and 5% of the Mo^{99} were reported to have escaped from the container, during the test, presumably from an opening.

The most comprehensive experimental program to investigate spray systems as engineered removal devices for fission products is that described by Row [140]. Removal characteristics of various solutions, stability of solutions, and effects of spray solutions on containment and equipment were investigated. The objectives of the program were to establish the feasibility of containment sprays as engineered safety features for nuclear reactors and to provide data that can be used by designers and operators of these systems. Laboratory-scale and large-scale engineering proof tests are included in the program.

4.2.2.1 Capacity of Solutions for Fission-Product Iodine Removal

A search for soluble additives reactive to molecular iodine and methyl iodide was conducted by Patterson. The equilibrium iodine capacity, K_d, of

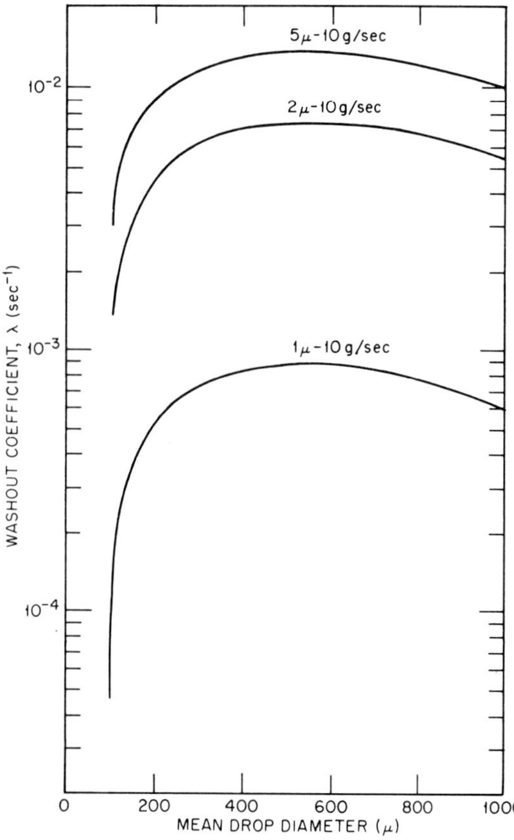

FIG. 4-2 Rates of removal of particles by water drops (from Griffiths, [126] Fig. 5).

various test solutions was determined using a gas scrubber system. The iodine capacity is defined as:

$$K_d = \frac{C_o}{C_g} = \frac{\text{total moles of iodine dissolved in the solution}}{\text{moles of iodine in the gas}} .$$

The iodine capacities for a number of the additives tested at 25° C are given in Table 4-4. The variation of K_d with temperature and concentration for some of the solutions of practical interest are given in Table 4-5. This search has revealed that there are several solutions in addition to sodium hydroxide and sodium thiosulfate which have potential for iodine removal.

Additional parameters on the effectiveness of various solutions for removing iodine were investigated in wind tunnel experiments. A spray solution drop was suspended in the tunnel gas containing I_2

TABLE 4-4.

Survey of Potential Additives to Sprays or Pools for I_2 Removal [140]

Additive	Concentration	K_d at 25°C
H_2O		92
$Na_2S_2O_3$	2.31×10^{-3} M	250
NaOH	2.52×10^{-3} M	250
KOH	2.39×10^{-3} M	202
Piperdine	$\sim 2.5 \times 10^{-3}$ M	369
Morpholine	$\sim 2.5 \times 10^{-3}$ M	277
N_2H_4	2.32×10^{-3} M	618
$HO\text{-}CH_2\text{-}CH_2\text{-}OH$	10 vol %	151
	45 vol %	428
Resorcinol	$\sim 2.5 \times 10^{-3}$ M	728
$NaNO_3$ (sodium nitrate)	9.56×10^{-2} M	103
CH_3CO_2H (acetic acid)	2.56×10^{-3} M	109
$NaClO_3$ (sodium chlorate)	9.8×10^{-2} M	112
H_2SO_4 (sulfuric acid)	$.252 \times 10^{-3}$ M	117
$Na_3C_6H_5O_7$ (sodium citrate)	2.41×10^{-3} M	122
NH_4OH (ammonium hydroxide)	1.775×10^{-4} M	131
LiOH (lithium hydroxide)	2.755×10^{-3} M	313
$Na_2B_4O_7 \cdot 10H_2O$ (sodium tetraborate or borax)	0.1 M	874
Creatinine	2.52×10^{-3} M	114
Glycine	2.37×10^{-3} M	125
Pyridine	2.5×10^{-3} M	197 (s)[a]
$AgNO_3$ (silver nitrate)	2.255×10^{-3} M	242 (s)
4-Aminopyridine	2.5×10^{-3} M	391 (s)
Quinoline	2.5×10^{-3} M	450 (s)
Piperazine	2.5×10^{-3} M	517 (s)
1, 4-Diazabicyclo (222) octane	2.76×10^{-3} M	702 (s)
1, 3, 5-Trihydroxybenzene	2.32×10^{-3} M	817 (s)
Pyrrole	2.5×10^{-3} M	1100 (s)
Urea	2.501×10^{-3} M	115
$CH_5N_3 \cdot HNO_3$ (guanidine nitrate)	2.32×10^{-3} M	119
KCNS (potassium thiocyanate)	2.46×10^{-3} M	368
Semicarbazide · HCl	2.29×10^{-3} M	690
Thiourea	2.48×10^{-3} M	767
Thiosemicarbazide	2.34×10^{-3} M	1015
Carbohydrazide	2.40×10^{-3} M	1512
Glycerine	30 vol %	173
Ethyl alcohol	30 vol %	203
Dipropylene glycol	20 vol %	215
Dioxane	20 vol %	235
Dimethylformamide	20 vol %	278
Triethylene glycol	30 vol %	342
Ethylene glycol	10 vol %	151
	45 vol %	428

[a] Results in solid product denoted by (s).

TABLE 4-5.

Variation with Temperature and Concentration (K_d) [140]

Additive	Concentration	K_d		
		25°C	45°C	66.2°C
H_2O		92	87	72
H_3BO_3 in H_2O	3000 ppm B	85	72	
$Na_2S_2O_3$ in H_2O	0.921×10^{-3} M	150		
	2.31×10^{-3} M	250	288	345
	4.62×10^{-3} M	380		
$Na_2S_2O_3 + H_3BO_3$	2.31×10^{-3} M	237		
NaOH in H_2O	1.012×10^{-3} M	155		
	2.52×10^{-3} M	250	219	155
	5.05×10^{-3} M	400		
$NaOH + H_3BO_3$	2.52×10^{-3} M	83		
$Na_2S_2O_3$ + NaOH	2.42×10^{-3} M (total: 50:50 mixture)	261		221

or CH3I and the uptake of these contaminants by the drop under controlled conditions was determined. The parameters that were investigated in the wind tunnel studies were humidity, temperature, drop size, contaminant concentration in the gas phase, rate of contaminant uptake, additive concentration in the solution, and pH of the solution. The effect of additive type and concentration on the mass transport (vt) for CH_3I at 26° C is shown in Fig. 4-3, and some data on the effect of pH and temperature are given in Fig. 4-4. The wind tunnel experimental data can be used in choosing the most appropriate spray solution composition.

4.2.2.2 Spray Removal Studies

A more direct evaluation of the feasibility of containment sprays for iodine removal and retention is provided by experiments at the Nuclear Safety Pilot Plant (NSPP) at Oak Ridge National Laboratory and at the Containment Systems Experiment (CSE) program at Battelle's Pacific Northwest Laboratory. The NSPP is a stainless steel model containment vessel with a volume of 1350 ft^3. It is a 10 ft-diam by 15 ft cylindrical vessel with ASME flanged and dished heads designed for 60 psig. The vessel contains several spray headers, and a number of nozzles have been used in tests, including the 7G3 nozzle supplied by the Spraying Systems Company. The facility used in the CSE program is a factor of 23 larger in gross volume than the NSPP. In this facility, the spray header is installed so that a drop falls 33.8 ft before striking the deck at the top of the dry well. If the hatch at the top of the dry well is open, an additional fall of 50.5 ft occurs before the drops strike the bottom of the dry well. The 7G3 nozzle is also used in this facility.

A summary of the results obtained in elemental iodine absorption experiments in the NSPP is given in Table 4-6 [140] and in the CSE experiments in Table 4-7 [141]. As can be seen in these tables, a number of factors, such as solution composition, solution flow rate, drop diameter, and temperature, have been investigated. In all cases, the sprays effectively removed I_2 from the containment atmosphere. The observed half-life for I_2 removal can be compared with the predicted half-life for removal based upon the Griffith model discussed earlier. The predicted values are given in the tables for both the NSPP and CSE experiments and a curve showing the agreement between the CSE experimental results and the model is given in Fig. 4-5 [142]. If the model is valid, all of the data should lie on the 45° line in Fig. 4-5. It can be concluded from the agreement shown in this figure that the Griffith model for elemental iodine removal by reactive sprays is valid and that it provides a means whereby the data from the NSPP and CSE tests can be applied to larger containment vessels.

Data on methyl iodide removal by spray systems in the NSPP are given in Table 4-8 [143]. As discussed previously, the removal of CH_3I by sprays involves simultaneous absorption and chemical reaction. The reaction of methyl iodide with thiosulfate and hydroxyl ions has been studied and

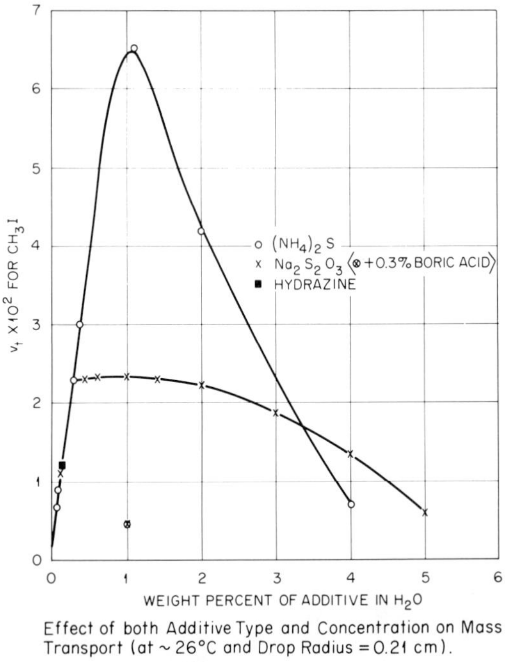

FIG. 4-3 Effect of both additive type and concentration on mass transport (at ~ 26°C and a drop radius of 0.21 cm) [140].

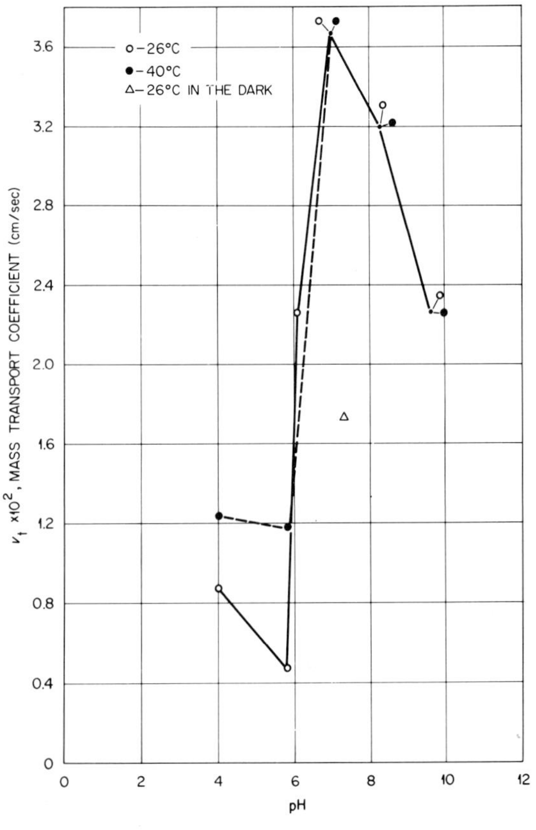

FIG. 4-4 Effect of pH and temperature on the mass transport of CH_3I [140].

TABLE 4-6.

Summary of Results Obtained in Elemental Iodine Absorption Experiments at the NSPP

	Run No.												
	21	22	26	27	28	30	31	32	33	37	38	42	44
Number of spray nozzles	12	12	1	1	1	1	1	1	1	1	3	3	3
Nozzle identification	J-140D[a]	J-140D	1713[a]	1713	1713	1713	1713	1713	1713	1713	7G3[b]	7G3	7G3
Solution composition													
13.7 g/liter boric acid	×	×	×		×	×	×	×	×	×	×	×	×
3.4 g/liter sodium hydroxide		×	×		×	×	×	×	×	×	×	×	×
8 g/liter sodium thiosulfate	×	×	×						×	×	×	×	×
Initial pressure (psig)	3	3	45	45	45	3	3	45	45	45	10	45	5
Initial atmosphere temperature (°C)	30	30	130	130	130	130	30	130	130	130	70	130	30
Solution temperature (°C)	30	30	30	30	120	30	30	30	30	120	100	30	30
Solution flow (gpm)	0.57	0.52	10.3	9.9	10.1	15.5	15.3	10.3	10.6	12	11	11	11
Observed removal half-life (sec)	37	38	31	48	48	24	38	41	21	20	11	12	14/80
Calculated removal half-life (sec)	3	3	98	107	102	36	37	98	94				

[a] Spray Engineering Co. catalog number.
[b] Spraying Systems Co. catalog number.

TABLE 4-7

Experimental Conditions and Results — CSE Spray Tests [141]

	Run A-3	Run A-4	Run A-6	Run A-7	Run A-8	Run A-9
Atmosphere	Air	Air	Steam-Air	Steam-Air	Steam-Air	Steam-Air
Temperature (°F)	80	80	250	250	250	250
Pressure (psia)	14.5	14.5	44	50	48	44
Nozzle type	(b)	(b)	(b)	(b)	(c)	(d)
Drop MMD (μ)[a]	1210	1210	1210	1210	770	1220
Geom. Std. Dev. (σ)	1.53	1.53	1.53	1.53	1.50	1.50
No. of nozzles	3	12	12	12	12	12
Spray rate (gal/min)	12.8	48.8	49	49	50.5	145
Total spray volume (gal)	510	1950	1960	1960	2020	2300
Spray solution	(e)	(e)	(f)	(g)	(f)	(f)
Initial removal half-life for I_2						
Predicted[h]	6.4	1.7	1.9	1.9	0.71	0.66
Observed	5.5	1.4	2.1	2.2	0.67	0.61

[a] Mass median diameter.
[b] Spraying Systems Co., 3/4 7G3, full cone.
[c] Spraying Systems Co., 3/8 A20, hollow cone.
[d] Spraying Systems Co., 3/4 A50, hollow cone.
[e] 525 ppm boron as H_3BO_3 in NaOH, pH 9.5.
[f] 3000 ppm boron as H_3BO_3 in NaOH, pH 9.5.
[g] 3000 ppm boron as H_3BO_3 in demineralized water, pH 5.
[h] Half-life based on drop MMD and terminal setting velocity of drops.

kinetic data for these reactions are available [144]. The results of the NSPP tests show that, at room temperature, the process occurs in the "slow reaction" regime. The overall removal rate is controlled by the rate at which methyl iodide is destroyed by chemical reaction. At 100° to 120° C, the half-lives are shorter than would be predicted on the basis of absorption into a rigid drop plus fast chemical reaction. Apparently the partition coefficient of methyl iodide between water and air above 100°C is higher than predicted, and internal circulation is occurring within the drop. Although the observed half-lives for CH_3I removal are considerably longer than those for I_2 removal, some reduction of dose from CH_3I can be achieved with spray solutions provided primarily for I_2 removal. Further searches for additives to enhance the CH_3I removal appear warranted.

Removal of CH_3I by spray solutions containing hydrazine has been investigated by Schwendiman [133] and Viles [145, 146]. Since the absorption of methyl iodide by sprays of aqueous solutions of hydrazine is controlled by liquid-phase mass transfer, reactions between CH_3I and hydrazine in the aqueous phase were extensively studied. Reaction rates were determined as a function of hydrazine concentration, sodium hydroxide concentration, salt concentration, and temperature. The reaction rate increased with increasing hydrazine concentration and was determined to be second-order. An activation energy for the reaction was calculated to be 20.2 kcal/mole and the rate was independent of hydroxyl concentration. Comparison of reaction rates with CH_3I in solution, based upon extrapolation of lower temperature data, indicates that hydrazine ranks approximately equal to sodium thiosul-

TABLE 4-8.

Summary of Methyl Iodide Removal Experiments at the NSPP [143]

Experimental conditions												
Spray nozzle installation												
Type of nozzle	1713[a]	1713	1713	7G3[b]	7G3	7G3	7G3	7G3	7G3	7G3	7G3	7G3
Number of nozzles	1	1	1	3	3	3	3	3	3	3	3	3
Pressure, temperature, etc.												
Initial vessel pressure (psig)	3	45	45	3	45	45	45	3	45	45	45	45
Initial temperature (°C)	30	130	130	30	130	130	130	30	130	130	130	30
Solution temperature (°C)	30	120	120	30	120	120	120	30	120	120	120	120
Mass flow (gpm/in^2)[c]	0.19	0.15	0.15	0.14	0.14	0.14	0.14	0.14	0.14	0.14	0.14	0.14
Spraying time (min)	153	120	157	180	180	180	180	180	180	180	180	180
Solution composition, moles/liter												
H_3BO_3	0.28	0.28	0.28	0.28	0.28	0.28	0.28	0.28	0.28	0.28	0.28	0.28
NaOH	0.17	0.17	0.17	0.17	0.17	0.17	0.17	0.17	0.17	0.17	0.17	0.17
$Na_2S_2O_3$	0.063	0.063	0.063		0.063	0.063	0.063		0.063		0.315	0.063
CTAB[d]					10^{-5}		10^{-4}	10^{-4}	10^{-4}			
PEI-1000[e]										2×10^{-5}		2×10^{-4}
HCHO										0.15		
Results												
Iodine remaining in U-tube (mCi)	3.07	6.67	6.43	0.003	0.045	0.003	0.004	0.05	0.001	<0.001	0.034	0
Iodine in spray solution	12.33	15.53	13.13	0.77	25.92	34.67	23.24	4.89	13.19	8.34	27.63	19.9
Iodine in MCV decontaminating solution	1.74	1.49	0.73	0.13	0.84	1.78	1.97	0.23	3.52	2.67	1.32	1.35
I_2 in purge (collected on Ag)	0.005	0.002	0.001	0.001	<0.001	<0.001	<0.001	0.001	<0.001	<0.001	<0.001	0
CH_3I in purge (collected on charcoal)	0.84	5.62	4.04	23.33	2.77	4.37	3.51	18.15	13.20	13.49	0.24	1.59
Overall decontamination factor for CH_3I	2.78	3.77	4.25	1.03	10.4	8.95	7.61	1.27	2.00	1.62	110	13
Mean half-life (min)	107	63	75		53	57	61	528	180	260	27	49

[a] Spray Engineering Co., Cat. No. 1713, Ramp bottom nozzle.
[b] Spraying Systems Co., Cat. No. 7G3.
[c] Typical reactor containment building design is 0.16 gpm/ft^2.
[d] Cetyl-trimethyl ammonium bromide.
[e] Dow Chemical Co., PEI-1000 (polyethyleneimine, mol wt assumed to be 60,000).

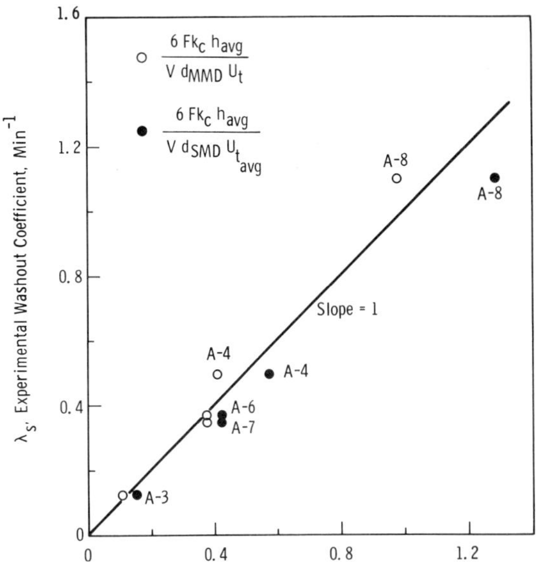

FIG. 4-5 Comparison of initial spray washout of elemental iodine with drop absorption models [142].

fate at 120° C on an equal wt. % basis. Computer solutions of a model for CH_3I removal based upon stagnant liquid drops and films flowing down the wall have predicted that a 5 w/o hydrazine solution should show a half-time of removal of about 1 hr in a large containment. Reactions between CH_3I and hydrazine at high concentrations have resulted in the formation of an aerosol which can be removed by filtration or by settling, diffusion, or solubility processes. While CH_3I removal can be obtained with hydrazine-containing solutions, this spray solution shows little promise of application to reactor containments due to lack of stability in a radiation environment. Decomposition of the reactive additive reduces its effectiveness for CH_3I removal and the production of hydrogen causes potential combustion problems in the containment. Also hydrazine can cause stress corrosion of copper-containing alloys. Hydrazine as a reagent in foams has been investigated and is discussed in Sec. 4.6.

Some data have been obtained on the removal of particles by sprays. The results of the CSE tests which have yielded data on aerosol removal are shown in Table 4-9. Although the data on particle removal are limited, some reduction in airborne particulates by spray systems appears to be possible.

4.2.2.3 Thermal and Radiation Stability of Spray Solutions

The thermal and radiation environment in the reactor containment during an accident must be considered in the selection of spray solutions for iodine removal. In a 1000 Mw(e) plant, the core fission-product inventory will reach 1.1×10^{10} curies. Noble gases represent 8.4×10^8 curies, and the halogens represent 7.5×10^8 curies. If the noble gases and halogens are released to the containment in the amounts defined by TID-14844, a high radiation environment will result. In addition to exposure to the radiation field in the containment, the containment spray solution will also be exposed to the radiation field in the core as a result of recirculation of the water from the sump. In a typical containment building of 10^6 ft^3 with recycle of the liquid through the reactor core, the accumulated dose in the spray solution will be about 10^8 rads in 8 1/2 days and will approach 10^9 rads in 120 days [147]. During an accident, temperature conditions which the spray solutions may experience can vary from slightly above room temperature in the containment to an excess of 150° C early in the accident and in the vicinity of the core. Since the spray solutions must retain the fission products for periods of time of up to several months, it is imperative to know the stability of the solutions over the postulated range of conditions.

Studies carried out by Zittel [148] have shown that a basic borate solution (0.15 N NaOH-3000 ppmB) is unaffected by the 150° C temperature, while the basic thiosulfate (1 w/o $Na_2S_2O_3$-0.15 N NaOH-3000 ppm B) shows acceptable stability (only 10% degradation after 72 hr at 140° C). The radiolytic stability of the two test solutions, as established by exposure to a Co^{60} gamma source, follows the same pattern (the basic borate being essentially unaffected by a dose of 10^8 rad), while the basic thiosulfate is additive, i.e., radiation at 140° C gives approximately the same degradation as the sum of the two processes carried out separately. Of greater concern is the generation of radiolytic H_2 from the spray solutions. Contrary to results that might be expected in a pure water system, the radiolytic gases do not reach a low overpressure equilibrium state in these solutions (Table 4-10). It can be seen from the $G(H_2)$ values that the systems are moving toward an equilibrium state. However, in neither case is an equilibrium state attained at any low overpressure. The data presented show clearly that the radiolytic reaction path differs for the two solutions. The basic borate solution produces both radiolytic O_2 and H_2, whereas the basic thiosulfate solution uses up O_2 and produces H_2. In both cases, it is evident that radiolytic H_2 is formed in amounts sufficient to be a problem.

Some radiation stability studies have been conducted on solutions containing additives for CH_3I removal. In general, these additives have been of a polyamine structure and show radiation stability sufficient for usage under conditions expected during an accident.

4.2.2.4. Effect of Spray Solutions on Containment and Equipment

Consideration must also be given to the possible effects of spray solutions on containment and the equipment within it. The use of chemical additives to a water spray generally increases the probability of corrosion of metals by the solution. If reactive sprays are to be satisfactory as engineered safety systems, they must not propagate the accident by corrosive action that would violate the containment or render useless any of the equipment that must operate during an accident. Long-term storage of

TABLE 4-9.

Aerosol Concentration Half-Lives Due to Operation of Sprays in CSE vessel — Runs A-3 and A-4

Containment conditions	Half-life (min)				
	Elemental Iodine	Particulate Iodine	CH_3I	Cesium	Uranium
First spray, run A-3	5.4	11.3	73	16	34
First spray, run A-4	1.64	2.6	53	3.5	7
Second spray, run A-3	6.2	13	170	16.7	32
Second spray, run A-4	10.4	8.7	240	10.3	21

TABLE 4-10.

Radiolytic Gas Generation

Gas/liquid ratio	ΔO_2 (cm^3/ml)	ΔH_2 (cm^3/ml)	$P(H_2)$ (Atm)	$P(H_2 + O_2)$ (Atm)	$G(H_2)$
Solution I. (0.15 N NaOH, 3000 ppm B) Dose = 1.2×10^8 rad					
25	0.05	1.06	0.04	0.25	0.43
5	0.48	1.40	0.28	0.57	0.57
1	0.46	1.14	1.14	1.79	0.46
0.5	0.36	0.32	1.66	2.53	0.33
0.1	0.15	0.39	3.92	5.58	0.16
Solution II. (1 wt% $Na_2S_2O_3$, 3000 ppm B, 0.15 N NaOH) Dose = 1.1×10^8 rad					
25	-3.26	0.97	0.04	0.01	0.40
5	-	1.21	0.24	0.24	0.50
1	-	1.17	1.16	1.16	0.48
0.5	-	1.12	2.24	2.24	0.46
0.1	-	0.73	7.40	7.40	0.30

the solutions must not affect the mixing capability of the system by corrosion of the storage tank components. Hence corrosion in a radiation environment must be considered. Another aspect to consider is the effect of the reactive spray solutions on protective coatings in the containment under accident conditions. The protective coatings must be able to withstand the accident transient coupled with the spray system operation. An unacceptable situation would result if larger quantities of the protective coatings were removed from surfaces and were introduced into the spray system and the reactor core. Conditions under which coatings may be exposed are discussed in Section 4.5 of this chapter.

The corrosion resistance of a variety of materials in iodine-absorbing spray solutions has been under study at ORNL [149], and a list of samples is given in Table 4-11. In general, only the aluminum alloy specimens exhibited marked weight losses when exposed to 0.15 M NaOH-0.28 M H_3BO_3 solutions. When exposed to sprays, all of the aluminum alloys showed corrosion rates of 140 to 200 mpy (mils/yr) and rates of 27 to 98 mpy when totally submerged. Of other materials showing weight losses, copper was the least resistant, where the corrosion rate was less than 1 mpy. There was no evidence of localized attack on any materials in the basic borate solution.

TABLE 4-11.

Corrosion Test Samples

Material	Application
1. Stainless steel, type 304, type 316	Piping
2. Carbon steel, A302B, A283 — grade C	Primary vessel, steam generator
Carbon steel, 212B	Containment building
3. Zircaloy-2	Used in clad of early plants
Zircaloy-4	Currently popular clad
4. Inconel-600	Steam generator tubing
Inconel-718	Core structural material
5. Aluminum — type varies: 1100, 5005, 5052, 6061, 3003, 3004	Wire-way, containment insulation, many miscellaneous items
6. Copper — type K	Water pipe
7. Copper/nickel alloy 90/10 and 70/30	Precoolers in air cleanup system
8. Brass — shim stock probably representative	Very small amounts used, motors
9. Monel — ASTM spec.	Small amounts used
10. Galvanized — commercial grade	Ventilation duct work

When the corrosion tests were performed with alkaline solutions containing thiosulfate and borate (0.15 M NaOH—0.28 M H_3BO_3—0.064 M $Na_2S_2O_3$), the aluminum specimens reduced some of the thiosulfate ions to sulfide ions which reacted with the copper corrosion products to form copper sulfide. The corrosion rates for the aluminum alloys were again high, but no higher than with basic borate solutions. Average weight losses of the copper specimens in the thiosulfate solution corresponded to a corrosion rate of 210 mpy, and the cupronickels showed lower corrosion rates, less than 2 mpy. In both the spray and solution tests, the carbon steel specimens developed random patches of rust under which pits as deep as 5 mils were found. All other materials showed negligible attack.

4.2.3 Typical Spray System Designs

Typical spray systems of the first generation are listed in Table 4-12 [3]. It should be noted that the primary purpose of many of these systems was to reduce containment pressure rapidly. All reactor systems which have such spray provisions must have all vital electrical equipment such as motors, valve operators, signal cables, etc., waterproofed. Some reactor installations have emergency bunkers or control centers which include remote pumps and systems to permit continuous operation of the spray system for long periods of time after an accident.

A few reactors provide spray systems external to the containment to spray the outside of the containment and thus help to reduce pressure. The BONUS and Indian Point-1 reactors have such provisions. Such a system is shown in Fig. 4-6 for Bonus [150]. Others, such as Yankee and San Onofre, which have steel containment shells with no concrete shell attached either internally or externally, could make use of their situation to provide emergency sprays even after the accident. For instance, fire hoses could be used.

In a few reactors, a ''caustic scrubber'' for fission-product removal—especially iodine—is used. Such a system used for the Oak Ridge Research Reactor is shown in Fig. 4-7 [151]. Iodine removal efficiencies of 95% to 99% are obtained. The ORNL unit was tested at an air flow rate of 5000 to 6000 cfm with an iodine concentration of 15 mg/ft^3; it achieved a removal efficiency of 99%. The efficiency appeared to be relatively independent of air-flow rate over quite a wide range, but increased with inlet iodine concentration. In at least four second-generation power reactors it is proposed to add borated basic (pH $\sim$9.0) sodium thiosulfate ($Na_2S_2O_3$) to containment sprays (Indian Point-2, Palisades, H. B. Robinson-2 and Diablo Canyon).

The Diablo Canyon plant [152] is typical of these spray protective systems and Fig. 4-8 shows a schematic diagram of the containment spray system. The plant is a four-loop design with a total heat output of 3250 Mw(t). The containment building is a reinforced concrete vertical cylinder with a flat base and hemispherical dome. The concrete vessel has a welded steel liner with a minimum thickness of 1/4 in. The side walls are approximately 142 ft high, and the inside diameter of the structure is 140 ft. The free gas volume contained in the building is 2.6×10^6 ft^3. The containment spray system is designed for heat and iodine removal. In the event of a loss-of-coolant accident, the spray system will be automatically actuated by a high containment pressure and possibly other system signals. These will cause the two containment spray pumps located in an auxiliary building to start and take suction directly from the refueling-water storage tank. The safety injection system (emergency cone-cooling system) also takes suction from the refueling-water storage. The spray solution chemical is added to the borated refueling water in the spray system and discharged into the spray head located in the dome of the containment building, at the design flow rate of 2600 gpm/pump.

The spray nozzles are located on two ring heads attached to the steel containment liner in the upper part of the containment building. The spray nozzles are arranged to provide maximum coverage of the free gas volume as well as wall washdown, and they have an average spray drop fall of 140 ft. The spray headers are protected from missiles by concrete shielding.

The spray system and safety injection will exhaust the 350,000 gal of refueling water in approximately 1/2 hr under design operating conditions. Pumps then begin drawing from the containment building sumps, which by now have accumulated enough spray solution for recirculation to the spray headers and the reactor vessel for shutdown cooling of the reactor. Fresh solution is injected through the spray headers during the initial 1/2 hr when refueling water storage is used. It has not encountered either high temperature, fission-product radiation, or fission-product contaminants before entering the vessel. The solution has the maximum theoretical iodine sequestering ability when sprayed during this period.

Operation of the safeguards equipment after the 1/2-hr switch to a recirculation mode will continue for some time. Spray cooling of the containment building interior may be terminated in the first 24 hr following the accident, but shutdown cooling of the reactor core will be required for months. This means that the spray solution must be circulated through the reactor core for an extended time period and therefore must demonstrate acceptable thermal and radiation stability.

Figure 4-9 [153] shows the pressure suppression containment spray system for Dresden-2 and -3. Note that the spray headers for each reactor are located in the dry well and are in duplicate. Note also that each ring and its associated piping and pumps are separate from the other two. The system is designed so that one spray system serves both reactors, thus economizing by eliminating one spray system. The diagram clearly shows that each reactor is doubly protected; it is not anticipated that both reactors would have an accident at the same time.

In each of these systems the pumps, coolers, and most valves are located outside of the containment. Since these piping systems carry water from the containment sump and recirculate it, the pipes and system components are part of the containment and must be assumed to be contaminated i.e., potential fission-product release zones.

TABLE 4-12

Building Spray Systems

Reactor	Addition or circulation rate	Spray Arrangement	Source and capacity information	Comments
Big Rock Point	1000 gpm maximum addition rate 400 gpm recirculation rate	There are two sets of spray nozzles available; one set is automatically put into service when the sphere pressure reaches 2 psig, and the other set is put into service manually from a location outside the sphere	100% coverage of the building with water supplied from Lake Michigan via the electric or diesel fire pumps until water level reaches a certain level in the sphere, at which time one of two recirculation pumps may be placed in service	Spray systems furnish pressure reduction after the accident; can be operated from a remote, shielded location
Elk River	1000 gpm	Spray headers located above main and basement floors	100% coverage; 30,000-gal water storage tank	Spray reduces the pressure after the accident
EGCR, external	2500 gpm	Ring headers with spray nozzles; most concentrated near the top of the dome	32×10^6 Btu/hr heat removal capacity; supplied from Melton Hill Lake	This is the primary method of energy removal and must operate after the accident
HWCTR	1000 gpm for 15 min; then 120 gpm	Sprinkler headers in the dome of the container and at all levels in the containment building	Initially from storage tanks in top of container, then by pumps from wells (pressure reduction from 23 to 10 psig in 15 min)	This furnishes pressure reduction after accident; can be operated from a remote location
Indian Point Internal	1000 gpm	Eight spray headers in top hemisphere	Installed in such a manner as to wash down surfaces	No credit taken for internal spray cooling in mca analysis
External	3000 gpm	Eight spray headers in top hemisphere	Recirculating type; the water is collected in a sump and pumped back to spray headers	
PRTR	500 gpm	Fog nozzles in two ring headers above main floor	Cooling power of 37.0×10^6 Btu/hr; 100% coverage of main floor	Can be operated from remote location (1/4 mile)
BONUS Internal	1000 gpm	Headers and nozzles over main floor and in basement	Water supply[c] (pressure reduction from 4.3 to 2.4 psig in 50 min)	No credit taken in mca analysis for either system; both can be operated from remote location
External	450 gpm	Two circular perforated ring headers around the upper dome	To be used to keep pressure down after initial pressure reduction by internal system; water supply[c]	
NPD	Reactor vault: 1500 Igpm	Fog nozzles suspended at two ends of reactor vault; actuated at 0.72 psig	Heavy water from moderator system	
	1500 Igpm	Fog nozzles (same as above)	Supplied from standby water system	
	8700 Igpm	Two perforated tanks suspended at two ends of reactor vault; actuated at 3.7 psig	Supplied from 250,000 imperial gal storage tank[a]	
	Boiler room (large leaks): 0 to 10 sec, 97,000 Igpm 10 to 20 sec, 75,000 Igpm 20 to 30 sec, 55,000 Igpm 30 to 40 sec, 36,000 Igpm 40 to 60 sec, 15,500 Igpm	Seven perforated tanks suspended in the boiler containment vessel; actuated at 1.5 psig	250,000 imperial gal storage tank[b]	

[a] 39,000 imperial gal unavailable because of internal arrangement of the tank.

[b] Initial 100,000 imperial gal for dousing; 150,000 imperial gal for light-water injection.

[c] Gravity fed (100 ft H_2O) from 100,000 gal storage which is shared by (1) core spray, (2) external building spray, (3) internal building spray, and (4) fire protection; additional makeup can be supplied from wells and pipeline.

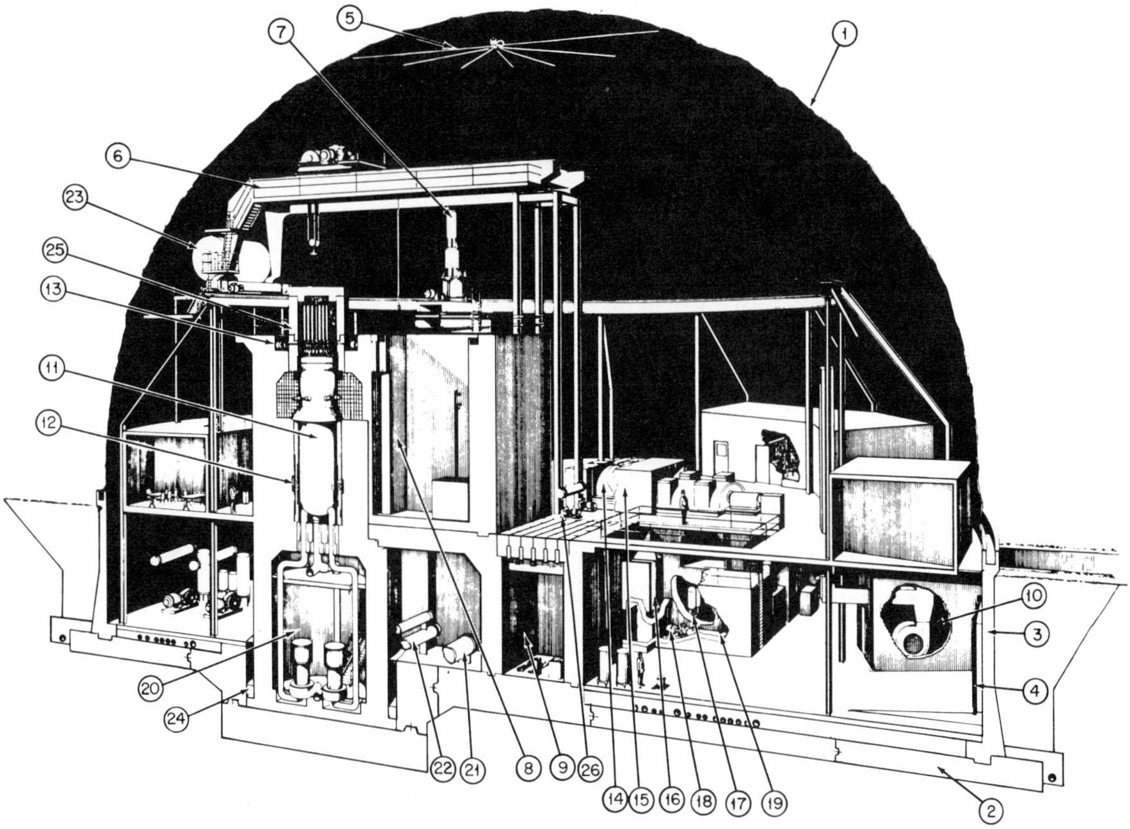

1. STEEL DOME
2. FOUNDATION MAT
3. RETAINER WALL
4. FREIGHT DOOR
5. BUILDING SPRAY
6. POLAR GANTRY CRANE
7. FUEL UNLOADING COFFIN
8. SPENT-FUEL STORAGE POOL
9. SOLID-RADIOACTIVE-WASTE STORAGE
10. BUILDING VENTILATION INTAKE FAN
11. REACTOR PRESSURE VESSEL
12. NEUTRON-SHIELD TANK
13. CONTROL-ROD-DRIVE MOTOR TRENCH
14. TURBINE-GENERATOR
15. TURBINE SHIELD
16. CONDENSER
17. CONDENSATE PUMPS
18. EVACUATOR PUMP
19. GLAND SEAL CONDENSER
20. REACTOR CIRCULATING-WATER PUMP ROOM
21. STARTUP HEATER
22. REACTOR WATER-PURIFICATION COOLERS
23. EMERGENCY CONDENSER
24. REACTOR PIT WATER MOAT
25. REMOVABLE CONCRETE SHIELD
26. FUEL POOL COOLING SYSTEM

FIG. 4-6 BONUS containment building (from ref. [150]).

4.3 Filter-Charcoal Adsorber Systems

4.3.1 General

Filters and charcoal adsorbers for the removal of iodine and particulates may be provided for cleaning the containment atmosphere on a recirculating basis before a substantial amount of radioactivity leaks out or before treated air is released to the environment. The filter-charcoal adsorber beds can conveniently be included as a part of the air-recirculation cooling systems in the containment and in many cases provide an engineered fission-product removal device with only a modest additional investment in terms of equipment. Much development and testing work has been done on the performance characteristics of filter and absorber media and these systems have been installed in many power plants. Filter-absorber systems can serve to complement the containment sprays or any other methods used for removing fission products during an accident, or can be used independently as an engineered removal system.

In order for these plants to claim any benefit from the filter-absorber systems, the reliability of the overall system must be demonstrated, as well as the efficiency of the filtering system under the expected accident atmosphere. Provision must be made for the unexpected, such as fires in charcoal filters initiated by decay heat of fission products; attrition of the charcoal beds, commonly called "dusting"; and blanketing of charcoal filters or metallic absorbers with water vapor, which may reduce the efficiency. The efficiency of any

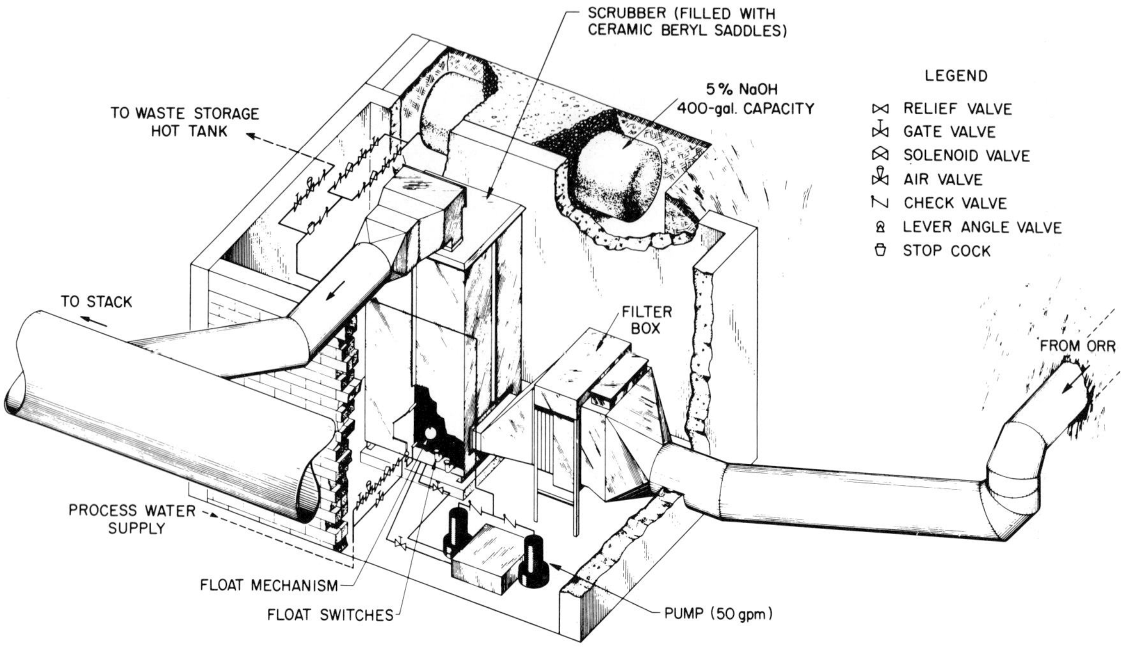

FIG. 4-7 Oak Ridge research reactor caustic scrubber (from ref. [151]).

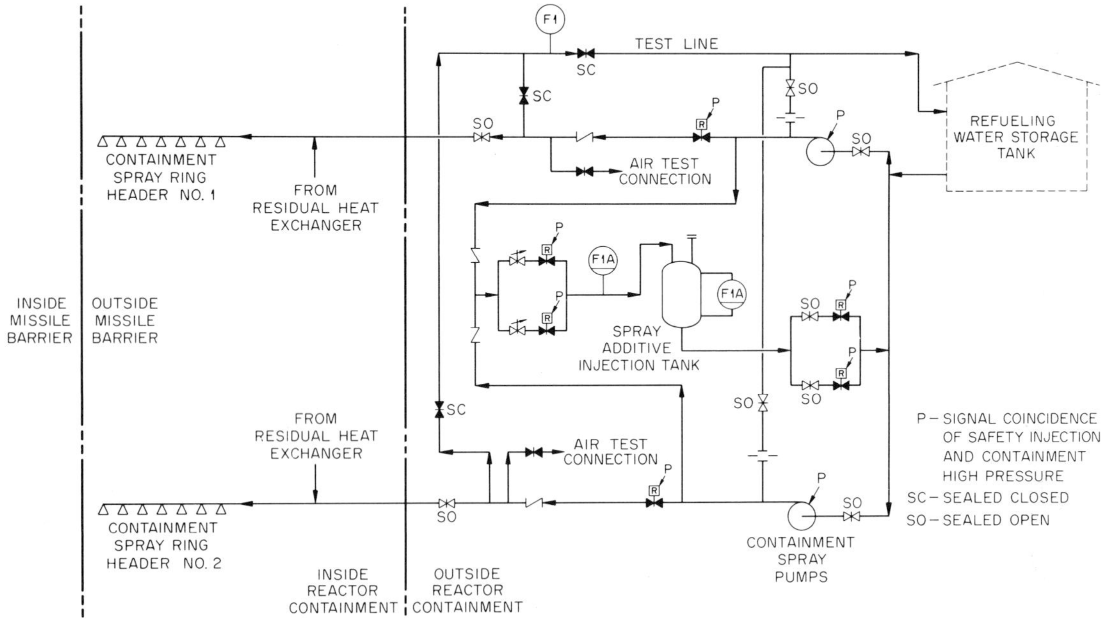

FIG. 4-8 Containment spray system [152].

iodine removal system also depends strongly upon the chemical form of the iodine.

Two important considerations are the general reliability of the blowers, filters, filter housings, seals, etc., and the relative invulnerability of the system to damage from particles, missiles, chemical reagents, vapors, shock waves, etc. Furthermore, adequate cooling must be provided for filters (as well as other absorbers) to remove decay heat from the fission products that are collected. It may also be necessary to shield the filters (or absorbers) if a significant amount of fission-product activity is expected.

Two types of filtering systems are widely included as engineered safety features in water-cooled power reactors at the present time (see Sec. 4.3.4): once-through filter-absorber systems in secondary containment exhaust lines, and recir-

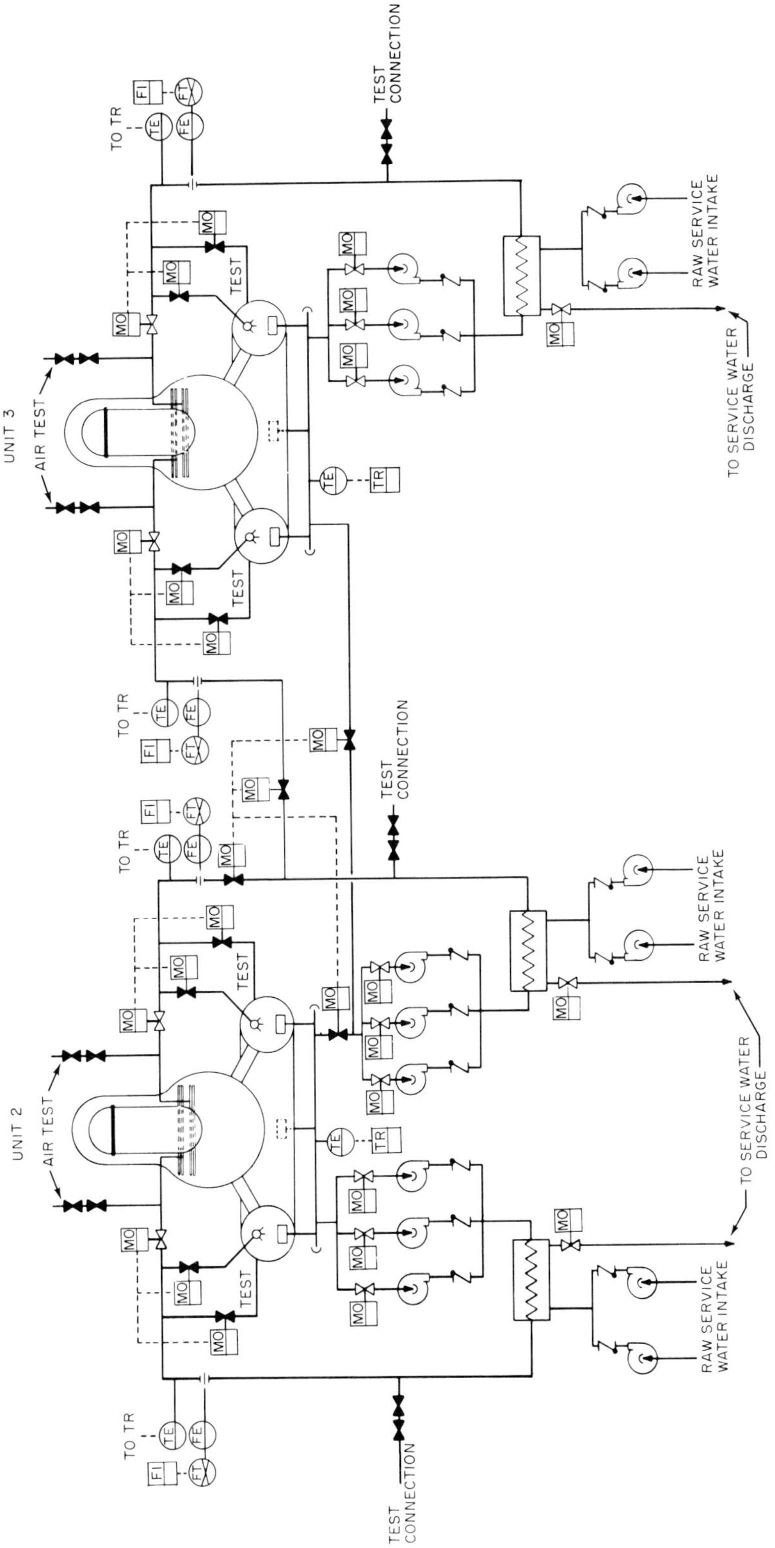

FIG. 4-9 Pressure-suppression containment spray system for Dresden-3.

culating filter-absorber systems within the containment shell. Both types have the same basic components: moisture deentrainers, prefilters, HEPA filters, and charcoal adsorbers. The major difference in these two systems is what is done with the filtered air. In the recirculating system, it is discharged back to the containment vessel, thus reducing the concentration of activity in the containment vessel by an amount that is a function of the filtering system performance. The filter-charcoal adsorber system is thus an alternate or complementary method to chemical sprays for a recirculation system such as is used in PWR confinements like Indian Point. The once-through system discharges the filtered air directly to the environment, usually through a stack. This discharged air is closely monitored for fission-product concentration.

In reactor installations as well as in laboratories and processing plants that handle radioactive materials, it is general practice to control contamination at the source. This requires utilization of special containments, such as work rooms, ventilation hoods, and glove boxes, for specific processes. In particularly hazardous programs it is necessary to consider the concentrations that could arise under accident conditions and to provide for control of such an emergency.

The practice of containing radioactivity at each source has several advantages. It reduces the spread of contamination, the deposition of activity in the ducts leading to large off-gas collection systems, the possibility that excess radioactivity from a large number of laboratories will accumulate to cause the sudden overloading of the main stack off-gas system, and the possibility that pressure surges somewhere in the system will force radioactivity into otherwise clean regions. Local removal systems can usually operate at low velocities and at higher efficiencies than the main off-gas system.

Where high concentrations of dust are involved, a large portion of the more massive particles may be removed locally by a prefilter or by a roughing-filter system. To avoid the deposition of airborne dust in the duct system before the prefilter, its design requires a suitably high air velocity. Deposition is particularly undesirable for combustible materials or for gamma-radioactive materials. If the contamination of the air after local treatment is still above the maximum allowable concentration and if high-efficiency (HEPA) filters are necessary, a precleaning device is often required to bring the dust loading down to a value suitable for economical operation of the high-efficiency filters. In a reactor containment for a water-cooled power reactor under accident conditions, premoisture removal is probably necessary and a recirculation cooling system may also be required. When very high efficiencies are required, several high-efficiency filters may be assembled in series; the effect of such cascade filtration on efficiency has been calculated and checked experimentally. In most reactor installations a stack is used to give a dilution effect. However, radioactivity concentration in the air from the air-cleaning system may be higher than the maximum allowable concentration at the site boundary if credit is allowed in the calculations for the stack dilution factor.

As dust collects on the filters during operation, the designer must be concerned with the economic balance between overcoming high pressure drop and frequently changing prefilters and main filters. Pressure-drop indicators, perhaps with alarms, must be installed and rules must be set up for their use to indicate filter charge frequency. The possibility of pressures high enough to produce a break in the filter integrity must be considered. Other physical conditions, such as humidity and temperature, must also be considered in relation to the performance of the filter. Also, certain physical conditions which may exist under abnormal or emergency conditions, as well as those existing in normal operation, must be anticipated.

The changing of filter units at frequent intervals necessitates careful planning of the initial installation and the methods used in the filter replacement process. This is particularly important in the case of radioactive materials, where there is danger of spreading the contamination around the area. Careful design of the installation with regard to removal and replacement operations will avoid unnecessary contamination of personnel and surrounding areas. Transportation and disposal of radioactive filters should be carefully planned.

The safety features of the system should be thoroughly considered in relation to the prevention of fires and the possible spread of radioactive materials under accident conditions. The location of the final filters in a system should always be remote from a possible source of fire. Protection from incandescent solid particles traveling along the ducting can be provided by a metal-screened glass-fiber prefilter which is placed well upstream from the final filters.

With the design and installation of a complete air-cleaning system, particular attention should be directed to improving the accessibility of individual units for periodic testing and for ease of maintenance and removal. This will be a convenient aid in testing and maintaining filters over a longer period of operation at lower cost. Instrumentation should be simple and reliable and should provide adequate warnings of any changes which might lead to hazards inside the plant or in outside facilities. Control panels should include information such as air-flow conditions at key positions in the system, pressure readings across filter banks, activity concentrations measured by air samplers, and indications from fire-detection systems.

The construction material for air-cleaning systems should be carefully selected according to the following safety considerations:

1. strength to withstand credible accident conditions in case of plant equipment failures;
2. corrosion resistance, particularly when associated with chemical and metallurgical processes;
3. surface finish as an aid in decontamination procedures;
4. fire resistance to protect against fires occurring either inside or outside the containment system;
5. long operating life to avoid early replacement of filters under hazardous conditions.

Plant, equipment, and materials must be reliable, since maintenance, repairs, and replacements must often be carried out under hazardous conditions. Particulate filter units and prefilters should be readily accessible and easily removable. Ease in changing a filter unit will reduce maintenance costs and releases of contamination into clean areas. Plant maintenance, repairs, or replacements and the arrangements for carrying out such work must be carefully planned in advance.

4.3.2 High Efficiency Particulate Air (HEPA) Filters

Although the final basis for acceptance of air-cleaning systems as engineered fission-product removal devices rests upon the performance of the system as a whole, each of the major components must perform its function. Thus it makes sense to discuss the components separately and gain a better appreciation of their capabilities as fission-product removal devices.

As explained in the previous section, high efficiency particulate filters are an integral part of most air-cleaning systems for power reactors. In this section, information pertaining to the installation, testing, and performance of these filters is discussed.

High efficiency particulate filters used in radioactive ventilation and exhaust systems are described by White and Smith [154], with reference to the developments in new filter media and in filter design during the 1950s. Newer designs are fully noncombustible and show greatly increased economy over older types. Again, consideration must be given to the use of primary separators where the dust burden is high.

Mulcaster [155] explains the requirements for high efficiency filters in the nuclear industries. This is followed by a survey of their development during the 1950s in England. Finally, the applications of high efficiency filters in nuclear industry, mechanical industry, medical research laboratories, and air-conditioning plants are described.

Smith and White [154] describe the different fibrous filters commercially available, outline the information required when designing a filtration plant, and report on a typical installation.

4.3.2.1 Preinstallation Efficiency Testing of HEPA Filters

The preinstalled efficiency of new high efficiency particulate air (HEPA) filters in the United States is generally tested following the procedures developed at Edgewood Arsenal [156]. The test consists of measuring the efficiency with which it removes thermally generated monodisperse dioctylphthalate (DOP) of 0.3μ particle diam from an air stream. The manufacturer tests the efficiency to be not less than 99.97% removal before acceptance. Most HEPA filters test at substantially higher removal values [157]. The British standards specify tests with particles of sodium chloride [158] or methylene blue [159]. The Dust Research Institute in Bonn, West Germany, uses a combination test with three kinds of solid and liquid particles [160]. There exist two USAEC Quality Assurance Stations with filter testing facilities for verifying the efficiency of new HEPA filters by subjecting them to a standard test procedure with carefully calibrated standard measuring instruments. This service is available on request or when specified by the purchaser [161].

The United States procedure calls for installation of the unit in a duct through which flows air containing concentrations of the DOP aerosol. Concentrations of the aerosol upstream and downstream from the filter are then measured in collected air samples. The AEC recommends testing at 20% and 100% of rated flow to test for pinholes [162] and leaks that cannot be found by visual inspection or at rated flow alone [163].

The pinhole effect occurs because the flow through the unharmed part of the filter is essentially stream line and the flow rate is proportional to the pressure drop, while the flow rate through the pinhole is turbulent and is proportional to the square root of the pressure drop. Therefore if pinhole defects exist in a filter, the observed efficiency of the filter increases with the flow rate, since the fraction of air passing through the pinhole is reduced at higher pressure drops. Thomas [164] and Adley and Anderson [165] have considered the theory of this effect. Of course, if the diameter of the leakage path increased in size with an increasing pressure drop, one would anticipate a lower filter efficiency with increasing flow and would search for a leaky flexible gasket or a filter loose in its frame. Tests of actual filters [166, 167] have demonstrated the validity of this theory.

A statistical analysis of filter test results showed a 95% probability of significant pinholing if the difference in penetrations between the full-flow test and the 20%-flow test exceeded 0.01% [168]. It is recommended that a requirement for two-flow testing be established with a maximum permissible penetration difference of 0.01% between the two flows for 500-cfm and larger filter units. It is generally considered that testing of smaller units at 20% operated flow is not practical as the low flow velocity results are too erratic.

These filters are quite delicate and must be handled, shipped, and installed with care. Tests by the manufacturer or a Quality Assurance Station does not ensure that the filter will perform as expected when installed.

4.3.2.2 Efficiency of HEPA Filters Under Simulated Accident Conditions

After an accident, the HEPA filter may be exposed to an aerosol of oxides of uranium and cladding together with fission products. If the reactor is water cooled, the atmosphere will be one of saturated steam at about 275° F. These conditions are quite different from those under which a new filter is tested.

Laboratory-scale tests have been run at ORNL on assemblies of three 1.5 in.-diam disks of HEPA filter media mounted in series, using aerosols of oxides of stainless steel and UO_2. The aerosol was generated in an electric-arc furnace. The tests were run as a function of gas velocity and humidity, and the amount of water in the filter media [169, 170, 171]. No full-scale tests have been carried out.

Six waterproofed HEPA filter media (Flanders-700, Flanders-800, AAF Type A57, AAF1, MSA Ultra HEPA and Cambridge 115EWP) and two nonwaterproof HEPA media (Flanders-600 and Cambridge 115E) were tested. The approximate linear velocity equivalent to the rated volumetric test flow for DOP testing was 5 ft/min. At this velocity in a dry atmosphere at room temperature, the stainless steel-uranium oxide aerosol mixture showed 99.97 ± 0.01% removal efficiency by the Flanders-700 media in 15 measurements.

In a water-saturated atmosphere at the same flow rate, the two nonwaterproofed HEPA filter media showed efficiencies of 99.84% and 99.80%. Two of the waterproofed media (Cambridge 115 EWP and AAF1) were also less efficient (99.93% and 99.92%) than the others which tested between 99.97% and 99.98%. Each of these efficiencies under saturated conditions is an average of 3 to 4 measurements.

Tests of all eight HEPA media in a water-saturated atmosphere were run at flow velocities of 3.5, 5, 7.5, and 10 ft/min. The results shown in Fig. 4-10 indicate that the efficiency was relatively constant up to 10 ft/min but dropped markedly at that value. The Flanders-800 and MSA Ultra HEPA, with measured efficiencies of $99.97^{+}\%$ when dry, had average efficiencies of 99.79% and 99.91% after storage in 100% humidity air at room temperature for 12 to 13 days, 99.31% and 99.46% after storage under the same conditions for 43 to 45 days, and 99.79% and 99.72% after storage in 100% humidity air at 80°C for 24 to 28 hr. Samples of one of the nonwaterproofed media, the Flanders-600, had an average efficiency of 98.89% after one drop of water was put on each 1.5-cm disk just before testing.

Measurements of the fiber diameters in four of the eight media showed two ranges of fiber diameter, one centered at about 2μ and one at about 0.15μ. It has been postulated and borne out by calculations that the moisture may condense on the 0.15μ fibers and cause a consequent increase in apparent fiber diameter. Also, high relative humidity has been shown to decrease the size of agglomerates of uranium and stainless-steel oxides by at least an order of magnitude and make them harder to filter [170-173]. Particle sizes were inferred indirectly by fibrous-filter analyzers [174] and by electron microscope examination. The dry agglomerates were chains of primary particles a few microns in length. At high aerosol concentrations, almost 80% of the agglomerates were greater than 10μ in diameter. In contrast, aerosols generated in 80% to 90% relative humidity air contained compact, approximately spherical agglomerates 2μ to 3μ in diameter. Passage of either wet- or dry-generated aerosols through water reduced the diameters to 0.1μ to 1μ. Thus it is conceivable that use of recirculating water sprays will result in reducing agglomerate size and causing a finer aerosol to be generated or recirculated than originally existed. Attention should be

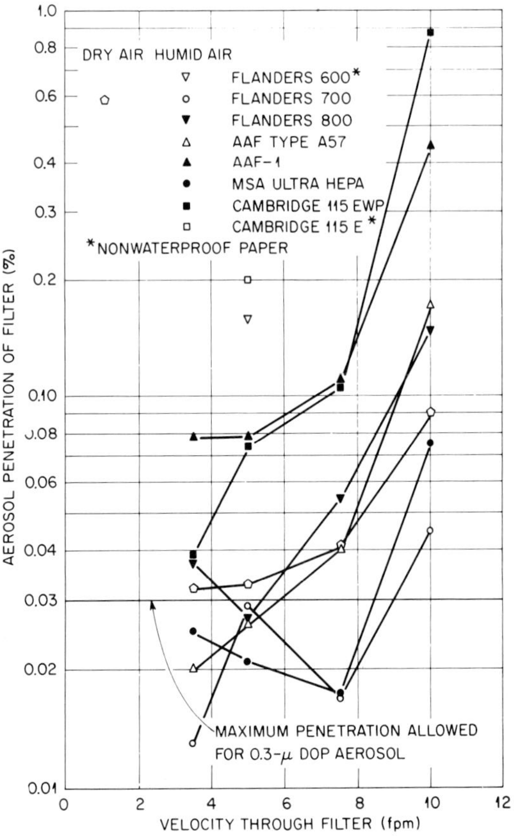

FIG. 4-10 Penetration of filter media by stainless-steel-UO_2 aerosols under dry and humid conditions (from ref. [169]).

given to the methods used to filter the water before recirculation to respray.

A dry aerosol of steel and uranium oxides was filtered with relatively high efficiency by roughing-filter packs of Dacron mats (fiber diameter ~ 11μ) in series. From 80% to 90% was retained on the first two mats. However, in a humid atmosphere, often over 50% of the aerosol penetrated completely through a pack of 8 to 12 mats. Thus such roughing filters would not appear to be very effective in postaccident conditions for water-cooled power reactors. Their use could lead to overloading of the HEPA filters which follow [172]. The filter efficiencies of the Dacron fibers could be affected by the melting of the fibers, but tests conducted to observe such effects showed none.

4.3.2.3 Filter Efficiency Under Operating Conditions

It is necessary to know at all times that the filter system is capable of carrying out its function properly. The filter media, separators, gaskets, seals, frames, and other components must be properly installed, properly operating, and must not have suffered deterioration. The effects of the accident itself must not harm the filter efficiency and postaccident conditions must be accounted for by properly pretreating the air before it reaches the HEPA filters. A manual on design and construction of high efficiency filtration systems [175] points out many gross errors made in the installation or maintenance of filter systems. In some cases, gaps were observed around the edges of sagging filter banks. In a number of cases, the HEPA filters were installed with the pleats horizontal instead of vertical, which increased stresses beyond design limits. Often no space was left for inspection or maintenance of the filters. Often the frames did not have enough members or had unwelded seams.

In order to prevent the HEPA filters from plugging or deteriorating due to excessive water, high efficiency moisture de-entrainers (HEMD) are now usually employed. All the second generation recirculation filtration systems in PWRs use such units. The second-generation BWR systems with vapor suppression also use HEMD units in their once-through secondary containment cleanup systems.

The Savannah River AEC Laboratories have carried out tests on such systems and have prepared specifications for moisture separators, water-repellent HEPA filters, and activated charcoal beds [176, 177, 178]. Each production reactor building at the Savannah River Plant has five parallel filter compartments, each containing in series a bank of moisture separators, a bank of HEPA filters, and a bank of activated charcoal beds. The moisture separators (York M321 Demisters) are mats of DuPont Teflon yarn woven on stainless steel wire and wrapped with stainless steel reinforcing wire in a form 2 ft × 2 ft × 2 in. thick and rated for 1600 scfm (standard cubic feet/minute) air at 0.95 in. H_2O-pressure. The HEPA filter banks are 1000-scfm open-face steel-cased units 24 × 24 × 11 1/2 in.^{3}, designed to remove more than 99% of all airborne particles greater than 0.3μ in diameter, even in a fog stream.

With the moisture separators, the HEPA filters met and exceeded specifications in passing flows of wet steam and fog mixtures for 10 days, even when the filters contained dust equivalent to 18 months of constant use [176]. In one test, wet steam was emitted for 30 sec at 7000 scfm/filter, and the subsequent evaporation caused mixtures of steam, air, and entrained water to be evolved at rates up to 1 lb of water/min per filter for ten days [177]. Partial plugging of the dusty filters by water that initially escaped the separator reduced the mixed flow to a minimum of 60% of the 1000 scfm-rated air flow 6 hr after the test started. The flow gradually increased to 900 scfm at the end of the tests. Filters tested without moisture separators upstream were almost completely plugged, and some actually ruptured during the fog test.

The service life of the moisture separator is indefinite. They are cleaned about once a year with steam to restore their original flow and pressure drop characteristics. The HEPA service life has been observed to be about two years at Savannah River.

A set of tests was run by the Clean Air Group Research Laboratory, American Air Filter Company, for the Connecticut Yankee Atomic Power Company. The moisture separator used consisted of waste-plate steel baffles followed by three successive 2-in.-thick fiberglass pads with a 1-in. space between the second and third pad [179]. HEPA filters were located downstream from the moisture separators and withstood test periods of up to 24 hr at predicted maximum conditions without measurable loss of performance. The test unit could circulate up to 1000 cfm of saturated air-steam mixtures at as high as 40 psig and 261° F. The moisture separators removed essentially all droplets under six test pressure conditions ranging from 10 to 40 psig.

The Underwriters' Laboratory has issued Laboratory Standard UL-586 [180] covering the specifications and tests for fire-resistant HEPA filters. Hot air at 700° ± 50° F is passed through the filter for five minutes at rated air flow. A spot flame at 1750° ± 50° F is directed against each of several points on the upstream face of the filter for five min while the filter is operating at rated flow. The hot air test checks the effect of hot air on filter performance and the spot flame test is a test of combustibility [181]. Only UL-labeled filters should be used in reactor installations.

It is necessary that the filter system not be damaged or ruptured by the overpressure due to shock waves. The Naval Research Laboratory and the Naval Ordnance Laboratory (NRL-NOL) have subjected HEPA filters of the sizes common to AEC programs to shock waves that produced a range of overpressures [182]. Small test charges were exploded in the chamber of a 6-in. gun. The muzzle of the gun was fastened to a tube 180 ft long and tapered uniformly from a diameter of 6 in. at the gun muzzle to 30 in. at the other end. The HEPA filter to be tested was mounted in a steel plate at the end of the tube. The positive duration of the shock wave (time for the overpressure to decay from peak pressure to zero) was observed to be almost constant, regardless of explosive charge, at about 50 msec. If the charge had been a spherical one, the positive duration would be expected to increase with charge size and could be 100 to 200 msec in a reactor accident. Variously sized filters from the same manufacturer were tested. Clean filters "failed" to pass a DOP test after the shock wave test with an efficiency of 99.97% at the overpressures shown in Table 4-13.

As can be seen, for a given thickness the ability to withstand shock increases with decreasing surface area. Pressures near the failure pressures caused cracks in the adhesive or small leaks at

TABLE 4-13.

Filter Failure Due to Shock Waves

Filter size (in.^{3})	Overpressure at failure (psi)
8 × 8 × 3	3.6
8 × 8 × 6	4.5
12 × 12 × 6	3.6
24 × 24 × 6	2.2
24 × 24 × 12	3.2

the media-adhesive bond. Pressures 0.5 to 1 psi greater than the failure pressure caused blowout slits in the downstream folds of the pleats. Pressures more than 2 psi greater than failure pressure caused extensive damage, including long cracks perpendicular to the pleats. Very high pressures, over 5 psi greater than the failure pressure, caused gross damage, in some cases blowing out all the media and separators.

NRL-NOL also conducted tests on "end of life" filters loaded with a test dust that had a number median diameter of about 0.5μ and a mass median diameter of $1.8\ \mu$. The loading resulted in a pressure drop of 4 in. H_2O at rated flow, though actual filters may be loaded much higher. When tested with the dust-loaded side towards the gun, a $8 \times 8 \times 6$ in.^{3} filter failed at 3.9 psi rather than 4.5 psi overpressure. A $24 \times 24 \times 12$ in.^{3} filter failed at 2.9 psi rather than 3.2 overpressure. The decrease was about 12% in both cases.

Some units of the $24 \times 24 \times 6$ in.^{3} size were obtained without the standard 1/4-in. wire mesh hardware cloth face guards. When tested, these failed at overpressures of about 1.4 psi rather than 2.2 psi, a decrease of about 40% in resistance. Burchsted [157] has plotted shock overpressure resistance of clean HEPA filters without face guards as a function of the ratio of thickness to face area. For three thicknesses, about 3, 6, and 12 in., he obtained a family of three straight lines of almost identical slope, showing the need for face guards and center support for filters with large face areas or small thickness.

It is clear that a loss-of-coolant accident can result in unbalanced overpressures on filters, even when they are part of a recirculation system fully within the containment. Reactor air cleanup system designers must consider this problem in the design as well.

It is obvious from the above discussions that the efficiency of the filter system must be tested after the initial installation is complete and must also be periodically tested thereafter. Filters to be used in a once-through system in the secondary containment exhaust lines are not likely to be subjected to large loadings of fission products or to steam water loadings, and DOP tests carried out on such systems in an ordinary atmospheric environment are likely to be indicative of the system efficiency in the event of an accident. The filter efficiency of these once-through systems must normally be higher than that of recirculating systems. On the other hand, the recirculating filter system must have high component and system integrity. A recirculating system can have lower filter efficiency since it is the efficiency/air-change multiplied by the rate of air change (usually changes/hour) that counts in such a system.

Inplace tests of HEPA filter systems are final proof tests carried out with the dual purpose of either establishing that the filter efficiency is within specification or showing that there is excessive leakage and locating its source(s). ORNL has described an inplace test using polydisperse DOP, produced by atomization with compressed air [166, 183, 184]. The DOP aerosol is introduced into any convenient air intake, far enough upstream of the HEPA filter section so that thorough mixing occurs before the filter bank is reached. Air samples are taken both upstream and downstream of the filter bank. The downstream sample may be taken after the exhaust blower to ensure thorough mixing. (In that case blower-shaft-seal leakage must be negligible.) Sampling lines should not be too long to avoid deposition problems in the line. In the inplace DOP tests at ORNL, no difference was found in the measured efficiency of a HEPA filter system when a 2-1/4 in.-thick (two 1-1/8-in. beds in series) bed of charcoal was interposed between the HEPA filter bank and the downstream sampling point [185]. If high leakage is observed, the leakage paths can usually be detected by passing the aerosol into the upstream flow and probing the downstream filter face with a probe directly connected to a photometer.

Similar procedures are also followed by LASL [186] for inplace testing of HEPA units at the UHTREX. A proposed standard, "Efficiency Testing of Air-Cleaning Systems Containing Devices for Removal of Particulates," was drafted by USASI Task Force Group N5.2.11, and will be a USA standard.

In place testing can be greatly facilitated by proper planning in the design stage. There should be built-in test facilities, room for inspection, adequate sized ducts, etc.

Although the compressed-air-induced DOP aerosol particulates are usually about 0.7μ in diameter while the standard thermally-induced aerosol is only 0.3μ in diameter, this particle-size difference has little or no effect on test results. Most particles detected on the downstream side pass through holes and continuous leakage paths which are very much larger than either of these particle sizes.

4.3.3 Charcoal Adsorber Systems

4.3.3.1 General Considerations

The removal of radioiodine from gases has been the subject of extensive reviews [151, 187, 188, 189, 190] which covered removal by a variety of methods and materials. The processes which have proved most practical are sorption on a special material and scrubbing with a reactive solution. Materials used included activated charcoal, beds of silver-plated wire or heated silver nitrate, silver or copper mesh, caustic scrubbers, and dry soda lime.

The literature [191-193] containing information on the iodine-trapping efficiency of these materials compares the effects of variations in iodine concentration, gas velocity, temperature, as well as the effects of the concentration of airborne impurities on the efficiencies for iodine removal from air, steam-air mixtures, helium, and carbon dioxide. Selection, design, testing, and efficiency of removal systems are also reviewed.

Activated charcoal beds have a high efficiency for trapping radioactive molecular iodine if (a) the charcoal has not lost its efficiency due to long exposure to moisture or impurities, (b) it has not settled or moved so that air bypasses the charcoal through leakage paths, and (c) waterlogging is prevented.

The high efficiency particulate filter and charcoal bed combination seems to be generally accepted as the best available trapping system for iodine released from irradiated fuel (an additional high efficiency particulate filter downstream of the charcoal to collect any charcoal dust is recommended). The principal shortcomings of this system are the susceptibility of charcoal to ignition and its rather poor retention of alkyl iodides in moist air. Organic iodides must be removed by use of activated charcoal impregnated with substances containing nonradioactive iodine which will exchange with the organic iodide. Partly because the biological effect of organic iodides has not been determined and partly because the fraction of released iodine converted to this form in reactor accidents is not yet well defined, this subject is still under active investigation. In any event, much progress has been made in understanding the behavior of alkyl iodides in trapping systems since the recognition of their presence in iodine sources and simulated reactor accidents. Hence a satisfactory solution to the problem of trapping these materials seems likely.

The possibility of ignition of charcoal beds in reactor accidents is difficult to assess. In situations where a significant probability of this type of failure exists, or where the resulting iodine release would have disastrous results, it seems prudent to consider the installation of a noncombustible heat sink, such as granular alumina, ahead of the charcoal bed. Beds of this type might also serve to delay sufficiently the arrival of iodine at the charcoal beds to allow at least partial decay of short-lived isotopes, thus alleviating the decay-heat problem. Automatically activated fire extinguishing systems offer an alternate method of minimizing ignition problems. What happens to the filter efficiency and the previously retained iodine in the event of a fire and a subsequent use of fire extinguishers is not clear.

Final testing of installed iodine trapping systems should be performed with the most realistic iodine sources available, but faults in the trap components or in their installation may be detected by other, less hazardous means, like the Freon test method. Further study of methods simulating iodine release from high burnup fuel is required in order to improve the realism of such tests with a minimum of hazard.

Chapter 4 of U.S. Reactor Containment Technology [3] reviews the problem of radioactive iodine removal from the containment. Keilholtz and Barton [8] discuss the information available through 1964 on the behavior of iodine in containment systems. Keilholtz, Guthrie, and Battle [9] discuss the performance and testing of charcoal adsorbers and air-cleaning systems in general up to mid-1968. The latter review has been used extensively in this section. The subsections below discuss the methods for efficiency testing of such systems, types of charcoal, and operational and accident considerations.

4.3.3.2 Efficiency Testing of Charcoal Beds

In testing the efficiency of a charcoal bed, it is necessary to first see that gas leakage through the bed due to channeling or settling is not excessive. Then it is necessary to test for the removal efficiency for both molecular iodine (it is assumed that the iodine on particulates was removed in an upstream HEPA filter) and organic iodides. In order to check that the bed is not near saturation, the total capacity of a representative sample of charcoal for both molecular iodine and methyl iodide should be measured. While no industrial standards yet exist, the general procedures are clear.

It is technologically and economically feasible to test the efficiency of representative samples of charcoal utilizing I_2^{131} or CH_3I^{131}. Such tests for inplace charcoal beds using small amounts of radioactive iodine or organic iodides have been made at ORNL [194]. Nonradioactive iodine methods have also been used for molecular iodine detection. Gukeisen and Malaby [195] have described such a test in which the amount of iodine or methyl-iodide passed is identified by neutron activation. Viles and Silverman [196] have described a method in which the iodine passed is identified by catalytic reduction of ceric ions by iodine. Thomson and Grossman [197] used the iodine as an oxidant.

The efficiency of impregnated charcoal for the removal of organic iodides can only be tested by using appropriate organic iodide test compounds. In order to remove radioactive organic iodides from the gas, the charcoal is impregnated with one or more substances containing nonradioactive iodine [198, 199, 200]. When the air containing radioactive organic iodides passes through the charcoal bed, the nonradioactive iodine exchanges with the radioactive iodine, thus removing it from the air stream. The exact mechanism by which this action occurs is not clear. Some five commercially available impregnated charcoals have been shown to be effective at humidities as high as 70%. Since these may include methyl, ethyl, butyl, and perhaps other iodides, and since the removal efficiency may be different for different iodides, a series of tests may be required.

The AEC Savannah River Laboratory routinely makes use of a leak-tight test utilizing DuPont Freon-112[201, 202]. Keilholtz, Guthrie, and Battle [9] suggest that the Freon leak test be used to provide quality assurance for all beds, whether new or recharged with charcoal. This test checks that no settling or channeling due to flooding or air passage velocities has reduced the bed efficiency. They then suggest that the ORNL-tracer radioactive-iodine tests for both molecular iodine and organic iodides be used on representative beds and/or samples to check iodine removal efficiency and bed capacity. This combination of tests is recommended by these authors as adequate.

Commercial equipment for testing charcoal beds and personnel to operate it may be available for on-site testing. A typical unit is available for testing with test gases in air-steam mixtures at a maximum velocity of 1000 cfm and a maximum temperature of 320° F (160°C). The facility can be used for radioactive or nonradioactive I_2, HI, or CH_3I, or with Freon or other compounds. It can carry out tracer analysis or gas chromatography.

As noted elsewhere, the relative humidity of HEPA filters and charcoal beds should be kept below 90%, using moisture deentrainers or other methods, because of the adverse effect of high humidity on the retention of methyl iodide by impregnated charcoals.

4.3.3.3 Unimpregnated Charcoal Efficiency

A number of tests on the efficiency of molecular iodine removal have been carried out on unimpregnated graphite charcoal under simulated accident conditions. For the N.S. Savannah Project, laboratory tests were made on charcoal units 11 in. square and 1-1/8 in. thick containing 12/30 mesh Pittsburgh BPL charcoal [203]. Iodine vapor (stable I^{127} labelled with I^{131}) was continuously injected into air 80% to 90% saturated with steam at 96° to 100° C in the concentrations predicted in a postulated reactor accident (10^{-2} mg/m^3 of steam-air). The mixture was passed through the charcoal units at 4.3 to 5.0 ft/min, giving a bed residence time of about 1.0 to 1.1 sec. The efficiency of the charcoal for removal of iodine was observed to be 99.186% ± 0.07% at the 95% confidence level. It was later discovered that what is now known to be methyl iodide may have been part of the iodine and the actual efficiency for molecular iodine may have been even higher than reported [204].

The two methods that were developed for inplace testing of adsorber systems, one using I^{131} with radioassay and the other using I^{127} with activation analysis, were applied on board the N.S. Savannah [203]. The I^{131} method was rapid and more sensitive but required precautions to avoid accidental release of I^{131}; the I^{127} method was slow and less sensitive but required fewer safety precautions. Fifty-six shipboard inplace tests, conducted under various circumstances, demonstrated the system efficiency to be greater than 99.9% and showed that the adsorbers had been installed so that their full efficiency could be realized. A method using Freon for testing adsorbers showed little promise for shipboard application because of interference by humidity at levels expected in the ship. An adsorber was developed for environmental iodine monitoring.

In 1964-65, four charcoals were tested for possible engineered safety system use in 1.5 in.-deep beds [205]. In eight molecular iodine removal tests, the air was from 56.3% to 90.7% saturated with steam at temperatures ranging from 95° to 180° C and linear gas velocities ranging from 27.9 to 39.6 ft/min. These velocities were much closer to those of current power reactor design (60-70 ft/min) for elemental iodine removal than those in the N.S. Savannah studies but showed no correlation of efficiency with velocities in the range investigated. In the eight tests, the average measured efficiency was about 98%, but the actual efficiency of molecular iodine removal may have been higher.

Some comprehensive work on the adsorption of elemental iodine on charcoal beds has been done at the Savannah River Laboratory [206]. The experiments were performed using activated carbon made from coconut shells supplied by the Barnebey-Cheney Co. and by "manufacturer A". Carbon from coconut shells has been shown to provide better retention of iodine and methyl iodide than carbon from bituminous coal or containing metallic impregnants (Whetlerite) [207]. These experiments use carbon beds 3-5/8 in. in diameter with depths of 5/8, 3/4, or 1 in. However, the edges of the beds were baffled to avoid bypass leakage, which resulted in only the 3 in.-diam core being exposed to the impinging gas stream. The experiments were performed as a function of flow rate through the bed, temperature, and humidity.

The results of these experiments showed that carbon adsorbers consistently removed 99.99% of the iodine vapor from (1) prefiltered mixtures of steam, air, and entrained liquid water particles, (2) unfiltered steam-air mixtures with rates of entrainment as high as 0.43 lb H_2O/min/filter (full-size carbon filter with 14 ft^2 of surface area), with the carbon bed initially saturated with water from the mixture. Iodine was not desorbed from the bed when the filters were exposed to the mixtures for periods as long as 2 hr. When the test filters were exposed to unfiltered steam-air mixtures with entrainment rates as high as 0.71 lb H_2O/min/filter, as much as 4.8% of the iodine was not removed. The tests were conducted with steam-air mixtures at temperatures of about 65° C and at a velocity of about 70 ft/min. This velocity was equivalent to a total volumetric flow of 1000 cfm through a full-size carbon filter.

Laboratory studies have shown that the removal efficiency of unimpregnated coconut charcoal at room temperature for methyl iodide was drastically reduced as the relative humidity was increased. The efficiency also tended to vary inversely with air velocity. The test time durations over which removal efficiencies greater than 90% were observed varied from about 100 hr at less than 3% relative humidity with a 10 ft/min air velocity and a charcoal bed depth of 3 in., to less than 0.2 hr at almost 100% relative humidity at 100 ft/min and a bed depth of 1 in. [208]. These conditions were still generally not as severe as might be encountered in a postaccident environment. The tests demonstrate the inadequacy of even unimpregnated coconut charcoals for organic iodide removals in most environments.

The charcoal unit on the EVESR at the General Electric Vallecitos Atomic Power Laboratory is said to have demonstrated a retention of 99.8% to 99.9% of organic halogens produced during power operation at a relative humidity of 10% to 15% [209].

The Savannah River Plant reactors make use of unimpregnated activated charcoal. It is believed that under the assumed accident conditions for the SRP reactors, the production of methyl iodide will be negligible [201, 210].

4.3.3.4 Impregnated Charcoal Efficiency

Since 1966 it has been known that activated coconut charcoals impregnated with one or more substances containing nonradioactive iodine have good efficiencies for the removal of radioactive methyl iodide at fairly high humidities, ambient pressures, and ambient or moderately high temperatures up to 115° F (46°C) [211, 212, 208]. More recently, tests [198, 213] of five commercial impregnated charcoals have shown effective methyl iodide removal at the temperature, pressure, and humidity conditions predicted for the containment after a loss-of-coolant accident. It is assumed that flooding of the charcoal beds can be prevented and that the humidity remain preferably below 90%. The five charcoals are identified as (1) Mine Safety Appliances Company, MSA-85851 and (2) MSA-24207, (3) Barnebey-Cheney, BC-727, and (4) BC-239, and (5) North American Carbon, Inc., G-601. While all the impregnants contain non-radioactive iodine, their exact nature is proprietary.

The tests conducted at about 270°F (132°C) and 55 psia with a steam-air velocity of 50 ft/min and with 3.5 mg of CH_3I injected per gram of charcoal. The 2 in.-deep beds of each of the five charcoals tested demonstrated efficiencies of 90% or higher for removal of I^{131} from CH_3I^{131}, provided the relative humidity of the charcoal did not exceed 90%.

The general efficiency observed is shown in Fig. 4-11. Note that if the relative humidity is less than 80% to 85%, the efficiencies observed are near 98%. Useful CH_3I^{131} trapping capability is attainable with the iodized charcoals at relative humidities higher than 90%, possibly even approaching 100%. However, at 100% relative humidity, bulk water may be associated with the charcoal and serious loss of efficiency could result [198].

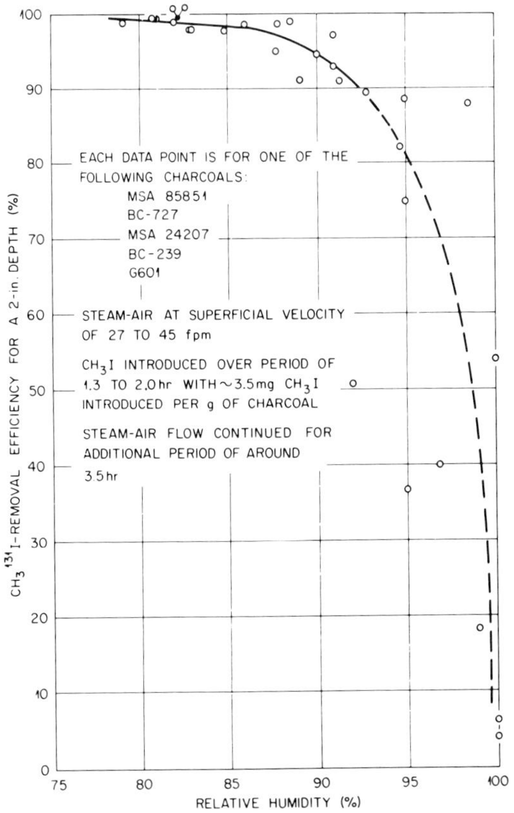

FIG. 4-11 Effect of relative humidity on the removal of I^{131} from CH_3I^{131} in flowing steam-air by commercial iodized charcoals at temperatures and pressures of around 270°F and 60 psia and bed depth of 2 in. (from ref. [213]).

These results were obtained on a once-through system, so higher total efficiencies would probably be obtained in a recirculating system. Weathering and poisoning could have adverse effects.

A heater has been added in order to reduce humidity in the charcoal beds of the Browns Ferry Plant [214, 215], a boiling-water reactor system utilizing a pressure suppression containment. The charcoal beds are located in the effluent duct of the secondary containment as indicated in Table 4-14. If perfect mixing is assumed, the humidity at an ambient temperature of 80°F (27°C) probably typical for the secondary refueling containment area, could be reduced from 100% to less than 70% by raising the temperature to about 90°F. With imperfect mixing, the temperature increase would have to be higher to insure sufficiently low humidity throughout the bed.

The UKAEA Reactor Development Laboratory at Windscale has carried out extensive tests of impregnated charcoal utilizing both organic and inorganic impregnants. Two coal-base (Sutcliffe Speakman and Company, Ltd., Type 207 B) charcoals gave the best results. The first, designated as 0.5% KI/207 B is impregnated with 0.5% potassium iodide. The other, designated 5% TEDA/207 B is impregnated with 5 wt% triethylene diamine.

The performance curve for the KI-impregnated charcoal was virtually flat over an impregnant concentration range from about 0.05 to 5.0 wt%, two orders of magnitude [216]. The use of 0.5 wt% KI lowered the ignition point of 207B charcoal from 500° C to about 350° C, approximately that of the unimpregnated charcoal. The best performance for this impregnated charcoal was considered to be in the range of 0.1 to 1 wt% KI, and 0.5 wt% was selected as the center of the range. The 0.5% KI/207B chosen was found to have an acceptable CH_3I loading of only slightly over 100 μg/g, whereas serious overloading of the 5% TEDA/207B did not occur until it was loaded to well over 1 mg/g. Because of the much larger capacity of 5% TEDA/207B, a mixture of the two impregnated charcoals was recommended. This recommendation was made in spite of the fact that TEDA was much more subject to loss by volatilization at high temperatures than was KI.

4.3.3.5 Prevention of Charcoal Ignition

In general, charcoal in an air stream will ignite at temperatures of 250° to 300° C. Once ignition starts, air flow accelerates the process. Small fires may be extinguished by cutting off the air supply, while large fires need fog sprays or water, coupled with continued air flow to cool the system and prevent reignition.

The principal source of heat is, of course, the radioactive iodine deposited in the charcoal bed. This heat can either cause combustion or may drive off the iodine that has previously been collected. Heat-resistant prefilters should be used to collect the initial load of fission products that generate excessive heat. An air precooler has been added in most of the second generation recirculation air cleanup systems to hold down the temperatures.

TABLE 4-14

Filter-Absorber Systems

Facility	Circulation or discharge rate (cfm)	Filter function	Filter description	Required performance at accident condition
CVTR	1000	Recirculation for particulate and iodine removal	Prefilter, absolute filter (99.9% removal of 0.3μ particles), and iodine-removal section of activated charcoal (1 in. deep, 90% iodine removal); two such units, one in standby	No credit taken in accident condition analysis
EGCR (1)	750 (max)	Once through; discharge to atmosphere	Prefilter (95% removal of 5μ particles), absolute filter (99% removal of 0.3μ particles), charcoal trap (95% iodine removal), and absolute filters (99% removal of 0.3μ particles)	95% removal of halogens and 99% of particulates
EGCR (2)	41,660	Recirculation system	Absolute (99.97% removal of 0.3μ particles); silver-plated copper mesh for iodine removal	No credit taken in accident condition analysis
HWCTR	Four units of 1000 each	Recirculation for particulate and iodine removal	A unit contains a moisture separator and a bed of activated (coconut shell) charcoal (56 lb) shown to remove iodine with an efficiency of > 99.9%	Credit is taken for one of the four units operating at stated efficiency (primarily for iodine removal)
Indian Point	10,000	Once through for particulate removal	Prefilters; airborne activity filters (99.97% removal of 0.3μ particles)	No credit taken in accident condition analysis
NS Savannah	Two units (1000 each), one in standby	Once through discharge to atmosphere; system for particulate and iodine removal	Prefilter, absolute filter (99.95% removal of 0.3μ particles), and iodine-removal section consisting of silver-plated copper mesh (6 in. thick), activated charcoal (1 in. effective thickness), silver-plated copper mesh (6 in. thick)	99% removal of iodine with infinite cloud release is assumed
Yankee	12,000, three units each of 4000 capacity	Recirculates container air to remove particulates 0.77 air changes/hr	Prefilters, airborne activity filters (99.95% removal of 0.3μ particles), fan	No credit taken in accident condition analysis
NPD (1)	32,000	Recirculation through reactor vault	D_2O sprays, filters, and ion exchange[a]	
NPD (2)	1000	Recirculation from boiler room to remove particulates	Filter (99.97% removal of 0.3μ particles)[a]	
Humboldt Bay (refueling building)	134	Once through; for particulates and iodine removal	Caustic scrubber and "absolute filters" (95% minimum removal of halogens and particulates)	
BONUS	50,000 (building ventilation)	Once through for particulate removal	Oil-treated prefilter and glass fiber filter	
Elk River	3500 (max)	Once through for particulate removal	Prefilter, absolute filter (99.95% removal of 0.3μ particles)	
Peach Bottom	Two units; 2000 cfm each; one in standby	Recirculation for strontium removal	Absolute (99.97% removal of 0.3μ particles)	50% removal of strontium

TABLE 4-14 (Continued)

Facility	Circulation or discharge rate (cfm)	Filter function	Filter description	Required performance at accident condition
Oconee-1, -2 (PWR)	2000[b]	Once through for secondary containment exhaust lines	In order: roughing filter, HEPA[c] filter, charcoal adsorber, fan, stack	Halogen removal efficiency of 90% for penetrations of room filters; 50% containment leakage assumed to pass through room
Millstone Point (typical BWR)	2000[b]	Once through for secondary containment pressure suppression	In order: HEMD[d], HEPA filter, charcoal adsorber, HEPA filter, fan, stack	Halogen removal efficiency of 90% (99% for Vermont Yankee and 95% for solids removal)
Browns Ferry-1, -2 (typical BWR)	2000[b]	Once through for secondary containment pressure suppression	In order: HEMD, heater, roughing filter, HEPA filter, charcoal adsorber, HEPA filter, fan, stack	Halogen removal efficiency of 90%
	4000[b]	Once through for secondary containment pressure suppression		
Conn. Yankee	50,000, four units in parallel	Recirculation, 5.37 air changes/hr	In order: HEMD, HEPA filter, charcoal absorber, cooler, fan efficiency of 50% (1 charcoal unit), enough to satisfy 10CFR100	Organic iodide removal in impregnated charcoal beds 80 to 90% and excellent efficiency for molecular iodine at 10% humidity
Indian Point -2	65,000, five units in parallel	Recirculation, 7.4 air changes/hr	In order: HEMD, cooler, roughing filter, HEPA filter, fan, HEMD, charcoal adsorber,	Elemental-iodine removal 90%, zero efficiency organic iodine assumed, efficiency of 45% enough to satisfy 10CFR100 (no credit for thiosulfate spray)
H.B. Robinson-2	100,000, four units in parallel	Recirculation, 11.4 air changes/hr	In order: HEMD, cooler, HEPA filter, fan	Uses thiosulfate spray to reduce iodine by factor 8.8/hr
Diablo Canyon	65,000, five units in parallel	Recirculation 7.5 air changes/hr	In order: Cooler, HEMD, roughing filter, HEPA filter, fan	

a In emergency, vented to stack through activated-charcoal and absolute filters.
b One spare system is provided.
c High-Efficiency Particle Air
d High-Efficiency Moisture Deentrainer

The Savannah River Laboratory (SRL) has made small-scale ignition tests on carbon beds with diameters and depths of up to 3 in. [217]. An ignition temperature of 340° to 350° C was allowed for a 1 in.-thick bed of unimpregnated activated coconut carbon (Barnebey-Cheney type 416) now used in the Savannah River Plant (SRP) confinement system. Air velocities from 10 to 125 ft/min had little effect on the ignition temperature. Exposure of the charcoal in the SRP confinement system for about two years increased the ignition temperature about 80° C. An ignition temperature of 530° C was measured at a velocity of 105 ft/min for samples of a high-temperature coconut carbon (Barnebey-Cheney type 592) packed in a 1 in.-thick bed. Tests indicate that the new carbon meets SRP specifications [201]. Activated carbon prepared from bituminous coal ignited at 480° C, coconut carbon ignited at 340° C, and Whetlerite carbon ignited at 250° C.

It has been conjectured that local "hot spots" in the graphite resulting from deposition of very radioactive particles or nonuniform iodine deposition might have the effect of lowering the ignition temperatures. However, results seemed to indicate no large changes due to such local effects [218-221].

Tests on a full-size prototype of SRP beds framed in stainless steel were carried out at the Lawrence Radiation Laboratory [201, 217]. Ignition temperatures of 340° C for type-416 carbon and 530° C for type-592 carbon were obtained which closely agreed with the small-scale tests at SRL. The airflow was held at 1000 cfm, corresponding to a face velocity of 70 ft/min.

At ignition, glowing particles of carbon were swept from the bed. Simultaneously, thermocouples in the bed suddenly indicated temperatures higher than the 540° C recorder limit. The burners and air flow were stopped immediately and carbon dioxide was sprayed on the upstream face, but it did not extinguish the fire. It was extinguished by restoring the air flow and spraying water on both faces of the bed for about 15 min.

4.3.3.6 Plant Operating Experience with Charcoal Adsorbers

The efficiency of charcoal beds for iodine removal from air as a function of length of service and complexity of the air composition has been discussed by McCormack [222]. Ten large-scale charcoal beds have been operated at Hanford for periods ranging from a few months to several years. From 6000 to 100,000 cfm of air pass through these beds. Eight of the beds are part of the reactor confinement system, one is installed in a high level radiometallurgy laboratory, and one was more recently installed at a chemical separation plant (Redox) downstream from a high efficiency filter. Efficiencies ranging from 99.9% to nearly zero were noted.

Beds in three types of installations were investigated. There seems to be a correlation between the retention of iodine-removal efficiency and the quality of the air streams. Charcoal in the first type of installation, whose air stream consisted of the reactor-building ventilation-system air (which can be thought of as clean air from a building that includes no chemical processing activities), showed no significant deterioration during one year of exposure. In the second type of system, charcoal in service for over two years in a radiometallurgy lab showed some loss of iodine retention, being only 58% efficient under test conditions. The air from the hot-cell exhaust contained impurities generated from physical testing of irradiated fuel elements and from a small amount of chemical treatment of the specimens. In the third installation, in the process-building ventilation air and vessel off-gas streams, charcoal lost its iodine removal efficiency in a period of a few weeks. This air stream contained organic and inorganic components arising from the routine processing of irradiated fuel elements.

The varying response of the charcoal behavior of these three filtered gas streams gives some indication of the complex nature of iodine and the care that should be used in applying iodine-adsorbing systems in real gas streams. In addition, because of the possible "poisoning" potential of trace components in the air being treated, the performance of charcoal beds should be periodically monitored.

The radioiodine collection efficiency of small activated charcoal cartridges was investigated by McConnon [223] under field conditions at a chemical processing plant stack. An average collection efficiency of 95% was obtained when the gas sample flow was limited to 0.1 cfm. Comparisons with results obtained with conventional caustic scrubber samplers indicate the need for further development of the charcoal cartridges. Recommendations are made to limit the use of the cartridges to situations where high I^{131} concentrations are expected, since a higher analytical precision is possible with increased I^{131} content in the cartridges.

The average collection efficiency increased from 68.7% at a superficial velocity (volumetric air flow divided by the cartridge cross-sectional area) of 480 ft/min to 96.6% at 96 ft/min. In addition, the range in sample results narrowed as the superficial velocity decreased. In other tests at Hanford (HW-73288), a standard deviation of 5.5% was achieved with samples simultaneously collected at a flow rate of 0.1 cfm.

A four-section charcoal column was installed in the ORNL Isotopes Development Center hot off-gas system to reduce atmospheric release of radioiodine from the I^{131} processing equipment [224, 225]. Since its installation, the column has removed approximately 98.7% of the radioiodine vented to the system, with an average individual section efficiency of 75%. The column contains four Dorex type H-42 charcoal cannisters enclosed in 6-in. pipe sleeves. Each cannister contains 1-1/2 lb of 8-14 mesh activated charcoal in a 3/4-in. bed in the annulus between two 10-1/2 in.-long perforated cylinders. The hot off-gas enters the inner cylinder at 3 cfm, passes through the inner wall, the charcoal bed, and the outer wall, and continues to the inner cylinder of the next section.

To obtain information on adsorption media other than charcoal, the fourth section of the column was replaced by a section heated to 100°C at the inlet and 220°C at the outlet and containing a 12 in.-high by 6 in.-diam bed of 1/4-in. beryl saddles coated with $AgNO_3$. Efficiencies ranging from 10% to 60% were found. These efficiencies were lower than in the charcoal section which had been replaced; however, the optimum operating temperature of 200°C was not maintained due to a malfunction of the section heater.

To compare the relative effects of polyethylene and stainless steel sample lines on iodine adsorption, a length of polyethylene tubing was compared with a length of stainless steel tubing of equal dimensions by connecting the two in parallel upstream of identical charcoal samplers. Equal flows were maintained. The sampler below the stainless tube had approximately 200 times as much I^{131} activity as the sampler below the polyethylene tubing.

A heated pipe was placed in the system with samplers on either end to determine the effect of temperature on the system. In several runs, no effect was seen at 260°C.

4.3.4 Nuclear Reactor Air-Cleaning Systems

As noted earlier, several types and combinations of air-cleaning systems can be used in nuclear reactors, depending upon the specific conditions of operation and the purposes in mind. Most of the systems are designed to meet the stringent requirements for fission-product cleanup imposed under serious accident conditions, but the air-cleaning systems also serve to provide maximum safety during normal operation. Table 4-15 describes the filter-absorber systems installed or planned for a number of typical reactor containments.

Once-through systems are used for the secondary containment of pressure-suppression systems for BWRs and for the secondary containments of PWRs such as the Indian Point-1 and Oconee-1 and -2 plants. Since the Oconee Plant secondary containment is unusual, Fig. 4-12 [226] shows a schematic diagram of the one symmetric half of the system for the two reactors. They are also used for systems such as Turkey Point-3 and -4, with provisions for a slow purge of containment contents. Power reactor once-through systems used HEPA filters and charcoal absorbers except the early Indian Point-1, and all pressure-suppression BWRs use moisture deentrainers. The Windscale Reactor installation also uses a once-through system in its air-cooled operation. Mossop [227] has outlined the restrictions and conditions for this installation and describes the system found most suitable for this application.

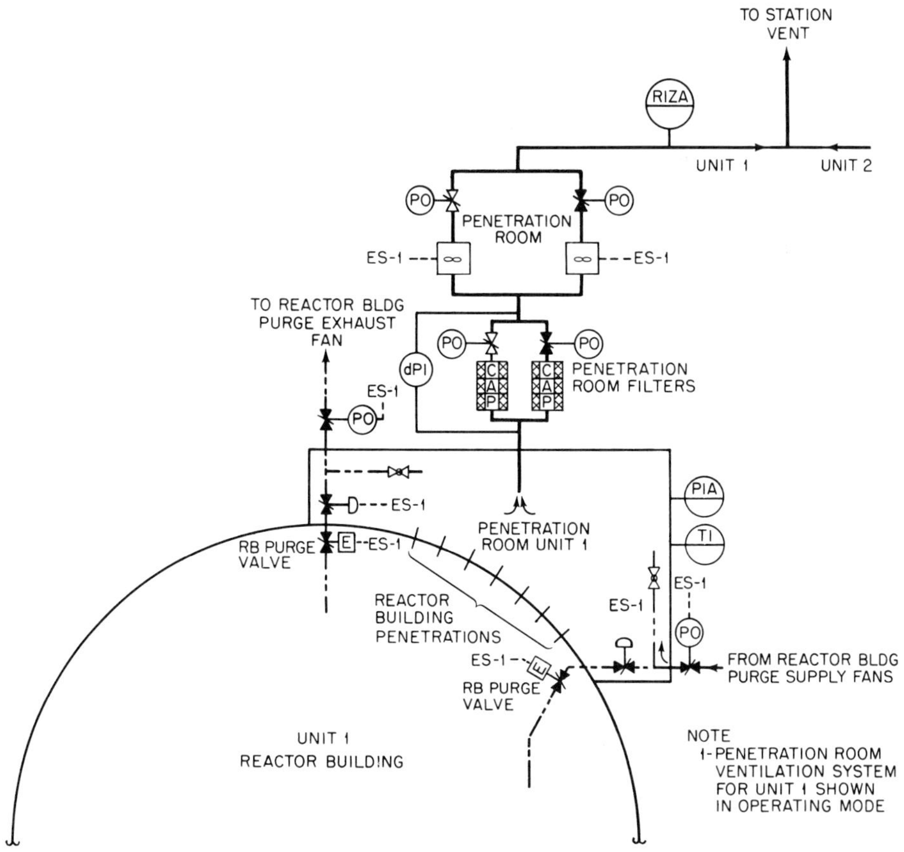

FIG. 4-12 Ventilation system for penetration room of Oconee Nuclear Station reactor building. The penetration room filters are designated C for charcoal adsorber, A for HEPA filter, and P for prefilter [226].

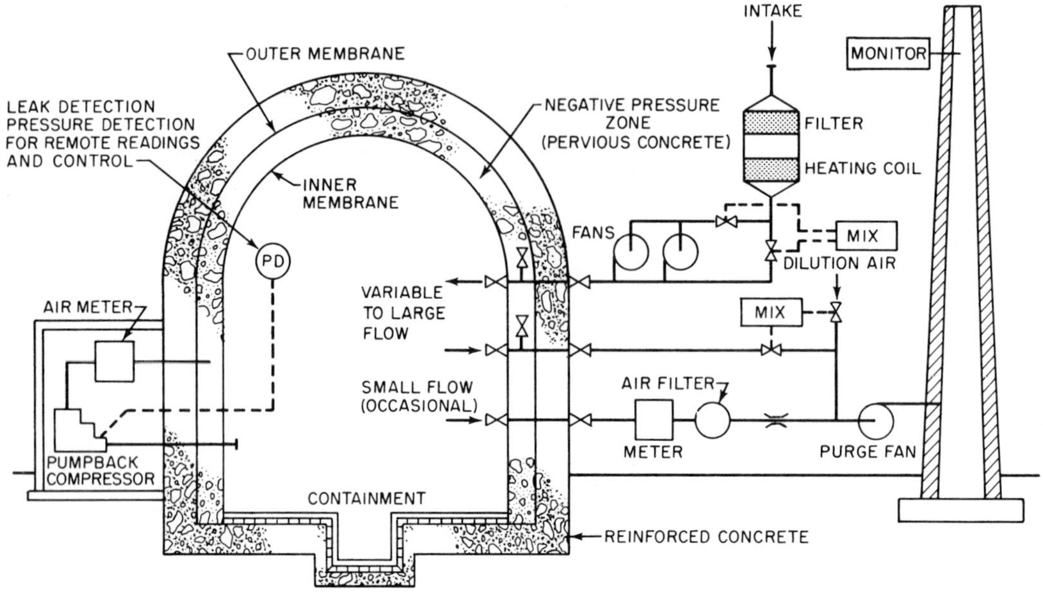

FIG. 4-13 Schematic diagram of air-cleaning system for proposed double containment concept (from ref. [228]).

A pump-back containment such as shown in Fig. 4-13 [288] was proposed for an early version of the Ravenswood Reactor and for the Malibu System [229]. In the Malibu design, the two domed steel cylinders are separated by a 20 in.-thick annulus filled with porous concrete. The annulus is kept at a pressure slightly lower than that of the outside air, and any leakage will be pumped back into the inner containment shell. Variable-to-large flow through the air-cleaning system is encountered under normal conditions and small flow under purge conditions. Recirculating air-cleaning systems, such as those designed for Connecticut Yankee, Indian Point-2, H. B. Robinson-2, and Diablo Canyon, are installed within the containment shell, where blowers induce air movement through the filter system. As may be seen in Table 4-15, the filter units usually consist of a prefilter followed by an absolute particulate (HEPA or high efficiency particulate air) filter which is in turn followed by an iodine-removal section consisting of activated charcoal or silver-plated wire mesh. Moisture deentrainers and coolers are present in most power reactor recirculator systems except the early Yankee reactor. The HWCTR, for example, used four separate systems, each with its own independent set of blowers, motors, and filters. When all are operating, they can filter 4000 cfm, or one containment volume every 1.3 hr.

A different approach to the use of filters is demonstrated by the Indian Point and N. S. Savannah systems. In these cases, the containment shell is housed in a compartment or building, and thus any leakage from the containment vessel is held up in the second compartment. It is the atmosphere of this second compartment that is filtered and fed to the stack. This concept is shown schematically in Fig. 4-14 [3].

In the New Production Reactor (NPR) constructed at Hanford, protection against fission-product release after a nuclear incident is provided by a confinement system [230]. The term confinement is used here to describe a system that permits the unfiltered release to the atmosphere of the initial burst of essentially nonradioactive steam, followed by sealing of the building and subsequent filtration of all building exhaust. This is in contrast to a containment system in which the entire release is retained within an essentially leaktight container. Either system provides adequate protection for the Hanford environs; however, the confinement approach was pursued for the NPR because of its inherently lower cost (estimated at 1/10 to 1/15 the cost of a containment system) and because of its greater versatility during construction and reaction operation. For the most serious accident, the time separation between closure of the building vents and potential fission product release is about

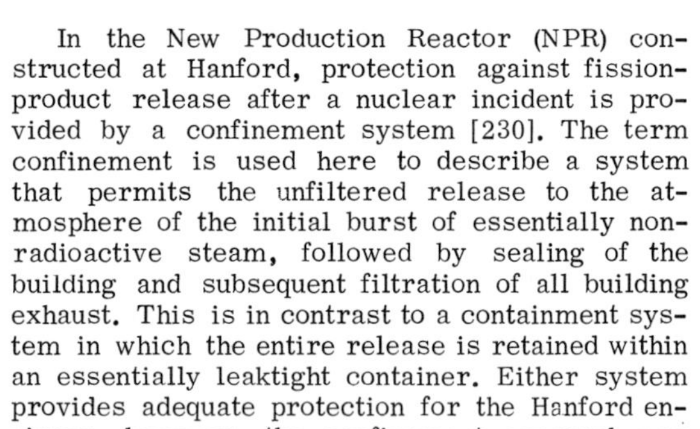

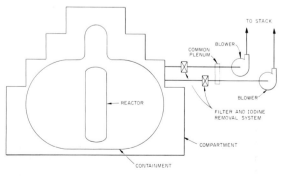

FIG. 4-14 The NS Savannah containment concept.

100 sec. Green, Thompson, and Copeland [231] describe the ventilation system, filter facility, instrumentation, and fog spray system of the Hanford reactor confinement system in some detail and discuss its testing and operational experience.

The Savannah River Laboratory [232] has developed a system to contain radioactive steam and steam-air mixtures (fog). The system filters moisture and radioactive materials from the ventilation exhaust stream and vents clean exhaust to the atmosphere. It is easily installed in existing reactor buildings that do not have containment shells, it is less expensive than other methods for reactor containment, and it may also be used with reactors in containment shells. The system consists of high efficiency moisture deentrainment devices (demisters), high efficiency particulate filters with water repellent medium, and activated carbon filters to remove radioactive iodine. For each 8000 scfm of capacity, the system contains five demisters in series with eight particulate filters followed by eight carbon filters. The pressure drop at 8000 scfm is 2.6 ± 0.1in. H_2O. The moisture separators and HEPA filter performance is described in more detail in Section 4.3.2.3.

The Piqua OMR (Organic Moderated Reactor) required processes to prevent atmospheric contamination from both radioactive and organic materials released during reactor operation [233]. Conventional air filters were demonstrated to be successful in the removal of organic coolant dusts from air streams. Activated carbon adsorbers are suitable for the reduction of organic coolant vapors to safe concentrations. Water scrubbers or steam jets precede the activated carbon adsorber in the waste-gas treatment system. Polyphenyl dusts can be removed from air streams by a coarse mesh prefilter.

The air-and-gas handling facilities at the High Flux Isotope Reactor (HFIR) at ORNL were specifically designed to minimize the spread of contamination inside and outside the building in the event of an accidental release of activity and to provide maximum safety during normal operation. Binford and Cramer [234] have described in detail the two separate systems which are provided for the disposal of gaseous waste: the special-building hot-exhaust system, which provides dynamic containment in the event of an abnormal release of activity, and the hot off-gas system, which handles the routine disposal of gaseous activity from the various system components. Certain areas such as the control room and offices, which are normally clean areas, are isolated from the other portions of the building and are served by a separate air-conditioning system.

The general concept of dynamic containment consists of maintaining an inward leakage of air into the contained region by creating a partial vacuum in that region. The air exhausted from the contained region is decontaminated by suitable filtering equipment and is discharged to the atmosphere in a manner designed to eliminate excessive environmental pollution.

The special-building hot-exhaust system is designed to provide a constant flow of air through those portions of the building which contain components capable of releasing significant quantities of activity. These areas include the primary heat exchanger cells and pipe tunnel, the reactor bay, the primary coolant deaerator and demineralizer cells, and the primary coolant demineralizer pump cell. Certain other areas, including the beam room, experiment room, and the cells containing the pool cleanup equipment, are also served by the system because of the configuration of the building and the convenience in locating the equipment. By maintaining appropriate pressure gradients, clean air flows into the areas served by the system, then through ducts to appropriate filtering equipment, and finally into the atmosphere from the top of a 250-ft stack.

The hot off-gas system is designed to handle low-volume high-concentration releases which normally accompany operations with radioactive material. It is connected directly to the components (deaerators, demineralizers, etc.) which release the gaseous activity. There are actually two such systems: the closed hot off-gas system, which collects gas from components which may be pressurized; and the open hot off-gas system, which collects gas only from unpressurized components. The hot off-gas systems are each constructed of type-304L stainless steel and schedule-5 pipe, except where heavier guage pipe was required for embedment. Piping within the pool structure has additional shielding where necessary to compensate for the concrete removed to allow pipe space. Embedded pipe is pressure-tested before concrete is poured to assure an airtight system.

Both charcoal and absolute filters, designed to reduce the potential activity in the effluent streams to tolerable discharge levels, are included in the filter assemblies of the two hot off-gas systems. Silver-plated copper-wool filters are located just upstream from the charcoal filters. The primary purpose of these silver filters is to retain the bulk of the iodine, thus preventing an excessive heat load in the charcoal.

Air from the special-building hot-exhaust system and the two hot off-gas systems is led from the various headers to separate underground filter pits which contain both high efficiency particulate filters and charcoal filters. After passing through the filters, the air is monitored and forced up a 250-ft stack so that any residual activity is dissipated by dispersion, diffusion, and decay. Separate filter banks and fans are provided for each of the three systems, although they discharge through a common header to the stack. Backdraft dampers on each fan prevent recirculation.

4.4 Pressure-Suppression Pools

Many power reactors utilize pressure-suppression pools to control the pressure buildup resulting from a loss-of-coolant accident. Usually such systems provide for blowdown of the contents of the primary system under water contained by an associated system. Energy released by the primary system must be dissipated in the water to prevent overpressurization of the containment. While the suppression pool is designed primarily for energy removal, it further offers a means to control fission products. Existing information on fission-product entrainment by suppression pools is somewhat limited at the present time, but the available data appear encouraging on the use of the pool for fission-product retention.

Five tests were conducted at the Moss Landing Power Plant specifically to determine the effectiveness of a water pool as a barrier to escaping fission products [235]. These were scaled-down, transient-blowdown tests simulating a maximum credible accident for the Humboldt Bay reactor. The simulated fission products were xenon, krypton, sodium iodide, iodine crystals, and zinc sulfide with a mean particle size of 2μ. The tests indicated that the water pool retained a very high proportion of impurities entering from the drywell. The measured separation factor between the drywell and the vapor space in the suppression chamber was of the order of 10^{-8} for solid particles, and 10^{-5} to 10^{-6} for the halogens and soluble salts. More than half of the noble gases were retained initially. Therefore the conclusion was drawn that if fission products are transported through the vent pipes, the water pool would permit the release of only minute fractions of the solids and halogens to its air space.

Some large-scale experiments have been reported which studied the removal of elemental iodine, methyl iodide, and lithium sulfate (representing soluble salts) in suppression pools [236]. Both transient-flow and steady-state-flow tests were run to simulate, respectively, the high flow rates encountered with maximum credible accidents and the lower flow rates corresponding to smaller ruptures which may exist for longer periods of time. The lutes, or vent pipes, were 13 in. in diameter and were immersed 2 ft into the large pool of water. The results from particle-removal experiments are given in Table 4-15. Water temperatures ranged between 32° and 176°F (0° to 80° C). The following was concluded: (1) Under transient flow conditions, the suppression pool gave a decontamination factor of at least 10 for elemental iodine and a soluble salt, but only poor removal of methyl iodide; (2) With steady-state-flow conditions, the decontamination factors for elemental iodine were from 14 to over 300, depending on the proportion of air (100% to 0%) present with the steam; iodine release time variations in depth of lute immersion between 1 and 2 ft had little effect on the decontamination factor and the iodine was mostly held permanently by the water; (3) Decontamination factors for small particles (0.06μ-diam) in the steady-state system were between 15 and 500. The relatively poor removal of methyl iodide as compared with that of the other forms of iodine may well control the overall iodine decontamination factor achieved by a suppression pool, depending on the proportion of the total iodine release as methyl iodide.

In small-scale tests, very similar to the large-scale steady-state tests described above, decontamination factors are reported as follows: 100 for elemental iodine at concentrations greater than 10^{-5} g-mole/liter, 1000 for elemental iodine at low concentrations and when sodium thiosulfate has been added to the pond water, 2 for methyl iodide, 100 for hydrogen iodide, and 50 for 0.06μ-diam particles [237]. These results are very similar to those for the large-scale tests of Ref. [236], even though the lutes for these small-scale tests were only 3 mm or 50 mm in diameter, and the depth of immersion in cold water was 20 and 10 lute diameters, respectively. Particle removal data are given in Table 4-15. Air-steam mixtures were used, and the total volume of air flow was the same as the volume of the pond water through which it passed.

Another article [238] reports experiments and calculations on trapping molecular iodine in laboratory and large-scale boiling water pools. Results show that, under certain conditions, superheated steam can be effectively decontaminated of iodine even by a boiling water pool with no net steam-condensation capacity. The parameters were pH (0 to 12), iodine concentration (0.001 to 1 ppm), and bubble size and time of interaction.

A modest exploratory program [239] has been established at ORNL for the purpose of gaining an understanding of the quantitative aspects of the basic mechanisms that control energy dissipation in pressure-suppression systems. Investigations of phenomena resulting from the injection of steam and steam-gas mixtures into water are being conducted to obtain data to permit confident extrapolation of results obtained with small-scale experi-

TABLE 4-15.

Nickel/Chromium Particle Collection in Suppression Pools

Reference	Total gas flow rate (lb/sec)	Steam (%)	Diameter of tube (mm)	Immersion depth of tube (mm)	Initial pond temperature (°C)	Particle size (μ)	Penetration (%)
[236]	2.6	3.7	330	600	0 - 10	0.06	0.42
	2.6	13.1	330	600	0 - 10	0.06	0.42
	4.3	42.0	330	600	0 - 10	0.06	0.13
	10.0	75.0	330	600	0 - 10	0.06	0.45
	10.0	75.0	330	600	0 - 10	0.06	0.12
	10.8	16.7	330	600	0 - 10	0.06	2.0
	16.4	45.0	330	600	0 - 10	0.06	0.55
[237]	6.2	90.0	50	500	50	0.06	0.25
	6.2	90.0	50	500	50	0.06	0.17
	6.2	90.0	50	500	50	0.06	0.15
	6.2	90.0	50	500	50	0.06	0.11
	6.2	90.0	3	60	50	0.06	0.09
	6.2	80.0	3	60	50	0.06	0.11
	6.2	60.0	3	60	50	0.06	0.24

mental equipment. An attempt is being made to develop appropriate scaling laws for the prediction of the behavior of full-size systems from studies of fission-product behavior in small models. The experimental equipment consists of an aluminum observation tank into which steam, water, and air can be provided. Full-length windows have been installed so that illumination and observation of phenomena can be made.

Although the investigation of steam-jet characteristics in degassed water has not been exhaustive, tests to date have produced consistent and plausible results. Analysis of the test results leads to the following conclusions:

1. In a water temperature range of 60° to 130°F (16° to 54°C) and a water pressure of approximately 16 psia at the point of steam injection, the ratio of the length of the steam jet to the inside diameter of the steam injection pipe varies as the natural logarithm of the mass velocity of the steam at the pipe exit (for vertical pipes discharging the steam in a downward direction).

2. The ratio of the jet length to the inside pipe diameter is apparently independent of the pipe size over the range tested. It is, however, a function of the water and steam conditions.

3. There are two distinct flow regimes. One of these spans the range of flow rates below which the linear flow velocity of the steam equals the sonic steam velocity under the existing conditions. The other flow regime involves flow rates above this range to an indefinitely high value. In both regions the ratios of jet length to inside pipe diameter are proportional to the natural logarithms of the mass velocities, but the constants determining the slopes are different.

4. There is a transition region between the two flow ranges that apparently spans the range between flows corresponding to that at which the center-line steam velocity becomes sonic and that at which the average steam velocity is sonic. The exact boundaries of this region have not been determined but probably are not greater than 5% to 10% of the flow rate at the initial point.

5. In the flow range below which the steam velocity becomes sonic, condensation of steam takes place more rapidly in degassed water than in nondegassed water, under otherwise similar conditions. Above this point there appears to be little difference in condensation rates, whether the water is degassed or saturated with air.

6. In relatively cool water (130°F and below), the condensation of saturated steam takes place very rapidly. The water currents generated by the jet continuously remove the heated water from the jet region and provide a source of cool water flowing downward around the jet. Thus there appears to be no problem in dissipating the steam energy with little pressure buildup. The largest jet-length-to-diameter ratio obtained in these tests was 2.33 at a mass velocity of approximately 300,000 lb/hr-ft^2. This mass flow rate is comparable with the maximum possible values postulated for most power reactor accident conditions.

Quantitative work with water saturated with air has not been as extensive as with degassed water. However, the data obtained indicate the following:

1. In the subsonic flow range, the length-to-diameter ratios for a saturated steam jet in nondegassed water are greater than in degassed water for the same steam flow rates and under otherwise similar water conditions. As in the case with degassed water, the ratios vary as the natural logarithm of the mass steam velocity, but with a lesser slope in the subsonic flow range.

2. At flow rates in the sonic range there appears to be no appreciable difference in the ratios for degassed and nondegassed water at comparable steam mass velocities.

3. In the subsonic flow range, saturated steam is condensed less rapidly in nondegassed water than in degassed water for the same flow rates and under otherwise similar water conditions.

4. Gas bubble generation occurred in all the steam injection tests performed in nondegassed water. Quantitative evaluation of this phenomenon will require considerable additional experimental work.

The General Electric Company [240], in a program complementary to the ORNL program, is conducting fission-product removal in a 1/10,000 scale model of a BWR primary containment system. The purpose of the scale-model tests is to identify and optimize those factors that would have a significant influence on the effectiveness of pressure suppression pools in trapping released fission products. A 2400Mw (t) General Electric BWR was selected for the design basis. Pressure vessel volume, drywell volume, suppression chamber volume, maximum break area, and downcomer area all were scaled down linearly. The experimental program is designed for parametric studies of methyl iodide concentration, suppression pool temperature, air-to-steam ratios in the downcomer, downcomer submission depth, downcomer radial location, suppression-pool chemical additives, and break size on fission-product retention. These tests, as well as those performed by ORNL, should provide much of the information required for development of analytical models of fission-product retention by suppression pools that can be used for the assessment of fission-product removal under accident conditions in large power reactors.

Only limited analyses of fission-product removal by suppression pools has been presented to date. Gieseke and Clark [241] have attempted to describe particle deposition from bubbles in a pressure-suppression pool. Several mechanisms including impaction, sedimentation, Brownian diffusion, thermophoresis, and diffusiophoresis can affect the particle deposition from bubbles. The theoretical aspects of deposition have been reviewed by Fuchs [54].

The motion of gas bubbles through liquids depends on the size of the bubbles, and Fuchs has used the analysis of Levich [242] for Reynolds numbers between 1 and 700 to predict particle deposition by several mechanisms. On the basis of the internal circulation predicted by Levich [242], the rate of deposition by impaction is

$$\phi = 6\pi V_b \tau n R, \qquad (4\text{-}21)$$

where ϕ = number of particles deposited/unit distance of bubble rise, V_b = velocity of bubble rise, n = particle concentration in gas, R = bubble radius, $\tau = 2r^2\rho/9\eta$, ρ = particle density, r = particle radius, and η = gas viscosity. The ratio of particles deposited by impaction to particles contained in the bubble is called the coefficient of inertial absorption (α_i) and is given by:

$$\alpha_i = \frac{9V_b\tau}{2R^2}. \qquad (4\text{-}22)$$

In a similar fashion the coefficient of sedimentation collection (α_s) is

$$\alpha_s = \frac{3g\tau}{4RV_b}. \qquad (4\text{-}23)$$

and the coefficient of diffusive collection (α_d) is

$$\alpha_d = 1.8\sqrt{\frac{D}{V_bR^3}}, \qquad (4\text{-}24)$$

where D = particle diffusivity.

In general, collection predicted by the coefficients defined above is quite small. For example, with R = 0.5 cm, V_b = 25 cm/sec, and r = 0.1 μ, the coefficient is $\alpha_d = 15 \times 10^{-4}$. A total collection efficiency can be calculated as follows. As given by Fuchs, the rate of diffusive removal is

$$-\frac{dn}{dx} = 2.4\pi n\sqrt{DV_bR^3}. \qquad (4\text{-}25)$$

Using the definition of α_d given above and integrating gives:

$$\frac{n_f}{n_0} = \exp\left(-\frac{4\pi V_bR^3\alpha_dX}{3}\right). \qquad (4\text{-}26)$$

n_0 is the number of particles initially present and n_f is the number remaining after a bubble has traveled a distance X. For the conditions noted above and a distance for bubble travel of 2 ft, the fraction of particles removed is about 69%. A distance for bubble travel of 8 ft would be required to give a collection efficiency of 99%.

The above analysis is expected to give an estimate of approximately an order of magnitude for particle removal from bubbles with the mechanisms considered. The effects of bubble collapse or condensation on particle collection have been neglected in this analysis. Condensing vapors are known to have a large influence on particle movement and it is expected that steam-air mixtures bubbling through pressure-suppression pools will result in considerable condensation and significantly increased rates of particle removal. There effects can be determined analytically only when condensation rates are known and condensation effects on particle motion are more quantitatively describable.

4.5 Reactive Coatings

Many surfaces within a containment shell are mild steel or concrete and are normally painted for protective purposes. If these surfaces could be painted with a reactive coating which would irreversibly sorb the iodine within the containment vessel under accident conditions, the potential for release to the environment due to leakage would be reduced. Laboratory experiments [243] have shown that iodine will deposit on most commercial painted surfaces, but the rate of deposition is low and often of little consequence in irreversibly immobilizing the iodine. In principle, it should be possible to develop a surface coating that would react rapidly and irreversibly with the chemical forms of iodine released under reactor accident conditions. What is required is a chemical compound or group of compounds that can be used either as a vehicle or a pigment and can react, when in the form of a surface coating, with vapor forms of iodine. Coatings of this type could provide a passive safety feature within the containment and supplement the active removal systems, such as sprays and filters, in contacting the fission products in a reactor accident.

Based upon the conditions that are postulated to exist during an accident in a light-water reactor, a number of criteria can be determined for reactive coatings [244]. The coatings must react irreversibly with fission-product iodine (primarily I_2 and CH_3I) from ambient temperature to 175°C. A minimum coating capacity for I_2 and CH_3I of 0.2 and 0.04 mg/cm^2 of geometric area, respectively, is required. High deposition rates of approximately 1 to 10 cm/sec are needed, and the coatings must function in steam and under condensing steam conditions. Thermal and radiation stability and compatibility with active safety systems such as containment spray solutions are required. Solubility in common solvents is also a desirable characteristic.

4.5.1 Coating Selection

An extensive screening program was undertaken at Battelle [246] to search for materials that were reactive with both I_2 and CH_3I and which could be used either as a coating vehicle or a pigment. Amines proved to be the most desirable class of compounds since they form charge-transfer complexes with I_2 and quaternary salts with CH_3I. The screening and evaluation program revealed that a compromise was needed between reactivity and resistance to loss-of-coolant accident environmental conditions. Candidate coatings that showed the highest reactivity for CH_3I showed the poorest resistance to steam and spray solutions, and vice versa. A compromise was affected by crosslinking the most reactive materials with an epoxy to improve their resistance to environment.

The coating recommended for carbon steel consists of 50-w/o binder and 50-w/o TiO_2 pigment with the binder containing 90-w/o DMAM and 10-w/o EPON 828 over a suitable primer. Poly (dimethylaminoethylmethacrylate), or DMAM, is a polymer containing tertiary amine groups. The epoxy EPON 828 is formed by reacting epichlorhydrin with bis (2, 2-hydroxphenyl) propane. Over concrete, a coating consisting of 50-w/o binder and 50-w/o TiO_2 with the binder containing 50-w/o Genamid 2000 and 50-w/o Epon 828 is recommended. Genamid 2000 is a polyamide resulting from reacting dimerized monobasic fatty acids with a molar excess of diamine.

The two recommended coatings were subjected to environmental testing as defined by the USAEC Committee on Standards for Protective Coatings. These tests included [245]: (1) exposure to steam in an autoclave (pressure decay from 60 psig down to 5 psig in 24 hr, hold at 5 psig for 12 days, followed by 2 psig for 14 days); (2) exposure to steam-air in an autoclave (pressure decay from 70 psig down to 15 psig in 24 hr, hold at 15 psig for 6 days); (3) exposure to containment spray solutions (1-month immersion at 90°C in each of the following solutions: 1.7-w/o H_3BO_3; 1.7-w/o H_3BO_3, 1.6-w/o $Na_2S_2O_3$:$5H_2O$, 0.6-w/o NaOH; 1.7-w/o H_3BO_3, 6.0-w/o NaOH); (4) simultaneous exposure to gamma radiation and steam (10^9 rads at 7×10^5 rads/hr in steam-air at 105°C); (5) exposure to a high-pressure steam jet (place specimen 1 in. from nozzle of a 1,000 psig steam jet for 3 min exposure at right angle to nozzle and inspect for failure); (6) repairability of coatings (age the specimen for 2 weeks at 65°C, remove a 1/2 in.-diam area down to the substrate, paint the base area with primer and the finish coat, age the new paint for 2 weeks at 65°C, autoclave for 3 days at 30 psig, and visually inspect for failure); (7) elongation test (perform test in accordance with ASTM-B522-60); and (8) weatherometer test (expose coatings for 500 hr in a xenon lamp weatherometer and inspect for failure). The only unsatisfactory performance under the recommended application was observed for the DMAM/Epon coating on steel in a strong sodium hydroxide spray solution. Since spray solutions of this pH are not being proposed, this should not be a major objection against the reactive coating.

Deposition experiments were performed on samples of the coatings to determine their capacities for and rates of reaction with fission-product iodine. The results of these experiments on the recommended coatings are shown in Table 4-16. The minimum capacity for elemental iodine was exceeded in 4 hr of experiments and the minimum capacity for CH_3I was nearly achieved in this time. On a theoretical basis, more than an adequate capacity is available in the coating. Minimum deposition velocities were on the order of 10^{-3} cm/sec for I_2 and 10^{-2} cm/sec for CH_3I. Most of the iodine and methyl iodide that was deposited on the coating reacted irreversibly with it.

4.5.3 Performance

The ability of the DMAM/10% Epon coating to remove airborne CH_3I from a containment atmosphere under simulated accident conditions was tested at Pacific Northwest Laboratories [246]. The 1.54-m^3 ADF tank was painted with the coating, and a simulated fission-product source of methyl iodide and iodine was injected into the tank. A 50-v/o steam-50-v/o air environment was maintained at a nearly constant temperature of 80°C and a pressure of 1 atm during the tests. Based upon laboratory deposition experiments, an airborne half-life for CH_3I of 22 min under these conditions was predicted.

A baseline experiment performed in the ADF tank prior to painting with the reactive coating yielded an airborne CH_3I half-life of at least 1000 min and an I_2 half-life of 2 min. In the experiments following painting with the reactive coating, the half-life for CH_3I was 16 min and the I_2 half-life was less than 0.5 min. The reactive coating appeared to perform as predicted. The tank surface appeared unaffected in continuity and luster following the two tests.

Use of a reactive coating as a passive safety feature in a reactor containment should materially decrease the half-life for airborne methyl iodide

TABLE 4-16

Laboratory Deposition Studies on Recommended Coatings

Coating	CH_3I				I_2	
	Temperature of deposition (°C)					
	37°	90°	115°	170°	90°	115°
	Capacity after 4 hr of deposition (mg/cm^2)					
GEN/EPON	2.2×10^{-3}	1.5×10^{-2}	1.4×10^{-2}	1.5×10^{-2}	1.8	1.5
DMAM/10% EPON	4.7×10^{-2}	6.7×10^{-2}	2.2×10^{-2}	9.0×10^{-2}	2.5	1.7
	Deposition velocity (cm/sec)					
GEN/EPON	8.7×10^{-4}	6.0×10^{-3}	5.4×10^{-3}	5.8×10^{-4}	7.0×10^{-1}	6.0×10^{-1}
DMAM/10% EPON	1.9×10^{-2}	2.7×10^{-2}	8.9×10^{-3}	3.6×10^{-2}	9.8×10^{-1}	6.7×10^{-1}
	Loss after 24-hr desorption at 170°C in flowing helium (%)					
GEN/EPON	2.70	6.84	3.70	0.23	9.71	9.89
DMAM/10% EPON	14.6	10.5	--	3.56	18.2	11.4

under natural effects conditions. This may be quantitatively estimated by considering the deposition occurring within the containment vessel as described by a well-mixed deposition model. If a puff release of methyl iodide is assumed, the airborne half-life is inversely proportional to the deposition velocity. Figure 4-15 shows the airborne half-life for CH_3I in large reactor containments if only the geometric area of a sphere with volume equivalent to the face volume of the containment is assumed available for deposition. It is expected that more surface area would be available for deposition, so these curves represent a conservative estimate of removal rates. The range of deposition velocities observed in laboratory experiments for the reactive coatings is shown. Since common containment coatings show little or no affinity for CH_3I, the reactive coatings appear to offer a means of controlling the airborne behavior of CH_3I during an accident if active removal systems are inoperable.

4.6 Foams and Encapsulations

4.6.1 General

The concept of using high expansion aqueous foams and possibly organic, plastic, mineral, or other foams for preventing environmental releases is a technique proposed originally by Silverman of the Harvard Air Cleaning Laboratory [247-250]. Foam is intended to contain and convert released radioactive gases and particulates to a form which would reduce their release rate. It has been evaluated experimentally with inert and radioactive iodine as well as with solid aerosols in the submicron range. Only aqueous foams have been studied to date because of their ease in disposal and their spray-down characteristics.

Encapsulation of contaminated air by reactive foams offers several potential advantages. (1) Large surfaces are available for liquid-phase reactions [250] with gaseous iodine and its compounds (I_2, HI, methyl and other alkyl iodides). This is shown in Table 4-17, which relates the liquid-surface/contained-gas-volume ratio for various foam bubble sizes to the total foam surface area in a 10^6 ft^3 spherical containment vessel, assuming the vessel is entirely filled with foam. A

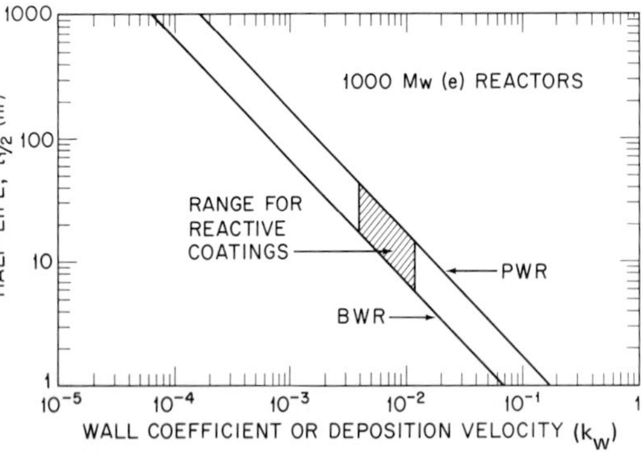

FIG. 4-15 CH_3I half-life vs. wall coefficient.

TABLE 4-17

Relative and Total Surface Areas of Single Spherical Bubbles (Foams)

Bubble diam.	Contained vol. (ft^3)	Surface-to-volume ratio (ft^{-1})	Total surface area in vol. of 10^6 ft^3 (ft^2)
0.25 in.	4.8×10^{-6}	288	2.88×10^8
0.75 in.	1.3×10^{-4}	96	9.6×10^7
1.5 in.	1.02×10^{-3}	48	4.8×10^7
124.1 ft.	10^6	.048	4.8×10^4

single foam bubble with an average diameter of 0.75 in. has a calculated surface-to-volume ratio 2000 times that of a spherical 10^6-ft^3 containment vessel, and the total surface area of the foam will be in the same proportion. (2) Aerosols may be formed within foams by gas-phase reactions between iodine (and iodine compounds) and volatile reactants in the foaming solution. When this occurs, the particles have relatively small distances to travel under the influence of settling and diffusional processes to make contact with a foam film and become trapped. (3) Long and intimate contact time is provided for slow mass-transfer reactions. (4) Foams permit the concentration of contaminants into a relatively small liquid waste volume. (5) A primary advantage of the use of foams is their potential for reducing containment leakage by trapping contaminated air inside foam bubbles which resist passage through cracks and porosities in the vessel.

A major disadvantage of foam application is incompatibility with spray cooling, as foams cannot be produced or maintained in the presence of copious liquid sprays. However, foams can be applied prior to or following cessation of spraying, although the time available for foam application would be reduced accordingly.

The foam generator was originally developed for use in fighting mine fires; it was intended to produce a rapid plugging of a mine shaft or airway by creating a foam plug (called "stemming" by British investigators [251]). The basic generator consists of an open-mesh nylon-weave cloth stretched across a pleated frame (for increased generating surface), on which an aqueous detergent solution (for example 2.5% lauryl ether sulfate) with lauryl alchohol stabilizer is sprayed as described by Davies and Rideal [252]. The detergent solution is pumped and sprayed through a low-pressure nozzle. Air flow is produced by an inexpensive centrifugal blower operating against a low static head system.

The foam obtained is characterized as a high expansion foam, because a thousand-fold ratio of volume of air to volume of water is obtained. This is in contrast to typical protein-base fire fighting foams in which a 30-to-1 ratio is commonly achieved. A high expansion ratio is necessary for containment purposes in order to obtain rapid generation and high surface area with minimal solution consumption and wetting down of walls and other nearby areas.

If the intake gas to the blower-bubble system is drawn from the top of the containment and discharged at the bottom of the gas face, the upper surface of the foam provides a barrier between the nontreated gas and the treated gas. The entire volume of the containment could be passed once through a moisture deentrainer-HEPA filter-charcoal bed train, and then through the blower-bubbler to ensure complete treatment of the contained gas. In the case of a normal iodine-entrapment recirculation system, the collection of iodine on the filter and charcoal beds, even if assumed to be 100% efficient, will be much slower according to the relation $(1-e^{-kt})$, and all of the iodine will never be removed.

It is necessary to provide a reactant in the solution to fix the iodine and methyl iodide and thus to prevent the remission effect of dissolved or volatile iodine. If this were not done, it is anticipated that iodine would diffuse out again when the foam bubbles collapsed. Sodium thiosulfate, silver nitrate, hydrazine, and unsymmetrical dimethyl hydrazine have been added to the foaming agents for reaction with elemental iodine. Hydrazine, thioacetamide, tributyl phosphine, and other additives have been investigated for methyl iodide removal.

4.6.2 Elemental Iodine Removal Studies

Silverman et al. [247, 248] reported on the use of two reactive foams, one containing 0.5% $Na_2S_2O_3$ and the other 0.5% $AgNO_3$, for removal of elemental iodine at a concentration of 13.6 ppm from air. Tests in a 1200-ft^3 chamber using $Na_2S_2O_3$ reactive foam showed a reduction in iodine concentration of 92% after 1/2 hr when the foam generator was shut off as soon as the chamber was filled with foam, and up to 96% removal efficiency when the foam generator operated continuously over the entire period, i.e., recirculated the chamber-air-foam mixture through the foam generator. Comparative studies of 0.5% $Na_2S_2O_3$ and 0.5% $AgNO_3$ in foam, using a small 53-ft^3 chamber, showed 1/2-hr elemental iodine removal efficiencies of 99% and 98%, respectively, when using continuous foam production with recycled air.

The removal of elemental iodine by nonreactive foams, with and without uranine particles in the air-iodine test atmosphere, showed greater elemental iodine removal when the aerosol was present, indicating significant absorption and reaction of iodine on the uranine particles. Both 0.2μ and 0.07μ uranine aerosols were generated and released into a 1500-liter (53-ft^3) chamber. The normal settling rate in each case was obtained and then the experiment was repeated using the high expansion foam. Uranine (disodium fluorescein) aerosols were prepared by atomizing and drying from an aqueous solution of 0.05% to 2.3% uranine. The dried particles thus may be more readily wetted but it is unlikely that other aerosols such as metal fumes or particulates with iodine or other gas films would be unable to stick readily to the bubble wall. The settling rate for the fine aerosol particles was quite low, with lower rates for the smaller particles, as would be expected. With the addition of foam, both aerosols were removed much more rapidly, with the large diameter aerosol being removed somewhat faster.

It is of interest to note, however, that Browning and Ackly [253], using radioactive-iodine-labelled aluminum oxide particles in the 14 to 65 Å size range found that 3-1/2 and 1-ft foam columns were capable of removing 90% to 95% of the suspended flowing aerosol on a dynamic basis. The smaller particles were removed with the greater efficiency. Results with a shorter foam column were equally impressive.

Tests made by Yoder, Fontana, and Silverman [254] at Oak Ridge National Laboratory in a 6000-ft^3 hot cell with continuous recycling of $Na_2S_2O_3$ reactive foams showed an I_2^{131} removal efficiency of 90% in 1 hr. Removal of I_2^{131} without foam (by plate-out and leakage from the dry cell) was found to be 70% during a like period, suggesting that there was a corrected iodine removal efficiency of only 67% by the foam alone. However, losses from the chamber by plate-out and leakage in the presence of foam are greatly reduced and most of the 90% reduction can be attributed solely to the foam used in the tests.

Iodine removal studies with foams containing 0.053 M and 0.156 M hydrazine and 0.108 M unsymmetrical dimethyl hydrazine (UDMH) were made in a 3260-ft^3 chamber by Viles [250]. A marked increase in iodine removal rates was observed in the presence of the reactive foams as compared to the plain foams, and UDMH was more reactive than hydrazine. Since hydrazine and UDMH rapidly form aerosols with iodine, it was believed that application of these reactants in foam for iodine removal involves the formation of an aerosol and the deposition on the bubble film by settling and diffusion processes.

4.6.3 Methyl Iodide Removal Studies

Viles et al. [250] have conducted screening tests of reactive foam solutions for the removal of gaseous methyl iodide. Static and dynamic (shaking) conditions were employed, and hydrazine, UDMH, piperazine, tri-n-butylphosphine (TBP), triethylenediamine, and thioacetamide were included as reactants. The latter three reactants were found to be the most effective for methyl iodide removal, and TBP vapors formed a water-soluble aerosol with methyl iodide. Water-foam encapsulation studies for air containing 0.8 to 2.0 ppm methyl iodide showed removal rates not significantly different from those found in the static screening tests for the same reactants, even though reactive surface areas were 20 to 100 times greater after foaming. This suggests that there is an absence of reactant in the foam bubble film, probably caused by displacement by surfactants in the foaming agent. If reactive foams containing liquid-phase reactants for methyl iodide are to be used successfully, a means must be found to obtain high concentrations of reactants in the surface layers of foam films.

4.6.4 Foam Stability

The critical experiments which are important if foam application is to be of general use are those designed to ascertain the effect of direct radiation and radioactive aerosols on the efficiency of foam suppression. Another important factor to ascertain is the height to which a foam column can be elevated before its own weight acts as a limiting or collapsing factor.

Radiation effects on foam were evaluated during the series of experiments by Yoder, Fontana, and Silverman [249] described earlier. In these experiments, beakers of foam were exposed to 150-kv X-rays at levels to 300r. No deleterious effects from the radiation were noted. Since the mass of the foam is small, radiation effects at these levels would not be expected. Higher radiation fields could produce radiolysis of the foam solution, and radiation effects such as those discussed for spray solutions in Sec. 4.2.2.3 must be considered. Hydrazine-containing solutions may be particularly sensitive to radiation effects.

Heat appears to be destructive to the bubbles, although the effect can be compensated for by continuous generation of the foam. The influence of steam was also observed and although its effects are likely to be the same as those of dry heat, it was possible to continuously generate foam in a steam environment. In the actual accident case, it is likely that the foam generator would not be turned on until the water sprays had reduced the steam pressure to avoid melting of the core.

The height factor of the foam column has been checked by actual tests with stabilized foam in a 90-ft tower at the Idaho NRTS. Due to a poor expansion mixture and a static-pressure limiting fan (3 in. H_2O), it was only possible to reach 57 ft of height, although based on the fan static value, over 100 ft should be possible. This height is sufficient, however, to permit complete immersion of a pressure vessel as well as submersion of a given volume surrounding it. Foam generators could be placed at various heights to cover zones where releases might take place.

It is likely that if the foam method were to be applied in practice, it would be expected to act as a backup safety system in a containment or confinement system. Foams have been shown to reduce by a factor of 7 to 10 the time required for iodine concentrations to be reduced to 1/2 their original value. The foam would probably be used after sprays had condensed the steam. Use of a foam generator at the outlet end of a moisture deentrainer-HEPA filter-charcoal bed chain would ensure that all of the containment atmosphere had passed through the air-cleaner system in a single pass. Definitive tests do not exist on the use of foams at high temperatures in an air-steam atmosphere.

Other experiments have shown that foam can be projected down ducts or tunnels to act as a sealant capable of absorbing contamination. It could thus be used as a valve or lock closure backup device for ventilation ducts or personnel locks.

4.7 The Diffusion Board Containment Concept

The ideal diffusion board is a cell-like structure which has the capability of resisting pressures encountered in postulated reactor accidents or releases. The structure must resist the inherent steam pressure developed by a sudden release from a severed coolant line and any shock wave that might be created. The objective is to design a diffusion structure that could be made strong enough to resist both forces. It could also act as an air cleaner and would replace the present containment pressure vessels made of steel plate. In other words, it would be a reactor confinement which would hold up any fission products released in an accident, with the exception of the noble gases and possible organic iodides. It could only be used at reactor sites with large exclusion areas, but might be useful in such situations. Such a structure would be able to function for both water- and gas-cooled reactors, and could handle many types of releases.

The diffusion board should have the following properties:

1. It should be noncombustible and unaffected by normal atmospheres, i.e., it should undergo no serious deterioration or corrosion.
2. It should resist steam pressures as high as 100 psi and temperatures of 250°F or more.
3. It should resist shock waves with overpressures not exceeding 3 in. Hg.
4. It should resist prolonged water-vapor contact and radiation exposure.
5. It should remove radioactive halogen gases with 99.9% efficiency.
6. It should remove 0.3μ and larger particulates with 99.5% to 99.9% efficiency.
7. It should adsorb rare gases to the maximum extent possible, and be permeable to air and water vapor.
8. It should have a gas flow-resistance below 6 in. H_2O so that diffusion can take place readily.
9. It should be inexpensive enough to show a substantial reduction in cost of materials and labor compared to a reactor containment or an off-gas cleaning unit.

The Harvard Air Cleaning Laboratory developed a honeycomb structure having two filter layers and a contained adsorbent within the interstices of a strong structure. Its performance on submicron particulate and halogen removal should be in excess of the above criteria, since its components have been subjected to tests of this type successfully. Its resistance to shock, steam, and static loading remain to be evaluated.

Other porous media have been studied, such as the U. S. Army Chemical Corps pilot-production diffusion board (a cellulose and carbon medium), porous carbon and impregnated refractory-fiber adsorbent combinations.

Samples of diffusion boards mounted in a 6-in. duct equipped with upstream and downstream sampling probes were evaluated with sublimed iodine vapor diluted to ppm levels and with the uranine aerosols described in Sec. 4.6. Resistance to air flow at three face velocities (0.1, 1.0, and 5.0 ft/min) was measured with manometers connected to duct taps placed before and after the media. Air at atmospheric pressure and temperature was used for efficiency and flow-resistance measurements.

Limited data are shown in Table 4-18 [3]. The efficiency data for the Chemical Corps pilot material is extremely favorable for both 0.05μ and 0.2μ aerosols, as well as for elemental iodine. Low-porosity carbon does not fare as well; for short runs, it exhibits a significant removal efficiency but it quickly becomes saturated.

Tests of the honeycomb media can almost be

TABLE 4-18

Diffusion Board Performance on Aerosols and Gases

Type of Diffusion Board	Aerosol or gas	Velocity (fpm)	Resistance to air flow (in. H_2O)	Penetration (%)	Efficiency (%)	Remarks
Chemical Corps pilot material	Uranine 0.05μ	0.1	0.72	0.37	99.63	Mean of 3 tests
Same	0.05μ	1.0	7.2	0.18	99.82	Same
Same	0.05μ	5.0	39.5	0.044	99.956	Same
Same	0.2μ	0.1	0.72	0.0026	99.9974	Mean of 2 tests
Same	0.2μ	1.0	7.2	0.0019	99.9981	Same
Same	0.2μ	5.0	39.5	0.0011	99.9989	Same
Same	I^{127}	0.1	0.68	0.16	99.840	Mean of 3 tests of 15 to 60 min each
Same	Same	1.0	6.8	0.052	99.948	Same
Same	Same	5.0	38.8	0.08 0.14 1.15	98.63	3 tests with progressively higher penetration because of saturation of media with iodine
Porous (1/4" thick) Carbon disc.	0.5μ	0.1	0.012	3.2	96.8	Mean of 3 tests
Porosity is that used for sewage	0.05μ	1.0	0.112	33.6	66.4	Mean of 2 tests
Aerator plates	0.05μ	5.0	0.56	37.4	62.6	Shows plugging on continuous exposure

predicted since the filter medium, all-glass web 1106 B, has been evaluated at well over 99.95% efficiency on 0.3μ, 0.07μ and 0.2μ solid particles and liquid droplets, and on many other aerosol sizes by the Harvard Air Cleaning Laboratory, Posner [256], and many others. The performance of the activated-carbon filling material can also be predicted from the Harvard results [257] and those of Adams and Browning [258]. The overall iodine efficiency will be only slightly increased by the presence of the glass-fiber filter medium on both sides of the honeycomb. On the other hand, an increase in efficiency can be expected due to the low velocity used for filtration as contrasted to the higher velocities in normal filter systems. Another beneficial effect of the lower velocity is to permit a reasonable life expectancy for shallow adsorbent beds.

For particle removal, it would be desirable to have a more efficient medium than 1106B on the downstream side, or a less efficient one on the upstream side. The former is preferred to avoid any significant deposition in the adsorbent layer. The 1106B on the upstream side, which would normally face the interior of the structure or flow path, can also be provided with a roughing filter such as spun-glass fiber mat (PF 316 or 105 Owens-Corning or equivalent; see reference [259]).

4.8 Retention of Noble Gases

The chemically inert character and the gaseous state of the elements of the periodic table which have full valence electron shells makes them particularly hard to handle when they are found among the fission products. In particular, the noble gases xenon and krypton cannot currently be removed by any practical means from the containment shell atmosphere during the high-temperature high-pressure stage of a loss-of-coolant accident. Their low concentrations, coupled with large air volumes, make such a concept unthinkable in terms of present day technology.*

The subject has been reviewed by Keilholtz [260]. Current considerations are devoted mainly to methods which might be used to clean up the containment atmosphere over several days. Obviously, the best current method of preventing release of any noble gas fission products to the atmosphere is the rapid reduction of containment pressure after the accident. This lowers the post-accident containment leakage rate rapidly and thus reduces noble-gas fission-product release. Such schemes as the partial vacuum containment method proposed by Stone and Webster, the pressure suppression system used by General Electric boiling water reactor designers, the containment sprays used in pressurized water reactors, and the ice condenser proposed by Westinghouse are all aimed at prompt reduction in containment pressure with the objective of reducing the leakage rate.

In order to remove xenon and krypton from the containment atmosphere, a method must be devised to extract them selectively from air. The simplest proven method is adsorption on activated charcoal. The charcoal can be poisoned or made less effective by standing under varying atmospheric conditions. Adsorption on charcoal at or

*If it were possible to trap them in a small volume, say at the top of the primary pressure vessel otherwise filled with water, they could be pumped out and compressed in a storage flask for retention, radioactive decay, and ultimate disposal by means of a high stack or transport to a remote site for gradual discharge there.

below liquid nitrogen temperature requires a smaller volume of charcoal, but the designer must add a liquid nitrogen system and arrange for its maintenance during and after an accident. Loss of liquid nitrogen or even heat-up due to radioactive energy release in the charcoal bed can cause re-release of the noble gases. Such systems have been proposed by the Air Reduction Company and the Babcock and Wilcox Company for the N. S. Savannah and certain other reactors [261].

Removal of the noble gases by solution of the gases in liquids has been proposed by J. M. Holmes of ORNL. While such a system has the advantage of being continuous, it requires large and complex equipment and ultimate disposal of the liquids. Absorption in liquid fluorocarbon has been studied [262].

The use of selective permeability in a cascade of appropriate membranes shows promise but has not been proven economically feasible [263]. The use of a cascade of membranes has been studied at ORNL [264].

A permselective membrane system consisting of thin sheets of methylphenyl silicone rubber has been developed to remove fission-product xenon and krypton from various gases, including oxygen, nitrogen, argon, and helium. The process is based on the greater solubility of xenon and krypton in a permselective membrane, thus leading to more rapid diffusion through it [265]. Possible applications of the process include removal of the noble fission-product gases from the air within a reactor containment, from the off-gas of a spent fuel reprocessing plant, or from the blanket gas in reactors that use vented or molten fuels.

Membranes have been successfully made and tested with permeabilities for nitrogen of about 0.27 ft^3/hr for one square foot of membrane. The membranes can withstand a pressure difference of 150 psi without rupture. A single-stage separation factor of at least 4 has been observed. The silicone rubber membranes have high radiation and chemical stabiltiy and tests indicate satisfactory functioning under normal operating conditions.

Permeability data for a mixture of less than 0.5% krypton in oxygen are shown in Table 4-19. The tests were conducted with pressure drops of 25 psi or 100 psi across the membrane; the first column shows the percent of total gas containing krypton which was allowed to pass through the membrane from the high-pressure side. Flow of the gases on the two sides of the membrane was either in the same direction (cocurrent) or in the opposite direction (countercurrent). The permeability, stage-separation factor, and material balance check for each test is shown. The data from the series

TABLE 4-19

Effects of Operating Pressure, Percentage of Gas Flowing Through the Membrane, and Mode of Flow on the Permeabilities and Separation Factors of Oxygen and Krypton[a]

Gas Flow Through Membrane (%)	Mode of Flow	Permeability[b]		Stage Separation Factor[c]	Material Balance (%)
		Oxygen	Krypton		
25-psi pressure drop					
18.7	Cocurrent	32.29	50.9	1.34	100.5
19.9	Cocurrent	32.25	52.1	1.39	96.1
19.9	Countercurrent	32.66	54.46	1.37	103.4
20.1	Cocurrent	33.10	52.11	1.33	102.5
42.0	Cocurrent	31.0	49.5	1.49	97.1
42.0	Cocurrent	32.7	51.2	1.40	100.0
43.0	Cocurrent	32.4	52.5	1.41	104.0
44.0	Countercurrent	32.7	61.9	1.54	107.1
60.4	Cocurrent	31.6	48.3	1.50	96.5
62.3	Countercurrent	32.9	51.0	1.46	101.8
62.5	Cocurrent	33.1	48.6	1.42	99.7
64.5	Cocurrent	31.6	46.9	1.46	98.2
80.6	Cocurrent	31.8	44.8	1.46	98.1
81.4	Cocurrent	32.2	46.3	1.51	97.5
85.0	Countercurrent	32.4	46.1	1.45	102.2
87.5	Cocurrent	32.4	46.8	1.45	104.9
100-psi pressure drop					
21.8	Cocurrent	25.51	43.2	1.71	96.7
22.5	Countercurrent	26.8	45.9	1.69	99.9
22.8	Cocurrent	27.3	44.9	1.64	99.7
40.0	Cocurrent	27.0	43.9	1.76	97.2
40.1	Cocurrent	26.8	43.9	1.68	104.0
40.3	Countercurrent	27.2	44.0	1.70	102.0
56.5	Cocurrent	26.9	43.4	1.74	106.6
56.6	Countercurrent	26.2	42.3	1.77	104.8
59.5	Cocurrent	27.2	41.5	1.88	93.6
79.7	Cocurrent	26.5	40.2	1.86	104.1
80.4	Cocurrent	26.4	39.6	1.97	99.4
80.7	Countercurrent	26.6	40.8	1.91	104.0

[a] Original gas mixture was 0.5% krypton in oxygen.

[b] Permeability is defined as cm^3/sec through a 1cm-thick membrane per square centimeter of surface times 10^9 divided by the difference in pressure (cm Hg) across the membrane.

[c] Ratio of concentrations of noble gas in product and raffinate streams at zero flow through the membrane.

of experiments were plotted and extrapolated back to zero passage of gas through the membrane. The ratios of krypton permeability to oxygen permeability at pressure differences of 25 and 100 psi and at zero flow were 1.65 and 1.74, respectively. Robb [266] reported a ratio of 1.63. The permeabilities reported in these experiments are about one-half those found by Robb. Evidently the Dacron mats used to support the membranes against the pressure differences reduced the permeabilities of the gases by about 50%, but did not affect the separation factors.

For a single stage, the separation factor (α^*), which is the ratio of the concentration of the noble gas in the product to its concentration in the raffinate streams at zero flow through the membrane, is given by [267]:

$$\alpha^* = \frac{\alpha_0}{1 + R(\alpha_0 - 1)}, \qquad (4\text{-}27)$$

where R is the ratio of pressure on the low-pressure side to that on the high-pressure side, and α_0 is the ratio of the permeabilities of the two pure gases at zero back pressure. From the α_0 values of the experiments with 0.5% krypton in oxygen, α^* values at 25 and 100 psi were calculated to be 1.33 and 1.59, respectively. There was no noticeable difference whether the flow was cocurrent or countercurrent.

Similar experiments with krypton in nitrogen gave calculated permeabilities at zero flow of 56 and 48 at 25 psi and 100 psi, respectively. These values agree well with the values of 54 and 47 for krypton in oxygen, since at higher flow rates through the membrane, the permeabilities of krypton in nitrogen are less than those for krypton in oxygen. The decrease in permeability of the dilute gas as the percentage of the gas flowing through the membrane is increased explains the observed effect of the carrier gas on the permeability of the gas as discussed by Rainey and Carter [264].

Calculations have been made [268] to determine the size, performance, and 1968 cost of a plant for processing argon cover gas for a 1000 Mw(e) sodium-cooled fast reactor with vented fuel. Babcock and Wilcox furnished the data shown in Table 4-20 for such a plant. Processing rates are relatively small (less than 10 scfm) and a cascade of membranes would be suitable for such an operation. The results of Table 4-21 were calculated for a membrane separation plant fitted in a cubic cell 15 ft on each side. The feed rate was 10 scfm, and the concentration factors of the noble gases were 10, 100, and 500. In each case, the radioactive krypton and xenon of the argon carrier gas was reduced to 5000 μCi/sec or less. The gas could then be either discharged or recycled for use in the reactor.

A similar set of calculations was carried out for the separation of the noble gases from a containment atmosphere (3×10^6 ft^3 volume) for a 3200 Mw(t) nuclear plant with an end-of-life (625 days at 3×10^{13} average neutron flux) core of 100 metric ton and a fuel enrichment of 2.3% U^{235} and average irradiation of 20,000 Mwd/metric ton. In such a case, the initial xenon and krypton source levels are about 6×10^8 and 3×10^8 curies, respectively. It was assumed that the containment shell atmosphere would be processed after a 30-day decay, a 60-day decay, or a 120-day decay. In each case the rate of processing was sufficient to reduce the krypton activity in the shell by a factor of 100 in a

TABLE 4-20.

Calculated Concentrations of Noble Gases Above the Sodium Coolant of a 1000-Mw(e) Reactor Designed by the Babcock & Wilcox Company

Blanket gas volume: 25,000 ft^3
Total krypton volume: 1 ft^3
Total xenon volume: 34 ft^3
Gas feed rate: 10 scfm

	Noble Gas Concentrations	
	Curies	Atoms/ft^3
^{83m}Kr	9,800	0.140135×10^{15}
^{85m}Kr	39,800	0.134638×10^{16}
^{85}Kr	43	0.311655×10^{17}
^{87}Kr	70,000	0.681579×10^{15}
^{88}Kr	108,000	0.232494×10^{16}
^{89}Kr	9,660	0.396253×10^{13}
Total Kr	237,303	
^{131m}Xe	16,300	0.360976×10^{17}
^{133m}Xe	89,400	0.379552×10^{17}
^{133}Xe	4,470,000	0.434665×10^{19}
^{135m}Xe	62,200	0.127519×10^{15}
^{135}Xe	356,000	0.251855×10^{17}
^{137}Xe	33,900	0.182444×10^{14}
^{138}Xe	258,000	0.562025×10^{15}
Total Xe	5,285,800	
Total Kr + Xe	5,523,103	
^{41}Ar	2,200	0.309506×10^{14}

TABLE 4-21

Performance and Cost of Several Cascades of Permselective Membranes for Separating Krypton and Xenon from Cover Gas of a Sodium-Cooled 1000-Mw(e) Reactor at Various Product Gas Flow Rates

Gas feed rate to cascade: 10 scfm			
Product gas flow rate, scfm	1	0.1	0.02
Cascade characteristics			
Number of enriching stages	9	15	20
Number of stripping stages	19	23	24
Membrane area (yd^2)	4630	5020	5140
Cascade volume (ft^3)	~9	~10	~11
Power requirement (kw)	126	136	140
Largest compressor (hp)	9.2	7.9	7.7
Pressure on membrane (psi)			
High-pressure side	150	150	150
Low-pressure side	0	0	0
Feed gas concentration and activity			
Concentration (at. %)			
Kr	0.004	0.004	0.004
Xe	0.136	0.136	0.136
Activity (Ci/ft^3)			
Kr	24.4	24.4	24.4
Xe	542.6	542.6	542.6
Recycle (or vented) gas concentration and activity			
Concentration (at.%)			
Kr	0.47×10^{-5}	0.49×10^{-5}	0.46×10^{-5}
Xe	0.53×10^{-9}	0.51×10^{-10}	0.25×10^{-10}
Kr + Xe activity			
Ci/ft^3	0.029	0.030	0.028
μ Ci/sec	4350	4950	4657
Product gas (to be stored) characteristics			
Concentration (at.%)			
Kr	0.04	0.4	2
Xe	1.4	14	68
Ar	98.6	85.0	30
Activity (Ci/ft^3)			
Kr	244	2440	12,200
Xe	5426	54,260	271,300
Concentration factor (product/feed)	10	100	500
Number of storage cylinders required per week[a]	50	5	1
Installed cost	$289,000	$342,000	$366,000

[a] Standard N_2 cylinders at a pressure of 2200 psi.

period of one week (see Table 4-22). During the 30- and 60-day decay periods the only significant activity is due to Kr^{85} and Xe^{133}. After about 60 days of decay, the Kr^{85} activity dominates. Previous studies of noble-gas removal using diffusion membranes [263] had postulated a removal time of seven days, resulting in high costs and large plants. The study outlined above postulated a removal time of 30

TABLE 4-22

Noble Gas Activity in the Shell and Irradiation Exposure[a] of Membrane During Removal of Noble Gases from Nuclear Containment Shell

Decay time before processing (days)	30	60	120
Activity in shell (Ci)			
Kr	9.0×10^5	8.95×10^5	8.86×10^5
Xe	2.0×10^7	8.0×10^5	1.0×10^3
Initial dose rate to membrane (rads/min)			
Kr	2.78×10^2	2.78×10^2	2.71×10^2
Xe	9.72×10^3	3.82×10^2	4.79×10^{-1}
Total	9.99×10^3	6.59×10^2	2.71×10^2
Integrated dose in processing period (rads)	1.22×10^7	1.07×10^6	5.89×10^5

[a] Measured in top stage of the cascade where the noble gases are most highly concentrated (and thus maximum exposure occurs).

days, and the cost of a plant to do this task is about the same as the cost of purifying the fast reactor cover gas, $300,000.

Robb has reported some results on the radiation stability of silicone rubber membranes at levels of 10^7 to 10^8 r. He checked the permeability of oxygen, nitrogen, and carbon dioxide. He found decreased permeabilities of 10% to 20%, but the separation factors of the gases increased by 2% to 20% as a result of the irradiation. Irradiation tends to increase linking and decrease the flexibility of the membranes. Additional studies are probably required in this area.

REFERENCES

1. J. R. Beattie, "An Assessment of Environmental Hazards from Fission Product Releases," British Report AhSB(S)R. 64 (1963).
2. J. O. Blomeke and M. F. Todd, "Uranium-235 Fission Product Production as a Function of Thermal Neutron Flux Irradiation Time and Decay Time. 1. Atomic Concentrations and Gross Totals," ORNL-2127 (TID-4500 rev.) (November, 1958).
3. W. B. Cottrell and A. W. Savolainen, Eds., "U. S. Reactor Containment Technology," ORNL-NSIC-5 (August, 1965).
4. J. J. DiNunno, F. D. Anderson, R. E. Baker, and R. L. Waterfield, "Calculation of Distance Factors for Power and Test Reactor Sites," USAEC Report TID-14844 (March, 1962).
5. D. A. Nitti, "A Study of Containment Sprays and Charcoal Filters for the Removal of Iodine Following a PWR Loss of Coolant Accident," IAEA Conference on the Treatment of Airborne Radioactive Wastes (1968).
6. F. R. Farmer, "Siting — A New Approach," IAEA Conference on Containment and Siting, SM-89/34, Vienna (April, 1967).
7. F. R. Farmer, "A Method of Assessing Fast Reactor Safety," Proceedings of the International Conference on the Safety of Fast Reactors, Aix-en-Provence, France (September 19-22, 1967), Vol. VI, part 2, pp. 1-9.
8. G. W. Keilholtz and C. J. Barton, "Behavior of Iodine in Reactor Containment Systems," ORNL-NSIC-4 (February, 1965).
9. G. W. Keilholtz, C. E. Guthrie, and G. C. Battle, Jr., "Air Cleaning as an Engineered Safety Feature in Light-Water Cooled Power Reactors," ORNL-NSIC-25 (September, 1968).
10. H. N. Culver, "Containment of Gas-Cooled Power Reactors," Nuclear Safety, 4 (4), 90 (June, 1963).
11. F. R. Farmer, "Safety Research in Support of a Power Reactor Program," Nuclear News, 6, 7 (August, 1963).
12. S. I. Kaplan, "Safety in High-Temperature Gas-Cooled Reactors," Nuclear Safety, 9 (1), 4 (January-February, 1968).
13. P. G. Gormley and M. Kennedy, "Diffusion from a Stream Flowing through a Cylindrical Tube," Proc. Royal Irish Academy, 52A, 163 (1949).
14. E. R. Gilliland, quote on p. 10 of T. K. Sherwood and R. L. Pigford, Absorption and Extraction, 2nd Ed., McGraw-Hill Book Company, New York (1952).
15. J. H. Perry, Chemical Engineers Handbook, 3rd Ed., McGraw-Hill Book Company, New York.
16. J. A. Gieseke, "Particle-Size Analysis with Diffusion Tubes," An Evaluation of the Applicability of Existing Data to the Analytical Description of a Nuclear Reactor Accident," Quarterly Progress Report for January-March, 1966, USAEC Report BMI-X-10163 (EURAEC-1641) (April, 1966), pp. 30-38.
17. N. A. Fuchs, I. B. Stechkina, and V. I. Starosselskii, "On the Determination of Particle Size Distribution in Polydisperse Aerosols by the Diffusion Method," Brit. J. Appl. Phys., 13, 208 (1962).
18. W. E. Browning, Jr., R. D. Ackley, and M. D. Silverman, "Characterization of Gas Borne Fission Products," Eighth AEC Air Cleaning Conference, ORNL, TID-7677 (1963), p. 155.
19. W. E. Browning, R. E. Adams, R. D. Ackley, M. E. Davis, and J. E. Attrill, "Identity, Character, and Chemical Behavior of Vapor Forms of Radioiodine," Proc. Inter. Symp. on Fission Product Release and Transport Under Accident Conditions, ORNL CONF-650407 (April, 1965).
20. A. W. Castleman, Jr. and I. N. Tang, "Vaporization of Fission Products from Irradiated Fuels - 1. Experimental Method and General Fission-Product Behavior," Nucl. Sci. and Engr., 29, 159 (1967).
21. W. J. Megaw and F. G. May, "The Behavior of Iodine Released in Reactor Containers," AERE-3781 (1961).
22. R. K. Hilliard, L. F. Coleman, and J. D. McCormack, "Comparisons of the Containment Behavior of a Simulant With Fission Products Released from Irradiated UO_2," USAEC Report BNWL-581 (March, 1968).
23. D. A. Collins, R. Taylor, and W. D. Yuill, "Sampling and Characterization Techniques Used in Study of Iodine Release from Irradiated Fuel Elements," Proc. Inter. Symp. on Fission Product Release and Transport Under Accident Conditions, ORNL CONF-650407 (April, 1965).
24. L. F. Parsly, Jr., "Experience With Maypacks in the Nuclear Safety Pilot Plant," ORNL-TM-2044.
25. R. L. Bennett, W. H. Hinds, and R. E. Adams, "Devolopment of Iodine Characteristic Sampler for Application in Humid Environments," USAEC Report ORNL-TM-2071 (May 1, 1968).
26. M. D. Silverman and W. E. Browning, Jr., "Fibrous Filters as Particle Size Analyzers," Science, 143, 572 (February, 1964).
27. L. Silverman, C. E. Billings, and M. W. First, AEC Monograph on Particle Size and Particle Size Analysis, USAEC (1966).
28. A. C. Chamberlain and R. C. Chadwick, Nucleonics 11, 22 (1963).
29. J. B. Morris, H. R. Diffy, B. Nicholls, and C. H. Rumary, "The Removal of Low Concentrations of Iodine from Air on a Plant Scale", J. Nucl. Energy, Parts A and B (Reactor Science and Technology), 16, 437-445 (1962).
30. A. E. J. Eggleton and D. H. F. Atkins, "Iodine Compounds Formed on Release of Carrier-Free Iodine-131," Trans. Amer. Med. Soc., 6, 129 (June, 1963).
31. J. Mishima, "Review of Methyl Iodide Behavior in Systems Containing Airborne Radioiodine," USAEC Report BNWL-319 (June, 1966).
32. W. S. Durant, R. C. Milham, D. R. Mulhbaier, and A. H. Peters, "Activity Confinement System of the Savannah River Plant Reactors," USAEC Report DP-1071 (August, 1966).
33. D. A. Collins et al., Experiments Relative to the Control of Fission-Product Release from Advanced Gas-Cooled Reactors, TRG Report 956 (W) United Kingdom Atomic Energy Authority, (May, 1965) p. 7.
34. R. D. Collins and H. Hillary, "Some Experiments Relating to the Behavior of Gas-Borne Iodine," International Symposium of Fission Product Release and Transport under Accident Conditions, Oak Ridge, Tennessee, April 5-7, 1965, CONF-650407 (1965) Vol. 2, p. 830.
35. R. K. Hilliard and J. W. McCormack, "Fission-Product Simulation in the Containment Systems Experiment," International Symposium on Fission Product Release and Transport under Accident Conditions, Oak Ridge, Tennessee, April 5-7, 1965, CONF-650407 (1965) Vol. 1, p. 588.
36. Nuclear Safety Program Semiannual Progress Report for Period Ending December 31, 1964, ORNL-3776, Oak Ridge National Laboratory, Oak Ridge, Tennessee (March, 1965).
37. Nuclear Safety Program Semiannual Progress Report for Period Ending June 30, 1964, ORNL-3691, Oak Ridge National Laboratory, Oak Ridge, Tennessee (November, 1964).
38. Nuclear Safety Program Semiannual Progress Report for Period Ending June 30, 1965, ORNL-3843, Oak Ridge National Laboratory, Oak Ridge, Tennessee, (September, 1965).
39. J. F. Croft and R. S. Iles, "Experimental Release of Radioiodine in the Zenith Reactor Containment," AEEW-R-172, VKAEA (September, 1962).
40. J. J. Hillary and J. C. Taylor, A High-Activity Iodine Release from Irradiated Fuel and to Subsequent Clean-up, TRG-888(W) United Kingdom Atomic Energy Authority (April, 1965).
41. I. N. Tang and A. W. Castleman, Jr., "Radiation Induced Decomposition of Methyl Iodide in Air," Trans. Amer. Nucl. Soc., 11, 67 (1968).
42. J. F. Kircher and R. S. Barnes, "Methyl Iodide Formation under Postulated Nuclear Reactor Accident Conditions," Treatment of Airborne Radioactive Wastes, International Atomic Energy Agency, Vienna (1968), pp. 137-162.
43. A. W. Castleman, Jr., I. N. Tang, and N. R. Munkelwitz, "The Chemical State of Fission-Product Iodine Emanating into a High Temperature Environment," J. Inorg. Nucl. Chem., 30, 5 (1968).
44. F. O. Cartan, H. R. Beard, F. A. Duce, and J. H. Keller, "Evidence for the Existence of Hypoiodous Acid as a Volatile Iodine Species Produced in Water-Air Mixtures," in Proceedings of the Tenth AEC Air Cleaning Conference, USAEC Report CONF-680821 (December, 1968), p. 342.
45. A. W. Castleman, Jr., "The Chemical and Physical Behavior of Released Fission Products," USAEC Report BNL-6415 (1962).
46. A. W. Castleman, Jr. and F. J. Salzano, "Current Studies of Fission Product Behavior at BNL," Eighth AEC Air Cleaning Conference, ORNL October 22-25, 1963, TID-7677 (1963) pp. 16-33.

47. A. W. Castleman, Jr. and I. N. Tang, "Vaporization of Fission Products from Irradiated Fuels - 1. Experimental Method and General Fission-Product Behavior," Nucl. Sci. and Engr. 29, 159 (1967).
48. A. W. Castleman, Jr., I. N. Tang, and R. A. Mackay, "Fission Product Behavior in Sodium Systems," USAEC Report BNL-10727 (1966).
49. B. C. Pollock, M. Silberberg, and R. L. Koontz, "Vaporization of Fission Products from Sodium," Proceedings of the International Conference on Sodium Technology and Large Fast Reactor Design, USAEC Report ANL-7520 (1968), p. 549.
50. A. P. Malinauskas, J. W. Gooch, Jr., and J. D. Redman, "High Temperature Behavior of Gasborne Fission Products," Nuclear Safety Program Annual Progress Report for Period Ending December 31, 1968, USAEC Report ORNL-4374 (June, 1969), pp. 309-316.
51. R. J. Davis, "A Nuclear Safety Particle Primer," USAEC Report ORNL-4337 (January, 1969).
52. T. W. Horst, "A Review of Particle Transport in a Condensing Steam Environment", USAEC Report BNWL-848 (June, 1968).
53. D. L. Morrison, J. M. Genco, J. A. Gieseke, R. L. Ritzman, C. T. Walters, and D. N. Sunderman, "An Evaluation of the Applicability of Existing Data to the Analytical Description of a Nuclear-Reactor Accident," USAEC Report BMI-1779 (EURAEC-1735) (August, 1966).
54. N. A. Fuchs, The Mechanics of Aerosols, The Macmillan Company, New York (1964).
55. H. L. Green and W. R. Lane, Particulate Clouds, Dusts, Smokes, and Mists, E. & T. N. Spon., Ltd., London (1957).
56. C. N. Davies, Aerosol Science, Academic Press, Inc., London (1966).
57. V. K. LeMer, "Nucleation in Phase Transitions," Ind. Eng. Chem., 44, 1270 (1952).
58. W. J. Dunning, "Nucleation Processes and Aerosol Formation," Discussions Faraday Soc., 30, 9 (1960).
59. M. Volmer and Weber, Z. Physik. Chem. (Leipzig), 119, 227 (1925).
60. R. Becker and W. Doring, Ann. Physik., 24, 719 (1935).
61. Y. B. Zeldovich, Experimental and Theoretical Physics, 12, Moscow (1942).
62. J. Frenkel, Kinetic Theory of Liquids, Clarendon Press, Oxford (1946).
63. Kuhrt, Z. Physik., 131, 185 (1952).
64. M. Volmer, Kinetik des Phasenbildung, Steinkopf, Dresden (1939).
65. J. P. Hirth and G. M. Pound, Condensation and Evaporation, Nucleation and Growth Kinetics, The Macmillan Company, New York (1963).
66. P. G. Hill, H. Witting, and E. P. Demetri, "Condensation of Metal Vapors During Rapid Expansion," J. Heat Transfer, 85, 303 (1963).
67. K. Stewart, "The Condensation of a Vapour to an Assembly of Droplets or Particles," Trans. Faraday Soc., 53, 161 (1956).
68. V. N. Lavrenchik, Global Fallout of the Products of Nuclear Explosions, State Press "Atomizdet," Moscow (1965).
69. J. Stockham and R. Snow, "Ablation Product Coalescence," Report IITRI C6007-12 (March 13, 1964).
70. J. M. Genco, W. E. Berry, H. S. Rosenberg, and D. L. Morrison, "Fission Product Deposition and its Enhancement under Accident Conditions: Deposition on Primary System Surfaces," USAEC Report BMI-1863 (March, 1969).
71. W. E. Browning, Jr., C. E. Miller, Jr., B. F. Roberts, R. P. Shedds, W. H. Montgomery, J. G. Wilhelm, and O. W. Thomas, "Simulated Loss-of-Coolant Accidents in the ORR," Nuclear Safety Program Semiannual Progress Report for Period Ending June 30, 1965, USAEC Report ORNL-3843 (September, 1965), pp. 3-39.
72. G. W. Parker, W. J. Martin, G. E. Couk, and C. J. Barton, "Behavior of Fission Products Released from Simulated Fuel in the Containment Mockup Facility," Nuclear Safety Program Semiannual Progress Report Ending June 30, 1965, USAEC Report ORNL-3843 (September, 1965), pp. 83-97.
73. W. A. Freeby, L. T. Lakey and D. E. Black, "Fission Product Behavior under Simulated Loss-of-Coolant Accident Conditions in the Contamination-Decontamination Experiment," USAEC Report IN-1172 (January, 1969).
74. D. L. Morrison, J. M. Genco, J. A. Gieseke, R. L. Ritzman, C. T. Walters, and R. O. Wooton, "An Evaluation of the Applicability of Existing Data to the Analytical Description of a Nuclear-Reactor Accident. Quarterly Progress Report for July-September, 1966," USAEC Report BMI-1777 (October, 1966).
75. A. C. Chamberlain, A. E. J. Eggleton, W. J. Megaw, and J. B. Morris, "Physical Chemistry of Iodine and Removal of Iodine from Gas Streams," Reactor Science and Technology. (J. Nucl. Energy, Parts A and B), 17, 519 (1963).
76. J. B. Morris and B. Nicholls, "The Deposition of Iodine Vapour on Surfaces," AERE-R 4502 (March, 1965).
77. S. Forberg, T. Westermark, and C. E. Holmquist, "Sorption of Fission Product Iodine on Different Materials with Application to Nuclear Reactor Accidents," Nucleonik, 3, 31 (1961).
78. J. F. Croft, R. E. Davies, and R. S. Iles, "Experiments on the Surface Deposition of Airborne Iodine of High Concentration," Health Phys., 11, 1 (1965).
79. L. G. Jarman, unpublished work, as reported by Morris and Nicholls, Ref. [76].
80. W. J. Megaw and F. G. May, "The Behavior of Iodine Released in Reactor Containers," Reactor Science and Technology (J. Nucl. Energy, Parts A and B), 16, 427 (1962).
81. H. S. Rosenberg, J. M. Genco, and D. L. Morrison, "Fission-Product Deposition and its Enhancement under Reactor Accident Conditions: Deposition on Containment System Surfaces," USAEC Report BMI-1865 (July, 1969).
82. J. M. Genco, H. S. Rosenberg, and D. L. Morrison, "Iodine Deposition and its Enhancement under Reactor Accident Conditions," Nuclear Safety, 9 (3), 226 (1968).
83. G. E. Creek, W. J. Martin, R. A. Lorenz, and G. W. Parker, "Behavior of Fission Products Released from Zircaloy-Clad High Burnup UO_2. In the Containment Mockup Facility," Nuclear Safety Program Annual Progress Report for Period Ending December 31, 1966, USAEC Report ORNL-4071 (March, 1967), pp. 75-78.
84. G. W. Parker, W. J. Martin, G. E. Creek, and R. A. Lorenz, "LOFT Support Studies—Fission-Product Release and Transport in Out-of-Pile Tests," Nuclear Safety Program Annual Progress Report for Period Ending December 31, 1968, USAEC Report ORNL-4374 (June, 1969), pp. 18-41.
85. J. L. Wantland and T. H. Row, "Behavior of Iodine in Air—A Resume of the First Seven Runs Conducted at the Nuclear Safety Pilot Plant," USAEC Report ORNL-4050 (April, 1967).
86. L. F. Parsly and T. H. Row, "Study of Fission Products Released from Trace-Irradiated UO_2 into Steam-Air Atmospheres (Nuclear Safety Pilot Plant Runs 8 and 9)," USAEC Report ORNL-TM-1588 (May, 1966).
87. L. F. Parsly and T. H. Row, "Behavior of Fission Products Released from Synthetic High Burnup UO_2 in Steam Atmospheres (Nuclear Safety Pilot Plant Runs 10-12)," USAEC Report ORNL-TM-1698 (February, 1968).
88. L. F. Parsly, J. K. Franzreb, P. P. Holz, I. Iyori, T. H. Row, and J. L. Wantland, "Fission-Product Transport in the Nuclear Safety Pilot Plant. Comparison of Aerosols from Simulants and Highly Irradiated UO_2 (Runs 12-15)," Nuclear Safety Program Annual Progress Report for Period Ending December 31, 1966, USAEC Report ORNL-4071 (March, 1967), pp. 78-87.
89. G. J. Rogers, "Containment Systems Experiment," Nuclear Safety Quarterly Report, August, September, October, 1968, for Nuclear Safety Branch of USAEC Division of Reactor Development and Technology, USAEC Report BNWL-926 (December, 1968), pp. 2.1-2.18.
90. R. K. Hilliard and J. G. Knudsen, "A Natural Convection Model for Predicting Elemental Iodine Transport in Containment Vessels," Trans. Amer. Nucl. Soc., 11 (2), 668 (1968).
91. J. F. Croft and R. S. Iles, "Experimental Release of Radioiodine in the Zenith Reactor Containment," AEEW-R-265 (June, 1963).
92. T. S. Kress and F. H. Neill, "A Model for Fission Product Transport and Deposition under Isothermal Conditions," ORNL-TM-1274 (October, 1965).
93. G. E. Raines, A. Abriss, D. L. Morrison, and R. A. Ewing, "Experimental and Theoretical Studies of Fission-Product Deposition in Flowing Helium," BMI-1688 (April, 1964).
94. L. F. Epstein and T. F. Evans, "Deposition of Matter from a Flowing Stream, Part I: General Relations and Equations," GEAP-4140 (1962).
95. M. N. Ozisik, "A Heat-Mass Analogy for Fission Product Deposition from Gas Streams," Nucl. Sci. Eng., 19 (2), 164 (June, 1964).
96. M. N. Ozisik, "Effects of Temperature on Fission Product Deposition," ORNL-3542 (March, 1964).
97. M. N. Ozisik, "A Transient Analysis of Fission Product Deposition," ORNL-TM-1650 (October, 1963).
98. E. R. Venerus and M. N. Ozisik, "Theoretical Investigations of Fission Product Deposition from Flowing Gas Streams," Trans. Amer. Nucl. Soc., 8 (2), 337 (1965).
99. M. N. Ozisik and F. H. Neill, "On the Theory of Diffusion Tubes," Trans. Amer. Nucl. Soc., 9 (2), 383 (1966).
100. M. N. Ozisik, "An Analytical Model for Fission-Product Transport and Deposition from Gas Streams," ORNL-3379 (1963).
101. C. N. Davies, "Deposition from Moving Aerosols," Aerosol Science, C. N. Davies, Editor, Academic Press, New York (1960), pp. 393-446.
102. G. M. Watson, R. B. Perez, and M. H. Fontana, "Effects of Containment Size on Fission Product Behavior," USAEC Report ORNL-4033 (January, 1967).
103. R. L. Ritzman, J. A. Gieseke, J. N. Blutreich, and D. L. Morrison, "Analytical Description of Fission Product Transport and Deposition in Containment Vessels," Trans. Amer. Nucl. Soc., 10 (2), 714 (1967).

104. W. A. Yuill, Jr. and V. F. Baston, "A Model for Fission Product Deposition under Natural Response," Nuclear Safety, 10 (6) (1969).
105. J. G. Knudsen and R. K. Hilliard, "Fission Product Transport by Natural Processes in Containment Vessels," USAEC Report BNWL-943 (January, 1969).
106. D. L. Morrison et al., "An Evaluation of the Applicability of Existing Data to the Analytical Description of a Nuclear-Reactor Accident. Quarterly Progress Report for January through March, 1967," USAEC Report BMI-1797 (April, 1967).
107. L. C. Richardson et al., "CONTEMPT - A Computer Program for Predicting the Containment Pressure - Temperature Response to a Loss-of-Coolant Accident," USAEC Report IDO-17220 (June, 1967).
108. W. H. McAdams, Heat Transmission, McGraw-Hill, 3rd Ed. (1954), p. 172.
109. R. K. Hilliard and J. G. Knudsen, "A Natural Convection Model for Predicting Elemental Iodine Transport in Containment Vessels," BNWL-SA-1994 (November, 1968).
110. R. L. Ritzman, paper presented at Amer. Nucl. Soc. Meeting, November, 1967.
111. M. N. Ozisik and D. Hughes, "Effects of Condensation on Removal of Fission Products from Steam-Air Mixtures," Trans. Amer. Nucl. Soc., 11 (2) 667 (1968).
112. P. C. Reist, "Review of the Probable Characteristics of Aerosols Resulting from Accident Loss of Sodium Coolant," USAEC Report NYO-841-18 (May, 1969).
113. P. Goldsmith, H. J. Delafield, and L. C. Cox, "The Role of Diffusiophoresis in the Scavenging of Radiative Particles from the Atmosphere," Quart. J. Roy. Meteorol. Soc., 89, 43 (1963).
114. L. Waldmann and K. H. Schmitt, "Thermophoresis and Diffusiophoresis of Aerosols," Aerosol Science, C. N. Davies, Editor, Academic Press, Inc., New York (1966), pp. 137-162.
115. S. Chapman and T. G. Cowling, The Mathematical Theory of Nonuniform Gases, 2nd Ed., Cambridge University Press, Cambridge, England (1953).
116. L. Waldmann, "Uber die Kraft eines Inhomogenen Gases auf Kleine Suspendierte Kugeln," Z. Naturforsch, A, 14 (1959), pp. 589-599.
117. S. P. Bakanov and B. V. Derjaguin, "The Motion of a Small Particle in a Nonuniform Gas Mixture," Discuss. Faraday Soc., 30, 130-138 (1960).
118. P. Goldsmith and F. G. May, "Diffusiophoresis and Thermophoresis in Water Vapor Systems," Aerosol Science, C. N. Davies, Ed., Academic Press, Inc., New York (1966), pp. 163-194.
119. K. H. Schmitt and L. Waldmann, "Untersuchungen an Schwebstoffteilchen im Diffundierenden Gasen," Z. Naturforsch, A, 15 (1960), pp. 843-851.
120. K. H. Schmitt, "Untersuchungen an Schwebstoffteilchen im Temperaturfeld," Z. Naturforsch, A, 14 (1959), pp. 870-881.
121. G. Zobel, Kollardzeitschrift, 157, 17 (1958).
122. P. Spiegler, J. G. Morgan, M. A. Greenfield, and R. L. Koontz, "Characterization of Aerosols Produced by Sodium Fires," USAEC Report NAA-SR-11997 (May, 1967).
123. U. S. A. Press Release G-293 (December 17, 1964).
124. K. Smith and G. R. Bainbridge, "Some Aspects of Aerosol Filtration in Stage I Civil Reactors," AERO-CONF-3 (November, 1960).
125. L. C. Schwendiman et al., "The Washout of Methyl Iodide by Hydrazine Sprays, Progress Report (through September, 1957)," USAEC Report BNWL-530 (1957).
126. V. Griffiths, "The Removal of Iodine from the Atmosphere by Sprays," British Report HHSB(S)R-45 (January 9, 1963).
127. C. R. Wilke and P. Chang, "Correlation of Diffusion Coefficients in Dilute Solutions," Amer. Inst. Chem. Eng. Jour. 1, 264 (1955).
128. F. H. Garner and P. J. Haycock, "Circulation in Liquid Drops," Proc. Roy. Soc. (London), 252A, 457, (1959).
129. T. D. Taylor and A. Acrivos, "On the Deformation and Drag of a Falling Viscous Drop at Low Reynolds Number," J. Fluid Mech., 18, 446 (1964).
130. F. Y. Pan and A. Acrivos, "Shape of a Drop or Bubble at Low Reynolds Number," Ind. Eng. Chem., Fundam., 7 (2), 227 (1968).
131. A. E. J. Eggleton, "A Theoretical Examination of Iodine-Water Partition Coefficient," AERE-R-4887 (1967).
132. M. A. Styrikovich, O. I. Marynova, K. Ya. Katkovskaya, I. Ya. Dubrovskii, and I. N. Smirnova, "Transfer of Iodine from Aqueous Solutions to Saturated Vapor," At. Energ. (USSR), 17 (1), 45 (1964).
133. L. C. Schwendiman, R. A. Hasty, and A. K. Postma, "The Washout of Methyl Iodide by Hydrazine Sprays. Final Report," BNWL-935 (November, 1968).
134. W. E. Ranz and W. R. Marshall, "Evaporation from Drops. Part II," Chemical Engineering Progress, 48(141): 173 (1952).
135. L. F. Parsly, Jr., "Removal of Elemental Iodine from Steam-Air Atmospheres by Reactive Sprays," ORNL-TM-1911. (Oct., 1967).
136. T. H. Row, L. F. Parsly, and H. E. Zittel, "Design Considerations of Reactor Containment Spray Systems - Part I," ORNL-TM-2412 (1969).
137. W. E. Clark and J. A. Gieseke, "Spray Droplet Deceleration in an Evaluation of the Applicability of Existing Data to the Analytical Description of a Nuclear-Reactor Accident," Quarterly Report for April through June, 1968, BMI-1844 (July, 1968), pp. 41-45.
138. L. F. Parsly, Jr. and J. L. Wantland, "Spray Studies at the Nuclear Safety Pilot Plant," Nuclear Safety Annual Progress Report for Period Ending December 31, 1968, ORNL-4374 (June, 1969), pp. 214-215.
139. J. W. Mausteller and J. J. Campana, "Activity Distribution from Simulated Pressurized Water Leaks," Proceedings of Second United Nations International Conference on the Peaceful Uses of Atomic Energy, 11, 153-156 (Geneva, 1958).
140. T. H. Row, "Spray and Pool Absorption Technology Program," ORNL-4360 (April, 1969).
141. G. J. Rogers, L. F. Coleman, R. K. Hilliard, and J. D. McCormack, "Removal of Airborne Fission Products by Containment Sprays," Trans. Am. Nucl. Soc. 12, 327 (1969).
142. L. F. Coleman, R. K. Hilliard, C. E. Linderoth, J. C. McCormack, and A. K. Postma, "Large-Scale Fission Product Transport Experiments," in BNWL-1009 (March, 1969), pp. 2.1-2.23.
143. L. F. Parsly and B. A. Soldano, "Removal of Methyl Iodide from Containment Atmospheres with Sprays," Trans. Am. Nucl. Soc. 12, 326 (1969).
144. A. Slater, J. Chem. Soc. 85, 1268 (1904); Moelwyn-Hughes, E. A., Proc. Royal Soc. (London) A-196, 540 (1949).
145. F. J. Viles, Jr. and L. Silverman, "Removal of Iodine and Methyl Iodide by Aerosol Formation with Hydrazines," Proc. of the 9th Air Cleaning Conf., CONF-660904 (1966), pp. 273-297.
146. F. J. Viles, Jr., E. Bulba, J. L. Lynch, and M. W. First, "Reactants for the Removal of Iodine and Methyl Iodine and their Application in Foams," Proc. of a Symp. on Treatment of Airborne Radioactive Wastes, IAEA (1968).
147. Sacramento Municipal Utility District to AEC, Docket 50-312 (April, 1968).
148. H. E. Zittel and T. H. Row, "Radiation and Chemical Stability of Spray Additions," Trans. ANS 12, 166 (1969).
149. "Nuclear Safety Program Annual Progress Report for Period ending December 31, 1968," USAEC Report ORNL-4374, June 1969, pp. 224-234.
150. "Boiling Nuclear Superheater (BONUS) Power Station Final Hazards Summary Report," Report GNEC-5, General Nuclear Engineering Corporation (February, 1962).
151. W. E. Browning, Jr., "Removal of Fission-Product Activity from Gases," Nuclear Safety 1, 3, (1960), pp. 40-46.
152. Preliminary Safety Analysis Report, Nuclear Plant - Diablo Canyon Site - Pacific Gas and Electric Company, USAEC Docket No. 50-275 (January, 1967).
153. Commonwealth Edison Company, "The Dresden Nuclear Power Station Unit Number 3, Preliminary Design and Analysis Report," 2, Docket 50-249 (February, 1966).
154. R. A. F. White and S. E. Smith, "Removal of Radioactive Particulates from Air," Research Applied in Industry 8 (6), 228-233 (June, 1960).
155. K. D. Mulcaster, "The Development and Application of High Efficiency Filters in the Nuclear Industry in England," Staub 21, 302-306 (July, 1961).
156. Quality Assurance Directorate, U. S. Army Edgewood Arsenal, Instruction Manual for Q76 DOP Filter Testing Penetrometer, Document No. 136-300 195A and Instruction Manual for Q107 DOP Filter Testing Penetrometer, Document No. 136-300-175A, Edgewood Arsenal, Maryland. (These manuals are replacing MIL-STD-282 and will eventually be replaced by a USA standard.)
157. C. A. Burchsted, "Requirements for Fire-Resistant High-Efficiency Particulate Air Filters," Proceedings of Ninth AEC Air Cleaning Conference, Boston, Mass., September, 1966, USAEC Report CONF-660904, 1, (January, 1967), pp. 62-74.
158. British Standard Method of Test for Low-Penetration Air Filters (other than for Air Supply to I. C. Engines and Compressors), B. S. 3928; 1965, British Standards Institution, London (1965).
159. British Standard Specification for Methods of Test for Air Filters Used in Air-Conditioning and General Ventilation, B. S. 2831; 1957, British Standard Institution, London, (1957).
160. D. Hasenclever, "The Testing of High Efficiency Filters for the Collection of Suspended Particles," International Symposium on Fission Product Release and Transport Under Accident Conditions, Oak Ridge, Tennessee, April 5-7, 1965, USAEC Report CONF-650407 (Jan. 1966), pp. 805-813.
161. USAEC Health and Safety Information Issue No. 253, Filter Unit Inspection and Testing Service, Fiscal Year 1968, Division of Operational Safety, USAEC Washington, D. C. (July 10, 1967); also, USAEC Health and Safety Information Issue No. 212, Minimal Specification for the Fire-Resistant High-Efficiency Filter Unit (June 25, 1965).

162. H. W. Knudsen and L. White, "Development of Smoke Penetration Meters," Report NRI-P-2642, Naval Research Lab. (Sept. 14, 1945).
163. G. Humphrey, Comment on p. 447 in Panel B, Round Table Session: Specifications, Maintenance and Monitoring of Filters, Proceedings of Eighth AEC Air Cleaning Conference, Oak Ridge National Laboratory, October 22-25, 1963, USAEC Report TID-7677 (March 1964), pp. 439-452.
164. J. W. Thomas, "Aerosol Penetration through Pinholed Filters," Health Physics, 11, 667-673 (1965).
165. F. E. Adley and D. E. Anderson, "The Effects of Holes on the Performance Characteristics of High-Efficiency Filters," Proceedings of Eighth AEC Air Cleaning Conference, Oak Ridge National Laboratory, October 22-26, 1963, USAEC Report TID-7677 (March 1964), pp. 494-507.
166. E. C. Parrish and R. W. Schneider, "Tests of High Efficiency Filters and Filter Installations at ORNL," USAEC Report ORNL-3442 (May 17, 1963).
167. E. Stafford and W. J. Smith, Ind. Eng. Chem., 43, 1346 (1951).
168. C. A. Burchsted, Oak Ridge National Laboratory, unpublished data (July, 1966).
169. R. E. Adams et al., "Filtration of Stainless Steel-UO_2 Aerosols," Nuclear Safety Program Annual Progress Report, December 31, 1967, USAEC Report ORNL-4228, Oak Ridge National Laboratory (April, 1968), pp. 133-148.
170. R. E. Adams et al., "Filtration of Particulate Aerosols under Reactor Accident Conditions," USAEC Report ORNL-TM-1707 (December, 1966).
171. R. J. Davis et al., "Filtration of Solid Aerosols," ORNL Nuclear Safety Research and Development Program Bimonthly Report. November-December, 1967, USAEC Report ORNL-TM-2095 (February, 1968). Further information contained under the same subtitle in preceding bimonthly reports: USAEC Reports ORNL-TM-2057 (September-October, 1967), ORNL-TM-1986 (July-August), ORNL-TM-1913 (May-June), etc.
172. W. D. Yuill and R. E. Adams, "Behavior of Oxide Aerosols of Uranium and Stainless Steel in Humid Atmospheres," USAEC Report ORNL-4198 (January, 1968).
173. W. D. Yuill and R. E. Adams, "Behavior of Aerosols in Humid Atmospheres," Nuclear Safety Program Annual Progress Report, December 31, 1967, USAEC Report ORNL-4228 (April, 1968), pp. 142-148.
174. M. D. Silverman et al., "Characterization of Radioactive Particulate Aerosols by the Fibrous Filter Analyzer," USAEC Report ORNL-4047 (March, 1967).
175. C. A. Burchsted and A. B. Fuller, "Design Construction and Testing of High-Efficiency Air Filtration Systems for Nuclear Applications," USAEC Report ORNL NSIC - 65 (Jan. 1970).
176. W. S. Durant et al., "Activity Confinement System of the Savannah River Plant Reactors," USAEC Report DP-1071 (August, 1966).
177. A. H. Peters, "Application of Moisture Separators and Particulate Filters in Reactor Containment," USAEC Report DP-812 (December, 1962).
178. J. W. Walker and A. H. Peters, "Filters for Reactor Containment," Mech. Eng., 85 (9), 46-50 (1963).
179. R. D. Rivers and J. L. Trinkle, "Moisture Separator Study," USAEC Report NYO-3250-6 (June, 1966).
180. Standard UL-586, High Efficiency Air Filter Units, Underwriters' Laboratories, Inc., Chicago, Illinois (June, 1964).
181. C. A. Burchsted, "Requirements for Fire-Resistant High-Efficiency Particulate Air Filters," Proceedings of Ninth AEC Air Cleaning Conference, Boston, Mass., September 13, 1966, USAEC Report CONF-660904, (January, 1967), Vol. 1, pp. 62-74.
182. W. L. Anderson and T. Anderson, Proceedings of Ninth AEC Air Cleaning Conference, Boston, Mass., September 13-16, 1966, USAEC Report CONF-660904 (January, 1967), Vol. 1, pp. 79-95.
183. R. W. Schneider, "In-Place Testing of High-Efficiency Filters," Nuclear Safety 4, 3, 56-68 (1963).
184. W. W. Goshorn and A. B. Fuller, "Particulate Filter Testing and Inspection Program," Nuclear Safety, 2, 2, 37-38 (1960).
185. E. C. Parrish, Oak Ridge National Laboratory, Private Communication (March, 1968).
186. J. D. DeField and H. J. Ettinger, "Efficiency Testing of the Air Cleaning System for a High Temperature Reactor (SM-110/32)," Treatment of Airborne Radiation Wastes, IAEA (1968), pp. 265-278.
187. W. E. Browning, Jr., "Removal of Fission-Product Activity from Gases", Nuclear Safety 4, 2, 83-86 (1962).
188. W. E. Browning, Jr., "Removal of Radioiodine from Gases," Nuclear Safety, 2, 3, 35-38, 43-44 (1961).
189. W. E. Browning, Jr., "Removal of Radioiodine from Gases," Nuclear Safety 4, 2, 83-86 (1962).
190. J. M. Holmes, "Removal of Radioiodine From Gases," Nuclear Safety 2, 4, 39-41 (1961).
191. W. B. Cottrell, W. E. Browning, Jr., G. W. Parker, A. W. Castleman, Jr., and R. L. Junkins, "U. S. Experience on Release and Transport of Fission Products within Containment Systems under Simulated and Actual Reactor Accident Conditions," Paper A/Conf 28 P/285, presented at the Third United Nations Inter. Conference on the Peaceful Uses of Atomic Energy, Geneva, 1964.
192. A. C. Chamberlain et al., "Report of Aerosol Group, Part I. Physical Chemistry of Iodine and Removal of Iodine from Gas Streams," AERE-R-4286, April 1963, J. Nuclear Energy, parts A and B. Reactor Science and Technology 17, 519 (1963).
193. J. B. Morris, H. R. Diffey, D. Nicholls and C. H. Rumary, "The Removal of Low Concentrations of Iodine from Air on a Plant Scale," AERE-R-3917, J. Nuclear Energy, parts A and B, Reactor Science and Technology 16, 437-445, (1962).
194. J. H. Swanks, "In-Place Iodine Filter Testing," Proceedings of Ninth AEC Air Cleaning Conference, September 1966, USAEC Report CONF-660904, 2 (January, 1967), pp. 1092-1104.
195. C. A. Gukeisen and K. L. Malaby, "In-Place Testing of Charcoal Filter Banks at Ames Laboratory Research Center (ALRC)," Proceedings of Ninth AEC Air Cleaning Conference, USAEC Report CONF-660904, 2 (January 1967), pp. 1063-1068.
196. F. J. Viles, Jr. and L. Silverman, "In-Place Iodine Removal Efficiency Test," Proceedings of Ninth AEC Air Cleaning Conference, USAEC Report CONF-660904, 2 (January 1967), pp. 1108-1132.
197. W. G. Thomson and R. E. Grossman, "In-Place Testing for Iodine Removal Efficiency Using an Electronic Detector," Proceedings of Ninth AEC Air Cleaning Conference, USAEC Report CONF-660904, (January, 1967), Vol. 2, pp. 1134-1149.
198. R. D. Ackley and R. E. Adams, "Removal of Radioactive Methyl Iodide from Steam-Air Systems (Test Series II)," USAEC Report ORNL-4180 (October, 1967).
199. Safety Evaluation by the Division of Reactor Licensing, USAEC, in the matter of Consolidated Edison Company of New York, Inc., Indian Point Nuclear Generating Unit No. 2, Peekskill, New York, Docket No. 50-247 (August 25, 1966), p. 64.
200. Safety Evaluation by the Division of Reactor Licensing, USAEC in the matter of Philadelphia Electric Company, Peach Bottom Atomic Power Station Unit Nos. 2 and 3, Peach Bottom Township, York County, Pennsylvania, Docket Nos. 50-277 and 50-278 (November 7, 1967), p. 34.
201. W. S. Durant et al., "Activity Confinement System of the Savannah River Plant Reactors," USAEC Report DP-1071, Savannah River Plant (August, 1966).
202. D. R. Muhlbaier, "Standardized Nondestructive Test of Carbon Beds for Reactor Confinement Applications," Final Progress Report, February-June 1966, USAEC Report DP-1082 (July, 1967).
203. R. E. Adams and W. E. Browning, Jr., "Iodine Vapor Adsorption Studies for the N. S. Savannah Project," ORNL-3726 (February, 1965), pp. 3-10.
204. R. E. Adams and W. E. Browning, Jr., "Iodine Vapor Adsorption Studies for the N. S. Savannah Project," ORNL-3726 (February, 1965), pp. 25-30.
205. R. E. Adams and W. E. Browning, Jr., "Removal of Iodine and Volatile Iodine Compounds from Air Systems by Activated Charcoal," International Symposium on Fission Product Release and Transport under Accident Conditions, April, 1965, USAEC Report CONF-650407, Vol. 2, pp. 869-884.
206. G. H. Prigge, "Application of Activated Carbon in Reactor Containment," DP-778 (September, 1962).
207. W. E. Browning, Jr. and R. E. Adams, "Removal of Iodine and Volatile Iodine Compounds from Air Streams by Activated Charcoal," Nuclear Safety Program Semiannual Progress Report for Period Ending June 30, 1964, ORNL-3691 (November, 1964), pp. 70-72.
208. R. D. Ackley et al., "Retention of Methyl Iodide by Charcoal under Accident Conditions," Nuclear Safety Program Semiannual Progress Report, December 31, 1965, USAEC Report ORNL-3915, pp. 61-80. (Mar. 1966)
209. Vermont Yankee Nuclear Power Station Plant Design and Analysis Report, Docket No. 50-271 (1966), pp. XIV-3-39.
210. Comments by G. W. Parker and A. H. Peters in Discussion, Proceedings of Ninth AEC Air Cleaning Conference, Boston, Mass., September 1966. USAEC Report CONF-660904 January, 1967, Vol. 1, p. 370.
211. R. E. Adams, W. E. Browning, Jr., W. B. Cottrell, and G. W. Parker, "The Release and Adsorption of Methyl Iodide in the HFIR Maximum Credible Accident," USAEC Report ORNL-TM-1291 (October, 1965).
212. G. W. Keilholtz, "Filters, Sorbents and Air Cleaning Systems as Engineered Safeguards in Nuclear Installations," USAEC Report ORNL-NSIC-13 (October, 1966).
213. R. E. Adams, R. D. Ackley and W. E. Browning, Jr., "Removal of Radioactive Methyl Iodide from Steam Air Systems," USAEC Report ORNL-4040 (January, 1967).
214. Tennessee Valley Authority, Design and Analysis Report for Browns Ferry Nuclear Power Station, Amendment H9-1, Docket Nos. 50-257 and 50-260.
215. Tennessee Valley Authority, Design and Analysis Report for Browns Ferry Nuclear Power Station Units 1 and 2, Dockets

50-259 and 260.

216. D. A. Collins, L. R. Taylor, and R. Taylor, "The Development of Impregnated Charcoals for Trapping Methyl Iodide at High Humidity," British TRG Report 1300(W) (1967).
217. R. C. Milham, "High Temperature Adsorbents for Iodine," Progress Report, January 1965-September 1966, USAEC Report DP-1075 (December, 1966).
218. W. E. Browning, Jr. et al., "Ignition of Charcoal Adsorbers by Fission-Product Decay Heat," Nuclear Safety Program Semiannual Progress Report, June 30, 1965, USAEC Report ORNL-3843 (September, 1965), pp. 156-167.
219. R. P. Shields and C. E. Miller, Jr., "The Effect of Fission Products on Charcoal Ignition (In-Pile Experiment)," USAEC Report ORNL-TM-1739 (January, 1967).
220. R. P. Shields and R. E. Adams, "The Effect of Fission Products on Ignition of Iodized Charcoal (In-Pile Experiment IGR-3)," USAEC Report ORNL-TM-2321 (December, 1968).
221. W. E. Browning, Jr. et al., "Ignition of Charcoal Adsorbers by Fission-Product Decay Heat," Nuclear Safety Program Semiannual Progress Report, December 31, 1965, USAEC Report, ORNL-3915 (March, 1966), pp. 81-85.
222. J. D. McCormack, "Some Observations on Iodine Removal from Plant Streams with Charcoal," HW-SA-3187 (1963), pp. 35-40.
223. D. McConnon, "Radioiodine Sampling with Activated Charcoal Cartridges," HW-77126 (April 19, 1963).
224. R. E. McHenry, "Fission Products," ORNL-TM-352 (June 16, 1953).
225. W. Cox, "Evaluation of Charcoal and Other Adsorbents for Removing Radioiodine from Hot Off-Gas Systems," TID-7688, 68-73 (1963).
226. Duke Power Company, Oconee Nuclear Station Units 1 and 2, Preliminary Safety Analysis Report, Dockets 50-269 and 50-270.
227. I. A. Mossop, "Filtration of the Gaseous Effluent of an Air-Cooled Reactor," Brit. Chem. Eng. 5, 420-424, 426 (June, 1960).
228. Preliminary Hazards Summary Report of Consolidated Edison Company of New York, Inc., Ravenswood Nuclear Generating Unit A, 1963, USAEC Report NP-12467 (1962).
229. City of Los Angeles Department of Water and Power, Malibu, Docket 50-214.
230. H. S. Davis, D. D. Stepnewski, G. E. Wade, and D. L. Condotta, "Fission Product Containment for the New Hanford Production Reactor," ANS Transactions, 6, 118-119 (June, 1963).
231. J. W. Green, W. V. Thompson, and H. Copeland, "Hanford Experience with Reactor Confinement," HW-SA-3213, (1963).
232. J. W. Walker and A. H. Peters, "The Use of Filters in Nuclear Reactor Containment," TID-17548, (December, 1962).
233. H. M. Gilroy and J. H. Wilson, "Piqua OMR Waste Gas Disposal," NAA-SR-4576 (July, 1960).
234. F. T. Binford and E. N. Cramer, "The High-Flux Isotope Reactor, A Functional Description," ORNL-3572 (May, 1964).
235. Pacific Gas and Electric Company, "Final Hazards Summary Report, Humboldt Bay Power Plant, Unit No. 3," Appendix IV, "Pressure Suppression Development Program," NP-14319 (September 1, 1961).
236. F. Abbey and J. J. Hillary (UKAEA), "Fission Product Removal for Reactors with Vented Containment," presented at IAEA Symposium on the Containment and Siting of Nuclear Power Plants, SM-89/36, NISC Accession No. 8827, (April, 1967).
237. H. R. Diffey, C. H. Rumary, M. J. S. Smith, and R. A. Stinchcombe, "Iodine Clean-Up in a Steam Suppression System," USAEC International Symposium on Fission Product Release and Transport under Accident Conditions, April 5-7, 1965, CONF-650407, vol. 2, pp. 776-804. (Jan. 1966).
238. L. Devell, R. Hesbol, and E. Bachofner, "Trapping of Iodine in Water Pools at 100°C," presented at IAEA Symposium on the Containment and Siting of Nuclear Power Plants, SM-89/15 NSIC Accession No. 18225 (April, 1967).
239. F. T. Binford, L. E. Stanford, and C. C. Webster, "Pressure-Supression Experiments," USAEC Report ORNL-4374 (June, 1969), pp. 234-250.
240. M. Siegler and D. P. Siegwarth, "Scale-Model Tests of Fission-Product Removal in Suppression Pools," USAEC Report ORNL-4374 (June, 1969), pp. 250-256.
241. J. A. Gieseke and W. E. Clark, "Particle Collection in Pressure-Suppression Pools," USAEC Report BMI-1835 (April, 1968), pp. 17-19.
242. V. G. Levich, Physicochemical Hydrodynamics, Prentice-Hall, Inc., Englewood Cliffs, New Jersey (1962), pp. 402, 404-409.
243. J. M. Genco, H. S. Rosenberg, D. A. Berry, G. E. Cremeans, W. E. Berry, and D. L. Morrison, "Fission-Product Deposition and its Enhancement under Reactor Accident Conditions," USAEC Report BMI-X-10213 (EURAEC-1968) (October, 1967).
244. J. M. Genco, D. A. Berry, H. S. Rosenberg, G. E. Cremeans, and D. L. Morrison, "Research and Development on Coatings for Retaining Fission Product Iodine," IAEA Conference on Treatment of Airborne Radioactive Wastes (1968).
245. J. M. Genco, H. S. Rosenberg, D. A. Berry, G. E. Cremeans, W. E. Berry, and D. L. Morrison, "Fission-Product Deposition and its Enhancement under Reactor Accident Conditions," Quarterly Progress Report for July-September, 1968, USAEC Report BMI-X-102044 (EURAEC-2073) (October, 1968).
246. H. S. Rosenberg, G. E. Cremeans, J. M. Genco, D. A. Berry, and D. L. Morrison, "Fission Product Deposition and its Enhancement under Reactor Accident Conditions: Reactive Coating Development," USAEC Report BMI (in press).
247. L. Silverman, "Foam and Diffusion Board Approaches to Containment of Reactor Releases," Third Gatlinburg Conference on Reactor Chemistry, TID-7641, 169-184 (1963).
248. L. Silverman, M. Corn, F. Stein, "Diffusion Board Containment Concepts and Foam Encapsulation Studies," 7th USAEC Air Cleaning Conference, TID-7627, (March 1962).
249. R. E. Yoder, M. H. Fontana, L. Silverman, "Foam Suppression of Radioactive Iodine and Particulates," NYO-9323, HACL-98 (February 25, 1964).
250. F. J. Viles, Jr., E. Bulba, J. L. Lynch, and M. W. First, "Reactants for the Removal of Iodine and Methyl Iodide and their Application in Foams," IAEA Conference (1968).
251. Safety in Mines Research Establishment, Buxton, England (1955).
252. J. T. Davis and E. K. Rideal, Interface Phenomena, Academic Press, New York (1962).
253. W. E. Browning, Jr., and R. D. Ackley, "Characterization of Millimicron Radioactive Aerosols and their Removal from Cases," Third Conference on Nuclear Reactor Chemistry, TID-7641, 130 (October, 1962).
254. R. E. Yoder, M. H. Fontana, and L. Silverman, "Foam Suppression of Radioactive Iodine and Particulates," USAEC Report NYO-9324 (1964).
255. R. J. Beers, Safety Engineers 100-USAEC, personal communication, 1963. Internal AEC report is available.
256. S. Posner, "Air Sampling Filter Paper Retention Studies Using Solid Particles," Ibid, p. 43.
257. R. Dennis, L. Silverman, and F. Stein, "Iodine Collection Studies, A Review," Ibid, p. 327.
258. R. E. Adams and W. E. Browning, Jr., "Removal of Iodine from Gas Stream, Ibid, p. 242.
259. S. K. Friedlander, L. Silverman, P. Drinker, and M. W. First, Handbook on Air Cleaning, U. S. Atomic Energy Commision (September, 1952).
260. G. W. Keilholtz, "Removal of Radioactive Noble Gases from Off-Gas Streams," Nuclear Safety, 8 (2), 155-160 (1966-67).
261. Air Reduction Company, Inc., "Noble Gas Recovery Study - Maritime Nuclear Ship Savannah, Final Report April 5, 1965," undocumented.
262. J. R. Merriman, J. H. Pashley, and S. H. Smiley, "Engineering Development of an Absorption Process for the Concentration and Collection of Krypton and Xenon, Summary of Progress through July 1, 1967," USAEC Report K-1725 (December, 1967).
263. S. Blumkin et al., "Priliminary Results of Diffusion Membrane Studies for the Separation of Noble Gases from Reactor Accident Atmospheres," presented at the Ninth AEC Air Cleaning Conference, Boston, Mass., September 13-16, 1966.
264. R. H. Rainey and W. L. Carter, "Separation of Noble Gases from Air by Permselective Membranes," Nuclear Safety Program Annual Progress Report, USAEC Report ORNL-4228 (December 31, 1967), pp. 173-182.
265. R. H. Rainey, "Criteria for Noble-Gas Removal Using Permselective Membranes," USAEC Report ORNL-TM-1822 (April 3, 1967).
266. W. L. Robb, "Thin Silicone Membranes - Their Permselective Properties and Some Applications," Report 65-C-031, General Electric Company (October, 1965).
267. S. Blumkin, "A Method for Calculating Cascade Gradients for Multicomponent Systems Involving Large Separation Factors," USAEC Report K-OA-1559, Oak Ridge Gaseous Diffusion Plant (January 15, 1968).
268. S. Blumkin, "Permselective Membrane Cascades for the Separation of Xenon and Krypton from Argon," USAEC Report K-OA-1622, Oak Ridge Gaseous Diffusion Plant (July 15, 1968).

CHAPTER 20

Radioactive Waste Management

W. A. RODGER
Nuclear Fuel Services, Inc.
Washington, D. C.

S. McLAIN
McLain Associates
Lafayette, Indiana

CHAPTER CONTENTS*

*Except for an appendix, this chapter is based on information in the literature or known to the authors prior to January 1964. An appendix summarizes relevant data up to mid-1969.

1 INTRODUCTION

The amount and type of nuclear fuel present in a reactor varies from a few kilograms of enriched uranium to hundreds of kilograms of fissionable material plus thousands of kilograms of source (fertile) material. During operation, large quantities of radioactive materials are formed in the fuel, the coolant, and the structural components of the reactor core. The amounts of the radioactive substances formed are a function of the power level, the operating schedules, and the materials used in the reactor core and adjacent complex. Thus, a reactor which has operated for any length of time at a high power level has a large inventory of potentially hazardous radioactive materials. Under ordinary operating conditions, essentially all these hazardous materials are maintained within the reactor.

This chapter discusses the sources and quantities of the wastes produced in conjunction with typical reactors and the methods and equipment used to control and manage the radioactive wastes. While emphasis is placed on experience in the United States, some discussion of foreign reactors and experience is also presented.

The primary emphasis has been placed upon a description of the wastes to be expected from various types of reactors, operating experience with relation to wastes, and descriptions of the methods used for controlling these wastes. This discussion (Sec. 3) together with the references will provide the designers of future reactors with tools which will allow them more easily to assess their waste source terms.

In Secs. 4, 5, and 6 a general discussion of methods for handling solid, gaseous, and liquid wastes is presented. Space is not avaliable to give a detailed discussion of all the possible methods for handling these wastes, and furthermore the authors believe that the methods used in any given situation must be chosen and designed to fit the specific local problem and that reactor designers should excercise discretion and ingenuity in the development of their waste-disposal methods. The time for waste treatment "standard methods" or "cook-books" is not yet here.

In an accident it is conceivable that a hazardous portion of the radioactive materials inventory could be released to the plant or to the surrounding environment. This possibility is properly considered in evaluating the hazards for any reactor. Handling these materials is not a primary consideration in designing the facilities that provide for collection and treatment of the small quantities of radioactive materials expected to escape from the primary coolant system during normal operations. The somewhat larger, but still small, quantities released by operating abnormalities such as fuel-element failures, leaking tubes and valves, decontamination of equipment, etc. must be considered during design of the waste-handling equipment.

Reactor designers, the Atomic Energy Commission and its advisory committees, the owners, the State and local authorities, and the various code- and standard-writing groups are all interested in preventing releases of appreciable quantities of radioactive materials from the normal operating systems. As a result, the plants are designed to prevent the accidental release of appreciable quantities of radioactive material. Such a release would put an overwhelming load on the normal waste-handling system. Rather than designing waste-handling facilities with sufficient capacity to handle this sort of eventuality, it has been the practice of reactor designers to provide containment or other protective procedures to handle accidental releases.

Since by their very nature accidents are not completely predictable, it has been necessary to devise decontamination and cleanup procedures specifically for each of the major radioactivity releases associated with incidents which have occurred. It is not within the scope of this chapter to discuss the effect of accidental releases on the environment. However, in Sec. 7 several incidents are described to illustrate the kinds of cleanup procedures that have been used, the relationship of waste management to the mitigation of release to the reactor environment, and the ways in which careful design of the reactor plant and its associated waste-handling facilities may make the control and cleanup of an incident less hazardous and troublesome.

2 GENERAL CONSIDERATIONS

Radioactive materials become associated with many wastes produced in nuclear reactor plants. These include gases, liquids, and solids. Small quantities of each of these materials must be removed from the reactor plant either continuously or intermittently.

One goal in the management of radioactive wastes is to design the plant so that the quantities of radioactive materials associated with each effluent gaseous and liquid stream are so low that they may be discharged to the environment without any treatment at all and still be below the maximum approved concentrations at the site boundaries. Except for reactors operating at very low power levels, this goal is seldom attainable. Some treatment of at least part of the gaseous and liquid waste streams is almost always required. Solid wastes are conveniently handled in sealed containers, stored, and shipped to approved disposal areas.

The treatment of a gaseous or liquid waste depends on its chemical content and the concentrations of radioactive materials. Isolation of the waste streams is normal. For example, the low-activity liquid wastes may be treated by ion exchange while the high-activity wastes require concentration usually by evaporation and calcination. Almost always the very low-level* waste streams are kept isolated from those containing larger quantities of active materials. The higher level waste streams generally have very small volumes, while the low-level* waste streams frequently have quite large volumes. Examples of large volume and very low-level wastes are the ventilation air and condenser cooling water which contain (during normal operation) only incidental quantities (if any) of added radioactive material; these and other low-level streams are simply monitored continuously, as they normally have no added radioactivity or, at most, quantities low enough to permit direct discharge. Examples of small-volume, high-level wastes are rejected coolant test samples and gases discharged from condensers of direct-cycle boiling-water reactors; these may require treatment and control.

Thus, the treatment of the various wastes usually consists of separating each of the gaseous and liquid waste streams into at least two fractions. The smaller streams contain the bulk of the radioactivity in more concentrated form, while the

*The terms "low level" and "high level" are far from standardized. They mean different things to different people and their meaning is strongly affected by the past experience of the user. Something which is "high level" to a research reactor operator may be "low level" to a chemical reprocessor. A set of working definitions used by the authors is given in Table 2-1. It is emphasized that these are the authors' definitions and are not generally accepted as "standard definitions."

TABLE 2-1

Authors' Definitions of Terms

Physical state	Class	Concentrations
All	Unrestricted	Background to 0.1 of limits in 10 CFR Part 20
	Low	0.1 of limits in 10 CFR Part 20 to 10^{-4} μc/cm^3
	High	10^{-4} μc/cm^3
Liquid	Low	0.1 of limits in 10 CFR Part 20 to 10^{-2} μc/cm^3
	High	10^{-2} μc/cm^3
Solid	Low	0.1 of limits in 10 CFR Part 20 50 mr/hr at contact or 10^3 μc of alpha material per container
	High	All solid wastes above 50 mr/hr at contact or containing over 10^3 μc of alpha material pre container

larger streams have sufficiently low radioactivity that they can be discarded after a minimum treatment without exceeding maximum permissible limits. The concentrated liquid streams must be reduced in volume so they can be economically stored until the activities decay to very low levels or until they can be discarded in some particular circumstance known to be nonhazardous.

A number of methods for treating various types of wastes are described very briefly herein. Because methods producing high degrees of separation are almost always more expensive than those producing more modest separation, it is uneconomical to choose methods which produce a higher degree of separation than is needed. Nevertheless, there has been a strong tendency to over-design the waste-disposal features of reactor plants.

The concentrations of radioactivity which may be discharged into the environment in either air or water are set by various governmental bodies, federal, state, and local. The fundamental regulations in the United States are contained in the Code of Federal Regulations, Title 10, Part 20 (usually abbreviated as "10 CFR20").

It will be noted from § 20.103 and § 20.106 of 10 CFR20, that the offsite allowable concentrations are generally 1/30 of those permitted for occupational exposure.

The discharge limits are values integrated over a period of time, usually a calendar quarter or year. Often instantaneous limits will also be set at a somewhat higher level. To assure that discharge limitations are being met, monitoring systems must be established within the controlled area and at the points of discharge. Frequently monitoring stations are established outside the controlled area to check that any increase in the off-site background, caused by either the gaseous or liquid wastes, does not exceed the permissible dose rates for the environs.

Nuclear power plants constructed near populated areas normally discharge such small quantities of wastes that the radioactivity is measurable only near the plants. Processes have been developed for handling those small waste streams that might be hazardous to the public. While these processes add to the cost of construction and operation of nuclear reactors, they are effective and result in discharge of only innocuous quantities of radioactive materials under normal and abnormal operating conditions. As noted above, a normal waste system cannot be expected to handle a release associated with a major accident. As a result, containment buildings are designed to retain all materials released in plausible accidents. (See Chapters 19, 21).

The experience with the power plants built in the United States has been excellent. The processes for handling the wastes have been effective. The actual wastes have been only a few percent of the maximum quantities which were expected by the designers. (Appendix 1 of this chapter summarizes data through mid-1969 on actual experience with radioactive waste systems at 8 operating power reactors.)

3 REACTOR WASTES AND THEIR HANDLING*

Most radioactive materials found in the coolant of a heterogeneous reactor are formed by neutron activation of the coolant and its impurities plus the corrosion and wear products removed from the cladding and core structures. Occasionally, metal-clad fuel elements are defective or fail, and small amounts of fission products escape to the coolant; and in a few cases traces of fuel have been left on the outside of fuel elements, and fission products from the "tramp" fuel have entered the coolant. In gas-cooled reactors using graphite-clad ceramic fuels, small quantities of gaseous or volatile fission products diffuse into the coolant.

---

*Several persons, including those listed below, assisted in preparation of portions of this section:
Frank Binford, Oak Ridge National Laboratory
Joel W. Chastain, Jr., Battelle Memorial Institute
John J. Hartig, Argonne National Laboratory
John R. Huffman, Phillips Petroleum Company, Idaho Falls
Howard A. McLain, Oak Ridge National Laboratory
Robert W. Powell, Brookhaven National Laboratory
R. F. S. Robertson, Chalk River
Charles A. Trilling, Atomics International
R. Kenneth Winkleblack, Atomics International
Clifford Zitek, Dresden Nuclear Power Station

As a result, all coolants in power reactors are continuously treated to maintain their purity.

In homogeneous reactors, the fuel is slurried with or dissolved in the coolant. Fission products in very large amounts are carried throughout the entire primary coolant system. The problems of handling these materials are similar to those encountered in irradiated fuel recovery plants. Because homogeneous reactors are still in early stages of development for power reactors, they are not emphasized in this chapter.

The various kinds of wastes encountered in several of the most highly developed types of heterogeneous reactors are discussed below.* The sources of the radioactive materials in each coolant are discussed along with the separation, collection, and handling procedures in use.

3.1 Water-Cooled Reactors

Most United States reactors are light-water-cooled; these include research, test, plutonium-production, power, and propulsion reactors. In addition, there are several research and production heavy-water-cooled reactors. The liquid and solid wastes produced by pressurized-water and boiling-water reactors are similar, but the gaseous wastes differ. Characteristics of light-water-cooled reactors, which are representative of several of the various technologies in use today, are presented in Table 3-1 [1, 2].

During operation, the major radioactive isotopes in the water are due to the activation of the water by neutrons and protons (recoil hydrogen nuclei) see Table 3-2 [3]. All of these radioisotopes are gases, most of them have short half-lives and none are discharged in any significant quantities from pressurized light-water-cooled reactors. Only N^{13} is discharged in appreciable quantities from boiling-water reactors. This is produced in such small amounts that dilution with air reduces its hazard to negligible quantities. F^{18} is produced but it is soluble in the condensate and remains with the liquid; so it is found on the condensate demineralizer resins. Very small

*The data used in this chapter were obtained in the late 1950s and early 1960s. More recent data are included in Appendix 1 of this chapter.

TABLE 3-1

Characteristics of Representative Water-Cooled Reactors [a]

Reactor	Power Mw(t)	Core			Control rods			Non-core materials	
		Fuel	Clad	Area (ft^2)	Poison	Clad	Area (ft^2)	Heat exchange area, (ft^2 of ss)	Other
Research									
Battelle [b] (BRR)	2	UAl	Al		B_4C	Al			Al, concrete
Oak Ridge (ORR)	20								
Argonne [c] (CP-5)	5	UAl	Al						
Test									
Materials Testing (MTR)	40	UAl	Al		Cd	Al			Cast iron pump
Engineering Test (ETR)	175	UAl	Al						
Power									
Boiling									
Exptl. Boiling (EBWR)	100	UZrNb	Zr2	2949 [d]	Boron	ss		---	
Dresden (DNPS)	700	UO_2	Zr2 [e] ss	35,000	B_4C	304 ss	1700	26,080	f
Pressurized Water									
Shippingport (PWR)	505 [g]	UZr UO_2	Zr4 Zr4	4,600 15,800	Hf	None	~320	34,800	
Yankee	540	UO_2	ss	15,500	Ag-In-Cd	None	~580	53,700	
Indian Point	585	UO_2-ThO_2	ss [h]	15,600	Hf	None	~394	54,100	
Production									
Hanford [i]	---	U	Al		---	---	---	None	Al
Savannah River	---	U	Al		---	Al	---	Large	
Propulsion									
Savannah	70	UO_2	ss	3,780	B-ss	ss	350 [j]	ss 6,360	

[a] Data taken from references [1] and [2]. Abbreviations: ss = stainless steel, Zr2 = Zircaloy-2. Conversion factor: 1 ft^2 = 929 cm^2.
[b] Pool type.
[c] Heavy-water-cooled and -moderated.
[d] The active surface of the first core was about 1680 ft^2 (156 m^2). In addition there were dummy fuel elements in the first core with an area of about 334 ft^2 (31 m^2) of aluminum.
[e] Zircaloy-2, Type I fuel and channels. Stainless steel, Type II fuel. Area of 35,000 ft^2 (3250 m^2) is for core 2.
[f] Cupro-nickel, 18,700 ft^2 (1740 m^2), monel, 19,500 ft^2 (1810 m^2), primary feedwater heaters after condensate demineralizers. 51,000 ft^2 (4750 m^2) of 304 L stainless steel in the primary system.
[g] Core 2.
[h] Zirconium channels.
[i] The New Production Reactor being built utilizes Zircaloy cladding and tubes. It is cooled by demineralized water with heat exchange by ss tubes. The older Hanford Reactors are all cooled by once-through purified Columbia River water.
[j] Plus ~220 ft^2 (20.4 m^2) of Zircaloy followers.

TABLE 3-2

Water Activation Products

Isotope	Reaction	Half-life	Most significant decay particles (Mev and percent): Beta	Positron	Gamma	Total decay energy (Mev)
H^3 [a]	$H^2 (n, \gamma)$ $Li^6 (n, \alpha)$ [b] Fission product	12.26 year	0.0186^{100}			0.0186
N^{13}	$O^{16} (p, \alpha)$	10.0 min		1.19^{100}		2.21
N^{16}	$O^{16} (n, p)$ $N^{15} (n, \gamma)$	7.4 sec	10.4^{26} 4.3^{68}		6.13^{73} 7.12^{5}	10.4
N^{17}	$O^{17} (n, p)$	4.14 sec	3.7			8.8
O^{19}	$O^{18} (n, \gamma)$	29 sec	3.25^{58} 4.60^{42}		0.200^{96} 1.36^{54}	4.80
F^{18}	$O^{18} (p, n)$	1.87 hr	EC^{3} [c]	0.65^{97}		1.67
Ar^{37} [d]	$Ar^{36} (n, \gamma)$	35 day	EC^{100}			0.82
Ar^{41} [d]	$Ar^{40} (n, \gamma)$	1.83 hr	$1.20^{99} 2.49^{1}$		1.29^{99}	2.49

[a] Tritium. While tritium is a gas, it usually is only present in readily detectable quantities as the oxide in liquid wastes discharged from light-water-cooled reactors and aqueous fuel reprocessing plants. It may be present in hazardous amounts in the coolant of heavy-water reactors.

[b] Lithium is sometimes added to coolants for pH control. Li^7OH has been in use at Shippingport.

[c] Electron capture. Gamma rays characteristic of the daughter are emitted.

[d] Ar^{37} and Ar^{41} may build up in small amounts if the water is not deaerated.

amounts of tritium are also present in the coolant as the oxide.

In the operating reactor complex the most significant activity is that of N^{16}. This determines the amount of shielding required for the piping in the primary coolant system external to the reactor. Upon reactor shutdown, this short-lived (7 sec) activity decays to a negligible level within a few minutes.

Radioactive products formed from neutron activation of impurities and corrosion products in the water interfere with maintenance operations when the reactor is shut down. The principal isotopes of interest are shown in Table 3-3 [3]. Reactor systems which contain aluminum have Na^{24} as the principal nongaseous activity. In systems containing stainless steel the principal long-life activities, after several days' decay, are Co^{58} and Co^{60}. Significant quantities of Cr^{51}, Fe^{59}, and a number of other activated corrosion products are also present. The quantities of these last are dependent on the materials used in construction of the primary reactor system, the coolant temperature, pH, etc. Cu^{64} and Mn^{56} are very significant in the first 24 to 48 hours after shutdown. The Co^{58} is formed from nickel and the Co^{60} from cobalt in the stainless steel. Cu^{64} is formed from small amounts of copper picked up in feedwater heaters. From the standpoint of discharge of this type of waste, the most significant nuclide is the Co^{60} because its 5.27-year half-life does not permit on-site decay. The principal airborne activities, other than the noble gases, are their daughters, especially Cs^{138} and the iodines [4].

As noted, fission products may be introduced into the coolant from activation and fission of "tramp" uranium on the surface of fuel elements, from the failure of fuel elements, and from low concentrations (ppm) of uranium in the structural materials. The amount introduced by fuel element failures markedly increases the concentration of the wastes but should not severely overload the equipment. However, the amount of shielding required, particularly for the waste-treatment equipment and storage tanks, may be affected.

The chemical and physical form of the impurities carried in the water is of significance in respect to the reactor design and operation as well as coolant maintenance. The word "crud" is commonly given to the materials that may be deposited from the water. The term refers particularly to the insoluble matter but may include absorbed soluble materials. Crud is responsible for the buildup of most of the activity in the primary coolant systems and it may cause fouling of heat-transfer surfaces and moving parts. The crud found in water-cooled reactors is composed primarily of corrosion products but it may also contain wear particles from pump bearings and control-rod drives plus miscellaneous debris such as dust and weld scale (see Fig. 3-1 [6]).

Since corrosion is the most significant source of crud, materials are chosen to minimize this. This is the main reason for the extensive use of stainless steels and stainless-steel-lined equipment [5, 6]. (See Sec. 2 of Chapter 13 and Sec. 2 of Chapter 17.)

Crud-formation rates are not always equivalent to crud-release rates since, in general, only part of the corrosion products enter the moving coolant.

TABLE 3-3

Impurity Activities in Water

Radioactive isotope	Half-life	Nuclear reaction	Most significant decay particles (Mev and Percent)			Total decay energy (Mev)
			Beta	Positron	Gamma	
Na^{24}	15.0 hr	Al^{27} (n, α) Na^{23} (n, γ)	$1.39^{\sim 100}$		2.75^{100}, 1.37^{100}	5.51
Al^{28}	2.30 min	Al^{27} (n, γ)	2.87^{100}		1.78^{100}	4.65
Si^{31}	2.62 hr	P^{31} (n, p)	1.48			1.48
Cl^{38}	37.3 min	Cl^{37} (n, γ)	4.8^{53}, 1.1^{31}		2.1^{47}, 1.6^{31}	4.8
Ar^{37}	35 day	Ar^{36} (n, γ)	EC^{100} [a]			0.82
Ar^{41}	1.83 hr	Ar^{40} (n, γ)	1.20^{99}, 2.49^{1}		1.29^{99}	2.49
Ca^{45}	160 day	Ca^{44} (n, γ)	0.25^{100}			0.25
Cr^{51}	27.8 day	Cr^{50} (n, γ)	EC		0.32^{10}	0.75
Mn^{54}	280 day	Fe^{54} (n, p)	EC^{100}		0.84^{100}	1.38
Fe^{55}	2.6 year	Fe^{54} (n, γ)	EC^{100}			0.22
Mn^{56}	2.58 hr	Fe^{56} (n, p) Mn^{55} (n, γ)	2.86^{60}, 1.05^{24}		0.84^{99}, 1.81^{23} 2.11^{14}	3.70
Co^{58}	71 day	Ni^{58} (n, p)	EC^{85}	0.48^{15}	0.81^{100}, $1.62^{0.5}$	2.31
Fe^{59}	45 day	Co^{59} (n, p)	0.46^{53}, 0.27^{46}		1.10^{56}, 1.29^{44}	1.56
Co^{60}	5.27 year	Co^{59} (n, γ)	0.31^{100}		1.33^{100}, 1.17^{100}	2.81
Cu^{64}	12.9 hr	Cu^{63} (n, γ)	0.57^{38}, EC^{43}	0.66^{19}	1.34^{1}	[c]
Ni^{65}	2.56 hr	Ni^{64} (n, γ)	2.10^{69}, 0.6^{23} 1.01^{8}		1.49^{20}, 1.11^{10}	2.10
Zn^{65}	245 day	Zn^{64} (n, γ)	EC^{98}	0.33^{2}	1.11^{50}	1.35
Cu^{66}	5.1 min	Cu^{65} (n, γ)	2.63^{91}, 1.59^{9}		1.04^{9}	2.63
Sr^{89}	51 day	Sr^{88} (n, γ)	1.46^{100}			1.46
Zr^{95}	65 day	Zr^{94} (n, γ)	0.40^{55}, 0.36^{43}		0.72^{55}, 0.76^{43}	1.12
Hf^{181}	45 day	Hf^{180} (n, γ)	CE [b] 0.41		0.137	1.02
Ta^{182}	115 day	Ta^{181} (n, γ)	0.51^{8}		0.10^{56}, 1.12^{33} 0.122^{28}, 1.19^{15}	1.73
Ta^{183}	5.0 day	Ta^{182} (n, γ)	0.61^{91}		0.108^{44}, 0.246^{35}	1.07
W^{187}	24.0 hr	W^{186} (n, γ)	0.63^{70}, 1.31^{20}		0.69, 0.48, 0.134	1.31

[a] Electron capture.
[b] Conversion electron.
[c] EC and positron, 1.68; beta, 0.57.

Breden has estimated that the corrosion product release rate of type-304 steel may be as low as 2 to 3 mg/dm^2-month, while the corrosion rate is 5 to 10 mg/dm^2-month in water under conditions normally found in primary coolant systems of boiling- and pressurized-water reactors [5]*.

Corrosion-product-release rates depend somewhat on the pH of the water. In neutral water loosely adherent films are formed, while at high pH the corrosion films on the stainless steels normally used are thicker and less crud is released. In addition to pH, oxygen concentration may also be a factor. In boiling-water reactors the 0.2 to 0.3 cc of oxygen per liter is credited with producing a very tenacious corrosion film. The corrosion-product films formed on zirconium are quite adherent. A large part of the corrosion products formed on aluminum surfaces slough off into the water, although a portion are soluble.

The materials in contact with the coolant include the fuel cladding, structural materials, and small quantities of various alloys. Aluminum is used for cladding and structural materials in low-temperature water-cooled reactors including research, test, and plutonium-production reactors. Stainless steel is used for most of the pumps, piping, etc. In power reactors stainless steel is used for cladding in many cases, as well as for

*For stainless steel, 10 mg/dm^2-month is equal to a corrosion rate of 0.06 mil/year.

structural and most other parts. Zirconium alloys are used for fuel cladding in the propulsion, the Shippingport, and Dresden Reactors, but stainless steel is used for other parts of these reactors (see Tables 3-1 and 3-4) [5, 7].

In reactors in which aluminum is used as the fuel cladding and structural material, the crud is composed of a mixture of iron and aluminum oxides. In the MTR the crud is apparent as a slight brown flocculent material [7]. Very little of this crud collects on the fuel and it does not interfere with heat transfer. Of interest is the EBWR which operates at 254°C (490°F). Originally it had aluminum dummy fuel assemblies. The total solids in the water of 70 ppb developed noticeable coatings on the active portions of the Zircaloy-clad fuel assemblies. The wet chemical analysis for these solids indicated that the assumed compounds had the following compositions by weight: $Al_2O_3 \cdot H_2O$, 70%; Fe_2O_3, 19%; and NiO, 11%. The aluminum was present primarily as colloidal boehmite, AlOOH. The presence of the crud increased the heat-transfer resistance and therefore the temperature of the fuel plates. For a 4-mil (0.1-mm) scale, calculations indicated a fuel centerline temperature increase to 665°C (1229°F) compared to the design value of 408°C (764°F) using the measured conductivity value for the scale of 0.0076 w/cm-°C.

It was expected that the uranium-alloy fuel material (U, 5 wt.% Zr, 1.5 wt.% Nb) used in the EBWR would grow at temperatures above 482°C (900°F); so a series of descaling efforts were carried out. These included treatment with dilute oxalic and nitric acids, dry heating to 454°C (849°F) for 25 hr, and slurry blasting. Only the last treatment was effective. The dry heating buckled the Zircaloy-clad fuel plates, but slurry

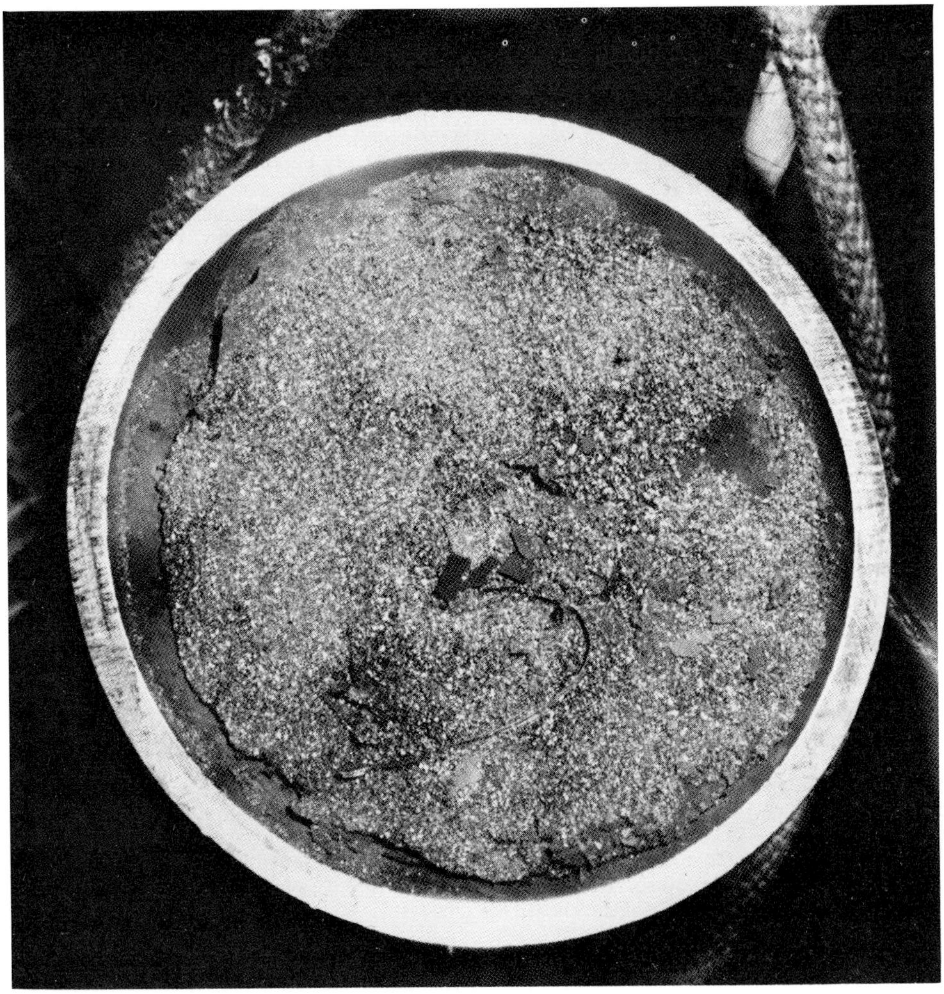

FIG. 3-1 Corrosion-product deposit in recirculation nozzle of the EBWR.

TABLE 3-4

Materials (Other than Cladding) Used in the Primary System and Typical Percentages of the Total Surface

Material	Typical percentages of total surface	
	Power	Research and test
AISI Type 304, or 347	>75	>80*
Aluminum	0	<20
Armco 17-4 PH	<10	0
Monel	<5	<1
Inconel	<5	0
AISI Type 410	<1	0
Hard chromium plate	<1	0
Stellited surfaces	<1	0
Inconel-X	<1	0
Cast iron	0	<1

*Some pool reactors are built in coated concrete pools and the stainless steel is decreased to near zero.

blasting removed some of the clad along with the scale. Actually, the scale cracked off at thicknesses of 4 to 8 mils (0.1 to 0.2 mm). This created large, highly activated, and hard particles that settled to the bottom of the reactor and fouled the control rod bushings and recirculator nozzles and produced a high radiation level in the subreactor room (see Fig. 3-1).

Crud deposition also occurred on the all-aluminum cores in the BORAX and ALPR (SL-1) reactors. The deposition was considerably less than that which occurred in the EBWR [5].

The insoluble crud in water-cooled power reactors which operate at 250 to 343°C (~500 to 650°F) consists largely of particles of black magnetite (Fe_3O_4) containing chromium and nickel oxides. The concentration varies from 0.05 to 3.0 ppm, with an average of 0.1 to 0.2 ppm. When oxygen is present in the water, a brown oxide, probably Fe_2O_3, is formed. Bursts or showers of crud occur in the water at intervals due to the effects of hydraulic instabilities or changes in oxygen concentration or pH of the water. Soluble materials are about 0.05 ppm.

Chromium is usually lower in the insoluble crud than in the parent steel, which indicates that it may be present in soluble form. Soluble chromate has been observed in the Naval Reactor Test Facility, perhaps due to the oxidation of Cr_2O_3 to chromate.

The factors influencing crud transport and deposition are not well understood. The insoluble products appear to be suspended and deposited repeatedly [8]. Dehydration of particles at high-temperature surfaces, precipitation by discharge of colloidal charges by ionizing radiation, and settling of particles in low-velocity areas have been suggested. This question has been extensively studied at Bettis [9].

The discussion above indicates that the impurities in the water exist as dissolved gases, as dissolved solids, at least partly, in both cationic and anionic forms, and as insoluble crud. These impurities may be removed in a variety of ways, but the most common methods used are degasification, ion exchange, precipitation, and filtration.

Degasification of the coolant occurs in the condenser of direct-cycle boiling-water reactors. Some of these gases may be recirculated. In pressurized-water reactors hydrogen may be added to limit oxygen production. When coolant is purged, it is frequently degasifed in the liquid-waste treatment system.

Ion-exchange systems vary in design in that some are operated at the pressure of the reactor coolant system while others operate at lower pressure. In all cases the water must be cooled to 49 to 66°C (120 to 150°F) before passage through the ion-exchange columns.

The resins are high-molecular-weight polymers containing special ionic groupings. The amount of cross-linkage is about 8%. This cross-linkage determines the porosity, stability, and solubility of the resin. The resins are available in 20 to 50-mesh screen sizes*. They are used in pure form or mixed and placed in tanks or large pipes through which the water is circulated. Frequently the tanks are used in parallel and in series so that the resins can be renewed during operation and the second bed can be a mixed bed for maximum efficiency [10, 11].

Because of the activity collected on the resins, the columns must be placed in shielded rooms or each column must be shielded. The fraction of the column that is saturated may be determined (approximately) by simply scanning the outside of the columns. This method of checking the resins is particularly useful in small research reactors.

The resins not only remove the ionic impurities but also are quite effective (when fresh) in removing the insoluble crud. However, experience has shown that the resins lose their crud-removal capability before they are saturated with ions. The resins used for purification of reactor coolant are almost never regenerated but rather are flushed to solid-waste storage and replaced.

Filters are frequently installed in the coolant-purification systems. These are high-pressure thick-mat or cartridge-type cotton or glass filters. When removed, they are sent to solid wastes.

The activity collected on the resins used with low-power reactors may be examined with a pulse-height gamma-ray analyzer to give information about difficulties. Water conductivity meters may also be used; these are very sensitive to changes in the reactor systems and may indicate operational difficulties. In power reactors, water analyses are used. For example, analysis for iodine in boiling-water reactors gives an excellent indication of fuel-element failures.

3.1.1 Research Reactors

Light-water-cooled research reactors may be pool or tank types, while all heavy-water research reactors are tank type. In pool reactors, the core is constructed near the bottom of an open pool of water which constitutes the moderator, reflector, coolant, and part of the shielding. All the pool

*Sieve openings are 33 mils (0.84 mm) for 20-mesh screen and 11.7 mils (0.30 mm) for 50-mesh screen.

water comes in contact with the reactor core and is, therefore, subject to contamination. The power level of pool-type reactors is limited to a few megawatts, due mainly to the escape of N^{16}, and secondarily, to formation of Na^{24} from aluminum. A pool reactor may be operated at higher power levels if the coolant flow is downward into a plenum with external storage (~ 2 min) for N^{16} decay. The coolant could be retained in tubes or a tank around the core.

In a typical pool reactor, Battelle Research Reactor, the pool-water purification system simply consists of the withdrawl of 20 gpm (1.26 liter/sec) of coolant from the pool with circulation through a filter and an ion-exchange unit. The liquid waste from the regeneration of the ion-exchange unit is concentrated by evaporation and disposed of as solid waste. The concentration of radioactivity before evaporation is about 1×10^{-4} μc/ml. The alpha activity in the ventilating air in the reactor room was found [12] to be only 5×10^{-14} μc/ml. The solid wastes consist of the ion-exchange resins that must be replaced periodically, cleanup rags, various dry materials, etc. The radioactivity of these wastes is negligible.

Many research reactors include graphite thermal columns that are opened to the air, and some of the test facilities are air-cooled. The Ar^{41} which forms in this air may contribute hazardous amounts of radioactivity to the test areas. These facilities should be vented to the stack with air inflow to the graphite thermal column and test facilities.

In the tank type of reactor, the core is constructed in a tank through which light or heavy water may be circulated as a coolant-moderator. The tank and reactor core may be located at the bottom of a large pool of water which is used to provide shielding and experimental space, or the core tank may be surrounded by graphite and concrete, steel and concrete, or simply concrete. Only the water inside the tank normally comes into contact with the core. The coolant flow and cleanup cycles for a reactor representative of this type, the Oak Ridge Research Reactor, are shown in Fig. 3-2 [13]. This reactor was selected because it is equipped with a tank which surrounds the core and contains experimental facilities, and this tank is placed in a pool. The reactor and pool have separate cooling circuits. Eighteen thousand gpm (1140 liter/sec) of water are pumped through the core; 400 gpm (25 liter/sec) are pumped through the experimental facility, cooled, and recirculated. Eighty gpm are purified by ion exchange, and 200 gpm (12.5 liter/sec) are filtered. A separate system purifies the water in the pool. The activity levels in the core and experimental-facility circuits are indicated in Table 3-5 [14].

The reactor-coolant purification system consists of cation and anion exchange columns that remove radioactive materials at the rate of ~ 2 gpm/cu ft of resin (~ 4.5 cm^3/sec per liter of resin). Each of these units consists of a cation resin column containing 10 ft^3 (283 liter) of Amberlite IR 120 resin and an anion column containing 37 ft^3 (1050 liter) of Amerlite IRA 401 resin. The total volume of the coolant is 65,000 gal (246,000 liter). The pool water is purified by a

TABLE 3-5

Activity Levels
in the Oak Ridge Research Reactor

[After eight (8) days operation at 30 Mw(t) and with 60 gpm (3.8 liter/sec) flow through the demineralizer and a flow rate of 18,000 gpm (1140 liter/sec) through the core]

Radioisotope	μc/ml
Na^{24}	$1.9 \cdot 10^{-2}$
Np^{239}	$2.8 \cdot 10^{-5}$
I^{133}	$1.4 \cdot 10^{-5}$
I^{131}	$1.4 \cdot 10^{-6}$

40 ft^3 (1130 liter) cation and 35 ft^3 (990 liter) anion column [14].

The Materials Testing, Engineering Test, and General Electric Test Reactors are tank type. They are similar to the Oak Ridge Reactor except that they are built to operate at higher power levels. The first two are not located in pools of water; the GETR is. The Advanced Engineering Test Reactor is similar except that it has five flux traps and is essentially several reactors built into one core.

The flow system for the Materials Testing Reactor* is shown in Fig. 3-3 [15]. The total flow through the core and reflector is about 23,000 gpm (1450 liter/sec). The water fed to the primary system is pretreated by ion exchange to reduce impurities to < 1 ppm. The major materials of construction exposed to the water are stainless steels, cast iron, beryllium, and aluminum alloys.

Originally the MTR was designed with continuous feed of 50 gpm (3.15 liter/sec) of deionized water with continuous overflow to waste. During the first two years of operation at 30 Mw(t), minor fission-product activities were released to the coolant system. The activity near the piping, after the N^{16} had decayed, was about 50 to 100 mr/hr during operation, and 10 to 20 mr/hr during shutdown. This mainly was due to the activated corrosion products. Activities included Al^{28}, Na^{24}, Co^{60}, and trace amounts of nine isotopes.

Beginning in June 1954, a series of fuel element ruptures occurred in the MTR. Fission products escaped and contaminated the piping, reactor tank, and core structural components. Activities were as high as 3 r/hr. In October 1955, a bypass cation resin bed with 125 cu ft (3540 liter) of resin was installed and a side stream of 1000 gpm (63 liter/sec) of coolant was passed through it. As the U^{235} and fission products were removed from the system, radiation levels slowly decreased to nearly their original values. The activities present in the water are indicated in Table 3-6 [16].

The MTR is equipped with flash evaporators for cooling the primary coolant. About 12 scfm (5.7 liter/sec at S.T.P.) of gas are produced at the present power level of 40 Mw(t). A typical analysis is 35.1 vol.% hydrogen, 30.5 vol.% oxygen, and 34.4 vol.% nitrogen [6]. This gas is diluted and discharged.

*The MTR is now (1971) in standby status.

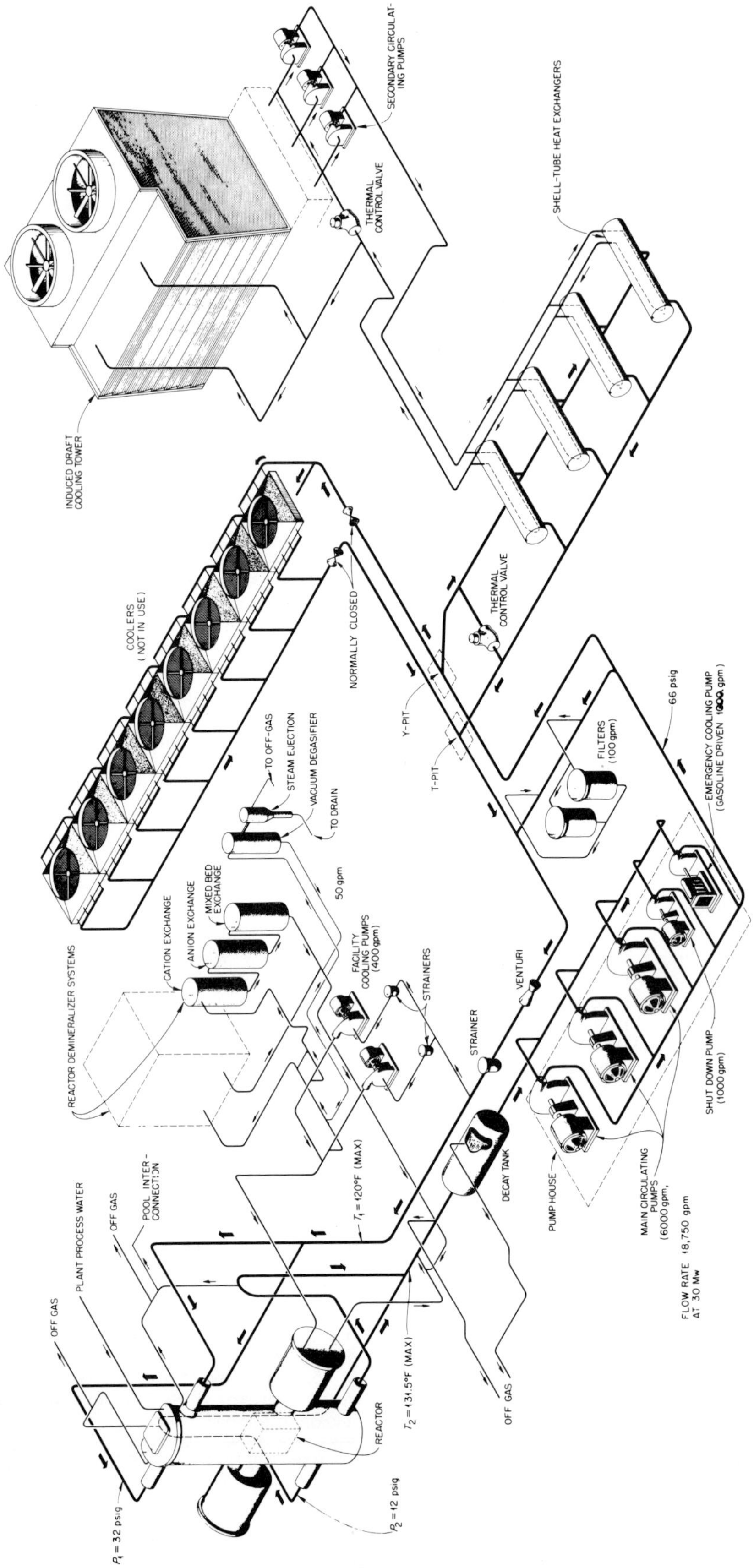

FIG. 3-2 Coolant flow diagram for the Oak Ridge Research Reactor.

TABLE 3-6

Normal Level of Radionuclides in the MTR Primary Water at 40 Mw(t)

Radionuclide	μc/ml
I^{131}	$4.5 \cdot 10^{-4}$
I^{133}	$1.8 \cdot 10^{-3}$
I^{135}	$1.8 \cdot 10^{-3}$
Ba^{139}	$1.8 \cdot 10^{-3}$
Ba^{140}	$4.5 \cdot 10^{-4}$
Sr^{91}	$1.8 \cdot 10^{-3}$
Sr^{92}	$1.4 \cdot 10^{-3}$
Ru^{103}	$9.0 \cdot 10^{-4}$
Ru^{105}	$4.5 \cdot 10^{-4}$
Np^{239}	$2.7 \cdot 10^{-3}$
Pa^{233}	$1.8 \cdot 10^{-5}$
Na^{24}	$1.8 \cdot 10^{-2}$
Zr^{95}	$2.7 \cdot 10^{-4}$
Zr^{97}	$2.7 \cdot 10^{-4}$
Cd^{115}	$9.0 \cdot 10^{-3}$

The thermal shield and graphite reflector are cooled with air drawn from the reactor room. About 340 curies of Ar^{41} are discharged each day through the stack.

3.1.2 Boiling-Water Reactors

Boiling-water reactors are characterized by bulk boiling in the core. The steam produced may be run directly to the turbine. The water in the core is either recirculated by density differences, as in the Experimental Boiling Water Reactor,* or by forced circulation. If forced circulation is used, the water may be passed through a heat exchanger-boiler to produce additional low-pressure steam, as in the Dresden Reactor. In any case, part or all of the condensate is filtered, deionized, and returned to the reactor. For example, in the EBWR the condensate is filtered but not deionized; while in the Dresden Reactor, the condensate which is returned to the reactor is passed through a full-flow deionizer. The condenser off-gases contain any gaseous fission products released to the coolant along with the gaseous activities released from the coolant.

The radioisotope N^{16} is the most significant activity in the water during operation when there are no defective fuel elements present. In the EBWR it is held in cationic, anionic, and gaseous forms, perhaps as NH_4^+, NO_3^-, and N_2. Although the N^{16} is not completely stripped from the reactor water, it occurs in curie quantities in steam and is a major radiation source, setting the shielding requirements for the steam equipment. Addition of nitrogen, oxygen, or helium in BORAX IV had no effect on its carry over with the steam; but addition of hydrogen caused a marked increase of nitrogen activity in the steam. By lowering the pH to 4 or 5 the nitrogen increased by a factor of 2 to 3. At still lower pH it was up by a factor of 5. It has also been shown that most of the N^{16} is removed by a mixed-bed resin column. N^{13} is also present and is significant in the half discharged gases because of its 10.0-min half-life. O^{19}, with a 29.5-sec half-life, is of significance in the off-gases from the main condenser at Dresden.

Most of the nonvolatile activated crud or soluble compounds are left in the reactor because of the water-to-steam decontamination factors of 10^4 to 10^5, depending on the quality of the steam.

3.1.2.1 Experimental Boiling Water Reactor. A flow diagram of the Experimental Boiling Water Reactor is shown in Fig. 3-4 [17]. The reactor was constructed to operate at 20 Mw(t), and later the equipment (to the right of the generator in Fig. 3-4) was added to permit operations at 100 Mw(t). During initial operation at 20 Mw(t), the predominant gaseous activities, after decay of the nitrogen, were Xe^{138} and Kr^{88}. Apparently these were produced in the uranium contamination, estimated as 0.040 g of 1.4% enriched uranium, on the outside of the fuel elements. The condenser gases are discharged to the atmosphere. The normal discharge rates of 20 Mw(t) operation were 0.2 to 0.4 curie/day of Xe^{138} and 0.005 to 0.01 curie/day of Kr^{88}. During a test period when a fuel element was deliberately defected, the discharge of these isotopes rose to 1.0 and 0.08 curie/day, respectively. The condenser air, on leaving the ejector, contained 7.4×10^{-3} μc/ml of Xe^{138}. Some of this is believed to have entered the containment room in the ventilation air where the Xe decayed to Cs^{138} to give a concentration of 1.2×10^{-7} μc/ml. The reactor has been reloaded and the amounts of Xe and Kr discharged are extremely small [18, 19].

The activities in the reactor water under various operating conditions are shown in Table 3-7 [19] and the corrosion-product activities in the coolant of the EBWR are indicated in Table 3-8 [20].

The main liquid wastes have been produced during shutdown. After each shutdown, the pressure vessel is filled with water and given a hydrostatic pressure and leak test. Following the test, the excess water, about 30,000 gal/year (113,550 liter/year), is bled to waste. This contains about 5×10^{-5} μc/ml. This water is added to the regular laboratory waste-disposal system of Argonne National Laboratory without difficulty.

Somewhat more than 100 ft^3 (2.8 m^3) of solid wastes containing about 10^5 μc of radioactive materials were collected during the first full year of operation. Ten to twenty percent of this volume was spent ion-exchange resins; the remainder consisted of wiping rags and other assorted debris. Radiation levels of the spent resin have been about 10 r/hr at contact. Power levels were 20 Mw(t) with intermittent operation.

3.1.2.2. Dresden Reactor. The Dresden Reactor is now licensed to operate at power levels up to 700 Mw(t). Primary steam is produced in the reactor and secondary steam is produced in four secondary steam generators with the forced recirculated coolant. A diagram of the sources and

*The EBWR is no longer in operation.

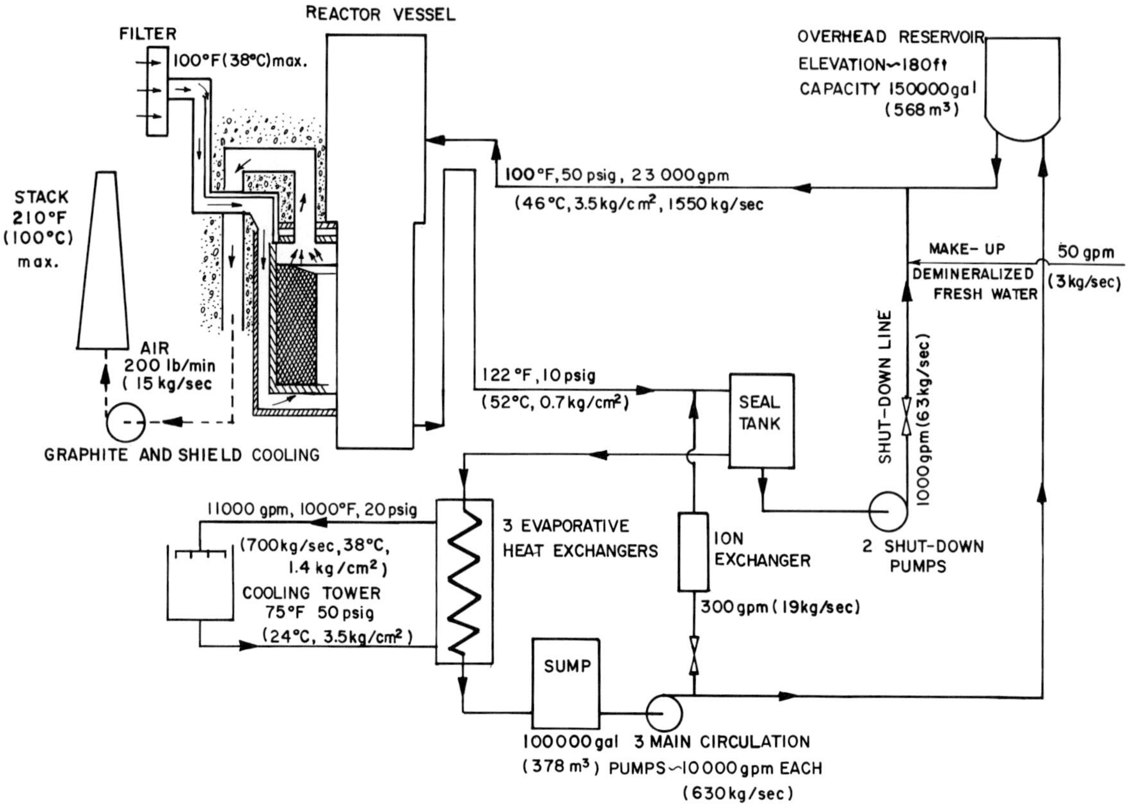

FIG. 3-3 Flow diagram for MTR.

TABLE 3-7

Activities Detected
in Experimental Boiling Water Reactor Water
During Operation under Several Conditions[a]

Isotope	Full flow through ionex[b] (μc/ml)	No flow through ionex[c] (μc/ml)	Partial flow through ionex and fuel defect[d] (μc/ml)
Ba^{140}	$7.2 \cdot 10^{-6}$	$5.4 \cdot 10^{-5}$	$1.4 \cdot 10^{-5}$
Cs^{138}	$6.8 \cdot 10^{-5}$		$2.3 \cdot 10^{-5}$
I^{131}	$2.7 \cdot 10^{-6}$	$2.3 \cdot 10^{-5}$	$9.0 \cdot 10^{-6}$
Mo^{99}	$4.1 \cdot 10^{-5}$	$6.3 \cdot 10^{-5}$	$3.6 \cdot 10^{-5}$
Sr^{89}	$2.3 \cdot 10^{-6}$	$2.3 \cdot 10^{-6}$	$2.7 \cdot 10^{-6}$
Na^{24}			$9.0 \cdot 10^{-2}$
Mn^{56}			$2.7 \cdot 10^{-2}$
Co^{58}			$1.8 \cdot 10^{-2}$

[a] Operating level—20 Mw(t).
[b] Ion-exchange flow rate of 10 gpm (0.63 liter/sec) with decontamination factors ranging from 250-5000.
[c] No circulation through the ion exchanger for 2.5 days.
[d] Ion-exchange flow rate of about 7 gpm (0.44 liter/sec) during first defect test.

design quantities of wastes is shown in Fig. 3-5 [21, 22]. Gaseous wastes come principally from three sources: the main condenser air ejector, leakage from the turbine gland seals, and ventilation air from various compartments in the reactor enclosure. Actual activity releases from the stack as compared with calculations for design are summarized in Table 3-9 [21, 23]. The short-lived activities in the steam from the reactor are N^{16} and O^{19}, with 10-min half-life N^{13} being the only significant activation gas discharged.

The total off-gas from the condenser is about 20-25 ft^3/min (9.5 to 11.8 liter/sec). The average holdup, in buried pipe, is about 20 min. The remaining N^{13} activity is about 500 μc/sec at full power. In addition, small amounts of Xe and Kr are released [2]. About 400 curies/sec of N^{16} are produced in the core; but only about 5.8 curies/sec reach the turbine, and negligible amounts are discharged from the stack [24].

The turbine gland seal off-gases and containment-vessel exhaust ventilation are not filtered. In the event of any automatic reactor scram signal, the reactor containment inlet and exhaust lines are closed by fast-operating valves. This prevents any accidental discharge to the environs.

The exhaust from the air ejector and other gas streams that may contain particulate matter

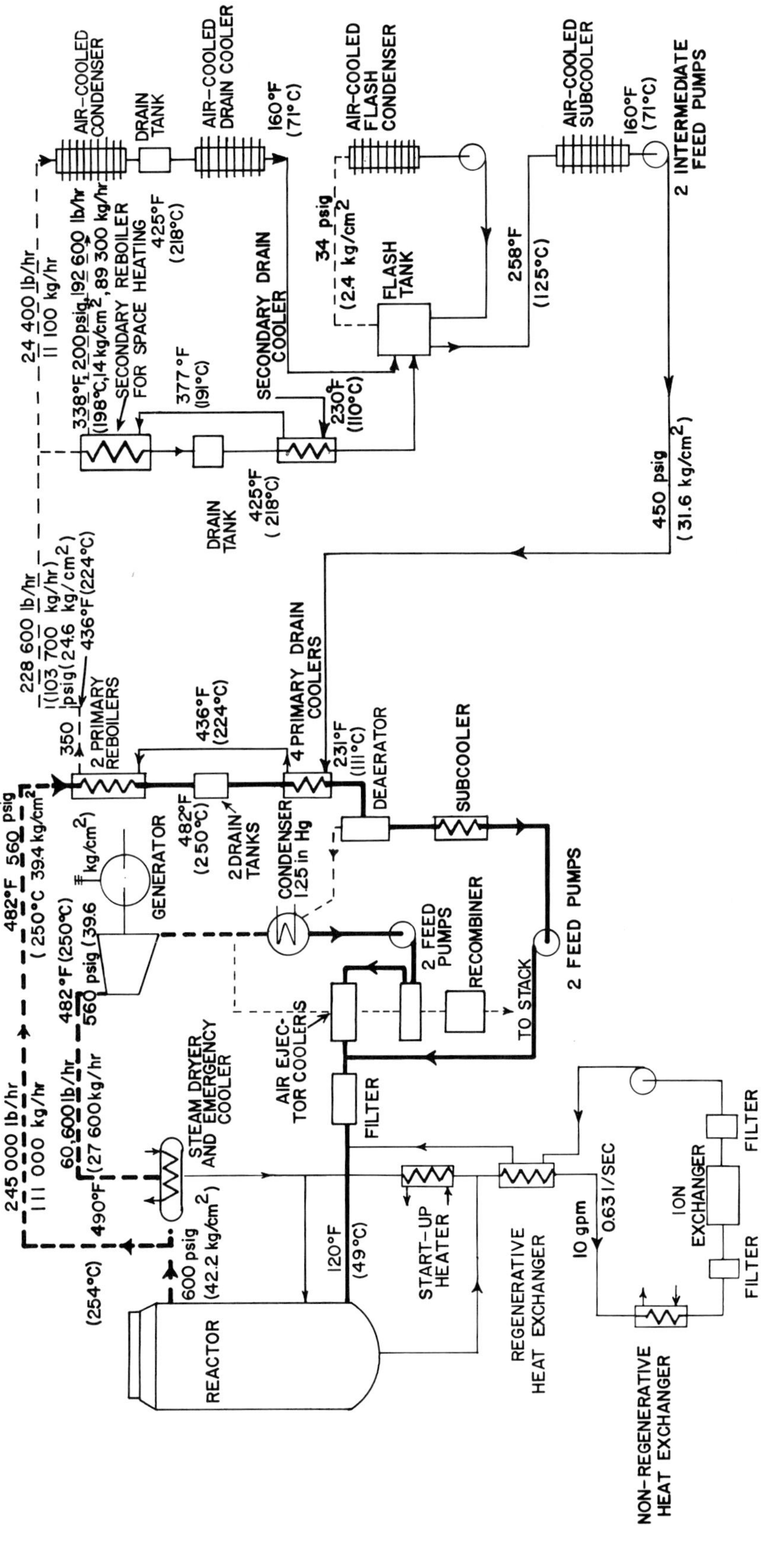

FIG. 3–4 Flow diagram for EBWR.

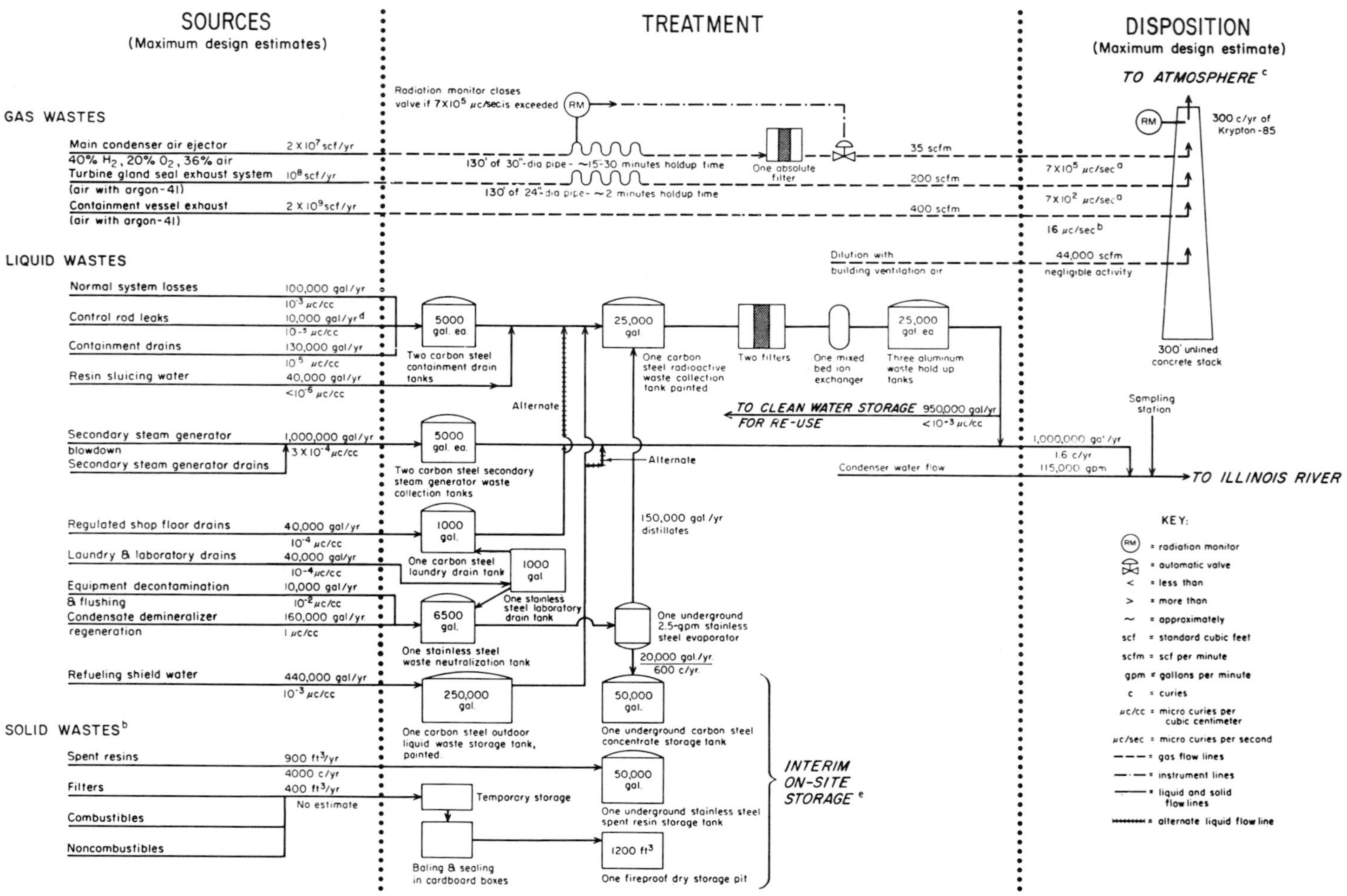

[a] *This activity, measured at discharge, is primarily from short-lived xenon and krypton isotopes with half-lives from ten seconds to 12 days. The only significant long-lived isotope is krypton-85, of which about 300 curies per year will be discharged on the maximum basis.*

[b] *This activity is primarily from argon-41.*

[c] *At maximum activity release rates it is estimated that dose rate to off-site personnel will not exceed maximum permissible rates of 0.5 rem per year.*

[d] *Most of the activity is from leakage during scrams--about 1000 gallons per year, containing one uc/cc. Daily leakage is less than 10^{-5} uc/cc.*

[e] *Initially there will be no off-site disposal of solid wastes.*

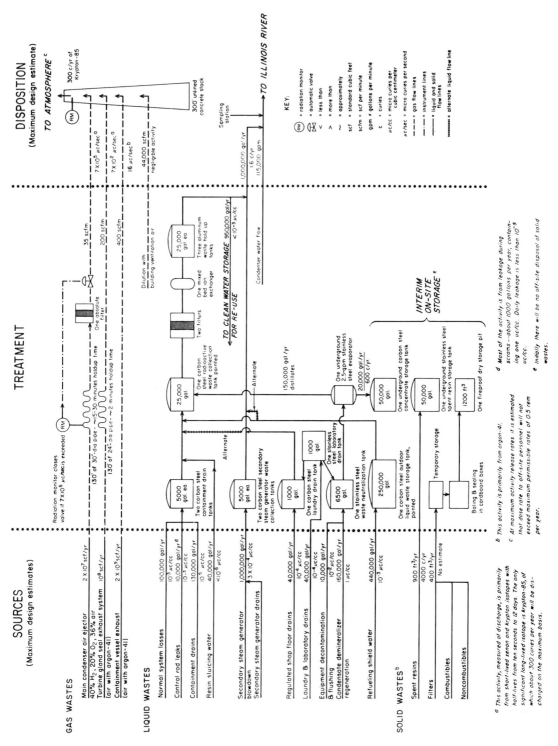

FIG. 3-5 Diagrammatic flow sheet for Dresden Nuclear Power Station waste plant.

TABLE 3-8

EBWR Corrosion-Product Activities

April-May 1957, Operation at 20 Mw(t)

Location	Isotope	Conc (μc/ml)
Reactor water	Mn^{56}	$3.6 \cdot 10^{-2}$
	Na^{24}	0.14
	Co^{58}	$3.6 \cdot 10^{-3}$
	Cu^{64}	$5.4 \cdot 10^{-3}$

are passed through high-efficiency filters. (See Chapter 19.) The ventilating air and waste gases are combined and discharged through a 300-ft (91-m) unlined concrete stack with an exit velocity of 50 ft/sec (15.2 m/sec) [22].

The main condensate is divided at the discharge of the condensate pumps. The primary feedwater passes through full flow deionizers while the secondary feedwater bypasses the demineralizer.

The steam piping systems are made of carbon steel, the condenser has admiralty tubing, and the feedwater heaters tubing are of copper-nickel and monel alloys. Type-304 stainless steel is used for the reactor system piping, the reactor vessel lining, and the steam drum lining. Type-304 and 400-series steels are used within the reactor. Type I fuel assemblies are clad with Zircaloy-2 and the Type II fuel assemblies have stainless steel cladding.

The condensate is deaerated in the hotwell to about 5 ppb oxygen; but radiolytic decomposition results in a concentration of 200 to 300 ppb (1 ppb = 10^{-9}) of oxygen in the reactor water. The reactor feedwater demineralizer removes 80 to 90% of the suspended materials, so that the total iron, copper, and nickel in the reactor feedwater is less than 15 ppb. The feedwater has an average conductivity of 0.10 micromho. However, it may pick up some impurities in the feedwater heaters and from the piping and reactor internals. The significant corrosion products in the reactor include Cu^{64}, Co^{58}, Co^{60}, Cr^{51}, and Mn^{56}.

Since the reactor water is circulated through the steam generators, activated corrosion products are deposited as an adherent film in the piping and tubing of the recirculation system. This activity has not unduly restricted maintenance work. It cannot be removed by simple flushing, but some crud, which settles out in pockets, can be flushed out [2]. After each startup the reactor water is cleaned up by blowdown to the waste-treatment system. In a few days the water has a turbidity of less than 10 American Public Health Association units [25]. Then an 80 ft^3 (2.26 m^3) mixed-resin bed is placed in operation at a flow rate of about 140 gpm (8.8 liter/sec). A new charge of resin reduces the turbidity further but continues to deionize the water long after it ceases to remove crud. On breakthrough the resin is pumped to a waste tank for storage. Blowdown is not needed after the reactor is at full load and the crud inventory is stabilized. The crud has a specific activity of 1.5×10^7 cpm/mg Fe.

Most of the liquid wastes (see Fig. 3-5) are intermittent in nature. From the startup of the reactor in 1959 through 1961, 17,700,000 gal (67,000 m^3) of water were passed through the radioactive waste demineralizer at a cost of about \$2.63/1000 gal (\$0.701/m^3). Of this, 1,040,000 gal (3900 m^3) were recycled, 16,650,000 gal (63,000 m^3) were reused, and only a small amount was added to the effluent to give a total discharge of 6,900,000 gal (26,000 m^3). These wastes are diluted with condenser cooling water to an activity level of about 1×10^{-8} μc/ml above average river background activity. This discharge is estimated to be greater than a factor of 1000 below the requirements of 10 CFR 20 based on isotope identification [24]. Tables 3-10a and 3-10b summarize, for the period October 1, 1959, to December 31, 1966, the volume and activity of radioactive liquid waste discharged and processed for reuse at the Dresden plant.

Solid wastes include the spent resins, worn-out and contaminated tools, and various trash. Some of the material is baled and stored; some of it is barrelled and shipped to burial sites; and the resins are sluiced to a permanent underground storage tank [26, 27].

3.1.3 Pressurized-Water Reactors

In pressurized-water reactors the primary coolant pressure is maintained high enough that bulk boiling does not occur; and the heat is transferred to boiling water in heat exchangers. The system is frequently pressurized in a separate tank by steam to permit slight change in volume of the coolant. Two or more coolant loops, consisting of appropriate piping, valves, pumps, and heat exchangers, are used. These external loops as well as the core supports and lining of the pressure vessels are almost all stainless steel or stainless-steel-lined. Only valve seats, pump bearings, etc.,

TABLE 3-9

Dresden Gaseous Activity Releases Compared to Calculated for Design Basis

Radioactive gases	Discharge rate from stack (μc/sec)	
	Calculated for design	Actual[a]
H^3	$1 \cdot 10^{-3}$	Not significant
Ar^{37}	$1 \cdot 10^{-3}$	Not significant
Ar^{41}	18	Less than 18
O^{19}	4	Less than 4
N^{16}	2	Less than 2
N^{13}		Up to about 500
Fission gases, primarily Xe and Kr		
No fuel-element leakage		100-200
With some fuel-element failures		200-70,000

[a] DPR 2 License Limit is $7 \cdot 10^5$ μc/sec of fission gases.

TABLE 3-10a

Radioactive Liquid Waste Discharged from Dresden Nuclear Power Station
Period October 1, 1959, to December 31, 1966

Period	To river											
	High purity			Moderate purity			Corrosive (waste neut)			Laundry		
	1000 gal	mc	$10^4 \mu\mu c/l$	1000 gal	mc	$10^5 \mu\mu c/l$	1000 gal	mc	$10^5 \mu\mu c/l$	1000 gal	mc	$10^4 \mu\mu c/l$
1959	250	< 1	< 1	339	< 1	< 1	26	< 1	< 1	31	< 1	< 1
1960	1,069	120	3.0	1,855	610	0.9	473	30	0.2	169	10	1.2
1961	1,088	130	3.1	1,296	810	1.7	114	70	1.6	292	30	3.2
1962	186	10	1.2	1,108	1,080	2.6	506	1,270	6.6	120	10	3.1
1963	194	40	5.8	790	1,740	5.8	576	980	4.5	184	10	1.6
1964	76	20	5.2	534	870	4.3	377	2,630	18.4	87	10	2.7
1965	0	0	0	41	1,590	100.0	263	6,630	67.0	160	40	7.0
1966	11	< 1	< 1	87	1,089	33.1	580	10,139	46.2	119	80	17.9
Total	2,874	320	2.9	6,050	7,788	3.4	2,915	21,749	19.7	1,162	190	4.3

are made of other materials. The fuel is usually clad with stainless steel, but Zircaloy has been used at Shippingport.

A side stream of the primary coolant is circulated through an ion-exchange — filtration cleanup system, and the oxygen concentration in the coolant is maintained at a low level by the introduction of hydrogen. In the Shippingport Reactor Li^7 OH is added to maintain a pH of 9.5 to 10.5. Boric acid is used as a soluble poison at Indian Point during reloading and at Yankee in extended tests.

The discussion of the MTR Reactor (see Sec. 3.1.1) is generally applicable to the pressurized-water reactors. Corrosion is somewhat higher be-

TABLE 3-10b

Radioactive Liquid Wastes Processed for Reuse
at Dresden Nuclear Power Station

Period October 1, 1959, to December 31, 1966

Period	To demineralizer			To storage		
	1000 gal	mc	$10^6 \mu\mu c/l$	1000 gal	mc	$10^4 \mu\mu c/l$
1959	1,665	< 1	< 1	1,665	< 1	< 1
1960	7,728	152,840	5.2	7,235	390	1.4
1961	7,764	379,910	12.9	7,753	3,180	10.8
1962	12,640	430,980	9.0	12,600	1,620	3.4
1963	9,900	545,140	14.5	9,740	540	1.5
1964	10,020	174,460	4.6	10,000	190	5.1
1965	8,260	659,320	21.0	8,140	4,620	15.0
1966	11,590	794,113	18.1	11,470	7,046	16.2
Total	69,567	3,136,763	11.9	68,603	17,586	6.7

cause the water temperatures are higher but it is still only a few tenths of a mil per year (~ microns/year). Even so, from 40 to 200 lb (18 to 90 kg) of corrosion products must be removed from a large power-reactor coolant system each year. Since gases are mainly retained in the primary system, only small quantities of gaseous wastes are released; thus it is the liquid wastes that are most significant. The concentrations of activities in the coolant and waste streams are discussed below.

The sources and assumed quantities of wastes utilized in design of the Shippingport Reactor are indicated in Fig. 3-6 [21]. Experience during the first two years at Shippingport indicated that about 80% of the radioactivity delivered to the waste-disposal system came from fission and corrosion products. The remaining 20% was due to the tritium that came from use of lithium, which was added as the hydroxide. The wastes produced and the activities released to the Ohio River during the first 2 years of operation are indicated in Tables 3-11 [28] and 3-12 [21, 28-30]. The tritium production has been reduced by using Li^7 rather than natural lithium. Improved fuel-manufacturing techniques have reduced the release of fission products in most pressurized-water reactors.

Data on the composition of the crud in the Shippingport Reactor are presented in Table 3-13 [6]. The main source of the activity in this reactor appears to arise from corrosion and wear products deposited and activated on surfaces exposed to the neutron flux. Part of the Co^{60} may result from corrosion products, which are sloughed off the control-rod mechanisms above the reactor.

The specific activity of the primary coolant (without degassing) of the Indian Point Reactor varies between 0.2 and 0.3 μc/ml at 75 to 100% of full power, 585 Mw(t). The major isotope is Mn^{56} (see Table 3-14) [31]. A coolant side stream is held up in a delay tank for several hours before deionization. As a result the feed to the demineralization system has decayed to about 10^{-2} μc/ml and this is reduced by cation exchange to about 3×10^{-3} μc/ml. Anions are then removed in a mixed-bed exchanger, thereby reducing the coolant activity to about 10^{-5} μc/ml. This latter step is accompanied by the pickup of Xe^{133} and Xe^{135} from the mixed-bed exchanger due to decay of I^{133} and I^{135} (see Fig. 3-7) [6, 21].

The volumes and activities of the liquid wastes discharged by the Yankee Atomic Power Plant over a period of 30 months are summarized in Table 3-15 [32], and the gaseous wastes are summarized in Table 3-16 [32]. The total numbers of drums of solid wastes shipped and the activities are summarized in Table 3-17 [32].

TABLE 3-12

Activity Discharged to the Ohio River by Shippingport Reactor during First Two Years Operation

Waste	Radioactivity					
	Daily volume[a] (gallons)		Non-Tritium (daily μc)		Tritium (daily μc)	
	Design	Actual	Design	Actual	Design	Actual
1958						
"Non active"	5630	1525	20	5	--	--
Service bldg.	2950	1060	1120	20	--	--
Reactor plant	750	2600	130	67	10	0.14
1959						
"Non active"	2630	2240	20	7.3	--	--
Service bldg.	2950	1580	1120	39	--	--
Reactor plant	750	680	130	65	10	0.20

[a] 1 gallon (U.S.) = 3.785 liters.

TABLE 3-11

Radioactive Wastes at PWR[a]

Type	Monthly amount	Special activity (μ c/ml)[b]
Reactor plant effluent	<16,000 gal (60,500 liter)	7.6[c] 1.7[d]
High-solid-content liquids	7500 gal (28,400 liter)	<0.27
Spent resin	36 ft^3 (1.02 m^3)	10^4
Combustible wastes:		
Average	3,000 lb (1360 kg)	Variable
Maximum	10,000 lb (4540 kg)	

[a] 225 Mw(t).
[b] Total activity of liquids and gases, ~20 curie/day prior to treatment.
[c] Volatile.
[d] Nonvolatile.

TABLE 3-13

Shippingport Crud Activities after 1440 Equivalent Full-Power Hours

Nuclide	Activity per mg crud (dis/min)
Co^{60}	$5.8 \cdot 10^6$
Co^{58}	$2.5 \cdot 10^6$
Fe^{59}	$1.8 \cdot 10^6$
Cr^{51}	$2.2 \cdot 10^6$
Mn^{54}	$0.8 \cdot 10^6$
Zr^{95}	$0.97 \cdot 10^6$
Hf^{181}	$0.76 \cdot 10^6$
Total	$1.5 \cdot 10^7$

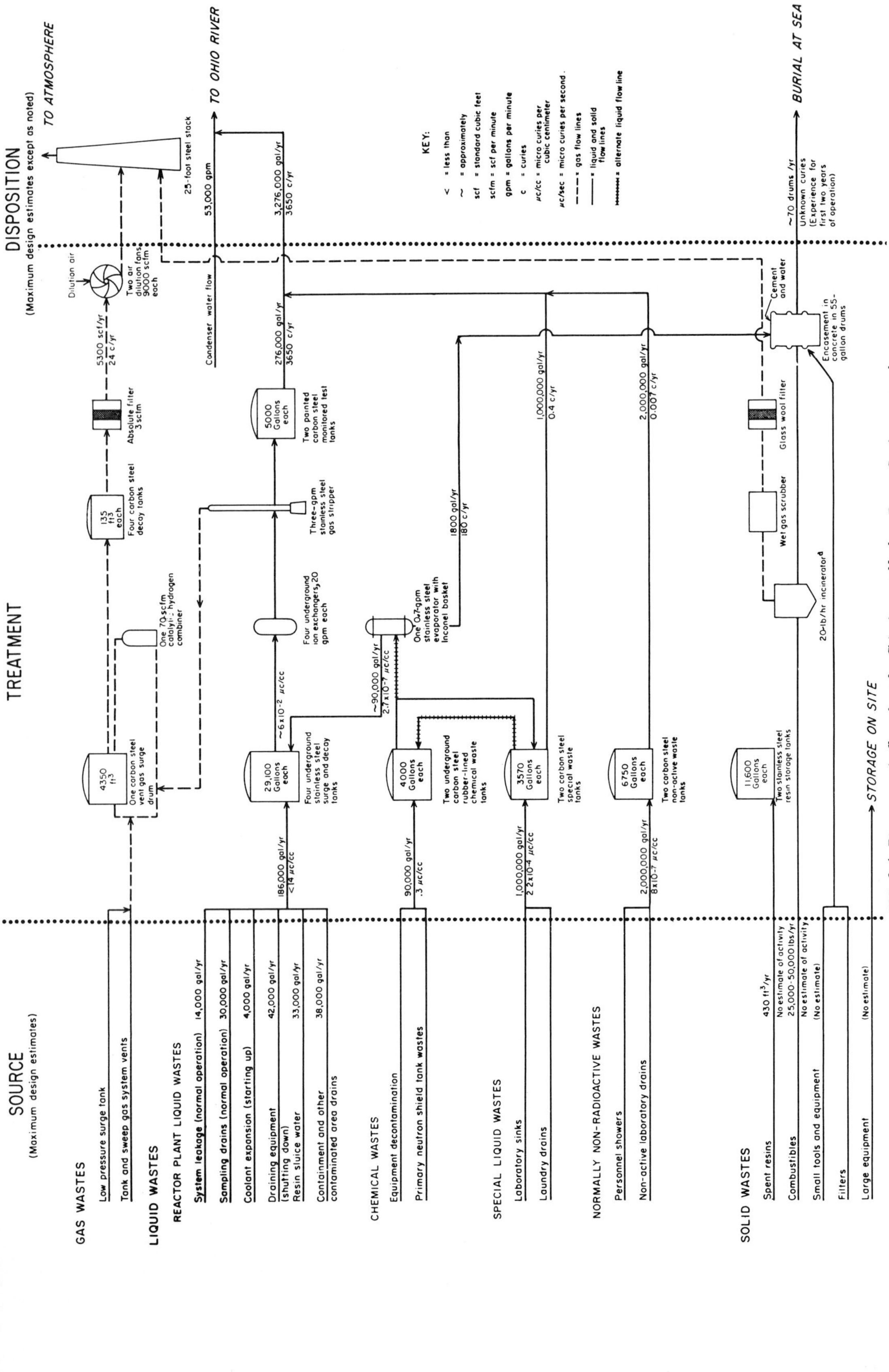

FIG. 3-6 Diagrammatic flow sheet for Shippingport Nuclear Power Station waste plant.

[a] *Burning rate is 40 pounds per hour.*

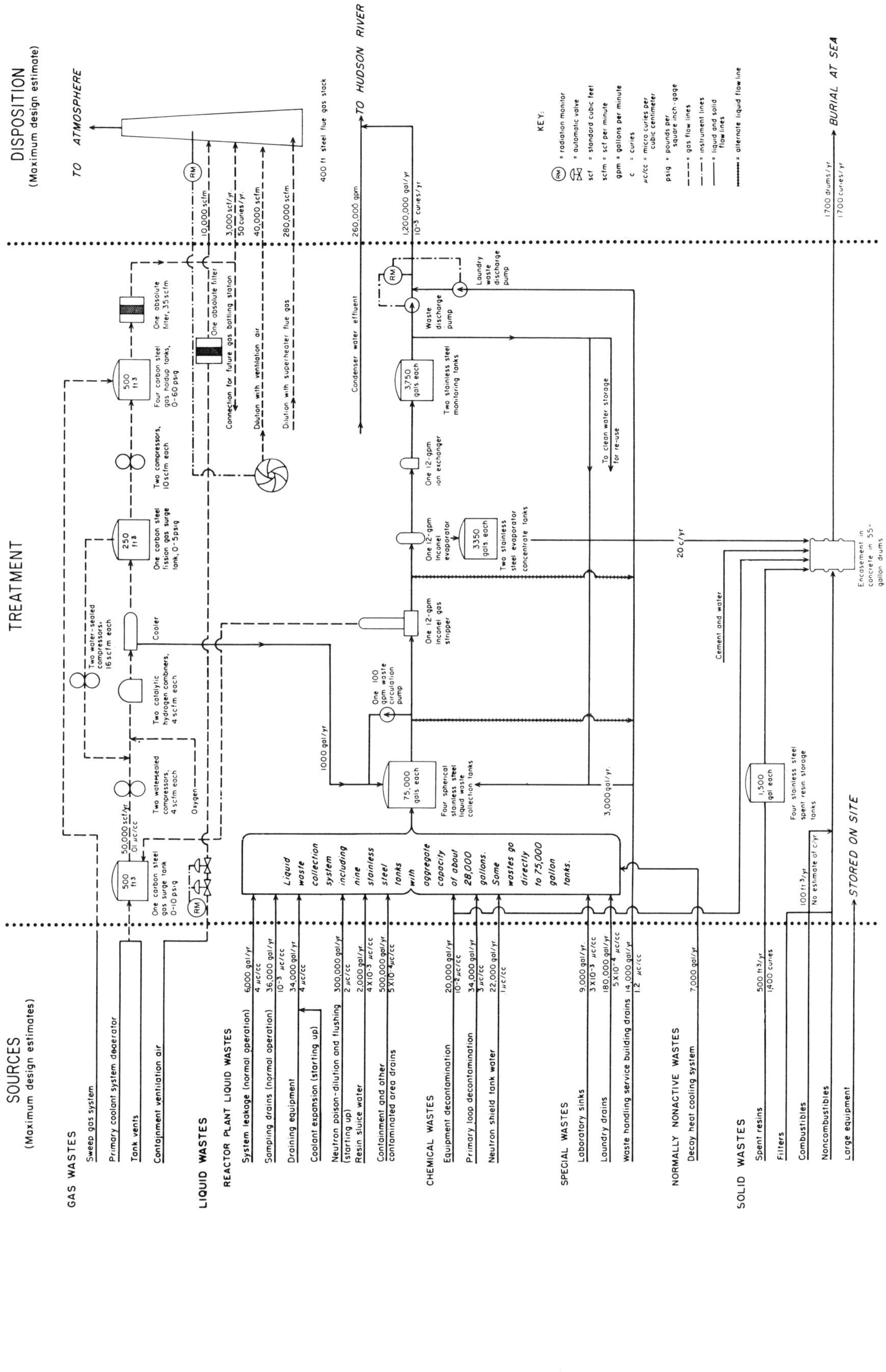

FIG. 3-7 Diagrammatic flow sheet (exclusive of tritium) for Indian Point Nuclear Power Station waste plant.

TABLE 3-14

Indian Point Station Primary Coolant Activities

Primary Coolant System Fission Fragment Activity (100% Reactor Power)

Isotope	Specific activity (μc/ml)[a]
Cs^{137}	$<10^{-5}$
Cs^{138}	$<10^{-5}$
I^{131}	$\sim 1 \cdot 10^{-3}$
I^{133}	$\sim 5 \cdot 10^{-3}$
Sr^{89}	$<10^{-5}$
Sr^{90}	$<10^{-5}$
Xe^{133}	$\sim 10^{-3}$
Xe^{135}	$\sim 10^{-3}$
Kr^{85}	Not observed
Kr^{87}	$\sim 10^{-4}$
Kr^{88}	$\sim 10^{-4}$

Primary Coolant System Undissolved Corrosion-Product Activity (100% Reactor Power)

Isotope	Concentration in total particulate (μc/mg)	Specific activity (μc/ml)[a]
Co^{58}	~5	$\sim 1 \cdot 10^{-4}$
Co^{60}	~1	$\sim 2 \cdot 10^{-5}$
Fe^{59}	~0.5	$\sim 1 \cdot 10^{-5}$
Mn^{54}	~0.2	$\sim 4 \cdot 10^{-6}$
Mn^{56}	Varies over wide range, depending on particulate level	
Cr^{51}	1	$\sim 2 \cdot 10^{-5}$

[a] These data were collected soon after the reactor started operation and the fuel had not reached the equilibrium burnout.

3.1.4 Propulsion Reactors

The nuclear submarines and the Nuclear Ship Savannah use pressurized-water reactors. The principal activities in the coolant of the Naval Reactor Test Facility are shown in Table 3-18 [29]. The submarines have Zircaloy-clad fuel, hafnium control-rods, and stainless-steel primary systems.

TABLE 3-15

Summary of Yankee Liquid-Waste Discharges

Year	Liquid discharges	
	gallons	millicuries
1960	262,000	2.47
1961	592,000	7.97
1962	867,000	7.59
1963	684,000	3.54
1964	663,000	2.05
1965	829,000	29.26
1966	1,284,000	36.41
Total	5,181,000	89.29

NOTES: Average liquid activity at discharge point: 5.5×10^{-11} μc/ml. Percent of 10 CFR 20 MPC (unrestricted-unknown mixture): 0.055%.

TABLE 3-16

Summary of Yankee Gaseous-Waste Discharges

Year	Gas discharges	
	ft^3	curies
1960	8,000	0.001
1961	9,983	0.002
1962	14,491	21.700
1963	500	7.377
1964	8,280	1.024
1965	1,466	1.290
1966	7,500	2.360
Total	50,220	33.754

NOTES:
Average gaseous activity discharge from stack: 2.4×10^{-11} μ c/ml;
Percent of 10 CFR 20 MPC (unrestricted Xe^{133}): within the detection limit of the laboratory instrumentation all identified activity was Xe^{133};
Percent of 10 CFR 20 MPC (unrestricted-unknown mixture): 24.0%.

Only small amounts of other materials and alloys are present.

About 500 gal (1900 liter) of coolant water are discharged each time a naval reactor plant is brought up to operating temperature [33]. This is normally discharged. The assumed dilution factor is 10^5 even in harbors. At one time the used resins were discharged at sea under conditions of very great dilution and away from other ships and from fishing banks. Reference [33] states that since the resins sink and the activity is rapidly exchanged with sea water, there is very great dilution of the activity discharged with the resins. Each discharge contains up to 10 curies of Co^{60}, 0.5 curie of Co^{58} and of Fe^{59}, 0.3 curie of Cr^{51}, 0.2 curie of Mn^{54}, and 1 curie of Hf^{175} [33].

The Nuclear Ship Savannah is a 70 Mw(t) pressurized-water reactor. The fuel is slightly enriched uranium dioxide clad with stainless steel. The reactor and heat exchangers are surrounded by

TABLE 3-17

Yankee Waste-Disposal-System Effectiveness

Year	Drums prepared*	Drum activity* (mc)	River discharge (mc)	Waste system removal (%)
1960	10	neg	2.47	-
1961	70	110	7.97	93.20
1962	517	1,480	7.59	99.50
1963	578	8,708	3.54	99.96
1964	465	9,324	2.05	99.98
1965	622	5,802	29.26	99.50
1966	441	2,886	36.41	99.10
Total	2703	28,310	89.29	99.70

*Includes that material prepared from the evaporator bottoms; spent resin, control rods and miscellaneous solid wastes are not included.

TABLE 3-18

Typical Radiochemistry of the Coolant in the Naval Reactor Test Facility

Group	Nuclide	Half-life	Specific activity (μc/ml)	Source
Very short-lived activity	N^{16}	7.3 sec	100	O^{16} in H_2O
	N^{17}	4.1 sec	800[a]	O^{17} in H_2O
Short-lived activity	K^{38}	7.7 min	$5 \cdot 10^{-2}$	?
	Ar^{41}	1.8 hr	$4 \cdot 10^{-2}$	Air in H_2O
	F^{18}	1.9 hr	$4 \cdot 10^{-2}$	
	Mn^{56}	2.6 hr	$0.5 \cdot 10^{-2}$	Steel
Intermediate-lived activity	Cu^{64}	12.8 hr	$3 \cdot 10^{-4}$	17-4 PH steel
	Na^{24}	15 hr	10^{-3}	Na in H_2O
	W^{187}	24 hr	$3 \cdot 10^{-3}$	Stellite
Long-lived activity	Fe^{59}	45 day	$1.1 \cdot 10^{-5}$	Steel
	Ta^{182}	111 day	$0.6 \cdot 10^{-5}$	Steel
	Co^{60}	5.3 year	$2.5 \cdot 10^{-5}$	Steel

[a] Neutrons/sec-ml.

a steel containment shell. There is some activation of the air; particularly there is formation of Ar^{41} to a maximum concentration of 4×10^{-7} μc/ml. Some fission gases also escape from the liquid wastes and are stripped from the primary coolant. These are ordinarily filtered and discharged. However, if they have a high concentration, they can be pumped into the containment vessel until they can be discharged.

Some 3200 gal (12,000 liter) of very low-level liquid wastes are produced each startup due to expansion. This can be held and reused or discharged. About 40 gal (150 liter) of liquid wastes due to leaks, etc., are collected daily. The main solid wastes are the ion exchange resins [34].

3.1.5 Plutonium-Production Reactors

The eight Hanford Production Reactors are designed for low-pressure, single-pass water cooling.* The large size and power of these reactors demands enormous quantities of cooling water (see Fig. 3-8 [35]). Treatment of the raw water, which is obtained from the Columbia River, consists of coagulation of solids and removal by settling and filtration by rapid sand filters, pH adjustment, corrosion inhibition, and chlorination.

Effluent water leaving the reactors flows by gravity to retention basins which provide a holdup from 1 to 3 hr to permit partial decay of the short-lived radioisotopes. The composition of the outfall from these retention basins varies considerably. Uncontaminated water from the filter plant, boiler house, and pump rooms is used for dilution. The effluent lines extend well into the main current of the river and discharge at the bottom. Should the reactor effluent water contain greater concentrations of radioactivity than normal (for example, after fuel-element ruptures), it may be pumped to a basin where radioactive materials are retained by percolation of the water into the ground.

The effect of a fuel-element defect is to permit water to enter the fuel through the cladding, react with the uranium, and form oxides and hydrides. Low-density materials are thus formed and the fuel element swells until the cladding breaks, releasing the oxide and hydride particles to the coolant stream*. Usually the defective fuel elements are found and removed before appreciable swelling of the cladding occurs and before appreciable radioactivity is released to the coolant streams. Even when the cladding breaks open and oxide particles are released, the oxide particles carry with them over 95% of the fission products. These particles are readily separated from the effluent by settling or filtration.

One procedure that is unique to the Hanford Reactors is that of "purging." In spite of the low turbidity in the treated river water, colloidal material carried into the reactor deposits on the aluminum surfaces of the fuel elements and coolant tubes. This material is removed at intervals by adding powdered diatomaceous earth to the coolant for a brief period. This results in a surge of radioactivity in the effluent up to six times normal. Because of the short time of the purge, the activity in the river is not increased over 0.5% [35].

The activity level of the effluent water is continuously monitored to make certain that the quantities of activity discharged are low in terms of environmental contamination. The radioactive isotopes discharged to the river are mainly activation products and decay in the first day after discharge to 10% of their discharge value. Mn^{56}, Cu^{64}, Na^{24}, Cr^{51}, Np^{239}, and As^{76} are the most prominent on discharge [35].

3.1.6 Heavy-Water Reactors

Recirculated heavy water is used in such reactors as the Argonne Research Reactor, CP-5, and the Savannah River Reactors as the moderator and coolant. The economics of such a system dictate the use of indirect heat exchange. Hence, waste-disposal problems for this type of reactor are alleviated, since the secondary cooling water is not exposed to the neutron flux, and, therefore, does not build up induced radioactivity. These reactors use aluminum fuel cladding and coolant tubes and operate at modest temperatures. Power reactors such as the CANDU Reactor use Zircaloy cladding and tubes with cold moderator in an aluminum calandria. These reactors are very similar to pressurized-water reactors except for greater buildup of tritium in the heavy water.

At Savannah River the heavy-water moderator-coolant is cooled in turn by river water in shell-and-tube heat exchangers. The river water is

*Only one of these is now (1971) in operation.

*See discussion of waterlogging in Chapter 13.

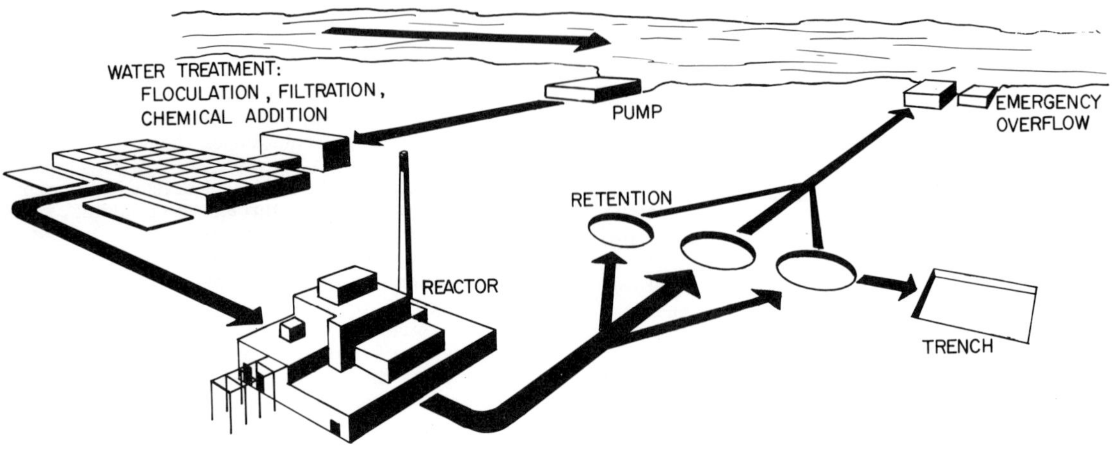

FIG. 3-8 Diagram of waste preparation and retention basin of a typical Hanford (plutonium-production) reactor.

monitored to detect moderator leakage and to ensure that radioactive materials accompanying a leak are not discharged directly to the river. The liquid wastes include auxiliary cooling water, decontamination, laboratory, and wash waters.

In addition to N^{16} the activities in the heavy-water moderator include N^{17}, O^{19}, and H^3. The heavy water also picks up radioactivity from the activation of corrosion products. Suspended solids and dissolved ions are removed by a filter and ion-exchange system.

Tritium continues to build up in the heavy water of the reactor for many years, so that eventually precautions must be taken to avoid exposure of personnel. It can be released to the atmosphere if heavy water leaks from the moderator or coolant loop. At Savannah River atmospheric discharge is through a 200-ft (60-m) stack. To date there has never been an occasion when the tritium concentration on or off the site has approached the permissible limit for continuous exposure [36].

A summary of the data available on tritium buildup in the CP-5 Reactor (see Figs. 3-9 and 3-10) shows the tritium concentration in millicuries of tritium per milliliter of heavy water as a function of the integrated power. The gap at 42,000 Mwh is due to dilution caused by the addition of 3300 lb (1500 kg) of fresh heavy water to the primary coolant system. Figure 3-10 represents the buildup of the total tritium in the system, where the total tritium is the product of the tritium concentration times the total volume of heavy water. No attempt has been made to correct for water losses. At the last point, 92,000 Mwh, the reactor contained 22,700 lb (10,300 kg) of heavy water and some 11,700 curies of tritium as liquid DTO or T_2O [37].

3.1.7 Homogeneous Reactors

Several small research reactors and several experimental and test reactors of the homogeneous type have been built. The research reactors have a maximum power of a few kilowatts and produce small amounts of fission products.

An example of a homogeneous test reactor is the Illinois Institute of Technology aqueous homogeneous reactor which operates at 75 kw(t). An atmosphere of oxygen is maintained over the coolant in the reactor. Ten liters per minute of hydrogen and oxygen are released, and these gases

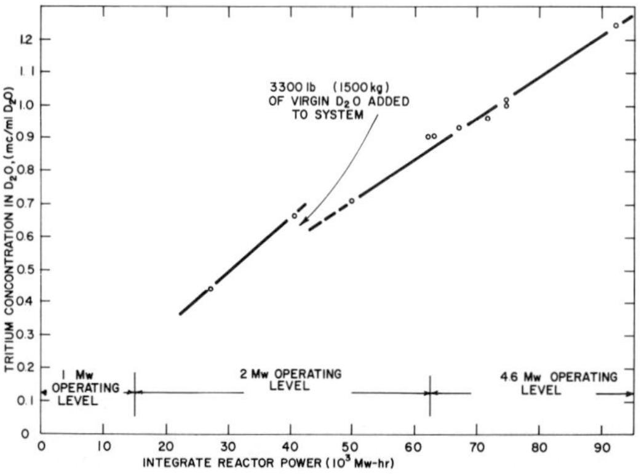

FIG. 3-9 Tritium concentration in D_2O of the CP-5 Reactor versus the integrated power.

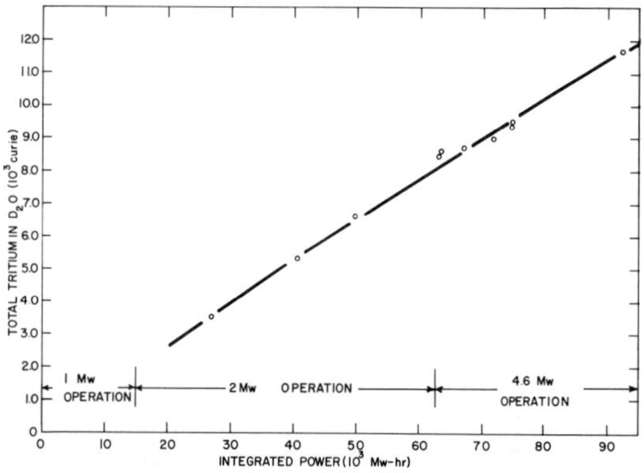

FIG. 3-10 Total tritium in D_2O of the CP-5 Reactor versus the integrated power.

are swept out of the reactor by oxygen and recombined. The fixed gases remain in the oxygen and are purged periodically. No liquid wastes are formed except during maintenance operations.

The Homogeneous Reactor Experiments 1 and 2 were built and operated for short periods at Oak Ridge. In HRE-2, which was designed for a core power of 5000 kw(t) and a blanket power of 220 kw(t), the core solution was UO_2SO_4 in heavy water and the blanket a slurry of thorium oxide in heavy water. The two streams were processed continuously by gas removal and total reprocessing of a small side stream.

About 1 ft^3/sec (~ 30 liter/sec) at STP of off-gases, mainly deuterium and oxygen, were produced. These carried the noble gases and some iodine. The gases were passed through a recombiner, cooler, cold traps, fission-gas absorbers, and blown to the stack. The fission-gas absorbers are of most interest since the data obtained can be used in design of other facilities (see Sec. 5) [38].

3.2 Sodium-Cooled Reactors

Sodium has been the most extensively investigated and used of the liquid metals considered as nuclear reactor coolants. Sodium-cooled reactors include the Sodium Reactor Experiment, the Hallam Nuclear Power Facility, the Experimental Breeder Reactors I and II, and the Enrico Fermi Reactor (see Table 3-19). The wastes from the Hallam and Fermi Reactors are discussed here as typical examples.

The corrosion rates of stainless steels and fuel materials by sodium at reactor operating temperatures of 425 to 540°C (800 to 1000°F) are very low. Consequently, there are only low levels of corrosion-activation products in the sodium coolant, and the only significant activities present come from activation of the sodium itself and from impurities in the sodium and the coolant system (see Table 3-20). There is about 4 ppm of uranium in commercial sodium. Apparently, this is rapidly removed in the cold traps, as fission products do not appear to be produced. Sodium is capable of causing metal transport but this is an operating difficulty and only affects the wastes indirectly.

The most significant impurities from an operational standpoint are oxygen and hydrogen, with oxygen being the most important. Because of its low solubility at low temperatures, the Na_2O formed tends to precipitate from the sodium and cause plugging of small lines and small openings. The solubility of the oxide increases from about 0.005 wt.% at 149°C (300°F) to about 0.1 wt.% at 482°C (900°F). However, there is considerable variation in reported results [39]. Na_2CO_3 and particularly NaH appear to increase the solubility of the oxide above 288°C (550°F) [40]. Sodium hydride forms rapidly at 204°C (400°F) but decomposes at about 427°C (800°F).

Sodium may be kept pure and both the oxides and hydrides removed by cold trapping. The sodium is cooled by an external NaK or boiling toluene loop and the precipitated materials removed by passage through wire mesh packed in the cold trap at about 149°C (300°F) (see Figs. 3-11a and 3-11b [41, 42]). The cold trap removes not only the sodium oxides and hydrides but also most of the other impurities, including essentially all the uranium.

The cold trap used in the EBR-II is about 3.5 ft (1.07 m) diam by 5 ft (1.52 m) high. To replace the cold trap requires cutting the sodium inlet and outlet pipes. It is expected that this cold trap will operate for the full life of the reactor. No plans have been made for its disposal. At the Enrico Fermi power plant it is planned to store any cold traps which are removed. Cold traps are about 50% efficient in respect to removal of most activated impurities in sodium reactor systems.

The sodium carried by the filters used to clean up the fresh sodium in the EBR-II and used during the cleanup of the reactor systems are recovered by distillation. It would be expected that sodium in the cold traps after a few years decay could be recovered by distillation and reused.

A hot trap is used to reduce the oxide composition to less than 10 ppm (see Fig. 3-12) [42]. The trap consists of a mesh of zirconium sheets at

TABLE 3-19

Sodium-Cooled Reactors

Type, name	Abbrev.	Power Mw(t)	Core		Control		Construction materials
			Fuel	Clad	Poison	Clad	
Thermal							
Sodium Reactor Experiment	SRE	20	U-Th	ss	B Ni	ss	ss
Hallam	HNPF	240	U-Mo	Zr	Gd Sm	Zr	ss
Fast							
Experimental Breeder Reactor	EBR II	62.5	U alloy	ss	(Fuel)	ss	ss
Fermi	---	200	U alloy	Zr	B_4C	ss	ss

TABLE 3-20

Activation Isotopes and Impurities in Sodium Coolant

Radioactive isotope	Half-life	Nuclear reaction
Na^{24}	15 hr	Na^{23} (n, γ)
Na^{22}	2.6 year	Na^{23} (n, 2n)
Rb^{86}	19.5 day	Rb^{85} (n, γ)
Sb^{124}	60 day	I^{127} (n, α)

649°C (1200°F). These react with the sodium oxide. The hot trap also removes hydrides by formation of zirconium hydride and removes carbon by reaction with the steel container.

Calcium, lithium, barium, or magnesium may be added to the sodium to form insoluble oxides. Calcium also reacts with any nitrogen present to form the insoluble nitride.

Sodium must be kept covered with an inert gas such as nitrogen, helium, or argon. Argon is purged frequently to keep the impurities, including Ar^{41}, to a minimum.

3.2.1 Hallam Facility*

The Hallam Nuclear Power Facility, HNPF, is a sodium-cooled graphite-moderated reactor fueled with slightly enriched uranium, 10 wt.% molybdenum alloy with stainless-steel-clad fuel and moderator. The normal power level is 240 Mw(t). The radioactivity of the coolant is almost all due to Na^{24}, the sodium activation being about 0.2 curie/g at full power [43].

*Since this was written, the reactor has been deactivated and dismantled.

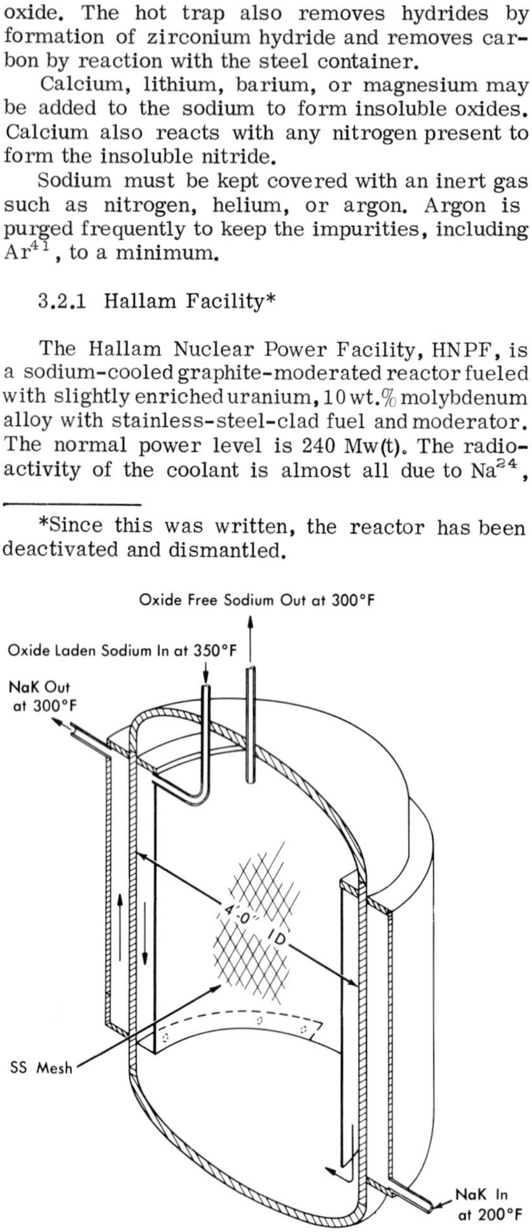

FIG. 3-11a NaK-cooled cold trap for removal of Na_2O from flowing Na or NaK.

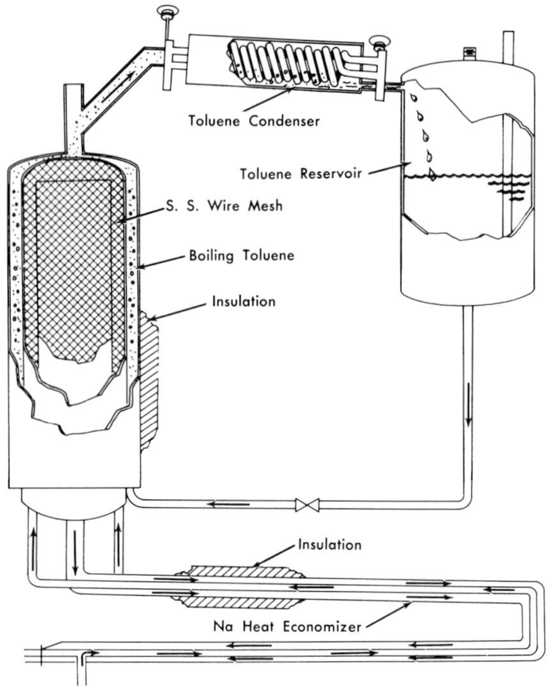

FIG. 3-11b Toluene-cooled cold trap for removal of Na_2O from flowing Na or NaK.

The gaseous wastes include the fuel cleaning, ventilation air, and helium and nitrogen purge gases. These are filtered and discharged through a 100-ft (30-m) stack (see Fig. 3-13) [21]. Some of the blanket gases escape and some Ar^{41} is formed in the air which surrounds the reactor. These are collected and stored for decay. Compressors of 25 ft^3/min capacity pump the gases into 150 psig tanks that hold 40,000 ft^3 at standard conditions. After decay the gases are filtered and discharged to the stack [44]. The main liquid wastes are those from the fuel cleaning cells, the decontamination of equipment and tools, and the laundry.

The main solid wastes are those produced in reloading and cleaning operations. Rejected core components represent the only significant quantities of activities. Since the activity is all short and intermediate life and it is physically fixed in the solid core components, there is no hazard except in handling and burial operations (see Fig. 3-13). The core components are expected to include 10 process tubes per year, containing 3×10^4 curies; one moderator can each four years, containing 4×10^3 curies; 3 shim safety rods per year, containing 9×10^3 curies; and 10 fuel assembly end fixtures per year, containing 4×10^3 curies; or a total of 4.4×10^4 curies/year.

3.2.2 Fermi Plant

The Enrico Fermi Fast Reactor at Lagoona Beach, Michigan, is designed for a maximum power level of 200 Mw(t). The uranium-alloy fuel is clad with Zircaloy and the blanket rods are clad with

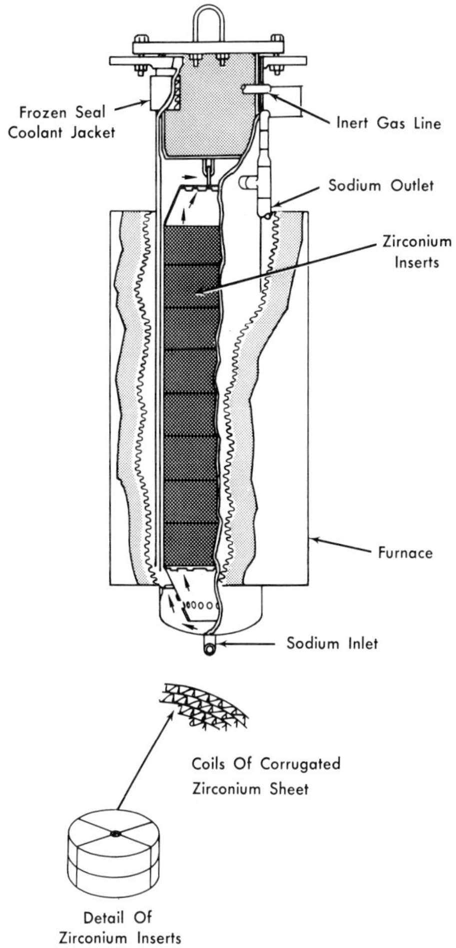

FIG. 3-12 Hot trap for further purification of Na or NaK.

stainless steel; stainless steel is used for most of the primary system.

The plant is expected to produce about 2000 ft^3 (56,000 liters) of waste gases per day. The bulk of this gas will be argon vented from the primary shield tank and lower containment vessel (see Table 3-21) [45]. When the gases contain more than 1×10^{-3} μc/ml, they are stored for decay. Lower concentration gases are diluted so that their concentration is $< 1.8 \times 10^{-6}$ μc/ml and discharged (see Fig. 3-14) [21]. When wind velocities are less than 4 miles/hr (~ 6.5 km/hr), discharge is limited and an automatic shutoff on the stack prevents discharge when the diluted gases have a higher concentration than 2×10^{-6} μc/ml. The shielded storage tanks have a capacity of 35,000 standard cubic feet of radioactive gases. All exit gases are filtered through high-efficiency filters.

The liquid wastes which are expected are indicated in Table 3-22 [45]. Wastes with higher activity than 4×10^{-3} μc/ml will be held for treatment or decay; those with less than 10^{-4} μc/ml will be discharged at the normal dilution rate of 130,000 gpm (8200 liter/sec); and wastes with intermediate amounts of activity will be discharged

TABLE 3-21

Fermi Gaseous Wastes

Source	Volume per month		Concentration (μc/ml)	Total activity per month (curies)
	(ft^3)	(liters)		
Primary gas discharge	650	18400	0.9 (Ar^{41})	17
Primary gas leakage	720	20400	146 (Ar^{41})	3000
System purge (intermittent)	$6 \cdot 10^4$	$1.7 \cdot 10^6$	0.06 (fission gases)	100
Subassembly cleaning	8500	$2.4 \cdot 10^5$	0.008 (fission gases)	1.9

under metered conditions. The diluted wastes will have a concentration less than 8×10^{-9} μc/ml at discharge. Tank storage capacity is 17,500 gal (66,300 liter).

Solid wastes will consist of about 50 ft^3/month (1.4 m^3/month) of filters, wastes, and ion-exchange resins containing a total of a maximum of 2.0×10^4 μc. After storage for decay they will be barrelled and shipped to regular burial sites.

The cold traps will be removed with the sodium and liners and stored in concrete vaults. No plans for burning the sodium or ultimate disposal of these wastes have been made.

3.3 Gas-Cooled Reactors*

Carbon dioxide is used as the coolant in most of the United Kingdom reactors and in the French reactors. Helium is used in the European Nuclear Energy Agency's DRAGON Reactor and in the two gas-cooled prototype power reactors being built in the United States. Once-through air is used in the graphite-moderated research reactors at Oak Ridge and Brookhaven and the French reactor, G-1; and it was used in the two Windscale Reactors.

A flow diagram of the Calder Hall Reactor is shown in Fig. 3-15 [46]. The amounts of activity

TABLE 3-22

Predicted Liquid Radioactive Wastes from Fermi Reactor at 630 Mw(t)

Source	Average volume per month		Average radioactivity (μc/ml)	Total activity per month (μc)
	(gal)	(liter)		
Subassembly cleaning	2600	9800	$8 \cdot 10^{-3}$	$8 \cdot 10^4$
Decontamination	350	1330	0.1	$1 \cdot 10^5$
Laundry	5000	18900	$1 \cdot 10^{-5}$	$2 \cdot 10^2$
Laboratory sinks	1500	5700	$1 \cdot 10^{-5}$	60

*See also Chapter 9.

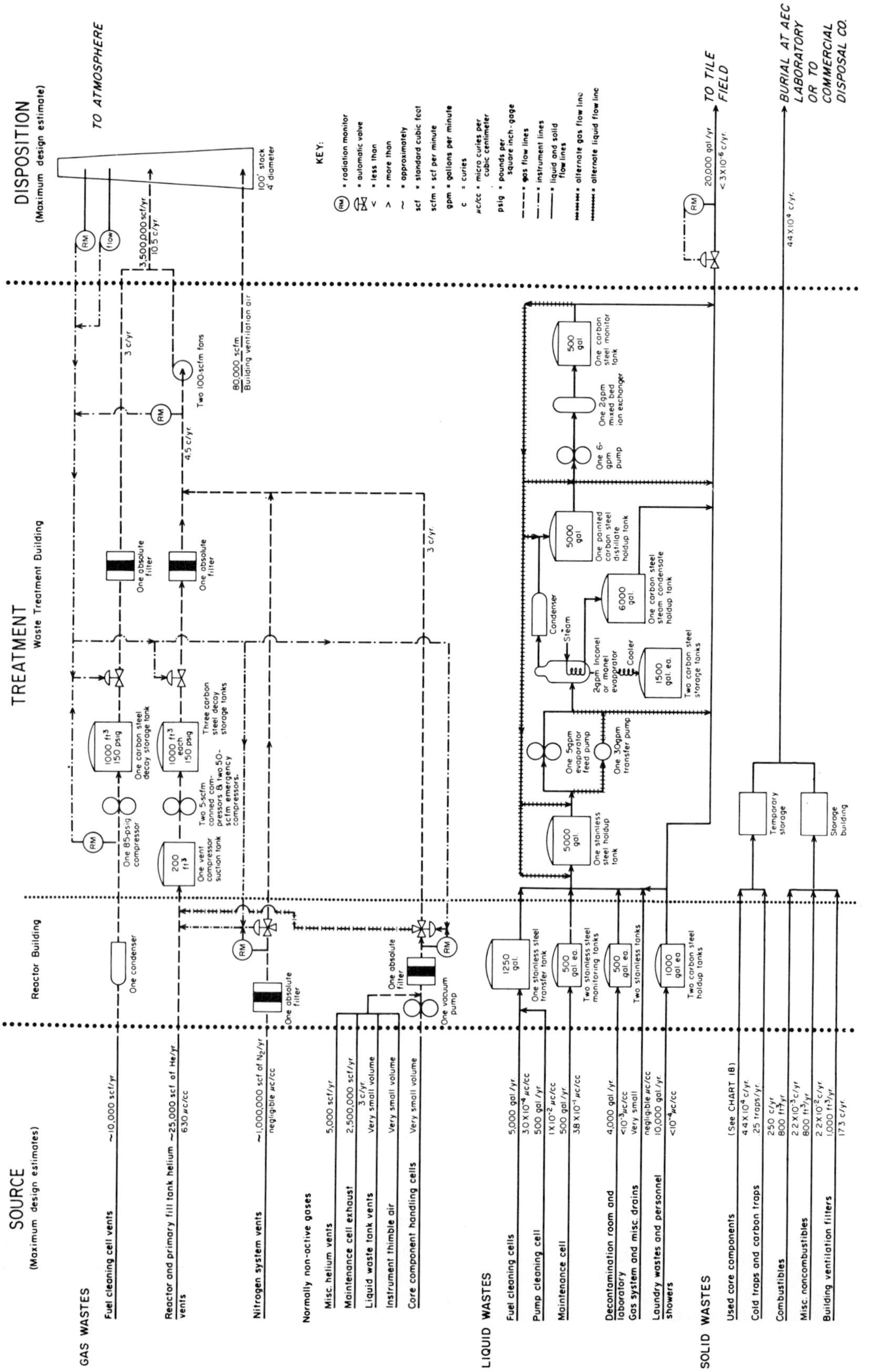

FIG. 3-13 Diagrammatic flow sheet for Hallam Nuclear Power Station waste plant.

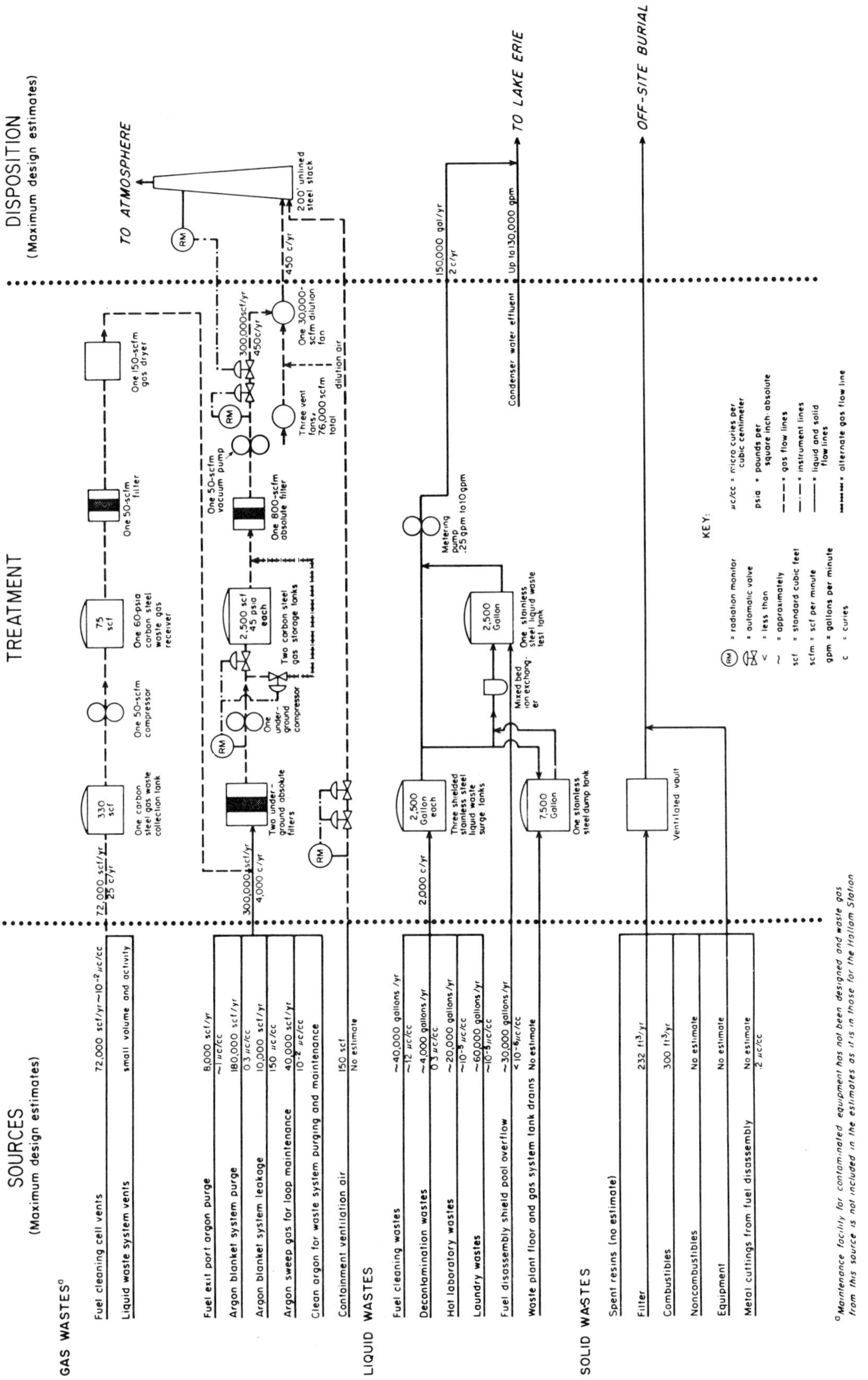

FIG. 3-14 Diagrammatic flow sheet for Enrico Fermi Nuclear Power Station waste plant.

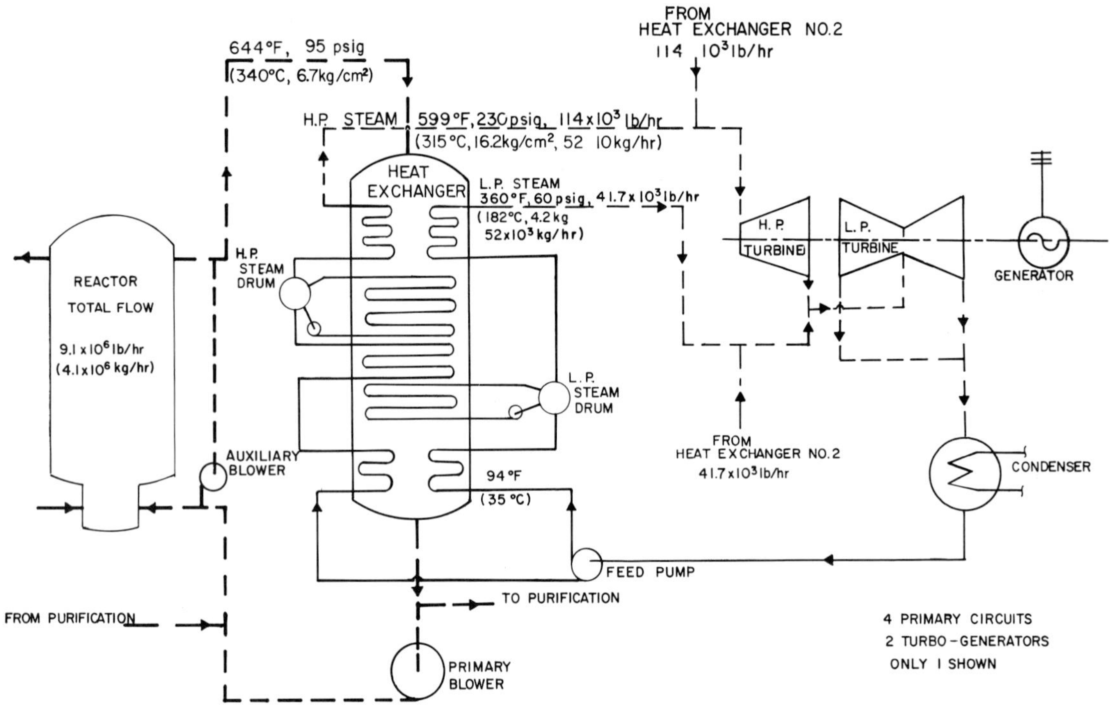

FIG. 3-15 Flow diagram of Calder Hall Reactor.

entrained in the gas stream are usually small. The presence of N^{16}, formed from the oxygen in the CO_2, requires shielding of the recirculating pumps and ducts. Small amounts of C^{14} are also formed from the CO_2. The argon concentration is about 1.5×10^{-3} μc/ml when the argon concentration is 3 ppm. An inspection at Calder Hall after a year's operation indicated that the deposited activity on ducts, pumps, etc., was very low [47].

An exhaust-air-filtration installation was built as a part of the once-through, air-cooled (1.13×10^6 lb/hr or 0.51×10^6 kg/hr), graphite-moderated Brookhaven Research Reactor.† This reactor was originally fueled with aluminum-clad natural uranium. Although a number of cladding failures occurred, none were considered serious. The reactor has since been reloaded with enriched uranium fuel using uranium-aluminum alloy techniques. The rated power level of the reactor was at that time reduced from 30 to 20 Mw(t). The airflow requirements were also reduced, and new cooling fans were installed which more appropriately matched the cooling requirements and system pressure losses. The inlet filters consist of American Air Filter Company, FG100, Deep Bed Mat, 1-in. thick. The exhaust filters* consist of a glass cloth which is efficient for particles larger than 1 μ (0.04 mil) [48]. About 17,000 curies of Ar^{41} are released each 24 hr.

The two Windscale Reactors were equipped with coarse glass filters which removed about 50% of the 5 μ particles. A fire occurred in one reactor, resulting in a release of appreciable activity. Details of the reactors and this accident are presented in Sec. 7. The reactors were shut down and closed up after the fire.

The Philadelphia Electric Company, with assistance from a large number of utilities, is constructing an all-ceramic-core gas-cooled reactor, known as the Peach Bottom plant. The fuel will consist of graphite-coated uranium and thorium dicarbide particles dispersed in graphite rings surrounding graphite rods. Outer sleeves made of impermeable graphite will be used to inhibit diffusion of gaseous fission products into the helium coolant.

A purge stream of helium will be pumped through grooves between the fuel rings and the outer sleeves to transport fission gases from the reactor as rapidly as they diffuse from the fuel particles. Purge gas will be continuously purified as will a stream taken from the main coolant stream. The purge gas passes through an internal fission trap, which is located in the lower reflector region of the fuel assembly. The trap consists of a slotted graphite cylinder containing treated charcoal in granular form. This trap retains most of the fission products. The gas passes through an external fission-product trap, where the bulk of the remaining fission products are removed. The clean helium returns to the main circuit [49].

*See discussion of filters in Chapter 19.

†Shut down in 1969.

The helium gas in the full-scale reactor is expected to become slightly contaminated with H^3 and Ar^{41}. Direct leakage is estimated to be 10^{-4} curie/day. In addition, an estimated 0.16 curie/yr will be released during the reloading and 2.5 curie/yr will be released as a result of maintenance operations. There will be some neutron activation of the containment atmosphere. The most important radioisotope is expected to be Kr^{85}. It will be adsorbed on activated carbon and shipped off the site.

Tests run in loops at Vallecitos have demonstrated that the gas diffusion from the fuel is low (see Fig. 3-16). At a heat-production rate of 240,000 Btu/ft^2-hr (18 cal/cm^2-sec) compared to the design rate of 102,000 Btu/ft^2-hr (7.7 cal/cm^2-sec), the activities in the purge and coolant gases were 25 μc/ml and 0.03 μc/ml compared to design values of 5×10^4 μc/ml and 5×10^2 μc/ml, respectively.

Some liquids will be produced from laundry, shielding cooling, etc. The molecular sieves in the helium cleanup system used to remove water and carbon dioxide will be regenerated weekly. The 75 gal of wastes are expected to contain 6.5×10^{-2} curie of Kr^{85}.

The high-level solid wastes will include the activated carbon with Kr^{85}. The wastes will be primarily filters which will be handled in shielded casks. Low-level wastes will include various cleaning rags, papers, and trash [50].

3.4 Organic-Cooled Reactors

The use of organic materials as reactor coolants is based on the use of multiring compounds such as the terphenyls. These deteriorate but they do not become appreciably radioactive in a neutron flux nor are they corrosive. One problem in their use is that of separation and disposal of the high-volatility gases and low-volatility or high-boiling-point compounds formed as a result of exposure to reactor radiations. These separated materials carry very little radioactivity. Attempts have been made to burn the gases with subsequent cooling and filtration of the combustion gases. This would remove the bulk of the radioactive materials as particulate matter. Burning proved to be difficult as the heavy tars do not burn readily and they are carried through to the filter which is quickly plugged. These difficulties have apparently been resolved in tests at Santa Susanna. A high degree of purity of the coolant must be maintained at all times to avoid fouling of the heat-transfer surfaces.

The Piqua Nuclear Power Facility at Piqua, Ohio, was designed to operate at 45.5 Mw(t). The fuel is slightly enriched U, 3.5 wt.% Mo, 0.1 wt.% Al alloy with finned aluminum-clad tubes. The coolant-moderator is a mixture of terphenyls.

The radioactive impurities in the coolant include traces of H^3, S^{35}, and C^{14}. Other materials include P^{32}, Cl^{36}, Mn^{54}, Fe^{59}, and Co^{60} [51, 52].

The gaseous wastes are distilled off under

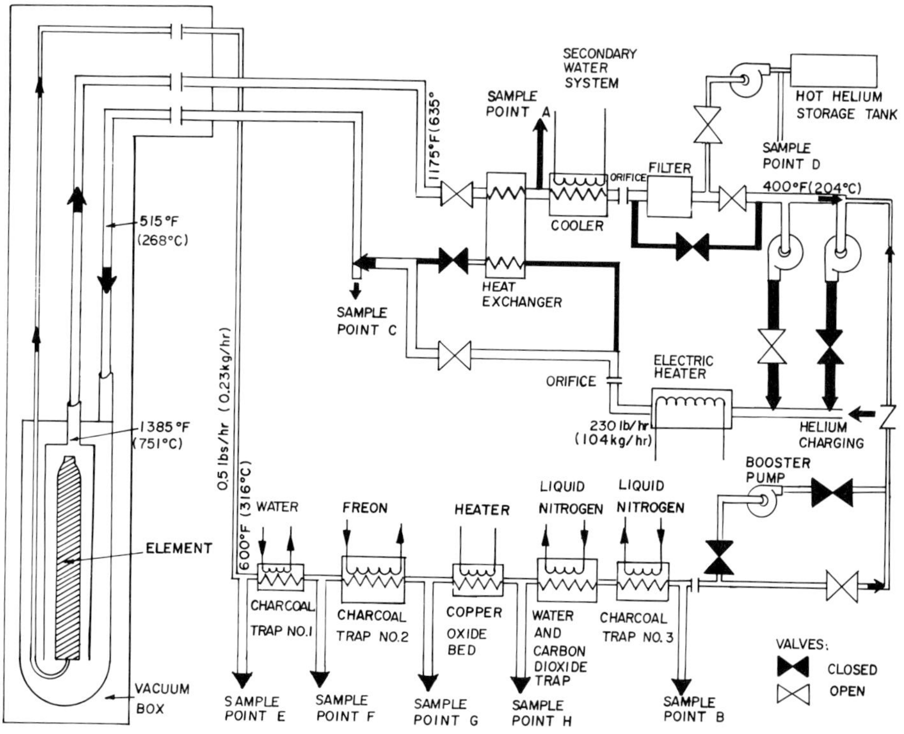

FIG. 3-16 Vallecitos test loop for the Peach Bottom Reactor.

vacuum and compressed by a steam ejector. After cooling, the diphenyl and heavier compounds are removed by cold traps and activated carbon at 71°C (160°F). Then the gas is cooled and the water condensed and sent to the liquid wastes. The residual gas is passed through an activated charcoal-decay train and to the stack. The high boiling residues are absorbed on vermiculite and shipped to solid-waste burial grounds. Solid wastes consist of the usual cleanup materials, filters, etc.

4 SOLID-WASTE SYSTEMS FOR REACTORS

4.1 General Considerations

As in any other nuclear activity, the operation of a nuclear reactor results in the contamination of a wide variety of solid materials. By far the largest volume of the contaminated solids falls into the category of low-level "trash." Included would be cleaning rags, Kleenex, sweeping compounds, paper towels, contaminated tools, and piping and equipment removed from the external circuits and from associated experiments during maintenance or dismantling operations. The amount of such material produced is quite variable depending on the size and type of the reactor and the degree of administrative control exercised in its operation. Generally, 1000 to 5000 ft^3/yr (30 to 140 m^3/yr) of this type of waste are produced in a power reactor plant. Low-level solids vary in activity level from just above background to about 50 mr/hr on contact. As an example, the Yankee Reactor has shipped off site a total of about 5000 ft^3 of this type of waste in 30 months. The activity content of this waste is estimated to be 2000 mc (see Table 3-17) [32].

Somewhat smaller quantities of waste — measured in the hundreds of cubic feet per year — containing higher concentrations of radioactivity are also formed. These include, for example, spent resins from the ion-exchange cleanup of reactor coolant, evaporator bottoms from liquid-waste treatment plants, incinerator ash, and hardware from fuel elements, control rods, and various reactor internals. These latter are removed during maintenance operations. Several thousands of curies of radioactivity will generally be associated with the yearly collection of this type of waste.

By far the greatest quantities of activity produced in a reactor are rejected from it in the form of spent fuel elements. Few of today's reactors produce more than 20 tons/yr ($\sim 10^4$ kg/yr) of these. From the standpoint of reactor operation the spent fuel elements may be thought of as a "waste." They contain, however, so much valuable material that they may not be treated as such; accordingly, they are shipped to a reprocessing plant for recovery of the valuable constituents. The associated high-level-waste problems are therefore transferred to other sites.

4.2 Waste Collection

The usual practice in handling solid wastes consists of accumulating the solids in suitable containers. Some degree of segregation of the wastes is required at the source -- activity level and combustibility are two common criteria. It is highly desirable to adopt measures which permit the separate collection of noncontaminated waste from that which is contaminated. If all the solid trash from an area handling radioactivity is assumed to be contaminated, the volume of material which must be handled and discarded as radioactive waste will be greatly increased. Separate collection systems must be set up and either routine or spot checks of the noncontaminated trash must be made, depending on the hazard potential of the possible contaminants.

In handling highly active solid wastes, provision must be made to safeguard personnel from radioactive hazards and to prevent the spread of contamination. Protective clothing is frequently required, protective respiratory devices such as masks, respirators, or supplied-air devices are used when inhalation hazards exist, and radiation surveys are made prior to and during handling.

Low-level wastes are handled directly and generally require no particular precaution. Collection practices for this type of waste are quite uniform and consist of distributing suitable containers throughout the work areas to receive and hold contaminated trash until collection. The containers range from cardboard cartons and kitchen-style garbage cans to 55-gal (208-liter) drums. They are plainly marked with radiation symbols to distinguish them from ordinary noncontaminated trash cans. These containers must be frequently and routinely surveyed for protection of workers in the areas and they should be surveyed before collection for the protection of the collectors. A frequently used limit for waste which may be collected by direct handling of unshielded containers is 50 mr/hr on contact. When collected, the low-level waste is usually taken to a central collection point where it may or may not receive further treatment, and it is then either stored on site or packaged for shipment off site.

The high-level wastes, such as the spent ion-exchange resins, require special treatment, since the wastes cannot be handled directly or without shielding. Two methods are available for handling such resins. Either they may be flushed into storage tanks for decay or they may be taken out of the exchange units in "cartridges" and placed in shielded containers for handling.

Hardware taken from within the reactor involves such a wide variety of shapes and sizes that each problem must be treated individually. About the only common denominator is that a shield of suitable size, shape and thickness must be available in order to accomplish the necessary transfers.

Irradiated fuel assemblies are removed from reactors by several different procedures, as discussed in Chapter 14. Regardless of the reloading procedure used, the fuel assemblies are stored in water or other liquid coolant for several months for decay of the short-lived fission products. The fuel assemblies are then either placed in shipping casks directly or disassembled. When they are disassembled, as much as possible of the non-fuel-bearing components are removed by cutting or sawing. The fuel portions of the assemblies or individual fuel elements are placed in shipping casks by various underwater grappling devices, and the shipping casks transferred from the underwater loading areas to railroad cars or trucks. The parts removed from the fuel assemblies are placed in storage tanks in the underwater canals for decay. They are later transferred to casks for removal to the solid-waste-treatment area, or placed in concrete-shielded containers suitable for direct burial.

4.3 Storage for Decay

Since all radioactive materials decay with time, it is sometimes desirable to provide a ''cooling'' period on site before attempting to move a radioactive solid to another place. This may be done to reduce either the amount of shielding or the amount of cooling that will be required, or both. By so doing it is often possible to reduce the cost of the transfer operations. While the high-level wastes generally do not contain sufficient activity that they have to be permitted to decay before shipment, storage for a few months to even years before shipment frequently simplifies the problems of handling and shipping.

In the case of the low-level trash nothing is gained by decay storage. The waste is low enough in activity level that it may be safely shipped at a minimum cost. However, intermediate storage is almost always provided to allow the accumulation of an optimum shipping load.

Spent fuel elements must be cooled 90 days before they can be shipped under current U.S. regulations [53]. This reduces the heat production and the I^{131} content to acceptable levels.

4.4 Incineration

Between 50 and 80% of the low-level trash from a reactor is combustible. This has led to the consideration of incineration as a means of volume reduction of solid waste. Several sites (e.g., Shippingport and Yankee) have installed incinerators for their solid wastes. A volume reduction of about 95% and a weight reduction of about 70% of the combustible wastes can be achieved.

Incinerators can be built very simply along the lines of a home unit or may be quite complicated as, for example, that shown in Fig. 4-1 [54]. Because some of the activity will be transferred to the combustion gases, a cleanup train for the off-gases is required. By adding units to this off-gas train

FIG. 4-1 Active waste incinerator at Argonne National Laboratory.

TABLE 4-1

Volume Reduction of Solid Wastes

	Volume as collected (ft^3)	Volume after baling (ft^3)	Volume after incineration (ft^3)
Paper, clothes, etc.	12,700	1,800	250
Filters	2,200	2,200	1,030
Evaporator bottoms	1,400	1,400	1,400
Miscellaneous	6,600	6,600	6,600
Total	22,900	12,000	9,280

NOTE: 1 $ft^3 = 0.02832\ m^3$.

nearly any necessary degree of decontamination of the exhaust gas can be accomplished.

Incinerators require a high degree of segregation of the wastes into combustible and noncombustible categories at the sources. It has been found difficult to obtain the required degree of cooperation from the producers of the waste.

In 1953 the figures shown in Table 4-1 [55] were obtained at Knolls Atomic Power Laboratory, indicating that the volume reduction obtained by incineration did not justify its cost [55]. Experience since then has generally borne out this conclusion [56].

4.5 Baling

Another technique which can be used to reduce the volume of the low-level trash to be stored is baling. This is a rather straightforward procedure which does not require any segregation of the waste other than by activity level. Simple paper balers, with the addition of a hooded enclosure, can be used. In the unit described in reference [10] a commercial hydraulic press operating at a pressure of 2100 psi (~ 150 atm) accomplished a reduction to about one-fourth the original volume. The cost is only about 15 to 20% of that of incineration.

4.6 Sampling and Monitoring

Solid wastes are almost never sampled, because samples are hard to obtain and are of little value. The material is so heterogeneous that the results obtained from samples are nearly meaningless. It is usually sufficient to determine the activity content of solid wastes by rough calculations based on readings taken with survey instruments.

Monitoring of solid-waste operations is required to assure the safety of personnel and to permit shipment under the various regulations (see Sec. 4.7). There are no particular problems associated with the monitoring of these operations, and the work is generally done as a part of the routine work of the Health Physics group.

4.7 Transportation and Final Disposal

Small amounts of solid wastes can be buried on site in soil under the restrictions of paragraph 20.304 of 10 CFR, Part 20. Ordinarily the solid waste from a reactor site are packaged for shipment to some other place for disposal.

Interstate shipment of radioactive material by land or water in the United States is subject to regulation of the Interstate Commerce Commission (ICC). The regulations applicable to radioactive materials are published as Title 49, Parts 71-78 of the "Code of Federal Regulations". Between revisions, annual supplements are issued and amendments may be published in the daily issues of the "Federal Register." These regulations may be obtained from the Superintendent of Documents, U. S. Government Printing Office, Washington, D. C., 20402. The ICC regulations are also published by the Bureau of Explosives [57] and by the American Trucking Association [58]. Transportation of radioactive materials by water is subject to regulation of the U.S. Coast Guard [59, 60].

In an effort to predict something about the problems of shipping radioactivity, the USAEC had for some time an Ad Hoc Committee on Transportation of Highly Radioactive Materials. This Committee has approved various shipping containers for radioactive materials. However, wastes are seldom shipped in reuseable containers but rather in steel barrels, in boxes made of various materials, and in concrete.

Dry low-level solid wastes, such as used clothing and paper, are nearly all packaged in cardboard containers, either cylindrical or cubical in shape. These containers, with their wastes, are buried directly.

High-level wastes are nearly always shipped in used steel barrels. These may be lined with one to several inches of concrete for shielding prior to filling. The barrels are then filled to within one to several inches (~ 10 cm) of the top and may be capped with concrete. The barrels are shipped to the burial site by truck or rail and buried directly. Very high level wastes that require over 1 ft (0.3 m) of concrete for shielding may be cast in large concrete blocks.

Rejected equipment which is impractical to decontaminate is sometimes "stored" in an open field (in an area that is fenced in and routinely surveyed by Health Physics personnel) and permitted to decay for several years. This permits later decontamination, disassembly, and repair for return to use.

Several reactors have been removed from service. One of these, the CP-3 Research Reactor at Argonne, was shut down in 1955 and the site returned to its original use. After the fuel, coolant, and useable equipment were removed and the shield filled with concrete, a hole was dug alongside the reactor shield large enough to permit its burial. After the hole was dug, the shield was undermined and a dynamite blast toppled it into the hole. The hole was then filled with dirt, leaving the site much as it had been before the reactor was built. It should be noted that the CP-3 was a very low power, 300 kw(t). It is not inferred that higher power reactors can be as easily disposed of.

The usual method for disposing of low-level solid wastes is by conventional sanitary land fill. This consists of digging a ditch of sufficient dimensions, putting the wastes in the ditchs, and backfilling to eliminate the spread of activity by wind and predatory animals. It is most safely practiced in isolated areas with favorable geological conditions. The AEC operates rather extensive land fills at Oak Ridge National Laboratory, Hanford, and the National Reactor Testing Station in Idaho. These have been, until the middle of 1963, available to private users. By the middle of 1963 two privately operated burial grounds had been established on state-owned land (Nevada and Kentucky). A third (in New York) was opened later in 1963. The AEC, therefore, has withdrawn from receiving commercial radioactive waste.

There are a number of commercial firms which have been picking up wastes and barging them to designated disposal sites off either coast. Although the amounts of activity which have been placed in the sea off the coast of this country are extremely small, this operation is one which has run into considerable political objection — local, state, and international. In addition, it seems to be becoming clearer that the economics of disposal favor land burial and it would seem safe to predict that, except for some specialized situations where location and transportation costs are unfavorable to land disposal, sea disposal of solid wastes will soon cease. Actually, less than 5% of the low-level solid wastes produced (up to 1963) in the United States have been disposed of at sea.

5 GASEOUS-WASTE SYSTEMS FOR REACTORS*

5.1 General Considerations

As indicated in Sec. 3, gaseous wastes are produced in heterogeneous nuclear reactor plants from a number of sources and they vary greatly. The ventilation air from reactor plants is usually monitored in potentially contaminated areas and prior to release to the stacks. The volumes of true radioactive gaseous wastes are small and the quantity of radioactive materials discharged are usually reduced to negligible quantities. The gaseous wastes produced by homogeneous reactors, however, include very active gases. Since there are only small homogeneous reactors in operation, emphasis is placed here on heterogeneous reactors.

The gaseous-waste systems are designed (1) to store, as necessary, and treat the small volumes of highly active off-gases such as purge gases from pressurizer tanks of pressurized-water reactors and the condenser off-gases of boiling-water reactors, (2) to filter the vent gases from irradiated fuel areas and the liquid-waste and solid-waste processing equipment, (3) to monitor all exit gas streams, and (4) to discharge the gases through 100- to 400-ft (30- to 120-m)-high stacks. As noted in Sec. 3, the coolant gases for the once-through air-cooled reactors such as the Brookhaven Graphite Research Reactor require filtration, as this air may become highly contaminated. The design of filter and purification equipment is discussed elsewhere, while the utilization of the equipment is discussed in this section.

As indicated in Table 3-2, most of the coolant-activation products are short-lived. Tritium is formed in readily detectable amounts in the coolant of light-water-cooled reactors, and very small amounts escape from heavy-water-moderated and -cooled reactors. The value of heavy water justifies the installation of leak-tight designs. For the other gases natural decay and dilution may be relied on to take care of the activated coolant gases and complex equipment is seldom required.

While gaseous-waste systems involve few new design principles, as most equipment is similar to that used in chemical industries for similar functions, the specifications for particulate removal, vapor removal, and leak rate are more stringent. These specifications indicate greater care in respect to details in design and construction. First, the radioactive-gas-transport systems must be essentially leak tight, and this means that vacuum techniques must be used and inspection must be made with halide or mass-spectrometer leak detectors. Seamless tubing and all-welded joints are sometimes specified. Second, the gas-transport systems are designed so that so far as possible leaks are into the system. Reactor areas are usually divided into areas of increasing hazard with pressure differences of 0.05 to 0.10 in. of water (0.1 to 0.2 mm Hg) between the areas. The pressure pattern is arranged so leaks are always from areas of low hazard to areas of higher hazard. Control rooms are under highest pressure with the waste-processing areas at the lowest pressure. Third, high-quality materials, sometimes stainless steel, are used to limit corrosion products and retention of particulate matter in the pipes. Fourth, greater precautions must be used to prevent blowouts or fires in the filters and to prevent release of trapped material. Fire-resistant filters must be used wherever fires appear possible. Fifth, valves installed in containment vents must be quick-operating and essentially leakproof. Sixth, various gas streams should be kept segregated until they have been monitored. After monitoring, the gases from several sources may be mixed before treatment or kept separate. In general, segregation is favored since not all the purification streams will be severely contaminated in case of an accidental release. Seventh, inspection and tests during operation must be frequent and be designed to actually test the emergency facilities.

Gases may be stored; scrubbed with various liquids to entrap or dissolve particulate matter, both liquid and solid, or to adsorb various radioactive gases such as iodine; passed over activated charcoal at room or subzero temperatures to remove fixed gases or high-molecular-weight organic compounds; and filtered to remove particulate matter. Filtration is used as the final treatment of the higher level gas streams before dispersal. And all streams are monitored separately and mixed and diluted at the bottom of the stack. A subcommittee of the American National Standards Association on Radioactive Waste Disposal is preparing a report covering current practices in respect to disposal of gaseous wastes [61].

*See also Section 4 of Chapter 19.

5.2 System Design

Several flowsheets of radioactive waste systems for various reactors are presented in Sec. 3. These are layouts that have been found to be practical and workable.

Since there are several sources of radioactive gas streams that may contain appreciable quantities of radioactive gases or particulates and since these should be monitored and treated separately, several monitors, pipe lines, storage tanks, and filters may be required in a single nuclear power plant. The systems are designed for in-leakage as much as possible, and pipe lines and equipment are located in underground or in concrete-walled areas. Exhaust fans are placed downstream from the treatment areas and just before the stack. Duplicate fans or cross piping just before the fans are used. Emergency power is always available to maintain fan operation; but the emergency fan operation may be at a small fraction of normal power.

5.2.1 Storage for Decay

Storage for decay is the simplest of operations in that the gases are cooled (if required for removal of excess water vapor) and stored in ferritic steel tanks. Pressures are less than atmospheric in vent tanks and up to 150 psig (~ 12 atm) in storage tanks. Atmospheric pressure decay tanks are frequently used in series in the exhaust line. If a high level of activity occurs as a result of operating difficulty, these tanks provide ample time to monitor the gases and decide if it is advisable to shut the reactor down before excessive dispersal of radioactive gases occurs.

The size of the storage system is based on the time required for decay of the expected radioactive isotopes in order to limit discharge under the worst assumed operating conditions to acceptable values. Usually, the decay times are in minutes to hours or from 60 to 120 days. The shorter period results in decay of gaseous activation products, except Ar^{41} and C^{14} and many of the gaseous fission products. I^{131} may limit the discharge of the gases until it has decayed. Kr^{85}, the only intermediate life radioisotope of importance, is always discharged.

There are no special design requirements for the high-pressure storage vessels except that they must be built in accordance with the ASME Boiler and Pressure Vessel Code. (See Chapter 14.) They may be built in accordance with Section VIII of the code or, if specified by the purchaser, in accordance with Section I or Section III, Nuclear Vessels. However, this is not considered necessary. Safety valves should be vented to other vessels or to the stack ahead of the final filters.

A typical decay storage system is the one installed at the Yankee pressurized-water reactor. Small amounts of hydrogen are released from the stripper used to degas the liquid wastes and from the vents of the liquid-waste tanks. This hydrogen is mixed with nitrogen and may contain some gaseous fission products. The gas is pumped into a 4160-ft^3 (118-m^3) surge drum. When sufficient gas is collected, it can be pumped into three higher pressure gas storage drums. The tanks have sufficient capacity for at least 60 days of decay. These gases are discharged after decay through a deep-bed high-efficiency filter and diluted with the ventilation air before discharge to the stack. In comparison, Dresden operates satisfactorily with an underground pipe which provides about 20 min for decay of the gases.

5.2.2 Recombiners

Various catalytic recombiners are used in pressurized-water and heavy-water reactor systems to reduce the hydrogen gas concentration below the possibility of an explosion. Or the hydrogen and oxygen mixture is diluted and dispersed through the stack. Palladium wire or platinum-clad stainless steel wire is in use in recombiners. A flame recombiner was used at Oak Ridge in the HRE-I.

Tests at Oak Ridge indicated that platinum black supported on aluminum pellets or on stainless steel wire mesh gave 100% recombination at reasonable rates.

Pressurized-water reactors are normally operated with excess hydrogen dissolved in the water; so the hydrogen reacts with oxygen as fast as formed. Tests are under way to determine if the mixed off-gases can be returned to boiling-water reactors for direct combination [62].

5.2.3 Compressors

Motor-driven positive displacement compressors of the Roots-Connersville type may be used, but any sealed compressor is satisfactory. The exhaust fans are sealed with external bearings.

5.2.4 Stacks

Concrete stacks have been most commonly used because of their low cost. For reactor wastes, these need have no special construction. In the case of the United Kingdom Windscale air-cooled reactors, the filters were installed in an enlarged section at the top of the stack. The practice in the United States is to locate monitors well up the stack or at the top of the stacks to indicate the total amount of beta and gamma activities released. The gaseous wastes are usually monitored prior to dilution with the ventilation air. The stacks may be designed for exit velocities of 50 to 80 ft/sec (15 to 25 m/sec) to give maximum rise and mixing with the air and to minimize downwash in the lee of the stacks.

Stacks cost from \$300 to \$1000/ft (\$1000 to \$3300/m) of height. For reactors, unlined concrete stacks of minimum cost are satisfactory. The height must be determined from a consideration of the location [48]. Some reactor plants, such as the EBWR, have been designed with vents or short stacks on the top of the containment shell. Under some atmospheric conditions relatively little mixing took place and air exhausted from the EBWR flowed downward around the containment shell and entered the air intakes. During the defected-fuel-element tests, this led to high air counts in the containment building.

Usually, rapid mixing occurs and dilution factors from 10^3 to 10^5 may be considered reasonable [63-69]. The MIT Reactor, which is located in a city and has very little controlled area (only 250 ft or 75 m to the boundary), uses a dilution factor of

3×10^3; this factor has always been exceeded. For larger areas with 1 mile (1.6 km) to the edge of the controlled area, higher factors may be attained. These factors are strongly dependent upon detailed local conditions, including structures and the local terrain, as well as on the stack height. The stack height for each location must be calculated on the basis of the local factors. It is usually high enough that local obstructions will not cause local down-currents.

5.2.5 Filter Installations*

The installations of filters for removal of particulate materials were discussed in Sec. 3, and specifications for various filters are discussed in Sec. 4 of Chapter 19. The purpose of this section is to discuss the installation and use of filter systems. The filters used are frequently the 24 × 24 in. (61 × 61 cm) size by 12 in. (30.5 cm) deep to handle 800 to 1000 ft^3/min (380 to 470 liters/sec) per filter at an initial pressure drop of 1 in. of water (1.9 mm Hg).

The filter housing may be stainless steel sheeting, all welded, with doors gasketed by O-rings. In this case the filters are sealed by gasket seals to the framework supports. O-rings that are automatically compressed may be used. However, seals of soft rubber have proven satisfactory and are commonly used. These seals have the advantage that they may be compressed from about 1/8 in. to 3/8 in. (3.2 to 9.5 mm); consequently the frame can be made of welded angles. The filters may be sealed by tape but rubber seals which are automatically compressed have given the best results. The filters are changed by opening the doors, removing the sealing tapes if present, and pushing the used filters out with new filters into cardboard containers or plastic bags that can be sealed without loss of particulate material. The new filters are sealed and the doors replaced. Or the ductwork may be concrete with mild or stainless steel frames that hold the filter frames. In an installation of this type, manholes are provided for entry into the filter room and the air flow is stopped or reversed slightly during the filter change.

High-efficiency filters have failed due to the effects of moisture absorption by the filter paper, by softening of the adhesive, and by moisture absorption by the corrugated paper separators. These defects were easily corrected by waterproofing the filter media and frames, using different adhesives, increasing the strength of filter paper, and using fiberglass media.

As a result of leaks and failures, inspection and testing stations have been established by the AEC at the Edgewood Chemical Center and the Hanford Works. The testing there has led to improvement in the filters. Initially rejections were high due to excessive penetration; later rejections were due to high pressure resulting from the use of humidity-resistant and heavier papers [70]. Rejections were several percent in 1960 and 1961.

In-plant handling resulted in many failures. To remedy this an installation manual was prepared [71].

Each filter installation should be tested in place. Portable dioctylphthalate aerosol generators are available. The NRL Model II generates aerosols with average particle sizes of about 0.8 μ. Detection measurement is done by the NRL E-3 light-scattering meter. The sensitivity of the instrument is 1 part in 10^5. A high-efficiency filter installation should give an efficiency greater than 99.95% [72].

Operation of the aerosol generator and detector depends upon complete mixing of the aerosol with the air and upon good sampling. Since these requirements are not likely to be completely met, several samples from different parts of the gas streams should be checked. This is necessary to locate leaks which may be present. Poor seals may occur around the frames and between the frames and housing.

Planning and design of a filter installation should include consideration of the following:

1. The system should include built-in probe holes for testing and monitoring.
2. Modular or unitized sealing of the individual filters has proven more effective than trying to seal large numbers of filters simultaneously.
3. Sealing should be at least semiautomatic and not depend upon a given tightness of screws or bolts.
4. Use of tape and caulking should be minimized. Gasket seals should be used as a part of the filter framework.
5. Use of excess capacity increases efficiency and filter life and may decrease costs.
6. The filter system should be equipped for testing.

During operation the efficiency of the filters increases. The pressure drops which can be tolerated are perhaps 10 in. of water (19 mm Hg) and the tendency today is to design for pressure drops of 8 in. of water (15 mm Hg) or more. Frequently, fan capacity proves to be the limiting factor.

No economic comparison between efficiency and velocity is available, but enough data are available to permit a reasonable estimate for a given installation. For small installations such as the various off-gas streams, a balance of capital versus operating costs could be made. A 1000 cfm (470 liter/sec) high-efficiency glass-fiber filter will cost about $40 plus installation costs of about $10, including labor [48].

5.2.6 Activated Charcoal Adsorbers

Extended experimental programs have been conducted using activated charcoal to remove iodine and rare gases from air streams. Actually, such complex systems are not required in most power reactor plants because only small amounts of these materials escape from the primary coolant systems. Complex systems are used to purify gaseous coolants and activated charcoal is

---

*Recent technology is summarized by C. A. Burchsted and A. B. Fuller, "Design, Construction and Testing of High-Efficiency Air Filtration Systems for Nuclear Application," USAEC Report ORNL-NSIC-65, Oak Ridge National Laboratory, January 1970.

used to remove tars from the organic-coolant combustion gases at Piqua. Iodine is removed from the chemical plant dissolver off-gases at Hanford and Savannah River. Iodine and the rare gases were also removed from the off-gases of the HRE-2 [73, 74, 75].

The design of the charcoal adsorbers for iodine and the rare gases involves consideration of the charcoal, its absorptive capacity, screen size, temperature, and gas velocity. A test conducted at Oak Ridge at 25°C (77°F) on iodine-containing air indicated that gas velocities in the range of 80 to 275 ft/min (41 to 140 cm/sec) and charcoal (Columbia SXC), sizes of 2 to 4 to 6 to 8 mesh*, had little effect on the efficiency of adsorption. The relative activity was reduced from 5×10^3 to 0.5 (the counting limit) in 4 in. (10 cm) of charcoal. The main loss was due to particulate charcoal that carried adsorbed material to the outlet filter [75].

Retention of iodine at 320 to 325°C (608 to 617°F) varied from 11 to 97% in short tests but with charcoal (Whetlerite manufactured by Pittsburgh BPL) impregnated with salts of silver, copper, and chromium, retention was over 99.99%. For final cleanup of gas streams the charcoal can be held at liquid-nitrogen temperatures, as indicated in Fig. 3-16.

Tests carried out in Great Britain have demonstrated that copper wire and charcoal beds give removal efficiencies for iodine of about 99.9%. Particulate matter is readily removed by absolute filters [76-79]. The iodine reacts with methane and other gases in the air possibly to form methyl iodide, which is filtered or removed with difficulty, and the compound, at least in part, passes through filters, metal sieves, charcoal adsorbers, etc [80]. While the amount of this iodine is very small, it may result in high concentrations, particularly after releases in postulated possible reactor accidents.

Charcoal traps may be ordinary welded steel or stainless steel pipes filled with crushed charcoal. In the HRE-2, because of the heat of decay of the radioactive materials and the necessity for shielding, the pipe consisted of two units of charcoal-filled pipe of 40 ft (12.2 m) of 1/2-in. (1.27-cm) pipe, 40 ft of 1-in. (2.54-cm) pipe, 40 ft of 2-in. (5.08-cm) pipe, and 60 ft (18.3 m) of 6-in. (15.2-cm) pipe in series. Each unit contained about 520 lb (236 kg) of Columbia-G activated charcoal. Most of the short-lived gases were allowed to decay by passage through 160 ft (49 m) of empty 3-in. (7.62-cm) pipe placed ahead of the charcoal-filled pipe. At an oxygen carrier gas flow rate of 250 ml/min (15 in.^{3}/min) per unit the average hold time of Kr^{85} was measured as 30 days. From this it was calculated the holdup time for krypton at higher temperature due to decay heat would be 23 days, and for xenon 700 days.

5.3 Monitoring

There are four general locations for monitors or instruments to measure and/or record the levels of radioactive materials in gaseous-waste-disposal areas. These are monitors for (1) individual pipe lines, storage tanks, or sources, (2) discharge to the stack, (3) discharge from the stack, and (4) the surrounding areas. Individual monitors are utilized to measure particular gaseous isotopes, gross gamma, beta, alpha, and mixed particulate matter. There are many types of detectors available as well as many direct reading and recording instruments. Several general papers on monitoring stacks have been presented [81-84].

Monitoring gases exhausted through a stack normally involves three measurements. First, the gross gamma activity may be measured high enough in the stack to obtain a representative sample or at the top of the stack by means of scintillation detectors on the outside of the stack. These are located to measure the gamma activity in the exhaust plume; detectors may be placed in the stack to determine the mr/hr; or samples may be taken and piped to a detector on the ground. Second, the gross beta-gamma and alpha may be measured by use of filter-charcoal cartridges which are changed daily and counted. Third, the particulate formed by decay of the radioactive gases may be trapped on suitable filters and counted. For most reactor installations only the mr/hr exhausted is measured and recorded continuously and the reactor is shut down when high levels occur. The reason for this is that the presence of alpha or particulate matter is so unlikely that they may be ignored in the stack exhaust. In some cases duplicate readings taken at several hours to several days are used to separate plant discharges from the background. Actually, the normal discharge of radioactive materials to the stack is monitored and the concentration in the stack discharge calculated from the quantities of radioactive materials metered to the stack. This is usually necessary in order to have a reading sufficiently above background to be significant and to reduce the error in measurement to a reasonable level. Nevertheless, the monitors at the top of the stack are necessary for emergency use and for obtaining permanent records.

The withdrawal probe, if used, must be carefully designed. The sampling linear velocity should be the same as that in the stack, and a minimum number of sharp turns should be present between the sampling point and the monitor. All lines should be the minimum length to avoid absorption of the radioactive isotopes on the walls. The gas may be drawn through a charcoal trap in which there is placed an ion chamber or the gas may be filtered and passed over a tape which moves continuously under an end-window GM tube for measurement of gross beta and gamma activity. Both the charcoal and tape are very sensitive and may be instrumented to give warning of higher-than-normal levels of activity.

In some cases additional dilution air may be added at the base of the stack. In this case the dilution fan should be controlled by the monitor on the gaseous radioactive gas line being discharged and the monitor on top of the stack should be used to check high levels.

*Note: Mesh or sieve numbers 2, 4, 6, 8 have openings of 6.20, 4.76, 3.36, 2.38 mm, respectively.

The activity levels in the surrounding areas may be measured by direct detectors or filters. Direct recorders are usually of little value due to the low readings. Filters which absorb the cesium and rubidium formed from decay of the xenon radioisotopes are useful.

6 LIQUID-WASTE SYSTEMS FOR REACTORS

6.1 General Considerations

It is apparent from the discussion of Sec. 3 that while many of the liquid wastes from water-cooled reactors are far too active to be discarded to the environment without any treatment, they are seldom of sufficient activity to be considered really high-level wastes. There is a need, therefore, for methods of waste treatment that have relatively modest decontamination factors and are reasonably priced. Because reactor water generally has a very low solids content, wastes produced therefrom readily lend themselves to treatment with ion-exchange resins. Processes based upon flocculation techniques may also be useful. Evaporation, while it is more expensive than the foregoing and tends to produce decontamination factors in excess of those which may be needed, also finds a place in many reactor waste-disposal systems. Conversion of liquids to solids, except for very special situations, does not have an important place in reactor liquid-waste disposal. Each of these methods will be discussed in this section.

6.2 Ion Exchange

Ion-exchange resins are high-molecular-weight polymers containing particular ionic groupings as an integral part of the structure. Anion exchangers contain amine groups with an equivalent amount of mobile anions such as chloride or hydroxyl ions. Cation resins contain phenolic, sulfonic, carboxylic, or phosphonic acid groups with an equivalent amount of cations such as sodium or hydrogen ions. The mobile ions can be exchanged for other ions of the same sign. This exchange is stoichiometric and generally reversible. The polymeric structure is sufficiently cross-linked to render it virtually insoluble. Commercially available resins are listed in reference [85].

Contact between a solution and an ion exchanger involves either a batch or a column process. In the batch process, the resin is mixed with the solution and agitated continuously or intermittently until equilibrium is reached. After equilibration the resin is removed by filtration. Column processes generally involve a fixed-bed operation. The resin is supported on a porous base while the solution flows through the resin. Moving-bed techniques have also been developed.

If a solution containing exchangeable ions is fed into the top of a fixed bed, it will exchange essentially all of these ions with the mobile ion of the same sign in the exchange resin in a comparatively narrow zone at the top of the bed. The solution will then pass through the rest of the column carrying the mobile ion of the exchanger, and there will be essentially no further change in composition as it does so. As the feed solution is continued, the upper layers of the bed will be completely converted to a form containing the ion in the feed solution, i.e., "exhausted," and no further exchange will take place. The zone in which exchange is taking place will be displaced farther down the column. Eventually this exchange zone will reach the bottom of the column and the exchanging ions in the feed stream will show up in quantity in the effluent. This is known as "breakthrough." The operation is usually discontinued before breakthrough and the column regenerated by reversing the exchange process, i.e. by treating the exhausted resin with a concentrated solution containing the original mobile ion that was incorporated into the resin [86].

It is possible to obtain nearly complete deionization by passing a solution through both an anion and a cation bed. Either bed may be used first. Or the resins may be mixed together and used as a single bed or mixed bed. Since the capacity of anion resins is only about half that of cation resins, it is customary to employ about two parts of anion resin to one of cation. To regenerate a mixed bed it is first necessary to classify the bed by backwashing with water, then to introduce acid into the cation portion of the bed and caustic into the anion portion, and to follow this with a water rinse. Finally the bed is remixed by blowing air through it.

Resin beds should not be less than 30 in. (76.2 cm) deep. Manufacturers recommend flow rates of about 2 gpm/ft^3 (4.5 liter/sec per m^3 of resin). It is wise to put a filter ahead of the resin bed to remove suspended solids and one following it to keep fine resin from being carried over into the system. Regenerant solutions usually employed are 6 N HCl and 4% NaOH.

A number of variables affect the removal of radioactivity by ion exchange resins. These include crosslinkage, resin-particle size, feed flow rate, pH of the feed, extraneous salts in the feed, colloidal material in the feed, resin-bed composition, resin-bed depth, and temperature.

The degree of cross linking in an ion-exchange resin determines its porosity, stability, and solubility. A change in the porosity of an ion-exchange resin affects its rate of ionic diffusion and therefore its rate of exchange [79].

Comparative tests using ordinary commercial-size* resins (20 to 60 mseh) and fines (40 to 80 mesh) indicate that the total ionic capacity is about the same, but that effluents with somewhat lower activity may be obtained with the smaller size. However, the use of the smaller particle size causes a substantial increase in pressure drop through the bed for a given flow rate. Since equivalent decontamination may be obtained with the larger particle size by reducing the flow rate or increasing the depth of the resin bed, the larger size is often selected to obtain maximum ease of operation.

---

*20 to 60 mesh corresponds to sieve openings 0.84 to 0.25 mm (0.0331 to 0.0098 in.); 40 to 80 mesh corresponds to 0.42 to 0.177 mm (0.0166 to 0.0070 in.).

It has also been shown that for the removal of gross fission-product activity, there is no appreciable difference in the decontamination obtained at 2 gpm/ft^3 resin (4.5 liter/sec per m^3 resin), the manufacturer's recommended flow rate, and at a rate five times as large [86].

The volume of feed which may be decontaminated per cycle by the resin bed is determined by the total ionic content of the feed. When the ionic capacity of the resin has been spent in the adsorption of ions, either active or inactive, good decontamination of the feed is no longer possible, although in some cases selective adsorption of the radioactive ions may continue. The concentration of total ions in the feed is therefore a very important factor in determining the volume of waste that may be treated per unit volume of resin per cycle, which in turn is vital in determining the cost of the operation.

If un-ionized materials such as organic colloids or iron or nickel compounds are present in the feed, the apparent capacity of the resin bed may be reduced substantially as active material may be carried through the bed by the colloid that is not absorbed. The unexpectedly low decontamination factors obtained during the cleanup of the Waltz Mill reactor (see Sec. 7.5) were explained on the basis of colloidal material.

Most of the applications of ion-exchange resins to the treatment of reactor wastes involve activity levels at which the effect of radiation upon the resin is negligible. For those situations in which higher levels of radioactivity are involved, the ability of the resins to withstand radiation damage is of some interest. Parker et al. have summarized the results of studies by several investigators on the effects of beta (absorbed), gamma, and X radiation on commercially available organic ion-exchange resins [87-90]. The data are shown in Table 6-1. In this tabulation the adsorption of 1 watt-hr of energy by 1 g of dry resin is equivalent to 3.8×10^8 r.

TABLE 6-1

Radiation Damage to Ion-Exchange Resins

Type of resin	% Capacity loss [a]	Total capacity	
		Milli-equiv./g	Milli-equiv./ml
Strong Acid Nuclear Sulfonic Polystyrene			
Dowex 50 (X-8 and X-12)	23, 10-20[b]	4.25	2.20
Nalcite HCR (X-8)	8	4.25	2.20
Amberlite IR-120	9, ~12[c]	4.20	2.15
Permutit Q	2		
Dowex 30	1[b]	4.00	1.35
Amberlite IR-105	1[b]	2.70	1.00
Amberlite IR-112	12[c]		
Weak Acid-Carboxylic			
Amberlite IRC-50	100	10.0	4.20
Permutit H70	100		
Strong Base-Quaternary Amine Polystyrene			
Dowex I	44	2.4	1.0
Nalcite SAR	37	2.3	0.9
Permutit S2	38		
Amberlite IRA-400	42	2.3	1.00
Amberlite IRA-410	40	2.5	1.0
Weak Base-Weakly Basic Amine Groups			
Nalcite WBR	20		
Amberlite XE-58	20		
Dowex 3	19		
Amberlite IR-4B	13	10.0	2.5
Amberlite IR-45	53	6.0	2.0
Permutit Deacidite	3	9.3	1.5

[a] Units are percent of capacity loss per watt-hour ($3.8 \cdot 10^8$r) of energy absorbed per gram of oven-dry resin. All data on capacity loss are from Wedemeyer [90] unless otherwise noted.
[b] Data from Higgins [88].
[c] Data from Fisher [89].

Examples of systems using ion exchange to keep reactor water at a low solids content and with a minimum of circulating radioactivity were cited in Sec. 3. Both full-flow and bypass systems were illustrated. When resins are used in power reactor systems, it is necessary to cool the water before putting it through the resins, since they will not stand the operating temperatures of reactor water. The water to be processed may either be cooled and then demineralized at pressure or it may be cooled and demineralized at a low pressure and then pumped back up to operating pressure.

Two techniques quite useful in operating an ion-exchange system are conductivity measurement and gamma scanning. The effluent from an ion-exchange unit can be put through a continuously recording conductivity cell; the cell readings can usually be counted on to signal approaching breakthrough. A simple gamma scan is also useful in determining the degree of resin-bed loading.

In some cases it may be more economical to dispose of the exhausted resin than to regenerate and reuse it. For such disposal it may be practical to mix the exhausted resin with Portland Cement and water and allow it to solidify. This method appears more attractive than incineration of the resin, which would require a specially built furnace and means for furnace-gas decontamination.

Whether or not regeneration is used should be strictly a matter of economics. For many reactor installations the cost of regeneration facilities and those necessary to handle the regenerant are simply not justifiable. At the Dresden plant, however, regeneration facilities were added at a later date since it was found that under the special conditions existing there, it was more economical to regenerate a portion of the resins than to discard them after a single use [91].

6.3 Evaporation

Even though evaporation often may be expected to give a decontamination factor higher than that needed in many reactor applications, many reactor systems do include an evaporator to provide an added safety factor or to handle special higher level wastes such as the regenerant solution from ion exchangers. A wide variety of evaporators have been tested for their usefulness in handling radioactive wastes. Some of the types used are forced-circulation evaporator with external horizontal heating surface, natural-circulation evaporator with external vertical heating surface, coil evaporator, vapor-compression evaporator, and double effect evaporator.

The results of pilot tests on the above types of evaporators carried on over a decade ago at a number of AEC sites are shown in Table 6-2 [92]. Essentially all types of evaporators which have been tried work reasonably well for concentrating radioactivity. The choice is usually made on the basis of special local requirements.

To minimize entrainment of radioactive contaminants in the condensate, various methods and pieces of equipment have been employed: centrifugal entrainment separators, internal and external types, filtration through a bed of fiberglass, and reflux.

The exact mechanism of entrainment formation is not known, but presumably it is a function of mechanical action of the surface of the boiling liquid and of bubble breakage. Particles formed by either of these mechanisms are carried by the vapor stream if their rate of fall (from Stokes' law) is less than the vapor velocity. Particulate formation, then, primarily should be affected by rate of boiling.

The resistance to bubble breakage—hence the particle-size distribution—is probably a function of the surface tension, viscosity, and salt concentration of the bubble film. Whatever mechanism projects liquid droplets into the vapor space may also project any suspended solids present. Although suspended solids (such as rust) normally are not considered noxious contaminants, in this case suspended solids are likely to have adsorbed radioactivity and thus are capable of imparting radioactive contamination to the condensate by desorption.

Initially the liquid droplets thrown into the vapor space probably have a normal size distribution. The distribution may be considerably altered by conditions in the vapor space, e.g., drops may evaporate until the droplet surface-tension forces are equalized by the vapor pressure. A difference in the particle-size distribution of the entrained droplets from boiling pure water and from boiling concentrated salt solution would also be expected.

Entrainment may be materially reduced by providing sufficient cross section and height above the boiling liquid surface that the larger droplets may fall back into the boiling liquid. In addition, some type of de-entrainer is usually needed if large decontamination factors are desired.

Many conventional devices have been used to reduce entrainment and to increase the overall decontamination factors. These include cyclones, packed columns, sprays, bubble-cap columns, impingement plates, baffles and settling domes. The decontamination contribution of the de-entrainment devices used varies from 10 to 100 [93].

In addition to de-entrainment, foaming during evaporation must also be controlled. While a small amount of foaming may have no adverse effect on evaporator performance, entrainment of foam may result in poor decontamination of distillates. Antifoam agents, baffles, pH control, and lowest possible operating liquid level are methods that have been used to control foaming.

6.4 Flocculation

Variations of water treatment have been studied exhaustively in an attempt to develop a waste-treatment process of general utility. While floc-

TABLE 6-2

Comparison of Various Pilot Evaporators

	Knolls	Oak Ridge	Brookhaven	Argonne	Mound
Capacity of equipment	400 gal/hr (0.42 l/sec)	285 gal/hr (0.30 l/sec)	600 gal/hr (0.63 l/sec)	150 gal/hr (0.158 l/sec)	100 gal/hr (0.105 l/sec)
Type of equipment	forced-feed flash	pot-type removable heating coils	vapor compression	vertical tube circulation	1st effect-vertical tube forced-feed
De-entrainment device	baffled separating column	cyclone-type separator	vapor dome, fiberglas bed	centrifugal separator, centrifix scrubber, (reflux if necessary)	2nd effect-vertical tube natural circulation
Typical feed (average activity)	$3 \cdot 10^{-2}$ $\mu c/cm^3$	(β) $2 \cdot 10^{-2}$ $\mu c/cm^3$	$1 \cdot 10^{-2}$ $\mu c/cm^3$	($\alpha + \beta$) 10^{-3} $\mu c/ml$	(β) 10^{-4} $\mu c/cm^3$
Typical feed (average solids)	0.3%	8.0%	0.5%	0.2%	3.0%
Overall decontamination factor	10^4-10^5	10^{3+}	10^6-10^7	10^4-10^5	$\sim 10^8$
Volume reduction factor	400	15	110	100	32
Slurry (average solids)	70%	70%	65%	20%	60%
Steam efficiency	85%	70%	100%	85%	92.5%

culation has not been used in reactor installations, it appears that such a system may have value in treating large volumes of slightly contaminated wastes, but as a method for handling high-level wastes, flocculation is not satisfactory. With the use of a wide range of flocculating agents, over-all decontamination factors of about 10 are obtained for mixed fission products. If it is possible to design the process for a specific single radioactive species, much better results can be obtained. Advantages of flocculation are rather low cost, the ability to handle a wide range of solid content in the feed, and the production of a waste floc volume which is relatively independent of feed solid content. Suitable storage or disposal facilities have to be provided for the resulting radioactive sludges.

A number of water-treatment processes were investigated at Oak Ridge to determine their removal efficiencies for Sr^{90}, Cs^{137}, and the rare earths from tap water. The characteristic efficiencies of five treatment processes are shown in Table 6-3 [94].

Sand filtration and chemical coagulation with aluminum and iron salts were unsatisfactory for the removal of strontium and cesium. Lime-soda softening and phosphate coagulation were found to be capable of removing more than 90% of the strontium.

Laboratory studies were extended to include actual process wastes at ORNL. Up to 90% removal of gross radioactivity could be obtained with excess lime-soda softening or phosphate coagulation when clay was added for the removal of cesium. Although phosphate coagulation was promising, efficient removals of strontium required accurate control of pH and of the ratio of phosphate and lime dosages.

Some earlier Oak Ridge work reported at Geneva in 1955 gave similar results [95]. It was shown that coagulation was most effective for the removal of radioactive ions of valence +3, +4, and +5. Data for 18 specific isotopes are given in Table 6-4 [95].

A really vast amount of work has been done on this type of processing. A number of representative references are given; but there are many more [96-102]. All of these agree generally with the conclusions stated herein.

7. NUCLEAR REACTOR INCIDENTS*

A few nuclear reactor incidents involving the release of fission products are reviewed in this section. The purpose is to indicate what design and operating changes might have been made that would have ameliorated the severity of the incidents—the causes of the accidents are not stressed. In these incidents significant quantities of radioactivity were released and spread through the reactor core or throughout a room, building, site, and even to the surrounding areas. The emergency measures adopted for cleanup of the radioactivity in these incidents affected the magnitude of the incidents. Failure to have prepared emergency plans prior to the incidents and to institute a rational plan of decontamination when an incident occurs can lead to unnecessary hazards and excessive costs.

7.1 Oak Ridge Graphite Reactor

In 1948 an incident occurred with the Oak Ridge Graphite Reactor which permitted the spread of potentially hazardous amounts of radioactive particulate matter over an area of several acres (2.5 acres = 1 hectare) near the reactor site. The reactor is graphite-moderated and once-through air-cooled. At that time it was fueled with unbonded natural uranium rods in aluminum cans. The fuel channels consist of 2-in. (5.08-cm)-square holes in the graphite oriented with the diagonals vertical and horizontal. The 1-in. (2.54-cm)-diam, 4-in. (10.16-cm)-long fuel elements were simply laid in these channels and cooled by air which entered from a plenum on the front face inside the shielding and left from the rear into a second plenum. In 1948 the air was passed through a roughing inlet filter but it was not filtered after passage through the reactor. The fuel was unloaded by shoving it manually with new fuel elements or wooden rods to the rear plenum where it dropped into a canal.

During routine operation a fuel-element failure occurred. By the time the defective element had been located and an attempt made to discharge it, it had swollen to the point where it could not be moved. Attempts to force the fuel element through by use of more pressure only aggravated the situation and resulted in additional fuel-element failures.

These fuel-element failures resulted in release of perhaps a pound of uranium oxide which was spread through the channel, rear plenum, concrete exhaust tunnel, fans, stack, and surrounding area. The beta and gamma activity, even near the stack, was low, but it was feared that particles of the oxide might be harmful.

Subsequent to the accident, the reactor was shut down and the area surveyed and partially decontaminated by simply planting grass in the clear areas to reduce continued dispersion due to the surface winds. A reactor filter installation was constructed prior to further reactor operation to eliminate further particulate discharge.

This incident indicated that measures for identification of failed fuel elements must be in continuous use, failed fuel elements must be removed with care that others are not damaged, and there is need for high-efficiency filters on all air exhaust streams that may contain radioactive particulate matter.

The cost of the incident involved about two months downtime for the reactor and the cost of special surveys and monitoring, plus a moderate expense for seeding of some surrounding areas. The installation of the filter building cost about $340,000. Perhaps a part of this should be charged to the emergency since the building was constructed under emergency conditions in about 30 days [103].

7.2 Windscale

The two Windscale Reactors were graphite-moderated, natural-uranium-fueled, and once-

*See also Chapter 11.

TABLE 6-3

Removal of Radionuclides from Water by Conventional Water-Treatment Processes

Radioisotope	Process waste stream composition (percent of gross beta 1954-1956)	Chemical[a] coagulation	Percent removal by treatment process			
			Chemical coagulation plus 100 ppm clay	Sand filtration	Lime-soda softening (150 ppm excess)	Phosphate coagulation (240 ppm dose)
Sr	19.6	3	0-51	4	97.3	97.8
Ce	15.2	91	85-96	-	-	99.9
Trivalent Rare Earths (including Y)	30.4	91	-	87	90.0	-
Cs	29.9	0.5	35-65	50	Not effective	-
Ru	1.9	77	-	-	-	-
	97.0					

[a] Coagulant includes alum, ferrous sulfate or ferric chloride, lime, soda ash or sodium silicate.

TABLE 6-4

Removal of Radioactive Materials by Conventional Water-Treatment Processes[a]

Isotope	Removal range in percent of initial activity		
	Chemical coagulation and settling	Sand filtration	Soda-ash softening
Cs^{137}-Ba^{137} (Cl)	0-37	10-70	<50
Sr^{89} (Cl)	0-15	1-13	50-95
Ba^{140}-La^{140} (Cl)	1-84	39-99	50-95
Cd^{115} (NO_3)		60-99	50-99
Sc^{46} (Cl)	62-99+	94-99	50-99
Y^{91} (Cl)	1-99+	84-99	50-95
Zr^{95}-Nb^{95} (oxalate complex)	2-99	91-96	50-99+
P^{32} (phosphate)	68-99+	91-96	50-99+
Cr^{51} (Cl)	0-60		
Mo^{90} (MoO_3)	0-60		
W^{185} (tungstate)	1-96	3-18	<50
Re^{186} (metal)	0-29		
I^{131} (iodide)	0-96		
Ru^{103} (Cl)	43-96		
Pr^{142} (Pr_2O_3)	83-99+		
Ce^{144}-Pr^{144} (Cl)	28-99+		
Pm^{147} (Cl)	4-99+		
Sm^{153} (Sm_2O_3)	44-99+		

[a] Variable chemical dose, coagulants pH conditions, activity concentrations, and waters.

through air-cooled. The design was similar in its general features to that of the Oak Ridge Graphite Reactor. The air was drawn through inlet roughing filters, blown through the core, and exhausted through a filter bank located near the top of the 410-ft (125-m) stack [104-105]. The exhaust filters were designed to retain the larger particles, which could severely contaminate the area near the reactors, but they were not intended to have a high efficiency for particles of all sizes.

In October 1957, an accident occurred during a planned release of stored (Wigner) energy from the graphite in one of the reactors. The reactor had been shut down and to heat the graphite it was brought to low power with reduced air flow. Due to errors in temperature readings, one reactor apparently was overheated. The first warning of trouble was given by an air-sampling instrument in operation in the open about 0.5 mile (~1 km) from the stack. The sample collected in 3.0 hr at midday indicated an air activation of 3000 beta disintegrations/min-m^3 or about 10 times background. Visual inspection through a plug hole on the charge face of the reactor revealed glowing fuel elements. Later it was found that there were about 150 channels containing uranium fuel elements glowing at red heat.

An attempt to discharge the thermally hot fuel elements was unsuccessful due to swelling. However, adjacent channels were discharged to create a fire break and to limit the affected zone. An attempt was then made to cool the affected zone by blowing carbon dioxide into the reactor core. This was ineffective. Some 10 hr after the incident was noted, it was decided to pump water into the reactor, and about 9 hr later (19 hr after the hot fuel elements were noted), water was injected. This was continued for 24 hr, by which time the reactor was cold.

The maximum radioactivity measured on the site varied from 2.3×10^{-8} to 4.5×10^{-7} μc/ml of beta with an average of about 2.0×10^{-8} μc/ml of beta. The highest level measured off the site was 4 mr/hr in the plume about 1 mile (~1.6 km) downwind. Due to wind variations there was considerable spread of the activity over very large areas.

The radioactive materials released were estimated from samples obtained on the filters and surrounding district. The estimated magnitude of the release was I^{131}-20,000 curies, Cs^{137}-600 curies, Sr^{89}-80 curies, and Sr^{90}-9 curies. The released I^{131} resulted in deposits on the pastures sufficient to make it necessary to condemn the milk in some areas until the I^{131} had decayed. The reactor has not been disassembled but simply closed up.

7.3 NRX

The Canadian Research Reactor, NRX, is located at Chalk River, Ontario. It is a 40 Mw(t) heavy-water-moderated, light-water-cooled, natural-uranium reactor with aluminum fuel cladding, coolant tubes, and calandria. The uranium fuel rods are 1 in. (2.54 cm) diam and 10 ft (3.05 m) long. There is an air gap between the coolant and calandria tubes to provide heat insulation, and these gaps were open at the bottom [106]. (This section is abstracted from Gilbert's excellent paper describing the incident and the decontamination of the plant.)

On December 12, 1952, after part of the core had been reloaded and during low-power experiments in which water flow had been shut off in some tubes, a power surge occurred. As a result of the power surge, portions of several uranium rods were melted along with portions of the cladding of about 10% of the fuel rods. The molten uranium and aluminum melted or reacted with the coolant tubes and portions of the calandria tubes. This permitted the light and heavy water to flow down the air space and out at the bottom of the tubes into the room in the basement under the reactor. The coolant water could not be shut off as it was necessary to provide shutdown cooling for the reactor and the failed rods had to be kept cool to prevent burning of the uranium. Before the situation was brought under control about 1.25×10^6 gal (4700 m^3) of water containing some 10,000 curies of radioactive isotopes were collected. After draining the water from the basement, the residual contamination of the concrete surfaces measured 10 r/hr on the average. Ceilings, walls, and floors throughout the reactor room measured about 50 mr/hr with many local spots much higher.

Immediately after the incident the area was evacuated due to the high radioactivity level; personnel were permitted to return within three days. The first problem in the decontamination of the plant was the disposal of the 1.25×10^6 gal of contaminated water. It was run into a process reservoir for temporary storage and then a pipeline was laid to a suitable waste-disposal area 1.25 miles (~2 km) from the reactor site. Laboratory tests, which were run quickly, indicated that the soil in the proposed waste-disposal area had a high ion-exchange capacity for the activity, particularly the strontium; so the water was pumped into the area and allowed to percolate into trenches in the ground. Eventually the water would reach a small lake and finally the Ottawa River. Subsequent checking of the lake and river has not shown significant leakage of the strontium into the river.

Since the pipeline was laid in winter and the ground was frozen as well as rocky, the pipe was placed above ground and surrounded by a continuous box or cover about 2 ft (60 cm) square. This was filled with sawdust to permit use of the pipe at -34°C (-30°F).

Next it was necessary to reduce the activity to levels where it would be possible to work in the building and decontaminate and disassemble the reactor. The areas were first flushed with large volumes of water which were pumped to the disposal area. This was followed by high-velocity streams of hot water with and without detergents.

Then all possible equipment was removed from the reactor building, wrapped in plastic covers, and sent to either the central decontamination area or the solid-waste disposal area. Then the areas were washed a third time. Finally, all areas were surveyed and hot spots either individually cleaned or shielded. The activity was thus reduced to the point where only the spots of highest activity read as much as 10 r/hr.

Because the high-activity levels permitted only brief personnel exposure, large numbers of personnel were used. These were loaned by the Canadian Army, the United States Navy, and United States AEC. The maximum exposure of any person was 17 r. During the early stages of cleanup some difficulty was experienced with ingestion of activity by personnel. This was traced to faulty use of the respirators by the personnel. A program of rigid inspection of the equipment and instruction of personnel was instituted; these steps corrected the difficulties.

Respirators were used at levels of airborne activity up to seven times tolerance. Above this masks and hoods having clean air supply were used. It was not found practical to use self-contained air-supply suits because of the difficulty in decontaminating the equipment and because such large numbers were needed.

A lead-shielded pinhole camera supplied by Knolls Atomic Power Laboratory proved very useful in the location of "hot spots." Two pictures were taken of various areas, one with visible light and one with a light shield. When the two negatives were superimposed, the high-activity spots were easily located.

Cleaning required a variety of techniques which had to be developed on the spot. Initial cleaning of the floors was done with the floors flooded with 2 ft (60 cm) of water to serve as shielding. Chipping removed the hottest spots. Later the concrete was decontaminated by flame priming, chipping, sand blasting, and grinding. Essentially all cracks in the floors had to be chipped out. All these operations were conducted using vacuum systems to remove the dust. An incident resulted in failure of one of the vacuum systems and a temporary but local air contamination of 1500 times tolerance occurred. When the activity of the concrete floors and walls had been reduced to 20 mr/hr, the remaining activity was often sealed in by pouring a fresh layer of concrete up to 6 in. (15 cm) thick.

Mild steel was cleaned with brushes and rust was removed with inhibited hydrochloric acid. Stainless steel was best cleaned by scrubbing with cotton wipers wet with detergents or acids. Lead was most easily decontaminated by removing, melting, and recasting, as most of the activity remained in the dross.

When the reactor room was clean enough to permit reasonable work periods, dismantling of the reactor was begun. Parts were removed, wrapped, and moved into a shielded trailer for transfer to burial or to a temporary ventilated structure built inside the reactor room wherein decontamination of large pieces was done. The most difficult job in manpower and time was the removal and disposal of the damaged calandria. After removal from the reactor in a plastic bag, the calandria was transferred to a truck and hauled to burial.

Once the calandria was removed, the lower shield plates were removed and the internal areas decontaminated. The reactor was then easily rebuilt. In putting the reactor back together the following measures were taken in order to facilitate cleanup should another such incident occur:

a. A permanent method of disposing of active water by pumping to the disposal area was installed.
b. All equipment possible was removed from the basement.
c. All surfaces which might become contaminated were sealed and painted. This was true for lead and stainless steel.
d. All cracks and crevices were filled and sealed.

These steps would be relatively simple, particularly if done during the original design. Not only do they offer protection in case of an accident, but they make maintenance easier during normal working conditions. Painting and sealing the concrete is particularly important; in the incident the activated water penetrated the concrete walls and floors to considerable depth.

The decontamination and rebuilding of the NRX Reactor cost about $2,500,000—roughly one-fourth the cost of a new reactor—and required one-fourth the time the replacement of the reactor would have required. The cleanup and repair of this reactor was a notable achievement. It showed that an almost catastrophic incident can occur to a reactor and yet the results can be handled without serious personnel exposure, albeit the cost was high. It was also notable as it was the first incident and undertaking of its type. The reactor operators handled the incident intelligently. The lessons were applied to the design of the NRU to an advantage.

7.4 NRU

On May 23, 1958, Chalk River had a second major reactor incident. (This section is abstracted from the discussion of the incident and the cleanup by Hughes and Greenwood [107]).

NRU is a 200 Mw(t) heavy-water-moderated and heavy-water-cooled research reactor which is fueled with natural uranium with aluminum cladding. The fuel is in the form of five vertical plates 0.171 and 0.177 in. (4.35 mm and 4.50 mm) thick, from 1.224 to 2.144 in. (3.11 to 5.45 cm) wide, and 10 ft (305 cm) long, containing 120 lb (54.5 kg) of uranium. The reactor is designed to be refueled during full operation and since the maximum thermal flux is 2.5×10^{14} n/cm^2-sec, the fuel is very active on removal, the very short-lived fission products being present in equilibrium quantities. Removal and transfer of the fuel assemblies is accomplished by means of a shielded transfer flask which has provision for cooling the assembly during transfer.

The incident occurred when a damaged fuel assembly was being unloaded. Difficulty was experienced in bringing the fuel assembly up into the flask. Cooling water was lost and the assembly stuck part way up into the flask. An attempt was made to transfer the damaged rod to the storage basin as quickly as possible; but the rod apparently melted, caught fire, and a piece about 3 ft (0.9 m) long dropped off and fell into a maintenance pit during the transfer. It set fire to paper and other materials in the maintenance pit and apparently continued to oxidize until it was quenched by subsequent action. The transfer cask with the remainder of the fuel assembly was successfully placed in the storage block.

Radiation fields around the maintenance pit were over 1000 r/hr. Notwithstanding, a team of men wearing respirators was able to extinguish the fire in about 15 min by covering the burning uranium with wet sand. Radiation exposures of 5 r were accepted during this operation.

The building was severely contaminated and considerable contamination resulted from the fire. It is the opinion of the Chalk River staff that little damage would have taken place had the damaged assembly been left in the reactor until steps could have been taken to guarantee cooling during transfer.

Once the fire was out, the immediate emergency was over. The most immediate requirement then was to remove the burned rod segment from the maintenance pit. A crew of 35 NRU staff men working 1 minute at a time, using rakes and hoes with 24 ft (7.3 m) handles, were able to get the segment into a wooden tray. This was then moved with considerable difficulty into a jury-rigged shielded trailer and taken to the disposal area. It was apparent that additional manpower would be required if individual exposure limits were to be maintained. The next day 40 Chalk River personnel who normally do not receive radiation in their work removed the sand and other debris from the pit. This was done by pairs of men working 1.5 min each and using long-handled shovels to put the sand into garbage cans. The cans, some of which read 200 r/hr on contact, had to be carried to an elevator. Fourteen of these men received more than the 5 r limit; one received 19 r.

At this point work was suspended until a complete survey of the extent of the contamination and activity levels could be made. Radiation levels of 1000 r/hr were found at the reactor deck and above the maintenance pit. Readings of 100 r/hr were common. Filter-paper smears taken on surfaces throughout the building gave the results shown in Table 7-1 [107]. Airborne activity at the time of the fire was too high to be measured. Even 12 days

TABLE 7-1

Contamination of NRU Reactor Building Surfaces

Location	Vertical surfaces (mr/hr)	Horizontal surfaces (max) (mr/hr)
Reactor room		
Walls		
North, south, and east	6-60[a]	450
West	20-400	2,500
Floor, spots		100,000[b]
Crane, and crane tracks		1,000
Offices, corridors, and change rooms	1-2	30

[a] A reading of 1 mr/hr corresponds to a GM counter reading of about 10,000 counts/min of mixed fission products.
[b] Measured with a high-range ion-chamber instrument sensitive to beta and gamma radiation.

after the incident air contamination in the reactor hall was about 4.5×10^{-8} μc/ml. Some contamination was found outside the building up to 5000 ft (1.5 km) downwind. But more contamination was spread on the roads used for transporting the debris to the burial ground than escaped directly during the incident. The roads were cleaned as quickly as possible, using vacuuming, washing with hoses, and, when necessary, removing some of the contaminated surfaces.

A plan of attack for cleanup was organized. Emergency headquarters and change rooms were set up in a nearby building. Closed-circuit television was set up between the reactor room and the headquarters. This was used to instruct workers before they went into the contaminated area and allowed the NRU personnel, who had used up their radiation allowance, to supervise the work.

Each man going into the area was completely clothed in a disposable plastic suit. Rubber gloves, respirator, and overshoes were sealed to the suit with tape. Film badges and dosimeters were worn. These suits cannot be ventilated and are uncomfortable. They can be worn only about 2 hr even when additional working time is permitted. In the early stages of the cleanup it was not unusual to find 200 mr/hr contamination on the outside of the suits. Consequently, it was necessary to remove the suits very carefully to avoid body contamination. The suits were discarded after a single use.

It was first necessary to remove the remainder of the sand and other contamination from the top of the reactor and from the maintenance pit. This was done largely by vacuuming. Methods had to be worked out to shield the vacuum filters and to change them semi-remotely. Considerable trouble was experienced by plugging of the vacuum system with assorted debris. In about a week radiation levels were reduced generally to below 1 r/hr. Nearly 300 men took part in this phase of the work with a maximum exposure of 3 r.

For the next phase of the work the armed forces provided 70 men per day to work three 8-hr shifts for the next 6 weeks. These men succeeded in reducing the radiation level to the point where a commercial firm of building cleaners and steeple jacks could be brought in to do the final cleaning of surfaces, some of which were 90 ft (~27 m) above the floor.

The surfaces were cleaned by vacuuming, wet mopping, and wiping with damp rags. Truly remarkable quantities of material were required. For example, during the 6-week period 15,000 mopheads were used and sent to the waste-disposal area. Chalk River has a central decontamination center which was put on a three-shift basis during the emergency. About 25,000 toe rubbers and 10,000 respirators were decontaminated during the month of July.

The decontamination of NRU was made simpler because of lessons learned in the NRX accident. All of the surfaces, horizontal and vertical, in the NRU building had been provided with smooth finishes which were easy to decontaminate. Lessons learned from the NRU incident include a decision to have available a portable change room that can be brought up to service any building. While the use of wet sand to smother the fire was a good improvisation, powdered graphite would have been better and this material is now stocked in quantity in every building in which there is a chance of this type of accident. Chalk River is also studying mechanized cleaning methods, remote-controlled monitors, and portable shielding walls.

7.5 Waltz Mill

On April 3, 1960, a fuel-element failure occurred in the Westinghouse Test Reactor at Waltz Mill, Pennsylvania. The WTR was a light-water-cooled and light-water-moderated tank-type test reactor fueled with aluminum-enriched uranium fuel assemblies. The fuel was in the form of 60 concentric cylindrical tubular assemblies. Each assembly contained about 200 g of U^{235}. Prior to the incident the reactor was operated at 20 Mw(t) but escalation tests were being conducted at the time of the incident to permit operation of 60 Mw(t) [108].

During the escalation tests the reactor was in operation at 38 Mw(t). Trouble developed in maintaining power levels and the activity began to rise; so the reactor was scrammed and the building evacuated. The coolant flow was left at full flow for a time but was soon placed on shutdown flow, and the reactor containment building was placed on recirculate, since the stack gas monitors were indicating high levels of activity.

It was later determined that one fuel element had partially melted. It is thought, although this is by no means certain, that the meltdown was caused by a defect in the fuel element rather than by malfunction of the reactor. It was also eventually established that about 5000 curies of total activity were released to the primary system, including 800 curies of radioactive gases which were primarily Xe and Kr. The initial burst of gaseous activity which was released to the atmosphere was about 250 curies. The total inventory of contaminated water initially consisted of that in the primary coolant system—about 180,000 gal (680 m^3). Initial activity levels in this water were about 5 μc/ml, of which about half appeared to be due to dissolved Xe^{133}.

Activity levels measured at various points 1 hr and 12 hr after the incident are given in Table 7-2 [108]. Immediately following the incident steps were taken to measure activities in the surrounding countryside. At distances of from 1 to 3 miles (1.6 to 5 km) no detectable activity was found. Readings at the main road, 750 ft (0.23 km) from the reactor, gave the following results: 20 mr/hr immediately after incident; 6 mr/hr, 1.5 hr after incident; 5 mr/hr, 4 hr after incident; and 3.5 mr/hr, 5 hr after incident. An environmental survey of the surrounding countryside showed that the incident had not contributed any measurable amount to the natural radioactive background of the area.

Initial cleanup efforts were directed at reducing the activity level in the primary coolant water using the primary coolant ion-exchange system and about 100 homemade auxiliary demineralizers. In addition the water was put through the bubble-cap tower of the waste-disposal-system evaporator in order to remove the xenon and krypton by degassing. After treatment the water was discharged to the main retention basin (capacity 240,000 gal or 910 m^3) at an activity level of approximately 10^{-2} to 10^{-3} $\mu c/ml$. Throughout the course of the cleanup, mixed-bed ion exchange gave a reported decontamination factor of only about 10^3. This is somewhat lower than might be expected for this operation and was accounted for by the WTR staff by the existence of colloids in the water.

After a week it was possible to raise the head of the reactor and wash it down sufficiently to permit its removal. The washing and scrubbing was done using automobile-washing brushes hooked up for continuous scrubbing. The general radiation background in the vicinity of the reactor head at that time was time about 1 r/hr. Installation of a 3-in. (7.6-cm) iron shield near the top of the reactor permitted the removal of all of the fuel elements except one. The one which had melted and caused the incident had to be drilled out. This operation took 5 days and, while it succeeded in getting the damaged element out, it also increased the activity of the water and spread radioactive chips and pieces throughout the system, which then had to be removed laboriously by a succession of flushing and vacuuming operations.

TABLE 7-2

Radiation Readings Following Incident at Waltz Mill

Location	Time after incident (hr)	Radiation level (mr/hr)
Health physics office	1	2
Reactor service building	1	5
Back of hot cells	1	200
Reactor control room	1	40
Main gate exclusion area	1	200
Pump room	1	10,000
Pump room (door)	12	200
Head tank	1	4,000
Head tank downcomer	1	40,000
Near foot of head tank	12	750
Reactor top	1	1,000
Reactor top	12	35-4,000
Pumps	12	up to 5,000

Building decontamination started at the top of the reactor. Tools and equipment were removed to an active storage area and waste paper and plastic were boxed for shipment. Solid active waste totaled about 2200 cu ft (~ 60 m^3) during the first 2 months after the incident. Surfaces were vacuumed and scrubbed with versene, cleanser, and absorbent pads which were discarded after a single use. When loose contamination (as indicated by smears) was reduced to less than 100 counts/min, the surfaces were repainted. Concrete surfaces were vacuumed and scrubbed in a similar manner but they were also scribbed with citric and muriatic acids. When the activities of these surfaces were reduced to 100 counts/min, they were covered with vinyl tile. Some asphalt roadway areas which became contaminated during the cleanup were vacuumed and scrubbed in a like manner. This did not always reduce the levels sufficiently and some roadway was dug up and drummed for disposal.

Decontamination of tools was accomplished by three methods: (1) hand scrubbing with a mixture of 20% Dowfax, 5% Stephan (LDA), 10% versene 100, 3% glycerine, 2% phosphoric acid; (2) wiping stainless-steel surfaces with absorbent pads wet with citric or nitric acids; (3) use of ultrasonic decontamination with a mixture of versene-Dowfax or with citric acid. The report of this incident states, "The problem of decontamination of tools was magnified perhaps tenfold by the surplus of tools and equipment in areas which became contaminated during spills or were engulfed by spread of contamination areas" [108].

The use of decontamination water in this cleanup was prodigious. By July 1960, the inventory of contaminated water was 1,600,000 gal (6×10^6 liter). Over 1,000,000 gal (3.8×10^6 liter) capacity of steel storage tanks were purchased. In addition a 2000 gal/hr (7,500 liter/hr) evaporative unit was also acquired.

It seems clear that this reactor was not well designed from the standpoint of handling a possible incident. It is equally clear from reading the account of the cleanup that operating practices before and after the incident contributed markedly to the severity and its cost. There are three recorded instances of spills or breakage which spread additional contamination. This plant had built into it nearly 500,000 gal (1.9×10^6 liter) of liquid storage capacity. It would seem that by the judicious use of water in the course of the decontamination, by reuse of processed or partially processed water, and by more rigorous control of the cleanup operation, it should have been possible to avoid the purchase of another million gallons of tankage.

7.6 SRE Core

The damage to the Sodium Reactor Experiment and its cleanup illustrates a different type of recovery operation. In this case the reactor was sealed and very little gaseous activity escaped into the reactor room at the time of the incident; and the rigid procedures followed in the disassembly and cleanup of the reactor prevented escape of appreciable gaseous radioactivity during these operations. The excellent, well-illustrated report prepared by Freede has been used for this discussion [109].

The SRE is a sodium-cooled graphite-moderated reactor fueled with slightly enriched uranium, with stainless-steel cladding. Seven 0.75-in. (1.90-mm)-diam rods were clustered into a single assembly with a Zircaloy sleeve. The hexagonal graphite moderator blocks were also Zircaloy-clad with the exception of the outer row of blocks, which were stainless-steel clad. The 119 moderator and reflector cans were 11 in. (27.9 cm) between flat faces and 10 ft (~3 m) long. The reactor operated at 21 Mw(t).

In mid-1959 there was a small leak of tetralin, an organic liquid, into the core. The tetralin decomposed, forming carbonaceous deposits which reduced sodium coolant flow, and 13 of the 43 uranium fuel assemblies were damaged. To clear up the damaged fuel and moderator components meant a complete disassembly of the core remotely and behind shielding.

Special equipment, including vacuum cleaners, optical scopes, lights, special fishing and grappling tools, cutters, and probes, as well as special casks, etc., were developed and used through the openings in the top plug. The equipment was all installed and used without escape of radioactive fission gases or particles and without admission of air to the reactor.

The disassembly was planned after tests by probes to determine the points of decreased flow and after removal of most of the sodium. This was followed by photographs of the top of the core. The photographs showed 81 fuel slug pieces (the original slugs were 6 in., or about 15 cm, long), and pieces of can and wire wrapping on top of the moderator cans. Subsequent examination showed high temperatures had resulted in formation of the uranium-iron alloy with destruction of the cladding and distortion of the fuel elements. Destruction of the moderator cans permitted sodium to enter the cans and to cause swelling of the graphite by about 1%.

The gas over the core contained about 10^{-3} $\mu c/ml$, but purging combined with decay reduced this to 10^{-6} $\mu c/ml$. This was mainly Kr^{85}; so very small gas leaks could be tolerated. However, all openings were made through two containments with an inert purge between the two.

During the disassembly, the 81 fuel slug pieces, numerous bits of stainless steel, and about 2 lb (~1 kg) of low-density, frothy, carbonaceous materials were removed from the top of the reactor core. Sixteen moderator cans were removed. Two of these contained whole fuel assemblies, two held substantial portions of broken assemblies, and four moderator cans had leaks.

The special equipment designed and built for the removal of the damaged fuel slugs and moderator cans worked well and no accidents or overexposures occurred. After cleanup the reactor was reloaded with new moderator assemblies as required and enriched uranium thorium fuel and returned to service.

7.7 <u>SL-1</u>

The Stationary Low Power Plant No. 1, or SL-1, underwent a rapid criticality with resultant pressure rise in the evening of January 3, 1961. The accident resulted in the immediate death of two persons, followed within about two hours by the death of the third man of the operating crew.

The SL-1 was a low-power reactor designed to be air transportable and to have a total power of 3 Mw(t). The core was of the MTR type in that aluminum-enriched uranium-alloy aluminum-clad fuel plates were used. The core was designed for very long fuel life and therefore had a large amount of excess reactivity held by five cruciform control rods and Al-Ni-B^{10} strips welded to the fuel-element side plates.

After construction the reactor was used for the usual series of criticality and control-rod worth measurements and given a 500-hr full-power run by Argonne National Laboratory. In February 1959, it was turned over to Combustion Engineering for operation under the Idaho Operations Office. Military personnel were assigned for training under the overall management and technical direction of Combustion Engineering. At the time of the accident three Army trainees, consisting of the shift supervisor, an operator-mechanic, and an apprentice-trainee were present.

The reactor had been shut down for 11 days for various types of maintenance work. At the time of the accident the three crew members were on top or near the top of the reactor, apparently connecting four of the five control rod drives (one had not been disconnected). The central rod apparently was withdrawn manually at less than maximum effort and this resulted in a rapid criticality.

As a result of the criticality, radioactivity was released to the extent that several days after the accident activity levels in the building were 1000 r/hr. One body, on removal, had readings as high as 400 r/hr. However, the building retained essentially all the activity, and activity levels outside the building were low. Some iodine escaped from the building but most of the particulate matter was retained. About 130 Mw-sec of energy had been released. About 5% of the gross fission products in the reactor prior to and produced during the criticality were released from the reactor vessel. Melting of about 20% of the core occurred. The fuel plates in the highest flux portions of the core were estimated to have reached 2060°C (3740°F) or the vaporization point. This resulted in destruction of these plates and production of additional steam due to reaction of aluminum and water. This was determined by the presence of alpha-phase alumina. This reaction was estimated to have released heat equal to less than 25% of the nuclear energy released [110].

The SL-1 was disassembled and the area decontaminated in 13 months, beginning in May 1961. This work is described in detail in the final report [111]. First, a control center was established for the various recovery operations. This area was utilized for briefing and for obtaining information from the working teams, for health physics, for assembly and maintenance of equipment, and for decontamination. Second, the radiation levels were reduced by removal of large pieces of contaminated equipment, by sweeping and vacuuming to remove the smaller pieces of debris, and by placing shielding over the top of the reactor. Part of this was done by remote operations and part directly by personnel. A burial site was opened near the reactor and all equipment buried that was not needed for examination to determine the history of the incident. This material was placed in casks and taken to the ANP area for examination.

After the building was partially cleaned, the upper portions were dismantled. With the area above the pressure vessel clean, the vessel was transferred to a large cask and transported to the ANP area for disassembly. Subsequently, the remainder of the building was disassembled and the site and remaining buildings decontaminated.

Detailed examinations of much of the equipment were necessary in analyzing the cause of the accident, its severity, and the sequence of events. The disassembly and examination required construction and use of special equipment. The entire procedure was conducted without serious exposure to personnel. Less than 6% of the 475 individuals involved received exposures over the "guide values" and the highest was only 16% over these values. Over 80,000 cu ft (~2300 m^3) of the equipment and building were buried at a burial ground opened near the site [112, 113].

7.8 EBR-1 Core Meltdown*

The following was abstracted from the papers presented by R. O. Brittan and by J. H. Kittel, et al., at the Second Geneva Conference [114, 115]. The EBR-1 is a 1400 Kw(t) experimental fast reactor located at the Reactor Testing Station. It was originally fueled with enriched-uranium bars in stainless steel tubes with a core about 8 1/2 in. (21.6 cm) high and 6 in. (15.1 cm) diam with 55 kg of uranium. The reactor contained natural-uranium rods 4 in. (10.2 cm) below and 8 in. (20.3 cm) above the enriched fuel. The reactor is cooled by downflow of NaK. At the time the incident occurred the reactor was being used in a power-rise study at a positive period of about 1 min. The coolant rate had been reduced to 1.5 gpm (94.6 cm^3/sec) at a temperature of 65°C (149°F). The period shortened as the temperature increased and a high temperature of the fuel occurred. The reactor was scrammed manually.

With a maximum heat production of about 8000 kw and a total energy of about 14,000 kw-sec the fuel was raised to a maximum of about 1030°C (1886°F). This was sufficient for the fuel and cladding to interact, forming a molten eutectic which partially dropped from the center of the core, leaving a hole.

The reactor was disassembled without hazard, as it was adequately shielded. A temporary cave was constructed over the top of the reactor and the core removed and placed in a coffin for shipment to Argonne. At Argonne the core was readily disassembled in a high-level cave without excessive contamination of the surrounding areas.

7.9 Other Reactors

The CP-3 at Argonne, Clementine at Los Alamos, HRE-I at Oak Ridge, and BORAX-1 have all been disassembled, the entire reactor complex decontaminated, and the equipment either recovered or buried. CP-3 was quite easily disassembled as mentioned earlier, and the central portion of the shield filled with concrete.

Clementine was the small fast reactor built at Los Alamos during World War II. It contained steel-clad plutonium fuel elements, which failed due to corrosion, and 1 to 10 g of plutonium contaminated the mercury coolant [116]. The mercury coolant was first drained down to the top level of the fuel rods, the top of the core pot cut open, the fuel elements removed, and the remainder of the mercury removed. Next the shielding, which was made of concrete blocks, was removed. Then the reflector, core pot, and shielding blocks above the reactor were removed as one unit. To do this, the radiation was reduced by welding 2-in. (5-cm) steel plates around the unit. It was then transferred by truck to a storage area for contaminated equipment. The mercury-coolant system was cut up by erecting small temporary dry boxes around the cutting points and cutting with tubing cutters rather than saws. The items were then placed in plastic tubing, which was cut and sealed for transfer to plutonium recovery areas. A large dry box was placed over the reactor for the fuel unloading. Some Pb turnings used to seal joints against gamma rays were alpha active. This was believed to be due to 138-day Po^{210} since the Pb contained 0.1% Bi. Only barely detectable quantities of plutonium escaped during the disassembly operations.

The HRE-I is of interest in that the entire primary system was decontaminated in about 30 days from a level of 1000 r/hr to the point that it could be disassembled with long-handled tools. The system was flushed alternately with 35% HNO_3 and 10% NaOH, 1.5% sodium tartrate, and 1.5% H_2O_2. About 1000 curies of Ce, Zr, Ba, La, Nb, and Ru were removed. Significant amounts of Zr and Nb remained in the oxide film. This film was not removed, as it was desired to measure the corrosion that had occurred. It was estimated that had the film been removed, the activity could have been reduced by another factor of 100 [117].

The BORAX-1 reactor was deliberately placed on a very short period as a part of an experimental program. This reactor was fueled with fully enriched uranium-aluminum fuel assemblies of the MTR type and light-water cooled and moderated. The core was located in an open vessel and the "control room" was located about 1/4 mile (~0.4 km) away. The reactor was operated, as are the SPERT reactors, only for short periods; so there was a minimum inventory of long-life fission products present.

During the destructive experiment some of the fuel was melted and the steam, estimated to have reached a pressure of over 6000 psi (~400 atm), ejected much of the core and control system, burst the thin-walled pressure vessel, and severely damaged the reactor components. Fifteen minutes after the experiment the total beta-gamma activity 0.8 mile (1.3 km) downwind from the reactor gave an external dose rate of about 5 mrem/hr. At 0.5 mile (0.8 km) the maximum dose rate due to the

*The problem of core meltdown in fast reactors is also discussed in Chapter 10.

explosion was 400 mrem/hr. Within a few days, recovery of the usable equipment was started and within a month the entire reactor area had been cleaned up. The usable equipment was readily decontaminated and the remainder buried. There were no excessive exposures.

7.10 Discussion*

The lessons that may be learned from a study of the incidents mentioned in previous paragraphs include the following:

a. A minimum amount of reactivity control should be incorporated in one control rod.
b. While it was desired to extend fuel life to very long periods, it perhaps would have been simpler in the case of the SL-1 to have provided additional fuel loading every 2 years.
c. There is a higher probability for incidents occurring when a reactor is being used for experimental measurements than when it is in full operation. The NRX, SL-1, and EBR-1 incidents occurred under these conditions and under partial or full manual control.
d. Nuclear reactor buildings should be designed so that any incident will be minimized and the reactor and building can be decontaminated. Ideally the reactor should be located so that coolant will flow into it in case of an incident (the SL-1 might have been worse in this case but the NRX would have been simplified), and there should be a minimum of components below the reactor that could become contaminated. All surfaces should be sealed and painted with an easily decontaminated paint. All equipment possible should be located outside the reactor building.
e. A disaster plan should be prepared for each reactor and a minimum of emergency equipment obtained. The disaster plan should emphasize putting the reactor in a safe shutdown condition in case of an emergency.
f. In case of an incident a recovery and decontamination program should be prepared prior to start of these operations.

*See also Chapter 11.

APPENDIX 1

RADIOACTIVE WASTE MANAGEMENT EXPERIENCE TO MID-1969

Since the material for this chapter was prepared (in the early 1960s), the use of nuclear energy for the generation of central-station electrical power has grown considerably. As a consequence, much of the data included in the chapter are now out-of-date. In this appendix more recent data are presented. The data were obtained in response to questionnaires sent in spring of 1969 to all utilities having operating reactors or reactors being designed or constructed.

Data on radioactive waste systems and experience were obtained from 8 operating reactors (see Table A-1) and from 18 reactors being designed or under construction (see Table A-2). The data are presented in the following tables:

Gaseous wastes:

From operating reactors . . . . Table A-3

From reactors under construction (design data) . . . . Table A-4

Liquid wastes:

From operating reactors . . . . Table A-5

From reactors under construction (design data) . . . . Table A-6

Solid wastes:

From operating reactors . . . . Table A-7

From reactors under construction (design data) . . . . Table A-8

The response to the questionnaires was gratifying. The authors wish to express their appreciation to all those who responded to the lengthy set of questions.

TABLE A-1

Reactors in operation

Reactor	Rating Mw(e)	Type	Nuclear steam system by	Date in operation	Gross electricity[a] generated (Mw-hr)	Approximate capacity factor (%)
Shippingport	68 (Core I) 100 (Core II)	PWR	Westinghouse	Dec. 18, 1957	3,752,000	Core I 27% Core II 45%
Dresden #1	200	BWR	General Electric	Aug. 1, 1960	8,950,000	--------
Yankee	185	PWR	Westinghouse	Aug. 19, 1960	9,609,000	75
Big Rock Point	75	BWR	General Electric	Sept. 1962	2,034,000	81 1967 68 1968
Elk River	23	BWR	Allis-Chalmers	Nov. 1962	511,000	53
Humboldt Bay	70	BWR	General Electric	Aug. 1, 1963	1,947,000	67
Connecticut Yankee	491	PWR	Westinghouse	Jan. 1, 1968	5,153,000	66
San Onofre	450	PWR	Westinghouse	Jan. 1, 1968	2,954,000	34 75 [b]

[a] Through May 30, 1969

[b] Between October 1968 and March 1969.

TABLE A-2

Reactors under Construction

Reactor	Rating Mw(e)	Type	Nuclear steam system by	Expected operation date
Oyster Creek #1	515	BWR	General Electric	July 1969
Monticello	545	BWR	General Electric	May 1970
Palisades	710	PWR	Combustion	May 1970
Millstone #1	652	BWR	General Electric	June 1970
Oconee #1	874	PWR	Babcock & Wilcox	May 1971
Oconee #2	874	PWR	Babcock & Wilcox	May 1972
Oconee #3	874	PWR	Babcock & Wilcox	June 1973
Salem #1	1050	PWR	Westinghouse	March 1972
Salem #2	1050	PWR	Westinghouse	March 1973
Prairie Island #1	550	PWR	Westinghouse	May 1972
Prairie Island #2	550	PWR	Westinghouse	May 1974
Diablo Canyon	1060	PWR	Westinghouse	1972
Donald C. Cook #1	1100	PWR	Westinghouse	1972
Donald C. Cook #2	1100	PWR	Westinghouse	1973
Kewaunee	527	PWR	Westinghouse	1972
Beaver Valley	800	PWR	Westinghouse	June 1973
Midland Unit #1	830	PWR	Babcock & Wilcox	Feb. 1974
Midland Unit #2	520[a]	PWR	Babcock & Wilcox	Feb. 1975

[a] Plus 4,000,000 lb/hr process steam.

TABLE A-3

Gaseous Wastes from Operating Reactors

Reactor	Type	Treatment			Stack		Average Annual Releases		
		Delay time	Filtration	Other	Height (ft)	Exhaust rate (cfm)	Activation & noble gases ($\mu c/sec$)	Halogens ($\mu c/sec$)	Particulates ($\mu c/sec$)
Shippingport	PWR	---	Yes	Storage Dilution H_2Gas Burner	25	9,000 (1 fan) 18,000 (2 fans)	2×10^{-5} to 1×10^{-2}	--------	--------
Dresden #1	BWR	20 min	Yes	--------	300	44,000	100 to 25,000	0.002 to 0.003	
Yankee	PWR	2 months	Deep bed	--------	127	15,000	3×10^{-9} $\mu c/cc$	--------	5×10^{-12} $\mu c/cc$
Big Rock Point	BWR	30 min	Yes	None	240	30,000	8000 to 12,000	2×10^{-10} $\mu c/cc$	2×10^{-11} $\mu c/cc$
Elk River	BWR	30 min	Prefilters Absolute filters	None	97	3,500	3.3×10^{-1} to 1.4×10^{2}	8×10^{-7} to 9.8×10^{-5}	7.9×10^{-4} to 7.5×10^{-3}
Humboldt Bay	BWR	40 min	Yes	Caustic scrubber for building air during refueling	250	12,000	35[a] 1600[b]	9×10^{-5} to 6×10^{-3} [a] 3×10^{-5} to 3×10^{-2} [b]	
Connecticut Yankee	PWR	100 days	Fiberglass	None	175	70,000	0.13	Not detectable	--------
San Onofre	PWR	---	Yes	None	100	40,000	1.65×10^{3} (gross)	--------	--------

[a] Without failed fuel in core

[b] With failed fuel in core

TABLE A-4

Design Data—Reactors under Construction—Gaseous Wastes

Reactor	Type	Treatment			Stack		Design or License Limits		
		Delay time	Filtration	Other	Height (ft)	Exhaust rate (cfm)	Activation & noble gases (μc/sec)	Halogens (μc/sec)	Particulates (μc/sec)
Oyster Creek #1	BWR	30 min	Absolute	None	368	190,000	3×10^5	4	13
Monticello	BWR	30 min	Yes	Dilution	328	4,150	4.8×10^5 [a] 1 to 5×10^4 [b]	3.3 < 0.5	a b
Palisades	PWR	10 min to 30 days	Yes	Holdup tanks	Top of contain-ment	60,000	10^5 total		b
Millstone #1	BWR	Air ejector- 30 min Gland seal- 1.75 min	Yes	None	375	180	1.4×10^6 [a] 5×10^3 [b]	12[a]	None[a]
Oconee #1, 2, 3	PWR	None routinely	Prefilter Absolute Charcoal in series	Capability for decay holdup	199	43,000	--------	-------	-------
Salem #1, 2	PWR	45 days	Absolute and/or iodine	-----	192	40,000 to 100,000	10 CFR20 off-site 240 [b]	~0[b]	a ~0[b]
Prairie Island #1, 2	PWR	45 days	Charcoal	None	Vent at top of contain-ment vessel	------	Kr^{85} 200[b] Xe^{133} 50[b]	~0[b]	~0[b]
Diablo Canyon	PWR	27-64 days	-----	-----	-----	-----	-----	-------	-------
Donald C. Cook #1	PWR	45 days	Yes	None	160	350,000	10 CFR20		a
Donald C. Cook #2	PWR	45 days	Yes	None	160	150,000	10 CFR20		a
Kewaunee	PWR	45 days	----	None	None	1.5 air changes/ hr for plant	10 CFR20		a
Beaver Valley	PWR	30 days	Yes	----	None	-------	10 CFR20		a
Midland Units #1, 2	PWR	-----	Yes	Holdup tanks	---	-------	-------	-------	-------

[a] Proposed license limit.
[b] Design estimate.

TABLE A-5

Liquid Wastes from Operating Reactors

Reactor	Type	Principal means of treatment	Volume (gal/yr)	Disposition: Treated (%)	Disposition: Reused (%)	Disposition: Discarded (%)	Discharge point	Activity before treatment (μc/cc)	Available dilution water (gpm)	Activity Discharged: $\beta - \gamma$[a] (curies/yr)	Activity Discharged: Tritium (curies/yr)	Activity Discharged: Concentration (μc/cc)
Shippingport	PWR	Evaporation Ion exchange Gas stripping Dilution	2×10^{6}	63	0	100	Condenser outlet to Ohio River	2×10^{-3} to 4×10^{-5}	118,000	0.07	35	1×10^{-4} to 7×10^{-6}
Dresden #1	BWR	Decay Ion exchange Filtration	11×10^{6}	---	84	---	Illinois River	10^{-2} to 10^{-5}	167,000	4	5 to 10	10^{-8} to 5×10^{-8}
Yankee	PWR	Evaporation Solidification with cement	500,000	100	0	100	With circulating water to Deerfield River	1×10^{-3}	140,000	6×10^{-4}	1000	3×10^{-7} [b] 7×10^{-11} [c]
Big Rock Point	BWR	Sock filter 25-μ filter Precipitation Ion exchange	4.2×10^{6}	100	90	10	Canal to Lake Michigan	1×10^{-2} to 4×10^{-4}	52,000	7	------	------
Elk River	BWR	Filtration Ion exchange	80,000	100	0	100	Condenser coolant discharge to Mississippi River	10 to 10^{-3}	58,000	0.22	8	1×10^{-6} to 1×10^{-8}
Humboldt Bay	BWR	Filtration Ion exchange	600,000	100	0	100	Condenser coolant discharge to Humboldt Bay	0.2 to 10^{-6}	100,000	2	~ 20	3.1×10^{-8} to 1.6×10^{-9}
Connecticut Yankee	PWR	Evaporation	1,200,000[d]	50	5	95	Connecticut River thru Mile-long Canal	10^{-4} to 10^{-6}	400,000	0.14	1700	10^{-7} to 10^{-10} β-γ 10^{-4} to 10^{-7} tritium
San Onofre	PWR	Ion exchange Deaeration Decay	$\sim 2\times10^{6}$	----	0	100	With condenser coolant to Pacific Ocean	----	350,000	~ 6	----	10^{-7} gross

a Exclusive of tritium.
b Before dilution.
c After dilution.
d Data from a typical year.

TABLE A-6

Design Data—Reactors under Construction—Liquid Wastes

Reactor	Type	Principal means of treatment	Expected volume (gal/yr)	Disposition			Discharge point	Activity before treatment (μc/cc)	Available dilution water (gpm)	Activity Discharged		
				Treated (%)	Reused (%)	Discarded (%)				β-γ[a] (curies/yr)	Tritium (curies/yr)	Concentration (μc/cc)
Oyster Creek #1	BWR	Filtration Ion exchange Evaporation	36×10^6	100	80	20	Barnegat Bay	---------	1.2×10^6	~ 25 (max)	not expected	$< 10^{-7}$
Monticello	BWR	Filtration Ion exchange	6.5×10^6	95	50 to 90	10 to 50	Discharge canal to Mississippi River	10^{-2} to 10^{-3}	16,000 to 300,000	0.25 to 4	7 to 15	$< 10^{-7}$
Palisades	PWR	Filtration Ion exchange Decay	1 to 3×10^6	100	0	100	Lake Michigan	10^{-3} to 10	400,000	-------	-------	------
Millstone #1	BWR	Filtration Ion exchange Centrifugation Evaporation	-------	----	----	----	Long Island Sound	--------	450,000	~ 1	-------	-------
Oconee #1, 2, 3	PWR	Ion exchange Decay Evaporation	-------	----	----	----	Tailrace of Keowee Hydro Station to Keowee River	--------	490,000	-------	-------	--------
Salem #1, 2	PWR	Evaporation Ion exchange	580,000	100	0	100	Delaware River	20[b]	1.1×10^6 ea	0.035 ea	4000 ea	2×10^{-11}
Prairie Island #1, 2	PWR	Ion exchange Evaporation	-------	----	----	----	Condenser discharge canal to Mississippi River	--------	300,000 ea	0.07	2200	0.01 mpc
Diablo Canyon	PWR	Ion exchange Filtration Evaporation	1.6×10^6	100	97	3	Pacific Ocean	10 to 10^{-2}	870,000	1.2 to 1.2×10^2	4400	1×10^{-7} to 1×10^{-9} β-γ 3×10^{-6} tritium
Donald C. Cook #1, 2	PWR	Ion exchange Decay Evaporation	480,000 ea	100	0	100	Condenser water to Lake Michigan	--------	750,000 ea	0.1	3400 ea	------
Kewaunee	PWR	Ion exchange Evaporation	400,000	----	----	----	Condenser water to Lake Michigan	--------	400,000	0.04	-------	-------
Beaver Valley	PWR	Ion exchange Evaporation	-------	----	----	----	Condenser cooling water to Ohio River	10^2	125,000	-------	-------	-------
Midland Units #1, 2	PWR	Evaporation Filtration Ion exchange	1.1×10^6	100	----	----	Tittabawassee River	--------	-------	-------	-------	-------

[a] Exclusive of tritium.
[b] Design maximum.

TABLE A-7

Solid Wastes from Operating Reactors

Reactor	Type	Resins			Other Solids		
		Volume (ft^3/yr)	Approximate activity (curies/ft^3)	Disposition	Volume (ft^3/yr)	Approximate activity (curies/ft^3)	Disposition
Shippingport	PWR	350	2.5×10^{-2}	Storage tanks	3500	2×10^{-3}	Off-site burial
Dresden #1	BWR	770	---	Burial	3200	~ 0.03	Burial
Yankee	PWR	70	~ 8	Burial	3700	10^{-3}	Burial
Big Rock Point	BWR	300	1	Off-site burial	700	0.03	Off-site burial
Elk River	BWR	20	0.8 to 8	Off-site burial	420	10^{-3}	Off-site burial
Humboldt Bay	BWR	50	---	Stored for later shipment	900	0.2	Off-site burial
Connecticut Yankee	PWR	45	---	Off-site burial	550	---	Off-site burial
San Onofre	PWR	---	---	Storage tanks for later burial	1000	6×10^{-4}	Off-site burial

TABLE A-8

Design Data—Reactors under Construction—Solid Wastes

Reactor	Type	Resins			Other Solids		
		Volume (ft^3/yr)	Approximate activity (curies/ft^3)	Disposition	Volume (ft^3/yr)	Approximate activity (curies/ft^3)	Disposition
Oyster Creek #1	BWR	1000	0.5	Off-site burial	25,000[a] 2,500[b]	0.005[a] 0.15[b]	Off-site burial
Monticello	BWR	2000	1	Off-site burial (storage vault)	------	------	------
Palisades	PWR	400	----	Off-site burial	------	------	------
Millstone #1	BWR	----	----	Off-site burial	------	------	Off-site burial
Oconee #1, 2, 3	PWR	100 ea	----	Off-site burial	------	------	Off-site burial
Salem #1, 2	PWR	240 ea	800 (max)	Off-site burial	~ 1000 ea	------	Off-site burial
Prairie Island #1, 2	PWR	----	----	------	------	------	------
Diablo Canyon	PWR	100	4 to 300	Off-site burial	4,800	5×10^{-2} to 5×10^{-1}	Off-site burial
Donald C. Cook #1, 2	PWR	200 ea	----	Off-site burial	------	------	Off-site burial
Kewaunee	PWR	----	----	------	------	------	------
Beaver Valley	PWR	----	----	Off-site shipment	------	------	Off-site burial
Midland Units #1, 2	PWR	800	----	Off-site burial	1,200	------	Off-site burial

[a] Evaporator concentrates.
[b] Filter sludge.

REFERENCES

1. Directory of Nuclear Reactors, Vols. II-IV, International Atomic Energy Agency, Vienna, Austria, 1959, 1962.
2. H. K. Hoyt, private communication, November 22, 1963, Commonwealth Edison Company, Dresden Station, Morris, Illinois.
3. J. F. Stehn, "Table of radioactive nuclides", Nucleonics, 18, 11 (1960) 186.
4. G. L. Redman, "Power Generation Statistics, Load Schedules, and Water Chemistry", talk presented at the 24th American Power Conference, Chicago, March 27-29, 1962.
5. C. R. Breden, "Behavior of Reactor Structural Materials from the Standpoint of Corrosion and Crud Formation", AEC-EURATOM Conference on Aqueous Corrosion of Reactor Materials, Brussels, Belgium, October 14-17, 1959. USAEC Report TID-5787, p. 48, 1960.
6. C. R. Breden, "Boiling Water Reactor Technology Status of the Art Report: Vol. II. Water Chemistry and Corrosion", USAEC Report ANL 6562, Argonne National Laboratory 1963.
7. L. P. Bupp, "Maintenance of Coolants", Chapter 7 in Vol. 4, "Engineering", Reactor Handbook, (2nd ed), Interscience Publishers, Inc., N.Y., 1964.
8. D. M. Wroughton and P. Cohen, "Radioactivity Levels in Pressurized Water Reactor Systems", Proceedings of the Second U.N. International Conference on Peaceful Uses of Atomic Energy, Geneva, 1958, Vol. 7, p. 427.
9. D. J. DePaul (Ed.), "Corrosion and Wear Handbook for Water-Cooled Reactors", USAEC Report TID-7006, 1957.
10. W. A. Rodger, "Radioactive Waste Disposal", USAEC Report ANL-6233, Argonne National Laboratory, 1960.
11. R. Kunin and R. J. Meyers, Ion Exchange Resins, John Wiley and Sons, Inc., N.Y., 1950.
12. J. N. Anno, A. M. Plummer and J. W. Chastain, "Experience with a 1-Megawatt Pool-Type Research Reactor", Proceedings of the Second U.N. International Conference on Peaceful Uses of Atomic Energy, Geneva, 1958, Vol. 10, p. 237.
13. ORNL-LR-DWG 28396 R6, in press.
14. T. E. Cole and J. A. Cox, "Design and Operation of the ORR", Proceedings of the Second U.N. International Conference on Peaceful Uses of Atomic Energy, Geneva, 1958, Vol. 10, p. 86.
15. Directory of Nuclear Reactors, Vol. II, p. 98, International Atomic Energy Agency, Vienna, Austria, 1959.
16. John R. Huffman, private communication, August 30, 1963, Phillips Petroleum Company, Idaho Falls, Idaho.
17. Directory of Nuclear Reactors, Vol. IV, p. 77, International Atomic Energy Agency, Vienna, Austria, 1962.
18. S. J. Goslovich et al., "Scintillation Counter Analysis of EBWR Radioactivity", Industrial Hygiene and Safety Division, Argonne National Laboratory, Internal Report, Undated. See also: Nucleonics, 16, 5 (1958) 94.
19. R. F. S. Robertson, "Tests of Defected Thoria-Urania Fuel Specimens in EBWR", USAEC Report ANL-6022, p. 30, Argonne National Laboratory, 1960.
20. R. F. S. Robertson, "First Defect Test in EBWR", Reactor Engineering Division Test Report 55, Argonne National Laboratory, Aug. 8, 1958. See also: V. M. Kolba, "EBWR Test Reports", USAEC Report ANL-6229, Argonne National Laboratory, 1960.
21. "Radioactive Waste Handling in the Nuclear Power Industry", Edison Electric Institute, 250 Third Ave., N.Y. 12, N.Y., 1960.
22. "Preliminary Hazards Survey Report for the Dresden Nuclear Power Station", USAEC Docket 50-10, 1957.
23. C. B. Zitek, private communication, Dresden Nuclear Power Station, Commonwealth Edison Co., Morris, Illinois.
24. E. R. Owen, "Performance Measurements of the Dresden Nuclear Power Station", Paper presented at the 24th American Power Conference, Chicago, Ill., March 27-29, 1962.
25. "Standard Methods for the Examination of Water, Sewage, and Industrial Wastes", American Public Health Association, Inc., 1790 Broadway, N.Y. 19, N.Y., (10th ed.), 1955, p. 207.
26. W. Kiedaisch, "Experience with Liquid Waste Handling at Dresden Nuclear Power Station", Paper presented at meeting of Am. Inst. Chem. Eng., September, 1961. See: Chem. Eng. Progr, 58, 1 (1962) 79.
27. H. K. Hoyt, "Plant Description", Paper presented at 24th American Power Conference, Chicago, Ill., March 27-29, 1962.
28. J. R. LaPointe, "How radioactive wastes will be handled at PWR", Nucleonics, 15, 5 (1957) 114.
29. I. H. Welinsky et al., "Chemistry of a Pressurized Water Nuclear Power Plant: A Review of Two Years Operating Experience", Proceedings of American Power Conference, 1956, Vol. 18, p. 559.
30. J. E. Gray, "Shippingport Experience Related to Release of Radioactivity to the Environment", Paper presented to 6th Engineering and Science Conference, N.Y., April 4-7, 1960.
31. W. J. Cahill, Jr. and R. H. Freyberg, "Operating Experience at the Indian Point Station", Proceedings of 25th American Power Conference, Chicago, Ill., March 26-28, 1963.
32. John Kaslow, private communication, July 11, 1963.
33. T. J. Iltis and M. E. Miles, "Radioactive Waste Disposal from U.S. Naval Nuclear Powered Ships", in "Industrial Radioactive Waste Disposal", Hearings of 86th Congress, Jan. 28-Feb. 3, 1959, U.S. Government Printing Office, Washington 25, D. C. 37457 O, p. 924.
34. A. W. Kramer, "Nuclear Propulsion for Nuclear Ships", U.S. Government Printing Office, Washington 25, D.C., 1962, p. 272.
35. H. M. Parker, "Radioactive Waste Management Operations at the Hanford Plant" in "Industrial Radioactive Waste Disposal", Hearings of 86th Congress, Jan. 28-Feb. 3, 1959, U.S. Government Printing Office, Washington, D.C. 37457 O, p. 171.
36. H. L. Butler, "Tritium hazards in heavy-water-moderated reactors", Nucl. Safety, 4, 3 (1963) 77.
37. J. J. Hartig, private communication, August 23, 1963, Argonne National Laboratory, Argonne, Illinois.
38. J. A. Lane, H. G. MacPherson and F. Maslan, Fluid Fuel Reactors, Addison-Wesley Publishing Co., Reading, Mass., 1958.
39. G. J. Barenborg, F. H. Haag, and H. F. Karnes, "Sodium and NaK", Section 6 of Chapter 7 in Vol. 4, "Engineering", of Reactor Handbook, 2nd ed., Interscience Publishers Inc., N. Y., 1964.
40. R. N. Lyon, "Liquid Metals", Chapter 49 in Vol. I. "Materials", of Reactor Handbook, Interscience Publishers, Inc., N.Y., 1960.
41. J. R. Dietrich and W. H. Zinn, Solid Fuel Reactors, p. 351, Addison-Wesley Publishing Co., Reading, Mass., 1958.
42. C. Starr and R. W. Dickenson, Sodium-Graphite Reactors, p. 167, Addison-Wesley Publishing Co., Reading, Mass., 1958.
43. J. D. Cochran, "Operating Experience at Hallam Nuclear Power Facility", Paper presented at 25th American Power Conference, Chicago, Ill., March 26-28, 1963.
44. T. L. Gershun (Ed.), "Preliminary Safeguards Report Based on Uranium-Molybdenum Fuel for the Hallam Nuclear Power Facility", Report NAA-SR-3379, North American Aviation, Inc., 1961.
45. "Part B. Revised License Application - Technical Information and Hazards Summary Report, Section 1: Reactor and Plant Design", USAEC Docket No. 50-16, Vol. 3.
46. Directory of Nuclear Reactors, Vol. IV, p. 187, International Atomic Energy Agency, Vienna, Austria, 1959.
47. H. G. Davey et al., "Operating Experience at Calder Hall", Proceedings of the Second U.N. International Conference on Peaceful Uses of Atomic Energy, Geneva, 1958, Vol. 8, p. 10.
48. L. Silverman, "Economic Aspects of Air and Gas Cleaning for Nuclear Energy Processes", in Disposal of Radioactive Wastes, International Atomic Energy Agency, Vienna, Austria, 1960, p. 181.
49. R. A. Meyer and C. L. Richard, "The Impact of Graphite Fuel Development on High Temperature Steam Generation", Paper presented at 25th American Power Conference, Chicago, Ill., March 26-28, 1963. See: Report GA-4051, General Atomic, Division of General Dynamics Corp., 1963.
50. "Application of Philadelphia Electric Company for Construction Period and Class 104 License. Part B. Preliminary Hazards Report, Peach Bottom Atomic Power Station", USAEC Docket 50-171.
51. H. C. Gilroy and J. H. Wilson, "OMR Piqua Waste Gas Treatment System, Prototype System Description and Experimental Program", Report NAA-SR-Memo-4593, North American Aviation, 1959.
52. "Final Safeguards Summary Report for the Piqua Nuclear Power Facility", Report NAA-SR-5608, North American Aviation, 1961.
53. 10 CFR, Part 72.
54. D. C. Hampson, E. H. Hykan, and W. A. Rodger, "Basic Operational Report on the Argonne Waste Incinerator", USAEC Report ANL-5067, Argonne National Laboratory, 1953.
55. R. C. Larson and R. H. Simon, "Solid Waste Disposal at KAPL", Report KAPL-936, Knolls Atomic Power Laboratory, 1953.
56. R. Dennis and L. Silverman, "Radioactive Incinerator Design and Operational Experience - A Review", in "Seventh AEC Air Cleaning Conference", Oct. 10-12, 1961. USAEC Report TID-7627, 1962.
57. Tariff No. 10, "Published ICC Regulations for the Transportation of Explosives and Other Dangerous Articles", Bureau of Explosives of the Association of American Railroads, H. A. Campbell, Agent, 30 Vesey Street, N.Y., 7, N.Y.
58. "Motor Carriers Explosives and Dangerous Articles Tariff No. 8", Tariff Bureau of the American Trucking Associations, F. G. Freund, Agent, 1424 16th St., N.W., Washington 6, D.C.
59. 46 CRF 146, Federal Register, July 17, 1952, p. 6460 ff; and December 31, 1952.
60. Water Carrier Tariff No. 6, Bureau of Explosives, H. A. Campbell, Agent, 30 Vesey Street, N.Y. 7, N.Y.

61. C. E. Stevenson et al., "Current Practices in Disposal of Waste Radioactive Gases from Nuclear Reactors", American Standards Association Subcommittee N 5.2, Working Group No. 4. (In preparation)
62. J. A. Lane, H. G. MacPherson, and F. Maslan, "Fluid Fuel Reactors", Addison-Wesley Publishing Co., Reading, Mass., 1958, p. 436.
63. O. G. Sutton, "A Theory of Eddy Diffusion in the Atmosphere", Proceedings of the Royal Society (London) A 135 (1932) 143.
64. C. H. Bosanquet and J. L. Pearson, "The Spread of Smoke and Gases from Chimneys", Trans. Faraday Soc., 32 (1936) 1249.
65. World Meteorological Organization, "Meteorological Aspects of the Peaceful Uses of Atomic Energy", Proceedings of Second U.N. International Conference on Peaceful Uses of Atomic Energy, Geneva, 1958, Vol. 18, p. 245, and following papers.
66. "Meteorology and Atomic Energy", U.S. Government Printing Office, Washington 25, D.C., 1955.
67. F. Pasquill, "Atmospheric Diffusion—The Dispersion of Windborne Material from Industrial and Other Sources." D. Van Nostrand Co., 120 Alexander St., Princeton, N.J.
68. D. H. Peck, "Meteorology and Air Cleaning", in "Seventh Air Cleaning Conference", Brookhaven National Laboratory, Oct. 10-12, 1961, USAEC Report TID-7627, p. 475, 1961.
69. F. A. Gifford, Jr., "Use of Routine Meteorological Observations for Estimating Atmospheric Dispersion", Nucl. Safety, 2, 4 (1961) 47.
70. H. Gilbert, "The Filter Test Program, An Installation Manual, and Filter Research" in "Seventh AEC Air Cleaning Conference", Brookhaven National Laboratory, Oct. 10-12, 1961, USAEC Report TID-7627, p. 74, 1962.
71. H. Gilbert and J. H. Palmer, "Inspection, Storage, Handling, and Installation of High-Efficiency Particulate Air Filter Units", USAEC Report TID-7023, 1961.
72. J. A. Young, "Evaluation of High-Efficiency Air Filter Systems", in "Seventh AEC Air Cleaning Conference", Brookhaven National Laboratory, Oct. 10-12, 1961, USAEC Report TID-7627, p. 84, 1962.
73. J. A. Lane, H. G. MacPherson, and F. Maslan, Fluid Fuel Reactors, Addison-Wesley Publishing Co., Reading, Mass., 1958.
74. R. Dennis, L. Silvermann, and F. Stein, "Iodine Collection Studies—A Review", in "Seventh AEC Air Cleaning Conference", Brookhaven National Laboratory, Oct. 10-12, 1961, USAEC Report TID-7627, p. 327, 1962.
75. R. E. Adams and W. E. Browning, Jr., "Removal of Iodine from Gas Streams", in "Seventh AEC Air Cleaning Conference", Brookkhaven National Laboratory, Oct. 10-12, 1961, USAEC Report TID-7627, p. 242, 1962.
76. "Removal of Radioiodine from Gases", Nuclear Safety, 4, 2 (1962) 83.
77. W. J. Megain and F. G. May, "The Behavior of Iodine Released in Reactor Containers", Reac. Sci. Technol., J. Nucl. Eng., Parts A and B, 16 (1962) 427.
78. J. B. Morris, et al., "The Removal of Low Concentrations of Iodine From Air on a Plant Scale," Reactor Sci. Technol., J. Nucl. Energy., Parts A and B, 16 (1962) 427.
79. J. B. Morris et al., "Removal of Iodine Vapour from Air by Metallic Copper", Reactor Sci. Technol., J. Nucl. Energy, Parts A and B, 17, 3 (1963) 70.
80. F. R. Farmer, "Reactor Safety Work in England," Talk presented to the Atomic Industrial Forum, New York, New York, 19 November 1963.
81. R. E. Tomlinson, "Release of Gases, Vapors, and Particles to the Atmosphere", in "Industrial Radioactive Waste Disposal", Hearings of 86th Congress, Jan. 26-Feb. 3, 1959, U.S. Government Printing Office, Washington 25, D. C.
82. J. F. Manneschmidt, "Equipment and Procedures for Stack Gas Monitoring at ORNL", in "Seventh AEC Air Cleaning Conference", Brookhaven National Laboratory, Oct. 10-12, 1961, USAEC Report TID-7627, p. 168, 1962.
83. F. N. Browder, "Radioactive Waste Management at Oak Ridge National Laboratory" in "Industrial Radioactive Waste Disposal", Hearings of 86th Congress, Jan. 28-Feb. 3, 1959, Government Printing Office, Washington 25, D. C., p. 461.
84. Staff of the Hanford Atomic Products Operation, "Radioactive Waste Management Operations at the Hanford Works" in "Industrial Radioactive Waste Disposal", Hearings of 86th Congress, Jan. 28-Feb. 3, 1959, U.S. Government Printing Office, Washington 25, D. C., p. 199.
85. R. Kunin and R. J. Myers, Ion Exchange Resins, John R. Wiley and Sons, N.Y., 1951.
86. H. G. Swope and E. Anderson, "Cation Exchange Removal of Radioactivity from Wastes", Ind. Eng. Chem., 47 (1955) 78.
87. G. W. Parker, I. R. Higgins, and J. T. Roberts, Table 16.8 in Ion Exchange Technology, F. C. Nachod and J. Schubert (Eds.), Academic Press, N.Y., 1956, p. 144.
88. I. R. Higgins, "Radiation Damage to Organic Ion Exchange Materials", USAEC Report ORNL-1325, Oak Ridge National Laboratory, 1953, (Confidential).
89. S. A. Fisher, "Effect of Gamma Radiation on Ion Exchange Resins", USAEC Report RMO-2528, USAEC Division of Raw Materials, 1954.
90. R. E. Wedemeyer, "The Stability of Ion Exchange Resins to X-Rays", PhD. Thesis, Vanderbilt University, 1953.
91. W. Kiedaisch, "Experience with Liquid Waste Handling at Dresden Nuclear Power Station", Paper Presented at Meeting of American Institute of Chemical Engineers, Lake Placid, N. Y., Sept. 1961. See: Chem. Eng. Progr., 58, 1 (1962) 79.
92. H. Etherington (Ed.), Nuclear Engineering Handbook, Section 11, McGraw-Hill Book Co., N.Y., 1958.
93. "Waste Processing 2. Evaporation", USAEC Report BNL-59 (C-12), Brookhaven National Laboratory, 1950.
94. K. E. Cowser and R. J. Morton, "Treatment Plant for Removal of Radioactive Contaminants from Process Waste Water. Part II: Evaluation of Performance", Statement for the Record, Hearings on Industrial Radioactive Waste Disposal, Joint Committee on Atomic Energy, 86th Congress, August 1959, U.S. Government Printing Office, Washington, D.C., p. 547.
95. C. P. Straub, W. J. Lacy, and R. J. Morton, "Methods for the Decontamination of Radioactive Liquid Wastes", Proceedings of the First U.N. International Conference on Peaceful Uses of Atomic Energy, Geneva, 1955, Vol. 9, p. 24.
96. R. F. McCauley, R. A. Lauderdale, and R. Eliassen, "A Study of the Lime-Soda Softening Process as a Method for Decontaminating Radioactive Waters", USAEC Report NYO-4439, New York Operations Office, 1953.
97. J. B. Nesbitt, W. J. Kaufman, R. F. McCauley, and R. Eliassen, "The Removal of Radioactive Strontium from Water by Phosphate Coagulation", USAEC Report NYO-4435, New York Operations Office, 1952.
98. W. L. Wilson, P. A. F. White, and J. G. Milton, "The Control, Conveyance, Treatment and Disposal of Radioactive Effluents from the Atomic Weapons Research Establishment, Aldermaston", J. Brit. Nucl. Energy Conf., 1 (1956) 149.
99. F. R. Farmer, "The Problem of Liquid and Gaseous Effluent Disposal at Windscale", J. Brit. Nucl. Energy Conference, 2 (1957) 26.
100. E. Glueckauf and T. V. Healy, "Chemical Processing of Fission Product Solutions", Proceedings of the First U.N. International Conference on Peaceful Uses of Atomic Energy, Geneva, 1955, Vol. 9, p. 635.
101. W. W. Schulz and T. R. McKenzie, "The Removal of Cesium and Strontium from Radioactive Waste Solutions", Sanitary Engineering Conference, Cincinnati, O., USAEC Report TID-7517, 1955.
102. R. E. Burns and M. J. Stedwell, "Volume Reduction of Radioactive Waste by Carrier Precipitation", Chem. Eng. Progr., 53 (1957) 93-F.
103. M. E. Ramsey and C. D. Cagle, "Ten Years' Operating Experience on the ORNL Graphite Moderator Normal-Uranium Reactor", Chem. Eng. Progr. Symposium Series, 50, 11 (1954) 149.
104. H. J. Dunster, H. Howells, and W. L. Templeton, "District Surveys Following the Windscale Incident, October 1957", Proceedings of the Second U.N. International Conference on Peaceful Uses of Atomic Energy, Geneva, 1958, Vol. 18, p. 296.
105. This write-up and several of the following ones are nearly identical with those published by W. J. Tyrrell, G. D. Dymmel, and W. A. Rodger, "Notes on Decontamination of Surfaces and Equipment", International Institute of Nuclear Science and Engineering, Argonne National Laboratory, April, 1961.
106. F. W. Gilbert, "Decontamination of the Canadian Reactor", Chem. Eng. Progr., 50, 5 (1954) 267.
107. E. O. Hughes and J. W. Greenwood, "Contamination and Cleanup of NRU", Nucleonics, 18, 1 (1960) 76.
108. "Report on WTR Fuel Element Failure, April 3, 1960", USAEC Report WTR-49, 1960.
109. W. J. Freede, "SRE Core Recovery Programs", Report NAA-SR-6359, North American Aviation, 1961.
110. C. A. Nelson et al., "Final Report on the SL-1 Incident, January 3, 1961, the General Manager's Board of Investigation", USAEC Information Release No. E-326, Sept. 24, 1962.
111. "Final Report of SL-1 Recovery Operation", Report IDO-19311, Phillips Petroleum Co., 1962.
112. W. J. Kann, D. H. Shaftman, and B. I. Spinrad, "Postincident Analysis of the SL-1 Design", Nucl. Safety, 4, 3 (1963) 39.
113. "SL-1 Final Report", Nucl. Safety, 4, 3 (1963) 83.
114. R. O. Brittan, "Analysis of the EBR-1 Core Meltdown", Proceedings of the Second U.N. International Conference on Peaceful Uses of Atomic Energy, Geneva, 1958, Vol. 12, p. 267.

115. J. H. Kittel et al., "Disassembly and Metallurgical Evaluation of the Melted-Down EBR-1 Core", Proceedings of the Second U.N. International Conference on Peaceful Uses of Atomic Energy, Geneva, 1958, Vol. 7, p. 472.
116. E. T. Jurney, "The Failure and Disassembly of the Los Alamos Fast Reactor", Chem. Eng. Progr. Symposium Series, No. 13, Vol. 5a, Part 30, 1954, p. 191.
117. J. A. Lane, H. G. MacPherson, and F. Maslan, Fluid Fuel Reactors, p. 358, Addison-Wesley Publishing Co., Inc., Reading, Mass., 1958.

CHAPTER 21

The Concepts of Reactor Containment

T. J. THOMPSON
Massachusetts Institute of Technology
Cambridge, Massachusetts

C. ROGERS McCULLOUGH
Southern Nuclear Engineering, Inc.
Washington, D.C.

1 GENERAL CONSIDERATIONS

1.1 Development of the Containment Concept

The first nuclear reactor, built under the West Stands at the University of Chicago under the direction of Enrico Fermi, was an experiment to prove that a self-sustaining nuclear chain reaction could be achieved. The workers there, well aware of the potential dangers, provided crude but effective shutdown measures to be used in the event of an accident. For example, they stationed a man with an axe to cut the rope holding a shutdown rod and assigned another man to break a large bottle of boric acid placed on top of the pile.

The emphasis in this first nuclear reactor was primarily to prevent an accident, not to ameliorate the consequences. As far as the present authors are aware, the possible extent of an excursion or the amount of environmental contamination that might have resulted were not given extensive consideration. However, curtains were provided to separate the region in which the pile was

1.2.5 Economics

It is worth reemphasizing that containment measures are taken solely to protect the general public by amelioration of the consequences of a "public-safety accident" as defined in Chapter 1, Sec. 2.1. The expenditures assignable to containment do not contribute to the generation of useful power but, instead, contribute to the public safety. In this sense, containment expenditures are to be contrasted with safety and protective measures designed to prevent reactor accidents. Expenditures for accident prevention are made in the best interests of the operator of the plant, as well as those of the general public, and cannot be said to constitute an undue financial burden on the facility operator. It is indeed fortunate that in only one area, containment, can the expenditures for safety be said to be nonproductive. However, even in this case, all reactors must have some protection against the weather, provisions for handling heavy equipment and radioactive fuel, protection for the operating crew, etc. Therefore, only the difference in the cost of two possible forms of housing (i.e., containment versus a minimum housing) should be considered as a true safety expense. In some instances there will be a substantial difference between the two kinds of housing, but in others the difference is not as great as might be imagined.

It is difficult, if not impossible, to make accurate estimates of the real costs of containment. Several attempts have been made [8,10,13]. Many of these include costs for reliable emergency electrical-supply systems and for emergency coolant-flow and other components or subsystems required to ensure that the plant will not be damaged under any credible circumstances. These expenditures cannot be divorced from adequate plant-protection requirements and are simply a part of the economics of nuclear power plants versus fossil-fueled power plants.

However, the containment cost studies do tend to show, for large plants, a general range of the order of \$10 to \$30/kw(e) for pressure-containment costs; for pressure-relief confinement methods the studies show even lower costs of \$2 to \$3/kw(e). At present, containment costs are by no means stabilized. Some of the present concepts may be providing less safety at higher costs than others. There are likely to be major developments in the next few years, which will lead to containment methods that may be less expensive and yet provide even greater safety.

1.3 Goals of Reactor Containment

1.3.1 The Primary Goal: Containing Fission Products

It may be of interest to set down briefly some of the principal goals that are sought by use of reactor containment. It must be emphasized that these are goals and that they have not been achieved to date in any containment design. In a sense, the success of the design can be measured by how close it is to achieving these goals.

The central goal may be stated as follows: A containment should prevent significant fission-product leakage of radioactive elements under all physically realizable circumstances and at all times. Obviously, if this goal is achieved, the purpose of containment is fulfilled and the containment is a success. The use of the term "physically realizable circumstances" defines a level of reliability above that meant by "credible." The former term implies that any situation that is possible under the laws of nature could conceivably happen. This goal is not likely ever to be fully achieved, although some present containment structures appear to be reasonably close to to it. In addition, there now appear to be ways to approach even closer without overly severe economic penalties.

The word "significant" in the stated goal is meant to imply that the levels of release set forth in whatever regulations are current are complied with under all physically realizable circumstances. At present in the United States, this implies compliance with 10 CFR 20 of the Federal Regulations under normal circumstances and with 10 CFR 100 under very severe and unusual accident conditions. Ultimately, as large reactors are sited near cities, "significant" will probably imply that even in a very severe accident no person anywhere outside the containment should receive a dose in excess of 25 rem whole-body or 300 rem thyroid in a period of about 2 hr (or the length of time considered necessary to evacuate him from the area).

1.3.2 Protecting from Missiles

The reactor containment should remain intact and leak-tight against the onslaught of any missiles from within or without. This goal implies that missiles of any sort should not be able to penetrate the containment or to lower its effectiveness. Several designs approach this goal. For instance, a heavy concrete containment with thick walls of concrete both inside and outside of a leak-tight steel shell should resist most missiles. An underground containment with the turbine-generator located in a separate vault and with the reactor vault lined with a concrete wall inside to protect the shell would come close to satisfying this goal.

Except during warfare, missiles that might be generated within the containment are generally considered far more dangerous to reactor safety than those from outside. Such internally generated missiles include plugs, valve bodies or parts, pipe sections, control rods or mechanisms, and even fragments of the pressure vessel or of the core itself. If the missiles are generated within the containment, there is a high probability that the primary system has been breached. If the rupture of the primary system also causes the containment to be breached, the containment is useless. Only adequate emergency core-cooling can then prevent the release of at least some fission products into the public environment. Containment sprays can reduce the overpressure and can wash down some fission products, but some finite fraction will certainly escape. Thus, protection of the containment against internal missiles conceivably generated by failures in the primary system is exceedingly important. Of lesser importance are missiles generated within the containment by the release of stored energy from regions not connected with the primary system. A typical example might be the failure of a secondary-system pipe or a pressurized-gas bottle stored in the containment.

Of still less importance would be the failure of the containment by a missile generated outside the containment. Since the reactor itself and the vital parts of the primary system are normally protected by heavy biological shielding, as well as by the containment shielding, it is unlikely that a missile from outside could cause damage to the primary system. Thus, the most serious consequence, if a missile strikes the containment from outside, is likely to be a containment rupture without harm to the reactor. However, wars and large aircraft could constitute serious missile hazards under certain circumstances. In order to avoid one possible source of external missiles, it is normal practice to align the turbine-generator so that failure of this rotating machinery cannot breach the containment or damage other vital equipment or instrumentation.

1.3.3 Containing Total Energy Released

The reactor containment should contain successfully any energy release of which the primary system is capable. As will be discussed in Sec. 4.3, the total energy release from any conceivable accident can have several components, including stored energy associated with the pressure, volume, and temperature of the primary system; nuclear-transient energy; chemical-reaction energy; and nuclear-decay heat. The stored energy and the nuclear-decay heat may be estimated fairly accurately. The other two components will be very much a function of the materials used in the system, the details of the design, and the possible methods of reactivity addition.

At present, a proven upper limit on the conceivable magnitude of the total energy release cannot be set, although for several types of reactors upper limits can be estimated with some confidence. Of the components of the total energy that may be released, the stored (thermodynamic) energy and the nuclear-decay heat appear to be the best known; for many reactors these are the largest components.

The wording of the goal implies that the containment should withstand the energy release independently of the rate of release. At present, this cannot be achieved for several commonly used systems.

1.3.4 Dissipating Afterheat

The containment design should include a completely reliable and adequately sized core- and containment-heat-removal system capable of continuous operation after an accident. In many containment designs the containment envelope acts as an insulating layer. In such designs, if there is an accident that releases energy and if there is no means for removal of heat from inside the containment, the pressure and temperature inside continue to rise, and rupture is inevitable. The heat-removal system is vital in such designs. In other designs, the containment envelope is essentially bare steel and may not require a separate heat-removal system.

An adequate emergency core-cooling system must function properly regardless of the location of the break in the primary system, regardless of the size of the break, and in spite of any other possible interfering complexities. It must function until the containment can be entered and the break in the primary system repaired. Residual pressure in the primary system, vapor evolution, siphons, or coolant pressure heads within pipes must not prevent operation of the emergency cooling system.

Containment-spray systems do not function to cool the core directly. Their purpose is to remove heat from the air-steam mixture within the containment by cooling and condensing the steam. In so doing, they reduce the pressure within the containment. Water used in such sprays must be cooled and recirculated in a heat-exchanger pumping system.

The reliability of such systems is very difficult to assess. Furthermore, it is difficult to devise tests that will demonstrate that effective reliability is being achieved.

1.3.5 Mitigating Core-Melting Effects

The system should be so designed that either (a) core melting cannot lead to violation of the containment or (b) gross melting cannot occur. The afterheat resulting from the decay of fission-product radioactivity is sufficiently great in some reactors that substantial portions of the core could melt and remain molten if coolant or coolant flow were lost for a sufficient interval of time. In turn, this molten mass could melt core internals and conceivably melt through the reactor vessel itself. It could even melt into the supporting structure beneath the reactor and thus penetrate the sealed-off base of the containment. In fast reactors, there may be problems of reassembly of the fuel into a supercritical mass.* Unless cooling can be supplied at some phase of this process, the melt-through will continue until a heat balance is established with the materials surrounding the molten region such that the fission-product-decay heat is dissipated into the surrounding solid materials without further melting of these materials.

The extent and consequences of such a melt-through are a function of the type of reactor, its detailed design, and the containment design. A combination of intensive and extensive variables is involved. One intensive variable is the heat generation per unit volume. The fraction of fission products per unit volume of core material can be greater in fast reactors than in thermal reactors, if the design objectives of high burnup and high power density are achieved. Hence, higher afterheat generation rates per unit volume are likely, and, consequently, a molten fast-reactor core might be expected to melt through to a greater depth than some other types. The total quantity of heat generated is a function of total reactor power, an extensive variable. A large total reactor power might cause melt-through even in other types of reactors, such as solid-moderated and water-cooled types, that lose their coolant in an accident.

It would appear possible to design a reactor so that even in the event of complete loss of core coolant, the molten fuel would not penetrate the containment. The molten pool will no longer increase in size when the rate of heat transfer per unit area from the outer surface of the molten mass to the solid surroundings, multiplied by the area of the outer surface, is equal to or greater than the rate of evolution of decay heat. It may well be possible to reach this equilibrium condition before the leak-tight containment has been breached.

On the other hand, it is much more desirable that no gross core melting occur at all. Melting can be prevented if it is ensured that adequate core cooling will always be available within the primary system. As indicated in Sec. 1.3.4, the design of a reliable system, which will add coolant and provide for its circulation indefinitely, and which can be tested, is most difficult to accomplish.

Water added by a containment spray may act as a heat sink for the containment as a whole but, with most current designs, will not provide much help in core cooling. Eventually, this heat must be removed by pumping the water through heat exchangers before it is returned to the containment system. The containment spray system may act to build up a volume of water in the bottom of the containment, which could provide some cooling for the pressure vessel or the molten core. However, the escape of the molten core through the primary-vessel bot-

*Core melt-through for fast reactors is discussed in Chapter 10, Sec. 4.3.

tom into the water may lead to unpredictable and violent results.

1.3.6 Providing for Containment Testing and Surveillance

The reactor containment must always be in a state of readiness to accept the consequences of all credible accidents without violating the containment integrity. It is necessary to provide assurance that the containment system will function properly if required to do so. For the system to operate correctly, it must be reasonably leak-tight, it must be able to withstand any internal pressures that may develop, it must be protected from or be able to withstand missiles, and it must have its engineered safeguards in a state of complete readiness.

It would appear to be necessary to verify that the containment can withstand the design pressure by an appropriate set of initial proof tests. In addition, leak tests should be conducted as a function of pressure when the reactor is ready to start up. Subsequently, a means should be provided to verify continuously or periodically that the leakage rates remain acceptably low. In general, the tests and monitoring systems now being used to check leak-tightness are not too far from the best that can be expected.

Surveillance methods must be provided to ensure that missile-protection systems and engineered safeguards are not allowed to deteriorate with time.

1.3.7 Making Containment "Passive"

The containment system that is always ready to accept the consequences of any accident is more likely to be effective than one that requires certain positive actions to be taken before it is ready.

A system capable of handling the consequences of any accident without requiring valves to close, emergency sprays to come on, or pumps to operate will be safer than one that has such requirements. A system that requires no action can be termed a "passive" system. One that requires certain actions can be termed an "active" or a "dynamic" system.

A containment system that permits air to circulate through the containment and back out must have ventilation valves that close in order to make the system effective. Such a system would appear to be somewhat less safe than one that is always sealed.

A containment that provides sufficient heat transfer through its walls so that no internal, recirculating containment-cooling system is required to dissipate the afterheat is more likely to prevent containment overpressure than one that requires the operation of a separate, dynamic heat-removal system.

It is likely that the ideal passive system can never be developed, since core cooling, at least, will probably require the functioning of one or more coolant-recirculation pumps (although the pumps could function from positions outside the containment and could make use of existing heat exchangers in some designs, thus reducing the number and extent of emergency actions required in case of an accident).

1.3.8 Incorporating Reliable Engineered Safeguards

Means must be available to ensure continuous operation of all vital active engineered-safeguard components for long periods and under the accident conditions that could exist. If the goal of a passive system is unattainable, then it is necessary to establish an essentially 100%-reliability criterion for certain components. In the event of a serious accident, it is necessary that those measures required to protect the general public function initially and continue to function indefinitely afterwards, at least for a year or so. It has become customary in the United States to identify certain special measures taken to ameliorate the consequences of an accident as engineered safeguards. Some of these play a vital role in the protection of the public.

Among the most important engineered safeguards are core-cooling, containment-atmosphere-cooling, and leak-tightness provisions. If core cooling can be maintained so that little or no core melting occurs, the consequences of an accident will not be too serious, either to the public or to the owners. Therefore, continuous and prolonged core cooling is the first essential. It is also essential that the design pressure of the containment not be exceeded. Therefore, heat-removal provisions, and perhaps containment sprays, must be capable of continuous operation for prolonged periods. Certain valves or other components may be required to function under a variety of conditions of temperature, pressure, and environment. If electrical motors or switches or seals are involved, they must be able to operate under the conditions that may follow an accident, including atmospheres of steam and air, high-level radiation, water, and perhaps corrosive atmospheres. Valves must function and often must be remotely operable, in spite of any damage that the accident may have caused (including primary-system rupture) and in spite of any conditions that may exist later. Components such as charcoal adsorbing beds must be able to operate in mixed steam-and-air atmospheres and must be able to dissipate the heat from the adsorbed fission products.

All these requirements mean that the engineered-safeguard systems must be maintained at the highest level of readiness at all times before any possible accidents. The reliability of such systems is difficult to assess or to test. Consequently, the achievement of this goal is one of the most difficult problems facing reactor designers today.

1.3.9 Providing for Reactor-System Surveillance

Normal surveillance of the reactor system within the containment must be possible. In all reactors it is necessary, or at least desirable, to enter the containment for a variety of reasons. This increases the risk to the personnel involved, but probably the risks are not greatly different from those of anyone who carries out similar tasks near any high-pressure, high-temperature system. In many reactor facilities, important mechanical devices, such as pumps and even turbines and electrical generators, are located within the containment. Often these devices may require servicing during operation. Certainly the system reliability will be increased if these devices, and the system as a whole, are frequently inspected.

One of the primary means by which accidents have been prevented to date in conventional plants, as well as in nuclear plants, is frequent and careful inspection. It is argued, and argued with a great deal of factual backing, that pipe breaks and failures in valves or other components can usually be detected before serious consequences develop if there are regular inspections. The size and complexity of nuclear plants make it virtually impossible to locate a sufficient number of sensors to detect all possible failures. Even listening devices cannot reveal the first small leak or, in any case, cannot show exactly where it is. There are no good "universal" locations for television monitors within containments. Many signs of impending trouble may be difficult to locate from inspections conducted outside the containment. There appears to be no adequate substitute for regularly scheduled inspections of the system in order to ensure that deteriorating or abnormal conditions are detected early.

Some systems require the containment to be filled with an inert gas (usually nitrogen). For instance, some reactor containments are filled with inert gas to ensure that no fires can develop.* Others can be "inerted" to reduce the consequences of a metal-water reaction in event of a serious accident. The existence of an inert atmosphere in the containment discourages frequent inspection within the containment, thereby increasing the vulnerability of the system to accident. Moreover, inspection within the inerted containment involves the possibility of accidental asphyxiation of employees who may enter the containment without adequate self-contained oxygen supplies. (Asphyxiation accidents have led to several deaths in the AEC programs to date.) In addition, if an employee wears proper self-contained breathing equipment, it is difficult for him to get into confined spaces where inspection may be desirable. The equipment also reduces the effectiveness of the inspector, as he has difficulty in communicating with his colleagues. All these factors (plus the real discomfort of prolonged use of breathing apparatus) constitute barriers to effective surveillance of the containment interior.

Some inspections of the primary system within the containment must be made with the system at operating temperature and pressure. The "heatup" of the system is accompanied by a variety of differential expansion effects and leaks may develop. These possibilities must be checked by inspection. To date at least, it has not been feasible to inspect all the key areas remotely. A number of facilities have installed sensitive microphones in the containment, with loudspeakers in the control room permitting the operators to hear any changes in the noise level of the system such as might be caused by steam leaks, increased bearing noises, etc.

Some reactor containments can be entered while the reactor is at full power (e.g., Dresden, Yankee), while others, because of the shielding design, can be entered only with the reactor at zero power but with full pressure and temperature (e.g., Indian Point-1.) Normally, construction of a personnel-access airlock is worthwhile for any power reactor, not only from the reactor-safety viewpoint, but also from the economic viewpoint. Without an airlock, it is necessary† to shut down and depressurize the primary system before entry—an expensive process.

The ideal arrangement, of course, would permit access to the containment for regular planned inspections and would, at the same time, provide for sufficiently large containment volume and containment design pressure to accommodate any possible energy releases. This ideal may be extremely difficult to realize.

*For example, N. S. Savannah (see Sec. 5.6 of Chapter 11) and the Peach Bottom Gas-Cooled Reactor.

†If the primary system is pressurized, an energy source exists that could cause primary-system rupture. Hence, the containment should be sealed off when the system is pressurized, just as it should be sealed off when positive reactivity changes could occur.

The use of an inert atmosphere in the containment has one definite advantage. If air is present in the containment, the chemical energy available during an accident may be increased by reactions of the oxygen with the fuel, with metals, or with hydrogen that may be released. If the system is "inerted," chemical reactions requiring air are no longer an important energy source. Thus, the decision of whether or not to fill the containment with a carbon dioxide or a nitrogen atmosphere is a difficult one.

1.3.10 Planning Postaccident Rehabilitation

Plans and means to return the entire system to normal after an accident or to neutralize it permanently must be available. Little thought has gone into this aspect of reactor accidents to date, largely because of the small probability of a major reactor accident; nonetheless, careful reflection shows that such provisions are necessary. The cleanup after the accidents at Chalk River (see Chapter 11, Secs. 3.3 and 4.2 and Chapter 20, Secs. 7.3 and 7.4) shows what is involved in the rehabilitation of a reactor. Appropriate efforts during design of a reactor system can substantially reduce the problem of cleanup.

Even the rehabilitation of a reactor with a relatively minor fuel-element break may involve serious problems. For instance, the disassembly of Clementine, the world's first fast reactor, was carried out because of a single fuel-pin rupture, which caused the plutonium fuel to disolve extensively in the mercury coolant. The resulting metallic solution was said by metallurgists to be pyrophoric on exposure to air, and its vapors were extremely toxic. Yet the reactor could not be abandoned intact, since its reactivity-control properties over a long period could have changed substantially. Prior plans for disassembly in such an eventuality would have helped.

The Windscale Reactor, which was ruined by a fire (see Sec. 3.7 of Chapter 11 and Sec. 7.2 of Chapter 20), was placed in a safe condition, sealed off, and abandoned without disassembly. The SL-1 Reactor was completely disassembled and removed. The cleanup after all of the major accidents was costly (in a number of instances, well over a million dollars) and time-consuming. In almost every case, planning during the design stage could have made the job much easier.

If a major accident were to occur in a large power plant, it would eventually be necessary somehow to remove, encapsulate, and dispose of radioactive material. The first steps would be taken from positions outside the containment. At some point, personnel would have to penetrate the containment and decontaminate the equipment sufficiently to permit continued operation of the plant, if that were economically feasible. In any event, the fuel, regardless of its condition, would have to be removed and stored or reprocessed. Such difficult processes could be carried out better if the designers anticipated the situation and made appropriate provisions. The extent of such provisions involves economic decisions at the design stage and, since many designers do not believe serious accidents will happen, the decisions tend to be made without considering postaccident rehabilitation.

2 GENERAL METHODS OF REACTOR CONTAINMENT AND CONFINEMENT

2.1 Introduction

Some facilities provide containment for the entire primary system. Examples include the Shippingport, Yankee, and San Onofre plants. Others permit major pipes of the primary system to penetrate the containment (often called partial containment). Usually, the pipes are for steam and condensate lines to and from the turbine. If the primary system fails external to the containment, isolation valves in the primary system must close to seal off the reactor section from that part of the primary system which is outside the containment. Examples are the Dresden-1, -2, -3, Humboldt Bay, and Quad Cities plants. Other systems can be said to be contained only after appropriate isolation valves have sealed off the containment openings. The NPR reactor plant is an example of this type. Numerous research reactors, such as the MITR, that circulate fresh air through the containment are also in this class.

From a practical viewpoint, it has proved necessary, even in the so-called fully-contained systems, for some small primary-system pipes, such as feedwater and sampling lines, to penetrate the containment. Thus, the terms full containment and partial containment must be viewed as relative. The variation from case to case in the fraction of the primary system contained and in the sizes and vulnerability of the pipes that penetrate the containment is great. In all primary-system lines that penetrate the containment, there must be automatically actuated and reliable isolation valves.

Some facilities have special provisions to reduce rapidly the pressure that would result in the containment if the primary system ruptured and released its pressurized coolant into the containment. Of particular note, since it has been incorporated into a number of reactor facilities, is the system that uses a pool into which escaping steam may be discharged and condensed. Some systems employ dousing sprays or other means to reduce pressure, including pressure-relief valves.

In order to reduce the leak rates, some systems employ various schemes that amount to double containment with or without "pump-back." (In pump-back, the leakage through the inner shell to the annular space is returned to the inner region, accompanied by any leakage into the annular space from outside.)

Confinement systems permit normal air circulation through the confined volume during routine operations, usually with the effluent passing through filters and scrubbers. The behavior of confinement schemes during accidents can be classified into two categories. In the first category, the pressurized coolant is released to the environment early in the accident (i.e., before fission products begin to be transported from their normal location within the fuel elements), and then the confinement zone is sealed off completely in order to retain the fission products that may be freed later during fuel melting. In the second category, the released effluent is filtered or otherwise treated at all times, both during the accident and afterwards, in order to remove at least some portion of the fission products. This type of confinement system would be likely to release substantially all the noble-gas fission products; hence, it is not generally as effective as systems in the first category.

The various containment and confinement concepts and their advantages and disadvantages are discussed in the following subsections. The subsections also discuss various independent or semi-independent concepts, which are often combined in various ways to fulfill specific requirements.

2.2 Full Containment of the Primary System

A reactor system is said to be fully contained if all the major components of the primary system and their interconnecting pipes are within the containment. The primary system consists of all pipes and components that contain or could contain the primary coolant.* In a pressurized- or boiling-water system, a sodium-cooled reactor, a gas-cooled reactor, or a fluid-fuel reactor, this definition would mean that the primary-coolant pumps, the reactor vessel, the primary-secondary heat exchangers, and the connecting pipes are located within the containment (see Fig. 2-1). (A liquid-metal-cooled reactor normally requires an additional secondary-tertiary nonradioactive sodium heat exchanger external to the containment.) The only major pipes leaving and entering the containment are those from the secondary side of the heat exchangers.

There are likely to be some lines that connect to the primary system and yet penetrate the containment. Such lines will normally include primary-coolant-purification lines, sampling lines, charge and discharge lines, and perhaps others. In a fully contained system, these pipes are normally small, so that rupture is not a serious accident. The pipes should be easily valved off and controlled. Since they are often used only occasionally, and then under careful surveillance, the question may be raised whether these auxiliary lines are major lines. One approach might be to designate as minor all those lines that are small enough in diameter so that a double-ended pipe break anywhere in the line would lead to a rate of loss of coolant sufficiently small to be compensated by readily available emergency coolant supplies and existing injection methods. In practice, auxiliary lines are usually made as small as possible, commensurate with their ability to carry out their intended function. Prudent design also provides automatically acting isolation valves near the envelope of the containment to ensure that these lines will not serve as a means of breaching the containment in the event of an accident. There are usually two valves in a series. The details of this design are discussed in the next section. In the event of the double-ended rupture of a minor pipe, properly operated isolation valve system should not permit the uncovering of the core, and, therefore, no fuel melting should occur. The pressure within the containment might rise to near its design pressure, but the released fluid would contain only a low concentration of fission products typical of normal operation.

In pressurized-water, steam-cooled, or gas-cooled reactors, the secondary coolant is normally identical to, or at least compatible with, the primary coolant so that an internal failure of the heat exchanger does not result in a chemical reaction between coolants. It is permissible, in these systems, to allow the secondary coolant to enter the containment, to flow through the secondary side of the heat exchanger, and to carry the heat removed back out of the containment, usually to a steam turbine. The case of two incompatible coolants, say sodium and water, is discussed later in this section.

There are three requirements that must be placed on the secondary-coolant sections that penetrate the containment. First, in the event of a rupture in the secondary system, the heat exchangers, which are the only points of contact between the two systems, must be able to withstand safely the full pressure of the primary system (in a pressurized-water system, usually 2,000 psi or 135 atm); conversely, the heat exchangers must withstand the full pressure of the secondary system in the event that the primary pressure is lost. Second, the volume of fluid in

*By definition, the primary coolant contacts the fuel during normal operation.

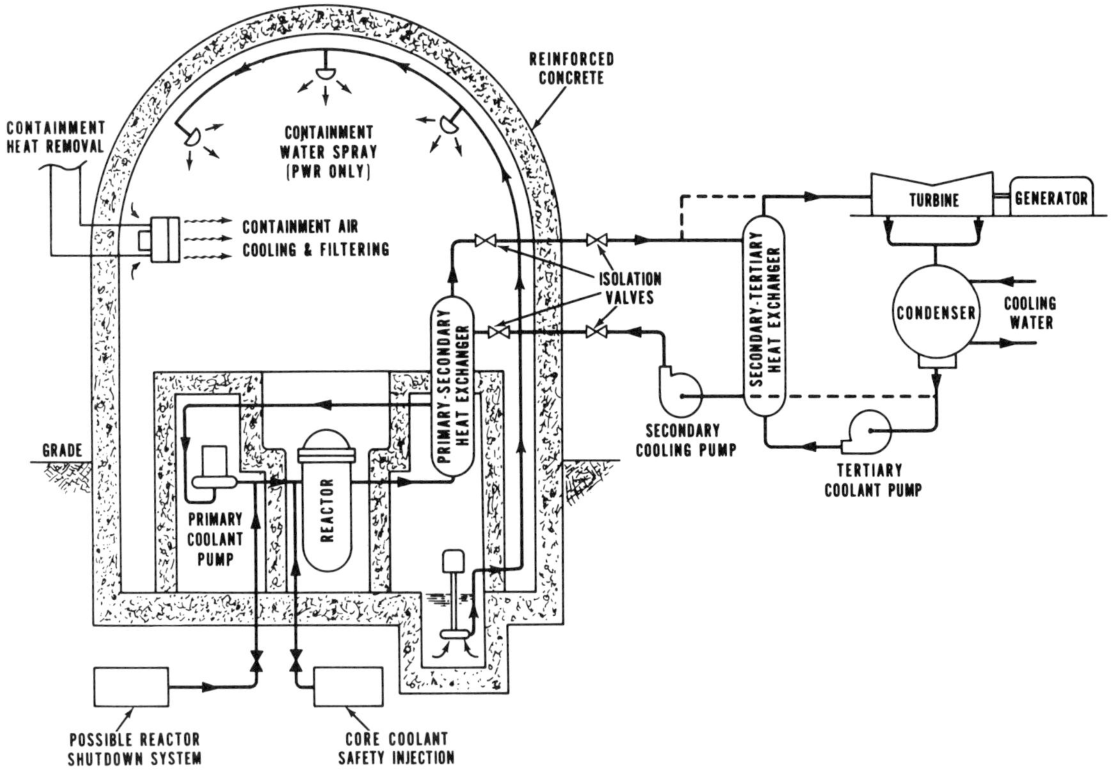

FIG. 2-1 Schematic of high-pressure full containment for a generalized power reactor. Broken lines, eliminating secondary-tertiary heat exchanger, apply to pressurized-water, gas-cooled, or organic-cooled reactors. The secondary-tertiary heat exchanger is necessary for liquid-metal-cooled reactors.

the secondary system and its maximum operating temperature and pressure should be such that, if a double-ended rupture of a major secondary-system pipe were to occur within the containment, the pressure within the containment would not be greater than the containment design pressure. In a sense, this second requirement is not as important as the first, since even if the rupture of a secondary pipe led in turn to the rupture of the containment, the primary system would presumably remain intact. The reactor can be shut down safely as long as the first requirement is met. In any event, the second requirement is not normally restrictive. For the most part, pressurized-water reactors use this type of containment. (One exception is the Elk River Reactor, a boiling-water reactor with a natural-circulation primary loop and a heat exchanger located within the containment, exchanging to a secondary steam system.) Third, the heat exchanger and all attached pipes must not fail because of any loading or combination of forces that could occur during any accident.

In liquid-metal-cooled reactors, it is necessary to include an intermediate nonradioactive heat-exchanger loop between the radioactive primary liquid-metal loop and the outer steam-generating loop. (Primary and intermediate coolants are usually identical.) As mentioned above, the intermediate loop is included because liquid metals and water are generally incompatible (see Chapter 17), and any leak in the heat-exchanger walls that separate the incompatible coolants could result in the release of considerable chemical energy and, perhaps, in the destruction of the heat exchanger. It is not prudent to have a heat exchanger with incompatible coolants within the containment because of its potential for releasing chemical energy in the event of an accident. At the same time, the radioactive liquid-metal coolant in the primary should not be allowed to pass outside the containment. The best solution proposed to date appears to be an intermediate (nonradioactive) coolant loop, where the intermediate coolant is compatible with the primary coolant. The heat exchanger between the intermediate coolant and water or steam is located outside the containment.

In a boiling-water or a gas-cooled reactor system, the primary coolant may also be the driving fluid for the turbine. If so, then full containment would require that the turbine, at least, and perhaps the electrical generator as well, be housed within the containment. To date, this system has been used on only one United States reactor, the BONUS Reactor in Puerto Rico. It has some advantages and some disadvantages. An advantage is that no special isolation valves are needed to close off the primary loop in the event of a pipe rupture. The system has all the advantages of a fully-contained primary system. Another advantage is that a containment designed to house both the reactor and the turbine-generator (with its requirements for an adequate overhaul space) is likely to be large. Its size will result in lower pressure in the containment in the event that all the primary coolant is released. (It is conceivable that other ingenious arrangements might change the containment size greatly for such a plant.) Since it has to withstand a lower pressure, the cost per unit volume of containment vessel will be lower; but since the volume is greater, it may have a higher total cost.

For the fully-contained reactor-plus-turbine system,

the presence of large rotating machines within the containment increases the vulnerability of the containment to penetration by missiles generated by any breakup of the rotating equipment. It is conceivable that a turbine could fly apart in such a manner as to rupture its housing and the containment as well, thus both causing a loss-of-coolant accident and providing the means by which fission products can escape the containment. The cost of adequate protection against this accident may be high. Unfortunately, the possibility of occurrence for this accident is high enough so that it cannot be ignored: roughly 1% of all large turbines have such accidents. The disadvantages just mentioned have been considered, in the past, to outweigh the advantages. As a consequence, it has been common practice in boiling-water-reactor power plants (except for BONUS), and in other plants where the coolant passes through the reactor core to drive a turbine, to have the turbine-generator set outside the containment. (Of course, the external location of the turbine and generator should not be such that a dynamic failure of these units would rupture the containment by means of an external missile, either.)

2.3 Partial Containment of the Primary System

A reactor is said to be partially contained if those portions of the primary system that normally retain the fission products (usually the reactor core with its clad fuel) are within the containment and can be isolated from the uncontained remainder of the primary system.

From a safety viewpoint, the principal reason for a partially contained primary system is to exclude from the containment any components of the primary system whose failure, when inside the containment, would greatly increase its vulnerability. Thus, in a direct-cycle boiling-water reactor, the turbine-generator set, with its probability, small as it may be, for failure under disruptive centrifugal forces, is such a component. If a liquid-metal-cooled reactor were to be designed without an intermediate nonradioactive liquid-metal loop, then the primary radioactive liquid-metal-water heat exchanger would be such a component. (In that case, a specially designed, very substantial, separate containment might be necessary for the heat exchanger).

Aside from the foregoing safety considerations, partial containment also has the advantage of lower costs than full containment.

The concept of partial containment requires that the fission products be retained in the portion of the system protected by the containment. Therefore, in the lines leaving the containment, valves or shutoff mechanisms must be located at the site of the penetration of the containment. The location of the isolation valves, in a sense, defines the extent of the containment. It is of the utmost importance that these valves be reliable, since, if a valve were to fail to operate during an accident, major quantities of fission products might escape, and the containment might be of little use in ameliorating the consequences of an accident.

Since a partial-containment system requires certain positive actions to ensure safety, whereas a fully contained system requires no action, it would seem logical to believe that the fully contained system might be safer, judged on this basis alone. Since certain essential safety actions are required in event of an accident, partially contained systems are dynamic or active containment systems. (See Sec. 1.3.7.)

To illustrate the problems involved in sealing off the part of a system containing the fission products in the event of an accident, consider Fig. 2-2. In the oversimplified figure, a single primary steam-line valve A is located at a distance a inside the containment wall; a return-line valve B is at the wall; and a turbine-stop valve C is at the turbine. Note that only two of these valves are specifically selected and designed for the containment isolation function. The turbine-stop valve C has a different function, but may be useful for sealing off parts of the system as well.

A failure between valve A and the reactor will release the primary coolant into the containment, and, if valve A closes properly, then valve A and the length a of pipe effectively become part of the containment. If valve A does not close, then the containment effectively includes all the external components of the system; it includes the connecting lines up to the turbine if valve C closes, and all the external system if valve C does not close. Since, in this concept, the turbine and connecting lines may be unshielded, they then become a potential gamma-ray source for the local environment, even though this part of the system is not breached. Further, if there are points in this section of the system where interchange occurs between the system and the outside world, such as off-gas lines, etc., these locations now become potential sources for the spread of radioactive contamination to the public. If none of the valves work, there may be convective circulation of fission products through this external loop, and, if there are leakage paths to the outside from this section, the release will be proportionally greater.

If the failure occurs between valve A and the containment wall, the containment is potentially breached; thus, unless at least valve A and either valve B or valve C work, the system may blow down directly to the containment by normal or reversed flow, and fission products can be spread throughout both parts of the system. (To prevent reversed flow, a check valve or valves can be added in the external system.) In fact, if valve A works and valves B and C do not, any fission-product release from the fuel to the containment must occur by reversed flow through the external part of the primary system where the off-gas section is normally located.

If the failure occurs between the containment wall and the turbine, the containment valve A and either valve B or valve C must close, in order to prevent a direct blowdown of the system to the atmosphere.

In one sense, three successively worse breaks have just been described. The first released coolant directly to the containment, but only one valve in three was required to

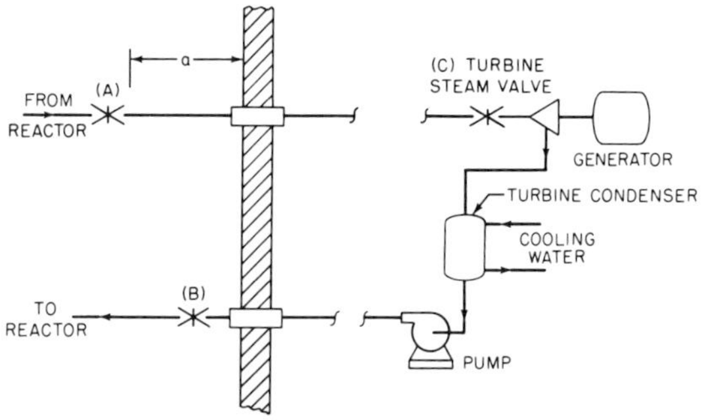

FIG. 2-2 A containment isolation valve system.

close in order to retain any fission products released from the fuel within the containment. The second case required at least the proper functioning of two valves to ensure that fission products released from the fuel were retained within the containment. The third case required that two valves function properly in order to prevent direct release of fission products to the atmosphere.

On the other hand, if the two valves function properly in the third case, the primary core vessel will not blow down at all (it is here assumed that there is an emergency means for recirculation of the coolant through the core to remove heat). In the second case, the primary core vessel will not blow down either. But in the first case, the core vessel will blow down, and emergency in-core cooling of some sort is essential if fuel is not to melt.

Obviously, the first and third cases are more important than the second, since the reactor designer can move valve A close to the containment wall, thus reducing the probability of a failure in that region.

It is clear that proper location of the check valves in the system shown in Fig. 2-2 help the situation materially. The rate of blowdown can also be reduced by addition of appropriate Venturi sections to limit exit flow. The system shown is illustrative and does not represent any particular reactor. In most large reactors now being designed and constructed, there are multiple steam and feed water supply lines, which complicate the problem greatly. However, an analysis of the type described above should be carried out for each system to see that no "unprotected" break can occur.

If a double-ended pipe break were to occur at any of the locations shown in Fig. 2-2, large hydraulic forces of action and reaction could cause violent pipe motion or whipping, which could bend or break the valves, sever the electrical lines that control or operate the valves, or even fracture the pipe-containment connection. For these reasons, it has become customary to locate two valves on each major containment pipe penetration as close as possible to the containment wall. To isolate the two valves, one from the other, and thus to minimize the possibility of an interaction that could prevent either from working, one valve should be located inside and one outside the containment. If there is an accident, the inside valve may be exposed to a severe environment that could cause malfunction. Moreover, the valve located inside the containment is less accessible for maintenance or hand manipulation, especially in the event of an accident. To avoid the possibility of inaccessibility, valve-operating motors and electrical connections specially protected against steam and heat may be required. If both valves must be capable of remote operation, the inside one will require additional electrical penetrations of the containment and hence will be more expensive.

The choice of locations for a valve pair is difficult to make. In some cases, e.g., where large amounts of hydraulic energy are involved, it may even be necessary to duplicate both the inside and the outside valves on one pipe in order to provide adequate protection. This "redundancy" in isolation valves is in keeping with the general control philosophy discussed in Secs. 1.4.3 and 4.2 of Chapter 6. It is also necessary to consider "diversity" (see Sec. 1.4.3 of Chapter 6). The valves themselves, the valve operators, and the power supplies must be located so that both of a redundant set of valves cannot be disabled by a common cause. This is a complex problem, and any proposed solution must be thoroughly studied and reviewed to be certain it is adequate under all conditions.

It should be pointed out that a partial-containment system may not afford adequate protection from an accidental reactor transient where major fuel melting occurs, whether or not a primary-system pressure buildup causes a primary-system break. In such an accident, gross amounts of fission products are released rapidly to the primary-coolant system, part of which is outside the containment. The design of the system and the normal off-gas radioactivity release rate in such a plant is based on a certain maximum allowable inventory in the primary coolant at any time. To exceed this inventory in a gross way because of a transient is not acceptable, and special precautions must be taken to ensure that such an accident cannot occur.

2.4 Pressure Suppression

In its broadest definition, a pressure-suppression containment system makes use of various heat-removal schemes to condense, or to reduce the pressure of, the effluent gases or vapors that might be emitted from a reactor primary system during an accident. Such a system is required to maintain the pressure in the containment within acceptable bounds for any credible accident where primary coolant is released into the free containment volume. In order to keep the initial pressure within the containment low, there must be a rapid condensation of the steam emerging from the primary system. Thus, only by intimately mixing the emerging steam with water (either in a pool or in a spray deluge) can one assure efficient condensation. Circulating fans and heat-exchanger combinations can be used for the long-term removal of decay heat, but not for lowering the containment pressure during the initial break.

Pressure-suppression techniques can be used in conjunction with either full or partial primary containment. There is no reason why such a system need be limited to boiling-water reactors. Pressure-suppression systems have been considered for pressurized-water reactors, although none have been built to date. Although pressure suppression is far more efficient when applied to systems that release condensable vapors in the event of a serious accident, it could be helpful even for high-temperature gas systems. The condensation or cooling system could conceivably involve coolant sprays, discharge of the vapor into a coolant-condensing medium such as water or finely divided ice, or heat exchangers.

One method of pressure suppression, first used in the Humboldt Bay Boiling Water Reactor Plant, is shown in Fig. 2-3. The contained part of the primary system is surrounded by a pressure vessel or "dry well," which is the primary containment barrier. In addition, there is a suppression chamber, which is partially filled with water. Connecting these two chambers is a set of vent pipes, with their outer open ends extending well below the liquid surface, as shown. If a primary pipe breaks within the dry well, steam is released, and the pressure there rises. Steam is forced through the vent pipes and bubbles out into the suppression pool, where it is condensed. If the original pipe break is small enough, the pressure observed within the dry well may never exceed by very much that due to the head of water maintained between the surface of the suppression pool and the vent pipe openings beneath the pool surface. For larger leaks, the pressure in the dry well is a function of leak size, number and size of vent pipes, pressure in the suppression chamber, and other factors dependent on the particular design.

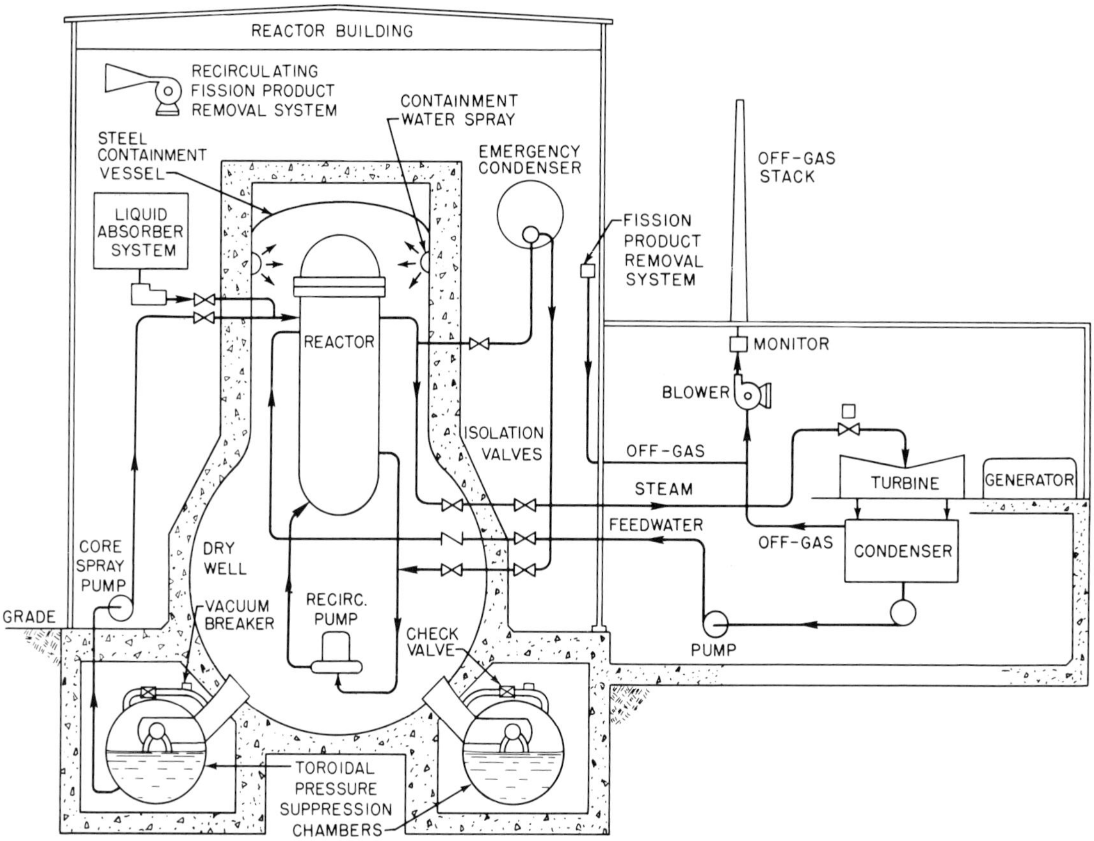

FIG. 2-3 Pressure-suppression containment for a typical BWR, including a superposed refueling region that uses venting confinement as a backup.

Any gases or air swept over with the steam collect in the suppression chamber above the pool surface and then can be retained at a relatively low pressure or released in a controlled manner. The water treatment of the steam and gases released in the accident may also be effective for removal of some of the condensable, particulate, and gaseous fission-products from the effluent steam. Addition of certain chemicals, such as sodium thiosulfate, may enhance the effectiveness.

The principal advantage of this type of containment is a reduction of the size and pressure requirements thus achieving a possible economic advantage over a simple partial- or full-containment system.

The suppression chamber is sized so that the maximum pressure reached is low. The pressure is a function of the volume of noncondensable gases in the dry well and suppression chamber, the thermal efficiency of the suppression system, the temperatures in the dry well and the pool, the transport rate from the dry well, and perhaps other parameters. For the system to be most effective, the volume within the dry well should be small; this minimizes the volume of entrained air or gas that can be carried over into the suppression chamber. Otherwise, the suppression chamber must be larger; or else the pressure in it will be high, and some of the advantages of the method will be lost.

Only that part of the primary-system piping within the dry well is protected.* In the example shown in Fig. 2-3, only those pipes between the pressure vessel and the containment-penetration points are protected, so that the pressure-suppression system is protecting only a few feet of primary-system piping against the consequences of pipe breaks. In most present installations, however, recirculating pipes, pumps, and other primary piping systems have been contained within the dry well, so the usefulness of the method is greatly increased. However, each reactor containment should be examined carefully, since it is always good design practice to understand clearly the degree of protection provided by any given safety system.

In common with other methods designed to permit release of the effluent vapors (steam-air mixtures) from the containment during the course of an accident, the pressure-suppression containment for a specific reactor must be designed so that the dry well can withstand successfully the overpressures due to any credible primary-

*In the United States, it has been considered incredible that a properly designed and constructed pressure vessel would rupture and breach the containment. Containments therefore, have not been designed to withstand a gross vessel failure.

system rupture or subsequent events such as metal-water reactions.

It is conceivable that the cooling of the dry-well section of the system after the accident could cause a partial vacuum in that zone. This vacuum could force water from the suppression pool back into the dry well, and another pressure surge could be developed in the dry well when the water contacted the hot primary system. Fission products might then be transported from the dry well to the suppression chamber, especially if the water trap were lost from the suppression pool because of these events. As a result, the maximum design pressure of the suppression chamber could be exceeded.

In some designs, a partial vacuum in the dry well might even cause the collapse and breaking of the containment vessel.* Then, if the core-injection systems were not effective in keeping the core covered, subsequent afterheating of the core could provide sufficient thermal driving force to move large quantities of fission products from the breached dry well to the region directly outside. In pressure-suppression designs to date, the region outside the dry well has been within a refueling building normally maintained at a slightly negative pressure relative to the atmosphere. The effluent gases and air from the refueling building are usually passed through a filter for iodine removal and thence up a stack. Thus, this outer region constitutes a coolant-venting containment as described in Sec. 1.2.6. Because of these possibilities, pressure-suppression designs generally have the pipes connecting the dry well and the suppression chamber equipped with check valves that prevent flow from the dry well to the suppression chamber but allow the reverse flow (see Fig. 2-3).

As indicated at the beginning of this section, other means than condensation pools can be provided to reduce containment pressures rapidly. Heavy deluge-type water sprays are installed at the Hanford Production Reactors to remove fission products from the air in certain postulated accident situations, and similar schemes are part of the plans for other reactors. However, such systems can be used effectively for pressure reduction only if it is known that the accidental release of steam from the primary system will be gradual. In the event of a sudden steam release, the inefficient interaction between steam and spray-water droplets might permit excessive pressure to develop in the containment.

There have been a number of tests carried out to demonstrate the feasibility of containment pressure-suppression systems of various types. In one experiment, the Sargent and Lundy engineers carried out a series of tests, reported in 1957 [14], in which steam was released into a steel pressure vessel. They showed that the cooling effect of the tank walls resulted in a maximum tank pressure considerably lower than that calculated with no such cooling. For example, a total quantity of 266 lb (120 kg) of 1,000 psig (69 atm) boiling water suddenly released into the test tank resulted in a maximum pressure of 24 psig (2.6 atm), rather than the 50 psig (4.3 atm) calculated. With a concrete-covered wall, the observed pressure for the same test was 38 psig (3.5 atm). When the system was arranged so that the boiling hot water was ejected directly into cold water within the containment, no pressure rise was observed.

Encouraged by these tests, a further set of experiments was run in 1959 by Sargent and Lundy [15]. The test equipment included a vertical containment shell, 14 ft (4.27m) in diameter and 32 ft (9.75m) high, designed for a pressure of 100 psig, and a hot-water-steam drum, 4-1/2 ft (1.37m) in diameter and 23 ft (7.01m) long, designed for a pressure of 700 psig. The containment shell was installed underground. The high-pressure drum had a number of 12-in. (30.5-cm) flanged openings along its vertical side, as well as one at the bottom. Before a test, all the side openings were closed by blind flanges except one. This one, covered by an explosion diaphragm designed to break at approximately 600 psig (42 atm), was mounted on a pipe elbow pointed downward, as shown in Fig. 2-4, which indicates the three locations actually used in the tests. Instrumentation, with readouts in an adjoining building for pressures and temperatures within both the containment shell and the drum, was installed. Up to 16 readings per second were obtained by photographic methods. The pressure instrumentation was capable of recording pressure changes to within 0.01 sec.

A summary of the results for 11 tests is shown in Fig. 2-4. Graphs of three typical tests are shown in Figs. 2-5, 2-6, and 2-7. The first test run, shown in Fig. 2-5, resulted in tearing loose the grating and some pressure-vessel insulation. The second test, carried out with the use of the rupture disk at the bottom of the drum, was equivalent to an energy release of approximately 500,000 kw of heat in 5 sec. The outlet was 1 ft (30.5 cm) from the surface of the cold water in this case. Subsequent tests were carried out with the outlet at successively greater distances from the cold-water surface in the bottom of the tank. As might be expected, the observed containment pressures were successively higher with greater distances between the hot-water release point and the cold-water surface.

Figure 2-7 shows the single experiment carried out with no cold water present. The observed maximum pressure of 95 psig (7.4 atm) is about 90% of the calculated theoretical pressure, the difference being due to heat absorption by the walls and other internal surfaces. Note also that there is evidence for shock-wave oscillations in this case, but not in those where cold water was present.

Two sets of tests of the pressure-suppression concept have been run by the Pacific Gas and Electric Company at their Moss Landing Power Station. One was run in 1959, primarily to demonstrate the design capability of the Humboldt Bay containment design, and the second, to demonstrate the capability of the Bodega Bay containment design. Since the second set put in a maximum of about 2.4 times as much energy per unit water volume as the first, only this will be discussed.

Basically, the system mocked up a full-scale 1/112 segment of the Bodega suppression chamber with one full-scale 24-in. (61.0-cm)-diam vent pipe and one set of model reactor and dry-well vessels with about 1/112 of the Bodega design volumes. The vertical reactor vessel was 27 in. (68.6 cm) ID by 21 ft (6.40m) long, designed for 1,250 psig (86 atm). It had a 10-in. (25.4-cm) discharge nozzle at the bottom end. The contained volume was 80 cu ft (2.27m^3), and most tests were run with

*Pressure tests of any containment vessel that has a relatively thin metallic skin lining the inside of a concrete or rock cavity can lead to difficulties. If there is a substantial void volume in the concrete or rock outside, then a leak in the liner will introduce high-pressure gas into the void outside during the test. Subsequent depressurization of the internal region can cause an inward collapse of the thin shell, just as a partial internal vacuum can.

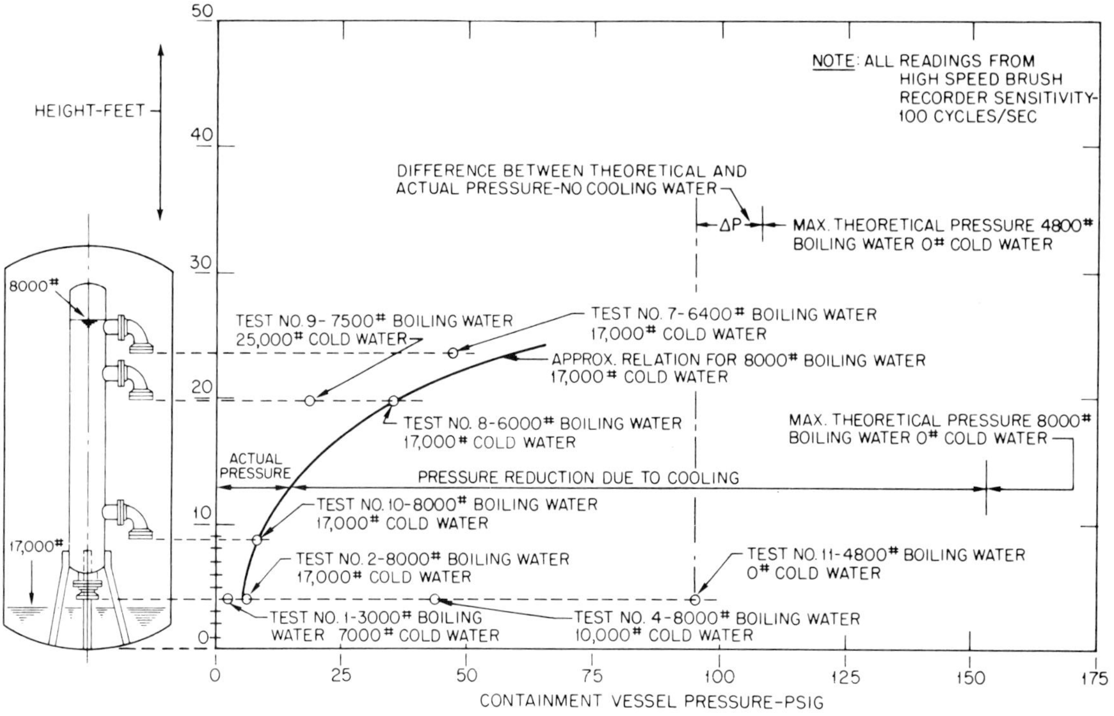

FIG. 2-4 Containment-vessel test, pressure versus height. (Sargent & Lundy Engineers.)

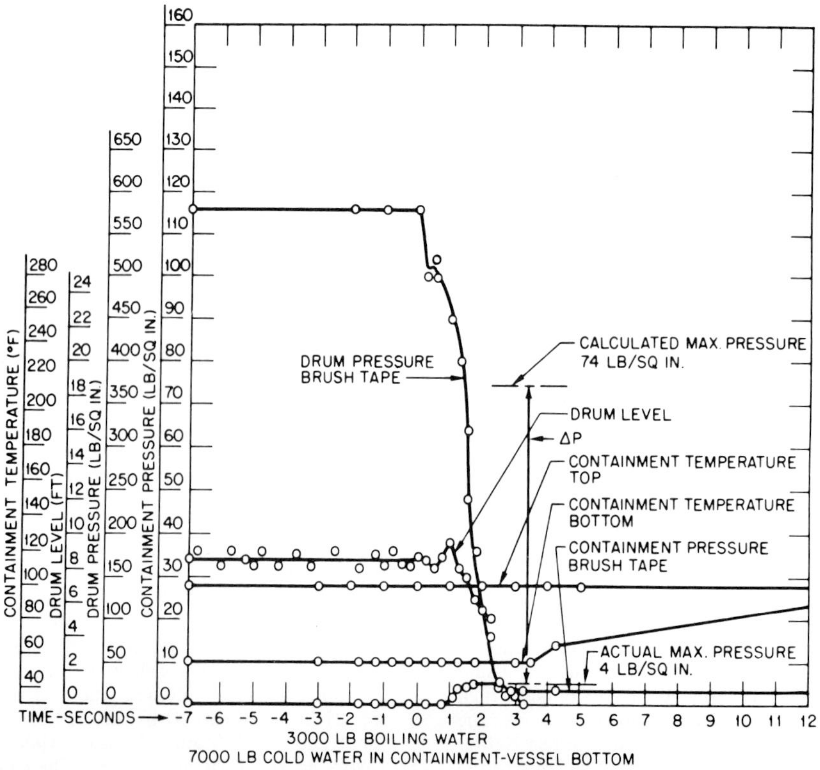

FIG. 2-5 Containment-vessel test (#1, June 19, 1959) showing pressure, temperature, etc. as a function of time.

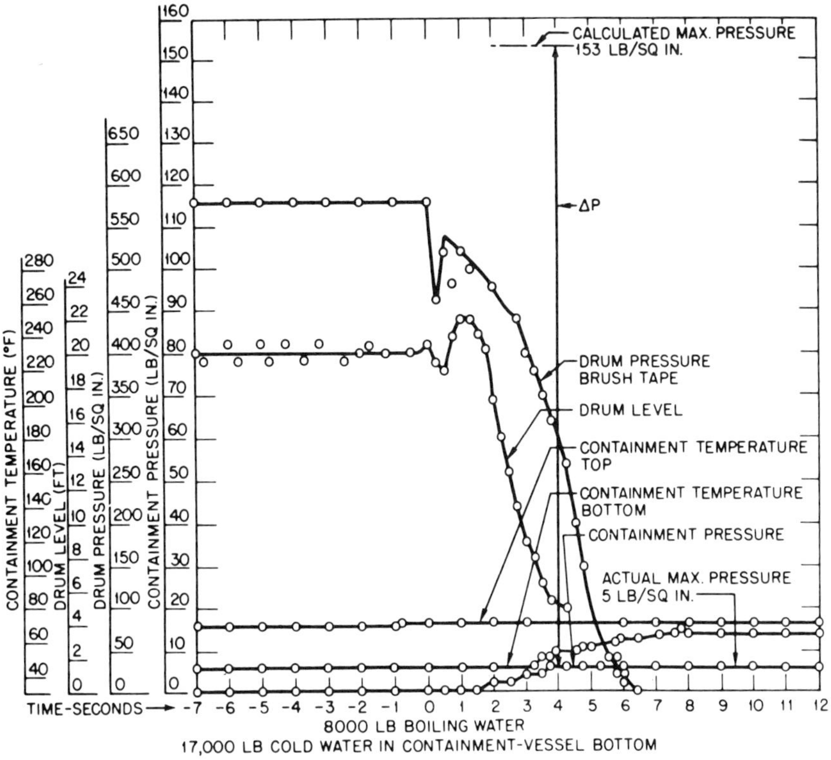

FIG. 2-6 Containment-vessel test (#2, June 19, 1959) showing pressure, temperature, etc. as a function of time.

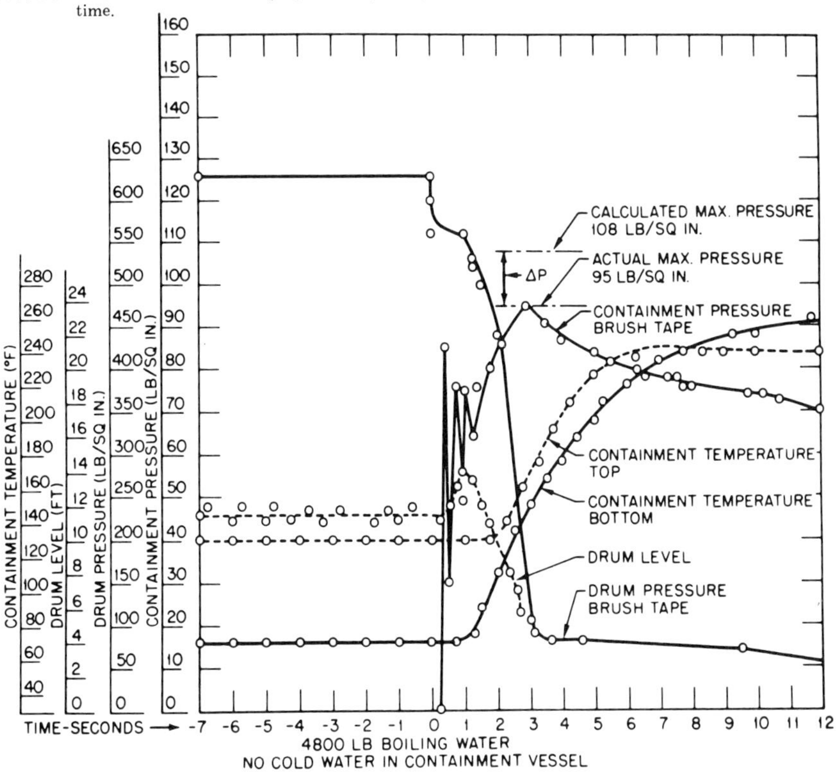

FIG. 2-7 Containment-vessel test (#11, August 13, 1959) showing pressure, temperature, etc. as a function of time.

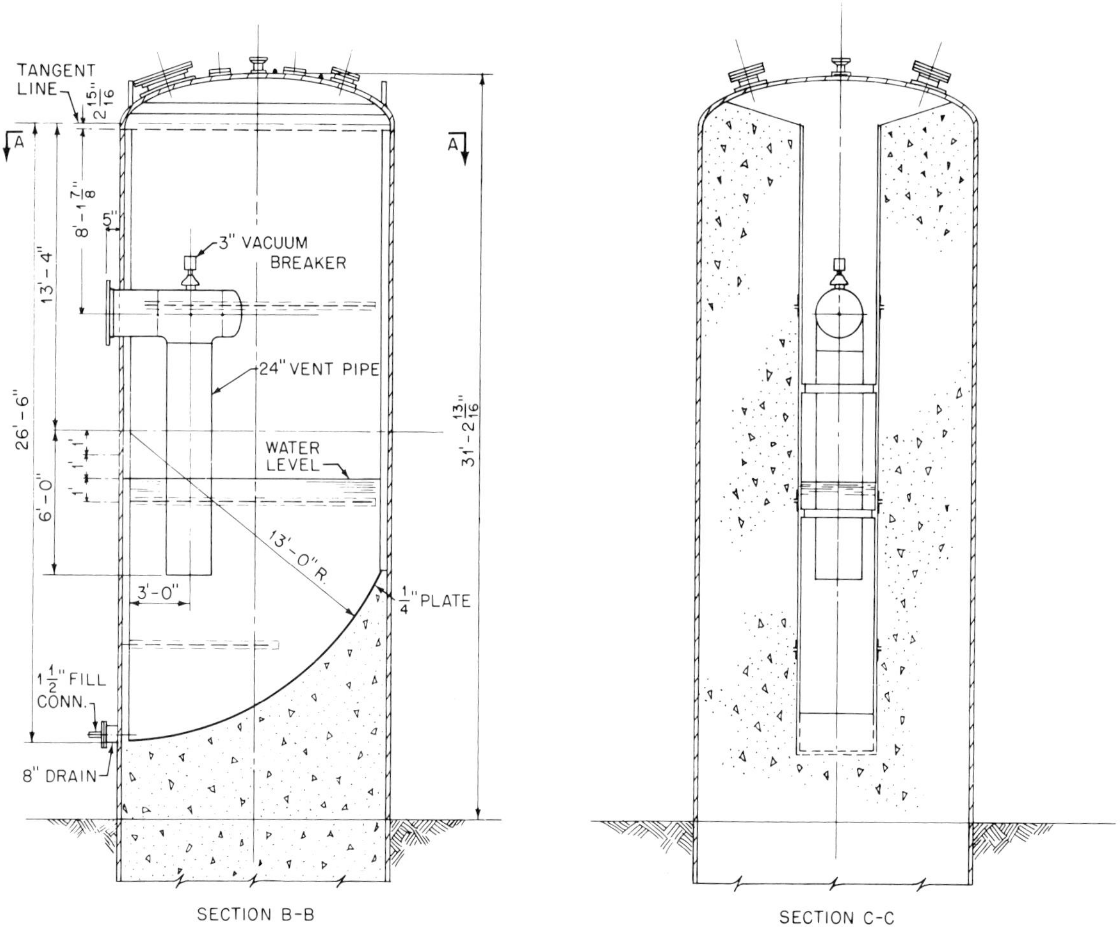

FIG. 2-8 Design of the Bodega suppression chamber.

about 54 cu ft (1.53m^3) of water present ($\sim$ 1/112 of the water in Bodega primary system). The 10-in. discharge line passed a double rupture disk and expanded to a 20-in. (50.8-cm) inlet nozzle to the 85-in. (2.16-m)-ID by 29-ft (8.84-m)-high simulated dry well. From a point on the side of the dry-well vessel about 1/5 of the way from the bottom, a 24-in. (61.0-cm) vent pipe 45 ft-(13.7m) long, containing two tees and one ell, led to the suppression pool. The end of the vent pipe was normally submerged 4 ft (1.22m) in the pool water (normally 11 ft (3.35m) deep). The suppression vessel was 12 ft (3.66m) ID and 49 ft (14.9m) long at the heads. Within this large tank, the suppression chamber was mocked up. The design of the suppression chamber is shown in Fig. 2-8.

The test results and values of the parameters are shown in Tables 2-1a and b. The first six tests were made with the suppression chamber open to the atmosphere. Tests 7 through 13 were run with the suppression chamber closed and water levels normal. However, suppression-chamber pressure was lower than predicted, presumably because all the air from the dry well was not carried over. To increase air carryover, a deflector plate was installed in front of the outlet nozzle in the dry well, so that the jet of steam-water mixture would be dispersed and mixed better in the dry well. Test 14 and those beyond were run with the deflector plate in place. Typical pressure-time traces are shown in Figs. 2-9 and 2-10. The maximum dry-well pressure observed was 63 psig (5.3 atm) for an orifice representing an accident 250% larger than the design-accident break size.

Tests 39 through 45 were run with the dry well pre-purged and with the reactor-vessel water subcooled from 25 to 110F° (14 to 61C°). In preparation, steam was allowed to flow through the dry well until the suppression-chamber pressure stopped rising. Presumably, all air then had been purged from the dry well. The results do not give any trend for these tests and probably indicate differing amounts of air present. These latter tests were designed to simulate accidents during rise-to-power situations.

In all cases when the disk ruptured, the reactor-vessel pressure dropped sharply, the drop increasing with orifice size and amount of initial subcooling. A short period of fairly steady pressure followed, and then a gradual and increasingly rapid pressure decrease. The initial drop was attributed to a brief delay in initiation of flow and start of flashing and to the need for a sufficient pressure loss to establish the rate of flashing that corresponds to the flow rate. The more rapid pressure drop at the end occurred probably because all the water had been expelled from the vessel and there was no source of further flashing to keep up the pressure.

The suppression vessel showed an initial sharp rise of 12 to 13 psig (pressure rise of $\sim$ 0.8 atm), followed by a slight drop and then a gradual rise to a maximum at

TABLE 2-1a

Bodega Suppression Chamber: Test Results and Parameter Values

	Test	Reactor Vessel			Orifice		Dry Well		Suppression Chamber	
		Pressure range during blow-down (psig)	Water volume (cu ft)	Water discharge time (sec)	Diameter (in.)	Water blowdown (lb/sec ft^2)	Maximum Pressure (psig)	Temp.(°F) before /max	Maximum Pressure (psig)	Temp.(°F) before / max
without deflector plate	1	1225/	14.4		0.453			70/	open	66/
	2	1250/	13.7		0.906			65/	"	67/
	3	1250/1050	13.7	5.3	1.66	7700		61/	"	71/
	4	1250/			3.24			64/	"	67/
	5	1250/725	54.8	7.97	3.24	5200	23	71/	"	79/
	6	1250/845	54.8	20.9	1.66	7450	9	65/	open	70/
	7	1250/845	54.8	20.6	1.66	7590	28	78/257	25	93/128
	8	1250/810	54.8	11.2	2.48	6270	24	70/240	21	79/111
	9	1250/720	54.8	7.65	3.24	5440	21	75/229	17	84/110
	10	1250/660	54.5	6.34	3.74	4910	31	62/222	19	76/108
	11	1250/630	54.5	4.72	4.50	4560	42	70/220	18	85/117
	12	1250/590	54.8	3.80	5.12	4420	55	83/230	18	101/131
	13	1250/900	54.8	53.2	0.906	9830	26	61/265	24	71/90
with deflector plate	14	1250/750	53.5	7.45	3.24	5440	37	65/248	27	70/97
	15	1250/710	52.8	5.99	3.74	5030	40	70/242	26	71/100
	16	1250/570	54.5	3.80	5.12	4400	63	77/239	26	82/110
	17	1250/730	54.5	7.88	3.24	5240	37	68/252	28	77/105
	18	1250/900	54.1	46.8	0.90[e]	11040	30	68/270	27	68/82
	19	1250/720	54.1	7.68	3.24[e]	5350	36	60/247	28	63/92
	20	1250/620	52.8	4.24	4.50[e]	4930	52	72/239	27	74/97
	21	1250/700	54.5	7.41	3.24[f]	5580	36	65/240	27	70/96
	22[a]	1210/740	54.8	7.65	3.24	5440	36	75/254	28	76/101
	23[b]	1250/690	54.8	8.20	3.24	5070	37	70/253	29	75/97
	24[c]	1250/705	54.5	7.37	3.24[f]	5600	34	150/261	22	80/110
	25[d]	1250/700	54.5	7.94	3.24	5210	36	67/251	28	70/102
	26	1250/830	54.8	15.75	2.00[e]	6850	33	65/268	29	81/102
	27	1250/690	54.8	7.47	3.24	5560	38	66/245	29	88/112

[a] Approximately 16°F initial subcooling
[b] 5 ft submergence
[c] Preheated dry well
[d] 3 ft submergence
[e] Nozzle
[f] Shortened nozzle

TABLE 2-1b

Bodega Suppression Chamber: Test Results with Dry Well Prepurged and Reactor-Vessel Water Subcooled

Test no.	Reactor Vessel				Orifice diameter (in.)	Dry Well		Suppression Chamber	
	Pressure range during blow-down (psig)	Initial subcooling (°F)	Water volume (cu ft)	Water discharge time (sec)		Maximum pressure (psig)	Temp.(°F) before /max	Press. (psig) before /max	Temp.(°F) before /max
39	1250/640	25	54.8	7.45	3.24 [a]	41	265/269	14/19	120/163
40	1250/600	35	54.8	7.57	3.24 [a]	52	255/271	22/30	65/113
41	1250/510	50	54.8	6.9	3.24 [a]	50	260/274	22/28	77/121
42	1250/620	50	54.8	7.35	3.24	47	262/277	23/29	89/126
43	1250/610	70	54.8	7.14	3.24 [a]	49	262/275	23/29	90/121
44	1250/460	90	54.8	6.8	3.24 [a]	50	260/269	20/25	83/126
45	1250/450	110	54.8	6.2	3.24 [a]	47	257/271	20/28	74/112

[a] Shortened nozzle

about the time when all the water and steam were expelled from the vessel. The sharp rise and drop were attributed to the rapid increase in water height as large quantities of air were first expelled into the water. The gradual rise was due to the purging of the remaining air from the dry well.

Calculations predicted that critical flow would not occur at the end of the vent pipe to the suppression chamber, as was the case in the Humboldt design (14-in. or 35.6-cm diam). Therefore, the dry-well pressure would then be the suppression-chamber pressure plus the vent-pipe-pressure drop. The calculations also predicted that the maximum dry-well pressure would not occur as the water was blown out of the submerged end of the vent pipe. These calculations were generally confirmed in the tests.

The Pacific Gas and Electric Company states that the test results show the following:

1) Condensation of steam in the suppression chamber is rapid and complete for flows associated with break areas, at least up to 250% of design-basis accident.
2) Variations in suppression-pool level of ± 1 ft (± 30.5 cm) did not affect performance of the containment system.
3) Moderate subcooling of reactor water and dry-well preheating did not affect performance significantly.
4) Use of nozzles instead of orifices to simulate a break resulted in an increase of rupture flow of less than 10%.

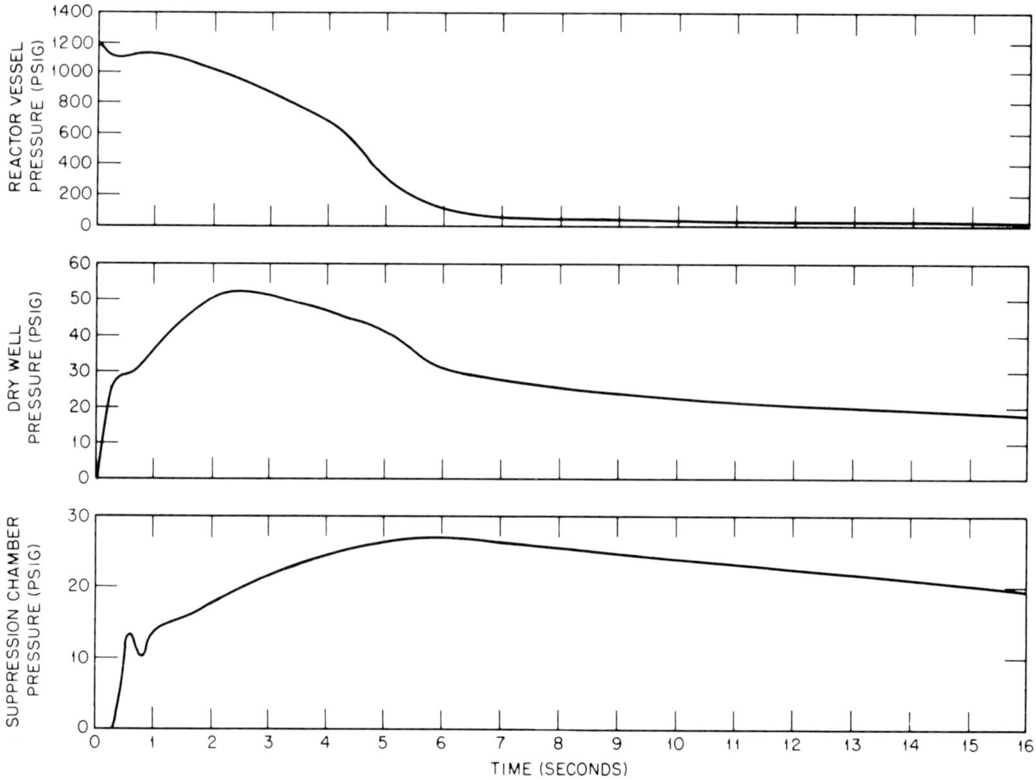

FIG. 2-9 Results of suppression-chamber test (#20) for a 4.50-in. (11.4-cm) nozzle (approximately twice the break area for a maximum credible operational accident).

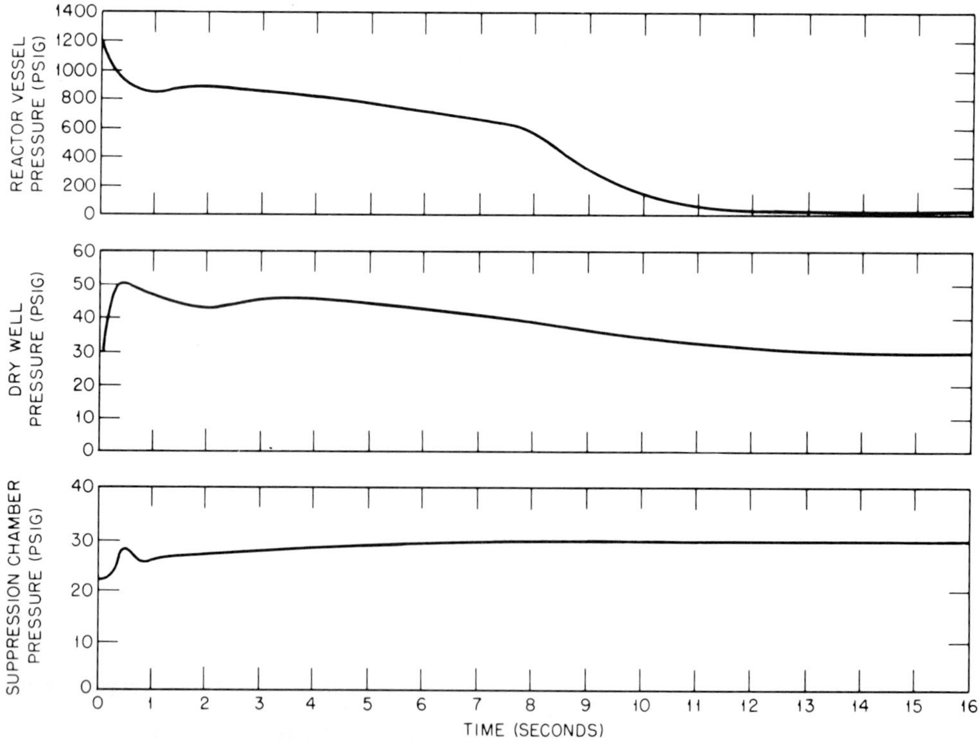

FIG. 2-10 Results of suppression-chamber test (#40) for a 3.24-in. (8.23-cm) nozzle with 35° F (19.4°C) initial subcooling and prepurged dry well.

5) Prepurging of the dry well, coincident with subcooled reactor water, resulted in a maximum dry-well peak pressure of 52 psig (~ 4.5 atm).

References [16], [17], [18], and [19] consider various aspects of pressure-suppression design. The maximum pressure in the suppression chamber is determined largely by the amount of air transferred from the dry well during the design-basis accident. To carry out the calculations completely, a detailed weight-and-energy balance must be considered for the given reactor. It is assumed, on the basis of the above tests, that the steam is condensed rapidly and completely. At least for the Bodega-type plant, the maximum dry-well pressure will occur under the quasi-steady vent-flow condition, because it is higher than the pressure initially needed to expel the water from the pipes. Figure 2-9 shows that the suppression-chamber pressure starts to rise at about 0.3 sec, a rise that indicates clearing of the vents; and the dry-well pressure reaches a maximum at about 2.5 sec (52 psig).

The Bodega Reactor Plant assumed a design-basis accident of a loss-of-coolant from the largest pipe (24 in. or 60.5 cm), coupled with the following pessimistic assumptions:

1) Dry-well air had been previously prepurged by a small undetected leak.
2) The large break occurred instantly and completely, before the reactor depressurized itself or was depressurized by operator action.
3) The large break occurred at a time when there was considerable subcooling, but the reactor had not yet lost pressure relative to the 1250 psig (~ 85 atm) overpressure assumed in the design-basis accident.

Figure 2-11 shows the results of calculations for varying rupture flow rates on dry-well pressure with these assumptions for a Bodega-type boiling-water reactor.

In 1966, the Westinghouse Electric Corporation advanced the ice-condenser reactor-containment concept. This is another method of pressure suppression that can be used in water-cooled and -moderated power reactors. As shown in Fig. 2-12, the primary reactor-coolant system is to be located below an operating deck within the containment. In event of a primary-system rupture, the steam that is ejected into the lower compartment increases the pressure there and opens the door panels at the bottom of the ice bed. The mixture of steam and air moves upward through the bed, where the steam is condensed.

Westinghouse has conducted engineering and full-scale-section tests of the concept. They believe that the pressure-suppression system has the following attractive features:

a. The peak containment pressure in the event of a loss-of-coolant accident is very low—of the order of 10 psig (~ 1.7 atm) or less—and is reduced to a few psi within minutes after the blowdown. These features have been demonstrated by tests duplicating large breaks, such as the maximum credible accident involving blowdowns as short as 9 seconds, and smaller breaks involving blowdowns of 4 hours or longer.

b. The performance of the system is relatively insensitive to large reductions in the ice heat transfer surface area. For example, tests have demonstrated that a reduction in heat transfer surface area by a factor of as much as 5 results in an increase in containment peak pressure of only 1 to 2 psi (~0.1 atm).

c. Storage tests demonstrate that the ice condenser designs under consideration are adequate to preserve the integrity of the ice for at least a number of years.

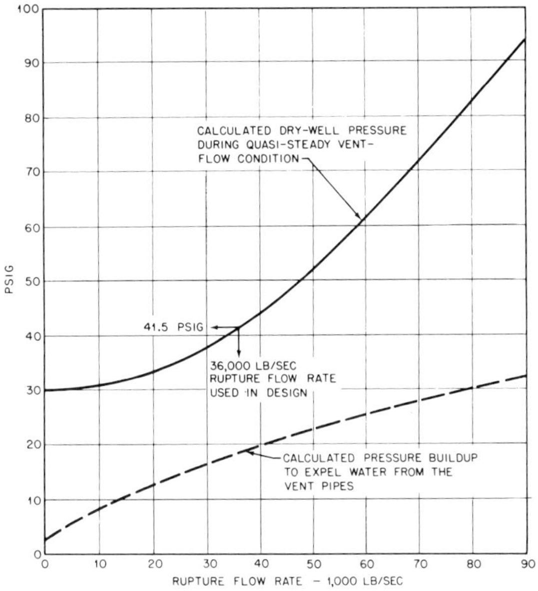

FIG. 2-11 Calculated dry-well pressure versus rupture flow rate for Bodega-type BWR.

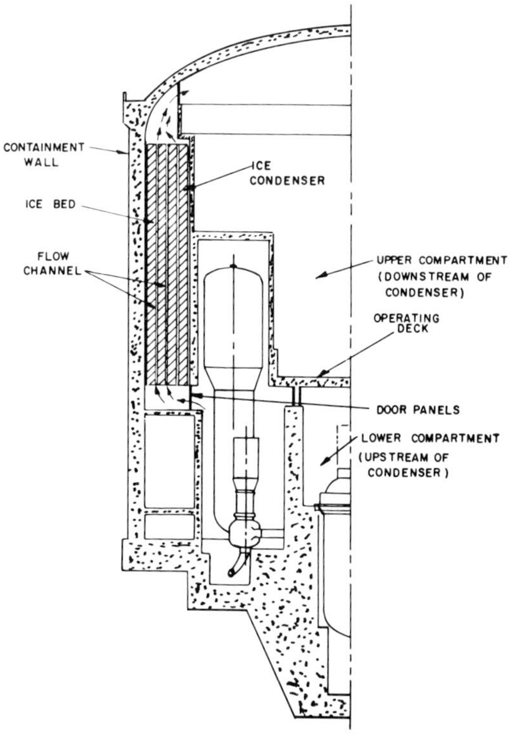

FIG. 2-12 Ice-condenser reactor-containment concept.

They also believe that this system is capable of absorbing a significant amount of reactor residual heat over a longer period of time. The melted ice (borated for pressurized-water reactors that use borated primary coolant) could be used as a water source for containment spray and core-cooling water. The containment could be arranged so that the cool water, together with normal spray water, would flood the reactor compartment to a level above the reactor-vessel nozzles, thus providing a continuing source of cooling water to the outside of the pressure vessel.

Results of the tests are summarized in Fig. 2-13. The results of the test receiver pressure have been corrected for differences in initial air-temperature and air-volume ratios. Ice beds with an equivalent unsupported height of up to 30 ft (9.1m) have been tested at storage temperatures of 10 to 25° F (−12.2 to −3.9° C) to determine the effect of bed depth on compaction rate. The 10° F tests indicate that the compaction rate of an unsupported depth of 30 ft of ice is of the order of 1 to 2 in. (2.5 to 5.0 cm) per year. Tests of such beds now exceed times of over 1 year. Westinghouse indicates that the present approximately 2.6×10^6 ft^3 (7.4×10^4 m^3) of free volume used on typical pressurized-water-reactor designs for a 1,000 Mw(e) plant can be reduced to about 800,000 ft^3 (2.3×10^4 m^3) by use of this method, while the maximum design accident pressure is still held at about 10 psig (~ 1.7 atm absolute). This should result in considerable savings.

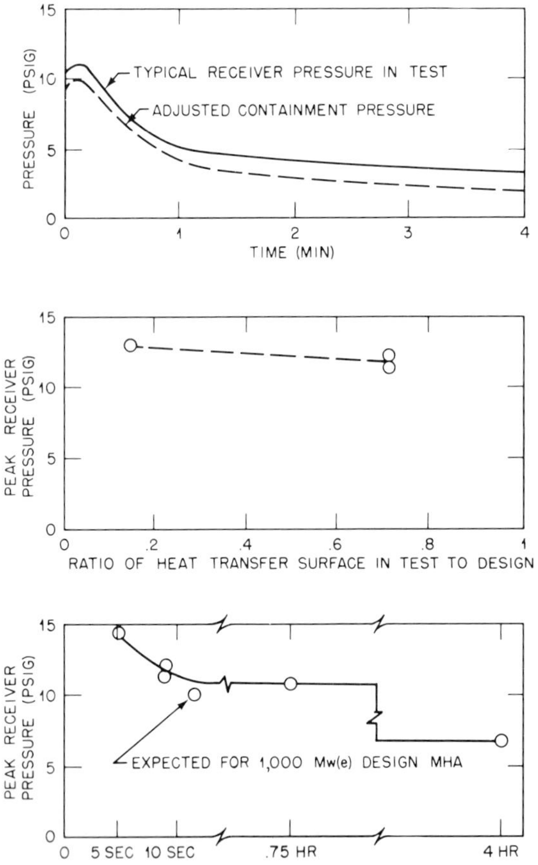

FIG. 2-13 Results of full-scale section tests of ice-condenser reactor containment.

Precautions must be taken to be sure that the system will function correctly. While tests seemed to indicate that melting, mass transfer, and compaction of ice are not serious problems in a suitably designed system, some provision for inspection of the units is probably necessary. Severe leakage of the operating deck seals could constitute a problem but Westinghouse tests seem to show little effect on the peak pressures. Westinghouse ran full-scale-section tests with operating deck leaks equivalent to an opening about 120 ft^2 (11.1m^2) with peak pressures still remaining "about 10 psig." In some areas with severe earthquake design requirements, the large mass of ice located high in the walls may create design problems. Provisions should be made to ensure that the flow of steam upward through the ice is well distributed over the entire ice volume in order to utilize it most effectively and prevent any chance of short circuits. Westinghouse proposes to solve this problem by requiring a certain minimum and uniform pressure on the ice-condenser door panels before they open. The smaller containment volume may affect the residence time of sprays and change the efficiency of iodine-removal sprays; it may also change the effect of any hydrogen generated by metal-water reactions in the containment. In general, this concept appears to be promising.

In 1967, the Stone and Webster Engineering Corporation introduced another means of pressure suppression [20]. If the reactor containment is maintained normally at a properly chosen subatmospheric pressure, the peak pressure is substantially reduced, and the time during which the containment pressure is above atmospheric is substantially shortened. For reference purposes, a pressure of 9.5 psia (0.64 atm) was selected. This pressure would allow access to the containment for inspection without the need of supplementary breathing apparatus. It is equivalent to the atmospheric pressure at about 10,000 ft (3.05 km) altitude. In addition, when the volume of in-leaking air during normal reactor operations is monitored, an indication of the leak-tightness of the system results.*

Figure 2-14 shows typical transient-pressure curves, resulting from a pipe rupture, for both an atmospheric and a subatmospheric containment. The containment volume for the vacuum-pressure transient has been reduced to give the same peak pressure as the atmospheric transient.

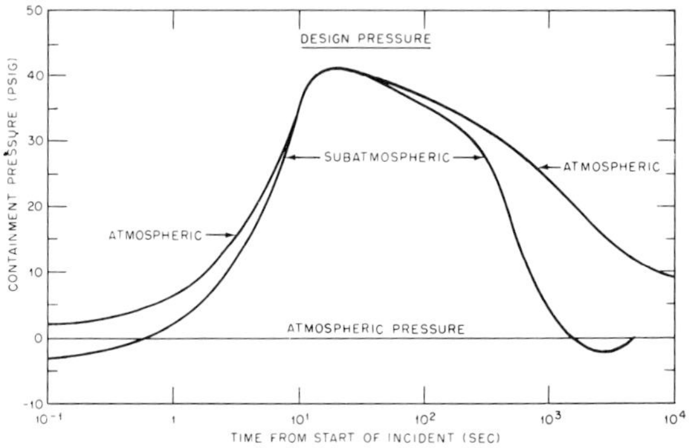

FIG. 2-14 Containment pressure transients following a loss-of-coolant accident.

*It should be noted, however, that the in-leakage characteristics of a system under partial vacuum may not be the same as those of out-leakage during accident overpressure.

TABLE 2-2

Subatmospheric Containment Conditions

	Subatmospheric	Atmospheric
Initial conditions, 105° F		
Initial pressure (psia)	10.0	15.5
Initial water partial pressure 80° F, D.P. (psia)	0.5	0.5
Initial air partial pressure (psia)	9.5	15.0
Postincident conditions, 140° F		
Air partial pressure (psia)	10.1	15.9
Water partial pressure saturated 140° F (psia)	2.9	2.9
Total pressure (psia)	13.0	18.8
Total pressure (psig)	-1.7	+4.1
Peak containment pressure 50 psig (64.7 psia)		
Temperature (°F)	285	273
Air partial pressure (psia)	12.4	19.6
Water partial pressure (psia)	52.3	45.1
Total pressure (psia)	64.7	64.7

Note that the out-leakage from the atmospheric containment continues indefinitely, while the out-leakage from the subatmospheric containment stops completely when the pressure drops to below atmospheric pressure. In the case shown, this occurs about 1/2 hr after the initiation of the accident. Typical subatmospheric and atmospheric conditions are shown in Table 2-2.

In order that the system operate correctly, vacuum pumps must be supplied. The engineered-safeguards system would include a containment-spray system and a recirculation-spray system to remove heat from the containment and to spray and recirculate the coolant for an indefinitely long time. A schematic diagram for this is shown in Fig. 2-15. Note that all components are supplied in duplicate. There are two "full-capacity" containment spray pumps located outside the containment, two half-capacity recirculation pumps inside the containment, and two half-capacity recirculation pumps outside the containment. The systems connected to regions outside the containment are sealed off by a double set of valves.

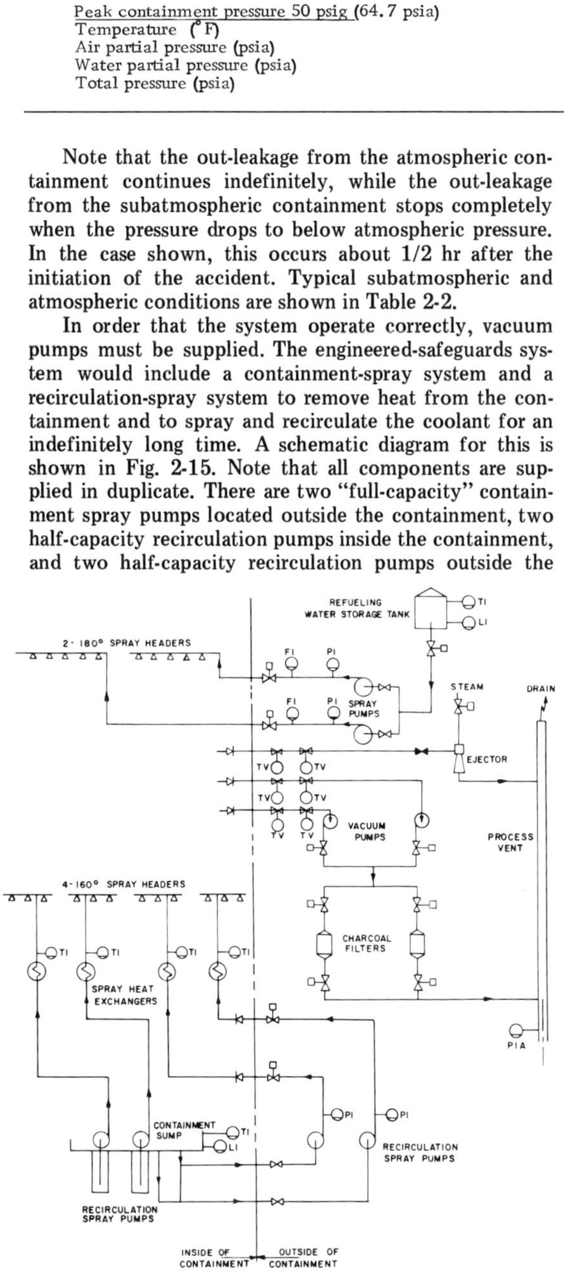

FIG. 2-15 Schematic of containment engineered safeguards.

In the long run, after an accident and the return of the system to a postincident condition, subatmospheric in-leakage of air would tend to permit the internal pressure to drift slowly up toward atmospheric pressure. One would not like to use the vacuum system during this time, as it would discharge radioactive effluents; but a special 100% filter and iodine-removal system could be installed at the discharge point if it proved necessary after an accident. The subatmospheric containment does exhibit a slightly higher postaccident temperature than atmospheric containment, as shown in Table 2-2.

The design shown eliminates the need for charcoal air-recirculation filters and fans operating in a steam environment. Instead a redundant spray system is relied on as shown in Fig. 2-15. Stone and Webster believes that this concept will reduce system costs. The structural design analysis must be carefully performed, since the steel inner shell envelope used to line the concrete containments of almost all the second-generation water-cooled power reactors must now withstand both vacuum and pressure requirements. However, that is not an insurmountable problem.

For the Shoreham Nuclear Power Station, the Stone and Webster Engineering Corporation has proposed a modified boiling-water reactor pressure-suppression containment, as shown in Fig. 2-16 [21]. The principles behind this design, developed by the Moss Landing tests, are identical with those used in the General Electric designs described earlier. However, the steel dry well and suppression pool have been replaced by concrete chambers with walls from 4-1/2 ft (1.37m) to 7 ft (2.13m) thick, lined with 3/8-in. (9.5-mm)-thick steel plate. Either reinforced or prestressed concrete can be used. The important parameters of the reference design are summarized in Table 2-3. The design anticipates that the steel lining will be backed up by reinforced concrete, rather than requiring an expansion space of several inches as in other boiling-water reactor pressure-suppression systems, such as the Millstone Point Reactor. Thus, a lighter steel-walled dry well would be required.

The containment described is 97 ft (29.56m) ID at the base and 152 ft (45.11m) high from the mat to the operating floor. The conic shape, with a low height-to-width ratio of 1.5, makes for structural stability. The 4-1/2 ft (121.9-cm) thickness at the top is determined by structural design requirements, while the increase to a 7-ft (213.4-cm) thickness opposite the core is for shield-

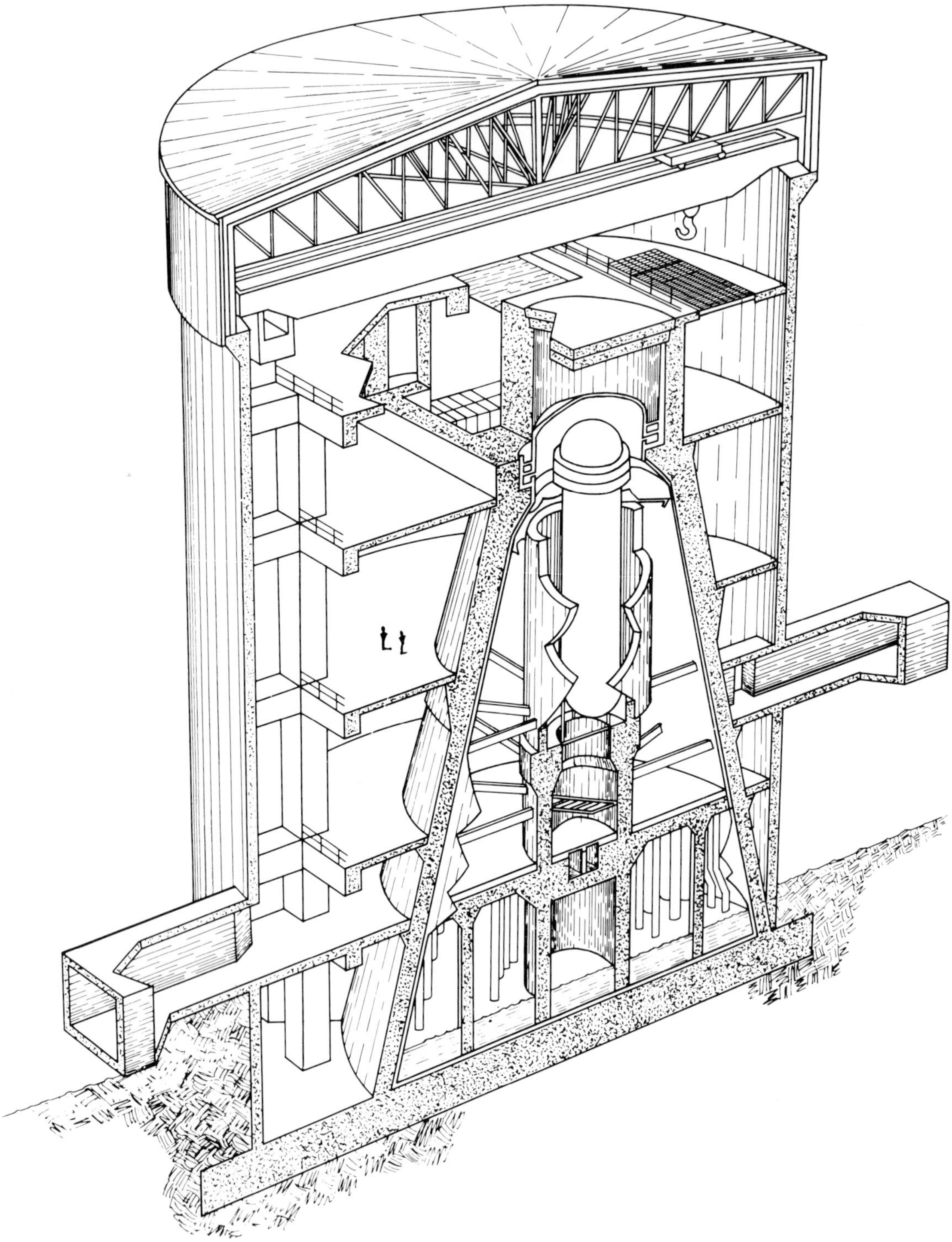

FIG. 2-16 Composite suppression containment for Shoreham Nuclear Power Station, Unit 1.

TABLE 2-3

Important Design Parameters, Shoreham Nuclear Power Station [21]

Drywell	
A. Internal design pressure (psig)	62
B. External design pressure (psig)	-10
C. Design temperature equivalent to 62 psig (°F)	310
D. Free volume, approximate (cu ft)	183,000
Suppression Chamber	
A. Internal design pressure (psig)	62
B. External design pressure (psig)	-10
C. Design temperature equivalent to 62 psig (°F)	310
D. Minimum net pool water volume at 110°F (cu ft)	58,900
E. Minimum net chamber gas volume, approximate (cu ft)	127,000
F. Break/area total vent area	0.0194
G. Initial water temperature rise (°F)	50
H. Initial pressure rise (psi)	35
I. Submergence of vent pipe below pool surface (ft)	3.0

ing purposes. Surrounding the containment is a circular concrete reactor building 133 ft (40.54m) in outside diameter and approximately 206 ft (62.79m) high above the concrete mat.

The containment-loading criteria are given as follows [21]:

1) Normal operating condition and hypothetical incident,
$(1.0 \pm 0.05)\ D + 1.0\ T_{1.5P}$;
2) Operating condition and hypothetical incident and design earthquake,
$(1.0 \pm 0.05)\ D + 1.0\ P + 1.0\ T + 1.5\ E$;
3) Operation condition and hypothetical incident and hypothetical earthquake,
$(1.0 \pm 0.05)\ D + 1.0\ P + 1.0\ T + 2.0\ E$;

where D = total dead-load of structure and equipment, P = design pressure, T = temperature associated with design pressure, $T_{1.5P}$ = temperature associated with 1.5 × design pressure and E = the design earthquake.

2.5 Initial-Pressure-Relief Confinement

An initial-pressure-relief confinement system vents to the atmosphere the effluent initially released by the primary system to the confinement in the event that an accident ruptures the primary-coolant system (see Fig. 2-17). After the initial release, the confinement system is sealed off, and an internal cooling system extracts continuously the fission-product after-heat developed within the sealed-off system. The normal operating inventory of low-level radioactivity in the primary coolant would be released to the atmosphere in the event of an accident. Therefore, the permissible limit on the quantity of radioactivity circulating in the coolant during normal operation is of fundamental importance in the safety analysis of any initial-pressure-relief system.*

*It should be noted that some fuel cladding may permit passage of certain fission products from the fuel into the coolant even though no cladding leak exists. For instance, in plate-type aluminum-clad fuel, MIT has found evidence of diffusion of certain noble-gas fission products. Evidence of leakage of tritium through stainless steel cladding has been observed at Yankee and other boiling- and pressurized-water-reactor plants.

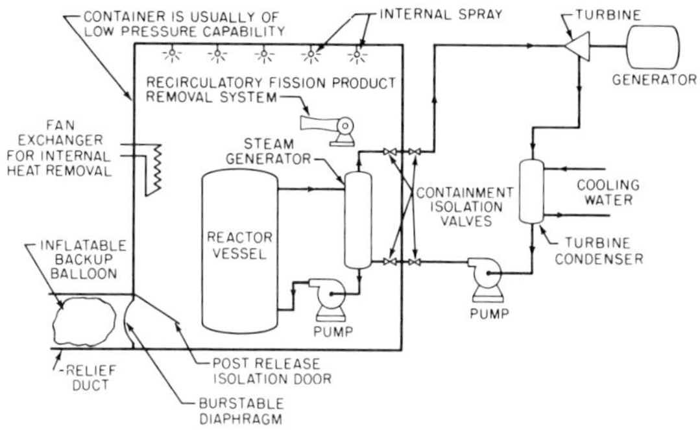

FIG. 2-17 Initial-pressure-relief confinement for a PWR, shown with internal sprays for initial steam quenching, a water-cooled fan heat-exchanger for continuous removal of internal heat, and an internal, recirculating fission-product removal system. In the event of an accident, the diaphragm bursts during the initial pressure release. The post-release gravity-operated isolation door, with its inflatable backup balloon, serves to seal off the passage.

In this confinement system, the coolant initially vented to the atmosphere must not be heavily contaminated with radioactivity. Because of this requirement, it is necessary to seek assurance that an accident cannot take place in such a way that fission products are released rapidly and simultaneously with the vented coolant. For instance, if the fuel elements were improperly designed, a sudden drop in pressure in the primary system could cause a large number of fuel elements to be ruptured by the buildup of fission-product gas pressure within the fuel cladding. In turn, the rupture of fuel elements could rapidly release fission-product gases stored within the fuel elements, and these could be carried out with the initial effluent discharge to the atmosphere. Similarly, a serious reactivity transient could cause fuel melting, as well as a transient pressure that might rupture the primary system. Such accidents are not protected against by initial-pressure-relief confinement.

A highly reliable means must be available for sealing off the confined volume after the initial coolant release has occurred. Some systems have made use of butterfly valves with pneumatic pressure-energy storage cylinders tripped by a fail-safe solenoid valve to ensure reliability. In one case (NPR), a backup system consists of a pressurized-gas cylinder that is designed to inflate a balloon on signal, thereby blocking the gas passage. Another scheme makes use of a water seal. This method was used in CP-5.* Because of the need for reliability in the closure, it is prudent, with this type of confinement, to follow the principles of redundancy and diversity (see Chapter 6, p. 296), providing a reliable containment-isolation method, then backing it up with another one as reliable as possible but based on a different principle.

In such schemes, means must be provided for continuously removing decay heat from the system. Obviously, a circulating coolant that keeps the core covered and prevents melting completely solves the problem. A spray or deluge system within the containment can be activated remotely or automatically to hold down the containment pressure, which would otherwise rise because of the afterheat. Other possibilities, such as heat exchangers, can be considered.

*In one sense, this design was the forerunner of pressure suppression.

The principal advantage of the initial-pressure-relief system is that the required design pressure is low, and hence the cost is low. The system may be used to advantage with primary systems that are purposely blown down to atmospheric pressure before emergency coolant is injected.† This use is particularly important, from an economic viewpoint, for reactors that are physically large or that have large inventories of primary coolant. There are several difficult problems encountered in establishing the design pressure of the containment. The rate of release of the coolant from the primary system, the coolant volume and conditions, the volume of the containment, and the orifice size of the containment pressure-relief pipe all enter into the calculations. The one factor that is difficult to estimate is the rate of release of the primary coolant. If it is underestimated, the containment could be ruptured because of the transient pressure.

An improperly designed containment system, which develops overpressures after it has been sealed off and after fission products have been released from the fuel, could be worse than no containment at all. By the time the overpressure bursts the system, gross amounts of fission products could be mixed in the vapor within the containment and could be released by the burst.

A pressure-relief system of this type is a dynamic system. If the containment is sealed off too soon and the dousing sprays then turned on, it is conceivable that a partial vacuum will develop within the containment, perhaps collapsing it. It is common practice to design many low-pressure structures to withstand internal pressures and yet to allow them to be vulnerable to small external overpressures. It may, therefore, be necessary to install vacuum-relief valves. This requirement introduces, in turn, an interesting problem in redundancy, since two valves in series, as required for normal containment seal-off, must both work to provide vacuum relief. Two single valves in parallel will increase the reliability of vacuum relief (only one needs to work) but will reduce containment reliability and, consequently, may be unacceptable from the standpoint of containment seal-off.

2.6 Continuous-Vapor-Venting Confinement

A continuous-vapor-venting confinement system provides a fixed venting path for coolant that might be released from the primary system to the containment during an accident (see refueling region of Fig 2-3). Normally, this path includes a means for treating the effluent mixture to extract from it as much as possible of the fission-product burden. This system is sometimes called confinement, vapor-venting, or negative-pressure containment. In general, it has the same advantages and problems as an initial-pressure-relief confinement. It has, in addition, two very important advantages and one important disadvantage.

The first advantage arises because the system is always connected to the outside atmosphere; therefore, there should be no problem of an overpressure or partial vacuum within the containment. The second advantage is that sequential release of coolant, followed by the bulk of the fission products, is not required in order that the system function as designed. If the system is properly designed so that it operates under all coolant-release conditions, effluent fission products can be removed with approximately the same efficiency no matter when or how they may be mixed with the coolant during the course of the accident. This is, however, a difficult task, as Chapter 19 makes clear. In general, this type of confinement aims at retention of all the fission products, except the noble gases, with relatively high efficiency.

Even in the case of the noble gases, there is now some hope of removal. Work at General Electric and ORNL has indicated that the permeability of noble gases through thin sheets (~ 1 mil or 0.025 mm) of dimethyl silicone rubber is greater by a factor of 3 to 10 than that of nitrogen or oxygen. This fact opens up the possibility that methods can be developed to separate the radioactive noble gases from air in appropriate process cascades. Since the principal disadvantage of the vapor-venting scheme arises from the fact that, to date, no practical means has been devised to extract the noble-gas fission products from the coolant, this development is of some importance. Until such a method is perfected and proved to work reliably, it must be assumed in an accident analysis that all the noble-gas fission products are released to the atmosphere. In addition, some other fission products or chemical forms of these products, such as methyl iodide, may not be picked up on charcoal adsorbers or filters or removed by chemical means. The merits and problems of such systems are discussed extensively in Chapters 18 and 19. Since the effluent-cleaning system must work no matter what conditions exist, all contingencies must be provided for. As an example, if the reactor is water-cooled, the clearning system must work for air alone, mixtures of air and steam, or steam alone. The filters and adsorbing beds (activated charcoal, for instance) must not catch fire because of overheating from deposited radioactive fission products. The cleaning system must not lose efficiency due to "poisoning" that may be accidentally or purposely added.

Systems of this type can be incorporated in reactor facilities for which the confinement is little more than an ordinary building of reasonably tight construction (see Fig. 2-3). By means of a blower or fan system, the interior of the confinement building is maintained at slightly below atmospheric pressure. Thus any air leakage is into the building. The air drawn in, together with any fission products and uncondensed primary-system effluent, is drawn through a high-efficiency filter and, perhaps, layers of adsorbing material, through a noble-gas-removal system (when one is perfected), and thence through the blower and a monitor; it is then released from the top of a stack. Notice in Fig. 2-3 that the filter and adsorber beds are on the suction side of the blower, so that external gas leakage in this region is into the filter system, not out. Radioactive-effluent monitors can be calibrated and, together with wind velocity and direction, provide a rough indication of the hazard of released activity to the general public. This type of system was used for the Hallam sodium-cooled system, with special filter provisions for the collection of sodium oxide in the event of a sodium fire.

A recirculation fission-product-removal system may be provided to augment the blower system already described. This system may make use of filters, adsorber beds, chemical absorbers or solutions, noble-gas diffusion

†Purposeful blowdowns in a reactor system in a planned manner, followed by the planned injection of coolant, may, in some cases, prove to be safer than attempts to inject emergency coolant into a pressurized system. In a pressurized system, some pipe breaks may starve the core of coolant and result in fuel melting, where a full, controlled blowdown and coolant injection could have prevented it.

separating systems, and even foams of various types. These methods are described in Chapter 19.

It is clear that a continuous-venting confinement must have an effluent path of sufficient capacity to prevent rupture of the building or overpressure escape of radioactivity through leaks in the building walls. Thus, the effectiveness of this form of containment again depends on a conservative accident evaluation.

Vapor-venting containment is inexpensive, as no more may be required than a steel-frame building with well-sealed exterior metal panels and gasketed doors. Normally, few if any exterior windows would be provided. This type of containment has been used at the Hallam Reactor, the Oak Ridge Research Reactor, the Armour Research Reactor, certain production reactors, and some swimming pool reactors. Some reactor plant facilities employ this type of confinement to lend added assurance to other kinds of containment. For instance, in the refueling buildings of a number of boiling-water reactors (Humboldt Bay, Dresden-2, etc.) vapor-venting systems are immediately superposed on the inner containments.

The use of a stack alone will be of considerable help in reducing the hazard of the effluent to the local population. The release of fission products from an elevated point rather than at ground level will provide additional air mixing and dispersion before any fission products can reach ground level. High stack velocities, high stack-gas temperatures, and mixture with a diluting air stream injected into the base of the stack can provide additional protection. At the same time, high stacks cannot provide much additional protection of regions more remote from the stack.

It is possible that a holdup tank of some sort can be provided to store, at least temporarily, the released fission products. Such an arrangement, including underground storage, holds promise of additional protection. The Armour Research Reactor containment has such a provision. Other possibilities have been considered, such as storage in a sealed-off underground mine.

2.7 Double or Multiple Containment

The types of containment described in the previous subsections can be combined. It has already been pointed out that pressure suppression has been combined with a peripheral region of continuous-vapor-venting confinement in a number of boiling-water reactors. In a sense, this is a double containment, although more logically it can be described as a combination of two systems: a pressure-suppression containment to cover that part of the primary system normally containing the fission-product burden, combined with a confinement system to cover all or part of the entire primary system. A similar combination of systems is provided for the N.S. Savannah. The reactor and its heat exchangers are within a conventional pressure-vessel containment (full primary containment); this containment is surrounded by an outer, sealed ship compartment, which is vented to a stack at the mast head (continuous-vapor-venting confinement).

The Consolidated Edison Indian Point-1 plant has a full-primary-system double containment that consists of a steel shell capable of containing all the effluent from a major pipe break. Surrounding this is a concrete outer shell. The space between the two shells is vented to the facility stack, thus providing a second confinement barrier to take care of any leakage from the primary system.

In several pressurized-water reactors (e.g., Malibu) it is proposed to use a thick and somewhat porous concrete (popcorn concrete) shell poured between an inner and outer leak-tight carbon steel shell. The region between these two shells is pumped to maintain a slightly negative pressure, and the leakage is pumped back into the inner shell containment. The region outside the two steel shells is covered with a thick layer of ordinary reinforced concrete to provide strength to back up the two steel shells and the porous concrete. This system provides extra protection against fission-product leakage and should have the lowest leakage rate of any scheme proposed to date. This is offset by the requirement that an adequate decay-heat-removal system must function if the containment is not to rupture. The thick-concrete-shell arrangement provides effective insulation that prevents conduction of the decay heat through the containment envelope.

All the multiple containments proposed to date afford what appears to be potentially more protection to the public than any of the single-component containment schemes. Yet all have some flaws. There are interactions between the two barrier systems. Both may be subject to rupture by a common missile or both may rupture in the event of a more severe pressure wave than anticipated. Any pumpback scheme will raise the pressure within the inner containment and, ultimately, if the leakage from the outside is severe, can conceivably rupture the containment because of internal overpressure. This has a distinct drawback since, as noted earlier, if it is necessary to vent the containment to prevent its rupture, it is better to do so before fuel melts and major quantities of fission products are released into the containment.

3 REACTOR—SYSTEM CHARACTERISTICS THAT AFFECT CONTAINMENT

3.1 General Approach

The containment designer must initially make a conservative estimate of the inventory of fission products that could conceivably be present in the vapor within the containment in case of a serious accident. The approximate quantity, chemical form, and radioactive characteristics of each biologically important fission-product isotope that could be present should be known.

To do so, the designer must first estimate the maximum inventory of fission products in the reactor system. Then he must estimate what fraction of that inventory, and its chemical and physical nature, might be present in the vapors within the containment in the event of a serious accident. Sufficient information is presented in this volume to make rough estimates of the inventory for most practical cases, but for more precise estimates basic references should be consulted [13b, 22, 23].

It is generally accepted that only a fraction of the fission products can escape from the primary system, even in the worst case, unless the entire core melts through the bottom of the vessel. Even then, not all the fission products will be localized. The estimation of this fraction is treated in Chapter 18 and discussed briefly in the next subsection. The containment may incorporate sprays, filters, or adsorber beds to remove fission products. The escape of fission products to the containment and beyond, and the measures that can be taken to reduce the containment inventory and the amount escaping, are the principal subjects of Chapter 19 and are discussed only briefly here.

The chemical nature and state of the reactor coolant, the containment atmosphere, and the reactor fuel with its cladding determine the rate of possible chemical reaction and the total energy that could be released due to chemical reactions during an accident. (This is discussed in Sec. 3.3 of this chapter and in Chapter 17.) The pressure, temperature, and volume of the primary coolant also are important parameters, since they determine the quantity of stored thermal energy that would be released to the containment in the event of a serious accident. (These factors are discussed briefly in Sec. 3.4)

The pressure and temperature requirements for containment design are calculated from estimates of the rate of energy release and of the maximum energy that might be made available from all sources in the event of a very serious accident. The energy sources are chemical, thermal, and nuclear (including core afterheat). The detailed nature of a postulated accident, or of any one of a set of such accidents, and often a postulated time sequence may have to be considered.

Once the maximum fission-product inventory, the fraction released to the containment, and the chronological sequence of energy release have been calculated from the reactor-system characteristics, the designer can set the specifications for the containment.

3.2 Fission-Product Inventory

The power level determines the rate at which the fission products are generated. Since the activity of fission products within the reactor is reduced by decay, by transmutation due principally to neutron capture, and, in some cases (e.g., Peach Bottom), by the physical removal of some species from the fuel, the fission-product inventory at any time depends on the detailed schedule of previous operations. A conservative maximum inventory is obtained by assuming a long period of steady-state power operation, followed by shutdown at the moment of the accident. As a generality, short-lived fission products (e.g, iodine) reach a saturation value (which depends on the power level) that results in a fixed activity value (kilocuries) per megawatt of thermal power as shown in Table 3-1 [23]. For Xe^{135} the equilibrium value at power saturates (in thermal reactors) and becomes independent of power level at sufficiently high neutron-flux values (because of the high value of its capture cross section). The activity due to the long-lived fission products that are important from a reactor-safety point of view increases with the total energy generated in the reactor. Typical values after 1 and 5 yr of irradiation are shown in Table 3-2 from Beattie [23]. The concentrations of certain of the hazardous alpha-emitting isotopes, such as Pu^{240}, also increase with time; their concentration depends on fuel composition and neutron-exposure history. Some isotopes formed by neutron capture, such as Pu^{239}, subsequently become part of the fissile fuel present. Among the common fissionable nuclei (Th^{232}, U^{235}, U^{238}, Pu^{239}), there are some differences in the fraction of the various fission-product isotopes produced, but these differences normally do not warrant use (in hazards calculations) of separate tables of properties.

It should not be overlooked that stable as well as radioactive isotopes accumulate in an operating reactor. Some of the stable isotopes are important because they contribute to the total amount of gas released, while others are important because they influence the life and efficiency of cleanup systems. Table 3-3 shows the relative amounts of stable and radioactive krypton, xenon, iodine, and cesium in a particular power reactor core after 3.2-yr of operation. The details of these processes are discussed in Chapter 18 and in reference [23].

It is difficult to estimate what fraction of the fission-product inventory can escape from the reactor fuel to reach and mix with the primary-system effluent vapor. This vapor, with its fission-product content, is retained within the containment and is the source of any leakage of radioactivity from the containment. To be conservative, one must assume that the noble gases will all escape from the melting fuel and mix in the containment vapor. The problem of what fraction to choose for the other gaseous or volatile fission products is the principal subject of Chapter 18. The reader is referred to that chapter for detailed considerations. In general, when no other reasoning has led to specific answers for a particular installation, designers in the United States use the guideline set forth in TID-14844 [24]: "It is assumed that the reactor is a pressurized-water type for which the maximum credible accident will release into the reactor building 100 percent of the noble gases, 50 percent of the halogens and 1 percent of the solids in the fission-product inventory. Such a release represents approximately 15 percent of the gross fission-product activity."

The reactor designer may modify the estimate of the fission-product inventory to some extent by installation of certain pieces of equipment or by methods ("engineered safeguards") that would act to reduce the inventory in a reliable manner (see Chapter 19). In summary, the objective of this section, and in part the objective of Chapters 18 and 19, is to set forth ways to establish a conservative yet realistic estimate of the fission-product inventory that might be present in the vapors held within the containment in the event of a serious accident.

3.3 Physical State and Chemical Composition of the Coolant

The physical state of the coolant is important in determining the containment design. For example, if the coolant is a gas, there is no simple way to reduce the containment pressure to atmospheric pressure at ambient temperature after an accident except to filter out particulates, to absorb the iodines, and to permit the gas to escape with the remainder of the unremoved fission products. Moveover, it is difficult to provide emergency coolant in case the primary system ruptures. If the coolant is a liquid or can be easily liquified by cooling, then, even if the liquid has partially vaporized into the containment, cooling will condense it and dramatically reduce the driving force that tends to expel vapors from the environment. If the coolant is water, reduction of pressure by condensation is well understood, and emergency supplies of water are relatively easy to obtain.

The chemical compositions of the coolant, the moderator, the containment atmosphere, and the fuel and its clad are important. The nature and extent of their chemical interactions are also important. If the coolant is a liquid metal, reactions with the containment air or the water in a heat exchanger must be considered; these are discussed in the next section and in Chapter 17. If an organic coolant is used, there is also a potential fire hazard when the containment atmosphere is air, and fire-fighting systems must be provided. Heavy water (D_2O) poses special problems because of its economic value, and because, after use in a reactor, it contains tritium.

It is desirable to have the fuel and its cladding, the

TABLE 3-1

Characteristics of Important Short-Half-Life Fission-Product Isotopes[a]

Isotope	Half-Life	Activity in kilocuries per megawatt of thermal power		Boiling point (°C)	Volatility	Health physics properties
		Shutdown	1 day after shutdown			
Br-83	2.3 h	3	0	59	Highly volatile	External whole-body radiation, moderate health hazard
-84	32 m	6	0	"	" "	
-85	3 m	8	0	"	" "	
-87	56 s	15	0	"	" "	
Kr-83m	114 m	3	0	-153	Gaseous	External radiation, slight health hazard
-85m	4.4 h	8	0.2	"	"	
-87	78 m	15	0	"	"	
-88	2.8 h	23	0.1	"	"	
-89	3 m	31	0	"	"	
-90	33 s	38	0	"	"	
I-131	8 d	25	23	185	Highly volatile	External radiation, internal irradiation of thyroid, high radiotoxicity
-132[b]	2.3 h	38	0	"	" "	
-133	21 h	54	25	"	" "	
-134	52 m	63	0	"	" "	
-135	6.7 h	55	4.4	"	" "	
-136	86 s	53	0	"	" "	
Xe-131m	12 d	0.3	0.3	-108	Gaseous	External radiation, slight health hazard
-133m	2.3 d	1	0.7	"	"	
-133	5.3 d	54	47	"	"	
-135m	15.6 m	16	0	"	"	
-135	9.2 h	25	4	"	"	
-137	3.9 m	48	0	"	"	
-138	17 m	53	0	"	"	
-139	41 s	61	0	"	"	
Te-127m	105 d	0.5	0.5	Released from oxidizing uranium		External radiation, moderate health hazard
-127	9.4 h	2.9	0.5	" " " "		
-129m	34 d	2.3	2.3	" " "		
-129	72 m	9.5	0	" " " "		
-131m	30 h	3.9	2.2	" " " "		
-131	25 m	26	0	" " " "		
-132	77 h	38	31	" " " "		Health hazard from I-132 daughter[b]
-133m	63 m	54	0	" " " "		External radiation, moderate health hazard
-133	2 m	54	0	" " " "		
-134	44 m	63	0	" " " "		
-135	2 m	55	0	" " " "		

[a] From Beattie [23], pp. 10-11.

[b] 38 Kilocuries of I-132 per megawatt of thermal power is generated in the reactor by decay of Te-132. Analyses that follow will also consider the I-132 formed outside the reactor by decay of Te-132 released from a reactor accident.

TABLE 3-2

Characteristics of Important Long-Half-Life Fission-Product Isotopes[a]

Isotope	Half-Life	Activity in kilocuries per megawatt of thermal power		Boiling point (°C)	Volatility	Health physics properties
		After 1 yr of irradiation	After 5 yr of irradiation			
Kr-85	10.4 y	0.12	0.62	-153	Gaseous	Slight health hazard
Sr-89	54 d	39	39	1366	Moderately volatile	Internal hazard to bone and lung
-90	28 y	1.2	6.0	1366	Moderately volatile	
Ru-106	1.0 y	5	10	4080*	Highly* volatile oxides, RuO3 and RuO4	Internal hazard to kidney and GI tract
Cs-137	33 y	1.1	5.3	670	Highly volatile	Internal hazard to whole body
Ce-144	282 d	30	50	3470*	Slightly* volatile	Internal hazard to bone, liver, and lung
Ba-140	12.8 d	53	53	1640*	Moderately volatile	Internal hazard to bone and lung

[a] From Beattie [23], pp. 10-11, with a few corrections indicated by asterisks [13a]

TABLE 3-3

Amounts of Some Fission Products in a Power Reactor Core [a]

Radioisotope	Amounts (g)	Total Activity (curie)	Half-life
Kr-85	966	4.12×10^5	10.27 y
Total Krypton (stable and radioactive)	14,000	-	-
Xe-133	255	4.7×10^7	5.27 d
Xe-135	3.6	9.2×10^6	9.13 h
Total Xenon (stable and radioactive)	157,000	-	-
I-129	5,620	0.90	1.72×10^7 y
I-131	168	20.7×10^6	8.05 d
I-133	42	4.7×10^7	20.8 h
I-135	14	4.3×10^7	6.68 h
Total Iodine (stable and radioactive)	7,230	-	-
Cs-135	7,390	6.5	3.0×10^6 y
Cs-137	34,800	3.41×10^6	26.6 y
Total Cesium (stable and radioactive)	80,100	-	-

[a] Calculated for an 843 Mw(t) reactor containing 1.86×10^{27} atoms U^{235}, running at full power continuously for 10^8 sec. $\phi = 2.5 \times 10^{13}$ n/cm^2 - sec. (From Blomeke and Todd [22]).

coolant, the moderator (if different from the coolant), and the containment atmosphere made of elements or compounds that cannot chemically interact with one another under the conditions that might be present. During normal operation, such a goal can be achieved. However, the very nature of chemical elements makes it virtually certain that no system of "noninteracting" components can exist in a reactor if the high temperatures associated with the fission process are approached. Many materials that are entirely compatible at low temperature are capable of exothermic reaction at high temperature.

The reactor designer must examine the accident conditions that might elevate the temperatures of reactor components and determine which chemical reactions are thermodynamically possible and which of those theoretically possible are credible under the conditions likely to exist during the accident. Some difficult choices are required. For instance, as pointed out in Sec. 1.3.9, metal-air or sodium-air reactions can be precluded by using an inert-gas atmosphere in the containment. But this has, as noted, serious disadvantages.

Containment design must also take into account the composition and physical state of the fuel. A reactor with fluid fuel (MSRE, HWR, Water Boiler) has all the fission products either dissolved or suspended in the fluid; there are no barriers to fission-product release such as are presented by a solid fuel and its cladding. The release of fission products from homogeneous systems may appear more likely because only the primary-system walls and the containment are barriers between the fission products and the public. On the other hand, in a heterogeneous reactor the rapid rate of heatup of the solid fuel in the event of loss of coolant causes thermal and pressure stresses or the formation of eutectics between fuel and clad that can rupture or melt cladding and permit the escape of gaseous fission products. In fact, there is likely to be less and less difference between homogeneous and heterogeneous systems as the accident becomes more severe. Moreover, with a fluid-fuel system, it should be possible to remove the long-life fission products, thereby minimizing the fission-product inventory. From a safety viewpoint, it is difficult to say with certainty whether a homogeneous reactor system is intrinsically less safe or more safe than a reactor that has solid fuel in a cladding. Much depends on the specifics of design. It seems likely that the relative safety of fluid-fuel reactors, compared with the conventional water reactors, has not received adequate consideration, since only small experimental homogeneous reactors have been built to date.

3.4 Coolant Pressure, Volume, and Temperature

The pressure, volume, and temperature of the reactor coolant are important in containment design primarily because they determine the amount of thermodynamic stored energy available in the coolant for release in case of a primary-system rupture. Stored energy often is a large fraction of the total available energy. Coolant pressure, volume, and temperature are also important from a safety standpoint in the overall reactor-system design since they determine the potential driving force in pipe or vessel failure and in leakage of radioactive material to the environment.

For example, if the pressure in the primary system is higher than that in the secondary system, as is usually the case, then any leak in the heat exchanger permits the passage of primary fluid (and any radioactive material that it might contain) into the secondary system. If the containment includes the heat exchanger (but does not include the rest of the secondary system), and if the heat exchanger leaks, then the pressure differential could drive radioactive material outside the containment. To prevent this, the heat exchanger is normally designed to withstand without failure the full primary-system pressure with no pressure on the secondary side; in addition, valves may be placed to isolate the secondary system from the heat exchanger. In a similar way, the heat exchanger must be designed to withstand the full pressure of the secondary system with no pressure on the primary system, since this situation can exist when coolant is lost in the primary system. In some Safety Analysis Reports (SAR), the analysis assumes that the volume of the secondary coolant in a single heat exchanger is added to the primary system by the accident. This situation could result from the catastrophic rupture of a single heat exchanger. Even in this case, it is assumed that the remainder of the secondary system can be isolated from the affected heat exchanger. The validity of these assumptions must be investigated for each design.

The volume of the primary system and the volume of coolant contained within it are both important. All other things being equal, the smaller the volume, the smaller the stored thermal energy. Hence, a small primary-system volume tends to permit a smaller containment volume or a lower design-accident pressure or both. Often, if the primary-system volume is small enough, the containment volume can be set by the space requirements for normal operation and for refueling. In this way, the extra cost of containment, above that required for normal operations, is minimized. A larger primary-system volume may be required to reduce the radiation exposure of the pressure vessel, to provide sufficient coolant to assure that core blowdown is not so rapid that core melting can start before initiation of emergency cooling, to avoid excessive coolant-flow velocity with its associated vibration or high pumping power, or to provide sufficient heat-transfer surface, etc. As is usually the case, a balance must be struck between conflicting requirements.

The quantity of the coolant within the primary system is independently important. This quantity determines the stored energy. In a direct-cycle boiling-water reactor, the coolant volume is appreciably less than the primary-system volume. If the actual quantity of coolant is strictly limited to some maximum amount, it would seem reasonable (in the accident analysis) to use that maximum amount in determining the energy that could be released.

The temperature of the coolant is important as an indication of the amount of stored energy present. If the temperature is above the boiling point of the liquid at atmospheric pressure, then expansion occurs on the release of pressure, and a driving force exists to expel coolant from any rupture of the primary system; any fission products in the coolant are expelled with it.

In addition, chemical-reaction rates are greatly increased with rise in temperature. This means not only that energy from the exothermic reactions accompanying a primary-system break may add to the pressure load on the containment (see Sec. 4.3.4), but also that chemical attack of the hot coolant on the containment structure itself (e.g., gaskets, packing, materials, insulation) may be severe.

Temperature gradients within the containment system and temperature variations with time create thermal stresses and fatigue. The higher the temperature, the greater the probability of temperature variations that could cause large thermal stresses. Again, these phenomena are well known and their effects can be estimated with sufficient accuracy in the design of the containment system, if reactor designers remember to consider them.

The physical properties of many materials are adversely affected by high temperature. For instance, improperly selected gasket and sealing materials used in a containment could fail at the elevated temperature resulting from an accident. Temperature, therefore, is a potential cause of failure of parts of the primary system and also of the containment structure. Obviously, in containment design, the possible consequences of the assault on the containment of hot coolant and other materials that might be released in an accident must be considered and appropriate countermeasures taken.

4 REQUIREMENTS FOR CONTAINMENT DESIGN

4.1 Containment selection

The process of selecting a suitable reactor containment is shown schematically in Fig. 4-1. This figure is intended to display the general logic involved; an actual selection may not be done in this manner, or in this detail or sequence, for any particular reactor. Often a given site and reactor requirement lead designers and owners directly to an acceptable answer. In many cases, the logical process shown has continued over many years, and more containment variants have been considered than those indicated in the figure.

The first steps in the process are to select a reactor and a site (steps 1 and 2 in Fig. 4-1). Then a containment type is tentatively selected (step 3) and determined as feasible and safe (step 4). As can be seen from Fig. 4-1, the processes of selection of site, reactor, and containment are so intertwined as to be virtually inseparable. The reactor designer must specify the reactor type, power level, core configuration (including materials and operating conditions), coolant pressure, coolant temperature, primary-coolant volume, and, in some cases, other items as well. The site must also be identified and a conservative set of meteorological conditions chosen. If no other set is available from observations made at the site, those mentioned in TID-14844 are sometimes used [24]. The seismic characteristics of the site may play an important role in site selection. If a site with poor seismic characteristics is chosen, the plant costs are likely to be increased.

It is necessary, next, to select a set of design accidents in order to carry out analyses of containment capability (step 5). The accidents selected depend on the type of reactor chosen and its particular characteristics; in Fig. 4-1 this dependence is indicated by the arrow from reactor design to design accident. Obviously, an accident that is physically impossible is not suitable. The most "conservative" containment is one that can contain all radioactive material under all physically possible conditions.

It should be emphasized that "design accidents" are postulated in order to provide a sound basis for probing the safety of the design and the adequacy of the containment. It is not expected that the most severe design accident will happen; in fact, such accidents are extremely unlikely. Every attempt is made to see that they cannot happen. Many reactor scientists and engineers outside the United States have overemphasized the importance that they believe this concept plays in the United States, perhaps without having fully understood that its intent was solely that of supplying a design basis for the containment.

The energy released in the design accident is determined in step 6. The energy released depends on the characteristics of the reactor and the nature of the accident and, in some cases, on the time sequence of the assumed accidents.

The energy release, in turn, plays a primary role in determining the pressure and temperature within the containment (step 7). In order to permit construction of a containment that satisfies code requirements, the volume within the containment may have to be larger than that needed for normal operation. If so, that adjustment in volume is made at this point in the analysis. The type of containment (step 3), together with the pressure and temperature during the design accident, in turn determines the leak rate (step 7).

The energy release (together with the type of containment, including engineered safeguards such as filters, etc.), the reactor, its power level, and the postulated accident determine the fraction and absolute amount of fission products available for release (step 8). The input into step 8 of Fig. 4-1 is discussed in detail in Chapters 18 and 19.

At some point, it is necessary to assess the suitability of the type of containment to the reactor type (step 4). For instance, a water-filled, pressure-suppression pool containment would be unsuitable for a sodium-cooled reactor. Other examples, not quite so clear, come to mind, including the problem of whether to place the steam turbine-generator for a boiling-water system inside or outside the containment, or whether to place the boilers inside or outside the primary vessel in a gas-cooled reactor. If the reactor type and its containment are not compatible, another containment must be chosen.

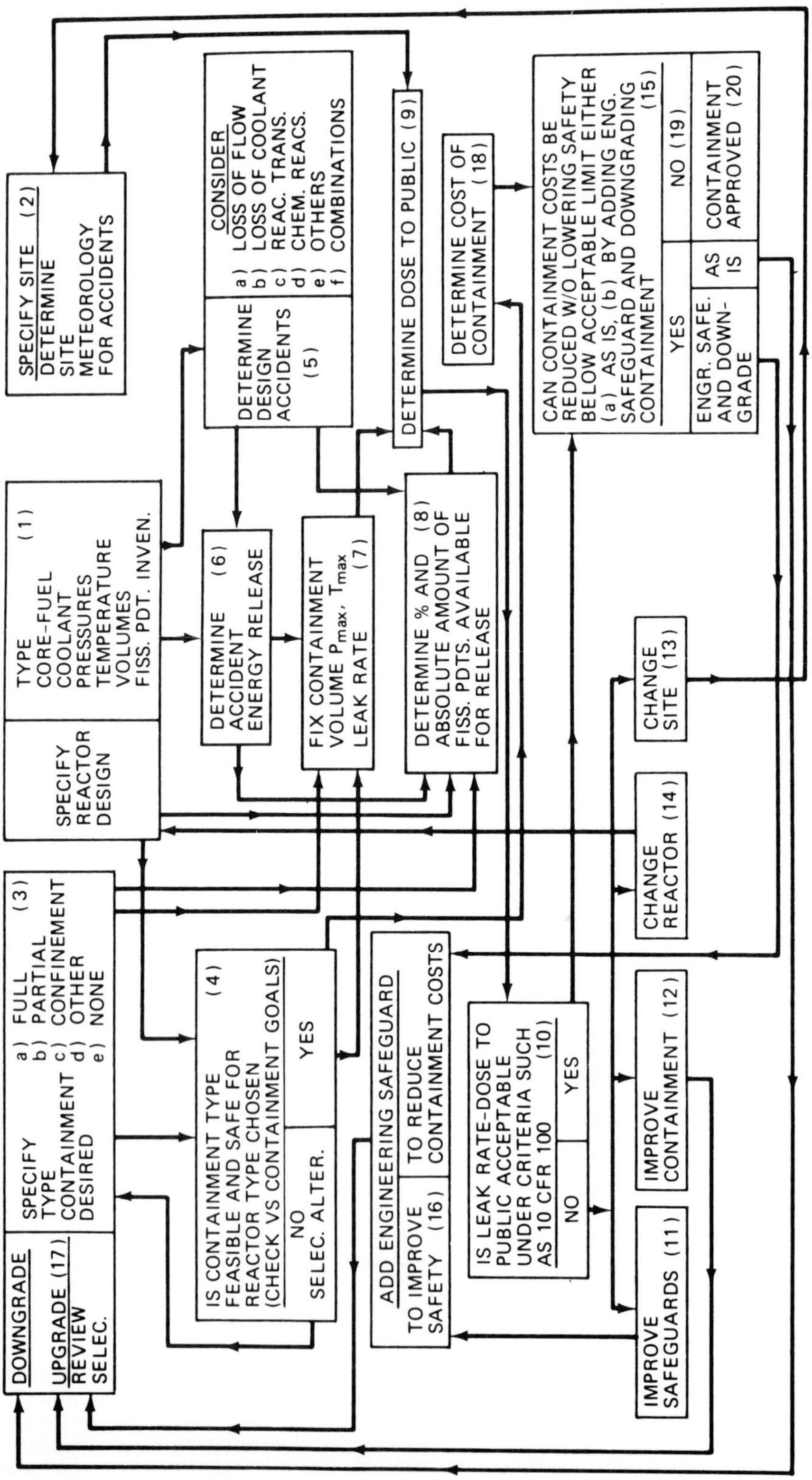

SPECIFY – SOME CHOICE INVOLVED; MAY NOT NECESSARILY BE DONE ON SAFETY GROUNDS.
DETERMINE – CALCULATE, FROM COMBINATIONS OF THE INPUTS AND PHYSICAL LAWS AND FROM INFORMATION AVAILABLE, A CONSERVATIVE AND LOGICAL ANSWER.
ACCEPTABLE OR ALLOWABLE – A LEVEL OF DOSE OR A CONDITION THAT IS NOT NORMALLY LIKELY AND IS USED ONLY FOR BASIC GUIDANCE IN EVENT OF A HIGLY IMPROBABLE ACCIDENT.

FIG. 4-1 Selection of a suitable reactor containment.

From the leak rate, the quantity of fission products available, and the meteorology, the possible exposure of the public in the event of a specific design accident can be estimated (step 9). This dose, in turn, can be compared (step 10) with such criteria as those of 10 CRF 20 (for normal operation) and 10 CFR 100 (for a calculated severe accident). If the design is found acceptable under both of these criteria, the entire process is reviewed again to see whether the system can be made more economical and just as safe, or more economical with added engineered safeguards (such as filters, adsorbers, sprays, etc.) to maintain the same level of safety (see steps 15, 16). This review process for reduced costs without loss of safety is shown symbolically in the diagram.

If the review at this point shows that the system does not meet the public-safety criteria and is therefore not acceptable, four choices remain. The designers can improve the safeguards and containment, locate the reactor at another site, change the reactor type or power output, or any combination of these that provides a sound and technically defensible answer. These steps are shown in the figure as steps 11 to 14, inclusive.

In step 3 of Fig. 4-1, five general choices are shown, including choice of no containment. Not many designers start out assuming no containment, although such an exercise can be valuable to provide an ultimate assurance of adequate protection of the public, even in the event of an "unprotected accident." The order of steps in the figure is subject to change and, in fact, the figure symbolically represents only one of many possibilities.

The variants are as many as there are reactor projects. However, the basic scheme of the containment selection process of Fig. 4-1, or most of the parts of it, is similar to that used for all reactors now being designed and constructed in the United States. Normally, the responsible reviewers of reactor safety are not concerned with the economic aspects, although they know that costs do play a part in choosing the reactor containment. To show how all parts of the scheme fit together, the economic factors have been included in Fig. 4-1.

The selection of a reactor type and the safety considerations involved are considered in most of the chapters of these volumes, especially in Chapters 2, 8, 9, 10, 12, 13 and 14. A complete elementary treatment of the relevant meteorology is given in references [25, 26]. A full review of the entire subject of containment, including detailed discussions of problems and useful calculational methods, is given in reference [13].

In the following sections, the topics are first discussed in terms of full containment and then modified to take into account the other possible types of containment. It is assumed that reactor type and power level have been chosen and that the principal characteristics of the reactor system relevant to safety analysis have been determined, as discussed in Sec. 3 of this chapter.

Sec. 4.2 discusses briefly the principal problems of the selection and analysis of "design accidents," that is, accidents hypothesized to probe the design. Section 4.3 discusses problems of estimation of the energy release that could occur during such accidents.

4.2 Design Accidents

The subject of conceivable types of accidents is considered in a number of places in these volumes. In Vol. 1, various types of primary accidents are outlined and discussed in Sec. 3 of Chapter 8. Causes of reactivity accidents and analyses of excursions for solid-moderator reactors are reviewed in Secs. 4 and 5 of Chapter 9. Fast-reactor reactivity accidents are discussed in Secs. 3 and 4 of Chapter 10. Accident experience and destructive tests are described and analyzed in Chapter 11. In Vol. 2, Chapter 16 discusses loss-of-coolant accidents. Finally, the radioactivity contamination of some accidents is considered in Chapter 20.

This section is concerned with hypothesized accidents severe enough to be considered design accidents, that is, severe enough to probe the design.

The choice of the design accident has a key effect on the containment design and can often materially affect the cost of containment. It is necessary that all concerned be sure that this accident is selected in a conservative manner. The accident, or accidents, that the containment must cope with successfully must be severe up to the point of total financial catastrophe for the primary system, and yet must be realistic. It cannot be an accident that is physically impossible.

Often in the past, the postulated accident (or accidents) used to establish the basis for containment design has been called the "maximum credible accident." This term has caused some misunderstandings and is now largely superseded, or should be, by the term "design accident" or other terms with a similar meaning.

It is worth pointing out that the dictionary defines "credible" as "believable" and "incredible" as "unbelievable." There is a difference between "incredible" and "impossible," which may be of great importance in reactor safety. Thus, a number of accidents may be postulated that are incredible but are, in no sense, impossible. Since in many power reactors the primary systems are at high pressure, any type of rupture of the primary system is theoretically possible or conceivable. However, the metallurgical and mechanical properties of the materials used, the fabrication and inspection of materials and components according to established codes, and the proper use of materials and components may make certain types of failure virtually impossible. Statistics for rare types of failure are not very meaningful. The design specifications may call for a reactor plant capable of withstanding, without danger to the public, an incredible, but still possible, accident. The accident may have a vanishingly small (but nonzero) probability of occurrence. Such an accident could be the complete rupture of the containment, with its disastrous possibilities [4]. The evidence must be reviewed especially carefully, with due regard to the site, and a conservative decision made.

There are problems with this approach, as with all other possible ones. For the most part, the problems arise from the credibility of the postulated accident. In pressurized-water, boiling-water, and gas-cooled reactors, it has been common United States practice to postulate at least one accident leading to the blowdown of the primary system. This accident tests the adequacy of the containment provisions to handle the steam or gas released, the possible fission-product release, and the continued core cooling. It is responsive to the objective of Sec. 1.3.3, i.e. to contain the total released energy. Thus, to explore this objective it is necessary to postulate such an accident, regardless of its credibility.

Reactor designers are understandably reluctant to postulate accidents that they consider incredible. To satisfy the objective of total energy containment, it might be better to postulate a standard series of hypothetical primary-system blowdowns, so that the consequences could be investigated without apparent reflection on the adequacy of the design. For instance, one might set up a standard procedure that would require consideration of a set of hypothetical blowdowns, due to openings of increasing size up to that equivalent to the open double-ended pipe break, for each major primary-system section between major valves and components. The assumption of the break conditions should be conservative but as realistic as possible. The double-ended instantaneous break is obviously conservative but also obviously unrealistic. The minimimum size opening to be considered could be that opening for which the normal coolant supplies could provide water faster than it can blow down from the system. Such a system would not require that failures be postulated directly.

Postulating a loss of coolant from failure of a major component also raises problems. Obviously, a major break in a primary-system component, such as the primary pressure vessel or a pump or blower, would be a serious accident. Its consequences would include not only the blowdown of the coolant, but also the effects of the missiles generated and the possibility of being unable to continue to cool the core. The most important component to consider is the reactor pressure vessel. Failure of the pressure vessel itself would directly involve the reactor core in the accident.

In all power reactors, it has generally been considered incredible that a nuclear transient would occur at the same time as, or would be the cause of, a primary-system rupture. This view must be reexamined for each case. It has also been considered incredible that the reactor pressure vessel would rupture. The basis for this latter view differs depending on the type of reactor.

The argument for the safety of reactor vessels for pressurized- and boiling-water reactors is based on the use of codes to assure control of vessel design, fabrication, testing, and use. To the best of the authors' knowledge, no pressure vessel designed, constructed, inspected, tested, and used in accordance with the appropriate sections of the ASME Unfired Pressure Vessel Code has failed catastrophically. Authorities generally agree that the failure of such a vessel is incredible. The designers of concrete vessels for gas-cooled reactors base their contention that pressure-vessel failure is incredible on the view that the many tendons that give the concrete vessel its strength are elastic enough to stretch and to relieve internal overpressures by minor venting, followed by reclosing and at least partial resealing.

In neither case is the proof of reliability absolutely certain. There remains still a very, very small but finite chance that such an accident might occur. In many power-reactor facilities, both in the United States and abroad, rupture of the pressure vessel would probably cause failure of the containment.

The reluctance to allow power reactors in metropolitan areas is due, in part, to the conservative approach taken with regard to the possibility of pressure vessel failure by those safety groups responsible for protection of the public. A complete solution of this problem would no doubt result in a relaxation of existing limitations on reactor sites. One solution might involve the use of underground and below-water-level installations. Other solutions are possible.

If codes are adequate to ensure the safety of primary pressure vessels, then it should be possible to write appropriate codes with a similar level of safety for connecting pipes and for other components in the primary system. Codes for the interactions between the members of the system could also be written. If one then had the same degree of faith in these codes and their use that now exists in the industry with regard to pressure-vessel codes, it might be that no containment would be necessary. However, the fact that containment is required, and that power reactors are not sited in metropolitan areas, still indicates that the industry and safety evaluators have not yet reached a point of total acceptance of the reliability of these codes and methods.

Chapter 11 discusses, in some detail, reactor accidents experienced to date. A review of that chapter may be of assistance in the provision of qualitative judgment as to the credibility of certain types of accidents. But for no type of accident experienced to date is there a sufficiently firm statistical base for elimination of that type of accident as a possibility. It can only be said that if accidents of a given type have occurred, or have even come close to occurring, it would be difficult to prove that type incredible. The only way to eliminate a particular type of accident with certainty is to eliminate the cause of the accident. The more conservative the design accident (i.e., the more severe the accident), and the more conservative the choice of fission-product-release fraction, the more conservative will be the containment design. In some cases, additional assurance of the safety of the facility can be obtained by investigation of the consequences of an uncontained but realistic fission-product release. A facility for which such a calculation shows nondisastrous consequences would seem to be adequately safe on any basis.

Although a single design accident is chosen as a basis for containment design, a series of different accidents must be investigated to confirm that the most conservative credible one has been chosen, and also that all variations of relevant parameters are accounted for. For instance, the double-ended rupture of a main coolant line may not be the worst accident for a particular reactor design. A smaller rupture of the same line might depressurize the system so slowly that core-cooling sprays could not be activated against the remaining pressure. After-heat might cause severe melting near the top of the core, even though the reactor were shut down. In this case, rapid blow-down of the system would permit the core-cooling sprays to be turned on more rapidly and, if the sprays worked as planned, would prevent core melting.

It is clear that, if several different types of design accidents seem credible or almost so, then the containment system and engineered safeguards must be so designed as to protect against all of them. Thus, in the previous example, the containment system should be designed to cope with a pressure surge, even a shock wave, that might be created by a sudden double-ended rupture, as well as with a slower release of pressure.

Even a single design accident may take several courses, resulting in different terminal conditions, depending upon the reactor-system operation in progress and the reactor configuration at the time of the accident. The varying causes of an accident may lead to different sets of requirements in different parts of the system.

The choice of design accident is often somewhat simplified if one source of energy may be responsible for a large fraction of the total prompt energy release in an accident. In this case, any accident that releases all the energy from this one cause may be an adequate design accident. For instance, it is estimated that the design accident in the Yankee pressurized-water reactor would release the energy amounts shown in Table 4-1 [11]. It can readily be seen that the stored thermodynamic energy far outweighs all other contributions to the released energy. The same situation probably exists in gas-

TABLE 4-1
Computed Energy Sources in the Yankee Pressurized-Water Reactor

Source of energy	Approximate magnitude (Btu)	Comments
Nuclear excursion	2×10^7	Vaporation of 20% of core
Chemical reaction	10^7	Metal-water reaction of all Zircaloy-2 and stainless steel in core
Decay heat	3×10^7	Integrated release over a 1-day shutdown period after infinite operation
Stored energy in coolant (water)	10^8	Includes all primary-system water

cooled power reactors. Fortunately, the thermodynamic release is the one energy source that can be most easily calculated in a realistic and yet completely conservative manner, as will be discussed in Sec. 4.3.2. In such cases, large uncertainties in other energy sources are less important. This happy state of affairs can provide assurance as to the adequacy of the containment design. It must be pointed out, however, that this situation may not exist for all water reactors and certainly does not exist for all other types of reactors.

If, for instance, a containment is designed so that the stress in a steel shell (or the steel-tensioned members in a concrete containment) does not exceed yield when the design accident occurs, and if the major portion of the energy released exhibits well-known and fully calculable behavior, then it may be possible to show that, even with ridiculously large energy-release values for the less well-known chemical and nuclear effects, the rupture strength of the steel would not be exceeded. This sort of analysis provides reassurance that the selected design accident is sufficiently conservative.

The selection of an appropriate design accident for those reactor types that have no large, well-defined potential source of energy release is more difficult. For instance, the fast reactor has virtually no stored pressurization energy, and the accident mechanisms that have been postulated are nuclear transients and chemical reactions. In this case, the entire sequence of events in the accident is important, and, in the final analysis, the estimate of the energy released may be subject to a large probable error. It may be that the containments of existing fast reactors are overdesigned, but the need for conservatism and the lack of knowledge of the course of the accident leave no other choice. The selection of this design accident and its course are principal topics of Chapter 10.

Once the choice of the cause of a design accident or accidents has been made, the course and the consequences must be traced out. Of primary importance is the determination of whether missiles such as valve parts, etc., can be released and breach the containment. If this is physically possible, then it must be made impossible by including missile shields. As discussed earlier in this section, it has been deemed incredible that the primary pressure vessel can rupture. However, as power reactors are located nearer to cities where the consequences of a rupture are worse, this viewpoint may change. Missiles consisting of parts of the pressure vessel may have to be considered and means devised to provide, even in this incredible event, adequate missile protection. Actually, in the design of a power-reactor facility, all credible missiles must be considered, and it must be determined that the containment is adequately protected against breaching by them. Otherwise the safety provided by the containment could be, in part at least, illusory.

The design accident must take into account the possibility of dangerous movements of pipes, or even of the pressure vessel itself, resulting from the reaction forces associated with a pipe break. The force of mechanical reaction to steam or gas release may be very large and may result in containment rupture, severance of core-cooling lines, severance or maloperation of electrical or mechanical connections for core control, or malfunction of safety valves.

To trace the sequence of events in an accident, it is necessary to know that nothing can occur that will alter the course or severity of the accident. To pursue the course of a design accident involves linking the reactor design with the containment design and provides a test of the adequacy of the two as a unit in their ability to withstand the consequences of the postulated accident. It is necessary to ascertain whether or not certain subsystems will operate as planned, whether certain subsystems can be isolated, and whether core melting will occur.

If emergency systems function as planned, no core melting should occur, and no serious consequences should result. As a part of the design-accident investigation, all subsystems, closures, and protective measures are probed to see whether each can function as planned. Important valves are installed in duplicate, with separate operators and with separate controls and primary and backup power supplies. Equipment for emergency power and controls are supplied. Backup pumps are provided for coolant-injection systems. Containment blowers, heat-removal systems, and fission-product filter and absorber banks are provided and checked for adequacy. Every reasonable effort is made to ensure containment of radioactive material, even from the fuel itself.

In order to pursue the hypothetical design accident to its most pessimistic conclusion, it is assumed that an engineered safeguard fails, core melting does occur, and a fraction of the fission products do gain access to the containment. Values for the released fraction are secured from the best available technical data.

Part of the process of investigation of the design accident is to ascertain that the operation of the emergency subsystems is as reliable as can be obtained. Also, the investigation should point up which subsystems are most important and, therefore, which subsystems need the most protection and backup. As an example, for some containments with insulating concrete walls, the post-accident heat-removal systems for the containment must be totally reliable, since otherwise the fission-product afterheat could rupture the containment.

Estimates of the percentage of the various fission products that are released and their absolute quantities must be made for each reactor design accident. Information developed in Chapter 18 and elsewhere may be used to make these estimates. A rough and generally conservative estimate of the percentage of the fission product released to the containment is given in Table 4-2 [24]. The absolute quantities may be obtained from the percentage release and the calculated fission-product inventory.

After estimating the fraction of the total fission-product burden that is released to the containment under

TABLE 4-2
Fission Products Released to Reactor Building [a]

Noble gases (Xe, Kr, etc.)	100%
Halogens (I^{131}, I^{132}, I^{133}, I^{134}, I^{135})	50%
"Solid" fission products	1%

[a] The total release represents approximately 15% of the gross fission-product activity.

the conditions (vapor composition, temperature, pressure, etc.) of the design accident, a fraction of the resulting contents of the containment is then considered to be released to the environment at a rate corresponding to the design leak rate.* The release from the containment is assumed to be at the worst position, generally at ground level, and under unfavorable meteorological conditions. The thyroid and whole-body doses are then calculated for the public in the most unfavorable direction. The correct meteorology for the specific reactor site should be used, and the meteorological practices set forth in references [25] and [26] should be followed.

In general, there are four mechanisms that could cause accidents leading to the release of large quantities of fission products from the fuel elements into the primary system. One mechanism is that of a nonnuclear energy release. The other three mechanisms involve directly the production of nuclear heat and its removal from the reactor core. They are reactivity excursions, loss of flow, and loss of coolant. A study of Table 3-1 of Chapter 2 will demonstrate that this simple classification of serious accidents covers those described in that more extensive table.

During the decade from 1965 to 1975, most of the world's power reactors will be pressurized-water reactors, boiling-water reactors, or gas-cooled reactors. Modern reactor technology provides assurance that, with reasonable care in design and operation, these reactors are unlikely to experience any accidents whose primary cause is a nonnuclear energy release and that the likelihood of a severe reactivity transient is smaller than that of either the loss-of-flow or loss-of-coolant accidents.

4.2.1 Nonnuclear-Energy Releases

Nonnuclear-energy releases, as a cause of a severe accident, might include such possibilities as the Wigner energy releases in graphite-moderated reactors and possible explosions involving hydrogen and oxygen formed by electrolysis or radiolysis of water. As long as large graphite-moderated gas-cooled power reactors operate with the graphite at high temperature, the possibility of Wigner energy release is not serious. Explosions of hydrogen-and-oxygen combinations could occur only in water-cooled or -moderated reactors. Since the cores are covered with water or steam in all cases, such explosions in the core would seem to be ruled out for normal operation. Explosions outside the core would seem a very remote possibility. Even in that unlikely case, the accident would result, at worst, in a primary-system rupture not involving the pressure vessel, which changes the accident to a loss-of-coolant accident. Thus it seems very unlikely that nonnuclear-energy releases can be considered as primary causes for design accidents in current power reactors. However, metal-water reactions between the cladding and the water, in boiling-water or pressurized-water reactors, or metal-air reactions in gas-cooled reactors, are potential sources of additional energy as an aftermath of loss-of-coolant accidents.

The burning of liquid-metal coolants in air is a possible accident in fast reactors. If the liquid metal is radioactive, radioactivity will be released from the primary system. Such an accident is more likely to involve simply the metal coolant than it is to involve the reactor core with its fission products. This accident is one that must be investigated for each fast-reactor design, but present and projected designs seem able to cope with it adequately.

4.2.2 Reactivity Excursions

Pressurized-water reactors are now operated with boron shim control, and, hence, the control rods act as safety rods. They are completely withdrawn from the core in normal power operation, except that one control group is slightly inserted. If care is exercised in the manner of going critical, and if any positive moderator effects are kept small, there is virtually no way in which a reactivity excursion can occur in such a system.

The boiling-water reactors have bottom-entry control rods, well-protected during operation, and operate in the power range where voids serve as a reasonably prompt shutoff mechanism. If care is exercised to ensure that control-rod mechanisms remain independent from one another, and if appropriate precautions are taken in start-up, there is virtually no way in which a reactivity excursion can occur in this system, either. Both these types of reactors tend to shut down on loss of coolant and loss of flow.

The gas-cooled natural-uranium reactors do not have large excess amounts of reactivity available, and hence cannot undergo fast nuclear transients. Gas-cooled reactors with fuel or higher enrichment do have enough excess reactivity to undergo reactor transients, but can be easily controlled by control rods and by backup control methods such as neutron-absorbing ball systems or gases. Further, properly designed control-rod mechanisms assure that rod blowout accidents are extremely unlikely. Since the coolant is not also the moderator in gas-cooled reactors, a loss-of-coolant accident will not shut off the reactor, and other means must be relied upon. Thus, possible interactions between loss-of-coolant accidents and even slow nuclear excursions, such as those resulting from moderator temperature coefficients or rod motions, must be investigated. However, in general, the possibility of a severe reactor transient in a gas-cooled reactor is very small. (See also Chapter 9.)

Other reactor types that are likely to be important in future power plants must be evaluated. If such systems have strong positive reactivity effects, then the possibility of reactivity transients is greatly increased. It appears possible to design large, fast reactors, and other power reactors of interest, in such a way as to reduce the likelihood of severe nuclear transients to an acceptable level, but economic penalties, such as limits in core size or shape, may be involved. For more details, Chapter 10 should be consulted.

*This leak rate is the limiting rate above which the reactor should not operate. The rate should be verified by continuous or spot tests during the entire period of reactor operation.

4.2.3 Loss of Coolant Flow

This accident can occur from two separate causes: loss of pumping or obstruction of flow. In general, large power reactors have at least two loops circulating coolant into and out of the core through common plenums. Each of these loops has a separate pump, and usually these pumps can be connected independently to one of two or more independent electrical power supplies. It thus appears unlikely that all pumps would lose their power simultaneously, although a large-area electrical blackout, such as occurred in November 1965 in New England, might cause such a loss. Usually, backup diesel engines are provided to take care of that contingency.

A flow obstruction could occur in one of at least two ways. Valves in all of the coolant loops could somehow be closed, thus shutting off the flow. This accident is very remote in all cases and impossible in systems which have no such valves. Obstructions could plug the loops or fuel-element inlet-flow orifices. It is extremely unlikely that obstructions could plug all the coolant loops or all the fuel elements.

Much more likely is the possibility that one or several fuel-element channels will be plugged by obstructions left in the system during construction or refueling, or by failure of a gasket or plate, which can act as an obstruction. Such accidents have happened on several occasions. Two are discussed in Secs. 4.3 and 4.4 of Chapter 11 (pp. 689-90). A third occurred on October 5, 1966, when an obstruction blocked the flow to one or two fuel elements in the Enrico Fermi sodium-cooled fast reactor. The operator observed a decrease in reactivity and, on checking thermocouple readings, an unusual distribution was observed which caused him to shut down the reactor [27-32]. Investigations showed that the fuel pins in one fuel element had melted quite completely and that two other elements had been damaged. A partially crumpled metal sheet about 8×10 in. was identified as the obstruction in this case.

Another type of loss of flow, with more potential for general core involvement, is given in the first item of Table 4-1 of Chapter 11. In that case, fuel-element swelling cut off coolant flow. Such a mechanism could be widespread in cores where the fuel is located in separate coolant passages. In any open-core reactor where cross-flow is permitted, it is difficult to see how a complete loss of coolant flow can occur, unless there is violent boiling.

In general, it would appear that loss-of-flow accidents involving the entire core are extremely unlikely, but accidents involving small portions of a core can, and no doubt will, occur. They are likely to be limited in extent and should not involve any serious danger to the public, unless strong positive reactivity effects are involved or unless melting can propagate. If such accidents should occur, local fuel melting is likely to be the worst possible consequence. Provisions to handle the consequences of loss-of-coolant accidents should be of value in this case as well. Thus, if redundancy and diversity in cooling-loop and pumping are provided, an overall loss-of-flow accident can be made extremely improbable. The loss-of-flow accident involving small fractions of the core—say, one element or so—cannot be eliminated so easily as an accident cause, but the methods used to cope with the consequences of local loss of flow do not differ from those required for the loss-of-coolant accident.

4.2.4 Loss of Coolant

The three types of power reactors most popular at present have pressurized coolants. Therefore, a failure in a primary-system pipe or component will lead to a loss-of-coolant accident. Even though the chain reaction may be shut down promptly after the initiation of such an accident, the decay of fission products will continue to develop heat, which must be removed by restoration of cooling if the fuel elements are to be prevented from melting. If fuel elements should melt, fission products would be released into the primary system, and a fraction of them would escape through the primary-system rupture into the containment.

Loss of coolant has been a safety concern since the earliest reactors. It has been thoroughly discussed, and provisions for preventing it have been thoroughly reviewed for every reactor. With the increasing size of power reactors now being built, the decay-heat-generation rate often amounts to values as high as 20 to 50 Mw(t). Largely because of the increased size and the complexity of these systems, the AEC Regulatory Staff and the Advisory Committee on Reactor Safeguards in October 1966 set up a task force to review certain aspects of these problems [33].

Among other considerations, the task force concluded that the description of the events "that could take place subsequent to a postulated meltdown of large portions of a core is at present indeterminate and quite speculative." Thus, they reaffirmed the earlier judgments of the industry and of the safety-evaluating groups that there is only one sure way to cope successfully with the loss-of-coolant accident: to get emergency cooling back onto the core before melting occurs and to maintain this cooling indefinitely.

4.3 Accident Energy Release

4.3.1 Energy Sources

It is necessary to know the energy release that might occur during an accident in order to determine the pressure and volume requirements for the containment. In the event of a serious accident, the energy may come from several sources, including stored thermodynamic energy in the primary system, nuclear-transient energy, the energy from chemical reactions, and fission-product-decay energy. All these potential sources may contribute energy in varying amounts and at varying rates, dependent upon the type of system, the design details, and, sometimes, the design accident considered.

The time scale for energy release varies considerably. The loss-of-coolant accident could happen in a fraction of a second (e.g., a brittle-fracture pressure-vessel rupture) or it could take a long time. While a nuclear transient could take a long time (i.e., slow ramp rod withdrawal) it is generally conceded that those transients likely to be of serious consequence have periods in the low millisecond range.* Chemical-energy release can be as slow as the corrosion process and as fast as a hydrogen-oxygen explosion. Of course, the release of fission-product-decay heat is a slow, orderly and well-known process.

*The very rapid nuclear transient is the subject of Chapter 7 in Vol. 1. Some aspects of it are also discussed in Chapters 8, 9, 10, and 11 for particular reactor types. Destructive-transients test experience is discussed in Chapter 11.

In most of today's water-cooled power reactors, the stored thermodynamic energy in the primary system is the dominant energy source. Some reactors, pressure-tube reactors, for example, have a relatively small volume at high pressure in the primary system, so that some other energy source may dominate at any time after the beginning of the accident. It is likely that the stored-energy source dominates as well, at least initially, in the gas-cooled reactors being designed and built now. For sodium-cooled fast reactors, the stored energy contributes almost nothing to the energy release in an accident.

As indicated in the preceding section, the increased understanding of the kinetics and control of large power reactors and the modified control methods now being used have reduced considerably the likelihood of a severe nuclear transient in large thermal power reactors from that believed possible 10 years ago. The problem of estimating the possible energy release from a severe transient is also beginning to be better understood. It may even be that something can be said about limits to the maximum amount of energy that could be released in a severe transient.

Energy release by chemical reaction is theoretically possible in any system in which an oxidation-reduction reaction is possible. For instance, metal-water reactions can occur whenever a coolant contains oxygen and a fuel consists of a cladding or fuel material that can be oxidized in an exothermic reaction. Thus there are many systems in which chemical reactions can release energy in an accident. The rate of reaction may be relatively slow or quite rapid. Usually, if the reaction is rapid, it occurs as the result of sudden heating, such as in a nuclear transient.

The release of fission-product afterheat or decay energy is a function of the reactor power, the method of reactor operation (i.e., "base-loaded" or "off-on" operation), the core lifetime, time after shutdown, degree of subcriticality of the shutdown core, and, sometimes, the specific characteristics of the reactor. The afterheat energy is well understood and predictable. Therefore, as long as the total contribution from stored thermal energy and fission-product afterheat is considerably larger than the total contribution from other, less predictable processes (such as nuclear transients or chemical reactions), the stored energy and afterheat play the fundamental role in establishment of containment requirements. As both are known and predictable, the total energy to be contained is also known.

In the subsections that follow, the four contributions, thermodynamic stored energy, nuclear-transient energy, chemical energy, and fission-product-decay energy, are discussed separately.

4.3.2 Primary-System Thermodynamic Energy Release

As mentioned in Sec. 4.3.1, the principal source of energy release in a pressurized- or boiling-water or a gas-cooled reactor is likely to be the energy stored in the primary coolant. If an accident ruptures the primary system, the expansion of the coolant (and the formation of steam, if it is a water-cooled power reactor) can provide the energy to accelerate projectiles that could penetrate the containment. But an even more important fact is that the energy released, and the resulting pressures, temperatures, and environment created within the containment, are often the principal considerations when the containment specifications are set up.

Since it is difficult to estimate accurately the rate of heat transfer to the containment walls and to other components external to the primary system, it is usually assumed that the release of the primary fluid to the containment is so rapid that heat transfer plays no part in the initial expansion. That is, the expansion is considered to be adiabatic. This is a conservative assumption, as it leads to a maximum initial pressure in the containment (as any heat transfer lowers the pressure), unless the core remains at or above critical in the nuclear sense.

Assuming adiabatic conditions, the internal energy of the system of pressurized high-temperature primary coolant plus atmospheric containment gas before the accident can be equated to the energy of the mixed system of coolant plus containment gas at some intermediate postaccident temperature. Since the problem is relatively straightforward for a gas-cooled reactor, requiring only the use of common properties of gases and the perfect gas law, a pressurized-water system will be considered here. The development of Pasqua will be followed [13c].

Equating the internal energies gives

$$M_a C_v T_{a1} + M_w U_{w1} = M_a C_v T_{a2} + M_w U_{w2}, \quad (4\text{-}1)$$

where M_a = mass of air in the system, M_w = mass of water in the system, C_v = specific heat of air at constant volume, T_{a1}= initial absolute temperature of air in the containment, T_{a2}= final absolute temperature of air in the containment, U_{w1}= initial specific internal energy of the water in the reactor, and U_{w2}= final specific internal energy of water in the containment.

The initial masses of air and water and all initial conditions are known. The final volume of the system is also known. The final temperature, T_{a2} of the system and the internal energy of the water in the containment are unknown. It is assumed that the final system is an equilibrium mixture of a gas, a saturated vapor, and a saturated liquid.

A quality of the mixture, X, may be defined. The volume of containment present per unit mass of water in the containment is V/M_w. Subtracting from this the volume per unit mass of saturated water at T_{a2} (which is V_{f2}, the specific volume at T_{a2}), gives the volume per unit mass occupied by vapor. This difference, divided by the specific volume change from saturated liquid to saturated vapor at T_{a2} (which would be the specific volume for just saturated vapor, V_{fg2}) gives the quality of the mixture:

$$X = \frac{(V/M_w) - V_{f2}}{V_{fg_2}} \quad (4\text{-}2)$$

Then Eq. (4-1) becomes

$$M_a C_v T_{a1} + M_w U_{w1} = M_a C_v T_{a2} + M_w (U_{f2} + X U_{fg_2}), \quad (4\text{-}3)$$

where U_{f2} is the specific internal energy of the saturated liquid and U_{fg_2} is the specific internal energy change from saturated vapor at temperature T_{a2}. This equation can be solved by using the steam tables, assuming an initial and final temperature, and then solving for X. By substituting the value of X obtained and the assumed T_{a2} Eq. (4-3) can be checked for equality. Repeated iteration then gives a satisfactory answer.

The final containment pressure, P_2, can then be found from

$$P_2 = P_{s2} + P_{a2}, \quad (4\text{-}4)$$

where P_{s2} is the partial pressure of the saturated steam at

T_{a2} and P_{a2} is the partial pressure of the containment gas (air) at T_{a2}. The perfect gas law can then be applied to obtain P_{a2}:

$$P_{a2} = P_{a1} (T_{a2}/T_{a1}),$$

where P_{a1} is the initial pressure of the gas in the containment vessel. The final pressure is

$$P_2 = P_{s2} + P_{a1} (T_{a2}/T_{a1}). \tag{4-5}$$

(Note that T_{a1} and T_{a2} must be in degrees Kelvin or degrees Rankine.)

This same problem may be solved by use of nomographs. The method and the nomographs are presented here in a short form; they are given in more detail (with examples) by Pasqua [13c]. If Eq. (4-3) is divided through by M_w, an equation for the internal energy per unit mass of water is obtained:

$$U_{a1} + U_{w1} = U_{a2} + U_{w2} = (U/M_w)$$

where $U_{a1} = (M_a/M_w)\, C_v T_{a1}$,

$$U_{a2} = (M_a/M_w)\, C_v T_{a2},$$

$$U_{w2} = U_{f2} + XU_{fg_2} \tag{4-6}$$

and X is given by Eq. (4-2).

The necessary nomographs for solution of these equations are shown in Figs. 4-2, 4-3, 4-4, and 4-5. If the initial air temperature in the containment and the mass ratio M_a/M_w are known, Fig. 4-2 can be used to obtain the initial internal energy of the air. When the saturation pressure or average temperature of the water coolant is known, the initial internal energy of the water can be obtained from Fig. 4-3. The sum of these two then gives the total specific internal energy of the system (Btu/lb H_2O). The problem is then to choose a final temperature, T_{a2}, that gives the same total specific internal energy. The ratio V/M_w is known, and so an appropriate curve or interpolated line between two curves can be chosen in Fig. 4-4. A value of U_{w2} must now be selected such that its value at a given T_{a2}, plus a corresponding value for U_{a2} at the same T_{a2}, from Fig. 4-4, gives agreement for the internal energy already obtained for the system. This may take several trials. The particular choices shown in the nomographs are for $T_{a1} = 200°F$, $M_w = 300$ lb, $P_{s1} = 1500$ psia (or $T_{s1} = 596°F$), V = 2000 cu ft, $U_{a1} = 16$ Btu/lb H_2O, $U_{w1} = 611$ Btu/lb H_2O, $T_{a2} = 245°F$, $P_2 = 33$ psia. Figure 4-5 gives final values of the containment pressure for various M_a/M_w ratios as a function of initial reactor pressure, calculated by these methods. Pasqua [13c] presents a FORTRAN program for obtaining solutions to Eq. (4-6).

The energy stored in the form of heat in the steel vessel core internals, primary-system piping, etc., must also be considered in a complete analysis of the behavior of the system. The heat content of these components will often be high enough to make this source of energy very important in the behavior of the system during the first few minutes after a loss-of-coolant accident.

4.3.3 Nuclear-Transient Energy

This subject is covered in a number of chapters of Vol. 1. Mathematical models for fast transients are discussed in Chapter 7. Solid-moderator reactor transients are discussed in Chapter 9. Fast-reactor transients are discussed in Chapter 10. Accidents or destructive tests resulting from reactivity additions are discussed in Chapter 11.

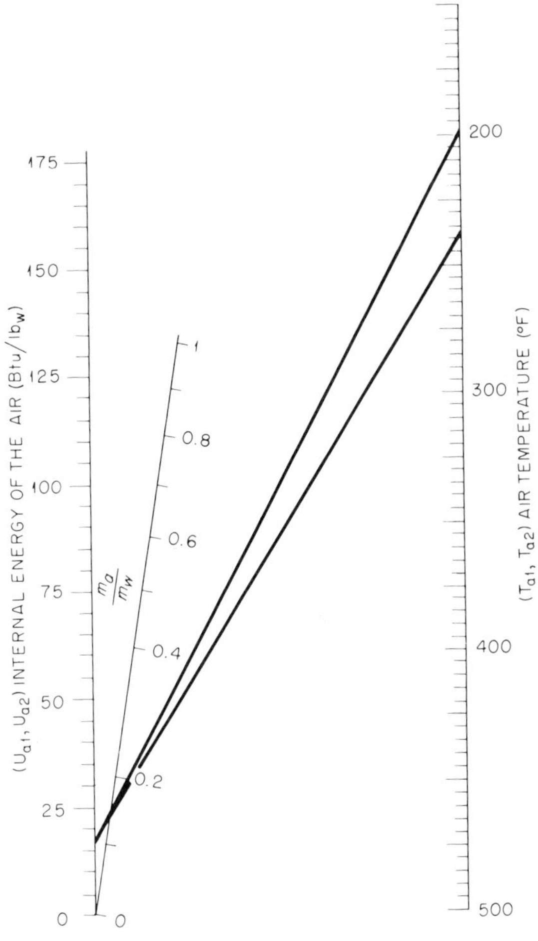

FIG. 4-2 Nomograph for determining the initial internal energy of the air from the initial air temperature and the mass ratio.

Some generalizations can be made. All reactors are designed to provide assurances that large reactivity additions cannot be made suddenly. Control-rod-withdrawal rates are limited electrically and mechanically. Cold-water accidents are precluded by slow-opening valves and certain interlocks. Neutron and temperature monitors give warnings and initiate automatic shutdown procedures in case of nuclear transients.

Perhaps the greatest assurances are given by inherent shutdown mechanisms, which exist within the core assembly and act regardless of the proper function of external controls. Almost all power reactors, except military ones, use natural-uranium fuel, or fuel only partially enriched by U^{235}. Therefore, most of them have a negative Doppler effect (see Chapter 2, Sec. 6.3, and Chapter 4) that tends to counteract promptly the addition of reactivity. (It should be noted that reactors do not necessarily have a negative Doppler effect.)

The boiling-water and the pressurized-water reactors, operating as they do at or near the coolant's boiling point, tend to shut down on loss of pressure; excessive and rapid pressure increases, which can cause positive reactivity effects, can usually be prevented by proper design. The boiling-water-reactor shutdown response to a

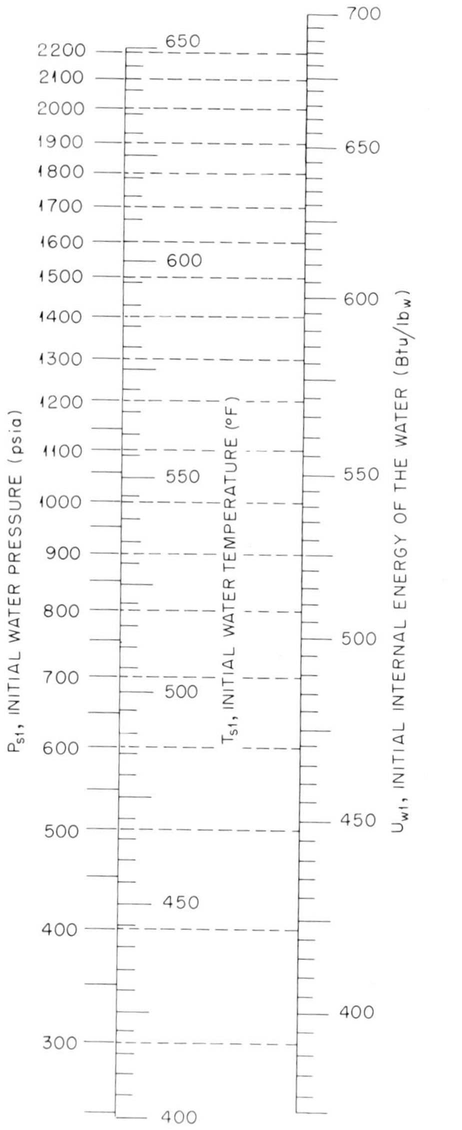

FIG. 4-3 Nomograph for determining the initial internal energy of the water from the saturation pressure or temperature.

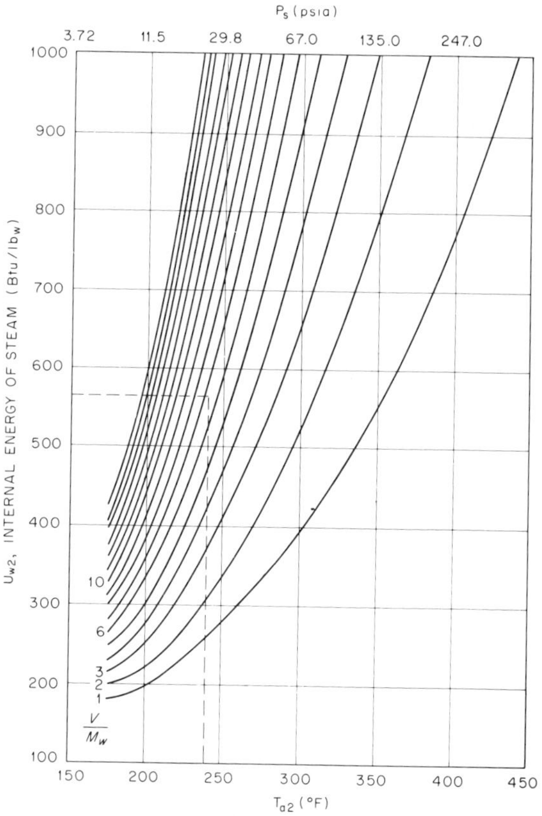

FIG. 4-4 Final internal energy of water mixture in containment vessel as a function of temperature and ratio of containment volume to mass of water mixture.

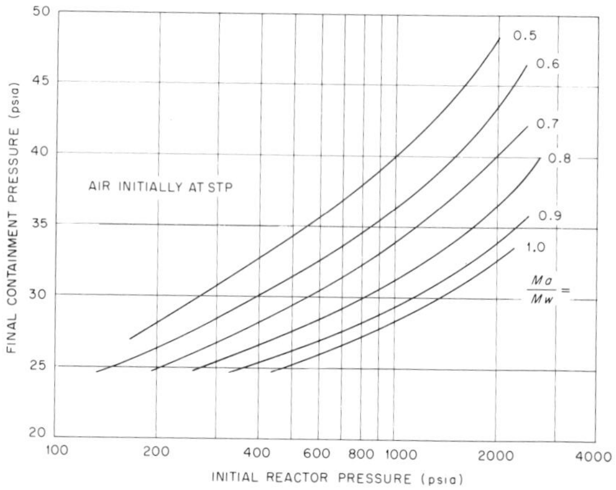

FIG. 4-5 Containment vessel conditions following reactor accident for various mass ratios.

nuclear transient is somewhat faster than a pressurized-water-reactor shutdown. Prompt gamma and fast-neutron transient heating of the water can rapidly add more voids to the already boiling water. The pressurized-water reactor using boron is able to operate with control rods essentially fully withdrawn, thus usually preventing large, rapid addition of reactivity by a rod-blowout accident; any positive moderator temperature or void coefficients that may result from use of boron can be limited by proper design. Gas-cooled power reactors that operate with natural uranium or very low enrichment generally do not have much excess reactivity and, hence cannot undergo rapid nuclear transients. (See, for instance, Chapter 9, Sec. 5.4, and note that the reactor periods, even in the worst cases shown, are of the order of 10 sec.)

In today's boiling-water, pressurized-water, and gas-cooled power reactors, the energy stored in the coolant far exceeds any reasonable estimate of the upper bound of the potential release of nuclear-transient energy. For example, in some of the older power reactors, a rough estimate of the upper bound was made by calculating the energy required to vaporize 20% of the core. Such an estimate for the Yankee Reactor has been made [13]; the essential data are as follows:

Core weight, W_F: 2.37×10^7g of UO_2.
Specific heat of solid UO_2, C_{Ps}: 18.45 + $(2.431 \times 10^{-3}T) - (2.272 \times 10^{-5}T^2)$ in cal/mole °C.
Reactor operating temperature, T_o : 268 °C (water) (assume 1000° C for average center UO_2).
Reactor operating pressure, P_o: 2000 psig.
UO_2 melting temperature, T_M: 2760°C.
UO_2 vaporization temperature, T_B:= 3180°C at 1 atm, 4380°C at P_o.
Heat of fusion, ΔH_F: 16 kcal/g mole.
Heat of vaporization, ΔH_V: 137 kcal/g mole.

$$Q_v = 0.2W_F C_{Ps} T_M - T_o) + \Delta H_F + C_{Pm}(T_B - T_M) + \Delta H_V$$

$$= (0.2)\,(2.37 \times 10^7)\,(0.081)\,(2760 - 1000) + \frac{16{,}000}{270}$$

$$+ (0.100)\,(4380 - 2760) + \frac{137{,}006}{270}$$

$$= 5.1 \times 10^9 \text{ cal} = 2.2 \times 10^4 \text{ Mw-sec} = 2.0 \times 10^7 \text{ Btu.}$$

This calculation assumes, somewhat arbitrarily, that 0.2 of the core is vaporized. Obviously, the calculation is not completely realistic. However, it should be noted that over half the total energy evolved goes into vaporizing the fuel. A release of about two times as much heat would be sufficient to bring the entire core (if a flat power distribution is assumed) to a temperature just below the vaporization temperature.

Since the actual ratio of maximum to average power in PWR cores is about 2 or 3, to raise the overall power of the core by adding reactivity uniformly throughout (as, for instance, by adding cold water) certainly would result in considerable fuel vaporization before all the fuel reached a temperature just below the vaporization temperature. In fact, some fuel is likely to vaporize even if the total heat added were just that of the calculation above, not twice that amount. It is somewhat fortuitous that this criterion is not too different from that given in Sec. 6.3 of Chapter 11, which, paraphrased, states that the energy evolved in the maximum nuclear transient is not likely to exceed the energy required to vaporize, say, 10% of the core, while the rest of the core is being heated as well. No proof for this statement exists at the moment, but certainly, in almost all cases, the core should disassemble by the time 10% of it is melted, thus reducing the reactivity.

Since a fast transient caused by local effects in large cores can vaporize fuel at a lower total heat input than considered above and can cause more rapid local disassembly, the full-core transient should be the upper limiting case. All reactor cores should be examined to ensure that during disassembly no more critical configuration is developed. Further, more experimental and theoretical work is needed to establish the validity of the concept discussed above.

A recent analysis that starts from the general hypotheses given in Sec. 6.3 of Chapter 11 lends additional support to the views expressed there [34]. In this work, a theoretical study was made of fast transients in water-moderated power reactors. In a fast transient, it is valid to assume no heat transfer across the cladding. The transient then proceeds in accordance with the well-developed reactor physics equations for such cores, until fuel vaporization occurs and high fuel internal pressure developes, which bursts the clad. The bursting of fuel intimately mixes molten fuel with water and causes prompt formation of a steam bubble, which helps terminate the excursion.

Figure 4-6 shows a comparison between the test results for a SPERT slightly enriched-UO_2, stainless-steel-clad core (see pp. 684-5 Vol. 1) and predictions for a shutdown (a) by Doppler effect alone and (b) by Doppler plus the negative reactivity effect of the steam bubble formed by bursting of two waterlogged pins. The agreement is much improved by including the steam-bubble mechanism.

Figures 4-7, 4-8, and 4-9 show the results of application of this same general type of theory to large boiling-water and pressurized-water reactors. Again, dotted lines indicate the predicted course of the transient if the only shutdown mechanism is the Doppler effect.

The first and most important feature to note in these graphs is that the total amount of energy released in the transients and the highest pressures developed increase very little, even for extremely large reactivity-addition rates.* Thus, there does appear to be almost an upper limit to the nuclear-transient energy that can be released in this type of reactor. The shutdown mechanism is aided by an augmented Doppler effect due to the larger fuel surface area exposed by fuel fragmentation.

Second, the energy developed in the transient appears to be less than that already stored in the system. Hence, it is possible that a containment could be developed that would be safe, even for a hypothetical transient that added an unrealistically large amount of reactivity. It is not clear whether such a transient could cause pressure-vessel rupture, but there appears to be a good chance that it might not. A more precise calculation, with exact core geometries and flux configurations, together with better defined fuel-particle sizes, is required. Such a calculation, to be meaningful, must use data from reactor designers and from tests of fuel bursting.

Third, the steam-bubble effect is more useful in terminating excursions in boiling-water reactors than in pressurized-water reactors, as the available free boiling surface permits easier expansion of the bubble when the clad bursts. However, the boiling-water reactor in such an accident might develop water-hammer effects similar to those observed in the SL-1 (see pp. 675-6 of Vol. 1), since the bubble would develop near the position of peak flux in the core, which is often well below the boiling region.

It is also likely that a few lower-melting fuel rods, properly spaced through the core, would have benefit as a shutdown mechanism. The mechanism should also help terminate fast-reactor transients in liquid-metal-cooled reactors.

In none of the cases just described have the effects of the shock wave formed been analyzed in any detail. If containment is to be effective, the action of shock waves and missiles that could be generated by such an excursion should be assessed. (Tests using explosive charges of various types are reported in the literature.)

In summary, for pressurized- and boiling-water reactors, it is unlikely that any conceivable transient can develop energy in excess of that necessary to vaporize a small part of the core, say 5% to 20%. For gas-cooled reactors utilizing natural uranium fuel, it is unlikely that any fast-reactivity transients can occur, since the available excess reactivity is small. (This assumption must be

*As pointed out in Sec. 4 of Chapter 7, the rate of activity addition, not the total amount of reactivity available, is the important parameter.

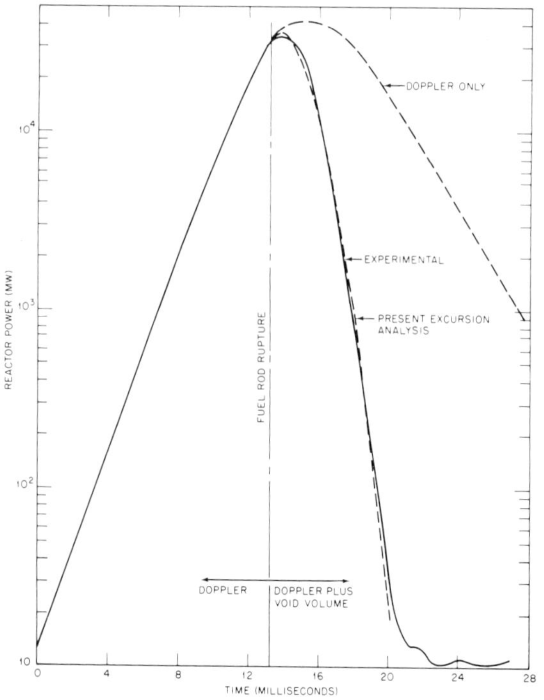

FIG. 4-6 Reactor power versus time, experimental and predicted results. (SPERT slightly-enriched UO_2, stainless-steel-clad core; test OC-599C, run 37.)

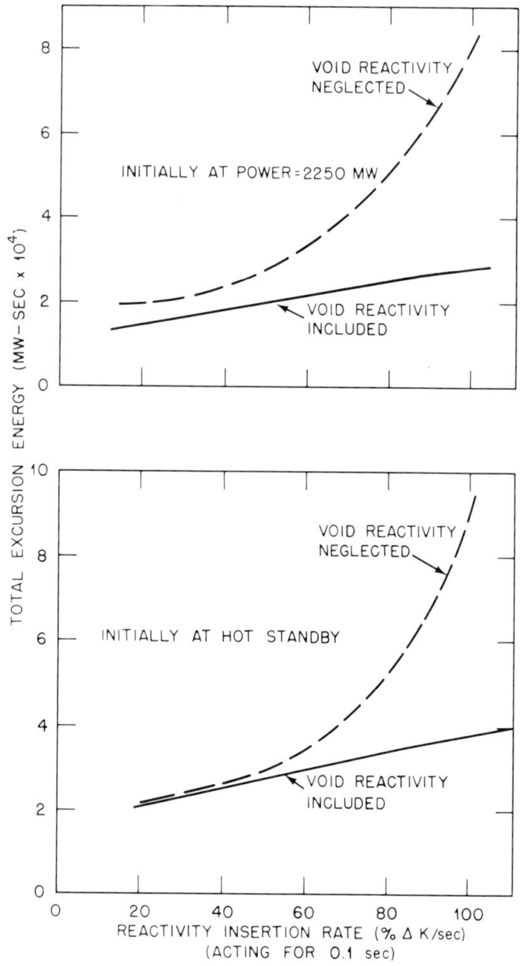

FIG. 4-7 Excursion energy versus reactivity insertion rate in a large BWR.

examined for each specific reactor.) For any of these reactor types, the energy release that could be expected is normally only a fraction, less than 30%, say, of the other sources of energy, particularly the stored thermodynamic energy and the fission-product afterheat. Accordingly, it is common practice for pressurized- and boiling-water reactors located outside metropolitan areas to exclude reactivity transients from the design accident. It is considered that, if the reactor containment is designed with a reasonable factor of safety, say 2 or 4, a nuclear transient, as well as other energy releases, cannot alter the effectiveness of the containment under accident conditions. However, this subject must be investigated for each reactor. In particular, the effects of a shock wave on the primary system or on the containment may alter this judgment.

Clearly, the ultimate hypothetical accident is one in which a nuclear transient creates a shock wave that ruptures the primary system, thus causing a loss-of-coolant accident, core melting, and exothermic chemical reactions. All that can be said with certainty for such a sequence of events is that the energy released by the nuclear transient for the reactor types discussed here is likely to be a relatively small fraction of the total energy release.

Estimation of the energy released in a severe nuclear transient in a liquid-metal-cooled fast reactor is difficult. (Gas-cooled or steam-cooled fast reactors are likely to have a relatively large store of thermodynamic energy in the primary system, thus modifying the following observations somewhat.) First, the nuclear transient is one of two important energy sources for this type of reactor, the other being chemical reaction. Second, the uncertainty involved in estimating either the nuclear transient or the chemical energy is large, since there is little experimental evidence that is directly applicable; the course of the accident must be mostly conjecture. Third, the entire containment design must be based on these estimates. This entire subject is discussed in detail in Sec. 4 of Chapter 10, Vol. 1, and is not dealt with here.

4.3.4 Chemical-Reaction Energy

The chemical-reaction energies that must be taken into account during a postulated nuclear accident are considered in Secs. 3 and 4 and the Addendum of Chapter 17. In Sec. 4.4 of Chapter 17, two types of accidents, loss-of-coolant and nuclear excursions, are discussed. In the loss-of-coolant accident, the heat source is fission-product decay, whose magnitude as a function of time is quite well known (see Sec. 4.3.5 of this chapter). Even for this accident, the containment problem is difficult. As pointed out in Chapter 17, Sec. 4.4, there may be autocatalytic effects even below the melting point of the cladding. Chemical rate equations and reliable estimates of the heat transfer between fuel and cladding and between cladding and steam are needed. Moreover, there

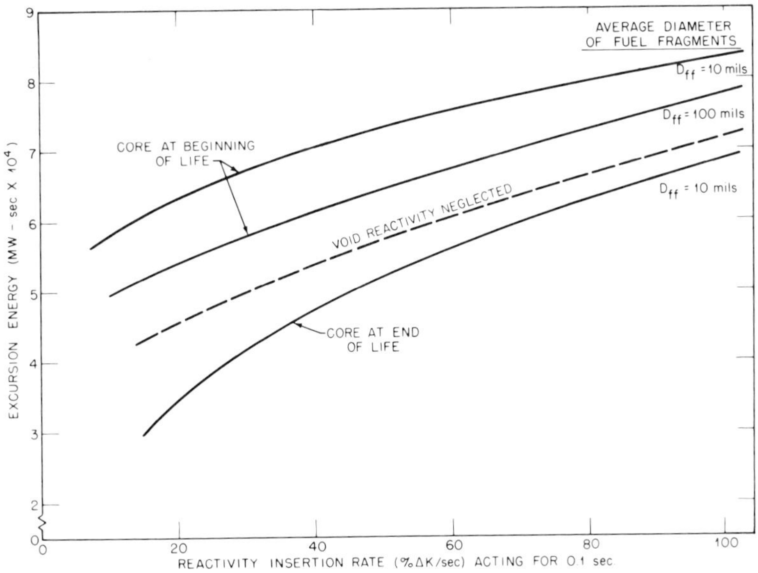

FIG. 4-8 Excursion energy versus reactivity insertion rate in a large PWR.

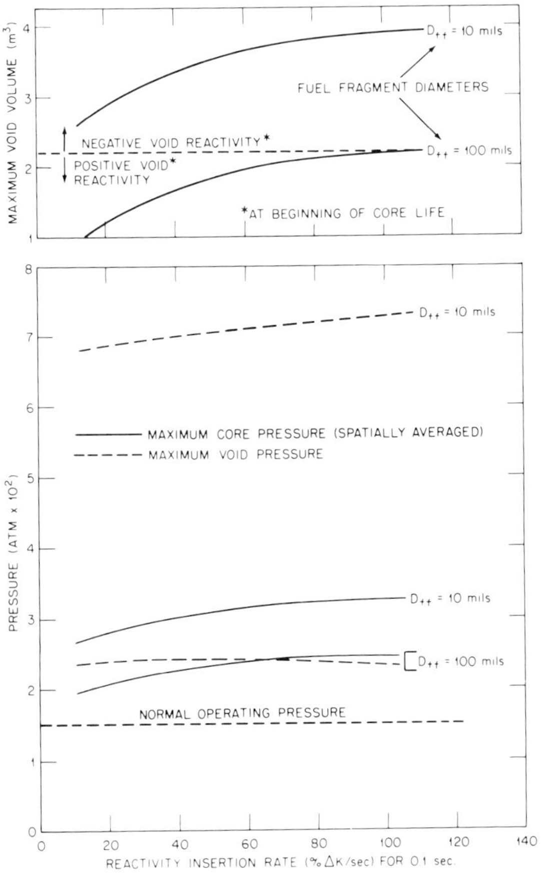

FIG. 4-9 Void volume and pressure during excursions in a large PWR.

is great uncertainty about the core fragmentation mechanism during a meltdown.

As pointed out in Sec. 4.4 of Chapter 17, the amount of metal-water reaction to be expected in a nuclear excursion is dependent on the magnitude of the nuclear excursion. If no fuel is melted, no significant metal-water reaction would be expected. In transients fast enough to be of concern, an adiabatic core-temperature calculation should be made. An average particle size and the extent of reaction constituting the energy release can then be estimated from Tables 4-3, 4-4, and 4-5, and from Fig. 4-2 of Chapter 17. The result is probably as good an estimate of chemical-energy release as can be obtained at this stage in the development of nuclear technology.

Only three nuclear transients have occurred that resulted in core shutdown by disassembly and core destruction. These were BORAX-1, SL-1, and the SPERT-1 destructive test. These transients are discussed in Chapter 11, and the metal-water-reaction aspects of them are discussed in more detail in Chapter 17. The fuel elements in these three reactors were all plate-type, with aluminum cladding and aluminum-uranium-alloy meat. (The SL-1 had 2% nickel in the cladding.) In all three, the metal-water reaction resulted in a release of chemical energy from 10 to 20% of the nuclear energy released. Thus, the metal-water reaction did not play a major role in these accidents.

It must be pointed out that this experience with aluminum-clad plate elements in a thermal reactor with little or no Doppler effect may not be typical of what can occur. Additional work is required in this area. If, as postulated in Sec. 4.3.3, severe nuclear transients always release about the same amount of energy, then one or two definitive tests of this entire mechanism should serve to establish its validity. The tests would obviously include verification of the shutdown mechanisms (Doppler and fuel-element vaporization), as well as the energy developed in metal-water reactions.

If it can be shown that severe nuclear transients are not possible, then metal-water reactions can result only from loss-of-coolant accidents followed by core melting.

TABLE 4-3

Constants for Total Power Emission Equation [a]

Applicable Time Interval (sec)	A	a	Maximum Positive Deviation	Maximum Negative Deviation
$10^{-1} \leq t < 10^{1}$	12.05	0.0639	4% at 10^{0} sec	3% at 10^{1} sec
$10^{1} \leq t \leq 1.5 \times 10^{2}$	15.31	0.1807	3% at 1.5×10^{2} sec	1% at 3×10^{1} sec
$1.5 \times 10^{2} < t < 4 \times 10^{6}$	26.02	0.2834	5% at 1.5×10^{2} sec	5% at 3×10^{3} sec
$4 \times 10^{6} \leq t \leq 2 \times 10^{8}$	53.18	0.3350	8% at 4×10^{7} sec	9% at 2×10^{8} sec

[a] From reference [13], p. 5.124.

TABLE 4-4

Constants for Total Gamma Power Emission Equation [a]

Applicable Time Interval (sec)	A	a	Maximum Positive Deviation	Maximum Negative Deviation
$10^{0} \leq t < 1.5 \times 10^{2}$	6.710	0.1316	7% at 1.5×10^{2} sec	7% at 10^{1} sec
$1.5 \times 10^{2} \leq t \leq 10^{6}$	14.77	0.2919	7% at 1.5×10^{2} sec	6% at 4×10^{3} sec
$10^{6} < t < 3.5 \times 10^{7}$	536.4	0.5500	4% at 2×10^{7} sec	5% at 4×10^{6} sec
$3.5 \times 10^{7} \leq t \leq 4 \times 10^{8}$	0.4285	0.1405	2% at 4×10^{8} sec	1% at 3.5×10^{7} sec

[a] From reference [13], p. 5.124.

This very much narrows the region of conjecture. In such cases, it could happen that molten fuel could fall into water, or safety-injection water could be sprayed into hot or molten fuel. It is relatively easy to investigate the mechanisms for this type of accident in laboratory or field tests.

4.3.5 Core Nuclear-Decay Heat

After a nuclear reactor has been shut down following power operation, it continues to generate heat. This heat arises from the continued presence of neutrons that cause further fissions, and from the emission of beta particles and gamma rays by the radioactive decay of the various fission-products. The first mechanism is much less important than the second and is usually evident only during the first few moments after shutdown.

Some neutrons evolved as a result of the fission process are emitted with varying half-lives as a part of certain fission-product-decay chains. (See Sec. 3.2 of Chapter 5.) These delayed neutrons amount to about 0.7% of the total neutrons from fission. Each delayed neutron can cause additional fissions. As the delayed neutrons die out with their characteristic half-lives, the fission power of the reactor dies out correspondingly. Since, by definition, a shutdown reactor is one whose k_{eff} is less than unity, any reactor that complies with this conditon is shutdown. However, as discussed in Sec. 1.1 of Chapter 5, the value of k_{eff} may still be close to unity, and the neutron multiplication may still be very high. In this case, the decay of the fission power is slower. Zubarev and Sokolov [35] have considered this problem and developed an equation on the basis of diffusion theory:

$$\frac{\phi(t)}{\phi_o} \cong \frac{\rho}{\rho+\beta} \exp -\left(\frac{\rho+\beta}{\ell} t\right) \times \sum_{i=1}^{m} \frac{\beta_i}{\rho+\beta} \exp\left(-\frac{\rho}{\rho+\beta} \lambda_i t\right), \qquad (4\text{-}7)$$

where $\beta = \sum_{i=1}^{m} \beta_i$, $\phi(t)$ = time-dependent thermal flux, ϕ_0 = constant flux up to start of shutdown $t \leqslant 0$, λ_i = decay constant for i^{th} group of delayed neutrons, m = number of delayed-neutron groups (six used in Fig. 4-10), ρ = reactivity = $(k_{eff}-1)/k_{eff}$. (See Sec. 2.1 of Chapter 2 for discussion of the significance of reactivity.)

This equation is plotted in Fig. 4-10. The curves are plotted for six groups of delayed neutrons and correspond to 1% and 25% shutdown. The delayed-neutron groups can include those neutrons resulting from energetic gamma-ray (for instance, above the 2.2-Mev threshold for deuterium) interaction with certain constituents of the core. This effect is particularly important for heavy-water-and beryllium-moderated cores. In such reactors, (γ, n) reactions may add to the afterheat and should be estimated. (Delayed-neutron characteristics for heavy water and beryllium are given by Bernstein [36] and by Keepin [37]). However, in most reactors the effect of post-shutdown delayed neutrons as a source of energy is of little importance; if it should be of importance, it will only be for a relatively short time, as indicated in Fig. 4-10.

A more important source of energy is from the beta and gamma rays emitted during the decay of fission products subsequent to the reactor shutdown. Normally fission-product decay is not an important source of energy during the initial stages of a postulated loss-of-coolant accident. The release of the stored thermal

TABLE 4-5

Energy Release Rate and Energy Release from Fission-Product Decay in a U^{235}-Fueled Reactor After Shutdown[a]

Operating Time, t_0 (hr)	Shutdown Time, t_s (sec)	Q_γ (%)[b]	Q_β (%)[b]	Q_t (%-sec)[c]
10^0	10^0	2.53	2.35	6
	10^1	1.95	1.70	55
	10^2	1.06	0.93	250
	10^3	0.34	0.29	1,180
	10^4	0.01	0.005	3,150
	10^5			4,950
10^1	10^0	2.90	2.68	6
	10^1	2.32	2.03	55
	10^2	1.42	1.24	400
	10^3	0.66	0.55	1,880
	10^4	0.20	0.15	7,400
	10^5	0.023	0.007	16,200
	10^6	0.0025		23,600
10^2	10^0	3.04	2.84	6
	10^1	2.46	2.19	55
	10^2	1.57	1.39	400
	10^3	0.81	0.71	2,480
	10^4	0.33	0.30	9,600
	10^5	0.088	0.060	33,500
	10^6	0.016	0.009	86,000
	10^7			200,000
	10^8			230,000
10^3	10^0	3.17	2.89	6
	10^1	2.59	2.24	55
	10^2	1.70	1.44	400
	10^3	0.93	0.76	2,480
	10^4	0.44	0.36	13,600
	10^5	0.185	0.123	51,500
	10^6	0.068	0.044	203,000
	10^7	0.007	0.005	481,000
	10^8			771,000
10^4	10^0	3.22	2.94	6
	10^1	2.64	2.29	55
	10^2	1.75	1.49	400
	10^3	0.98	0.81	2,480
	10^4	0.492	0.412	13,600
	10^5	0.236	0.172	68,500
	10^6	0.110	0.088	260,000
	10^7	0.022	0.028	1,090,000
	10^8			2,270,000
10^5	10^0	3.22	2.97	6
	10^1	2.64	2.32	55
	10^2	1.75	1.52	400
	10^3	0.99	0.83	2,480
	10^4	0.497	0.436	13,600
	10^5	0.240	0.197	68,500
	10^6	0.113	0.113	370,000
	10^7	0.025	0.049	1,420,000
	10^8	0.003	0.012	4,150,000
∞	10^0	3.24	2.99	6
	10^1	2.66	2.34	55
	10^2	1.77	1.54	400
	10^3	1.00	0.87	2,480
	10^4	0.510	0.465	13,600
	10^5	0.254	0.225	68,500
	10^6	0.127	0.142	370,000
	10^7	0.038	0.079	1,822,500
	10^8	0.015	0.040	8,052,500

[a] From reference [13], p. 125 and 126.
[b] Multiply tabulated values by (operating power/100) to obtain units of power.
[c] Multiply tabulated values by (operating power/100) to obtain units of energy, i.e., power-sec.

energy far outweighs the fission-product-decay heat. The decay-heat rate from this source represents at most 6 to 8% of the original power level of the reactor in operation. However, it may become very important later. If the fission-product-decay heat output exceeds the rate at which heat can be removed from the containment system, temperatures and pressures will rise within the containment, and a containment rupture could occur.

The energy released in fission-product decay depends upon the beta-and gamma-ray energies of the decay steps in all the decay chains, together with their half-lives and relative abundance in the core. The fission-product mass

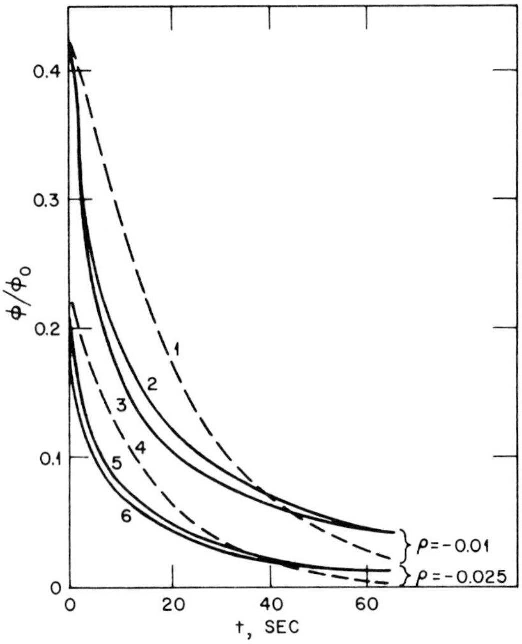

FIG. 4-10 Flux decrease according to Eq. (4-7).

distributions for thermal-neutron fission are relatively independent of the fissile isotopes involved; the range of beta rays is short; and power reactors, at least, are large enough so that no serious error is made if it is assumed that all beta- and gamma-ray energy is deposited in the core and that all fission-product decays proceed in the same manner in all reactors. In fast reactors there might be reason to consider the fission-product yields in more detail.) Calculations of the energy release after fission, suitable for essentially all reactors, can then be made. A definitive estimate was first made by Way and Wigner [38]. It may be stated in the form of an equation:

$$\frac{P}{P_o} = 6.22 \times 10^{-2} \, [t^{-0.2} - (T_o + t)^{-0.2}],$$

where P is the power generation due to beta and gamma rays (essentially equal contributions), P_o is the reactor power before shutdown, T_o is the power-operation time, in seconds, before shutdown, and t is the time of shutdown. This equation is said to give results correct to within a factor of two for decay times between 10 sec and 100 days.

Untermyer and Weills [39] developed an empirical formula to fit the experimental data for irradiated natural uranium (including U^{239} and Np^{239}) as follows:

$$\frac{P}{P_o} = 0.1 \left\{ (t + 10)^{-0.2} - (t + T_o + 10)^{-0.2} - 0.87 \, [(t + 2 \times 10^7)^{-0.2} - (t + T_o + 2 \times 10^7)^{-0.2}] \right\}$$

They estimated their errors as follows:

t under 1 sec, large error;
$1 \leqslant t \leqslant 10^2$ sec, within ±50%;
$10^2 \leqslant t \leqslant 10^4$ sec, within ±30%;
$10^4 \leqslant t \leqslant 10^6$ sec, within ±10%;
$10^6 \leqslant t \leqslant 10^8$ sec, within ±50%.

Shure [40] has combined what he believes to be the most reliable information (see also pp. 139-143, Sec. 5 of ref. [13]) into one set of graphs, which are presented here in Figures 4-11 through 4-14. The data are for an

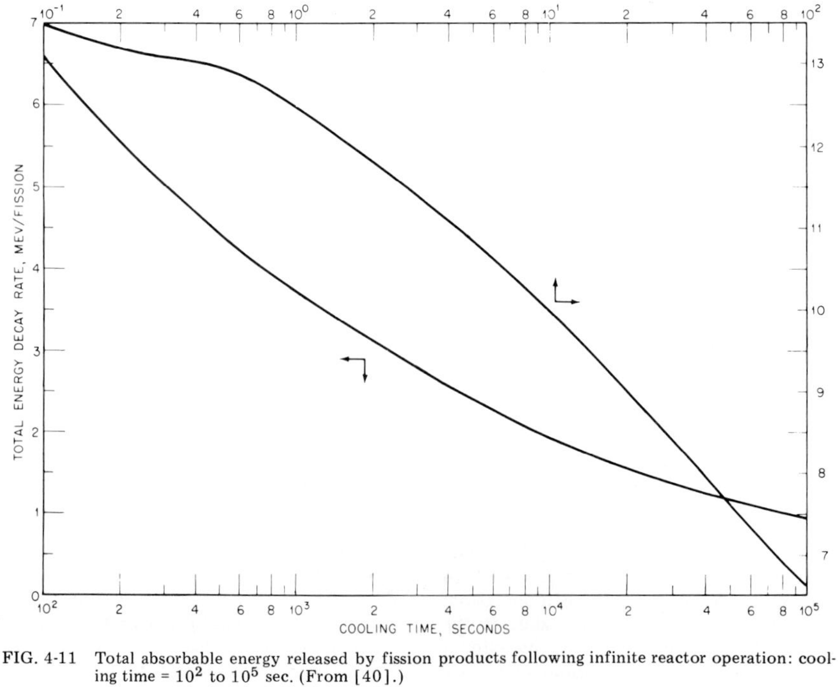

FIG. 4-11 Total absorbable energy released by fission products following infinite reactor operation: cooling time = 10^2 to 10^5 sec. (From [40].)

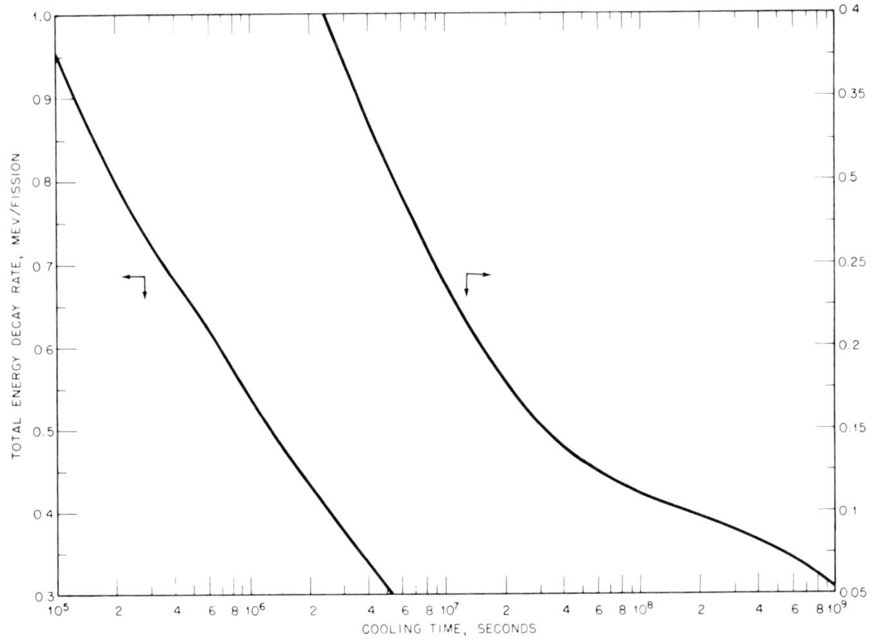

FIG. 4-12 Total absorbable energy released by fission products following infinite reactor operation: cooling time = 10^5 to 10^9 sec. (From [40].)

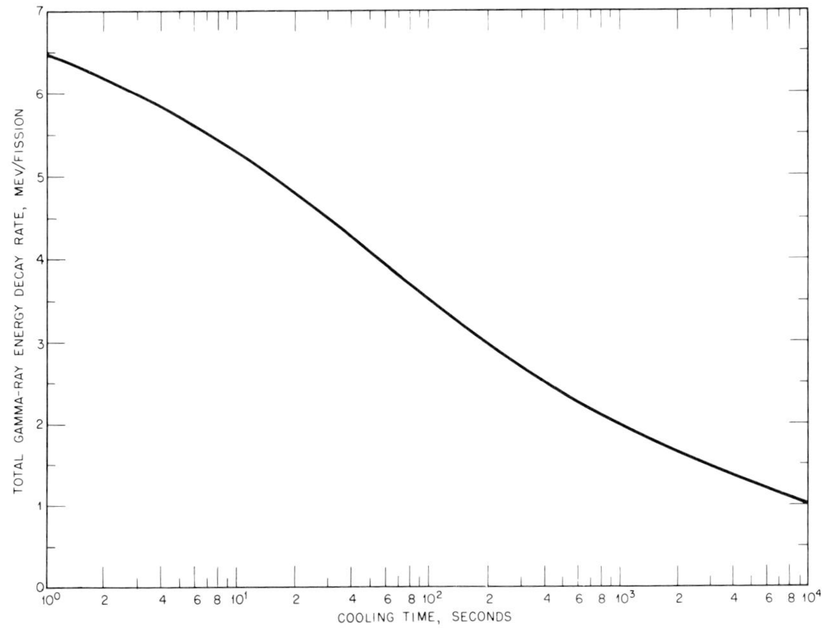

FIG. 4-13 Total gamma-ray energy released by fission products following infinite reactor operation: cooling time = 10^0 to 10^4 sec. (From [40].)

assumed infinite reactor-operating time, followed by a cooling time, t_s sec. This can be converted to finite operating times t_o sec for the total-energy-release curves by the equation:

$$M(t_o, t_s) = M(\infty, t_s) - M(\infty, t_o + t_s) \text{ Mev/fission.}$$

Gamma-energy release can be obtained from the curves for $G(\infty, t_s)$, in a similar way. If desired, these gamma-energy-release curves can be subtracted from the total energy release, $M(\infty, t_s)$, to give beta-ray energy releases separately. The total heat-release rate due to both beta rays and gamma rays is given by

$$Q = \frac{PM(t_o, t_s)}{200} \text{ watts} = 49.8\ PM(t_o, t_s) \text{ cal/sec,}$$

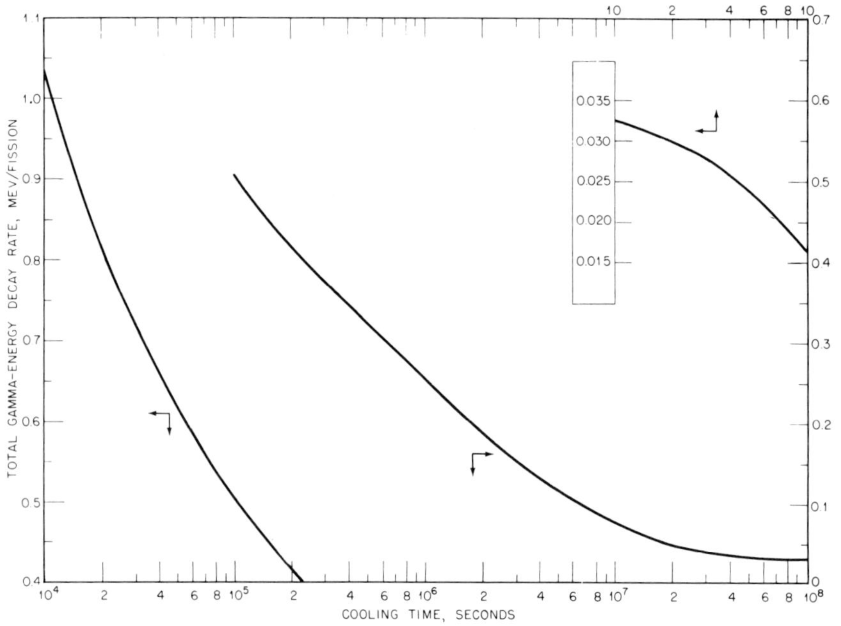

FIG. 4-14 Total gamma-ray energy released by fission products following infinite reactor operation: cooling time = 10^4 to 10^8 sec. (From [40].)

where it has been assumed that 200 Mev are released per fission, and where P is the operating power in watts.

Shure has represented and M(∞, t) and G(∞, t), as given by these graphs, analytically by an exponential form At^{-a}. Tables 4-3 and 4-4 [13] list values for A and a, together with the maximum deviations to be expected if this equation is used. In order to obtain the total energy released after shutdown, $M(t_o, t_s)$ must be integrated over t_s. Table 4-5 [13] presents the energy-release rates Q_γ and Q_β for gamma and beta rays. To obtain the total release rate, add Q_γ to Q_β. The cumulative energy release (Q_t) after the shutdown time listed (t_s sec), for the operating time listed (t_o hr), is also given. These are all given as the percentage of the operating power of the reactor.

In general, of the methods listed, that of Untermyer and Weills [39] appears to give the most conservative, i.e., highest, values of heat release for the times of interest. A more detailed discussion, with numerical examples, is given in Sec. 5 of reference [13]. The accuracy of the estimates is certainly sufficient for estimating reactor-accident energy-release values. Earlier reviews of the subject are presented in references [41-45].

REFERENCES

1. R. G. Hewlett and O. E. Anderson, Jr., "The New World, 1939/46," Vol. I, *A History of the United States Atomic Energy Commission*, The Pennsylvania State University Press, 1962.
2. H. Etherington (Ed.), Chapter 13, pp. 166-167, "Mechanical Design and Operation of Reactors," *Nuclear Engineering Handbook*, McGraw-Hill Book Co., New York, 1958.
3. J. S. Zizzi and E. P. Diehl, "Leak Rate Test Power Plant Building, West Milton Site", USAEC Report KAPL-Memo-SIR-35-53, General Electric Co., Knolls Atomic Power Laboratory, 1953.
4. Reactor Safeguard Committee "Summary Report of Reactor Safeguard Committee", USAEC Report WASH-3 (rev.), March 31, 1950.
5. "PWR Hazards Summary Report", Report WAPD-SC-541, Westinghouse Electric Corp., Bettis Plant, Pittsburgh, Pa., Sept. 1957.
6. "Theoretical Possibilities and Consequences of Major Accidents in Large Nuclear Power Plants," USAEC Report WASH-740, March 1957.
8. R. L. Koontz, et al, "Low Pressure Containment Buildings, Component Tests and Design Data", USAEC Report NAA-SR-7234, p. II-1, Atomics International, March 15, 1963.
9. "Report of Subcommittee on Pathologic Effects of Thyroid Radiation to the National Research Council", Jan. 1967.
10. "Guide to Nuclear Power Cost Evaluation", USAEC Report TID-7025, March 15, 1962.
11. R. A. Johnson and I. Nelson, "Reactor Containment Design Study", USAEC Report SL-1868, Sargent and Lundy, Chicago, Illinois, May 18, 1961. See also USAEC Report SL-1868, Supp. 1, April 23, 1963.
12. D. Kallman and O. G. Hanson, "Some Economic Aspects of Reactor Safety", Reactor Safety Conference, New York, N.Y., Oct. 31, 1957 in USAEC Report TID-7549, Part 2, pp. 1-6, April 1958.
13. W. B. Cottrell and A. W. Savolainen (Eds.), "U.S. Reactor Containment Technology. A compilation of Current Practice in Analysis, Design, Construction, Test and Operation", USAEC Report ORNL-NSIC-5, Aug. 1965. (a) Chapter 11, (b) Chapter 4, (c) Chapter 6.3.
14. Proceedings of the American Power Conference, Chicago, March 1957, reported in *Electrical World*, July 1957.
15. "Preliminary Report 1959 Containment Test" Sargent and Lundy Engineers, Chicago, Ill.
16. Pacific Gas & Electric Co. "Humboldt Bay Power Plant, Unit No. 3. Final Hazards Summary Report", Docket 50-133, Sept. 1, 1961.
17. C. P. Ashworth, D. B. Barton, and C. H. Robbins, "Pressure Suppression", *Nucl. Eng.*, 7:313-21 (Aug. 1962).
18. C. P. Ashworth, D. B. Barton, E. Janssen, and C. H. Robbins, "Predicting Maximum Pressures in Pressure Suppression Containment", ASME Paper No. 61-WA-222. Abstract in *Mech. Eng.* 84(1), 64 (Jan. 1962)
19. Pacific Gas & Electric Co. "Bodega Bay Atomic Park, Unit No. 1. Preliminary Hazards Summary Report", Dec. 28, 1962.
20. J. H. Noble, "Subatmospheric Containment", *Nucl. Eng. Design 6*, 489, 1967.
21. Long Island Lighting Company, "Shoreham Nuclear Power Station, Preliminary Safety Analysis Report", 1967.
22. J. O. Blomeke and M. F. Todd, "Uranium-235 Fission-Product Production as a Function of Thermal Neutron

Flux, Irradiation Time, and Decay Time", USAEC Report ORNL-2127, Oak Ridge National Laboratory, Nov. 1958.
23. J. R. Beattie, "An Assessment of Environmental Hazard from Fission-Product Releases", British Report AHSB(5)R-64, UKAEA, Authority Health & Safety Branch, Risley, Lancs., England, May 16, 1961.
24. J. J. DiNunno, F. D. Anderson, R. E. Baker, R. L. Waterfield, "Calculations of Distance Factors for Power and Test Reactor Sites", USAEC Report TID-14844, March 1962.
25. U.S. Dept. Commerce (Weather Bureau)," Meteorology and Atomic Energy" U.S. Govt. Printing Office, July 1955, (Designated as USAEC Report AECU-3066.)
26. D. H. Slade (Ed.), "Meteorology and Atomic Energy, 1968", USAEC Report TID-24190, July 1968.
27. J. G. Duffy, W. H. Jens, J. G. Feldes, K. P. Johnson, and W. J. McCarthy, Jr., "Investigation of the Fuel Melting Incident at the Enrico Fermi Atomic Power Plant", p. 2.15-37 of *Fast Reactors National Topical Meeting, San Francisco, Calif. April 10-12, 1967,* American Nuclear Society, Hinsdale, Illinois, 1967.
28. E. L. Alexanderson, J. A. Ford, J. B. Nims, and W. R. Olson, "Enrico Fermi Atomic Power Plant Operating Experiences through 100 Mwt", p. 1.21-45 of *Fast Reactors National Topical Meeting, San Francisco, Calif. April 10-12, 1967,* American Nuclear Society, Hinsdale, Illinois, 1967.
29. A. J. Friedland, "Relation Between Maximum Core Coolant Temperatures and Outlet Thermocouple Readings for Plugged Subassemblies, Proc. of International Conference on the Safety of Fast Reactors, Aix-en-Provence, Sept 19-22, 1967. Denielou, G., ed., Paris, Commissariat à l'Energie Atomique, 1967, pp. Va. 3.1-13.
30. R. C. Callen, et al, "Response of Fermi Fuel Outlet Thermocouples Under Normal and Fuel Failure Conditions", *ANS Trans.* Vol. 10, No. 2, November 1967.
31. E. R. Volk, W. H. Jens, K. Kishida, J. B. Nims, "Malfunction Detection in Fast Reactors," Proc. of International Conference on the Safety of Fast Reactors. pp. Va. 2.1-13
32. W. J. McCarthy, Jr., W. H. Jens, "A Review of the Fermi Reactor Fuel Damage Incident and a Preliminary Assessment of Its Significance to the Design and Operation of Sodium Cooled Fast Reactors", Power Reactor Development Corp. Detroit, Mich. 1967.
33. "Emergency Core Cooling", Report of Advisory Task Force on Power Reactor Emergency Cooling, USAEC Report TID-24226 (1967).
34. G. J. Elbaum and T. J. Thompson, "Rapid Excursions on Water Reactors Involving Fuel Element Rupture", *ANS Trans.* Vol. 10, No. 2, 707, November 1967.
35. T. N. Zubarev, A. R. Sokolov, "Calculation of the Heat Release in a Shutdown Reactor", (letter to the editor) *Jour. Nucl. Energy,* Part A, *13,* 72, 1960.
36. S. Bernstein et al, *Phys. Rev., 71,* 573, 1947.
37. G. R. Keepin, "Neutron Data for Reactor Kinetics. 1.) Dealyed Neutrons from Fission. 2.) Photoneutrons from D_2O and Beryllium", *Nucleonics, 20*(8), 150, 1962.
38. K. Way, E. P. Wigner, *Phys. Rev., 70,* 1318, 1948.
39. S. Untermyer and J. T. Weills, "Heat Generated in Irradiated Uranium", USAEC Report ANL-4790, Argonne National Laboratory, February 25, 1962.
40. K. Shure, "Fission Product Decay Energy", pp. 1-17, *Bettis Technical Review. Reactor Technology,* Report WAPD-BT-24, Westinghouse Electric Corp., Bettis Atomic Power Lab. December 1961.
41. J. R. Stehn and E. F. Clancy, "Fission-Product Radioactivity and Heat Generation," Proceedings of the Second U.N. International Conference on the Peaceful Uses of Atomic Energy, Geneva, 1958, Vol. 13, p. 49.
42. I. J. MacBean, "Energy Release from Fission Product Decay", British Report AERE-R-3033, United Kingdom Atomic Energy Authority, Research Group. Atomic Energy Research Establishment, Harwell, Berks, Eng., September 1959.
43. J. F. Perkins and R. W. King, "Energy Release from the Decay of Fission Products", *Nucl. Sci. Eng., 3,* 726, 1958.
44. F. C, Maienschein et al, "Gamma Rays Associated with Fission", Proceedings of the Second U.N. International Conference on the Peaceful Uses of Atomic Energy, Geneva 1958, Vol. 15, p. 366.
45. P. Spiegler, "Energy Release from the Decay of Fission Products," Report NAA-SR-Memo-4126, Atomics International, Canoga Park, Calif., July 31, 1959.

APPENDIX 1

Tabulation of Parameters Relevant to Safety for Five Typical Power Reactors

Appendix 1 of Volume 1 (pages 709 to 733, inclusive) consisted of a tabulation of basic parameters of a number of power reactors, all of which operate at well below 800 Mw(t). A similar tabulation is presented here, but only for reactors operating above 800 Mw(t). The parameters of two pressurized-water reactors, two boiling-water reactors, and one high-temperature gas-cooled reactor are given.

As in the earlier tabulation, the information is organized under four headings:

Table 1

General Description (reactor name, location; identification of fuel, moderator, coolant; operating temperature, pressure; effective core size; thermal and electrical power output)

Table 2

Nuclear Parameters (effective multiplication constant, fuel burnup, Doppler coefficient, moderator void coefficient, moderator temperature coefficient, coolant temperature coefficient, pressure coefficient, power coefficient, other reactivity coefficients, peaking factors and relative fuel assembly power, effective delayed neutron fraction, prompt neutron lifetime)

Table 3

Control (rod description, other control methods, rod deceleration, rod withdrawal, rod insertion, method of use of control devices)

Table 4

Reactivity Inventory (temperature defect, cold to hot zero power; power defect, zero to full power; equilibrium Xe and Sm; fuel depletion and gross fission products; total installed reactivity, and shutdown margin).

TABLE 1

General Description

Abbreviated Name	Full Name; Location	Fuel	Moderator; Coolant	Operating		Effective Core Size[a]	Power	
				Temp. (°F)	Press. (psia)		Thermal Mw (t)	Elec. Mw (e)
Indian Point-2	Consolidated Edison Company of New York Inc. Village of Buchanan, Westchester County, New York	215, 319 lb 2.92% enrich UO_2 (equilibrium) Feed enrichments, w/o Region 1 2.23 Region 2 2.38 Region 3 2.68	H_2O / H_2O	T_{in} 543 T_{out} 596	2235 PWR	h = 12 ft d = 11.14 ft v = 1170 ft^3	2758	873 (net)
Connecticut Yankee	Connecticut Yankee Atomic Power Company, Haddam Neck Plant, Haddam, Conn.	170,000 lb UO_2 Feed enrichment w/o Region 1 3.4 Region 2 3.8 Region 3 4.2	H_2O / H_2O	T_{in} 546 T_{out} 587	2050 PWR	h = 10 ft d = 9.92 ft v = 772.9 ft^3	1473	462 (net)
Dresden-3	Commonwealth Edison Company, Dresden Nuclear Power Station, Unit 3. Grunog, County, Illinois	UO_2, 2% (average fuel enrichment) Cladding–Zircaloy-2	H_2O / H_2O	T_{in} 530.4	1000 BWR	h = 12 ft d = 15.8 ft v = 2355.3 ft^3	2255	752 (gross) 715 (net)
Browns Ferry	Browns Ferry Nuclear Plant Tennesse Valley Authority Limestone Co., Alabama	UO_2, 2.19% (initial average fuel enrichment) Cladding–Zircaloy-2	H_2O / H_2O	T_{in} 530.4	1000 BWR	h = 12 ft d = 16.6 ft v = 2581.5 ft^3	3293	1098 (gross)
Fort St. Vrain	Fort St. Vrain Nuclear Generating Station for Public Service Company of Colorado.	UC, 93% U^{235} and Thorium. Fertile Elements—100% thorium. Cladding–graphite	Carbon / He	T_{in} 760 T_{out} 1430	Pin 700 GCR	h = 15.5 ft d = 19.5 ft v = 4629 ft^3	837.5	330 (net)

[a] cylindrical unless otherwise noted; h = height, d = diameter, v = volume.

TABLE 2

Nuclear Parameters

Item	Indian Point-2	Connecticut Yankee	Dresden-3	Browns Ferry	Fort St. Vrain
k_{eff}	1.275 — Cold, no power, clean 1.225 — Hot, no power, clean 1.170 — Hot, full power, Xe and Sm equal.	1.257 — Cold, no power, clean 1.205 — Hot, no power, clean 1.185 — Hot, full power, Xe and Sm equal.	1.26 — Cold, no power, clean 0.96 — With all control rods in <0.99 — With strongest control rod out	1.25 — Cold, no power, clean 0.96 — With all control rods in <0.99 — With strongest control rod out	1.143 — 80°F Initial core 1.110 — 400°F Initial core 1.03 — Operating 0.923 — All control rods in (80°F) 0.934 — Max. worth rod pair stuck out
Fuel Burnup MWD/MTU	Avg. first cycle 12,000 Avg. first core 21,800 Equil. core avg. 27,000	Avg. first cycle 14,100 Equil. core avg. 21,800	Avg. first core 15,000	Avg. first core 19,000	100,000
Doppler coefficient (Δ k/k/°F)	-1×10^{-5} to -2×10^{-5}	-0.5×10^{-5} to -2.1×10^{-5}	Cold: -1.3×10^{-5} Hot: -1.2×10^{-5} Operating: -1.3×10^{-5}	Cold: -1.3×10^{-5} Hot: -1.2×10^{-5} Operating: $\leq 1.3 \times 10^{-5}$	Equil. core, beginning of cycle, with Xe and Sm -2.8×10^{-5} (80°F) -2.3×10^{-5} (400°F) -1.1×10^{-5} operating
Moderator void coefficient (Δ k/k/% void)	1×10^{-3} to -3×10^{-3}	0 to -2×10^{-3}	Hot: -1.0×10^{-3} Operating: -1.5×10^{-3}	Hot: -1.0×10^{-3} Operating: -1.6×10^{-3}	- - - - -
Moderator Temperature coefficient (Δ k/k/°F)	$+1 \times 10^{-4}$ to -3×10^{-4}	0 to $+2.4 \times 10^{-4}$	Cold: -5×10^{-5} Hot: -39×10^{-5}	Cold: -5×10^{-5} Hot: -39×10^{-5}	Isothermal coeff. equil. core, BOC[4], with Xe and Sm -2.9×10^{-5} (80°F) -2.3×10^{-5} (400°F) -5.7×10^{-6} operating
Coolant temperature coefficient (per °F)	Same as above	Same as above	Same as above	Same as above	$\Delta\rho$ due to the loss of all the helium from the reactor = 5×10^{-5} Δ k/k

Table 2 (Continued)

Pressure coefficient ($\Delta k/k/°F$)	-1×10^{-6} to $+3 \times 10^{-6}$ BOC[4] EOL[5] 2,170 ppm boron no boron	0 to $+2.4 \times 10^{-6}$ BOC EOL	-----	-----	Negligible
Power coefficient	$-0.02 \frac{\Delta k/k}{\Delta p/p}$ at the beginning of core life	$-0.015 \frac{\Delta k/k}{\Delta p/p}$ at the beginning of core life	$-0.04 \frac{\Delta k/k}{\Delta p/p}$ at the beginning of core life; $-0.02 \frac{\Delta k/k}{\Delta p/p}$ at end of core life	$-0.04 \frac{\Delta k/k}{\Delta p/p}$ at the beginning of core life; $-0.02 \frac{\Delta k/k}{\Delta p/p}$ at end of core life	$-0.037 \frac{\Delta k/k}{\Delta p/p}$
Other reactivity coefficients	Boron worth; Hot → 1% $\frac{\Delta k}{k}$ / 150 ppm Cold → 1% $\frac{\Delta k}{k}$ / 120 ppm	Boron worth; Hot → 1% $\frac{\Delta k}{k}$ / 150 ppm Cold → 1% $\frac{\Delta k}{k}$ / 120 ppm	-----	-----	-----
Max. rel. ass'y power[1] Local peaking factor[2] Axial peaking factor[3] Total peaking factor	1.88 --- --- 3.25	1.61 --- --- 3.09	1.47 1.30 1.57 3.00	1.4 1.24 1.5 2.60	--- --- 1.5 2.67
Effective delayed neutron fraction (sec)		0.005 to 0.0007			0.00517 with Xe, BOC[4], equil. core. 0.00461 end of cycle, equil. core.
Prompt neutron lifetime (sec)		16.0×10^{-6} to 18.0×10^{-6}			2.39×10^{-4} sec with Xe, BOC, equil. core. 2.63×10^{-4} sec end of cycle, equil. core.

1. The power of a fuel assembly divided by the core average assembly power.
2. Maximum fuel rod average heat flux in an assembly divided by the assembly average heat flux.
3. Maximum heat flux on a given fuel rod divided by the average heat flux on that rod.
4. BOC = Beginning of cycle.
5. EOL = End of life.

TABLE 3

Control

Item	Indian Point-2	Connecticut Yankee	Dresden-3	Browns Ferry	Fort St. Vrain
Rods	53 RCC Assemblies (5% Cd-15% In-80% Ag) 20 Absorber rods per RCC. Total rod worth 7%	45 RCC Assemblies (5% Cd-15% In-80% Ag) 16 Absorber rods per RCC. Total rod worth 7 1/2%	177 cruciform rods, B_4C granules; avg. worth/rod = 0.005 Δk; max. worth/rod= 0.01Δk total worth of control rods: -0.18 Δk	185 cruciform rods, B_4C granules; avg. worth/rod = 0.005 Δk; max. worth/rod= 0.01 Δk; total worth of control rods: -0.17 Δk	37 pairs of control rods; 30% wt–boron carbide. Worth of all control rods inserted: (equil. core) 0.203 Δk (80°F) 0.207 Δk (400°F) 0.212 Δk operating
Other control methods	Boron concentration: To shut reactor down no rods inserted, clean cold/hot-3400 ppm/2300 ppm; control with no rods, clean/ eq. Xe and Sm —2800 ppm/2300 ppm	Boron concentration: To shut reactor down no rods inserted, clean cold/hot-3360 ppm/3250 ppm	324 temporary control curtains of natural boron and stainless steel: total worth -0.12 Δk. Liquid control is provided as a redundant shut down (standby system)	356 temporary control curtains of natural boron and stainless steel; total worth -0.12 Δk. Standby liquid control is provided as a redundant shut down	Reserve shutdown system. It will use B_4C in granular form mixed with graphite granules. They will be stored in a hopper in each refueling penetration from which they can be released and allowed to fall into channel in the core
Deceleration	The bottom plug is made bullet–nose to guide the RCC smoothly into the dash-pot section of the guide thimble	The bottom of the guide thimble is of reduced inside diameter to perform a dash-pot action when the rods are dropped	A stack of Inconel-750 spring washers helps absorb the final mechanical shock at the end of travel	A stack of Inconel-750 spring washers helps absorb the final mechanical shock at the end of travel	The lower end of the rod is attached to an end section which incorporates a tubular type crushable shock absorbor
Withdrawal	The latch assembly moves the control rod up or down by 3/8-inch steps. It develops a lifting force of 400 lbs. rate: 15 in/min	The latch assembly moves up or down to raise or lower the drive rod by 3/8 inch (one step). The mechanisms develop a lifting force of 400 lbs. Rate 15 in/min	Notch increments 6"; worth of notch increments $\leq 0.002 \Delta k$	Notch increments 6"; worth of notch increments $\leq 0.002 \Delta k$	Rod withdrawal speed: 3.6 in/sec minimum rod withdrawal time: 58 sec. Max. rod pair worth upon withdrawal = 1.5% Δk
Insertion	Fast total insertion (fall by gravity): 2 to 3 secs	Fast total insertion (fall by gravity): 2 to 3 secs	Velocity = 5 ft/sec; -3% Δk/sec for the first 10% of the rod; -4% Δk/sec to the 90% insertion point	Velocity = 5 ft/sec; -3% Δk/sec for the first 10% of the rod; -4% Δk/sec to the 90% insertion point	Rod insertion speed: 3.7 in/sec. Maximum rod insertion time: Shim -57.0 sec; scram—48 sec
Method of use	Chemical shim rods: regulating and safety	Chemical shim control rods: regulating and safety	Shim, regulating and safety	Shim, regulating and safety	Shim, regulating and safety

TABLE 4

Reactivity Inventory

Item	Indian Point-2	Connecticut Yankee	Dresden-3	Browns Ferry	Fort St. Vrain
Temperature defect ($\Delta\rho$); cold to hot zero power	0.050	0.036	0.028	0.028	0.028
Power defect ($\Delta\rho$); zero power to full power	0.020	0.015	0.040	0.040	0.037
Equilibrium xenon and samarium ($\Delta\rho$)	0.035	0.028	0.062	0.062	0.026
Fuel depletion and gross fission products ($\Delta\rho$)	0.170	0.140	0.120	0.120	0.055
Total installed reactivity ($\Delta\rho$)	0.275	0.219	0.250	0.250	0.146
Shutdown margin	1% with the highest worth rod stuck	1% with the highest worth rod stuck; 3% with all rods in	1% with the highest worth rod stuck; 4% with all rods in	1% with the highest worth rod stuck; 4% with all rods in	0.077 Δk (80° F); all control rods inserted; 0.111 Δk (400° F), all rods in

APPENDIX 2

Abbreviations Used in Text

Listed below are brief identifications of abbreviations referred to in the text of this volume. For identification of abbreviations used in United States nuclear reactor projects, the reader is referred to AEC Report TID-8200 (26th Rev.) "Nuclear Reactors Being Built or planned in the United States as of June 30, 1972." For other abbreviations, the "Directory of Nuclear Reactors" published by the International Atomic Energy Agency, Vienna and the USAEC Report TID-7031 "A Handbook of Abbreviations and Nicknames" C. B. Yulish (Ed.) (May 1964) should be consulted.

ACRS	Advisory Committee on Reactor Safeguards
ADF	Aerosol Development Facility
AGCR	Army Gas-Cooled Reactor
AGR	Advanced Gas-Cooled Reactor
AIME	American Institute of Mining, Metallurgical and Petroleum Engineers
AISI	American Iron and Steel Institute
ALPR	see SL-1
ANL	Argonne National Laboratory
ANSI	American National Standards Institute (formerly American Standards Association)
ANP	Aircraft Nuclear Propulsion
APDA	Atomic Power Development Associates
APPR	Army Power Package Reactor, see SM-1
ARE	Aircraft Reactor Experiment
ASA	see ANSI
ASME	American Society of Mechanical Engineers
ASTM	American Society for Testing and Materials
ATR	Advanced Test Reactor
AVR	Arbeitsgemeinschaft Versuchs-Reaktor (Jülich, W. Germany)
BGRR	Brookhaven Graphite Research Reactor
BMI	Battelle Memorial Institute
BNL	Brookhaven National Laboratory
BNW	Battelle-Northwest
BONUS	Boiling Nuclear Superheat Reactor
BORAX	Boiling Reactor Experiment
BPVC	Boiler and Pressure Vessel Committee (ASME)
BWR	boiling water reactor
CANDU	Douglas Point Nuclear Power Station (Canada)
CAT	crack arrest temperature
CC	cold, clean condition
CDE	Contamination-Decontamination Experiment
CFR	Code of Federal Regulations
CISE	Centro Informazioni Studie d'Experienze (Milan, Italy)
CMF	Containment Mockup Facility
CP-5	Argonne Research Reactor
CRI	Containment Research Installation
CSE	Containment Systems Experiment
CVTR	Carolinas-Virginia Tube Reactor
DBA	design-basis accident
DMAM	dimethylaminoethylmethacrylate
DOP	dioctylphthalate
EBOR	Experimental Beryllium Oxide Reactor
EBR	Experimental Breeder Reactor
EBWR	Experimental Boiling Water Reactor
ECCS	emergency core-cooling system
EFFBR	Enrico Fermi Fast Breeder Reactor
EGCR	Experimental Gas-Cooled Reactor
EL-2, EL-3	Reactors at Saclay, France
ESADA	Empire State Atomic Development Authority
ETR	Engineering Test Reactor
EVESR	Esada-Vallecitos Experimental Superheat Reactor

FCR	Fast Ceramic Reactor Program
FR-2	heavy water research reactor at Karlsruhe, W. Germany
FSAR	final safety analysis report
FTE	fracture transition for elastic loading
FTP	fracture transition for plastic loading
FWCNP	Florida West Coast Nuclear Power Plant
G-2, G-3	plutonium production reactors at Marcoule, France
GCRE	Gas-Cooled Reactor Experiment
GETR	General Electric Test Reactor
HATR	Highly Active Test Facility (UKAEA)
HC	hot, clean condition
HEMD	high-efficiency moisture de-entrainers
HEPA	high-efficiency particulate air
HFIR	High Flux Isotope Reactor
HRE	Homogeneous Reactor Experiment
HRT	Homogeneous Reactor Test
HTGR	High-Temperature Gas-Cooled Reactor
HTRE	Heat Transfer Reactor Experiment
HWCTR	Heavy Water Components Test Reactor
HWGCR	Heavy Water Gas-Cooled Reactor (Czechoslovakia)
HWOCR	Heavy Water Organic-Cooled Reactor
ICC	Interstate Commerce Commission
LAMPRE	Los Alamos Molten Plutonium Reactor Experiment
LMFBR	Liquid-Metal Fast-Breeder Reactor
LOCA	loss-of-coolant accident
LOFT	Loss of Fluid Test
LRL	Lawrence Radiation Laboratory
MGCR	Maritime Gas-Cooled Reactor
MITR	Massachusetts Institute of Technology Reactor
MSRE	Molten-Salt Reactor Experiment
MTR	Materials Testing Reactor
NDT	nil-ductility transition
NOL	Naval Ordnance Laboratory
NPD	Nuclear Power Demonstration Station (Canada)
NPR	New Production Reactor
NRL	Naval Research Laboratory
NRTS	National Reactor Testing Station
NRU, NRX	reactors at Chalk River (Ontario, Canada)
NSPP	Nuclear Safety Pilot Plant
OMRE	Organic-Moderated Reactor Experiment
ORNL	Oak Ridge National Laboratory
ORR	Oak Ridge Research Reactor
PAT	painted aerosol tank
PCVR	prestressed concrete reactor vessel
PEGCPR	Partially Enriched Gas-Cooled Power Reactor
PH	precipitation-hardened
PLUTO	tank reactor for plutonium tests (United Kingdom)
PM-2A	Portable Medium Power Plant No. 2A
PRTR	Plutonium Recycle Test Reactor
PSAR	preliminary safety analysis report
PTFE	polytrifluorethylene
PWR	pressurized water reactor
RA-1	Argentine Reactor No. 1 (Buenos Aires)
S1G	Submarine Intermediate Reactor Mark A (formerly SIR)
SAE	Society of Automotive Engineers
SAP	sintered aluminum powder
SAR	safety analysis report
SAT	stainless steel aerosol tank
SETOR	Southwest Experimental Test Oxide Reactor
SL-1	Stationary Low-Power Plant No. 1 (formerly ALPR)
SM-1	Stationary Medium-Power Plant No. 1 (formerly APPR-1)
S-N	stress versus number of cycles
SNAP	systems for nuclear auxiliary power
SPERT	Special Power Excursion Reactor Test
SRE	Sodium Reactor Experiment
SRL	Savannah River Laboratory
SRP	Savannah River plant
TBP	tri-n-butylphosphine
TIG	tungsten-inert gas (welding)
TREAT	Transient Reactor Test Facility
TTT	time-temperature-transformation
UDMH	unsymmetrical dimethyl hydrazine
UHTREX	Ultra-High Temperature Reactor Experiment
UKAEA	United Kingdom Atomic Energy Authority
VBNR	Vallecitos Boiling Water Reactor
WQ	water-quenched
WR-1	Whiteshell Reactor No. 1 (Pinawa, Manitoba, Canada)
WTR	Westinghouse Test Reactor
X-10	an area (original reactor site) at ORNL

APPENDIX 3

Contents of Volume 1

CHAPTER 3

General Reactor Dynamics

E. P. Gyftopoulos

CHAPTER 4

The Doppler Coefficient

L. W. Nordheim

CHAPTER 5

Criticality

H. C. Paxton, G. R. Keepin

CHAPTER 6

Sensing and Control Instrumentation

A. Pearson and C. G. Lennox

CHAPTER 7

Mathematical Models of Fast Transients

W. E. Nyer

CHAPTER 8

Water Reactor Kinetics

J. A. Thie

CHAPTER 9

Kinetics of Solid-Moderator Reactors

H. B. Stewart and M. H. Merrill

CHAPTER 10

Fast Reactor Kinetics

W. J. McCarthy, Jr. and D. Okrent

CHAPTER 11

Accidents and Destructive Tests

T. J. Thompson

INDEX